Principles and Practice of Clinical Bacteriology
Second Edition

Principles and Practice of Clinical Bacteriology
Second Edition

Editors

Stephen H. Gillespie

Royal Free and University College Medical School, London, UK

and

Peter M. Hawkey

University of Birmingham, Birmingham, UK

John Wiley & Sons, Ltd

Other Wiley Editorial Offices

John Wiley & Sons Inc., 111 River Street, Hoboken, NJ 07030, USA

Jossey-Bass, 989 Market Street, San Francisco, CA 94103-1741, USA

Wiley-VCH Verlag GmbH, Boschstr. 12, D-69469 Weinheim, Germany

John Wiley & Sons Australia Ltd, 33 Park Road, Milton, Queensland 4064, Australia

John Wiley & Sons (Asia) Pte Ltd, 2 Clementi Loop #02-01, Jin Xing Distripark, Singapore 129809

John Wiley & Sons Canada Ltd, 22 Worcester Road, Etobicoke, Ontario, Canada M9W 1L1

Wiley also publishes its books in a variety of electronic formats. Some content that appears in print may not be available in electronic books.

Library of Congress Cataloging in Publication Data

Principles and practice of clinical bacteriology / editors, Stephen H.
   Gillespie and Peter M. Hawkey. — 2nd ed.
     p.   ;   cm.
   Includes bibliographical references and index.
   ISBN-13: 978-0-470-84976-7 (cloth : alk. paper)
   ISBN-10: 0-470-84976-2 (cloth : alk. paper)
   1. Medical bacteriology.
   [DNLM: 1. Bacteriological Techniques. QY 100 P957 2005]   I. Gillespie, S. H.
   II. Hawkey, P. M. (Peter M.)
 QR46.P84   2005
 616.9′041—dc22

                      2005013983

British Library Cataloguing in Publication Data

A catalogue record for this book is available from the British Library

ISBN-13 978-0-470-84976-7 (cloth: alk. paper)
ISBN-10 0-470-84976-2 (cloth: alk. paper)

Typeset in 9/10pt Times by Integra Software Services Pvt. Ltd, Pondicherry, India
Printed and bound in Great Britain by Antony Rowe Ltd, Chippenham, Wiltshire
This book is printed on acid-free paper responsibly manufactured from sustainable forestry in which at least two trees are planted for each one used for paper production.

Contents

List of Contributors

Dlawer A. A. Ala'Aldeen Molecular Bacteriology and Immunology Group, Division of Microbiology, University Hospital, Nottingham NG7 2UH, UK

Indran Balakrishnan Department of Medical Microbiology, Royal Free Hospital, Pond Street, London NW3 2QG, UK

Christopher L. Baylis Campden & Chorleywood Food Research Association (CCFRA), Chipping Campden, Gloucestershire GL55 6LD, UK

Cécile M. Bébéar Laboratoire de Bactériologie, Université Victor Segalen Bordeaux 2, 146 rue Léo Saignat, 33076 Bordeaux, France

Christiane Bébéar Laboratoire de Bactériologie, Université Victor Segalen Bordeaux 2, 146 rue Léo Saignat, 33076 Bordeaux, France

Alex van Belkum Department of Medical Microbiology & Infectious Diseases, Erasmus University Medical Center Rotterdam EMCR, Dr Molewaterplein 40, 3015 GD Rotterdam, The Netherlands

Eugenie Bergogne-Berezin University Paris 7, Faculty of Medicine Bichat, 100 bis rue du Cherche-Midi, Paris 75006, France

Patrick J. Blair US Embassy Jakarta, US NAMRU 2, FPO AP 96520-8132

Alpana Bose Centre for Medical Microbiology, Royal Free and University College Medical School, Hampstead Campus, Rowland Hill Street, London NW3 2PF, UK

Tom Cheasty HPA Colindale, Laboratory of Enteric Pathogens, Centre for Infection, 61 Colindale Avenue, London NW9 5HT, UK

Derrick W. Crook Department of Clinical Microbiology, John Radcliffe Hospital, Level 7, Headington, Oxford OX3 9DU, UK

Aruni De Zoysa Health Protection Agency, Centre for Infection, Department of Respiratory and Systemic Infections, 61 Colindale Avenue, London NW9 5HT, UK

Brian I. Duerden Anaerobe Reference Laboratory, Department of Medical Microbiology, University of Wales College of Medicine, Heath Park, Cardiff CF14 4XN, UK

Androulla Efstratiou Respiratory and Systemic Infection Laboratory, Health Protection Agency, Centre for Infection, 61 Colindale Avenue, London NW9 5HT, UK

Florence Fenollar Unité des Rickettsies, CNRS UMR 6020A, Faculté de Médecine, Université de la Méditerranée, 27 Bd Jean Moulin, 13385 Marseille Cedex 05, France

Roger G. Finch Department of Microbiology and Infectious Diseases, The Nottingham City Hospital NHS Trust, The Clinical Sciences Building, Hucknall Road, Nottingham NG5 1PB, UK

Sarah J. Furrows Department of Clinical Parasitology, The Hospital of Tropical Diseases, Mortimer Market off Tottenham Court Road, London WC1E 6AU, UK

Stephen H. Gillespie Centre for Medical Microbiology, Royal Free and University College Medical School, Hampstead Campus, Rowland Hill Street, London NW3 2PF, UK

Richard L. Guerrant Division of Infectious Diseases, University of Virginia School of Medicine, Charlottesville, VA 22908, USA

Val Hall Anaerobe Reference Laboratory, National Public Health Service for Wales, Microbiology Cardiff, University Hospital of Wales, Heath Park, Cardiff CF4 4XW, UK

T. G. Harrison Respiratory and Systemic Infection Laboratory, Health Protection Agency, Centre for Infection, 61 Colindale Avenue, London NW9 5HT, UK

C. Anthony Hart Medical Microbiology Department, Royal Liverpool Hospital, Duncan Building, Daulby Street, Liverpool L69 3GA, UK

Peter M. Hawkey Division of Immunity and Infection, The Medical School, Edgbaston, Birmingham B15 2TT, UK

Qiushui He Pertussis Reference Laboratory, National Public Health Institute, Kiinamyllynkatu 13, 20520 Turku, Finland

Brian Henderson Division of Infection and Immunity, Eastman Dental Institute, 256 Gray's Inn Road, London WC1X 8LD, UK

Derek W. Hood Molecular Infectious Diseases, Department of Paediatrics, Weatherall Institute of Molecular Medicine, John Radcliffe Hospital, Headington, Oxford OX3 9DS, UK

Catherine Ison Sexually Transmitted Bacteria Reference Laboratory, Centre for Infection, Health Protection Agency, 61 Colindale Avenue, London NW9 5HT, UK

Claire Jenkins Department of Medical Microbiology, Royal Free Hospital NHS Trust, Rowland Hill Street, London NW3 2PF, UK

Peter J. Jenks Department of Microbiology, Plymouth Hospitals NHS Trust, Derriford Hospital, Plymouth PL6 8DH, UK

Franca R. Jones National Naval Medical Center, Microbiology Laboratory, 8901 Wisconsin Avenue, Bethesda, MD 20889, USA

Monika Keelan Department of Laboratory Medicine & Pathology, Division of Medical Laboratory Science, B-117 Clinical Sciences Building, University of Alberta, Edmonton, Alberta, Canada T6G 2G3

Kevin G. Kerr Department of Microbiology, Harrogate District Hospital, Lancaster Park Road, Harrogate HG2 7SX, UK

Theresa Lamagni Healthcare Associated Infection & Antimicrobial Resistance Department, Health Protection Agency, Centre for Infection, 61 Colindale Avenue, London NW9 5EQ, UK

Paul N. Levett Provincial Laboratory, Saskatchewan Health, 3211 Albert Street, Regina, Saskatchewan, Canada S4S 5W6

Niall A. Logan Department of Biological and Biomedical Sciences, Glasgow Caledonian University, Cowcaddens Road, Glasgow G4 0BA, UK

Benjamin J. Luft Department of Medicine, Division of Infectious Diseases, SUNY at Stony Brook, New York, NY 11794-8160, USA

Roderick McNab Consumer Healthcare, GlaxoSmithKline, St. George's Avenue, Weybridge, Surrey KT13 0DE, UK

Jussi Mertsola Department of Pediatrics, University of Turku, Kiinamyllynkatu 4-8, 20520 Turku, Finland

David A. Murdoch Department of Medical Microbiology, Royal Free Hospital NHS Trust, Rowland Hill Street, London NW3 2PF, UK

Barbara E. Murray Division of Infectious Diseases, University of Texas Health Science Center, Houston Medical School, Houston, TX 77225, USA

Esteban C. Nannini Division of Infectious Diseases, University of Texas Health Science Center, Houston Medical School, Houston, TX 77225, USA

James G. Olson Virology Department, U.S. Naval Medical Research Center Detachment, Unit 3800, American Embassy, APO AA 34031

Petra C. F. Oyston Defence Science and Technology Laboratory, Porton Down, Salisbury, Wiltshire SP4 0JQ, UK

Sudha Pabbatireddy Department of Medicine, Division of Infectious Diseases, SUNY at Stony Brook, New York, NY 11794-8160, USA

Sheila Patrick Department of Microbiology and Immunobiology, School of Medicine, Queen's University of Belfast, Grosvenor Road, Belfast BT12 6BN, UK

Sharon Peacock Nuffield Department of Clinical Laboratory Sciences, Department of Microbiology, Level 7, The John Radcliffe Hospital, Oxford OX3 9DU, UK

Charles W. Penn Professor of Molecular Microbiology, School of Biosciences, University of Birmingham, Edgbaston, Birmingham B15 2TT, UK

Sabine Pereyre Laboratoire de Bactériologie, Université Victor Segalen Bordeaux 2, 146 rue Léo Saignat, 33076 Bordeaux, France

Tyrone L. Pitt Centre for Infection, Health Protection Agency, 61 Colindale Avenue, London NW9 5HT, UK

Ian R. Poxton Medical Microbiology, University of Edinburgh Medical School, Edinburgh EH8 9AG, UK

Michael B. Prentice Department of Microbiology University College Cork, Cork, Ireland

Didier Raoult Unité des Rickettsies, CNRS UMR 6020A, Faculté de Médecine, Université de la Méditerranée, 27 Bd Jean Moulin, 13385 Marseille Cedex 05, France

Derren Ready Eastman Dental Hospital, University College London Hospitals NHS Trust, 256 Gray's Inn Road, London WC1X 8LD, UK

John Richens Centre for Sexual Health and HIV Research, Royal Free and University College Medical School, The Mortimer Market Centre, Mortimer Market, London WC1E 6AU, UK

Geoff L. Ridgway Pathology Department, The London Clinic, 20 Devonshire Place, London W1G 6BW, UK

Marina Rodríguez-Díaz Department of Biological and Biomedical Sciences, Glasgow Caledonian University, Cowcaddens Road, Glasgow G4 0BA, UK

J. M. Rolain Unité des Rickettsies, CNRS UMR 6020A, Faculté de Médecine, Université de la Méditerranée, 27 Bd Jean Moulin, 13385 Marseille Cedex 05, France

Andrew J. H. Simpson Defence Science and Technology Laboratory, Biological Sciences Building 245, Salisbury, Wiltshire SP4 0JQ, UK

Shiranee Sriskandan Gram Positive Molecular Pathogenesis Group, Department of Infectious Diseases, Faculty of Medicine, Imperial College London, Hammersmith Hospital, Du Cane Road, London W12 0NN, UK

Diane E. Taylor Department of Medical Microbiology and Immunology, 1-41 Medical Sciences Building, University of Alberta, Edmonton AB T6G 2H7, Canada

Nathan M. Thielman Division of Infectious Diseases, University of Virginia School of Medicine, Charlottesville, VA 22908, USA

Andrew J. L. Turner Manchester Medical Microbiology Partnership, Department of Clinical Virology, 3rd Floor Clinical Sciences Building 2, Manchester Royal Infirmary, Oxford Road, Manchester M13 9WL, UK

David P. J. Turner Molecular Bacteriology and Immunology Group, Division of Microbiology, University Hospital of Nottingham, Nottingham NG7 2UH, UK

Cees M. Verduin PAMM, Laboratory of Medical Microbiology, P.O. Box 2, 5500 AA Veldhoven, The Netherlands

Matti K. Viljanen Department of Medical Microbiology, University of Turku, Kiinamyllynkatu 13, 20520 Turku, Finland

Adrian Whatmore Department of Statutory and Exotic Bacterial Diseases, Veterinary Laboratory Agency – Weybridge, Woodham Lane, Addlestone, Surrey KT15 3NB, UK

Mark H. Wilcox Department of Microbiology, Leeds General Infirmary & University of Leeds, Old Medical School, Leeds LS1 3EX, UK

Edward J. Young Department of Internal Medicine, Baylor College of Medicine, Veterans Affairs Medical Center, One Baylor Place, Houston, TX 77030, USA

Preface

Change is inevitable and nowhere more than in the world of bacteriology. Since the publication of the first edition of this book in 1997 the speed of change has accelerated. We have seen the publication of whole genome sequences of bacteria. The first example, *Mycoplasma genitalium*, was heralded as a breakthrough, but this was only the first in a flood of sequences. Now a wide range of human, animal and environmental bacterial species sequences are available. For many important organisms, multiple genome sequences are available, allowing comparative genomics to be performed. This has given us a tremendous insight into the evolution of bacterial pathogens, although this is only a part of the impact of molecular biological methods on bacteriology. Tools have been harnessed to improve our understanding of the pathogenesis of bacterial infection, and methods have been developed and introduced into routine practice for diagnosis. Typing techniques have been developed that have begun unravelling the routes of transmission in human populations in real time. New pathogens have been described, and some species thought to have been pathogenic have been demonstrated only to be commensals. Improved classification, often driven by molecular methods, has increased our ability to study the behaviour of bacteria in their interaction with the human host. New treatments have become available, and in other instances, resistance has developed, making infections with some species more difficult to manage.

In this revision the authors and editors have endeavoured to provide up-to-date and comprehensive information that incorporates this new knowledge, while remaining relevant to the practice of clinical bacteriology. We hope that you find it helpful in your daily practice.

Stephen H. Gillespie
Peter M. Hawkey

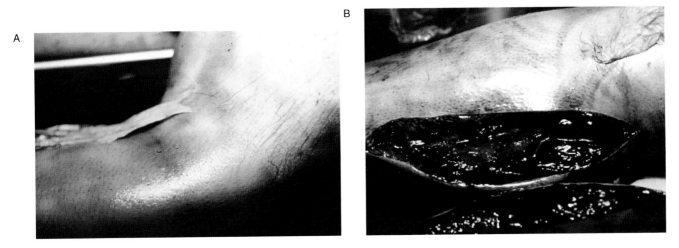

Plate 1 Appearances of group A streptococcal cellulitis and necrotising fasciitis (A). The mottled, tense appearances of the skin are an ominous predictor of underlying changes of advanced necrotising fasciitis. Necrosis and liquefaction of adjacent muscle may accompany advanced group A streptococcus (GAS) necrotising fasciitis (B). Note the blanching erythema of the skin, consistent with a diagnosis of streptococcal toxic shock syndrome (STSS). Photographs: S. Sriskandan, courtesy of Science Press.

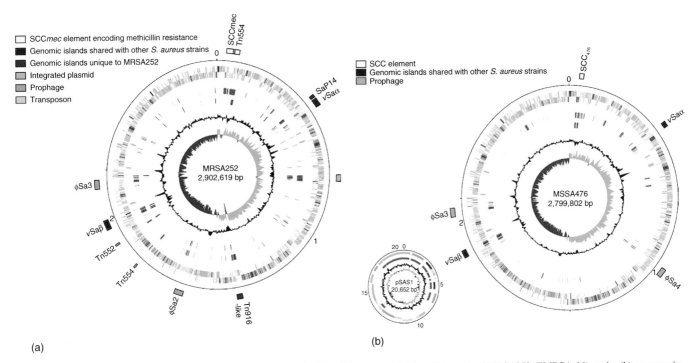

(a)   (b)

Plate 2 Schematic circular diagrams of the *S. aureus* MRSA252 and MSSA476 genomes. (a), hospital-acquired MRSA252 (EMRSA-16) strain; (b), community-acquired MSSA476 strain. Key to the chromosomal circular diagrams (outside to inside): scale (in Mb), annotated CDS (coloured according to predicted function), tRNA and rRNA (green), additional DNA compared to the other *S. aureus* strain described here (MSSA476 or MRSA252 where appropriate) (red), additional DNA compared to other sequenced *S. aureus* strains [N315(5), Mu50(5), MW2(6)] (blue), %G + C content, G + C deviation (>0% olive, <0% purple). Colour coding for CDSs: dark blue, pathogenicity/adaptation; black, energy metabolism; red, information transfer; dark green, surface associated; cyan, degradation of large molecules; magenta, degradation of small molecules; yellow, central/intermediary metabolism; pale green, unknown; pale blue, regulators; orange, conserved hypothetical; brown, pseudogenes; pink, phage + IS elements; grey, miscellaneous. Key to the MSSA476 plasmid circular diagram (outside to inside): scale (in kb), annotated CDS (coloured according to predicted function, see above), DNA with similarity to the integrated plasmid of MRSA252 (purple), %G + C content, G + C deviation (>0% olive, <0% purple).

(a)

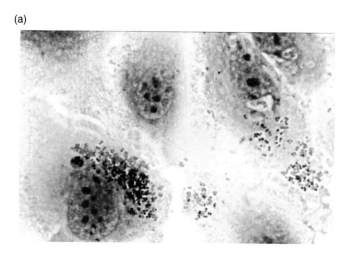

(b)

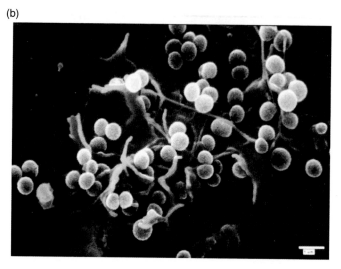

Plate 3 (a) Human endothelial cell monolayer following adherence assay with *S. aureus* in vitro. Light microscopy view, ×100 magnification. Bacteria are associated with the host cells. (b) Human endothelial cell monolayer following adherence assay with *S. aureus* in vitro. Scanning electron microscopy, colour enhanced view ×8800 magnification. Bacteria appear to be adherent to the monolayer and are undergoing a process of uptake by the endothelial cell.

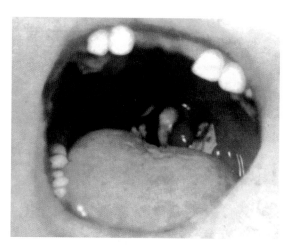

Plate 4 Faucial diphtheria. The characteristic thick membrane of diphtheria infection in the posterior pharynx. Reproduced by permission of Dr A. Sukonthaman, Bangkok, Thailand.

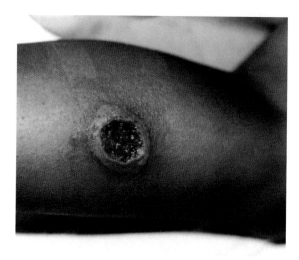

Plate 5 Cutaneous diphtheria infection. A tender pustule has enlarged to form an oval punched-out ulcer with a membrane and an oedematous rolled border. Reproduced by permission of Dr A. Sukonthaman, Bangkok, Thailand.

Plate 6 Zhiel Neelsen stain of *M. tuberculosis* illustrating the cording characteristic of this organism.

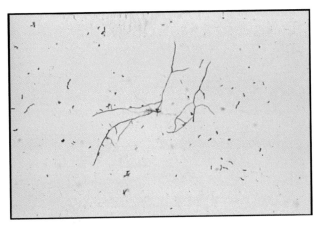

Plate 7 Gram stain of an isolate of Nocardia obtained from barin abscess pus demonstrating long branched filaments and coocal forms.

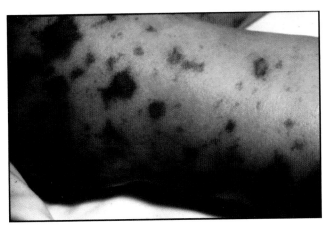

Plate 8 Characteristic meningococcal rash in a child with meningococcal speticaemia. Note petechiae coalescing in places. (Courtesy of Dr H. Vyas, Nottingham.)

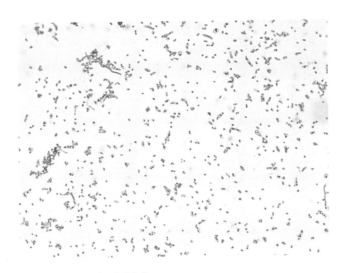

Plate 9 Gram stain of *H. influenzae*.

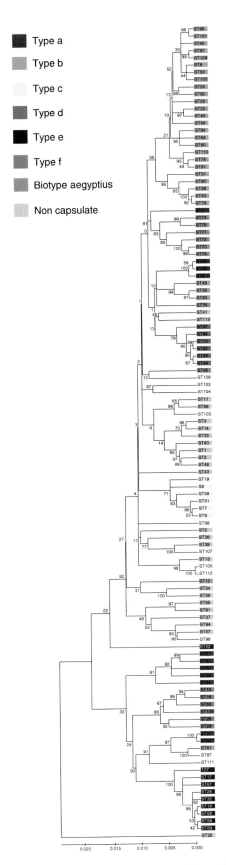

Plate 10 A neighbourhood joining tree with bootstrap values at the nodes of MLST data for all types available from http://haemophilus.mlst.net. A limited number of Nc-hi have been MLST typed which does not give a view of the wide diversity within that population. These data illustrate those strains of each capsular type clustering into tight closely related groups or clonal complexes, reflecting the clonal structure of capsulate *H. influenzae*.

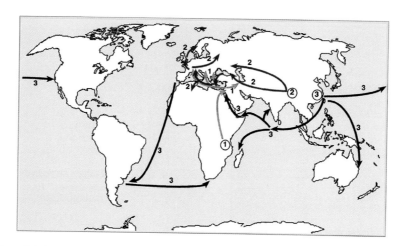

Plate 11 Pandemic spread of *Y. pestis* in human history. Routes followed by the three plague pandemic waves, labelled 1, 2 and 3. Circled numbers indicate the regions thought to be the origins of these pandemic waves. From Achtman M., Zurth K., Morelli G., Torrea G., Guiyoule A., Carniel E. (1999). Proc Natl Acad Sci USA 2001 **96**(24):14043–8, reproduced with permission. Copyright (1999) National Academy of Sciences, USA.

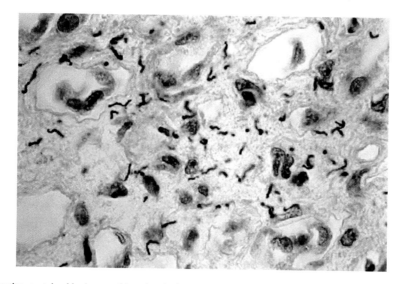

Plate 12 Leptospires in kidney tissue, stained by immunohistochemical stain (Layne, Prussia and Levett, in preparation). Reproduced from Principles and Practice of Infectious Diseases, 6th Edition, Mandell *et al.*, Figure 237–56, page 2793, copyright (2005), with permission from Elsevier.

Section One

Gram-Positive Cocci

1

β-Haemolytic Streptococci

Androulla Efstratiou[1], Shiranee Sriskandan[2], Theresa Lamagni[1] and Adrian Whatmore[3]

[1]Health Protection Agency, Centre for Infections, London; [2]Department of Infectious Diseases, Imperial College School of Medicine, London; and [3]Veterinary Laboratory Agency, Addlestone, Surrey, UK

GENERAL INTRODUCTION

Streptococcal diseases such as scarlet fever, erysipelas and puerperal fever were recognised as major problems for centuries before characterisation of the causative organisms. The first attempts to differentiate streptococci were probably made by Schottmüller in 1903, who used haemolysis to distinguish them. The β-haemolytic streptococci, characterised by the production of clear zones of haemolysis around colonies following overnight incubation on blood agar, turned out to contain some of the most ubiquitous bacterial colonisers of humans. Although there are now numerous additional approaches for differentiating streptococci, both haemolytic activity and another early approach, Lancefield typing (based on group-specific carbohydrate antigens), remain useful for differentiating them in the modern clinical microbiology laboratory. Table 1.1 lists the major β-haemolytic streptococci currently recognised. Interestingly, the accumulation of molecular data in recent years has revealed that the β-haemolytic streptococci do largely correspond to a phylogenetic grouping. All the species listed in Table 1.1 fall within the pyogenic grouping recognised by Kawamura *et al.* (1995) on the basis of 16S rRNA sequence, although this grouping also contains non-β-haemolytic streptococci. Moreover, many other streptococci, particularly *anginosus* species, could on occasion show β-haemolytic activity. However, these are not considered primarily β-haemolytic organisms and are discussed by Roderick McNab and Theresa Lamagni in Chapter 2, under nonhaemolytic streptococci.

The most important β-haemolytic streptococci are *Streptococcus pyogenes* and *Streptococcus agalactiae*. *Streptococcus pyogenes*, also known as the group A streptococcus (GAS) – considered the most pathogenic of the genus – appears to be essentially confined to humans and is associated with a wide spectrum of diseases. These range from mild, superficial and extremely common diseases such as pharyngitis (strep throat) and impetigo to severe invasive disease from which the organism has gained public notoriety as the flesh-eating bacterium and immune-mediated sequelae such as rheumatic fever (RF) and poststreptococcal glomerulonephritis. Although well adapted to asymptomatic colonisation of adults, *S. agalactiae*, or the group B streptococcus (GBS), is a major cause of neonatal sepsis and an increasingly common pathogen of the elderly and immunocompromised as well as an important cause of mastitis in cattle. *Streptococcus dysgalactiae* subsp. *equisimilis* is considered a less pathogenic species than GAS but is a relatively common cause of similar superficial disease and may also be associated with invasive disease of the elderly or immunocompromised. A further four members of the β-haemolytic streptococci are considered to be primarily non-human pathogens, although they have occasionally been associated with zoonotic human infection and will be discussed only briefly in this chapter. *Streptococcus equi* subsp. *zooepidemicus* infections are generally associated with the consumption of unpasteurised dairy products and may induce a broad disease spectrum. *Streptococcus equi* subsp. *zooepidemicus* has been associated with outbreaks of nephritis (Francis *et al.* 1993). There have been some reports (although rare) of the isolation of *Streptococcus canis* from human blood cultures (the microbiological identification of the species does remain questionable). *Streptococcus iniae* has been associated with invasive disease in fish-handlers, and *Streptococcus porcinus* has been occasionally isolated from the human genitourinary tract, although the pathogenic potential of these isolates is unclear. A further three β-haemolytic streptococcal species have not been reported as having been isolated from humans (*Streptococcus equi* subsp. *equi*, *Streptococcus didelphis* and *Streptococcus phoacae*) – *S. phoacae* and *S. didelphis* have been isolated from seals and opossums. *Streptococcus iniae* is also primarily animal pathogen, classically associated with infections amongst freshwater dolphins. The review by Facklam (2002) provides a more detailed description of these streptococci.

Table 1.1 Features of β-haemolytic streptococci

Species	Common Lancefield group(s)	Major host(s)
S. pyogenes	A	Human
S. agalactiae	B	Human, bovine
S. dysgalactiae subsp. *equisimilis*	C, G	Human, animals
S. equi subsp. *equi*	C	Animals
S. equi subsp. *zooepidemicus*	C	Animals (human[a])
S. canis	G	Dog (human[a])
S. porcinus	E, P, U and V	Swine (human[a])
S. iniae	None	Marine life (human[a])
S. phoacae	C and F	Seal
S. didelphis	None	Opossum

[a] Rare causes of human infection.

PATHOGENESIS AND VIRULENCE FACTORS

GAS Pathogenesis

The status of GAS as one of the most versatile of human pathogens is reflected in the astonishing array of putative virulence determinants associated with this organism. Lack of space and the remit of this chapter permit only a cursory discussion of the extensive and ever-expanding

Table 1.2 Putative virulence factors of group A streptococcus (GAS)

Cell-surface molecules	Major function(s)/role
M-protein family (Emm, Mrp, Enn)	Resistance to phagocytosis
	Multiple ligand-binding capacities
Fibronectin-binding proteins (multiple)	Adhesion, invasion?
C5a peptidase	Inhibits leukocyte recruitment
Hyaluronic acid capsule	Resistance to phagocytosis
	Adherence?

Excreted molecules	Probable function/role
Streptolysin O/streptolysin S	Haemolytic toxins
Streptokinase	Lysis of blood clots and tissue spread
Superantigens (multiple)	Nonspecific immune stimulation
Protein Sic	Inhibitor of complement
SpeB (cysteine protease)	Generates biologically active molecules
	Inflammation, shock and tissue damage

literature in this area. For a more encompassing view, readers are referred to the many excellent reviews covering the role of putative virulence factors identified in GAS (Cunningham 2000; Nizet 2002; Bisno, Brito and Collins 2003; Collin and Olsén 2003). Throughout the ensuing section the reader should bear in mind that the repertoire of putative virulence factors is now known to vary between strains and that even when the same factors are present there may be extensive genotypic and phenotypic diversity. It is thus increasingly apparent that GAS pathogenesis involves a complex and interacting array of molecules and that the fine details may vary between different strains. The best-studied virulence factors are listed in Table 1.2, though this table is far from exhaustive as there are numerous less-well-characterised molecules that could be involved in pathogenesis.

Cell-Associated Molecules

The M protein has long been considered the major cell-surface protein of GAS. M protein was originally defined as an antiphagocytic protein that determines the serotype of a strain and evokes serotype-specific antibody that promotes phagocytosis by neutrophils in fresh human blood. Despite extensive antigenic variation all M proteins have an α-helical coiled–coil conformation and possess an array of properties that may be important in their antiphagocytic role. The best-characterised activities are binding of factor H, a regulatory component of complement control system, and binding of fibrinogen that diminishes alternate complement pathway-mediated binding of C3b to bacterial cells, thus impeding recognition by polymorpholeukocytes (PMLs). The M-protein-encoding gene (*emm*) is now known to belong to a family of so-called M-like genes linked in a cluster that appear to have evolved by gene duplication and intergenic recombination (Whatmore and Kehoe 1994). A variety of genomic arrangements of this family has been identified, the simplest of which consists of only an *emm* gene flanked by a regulatory gene *mga* (multiple gene activator) and *scpA* (see *Virulence Gene Regulators*). However, many isolates contain an upstream gene (usually designated *mrp*) and/or a downstream gene (usually designated *enn*), and several members of the M-like gene family have been found to bind a whole array of moieties including various immunoglobulin A (IgA) and IgG subclasses, albumin, plasminogen and the factor H-like protein C4bBP. In recent years it has become clear that resistance to phagocytosis is not solely dependent on the M protein itself. For example, inactivation of *emm49* has little impact on resistance to phagocytosis, with *mrp49*, *emm49* and *enn49* all appearing to be required (Podbielski *et al*. 1996; Ji *et al*. 1998).

The capacity of some strains (notably M18) to resist phagocytosis is virtually entirely dependent on the extracellular nonantigenic hyaluronic acid capsule (Wessels and Bronze 1994). Many clinical isolates, particularly those epidemiologically associated with severe disease or RF, produce large mucoid capsules. These may act to mask streptococcal antigens or may act as a physical barrier preventing access of phagocytes to opsonic complement proteins at the bacterial cell surface, as well as being involved in adherence.

In addition to the antiphagocytic molecules described above many other cellular components have postulated roles in pathogenesis. All GAS harbour *scpA* mentioned above that encodes a C5a peptidase. This protein destroys C5a complement protein that is a chemotactic signal that initially attracts neutrophils to sites of infection. GAS have been shown to possess an array of fibronectin-binding proteins, indicating the importance of interactions with this molecule. Fibronectin is a Principal glycoprotein in plasma, body fluids and the extracellular matrix. Perhaps the best-characterised GAS fibronectin-binding protein is protein F (SfbI) that has roles in adherence and invasion (see *Adherence, Colonisation and Internalisation*). However, an ever-expanding array of other molecules has been shown to possess this activity, including proteins such as SfbII, SOF, proteinF2, M3, SDH, FBP-51, PFBP, Fba and glyceraldehyde-3-phosphate dehydrogenase.

With the recent completion of genome sequencing projects many novel cell-surface proteins have been identified. Most remain to be fully characterised, but their surface location facilitates potential roles in host interaction. Examples include Slr, a leucine-rich repeat protein, isogenic mutants of which are less virulent in an intraperitoneal mouse model and more susceptible to phagocytosis by human PMLs (Reid *et al*. 2003). Two novel collagen-like proteins, SclA and SclB, have also been identified. Their function is unclear, but they may be important in adherence and are of potential interest in the pathogenesis of autoimmune sequelae (Lukomski *et al*. 2000; Rasmussen, Eden and Bjorck 2000; Whatmore 2001). A further potential virulence factor identified from the genome sequence is protein GRAB, present in most GAS and possessing high-affinity binding capacity for the dominant proteinase inhibitor of human plasma α2-macroglobulin. Again, mutants are attenuated in mice and it has been suggested that α2-macroglobulin bound to the bacterial surface via protein GRAB entraps and inhibits the activity of GAS and host proteinases, thereby protecting other virulence determinants from proteolytic degradation (Rasmussen, Muller and Bjorck 1999).

Excreted Products

Most GAS produce two distinct haemolysins, streptolysin O (SLO) and streptolysin S (SLS). SLO is an oxygen-labile member of the cholesterol-binding thiol-activated toxin family that elicits antibodies useful for documenting recent exposure and is toxic to a variety of cells. SLS belongs to the bacteriocin family of microbial toxins and is not immunogenic in natural infection but shares similar toxic activities to SLO. Streptokinase facilitates the spread of organisms by promoting the lysis of blood clots. Streptokinase binds to mammalian plasminogen, and this complex then converts other plasminogen molecules to the serum protease plasmin that subsequently acts on a variety of proteins including fibrin. GAS have cell-surface receptors capable of binding plasminogen, which following conversion to plasmin in the presence of streptokinase may generate cell-associated proteolytic enzyme capable of causing tissue destruction.

GAS produce an array of superantigens. These protein exotoxins have the ability to trigger excessive and aberrant activation of T cells by bypassing conventional major histocompatibility complex (MHC)-restricted antigen processing (Llewelyn and Cohen 2002). Subsequent excessive and uncoordinated release of inflammatory cytokines such as tumour necrosis factor-α (TNF-α), interleukin-2 (IL-2) and interferon-γ (IFN-γ) results in many of the symptoms consistent with toxic shock syndrome. Over a dozen distinct superantigens have now been recognised in GAS, and many appear to be located on mobile genetic elements. Different isolates possess distinct arrays of superantigens with subtly different specificities and activities. SpeB was initially thought to be a superantigen, but it is now thought that its disease contribution is related to protease activity. SpeB is a chromosomally encoded cysteine proteinase that acts on many targets including various Ig classes and generates biologically active molecules by activities that include cleaving IL-1β precursor to an active form and releasing bradykinin from a precursor form. These activities generate reactive molecules with roles in inflammation, shock and tissue destruction.

A further proposed secreted virulence factor is the streptococcal inhibitor of complement (Sic) protein found predominantly in M1 strains which binds to components of the complement membrane attack complex (C5b-C9) and thus inhibits complement-mediated lysis *in vitro*. The *in vivo* function of Sic is not clear, but it appears to be rapidly internalised by human epithelial cells and PMLs where specific interactions result in the paralysis of the actin cytoskeleton and significantly decreased opsonophagocytosis and killing of GAS. Sic is one of the most variable bacterial proteins known, and variants arise rapidly *in vivo* (Matsumoto *et al.* 2003), suggesting an important role for this molecule in M1 isolates.

Adherence, Colonisation and Internalisation

The GAS infectious process entails several steps proceeding from initial adherence to mucosal cells to subsequent invasion of deeper tissue, leading to occasional penetration of the bloodstream. The strategies by which GAS adhere and invade are multiple and complex and vary between strains and between host cell types, with the potential involvement of multiple adhesins and invasins. An astonishing array of putative adhesins has been described (reviewed by Courtney, Hasty and Dale 2002), although only the roles of lipoteichoic acid (LTA), M protein and some fibronectin-binding proteins have been studied in any great detail.

LTA adheres to fibronectin on buccal epithelial cells, an interaction that is blocked by excess LTA or anti-LTA antibody, and this serves as a first step in adhesion with secondary adhesins, thus facilitating stronger and more specific binding. Many studies have implicated M protein in adhesion, although observations vary greatly from strain to strain and depend on the *in vitro* model and tissue type studied. However, there is evidence that M protein mediates adherence to skin keratinocytes via CD46. There is substantial evidence that fibronectin-binding proteins such as SfbI and related proteins are important in adhesion. SfbI mediates adherence to respiratory epithelial cells and cutaneous Langerhans cells. Expression is environmentally regulated, being enhanced in oxygen-rich atmospheres, in contrast to that of M protein.

Following adhesion GAS must maintain itself on the pharynx. The capsule may be important at this stage as encapsulation facilitates more persistent colonisation and leads to higher mortality in animal models than is seen in isogenic acapsular mutants. Although not generally considered intracellular pathogens, GAS have been shown to penetrate and, in some cases, multiply within a variety of epithelial cell lines *in vitro*. Both M protein and SfbI are implicated in internalisation because of their fibronectin-binding capacities that bridge GAS to integrin receptors, promoting actin rearrangement by independent mechanisms and leading to invasion. The biological significance of these interactions is unclear, but they may facilitate deep-tissue invasion or be important in the persistence of GAS in the face of host defences or therapy.

Virulence Gene Regulators

In recent years it has become clear that complex regulatory circuits control virulence factor expression in GAS, allowing it to respond to environmental signals and adapt to various niches (reviewed by Kreikemeyer, McIver and Podbielski 2003). The multiple gene regulator Mga is located upstream of the *emm* gene cassette and is thought to form part of a classic two-component regulatory signal transduction system, though the sensor remains unidentified. It controls transcription of itself and a variety of other genes including those encoding M and M-like proteins, C5a peptidase, SclA and Sic. Mga positively regulates itself and binds to a consensus sequence upstream of the gene it regulates, increasing expression in response to increased carbon dioxide concentrations, increased temperature and iron limitation. A further two-component system known as CsrRS (capsule synthesis regulator) or CovRS (control of virulence genes) has also been identified. This system represses the synthesis of capsule and several other virulence factors including streptokinase, SpeB and SLO and the CovRS operon itself. Recent microarray studies have indicated that this system may influence transcription directly or indirectly of as many as 15% of GAS genes (Graham *et al.* 2002). A further regulator, RofA, was first identified in an M6 strain as a positive regulator of SfbI in response to reduced oxygen, but it also negatively regulates SLS, SpeB and Mga. Nra, identified in an M49 strain shares 62% identity with RofA and shares many of its activities. A global regulator Rgg or RopB homologous to transcriptional regulators in other Gram-positive organisms has been identified and appears to affect the transcription of multiple virulence genes genome-wide through modulation of existing regulatory networks (Chaussee *et al.* 2003). Studies carried out to date have highlighted the complexity and interdependence of GAS regulatory circuits. The understanding of these is clearly in its infancy, and it will be crucial for a fuller understanding of GAS pathogenesis.

Host Immune Response to Infection

Protective immunity against GAS correlates with the presence of opsonising antibody against type-specific M protein. However, the paucity of infection in adults relative to children suggests that other mechanisms help to protect against infection. Secretory IgA against nonserotype-specific regions of M protein plays a role in host protection by preventing adhesion to mucosal surfaces. In addition, it is likely that immune responses to other streptococcal molecules have roles in protection. Thus, for example, a novel antigen generating opsonising protective antibody (Dale *et al.* 1999) has been reported in an M18 isolate, and surface molecules such as the C5a peptidase, SOF and the group A carbohydrate induce protective immune responses. Furthermore, antibodies against the GAS superantigens may be important in neutralising the toxic activity of these molecules.

Pathogenesis of Sequelae

Although a close link between GAS and rheumatic fever (RF) has been established for many years, the exact mechanism by which GAS evokes RF remains elusive. Molecular mimicry is assumed to be the reason for RF, and antigenic similarity between various GAS components, notably M protein, and components of human tissues such as heart, synovium, and the basal ganglia of brain could theoretically account for most manifestations of RF (Ayoub, Kotb and Cunningham 2000). Only certain strains of GAS appear capable of initiating the immune-mediated inflammatory reaction, related either to cross-reacting antibodies generated against streptococcal components or to the stimulation of cell-mediated immunity that leads to disease in susceptible hosts.

Like RF, acute poststreptococcal glomerulonephritis (APSGN) is associated with only some GAS strains and certain susceptible hosts. Again, the precise causative mechanisms are not clear. Renal injury associated with APSGN appears to be immunologically mediated, and antigenic similarities between human kidney and various streptococcal constituents (particularly M protein and fragments of the streptococcal cell membrane) that could generate cross-reactive antibodies have been investigated. However, other nonmutually exclusive mechanisms such as immune complex deposition, interactions of streptococcal constituents such as SpeB and streptokinase with glomerular tissues and direct complement activation following the deposition of streptococcal antigens in glomeruli have also been proposed.

GBS Pathogenesis

Despite GBS being the predominant cause of invasive bacterial disease in the neonatal period, little is known about the molecular events that lead to invasive disease. The capsule is considered the most important virulence factor, and most isolates from invasive disease are encapsulated. The capsular polysaccharide inhibits the binding of the activated complement factor C3b to the bacterial surface, preventing the activation of the alternative complement pathway (Marques et al. 1992). GBS also produce a haemolysin (cytolysin) (Nizet 2002), and a surface protein, the C protein, has been extensively characterised. C protein is a complex of independently expressed antigens α and β, both of which elicit protective immunity in animal models. The function of the C protein remains unclear, although the β component binds nonspecifically to IgA and is presumed to interfere with opsonophagocytosis. Like GAS, S. agalactiae also harbours a C5a peptidase that interferes with leukocyte recruitment to sites of infection and hyaluronidase, protease and nuclease activities, though the roles, if any, of these molecules in pathogenesis are unclear. Recently completed GBS genome sequences have revealed an array of uncharacterised cell-wall-linked proteins and lipoproteins, many of which may be important in host interaction and pathogenesis.

Pathogenesis of Other β-Haemolytic Streptococci

Infection of humans with S. dysgalactiae subsp. equisimilis [human group G streptococcus (GGS)/group C streptococcus (GCS)] causes a spectrum of disease similar to that resulting from GAS infection. These organisms have been found to share many virulence determinants such as streptokinase, SLO, SLS, several superantigens, C5a peptidase, fibronectin-binding proteins, M proteins and hyaluronidase. An emerging theme in the field is the occurrence of horizontal gene transfer from human GGS/GCS to GAS and vice versa. The presence of GGS/GCS DNA fragments in GAS in genes such as those encoding hyaluronidase, streptokinase and M protein was observed some years ago, and it has recently become clear that many superantigen genes are also found in subsets of GGS/GCS (Sachse et al. 2002; Kalia and Bessen 2003). It is likely that this horizontal gene transfer plays a crucial role in the evolution and biology of these organisms with the potential to modify pathogenicity and serve as a source of antigenic variation.

Most work with the two S. equi subspecies has concerned itself with S. equi subsp. equi, resulting in the characterisation of many virulence factors equivalent to those seen in other β-haemolytic streptococci. Characterisation of S. equi subsp. zooepidemicus remains incomplete; however, the organisms are known to possess a fibronectin-binding protein and an M-like protein (SzP). In contrast to S. equi subsp. equi, the subspecies zooepidemicus appears to lack superantigens (Harrington, Sutcliffe and Chanter 2002). Indeed, this has been postulated as an explanation for the difference in disease severity observed between the subspecies.

Reports of streptococcal toxic shock-like syndrome in dogs have led to searches for GAS virulence gene equivalents in S. canis (DeWinter, Low and Prescott 1999), with little success to date. The factors involved in the pathogenesis of disease caused by S. canis consequently remain poorly understood.

Genomes, Genomics and Proteomics of β-Haemolytic Streptococci

At the time of writing, the genomic sequences of four GAS strains (one M1 isolate, two M3 isolates and one M18 isolate) and two GBS strains (serotype III and V isolates) are completed and publicly available, with genomic sequencing of others likely to be completed shortly.

The first GAS sequence completed (M1) confirmed that the versatility of this pathogen is reflected in the presence of a huge array of putative virulence factors in GAS (>40) and identified the presence of four different bacteriophage genomes (Ferretti et al. 2001). These prophage genomes encode at least six potential virulence factors and emphasise the importance of bacteriophage in horizontal gene transfer and as a possible mechanism for generating new strains with increased pathogenic potential. Subsequent sequencing of other genomes confirmed that phage, phage-like elements and insertion sequences are the major sources of variation between genomes (Beres et al. 2002; Smoot et al. 2002; Nakagawa et al. 2003) and that the creation of distinct arrays of potential virulence factors by phage-mediated recombination events contributes to localised bursts of disease caused by strains marked by particular M types. A potential example of this is the increase in severe invasive disease caused by M3 isolates in recent years. Integration of the M3 genome sequence data with existing observations provided an insight into this – contemporary isolates of M3 express a particularly mitogenic superantigen variant (SpeA3) that is 50% more mitogenic in vitro than SpeA1 made by isolates recovered before the 1980s. They also harbour a phage encoding the superantigen SpeK and the extracellular phospholipase Sla not present in M3 isolates before 1987 and many also have the superantigen Ssa, again, not present in older isolates (Beres et al. 2002). The two S. agalactiae genomes also reveal the presence of potential virulence factors associated with mobile elements including bacteriophage, transposons and insertion sequences, again suggesting that horizontal gene transfer may play a crucial role in the emergence of hypervirulent clones (Glaser et al. 2002; Tettelin et al. 2002).

The use of comparative genomics, proteomics and microarray-based technologies promises to add substantially to current understanding of the β-haemolytic streptococci and provide means of rapidly identifying novel bacterial proteins that may participate in host–pathogen interactions or serve as therapeutic targets. For example, as described already, many new surface proteins have been identified from genome sequences (Reid et al. 2002). In addition, microarray technology is being used in GAS comparative genomic studies (Smoot et al. 2002) and in the analysis of differential gene expression (Smoot et al. 2001; Graham et al. 2002; Voyich et al. 2003), and proteomic approaches have been used to identify major outer surface proteins of GBS (Hughes et al. 2002).

EPIDEMIOLOGY OF β-HAEMOLYTIC STREPTOCOCCUS INFECTION

Our understanding of the epidemiology of β-haemolytic streptococcal infections and their related diseases is relatively poor compared with that of many other infectious diseases. Many countries with established infectious disease surveillance programmes undertake relatively little surveillance of diseases caused by these streptococci. However, most countries are expanding or modifying their surveillance programmes to include β-haemolytic streptococcal diseases, not least in the light of recent worrying trends in incidence.

To fully understand the epidemiology of these diseases in terms of how these organisms spread, host and strain characteristics of importance to onward transmission, disease severity and inter- and intraspecies competition for ecological niches, one would need to undertake the most comprehensive of investigations following a large cohort for a substantial period of time. Understanding these factors would allow us to develop effective prevention strategies. An important such measure, discussed in the section Vaccines for β-haemolytic streptococcal disease, is the introduction of a multivalent vaccine. Although existing M-typing data allow us to predict what proportion of disease according to current serotype distribution could be prevented, the possibility of serotype replacement occurring is something which dramatically limits the impact of such a measure (Lipsitch 1999), a phenomenon witnessed for pneumococcal infection and suggested in at least one GAS carriage study (Kaplan, Wotton and Johnson 2001).

The Burden of Disease Caused by β-Haemolytic Streptococci

Superficial Infections

For the reasons specified above the bulk of incidence monitoring is focused at the severe end of streptococcal disease. Our understanding of the epidemiology of less severe but vastly common superficial infections is severely limited despite the substantial burden represented by these diseases, especially the ubiquitous streptococcal pharyngitis. Such diseases represent a significant burden on healthcare provision and also provide a constant reservoir for deeper-seated infections. Pharyngitis is one of the most common reasons for patients to consult their family practitioner. Acute tonsillitis and pharyngitis account for over 800 consultations per 10 000 patients annually, in addition to the economic impact of days missed from school or work (Royal College of General Practitioners 1999).

Since the end of the nineteenth century, clinicians in the United Kingdom have been required by law to notify local public health officials of the incidence of scarlet fever, among other conditions. Reports of scarlet fever plummeted dramatically since the mid-1900s, with between 50 000 and 130 000 reports per year being typically reported until the 1940s. Annual reports dropped dramatically, with around 2000 cases still reported each year (HPA 2004a).

Severe Disease Caused by β-Haemolytic Streptococci

Data from the United Kingdom indicate streptococci to be the fourth most common cause of septicaemia caused by bacterial or fungal pathogens, 15% of monomicrobial and 12% of polymicrobial positive blood cultures isolating a member of the *Streptococcus* genus (HPA 2003a). β-Haemolytic streptococci themselves accounted for 5% of blood culture isolations (HPA 2004b), a similar estimate to that from the United States (Diekema, Pfaller and Jones 2002), with rates of infection being highest for GBS bacteraemia at 1.8 per 100 000 population, substantially lower, however, than estimates in the United States of around 7.0 per 100 000 (Centers for Disease Control and Prevention 2003), although this includes other sterile-site isolations. Recent estimates of early-onset GBS disease incidence in countries without national screening programmes range from 0.48 in the United Kingdom to 1.95 in South Africa (Madhi *et al.* 2003; Heath *et al.* 2004), with current estimates in the United States of around 0.4 per 1000 live births (Centers for Disease Control and Prevention 2003).

In contrast, invasive GAS disease rates in the United States and United Kingdom are more similar, around 3.8 per 100 000 in 2003 (Centers for Disease Control and Prevention 2004a; HPA 2004c), the UK data being derived from a pan-European enhanced surveillance programme (Strep-EURO). Current estimates from Netherlands are also reasonably similar at 3.1 per 100 000 in 2003 (Vlaminckx *et al.* 2004).

Less widely available than for GAS and GBS, data for GGS from the United Kingdom suggest an incidence of GGS bacteraemia of just over 1 case per 100 000 population, with GCS being reported in 0.4 per 100 000.

Global Trends in Severe Diseases Caused by β-Haemolytic Streptococci

Since the mid-1980s, a proliferation of global reports has suggested an upturn in the incidence of severe diseases caused by β-haemolytic streptococci. Many possible explanations could account for this rise, including a proliferation of strains with additional virulence properties or a decrease in the proportion of the population with immunological

protection. A further factor fuelling the rise in opportunistic infections generally is the enlargement of the pool of susceptible individuals resulting from improved medical treatments for many life-threatening conditions (Cohen 2000). This has been borne out in the rises in incidence seen for many nonvaccine-preventable diseases caused by a range of bacterial and fungal pathogens.

Trends in Invasive GAS Disease

Many reports published during the 1980s and 1990s suggested resurgence of invasive diseases caused by GAS. The United States was among the first to document this upturn, although some estimates were based on restricted populations. Subsequent findings from multistate sentinel surveillance conducted in the second half of the 1990s failed to show any clear trends in incidence (O'Brien *et al.* 2002). Surveillance activities in Canada during the early 1990s identified such an increase, although rates of disease were relatively low (Kaul *et al.* 1997). Norway was among the first European countries to report an increase in severe GAS infections during the late 1980s (Martin and Høiby 1990). Several other European countries subsequently stepped up surveillance activities and began to show similar trends. Among these was Sweden, where the incidence of invasive GAS increased at the end of the 1980s and again during the mid-1990s (Svensson *et al.* 2000). The United Kingdom also saw rises of GAS bacteraemia, from just over 1 per 100 000 in the early 1990s to a figure approaching 2 per 100 000 in 2002 (Figure 1.1). Other European countries reported more equivocal changes – surveillance of bacteraemia in Netherlands showing a fall in invasive GAS disease from the late 1990s to 2003 (Vlaminckx *et al.* 2004). National incidence estimates from Denmark have shown an unclear pattern, although recent upturns have been reported (Statens Serum Institut 2002).

Trends in Invasive GBS Disease

The introduction of neonatal screening programmes has had a substantial impact on the incidence of neonatal GBS disease in the United States (Schrag *et al.* 2000, 2002). Centers for Disease Control and Prevention's (CDC) Active Bacterial Core (ABC) programme showed a dramatic fall in the incidence of early-onset disease between 1993 and 1997 as increasing numbers of hospitals implemented antimicrobial prophylaxis, with little change in late-onset disease (Schuchat 1999).

Neonatal disease aside, trends in overall GBS infection point to a general rise in incidence, including in the United States (Centers for Disease Control and Prevention 1999, 2004b). Surveillance of bacteraemia in England and Wales shows a 50% rise in reporting rates from the early 1990s to the early 2000s (Figure 1.1).

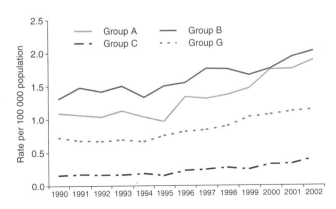

Figure 1.1 Annual rate of reports of bacteraemia caused by β-haemolytic streptococci – England and Wales. Source: Health Protection Agency.

Trends in Invasive GGS and GCS Disease

Few countries undertake national surveillance of invasive GGS or GCS disease. Population-based surveillance from England and Wales has shown some interesting trends in GCS in particular, which has shown a steeper rise in rates of reports than any of the other β-haemolytic streptococci, from 0.16 per 100 000 in 1990 to 0.42 per 100 000 in 2002 (Figure 1.1). Rates of GGS bacteraemia have showed a similar magnitude of rise to that for GAS, from 0.73 to 1.15 per 100 000 over the same time period.

CLINICAL ASPECTS OF β-HAEMOLYTIC STREPTOCOCCAL INFECTION

Clinical Presentation of Disease

Superficial Infections

All β-haemolytic streptococci cause colonisation and superficial infection of epithelial and mucosal surfaces. The clinical scenarios are summarised in Table 1.3 and, for the purposes of this description, include all conditions affecting the upper respiratory tract. These infections are only rarely associated with bacteraemia. The occurrence of streptococcal bacteraemia in a patient with streptococcal pharyngitis should prompt a search for additional, more deep-seated foci of infection.

Streptococcus pyogenes accounts for 15–30% of pharyngitis cases (Gwaltney and Bisno 2000). The classical presentation is of

Table 1.3 Superficial infections associated with β-haemolytic streptococci

	Group A	Group B	Group C	Group G
Pharyngitis	+++	–	+	(+)
Otitis externa	+	–	–	–
Otitis media	(+)	–	–	–
Laryngitis	+	–	(+)	–
Vulvovaginitis/perianal	+++	+[a]	–	–
Impetigo	+++	(+)	(+)[b]	+[b]

+++, common identified bacterial cause; +, recognised but uncommon pathogen; (+), rare or poorly described; –, not known to be associated.
[a] Association with diabetes mellitus.
[b] Reported in tropical climates.

bilateral acute purulent tonsillitis with painful cervical lymphadenopathy and fever. With prompt antibiotic treatment the illness normally lasts less than 5 days. Outbreaks have occurred in institutionalised settings such as military recruitment centres. Illness is common in the young, and the incidence declines dramatically with age. In children *S. pyogenes* also is a recognised cause of perianal disease and vulvovaginitis, presenting as pruritus with dysuria and discharge in the latter case (Mogielnicki, Schwartzman and Elliott 2000; Stricker, Navratil and Sennhauser 2003). Most sufferers have throat carriage of *S. pyogenes* or might have had recent contact with throat carriers. Outbreaks of both pharyngitis and perineal disease are common in the winter months in temperate regions.

Impetigo is the other most common superficial manifestation of β-haemolytic streptococcal infection. The condition presents with small-to-medium-sized reddish papules, which when de-roofed reveal a crusting yellow/golden discharge. Often complicated by coexisting *Streptococcus aureus* infection, the condition is contagious and spreads rapidly amongst children living in proximity to each other. Although group C streptococci (GCS) and G group streptococci (GGS) are recognised causes of impetigo in tropical climes, this would be unusual in temperate areas (Belcher *et al.* 1977; Lawrence *et al.* 1979).

Complications of Superficial Disease

In addition to the direct effects of clinical infection several major immunological syndromes are associated with or follow superficial streptococcal infection, such as scarlet fever, RF, poststreptococcal reactive arthritis, poststreptococcal glomerulonephritis and guttate psoriasis. It is extremely rare for true superficial streptococcal disease to be associated with streptococcal toxic shock syndrome (STSS). Most cases where pharyngitis is diagnosed in association with STSS often also have bacteraemia or additional foci of infection.

Invasive GAS Infections (Figure 1.2)

Invasive GAS disease has been the subject of several studies since the late 1980s and early 1990s because of its apparent reemergence during this period. Although the common manifestation of invasive GAS disease is cellulitis and other types of skin and soft-tissue infections (almost 50% of invasive disease cases), it is notoriously difficult to diagnose microbiologically because bacterial isolates are seldom recovered from infected tissue. Most invasive infections are confirmed because of bacteraemia.

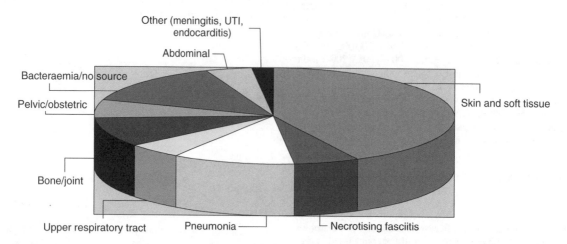

Figure 1.2 Clinical syndromes associated with invasive GAS disease. Adapted from data in Kristensen and Schonheyder (1995), Davies *et al.* (1996) and Zurawski *et al.* (1998).

Cellulitis, which spreads along the long axis of a limb, rather than in a centrifugal pattern, is more likely to be streptococcal than staphylococcal. Furthermore, cellulitis occurring in a butterfly distribution on the face (erysipelas) is almost always due to the GAS. Occasionally, cracks between digits, or behind the ears, reveal likely portals of entry. Large-colony GCS and GGS, as well as GBS, cause cellulitis, although the incidence of premorbid illness or advanced age is more common in these latter groups.

In some circumstances bacteria spread between the fascial planes, along fibrous and fatty connective tissues, which separate muscle bundles. This results in necrotising fasciitis (Plate 1), where connective tissues become inflamed and rapidly necrotic. In several recent series necrotising fasciitis has accounted for approximately 6% of all invasive GAS cases (Davies *et al.* 1996; Ben-Abraham *et al.* 2002). Streptococcal necrotising fasciitis often arises because of bacteraemic seeding into fascia underlying otherwise quite normal-looking skin. In some cases there is history of prior blunt trauma to an infected region. In other cases there is an obvious portal of entry, such as recent surgery or varicella. It is useful to distinguish streptococcal necrotising fasciitis from synergistic or mixed-infection necrotising infections. Streptococcal necrotising fasciitis often occurs as a pure infection and may arise in previously healthy individuals, often affecting the limbs, resulting in extreme pain in a febrile or septic patient. In contrast, synergistic necrotising fasciitis tends to follow surgery or debilitation and occurs in areas where enteric and anaerobic bacteria have opportunity to mix with normal skin flora, such as abdominal and groin wounds.

Although less common than skin and soft-tissue infection, puerperal sepsis with endometritis and pneumonia due to *S. pyogenes* still occur and should not go unrecognised (Zurawski *et al.* 1998; Drummond *et al.* 2000). Epiglottitis because of GAS infection is also well recognised in adults (Trollfors *et al.* 1998). Importantly, a significant proportion of invasive GAS disease occurs as isolated bacteraemia in a patient without other apparent focus. This does not obviate the need for careful inspection of the whole patient in an effort to rule out areas of tissue necrosis.

Predisposition to Invasive GAS Disease

Although capable of occurring during any stage of life, invasive GAS disease is more common in early (0–4 years) and later life (>65 years). As for the other β-haemolytic streptococci, incidence in males generally exceeds that in females (HPA 2003b).

Many underlying medical conditions have been associated with increased risk of invasive GAS disease (Table 1.4). Some, such as varicella, provide obvious portals of entry for the bacterium, whilst

Table 1.4 Risk factors for acquiring invasive group A streptococcus (GAS) disease

Predisposing skin conditions
Surgery
Intravenous drug use
Varicella
Insect bites

Predisposing medical conditions
Cardiovascular disease (16–47%)
Diabetes (6–19%)
Alcoholism (10–15%)
Malignancy (8–10%)
Lung disease (4–9%)
HIV infection (2–5%)
Recent childbirth
None (30%)

Data adapted from Kristensen and Schonheyder (1995), Davies *et al.* (1996) and Eriksson *et al.* (1998).

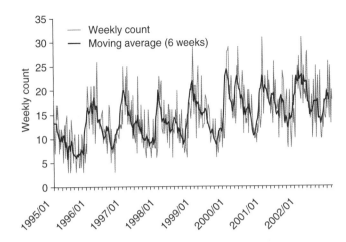

Figure 1.3 Weekly distribution of *Streptococcus pyogenes* bacteraemia – England, Wales, Northern Ireland and Channel Islands, 1995–2003. Source: Health Protection Agency.

others appear to result in more subtle immunoparesis. Injecting-drug users are now recognised as a major risk group for invasive GAS disease, accounting for up to a fifth of cases in some countries (Factor *et al.* 2003; Lamagni *et al.* 2004). Most infections are thought to be sporadic community-acquired infections, around 5% resulting from nosocomial infection. However, importantly, in approximately 30% of invasive disease cases there is no underlying medical or other predisposing condition. This distinguishes the GAS from other β-haemolytic streptococci.

Invasive GAS infections exhibit a distinct seasonal pattern, with striking late winter/early spring peaks (Figure 1.3). How much of this can be attributed to winter vulnerability to GAS pharyngitis and carriage and how much is likely to be the result of postviral infection susceptibility to bacterial infections is unclear.

Complications of Invasive GAS Disease

The mortality from invasive GAS disease is high (14–27%) and rises considerably with patient age and number of organ system failures, reaching 80% in the presence of STSS (Davies *et al.* 1996; Efstratiou 2000). STSS complicates roughly 10–15% of all cases of invasive GAS disease, accounting for most patients with invasive GAS disease who require admission to an intensive care facility (Eriksson *et al.* 1998).

Streptococcal Toxic Shock Syndrome

In the late 1980s it was recognised that invasive GAS disease could be associated with a syndrome of multiorgan system failure coupled with erythematous blanching rash. The similarity to staphylococcal toxic shock syndrome (described a decade earlier) was striking, particularly as both conditions were associated with a rash that subsequently desquamates and as both conditions were associated with bacteria that produced superantigenic exotoxins. Streptococcal pyrogenic exotoxin A (SPEA) was particularly implicated in GAS infections (Cone *et al.* 1987; Stevens *et al.* 1989).

Scarlet fever: S. pyogenes produces many phage-encoded classical superantigens, in particular, SPEA and SPEC. Intriguingly, isolates producing these toxins historically were linked to the development of scarlet fever, a disease accompanying streptococcal pharyngitis in children that was characterised by fever and widespread blanching erythema. A similar, more feared syndrome known as septic scarlet fever could accompany invasive streptococcal infection, classically

Table 1.5 Clinical syndrome of STSS (from Working Group on Severe Streptococcal Infections 1993)

Symptoms
Renal impairment (raised creatinine)
Coagulopathy or low platelets
Liver dysfunction (raised transaminases)
Respiratory failure/hypoxaemia
Erythematous rash (which desquamates)
Soft-tissue necrosis

Identification of a GAS from a normally sterile site
(definite case)
or
Identification of a GAS from a nonsterile site (probable case)
and
Systolic blood pressure <90 mmHg
plus
At least two of the above symptoms

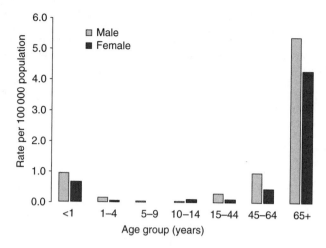

Figure 1.4 Age-specific rates of GGS bacteraemia reports – England, Wales and Northern Ireland, 2002. Source: Health Protection Agency.

puerperal fever, and often resulted in death from multiorgan failure (Katz and Morens 1992). Although the incidence of scarlet fever regressed markedly during the twentieth century, it seems likely that STSS represents the reemergence (or re-recognition) of septic scarlet fever.

Modern STSS: In 1990 a consensus definition for the diagnosis of STSS was established (Working Group on Severe Streptococcal Infections 1993) (Table 1.5). Epidemiological studies suggested that 10–15% of all invasive disease cases were complicated by STSS. The development of STSS became an important marker of subsequent morbidity and mortality, and many therapeutic approaches were developed specifically for the management of STSS cases. Although a disproportionately large number of STSS isolates are found to be positive for the *speA* gene, this may be caused by a preponderance of certain clones sharing common M serotypes (Cleary *et al.* 1992; Musser *et al.* 1993). Cases of STSS caused by *speA* and *speC* strains are recognised (Hsueh *et al.* 1998). Recent advances in bacterial genomics have led to the identification of many novel *S. pyogenes* superantigen genes, several of which appear to be chromosomal rather than phage derived (Proft and Fraser 2003). It seems plausible that many cases of STSS are attributable to superantigens other than SPEA or SPEC.

Invasive Disease due to GCS and GGS

Recognition of invasive group C and G β-haemolytic streptococcal disease is rising, though dependent largely on the identification of bacteraemic cases (Kristensen and Schonheyder 1995; Hindsholm and Schonheyder 2002; Sylvetsky *et al.* 2002). Unlike GAS disease, diseases due to GCS and GGS usually have comorbidity from other medical conditions, with cases being common in those over 75 years of age and in males (Figure 1.4).

The reported mortality from invasive GCS or GGS bacteraemia is 15–23%, slightly lower than that reported for GAS disease. There have been sporadic reports of toxic shock-like syndromes associated with group C and G infections, and very recently a novel set of superantigen toxin genes has been identified in GCS (Wagner *et al.* 1996; Hirose *et al.* 1997; Proft and Fraser 2003).

Invasive Disease due to GBS

Although predominantly feared as a neonatal disease, GBS can also cause invasive infection in adults.

Neonatal GBS Disease

A full discussion of the neonatal presentation of GBS disease is beyond the scope of this chapter. Between 65 and 80% of cases are early onset and arise from colonisation during or at the time of birth (Kalliola *et al.* 1999; Madhi *et al.* 2003; Heath *et al.* 2004). Early-onset GBS disease is defined as that occurring within the first 7 days of life, although most cases are diagnosed within the first 48 hours of life. Because of this several countries have instituted intrapartum antibiotic prophylaxis for women known to be colonised with GBS or those assessed to be at high risk. Though still under debate in many countries, the implementation of intrapartum antibiotic prophylaxis in the United States was followed by a 65% reduction in early-onset neonatal GBS infection, which is otherwise associated with a mortality rate of 28% (Schrag *et al.* 2000). Most infections present as bacteraemia, without identified focus of infection, though meningitis and pneumonia are also frequent (Kalliola *et al.* 1999). Late-onset disease (7–90 days) is thought to occur because of postnatal transmission of GBS, potentially from within hospital sources.

Many maternal and pregnancy-related factors have been associated with increased risk of neonatal GBS infection, including preterm birth and prolonged rupture of membranes (Heath *et al.* 2004). Black ethnicity is also associated with increased risk of GBS infection, as is multiple births, although it is unclear whether these are acting as risks independent of each other.

Adult GBS Disease

GBS also cause adult invasive disease, though mainly in those with predisposing factors, in particular diabetes (11–49%), cancer, alcoholism, HIV infection, the bedridden/elderly or during pregnancy. The most common sources of bacteraemia in nonpregnant adults are pneumonia, soft-tissue infections (including wound infections) and urinary tract infection, although bacteraemia without identified source is also common (Farley *et al.* 1993; Trivalle *et al.* 1998; Larppanichpoonphol and Watanakunakorn 2001). Amongst pregnant women the risk of invasive GBS bacteraemia is also increased – pregnant women account for approximately 20% of all cases of GBS bacteraemia amongst adults (Farley *et al.* 1993). The introduction of intrapartum antibiotic prophylaxis has reduced the incidence of pregnancy-related invasive GBS disease by 21% in some parts of the United States (Schrag *et al.* 2000). Aside from diseases in women of childbearing age, adult and childhood infection is more common in males than females (HPA 2003b).

Immunologically Mediated Disease: Nonsuppurative Sequelae of β-Haemolytic Streptococcal Disease

In addition to toxin-mediated complications, such as scarlet fever and STSS, GAS infection can also be complicated by defined postinfectious sequelae. Although the aetiology of these heterogeneous conditions is poorly understood, it is likely that some, if not all, result from cross-reactive T- and B-cell epitopes shared by streptococcal cell wall and human matrix protein antigens. It is also possible that bacterial superantigens drive some of these conditions in susceptible individuals.

A full description of all these conditions is beyond the scope of this chapter. The references in Table 1.6 provide good overviews of each subject.

Rheumatic Fever

RF affects 1–4% of children with untreated streptococcal pharyngitis and is a major cause worldwide of acquired cardiovascular disease, with an incidence of 100–200 per 100 000 children in countries where RF is endemic (Guilherme *et al*. 2000; Olivier 2000). Although the incidence of RF in developed countries is usually low (1–2 per 100 000), the disease occurs with marked frequency amongst the economically deprived native communities of New Zealand and Australia. RF can also cause outbreaks in affluent communities: an eightfold increase in RF incidence occurred across Utah and other states within the United States in the late 1980s, leading to carditis in over 90% of cases (Veasy *et al*. 1987; Bisno 1991). Rheumatic heart disease follows acute RF in 30–50% of cases (Majeed *et al*. 1992; Guilherme *et al*. 2000). The so-called rheumatogenic *S. pyogenes* serotypes have been associated with outbreaks of RF (M1 and M18 in the United States and M1, M3 and M5 in the United Kingdom), although RF can be associated with a broad range of strains. Epidemiological study is hampered because isolates are rarely available for typing (Veasy *et al*. 1987; Colman *et al*. 1993). Study of the pathogenesis of RF has focused on the role of streptococcal M protein. Cases with arthritis not meeting the acknowledged criteria for RF are often designated as poststreptococcal reactive arthritis. Here, the risk of cardiac involvement appears low (Shulman and Ayoub 2002). The diagnosis of RF requires clinical recognition of a collection of signs coupled with laboratory evidence of recent streptococcal infection (Table 1.7). Although numerous cases suggest that acute RF follows GAS infection of the skin, best-documented cases and outbreaks have followed episodes of pharyngitis.

Table 1.6 Spectrum of poststreptococcal immunologically mediated disease

Recognised poststreptococcal sequelae
Rheumatic fever
Sydenham's chorea
Poststreptococcal reactive arthritis
Glomerulonephritis
Guttate psoriasis
Proposed poststreptococcal sequelae
PANDAS
Kawasaki disease

PANDAS, paediatric autoimmune disorders associated with streptococcal infections, overlap with chorea and tic disorders. For reviews on Sydenham's chorea and PANDAS, see Snider and Swedo (2003), Simckes and Spitzer (1995) and Shulman and Ayoub (2002). For reviews on guttate psoriasis, see Valdimarsson *et al*. (1995). For reviews on Kawasaki disease, see Curtis *et al*. (1995).

Table 1.7 Revised Jones Criteria for diagnosing acute rheumatic fever (Adapted from American Heart Association 1992)

Major
Carditis
Polyarthritis
Erythema marginatum
Chorea
Subcutaneous nodules
Minor
Arthralgia
Fever
Prolonged PR interval
Raised C-reactive protein/erythrocyte sedimentation rate

Evidence of recent streptococcal throat infection
plus
One major and two minor criteria
or
Two major criteria

It is difficult to predict how RF may evolve in the coming years. It is not clear whether the epidemic outbreaks or the observed rare cases are aberrations in the generally declining profile currently observed in developed countries or whether there is a true risk of resurgence of the disease. Consequently, accurate identification of GAS pharyngitis and follow-up of rheumatogenic strains with their appropriate treatment are still important (Olivier 2000).

LABORATORY DIAGNOSIS AND IDENTIFICATION

Serological grouping provides the first precise method for identifying the β-haemolytic streptococci, in particular the human pathogens of Lancefield groups A, B, C and G. The main characteristics of these β-haemolytic streptococci are summarised in Table 1.8. Isolates from primary culture are traditionally identified by colony morphology, Gram stain, catalase test and Lancefield grouping. If Lancefield grouping does not provide sufficient identification, full biochemical identification may be obtained usually using a commercial identification system. *Streptococcus pyogenes* is identified in a diagnostic laboratory from its polysaccharide group A antigen, using commercial grouping systems. Presumptive identification is usually made by bacitracin susceptibility or pyrrolidonylarylamidase activity. *Streptococcus pyogenes* is the only species within the genus that is positive for both tests (Facklam 2002). *Streptococcus agalactiae* is the most common cause of neonatal sepsis and is the only species that carries the Lancefield group B antigen. Lancefield grouping usually identifies the organism, but the organism can also be identified presumptively by the CAMP test and hippurate hydrolysis. *Streptococcus dysgalactiae* subsp. *equisimilis* is the revised taxonomic species for the human streptococci of Lancefield group C (*S. equisimilis*) and group G (nameless). Detection of the group antigen in combination with the phenotypic characteristics will subdivide these organisms into the appropriate species (Table 1.8). Studies have also revealed that strains of Lancefield group L should also be included within this species. To complicate this even further some strains of *S. dysgalactiae* subsp. *equisimilis* possess the group A antigen. Therefore, the group antigen can only be used as an aid in species identification. The phenotypic tests listed in Table 1.8 should be used, together with the haemolytic reaction and group antigen, to identify the species.

Streptococcus equi subsp. *equi* is the classic cause of strangles in horses. Identification of the species is again based upon the phenotypic test reactions and demonstration of the group C antigen. *Streptococcus equi* subsp. *zooepidemicus* is usually identified according to Lancefield group and phenotypic profile. The animal pathogen *S. canis* carries

Table 1.8 Biochemical characteristics of the β-haemolytic streptococci of Lancefield groups A, B, C and G

Streptococcus sp.	Group	Bacitracin[a]	Pyrrolidonylarylamidase	CAMP reaction	Arginine hydrolysis	Voges–Proskauer reaction	β-Galactosidase activity	Hippurate	Glycogen	Ribose	Sorbitol	Trehalose
S. pyogenes	A	+	+	–	+	–	–	–	v	–	–	+
S. agalactiae	B	–	–	+	+	–	–	+	–	+	–	+
S. dysgalactiae subsp. equisimilis	C	–	v	–	+	–	–	–	v	+	–	+
S. dysgalactiae subsp. equisimilis	G	–	v	–	+	–	–	–	–	+	–	+
S. equi subsp. equi	C	–	–	–	+	–	v	–	+	–	–	v
S. equi subsp. zooepidemicus	C	–	–	–	+	–	–	–	+	+	+	–
S. canis	G	–	–	–	+	–	+	–	–	+	–	–

Lancefield group, group carbohydrate antigen. v, variable reaction + or –.
[a] Sensitivity to 0.1 U of bacitracin.

the group G antigen. *Streptococcus porcinus* may cause difficulties in identification as it carries the group E, P, U or V antigen and also cross-reacts with group B sera in commercial kit systems.

Rapid Identification Assays for β-Haemolytic Streptococci

The signs and symptoms of pharyngitis caused in particular by GAS can be nonspecific, hampering accurate clinical diagnosis. Microbiological investigation offers the most robust diagnostic route, although not always feasible. A significant limitation of culturing throat swabs is the delay in obtaining the results. In the early 1980s commercial rapid antigen tests were developed for detecting GAS directly from throat swabs. These became quite widely used in the United States in particular (Gerber and Shulman 2004). Most of these kits exhibit high specificities; however, there is considerable variability in sensitivity. Current published data are inadequate to allow definitive conclusions to be made regarding the relative performance characteristics of these kits, in particular for GAS pharyngitis. Also, the rarity of serious consequences of acute GAS pharyngitis also suggests the need for careful cost–benefit and risk–benefit analyses of different diagnostic strategies in various clinical settings.

Given the importance of GBS as a cause of neonatal sepsis, point-of-care or rapid laboratory tests for GBS have been developed in recent years primarily for identifying GBS-colonised women in labour. These tests are used in the United States; however, their use in Europe is somewhat limited. Identification of GBS-colonised women is critical to the prevention of GBS neonatal infections. Currently, antenatal screening culture using broth culture in selective medium is the gold standard for detecting anogenital GBS colonisation. Rapid tests have been developed, but again they lack specificity and sensitivity and have yet to offer a potential substitute for culture. GBS-specific polymerase chain reaction (PCR)-based assays have demonstrated better sensitivity than those for GAS, but they require complicated procedures that are not applicable to clinical use. More promising results have emerged from the development of a rapid assay using the LightCycler technology (Bergeron *et al.* 2000). It was found that GBS could be detected rapidly and reliably by a PCR assay of combined vaginal and anal secretions from pregnant women at the time of delivery. The use of such technology at the time of delivery is, however, not entirely practical, and simpler assays are currently under development using technologies such as microfluidics.

Typing of β-Haemolytic Streptococci

Many methods have been developed for the serological classification of the β-haemolytic streptococci, in particular for GAS, GBS, GCS

and GGS, based primarily on the characterisation of cell-wall protein and polysaccharide antigens.

Typing of GAS

The best-known scheme for GAS is that based upon the characterisation of the T- and M-protein antigens, as developed by Griffith (1934) and Lancefield (1962). The variable sequences of the surface-exposed amino terminal end of the M protein provide the basis for the classic M typing scheme. Thus far, there are more than 100 validated M proteins amongst GAS and 30 T-protein antigens. The correlation between M and T type is not always clear. For instance, for each T type more than one M type can exist, and to further complicate matters, an M-type strain can express one or more different T-type antigens (Johnson and Kaplan 1993). Approximately 50% of GAS strains also produce an apoproteinase, an enzyme that causes mammalian serum to increase in opacity. This reaction is called the serum opacity factor (OF), and the enzyme responsible is referred to as OF. The OF has proved useful for the serological identification of GAS. Antibody to OF is specific in its inhibition of the opacity reaction of the M type producing it, and this characteristic has proved extremely useful as a supplementary aid to the classical typing scheme. Table 1.9 summarises the common reactions observed amongst GAS currently isolated within the United Kingdom and the relationship of T, M and OF antigens. Most epidemiological studies classify the organisms by M type, but it is sometimes difficult to identify the M-protein types in this way because of the unavailability of typing reagents and difficulties in their preparation and maintenance.

Molecular typing is now increasingly replacing classical serological methods for typing GAS, with few national reference centres still using the classical schemes. Sequencing of the *emm* gene, which encodes the M protein, has now become the key method for typing

Table 1.9 Some common group A streptococcus (GAS) M types (in boldface) within the United Kingdom and Europe and their related T and OF antigens

T pattern	Opacity factor	
	Positive	Negative
1	68	**1**
3/13/B3264	73, 77, **82, 83, 89**	**3, 33, 52, 53**
4	**4, 48, 60, 63**	24, 26
5/27/44	**82**	5
12	22, 62, 66, 76	**12**
28	**28, 87**	–

GAS. Currently, more than 150 validated *emm* types exist. The extensive variability within the N-terminus of *emm* genes forms the basis for distinguishing between the different *emm* types. Depending on the *emm* genotype between 160 and 660 nucleotide bases are sequenced from the 5′-terminal end and matched with available sequences in the *emm* gene database (http://www.cdc.gov/ncidod/biotech/strep/strepindex.html). Two strains are regarded to represent the same *emm* type if they share greater than or equal to 95% identity in the N-terminal end (Beall, Facklam and Thompson 1996; Facklam *et al.* 2002).

There are primarily four major subfamilies of *emm* genes that are defined by sequence differences within the 3′ end encoding the peptidoglycan-spanning domain (Hollingshead *et al.* 1993). The chromosomal arrangement of *emm* subfamily genes reveals five major *emm* patterns, denoted *emm* A to E (Bessen *et al.* 2000). Patterns B and C are rare and are currently grouped with *emm* pattern A strains. A GAS isolate has one, two or three *emm* genes lying in tandem on the chromosome, and each gene differs in sequence from the others. In strains having three *emm* genes the determinants of *emm* type lie within the central *emm* locus. The *emm* pattern serves as a genotypic marker for tissue-site preferences amongst GAS strains.

Most GAS *emm* sequence types represent M-protein types that are widely distributed according to geographic region. This information is vital to researchers currently formulating multivalent M-protein vaccines representative of common circulating GAS isolates.

Several molecular subtyping methods have been used to characterise and measure the genetic diversity of these organisms. These include multilocus sequence typing (MLST), restriction endonuclease analysis of genomic DNA, ribotyping, random-amplified polymorphic DNA (RAPD) fingerprinting, PCR–restriction fragment length polymorphism (PCR-RFLP) analysis of the *emm* gene, pulsed-field gel electrophoresis (PFGE) and fluorescent amplified fragment length polymorphism (FAFLP) analysis. The most reproducible of these methods, PFGE, is time consuming and labour intensive. The PCR-based methods are poorly reproducible. FAFLP is a modification of the amplified fragment length polymorphism analysis (AFLP) methodology using a fluorescently labelled primer and has been used to characterise GAS isolates from an outbreak of invasive disease and was found to be an effective and more discriminatory method than PFGE. As well as being able to distinguish between different M types, the method was also able to subtype isolates within an M type (Desai *et al.* 1999).

MLST is a relatively new tool for the molecular typing of GAS strains (Enright *et al.* 2001). The main advantage of MLST over gel-based methods is that the sequence data, which are generated for a series of neutral housekeeping loci, are unambiguous, electronically portable and readily queried via the Internet (http://www.mlst.net). A recent study by McGregor *et al.* (2004) has documented a comprehensive catalogue of MLST sequence types (STs) and *emm* patterns for most of the known *emm* types and thus has provided the basis for addressing questions on the population structure of GAS. In brief *emm* pattern D and E strains account for greater than 80% of *emm* types, and therefore, from a global perspective, these strains are said to be of medical importance. The classic throat strains (*emm* patterns A–C) displayed the least diversity.

Typing of GCS and GGS

Early typing methods for group C and G streptococci were based upon bacteriophage typing (Vereanu and Mihalcu 1979) and bacteriocins (Tagg and Wong 1983). A serological typing scheme described in the 1980s (Efstratiou 1983) can subdivide GCS and GGS of human origin into more than 20 different T-protein types. M proteins described amongst these isolates have also been used in a complementary typing scheme and are clinically relevant in view of their virulence characteristics (Efstratiou *et al.* 1989). However, these schemes have only been used in specialist reference centres (e.g. in the United Kingdom and also in Thailand where GCS and GGS infections exceed those of GAS). Other typing methods for these streptococci have also been explored, but M typing and molecular typing (PFGE and ribotyping) appeared to be the most discriminating (Efstratiou 1997; Ikebe *et al.* 2004). Ribotyping was useful for examining representative isolates from three different outbreaks caused by one serotype T204 in the United Kingdom and overseas (Efstratiou 1997). This technique has also been successfully applied to the outbreaks caused by *S. equi* subsp. *zooepidemicus* in humans (Francis *et al.* 1993). In this instance ribotyping was useful because phenotypic typing schemes for *S. equi* subsp. *zooepidemicus* are unavailable.

Because the *emm* genes of GAS and GCS/GGS share some homology *emm*-based sequence typing has also been described for these organisms and is based upon the heterogeneity of the 5′ ends of the gene, which give rise to different STs. Approximately 40 *emm* types of GGS and GCS have thus far been identified (http://www.cdc.gov/ncidod/biotech/strep/emmtypes.htm). However, in contrast to GAS, little is known about which *emm* types are dominant in the GGS/GCS population and genetic diversity within an *emm* type, whether *emm* types represent closely related groups as reported for GAS (Enright *et al.* 2001) or whether particular *emm* types are associated with invasive disease. Recent typing studies have shown the distribution of these STs amongst strains from invasive disease, with the emergence of two novel types (Cohen-Poradosu *et al.* 2004).

An important distinguishing feature of GAS and GCS/GGS lies in the relationship between *emm* type and clone (Kalia *et al.* 2001). Among GAS most isolates of a given *emm* type are clones or form a clonal complex, as defined by MLST STs with greater than or equal to five housekeeping alleles in common (Enright *et al.* 2001). In contrast, GCS and GGS have distinct allelic profiles and only one pair of isolates bearing the same *emm* type has similar genotypes (Kalia *et al.* 2001).

Typing of GBS

Serotyping has been traditionally the predominant method used to classify GBS. Type-specific capsular polysaccharide antigens allow GBS to be classified into nine serotypes – Ia, Ib, II–VIII – and there are also three protein antigens that are serologically useful – c, R and X. Serotype III strains are of particular importance as they are responsible for most infections, including meningitis in neonates worldwide (Kalliola *et al.* 1999; Davies *et al.* 2001; Dahl, Tessin and Trollfors 2003; Weisner *et al.* 2004). Thus, serotyping offers low discrimination for these organisms. Bacteriophage typing was developed in the 1970s primarily to enhance the power of serotyping. But again phage typing is also somewhat limited in that there is a high incidence of nontypability. The method itself is cumbersome and requires rigid quality control with the use of specific reagents (Colman 1988).

In recent years there has been a shift towards molecular methods for typing GBS. Diverse lineages of, for example, serotype III strains can be distinguished with multilocus enzyme electrophoresis (MLE), PFGE and restriction digest pattern analysis, ribotyping. MLE was found not to be useful for differentiating serotype III because most (particularly from patients with invasive disease) grouped into a single MLE type. More recently, MLST has been applied to a collection of globally and ecologically diverse human GBS strains that included representatives of the capsular serotypes Ia, Ib, II, III, V, VI and VIII. A website was also established for the storage of GBS allelic profiles from across the globe and can be accessed at http://sagalactiae.mlst.net. Seven housekeeping genes are amplified, and the combination of alleles at the seven loci provides an allelic profile or ST for each strain. Most of the strains (66%) are assigned to four major STs. ST-1 and ST-19 are significantly associated with asymptomatic carriage, whereas ST-23 includes both carriage and invasive strains. All isolates of ST-17 are serotype III clones, and this ST apparently defines a homogenous clone that had a strong association with neonatal invasive infections. The variation in serotype with a single ST and the presence of genetically diverse isolates with the same serotype suggested that the capsular biosynthesis genes of GBS are subjected to relatively frequent horizontal gene transfer, as observed with *S. pneumoniae* (Coffey *et al.* 1998; Jones *et al.* 2003). A single gene therefore confers serotype specificity in GBS of capsular types III and Ia (Chaffin *et al.* 2000), and recombinational replacement of this gene with that from an isolate of a different type results in a change of capsular type. Therefore, confirmation of serotype identity is possible when a DNA sequence-based serotyping method is used, such as the one described recently by Kong *et al.* (2003). They used three sets of markers – the capsular polysaccharide synthesis (*cps*) gene cluster, surface protein antigens and mobile genetic elements. The use of three sets of markers resulted in a highly discriminatory typing scheme for GBS as it provides useful phenotypic data, including antigenic composition, thus important for epidemiological surveillance studies especially in relation to potential GBS vaccine use. Further studies are needed on the distribution of these mobile genetic elements and their associations with virulence and pathogenesis of GBS disease.

International Quality Assurance Programmes for GAS Characterisation

The various reported additions to the classical serotyping methods for these organisms have increased the likelihood of differences both in the determination of previously unknown types and in the confirmation of previously unrecognised serotypes. Although international reference laboratories have informally exchanged strains and confirmed the identification of unique isolates for many years, the 'new' molecular techniques have made precise confirmation and agreement even more important to clinical and epidemiological laboratory research. The first international quality assurance programme was established amongst six major international streptococcal reference centres (two in United States and one each in United Kingdom, Canada, New Zealand and Czech Republic). Five distributions were made, and both traditional and genotypic methods were used. The correlation of results between centres was excellent, albeit with a few differences noted. This continuing self-evaluation demonstrated the importance of comparability, verification, standardisation and agreement of methods amongst reference centres in identifying GAS M and *emm* types. The results from this programme also demonstrate and emphasise the importance of precisely defined criteria for validating recognised and accepted types of GAS (Efstratiou *et al.* 2000). There have been more recent distributions amongst European and other typing centres (20 centres), within the remit of a pan-European programme 'Severe *Streptococcus pyogenes* infections in Europe' (Schalén 2002).

ANTIMICROBIAL RESISTANCE

β-Haemolytic streptococci of Lancefield groups A, B, C and G remain exquisitely sensitive to penicillin and other β-lactams, but resistance to sulphonamides and tetracyclines has been recognised since the Second World War. Penicillin has, therefore, always been the drug of choice for most infections caused by β-haemolytic streptococci, but macrolide compounds, including erythromycin, clarithromycin, roxithromycin and azithromycin, have been good alternatives in penicillin-allergic patients.

β-Lactam Resistance

In striking contrast to erythromycin no increase in penicillin resistance has been observed *in vitro* among clinical isolates of β-haemolytic streptococci, in particular GAS. Penicillin tolerance, however, amongst GAS, defined as an increased mean bactericidal concentration (MBC) : minimum inhibitory concentration (MIC) ratio (usually >16), has been reported at a higher frequency in patients who were clinical treatment failures than in those successfully treated for pharyngitis (Holm 2000). The presence of β-lactamase-producing bacteria in the throat has been assumed to be another significant factor in the high failure rate of penicillin V therapy. Although many β-lactamase-producing bacteria are not pathogenic to this particular niche, they can act as indirect pathogens through their capacity to inactivate penicillin V administered to eliminate β-haemolytic streptococci.

Macrolide Resistance

Macrolide resistance amongst GAS has increased between the 1990s and early 2000s in many countries (Seppala *et al.* 1992; Cornaglia *et al.* 1998; Garcia-Rey *et al.* 2002). The resistance is caused by two different mechanisms: target-site modification (Weisblum 1985) and active drug efflux (Sutcliffe, Tait-Kamradt and Wondrack 1996). Target-site modification is mediated by a methylase enzyme that reduces binding of macrolides, lincosamides and streptogramin B antibiotics (MLS$_B$ resistance) to their target site in the bacterial ribosome, giving rise to the constitutive (CR) and inducible (IR) MLS$_B$ resistance phenotypes (Weisblum 1985). Another phenotype called noninducible (NI) conferring low-level resistance to erythromycin (MIC <16 mg/L) and to other 14- or 15-membered compounds but sensitive to the 16-membered macrolides has also been described (Seppala *et al.* 1993). This phenotype is invariably susceptible to clindamycin and has been designated as NI because of an efflux mechanism.

Since the identification of plasmid-mediated MLS resistance in the *Streptococcus* sp. in the early 1970s it has been shown that conjugation is the most common dissemination mode of resistance. Transduction may also contribute to the spread of antibiotic resistance amongst GAS but has not been reported for other streptococci (Courvalin, Carlier and Chabbert 1972). Transposon-mediated MLS resistance in GAS has been described, and chromosomally integrated resistance genes of presumed plasmid origin have been found in many plasmid-free strains (Le Bouguenec, de Cespedes and Horaud 1990).

The genes encoding the methylases have been designated *erm* (erythromycin ribosome methylation). The methylases of the various species have been characterised at the molecular level and are subdivided into several classes, named by letters (Levy *et al.* 1999). In GAS the so-called *erm*B (*erm*AM) gene is among the most common, found to be present in 31% of 143 macrolide-resistant GAS lower respiratory tract infection strains obtained from 25 countries participating in the Prospective Resistant Organism Tracking and Epidemiology for the Ketolide Telithromycin (PROTEKT) study during 1999–2000 (Farrell *et al.* 2002). A closely related gene, *erm*TR (*erm*A) (Seppala *et al.* 1998), was found in 23% of these isolates, whilst *mef*A, the resistance gene due to macrolide efflux (Levy *et al.* 1999), was detected in 46% of these strains.

The increase in erythromycin resistance amongst GAS has been shown to be caused by isolates with the M phenotype, known to be associated with active efflux (Tait-Kamradt *et al.* 1997). The increase in the prevalence of erythromycin-resistant GAS carrying the *erm*A, *erm*B and/or *mef*A genes has been the subject of many recent reports (Giovanetti *et al.* 1999; Kataja, Huovinen and Seppala 2000; Leclercq 2002). In contrast, less work has been undertaken to characterise the mechanism of tetracycline resistance amongst β-haemolytic streptococci. For GAS, only the *tet*M gene has been the commonly identified gene. Other streptococcal species that carry the *tet*(O) gene, which codes for another tetracycline resistance ribosomal protection protein, or the *tet*(K) and the *tet*(L) genes, which code for efflux-mediated tetracycline resistance, have also been identified (Chopra and Roberts 2001).

Current estimates of macrolide resistance in GAS strains within Europe show some variation, from 1.8% in a collection of invasive and noninvasive strains from Denmark (Statens Serum Institut, Danish Veterinary and Food Administration, Danish Medicines Agency and Danish Institute for Food and Veterinary Research 2004), 4% of bacteraemic strains in the United Kingdom (HPA 2004b) to 20% of noninvasive strains from across Spain (Perez-Trallero *et al.* 2001). In a mixed collection of GAS strains obtained from Russia 11% were found to be erythromycin resistant (Kozlov *et al.* 2002). Among the β-haemolytic streptococcal strains collected as part of the SENTRY worldwide antimicrobial susceptibility monitoring programme between 1997 and 2000, 10% of European isolates from invasive and noninvasive disease were found to be erythromycin resistant and 11% of strains collected from Asia-Pacific region, 19% from North American and only 3% from Latin American exhibited erythromycin resistance (Gordon *et al.* 2002).

Resistance to Other Therapeutic Agents

Rifampicin is a particularly active agent against Gram-positive bacteria and mycobacteria. There have been few reports about the susceptibility of *S. pyogenes* to rifampicin; however, resistance has been estimated to be approximately 0.3% amongst isolates from Spain. Resistance is caused by an alteration of one or more regions of the target site, the β-subunit of RNA polymerase. Most of the mutations associated with rifampicin resistance are located in a segment in the centre of the *rpo*B gene cluster (Herrera *et al.* 2002).

TREATMENT STRATEGIES

Antimicrobial Therapy for β-Streptococcal Disease

Superficial Disease

β-Haemolytic streptococci have remained exquisitely sensitive to penicillin. It is normal to give 10 days of treatment with penicillin V to fully treat streptococcal pharyngitis and prevent RF, although there is some preliminary evidence to suggest that 5 days of azithromycin will suffice (Bisno *et al.* 2002). Up to 30% of streptococcal sore throats treated with penicillin may relapse, possibly because of the internalisation of bacteria into buccal or tonsillar epithelial cells (Sela *et al.* 2000; Kaplan and Johnson 2001). Macrolides have a theoretical advantage over β-lactams because of intracellular penetration. Such compounds may afford the best eradication rate. Although macrolides have traditionally been considered as alternatives to β-lactams in patients with penicillin allergy, it is noteworthy that up to 45% of GAS isolates are reported to be resistant to macrolides in certain European countries. In the United Kingdom the incidence of erythromycin resistance is lower, at around 5–10%.

For impetigo initial empiric oral systemic treatment directed against both *S. aureus* and *S. pyogenes* (e.g. flucloxacillin) is normally combined with agents to clear carriage of the organisms. Notably, the use of topical fusidic acid may be of little benefit, as only approximately half of GAS are sensitive to this drug, and many *S. aureus* strains are now demonstrating fusidin resistance.

Invasive β-Haemolytic Disease

For invasive β-haemolytic disease intravenous treatment with penicillin (or ceftriaxone) is mandatory, at least until an initial response is observed. Most authorities now recommend the use of clindamycin alongside penicillin, because it has been shown to improve outcome both in animal models of invasive streptococcal disease and in clinical studies (Stevens *et al.* 1988; Stevens, Bryant and Yan 1994; Zimbelman, Palmer and Todd 1999). Clindamycin appears to inhibit the synthesis or release of many streptococcal virulence factors such

as superantigen toxins and cell-wall components (Gemmell *et al.* 1981; Brook, Gober and Leyva 1995; Sriskandan *et al.* 1997).

In some cases of invasive GBS disease an aminoglycoside is added to β-lactam therapy for the first 2 weeks. This is particularly true for neonatal GBS disease.

Surgery for β-Streptococcal Disease

In addition to appropriate antibiotic therapy it is essential that cases of invasive disease be examined for evidence of tissue necrosis or gangrene. This may present subtly, and many cases will require surgical exploration. All necrotic tissue must be removed promptly, and reexploration at daily intervals may be necessary (Stamenkovic and Lew 1984; Heitmann *et al.* 2001). Early liaison with plastic surgery specialists can facilitate appropriate debridement. Following extensive surgery the patient remains at risk of nosocomial infection.

Immunological Therapies

Intravenous immunoglobulin (IVIG) is recommended as an adjunct to treatment of severe invasive GAS infection, in particular those cases complicated by STSS. Although there are no controlled trials to support its use in treatment, anecdotal reports and a retrospective clinical study suggest that there are some benefits (Lamothe *et al.* 1995; Kaul *et al.* 1999). A European clinical trial was abandoned because of difficulties in patient recruitment, though initial responses were promising (Darenberg *et al.* 2003). Although the cost of the proposed doses of IVIG is likely to be high, surprisingly there are no preclinical studies that have yet demonstrated the efficacy of IVIG.

There are many mechanisms that may explain the apparent efficacy of IVIG. Firstly, IVIG-mediated neutralisation of superantigenic toxins leads to recovery from STSS, and secondly, it has been shown that IVIG can assist in the opsonisation of GAS in nonimmune hosts (Skansen-Saphir *et al.* 1994; Norrby-Teglund *et al.* 1996; Basma *et al.* 1998). Of course, IVIG is likely to contain an abundance of polyclonal antistreptococcal antibodies, including antibodies to a wide range of secreted virulence factors, which may be of importance in disease progression.

Phage Therapy

Although antibiotic resistance has not yet affected therapy for β-haemolytic streptococcal disease substantially, the identification of novel compounds that can specifically target pathogens is welcome. Recently, purified recombinant phage lytic enzymes (lysins) have been demonstrated to have excellent rapid bactericidal activity. Although not developed to a degree that would allow systemic use, phage lytic enzyme has been recently used to successfully eradicate pharyngeal carriage of *S. pyogenes* (Nelson, Loomis and Fischetti 2001).

PREVENTION AND CONTROL

Vaccines for β-Haemolytic Streptococcal Disease

GAS opsonic antibodies directed against the N-terminal sequences of the streptococcal M protein are known to confer strain-specific immunity to *S. pyogenes*. There is therefore an obvious advantage to exploiting the M protein as a vaccine candidate. Unfortunately, there are in excess of 100 M serotypes and in excess of 150 *emm* gene types, impeding the process of choosing vaccine candidates. Indeed, the prevalent M serotypes in different communities differ significantly. To counter this multivalent vaccines have been designed that incorporate several dominant M serotypes. A recombinant multivalent vaccine has recently undergone phase I clinical trial evaluation

(Kotloff *et al.* 2004), which has provided the first evidence that a hybrid fusion protein is a feasible strategy for evoking type-specific opsonic antibodies against multiple serotypes of GAS without eliciting antibodies that cross-react with host tissues. This represents a critical step in GAS vaccine development.

The conserved C-terminus of the M protein has also been considered as a vaccine candidate. Conserved T-cell epitopes, which are thought to be rheumatogenic, reside in the C-terminus of the protein (Robinson, Case and Kehoe 1993; Pruksakorn *et al.* 1994), though rheumatogenic T-cell epitopes may also exist in the variable N-terminus (Guilherme *et al.* 2000). These considerations have greatly complicated GAS vaccine development because of fears of precipitating RF in vaccinated individuals (Pruksakorn *et al.* 1994; Brandt *et al.* 2000).

Recent studies suggest that opsonic antibodies to M protein alone are insufficient for opsonophagocytosis of *S. pyogenes*, emphasising the need to seek additional vaccine candidates. Certain non-M-protein antigens have already been evaluated in preclinical models, including the streptococcal C5a peptidase, the streptococcal cysteine protease (SPEB) and the GAS carbohydrate.

Although maternal intrapartum antibiotic prophylaxis is clearly effective and has reduced the incidence of early-onset GBS neonatal disease substantially in the United States, it cannot prevent late-onset GBS disease. Vaccination of women of childbearing age against GBS could theoretically prevent both early-onset and late-onset GBS disease in the neonate, in addition to preventing GBS disease in pregnant women. Opsonic antibodies directed against the capsular polysaccharide of GBS confer serotype-specific protection. Initial vaccines developed focused on capsular type III isolates, but the emergence of type V isolates in recent years has prompted the development of a polyvalent GBS vaccine (Baker and Edwards 2003). Vaccines have been developed by coupling purified capsular polysaccharide antigen of GBS with an immunogenic protein carrier. Glycoconjugate vaccines against all nine currently identified GBS serotypes have been synthesised and shown to be immunogenic in animal models and in human phase I and II trials. The recent results strongly suggest that GBS conjugate vaccines may be effective in the prevention of GBS disease (Paoletti and Kasper 2003).

REFERENCES

American Heart Association (1992) Guidelines for the diagnosis of rheumatic fever. Jones Criteria, 1992 update. Special Writing Group of the Committee on Rheumatic Fever, Endocarditis, and Kawasaki Disease of the Council on Cardiovascular Disease in the Young of the American Heart Association. *Clinical Infectious Diseases*, **268**, 2069–2073.

Ayoub EM, Kotb M, Cunningham MW (2000) Rheumatic fever pathogenesis. In *Streptococcal Infections: Clinical Aspects, Microbiology and Molecular Pathogenesis*. Stevens DL, Kaplan EL (eds). Oxford University Press, pp. 102–132.

Baker CJ, Edwards MS (2003) Group B streptococcal conjugate vaccines. *Archives of Disease in Childhood*, **88**, 375–378.

Basma H, Norrby-Teglund A, McGeer A *et al.* (1998) Opsonic antibodies to the surface M protein of group A streptococci in pooled normal immunoglobulins (IVIG): potential impact on the clinical efficacy of IVIG therapy for severe invasive group A streptococcal infections. *Infection and Immunity*, **66**, 2279–2283.

Beall B, Facklam R, Thompson T (1996) Sequencing *emm*-specific PCR products for routine and accurate typing of group A streptococci. *Journal of Clinical Microbiology*, **34**, 953–958.

Belcher DW, Afoakwa SN, Osei-Tutu E *et al.* (1977) Endemic pyoderma in Ghana: a survey in rural villages. *Transactions of the Royal Society of Tropical Medicine and Hygiene*, **71**, 204–209.

Ben-Abraham R, Keller N, Vered R *et al.* (2002) Invasive group A streptococcal infections in a large tertiary center: epidemiology, characteristics and outcome. *Infection*, **30**, 81–85.

Beres SB, Sylva GL, Barbian KD *et al.* (2002) Genome sequence of a serotype M3 strain of group A streptococcus: phage-encoded toxins, the high-virulence phenotype, and clone emergence. *Proceedings of the National Academy of Sciences of the United States of America*, **99**, 10078–10083.

Bergeron MG, Ke D, Menard C *et al.* (2000) Rapid detection of group B streptococci in pregnant women at delivery. *The New England Journal of Medicine*, **343**, 175–179.

Bessen DE, Carapetis JR, Beall B *et al.* (2000) Contrasting molecular epidemiology of group A streptococci causing tropical and nontropical infections of the skin and throat. *The Journal of Infectious Diseases*, **182**, 1109–1116.

Bisno AL (1991) Group A streptococcal infections and acute rheumatic fever. *The New England Journal of Medicine*, **325**, 783–793.

Bisno AL, Brito MO, Collins CM (2003) Molecular basis of group A streptococcal virulence. *The Lancet Infectious Diseases*, **3**, 191–200.

Bisno AL, Gerber MA, Gwaltney JM Jr *et al.* (2002) Practice guidelines for the diagnosis and management of group A streptococcal pharyngitis. Infectious Diseases Society of America. *Clinical Infectious Diseases*, **35**, 113–125.

Brandt ER, Sriprakash KS, Hobb RI *et al.* (2000) New multi-determinant strategy for a group A streptococcal vaccine designed for the Australian Aboriginal population. *Nature Medicine*, **6**, 455–459.

Brook I, Gober AE, Leyva F (1995) *In vitro* and *in vivo effects* of penicillin and clindamycin on expression of group A beta-hemolytic streptococcal capsule. *Antimicrobial Agents and Chemotherapy*, **39**, 1565–1568.

Centers for Disease Control and Prevention (1999) Active Bacterial Core Surveillance (ABCs) Report. Emerging Infections Program Network: group B Streptococcus, 1997. www.cdc.gov/ncidod/dbmd/abcs/survreports/gbs97.pdf (25 August 2004).

Centers for Disease Control and Prevention (2003) Active Bacterial Core Surveillance (ABCs) Report. Emerging Infections Program Network: Group B Streptococcus, 2002. www.cdc.gov/ncidod/dbmd/abcs/survreports/ gbs02.pdf (25 August 2004).

Centers for Disease Control and Prevention (2004a) *Active Bacterial Core Surveillance Report, Emerging Infections Program Network, Group A Streptococcus, 2003*. CDC, Atlanta, GA.

Centers for Disease Control and Prevention (2004b) *Active Bacterial Core Surveillance Report, Emerging Infections Program Network, Group B Streptococcus, 2003*. CDC, Atlanta, GA.

Chaffin DO, Beres SB, Yim HH, Rubens CE (2000) The serotype of type Ia and III group B streptococci is determined by the polymerase gene within the polycistronic capsule operon. *Journal of Bacteriology*, **182**, 4466–4477.

Chaussee MS, Somerville GA, Reitzer L, Musser JM (2003) Rgg coordinates virulence factor synthesis and metabolism in *Streptococcus pyogenes*. *Journal of Bacteriology*, **185**, 6016–6024.

Chopra L, Roberts M (2001) Tetracycline antibiotics: mode of action, applications, molecular biology, and epidemiology of bacterial resistance. *Microbiology and Molecular Biology Reviews*, **65**, 232–260.

Cleary PP, Kaplan EL, Handley JP *et al.* (1992) Clonal basis for resurgence of serious *Streptococcus pyogenes* disease in the 1980s. *Lancet*, **339**, 518–521.

Coffey TJ, Enright MC, Daniels M *et al.* (1998) Recombinational exchanges at the capsular polysaccharide biosynthetic locus lead to frequent serotype changes among natural isolates of *Streptococcus pneumoniae*. *Molecular Microbiology*, **27**, 73–83.

Cohen ML (2000) Changing patterns of infectious disease. *Nature*, **406**, 762–767.

Cohen-Poradosu R, Jaffe J, Lavi D *et al.* (2004) Group G streptococcal bacteremia in Jerusalem. *Emerging Infectious Diseases*, **10**, 1455–1460.

Collin M, Olsén A (2003) Extracellular enzymes with immunomodulating activities: variations on a theme in *Streptococcus pyogenes*. *Infection and Immunity*, **71**, 2983–2992.

Colman G (1988) Typing of *Streptococcus agalactiae* (Lancefield group B). *European Journal of Clinical Microbiology & Infectious Diseases*, **7**, 226–231.

Colman G, Tanna A, Efstratiou A, Gaworzewska ET (1993) The serotypes of *Streptococcus pyogenes* present in Britain during 1980–1990 and their association with disease. *Journal of Medical Microbiology*, **39**, 165–178.

Cone LA, Woodard DR, Schlievert PM, Tomory GS (1987) Clinical and bacteriologic observations of a toxic shock-like syndrome due to *Streptococcus pyogenes*. *The New England Journal of Medicine*, **317**, 146–149.

Cornaglia G, Ligozzi M, Mazzariol A *et al.* (1998) Resistance of *Streptococcus pyogenes* to erythromycin and related antibiotics in Italy. The Italian Surveillance Group for Antimicrobial Resistance. *Clinical Infectious Diseases*, **27**(Suppl. 1), S87–S92.

Courtney HS, Hasty DL, Dale JB (2002) Molecular mechanisms of adhesion, colonization, and invasion of group A streptococci. *Annals of Medicine*, **34**, 77–87.

Courvalin PM, Carlier C, Chabbert YA (1972) Plasmid-linked tetracycline and erythromycin resistance in group D "streptococcus". *Annales de l'Institut Pasteur*, **123**, 755–759.

Cunningham MW (2000) Pathogenesis of group A streptococcal infections. *Clinical Microbiology Reviews*, **13**, 470–511.

Curtis N, Zheng R, Lamb JR, Levin M (1995) Evidence for a superantigen mediated process in Kawasaki disease. *Archives of Disease in Childhood*, **72**, 308–311.

Dahl MS, Tessin I, Trollfors B (2003) Invasive group B streptococcal infections in Sweden: incidence, predisposing factors and prognosis. *International Journal of Infectious Diseases*, **7**, 113–119.

Dale JB, Chiang EY, Liu S, Courtney HS, Hasty DL (1999) New protective antigen of group A streptococci. *The Journal of Clinical Investigation*, **103**, 1261–1268.

Darenberg J, Ihendyane N, Sjolin J *et al.* (2003) Intravenous immunoglobulin G therapy in streptococcal toxic shock syndrome: a European randomized, double-blind, placebo-controlled trial. *Clinical Infectious Diseases*, **37**, 333–340.

Davies HD, McGeer A, Schwartz B *et al.* (1996) Invasive group A streptococcal infections in Ontario, Canada. Ontario Group A Streptococcal Study Group. *The New England Journal of Medicine*, **335**, 547–554.

Davies HD, Raj S, Adair C *et al.* (2001) Population-based active surveillance for neonatal group B streptococcal infections in Alberta, Canada: implications for vaccine formulation. *The Pediatric Infectious Disease Journal*, **20**, 879–884.

Desai M, Efstratiou A, George R, Stanley J (1999) High-resolution genotyping of *Streptococcus pyogenes* serotype M1 isolates by fluorescent amplified-fragment length polymorphism analysis. *Journal of Clinical Microbiology*, **37**, 1948–1952.

DeWinter LM, Low DE, Prescott JF (1999) Virulence of *Streptococcus canis* from canine streptococcal toxic shock syndrome and necrotizing fasciitis. *Veterinary Microbiology*, **70**, 95–110.

Diekema DJ, Pfaller MA, Jones RN, SENTRY Participants Group (2002) Age-related trends in pathogen frequency and antimicrobial susceptibility of bloodstream isolates in North America: SENTRY Antimicrobial Surveillance Program, 1997–2000. *International Journal of Antimicrobial Agents*, **20**, 412–418.

Drummond P, Clark J, Wheeler J *et al.* (2000) Community acquired pneumonia—a prospective UK study. *Archives of Disease in Childhood*, **83**, 408–412.

Efstratiou A (1983) The serotyping of hospital strains of streptococci belonging to Lancefield group C and group G. *The Journal of Hygiene*, **90**, 71–80.

Efstratiou A (1989) Outbreaks of human infection caused by pyogenic streptococci of Lancefield groups C and G. *Journal of Medical Microbiology*, **29**, 207–219.

Efstratiou A (1997) Pyogenic streptococci of Lancefield groups C and G as pathogens in man. *Journal of Applied Microbiology Symposium Supplement*, **83**, 72S–79S.

Efstratiou A (2000) Group A streptococci in the 1990s. *The Journal of Antimicrobial Chemotherapy*, **45**(Suppl.), 3–12 [topic T1].

Efstratiou A, Facklam RR, Kaplan EL *et al.* (2000) Report on an international quality assurance programme for group A streptococcal characterization. In *Streptococci and Streptococcal Diseases: Entering the New Millennium. Proceedings of the XIV Lancefield International Symposium on Streptococci and Streptococcal Diseases.* Martin DR, Tagg JR (eds). Auckland, New Zealand: Securacopy, pp. 797–803.

Efstratiou A, Teare EL, McGhie D, Colman G (1989) The presence of M proteins in outbreak strains of *Streptococcus equisimilis* T-type 204. *The Journal of Infection*, **19**, 105–111.

Enright MC, Spratt BG, Kalia A *et al.* (2001) Multilocus sequence typing of *Streptococcus pyogenes* and the relationships between *emm* type and clone. *Infection and Immunity*, **69**, 2416–2427.

Eriksson BK, Andersson J, Holm SE, Norgren M (1998) Epidemiological and clinical aspects of invasive group A streptococcal infections and the streptococcal toxic shock syndrome. *Clinical Infectious Diseases*, **27**, 1428–1436.

Facklam R (2002) What happened to the streptococci: overview of taxonomic and nomenclature changes. *Clinical Microbiology Reviews*, **15**, 613–630.

Facklam RF, Martin DR, Lovgren M *et al.* (2002) Extension of the Lancefield classification for group A streptococci by addition of 22 new M protein gene sequence types from clinical isolates: *emm*103 to *emm*124. *Clinical Infectious Diseases*, **34**, 28–38.

Factor SH, Levine OS, Schwartz B *et al.* (2003) Invasive group A streptococcal disease: risk factors for adults. *Emerging Infectious Diseases*, **9**, 970–977.

Farley MM, Harvey RC, Stull T *et al.* (1993) A population-based assessment of invasive disease due to group B streptococcus in nonpregnant adults. *The New England Journal of Medicine*, **328**, 1807–1811.

Farrell DJ, Morrissey I, Bakker S, Felmingham D (2002) Molecular characterization of macrolide resistance mechanisms among *Streptococcus pneumoniae* and *Streptococcus pyogenes* isolated from the PROTEKT 1999–2000 study. *The Journal of Antimicrobial Chemotherapy*, **50**(Suppl. S1), 39–47.

Ferretti JJ, McShan WM, Ajdic D *et al.* (2001) Complete genome sequence of an M1 strain of *Streptococcus pyogenes*. *Proceedings of the National Academy of Sciences of the United States of America*, **98**, 4658–4663.

Francis AJ, Nimmo GR, Efstratiou A *et al.* (1993) Investigation of milk-borne *Streptococcus zooepidemicus* infection associated with glomerulonephritis in Australia. *The Journal of Infection*, **27**, 317–323.

Garcia-Rey C, Aguilar L, Baquero F *et al.* (2002) Pharmacoepidemiological analysis of provincial differences between consumption of macrolides and rates of erythromycin resistance among *Streptococcus pyogenes* isolates in Spain. *Journal of Clinical Microbiology*, **40**, 2959–2963.

Gemmell CG, Peterson PK, Schmeling D *et al.* (1981) Potentiation of opsonization and phagocytosis of *Streptococcus pyogenes* following growth in the presence of clindamycin. *The Journal of Clinical Investigation*, **67**, 1249–1256.

Gerber MA, Shulman ST (2004) Rapid diagnosis of pharyngitis caused by group A streptococci. *Clinical Microbiology Reviews*, **17**, 571–580.

Giovanetti E, Montanari MP, Mingoia M, Varaldo PE (1999) Phenotypes and genotypes of erythromycin-resistant *Streptococcus pyogenes* strains in Italy and heterogeneity of inducibly resistant strains. *Antimicrobial Agents and Chemotherapy*, **43**, 1935–1940.

Glaser P, Rusniok C, Buchrieser C *et al.* (2002) Genome sequence of *Streptococcus agalactiae*, a pathogen causing invasive neonatal disease. *Molecular Microbiology*, **45**, 1499–1513.

Gordon KA, Beach ML, Biedenbach DJ *et al.* (2002) Antimicrobial susceptibility patterns of beta-hemolytic and viridans group streptococci: report from the SENTRY Antimicrobial Surveillance Program (1997–2000). *Diagnostic Microbiology and Infectious Disease*, **43**, 157–162.

Graham MR, Smoot LM, Migliaccio CA *et al.* (2002) Virulence control in group A streptococcus by a two-component gene regulatory system: global expression profiling and *in vivo* infection modeling. *Proceedings of the National Academy of Sciences of the United States of America*, **99**, 13855–13860.

Griffith F (1934) The serological classification of *Streptococcus pyogenes*. *The Journal of Hygiene*, **34**, 542–584.

Guilherme L, Dulphy N, Douay C *et al.* (2000) Molecular evidence for antigen-driven immune responses in cardiac lesions of rheumatic heart disease patients. *International Immunology*, **12**, 1063–1074.

Gwaltney JM, Bisno AL (2000) Pharyngitis. In *Mandell, Douglas, and Bennett's Principles & Practice of Infectious Diseases, 5th edition.* Mandell GL, Bennett JE, Dolin R (eds). New York: Churchill Livingstone, vol. 1, pp. 656–662.

Harrington DJ, Sutcliffe IC, Chanter N (2002) The molecular basis of *Streptococcus equi* infection and disease. *Microbes and infection/Institut Pasteur*, **4**, 501–510.

Heath PT, Balfour G, Weisner AM *et al.* (2004) Group B streptococcal disease in UK and Irish infants younger than 90 days. *Lancet*, **363**, 292–294.

Heitmann C, Pelzer M, Bickert B *et al.* (2001) [Surgical concept and results in necrotizing fasciitis]. *Chirurg*, **72**, 168–173.

Herrera L, Salcedo C, Orden B *et al.* (2002) Rifampin resistance in *Streptococcus pyogenes*. *European Journal of Clinical Microbiology & Infectious Diseases*, **21**, 411–413.

Hindsholm M, Schonheyder HC (2002) Clinical presentation and outcome of bacteraemia caused by beta-haemolytic streptococci serogroup G. *APMIS*, **110**, 554–558.

Hirose Y, Yagi K, Honda H *et al.* (1997) Toxic shock-like syndrome caused by non-group A beta-hemolytic streptococci. *Archives of Internal Medicine*, **157**, 1891–1894.

Hollingshead SK, Readdy TL, Yung DL, Bessen DE (1993) Structural heterogeneity of the *emm* gene cluster in group A streptococci. *Molecular Microbiology*, **8**, 707–717.

Holm SE (2000) Treatment of recurrent tonsillopharyngitis. *The Journal of Antimicrobial Chemotherapy*, **45**(Suppl.), 31–35 [topic T1].

HPA (2003a) Candidaemia and polymicrobial bacteraemias: England, Wales, and Northern Ireland, 2002. *CDR Weekly*, **13**(42), p. Bacteraemia. www.hpa.org.uk/cdr/PDFfiles/2003/cdr4203.pdf (25 August 2004).

HPA (2003b) Pyogenic and non-pyogenic streptococcal bacteraemias, England, Wales, and Northern Ireland: 2002. *CDR Weekly*, **13**(16). www.hpa.org.uk/cdr/PDFfiles/2003/1603strep.pdf (25 August 2004).

HPA (2004a) Statutory Notifications of Infectious Diseases (NOIDs)—Annual Totals 1991 to 2003—England and Wales. Health Protection Agency. www.hpa.org.uk/infections/topics_az/noids/annualtab.htm (25 August 2004).

HPA (2004b) Laboratory reports of bacteraemias, England, Wales, and Northern Ireland: 2002 and 2003. *CDR Weekly*, **14**(3). www.hpa. org.uk/cdr/archives/archive04/bacteraemia04.htm#bact_02/03 (25 August 2004).

HPA (2004c) Pyogenic and non-pyogenic streptococcal bacteraemias, England, Wales, and Northern Ireland: 2003. *CDR Weekly*, **14**(16). www.hpa.org.uk/cdr/PDFfiles/2004/bact_1604.pdf (25 August 2004).

Hsueh PR, Wu JJ, Tsai PJ *et al.* (1998) Invasive group A streptococcal disease in Taiwan is not associated with the presence of streptococcal pyrogenic exotoxin genes. *Clinical Infectious Diseases*, **26**, 584–589.

Hughes MJ, Moore JC, Lane JD *et al.* (2002) Identification of major outer surface proteins of *Streptococcus agalactiae*. *Infection and Immunity*, **70**, 1254–1259.

Ikebe T, Murayama S, Saitoh K *et al.* (2004) Surveillance of severe invasive group-G streptococcal infections and molecular typing of the isolates in Japan. *Epidemiology and Infection*, **132**, 145–149.

Ji Y, Schnitzler N, DeMaster E, Cleary P (1998) Impact of M49, Mrp, Enn, and C5a peptidase proteins on colonization of the mouse oral mucosa by *Streptococcus pyogenes*. *Infection and Immunity*, **66**, 5399–6405.

Johnson DR, Kaplan EL (1993) A review of the correlation of T-agglutination patterns and M-protein typing and opacity factor production in the identification of group A streptococci. *Journal of Medical Microbiology*, **38**, 311–315.

Jones N, Bohnsack JF, Takahashi S *et al.* (2003) Multilocus sequence typing system for group B streptococcus. *Journal of Clinical Microbiology*, **41**, 2530–2536.

Kalia A, Bessen DE (2003) Presence of streptococcal pyrogenic exotoxin A and C genes in human isolates of group G streptococci. *FEMS Microbiology Letters*, **219**, 291–295.

Kalia A, Enright MC, Spratt BG, Bessen DE (2001) Directional gene movement from human-pathogenic to commensal-like streptococci. *Infection and Immunity*, **69**, 4858–4869.

Kalliola S, Vuopio-Varkila J, Takala AK, Eskola J (1999) Neonatal group B streptococcal disease in Finland: a ten-year nationwide study. *The Pediatric Infectious Disease Journal*, **18**, 806–810.

Kaplan EL, Johnson DR (2001) Unexplained reduced microbiological efficacy of intramuscular benzathine penicillin G and of oral penicillin V in eradication of group A streptococci from children with acute pharyngitis. *Pediatrics*, **108**, 1180–1186.

Kaplan EL, Wotton JT, Johnson DR (2001) Dynamic epidemiology of group A streptococcal serotypes associated with pharyngitis. *Lancet*, **358**, 1334–1337.

Kataja J, Huovinen P, Seppala H (2000) Erythromycin resistance genes in group A streptococci of different geographical origins. The Macrolide Resistance Study Group. *The Journal of Antimicrobial Chemotherapy*, **46**, 789–792.

Katz AR, Morens DM (1992) Severe streptococcal infections in historical perspective. *Clinical Infectious Diseases*, **14**, 298–307.

Kaul R, McGeer A, Low DE *et al.* (1997) Population-based surveillance for group A streptococcal necrotizing fasciitis: clinical features, prognostic indicators, and microbiologic analysis of seventy-seven cases. Ontario Group A Streptococcal Study. *The American Journal of Medicine*, **103**, 18–24.

Kaul R, McGeer A, Norrby-Teglund A *et al.* (1999) Intravenous immunoglobulin therapy for streptococcal toxic shock syndrome—a comparative observational study. The Canadian Streptococcal Study Group. *Clinical Infectious Diseases*, **28**, 800–807.

Kawamura Y, Hou XG, Sultana F *et al.* (1995) Determination of 16S rRNA sequences of *Streptococcus mitis* and *Streptococcus gordonii* and phylogenetic relationships among members of the genus *Streptococcus*. *International Journal of Systematic Bacteriology*, **45**, 406–408.

Kong F, Martin D, James G, Gilbert GL (2003) Towards a genotyping system for *Streptococcus agalactiae* (group B streptococcus): use of mobile genetic elements in Australasian invasive isolates. *Journal of Medical Microbiology*, **52**, 337–344.

Kotloff KL, Corretti M, Palmer K *et al.* (2004) Safety and immunogenicity of a recombinant multivalent group A streptococcal vaccine in healthy adults: phase 1 trial. *The Journal of the American Medical Association*, **292**, 709–715.

Kozlov RS, Bogdanovitch TM, Appelbaum PC *et al.* (2002) Antistreptococcal activity of telithromycin compared with seven other drugs in relation to macrolide resistance mechanisms in Russia. *Antimicrobial Agents and Chemotherapy*, **46**, 2963–2968.

Kreikemeyer B, McIver KS, Podbielski A (2003) Virulence factor regulation and regulatory networks in *Streptococcus pyogenes* and their impact on pathogen–host interactions. *Trends in Microbiology*, **11**, 224–232.

Kristensen B, Schonheyder HC (1995) A 13-year survey of bacteraemia due to beta-haemolytic streptococci in a Danish county. *Journal of Medical Microbiology*, **43**, 63–67.

Lamagni TL, Neal S, Alhaddad N, Efstratiou A (2004) Results from the first six months of enhanced surveillance of severe *Streptococcus pyogenes* disease in England and Wales. 14th European Congress of Clinical Microbiology and Infectious Diseases, Prague, Czech Republic, 1–4 May 2004. *Clinical Microbiology and Infection*, **10**(Suppl. 3), 1–86 [abstract O199].

Lamothe F, D'Amico P, Ghosn P *et al.* (1995) Clinical usefulness of intravenous human immunoglobulins in invasive group A streptococcal infections: case report and review. *Clinical Infectious Diseases*, **21**, 1469–1470.

Lancefield RC (1962) Current knowledge of type specific M antigen of group A streptococci. *Journal of Immunology*, **89**, 307–313.

Larppanichpoonphol P, Watanakunakorn C (2001) Group B streptococcal bacteremia in nonpregnant adults at a community teaching hospital. *Southern Medical Journal*, **94**, 1206–1211.

Lawrence DN, Facklam RR, Sottnek FO *et al.* (1979) Epidemiologic studies among Amerindian populations of Amazonia. I. Pyoderma: prevalence and associated pathogens. *The American Journal of Tropical Medicine and Hygiene*, **28**, 548–558.

Le Bouguenec C, de Cespedes G, Horaud T (1990) Presence of chromosomal elements resembling the composite structure Tn3701 in streptococci. *Journal of Bacteriology*, **172**, 727–734.

Leclercq R (2002) Mechanisms of resistance to macrolides and lincosamides: nature of the resistance elements and their clinical implications. *Clinical Infectious Diseases*, **34**, 482–492.

Levy SB, McMurry LM, Barbosa TM *et al.* (1999) Nomenclature for new tetracycline resistance determinants. *Antimicrobial Agents and Chemotherapy*, **43**, 1523–1524.

Lipsitch M (1999) Bacterial vaccines and serotype replacement: lessons from *Haemophilus influenzae* and prospects for *Streptococcus pneumoniae*. *Emerging Infectious Diseases*, **5**, 336–345.

Llewelyn M, Cohen J (2002) Superantigens: microbial agents that corrupt immunity. *The Lancet Infectious Diseases*, **2**, 156–162.

Lukomski S, Nakashima K, Abdi I *et al.* (2000) Identification and characterization of the *scl* gene encoding a group A streptococcus extracellular protein virulence factor with similarity to human collagen. *Infection and Immunity*, **68**, 6542–6553.

Madhi SA, Radebe K, Crewe-Brown H *et al.* (2003) High burden of invasive *Streptococcus agalactiae* disease in South African infants. *Annals of Tropical Paediatrics*, **23**, 15–23.

Majeed HA, Batnager S, Yousof AM *et al.* (1992) Acute rheumatic fever and the evolution of rheumatic heart disease: a prospective 12 year follow-up report. *Journal of Clinical Epidemiology*, **45**, 871–875.

Marques MB, Kasper DL, Pangburn MK, Wessels MR (1992) Prevention of C3 deposition by capsular polysaccharide is a virulence mechanism of type III group B streptococci. *Infection and Immunity*, **60**, 3986–3993.

Martin PR, Høiby EA (1990) Streptococcal serogroup A epidemic in Norway 1987–1988. *Scandinavian Journal of Infectious Diseases*, **22**, 421–429.

Matsumoto M, Hoe NP, Liu M *et al.* (2003) Intrahost sequence variation in the streptococcal inhibitor of complement gene in patients with human pharyngitis. *The Journal of Infectious Diseases*, **187**, 604–612.

McGregor KF, Spratt BG, Kalia A *et al.* (2004) Multilocus sequence typing of *Streptococcus pyogenes* representing most known *emm* types and distinctions among subpopulation genetic structures. *Journal of Bacteriology*, **186**, 4285–4294.

Mogielnicki NP, Schwartzman JD, Elliott JA (2000) Perineal group A streptococcal disease in a pediatric practice. *Pediatrics*, **106**, 276–281.

Musser JM, Kapur V, Kanjilal S *et al.* (1993) Geographic and temporal distribution and molecular characterization of two highly pathogenic clones of *Streptococcus pyogenes* expressing allelic variants of pyrogenic exotoxin A (scarlet fever toxin). *The Journal of Infectious Diseases*, **167**, 337–346.

Nakagawa I, Kurokawa K, Yamashita A *et al.* (2003) Genome sequence of an M3 strain of *Streptococcus pyogenes* reveals a large-scale genomic rearrangement in invasive strains and new insights into phage evolution. *Genome Research*, **13**, 1042–1055.

Nelson D, Loomis L, Fischetti VA (2001) Prevention and elimination of upper respiratory colonization of mice by group A streptococci by using a bacteriophage lytic enzyme. *Proceedings of the National Academy of Sciences of the United States of America*, **98**, 4107–4112.

Nizet V (2002) Streptococcal β-hemolysins: genetics and role in disease pathogenesis. *Trends in Microbiology*, **10**, 575–580.

Norrby-Teglund A, Kaul R, Low DE *et al*. (1996) Plasma from patients with severe invasive group A streptococcal infections treated with normal polyspecific IgG inhibits streptococcal superantigen-induced T cell proliferation and cytokine production. *Journal of Immunology*, **156**, 3057–3064.

O'Brien KL, Beall B, Barrett NL *et al*. (2002) Epidemiology of invasive group A streptococcus disease in the United States, 1995–1999. *Clinical Infectious Diseases*, **35**, 268–276.

Olivier C (2000) Rheumatic fever—is it still a problem? *The Journal of Antimicrobial Chemotherapy*, **45**(Suppl.), 13–21 [topic T1].

Paoletti LC, Kasper DL (2003) Glycoconjugate vaccines to prevent group B streptococcal infections. *Expert Opinion on Biological Therapy*, **3**, 975–984.

Perez-Trallero E, Fernandez-Mazarrasa C, Garcia-Rey C *et al*. (2001) Antimicrobial susceptibilities of 1,684 *Streptococcus pneumoniae* and 2,039 *Streptococcus pyogenes* isolates and their ecological relationships: results of a 1-year (1998–1999) multicenter surveillance study in Spain. *Antimicrobial Agents and Chemotherapy*, **45**, 3334–3340.

Podbielski A, Schnitzler N, Beyhs P, Boyle MD (1996) M-related protein (Mrp) contributes to group A streptococcal resistance to phagocytosis by human granulocytes. *Molecular Microbiology*, **19**, 429–441.

Proft T, Fraser JD (2003) Bacterial superantigens. *Clinical and Experimental Immunology*, **133**, 299–306.

Pruksakorn S, Currie B, Brandt E *et al*. (1994) Identification of T cell autoepitopes that cross-react with the C-terminal segment of the M protein of group A streptococci. *International Immunology*, **6**, 1235–1244.

Rasmussen M, Eden A, Bjorck L (2000) SclA, a novel collagen-like surface protein of *Streptococcus pyogenes*. *Infection and Immunity*, **68**, 6370–6377.

Rasmussen M, Muller HP, Bjorck L (1999) Protein GRAB of *Streptococcus pyogenes* regulates proteolysis at the bacterial surface by binding alpha2-macroglobulin. *The Journal of Biological Chemistry*, **274**, 15336–15344.

Reid SD, Green NM, Sylva GL *et al*. (2002) Postgenomic analysis of four novel antigens of group A streptococcus: growth phase-dependent gene transcription and human serologic response. *Journal of Bacteriology*, **184**, 6316–6324.

Reid SD, Montgomery AG, Voyich JM *et al*. (2003) Characterization of an extracellular virulence factor made by group A streptococcus with homology to the *Listeria monocytogenes* internalin family of proteins. *Infection and Immunity*, **71**, 7043–7052.

Robinson JH, Case MC, Kehoe MA (1993) Characterization of a conserved helper-T-cell epitope from group A streptococcal M proteins. *Infection and Immunity*, **61**, 1062–1068.

Royal College of General Practitioners (1999) General practitioner workload. *RCGP Information Sheet 3*, 1–2.

Sachse S, Seidel P, Gerlach D *et al*. (2002) Superantigen-like gene(s) in human pathogenic *Streptococcus dysgalactiae*, subsp. *equisimilis*: genomic localisation of the gene encoding streptococcal pyrogenic exotoxin G (*speG*(dys)). *FEMS Immunology and Medical Microbiology*, **34**, 159–167.

Schalén C (2002) European surveillance of severe group A streptococcal disease. *Eurosurveillance Weekly*, **6**(35), 1–4. www.eurosurveillance.org/ew/2002/020829.asp (25 August 2004).

Schrag S, Gorwitz R, Fultz-Butts K, Schuchat A (2002) Prevention of perinatal group B streptococcal disease. Revised guidelines from CDC. *Morbidity and Mortality Weekly Report. Recommendations and Reports/Centers for Disease Control*, **51**(RR-11), 1–22.

Schrag SJ, Zywicki S, Farley MM *et al*. (2000) Group B streptococcal disease in the era of intrapartum antibiotic prophylaxis. *The New England Journal of Medicine*, **342**, 15–20.

Schuchat A (1999) Group B streptococcus. *Lancet*, **353**, 51–56.

Sela S, Neeman R, Keller N, Barzilai A (2000) Relationship between asymptomatic carriage of *Streptococcus pyogenes* and the ability of the strains to adhere to and be internalised by cultured epithelial cells. *Journal of Medical Microbiology*, **49**, 499–502.

Seppala H, Nissinen A, Jarvinen H *et al*. (1992) Resistance to erythromycin in group A streptococci. *The New England Journal of Medicine*, **326**, 292–297.

Seppala H, Nissinen A, Yu Q, Huovinen P (1993) Three different phenotypes of erythromycin-resistant *Streptococcus pyogenes* in Finland. *The Journal of Antimicrobial Chemotherapy*, **32**, 885–891.

Seppala H, Skurnik M, Soini H, Roberts MC, Huovinen P (1998) A novel erythromycin resistance methylase gene (*ermTR*) in *Streptococcus pyogenes*. *Antimicrobial Agents and Chemotherapy*, **42**, 257–262.

Shulman ST, Ayoub EM (2002) Poststreptococcal reactive arthritis. *Current Opinion in Rheumatology*, **14**, 562–565.

Simckes AM, Spitzer A (1995) Poststreptococcal acute glomerulonephritis. *Pediatrics in Review/American Academy of Pediatrics*, **16**, 278–279.

Skansen-Saphir U, Andersson J, Bjork L, Andersson U (1994) Lymphokine production induced by streptococcal pyrogenic exotoxin-A is selectively down-regulated by pooled human IgG. *European Journal of Immunology*, **24**, 916–922.

Smoot JC, Barbian KD, Van Gompel JJ *et al*. (2002) Genome sequence and comparative microarray analysis of serotype M18 group A streptococcus strains associated with acute rheumatic fever outbreaks. *Proceedings of the National Academy of Sciences of the United States of America*, **99**, 4668–4673.

Smoot LM, Smoot JC, Graham MR *et al*. (2001) Global differential gene expression in response to growth temperature alteration in group A *Streptococcus*. *Proceedings of the National Academy of Sciences of the United States of America*, **98**, 10416–10421.

Snider LA, Swedo SE (2003) Post-streptococcal autoimmune disorders of the central nervous system. *Current Opinion in Neurology*, **16**, 359–365.

Sriskandan S, McKee A, Hall L, Cohen J (1997) Comparative effects of clindamycin and ampicillin on superantigenic activity of *Streptococcus pyogenes*. *The Journal of Antimicrobial Chemotherapy*, **40**, 275–277.

Stamenkovic I, Lew PD (1984) Early recognition of potentially fatal necrotizing fasciitis. The use of frozen-section biopsy. *The New England Journal of Medicine*, **310**, 1689–1693.

Statens Serum Institut (2002) A rise in invasive group A streptococcal infections. *Epidemiology News*, 37. www.ssi.dk/sw1875.asp (21 May 2005).

Statens Serum Institut, Danish Veterinary and Food Administration, Danish Medicines Agency and Danish Institute for Food and Veterinary Research (2004) DANMAP 2003 – Use of antimicrobial agents and occurrence of antimicrobial resistance in bacteria from food animals, foods and humans in Denmark. Statens Serum Institut, Denmark. www.dfvf.dk/Files/Filer/Zoonosecentret/Publikationer/Danmap/Danmap_2003.pdf (21 May 2005).

Stevens DL, Bryant AE, Yan S (1994) Invasive group A streptococcal infection: new concepts in antibiotic treatment. *International Journal of Antimicrobial Agents*, **4**, 297–301.

Stevens DL, Gibbons AE, Bergstrom R, Winn V (1988) The Eagle effect revisited: efficacy of clindamycin, erythromycin, and penicillin in the treatment of streptococcal myositis. *The Journal of Infectious Diseases*, **158**, 23–28.

Stevens DL, Tanner MH, Winship J *et al*. (1989) Severe group A streptococcal infections associated with a toxic shock-like syndrome and scarlet fever toxin A. *The New England Journal of Medicine*, **321**, 1–7.

Stricker T, Navratil F, Sennhauser FH (2003) Vulvovaginitis in prepubertal girls. *Archives of Disease in Childhood*, **88**, 324–326.

Sutcliffe J, Tait-Kamradt A, Wondrack L (1996) *Streptococcus pneumoniae* and *Streptococcus pyogenes* resistant to macrolides but sensitive to clindamycin: a common resistance pattern mediated by an efflux system. *Antimicrobial Agents and Chemotherapy*, **40**, 1817–1824.

Svensson N, Oberg S, Henriques B *et al*. (2000) Invasive group A streptococcal infections in Sweden in 1994 and 1995: epidemiology and clinical spectrum. *Scandinavian Journal of Infectious Diseases*, **32**, 609–614.

Sylvetsky N, Raveh D, Schlesinger Y *et al*. (2002) Bacteremia due to betahemolytic streptococcus group G: increasing incidence and clinical characteristics of patients. *The American Journal of Medicine*, **112**, 622–626.

Tagg JR, Wong HK (1983) Inhibitor production by group-G streptococci of human and of animal origin. *Journal of Medical Microbiology*, **16**, 409–415.

Tait-Kamradt A, Clancy J, Cronan M *et al*. (1997) mefE is necessary for the erythromycin-resistant M phenotype in *Streptococcus pneumoniae*. *Antimicrobial Agents and Chemotherapy*, **41**, 2251–2255.

Tettelin H, Masignani V, Cieslewicz MJ *et al*. (2002) Complete genome sequence and comparative genomic analysis of an emerging human pathogen, serotype V *Streptococcus agalactiae*. *Proceedings of the National Academy of Sciences of the United States of America*, **99**, 12391–12396.

Trivalle C, Martin E, Martel P *et al*. (1998) Group B streptococcal bacteraemia in the elderly. *Journal of Medical Microbiology*, **47**, 649–652.

Trollfors B, Nylen O, Carenfelt C *et al*. (1998) Aetiology of acute epiglottitis in adults. *Scandinavian Journal of Infectious Diseases*, **30**, 49–51.

Valdimarsson H, Baker BS, Jonsdottir I *et al.* (1995) Psoriasis: a T-cell-mediated autoimmune disease induced by streptococcal superantigens. *Immunology Today*, **16**, 145–149.

Veasy LG, Wiedmeier SE, Orsmond GS *et al.* (1987) Resurgence of acute rheumatic fever in the intermountain area of the United States. *The New England Journal of Medicine*, **316**, 421–427.

Vereanu A, Mihalcu F (1979) Improved lysotyping scheme for group C streptococci with new phage preparations. *Archives roumaines de pathologie experimentales et de microbiologie*, **38**, 265–272.

Vlaminckx BJM, van Pelt W, Schouls LM *et al.* (2004) Long-term surveillance of invasive group A streptococcal disease. 14th European Congress of Clinical Microbiology and Infectious Diseases, Prague, Czech Republic, 1–4 May 2004. *Clinical Microbiology and Infection*, **10**(Suppl. 3), 1–86 [abstract P1411].

Voyich JM, Sturdevant DE, Braughton KR *et al.* (2003) Genome-wide protective response used by group A streptococcus to evade destruction by human polymorphonuclear leukocytes. *Proceedings of the National Academy of Sciences of the United States of America*, **100**, 1996–2001.

Wagner JG, Schlievert PM, Assimacopoulos AP *et al.* (1996) Acute group G streptococcal myositis associated with streptococcal toxic shock syndrome: case report and review. *Clinical Infectious Diseases*, **23**, 1159–1161.

Weisblum B (1985) Inducible resistance to macrolides, lincosamides and streptogramin type B antibiotics: the resistance phenotype, its biological diversity, and structural elements that regulate expression—a review. *The Journal of Antimicrobial Chemotherapy*, **16**(Suppl. A), 63–90.

Weisner AM, Johnson AP, Lamagni TL *et al.* (2004) Characterization of group B streptococci recovered from infants with invasive disease in England and Wales. *Clinical Infectious Diseases*, **38**, 1203–1208.

Wessels MR, Bronze MS (1994) Critical role of the group A streptococcal capsule in pharyngeal colonization and infection in mice. *Proceedings of the National Academy of Sciences of the United States of America*, **91**, 12238–12242.

Whatmore AM (2001) *Streptococcus pyogenes sclB* encodes a putative hypervariable surface protein with a collagen-like repetitive structure. *Microbiology*, **147**, 419–429.

Whatmore AM, Kehoe MA (1994) Horizontal gene transfer in the evolution of group A streptococcal *emm*-like genes: gene mosaics and variation in *Vir* regulons. *Molecular Microbiology*, **11**, 363–374.

Working Group on Severe Streptococcal Infections (1993) Defining the group A streptococcal toxic shock syndrome. Rationale and consensus definition. *The Journal of the American Medical Association*, **269**, 390–391.

Zimbelman J, Palmer A, Todd J (1999) Improved outcome of clindamycin compared with beta-lactam antibiotic treatment for invasive *Streptococcus pyogenes* infection. *The Pediatric Infectious Disease Journal*, **18**, 1096–1100.

Zurawski CA, Bardsley M, Beall B *et al.* (1998) Invasive group A streptococcal disease in metropolitan Atlanta: a population-based assessment. *Clinical Infectious Diseases*, **27**, 150–157.

2

Oral and Other Non-β-Haemolytic Streptococci

Roderick McNab[1] and Theresa Lamagni[2]

[1]*GlaxoSmithKline, Surrey; and* [2] *Health Protection Agency, Centre for Infection, London, UK*

INTRODUCTION

The genus *Streptococcus* consists of Gram-positive cocci that are facultatively anaerobic, nonsporing and catalase negative, with cells arranged in chains or pairs. Their nutritional requirements are complex, and carbohydrates are fermented to produce L-(+)-lactic acid as the main end product.

All streptococci are obligate parasites of mammals, and within the genus are many species that are important pathogens of humans and domestic animals as well as many well-known opportunistic pathogens.

The determination of haemolysis on blood agar plates is a useful characteristic for identifying streptococci. Members of this genus may be broadly divided into β-haemolytic (pyogenic) and non-β-haemolytic groups. The latter group comprises species that are α-haemolytic on blood agar (the so-called viridans streptococci that produce a greenish discolouration surrounding colonies on blood agar plates) as well as nonhaemolytic (γ-haemolytic) strains.

This chapter focuses on these non-β-haemolytic streptococci, with the exception of *Streptococcus pneumoniae*, a close relative of the mitis group of oral streptococci, which is given its own chapter (Chapter 3) as befits its more pathogenic status. For a list of non-β-haemolytic streptococci, see Tables 2.1 and 2.2.

DESCRIPTION OF THE ORGANISMS

Current Taxonomic Status

Since the 1980s the genus *Streptococcus* has undergone considerable upheaval including the removal of lactic and enteric species into separate genera (*Lactococcus* and *Enterococcus*, respectively). Additionally, many new genera of Gram-positive cocci that grow in pairs or chains were established, primarily through reclassification from the genus *Streptococcus* by genotypic and phenotypic characteristics. For reviews, see Whiley and Beighton (1998) and Facklam (2002).

Classification of species that remain within the genus *Streptococcus* has been fraught with difficulty partly due to the early overreliance on serological and haemolytic reactions. Although Lancefield serological grouping, based on carbohydrate antigens present in the cell walls of streptococci, has proved a very useful tool for the more pathogenic β-haemolytic streptococci (see Chapter 1), its application to non-β-haemolytic streptococci is of little value for identification where group-specific antigens may be absent or shared by several distinct taxa, for example. Similarly, although the type of haemolysis demonstrated by streptococci is a useful marker in the initial examination of clinical isolates, the distinction between α- and γ-haemolytic reactions on blood agar plates may be of limited taxonomical value. β-Haemolysis by

Table 2.1 Oral streptococci

Group and species	Source	Comments
Anginosus group		
S. anginosus	Human	
S. constellatus	Human	
S. intermedius	Human	
Mitis group		
S. mitis	Human	
S. oralis	Human	
S. crista	Human	*S. cristatus*[a]
S. infantis	Human	
S. peroris	Human	
S. orisratti	Rat	
Sanguis group[b]		
S. sanguis	Human	*S. sanguinis*[a]
S. parasanguis	Human	*S. parasanguinis*[a]
S. gordonii	Human	Formerly *S. sanguis*
Salivarius group		
S. salivarius	Human	
S. vestibularis	Human	
S. thermophilus	Dairy product	
Mutans group		
S. mutans	Human	Serotype c, e, f
S. sobrinus	Human, rat	Serotype d, g
S. rattus	Rat, (human)	Serotype b, *S. ratti*[a]
S. downei	Monkey	Serotype h
S. cricetus	Hamster, (human)	Serotype a
S. macacae	Monkey	Serotype c
S. hyovaginalis	Swine	

The generic term oral streptococcus is used with caution since many of the species listed may also be found in the gastrointestinal and genitourinary tracts. (human), rarely found in humans.
[a] Proposed correction to epithet (Trüper and de'Clari, 1997, 1998).
[b] *S. sanguis* group separated from larger *S. mitis* group by Facklam (2002).

streptococci is caused by well-characterized haemolysins (see Chapter 1), whereas the α-haemolytic reaction on blood agar plates results from the production and release of hydrogen peroxide by streptococcal colonies grown under aerobic conditions that causes oxidation of the haeme iron of haemoglobin in erythrocytes (Barnard and Stinson, 1996, 1999). Differences exist in the haemolytic reaction within species, and the reaction is dependent on culture conditions and the origin of the blood incorporated into the agar.

From the 1980s onwards the taxonomy of the streptococci has been developed using a wide range of approaches, thereby reducing the emphasis previously placed on serology and haemolysis. Significant contributions have included the analysis of cell-wall composition,

Principles and Practice of Clinical Bacteriology Second Edition Editors Stephen H. Gillespie and Peter M. Hawkey

Table 2.2 Lancefield group D *S. bovis/S. equinus* complex as proposed by Schlegel *et al.* (2003a)

Species	Previous designation	Comments
S. bovis/S. equinus[a]	Bovis II.1 (mannitol- and β-glucuronidase-negative and α-galactosidase-positive)	Bovine, equine, human[b]
S. gallolyticus subsp. *gallolyticus*	Bovis I, II.2 (mannitol-negative and β-glucuronidase- and β-mannosidase-positive) isolates	Human, bovine, environmental
subsp. *pasteurianus* subsp. *macedonicus*		Biotype I strains more commonly associated with colonic cancer patients than *S. bovis* II.1 (Rouff *et al.*, 1989)

S. infantarius
  subsp. *infantarius*
  subsp. *coli*
	New species designation, Schlegel *et al.* (2000)	Human, environmental

S. alactolyticus		Porcine

[a] The epithet *equinus* has nomenclatorial priority; however, the epithet *bovis* remains widely used in the medical literature.
[b] Rarely found in humans.

metabolic studies, numerical taxonomy and genotypic analysis such as DNA–DNA hybridization, DNA base composition [mol% guanine + cytosine (G+C)] and comparative sequence analysis of small subunit (16S) ribosomal DNA (reviewed by Facklam, 2002). This latter technique was used to subdivide the genus *Streptococcus* into six major clusters that broadly conform to the results of previous taxonomic studies (Kawamura *et al.*, 1995a) (Figure 2.1).

To further complicate matters, a taxonomic note by Trüper and de'Clari (1997, 1998) called for grammatical corrections in the Latin epithets of four streptococcal species, among 20 species. For the sake of clarity, where the epithets *sanguis*, *rattus* and *crista* have been used in the scientific literature for over 30 years, we have chosen to retain the historical nomenclature throughout this chapter (Kilian, 2001).

Anginosus Group Streptococci

These streptococci are common members of the oral flora and are found in the gastrointestinal and genital tracts and are clinically significant owing to their association with purulent infections in humans. The classification of these streptococci was confused for many years, and the term *Streptococcus milleri* has been used for a biochemically and serologically diverse collection of streptococci, despite not being accepted by taxonomists as a confirmed taxonomic entity. The weight of evidence from nucleic acid studies currently supports the recognition of three separate, albeit closely related, species – *Streptococcus anginosus*, *Streptococcus constellatus* and *Streptococcus intermedius* – and these have been published with amended species descriptions (Whiley and Beighton, 1991; Whiley *et al.*, 1993). Isolates of *S. intermedius* rarely have Lancefield group antigens, whereas isolates of *S. anginosus* and *S. constellatus* may have Lancefield F, C, A or G antigens. There are β-haemolytic strains of each of the three species; however, these are outnumbered by the non-β-haemolytic strains.

Mitis Group Streptococci

The mitis group of streptococci are prominent components of the human oropharyngeal microflora and are among the predominant

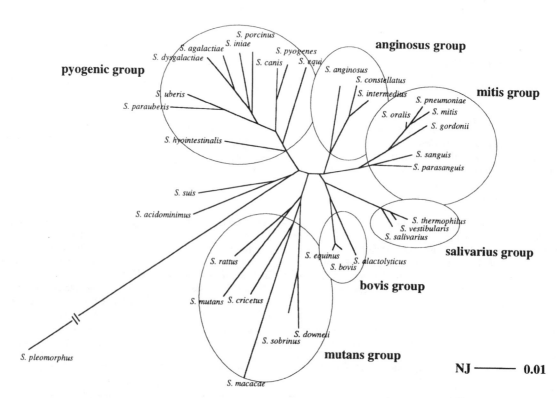

Figure 2.1 Phylogenetic relationships among 34 *Streptococcus* species. Distances were calculated from the neighbour-joining (NJ) method. Reproduced from Kawamura *et al.* (1995a) with permission of the International Union of Microbiological Societies.

causes of bacterial endocarditis. The application of nucleic acid studies has enabled some clarity to be derived for this taxonomically refractory group of oral streptococci. The mitis group as originally described by Kawamura *et al.* (1995a) included *Streptococcus sanguis*, *Streptococcus parasanguis*, *Streptococcus gordonii*, *Streptococcus crista*, *Streptococcus oralis*, *Streptococcus mitis* and *S. pneumoniae* (considered separately in Chapter 3). For several species, notably *S. sanguis*, *S. oralis* and *S. mitis*, combined taxonomic and nomenclatorial changes have been the cause of considerable confusion. Part of this confusion arose from the emphasis placed on the possession of Lancefield group H antigen, which was not well defined. Additionally, alternative nomenclature schemes have added to the confusion, particularly the use of *S. mitior*, a name traditionally assigned to an ill-defined species from the early days of streptococcal classification. The legacy of confusion that stems from a poor definition of the Lancefield group H antigen and the emphasis placed on its possession highlights the difficulties associated with the application of Lancefield grouping as a taxonomic tool for non-β-haemolytic streptococci.

Three additional mitis group species have been described: *Streptococcus peroris* and *Streptococcus infantis* were isolated from the human oral cavity (Kawamura *et al.*, 1998), whereas *Streptococcus orisratti* was isolated from the oral cavity of rats (Zhu, Willcox and Knox, 2000). Continued refinement and redefining of the mitis group of streptococci seem likely as the application of molecular taxonomic tools expands, and consequently it may be some time before we can fully correlate new and revised species in this group with human infections. The classification system proposed by Facklam (2002) and used in epidemiology reporting by the US Centers for Disease Control and by the Health Protection Agency in the UK has split *S. sanguis*, *S. parasanguis* and *S. gordonii* from the mitis group and placed them in a separate sanguis group on the basis of phenotypic characteristics. The division is used here for epidemiology reporting purposes, although 16S rRNA sequencing supports the taxonomic position described in Figure 2.1 (Kawamura *et al.*, 1995a; Whiley and Beighton, 1998).

Salivarius Group Streptococci

Streptococcus salivarius is commonly isolated from most areas within the human oral cavity, particularly from the tongue and other mucosal surfaces and from saliva, though it is rarely associated with disease. This species is considered to be an important component of the oral ecosystem because of the production of an extracellular fructose polysaccharide (levan) from dietary sucrose and the production of urease by some strains. DNA–DNA reassociation experiments have revealed that *S. salivarius* is closely related to *Streptococcus thermophilus*, a nonoral bacterium from dairy sources, and also to an oral species *Streptococcus vestibularis* which is an α-haemolytic, urease-producing streptococcus that is found predominantly on the vestibular mucosa of the human oral cavity (Hardie and Whiley, 1994). *Streptococcus vestibularis* does not produce extracellular polysaccharide. The formation of a distinct species group by these three streptococci was confirmed by 16S rDNA sequence comparisons (Bentley, Leigh and Collins, 1991; Kawamura *et al.*, 1995a).

Mutans Group Streptococci

The mutans group of streptococci exclusively colonize the tooth surfaces of humans and certain animals and are associated with dental caries, one of the most common infectious diseases of humans. Their characteristics include the ability to produce both soluble and insoluble extracellular polysaccharides from sucrose which are important in the formation of dental plaque and in the pathogenicity (cariogenicity) of the organism. Following the original species description of *Streptococcus mutans* from carious teeth by Clarke (1924) and its isolation from a case of bacterial endocarditis shortly afterwards, little attention was

paid to this species until the 1960s when it was demonstrated that caries could be experimentally induced and transmitted in animals. Similar caries-inducing streptococci were found in the dental plaque of humans, and from this point on, *S. mutans* became the focus of considerable attention. Studies revealed that isolates identified as *S. mutans* comprised a phenotypically and genetically heterogeneous taxon. Eight serotypes or serovars (designated a–h) have so far been identified, and six distinct species are currently recognized within the mutans group: *S. mutans* (serotypes c, e and f), *Streptococcus sobrinus* (serotypes d and g), *Streptococcus rattus* (serotype b), *Streptococcus downei* (serotype h), *Streptococcus cricetus* (serotype a) and *Streptococcus macacae* (serotype c). A seventh species initially included in the mutans group, *Streptococcus ferus* (serotype c), was shown to be only distantly related to all currently described streptococci, with no strong evidence to support its inclusion in this group (Whatmore and Whiley, 2002). Of these streptococci, *S. mutans* and *S. sobrinus* are commonly isolated from humans, whereas *S. cricetus* and *S. rattus* are occasionally recovered. *Streptococcus hyovaginalis*, isolated from the genital tract of pigs, is included in the mutans group because of similar phenotypic characteristics; however, identification from human sources has not yet been documented.

Bovis Group Streptococci

Streptococcus bovis was originally described as a bovine bacterium causing mastitis. Strains described as *S. bovis* have been isolated from patients with endocarditis, and this species is also reported to be associated with colon cancer and inflammatory bowel disease in humans. Human *S. bovis* strains are divided into two biotypes based on their ability (biotype I) or inability (biotype II) to ferment mannitol. Despite the recognition of the clinical significance of this organism, its classification remains confused. Schlegel *et al.* (2000) described a new species isolated from human and environmental sources, *Streptococcus infantarius*, and have proposed that the *S. bovis/Streptococcus equinus* complex be reorganized into seven species or subspecies (Table 2.2) based on a combination of phenotypic, genetic and phylogenetic methods (Schlegel *et al.*, 2003a). In this classification system, *S. bovis* biotype II.1 and *S. equinus* form a single genospecies. The epithet *equinus* has nomenclatorial priority; however, the epithet *bovis* remains widely used in the medical literature. *Streptococcus bovis* biotype II.2 isolates were reclassified along with *Streptococcus macedonicus* as *Streptococcus gallolyticus*.

Nutritionally Variant Streptococci

This clinically important group of bacteria, so named because of the need to supplement complex growth medium with cysteine or one of the active forms of vitamin B6 (pyroxidal hydrochloride or pyridoxamine hydrochloride) to obtain growth, forms part of the normal flora of the human throat and urogenital and intestinal tracts. They are of clinical interest because of their association with infective endocarditis and other conditions including otitis media, abscesses of the brain and pancreas, pneumonia and osteomyelitis (Bouvet, Grimont and Grimont, 1989). The names *Streptococcus adjacens* and *Streptococcus defectivus* were first proposed by Bouvet, Grimont and Grimont in 1989 but were subsequently reclassified on the basis of 16S rRNA gene sequence data and other phylogenetic analysis within a new genus, *Abiotrophia*, as *Abiotrophia adiacens* and *Abiotrophia defectiva*, respectively (Kawamura *et al.*, 1995b). *Abiotrophia adiacens* was subsequently transferred to *Granulicatella adiacens* (Collins and Lawson, 2000).

INFECTIONS CAUSED BY NON-β-HAEMOLYTIC STREPTOCOCCI

Non-β-haemolytic streptococci comprise a significant proportion of the normal commensal flora of the human body. However, they are involved in many types of infections, in which the source of the infection is almost invariably endogenous, being derived from the host's micro-flora. The streptococcal species themselves are generally thought to be of relatively low virulence, not normally associated with acute, rapidly spreading infections such as those caused by *Streptococcus pyogenes*, although they clearly have phenotypic features that result in the production of disease under appropriate circumstances.

Since many of these streptococci are present in the mouth, upper respiratory tract, genitourinary tract and, to a lesser extent, gastrointestinal tract, they are sometimes involved in pathological processes at these sites, possibly following some local or systemic change in host susceptibility or an alteration in local environmental conditions. A classic example is the manifestation of dental caries that arises following excessive consumption of dietary sugars, particularly sucrose. Alternatively, the streptococci at a mucosal site may gain access to the blood stream because of some local traumatic event and set up an infection at a distant location, such as the heart valve in endocarditis or in the brain or liver, giving rise to an abscess. The key event for infections at distant body sites is bacteraemia.

Epidemiology

Our understanding of the importance of non-β-haemolytic streptococci as bacteraemic pathogens has been hampered by a dearth of surveillance activity in many countries. Robust estimates of incidence have been, and to a degree remain, few and far between. Most quantitative studies undertaken have been based on case series originating from localized areas whose catchment populations are often difficult to enumerate and, therefore, difficult to translate into estimates of incidence.

In England and Wales a comprehensive laboratory-based surveillance network gathers reports of bacteraemia caused by all pathogens. Data from this surveillance on non-β-haemolytic streptococci indicate an incidence of 3.8 per 100 000 population in 2002 (HPA, 2003a). Non-β-haemolytic streptococci comprised approximately 3% of all bacteraemia reported through this system (HPA, 2002), broadly in line with estimates from other European countries and the Americas, which range from 1.5% to 5.9% (Jacobs *et al.*, 1995; Casariego *et al.*, 1996; Diekema *et al.*, 2000; Fluit *et al.*, 2000).

Of the non-β-haemolytic streptococci, mitis group organisms appear to be the most common cause of bacteraemia overall (Venditti *et al.*, 1989; Doern *et al.*, 1996; HPA, 2003a), the rate of reports in England and Wales being 1.4 per 100 000 population in 2002, with *S. oralis* being the most common single species identified (12% of all non-β-haemolytic streptococci). A smaller study from The Netherlands similarly found *S. oralis* to be the most common non-β-haemolytic streptococci causing bacteraemia, particularly associated with infection in haematology patients (Jacobs *et al.*, 1995). Studies focusing on neutropenic patients, one of the most vulnerable patient groups, suggest that between 13% and 18% of bacteraemias are due to non-β-haemolytic streptococci (Wisplinghoff *et al.*, 1999; Marron *et al.*, 2001), with *S. mitis* featuring as one of the most prominent species (Alcaide *et al.*, 1996).

Interestingly, the relative contribution of each non-β-haemolytic streptococcal group to bacteraemic episodes in England and Wales has changed over 1990–2000. The incidence of the formerly dominant sanguis group (*S. sanguis*, *S. parasanguis* and *S. gordonii*) has declined by half, whereas those of mitis and anginosus group strepto-cocci have increased dramatically by three- and twofold, respectively (Figure 2.2).

Estimates of the relative importance of healthcare exposure in the aetiology of non-β-haemolytic streptococcal bacteraemia have differed between studies. Two studies of *S. bovis* bacteraemia in Hong Kong (Lee *et al.*, 2003) and Israel (Siegman-Igra and Schwartz, 2003) found none and 10% of cases, respectively, to have been hospital acquired, whereas 6 of 31 (19%) *S. milleri* bacteraemias identified in a case series from Spain were considered to be hospital acquired

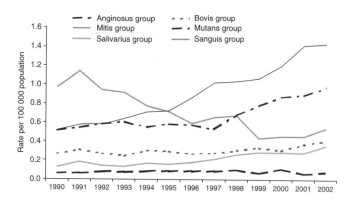

Figure 2.2 Annual rate of laboratory reports of bacteraemia caused by non-β-haemolytic streptococci, England and Wales. Source: Health Protection Agency (Communicable Disease Surveillance Centre).

(Casariego *et al.*, 1996). Nosocomial infection surveillance from the United States (Emori and Gaynes, 1993; Pfaller *et al.*, 1997) and England (NINSS, 2003) identified between 1% and 3% of hospital-acquired bacteraemia to involve non-β-haemolytic streptococci.

Being opportunistic pathogens, bacteraemia involving non-β-haemolytic streptococci tend to be concentrated in vulnerable individuals, namely the very young and older age groups (Figure 2.3). Much like many other blood stream infections, non-β-haemolytic streptococcus infection tends to have rates that are higher in males than in females, this pattern being seen across all non-β-haemolytic streptococcal groups (HPA, 2003b) and often also seen in studies of infective endocarditis (Mylonakis and Calderwood, 2001; Hoen *et al.*, 2002; Mouly *et al.*, 2002), with some exceptions (Hogevik *et al.*, 1995).

Infective Endocarditis

Infective endocarditis involving non-β-haemolytic streptococci usually occurs in patients with preexisting valvular lesions and is typically subacute, whereas the acute form of endocarditis which can occur in those with previously undamaged heart valves is associated with more virulent bacteria such as *Staphylococcus aureus*, *S. pyogenes* or *S. pneumoniae*. Patients at particular risk of developing subacute endocarditis include those with congenital heart defects affecting valves, those with acquired cardiac lesions following rheumatic fever

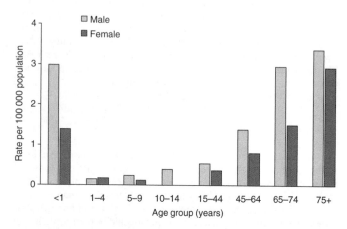

Figure 2.3 Age- and sex-specific incidence of anginosus group streptococcus laboratory reports in 2002, England and Wales. Source: Health Protection Agency (Communicable Disease Surveillance Centre).

and those who have undergone valve replacement therapy. These predisposing conditions may cause the development of noninfected platelet–fibrin vegetations (nonbacterial thrombotic vegetations) on the endocardium that may subsequently become infected by circulating microorganisms in the bloodstream during a transient bacteraemia.

It is often difficult to pinpoint the exact precipitating event that gives rise to the development of endocarditis, and some controversy exists over the incubation period between the bacteraemia that results in infection and the first symptoms of disease. An important potential source of the organisms that cause infectious bacteraemia is the oral cavity, although streptococci and other causative organisms may also enter the bloodstream from other body sites. Most invasive dental procedures (e.g. tooth extraction) may cause transient bacteraemia (Hall, Heimdahl and Nord, 1999). However, formal epidemiological studies suggest that dental procedures cause relatively few cases (van der Meer *et al.*, 1992; Strom *et al.*, 1998), and the value of antibiotic prophylaxis has been questioned (Durack, 1998; Morris and Webb, 2001). Considered more relevant is bacteraemia associated with poor dental hygiene and with routine oral procedures including brushing, flossing and chewing (Lockhart and Durack, 1999). Nevertheless, antibiotic prophylaxis is still recommended before certain invasive dental procedures on patients known to be at risk of developing bacterial endocarditis (see *Prevention and Control*).

Pathogenesis

The pathogenesis of infectious endocarditis is highly complex, involving numerous host–pathogen interactions. Review of the literature has indicated that many of the oral species are cultured from blood following oral procedures (Lockhart and Durack, 1999). Nevertheless, the non-β-haemolytic streptococci account for many cases of infective endocarditis (see *Epidemiology*), indicating that these species may possess specific pathogenic features. Some factors thought to be important are listed in Table 2.3, although only a limited number of these postulated pathogenic features have been studied in experimental models of infective endocarditis. The pathogenesis of infective endocarditis may be divided into several distinct and sequential steps. The first event is bacterial adhesion to the damaged tissue following an episode of bacteraemia. The second step is the establishment and persistence of streptococci, followed, thirdly, by bacterial growth and local tissue damage.

Bacterial attachment to damaged tissue is critical to disease, and bacterial adhesins that recognize and bind to host tissue components such as fibronectin and fibrinogen have been implicated in the disease process. Bacterial adhesion to fibronectin is demonstrated by numerous non-β-haemolytic streptococci (Table 2.3). Fibronectin binding was shown to be important for *S. sanguis* induction of experimental endocarditis in rats, and a mutant lacking fibronectin-binding activity was significantly less infective than the parent strain (Lowrance, Baddour and Simpson, 1990). Numerous fibronectin-binding proteins have been characterized among non-β-haemolytic streptococci, including CshA of *S. gordonii* that is expressed by most members of the mitis and sanguis groups of streptococci (McNab *et al.*, 1996; Elliott *et al.*, 2003).

The development of valvular vegetations in *S. sanguis*-induced endocarditis in rabbits has been shown to depend, at least in part, on the availability of fibrin (Yokota *et al.*, 2001). FimA protein of *S. parasanguis* has been identified as an adhesin mediating the attachment of bacteria

Table 2.3 Streptococcal virulence properties in infective endocarditis

Pathogenic feature	Determinant	Comment	Reference
Adherence to fibronectin			
S. sanguis[a]	Unidentified	Rat model	Lowrance, Baddour and Simpson (1990)
S. gordonii	CshA	Expressed by most sanguis and mitis group streptococci	Elliott *et al.* (2003), McNab *et al.* (1996)
S. gordonii	FbpA	Vaccination with homologous protein FBP54 of *S. pyogenes* protects against *S. pyogenes* challenge in mice	Christie, McNab and Jenkinson (2002), Kawabata *et al.* (2001)
S. intermedius	Antigen I/II	Expressed widely by members of the mitis, mutans and anginosus groups of streptococci	Petersen *et al.* (2001, 2002), Sciotti *et al.* (1997)
S. anginosus	14-kDa protein		Willcox *et al.* (1995)
Adherence to fibrin			
S. parasanguis	FimA	Vaccination studies demonstrate cross-protection afforded in rat model	Burnette-Curley *et al.* (1995), Viscount *et al.* (1997), Kitten *et al.* (2002)
S. mutans	Glucosyl transferases	Rat model	Munro and Macrina (1993)
Adherence to platelets			
S. mitis	PblA, PblB, PblT	Genes encoding these putative adhesins are carried on a prophage	Bensing, Rubens and Sullam (2001), Bensing, Siboo and Sullam (2001)
Platelet aggregation			
S. sanguis	PAAP	Rabbit model. *S. sanguis* strains expressing PAAP caused endocarditis with a more severe clinical course	Herzberg *et al.* (1992)
Anginosus group streptococci	Unidentified	Rat model. No strong correlation between endocardial infectivity and platelet-aggregating activity	Willcox *et al.* (1994), Kitada, Inoue and Kitano (1997)
Resistance to host defence mechanisms			
S. oralis, S. mitis	Unidentified	Rabbit model. Platelet releasate-resistant *S. oralis* persisted in vegetations	Dankert *et al.* (1995, 2001)
Other			
S. gordonii	MsrA	Repair of oxidized proteins	Kiliç *et al.* (1999); Vriesema *et al.* (2000)
S. sanguis	SOD, unidentified 190-kDa protein	Induction of proinflammatory cytokines	Banks *et al.* (2002)

[a] *S. sanguis* strain Challis used in this study was reclassified as *S. gordonii* Challis.

to fibrin clots, and inactivation of the *fimA* gene in *S. parasanguis* abrogates the ability of cells to bind to fibrin and to cause endocarditis in rats (Burnette-Curley *et al.*, 1995). Further, immunization of rats with *S. parasanguis* FimA conferred protection against subsequent challenge with *S. parasanguis* (Viscount *et al.*, 1997). FimA homologues in diverse streptococcal species share significant antigenic similarity, and Kitten *et al.* (2002) have demonstrated that vaccination with *S. parasanguis* FimA protected rats from endocarditis caused by other oral streptococci, raising the possibility of FimA being used as a vaccine for at-risk individuals. Studies have demonstrated that the FimA-like family of proteins function in manganese transport (Kolenbrander *et al.*, 1998) and so may also contribute to infective endocarditis through acquisition of this essential growth factor.

Many of the oral streptococcal species produce high-molecular mass glucans, through the enzymatic activity of glucosyltransferase (GTF), when grown in the presence of sucrose, a property that has been implicated in the pathogenesis of infective endocarditis. Thus, rats inoculated with an isogenic mutant of *S. mutans* lacking GTF activity developed endocarditis less frequently than those inoculated with the parental strain, and additionally, the isogenic mutant adhered in lower numbers to fibrin *in vitro* (Munro and Macrina, 1993). In contrast, there was no difference in virulence between sucrose-grown wild-type *S. gordonii* and its isogenic GTF-negative mutant (Wells *et al.*, 1993), highlighting species differences and indicating that multiple virulence factors may be involved in pathogenesis.

Bacterial interaction with platelets is considered a major factor in endocarditis (reviewed by Herzberg, 1996; Herzberg *et al.*, 1997). As well as the direct adhesion of bacteria to platelets (Table 2.3), many streptococci, particularly *S. sanguis* strains, induce the aggregation of platelets *in vitro*. These aggregates show densely compacted and degranulated platelets and contain entrapped streptococcal cells. Thus, bacterial aggregation of platelets has been proposed to contribute to the establishment and persistence of adherent bacteria through the creation of a protective thrombus. The interactions between *S. sanguis* and platelets have been studied in some detail (Herzberg, 1996). Interactions are complex and involve at least three streptococcal sites, and these interactions result in platelet activation and the release of ATP-rich dense granules. Platelet aggregation-associated protein (PAAP) plays an important role through interaction with a signal-transducing receptor on the platelet surface, inducing platelet activation and aggregation. Platelet-aggregating strains of *S. sanguis* induce significantly larger vegetations than nonaggregating strains in a rabbit model of endocarditis, and antibodies to PAAP can ameliorate the clinical severity of disease (Herzberg *et al.*, 1992). Nevertheless, non-platelet-aggregating *S. sanguis* isolates can still cause endocarditis, again highlighting the multiplicity of bacterial–host interactions that are involved in the disease process (Douglas, Brown and Preston, 1990; Herzberg *et al.*, 1992).

In addition to contributing to the disease process, platelet activation and aggregation are also a key aspect of host defence against disease, and the persistence of adherent streptococci is related, at least in part, to their resistance to microbicidal factors released from activated platelets as well as to circulating defence mechanisms (Dankert *et al.*, 1995). In one study over 80% of the oral streptococcal strains isolated from the blood of patients with infective endocarditis were resistant to platelet microbicidal activity compared with only 23% of streptococcal strains isolated from the blood of neutropenic patients without infective endocarditis (Dankert *et al.*, 2001). Further, immunizing rabbits to produce neutralizing antibodies against platelet microbicidal components rendered these animals more susceptible to infective endocarditis when challenged with a sensitive *S. oralis* strain (Dankert *et al.*, 2001).

Two studies have applied modern molecular screens to identify streptococcal genes encoding potential virulence factors for endocarditis. Using *in vivo* expression technology (IVET) in conjunction with a rabbit model of endocarditis, Kiliç *et al.* (1999) identified 13 genes whose expression was required for *S. gordonii* growth within valvular vegetations. One of these genes encoded methionine sulphoxide reductase (MSR), a protein involved in the repair of oxidized proteins. MSR has been proposed to play a role in the maintenance of the structure and function of proteins, including adhesins, where sulphur groups of methionine residues are highly sensitive to damage by oxygen radicals. In a separate study a shift in pH from 6.2 to 7.3 was used to mimic the environmental clues experienced by *S. gordonii* entering the blood stream from their normal habitat of the mouth (Vriesema, Dankert and Zaat, 2000). Among five genes induced by this pH shift was one encoding the MSR described above. Further application of molecular techniques should help dissect the complex host–bacterial interactions required for disease caused by these otherwise nonpathogenic bacteria.

Epidemiology

It has been estimated that about 20 cases of infective endocarditis per million of the population per year can be expected in England and Wales, with an associated mortality rate of approximately 20% (Young, 1987). Between 1975 and 1987 around 200 (±30) deaths per year were recorded in these countries. There have been numerous surveys of the causative agents in infective endocarditis over the years. Streptococci remain the predominant cause of infective endocarditis, accounting for up to 50% of cases in most published series, despite reported changes in the spectrum of disease due, in part, to an increased frequency of intravenous drug abuse, more frequent use of invasive procedures involving intravenous devices and heart surgery with prosthetic valves and a decrease in the incidence of rheumatic fever-associated cardiac abnormalities. It has sometimes been difficult to equate the identity of streptococci reported in earlier studies with the currently accepted classification and nomenclature, particularly with respect to the continued use of the term *Streptococcus viridans* as a catch-all for streptococci that are α-haemolytic on blood agar. Nevertheless, a study by Douglas *et al.* (1993) of 47 streptococcal isolates from 42 confirmed cases of infective endocarditis revealed that members of the mitis group of streptococci, particularly *S. sanguis*, *S. oralis* and *S. gordonii*, comprised the most commonly isolated oral streptococci (Table 2.4). Nonoral *S. bovis* is also a significant aetiological agent of both native and prosthetic heart valve infections (Duval *et al.*, 2001; Siegman-Igra and Schwartz, 2003). In a French study (390 patients) the incidence of infective endocarditis was 31 cases per million of the population (Hoen *et al.*, 2002). Streptococci were isolated in 48% of these cases, with the second

Table 2.4 Streptococci isolated from infective endocarditis

Species	Isolation frequency (% of streptococcal isolates)	
	Douglas *et al.* (1993)	Hoen *et al.* (2002)
Mitis and sanguis groups		
S. sanguis	31.9	4
S. oralis	29.8	5.8
S. gordonii	12.7	1.3
S. parasanguis	4.2	0.9
S. mitis	4.2	9.8
S. mutans	4.2	4.0
Salivarius group	4.2	1.8
Anginosus group	Not detected	0.9
Group D streptococci	Not detected	43.5
Nutritionally variant streptococci	Not detected	1.3
Enterococci	Not detected	12.9
Pyogenic group	Not detected	9.8
S. pneumoniae	Not detected	1.8
Unidentified/other	2.1	2.2

The Douglas *et al.* (1993) study focused on oral viridans streptococci.

most frequently isolated genus being *Staphylococcus* (29% of cases). Among the streptococci, oral (30%) and group D species (43.5%, the bulk of which were identified as *S. gallolyticus*) comprised most isolates, with the remainder being predominantly pyogenic streptococci (9.8% of streptococci) or enterococci (12.9%) (Table 2.4). In comparison with a similar study conducted in 1991 the incidence of infective endocarditis due to oral streptococci decreased slightly (Hoen *et al.*, 2002).

Clinical Features

Infective endocarditis due to non-β-haemolytic streptococci is usually subacute and may be difficult to diagnose in the earlier stages because of the vagueness and nonspecificity of the signs and symptoms. The patient often presents initially with fever, general malaise and a heart murmur, and other symptoms such as emboli, cardiac failure, splenomegaly, finger clubbing, petechial haemorrhages and anaemia may also be observed at some stage. In addition to fever, rigours and malaise, patients may also suffer from anorexia, weight loss, arthralgia and disorientation. Because of the decline in the number of cases of rheumatic heart disease in some countries and the increase in other predisposing causes, the clinical presentation of infective endocarditis may vary considerably and may not conform to classical descriptions of the disease. This variability in the clinical presentation of infective endocarditis presents a continuing challenge to diagnostic strategies that must be sensitive for disease and specific for its exclusion across all forms of the disease. The Duke Criteria were developed by Durack and colleagues from Duke University in a retrospective study (Durack, Lukes and Bright, 1994) and validated in a prospective cohort (Bayer *et al.*, 1994). The Duke Criteria have proven to be more sensitive at establishing definitive diagnoses than previous systems (reviewed by Bayer *et al.*, 1998).

Laboratory Diagnosis

Positive blood cultures are a major diagnostic criterion for infective endocarditis and are key in identifying aetiological agent and its antimicrobial susceptibility. Therefore, whenever there is a clinical suspicion of infective endocarditis it is important to take blood cultures as soon as possible, before antibiotic treatment is started. At least 20 ml of blood should be taken from adults on each sampling occasion, and it is usually recommended that three separate samples be collected during the 12–24-h period following the initial provisional diagnosis. Most positive cultures are obtained from the first two sets of blood cultures. Administration of antimicrobial agents to patients with infective endocarditis before blood cultures are obtained may reduce the recovery rate of bacteria by 35–40%. Thus, if antibiotic therapy has already been commenced, it may be necessary to collect several more blood cultures over a few days to increase the likelihood of obtaining positive cultures.

Aseptically collected blood samples should, ideally, be inoculated into at least two culture bottles to allow reliable isolation of both aerobic and anaerobic bacteria. The nutritionally variant streptococci (NVS) (see *Nutritionally Variant Streptococci*) are fastidious and require the addition of pyridoxal and/or L-cysteine to the medium for successful identification. Many laboratories now use highly sensitive (semi)automated systems for processing blood culture specimens, but both these and conventional methods sometimes yield culture-negative results from patients with suspected endocarditis. Such negative results may be due to previous antibiotic therapy, the presence of particularly fastidious bacteria, use of poor culture media or isolation techniques or infection due to microorganisms other than bacteria. If blood cultures remain negative after 48–72 h, they should be incubated for a more prolonged period (2–3 weeks) and should be examined microscopically on day 7, day 14 and at the end of the incubation period. Additionally, aliquots should be subcultured on chocolate agar for further incubation in an atmosphere with increased CO_2 levels.

Molecular techniques may be used for the detection and identification of bacteria in blood or on cardiac tissue taken from patients with suspected endocarditis, but that otherwise remains blood culture negative. Techniques include polymerase chain reaction (PCR) amplification of target sequences, and routine DNA sequencing of the amplified DNA may additionally allow for rapid identification of the causative organism (reviewed by Lisby, Gutschik and Durack, 2002).

Management

Effective management of infective endocarditis includes both treatment to control and eliminate the causative infectious agent and other measures to maintain the patient's life and well-being. Increasingly, cardiac surgery is carried out at a relatively early stage to replace damaged and ineffective heart valves.

The antimicrobial treatment depends upon maintaining sustained, high-dose levels of appropriate bactericidal agents, usually administered parenterally, at least in the early stages. It is vital that the aetiological agent, once isolated from repeated blood cultures, be fully identified and tested for antibiotic sensitivity. In addition to determining which antibiotics are most likely to be effective against the particular organisms isolated, the microbiology laboratory will also be required to periodically monitor whether bactericidal levels of the selected drug(s) are being maintained in the patient's blood.

Clearly, the choice of antimicrobial agent depends upon the identity of the causative agent. Most of the oral streptococcal and *S. bovis* strains remain sensitive to penicillin [defined as minimum inhibitory concentrations (MIC) = 0.1 mg/l, although see *Antibiotic Susceptibility*] and to glycopeptides; consequently, a combination of benzylpenicillin and gentamicin is recommended (Table 2.5). Penicillin-allergic patients should be treated with a combination of vancomycin and gentamicin.

Table 2.5 Endocarditis treatment regimens

Treatment regimens for adults not allergic to the penicillins
Viridans streptococci and *S. bovis*

(A) Fully sensitive to penicillin (MIC ≤ 0.1 mg/l)
      Benzylpenicillin 7.2 g daily in six divided doses by intravenous bolus injection for 2 weeks plus intravenous gentamicin 80 mg twice daily for 2 weeks[a]
(B) Reduced sensitivity to penicillin (MIC > 0.1 mg/l)
      Benzylpenicillin 7.2 g daily in six divided doses by intravenous bolus injection for 4 weeks plus intravenous gentamicin 80 mg twice daily for 4 weeks[a]

Treatment regimens for adults allergic to the penicillins
Viridans streptococci, *S. bovis* and enterococci

- Initially either vancomycin 1 g by intravenous infusion given over at least 100 min twice daily (determine blood concentrations and adjust dose to achieve 1-h postinfusion concentrations of about 30 mg/l and trough concentrations of 5–10 mg/l) or teicoplanin 400 mg by intravenous bolus injection 12 hourly for three doses and then a maintenance intravenous dose of 400 mg daily
- Give vancomycin or teicoplanin for 4 weeks plus intravenous gentamicin 80 mg twice daily. Viridans streptococcal and *S. bovis* endocarditis should be treated with gentamicin for 2 weeks and enterococcal endocarditis for 4 weeks[a]

Conditions to be met for a 2-week treatment regimen for viridans streptococcal and S. bovis *endocarditis*

- Penicillin-sensitive viridans streptococcus or *S. bovis* (penicillin MIC ≤ 0.1 mg/l)
- No cardiovascular risk factors such as heart failure, aortic insufficiency or conduction abnormalities
- No evidence of thromboembolic disease
- Native valve infection
- No vegetations more than 5 mm in diameter on echocardiogram
- Clinical response within 7 days. Temperature should return to normal, and patient should feel well and appetite should return

---

Table adapted from Simmons *et al.* (1998), based on the Endocarditis Working Party of the British Society for Antimicrobial Chemotherapy.
[a] Gentamycin blood levels must be monitored.

Under certain conditions (Table 2.5) a 2-week course of treatment with penicillin and gentamycin is adequate for endocarditis caused by fully sensitive organisms.

Infective endocarditis occurs less frequently in children than in adults, although the frequency of disease in children appears to be increasing since the 1980s despite a decline in the incidence of rheumatic fever, a major predisposing factor in the past. This is due, at least in part, to improved survival of children who are at risk, such as those with congenital heart disease. In children over 1 year old with endocarditis, the oral streptococci are the most frequently isolated organisms (32–43% of patients; reviewed by Ferrieri *et al.*, 2002). In general, the principles of antibiotic treatment of paediatric endocarditis are similar to those for the treatment of adults.

Prevention and Control

When patients are known to have predisposing cardiac abnormalities, great care should be taken to protect them from the risk of endocarditis when undergoing any dental, surgical or investigational procedures which might induce a transient bacteraemia. However, even if carried out perfectly, this approach is not likely to prevent all episodes of endocarditis since up to 50% of cases occur in individuals without previously diagnosed cardiac abnormalities (Hoen *et al.*, 2002). For the identified at-risk group, the appropriate antibiotic prophylaxis is summarized in Table 2.6.

The main principle governing these prophylactic regimens is that a high circulating blood level of a suitable bactericidal agent should be achieved at a time when the bacteraemia would occur. For bacteraemia arising during dental surgery, additional protection may be achieved by supplementing the use of systemic antibiotics with locally applied chlorhexidine gluconate gel (1%) or chlorhexidine gluconate mouthwash (0.2%) 5 min before the procedure. Dental procedures that require antibiotic prophylaxis include extractions, scaling and surgery involving gingival tissues. A most important consideration for patients who are at risk of endocarditis is that their dental treatment be planned in such a way that the need for frequent antibiotic prophylaxis and the consequent selection for resistant bacteria among the resident flora is avoided. For multistage dental procedures, a maximum of two single doses of penicillin may be given in a month and alternative drugs should be used for further treatment, and penicillin should not be used again for 3–4 months.

Abscesses Caused by Non-β-Haemolytic Streptococci

Streptococci are frequently isolated from purulent infections in various parts of the body, including dental, central nervous system (CNS), liver and lung abscesses. Commonly, there is a mixture of several organisms in the pus, which may contain obligate anaerobes as well as streptococci and other facultative anaerobes. Consequently, it may be difficult to determine the contribution that any single strain or species is making to the infectious process. The source of these bacteria is usually the patient's own commensal microflora and may be derived from the mouth, upper respiratory tract, gastrointestinal tract or genitourinary tract.

Members of the anginosus group of streptococci (previously known as *S. milleri* group, or SMG, streptococci) have a particular propensity to cause abscesses. The development of reliable identification and speciation methods for this group of streptococci has allowed epidemiological analysis to determine whether there is a correlation between species and anatomical site of infection (reviewed by Belko *et al.*, 2002). Studies indicate that *S. intermedius* was more frequently isolated from infections of the CNS and liver, *S. constellatus* was more frequently isolated from lung infections, whereas *S. anginosus* was more frequently associated with infections of the gastrointestinal and genitourinary tracts and with soft-tissue infections. Although studies have identified several virulence determinants of anginosus

Table 2.6   Recommended antibiotic prophylaxis for endocarditis

Dental procedures under local or no anaesthesia
Patients who have not received more than a single dose of a penicillin in the previous month, including those with a prosthetic valve (but not those who have had endocarditis)
- Oral amoxicillin 3 g 1 h before procedure (children under 5 years, quarter adult dose; 5–10 years, half adult dose)

Patients who are penicillin-allergic or have received more than a single dose of a penicillin in the previous month
- Oral clindamycin 600 mg 1 h before procedure (children under 5 years, clindamycin quarter adult dose or azithromycin 200 mg; 5–10 years, clindamycin half adult dose or azithromycin 300 mg)

Patients who have had endocarditis
- Amoxicillin plus gentamycin, as under general anaesthesia

Dental procedures under general anaesthesia
No special risk (including patients who have not received more than a single dose of a penicillin in the previous month)
Either
- Intravenous amoxicillin 1 g at induction, then oral amoxicillin 500 mg 6 hours later (children under 5 years, quarter adult dose; 5–10 years, half adult dose)
Or
- Oral amoxicillin 3 g 4 h before induction, then oral amoxicillin 3 g as soon as possible after procedure (children under 5 years, quarter adult dose; 5–10 years, half adult dose)

Special risk (patients with a prosthetic valve or who have had endocarditis)
- Intravenous amoxicillin 1 g plus intravenous gentamicin 120 mg at induction, then oral amoxicillin 500 mg 6 h later (children under 5 years, amoxicillin quarter adult dose, gentamicin 2 mg/kg; 5–10 years, amoxicillin half adult dose, gentamicin 2 mg/kg)

Patients who are penicillin-allergic or who have received more than a single dose of a penicillin in the previous month
- Intravenous vancomycin 1 g over at least 100 min, then intravenous gentamycin 120 mg at induction or 15 min before procedure (children under 10 years, vancomycin 20 mg/kg, gentamicin 2 mg/kg)
Or
- Intravenous teicoplanin 400 mg plus gentamicin 120 mg at induction or 15 min before procedure (children under 14 years, teicoplanin 6 mg/kg, gentamicin 2 mg/kg)
Or
- Intravenous clindamycin 300 mg over at least 10 min at induction or 15 min before procedure, then oral or intravenous clindamycin 150 mg 6 h later (children under 5 years, quarter adult dose; 5–10 years, half adult dose)

Table adapted from the British National Formulary (2003), based on Recommendations of the Endocarditis Working Party of the British Society for Antimicrobial Chemotherapy.

group streptococci (see *Pathogenesis*), we have little understanding of the bacterial–host interactions that define the body-site specificity identified by these epidemiological studies.

Pathogenesis

Members of the anginosus group of streptococci possess multiple pathogenic properties that may contribute to disease. These include adhesion to host tissue components such as fibronectin, fibrinogen and fibrin–platelet clots (Willcox, 1995; Willcox *et al.*, 1995) and the aggregation of platelets (Willcox *et al.*, 1994; Kitada, Inoue and Kitano, 1997) (Table 2.7).

Abscesses are frequently polymicrobial in nature, and studies suggest that coinfection of streptococci with strict anaerobes such as *Fusobacterium nucleatum* and *Prevotella intermedia*, both common oral isolates, enhanced pathology. Thus, coinfection of anginosus group streptococci with *F. nucleatum* in a mouse subcutaneous abscess model resulted in increased bacterial survival and increased abscess size compared with abscesses formed following monoculture inoculation of streptococcus or *F. nucleatum* alone (Nagashima,

Table 2.7 Pathogenic properties of anginosus group streptococci

Property	Reference
Adhesion to host components	Willcox (1995), Willcox et al. (1995)
Platelet aggregation	Willcox et al. (1994), Kitada, Inoue and Kitano (1997)
Synergistic interactions with anaerobes	Shinzato and Saito (1994, 1995), Nagashima, Takao and Maeda (1999)
Production of hydrolytic enzymes	Unsworth (1989), Jacobs and Stobberingh (1995), Shain, Homer and Beighton (1996)
Intermedilysin-specific haemolytic activity	Nagamune et al. (1996)

Takao and Maeda, 1999). In a second study, coinfection of *S. constellatus* with *P. intermedia* resulted in an increased mortality rate in mouse pulmonary infection model (Shinzato and Saito, 1994). In both cases *in vitro* studies indicated that coculture enhanced the growth of the streptococcus and suggested that secreted factors produced by the anaerobic bacteria suppressed host bactericidal activity.

The streptococci themselves may secrete enzymes capable of degrading host tissue components including chondroitin sulphate and hyaluronic acid (Unsworth, 1989; Jacobs and Stobberingh, 1995; Shain, Homer and Beighton, 1996). In one clinical study most (85%) of the anginosus group streptococci isolated from abscesses produced hyaluronidase compared with only 25% of strains isolated from the normal flora of healthy sites (Unsworth, 1989). Despite such epidemiological evidence, the role of hydrolytic enzymes in the disease process has not yet been investigated in any detail.

In 1996, Nagamune and coworkers reported on a human-specific cytolysin, intermedilysin, that was secreted by a strain of *S. intermedius* isolated from a human liver abscess and shown to be able to directly damage host cells including polymorphonuclear cells (Nagamune *et al.*, 1996; Macey *et al.*, 2001). Intermedilysin-specific haemolytic activity is distinct from the haemolytic activity found on blood agar with some anginosus group streptococci. The 54-kDa protein is a member of cholesterol-binding cytolysin family of pore-forming toxins expressed by a range of Gram-positive bacteria (Billington, Jost and Songer, 2000) including *S. pneumoniae* (pneumolysin). PCR amplification studies demonstrated that the intermedilysin gene, *ily*, is restricted to typical *S. intermedius* isolates and is absent from *S. anginosus* and *S. constellatus* strains (Nagamune *et al.*, 2000). The occurrence of human-specific haemolysis and/or the presence of the *ily* gene consequently offers a useful marker of *S. intermedius* (Jacobs, Schot and Schouls, 2000; Nagamune *et al.*, 2000). In the study by Nagamune *et al.* (2000), intermedilysin-specific haemolytic activity was approximately 6–10-fold higher in brain abscess or abdominal infection strains compared with *S. intermedius* isolates from dental plaque. In contrast, in this study no apparent association was observed between the degree of hydrolytic activity and site of infection, and dental plaque isolates demonstrated comparable chondroitin sulphatase, hyaluronidase and sialidase (neuraminidase) activity (Nagamune *et al.*, 2000).

Clinical and Laboratory Considerations

There are no particular features that clearly distinguish abscesses associated with non-β-haemolytic streptococci from those caused by other microorganisms. The actual presentation depends upon the site and extent of the abscess as well as upon the nature of the causative organism(s). Since the infections are often polymicrobial, sometimes with obligate anaerobes, streptococci may be isolated from foul-smelling, apparently anaerobic pus. Successful diagnosis depends to a large extent on obtaining adequate clinical material that has not been contaminated with normal commensal bacteria from the skin or mucosal surfaces. Whenever possible, aspirated pus samples should be collected and inoculated into appropriate culture media for aerobic,

microaerophilic and anaerobic incubation. Isolates of presumptive streptococci should be identified, as described in the section *Laboratory Diagnosis*, and tested for antibiotic sensitivity.

The clinical management of all abscesses, whether or not streptococci are involved, requires both surgical drainage and antimicrobial chemotherapy. Since the infection is frequently mixed, a combination of agents may be indicated to combat the different species present. For example, a combination of penicillin and metronidazole may be appropriate for abscesses caused by streptococci in conjunction with one or more strict anaerobes, as is often the case with infections around the head and neck.

Streptococcal Infections in the Immunocompromised

Advances in organ transplantation and the treatment of patients with cancer have resulted in increased numbers of immunocompromised individuals at risk of infection from endogenous organisms. Between 1970 and 2000, the non-β-haemolytic streptococci have emerged as significant pathogens in these patients, capable of causing septicaemia, acute respiratory distress syndrome and pneumonia. Non-β-haemolytic streptococci are also implicated in neonatal septicaemia and meningitis. Centres in North America and Europe have reported variable experiences with infection by these species in cancer and transplant patients, with a mortality rate of up to 50% (reviewed by Shenep, 2000). The source of infection is generally the oral cavity or gastrointestinal tract, and risk factors, in addition to profound neutropenia, include chemotherapy-induced mucositis, use of cytosine arabinoside (over and above its association with mucositis) and the use of certain prophylactic antimicrobial therapies (quinolones and cotrimoxazole) with reduced activity against the non-β-haemolytic streptococci (Kennedy and Smith, 2000; Shenep, 2000).

The species which are most commonly associated with infection in immunocompromised individuals are *S. oralis*, *S. mitis* and *S. sanguis*; however, little is known regarding the virulence determinants or pathogenic mechanisms involved in these infections beyond the ability of the streptococci to induce the production of proinflammatory cytokines (Vernier *et al.*, 1996; Scannapieco, Wang and Shiau, 2001).

Clinical and Laboratory Considerations

The initial clinical feature of septicaemia is typically fever, which is generally at least 39°C and may persist for several days despite clearance of cultivable organisms from the blood. Most patients respond to appropriate antibiotic therapy; however, in few cases there is progression to fulminant septicaemia associated with prolonged fever and severe respiratory distress, some 2–3 days after the initial bacteraemia. Identification of the causative organism is routinely by culture of blood or other normally sterile tissue.

Empiric antimicrobial therapy for episodes of febrile neutropenia should ideally have activity against both Gram-negative and Gram-positive bacteria, and consequently, combination therapy or use of a broad-spectrum antibiotic is recommended. As discussed in the section *Antibiotic Susceptibility*, some streptococcal isolates from neutropenic patients are found to be relatively resistant to penicillin and other antibiotics, and the choice of empiric therapy where infection by non-β-haemolytic streptococci is suspected should take into account the local pattern of antibiotic susceptibilities among recent isolates. For endocarditis prophylaxis, the establishment and maintenance of good oral hygiene is an important preventative measure both before and during the period of neutropenia.

Caries

Dental caries is a disease that destroys the hard tissues of the teeth. If unchecked, the disease may progress to involve the pulp of the tooth and, eventually, the periapical tissues surrounding the roots. Once the process has reached the periapical region, infection either may remain localized as an acute dental abscess or as a chronic granuloma or may spread more widely in various directions depending on its anatomical position. In some cases such spreading infections may cause a life-threatening situation, for example, if the airway is obstructed by submandibular swelling (Ludwig's angina).

The disease is initiated by acid demineralization of the teeth because of the metabolic activities of saccharolytic bacteria, including streptococci, which are situated on the tooth surface as part of the complex microbial community known as dental plaque. Dental plaque accumulates rapidly on exposed tooth surfaces in the mouth and consists of a complex mixture of bacteria and their products. Many of the oral streptococci listed in Table 2.1 are prominent components of dental plaque. When an external source of carbohydrate becomes available, in the form of dietary carbohydrate (particularly sucrose), streptococci and other plaque bacteria rapidly utilize the fermentable sugars and release acidic metabolic end products such as lactic acid. This can result in a rapid drop in pH in the vicinity of teeth that, if sufficiently low (pH 5.5 or less), in turn results in the demineralization of the dental enamel.

Most detailed studies on the pathogenesis of dental caries have focused on the mutans streptococci since these are widely regarded as the most significant initiators of the disease. Properties of these streptococci that are considered to be important in caries include ability to survive and grow at relatively low pH, production of extracellular polysaccharides from glucose and production of acid from carbohydrates, and mutant strains lacking in one or more of these attributes have been shown to be less cariogenic in experimental animals (reviewed by Kuramitsu, 2003). The frequent consumption of sugar is proposed to shift the plaque population in favour of aciduric species able to grow and survive in low pH conditions, which in turn results in greater plaque acidification and, consequently, greater enamel dissolution (Marsh, 2003). However, oral species other than mutans group streptococci are thought to contribute to the disease process.

The main approaches to caries prevention include the control of dietary carbohydrates, particularly by reducing the frequency of sugar intakes, the use of fluorides (both topically and systemically), maintenance of good oral hygiene and plaque control, application of fissure sealants and regular dental check-ups. With a better understanding of mucosal immunity and using new technologies available in molecular biology, researchers have developed novel caries preventative measures. These include local passive and active immunization, replacement therapy (the use of engineered noncariogenic *S. mutans* strains to replace cariogenic species within dental plaque) and the use of anti-adhesive peptides (Kelly *et al.*, 1999; Hillman, 2002; Koga *et al.*, 2002). Despite growing evidence from laboratory animal and human clinical studies of the ability of these approaches to control *S. mutans* numbers, their potential as anti-caries treatments has yet to be fully realized.

LABORATORY DIAGNOSIS

Specimens and Growth Media

Non-β-haemolytic streptococci are isolated from a wide range of clinical specimens such as blood, pus, wounds, skin swabs and biopsies as well as from dental plaque, saliva and other oral sites. Obtaining pus from an oral abscess is best done by direct aspiration using a hypodermic syringe rather than by swabbing, to reduce the risk of contaminating the sample with the oral flora, although in some instances (e.g. with infants), swab samples may be the only option available, unless the patient is undergoing a general anaesthesia. When sampling oral sites for ecological studies, it is necessary to ensure that the area sampled is small enough to be representative of a discrete site to avoid the risk of obscuring the differences between sites by sampling too big an area.

Where the laboratory processing of a specimen may be delayed, the clinical sample is best held in a suitable reduced transport fluid such as the one described by Hardie and Whiley (1992).

The non-β-haemolytic streptococci are fastidious organisms, with a need for a carbohydrate source, amino acids, peptides and proteins, fatty acids, vitamins, purines and pyrimidines. These bacteria therefore need complex growth media commonly containing meat extract, peptone and blood or serum. Non-β-haemolytic streptococci are best isolated from clinical samples on a combination of nonselective and selective agar media. Nonselective examples include blood agar 2 (Oxoid, Hampshire, UK), Columbia agar (Gibco BRL, Life Technologies, Paisley, UK), fastidious anaerobic agar (Laboratory M, Amersham, UK) and brain–heart infusion agar (Oxoid) supplemented with 5% defibrinated horse or sheep blood. Several selective media are available for the non-β-haemolytic streptococci. The two most commonly used agars are trypticase–yeast–cystine (TYC) and mitis–salivarius (MS) agars that contain 5% sucrose to promote the production of extracellular polysaccharides, resulting in the production of characteristic colonial morphologies as an aid to identification. TYC is available commercially from Laboratory M; MS agar is available from Oxoid and from Difco (Detroit, MI, USA). Other selective media commonly used for the isolation of mutans streptococci are based on TYC or MS agars and have the addition of bacitracin (0.1–0.2 U/ml) and an increased amount of sucrose (20%). Nalidixic acid–sulphamethazine (NAS) agar is a selective medium for the anginosus group streptococci and uses 40 g/l of sensitivity agar supplemented with 30 μg/ml of nalidixic acid, 1 mg/ml of sulphamethazine (4-amino-*N*-[4,6-dimethyl-2-pyrimidinyl]benzene sulphonamide) and 5% defibrinated horse blood. This medium is also selective for *S. mutans*.

As streptococci are facultative anaerobes, incubation is best carried out routinely in an atmosphere of air plus 10% carbon dioxide or in an anaerobic gas mix containing nitrogen (70–80%), hydrogen (10–20%) and carbon dioxide (10–20%). Some strains have an absolute requirement for carbon dioxide, particularly on initial isolation. Colonies on blood agar are typically 1 mm or less in diameter after incubation for 24 h at 37 °C, are nonpigmented and often appear translucent. On blood agar the streptococci discussed in this chapter usually produce α-haemolysis or are non-(γ)-haemolytic. However, as mentioned previously, the haemolytic reactions of different strains within a species may vary and are sometimes influenced by the source of blood (e.g. horse, sheep) and by the incubation conditions. Some examples of colonial morphology on different culture media are illustrated in Figure 2.4.

Liquid culture of these streptococci may be carried out in a commercial broth such as Todd–Hewitt broth or brain–heart infusion broth (Oxoid) with or without supplementation with yeast extract (0.5%). The growth obtained in broth cultures varies from a diffuse turbidity to a granular appearance with clear supernatant depending on strain and species.

Initial Screening Tests

Streptococci are usually spherical, with cells of approximately 1-μm diameter arranged in chains or pairs. The length of the chains may vary from only a few cells to over 50 cells, depending on the strain and cultural conditions (longer chains are produced when the organisms are grown in broth culture). Streptococci stain positive in the Gram stain, although older cultures may appear Gram variable. Some strains may appear as short rods under certain cultural conditions. Isolates should be tested for catalase reaction, and only catalase-negative strains should be put through further streptococcal identification tests.

The clinical microbiologist should be aware that, because of developments in taxonomic studies, several other genera of facultative anaerobic Gram-positive cocci that grow in pairs or chains have been proposed that may superficially resemble streptococci. The scheme described in Table 2.8 should allow the differentiation of these genera on the basis of a few cultural and biochemical tests, with great caution to be exercised in the interpretation of morphological observations.

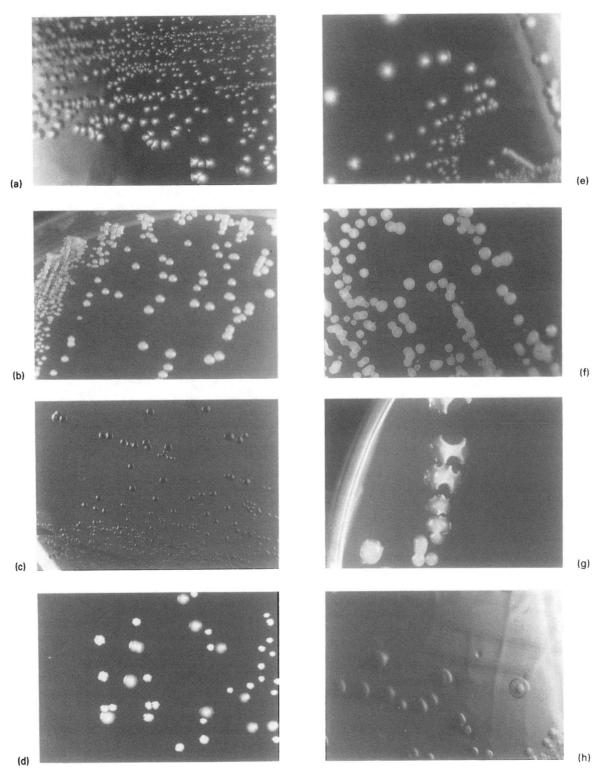

Figure 2.4 Colonial appearance of non-β-haemolytic streptococci on blood agar (BA) or 5% sucrose containing agars [trypticase–yeast–cystine (TYC) and mitis–salivarius (MS)]: (a) *S. anginosus* (BA); (b) *S. anginosus* (TYC); (c) *S. anginosus* (MS); (d) *S. intermedius* (BA) showing rough and smooth colony variants; (e) *S. salivarius* (BA); (f and g) *S. salivarius* (TYC) showing extracellular polysaccharide (fructan) production; (h) *S. salivarius* (MS); (i) *S. sanguis* (BA); (j) *S. sanguis* (TYC) showing extracellular polysaccharide (glucan) production; (k) *S. sanguis* (MS); (l) *S. mutans* (BA); (m) *S. mutans* (TYC) showing extracellular polysaccharide (glucan) production; (n) *S. mutans* (MS); (o) *S. bovis* (BA); (p) *S. bovis* (MS) showing extracellular polysaccharide (glucan) production.

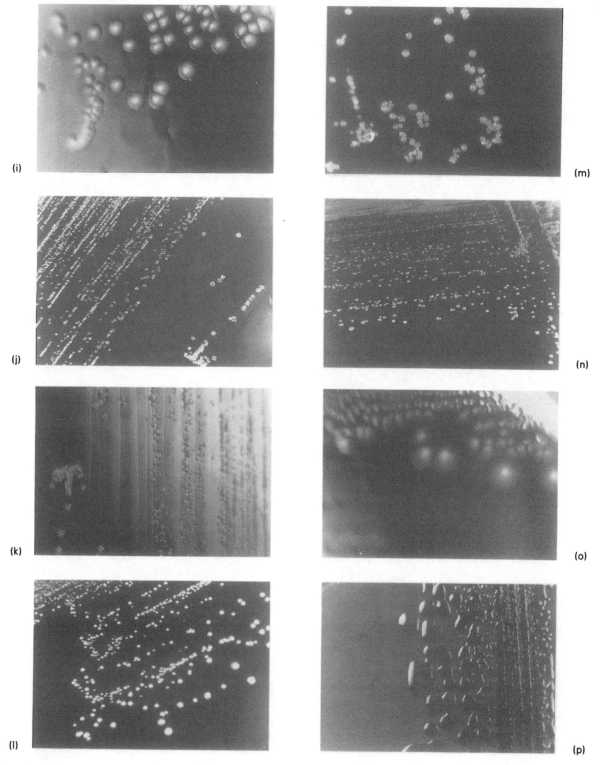

Figure 2.4 (Continued).

Identification to Species Level

The identification of isolates to species level may be carried out using either laboratory-based biochemical test schemes or commercial identification kits with accompanying databases. Although species-level identification of non-β-haemolytic streptococci is not routinely carried out in many clinical laboratories, this may be deemed necessary for suspected pneumococcal infection, isolates obtained from deep-seated abscesses, specimens from endocarditis or where pure cultures have been grown. Differential tests for the identification of non-β-haemolytic streptococci are summarized in Tables 2.9–2.13.

Table 2.8 Differential characteristics of *Streptococcus* and other Gram-positive, catalase-negative, facultatively anaerobic coccus genera that grow in pairs or chains

Genus	Pyrrolidonyl arylamidase	Leucine aminopeptidase production	6.5% NaCl	BE	Growth at 10 °C	Growth at 45 °C	Susceptibility to vancomycin	Haemolysis
Streptococcus[a]	–	+	–[b]	–[c]	–	V	S	α, β, γ
Enterococcus[d]	+	+	+	+	+	+	S[e]	α, β, γ
Lactococcus	+	+	V	+	+	–[f]	S	α, γ
Vagococcus[d]	+	+	+	+	+	–	S	α, γ
Globicatella	+	–	+	–	–	–	S	α
Leuconostoc[g]	–	–	V	V	+	V	R	α, γ

BE, reaction on bile–aesculin medium. +, ≥95% of strains are positive; –, ≥5% of strains are negative; V, variable reaction; R, resistant; S, susceptible.

[a] *Abiotrophia* and *Granulicatella* (formerly nutritionally variant streptococci strains) are weakly pyrrolidonyl arylamidase and leucine aminopeptidase production positive and grow in 6.5% NaCl broth. Some group A streptococci are pyrrolidonyl arylamidase positive.
[b] Some β-haemolytic streptococci grow in 6.5% NaCl broth.
[c] 5–10% of viridans streptococci are bile–aesculin positive.
[d] Some enterococci are motile; *Vagococcus* is motile (formerly called group N streptococci).
[e] Some strains are vancomycin resistant.
[f] Some strains grow very slowly at 45 °C.
[g] Gas is produced by *Leuconostoc* from glucose in Mann, Rogosa, Sharpe (MRS) *Lactobacillus* broth.

Table 2.9 Differential biochemical characteristics of the anginosus group

Test	S. anginosus	S. constellatus	S. intermedius
Acid from			
N-Acetyl-D-glucosamine	–/+	–/+	+
Amygdalin	+	–/+	+/–
Lactose	+	+/–	+
Mannitol	–/+[a]	–	–
Raffinose	–/+[a]	–	–
Production of			
β-D-Galactosidase	–[b]	–	+
α-D-Glucosidase	–/+	+	+
β-D-Glucosidase	+	–	–/+
β-D-Fucosidase	–	–	+
β-D-N-Acetylglucosaminidase	–	–	+
β-D-N-Acetylgalactosaminidase	–	–	+
Sialidase	–	–	+
Hydrolysis of			
Esculin	+	+[c]	+
Production of			
H₂O₂	–/+	–	–
Hyaluronidase	–	+/–	+

+, ≥90% of strains are positive; +/–, 50–89% of strains are positive; –/+, 11–49% of strains are positive; –, ≤10% of strains are positive. Table taken from Whiley and Beighton (1998), with permission of Blackwell Publishing Ltd.
[a] Strains of *S. anginosus* isolated from the genitourinary tract are frequently positive for acid production from this carbohydrate.
[b] Ahmet, Warren and Huoang (1995) reported >50% of strains as positive when strains were incubated under hypercapnic conditions.
[c] Weak or slow reactions given by some strains.

The rapid and extensive taxonomic revision of the genus *Streptococcus*, including the demonstration of several new species and amended descriptions of others, has, to some extent, outpaced the availability of comprehensive schemes for routine identification. However, significant improvements to this situation have come about through the incorporation of fluorogenic and chromogenic substrates into identification schemes for the rapid detection of preformed enzyme activities. These test schemes combine tests for the detection of glycosidase and arylamidase reactions with more traditional tests that detect carbohydrate fermentation, arginine dihydrolase and acetone production. A further development has been the inclusion of these tests in a more standardized format in some commercially available test kits, which helps reduce the degree of discrepancy between biochemical test results from different laboratories. Currently, a 32 Test Kit for identifying streptococci is available, which takes into account most of the currently recognized species (Rapid ID32 Strep system; bioMérieux Vitek, bioMérieux, Durham, NC, USA). In an independent evaluation the kit gave correct identification for 95.3% (413/433) of strains examined, including 109 strains that required some additional tests for complete identification. Sixteen strains remained unidentified and four were misidentified (Freney *et al.*, 1992). In another study the test kit gave correct identification for 87% of strains examined. However, *S. mitis* and *S. oralis* were not easily differentiated using this method (Kikuchi *et al.*, 1995). Other phenotypic markers that have been reported to be potentially useful, particularly for the oral streptococci, include acid and alkaline phosphatases, neuraminidase (sialidase), IgA1 protease production, salivary amylase binding, extracellular polysaccharide production and the detection of hyaluronidase and chondroitin sulphate depolymerase activities. Several other approaches to species identification have been investigated such as pyrolysis–mass spectrometry, whole-cell–derived polypeptide patterns by sodium dodecyl sulphate–polyacrylamide gel electrophoresis (SDS-PAGE) or other electrophoretic separation techniques and the analyses of long-chain fatty acids and cell-wall components. However, these strategies have not been widely adopted and remain a part of the repertoire of the specialist research laboratory.

Significant progress has been already made in developing genetic-based approaches to species identification. DNA probes for this purpose have been described that utilize whole chromosomal preparations or cloned DNA fragments. However, as with most other groups of bacteria, by far the greatest promise for a genetically based identification approach lies in the analysis of nucleotide sequence data derived from the 16S rRNA gene. Published studies have provided 16S rRNA gene sequence data for most, but not all, non-β-haemolytic streptococci, which have been found to be distinct for each of the species so far examined (Bentley, Leigh and Collins, 1991; Kawamura *et al.*, 1995a). Other approaches have been described including restriction fragment length polymorphism (RFLP) analysis of amplicons of 16S–23S rDNA and intergenic spacer regions (Rudney and Larson, 1993; Schlegel *et al.*, 2003b) or tDNA intergenic spacer regions (De Gheldre *et al.*, 1999) and PCR amplification and sequence analysis of the gene encoding manganese-dependent superoxide dismutase (Poyart, Quesne and Trieu-Cuot, 1998).

Serology

The success obtained in producing a serological classification for the pyogenic streptococci was not extended to the non-β-haemolytic species. Several early attempts to this end were unsatisfactory and failed to produce an all-encompassing scheme for these streptococci.

Table 2.10 Differential biochemical characteristics of the mitis group

Test	S. sanguis	S. gordonii	S. mitis	S. oralis	S. parasanguis	S. crista	S. pneumon
Acid from							
Amygdalin	–/+	+	–	–	–/+	–	–
Arbutin	+/–	+	–	–	–/+	+	–
Inulin[b]	–/+	+	–/+	–	–	–	–
Lactose	+	+	+/–	+	+	+/–	+
Melibiose	+/–	–/+	+/–	+/–[a]	+/–	–	–
Raffinose	+/–	–/+	+/–	+/–	+/–	–	+
Sorbitol	–/+	–	–/+	–	–/+	–	–
Trehalose	+	+	–/+	–/+	+/–	+	+
Salicin	+	+	–/+	–	+/–	Not determined	+/–
Production of							
α-D-Galactosidase	+/–	–/+	+/–	–/+	+	–	+
β-D-Galactosidase[c]	–/+[b]	–/+[b]	–/+[b]	+/–	+	–/+	+
α-D-Glucosidase[c]	–	–/+[b]	+/–[b]	+	+	–	+
β-D-Glucosidase[c]	+/–	+	+/–[b]	–	–/+	–	–/+
α-L-Fucosidase[c]	–	+	–	–	–/+	+	–/+
β-D-Fucosidase	+/–[b]	–	+/–[b]	–[b]	–/+	–	–/+
β-D-N-Acetylgalactosaminidase	–	+/–	–	+	Not determined	+	+
β-D-N-Acetylglucosaminidase	–/+[b]	+	–[b]	+	+	+	+/–
Sialidase	–	–	–/+	+	–	–	+
Hydrolysis of							
Arginine	+	+	–[b]	–	+	+/–	–
Esculin	+/–	+	–	–/+	–/+	–	–/+
Production of							
Hyaluronidase	–	–	–	–	–	–	+
Alkaline phosphatase	–	+	+/–	+	+	–	–
Extracellular polysaccharide	+	+	–	+/–	–	v[d]	–
IgA protease	+	–	–/+	+	Not determined	Not determined	+
Amylase binding	–	+	+	–	+	+	+

Mitis group presented in this table comprises both mitis and sanguis groups reported in Facklam (2002). Refer to Table 2.9 footnote for the description of symbols. Table taken from Whiley and Beighton (1998), with permission of Blackwell Publishing Ltd.
[a] Most of the fermentation reactions are weaker than with other carbohydrates (pH 5.6–6.2).
[b] The percentage of strains reported positive varies between studies.
[c] The percentage of strains giving a positive test reaction may vary depending on the exact test method used (Kilian et al., 1989).
[d] Reported as 'variable'.

Table 2.11 Differential biochemical characteristics of the salivarius group

Test	S. salivarius	S. vestibularis	S. thermophilus[c]
Acid from			
N-Acetyl-D-glucosamine	–/+[a]	+/–	–
Amygdalin	–/+[a]	+/–	–
Arbutin	+	–/+	v[b]
Inulin	+/–	–	–
Lactose	+	+/–	+
Melibiose	–/+	–	v[b]
Raffinose	+/–[a]	–	v[b]
Salicin	+	+	–
Production of			
α-L-Arabinosidase	+[a]	+	Not determined
α-D-Galactosidase	–/+	–	–
α-D-Glucosidase	+/–[b]	+/–	–
β-D-Glucosidase	+/–[b]	–	–
β-D-Fucosidase	+/–[b]	–	Not determined
Hydrolysis of			
Esculin	+	+/–	–
Urea	+/–	+	–
Production of			
Acetoin (VP)	+/–	+/–	+
H_2O_2	–	+/–	Not determined
Alkaline phosphatase	+/–	–	–
Extracellular polysaccharide	+	–	–

Refer to Table 2.9 footnote for the description of symbols. Table taken from Whiley and Beighton (1998), with permission of Blackwell Publishing Ltd.
[a] Number of strains reported as positive varies between studies.
[b] Results reported as 'variable'.
[c] S. thermophilus strains frequently grow at 50 °C and survive heating at 65 °C for 30 min.

In retrospect, these studies were probably frustrated by the unsatisfactory classification of the viridans streptococci at the time, as well as by the numerous serological cross-reactions that characterize these species. Serological studies have also been undertaken on the S. milleri group (anginosus group) with the aim of developing a useful scheme for serotyping clinical isolates. In one study (Kitada et al., 1992) 91 clinical isolates were tested for the possession of a Lancefield group antigen and/or one of eight cell-surface carbohydrate serotyping antigens. Unfortunately, 19 of 91 isolates (21%) failed to react against any of the antisera, added to which the identity of the streptococci examined cannot be related to currently accepted species with any certainty. However, serological analysis of some of these streptococci has provided useful data with important consequences. Serological subdivisions within the mutans streptococci together with biochemical and genetic data led to the recognition of several distinct species of acidogenic oral bacteria, and this is of considerable significance in studies of dental caries. There are currently eight serotypes (serovars) recognized within these streptococci (a–h) based on the possession of serotype-specific cell-wall polysaccharide antigens (Hamada and Slade, 1980) (Table 2.1).

ANTIBIOTIC SUSCEPTIBILITY

Although antibiotic activity against non-β-haemolytic streptococci has long since been known to be poorer than for their β-haemolytic counterparts, the 1990s has witnessed some alarming changes in the susceptibility of non-β-haemolytic streptococcal strains to key therapeutic agents. Early indications of emerging β-lactam resistance came from the United States, Italy and the United Kingdom (Wilson et al., 1978; Bourgault, Wilson and Washington, 1979; Venditti et al., 1989;

Table 2.12 Differential biochemical characteristics of the mutans group

Test	S. mutans	S. sobrinus	S. cricetus	S. rattus	S. macacae	S. downei
Acid from						
N-Acetyl-D-glucosamine	+	−	+	+	+	Not determined
Amygdalin	+/−[a]	−	+	+	+	Not determined
Arbutin	+[b]	−	+	+	Not determined	Not determined
Dextrin	−	−	−	−	−	Not determined
Inulin	+	+	+	+	−	+
Lactose	+	+/−	+/−	+/−	Not determined	+
Mannitol	+	+/−	+	+	+	+
Melibiose	+/−[b]	−	+	+	−	+
Raffinose	+	−	+	+	+	−
Sorbitol	+	−/+	+	+	+	−
Trehalose	+	+	+/−	+	+	+
Salicin	+	−	+	+	Not determined	+
Glycogen	−	−	−	−	Not determined	−
Starch	−	−	−	−	−	−
Production of						
α-D-Galactosidase	+[b]	−	Not determined	Not determined	Not determined	Not determined
β-D-Glucosidase	+	−	Not determined	Not determined	Not determined	Not determined
Hydrolysis of						
Arginine	−	−	−	+	−	−
Esculin	+	+/−	+/−	+	+	−
Production of						
Acetoin (VP)	+	+	+	+	+	+
H₂O₂	−	+	−	−	−	−
Resistance to bacitracin	+	+	−	+	−	−

Refer to Table 2.9 footnote for the description of symbols. Table taken from Whiley and Beighton (1998), with permission of Blackwell Publishing Ltd.
[a] Proportion of strains reported as positive varies between studies.
[b] Non-melibiose-fermenting strains ferment amygdalin and raffinose less often and do not produce α-galactosidase or α-glucosidase.

Table 2.13 Differential biochemical characteristics of the S. bovis/S. equinus complex

Test	S. bovis/S. equinus	S. gallolyticus		S. infantarius			S. alactolyticu
		subsp. gallolyticus	subsp. pasteurianus	subsp. macedonicus	subsp. infantarius	subsp. coli	
Hydrolysis of							
Aesculin	+	+	+	−	+/−	+	+
Gallate	−	+	−	−	−	−	−
Production of							
β-Glucosidase	+	+	+	−	+/−	+	+
β-Glucuronidase	−	−	+	−	−	−	−
α-Galactosidase	v[a]	+	+/−	+/−	+	+	+
β-Galactosidase (β-GAR test)	−	−	−	+	−	−	−
β-Galactosidase (β-GAL test)	−	−	+	+/−	−	−	−
β-Mannosidase	−	+/−	+	−	−	−	−
Acid from							
Starch	v[a]	+	−	+	+	+/−	−
Glycogen	v[a]	+	−	−	+	−	−
Inulin	v[a]	+	−	−	−	−	−
Lactose	v[a]	+	+	+	+	+	−
Mannitol	−	+	−	−	−	−	−
Methyl-β-D-glucopyranoside	+	+	+	−	−	+	+/−
Raffinose	v[a]	+	+/−	−	+	−	−
Trehalose	+/−	+	+	−	−	−	−

+, ≥80% of strains are positive; +/−, 21–79% of strains are positive; −, ≤20% of strains are positive. Table adapted from Schlegel et al. (2003a).
[a] Different reactions were observed for isolates formerly assigned to S. equinus or S. bovis.

McWhinney et al., 1993; Renneberg, Niemann and Gutschik, 1997), with further reports emerging throughout the 1990s from almost all continents, indicating emerging penicillin resistance.

Within Europe, many reports from the late 1980s, and continuing throughout the 1990s, observed an increase in the numbers of antibiotic-resistant non-β-haemolytic streptococcal strains implicated in severe disease in neutropenic and other patient groups. Research from Barcelona, Spain, based on non-β-haemolytic streptococcal strains

causing bacteraemia in neutropenic patients collected between 1987 and 1992 documented such an emergence of penicillin resistance (Carratala et al., 1995). The strain collection was expanded to span from 1986 to 1996, and microbiological analysis revealed a remarkably high number of strains to have reduced susceptibility to penicillin (39%), with resistance to erythromycin and ciprofloxacin being found in just over a third of isolates (Marron et al., 2001). Susceptibility to fourth-generation cephalosporins was also tested in 35 of these isolates, 5 of which were found to be resistant to cefpirome and 12 to cefepime. How widespread this phenomenon of penicillin resistance was is difficult to ascertain, since a study from the opposite side of the Iberian Peninsula (Lugo), although admittedly based on a small collection of anginosus group bacteraemic strains (1988–1994), found all to be sensitive to penicillin and erythromycin (Casariego *et al.*, 1996). More recent evidence from central Spain (Madrid) also reported a high proportion of non-β-haemolytic streptococci isolated from blood culture between 1998 and 2001 to be to erythromycin (42%), tetracycline (35%) and clindamycin (25%) resistant (Rodriguez-Avial *et al.*, 2003). Researchers also found evidence of cross-resistance between erythromycin and tetracycline, with resistance to one agent being associated with the occurrence of the other. Also of note is that 3 of the 111 isolates tested by Rodriguez-Avial *et al.* (2003) were found to be resistant to the streptogramin quinupristin/dalfopristin.

National surveillance data from England and Wales also point to a similar increase in resistant phenotypes (PHLS, 2001, 2002; HPA, 2003a). The numbers of bacteraemic strains reported through routine surveillance as having reduced susceptibility to penicillin increased between 2000 and 2002 for all non-β-haemolytic streptococcal groups, with yearly increases for *S. mitis* group being particularly pronounced, stepping from 9 reports in 2000 to 33 in 2001 to 73 in 2003. Of the multitude of species that constitute the non-β-haemolytic streptococci, *S. mitis*, *S. sanguis* and *S. oralis*, have been described as most frequently exhibiting reduced susceptibility to penicillin (Alcaide *et al.*, 1995; Mouton *et al.*, 1997; de Azavedo *et al.*, 1999; Wisplinghoff *et al.*, 1999; Gershon *et al.*, 2002). Analysis of streptococcal isolates submitted to the UK national reference laboratories between 1996 and 2000, originating from patients with endocarditis, found one in four of *S. mitis* isolates to have reduced susceptibility to penicillin (MIC >0.125 mg/l), with one in seven of *S. sanguis* and one in eight of *S. oralis* isolates also showing reduced penicillin susceptibility (Johnson *et al.*, 2001). Tetracycline and erythromycin resistance similarly increased over this period, with *S. mitis* group showing highest frequency of erythromycin resistance and *S. bovis* group showing the highest frequency of tetracycline resistance (HPA, 2003a). In the light of the common target of activity for streptogramins and macrolides, studies have found evidence of cross-resistance between erythromycin and quinupristin/dalfopristin (Mouton *et al.*, 1997).

Similarly worrying reports of antibiotic resistance in non-β-haemolytic streptococci causing bacteraemia and/or endocarditis emerged from other parts of Europe during this period. Although frequently based on small collections of strains, studies in France, Germany, Sweden and Denmark documented substantial proportions of non-β-haemolytic streptococcal strains as having decreased sensitivity to erythromycin and/or penicillin (Wisplinghoff *et al.*, 1999; Lefort *et al.*, 2002; Westling *et al.*, 2002). Pooled data for 1997–1998 from European countries participating in the SENTRY Antimicrobial Surveillance Programme found 39% of non-β-haemolytic streptococci to be nonsusceptible to penicillin, whereas 30% exhibited reduced susceptibility to erythromycin (Fluit *et al.*, 2000).

Interestingly, two studies from the Far East both showed high levels of macrolide resistance in bacteraemic isolates. Sixty-three per cent of non-β-haemolytic streptococcal strains collection from Taiwan were reported as having reduced susceptibility to erythromycin and clarithromycin, although all strains were sensitive to penicillin, cefotaxime, imipenem, vancomycin and teicoplanin (Teng *et al.*, 2001). Similarly, in Hong Kong, characterization of bacteraemic *S. bovis* isolates showed all to be sensitive to penicillin, cephalothin and vancomycin, whereas 65% had reduced susceptibility to erythromycin. Forty-one per cent also showed reduced susceptibility to clindamycin (Lee *et al.*, 2003). Subsequent reports from participants in the Asia-Pacific region SENTRY Antimicrobial Surveillance Programme, which includes Hong Kong and Taiwan, documented a frequency of penicillin and erythromycin resistance in non-β-haemolytic streptococci more akin to that documented in Europe and the Americas,

possibly due to the effect of pooling data from many countries, including Australia and South Africa (Gordon *et al.*, 2002).

Research from North America began to show a similar pattern of emerging penicillin resistance as that seen in Europe among non-β-haemolytic streptococci. Studies conducted in the United States during the mid-1990s began to report levels of penicillin tolerance (nonsusceptibility) at between 39% and 44% of isolates (Doern *et al.*, 1996; Pfaller *et al.*, 1997). Cases of quinupristin/dalfopristin-resistant *S. mitis* have also been uncovered in the United States through the SENTRY Antimicrobial Surveillance Programme (Kugler *et al.*, 2000). Similarly, research from the Canadian Bacterial Surveillance Network found levels of penicillin resistance increasing from the mid-1990s (28% of blood culture isolates) to 2000 (37%) (de Azavedo *et al.*, 1999; Gershon *et al.*, 2002). Bigger increases still were seen in the frequency of isolates nonsusceptible to erythromycin, from 29% to 42%, with frequency of reduced clindamycin susceptibility also increasing from 4% to 10%.

Following the mottled array of reports describing the emergence of resistant phenotypes over 1990s, more recent descriptions from the SENTRY global antimicrobial surveillance network suggest that some degree of uniformity of antibiotic resistance may now exist among non-β-haemolytic streptococci causing blood stream infection (Gordon *et al.*, 2002). The increase in resistance in these important pathogens presents a challenge to the treatment and control of infections and warrants further vigilance, especially in the light of the first description of a naturally occurring vancomycin-resistant *S. bovis* strain (Poyart *et al.*, 1997).

REFERENCES

Ahmet, Z., Warren, M. and Huoang, E.T. (1995) Species identification of members of the *Streptococcus milleri* group isolated from the vagina by ID 32 Strep System and differential phenotypic characteristics. *Journal of Clinical Microbiology*, **33**, 1592–1595.

Alcaide, F., Linares, J., Pallares, R. *et al.* (1995) *In vitro* activities of 22 beta-lactam antibiotics against penicillin-resistant and penicillin-susceptible viridans group streptococci isolated from blood. *Antimicrobial Agents and Chemotherapy*, **39**, 2243–2247.

Alcaide, F., Carratala, J., Linares, J. *et al.* (1996) *In vitro* activities of eight macrolide antibiotics and RP-59500 (quinupristin-dalfopristin) against viridans group streptococci isolated from blood of neutropenic cancer patients. *Antimicrobial Agents and Chemotherapy*, **40**, 2117–2120.

Banks, J., Poole, S., Nair, S.P. *et al.* (2002) *Streptococcus sanguis* secretes CD14-binding proteins that stimulate cytokine synthesis: a clue to the pathogenesis of infective (bacterial) endocarditis? *Microbial Pathogenesis*, **32**, 105–116.

Barnard, J.P. and Stinson, M.W. (1996) The alpha-hemolysin of *Streptococcus gordonii* is hydrogen peroxide. *Infection and Immunity*, **64**, 3853–3857.

Barnard, J.P. and Stinson, M.W. (1999) Influence of environmental conditions on hydrogen peroxide formation by *Streptococcus gordonii*. *Infection and Immunity*, **67**, 6558–6564.

Bayer, A.S., Ward, J.I., Ginzton, L.E. and Shapiro, S.M. (1994) Evaluation of new clinical criteria for the diagnosis of infective endocarditis. *The American Journal of Medicine*, **96**, 211–219.

Bayer, A.S., Bolger, A.F., Taubert, K.A. *et al.* (1998) Diagnosis and management of infective endocarditis and its complications. *Circulation*, **98**, 2936–2948.

Belko, J., Goldmann, D.A., Macone, A. and Zaidi, A.K. (2002) Clinically significant infections with organisms of the *Streptococcus milleri* group. *The Pediatric Infectious Disease Journal*, **21**, 715–723.

Bensing, B.A., Rubens, C.E. and Sullam, P.M. (2001) Genetic loci of *Streptococcus mitis* that mediate binding to human platelets. *Infection and Immunity*, **69**, 1373–1380.

Bensing, B.A., Siboo, I.R. and Sullam, P.M. (2001) Proteins PblA and PblB of *Streptococcus mitis*, which promote binding to human platelets, are encoded within a lysogenic bacteriophage. *Infection and Immunity*, **69**, 6186–6192.

Bentley, R.W., Leigh, J.A. and Collins, M.D. (1991) Intrageneric structure of *Streptococcus* based on comparative analysis of small-subunit rRNA sequences. *International Journal of Systematic Bacteriology*, **41**, 487–494.

Billington, S.J., Jost, B.H. and Songer, J.G. (2000) Thiol-activated cytolysins: structure, function and role in pathogenesis. *FEMS Microbiology Letters*, **182**, 197–205.

Bourgault, A.M., Wilson, W.R. and Washington, J.A. (1979) Antimicrobial susceptibilities of species of viridans streptococci. *The Journal of Infectious Diseases*, **140**, 316–321.

Bouvet, A., Grimont, F. and Grimont, P.A.D. (1989) *Streptococcus defectivus* sp. nov. and *Streptococcus adjacens* sp. nov. nutritionally variant streptococci from human clinical specimens. *International Journal of Systematic Bacteriology*, **39**, 290–294.

Burnette-Curley, D., Wells, V., Viscount, H. *et al.* (1995) FimA, a major virulence factor associated with *Streptococcus parasanguis* endocarditis. *Infection and Immunity*, **63**, 4669–4674.

Carratala, J., Alcaide, F., Fernandez-Sevilla, A. *et al.* (1995) Bacteremia due to viridans streptococci that are highly resistant to penicillin: increase among neutropenic patients with cancer. *Clinical Infectious Diseases*, **20**, 1169–1173.

Casariego, E., Rodriguez, A., Corredoira, J.C. *et al.* (1996) Prospective study of *Streptococcus milleri* bacteremia. *European Journal of Clinical Microbiology & Infectious Diseases*, **15**, 194–200.

Christie, J., McNab, R. and Jenkinson, H.F. (2002) Expression of fibronectin-binding protein FbpA modulates adhesion in *Streptococcus gordonii*. *Microbiology*, **148**, 1615–1625.

Clarke, J.K. (1924) On the bacterial factor in the aetiology of dental caries. *British Journal of Experimental Pathology*, **5**, 141–147.

Collins, M.D., Lawson, P.A. (2000) The genus *Abiotrophia* (Kawamura et al.) is not monophyletic: proposal of *Granulicatella* gen. nov., *Granulicatella adiacens* comb. nov., *Granulicatella elegans* comb. nov. and *Granulicatella balaenopterae* comb. nov. *International Journal of Systematic and Evolutionary Microbiology*, **50**, 365–369.

Dankert, J., van der Werff, J., Zaat, S.A. *et al.* (1995) Involvement of bactericidal factors from thrombin-stimulated platelets in clearance of adherent viridans streptococci in experimental infective endocarditis. *Infection and Immunity*, **63**, 663–671.

Dankert, J., Krijgsveld, J., van Der Werff, J. *et al.* (2001) Platelet microbicidal activity is an important defense factor against viridans streptococcal endocarditis. *The Journal of Infectious Diseases*, **184**, 597–605.

de Azavedo, J.C., Trpeski, L., Pong-Porter, S. *et al.* (1999) *In vitro* activities of fluoroquinolones against antibiotic-resistant blood culture isolates of viridans group streptococci from across Canada. *Antimicrobial Agents and Chemotherapy*, **43**, 2299–2301.

De Gheldre, Y., Vandamme, P., Goossens, H. and Struelens, M.J. (1999) Identification of clinically relevant viridans streptococci by analysis of transfer DNA intergenic spacer length polymorphism. *International Journal of Systematic Bacteriology*, **49**, 1591–1598.

Diekema, D.J., Pfaller, M.A., Jones, R.N. *et al.* (2000) Trends in antimicrobial susceptibility of bacterial pathogens isolated from patients with bloodstream infections in the USA, Canada and Latin America. SENTRY Participants Group. *International Journal of Antimicrobial Agents*, **13**, 257–271.

Doern, G.V., Ferraro, M.J., Brueggemann, A.B. and Ruoff, K.L. (1996) Emergence of high rates of antimicrobial resistance among viridans group streptococci in the United States. *Antimicrobial Agents and Chemotherapy*, **40**, 891–894.

Douglas, C.W., Brown, P.R. and Preston, F.E. (1990) Platelet aggregation by oral streptococci. *FEMS Microbiology Letters*, **72**, 63–68.

Douglas, C.W., Heath, J., Hampton, K.K. and Preston, F.E. (1993) Identity of viridans streptococci isolated from cases of infective endocarditis. *Journal of Medical Microbiology*, **39**, 179–182.

Durack, D.T. (1998) Antibiotics for prevention of endocarditis during dentistry: time to scale back? *Annals of Internal Medicine*, **129**, 829–831.

Durack, D.T., Lukes, A.S. and Bright, D.K. (1994) New criteria for diagnosis of infective endocarditis: utilization of specific echocardiographic findings. Duke Endocarditis Service. *The American Journal of Medicine*, **96**, 200–209.

Duval, X., Papastamopoulos, V., Longuet, P. *et al.* (2001) Definite *Streptococcus bovis* endocarditis: characteristics in 20 patients. *Clinical Microbiology and Infection*, **7**, 3–10.

Elliott, D., Harrison, E., Handley, P.S. *et al.* (2003) Prevalence of Csh-like fibrillar surface proteins among mitis group oral streptococci. *Oral Microbiology and Immunology*, **18**, 114–120.

Emori, T.G. and Gaynes, R.P. (1993) An overview of nosocomial infections, including the role of the microbiology laboratory. *Clinical Microbiology Reviews*, **6**, 428–442.

Facklam, R. (2002) What happened to the streptococci: overview of taxonomic and nomenclature changes. *Clinical Microbiology Reviews*, **15**, 613–630.

Ferrieri, P., Gewitz, M.H., Gerber, M.A. *et al.* (2002) Unique features of infective endocarditis in childhood. *Circulation*, **105**, 2115–2127.

Fluit, A.C., Jones, M.E., Schmitz, F.J. *et al.* (2000) Antimicrobial susceptibility and frequency of occurrence of clinical blood isolates in Europe from the SENTRY antimicrobial surveillance program, 1997 and 1998. *Clinical Infectious Diseases*, **30**, 454–460.

Freney, J., Bland, S., Etienne, J. *et al.* (1992) Description and evaluation of the semiautomated 4-hour Rapid ID 32 Strep method for identification of streptococci and members of related genera. *Journal of Clinical Microbiology*, **30**, 2657–2661.

Gershon, A.S., de Azavedo, J.C., McGeer, A. *et al.* (2002) Activities of new fluoroquinolones, ketolides, and other antimicrobials against blood culture isolates of viridans group streptococci from across Canada, 2000. *Antimicrobial Agents and Chemotherapy*, **46**, 1553–1556.

Gordon, K.A., Beach, M.L., Biedenbach, D.J. *et al.* (2002) Antimicrobial susceptibility patterns of beta-hemolytic and viridans group streptococci: report from the SENTRY Antimicrobial Surveillance Program (1997–2000). *Diagnostic Microbiology and Infectious Disease*, **43**, 157–162.

Hall, G., Heimdahl, A. and Nord, C.E. (1999) Bacteremia after oral surgery and antibiotic prophylaxis for endocarditis. *Clinical Infectious Diseases*, **29**, 1–8.

Hamada, S. and Slade, H.D. (1980) Biology, immunology, and cariogenicity of *Streptococcus mutans*. *Microbiological Reviews*, **44**, 331–384.

Hardie, J.M. and Whiley, R.A. (1992) The genus *Streptococcus* – oral. In *The Prokaryotes*, 2nd edn, Vol. II (eds A. Balows, H.G. Trüper, M. Dworkin, W. Harder and K.-H. Schleifer), pp. 1421–1449. Springer-Verlag, New York.

Hardie, J.M. and Whiley, R.A. (1994) Recent developments in streptococcal taxonomy: their relation to infections. *Reviews in Medical Microbiology*, **5**, 151–162.

Herzberg, M.C. (1996) Platelet–streptococcal interactions in endocarditis. *Critical Reviews in Oral Biology and Medicine*, **7**, 222–236.

Herzberg, M.C., MacFarlane, G.D., Gong, K. *et al.* (1992) The platelet interactivity phenotype of *Streptococcus sanguis* influences the course of experimental endocarditis. *Infection and Immunity*, **60**, 4809–4818.

Herzberg, M.C., Meyer, M.W., Kiliç, A.O. and Tao, L. (1997) Host–pathogen interactions in bacterial endocarditis: streptococcal virulence in the host. *Advances in Dental Research*, **11**, 69–74.

Hillman, J.D. (2002) Genetically modified *Streptococcus mutans* for the prevention of dental caries. *Antonie van Leeuwenhoek*, **82**, 361–366.

Hoen, B., Alla, F., Selton-Suty, C. *et al.* (2002) Changing profile of infective endocarditis: results of a 1-year survey in France. *Journal of the American Medical Association*, **288**, 75–81.

Hogevik, H., Olaison, L., Andersson, R. *et al.* (1995) Epidemiologic aspects of infective endocarditis in an urban population. A 5-year prospective study. *Medicine (Baltimore)*, **74**, 324–339.

HPA (2002) Candidaemia and polymicrobial bacteraemias, England, Wales, and Northern Ireland 2002. *CDR Weekly* [Serial Online], 13. http://www.hpa.org.uk/cdr/PDFfiles/2003/cdr4203.pdf (10 February 2004).

HPA (2003a) Pyogenic and non-pyogenic streptococcal bacteraemias, England, Wales, and Northern Ireland 2002. *CDR Weekly*, **13**(16). www.hpa.org.uk/cdr/PDFfiles/2003/cdr1603.pdf (10 February 2004).

HPA (2003b) *Klebsiella, Enterobacter, Serratia*, and *Citrobacter* spp bacteraemias, England, Wales, and Northern Ireland 2002. *CDR Weekly*, **13**(20). www.hpa.org.uk/cdr/archives/2003/cdr2003.pdf (10 February 2004).

Jacobs, J.A., Schouten, H.C., Stobberingh, E.E. and Soeters, P.B. (1995) Viridans streptococci isolated from the bloodstream. Relevance of species identification. *Diagnostic Microbiology and Infectious Disease*, **22**, 267–273.

Jacobs, J.A., Schot, C.S. and Schouls, L.M. (2000) Haemolytic activity of the 'Streptococcus milleri group' and relationship between haemolysis restricted to human red blood cells and pathogenicity in *S. intermedius*. *Journal of Medical Microbiology*, **49**, 55–62.

Jacobs, J.A. and Stobberingh, E.E. (1995) Hydrolytic enzymes of *Streptococcus anginosus*, *Streptococcus constellatus* and *Streptococcus intermedius* in relation to infection. *European Journal of Clinical Microbiology & Infectious Diseases*, **14**, 818–820.

Johnson, A.P., Warner, M., Broughton, K. *et al.* (2001) Antibiotic susceptibility of streptococci and related genera causing endocarditis: analysis of UK reference laboratory referrals, January 1996 to March 2000. *BMJ (Clinical Research Ed.)*, **322**, 395–396.

Kawabata, S., Kunitomo, E., Terao, Y. *et al.* (2001) Systemic and mucosal immunizations with fibronectin-binding protein FBP54 induce protective immune responses against *Streptococcus pyogenes* challenge in mice. *Infection and Immunity*, **69**, 924–930.

Kawamura, Y., Hou, X.G., Sultana, F. *et al.* (1995a) Determination of 16S rRNA sequences of *Streptococcus mitis* and *Streptococcus gordonii* and phylogenetic relationships among members of the genus *Streptococcus*. *International Journal of Systematic Bacteriology*, **45**, 406–408.

Kawamura, Y., Hou, X.G., Sultana, F. *et al.* (1995b) Transfer of *Streptococcus adjacens* and *Streptococcus defectivus* to *Abiotrophia* gen. nov. as *Abiotrophia adiacens* comb. nov. and *Abiotrophia defectiva* comb. nov., respectively. *International Journal of Systematic Bacteriology*, **45**, 798–803.

Kawamura, Y., Hou, X.G., Todome, Y. *et al.* (1998) *Streptococcus peroris* sp. nov. and *Streptococcus infantis* sp. nov., new members of the *Streptococcus mitis* group, isolated from human clinical specimens. *International Journal of Systematic Bacteriology*, **48**, 921–927.

Kelly, C.G., Younson, J.S., Hikmat, B.Y. *et al.* (1999) A synthetic peptide adhesion epitope as a novel antimicrobial agent. *Nature Biotechnology*, **17**, 42–47.

Kennedy, H.F. and Smith, A.J. (2000) Viridans streptococcal infection in the medically compromised. *Reviews in Medical Microbiology*, **11**, 77–86.

Kikuchi, K., Enari, T., Totsuka, K. and Shimizu, K. (1995) Comparison of phenotypic characteristics, DNA–DNA hybridization results, and results with a commercial rapid biochemical and enzymatic reaction system for identification of viridans group streptococci. *Journal of Clinical Microbiology*, **33**, 1215–1222.

Kilian, M. (2001) Recommended conservation of the names *Streptococcus sanguis*, *Streptococcus rattus*, *Streptococcus cricetus*, and seven other names included in the Approved Lists of Bacterial Names. Request for an opinion. *International Journal of Systematic and Evolutionary Microbiology*, **51**, 723–724.

Kilian, M., Mikkelsen, L. and Henrichsen, J. (1989) Taxonomic study of viridans streptococci: description of *Streptococcus gordonii* sp. nov. and emended descriptions of *Streptococcus sanguis* (White and Niven 1946), *Streptococcus oralis* (Bridge and Sneath 1982), and *Streptococcus mitis* (Andrewes and Horder, 1906). *International Journal of Systematic Bacteriology*, **39**, 471–484.

Kiliç, A.O., Herzberg, M.C., Meyer, M.W. *et al.* (1999) Streptococcal reporter gene-fusion vector for identification of *in vivo* expressed genes. *Plasmid*, **42**, 67–72.

Kitada, K., Nagata, K., Yakushiji, T. *et al.* (1992) Serological and biological characteristics of "Streptococcus milleri" isolates from systemic purulent infections. *Journal of Medical Microbiology*, **36**, 143–148.

Kitada, K., Inoue, M. and Kitano, M. (1997) Experimental endocarditis induction and platelet aggregation by *Streptococcus anginosus*, *Streptococcus constellatus* and *Streptococcus intermedius*. *FEMS Immunology and Medical Microbiology*, **19**, 25–32.

Kitten, T., Munro, C.L., Wang, A. and Macrina, F.L. (2002) Vaccination with FimA from *Streptococcus parasanguis* protects rats from endocarditis caused by other viridans streptococci. *Infection and Immunity*, **70**, 422–425.

Koga, T., Oho, T., Shimazaki, Y. and Nakano, Y. (2002) Immunization against dental caries. *Vaccine*, **20**, 2027–2044.

Kolenbrander, P.E., Andersen, R.N., Baker, R.A. and Jenkinson, H.F. (1998) The adhesion-associated *sca* operon in *Streptococcus gordonii* encodes an inducible high-affinity ABC transporter for Mn^{2+} uptake. *Journal of Bacteriology*, **180**, 290–295.

Kugler, K.C., Denys, G.A., Wilson, M.L. and Jones, R.N. (2000) Serious streptococcal infections produced by isolates resistant to streptogramins (quinupristin/dalfopristin): case reports from the SENTRY antimicrobial surveillance program. *Diagnostic Microbiology and Infectious Disease*, **36**, 269–272.

Kuramitsu, H.K. (2003) Molecular genetic analysis of the virulence of oral bacterial pathogens: an historical perspective. *Critical Reviews in Oral Biology and Medicine*, **14**, 331–344.

Lee, R.A., Woo, P.C., To, A.P. *et al.* (2003) Geographical difference of disease association in *Streptococcus bovis* bacteraemia. *Journal of Medical Microbiology*, **52**, 903–908.

Lefort, A., Lortholary, O., Casassus, P. *et al.* (2002) Comparison between adult endocarditis due to beta-hemolytic streptococci (serogroups A, B, C, and G) and *Streptococcus milleri*: a multicenter study in France. *Archives of Internal Medicine*, **162**, 2450–2456.

Lisby, G., Gutschik, E. and Durack, D.T. (2002) Molecular methods for diagnosis of infective endocarditis. *Infectious Disease Clinics of North America*, **16**, 393–412.

Lockhart, P.B. and Durack, D.T. (1999) Oral microflora as a cause of endocarditis and other distant site infections. *Infectious Disease Clinics of North America*, **13**, 833–850.

Lowrance, J.H., Baddour, L.M. and Simpson, W.A. (1990) The role of fibronectin binding in the rat model of experimental endocarditis caused by *Streptococcus sanguis*. *The Journal of Clinical Investigation*, **86**, 7–13.

Macey, M.G., Whiley, R.A., Miller, L. and Nagamune, H. (2001) Effect on polymorphonuclear cell function of a human-specific cytotoxin, intermedilysin, expressed by *Streptococcus intermedius*. *Infection and Immunity*, **69**, 6102–6109.

Marron, A., Carratala, J., Alcaide, F. *et al.* (2001) High rates of resistance to cephalosporins among viridans-group streptococci causing bacteraemia in neutropenic cancer patients. *The Journal of Antimicrobial Chemotherapy*, **47**, 87–91.

Marsh, P.D. (2003) Are dental diseases examples of ecological catastrophes? *Microbiology*, **149**, 279–294.

McNab, R., Holmes, A.R., Clarke, J.M. *et al.* (1996) Cell surface polypeptide CshA mediates binding of *Streptococcus gordonii* to other oral bacteria and to immobilized fibronectin. *Infection and Immunity*, **64**, 4204–4210.

McWhinney, P.H., Patel, S., Whiley, R.A. *et al.* (1993) Activities of potential therapeutic and prophylactic antibiotics against blood culture isolates of viridans group streptococci from neutropenic patients receiving ciprofloxacin. *Antimicrobial Agents and Chemotherapy*, **37**, 2493–2495.

Morris, A.M. and Webb, G.D. (2001) Antibiotics before dental procedures for endocarditis prophylaxis: back to the future. *Heart*, **86**, 3–4.

Mouly, S., Ruimy, R., Launay, O. *et al.* (2002) The changing clinical aspects of infective endocarditis: descriptive review of 90 episodes in a French teaching hospital and risk factors for death. *The Journal of Infection*, **45**, 246–256.

Mouton, J.W., Endtz, H.P., den Hollander, J.G. *et al.* (1997) *In-vitro* activity of quinupristin/dalfopristin compared with other widely used antibiotics against strains isolated from patients with endocarditis. *The Journal of Antimicrobial Chemotherapy*, **39**(Suppl. A), 75–80.

Munro, C.L. and Macrina, F.L. (1993) Sucrose-derived exopolysaccharides of *Streptococcus mutans* V403 contribute to infectivity in endocarditis. *Molecular Microbiology*, **8**, 133–142.

Mylonakis, E. and Calderwood, S.B. (2001) Infective endocarditis in adults. *The New England Journal of Medicine*, **345**, 1318–1330.

Nagamune, H., Ohnishi, C., Katsuura, A. *et al.* (1996) Intermedilysin, a novel cytotoxin specific for human cells secreted by *Streptococcus intermedius* UNS46 isolated from a human liver abscess. *Infection and Immunity*, **64**, 3093–3100.

Nagamune, H., Whiley, R.A., Goto, T. *et al.* (2000) Distribution of the intermedilysin gene among the anginosus group streptococci and correlation between intermedilysin production and deep-seated infection with *Streptococcus intermedius*. *Journal of Clinical Microbiology*, **38**, 220–226.

Nagashima, H., Takao, A. and Maeda, N. (1999) Abscess forming ability of *Streptococcus milleri* group: synergistic effect with *Fusobacterium nucleatum*. *Microbiology and Immunology*, **43**, 207–216.

NINSS (2003) *Surveillance of Hospital-Acquired Bacteraemia in English Hospitals, 1997–2002*. PHLS, London. www.hpa.org.uk/infections/publications/ninns/hosacq_HAB_2002.pdf (10 February 2004).

Petersen, F.C., Pasco, S., Ogier, J. *et al.* (2001) Expression and functional properties of the *Streptococcus intermedius* surface protein antigen I/II. *Infection and Immunity*, **69**, 4647–4653.

Petersen, F.C., Assev, S., van der Mei, H.C. *et al.* (2002) Functional variation of the antigen I/II surface protein in *Streptococcus mutans* and *Streptococcus intermedius*. *Infection and Immunity*, **70**, 249–256.

Pfaller, M.A., Jones, R.N., Marshall, S.A. *et al.* (1997) Nosocomial streptococcal blood stream infections in the SCOPE Program: species occurrence and antimicrobial resistance. The SCOPE Hospital Study Group. *Diagnostic Microbiology and Infectious Disease*, **29**, 259–263.

PHLS (2001) Non-pyogenic and group B streptococcal bacteraemias, England and Wales: 1999 and 2000. *CDR Weekly*, **11**(42). www.hpa.org.uk/cdr/PDFfiles/2001/cdr4201.pdf (10 February 2004).

PHLS (2002) Pyogenic and non-pyogenic streptococcal bacteraemias, England and Wales: 2001. *CDR Weekly*, **12**(16). www.hpa.org.uk/cdr/PDFfiles/2002/cdr1602.pdf (10 February 2004).

Poyart, C., Pierre, C., Quesne, G. *et al.* (1997) Emergence of vancomycin resistance in the genus *Streptococcus*: characterization of a *vanB* transferable determinant in *Streptococcus bovis*. *Antimicrobial Agents and Chemotherapy*, **41**, 24–29.

Poyart, C., Quesne, G. and Trieu-Cuot, P. (1998) Taxonomic dissection of the *Streptococcus bovis* group by analysis of manganese-dependent superoxide dismutase gene (*sodA*) sequences: reclassification of '*Streptococcus infantarius* subsp. *coli*' as *Streptococcus lutetiensis* sp. nov. and of

Streptococcus bovis biotype II.2 as *Streptococcus pasteurianus* sp. nov. *International Journal of Systematic and Evolutionary Microbiology*, **52**, 1247–1255.

Renneberg, J., Niemann, L.L. and Gutschik, E. (1997) Antimicrobial susceptibility of 278 streptococcal blood isolates to seven antimicrobial agents. *The Journal of Antimicrobial Chemotherapy*, **39**, 135–140.

Rodriguez-Avial, I., Rodriguez-Avial, C., Culebras, E. and Picazo, J.J. (2003) Distribution of tetracycline resistance genes *tet*(M), *tet*(O), *tet*(L) and *tet*(K) in blood isolates of viridans group streptococci harbouring *erm*(B) and *mef*(A) genes. Susceptibility to quinupristin/dalfopristin and linezolid. *International Journal of Antimicrobial Agents*, **21**, 536–541.

Rouff, K.L., Miller, S.I., Garner, C.V. *et al.* (1989) Bacteremia with *Streptococcus bovis* and *Streptococcus salivarius*: clinical correlates of more accurate identification of isolates. *Journal of Clinical Microbiology*, **27**, 305–308.

Rudney, J.D. and Larson, C.J. (1993) Species identification of oral viridans streptococci by restriction fragment polymorphism analysis of rRNA genes. *Journal of Clinical Microbiology*, **31**, 2467–2473.

Scannapieco, F.A., Wang, B. and Shiau, H.J. (2001) Oral bacteria and respiratory infection: effects on respiratory pathogen adhesion and epithelial cell proinflammatory cytokine production. *Annals of Periodontology*, **6**, 78–86.

Schlegel, L., Grimont, F., Collins, M.D. *et al.* (2000) *Streptococcus infantarius* sp. nov., *Streptococcus infantarius* subsp. *infantarius* subsp. nov. and *Streptococcus infantarius* subsp. *coli* subsp. nov., isolated from humans and food. *International Journal of Systematic and Evolutionary Microbiology*, **50**, 1425–1434.

Schlegel, L., Grimont, F., Ageron, E. *et al.* (2003a) Reappraisal of the taxonomy of the *Streptococcus bovis/Streptococcus equinus* complex and related species: description of *Streptococcus gallolyticus* subsp. *gallolyticus* subsp. nov., *S. gallolyticus* subsp. *macedonicus* subsp. nov. and *S. gallolyticus* subsp. *pasteurianus* subsp. nov. *International Journal of Systematic and Evolutionary Microbiology*, **53**, 631–645.

Schlegel, L., Grimont, F., Grimont, P.A. and Bouvet, A. (2003b) Identification of major Streptococcal species by rrn-amplified ribosomal DNA restriction analysis. *Journal of Clinical Microbiology*, **41**, 657–666.

Sciotti, M.A., Yamodo, I., Klein, J.P. and Ogier, J.A. (1997) The N-terminal half part of the oral streptococcal antigen I/IIf contains two distinct binding domains. *FEMS Microbiology Letters*, **153**, 439–445.

Shain, H., Homer, K.A. and Beighton, D. (1996) Degradation and utilisation of chondroitin sulphate by *Streptococcus intermedius*. *Journal of Medical Microbiology*, **44**, 372–380.

Shenep, J.L. (2000) Viridans-group streptococcal infections in immunocompromised hosts. *International Journal of Antimicrobial Agents*, **14**, 129–135.

Shinzato, T. and Saito, A. (1994) A mechanism of pathogenicity of "Streptococcus milleri group" in pulmonary infection: synergy with an anaerobe. *Journal of Medical Microbiology*, **40**, 118–123.

Shinzato, T. and Saito, A. (1995) The *Streptococcus milleri* group as a cause of pulmonary infections. *Clinical Infectious Diseases*, **21**, S238–S243.

Siegman-Igra, Y. and Schwartz, D. (2003) *Streptococcus bovis* revisited: a clinical review of 81 bacteremic episodes paying special attention to emerging antibiotic resistance. *Scandinavian Journal of Infectious Diseases*, **35**, 90–93.

Simmons, N.A., Ball, A.P., Eykyn, S.J. *et al.* (1998) Antibiotic treatment of streptococcal, enterococcal, and staphylococcal endocarditis. *Heart*, **79**, 207–210.

Strom, B.L., Abrutyn, E., Berlin, J.A. *et al.* (1998) Dental and cardiac risk factors for infective endocarditis. A population-based, case-control study. *Annals of Internal Medicine*, **129**, 761–769.

Teng, L.J., Hsueh, P.R., Ho, S.W. and Luh, K.T. (2001) High prevalence of inducible erythromycin resistance among *Streptococcus bovis* isolates in Taiwan. *Antimicrobial Agents and Chemotherapy*, **45**, 3362–3365.

Trüper, H.G. and de'Clari, L. (1997) Taxonomic note: necessary correction of specific epithets formed as substantives (nouns) "in apposition". *International Journal of Systematic Bacteriology*, **47**, 908–909.

Trüper, H.G. and de'Clari, L. (1998) Taxonomic note: erratum and correction of further specific epithets formed as substantives (nouns) "in apposition". *International Journal of Systematic Bacteriology*, **48**, 615.

Unsworth, P.F. (1989) Hyaluronidase production in *Streptococcus milleri* in relation to infection. *Journal of Clinical Pathology*, **42**, 506–510.

van der Meer, J.T., Thompson, J., Valkenburg, H.A. and Michel, M.F. (1992) Epidemiology of bacterial endocarditis in The Netherlands. II. Antecedent procedures and use of prophylaxis. *Archives of Internal Medicine*, **152**, 1869–1873.

Venditti, M., Baiocchi, P., Santini, C. *et al.* (1989) Antimicrobial susceptibilities of *Streptococcus* species that cause septicemia in neutropenic patients. *Antimicrobial Agents and Chemotherapy*, **33**, 580–582.

Vernier, A., Diab, M., Soell, M. *et al.* (1996) Cytokine production by human epithelial and endothelial cells following exposure to oral viridans streptococci involves lectin interactions between bacteria and cell surface receptors. *Infection and Immunity*, **64**, 3016–3022.

Viscount, H.B., Munro, C.L., Burnette-Curley, D. *et al.* (1997) Immunization with FimA protects against *Streptococcus parasanguis* endocarditis in rats. *Infection and Immunity*, **65**, 994–1002.

Vriesema, A.J., Dankert, J. and Zaat, S.A. (2000) A shift from oral to blood pH is a stimulus for adaptive gene expression of *Streptococcus gordonii* CH1 and induces protection against oxidative stress and enhanced bacterial growth by expression of *msrA*. *Infection and Immunity*, **68**, 1061–1068.

Wells, V.D., Munro, C.L., Sulavik, M.C. *et al.* (1993) Infectivity of a glucan synthesis-defective mutant of *Streptococcus gordonii* (Challis) in a rat endocarditis model. *FEMS Microbiology Letters*, **112**, 301–305.

Westling, K., Ljungman, P., Thalme, A. and Julander, I. (2002) *Streptococcus viridans* septicaemia: a comparison study in patients admitted to the departments of infectious diseases and haematology in a university hospital. *Scandinavian Journal of Infectious Diseases*, **34**, 316–319.

Whatmore, A.M. and Whiley, R.A. (2002) Re-evaluation of the taxonomic position of *Streptococcus ferus*. *International Journal of Systematic and Evolutionary Microbiology*, **52**, 1783–1787.

Whiley, R.A. and Beighton, D. (1991) Emended descriptions and recognition of *Streptococcus constellatus*, *Streptococcus intermedius*, and *Streptococcus anginosus* as distinct species. *International Journal of Systematic Bacteriology*, **41**, 1–5.

Whiley, R.A. and Beighton, D. (1998) Current classification of the oral streptococci. *Oral Microbiology and Immunology*, **13**, 195–216.

Whiley, R.A., Freemantle, L., Beighton, D. *et al.* (1993) Isolation, identification and prevalence of *Streptococcus anginosus*, *S. intermedius*, and *S. constellatus* from the human mouth. *Microbial Ecology in Health and Disease*, **6**, 285–291.

Willcox, M.D. (1995) Potential pathogenic properties of members of the "*Streptococcus milleri*" group in relation to the production of endocarditis and abscesses. *Journal of Medical Microbiology*, **43**, 405–410.

Willcox, M.D., Loo, C.Y., Harty, D.W. and Knox, K.W. (1995) Fibronectin binding by *Streptococcus milleri* group strains and partial characterisation of the fibronectin receptor of *Streptococcus anginosus* F4. *Microbial Pathogenesis*, **19**, 129–137.

Willcox, M.D., Oakey, H.J., Harty, D.W. *et al.* (1994) Lancefield group C *Streptococcus milleri* group strains aggregate human platelets. *Microbial Pathogenesis*, **16**, 451–457.

Wilson, W.R., Geraci, J.E., Wilkowske, C.J. and Washington, J.A. (1978) Short-term intramuscular therapy with procaine penicillin plus streptomycin for infective endocarditis due to viridans streptococci. *Circulation*, **57**, 1158–1161.

Wisplinghoff, H., Reinert, R.R., Cornely, O. and Seifert, H. (1999) Molecular relationships and antimicrobial susceptibilities of viridans group streptococci isolated from blood of neutropenic cancer patients. *Journal of Clinical Microbiology*, **37**, 1876–1880.

Yokota, M., Basi, D.L., Herzberg, M.C. and Meyer, M.W. (2001) Anti-fibrin antibody binding in valvular vegetations and kidney lesions during experimental endocarditis. *Microbiology and Immunology*, **45**, 699–707.

Young, S.E.J. (1987) Aetiology and epidemiology of infective endocarditis in England and Wales. *The Journal of Antimicrobial Chemotherapy*, **20**(Suppl. A), 7–14.

Zhu, H., Willcox, M.D. and Knox, K.W. (2000) A new species of oral *Streptococcus* isolated from Sprague-Dawley rats, *Streptococcus orisratti* sp. nov. *International Journal of Systematic and Evolutionary Microbiology*, **50**, 55–61.

3

Streptococcus pneumoniae

Indran Balakrishnan

Department of Medical Microbiology, Royal Free Hospital, London, UK

INTRODUCTION

The pneumococcus was discovered in 1881 simultaneously by Sternberg in the United States, who named it *Micrococcus pasteuri*, and by Pasteur in France, who named it *Microbe septicèmique du salive* (Watson *et al.*, 1993). Within 10 years, the great propensity of this organism to cause serious infection was recognised and its causative role in pneumonia, meningitis, otitis media, arthritis and bacteraemia described. The identification of this organism as the most frequent cause of lobar pneumonia led to common use of the term pneumococcus, and its microscopic morphology in Gram-stained sputum led to it being renamed *Diplococcus pneumoniae* in 1926. Based on its ability to form short chains in liquid media, the organism was subsequently reclassified into the genus *Streptococcus* in 1974 and its name changed to *Streptococcus pneumoniae*.

Research performed during the first half of the twentieth century revealed that the pneumococcus could be divided into numerous serotypes using antisera against its capsular polysaccharides.

The history of *S. pneumoniae* is inextricably intertwined with that of microbiology. The pneumococcus played a pivotal role in the elucidation of the process of opsonisation in the 1890s, providing the basis for our understanding of humoral immunity. These principles soon found application when, in the early twentieth century, vaccination with killed pneumococci was successful in reducing the incidence of lobar pneumonia amongst African miners (reviewed in Musher, Watson and Dominguez, 1990). The first pneumococcal capsular polysaccharide vaccine was shown to be effective in 1938 (Smillie, Wornock and White, 1938).

The most significant role played by the pneumococcus in expanding our understanding of science, however, was in the discovery of DNA. The pneumococcus led to the discovery of the process of transformation in the 1920s by Griffith (1928). This process remained somewhat mysterious until the 1940s, when Avery, MacLeod and McCarty (1944) demonstrated that transformation is mediated by the transfer of DNA and that DNA is in fact the carrier of genetic information.

DESCRIPTION OF THE ORGANISM

Microscopic and Colonial Appearance

Streptococcus pneumoniae is a small (0.5–1.0 μm diameter) lanceolate Gram-positive coccus (Figure 3.1). It replicates in chains in liquid media but forms pairs joined along the horizontal axis on solid media. Pneumococci grow better in the presence of a source of catalase and form small (0.5–2 mm diameter) colonies that are dome-shaped initially but collapse centrally owing to autolysis after longer incubation, producing the classic draughtsman shape. Pneumococci breakdown haemoglobin by production of pneumolysin, resulting in a greenish

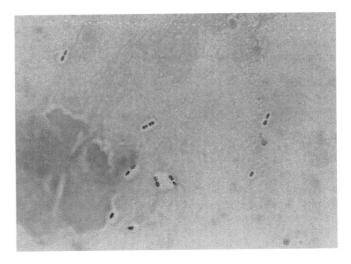

Figure 3.1 Gram stain appearance of *S. pneumoniae*.

discolouration of blood agar surrounding the colonies – this characteristic continues to be termed α-haemolysis, although lysis of erythrocytes is not a feature.

Physiology

Streptococcus pneumoniae is a facultative anaerobe. It is capnophilic, preferring atmospheres of 5–10% carbon dioxide. Pneumococci exhibit fermentative metabolism, utilising glucose via the hexose monophosphate pathway.

Four characteristics are especially useful in laboratory identification. The organism is

1. catalase negative;
2. inhibited by ethyl hydrocupreine (optochin);
3. bile soluble (bile salts activate autolytic activity, producing cell lysis);
4. α-haemolytic.

PATHOGENICITY

Streptococcus pneumoniae has a collection of characteristics that permit it to colonise a variety of niches. Most pneumococci are found in the nasopharynx, a location where transformation can readily take place. This location also provides a convenient niche from which invasive disease can occur, and a complex interplay of factors determines whether the colonising organisms go on to establish invasive infection. Severe pneumococcal infection and its complications result partly from the direct actions of pneumococcal virulence determinants and partly from the host immune response to various pneumococcal components. Pneumococcal virulence determinants and the corresponding immune responses produce four key effects: adhesion, invasion, inflammation and shock.

Adhesion

Opacity

Changes in colonial opacity of mucosal pathogens that alter their ability to colonise these surfaces have been described (James and Swanson, 1978; Weiser, 1993). Opacity variants have been described after culture on tryptose soya agar supplemented with catalase. The organisms spontaneously revert between transparent and opaque phases at a rate of between 10^{-3} and 10^{-6}. Transparent variants are able to colonise the nasopharynx in the infant rat model more easily than opaque variants, and this is in keeping with their greater capacity for adhering to a variety of cells (e.g. human buccal epithelial cells, type II pneumocytes and vascular endothelial cells) (Cundell et al., 1995b). The adherence of transparent variants is enhanced by treatment with interleukin-1 (IL-1) and tumour necrosis factor-α (TNF-α). This is probably due to cytokine-induced expression of cellular platelet-activating factor (PAF) receptors, to which the transparent phenotype binds via a phosphorylcholine (PC) ligand (see *Attachment and Phase Variation*) (Cundell et al., 1995a).

The variants do not change the physical characteristics of the peptidoglycan stem peptides, but opaque variants do not undergo autolysis as quickly as transparent variants. The opaque phenotype is associated with enhanced production of capsular polysaccharide and PspA and the transparent phenotype with increased amounts of C-polysaccharide (Kim and Weiser, 1998). The choline-binding protein CbpA is present in higher concentrations in transparent phenotypes, and this protein is required for efficient colonisation in the infant rat model [see *Choline-Binding Protein A (CbpA, PspC, SpsA and PbcA)*] (Rosenow et al., 1997). Opaque variants have been shown to be more virulent in the mouse intraperitoneal model of infection, with a reduction of 2–3 days in the time to death (Kim and Weiser, 1998). Mice dying after inoculation with opaque phenotypes have organisms of the same phenotype recovered from their spleen, whereas those mice inoculated with a transparent phenotype had organisms of a more opaque phenotype recovered from their spleen. Table 3.1 summarises the differences between opaque and transparent phenotypes.

Attachment and Phase Variation

Streptococcus pneumoniae binds to oral epithelial cells via *N*-acetyl galactosamine-β1–3-galactose (Andersson et al., 1983, 1988). Resting pneumocytes express two classes of receptors: *N*-acetyl-D-galactosamine linked by either β1,3 or β1,4 to galactose. Transparent and opaque phenotypes bind equally well to these receptors. Activated type II pneumocytes express a PAF receptor when stimulated by cytokines such as IL-1. Transparent, but not opaque, colonial variants of

pneumococci adhere to the PAF receptor via a PC ligand. It has been proposed that the pneumococcus binds to the PAF receptor by mimicking PAF, which, like the pneumococcal cell surface, contains PC – transparent phenotypes harbour more PC-containing cell-wall teichoic acid components, and binding is reduced by PAF receptor antagonists and exogenous choline (Cundell et al., 1995a).

Pneumococci cannot enter resting endothelial cells, but experiments with activated cells show that they can allow internalisation of 2–3% of the inoculum (Geelen, Bhattacharyya and Tuomanen, 1993). Activated endothelial cells upregulate PAF receptor expression and allow pneumococcal entry using the PAF receptor recycling pathway – this may provide a route of transition across the alveolar space into the bloodstream and across the blood–brain barrier into the meninges. It is thought that either viral infections or cell-wall components released from dividing pneumococci could provide the initial stimulus for cytokine production (Tuomanen, 1999).

Pneumococcal Surface Adhesin A

Pneumococcal surface adhesin A (PsaA) is a recently identified 37-kDa surface lipoprotein essential for pneumococcal virulence. Sequencing studies have shown homology between PsaA and lipoprotein adhesins found in other viridans-type streptococci, such as FimA in *Streptococcus parasanguis* and SsaB in *Streptococcus sanguis* (Fenno, LeBlanc and Fives-Taylor, 1989; Ganeshkumar et al., 1991). PsaA is a component of an ABC-type manganese permease membrane transport system – many such transporters exist in the pneumococcal genome, and some, such as Adc and PsaA, are particularly important for virulence.

PsaA is encoded by the *psaA* gene, which is part of the *psa* locus. This locus contains three other open-reading frames, encoding an ATP-binding protein (PsaB), a hydrophobic membrane protein (PsaC) and another protein (PsaD) which has homology with a periplasmic thiol peroxidase of *E. coli*.

In keeping with the actions of other pneumococcal permeases, the *psa* operon appears to have a regulatory role in adhesion. This operon affects the expression of choline-binding proteins on the pneumococcal surface, *psa* mutants demonstrating a complete absence of CbpA (Novak et al., 1998). PsaA and PsaD mutants have been shown to display reduced virulence in mice and reduced adherence to lung cells *in vitro* (Berry and Paton, 1996). *psaB*, *psaC* and *psaD* mutants have been noted to have a reduced capacity for transformation. Mutations within the *Psa* operon have also been shown to confer penicillin tolerance, mutants being able to survive 10× minimum inhibitory concentration (MIC) of penicillin (Novak et al., 1998).

Choline-Binding Protein A (CbpA, PspC, SpsA and PbcA)

Choline-binding protein A was the first pneumococcal surface protein adhesin to be described (Rosenow et al., 1997). It is composed of 663 amino acids and has a molecular mass of 75 kDa. The protein has two distinct domains: a C-terminal that shares >95% homology with PspA and a unique N-terminal that, although different in primary sequence, has an α-helical coiled-coil structure like PspA. Like PspA, the choline-binding C-terminal of CbpA has a proline-rich region and the ten tandem, direct repetitive sequences of 20 amino acids that are characteristic of choline-binding proteins.

CbpA is thought to play an important role in the adhesion of pneumococci to human cells – its expression is upregulated in the transparent phenotype, and it has been shown that CbpA is essential for pneumococcal penetration across the blood–brain barrier (see *Meningitis*) (Ring, Weiser and Tuomanen, 1998). It has been suggested that pneumococci first adhere to cell-surface glycoconjugates. Cytokine activation of the cells leads to a change in the expression of glycoprotein ligands, allowing increased pneumococcal adherence, and it is in this second stage, the change from colonisation to invasion, that CbpA has a function (Cundell, Masure and Tuomanen, 1995).

Table 3.1 Summary of differences between opaque and transparent phenotypes

Characteristic	Transparent	Opaque
Capsule production	+	+++
Teichoic acid	++++	+
Surface expression *lytA*	+++	+
PspA	+	++
Colonisation	+++	+
Virulence	+	+++

CbpA may act as a bridge between pneumococcal PC and activated human cell glycoconjugates. It could also mask PC by preventing it from binding the PAF receptor. It is apparent that the adhesive/invasive process is a complex interplay between CbpA and PC, and their precise relationship and relative importance remain unclear.

The importance of the interactions between CbpA and both secretory component (SC) and the polymeric immunoglobulin receptor (pIgR) in adhesion and invasion is being increasingly recognised (Zhang *et al.*, 2000; Kaetzel, 2001). SC is a peptide found free in several secretions including respiratory secretions, milk and bile as well as in association with pIgR and human secretory IgA (sIgA) (Zhang *et al.*, 2000; Kaetzel, 2001). The polymeric immunoglobulin receptor is an integral membrane protein of 764 amino acids in humans that is initially targeted to the basolateral surface of epithelial cells. An important function of pIgR is the transcytosis of polymeric immunoglobulins synthesised by plasma cells in the lamina propria. These immunoglobulins bind covalently to SC, an extracellular ligand-binding domain of about 550 amino acids at the N-terminal of pIgR. Internalisation signals cause pIgR (with or without pIg) to be endocytosed and transported from the basolateral epithelial surface via a series of endosomal compartments to the mucosal surface. Here, SC is proteolytically cleaved from pIgR and released into mucosal secretions, either free or as a pIg–SC complex termed secretory immunoglobulin (Mostov *et al.*, 1999). Uncleaved pIgR on the mucosal surface can be internalised and transcytosed back to the basolateral surface of the cell. The major consequences of the interaction between CbpA and pIgR are that it provides both a mechanism for adherence and, therefore, colonisation of mucosal surfaces and a pathway for the translocation of pneumococci across epithelial cells by co-opting the transcytosis machinery.

SC-binding domains of 107 amino acids each have been identified at two sites in CbpA, with a hexapeptide motif (Y(H/R)RNYPT) between amino acids 198 and 203 being the SC/sIgA-binding motif (Hammerschmidt *et al.*, 1997, 2000; Zhang *et al.*, 2000). CbpA has also been shown to directly bind domains D3 and D4 of pIgR (Lu *et al.*, 2003).

Several other functions have also been ascribed to CbpA under different names, including C3-binding activity and inhibition of the complement cascade by factor H-binding activity (Cheng, Finkel and Hostetter, 2000; Janulczyk *et al.*, 2000).

Invasion

The most important event in pneumococcal infection is invasion of the lower respiratory tract. Pneumococcal pneumonia is in itself a life-threatening illness, and it also serves as a focus for bloodstream invasion. Paradoxically, pneumococci also have an ability to establish systemic invasion in the absence of a clinically evident focus of infection.

At the alveolar level, the factors that determine whether or not invasive disease is established could be viewed as a dynamic balance between adherence and uptake into pneumocytes versus clearance by alveolar macrophages. Table 3.2 summarises the various scenarios and their likely outcomes.

Table 3.2 Factors determining risk of pneumococcal pulmonary invasion

Situation	Pneumocyte uptake	Macrophage clearance	Outcome
Resting pneumocytes and opsonic antibody absent	Low	Low	Low risk of invasion
Resting pneumocytes and opsonic antibody present	Low	High	Very low risk of invasion
Activated pneumocytes and opsonic antibody absent	High	Low	Invasion likely
Activated pneumocytes and opsonic antibody present	High	High	Invasion risk uncertain

The global population of pneumococci consists of more than 90 serotypes, but human infection/colonisation is dominated by a relatively small number of these. Multilocus sequence typing has shown that some sequence types define strains with an increased capacity to cause invasive disease (Enright and Spratt, 1998). These apparently successful clones may have gathered a collection of genes (or have enhanced expression of a group of genes) that confers a selective advantage that shifts the host–pathogen balance, facilitating the transition from colonisation to invasion. Some of the proteins encoded by these genes are discussed below.

Hyaluronidase

This enzyme could facilitate pneumococcal invasion by degrading connective tissue. The importance of this enzyme has been illustrated by the demonstration that pneumococcal strains with higher hyaluronidase activity breach the blood–brain barrier and disseminate more effectively (Kostyukova *et al.*, 1995). Studies have shown that pneumococcal strains isolated from patients with meningitis and meningoencephalitis have significantly higher hyaluronidase activity than strains causing purulent otitis media, indicating the importance of hyaluronidase in the pathogenesis of human pneumococcal meningitis (Volkova, Kostiukova and Kvetnaia, 1994).

Neuraminidase

These enzymes cleave sialic acid residues from a wide variety of molecules, damaging host tissues. This action may serve to expose receptors for pneumococcal adhesins, thus facilitating both colonisation and invasion. The role of neuraminidase in adhesion has been demonstrated in the chinchilla respiratory tract – neuraminidase treatment of the trachea increased pneumococcal adherence and reversed the inhibition produced by prior incubation with glycoconjugate analogues of known pneumococcal receptors. These data suggest that neuraminidase facilitates adhesion by increasing the number of adhesins available for pneumococcal binding (Tong *et al.*, 1999).

The importance of neuraminidase activity in pneumococcal virulence has been suggested by the finding that patients with pneumococcal meningitis with high levels of *N*-acetyl neuraminic acid in their cerebrospinal fluid (CSF) were more likely to develop coma and bacteraemia (O'Toole, Goode and Howe, 1971).

Pneumococcal Surface Protein A

Pneumococcal surface protein A (PspA) has been confirmed as a pneumococcal virulence determinant. Immunisation of X-linked immunodeficient mice with recombinant PspA conferred protective immunity to challenge with three of four virulent pneumococcal strains (McDaniel *et al.*, 1991), and monoclonal anti-PspA provided passive protection (McDaniel *et al.*, 1984).

Hammerschmidt *et al.* (1999) have shown that PspA functions as a specific receptor for lactoferrin, the N-terminal being responsible for lactoferrin binding. PspA, hence, plays an essential role in enabling iron acquisition by *S. pneumoniae*. Mutant strains lacking PspA expression fix more complement than isogenic parent strains expressing PspA (Fenno, LeBlanc and Fives-Taylor, 1989) – this interference with complement activation serves to facilitate pneumococcal survival and host invasion.

Pneumolysin

The importance of pneumolysin as a virulence determinant has been recognised following active and passive immunisation studies of mice (Paton, Lock and Hansman, 1983). This finding has since been confirmed by studies demonstrating considerable loss of virulence in isogenic mutants of *S. pneumoniae* in which the gene for pneumolysin had been disrupted (Berry *et al.*, 1989a).

Pneumolysin is a cytoplasmic toxin released by autolysis of the cell. It belongs to the family of thiol-activated cytolysins, the mechanism of action of which is thought to follow a two-stage common pathway (Bhakdi and Tranum-Jensen, 1986). The first stage entails binding of the monomeric toxin to the target-cell membrane. The second stage comprises lateral movement and oligomerisation of the monomers to form a high-molecular-weight transmembrane pore. This results in the leakage of intracellular solutes and an influx of water, resulting in colloid–osmotic lysis of the cell. By this mechanism, pneumolysin is able to damage a wide range of eukaryotic cells. Pneumolysin also has modulatory effects on cells at sublytic concentrations, e.g. stimulation of TNF-α and IL-1β production by human monocytes (Houldsworth, Andrew and Mitchell, 1994).

Pneumolysin is a 52.8-kDa polypeptide consisting of 470 amino acids. Although much of its structure–function relationship remains unclear, the functional relevance of some areas has been elucidated. Pneumolysin contains eight histidine residues, one of which (position 367 in pneumolysin) is conserved in all thiol-activated toxins. Mutagenesis of this residue seriously impairs oligomerisation, causing a dramatic loss of activity (Mitchell *et al.*, 1992). Also in common with other thiol-activated toxins, pneumolysin contains only a single cysteine residue (at position 428 in a domain towards the carboxyl terminus). Substitution of this residue has been shown to greatly reduce cytolytic activity without affecting binding or oligomerisation (Saunders *et al.*, 1989), implying that there is a crucial action required for target-cell lysis beyond oligomerisation, in which this residue plays an essential role. Deletion studies have shown that the C-terminus is involved in membrane binding via cholesterol (Alouf and Geoffroy, 1991; Owen *et al.*, 1994).

The secondary structure of pneumolysin consists of 36% β-sheet and 31% α-helix (Morgan *et al.*, 1993). Hydropathy plots suggest that it is a highly hydrophilic molecule with no significant regions of hydrophobicity; hence, oligomerisation is essential for the formation of channels. Electron micrographs show that pneumolysin monomers are asymmetric molecules composed of four domains (Morgan *et al.*, 1994). Three of the domains lie side by side, forming the stalk of the oligomer, from which domain 4 projects out as a flexible flange.

Oligomeric pneumolysin exists as large arc-and-ring structures in cell membranes, measuring 30–40 nm in diameter (Morgan *et al.*, 1995). The problem posed by this is the apparent absence of hydrophobic areas both on the surface of pneumolysin and at the domain interfaces, making membrane insertion impossible. There is, however, a hydrophobic pocket that could act as a cholesterol-binding site. This pocket is blocked by a tryptophan-rich region at the tip of domain 4, of which one residue (Trp 433) is crucial for maintaining a loop conformation. It is hypothesised that cholesterol could displace the tryptophan-rich region to reveal the hydrophobic pocket, which could then initiate membrane penetration (Rossjohn *et al.*, 1998).

The cytolytic properties of pneumolysin enable it to inflict damage on a wide variety of eukaryotic cells, including bronchial epithelial cells, alveolar epithelial cells and pulmonary endothelium. Its detrimental effects on bronchial epithelium result in the slowing of the ciliary beat, impairing the ability of the mucociliary escalator to effectively clear particles. Pneumolysin also induces separation of the tight junctions of the alveolar epithelial cells (Rayner *et al.*, 1995). This, together with its direct cytotoxic effect on the cells of the alveolar capillary barrier, serves to facilitate bloodstream invasion by pneumococci. Pneumolysin also inhibits the respiratory burst and chemotaxis of polymorphs at doses as low as 1 ng/mL, hence facilitating pneumococcal survival and invasion (Paton, Lock and Hansman, 1983).

Opsonophagocytosis by the complement system is essential for the clearance of pneumococci from the lung (Winkelstein, 1984). The ability of pneumolysin to activate the classical complement pathway may allow pneumococci to evade opsonophagocytosis by consuming the limited supply of complement factors in the alveoli, facilitating invasion.

Inflammation and Shock

Pneumolysin

An interesting mechanism of injury that mediates both the activation of the inflammatory response and direct lung damage is suggested by the finding that phospholipase A in pulmonary artery endothelium is strongly activated by pneumolysin and, once activated, breaks down a wide variety of cell-membrane phospholipids (Rubins *et al.*, 1994). This results in the release of free fatty acids and lysophosphatides, which are directly cytotoxic. Arachidonic acid released by phospholipase activity, together with its eicosanoid cascade metabolites, is a potent neutrophil chemotaxin. Recruitment and activation of neutrophils further contribute to lung damage as well as to the inflammatory response. Inflammation would be enhanced by the ability of pneumolysin to activate the classical complement pathway without the need for specific antibody (Paton, Rowan-Kelly and Ferrante, 1984).

Autolysin

The major enzyme responsible for cell-wall turnover is the autolysin of *S. pneumoniae*, *N*-acetyl muramoyl-L-alanine amidase. This enzyme is responsible for cellular autolysis that occurs at the end of log phase in pneumococci and for lysis of the cell wall in response to treatment with deoxycholate or with β-lactam antibiotics. Autolysin is one of about a dozen proteins that is held on the pneumococcal surface by noncovalently docking with the choline of the pneumococcal cell wall, instead of the usual method of anchorage in Gram-positive bacteria using the LPXTGE motif. These choline-binding proteins have a common choline-binding carboxy terminal, their functional diversity being produced by the amino terminal.

Autolysin contributes to pneumococcal virulence in several ways. Its role in cell-wall turnover means that autolysin activity generates cell-wall breakdown products that are highly inflammatory. The cell-wall-degrading activity of autolysin also allows the release of intracellular toxins (e.g. pneumolysin) into the external medium. Many experimental studies have demonstrated the role of autolysin in virulence. Reduced virulence has been demonstrated when normally virulent strains are transformed with inactivated *lytA* (the gene that encodes pneumococcal autolysin), and the deficit has been shown to be reversible by back transformation (Berry *et al.*, 1989b).

C-Polysaccharide (PnC)

The cell wall of *S. pneumoniae* has an important role in generating inflammatory mediators and consequent pathological changes. Injection of cell-wall preparations but not heat-killed capsulate strains or isolated capsular antigen has been shown to generate a strong inflammatory response in chinchilla rabbits (Tuomanen *et al.*, 1985a,b). The teichoic acid C-polysaccharide is important as removal of this antigen diminishes complement activation by cell-wall components (Tomasz and Saukkonen, 1989). The components of the complement cascade generated by interaction with PnC are thought to be crucial for the generation of an inflammatory reaction in the alveoli, the meninges and the middle ear (Tuomanen *et al.*, 1985a,b; Tuomanen, Rich and Zak, 1987). PnC activates the alternate complement pathway via its PC component (Winkelstein, Abramovitz and Tomasz, 1980; Hummell *et al.*, 1985). The pneumococcal plasma membrane is also a weak activator of the alternate pathway through the PnC analogue lipoteichoic acid (F antigen) (Hummell *et al.*, 1985). In the chinchilla model of otitis media, death of the organism induced by penicillin was responsible for a rapid increase in the inflammatory response. This was thought to be due to the release of cell-wall products and pneumolysin (Sato *et al.*, 1996). Organisms with high cell-wall turnover, resulting in a greater release of teichoic acid components, induce a greater inflammatory response in rabbit meninges (Schmidt and Täuber, 1993).

PC is of considerable functional relevance in organisms colonising mucosal surfaces, particularly the respiratory tract. The expression of this epitope is subject to phase variation, there being a selection for PC+ variants during nasopharyngeal colonisation. PC has been shown to protect organisms against clearance, hence contributing to persistence in the respiratory tract. However, PC+ variants exhibit increased serum sensitivity. Hence, it appears that there is differential expression of PC, with upregulation occurring in the initial colonisation phase and downregulation in the subsequent invasive phase to avoid anti-PC antibody, C-reactive protein (CRP) binding and complement activation (Weiser, 1998).

Immunopathology

Pneumonia

The interactions that occur between the pneumococcus and host during the evolution of infection are complex and remain to be fully elucidated. Using a mouse model, Bergeron et al. (1998) have broadly divided pulmonary pneumococcal infection into five stages. In the first stage (0–4 h), infection is established – there is evidence of ineffective phagocytosis by alveolar macrophages and cytokine release. The major pneumococcal antiphagocytic components are its polysaccharide capsule and pneumolysin (see *Invasion*). The predominant cytokines vary between sites, with TNF-α, IL-6 and nitric oxide (NO) predominating in bronchial lavage (BAL) fluid; TNF, IL-6 and IL-1 in lung tissue; and IL-6 in serum. The second stage (4–24 h) is marked by the multiplication of pneumococci in the alveoli and neutrophil chemotaxis into lung tissue – TNF, IL-6, IL-1 and leukotriene B4 levels all increase in both lung tissue and BAL fluid. Serum IL-1 levels also rise transiently. Neutrophil chemotaxis is mediated by the complement-activating properties of pneumolysin and C-polysaccharide as well as the cytokines listed above. Lung damage is seen in the third stage (24–48 h) – there is evidence of alveolar injury and interstitial oedema, reflecting the cytolytic actions of pneumolysin. There is also evidence of regeneration during this stage: type II pneumocytes proliferate to regenerate both type I and type II cells. This stage sees the progression of infection from the presepticaemic to the septicaemic phase, bacteria shifting from the alveoli through lung tissue into the bloodstream and there being a reduction in IL-1 and TNF in the lung tissue. Both CbpA and the changes made to the alveolar–capillary barrier made by pneumolysin (see *Invasion*) would facilitate bloodstream invasion. The fourth stage (48–72 h) is characterised by a sharp increase in alveolar monocyte and lymphocyte activity, together with NO release in lung tissue and alveolar spaces. The effects of pneumococcal bacteraemia are manifest – leucopenia and thrombocytopenia are seen, together with an increase in blood TNF and IL-6. The final stage (72–96 h) is characterised by further bacterial proliferation, high NO levels, massive tissue damage, lipid peroxidation and high mortality.

Cytokine production is stimulated by both CD14-dependent and -independent pathways, with as many as two receptors other than CD14 being involved (Cauwels et al., 1997). Although the precise roles and effects of the various cytokines produced at the different stages of the pathogenesis of pneumococcal pneumonia remain to be fully elucidated, some progress has been made towards unravelling this maze.

There is considerable evidence to suggest that IL-1 plays a far more dominant role in Gram-positive inflammation than it does in Gram-negative sepsis (Riesenfeld-Orn et al., 1989). Work indicates that TNF, although pivotal in the pathogenesis of Gram-negative sepsis syndrome, plays a protective role in pneumococcal pneumonia, inhibiting the onset of bacteraemia by stimulating antibody production, neutrophil recruitment and granulocyte–macrophage colony-stimulating factor production by endothelial cells (Takashima et al., 1997). IL-10, on the other hand, demonstrates a protective role in Gram-negative

sepsis, but elevated lung levels have been shown to be detrimental in a murine model of pneumococcal pneumonia (van der Poll et al., 1996). There is also evidence that IL-6 plays an important role as a mediator of inflammation, controlling the production of both pro- and anti-inflammatory cytokines. Its net effect has, however, been shown to be protective (van der Poll et al., 1997).

Meningitis

There has been considerable advance in our understanding of pneumococcal meningitis. Whether meningeal invasion is the result of specific targeting or purely a chance process remains unclear. However, pneumococci have been shown to be nearly twice as efficient at invading cerebral as compared with peripheral endothelium. Transparent phenotypes, with less capsular polysaccharide, more teichoic acid and more CbpA, are three to five times more efficient at invading cerebral microvascular endothelia than are opaque phenotypes, which are killed intracellularly. This is in contrast to the bloodstream, where opaque phenotypes are more invasive. Inflammatory activation is a prerequisite for the migration of pneumococci through cerebral microvascular endothelial cells. It is thought that transparent phenotypes are able to enter by PAF receptor-mediated endocytosis and corrupt the PAF receptor recycling pathway, producing a greater degree of vesicular transcytosis to the basolateral surface of the cell, with some recycling to the apical surface. Opaque variants, in contrast, appear to be killed intracellularly (Ring, Weiser and Tuomanen, 1998).

Streptococcus pneumoniae does not directly mediate cytotoxicity. Rather, tissue damage is the result of inflammation and its sequelae. The critical stimuli for the inflammatory response are the major components of the pneumococcal cell wall (teichoic acid and peptidoglycan) and pneumolysin, and not the polysaccharide capsule, which lacks inflammatory potential and acts mainly to protect pneumococci against phagocytosis in the absence of anticapsular antibody (Tuomanen et al., 1985a,b). Although different inflammatory responses have been found with strains of different capsular serotypes, the strains studied were not otherwise isogenic, and therefore, it is not possible to ascribe these differences to any one parameter. Cell-wall components released by autolytic activity induce the release of proinflammatory cytokines (e.g. TNF-α, IL-1, IL-6) from mononuclear cells within the central nervous system (CNS) such as ependymal cells, astrocytes and macrophages/microglia. These cytokines trigger a complex cascade of inflammatory mediators, which in concert regulate the various arms of the inflammatory response (Tuomanen et al., 1986).

The cytokines released induce an opening of the tight junctions between brain capillary endothelial cells associated with enhanced pinocytotic activity, allowing an influx of serum components, notably chemotactic complement factors (Ernst et al., 1984; Quagliarello, Long and Scheld, 1986). They also upregulate adhesion molecules [e.g. intercellular adhesion molecule-1 (ICAM-1)] on cerebral vascular endothelial cells and attract neutrophils across the blood–brain barrier (Täuber, 2000). In consequence, a neutrophil infiltrate is evident in the CSF within 8–12 h of bacterial inoculation (Sande and Täuber, 2000). However, the neutrophils have no effect on pneumococcal multiplication, reflecting the low concentration of opsonins in the CSF (Propp, Jabbari and Barron, 1973; Smith and Bannister, 1973; Ernst, Decazes and Sande, 1983).

The cytokine milieu may also have more direct neurotoxic effects by contributing to a metabolic shift to anaerobic glycolysis, astrocyte dysfunction and neuronal damage (Täuber, 2000).

Probably the most important consequence of this inflammatory reaction in the CSF is increased intracranial pressure, which results from cerebral oedema, increased cerebral blood volume and alterations of CSF hydrodynamics (in particular, reduced CSF resorption) (Scheld et al., 1980; Täuber, 1989; Tureen, Täuber and Sande, 1992; Tureen, Liu and Chow, 1996). This increased intracranial pressure, together with the impairment of autoregulation of cerebral blood flow seen in pneumococcal meningitis, plays a significant role in reducing cerebral blood flow, resulting in ischaemic necrosis and neuronal loss from energy failure. Several vasoconstrictive mediators, such as endothelin and reactive oxygen intermediates, counteract the vasodilatory effects of the increased NO production seen in meningitis and further contribute to cerebral ischaemia. These agents also have a variety of deleterious effects on cells and macromolecules, including lipid peroxidation, DNA damage and protein oxidation (Sande and Täuber, 2000).

Direct neurotoxicity is another important mechanism by which brain damage occurs. It is mediated by pneumolysin and several other agents released during pneumococcal meningitis. NO can be directly neurotoxic or can combine with superoxide to form peroxynitrite, a highly cytotoxic agent (Sande and Täuber, 2000). The production of NO by neuronal oxide synthase (NOS) and oxidative radical release is increased by excitatory amino acids (e.g. glutamate), which are released at higher levels from neurones under stress. A commonly affected site is the dentate gyrus in the hippocampus, cell death occurring by apoptosis (Täuber, 2000).

EPIDEMIOLOGY

High-Risk Groups

Both the rate of nasopharyngeal carriage and the rate of pneumococcal infection are highest at the extremes of age. Invasive pneumococcal infection is relatively common in neonates and children below the age of 2 years (160 per 100 000), the incidence falling sharply in adolescent and young adult years (5 per 100 000), to rise again in populations over the age of 65 years (70 per 100 000) (Breiman *et al.*, 1990). Rates of both colonisation and invasive infection show marked seasonal variation, with considerable increase in the colder months (Balakrishnan *et al.*, 2000).

Although pneumococci undoubtedly attack previously healthy people, several anatomical (dural tears and basal skull fracture) and physiological (mucociliary escalator dysfunction due to pollution, cigarette smoke or viral infection) defects predispose to pneumococcal infection. Defects of the reticuloendothelial system (hyposplenism due to any cause) and humoral immune system (hypogammaglobulinaemia and complement defects) and AIDS also greatly increase host susceptibility. Other conditions that predispose to pneumococcal infection include alcoholism, diabetes mellitus and chronic cardiac, respiratory, liver and renal disease.

Host-to-host transmission occurs from extensive, close contact. Daycare centre attendance and crowded living conditions (prisons, nursing homes and homeless shelters) are associated with an increased transmission rate, but casual contact is usually not (Rauch *et al.*, 1990; Hoge *et al.*, 1994).

IgG2 has a pivotal role in defence against pneumococcal infection. Human FcγRIIa has two codominantly expressed allotypes, which greatly differ in their ability to ligate IgG2, FcγRIIa–R131 binding only weakly. Studies have shown that half of all patients with bacteraemic pneumococcal pneumonia were homozygous for FcγRIIa–R131, which was significantly higher than in other groups. Moreover, all bacteraemic patients who died within 1 week of admission were homozygous for this allele (Yee *et al.*, 2000).

Serotype Distribution

More than 90 known serotypes of pneumococci vary in their distribution and propensity to cause different infections. Seven serotypes (6A, +6B, 7F, 14, 18C, 19F and 23F) account for 70% of isolates from blood and CSF amongst children in North America and Western Europe (Klein, 1981; Nielsen and Henrichsen, 1992). Together with these serotypes, type 3 is commonly implicated in otitis media in children.

CLINICAL SYNDROMES

Streptococcus pneumoniae causes infection of the bronchi, lungs, sinuses and middle ear by direct spread from the site of nasopharyngeal colonisation and of the CNS, peritoneal cavity, bones, joints and heart valves by haematogenous spread. Local extension to the CNS may also result in recurrent meningitis in patients who have a dural defect. Primary bacteraemia (i.e. one without identifiable source) is particularly common in children aged 6–24 months – patients with this syndrome are often only moderately ill and may recover swiftly and spontaneously; if left untreated, however, overt foci of infection may subsequently develop.

Pneumonia

Streptococcus pneumoniae is the most common cause of pneumonia worldwide, and pneumonia is the most common manifestation of pneumococcal infection, being present in 68% of cases of pneumococcal bacteraemia in a study by Balakrishnan *et al.* (2000). Pneumococcal pneumonia is estimated to affect 1 in 1000 adults each year (World Health Organization, 1999). Disease occurs when body defences fail to prevent pneumococcal access to and subsequent replication in the alveoli.

The most frequent clinical picture is an ill patient who reports sudden onset of fever in association with pleuritic chest pain and other nonspecific symptoms such as headache, vomiting and diarrhoea. Cough soon develops, which is initially dry but often followed by haemoptysis. The presentation may be much more subtle in elderly and immunocompromised patients. Physical examination reveals signs consistent with consolidation. Pleural effusion may also be detectable. These findings are confirmed by chest X-ray, which usually shows consolidation filling all or most of a lobe. *S. pneumoniae* is not a classic abscess-producing organism, and lung abscess should raise suspicion of other pathology such as bronchial obstruction or pulmonary infarction. Most patients with pneumococcal pneumonia have a leucocytosis. Some patients, however, have a low leucocyte count, which indicates a very poor prognosis (Balakrishnan *et al.*, 2000). Abnormal liver function tests are also a common finding.

The presence of abundant neutrophils together with lanceolate Gram-positive diplococci in sputum strongly suggests the diagnosis, which is usually confirmed by microbiological culture of respiratory tract specimens (most often sputum) and blood.

The case fatality of pneumococcal pneumonia approximates 5–10% and increases to 20% in the presence of bacteraemia. Besides the bloodstream, infection can rarely extend to involve the pericardial and pleural cavities, resulting in empyema.

Meningitis

Streptococcus pneumoniae causes meningitis in all age groups and is the most common cause of bacterial meningitis in middle-aged and elderly populations and in patients with dural defects. Meningitis accounts for about 10% of invasive pneumococcal infections. Its incidence in developed countries is about 1.5 per 100 000, rising to 8 per 100 000 in children below the age of 5 years (Spanjaard *et al.*, 2000).

Onset is usually over 1–2 days, although it may be more gradual. Meningism is usually obvious at presentation, and confusion and obtundation are not uncommon. Rash is not a typical feature and should prompt investigation for an alternative aetiology. A petechial rash may, however, be seen in overwhelming pneumococcal sepsis. The patient may have evidence of other foci of pneumococcal infection.

Peripheral neutrophil leucocytosis ($>20 \times 10^9$/L) is usual, a low white count being associated with a poor prognosis (Balakrishnan *et al.*, 2000). CSF examination reveals a high neutrophil count, accompanied by other features of bacterial meningitis (high CSF protein and low CSF: plasma glucose ratio). Pneumococci are often evident on Gram stain examination of CSF (organisms are seen in 75–90% of culture-positive specimens) and are readily cultured from CSF and often from blood. Antigen detection in CSF provides a rapid adjunct to diagnosis and is of particular benefit in specimens taken after commencement of chemotherapy. Molecular amplification techniques may also be applied.

Pneumococcal meningitis is particularly prone to complications, common sequelae being subdural empyema, cranial nerve palsies and hydrocephalus. Seizures occur in about 30% of cases and are associated with higher mortality. Brain infarcts due to cortical thrombophlebitis or arteritis occur in about 15%. It carries a case fatality of about 30% (Immunization Practices Advisory Committee, 1981), although mortality may be much higher at the extremes of age.

Otitis Media

Streptococcus pneumoniae is recognised as being the most common cause of otitis media, being implicated in 30–40% of cases. Pneumococcal otitis media has a high prevalence the world over, nearly every child suffering at least one episode of this infection before the age of 5 years. Pneumococcal otitis media usually follows viral upper respiratory tract infection, which plays a contributory role by causing congestion of the meatus of the eustachian tube. Infection usually follows soon after the host is colonised with a novel serotype, serotypes 6, 14, 19F and 23F being predominant.

Sinusitis

Streptococcus pneumoniae is a leading cause of sinusitis. Pneumococcal sinusitis is usually secondary to mucosal congestion by viral infection or allergy, resulting in impaired bacterial clearance.

Other Syndromes

Streptococcus pneumoniae is recognised as the most common cause of primary bacterial peritonitis in prepubertal females. Pneumococcal endocarditis and osteomyelitis have been described rarely, pneumococci having a vertebral predilection. Pneumococcal septic arthritis is also well recognised. Unusual or recurrent pneumococcal infection without obvious predisposing cause should prompt exclusion of HIV infection.

LABORATORY DIAGNOSIS

Microscopy

Streptococcus pneumoniae is readily recognised in clinical specimens by its distinctive appearance. The presence of Gram-positive lanceolate diplococci (or short chains) in clinical specimens is, in most cases, diagnostic of pneumococcal infection – this diagnosis is supported by the presence of pus cells. An interesting adjunct to the Gram stain is the demonstration of capsular swelling (visible under phase-contrast or bright-field microscopy) on addition of polyvalent antisera – the quellung reaction.

Culture

Although a fastidious organism, pneumococci are readily cultured from appropriate clinical specimens on blood and chocolate agar, usually after overnight incubation at 37 °C. These organisms grow better under capnophilic (5% CO_2) and anaerobic conditions than they do aerobically. The typical draughtsman-shaped colonial appearance is often diagnostic. Confirmation may be obtained by demonstrating either inhibition of growth around a 5-μg optochin disc after overnight incubation or bile solubility.

Antigen Detection

Detection of antigen in body fluids enables rapid diagnosis of pneumococcal infection and is of particular advantage in patients who have received chemotherapy before specimens were obtained. Counterimmunoelectrophoresis was the earliest rapid antigen detection employed in clinical settings. This has now been largely replaced by latex agglutination and by immunochromatography.

Latex agglutination and coagglutination detect soluble pneumococcal capsular polysaccharide by using latex particles (or *Staphylococcus aureus* cells) coated with polyvalent antisera. These techniques are more sensitive and more rapid than counterimmunoelectrophoresis and have the added advantage of requiring no specialised equipment. Latex agglutination has a sensitivity of 90–100% and specificity of more than 98% in CSF but is considerably less reliable in other specimens such as sputum, urine (even after concentration) and pleural fluid. In sputum, sensitivity remains high (>90%), but specificity is considerably lower (80–85%).

The rapid diagnosis of pneumococcal infection has been greatly facilitated by immunochromatographic detection of C-polysaccharide in urine and CSF. This test takes only 15 min, requires no specialised equipment and is very reliable, producing sensitivities of 82% and 97% and specificities of 97% and 99% in urine and CSF, respectively (Smith *et al.*, 2003).

Nucleic Acid Amplification

Amplification of specific nucleic acid sequences by using the polymerase chain reaction (PCR) has provided a rapid and sensitive diagnostic technique for detecting pneumococcal infection that is virtually unaffected by prior chemotherapy. This technology has been applied to a variety of clinical samples, including CSF, blood and sputum. Several targets have been examined for suitability in this technique, the most popular being autolysin and pneumolysin. Studies using the autolysin gene as an amplification target in sputum samples have shown sensitivity and specificity of 95% and 100%, respectively (Gillespie *et al.*, 1994). Quantitative real-time PCR targeting the pneumolysin gene by using fluorescence resonance energy transfer (FRET) probes was shown to be highly sensitive and specific in CSF and whole-blood specimens – sensitivity was determined to be between one and ten target copies per reaction, and specificities of 97% and 100% were obtained for CSF and whole blood, respectively (van Haeften *et al.*, 2003). Another new development using molecular methodology is the evolution of molecular typing techniques such as multilocus sequence typing (MLST), field inversion gel electrophoresis, ribotyping, restriction fragment end labelling, BOX PCR and random amplification of polymorphic DNA, which surpass the quellung reaction in facilitating the analysis of the way in which this organism interacts with human populations (Enright and Spratt, 1998; Gillespie, 1999).

Susceptibility Testing

Disc Susceptibility Testing

Historically, clinical laboratories in the United Kingdom and Ireland have used a comparative method of disc testing to interpret susceptibility, rather than one based on a correlation between MIC and zone of inhibition (which is used widely in Europe). This latter methodology is now recommended by the British Society for Antimicrobial Chemotherapy (BSAC), which publish a standardised method of disc testing, together with zone limits that correlate with MIC breakpoints (Andrews, 2001). Testing is performed on Isosensitest agar supplemented with 5% defibrinated horse blood. Incubations are carried out for 18–20 h at 35–37 °C in air plus 4–6% CO_2. The inoculum used should match the density of the 0.5 McFarland standard. Susceptibility testing using penicillin discs is not a reliable indicator of resistance, and it is recommended that penicillin resistance be detected with an oxacillin 1 μg disc.

MIC Determination

The method of MIC determination generally used in North America is National Committee for Clinical Laboratory Standards (NCCLS) broth microdilution, using cation-adjusted Mueller–Hinton broth supplemented with 5% defibrinated lysed horse blood and inoculated with standardised inocula of 5×10^4 CFU/mL (National Committee for Clinical Laboratory Standards, 2000). Incubations are carried out in air at 35–37 °C for 18–20 h. The BSAC recommend either broth or agar dilution. The latter methodology, which was employed in the BSAC Surveillance Programme, is carried out using Isosensitest agar supplemented with 5% defibrinated horse blood and inoculated with standardised inocula of 10^4 CFU/mL. Incubations are carried out for 18–20 h at 35–37 °C in air plus 4–6% CO_2. Reynolds *et al.* (2003a) reported very good agreement between BSAC agar dilution and NCCLS broth microdilution methodologies.

A more rapid method of MIC determination better suited to clinical laboratories is E-testing, which involves application of a calibrated, antibiotic-impregnated plastic strip to the surface of an inoculated plate.

Nucleic Acid Amplification Techniques

Molecular techniques for susceptibility testing are being developed. The effectiveness of a PCR-restriction fragment polymorphism strategy directed against the *pbp2b* gene in identifying penicillin-nonsusceptible pneumococci has been demonstrated (O'Neill, Gillespie and Whiting, 1999), and Kearns *et al.* (2002) have shown that a real-time PCR hybridisation assay directed against the same gene can be used to rapidly (<1 h) determine penicillin susceptibility in pneumococcal meningitis, even in culture-negative cases.

Susceptibility Profiles

Data obtained by the BSAC Respiratory Resistance Surveillance Programme in Great Britain, which collected 1154 pneumococcal isolates, are summarised in Table 3.3 (Reynolds *et al.*, 2003b). Penicillin susceptibility (MIC ≤ 0.06 mg/L) overall was 89.5%, there being 92.8% and 67.8% susceptibility rates in Great Britain and Ireland, respectively (Table 3.4). Clinically relevant penicillin

Table 3.4 Per cent susceptible (intermediate) and MIC_{90} values for *S. pneumoniae* in Great Britain and Ireland

Antibiotic	%S (%I)		MIC_{90} (mg/L) (MIC_{max})	
	Great Britain	Ireland	Great Britain	Ireland
Penicillin	92.8 (7.0)	67.8 (28.7)	0.03 (2)	1 (4)
Amoxicillin	99.2	95.4	0.03 (2)	1 (8)
Cefaclor	91.9	67.8	1 (1)	128 (128)
Cefotaxime	100	98.9	0.06 (1)	1 (2)
Cefuroxime	95.9	71.8	0.12 (4)	4 (8)
Erythromycin	89.2	78.2	4 (4)	≥32 (32)
Clindamycin	95.9	87.9	0.25 (4)	≥4 (4)
Tetracycline	93.1	81.6	0.25 (32)	≥32 (32)

I, intermediate resistance to penicillin; MIC_{90}, minimum inhibitory concentration of 90% of isolates; MIC_{max}, maximum minimum inhibitory concentration; S, susceptibility to penicillin.

resistance (MIC ≥ 2 mg/L) remains very rare (0.3% overall). These rates mirror those reported from the European Antimicrobial Resistance Surveillance System (EARSS) and Nearchus programmes (http://www.earss.rivm.nl; Grüneberg *et al.*, 1998). The data show near-universal susceptibility to cefotaxime, moxifloxacin and levofloxacin and universal resistance to trimethoprim. This high level of susceptibility to fluoroquinolones contrasts with that seen in some other areas, notably in Hong Kong, where the prevalence of fluoroquinolone nonsusceptibility (MIC ≥ 4 mg/L) in 2000 was reported as 13.3% (Ho *et al.*, 2001). There was considerable regional variation in other β-lactams, erythromycin, clindamycin and tetracycline, as shown in Table 3.4.

As compared with British isolates, Irish isolates had MIC_{90}s that were five to six doubling dilutions higher for all β-lactams tested including cefotaxime, six doubling dilutions higher for tetracycline and at least three doubling dilutions higher for erythromycin and clindamycin. Logistic regression analysis of the data shows that for all agents tested except ciprofloxacin, isolates from Ireland were less likely to be susceptible than were those from England. The reasons for these differences remain unclear. Outpatient prescribing data do not suggest significant differences in prescribing practices (Cars, Molstad and Melander, 2001). For any given country and antibiotic, isolates from patients aged 20–49 years were more likely to be susceptible than those from patients aged 50–69 years.

All 17 amoxicillin-resistant isolates (1.3%) were nonsusceptible to penicillin; 123 isolates (9.3%) were resistant to two or more unrelated antimicrobials. The most common linked nonsusceptibilities were to erythromycin/tetracycline (6.1%), penicillin/tetracycline (5.3%) and penicillin/erythromycin (5.2%). Fifty-five isolates (4.1%) were resistant to three or more unrelated antimicrobials, 51 (3.8%) being nonsusceptible to penicillin, erythromycin and tetracycline. Three isolates (0.2%) were resistant to penicillin, erythromycin and ciprofloxacin, and one isolate (0.1%) was resistant to erythromycin, ciprofloxacin and tetracycline. No isolate was resistant to amoxicillin, erythromycin, tetracycline and ciprofloxacin. Resistance to amoxicillin, penicillin, erythromycin and tetracycline showed positive associations in all possible pairwise combinations. There was no evidence of association between ciprofloxacin resistance and resistance to any of these antibiotics.

Resistance to β-Lactams

Activity against pneumococci varies between β-lactams. Amoxicillin remains the most active oral agent, followed by cefditoren and cefpodoxime. Cefuroxime and cefprozil are notably less active. Amongst parenteral cephalosporins, cefotaxime, ceftriaxone, cefepime and cefpirome have most activity. Other third-generation cephalosporins (ceftazidime and ceftizoxime) are much less active.

Table 3.3 Breakpoints, per cent susceptible (intermediate) and MIC_{90} values for *S. pneumoniae* in Great Britain

Antibiotic	Low breakpoint (mg/L): (S≤)	High breakpoint (mg/L): (R≥)	%S (%I)	MIC_{90} (mg/L)
Penicillin	0.06	2	92.8 (7.0)	0.03 (MIC_{max} = 2)
Amoxicillin	1	2	99.2	0.03
Cefaclor	1	2	91.9	1
Cefuroxime	1	2	95.9	0.12
Cefotaxime	1	2	100	0.06
Erythromycin	0.5	1	89.2	4
Clindamycin	0.5	1	95.9	0.25
Ciprofloxacin	NA	4	0 (94.9)	2
Moxifloxacin	1	2	99.7	0.25
Levofloxacin	2	4	99.6	1
Tetracycline	1	2	93.1	0.25
Trimethoprim	0.5	1	0	8

I, intermediate resistance to penicillin; MIC_{90}, minimum inhibitory concentration of 90% of isolates; MIC_{max}, maximum minimum inhibitory concentration; NA, not applicable; R, resistance to penicillin; S, susceptibility to penicillin.

Table 3.5 Antimicrobial resistance rates for *S. pneumoniae* in the United States

Antibiotic	All strains	% Resistance		
		Susceptibility to penicillin (MIC ≤ 0.16 mg/L)	Intermediate resistance to penicillin (MIC = 0.12–0.1 mg/L)	Resistance to penicillin (MIC ≥ 2 mg/L)
Macrolides	25.7	5.6	42.3	77.8
Clindamycin	8.9	1.3	19.6	25.8
Tetracycline	16.3	3.1	31.4	48.0
Co-trimoxazole	30.3	7.6	39.2	94.5
Chloramphenicol	8.3	1.0	13.4	27.7

MIC, minimum inhibitory concentration.

The first clinical isolate with reduced susceptibility to penicillin was reported in 1967 (Hansman and Bullen, 1967). The prevalence of resistance increased sharply in the 1990s, and up to 34% of American isolates of the early 2000 have reduced susceptibility to penicillin (MIC ≥ 0.12 mg/L), between half and two-thirds of which are considered resistant (MIC ≥ 2 mg/L). The levels of reduced susceptibility and resistance to penicillin seen in several other countries (Hong Kong, Spain, Hungary and South Africa) are far greater, being in the range of 40–70%. As happens for several other pathogens, this situation is exacerbated by co-resistance to other antibiotics. Data from a large American survey conducted during 1999–2000 are summarised in Table 3.5 (Doern *et al.*, 2001).

Two factors have been identified as being strongly associated with the development of β-lactam resistance: β-lactam exposure and contact with carriers. Hence, treatment with β-lactams in the last 3 months, age, hospitalisation, institutionalisation, overcrowding, daycare attendance (or exposure to children in day care) and immunosuppression have all been associated with carriage of β-lactam-resistant strains (Clavo-Sanchez *et al.*, 1997).

β-Lactams kill pneumococci by binding and inhibiting the enzymes involved in peptidoglycan biosynthesis, namely the transpeptidases and carboxypeptidases, forming a serine ester-linked, enzymatically inactive penicilloyl complex. Six such enzymes [also called penicillin-binding proteins (PBPs)] have been identified: 1A, 1B, 2A, 2B, 2X and 3. Resistance is a consequence of minor alterations in the genes encoding PBPs, resulting in a decreased affinity of the PBP–β-lactam interaction. Genomic studies suggest that many of these alterations are the result of transformation and homologous recombination with *PBP* genes from related species, notably *Streptococcus oralis* and *Streptococcus mitis* in the oropharynx with which pneumococci share a niche, resulting in the formation of a mosaic gene structure (Dowson *et al.*, 1993). Point mutations have also contributed to the alterations in some strains (Coffey *et al.*, 1995). The altered PBPs seem to synthesise a chemically distinct cell wall characterised by abundant branched muropeptides in contrast to the linear muropeptide structure of penicillin-susceptible strains (Garcia-Bustos and Tomasz, 1990).

Stepwise accumulation of PBP genetic alterations leading to progressive loss of β-lactam affinity has meant that pneumococcal β-lactam susceptibility is concentration dependent and is not an all-or-none phenomenon. The efficacy of a given β-lactam in a particular infection therefore depends both on the MIC of the isolate and on the drug concentration that can be attained within the site of infection, antibacterial activity being maximal at four times MIC and being a function of the time above MIC (optimal efficacy being obtained when drug concentration exceeds MIC for 30%, 40% and 50% of the dosing interval for the carbapenems, penicillins and cephalosporins, respectively) (Craig, 1998). This has two important consequences. Firstly, the different permeabilities of different tissues mean that it is necessary to adjust breakpoints accordingly when interpreting MICs. Hence, the excellent lung penetration of intravenous benzylpenicillin (concentrations obtained are similar to those in serum) means that it remains an effective treatment for pneumococcal pneumonia caused by strains with MICs up to 4 mg/L. In contrast, its vastly inferior

CSF penetration makes it unsuitable for pneumococcal meningitis caused by strains with MICs ≥ 0.25 mg/L, and current British recommendations are that such CSF isolates be reported as resistant. In view of these considerations the NCCLS has published revised breakpoints for nonmeningeal pneumococcal infection (National Committee for Clinical Laboratory Standards, 2001a,b) (Table 3.6). Secondly, it is often possible to overcome loss of susceptibility either by increasing the dosage of the β-lactam or by choosing an agent with better penetration for the tissue concerned. This principle has led to an increased dosage of amoxicillin being recommended for empiric treatment of otitis media and third-generation cephalosporins being recommended as empiric treatment for pneumococcal meningitis.

The common target site for all β-lactams has resulted in cross-resistance to varying degrees within this group of antibiotics. MICs to second- and third-generation cephalosporins have been found to rise with that to penicillin for many strains (Doern *et al.*, 2001).

Resistance to Macrolides

Macrolide resistance has increased markedly in many areas over the 1990s and is particularly prevalent amongst penicillin-resistant isolates. Whereas in the United Kingdom resistance rates are only around 10%, Ireland and the United States have between two and three times this prevalence. In Spain, erythromycin resistance rose from 0% in 1979 to 9.4% in 1990. By 1997 one-third of isolates were macrolide resistant. Italian studies show a similar pattern (Lynch and Martinez, 2002). By the mid-1990s macrolide resistance rates in Greece, Slovakia, Bulgaria, Romania, Uruguay and some Asian countries were in excess of 20%. By the late 1990s 45% of isolates in southwest France and more than two-thirds of Chinese and Japanese isolates were macrolide resistant. Macrolide resistance has, however, remained low (0–7%) in many parts of the world, notably South Africa, Scandinavia, Germany, Austria, Portugal and Israel.

Previous macrolide exposure, in particular to the newer longer-acting agents that have protracted sub-MIC periods (notably azithromycin), is strongly associated with resistant isolates. Nasopharyngeal carriage allows dissemination of macrolide-resistant clones, children being the predominant reservoir. The pace of transmission is accelerated by selection pressure imposed by antibiotic use. An increased prevalence

Table 3.6 National Committee for Clinical Laboratory Standards (NCCLS) breakpoints for *S. pneumoniae* – nonmeningeal infection

	Old breakpoint (mg/L)			New breakpoint (mg/L)		
	S	I	R	S	I	R
Amoxicillin	<1	1	>1	<4	4	>4
Cefuroxime	<1	1	>1	<2	2	>2
Ceftriaxone or cefotaxime	<1	1	>1	<2	2	>2

I, intermediate resistance to penicillin; R, resistance to penicillin; S, susceptibility to penicillin.

of resistant isolates is also associated with day care, institutionalisation, penicillin resistance, a persistent source of infection (middle ear or sinus), age below 5 years or above 65 years and hospitalisation (Gay et al., 2000).

Macrolides, lincosamides, streptogramins and ketolides exert their antimicrobial properties by binding bacterial ribosomal target sites, causing premature dissociation of the peptidyl tRNA from the 50S ribosome and inhibition of protein synthesis.

Target-site alteration disrupts the ribosomal tertiary structure, resulting in the loss of binding affinity and drug resistance. This alteration is most commonly mediated by a methyltransferase, which methylates a highly conserved region (A2058) of the peptidyl transferase loop in domain V of 23S mRNA. This enzyme is encoded by the ermAM (erythromycin ribosome methylation) gene, which may be either chromosomal or borne on plasmids or conjugative transposons. Expression may be either constitutive or inducible. The most common variant of ermAM in pneumococci is ermB – this gene is usually transposon-borne, allowing rapid dissemination. This mechanism, which is the predominant cause of macrolide resistance in Europe and South Africa, confers high-level resistance (MIC to erythromycin >64 mg/L) to macrolides, lincosamides and streptogramins – the MLS$_B$ phenotype. Telithromycin, a new ketolide, retains activity against strains of the MLS$_B$ phenotype where ermB expression is inducible because of higher affinity and weaker induction. Activity is reduced, however, when expression is constitutive. Less commonly, target-site alteration may occur by point mutation, either in ribosomal protein or in the 23S rRNA – this is responsible for only 1–3% of macrolide resistance but has led to the emergence of novel phenotypes including high-grade telithromycin resistance (Tait-Kamradt et al., 2000). Fortunately, ketolides and oxazolidinones remain active against >99% of macrolide-resistant strains, and the glycopeptides remain universally active (Lynch and Martinez, 2002).

A very different mechanism by which resistance may occur is by active efflux of the antibiotic. Macrolide efflux is mediated by a 405-amino acid hydrophobic protein encoded by the chromosomal mefE (macrolide efflux) gene, which confers resistance to 14- and 15-membered macrolides (erythromycin, clarithromycin and azithromycin) but not to 16-membered macrolides (josamycin), lincosamides, streptogramins or ketolides – the so-called M phenotype. This is the predominant mechanism of resistance in North America and Japan, being detected in more than two-thirds of resistant strains. Resistance is relatively low level (MIC to erythromycin 1–32 mg/L) and transferable by conjugation (Bellido et al., 2002). The M phenotype remains rare in Europe and South Africa, being found in <20% of macrolide-resistant isolates (Tait-Kamradt et al., 2000).

Any combination of these various mechanisms of macrolide resistance may coexist, resulting in a wide range of MICs. The extent to which in vitro macrolide resistance translates into therapeutic failure may be significantly affected by the mechanism(s) of resistance involved.

Current NCCLS guidelines state that pneumococci with MICs to erythromycin and clarithromycin of 0.5 mg/L (1 mg/L for azithromycin) should be considered as being of intermediate sensitivity and be regarded as resistant above this MIC (National Committee for Clinical Laboratory Standards, 2001b). The most appropriate pharmacodynamic parameters that best correlate with efficacy remain controversial – it appears that, as for β-lactams, duration above MIC is the best correlate for erythromycin and clarithromycin (optimal when >40% of the dosing interval) and area under the curve (AUC)/MIC is the best correlate for azithromycin (optimal when >25%) (Craig, 2001).

The extent to which in vitro macrolide resistance translates into therapeutic failure is influenced by several pharmacokinetic and pharmacodynamic variables and remains poorly understood. Studies indicate that, as for β-lactams, an in vivo–in vitro paradox exists, owing largely to the excellent tissue-penetration properties and long tissue-elimination half-lives of these agents which allow target attainment despite relatively high MICs. As a result, good clinical outcomes are usually achieved in pneumococcal pneumonia and otitis

media caused by isolates with MIC values as high as 4 μg/mL or even 8–16 μg/mL. There are, however, increasing numbers of reports of treatment failure attributed to bacteraemic infection with macrolide-resistant pneumococci (where outcome is related to serum macrolide levels rather than pharmacodynamics) (Garau, 2002).

Resistance to Fluoroquinolones

The tide of fluoroquinolone resistance (MIC ≥ 4 mg/L) in pneumococci began in Canada, with an increase in ciprofloxacin resistance from 0% in 1993 to 1.7% in 1997–1998. This sharp increase was preceded by an increase in the use of fluoroquinolones, principally ciprofloxacin, and was associated with older age (>65) and residence in Ontario, both of which correlated with increased fluoroquinolone use (Chen et al., 1999). Resistance was also more common in penicillin-resistant isolates. A similar rate of resistance (1.4%) was reported in 2000, but there was now evidence of cross-resistance with the latest, most potent fluoroquinolones: 62% of ciprofloxacin-resistant isolates were cross-resistant to levofloxacin, 7% to gatifloxacin and 3% to moxifloxacin (Low et al., 2002).

A survey of the UK isolates from pneumococcal invasive infections showed resistance rates of 1% to moxifloxacin (Johnson et al., 2001). Similarly low rates of fluoroquinolone resistance have been found in the United States. However, increasing resistance to the newer drugs has been noted – levofloxacin resistance rates increased from 0.2% in 1999–2000 to 0.7% in 2000–2001 and gatifloxacin from 0.3% to 0.4% during the same period (Doern et al., 2001). Studies also suggest an increase in the number of strains containing single parC mutations [from 0.4% in 1992–1996 to 4.5% in 1999–2000: these strains would be classified as susceptible by routine susceptibility testing but are the progenitors of fully resistant dual mutants (parC and gyrA) (Davies et al., 2002)].

Higher rates of fluoroquinolone resistance have been reported in other areas. Hong Kong has witnessed a dramatic rise in levofloxacin resistance (<0.5% in 1995, 5.5% in 1998 and 13.3% in 2000) (Ho et al., 2001). Typing studies (pulse-field gel electrophoresis and MLST) indicate that this increase is largely due to clonal dissemination of a newly emerged fluoroquinolone-resistant variant of the widely distributed multidrug-resistant Spanish 23F clone. Fluoroquinolone resistance rates of 5.3% have been reported from Spain. Most resistant isolates were polyclonal in origin, but 30% of strains were fluoroquinolone-resistant variants of the multiresistant internationally epidemic Spanish 23F and French 9V clones (Alou et al., 2001).

Since paediatric fluoroquinolone use is exceptional, fluoroquinolone-resistant pneumococci are largely limited to adult populations. Besides older age, other risk factors for acquisition include chronic obstructive pulmonary disease, nosocomial acquisition, institutionalisation and prior fluoroquinolone exposure.

Fluoroquinolones inhibit pneumococcal DNA replication by binding vital enzymes, namely topoisomerase IV (which acts to separate interlocked DNA strands to allow segregation of daughter chromosomes) and DNA gyrase (which alters DNA supercoiling). Different fluoroquinolones have different affinities for the two enzymes, each of which is composed of two subunits: ParC and ParE in topoisomerase IV and Gyr A and GyrB in DNA gyrase. Ciprofloxacin, levofloxacin, norfloxacin and trovafloxacin preferentially target ParC, whereas GyrA seems to be the preferred target for moxifloxacin, gatifloxacin and gemifloxacin (Fukuda and Hiramatsu, 1999). The rank order of activity against pneumococci, from most to least active, is gemifloxacin, trovafloxacin, sparfloxacin, moxifloxacin, grepafloxacin, gatifloxacin, levofloxacin and ciprofloxacin (Jones et al., 2000).

Resistance to fluoroquinolones may occur by two mechanisms. Target-site alteration most commonly occurs because of stepwise accumulation of chromosomal mutations, particularly in the quinolone-resistance–determining region (QRDRs) of gyrA or parC. Reports of mosaic genes in fluoroquinolone-resistant strains suggest that horizontal transfer of altered type II topoisomerase genes from viridans-type streptococci by recombination and transformation also contributes significantly to resistance by target-site alteration (Janoir *et al.*, 1999). The second mechanism by which fluoroquinolone resistance occurs is active drug efflux from the bacterial cell by overexpression of the multidrug resistance efflux pump PmrA.

QRDR mutations occur at a rate of about 1 in 10^{-6} to 1 in 10^{-9} (Fukuda and Hiramatsu, 1999). The first QRDR mutation to confer a selective advantage has to occur in the preferred subunit of the quinolone, imposing a selection pressure. This has the effect of raising the MIC of the organism to all quinolones with the same preference by four- to eightfold. Given time, this population acquires a mutation in the second subunit, which further increases its MIC and also renders it resistant to quinolones that target either subunit (or both). Continued or repeated fluoroquinolone exposure augments the selective advantage, allowing further mutations and higher levels of resistance to build up. As the pulmonary bacterial load in a patient with pneumonia often exceeds 10^9, the risks of QRDR mutants arising are great – fluoroquinolones further increase this risk because of their ability to damage bacterial DNA and trigger the error-prone SOS DNA repair system (Hooper, 2002).

MDR efflux pumps are normally present even in susceptible wildtype bacteria. Low-level resistance to fluoroquinolones (MIC 4–8 mg/L) can occur by increased expression of these pumps, usually because of mutations of the structural gene *pmrA*. Fluoroquinolones differ in the extents to which they are affected by efflux mechanisms. Studies suggest that sparfloxacin, levofloxacin, grepafloxacin, gatifloxacin and moxifloxacin are poorer substrates for PmrA than hydrophilic fluoroquinolones such as ciprofloxacin and norfloxacin and are hence less susceptible to this mechanism of resistance. There is also preliminary evidence of other as-yet-unidentified fluoroquinolone efflux mechanisms (Piddock *et al.*, 2002). Efflux-mediated resistance may be additive to that produced by target-site alteration.

Our understanding of the mechanism of action of fluoroquinolones and the mechanisms by which resistance arises can be extrapolated to devise strategies that minimise the risks of resistance. A fluoroquinolone with equal affinities for both topoisomerase II enzymes should confer no selective advantage on single mutants – double mutants, which are much less frequent (1 in 10^{12}), would be needed before resistance could emerge. Gemifloxacin, a new fluoroquinolone, is active against both single mutants (either *gyrA* or *parC*) and efflux mutants. There is some evidence that it remains effective even against double mutants (Nagai *et al.*, 2000). An alternative strategy is to design drugs and dosing regimens to achieve a tissue antibiotic concentration that is inhibitory to single mutants [so that mutant prevention concentration (MPC) about 8× MIC is achieved]. MPC can be safely achieved with moxifloxacin and gatifloxacin and high-dose levofloxacin regimens. Conversely, the use of weaker fluoroquinolones may facilitate the stepwise selection of mutants and, therefore, should be avoided.

The susceptibility breakpoints for pneumococci in accordance with current NCCLS guidelines are summarised in Table 3.7 (National Committee for Clinical Laboratory Standards, 2001b).

Table 3.7 National Committee for Clinical Laboratory Standards (NCCLS) breakpoints for *S. pneumoniae*

Antibiotic	Breakpoint (mg/L)		
	S	I	R
Levofloxacin	≤2	4	≥8
Moxifloxacin or gatifloxacin	≤1	2	≥4

I, intermediate resistance to penicillin; R, resistance to penicillin; S, susceptibility to penicillin.

The bactericidal activity of fluoroquinolones is related to AUC: MIC, being maximal when the ratio exceeds 25–40 (Wright *et al.*, 2000). Target attainment is more easily achieved with moxifloxacin and gatifloxacin than with levofloxacin (Nuermberger and Bishai, 2004). Moxifloxacin and gatifloxacin have the added advantage of requiring mutations in both *parC* and *gyrA* before resistance becomes clinically significant, because their peak serum levels exceed the MICs produced by single-step mutants (Blondeau *et al.*, 2001). Hence, these newer agents have lower MPCs (1 mg/L) as compared with levofloxacin (4 mg/L) and ciprofloxacin (8 mg/L). One should, however, be wary of using these agents to treat a population of single-step mutants (such as might be present in the alveoli of a patient with pneumonia who has been recently exposed to ciprofloxacin), as dual mutants will emerge at a much higher frequency (1 in 10^6) than in a wildtype population and could lead to clinical failure (Gillespie *et al.*, 2002, 2003).

Resistance to Rifampicin

The incidence of rifampicin nonsusceptibility remains low, at <3%. Rifampicin exerts its bactericidal action by binding the β-subunit of DNA-dependent RNA polymerase, hence preventing chain initiation. Highly resistant isolates result from mutations within the *rpo*B gene.

Resistance to Vancomycin

Vancomycin resistance remains almost entirely a hypothetical issue; 99.7–100% of isolates remain susceptible. However, vancomycin tolerance (MBC: MIC > 32) was described in a survey of clinical isolates – all these strains were serotype 9V, penicillin resistant and had a unique valine-for-alanine substitution at position 440 of the *vnc*S gene (Novak *et al.*, 1999). This mutation results in the loss of function of the *vnc*S histidine kinase, producing a reduction in autolytic activity. Tolerance could lead to clinical failure; however, there are no human data to support this. By allowing survival in the presence of the drug, tolerance could also facilitate the acquisition of resistance-mutations and hence accelerate the development of resistance.

Resistance to Other Drugs

Tetracycline resistance in *S. pneumoniae* is recognised in about 7% of isolates in the United Kingdom. It is mediated by a soluble protein encoded by the *tet*M gene, which shares N-terminal homology with elongation factors involved in translation. The tetM protein is hence able to bind to the ribosome, thereby protecting it from the inhibitory activity of tetracycline. Chloramphenicol resistance is found in about 8% of UK isolates. It is mediated by an inducible chloramphenicol acetyltransferase (CAT). Both tetM and CAT are most commonly encoded by genes on conjugative transposon Tn*1545*. Other conjugative elements bearing these resistance genes have also been described.

Multidrug-Resistant S. pneumoniae

Multidrug-resistant *S. pneumoniae* (MDRSP) (defined as resistance to penicillin and at least two other antibiotic classes) was first reported from South Africa in 1978. The isolate was reported to be resistant to penicillin, erythromycin, tetracycline, clindamycin, co-trimoxazole and chloramphenicol (Jacobs *et al.*, 1978). Global dissemination has since occurred, and multidrug-resistant isolates are now recognised in every continent. Their prevalence in the United States has been estimated at 14–16% (Whitney *et al.*, 2000). Most of these strains are derived from a few clones, notably the Spanish clone (serotype 23F) and its variants (serotypes 19A, 19B and 19F) that are associated with resistance to penicillins, chloramphenicol (cat+), tetracycline (tetM+), co-trimoxazole and macrolides (75% are *mef*E+ or *erm*B+) and the more sensitive Spanish–French clones (serotypes 6B, 9 and 14) that are resistant to penicillin and co-trimoxazole but only occasionally show resistance to macrolides, tetracycline and chloramphenicol (Corso *et al.*, 1998). The French clone is prevalent in Europe and South America. The Spanish clone has a broader distribution, being prevalent in North America, Europe, Asia, South America and South Africa. Patterns of drug resistance and the relative prevalences of different clones show considerable regional and international variation.

Fluoroquinolone-resistant variants of both the Spanish 23F and French 9V clones have emerged. Selection pressures imposed by a variety of unrelated antibiotics could fuel a rapid increase in fluoroquinolone resistance, rendering useless the agents that are currently most active against MDRSP (Doern, 2001).

MANAGEMENT

The treatment of choice for severe penicillin-susceptible pneumococcal infection is benzylpenicillin. The major reasons for departure from this therapy are allergy and possible or confirmed resistance. The treatment approaches in the face of these issues and the nuances of different infections are addressed below.

Pneumonia

The incidence of pneumococci that are highly resistant to penicillin (MIC≥4μg/mL) in the United Kingdom remains low, and documented clinical failures among penicillin-resistant pneumococcal pneumonia treated with adequate doses of penicillin are rare.

The optimal duration of therapy for pneumococcal pneumonia is uncertain. Most physicians now treat for 10–14 days.

Oral Therapy

Amoxicillin is the preferred agent, at a dose of 0.5–1 g three times daily. Clarithromycin 500 mg twice daily or the newer fluoroquinolones (e.g. levofloxacin 500 mg once daily) are alternatives for β-lactam-intolerant patients or, much more rarely, those infected with strains that are highly resistant to penicillin.

Intravenous Therapy

Amoxicillin 0.5–1 g three times daily or benzylpenicillin 1.2 g 4–6 hourly is the treatment of choice for pneumococcal pneumonia caused by strains that are either sensitive or intermediately resistant to penicillin. Third-generation cephalosporins (ceftriaxone 1 g once daily or cefotaxime 1 g three times daily) are suitable alternatives in 90% of penicillin-intolerant patients. There are few data on whether treatment should be altered in infection with penicillin-resistant strains. Benzylpenicillin at a dose of 18–24 MU per day in four to six divided doses remains a viable option for isolates with MICs up to 4 mg/L. At the time of writing, levels of cefotaxime or ceftriaxone achievable in the pulmonary tissues are adequate in most cases, and even critically ill patients are treated with ceftriaxone 1–2 g 12–24 hourly. However, the macrolides, newer fluoroquinolones, glycopeptides, linezolid and carbapenems present alternatives in the rare cases of exceptional resistance. All these agents except the carbapenems could also be used when faced with β-lactam intolerance.

Meningitis

Pneumococcal meningitis caused by penicillin-susceptible strains should ideally be treated with benzylpenicillin 1.2 g 4 hourly for 10–14 days. The preferred treatment for infection caused by strains in which penicillin MIC levels are ≥0.25 mg/L is ceftriaxone 2–4 g once daily. Penicillin is ineffective against these resistant strains because CSF concentrations are unlikely to exceed 2–10% of serum levels.

The efficacy of third-generation cephalosporins in pneumococcal meningitis diminishes once their MIC ≥ 0.5 mg/L, and isolates with MIC 1 mg/L are defined as partially resistant (highly resistant if MIC > 1 mg/L). However, there have been several reports of success in using cefotaxime therapy in partially cefotaxime-resistant pneumococcal meningitis (Viladrich *et al.*, 1988; Tan *et al.*, 1994). High-dose regimens (300–350 mg/kg/day to a maximum of 24 g) have been found to be safe and effective for meningitis caused by cefotaxime-resistant strains (MICs of cefotaxime up to 2 mg/L) (Viladrich, 2004).

However, high-dose cefotaxime therapy might sometimes fail, either because the isolate is particularly resistant (MIC ≥ 4 mg/L) or because CSF penetration is particularly low. In these cases, several possible antibiotic regimens may be employed:

1. Third-generation cephalosporin plus vancomycin: The main disadvantage of this regimen is that vancomycin penetration into the CSF is unreliable, particularly if dexamethasone is co-administered. Clinical experience is lacking.
2. Third-generation cephalosporin plus rifampicin: CSF penetration of rifampicin is unaffected by dexamethasone administration, and CSF levels of 1 mg/L can be attained. Nearly all pneumococcal isolates remain susceptible, with MICs of about 0.12 mg/L. However, clinical experience with this regimen is lacking, and concerns have been raised that rifampicin might antagonise the β-lactam.
3. Vancomycin plus rifampicin.
4. Addition of intrathecal vancomycin to the intravenous regimen (10–20 mg every 24–48 h).
5. Imipenem ± rifampicin: Imipenem MICs of penicillin-resistant pneumococci tend to be lower than those of third-generation cephalosporins, and there are documented cases of cure (Asensi *et al.*, 1989; Guibert *et al.*, 1995). Imipenem-induced neurotoxicity is an obvious hazard in this setting. MICs to meropenem tend to be two- to fourfold higher, and this agent is therefore unlikely to confer any significant advantage.
6. New antibiotics, such as levofloxacin and linezolid, present hope for the future.

It has been shown that early treatment with dexamethasone (i.e. before or with the first dose of antibiotic, and then continued for 2–4 days) improves outcome in pneumococcal meningitis (de Gans and van de Beek, 2002). This adjuvant therapy is now unequivocally recommended in both adults and children (Chaudhuri, 2004).

Otitis Media and Sinusitis

In a report from the Drug-resistant *S. pneumoniae* Therapeutic Working Group it was recommended that oral amoxicillin should remain the first-line antimicrobial agent for treating pneumococcal otitis media (Dowell *et al.*, 1999). However, in view of the increasing prevalence of penicillin resistance, a higher dose (an increase from 40–45 mg/kg/day to 80–90 mg/kg/day) was recommended for empiric treatment. For patients with defined treatment failure after 3 days, alternative regimens such as oral cefpodoxime, a single parenteral dose of ceftriaxone followed by oral cefpodoxime for 5 days or co-amoxiclav are recommended to cover infection caused either by a penicillin-resistant pneumococcus or by a β-lactamase-producing organism such as *Moraxella catarrhalis* or *Haemophilus influenzae*. The macrolides provide a viable alternative for the β-lactam-allergic patient.

The same therapeutic considerations apply for sinusitis, which has a similar pathogenesis to otitis media.

PREVENTION AND CONTROL

Pneumococcal Vaccines

There are two types of pneumococcal vaccines: a 23-valent pneumococcal polysaccharide vaccine (PPV) and a 7-valent pneumococcal conjugate vaccine (PCV).

Pneumococcal Polysaccharide Vaccine

This is a polyvalent vaccine containing 25 µg of purified capsular polysaccharide from each of 23 capsular types of pneumococci that together account for about 96% of the pneumococcal isolates causing serious infection in Great Britain (World Health Organization, 1999).

Most healthy adults develop a good antibody response to a single dose of the vaccine by the third week following immunisation. Protection is much less reliable in the immunocompromised and in children below the age of 2 years.

Pneumococcal Conjugate Vaccine

This vaccine contains polysaccharide antigens from seven common serotypes (4, 6B, 9V, 14, 18C, 19F and 23F) conjugated to a protein carrier (currently CRM197 protein). The selected serotypes accounted for 65.6% of pneumococcal invasive infections in England and Wales in 2000 (82.3% in children <5 years) (World Health Organization, 1999). The main advantage of this vaccine is that it is protective in children over 2 months of age. Conjugation of polysaccharide antigen to a protein carrier has the effect of converting a T-independent immune response to a T-dependent one, hence achieving better immunogenicity.

Vaccine Effectiveness

Studies on the polysaccharide vaccine have produced variable results, but overall protective efficacy in preventing bacteraemic infection ranges from 50% to 70% (Mangtani, Cutts and Hall, 2003). The main drawbacks of this vaccine are its ineffectiveness at preventing nonbacteraemic pneumococcal pneumonia, otitis media and exacerbations of chronic bronchitis and inability to protect children below the age of 2 years and the immunocompromised. In addition, data from several Asian countries suggest that the 23 vaccine serotypes account for only 63% of infections, and serotype data from many parts of the world are scarce (Lee, Banks and Li, 1991).

The PCV has been shown to protect against meningitis, pneumonia, bacteraemia and otitis media in children vaccinated at 2, 4, 6 and 15 months of age. Serotype-specific efficacy was 94% after the first dose of vaccine and 97% after the fourth (Black et al., 2000).

The potential of pneumococcal proteins, such as pneumolysin, PspA and CbpA, as vaccine candidates is being examined. These antigens should circumvent the issues of serotype-specific immunity and low immunogenicity that hamper polysaccharide-based vaccines, but many factors require further study, including allelic variation and its effects on cross-reactivity.

Recommendations

Vaccination is recommended for all those in whom pneumococcal infection is likely to be common and/or serious, i.e.

1. all adults aged 65 and over;
2. asplenia and severe splenic dysfunction [ideally 4–6 weeks (no less than 2 weeks) before splenectomy. If not possible, vaccination should be delayed until after recovery. Vaccination against H. influenzae type b, influenza and Neisseria meningitidis group C is additionally recommended, together with antibiotic prophylaxis (usually with phenoxymethyl penicillin) until at least age 16 years];
3. chronic renal disease;
4. chronic heart disease;
5. chronic liver disease including cirrhosis;
6. diabetes mellitus;
7. immunodeficiency or immunosuppression [ideally 4–6 weeks (no less than 2 weeks) before chemotherapy or radiotherapy. If not given, vaccination should be delayed until at least 3 months after treatment];
8. HIV infection at all stages. A report of increased risk of pneumococcal disease in severely immunocompromised HIV-infected Ugandan adult vaccinees requires further confirmatory studies and additional studies to assess its relevance to the United Kingdom (French et al., 2000). However, current UK recommendations are to vaccinate all HIV-infected individuals;
9. patients with cochlear implants.

The vaccine routinely used is the 23-valent polysaccharide vaccine, the conjugate vaccine only being recommended in children belonging to the groups above, who are below the age of 2 years. These children should be subsequently vaccinated with the polysaccharide vaccine after they reach the age of 2 years, to broaden the spectrum of serotype coverage.

Dosage and Administration

The polysaccharide vaccine may be administered either subcutaneously or intramuscularly as a single dose. It may be given simultaneously with the influenza vaccine, at a different site. Revaccination is not routinely recommended, except at 5-yearly intervals in patients in whom antibody levels are more likely to have declined rapidly, such as those with nephrotic syndrome, asplenia or splenic dysfunction.

The conjugate vaccine is given intramuscularly. It may be administered at the same time as other childhood immunisations but must be given at a separate site in a separate limb. The dosage regimen varies depending on the age of the child (Table 3.8).

The polysaccharide vaccine should be administered at least 2 months following the last dose of conjugate vaccine.

Adverse Reactions

Both vaccines are generally well tolerated but have been associated with mild erythema and induration that last 1–3 days at the site of injection. Systemic reactions are rare.

A transient increase in viral load usually follows vaccination in HIV-infected patients, the clinical significance of which remains to be ascertained.

Table 3.8 Dosage regimen for pneumococcal conjugate vaccine

Age	Regimen
<6 months	Three doses separated by at least 1 month Fourth dose after first birthday
7–11 months	Two doses separated by at least 1 month Third dose after first birthday
12–23 months	Two doses separated by at least 2 months

Contraindications

1. hypersensitivity to any component of vaccine – an allergic reaction following administration of diphtheria toxoid or CRM197-containing vaccines is a contraindication to vaccination with PCV;
2. acute infection;
3. pregnancy/breastfeeding – safety not assessed;
4. revaccination with PPV within 3 years of a previous dose of polysaccharide vaccine.

REFERENCES

Alou L, Ramirez M, García-Rey C *et al.* (2001) *Streptococcus pneumoniae* isolates with reduced susceptibility to ciprofloxacin in Spain: Clonal diversity and appearance of ciprofloxacin-resistant epidemic clones. *Antimicrob Agents Chemother*, **45**, 2955–2957.

Alouf JE, Geoffroy C (1991) The family of the antigenically related cholesterol binding (sulphydryl activated) cytolytic toxins. In *Source of Bacterial Protein Toxins* (eds JE Alouf and JH Freer), pp. 147–186. Academic Press, London.

Andersson B, Dahmen J, Frejd T *et al.* (1983) Identification of an active disaccharide unit of a glycoconjugate receptor for pneumococci attaching to human pharyngeal epithelial cells. *J Exp Med*, **158**, 559–570.

Andersson B, Beachey EH, Tomasz A *et al.* (1988) A sandwich adhesion on *Streptococcus pneumoniae* attaching to human oropharyngeal epithelial cells *in vitro*. *Microb Pathog*, **4**, 267–278.

Andrews JM; BSAC Working Party On Susceptibility Testing ft (2001) BSAC standardized disc susceptibility testing method. *J Antimicrob Chemother*, **48(Suppl. 1)**, 43–57.

Asensi F, Pérez-Tamarit D, Otero MC *et al.* (1989) Imipenem-cilastatin therapy in a child with meningitis caused by a multiply resistant pneumococcus. *Pediatr Infect Dis J*, **8**, 895.

Avery OT, MacLeod CM, McCarty M (1944) Studies on the chemical nature of the substance inducing transformation of pneumococcal types. *J Exp Med*, **79**, 137–157.

Balakrishnan I, Crook P, Morris R, Gillespie SH (2000) Early predictors of mortality in pneumococcal bacteraemia. *J Infect*, **40**, 256–261.

Bellido JLM, Guirao GY, Zufiaurre NG, Manzanares AA (2002) Efflux-mediated antibiotic resistance in Gram positive bacteria. *Rev Med Microbiol*, **13**, 1–13.

Bergeron Y, Ouellet N, Deslauriers AM *et al.* (1998) Cytokine kinetics and other host factors in response to pneumococcal pulmonary infection in mice. *Infect Immun*, **66**, 912–922.

Berry AM, Paton JC (1996) Sequence heterogeneity of PsaA, a 37-kilodalton putative adhesin essential for virulence of *Streptococcus pneumoniae*. *Infect Immun*, **64**, 5255–5262.

Berry AM, Lock RA, Hansman D, Paton JC (1989a) Contribution of autolysin to virulence of *Streptococcus pneumoniae*. *Infect Immun*, **57**, 2324–2330.

Berry AM, Yother J, Briles DE *et al.* (1989b) Reduced virulence of a defined pneumolysin-negative mutant of *Streptococcus pneumoniae*. *Infect Immun*, **57**, 2037–2042.

Bhakdi S, Tranum-Jensen J (1986) Membrane damage by pore-forming bacterial cytolysins. *Microb Pathog*, **1**, 5–14.

Black S, Shinefield H, Fireman B *et al.* (2000) Efficacy, safety and immunogenicity of heptavalent pneumococcal conjugate vaccine in children. Northern California Kaiser Permanente Vaccine Study Center Group. *Pediatr Infect Dis J*, **19**, 187–195.

Blondeau JM, Zhao X, Hansen G, Drlica K (2001) Mutant prevention concentrations of fluoroquinolones for clinical isolates of *Streptococcus pneumoniae*. *Antimicrob Agents Chemother*, **45**, 433–438.

Breiman RF, Spika JS, Navarro VJ *et al.* (1990) Pneumococcal bacteremia in Charleston County, South Carolina. A decade later. *Arch Intern Med*, **150**, 1401–1405.

Cars O, Molstad S, Melander A (2001) Variation in antibiotic use in the European Union. *Lancet*, **357**, 1851–1853.

Cauwels A, Wan E, Leismann M, Tuomanen E (1997) Coexistence of CD14-dependent and independent pathways for stimulation of human monocytes by Gram-positive bacteria. *Infect Immun*, **65**, 3255–3260.

Chaudhuri A (2004) Adjunctive dexamethasone treatment in acute bacterial meningitis. *Lancet Neurol*, **3**, 54–62.

Chen DK, McGeer A, de Azavedo JC, Low DE (1999) Decreased susceptibility of *Streptococcus pneumoniae* to fluoroquinolones in Canada. Canadian Bacterial Surveillance Network. *N Engl J Med*, **341**, 233–239.

Cheng Q, Finkel D, Hostetter MK (2000) Novel purification scheme and functions for a C3-binding protein from *Streptococcus pneumoniae*. *Biochemistry*, **39**, 5450–5457.

Clavo-Sanchez AJ, Giron-Gonzalez JA, Lopez-Prieto D *et al.* (1997) Multivariate analysis of risk factors for infection due to penicillin-resistant and multidrug-resistant *Streptococcus pneumoniae*: A multicenter study. *Clin Infect Dis*, **24**, 1052–1059.

Coffey TJ, Daniels M, McDougal LK *et al.* (1995) Genetic analysis of clinical isolates of Streptococcus pneumoniae with high-level resistance to expanded-spectrum cephalosporins. *Antimicrob Agents Chemother*, **39**, 1306–1313.

Corso A, Severina EP, Petruk VF *et al.* (1998) Molecular characterization of penicillin-resistant *Streptococcus pneumoniae* isolates causing respiratory disease in the United States. *Microb Drug Resist*, **4**, 325–327.

Craig WA (1998) Pharmacokinetic/pharmacodynamic parameters: Rationale for antibacterial dosing of mice and men. *Clin Infect Dis*, **26**, 1–10.

Craig WA (2001) The hidden impact of antibacterial resistance in respiratory tract infection. Re-evaluating current antibiotic therapy. *Respir Med*, **95(Suppl. A)**, S12–S19.

Cundell D, Masure HR, Tuomanen EI (1995) The molecular basis of pneumococcal infection: A hypothesis. *Clin Infect Dis*, **21**, S204–S211.

Cundell DR, Weiser JN, Shen J *et al.* (1995a) Relationship between colonial morphology and adherence of *Streptococcus pneumoniae*. *Infect Immun*, **63**, 757–761.

Cundell DR, Gerard NP, Gerard C *et al.* (1995b) *Streptococcus pneumoniae* anchor to activated human cells by the receptor for platelet-activating factor. *Nature*, **377**, 435–438.

Davies TA, Evangelista A, Pfleger S *et al.* (2002) Prevalence of single mutations in topoisomerase type II genes among levofloxacin-susceptible clinical strains of *Streptococcus pneumoniae* isolated in the United States in 1992 to 1996 and 1999 to 2000. *Antimicrob Agents Chemother*, **46**, 119–124.

de Gans J, van de Beek D. European Dexamethasone in Adulthood Bacterial Meningitis Study Investigators (2002) Dexamethasone in adults with bacterial meningitis. *N Engl J Med*, **347**, 1549–1556.

Doern GV (2001) Antimicrobial use and the emergence of antimicrobial resistance with *Streptococcus pneumoniae* in the United States. *Clin Infect Dis*, **33(Suppl. 3)**, S187–S192.

Doern GV, Heilmann KP, Huynh HK *et al.* (2001) Antimicrobial resistance among clinical isolates of *Streptococcus pneumoniae* in the United States during 1999–2000, including a comparison of resistance rates since 1994–1995. *Antimicrob Agents Chemother*, **45**, 1721–1729.

Dowell SF, Butler JC, Giebink GS *et al.* (1999) Acute otitis media: Management and surveillance in an era of pneumococcal resistance – a report from the Drug-resistant *Streptococcus pneumoniae* Therapeutic Working Group. *Pediatr Infect Dis J*, **18**, 1–9.

Dowson CG, Coffey TJ, Kell C, Whiley RA (1993) Evolution of penicillin resistance in *Streptococcus pneumoniae*; the role of *Streptococcus mitis* in the formation of a low affinity PBP2B in *S. pneumoniae*. *Mol Microbiol*, **9**, 635–643.

Enright MC, Spratt BG (1998) A multilocus sequence typing scheme for *Streptococcus pneumoniae*: Identification of clones associated with serious invasive disease. *Microbiology*, **144**, 3049–3060.

Ernst JD, Decazes JM, Sande MA (1983) Experimental pneumococcal meningitis: Role of leukocytes in pathogenesis. *Infect Immun*, **41**, 275–279.

Ernst JD, Hartiala KT, Goldstein IM, Sande MA (1984) Complement (C5)-derived chemotactic activity accounts for accumulation of polymorphonuclear leukocytes in cerebrospinal fluid of rabbits with pneumococcal meningitis. *Infect Immun*, **46**, 81–86.

Fenno JC, LeBlanc DJ, Fives-Taylor P (1989) Nucleotide sequence analysis of a type 1 fimbrial gene of *Streptococcus sanguis* FW213. *Infect Immun*, **57**, 3527–3533.

French N, Nakiyingi J, Carpenter LM *et al.* (2000) 23-valent pneumococcal polysaccharide vaccine in HIV-1-infected Ugandan adults: Double-blind, randomised and placebo controlled trial. *Lancet*, **355**, 2106–2111.

Fukuda H, Hiramatsu K (1999) Primary targets of fluoroquinolones in *Streptococcus pneumoniae*. *Antimicrob Agents Chemother*, **43**, 410–412.

Ganeshkumar N, Hannam PM, Kolenbrander PE, McBride BC (1991) Nucleotide sequence of a gene coding for a saliva-binding protein (SsaB) from *Streptococcus sanguis* 12 and possible role of the protein in coaggregation with actinomyces. *Infect Immun*, **59**, 1093–1099.

Garau J (2002) Treatment of drug-resistant pneumococcal pneumonia. *Lancet Infect Dis*, **2**, 404–415.

Garcia-Bustos J, Tomasz A (1990) A biological price of antibiotic resistance: Major changes in the peptidoglycan structure of penicillin-resistant pneumococci. *Proc Natl Acad Sci U S A*, **87**, 5415–5419.

Gay K, Baughman W, Miller Y *et al.* (2000) The emergence of *Streptococcus pneumoniae* resistant to macrolide antimicrobial agents: A 6-year population-based assessment. *J Infect Dis*, **182**, 1417–1424.

Geelen S, Bhattacharyya C, Tuomanen E (1993) The cell wall mediates pneumococcal attachment to and cytopathology in human endothelial cells. *Infect Immun*, **61**, 1538–1543.

Gillespie SH (1999) The role of the molecular laboratory in the investigation of *Streptococcus pneumoniae*. *Semin Respir Infect*, **14**, 269–275.

Gillespie SH, Ullman C, Smith MD, Emery V (1994) Detection of *Streptococcus pneumoniae* in sputum samples by PCR. *J Clin Microbiol*, **32**, 1308–1311.

Gillespie SH, Voelker LL, Dickens A (2002) Evolutionary barriers to quinolone resistance in *Streptococcus pneumoniae*. *Microb Drug Resist*, **8**, 79–84.

Gillespie SH, Voelker LL, Ambler JE *et al.* (2003) Fluoroquinolone resistance in *Streptococcus pneumoniae*: Evidence that gyrA mutations arise at a lower rate and that mutation in gyrA or parC predisposes to further mutation. *Microb Drug Resist*, **9**, 17–24.

Griffith F (1928) The significance of pneumococcal types. *J Hyg*, **27**, 113–159.

Grüneberg RN, Felmingham D, Harding I *et al.* (1998) The Nearchus project: Antibiotic susceptibility of respiratory pathogens and clinical outcome in lower respiratory tract infections at 27 centres in the UK. *Int J Antimicrob Agents*, **10**, 127–133.

Guibert M, Chahime H, Petit J *et al.* (1995) Failure of cefotaxime treatment in two children with meningitis caused by highly penicillin-resistant *Streptococcus pneumoniae*. *Acta Paediatr*, **84**, 831–833.

Hammerschmidt S, Talay SR, Brandtzaeg P, Chhatwal GS (1997) SpsA, a novel pneumococcal surface protein with specific binding to secretory Immunoglobulin A and secretory component. *Mol Microbiol*, **25**, 1113–1124.

Hammerschmidt S, Bethe G, Remane PH, Chhatwal GS (1999) Identification of pneumococcal surface protein A as a lactoferrin-binding protein of *Streptococcus pneumoniae*. *Infect Immun*, **67**, 1683–1687.

Hammerschmidt S, Tillig MP, Wolff S *et al.* (2000) Species-specific binding of human secretory component to SpsA protein of *Streptococcus pneumoniae* via a hexapeptide motif. *Mol Microbiol*, **36**, 726–736.

Hansman D, Bullen M (1967) A resistant pneumococcus. *Lancet*, **2**, 264–265.

Ho PL, Yung RW, Tsang DN *et al.* (2001) Increasing resistance of *Streptococcus pneumoniae* to fluoroquinolones: Results of a Hong Kong multicentre study in 2000. *J Antimicrob Chemother*, **48**, 659–665.

Hoge CW, Reichler MR, Dominguez EA *et al.* (1994) An epidemic of pneumococcal disease in an overcrowded, inadequately ventilated jail. *N Engl J Med*, **331**, 643–648.

Hooper DC (2002) Fluoroquinolone resistance among Gram-positive cocci. *Lancet Infect Dis*, **2**, 530–538.

Houldsworth S, Andrew PW, Mitchell TJ (1994) Pneumolysin stimulates production of tumor necrosis factor alpha and interleukin-1 beta by human mononuclear phagocytes. *Infect Immun*, **62**, 1501–1503.

Hummell DS, Swift AJ, Tomasz A, Winkelstein JA (1985) Activation of the alternative complement pathway by pneumococcal lipoteichoic acid. *Infect Immun*, **47**, 384–387.

Immunization Practices Advisory Committee (1981) Pneumococcal polysaccharide vaccine. *MMWR Morb Mortal Wkly Rep*, **30**, 410–412, 417–419.

Jacobs MR, Koornhof HJ, Robins-Browne RM *et al.* (1978) Emergence of multiply resistant pneumococci. *N Engl J Med*, **299**, 735–740.

James JF, Swanson J (1978) Studies on gonococcus infection. XIII. Occurrence of color/opacity colonial variants in clinical cultures. *Infect Immun*, **19**, 332–340.

Janoir C, Podglajen I, Kitzis MD *et al.* (1999) *In vitro* exchange of fluoroquinolone resistance determinants between *Streptococcus pneumoniae* and viridans streptococci and genomic organization of the *parE-parC* region in *S. mitis*. *J Infect Dis*, **180**, 555–558.

Janulczyk R, Iannelli F, Sjoholm AG *et al.* (2000) Hic, a novel surface protein of *Streptococcus pneumoniae* that interferes with complement function. *J Biol Chem*, **275**, 37257–37263.

Johnson AP, Warner M, George RC, Livermore DM (2001) Activity of moxifloxacin against clinical isolates of *Streptococcus pneumoniae* from England and Wales. *J Antimicrob Chemother*, **47**, 411–415.

Jones ME, Sahm DF, Martin N *et al.* (2000) Prevalence of *gyrA*, *gyrB*, *parC*, and *parE* mutations in clinical isolates of *Streptococcus pneumoniae* with decreased susceptibilities to different fluoroquinolones and originating

from Worldwide Surveillance Studies during the 1997–1998 respiratory season. *Antimicrob Agents Chemother*, **44**, 462–466.

Kaetzel CS (2001) Polymeric Ig receptor: Defender of the fort or Trojan horse? *Curr Biol*, **11**, R35–R38.

Kearns AM, Graham C, Burdess D *et al.* (2002) Rapid real-time PCR for determination of penicillin susceptibility in pneumococcal meningitis, including culture-negative cases. *J Clin Microbiol*, **40**, 682–684.

Kim JO, Weiser JN (1998) Association of intrastrain phase variation in quantity of capsular polysaccharide and teichoic acid with the virulence of *Streptococcus pneumoniae*. *J Infect Dis*, **177**, 368–377.

Klein JO (1981) The epidemiology of pneumococcal disease in infants and children. *Rev Infect Dis*, **3**, 246–253.

Kostyukova NN, Volkova MO, Ivanova VV, Kvetnaya AS (1995) A study of pathogenic factors of *Streptococcus pneumoniae* strains causing meningitis. *FEMS Immunol Med Microbiol*, **10**, 133–137.

Lee CJ, Banks SD, Li JP (1991) Virulence, immunity, and vaccine related to *Streptococcus pneumoniae*. *Crit Rev Microbiol*, **18**, 89–114.

Low DE, de Azavedo J, Weiss K *et al.* (2002) Antimicrobial resistance among clinical isolates of *Streptococcus pneumoniae* in Canada during 2000. *Antimicrob Agents Chemother*, **46**, 1295–1301.

Lu L, Lamm ME, Li H *et al.* (2003) The human polymeric immunoglobulin receptor binds to *Streptococcus pneumoniae* via domains 3 and 4. *J Biol Chem*, **278**, 48178–48187.

Lynch III JP, Martinez FJ (2002) Clinical relevance of macrolide-resistant *Streptococcus pneumoniae* for community-acquired pneumonia. *Clin Infect Dis*, **34(Suppl. 1)**, S27–S46.

Mangtani P, Cutts F, Hall AJ (2003) Efficacy of polysaccharide pneumococcal vaccine in adults in more developed countries: The state of the evidence. *Lancet Infect Dis*, **3**, 71–78.

McDaniel LS, Scott G, Kearney JF, Briles DE (1984) Monoclonal antibodies against protease-sensitive pneumococcal antigens can protect mice from fatal infection with *Streptococcus pneumoniae*. *J Exp Med*, **160**, 386–397.

McDaniel LS, Sheffield JS, Delucchi P, Briles DE (1991) PspA, a surface protein of *Streptococcus pneumoniae*, is capable of eliciting protection against pneumococci of more than one capsular type. *Infect Immun*, **59**, 222–228.

Mitchell TJ, Andrew PW, Boulnois GJ *et al.* (1992) Molecular studies of pneumolysin as an aid to vaccine design. *Zentralbl Bakteriol*, **23**, 429–438.

Morgan PJ, Varley PG, Rowe AJ *et al.* (1993) Characterization of the solution properties and conformation of pneumolysin, the membrane-damaging toxin of Streptococcus pneumoniae. *Biochem J*, **296**, 671–674.

Morgan PJ, Hyman SC, Byron O *et al.* (1994) Modeling the bacterial protein toxin, pneumolysin, in its monomeric and oligomeric form. *J Biol Chem*, **269**, 25315–25320.

Morgan PJ, Hyman SC, Rowe AJ *et al.* (1995) Subunit organisation and symmetry of pore-forming, oligomeric pneumolysin. *FEBS Lett*, **371**, 77–80.

Mostov K, ter Beest MB, Chapin SJ (1999) Catch the mu1B train to the basolateral surface. *Cell*, **2**, 121–122.

Musher DM, Watson DA, Dominguez E (1990) Pneumococcal vaccination: Work to date and future prospects. *Am J Med Sci*, **300**, 45–52.

Nagai K, Davies TA, Dewasse BE *et al.* (2000) *In Vitro Selection of Resistance to Gemifloxacin, Trovafloxacin, Ciprofloxacin, Gatifloxacin and Moxifloxacin in* Streptococcus pneumoniae. 40th Interscience Conference on Antimicrobial Agents and Chemotherapy, September 17–20, 2000, Toronto, Ontario, Canada. Abstract 744.

National Committee for Clinical Laboratory Standards (2000) *Methods for Dilution Antimicrobial Susceptibility Tests for Bacteria That Grow Aerobically – Fifth Edition: Approved Standard M7-A5*. National Committee for Clinical Laboratory Standards, Wayne, PA.

National Committee for Clinical Laboratory Standards (2001a) *Performance Standards for Antimicrobial Susceptibility Testing. 10th Information Supplement, M100-S11*. National Committee for Clinical Laboratory Standards, Wayne, PA.

National Committee for Clinical Laboratory Standards (2001b) *Performance Standards for Antimicrobial Susceptibility Testing. 11th Information Supplement M100-S13*. National Committee for Clinical Laboratory Standards, Wayne, PA.

Nielsen SV, Henrichsen J (1992) Capsular types of *Streptococcus pneumoniae* isolated from blood and CSF during 1982–1987. *Clin Infect Dis*, **15**, 794–798.

Novak R, Braun JS, Charpentier E, Tuomanen E (1998) Penicillin tolerance genes of *Streptococcus pneumoniae*: The ABC-type manganese permease complex Psa. *Mol Microbiol*, **29**, 1285–1296.

Novak R, Henriques B, Charpentier E *et al.* (1999) Emergence of vancomycin tolerance in *Streptococcus pneumoniae*. *Nature*, **399**, 590–593.

Nuermberger EL, Bishai WR (2004) Management of community-acquired pneumonia caused by drug-resistant *Streptococcus pneumoniae*. In *Management of Multiple Drug Resistant Infections* (ed. SH Gillespie), pp. 3–29. Humana Press, Totowa, NJ.

O'Neill AM, Gillespie SH, Whiting GC (1999) Detection of penicillin susceptibility in *Streptococcus pneumoniae* by pbp2b PCR-restriction fragment length polymorphism analysis. *J Clin Microbiol*, **37**, 157–160.

O'Toole RD, Goode L, Howe C (1971) Neuraminidase activity in bacterial meningitis. *J Clin Invest*, **50**, 979–985.

Owen RH, Boulnois GJ, Andrew PW, Mitchell TJ (1994) A role in cell-binding for the C-terminus of pneumolysin, the thiol-activated toxin of *Streptococcus pneumoniae*. *FEMS Microbiol Lett*, **121**, 217–221.

Paton JC, Lock RA, Hansman DJ (1983) Effect of immunization with pneumolysin on survival time of mice challenged with *Streptococcus pneumoniae*. *Infect Immun*, **40**, 548–552.

Paton JC, Rowan-Kelly B, Ferrante A (1984) Activation of human complement by the pneumococcal toxin pneumolysin. *Infect Immun*, **43**, 1085–1087.

Piddock LJ, Johnson MM, Simjee S, Pumbwe L (2002) Expression of efflux pump gene *pmr*A in fluoroquinolone-resistant and -susceptible clinical isolates of *Streptococcus pneumoniae*. *Antimicrob Agents Chemother*, **46**, 808–812.

Propp RP, Jabbari B, Barron K (1973) Measurement of the third component of complement in cerebrospinal fluid by modified electroimmunodiffusion. *J Lab Clin Med*, **82**, 154–157.

Quagliarello VJ, Long WJ, Scheld WM (1986) Morphologic alterations of the blood–brain barrier with experimental meningitis in the rat. Temporal sequence and role of encapsulation. *J Clin Invest*, **77**, 1084–1095.

Rauch AM, O'Ryan M, Van R, Pickering LK (1990) Invasive disease due to multiply resistant *Streptococcus pneumoniae* in a Houston, Tex, day-care center. *Am J Dis Child*, **144**, 923–927.

Rayner CF, Jackson AD, Rutman A et al. (1995) Interaction of pneumolysin-sufficient and -deficient isogenic variants of *Streptococcus pneumoniae* with human respiratory mucosa. *Infect Immun*, **63**, 442–447.

Reynolds R, Shackcloth J, Felmingham D, MacGowan A. BSAC Extended Working Party on Respiratory Resistance Surveillance (2003a) Comparison of BSAC agar dilution and NCCLS broth microdilution MIC methods for *in vitro* susceptibility testing of *Streptococcus pneumoniae*, *Haemophilus influenzae* and *Moraxella catarrhalis*: The BSAC Respiratory Resistance Surveillance Programme. *J Antimicrob Chemother*, **52**, 925–930.

Reynolds R, Shackcloth J, Felmingham D, MacGowan A. BSAC Extended Working Party on Respiratory Resistance Surveillance (2003b) Antimicrobial susceptibility of lower respiratory tract pathogens in Great Britain and Ireland 1999–2001 related to demographic and geographical factors: The BSAC Respiratory Resistance Surveillance Programme. *J Antimicrob Chemother*, **52**, 931–943.

Riesenfeld-Orn I, Wolpe S, Garcia-Bustos JF et al. (1989) Production of interleukin-1 but not tumor necrosis factor by human monocytes stimulated with pneumococcal cell surface components. *Infect Immun*, **57**, 1890–1893.

Ring A, Weiser JN, Tuomanen EI (1998) Pneumococcal trafficking across the blood–brain barrier. Molecular analysis of a novel bidirectional pathway. *J Clin Invest*, **102**, 347–360.

Rosenow C, Ryan P, Weiser JN et al. (1997) Contribution of novel choline-binding proteins to adherence, colonization and immunogenicity of *Streptococcus pneumoniae*. *Mol Microbiol*, **25**, 819–829.

Rossjohn J, Gilbert RJ, Crane D et al. (1998) The molecular mechanism of pneumolysin, a virulence factor from *Streptococcus pneumoniae*. *J Mol Biol*, **284**, 449–461.

Rubins JB, Mitchell TJ, Andrew PW, Niewoehner DE (1994) Pneumolysin activates phospholipase A in pulmonary artery endothelial cells. *Infect Immun*, **62**, 3829–3836.

Sande MA, Täuber MG (2000) Pneumococcal meningitis: Current pathophysiologic concepts. In *Streptococcus Pneumoniae: Molecular Biology and Mechanisms of Disease* (ed. A Tomasz), pp. 315–320. Mary Ann Liebert, New York.

Sato K, Quartey MK, Liebeler CL, Le CT, Giebink GS (1996) Roles of autolysin and pneumolysin in middle ear inflammation caused by a type 3 *Streptococcus pneumoniae* strain in the chinchilla otitis media model. *Infect Immun*, **64**, 1140–1145.

Saunders FK, Mitchell TJ, Walker JA et al. (1989) Pneumolysin, the thiol-activated toxin of *Streptococcus pneumoniae*, does not require a thiol group for *in vitro* activity. *Infect Immun*, **57**, 2547–2552.

Scheld WM, Dacey RG, Winn HR et al. (1980) Cerebrospinal fluid outflow resistance in rabbits with experimental meningitis. Alterations with penicillin and methylprednisolone. *J Clin Invest*, **66**, 243–253.

Schmidt T, Täuber MG (1993) Pharmacodynamics of antibiotics in the therapy of meningitis: Infection model observations. *J Antimicrob Chemother*, **31(Suppl. D)**, 61–70.

Smillie WG, Wornock GH, White HJ (1938) A study of a type I pneumococcus epidemic at the State Hospital at Worcester, Mass. *Am J Public Health*, **28**, 293–302.

Smith H, Bannister B (1973) Cerebrospinal-fluid immunoglobulins in meningitis. *Lancet*, **2**, 591–593.

Smith MD, Derrington P, Evans R et al. (2003) Rapid diagnosis of bacteremic pneumococcal infections in adults by using the Binax NOW *Streptococcus pneumoniae* urinary antigen test: A prospective, controlled clinical evaluation. *J Clin Microbiol*, **41**, 2810–2813.

Spanjaard L, van der Ende A, Rumke H et al. (2000) Epidemiology of meningitis and bacteraemia due to *Streptococcus pneumoniae* in The Netherlands. *Acta Paediatr Suppl*, **89**, 22–26.

Tait-Kamradt A, Davies T, Appelbaum PC et al. (2000) Two new mechanisms of macrolide resistance in clinical strains of *Streptococcus pneumoniae* from eastern Europe and North America. *Antimicrob Agents Chemother*, **44**, 3395–3401.

Takashima K, Tateda K, Matsumoto T et al. (1997) Role of tumor necrosis factor alpha in pathogenesis of pneumococcal pneumonia in mice. *Infect Immun*, **65**, 257–260.

Tan TQ, Schutze GE, Mason EO Jr, Kaplan SL (1994) Antibiotic therapy and acute outcome of meningitis due to *Streptococcus pneumoniae* considered intermediately susceptible to broad-spectrum cephalosporins. *Antimicrob Agents Chemother*, **38**, 918–923.

Täuber MG (1989) Brain edema, intracranial pressure and cerebral blood flow in bacterial meningitis. *Pediatr Infect Dis J*, **8**, 915–917.

Täuber MG (2000) *Pathophysiology of Pneumococcal Meningitis*. Second International Symposium of Pneumococci and Pneumococcal Diseases, March 19–23, 2000, Sun City, South Africa.

Tomasz A, Saukkonen K (1989) The nature of cell wall-derived inflammatory components of pneumococci. *Pediatr Infect Dis J*, **8**, 902–903.

Tong HH, McIver MA, Fisher LM, DeMaria TF (1999) Effect of lacto-N-neotetraose, asialoganglioside-GM1 and neuraminidase on adherence of otitis media-associated serotypes of *Streptococcus pneumoniae* to chinchilla tracheal epithelium. *Microb Pathog*, **26**, 111–119.

Tuomanen E (1999) Molecular and cellular biology of pneumococcal infection. *Curr Opin Microbiol*, **2**, 35–39.

Tuomanen E, Liu H, Hengstler B et al. (1985a) The induction of meningeal inflammation by components of the pneumococcal cell wall. *J Infect Dis*, **151**, 859–868.

Tuomanen E, Tomasz A, Hengstler B, Zak O (1985b) The relative role of bacterial cell wall and capsule in the induction of inflammation in pneumococcal meningitis. *J Infect Dis*, **151**, 535–540.

Tuomanen E, Hengstler B, Zak O, Tomasz A (1986) Induction of meningeal inflammation by diverse bacterial cell walls. *Eur J Clin Microbiol*, **5**, 682–684.

Tuomanen E, Rich R, Zak O (1987) Induction of pulmonary inflammation by components of the pneumococcal cell surface. *Am Rev Respir Dis*, **135**, 869–874.

Tureen JH, Täuber MG, Sande MA (1992) Effect of hydration status on cerebral blood flow and cerebrospinal fluid lactic acidosis in rabbits with experimental meningitis. *J Clin Invest*, **89**, 947–953.

Tureen J, Liu Q, Chow L (1996) Near-infrared spectroscopy in experimental pneumococcal meningitis in the rabbit: Cerebral hemodynamics and metabolism. *Pediatr Res*, **40**, 759–763.

van der Poll T, Marchant A, Keogh CV et al. (1996) Interleukin-10 impairs host defense in murine pneumococcal pneumonia. *J Infect Dis*, **174**, 994–1000.

van der Poll T, Keogh CV, Guirao X et al. (1997) Interleukin-6 gene-deficient mice show impaired defense against pneumococcal pneumonia. *J Infect Dis*, **176**, 439–444.

van Haeften R, Palladino S, Kay I et al. (2003) A quantitative LightCycler PCR to detect *Streptococcus pneumoniae* in blood and CSF. *Diagn Microbiol Infect Dis*, **47**, 407–414.

Viladrich PF (2004) Management of meningitis caused by resistant *Streptococcus pneumoniae*. In: *Management of Multiple Drug Resistant Infections* (ed. SH Gillespie), pp. 31–48. Humana Press, Totowa, NJ.

Viladrich PF, Gudiol F, Liñares J et al. (1988) Characteristics and antibiotic therapy of adult meningitis due to penicillin-resistant pneumococci. *Am J Med*, **84**, 839–846.

Volkova MO, Kostiukova NN, Kvetnaia AS (1994) [The role of hyaluronidase in the occurrence of a generalized pneumococcal infection]. *Zh Mikrobiol Epidemiol Immunobiol*, **Aug-Sep(Suppl. 1)**, 118–122 [Russian].

Watson DA, Musher DM, Jacobson JW, Verhoef J (1993) A brief history of the pneumococcus in biomedical research: A panoply of scientific discovery. *Clin Infect Dis*, **17**, 913–924.

Weiser JN (1993) Relationship between colony morphology and the life cycle of *Haemophilus influenzae*: The contribution of lipopolysaccharide phase variation to pathogenesis. *J Infect Dis*, **168**, 672–680.

Weiser JN (1998) Phase variation in colony opacity by *Streptococcus pneumoniae*. *Microb Drug Resist*, **4**, 129–135.

Whitney CG, Farley MM, Hadler J *et al.* (2000) Increasing prevalence of multidrug-resistant *Streptococcus pneumoniae* in the United States. *N Engl J Med*, **343**, 1917–1924.

Winkelstein JA (1984) Complement and the host's defence against the pneumococcus. *CRC Crit Rev Microbiol*, **11**, 187–208.

Winkelstein JA, Abramovitz AS, Tomasz A (1980) Activation of C3 via the alternative complement pathway results in fixation of C3b to the pneumococcal cell wall. *J Immunol*, **124**, 2502–2506.

World Health Organization (1999) Pneumococcal vaccines. WHO position paper. *Wkly Epidemiol Rec*, **74**, 177–183.

Wright DH, Brown GH, Peterson ML, Rotschafer JC (2000) Application of fluoroquinolone pharmacodynamics. *J Antimicrob Chemother*, **46**, 669–683.

Yee AM, Phan HM, Zuniga R *et al.* (2000) Association between FcgammaRIIa-R131 allotype and bacteremic pneumococcal pneumonia. *Clin Infect Dis*, **30**, 25–28.

Zhang J-R, Mostov KE, Lamm ME *et al.* (2000) The polymeric immunoglobulin receptor translocates pneumococci across human nasopharyngeal epithelial cells. *Cell*, **102**, 827–837.

4

Enterococcus spp.

Esteban C. Nannini[1,2] and Barbara E. Murray[2]

[1]*Department of Infectious Diseases, Sanatorio Parque, Rosario, Argentina; and* [2]*University of Texas Health Science Center, Houston Medical School, Houston, TX, USA*

INTRODUCTION

The term enterocoque, emphasizing the intestinal origin of a newly recognized Gram-positive coccus, was encountered in a French paper published in 1899 (Murray, 1990). Because of their cell shape, staining characteristics and lack of catalase, enterococci were considered members of the genus *Streptococcus* until the early 1990s. The name *Streptococcus faecalis*, which came to be used for the most common species, emphasized the relationship of these organisms to feces. In Sherman's classification scheme of streptococci proposed in 1937, the term enterococcus was used for those streptococci that had the ability to grow at 10 °C and 45 °C, in sodium chloride concentrations of 6.5% and at pH 9.6, to resist 60 °C for 30 min and to split esculin (Sherman, 1937). Some enterococci were also noted to be β-hemolytic, at least on some types of blood. Based on Lancefield's serological classification, enterococci were considered salt-tolerant group D streptococci. Biochemical characteristics that distinguished *Streptococcus faecium* from *S. faecalis* were clarified in the 1940s, and the species status of *S. faecium* was formally accepted in the mid-1960s. In 1984 the genus *Enterococcus*, with inclusion of the species *Enterococcus* (*S.*) *faecalis*, *Enterococcus* (*S.*) *faecium* and others, was proposed based on the low nucleic acid homology between enterococci and streptococci (Schleifer and Kilpper-Bälz, 1984). Many new species of *Enterococcus* have also been described in recent years.

DESCRIPTION OF THE GENUS

Biochemical Characteristics

Enterococci are Gram-positive, facultatively anaerobic cocci that typically grow in pairs or short chains. They are catalase negative, although a weak pseudocatalase is occasionally seen. Most strains have the ability to grow in the presence of 6.5% sodium chloride, at 10 °C and at 45 °C and at pH 9.6 and can survive at 60 °C for 30 min. Enterococci are capable of hydrolyzing esculin in the presence of bile, and most strains react with group D antisera. Some react also with group Q antisera. Detection of the group D antigen, which is a cell-wall-associated glycerol teichoic acid antigen, is not specific to enterococci, and group D antigen may be found in other Gram-positive bacteria such as *Streptococcus bovis*, *Streptococcus equines*, *Streptococcus suis*, *Pediococcus* spp. and *Leuconostoc* spp. Genera of Gram-positive, catalase-negative, facultatively anaerobic bacteria whose members may also show some of the other phenotypic characteristics of enterococci include *Aerococcus*, *Gemella*, *Streptococcus*, *Lactococcus*, *Pediococcus* and *Leuconostoc* (Table 4.1). The emergence of vancomycin resistance in enterococci has increased the importance and difficulty of differentiating enterococci from intrinsically vancomycin-resistant Gram-positive cocci, for example, *Leuconostoc* spp. or *Pediococcus* spp., which also may be isolated from human specimens. Most enterococci possess the enzyme

Table 4.1 Useful tests to differentiate selected Gram-positive organisms from enterococci

	Gas	Vancomycin resistance	Group D antigen	Bile–esculin	PYR	Leucine aminopeptidase	Hemolysis[a]	Growth in		
								6.5% NaCl	45 °C	10 °C
Enterococci	−[a]	±[b]	V+	+	+	+	α/γ	+	+	V+
Viridans streptococci	−	−	−	V−	−[a]	+	α/γ	−	V−	−[a]
Lactococci	−	−	−	V+	V+	+	α/γ	V+	V−	+
Aerococci	−	−	−	V+	+	−	α	+	−	−
Pediococci	−	+	+	+	−	+	α	V−	V+	−
Leuconostocs	+	+	V−	V+	−	−	α/γ	V+	−	V+
Lactobacilli	V+	V+	V−	V+	−	−	α/γ	V−	V+	+
Gemella	−	−	−	−	+	+	γ	−	−	−

−, most (>90%) strains negative; +, most (>90%) strains positive; ±[b], depending on the species could be intrinsic or acquired vancomycin resistance (see *Glycopeptide Resistance*); −[a], occasionally positive; Gas, gas production from glucose; PYR, L-pyrrolidonyl-β-naphthylamide; V−, more than half of the strains negative; V+, more than half of the strains positive. Adapted from Facklam and Collins (1989) and Facklam, Carvalho and Teixeira (2002).
[a] Hemolysis on 5% sheep blood agar.

Principles and Practice of Clinical Bacteriology Second Edition   Editors Stephen H. Gillespie and Peter M. Hawkey

pyrrolidonylarylamidase and are able to hydrolyze L-pyrrolidonyl-β-naphthylamide (PYR) (Facklam, Hollis and Collins, 1989), although some species are negative (*Enterococcus cecorum*, *Enterococcus columbae*, *Enterococcus pallens* and *Enterococcus saccharolyticus*) (Facklam, Carvalho and Teixeira, 2002). This reaction is very useful for differentiating enterococci from most streptococci, *Leuconostoc* spp. and pediococci. Other Gram-positive, catalase-negative cocci that test positive for PYR include group A streptococci, *Abiotrophia* and *Granulicatella* (both formerly known as nutritionally variant streptococci), *Helcococcus*, *Aerococcus* spp. and *Gemella*. *Leuconostoc* spp. and pediococci can be group D positive, but both are PYR negative, and leuconostocs produce gas from glucose in Mann–Rogosa–Sharpe (MRS) *Lactobacillus* broth and do not produce leucine aminopeptidase (Facklam and Collins, 1989). The ability to grow on bile-containing media and hydrolyze esculin [bile–esculin (BE) test] is characteristic of group D streptococci, but growth in 6.5% sodium chloride broth distinguishes enterococci from the other group D-reactive species (Facklam *et al.*, 1974). Table 4.1 summarizes the reactions useful for differentiating selected Gram-positive organisms. Some investigators have used molecular methods to differentiate enterococci from other genera (see *Laboratory Diagnosis*).

Growth Characteristics

It has been known for decades that enterococci have complex growth requirements. *E. faecalis* has been shown to grow well on Davis minimal medium (consisting of salts, citrate, thiamine, glucose and agar) supplemented with additional vitamins (biotin, calcium pantothenic acid, pyridoxine, nicotinic acid, riboflavin and folic acid) plus 20 amino acids (Murray *et al.*, 1993). By deleting individual amino acids from this medium, it was shown that most of 23 *E. faecalis* strains tested were prototrophic for purines and pyrimidines and eight amino acids (Ala, Asn, Asp, Gln, Lys, Pag, Pro and Tyr) and auxotrophic (or almost so) for Arg, Glu, Gly, His, Ile, Leu, Met, Trp and Val. Cys, Ser and Thr were stimulatory for the growth of some *E. faecalis* strains (Murray *et al.*, 1993). Availability of the genome sequence from a clinical isolate of vancomycin-resistant *E. faecalis* (V583) has allowed the determination of the location of purine, pyrimidine and other biosynthesis genes (Paulsen *et al.*, 2003).

Habitat

Enterococci are usual inhabitants of the gastrointestinal (GI) tracts of warm-blooded animals and are also found in insects and plants. *E. faecalis* is usually more common in human feces than *E. faecium*, averaging 10^5–10^7 and 10^4–10^5 colony-forming units (CFU) per gram, respectively, but *E. faecium* predominates especially in the hospital setting (Murray, 1990; Suppola *et al.*, 1996). Enterococci are less frequently found at other sites such as vagina, skin, oral cavity and dental plaque. Enterococci are also found routinely in sewage, and their presence has been used to monitor fecal contamination. Because of their production of lactic acid, enterococci have been used as starters in the manufacture of cheese, and they have been isolated from cheese products as well as from certain meats and other foods.

PATHOGENICITY

Virulence Factors in Enterococci

It was recognized early on that enterococci were less virulent than staphylococci, pneumococci and group A streptococci and were even called facultative parasites. The introduction and widespread use of antibiotics, many of which had no or poor activity against enterococci (i.e. cephalosporins), was soon followed by an increase in the incidence of enterococcal infections. However, there were many reports of infections before the antibiotic era, which illustrates that these organisms are able to produce infection in normal hosts, particularly endocarditis and pelvic, intra-abdominal and urinary infections.

Virulence factors associated with the pathogenesis of enterococcal infections in humans are currently under active investigation. Some have been identified, and their role has been tested in diverse animal models of enterococcal infection, although the degree of correlation with human disease is uncertain for most of the models. These virulence factors may be divided into secreted factors (cytolysin/hemolysin, gelatinase and serine protease), cell-surface-located proteins or adhesins [aggregation substance, enterococcal surface protein (Esp) and the adhesin of collagen from *E. faecalis* (Ace)] and cell-wall polysaccharides.

Secreted Factors

Cytolysin is a bacterial toxin produced by some strains of *E. faecalis* that is often encoded within pheromone-responsive plasmids (Ike *et al.*, 1990) and also the chromosome (Ike and Clewell, 1992). It has β-hemolytic activity against some types of blood, including human, rabbit and horse blood. Diverse animal studies have shown that cytolysin-producing strains are associated with higher lethality rates (or increased organ destruction in the rabbit endophthalmitis model) than nonproducing isogenic strains; however, the exact mechanism by which this enzyme contributes to the pathogenesis of *E. faecalis* is still unknown.

Early studies identified certain enterococcal strains with the ability to liquefy gelatin, therefore called *S. faecalis* var. *liquefaciens* or *S. faecalis* var. *zymogenes*, if they were also hemolytic strains. The enzyme thought to be responsible for this reaction was named gelatinase, a metalloprotease which hydrolyzes gelatin, collagen, casein, lactoglobulin and other peptides, although it is now known that gelatinase producers also produce a co-transcribed serine protease, and both are regulated by the *fsr* system, a homologue of the very-well-studied *agr* system of *Staphylococcus aureus*. An experimental endocarditis model from the late 1970s reported a more aggressive course of disease (higher degree of bacteremia, earlier mortality and more embolic phenomena) in animals infected with proteolytic strains than in those infected with the nonproteolytic ones; however, these were not isogenic strains, which limits conclusions from this study (Gutschik, Moller and Christensen, 1979). Using a mouse peritonitis model, a lower lethal dose 50% (LD_{50}) and a delayed mortality were observed when isogenic strains not producing the proteases were compared with gelatinase- and serine protease-producing parental strains (Singh *et al.*, 1998b), and other models have reported similar findings.

Surface-Located Proteins

Aggregation substance is a surface-located protein encoded by pheromone-responsive plasmids that may play a role in binding to human cells, in resistance to neutrophil-mediated killing and in promoting internalization by intestinal cells. It is proposed that aggregation substance, by mediating clumping and/or binding between enterococcal cells and target cells, leads to increased local density of cells, necessary to induce cytolysin/hemolysin production and protease expression via quorum-sensing mechanisms.

The Esp is another surface protein and has been shown to promote colonization and persistence in bladder (but not kidneys) of mice (Shankar *et al.*, 2001). Esp also influences, at least in some strains, biofilm formation *in vitro* (Toledo-Arana *et al.*, 2001). A collagen adhesin called Ace has also been described, which has significant similarity to the A domain of the collagen-binding protein (Cna) of *S. aureus* and the *E. faecium* adhesin Acm (Nallapareddy *et al.*, 2000). The *ace* gene is apparently well conserved and specifically present in *E. faecalis* isolates (Duh *et al.*, 2001).

Another factor that seems to be specific to *E. faecalis* isolates is the *E. faecalis* antigen A (EfaA), which encodes a putative ABC transporter that is regulated by manganese and appears to be important for full virulence (Singh *et al.*, 1998a; Low *et al.*, 2003).

Cell-Wall Polysaccharides

Polysaccharides on the surface of enterococci may represent an effective way to prevent phagocytosis. Some of these appear to be variable capsular carbohydrates, and several have been described in *E. faecalis* and *E. faecium* (Arduino *et al.*, 1994; Huebner *et al.*, 1999; Rakita *et al.*, 2000; Hancock and Gilmore, 2002). The Epa, enterococcal polysaccharide antigen, is a common *E. faecalis* cell-wall polysaccharide, and mutants with an insertion in genes of the *epa* cluster showed attenuation in the mouse peritonitis model and reduced biofilm formation and are more susceptible to neutrophil killing (Teng *et al.*, 2002).

Although adhesins (Acm and SagA) and a variant of Esp have been described in *E. faecium*, virulence factors of this species remain poorly understood, and the traits typically described in *E. faecalis*, such as cytolysin, gelatinase, serine protease and aggregation substance, are rarely, if ever, found in *E. faecium*. The availability of the genome sequence of an *E. faecium* strain (partial sequence available at http://www.hgsc.bcm.edu/microbial/efaecium) should aid in the study of enterococcal pathogenesis.

CLINICAL INFECTIONS

Historically, *E. faecalis*, *E. faecium* and other species (e.g. *Enterococcus gallinarum*, *Enterococcus avium*, *Enterococcus casseliflavus* and *Enterococcus raffinosus*) accounted for 80–90%, 5–20% and 2–4% of all enterococcal infections, respectively (Maki and Agger, 1988; Lewis and Zervos, 1990; Patterson *et al.*, 1995). However, in the 1990s, a shift to a higher incidence of *E. faecium* infections has been observed in the United States, coincident with increased antibiotic pressure and development of vancomycin resistance by this species.

Urinary Tract Infections

Since the early 1900s enterococci have been known to cause urinary tract infections (UTIs). Elderly men, the presence of an indwelling bladder catheter, structural abnormalities of the urinary tract and recent urologic instrumentation are recognized risk factors for the isolation of enterococci from UTIs, which has been reported in ~15% of nosocomial cases (Murray, 1990). In contrast, enterococci cause less than 5% of uncomplicated UTIs in young women. Despite the associated low morbidity and mortality, enterococcal UTIs have clinical importance because of additional costs of hospitalization and therapy.

Intra-abdominal and Pelvic Infections

Enterococci are frequently found in intra-abdominal and pelvic infections, including salpingitis, endometritis and abscesses after cesarean section, but very rarely as the sole agent. The role of enterococci in the early phases of these infections remains controversial, since most of these resolve with antibiotics that do not specifically target enterococci, although occasionally breakthrough enterococcal bacteremia does occur. Animal experiments show that enterococci act synergistically with other bacteria in polymicrobial intra-abdominal infections (Montravers *et al.*, 1997). Enterococci are also a cause of spontaneous bacterial peritonitis in cirrhotic and nephrotic patients and peritonitis in patients on continuous ambulatory peritoneal dialysis.

Endocarditis

Enterococci are the third most common cause of infective endocarditis, following viridans streptococci and *S. aureus*, and account for 5–20% of cases (Megran, 1992). *E. faecalis* is most often the species isolated, but other enterococcal species also cause endocarditis. Enterococcal endocarditis has an acute or, more frequently, subacute onset, affects men (especially elderly men) more frequently than women and often has a genitourinary origin, although GI or biliary tract sources are also probable sources. A high percentage of women with enterococcal endocarditis have a recent history of abortion, cesarean section or genitourinary manipulation. Even though subjects with prior valvular heart disease are common in the series of enterococcal endocarditis, enterococci also affect normal valves (mainly aortic and mitral valves) and prosthetic valves, for which enterococci have been implicated in ~7% of the cases (Rice *et al.*, 1991; Megran, 1992). Enterococci are also among the organisms to consider in intravenous drug users with endocarditis, although, interestingly, they usually do not affect the tricuspid valve. Studies have shown the relatively high propensity of enterococci to adhere to normal and damaged valvular endothelium, comparable to that observed with viridans streptococci and *S. aureus* (Gould *et al.*, 1975). As with other enterococcal infections, the mortality rate (15–20%) is probably affected by the seriousness of the underlying disease present in most of these patients (Megran, 1992).

Bacteremia

Enterococcal bacteremia without endocarditis is a much more common event than endocarditis. The source of the bacteremia in the absence of endocarditis is often the urinary tract, the biliary tract, intra-abdominal or pelvic collections, intravascular catheters or wound infections, situations in which the finding of polymicrobial bacteremia is not uncommon (Maki and Agger, 1988). The features of enterococcal bacteremia that are associated with endocarditis include community-onset, the presence of preexisting valvular heart disease or audible heart murmur and no primary site of infection to explain the bacteremia. On the other hand, nosocomial and/or polymicrobial bacteremia is typically observed in the absence of endocarditis (Maki and Agger, 1988). When there are no signs of intra-abdominal, genitourinary tract or intravascular catheter-related infections, the source of the bacteremia is usually presumed to be translocation of enterococci from the GI tract. Enterococcal bacteremia is frequently observed in patients with serious underlying disease and in those who have undergone major surgery, have received antibiotic therapy or have urethral or intravascular catheters (Wells and von Graevenitz, 1980; Shlaes, Levy and Wolinsky, 1981; Maki and Agger, 1988; Pallares *et al.*, 1993). The severity of the underlying disease, the invasiveness of in-hospital procedures (surgeries, intravascular and bladder catheter placements, etc.), the length of intensive care unit (ICU) stay and the suppression of the normal gut flora by antibiotic pressure (especially antibiotics with antianaerobic effect) are factors that have been associated with enterococcal infections, particularly with *E. faecium* strains. When these factors are prolonged, resistant strains acquired from the hospital environment [i.e. ampicillin-resistant *E. faecium*, *E. faecalis* with high-level resistance (HLR) to gentamicin, vancomycin-resistant enterococci (VRE) and, possibly, in the future, linezolid- and quinupristin–dalfopristin-resistant *E. faecium* strains] are allowed to persist and proliferate in the GI tract, predisposing to subsequent disseminated enterococcal infection.

It has been speculated that the presence of enterococcal bacteremia by itself is a marker of a serious underlying condition. Of note, patients with bacteremia due to *E. faecium* are more often seriously ill or are receiving immunosuppressors or had been previously treated with antibiotics than patients with *E. faecalis* bacteremia (Noskin, Peterson and Warren, 1995; Suppola *et al.*, 1998), which may explain, at least in part, the higher mortality associated with *E. faecium* observed in some series (up to 50%) (Noskin, Peterson and Warren, 1995). Thus, the difference in mortality rates observed in infections caused by these two species may be related to host factors (and, perhaps, resistance) rather than to existing differences in virulence.

Nosocomial Infections

In the United States, enterococci have been ranked as the third most common agent recovered from nosocomial bloodstream infections (Edmond *et al.*, 1999). The high incidence of enterococcal infections has been attributed to the widespread use of antimicrobial agents to which enterococci are resistant, the high proportion of patients who are immunocompromised and with invasive devices and the nosocomial spread of resistant strains. Of note, a dramatic increase in the proportion of VRE among nosocomial isolates has been observed over the past 15 years, reaching 26% of the ICU isolates collected in 2000 (NNIS, 2001). Enterococci are important pathogens in nosocomial-acquired UTIs and are the second most common agent isolated, both in Europe (Bouza *et al.*, 2001) and in US hospitals (Mathai, Jones and Pfaller, 2001). Enterococci are also increasingly reported as a cause of catheter-related bloodstream infection (Sandoe *et al.*, 2002).

Neonatal Infections

Enterococci cause neonatal sepsis and/or meningitis, and it was reported as the second most common isolate in a point prevalence study of nosocomial infections in neonatal ICUs in the United States (Sohn *et al.*, 2001). Although this may occur in normal-term infants, low-birth-weight infants and premature infants with severe underlying conditions are a high-risk group for enterococcal infections (Das and Gray, 1998). Use of a central catheter, the time the central line was in place, GI surgery, prior antibiotic use and intubation have been identified as risk factors for neonatal enterococcal disease (Luginbuhl *et al.*, 1987; Sohn *et al.*, 2001). As with adults, VRE may also spread rapidly among hospitalized neonates (Malik *et al.*, 1999).

Other Less Common Infections

Besides neonatal meningitis, enterococci have also been cited as a cause of shunt infections and meningitis in older children and adults (Durand *et al.*, 1993). In most of these patients, invasive procedures of the central nervous system, underlying diseases, previous antibiotic therapy and even disseminated strongyloidiasis were the predisposing factors for infection. Enterococci have also been found, usually with other microorganisms, in diabetic foot infections, decubitus ulcers, burns and postsurgical abdominal wounds (Lewis and Zervos, 1990).

LABORATORY DIAGNOSIS

Isolation Procedures

Enterococci grow well on blood agar base media containing 5% animal blood and on many other media [e.g., Mueller–Hinton, brain–heart infusion (BHI), dextrose phosphate, chocolate] that are not selective for Gram-negative bacteria. Some strains of *E. faecalis* produce β-hemolysis on media containing rabbit, human or horse blood but not sheep blood. Bile–esculin azide (Enterococcosel agar) or other commercially available azide-containing media, Columbia colistin–nalidixic acid (CNA) agar, phenylethyl alcohol (PEA) agar, SF agar, triphenyl tetrazolium chloride azide agar (m-Enterococcus agar), among others, may be used to isolate enterococci from mixed samples containing Gram-negative bacteria.

Species Identification

Enterococcus faecalis accounts for more than 80% of the clinical isolates, followed by *E. faecium* and then by other enterococcal species (e.g. *E. gallinarum*, *E. avium*, *E. casseliflavus* and *E. raffinosus*), although in settings with a high rate of VRE an increased proportion of *E. faecium* strains is usually found. Some other species are encountered less often (*Enterococcus mundtii*, *Enterococcus durans* and *Enterococcus hirae*), and some are mainly isolated from non-human sources (*Enterococcus malodoratus*, *Enterococcus pseudoavium*, *Enterococcus sulfureus*, *E. cecorum*, *E. columbae*, *E. saccharolyticus*, *Enterococcus asini*, *Enterococcus ratti* and *Enterococcus phoeniculicola*). Differentiation of non-faecalis enterococci may be helpful in serious infections for therapeutic (differences in susceptibilities to antibiotics exist) and epidemiological purposes. The genus *Enterococcus* now contains over 20 species, some of which have been isolated only from plants and animals (Facklam, Carvalho and Teixeira, 2002). Based on DNA reassociation and/or 16S rRNA sequencing studies, the proposed species *Enterococcus solitarius* and *Enterococcus seriolicida* no longer belong to the *Enterococcus* genus (Williams, Rodrigues and Collins, 1991) and *Enterococcus flavescens* and *E. casseliflavus* constitute a single species (hereafter called *E. casseliflavus*). New species have been recently described, namely, *Enterococcus haemoperoxidus* and *Enterococcus moraviensis* from surface waters (Svec *et al.*, 2001), *Enterococcus villorum* (also named *Enterococcus porcinus*) (Vancanneyt *et al.*, 2001), *E. phoeniculicola* (Law-Brown and Meyers, 2003) and *E. ratti* (Teixeira *et al.*, 2001) from animals and the pigmented *E. pallens* and *Enterococcus gilvus* from human clinical specimens (Tyrrell *et al.*, 2002).

Conventional Tests

Historically, enterococci have been separated into three groups on the basis of acid formation in conventional tubes of mannitol, sorbitol and sorbose broths and hydrolysis of arginine (Facklam and Collins, 1989). According to this scheme, group I species (*E. avium*, *E. malodoratus*, *E. raffinosus*, *E. pseudoavium*, *E. saccharolyticus*, *E. pallens* and *E. gilvus*) form acid in mannitol, sorbitol and sorbose broths but do not hydrolyze arginine. Group II species (*E. faecalis*, *E. faecium*, *E. casseliflavus*, *E. mundtii* and *E. gallinarum* as well as lactococci) produce acid in mannitol broth but not in sorbose broth and are able to hydrolyze arginine. Variable reactions are seen in sorbitol broth. Group III species (*E. durans*, *E. hirae*, *Enterococcus dispar*, *E. ratti* and asaccharolytic variants of *E. faecium* and *E. faecalis*) hydrolyze arginine but fail to form acid in these three carbohydrate broths. More recently, two groups have been added: group IV (*E. asini*, *E. sulfureus* and *E. cecorum*), which do not produce acid in mannitol and sorbose broths and fail to hydrolyze arginine, and group V (mostly variant strains of *E. faecalis*, *E. casseliflavus* and *E. gallinarum*), which are not able to hydrolyze arginine (Facklam, Carvalho and Teixeira, 2002). The species within each group may be identified by the reactions summarized in Table 4.2. *Enterococcus faecalis* strains are usually tellurite tolerant when tested on agar medium containing 0.04% potassium tellurite and form black colonies. A few *E. gallinarum*, *E. casseliflavus* and *E. mundtii* strains may also be tellurite tolerant. Motility and yellow pigmentation are helpful for differentiating three species in the group II. *Enterococcus casseliflavus* is motile and produces yellow pigment; *E. mundtii* produces yellow pigment, but it is not motile; *E. gallinarum* is motile but does not produce yellow pigment. Occasionally, pigment production and motility are unreliable features on which to base identification of *E. gallinarum* and *E. casseliflavus* (Vincent *et al.*, 1991). Other investigators found discrepancies in conventional tube test results described by Facklam and Collins, particularly in arginine and sorbitol tests (Knudtson and Hartman, 1992). *Enterococcus sulfureus* belongs to group IV and is a nonmotile pigment producer that does not possess

Table 4.2 Tests for identifying *Enterococcus* species

	Mannitol	Sorbitol	Sorbose	Arginine	Arabinose	Raffinose	Tellurite	Motility	Pigmentation	Sucrose	Pyruvate	Methyl-α-D-glucopyranoside
Group I												
E. avium	+	+	+	−	+	−	−	−	−	+	+	V
E. malodoratus	+	+	+	−	−	+	−	−	−	+	+	V
E. raffinosus	+	+	+	−	+	+	−	−	−	+	+	V
E. pseudoavium	+	+	+	−	−	−	−	−	−	+	+	+
E. saccharolyticus	+	+	+	−	−	+	−	−	−	+	−	+
E. pallens	+	+	+	−	−	+	−	−	+	+	−	+
E. gilvus	+	+	+	−	−	+	−	−	+	+	+	−
Group II												
E. faecalis	+	+	−	+	−	−	+	−	−	+[a]	+	−
E. faecium	+	−	−	+	+	−	−	−	−	+[a]	−	−
E. casseliflavus/ E. flavescens	+	V	−	+	+	+	−[a]	+	+	+	V	+
E. mundtii	+	V	−	+	+	+	−	−	+	+	−	−
E. gallinarum	+	−	−	+	+	+	−	+[a]	−	+	−	+
Group III												
E. durans	−	−	−	+	−	−	−	−	−	−	−	−
E. hirae	−	−	−	+	−	+	−	−	−	+	−	−
E. dispar	−	−	−	+	−	+	−	−	−	+	+	+
E. ratti	−	−	−	+	−	−	−	−	−	−	−	−
E. faecalis (var)	−	−	−	+	−	−	+	−	−	−	+[a]	
Group IV												
E. asini	−	−	−	−	−	−	−	−	−	+	−	−
E. sulfureus	−	−	−	−	−	+	−	−	+	+	−	+
E. cecorum	−	+	−	−	−	+	−	−	−	+	+	−

+, most (>90%) strains positive; −, most (>90%) strains negative; V, variable. Adapted from Facklam and Collins (1989) and Facklam, Carvalho and Teixeira (2002).
[a] Occasional exceptions reported.

group D antigen, and mannitol, inulin, arabinose and arginine tests are negative (Martinez-Murcia and Collins, 1991). The two recently recognized nonmotile pigmented species, *E. pallens* and *E. gilvus*, may be differentiated from other pigmented species because they fit into group I, and therefore, mannitol, sorbitol and sorbose tests are positive and arginine hydrolysis is negative. PYR and methyl-α-D-glucopyranoside (MGP) reactions are negative for *E. pallens* and positive for *E. gilvus*. Owing to phenotypic similarities to enterococci, some authors have incorporated *Lactococcus* spp. and *Vagococcus* spp. into groups II and V, respectively (Facklam, Carvalho and Teixeira, 2002).

Because of the increased prevalence of VRE and its epidemiological and probably clinical implications, proper identification of these enterococci is important. The presence of Gram-positive cocci obtained from the growth of organisms on agars containing vancomycin (usually 6 μg/ml), used for surveillance of VRE strains, must be confirmed, and then, the organisms should be subcultured onto blood agar (Willey *et al.*, 1999). The presence of yellow pigment (observed after picking up some colonies with a cotton swab) and the negative PYR reaction confirms the presence of *E. casseliflavus* and *Leuconostoc* or *Pediococcus* spp., respectively. If the strain is nonpigmented and PYR positive, then *E. gallinarum* may be identified by the motility test (although ~8% of the strains may test negative) or, more specifically, by the xylose test, which can be performed in 2 h (Willey *et al.*, 1999). Overnight fermentation tests using arabinose, pyruvate and sorbitol may further differentiate between *E. faecalis* and *E. faecium*. The results of the vancomycin minimal inhibitory concentration (MIC) might also be considered since the intrinsically vancomycin-resistant strains *E. casseliflavus* and *E. gallinarum* display intermediate-to-low level of resistance, whereas VanA or VanB type of VRE usually shows high-level vancomycin resistance.

Other Laboratory Tests

A method that has been used by several laboratories involves whole-cell protein analysis, which compares patterns generated by different

species (Teixeira *et al.*, 2001; Devriese *et al.*, 2002). Several methods based on genotypic rather than phenotypic characteristics have been useful for discriminating enterococcal species. Some of these tests include sequencing of *ddl* genes encoding D-alanine:D-alanine ligases (Ozawa, Courvalin and Gaiimand, 2000), the *sodA* gene coding for manganese-dependent superoxide dismutase (Poyart, Quesnes and Trieu-Cuot, 2000), the *tuf* gene (Ke *et al.*, 1999), PCR amplification of the intergenic spacer between 16S and 23S rRNA (Tyrrell *et al.*, 1997) and 16S rRNA sequencing (Angeletti *et al.*, 2001; Devriese *et al.*, 2002; Facklam, Carvalho and Teixeira, 2002; Ennahar, Cai and Fujita, 2003). Other methods have focused on the rapid identification of *E. faecium* strains with a PCR assay using specific inserts (Cheng *et al.*, 1997) or through amplification of or hybridization to species-specific genes, such as the *aac(6′)Ii* gene (encoding the *E. faecium*-specific aminoglycoside-modifying enzyme 6′-*N*-acetyltransferase) (Costa *et al.*, 1993), the *msrC* gene (associated with low-level macrolide resistance of *E. faecium*) (Singh, Malathum and Murray, 2001), or detection using hybridization for 16S rDNA (Manero and Blanch, 2002), *efaA*, *ace* (Duh *et al.*, 2001) and *acm* genes (Nallapareddy *et al.*, 2000). The *lsa* gene [responsible for the natural resistance of *E. faecalis* to quinupristin–dalfopristin (Singh, Weinstock and Murray, 2002)] has also been used for identification of *E. faecalis* strains.

Commercial Identification Systems

Several commercially available devices are available for identifying *Enterococcus* species. The API 20S and the API Rapid ID 32 systems appear to work relatively well for *E. faecalis*, but supplemental tests are frequently needed for proper identification of *E. faecium*, *E. gallinarum*, *E. casseliflavus* and *E. durans* (Sader, Biedenbach and Jones, 1995; Hamilton-Miller and Shah, 1999). The Vitek 2 system shows better identification of enterococci than previous a version, although motility and pigmentation tests are frequently required to avoid misidentification of *E. gallinarum* or *E. casseliflavus* as *E. faecium* (Garcia-Garrote, Cercenado and Bouza, 2000; Ligozzi *et al.*, 2002). Similar findings were reported when the automated MicroScan WalkAway 96 system was evaluated (d'Azevedo *et al.*, 2001).

Susceptibility Testing

Routine Testing (Kirby–Bauer)

Since enterococcal susceptibility to antimicrobial agents is unpredictable, the site of infection and/or the significance of a particular isolate determines which antimicrobials should be included in susceptibility testing. Drugs to which enterococci are intrinsically resistant, such as cephalosporins, oxacillin, trimethoprim–sulfamethoxazole (TMP–SMX) (*in vivo* resistance), clindamycin and aminoglycosides (when used as monotherapy), should not be tested. On the other hand, susceptibility to penicillin or ampicillin and vancomycin should be determined routinely. For urine isolates, fluoroquinolones, nitrofurantoin and fosfomycin may be added. Using the Kirby–Bauer technique, a 16 mm or less zone diameter of inhibition around ampicillin 10 μg disks and a zone of 14 mm or less around penicillin 10 unit disks are considered resistant. A vancomycin (30 g disk) zone of inhibition of 14 mm or less is reported as resistant, 15–16 mm as intermediate and 17 mm or more as susceptible (plates should be held for a full 24-h period). It is recommended that any haze or colonies within the zone should be taken into account and that an MIC test be performed for strains with intermediate-susceptibility zones if vancomycin is to be used for treatment (Swenson *et al.*, 1992).

Agar Dilution MICs

For agar dilution testing, interpretative criteria of the National Committee for Clinical Laboratory Standards (NCCLS) for ampicillin and penicillin are the following: MIC ≤ 8 μg/ml, susceptible; MIC ≥ 16 μg/ml, resistant (NCCLS, 2003). For vancomycin, MICs of ≤4 μg/ml and 8–16 μg/ml are considered susceptible and intermediate, respectively (the latter are considered resistant in some countries), whereas an MIC of vancomycin of ≥32 μg/ml is considered resistant (NCCLS, 2003). For the screening of VRE, NCCLS recommends the use of a BHI agar with 6 μg/ml of vancomycin and an inoculum of 1–10 μl of a 0.5 McFarland standard suspension. Growth after 24 h of incubation at 35 °C is interpreted as resistant.

Aminoglycoside HLR and β-Lactamase Screening

For serious infections, particularly endocarditis and possibly for meningitis and deep-seated infections in immunocompromised patients, HLR to aminoglycosides and β-lactamase testing should be performed. HLR to gentamicin indicates resistance to synergism with all currently available aminoglycosides except streptomycin and, occasionally, arbekacin. MICs of arbekacin (available in Japan) are variable for strains with HLR to gentamicin. High-level aminoglycoside resistance can be detected by agar or single-tube broth screening with 500 μg/ml of gentamicin. For streptomycin, 2000 μg/ml and 1000 μg/ml is used for agar and broth screening, respectively. Disks containing 300 μg of streptomycin and 120 μg of gentamicin may also be used to predict synergy or the lack of synergy. E-Test also demonstrates concordance in the detection of HLR to aminoglycosides among enterococci when compared with agar dilution screening. Broth microdilution systems appear reliable for HLR to gentamicin, but they have missed some strains with HLR to streptomycin in the past. New versions of automated systems for susceptibility testing of enterococci display a high degree of correlation with standard methods, including VRE strains and strains with HLR to aminoglycosides (d'Azevedo *et al.*, 2001; Ligozzi *et al.*, 2002).

Molecular Typing Systems

Among various methods used for molecular epidemiology studies, total plasmid content, plasmid DNA digestion patterns, amplified fragment length polymorphism, ribotyping and conventional as well as pulsed-field gel electrophoresis (PFGE) of chromosomal DNA have been used to evaluate nosocomial enterococcal infections. PFGE is considered the gold standard and performs better at identifying clonality, at least during prolonged outbreaks. It is reproducible and generally available. A pilot study using multilocus sequence typing suggests that this method may also be used.

MECHANISMS OF RESISTANCE TO ANTIMICROBIAL AGENTS

The presence of tolerance and/or resistance to different classes of antibiotics is one of the typical characteristics of enterococci. Antimicrobial resistance among enterococci may be divided into two types: intrinsic (or inherent) and acquired (Table 4.3). Intrinsic resistance refers to naturally occurring, chromosomally encoded characteristics encountered in all or almost all of the strains of a particular species. Acquired resistance is caused by acquisition of new DNA or mutations in the existing DNA.

Intrinsic Resistance

Inherent resistance characteristics include resistance to semisynthetic penicillinase-resistant penicillins (e.g., methicillin), cephalosporins, monobactams, quinupristin–dalfopristin (*E. faecalis*), low levels of aminoglycosides and clindamycin and TMP–SMX (*in vivo*) and only moderate susceptibility to most fluoroquinolones. Vancomycin resistance in *E. gallinarum* and *E. casseliflavus* is also intrinsic and is discussed in the *Glycopeptide Resistance* section. Relative intrinsic resistance of enterococci to β-lactams is a characteristic feature of these organisms because of the presence of penicillin-binding proteins (PBPs) with low affinity for penicillin (Fontana *et al.*, 1983). MICs of penicillin for *E. faecalis* are generally between 1 μg/ml and 4 μg/ml, approximately 10–1000 times greater than those for streptococci such as *Streptococcus pyogenes*. Generally, MICs of ampicillin and ureidopenicillins are one dilution lower than those of penicillin. *Enterococcus faecium* is even more resistant to β-lactams with typical MICs of penicillin of ≥8 μg/ml. *Enterococcus faecium* strains with higher MICs of penicillin (64 to >256 μg/ml) are increasingly reported in recent years and are discussed in the section *Acquired Resistance*. None of the available cephalosporins inhibits enterococci sufficiently to be used clinically, and enterococcal superinfection may in fact occur in patients receiving cephalosporins. As with the penicillins, carbapenems are more active against *E. faecalis* than *E. faecium*. Low-level resistance to clindamycin is another characteristic feature of enterococci, particularly *E. faecalis*. A gene called *lsa*, which encodes a putative ABC transporter, accounts for the intrinsic resistance of *E. faecalis* to both clindamycin and quinupristin–dalfopristin (Singh, Weinstock and Murray, 2002). As a corollary, superinfection with *E. faecalis* isolates has occurred in patients receiving quinupristin–dalfopristin (Moellering *et al.*, 1999), and acquired resistance also occurs in *E. faecium*. Another example of intrinsic resistance is the low-level resistance to aminoglycosides (4–64 μg/ml for gentamicin and 32–500 μg/ml for streptomycin). This resistance appears to be related to limited drug uptake. Combining aminoglycosides with cell-wall–active agents such as penicillins or vancomycin results in markedly enhanced uptake of the aminoglycoside, leading to the well-known synergistic effect of cell-wall synthesis inhibitors plus aminoglycosides (defined as ≥2 log$_{10}$ enhanced killing relative to the effect of the cell-wall–active agent alone when the aminoglycoside effect is subinhibitory). Strains of *E. faecium* normally contain a chromosomally located gene, named *aac(6′)Ii*, encoding the aminoglycoside-modifying enzyme

Table 4.3 Common resistances in enterococci

Pattern of resistance	Mechanisms/Comments
Intrinsic resistances	
Aminoglycosides (low level)	Limited drug uptake
Aminoglycosides (tobramycin, kanamycin, netilmicin and sisomicin) (MICs usually ≤ 2000 µg/ml, but synergism with penicillin is abolished)	Chromosomally encoded AAC(6′)-Ii enzyme present in *E. faecium* strains
β-Lactams (semisynthetic penicillinase-resistant penicillins, cephalosporins and aztreonam)	Low-affinity PBPs (*E. faecium* more so than *E. faecalis*)
Trimethoprim–sulfamethoxazole	*In vivo* resistance (MIC may fall within the susceptible range)
Quinupristin–dalfopristin	ABC transporter for lincosamides, and dalfopristin present in *E. faecalis* strains
Glycopeptides	VanC-type in *E. casseliflavus* and *E. gallinarum* isolates (peptidoglycan precursor ending in D-Ala-D-serine)
Acquired resistances	
Aminoglycosides (high level, MICs usually ≥ 2000 µg/ml)	Aminoglycoside-modifying enzymes (AAC, ANT and APH)
	Ribosomal subunit mutation (described for streptomycin)
Chloramphenicol	Chloramphenicol acetyltransferase
Tetracyclines	Efflux system and ribosomal protection
Penicillins	Overproduction of low-affinity PBPs
	Mutations in low-affinity PBPs (*E. faecium* isolates)
	Production of β-lactamase (*E. faecalis* isolates)
Quinolones	Mutations in *gyrA* and *parC* genes, coding for submit of DNA gyrase and topoisomerase IV, respectively
Macrolides, lincosamides and streptogramins B (MLS$_B$ group)	Reduced drug binding through methylation of an adenine residue of the 23S rRNA (most common is ErmB)
Glycopeptides	Production of low-affinity peptidoglycan precursors ending in D-Ala-D-lactate (VanA, VanB and VanD) or D-Ala-D-serine (VanC and VanE) instead of the normal D-Ala-D-alanine)
Oxazolidinones (linezolid)	Mutations in the 23S rRNA

AAC, aminoglycoside acetyltransferase; ANT, aminoglycoside nucleotidyltransferase; APH, aminoglycoside phosphotransferase; MIC, minimum inhibitory concentration; PBP, penicillin-binding protein.

6′-*N*-acetyltransferase, which is specific to this species (Costa *et al.*, 1993). Although the low-level production of this enzyme does not confer HLR to aminoglycosides, MICs are higher than those of *E. faecalis* and synergism between cell-wall–active agents and the aminoglycosides carrying an unprotected amino group at the 6′ position (tobramycin, kanamycin, netilmicin and sisomicin) is not observed. Another apparently intrinsic property of enterococci is that of *in vivo* resistance to TMP–SMX. Even though TMP–SMX may test active *in vitro*, it has been found to be ineffective as monotherapy in both animal and clinical case report studies.

Acquired Resistance

Enterococci possess several different types of mobile genetic elements that may be transferred from one strain to another. One class consists of broad-host-range plasmids, which may be transferred between enterococci and to various other Gram-positive bacteria such as staphylococci, streptococci, *Bacillus* spp. and lactobacilli. These plasmids transfer during filter mating but not usually in broth. Another type of plasmid, referred to as narrow-host-range plasmids because they appear to transfer primarily to *E. faecalis*, responds to sex pheromones produced by the recipient cells. There are many different known pheromones that act on specific pheromone-responsive plasmids. In response to pheromones, these plasmids initiate the production of aggregation substance (clumping factor) which leads to sticking together of donor and recipient cells. The transfer frequency of plasmids is markedly increased by this mechanism, and transfer occurs in broth as well as during filter matings (Clewell and Weaver, 1989). Transposons are also common in enterococci.

Macrolide, Lincosamide and Streptogramin Resistance

The erythromycin resistance determinant (*ermB*), which is commonly found in human enterococcal isolates, is often encoded on a transposon, exemplified by the well-studied Tn*917*. This gene codes for an enzyme that methylates a specific adenine residue in the 23S ribosomal RNA within the 50S ribosomal subunit (Roberts *et al.*, 1999), resulting in reduced binding affinity of macrolides, lincosamides (clindamycin) and streptogramins B (quinupristin), the so called MLS$_B$-type resistance. This same determinant also reduces the bactericidal activity of quinupristin–dalfopristin (Synercid®). A gene, *vgb*(A), responsible for the hydrolysis of streptogramins B has also been identified in enterococci (*E. faecium* isolates) (Jensen *et al.*, 1998), as have acetyltransferases encoded by the genes *vat*(D) and *vat*(E) that inactivate streptogramin compounds (Soltani *et al.*, 2000). It generally appears that high MICs of quinupristin–dalfopristin require the presence of more than one gene conferring resistance to the individual components (Bozdogan and Leclercq, 1999).

Chloramphenicol, Tetracycline and Fluoroquinolone Resistance

Chloramphenicol resistance, found in 20–40% of enterococci, is often mediated by chloramphenicol acetyltransferase (Murray, 1990). Tetracycline resistance, which is found in 60–80% of enterococci, is mediated by different genes that code for energy-dependent efflux proteins, including *tet*(K), *tet*(L) or *tet*(A). Other genes associated with tetracycline resistance include *tet*(M), *tet*(O), *tet*(Q) and *tet*(S) [with *tet*(M) by far the most common], which encode cytoplasmic proteins known as ribosomal protection proteins that confer cross-resistance to all currently available agents of this class (Roberts, 1996). The well-studied conjugative transposon Tn*916* carries the tetracycline resistance gene *tet*(M), the same gene found in tetracycline-resistant *Neisseria gonorrhoeae*. This and other conjugative transposons differ from ordinary or nonconjugative transposons (e.g., Tn*917*) because they initiate bacterial mating (conjugation), which allows them to self-transfer from one bacterial cell to another. Of note, the recently developed derivatives of minocycline, called glycylcyclines, remain active against tetracycline- and minocycline-resistant enterococcal isolates (Rasmussen, Gluzman and Tally, 1994).

Fluoroquinolones are considered agents with rather poor intrinsic activity against enterococci. Mutations in the *gyrA* and the *parC* genes, coding for subunits of DNA gyrase and topoisomerase IV, respectively, have been related to increasing quinolone MICs (Korten, Huang and Murray, 1994; Kanematsu *et al.*, 1998). Most of the quinolone-resistant enterococcal isolates studied had mutation in both genes.

HLR to Aminoglycosides

Besides intrinsic low-level resistance to aminoglycosides, many enterococcal strains have acquired HLR (MIC > 2000 µg/ml) to an aminoglycoside, which causes resistance to the synergism otherwise seen between cell-wall–active agents and the involved aminoglycoside. HLR to aminoglycosides is most often due to the production of one or more aminoglycoside-modifying enzymes: aminoglycoside phosphotransferases (APH), aminoglycoside nucleotidyltransferases (ANT) or aminoglycoside acetyltransferases (AAC). *Enterococcus faecalis* strains with HLR to gentamicin were first reported in 1979, and strains highly resistant to all aminoglycosides (including gentamicin and streptomycin) were first reported in 1983 (Murray, 1990; Patterson and Zervos, 1990). Such strains have now been reported around the world, and other enterococcal species with this trait have been increasingly reported in recent years. In some tertiary centers, most of the enterococci are resistant to penicillin and gentamicin synergism. More than 90% of the enterococci with HLR to gentamicin contain plasmids with a gene coding for a bifunctional enzyme 2″-phosphotransferase-6′-acetyltransferase (APH(2″)-AAC(6′)) activity, identical to the gene found in gentamicin-resistant staphylococci (Ferretti, Gilmore and Courvalin, 1986). This gene may be located on a transposon probably inherited from *S. aureus* (Hodel-Christian and Murray, 1991). This enzyme causes resistance to synergy between cell-wall–active agents and all currently available aminoglycosides, except streptomycin (and occasionally arbekacin). More recently, other genes such as *aph(2″)Ib*, *aph(2″)-Ic* and *aph(2″)-Id* encoding phosphotransferases that confer HLR to gentamicin and some, but not all, other aminoglycosides have been reported (Chow, 2000). However, because the bifunctional APH(2″)-AAC(6′) enzyme accounts for the vast majority of HLR to gentamicin and confers cross-resistance to all clinically available aminoglycosides except streptomycin, screening using high concentrations of gentamicin and streptomycin is still recommended. A different approach may arise if other genes conferring resistance to gentamicin, but not to other aminoglycosides, are found more frequently in clinical isolates (Chow, 2000).

HLR to streptomycin may result by enzymatic modification or secondary to mutations in the 30S ribosomal subunit. Strains with ribosomal resistance have very high MICs of streptomycin (>32 000 µg/ml), whereas the ANT(6′) enzyme producers usually have MICs of streptomycin between 4000 µg/ml and 16 000 µg/ml. HLR to kanamycin without HLR to gentamicin is caused by the production of the widespread APH(3′) or, rarely, ANT(4′). The latter enzyme also causes HLR to tobramycin. These enzymes are important because they also prevent penicillin–amikacin synergy, even though they do not confer HLR to amikacin (Leclercq *et al.*, 1992).

β-Lactamase Production and High-Level Penicillin Resistance due to Other Mechanisms

The first β-lactamase-producing enterococcus was an *E. faecalis* isolate recovered from a urine culture in 1981. Since then, such strains have been isolated from different geographical locations across the United States and several other countries (Murray, 1992). The *blaZ* gene in enterococci is identical to that of the type A β-lactamase found in *S. aureus* (Murray, 1992) and usually has been found on conjugative plasmids, although a chromosomal location has also been described. In one instance, the gene appears to be incorporated into a transposon-like element derived from staphylococci (Rice and Marshall, 1992). Since the amount of enzyme produced by enterococci is lower than that produced by *S. aureus*, β-lactamase-producing strains may be missed by routine susceptibility-testing methods unless a high inoculum or a detection system such as the chromogenic cephalosporin nitrocefin is used. Although β-lactamase-producing enterococci are still rare among clinical isolates, they may cause outbreaks and severe infections. For this reason, it is advisable to test enterococci isolated from severe infections, particularly endocarditis, for β-lactamase production. With few exceptions, β-lactamase-producing enterococci are also highly resistant to gentamicin. The only β-lactamase-producing *E. faecium* reported was isolated in a hospital having an outbreak due to β-lactamase-producing *E. faecalis*, suggesting interspecies spread of this resistance determinant (Coudron, Markowitz and Wong, 1992).

Although it has been known for many years that strains of *E. faecium* are more resistant to β-lactams than *E. faecalis*, strains of *E. faecium* with much higher penicillin MICs (>64 µg/ml) have become increasingly common. Mechanisms that have been implicated in causing high levels of penicillin resistance include overproduction of PBPs with low affinity for β-lactams and mutations in the penicillin-binding domain of the PBP, resulting in further reduction in the affinity for β-lactams (Fontana *et al.*, 1996).

Glycopeptide Resistance

Glycopeptide resistance, the most worrisome acquired resistance in enterococci, was first described in England and France in the mid–late 1980s, with a rapidly increasing number of reports from the United States and other countries. In recent years, horizontal transfer of the enterococcal *vanA* gene associated with this type of resistance to *S. aureus* has been reported (CDC, 2002a,b).

Six types of glycopeptide resistances in enterococci have been described, named for their specific ligase genes (e.g., *vanA*, *vanB*), which may be intrinsic (e.g., *vanC*, as a species characteristic) or acquired. Within the acquired types of glycopeptide resistance, *E. faecium* accounts for most of the isolates, followed by *E. faecalis* and, less often, by other enterococcal species. These types of glycopeptide resistances share the capability of producing a peptidoglycan precursor with decreased affinity for glycopeptides. These peptidoglycan precursors end in the depsipeptide D-alanyl-D-lactate in VanA, VanB and VanD strains and in the dipeptide D-alanyl-D-serine in VanC and VanE isolates, instead of the normal D-alanyl-D-alanine ending. The VanA and VanB types are the most commonly found. High MICs of vancomycin and teicoplanin are usually observed in VanA isolates, whereas VanB strains typically display low teicoplanin MICs and intermediate-to-high MICs of vancomycin.

The *vanA* gene cluster is typically found on the transposon Tn*1546* or related elements (e.g., Tn*5482*) (Arthur *et al.*, 1993; Handwerger and Skoble, 1995), which in turn may be carried on plasmids as well as on bacterial chromosomes. The three main, necessary and co-transcribed, genes of the *vanA* gene cluster are *vanH*, *vanA* and *vanX* (Arthur, Molinas and Courvalin, 1992), which encode a dehydrogenase (VanH), a ligase (VanA) and a D,D-dipeptidase (VanX), respectively. The ligase synthesizes D-Ala-D-Lac using D-Lac obtained from the reduction of the pyruvate by the dehydrogenase (VanH) (Arthur and Courvalin, 1993), which is then incorporated into cell-wall precursors. These D-Ala-D-Lac-ending precursors have less affinity for glycopeptides than D-Ala-D-Ala precursors; therefore, cell-wall synthesis continues in the presence of vancomycin. The D,D-dipeptidase VanX hydrolyzes free D-Ala-D-Ala dipeptides, decreasing the production of the normal D-Ala-D-Ala-containing precursor. Two-component regulatory systems composed of a membrane-associated sensor kinase (e.g., VanS) and a cytoplasmic regulator (e.g., VanR) regulate the expression of the *vanA* and *vanB* gene clusters. The presence of glycopeptides is sensed by VanS, which activates VanR in the cytoplasm, leading to the transcriptional activation of the resistance genes and regulatory genes (Arthur and Quintiliani, 2001). The sensor kinase in VanB strains (VanS$_B$) does not recognize teicoplanin in the medium, which may explain the low teicoplanin MIC usually expressed by these strains (Evers and Courvalin, 1996).

The motile species *E. gallinarum* and *E. casseliflavus* have, as an intrinsic characteristic, the genes *vanC-1* and *vanC-2*, respectively (Navarro and Courvalin, 1994). Since these species synthesize both normal ending precursors (D-Ala-D-Ala) and low-glycopeptide-affinity precursors (D-Ala-D-Ser), variable MICs of vancomycin, even within the susceptible range, are usually observed in these enterococcal species (Dutka-Malen *et al.*, 1992).

The overall epidemiological impact of VRE strains is different in Europe and the United States. In the former, VRE strains that generally have low MICs of ampicillin are found relatively often in food animals and healthy volunteers, probably secondary to the prior use of the glycopeptide avoparcin as a growth enhancer to feed animals. In the United States, avoparcin was never approved for use in animals and, instead, a high prevalence of VRE is typically observed in hospitalized patients, probably because of the widespread use of vancomycin within the hospital setting.

Some isolates of vancomycin-resistant *E. faecalis* and *E. faecium* show growth only in the presence of vancomycin, thereby called vancomycin-dependent enterococci (VDE) (Fraimow *et al.*, 1994). These strains cannot produce D-Ala-D-Ala-ending precursors because they carry an inactive D-alanyl:D-alanine ligase, but they can continue cell-wall synthesis using products from their *vanA* or *vanB* gene cluster if they are under vancomycin-inducing conditions. Spontaneous revertants (vancomycin resistant but not dependent) arise because of restoration of endogenous D-alanyl:D-alanine ligase or mutations that cause constitutive (rather than inducible) expression of *van* resistance genes.

Linezolid Resistance

In vitro studies have indicated that the presence of a mutation in multiple copies of the 23S rRNA genes in *E. faecalis* and *E. faecium* strains is needed to cause linezolid resistance (Prystowsky *et al.*, 2001; Marshall *et al.*, 2002). Some patients receiving linezolid for infections caused by VRE (*E. faecium*) have failed treatment because of the development of resistance to linezolid (Gonzales *et al.*, 2001). These patients have had infections associated with high bacterial inoculum, such as empyema and liver abscesses, and usually received prolonged courses of linezolid therapy (Gonzales *et al.*, 2001).

MANAGEMENT OF ENTEROCOCCAL INFECTIONS

Most of the infections caused by enterococci in non-immunocompromised patients, such as UTIs, soft-tissue infections and intra-abdominal abscesses, do not require therapy using a regimen with bactericidal activity. Single-drug treatment with ampicillin, penicillin or vancomycin (in case of allergy to β-lactams) is usually sufficient. Ureidopenicillins, such as piperacillin and mezlocillin, with broader spectrum might also be used, especially when mixed infection is suspected. Ticarcillin and carbenicillin are not regarded as active anti-enterococcal agents and should not be prescribed for that purpose. Fluoroquinolones and nitrofurantoin alone are only marginally active against enterococci and have been successfully used only in lower UTIs (Lefort *et al.*, 2000). Erythromycin and rifampin have had little clinical use against enterococci, whereas tetracyclines and chloramphenicol had some use for VRE before the approval of quinupristin–dalfopristin and linezolid. Despite the fact that enterococci are frequently isolated from mixed intra-abdominal infections, most clinical trials with antibiotics without specific anti-enterococcal coverage do not show clinical failure or persistent isolation of these organisms. For this reason, many authorities do not recommend specific anti-enterococcal therapy initially (Gorbach, 1993). However, in some cases, enterococci may be important pathogens and patients with persistent positive cultures in the absence of clinical improvement should be treated with specific anti-enterococcal therapy.

Endocarditis

Therapy of endocarditis and other serious systemic infections has been a challenge to clinicians since the beginning of the antibiotic era. Many patients with enterococcal endocarditis failed even with high doses of penicillin, presumably because a bactericidal effect is required in the treatment of such infections and because cell-wall agents alone are often not bactericidal for enterococci. With the introduction of streptomycin, successful outcome with penicillin–streptomycin therapy was reported and bactericidal synergism between penicillin and streptomycin could be demonstrated by time-kill techniques. The usual choices for the aminoglycoside component of combined therapy are streptomycin and gentamicin. The traditional treatment of enterococcal endocarditis has consisted of penicillin or ampicillin (or vancomycin for patients with β-lactam intolerance) combined with gentamicin or streptomycin administered for 4–6 weeks. (Because of its safety profile and the ability to monitor serum concentrations, gentamicin is preferred by many to streptomycin.) A recent study from Sweden suggested that shorter courses may suffice for some patients (Olaison *et al.*, 2002). Low-dose gentamicin (3 mg/kg/day, divided in three doses) plus high-dose penicillin (20×10^6 U/day divided in six doses) or ampicillin (12 g/day divided in six doses) is the recommended regimen.

Bacteremia

Enterococcal bacteremia without endocarditis and with a known extracardiac source generally responds well to 10–14 days of monotherapy. However, Maki and Agger (1988) recommend 4 weeks of bactericidal therapy for patients in whom an extracardiac source cannot be identified, especially if the infection is community acquired and/or if the patient has known valvular heart disease.

HLR Strains

HLR to either streptomycin or gentamicin abolishes the bactericidal synergism usually seen when these drugs are combined with cell-wall–active antibiotics. HLR to gentamicin predicts lack of synergistic killing with all other generally available aminoglycosides, except streptomycin. For this reason, all high-level gentamicin-resistant strains should be screened for high-level streptomycin resistance. Because the normally present chromosomally encoded aminoglycoside-modifying enzyme AAC(6′) of *E. faecium*, synergism is not observed with the combination of penicillin and tobramycin for this species. Because both identification to species and screening for HLR to tobramycin [for detecting APH(2″)-AAC(6′) or ANT(4′)] would need to be performed, most clinicians avoid using tobramycin for enterococci. Amikacin is also generally avoided because of the need to perform screening for HLR to kanamycin or to perform time-kill synergy studies, which many laboratories are not prepared to do. Some investigators have used 8–10 weeks of continuous intravenous ampicillin or high doses of vancomycin for patients with enterococcal endocarditis caused by isolates with HLR to all aminoglycosides, but the optimal therapy for this condition is uncertain (Eliopoulos, 1993). In some cases, cardiac valve replacement, as an adjunct to medical therapy, may need to be considered if there is an inadequate initial response or relapse.

Multiple Drug-Resistant Isolates

The recognition of isolates resistant to clinically achievable concentrations of penicillin or glycopeptides or both further complicates the therapy of enterococcal infections. For β-lactamase-producing strains of *E. faecalis*, the combination of penicillin and a β-lactamase inhibitor (ampicillin–sulbactam, amoxicillin–clavulanic acid and piperacillin–tazobactam), vancomycin or imipenem is an alternative option.

Although vancomycin may be used for strains that are highly resistant to penicillins or in penicillin-intolerant patients, the optimal therapy for VRE strains is still unclear. When the infecting VRE strain is *E. faecalis*, ampicillin will almost always be active. Even when a vancomycin-resistant *E. faecium* isolate is reported as ampicillin resistant, based on NCCLS breakpoints, if the MIC is ≤64 µg/ml, high-dose ampicillin (20–24 g/day) might still be effective (Murray, 1997), especially if there is an active aminoglycoside to use in combination. If the VRE isolate is motile, the involved species is either *E. gallinarum* or *E. casseliflavus* with the intrinsic *vanC* gene, and the treatment of choice is ampicillin or penicillin. (Glycopeptides should be avoided even if the MICs are within the susceptible range). VRE strains carrying the *vanB* gene cluster typically display low MIC of teicoplanin; however, emergence of resistance to this agent has been described in animal models (Lefort *et al.*, 1999) and in clinical practice (Hayden *et al.*, 1993). If the VRE infection is limited to the lower urinary tract, nitrofurantoin and fosfomycin tromethamine salt appear to be useful options.

Quinupristin–Dalfopristin

A little more than a decade after the first VRE strain was isolated, two antibiotics were approved for the treatment of VRE (*E. faecium*) infections. The first one was quinupristin–dalfopristin (a 30:70 mixture of two streptogramins). Dalfopristin, a streptogramin A, and quinupristin, a streptogramin B, act synergistically to inhibit bacterial protein synthesis, leading to bacteriostatic to modest bactericidal activity against enterococci (eliminated by the expression of the *ermB* gene found in most VRE isolates), except for *E. faecalis*. This enterococcal species is naturally resistant to quinupristin–dalfopristin because of an efflux mechanism (Singh, Weinstock and Murray, 2002). Clinical trials have demonstrated the usefulness of quinupristin–dalfopristin in the treatment of diverse VR *E. faecium* infections (Moellering *et al.*, 1999; Winston *et al.*, 2000); however, side effects, cost and the need for central venous catheter limit its use.

Linezolid

The second agent with Food and Drug Administration (FDA) approval for the treatment of VRE infections is linezolid, the first antibiotic from the oxazolidinone class. Like quinupristin–dalfopristin, the oxazolidinones also inhibit bacterial protein synthesis, but at an early step, binding to the 23S ribosomal RNA of the 50S subunit to prevent the formation of the initiation complex (Shinabarger *et al.*, 1997). Published studies including patients with various infections caused by VRE strains have reported relatively good clinical and microbiological cure rates (Chien, Kucia and Salata, 2000; Leach *et al.*, 2000).

Other Agents

Occasional reports of successful treatment of VRE using chloramphenicol, tetracyclines or a combination of both exist (Lautenbach *et al.*, 1998; Safdar *et al.*, 2002). Even though quinupristin–dalfopristin and linezolid have brought therapeutic alternatives for infections that were very difficult to eradicate, these agents still have negative aspects, including the lack of bactericidal effect, the high rate of side effects and the already reported development of resistance. These considerations highlight the need to continue the search for new active drugs against VRE. In fact, new agents with promising activity against VRE strains are already in advanced phase of clinical development, including lipopeptide–daptomycin (recently approved by the FDA for

complicated skin and skin structure infections due to staphylococci or streptococci), the semisynthetic glycopeptide oritavancin and the glycylcycline tigecycline (a derivative of minocycline).

PREVENTION AND CONTROL

After the description of the first VRE strain in the late 1980s, many outbreaks of VRE have been reported. Initial outbreaks of VRE represented clonal dissemination of a single strain, which could often be abolished with infection control measures (Boyce *et al.*, 1994). Later, VRE infection and colonization were often associated with more than one strain and were more difficult to eliminate (Morris *et al.*, 1995). The capability of VRE to cause widespread dissemination has been demonstrated by the finding of the same strain in different hospitals from the same region and even from different states. In this context the role of health care workers (HCWs), the environment and VRE colonized-patients as reservoirs and agents of transmission must be emphasized. The actual proportion of VRE-colonized patients, known as colonization pressure, is another important factor that influences the rate of new VRE-colonized or -infected patient (Bonten *et al.*, 1998). Since the carriage of VRE (and any enterococcal strain) on the hands of HCW has been implicated as the main form of person-to-person transmission, the use of gloves has been recommended when taking care of a VRE-colonized or -infected patient (CDC, 1995), to be followed by handwashing or antisepsis after glove removal. Inanimate objects from the rooms of VRE-colonized patients are frequently contaminated (especially if the patient has diarrhea), explained by the capacity of enterococci to survive on dry surfaces, facilitating the transmission of the resistant strain to other hospitalized patients. It is worth noting that the recovery of the species *E. gallinarum* and *E. casseliflavus* (which carry the nontransferable *vanC* gene) does not necessarily trigger the application of isolation procedures, although highly vancomycin-resistant strains of these species with the transferable *vanA* or *vanB* genes have been reported (Dutka-Malen *et al.*, 1994).

Because control of methicillin-resistant *S. aureus* (MRSA) within the hospital may help in the control of VRE by reducing the selective pressure of vancomycin use (Herwaldt, 1999), the Hospital Infection Control Practices Advisory Committee (HICPAC) recommends restricted use of vancomycin in several situations, the most important being routine surgical prophylaxis, treatment of single positive blood culture for coagulase-negative staphylococci, prophylaxis for intravascular catheter placement, selective decontamination of the digestive tract, attempts to eradicate MRSA colonization and treatment of *Clostridium difficile* colitis (CDC, 1995). The overall decrease of antibiotic selective pressure may also result in better control of VRE in the hospital setting, as suggested by a study showing that antibiotics with antianaerobic activity were associated with higher density of VRE colonization (Donskey *et al.*, 2000). Some of the HICPAC recommendations for controlling nosocomial spread of VRE include the placement of VRE-colonized or -infected patients in private rooms or cohorting with other subjects who are also VRE positive, use of clean nonsterile gloves and gown (the latter if substantial patient contact is anticipated), strict handwashing with antiseptic soap or a waterless antiseptic agent after glove removal and avoiding sharing inanimate objects between patients. Other important measures should include adequate educational programs for personnel and training of the microbiology laboratory staff in the detection and reporting of isolated VRE strains. In settings such as transplant units and ICUs, more aggressive infection control practices may need to be implemented, which will have greater effect when the first VRE strains are isolated. In practical terms, once a patient becomes colonized or infected with VRE it is very difficult to discontinue the isolation precautions described above. Of note, if the patient is readmitted, precautions should resume. Units in an endemic situation might also consider the use of staff cohorting to decrease HCW's exposure to

VRE-positive patients, detect VRE carriers among HCWs and perform environmental cultures to ensure adequate disinfection of the room.

An interesting approach is to attempt to suppress or eradicate VRE from the GI tract to decrease the risk of bacteremia caused by this microorganism in patients who are particularly susceptible to this event (e.g., colonized neutropenic patients with mucositis). Initially, oral bacitracin and doxycycline were tried without much success (Weinstein *et al.*, 1999), but more recently, ramoplanin, a nonabsorbable glycolipodepsipeptide active against VRE, was shown to have efficacy, in a phase II trial, in decreasing counts of VRE, in stool, although only while subjects were taking the drug (Wong *et al.*, 2001).

REFERENCES

Angeletti, S., Lorino, G., Gherardi, G. *et al.* (2001) Routine molecular identification of enterococci by gene-specific PCR and 16S ribosomal DNA sequencing. *J Clin Microbiol* **39**, 794–797.

Arduino, R. C., Jacques-Palaz, K., Murray, B. E. and Rakita, R. M. (1994) Resistance of *Enterococcus faecium* to neutrophil-mediated phagocytosis. *Infect Immun* **62**, 5587–5594.

Arthur, M. and Courvalin, P. (1993) Genetics and mechanisms of glycopeptide resistance in enterococci. *Antimicrob Agents Chemother* **37**, 1563–1571.

Arthur, M. and Quintiliani, R. Jr (2001) Regulation of VanA- and VanB-type glycopeptide resistance in enterococci. *Antimicrob Agents Chemother* **45**, 375–381.

Arthur, M., Molinas, C. and Courvalin, P. (1992) The VanS-VanR two-component regulatory system controls synthesis of depsipeptide peptidoglycan precursors in *Enterococcus faecium* BM4147. *J Bacteriol* **174**, 2582–2591.

Arthur, M., Molinas, C., Depardieu, F. and Courvalin, P. *et al.* (1993) Characterization of Tn*1546*, a Tn*3*-related transposon conferring glycopeptide resistance by synthesis of depsipeptide peptidoglycan precursors in *Enterococcus faecium* BM4147. *J Bacteriol* **175**, 117–127.

Bonten, M. J., Slaughter, S., Ambergen, A. W. *et al.* (1998) The role of "colonization pressure" in the spread of vancomycin-resistant enterococci: an important infection control variable. *Arch Intern Med* **158**, 1127–1132.

Bouza, E., San Juan, R., Munoz, P. *et al.* (2001) A European perspective on nosocomial urinary tract infections I. Report on the microbiology workload, etiology and antimicrobial susceptibility (ESGNI-003 study). European Study Group on Nosocomial Infections. *Clin Microbiol Infect* **7**, 523–531.

Boyce, J. M., Opal, S. M., Chow, J. W. *et al.* (1994) Outbreak of multidrug-resistant *Enterococcus faecium* with transferable vanB class vancomycin resistance. *J Clin Microbiol* **32**, 1148–1153.

Bozdogan, B. and Leclercq, R. (1999) Effects of genes encoding resistance to streptogramins A and B on the activity of quinupristin–dalfopristin against *Enterococcus faecium*. *Antimicrob Agents Chemother* **43**, 2720–2725.

CDC (1995) Recommendations for preventing the spread of vancomycin resistance. Recommendations of the Hospital Infection Control Practices Advisory Committee (HICPAC). *MMWR Recomm Rep* **44**, 1–13.

CDC (2002a) *Staphylococcus aureus* resistant to vancomycin – United States, 2002. *MMWR Morb Mortal Wkly Rep* **51**, 565–567.

CDC (2002b) Vancomycin-resistant *Staphylococcus aureus* – Pennsylvania, 2002. *MMWR Morb Mortal Wkly Rep* **51**, 902.

Cheng, S., McCleskey, F. K., Gress, M. J. *et al.* (1997) A PCR assay for identification of *Enterococcus faecium*. *J Clin Microbiol* **35**, 1248–1250.

Chien, J. W., Kucia, M. L. and Salata, R. A. (2000) Use of linezolid, an oxazolidinone, in the treatment of multidrug-resistant gram-positive bacterial infections. *Clin Infect Dis* **30**, 146–151.

Chow, J. W. (2000) Aminoglycoside resistance in enterococci. *Clin Infect Dis* **31**, 586–589.

Clewell, D. B. and Weaver, K. E. (1989) Sex pheromones and plasmid transfer in *Enterococcus faecalis*. *Plasmid* **21**, 175–184.

Costa, Y., Galimand, M., Leclercq, R. *et al.* (1993) Characterization of the chromosomal *aac(6')-Ii* gene specific for *Enterococcus faecium*. *Antimicrob Agents Chemother* **37**, 1896–1903.

Coudron, P. E., Markowitz, S. M. and Wong, E. S. (1992) Isolation of a beta-lactamase-producing, aminoglycoside-resistant strain of *Enterococcus faecium*. *Antimicrob Agents Chemother* **36**, 1125–1126.

d'Azevedo, P. A., Dias, C. A., Goncalves, A. L. *et al.* (2001) Evaluation of an automated system for the identification and antimicrobial susceptibility testing of enterococci. *Diagn Microbiol Infect Dis* **40**, 157–161.

Das, I. and Gray, J. (1998) Enterococcal bacteremia in children: a review of seventy-five episodes in a pediatric hospital. *Pediatr Infect Dis J* **17**, 1154–1158.

Devriese, L. A., Vancanneyt, M., Descheemaeker, P. *et al.* (2002) Differentiation and identification of *Enterococcus durans*, *E. hirae* and *E. villorum*. *J Appl Microbiol* **92**, 821–827.

Donskey, C. J., Chowdhry, T. K., Hecker, M. T. *et al.* (2000) Effect of antibiotic therapy on the density of vancomycin-resistant enterococci in the stool of colonized patients. *N Engl J Med* **343**, 1925–1932.

Duh, R. W., Singh, K. V., Malathum, K. and Murray, B. E. (2001) *In vitro* activity of 19 antimicrobial agents against enterococci from healthy subjects and hospitalized patients and use of an *ace* gene probe from *Enterococcus faecalis* for species identification. *Microb Drug Resist* **7**, 39–46.

Durand, M. L., Calderwood, S. B., Weber, D. J. *et al.* (1993) Acute bacterial meningitis in adults. A review of 493 episodes. *N Engl J Med* **328**, 21–28.

Dutka-Malen, S., Blaimont, B., Wauters, G. and Courvalin, P. (1994) Emergence of high-level resistance to glycopeptides in *Enterococcus gallinarum* and *Enterococcus casseliflavus*. *Antimicrob Agents Chemother* **38**, 1675–1677.

Dutka-Malen, S., Molinas, C., Arthur, M. and Courvalin, P. (1992) Sequence of the *vanC* gene of *Enterococcus gallinarum* BM4174 encoding a D-alanine:D-alanine ligase-related protein necessary for vancomycin resistance. *Gene* **112**, 53–58.

Edmond, M. B., Wallace, S. E., McClish, D. K. *et al.* (1999) Nosocomial bloodstream infections in United States hospitals: a three-year analysis. *Clin Infect Dis* **29**, 239–244.

Eliopoulos, G. M. (1993) Aminoglycoside resistant enterococcal endocarditis. *Infect Dis Clin North Am* **7**, 117–133.

Ennahar, S., Cai, Y. and Fujita, Y. (2003) Phylogenetic diversity of lactic acid bacteria associated with paddy rice silage as determined by 16S ribosomal DNA analysis. *Appl Environ Microbiol* **69**, 444–451.

Evers, S. and Courvalin, P. (1996) Regulation of VanB-type vancomycin resistance gene expression by the VanS(B)-VanR(B) two-component regulatory system in *Enterococcus faecalis* V583. *J Bacteriol* **178**, 1302–1309.

Facklam, R. R. and Collins, M. D. (1989) Identification of *Enterococcus* species isolated from human infections by a conventional test scheme. *J Clin Microbiol* **27**, 731–734.

Facklam, R. R., Padula, J. F., Thacker, L. G. *et al.* (1974) Presumptive identification of group A, B, and D streptococci. *Appl Microbiol* **27**, 107–113.

Facklam, R., Hollis, D. and Collins, M. D. (1989) Identification of gram-positive coccal and coccobacillary vancomycin-resistant bacteria. *J Clin Microbiol* **27**, 724–730.

Facklam, R. R., Carvalho, M. G. and Teixeira, L. M. (2002) History, taxonomy, biochemical characteristics, and antibiotic susceptibility testing of enterococci. In *The Enterococci: Pathogenesis, Molecular Biology, and Antibiotic Resistance*. M. S. Gilmore, D. B. Clewell, P. M. Courvalin *et al.* Washington, DC: ASM Press: 1–54.

Ferretti, J. J., Gilmore, K. S. and Courvalin, P. (1986) Nucleotide sequence analysis of the gene specifying the bifunctional 6'-aminoglycoside acetyltransferase 2"-aminoglycoside phosphotransferase enzyme in *Streptococcus faecalis* and identification and cloning of gene regions specifying the two activities. *J Bacteriol* **167**, 631–638.

Fontana, R., Cerini, R., Longoni, P. *et al.* (1983) Identification of a streptococcal penicillin-binding protein that reacts very slowly with penicillin. *J Bacteriol* **155**, 1343–1350.

Fontana, R., Ligozzi, M., Pittaluga, F. and Satta, G. (1996) Intrinsic penicillin resistance in enterococci. *Microb Drug Resist* **2**, 209–213.

Fraimow, H. S., Jungkind, D. L., Lander, D. W. *et al.* (1994) Urinary tract infection with an *Enterococcus faecalis* isolate that requires vancomycin for growth. *Ann Intern Med* **121**, 22–26.

Garcia-Garrote, F., Cercenado, E. and Bouza, E. (2000) Evaluation of a new system, VITEK 2, for identification and antimicrobial susceptibility testing of enterococci. *J Clin Microbiol* **38**, 2108–2111.

Gonzales, R. D., Schreckenberger, P. C., Graham, M. B. *et al.* (2001) Infections due to vancomycin-resistant *Enterococcus faecium* resistant to linezolid. *Lancet* **357**, 1179.

Gorbach, S. L. (1993) Treatment of intra-abdominal infections. *J Antimicrob Chemother* **31 (Suppl. A)**, 67–78.

Gould, K., Ramirez-Ronda, C. H., Holmes, R. K. and Sanford, J. P. (1975) Adherence of bacteria to heart valves *in vitro*. *J Clin Invest* **56**, 1364–1370.

Gutschik, E., Moller, S. and Christensen, N. (1979) Experimental endocarditis in rabbits. 3. Significance of the proteolytic capacity of the infecting strains of *Streptococcus faecalis*. *Acta Pathol Microbiol Scand [B]* **87**, 353–362.

Hamilton-Miller, J. M. and Shah, S. (1999) Identification of clinically isolated vancomycin-resistant enterococci: comparison of API and BBL Crystal systems. *J Med Microbiol* **48**, 695–696.

Hancock, L. E. and Gilmore, M. S. (2002) The capsular polysaccharide of *Enterococcus faecalis* and its relationship to other polysaccharides in the cell wall. *Proc Natl Acad Sci U S A* **99**, 1574–1579.

Handwerger, S. and Skoble, J. (1995) Identification of chromosomal mobile element conferring high-level vancomycin resistance in *Enterococcus faecium*. *Antimicrob Agents Chemother* **39**, 2446–2453.

Hayden, M. K., Trenholme, G. M., Schultz, J. E. and Sahm, D. F. (1993) *In vivo* development of teicoplanin resistance in a VanB *Enterococcus faecium* isolate. *J Infect Dis* **167**, 1224–1227.

Herwaldt, L. A. (1999) Control of methicillin-resistant *Staphylococcus aureus* in the hospital setting. *Am J Med* **106**, 11S–18S; discussion 48S–52S.

Hodel-Christian, S. L. and Murray, B. E. (1991) Characterization of the gentamicin resistance transposon Tn*5281* from *Enterococcus faecalis* and comparison to staphylococcal transposons Tn*4001* and Tn*4031*. *Antimicrob Agents Chemother* **35**, 1147–1152.

Huebner, J., Wang, Y., Krueger, W. A. *et al.* (1999) Isolation and chemical characterization of a capsular polysaccharide antigen shared by clinical isolates of *Enterococcus faecalis* and vancomycin-resistant *Enterococcus faecium*. *Infect Immun* **67**, 1213–1219.

Ike, Y. and Clewell, D. B. (1992) Evidence that the hemolysin/bacteriocin phenotype of *Enterococcus faecalis* subsp. *zymogenes* can be determined by plasmids in different incompatibility groups as well as by the chromosome. *J Bacteriol* **174**, 8172–8177.

Ike, Y., Clewell, D. B., Segarra, R. A. and Gilmore, M. S. (1990) Genetic analysis of the pAD1 hemolysin/bacteriocin determinant in *Enterococcus faecalis*: Tn*917* insertional mutagenesis and cloning. *J Bacteriol* **172**, 155–163.

Jensen, L. B., Hammerum, A. M., Aerestrup, F. M. *et al.* (1998) Occurrence of *satA* and *vgb* genes in streptogramin-resistant *Enterococcus faecium* isolates of animal and human origins in the Netherlands. *Antimicrob Agents Chemother* **42**, 3330–3331.

Kanematsu, E., Deguchi, T., Yasuda, M. *et al.* (1998) Alterations in the GyrA subunit of DNA gyrase and the ParC subunit of DNA topoisomerase IV associated with quinolone resistance in *Enterococcus faecalis*. *Antimicrob Agents Chemother* **42**, 433–435.

Ke, D., Picard, F. J., Martineau, F. *et al.* (1999) Development of a PCR assay for rapid detection of enterococci. *J Clin Microbiol* **37**, 3497–3503.

Knudtson, L. M. and Hartman, P. A. (1992) Routine procedures for isolation and identification of enterococci and fecal streptococci. *Appl Environ Microbiol* **58**, 3027–3031.

Korten, V., Huang, W. M. and Murray, B. E. (1994) Analysis by PCR and direct DNA sequencing of *gyrA* mutations associated with fluoroquinolone resistance in *Enterococcus faecalis*. *Antimicrob Agents Chemother* **38**, 2091–2094.

Lautenbach, E., Schuster, M. G., Bilker, W. B. and Brennan, P. J. (1998) The role of chloramphenicol in the treatment of bloodstream infection due to vancomycin-resistant *Enterococcus*. *Clin Infect Dis* **27**, 1259–1265.

Law-Brown, J. and Meyers, P. R. (2003) *Enterococcus phoeniculicola* sp. nov., a novel member of the enterococci isolated from the uropygial gland of the Red-billed Woodhoopoe, Phoeniculus purpureus. *Int J Syst Evol Microbiol* **53**, 683–685.

Leach, T. S., Schaser, R. J., Hempsall, K. A. *et al.* (2000) Clinical efficacy of linezolid for infections caused by vancomycin-resistant enterococci (VRE) in a compassionate-use program. *Clin Infect Dis* **31**, 224 [abstract, 66].

Leclercq, R., Dutka-Malen, S., Brisson-Noel, A. *et al.* (1992) Resistance of enterococci to aminoglycosides and glycopeptides. *Clin Infect Dis* **15**, 495–501.

Lefort, A., Baptista, M., Fantin, B. *et al.* (1999) Two-step acquisition of resistance to the teicoplanin-gentamicin combination by VanB-type *Enterococcus faecalis in vitro* and in experimental endocarditis. *Antimicrob Agents Chemother* **43**, 476–482.

Lefort, A., Mainardi, J. L., Tod, M. and Lortholary, O. (2000) Antienterococcal antibiotics. *Med Clin North Am* **84**, 1471–1495.

Lewis, C. M. and Zervos, M. J. (1990) Clinical manifestations of enterococcal infection. *Eur J Clin Microbiol Infect Dis* **9**, 111–117.

Ligozzi, M., Bernini, C., Bonora, M. G. *et al.* (2002) Evaluation of the VITEK 2 system for identification and antimicrobial susceptibility testing of medically relevant gram-positive cocci. *J Clin Microbiol* **40**, 1681–1686.

Low, Y. L., Jakubovics, N. S., Flatman, J. C. *et al.* (2003) Manganese-dependent regulation of the endocarditis-associated virulence factor EfaA of *Enterococcus faecalis*. *J Med Microbiol* **52**, 113–119.

Luginbuhl, L. M., Rotbart, H. A., Facklam, R. R. *et al.* (1987) Neonatal enterococcal sepsis: case-control study and description of an outbreak. *Pediatr Infect Dis J* **6**, 1022–1026.

Maki, D. G. and Agger, W. A. (1988) Enterococcal bacteremia: clinical features, the risk of endocarditis, and management. *Medicine (Baltimore)* **67**, 248–269.

Malik, R. K., Montecalvo, M. A., Reale, M. R. *et al.* (1999) Epidemiology and control of vancomycin-resistant enterococci in a regional neonatal intensive care unit. *Pediatr Infect Dis J* **18**, 352–356.

Manero, A. and Blanch, A. R. (2002) Identification of *Enterococcus* spp. based on specific hybridisation with 16S rDNA probes. *J Microbiol Methods* **50**, 115–121.

Marshall, S. H., Donskey, C. J., Hutton-Thomas, R. *et al.* (2002) Gene dosage and linezolid resistance in *Enterococcus faecium* and *Enterococcus faecalis*. *Antimicrob Agents Chemother* **46**, 3334–3336.

Martinez-Murcia, A. J. and Collins, M. D. (1991) *Enterococcus sulfureus*, a new yellow-pigmented *Enterococcus* species. *FEMS Microbiol Lett* **64**, 69–74.

Mathai, D., Jones, R. N. and Pfaller, M. A., SENTRY Participant Group North America (2001) Epidemiology and frequency of resistance among pathogens causing urinary tract infections in 1,510 hospitalized patients: a report from the SENTRY Antimicrobial Surveillance Program (North America). *Diagn Microbiol Infect Dis* **40**, 129–136.

Megran, D. W. (1992) Enterococcal endocarditis. *Clin Infect Dis* **15**, 63–71.

Moellering, R. C., Linden, P. K., Reinhardt, J. *et al.* (1999) The efficacy and safety of quinupristin/dalfopristin for the treatment of infections caused by vancomycin-resistant *Enterococcus faecium*. Synercid Emergency-Use Study Group. *J Antimicrob Chemother* **44**, 251–261.

Montravers, P., Mohler, J., Saint Julien, L. and Carbon, C. (1997) Evidence of the proinflammatory role of *Enterococcus faecalis* in polymicrobial peritonitis in rats. *Infect Immun* **65**, 144–149.

Morris, J. G. Jr, Shay, D. K., Hebden, J. N. *et al.* (1995) Enterococci resistant to multiple antimicrobial agents, including vancomycin. Establishment of endemicity in a University medical center. *Ann Intern Med* **123**, 250–259.

Murray, B. E. (1990) The life and times of the *Enterococcus*. *Clin Microbiol Rev* **3**, 46–65.

Murray, B. E. (1992) Beta-lactamase-producing enterococci. *Antimicrob Agents Chemother* **36**, 2355–2359.

Murray, B. E. (1997) Vancomycin-resistant enterococci. *Am J Med* **102**, 284–293.

Murray, B. E., Singh, K. V., Ross, R. P. *et al.* (1993) Generation of restriction map of *Enterococcus faecalis* OG1 and investigation of growth requirements and regions encoding biosynthetic function. *J Bacteriol* **175**, 5216–5223.

Nallapareddy, S. R., Qin, X., Weinstock, G. M. *et al.* (2000) *Enterococcus faecalis* adhesin, ace, mediates attachment to extracellular matrix proteins collagen type IV and laminin as well as collagen type I. *Infect Immun* **68**, 5218–5224.

Navarro, F. and Courvalin, P. (1994) Analysis of genes encoding D-alanine–D-alanine ligase-related enzymes in *Enterococcus casseliflavus* and *Enterococcus flavescens*. *Antimicrob Agents Chemother* **38**, 1788–1793.

NCCLS (2003) *MIC Testing – Supplemental Tables. NCCLS Document M100-S13 (M7)*. Wayne, PA: National Committee for Clinical Laboratory Standards.

NNIS (2001) National Nosocomial Infections Surveillance (NNIS) System Report, Data Summary from January 1992–June 2001, issued August 2001. *Am J Infect Control* **29**, 404–421.

Noskin, G. A., Peterson, L. R. and Warren, J. R. (1995) *Enterococcus faecium* and *Enterococcus faecalis* bacteremia: acquisition and outcome. *Clin Infect Dis* **20**, 296–301.

Olaison, L. and Schadewitz, K., Swedish Society of Infectious Diseases Quality Assurance Study Group for Endocarditis (2002) Enterococcal endocarditis in Sweden, 1995–1999: can shorter therapy with aminoglycosides be used? *Clin Infect Dis* **34**, 159–166.

Ozawa, Y., Courvalin, P. and Gaiimand, M. (2000) Identification of enterococci at the species level by sequencing of the genes for D-alanine:D-alanine ligases. *Syst Appl Microbiol* **23**, 230–237.

Pallares, R., Pujol, M., Pena, C. *et al.* (1993) Cephalosporins as risk factor for nosocomial *Enterococcus faecalis* bacteremia. A matched case-control study. *Arch Intern Med* **153**, 1581–1586.

Patterson, J. E. and Zervos, M. J. (1990) High-level gentamicin resistance in *Enterococcus*: microbiology, genetic basis, and epidemiology. *Rev Infect Dis* **12**, 644–652.

Patterson, J. E., Sweeney, A. H., Simms, M. *et al.* (1995) An analysis of 110 serious enterococcal infections. Epidemiology, antibiotic susceptibility, and outcome. *Medicine (Baltimore)* **74**, 191–200.

Paulsen, I. T., Banerjei, L., Myers, G. S. *et al.* (2003) Role of mobile DNA in the evolution of vancomycin-resistant *Enterococcus faecalis*. *Science* **299**, 2071–2074.

Poyart, C., Quesnes, G. and Trieu-Cuot, P. (2000) Sequencing the gene encoding manganese-dependent superoxide dismutase for rapid species identification of enterococci. *J Clin Microbiol* **38**, 415–418.

Prystowsky, J., Siddiqui, F., Chosay, J. *et al.* (2001) Resistance to linezolid: characterization of mutations in rRNA and comparison of their occurrences in vancomycin-resistant enterococci. *Antimicrob Agents Chemother* **45**, 2154–2156.

Rakita, R. M., Quan, V. C., Jacques-Palaz, K. *et al.* (2000) Specific antibody promotes opsonization and PMN-mediated killing of phagocytosis-resistant *Enterococcus faecium*. *FEMS Immunol Med Microbiol* **28**, 291–299.

Rasmussen, B. A., Gluzman, Y. and Tally, F. P. (1994) Inhibition of protein synthesis occurring on tetracycline-resistant, TetM-protected ribosomes by a novel class of tetracyclines, the glycylcyclines. *Antimicrob Agents Chemother* **38**, 1658–1660.

Rice, L. B. and Marshall, S. H. (1992) Evidence of incorporation of the chromosomal beta-lactamase gene of *Enterococcus faecalis* CH19 into a transposon derived from staphylococci. *Antimicrob Agents Chemother* **36**, 1843–1846.

Rice, L. B., Calderwood, S. B., Eliopoulos, G. M. *et al.* (1991) Enterococcal endocarditis: a comparison of prosthetic and native valve disease. *Rev Infect Dis* **13**, 1–7.

Roberts, M. C. (1996) Tetracycline resistance determinants: mechanisms of action, regulation of expression, genetic mobility, and distribution. *FEMS Microbiol Rev* **19**, 1–24.

Roberts, M. C., Sutcliffe, J., Courvalin, P. *et al.* (1999) Nomenclature for macrolide and macrolide-lincosamide-streptogramin B resistance determinants. *Antimicrob Agents Chemother* **43**, 2823–2830.

Sader, H. S., Biedenbach, D. and Jones, R. N. (1995) Evaluation of Vitek and API 20S for species identification of enterococci. *Diagn Microbiol Infect Dis* **22**, 315–319.

Safdar, A., Bryan, C. S., Stinson, S. and Saunders, D. E. (2002) Prosthetic valve endocarditis due to vancomycin-resistant *Enterococcus faecium*: treatment with chloramphenicol plus minocycline. *Clin Infect Dis* **34**, E61–E63.

Sandoe, J. A., Witherden, I. R., Au-Yeung, H. K. *et al.* (2002) Enterococcal intravascular catheter-related bloodstream infection: management and outcome of 61 consecutive cases. *J Antimicrob Chemother* **50**, 577–582.

Schleifer, K. H. and Kilpper-Bälz, R. (1984) Transfer of *Streptococcus faecalis* and *Streptococcus faecium* to the genus *Enterococcus* nom. rev. as *Enterococcus faecalis* comb. nov. and *Enterococcus faecium* comb. nov. *Int J Syst Bacteriol* **34**, 31–34.

Shankar, N., Lockatell, C. V., Baghdayan, A. S. *et al.* (2001) Role of *Enterococcus faecalis* surface protein Esp in the pathogenesis of ascending urinary tract infection. *Infect Immun* **69**, 4366–4372.

Sherman, J. M. (1937) The streptococci. *Bacteriol Rev* **1**, 3–97.

Shinabarger, D. L., Marotti, K. R., Murray, R. W. *et al.* (1997) Mechanism of action of oxazolidinones: effects of linezolid and eperezolid on translation reactions. *Antimicrob Agents Chemother* **41**, 2132–2136.

Shlaes, D. M., Levy, J. and Wolinsky, E. (1981) Enterococcal bacteremia without endocarditis. *Arch Intern Med* **141**, 578–581.

Singh, K. V., Coque, T. M., Weinstock, G. M. and Murray, B. E. (1998a) *In vivo* testing of an *Enterococcus faecalis* efaA mutant and use of efaA homologs for species identification. *FEMS Immunol Med Microbiol* **21**, 323–331.

Singh, K. V., Qin, X., Weinstock, G. M. and Murray, B. E. (1998b) Generation and testing of mutants of *Enterococcus faecalis* in a mouse peritonitis model. *J Infect Dis* **178**, 1416–1420.

Singh, K. V., Malathum, K. and Murray, B. E. (2001) Disruption of an *Enterococcus faecium* species-specific gene, a homologue of acquired macrolide resistance genes of staphylococci, is associated with an increase in macrolide susceptibility. *Antimicrob Agents Chemother* **45**, 263–266.

Singh, K. V., Weinstock, G. M. and Murray, B. E. (2002) An *Enterococcus faecalis* ABC homologue (Lsa) is required for the resistance of this species to clindamycin and quinupristin–dalfopristin. *Antimicrob Agents Chemother* **46**, 1845–1850.

Sohn, A. H., Garrett, D. O., Sinkowitz-Cochran, R. L. *et al.* (2001) Prevalence of nosocomial infections in neonatal intensive care unit patients: results from the first national point-prevalence survey. *J Pediatr* **139**, 821–827.

Soltani, M., Beighton, D., Philpott-Howard, J. and Woodford, N. (2000) Mechanisms of resistance to quinupristin–dalfopristin among isolates of *Enterococcus faecium* from animals, raw meat, and hospital patients in Western Europe. *Antimicrob Agents Chemother* **44**, 433–436.

Suppola, J. P., Volin, L., Valtonen, V. V. and Vaara, M. (1996) Overgrowth of *Enterococcus faecium* in the feces of patients with hematologic malignancies. *Clin Infect Dis* **23**, 694–697.

Suppola, J. P., Kuikka, A., Vaara, M. and Valtonen, V. V. (1998) Comparison of risk factors and outcome in patients with *Enterococcus faecalis* vs *Enterococcus faecium* bacteraemia. *Scand J Infect Dis* **30**, 153–157.

Svec, P., Devriese, L. A., Sedlacek, I. *et al.* (2001) *Enterococcus haemoperoxidus* sp. nov. and *Enterococcus moraviensis* sp. nov., isolated from water. *Int J Syst Evol Microbiol* **51**, 1567–1574.

Swenson, J. M., Ferraro, M. J., Sahm, D. F. *et al.* (1992) New vancomycin disk diffusion breakpoints for enterococci. The National Committee for Clinical Laboratory Standards Working Group on Enterococci. *J Clin Microbiol* **30**, 2525–2528.

Teixeira, L. M., Carvalho, M. G., Espinola, M. M. *et al.* (2001) *Enterococcus porcinus* sp. nov. and *Enterococcus ratti* sp. nov., associated with enteric disorders in animals. *Int J Syst Evol Microbiol* **51**, 1737–1743.

Teng, F., Jacques-Palaz, K. D., Weinstock, G. M. and Murray, B. E. (2002) Evidence that the enterococcal polysaccharide antigen gene (*epa*) cluster is widespread in *Enterococcus faecalis* and influences resistance to phagocytic killing of *E. faecalis*. *Infect Immun* **70**, 2010–2015.

Toledo-Arana, A., Valle, J., Solano, C. *et al.* (2001) The enterococcal surface protein, Esp, is involved in *Enterococcus faecalis* biofilm formation. *Appl Environ Microbiol* **67**, 4538–4545.

Tyrrell, G. J., Bethune, R. N., Willey, B. and Low, D. E. (1997) Species identification of enterococci via intergenic ribosomal PCR. *J Clin Microbiol* **35**, 1054–1060.

Tyrrell, G. J., Turnbull, L., Teixeira, L. M. *et al.* (2002) *Enterococcus gilvus* sp. nov. and *Enterococcus pallens* sp. nov. isolated from human clinical specimens. *J Clin Microbiol* **40**, 1140–1145.

Vancanneyt, M., Snauwaert, C., Cleenwerck, I. *et al.* (2001) *Enterococcus villorum* sp. nov., an enteroadherent bacterium associated with diarrhoea in piglets. *Int J Syst Evol Microbiol* **51**, 393–400.

Vincent, S., Knight, R. G., Green, M. *et al.* (1991) Vancomycin susceptibility and identification of motile enterococci. *J Clin Microbiol* **29**, 2335–2337.

Weinstein, M. R., Dedier, H., Brunton, J. *et al.* (1999) Lack of efficacy of oral bacitracin plus doxycycline for the eradication of stool colonization with vancomycin-resistant *Enterococcus faecium*. *Clin Infect Dis* **29**, 361–366.

Wells, L. D. and von Graevenitz, A. (1980) Clinical significance of enterococci in blood cultures from adult patients. *Infection* **8**, 147–151.

Willey, B. M., Jones, R. N., McGeer, A. *et al.* (1999) Practical approach to the identification of clinically relevant *Enterococcus* species. *Diagn Microbiol Infect Dis* **34**, 165–171.

Williams, A. M., Rodrigues, U. M. and Collins, M. D. (1991) Intrageneric relationships of Enterococci as determined by reverse transcriptase sequencing of small-subunit rRNA. *Res Microbiol* **142**, 67–74.

Winston, D. J., Emmanouilides, C., Kroeber, A. *et al.* (2000) Quinupristin/dalfopristin therapy for infections due to vancomycin-resistant *Enterococcus faecium*. *Clin Infect Dis* **30**, 790–797.

Wong, M. T., Kauffman, C. A., Standiford, H. C. *et al.* (2001) Effective suppression of vancomycin-resistant *Enterococcus* species in asymptomatic gastrointestinal carriers by a novel glycolipodepsipeptide, ramoplanin. *Clin Infect Dis* **33**, 1476–1482.

5

Staphylococcus aureus

Sharon Peacock

Nuffield Department of Clinical Laboratory Sciences, The John Radcliffe Hospital, Oxford, UK

DESCRIPTION OF THE ORGANISM

Taxonomic History

Rosenbach provided the first taxonomic description of *Staphylococcus* in 1884 when he divided the genus into *Staphylococcus aureus* and *Staphylococcus albus* (Rosenbach 1884), although Pasteur and Ogston had observed spherical bacteria in abscess pus 4 years earlier (Ogston 1880; Pasteur 1880). The staphylococci and a group of saprophytic, tetrad-forming micrococci were placed into the genus *Micrococcus* by Zopf in 1885 (Zopf 1885) and subsequently separated again by Flügge. Evans, Bradford and Niven (1955) divided facultative anaerobic cocci and obligate aerobes into the genus *Staphylococcus* and *Micrococcus*, respectively, based on the oxidation–fermentation (OF) test for glucose fermentation. A major advance occurred when DNA base composition was compared between *Staphylococcus* and *Micrococcus* (Silvestri and Hill 1965). This demonstrated that micrococci have a G + C content of 63–73 mol%, compared with staphylococci which have a G + C content of DNA of 30–39 mol%, indicating that they are not significantly related. More recent systematic studies have distinguished staphylococci from micrococci and other bacteria by using a range of factors including cell-wall composition (Schleifer and Kandler 1972), cytochromes and menaquinones (Faller, Götz and Schleifer 1980; Collins and Jones 1981), cellular fatty acids and polar lipids (Nahaie *et al.* 1984), DNA–rRNA hybridization (Kilpper, Buhl and Schleifer 1980) and comparative oligonucleotide cataloguing of 16S rRNA (Ludwig *et al.* 1981). The genus *Staphylococcus* now belongs to the broad *Bacillus–Lactobacillus–Streptococcus* cluster consisting of Gram-positive bacteria that have a low G + C content of DNA. The closest relatives of staphylococci are the macrococci (Kloos *et al.* 1998a). They are also related to salinicocci, enterococci, planococci, bacilli and listerias on the basis of partial oligonucleotide sequencing of 16S rRNA and rDNA (Ludwig *et al.* 1985; Stackebrandt *et al.* 1987).

Cell Morphology and Cultural Characteristics

Light Microscopy and Staining Reactions

Gram-stained cells of staphylococci are uniformly Gram positive in young (18–24 h) cultures and appear spherical with an average diameter of 0.5–1.5 μm on light microscopy. Cells divide in more than one plane to form irregular clusters. This is the commonest appearance but is not absolute. The possibility of *S. aureus* should not be discounted when organisms are seen as singles, pairs or other configurations, particularly when being observed on direct stain of clinical material. Cell-wall–defective or –deficient (L-form) cells have been described

(Kagan 1972). These fail to take Gram strain, are osmotically sensitive and are not easily cultured on the usual isolation media. As a result, their frequency and clinical relevance are unknown.

Growth Characteristics, Colony Morphology and Metabolism

Staphylococcus aureus is aerobic, facultatively anaerobic and nonmotile. It grows well in a variety of commercial broth media, including trypticase soy broth, brain–heart infusion broth and tryptose phosphate broth, with or without the addition of blood. Commonly used selective media include mannitol salt agar, lipase salt mannitol agar, Columbia colistin–nalidixic acid (CNA) agar and Baird–Parker agar base supplemented with egg yolk tellurite enrichment. These media inhibit the growth of Gram-negative bacteria but allow the growth of staphylococci and certain other Gram-positive bacteria. Staphylococci produce distinctive colonies on a variety of selective and nonselective agar media, but accurate distinction between *S. aureus* and other staphylococci requires further testing. Colonies of the same strain generally exhibit similar features of size, consistency, edge, profile, luster and pigment, but some strains may produce two or more colony morphotypes.

Small-colony variants (SCVs) of *S. aureus* have been described. These are sometimes identified in association with cases of *S. aureus* infection that prove resistant to antibiotic cure as manifest by persistence or relapse (Proctor *et al.* 1995; Kahl *et al.* 1998; Abele-Horn *et al.* 2000). Colony size is around one-tenth or less of that of wild type, indicative of slow growth. These may be mixed with the wild-type morphology, often initially giving the impression of a mixed specimen, or may be the only morphotype observed. SCVs may be stable or may revert to wild type after serial passage (McNamara and Proctor 2000). Reversion within a single colony may also give rise to the appearance of colony sectoring. SCVs lack pigment, are often auxotrophic for menadione or hemin, have a reduced range of carbohydrate utilization, may fail to express several putative virulence factors and are resistant to gentamicin owing to poor drug uptake (Wilson and Sanders 1976; Proctor and Peters 1998). SCVs have been reported to arise following exposure to gentamicin in broth culture, after uptake of bacteria by endothelial cells *in vitro* (Vesga *et al.* 1996), in experimental animal models and during human infection (Proctor *et al.* 1995; von Eiff *et al.* 1997; Kahl *et al.* 1998; Abele-Horn *et al.* 2000; von Eiff *et al.* 2001a). SCVs also persist within endothelial cells *in vitro* (Balwit *et al.* 1994). It has been proposed that SCVs represent a small subpopulation of organisms that become auxotrophs following the acquisition of a mutation in genes involved in the electron transport chain. However, reversible phenotypic switching may occur between *S. aureus* SCV and the parental phenotype (Massey, Buckling and Peacock 2001; Massey and Peacock 2002). It is possible that switching between gentamicin-resistant SCV and gentamicin-sensitive wild type circumvents the lasting fitness cost of antibiotic resistance associated with permanent genetic mutation.

A rabbit endocarditis model has been used to examine *menD* or *hemB* mutants (resulting in an SCV phenotype and menadione and hemin auxotrophy, respectively). No differences were observed in infectious dose for either mutant compared with wild type. The response to oxacillin therapy was the same for the *hemB* mutant and wild type, but oxacillin therapy failed to reduce bacterial densities of the *menD* mutant in kidney or spleen (Bates *et al.* 2003). A murine model of septic arthritis has also been used to compare a *hemB* mutant with the isogenic wild-type strain. The mutant was more virulent on a per-organism basis compared with wild type (Jonsson *et al.* 2003). This was considered to be due in part to the increase in protease production by the mutant. The similarity between an SCV that arises during human infection and site-directed bacterial mutants with an SCV phenotype is unclear.

Staphylococci are capable of using a variety of carbohydrates as carbon and energy sources. The Embden–Meyerhof–Parnas (EMP) (glycolytic) pathway and the oxidative hexose monophosphate pathway (HMP) are the two central pathways used by staphylococci for glucose metabolism (Strasters and Winkler 1963; Blumenthal 1972). *Staphylococcus aureus* metabolizes glucose mainly by glycolysis and to a limited extent by the HMP (Sivakanesan and Dawes 1980). The major end product of anaerobic glucose metabolism in *S. aureus* is lactate, while only 5–10% of glucose appears as lactate under aerobic conditions. Most staphylococcal species are capable of synthesizing many different amino acids needed for growth. The genetic control of histidine (Kloos and Pattee 1965), isoleucine and valine (Smith and Pattee 1967), lysine (Barnes, Bondi and Fuscaldo 1971), tryptophan (Proctor and Kloos 1970), leucine (Pattee *et al.* 1974) and alanine, threonine, tyrosine and methionine (Schroeder and Pattee 1984; Mahairas *et al.* 1989) biosynthesis has been studied extensively in *S. aureus*. Naturally occurring strains may have mutations in one or more of the amino acid biosynthesis genes, resulting in certain amino acid requirements.

CELLULAR AND SECRETED COMPONENTS

Cellular and Secreted Components

Cell Membrane and Cell Wall

The cell membrane is a typical lipid–protein bilayer composed mainly of phospholipids and proteins. The cytochromes and menaquinones bound to cell membranes are important components of the electron transport system. Proteins isolated from the membranes of *S. aureus* include several penicillin-binding proteins (PBPs), which catalyse terminal reactions of peptidoglycan biosynthesis (Hartman and Tomasz 1984). Iron-regulated cell-membrane proteins are expressed under iron limitation (Domingue, Lambert and Brown 1989). A 42-kDa cell-wall protein that binds human transferrin, the major iron-binding protein in serum, has been detected (Modun, Kendall and Williams 1994). A protein that binds human lactoferrin, an iron-binding protein found in milk, tears, saliva and some other body fluids, has also been identified (Naidu, Andersson and Forsgren 1992).

Peptidoglycan and teichoic acid are the major components of the staphylococcal cell wall. Peptidoglycan makes up 50–60% of the dry weight (Schleifer and Kandler 1972; Schleifer 1983). It is the main structural polymer in the wall and consists of a heteropolymer of glycan chains cross-linked by short peptides. The glycan moiety is made up of alternating β-1,4-linked units of *N*-acetylglucosamine and *N*-acetylmuramic acid. The carboxyl group of muramic acid is substituted by an oligopeptide containing alternating L- and D-amino

acids. These peptide subunits are cross-linked by the insertion of an interpeptide oligoglycine bridge that extends from the COOH-terminal D-alanine in position 4 of one peptide subunit to the ε-amino group of L-lysine in position 3 of an adjacent peptide subunit (Schleifer 1973). Biological activities of peptidoglycans include endotoxin-like properties (pyrogenicity, complement activation, generation of chemotactic factors and aggregation and lysis of animal blood platelets), inflammatory skin reactions, inhibition of leucocyte migration and adjuvant activity (Schleifer 1983). Staphylococcal cell-wall teichoic acid is a water-soluble polymer that is covalently linked to peptidoglycan acid and amounts to about 30–50% dry weight. *Staphylococcus aureus* lipoteichoic acid (LTA) activates immune cells via toll-like receptor 2 (TLR-2), lipopolysaccharide (LPS)-binding protein (LBP) and CD14 (Morath *et al.* 2002; Schroder *et al.* 2003). *Staphylococcus aureus* peptidoglycan is also recognized by TLR-2 (Yoshimura *et al.* 1999; Kyburz *et al.* 2003).

Cell-Surface–Associated and Secreted Adhesins

Adherence to soluble plasma components and/or extracellular matrix proteins is promoted by a range of *S. aureus* cell-wall–associated proteins (Patti *et al.* 1994a) (Table 5.1). These adhesins may play a central role in the colonization of the host and during invasive disease. *In silico* analysis of six *S. aureus* genome sequences has identified 21 genes predicted to encode surface proteins anchored to the peptidoglycan cell wall. The best characterized of these are protein A SpA (Forsgren *et al.* 1983; Uhlén *et al.* 1984; Hartleib *et al.* 2000), two fibronectin-binding proteins FnBPA and FnBPB (Rydén *et al.* 1983; Switalski *et al.* 1983; Signas *et al.* 1989; Jönsson *et al.* 1991; Greene *et al.* 1995), the fibrinogen-binding proteins clumping factors A and B (ClfA and ClfB) (McDevitt *et al.* 1994; Ni Eidhin *et al.* 1998) and a collagen-binding protein Cna (Patti *et al.* 1992). Other proteins include three members of the serine aspartate multigene family (SdrC, SdrD and SdrE) (Josefsson *et al.* 1998), the bone sialoprotein-binding protein Bbp (an allelic variant of SdrE) (Tung *et al.* 2000), the plasmin-sensitive protein Pls (Savolainen *et al.* 2001; Huesca *et al.* 2002) and biofilm-associated protein (Bap) (Cucarella *et al.* 2001, 2002). These proteins have features characteristic of Gram-positive bacterial surface-expressed proteins, including a secretory signal sequence at the N-terminus and a positively charged tail, a hydrophobic transmembrane domain and a wall-spanning region with an LPXTG (Leu-Pro-X-Thr-Gly) motif at the C-terminus. The LPXTG motif represents the target for sortase (SrtA), a membrane-associated enzyme that cleaves the motif between the threonine and glycine residues. The protein is then covalently anchored to the peptidoglycan cell wall (Schneewind, Fowler and Faull 1995; Mazmanian *et al.* 1999; Ton-That *et al.* 1999; Ilangovan *et al.* 2001). A second sortase (SrtB) is required for anchoring a surface protein with an NPQTN motif. The gene encoding SrtB is part of an iron-regulated locus called iron-responsive surface determinants (*isd*), which also contains surface proteins and a ferrichrome transporter with NPQTN and LPXTG motifs (Mazmanian *et al.* 2002). Two further cell-wall proteins (FrpA and FrpB) are expressed under iron-restricted growth conditions. FrpA may be involved in the adhesion of *S. aureus* to plastic *in vitro*, and both are regulated by the ferric uptake regulator Fur (Morrissey *et al.* 2002).

Staphylococcus aureus also expresses bacterial adhesins that lack an LPXTG motif (Table 5.1). These include the major histocompatibility complex class II analogue Map and the highly similar extracellular adhesin protein Eap (McGavin *et al.* 1993; Jönsson *et al.* 1995; Palma, Haggar and Flock 1999), Efb (formally Fib) (Boden and Flock 1994), the elastin-binding protein EbpS (Park *et al.* 1996; Downer *et al.* 2002) and Ehb, which is a very large fibronectin-binding protein that appears on localization studies to be cell envelope associated (Clarke *et al.* 2002).

Table 5.1 *S. aureus* cell-surface–associated and secreted proteins

Bacterial determinant	Putative function	References
Protein A (SpA)	Binds Fc domain of Ig and von Willebrand factor	Forsgren *et al.* (1983), Uhlén *et al.* (1984), Hartleib *et al.* (2000)
Fibronectin-binding proteins A and B (FnBPA and FnBPB)	Adhesins for fibronectin FnBPA recognizes fibrinogen	Rydén *et al.* (1983), Switalski *et al.* (1983), Signas *et al.* (1989), Jönsson *et al.* (1991), Greene *et al.* (1995), Wann, Gurusiddappa and Hook (2000)
Clumping factor A and B (ClfA and ClfB)	Adhesins for fibrinogen ClfA and ClfB mediate platelet activation and aggregation ClfB mediates binding to epidermal cytokeratin	McDevitt *et al.* (1994), Ni Eidhin *et al.* (1998), Siboo *et al.* (2001), O'Brien *et al.* (2002a, 2002b)
Collagen-binding protein (Cna)	Adhesin for collagen	Patti *et al.* (1992)
SdrC, SdrD and SdrE – members of serine–aspartate multigene family	Function of SdrC and SdrD unknown; putative adhesins SdrE mediates platelet activation and aggregation	Josefsson *et al.* (1998)
Bone sialoprotein-binding protein (Bbp)	Adhesin for bone sialoprotein Member of the Sdr family; allelic variant of SdrE	Rydén *et al.* (1989), Tung *et al.* (2000)
Plasmin-sensitive protein (Pls)	Associated with mecA in some strains Large surface protein that may interfere with bacterial interactions with host proteins such as fibronectin and immunoglobulins	Savolainen *et al.* (2001)
Biofilm-associated protein (Bap)	May interfere with bacterial attachment to host tissues and cellular internalization. Present in 5% of bovine strains but absent from 75 human isolates tested	Cucarella *et al.* (2001, 2002)
Elastin-binding protein (EbpS)	Adhesin for elastin	Park *et al.* (1996), Downer *et al.* (2002)
Map/Eap	Map (cell associated) described as MHC class II analogue protein Eap (secreted protein) has a broad binding specificity, involved in adherence to fibroblasts, enhances internalization into eukaryotic cells *in vitro* and serves as antiinflammatory factor by inhibiting the recruitment of host leucocytes	McGavin *et al.* (1993), Jönsson *et al.* (1995), Palma, Haggar and Flock (1999), Flock and Flock (2001), Hussain *et al.* (2001, 2002), Chavakis *et al.* (2002), Haggar *et al.* (2003)
Extracellular matrix-binding protein homologue (Ehb)	Very large cell-wall–associated protein; binds human fibronectin	Clarke *et al.* (2002)

Extracellular Toxins

Staphylococcus aureus produces a group of functionally related pyrogenic toxins that cause fever and shock in its hosts (Table 5.2). These toxins are classified as superantigens (SAgs) and include the staphylococcal enterotoxins (SEs) and toxic shock syndrome toxin-1 (TSST-1), which have, in common, a potent mitogenic activity for T lymphocytes of several host species (Iandolo 1989; Fleischer 1994). Exfoliative toxins are also reported to have superantigenic activity. Positivity of human isolates for the genes encoding enterotoxins varies between toxins. For example, 23%, 8% and 60% of over 300 human isolates tested in one study were positive for *sea*, *seb* and *seg*, respectively (Peacock *et al.* 2002b). TSST-1 causes approximately 75% of all staphylococcal toxic shock syndrome (TSS) cases (Bergdoll *et al.* 1981; Schlievert *et al.* 1981), although about 90% of *S. aureus* isolated from the vagina of patients with TSS produce TSST-1. The organisms associated with non-menstrual TSS are positive for TSST-1 in around half of cases, the remainder being positive for staphylococcal enterotoxin B (SEB) and staphylococcal enterotoxin C (SEC) in 47% and 3%, respectively. The incidence of TSST-1 is much lower (in the region of 25%) in *S. aureus* strains isolated from other types of patients or healthy individuals, and it has not been detected in other staphylococcal species. The *S. aureus* epidermolytic toxins (ETs) known as exfoliatin A (ETA) and exfoliatin B (ETB) are recognized as the cause of staphylococcal scalded skin syndrome (SSSS), a disease associated with severe blistering of the skin, especially in young children (Rogolsky 1979). ETs are produced by a low percentage of *S. aureus* strains (14% and 2% for *eta* and *etb*, respectively, in one study; Peacock *et al.* 2002b). *Staphylococcus aureus* SAgs are encoded by accessory genetic elements (reviewed in McCormick, Yarwood and Schlievert 2001; Novick, Schlievert and Ruzin 2001). The first five to be described (SapI 1–4 and SapI bov) are pathogenicity islands that each encode one or more of the SAgs (Novick, Schlievert and Ruzin 2001). A further six novel pathogenicity islands have been identified in the genome sequences of two *S. aureus* isolates (N315 and Mu50) (Kuroda *et al.* 2001).

Staphylococcus aureus also produces four types of haemolysins known as α-, β-, δ- and γ-toxin (Table 5.2). These attack the membranes of erythrocytes as well as some other cell types from several host species. A high percentage (86–95%) of human *S. aureus* strains produce α-toxin, although many produce only small amounts. α-Toxin-defective mutants have reduced virulence in animal models compared with the wild-type parental strain (Foster *et al.* 1990). β-Toxin has species-dependent activity, with goat, cow and sheep erythrocytes being most sensitive, human erythrocytes having intermediate sensitivity and dog and mouse erythrocytes being resistant (Bohach and Foster 2000). Around 85% of human *S. aureus* strains are positive for β-toxin. δ-Toxin is produced by a high percentage (86–97%) of *S. aureus* strains of human and animal origin. γ-Toxin and Panton–Valentine leukocidin (PVL) are bicomponent toxins encoded by the *hlg* and *luk-PV* loci, respectively (Bohach and Foster 2000). The toxins contain two synergistically acting

Table 5.2 *S. aureus* toxins

Protein	Biological activities	References
Toxic shock syndrome toxin-1 (TSST-1)	Causes toxic shock syndrome Superantigen with mitogenic activity for T cells	Schlievert *et al.* (1981), Kreiswirth *et al.* (1983), Chu *et al.* (1988)
Exfoliative toxin A (ETA) and Exfoliative toxin B (ETB)	Staphylococcal scalded skin syndrome Causes intraepidermal blisters Has superantigen activity	Melish and Glasgow (1970), Rogolsky (1979), Dancer *et al.* (1990), Jackson and Iandolo (1986), O'Toole and Foster (1986)
Enterotoxins SEA-SEE and SEG-SEQ	Ingestion of preformed toxin in food results in nausea, vomiting and diarrhoea	Reviewed by Dinges, Orwin and Schlievert (2000)
α-Toxin	Cytolytic pore-forming toxin	Bhakdi and Tranum-Jensen (1991)
β-Toxin	Neutral sphingomyelinase Acts as a type C phosphatase, hydrolyzing sphingomyelin to phosphorylcholine and ceramide	Doery *et al.* (1965), Smyth, Möllby and Wadström (1975), Coleman *et al.* (1989)
δ-Toxin	Surfactant on various cells: erythrocytes, leucocytes and bacterial protoplasts	Fitton, Dell and Shaw (1980)
γ-Toxin	Bicomponent toxin Cytolytic for erythrocytes and leucocytes Can form active combinations with Panton–Valentine leukocidin components	Fackrell and Wiseman (1976), Cooney *et al.* (1988), Prevost *et al.* (1995)
Panton–Valentine leukocidin	Cytolytic for leucocytes	Noda *et al.* (1980), Rahman *et al.* (1991), Prevost *et al.* (1995)

proteins, one S component (HlgA, HlgC or LukS-PV) and one F component (HlgB or LukF-PV). Any S component may combine with either F component, leading to six combinations in strains that carry both loci (currently reported to be <2% of human isolates) or two combinations in strains positive for *hlg* alone (>95%). γ-Toxin is strongly haemolytic but weakly leukocytic, whereas PVL may lyse neutrophils and macrophages. However, the effect of a given toxin will be modulated depending on the component pairing. For example, Luk-PV paired with HlgA or HlgC results in a toxin that is weakly haemolytic but which has strong leukocytic activity. The recently described community-acquired methicillin-resistant *S. aureus* (CA-MRSA) strains commonly carry the PVL locus (Vandenesch *et al.* 2003). This may result in a rise in the proportion of *S. aureus* positive for PVL.

Other Secreted Extracellular Components

Staphylococcus aureus produces four major extracellular proteases (Karlsson and Arvidson 2002), namely staphylococcal serine protease (V8 protease; SspA), cysteine protease (SspB), metalloprotease (aureolysin; Aur) and staphopain (Scp). These are secreted as proenzymes that are cleaved by proteolysis to become mature enzymes. A survey of 92 clinical isolates demonstrated that protease genes appear to be conserved (Karlsson and Arvidson 2002). Their role in human disease pathogenesis is unknown.

Several other extracellular components are secreted by *S. aureus*, including coagulase, staphylokinase, lipase, urease and hyaluronidase. Coagulase clots plasma in the absence of Ca^{2+} but requires a plasma constituent known as coagulase-reacting factor (CRF). Interaction between the two results in a complex (staphylothrombin) that converts fibrinogen to fibrin (Drummond and Tager 1963). Most *S. aureus* strains (98–99%) exhibit coagulase activity. Staphylokinase binds to plasminogen and activates it to become the fibrinolytic enzyme plasmin. This is produced by a relatively high percentage (60–95%) of human *S. aureus* strains. Hyaluronate lyase (hyaluronidase) is a glycoprotein that hydrolyses the mucopolysaccharide hyaluronate (Rautela and Abramson 1973). Staphylococcal lipases hydrolyse a

wide range of substrates, and a large percentage (95–100%) of strains produce detectable levels of the heat-stable nuclease (or thermonuclease).

Capsule and Biofilm

Staphylococcus aureus strains may be assigned to one of at least 11 capsular serotypes by using polyclonal rabbit antiserum specific for their associated capsular polysaccharides. Human isolates are usually microencapsulated with either serotype 5 or serotype 8 (Arbeit *et al.* 1984; Hochkeppel *et al.* 1987; Albus *et al.* 1988). Capsular polysaccharides have been purified from strains with serotypes 1, 2, 5 and 8, and the biochemical structure has been defined. Readers seeking a review of this area are referred to O'Riordan and Lee (2004).

Staphylococcus aureus produces an extracellular material called slime termed biofilm. The mechanism by which staphylococci attach to prosthetic material and elaborate biofilm is being increasingly understood and is clearly a complex and multistep process (reviewed in Mack 1999; von Eiff, Heilmann and Peters 1999; von Eiff, Peters and Heilmann 2002). One important element in this pathway is the *ica* operon, a gene cluster present in *S. aureus* (and *Staphylococcus epidermidis*) that encodes the production of polysaccharide intercellular adhesin (PIA). This mediates intercellular adherence of bacteria and the accumulation of multilayered biofilm (Heilmann *et al.* 1996). Biofilm production is likely to be relevant to human disease, since it embeds sessile bacteria adherent to prosthetic surfaces, making them more resistant to the effect of antibiotics and inaccessible to the immune response.

Regulation of Gene Expression

Staphylococcus aureus gene expression is under the control of regulatory systems that respond to changes in environmental conditions. As a broad generalization and largely based on experimental data from *in vitro* systems, syntheses of surface proteins occur very early in exponential phase and are downregulated at the transcriptional level later in the exponential phase. This is followed in the post-exponential phase by the transcription of many genes encoding extracellular proteins. These observations may be explained by the coordinated action of several regulatory systems. The best characterized of these is Agr (accessory global regulator) and Sar (staphylococcal accessory regulator), with a third regulator termed Sae (*S. aureus* exoprotein expression). Readers are referred to the large bodies of published work by Richard Novick and colleagues and Ambrose Cheung and colleagues for more in-depth details on Agr and Sar, respectively. Relevant literature on Sae includes Giraudo *et al.* (1994, 1996, 1999, 2003), Giraudo, Cheung and Nagel (1997), Goerke *et al.* (2001), Novick and Jiang (2003) and Steinhuber *et al.* (2003). Regulation *in vivo* is poorly defined overall, but studies addressing this area indicate that regulation does not always mirror that seen under laboratory conditions (Goerke *et al.* 2000).

Other factors may also affect gene expression. For example, FnBP expression is altered *in vitro* on exposure to subinhibitory ciprofloxacin (Bisognano *et al.* 2000). Clinical isolates with grlA and gyrA mutations (encoding altered topoisomerase IV and DNA gyrase, respectively) exhibited significant increases in attachment to fibronectin-coated surfaces after growth in the presence of one-quarter the minimal inhibitory concentration (MIC) of ciprofloxacin. This phenomenon may be explained by the observation that ciprofloxacin activates the *fnbB* promoter (Bisognano *et al.* 2000). It is also associated with the activation of *recA* and derepression of *lexA* (an SOS repressor)-regulated genes (Bisognano *et al.* 2004). A *hemB* mutant defective in hemin biosynthesis and exhibiting an SCV phenotype has also been reported to have increased expression of genes encoding ClfA and fibronectin-binding proteins (Vaudaux *et al.* 2002).

GENETIC EXCHANGE

Antimicrobial Resistance of *S. aureus*

The most significant events in the history of *S. aureus* antibiotic resistance are the emergence of resistance to penicillin (which is now almost ubiquitous), the emergence of and epidemic rise in methicillin resistance, the recognition of strains with intermediate resistance to vancomycin and the recent emergence of *S. aureus* that is fully resistant to vancomycin. The mechanisms of resistance to other antimicrobial agents are summarized in Table 5.3.

Resistance to β-Lactam Antibiotics

Penicillin-resistant *S. aureus* strains emerged in the early 1940s, shortly after the introduction of penicillin into clinical practice. Resistance to methicillin and other β-lactamase-resistant penicillins was likewise observed soon after methicillin was introduced into clinical use in Britain (Jevons 1961). At this time the methicillin-resistant strains isolated in Britain demonstrated heterogeneous resistance to methicillin (i.e. affecting only a minority of the cell population), were multiply antibiotic resistant and were isolated from hospitalized patients (Barber 1961). After the mid-1970s, large outbreaks of infection caused by MRSA were reported in many hospitals in Britain (Shanson, Kensit and Duke 1976; Cookson and Phillips 1988), Ireland (Cafferkey *et al.* 1985), the United States (Schaefler *et al.* 1981) and Australia (Pavillard *et al.* 1982). Many of these early MRSA epidemics were caused by a single epidemic strain that was transferred between hospitals (Duckworth, Lothian and Williams 1988). Since then, many clones of MRSA associated with epidemic spread or sporadic infections have been described throughout the world.

Approximately one-third of serious *S. aureus* infections in the United Kingdom are now caused by MRSA, although the figure varies considerably worldwide. Until recently, MRSA was mostly confined to the hospital setting and MRSA colonization of those discharged from the community rarely persisted long term except when associated with

defects in the skin integrity or the presence of prosthetic material. However, community-acquired MRSA associated with both colonization and infection is being increasingly recognized (Daum *et al.* 2002; Okuma *et al.* 2002). These strains are resistant to fewer non-β-lactam antibiotics than most of the previously defined MRSA strains.

The gene encoding methicillin resistance (*mecA*) is carried by the chromosome of MRSA and methicillin-resistant *S. epidermidis* (MRSE), the mechanism for which is the synthesis of an altered low-affinity PBP termed PBP2a. *mecA* is part of a mobile genetic element termed staphylococcal cassette chromosome mec (SCC*mec*) (Katayama, Ito and Hiramatsu 2000). Four types of SCC*mec* have been defined based on sequence analysis. Comparison of SCC*mec* types I–III has demonstrated substantial differences in both size and nucleotide sequence but shared conserved terminal inverted repeats and direct repeats at the integration junction points, conserved genetic organization around the *mecA* gene and the presence of *ccr* genes, which are responsible for the movement of the island (Ito *et al.* 2001). Evaluation of 38 epidemic MRSA strains isolated in 20 countries showed that the majority possessed one of the three typical SCC*mec* elements on the chromosome (Ito *et al.* 2001). SCC*mec* type IV was subsequently defined in community-acquired MRSA (Daum *et al.* 2002). This element is smaller and lacks non-β-lactam genetic-resistance determinants. Multiple MRSA clones carrying type IV SCC*mec* have been identified in community-acquired MRSA strains in the United States and Australia (Okuma *et al.* 2002). An evaluation of SCC*mec* types (as defined by the PCR) present in 202 MRSA isolates from two hospitals in Florida reported that four isolates failed to give an amplification product, indicating the possible existence of additional types (Chung *et al.* 2004).

Genetic relationships of 254 MRSA isolates recovered between 1961 and 1992 from nine countries on four continents have been analysed using electrophoretic mobility of enzymes (multilocus enzyme electrophoresis) (Musser and Kapur 1992). Fifteen distinctive electrophoretic types (clones) were identified, and the *mec* gene was found in divergent phylogenetic lineages. This was interpreted as evidence that multiple episodes of horizontal transfer and recombination have contributed to the spread of methicillin resistance. MLST has subsequently been used to define the population genetic structure of MRSA and methicillin-susceptible strains (Enright *et al.* 2002). Eleven major MRSA clones were identified within five groups of related genotypes. The major MRSA clones appeared to have arisen repeatedly from successful epidemic MSSA strains, and isolates with decreased susceptibility to vancomycin have arisen from some of these major MRSA clones (Enright *et al.* 2002).

Possible precursors and reservoirs of the genetic determinants of methicillin resistance include *Staphylococcus sciuri* (Wu, de Lencastre and Tomasz 1998) and *Staphylococcus hominis* subsp. *novobiosepticus* (Kloos *et al.* 1998b). A genetic element that is structurally very similar to SCC*mec* except for the absence of *mecA* has been defined in a methicillin-susceptible strain of *S. hominis* (Katayama *et al.* 2003). An SCC element that lacks *mecA* has also been defined during sequencing of *S. aureus* MSSA252 (Holden *et al.* 2004). The role of this element in horizontal transfer of genes encoding antibiotic resistance and other determinants that may endow a biological fitness requires further investigation.

Some *mecA S. aureus* strains exhibit borderline susceptibility to methicillin (oxacillin MIC, 1–2 µg/ml) and other β-lactamase-resistant penicillins (McMurray, Kernodle and Barg 1990; Barg, Chambers and Kernodle 1991). This is partly due to the hyperproduction of β-lactamase type A (McDougal and Thornsberry 1986; Chambers, Archer and Matsuhashi 1989).

Resistance to Vancomycin

The glycopeptide antibiotics vancomycin and teicoplanin prevent the transglycosylation and transpeptidation steps of cell-wall peptidoglycan synthesis by binding to the peptidyl-D-alanyl-D-alanine termini

78

Table 5.3 Antibiotic resistance mechanisms for *S. aureus* (see text for β-lactams and glycopeptides)

Responsible protein	Major activity or mechanism	Gene	Gene location	References
Macrolide, lincosamide and streptogramin B (MLS)				
23S rRNA methylases A–C	Methylases cause a conformational change in the ribosome, resulting in a reduced binding of MLS antibiotics to target site in 50s subunit	*ermA*	Chromosome: on transposon Tn554 which prefers a single *att* site	Murphy (1985a), Murphy, Huwyler and de Freire Bastos Mdo (1985), Thakker-Varia *et al.* (1987), Lim *et al.* (2002), Strommenger *et al.* (2003)
		ermB	Plasmids and chromosome: on transposon Tn551 which can transpose to many sites	Khan and Novick (1980), Luchansky and Pattee (1984)
	Require induction by erythromycin or are made constitutively	*ermC*	Plasmids (mainly class I) > chromosome	Horinouchi and Weisblum (1982), Weisblum (1985)
Macrolide and streptogramin B (MS)				
ATP-dependent efflux protein MsrA	ATP-dependent efflux pump to remove MS antibiotics	*msrA*	Plasmid	Ross *et al.* (1989, 1990), Lina *et al.* (1999)
Streptogramin A				
Acetyltransferase	Inactivation of drug by enzymatic modification	*vatA* *vatB*	Plasmid	Allignet *et al.* (1993, 1996), Allignet, Loncle and el Sohl (1992), Allignet and el Solh (1995), Allignet and El Solh (1997), Haroche *et al.* (2000, 2003), Haroche, Allignet and El Solh (2002), Strommenger *et al.* (2003)
ATP-binding proteins	Mechanism not fully elucidated	*vgaA* *vgaB* *vgaAv* (variant of A)	Plasmid; *vgaAv* is carried by Tn5406 located on plasmid and/or chromosome	
Streptogramin B				
Hydrolase	Hydrolysis of drug	*vgbA*	Plasmid	Allignet *et al.* (1988), Allignet, Liassine and el Solh (1998), Haroche *et al.* (2003)
		vgbB	Located on plasmids conferring resistance to A compounds	
Lincosamide				
3-Lincomycin, 4-clindamycin O-nucleotidyltransferase	Inactivates lincosamides	*linA'*	Plasmid	Brisson-Noël *et al.* (1988), Lina *et al.* (1999)
Tetracycline				
Tetracycline export proteins	Energy-dependent efflux pump removes tetracycline	*tetK* *tetL*	Plasmid	Khan and Novick (1983), Mojumdar and Khan (1988), Guay, Khan and Rothstein (1993), Strommenger *et al.* (2003)
Tetracycline and minocycline				
Ribosomal protection protein		*tetM*	Chromosome: Tn916- and Tn916-like transposons	Nesin *et al.* (1990), Burdett (1993), Strommenger *et al.* (2003)
Aminoglycosides				
Aminoglycoside-modifying enzymes				
Aminoglycoside acetyltransferase (AAC)	Acetylation of an amino group: tobramycin, netilmicin, amikacin and gentamicin	*aac(6')-aph(2")*	Plasmids (mainly class III) and chromosome on transposons (Tn4001 and Tn4001-like, Tn4031, Tn3851)	Townsend, Grubb and Ashdown (1983), Townsend *et al.* (1984), Rouch *et al.* (1987)

Aminoglycoside phosphotransferase (APH)	Phosphorylation of a hydroxyl group: kanamycin, neomycin, paromomycin, amikacin and gentamicin	*aph(3′)-IIIa* (*aphA*)	Plasmids and chromosome	Gray and Fitch (1983), Coleman *et al.* (1985), el Solh, Moreau and Ehrlich (1986)
Aminoglycoside adenylyltransferase (ANT)	Adenylation of a hydroxyl group: tobramycin, amikacin, paromomycin, kanamycin, neomycin, gentamicin and dibekacin	*ant(4′)-Ia* (*aadD*)	Plasmids	Schwotzer, Kayser and Schwotzer (1978), Thomas and Archer (1989)
Aminoglycoside adenyltransferase	Adenylates spectinomycin	*ant(9)-Ia* (*spc*)	Chromosome: on transposon Tn554	Murphy (1985b)
Trimethoprim Dihydrofolate reductase (DHFR)	Single amino acid substitution of *dhfr* gene leads to marked reduced affinity for trimethoprim	*drfB*	Plasmid > chromosome	Coughter, Johnston and Archer (1987), Tennent *et al.* (1988), Rouch *et al.* (1989), Dale, Then and Stuber (1993)
	Acquisition of second gene via plasmid that encodes trimethoprim-resistant DHFR	*dfrA*		
Chloramphenicol Chloramphenicol acetyltransferase (CAT)	Acetylates chloramphenicol via acetyl coenzyme A to yield derivatives that are unable to bind to the ribosome	*cat*	Plasmids (class I)	Shaw (1983), Projan and Novick (1988), Cardoso and Schwarz (1992)
			Chromosome: with cat plasmid integrated	
Fluoroquinolones DNA gyrase subunits GyrA and GyrB	Mutation in genes encoding two subunits of DNA gyrase	*gyrA* *gyrB*	Chromosome	Sreedharan, Peterson and Fisher (1991), Goswitz *et al.* (1992), Margerrison, Hopewell and Fisher (1992), Brockbank and Barth (1993), Ito *et al.* (1994)
DNA topoisomerase IV	Mutation in genes encoding two subunits of DNA gyrase Reduced target enzyme expression (ParE)	*grlA* *grlB* (also termed *parC* and *parE*)	Chromosome	Trucksis, Wolfson and Hooper (1991), Ferrero *et al.* (1994), Ferrero, Cameron and Crouzet (1995), Hooper (2001), Ince and Hooper (2003)
Multidrug efflux protein NorA	Energy-dependent efflux pump removes hydrophilic fluoroquinolones; associated with alterations in gene or increased transcription of norA	*norA* Regulation of norA affected by *flqB*	Chromosome	Ohshita, Hiramatsu and Yokota (1990), Yoshida *et al.* (1990), Kaatz, Seo and Ruble (1993), Ng, Trucksis and Hooper (1994)
Rifampin DNA-dependent RNA polymerase β-subunit	Single base pair change in *rpoB* resulting in an amino acid substitution in the β subunit of RNA polymerase	*rpoB*	Chromosome	Aubry-Damon, Soussy and Courvalin (1998)
Mupirocin Isoleucyl-tRNA synthetase	Low-level mupirocin resistance (MIC 8–256 mg/l) because of mutation in gene for target enzyme	*ileS*	Chromosome Val-to-Phe changes at either residue 588 (V5 88F) or residue 631 (V63IF)	Gilbart, Perry and Slocombe (1993), Hodgson *et al.* (1994), Cookson (1998), Antonio, McFerran and Pallen (2002)
	High-level resistant (MIC > 512 mg/l) mediated by *ileS-2* gene, which encodes an additional isoleucyl-tRNA synthetase	*MupA*	Plasmid	

of peptidoglycan precursors. Glycopeptides are important in clinical practice because they are used to treat severe infections caused by MRSA. Three categories of *S. aureus* resistances to vancomycin have been described since 1997:

1. *S. aureus* with intermediate-level resistance to vancomycin (VISA);
2. *S. aureus* with heteroresistance to vancomycin (hVISA);
3. vancomycin-resistant *S. aureus* (VRSA).

Criteria have been developed by the CDC for the identification of VISA. These are the following:

1. growth within 24 h on commercial brain–heart infusion agar screen plates containing 6 µg/ml of vancomycin;
2. E-test vancomycin MIC >6 µg/ml;
3. broth microdilution MIC of 8–16 µg/ml (Tenover *et al*. 1998).

A variety of screening methods has been described to detect hVISA, but the optimal method is still under debate (Liu and Chambers 2003). VRSA is defined by an MIC >32 µg/ml (National Committee for Clinical Laboratory Standards 2002a).

Staphylococcus aureus with intermediate-level resistance to vancomycin (VISA) was first detected in Japan in 1996 (Hiramatsu *et al*. 1997b). The strain (Mu50) was isolated from an infant who developed a sternal wound infection that was refractory to treatment following surgery to correct a congenital cardiac defect. This strain was reported to have an MIC of 8 mg/l (Hiramatsu *et al*. 1997b). VISA strains have since been identified worldwide, although they are currently rare in clinical practice, and most appear to evolve from MRSA strains in patients who have received prolonged vancomycin treatment. The first hVISA (Mu3) was identified in Japan from the sputum of a patient with MRSA pneumonia following surgery (Hiramatsu *et al*. 1997a). This strain was reported to have an MIC of 3 µg/ml (Hiramatsu *et al*. 1997a). Serial passage of Mu5 in increasing concentrations of vancomycin gave rise to subpopulations with levels of resistance comparable to those of Mu50, and typing showed that Mu3 and Mu50 had the same PFGE pattern (Hiramatsu *et al*. 1997a). hVISA has since been reported from around the world and appears to be more common than VISA (Hiramatsu *et al*. 1997a; Hiramatsu 2001; Liu and Chambers 2003). The first VRSA was reported in Michigan in July 2002, with a second apparently unrelated case in Pennsylvania 2 months later (Centers for Disease Control and Prevention 2002a, 2002b).

The mechanisms of vancomycin resistance for VISA and hVISA are not fully elucidated. These strains do not carry the enterococcal vancomycin-resistance gene *vanA*, *vanB* or *vanC* 1–3 (Hanaki *et al*. 1998a). Findings at the time of writing indicate that changes in the cell wall are important in this process. Strain Mu50 has increased amounts of glutamine-non-amidated muropeptides and decreased cross-linking of peptidoglycan compared with Mu3, together with a decreased dimer-to-monomer ratio of muropeptides (Hanaki *et al*. 1998b). The peptidoglycan of Mu50 binds 1.4 times more vancomycin than that of Mu3 (Hanaki *et al*. 1998b). Both strains produce three to five times the amount of PBPs when compared with vancomycin-susceptible *S. aureus* control strains with or without methicillin resistance, and transmission electron microscopy has shown a doubling in the cell-wall thickness in Mu50 compared with control strains (Hanaki *et al*. 1998a). An increase in the consumption of vancomycin by the cell wall and reduction in the amount of vancomycin reaching the cytoplasmic membrane may both contribute to vancomycin resistance. Microarray transcription analysis of two clinical VISA isolates was compared with two derivatives with MIC of 32 µg/ml after passage in the presence of vancomycin (Mongodin *et al*. 2003). Of 35 genes with increased transcription, 15 involved purine biosynthesis or transport, and the regulator of the major purine biosynthetic operon was mutant (Mongodin *et al*. 2003). These genes may be involved in the metabolic

processes required for the production of the thicker cell wall. The genome sequences of Mu50 and vancomycin-susceptible MRSA strains N315, EMRSA-16 and COL have been compared (Avison *et al*. 2002). Several mutations affecting important cell-wall biosynthesis and intermediary metabolism genes were identified in Mu50.

Vancomycin-resistant enterococci were reported in 1988. Transfer from *Enterococcus faecalis* to *S. aureus* of high-level resistance to both vancomycin and teicoplanin was performed in the laboratory in the early 1990s through the interstrain transfer of *vanA* (Noble, Virani and Cree 1992). This gene encodes a ligase that causes an alteration in the composition of the terminal dipeptide in muramyl pentapeptide cell-wall precursors, leading to decreased binding affinity to glycopeptides (Fraimow and Courvalin 2000). This represents the mechanism of resistance in the two VRSA isolated in the United States that contain the *vanA* gene (Weigel *et al*. 2003; Tenover *et al*. 2004).

Resistance to Other Antibiotics

Table 5.3 summarizes the antibiotic-resistant mechanisms for *S. aureus* to the other major antibiotic groups.

Plasmids and Bacteriophages

Staphylococcus aureus plasmids have been classified into three general classes. Class I plasmids are small (1–5 kbp), have a high copy number (10–55 copies per cell) and are either cryptic or encode a single antibiotic resistance. These plasmids are the most widespread throughout the genus *Staphylococcus*. For example, the pT181 family comprises a group of small (4–4.6 kbp) plasmids that usually encode tetracycline or chloramphenicol resistance, and the pSN2 family plasmids often encode erythromycin resistance. Class II plasmids, commonly called penicillinase or β-lactamase plasmids, are relatively large (15–33 kbp), have a low copy number (4–6 per cell) and carry several combinations of antibiotic and heavy-metal resistance genes, many of which are located on transposons (Shalita, Murphy and Novick 1980; Lyon and Skurray 1987). Class III plasmids appear to be assemblages of transposons and transposon remnants (Gillespie *et al*. 1987).

Most strains of *S. aureus* are multiply lysogenic (Verhoef, Winkler and van Boven 1971; Pulverer, Pillich and Haklová 1976). The temperate phages of *S. aureus* may be subdivided into three main serological groups termed A, B and C. The relevance of bacteriophage is threefold:

1. these are commonly used in experimental transduction;
2. bacterial susceptibility to phage has been used for many years as the basis for a typing scheme;
3. phage may influence the gene complement or gene expression of an organism.

Alterations in gene complement or expression may occur either via carriage of genes into the organism [lysogenic conversion by prophage that carries genes encoding staphylokinase and staphylococcal enterotoxin A (SEA)] or via negative lysogenic conversion in which genes that contain phage-attachment sites within their sequence are disrupted (e.g. *hlb* encoding β-toxin; Coleman *et al*. 1986).

Genetic Elements and the S. aureus Genome Sequence

The genome of *S. aureus* is composed of a single chromosome of around 2.8 Mb, which is predicted to encode approximately 2500 genes. The genome sequence of two related *S. aureus* strains (N315 and Mu50) was published in 2001 (Kuroda *et al*. 2001). N315 is an MRSA strain isolated in Japan in 1982, and Mu50 is an MRSA strain with intermediate-level resistance to vancomycin, isolated in Japan in 1997. It was observed that most of the antibiotic resistance genes were carried either by plasmids or by mobile genetic elements including a unique resistance island and that many putative virulence genes seem to have been acquired by lateral gene transfer. Three classes of new pathogenicity islands were identified: a TSST island family, an exotoxin island and an enterotoxin island. Sequencing of strain MW2 was subsequently undertaken (Baba *et al*. 2002). This strain is a community-acquired MRSA isolated from a 16-month-old girl in the United States with fatal septicaemia and septic arthritis. The genome of this strain was compared with that of N315 and Mu50, including a comparison between the staphylococcal cassette chromosome *mec* (SCC*mec*) that carries the *mec* gene encoding methicillin resistance. Nineteen additional virulence genes were found in the MW2 genome, all but two of which were carried by one of the seven genomic islands of MW2. Sequencing of two further isolates by The Wellcome Trust Sanger Institute, UK, has provided an opportunity to undertake comparative genomics for five genomes (Holden *et al*. 2004). The strains sequenced were a methicillin-susceptible *S. aureus* (MSSA) strain (MSSA476) that is phylogenetically close to MW2 and a hospital-acquired representative of the epidemic methicillin-resistant EMRSA-16 clone (MRSA252) that is phylogenetically divergent from any other sequenced strain. Schematic circular diagrams of the *S. aureus* MRSA252 and MSSA476 genomes are shown in Plate 2. Comparison of the five *S. aureus* whole-genome sequences showed that 166 genes (6%) in the phylogenetically divergent MRSA252 genome were not found in the other published genomes. Around 80% of the genome was highly conserved between strains, the remaining 20% of each consisting of highly variable genetic elements that spread horizontally between strains at high frequency. This is consistent with the findings of a study that used a DNA microarray representing more than 90% coverage of the *S. aureus* genome (Fitzgerald *et al*. 2001). Comparison of the genomes of 36 strains from divergent clonal lineages showed that genetic variation was very extensive, with approximately 22% of the genome being comprised of dispensable genetic material.

Publication of the genome sequence of two further strains (COL and 8325) is imminent, with genome sequence of additional strains reportedly in progress. Comparison of the *S. aureus* genome with the genome sequence of *S. epidermidis* (ATCC 12228) has demonstrated that *S. epidermidis* contains a lower number of putative virulence determinants (Zhang *et al*. 2003).

Sequencing of smaller fragments of the genome in populations of isolates has provided interesting insights into rates of genetic mutation. Examination of the sequence changes at multilocus sequence typing (MLST) loci during clonal diversification has shown that point mutations give rise to new alleles at least 15-fold more frequently than by recombination (Feil *et al*. 2003). This contrasts with the naturally transformable species *Neisseria meningitidis* and *Streptococcus pneumoniae*, in which alleles change between five- and tenfold more frequently by recombination than by mutation.

STAPHYLOCOCCUS AUREUS AND THE HUMAN HOST

Staphylococcus aureus Carriage

Three *S. aureus* carriage patterns have been described in the healthy adult population, with approximately 20% of individuals being persistent *S. aureus* carriers, about 60% intermittent carriers and 20% persistent non-carriers (Kluytmans, van Belkum and Verbrugh 1997). This represents a useful framework during the study of *S. aureus* carriage (Peacock, de Silva and Lowy 2001) and has become accepted and widely quoted for many years. A review of the few longitudinal studies that are to be found in the older published literature suggests that this may be an oversimplification. However, there is sufficient data to support the idea that there is considerable heterogeneity within the population in which some individuals never or hardly ever carry *S. aureus*, others are usually carriers and the remainder carry the organism for variable periods of time. Rates of *S. aureus* carriage are

higher in those infected with human immunodeficiency virus compared with both healthcare workers and patients with a range of chronic diseases (Weinke *et al*. 1992). Carriage rates are also higher in individuals with insulin-dependent diabetes, in patients on continuous ambulatory peritoneal dialysis (CAPD) and haemodialysis and in intravenous drug abusers when compared with the healthy population (Kluytmans, van Belkum and Verbrugh 1997). Carriage rates vary between different ethnic groups (Noble 1974), and a community-based carriage study has reported a family predisposition to nasal carriage of *S. aureus* (Noble, Valkenburg and Wolters 1967). Several small genetic studies have been performed, but their findings are inconclusive (Hoeksma and Winkler 1963; Aly *et al*. 1974; Kinsman, McKenna and Noble 1983).

The nose is the dominant ecological niche for *S. aureus*, as demonstrated by loss of carriage from other sites following nasal decolonization using a topical antibiotic (assuming the absence of prosthetic material, wounds or skin defects). *Staphylococcus aureus* resides in the anterior nares, an area covered by stratified, keratinized, non-ciliated epithelium. Other sites of colonization include the axilla, perineum and hairline. Throat carriage also occurs in some individuals, sometimes in the context of a negative nasal swab.

The cellular and molecular basis of the interaction between bacterium and host that facilitates carriage and host tolerance is poorly understood. It seems likely that bacteria recognize one or more host receptors during attachment and colonization. Epithelial cells express glycoproteins, glycolipids and proteoglycans, and the presence/absence of heterogeneity in one or more of these molecules could be involved. Mucin may enhance adherence since *S. aureus* has been shown to adhere to ferret airway mucus, bovine submaxillary gland mucin and purified human mucin *in vitro* (Sanford, Thomas and Ramsay 1989; Shuter, Hatcher and Lowy 1996). An association has been described between adhesion of *S. aureus* to buccal epithelial cells and expression of the Lewis[a] blood group antigen (Saadi *et al*. 1993). Asialoganglioside 1 has been identified as a receptor for *S. aureus* in a cystic fibrosis bronchial epithelial cell line (Imundo *et al*. 1995). *Staphylococcus aureus* also adheres to glycolipids (fucosylasialo-GM1, asialo-GM1 and asialo-GM2) extracted from lung tissue (Krivan, Roberts and Ginsburg 1988). Adherence of *S. aureus* to gangliosides and asialo-GM1 extracted from corneal epithelial cells has also been described (Schwab *et al*. 1996). Studies are required to examine the role of these and other host receptors in the process of *S. aureus* carriage in the nose. The number or nature of host receptors in the nares may vary, since one study demonstrated significantly greater adherence of *S. aureus* to desquamated epithelial cells from carriers compared with non-carriers (Aly *et al*. 1977). The surface-expressed *S. aureus* protein clumping factor B (ClfB) is a multifunctional adhesin that interacts with epidermal cytokeratins and desquamated nasal epithelial cells *in vitro* (O'Brien *et al*. 2002b). The *S. aureus* surface protein SasG, identified previously by *in silico* analysis of genome sequences, and two homologous proteins, Pls of *S. aureus* and AAP of *S. epidermidis*, also promote bacterial adherence to desquamated nasal epithelial cells *in vitro* (Roche, Meehan and Foster 2003). Variation in host phenotype in the nares has been demonstrated by a study in which nasal secretions from three subjects colonized with *S. aureus* lacked antimicrobial activity against *S. aureus in vitro*, whereas fluid from non-carriers was bactericidal (Cole, Dewan and Ganz 1999). The host may modulate carriage through immune regulation. In a human experimental inoculation study, those susceptible to the smallest initial inoculum were colonized for the greatest period of time (Ehrenkranz 1966). After spontaneous nasal elimination, survival of the same strain on repeated reinoculation was uniformly brief, but duration of colonization of a second strain of *S. aureus* mirrored that of the primary inoculation. These data suggest an acquired immune response.

Colonization resistance caused by bacterial interference between different strains or species is likely to influence the strain of *S. aureus* carried. During the late 1950s it was recognized that colonization of newborn infants with *S. aureus* of the 52/52A/80/81 phage group complex was often followed by disease in infants and their contacts (Shaffer *et al.* 1957) and that the presence of *S. aureus* in the nose before contact with phage type 80/81 strains prevented colonization with this strain. Newborn infants inoculated with the nonpathogenic *S. aureus* strain 502A demonstrated reduction in colonization with a second strain (Shinefield *et al.* 1963), but recognition that 502A could cause both minor and major *S. aureus* disease led to a rapid halt to this practice. The interaction between *S. aureus* and other microbial species has been examined by a small number of studies. The presence of viridans streptococci has been reported to prevent colonization with MRSA in newborns, perhaps because of streptococcal production of H_2O_2 (Uehara *et al.* 2001a, 2001b). A low incidence of *S. aureus* carriage was found in individuals positive on nasal culture for *Corynebacterium* sp., and *S. aureus* carriage was interrupted in 12 of 17 persistent carriers after an average of 12 inoculations of *Corynebacterium* sp. (Uehara *et al.* 2000). The bacterial nasal flora of 216 healthy volunteers was defined to identify possible interactions between different species (Lina *et al.* 2003). The *S. aureus* colonization rate correlated negatively with the rate of colonization by *Corynebacterium* spp. and non-aureus staphylococci, especially *S. epidermidis*, suggesting that both *Corynebacterium* spp. and *S. epidermidis* antagonize *S. aureus* colonization. The effect of other components of the normal nasal flora on the development of *S. aureus* carriage has also been examined in 157 consecutive infants over the first 6 months of life (Peacock *et al.* 2003). Putative bacterial interference between *S. aureus* and other species occurred early in infancy but was not sustained. An increasing anti-staphylococcal effect was observed over time, but this was not attributable to bacterial interference.

Determinants of *S. aureus* Infection

Determinants of infection may be divided into host and bacterial factors. Whether a given individual develops *S. aureus* infection is likely to depend on a complex interplay between the two. Little is currently understood regarding the dominant players.

Host Factors

There is a paucity of studies examining host genetic susceptibility to *S. aureus* carriage or disease. The increased susceptibility for *S. aureus* infection seen in individuals with disorders of the immune system such as leucocyte adhesion deficiency syndromes, Job's syndrome (hyperimmunoglobulin E syndrome) and chronic granulomatous disease (Nauseff and Clark 2000) indicates the importance of host factors. The role of the innate immune system in *S. aureus* disease is undergoing investigation. *Staphylococcus aureus* is recognized by TLR-2. The role of TLR-2 and MyD88 (an adaptor molecule that is essential for TLR family signalling) in *S. aureus* disease has been examined using TLR-2- or MyD88-deficient mice. Both mutations led to an increased susceptibility to *S. aureus* infection (Takeuchi, Hoshino and Akira 2000).

The involvement of host factors may also be inferred from known risk factors for *S. aureus* disease. For example, the onset of septic arthritis after closed joint trauma is well recognized, and there is an increased risk of *S. aureus* septic arthritis in those with inflammatory arthritides such as rheumatoid arthritis. Diabetes mellitus is associated with a higher risk of *S. aureus* infection, but the relative importance of factors such as higher *S. aureus* carriage rate, relative immunodeficiency and repeated needle puncture is uncertain. Immune dysfunction may also be an important contributor to the higher rate of *S. aureus* sepsis in patients undergoing chemotherapy for malignant disease or haemodialysis. The presence of surgical incisions and prosthetic devices that breach the normal host defences is an important risk factor for nosocomial disease. Prosthetic material also

provides a site of colonization. The importance of foreign material as a determinant of infection is demonstrated by the variable rates of sepsis associated with different types of haemodialysis vascular accesses. Endogenous arteriovenous fistulae are less prone to infection than prosthetic shunts (Bonomo *et al.* 1997), and intravenous haemodialysis catheters have the greatest associated risk and are a leading cause of *S. aureus* bacteraemia in patients requiring renal replacement therapy (Hoen *et al.* 1998).

Bacterial Factors

Bacterial determinants of infection have undergone extensive study using *in vitro*, *ex vivo* and animal models. Fibrinogen and fibronectin-binding proteins facilitate adherence of *S. aureus* to host proteins deposited on artificial surfaces (Vaudaux *et al.* 1984, 1989, 1993, 1995; Herrmann *et al.* 1988) and are likely to be important in *S. aureus* device-related infection in humans. Experimental *S. aureus* endocarditis has been used to examine the role of several bacterial adhesins. ClfA appears to be involved in the disease process (Moreillon *et al.* 1995; Que *et al.* 2001), whereas ClfB has a limited effect in this model (Entenza *et al.* 2000). Rats inoculated with a transposon mutant defective in fibronectin binding had fewer bacteria on the heart valve (Kuypers and Proctor 1989). A second study that compared wild-type *S. aureus* strain 8325–4 and an isogenic mutant defective in fibronectin binding did not find a difference between the two (Flock *et al.* 1996). However, the nonpathogenic *Lactococcus lactis* engineered to express FnBPA required a lower infective dose to induce endocarditis (Que *et al.* 2001). Experimental septic arthritis has also been used widely to study disease pathogenesis. ClfA and Cna are determinants of experimental septic arthritis (Patti *et al.* 1994b; Josefsson *et al.* 2001), and Cna has been implicated in osteomyelitis (Elasri *et al.* 2002). FnBP has been shown to be involved in adherence to human airway epithelium (Mongodin *et al.* 2002), although FnBP expression was associated with decreased virulence in a rat model of pneumonia (McElroy *et al.* 2002). A study comparing the presence of 33 putative virulence determinants in a large number of *S. aureus* isolates associated with carriage and disease showed that the genes for three surface proteins (*fnbA, can* and *sdrE*) were more commonly detected in invasive isolates (Peacock *et al.* 2002b).

The fibronectin-binding proteins confer bacterial attachment to and invasion of a range of host cells *in vitro* (Vercellotti *et al.* 1984; Ogawa *et al.* 1985; Dziewanowska *et al.* 1999; Lammers, Nuijten and Smith 1999; Peacock *et al.* 1999; Sinha *et al.* 1999, 2000; Fowler *et al.* 2000; Massey *et al.* 2001). Plate 3 shows *S. aureus* adhering to and being taken up by endothelial cells *in vitro*. Cell invasion is an active process and appears to involve host cell integrins (Sinha *et al.* 1999; Dziewanowska *et al.* 1999; Fowler *et al.* 2000; Massey *et al.* 2001). The relevance of these observations to human disease is currently unknown.

Signature-tagged mutagenesis has been used to identify *S. aureus* genes required for virulence in a murine model of bacteraemia. Gene mutations associated with reduced virulence were of unknown function in half of mutants identified, the remaining genes being mainly involved in nutrient biosynthesis and cell-surface metabolism (Mei *et al.* 1997). Signature-tag transposon mutants have also been screened in four animal infection models (mouse abscess, wound, bacteraemia and rabbit endocarditis) (Coulter *et al.* 1998). The largest gene class identified encoded peptide and amino acid transporters, some of which were important for *S. aureus* survival in all four animal infection models.

CLINICAL SYNDROMES OF *S. AUREUS* DISEASE

Staphylococcus aureus causes a wide range of infections that may be broadly divided into community and hospital acquired.

Community-acquired infections include toxin-mediated diseases such as TSS and food poisoning, infections affecting the skin and soft tissues, infection of bones and joints, infection relating to other deep sites (e.g. endocarditis, abscess formation in liver and spleen) and infection of lung and urinary tract. This wide range of manifestations occurs both in predisposed and in previously healthy individuals. The commonest deep-site infections are endocarditis and bone and joint infection (Fowler *et al.* 2003).

Toxin-Mediated Staphylococcal Diseases

An epidemic of TSS in young women in the early 1980s associated with menstruation and the use of high-absorbancy tampons was caused by TSST-1-producing organisms (Dinges, Orwin and Schlievert 2000). Modification in tampon manufacture resulted in a fall in the number of cases in this group. Non-menstrual TSS is associated with a range of *S. aureus* disease types including invasive *S. aureus* infection. TSS may also occur in association with influenza and in patients with a wound that is colonized by TSST-1-producing organisms, but where clinical features of wound sepsis are absent.

TSS is a severe disease characterized by high fever, hypotension, diffuse erythematous rash with subsequent desquamation 1–2 weeks later and involvement of three or more organ systems (Dinges, Orwin and Schlievert 2000). Clinical features may include mucous membrane hyperaemia, myalgia, vomiting and diarrhoea, renal and hepatic impairment, altered level of consciousness, coagulopathy and low platelet count. The diagnosis is made on clinical grounds, and the Centers for Disease Control and Prevention (CDC) have devised a clinical case definition (Centers for Disease Control and Prevention 1997). Not all patients with clinical features consistent with TSS meet the CDC diagnostic criteria (Dinges, Orwin and Schlievert 2000). The results of the following laboratory tests may be helpful in reaching a diagnosis in such cases (Parsonnet 1998):

1. isolation of *S. aureus* from a mucosal or normally sterile body site;
2. production by this organism of TSST-1 or other SAg known to be associated with TSS;
3. absence of antibodies to the relevant toxin at the time of illness;
4. seroconversion to this toxin with detection of antibodies during convalescence.

SSSS primarily affects neonates and young children. This disease results from the action of *S. aureus* exfoliative toxins on skin epidermidis. Mid-epidermal splitting leads to the development of fragile, thin roofed blisters that rapidly rupture to leave denuded areas (Ladhani *et al.* 1999). SSSS often follows a localized infection in sites such as the umbilical stump, ear or conjunctiva. Presenting clinical features include fever, lethargy, poor feeding and irritability, followed by a tender red rash that spreads to involve the entire body within a few days. Patients have poor temperature control, suffer large fluid losses and may develop secondary infection. Diagnosis is made on the basis of clinical features, including Nikolsky's sign in which the skin wrinkles on gentle pressure. Culture of blister fluid in SSSS is usually negative since the blisters have developed because of toxin formed by *S. aureus* at a distant site. Bullous impetigo has been described as a less severe form of disease with localized skin blistering. Culture of blister fluid is often positive for *S. aureus*. This might indicate that features are caused by local production of bacterial factors.

Staphylococcal food poisoning results from ingestion of preformed SEs (Dinges, Orwin and Schlievert 2000). This occurs when *S. aureus* is inoculated into food which is then left in conditions that are permissive for bacterial multiplication and toxin secretion before consumption. The mechanism of action of the toxin is still under study, but the initiation of the emetic response is thought to be due to interactions with the emetic reflex located in the abdominal viscera, with subsequent activation of the medullary emetic centre in the brain stem that is stimulated via the vagus and sympathetic nerves (Dinges, Orwin and Schlievert 2000). Nausea and vomiting occur after an incubation period of 2 and 6 h. Abdominal pain and diarrhoea are also common features. Diagnosis is made on clinical grounds. Suspected food may be cultured for the presence of *S. aureus*.

Bacteraemia and Endocarditis

Around two-thirds of *S. aureus* bacteraemia cases are attributable to nosocomial sepsis, most of which relate to an intravenous device. The remainder represent community-acquired bacteraemia. Data from the National Nosocomial Infections Surveillance System indicate that rates of *S. aureus* bacteraemia are increasing (Banerjee *et al.* 1991). Rates of *S. aureus* endocarditis have also increased (Sanabria *et al.* 1990; Watanakunakorn and Burkert, 1993). The use of intravenous catheters may have contributed to this rise. These figures pertain to settings with systems of advanced medical care. Rates of nosocomial *S. aureus* sepsis are likely to vary considerably depending on the global setting, although there is a paucity of data from many parts of the world.

Readers are referred to an excellent review on *S. aureus* bacteraemia and endocarditis written by Cathy Petti and Vance Fowler (Petti and Fowler 2003). The general medical condition of cases presenting with *S. aureus* bacteraemia ranges from the febrile but haemodynamically stable patient, to one who is profoundly shocked and has multiorgan failure with respiratory distress and coagulopathy. All patients with *S. aureus* bacteraemia require assessment to distinguish between those with uncomplicated versus complicated bacteraemia. Uncomplicated bacteraemia may be defined as *S. aureus* bacteraemia but no evidence of seeding of bacteria from the bloodstream to other sites (such as heart valves, bone and joint), and complicated bacteraemia is defined as the presence of one or more foci. A study of 724 adult patients presenting to a hospital with *S. aureus* bacteraemia identified complications in 43% (Fowler *et al.* 2003). This distinction is important since the presence of complications will determine the length of treatment and the need for investigations and additional intervention. Clinical assessment should be repeated during admission since metastatic complications of bacteraemia may only become evident some time after the initial bacteraemia. However, clinical examination may be inadequate in identifying the complicated case. For example, the absence of signs of endocarditis does not exclude the diagnosis. A study of *S. aureus* endocarditis over a 10-year period in Denmark reported that the diagnosis was not suspected and was first detected at autopsy in 32% of patients (Roder *et al.* 1999). Underlying cardiac valvular disease or prosthetic heart valves are risk factors for endocarditis in patients with bacteraemia and should increase clinical suspicion. The Duke criteria should be used in the clinical evaluation of patients suspected of having endocarditis (Durack, Lukes and Bright 1994; Bayer *et al.* 1998). Cardiac echocardiography has been used widely in many centres because of the clinical difficulty in identifying patients with endocarditis. The sensitivity of transthoracic and transoesophageal echocardiography is around 60% and 90%, respectively (Chamis *et al.* 1999).

A study of clinical identifiers of complicated *S. aureus* bacteraemia represents significant progress (Fowler *et al.* 2003). A scoring system based on the presence or absence of four risk factors (community acquisition, skin examination findings suggesting acute systemic infection, persistent fever at 72 h and positive follow-up blood culture results at 48–96 h) was able to identify accurately complicated *S. aureus* bacteraemia. The findings of this study allow readily available clinical variables to be used in the identification of patients requiring further investigation and prolonged therapy. Serological testing in its present form is not useful for confirming the diagnosis of *S. aureus* sepsis or for defining complicated bacteraemia or the extent of disease.

Bone and Joint Sepsis

Staphylococcus aureus is the leading cause of primary septic arthritis and osteomyelitis in all ages except neonates (Baker and Schumacher 1993; Lew and Waldvogel 1997). The route of infection is obvious for individuals with conditions that allow direct bacterial access, including open fractures (where the bone is exposed), surgery involving the bone (including joint replacement or internal fracture fixation) and chronic ulceration down to bone or joint (e.g. in the diabetic foot or in a bed-bound patient with a pressure sore). However, many patients have no obvious predisposing factor and do not have identifiable skin lesions or any other primary focus. In such cases, bacteria presumably gain access to bone and joint because of seeding from the bloodstream. Once present in the bone or joint, they multiply, which triggers a host inflammatory response, the features of which are increased blood flow, increased capillary permeability to fluids and cells and the recruitment of inflammatory cells, initially neutrophil polymorphs. If this immune response is unsuccessful in controlling infection, the result is intense inflammation surrounding an area with high densities of bacteria and the development of an abscess. Figure 5.1 shows the appearance, on magnetic resonance imaging, of osteomyelitis of the femur, with oedema in the bone marrow and a large subperiosteal abscess. Secondary bacteraemia is common at this point in the absence of antibiotic therapy. On presentation, fever is common, but once again the overall condition of the patient may be highly variable, and signs and symptoms will depend on the site affected.

Pulmonary Infections

Infection arises because of inhalation of the organism from a site of colonization (community acquired), via an endotracheal tube in ventilated patients or, more rarely, because of haematogenous spread in a bacteraemic patient. *Staphylococcus aureus* infection may involve the lung parenchyma (localized abscess or more diffuse pattern), pleural cavity (empyema) or both. No clinical or radiological features are typical of *S. aureus* pneumonia. Microbiological diagnosis is complicated in that *S. aureus* is a normal commensal of the upper respiratory tract (predominantly the nose) in a proportion of individuals. As a result, interpretation of *S. aureus* in sputum culture is difficult, and more invasive procedures such as bronchoscopy and lavage may be necessary. A syndrome of rapidly progressive *S. aureus* necrotizing pneumonia affecting previously healthy children and young adults has been described (Gillet *et al.* 2002). The strains responsible were positive for PVL, and pneumonia was often preceded by influenza-like symptoms.

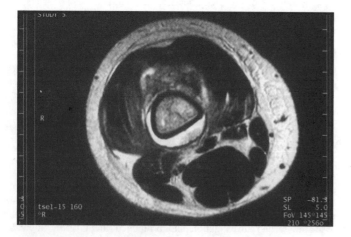

Figure 5.1 Osteomyelitis of the femur. Magnetic resonance imaging demonstrating features indicative of oedema in the bone marrow and a large subperiosteal abscess.

Other Clinical Manifestations

As a generalization, *S. aureus* may infect any organ or region of the body. Additional deep-site infections such as abscess formation in liver, spleen and kidney usually occur in the setting of a bacteraemic patient who develops seeding to multiple sites. Meningitis is an uncommon primary manifestation and usually occurs because of septic embolization associated with endocarditis.

Staphylococcus aureus is a common cause of skin and soft-tissue infections such as boils, impetigo and cellulitis. Primary infection of muscle is unusual in temperate climatic regions but is reported to be common in the tropical setting where paediatric centres may see several cases requiring surgical drainage each week. The reason for this difference is poorly understood. One possible explanation is that parasitic infection with muscle involvement provides a foreign body nidus that predisposes to infection. *Staphylococcus aureus* may cause acute otitis media, although is much less commonly found than other pathogens such as *S. pneumoniae* and *Haemophilus influenzae*. *Staphylococcus aureus* may cause urinary tract infection. Patients with *S. aureus* bacteraemia may also have viable organisms in the urine (Lee, Crossley and Gerding 1978). Whilst the overwhelming majority of patients with *S. aureus* cultured from the urine will not have bacteraemia, it is useful to bear this in mind as a possibility in those patients presenting unwell to hospital.

Nosocomial Infection

The range of nosocomial or hospital-acquired infections includes surgical wound infection, ventilator-associated pneumonia, bacteraemia associated with intravenous devices and infection associated with other types of prosthetic materials such as cerebrospinal fluid (CSF) shunts, prosthetic joints and vascular grafts. However, any disease manifestation may occur in the hospital setting. Investigation will depend on the nature and severity of infection and requires individual assessment for a given patient. The distinction between community and hospital infection is becoming less clear with the rise in home intravenous therapy and the intravenous device-related bacteraemia that occurs in a proportion of cases.

LABORATORY ISOLATION AND IDENTIFICATION OF *S. AUREUS*

This section provides an overview of the laboratory techniques used to isolate and identify *S. aureus*. Readers wishing to undertake this in practice are referred to dedicated texts detailing Standard Operating Procedures.

Direct microscopic examination of normally sterile fluids may provide a rapid, presumptive report of Gram-positive cocci resembling staphylococci. Isolation of *S. aureus* from primary clinical specimens is performed using 5% blood agar following an incubation period of 18–24 h in air at 35–37 °C. Screening for the presence of *S. aureus* in mixed cultures requires the use of selective agar. One example of a medium in common use is mannitol salt agar that is often used for nose swabs. *Staphylococcus aureus* ferments mannitol, resulting in a change in the colour of the medium from pink to yellow. Blood agar containing antibiotics that inhibit Gram-negative organisms should be used for specimens containing mixed flora from sites such as wounds.

Colony morphology may be used by the experienced observer to define presumptive staphylococci. Each discernible staphylococcal morphotype in a given specimen should be analysed. A Gram stain appearance of cocci in clusters and a positive catalase test provide rapid indicators of staphylococci. Distinguishing between *S. aureus* and the remaining staphylococcal species requires further testing. This is commonly performed using two or three simple tests. The coagulase test detects the production of coagulase by *S. aureus*. One colony is mixed with plasma, incubated at 37 °C for 4 h and observed for clot formation. Samples that are negative at 4 h are incubated and observed again for clotting at 24 h. The slide agglutination test detects clumping factor (ClfA). This is performed by making a heavy homogenous suspension of cells in distilled water on a glass slide to which a drop of plasma is added. The mixture should be examined for clumping within about 10 s. The DNase test detects the production of deoxyribonuclease. The test isolate is streaked onto solid agar containing DNA and incubated for 24 h. Hydrochloric acid is added to the surface of the plate that precipitates unhydrolysed DNA to produce a white opacity. A zone of clearing around the bacterial growth indicates the production of deoxyribonuclease that results in the hydrolysis of DNA. Most *S. aureus* isolates are positive for all three tests, but rare strains may be negative for one or more. Furthermore, some strains of coagulase-negative staphylococci are positive for one or more of these tests. For example, about 10–15% of *Staphylococcus intermedius* strains are clumping factor positive, and *Staphylococcus lugdunensis* and *Staphylococcus schleiferi* subsp. *schleiferi* are also positive if human plasma is used in the slide test (Freney *et al.* 1988). Many commercial tests are available that detect one or more of clumping factor, protein A, capsular polysaccharide and group-specific antigen by using latex agglutination (van Griethuysen *et al.* 2001). These are easy to use, reliable and popular in diagnostic laboratories. Other commercial rapid identification tests are available, such as the AccuProbe identification test for *S. aureus* (General-Probe, San Diego, CA, USA) and an immunoenzymatic assay based on a monoclonal antibody prepared against the *S. aureus* endo-β-*N*-acetylglucosaminidase (Guardati *et al.* 1993). These do not replace the inexpensive and reliable tests described above but may find a role in identifying coagulase-negative, protein A-negative, and/or clumping factor-negative mutants of *S. aureus* that would otherwise be misidentified by coagulase and latex agglutination tests. Any persisting uncertainty may usually be resolved using a commercially available rapid identification kit such as API STAPH. There are also a range of fully automated identification systems currently available.

Antibiotic susceptibility testing should be performed using guidelines and Standard Operating Procedures prepared by the National Committee for Clinical Laboratory Standards (NCCLS) or British Society for Antimicrobial Chemotherapy (BSAC). The method used will be strongly influenced by the global setting of the laboratory. Well-funded laboratories often use fully automated systems for identification and/or susceptibility testing. Many laboratories in the United Kingdom continue to use disc susceptibility testing for most antimicrobial agents. Accurate detection of MRSA is an essential service provided by diagnostic laboratories. The oxacillin agar screen test has been a mainstay for the detection of MRSA in diagnostic laboratories for many years. Readers are referred to NCCLS (National Committee for Clinical Laboratory Standards 2002a, 2002b) or BSAC recommendations for a detailed description of methodology for the reference methods (broth microdilution, disk diffusion and oxacillin agar screen).

A wide range of other tests are available for identifying methicillin resistance. Genotypic testing for detecting the presence of *mecA* is performed using the polymerase chain reaction (PCR) (Swenson *et al.* 2001; Grisold *et al.* 2002; Jonas *et al.* 2002; Fang and Hedin 2003). This is not in routine use in most clinical diagnostic laboratories but has become a gold standard for comparing sensitivity and specificity of other tests. Some investigators include PCR detection of a second gene such as *femA*, *femB* or the *S. aureus*-specific marker gene *nuc*, and multiplex PCR may be performed to detect several genes associated with resistance to different antibiotic groups (Perez-Roth *et al.* 2001). PCR methodology may also be used to detect MRSA directly from swabs (Jonas *et al.* 2002), although this involves an enrichment step for *S. aureus* by using immunomagnetic separation (Francois *et al.* 2003). Real-time PCR has also been developed for detecting MRSA (Grisold *et al.* 2002; Fang and Hedin 2003). Gene probe hybridization assays are commercially available, such as the EVIGENE MRSA detection kit that is accurate when valid tests are

produced (Levi and Towner 2003; Poulsen, Skov and Pallesen 2003). Many commercial automated systems are also available. Detection of PBP2a may be performed using a latex agglutination test (e.g. MRSA-Screen test; Horstkotte *et al.* 2001), although this has a lower sensitivity and specificity compared with PCR (Jureen *et al.* 2001; Yamazumi *et al.* 2001). A study using a set of 55 *S. aureus* challenge organisms including several borderline oxacillin-resistant strains that were *mecA* positive compared six routine methods [three reference methods (broth microdilution, disk diffusion and oxacillin agar screen), two commercial automated systems (MicroScan conventional panels and Vitek cards) and the MicroScan rapid panel] and two new rapid methods, Velogene (cycling probe assay) and the MRSA-Screen (latex agglutination). PCR for *mecA* was used as the gold standard (Swenson *et al.* 2001). Results (% sensitivity/% specificity) were reported as follows: broth microdilution, 100/100; Velogene, 100/100; Vitek, 95/97; oxacillin agar screen, 90/92; disk diffusion, 100/89; MicroScan rapid panels, 90/86; MRSA-Screen, 90/100; and MicroScan conventional, 74/97. The MRSA-Screen sensitivity improved to 100% if agglutination reactions were read at 15 min.

Clinical Interpretation of Cultures Positive for *S. aureus*

Isolation of *S. aureus* from a normally sterile site (such as CSF, pus from joints or deep sites obtained by aspiration or during theatre procedures and blood cultures) should be considered clinically significant until clearly proven otherwise. Unlike the case for coagulase-negative staphylococci, *S. aureus* are rare contaminants in good-quality specimens. Patients with invasive community-acquired *S. aureus* disease frequently have a relevant history and clinical signs and symptoms of disease, making the interpretation of a positive culture straightforward. Review of individuals with nosocomial infection should take account of the clinical setting, including recent medical procedures and the presence of intravenous devices and other prosthetic materials. Interpretation of *S. aureus* isolated from samples such as urine and bronchial lavage that have been taken from sites colonized with normal flora may be more difficult and will only approach accuracy if interpreted at the bedside with the full knowledge of the patient's clinical features. Isolation of *S. aureus* in swabs from wounds, ulcers and other skin defects may reflect colonization or infection, and the diagnosis of superficial tissue infection is best judged using clinical criteria. All clinically important positive cultures are best interpreted by a team including primary clinician and microbiologist.

Staphylococcus aureus Typing in the Clinical and Research Setting

Typing techniques performed in diagnostic or reference laboratories for clinical purposes usually aim to examine one of a limited number of situations. A cluster of *S. aureus* cases in a single ward or unit may be investigated to define whether strains are the same (indicating an outbreak) or different. This may be extended to tracking of the spread of strains between units or hospitals and to defining the relatedness of strains giving rise to outbreaks in more than one location. This type of information is important to infection control that aims to prevent further spread of disease. Typing may also be used for isolates cultured from the same patient over time to help define whether a second episode of *S. aureus* infection is due to relapse or reinfection. If isolates from a first and second episode have different genotypes, this is indicative of reinfection. If organisms have the same genotype, this is suggestive of relapse, although it could also represent a second episode of infection with a stable carriage strain. Typing in the research setting is most often applied to address questions of microepidemiology. Other applications include the evaluation of global population genetic structure, genetic evolution, genetic diversity and pathogenicity.

A wide range of typing techniques has been described. These differ in their reproducibility, portability and discriminatory ability, factors that will influence the choice of method for a given situation. The commonest method currently used worldwide is pulsed-field gel electrophoresis (PFGE). This technique is particularly suitable for use in outbreak situations. Problems of reproducibility exist both within and between laboratories, and large collaborative studies have been undertaken in Europe and Canada in an attempt to standardize methodology for the typing of MRSA (Mulvey et al. 2001; Murchan et al. 2003). MLST is a sequence-based technique that is both reproducible and portable. It promises to be useful for the study of S. aureus in a wide range of settings, particularly as sequencing technology becomes more widespread and easily accessible. This technique is based on the principles of multilocus enzyme electrophoresis. Fragments (around 500 bp) of seven unlinked housekeeping genes are amplified by PCR and sequenced. The sequences obtained are assigned allele numbers following comparison of the DNA sequence with results from previously typed strains using the MLST website (http://www.mlst.net). For each isolate, the allele numbers at each of the seven loci defines the allelic profile or sequence type (ST) (Enright et al. 2000). This technique may be used for studies of both micro- and macroepidemiology. Comparison of PFGE and MLST in a microepidemiological setting found them to have very similar discriminatory abilities (Peacock et al. 2002a). MLST is also well suited for the study of S. aureus population genetic structure and has been used to examine and compare isolates from around the world.

Other typing techniques that have been used include ribotyping, plasmid profiling (reviewed by Kloos and Bannerman 1994; Arbeit 1995), random amplified polymorphic DNA (RAPD) (Welsh and McClelland 1991; Marquet-Van Der Mee et al. 1995), PCR amplification and restriction digest of coa (Goh et al. 1992) and sequence analysis of coa (encoding coagulase) (Shopsin et al. 2000) and spa (encoding protein A) (Koreen et al. 2004). The PCR-based methodology applied to multiple-locus variable-number tandem repeat (VNTR) analysis has been applied to MRSA. A scheme using five VNTR loci (sdr, clfA, clfB, ssp and spa) was reported to be equivalent to PFGE in terms of discriminatory power and reproducibility (Sabat et al. 2003). Its utility in clinical practice and in the research setting awaits further validation. Matrix-assisted laser desorption/ionization-time of flight mass spectrometry (MALDI-TOF MS) has been used to examine the bacterial fingerprints of two S. aureus strains over time and to determine whether a profile for MRSA could be defined (Bernardo et al. 2002). Bacterial fingerprints obtained were specific for any given strain, but a uniform signature profile for MRSA could not be identified. Studies that define the utility of this technique in epidemiological studies are ongoing. Although phage typing was the mainstay of typing methodology for many years following its inception in 1952–1953 (Williams and Rippon 1952), this has largely been replaced by newer methods that are more discriminatory and/or reproducible.

TREATMENT OF S. AUREUS DISEASE

Toxin-Mediated Disease

Management of TSS often includes the need for a high-dependency setting. The presence of tampons should be determined and removed where appropriate, and possible sources of infection should be carefully considered in non-menstrual cases. Cultures should be taken for the isolation of S. aureus and the organism sent to a reference laboratory to determine TSST-1 production if the diagnosis is in doubt. Acute and convalescent sera should be taken and stored and tested for evidence of seroconversion to TSST-1 in doubtful cases. The need for other investigations should be guided by clinical need. A β-lactamase-resistant anti-staphylococcal antibiotic should be given to eradicate the toxin-producing strain of S. aureus. Clindamycin has been advocated for use as an adjunct to penicillin for the treatment of streptococcal TSS because of its ability to reduce toxin production in vitro (Sriskandan et al. 1997; Mascini et al. 2001). Evidence for the value of clindamycin in TSS caused by S. aureus is laboratory based. The effect of clindamycin, flucloxacillin or flucloxacillin plus gentamicin has been defined in terms of S. aureus growth and toxin production in vitro. Flucloxacillin alone or in combination with gentamicin was rapidly cidal, but clindamycin was bacteriostatic for logarithmic phase cultures. All three inhibited toxin production during logarithmic growth, with inhibition of 75%, 30% and 75% of the control, respectively, during stationary phase growth (van Langevelde et al. 1997). The relevance of these data to clinical practice is unclear, but many centres commonly give combination therapy using a β-lactamase-resistant anti-staphylococcal antibiotic plus clindamycin. The case fatality rate is reported to be 3%, although this reflects the outcome of individuals treated in affluent nations.

Fluid and electrolyte replacement are the mainstay of the treatment for scalded skin syndrome. Antibiotics are given to eradicate the toxin-producing S. aureus isolate. Fever usually settles after 2 or 3 days, and no new bullae form, although it takes 2–3 weeks for the skin lesions to resolve completely. Affected neonates should be isolated and strict barrier precautions applied to prevent cross-infection. Management of food poisoning requires fluid and electrolyte replacement, but antibiotic treatment is not necessary. The disease is self-limiting and usually resolves within 24 h.

Invasive Disease

Invasive disease includes bacteraemia, endocarditis, bone and joint infection and deep-site abscesses. One or more investigations may be required to delineate the extent of disease. Collections of pus should be drained where technically possible. Infected joints should be washed out by an orthopaedic surgeon in theatre to reduce joint damage and concomitant poor function. Temporary prosthetic material such as intravenous devices should be removed without delay.

Infection caused by methicillin-susceptible strains should be treated with a β-lactamase-resistant penicillin such as flucloxacillin. The current parenteral treatment of choice for methicillin-resistant strains is vancomycin. The choice of antibiotic for empiric treatment of suspected S. aureus, or for culture-proven disease before susceptibility test results becoming available, will depend on the rate of MRSA endemicity within a given unit and/or recent significant exposure to another patient either carrying or infected with MRSA. Strains with reduced susceptibility to vancomycin are currently rare in clinical practice, and treatment of such patients should be guided by a team including clinicians, microbiologists and the hospital infection control officer.

The broad range of clinical manifestations of S. aureus disease makes it impossible to present a suitable treatment regimen and duration in all cases. The optimal duration of treatment for patients with bacteraemia but no evidence of seeding to distant sites or heart valves is unclear. A meta-analysis of 11 studies in which patients were treated with 2 weeks of therapy reported an average late complication rate of 6.1% (Jernigan and Farr 1993). This study has been used to support the use of 2 weeks of intravenous antibiotics for uncomplicated S. aureus bacteraemia. Many centres are keen to treat for shorter periods or to change from intravenous to oral treatment after signs and symptoms have resolved. A switch to oral therapy for patients considered to be at low risk of complicated bacteraemia would reduce the potential problems associated with an extended hospital stay and the risk of (further) intravenous device-related infection. However, efficacy of shorter treatment has not been validated by clinical trial. It is recommended that management follow published guidelines and existing recommendations (Mermel et al. 2001).

Guidelines have been published for the treatment of native or prosthetic valve *S. aureus* endocarditis (Wilson *et al.* 1995; Working Party of the British Society for Antimicrobial Chemotherapy 1998; Horstkotte *et al.* 2004). An overview is given below.

Native Valve Endocarditis

Left-sided native valve *S. aureus* endocarditis requires 4 weeks of parenteral antibiotic therapy. MSSA infection is treated with a β-lactamase-resistant penicillin such as flucloxacillin. Vancomycin is the treatment of choice for patients with MRSA native valve endocarditis or for those with an anaphylactic penicillin allergy, although this is associated with a higher rate of treatment failure (Small and Chambers 1990; Levine, Fromm and Reddy 1991). Evidence for the benefit of combination therapy with gentamicin is limited. One study demonstrated that patients receiving nafcillin plus gentamicin had a more rapid rate of bacteraemia clearance, a higher rate of nephrotoxicity but no difference in mortality compared with those treated with nafcillin alone (Korzeniowski and Sande 1982). Gentamicin is currently added for the first 3–5 or 7 days (the duration of gentamicin differing between US and UK guidelines, respectively). The benefit of adding rifampicin to a penicillinase-resistant penicillin or vancomycin is also uncertain. Individuals with right-sided native valve endocarditis caused by MSSA (usually intravenous drug users) may be suitable for a shorter course of therapy (as little as 2 weeks) (Korzeniowski and Sande 1982; Chambers, Miller and Newman 1988; DiNubile 1994; Ribera *et al.* 1996).

Prosthetic Valve Endocarditis

Treatment requires a combined medical and surgical approach. Recommended antimicrobial therapy is 6 weeks of a penicillinase-resistant penicillin (or vancomycin if methicillin resistant) plus rifampicin and gentamicin for the first 2 weeks. Surgery is often required, and timely valve replacement surgery may be life saving. Mortality is high at 40% (Petti and Fowler 2003).

The decision to undertake valve replacement during the management of patients with both native and prosthetic valve *S. aureus* endocarditis should be taken on an individual patient basis. Clinical indications for surgery include uncontrolled infection, congestive cardiac failure, paravalvular abscess or haemodynamically significant valvular dysfunction (Petti and Fowler 2003). The American Heart Association Committee on infective endocarditis has identified echocardiographic features associated with a potential need for surgery (Bayer *et al.* 1998).

Other Invasive Disease

Infection of other deep sites often requires 4–6 weeks or more of treatment, depending on the nature and extent of infection, and treatment should be guided by experienced medical staff. Microbiologists in many hospitals in the United Kingdom provide bedside consultations and advice, while some hospitals have a more extensive consult service of microbiologists and virologists and general and infectious disease physicians.

There is a lack of consensus about the management of *S. aureus* meningitis (Norgaard *et al.* 2003). The rare community-acquired cases may be treated with high-dose β-lactamase-resistant penicillin assuming that the isolate is methicillin susceptible (Roos and Scheld 1997). Empiric treatment for community-acquired meningitis before culture results often includes the use of ceftriaxone, which is a suitable alternative for MSSA. Community-acquired meningitis caused by MRSA is most unusual given the current low rate of MRSA carriage in the community and usually occurs as a complication of neurosurgery, often in the presence of a ventricular shunt. Entry of vancomycin into the CSF is poor in the absence of meningeal inflammation, and penetration in the presence of inflammation is not well documented in human infection. Vancomycin is used for this infection, but intrathecal instillation may be used in the presence of a shunt or drain, and a second antibiotic is usually added (Pintado *et al.* 2002). This highly specialized area is best managed by experienced medical staff.

Management of Other *S. aureus* Infections

Individuals with mild skin and soft-tissue sepsis are often treated in the community with oral antibiotics. Duration of treatment will depend on the extent of infection but is generally around 1 week. Soft-tissue sepsis such as cellulitis that is severe enough to warrant hospital admission should be treated with a period of parenteral therapy. As a broad generalization, patients in this category who do not have documented bacteraemia may be treated by parenteral therapy until the fever has come under control. Oral antibiotics are then continued until the infection has clinically resolved. There is currently no good test to determine the causative pathogen associated with cellulitis, but most are caused by either *S. aureus* or group A streptococci. Community-acquired *S. aureus* pneumonia is uncommon. Patients presenting to hospital may be extremely unwell and require treatment as for invasive disease. The approach to the treatment of community-acquired urinary tract infection is no different from that associated with other pathogens.

Treatment of patients with nosocomial infection will depend on the nature of the infection. Superficial wound infection is usually treated for about 1 week or until clinical features of infection resolve. Intravenous catheter-related bacteraemia is treated as for other bacteraemias, current recommendations being 2 weeks of therapy. Infected permanent prosthetic material such as vascular grafts may be salvaged by long courses of antibiotics, but this should ideally be undertaken by a unit with specialist experience. Infected joint prostheses may be left *in situ* or removed in a one- or two-stage procedure. Again, such complicated disease should be treated in a unit with an interest in bone and joint infection.

PREVENTION AND CONTROL

Prevention of *S. aureus* disease falls into three main categories:

1. hospital infection control measures to prevent nosocomial infection, including perioperative antibiotic prophylaxis;
2. eradication of *S. aureus* carriage;
3. vaccination strategies.

Hospital Infection Control and Antibiotic Prophylaxis

Hospital infection control measures may prevent a proportion of nosocomial infections. Handwashing plays a central role, reducing transmission of pathogens between individuals and from the hands of a given individual to vulnerable sites such as wounds and dialysis catheters. Perioperative antibiotic prophylaxis is also important in preventing surgical sepsis, together with good skin preparation before surgery and aseptic and surgical techniques. There is a wide range of other possible measures, implementation of which will be dictated by the global setting and healthcare infrastructure. Affluent nations can implement hospital infection control through an infection control team who devise policies, monitor hospital infections from a diagnostic microbiology laboratory or by active ward-based surveillance and implement outbreak procedures where necessary. An infection control manual containing policies detailing the approach to a wide range of issues from antibiotic use, to dealing with patients colonized or infected with MRSA, waste disposal and disinfection is usually produced. Medical staff are educated and actively encouraged to follow guidelines that reduce infection rates in given settings. For example, high-quality care of intravenous devices includes training of staff on line insertion and after care, the use of protocols and good mechanisms for recording line insertion and removal and infective complications. Hospital infection control in resource-poor areas of the world, where even the cheapest of antibiotics are barely affordable, is low on the priority list. The extent to which nosocomial infection occurs in such settings is little studied.

Eradication of *S. aureus* Carriage

Rates of *S. aureus* infection are higher in carriers than in non-carriers in a range of clinical settings (Weinstein 1959; Yu *et al*. 1986; Luzar *et al*. 1990; Weinke *et al*. 1992). This is consistent with the finding that individuals are usually infected with their own carriage isolate (Yu *et al*. 1986; Luzar *et al*. 1990; Nguyen *et al*. 1999; von Eiff *et al*. 2001b). Temporary eradication of carriage has been reported to result in a reduction in nosocomial infection in several patient groups and has been the focus of much recent interest and research. Eradication of *S. aureus* carriage is usually achieved by the topical application of antibiotic to the anterior nose. The most common agent in use is mupirocin, which has also been applied to exit sites of prosthetic devices such as intravenous and peritoneal dialysis catheters. A potentially promising agent currently under trial is lysostaphin, a peptidoglycan hydrolase secreted by *Staphylococcus simulans* that cleaves the polyglycine interpeptide bridges of the *S. aureus* cell wall. This agent is undergoing evaluation in both treatment and carriage eradication trials. Lysostaphin has been shown to have efficacy in the treatment of experimental *S. aureus* keratitis (Dajcs *et al*. 2000) and endocarditis (Climo *et al*. 1998; Patron *et al*. 1999). A single application of 0.5% lysostaphin formulated in a petrolatum-based cream eradicates *S. aureus* nasal colonization in 93% of animals in a cotton rat model. A single dose of lysostaphin cream was more effective than a single dose of mupirocin ointment at eradicating *S. aureus* nasal colonization in this animal model (Kokai-Kun *et al*. 2003). Exposure of *S. aureus* to lysostaphin alone results in the emergence of lysostaphin-resistant mutants *in vitro* and *in vivo* (Climo, Ehlert and Archer 2001), but it is unclear whether this will prove to be a limiting factor in clinical practice. Publication of study data pertaining to human use is awaited.

Many studies have been conducted to examine the effect of temporary carriage eradication, most of which have been in the dialysis population. Eradication of carriage by topical application of mupirocin to the nose in a cohort of haemodialysis patients led to a reduction in the number of episodes of *S. aureus* infection per patient year (Boelaert *et al*. 1989, 1993; Holton *et al*. 1991; Kluytmans *et al*. 1996). Efficacy of nasal *S. aureus* carriage eradication has also been evaluated for patients received peritoneal dialysis, but the evidence for efficacy of nasal mupirocin in preventing morbidity in CAPD patients is not convincing. Topical application of mupirocin to the peritoneal dialysis catheter exit site has been shown to reduce exit-site infection and peritonitis (Bernardini *et al*. 1996; Thodis *et al*. 1998). However, routine application of mupirocin to the exit site of all patients is likely to result in the rapid emergence of drug resistance. A double-blind, placebo-controlled trial of patients undergoing surgery was conducted to assess the effect of mupirocin on rates of postoperative *S. aureus* infection. Mupirocin did not significantly reduce the rate of *S. aureus* surgical-site infections overall, but it did significantly decrease the rate of all nosocomial *S. aureus* infections among patients who were *S. aureus* carriers (Perl *et al*. 2002). An evidence-based review has been performed to examine the efficacy of intranasal mupirocin in the eradication of *S. aureus* nasal carriage and in the prophylaxis of infection. Sixteen randomized, controlled trials were appraised, of which seven assessed the reduction in the rate of infection and nine trials assessed eradication of colonization as a primary outcome measure. Mupirocin was highly effective at eradication of nasal carriage in the short term, although 2–14-day courses did not result in long-term

eradication. This did not convert into clinical benefit overall, although subgroup analyses and several small studies revealed lower rates of *S. aureus* infection among selected populations of patients with nasal carriage treated with mupirocin (Laupland and Conly 2003). The authors concluded that the available literature does not support routine use of topical intranasal mupirocin to prevent subsequent infections, but suggested that there are some patient groups who would benefit, particularly those with acute disease (cardiac surgery and head injury). Further studies are required to clarify this area. A Cochrane review has also been performed to examine the effects of topical and systemic antimicrobial agents on nasal and extranasal MRSA carriage, adverse events and incidence of subsequent MRSA infections. This study concluded that there is insufficient evidence to support the use of topical or systemic antimicrobial therapy for eradicating nasal or extranasal MRSA (Loeb *et al*. 2003).

Vaccination

A considerable body of experimental vaccine-related work has been published, although this has not yet translated into a licensed vaccine preparation. The most significant human anti-staphylococcal vaccine study published to date evaluated a single dose of a conjugate vaccine comprising *S. aureus* types 5 and 8 capsular polysaccharides conjugated to nontoxic recombinant *Pseudomonas aeruginosa* exotoxin A. Evaluation by randomized trial of 1804 adult patients at 73 hemodialysis centres demonstrated partial immunity against *S. aureus* bacteremia for approximately 40 weeks, after which protection waned as antibody levels decreased (Shinefield *et al*. 2002).

Staphylococcal vaccine research may be divided into studies that target bacterial determinants whose genes are always or almost always present (such as capsule, fibronectin-binding protein and clumping factor A) and those that are variably present in populations of clinical isolates (including collagen-binding protein, enterotoxins and PBP2a, the altered PBP encoded by *mecA* and expressed by methicillin-resistant strains).

Staphylococcus aureus synthesizes poly-*N*-succinyl-β-1–6-glucosamine (PNSG) as a surface polysaccharide during human and animal infection, although few strains expressed PNSG *in vitro*. All *S. aureus* strains examined in one study carried genes for PNSG synthesis, and immunization protected mice against death from strains that produced little PNSG *in vitro* (McKenney *et al*. 1999). A considerable body of work has been undertaken to examine fibronectin-binding protein as a potential vaccine target, but a complicated story has emerged. The presence of antibodies to fibronectin-binding protein has been reported to lead to the following:

1. more rapid bacterial clearance from mouse peritoneum and liver after peritoneal inoculation (Rozalska and Wadstrom 1993);
2. a reduction in bacteria count on injured heart valve in an animal model of endocarditis (Schennings *et al*. 1993);
3. lowered bacterial count in rabbit blood and excised organs compared with controls following intravenous inoculation (Park *et al*. 1999).

FnBP vaccination also reduced disease severity in a mouse mastitis model (Mamo *et al*. 1994). However, blocking of the interaction between FnBPs and fibronectin appears to be limited to 50% inhibition *in vitro*. This may be because antibodies raised *in vivo* are predominantly to epitopes that are presented by the ligand–FnBP complexes (i.e. after binding has occurred). These have been termed ligand-induced binding sites. Mapping of immunodominant epitopes has been performed (Rozalska, Sakata and Wadstrom 1994; Sun *et al*. 1997), and current research directions include the generation of antibodies by using recombinant FnBP protein with preserved antigenic activity but reduced ligand-binding activity (Brennan *et al*. 1999; Huesca *et al*. 2000; Rennermalm *et al*. 2001). Vaccination against

ClfA has had mixed results. A DNA vaccine directed against ClfA (injection of plasmids expressing the fibrinogen-binding region A of ClfA) induced a strong and specific antibody response but did not protect mice against an intraperitoneal challenge of *S. aureus* (Brouillette *et al.* 2002). A second group reported that mice immunized with recombinant ClfA and challenged with *S. aureus* developed less severe arthritis than controls and that passive immunization of mice with rat and rabbit anti-ClfA antibodies protected against *S. aureus* arthritis and sepsis-induced death (Josefsson *et al.* 2001).

Passive transfer of collagen adhesin-specific antibodies or vaccination using a recombinant fragment of the *S. aureus* collagen adhesin has been shown to protect mice against sepsis-induced death following inoculation of the Cna-positive strain Phillips (Nilsson *et al.* 1998). Only around half of clinical isolates carry the gene encoding Cna, and therefore, clinical application of a vaccine to Cna is only realistic if this is just one of several targets. Vaccination of mice against PBP2a followed by sublethal dose of MRSA reduced the *S. aureus* counts in kidney compared with non-immunized controls (Ohwada *et al.* 1999; Senna *et al.* 2003). It is unclear whether there is cross-protection against the PBP2 of MSSA, or whether this would only be effective against MRSA, thereby limiting its usefulness.

Toxins have also been evaluated as target vaccine candidates. Immunization of mice with recombinant SEA devoid of superantigenic properties provided good protection against *S. aureus* sepsis (Nilsson *et al.* 1999), and vaccination against SEB provided protection from the effect of subsequent toxin challenge given intraperitoneally or mucosally (Stiles *et al.* 2001). However, many clinical isolates do not express SEA or SEB.

An alternative approach to the identification of *S. aureus* vaccine candidates has been reported, in which a comprehensive list of immunologically relevant proteins is identified. In brief, *S. aureus* proteins are displayed *in vitro* using an expression library and screened using antibodies in sera from patients. Many immunogenic antigens have been identified for further evaluation (Etz *et al.* 2002; Weichhart *et al.* 2003).

REFERENCES

Abele-Horn, M., Schupfner, B., Emmerling, P. *et al.* (2000) Persistent wound infection after herniotomy associated with small-colony variants of *Staphylococcus aureus*. *Infection* **28**, 53–4.

Albus, A., Fournier, J.M., Wolz, C. *et al.* (1988) *Staphylococcus aureus* capsular types and antibody response to lung infection in patients with cystic fibrosis. *Journal of Clinical Microbiology* **26**, 2505–9.

Allignet, J., el Solh, N. (1995) Diversity among the gram-positive acetyltransferases inactivating streptogramin A and structurally related compounds and characterization of a new staphylococcal determinant, *vatB*. *Antimicrobial Agents and Chemotherapy* **39**, 2027–36.

Allignet, J., El Solh, N. (1997) Characterization of a new staphylococcal gene, *vgaB*, encoding a putative ABC transporter conferring resistance to streptogramin A and related compounds. *Gene* **202**, 133–8.

Allignet, J., Loncle, V., Mazodier, P., el Solh, N. (1988) Nucleotide sequence of a staphylococcal plasmid gene, *vgb*, encoding a hydrolase inactivating the B components of virginiamycin-like antibiotics. *Plasmid* **20**, 271–5.

Allignet, J., Loncle, V., el Sohl, N. (1992) Sequence of a staphylococcal plasmid gene, *vga*, encoding a putative ATP-binding protein involved in resistance to virginiamycin A-like antibiotics. *Gene* **117**, 45–51.

Allignet, J., Loncle, V., Simenel, C. *et al.* (1993) Sequence of a staphylococcal gene, *vat*, encoding an acetyltransferase inactivating the A-type compounds of virginiamycin-like antibiotics. *Gene* **130**, 91–8.

Allignet, J., Aubert, S., Morvan, A., el Solh, N. (1996) Distribution of genes encoding resistance to streptogramin A and related compounds among staphylococci resistant to these compounds. *Antimicrobial Agents and Chemotherapy* **40**, 2523–8.

Allignet, J., Liassine, N., el Solh, N. (1998) Characterization of a staphylococcal plasmid related to pUB110 and carrying two novel genes, vatC and vgbB, encoding resistance to streptogramins A and B and similar antibiotics. *Antimicrobial Agents and Chemotherapy* **42**, 1794–8.

Aly, R., Maibach, H.I., Shinefield, H.R., Mandel, A.D. (1974) *Staphylococcus aureus* carriage in twins. *American Journal of Diseases of Children* **127**, 486–8.

Aly, R., Shinefield, H.I., Strauss, W.G., Maibach, H.I. (1977) Bacterial adherence to nasal mucosal cells. *Infection and Immunity* **17**, 546–9.

Antonio, M., McFerran, N., Pallen, M.J. (2002) Mutations affecting the Rossman fold of isoleucyl-tRNA synthetase are correlated with low-level mupirocin resistance in *Staphylococcus aureus*. *Antimicrobial Agents and Chemotherapy* **46**, 438–42.

Arbeit, R.D. (1995) Laboratory procedures for the epidemiologic analysis of microorganisms. In *Manual of Clinical Microbiology*, 6th edn, eds. Murray, P.R., Baron, E.J. *et al.* ASM Press, Washington, DC, 190–208.

Arbeit, R.D., Karakawa, W.W., Vann, W.F., Robbins, J.B. (1984) Predominance of two newly described capsular polysaccharide types among clinical isolates of *Staphylococcus aureus*. *Diagnostic Microbiology and Infectious Disease* **2**, 85–91.

Aubry-Damon, H., Soussy, C.J., Courvalin, P. (1998) Characterization of mutations in the rpoB gene that confer rifampin resistance in *Staphylococcus aureus*. *Antimicrobial Agents and Chemotherapy* **42**, 2590–4.

Avison, M.B., Bennett, P.M., Howe, R.A., Walsh, T.R. (2002) Preliminary analysis of the genetic basis for vancomycin resistance in *Staphylococcus aureus* strain Mu50. *The Journal of Antimicrobial Chemotherapy* **49**, 255–60.

Baba, T., Takeuchi, F., Kuroda, M. *et al.* (2002) Genome and virulence determinants of high virulence community-acquired MRSA. *Lancet* **359**, 1819–27.

Baker, D.G., Schumacher, H.R. Jr (1993) Acute monoarthritis. *The New England Journal of Medicine* **329**, 1013–20.

Balwit, J.M., van Langevelde, P., Vann, J.M., Proctor, R.A. (1994) Gentamicin-resistant menadione and hemin auxotrophic *Staphylococcus aureus* persist within cultured endothelial cells. *The Journal of Infectious Diseases* **170**, 1033–7.

Banerjee, S.N., Emori, T.G., Culver, D.H. *et al.* (1991) Secular trends in nosocomial primary bloodstream infections in the United States, 1980–1989. National Nosocomial Infections Surveillance System. *American Journal of Medicine* **91**, 86S–89S.

Barber, M. (1961) Methicillin-resistant staphylococci. *Journal of Clinical Pathology* **14**, 385–93.

Barg, N., Chambers, H., Kernodle, D. (1991) Borderline susceptibility to antistaphylococcal penicillins is not conferred exclusively by the hyperproduction of β-lactamase. *Antimicrobial Agents and Chemotherapy* **35**, 1975–9.

Barnes, I.J., Bondi, A., Fuscaldo, K.E. (1971) Genetic analysis of lysine auxotrophs of *Staphylococcus aureus*. *Journal of Bacteriology* **105**, 553–5.

Bates, D.M., von Eiff, C., McNamara, P.J. *et al.* (2003) *Staphylococcus aureus menD* and *hemB* mutants are as infective as the parent strains, but the menadione biosynthetic mutant persists within the kidney. *The Journal of Infectious Diseases* **187**, 1654–61.

Bayer, A.S., Bolger, A.F., Taubert, K.A. *et al.* (1998) Diagnosis and management of infective endocarditis and its complications. *Circulation* **98**, 2936–48.

Bergdoll, M.S., Crass, B.A., Reiser, R.F. *et al.* (1981) A new staphylococcal enterotoxin, enterotoxin F, associated with toxic-shock-syndrome *Staphylococcus aureus* isolates. *Lancet* **1**, 1017–21.

Bernardini, J., Piraino, B., Holley, J. *et al.* (1996) A randomized trial of *Staphylococcus aureus* prophylaxis in peritoneal dialysis patients: mupirocin calcium ointment 2% applied to the exit site versus cyclic oral rifampin. *American Journal of Kidney Diseases* **27**, 695–700.

Bernardo, K., Pakulat, N., Macht, M. *et al.* (2002) Identification and discrimination of *Staphylococcus aureus* strains using matrix-assisted laser desorption/ionization-time of flight mass spectrometry. *Proteomics* **2**, 747–53.

Bhakdi, S., Tranum-Jensen, J. (1991) Alpha-toxin of *Staphylococcus aureus*. *Microbiological Reviews* **55**, 733–51.

Bisognano, C., Vaudaux, P., Rohner, P. *et al.* (2000) Induction of fibronectin-binding proteins and increased adhesion of quinolone-resistant *Staphylococcus aureus* by subinhibitory levels of ciprofloxacin. *Antimicrobial Agents and Chemotherapy* **44**, 1428–37.

Bisognano, C., Kelley, W.L., Estoppey, T. *et al.* (2004) A recA-LexA-dependent pathway mediates ciprofloxacin-induced fibronectin binding in *Staphylococcus aureus*. *The Journal of Biological Chemistry* **279**, 9064–71.

Blumenthal, H.J. (1972) Glucose catabolism in staphylococci. In *The Staphylococci*, ed. Cohen, J.O. Wiley, New York, 111–35.

Boden, M.K., Flock, J.I. (1994) Cloning and characterization of a gene for a 19 kDa fibrinogen-binding protein from *Staphylococcus aureus*. *Molecular Microbiology* **12**, 599–606.

Boelaert, J.R., De Smedt, R.A., De Baere, Y.A. *et al.* (1989) The influence of calcium mupirocin nasal ointment on the incidence of *Staphylococcus aureus* infections in haemodialysis patients. *Nephrology, Dialysis, Transplantation* **4**, 278–81.

Boelaert, J.R., Van Landuyt, H.W., Godard, C.A. *et al.* (1993) Nasal mupirocin ointment decreases the incidence of *Staphylococcus aureus* bacteraemias in haemodialysis patients. *Nephrology, Dialysis, Transplantation* **8**, 235–9.

Bohach, G.A., Foster, T.J. (2000) *Staphylococcus aureus* exotoxins. In *Gram-Positive Pathogens*, eds. Fischetti, V.A., Novick, R.P., Ferretti, J.J., Portnoy, D.A., Rood, J.I. ASM Press, Washington, DC, 367–91.

Bonomo, R.A., Rice, D., Whalen, C. *et al.* (1997) Risk factors associated with permanent access-site infections in chronic hemodialysis patients. *Infection Control and Hospital Epidemiology* **18**, 757–61.

Brennan, F.R., Jones, T.D., Longstaff, M. *et al.* (1999) Immunogenicity of peptides derived from a fibronectin-binding protein of *S. aureus* expressed on two different plant viruses. *Vaccine* **17**, 1846–57.

Brisson-Noël, A., Delrieu, P., Samain, D., Courvalin, P. (1988) Inactivation of lincosaminide antibiotics in Staphylococcus. Identification of lincosaminide *O*-nucleotidyltransferases and comparison of the corresponding resistance genes. *The Journal of Biological Chemistry* **263**, 15880–7.

Brockbank, S.M., Barth, P.T. (1993) Cloning, sequencing, and expression of the DNA gyrase genes from *Staphylococcus aureus*. *Journal of Bacteriology* **175**, 3269–77.

Brouillette, E., Lacasse, P., Shkreta, L. *et al.* (2002) DNA immunization against the clumping factor A (ClfA) of *Staphylococcus aureus*. *Vaccine* **20**, 2348–57.

Burdett, V. (1993) tRNA modification activity is necessary for Tet(M)-mediated tetracycline resistance. *Journal of Bacteriology* **175**, 7209–15.

Cafferkey, M.T., Hone, R., Coleman, D. *et al.* (1985) Methicillin-resistant *Staphylococcus aureus* in Dublin 1971–84. *Lancet* **2**, 705–8.

Cardoso, M., Schwarz, S. (1992) Nucleotide sequence and structural relationships of a chloramphenicol acetyltransferase encoded by the plasmid pSCS6 from *Staphylococcus aureus*. *Journal of Applied Bacteriology* **72**, 289–93.

Centers for Disease Control and Prevention (1997) Case definitions for infectious conditions under public health surveillance. *MMWR. Recommendations and Reports* **46**(RR-10), 1–55.

Centers for Disease Control and Prevention (2002a) *Staphylococcus aureus* resistant to vancomycin—United States, 2002. *MMWR.* **51**, 565–7.

Centers for Disease Control and Prevention (2000b) Vancomycin-resistant *Staphylococcus aureus*—Pennsylvania, 2002. *MMWR.* **51**, 902.

Chambers, H.F., Miller, R.T., Newman, M.D. (1988) Right-sided *Staphylococcus aureus* endocarditis in intravenous drug abusers: two-week combination therapy. *Annals of Internal Medicine* **109**, 619–24.

Chambers, H.F., Archer, G., Matsuhashi, M. (1989) Low-level methicillin resistance in strains of *Staphylococcus aureus*. *Antimicrobial Agents and Chemotherapy* **33**, 424–8.

Chamis, A.L., Gesty-Palmer, D., Fowler, V.G., Corey, G.R. (1999) Echocardiography for the diagnosis of *Staphylococcus aureus* infective endocarditis. *Current Infectious Disease Reports* **1**, 129–35.

Chavakis, T., Hussain, M., Kanse, S.M. *et al.* (2002) *Staphylococcus aureus* extracellular adherence protein serves as anti-inflammatory factor by inhibiting the recruitment of host leukocytes. *Nature Medicine* **8**, 687–93.

Chu, M.C., Kreiswirth, B.N., Pattee, P.A. *et al.* (1988) Association of toxic shock toxin-1 determinant with a heterologous insertion at multiple loci in the *Staphylococcus aureus* chromosome. *Infection and Immunity* **56**, 2702–8.

Chung, M., Dickinson, G., De Lencastre, H., Tomasz, A. (2004) International clones of methicillin-resistant *Staphylococcus aureus* in two hospitals in Miami, Florida. *Journal of Clinical Microbiology* **42**, 542–7.

Clarke, S.R., Harris, L.G., Richards, R.G., Foster, S.J. (2002) Analysis of Ebh, a 1.1-megadalton cell wall-associated fibronectin-binding protein of *Staphylococcus aureus*. *Infection and Immunity* **70**, 6680–7.

Climo, M.W., Patron, R.L., Goldstein, B.P., Archer, G.L. (1998) Lysostaphin treatment of experimental methicillin-resistant *Staphylococcus aureus* aortic valve endocarditis. *Antimicrobial Agents and Chemotherapy* **42**, 1355–60.

Climo, M.W., Ehlert, K., Archer, G.L. (2001) Mechanism and suppression of lysostaphin resistance in oxacillin-resistant *Staphylococcus aureus*. *Antimicrobial Agents and Chemotherapy* **45**, 1431–7.

Cole, A.M., Dewan, P., Ganz, T. (1999) Innate antimicrobial activity of nasal secretions. *Infection and Immunity* **67**, 3267–75.

Coleman, D.C., Pomeroy, H., Estridge, J.K. *et al.* (1985) Susceptibility to antimicrobial agents and analysis of plasmids in gentamicin- and methicillin-resistant *Staphylococcus aureus* from Dublin hospitals. *Journal of Medical Microbiology* **20**, 157–67.

Coleman, D.C., Arbuthnott, J.P., Pomeroy, H.M., Birkbeck, T.H. (1986) Cloning and expression in Escherichia coli and *Staphylococcus aureus*

of the beta-lysin determinant from *Staphylococcus aureus*: evidence that bacteriophage conversion of beta-lysin activity is caused by insertional inactivation of the beta-lysin determinant. *Microbial Pathogenesis* **1**, 549–64.

Coleman, D.C., Sullivan, D.J., Russell, R.J. *et al.* (1989) *Staphylococcus aureus* bacteriophages mediating the simultaneous lysogenic conversion of β-lysin, staphylokinase and enterotoxin A: molecular mechanism of triple conversion. *Journal of General Microbiology* **135**, 1679–97.

Collins, M.D., Jones, D. (1981) Distribution of isoprenoid quinone structural types in bacteria and their taxonomic implication. *Microbiological Reviews* **45**, 316–54.

Cookson, B.D. (1998) The emergence of mupirocin resistance: a challenge to infection control and antibiotic prescribing practice. *The Journal of Antimicrobial Chemotherapy* **41**, 11–8.

Cookson, B.D., Phillips, I. (1988) Epidemic methicillin-resistant *Staphylococcus aureus*. *The Journal of Antimicrobial Chemotherapy* **21**(Suppl. C), 57–65.

Cooney, J., Mulvey, M., Arbuthnott, J.P., Foster, T.J. (1988) Molecular cloning and genetic analysis of the determinant for gamma-lysin, a two-component toxin of *Staphylococcus aureus*. *Journal of General Microbiology* **134**, 2179–88.

Coughter, J.P., Johnston, J.L., Archer, G.L. (1987) Characterization of a staphylococcal trimethoprim resistance gene and its product. *Antimicrobial Agents and Chemotherapy* **31**, 1027–32.

Coulter, S.N., Schwan, W.R., Ng, E.Y. *et al.* (1998) *Staphylococcus aureus* genetic loci impacting growth and survival in multiple infection environments. *Molecular Microbiology* **30**, 393–404.

Cucarella, C., Solano, C., Valle, J. *et al.* (2001) Bap, a *Staphylococcus aureus* surface protein involved in biofilm formation. *Journal of Bacteriology* **183**, 2888–96.

Cucarella, C., Tormo, M.A., Knecht, E. *et al.* (2002) Expression of the biofilm-associated protein interferes with host protein receptors of *Staphylococcus aureus* and alters the infective process. *Infection and Immunity* **70**, 3180–6.

Dajcs, J.J., Hume, E.B., Moreau, J.M. *et al.* (2000) Lysostaphin treatment of methicillin-resistant *Staphylococcus aureus* keratitis in the rabbit. *Investigative Ophthalmology & Visual Science* **41**, 1432–7.

Dale, G.E., Then, R.L., Stuber, D. (1993) Characterization of the gene for chromosomal trimethoprim-sensitive dihydrofolate reductase of *Staphylococcus aureus* ATCC 25923. *Antimicrobial Agents and Chemotherapy* **37**, 1400–5.

Dancer, S.J., Garratt, R., Saldanha, J. *et al.* (1990) The epidermolytic toxins are serine proteases. *FEBS Letters* **268**, 129–32.

Daum, R.S., Ito, T., Hiramatsu, K. *et al.* (2002) A novel methicillin-resistance cassette in community-acquired methicillin-resistant *Staphylococcus aureus* isolates of diverse genetic backgrounds. *The Journal of Infectious Diseases* **186**, 1344–7.

Dinges, M.M., Orwin, P.M., Schlievert, P.M. (2000) Exotoxins of *Staphylococcus aureus*. *Clinical Microbiology Reviews* **13**, 16–34.

DiNubile, M.J. (1994) Short-course antibiotic therapy for right-sided endocarditis caused by *Staphylococcus aureus* in injection drug users. *Annals of Internal Medicine* **121**, 873–6.

Doery, H.M., Magnusson, B.J., Gulasekharam, J., Pearson, J.E. (1965) The properties of phospholipase enzymes in staphylococcal toxins. *Journal of General Microbiology* **40**, 283–96.

Domingue, P.A., Lambert, P.A., Brown, M.R. (1989) Iron depletion alters surface-associated properties of *Staphylococcus aureus* and its association to human neutrophils in chemiluminescence. *FEMS Microbiology Letters* **50**, 265–8.

Downer, R., Roche, F., Park, P.W. *et al.* (2002) The elastin-binding protein of *Staphylococcus aureus* (EbpS) is expressed at the cell surface as an integral membrane protein and not as a cell wall-associated protein. *The Journal of Biological Chemistry* **277**, 243–50.

Drummond, M.C., Tager, M. (1963) Fibrinogen clotting and fibrino-peptide formation by staphylocoagulase and the coagulase-reacting factor. *Journal of Bacteriology* **85**, 628–35.

Duckworth, G.J., Lothian, J.L., Williams, J.D. (1988) Methicillin-resistant *Staphylococcus aureus*: report of an outbreak in a London teaching hospital. *The Journal of Hospital Infection* **11**, 1–15.

Durack, D.T., Lukes, A.S., Bright, D.K. (1994) New criteria for diagnosis of infective endocarditis: utilization of specific echocardiographic findings. Duke Endocarditis Service. *American Journal of Medicine* **96**, 200–9.

Dziewanowska, K., Patti, J.M., Deobald, C.F. *et al.* (1999) Fibronectin binding protein and host cell tyrosine kinase are required for internalization of

Staphylococcus aureus by epithelial cells. *Infection and Immunity* **67**, 4673–8.

Ehrenkranz, N.J. (1966) Nasal rejection of experimentally inoculated *Staphylococcus aureus*: evidence for an immune reaction in man. *Journal of Immunology* **96**, 509–17.

el Solh, N., Moreau, N., Ehrlich, S.D. (1986) Molecular cloning and analysis of *Staphylococcus aureus* chromosomal aminoglycoside resistance genes. *Plasmid* **15**, 104–18.

Elasri, M.O., Thomas, J.R., Skinner, R.A. *et al.* (2002) *Staphylococcus aureus* collagen adhesin contributes to the pathogenesis of osteomyelitis. *Bone* **30**, 275–80.

Enright, M.C., Day, N.P., Davies, C.E. *et al.* (2000) Multilocus sequence typing for characterization of methicillin-resistant and methicillin-susceptible clones of *Staphylococcus aureus*. *Journal of Clinical Microbiology* **38**, 1008–15.

Enright, M.C., Robinson, D.A., Randle, G. *et al.* (2002) The evolutionary history of methicillin-resistant *Staphylococcus aureus* (MRSA). *Proceedings of the National Academy of Sciences of the United States of America* **99**, 7687–92.

Entenza, J.M., Foster, T.J., Ni Eidhin, D. *et al.* (2000) Contribution of clumping factor B to pathogenesis of experimental endocarditis due to *Staphylococcus aureus*. *Infection and Immunity* **68**, 5443–6.

Etz, H., Minh, D.B., Henics, T. *et al.* (2002) Identification of in vivo expressed vaccine candidate antigens from *Staphylococcus aureus*. *Proceedings of the National Academy of Sciences of the United States of America* **99**, 6573–8.

Evans, J.B. Jr, Bradford, W.L., Niven, C.F. (1955) Comments concerning the taxonomy of the genera *Micrococcus* and *Staphylococcus*. *International Bulletin of Bacteriological Nomenclature and Taxonomy* **5**, 61–6.

Fackrell, H.B., Wiseman, G.M. (1976) Properties of the gamma haemolysin of *Staphylococcus aureus* 'Smith 5R'. *Journal of General Microbiology* **92**, 11–24.

Faller, A.H., Götz, F., Schleifer, K.H. (1980) Cytochrome patterns of staphylococci and micrococci and their taxonomic implications. *Zentralbl Bakteriol Parasitenkd Infektionskr Hyg Abt 1 Orig C* **1**, 26–39.

Fang, H., Hedin, G. (2003) Rapid screening and identification of methicillin-resistant *Staphylococcus aureus* from clinical samples by selective-broth and real-time PCR assay. *Journal of Clinical Microbiology* **41**, 2894–9.

Feil, E.J., Cooper, J.E., Grundmann, H. *et al.* (2003) How clonal is *Staphylococcus aureus*? *Journal of Bacteriology* **185**, 3307–16.

Ferrero, L., Cameron, B., Manse, B. *et al.* (1994) Cloning and primary structure of *Staphylococcus aureus* DNA topoisomerase IV: a primary target of fluoroquinolones. *Molecular Microbiology* **13**, 641–53.

Ferrero, L., Cameron, B., Crouzet, J. (1995) Analysis of *gyrA* and *grlA* mutations in stepwise-selected ciprofloxacin-resistant mutants of *Staphylococcus aureus*. *Antimicrobial Agents and Chemotherapy* **39**, 1554–8.

Fitton, J.E., Dell, A., Shaw, W.V. (1980) The amino acid sequence of the delta haemolysin of *Staphylococcus aureus*. *FEBS Letters* **115**, 209–12.

Fitzgerald, J.R., Sturdevant, D.E., Mackie, S.M. *et al.* (2001) Evolutionary genomics of *Staphylococcus aureus*: insights into the origin of methicillin-resistant strains and the toxic shock syndrome epidemic. *Proceedings of the National Academy of Sciences of the United States of America* **98**, 8821–6.

Fleischer, B. (1994) Lymphocyte stimulating 'superantigens' of *Staphylococcus aureus* and *Streptococcus pyogenes*. In *Molecular Pathogenesis of Surgical Infections*, eds. Wadström, T., Holder, I.A., Kronvall, G. Gustav Fischer Verlag, Stuttgart, 171–81.

Flock, J.I., Hienz, S.A., Heimdahl, A., Schennings, T. (1996) Reconsideration of the role of fibronectin binding in endocarditis caused by *Staphylococcus aureus*. *Infection and Immunity* **64**, 1876–8.

Flock, M., Flock, J.I. (2001) Rebinding of extracellular adherence protein Eap to *Staphylococcus aureus* can occur through a surface-bound neutral phosphatase. *Journal of Bacteriology* **183**, 3999–4003.

Forsgren, A., Ghetie, V., Lindmark, R., Sjoquist, J. (1983) Protein A and its exploitation. In *Staphylococci and Staphylococcal Infections*, vol. 2, eds. Easmon, C.S.F., Adlam, C. Academic Press, London, 429–80.

Foster, T.J., O'Reilly, M. *et al.* (1990) Genetic studies of virulence factors of *Staphylococcus aureus*. Properties of coagulase and gamma-toxin and the role of alpha-toxin, beta-toxin and protein A in the pathogenesis of *S. aureus* infections. In *Molecular Biology of the Staphylococci*, ed. Novick, R.P. VCH Publishers, Cambridge, UK.

Fowler, T., Wann, E.R., Joh, D. *et al.* (2000) Cellular invasion by *Staphylococcus aureus* involves a fibronectin bridge between the bacterial fibronectin-binding MSCRAMMs and host cell beta1 integrins. *European Journal of Cell Biology* **79**, 672–9.

Fowler, V.G. Jr, Olsen, M.K., Corey, G.R. *et al.* (2003) Clinical identifiers of complicated *Staphylococcus aureus* bacteremia. *Archives of Internal Medicine* **163**, 2066–72.

Fraimow, H.S., Courvalin, P. (2000) Resistance to glycopeptides in Gram-positive pathogens. In *Gram-Positive Pathogens*, eds. Fischetti, V.A., Novick, R.P., Ferretti, J.J., Portnoy, D.A., Rood, J.I. ASM Press, Washington, DC.

Francois, P., Pittet, D., Bento, M. *et al.* (2003) Rapid detection of methicillin-resistant *Staphylococcus aureus* directly from sterile or nonsterile clinical samples by a new molecular assay. *Journal of Clinical Microbiology* **41**, 254–60.

Freney, J., Brun, Y., Bes, M. *et al.* (1988) *Staphylococcus lugdunensis* sp. nov. and *Staphylococcus schleiferi* sp. nov., two species from human clinical specimens. *International Journal of Systematic Bacteriology* **38**, 168–72.

Gilbart, J., Perry, C.R., Slocombe, B. (1993) High-level mupirocin resistance in *Staphylococcus aureus*: evidence for two distinct isoleucyl-tRNA synthetases. *Antimicrobial Agents and Chemotherapy* **37**, 32–8.

Gillespie, M.T., Lyon, B.R., Messerotti, L.J., Skurray, R.A. (1987) Chromosome- and plasmid-mediated gentamicin resistance in *Staphylococcus aureus* encoded by Tn*4001*. *Journal of Medical Microbiology* **24**, 139–44.

Gillet, Y., Issartel, B., Vanhems, P. *et al.* (2002) Association between *Staphylococcus aureus* strains carrying gene for Panton–Valentine leukocidin and highly lethal necrotising pneumonia in young immunocompetent patients. *Lancet* **359**, 753–9.

Giraudo, A.T., Raspanti, C.G., Calzolari, A., Nagel, R. (1994) Characterization of a Tn551-mutant of *Staphylococcus aureus* defective in the production of several exoproteins. *Canadian Journal of Microbiology* **40**, 677–81.

Giraudo, A.T., Rampone, H., Calzolari, A., Nagel, R. (1996) Phenotypic characterization and virulence of a *sae- agr-* mutant of *Staphylococcus aureus*. *Canadian Journal of Microbiology* **42**, 120–3.

Giraudo, A.T., Cheung, A.L., Nagel, R. (1997) The *sae* locus of *Staphylococcus aureus* controls exoprotein synthesis at the transcriptional level. *Archives of Microbiology* **168**, 53–8.

Giraudo, A.T., Calzolari, A., Cataldi, A.A. *et al.* (1999) The sae locus of *Staphylococcus aureus* encodes a two-component regulatory system. *FEMS Microbiology Letters* **177**, 15–22.

Giraudo, A.T., Mansilla, C., Chan, A. *et al.* (2003) Studies on the expression of regulatory locus sae in *Staphylococcus aureus*. *Current Microbiology* **46**, 246–50.

Goerke, C., Campana, S., Bayer, M.G. *et al.* (2000) Direct quantitative transcript analysis of the agr regulon of *Staphylococcus aureus* during human infection in comparison to the expression profile *in vitro*. *Infection and Immunity* **68**, 1304–11.

Goerke, C., Fluckiger, U., Steinhuber, A. *et al.* (2001) Impact of the regulatory loci agr, sarA and sae of *Staphylococcus aureus* on the induction of alpha-toxin during device-related infection resolved by direct quantitative transcript analysis. *Molecular Microbiology* **40**, 1439–47.

Goh, S.H., Byrne, S.K., Zhang, J.L., Chow, A.W. (1992) Molecular typing of *Staphylococcus aureus* on the basis of coagulase gene polymorphisms. *Journal of Clinical Microbiology* **30**, 1642–5.

Goswitz, J.J., Willard, K.E., Fasching, C.E., Peterson, L.R. (1992) Detection of gyrA gene mutations associated with ciprofloxacin resistance in methicillin-resistant *Staphylococcus aureus*: analysis by polymerase chain reaction and automated direct DNA sequencing. *Antimicrobial Agents and Chemotherapy* **36**, 1166–9.

Gray, G.S., Fitch, W.M. (1983) Evolution of antibiotic resistance genes: the DNA sequence of a kanamycin resistance gene from *Staphylococcus aureus*. *Molecular Biology and Evolution* **1**, 57–66.

Greene, C., McDevitt, D., Francois, P. *et al.* (1995) Adhesion properties of mutants of *Staphylococcus aureus* defective in fibronectin-binding proteins and studies on the expression of *fnb* genes. *Molecular Microbiology* **17**, 1143–52.

Grisold, A.J., Leitner, E., Muhlbauer, G. *et al.* (2002) Detection of methicillin-resistant *Staphylococcus aureus* and simultaneous confirmation by automated nucleic acid extraction and real-time PCR. *Journal of Clinical Microbiology* **40**, 2392–7.

Guardati, M.C., Guzmán, C.A., Piatti, G., Pruzzo, C. (1993) Rapid methods for identification of *Staphylococcus aureus* when both human and animal staphylococci are tested: comparison with a new immunoenzymatic assay. *Journal of Clinical Microbiology* **31**, 1606–8.

Guay, G.G., Khan, S.A., Rothstein, D.M. (1993) The tet(K) gene of plasmid pT181 of *Staphylococcus aureus* encodes an efflux protein that contains 14 transmembrane helices. *Plasmid* **30**, 163–6.

Haggar, A., Hussain, M., Lonnies, H. *et al.* (2003) Extracellular adherence protein from *Staphylococcus aureus* enhances internalization into eukaryotic cells. *Infection and Immunity* **71**, 2310–7.

Hanaki, H., Kuwahara-Arai, K., Boyle-Vavra, S. *et al.* (1998a) Activated cell-wall synthesis is associated with vancomycin resistance in methicillin-resistant *Staphylococcus aureus* clinical strains Mu3 and Mu50. *The Journal of Antimicrobial Chemotherapy* **42**, 199–209.

Hanaki, H., Labischinski, H., Inaba, Y. *et al.* (1998b) Increase in glutamine-non-amidated muropeptides in the peptidoglycan of vancomycin-resistant *Staphylococcus aureus* strain Mu50. *The Journal of Antimicrobial Chemotherapy* **42**, 315–20.

Haroche, J., Allignet, J., Buchrieser, C., El Solh, N. (2000) Characterization of a variant of vga(A) conferring resistance to streptogramin A and related compounds. *Antimicrobial Agents and Chemotherapy* **44**, 2271–5.

Haroche, J., Allignet, J., El Solh, N. (2002) Tn5406, a new staphylococcal transposon conferring resistance to streptogramin A and related compounds including dalfopristin. *Antimicrobial Agents and Chemotherapy* **46**, 2337–43.

Haroche, J., Morvan, A., Davi, M. *et al.* (2003) Clonal diversity among streptogramin A-resistant *Staphylococcus aureus* isolates collected in French hospitals. *Journal of Clinical Microbiology* **41**, 586–91.

Hartleib, J., Kohler, N., Dickinson, R.B. *et al.* (2000) Protein A is the von Willebrand factor binding protein on *Staphylococcus aureus*. *Blood* **96**, 2149–56.

Hartman, B.J., Tomasz, A. (1984) Low-affinity penicillin-binding protein associated with beta-lactam resistance in *Staphylococcus aureus*. *Journal of Bacteriology* **158**, 513–6.

Heilmann, C., Schweitzer, O., Gerke, C. *et al.* (1996) Molecular basis of inter-cellular adhesion in the biofilm-forming *Staphylococcus epidermidis*. *Molecular Microbiology* **20**, 1083–91.

Herrmann, M., Vaudaux, P.E., Pittet, D. *et al.* (1988) Fibronectin, fibrinogen, and laminin act as mediators of adherence of clinical staphylococcal isolates to foreign material. *The Journal of Infectious Diseases* **158**, 693–701.

Hiramatsu, K. (2001) Vancomycin-resistant *Staphylococcus aureus*: a new model of antibiotic resistance. *Lancet Infectious Diseases* **1**, 147–55.

Hiramatsu, K., Aritaka, N., Hanaki, H. *et al.* (1997a) Dissemination in Japanese hospitals of strains of *Staphylococcus aureus* heterogeneously resistant to vancomycin. *Lancet* **350**, 1670–3.

Hiramatsu, K., Hanaki, H., Ino, T. *et al.* (1997b) Methicillin-resistant *Staphylococcus aureus* clinical strain with reduced vancomycin susceptibility. *The Journal of Antimicrobial Chemotherapy* **40**, 135–6.

Hochkeppel, H.K., Braun, D.G., Vischer, W. *et al.* (1987) Serotyping and electron microscopy studies of *Staphylococcus aureus* clinical isolates with monoclonal antibodies to capsular polysaccharide types 5 and 8. *Journal of Clinical Microbiology* **25**, 526–30.

Hodgson, J.E., Curnock, S.P., Dyke, K.G. *et al.* (1994) Molecular characterization of the gene encoding high-level mupirocin resistance in *Staphylococcus aureus* J2870. *Antimicrobial Agents and Chemotherapy* **38**, 1205–8.

Hoeksma, A., Winkler, K.C. (1963) The normal flora of the nose in twins. *Acta Leidensia* **32**, 123–33.

Hoen, B., Paul-Dauphin, A., Hestin, D., Kessler, M. (1998) EPIBACDIAL: a multicenter prospective study of risk factors for bacteremia in chronic hemodialysis patients. *Journal of the American Society of Nephrology* **9**, 869–76.

Holden, M.T., Feil, E.J., Lindsay, J.A. *et al.* (2004) Complete genomes of two clinical *Staphylococcus aureus* strains: evidence for the rapid evolution of virulence and drug resistance. *Proceedings of the National Academy of Sciences of the United States of America* **101**, 9786–91.

Holton, D.L., Nicolle, L.E., Diley, D., Bernstein, K. (1991) Efficacy of mupirocin nasal ointment in eradicating *Staphylococcus aureus* nasal carriage in chronic haemodialysis patients. *The Journal of Hospital Infection* **17**, 133–7.

Hooper, D.C. (2001) Mechanisms of action of antimicrobials: focus on fluoro-quinolones. *Clinical Infectious Diseases* **32**(Suppl. 1), S9–S15.

Horinouchi, S., Weisblum, B. (1982) Nucleotide sequence and functional map of pE194, a plasmid that specifies inducible resistance to macrolide, lincosamide, and streptogramin type B antibiotics. *Journal of Bacteriology* **150**, 804–14.

Horstkotte, D., Follath, F., Gutschik, E. *et al.* (2004) Guidelines on prevention, diagnosis and treatment of infective endocarditis executive summary; the task force on infective endocarditis of the European society of cardiology. *European Heart Journal* **25**, 267–76.

Horstkotte, M.A., Knobloch, J.K., Rohde, H., Mack, D. (2001) Rapid detection of methicillin resistance in coagulase-negative staphylococci by a penicillin-binding protein 2a-specific latex agglutination test. *Journal of Clinical Microbiology* **39**, 3700–2.

Huesca, M., Sun, Q., Peralta, R. *et al.* (2000) Synthetic peptide immunogens elicit polyclonal and monoclonal antibodies specific for linear epitopes in the D motifs of *Staphylococcus aureus* fibronectin-binding protein, which are composed of amino acids that are essential for fibronectin binding. *Infection and Immunity* **68**, 1156–63.

Huesca, M., Peralta, R., Sauder, D.N. *et al.* (2002) Adhesion and virulence properties of epidemic Canadian methicillin-resistant *Staphylococcus aureus* strain 1: identification of novel adhesion functions associated with plasmin-sensitive surface protein. *The Journal of Infectious Diseases* **185**, 1285–96.

Hussain, M., Becker, K., von Eiff, C. *et al.* (2001) Identification and character-ization of a novel 38.5-kilodalton cell surface protein of *Staphylococcus aureus* with extended-spectrum binding activity for extracellular matrix and plasma proteins. *Journal of Bacteriology* **183**, 6778–86.

Hussain, M., Haggar, A., Heilmann, C. *et al.* (2002) Insertional inactivation of Eap in *Staphylococcus aureus* strain Newman confers reduced staphylococcal binding to fibroblasts. *Infection and Immunity* **70**, 2933–40.

Iandolo, J.J. (1989) Genetic analysis of extracellular toxins of *Staphylococcus aureus*. *Annual Review of Microbiology* **43**, 375–402.

Ilangovan, U., Ton-That, H., Iwahara, J. *et al.* (2001) Structure of sortase, the transpeptidase that anchors proteins to the cell wall of *Staphylococcus aureus*. *Proceedings of the National Academy of Sciences of the United States of America* **98**, 6056–61.

Imundo, L., Barasch, J., Prince, A., Al-Awqati, Q. (1995) Cystic fibrosis epithelial cells have a receptor for pathogenic bacteria on their apical surface. *Proceedings of the National Academy of Sciences of the United States of America* **92**, 3019–23.

Ince, D., Hooper, D.C. (2003) Quinolone resistance due to reduced target enzyme expression. *Journal of Bacteriology* **185**(23), 6883–92.

Ito, H., Yoshida, H., Bogaki-Shonai, M. *et al.* (1994) Quinolone resistance mutations in the DNA gyrase gyrA and gyrB genes of *Staphylococcus aureus*. *Antimicrobial Agents and Chemotherapy* **38**, 2014–23.

Ito, T., Katayama, Y., Asada, K. *et al.* (2001) Structural comparison of three types of staphylococcal cassette chromosome mec integrated in the chromosome in methicillin-resistant *Staphylococcus aureus*. *Antimicrobial Agents and Chemotherapy* **45**, 1323–36.

Jackson, M.P., Iandolo, J.J. (1986) Sequence of the exfoliative toxin B gene of *Staphylococcus aureus*. *Journal of Bacteriology* **167**, 726–8.

Jernigan, J.A., Farr, B.M. (1993) Short-course therapy of catheter-related *Staphylococcus aureus* bacteremia: a meta-analysis. *Annals of Internal Medicine* **119**, 304–11.

Jevons, M.P. (1961) 'Celbenin'-resistant staphylococci. *British Medical Journal* **1**, 124–5.

Jonas, D., Speck, M., Daschner, F.D., Grundmann, H. (2002) Rapid PCR-based identification of methicillin-resistant *Staphylococcus aureus* from screening swabs. *Journal of Clinical Microbiology* **40**, 1821–3.

Jonsson, I.M., von Eiff, C., Proctor, R.A. *et al.* (2003) Virulence of a hemB mutant displaying the phenotype of a *Staphylococcus aureus* small-colony variant in a murine model of septic arthritis. *Microbial Pathogenesis* **34**, 73–9.

Jönsson, K., Signäs, C., Muller, H.P., Lindberg, M. (1991) Two different genes encode fibronectin binding proteins in *Staphylococcus aureus*. The complete nucleotide sequence and characterization of the second gene. *European Journal of Biochemistry* **202**, 1041–8.

Jönsson, K., McDevitt, D., McGavin, M.H. *et al.* (1995) *Staphylococcus aureus* expresses a major histocompatibility complex class II analog. *The Journal of Biological Chemistry* **270**, 21457–60.

Josefsson, E., McCrea, K.W., Ni Eidhin, D. *et al.* (1998) Three new members of the serine-aspartate repeat protein multigene family of *Staphylococcus aureus*. *Microbiology* **144**, 3387–95.

Josefsson, E., Hartford, O., O'Brien, L. *et al.* (2001) Protection against experimental *Staphylococcus aureus* arthritis by vaccination with clumping factor A, a novel virulence determinant. *The Journal of Infectious Diseases* **184**, 1572–80.

Jureen, R., Bottolfsen, K.L., Grewal, H., Digranes, A. (2001) Comparative evaluation of a commercial test for rapid identification of methicillin-resistant *Staphylococcus aureus*. *APMIS* **109**, 787–90.

Kaatz, G.W., Seo, S.M., Ruble, C.A. (1993) Efflux-mediated fluoroquinolone resistance in *Staphylococcus aureus*. *Antimicrobial Agents and Chemotherapy* **37**, 1086–94.

Kagan, B.M. (1972) L-forms. In *The Staphylococci*, ed. Cohen, J.O. Wiley-Interscience, New York, 65–74.

Kahl, B., Herrmann, M., Everding, A.S. *et al.* (1998) Persistent infection with small-colony variant strains of *Staphylococcus aureus* in patients with cystic fibrosis. *The Journal of Infectious Diseases* **177**, 1023–9.

Karlsson, A., Arvidson, S. (2002) Variation in extracellular protease production among clinical isolates of *Staphylococcus aureus* due to different levels of expression of the protease repressor *sarA*. *Infection and Immunity* **70**, 4239–46.

Katayama, Y., Ito, T., Hiramatsu, K. (2000) A new class of genetic element, staphylococcus cassette chromosome *mec*, encodes methicillin resistance in *Staphylococcus aureus*. *Antimicrobial Agents and Chemotherapy* **44**, 1549–55.

Katayama, Y., Takeuchi, F., Ito, T. *et al.* (2003) Identification in methicillin-susceptible *Staphylococcus hominis* of an active primordial mobile genetic element for the staphylococcal cassette chromosome mec in methicillin-resistant *Staphylococcus aureus*. *Journal of Bacteriology* **185**, 2711–22.

Khan, S.A., Novick, R.P. (1980) Terminal nucleotide sequences of Tn551, a transposon specifying erythromycin resistance in *Staphylococcus aureus*: homology with Tn3. *Plasmid* **4**, 148–54.

Khan, S.A., Novick, R.P. (1983) Complete nucleotide sequence of pT181, a tetracycline-resistance plasmid from *Staphylococcus aureus*. *Plasmid* **10**, 251–9.

Kilpper, R., Buhl, U., Schleifer, K.H. (1980) Nucleic acid homology studies between *Peptococcus saccharolyticus* and various anaerobic and facultative anaerobic Gram-positive cocci. *FEMS Microbiology Letters* **8**, 205–10.

Kinsman, O.S., McKenna, R., Noble, W.C. (1983) Association between histocompatability antigens (HLA) and nasal carriage of *Staphylococcus aureus*. *Journal of Medical Microbiology* **16**, 215–20.

Kloos, W.E., Bannerman, T.L. (1994) Update on clinical significance of coagulase-negative staphylococci. *Clinical Microbiology Reviews* **7**, 117–40.

Kloos, W.E., Pattee, P.A. (1965) Transduction analysis of the histidine region in *Staphylococcus aureus*. *Journal of General Microbiology* **39**, 195–207.

Kloos, W.E., Ballard, D.N., George, C.G. *et al.* (1998a) Delimiting the genus *Staphylococcus* through description of *Macrococcus caseolyticus* gen. nov., comb. nov., *M. equipercicus* sp. nov., *M. bovicus* sp. nov., and *M. carouselicus* sp. nov. *International Journal of Systematic Bacteriology* **47**, 859–77.

Kloos, W.E., George, C.G., Olgiate, J.S. *et al.* (1998b) *Staphylococcus hominis* subsp. *novobiosepticus* subsp. nov., a novel trehalose- and N-acetyl-D-glucosamine-negative, novobiocin- and multiple-antibiotic-resistant subspecies isolated from human blood cultures. *International Journal of Systematic Bacteriology* **48**, 799–812.

Kluytmans, J., van Belkum, A., Verbrugh, H. (1997) Nasal carriage of *Staphylococcus aureus*: epidemiology, underlying mechanisms, and associated risks. *Clinical Microbiology Reviews* **10**, 505–20.

Kluytmans, J.A., Manders, M.J., van Bommel, E., Verbrugh, H. (1996) Elimination of nasal carriage of *Staphylococcus aureus* in hemodialysis patients. *Infection Control and Hospital Epidemiology* **17**, 793–7.

Kokai-Kun, J.F., Walsh, S.M., Chanturiya, T., Mond, J.J. (2003) Lysostaphin cream eradicates *Staphylococcus aureus* nasal colonization in a cotton rat model. *Antimicrobial Agents and Chemotherapy* **47**, 1589–97.

Koreen, L., Ramaswamy, S.V., Graviss, E.A. *et al.* (2004) *spa* typing method for discriminating among *Staphylococcus aureus* isolates: implications for use of a single marker to detect genetic micro- and macrovariation. *Journal of Clinical Microbiology* **42**, 792–9.

Korzeniowski, O., Sande, M.A. (1982) Combination antimicrobial therapy for *Staphylococcus aureus* endocarditis in patients addicted to parenteral drugs and in nonaddicts: a prospective study. *Annals of Internal Medicine* **97**, 496–503.

Kreiswirth, B.N., Lofdahl, S., Betley, M.J. *et al.* (1983) The toxic shock syndrome exotoxin structural gene is not detectably transmitted by a prophage. *Nature* **305**, 709–12.

Krivan, H.C., Roberts, D.D., Ginsburg, V. (1988) Many pulmonary pathogenic bacteria bind specifically to the carbohydrate sequence GalNAc beta 1–4Gal found in some glycolipids. *Proceedings of the National Academy of Sciences of the United States of America* **85**, 6157–61.

Kuroda, M., Ohta, T., Uchiyama, I. *et al.* (2001) Whole genome sequencing of meticillin-resistant *Staphylococcus aureus*. *Lancet* **357**, 1225–40.

Kuypers, J.M., Proctor, R.A. (1989) Reduced adherence to traumatized rat heart valves by a low-fibronectin-binding mutant of *Staphylococcus aureus*. *Infection and Immunity* **57**, 2306–12.

Kyburz, D., Rethage, J., Seibl, R. *et al.* (2003) Bacterial peptidoglycans but not CpG oligodeoxynucleotides activate synovial fibroblasts by toll-like receptor signaling. *Arthritis and Rheumatism* **48**, 642–50.

Ladhani, S., Joannou, C.L., Lochrie, D.P. *et al.* (1999) Clinical, microbial, and biochemical aspects of the exfoliative toxins causing staphylococcal scalded-skin syndrome. *Clinical Microbiology Reviews* **12**, 224–42.

Lammers, A., Nuijten, P.J., Smith, H.E. (1999) The fibronectin binding proteins of *Staphylococcus aureus* are required for adhesion to and invasion of bovine mammary gland cells. *FEMS Microbiology Letters* **180**, 103–9.

Laupland, K.B., Conly, J.M. (2003) Treatment of *Staphylococcus aureus* colonization and prophylaxis for infection with topical intranasal mupirocin: an evidence-based review. *Clinical Infectious Diseases* **37**, 933–8.

Lee, B.K., Crossley, K., Gerding, D.N. (1978) The association between *Staphylococcus aureus* bacteremia and bacteriuria. *American Journal of Medicine* **65**, 303–6.

Levi, K., Towner, K.J. (2003) Detection of methicillin-resistant *Staphylococcus aureus* (MRSA) in blood with the EVIGENE MRSA detection kit. *Journal of Clinical Microbiology* **41**, 3890–2.

Levine, D.P., Fromm, B.S., Reddy, B.R. (1991) Slow response to vancomycin or vancomycin plus rifampin in methicillin-resistant *Staphylococcus aureus* endocarditis. *Annals of Internal Medicine* **115**, 674–80.

Lew, D.P., Waldvogel, F.A. (1997) Osteomyelitis. *The New England Journal of Medicine* **336**, 999–1007.

Lim, J.A., Kwon, A.R., Kim, S.K. *et al.* (2002) Prevalence of resistance to macrolide, lincosamide and streptogramin antibiotics in Gram-positive cocci isolated in a Korean hospital. *The Journal of Antimicrobial Chemotherapy* **49**, 489–95.

Lina, G., Quaglia, A., Reverdy, M.E. *et al.* (1999) Distribution of genes encoding resistance to macrolides, lincosamides, and streptogramins among staphylococci. *Antimicrobial Agents and Chemotherapy* **43**, 1062–6.

Lina, G., Boutite, F., Tristan, A. *et al.* (2003) Bacterial competition for human nasal cavity colonization: role of Staphylococcal agr alleles. *Applied and Environmental Microbiology* **69**, 18–23.

Liu, C., Chambers, H.F. (2003) *Staphylococcus aureus* with heterogeneous resistance to vancomycin: epidemiology, clinical significance, and critical assessment of diagnostic methods. *Antimicrobial Agents and Chemotherapy* **47**, 3040–5.

Loeb, M., Main, C., Walker-Dilks, C., Eady, A. (2003) Antimicrobial drugs for treating methicillin-resistant *Staphylococcus aureus* colonization. *Cochrane Database of Systematic Reviews* **4**, CD003340.

Luchansky, J.B., Pattee, P.A. (1984) Isolation of transposon Tn551 insertions near chromosomal markers of interest in *Staphylococcus aureus*. *Journal of Bacteriology* **159**, 894–9.

Ludwig, W., Schleifer, K.H., Fox, G.E. *et al.* (1981) A phylogenetic analysis of staphylococci, *Peptococcus saccharolyticus* and *Micrococcus mucilaginosus*. *Journal of General Microbiology* **125**, 357–66.

Ludwig, W., Seewaldt, E., Kilpper-Balz, R. *et al.* (1985) The phylogenetic position of *Streptococcus* and *Enterococcus*. *Journal of General Microbiology* **131**, 543–51.

Luzar, M.A., Coles, G.A., Faller, B. *et al.* (1990) *Staphylococcus aureus* nasal carriage and infection in patients on continuous ambulatory peritoneal dialysis. *The New England Journal of Medicine* **322**, 505–9.

Lyon, B.R., Skurray, R. (1987) Antimicrobial resistance of *Staphylococcus aureus*: genetic basis. *Microbiological Reviews* **51**, 88–134.

Mack, D. (1999) Molecular mechanisms of *Staphylococcus epidermidis* biofilm formation. *The Journal of Hospital Infection* **43**(Suppl.), S113–25.

Mahairas, G.G., Lyon, B.R., Skurray, R.A., Pattee, P.A. (1989) Genetic analysis of *Staphylococcus aureus* with Tn4001. *Journal of Bacteriology* **171**, 3968–72.

Mamo, W., Jonsson, P., Flock, J.I. *et al.* (1994) Vaccination against *Staphylococcus aureus* mastitis: immunological response of mice vaccinated with fibronectin-binding protein (FnBP-A) to challenge with *S. aureus*. *Vaccine* **12**, 988–92.

Margerrison, E.E., Hopewell, R., Fisher, L.M. (1992) Nucleotide sequence of the *Staphylococcus aureus* gyrB–gyrA locus encoding the DNA gyrase A and B proteins. *Journal of Bacteriology* **174**, 1596–603.

Marquet-Van Der Mee, N., Mallet, S., Loulergue, J., Audurier, A. (1995) Typing of *Staphylococcus epidermidis* strains by random amplification of polymorphic DNA. *FEMS Microbiology Letters* **128**, 39–44.

Mascini, E.M., Jansze, M., Schouls, L.M. *et al.* (2001) Penicillin and clindamycin differentially inhibit the production of pyrogenic exotoxins A and B by group A streptococci. *International Journal of Antimicrobial Agents* **18**, 395–8.

Massey, R.C., Buckling, A., Peacock, S.J. (2001) Phenotypic switching of antibiotic resistance circumvents permanent costs in *Staphylococcus aureus*. *Current Biology* **11**, 1810–4.

Massey, R.C., Kantzanou, M.N., Fowler, T. *et al.* (2001) Fibronectin-binding protein A of *Staphylococcus aureus* has multiple, substituting, binding regions that mediate adherence to fibronectin and invasion of endothelial cells. *Cellular Microbiology* **3**, 839–51.

Massey, R.C., Peacock, S.J. (2002) Antibiotic-resistant sub-populations of the pathogenic bacterium *Staphylococcus aureus* confer population-wide resistance. *Current Biology* **12**, R686–7.

Mazmanian, S.K., Liu, G., Ton-That, H., Schneewind, O. (1999) *Staphylococcus aureus* sortase, an enzyme that anchors surface proteins to the cell wall. *Science* **285**, 760–3.

Mazmanian, S.K., Ton-That, H., Su, K., Schneewind, O. (2002) An iron-regulated sortase anchors a class of surface protein during *Staphylococcus aureus* pathogenesis. *Proceedings of the National Academy of Sciences of the United States of America* **99**, 2293–8.

McCormick, J.K., Yarwood, J.M., Schlievert, P.M. (2001) Toxic shock syndrome and bacterial superantigens: an update. *Annual Review of Microbiology* **55**, 77–104.

McDevitt, D., Francois, P., Vaudaux, P., Foster, T.J. (1994) Molecular characterization of the clumping factor (fibrinogen receptor) of *Staphylococcus aureus*. *Molecular Microbiology* **11**, 237–48.

McDougal, L.K., Thornsberry, C. (1986) The role of β-lactamase in staphylococcal resistance to penicillinase-resistant penicillins and cephalosporins. *Journal of Clinical Microbiology* **23**, 832–9.

McElroy, M.C., Cain, D.J., Tyrrell, C. *et al.* (2002) Increased virulence of a fibronectin-binding protein mutant of *Staphylococcus aureus* in a rat model of pneumonia. *Infection and Immunity* **70**, 3865–73.

McGavin, M.H., Krajewska-Pietrasik, D., Ryden, C., Hook, M. (1993) Identification of a *Staphylococcus aureus* extracellular matrix-binding protein with broad specificity. *Infection and Immunity* **61**, 2479–85.

McKenney, D., Pouliot, K.L., Wang, Y. *et al.* (1999) Broadly protective vaccine for *Staphylococcus aureus* based on an in vivo-expressed antigen. *Science* **284**, 1523–7.

McMurray, L.W., Kernodle, D.S., Barg, N.L. (1990) Characterization of a widespread strain of methicillin-susceptible *Staphylococcus aureus* associated with nosocomial infections. *The Journal of Infectious Diseases* **162**, 759–62.

McNamara, P.J., Proctor, R.A. (2000) *Staphylococcus aureus* small colony variants, electron transport and persistent infections. *International Journal of Antimicrobial Agents* **14**, 117–22.

Mei, J.M., Nourbakhsh, F., Ford, C.W., Holden, D.W. (1997) Identification of *Staphylococcus aureus* virulence genes in a murine model of bacteraemia using signature-tagged mutagenesis. *Molecular Microbiology* **26**, 399–407.

Melish, M.E., Glasgow, L.A. (1970) The staphylococcal scalded-skin syndrome. *The New England Journal of Medicine* **282**, 1114–9.

Mermel, L.A., Farr, B.M., Sherertz, R.J. *et al.* (2001) Guidelines for the management of intravascular catheter-related infections. *Clinical Infectious Diseases* **32**, 1249–72.

Modun, B., Kendall, D., Williams, P. (1994) Staphylococci express a receptor for human transferrin: identification of a 42-kilodalton cell wall transferrin-binding protein. *Infection and Immunity* **62**, 3850–8.

Mojumdar, M., Khan, S.A. (1988) Characterization of the tetracycline resistance gene of plasmid pT181 of *Staphylococcus aureus*. *Journal of Bacteriology* **170**, 5522–8.

Mongodin, E., Bajolet, O., Cutrona, J. *et al.* (2002) Fibronectin-binding proteins of *Staphylococcus aureus* are involved in adherence to human airway epithelium. *Infection and Immunity* **70**, 620–30.

Mongodin, E., Finan, J., Climo, M.W. *et al.* (2003) Microarray transcription analysis of clinical *Staphylococcus aureus* isolates resistant to vancomycin. *Journal of Bacteriology* **185**, 4638–43.

Morath, S., Stadelmaier, A., Geyer, A. *et al.* (2002) Synthetic lipoteichoic acid from *Staphylococcus aureus* is a potent stimulus of cytokine release. *Journal of Experimental Medicine* **195**, 1635–40.

Moreillon, P., Entenza, J.M., Francioli, P. *et al.* (1995) Role of *Staphylococcus aureus* coagulase and clumping factor in pathogenesis of experimental endocarditis. *Infection and Immunity* **63**, 4738–43.

Morrissey, J.A., Cockayne, A., Hammacott, J. *et al.* (2002) Conservation, surface exposure, and in vivo expression of the Frp family of iron-regulated cell wall proteins in *Staphylococcus aureus*. *Infection and Immunity* **70**, 2399–407.

Mulvey, M.R., Chui, L., Ismail, J. *et al.* (2001) Development of a Canadian standardized protocol for subtyping methicillin-resistant *Staphylococcus aureus* using pulsed-field gel electrophoresis. *Journal of Clinical Microbiology* **39**, 3481–5.

Murchan, S., Kaufmann, M.E., Deplano, A. *et al.* (2003) Harmonization of pulsed-field gel electrophoresis protocols for epidemiological typing of strains of methicillin-resistant *Staphylococcus aureus*: a single approach developed by consensus in 10 European laboratories and its application for tracing the spread of related strains. *Journal of Clinical Microbiology* **41**, 1574–85.

Murphy, E. (1985a) Nucleotide sequence of ermA, a macrolide-lincosamide-streptogramin B determinant in *Staphylococcus aureus*. *Journal of Bacteriology* **162**, 633–40.

Murphy, E. (1985b) Nucleotide sequence of a spectinomycin adenyltransferase AAD(9) determinant from *Staphylococcus aureus* and its relationship to AAD(3″) (9). *Molecular & General Genetics* **200**, 33–9.

Murphy, E., Huwyler, L., de Freire Bastos Mdo, C. (1985) Transposon Tn554: complete nucleotide sequence and isolation of transposition-defective and antibiotic-sensitive mutants. *EMBO Journal* **4**, 3357–65.

Musser, J.M., Kapur, V. (1992) Clonal analysis of methicillin-resistant *Staphylococcus aureus* strains from intercontinental sources: association of the mec gene with divergent phylogenetic lineages implies dissemination by horizontal transfer and recombination. *Journal of Clinical Microbiology* **30**, 2058–63.

Nahaie, M.R., Goodfellow, M., Minnikin, D.E., Hajek, V. (1984) Polar lipid and isoprenoid quinone composition in the classification of *Staphylococcus*. *Journal of General Microbiology* **130**, 2427–37.

Naidu, A.S., Andersson, M., Forsgren, A. (1992) Identification of a human lactoferrin-binding protein in *Staphylococcus aureus*. *Journal of Medical Microbiology* **36**, 177–83.

National Committee for Clinical Laboratory Standards (2002a) *Methods for Dilution Antimicrobial Susceptibility Tests for Bacteria That Grow Aerobically. Approved Standard—Fifth Edition.* Document M7-A5. NCCLS, Wayne, PA.

National Committee for Clinical Laboratory Standards (2002b) *Performance Standards for Antimicrobial Disk Susceptibility Tests. Approved Standard—Seventh Edition.* Document M2-A7. NCCLS, Wayne, PA.

Nauseff, W.M., Clark, R.A. (2000) Granulocytic phagocytes. In *Mandell, Douglas, and Bennett's Principles & Practice of Infectious Diseases, 5th edition*, eds. Mandell, G.L., Bennett, J.E., Dolin, R. Churchill Livingstone, New York, 89–112.

Nesin, M., Svec, P., Lupski, J.R. *et al.* (1990) Cloning and nucleotide sequence of a chromosomally encoded tetracycline resistance determinant, tetA(M), from a pathogenic, methicillin-resistant strain of *Staphylococcus aureus*. *Antimicrobial Agents and Chemotherapy* **34**, 2273–6.

Ng, E.Y., Trucksis, M., Hooper, D.C. (1994) Quinolone resistance mediated by norA: physiologic characterization and relationship to flqB, a quinolone resistance locus on the *Staphylococcus aureus* chromosome. *Antimicrobial Agents and Chemotherapy* **38**, 1345–55.

Nguyen, M.H., Kauffman, C.A., Goodman, R.P. *et al.* (1999) Nasal carriage of and infection with *Staphylococcus aureus* in HIV-infected patients. *Annals of Internal Medicine* **130**, 221–5.

Ni Eidhin, D., Perkins, S., Francois, P. *et al.* (1998) Clumping factor B (ClfB), a new surface-located fibrinogen-binding adhesin of *Staphylococcus aureus*. *Molecular Microbiology* **30**, 245–57.

Nilsson, I.M., Patti, J.M., Bremell, T. *et al.* (1998) Vaccination with a recombinant fragment of collagen adhesin provides protection against *Staphylococcus aureus*-mediated septic death. *Journal of Clinical Investigation* **101**, 2640–9.

Nilsson, I.M., Verdrengh, M., Ulrich, R.G. *et al.* (1999) Protection against *Staphylococcus aureus* sepsis by vaccination with recombinant staphylococcal enterotoxin A devoid of superantigenicity. *The Journal of Infectious Diseases* **180**, 1370–3.

Noble, W.C. (1974) Carriage of *Staphylococcus aureus* and beta haemolytic streptococci in relation to race. *Acta Dermato-venereologica* **54**, 403–5.

Noble, W.C., Valkenburg, H.A., Wolters, C.H. (1967) Carriage of *Staphylococcus aureus* in random samples of a normal population. *The Journal of Hygiene* **65**, 567–73.

Noble, W.C., Virani, Z., Cree, R.G. (1992) Co-transfer of vancomycin and other resistance genes from *Enterococcus faecalis* NCTC 12201 to *Staphylococcus aureus*. *FEMS Microbiology Letters* **72**, 195–8.

Noda, M., Hirayama, T., Kato, I., Matsuda, F. (1980) Crystallization and properties of staphylococcal leukocidin. *Biochimica et Biophysica Acta* **633**, 33–44.

Norgaard, M., Gudmundsdottir, G., Larsen, C.S., Schonheyder, H.C. (2003) *Staphylococcus aureus* meningitis: experience with cefuroxime treatment during a 16 year period in a Danish region. *Scandinavian Journal of Infectious Diseases* **35**, 311–4.

Novick, R.P., Jiang, D. (2003) The staphylococcal saeRS system coordinates environmental signals with agr quorum sensing. *Microbiology* **149**, 2709–17.

Novick, R.P., Schlievert, P., Ruzin, A. (2001) Pathogenicity and resistance islands of staphylococci. *Microbes and Infection* **3**, 585–94.

O'Brien, L., Kerrigan, S.W., Kaw, G. *et al.* (2002a) Multiple mechanisms for the activation of human platelet aggregation by *Staphylococcus aureus*: roles for the clumping factors ClfA and ClfB, the serine-aspartate repeat protein SdrE and protein A. *Molecular Microbiology* **44**, 1033–44.

O'Brien, L.M., Walsh, E.J., Massey, R.C. *et al.* (2002b) *Staphylococcus aureus* clumping factor B (ClfB) promotes adherence to human type I cytokeratin 10: implications for nasal colonization. *Cellular Microbiology* **4**, 759–70.

O'Riordan, K., Lee, J.C. (2004) *Staphylococcus aureus* capsular polysaccharides. *Clinical Microbiology Reviews* **17**, 218–34.

O'Toole, P.W., Foster, T.J. (1986) Epidermolytic toxin serotype B of *Staphylococcus aureus* is plasmid-encoded. *FEMS Microbiology Letters* **36**, 311–4.

Ogawa, S.K., Yurberg, E.R., Hatcher, V.B. *et al.* (1985) Bacterial adherence to human endothelial cells in vitro. *Infection and Immunity* **50**, 218–24.

Ogston, A. (1880) Ueber Abscesse. *Archiv fur klinische Chirurgie* **25**, 588.

Ohshita, Y., Hiramatsu, K., Yokota, T. (1990) A point mutation in *norA* gene is responsible for quinolone resistance in *Staphylococcus aureus*. *Biochemical and Biophysical Research Communications* **172**, 1028–34.

Ohwada, A., Sekiya, M., Hanaki, H. *et al.* (1999) DNA vaccination by *mecA* sequence evokes an antibacterial immune response against methicillin-resistant *Staphylococcus aureus*. *The Journal of Antimicrobial Chemotherapy* **44**, 767–74.

Okuma, K., Iwakawa, K., Turnidge, J.D. *et al.* (2002) Dissemination of new methicillin-resistant *Staphylococcus aureus* clones in the community. *Journal of Clinical Microbiology* **40**, 4289–94.

Palma, M., Haggar, A., Flock, J.I. (1999) Adherence of *Staphylococcus aureus* is enhanced by an endogenous secreted protein with broad binding activity. *Journal of Bacteriology* **181**, 2840–5.

Park, H.M., Yoo, H.S., Oh, T.H. *et al.* (1999) Immunogenicity of alpha-toxin, capsular polysaccharide (CPS) and recombinant fibronectin-binding protein (r-FnBP) of *Staphylococcus aureus* in rabbit. *The Journal of Veterinary Medical Science* **61**, 995–1000.

Park, P.W., Rosenbloom, J., Abrams, W.R. *et al.* (1996) Molecular cloning and expression of the gene for elastin-binding protein (ebpS) in *Staphylococcus aureus*. *The Journal of Biological Chemistry* **271**, 15803–9.

Parsonnet, J. (1998) Case definition of staphylococcal TSS: A proposed revision incorporating laboratory findings, p15. In *European Conference on Toxic Shock Syndrome International Congress and Symposium Series 299 Royal*, eds. Arbuthnott, J., Furman, B. Royal Society of Medicine Press, New York.

Pasteur, L. (1880) De l'extension de la thèorie des germes á l'étiologie de quelques maladies communes. *CR Acad Sci* **90**, 1033.

Patron, R.L., Climo, M.W., Goldstein, B.P., Archer, G.L. (1999) Lysostaphin treatment of experimental aortic valve endocarditis caused by a *Staphylococcus aureus* isolate with reduced susceptibility to vancomycin. *Antimicrobial Agents and Chemotherapy* **43**, 1754–5.

Pattee, P.A., Schutzbank, T., Kay, H.D., Laughlin, M.H. (1974) Genetic analysis of the leucine biosynthetic genes and their relationship to the *ilv* gene cluster. *Annals of the New York Academy of Sciences* **236**, 175–86.

Patti, J.M., Allen, B.L., McGavin, M.J., Hook, M. (1994a) MSCRAMM-mediated adherence of microorganisms to host tissues. *Annual Review of Microbiology* **48**, 585–617.

Patti, J.M., Bremell, T., Krajewska-Pietrasik, D. *et al.* (1994b) The *Staphylococcus aureus* collagen adhesin is a virulence determinant in experimental septic arthritis. *Infection and Immunity* **62**, 152–61.

Patti, J.M., Jönsson, H., Guss, B. *et al.* (1992) Molecular characterization and expression of a gene encoding a *Staphylococcus aureus* collagen adhesin. *The Journal of Biological Chemistry* **267**, 4766–72.

Pavillard, R., Harvey, K., Douglas, D. *et al.* (1982) Epidemic of hospital-acquired infection due to methicillin-resistant *Staphylococcus aureus* in major Victorian hospitals. *Medical Journal of Australia* **1**, 451–4.

Peacock, S.J., Foster, T.J., Cameron, B.J., Berendt, A.R. (1999) Bacterial fibronectin-binding proteins and endothelial cell surface fibronectin mediate adherence of *Staphylococcus aureus* to resting human endothelial cells. *Microbiology* **145**, 3477–86.

Peacock, S.J., de Silva, I., Lowy, F.D. (2001) What determines nasal carriage of *Staphylococcus aureus*? *Trends in Microbiology* **9**, 605–10.

Peacock, S.J., de Silva, G.D., Justice, A. *et al.* (2002a) Comparison of multilocus sequence typing and pulsed-field gel electrophoresis as tools for typing *Staphylococcus aureus* isolates in a microepidemiological setting. *Journal of Clinical Microbiology* **40**, 3764–70.

Peacock, S.J., Moore, C.E., Justice, A. *et al.* (2002b) Virulent combinations of adhesin and toxin genes in natural populations of *Staphylococcus aureus*. *Infection and Immunity* **70**, 4987–96.

Peacock, S.J., Justice, A., Griffiths, D. *et al.* (2003) Determinants of acquisition and carriage of *Staphylococcus aureus* in infancy. *Journal of Clinical Microbiology* **41**, 5718–25.

Perez-Roth, E., Claverie-Martin, F., Villar, J., Mendez-Alvarez, S. (2001) Multiplex PCR for simultaneous identification of *Staphylococcus aureus* and detection of methicillin and mupirocin resistance. *Journal of Clinical Microbiology* **39**, 4037–41.

Perl, T.M., Cullen, J.J., Wenzel, R.P. *et al.* (2002) Intranasal mupirocin to prevent postoperative *Staphylococcus aureus* infections. *The New England Journal of Medicine* **346**, 1871–7.

Petti, C.A., Fowler, V.G. Jr (2003) *Staphylococcus aureus* bacteremia and endocarditis. *Cardiology Clinics* **21**, 219–33.

Pintado, V., Meseguer, M.A., Fortun, J. *et al.* (2002) Clinical study of 44 cases of *Staphylococcus aureus* meningitis. *European Journal of Clinical Microbiology & Infectious Diseases* **21**, 864–8.

Poulsen, A.B., Skov, R., Pallesen, L.V. (2003) Detection of methicillin resistance in coagulase-negative staphylococci and in staphylococci directly from simulated blood cultures using the EVIGENE MRSA Detection Kit. *The Journal of Antimicrobial Chemotherapy* **51**, 419–21.

Prevost, G., Cribier, B., Couppie, P. *et al.* (1995) Panton–Valentine leucocidin and gamma-hemolysin from *Staphylococcus aureus* ATCC 49775 are encoded by distinct genetic loci and have different biological activities. *Infection and Immunity* **63**, 4121–9.

Proctor, A.R., Kloos, W.E. (1970) The tryptophan gene cluster of *Staphylococcus aureus*. *Journal of General Microbiology* **64**, 319–27.

Proctor, R.A., Peters, G. (1998) Small colony variants in staphylococcal infections: diagnostic and therapeutic implications. *Clinical Infectious Diseases* **27**, 419–22.

Proctor, R.A., van Langevelde, P., Kristjansson, M. *et al.* (1995) Persistent and relapsing infections associated with small-colony variants of *Staphylococcus aureus*. *Clinical Infectious Diseases* **20**, 95–102.

Projan, S.J., Novick, R. (1988) Comparative analysis of five related Staphylococcal plasmids. *Plasmid* **19**, 203–21.

Pulverer, G., Pillich, J., Haklová, M. (1976) Phage-typing set for the species *Staphylococcus albus*. In *Staphylococci and Staphylococcal Diseases*, ed. Jeljaszewicz, J. Gustav Fischer Verlag, Stuttgart, 153–7.

Que, Y.A., Francois, P., Haefliger, J.A. *et al.* (2001) Reassessing the role of *Staphylococcus aureus* clumping factor and fibronectin-binding protein by expression in *Lactococcus lactis*. *Infection and Immunity* **69**, 6296–302.

Rahman, A., Izaki, K., Kato, I., Kamio, Y. (1991) Nucleotide sequence of leukocidin S-component gene (*lukS*) from methicillin resistant *Staphylococcus aureus*. *Biochemical and Biophysical Research Communications* **181**, 138–44.

Rautela, G.S., Abramson, C. (1973) Crystallization and partial characterization of *Staphylococcus aureus* hyaluronate lyase. *Archives of Biochemistry and Biophysics* **158**, 687–94.

Rennermalm, A., Li, Y.H., Bohaufs, L. *et al.* (2001) Antibodies against a truncated *Staphylococcus aureus* fibronectin-binding protein protect against dissemination of infection in the rat. *Vaccine* **19**, 3376–83.

Ribera, E., Gomez-Jimenez, J., Cortes, E. *et al.* (1996) Effectiveness of cloxacillin with and without gentamicin in short-term therapy for right-sided *Staphylococcus aureus* endocarditis. A randomized, controlled trial. *Annals of Internal Medicine* **125**, 969–74.

Roche, F.M., Meehan, M., Foster, T.J. (2003) The *Staphylococcus aureus* surface protein SasG and its homologues promote bacterial adherence to human desquamated nasal epithelial cells. *Microbiology* **149**, 2759–67.

Roder, B.L., Wandall, D.A., Frimodt-Moller, N. *et al.* (1999) Clinical features of *Staphylococcus aureus* endocarditis: a 10-year experience in Denmark. *Archives of Internal Medicine* **159**, 462–9.

Rogolsky, M. (1979) Nonenteric toxins of *Staphylococcus aureus*. *Microbiological Reviews* **43**, 320–60.

Roos, K.L., Scheld, W.M. (1997) Central nervous system infections. In *The Staphylococci in Human Disease*, vol. 1, eds. Crossley, K.B., Archer, G.L. Churchill Livingstone, New York, 413–39.

Rosenbach, F.J. (1884) *Mikro-organismen bei den Wund-Infections-Krankheiten des Menschen*. Bergmann, Weisbaden.

Ross, J.I., Farrell, A.M., Eady, E.A. *et al.* (1989) Characterisation and molecular cloning of the novel macrolide-streptogramin B resistance determinant from *Staphylococcus epidermidis*. *The Journal of Antimicrobial Chemotherapy* **24**, 851–62.

Ross, J.I., Eady, E.A., Cove, J.H. *et al.* (1990) Inducible erythromycin resistance in staphylococci is encoded by a member of the ATP-binding transport super-gene family. *Molecular Microbiology* **4**, 1207–14.

Rouch, D.A., Byrne, M.E., Kong, Y.C., Skurray, R.A. (1987) The *aacA–aphD* gentamicin and kanamycin resistance determinant of Tn*4001* from *Staphylococcus aureus*: expression and nucleotide sequence analysis. *Journal of General Microbiology* **133**, 3039–52.

Rouch, D.A., Messerotti, L.J., Loo, L.S. *et al.* (1989) Trimethoprim resistance transposon Tn*4003* from *Staphylococcus aureus* encodes genes for a dihydrofolate reductase and thymidylate synthetase flanked by three copies of IS*257*. *Molecular Microbiology* **3**, 161–75.

Rozalska, B., Wadstrom, T. (1993) Protective opsonic activity of antibodies against fibronectin-binding proteins (FnBPs) of *Staphylococcus aureus*. *Scandinavian Journal of Immunology* **37**, 575–80.

Rozalska, B., Sakata, N., Wadstrom, T. (1994) *Staphylococcus aureus* fibronectin-binding proteins (FnBPs). Identification of antigenic epitopes using polyclonal antibodies. *APMIS* **102**, 112–8.

Rydén, C., Rubin, K., Speziale, P. *et al.* (1983) Fibronectin receptors from *Staphylococcus aureus*. *The Journal of Biological Chemistry* **258**, 3396–401.

Rydén, C., Yacoub, A.I., Maxe, I. *et al.* (1989) Specific binding of bone sialoprotein to *Staphylococcus aureus* isolated from patients with osteomyelitis. *European Journal of Biochemistry* **184**, 331–6.

Saadi, A.T., Blackwell, C.C., Raza, M.W. *et al.* (1993) Factors enhancing adherence of toxigenic *Staphylococcus aureus* to epithelial cells and their possible role in sudden infant death syndrome. *Epidemiology and Infection* **110**, 507–17.

Sabat, A., Krzyszton-Russjan, J., Strzalka, W. *et al.* (2003) New method for typing *Staphylococcus aureus* strains: multiple-locus variable-number tandem repeat analysis of polymorphism and genetic relationships of clinical isolates. *Journal of Clinical Microbiology* **41**, 1801–4.

Sanabria, T.J., Alpert, J.S., Goldberg, R. *et al.* (1990) Increasing frequency of staphylococcal infective endocarditis. Experience at a University hospital, 1981 through 1988. *Archives of Internal Medicine* **150**, 1305–9.

Sanford, B.A., Thomas, V.L., Ramsay, M.A. (1989) Binding of staphylococci to mucus in vivo and in vitro. *Infection and Immunity* **57**, 3735–42.

Savolainen, K., Paulin, L., Westerlund-Wikstrom, B. *et al.* (2001) Expression of *pls*, a gene closely associated with the *mecA* gene of methicillin-resistant *Staphylococcus aureus*, prevents bacterial adhesion in vitro. *Infection and Immunity* **69**, 3013–20.

Schaefler, S., Jones, D., Perry, W. *et al.* (1981) Emergence of gentamicin- and methicillin-resistant *Staphylococcus aureus* strains in New York City hospitals. *Journal of Clinical Microbiology* **13**, 754–9.

Schennings, T., Heimdahl, A., Coster, K., Flock, J.I. (1993) Immunization with fibronectin binding protein from *Staphylococcus aureus* protects against experimental endocarditis in rats. *Microbial Pathogenesis* **15**, 227–36.

Schleifer, K.H. (1973) Chemical composition of staphylococcal cell walls. In *Staphylococci and Staphylococcal Infections*, ed. Jeljaszewicz, J. Polish Medical Publishers, Warsaw, 13–23.

Schleifer, K.H. (1983) The cell envelope. In *Staphylococci and Staphylococcal Infections*, vol. 2, eds. Easmon, C.S.F., Adlam, C. Academic Press, London, 358–428.

Schleifer, K.H., Kandler, O. (1972) Peptidoglycan types of bacterial cell walls and their taxonomic implications. *Bacteriological Reviews* **36**, 407–77.

Schlievert, P.M., Shands, K.N., Dan, B.B. *et al.* (1981) Identification and characterization of an exotoxin from *Staphylococcus aureus* associated with toxic-shock syndrome. *The Journal of Infectious Diseases* **143**, 509–16.

Schneewind, O., Fowler, A., Faull, K.F. (1995) Structure of the cell wall anchor of surface proteins in *Staphylococcus aureus*. *Science* **268**, 103–6.

Schroder, N.W., Morath, S., Alexander, C. *et al.* (2003) Lipoteichoic acid (LTA) of *Streptococcus pneumoniae* and *Staphylococcus aureus* activates immune cells via Toll-like receptor (TLR)-2, lipopolysaccharide-binding protein (LBP), and CD14, whereas TLR-4 and MD-2 are not involved. *The Journal of Biological Chemistry* **278**, 15587–94.

Schroeder, C.J., Pattee, P.A. (1984) Transduction analysis of transposon Tn*551* insertions in the *trp–thy* region of the *Staphylococcus aureus* chromosome. *Journal of Bacteriology* **157**, 533–7.

Schwab, U., Thiel, H.J., Steuhl, K.P., Doering, G. (1996) Binding of *Staphylococcus aureus* to fibronectin and glycolipids on corneal surfaces. *German Journal of Ophthalmology* **5**, 417–21.

Schwotzer, U., Kayser, F.H., Schwotzer, W. (1978) R-plasmid mediated aminoglycoside resistance in *Staphylococcus epidermidis*: Structure determination of the products of an enzyme nucleotidylating the 4'- and 4″-hydroxyl group of aminoglycoside antibiotics. *FEMS Microbiology Letters* **3**, 29–33.

Senna, J.P., Roth, D.M., Oliveira, J.S. *et al.* (2003) Protective immune response against methicillin resistant *Staphylococcus aureus* in a murine model using a DNA vaccine approach. *Vaccine* **21**, 2661–6.

Shaffer, T.E., Sylvester, R.F. Jr, Baldwin, J.N., Rheins, M.S. (1957) Staphylococcal infections in newborn infants. II. Report of 19 epidemics caused by an identical strain of *Staphylococcus pyogenes*. *American Journal of Public Health* **47**, 990–4.

Shalita, Z., Murphy, E., Novick, R.P. (1980) Penicillinase plasmids of *Staphylococcus aureus*: structural and evolutionary relationships. *Plasmid* **3**, 291–311.

Shanson, D.C., Kensit, J.C., Duke, R. (1976) Outbreak of hospital infection with a strain of *Staphylococcus aureus* resistant to gentamicin and methicillin. *Lancet* **2**, 1347–8.

Shaw, W.V. (1983) Chloramphenicol acetyltransferase: enzymology and molecular biology. *Critical Reviews in Biochemistry* **14**, 1–46.

Shinefield, H., Black, S., Fattom, A. *et al.* (2002) Use of a *Staphylococcus aureus* conjugate vaccine in patients receiving hemodialysis. *The New England Journal of Medicine* **346**, 491–6.

Shinefield, H.R., Ribble, J.C., Boris, M., Eichenwald, H.F. (1963) Bacterial interference: its effect on nursery-acquired infection with *Staphylococcus aureus*. I. Preliminary observations on artificial colonzation of newborns. *American Journal of Diseases of Children* **105**, 646–54.

Shopsin, B., Gomez, M., Waddington, M. *et al.* (2000) Use of coagulase gene (*coa*) repeat region nucleotide sequences for typing of methicillin-resistant *Staphylococcus aureus* strains. *Journal of Clinical Microbiology* **38**, 3453–6.

Shuter, J., Hatcher, V.B., Lowy, F.D. (1996) *Staphylococcus aureus* binding to human nasal mucin. *Infection and Immunity* **64**, 310–8.

Siboo, I.R., Cheung, A.L., Bayer, A.S., Sullam, P.M. (2001) Clumping factor A mediates binding of *Staphylococcus aureus* to human platelets. *Infection and Immunity* **69**, 3120–7.

Signas, C., Raucci, G., Jonsson, K. *et al.* (1989) Nucleotide sequence of the gene for a fibronectin-binding protein from *Staphylococcus aureus*: use of this peptide sequence in the synthesis of biologically active peptides. *Proceedings of the National Academy of Sciences of the United States of America* **86**, 699–703.

Silvestri, L.G., Hill, L.R. (1965) Agreement between deoxyribonucleic acid base composition and taxonomic classification of Gram-positive cocci. *Journal of Bacteriology* **90**, 136–40.

Sinha, B., Francois, P.P., Nusse, O. *et al.* (1999) Fibronectin-binding protein acts as *Staphylococcus aureus* invasin via fibronectin bridging to integrin a5β1. *Cellular Microbiology* **1**, 101–17.

Sinha, B., Francois, P., Que, Y.A. *et al.* (2000) Heterologously expressed *Staphylococcus aureus* fibronectin-binding proteins are sufficient for invasion of host cells. *Infection and Immunity* **68**, 6871–8.

Sivakanesan, R., Dawes, E.A. (1980) Anaerobic glucose and serine metabolism in *Staphylococcus epidermidis*. *Journal of General Microbiology* **118**, 143–57.

Small, P.M., Chambers, H.F. (1990) Vancomycin for *Staphylococcus aureus* endocarditis in intravenous drug users. *Antimicrobial Agents and Chemotherapy* **34**, 1227–31.

Smith, C.D., Pattee, P.A. (1967) Biochemical and genetic analysis of isoleucine and valine biosynthesis in *Staphylococcus aureus*. *Journal of Bacteriology* **93**, 1832–8.

Smyth, C.J., Möllby, R., Wadström, T. (1975) Phenomenon of hot–cold hemolysis: chelator-induced lysis of sphingomyelinase-treated erythrocytes. *Infection and Immunity* **12**, 1104–11.

Sreedharan, S., Peterson, L.R., Fisher, L.M. (1991) Ciprofloxacin resistance in coagulase-positive and -negative staphylococci: role of mutations at serine 84 in the DNA gyrase A protein of *Staphylococcus aureus* and *Staphylococcus epidermidis*. *Antimicrobial Agents and Chemotherapy* **35**, 2151–4.

Sriskandan, S., McKee, A., Hall, L., Cohen, J. (1997) Comparative effects of clindamycin and ampicillin on superantigenic activity of *Streptococcus pyogenes*. *The Journal of Antimicrobial Chemotherapy* **40**, 275–7.

Stackebrandt, E., Ludwig, W., Weizenegger, M. *et al.* (1987) Comparative 16S rRNA oligonucleotide analyses and murein types of round-spore-forming bacilli and non-spore-forming relatives. *Journal of General Microbiology* **133**, 2523–9.

Steinhuber, A., Goerke, C., Bayer, M.G. *et al.* (2003) Molecular architecture of the regulatory locus *sae* of *Staphylococcus aureus* and its impact on expression of virulence factors. *Journal of Bacteriology* **185**, 6278–86.

Stiles, B.G., Garza, A.R., Ulrich, R.G., Boles, J.W. (2001) Mucosal vaccination with recombinantly attenuated staphylococcal enterotoxin B and protection in a murine model. *Infection and Immunity* **69**, 2031–6.

Strasters, K.C., Winkler, K.C. (1963) Carbohydrate metabolism of *Staphylococcus aureus*. *Journal of General Microbiology* **33**, 213–29.

Strommenger, B., Kettlitz, C., Werner, G., Witte, W. (2003) Multiplex PCR assay for simultaneous detection of nine clinically relevant antibiotic resistance genes in *Staphylococcus aureus*. *Journal of Clinical Microbiology* **41**, 4089–94.

Sun, Q., Smith, G.M., Zahradka, C., McGavin, M.J. (1997) Identification of D motif epitopes in *Staphylococcus aureus* fibronectin-binding protein for the production of antibody inhibitors of fibronectin binding. *Infection and Immunity* **65**, 537–43.

Swenson, J.M., Williams, P.P., Killgore, G. *et al.* (2001) Performance of eight methods, including two new rapid methods, for detection of oxacillin resistance in a challenge set of *Staphylococcus aureus* organisms. *Journal of Clinical Microbiology* **39**, 3785–8.

Switalski, L.M., Rydén, C., Rubin, K. *et al.* (1983) Binding of fibronectin to *Staphylococcus* strains. *Infection and Immunity* **42**, 628–33.

Takeuchi, O., Hoshino, K., Akira, S. (2000) Cutting edge: TLR2-deficient and MyD88-deficient mice are highly susceptible to *Staphylococcus aureus* infection. *Journal of Immunology* **165**, 5392–6.

Tennent, J.M., Young, H.K., Lyon, B.R. *et al.* (1988) Trimethoprim resistance determinants encoding a dihydrofolate reductase in clinical isolates of *Staphylococcus aureus* and coagulase-negative staphylococci. *Journal of Medical Microbiology* **26**, 67–73.

Tenover, F.C., Lancaster, M.V., Hill, B.C. *et al.* (1998) Characterization of staphylococci with reduced susceptibilities to vancomycin and other glycopeptides. *Journal of Clinical Microbiology* **36**, 1020–7.

Tenover, F.C., Weigel, L.M., Appelbaum, P.C. *et al.* (2004) Vancomycin-resistant *Staphylococcus aureus* isolate from a patient in Pennsylvania. *Antimicrobial Agents and Chemotherapy* **48**, 275–80.

Thakker-Varia, S., Jenssen, W.D., Moon-McDermott, L. *et al.* (1987) Molecular epidemiology of macrolides-lincosamides-streptogramin B resistance in *Staphylococcus aureus* and coagulase-negative staphylococci. *Antimicrobial Agents and Chemotherapy* **31**, 735–43.

Thodis, E., Bhaskaran, S., Pasadakis, P. *et al.* (1998) Decrease in *Staphylococcus aureus* exit-site infections and peritonitis in CAPD patients by local application of mupirocin ointment at the catheter exit site. *Peritoneal Dialysis International* **18**, 261–70.

Thomas, W.D. Jr, Archer, G.L. (1989) Mobility of gentamicin resistance genes from staphylococci isolated in the United States: identification of Tn4031, a gentamicin resistance transposon from *Staphylococcus epidermidis*. *Antimicrobial Agents and Chemotherapy* **33**, 1335–41.

Ton-That, H., Liu, G., Mazmanian, S.K. *et al.* (1999) Purification and characterization of sortase, the transpeptidase that cleaves surface proteins of *Staphylococcus aureus* at the LPXTG motif. *Proceedings of the National Academy of Sciences of the United States of America* **96**, 12424–9.

Townsend, D.E., Grubb, W.B., Ashdown, N. (1983) Gentamicin resistance in methicillin-resistant *Staphylococcus aureus*. *Pathology* **15**, 169–74.

Townsend, D.E., Ashdown, N., Greed, L.C., Grubb, W.B. (1984) Analysis of plasmids mediating gentamicin resistance in methicillin-resistant *Staphylococcus aureus*. *The Journal of Antimicrobial Chemotherapy* **13**, 347–52.

Trucksis, M., Wolfson, J.S., Hooper, D.C. (1991) A novel locus conferring fluoroquinolone resistance in *Staphylococcus aureus*. *Journal of Bacteriology* **173**, 5854–60.

Tung, H., Guss, B., Hellman, U. *et al.* (2000) A bone sialoprotein-binding protein from *Staphylococcus aureus*: a member of the staphylococcal Sdr family. *Biochemical Journal* **345**, 611–9.

Uehara, Y., Nakama, H., Agematsu, K. *et al.* (2000) Bacterial interference among nasal inhabitants: eradication of *Staphylococcus aureus* from nasal cavities by artificial implantation of *Corynebacterium* sp. *The Journal of Hospital Infection* **44**, 127–33.

Uehara, Y., Kikuchi, K., Nakamura, T. *et al.* (2001a) H_2O_2 produced by viridans group streptococci may contribute to inhibition of methicillin-resistant *Staphylococcus aureus* colonization of oral cavities in newborns. *Clinical Infectious Diseases* **32**, 1408–13.

Uehara, Y., Kikuchi, K., Nakamura, T. *et al.* (2001b) Inhibition of methicillin-resistant *Staphylococcus aureus* colonization of oral cavities in newborns by viridans group streptococci. *Clinical Infectious Diseases* **32**, 1399–407.

Uhlén, M., Guss, B., Nilsson, B. *et al.* (1984) Complete sequence of the staphylococcal gene encoding protein A. A gene evolved through multiple duplications. *The Journal of Biological Chemistry* **259**, 1695–702.

van Griethuysen, A., Bes, M., Etienne, J. *et al.* (2001) International multicenter evaluation of latex agglutination tests for identification of *Staphylococcus aureus*. *Journal of Clinical Microbiology* **39**, 86–9.

van Langevelde, P., van Dissel, J.T., Meurs, C.J. *et al.* (1997) Combination of flucloxacillin and gentamicin inhibits toxic shock syndrome toxin 1 production by *Staphylococcus aureus* in both logarithmic and stationary phases of growth. *Antimicrobial Agents and Chemotherapy* **41**, 1682–5.

Vandenesch, F., Naimi, T., Enright, M.C. *et al.* (2003) Community-acquired methicillin-resistant *Staphylococcus aureus* carrying Panton–Valentine leukocidin genes: worldwide emergence. *Emerging Infectious Diseases* **9**, 978–84.

Vaudaux, P., Suzuki, R., Waldvogel, F.A. *et al.* (1984) Foreign body infection: role of fibronectin as a ligand for the adherence of *Staphylococcus aureus*. *The Journal of Infectious Diseases* **150**, 546–53.

Vaudaux, P., Pittet, D., Haeberli, A. *et al.* (1989) Host factors selectively increase staphylococcal adherence on inserted catheters: a role for fibronectin and fibrinogen or fibrin. *The Journal of Infectious Diseases* **160**, 865–75.

Vaudaux, P., Pittet, D., Haeberli, A. *et al.* (1993) Fibronectin is more active than fibrin or fibrinogen in promoting *Staphylococcus aureus* adherence to inserted intravascular catheters. *The Journal of Infectious Diseases* **167**, 633–41.

Vaudaux, P., Francois, P., Bisognano, C. *et al.* (2002) Increased expression of clumping factor and fibronectin-binding proteins by hemB mutants of *Staphylococcus aureus* expressing small colony variant phenotypes. *Infection and Immunity* **70**, 5428–37.

Vaudaux, P.E., Francois, P., Proctor, R.A. *et al.* (1995) Use of adhesion-defective mutants of *Staphylococcus aureus* to define the role of specific plasma proteins in promoting bacterial adhesion to canine arteriovenous shunts. *Infection and Immunity* **63**, 585–90.

Vercellotti, G.M., Lussenhop, D., Peterson, P.K. *et al.* (1984) Bacterial adherence to fibronectin and endothelial cells: a possible mechanism for bacterial tissue tropism. *The Journal of Laboratory and Clinical Medicine* **103**, 34–43.

Verhoef, J., Winkler, K.C., van Boven, C.P. (1971) Characters of phages from coagulase-negative staphylococci. *Journal of Medical Microbiology* **4**, 413–24.

Vesga, O., Groeschel, M.C., Otten, M.F. *et al.* (1996) *Staphylococcus aureus* small colony variants are induced by the endothelial cell intracellular milieu. *The Journal of Infectious Diseases* **173**, 739–42.

von Eiff, C., Bettin, D., Proctor, R.A. *et al.* (1997) Recovery of small colony variants of *Staphylococcus aureus* following gentamicin bead placement for osteomyelitis. *Clinical Infectious Diseases* **25**, 1250–1.

von Eiff, C., Heilmann, C., Peters, G. (1999) New aspects in the molecular basis of polymer-associated infections due to staphylococci. *European Journal of Clinical Microbiology & Infectious Diseases* **18**, 843–6.

von Eiff, C., Becker, K., Metze, D. *et al.* (2001a) Intracellular persistence of *Staphylococcus aureus* small-colony variants within keratinocytes: a cause for antibiotic treatment failure in a patient with Darier's disease. *Clinical Infectious Diseases* **32**, 1643–7.

von Eiff, C., Becker, K., Machka, K. *et al.* (2001b) Nasal carriage as a source of *Staphylococcus aureus* bacteremia. Study Group. *The New England Journal of Medicine* **344**, 11–6.

von Eiff, C., Peters, G., Heilmann, C. (2002) Pathogenesis of infections due to coagulase-negative staphylococci. *Lancet Infectious Diseases* **2**, 677–85.

Wann, E.R., Gurusiddappa, S., Hook, M. (2000) The fibronectin-binding MSCRAMM FnbpA of *Staphylococcus aureus* is a bifunctional protein that also binds to fibrinogen. *The Journal of Biological Chemistry* **275**, 13863–71.

Watanakunakorn, C., Burkert, T. (1993) Infective endocarditis at a large community teaching hospital, 1980–1990. A review of 210 episodes. *Medicine (Baltimore)* **72**, 90–102.

Weichhart, T., Horky, M., Sollner, J. *et al.* (2003) Functional selection of vaccine candidate peptides from *Staphylococcus aureus* whole-genome expression libraries in vitro. *Infection and Immunity* **71**, 4633–41.

Weigel, L.M., Clewell, D.B., Gill, S.R. *et al.* (2003) Genetic analysis of a high-level vancomycin-resistant isolate of *Staphylococcus aureus*. *Science* **302**, 1569–71.

Weinke, T., Schiller, R., Fehrenbach, F.J., Pohle, H.D. (1992) Association between *Staphylococcus aureus* nasopharyngeal colonization and septicemia in patients infected with the human immunodeficiency virus. *European Journal of Clinical Microbiology & Infectious Diseases* **11**, 985–9.

Weinstein, H.J. (1959) The relation between the nasal-staphylococcal-carrier state and the incidence of postoperative complications. *The New England Journal of Medicine* **260**, 1303–8.

Weisblum, B. (1985) Inducible resistance to macrolides, lincosamides and streptogramin type B antibiotics: the resistance phenotype, its biological diversity, and structural elements that regulate expression – a review. *The Journal of Antimicrobial Chemotherapy* **16**(Suppl. A), 63–90.

Welsh, J., McClelland, M. (1991) Genomic fingerprints produced by PCR with consensus tRNA gene primers. *Nucleic Acids Research* **19**, 861–6.

Williams, R.E., Rippon, J.E. (1952) Bacteriophage typing of *Staphylococcus aureus. The Journal of Hygiene* **50**, 320–53.

Wilson, S.G., Sanders, C.C. (1976) Selection and characterization of strains of *Staphylococcus aureus* displaying unusual resistance to aminoglycosides. *Antimicrobial Agents and Chemotherapy* **10**, 519–25.

Wilson, W.R., Karchmer, A.W., Dajani, A.S. *et al.* (1995) Antibiotic treatment of adults with infective endocarditis due to streptococci, enterococci, staphylococci, and HACEK microorganisms. *The Journal of the American Medical Association* **274**, 1706–13.

Working Party of the British Society for Antimicrobial Chemotherapy (1998) Antibiotic treatment of streptococcal, enterococcal, and staphylococcal endocarditis. Working Party of the British Society for Antimicrobial Chemotherapy. *Heart* **79**, 207–10.

Wu, S., de Lencastre, H., Tomasz, A. (1998) Genetic organization of the *mecA* region in methicillin-susceptible and methicillin -resistant strains of *Staphylococcus sciuri. Journal of Bacteriology* **180**, 236–42.

Yamazumi, T., Furuta, I., Diekema, D.J. *et al.* (2001) Comparison of the Vitek gram-positive susceptibility 106 card, the MRSA-Screen latex agglutination test, and mecA analysis for detecting oxacillin resistance in a geographically diverse collection of clinical isolates of coagulase-negative staphylococci. *Journal of Clinical Microbiology* **39**, 3633–6.

Yoshida, H., Bogaki, M., Nakamura, S. *et al.* (1990) Nucleotide sequence and characterization of the *Staphylococcus aureus* norA gene, which confers resistance to quinolone. *Journal of Bacteriology* **172**, 6942–9.

Yoshimura, A., Lien, E., Ingalls, R.R. *et al.* (1999) Cutting edge: recognition of Gram-positive bacterial cell wall components by the innate immune system occurs via Toll-like receptor 2. *Journal of Immunology* **163**, 1–5.

Yu, V.L., Goetz, A., Wagener, M. *et al.* (1986) *Staphylococcus aureus* nasal carriage and infection in patients on hemodialysis. Efficacy of antibiotic prophylaxis. *The New England Journal of Medicine* **315**, 91–6.

Zhang, Y.Q., Ren, S.X., Li, H.L. *et al.* (2003) Genome-based analysis of virulence genes in a non-biofilm-forming *Staphylococcus epidermidis* strain (ATCC 12228). *Molecular Microbiology* **49**, 1577–93.

Zopf, W. (1885) *Die Spaltpilze*, 3rd edn. Edward Trewendt, Breslau.

6

Coagulase-Negative Staphylococci

Roger G. Finch

Department of Microbiology and Infectious Diseases, The University of Nottingham and Nottingham City Hospital NHS Trust, Clinical Sciences Building, Hucknall Road, Nottingham NG5 1PB, UK

INTRODUCTION

Coagulase-negative staphylococci (CONS) comprise an ever-expanding group of bacteria whose medical importance has emerged in the past three decades. They now count among the most frequent of nosocomial pathogens featuring prominently among blood culture isolates, often in association with intravascular devices, and as a cause of infection of more deep-seated prosthetic implants. Clinically, infection may be silent, overt and occasionally fulminant, and this reflects the diverse pathogenic profile of this group of organisms. CONS are also characterized by an unpredictable pattern of susceptibility to commonly used antibiotics. Multiple drug resistance is common and adds to the difficulties of treating these infections.

HISTORICAL ASPECTS

Staphylococcus aureus was first described by Rosenbech in 1884. Its pathogenic profile includes local invasion, systemic spread, toxin-mediated disease and increasing resistance to commonly used antibiotics. It was included among the group of pyogenic cocci, and hence, its former description, *S. pyogenes*. In contrast, CONS were considered for many years to be non-pathogenic commensal organisms of the skin. *Staphylococcus albus* was widely used to describe all CONS as distinct from the colonial appearance of the golden pigmented *S. aureus*. The first widely accepted pathogenic role of CONS was the association of *S. saprophyticus* (novobiocin-resistant CONS) with urinary tract infections (UTIs) in women (Pereira, 1962).

The widely held view that CONS were largely commensals and indeed contaminants of clinical specimens frustrated recognition of the pathogenic potential of this group of organisms for many years. However, by the 1980s CONS had clearly been identified with a wide variety of clinical problems such as bacteremia, endocarditis of both prosthetic and native heart valves, septic arthritis, peritonitis complicating continuous ambulatory peritoneal dialysis (CAPD), mediastinitis, pacemaker associated infections, cerebrospinal fluid (CSF) shunt device infections, prosthetic joint and other orthopaedic device infections, osteomyelitis, UTI and prostatitis (Kloos and Bennerman, 1994). While *S. saprophyticus* was clearly associated with the urinary tract, the predominant species among the remaining infections was *S. epidermidis*. However, many hospital diagnostic laboratories have used the species *S. epidermidis* description loosely to encompass all CONS, further frustrating recognition of the diverse microbial nature of CONS infections. This has changed in recent years largely as a result of increased awareness of the importance of CONS infections, together with the availability of commercial identification systems.

DESCRIPTION OF THE ORGANISM

Taxonomically, CONS, together with *S. aureus*, are members of the family Micrococcaceae. They are Gram-positive facultative anaerobes which appear in clusters, are non-motile, non-spore forming and catalase positive and in general do not produce the enzyme coagulase. A thin capsule may be detected in some strains.

CONS are divided into more than 30 species (Table 6.1) and more than a dozen subspecies, of which approximately half have been associated with humans (Kloos and Bennerman, 1994). The remainder are associated with domestic and other species of mammals. The relatedness of these species has been confirmed by guanine + cytosine ratios. DNA sequence homology of >50% has been used to group the species, although a number of species are too distantly related to fit into this arrangement.

Coagulase production is generally absent among CONS, although some strains of *S. intermedius* and *S. hyicus* are weak producers. Heat-stable thermonuclease is produced by *S. intermedius*, *S. hyicus*, *S. schleiferi* and some strains of *S. carnosus*, *S. epidermidis* and *S. simulans* and permits cleavage of nucleic acids.

Classification

CONS have been divided into various species based on a variety of characteristics including colonial morphology, coagulase and phosphatase production, acid formation from maltose, sucrose, D-mannitol, D-trehalose and D-xylose as well as susceptibility to novobiocin using

Table 6.1 Coagulase-negative staphylococci associated with humans

Staphylococcus epidermidis
S. auricularis
S. capitis
S. caprae
S. cohnii
S. haemolyticus
S. hominis
S. lugdunensis
S. pasteuri
S. saccarolyticus
S. saprophyticus
S. schleiferi
S. simulans
S. warneri
S. xylosus

Principles and Practice of Clinical Bacteriology Second Edition   Editors Stephen H. Gillespie and Peter M. Hawkey
© 2006 John Wiley & Sons, Ltd

Table 6.2 Laboratory characteristics of the genus *Staphylococcus*

	S. aureus	*S. intermedius*	*S. hyicus*	*S. chromogenes*	*S. epidermidis*	*S. capitis* subsp. *capitis*	*S. auricularis*	*S. saccharolyticus*	*S. haemolyticus*	*S. hominis* subsp. *hominis*	*S. warneri*	*S. simulans*	*S. saprophyticus* subsp. *saprophyticus*
Growth anaerobically	+	+	+	+	+	w	w	+[a]	+	w	+	+	w
Oxidase	–	–	–	–	–	–	–	–	–	–	–	–	–
VP[e]	+	–	–	–	+	+	d	?	+	+	+	–	+
Coagulase[e]	+	+	d	–	–	–	–	–	–	–	–	–	–
Acid from													
Lactose	+	+	+	+	D		–	–	D	+	–	+	+
Maltose[e]	+	–	–	d	+	–	d	–	+[b]	+	d	–	+
Mannitol	+	+	+	d	–	+	–	–	+[b]	–	d	+	+
Fructose	+	+	+	+	+	+	+	+	d	+	+	+	+
Sucrose	+	+	+	+	+	+	d	–	+	+	+	+	+
Trehalose[e]	+	+	+	+	–	+	+	–	+	+	+	+	+
Xylose	–	–	–	–	–	–	–	–	–	–	–	–	–
Cellobiose	–	–	–	–	–	–	–	?	–	–	–	–	–
Raffinose	–	–	–	–	–	–	–	–	–	–	–	–	–
Mannose	+	+	+	+	+	+	–	+	–	–	–	d	–
Phosphatase[e]	+	+	+	+	+	–	–	?	d	–	–	w	–
Nitrate	+	+	+	+	+	+	d	+	+	–	–	+	–
Arginine[e]	+	+	+	+	+	–	–	+	+	+[d]	+[d]	+	–
Urea	d	D	+	+	+	–	–	?	–	+	+	+	d
Protease	+	+	+	+	w	w	–	–	–	–	–	–	+
Novobiocin[e]	s	s	s	s	s	s	s	s	s	s	s	s	r

Table 6.2 (Continued)

	S. cohnii subsp. cohnii	S. xylosus	S. caprae	S. carnosus	S. caseolyticus	S. arlettae	S. equorum	S. gallinarum	S. kloosii	S. lentus	S. sciuri	S. lugdunensis	S. schleiferi subsp. schleiferi
Growth anaerobically	w	w	+	+	w	–	–	w	–	–	w	+	+
Oxidase	–	–	–	–	+	–	–	–	D	w	+	–	–
VP[e]	+	–	+	+	–	–	–	–	–	–	–	+	+
Coagulase[e]	–	–	–	–	–	–	–	–	–	–	–	–	–
Acid from													
Lactose	–	+	+	d	+	+	+	d	d	+	–	–	–
Maltose[e]	+	+	d	–	+	+	+	+	+	d	+	+	–
Mannitol	+	+	–	+	–	+	+	+	+	+	+	–	–
Fructose	+	+	–	+	+	+	+	+	+	+	+	+	–
Sucrose	–	+	–	d	d	+	+	+	–	+	+	+	d
Trehalose[e]	+	+	+	d	d	+	+	+	+	+	+	+	–
Xylose	–	+	–	–	–	+	+	+	–	–	–	–	–
Cellobiose	–	–	–	–	?	–	–	+	–	+	+	–	–
Raffinose	–	–	–	–	?	+	+	+	–	+	–	–	–
Mannose	+	+	+	+	–	d	+	+	–	+	d	+	+
Phosphatase[e]	+	+	+	?	?	+	+	+	+	+	+	–	+
Nitrate	–	+	+	+	+	–	+	+	–	+	+	+	+
Arginine[e]	–	–	+	?	?	–	–	?	–	?	?	–[c]	+
Urea[e]	d	+	+	–	?	–	+	+	–	d	–	?	?
Protease	–	–	–	–	+	–	–	–	–	w	+	?	?
Novobiocin[e]	r	r	s	s	s	r	r	r	r	r	r	s	s

+, 85–100% strains are positive (all, most, many, usually); –, 0–15% strains positive (none, one, few, some); ?, not known or insufficient information; d, 16–84% strains positive (many, several, some); w, weak reaction or growth; D, different reactions given by lower taxa (genera, species, varieties); s, sensitive; r, resistant.

[a] No growth anaerobically.

[b] Usual reaction.

[c] Omithine decarboxylated.

[d] Inferred reaction.

[e] These tests are usually sufficient to identify the species that may infect humans.

Reproduced from Barrow and Feltham (1993) with permission of Cambridge University Press.

a 5-μg disc (Pfaller and Herwald, 1988). More extensive biochemical testing is necessary to speciate less common strains such as *S. warneri*, *S. capitis*, *S. simulans* and *S. hominis*, although little call is made for this outside reference or research laboratories (Kloos, Schleifer and Götz, 1991). Table 6.2 summarizes the major differentiating biochemical features (Barrow and Feltham, 1993). *Staphylococcus intermedius* is coagulase positive as are many strains of *S. hyicus*.

PATHOGENICITY

Infection associated with implanted medical devices is currently the most important pathogenic consequence of CONS infection. This is recognized as a two-stage process initially of bacterial attachment, followed by cell division within an extracellular matrix to produce a biofilm. The process of bacterial attachment to cells and inanimate surfaces has been subject to much investigation (Tenney *et al.*, 1986). In the case of CONS, it is clear that the mechanism is complex, strain variable and affected by the nature of the solid surface and the environment in which attachment occurs (Figure 6.1).

Biomaterials are largely synthetic polymers but may occasionally be natural substances. Physicochemical factors affecting attachment include electrostatic forces and hydrophobicity. While more hydrophobic strains in general attach more readily to these surfaces, there is considerable variation which may be further affected by nutrient limitation, pH and variation in carbon dioxide tension (pCO_2) (Denyer *et al.*, 1990).

In addition to mechanisms surrounding device-associated infection, CONS are also known to express an increasing number of other virulence factors to varying degrees among the different species (Gemmel, 1986). These include hemolysins, phosphatases, thermonuclease, lipase, galactosidase, pyrrolidonyl arylamidase and various decarboxylases.

Attachment

The adherence of microorganisms to biomaterials varies in relation to the nature of the material and the surface characteristics of the microorganism. In addition to physiochemical properties, including hydrophobicity and polarity, the implanted medical device readily becomes coated with host substances such as collagen, fibrinogen, fibronectin, vitronectin and laminin (von Eiff, Peters and Heilmann, 2002). Variation in binding affinities to these proteins can be demonstrated; *S. haemolyticus* binds more strongly than strains of *S. epidermidis* (Paulsson, Ljungh and Wadström, 1992). The binding of *S. epidermidis* to fibronectin may be linked to their ability to colonize damaged tissues (Wadström, 1989). In contrast, strains of *S. saprophyticus* associated with urine infections have a higher capacity to adhere to laminin. However, the importance of polymer conditioning with these various host proteins in promoting attachment still remains a source of controversy (Mack, 1999).

The search for a specific adhesin continues. A capsular immuno-dominant polysaccharide adhesin (PS/A) with a molecular weight of >500 000 has been described (Tojo *et al.*, 1988; Mack *et al.*, 1994). Antibodies to PS/A have been shown to block adherence of *S. epidermidis* (strain RP-62A) to silastic catheter surfaces. The same adhesin has been demonstrated in other species such as *S. capitis* subsp. *ureolyticus*. Anti-adhesin antibodies have also been shown to enhance opsonophago-cytosis by polymorphonuclear leukocytes. Another protein that may be associated with attachment is a 220-kDa surface antigen of *S. epidermidis*, which immunogold electron microscopy suggests could be fimbrial in nature (Timmerman *et al.*, 1991). In general adhesin-positive isolates have been shown *in vitro* to be more adherent to medical devices (Muller *et al.*, 1993). Autolysin-mediated attachment has also been described (Heilmann *et al.*, 1997). Cellular accumulation follows as a result of polysaccharide intercellular adhesin (PIA) (Heilmann *et al.*, 1996). Molecular and genetic studies have indicated that PIA, and possibly capsular PS/A, are encoded by the *ica* operon (Heilmann *et al.*, 1996; McKenney *et al.*, 1998).

The presence of *ica* gene appears to correlate with pathogenicity. Blood culture and device-associated isolates are significantly more likely to contain this gene compared with non-pathogenic controls (Ziebutr *et al.*, 1997; Fitzpatrick *et al.*, 2002). Such isolates are also more likely to possess the *mecA* gene (Frebourg *et al.*, 2000).

Slime

Staphylococci, including CONS, produce variable amounts of capsular material. External to this is extracellular slime material produced under certain circumstances of growth (Bayston and Rodgers, 1990), usually following attachment to a foreign surface.

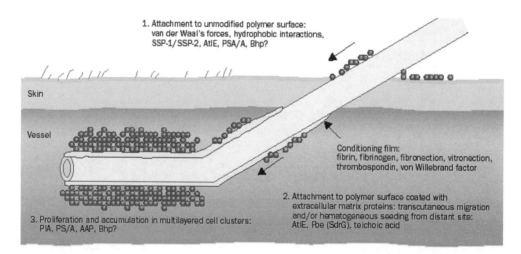

Figure 6.1 Schematic model of the phases involved in *Staphylococcus epidermidis* biofilm formation and bacterial factors involved. Biofilm-associated (Bap) homologous protein (Bhp) whose homologous counterpart in *S. aureus* has been shown to be involved in *S. aureus* biofilm formation is labeled with a question mark. AAP = accumulation-associated protein; AtlE = autolysin; Fbe = fibrinogen-binding protein; PIA = polysaccharide intercellular adhesin; PS/A = polysaccharide/adhesin; SdrG = serine-aspartate-repeat-containing protein G; SSP-1/SSP-2 = staphylococcal surface proteins. Reproduced with permission of The Lancet Infectious Diseases.

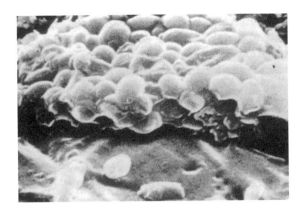

Figure 6.2 Microcolony of coagulase-negative staphylococci growing within a biofilm on the internal surface of a central venous catheter. Reprinted with permission from Elsevier (The Lancet, 2002, Vol 2, pp 677–685).

The term glycocalyx encompasses both capsular and extracellular slime material within which the microorganisms grow to produce a biofilm. The importance of extracellular slime material and biofilm formation lies in the relationship of CONS to medical device-associated infections and colonization of the uroepithelium. Extracellular slime production appears to occur more readily with certain strains of CONS associated with foreign bodies and in particular indwelling medical devices (Figure 6.2). The function of slime has been much debated, but it is believed to act as an ion-exchange resin for nutritional purposes. Bacterial growth within a biofilm differs markedly from the planktonic state.

Chemical analysis of slime indicates that it is a glycoconjugate composed of glycerol phosphate, D-alanine, N-acetylglucosamine and usually glucose (Kotilainen et al., 1990; Hussain, Hastings and White, 1991). It is now described as PIA. Production is strain variable and affected by circumstances of growth (Hussain et al., 1992a). Class I and class II phenotypes have been described (Barker, Simpson and Christensen, 1990). The former demonstrate good slime production under aerobic conditions but not in anaerobic states, whereas class II produce little or no slime under either condition. Slime production has been measured qualitatively and quantitatively. The Congo Red agar test has been widely used for individual colony testing, while radiochemical analyses have also been described (Hussain et al., 1992b). Slime production is now known to be under the genetic control of the *ica* gene cluster (Fey et al., 1999).

Biofilm-associated CONS appear to be protected from a number of control mechanisms. Slime from CONS interferes with various components of phagocytosis (Johnson et al., 1986). Leukocyte migration is inhibited, as is degranulation. Oxygen-dependent intracellular killing, as measured by chemiluminescence, is also impaired. In addition, complement and immunoglobulin G (IgG) opsonic phagocytosis and killing have also been shown to be inhibited (Noble, Grant and Hajen, 1990). It is thought that surface-exposed proteins may be important in this regard (Shiro et al., 1994). Human T-cell lymphocyte proliferation is inhibited (Gray et al., 1984). This may be a direct inhibitory effect or may be mediated by enhanced production of prostaglandin E2 (Stout et al., 1992). In addition, slime production has been associated with increased production of interleukin I and tumor necrosis factor α (Beutler and Cerami, 1989; Dinarello, 1989) as have changes in Ig production as a result of impaired B-cell blastogenesis. While these phenomena are of interest, they do not clarify the exact molecular and genetic nature whereby slime production is inhibitory to host defenses.

Most reports have focused on *S. epidermidis* and have generally, but not universally, shown slime production to be linked with pathogenic potential in relation to device-associated infections. *Staphylococcus saprophyticus* can also produce slime; urea appears to be an essential nutrient and is associated with biofilm production and the formation of urinary struvite and apatite stones (McLean *et al.*, 1985).

Iron-Scavenging Systems

Many bacteria have developed sophisticated iron-scavenging systems, as this element is essential to survival. Such systems have been well characterized for Gram-negative bacteria; however, their significance among Gram-positive pathogens is now emerging. *Staphylococcus epidermidis*-associated CAPD catheter-associated infections express two iron-regulated proteins in the cytoplasmic membrane when grown in used dialysate (Williams, Denyer and Finch, 1988). These have been shown to be antigenic and in a rat chamber model are immunodominant (Modun, Kendall and Williams, 1994). Furthermore, using lectin-binding studies human transferrin, but not transferrin from other species, has been shown to bind to a 42-kDa transferring receptor protein (Modun and Williams, 1999). This suggests that *S. epidermidis* has developed sophisticated iron-scavenging systems which include not only siderophores but also receptor and transporter systems to internalize this essential element (Cockayne *et al.*, 1998). Staphyloferrin A and B are siderophores which have been identified in *S. hyicus* and *S. epidermidis* (Lindsey, Riley and Mee, 1994).

EPIDEMIOLOGY

Normal Habitat

The distribution and concentration of CONS on the human skin and surface mucosae varies. The density of organisms ranges from 10^3 to 10^6 CFU/cm^2, with the lowest counts in the dry areas of the skin. They are particularly concentrated in the perineum, inguinal region, axillae and anterior nares. The distribution pattern is reflected in the relative frequency with which the various species are isolated from pathological specimens. For example, *S. capitis* subsp. *capitis* generally prefers the scalp and face. *Staphylococcus epidermidis* is the most widely distributed species on the skin and achieves the highest concentrations in moist skin areas. *Staphylococcus haemolyticus* and *S. hominis* are most numerous in the perineum, axillae and inguinal regions in association with apocrine glands. *Staphylococcus auricularis* is largely confined to the external auditory meatus. *Staphylococcus warneri* and *S. lugdunensis* are less frequent commensals than are other species, but when present they are widely distributed (Kloos and Bennerman, 1994).

The increasing importance of CONS as human pathogens has been recognized over the past 25 years. Prior to this, *S. epidermidis* was an uncommon pathogen, although *S. saprophyticus* was recognized as an occasional cause of community-acquired UTI in sexually active young women. However, community-acquired CONS infections are rare but may arise in patients who have been discharged from hospital following the insertion of prosthetic implants such as joints, heart valves and intracardiac patches. Infection by these organisms is generally a delayed expression of hospital-acquired infection. However, the recent growth in CAPD, earlier discharge of patients from hospital and interest in home intravenous drug therapy suggest that CONS infections will increase within the community.

Hospital

It is in the hospital setting that CONS infections have made their greatest impact. Many studies confirm the rising frequency of CONS infections (Sehaberg, Culver and Gaynes, 1991). This is demonstrated in surveys of nosocomial bacteremia; rates of 5–10% in the early 1980s increased to 25–30% in the 1990s for all nosocomial bacteremias (Bannerjee et al., 1991). Currently CONS are now the leading cause of hospital-acquired bacteremia as demonstrated in the National

Nosocomial Infection Survey in the USA (Sehaberg, Culver and Gaynes, 1991). These surveys suggest infection rates of 5.2–38.6 cases per 10 000 admissions in one institution (Martin, Pfaller and Wenzel, 1989). This increase is greatest in patients within critical care areas such as the intensive care unit (ICU), hematology and oncology facilities, dialysis and other high-dependency units, although no service is entirely free of this problem. CONS bacteremia among neonatal ICUs has increased and is more common in the low birth weight infant of <1000 g (Freeman et al., 1987). Nosocomial CONS bacteremia often prolongs hospitalization. Furthermore, there is an overall mortality of 30.5% and an attributable mortality of 13.6%, with a relative risk ratio for dying of 1.8 in comparison with controls (Martin, Pfaller and Wenzel, 1989).

One population of patients in whom CONS infections have figured prominently are those with malignant disease undergoing cytotoxic chemotherapy. Gram-positive infections are an increasing cause of neutropenic febrile episodes, among which CONS predominate (EORTC International Antimicrobial Therapy Cooperative Group, 1990). This not only reflects the use of intravascular devices such as Hickman catheters and dialysis lines (Quinn, Counts and Meyers, 1986) but also an absolute reduction in Gram-negative bacteremia as a result of improved chemoprophylactic regimens. More aggressive cytotoxic regimens also increase the risks of infection as a result of their effects on skin and mucous membranes.

While most nosocomial CONS infections are device associated, others have been clearly linked to infections at other sites such as skin and soft tissue, septic arthritis, mediastinitis, native valve endocarditis, ophthalmitis, pneumonia and meningitis (Kloos and Bennerman, 1994). In many situations CONS infection has been linked to surgical wound infection or mechanical ventilation. Although the possibility of contamination is high, it is important not to dismiss the isolation of CONS from a range of non-device-associated situations without assessing carefully the full aspects of a case.

Although most CONS infections arise from the host's normal flora, there is increasing evidence to suggest that some are the result of nosocomial spread. Studies have indicated transmission from staff to patients, while in other reports epidemic strains have been found circulating within high-dependency units such as a neonatal ICU (Huebner et al., 1994). Part of the difficulty in defining the epidemiology of nosocomial CONS infections accurately has been the lack of good data concerning normal carriage rates of the various CONS species, and the factors that affect these. The increasing availability of various molecular typing systems has greatly improved this epidemiological data (Bingen, Denamur and Elion, 1994).

CLINICAL FEATURES

Infection with CONS rarely exhibits characteristic clinical symptoms or signs. Patients may have a fever which is usually low grade. Occasionally, infection may have more striking systemic features of the sepsis syndrome with multiorgan involvement. Other signs are those of infection at the affected site and are described below.

Intravascular Line Infections

CONS are the commonest cause of intravascular line infections (Goldman and Pier, 1993). Clinical evidence for sepsis includes inflammation, with or without a purulent discharge at the insertion site. Evidence of systemic infection and occasionally metastatic sepsis at other sites may be present. Endocarditis is always a concern with regard to infected, centrally placed vascular catheters.

Peripheral blood cultures are often collected simultaneously with cultures collected through the catheter and compared in an attempt to define a pathogenic role. As peripheral devices are more readily removed, they have been sampled in various ways.

Some recommend quantitative blood cultures in this situation, although this can only increase the risk of contamination. Semiquantitative cultures of the distal (intravascular) section of peripheral venous catheters, cultured by rolling the cut off tip across an agar plate, have provided guidance in differentiating between contamination and true colonization; counts of >15 CFU have been used to indicate catheter-related sepsis (Maki, Weise and Sarafin, 1977). However, this approach is not usually of great practical value in the clinical management of such patients. Alternative approaches have included intraluminal brushings of the catheter in situ (Markus and Buday, 1989), semiquantitative cultures of the hub and skin (Cercenado et al., 1990) and flushing of removed catheter systems, a section of which is then incubated in broth media; the ability to distinguish infection from contamination is poor and may therefore lead to confusion in the management of such patients.

Bacteremia and Endocarditis

CONS are now among the most frequent of blood culture isolates, especially in hospitalized patients. However, the risk of contamination with commensal CONS, either during collection or as a result of laboratory manipulations, cannot be entirely eliminated, despite the use of automated blood culture systems. As a result it is generally considered desirable to have evidence that a similar isolate be present in at least two sets of blood cultures collected from different sites at different times. In routine diagnostic laboratories, evidence of similarity is often restricted to colonial morphology, coagulase activity and antibiogram with or without biochemical testing.

The microbiological diagnosis of infective endocarditis affecting a prosthetic valve or implant, or indeed a native valve, presents a particular challenge. Here repeated isolation of a similar strain from samples collected from different venepuncture sites over a period of time and viewed in the context of the overall clinical picture provides the best evidence for active infection (Caputo et al., 1987). Among CONS causing endocarditis, S. epidermidis, S. capitis, S. warneri and more recently S. lugdunensis predominate (Herchline and Ayers, 1991).

CSF Shunts/Valves

CSF-shunting systems include valves, ventriculoatrial and ventriculoperitoneal systems used in the management of hydrocephalus. Ventriculoperitoneal shunts are currently preferred. Unfortunately these are subject to microbial contamination, often at the time of insertion or subsequent manipulations, and they cause considerable difficulties with regard to early diagnosis and management. Those inserted in the first few months of life are at greatest risk of infection with rates of 10–20%, of which CONS accounts for approximately three-quarters of the infecting organisms (Bayston, 1994). Shunt infections may be internal when they affect the lumen of the shunt or less commonly external when the surgical insertion site and surrounding tissues become infected. Ventriculitis is common and requires examination of CSF for its diagnosis. Aspiration of CSF from the valve chamber of the shunt must be carried out carefully to avoid introducing infection; adequate skin disinfection is essential. Ventriculoperitoneal shunts may be associated with low-grade peritonitis. Bacteremia is a rare complication of ventriculoarterial shunt infection.

Dialysis Infections

CONS are the most common cause of dialysis-associated infections and together with S. aureus comprise the leading bacteria responsible for infection of either short-term or chronic dialysis catheter devices (Cheesbrough, Finch and Burden, 1986). The temporary use of double-lumen catheters for hemodialysis is often complicated by infection, usually with microorganisms of skin origin including CONS. These infections may be clinically evident but are often covert and are only recognized once the catheter is removed and cultured. Microbial attachment, colonization and biofilm production are common and are particularly associated with CONS (Baddour et al., 1986). The diagnosis is usually made clinically and supported by positive blood cultures which may be collected from a peripheral vein as well as through the device.

Arteriovenous fistulae created for hemodialysis are prone to infection during needle insertion. Staphylococci are the leading pathogens; S. aureus predominates but CONS may also occur. The risks from both these dialysis catheter-associated infections are recurrence and metastatic spread of infection. Diagnosis is largely based on clinical evidence of sepsis and positive blood cultures. Local evidence of inflammation with exudate formation may be present. Removal of dialysis catheters provides an opportunity to culture the endovascular and subcutaneous portions through techniques such as catheter rolling and flushing (Maki, Weise and Sarafin, 1977). Examination by scanning electron microscopy to visualize microcolony formation has been described within the research environment.

CONS are the most frequent cause of peritonitis in patients managed by CAPD (Working Party of the BSAC, 1987). The microbiological diagnosis is partly hampered by the low colony counts present in spent dialysate, which may be as few as <1 CFU/ml. Furthermore, the risk of contaminating samples at the time of collection or during processing, often with skin microorganisms, presents an added problem. The diagnosis is therefore based on clinical and microbiological evidence of peritonitis. The latter should include evidence of a minimum of 100 white blood cells per microliter. The cellular effluent is made up of polymorphonuclear cells and peritoneal macrophages (Verbrugh et al., 1983). Direct Gram staining of spun dialysate yields a positive result in only about 15% of instances and thus lacks sensitivity. Because of the low bacterial counts, techniques aimed at improving the yield have been adopted. Centrifuged deposits and Millipore-filtered samples are used in some diagnostic laboratories. However, broth enrichment culture is now widely adopted as the simplest diagnostic approach, often using conventional blood culture systems (Working Party of the BSAC, 1987). Isolation, identification and antibiotic susceptibility should follow. Recurrent infections are not uncommon and reflect either persistence of the original infection, usually from continued colonization of the implanted Tenkhoff catheter, or reinfection with a different microorganism. An interval of 6 weeks between two episodes of infection is often used to distinguish recurrence from reinfection. However, this is an arbitrary separation which does not always stand up to close microbiological scrutiny. Hence it is important to retain isolates in case more detailed microbiological analysis becomes necessary.

Urine

Midstream specimens of urine (MSU) are generally adequate for the diagnosis of UTI. CONS count for approximately 10% of UTIs in young women. Staphylococcus saprophyticus and S. epidermidis are the most frequent CONS isolates (Latham, Running and Stamm, 1983). The risk of contamination of the sample can present difficulties when interpreting culture findings, particularly in the absence of pyuria. Standard quantitative microbiological techniques are satisfactory in indicating levels of significance. Counts of 10^5 CFU/ml are considered significant, but in patients with dysuria and frequency counts as low as 10^3 CFU/ml can also be found. CONS grows satisfactorily on standard laboratory media used in the diagnosis of UTI such as cysteine lactose electrolyte-deficient and MacConkey agar. Colonial morphology, coagulase negativity and resistance to a 5-μg novobiocin disk is generally satisfactory for the routine identification of S. saprophyticus. Isolation in pure culture is usual, although repeat mixed cultures may occur. It is rarely necessary to resort to suprapubic aspiration of urine.

Soft Tissue and Deep Wound Samples

In the case of soft tissue and wound infections, aspirated material for culture on solid media is preferred. Surface swabs will almost certainly be contaminated by normal flora. Gram stain of infected material is valuable as a means of initial assessment and as an aid to interpretation of subsequent culture findings. In the case of infected medical implants such as joint prostheses, arthrocentesis and, when appropriate, surgical exploration, allow sampling of synovial fluid, the surrounding soft tissue as well as the bone cement interface. This permits a full range of cultures including tissue homogenates to be set up. When infected prosthetic valves are removed, vegetations and infected valve rings should be processed carefully to avoid laboratory contamination. Culturing tissue homogenates may improve the yield. Likewise, an infected joint prosthesis should also be subject to microbiological sampling and culture in order to provide the best information for eradicating deep-seated infection.

LABORATORY DIAGNOSIS

Specimens

CONS may be isolated from a wide range of laboratory specimens submitted for bacteriological analysis. However, difficulties in defining their pathogenic role are common. Repeated isolation from a normally sterile body site presents few problems when ascribing aetiology. However, the fact that CONS are ubiquitous members of the normal flora of the skin and surface mucosae inevitably leads to concerns with regard to contamination of samples collected from these sites or from regions contiguous or deep to such areas. Furthermore, the association of CONS with deeply situated medical device implants presents difficulties with sampling, processing and interpretation of cultures, should these become infected. In many clinical situations the risks associated with the removal of implanted medical devices, such as prosthetic hip joints and heart valves or CSF-shunting systems, present a major clinical dilemma in which the risks of removal must be carefully weighed against the failure to control the infectious process. Various procedures have been developed, aimed at increasing the sensitivity and specificity of laboratory diagnosis.

Media

CONS can be isolated using a variety of laboratory media such as blood agar, nutrient agar, tryptic soya agar, brain heart infusion agar and their broth equivalents. Colonies vary in diameter from 1 to 3 mm following overnight incubation in air. Morphological variation provides some assistance in identification and to some extent speciation. Colonies of CONS tend to be smaller than those of S. aureus, usually lack pigment and were formerly called S. albus because of their white appearance. Staphylococcus lugdunensis often appears creamy or yellow in colour. Occasional strains are sticky, reflecting slime production. In general the colonies are entire, complex and smooth. Colonial variation occurs; for example, those of S. haemolyticus are usually larger than S. epidermidis.

Speciation

Another laboratory dilemma includes the desirability of speciating CONS isolates accurately for diagnostic and epidemiological purposes. The lack of widely available, sensitive, reliable and economical typing systems creates difficulties in recognizing outbreaks of CONS infections.

A number of commercial miniaturized rapid identification systems have been produced. The API-STAPH and Vitek systems among others are increasingly widely used and have permitted ready speciation by routine diagnostic laboratories of many commonly isolated species. These diagnostic kits, in conjunction with novobiocin susceptibility testing, provide speciation accuracy of 70–90% (Kloos and Bennerman, 1994). Occasional difficulties arise in speciating *S. warneri* and *S. haemolyticus*. Other systems are based on fatty acid analysis, high-resolution gas chromatography and ribotyping. However, for routine diagnostic purposes, such systems are generally unnecessary and relatively expensive, especially those which are combined with automated reading systems. Speciation still has limitations in establishing the epidemiology of clusters of CONS infections. However, they are useful in defining the nature of repeat isolates from blood, implanted medical devices and other sites when distinguishing contaminant from pathogen.

Finally the increasing problem of multi-drug resistance among CONS has implications for the range of susceptibility tests that should be provided. In turn, the necessity for therapeutic drug monitoring of vancomycin adds an additional complexity to management.

Antibiotic Susceptibility

The growing importance of CONS as a cause of human infection has been matched by an increasing problem of drug resistance. Multiply antibiotic-resistant strains of CONS are now commonly recognized either as colonizers or as pathogens. Evidence for cross-infection between patients and staff is accumulating and CONS, as part of the normal human flora, provide a reservoir in which antibiotic resistance can be spread to other CONS and other potentially pathogenic bacteria.

The majority of CONS are able to elaborate an inducible β-lactamase and hence are resistant to penicillin, while the production of a low-affinity penicillin-binding protein, PB2a, is responsible for resistance to methicillin and related anti-staphylococcal drugs (Chambers, 1987). The *mec*A gene is common to all staphylococci exhibiting methicillin resistance and is expressed heterotypically among a variable minority of isolates within a bacterial population. Recognition is sometimes difficult to detect, as the more resistant subpopulation is in the minority and grows more slowly. Detection is improved by using a higher inoculum and incubation at lower temperatures (30 °C); 2% NaCl and prolonged incubation also improve detection, although mecA gene probe is probably the most sensitive method (Archer and Pennell, 1990); however, detection does not necessarily correlate with gene expression. More than 80% of hospital-acquired CONS are methicillin resistant (Schulin and Voss, 2001), and of these, more than half will be resistant to other agents such as gentamicin, trimethoprim, erythromycin and clindamycin (Diekema *et al.*, 2001). Tetracycline, chloramphenicol and quinolone resistance is more variable. Methicillin resistant CONS should be considered to be multi-resistant unless there is *in vitro* evidence to the contrary (Archer and Climo, 1994).

Gentamicin resistance among CONS is usually enzymatic as a result of either acetylation [AAC (6′)] or phosphorylation (APH 2′) (Thomas and Archer, 1989). Such resistance may be plasmid or chromosomally mediated and is common to most aminoglycosides. This is of particular concern when treating infections such as infective endocarditis for which a β-lactam–aminoglycoside regimen is widely used.

Quinolone resistance was recognized shortly after the introduction of agents such as ciprofloxacin and ofloxacin (Kotilainen, Nikoskelainen and Huovinen, 1990). The widespread use of these agents within hospital populations led to a rapid decline in the susceptibility of CONS as a result of DNA gyrase (*gyr*A) mutation (Piddock, 1994). More recently, alterations to the *nor*A gene have been linked to increased efflux of quinolones and other antibiotics (Piddock, 1994). Resistance rates vary throughout the world but are of the order of 7–22%. Cross-resistance to other quinolones is common and includes the more recently developed agents active against Gram-positive cocci.

Tetracycline resistance has declined owing to less frequent use in hospitals. Resistance rates of about 25–35% (Archer and Scott, 1991) occur and are related to either reflux or ribosomal resistance; while the

tet (K) gene is responsible for plasmid-mediated resistance (Bismuth *et al.*, 1990), the *tet* M gene is chromosomally carried and is responsible for resistance in minocycline, tetracycline and doxycycline, although minocycline often retains activity against other tetracycline-resistant strains.

Macrolide resistance is often linked to clindamycin and streptogramin resistance, as they share a similar target on the 50S ribosomal subunit, which is important in translocation and transpeptidation. The majority of methicillin-resistant CONS and even some methicillin-sensitive strains will be resistant to erythromycin and related compounds (Archer and Climo, 1994). Dissociated resistance between erythromycin and clindamycin is the result of differential induction of the *erm* gene (Jenssen *et al.*, 1987), although mutational resistance can occur during treatment with clindamycin. New macrolides – clarithromycin and azithromycin – appear to offer no advantage with regard to CONS, as *in vitro* activity is comparable and cross-resistance occurs (Hamilton-Miller, 1992).

Rifampicin is increasingly used to treat deep-seated CONS infection in combination with other antibiotics. CONS are usually highly sensitive, but rifampicin-resistant mutants occur spontaneously at a rate of approximately 1 in 10^6 as a result of genetic mutation in coding for the subunit of DNA-dependent RNA polymerase. It is therefore important that use of this agent be limited and should be based on *in vitro* evidence of susceptibility.

Vancomycin is currently the glycopeptide of choice for the treatment of most serious CONS infections. To date, most CONS remain susceptible to 4 mg/l of vancomycin or less, although occasional strains of *S. haemolyticus* are resistant (Srinivasan, Dick and Perl, 2002). Vancomycin resistance among *S. epidermidis* isolates is even more uncommon, but concerns have been heightened following the detection of vancomycin-resistant *S. aureus*.

Teicoplanin is being increasingly used as an alternative to vancomycin in the treatment of a wide range of staphylococcal infections. However, CONS of reduced susceptibility (minimal inhibitory concentration >8 mg/l) among *S. epidermidis* and *S. haemolyticus* are now recognized (Bannerman, Wadiak and Kloos, 1991). Resistance to *S. haemolyticus* can be readily reduced *in vitro* and among *S. epidermidis* by stepwise exposure (Biavasco *et al.*, 1991). However, such strains are currently uncommon. Quinupristin–dalfopristin (von Eiff *et al.*, 2000) and linezolid (Henwood *et al.*, 2000) are two new agents licensed for the treatment of Gram-positive infections including those caused by susceptible CONS. Experience to date remains limited but is encouraging in the case of linezolid. However, despite *in vitro* susceptibility of many isolates, there is little evidence to suggest that the antibiotic treatment of medical device-associated infections has been improved. While *in vitro* susceptibility to various antibiotics can be demonstrated, biofilm-embedded microorganisms are less susceptible to the effects of antibiotic. Originally it was felt that the biofilm prevented drug penetration, but this is now known to be incorrect. Diffusion is often rapid, and concentrations are high in the case of vancomycin and rifampicin (Schulin and Voss, 2001). *In vitro* biofilm models of infection have shed some light on the activity of established drugs (Widmer *et al.*, 1990) and new agents such as quinupristin–dalfopristin (Gander and Finch, 2000) and linezolid (Gander *et al.*, 2002).

MANAGEMENT OF INFECTION

The clinical presentations of CONS infection vary widely according to the circumstances in which they express disease. In patients with deep-seated infections, such as those complicating a prosthetic hip implant, clinical features are usually localized to the joint with progressive pain, and eventually instability; constitutional symptoms are often absent. This contrasts with the patient in an ICU with multiple endovascular lines, in whom fever and persistent bacteremia should rapidly raise the suspicion of line-associated sepsis. Peritonitis complicating CAPD is usually accompanied by a cloudy effluent dialysate and may precede clinical symptoms of peritonitis.

The diagnosis and management of CONS-associated infections presents major challenges. There is often difficulty in establishing a microbiological diagnosis, particularly in situations associated with deep-seated infection. Removal of an infected prosthetic device presents specific and individual problems; an implanted heart valve is less easily dealt with than an endovascular catheter which may be readily removed. The diagnostic problems are compounded by the occasional difficulties in confidently ascribing pathogenicity to organisms which are the most common laboratory or sampling contaminant. Furthermore, antibiotic choice is restricted owing to the frequency of multi-drug resistance. Finally, there may be uncertainty with regard to duration of therapy and the likelihood of success if prescribed in the continued presence of an infected medical device.

Urinary Tract Infection

Staphylococcus saprophyticus, and less commonly *S. epidermidis* and other species of CONS, account for about 10% of cases of UTI. *Staphylococcus saprophyticus* has a seasonal pattern peaking during the summer months (Hovelius and Mårdh, 1984). Although most commonly associated with acute cystitis in young women, the upper urinary tract or prostate may occasionally be involved. Acute cystitis can usually be treated with an appropriate antibiotic, although susceptibility guidance is desirable. Trimethoprim, nitrofurantoin or a quinolone such as norfloxacin are all appropriate choices, provided the organism is sensitive. Short-course therapy is now widely practised using a 3-day regimen with arrangements for checking MSU 6 weeks after treatment or earlier should symptoms fail to resolve.

Intravascular Line Infection

The range and complexity of intravascular lines in current medical use are reflected in the diverse nature of the infectious complications. One of the most widely used devices is the intravenous cannula of the Venflon variety. These may or may not have an integral injection port. Most are sited in peripheral veins for short-term vascular access. The complications include phlebitis and infection which are often inter-related and vary with the type of plastic, the length of the device and the site of insertion (Cheesbrough and Finch, 1985). Infection rates average 0.6 per 1000 device days (Safdar, Kluger and Maki, 2001). Evidence of infection at the exit site is usually managed by device removal with spontaneous resolution. Occasionally, cellulitis or septic phlebitis may occur, with the necessity for antibiotic therapy. Most respond to an anti-staphylococcal penicillin, such as flucloxacillin, although occasionally metastatic infection may arise with positive blood cultures. Antibiotic selection should then be governed by sensitivity testing but will usually include either a penicillase-stable penicillin or a glycopeptide such as vancomycin or teicoplanin.

Central venous catheters are used for a variety of purposes including pressure monitoring, infusion of fluids, blood or drugs and parenteral nutrition. Infectious complications are much more frequent, usually because of the multiple uses these systems are put to. Hickman, Broviac, Swan-Ganz and hemodialysis catheters are particularly subject to infectious complications from skin microorganisms (Cheesbrough, Finch and Burden, 1986; Goldman and Pier, 1993). Staphylococci including CONS predominate (Mermel *et al.*, 1991). There should be a low threshold for suspecting such an infection in any patient running a fever with a central line in place. Blood cultures should be obtained, and the exit site inspected and sampled (Linares *et al.*, 1985). It is occasionally helpful to draw blood cultures through the catheter (Moyer, Edwards and Farley, 1983).

The principles of treatment are the same as for peripheral devices (Bouza, Burillo and Muñoz, 2002). However, these devices are often essential to patient management, particularly to those patients in a high-dependency unit, so that a trial of antibiotic therapy is often conducted before a decision to remove the device is made; ideally the catheter should be removed and resited. This approach is unwise where there is clinical evidence of local or disseminated sepsis. Antibiotic choice should again be guided by laboratory susceptibility testing. In the case of CONS infections, greatest reliance is now placed on vancomycin (Mermel *et al.*, 2001), with teicoplanin as an alternative choice (Graninger *et al.*, 2002). Another approach has been to combine systemic with locally administered antibiotic using relatively high concentrations of drug which is locked into the lumen of the catheter for several hours (Capdevila, 1998). Where drug intolerance or glycopeptide resistance is a problem, alternative agents include linezolid and rifampicin (in combination with another agent) according to *in vitro* susceptibility testing. This strategy will occasionally prove successful, but as stated, line removal is often necessary to eradicate infection. There is the risk of developing metastatic infection or endovascular infection such as right-sided endocarditis when the catheter tip is sited within the heart.

Infective Endocarditis

Staphylococci are the most frequent cause of prosthetic valve endocarditis (PVE) and, less commonly, the cause of native valve endocarditis (Caputo *et al.*, 1987; Tornos *et al.*, 1997). Although predominantly arising within 2 months following valve insertion, it has become increasingly apparent that endocarditis beyond this period may also be staphylococcal in nature. CONS figure prominently in both situations (Karchmer, 2000). The key clinical features of endocarditis include low-grade fever, a murmur and systemic embolization. There should be a low threshold for suspecting infective endocarditis in patients with prosthetic heart valves, intracardiac patches or ventricular support systems. The diagnosis is normally confirmed microbiologically by evidence of persistent bacteremia on blood culture.

The management of PVE includes prolonged high-dose intravenous bactericidal antibiotic and careful monitoring of clinical response supported by investigations such as echocardiography. Failure to control bacteremia, a changing murmur, conduction abnormalities suggesting a septal abscess, progressive heart failure and major embolic phenomena are all indications for urgent valve replacement.

The antibiotic regimen should be guided by laboratory information on the nature and susceptibility of the infecting organism. In the case of CONS endocarditis an anti-staphylococcal penicillin such as flucloxacillin in combination with gentamicin, with or without the addition of rifampicin, is commonly prescribed for a period of 6 weeks (Wilson *et al.*, 1995). In those allergic to penicillin an appropriate cephalosporin or vancomycin is widely used, again in association with gentamicin. Teicoplanin provides an alternative to vancomycin (Galetto *et al.*, 1986), although failures have been reported (Gilbert, Wood and Kimbrough, 1991). The latter will require careful monitoring of blood levels to avoid drug toxicity, especially to the kidney. Treatment of PVE requires a minimum of 6 weeks of therapy with careful follow-up to detect relapse. PVE complicates approximately 2–3% of procedures. Prevention requires careful surgical technique with practices similar to those adopted for prosthetic joint insertion (q.v.). Perioperative antibiotic prophylaxis regimens vary, but a broad-spectrum cephalosporin such as cefuroxime is generally suitable. If surgery is prolonged, a second dose may need to be administered. Where there are concerns about possible MRSA infection complicating surgery, some centers are including a glycopeptide (Maki *et al.*, 1992).

CAPD Peritonitis

Peritonitis is the commonest complication of CAPD. The infecting organisms often arise from the skin of the patient and gain access to the Tenkhoff catheter during bag changes. Less commonly the catheter exit site becomes infected. There is considerable individual variation in the frequency of such infections. Patients need to be educated in the practice of sterile technique and observed to ensure good practices, as repeated episodes of peritonitis can alter the efficiency of the peritoneal surface as a dialysis membrane.

The principles of management have evolved over the past decade or so. Systemic antibiotics are less able to achieve sufficiently high therapeutic concentrations within the infected peritoneal cavity than those administered directly with the dialysate. The latter is the preferred route of administration, although in patients with systemic features of sepsis one or two doses of antibiotic are given intravenously at the start of therapy (Report of BSAC, 1987; Vas, 1994).

The choice of agent should be guided by microbiological information (Keane et al., 2000). However, the initial empirical regimen takes account of the importance of skin staphylococci and in particular CONS. Vancomycin plus gentamicin is the most widely adopted combination, although caution may be necessary in those patients with some residual renal function (Shemin et al., 1999). Alternatives have included cefuroxime, quinolones, other cephalosporins and carbapenems. While the drug has frequently been given into each dialysate bag, it is now recognized that alternate bag therapy, or even the overnight dwell bag, may be adequate for the control of such infections (Ad Hoc Advisory Committee, 1993; Keane et al., 2000). However, before varying the regimen it is important to monitor response to treatment clinically by observing the clearing of the dialysate bag. Microbiological information may permit subsequent simplification of the regimen to a single drug. Treatment is usually continued for a period of 7–10 days when managing CONS infections, but longer periods for infections caused by yeasts or Gram-negatives such as *Pseudomonas aeruginosa* will be required (Report of BSAC, 1987).

Patients should be monitored to detect early relapse. Should this occur within a 6-week period, then careful consideration should be given to whether the Tenkhoff catheter is heavily colonized with bacteria within a biofilm. This makes eradication of infection extremely difficult, and catheter change is often necessary for its clearance. Traditionally this required a period of hemodialysis following peritoneal catheter removal and the subsequent insertion of a new Tenkhoff device. However, removal and insertion of a new catheter as a single procedure has proved successful under cover of appropriate antibiotic therapy and has the advantage of avoiding unnecessary hemodialysis and a further surgical procedure to insert a new catheter as a delayed procedure (Ad Hoc Advisory Committee, 1993).

CSF-Shunt Infections

The management of CSF shunt infections is a specialized problem requiring close collaboration between experienced neurosurgical and microbiological services and an individualized decision as to whether this should be managed surgically or non-surgically (Bayston, 2001). Surgical treatment requires shunt removal and reshunting, with or without a period of external ventricular CSF drainage. An alternative approach includes shunt removal and regular ventricular taps in selected patients. However, both approaches run the risk of either ventriculitis from retrograde spread of organisms through the external ventricular drainage system or from direct puncture, which of itself may result in porencephalic cysts (Mayhall et al., 1984). Another approach has been to remove the infected shunt and replace it immediately with a fresh system. However, this can be technically difficult and the risk of infecting the new shunt, despite appropriate antibiotic cover, may occur.

The antibiotic management of CSF-shunt infections depends upon the causative microorganisms, their known susceptibility to antibiotics and the ability to deliver the antibiotic in sufficient concentrations to the site of the infection (Bayston et al., 1995). Furthermore, the ability of many microorganisms to form microcolonies and exist within a biofilm presents further difficulties with regard to eradicating these infections.

Various antibiotics have been used, such as flucloxacillin, cephalosporins such as cefuroxime, fusidic acid and erythromycin and to a lesser extent clindamycin. The modest inflammatory response in the CSF produced by infections such as CONS presents further difficulties in achieving adequate concentrations of β-lactam agents within the CSF. Furthermore, these drugs are only active against dividing cells, and within an infected CSF shunt, cells are often dividing extremely slowly. Greatest reliance is therefore currently placed on the use of vancomycin, with or without rifampicin (Bayston, 2001). It is important that the infecting organism be identified carefully, and in the case of CONS the laboratory evidence should suggest a true infection rather than the presence of a contaminating organism.

Intraventricular administration of antibiotics is usually essential to achieve sufficient bactericidal concentrations to sterilize an infected shunt and the ventricular CSF (Wen et al., 1992). In the case of ventriculoperitoneal shunts, infection is also present within the peritoneal cavity (although often clinically inapparent) and systemic antibiotics should also be given.

Vancomycin has achieved greater importance in the management of multiple antibiotic-resistant staphylococcal shunt infection. A daily dose of 20 mg of vancomycin administered intraventricularly has proved effective and safe (Bayston, Hart and Barniecoat, 1987; Bayston, 2001). CSF concentrations of between 200 and 300 mg/l are achieved. Systemic antibiotic is also given, and oral rifampicin is preferred when dealing with staphylococcal infections. The response to treatment is measured clinically and by repeated CSF examinations. CONS infections generally disappear within 4–5 days. CSF levels of antibiotic should be measured preferably to ensure adequate bactericidal concentrations. A decision whether to attempt medical eradication of infection or to reshunt requires assessment of all factors relevant to the case.

Prosthetic Joint Infections

Many thousands of total hip arthroplasties and total knee procedures are performed annually. The majority give rise to a few complications. However, the development of a wound infection, and more seriously infection of the prosthesis, can result in failure of the prosthetic implant (Fitzgerald et al., 1977; Norden, 1994). Infection rates for total hip replacements vary between 0.6 and 2%. The rate for total knee arthroplasties is approximately twofold higher. Underlying problems such as diabetes mellitus, rheumatoid arthritis, advanced age and infection remote from the operation site are all recognized risk factors of infection which may arise by direct implantation through the open wound, or as a result of hematogenous spread, or less commonly as a result of the reactivation of latent infection in a previously infected operative site (Berbari et al., 1998). Infections are reduced in frequency by the use of prophylactic antibiotics in the perioperative period and by high-efficiency particulate air (HEPA) exhaust ventilation systems in the operating theatre suite (Lidwell et al., 1982; Salvati et al., 1982). The use of exhausted suits for the operating team are in additional refinement but have not been widely adopted.

The diagnosis of joint infection can be difficult and may be delayed, as local symptoms often occur late (Inman et al., 1984). Loosening of the prosthesis is a late sign. Systemic signs of infection are often lacking in the early stages. Monitoring the erythrocyte sedimentation rate (ESR) can be valuable. The role of isotope scans for diagnosis remains controversial (Owen et al., 1995). Early acute infections are more likely to be caused by *S. aureus* and associated with poor wound healing and local evidence of infection. Infections delayed beyond a 3-month period are more usually associated with low-grade pathogens such as CONS and other skin bacteria such as *Propionibacterium* spp. Infection should be suspected if there is persistent joint pain and a raised ESR.

The management of prosthetic joint infections needs to be individualized and is a balance between medical and surgical treatment.

Prolonged antibiotic therapy for periods ranging from 1 to 12 months according to response is widely practised. Alternatively, this is often given following removal of the infected prosthesis and wound debridement. The choice of agent should be guided by microbiological data but should be directed at Gram-positive cocci including CONS in the absence of any laboratory information. Drugs such as flucloxacillin, co-amoxiclav or a cephalosporin are commonly used. These may be given parenterally for the first few weeks until there is clinical evidence of response including a falling ESR. Surgical drainage of any infected material should be part of the early management, while early resection of an obviously infected arthroplasty should not be delayed. A replacement implant should only be considered once all evidence of local infection has disappeared.

Prevention of prosthetic joint infections requires careful consideration of the operative environment, which should include a satisfactory ventilation system, preferably with a HEPA filtration arrangement (Lidwell et al., 1984). Meticulous preparation of the operative site and observance of all aspects of sterile technique are essential. Double glove techniques and operator isolator systems have their devotees. Patients should be free of infection at other sites and in particular the skin.

The use of antibiotic prophylaxis is now well established and in general when given as a short perioperative course is effective in encouraging a sterile operative site. The combined benefits of clean air within the operating room and short-course perioperative antibiotic prophylaxis have been associated with the lowest complication rates of <1%. Among the agents selected, flucloxacillin and the cephalosporins have proved the most popular (Hill et al., 1981; Norden, 1991). Cefuroxime, cephradine and cefazolin possess anti-staphylococcal activity, which is a primary consideration, although the increasing frequency of antibiotic resistance has led to alternative choices such as co-amoxiclav (Nelson, 1987). Other techniques such as joint irrigation and antibiotic-impregnated bone cement have been used by some but have not been universally adopted (Espehaug et al., 1997).

REFERENCES

Ad Hoc Advisory Committee on Peritonitis Management. (1993) Peritoneal dialysis-related peritonitis treatment recommendations: 1993 update. Peritoneal Dialysis Internation, 13, 14–28.

Archer, G.L. and Climo, M.W. (1994) Antimicrobial susceptibility of coagulase-negative staphylococci. Antimicrobial Agents and Chemotherapy, 38, 2231–2235.

Archer, G.L. and Pennell, E. (1990) Detection of methicillin resistance in staphylococci by using a DNA probe. Antimicrobial Agents and Chemotherapy, 34, 1720–1724.

Archer, G.L. and Scott, J. (1991) Conjugative transfer genes in staphylococcal isolates from the United States. Antimicrobial Agents and Chemotherapy, 35, 2500–2504.

Baddour, L.M., Smalley, D.L., Kraus, A.P. Jr et al. (1986) Comparison of microbiologic characteristics of pathogenic and saprophytic coagulase-negative staphylococci from patients on continuous ambulatory peritoneal dialysis. Diagnostic Microbiology and Infectious Disease, 5, 197–205.

Bannerjee, S.N., Emori, T.G., Culver, D.H. et al. (1991) Secular trends in Nosocomial primary blood-stream infections in the United States, 1980–1989. The American Journal of Medicine, 91 (Suppl. 3B), 86S–89S.

Bannerman, T.L., Wadiak, D.L. and Kloos, W.E. (1991) Susceptibility of staphylococcus species and subspecies to teicoplanin. Antimicrobial Agents and Chemotherapy, 35, 1919–1922.

Barker, L.P., Simpson, W.A. and Christensen, G.D. (1990) Differential production of slime under aerobic and anaerobic conditions. Journal of Clinical Microbiology, 28, 2578–2579.

Barrow, G.I. and Feltham, R.K.A. (1993) Cowan and Steel's Manual for the Identification of Medical Bacteria, Cambridge University Press, Cambridge.

Bayston, R. (1994) Hydrocephalus shunt infections. The Journal of Antimicrobial Chemotherapy (Suppl. A), 75–84.

Bayston, R. (2001) Epidemiology, diagnosis, treatment and prevention of cerebrospinal fluid shunt infections. Neurosurgery Clinics of North America, 12, 703–708.

Bayston, R., Hart, C.A. and Barniecoat, N. (1987) Intraventricular vancomycin the treatment of ventriculitis associated with cerebrospinal shunting and draining. Journal of Neurology, Neurosurgery Psychiatry, 50, 1419–1423.

Bayston, R. and Rodgers, J. (1990) Production of extracellular slime by Staphylococcus epidermidis, during stationary phase of growth: its association with adherence to implantable devices. Journal of Clinical Pathology, 43, 866–870.

Bayston, R., de Louvois, J., Brown, E.M. et al. (1995) Treatment of infections associated with shunting for hydrocephalus. British Journal of Hospital Medicine, 53, 368–373.

Berbari, E.F., Hanssen, A.D., Duffy, M.C. et al. (1998) Risk factors for prosthetic joint infection: case-control study. Clinical Infectious Diseases, 27, 1247–1254.

Beutler, B. and Cerami, A. (1989) The biology of cachectin/TNFα: a primary mediator of the host response. Annual Review of Immunology, 7, 625–656.

Biavasco, F., Giovanetti, E., Montanari, M.P. et al. (1991) Development of in-vitro resistance to glycopeptide antibiotics: assessment in staphylococci of different species. The Journal of Antimicrobial Chemotherapy, 27, 71–79.

Bingen, E.H., Denamur, E. and Elion, J. (1994) Use of ribotyping in epidemiological surveillance of nosocomial outbreaks. Clinical Microbiology Reviews, 7, 311–327.

Bismuth, R., Zilhao, R., Sakamoto, H. et al. (1990) Gene heterogeneity for tetracycline resistance in Staphylococcus spp. Antimicrobial Agents and Chemotherapy, 34, 1611–1614.

Bouza, E., Burillo, A. and Muñoz, P. (2002) Catheter-related infections: diagnosis and intravascular treatment. Clinical Microbiological Infection, 8, 265–274.

Capdevila, J.A. (1998) Catheter-related infection: an update on diagnosis, treatment, and prevention. International Journal of Infectious Diseases, 2 (4), 230–236.

Caputo, G.M., Archer, G., Calderwood, S.B. et al. (1987) Native valve endocarditis due to coagulase-negative staphylococci: clinical and microbiologic features. The American Journal of Medicine, 83, 619–625.

Cercenado, E., Ena, J., Rodriguez-Creixems, M. et al. (1990) A conservative procedure for the diagnosis of catheter-related infections. Archives of Internal Medicine, 150 (7), 1417–1420.

Chambers, H.F. (1987) Coagulase-negative staphylococci resistant to β-lactam antibiotics in-vivo produce penicillin-binding protein 2a. Antimicrobial Agents and Chemotherapy, 31, 1919–1924.

Cheesbrough, J.S. and Finch, R.G. (1985) Studies on the microbiological safety of the valved side-port of the 'Venflon' cannula. The Journal of hospital infection, 2, 201–208.

Cheesbrough, J.S., Finch, R.G. and Burden, R.P. (1986) A prospective study of the mechanisms of infection associated with hemodialysis catheters. The Journal of Infectious Diseases, 154, 579–589.

Cockayne, A., Hill, P.J., Powell, N.B. et al. (1998) Molecular cloning of a, 32-kilodalton lipoprotein component of a novel iron-regulated Staphylococcus epidermidis ABC transporter. Infection and Immunity, 66, 3767–3774.

Denyer, S.P., Davies, M.C., Evans, J.A. et al. (1990) Influence of carbon dioxide on the surface characteristics and adherence potential of coagulase-negative staphylococci. Journal of Clinical Microbiology, 28, 1813–1817.

Diekema, D.J., Pfaller, M.A., Schmitz, F.J. et al. (2001) Survey of infections due to Staphylococcus species: frequency of occurrence and antimicrobial susceptibility of isolates collected in the United States, Canada, Latin America, Europe and the Western Pacific region for SENTRY antimicrobial surveillance programme, 1997–1999. Clinical Infectious Diseases, 32 (Suppl. 2), S114–S132.

Dinarello, C.A. (1989) Interleukin 1 and its biologically related cytokines. Advances in Immunology, 44, 153–206.

EORTC International Antimicrobial Therapy Cooperative Group. (1990) Gram-positive bacteremia in granulocytopenic cancer patients: results of a prospective randomized therapeutic trial. European Journal of Cancer and Clinical Oncology, 26 (5), 569–574.

Espehaug, B., Engesaeter, L.B., Vollset, S.E. et al. (1997) Antibiotic prophylaxis in total hip arthroplasty. Review of 10,905 primary cemented total hip replacements reported to the Norwegian arthroplasty register, 1987 to 1995. The Journal of Bone and Joint Surgery. British Volume, 79, 590–595.

Fey, P.D., Ulphani, J.S., Gotz, F. et al. (1999) Characterization of the relationship between polysaccharide intercellular adhesin and hemagglutination in Staphylococcus aureus. The Journal of Infectious Diseases, 179, 1561–1564.

Fitzgerald, R.H., Nolan, D.R., Illstrup, D.M. et al. (1977) Deep wound sepsis following total hip arthroplasty. The Journal of Bone and Joint Surgery, 59A, 847–855.

Fitzpatrick, F., Humphreys, H., Smyth, E. et al. (2002) Environmental regulation of biofilm formation in intesive care unit isolates of Staphylococcus epidermidis. The Journal of hospital infection, 42, 212–218.

Frebourg, N.B., Lefebvre, S., Baert, S. and Lemeland, J.F. (2000) PCR-based assay for discrimination between invasive and contaminating *Staphylococcus epidermidis* strains. *Journal of Clinical Microbiology*, **38**, 877–880.

Freeman, J., Platt, R., Sidebottom, D.G. *et al.* (1987) Coagulase-negative staphylococcal bacteremia in the changing neonatal intensive care population. *Journal of the American Medical Association*, **258**, 2548–2552.

Galetto, D.W., Boscia, J.A., Kobasa, W.D. and Kaye, D. (1986) Teicoplanin compared with vancomycin for treatment of experimental endocarditis due to methicillin-resistant *Staphylococcus epidermidis*. *The Journal of Infectious Diseases*, **154**, 69–75.

Gander, S. and Finch, R. (2000) The effects of exposure at constant (1 h) or exponentially decreasing concentrations of quinupristin/dalfopristin on biofilms of Gram-positive bacteria. *The Journal of Antimicrobial Chemotherapy*, **46**, 61–67.

Gander, S., Hayward, K., Finch, R.G. *et al.* (2002) An investigation of the antimicrobial effects of linezolid on bacterial biofilms utilising an in vitro pharmacokinetic model. *The Journal of Antimicrobial Chemotherapy*, **49**, 301–308.

Gemmel, C.G. (1986) Virulence characteristics of *Staphylococcus epidermidis*. *Journal of Medical Microbiology*, **22**, 287–289.

Gilbert, D.N., Wood, C.A. and Kimbrough, R.C. (1991) Failure of treatment with teicoplanin at 6 milligrams/kilogram/day in patients with *Staphylococcus aureus* intravascular infection. *Antimicrob Agents Chemother*, **35**, 79–87.

Goldman, D.A. and Pier, G.B. (1993) Pathogenesis of infections related to intravascular catheterisation. *Clinical Microbiology Reviews*, **6**, 176–192.

Graninger, W., Assadian, O., Lagler, H. and Ramharter, M. (2002) The role of glycopeptides in the treatment of intravascular catheter-related infections. *Clinical Microbiological Infection*, **8**, 310–315.

Gray, E.D., Peters, G., Verstegen, M. *et al.* (1984) Effects of extracellular slime substance from *Staphylococcus epidermidis* on the cellular immune response. *Lancet*, **i**, 365–367.

Hamilton-Miller, J. (1992) In-vitro activities of 14-,15- and 16-membered macrolides against Gram-positive cocci. *The Journal of Antimicrobial Chemotherapy*, **29**, 141–147.

Heilmann, C., Hussain, M., Peters, G. and Gotz, F. (1997) Evidence for autolysin-mediated primary attachment of *Staphylococcus epidermidis* to a polystyrene surface. *Molecular Microbiology*, **24**, 1013–1024.

Heilmann, C., Schweitzer, O., Gerke, C. *et al.* (1996) Mollecular basis of intercellular adhesion in the biofilm-forming *Staphylococcus epidermidis*. *Molecular Microbiology*, **20**, 1083–1091.

Henwood, C.J., Livermore, D.M., Johnson, A.P. *et al.* (2000) Susceptibility of Gram-positive cocci from 25 UK hospitals to antimicrobial agents including linezolid. *The Journal of Antimicrobial Chemotherapy*, **46**, 931–940.

Herchline, T.E. and Ayers, L.W. (1991) Occurrence of *Staphylococcus lugdenensis* in consecutive clinical cultures and relationship of isolation to infection. *Journal of Clinical Microbiology*, **29**, 419–421.

Hill, C., Flamant, R., Mazas, F. *et al.* (1981) Prophylactic cefazolin versus placebo in total hip replacement. *Lancet*, **i**, 795–797.

Hovelius, B. and Mårdh, P.A. (1984) *Staphylococcus saprophyticus* as a common cause of urinary tract infections. *Reviews of Infectious Diseases*, **6**, 328–337.

Huebner, J., Pier, G.B., Maslow, J.N. *et al.* (1994) Endemic nosocomial transmission of *Staphylococcus epidermidis* bacteraemia isolates in a neonatal intensive care unit over, 10 years. *The Journal of Investigative Dermatology*, **169**, 526–531.

Hussain, M., Collins, C., Hastings, J.G.M. *et al.* (1992b) Radiochemical assay to measure the biofilm produced by coagulase-negative staphylococci on solid surfaces and its use to quantitate the effects of various antibacterial compounds on the formation of the biofilm. *Journal of Medical Microbiology*, **37**, 62–69.

Hussain, M., Hastings, J.G.M. and White, P.J. (1991) Isolation and composition of the extracellular slime made by coagulase-negative staphylococci in a chemically defined medium. *The Journal of Infectious Diseases*, **163**, 534–541.

Hussain, M.A., Wilcox, M.H., White, P.J. *et al.* (1992a) Importance of medium and atmosphere type to both slime production and adherence by coagulase-negative staphylococci. *The Journal of hospital infection*, **20**, 173–184.

Inman, R.D., Gallegos, K.V., Brause, B.D. *et al.* (1984) Clinical and microbial features of prosthetic joint infection. *The American Journal of Medicine*, **77**, 47–53.

Jenssen, W.D., Thakker-Varia, S., Dublin, D.T. *et al.* (1987) Prevalence of macrolide-licosamides-streptogramin B resistance and *erm* gene classes among clinical strains of staphylococci and streptococci. *Antimicrobial Agents and Chemotherapy*, **31**, 883–888.

Johnson, G.M., Lee, D.A., Regelmann, W.E. *et al.* (1986) Interference with granulocyte function by *Staphylococcus epidermidis* slime. *Infection and Immunity*, **54**, 13–20.

Karchmer, A.W. (2000) Infections of prosthetic heart valves, in *Infections Associated with Indwelling Devices*, 3rd edn (eds A.L. Bisno and F.A. Waldvogel), ASM Press, Washington, DC, pp. 145–172.

Keane, W.F., Bailie, G.R., Boeschoten, E. *et al.* (2000) Adult peritoneal dialysis-related peritonitis treatment recommendations. *Peritoneal Dialysis International*, **20** (4), 396–411.

Kloos, W.E. and Bennerman, T.L. (1994) Update on clinical significance of coagulase-negative staphylococci. *Clinical Microbiology Reviews*, **7**, 117–140.

Kloos, W.E., Schleifer, K.H. and Götz, F. (1991) The genus *Staphylococcus*, in *The Prokaryotes* (eds A. Balows, H.G. Trüper, M. Dworkin *et al.*), Springer-Verlag, New York, pp. 1369–1420.

Kotilainen, P., Maki, J., Oksman, P. *et al.* (1990) Immunochemical analysis of the extracellular slime substance of *Staphylococcus epidermidis*. *European Journal of Clinical Microbiology and Infectious Diseases*, **9**, 262–270.

Kotilainen, P., Nikoskelainen, J. and Huovinen, P. (1990) Emergence of ciprofloxacin-resistant coagulase-negative staphylococcal skin flora in immunocompromised patients receiving ciprofloxacin. *The Journal of Infectious Diseases*, **161**, 41–44.

Latham, R.H., Running, K. and Stamm, W.E. (1983) Urinary tract infections in young adult women caused by *Staphylococcus saprophyticus*. *Journal of the American Medical Association*, **250**, 3063–3066.

Lidwell, O.M., Lowbury, E.J., Whyte, W. *et al.* (1982) Effect of ultra-clean air in operating rooms on deep sepsis in the joint after total hip or knee replacement: a randomised study. *British Medical Journal (Clinical Research Edition)*, **285**, 10–14.

Lidwell, O.M., Lowbury, E.J.L., Whyte, W. *et al.* (1984) Infection and sepsis after operations for total hip or knee-joint replacement: influence of ultra-clean air, prophylactic antibiotics and other factors. *Journal of Hygiene (Cambridge)*, **93**, 505–529.

Linares, J., Sitges-Serra, A., Garau, J. *et al.* (1985) Pathogenesis of catheter sepsis: a prospective study with quantitative and semiquantitative cultures of catheter hub and segments. *Journal of Clinical Microbiology*, **25**, 357–360.

Lindsey, J.A., Riley, T.V. and Mee, B.J. (1994) Production of siderophore by coagulase-negative staphylococci and its relation to virulence. *European Journal of Clinical Microbiological Infection and Disease*, **13** (12), 1063–1066.

Mack, D. (1999) Indwelling devices and prosthesis: molecular mechanisms of *Staphylococcus epidermidis* biofilm formation. *Journal of Hospital Infection*, **43** (Suppl.), S113–S125.

Mack, D., Nedelmann, M., Krokotsch, A. *et al.* (1994) Characterization of transposon mutants of biofilm-producing *Staphylococcus epidermidis* impaired in the accumulative phase of biofilm production: genetic identification of a hexosamine-containing polysaccharide intercellular adhesin. *Infection and Immunity*, **62**, 3244–3253.

Maki, D.G., Bohn, M.J., Stolz, S.M. *et al.* (1992) Comparative study of cefazolin, cefamandole and vancomycin for surgical prophylaxis in cardiac and vascular operations. A double-blind randomized trial. *The Journal of Thoracic and Cardiovascular Surgery*, **104**, 1423–1434.

Maki, D.G., Weise, C.E. and Sarafin, H.W. (1977) A semiquantitative culture method for identifying intravenous catheter-related infection. *The New England Journal of Medicine*, **296**, 1305–1309.

Markus, S. and Buday, S. (1989) Culturing indwelling central venous catheters in situ. *Infections in Surgery*, **8**, 157–162.

Martin, M.A., Pfaller, M.A. and Wenzel, R.P. (1989) Coagulase-negative staphylococcal bacteraemia. *Annals of Internal Medicine*, **110**, 9–16.

Mayhall, C.G., Archer, N.H., Lamb, A. *et al.* (1984) Ventriculostomy-related infections: a prospective epidemiologic study. *The New England Journal of Medicine*, **310**, 553–559.

McKenney, D., Hubner, J., Muller, E. *et al.* (1998) The *ica* locus of *Staphylococcus epidermidis* encodes production of the capsular polysaccharide/adhesin. *Infection and Immunity*, **66**, 4711–4720.

McLean, R.J.C., Nickel, J.C., Noakes, V.C. *et al.* (1985) An in-vitro ultrastructural study of infectious kidney stone genesis. *Infection and Immunity*, **49**, 805–811.

Mermel, L., Farr, B., Sherertz, R. *et al.* (2001) Guidelines for the management of intravascular catheter-related infections. *Clinical Infectious Diseases*, **32**, 1249–1272.

Mermel, L.A., McCormick, R.D., Springman, S.R. *et al.* (1991) The pathogenesis and epidemiology of catheter-related infection with pulmonary artery

Swan-Ganz catheters: a prospective study utilizing molecular subtyping. *The American Journal of Medicine*, **91**, S197–S205.

Modun, B. and Williams, P. (1999) The staphylococcal transferrin-binding protein is a cell wall glyceraldehyde-3 phosphate dehydrogenase. *Infection and Immunity*, **67**, 1086–1092.

Modun, B., Kendall, D. and Williams, P. (1994) Staphylococci express a receptor for human transferrin: identification of a 42-kilodalton cell wall transferrin binding protein. *Infection and Immunity*, **62**, 3850–3958.

Moyer, M.A., Edwards, L.D. and Farley, L. (1983) Comparative culture methods on 101 intravenous catheters: routine, semiquantitative and blood cultures. *Archives of Internal Medicine*, **143**, 66–69.

Muller, E., Takeda, S., Shiro, H. *et al.* (1993) Occurrence of capsular polysaccharide/adhesin among clinical isolates of coagulase-negative staphylococci. *Journal of Infectious Diseases*, **168**, 1211–1218.

Nelson, C.L. (1987) The prevention of infection in total joint replacement surgery. *Reviews of Infectious Diseases*, **9**, 613–618.

Noble, M.A., Grant, S.K. and Hajen, E. (1990) Characterization of a neutrophil-inhibitory factor from clinically significant *Staphylococcus epidermidis*. *The Journal of Infectious Diseases*, **162**, 909–913.

Norden, C. (1994) Infections in total joint replacement, *Infections in Bones and Joints* (eds C. Norden, W.J. Gillispie and S. Nade), Blackwell Scientific Publications, Boston, pp. 291–319.

Norden, C.W. (1991) Antibiotic prophylaxis in orthopedic surgery. *Reviews of Infectious Diseases*, **13** (Suppl. 10), S842–S846.

Owen, R.J., Harper, W.M., Finlay, D.B. and Belton, I.P. (1995) Isotope bone scans in patients with painful knee replacements: do they alter management? *The British Journal of Radiology*, **68**, 1204–1207.

Paulsson, M., Ljungh, A. and Wadström, T. (1992) Rapid identification of fibronectin, vitronectin, laminin, and collagen cell surface binding proteins on coagulase-negative staphylococci by particle agglutination assays. *Journal of Clinical Microbiology*, **30**, 2006–2112.

Pereira, A.T. (1962) Coagulase-negative strains of *Staphylococcus* possessing antigen 51 as agents of urinary infections. *Journal of Clinical Pathology*, **15**, 252–259.

Pfaller, M.A. and Herwald, L.A. (1988) Laboratory, clinical and epidemiological aspects of coagulase-negative staphylococci. *Clinical Microbiology Reviews*, **1**, 281–299.

Piddock, L.J.V. (1994) New quinolones and Gram-positive bacteria. *Antimicrobial Agents and Chemotherapy*, **38**, 163–169.

Quinn, J.P., Counts, G.W. and Meyers, J.D. (1986) Intracardiac infections due to coagulase-negative staphylococcus associated with Hickman catheters. *Cancer*, **57**, 1079–1082.

Report of a Working Party of the British Society for Antimicrobial Chemotherapy. (1987) Peritonitis complicating continuous peritoneal dialysis: recommendations for its diagnosis and management. *The Lancet*, **1**, 845–849.

Safdar, N., Kluger, D.M. and Maki, D.G. (2001) A review of risk factors for catheter-related bloodstream infection caused by percutaneously inserted, noncuffed central venous catheters. *Medicine*, **8** (6), 466–479.

Salvati, E.A., Robinson, R.P., Zeno, S.M. *et al.* (1982) Infection rates after 3175 total hip and total knee replacements performed with and without a horizontal unidirectional filtered air-flow system. *The Journal of Bone and Joint Surgery*, **64A**, 525–535.

Schulin, T. and Voss, A. (2001) Coagulase-negative staphylococci as a cause of infections related to intravascular prosthetic devices: limitations of present therapy. *Clinical Microbiological Infection*, **7** (Suppl. 24), 1–7.

Sehaberg, D.R., Culver, D.H. and Gaynes, R.P. (1991) Major trends in the microbial etiology of nosocomial infection. *The American Journal of Medicine*, **91** (S3B), 72–75.

Shemin, D., Maaz, D., St. Pierre, D. *et al.* (1999) Effect of aminoglycoside use on residual renal function in peritoneal dialysis patients. *American Journal of Kidney Diseases*, **34**, 14–20.

Shiro, H., Muller, E., Gutierrez, N. *et al.* (1994) Transposon mutants of *Staphylococcus epidermidis* deficient in elaboration of capsular polysaccharide/

adhesin and slime are avirulent in a rabbit model of endocarditis. *The Journal of Infectious Diseases*, **169**, 1042–1049.

Srinivasan, A., Dick, J.D. and Perl, T.M. (2002) Vancomycin resistance in Staphylococci. *Clinical Microbiology Reviews*, 15 (3), 430–438.

Stout, R.D., Ferguson, K.P., Li, Y. *et al.* (1992) Staphylococcal exopolysaccharides inhibit lymphocyte proliferative responses by activation of monocyte prostaglandin production. *Infection and Immunity*, **60**, 922–927.

Tenney, J.H., Moody, M.R., Newman, K.A. *et al.* (1986) Adherent microorganisms on luminal surfaces of long-term intravenous catheters. *Archives of Internal Medicine*, **146**, 1949–1954.

Thomas, W.D. Jr and Archer, G.L. (1989) Mobility of gentamicin resistance genes from staphylococci isolated in the United States: identification of Tn*4031*, a gentamicin resistance transposon from *Staphylococcus epidermidis*. *Antimicrobial Agents and Chemotherapy*, **33**, 1335–1341.

Timmerman, C.P., Fleer, A., Besnier, J.M. *et al.* (1991) Characterization of a proteinaceous adhesin of *Staphylococcus epidermidis* which mediates attachment to polystyrene. *Infection and Immunity*, **59**, 4187–4192.

Tojo, M., Yamashita, N., Goldman, D.A. *et al.* (1988) Isolation and characterization of a capsular polysaccharide/adhesin from *Staphylococcus epidermidis*. *The Journal of Infectious Diseases*, **157**, 713–722.

Tornos, P., Almirante, B., Olona, M. *et al.* (1997) Clinical outcome and long-term prognosis of late prosthetic valve endocarditis: a 20-year experience. *Clinical Infectious Diseases*, **24**, 381–386.

Vas, S.L. (1994) Infections associated with the peritoneum and hemodialysis, in *Infections Associated with Indwelling Medical Devices*, 2nd edn (eds A.L. Bisno and F.A. Waldvogel), ASM Press, Washington, DC, pp. 309–346.

Verbrugh, H.A., Keane, W.F., Hoidal, J.R. *et al.* (1983) Peritoneal macrophages and opsonins: antibacterial defense in patients undergoing chronic peritoneal dialysis. *The Journal of Infectious Diseases*, **147**, 1018–1029.

von Eiff, C., Peters, G. and Heilmann, C. (2002) Pathogenesis of infections due to coagulase-negative staphylococci. *Lancet Infectious Diseases*, **2**, 677–685.

von Eiff, C., Reinert, R.R., Kresken, M. *et al.* (2000) Nationwide German multicenter study on prevalence of antibiotic resistance in staphylococcal bloodstream isolates and comparative in vitro activities of quinupristin–dalfopristin. *Journal of Clinical Microbiology*, **38**, 2819–2823.

Wadström, T. (1989) Molecular aspects of bacterial adhesion, colonization, and development of infections associated with biomaterials. *Journal of Investigative Surgery*, **2**, 353–360.

Wen, D.Y., Bottini, A.G., Hall, W.A. and Haines, S.J. (1992) Infections in neurologic surgery. The intraventricular use of antibiotics. *Neurosurgery Clinics of North America*, **3**, 343–354.

Widmer, A.F., Frei, R., Rajacic, Z. and Zimmerli, W. (1990) Correlation between in vivo and in vitro efficacy of antimicrobial agents against foreign body infections. *Journal of Infectious Diseases*, **162**, 96–102.

Williams, P., Denyer, S.P. and Finch, R.G. (1988) Protein antigens of *Staphylococcus epidermidis* grown under iron-restricted conditions of human peritoneal dialysate. *FEMS Microbiology Letters*, **50**, 29–33.

Wilson, W.R., Karchmer, A.W., Dajani, A.S. *et al.* (1995) Antibiotic treatment of adults with infective endocarditis due to streptococci, enterococci, staphylococci and HACEK microorganisms. American Heart Association [see comments]. *Journal of the American Medical Association*, **274**, 1706–1713.

Working Party of the British Society for Antimicrobial Chemotherapy (1987) Diagnosis and management of peritonitis in continuous ambulatory peritoneal dialysis. *Lancet*, **i**, 845–849.

Ziebutr, W., Heilmann, C., Gotz, F. *et al.* (1997) Detection of the intercellular adhesion gene cluster (*ica*) and phase variation in *Staphylococcus epidermidis* blood culture strains and mucosal isolates. *Infection and Immunity*, **65**, 890–896.

Section Two

Gram-Positive Bacilli

7

Corynebacterium spp.

Aruni De Zoysa and Androulla Efstratiou

Health Protection Agency, Centre for Infections, Department of Respiratory and Systemic Infections, London, UK

INTRODUCTION

Diphtheria is one of the most important diseases of bacterial aetiology. The causative organism is toxin-producing *Corynebacterium diphtheriae*, which infects the tissues of the pharynx (Lehmann and Neumann, 1896). Diphtheria is an ancient disease that probably existed even in the times of Hippocrates (Andrews *et al.*, 1923; Harries, Mitman and Taylor, 1951; Lloyd, 1978; English, 1985). The first modern clinical descriptions were made by Joost van Lom (Lommius) in 1560 and Baillou (Ballonius) in 1576 (Andrews *et al.*, 1923). They referred to the disease as 'quinsy' and 'croup', respectively. The first epidemic of the disease was described in Spain, where it was called 'morbus suffocans' or 'garrotillo' as suffocation brought by the disease resembled garrotting, the Spanish method of executing criminals (Rosen, 1954). The disease occurred in distinct waves: 1583–1618, 1630, 1645 and 1666. From Spain the disease spread to Italy in 1618, where it was known as 'male in canna' or 'gullet disease'. It first broke out in Naples and then spread to southern Italy including Sicily. Towards the end of the seventeenth century the disease died down; however, as the eighteenth century evolved, diphtheria again became prevalent in Europe and broke out in Britain and America. The first major epidemic of the eighteenth century occurred between 1735 and 1740 in New England, where it was commonly known as 'throat distemper' (Douglass, 1736; Andrews *et al.*, 1923; Rosen, 1954). In 1735, New Hampshire lost a thousand citizens (5% of the total population) to the disease and in New England there were about 5000 deaths (2.5% of the population) (English, 1985).

Diphtheria was first recognised as a distinct clinical entity by the French pathologist and clinician Pierre Bretonneau in 1826 (Semple, 1859; Rosen, 1954). Bretonneau coined the term diphthérite to cover a contagious condition, characterised by the presence of a false membrane in the pharynx or larynx. The word diphthérite is derived from the Greek word *diphthera* (διφθέρα), meaning leather or hide. Following Bretonneau's treatise, there was much speculation about the cause of the disease. In 1883, Edwin Klebs reported that two types of microorganisms existed in the diphtheritic membrane, chain-producing cocci and bacilli, and Klebs was unable to find the bacilli in the blood or organs of individuals who had died from the disease. He reported that they were confined to the membrane (Andrews *et al.*, 1923; Dolman, 1973). In 1884, Friedrich Loeffler proved that it was the bacillus and not the cocci, which were described previously by Klebs, that caused diphtheria (Loeffler *et al.*, 1908; Andrews *et al.*, 1923). He developed a new stain (methylene blue) and a new culture medium (blood agar) and used this to demonstrate the presence of the organism in post-mortem material from cases of diphtheria. He demonstrated that cultures of the organism were highly pathogenic in

laboratory animals by showing that it killed guinea pigs and small birds. Loeffler showed that the organism retained its virulence after repeated passage from guinea pig to guinea pig. He found that the organism was confined to the membrane and that it did not invade the body (confirmed Klebs findings). He postulated that the bacillus released a toxin capable of causing tissue damage away from the respiratory mucosa (Loeffler *et al.*, 1908). In 1988, Roux and Yersin, working at the Pasteur Institute, proved him correct by demonstrating that, when a sterile filtrate derived from broth cultures of the bacillus was injected into animals it caused all the symptoms, which would be produced if a live culture were injected. In 1890, von Behring and Kitasato produced serum antibodies in animals against the toxin and demonstrated that the serum could protect susceptible animals against the disease (von Behring and Kitasato, 1890). In 1901, von Behring won the Nobel Prize in medicine for his work on the development of antitoxin.

The first use of toxin–antitoxin mixtures to immunise children against diphtheria was reported by von Behring in 1913. Also in 1913, Bela Schick introduced an intradermal test, which was able to detect the presence or absence of circulating antitoxin (MacGregor, 2000). In 1923, Glenny and Hopkins showed that formalin treatment of diphtheria toxin (DT) abolished its toxicity but left its antigenicity intact. They coined the term toxoid for the non-toxic, antigenic material. It was useful as an immunising agent in animals in combination with antitoxin (Glenny and Hopkins, 1923). Around the same time, a stable, non-toxic diphtheria toxoid, which could be used for immunising humans, was finally produced by Ramon by the action of formalin on crude toxin, followed by incubation at 37 °C for several weeks (Relyveld, 1995). The initial clinical trials were extremely successful as it reduced the fatalities among those immunised to nearly zero (English, 1985). The observation by Glenny and others that the immunogenicity of the toxoid could be improved by precipitation with aluminium compounds led to a new generation of toxoids (Dixon, Noble and Smith, 1990). At first, alum-precipitated toxoid was used, but this has now been replaced by preparations in which purified toxoid is precipitated onto aluminium phosphate (Dixon, Noble and Smith, 1990).

DESCRIPTION OF THE ORGANISM

Genus *Corynebacterium*

The genus *Corynebacterium* was first proposed by Lehmann and Neumann in 1896 (Skerman, McGowan and Sneath, 1980) to include the diphtheria bacillus and other morphologically similar organisms.

Principles and Practice of Clinical Bacteriology Second Edition   Editors Stephen H. Gillespie and Peter M. Hawkey

The genus name is derived from the Greek words *koryne*, meaning club, and *bacterion*, little rod (Schwartz and Wharton, 1997). The genus *Corynebacterium* is presently composed of 46 species (and two taxon groups), 31 of which are medically relevant (Funke and Bernard, 1999). *Corynebacterium* species are commonly found in soil and water and reside on the skin and mucous membranes of humans and animals. *Corynebacterium diphtheriae* is the major human pathogen in this genus. Other potentially toxigenic corynebacterial species include *Corynebacterium ulcerans* and *Corynebacterium pseudotuberculosis*.

The cell wall of corynebacteria contains meso-diaminopimelic acid and short-chain mycolic acids with 22–36 carbon atoms (Collins, Goodfellows and Minnikin, 1982; Collins and Cummins, 1986). The *Corynebacterium* species *Corynebacterium amycolatum* and *Corynebacterium kroppenstedtii* are the only species that do not possess mycolic acids in the cell wall (Collins, Burton and Jones, 1988; Collins *et al.*, 1998). Palmitic ($C_{16:0}$), oleic ($C_{18:1\omega9c}$) and stearic ($C_{18:0}$) acids are the main cellular fatty acids in all corynebacteria. The G+C content of *Corynebacterium* spp. varies from 46 mol% in *Corynebacterium kutscheri* to 74 mol% in *Corynebacterium auris* (Funke *et al.*, 1997). Other microscopic characteristic features of the genus *Corynebacterium* include morphology, showing straight-to-slightly-curved rods with tapered ends and sometimes club-shaped forms; snapping division producing angular and palisade arrangements (Chinese lettering); Gram-positive staining (some cells stain unevenly); formation of metachromatic granules (which is visualised with Loeffler's methylene blue stain or Albert's stain); non-motile, non-capsulate and non-sporing; facultatively anaerobic growth (some are aerobic); and catalase production (Collins and Cummins, 1986). In 1995 Pascual *et al.* and Ruimy *et al.* outlined the phylogenetic relationships within the genus *Corynebacterium*, creating an extensive database for 16S rRNA gene sequences for the delineation of new species (Pascual *et al.*, 1995; Ruimy *et al.*, 1995).

Corynebacterium diphtheriae

Corynebacterium diphtheriae is the causative agent of diphtheria. Humans are its only host, and there is no known environmental reservoir of infection (MacGregor, 2000). *Corynebacterium diphtheriae*, when lysogenised by certain bacteriophages, produces DT. Colonies of *C. diphtheriae* grown on Hoyle's tellurite medium appear black or grey because of the reduced tellurite. However, other bacteria (e.g. streptococci and staphylococci) also grow as black colonies. When grown on a medium described by Tinsdale (1947) and modified by Moore and Parsons (1958), *C. diphtheriae*, *C. ulcerans* and *C. pseudotuberculosis* produce a well-delineated brown halo around the greyish-black colonies (only rarely may streptococci, staphylococci and some other bacteria show this phenomenon). A brown halo results when H_2S is produced by the action of cysteinase on L-cystine and potassium tellurite is reduced to metallic tellurium. Halo formation also depends on the organism acidifying the medium. Another feature that sets *C. diphtheriae*, *C. ulcerans* and *C. pseudotuberculosis* apart from most other corynebacteria that might be isolated from humans is their inability to hydrolyse pyrazinamidase.

There are four biotypes of *C. diphtheriae*: var. *gravis*, var. *mitis*, var. *belfanti* and var. *intermedius* (Anderson *et al.*, 1933; Chang, Laughren and Chalvardjian, 1978; Funke *et al.*, 1997), differentiated on the basis of cultural and biochemical differences (including colony morphology) (Krech, 1994). In practice, *C. diphtheriae* var. *intermedius* is the only biotype that is distinguished easily on the basis of colony morphology. *Corynebacterium diphtheriae* strains of the biotype *intermedius* are lipophilic and require lipids (e.g. hydrolysis on Tween 80 medium) for optimal growth. On routine medium, *intermedius* strains produce tiny (0.5–1 mm in diameter), grey, discrete, translucent colonies (Ward, 1948; Efstratiou and Maple, 1994; Clarridge, Popovic and Inzana, 1998). Strains of biotype *gravis* and *mitis* are

non-lipophilic, and they produce larger (1.5–2 mm in diameter), denser, grey, opaque colonies that are indistinguishable. A weak β-haemolysis is observed with *mitis* strains and a few *gravis* strains. Differences in the Gram stain morphology for each biotype have been described by Noble and Dixon in 1990: strains of the *mitis* biotype are usually long, pleomorphic, but rigid club-shaped rods; biotype *gravis* are usually short, coccoid or pyriform rods; and *intermedius* strains are highly pleomorphic, ranging from very long to very short rods (Noble and Dixon, 1990).

Strains of *C. diphtheriae* biotype *intermedius* are distinguished from biotype *mitis* strains because the former is dextrin positive. Biotypes *mitis* and *gravis* are distinguished by their glycogen and starch reactions (*gravis* strains are positive for both). *Corynebacterium diphtheriae* biotype *belfanti* is nitrate negative; otherwise, they resemble *mitis* strains. The *intermedius* biotype seems to be the least common biotype.

PATHOGENESIS

Diphtheria Toxin and its Mode of Action

The major virulence determinant of *C. diphtheriae* is diphtheria toxin (DT), the cause of the systemic complications seen with diphtheria. The DT gene is carried by a family of closely related bacteriophages (corynebacteriophages) that integrate into the bacterial chromosome and convert non-toxigenic, non-virulent *C. diphtheriae* strains into toxigenic, highly virulent species (Freeman, 1951; Uchida, Gill and Pappenheimer, 1971). Therefore, DT is only secreted by strains of *C. diphtheriae* lysogenically infected with a temperate bacteriophage (Freeman, 1951). Several different types of lysogenic corynephages have been recognised (β *tox*[+], γ *tox*[+] and ω *tox*[+]), with the β-corynephage being the most common. Some of these phages are also capable of lysogenising *C. ulcerans* and *C. pseudotuberculosis*.

DT is a protein molecule with a molecular weight (M_r) of 58 350 Da (Pappenheimer, 1977; Collier, 1982). The toxin is synthesised as a single polypeptide chain consisting of 535 amino acid residues. It contains four cysteine (C) residues and has two internal disulphide bonds that link C186 to C201 (links fragment A to B) and C461 to C471 (contained within fragment B). Treatment of DT with trypsin selectively cleaves the peptide bond at arginine (R) residue R190, R192 or R193, which are positioned in the protease-sensitive loop, to generate an amino-terminal fragment A (21 150 Da) and a carboxyl-terminal fragment B (37 200 Da) which remain covalently linked by the disulphide bond between C186 and C201 (Drazin, Kandel and Collier, 1971; Gill and Pappenheimer, 1971), and reduction of this disulphide bridge separates the toxin into fragments A and B. Studies performed on the three-dimensional structure of the molecule (Choe *et al.*, 1992; Bennett and Eisenberg, 1994) have shown that the toxin may be divided into three separate domains:

1. the amino-terminal C (catalytic) domain, responsible for ADP-ribosylation of elongation factor 2 (EF-2) which blocks protein synthesis;
2. the centrally located transmembrane or translocation (T) domain [composed of nine α-helices (TH1–TH9)], responsible for insertion into membranes at acidic pH, formation of channels and translocation of the C-domain across endosomal membranes for delivery to the cytosols;
3. the carboxyl-terminal R-domain, responsible for binding of DT to the heparin-binding epidermal growth factor (HB-EGF) precursor that functions as the DT receptor on susceptible cells (Morris *et al.*, 1985; Naglich *et al.*, 1992; Naglich, Rolf and Eidels, 1992).

The HB-EGF receptor is a protein on the surface of many types of cells. The occurrence and distribution of the HB-EGF receptor on cells determine the susceptibility of an animal species, and certain cells of an animal species, to the DT. More receptors are found on the heart and nerve cells than other cell types; hence, in severe forms of the disease, cardiac failure and neurological damage occurs (Efstratiou et al., 1998).

The intoxication of a single eukaryotic cell by DT involves at least four steps. First, the toxin binds to a specific receptor (HB-EGF receptor) on susceptible cells and enters the cell by receptor-mediated endocytosis (Morris et al., 1985; Naglich et al., 1992). The pH within the endosome becomes more acidic and triggers the partial unfolding of the protein, which exposes the hydrophobic regions of the B fragment (helices 8 and 9, the innermost layer of the transmembrane domain), and the translocation domain of DT inserts into the endosomal membrane and forms a channel through which translocation of the A fragment (C-domain) occurs [the A fragment is cleaved from the B fragment (transmembrane and receptor-binding domains), and it translocates into the cytoplasm[(Kagan, Finkelstein and Colombini, 1981; Hu and Holmes, 1984; Moskaug, Sandvig and Olsnes, 1988; Papini et al., 1993). Fragment A is a NAD^+-binding enzyme that catalyses the transfer of the ADP-ribosyl group to a post-translationally modified histidine residue termed diphthamide, present in the cytoplasmic EF-2 of eukaryotic cells (Brown and Bodley, 1979; Van Ness, Howard and Bodley, 1980). The resulting EF-2–ADP-ribose complexes are inactive and therefore cause inhibition of protein synthesis and, ultimately, cell death. In the absence of fragment B, the A fragment is non-toxic, as it cannot cross the plasma membrane to reach the cytoplasm. It has been demonstrated in vitro that a single molecule of fragment A in the cytoplasm is enough to kill one eukaryotic cell (Yamaizumi et al., 1978). The minimal lethal dose for humans and sensitive animals (monkeys, rabbits and guinea pigs) is below 0.1 µg/kg of body weight (Pappenheimer, 1984).

Regulation of DT Production

Production of DT by toxigenic strains of C. diphtheriae is affected greatly by the composition of the growth medium (Mueller, 1940, 1941; Mueller and Miller, 1941). Studies carried out during the 1930s showed that optimal yields of DT are obtained when C. diphtheriae is grown under conditions of iron limitation and that DT production is inhibited severely when grown under high iron conditions (Pappenheimer and Johnson, 1936).

Studies on the regulation of DT production began in the 1970s. In 1974, Murphy and colleagues reported that the DNA from tox⁺ β-corynephage directed the synthesis of DT and other phage-encoded proteins and that addition of C. diphtheriae extracts to this system inhibited the synthesis of DT, but not the production of other phage-encoded proteins (Murphy et al., 1974). This led to the hypothesis that C. diphtheriae contained a factor that acted as a negative controlling element in the regulation of toxin production. Fourel et al. in 1989 provided evidence for the presence of a chromosome-encoded DT repressor (Fourel, Phalipon and Kaczorek, 1989). These investigators observed that proteins in crude extracts from C. diphtheriae grown under high iron conditions were able to bind specifically to the tox operator and protect it from DNase I digestion. As protection of the tox operator required the presence of iron in the reaction, it was assumed that the binding factor was the DT repressor. In 1992, Krafft et al. demonstrated that an iron-dependent repressor, which inhibits the transcription of the tox gene of the β-corynephage under high iron growth conditions, was encoded by the C. diphtheriae genome (Holmes, 1975; Murphy, Skiver and McBride, 1976; Murphy, Bacha and Holmes, 1979; Krafft et al., 1992).

The gene for the DT repressor (dtxR) was cloned by Boyd et al. in 1990 and Schmitt and Holmes in 1991. DNA sequence analysis indicated that the dtxR gene encodes a M_r 25 316 polypeptide consisting of 226 amino acids (Boyd, Oza and Murphy, 1990; Schmitt and Holmes, 1991a,b). Each DtxR monomer has three domains. Domain 1 (residues 1–73) is located at the amino terminal and contains

a helix-turn-helix motif, which is required for binding DNA. Domain 2 (residues 74–144) is responsible for dimerisation and possesses two metal-binding sites (site 1, the ancillary site is composed of ligands H79, E83, H98, E170 and Q173, and site 2, the primary site consists of M10, C102, E105 and H106 and the main chain carbonyl oxygen of C102). Domain 3 (residues 145–226) is located at the carboxyl terminal. It exhibits an SH3-like fold and has greater flexibility than domains 1 and 2 (Qiu et al., 1995). Domain 3 is not resolved in most crystal structures of DtxR, and its function is unknown.

EPIDEMIOLOGY

Overview of Diphtheria since 1940

In the prevaccine era, diphtheria was a much feared childhood disease found in temperate climates. Widespread implementation of childhood immunisation in the 1940s and 1950s led to a marked decrease in the incidence of diphtheria in Western Europe, the United States and Canada (Galazka, Robertson and Oblapenko, 1995). However, since 1989, there has been a major resurgence of the disease in Europe, mainly within countries of the former Soviet Union. The diphtheria epidemic in the Newly Independent States (NIS) of the former USSR began in the Russian Federation in 1990 and affected all 15 NIS countries by the end of 1994. In Russia, the highest incidences were seen in Moscow and Saint Petersburg, where the outbreak first began. (Hardy, Dittman and Sutter, 1996; Efstratiou, Engler and De Zoysa, 1998). In 1990, 1431 diphtheria cases were reported in the NIS (1211 of these cases were from the Russian Federation). The number of cases reported increased to 5729 by 1992 and to 19 453 by 1993. In 1994, 47 324 cases were reported, and it reached a peak of 49 994 in 1995 (WHO, 2002).

In 1995, aggressive diphtheria control strategies were implemented (i.e. mass immunisation of the whole population), and as a result, the number of cases decreased by 60% in 1996 to 20 078. The decreasing trend continued in 1997 (7149 cases were reported in the NIS), and in 1998, 2610 cases were reported in the NIS. There has been excellent progress in the control of diphtheria in Armenia, Azerbaijan, Estonia, Kazakhstan, Lithuania, Moldova and Uzbekistan. Diphtheria control has improved in Belarus, Georgia, Kyrgyzstan, Russian Federation, Tajikistan, Turkmenistan and Ukraine. However, continued efforts and close monitoring are still required. Diphtheria still remains a major concern in Latvia, as there has been an increase in the number of reported cases from 42 in 1997 to 81 in 1999 and to 264 cases in 2000. The total number of cases reported to the World Health Organization (WHO) from the NIS in 2000 was 1573 (WHO, 2002). This outbreak has been the largest diphtheria outbreak in the developed world in recent years. Since the epidemic began, more than 170 000 cases and 4000 deaths have been reported to the WHO Regional Office for Europe (Roure, 1998). Also, importations from the NIS to other European countries such as Finland, Estonia, Germany, Norway, Poland, Bulgaria, Belgium, Greece and the United Kingdom have been reported. Over 20 imported cases of diphtheria in adults have been reported in the above countries (De Zoysa et al., 1993; Efstratiou, 1995; Rey, Patey and Vincent-Ballereau, 1996; Dittman, 1997; Clarridge, Popovic and Inzana, 1998; Wharton et al., 1998).

Figure 7.1 illustrates the diphtheria global annual reported incidence and DTP3 coverage between 1980 and 2001. The decline in reported diphtheria cases in the 1980s is consistent with the reported increasing DTP3 coverage. The sudden increase in incidence during the 1990s is due to the massive epidemic that occurred in the former Soviet Union. During the 1990s, outbreaks were also reported in Algeria, Iraq, Laos, Mongolia, Sudan and Thailand. Figure 7.1 shows only those cases reported to the WHO, and therefore, these figures probably underestimate the true incidence of the disease, as for many countries figures are unavailable.

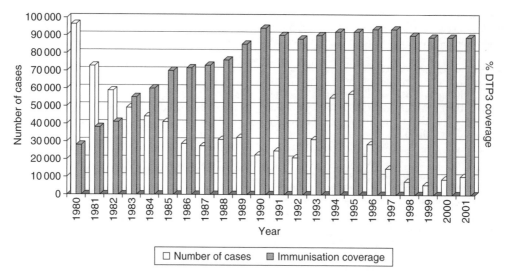

Figure 7.1 Diphtheria global annual reported incidence and DTP3 coverage, 1980–2001.

Molecular Epidemiology

Limitations of the phenotypic methods led to the development of molecular typing schemes for *C. diphtheriae*. In 1983, Pappenheimer and Murphy demonstrated the resolving power of molecular methods when they analysed *C. diphtheriae* isolates using DNA restriction fragment length polymorphism (RFLP) with different DNA probes for different areas of the toxin gene (Pappenheimer and Murphy, 1983). In 1987, Rappuoli *et al.* used an insertion element of *C. diphtheriae* as a DNA probe to characterise Swedish strains from a diphtheria outbreak in an alcoholic and drug-abusing population (Rappuoli, Perugini and Ratti, 1987), and in 1989, Coyle *et al.* studied the molecular epidemiology of the three biotypes of *C. diphtheriae* in the Seattle outbreak by analysing RFLPs with three DNA probes (Coyle *et al.*, 1989).

Since then, several new molecular subtyping methods have been developed and successfully applied to epidemiological investigations of diphtheria (De Zoysa *et al.*, 1995; De Zoysa and Efstratiou, 1999, 2000). Currently, ribotyping is the gold standard for typing *C. diphtheriae*, and molecular subtyping of *C. diphtheriae* by using ribotyping has identified significant genetic diversity within the species and has led to the identification of a unique clonal group that emerged in Russia in 1990 at the beginning of the Eastern European epidemic (De Zoysa *et al.*, 1995; Popovic *et al.*, 2000).

As a result of extensive collaborative typing studies amongst the European Laboratory Working Group on Diphtheria (ELWGD), a database of ribotype patterns has been set up at the Institut Pasteur, Paris, France, using the software Taxotron® (Grimont *et al.*, 2004). This database allows international exchange and dissemination of typing data, specifically on the origin of outbreak strains and their relationship to endemic and epidemic strains worldwide (Efstratiou *et al.*, 2000).

The Diphtheria European and Global Network

The ELWGD was formed at the initiative of the WHO European Office in July 1993 as a result of the epidemic situation in Eastern Europe. The main objectives of the ELWGD were, firstly, to establish a network of collaborating laboratories to monitor the microbiological surveillance of diphtheria, secondly, to standardise methodologies for laboratory diagnosis in epidemic areas and, thirdly, to standardise the

typing techniques, leading to the establishment of a microbiological surveillance system for all isolates of *C. diphtheriae* in the European Region and beyond.

In 1994, an international microbiological surveillance for all *C. diphtheriae* isolates from sporadic cases within the European Region was initiated with collaboration from the members of the ELWGD. Typing results obtained essentially by the laboratory in the United Kingdom facilitated the establishment of a standard technique (ribotyping) for typing *C. diphtheriae*, and the ribotyping results obtained were used for the establishment of a database of ribotype patterns.

The ELWGD plays a major role in the strengthening of laboratory diagnostics. The group provides advice to WHO and other agencies on laboratory diagnosis and molecular epidemiology of *C. diphtheriae*. In response to the diphtheria epidemics, a WHO manual on laboratory diagnosis of diphtheria has been produced, which complements the WHO manual on control and case management (Efstratiou *et al.*, 2000). The Diphtheria Surveillance Network (DIPNET) established in 2001 (Kelly and Efstratiou, 2003) is the definitive expansion of the ELWGD and encompasses key epidemiologists and microbiologists from most of the European Union (EU) member states, associated countries and beyond. The integration of ELWGD and DIPNET has established rapid communication between the laboratory network, public health specialists and WHO, which should lead to a standardised surveillance system for diphtheria and related infectious diseases in Europe.

CLINICAL MANIFESTATIONS

Diphtheria is usually classified according to the site of manifestation (i.e. tonsillar, pharyngeal, nasal and non-respiratory). Humans are the only natural host for *C. diphtheriae*, and the organism spreads by airborne upper respiratory tract droplets or direct contact with respiratory secretions of symptomatic individuals or exudate from infected skin. Dust and clothing also contribute to transmission as the organism may survive up to 6 months in dust and fomites. The incubation period is 2–5 days, occasionally longer. Infection is usually associated with low-grade fever, and systemic complaints early in the infection are minimal. Symptoms depend on the location and duration of the infection before treatment. Concentration of bacteria increases over time, resulting in the release of increasing amounts of DT. Therefore, early recognition and treatment are critical to reduce complications associated with toxin dissemination. Classically, diphtheria affects the respiratory tract, although diphtheritic infections of several other sites (skin, eyes, ears, vagina and endocardium) have been described (MacGregor, 2000).

Faucial Diphtheria

Faucial diphtheria is the most common clinical presentation and involves infection of the posterior structures of the mouth, tonsils and proximal pharynx. Symptoms may be mild in vaccinated patients, whereas unvaccinated patients tend to have more severe disease. The onset is usually abrupt, with low-grade fever (rarely >39 °C), malaise, sore throat, mild pharyngeal injection and the development of a membrane typically on one or both tonsils (Plate 4). The membrane gradually thickens and extends over the tonsillar pillars, uvula, soft palate, oropharynx and nasopharynx. Initially, the membrane is white and smooth but later becomes grey with patches of green and black necrosis. The extent of the membrane correlates with the severity of symptoms: localised tonsillar disease is often mild, but involvement of the posterior pharynx, soft palate and periglottic areas is associated with profound malaise, weakness, prostration, cervical adenopathy and swelling. Gross swelling of the cervical lymph nodes gives rise to a bullneck appearance, causing respiratory stridor (Dobie and Tobey, 1979; MacGregor, 2000; Demirci, 2001; Singh and Saba, 2001).

Anterior Nasal Diphtheria

Infection of the anterior nares results in mucoid or serosanguineous nasal discharge accompanied by few constitutional symptoms. Shallow ulceration of the external nares and upper lip are characteristic. This form of diphtheria is most common in infants (MacGregor, 2000; Demirci, 2001; Singh and Saba, 2001).

Laryngeal and Tracheobronchial Diphtheria

Laryngeal diphtheria occurs as a result of membrane extension. Symptoms include hoarseness, dyspnoea, respiratory stridor and a brassy cough. Rapid diagnosis, intubation or tracheostomy is required to relieve airway obstruction; otherwise, or the patient may rapidly die of asphyxiation. The disease may progress, if enough toxin enters the blood stream, causing severe prostration, striking pallor, rapid pulse, stupor and coma. These effects may result in death within a week of onset of symptoms (Clarridge, Popovic and Inzana, 1998).

Systemic Complications of Diphtheria

Systemic complications occur because of DT. The absorbed toxin causes delayed damage at distant sites. Although the toxin is toxic to all tissues, it has its most striking effects on the heart (myocarditis) and the cranial nerves. These complications may occur from 1 to 12 weeks after disease onset.

Cardiac Toxicity

Evidence of myocarditis is found in up to two-thirds of patients with respiratory diphtheria (Boyer and Weinstein, 1948). Clinically significant cardiac dysfunction is observed in 10–20 of patients (Morgan, 1963; Ledbetter, Cannon and Costa, 1964). Cardiac toxicity correlates directly with the extent and severity of local disease and generally occurs within 1–2 weeks after the onset of the illness. The DT damages the Purkinje system, and the toxin also causes necrosis of cardiac muscle cells, which results in acute congestive heart failure and circulatory collapse (MacGregor, 2000).

Neurologic Toxicity

Neurologic toxicity is also proportional to the severity of the primary infection. Mild disease only occasionally produces neurotoxicity, but up to 75% of patients with severe disease develop neuropathy. Symptoms develop 10–28 days after the onset of respiratory disease. Two types of neuropathies are seen. The first, cranial nerve involvement, is usually limited to the ninth and tenth nerves, resulting in aspiration, difficulty in swallowing, nasal regurgitation and loss of the gag reflex (Dobie and Tobey, 1979). Less often, oculomotor and facial nerves become impaired. This type of neuropathy tends to occur later in the course of illness and resembles Guillain–Barré syndrome. Patients have quadriparesis associated with hyporeflexia. Microscopic examination of affected nerves shows degeneration of myelin sheaths and axon cylinders. After treatment of the infection, slow but complete neurologic recovery follows (MacGregor, 2000). The frequency of complications such as myocarditis and neuritis was directly related to the time between the onset of symptoms and administration of anti-toxin and to the extent of membrane formation.

Cutaneous Diphtheria

Corynebacterium diphtheriae causes clinical skin infections characterised by chronic non-healing ulcers with a dirty grey membrane particularly in the tropics (Livingood, Perry and Forrester, 1946; Höfler, 1991). Primary infection of the skin with C. diphtheriae does occur; however, more commonly, the organism superinfects pre-existing skin lesions (Liebow et al., 1946), including underlying dermatoses, lacerations, burns, insect bites and impetigo. Primary cutaneous diphtheria begins as a tender pustule and enlarges to an oval punched-out ulcer with a membrane and oedematous rolled borders (Liebow et al., 1946; Höfler, 1991) (Plate 5). Group A streptococci and Staphylococcus aureus are often isolated concurrently. It has also been reported that toxigenic strains of C. ulcerans have been associated with mimicking classical cutaneous diphtheria (Wagner et al., 2001). Skin infections may vary in severity, but toxin-induced complications are usually uncommon. Cutaneous infections induce high levels of antitoxin antibody and, therefore, appear to act as natural immunising events (Gunatillake and Taylor, 1968; Bray et al., 1972; Hewlett, 1985). They also serve as a reservoir for the organism under conditions of both endemic and epidemic respiratory tract diphtheria. Cutaneous infections have been shown to contaminate the environment and to induce throat infections more efficiently than pharyngeal colonisation. Also, bacterial shedding from cutaneous infections continues longer than from the respiratory tract (Belsey et al., 1969; Koopman and Campbell, 1975; Belsey and LeBlanc, 1975).

Other Sites of Infection

The organism may uncommonly infect other sites such as the eye, genitalia and the ear. The eye may be infected through the nasolacrimal duct from the nose or directly by droplets during the performance of tracheotomy (Harries, Mitman and Taylor, 1951). Patients may develop conjunctival diphtheria as the only manifestation of diphtheria. An acute catarrhal condition of the eye may occur; a filmy exudate or a membrane may also develop inside the lower palpebral conjunctiva and rapidly spread over the eye (Hogan, 1947), which may result in the necrosis of the conjunctiva and severe damage to the cornea. Diphtheritic infections of the vulva and vagina have also been reported (Eigen, 1932; Wallfield and Litvak, 1933; Grant, 1934).

Infections Caused by Non-toxigenic C. diphtheriae

Non-toxigenic C. diphtheriae has been known to cause significant local disease and, since 1922, some times systemic infection even in the absence of toxin production (Jordan, Smith and Neave Kingsbury, 1922). There have been reports of classical diphtheria associated with non-toxigenic *C. diphtheriae* (Edward and Allison, 1951; Rakhmanova *et al.*, 1997). However, these reports are doubtful, as patients carry toxigenic and non-toxigenic strains at the same time (Simmons *et al.*, 1980). Therefore, it is likely that a toxigenic strain was present, but not identified from patient samples. There have also been several reports of endocarditis due to non-toxigenic *C. diphtheriae* (see below). Two cases who accidentally ingested non-toxigenic *C. diphtheriae* var. *mitis* in a laboratory developed sore throat with tonsillar membrane (Barksdale, Garmise and Horibata, 1960). A non-toxigenic strain of *C. diphtheriae* var. *gravis* was also documented as the causative organism in an outbreak of infection in a boy's residential school in 1975. The organism was isolated from 7 of 19 boys and from 24 asymptomatic carriers (Jephcott *et al.*, 1975).

Non-toxigenic isolates are often isolated from cases of pharyngitis. Wilson *et al.* in 1992 reported a cluster of cases of pharyngitis in homosexual men attending a genitourinary medicine clinic (GUM) in London. However, their role here is uncertain, as the pharyngitis may have been caused by a concurrent viral or streptococcal infection (578 homosexual men were screened for pharyngeal isolates, *C. diphtheriae* was isolated from 6 patients and streptococci were isolated from 55 patients). Four reports (Jephcott *et al.*, 1975; Wilson *et al.*, 1990, 1992; Sloss and Faithfull-Davies, 1993) that examined the occurrence of non-toxigenic *C. diphtheriae* in throat swabs showed that it was not possible to state whether non-toxigenic *C. diphtheriae* was the cause of pharyngitis or whether it was a mere coloniser, especially in the absence of a control group. A study carried out on 238 throat isolates (Reacher *et al.*, 2000) showed that non-toxigenic *C. diphtheriae* was reported as the predominant organism in 72% of the cases. β-Haemolytic streptococci were also isolated from 28% of the cases, and viral culture of the throat was undertaken in only 4% of the cases. Sore throat was reported in 211 of the 238 cases. The data indicated that non-toxigenic *C. diphtheriae* is frequently perceived as a pathogen. However, very little is known about the pathogenicity of non-toxigenic *C. diphtheriae*. It has been suggested that the organism may have an affinity for attaching to vascular endothelium (Belko, Wessel and Malley, 2000).

Corynebacterium diphtheriae Endocarditis and Septicaemia

Endocarditis due to *C. diphtheriae* was first reported by Howard in 1893 (Howard, 1893). Endocarditis and septicaemia caused by *C. diphtheriae* are often considered a rarity. Toxin production does not appear to be important in these invasive infections as most of the cases reported were caused by non-toxigenic strains, and immunisation with the toxoid offers no protection. Endocarditis caused by non-toxigenic strains of *C. diphtheriae* commonly occurs in previously healthy patients who do not have any predisposing factors for endocarditis. Endocarditis can be a devastating disease, with a high incidence of complications and mortality. Most of the cases do not have a coexisting respiratory infection. Therefore, the route of infection is usually obscure. Possible predisposing factors include pre-existing heart disease including mitral insufficiency, and prosthetic valves, intravenous drug use and malignancy. *Corynebacterium* spp. isolated in blood cultures are often classified as contaminants, and this may delay the diagnosis of endocarditis. Also, because of the absence of underlying cardiac risk factors in many of these patients, physicians may not initially consider the possibility of endocarditis. These factors may explain the high morbidity and case fatality rate associated with this disease.

LABORATORY DIAGNOSIS OF DIPHTHERIA

Diphtheria is no longer easily diagnosed on clinical grounds because of the lack of familiarity with the clinical features. Mild cases of the disease resemble streptococcal pharyngitis, and the classical pseudo-membrane of the pharynx may not develop, particularly in people who have been vaccinated (Bonnet and Begg, 1999). Because the disease is rare, many clinicians may never encounter a case and therefore miss clinical diagnosis (Bowler *et al.*, 1988). Therefore, diagnosis relies on the laboratory isolation and identification of toxigenic *C. diphtheriae*. It is therefore of utmost importance for the laboratory to provide accurate and rapid information to confirm or eliminate a suspected case of diphtheria. However, bacteriological diagnosis should be regarded as complementary to, and not as a substitute for, clinical diagnosis of diphtheria (Efstratiou and Maple, 1994).

Identification of Potentially Toxigenic *C. diphtheriae*

The currently recommended microbiological procedure for the laboratory diagnosis of diphtheria is shown in Figure 7.2.

Specimen Collection and Transport

Because diphtheria is primarily an upper respiratory tract infection, ear, throat, nasopharyngeal and nasal swabs should be taken. Also, if membranous material is present, it should be submitted for examination. The only other common form of diphtheria is cutaneous diphtheria, which is clinically indistinguishable from pyodermas. Swabs should be taken from the affected area of the skin. Any crusted material should be removed and the swabs taken from the base of the lesion (Efstratiou and Maple, 1994; Clarridge, Popovic and Inzana, 1998; Efstratiou and George, 1999; Efstratiou *et al.*, 2000).

Once swabs have been taken, they should be transported to the laboratory immediately. The laboratory should be notified ahead of time of a suspected diagnosis of diphtheria. If the specimen cannot be transported to the laboratory immediately, the use of a transport medium such as Amies (Amies, 1976), which was described in 1976, is recommended (Efstratiou and Maple, 1994; Efstratiou and George, 1999).

Primary Isolation

Culture specimens should be initially plated out on blood agar and Hoyle's tellurite agar and incubated at 35–37 °C. Tellurite-containing media suppress the growth of commensals and allow the growth of *C. diphtheriae* and some other corynebacteria as well as staphylococci and yeasts. They have the rare ability to reduce tellurite salts, producing dull black colonies (Efstratiou *et al.*, 2000). *Corynebacterium diphtheriae* grows most rapidly on Loeffler medium, which is a lipid-rich medium, and Gram stains from colonies grown on Loeffler are the best for demonstrating metachromatic granules. However, diagnosis of diphtheria based solely upon direct microscopy of a smear is unreliable. Potentially toxigenic species of corynebacteria (*C. diphtheriae*, *C. ulcerans* and *C. pseudotuberculosis*) are differentiated by the use of rapid screening tests such as detection of the enzyme cysteinase (using Tinsdale medium) and the absence of pyrazinamidase activity (Colman, Weaver and Efstratoiu, 1992; Efstratiou and Maple, 1994). On Tinsdale medium, the three potentially toxigenic species of corynebacteria produce black colonies surrounded by a brown halo. The use of Tinsdale medium for primary culture is not recommended, because the medium is highly selective and, therefore, increases the probability of false negatives, mainly with specimens that contain small numbers of organisms (Efstratiou and Maple, 1994; Efstratiou and George, 1999).

After isolation of pathogenic strains, identification is carried out by the use of a series of simple biochemical tests. The biochemical characteristics of the potentially toxigenic corynebacteria (four biotypes of *C. diphtheriae*, *C. ulcerans* and *C. pseudotuberculosis*) and other coryneforms that may be present in throat or wound swabs are described in Table 7.1. Commercial identification kits such as the

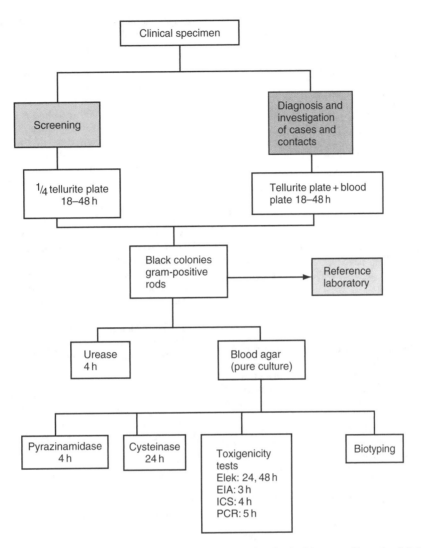

Figure 7.2 A schematic diagram of the currently recommended microbiological procedure for the laboratory diagnosis of diphtheria. Taken from Efstratiou *et al.* (2000). Reproduced by permission of University of Chicago Press, copyright (2000).

Table 7.1 Biochemical identification of clinically significant corynebacteria (Efstratiou and George, 1999) Reproduced by permission of the Health Protection Agency

Species	CYS[a]	PYZ[b]	Nitrate	Urea	Fermentation of			
					Glucose	Maltose	Sucrose	Starch
C. diphtheriae								
Var. *gravis*	+	−	+	−	+	+	−	+
Var. *mitis*	+	−	+	−	+	+	−	−
Var. *intermedius*	+	−	+	−	+	+	−	−
Var. *belfanti*	+	−	−	−	+	+	−	−
C. ulcerans	+	−	−	+	+	+	−	+
C. pseudotuberculosis	+	−	−	+	+	+	−	+
C. amycolatum	−	+	v	v	+	v	v	−
C. imitans	−	±	−	−	+	+	±	−
C. pseudodiphtheriticum	−	v	+	+	−	−	−	−
C. striatum	−	+	+	−	+	−	v	−

+, positive; −, negative; ± , weak; v, variable.

[a] Cysteinase production on Tinsdale medium.

[b] Pyrazinamidase activity.

API Coryne system (API bioMérieux, Marcy-l'Etoile, France) and Rosco Diagnostica tests (Rosco Diagnostica, Taastrup, Denmark) are readily available, and the results obtained are highly accurate and reproducible. All tests are described in detail in a WHO manual (Efstratiou and Maple, 1994). The four biotypes of *C. diphtheriae* (var. *gravis*, var. *mitis*, var. *belfanti* and var. *intermedius*) are classified on the basis of morphological and biochemical properties (Table 7.2). The biochemical tests that distinguish *C. diphtheriae* from other corynebacteria and coryneforms are as

Table 7.2 Differences in the morphological appearances of typical colonies of *C. diphtheriae* on tellurite blood agar (after 48 h of incubation aerobically)

Biotype	Colony morphology
C. diphtheriae var. *gravis*	Dull, dry grey, opaque colonies, 1.5–2 mm in diameter with a matte surface. Colonies are friable, they tend to break into small segments when touched with a straight wire. They can also be pushed across the surface of the medium without breaking. Non-haemolytic usually
C. diphtheriae var. *mitis*	Grey, opaque colonies, 1.5–2 mm in diameter with an entire edge and smooth surface. Variation in size is a common characteristic. They tend to exhibit a small zone of β-haemolysis on blood agar
C. diphtheriae var. *intermedius*	Small, grey, discrete, translucent colonies, 0.5–1 mm in diameter

follows: production of catalase and cysteinase, reduction of nitrate (except biotype *belfanti*, which is nitrate negative), absence of urease and pyrazinamidase and fermentation of glucose, maltose and starch (only biotype *gravis* ferments starch).

Detection of Toxigenicity

The test for toxigenicity is the most important test for the microbiological diagnosis of diphtheria and should be performed without delay on any suspect isolate that is found by routine screening or while investigating a suspect case of diphtheria (Brooks and Joynson, 1990). The *in vivo* toxin test using guinea pigs was the gold standard for testing toxin production since the end of the last century. However, the expense, the risk of accidental self-inoculation, slowness of the test in producing a result and the increasing unacceptability of *in vivo* tests in many countries led to a decline in the use of this test. An alternative gold standard for the *in vivo* assay is the *in vitro* Vero cell assay, which is based on the cytotoxicity of DT to cultured Vero cells (Efstratiou *et al.*, 1998). Engler *et al.* (1997) and Reinhardt, Lee and Popovic (1998) developed modified Elek tests, which produce results in 16–24 h. Other *in vitro* methods that are currently used include a polymerase chain reaction (PCR) assay for detecting the 'A' portion of the toxin gene (Pallen, 1991; Hauser *et al.*, 1993; Pallen *et al.*, 1994; Aravena-Roman, Bowman and O'Neill, 1995; Mikhailovich *et al.*, 1995; Nakao and Popovic, 1997), a rapid enzyme immunoassay (EIA) (Engler and Efstratiou, 2000) and a rapid immunochromatographic strip (ICS) test (Engler *et al.*, 2002).

MANAGEMENT, PREVENTION AND CONTROL OF *C. DIPHTHERIAE* INFECTIONS

Antitoxin has been used to treat diphtheria for over 100 years (MacGregor, 2000). Diphtheria antitoxin is a hyperimmune serum produced in horses. The antitoxin only neutralises circulating toxin that has not bound to the tissues, and therefore, it is critical that diphtheria antitoxin be administered as soon as a presumptive diagnosis has been made, without waiting for bacteriological confirmation.

Delayed administration increases the risk of late effects such as myocarditis and neuritis (Begg, 1994).

Because antitoxin is derived from horse serum, tests with a trial dose to exclude hypersensitivity to horse proteins should be carried out before administration. All patients should be tested by applying a 1:100 dilution of diphtheria antitoxin onto the underside of the forearm by scratching or pricking. If the result is negative, 0.02 ml of antitoxin diluted 1:1000 in saline is injected intracutaneously, with adrenaline available for immediate administration. If an immediate reaction occurs, the patient should be desensitised progressively with higher doses of antiserum (MacGregor, 2000).

The dose of antitoxin depends upon the severity of the disease and the site and the extent of the membrane. Table 7.3 summarises the dose range for various clinical situations (Bonnet and Begg, 1999). Many countries use this scheme; however, there may be variations, which have been recommended by manufacturers of antitoxin or the national health authorities. Antitoxin therapy is probably of no value for cutaneous disease; however, some authorities recommend using 20 000–40 000 units because there have been reports of toxic sequelae (Farizo *et al.*, 1993).

Antibiotics are necessary to eliminate the bacterium and prevent spread, but it is not a substitute for antitoxin treatment. Killing the organism by antibiotic therapy has three benefits:

1. toxin production is terminated;
2. local infection is improved;
3. spread of organism to uninfected contacts is prevented (MacGregor, 2000).

Antibiotics such as penicillin, erythromycin, clindamycin, rifampicin and tetracycline are effective. Penicillin and erythromycin appear to be the agents of choice. Intramuscular administration of procaine penicillin G (25 000–50 000 U/kg·day for children and 1.2 million units/day for adults, in two divided doses) is recommended or parenteral course of erythromycin (40–50 mg/kg·day, with a maximum of 2 g/day) until the patient can swallow comfortably, at which point erythromycin in four divided doses or oral penicillin V (125–250 mg four times daily) may be substituted. Antibiotic treatment should be continued for 14 days (Bonnet and Begg, 1999).

Prevention of diphtheria depends primarily on widespread routine vaccination with diphtheria toxoid (MacGregor, 2000). Control and prevention of diphtheria epidemics require the following:

1. *High vaccine coverage in target groups.* Achieving and maintaining high coverage in children through routine immunisation. Furthermore, mass immunisation of adults older than 25 years, belonging to high-risk groups (alcoholics, homeless, armed forces and health care workers) with diphtheria toxoid-containing vaccines (preferably Td, diphtheria–tetanus vaccine with reduced amount of diphtheria toxoid).
2. *Early diagnosis and proper management (immediate treatment and hospitalisation) of diphtheria cases.* Prompt recognition of suspected diphtheria cases; prompt and qualified collection and shipment of specimens for laboratory examination; prompt standard treatment with diphtheria antitoxin antibiotics; and notification of the case to the local health authorities.

Table 7.3 Dose of antitoxin recommended for various types of diphtheria

Type of diphtheria	Dose (units)	Route
Nasal	10 000–20 000	Intramuscular
Tonsillar	15 000–25 000	Intramuscular or intravenous
Pharyngeal or laryngeal	20 000–40 000	Intramuscular or intravenous
Combined types or delayed diagnosis	40 000–60 000	Intravenous
Severe diphtheria, e.g. with extensive membrane and/or severe oedema (bullneck diphtheria)	40 000–100 000	Intravenous or part intravenous and part intramuscular

3. *Rapid investigation of close contacts of people with diphtheria and their effective standardised treatment to prevent secondary cases.* Clinical surveillance for 7 days from the last date of contact with the case; treatment with antibiotics for 7 days; and immunisation with booster dose of diphtheria/tetanus toxoid if their immunisation is incomplete or unknown (Dittmann and Roure, 1994).

Immunity to Diphtheria

Acquired immunity to diphtheria is primarily due to toxin-neutralising antibody (antitoxin). Antibodies to fragment B of the DT are able to neutralise the effect of toxin and protect against disease. Diphtheria antitoxin production [primarily of immunoglobulin G (IgG) antibody] may be induced by the natural toxin during clinical infection or in the carrier state or by immunisation with diphtheria toxoid (Todar, 1997; Canada CDR, 1998). Diphtheria toxoid is a formaldehyde-inactivated preparation of DT, adsorbed onto aluminium salts to increase its antigenicity (WHO, 1998). DTP vaccine (diphtheria toxoid combined with pertussis vaccine and tetanus toxoid) is the core vaccine in childhood immunisation programmes. The vaccine is used in preschool children for primary and reinforcing immunisation, whereas a combined diphtheria–tetanus vaccine in the form DT is used for booster doses in preschool children (7 years and younger) and its adult form Td (which has a reduced amount of diphtheria toxoid) is used for booster doses in older children, adolescents and adults. There are three main DTP immunisation schedules in widespread use in different countries:

1. three doses: three primary doses of DTP vaccine given during the first year of life;
2. four doses: primary series of three doses reinforced with one booster dose administered usually at the age of 2–3;
3. five doses: primary series of three doses reinforced with the first booster at the age of 2 and second booster given before entering school at the age of 4–6 (WHO, 1998).

The three-dose schedule is mainly used in the African region, Southeast Asia and in the Western Pacific region. The four-dose schedule is used in most European countries, and the five-dose schedule is used mainly in the American region (WHO, 1998). In the United Kingdom, five doses are currently recommended: a primary series of three doses at 2, 3 and 4 months of age and a booster dose at school entry or 3 years after primary immunisation, and since 1994, a booster dose at 15–19 years has been recommended (PHLS CDR, 1994; Begg and Balraj, 1995).

In the prevaccine era, circulation of *C. diphtheriae* strains was common and the occurrence of diphtheria case high. This enabled adults to acquire natural immunity by subclinical infection. Most newborn infants passively acquire antibodies from their mothers via the placenta, which provided protection during the first 6 months after birth (Zingher, 1923; Hardy, 1998; Galazka, 2000). However, after widespread childhood immunisation, diphtheria became rare and exposure to these organisms was uncommon. Immunity induced by childhood immunisation usually wanes, and if adults are not exposed to toxigenic organisms to acquire natural immunity and if they do not receive booster doses of diphtheria toxoid, they become susceptible to the disease (Brainerd *et al.*, 1952; Galazka and Robertson, 1996; Galazka and Tomaszunas-Blaszczyk, 1997). Several studies have shown that many adults in industrialised countries have low antibody levels and are susceptible to diphtheria (Weiss, Strassburg and Feeley, 1983; Christenson and Böttiger, 1986; Simonsen *et al.*, 1987; Cellesi *et al.*, 1989; Maple *et al.*, 1995; Edmunds *et al.*, 2000; Egemen *et al.*, 2000; Marlovits *et al.*, 2000; von Hunolstein *et al.*, 2000). The age group with the lowest levels of immunity varies from country to country and probably depends on the year that childhood immunisation programme was implemented on a routine basis (Galazka and Robertson, 1996; Edmunds *et al.*, 2000).

According to internationally accepted definitions, an antibody titre of >0.1 IU/ml gives long-term protection against diphtheria, a titre of 0.1 IU/ml gives full protection, a titre between 0.01 IU/ml and 0.09 IU/ml gives basic protection and a titre of less than 0.01 IU/ml indicates no protection against diphtheria. Several assays have been developed to determine the diphtheria antitoxin level in serum (Efstratiou and Maple, 1994), and among these, the *in vivo* toxin neutralisation test is regarded as the gold standard. However, it is laborious, time consuming and expensive; therefore, equally accurate *in vitro* cell-culture methods have been developed. These assays measure the ability of serum from an individual to neutralise the cytopathic effects of pure DT in a tissue-culture system and thus measure functional antibodies. Assays that are cheaper and simpler to perform than the neutralisation tests include enzyme-linked immunosorbent assay (ELISA) and passive haemagglutination tests. However, they lack sensitivity for serum samples containing <0.1 IU/ml of antitoxin (Efstratiou and George, 1999).

OTHER POTENTIALLY TOXIGENIC CORYNEBACTERIA

Corynebacterium ulcerans

Corynebacterium ulcerans was first isolated in 1926 from throat lesions by Gilbert and Stewart (Gilbert and Stewart, 1926). Colonies of *C. ulcerans* resemble those of *C. diphtheriae* biotype *gravis*. They are dry, waxy and grey-white, with light haemolysis, and are 1–2 mm in diameter after 24 h. Microscopically, the organism is pleomorphic, with predominantly coccoid forms and some rod forms. On Tinsdale medium, it produces brownish-black colonies with distinct halos, which are indistinguishable from *C. diphtheriae* colonies. *Corynebacterium ulcerans* is urease positive and nitrate negative and ferments glucose, maltose and glycogen (von Graevenitz, Coyle and Funke, 1998; Efstratiou and George, 1999; Funke and Bernard, 1999). The ability to hydrolyse urea and the lack of nitrate reduction distinguish *C. ulcerans* from *C. diphtheriae* (Table 7.1). *Corynebacterium ulcerans* when lysogenic for a *tox*+ carrying phage, produces two exotoxins in varying proportions (Petrie and McLean, 1934; Carne and Onon, 1982). One is identical to *C. diphtheriae* toxin, which is neutralised by diphtheria antitoxin and can be detected by the Elek test. The other is identical to *C. pseudotuberculosis* toxin, phospholipase D (PLD), which is unaffected by diphtheria antitoxin. *Corynebacterium pseudotuberculosis* toxin PLD is discussed in more detail in the section *Corynebacterium pseudotuberculosis*.

Corynebacterium ulcerans is a veterinary pathogen and causes mastitis in cattle and other domestic and wild animals (Fox and Frost, 1974; Hommez , 1999). Toxigenic strains of *C. ulcerans* have been associated with classical diphtheria (Fakes and Downham, 1970; Meers, 1979; CDC, 1997; Ahmad, Gainsborough and Paul, 2000) as well as milder symptoms (Bostock *et al.*, 1984; Hart, 1984; Barrett, 1986; Pers, 1987; Kisely *et al.*, 1994). At least three deaths have been attributed to such infection (Leek, 1990; Anonymous, 2000; Wellinghausen *et al.*, 2002). Usually, human infections are acquired through contact with animals or by ingestion of unpasteurised dairy products (Bostock *et al.*, 1984; Hart, 1984; Barrett, 1986). However, such risk factors were not identified for three cases of classical diphtheria in the United Kingdom, caused by *C. ulcerans*, which suggests that there may be other routes of infection (White *et al.*, 2001). Recently, in the United Kingdom, toxigenic *C. ulcerans* was isolated from domestic cats with bilateral nasal discharge (Anonymous, 2002a; Taylor *et al.*, 2002).

Corynebacterium pseudotuberculosis

Corynebacterium pseudotuberculosis (previously known as *Corynebacterium ovis*) was called the Preisz–Nocard bacillus in honour of the researchers who first isolated the organism in the early 1890s from necrotic kidney of a sheep (Lipsky *et al.*, 1982). Colonies are yellowish white, opaque and convex with a matted surface and are about 1–2 mm in diameter after 48 h and produce slight β-haemolysis. Like *C. ulcerans*, *C. pseudotuberculosis* is urease positive. The organism is similar to both *C. diphtheriae* and *C. ulcerans* in its production of halos on Tinsdale media and failure to hydrolyse pyrazinamidase (Table 7.1). *Corynebacterium pseudotuberculosis* comprises two biovariants, *equi* and *ovis*, that are distinguished by the ability of biovariant *equi* to reduce nitrate (Songer *et al.*, 1988). The most discriminating biochemical tests for distinguishing *C. pseudotuberculosis* from *C. ulcerans* are the hydrolysis of glycogen by *C. ulcerans* and the lipid-stimulated growth of *C. pseudotuberculosis*.

Corynebacterium pseudotuberculosis also harbours the DT, and nearly all strains produce dermonecrotic toxin or PLD. PLD is a potential virulence factor of *C. pseudotuberculosis*. It is a protein much smaller than the DT (molecular weigh 14 000). Over 10% of the *C. pseudotuberculosis* isolates produce the DT (Maximescu *et al.*, 1974); however, there have not been any clinical cases of diphtheria attributed to infection with *C. pseudotuberculosis* (MacGregor, 2000).

Corynebacterium pseudotuberculosis causes caseous lymphadenitis in sheep and goats, ulcerative lymphangitis in horses and cutaneous abscesses and lesions in many other animal species (Yeruham *et al.*, 1997; Moller *et al.*, 2000; Watson and Preece, 2001). Infections of humans have also been recorded in farm workers and veterinarians exposed to infected animals, usually manifesting as lymphadenitis (Goldberger *et al.*, 1981; Richards *et al.*, 1985; House *et al.*, 1986; Mills *et al.*, 1997; Peel *et al.*, 1997).

OTHER CORYNEBACTERIA

Corynebacterium pseudodiphtheriticum

Corynebacterium pseudodiphtheriticum, previously known as *Bacillus pseudodiphtheriticum*, was probably first isolated by von Hoffman in 1888 from throats of humans (MacGregor, 2000). The current name *C. pseudodiphtheriticum* was given by Bergey *et al.* in 1925 (Lipsky *et al.*, 1982). Colonies of *C. pseudodiphtheriticum* are non-haemolytic, white, slightly dry with entire edges and 1–2 mm in diameter after 48 h of incubation at 37 °C. The organism is non-fermentative, reduces nitrate and hydrolyses urea.

Corynebacterium pseudodiphtheriticum is part of the oropharyngeal bacterial flora and is associated with respiratory disease mainly in immunosuppressed patients (Cimolai *et al.*, 1992; Freeman *et al.*, 1994; Ahmed *et al.*, 1995; Manzella *et al.*, 1995). It also causes pneumonia or bronchitis in non-immunocompromised hosts (Miller *et al.*, 1986; Ahmed *et al.*, 1995; Manzella *et al.*, 1995). Most of the cases reported have been in patients with endotracheal intubation. Therefore, there is a possibility that the intubation procedure introduces the organism into the lower respiratory tract (Freeman *et al.*, 1994). The organism also causes endocarditis and has been isolated from draining wounds (Lindner, Hardy and Murphy, 1986; Hatch, 1991; Morris and Guild, 1991; Wilson *et al.*, 1992).

Corynebacterium xerosis

Corynebacterium xerosis is a commensal of human skin and mucous membranes (von Graevenitz, Coyle and Funke, 1998). It has been reported that, in the past, *Corynebacterium striatum* strains were misidentified as *C. xerosis* (Coyle *et al.*, 1993). In the past, differentiation of *C. xerosis* from *C. striatum* relied on a few biochemical tests. Both organisms are positive for glucose and sucrose fermentation and are negative for urease production, esculin hydrolysis and mannitol and xylose fermentation. They are distinguished primarily by fermentation of maltose by *C. xerosis* but not *C. striatum*. True *C. xerosis* colonies appear yellowish, dry and granular with irregular edges,

whereas *C. striatum* colonies are white, moist and smooth. Other characteristics that may be helpful in differentiating *C. xerosis* from *C. striatum* are the following: hydrolysis of tyrosine and rapid fermentation of sugars (24–36 h) by strains of *C. striatum*. *Corynebacterium xerosis* strains do not hydrolyse tyrosine and ferment sugars slowly (72–96 h) (Martinez-Martinez *et al.*, 1995).

Furthermore, Funke *et al.* in 1996 demonstrated that nearly all isolates identified as *C. xerosis* in the routine clinical laboratory in the past in fact represent *C. amycolatum* strains. They concluded that *C. xerosis* strains are very rare and that nearly all *C. xerosis* strains that appeared in the literature before 1996 may have been misidentified *C. amycolatum* strains. Characteristics that differentiate *C. amycolatum* from *C. xerosis* are the following: *C. amycolatum* strains produce white-grey colonies, lack mycolic acids, show weak or no leucine arylamidase activity, produce large amounts of propionic acid from glucose fermentation and are resistant to the vibriocidal agent 0/129, whereas *C. xerosis* strains, as mentioned above, produce yellowish colonies, express mycolic acids and leucine arylamidase activity, produce lactic acid from glucose fermentation and are susceptible to 0/129 (Funke *et al.*, 1996a). Therefore, as *C. xerosis*, *C. striatum* and *C. amycolatum* strains have been misidentified in the past, the pathogenic role of these organisms is unclear.

Corynebacterium imitans

Corynebacterium imitans is a new species, and the only strain described to date was isolated in 1997 from a nasopharyngeal specimen of a child suspected of having throat diphtheria, and the same strain was isolated from three adult contacts (Funke *et al.*, 1997). This was the first well-documented case of person-to-person transmission of a *Corynebacterium* other than *C. diphtheriae* in a non-hospital environment. Colonies of *C. imitans* are 1–2 mm in diameter, white-grey in colour and shiny and creamy with entire edges after 24 h incubation at 37 °C. The organism is cysteinase negative and tellurite positive. It is catalase positive and nitrate and α-glucosidase negative, and pyrazinamidase activity is weak. Urea, esculin and tyrosine are not hydrolysed. Acid is produced from D-arabinose, ribose, D-glucose, D-fructose, D-mannose, maltose, lactose, L-fucose and weakly from sucrose. *Corynebacterium imitans* is be differentiated from *C. diphtheriae* by the following tests: negative nitrate reduction, weak pyrazinamidase activity, positive alkaline phosphatase, a positive CAMP test and susceptibility to 0/129. The strain isolated did not produce DT, and the toxin gene was not detectable.

Corynebacterium jeikeium

Corynebacterium jeikeium was first described by Hande *et al.* in 1976, as the cause of four cases of septicaemia in immunocompromised patients, and it consists of the isolates previously designated as CDC coryneform group JK bacteria (Jackman *et al.*, 1987).

Corynebacterium jeikeium exists as part of the normal skin flora in hospitalised patients (Gill *et al.*, 1981; Larson *et al.*, 1986; Soriano and Fernadez-Roblas, 1988), and it may survive in the hospital environment for long periods of time (Telander *et al.*, 1988). *Corynebacterium jeikeium* isolates are pleomorphic, club-shaped Gram-positive rods, arranged in V forms or palisades. On blood agar, these organisms form discrete pinpoint colonies that are greyish-white, glistening and non-haemolytic. The biochemical reactions that differentiate *C. jeikeium* from other corynebacteria are the inability to produce urease, reduce nitrate or readily ferment most carbohydrates (ferments glucose, galactose and, sometimes, maltose). The API Coryne identification system may incorrectly identify *C. jeikeium* because of insufficient acid formation from glucose and maltose (von Graevenitz, Coyle and Funke, 1998). The organism is catalase positive and oxidase negative, and no halo is produced on Tinsdale medium (Coyle, Hollis and Groman, 1985; Jackman *et al.*, 1987). It is a strict aerobe, with no growth occurring anaerobically (Jackman *et al.*, 1987).

Most infections have occurred in immunocompromised patients with malignancies, patients with medical devices (vascular or implanted access catheters), patients who are hospitalised for long periods, those who have breaks in their integument and those who have had therapy with broad-spectrum antibiotics (Coyle and Lipsky, 1990). Cutaneous infection by *C. jeikeium* was reported by Dan *et al.* in 1988. The following have been reported as being associated with *C. jeikeium* infections: neutropenia (Rozdzinski *et al.*, 1991), prosthetic and native heart valve endocarditis (Vanbosterhaut *et al.*, 1989; Ross *et al.*, 2001), osteomyelitis (Weller, McLardy-Smith and Crook, 1994), meningitis with transverse myelitis (Johnson, Hulse and Oppenheim, 1992) and cerebral ventriculitis (Knudsen, Nielsen and Espersen, 1994).

Corynebacterium urealyticum

The name *Corynebacterium urealyticum* was proposed in 1986 to describe *Corynebacterium* group D2 (Soriano *et al.*, 1986); however, this name was not validated until 1992 (Pitcher *et al.*, 1992). *Corynebacterium urealyticum* is Gram-positive rods that may appear coccoidal after prolonged incubation and are arranged in V forms or palisades. Colonies are non-haemolytic, pinpoint, whitish, smooth and convex. The organism is a strict aerobe, with no growth occurring on blood agar incubated anaerobically (Pitcher *et al.*, 1992). *Corynebacterium urealyticum* strains are catalase positive, produce pyrazinamidase, have very strong urease activity and are oxidase negative, nitrate negative and asaccharolytic. They may be differentiated from *C. jeikeium* by their inability to oxidise sugars and by their urease production.

Corynebacterium urealyticum has been documented to cause acute or chronic (encrusted) infections of the lower urinary tract if an underlying renal or bladder disease is present (Soriano *et al.*, 1990). The organism may also cause infection in the upper part of the urinary tract, resulting in pyelonephritis or pyeloureteritis (Aguado, Ponte and Soriano, 1987), particularly in immunocompromised or post–renal-transplant patients with surgical complications (Nadal, Schwöbel and von Graevenitz, 1988; Aguado *et al.*, 1993). There have been reports of *C. urealyticum* causing infections other than urinary tract infections, such as endocarditis (Ena *et al.*, 1991), bacteremia (Marshall, Routh and MacGowan, 1987; Soriano *et al.*, 1993; Wood and Pepe, 1994), osteomyelitis (Chomarat Breton and Dubost, 1991), soft-tissue infection (Saavedra *et al.*, 1996) and wound infection (Soriano *et al.*, 1993). Most *C. urealyticum* strains are multiresistant to antibiotics. *Corynebacterium urealyticum* is often resistant to β-lactams and aminoglycosides and susceptible to vancomycin.

REFERENCES

Ahmad, N., Gainsborough, N. & Paul, J. (2000) An unusual case of diphtheria and its complications. *Hospital Medicine* **61**: 436–437.

Amies, C.R. (1976) A modified formula for the preparation of Stuart's transport medium. *Canadian Journal of Public Health* **58**: 296–300.

Anderson, J.S., Cooper, K.E., Happold, F.C. & McLeod, J.W. (1933) Incidence and correlation with clinical severity of *gravis, mitis*, and *intermedius* types of diphtheria bacillus in a series of 500 cases at Leeds. *The Journal of Pathology and Bacteriology* **36**: 169–182.

Andrews, F.W., Bulloch, W., Douglas, S.R. *et al.* (1923) In *Diphtheria—Its Bacteriology, Pathology and Immunology*. Chapter 1. London: His Majesty's Stationery Office.

Aravena-Roman, M., Bowman, R. & O'Neill, G. (1995) Polymerase chain reaction for the detection of toxigenic *Corynebacterium diphtheriae*. *Pathology* **27**: 71–73.

Barksdale, L., Garmise, L. & Horibata, K. (1960) Virulence, toxinogeny, and lysogeny in *Corynebacterium diphtheriae*. *Annals of the New York Academy of Sciences* **88**: 1095–1096.

Begg, N. (1994) Diphtheria. Manual for the management and control of diphtheria in the European Region. *The Expanded Programme on Immunisation in the European Region*. Copenhagen. 1–18.

Begg, N. & Balraj, V. (1995) Diphtheria: are we ready for it? *Archives of Disease in Childhood* **73**: 568–572.

Belko, J., Wessel, D.L. & Malley, R. (2000) Endocarditis caused by *Corynebacterium diphtheriae*: case report and review of the literature. *The Pediatric Infectious Disease Journal* **19**: 159–163.

Belsey, M.A. & LeBlanc, D.R. (1975) Skin infections and the epidemiology of diphtheria. Acquisition and persistence of *C. diphtheriae* infections. *American Journal of Epidemiology* **102**: 179–184.

Belsey, M.A., Sinclair, M., Roder, M.R. & LeBlanc, D.R. (1969) *Corynebacterium diphtheriae* skin infections in Alabama and Louisiana. *The New England Journal of Medicine* **280**: 135–141.

Bennett, M.J. & Eisenberg, D. (1994) Refined structure of monomeric diphtheria toxin at 2.3 Å resolution. *Protein Science* **3**: 1464–1475.

Bonnet, J.M. & Begg, N.T. (1999) Control of diphtheria: guidance for consultants in communicable disease control. *Communicable Disease and Public Health* **4**: 242–249.

Bowler, I.C.J., Mandal, B.K., Schlecht, B. & Riordan, T. (1988) Diphtheria – the continuing hazard. *Archives of Disease in Childhood* **63**: 194–210.

Boyd, J., Oza, M.N. & Murphy, J.R. (1990) Molecular cloning and DNA sequence analysis of a diphtheria *tox* iron-dependent regulatory element (*dtx*R) from *Corynebacterium diphtheriae*. *Proceedings of the National Academy of Sciences of the United States of America* **87**: 5968–5972.

Boyer, N.H. & Weinstein, L. (1948) Diphtheritic myocarditis. *The New England Journal of Medicine* **239**: 913.

Brainerd, H., Kiyasu, W., Scaparone, M. & O'Gara, L. (1952) Susceptibility to diphtheria among elderly persons. Immunisation by the intracutaneous administration of toxoid. *The New England Journal of Medicine* **247**: 550–554.

Bray, J.P., Burt, E.G., Potter, E.V. *et al.* (1972) Epidemic diphtheria and skin infections in Trinidad. *The Journal of Infectious Diseases* **126**: 34–40.

Brooks, R. & Joynson, D.H. (1990) Bacteriological diagnosis of diphtheria (ACP Broadsheet no 125). *Journal of Clinical Pathology* **43**: 576–580.

Brown, B.A. & Bodley, J.W. (1979) Primary structure at the site in beef and wheat elongation factor 2 of ADP-ribosylation by diphtheria toxin. *FEBS Letters* **103**: 253–255.

Canada CDR (1998) Guidelines for the control of diphtheria in Canada. *Canada Communicable Disease Report Supplement* **24S3**. www.phac-aspc. gc.ca/publicat/ccdr-rmtc/98vol24/24s3/24s3g_e.html (February 2004).

Carne, H.R. & Onon, E.O. (1982) The exotoxins of *Corynebacterium ulcerans*. *The Journal of Hygiene* **88**: 173–191.

CDC (1997) Respiratory diphtheria caused by *Corynebacterium ulcerans* – Terre Haute, Indiana. *Morbidity and Mortality Weekly Report* **46**: 330–332.

Cellesi, C., Zanchi, A., Michelangeli, C. *et al.* (1989) Immunity to diphtheria in a sample of adult population from central Italy. *Vaccine* **7**: 417–420.

Chang, D.N., Laughren, G.S. & Chalvardjian, N.E. (1978) Three variants of *Corynebacterium diphtheriae* subsp. *mitis* (*belfanti*) isolated from a throat specimen. *Journal of Clinical Microbiology* **8**: 767–768.

Choe, S., Bennett, M.J., Fujii, G. *et al.* (1992) The crystal structure of diphtheria toxin. *Nature* **357**: 216–222.

Christenson, B. & Böttiger, M. (1986) Serological immunity to diphtheria in Sweden in 1978 and 1984. *Scandinavian Journal of Infectious Diseases* **18**: 227–233.

Clarridge, J.E., Popovic, T. & Inzana, T.J. (1998) Diphtheria and other corynebacterial and coryneform infections. In *Bacterial Infections: Topley & Wilson's Microbiology and Microbial Infections, Vol. 3, 9th ed.* Hausler, W.J. Jr & Sussman, M. (eds). New York: Oxford University Press. pp. 347–371.

Collier, R.J. (1982) Structure and activity of diphtheria toxin. In *ADP—Ribosylation Reactions*. Hayashi, D. & Ueda, K. (eds). New York: Academic Press. p. 575.

Collins, M.D. & Cummins, C.S. (1986) Genus *Corynebacterium* Lehmann and Neumann. In *Bergey's Manual of Systemic Bacteriology, Vol. 2.* Sneath, P.H.A., Mair, N.S., Sharpe, M.E. & Holt, J.G. (eds). Baltimore, MD: Williams & Wilkins. pp. 1266–1276.

Collins, M.D., Goodfellows, M. & Minnikin, D.E. (1982) A survey of the structures of mycolic acids in *Corynebacterium* and related taxa. *Journal of General Microbiology* **128**: 129–149.

Collins, M.D., Burton, R.A. & Jones, D. (1988) *Corynebacterium amycolatum* sp. nov., a new mycolic acid-less *Corynebacterium* species from human skin. *FEMS Microbiology Letters* **49**: 349–352.

Collins, M.D., Falsen, E., Akervall, E. *et al.* (1998) *Corynebacterium kroppenstedtii* sp. nov., a novel *Corynebacterium* that do not contain mycolic acids. *International Journal of Systematic Bacteriology* **48**: 1449–1454.

Colman, G., Weaver, E. & Efstratoiu, A. (1992) Screening tests for pathogenic corynebacteria. *Journal of Clinical Pathology* **45**: 46–48.

Coyle, M.B., Groman, N.B., Russell, J.Q. *et al.* (1989) The molecular epidemiology of three biotypes of *C. diphtheriae* in the Seattle outbreak 1972–1982. *The Journal of Infectious Diseases* **159**: 670–679.

De Zoysa, A. & Efstratiou, A. (1999) PCR typing of *Corynebacterium diphtheriae* by random amplification of polymorphic DNA. *Journal of Medical Microbiology* **48**: 335–340.

De Zoysa, A. & Efstratiou, A. (2000) Use of amplified fragment length polymorphisms for typing *Corynebacterium diphtheriae*. *Journal of Clinical Microbiology* **38**: 3843–3845.

De Zoysa, A., Efstratiou, A., George, R.C. *et al.* (1993) Diphtheria and travel. *Lancet* **342**: 446.

De Zoysa, A., Efstratiou, A., George, R.C. *et al.* (1995) Molecular epidemiology of *Corynebacterium diphtheriae* from northwestern Russia and surrounding countries studied by using ribotyping and pulsed-field gel electrophoresis. *Journal of Clinical Microbiology* **33**: 1080–1083.

Demirci, C.S. (2001) *Diphtheria*. www.emedicine.com/ped/topic596.htm (February 2004)

Dittman, S. (1997) Epidemic diphtheria in the Newly Independent States of the former USSR – situation and lessons learned. *Biologicals* **25**: 179–186.

Dittmann, S. & Roure, C. (1994) Plan of action for the prevention and control of diphtheria in the European Region. *The Expanded Programme on Immunisation in the European Region*. Copenhagen. 8–10.

Dixon, J., Noble, W. & Smith, G. (1990) Diphtheria; other corynebacterial and coryneform infections. In *Bacterial Diseases: Topley & Wilson's Principles of Bacteriology, Virology and Immunity, Vol. 3, 8th ed.* Smith, G. & Easmon, C. (eds). London: Edward Arnold (Parker, M. & Collier, L. (eds)). pp. 55–80.

Dobie, R.A. & Tobey, D.N. (1979) Clinical features of diphtheria in the respiratory tract. *Journal of the American Medical Association* **242**: 2197.

Dolman, C.E. (1973) Landmarks and pioneers in the control of diphtheria. *Canadian Journal of Public Health* **64**: 317–336.

Douglass, W. (1736) The practical history of a new epidemical eruptive miliary fever, with an Angina ulcusculosa, which prevailed in Boston, New England in the years 1735 and 1736. Reprinted in 1825. *New England Journal of Medicine and Surgery* **14**: 1–13.

Drazin, R., Kandel, J. & Collier, R.J. (1971) Structure and activity of diphtheria toxin. II. Attack by trypsin at a specific site within the intact toxin molecule. *The Journal of Biological Chemistry* **246**: 1504–1510.

Edmunds, W.J., Pebody, R.G., Aggerback, H. *et al.* (2000) The sero-epidemiology of diphtheria in Western Europe. *Epidemiology and Infection* **125**: 113–125.

Edward, D.G. & Allison, V.D. (1951) Diphtheria in the immunised with observations on a diphtheria-like disease associated with non-toxigenic strains of *Corynebacterium diphtheriae*. *The Journal of Hygiene* **49**: 205–219.

Efstratiou, A. (1995) *Corynebacterium diphtheriae*: molecular epidemiology and characterisation studies on epidemic and sporadic isolates. *Microecology and Therapy* **25**: 63–71.

Efstratiou, A. & George, R.C. (1999) Laboratory guidelines for the diagnosis of infections caused by *Corynebacterium diphtheriae* and *C. ulcerans*. *Communicable Disease and Public Health* **2**: 250–257.

Efstratiou, A. & Maple, P.A.C. (1994) Manual for the laboratory diagnosis of diphtheria. *The Expanded Programme on Immunization in the European Region of WHO*. ICP/EPI038 (C), Copenhagen.

Efstratiou, A., Engler, K.H. & De Zoysa, A. (1998) Diagnosis and epidemiology of diphtheria. In *Molecular Bacteriology*. Woodford, N. & Johnson, A.P. (eds). *Methods in Molecular Medicine Series*. Humana Press. pp. 191–212.

Efstratiou, A., Engler, K.H., Dawes, C.S. & Sesardic, D. (1998) Comparison of phenotypic and genotypic methods for detection of diphtheria toxin among isolates of pathogenic corynebacteria. *Journal of Clinical Microbiology* **36**: 3173–3177.

Efstratiou, A., Engler, K.H., Mazurova, I.K. *et al.* (2000) Current approaches to the laboratory diagnosis of diphtheria. *The Journal of Infectious Diseases* **181**: S138–S145.

Egemen, A., Kurugol, Z., Aksit, S. *et al.* (2000) Immunity to diphtheria in Izmir, Turkey. *European Journal of Epidemiology* **16**: 1039–1042.

Eigen, L.A. (1932) Vaginal diphtheria. Review of the literature to date and report of a case. *Journal of Medical Society of New Jersey* **29**: 778.

Engler, K.H. & Efstratiou, A. (2000) Rapid enzyme immunoassay for determination of toxigenicity among clinical isolates of corynebacteria. *Journal of Clinical Microbiology* **38**: 1385–1389.

Engler, K.H., Glushkevich, T., Mazurova, I.K. *et al.* (1997) A modified Elek test for detection of toxigenic corynebacteria in the diagnostic laboratory. *Journal of Clinical Microbiology* **35**: 495–498.

Engler, K.H., Efstratiou, A.E., Norn, D. *et al.* (2002) Immunochromatographic strip test for rapid detection of diphtheria toxin: description and multicentre evaluation in areas of low and high prevalence of diphtheria. *Journal of Clinical Microbiology* **40**: 80–83.

English, P.C. (1985) Diphtheria and theories of infectious disease: centennial appreciation of the critical role of diphtheria in the history of medicine. *Pediatrics* **76**: 1–9.

Fakes, R.W. & Downham, M. (1970) Toxic reaction to *Corynebacterium ulcerans*. *Lancet* **i**: 298.

Farizo, K.M., Strebel, P.M., Chen, R.T. *et al.* (1993) Fatal respiratory disease due to *Corynebacterium diphtheriae*: case report and review of guidelines for management, investigation and control. *Clinical Infectious Diseases* **16**: 59–68.

Fourel, G., Phalipon, A. & Kaczorek, A. (1989) Evidence for direct regulation of diphtheria toxin gene transcription by an Fe^{2+} dependent DNA-binding repressor, DtoxR, in *Corynebacterium diphtheriae*. *Infection and Immunity* **57**: 3221–3225.

Fox, J.G. & Frost, W.W. (1974) *Corynebacterium ulcerans* mastitis in a bonnet macaque (*Macaca radiata*). *Laboratory Animal Science* **24**: 820–822.

Freeman, V.J. (1951) Studies on the virulence of bacteriophage-infected strains of *C. diphtheriae*. *Journal of Bacteriology* **61**: 675–688.

Funke, G. & Bernard, K.A. (1999) Coryneform gram-positive rods. In *Manual of Clinical Microbiology, 7th ed.* Murray, P.R., Baron, E.J., Pfaller, M.A., Tenover, F.C. & Yolken, R.H. (eds). Washington, DC: American Society for Microbiology. pp. 319–345.

Funke, G., von Graevenitz, A., Clarridge, J.E. III & Bernard, A. (1997) Clinical microbiology of coryneform bacteria. *Clinical Microbiology Reviews* **10**: 125–159.

Galazka, A. (2000) The changing epidemiology of diphtheria in the vaccine era. *The Journal of Infectious Diseases* **181**: S2–S9.

Galazka, A. & Tomaszunas-Blaszczyk, J. (1997) Why do adults contract diphtheria? *Eurosurveillance* **2**: 60–63.

Galazka, A.M. & Robertson, S.E. (1996) Immunisation against diphtheria with special emphasis on immunisation of adults. *Vaccine* **14**: 845–857.

Galazka, A.M., Robertson, S.E. & Oblapenko, G.P. (1995) Resurgence of Diphtheria. *European Journal of Epidemiology* **11**: 95–105.

Gilbert, R. & Stewart, F.C. (1926) *Corynebacterium ulcerans*: a pathogenic microorganism resembling *C. diphtheriae*. *The Journal of Laboratory and Clinical Medicine* **12**: 756–761.

Gill, D.M. & Pappenheimer, J.M. Jr (1971) Structure–activity relationships in diphtheria toxin. *The Journal of Biological Chemistry* **246**: 1492–1495.

Glenny, A.T. & Hopkins, B.E. (1923) Diphtheria toxoid as an immunising agent. *British Journal of Experimental Pathology* **4**: 283–288.

Grant, I. (1934) Unusual case of diphtheria. *British Medical Journal* **1**: 1074.

Grimont, P.A.D., Grimont, F., Efstratiou, A. *et al.* (2004) International nomenclature for *Corynebacterium diphtheriae* ribotypes. *Research in Microbiology* **155**: 162–166.

Gunatillake, P.D.P. & Taylor, G. (1968) The role of cutaneous diphtheria in the acquisition of immunity. *The Journal of Hygiene* **66**: 83–8.

Hardy, I.R.B. (1998) Diphtheria. In *Bacterial Infections of Humans Epidemiology and Control, 3rd ed.* Evans, A.S. & Brachman, P.S. (eds). New York: Plenum Publishing.

Hardy, I.R.B., Dittman, S. & Sutter, R.W. (1996) Current situation and control strategies for resurgence of diphtheria in newly independent states of the former Soviet Union. *Lancet* **347**: 1739–1744.

Harries, E.H.R., Mitman, M. & Taylor, I. (1951) Diphtheria. In *Clinical Practice in Infectious Diseases, Fourth Edition*. Edinburgh: E & S Livingstone.

Hauser, D., Popoff, M.R., Kiredjian, M. *et al.* (1993) Polymerase chain reaction assay for diagnosis of potentially toxigenic *Corynebacterium diphtheriae* strains correlation with ADP-ribosylation activity assay. *Journal of Clinical Microbiology* **31**: 2720–2723.

Hewlett, E.L. (1985) Selective primary health care: strategies for control of disease in the developing world. XVII. Pertussis and diphtheria. *Reviews of Infectious Diseases* **7**: 426–433.

Höfler, W. (1991) Cutaneous diphtheria. *International Journal of Dermatology* **30**: 845–847.

Hogan, M.J. (1947) Conjunctivitis with membrane formation. *American Journal of Ophthalmology* **30**: 1495.

Holmes, R.K. (1975) Genetic aspects of toxinogenesis in bacteria. In *Microbiology*. Schlessinger, D. (ed.). Washington, DC: American Society for Microbiology. pp. 296–301.

Hommez, J., Devriese, L.A., Vaneechoutte, M. *et al.* (1999) Identification of nonlipophilic corynebacteria isolated from dairy cows with mastitis. *Journal of Clinical Microbiology* **37**: 954–957.

Howard, W.T. (1893) Acute ulcerative endocarditis due to the bacillus diphtheriae. *The Johns Hopkins Hospital Bulletin* **4**: 32–34.

Hu, V.W. & Holmes, R.K. (1984) Evidence for direct insertion of fragments A and B of diphtheria toxin into model membranes. *The Journal of Biological Chemistry* **259**: 12226–12233.

Jephcott, A.E., Gillespie, E.H., Davenport, C. *et al.* (1975) Non-toxigenic *Corynebacterium diphtheriae* in a boarding school. *Lancet* **1**: 1025–1026.

Jordan, J.H., Smith, F. & Neave Kingsbury, A. (1922) Pathogenicity of the diphtheria group. *Lancet* **2**: 1052–1056.

Kagan, B.L., Finkelstein, A. & Colombini, M. (1981) Diphtheria toxin fragment forms large pores in phospholipid bilayer membranes. *Proceedings of the National Academy of Sciences of the United States of America* **78**: 4950–4954.

Kelly, C., Efstratiou, A., European Diphtheria Surveillance Network (2003) Seventh International Meeting of the European Laboratory Working Group on Diphtheria – Vienna, June 2002. *Eurosurveillance* **10**: 189–195.

Koopman, J.S. & Campbell, J. (1975) The role of cutaneous diphtheria infections in a diphtheria epidemic. *The Journal of Infectious Diseases* **131**: 239–244.

Krafft, A.E., Tai, S.P., Coker, C. & Holmes, R.K. (1992) Transcription analysis and nucleotide sequence of tox promoter/operator mutants of corynebacteriophage beta. *Microbial Pathogenesis* **13**: 85–92.

Krech, T. (1994) Epidemiological typing of *Corynebacterium diphtheriae*. *Medical Microbiology Letters* **3**: 1–8.

Ledbetter, M.K., Cannon, A.B. & Costa, A.F. (1964) The electrocardiogram in diphtheritic myocarditis. *American Heart Journal* **68**: 599.

Lehmann, K.B. & Neumann, R.O. (1896) *Atlas und Grundriss der Bakteriologie und Lehrbuch der Speciellen Bakteriologischen Diagnostik, 1st edn.* Munich: Lehmann.

Liebow, A.A., MacLean, P.D., Bumstead, J.H. & Welt, L.G. (1946) Tropical ulcers and cutaneous diphtheria. *Archives of Internal Medicine* **78**: 255–295.

Livingood, C.S., Perry, D. & Forrester, J.S. (1946) Cutaneous diphtheria: Report of 140 cases. *The Journal of Investigative Dermatology* **7**: 341–364.

Lloyd, G.E.R. (ed.) (1978) Chadwick, J. & Mann, W.N. (trans.). New York: Penguin Books.

Loeffler, F., Newsholme, A., Mallory, F.B. *et al.* (1908) In *The Bacteriology of Diphtheria*. Nuttall, G.H.F. & Graham-Smith, G.S. (eds). London: Cambridge University Press.

MacGregor, R.R. (2000) *Corynebacterium diphtheriae*. In *Principles and Practice of Infectious Diseases, 5th ed.* Mandell, G.L., Bennett, J.E. & Dolin, R. (eds). Philadelphia: Churchill Livingstone. pp. 2190–2198, Chapter 193.

Maple, P.A., Efstratiou, A., George, R.C. *et al.* (1995) Diphtheria immunity in UK blood donors. *Lancet* **345**: 963–965.

Marlovits, S., Stocker, R., Efstratiou, A. *et al.* (2000) Seroprevalence of diphtheria immunity among injured adults in Austria. *Vaccine* **19**: 1061–1067.

Meers, P.D. (1979) A case of classical diphtheria, and other infections due to *Corynebacterium ulcerans*. *The Journal of Infection* **1**: 139–142.

Mikhailovich, V.M., Melnikov, M.G., Mazurova, I.K. *et al.* (1995) Application of PCR for detection of toxigenic *Corynebacterium diphtheriae* strains isolated during the Russian diphtheria epidemic, 1990 through 1994. *Journal of Clinical Microbiology* **33**: 3061–3063.

Moore, M.S. & Parsons, E.I. (1958) A study of modified Tinsdale's medium for the primary isolation of *Corynebacterium diphtheriae*. *The Journal of Infectious Diseases* **102**: 88–93.

Morgan, B.C. (1963) Cardiac complications of diphtheria. *Pediatrics* **32**: 549–557.

Morris, R.E., Gerstein, A.S., Bonventre, P.F. & Saelinger, C.B. (1985) Receptor-mediated entry of diphtheria toxin into monkey kidney (vero) cells: electron microscopic evaluation. *Infection and Immunity* **50**: 721–727.

Moskaug, J.O., Sandvig, K. & Olsnes, S. (1988) Low pH-induced release of diphtheria toxin A-fragment in Vero cells. Biochemical evidence for transfer to the cytosol. *The Journal of Biological Chemistry* **263**: 2518–2525.

Mueller, J.H. (1940) Nutrition of the diphtheria bacillus. *Bacteriological Reviews* **4**: 97–134.

Mueller, J.H. (1941) The influence of iron on the production of diphtheria toxin. *Journal of Immunology* **42**: 343–351.

Mueller, J.H. & Miller, P.A. (1941) Production of diphtheria toxin of high potency (100Lf) on a reproducible medium. *Journal of Immunology* **40**: 21–32.

Murphy, J.R., Bacha, P. & Holmes, R.K. (1979) Regulation of diphtheria toxin production. In *Microbiology*. Schlessinger, D. (ed.). Washington, DC: American Society for Microbiology. pp. 181–186.

Murphy, J.R., Pappenheimer, A.M. Jr & de Borms, S.T. (1974) Synthesis of diphtheria *tox*-gene products in *Escherichia coli* extracts. *Proceedings of the National Academy of Sciences of the United States of America* **71**: 11–15.

Murphy, J.R., Skiver, J. & McBride, G. (1976) Isolation and partial characterisation of a corynebacterium β *tox* operator constitutive-like mutant lysogen of *Corynebacterium diphtheriae*. *Journal of Virology* **18**: 235–244.

Naglich, J.G., Metherall, J.E., Russell, D.W. & Eidels, L. (1992) Expression cloning of a diphtheria toxin receptor: identity with a heparin-binding EGF-like growth factor precursor. *Cell* **69**: 1051–1061.

Naglich, J.G., Rolf, J.M. & Eidels, L. (1992) Expression of functional diphtheria toxin receptors on highly toxin-sensitive mouse cells that specifically bind radioiodinated toxin. *Proceedings of the National Academy of Sciences of the United States of America* **89**: 2170–2174.

Nakao, H. & Popovic, T. (1997) Development of a direct PCR for detection of the diphtheria toxin gene. *Journal of Clinical Microbiology* **35**: 1651–1655.

Noble, W.C. & Dixon, J.M.S. (1990) *Corynebacterium* and other coryneform bacteria. In *Topley & Wilson's Principles of Bacteriology, Virology and Immunity, Vol. 2*. Parker, M.T. & Duerden, B.I. (eds). London, UK: Edward Arnold. pp. 103–108.

Pallen, M.J. (1991) Rapid screening for toxigenic *Corynebacterium diphtheriae* by the polymerase chain reaction. *Journal of Clinical Pathology* **44**: 1025–1026.

Pallen, M.J., Hay, A.J., Puckey, L.H. & Efstratiou, A. (1994) Polymerase chain reaction for screening clinical isolates of corynebacteria for the production of diphtheria toxin. *Journal of Clinical Pathology* **47**: 353–356.

Papini, E., Rappuoli, R., Murgia, M. & Montecucco, C. (1993) Cell penetration of diphtheria toxin. Reduction of the interchain disulfide bridge is the rate limiting step of translocation in the cytosol. *The Journal of Biological Chemistry* **268**: 1567–1574.

Pappenheimer, A.M. Jr (1977) Diphtheria toxin. *Annual Review of Biochemistry* **46**: 69–94.

Pappenheimer, A.M. Jr (1984) Diphtheria. In *Bacterial Vaccines*. Germanier, R. (ed.). New York: Academic Press. pp. 1–16.

Pappenheimer, A.M. & Murphy, J.R. (1983) Studies on the molecular epidemiology of diphtheria. *Lancet* **2**: 923–926.

Pappenheimer, A.M. Jr & Johnson, S. (1936) Studies on diphtheria toxin production. I. The effect of iron and copper. *British Journal of Experimental Pathology* **17**: 335–341.

Pascual, C., Lawson, P.A., Farrow, J.A.E. *et al.* (1995) Phylogenetic analysis of the genus *Corynebacterium* based on 16S rRNA gene sequences. *International Journal of Systematic Bacteriology* **45**: 724–728.

Petrie, G.F. & McLean, D. (1934) The inter-relations of *Corynebacterium ovis*, *Corynebacterium diphtheriae*, and certain diphtheria strains derived from the human nasopharynx. *The Journal of Pathology and Bacteriology* **39**: 635–663.

PHLS CDR (1994) Diphtheria booster dose for school leavers. *CDR Weekly* **4**(17): 77. www.hpa.org.uk/cdr/archives/cdr94/cdr1794.pdf (25 June 2005).

Popovic, T., Mazurova, I.K., Efstratiou, A. *et al.* (2000) Molecular epidemiology of diphtheria. *The Journal of Infectious Diseases* **181S**: 168–177.

Rakhmanova, A.G., Lumio, J., Groundstroem, K.W. *et al.* (1997) Fatal respiratory tract diphtheria apparently caused by nontoxigenic strains of *Corynebacteria diphtheriae*. *European Journal of Clinical Microbiology & Infectious Diseases* **16**: 816–820.

Rappuoli, R., Perugini, M. & Ratti, G. (1987) DNA element of *Corynebacterium diphtheriae* with properties of an insertion sequence and usefulness for epidemiological studies. *Journal of Bacteriology* **169**: 308–312.

Reacher, M., Ramsay, M., White, J. *et al.* (2000) Non-toxigenic *Corynebacterium diphtheriae*: an emerging pathogen in England and Wales? *Emerging Infectious Diseases* **6**: 640–645.

Reinhardt, D.J., Lee, A. & Popovic, T. (1998) Antitoxin-in-membrane and antitoxin-in-well assays for detection of toxigenic *Corynebacterium diphtheriae*. *Journal of Clinical Microbiology* **36**: 207–210.

Relyveld, E.H. (1995) A history of toxoids. *Proceedings from the International Meeting on the History of Vaccinology*, 6–8 December, Paris, France.

Rey, M., Patey, O. & Vincent-Ballereau, F. (1996) Diphtheria's European come back. *Eurosurveillance* **1**: 14–16.

Rosen, G. (1954) Acute communicable diseases. In *The History and Conquest of Common Diseases*. Bett, W. (ed.). University of Oklahoma Press. pp. 6–25.

Roure, C. (1998) Overview of diphtheria activities within the European region. Report on the Planning Meeting for EU Concerted Action for the Microbiological Surveillance of Diphtheria in Europe, London, UK, 6–7 April 1998.

Ruimy, R., Riegel, P., Boiron, H. *et al.* (1995) Phylogeny of the genus *Corynebacterium* deduced from analyses of small-subunit ribosomal DNA sequences. *International Journal of Systematic Bacteriology* **45**: 740–746.

Schmitt, M.P. & Holmes, R.K. (1991a) Characterisation of a defective diphtheria toxin repressor (*dtx*R) allele and analysis of *dtx*R transcription in wild type and mutant strains of *Corynebacterium diphtheriae*. *Infection and Immunity* **59**: 3903–3908.

Schmitt, M.P. & Holmes, R.K. (1991b) Iron-dependent regulation of diphtheria toxin and siderophore expression by the cloned *Corynebacterium diphtheriae* repressor gene *dtx*R in *C. diphtheriae* C7 strains. *Infection and Immunity* **59**: 1899–1904.

Schwartz, D.A. & Wharton, M. (1997) Diphtheria. In *Pathology of Emerging Infections 2*. Nelson, A.N. & Horsburgh, C.R. Jr (eds). Washington, DC: ASM Press. pp. 145–166.

Semple, H.R. (1859) Memoirs on diphtheria. From the writings of Bretonneau, Guersant, Trousseau, Bouchut, Empis & Daviot. The New Sydenham Society, London.

Simmons, L.E., Abbott, J.D., Macaulay, M.E. *et al.* (1980) Diphtheria carriers in Manchester: simultaneous infection with toxigenic and non-toxigenic *mitis* strains. *Lancet* **1**: 304–305.

Simonsen, O., Kjeldsen, K., Bentzon, M.W. & Heron, I. (1987) Susceptibility to diphtheria in populations vaccinated before and after elimination of indigenous diphtheria in Denmark. *Acta Pathologica et Microbiologica Scandinavica. Section C, Immunology* **95**: 225–231.

Singh, M.K. & Saba, P.Z. (2001) *Diphtheria*. www.emedicine.com/emerg/topic138.htm (February 2004).

Skerman, V.B.D., McGowan, V. & Sneath, P.H.A. (eds) (1980) Approved lists of bacterial names. *International Journal of Systematic Bacteriology* **30**: 225–420.

Sloss, J.M. & Faithfull-Davies, D.N. (1993) Non-toxigenic *Corynebacterium diphtheriae* in military personnel. *Lancet* **341**: 1021.

Todar, K. (1997) *Diphtheria*. Available at: http://textbookofbacteriology.net/diphtheria.html.

Uchida, T., Gill, D.M. & Pappenheimer, A.M. Jr (1971) Mutation in the structural gene for diphtheria toxin carried by temperate phage β. *Nature: New Biology* **233**: 8–11.

Van Ness, B.G., Howard, J.B. & Bodley, J.W. (1980) ADP-ribosylation of elongation factor 2 by diphtheria toxin. NMR spectra and proposed structures of riboslyl-diphthamide and its hydrolysis products. *The Journal of Biological Chemistry* **255**: 10717–10720.

von Behring, E. & Kitasato, S. (1890) Ueber das Zustandekommen der Diphtherie-Immunität bei Thieren. *Deutsche Medizinische Wochenschrift*. In *Milestones in Microbiology 1546 to 1940*. Brock, T.D. (ed. & trans.). Washington, DC: ASM Press. **16**: 1113–1114.

von Graevenitz, A.W.C., Coyle, M.B. & Funke, G. (1998) Corynebacteria and rare coryneforms. In *Systemic Bacteriology: Topley & Wilson's Microbiology and Microbial Infections, Vol. 2, 9th ed.* Balows, A. & Duerden, B.I. (eds). New York: Oxford University Press. pp. 533–548.

von Hunolstein, C., Rota, M.C., Alfarone, G. *et al.* (2000) Diphtheria antibody levels in the Italian population. *European Journal of Clinical Microbiology & Infectious Diseases* **19**: 433–437.

Wagner, J., Ignatius, R., Voss, S. *et al.* (2001) Infection of the skin caused by *Corynebacterium ulcerans* and mimicking classical cutaneous diphtheria. *Clinical Infectious Diseases* **33**: 1598–1600.

Wallfield, M.J. & Litvak, A.M. (1933) Vulvo-vaginal diphtheria. *The Journal of Pediatrics* **3**: 756.

Ward, K.W. (1948) Effect of Tween 80 on certain strains of *C. diphtheriae*. *Proceedings of the Society for Experimental Biology and Medicine* **67**: 527–528.

Weiss, B.P., Strassburg, M.A. & Feeley, J.C. (1983) Tetanus and diphtheria immunity in an elderly population in Los Angeles County. *American Journal of Public Health* **73**: 802–804.

Wharton, M., Hardy, I.R.B., Vitek, C., Popovic, T. & Sutter, R.W. (1998) Epidemic diphtheria in the newly independent states of the former Soviet Union. In *Emerging Infections, 1st ed.* Scheld, W.M., Armstrong, D. & Hughes, J.M. (eds). Washington, DC: ASM Press. pp. 165–176.

WHO (1998) *Diphtheria*. www.who.int/vaccines-diseases/diseases/diphtheria_dis.shtml (February 2004).

WHO (2002) *Diphtheria control*. www.euro.who.int/vaccine/20030724_3. (6 July 2005).

Wilson, A.P.R., Ridgway, G.L., Gruneberg, R.N. *et al.* (1990) Routine screening for *Corynebacterium diphtheriae*. *Lancet* **336**: 1119.

Wilson, A.P.R., Efstratiou, A., Weaver, E. *et al.* (1992) Unusual non-toxigenic *Corynebacterium diphtheriae* in homosexual men. *Lancet* **339**: 998.

Yamaizumi, M., Mekada, E., Uchida, T. & Okada, Y. (1978) One molecule of diphtheria toxin fragment A introduced into a cell can kill the cell. *Cell* **15**: 245–250.

Zingher, A. (1923) A Schick test performed on more than 150,000 children in public and parochial schools in new York (Manhattan and the Bronx). *American Journal of Diseases of Children* **25**: 392–405.

8

Listeria and *Erysipelothrix* spp.

Kevin G. Kerr

Department of Microbiology, Harrogate District Hospital, Harrogate, UK

INTRODUCTION

In 1918 a bacterium resembling *Erysipelothrix rhusiopathiae* was isolated from the cerebrospinal fluid of a soldier serving in France. This isolate was deposited in the Institut Pasteur, Paris, but it was to be many years before this bacterium was recognised as *Listeria monocytogenes*.

In 1924 Murray and colleagues, during the investigation of an epizootic among experimental animals, isolated a bacterium from the lymph glands of affected animals which, when injected into healthy animals, induced a monocytosis and was accordingly given the name *Bacillus monocytogenes*. Three years later in South Africa, Pirie who was studying carriage of *Yersinia pestis* by rodents isolated a bacterium from the livers of gerbils, which he named *Listerella hepatolytica*. This bacillus was found to be indistinguishable from *B. monocytogenes*, and Pirie later proposed the name *L. monocytogenes*.

In 1929 Nyfeldt isolated *L. monocytogenes* from blood cultures of a patient with glandular fever and was convinced he had isolated the causative agent of infectious mononucleosis. Not surprisingly, other investigators were unable to substantiate his claim. Over the ensuing years, anecdotal reports of human listeriosis in both neonates and adults continued to appear in the literature, but the consensus was, however, that listeriosis was an extremely rare zoonotic pathogen. The pioneering work of Reiss and colleagues, and of Heinz Seeliger in the 1950s, encompassed the first truly systematic studies of human listeriosis; but it was not until three decades later, following a series of well-publicised outbreaks, that *L. monocytogenes* was catapulted from relative obscurity into the microbiological limelight. Since that time it has become the focus of much attention from a number of scientific disciplines, not only because of the public health importance of *L. monocytogenes* but also because the bacterium has become to be regarded as an important experimental model for facultative intracellular parasitism.

THE GENUS LISTERIA

General Description of the Genus

The intrageneric and intergeneric relatedness of the genus *Listeria* has been the subject of much debate. At first considered as a member of the family Corynebacteriaceae, primarily because of morphological characteristics, later chemotaxonomic studies demonstrated that *Listeria* spp. are quite distinct from the corynebacteria. It was not until 1984, however, that partial sequencing of 16S rRNA unambiguously confirmed the phylogenetic place of the genus *Listeria* in the *Clostridium-Bacillus-Lactobacillus* group, with *Brochothrix thermosphacta* as its nearest neighbour.

The genus *Listeria* as originally described was monotypic, with *L. monocytogenes* the sole member. In the ensuing years a further seven species were recognised: *L. ivanovii* (previously *L. monocytogenes* serovar 5), *L. innocua*, *L. welshimeri*, *L. seeligeri*, *L. murrayi*, *L. grayi* and *L. denitrificans*. *Listeria denitrificans* has been since reclassified as *Jonesia denitrificans*. *Listeria murrayi* and *L. grayi* were once considered sufficiently distinct from other *Listeria* spp. to warrant the creation of a new genus, *Murraya*, but have now been assigned to a single species *L. grayi*. Most recently, *L. ivanovii* has been divided into two subspecies: subsp. *ivanovii* and subsp. *londoniensis*. Of the non-*monocytogenes* species, only *L. ivanovii* is recognised as a human pathogen (Cummins, Fielding and McGlauchlin, 1994).

The genomes of *L. monocytogenes* and *L. innocua* have been sequenced with the former having a genome 2.94 Mbp long with 2583 open-reading frames and a mol% G + C of 39. *Listeria innocua* has a genome 3.01 Mbp in length with 2973 open-reading frames and a mol% G + C content of 37 (Glaser *et al.*, 2001). Of particular note is the fact that many of the proteins encoded by these bacteria bear a striking similarity to *Bacillus subtilis*. *Listeria monocytogenes* is found in an extraordinarily diverse range of habitats, and it is thus unsurprising that 331 genes encoding different transport proteins (11.6% of all predicted genes) and 209 transcriptional regulators have been identified in the bacterium (Glaser *et al.*, 2001).

Between 0 and 79% of *L. monocytogenes* isolates carry plasmids of various sizes. Although most of these are cryptic, cadmium resistance is plasmid associated and plasmids specifying for resistance to single or multiple antimicrobials have also been reported (Poyart-Salmeron *et al.*, 1992; Hadorn *et al.*, 1993). Plasmid carriage is much more predominant in isolates of *L. monocytogenes* belonging to serogroup 1 (Harvey and Gilmour, 2001).

Listeria spp. are short (0.4–0.5×0.5–2.0 µm) Gram-positive rods. They are nonacid fast and do not produce spores. At 20 °C they are motile by means of peritrichous flagella, but motility is not observed in cultures incubated at 37 °C. They are facultatively anaerobic and grow over a wide temperature range of 0–45 °C (optimum 30–36 °C), although growth at very low temperatures is slow. Growth over a wide pH range occurs, with some strains growing at pH 9.6, but is optimal at neutral to slightly alkaline pH. Key biochemical and other characteristics are listed in Tables 8.2 and 8.3.

The major antigenic determinants of *Listeria* spp. are the somatic (O) and flagellar (H) antigens. The current serotyping scheme, based on these antigens, distinguishes six serogroups (1/2, 3, 4, 5, 6 and 7) further subdividable into serovars for all species except *L. grayi* and *L. ivanovii*. While *L. ivanovii* and *L. welshimeri* can be readily differentiated from *L. monocytogenes* on serotyping, other species may share one or more antigens with *L. monocytogenes*. For example, some isolates of *L. seeligeri* may be antigenically indistinguishable from those belonging to *L. monocytogenes* to serogroup 1/2. It should also be noted that, although 13 serovars of L. monocytogenes are recognised currently, in excess of 90% of clinical isolates belong to only three serovars –1/2a, 1/2b and 4b, thereby limiting the value of this technique in epidemiological studies.

PATHOGENESIS

Apart from the very rare exception of direct inoculation via skin or conjunctiva, it is most likely that in the adult the bacteria gain access to the host via the gastrointestinal tract. In the case of the fetus, infection in the majority of cases arises from haematogenous seeding of the placenta and only a few cases are thought to be acquired intra partum as the neonate exits the birth canal. Late onset neonatal listeriosis is acquired through direct or indirect person-to-person transmission.

Although it has been suggested that co-infection with another gastrointestinal pathogen may predispose to invasion of the gut by L. monocytogenes, it is now known that the bacterium can specify its own uptake by mammalian cells, including those which are nonprofessional phagocytes. The expression of cell-surface-associated proteins, of which internalin (Inl) A and B are the best characterised, are of key importance in this process. Another protein (p60), the product of the iap gene, recently renamed cwhA (cell-wall hydrolase) was originally thought to play a significant role in this context, but its contribution to invasiveness has since been downplayed (Pilgrim et al., 2003).

Following internalisation by the host cell, the bacterium lyses the membranes of the phagosomal vacuole. Lysis is mediated by listeriolysin O, a haemolysin which is related to other cholesterol-dependent pore-forming toxins (CDTX) such as streptolysin O. Two phospholipases, PlcA and B, are also deemed to be important in allowing L. monocytogenes to escape from phagosomal vacuoles. Once free in the host cell cytoplasm, the protein ActA promotes actin accumulation and polymerisation resulting in the formation of an actin tail at a pole of the bacterium, and it is propelled in a pseudopod-like structure into an adjacent cell whereupon it is then phagocytosed. The bacterium escapes from the resulting double-membrane vacuole through production of listeriolysin and PlcA and B. Key virulence genes of L. monocytogenes are located on a 8.2-kb pathogenicity island, LIPI, and are transcriptionally regulated by the product of the prfA gene (PrfA). PrfA also partially regulates expression of internalin A and B which are located on a separate pathogenicity island (Roberts and Wiedmann, 2003).

The minimum infectious dose of L. monocytogenes for humans remains to be determined, although counts of the bacterium from foodstuffs implicated in most sporadic and epidemic cases are typically in excess of 10^6 CFU g^{-1}. Putative interstrain differences in virulence have been identified – for example, human strains of L. monocytogenes associated with asymptomatic carriages express a truncated InlA (Olier et al., 2003). Comparative genomics have also revealed serovar-specific genes which will permit further investigation of differences in virulence and may help to explain why most cases of human listeriosis are caused by serovar 4b (Doumith et al., 2004).

EPIDEMIOLOGY

The Environment

Listeria monocytogenes has been isolated from a very wide range of environmental sources including dust soil and sewage. Although it is often considered that L. monocytogenes in soil and vegetation results from faecal contamination by animals, Weis and Seeliger (1976) noted that the bacterium was isolated in greatest numbers from plants and soil from uncultivated areas, whereas the lowest numbers were obtained from agricultural areas.

Listeria monocytogenes can survive for extended periods in the environment. It can survive for up to 1500 days in certain types of soil and can persist and multiply in water, being isolated from surface water of canals and lakes as well as estuarine environments. The bacterium has also been isolated from raw and treated sewage. When sprayed on agricultural land, numbers of L. monocytogenes do not significantly decrease over time, leading to concern that crops grown in soil treated with sewage sludge may become contaminated with the bacterium (Garrec, Picard-Bonnaud and Pourcher, 2003). Similarly, aquatic environments may become contaminated by the addition of animal manure or human soil as occurs in several countries.

Listeria monocytogenes can also be found in the environment of food production facilities and retail outlets. The formation of biofilm, which is strain dependent, may be important in this respect (Borucki et al., 2003). The bacterium has also been isolated from the domestic environment, but the significance of this in relation to human listeriosis is uncertain (Beumer et al., 1996).

Food

Following the recognition that listeriosis is frequently a food-borne infection there have been few foods which have not escaped the attention of microbiologists seeking L. monocytogenes. Studies to determine the extent of listerial contamination of a given food product often yield different results. The reasons for this are complex and reflect, at least in part, geographic, seasonal and other differences as well as variations in isolation protocols. Marked differences in experimental technique also account for the discrepancies in the results of investigations into the growth, survival and heat tolerance of L. monocytogenes in foods.

Listeria monocytogenes has been associated with a wide variety of food products, particularly unpasteurised milk and dairy products, principally surface-ripened (soft) cheese. Hard cheese, because of the decrease in pH and a_w, which occur during ripening is much less likely to become contaminated during manufacture. It should be noted that soft cheeses made from pasteurised milk can become contaminated during production (Bind, 1989).

Listeriosis has been associated with the consumption of poultry, and recent surveys have shown that up to 62% of uncooked chicken specimens sampled at retail level may be contaminated with the bacterium (Miettinen et al., 2001). The prevalence of L. monocytogenes in cooked chicken, as might be expected, is lower than that for raw products; nevertheless, both anecdotal reports and case-control studies have implicated cooked chicken in human listeriosis. Other meat products, both raw and cooked, may also be contaminated with the bacterium.

Listeria monocytogenes has also been detected in a wide range of other foodstuffs including pâté, vegetables (including prepacked salads), seafood – particularly crab meat and smoked fish – and both commercial and institutional cook-chill products. Contamination of ready-to-eat foods may also occur at the retail level (Gombas et al., 2003).

Animals and Human Carriers

At least 50 species of wild, zoo and domestic animals and birds have been shown to harbour L. monocytogenes. Rates of faecal carriage in cattle, sheep and goats vary widely but may be very high, even in listeriosis-free herds.

Early studies investigating faecal carriage in humans suggested that up to 70% of selected populations may excrete L. monocytogenes for short periods, but these reports must be viewed with caution, as many of the isolates, given contemporary identification criteria, would not now be recognised as L. monocytogenes. More recent reports suggest that faecal carriage is much less frequent, occurring in approximately 3% of healthy individuals. Several studies have attempted to identify factors associated with carriage of the bacterium, such as H2 receptor antagonist therapy, but results from investigations should be interpreted with care, as the only long-term prospective study (Grif *et al.*, 2003) has indicated that faecal carriage may not only be intermittent but also short-lived (maximum of 4 days).

Carriage of *Listeria* spp. in the vagina has been investigated but does not seem to occur outside the setting of recent maternofetal listeriosis.

Human Infection

For many years, *L. monocytogenes* was considered a rare zoonotic pathogen. The first cases of presumed food-borne listeriosis were reported as long ago as 1951. Similar reports continued to appear in the ensuing two decades but failed to make an impact in the wider scientific and medical communities, presumably because the East German and Czechoslovak researchers had little contact with fellow investigators in the West. It was not until the large outbreaks in the 1980s that systematic studies of the epidemiology of *L. monocytogenes* infection were initiated. These studies resulted in the perception in many countries that the incidence of listeriosis was increasing. However, the question as to whether this represented a true increase or merely greater awareness of the disease among clinicians, bacteriologists and public health agencies remains unanswered. While there can be no doubt that improved diagnostic techniques and surveillance and reporting systems had improved, most authorities now believe that there was a real increase in the number of cases. The reasons for this are complex but reflect, at least in part, the growing number of individuals with conditions which render them vulnerable to *L. monocytogenes* infection as well as major changes in food production, distribution, retailing (such as the introduction of refrigerated products with extended shelf lives) and dietary habits.

Recognition of these factors leading to improved food production and retailing practices, increased microbiological and public health surveillance coupled with changes in consumer awareness relating to high-risk foods, especially in the most vulnerable patient groups, may well be responsible for the decrease in the incidence of listeriosis reported by several countries, although outbreaks continue to occur.

Listeria monocytogenes as a Food-Borne Pathogen

Evidence that contaminated food is the principal route of transmission of *L. monocytogenes* comes from three main sources: investigation of outbreaks (Table 8.1), case-control studies and individual case reports.

Even before the well-publicised outbreaks in North America and Europe in the 1980s, epidemics of listeriosis were well recognised. Several of these involved large numbers of individuals (in the 1966 outbreak in Halle, Germany, there were 279 cases), but insufficient data were collected to allow precise identification of the source of these outbreaks. Recognition of outbreaks is hampered by a number of

factors including the long incubation period in some cases of listeriosis, geographically wide distribution of implicated food products or food products not sold through conventional retail channels (CDC, 2001) as well as a failure to recognise that food-borne listeriosis may manifest with a clinical picture of gastroenteritis similar to that of salmonellosis or campylobacteriosis (as discussed later). Indeed, it is possible that some outbreaks have only come to light because of fortuitous circumstances (Linnan *et al.*, 1988) or through molecular epidemiological surveillance (Anon., 2003).

Case–control studies and anecdotal reports of human listeriosis have incriminated a variety of contaminated foodstuffs, such as soft cheese and undercooked poultry, as the source of infection. The investigation of sporadic cases is often difficult, as the incubation time for food-borne listeriosis may be very long compared with that for other food-borne pathogens (mean 3–4 weeks).

In contrast to classical food-borne listeriosis, there is increasing recognition that *L. monocytogenes* can cause a self–limited febrile gastroenteritis syndrome with a short incubation time (typically 24 h) (Frye *et al.*, 2002). Foods implicated in this form of listeriosis include turkey meat, sweet corn and tuna salad, chocolate milk, smoked fish and corned beef (Sim *et al.*, 2002).

Person-to-person transmission outside the hospital setting has never been convincingly demonstrated, but nosocomial spread of *L. monocytogenes* is well recognised, particularly on neonatal units, typically following introduction of a case of early onset sepsis onto the ward, which is followed by late onset cases several days later (Colodner *et al.*, 2003). Routes of transmission include contaminated resuscitation equipment, rectal thermometers and mineral oil applied to the skin of newborns. In other outbreaks, transmission of the bacterium by health care personnel or mothers appears to have occurred. Nosocomial listeriosis may also be food borne in origin (Graham *et al.*, 2002).

Zoonotic cases of listeriosis are uncommon and have presentations which differ from the more commonly recognised syndromes and include cutaneous lesions, conjunctivitis and pneumonia. There is no convincing evidence that *L. monocytogenes* can be transmitted by sexual intercourse. Transmission via blood transfusion, although theoretically possible given the psychrotrophic nature of the bacterium, has never been reported.

Molecular Epidemiology of *L. monocytogenes*

The limitations of conventional typing systems such as serotyping, which is poorly discriminative, and bacteriophage typing, available at only a few centres and limited by the non-typeability of many isolates, have long been known. Accordingly, molecular techniques have been applied and have contributed much to the understanding of *L. monocytogenes* infection. Methods include restriction fragment length polymorphism analysis, PCR-based techniques, such as random amplification of polymorphic DNA (RAPD), multilocus enzyme electrophoresis, amplified fragment length polymorphism and ribotyping. Pulsed-field gel electrophoresis is generally regarded as the gold standard typing method for *L. monocytogenes*, although

Table 8.1 Examples of outbreaks of listeriosis with a food-borne vector

Location	Date	Number of cases	% fatal[a]	Foodstuff	Reference
Sweden	1996	9	22	Gravadlax	Ericsson *et al.* (1997)
Finland	1998/1999	25	24	Butter	Lyytikäinen *et al.* (2000)
France	1999/2000	32	31	Pork tongue in aspic	de Valk *et al.* (2001)
France	1999/2000	10	33	Rillettes (paté-like product)	de Valk *et al.* (2001)
USA	2000/2001	12	41	Home-made cheese	CDC (2001)[b]
USA	2002	46	21	Turkey	CDC (2002)

[a] Includes stillbirths/abortions.
[b] Centers for Disease Control.

multilocus sequence typing may prove to be a more discriminative technique (Revazishvili *et al.*, 2004). In addition, surveillance using molecular techniques has been used to identify hitherto unrecognised outbreaks and clusters of infection (Anon., 2003; Sauders, Fortes and Morse, 2003) and to distinguish between relapsing and recurring listeriosis.

CLINICAL MANIFESTATIONS OF *L. MONOCYTOGENES* INFECTION

Listeriosis in Pregnancy

This syndrome accounts for approximately one-third of all cases of listeriosis, although the proportion may be much higher in epidemics. Although most frequently documented in the third trimester, listeriosis in earlier stages of pregnancy has also been reported. However, as early fetal loss is very often incompletely investigated, the number of cases occurring in the first trimester of pregnancy may be underestimated.

Bacteraemia is the most common manifestation of listeriosis in pregnancy and is classically accompanied by an episode of flu-like illness with fever, headache and myalgia. Associated low back pain may suggest a urinary tract infection. In up to 45% of cases, however, the pregnant woman may be asymptomatic with the first indication of infection being the abortion or stillbirth of the foetus or neonatal listeriosis in a live-born child.

The idea that maternal genital listeriosis is a cause of repeated abortion was first proposed over 40 years ago, but there is no compelling evidence to uphold it.

In contrast to adult listerial infections, complications in pregnant women are very rare and are usually seen in the context of underlying illness such systemic lupus erythematosis or AIDS and, in the overwhelming majority of women, infection is self-limiting.

Evidence that untreated *L. monocytogenes* bacteraemia in the gravid female does not inevitably lead to transmission of infection to the foetus is accumulating, especially if appropriate antimicrobial chemotherapy is administered (Mylonakis, Paliou and Hohmann, 2002).

Neonatal Listeriosis

Somewhat analogous to neonatal sepsis with Lancefield group B streptococci, neonatal listeriosis can be divided in to early and late onset groups with the former being defined as sepsis manifesting within 5 days of birth, but typically presenting within 48 h post partum. Late onset disease develops 5 or more days after birth, with a mean age of onset of 14 days. Mortality is high, up to 60 and 25% for the early- and late-onset forms, respectively.

Early onset disease results, in the most cases, from *in utero* transmission of *L. monocytogenes* from a bacteraemic mother. Ascending vaginal infection is very uncommon. The classical manifestation of early onset disease is granulomatosis infantisepticum characterised at autopsy by widely disseminated granulomas in several organs. Although noted frequently in early reports, it has been much less commonly observed by later investigators. More typical of this form of the disease is septicaemia often accompanied by pneumonic involvement or meningitis (Mylonakis, Paliou and Hohmann, 2002). Maculopapular or papulovesicular rashes are also seen. In practice, however, it is often difficult to distinguish between early onset listeriosis and other forms of neonatal sepsis on clinical grounds alone.

In contrast to the early onset form of the disease, infants who manifest the late onset form are usually healthy at birth. It has been suggested that affected infants are colonised at birth, but as a result of unexplained factors, the onset of infection is delayed. Much more likely, however, is that nosocomial acquisition has occurred as is borne out by

numerous reports of clusters of late onset listeriosis. Late onset disease is characterised by meningitis or meningoencephalitis, seen in up to 90% of cases. Onset may be more insidious than with the early form of the infection, with poor feeding, irritability and fever common presenting symptoms.

Listeriosis in Adults

Listeriosis in adults is frequently seen against a background of immune suppression, particularly where cell-mediated immunity is affected, e.g. lymphoreticular neoplasms, antirejection regimens in solid organ transplantation and autologous and allogeneic bone marrow transplantation in which the disease occurs typically as a late complication (Rivero *et al.*, 2003). Listeriosis may follow therapy with corticosteroids, chemotherapeutic agents, such as fludarabine, and the anti-tumour necrosis factor-α agents infliximab and etanercept (Slifman *et al.*, 2003). Early in the AIDS pandemic, *L. monocytogenes* was not considered to be a significant pathogen in HIV-infected individuals, and several plausible explanations for this, including the antilisterial activity of co-trimoxazole used for *Pneumocystis jiroveci* prophylaxis, were advanced. Later studies, however, demonstrated that listeriosis is 150–300 times more common in HIV-positive patients than in matched control groups from the general population. Other chronic diseases such as alcoholism and diabetes mellitus also predispose to *L. monocytogenes* infection.

The fact that invasive listeriosis can develop in previously apparently healthy individuals is often overlooked. Indeed, in several published series, approximately one-third of cases had no apparent predisposing factor (e.g. Skogberg *et al.*, 1992). Immunocompetent individuals are also susceptible to the febrile gastroenteritis syndrome associated with consumption of *L. monocytogenes*-contaminated foodstuffs.

In invasive listeriosis, reported mortality rates vary widely according to age, sex, underlying illnesses and it is thus extremely difficult to useful comparisons between published studies. Nevertheless, overall mortality rates of 35–45% are typical. Several investigators have reported decreasing mortality rates, perhaps because of more increased awareness of the infection leading to more vigorous bacteriological investigations and antimicrobial therapy in immunocompromised patients.

Adult invasive listeriosis can be categorised into three groups according to clinical manifestations: central nervous system (CNS) disease which accounts for as much as 75% of all cases, primary bacteraemia and a miscellaneous group which embraces an ever-widening spectrum of localised infections.

Neurological Infection

Meningitis, with or without focal neurological signs, is the most common form of CNS listeriosis. As with other pyogenic meningitides, listerial meningitis is usually acute in onset with high fever, nuchal rigidity and photophobia, but there are other additional features such as movement disorders, including tremors and/or ataxia. Seizures also appear to be more common than in other types of meningitis.

The most common non-meningitic form of the CNS listeriosis is encephalitis, most frequently affecting the rhombencephalon. This has a characteristic biphasic course: a nonspecific prodrome of headache, nausea, vomiting and fever, followed by progressive asymmetrical cranial nerve palsies, cerebellar signs or hemiparesis and impairment of consciousness. Prognosis is poor with significant rates of neurological sequelae in survivors. Given the frequency of CNS involvement, it is surprising that listerial brain abscesses, which can be single or multiple, are rare, accounting for only 1% of cases of CNS disease (Cone *et al.*, 2003). All cases reported thus far appear to be haematogenous in origin with no evidence of spread from contiguous sites.

Bacteraemia

Primary bacteraemia is the second most common manifestation of invasive listeriosis in nonpregnant adults, occurring in 25–50% cases. There are no distinguishing clinical features associated with this form of listeriosis. Patients may be profoundly hypotensive with an overall clinical picture which mimics Gram-negative endotoxic shock which may be accompanied by the adult respiratory distress syndrome.

Miscellaneous Syndromes

A variety of other clinical syndromes are associated with invasive adult listeriosis. Infections of the locomotor system include osteomyelitis and septic arthritis which may affect both native and prosthetic joints. Pulmonary infection, in contradistinction to neonatal listeriosis, is rare, although cases of pneumonia or empyaema have been reported. Hepatic involvement in listeriosis is also uncommon, although hepatic abscesses, both single and multiple, are recognised with the former almost exclusively occurring in patients with insulin-dependent diabetes (Bronniman *et al.*, 1998). Hepatitis in liver transplant recipients has also been described (Vargas *et al.*, 1998). Spontaneous bacterial peritonitis caused by *L. monocytogenes* has been reported in both cirrhotic and non-cirrhotic patients (Jammula and Gupta, 2002) as has peritonitis in patients undergoing continuous peritoneal dialysis (Tse *et al.*, 2003).

Listeria endocarditis occurs on both native and prosthetic valves, and although often an aggressive infection characterised by the development of complications, such as paravalvular abscess formation, some cases respond to medical therapy alone (Fernández Guerrero *et al.*, 2004). Infections of arteriovenous shunts, intravenous cannulae and prosthetic vascular grafts have also been recorded.

Ocular infections, although rare, have been described by several investigators. These infections are difficult to manage and are often sight threatening, although prompt intervention with appropriate antimicrobials may be associated with an improved outcome (Lohmann *et al.*, 1999). Primary cutaneous listeriosis is infrequently reported and is usually seen in veterinarians or farm workers after accidental inoculation of the skin following exposure to animals.

LABORATORY DIAGNOSIS

Diagnosis of listeriosis on clinical grounds alone is problematic, partly because there are few, if any pathognomic signs, and partly because it is such a rare infection and is often not considered in the differential diagnosis.

Direct Detection of *L. monocytogenes* in Clinical Material

Gram staining of specimens from normally sterile sites may suggest a diagnosis of listeriosis. Short Gram-positive rods may be seen in both intracellular and extracellular locations, but it should be remembered that the bacteria may appear filamentous, coccoid or even Gram negative, especially in cerebrospinal fluid from patients with partially treated meningitis. It should also be noted that analysis of CSF specimens may not yield a typical pyogenic picture; in up to 30% of cases there is no hypoglycorrachia and there may be no predominance, or even an absence, of polymorphs (Niemann and Lorber, 1980).

A large number of systems which detect listerial antigens, such as enzyme-linked immunosorbent assay or reversed passive latex agglutination or DNA using hybridisation of PCR-based techniques, are available commercially but these have been developed for the identification of *Listeria* spp. in foodstuffs (USDA/CFSAN, 2003a) and there is only very limited experience of these methodologies in the clinical laboratory (Lohmann *et al.*, 1999).

Serological Diagnosis of Listerial Infection

Detection of antibodies to listeriolysin has on occasion proved useful in the investigation of outbreaks as well as seroprevalence studies (Dalton *et al.*, 1997), but the technique is not widely available and has yet to find a place in the diagnostic laboratory for the investigation of sporadic cases of listeriosis.

Isolation of *L. monocytogenes* from Clinical Material

Listeria monocytogenes is easily isolated from specimens obtained from normally sterile sites, and the bacterium will grow on most commonly used nonselective media after 24–48 h of incubation at 37 °C. Blood agar is preferred as *L. monocytogenes* is β-haemolytic, although it should be noted that rare nonhaemolytic strains occur (USDA/CFSAN, 2003b). Isolation of *L. monocytogenes* from clinical (or food/environmental) specimens likely to be contaminated with competing microflora requires the use of selective differential media such as Oxford or PALCAM (polymyxin acriflavine lithium chloride ceftazidime individual aesculin mannitol) agars, both of which are commercially available. On these media, aesculin-hydrolysing colonies such as *Listeria* spp. are surrounded by a black aesculetin pigment (which may diffuse widely throughout the medium, making it difficult to identify aesculin-hydrolysing colonies). With the exception of the uncommonly encountered *L. grayi*, *Listeria* spp. do not ferment mannitol and this property is made use of in PALCAM agar which contains a mannitol/phenol red indicator system. Several commercially produced chromogenic agars for the isolation of *Listeria* spp. are available but experience with these products outside of food microbiology laboratories is limited. Food or contaminated clinical material may also be incubated in a one- or two-step enrichment process, using an enrichment broth, such as Fraser broth or buffered *Listeria* enrichment broth, which again are commercially available, before subculture onto a selective/differential medium. Cold enrichment of contaminated specimens at 4 °C is now regarded as obsolete.

Identification of *L. monocytogenes*

There are several reasons as to why *L. monocytogenes* may be misidentified in the clinical laboratory other than the well-recognised tendency of the bacterium to be labelled as a diphtheroid which may lead to an isolate being discarded as a contaminant. Zones of β-haemolysis may not be obvious on first inspection and may require removal of a colony from the plate for this to become apparent. In addition, *L. monocytogenes* may cross-react with antigens in some, but not all, commercially available kits for the grouping of β-haemolytic streptococci (Farrington *et al.*, 1991). Furthermore, tests for motility are sometimes carried out only at 20 °C without simultaneously testing for absence of motility at 37 °C. It should also be noted that only a minority of the motile population may exhibit the characteristic of the so-called tumbling leaf motility when viewed in hanging drop preparation. Very rarely, tests for catalase may be negative (Elsner *et al.*, 1996).

Although biochemical identification of *Listeria* spp., using traditional biochemical and other techniques as outline in Tables 8.2 and 8.3, should be well within the capability of most clinical laboratories, a number of commercially produced identification systems can be used, such as the API Listeria (Bille *et al.*, 1992). These are, however, primarily designed for the food microbiology laboratory, and it is unlikely, given the infrequency of isolation of *L. monocytogenes*

Table 8.2   Differentiation of *Listeria* species

Test	L. monocytogenes	L. innocua	L. seeligeri	L. ivanovii	L. welshimeri	L. grayi subsp. grayi	L. grayi subsp. murrayi
β-Haemolysis	+	−	+	+	−	−	−
CAMP test							
S. aureus	+	−	W	−	−	−	−
R. equi	w	−	−	+	−	−	−
Nitrate reduction	−	−	−	−	−	−	+
Acid from:							
D-Mannitol	−	−	−	−	−	+	+
L-Rhamnose	+	V	−	−	−	−	v
D-Xylose	−	−	+	+	+	−	−
α-Methyl-D-glucoside	+	+	+	+	+	v	v
α-Methyl-D-mannoside	+	+	−	V[a]	+	v	v
D-Ribose	−	−	−	V[a, b]	−	+	+
D-Tagatose	−	−	−	−	v	−	−

+, ≥90% strains positive; −, ≤10% strains positive; v, 11–89% strains positive; w, weakly positive.
[a] subsp. *ivanovii* is negative, subsp. *londoniensis* is positive.
[b] subsp. *ivanovii* is positive, subsp. *londoniensis* is negative.

Table 8.3   Differentiation of *Listeria* species and *Erysipelothrix rhusiopathiae*

Test	Listeria spp.	E. rhusiopathiae
Catalase	+	−
Oxidase	−	−
Motility		
20 °C	+	−
37 °C	−	−
Growth at 4 °C	+	−
Haemolysis on blood agar	β/Nonhaemolytic	α/Nonhaemolytic
H₂S production	−	+
Aesculin hydrolysis	+	−
Indole production	−	−
Vosges-Proskauer test	+	−
Urease	−	−
Vancomycin susceptibility	S	R

S, susceptible; R, resistant.

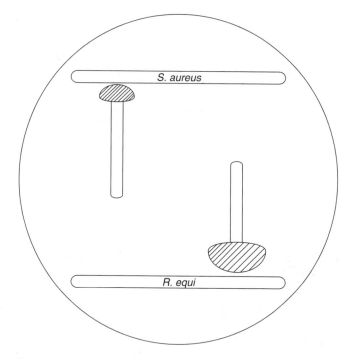

Figure 8.1   CAMP test with *Listeria monocytogenes* (top) and *L. ivanovii* (bottom). Hatched areas represent zones of haemolysis.

in the clinical setting, that the routine diagnostic laboratory would find it cost effective to employ these methods. However, many commonly used manual (e.g. API CORYNE) or automated identification systems (VITEK 2) likely to be in use in clinical laboratories are capable of identifying putative *Listeria* isolates to species level, although additional tests may be required to complete the identification process.

The CAMP (Christie Atkins and Munch-Petersen) test, originally employed in the identification of *Streptococcus agalactiae* has been modified for use in the characterisation of *Listeria* spp. The test relies upon synergism of haemolysis between cultures of *L. monocytogenes* and *Staphylococcus aureus* and *L. ivanovii* and *Rhodococcus equi* (*L. seeligeri* may also give a weakly positive CAMP test with *S. aureus*) (Figure 8.1). In the test *S. aureus* or *R. equi* is streaked across a 5% (v/v) sheep blood agar plate and the β-haemolytic *Listeria* isolate is streaked at 90° to the latter. Plates are incubated in air at 37 °C for 24 and 48 h. With *L. ivanovii* and *R. equi*, an easily identifiable arrowhead zone of haemolysis is observed. With *S. aureus* and *L. monocytogenes* (and also *L. seeligeri*), synergistic haemolysis also occurs, but this is less obvious and a more blunt zone is observed. It must be emphasised that, for the test to be successful, freshly prepared sheep blood agar should be used. The red cells should be washed twice in phosphate-buffered saline to remove inhibitors of haemolysis before use. Not all strains of *S. aureus* and *R. equi* are suitable for the CAMP test, and culture collection strains are recommended for this purpose [NCTC 1803 or NCTC 7428 (ATCC 49444) and NCTC 1621

(ATCC 6939), respectively]. Similarly, known isolates of *L. monocytogenes* (and other species if available) should be included as controls. Finally, it should be acknowledged that some strains of *L. monocytogenes* may also give a positive CAMP reaction with both *S. aureus* and *R. equi*, although the synergistic haemolysis seen with the latter is usually less pronounced than that seen with *L. ivanovii*.

MANAGEMENT OF INFECTION

Despite *in vitro* susceptibility to a wide range of agents (except the cephalosporins and fosfomycin), the outcome of antimicrobial therapy in invasive listeriosis is often disappointing, partly because of delays in the initiation of appropriate therapy as the diagnosis is not considered sufficiently early in the course of the illness and partly because so many patients with listeriosis have predisposing conditions such as immunosuppression. Moreover, *L. monocytogenes* is a facultative intracellular parasite, and many agents have poor intracellular penetration. Even if penetration does occur, compounds may be unstable or exhibit reduced activity in this milieu. Information on the most effective regimens for the management of listeriosis are derived from *in vitro* data, animal models and clinical reports, but methodologies vary greatly between studies and results are often conflicting. There remain no controlled trials in this field.

Despite poor intracellular penetration and a widely held view that β-lactams are only bacteriostatic against *L. monocytogenes* in intracellular locations, the aminopenicillins, amoxicillin and ampicillin are regarded as the drugs of choice in invasive listeriosis, albeit at higher than usual doses of 8–12 g per day in divided doses. Many authorities recommend the addition of an aminoglycoside, such as gentamicin at 5–7 mg/kg as a single daily dose, to this regimen, as *in vitro* data suggest this is a synergistic combination and also because of favourable clinical experience of this combination as cited in published series (Safdar and Armstrong, 2003a). Animal models using liposomal or nanoparticle-bound ampicillin have yielded encouraging results, although these have yet to be evaluated in the clinical setting.

The optimal duration of antimicrobial therapy for listeriosis remains to be determined; this is not altogether surprising, given the wide range of clinical manifestations and large number of predisposing conditions, but a general recommendation would be at least 2 weeks for primary bacteraemia and listeriosis in pregnancy and at least 3 weeks for meningitis and neonatal infection. Shorter courses are associated with relapsing infection. Endocarditis may require up to 6 weeks of therapy.

For patients unable to tolerate an ampicillin ± aminoglycoside regimen, co-trimoxazole is considered acceptable second-line therapy, on account of both *in vitro* synergy of the combination at concentrations readily achievable in serum and CSF and encouraging clinical outcomes as reported in the literature.

The choice of antimicrobial for patients where neither ampicillin nor co-trimoxazole can be used is problematic. Experience with vancomycin-based regimens has been disappointing (Safdar and Armstrong, 2003a), and although other agents such as carbapenems, newer fluoroquinolones, linezolid and quinupristin/dalfopristin have good *in vitro* activity against *L. monocytogenes*, clinical experience with these compounds remains limited (Weston *et al.*, 1998; Grumbach, Mylonakis and Wing, 1999).

Acquired antimicrobial resistance in clinical isolates is relatively uncommon (Safdar and Armstrong, 2003b), although tetracycline resistance in isolates carrying the *tetM* gene has been reported as has plasmid-mediated resistance (conferred by the *dfrD*) gene (Charpentier *et al.*, 1995; Charpentier, Gerbaud and Courvalin, 1999). Carriage of plasmid conferring resistance to multiple antimicrobials has also been reported (Hadorn *et al.*, 1993).

ERYSIPELOTHRIX RHUSIOPATHIAE

Introduction

Although Koch cultured a bacterium *Erysipelothrix muriseptica* from the blood of septicaemic mice in 1880, it was not until 8 years later that Löffler published the first accurate description of *E. rhusiopathiae*, which he isolated from a pig which had died from swine erysipelas. In the intervening period, Pasteur and Dumas (and later Thullier) used a strain of the bacterium which had been passaged through rabbits to immunise swine against erysipelas – the first time active immunisation by a live attenuated vaccine had been demonstrated. The bacterium was first identified by Rosenbach as a human pathogen in 1909 when he reported its isolation from a patient with localised cutaneous lesions to which he gave the name erysipeloid to distinguish them from those of human erysipelas.

Description of the Organism

Erysipelothrix spp. are facultative anaerobes which grow over a wide temperature range, with an optimum of 30–37 °C. They are nonmotile, asporogenous, nonacid fast, short Gram-positive rods (0.2–0.4 μm in diameter and 0.8–2.5 μm in length). They occur singly but, according to growth conditions, may be observed as chains, clusters or filamentous forms. Growth is favoured under alkaline conditions (range pH 6.7–9.2, optimum 7.2–7.6). Biochemical characteristics are discussed later and in Table 8.3. Until recently the genus *Erysipelothrix* contained only one species, *E. rhusiopathiae*, but chemotaxonomic and DNA studies have identified two other species, *E. tonsillarum* and *E. inopinata*, neither of which have been implicated as human pathogens.

For many years serotyping was used both as a tool for taxonomic and epidemiological analysis, but this has been rendered obsolescent with the introduction of molecular typing techniques such as ribotyping and RAPD-PCR.

Pathogenesis of *Erysipelothrix rhusiopathiae* Infection

Virulence factors of *E. rhusiopathiae* remain underinvestigated, particularly those which may be of importance in human infection. Putative virulence factors have included neuraminidase, hyaluronidase and adherence to mammalian cells, but the ability to survive intracellularly within phagocytic cells is probably of most importance (Shimoji, 2000).

EPIDEMIOLOGY

Ecology of *Erysipelothrix rhusiopathiae*

Erysipelothrix rhusiopathiae is distributed widely and has been isolated from a number of environmental sources such as soil and sewage effluent. The primary reservoir for the bacterium is, however, mammals and saltwater fish and crustaceans and, to a lesser extent, freshwater fish and avian species in which it can behave as either commensal or pathogen. Particularly important in this respect are pigs in which *E. rhusiopathiae* infection is associated with a number of syndromes including swine erysipelas, septicaemia and polyarthritis (which resembles rheumatoid arthritis in humans). Infection in pigs as well as other farmed animals and birds is of considerable economic importance to the agricultural industry.

Human Infection

There is no carrier state in humans. Although infections associated in humans are apparently rare, difficulties in the microbiological diagnosis, coupled with the rapid response of this condition to many antimicrobials and the fact that the disease is not notifiable, may lead to an underestimation of the true number of cases.

Given the reservoirs of this bacterium, it is not surprising that individuals who have frequent contact with animals, poultry, fish and their related products, such as farmers, abattoir workers, fishermen and butchers, are the most at risk of *Erysipelothrix* infection. In most instances of infection it is likely that the bacterium is acquired through cutaneous inoculation either through direct penetration of the skin with a sharp object or through contamination of a pre-existing skin lesion.

In some cases, no history of animal or fish contact can be elicited, while in others consumption of raw fish, shellfish or undercooked pork has been associated with subsequent systemic infection. There is no evidence that person-to-person transmission occurs.

CLINICAL FEATURES

The most common form of E. rhusiopathiae infection in humans, accounting for around 90% of cases, reported is the cellulitis known as erysipeloid. Most cases are occupationally related and follow puncture wounds or inoculation of cuts or abrasions. Very uncommonly, infection may follow dog bites. There is no convincing evidence to support earlier assertions that the bacterium can penetrate intact skin. After an incubation of 2–7 days, the lesions which are usually (but not exclusively) located on the hands or forearms develop. These typically have a blue or purplish colouration and a sharply defined edge and are often described as being intensely itchy or burning. Lesions may spread proximally with central clearing. Systemic symptoms are uncommon and, although there may be swelling and complaints of stiffness in adjacent joints, aspirates are sterile on culture. The differential diagnosis includes erysipelas caused by S. pyogenes and staphylococcal cellulites, although in these syndromes the incubation time is shorter and constitutional signs and lymphadenopathy are more frequent.

A more severe form of cutaneous infection, in which the cellulitis spreads more widely, often in conjunction with the appearance of lesions at other sites, occurs rarely. Despite the frequency of systemic symptoms in these cases, blood cultures are rarely positive when traditional manual systems are used. A case of necrotising fasciitis caused by E. rhusiopathiae has also been reported (Simoniescu et al., 2003).

The most serious manifestation of E. rhusiopathiae infection is endocarditis (Gorby and Peacock, 1988). Approximately 40% of patients give a history of preceding or coexisting skin lesions. Most cases occur on native valves in individuals without a history of valvular disease. Very rarely consumption of raw or undercooked seafood or pork has anteceded endocarditis. Males are more likely to be affected than females, presumably reflecting occupational exposure to the bacterium, and there is a predilection for the aortic valve. Presentation can be acute but is more likely to have an insidious onset (mean duration to diagnosis 6.6 weeks). The course of infection is often characterised by valve destruction (30% cases require valve replacement) which along with embolic events may account for the higher mortality rate of this infection when compared with endocarditis caused by other bacteria. Bacteraemia without endocarditis also occurs usually, but not exclusively, in immunocompromised individuals, particularly those with chronic liver disease and in patients receiving glucocorticosteroids (Fakoya et al., 1995).

Erysipelothrix rhusiopathiae is also associated with a range of other infections including septic arthritis in both immunocompromised and immunocompetent (Ruiz et al., 2003; Vallianatos, Tilentzoglou and Koutsoukou, 2003) and peritonitis in chronic ambulatory peritoneal dialysis (Dunbar and Clarridge, 2000).

LABORATORY DIAGNOSIS

PCR-based methods have been described but have found only limited use in the clinical laboratory (Brooke and Riley, 1999). Similarly, serological tests and antibody detection techniques have been little employed outside the veterinary setting. Accordingly, diagnosis of E. rhusiopathiae infection in humans remains dependent on the isolation and identification of the bacterium from clinical material.

The bacterium will grow readily on standard blood culture media in cases of septicaemia or endocarditis. Swabs are unsuitable for the diagnosis of erysipeloid, and aspirates, or preferably skin biopsies, from the edge of the lesion should be obtained. As the number of bacteria in these lesions may be small, enrichment for up to 2 weeks in glucose serum broth is recommended with frequent subcultures to blood agar. Primary and subcultures should be incubated at 37 °C in 5–10% CO_2. For isolation from heavily contaminated specimens, selective media are available (Bratberg, 1981).

After 24 h of incubation, colonies are pinpoint, but extended incubation permits the observation of two colonial types. Smooth (S) colonies are 0.5–1.5 mm in diameter, are convex and have a transparent bluish appearance. These are traditionally associated with isolates from acute infection, whereas rough (R) colonies, which are larger, flatter, opaque and with a rough irregular edge, are classically associated with chronic conditions. However, intermediate forms occur and both S and R forms can be isolated from a single specimen (Dunbar and Clarridge, 2000). On blood agar a narrow zone of α-haemolysis may be observed on prolonged incubation, and as E. rhusiopathiae is catalase negative, misidentification of these colonies as viridans group streptococci may occur. In addition, granules in the bacilli may resemble those of diphtheroids and smears made from older colonies may appear Gram negative.

The biochemical and other tests outlined in Table 8.3 can be used to identify the bacterium in the clinical laboratory, although this may be more easily achieved using commercially available systems, e.g. the API CORYNE kit.

When identifying E. rhusiopathiae and L. monocytogenes from environmental and food specimens, it is important to distinguish these bacteria from the Gram-positive genera Brochothrix and Kurthia. In practice this is not difficult as Brochothrix does not grow above 30 °C and Kurthia spp. are obligate aerobes and oxidase positive.

In clinical material, E. rhusiopathiae can be distinguished from Lactobacillus spp. by hydrogen sulphide production and from Corynebacterium spp. by failure to produce catalase and resistance to vancomycin. Listeria spp. can be distinguished from the former on the grounds of motility and catalase test and from the latter on testing for motility. [Leifsonia aquatica (formerly Corynebacterium aquaticum) is motile but, unlike Listeria spp., is negative in the Voges–Proskauer test.]

MANAGEMENT OF ERYSIPELOTHRIX INFECTION

Although the lesions of erysipeloid may resolve spontaneously, the time to resolution may be prolonged and relapses may occur and antimicrobial therapy is therefore recommended. Additionally, this may prevent progression to bacteraemia and/or endocarditis.

Given the relative infrequency of E. rhusiopathiae infection, it is not surprising that there have been no clinical trials to determine optimal antimicrobial therapy. Although there are several reports of the in vitro antimicrobial susceptibility of the bacterium, these have often involved small numbers of strains and different testing methods and there is a paucity of information regarding the clinical relevance of these data. However, the majority of isolates appear susceptible to penicillin, cephalosporins and imipenem as well as fluoroquinolones. The bacterium remains exquisitely susceptible to penicillin (MIC ≤ 0.01 µg ml^{-1}) which remains the drug of choice. Of the other β-lactams, imipenem appears the most active agent. In vitro susceptibility to erythromycin, tetracycline and chloramphenicol is variable. Although generally regarded as resistant to aminoglycosides, Schuster, Brennan and Edelstein (1993) report an isolate with an MIC to gentamicin of 1 µg ml^{-1}. Of particular note is the resistance of E. rhusiopathiae to vancomycin (but not necessarily teicoplanin), as this agent may be used in the empiric therapy of prosthetic valve endocarditis. There is no evidence that the genes for glycopeptide resistance in enterococci have been acquired from E. rhusiopathiae (Patel, 2000).

Although penicillin is regarded as the agent of choice for E. rhusiopathiae infection, there is very limited reported experience of the management of penicillin-hypersensitive patients, but ciprofloxacin appears highly active against the bacterium in vitro and has been used with the management of E. rhusiopathiae endocarditis in a patient who could not tolerate penicillin (MacGowan et al., 1991).

Optimal dosing regimens and duration of therapy remain ill defined. Benzylpenicillin for 4–6 weeks has been recommended for therapy of serious E. rhusiopathiae infections, although shorter courses of 2 weeks of intravenous therapy followed by 2–4 weeks of oral therapy have been used with success (Reboli and Farrar, 1989).

REFERENCES

Anon. (2003) First documented outbreak of *Listeria monocytogenes* in Quebec, 2002. *Canada Communicable Disease Report*, **29**, 181–186. www.hc-sc.gc.ca/pphb-dgspsp/publicat/ccdr-rmtc/03vol29/dr2921ea.html (4 February 2004).

Beumer, R., te Giffel, M., Spoorenberg, E. *et al.* (1996) *Listeria* spp. in domestic environments. *Epidemiology and Infection*, **117**, 437–442.

Bille, J., Catimel, B., Bannerman, E. *et al.* (1992) API *Listeria*, a new and promising one-day system to identify *Listeria* isolates. *Applied and Environmental Microbiology*, **58**, 1857–1860.

Bind, J.L. (1989) International aspects of the control of animal listeriosis. *Acta Microbiologica Hungarica*, **30**, 91–94.

Borucki, M.K., Peppin, D.J., White, D. *et al.* (2003) Variation in biofilm formation among strains of *Listeria monocytogenes*. *Applied and Environmental Microbiology*, **69**, 7336–7342.

Bratberg, A.M. (1981) Observations on the utilisation of a selective medium for the isolation of *Erysipelothrix rhusiopathiae*. *Acta Veterinaria Scandinavica*, **2**, 55–59.

Bronniman, S., Baer, H.U., Malinverni, R. and Buchler, M.W. (1998) *Listeria monocytogenes* causing solitary liver abscess. Case report and review of the literature. *Listeria monocytogenes*. *Digestive Surgery*, **15**, 364–368.

Brooke, C.J. and Riley, T.V. (1999) *Erysipelothrix rhusiopathiae*: bacteriology, epidemiology and clinical manifestations of an occupational pathogen. *Journal of Medical Microbiology*, **48**, 789–799.

Centers for Disease Control. (2001) Outbreak of listeriosis associated with homemade Mexican-style cheese. *Morbidity and Mortality Weekly Report*, **50**, 560–562.

Centers for Disease Control. (2002) Outbreak of listeriosis – Northeastern United States, 2002. *Morbidity and Mortality Weekly Report*, **51**, 950–951.

Charpentier, E., Gerbaud, G. and Courvalin, P. (1999) Conjugative mobilization of the rolling-circle plasmid pIP823 from *Listeria monocytogenes* BM4293 among Gram-positive and Gram-negative bacteria. *Journal of Bacteriology*, **181**, 3368–3374.

Charpentier, E., Gerbaud, G., Jacquet, J. *et al.* (1995) Incidence of antibiotic resistance in *Listeria* species. *The Journal of Infectious Diseases*, **172**, 277–281.

Colodner, R., Sakran, W., Miron, D. *et al.* (2003) *Listeria monocytogenes* cross-contamination in a nursery. *American Journal of Infection Control*, **31**, 322–324.

Cone, L.A., Leung, M.M., Byrd, R.G. *et al.* (2003) Multiple cerebral abscesses because of *Listeria monocytogenes*: three cases reports and a literature review of supratentorial listerial brain abscess(es). *Surgical Neurology*, **59**, 320–328.

Cummins, A.J., Fielding, A.K. and McGlauchlin, J. (1994) *Listeria ivanovii* infection in a patient with AIDS. *The Journal of Infection*, **28**, 89–91.

Dalton, C.B., Austin, C.C., Sobel, J. *et al.* (1997) An outbreak of gastroenteritis and fever due to *Listeria monocytogenes* in milk. *The New England Journal of Medicine*, **336**, 100–105.

de Valk, H., Vaillant, V., Jacquet, C. *et al.* (2001) Two consecutive nationwide outbreaks of listeriosis in France, October, 1999–February 2000. *American Journal of Epidemiology*, **154**, 944–950.

Doumith, M., Cazalet, C., Simoes, N. *et al.* (2004) New aspects regarding evolution and virulence of *Listeria monocytogenes* revealed by comparative genomics and DNA arrays. *Infection and Immunity*, **72**, 1072–1083.

Dunbar, S.A. and Clarridge, J.E. III. (2000) Potential errors in the recognition of *Erysipelothrix rhusiopathiae*. *Journal of Clinical Microbiology*, **38**, 1302–1304.

Elsner, H.-A., Sobottka, I., Bubert, A. *et al.* (1996) Catalase-negative *Listeria monocytogenes* causing lethal sepsis and meningitis in an adult hematologic patient. *European Journal of Clinical Microbiology and Infectious Disease*, **15**, 965–967.

Ericsson, H., Elköw, A., Danielsson-Tham, M.-L. *et al.* (1997) An outbreak of listeriosis suspected to have been caused by rainbow trout. *Journal of Clinical Microbiology*, **35**, 2904–2907.

Fakoya, A., Bendall, R.P., Churchill, D.R. *et al.* (1995) *Erysipelothrix rhusiopathiae* bacteraemia in a patient without endocarditis. *The Journal of Infection*, **30**, 180–181.

Farrington, M., Winters, S.M., Rubenstein, D. *et al.* (1991) Streptococci from primary isolation plates grouped by reverse passive haemagglutination. *Journal of Clinical Pathology*, **44**, 670–675.

Fernández Guerrero, M.L., Rivas, P., Rábago, R. *et al.* (2004) Prosthetic valve endocarditis due to *Listeria monocytogenes*. *International Journal of Infectious Diseases*, **8**, 97–102.

Frye, D.M., Zweig, R., Sturgeon, J. *et al.* (2002) An outbreak of febrile gastroenteritis associated with delicatessen meat contaminated with *Listeria monocytogenes*. *Clinical Infectious Diseases*, **35**, 943–949.

Garrec, N., Picard-Bonnaud, F. and Pourcher, A.M. (2003) Occurrence of *Listeria* sp. and *Listeria monocytogenes* in sewage sludge used for land application: effect of dewatering, liming and storage in tank on survival of *Listeria* species. *FEMS Immunology and Medical Microbiology*, **35**, 275–283.

Glaser, P., Frangeul, L., Buchrieser, C. *et al.* (2001) Comparative genomics of *Listeria* species. *Science*, **294**, 849–852.

Gombas, D.E., Chen, H.Y., Clavero, C.S. and Scott, V.N. (2003) Survey of *Listeria monocytogenes* in ready-to-eat foods. *Journal of Food Protection*, **66**, 559–569.

Gorby, G.L. and Peacock, J.E. (1988) *Erysipelothrix rhusiopathiae* endocarditis: microbiologic, epidemiologic, and clinical features of an occupational disease. *Review of Infectious Diseases*, **10**, 317–325.

Graham, J.C., Lanser, S., Bignardi, G. *et al.* (2002) Hospital-acquired listeriosis. *The Journal of Hospital Infection*, **51**, 136–139.

Grif, K., Patscheider, G., Dierich, M.P. and Allerberger, F. (2003) Incidence of fecal carriage of *Listeria monocytogenes* in three healthy volunteers: a one-year prospective stool survey. *European Journal of Clinical Microbiology and Infectious Diseases*, **22**, 16–20.

Grumbach, N.M., Mylonakis, E. and Wing, E.J. (1999) Development of listerial meningitis during ciprofloxacin treatment. *Clinical Infectious Diseases*, **29**, 1340–1341.

Hadorn, K., Hachler, H., Schaffner, A. and Kayser, F.H. (1993) Genetic characterization of plasmid-encoded multiple antibiotic resistance in a strain of *Listeria monocytogenes* causing endocarditis. *European Journal of Clinical Microbiology and Infectious Diseases*, **129**, 28–37.

Harvey, J. and Gilmour, A. (2001) Characterization of recurrent and sporadic *Listeria monocytogenes* isolates from raw milk and nondairy foods by pulsed-field gel electrophoresis, monocin typing, plasmid profiling and cadmium and antibiotic resistance determination. *Applied and Environmental Microbiology*, **67**, 840–847.

Jammula, P. and Gupta, R. (2002) *Listeria monocytogenes*-induced monomicrobial non-neutrocytic bacterascites. *Southern Medical Journal*, **95**, 1204–1206.

Linnan, M.J., Mascola, L., Lou, X.D. *et al.* (1988) Epidemic listeriosis associated with Mexican-style soft cheese. *The New England Journal of Medicine*, **319**, 823–828.

Lohmann, C.P., Gabel, V.P., Heep, M. *et al.* (1999) *Listeria monocytogenes*-induced endogenous endophthalmitis in an otherwise healthy individual: rapid PCR diagnosis as the basis for effective treatment. *European Journal of Ophthalmology*, **9**, 53–57.

Lyytikäinen, O., Autio, T., Maijala, R. *et al.* (2000) An outbreak of Listeria monocytogenes serotype 3a infections from butter in Finland. *Journal of Infectious Disease*, **181**, 1838–1841.

MacGowan, A.P., Reeves, D.S., Wright, C. and Glover, S.C. (1991) Tricuspid valve infective endocarditis and pulmonary sepsis due to *Erysipelothrix rhusiopathiae* successfully treated with high doses of ciprofloxacin but complicated by gynaecomastia. *The Journal of Infection*, **22**, 100–101.

Miettinen, M.K., Palmu, L., Bjorkroth, K.J. and Korkeala, H. (2001) Prevalence of *Listeria monocytogenes* in broilers at the abattoir, processing plant and retail level. *Journal of Food Protection*, **64**, 994–999.

Mylonakis, E., Paliou, M. and Hohmann, E.L. (2002) Listeriosis during pregnancy. *Medicine (Baltimore)*, **81**, 260–269.

Niemann, R.E. and Lorber, B. (1980) Listeriosis in adults: a changing pattern. Report of eight cases and a review of the literature. *Reviews of Infectious Diseases*, **2**, 207–227.

Olier, M., Pierre, F., Rousseaux, S. *et al.* (2003) Expression of truncated internalin A is involved in impaired internalization of some *Listeria monocytogenes* isolates carried asymptomatically by some humans. *Infection and Immunity*, **71**, 1217–1224.

Patel, R. (2000) Enterococcal-type glycopeptide resistance genes in non-enterococcal organisms. *FEMS Microbiology Letters*, **185**, 1–7.

Pilgrim, S., Kolb-Mäurer, A., Gentschev, I. *et al.* (2003) Deletion of the gene encoding p60 in *Listeria monocytogenes* leads to abnormal cell division and loss of actin-based motility. *Infection and Immunity*, **71**, 3473–3484.

Poyart-Salmeron, C., Carlier, C., Trieu-Cuot, P. *et al.* (1992) Transferable plasmid-mediated antibiotic resistance in *Listeria monocytogenes*. *Lancet*, **i**, 1422–1426.

Reboli, A.C. and Farrar, W.E. (1989) *Erysipelothrix rhusiopathiae*: an occupational pathogen. *Clinical Microbiology Reviews*, **2**, 354–359.

Revazishvili, T., Kotetishvili, M., Stine, O.C. *et al.* (2004) Comparative analysis of multilocus sequence typing and pulsed-field gel electrophoresis for characterizing *Listeria monocytogenes* strains isolated from environmental and clinical sources. *Journal of Clinical Microbiology*, **42**, 276–285.

Rivero, G.A., Torres, H.A., Rolston, K.V. and Kontoyiannis, D.P. (2003) *Listeria monocytogenes* infection in patients with cancer. *Diagnostic Microbiology and Infectious Disease*, **47**, 393–398.

Roberts, A.J. and Wiedmann, M. (2003) Pathogen, host and environmental factors contributing to the pathogenesis of listeriosis. *Cellular and Molecular Life Sciences*, **60**, 904–918.

Ruiz, M.E., Richards, J.S., Kerr, G.S. and Kan, V.L. (2003) Erysipelothrix rhusiopathiae septic arthritis. *Arthritis and Rheumatism*, **48**, 1156–1157.

Safdar, A. and Armstrong, D. (2003a) Antimicrobial activities against, 84 *Listeria monocytogenes* isolates from patients with systemic listeriosis at a comprehensive cancer center (1955–1997). *Journal of Clinical Microbiology*, **41**, 483–485.

Safdar, A. and Armstrong, D. (2003b) Listeriosis in recipients of allogeneic blood and bone marrow transplantation: thirteen year review of disease characteristics, treatment outcomes and a new association with human cytomegalovirus infection. *Bone Marrow Transplantation*, **29**, 913–916.

Sauders, B.D., Fortes, E.D. and Morse, D.L. (2003) Molecular subtyping to detect human listeriosis clusters. *Emerging Infectious Diseases*, **9**, 672–680.

Schuster, M.G., Brennan, P.J. and Edelstein, P. (1993) Persistent bacteremia in a patient with *Erysipelothrix rhusiopathiae* in a hospitalized patient. *Clinical Infectious Diseases*, **17**, 783–784.

Shimoji, Y. (2000) Pathogenicity of *Erysipelothrix rhusiopathiae*: virulence factors and protective immunity. *Microbes and Infection*, **2**, 965–972.

Sim, J., Hood, D., Finnie, L. *et al.* (2002) Series of incidents of *Listeria monocytogenes* non-invasive febrile gastroenteritis involving ready-to-eat meats. *Letters in Applied Microbiology*, **35**, 409–413.

Simoniescu, R., Grover, S., Shekar, R. and West, B.C. (2003) Necrotizing fasciitis caused by *Erysipelothrix rhusiopathiae*. *Southern Medical Journal*, **96**, 937–939.

Skogberg, K., Syrjanen, J., Jakhola, M. *et al.* (1992) Clinical presentation and outcome of listeriosis in patients with and without immunosuppressive therapy. *Clinical Infectious Diseases*, **14**, 815–821.

Slifman, N.R., Gershon, S.K., Lee J.H. *et al.* (2003) *Listeria monocytogenes* infection as a complication of treatment with tumor necrosis factor α-neutralizing factor agents. *Arthritis and Rheumatism*, **48**, 319–324.

Tse, K.C., Li, F.K., Chan, T.M. and Lai, K.N. (2003) *Listeria monocytogenes* peritonitis complicated by septic shock in a patient on continuous ambulatory peritoneal dialysis. *Clinical Nephrology*, **60**, 61–62.

USDA/CFSAN. (2003a) *Bacteriological Analytical Manual Online – Detection and Enumeration of Listeria monocytogenes in Foods*. http://vm.cfsan.fda.gov/~ebam/bam-10.html (4 February 2004).

USDA/CFSAN. (2003b) *Bacteriological Analytical Manual Online – Guideline on Identification of Atypical Hemolytic Listeria Isolates*. http://vm.cfsan.fda.gov/~ebam/bam-10s5.html (4 February 2004).

Vallianatos, P.G., Tilentzoglou, A.C. and Koutsoukou, A.D. (2003) Septic arthritis caused by *Erysipelothrix rhusiopathiae* infection after arthroscopically assisted anterior cruciate ligament reconstruction. *Arthroscopy*, **19**, 26–29.

Vargas, V., Aleman, C., de Torres, I. *et al.* (1998) *Listeria monocytogenes* associated acute hepatitis in a liver transplant recipient. *Liver*, **18**, 213–215.

Weis, J. and Seeliger, H.P.R. (1976) Incidence of *Listeria monocytogenes* in nature. *Applied Microbiology*, **30**, 29–32.

Weston, V.C., Punt, J., Vloebeghs, M. *et al.* (1998) *Listeria monocytogenes* meningitis in a penicillin-allergic paediatric renal transplant patient. *The Journal of Infection*, **37**, 77–78.

9

Bacillus spp. and Related Genera

Niall A. Logan and Marina Rodríguez-Díaz

Department of Biological and Biomedical Sciences, Glasgow Caledonian University, Glasgow, UK

HISTORICAL INTRODUCTION

The organism now known as *Bacillus subtilis* was first described as '*Vibrio subtilis*' by Christian Ehrenberg in 1835. Nearly 30 years later Casimir Davaine (1864) gave the name '*Bacteridium*' to the organism associated with anthrax infection – *la maladie charbonneuse* – but the name *Bacillus* was first applied to these organisms by Ferdinand Cohn (1872), who included the three species of rod-shaped bacteria, *B. subtilis* (type species), *B. anthracis* and the now unidentifiable *B. ulna*. Cohn (1876), Koch (1876) and Tyndall (1877) independently discovered that certain bacteria could spend part of their lives as the dormant cellular structures now known as endospores. The first two of these authors recognized the significance of these structures in the epidemiology of anthrax, and Koch's study of the life history of *B. anthracis* proved the germ theory of disease and so marked the genesis of clinical bacteriology. Winter (1880) was the first to include 'propagation through spores' in the description of the genus, and Prazmowski (1880) was the first to use sporulation as a differential (i.e., taxonomic) characteristic; he proposed the genus name *Clostridium* for organisms that differed from *Bacillus* in having spindle-shaped sporangia. Over the next 40 years at least 30 keys for the classification of bacteria were produced, and although many of these used the presence of spores as a character, there were many that did not. It was not until the Committee of the Society of American Bacteriologists on Characterization and Classification of Bacterial Types reported in the early volumes of the *Journal of Bacteriology* (Winslow *et al.*, 1917, 1920) that satisfactory and largely uncontested definitions of the bacterial groups emerged. The defining feature of the family *Bacillaceae* was 'Rods producing endospores', and the Committee used the requirement of oxygen and sporangial shape for differentiation between *Bacillus* and *Clostridium*, the other genus in the family. This description was applied in the first and second editions of *Bergey's Manual of Determinative Bacteriology* (Bergey *et al.*, 1923, 1925), and so the definition of the genus *Bacillus* as aerobic endospore-forming rods had become widely established by the 1920s.

The first major study of *Bacillus* was undertaken by Ford and colleagues (Laubach, Rice and Ford, 1916; Lawrence and Ford, 1916). They allocated 1700 strains to 27 species but produced an incomplete classification, particularly owing to the absence of adequate differential tests. The next scheme appeared 30 years later when Smith, Gordon and Clark (1946) published a monograph that dealt with 625 strains. Subsequently this work was expanded to include 1134 strains and to cover all known aerobic spore formers (Smith, Gordon and Clark, 1952). The work of these authors was thorough; they collected strains widely and studied rod, spore and sporangial morphologies as well as many physiological characters. They placed their organisms in three groups according to their spore and sporangial shapes. Their approach contrasted with the tendency at that time to describe new species on the basis of a single strain differing from the description of an existing taxon by a single character, and the soundness of the work was established when the scheme was successfully applied to 246 strains isolated by Knight and Proom (1950) from soil and to 400 isolates from clinical specimens, laboratory dust and soil (Burdon, 1956). An updated and expanded version of the Smith *et al.* scheme was published by Gordon, Haynes and Pang (1973), and this classification, despite attempts to split the genus (Gibson and Gordon, 1974), was widely followed until the early 1990s.

During this period the number of recognized species in the genus declined from its peak number of 146, in the 1939 (fifth) edition of *Bergey's Manual of Determinative Bacteriology*, to 22, with 26 of uncertain status in the 1974 (eighth) edition and to only 31 species in the ICSB Approved Lists of Bacterial Names (Skerman, McGowen and Sneath, 1980). Despite this relatively small number of species, however, it was generally agreed that the genus *Bacillus* was heterogeneous. The range of DNA base composition acceptable in a homogeneous genus is widely agreed to be no more than 10%, whereas that found in *Bacillus* was over 30% (Claus and Berkeley, 1986). This gave confirmation of heterogeneity, and numerical studies such as that by Logan and Berkeley (1981) also indicated that the genus might be divided into five or six genera.

Accordingly, *Bacillus* has now been divided into more manageable and better-defined groups on the basis of 16S rDNA sequencing studies, but this and other molecular characterization techniques have also allowed the recognition of many new species. Ten new genera have been proposed so far: *Alicyclobacillus* (Wisotzkey *et al.*, 1992), which contains six species of thermoacidophiles; *Paenibacillus* (Ash, Priest and Collins, 1993), containing 41 species and including organisms formerly called *B. polymyxa*, *B. macerans* and *B. alvei* and the honey bee pathogens *B. larvae* and *B. pulvifaciens* (now both subspecies of *P. larvae*); *Brevibacillus* (Shida *et al.*, 1996), containing 11 species and including organisms formerly called *B. brevis* and *B. laterosporus*; *Aneurinibacillus* (Shida *et al.*, 1996), with *A. aneurinilyticus* and two other species; *Virgibacillus* (Heyndrickx *et al.*, 1998) with *V. pantothenticus* and four other species; *Gracilibacillus* and *Salibacillus* (Wainø *et al.*, 1999), which each contain two species of halophiles; *Geobacillus* (Nazina *et al.*, 2001) with ten species of thermophiles including *B. stearothermophilus*; *Ureibacillus* (Fortina *et al.*, 2001) with two round-spored, thermophilic species and single-membered *Marinibacillus* (Yoon *et al.*, 2001b). *Salibacillus* has subsequently been merged with *Virgibacillus* (Heyrman *et al.*, 2003).

Sporosarcina contains the motile, spore-forming coccus *S. ureae*, which is closely related to *B. sphaericus*, and five rod-shaped, round-spored species, including *B. pasteurii*, that have now been placed in this genus (Yoon *et al.*, 2001a). Other genera of aerobic endospore formers, not derived from *Bacillus*, are: *Sulfobacillus*, *Amphibacillus*,

Halobacillus, Ammoniphilus, Thermobacillus, Filobacillus, Jeotgalibacillus and *Lentibacillus*, but these and *Alicyclobacillus, Gracilibacillus*, and *Ureibacillus* strains are unlikely to be encountered in a clinical laboratory, and clinical isolates of *Geobacillus* and *Sporosarcina* have not been reported; so these genera will not be considered further.

Bacillus continues to accommodate the best-known species such as *B. subtilis* (the type species), *B. anthracis, B. cereus, B. licheniformis, B. megaterium, B. pumilus, B. sphaericus* and *B. thuringiensis*. It still remains a large genus, with nearly 80 species, as losses of species to other genera have been balanced by proposals for new *Bacillus* species (six species await transfer to *Geobacillus* and *Paenibacillus*). Members of the *B. cereus* group, *B. anthracis, B. cereus* and *B. thuringiensis*, are really pathovars of a single species (Turnbull *et al.*, 2002).

Unfortunately, taxonomic progress has not revealed readily determinable features characteristic of each genus. They show wide ranges of sporangial morphologies and phenotypic test patterns. Many recently described species represent genomic groups disclosed by DNA–DNA pairing experiments, and routine phenotypic characters for distinguishing some of them are very few and of unproven value. New species and subspecies of aerobic endospore formers are regularly described (47 in the 3 years January 2000 to January 2003, during which period no proposals for merging species had been made), but often on the basis of very few strains. The habitats and physiologies of most of these recently described organisms are such that only in extraordinary circumstances would they be of clinical interest, but several of the remainder, in addition to the well-established pathogens *B. anthracis* and *B. cereus*, must be considered because they do occasionally cause clinical problems and the frequency of recognition of these problems is increasing.

DESCRIPTIONS OF ORGANISMS

Although the production of resistant endospores in the presence of oxygen remains the defining feature for *Bacillus* and the new genera derived from it, the definition was undermined by the discoveries of *B. infernus* and *B. arseniciselenatis*, which are strictly anaerobic (Boone *et al.*, 1995; Switzer Blum *et al.*, 1998), and spores have not been observed in *B. halodenitrificans, B. infernus, B. selenitireducens, B. subterraneus* and *B. thermoamylovorans*.

Nonetheless, those species of *Bacillus* and related genera likely to be isolated in a clinical laboratory are rod-shaped, endospore-forming organisms which may be aerobic or facultatively anaerobic. They are usually Gram positive (in young cultures), but sometimes Gram variable or frankly Gram negative. They are mostly catalase positive and may be motile by means of peritrichous flagella. Most species are mesophilic, but *Bacillus* contains some thermophiles and psychrophiles, *Aneurinibacillus* and *Brevibacillus* both contain thermophilic species, and *Paenibacillus* contains one psychrophilic species.

Although the aerobic endospore-forming genera are physiologically heterogeneous, containing some species that do not sporulate readily, all the clinically significant isolates reported to date are of species that grow, and often sporulate, on routine laboratory media such as blood agar or nutrient agar, at 37 °C. It is unlikely that clinically important, but more fastidious, strains are being missed for want of appropriate growth conditions. Occasional isolates are sent to reference laboratories as *Bacillus* species because they are large Gram-positive rods, even though spores have not been seen, or sometimes because poly-β-hydroxybutyrate granules or other inclusion bodies have been mistaken for spores.

Bacillus anthracis **and** B. cereus

These are the two *Bacillus* species of greatest medical and veterinary importance, and together with *B. mycoides, B. thuringiensis, B. weihenstephanensis* (a psychrophile) and their more distant relative *B. pseudomycoides*, they form the *B. cereus* group.

The colonies of this group are very variable (Figure 9.1a–c) but usually recognizable nonetheless. They are characteristically large (2–7 mm in diameter, and usually near the top end of this range) and vary in shape from circular to irregular, with entire to undulate, crenate or fimbriate (medusa head) edges; they have matt or granular textures and have been likened to little heaps of ground glass. Smooth and moist colonies are not uncommon, however.

Although colonies of *B. anthracis* and *B. cereus* can be similar in appearance (Figure 9.1a,b), those of the former are generally smaller, are nonhemolytic, may show more spiking or tailing along the lines of inoculation streaks and are very tenacious as compared with the usually more butyrous consistency of *B. cereus* and *B. thuringiensis* colonies so that they may be pulled into standing peaks with a loop. The edges of *B. anthracis* colonies are often described as medusa head, but this is a character to be found throughout the *B. cereus* group. *Bacillus cereus* itself has colonies that are butyrous and cream to whitish in colour. The white colonies often contain spores. On blood agar, which is usually vigorously hemolysed, the colonies may be tinged with pigments derived from hemoglobin degradation. Fresh cultures characteristically smell mousy. *Bacillus mycoides* colonies are characteristically rhizoid or hairy looking, adherent and readily cover the whole surface of an agar plate.

Bacillus anthracis cells are Gram-positive, nonmotile rods, 1.0–1.2×3.0–5.0 μm. They commonly appear as long, entangled chains, but some strains of *B. cereus* and *B. mycoides* (which is nonmotile) may look similar. Fresh isolates of *B. anthracis*, and laboratory strains grown on serum-containing agar in the presence of carbon dioxide, produce a poly-γ-D-glutamic acid capsule. Spores are ellipsoidal, central or nearly so and do not cause the sporangium to swell. This organism is a chemoorganotroph and grows above 15–20 °C and below 40 °C; its optimum temperature is about 37 °C.

Bacillus cereus cells are Gram positive and similar in size to those of *B. anthracis* but are usually motile, and single organisms, pairs and short chains are more common than in *B. anthracis* cultures. Capsules are not formed, but spore and sporangial morphology are similar to those of *B. anthracis* (Figure 9.2a). *Bacillus cereus* is chemoorganotrophic and grows above 10–20 °C and below 35–45 °C with an optimum temperature of about 37 °C.

Bacillus thuringiensis can be distinguished from *B. cereus* by the production of parasporal crystals that may be toxic for insects and other invertebrates. These crystals may be observed in sporulated cultures (after 2–5 days incubation) by phase-contrast microscopy (Figure 9.2b) or by staining with malachite green. The spores may lie obliquely in the sporangium, as might happen in *B. cereus*.

Vegetative cells of all members of the *B. cereus* group may, when grown on a carbohydrate medium, accumulate poly-β-hydroxybutyrate, and their cytoplasm appears vacuolate or foamy.

Other Species

Outside the *B. cereus* group, aerobic endosporeformers show a very wide range of colonial morphologies, both within and between species (Figure 9.1). Colonial appearance varies from moist and glossy, through granular, to wrinkled. Shape varies from round to irregular, sometimes spreading, with entire, through undulate or crenate, to fimbriate edges. Diameters range from 1 to 5 mm. Colours commonly range from buff or creamy grey to off-white, and strains of some species may produce yellow, orange or pink pigment. Hemolysis may be absent, slight or marked, partial or complete. Elevations range from effuse, through raised, to convex. Consistency is usually butyrous, but mucoid and dry; adherent colonies are not uncommon. Despite this diversity, colonies of these organisms are not generally difficult to recognize, and some species have characteristic, yet seemingly infinitely variable, colonial morphologies, as with the *B. cereus* group. *Bacillus subtilis* (Figure 9.2h) and *B. licheniformis* produce similar colonies which are exceptionally variable in appearance and often

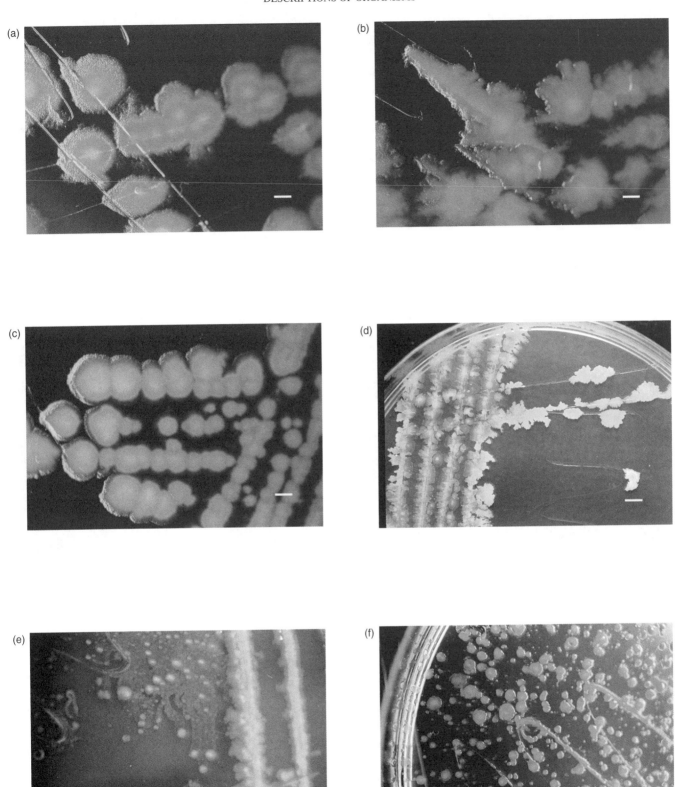

Figure 9.1 Colonies of aerobic endospore-forming bacteria on blood agar (a–c and g–l) and nutrient agar (d–f) after 24–36 h at 37 °C. Bar markers represent 2 mm. (a) *Bacillus anthracis*; (b) *B. cereus*; (c) *B. thuringiensis*; (d) *B. subtilis*; (e) motile microcolonies formed by some *Paenibacillus* strains formerly allocated to *B. circulans*; (f) *Paenibacillus alvei*; (g) *B. megaterium*; (h) *B. pumilus*; (i) *B. sphaericus*; (j) *Brevibacillus brevis*; (k) *B. laterosporus*; (l) *P. polymyxa*.

(g)

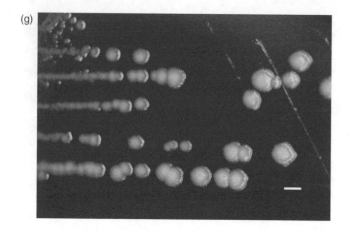

(h)

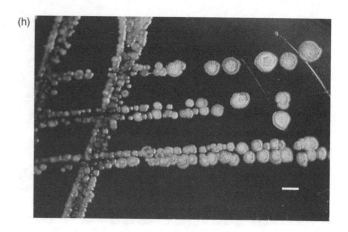

(i)

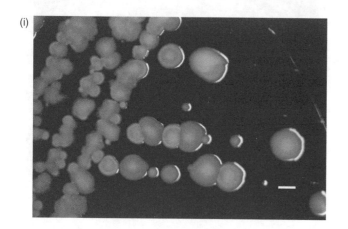

(j)

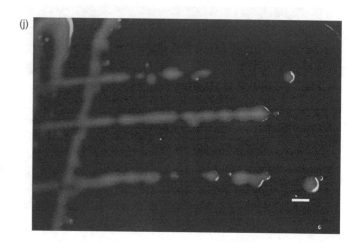

(k)

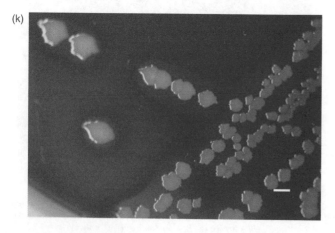

(l)

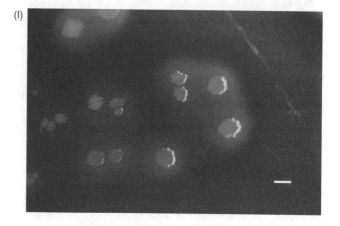

Figure 9.1   (Continued).

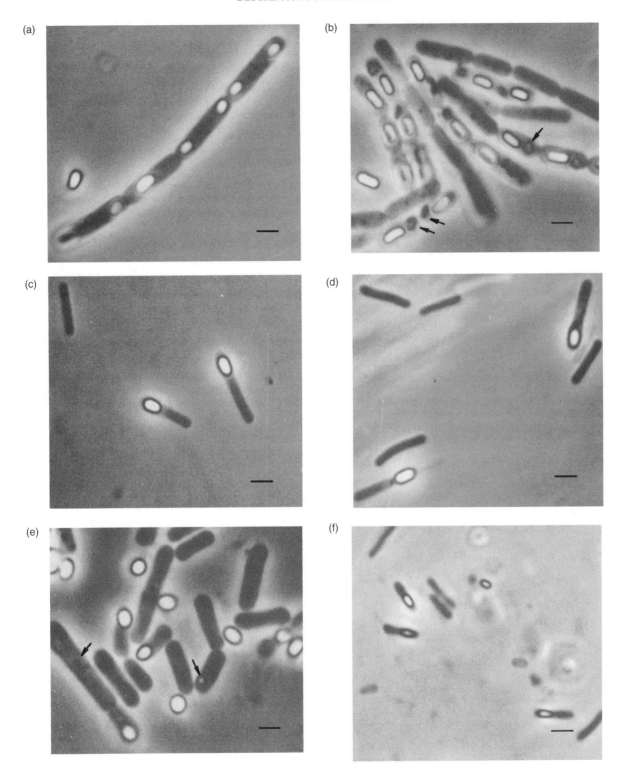

Figure 9.2 Photomicrographs of aerobic endospore-forming bacteria viewed by phase-contrast microscopy. Bar markers represent 2 μm. (a) *Bacillus cereus*, broad cells with ellipsoidal subterminal spores, not swelling the sporangia; (b) *B. thuringiensis*, broad cells with ellipsoidal subterminal spores, not swelling the sporangia and showing parasporal crystals of insecticidal toxin (arrowed); (c) *B. circulans*, ellipsoidal subterminal spores, swelling the sporangia; (d) *B. coagulans*, ellipsoidal subterminal spores, swelling the sporangia; (e) *B. megaterium*, broad cells with ellipsoidal and spherical subterminal and terminal spores, not swelling the sporangia and showing polyhydroxybutyrate inclusions (arrowed); (f) *B. pumilus*, slender cells with cylindrical subterminal spores not swelling the sporangia; (g) *B. sphaericus*, spherical terminal spores, swelling the sporangia; (h) *B. subtilis*, ellipsoidal central and subterminal spores, not swelling the sporangia; (i) *Brevibacillus brevis*, ellipsoidal subterminal spores, one swelling its sporangium slightly; (j) *B. laterosporus*, ellipsoidal central spores with thickened rims on one side (arrowed), swelling the sporangia; (k) *Paenibacillus alvei*, cells with tapered ends, ellipsoidal paracentral to subterminal spores, not swelling the sporangium; (l) *P. polymyxa*, ellipsoidal paracentral to subterminal spores, slightly swelling the sporangia.

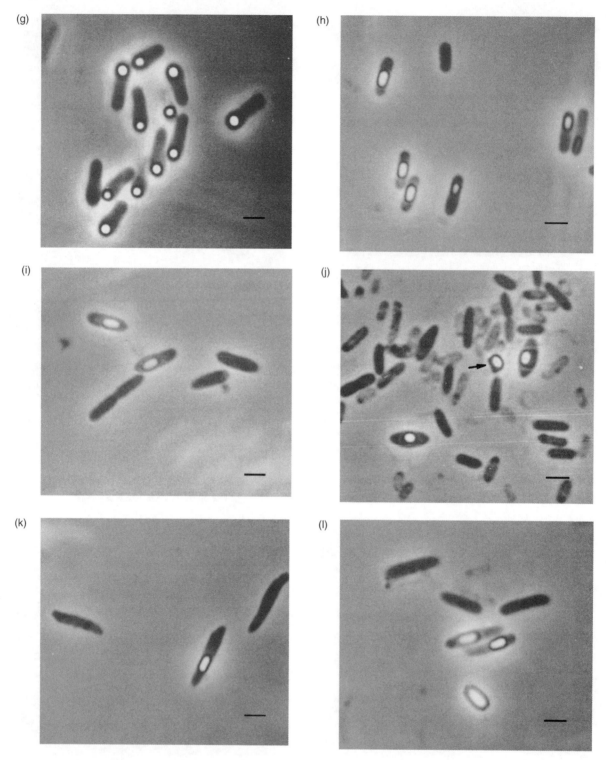

Figure 9.2 (Continued).

appear to be mixed cultures; colonies (Figure 9.1d) are irregular in shape, are of moderate (2–4 mm) diameter, range from moist and butyrous or mucoid, with margins varying from undulate to fimbriate, through membranous with an underlying mucoid matrix, with or without mucoid heading at the surface, to rough and are crusted as they dry. The licheniform colonies of *B. licheniformis* tend to be quite adherent. *Bacillus circulans* was so named because a protoplasm-like, circular motion of the interior of its colonies was observable by

low-power microscopy. Rotating and migrating microcolonies showing spreading growth (Figure 9.1e) were observed macroscopically in about 13% of strains received as *B. circulans* (Logan, Old and Dick, 1985), but the species is exceptionally heterogeneous and the strains with motile colonies have now been allocated to *Paenibacillus glucanolyticus*, *P. lautus*, and some unidentified *Paenibacillus* species. Clinical isolates previously identified as strains of *B. circulans* (Figure 9.2c) may therefore be *Paenibacillus* species. Motile microcolonies with spreading growth are also characteristic of *P. alvei* (Figures 9.1f and 9.2k); plate cultures of this species will readily grow across the entire agar surface and commonly have an unpleasant smell.

Other species that have been encountered in the clinical laboratory include *B. coagulans* (Figure 2d), *B. megaterium* (Figures 9.1g and 9.2e), *B. pumilus* (Figures 9.1h and 9.2f), *B. sphaericus* (Figures 9.1i and 9.2g), *B. brevis* (Figures 9.1j and 9.2i), *B. laterosporus* (Figures 9.1k and 9.2j), *P. macerans* and *P. polymyxa* (Figures 9.1l and 9.2l), and they do not produce particularly distinctive growth; it must be appreciated that few species have colonies sufficiently characteristic or invariant to allow tentative identification, even by the experienced worker.

Microscopic morphologies, particularly of sporangia (Figure 9.2), are much more helpful than colonial characters in distinguishing between all these non-*B. cereus* group species. Vegetative cells are usually round ended, but cells of *P. alvei* may have tapered ends (Figure 9.2k). The large cells of *B. megaterium* (Figure 9.2e) may accumulate poly-β-hydroxybutyrate and appear vacuolate or foamy when grown on glucose nutrient agar. Overall, cell widths vary from about 0.5 to 1.5 μm and lengths from 1.5 to 5 μm. Most strains of these species are motile. Spore shapes vary from cylindrical (Figure 9.2f,k), through ellipsoidal (Figure 9.2a–d,h–j,l), to spherical (Figure 9.2e,g), and bean- or kidney-shaped, curved cylindrical and pear-shaped spores are occasionally seen. Spores may be terminally (Figure 9.2g,d,e), subterminally (Figure 9.2a–c,f,h,i,k,l) or centrally positioned (Figure 9.2f,j) within sporangia and may distend it (Figure 9.2c,d,g,i,j,l). Although there is some within-species and within-strain variation, spore shape, position and size tend to be characteristic of species and may allow tentative identification by the experienced worker. One species, *B. laterosporus*, produces very distinctive ellipsoidal spores, which have thickened rims on one side, so that they appear to be laterally displaced in the sporangia (Figure 9.2j).

All these species are mesophilic and will grow well between 30 and 37 °C; minimum temperatures for growth lie mostly between 5 and 20 °C, maxima mostly between 35 and 50 °C. For more information on growth temperatures, see Claus and Berkeley (1986).

NORMAL HABITATS

Bacillus anthracis

Long before this organism was recognized as the cause of anthrax, it was understood that putting animals to graze in certain places was likely to lead to an outbreak of the disease. At one time this was attributed to the persistence of spores in the soil, introduced from infected animals. This does not, however, explain the regular, seasonal outbreaks in some areas (anthrax districts) and the long intervals between infections in others. The spores remain viable in soil for many years, and their persistence does not depend on animal reservoirs so that *B. anthracis* is exceedingly difficult to eradicate from an endemic area. *Bacillus anthracis* continues to be generally regarded as an obligate pathogen. Its continued existence in the ecosystem appears to depend on a periodic multiplication phase within an animal host, with its environmental presence reflecting contamination from an animal source at some time (Lindeque and Turnbull, 1994); however, some authorities believe that self-maintenance may occur within certain soil environments (Cherkasskiy, 1999).

Other Species

Most aerobic endospore forming species are saprophytes widely distributed in the natural environment, but some species are opportunistic or obligate pathogens of animals, including humans, other mammals and insects. *Bacillus anthracis* is, to all intents and purposes, an obligate

pathogen of animals and humans. The habitats of most species are soils of all kinds, ranging from acid through neutral to alkaline, hot to cold and fertile to desert, and the water columns and bottom deposits of fresh and marine waters. Their endospores readily survive distribution in soils, dusts and aerosols from these natural environments to a wide variety of other habitats, and Nicholson *et al.* (2000) have considered the roles of *Bacillus* spores in the natural environment. Some species appear to be ubiquitous contaminants of man, other animals, their foodstuffs, water and environments, natural, domestic, industrial and hospital. Their wide distribution is in part owing to the extraordinary longevity of their endospores, which show much greater resistance to physical and chemical agents, such as heat, cold, desiccation, radiation, disinfectants, antibiotics and other toxic agents, than their counterpart vegetative cells. Endospores are typically more resistant to heat than vegetative cells by a factor of 10^5 or more, while resistance to UV and ionizing radiation may be 100-fold or more. If protected from radiation, spores may survive for very long periods. Striking claims include the isolation of a viable strain of *B. sphaericus* from an extinct bee preserved in 25- to 40-million-year-old amber (Cano and Borucki, 1995), and the recovery of an aerobic endospore former related to *Salibacillus* and *Virgibacillus* from a water droplet trapped in a Permian salt crystal for an estimated 250 million years (Vreeland, Rosenzweig and Powers, 2000); both isolations need to be confirmed by independent laboratories. *Bacillus infernus* was isolated from Triassic shales, lying at depths of around 2.7 km below the land surface, that may have been hydrologically isolated for 140 million years (Boone *et al.*, 1995). Another organism from a deep subsurface environment is *B. subterraneus*, which was isolated from the Great Artesian Basin of Australia (Kanso, Greene and Patel, 2002).

The commonly isolated species, such as *B. subtilis* and *B. cereus*, are very widely distributed worldwide; *B. thuringiensis* has been isolated from all continents, including Antarctica (Forsyth and Logan, 2000). In studies of indoor and outdoor air and dusts, *Bacillus* species commonly dominate the cultivable flora or form a large part of it. The presence of *B. fumarioli* strains showing similar phenotypic behavior and substantial genotypic similarity from Candlemas Island in the South Sandwich archipelago and from volcanoes some 5600 km distant on continental Antarctica is most convincingly explained by their carriage in the air as free spores or spores attached to plant propagules, as no birds are known to visit the latter sites (Logan *et al.*, 2000). Strains indistinguishable from these Antarctic isolates have been cultivated from gelatin production plants in Belgium, France and the USA.

Aerobic endospore formers are often isolated following heat treatment of specimens in order to select for spores, and presence of spores in a particular environment does not necessarily indicate that the organism is metabolically active there; however, it is reasonable to assume that large numbers of endospores in a given environment reflects former or current activity of vegetative cells there. Isolation of organisms showing special adaptations to the environments in which they are found, such as acidophily, alkaliphily, halophily, psychrophily and thermophily, suggests that these organisms must be metabolically active in these niches, but it tells us little about the importances of their roles in the ecosystems, and nothing about their interactions with other members of the flora.

Many species will degrade biopolymers, with versatilities varying according to species, and it is therefore assumed that they have important roles in the biological cycling of carbon and nitrogen; it is further assumed that their activities in food spoilage and biodegradation reflect the contamination of these materials by endospores derived from dusts and other vehicles. Valid though these assumptions may be, the ever-increasing diversity of known species and their apparent primary habitats imply that such generalizations may deserve reconsideration in some cases and that certain species may have quite specialized activities. Although isolates of many of the established species have been derived from soil, or from environments that may have been contaminated directly or indirectly by soil, the range of isolation sources is very wide.

PATHOGENICITY

Bacillus anthracis

The spore is the primary infectious form of the organism, and infection occurs when endospores enter the body from the environment (Hanna and Ireland, 1999). Spores are rapidly phagocytosed by macrophages, some of which undergo lysis. In cases of inhalational anthrax the inhaled spores are carried, from the lungs (where there is no overt infection; thus pulmonary anthrax is a misnomer), by the surviving macrophages along the lymphatics toward the mediastinal lymph nodes, where the infection progresses. Germination and initial multiplication that begin within the macrophages may occur while in transit to the lymph nodes (Hanna and Ireland, 1999), but phagocytosed spores may not germinate for up to 60 days, and so incubation of the inhalational form of the disease may take between 2 days and 6–8 weeks; this latency does not appear to occur in the cutaneous form of the disease. By analogy with other *Bacillus* species, germination is presumed to be triggered by a specific chemical germinant, but this remains unidentified. The elevated CO_2 level and body temperature of the host cause the organism to transcriptionally activate the capsule and toxin genes. These genes are carried on two plasmids: plasmid pX01 encodes the toxin genes and plasmid pX02 encodes the capsule genes. Loss of either plasmid effectively renders the organism avirulent. The capsule of poly-γ-D-glutamic acid is purported to resist phagocytosis by virtue of its negative charge (Ezzell and Welkos, 1999). The anthrax toxin complex comprises three components: edema factor (EF), protective antigen (PA) and lethal factor (LF), none of which is toxic alone. EF and LF are active in binary combinations with PA and have different activities. PA molecules bind to molecules of a particular host cell membrane protein (anthrax toxin receptor or ATR) (Bradley *et al.*, 2001) and form ring-shaped prepores of heptameric oligomers (aggregates of seven); PA is then cleaved and hence activated by a furin-like protease on the surface of the cell under attack. An active PA heptamer can then bind one or more molecules of EF, LF or both. The complex passes into the cell by receptor-mediated endocytosis and into an acidified endosome; following conformational change of the heptamer in the low-pH environment, the complex escapes directly to the cytosol by insertion of the heptamer into the endosomal membrane. The PA–EF binary toxin interacts with the abundant host protein calmodulin (CaM; the major intracellular calcium receptor) and becomes an active adenylyl cyclase in most cell types; this elevates levels of the secretogogue cAMP and leads to hypovolemic shock. The crystal structure of EF in complex with CaM has been elucidated (Drum *et al.*, 2002). The PA–LF binary toxin is a zinc metalloprotease which cleaves members of the mitogen-activated protein kinase kinase family, thereby affecting certain signaling pathways, and levels of shock-inducing cytokines. This toxin primarily affects macrophages, and removal of macrophages from mice using silica renders them insensitive to the toxin (Hanna, 1999); however, the process which leads to macrophage lysis is unclear (Pannifer *et al.*, 2001). The importance of any interaction between EF and LF awaits clarification. Depending on the host, there is a rapid build up of the bacteria in the blood over the last few hours to terminal levels of 10^7–10^9/ml in the most susceptible species. The molecular pathogenesis of infection with *B. anthracis* was reviewed by Little and Ivins (1999).

Bacillus cereus

Bacillus cereus is next in importance to *B. anthracis* as a pathogen of humans (and other animals), causing food-borne illness and opportunistic infections, and its ubiquity ensures that cases are not uncommon. *Bacillus cereus* causes two distinct food-poisoning syndromes, a diarrheal type and an emetic type. Both syndromes arise as a direct result of the fact that *B. cereus* spores can survive normal cooking procedures. Under improper storage conditions after cooking, the spores germinate and the vegetative cells multiply. In diarrheal illness, the toxin(s) responsible is produced by organisms in the small intestine (infective doses 10^4–10^9 cells per gram of food), while the emetic toxin is preformed and ingested in food (about 10^5–10^8 cells per gram in order to produce sufficient toxin). Variations in infective dose of the diarrheal illness are reflections of the proportion of ingested cells that are sporulated and so able to survive the acid barrier of the stomach. The capacity of the strain concerned to produce toxin(s) will, of course, influence the infective or intoxicating dose in both types of illness. It is likely that cases showing both diarrheal and emetic symptoms are caused by organisms producing both diarrheal and emetic toxins. Strains of *B. thuringiensis* commonly carry genes for *B. cereus* enterotoxins (Rivera, Granum and Priest, 2000). The safety of using *B. thuringiensis* as a biopesticide on crop plants has been reviewed by Bishop (2002); Bishop, Johnson and Perani (1999) found that the main pesticide strains that they assayed produced low titers of enterotoxin.

Toxins

The toxigenic basis of *B. cereus* food poisoning and other *B. cereus* infections has begun to be elucidated, and a complex picture is emerging (Beecher, 2001; Granum, 2002). *Bacillus cereus* is known to produce six toxins: five enterotoxins and the emetic toxin. The enterotoxins are: (i) hemolysin BL (Hbl), a 3-component proteinaceous toxin which also has dermonecrotic and vascular permeability activities and causes fluid accumulation in ligated rabbit ileal loops; Hbl is produced by about 60% of strains tested (Granum, 2002), and it has been suggested that it is a primary virulence factor in *B. cereus* diarrhea, but the mechanism of its enterotoxic activity is unclear (Granum, 2002); (ii) nonhemolytic enterotoxin (Nhe) is another 3-component proteinaceous toxin which is produced by most strains tested (Granum, 2002) and whose components show some similarities to Hbl; (iii) and (iv) enterotoxin T (BceT) and enterotoxin FM (EntFM) are single-component proteinaceous toxins whose roles and characteristics are not known; (v) enterotoxin K (EntK) is similar to the β-toxin of *Clostridium perfringens* and was associated with a French outbreak of necrotic enteritis in which three people died (Lund, De Buyser and Granum, 2000). The genetics of toxin production are summarized by Granum (2002).

The emetic toxin, cereulide, is a dodecadepsipeptide comprising a ring of four amino- and oxy-acids: [D-*O*-Leu-L-Ala-L-*O*-Val-L-Val] thrice repeated; it is closely related to the potassium ionophore valinomycin (Agata *et al.*, 1994). It is resistant to heat, pH and proteolysis, but it is not antigenic (Kramer and Gilbert, 1989). Cereulide is probably an enzymatically synthesized peptide rather than a direct genetic product; it is produced in larger amounts at lower incubation temperatures. Its production does not appear to be connected with sporulation (Finlay, Logan and Sutherland, 2000), and it is produced in aerobic and microaerobic but not anaerobic conditions (Finlay, Logan and Sutherland, 2002). Its mechanism of action is unknown, but it has been shown to stimulate the vagus afferent through binding to the 5-HT_3 receptor (Agata *et al.*, 1995). The earliest detection system for emetic toxin involved monkey-feeding tests (Logan *et al.*, 1979), but a semi-automated metabolic staining assay has now been developed (Finlay, Logan and Sutherland, 1999).

Molecular studies are giving valuable insights into genetic profiles among the *B. cereus* group (Keim *et al.*, 2000; Turnbull *et al.*, 2002), and the genomes of *B. anthracis* and *B. cereus* have been sequenced (Ivanova *et al.*, 2003; Read *et al.*, 2003).

EPIDEMIOLOGY

Bacillus anthracis

Life Cycle

The natural cycle of infection is as follows: a grazing animal ingests the spores which may then gain access to the lymphatics, and so to the spleen, through abrasions in the alimentary canal. After several days of the organism multiplying and producing toxin in the spleen, the animal suffers a sudden and fatal septicemia and collapses; hemorrhagic exudates escape from the mouth, nose and anus and contaminate the soil, where the vegetative cells sporulate in the air. In an infected herd of herbivores, not all the animals develop clinical disease, and there is serological evidence that mild and symptomless infections occur (Turnbull *et al.*, 1992a). The spores may remain viable in soil for many years and so the persistence of the organism does not depend on animal reservoirs; *B. anthracis* is therefore exceedingly difficult to eradicate from an endemic area. The organism continues to be generally regarded as an obligate pathogen, with its continued existence in the ecosystem appearing to depend on a periodic multiplication phase within an animal host and its environmental presence reflecting contamination from an animal source at some time. Excepting cases of carnivorous scavenging of meat from anthrax carcasses, direct animal-to-animal transmission within a species is very rare; however, such scavengers can show transient carriage of the organism (Lindeque and Turnbull, 1994).

Natural Epidemics

Anthrax is a disease of great antiquity, and descriptions are found in ancient literature such as Virgil's *Georgics* (36–29 BC). Although it is primarily a disease of herbivores, the disease appears to derive its name from the black eschar associated with human cutaneous infections, *anthrax* being the Greek word for coal. Before an effective veterinary vaccine became available in the late 1930s, it was one of the foremost causes of mortality in cattle, sheep, goats and horses worldwide. An outbreak in Iraq in 1945 killed 1 million sheep. The development and application of veterinary and human vaccines together with improvements in factory hygiene and sterilization procedures for imported animal products, and the increased use of artificial alternatives to animal hides or hair, have resulted over the past half century in a marked decline in the incidence of the disease in both animals and humans. Sites where these natural materials were formerly handled, such as disused tanneries, may be sources of infection when they are disturbed during redevelopment. Likewise, anthrax carcasses can remain infectious for many years, even when buried with quicklime. Nonendemic regions must be constantly on the alert for the arrival of *B. anthracis* in imported products of animal origin; the disease continues to be endemic in several countries of Africa, Asia and central and southern Europe, particularly those that lack an efficient vaccination policy.

Human Infection

Apart from artificial attacks, humans almost invariably contract anthrax directly or indirectly from animals. It is a point-source type of disease, and direct human-to-human transmission is exceptionally rare. Indeed, circumstantial evidence shows that humans are moderately resistant to anthrax as compared with obligate herbivores; infectious doses in the human inhalational and intestinal forms are generally very high (LD$_{50}$ 2500–55 000 spores). Naturally-acquired human anthrax may result from close contact with infected animals or their carcasses after death from the disease or be acquired by those employed in processing wool, hair, hides, bones or other animal products. Most cases (about 99%) are cutaneous infections, but

B. anthracis meningitis and gastrointestinal anthrax are occasionally reported. In industrial settings, inhalation of spore-laden dust used to occur, but cases of clinical disease were rare.

Bioterrorism

Anthrax weapons are normally intended to cause the inhalational form but are likely to cause cutaneous cases as well.

In public consciousness, *B. anthracis* is associated more with warfare and terrorism than with a disease of herbivores, and it is feared accordingly. It has long been considered a potential agent for biological warfare or bioterrorism. Its first use is believed to have been against livestock during World War I (Christopher *et al.*, 1997; Barnaby, 2002). It has been included in various development and offensive programmes in several countries since (Zilinskas, 1997; Mangold and Goldberg, 1999; Miller, Engelberg and Broad, 2001; Barnaby, 2002; Alibek, 2003), and it has also been used in terrorist attacks (Christopher *et al.*, 1997; Lane and Fauci, 2001). There have been a few reports of laboratory-acquired infections, none of them recent (Collins, 1988), but a major outbreak of anthrax occurred in April 1979 in the city of Sverdlovsk (now Yekaterinburg) in the Urals as a result of the accidental release of spores from a military production facility; 77 cases were recorded, and 66 patients died (Meselson *et al.*, 1994).

Bacillus cereus

The diarrheal type of food-borne illness is characterized by abdominal pain with diarrhea 8–16 h after ingestion of the contaminated food and is associated with a diversity of foods from meats and vegetable dishes to pastas, desserts, cakes, sauces and milk. The emetic type of illness, a true food poisoning, is characterized by nausea and vomiting 1–5 h after eating the offending food, predominantly oriental rice dishes, although occasionally other foods such as pasteurized cream, milk pudding, pastas and reconstituted infant-feed formulas have been implicated. One outbreak followed the mere handling of contaminated rice in a children's craft activity (Briley, Teel and Fowler, 2001). Cases may occur at any time of year – as would be expected, as they are caused by improper food handling – and without any particular distribution pattern within a country. Between countries, however, there are differences in the proportions of all outbreaks of food-borne disease that are caused by *B. cereus*, but it is probable cases are underreported.

Other *B. cereus* infections occur mainly, though not exclusively, in persons predisposed by extremes of age, trauma, neoplastic disease, immunosuppression, alcoholism and other drug abuse, or some other underlying condition, and fatalities occasionally result. In the case of posttraumatic endophthalmitis, *Bacillus* species, particularly *B. cereus*, may be the second most commonly isolated organisms after *Staphylococcus epidermidis*. Endophthalmitis may follow penetrating trauma of the eye or hematogenous spread, evolving very rapidly. Loss of both vision and the eye is likely if appropriate treatment is instituted too late (Davey and Tauber, 1987).

Other Species

Opportunistic infections with aerobic endospore formers have been reported since the late nineteenth century. Given the ubiquity of these organisms, infections with them might be commonplace were it not for their low invasiveness and virulence. It is certain that many clinically significant isolates of *Bacillus* were disregarded in the past, and the perceived difficulties of identification may have contributed to this. Although awareness is now better than ever, the problem remains one of recognizing what is a contaminant and what is not. A common feature of case reports, right up to the present, has been emphasis on the importance of interpreting spore-forming isolates in light of any other species cultured and the clinical context and the danger of dismissing them as mere contaminants. The fact that individual case histories of infections with these organisms continue to warrant published reports indicates that incidents are rare and that authors feel that awareness needs to be improved. The earlier literature was reviewed by Norris *et al.* (1981) and Logan (1988). The factors contributing to the increasing frequency with which opportunistic infections are encountered are well known: host predisposition by suppressed or compromised immunity; malignant disease or metabolic disorder; exposure by accidental trauma and clinical and surgical procedures; drug abuse and advances in bacteriological technique, interpretation and awareness. It seems unlikely that these organisms have changed in virulence over the last century. Although reports of infections with non-*B. cereus* group species are comparatively rare, they are very diverse, and there have been several hospital pseudoepidemics associated with contaminated blood culture systems.

CLINICAL FEATURES

Bacillus anthracis

Cutaneous Infection

The organism is not invasive, and cutaneous infections occur through breaks in the skin; so the lesions generally occur on exposed regions of the body. This may include the eyelids. Before the availability of antibiotics and vaccines, 10–20% of untreated cutaneous cases were fatal, and the rare fatalities seen today are due to obstruction of the airways by the edema that accompanies lesions on the face or neck, and sequelae of secondary cellulitis or meningitis. The incubation period in cutaneous anthrax is generally 2–3 days (with extremes of approximately 12 h to 2 weeks). Results from a study of cutaneous anthrax cases in Turkey indicate that these occur in decreasing order of frequency on hands and fingers, eyelids and face, wrists and arms, feet and legs and neck (Doganay, 1990). A small pimple or papule appears, and over the next 24 h, a ring of vesicles develops around it, and it ulcerates, dries and blackens into the characteristic necrotic eschar (Figure 9.3a). Differential diagnoses include other cutaneous and subcutaneous infections (Hart and Beeching, 2002). The eschar enlarges, becoming thick and adherent to underlying tissues over the ensuing week, and it is surrounded by edema, which may be very extensive (Figure 9.3b). Pus and pain are normally absent; their presence, or the presence of marked lymphangitis and fever, probably indicates secondary bacterial infection. Historical records show mortality rates of approximately 20% in untreated cutaneous cases.

Inhalational Infection

Inhalational (formerly called pulmonary) anthrax cases are more often fatal than cutaneous cases, because they go unrecognized until too late for effective therapy, but undiagnosed, low-grade infections with recovery may occur. Analysis of ten of the cases associated with the bioterrorist events of 2001, in which spores were delivered in mailed letters and packages (Jernigan *et al.*, 2001), revealed a median incubation period of 4 days (range 4–6 days) and a variety of symptoms at initial presentation including fever or chills, sweats, fatigue or malaise, minimal or nonproductive cough, dyspnea and nausea or vomiting. All patients had abnormal chest X-rays with infiltrates, pleural effusion and mediastinal widening. Mediastinal lymphadenopathy was present in seven cases. The number of recorded cases of inhalational anthrax is lower than that might be expected from the high profile given to this condition. In the 20th century there were just 18 reported cases (two of them being laboratory acquired) in the USA, 16 (88.9%) fatal (Brachman and Kaufmann, 1998); figures in the UK

showed a similar picture. In 11 confirmed cases of inhalational anthrax that followed the bioterrorist attack in the USA, early recognition and treatment helped a survival level of 55% to be achieved (Jernigan *et al.*, 2001; Bell, Kozarsky and Stephens, 2002). It is possible that undiagnosed, low-grade inhalational infections with recovery may occur.

Gastrointestinal Infection

Oropharyngeal and gastrointestinal anthrax are not uncommon in endemic regions of the world where socioeconomic conditions are poor, and people eat the meat of animals that have died suddenly; such cases are greatly underreported. They are essentially cutaneous anthrax occurring on the oropharyngeal and gastrointestinal mucosas. Symptoms of gastroenteritis may occur prior to onset of systemic symptoms. Gastrointestinal infections are mainly characterized by gastroenteritis, and asymptomatic infections and symptomatic infections with recovery are not uncommon. The symptoms of oropharyngeal infections are fever, toxemia, inflammatory lesions in the oral cavity and oropharynx, cervical lymph node enlargement and edema, and there is a high case-fatality rate (Sirisanthana and Brown, 2002).

Meningitis

Meningitis can develop from any of the forms of anthrax. The emergence of clinical signs is rapidly followed by unconsciousness, and the prognosis is poor. Outbreaks of primary anthrax meningoencephalitis have been reported from India and elsewhere (George *et al.*, 1994; Kwong *et al.*, 1997).

In fatal cases of any of the human forms of anthrax, the generalized symptoms may be mild (fatigue, malaise, fever and/or gastrointestinal symptoms) and are followed by sudden onset of acute illness with dyspnoea, cyanosis, severe pyrexia and disorientation followed by circulatory failure, shock, coma and death, all within a few hours.

Bacillus cereus

Enteric Disease

The diarrheal, or long incubation, form of *B. cereus* food poisoning has an onset period of 8–16 h, followed by abdominal cramps, profuse watery diarrhea and rectal tenesmus. Occasionally there is fever and vomiting. Thus the symptoms are similar to those of *C. perfringens* food poisoning. Recovery is usually complete in 24 h, although there have been two reports of death (Logan, 1988). Emetic food poisoning due to *B. cereus* is characterized by a shorter onset period of 1–6 h, and the symptoms resemble those of *S. aureus* food poisoning. Nausea, vomiting and malaise, occasionally with diarrhea, characterize the disease. There are rarely complications, and recovery within 24 h is usual, but fulminant liver failure associated with the emetic toxin has been reported (Mahler *et al.*, 1997).

Ocular Disease

Bacillus cereus is one of the most virulent and destructive of ocular pathogens. The most serious of these conditions is panophthalmitis, a rapidly developing infection which may follow penetrating trauma of the eye (commonly by a metal fragment in an environment such as a farm or a garage), intraocular surgery, or hematogenous dissemination of the organism from another site (typically in intravenous drug abusers). Either way, the condition usually evolves so rapidly that irreversible damage occurs before effective treatment can be started; vision is therefore lost, and loss of the eye is normal. *Bacillus cereus* keratitis associated with contact lens wear has also been reported (Pinna *et al.*, 2001).

(a)

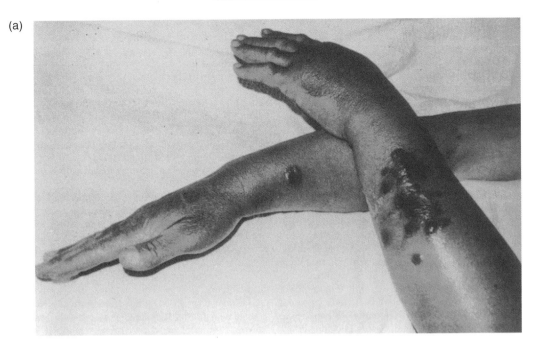

(b)

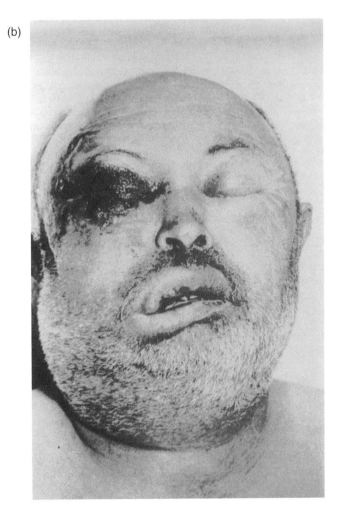

Figure 9.3   (a) Anthrax infection of the forearms. (Photograph kindly supplied by Professor Mehmet Doganay, Erciyes Universitesi, Kaysari, Turkey.) (b) Anthrax infection of the eye. The characteristic black eschar on the upper lid, particularly, is evident. The eye itself is usually not damaged. Edema of the right side of the face can also be seen. (Photograph kindly supplied by Professor Mehmet Doganay, Erciyes Universitesi, Kaysari, Turkey.)

Other Conditions

Other reported conditions associated with *B. cereus* infection include bacteremia, septicemia, fulminant sepsis with hemolysis, meningitis, brain hemorrhage, ventricular shunt infections, infections associated with central venous catheters, endocarditis (which may be associated with prosthetic valves or intravenous drug abuse), pseudomembranous tracheobronchitis, pneumonia, empyema, pleurisy, lung abscess, brain abscess, osteomyelitis, salpingitis, urinary tract infection and primary cutaneous infections. Wound infections, mostly in otherwise healthy persons, have been reported following surgery, road traffic and other accidents, scalds, burns, plaster fixation, drug injection (including a case associated with contaminated heroin) (Dancer *et al.*, 2002) and close-range gunshot and nail-bomb injuries; some became necrotic and gangrenous. A fatal inflammation was caused by a blank firearm injury; blank cartridge propellants are commonly contaminated with the organism (Rothschild and Leisenfeld, 1996). Neonates also appear to be particularly susceptible to *B. cereus*, especially with umbilical stump infections; respiratory tract infections associated with contaminated ventilation systems have also occurred (Van der Zwet *et al.*, 2000); other infections reported in neonates include intestinal perforation, meningoencephalitis and bacteremia refractory to therapy. There have been reports of wound, burn and ocular infections with *B. thuringiensis* (Damgaard *et al.*, 1997), but there is as yet no evidence of infections associated with the use of this organism as an insecticide. *Bacillus cereus* also causes infections in domestic animals. It is a well-recognized agent of mastitis and abortion in cattle and can cause these conditions in other livestock (Blowey and Edmondson, 1995).

Bacillus thuringiensis is a close relative of *B. cereus* and may produce the *B. cereus* diarrheal toxin, and it has indeed been implicated in cases of gastroenteritis (Damgaard *et al.*, 1997). There have also been reports of wound, burn and ocular infections with *B. thuringiensis* (Damgaard *et al.*, 1997). Cases of illness caused by *B. thuringiensis* may have been diagnosed as caused by *B. cereus*, as the former may not produce its characteristic insecticidal toxin crystals when incubated at 37 °C, owing to the loss of the plasmids carrying the toxin genes.

Other Species

In addition to *B. cereus*, a wide range of other aerobic endospore-forming bacteria have been isolated in association with endophthalmitis (Das *et al.*, 2001). *Bacillus licheniformis* has been reported from ventriculitis following the removal of a meningioma, cerebral abscess after penetrating orbital injury (Jones, Hanson and Logan, 1992), septicemia following arteriography, bacteremia associated with indwelling central venous catheters (Blue, Singh and Saubolle, 1995), bacteremia during pregnancy with eclampsia and acute fibrinolysis, peritonitis in a CAPD patient and in a patient with volvulus and small-bowel perforation, ophthalmitis and corneal ulcer after trauma. There have also been reports of L-form organisms, phenotypically similar to *B. licheniformis*, occurring in blood and other body fluids of patients with arthritis, patients with neoplasms, clinically normal persons, and in association with infectious synovitis in birds. Although some authors have claimed a relationship between these organisms and diseases with postulated immunological elements, and higher isolations from the synovial fluids and membranes of arthritic patients have been reported, Bartlett and Bisset (1981) were unable to confirm the latter association. *Bacillus licheniformis* can cause food-borne diarrheal illness and has been associated with an infant fatality (Mikkola *et al.*, 2000). A toxin possibly associated with *B. licheniformis* food poisoning has been identified (Mikkola *et al.*, 2000), but in general, toxins or virulence factors widely accepted as responsible

for symptoms periodically associated with *Bacillus* species other than those in the *B. cereus* group have not been identified. *Bacillus licheniformis* is frequently associated with bovine abortion and has been reproduced by experimental infection of cows, which demonstrated the tropism of the organism for the bovine placenta (Agerholm *et al.*, 1997); this species has also been associated with abortion in water buffalo (Galiero and De Carlo, 1998) and is occasionally associated with bovine mastitis (Blowey and Edmondson, 1995). Many of these types of *B. licheniformis* and *B. cereus* infections are associated with wet and dirty conditions during winter housing, particularly when the animals lie in spilled silage (Blowey and Edmondson, 1995); in one outbreak, a water tank contaminated with *B. licheniformis* was implicated (Parvanta, 2000).

The name *B. subtilis* was often used to mean any aerobic, endospore-forming organism, but since 1970, there have been reports of infection in which identification of this species appears to have been made accurately. Cases associated with neoplastic disease include fatal pneumonia and bacteremia, a septicemia and an infection of a necrotic axillary tumor in breast cancer patients. Breast prosthesis and ventriculoatrial shunt infections, endocarditis in a drug abuser, meningitis following a head injury, cholangitis associated with kidney and liver disease, isolations from dermatolymphangioadenitis associated with filarial lymphedema (Olszewski *et al.*, 1999) and surgical wound drainage sites have also been reported. A probiotic preparation of this species led to a fatal septicemia in an immunocompromised patient (Oggioni *et al.*, 1998). *Bacillus subtilis* has also been implicated in food-borne illness and cases of bovine mastitis and of ovine abortion (Logan, 1988).

Organisms identified as *B. circulans* have been isolated from cases of meningitis, a cerebrospinal fluid shunt infection, endocarditis, endophthalmitis (Tandon *et al.*, 2001), a wound infection in a cancer patient and a bite wound. Roy *et al.* (1997) reported epidemic endophthalmitis associated with isolates identified as *B. circulans* that contaminated a product used during cataract surgery. It must be noted, however, that many isolates previously identified as *B. circulans* might have been misallocated (see comments on this species in *Descriptions of Organisms* above). *Bacillus coagulans* has been isolated from corneal infection, bacteremia and bovine abortion. *Bacillus pumilus* has been found in cases of pustule and rectal fistula infection, and in association with bovine mastitis. *Bacillus sphaericus* has been implicated in a fatal lung pseudotumor and a case of meningitis and has been associated with food-borne illness (Turnbull and Kramer, 1995). Among 18 cancer patients with 24 bacteremic episodes, Banerjee *et al.* (1988) isolated *B. cereus* (eight cases), *B. circulans* (three cases), *B. subtilis* (two cases), *B. coagulans* (one case), *B. licheniformis* (one case), *B. pumilus* (one case), *B. sphaericus* (one case) and six unidentified aerobic endospore formers.

Bacillus brevis has been isolated from corneal infection and has been implicated in several incidents of food poisoning; since these reports, the species was split and transferred to the new genus *Brevibacillus*. Strains of the new species, *B. agri*, have been isolated in association with an outbreak of water-borne illness in Sweden (Logan *et al.*, 2002), and other *Brevibacillus* species have been found in human blood and bronchioalveolar lavage specimens. *Brevibacillus laterosporus* has been reported in association with a severe case of endophthalmitis.

Paenibacillus alvei has been isolated from cases of meningitis, endophthalmitis, a prosthetic hip infection in a patient with sickle cell anemia, a wound infection and, in association with *C. perfringens*, a case of gas gangrene. *Paenibacillus popilliae* has been reported from endocarditis (Wu *et al.*, 1999). *Paenibacillus macerans* has been isolated from a brain abscess following penetrating periorbital injury, a wound infection following removal of a malignant melanoma, catheter-associated infection in a leukemic patient and from bovine abortion, and *P. polymyxa* has been isolated from ovine abortion.

LABORATORY DIAGNOSIS

Safety

Bacillus anthracis

Anthrax is not highly contagious. Cutaneous anthrax is readily treated and is only life threatening in exceptional cases, and the infectious doses in the human inhalational and intestinal forms (also treatable if recognized early) are generally very high. The LD_{50} is of the order of 10 000 spores, although in the bioterrorist events of 2001 it would appear that at least some of the 11 inhalational cases resulted from exposures to considerably less than this figure. In general, precautions need to be sensible, not extreme. When collecting specimens related to suspected anthrax, disposable gloves, disposable apron or overalls and boots which can be disinfected after use should be worn. For dusty samples that might contain many spores, the use of headgear and dust masks should be considered.

Disposable items should be discarded into suitable containers for autoclaving followed by incineration. Nonautoclavable items should be immersed overnight in 10% formalin (5% formaldehyde solution; 5% glutaraldehyde is also effective). Items that cannot be immersed should be bagged and sent for formaldehyde fumigation. Ethylene oxide and hydrogen peroxide vapor are also effective fumigants, but the latter is inappropriate if organic matter is being treated. The best disinfectant for specimen spillages is again formalin, but where this is considered impractical, 10% hypochlorite solution can be used, but it is rapidly neutralized by organic matter and it corrodes metals. Other strong oxidising agents, such as hydrogen peroxide (5%) and peracetic acid (1%), are also effective but with the same limitations with respect to organic matter. For further discussion, see Turnbull et al. (1998). Isolation and presumptive identification of B. anthracis can be performed safely in the routine clinical microbiology laboratory, provided that normal good laboratory practice is observed; vaccination is not required for minimal handling of the organism. If aerosols are likely to be generated, the work should be performed in a safety cabinet.

In the case of suspected bioterrorism the laboratory examination of clinical specimens will be as described above for cases of naturally-acquired anthrax, but nonclinical materials may be very hazardous and no attempt to process them should be made without the appropriate instructions from the correct authorities.

Other Species

Clinical specimens for isolation of other aerobic endospore-forming species need no special precautions and can be collected, transported and cultivated in the normal way and handled safely on the open bench.

Specimens and Isolation

Bacillus anthracis

In specimens submitted for isolation of B. anthracis from old carcasses, animal products or environmental specimens, the organisms will mostly be present as spores, and these may be selected for by heat treatment (see Heat treatment below). There is no effective enrichment method for B. anthracis in old animal specimens or environmental samples; isolation from these is best done with polymyxin-lysozyme EDTA-thallous acetate (PLET) agar (Knisely, 1966). Aliquots (250 μl) of the undiluted and 1:10 and 1:100 dilutions of heat-treated suspension of the specimen are spread across PLET plates which are read after incubation for 36–40 h at 37 °C. Colonies of B. anthracis are smaller and smoother than on plain heart infusion agar, and B. cereus is generally inhibited by the medium. To make the PLET medium, heart infusion agar (or heart infusion broth with other agar base) is prepared according to the manufacturer's instructions. EDTA (0.3 g/l) and thallous acetate (0.04 g/l) are added before autoclaving. (Note: thallous acetate is poisonous; avoid skin contact and inhalation by wearing gloves and weighing out the powder in a fume cupboard.) After autoclaving at 121 °C for 15 min, the agar is cooled to 50 °C and polymyxin (30 000 units/l) and lysozyme (300 000 units/l) are added in filter-sterilized solutions. It is important that the agar is left at 50 °C for long enough to ensure that this temperature has been reached throughout the medium before the polymyxin and lysozyme are added. After swirling to ensure even suspension of the ingredients, the agar is poured into Petri dishes.

Roughly circular, creamy-white colonies, 1–3 mm diameter, with a ground-glass texture are subcultured (i) on blood agar plates to test for γ phage and penicillin susceptibility, and for hemolysis, and (ii) directly or subsequently in blood (or bicarbonate agar, see subsection Bacillus anthracis, under Diagnosis, below) to look for capsule production using M'Fadyean's stain or by indirect fluorescent antibody capsule staining. PCR-based methods are being used increasingly for confirming the identity of isolates (see subsection Bacillus anthracis, under Diagnosis, below).

In all human anthrax cases, specimens from possible sources of the infection (carcass, hides, hair, bones, etc.) should be sought in addition to specimens from the patients themselves. Swabs are appropriate for collecting vesicular fluid from cutaneous cases. Adequate material should be submitted for both culture and a smear for visualizing the capsule. Intestinal anthrax will be suspected only if an adequate history of the patient is known. If the patient is not severely ill, a fecal specimen may be collected, but isolation of B. anthracis may not be successful. If the patient is severely ill, blood should also be cultured, although isolation may not be possible after antimicrobial treatment, and such treatment should not await laboratory results. A blood smear may reveal the capsulated bacilli or, if treatment has started, capsule ghosts.

Postmortem blood may be collected by venipuncture, because it is characteristic of anthrax that the blood does not clot at death, and it should be examined by smear (for capsule) and culture. Any hemorrhagic fluid from nose, mouth or anus should be cultured, and if these are positive, no further specimens are required; if they are negative, specimens of peritoneal fluid, spleen and/or mesenteric lymph nodes, aspirated by techniques avoiding spillage of fluids, may be collected.

Inhalational anthrax will only be suspected if the patient's history suggests it. Unless the patient is severely ill, immediate specimen collection is likely to be unfruitful and the person should be treated and simply observed. Paired sera (when first seen and >10 days later) may be collected for subsequent confirmation of diagnosis. In the severely ill patient, blood smear and culture are necessary, but results will depend somewhat on any previous antibiotic treatment. The approach given for intestinal anthrax should be followed for postmortem cases.

Other Species

All the clinically significant isolates reported to date are of species that grow, and often sporulate, on routine laboratory media at 37 °C. It seems unlikely that many clinically important, but more fastidious, strains are being missed for the want of special media or growth conditions.

Enrichment procedures are generally inappropriate for isolations from clinical specimens, but when searching for B. cereus in stools ≥3 days after a food-poisoning episode, nutrient or tryptic soy broth with polymyxin (100 000 U/l) may be added to the heat-treated specimen.

Several media have been designed for isolation, identification and enumeration of B. cereus, and some of these are available commercially. They exploit the organism's egg-yolk reaction positivity and acid-from-mannitol negativity; pyruvate and polymyxin may be included for selectivity. Three satisfactory formulations are MEYP (mannitol, egg yolk, polymyxin B agar), PEMBA (polymyxin B, egg yolk, mannitol, bromthymol blue Agar) and BCM (*B. cereus* medium) (van Netten and Kramer, 1992). There are no selective media for other bacilli, but spores can be selected for by heat treating part of the specimen, as described below; the vegetative cells of both spore formers and nonsporeformers will be killed, but the heat-resistant spores will not only survive but may be heat-shocked into subsequent germination. The other part of the specimen is cultivated without heat treatment in case spores are very heat sensitive or absent. Heat treatment is not appropriate for fresh clinical specimens, where spores are usually sparse or absent.

Heat Treatment

For specimens with mixed microflora, heating at 62.5 °C for 15 min will both heat-shock the spores and effectively destroy nonspore-forming contaminants [solid samples should first be emulsified in sterile, deionized water, 1:2 (w/v)]. Direct plate cultures are made on blood, nutrient or selective agars, as appropriate, by spreading up to 250 µl volumes from undiluted and tenfold and 100-fold dilutions of the treated sample.

Maintenance

Maintenance is simple if spores can be obtained, but it is a mistake to assume that a primary culture or subculture on blood agar will automatically yield spores if stored on the bench or in the incubator. It is best to grow the organism on nutrient agar or trypticase soy agar containing 5 mg/l of manganese sulphate for a few days and refrigerate when microscopy shows that most cells have sporulated. For most species, sporulated cultures on slants of this medium, sealed after incubation, can survive in a refrigerator for years. Alternatively, cultures (preferrably sporulated) can be frozen at −70 °C or lyophilized.

Diagnosis

It is important to appreciate that these organisms do not always stain Gram positive. Although Gram staining is considered to be of limited value in anthrax diagnosis, because it does not reveal the capsule, it was found to be useful during the American bioterrorism attack in 2001. In a well-developed country it is unlikely that large numbers of Gram-positive bacteria in the blood at death are going to be anything but *B. anthracis*, particularly when this is supported by the recent history of events. Before attempting to identify to species level it must be confirmed that the isolate really is an aerobic endospore former and that other inclusions are not being mistaken for spores. A Gram-stained smear showing cells with unstained areas suggestive of spores can be stripped of oil with acetone/alcohol, washed, and then stained for spores. Spores are stained in heat-fixed smears by flooding with 10% aqueous malachite green for up to 45 min (without heating), followed by washing and counterstaining with 0.5% aqueous safranin for 30 s; spores are green within pink or red cells at ×1000 magnification. Phase contrast (at ×1000 magnification) should be used if available, as it is superior to spore staining and more convenient. Spores are larger, more phase-bright and more regular in shape, size and position than other kinds of inclusion such as polyhydroxybutyrate (PHB) granules (Figure 2e), and sporangial appearance is valuable in identification. Details of colonial and sporangial morphologies of the more frequently encountered species are given above (see *Descriptions of Organisms*).

Members of the *B. cereus* group and *B. megaterium* will produce large amounts of storage material when grown on carbohydrate media, but on routine media this vacuolate or foamy appearance is rarely sufficiently pronounced to cause confusion. Isolates of other organisms have often been submitted to reference laboratories as *Bacillus* species because they were large, aerobic Gram-positive rods, even though sporulation had not been observed, or because PHB granules or other storage inclusions had been mistaken for spores.

The most widely used diagnostic schemes use traditional phenotypic tests (Gordon, Haynes and Pang, 1973) or miniaturized tests in the API 20E and 50CHB Systems (Logan and Berkeley, 1984; Logan, 2002; bioMérieux, Marcy l'Etoile, France). The API 20E/50CHB kits can be used for the presumptive distinction of *B. anthracis* from other members of the *B. cereus* group within 48 h. bioMérieux also offers a *Bacillus* card for the Vitek automated identification system. As many new species have been proposed since these schemes were established, updated API and Vitek databases need to be prepared. Biolog (Biolog Inc., Hayward, CA, USA) also offers a *Bacillus* database. The effectiveness of such kits can vary with the genera and species of aerobic endospore formers concerned, but they are improving with continuing development and increased databases (Logan, 2002). It is stressed that their use should always be preceded by the basic characterization tests described below.

Other approaches include chemotaxonomic fingerprinting by fatty acid methyl ester (FAME) profiling, polyacrylamide gel electrophoresis (PAGE) analysis, pyrolysis mass spectrometry and Fourier-transform infrared spectroscopy. All these approaches have been successfully applied either across the genera or to small groups. As with genotypic profiling methods, large databases of authentic strains are necessary; some of these are commercially available, such as the Microbial Identification System software (Microbial ID. Inc., Newark, DE, USA) database for FAME analysis.

For diagnostic purposes, the aerobic endospore formers comprise two groups, the reactive ones that will give positive results in various routine biochemical tests, and which are therefore easier to identify, and the nonreactive ones which give few if any positive results in such tests. Nonreactive isolates may belong to the genera *Brevibacillus* or *Aneurinibacillus*, or be members of the *B. sphaericus* group or *B. badius*.

Bacillus anthracis

It is generally easy to distinguish virulent *B. anthracis* from other members of the *B. cereus* group. An isolate with the correct colonial morphology, white or grey in colour, nonhemolytic or only weakly hemolytic, nonmotile, susceptible to the diagnostic γ phage and penicillin and able to produce the characteristic capsule (as shown by M'Fadyean staining or indirect fluorescent antibody staining) is *B. anthracis*. (Enquiries about γ phage and indirect fluorescent antibody capsule staining should be addressed to the Diagnostics Systems Division, USAMRIID, Fort Detrick, Frederick, MD 21702–5011, USA.)

An isolate with the characteristic phenotype but unable to produce capsules may be an avirulent form lacking either its capsule or toxin genes or both (Turnbull *et al.*, 1992b) and should be referred to a specialist laboratory; such isolates are generally found in environmental samples, but they may be identified as *B. cereus* and discarded by routine laboratories.

The capsule of virulent *B. anthracis* can be demonstrated on nutrient agar containing 0.7% sodium bicarbonate incubated overnight under 5–7% CO_2 (candle jars perform well). Colonies of the capsulated *B. anthracis* appear mucoid, and the capsule can be visualized by staining smears with M'Fadyean's polychrome methylene blue or India ink (Turnbull *et al.*, 1998) or by indirect fluorescent antibody staining. More simply, 2.5 ml of blood (defibrinated horse blood seems best; horse or fetal calf serum are quite good) can be inoculated with a pinhead quantity of growth from the suspect colony, incubated statically for 6–18 h at 37 °C and M'Fadyean stained. For M'Fadyean staining, make a thin smear from the specimen, and also from a positive control, on a clean slide and allow to dry. Fix by immersion in

95% or absolute alcohol for 30–60 s. Put a large drop (approximately 50 μl) of polychrome methylene blue (M'Fadyean stain) on the smear, and ensure all the smear is covered by spreading the stain with an inoculating loop (flooding the slide is wasteful, unnecessary and ecologically undesirable). Leave for 1 min and wash the stain off with water (into 10% hypochlorite solution). Blot the slide and allow to dry. At ×100–400 magnification, the organisms can be seen as fine short threads; at ×1000 magnification (oil immersion), if virulent *B. anthracis* is present, the capsule should be seen as a clearly demarcated zone around the blue-black, often square-ended rods which lie in short chains of two to a few cells in number. The positive control can be prepared by culturing a virulent strain in defibrinated horse blood as described above. The M'Fadyean stain is preferable to other capsule staining methods, as it is more specific for *B. anthracis* capsules. It dates from 1903 and has proved a remarkably successful rapid diagnostic test since. However, reliable stain and adequate quality control of its performance are becoming hard to guarantee.

Primer sequences are now available for confirming the presence of the toxin and capsule genes (Jackson *et al.*, 1998) and hence the virulence of an isolate. Real-time PCR assays, which use primer and fluorescently-labeled probe systems to allow the rapid and sensitive detection of genes specific for *B. anthracis*, have been developed in several laboratories. Makino and Cheun (2003) described an assay that targeted genes for capsule and PA and allowed a single spore to be detected in 100 l of air in 1 h. Drago *et al.* (2002) outlined an assay targeting fragments of a chromosomal gene (*rpo*) for detecting the organism in clinical samples. Hoffmaster *et al.* (2002) evaluated and validated a three-target assay, with primers for capsule, PA and *rpo*, in order to test suspect isolates and to screen environmental samples during the outbreak that followed the 2001 bioterrorist attack in the USA.

Monoclonal antibody tests based on specific spore cortex or cell wall epitopes, for reliable differentiation of *B. anthracis* from close relatives, are not yet available. King *et al.* (2003) assessed three commercially available immunoassays for the direct detection of *B. anthracis* spores but found that they were limited by their sensitivities, needing at least 10^5–10^6 spores per sample. Breakthroughs have recently been made in the long-standing problem of differentiating strains of *B. anthracis* for epidemiological and strategic purposes using amplified fragment length polymorphism (AFLP) analysis (Jackson *et al.*, 1998; Keim *et al.*, 2000).

The three protein components of anthrax toxin [protective antigen (PA), lethal factor (LF) and edema factor (EF)] and antibodies to them can be used in enzyme immunoassay systems. For routine confirmation of anthrax infection, or for monitoring response to anthrax vaccines, antibodies against protective antigen alone appear to be satisfactory; they have proved useful for epidemiological investigations in humans and animals. In human anthrax, however, early treatment sometimes prevents antibody development (Turnbull *et al.*, 1992a).

Schuch, Nelson and Fischetti (2002) found that the PlyG lysin of γ phage may be used to detect *B. anthracis* by luminescence and that the same lysin could kill vegetative cells and germinating spores.

Bacillus cereus

A strain differentiation system for *B. cereus* based on flagellar (H) antigens is available at the Food Hygiene Laboratory, Central Public Health Laboratory, Colindale, London, UK, for investigations of food-poisoning outbreaks or other *B. cereus*-associated clinical problems. *Bacillus thuringiensis* strains are also classified on the basis of their flagellar antigens; 82 serovars have been recognized. This is done at the Pasteur Institute, Paris, France and at Abbott Laboratories, North Chicago, IL, USA.

The enterotoxin complex responsible for the diarrheal type of *B. cereus* food poisoning has been increasingly well characterized (Granum, 2002). Two commercial kits are available for its detection

in foods and feces, the Oxoid BCET-RPLA (Oxoid Ltd, Basingstoke, UK; Product Code TD950) and the TECRA VIA (TECRA Diagnostics, Roseville, NSW, Australia; Product Code BDEVIA48). However, these kits detect different antigens, and there is some controversy about their reliabilities. Other assays, based on tissue culture, have also been developed (Fletcher and Logan, 1999). The emetic toxin of *B. cereus* has been identified, and it may be assayed in food extracts or culture filtrates using HEp-2 cells (Finlay, Logan and Sutherland, 1999). A commercial kit for detecting the emetic toxin is under development. Restriction fragment length polymorphism (RFLP) analysis of *B. cereus* toxin genes or their fragments is of potential value in epidemiological investigations (Schraft *et al.*, 1996; Mäntynen and Lindström, 1998; Ripabelli *et al.*, 2000).

ANTIMICROBIAL SENSITIVITY AND MANAGEMENT OF INFECTION

Bacillus anthracis

Most strains of *B. anthracis* are susceptible to penicillin, and until recently there were few authenticated reports of resistant isolates (Lalitha and Thomas, 1997); consequently this antibiotic was the mainstay of treatment, and there were few studies on the organism's sensitivity to other antibiotics. Mild and uncomplicated cutaneous infections may be treated with oral penicillin V, but the treatment usually recommended is intramuscular procaine penicillin or benzyl penicillin (penicillin G). In severe cases, and gastrointestinal and inhalational infections, the recommended therapy has been penicillin G by slow intravenous injection or infusion until the fever subsides, followed by intramuscular procaine penicillin; the organism is normally susceptible to streptomycin, which may act synergistically with penicillin (Turnbull *et al.*, 1998). The use of an adequate dose of penicillin is important, as Lightfoot, Scott and Turnbull (1990) found that strains grown in the presence of subinhibitory concentrations of flucloxacillin *in vitro* became resistant to penicillin and amoxicillin. The study of Lightfoot, Scott and Turnbull (1990) on 70 strains and that of Doganay and Aydin (1991) on 22 isolates found that most strains were sensitive to penicillins, with minimal inhibitory concentrations of 0.03 mg/l or less; however, the former authors found that two resistant isolates from a fatal case of inhalational infection had MICs in excess of 0.25 mg/l. *Bacillus anthracis* is resistant to many cephalosporins. Coker, Smith and Hugh-Jones (2002) found that of 25 genetically diverse mainly animal and human isolates from around the world, five strains were resistant to the second generation cephalosporin cefuroxime, and 19 strains showed intermediate susceptibility to this agent; all strains were sensitive to the first generation cephalosporin cephalexin, and to the second generation cefaclor, and three were resistant to penicillin but were negative for β-lactamase production. Mohammed *et al.* (2002) studied 50 historical isolates from humans and animals and 15 clinical isolates from the 2001 bioterrorist attack in the USA; the majority of their strains could be regarded as nonsusceptible to the third generation cephalosporin ceftriaxone, and three strains were resistant to penicillin. Genomic sequence data indicate that *B. anthracis* possesses two β-lactamases: a potential penicillinase (class A) and a cephalosporinase (class B) which is expressed (Bell, Kozarsky and Stephens, 2002). Tetracyclines, chloramphenicol and gentamicin are suitable for the treatment of patients allergic to penicillin; tests in primates showed doxycycline to be effective, a finding confirmed by Coker, Smith and Hugh-Jones (2002) and indicated the suitability of ciprofloxacin (Turnbull *et al.*, 1998). Mohammed *et al.* (2002) found that most of their strains showed only intermediate susceptibility to erythromycin Esel, Doganay and Sumerkan (2003) found that ciprofloxacin and the newer quinolone gatifloxacin had a good *in vitro* activity against 40 human isolates collected in Turkey but that for another new quinolone, levofloxacin, high MICs were observed for ten strains.

Because human cases tend to be sporadic, clinical experience of alternative treatment strategies was sparse until the bioterrorist attack occurred in the US in late 2001. The potential and actual use of *B. anthracis* as a bioweapon has also emphasized the need for postexposure prophylaxis; recommendations include ciprofloxacin or doxycycline, with amoxycillin as an option for the treatment of children and pregnant or lactating women, given the potential toxicity of quinolones and tetracyclines; however, β-lactams do not penetrate macrophages well, and these are the sites of spore germination (Bell, Kozarsky and Stephens, 2002). Combination therapy, begun early, with a fluoroquinolone such as ciprofloxacin and at least one other antibiotic to which the organism is sensitive, appears to improve survival (Jernigan *et al.*, 2001). Following the 2001 outbreak in the US, the recommendation for initial treatment of inhalational anthrax is ciprofloxacin or doxycycline along with one or more agents to which the organism is normally sensitive; given supportive sensitivity testing, a penicillin may be used to complete treatment. The same approach is recommended for cutaneous infections (Bell, Kozarsky and Stephens, 2002). Doxycycline does not penetrate the central nervous system well and so is not appropriate for the treatment of meningitis.

Bacillus cereus

Despite the well-established importance of *B. cereus* as an opportunistic pathogen, there have been rather few studies of its antibiotic sensitivity, and most information has to be gleaned from the reports of individual cases or outbreaks. *Bacillus cereus* and *B. thuringiensis* produce a broad-spectrum β-lactamase and are thus resistant to penicillin, ampicillin and cephalosporins; they are also resistant to trimethoprim (an Antarctic *B. thuringiensis* isolate was penicillin sensitive, however) (Forsyth and Logan, 2000). An *in vitro* study of 54 isolates from blood cultures by disk diffusion assay found that all strains were susceptible to imipenem and vancomycin and that most were sensitive to chloramphenicol, ciprofloxacin, erythromycin and gentamicin (with 2, 2, 6 and 7% strains, respectively, showing moderate or intermediate sensitivities), while 22 and 37% of strains showed only moderate or intermediate susceptibilities to clindamycin and tetraclycine, respectively (Weber *et al.*, 1988); in the same study, microdilution tests showed susceptibility to imipenem, vancomycin, chloramphenicol, gentamicin and ciprofloxacin with MICs of 0.25–4, 0.25–2, 2.0–4.0, 0.25–2 and 0.25–1.0 mg/l, respectively. A plasmid carrying resistance to tetracycline in *B. cereus* has been transferred to a strain of *B. subtilis* and stably maintained (Bernhard, Schrempf and Goebel, 1978).

Although strains are almost always susceptible to clindamycin, erythromycin, chloramphenicol, vancomycin, and the aminoglycosides and are usually sensitive to tetracycline and sulphonamides, there have been several reports of treatment failures with some of these drugs: a fulminant meningitis which did not respond to chloramphenicol (Marshman, Hardwidge and Donaldson, 2000); a fulminant infection in a neonate which was refractory to treatment that included vancomycin, gentamicin, imipenem, clindamycin and ciprofloxacin (Tuladhar *et al.*, 2000); failure of vancomycin to eliminate the organism from cerebrospinal fluid in association with a fluid-shunt infection (Berner *et al.*, 1997); persistent bacteremias with strains showing resistance to vancomycin in two hemodialysis patients (A. von Gottberg and W. van Nierop, personal communication). Oral ciprofloxacin has been used successfully in the treatment of *B. cereus* wound infections. Clindamycin with gentamicin, given early, appears to be the best treatment for ophthalmic infections caused by *B. cereus*, and experiments with rabbits suggest that intravitreal corticosteroids and antibiotics may be effective in such cases (Liu *et al.*, 2000).

Other Species

Information is sparse on treatment of infections with species outside the *B. cereus* group. Gentamicin was effective in treating a case of

B. licheniformis ophthalmitis, and cephalosporin was effective against *B. licheniformis* bacteremia and septicemia. Resistance to macrolides appears to occur naturally in *B. licheniformis* (Docherty *et al.*, 1981). *Bacillus subtilis* endocarditis in a drug abuser was successfully treated with cephalosporin, and gentamicin was successful against a *B. subtilis* septicemia. Penicillin, or its derivatives, or cephalosporins probably form the best first choices for treatment of infections attributed to other *Bacillus* species. In the study by Weber *et al.* (1988), among the isolates of *B. megaterium* (13 strains), *B. pumilus* (four strains), *B. subtilis* (four strains), *B. circulans* (three strains), *B. amyloliquefaciens* (two strains) and *B. licheniformis* (one strain), along with five strains of *B.* (now *Paenibacillus*) *polymyxa* and three unidentified strains from blood cultures, over 95% of isolates were susceptible to imipenem, ciprofloxacin and vancomycin, while between 75 and 90% were susceptible to penicillins, cephalosporins and chloramphenicol. Isolates of *B. polymyxa* and *B. circulans* were more likely to be resistant to the penicillins and cephalosporins than strains of the other species – it is possible that some or all of the strains identified as *B. circulans* might now be accommodated in *Paenibacillus*, along with *B. polymyxa*. An infection of a human-bite wound with an organism identified as *B. circulans* did not respond to treatment with amoxicillin and flucloxacillin but was resolved with clindamycin (Goudswaard, Dammer and Hol, 1995). A recurrent septicemia with *B. subtilis* in an immunocompromised patient yielded two isolates, both of which could be recovered from the probiotic preparation that the patient had been taking; one isolate was resistant to penicillin, erythromycin, rifampin and novobiocin, while the other was sensitive to rifampin and novobiocin but resistant to chloramphenicol (Oggioni *et al.*, 1998).

A strain of *B. circulans* showing vancomycin resistance has been isolated from an Italian clinical specimen (Ligozzi, Cascio and Fontana, 1998). Vancomycin resistance was reported for a strain of *B.* (now *Paenibacillus*) *popilliae* in 1965, and isolates of this species dating back to 1945 have been shown to carry a *vanA*- and *vanB*-like gene, a gene resembling those responsible for high-level vancomycin resistance in enterococci. Vancomycin resistant enterococci (VRE) were first reported in 1986, and so it has been suggested that the resistance genes in *B. popilliae* and VRE may share a common ancestor, or even that the gene in *B. popilliae* itself may have been the precursor of those in VRE; *B. popilliae* has been used for over 50 years as a biopesticide, and no other potential source of *vanA* and *vanB* has been identified (Rippere *et al.*, 1998). Of two South African vancomycin-resistant clinical isolates, one was identified as *P. thiaminolyticus* and the other was unidentified but considered to be related to *B. lentus* (Forsyth and Logan, unpublished); the latter was isolated from a case of neonatal sepsis and has been shown to have inducible resistance to vancomycin and teicoplanin; this is in contrast to the *B. circulans* and *P. thiaminolyticus* isolates mentioned above, in which expression of resistance was found to be constitutive (A. von Gottberg and W. van Nierop, personal communication).

PREVENTION AND CONTROL

Bacillus anthracis

Cross-infection is not a problem with this organism, as person-to-person transmission is very rare and direct infections from soil are few. Soil is, however, a main reservoir from which animals and so man are infected. In countries free of the disease, disinfection of animal products imported from epizootic areas of the world is necessary to prevent its reintroduction. Hair and wool can effectively be treated with formaldehyde, as can hides and bone meal by heat (Brachman, 1990).

The Sterne attenuated live spore vaccine, based on a toxigenic but noncapsulate strain, was introduced for animal vaccination in the late 1930s, and spores of this strain remain in use as the basis of livestock vaccines in most parts of the world today. As this vaccine can show some slight virulence for certain animals, it is not considered suitable for human protection in the West, but live spore preparations are used in China and Russia. The former USSR vaccine was developed in the 1930s and 1940s and licensed for administration by injection in 1959. The UK vaccine is an alum-precipitated culture filtrate of the Sterne strain; it was first formulated in 1954, introduced for workers at risk in 1965 and licensed for human use in 1979. The current human vaccine in the USA is an aluminium hydroxide-adsorbed vaccine strain culture filtrate containing a relatively high proportion of PA and relatively low amounts of LF and EF; it was licensed in 1972 (Turnbull, 2000). Concerns about the lack of efficacy and safety data on the long-established UK and US vaccines, especially following the Sverdlovsk incident, and allegations that anthrax vaccination contributed to Gulf War syndrome in military personnel, have led to demands for new vaccines that would necessarily undergo stricter testing than was customary in the past. Favored active ingredients of these next-generation vaccines are whole-length recombinant PA or a mutant (nontoxic) portion of this molecule (Turnbull, 2000).

Bacillus cereus

Diarrheal and vomiting intoxications by this organism are readily preventable by appropriate food-handling procedures. Meat and vegetables should not be held at temperatures between 10 and 45 °C for long periods, and rice held overnight after cooking should be refrigerated and not held at room temperature. Prevention of infection by this organism, and by other *Bacillus* species, in patients following surgery or in those who are immunocompromised or who are otherwise predisposed to infection, depends on good practice.

ACKNOWLEDGEMENTS

We warmly acknowledge the generous assistance given by Dr Peter Turnbull; he kindly gave NAL freedom to use his material on anthrax that had been prepared for another publication for which they were co-authors. We also thank Professor Mehmet Doganay for material reused from the previous edition of this book and Kirsteen Barclay for assistance with the manuscript.

REFERENCES

Agata N, Mori M, Ohta M *et al.* (1994) A novel dodecadepsipeptide, cereulide, isolated from *Bacillus cereus* causes vacuole formation in Hep-2 cells. *FEMS Microbiology Letters*, **121**, 31–34.

Agata N, Ohta M, Mori M and Isobe M. (1995) A novel dodecadepsipeptide, cereulide, is an emetic toxin of *Bacillus cereus*. *FEMS Microbiology Letters*, **129**, 17–20.

Agerholm JS, Jensen NE, Dantzer V *et al.* (1997) Experimental infection of pregnant cows with *Bacillus licheniformis* bacteria. *Veterinary Pathology*, **36**, 191–201.

Alibek K. (2003) *Biohazard*, Hutchinson, London.

Ash C, Priest FG and Collins MD. (1993) Molecular identification of rRNA group, 3 bacilli (Ash, Farrow, Wallbanks and Collins) using a PCR probe-test: proposal for the creation of a new genus *Paenibacillus*. *Antonie van Leeuwenhoek*, **64**, 253–260.

Banerjee C, Bustamante CI, Wharton R *et al.* (1988) *Bacillus* infections in patients with cancer. *Archives of Internal Medicine*, **148**, 1769–1774.

Barnaby W. (2002) *The Plague Makers: the Secret World of Biology Warfare*, Vision, London.

Bartlett R and Bisset KA. (1981) Isolation of *Bacillus licheniformis* var. *endoparasiticus* from the blood of rheumatoid arthritis patients and normal subjects. *Journal of Medical Microbiology*, **14**, 97–105.

Beecher DJ. (2001) The *Bacillus cereus* group, in *Molecular Medical Microbiology* (ed. M Sussman), Academic Press, London, UK, pp. 1161–1190.

Bell DM, Kozarsky PE and Stephens DS. (2002) Clinical issues in the prophylaxis, diagnosis, and treatment of anthrax. *Emerging Infectious Diseases*, **8**, 222–225.

Bergey DH, Harrison FC, Breed RS *et al.* (1923) *Bergey's Manual of Determinative Bacteriology*, 1st edn, The. Williams & Wilkins Co, Baltimore, pp. 422.

Bergey DH, Harrison FC, Breed RS *et al.* (1925) *Bergey's Manual of Determinative Bacteriology*, 2nd edn, The. Williams & Wilkins Co, Baltimore, pp. 462.

Berner R, Heinen F, Pelz K *et al.* (1997) Ventricular shunt infection and meningitis due to *Bacillus cereus*. *Neuropediatrics*, **28**, 333–334.

Bernhard K, Schrempf H and Goebel W. (1978) Bacteriocin and antibiotic resistance plasmids in *Bacillus cereus* and *Bacillus subtilis*. *Journal of Bacteriology*, **133**, 897–903.

Bishop AH. (2002) *Bacillus thuringiensis* insecticides, in *Applications and Systematics of* Bacillus *and Relatives* (eds RCW Berkeley, M Heyndrickx, NA Logan and P De Vos), Blackwell Science, Oxford, pp. 160–175.

Bishop AH, Johnson J and Perani M. (1999) The safety of *Bacillus thuringiensis* to mammals investigated by oral and subcutaneous dosage. *World Journal of Biotechnology*, **15**, 375–380.

Blowey R and Edmondson P. (1995) *Mastitis Control in Dairy Herds. An Illustrated and Practical Guide*, Farming Press Books, Ipswich.

Blue SR, Singh VR and Saubolle MA. (1995) *Bacillus licheniformis* bacteremia: five cases associated with indwelling central venous catheters. *Clinical Infectious Diseases*, **20**, 620–633.

Boone DR, Liu Y, Zhao Z-J *et al.* (1995) *Bacillus infernus* sp. nov., an Fe(III)- and Mn(IV)-reducing anaerobe from the deep terrestrial subsurface. *International Journal of Systematic Bacteriology*, **45**, 441–448.

Brachman PS. (1990) Introductory comments on prophylaxis, therapy and control in relation to anthrax. *Salisbury Medical Bulletin*, **60** (Special Suppl.), 84–85.

Brachman P and Kaufmann A. (1998) Anthrax, in *Bacterial Infections of Humans* (eds Evans and Brachman), Plenum Medical Book Company, New York, NY, pp. 95–107.

Bradley KA, Mogridge J, Mourez M *et al.* (2001) Identification of the cellular receptor for anthrax toxin. *Nature*, **414**, 225–229.

Briley RT, Teel JH and Fowler JP. (2001) Nontypical *Bacillus cereus* outbreak in a child care center. *Journal of Environmental Health*, **63**, 9–11.

Burdon KL. (1956) Useful criteria for the identification of *Bacillus anthracis* and related species. *Journal of Bacteriology*, **71**, 25–42.

Cano RJ and Borucki MK. (1995) Revival and identification of bacterial spores in 25- to 40-million-year-old Dominican amber. *Science*, **268**, 1060–1064.

Cherkasskiy BL. (1999) A national register of historic and contemporary anthrax foci. *Journal of Applied Microbiology*, **87**, 192–195.

Christopher GW, Cieslak TJ, Pavlin JA and Eitzen EM. (1997) Biological warfare; a historical perspective. *The Journal of the American Medical Association*, **278**, 412–417.

Claus D and Berkeley RCW. (1986) Genus *Bacillus* Cohn 1872, 174[AL], in Bergey's Manual of Systematic Bacteriology, Vol. 2 (eds Sneath PHA, Mair NS, Sharpe ME and Holt JG), Williams & Wilkins, Baltimore, pp. 1105–1139.

Cohn F. (1872) Untersuchungen über Bakterien. *Beitrage zur Biologie der Pflanzen*, **1** (2), 127–224.

Cohn F (1876) Untersuchungen über Bakterien. IV. Beitrage zur Biologie der Bacillen. *Beitrage zur Biologie der Pflanzen*, **2**, 249–276.

Coker PR, Smith KL and Hugh-Jones ME. (2002) Antimicrobial susceptibilities of diverse *Bacillus anthracis* isolates. *Antimicrobial Agents and Chemotherapy*, **46**, 3843–3845.

Collins CH. (1988) *Laboratory Acquired Infections*, 2nd edn, Butterworths, London, UK, p. 16.

Damgaard PH, Granum PE, Bresciani J *et al.* (1997) Characterization of *Bacillus thuringiensis* isolated from infections in burn wounds. *FEMS Immunology and Medical Microbiology*, **18**, 47–53.

Dancer SJ, McNair D, Finn P and Kolstø A-B. (2002) *Bacillus cereus* cellulitis from contaminated heroin. *Journal of Medical Microbiology*, **51**, 278–281.

Das T, Choudhury K, Sharma S *et al.* (2001) Clinical profile and outcome in *Bacillus* endophthalmitis. *Ophthalmology*, **108**, 1819–1825.

Davaine MC. (1864) Nouvelles recherches sur la nature de la maladie charbonneuse connue sous ie nom de sang de rate. *Comptes Rendues de L' Academic des Sciences, Paris*, **59**, 393–396.

Davey RT Jr and Tauber WB. (1987) Posttraumatic endophthalmitis: the emerging role of *Bacillus cereus* infection. *Reviews of Infectious Diseases*, **9**, 110–123.

Docherty A, Grandi G, Grandi R *et al.* (1981) Naturally occurring macrolide-lincosamide-streptogramin B resistance in *Bacillus licheniformis*. *Journal of Bacteriology*, **145**, 129–137.

Doganay M. (1990) Human anthrax in Sivas, Turkey. *Salisbury Medical Bulletin*, **60** (Special Suppl.), 13.

Doganay M and Aydin N. (1991) Antimicrobial susceptibility of *Bacillus anthracis*. *Scandinavian Journal of Infectious Diseases*, **23**, 333–335.

Drago L, Lombardi A, De Vecchi E and Gismondo MR. (2002) Real-time PCR assay for rapid detection of *Bacillus anthracis* spores in clinical samples. *Journal of Clinical Microbiology*, **40**, 4399.

Drum CL, Yan S-Z, Bard J *et al*. (2002) Structural basis for the activation of anthrax adenylyl cyclase exotoxin by calmodulin. *Nature*, **415**, 396–402.

Ehrenberg CG. (1835) Dritter Beitrag zur Erkenntnis grosser Organisation in der Richtung des kleinsten Raumes. *Abhandlungen der Koniglichen Akademie der Wissenschaften zu Berlin Aus Den Jahre 1833–1835*, 145–336.

Esel D, Doganay M and Sumerkan B. (2003) Antimicrobial susceptibilities of 40 isolates of *Bacillus anthracis* isolated in Turkey. *International Journal of Antimicrobial Agents*, **22**, 70–72.

Ezzell JW and Welkos SL. (1999) The capsule of *Bacillus anthracis*, a review. *Journal of Applied Microbiology*, **87**, 250.

Finlay WJJ, Logan NA and Sutherland AD. (1999) Semiautomated metabolic staining assay for *Bacillus cereus* emetic toxin. *Applied and Environmental Microbiology*, **65**, 1811–1812.

Finlay WJJ, Logan NA and Sutherland AD. (2000) *Bacillus cereus* produces most emetic toxin at lower temperatures. *Letters in Applied Microbiology*, **31**, 385–389.

Finlay WJJ, Logan NA and Sutherland AD. (2002) *Bacillus cereus* emetic toxin production in relation to dissolved oxygen tension and sporulation. *Food Microbiology*, **19**, 423–430.

Fletcher P and Logan NA. (1999) Improved cytotoxicity assay for *Bacillus cereus* diarrhoeal enterotoxin. *Letters in Applied Microbiology*, **28**, 394–400.

Forsyth G and Logan NA. (2000) Isolation of *Bacillus thuringiensis* from northern Victoria Land, Antarctica. *Letters in Applied Microbiology*, **30**, 263–266.

Fortina MG, Pukall R, Schumann P *et al*. (2001) *Ureibacillus* gen. nov., a new genus to accommodate *Bacillus thermosphaericus* (Andersson *et al*. 1995), emendation of *Ureibacillus thermosphaericus* and description of *Ureibacillus terrenus* sp. nov. *International Journal of Systematic and Evolutionary Microbiology*, **51**, 447–455.

Galiero G and De Carlo E. (1998) Abortion in water buffalo (*Bubalis bubalis*) associated with *Bacillus licheniformis*. *The Veterinary Record*, **143**, 640.

George S, Mathai D, Balraj V *et al*. (1994) An outbreak of anthrax meningoencephalitis. *Transactions of the Royal Society of Tropical Medicine and Hygiene*, **88**, 206–207.

Gibson T and Gordon RE. (1974) Genus *Bacillus* Cohn, in *Bergey's Manual of Determinative Bacteriology*, 8th edn (eds Buchanan RE and Gibbons NE), Williams & Wilkins, Baltimore, pp. 529–550.

Gordon RE, Haynes WC and Pang CH-N. (1973) The genus *Bacillus*. Agriculture Handbook No. 427. US Department of Agriculture, Washington, DC.

Goudswaard WB, Dammer MH and Hol C. (1995) *Bacillus circulans* infection of a proximal interphalangeal joint after a clenched-fist injury caused by human teeth. *European Journal of Clinical Microbiology and Infectious Disease*, **14**, 1015–1016.

Granum PE. (2002) *Bacillus cereus* and food poisoning, in *Applications and Systematics of Bacillus and Relatives* (eds Berkeley RCW, Heyndrickx M, Logan NA and De Vos P), Blackwell Science, Oxford, pp. 37–46.

Hanna PC. (1999) Lethal toxin actions and their consequences. *Journal of Applied Microbiology*, **87**, 285–287.

Hanna PC and Ireland JAW. (1999) Understanding *Bacillus anthracis* pathogenesis. *Trends in Microbiology*, **7**, 180–182.

Hart CA and Beeching NJ. (2002) A spotlight on anthrax. *Clinics in Dermatology*, **20**, 365–375.

Heyndrickx M, Lebbe L, Kersters K *et al*. (1998) *Virgibacillus*: a new genus to accommodate *Bacillus pantothenticus* (Proom and Knight 1950). Emended description of *Virgibacillus pantothenticus*. *International Journal of Systematic Bacteriology*, **48**, 99–106.

Heyrman J, Balcaen A, Lebbe L *et al*. (2003) *Virgibacillus carmonensis* sp. nov., *Virgibacillus necropolis* sp. nov. and *Virgibacillus picturae* sp. nov., three new species isolated from deteriorated mural paintings, transfer of the species of the genus *Salibacillus* to *Virgibacillus*, as *Virgibacillus marismortui* comb. nov. and *Virgibacillus salexigens* comb. nov. and emended description of the genus *Virgibacillus*. *International Journal of Systematic and Evolutionary Microbiology*, **53**, 501–511.

Hoffmaster AR, Meyer RF, Bowen MP *et al*. (2002) Evaluation and validation of a real-time polymerase chain reaction assay for rapid identification of *Bacillus anthracis*. *Emerging Infectious Diseases*, **8**, 1178–1181.

Ivanova N, Sorokin A, Anderson I *et al*. (2003) Genome sequence of *Bacillus cereus* and comparative analysis with *Bacillus anthracis*. *Nature*, **423**, 87–91.

Jackson PJ, Hugh-Jones ME, Adair DM *et al*. (1998) PCR analysis of tissue samples from the, 1979 Sverdlovsk anthrax victims: The presence of multiple *Bacillus anthracis* strains in different victims. *Proceedings of the National Academy of Sciences of the United States of America*, **95**, 1224–1229.

Jernigan JA, Stephens DS, Ashford DA *et al*. (2001) Bioterrorism-related inhalational anthrax: the first 10 cases reported in the United States. *Emerging Infectious Diseases*, **7**, 933–944.

Jones BL, Hanson MF and Logan NA. (1992) Isolation of *Bacillus licheniformis* from a brain abscess following a penetrating orbital injury. *The Journal of Infection*, **24**, 103–105.

Kanso S, Greene AC and Patel BKC. (2002) *Bacillus subterraneus* sp. nov., an iron- and manganese-reducing bacterium from a deep subsurface Australian thermal aquifer. *International Journal of Systematic and Evolutionary Microbiology*, **52**, 869–874.

Keim P, Price LB, Klevytska AM *et al*. (2000) Multiple-locus variable-number tandem repeat analysis reveals genetic relationships within *Bacillus anthracis*. *Journal of Bacteriology*, **182**, 2928–2936.

King D, Luna V, Cannons A *et al*. (2003) Performance assessment of the three commercial assays for direct detection of *Bacillus anthracis* spores. *Journal of Clinical Microbiology*, **41**, 3454–3455.

Knight BCJ and Proom H. (1950) A comparative survey of the nutrition and physiology of mesophilic species in the genus *Bacillus*. *Journal of General Microbiology*, **4**, 508–538.

Knisely RF. (1966) Selective medium for *Bacillus anthracis*. *Journal of Bacteriology*, **92**, 784–786.

Koch R. (1876) Untersuchungen über Bakterien. V. Die Aetiologi der Milzbrand Krankheit, begrandet auf Entwicklurgsgeschichte des *Bacillus anthracis*. *Beitrage zur Biologie der Pflanzen*, **2**, 277–308.

Kramer JM and Gilbert RJ. (1989) *Bacillus cereus* and other *Bacillus* species, in *Foodborne Bacterial Pathogens* (ed. Doyle), Marcel Dekker, New York and Basel, pp. 21–70.

Kwong KL, Que TL, Wong SN and So KT. (1997) Fatal meningoencephalitis due to *Bacillus anthracis*. *Journal of Paediatrics and Child Health*, **33**, 539–541.

Lalitha MK and Thomas MK. (1997) Penicillin resistance in *Bacillus anthracis*. *Lancet*, **349**, 1522.

Lane HC and Fauci AS. (2001) Bioterrorism on the home front: a new challenge for American medicine. *The Journal of the American Medical Association*, **286**, 2595.

Laubach CA, Rice JL and Ford WW. (1916) Studies on aerobic spore-bearing non-pathogenic bacteria. Part II. *Journal of Bacteriology*, **1**, 493–533.

Lawrence JS and Ford WW. (1916) Studies on aerobic spore-bearing non-pathogenic bacteria. Part I. *Journal of Bacteriology*, **1**, 273–320.

Lightfoot NF, Scott RJD and Turnbull PCB. (1990) Antimicrobial susceptibility of *B. anthracis*. *Salisbury Medical Bulletin*, **60** (Special Suppl.), 95–98.

Ligozzi M, Cascio GL and Fontana R. (1998) *vanA* gene cluster in a vancomycin-resistant clinical isolate of *Bacillus circulans*. *Antimicrobial Agents and Chemotherapy*, **42**, 2055–2059.

Lindeque PM and Turnbull PCB. (1994) Ecology and epidemiology of anthrax in the Etosha National Park, Namibia. *The Onderstepoort Journal of Veterinary Research*, **61**, 71–83.

Little SF and Ivins BE. (1999) Molecular pathogenesis of *Bacillus anthracis* infection. *Microbes and Infection*, **2**, 131–139.

Liu SM, Way T, Rodrigues M and Steidl SM. (2000) Effects of intravitreal corticosteroids in the treatment of *Bacillus cereus* endophthalmitis. *Archives of Ophthalmology*, **118**, 803–806.

Logan NA. (1988) *Bacillus* species of medical and veterinary importance. *Journal of Medical Microbiology*, **25**, 157–165.

Logan NA. (2002) Modern identification methods, in *Applications and Systematics of Bacillus and Relatives* (eds Berkeley RCW, Heyndrickx M, Logan NA and De Vos P), Blackwell Science, Oxford, UK, pp. 123–140.

Logan NA and Berkeley RCW. (1981) Classification and identification of members of the genus *Bacillus* using API tests, in *The Aerobic Endospore-Forming Bacteria* (eds Berkeley RCW and Goodfellow M), Academic Press, London, pp. 105–140.

Logan NA and Berkeley RCW. (1984) Identification of *Bacillus* strains using the API system. *Journal of General Microbiology*, **130**, 1871–1882.

Logan NA, Capel BJ, Melling J and Berkeley RCW. (1979) Distinction between emetic and other strains of *Bacillus ceeus* using the API System and numerical methods. *FEMS Microbiology Letters*, **5**, 373–375.

Logan NA, Forsyth G, Lebbe L et al. (2002) Polyphasic identification of Bacillus and Brevibacillus strains from clinical, dairy, and industrial specimens and proposal of Brevibacillus invocatus, sp. nov. International Journal of Systematic and Evolutionary Microbiology, 52, 953–966.

Logan NA, Lebbe L, Hoste B et al. (2000) Aerobic endospore-forming bacteria from geothermal environments in northern Victoria Land, Antarctica, and Candlemas Island, South Sandwich archipelago, with the proposal of Bacillus fumarioli sp. nov. International Journal of Systematic and Evolutionary Microbiology, 50, 1741–1753.

Logan NA, Old DC and Dick HM. (1985) Isolation of Bacillus circulans from a wound infection. Journal of Clinical Pathology, 38, 838–839.

Lund T, De Buyser ML and Granum PE. (2000) A new enterotoxin from Bacillus cereus that can cause necrotic enteritis. Molecular Microbiology, 38, 254–261.

Mahler H, Pasi A, Kramer JM et al. (1997) Fulminant liver failure in association with the emetic toxin of Bacillus cereus. The New England Journal of Medicine, 336, 1142–1148.

Makino S-I and Cheun HI. (2003) Application of the real-time PCR for the detection of airborne microbial pathogens in reference to the anthrax spores. Journal of Microbiological Methods, 53, 141–147.

Mangold T and Goldberg J. (1999) Plague Wars: A True Story of Biology Warfare, Macmillan, London.

Mäntynen V and Lindström K. (1998) A rapid PCR-based DNA test for enterotoxic Bacillus cereus. Applied and Environmental Microbiology, 64, 1634–1639.

Marshman LAG, Hardwidge C and Donaldson PMW. (2000) Bacillus cereus meningitis complicating cerebrospinal fluid fistula repair and spinal drainage. British Journal of Neurosurgery, 14, 580–582.

Meselson M, Guillemin J, Hugh-Jones ME et al. (1994) The Sverdlovsk anthrax outbreak of 1979. Science, 266, 1202–1208.

Mikkola R, Kolari M, Andersson MA et al. (2000) Toxic lactonic lipopeptide from food poisoning isolates of Bacillus licheniformis. European Journal of Biochemistry, 267, 4068–4074.

Miller J, Engelberg S and Broad W. (2001) Germs, The Ultimate Weapon. Simon & Schuster, London and New York.

Mohammed MJ, Marston CH, Popovic T et al. (2002) Antimicrobial susceptibility testing of Bacillus anthracis: comparison of results obtained by using the National Committee for Clinical Laboratory Standards broth microdilution reference and Etest agar gradient diffusion methods. Journal of Clinical Microbiology, 40, 1902–1907.

Nazina TN, Tourova TP, Poltaraus AB et al. (2001) Taxonomic study of aerobic thermophilic bacilli: descriptions of Geobacillus subterraneus gen nov, sp. nov. and Geobacillus uzenensis sp. nov. from petroleum reservoirs and transfer of Bacillus stearothermophilus, Bacillus thermocatenulatus, Bacillus thermoleovorans, Bacillus kaustophilus, Bacillus thermoglucosidasius, Bacillus thermodenitrificans to Geobacillus as Geobacillus stearothermophilus, Geobacillus thermocatenulatus, Geobacillus thermoleovorans, Geobacillus kaustophilus, Geobacillus thermoglucosidasius, Geobacillus thermodenitrificans. International Journal of Systematic and Evolutionary Microbiology, 51, 433–446.

van Netten P and Kramer M. (1992) Media for the detection and enumeration of Bacillus cereus in foods: a review. International Journal of Food Microbiology, 17, 85–99.

Nicholson WL, Munakata N, Horneck G et al. (2000) Resistance of Bacillus endospores to extreme terrestrial and extraterrestrial environments. Microbiology and Molecular Biology Reviews, 64, 548–572.

Norris JR, Berkeley RCW, Logan NA and O'Donnell AG. (1981) The genera Bacillus and Sporolactobacillus, in The Prokaryotes: A Handbook on Habitats, Isolation and Identification of Bacteria, Vol. 2 (eds Starr MP, Stolp H, Trüper HG, Balows A and Shlegel HG), Springer-Verlag, Berlin, pp. 1711–1742.

Oggioni M, Pozzi G, Valensis PE et al. (1998) Recurrent septicemia in an immunocompromised patient due to probiotic strains of Bacillus subtilis. Journal of Clinical Microbiology, 36, 325–326.

Olszewski WL, Jamal S, Manokaran G et al. (1999) Bacteriological studies of blood, tissue fluid, lymph and lymph nodes in patients with acute dermatolymphangioadenitis (DLA) in course of 'filarial' lymphedema. Acta Tropica, 73, 217–224.

Pannifer AD, Wong TY, Schwarzenbacher R et al. (2001) Crystal structure of the anthrax lethal toxin. Nature, 414, 229–233.

Parvanta MF. (2000) Abortion in a dairy herd associated with Bacillus licheniformis. Tierarztliche Umschau, 55, 126.

Pinna A, Sechi LA, Zanetti S et al. (2001) Bacillus cereus keratitis associated with contact lens wear. Ophthalmology, 108, 1830–1834.

Prazmowski A. (1880) Untersuchung über die Entwickelungsgeschichte und Fermentwirkung einiger Bacterien-Arten, Hugo Voigt, Leipzig, p. 23.

Read TD, Peterson SN, Tourasse N et al. (2003) The genome sequence of Bacillus anthracis Ames and comparison to closely related bacteria. Nature, 423, 81–86.

Ripabelli G, McLauchlin J, Mithani V and Threlfall EJ. (2000) Epidemiological typing of Bacillus cereus by amplified fragment length polymorphism. Letters in Applied Microbiology, 30, 358–363.

Rippere K, Patel R, Uhl JR et al. (1998) DNA sequence resembling vanA and vanB in the vancomycin-resistant biopesticide Bacillus popilliae. The Journal of Infectious Diseases, 178, 584–588.

Rivera AMG, Granum PE and Priest FG. (2000) Common occurrence of enterotoxin genes and enterotoxicity in Bacillus thuringiensis. FEMS Microbiology Letters, 190, 151–155.

Rothschild MA and Leisenfeld O. (1996) Is the exploding powder from blank cartridges sterile? Forensic Science International, 83, 1–13.

Roy M, Chen JC, Miller M et al. (1997) Epidemic Bacillus endophthalmitis after cataract surgery. Ophthalmology, 104, 1768–1772.

Schraft H, Steele M, McNab B et al. (1996) Epidemiological typing of Bacillus spp. isolated from food. Applied and Environmental Microbiology, 62, 4229–4232.

Schuch R, Nelson D and Fischetti VA. (2002) A bacteriolytic agent that detects and kills Bacillus anthracis. Nature, 418, 884–889.

Shida O, Takagi H, Kadowaki K and Komagata K. (1996) Proposal for two new genera, Brevibacillus gen. nov. and Aneurinibacillus gen. nov. International Journal of Systematic Bacteriology, 46, 939–946.

Sirisanthana T and Brown AE. (2002) Anthrax of the gastrointestinal tract. Emerging Infectious Diseases, 8, 649–651.

Skerman VBD, McGowen V and Sneath PHA. (1980) Approved lists of bacterial names. International Journal of Systematic Bacteriology, 30, 225–420.

Smith NR, Gordon RE and Clark FE. (1946) Aerobic Mesophilic Sporeforming Bacteria: Miscellaneous Publication No. 559, US Department of Agriculture, Washington, DC.

Smith NR, Gordon RE and Clark FE. (1952) Aerobic Endosporeforming Bacteria: Agriculture Monograph No. 16, US Department of Agriculture, Washington, DC.

Switzer Blum J, Burns Bindi A, Buzzelli J et al. (1998) Bacillus arsenicoselenatis, sp. nov. and Bacillus selenitireducens, sp. nov.: two haloalkaliphiles from Mono Lake, California that respire oxyanions of selenium and arsenic. Archives of Microbiology, 171, 19–30.

Tandon A, Tay-Kearney ML, Metcalf C and McAllister I. (2001) Bacillus circulans endophthalmitis. Clinical and Experimental Ophthalmology, 29, 92–93.

Tuladhar R, Patole SK, Koh TH et al. (2000) Refractory Bacillus cereus infection in a neonate. International Journal of Clinical Practice, 54, 345–347.

Turnbull PCB. (2000) Current status of immunization against anthrax: old vaccines may be here to stay for a while. Current Opinion in Infectious Disease, 13, 113–120.

Turnbull PCB, Böhm R, Cosivi O et al. (1998) Guidelines for the Surveillance and Control of Anthrax in Humans and Animals, World Health Organization, Geneva, WHO/EMC/ZDI/98.6.

Turnbull PCB, Doganay M, Lindeque PM et al. (1992a) Serology and anthrax in humans, livestock and Etosha National Park wildlife. Epidemiology and Infection, 108, 299–313.

Turnbull PCB, Hutson RA, Ward MJ et al. (1992b) Bacillus anthracis but not always anthrax. Journal of Applied Bacteriology, 72, 21–28.

Turnbull PCB, Jackson PJ, Hill KK et al. (2002) Longstanding taxonomic enigmas within the 'Bacillus cereus group' are on the verge of being resolved by far-reaching molecular developments: forecasts on the possible outcome by an ad hoc team, in Applications and Systematics of Bacillus and Relatives (eds Berkeley RCW, Heyndrickx M, Logan NA and De Vos P), Blackwell Science, Oxford, pp. 23–36.

Turnbull PCB and Kramer JM. (1995) Bacillus, in Manual of Clinical Microbiology, 6th edn (eds Murray PR, Baron EJ and Pfaller MA), ASM Press, Washington, DC, pp. 349–350.

Tyndall J. (1877) Further researches on the department and vital persistence of putrefactive and infective organisms from a physical point of view. Philosophical Transactions of the Royal Society of London, 167, 149–206.

Van der Zwet WC, Parlevliet GA, Savelkoul PH et al. (2000) Outbreak of Bacillus cereus infections in a neonatal intensive care unit traced to balloons used in manual ventilation. Journal of Clinical Microbiology, 38, 4131–4136.

Vreeland RH, Rosenzweig WD and Powers DW. (2000) Isolation of a 250 million-year-old halotolerant bacterium from a primary salt crystal. Nature, 407, 897–900.

Wainø M, Tindall BJ, Schumann P and Ingvorsen K. (1999) *Gracilibacillus* gen. nov., with description of *Gracilibacillus halotolerans* gen. nov., sp. nov.: transfer of *Bacillus dipsosauri* to *Gracilibacillus dipsosauri* comb. nov., and *Bacillus salexigens* to the genus *Salibacillus* gen. nov., as *Salibacillus salexigens* comb. nov. *International Journal of Systematic Bacteriology*, **49**, 821–831.

Weber DJ, Saviteer SM, Rutala WA and Thomann CA. (1988) In vitro susceptibility of *Bacillus* spp. to selected antimicrobial agents. *Antimicrobial Agents and Chemotherapy*, **32**, 642–645.

Winslow C-EA, Broadhurst J, Buchanan RE *et al.* (1917) The families and genera of the bacteria. Preliminary report of the committee of the society of American bacteriologists on characterization and classification of bacterial types. *Journal of Bacteriology*, **2**, 505–566.

Winslow C-EA, Broadhurst J, Buchanan RE *et al.* (1920) The families and genera of the bacteria. Final report of the committee of the society of American bacteriologists on characterization and classification of bacterial types. *Journal of Bacteriology*, **5**, 191–229.

Winter G. (1880) Die Pilze, in *Dr L. Rabenhorst's Kryptogamen-Flora*, 2nd edn, Vol. 1, Eduard Kummer, Leipzig, p. 38.

Wisotzkey JD, Jurtshuk P, Fox GE *et al.* (1992) Comparative analyses on the 16S rRNA (rDNA) of *B. acidocaldarius, B. acidoterrestris* and *B. cycloheptanicus* and proposal for creation of a new genus, *Alicyclobacillus* gen. nov. *International Journal of Systematic Bacteriology*, **42**, 263–269.

Wu YJ, Hong TC, Hou CJ *et al.* (1999) *Bacillus popilliae* endocarditis with prolonged complete heart block. *American Journal of Medical Science*, **317**, 263–265.

Yoon J-H, Lee K-C, Weiss N *et al.* (2001a) *Sporosarcina aquimarina* sp. nov., a bacterium isolated from seawater in Korea, and transfer of *Bacillus globisporus* (Larkin and Stokes, 1967), *Bacillus psychrophilus* (Nakamura, 1984), and *Bacillus pasteurii* (Chester, 1898) to the genus *Sporosarcina* as *Sporosarcina globispora* comb. nov., *Sporosarcina psychrophila* comb. nov. and *Sporosarcina pasteurii* comb. nov. and emended description of the genus *Sporosarcina*. *International Journal of Systematic and Evolutionary Microbiology*, **51**, 1079–1086.

Yoon J-H, Weiss N, Lee K-C *et al.* (2001b) *Jeotgalibacillus alimentarius* gen. nov., sp. nov., a novel bacterium isolated from jeotgal with L-lysine in the cell wall, and reclassification of *Bacillus marinus* Rüger, 1983 as *Marinibacillus marinus* gen. nov., comb. nov. *International Journal of Systematic and Evolutionary Microbiology*, **51**, 2087–2093.

Zilinskas RA. (1997) Iraq's biological weapons – the past as future? *The Journal of the American Medical Association*, **278**, 418–424.

10

Mycobacterium tuberculosis

Stephen H. Gillespie

Centre for Medical Microbiology, Royal Free and University College Medical School, Hampstead Campus, London, UK

HISTORY

Described by John Bunyan as the Captain of the men of death and by Oliver Wendell Holmes as the White Plague, tuberculosis has been and remains a major threat to human health (Dye, 2000). It is a disease of great antiquity thought to have evolved as a human pathogen during the Neolithic period, and its rise has been associated with the change from a hunter-gatherer lifestyle to pastoralism and farming. Skeletons from the Neolithic period and Egyptian and Peruvian mummies have evidence of tuberculosis (Salo *et al.*, 1994). Chinese records from more than 4000 years ago describe a respiratory disease with cough fever and wasting suggestive of tuberculosis. The disease was recognised by Hippocrates who made an accurate clinical description and probably gave it the name by which it was known for more than 2000 years: phthysis, which recognised the wasting characteristic of the disease. Aristotle and Galen recognised that tuberculosis was transmissible, a point that remained in dispute until the description of the tubercle bacillus by Koch in 1882 and for some years afterwards (Dormandy, 1999).

Tuberculosis has been a major cause of death as evidenced by the seventeenth-century bills of mortality from London which indicate that approximately 20% of deaths were caused by tuberculosis. In Massachusetts during 1768–1773 pulmonary tuberculosis accounted for 18% of all deaths rising to 25% at the turn of the twentieth century (Holmberg, 1990). In Europe the peak of the tuberculosis epidemic was reached at the end of the eighteenth and early nineteenth centuries, whereas in Eastern Europe the peak was delayed several decades and only occurred in Asia in the late nineteenth and early twentieth centuries (Dubos and Dubos, 1952). In industrialised countries it is thought that the impact of social improvement brought about a steady fall in the incidence of tuberculosis but was confounded by a steep rise during and after the two world wars.

Tuberculosis is a disease of overcrowding poverty and poor nutrition. Modern day rates could be correlated with economic indicators such as per capita income and unemployment. More recently, HIV has brought about an enormous growth in the incidence of tuberculosis especially in sub-Saharan Africa (Corbett *et al.*, 2003). In addition, population migration through war and economic factors has resulted in individuals moving from countries of high endemicity to industrialised countries (Maguire *et al.*, 2002). There have been many outbreaks of multiple drug-resistant tuberculosis (MDRTB) in industrialised countries that have resulted in extensive press coverage and public anxiety (Alland *et al.*, 1994; Moss *et al.*, 1997). This has prompted lurid headlines announcing the 'return of the White Plague'. However, it is more accurate to say that for a short time tuberculosis rates in developing countries were falling rapidly and the possibility of eradication was ahead but for most of the world tuberculosis has remained a common threat with an estimated 8.3 million cases and 1.8 million deaths (Corbett *et al.*, 2003).

DESCRIPTION OF THE ORGANISM

The mycobacteria grouped in the tuberculosis complex are very closely related with >99.9% genetic similarity and identical 16S ribosomal RNA gene sequences. Restricted structural gene polymorphism suggests that the organism has disseminated globally relatively recently (in evolutionary terms) (Sreevatsan *et al.*, 1997b).

The genome of *Mycobacterium tuberculosis* was one of the first to be reported (Cole *et al.*, 1998). The genome size is 4 411 532 bp with 3959 protein coding genes and six pseudogenes. Function has been attributed to 2441 genes, and 912 are conserved hypothetical genes. A total of 606 have unknown function, and 129 genes are absent from *M. bovis* BCG (Cole *et al.*, 1998).

In *M. tuberculosis*, single-gene polymorphisms are rare and main mechanism for genetic plasticity is through the action of insertion sequences leading to gene insertion and deletion (Brosch *et al.*, 2001). Comparative genomics suggests that *M. tuberculosis* is undergoing successive gene deletion, as it adapts to its obligate pathogen lifestyle.

Previous authors had presumed that *M. bovis* was the ancestor of *M. tuberculosis*, but recent deletion analysis studies suggest that either *M. africanum* or *M. canetti* are the ancestral member of the genus with other species emerging from this. With this scheme, *M. canetti* diverged first from the ancestral strain and then successive branches led to the emergence of *M. tuberculosis*, *M. microti* and *M. bovis*, respectively (Brosch *et al.*, 2002). The population genetics of the organism is described below in *Epidemiology*.

PATHOGENESIS

Response to Infection

Classic animal studies suggest that there are four stages of infection (Lurie, 1931; Lurie, 1964). The first stage is contact with macrophage in which some of the organisms are destroyed and some are able to continue to multiply. Pathogenic mycobacteria are able to survive and multiply within macrophages because they inhibit the acidification of the phagolysosome stabilising at pH 6.2–6.3 and are retained within the endosomal recycling pathway, giving the organism access to essential nutrients such as iron (Kaufmann, 2001). Cord factor is thought to play an important role in preventing acidification (Indrigo, Hunter and Actor, 2003). The phagosomes also fail to fuse with lysosomes, although they do acquire some lysosomal proteins (Russell, 2001). During the second stage, blood monocytes and other immune cells are attracted to the site of infection. The monocytes differentiate into macrophages which are unable to kill the mycobacteria. It is only a few weeks later, the third stage, when the influx of antigen-specific T cells, that myco-bacterial multiplication is controlled. This happens because of secretion of IFN-γ which activates the macrophages and releases the block on phagosomal maturation (Russell, 2001). Efficient killing of mycobacteria depends on nitric oxide produced by nitric oxide synthase and by other related radicals (Kaufmann, 2001). This process is enhanced by acidification of the mycobacteria-containing vacuole which can now occur because the block is now lifted. The final stage infection may become dormant or, alternatively, if the immune cells fail to control the multiplication of the bacteria dissemination occurs and active disease develops.

Infected macrophages produce large amounts of proinflammatory mediators such as TNF-α, IFN-γ, IL-6, IL-1β and IL-12. A range of anti-inflammatory cytokines is produced including IL-4, TGF-β and IL-10 (van Crevel, Ottenhoff and van der Meer, 2002). The balance between proinflammatory and anti-inflammatory cytokines defines the expression of tuberculosis (van Creval et al., 2001). Uncontrolled cytokine release is responsible for many of the symptoms and signs of tuberculosis such as fever and wasting.

Histopathology

Another way of looking at the pathogenesis of tuberculosis is through histopathology. A complete understanding of the disease requires integration of information from both approaches. There are three main histological lesions described in tuberculosis: exudative, productive and caseous (Canetti, 1955). Although there is some disagreement about the order in which they occur, it is likely that exudative or productive lesions occur first and lead to caseating lesions. The caseous lesions have relatively few organisms present. Following formation of the caseum the lesion may remain solid and evolve toward peripheral tubercle formation and sclerosis. This takes place in all lesions of latent infection as well as in the great majority of active tuberculosis lesions. On the other hand the caseous lesion may soften which, when the lesion communicates with a bronchus, drains leading to the formation of a cavity, a development which has traditionally been associated with the onset of clinical tuberculosis. During the process of cavity formation alveoli are destroyed and neighboring cavities may unite. Cavities are especially difficult to manage because it provides excellent conditions for the multiplication of tubercle bacilli in a site where the immune system and antibiotic therapy may not be effective (see below). Additionally, draining cavities are a source of bacteria that spread to other parts of the lung. Histopathologically the cavity is surrounded by a fibrous capsule inside which there is a zone of intense round-cell infiltration made up of macrophages, CD4$^+$ and CD8$^+$ cells and no bacteria. Beyond this there is a zone of necrosis in which there are few immune cells and no bacteria beyond which at the air interface of the cavity there is a zone in which bacteria are present in large numbers, in association with macrophages and T cells that are CD4$^-$CD8$^-$. The bacteria are found within the macrophages and this indicates that, in this outer zone, the lack of T-cell help means that the macrophages are unable to control the growth of bacteria (Kaplan et al., 2003).

EPIDEMIOLOGY

Tuberculosis is spread by the respiratory route. Traditionally patients who are sputum smear positive are considered to be infectious and those who have extrapulmonary disease or who are smear negative are not. Several factors can influence transmission rates, including the infective dose and the energy with which they are expelled into the air. Several epidemiological models have been described to describe this pheno-menon (Sultan et al., 1960; Riley et al., 1962; Beggs et al., 2003).

Patients who are heavily infected and those with laryngeal tubercu-losis are more likely to transmit infection due to increased coughing (Sultan et al., 1960; Riley et al., 1962; Braden, 1995). Bronchoscopy and autopsy exposure is also more likely to result in infections (Beggs et al., 2003). Transmission is more likely to occur in overcrowded conditions (Kuemmerer and Comstock, 1967; Beggs et al., 2003). Children and individuals who have immunodeficiency, for example HIV infection, are especially vulnerable to infection and outbreaks in schools can be extensive (Hoge et al., 1994). When tuberculosis was introduced into isolated island populations during the nineteenth century, epidemics developed associated with high mortality (Holmberg, 1990).

In immunocompetent individuals infected with M. tuberculosis, there is a 5–10% risk of developing active tuberculosis within 2 years of primary disease. Later there is a 5–10% lifetime risk of developing acute disease – reactivation. In countries with high prevalence the disease is found in the young and very young adults (Rieder, 1999). As social and economic conditions improve, clinical tuberculosis is dominated by reactivation disease and is found in increasingly older individuals. Reactivation is associated with failing immunity, but in the absence of HIV or immunosuppressive therapy, it is rare for an identifiable defect to be identified. Individuals who have migrated from high endemicity countries to low endemicity countries appear especially vulnerable to reactivation within the first 2 years (Tocque et al., 1998). Throughout the twentieth century the average age of tuberculosis patients increased and a similar process was occurring in developing countries until the appearance of the HIV epidemic which reversed this trend.

The epidemiology of tuberculosis is quite different from most other infectious diseases. This is due to the relatively long incubation period and the presence of both primary and reactivation forms of disease. Mathematical modeling of the epidemiology suggests that epidemics of tuberculosis are slow to develop peaking between 50 and 200 years after introduction of the disease (Blower et al., 1995). However, the interaction between HIV and tuberculosis serves to facilitate the spread of the disease (Corbett et al., 2003). In many countries the interaction between HIV and tuberculosis has been very severe with co-infection rates greater that 5000/100 000 in Southern Africa. HIV makes the individual more susceptible to infection, and they also have a 10% risk of reactivation disease per year (Glynn, 1998).

A number of molecular techniques have been applied to M. tuber-culosis including IS6110 RFLP analysis (van Embden et al., 1993; Maguire et al., 2002), gene deletion arrays (Kato-Maeda et al., 2001) and mycobacterial intergenic repeat sequence (MIRU) analysis (see below) (Supply et al., 2000; Supply et al., 2003) which have shed light on the population genetics of this organism. Using these techniques it has been possible to identify groups of related strains that are spreading widely. An example of this is the Beijing lineage which has spread widely in Asia (Bifani et al., 2002). A branch of this lineage described as strain W was responsible for extensive outbreaks in the United States that centered on New York City (Alland et al., 1994; Moss et al., 1997). Recent studies have suggested that ability of this lineage to spread is due to an increased mutation rate caused by alteration in the mutT genes; it is suggested that this makes strains of the Beijing lineage more adaptable (Rad et al., 2003). This may also explain the apparent ease with which strain W developed resistance and spread widely.

CLINICAL FEATURES

Pulmonary Disease

Pulmonary disease can be part of primary of post-primary phases of the infection. Many cases are detected through screening radiology, something which played a part in the decline in infection rates in the second half of the twentieth century. Cough is the commonest presenting symptom (Banner, 1979) and is typically constant and irritating. It is usually productive and can be associated with hemoptysis. Fever and malaise are common but often low grade. Weight loss develops over a number of months and may cause clinical confusion with malignancy. Breathlessness is a late manifestation of disease.

Pulmonary disease can be complicated by extension to local structures resulting in empyema and pericarditis. Severe disease may result in significant loss of lung capacity further complicated by fibrosis.

The chest X-ray has a vital role in diagnosing pulmonary disease. In addition to identifying patients with changes suspicious of tuberculosis, it can demonstrate the presence of cavities, pleural effusions, empyema and pericardial effusions. CT scanning is also gaining a place in diagnosis and evaluation of extent of disease. Where facilities allow endoscopy with a flexible bronchoscope can aid diagnosis by allowing the abnormal area of lung to be visualised and specimens taken.

Tuberculosis Meningitis

Although relatively rare, tuberculosis of the CNS has a serious impact due to the relatively poor prognosis and difficulty in making the diagnosis. Most CNS disease takes the form of meningitis, but cases of localised tuberculosis, tuberculomas, do occur. Spread to the meningitis is usually hematogenous, and coexistent miliary disease is diagnosed in approximately 20% of cases.

Symptoms are typically nonspecific in the early stages of the disease characterised by fever, anorexia and weight loss. Later, headache and neck-stiffness appear, and these may be accompanied by drowsiness which will develop into coma if effective treatment is not instituted. The severity of disease is classified by the Medical Research Council scale: stage I, no disturbance of consciousness or neurological signs and stage II, where there is some clouding of consciousness but coma has not intervened and focal neurological signs may be found. In stage III, patients are comatose or stuporous.

Diagnosis is difficult, as the number of organisms present in cerebrospinal fluid (CSF) is relatively low; so samples are often falsely negative. The application of molecular techniques has improved diagnostic sensitivity. It is the CSF cell count and biochemistry that are most important in making the diagnosis.

Lymph Node Tuberculosis

Tuberculosis lymphadenitis is one of the commonest presentations of extrapulmonary disease, being responsible for over 30% of extrapulmonary disease in some areas (Mehta et al., 1991). In areas where tuberculosis is common the disease usually presents early in life, whereas in low prevalence areas tuberculosis lymphadenitis presents later, usually between 20 and 50 years of age. In many industrialised countries nontuberculosis species are the most common etiological agents of mycobacterial lymphadenitis. The cervical lymph nodes are the commonest site of infection, although mediastinal and abdominal lymphadentitis is found in association with pulmonary and gastrointestinal disease. In some cases the diagnosis of tuberculosis is made by biopsy of mediastinal or abdominal nodes.

Infected lymph nodes may show mild reactive hyperplasia or granulomata with caseation and necrosis. Tuberculous lymphadenitis must be distinguished from lymphoma, metastatic carcinoma, sarcoidosis and other alternative causes of noncaseating granulomata. The diagnosis can be made by fine-needle aspiration or open biopsy followed by bacteriological and molecular examination.

Bone and Joint Tuberculosis

Tuberculosis has a predilection for the spine, and examples of this form of disease have been identified in mummified human remains from Egyptian tombs. Infection in large joints also occurs. Spinal tuberculosis is characterised by local tenderness usually in the lumbosacral area (>40% of all bony tuberculosis), and this is often associated with the systemic signs of the disease. Collapse of the vertebra cause kyphosis and may produce a spinal cord compression resulting in the most serious complication. Rupture of pus from the vertebra may cause the pus to enter the paraspinal space and into the psoas, which may lead to a cold abscess in the groin. Diagnosis is usually made radiologically, and the availability of CT and MRI scanning has made diagnosis much easier in recent years.

Gastrointestinal Disease

Gastrointestinal tuberculosis is now a rare condition in developed countries but remains common in developing countries. Infection can occur in any part of the gastrointestinal tract. Abdominal disease can present acutely (one-third of cases) as an abdominal emergency with pain. More commonly it presents with a slow deterioration characterised by weight loss, fever, malaise and ascites. There are no characteristic diagnostic signs, and the diagnosis is usually made with a combination of clinical suspicion and culture from biopsy or ascites fluid. Radiology using barium examination, CT and ultrasound may be suggestive.

Cutaneous Tuberculosis

Mycobacterium tuberculosis can affect the skin in a number of ways, but cutaneous disease is rare. The organism is capable of causing a primary infection which usually occurs in children or young adults who are inoculated by skin trauma. Typically a nodule develops at the site of the lesion associated with regional lymphadenopathy, the primary cutaneous complex. Lupus vulgaris is chronic form of cutaneous tuberculosis disease where dull red lesions appear on the face, head or extremities. It is usually seen in older patients and often follows a long history before a diagnosis is made. It is a progressive disease that can cause severe facial scarring resulting in considerable deformity. A miliary form of tuberculosis has been described in the very young or in AIDS patients. It is characterised by small purplish papules which can be found over wide areas of the skin.

Tuberculides are fascinating and controversial syndromes that are believed to be cutaneous manifestations of immune responses to tuberculosis. One relatively common example of this is erythema nodosum. Others that may cause considerable diagnostic problems include Bazin's erythema induratum.

Genitourinary Tuberculosis

Renal tuberculosis is one of the less common presentation of tuberculosis. It arises as a consequence of miliary spread of the bacteria in primary infection. Tuberculosis infection is often asymptomatic in the early stages of infection, but later dysuria, hematuria and in advanced cases flank pain may be present. Renal infection can be complicated by fibrosis and renal obstruction which may lead to renal failure. In almost all cases the intravenous pyelogram is abnormal with signs of fibrosis in the ureters.

Endocrine Consequences of Tuberculosis

Mycobacterium tuberculosis has the ability to infect almost all organs of the body. When this involves endocrine tissue the effects can be devastating. Destruction of the adrenal glands produces failure of corticosteroid production and can be rapidly fatal if the diagnosis is not made quickly. This syndrome can be undiagnosed when pulmonary tuberculosis treatment is commenced, but the patient may suffer an adrenal crisis 2–4 weeks into treatment. This phenomenon may be an explanation for some of the cases of sudden death in patients early in their course of tuberculosis.

The pituitary gland may also be affected by *M. tuberculosis* causing failure of endocrine functions. In females this may present as amenorrhea. In men it is corticosteroid axis that is the first presenting system.

LABORATORY DIAGNOSIS

Organisation of Laboratories for the Diagnosis of *M. tuberculosis* and Other Mycobacteria

In industrialised countries the incidence of tuberculosis can vary significantly between areas where the disease is rare to towns and cities where it is common. For some laboratories the number of isolates is low, and this poses difficulties in maintaining the skill required to perform susceptibilities and identifications. To overcome this potential problem most countries in the industrialised world divide adopt a three-level structure. For example, in the UK, smears and cultures are performed in most laboratories sending isolates to regional reference centers for identification and susceptibility testing for first-line agents (Drobniewski *et al.*, 2003). A national reference center acts as a reference for the regional centers and provides a comprehensive susceptibility and identification service. Additionally such national centers should perform research and organise national quality control programs. Technological changes have made the diagnosis of mycobacterial disease considerably easier and the increase in the number of specimens and positives has increased the throughput for most urban laboratories. Additionally, many patients are at risk of MDRTB and are immunocompromised. Such patients may deteriorate rapidly if the diagnosis is not made quickly. This means that there is increasing need for mycobacterial diagnostic facilities to be placed at centers where tuberculosis has re-emerged and in hospital where there are many immunocompromised patients (Davies *et al.*, 1999). Those laboratories that isolate few mycobacteria should refer samples where definitive diagnosis can be made reliably. Such laboratories should aim to provide identification and susceptibility within 21 days of isolation (Association of State and Territorial Public Health Directors and Centers for Disease Control, 1995; Drobniewski *et al.*, 2003).

In developing countries, direct sputum smear examination should be provided at all district hospital laboratories. At least a proportion of all isolates should be sent to reference centers for conformation of identification and susceptibility testing. With the increasing burden of HIV infection and MDRTB, it is desirable that more specimens are cultured and susceptibility testing is performed.

In laboratories that process specimens for the diagnosis of tuberculosis and other mycobacteria, the processing should be performed in a sealable room with negative pressure ventilation, and these laboratories should provide protection against aerosol transmission through an exhaust protective cabinet. The laboratory should have access to an autoclave within or near it. Most industrialised countries have detailed requirements for such containment level 3 or P3 laboratories, the description of which is beyond the scope of this chapter (Collins, Grange and Yates, 1997; Advisory Committee on Dangerous Pathogens, 2001).

Specimens

Almost any tissue or body fluid is suitable for examination for mycobacteria, but the investigation of each disease syndrome requires a different protocol. Specimens should be collected into clean sterile containers and transported to the laboratory with minimal delay. When unusual or precious samples, e.g. biopsy specimens, are taken, the laboratory should be contacted to coordinate specimen transport and processing. Additionally, it is essential that the laboratory provides advice on the correct way of sending potentially infected samples.

When pulmonary tuberculosis is suspected the principal specimen is sputum, but not all patients are capable of producing a satisfactory sample. This may be due to physical weakness or failure to coordinate the effort required for effective specimen production, and some patients may have a nonproductive cough. Children are very rarely capable of producing an adequate sputum specimen. When sputum is not available, an alternative is an early morning gastric fluid. Fluid aspirated from the stomach contains respiratory secretions coughed up during the night and swallowed. Another moderately invasive method is induced sputum, but this method has been associated with transmission of infection and must be perform in suitable facilities (Breathnach *et al.*, 1998). A more invasive technique is a bronchoalveolar lavage (BAL), and this provides a better specimen when the facilities for this technique are available and the patient is able to tolerate the procedure (Anon., 2000).

Specimens should be considered in two groups, as the protocol for processing them is significantly different. The first group are the specimens which contain a normal flora such as sputum or BAL and these will require decontamination. The second group of specimens are those which come from a normally sterile site including CSF, blood, bone marrow and pleural, pericardial and joint fluid. In addition, aspirated pus and tissue biopsies should be considered in this group.

An accurate white cell count and differential should be performed on CSF and the protein, glucose concentrations should be measured. Tuberculosis meningitis is associated with a high protein and low glucose concentrations. The raised white cell count usually has a lymphocyte predominance, although rarely a neutrophilia may be found.

Specimens of urine are required for the diagnosis of renal tuberculosis. As 24-h collections are likely to become contaminated with other bacteria, three first void samples should be examined. Tissue samples may be obtained in cases of diagnostic difficulty when other less invasive tests have proved negative and the suspicion of tuberculosis or other mycobacterial disease remains. Open liver, lymph node and brain biopsies are commonly taken, but fine-needle aspirates and transbronchial biopsy should be considered to reduce the need for a surgical procedure. When such specimens are contemplated, the laboratory should be contacted so that they can be processed with the minimum of delay. It is particularly important to ensure that specimens are not placed in formalin for histopathological examination before culture can be performed. Additionally, pleural and pericardial fluids can be examined, although a higher diagnostic yield is obtained if biopsy specimens are collected.

Blood should be collected in heparin anticoagulant (not EDTA) or directly inoculated into commercially available mycobacterial liquid culture medium (e.g. MBBacT or BACTEC) (Vetter *et al.*, 2001; von Gottberg *et al.*, 2001). Feces may be examined by microscopy and culture for the presence of *M. avium* intracellulare.

Microscopy

One of the most successful tests for the diagnosis of mycobacterial disease is the direct microscopic examination of specimens for the presence of acid-fast bacteria. Although it requires staff who are experienced in reading smears, it has the advantage of being rapid and inexpensive. For laboratories in developing countries it may be the only diagnostic test that is available. It has a lower limit of detection of approximately 10 000 organisms per milliliter (Hobby *et al.*, 1973). This is 100 times less sensitive than culture, and consequently a negative smear does not preclude the diagnosis of tuberculosis (Cruickshank, 1952; Yeager *et al.*, 1967). Patients who are sputum smear negative are thought to pose a significantly lower risk of transmitting their infection than smear-positive cases. Thus a diagnostic system with limited resources which only uses sputum smear for diagnosis will identify the most important patients: those likely to contribute further new cases.

There are two main staining methods: the Ziehl–Neelsen and related methods and fluorochrome techniques that use auramine or auramine–rhodamine dyes. Smears are semi-quantitative and may provide the clinician an impression of the severity of the infection or progress of a patient on treatment (Anon., 2000). However, the excretion of bacteria into the sputum can be intermittent, and organisms that have been killed by therapy remain in the sputum for several months (see below). Thus at least three good quality samples of sputum should be examined before the diagnosis is rejected.

Smears can performed directly from almost all clinical specimens. Exceptions to this rule are urine, and gastric aspirates, which may be contaminated with commensal mycobacteria. This means that there is a significant risk of generating a false-positive result. To assist infection control procedures, laboratories should provide results of sputum examination within 24 h (Association of State and Territorial Public Health Directors and Centers for Disease Control, 1995). This is especially important when patients at risk of MDRTB are examined or there is the possibility of exposure of immunocompromised individuals to infectious tuberculosis cases.

Decontamination

Mycobacteria grow more slowly than other bacteria and fungi with which the specimen may be contaminated, and thus cultures can be overwhelmed with more rapidly growing species. This can be overcome by decontaminating the specimen or by inoculating it into medium that contains a cocktail of antibiotics that will inhibit faster growing bacteria and fungi.

Decontamination methods have a greater effect on rapidly growing bacteria and fungi, but they do kill mycobacteria too. Thus, it is important that the methods are chosen to be sufficiently stringent to reduce the risk of rapid growing organisms but not so stringent that the mycobacterial recovery rate suffers (Collins, Grange and Yates, 1997). The usual decontamination procedure is to use 2% sodium hydroxide. Because of its effect on mycobacterial viability, the efficacy of decontamination should be monitored closely. Not more than 5% or less than 2% of cultures should fail due to contamination. If the level exceeds 5% then too many cultures will be lost leading to diagnostic delay, whereas if less than 2% are contaminated it suggests that the decontamination is too stringent and mycobacterial viability is being adversely affected with an inevitable reduction in the sensitivity of culture. Similar considerations apply to cultures inoculated into medium containing an antibiotic cocktail (Collins, Grange and Yates, 1997).

Culture of M. tuberculosis

Culture of mycobacteria remains the cornerstone of microbiological diagnosis of tuberculosis, as it is still the most sensitive diagnostic techniques available, and the isolation of the organisms allows definitive identification, susceptibility and molecular epidemiological tests to be performed. The slow growth of *M. tuberculosis* is a challenge for the diagnostic laboratory to produce a result in as timely a fashion as possible. There are three main types of isolation media available for primary diagnosis: egg based such as Lowenstein–Jenssen, agar based such as Middlebrook 7H10 and finally liquid growth media, e.g. Middlebrook 7H9 reviewed by Collins, Grange and Yates (1997). Growth on liquid media tends to be faster, and automatic systems can be used to monitor growth and identify positive samples rapidly. Many commercial liquid culture systems have been marketed from manual growth detection to continuously monitored systems (Scarparo *et al.*, 2002; Bemer *et al.*, 2004). Almost all of the currently available automated products identify mycobacterial growth by detecting carbon dioxide production. In addition, semi-automated systems are available that provide the advantages of liquid culture without the large capital investment required for a fully automated system (Adjers-Koskela and Katila, 2003). Agar-based solid and liquid media can be used for primary isolation of organisms in specimens from sterile sites, but contaminated samples must be inoculated into liquid medium supplemented with an antibiotic cocktail. Positive specimens inoculated into liquid broth media will usually signal positive within 3 weeks, and samples can usually be discarded after 6 weeks as negative. Egg-based solid medium must be inspected at least weekly, and growth can usually be

seen after 4 weeks. Cultures can be discarded as negative after 8 weeks, although some precious samples, e.g. CSF, may be incubated for a further 4 weeks (Collins, Grange and Yates, 1997).

Cross-Contamination

Care must be taken to ensure that cross-contamination does not take place in the clinical laboratory (Bhattacharya *et al.*, 1998). With careful practices, the rate should be below 1% (Ruddy *et al.*, 2002), but in some circumstances when these methods have broken down, cross-contamination has been much more common (de C. Ramos *et al.*, 1999). Methods of prevention of cross-contamination is reviewed in Bhattacharya *et al.* (1998) and Ruddy *et al.* (2002).

Identification

Identifying mycobacteria can impose a significant delay in the diagnostic process. There are three main approaches, phenotypic tests, DNA hybrisation/amplification and direct sequencing of the 16S ribosomal gene. Phenotypic tests are the longest established methods but require considerable skill and experience to perform (Collins, Grange and Yates, 1997). Once growth has been detected, a ZN film should be examined and the morphology of the bacteria should be determined. *Mycobacterium tuberculosis* exhibits a characteristic cording which is almost diagnostic (Plate 6). The isolate can then be inoculated onto a medium such as para-aminobenzoic acid upon which *M. tuberculosis* will not grow. These two features together with the appearance of the colonies on solid medium make a presumptive identification of *M. tuberculosis* usually within 2 weeks. Definitive identification may be performed using the tests described above.

DNA hybridisation and amplification methods have transformed the speed of mycobacterial identification. These rapid techniques, many of which are produced commercially, use a direct hybridisation method such as Accuprobe which can identify culture isolates within a few hours (Scarparo *et al.*, 2001; Lebrun *et al.*, 2003). The species which are included are *M. tuberculosis*, *M. avium*, *M. intracellulare*, *M. kansasii* and *M. gordoni*. Alternatively, a DNA amplification step with or without reverse phase hybridisation to detect specific *M. tuberculosis* genes and several commercial systems such as Probtec or INNO-LIPA are available (Lebrun *et al.*, 2003; Wang, Sng and Tay, 2004). When incorporated into the routine workflow of a laboratory, they permit *M. tuberculosis* to be identified in 2–3 days (Davies *et al.*, 1999; Drobniewski *et al.*, 2003). Direct amplification of the 16S ribosomal gene and sequencing enable a definitive identification of any mycobacterial species. This approach is now more widely applied and has resulted in significant time saving to identification and the description of a large number of slow-growing mycobacteria which may be pathogens in humans (Tortoli, 2003). In addition to the 16S gene, the DNA gyrase gene *gyrA* and RNA-dependent polymerase *rpoB* amplicons can be used for speciation (Kim *et al.*, 1999; Dauendorffer *et al.*, 2003).

DNA Amplification Techniques

Increasingly, nucleic acid amplification methods are integrated into routine mycobacterial diagnosis. They are principally approved for the rapid diagnosis of tuberculosis from respiratory samples. Several techniques have now been licensed for use on smear-positive samples, although studies show that these can be useful in examining smear-negative samples (McHugh, personal communication). The application of NAA methods to diagnosis of nonrespiratory samples is more problematical, as biopsy pleural fluid samples often have low numbers of bacteria present making the test less sensitive. However, it is in these difficult cases that the test has the greatest potential to influence diagnosis and alter treatment plans. Great care must be taken in selecting specimens for processing and in reporting the results (Conaty *et al.*, 2005).

An integrated conventional and
molecular diagnostic service for
tuberculosis

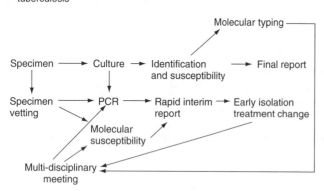

Figure 10.1 An integrated conventional and molecular diagnostic service for tuberculosis.

Molecular Susceptibility

The molecular mechanism of resistance is complex for most anti-tuberculosis agents, making reliable molecular susceptibility testing difficult. However, for rifampicin over 90% of resistance mutations are found in a small section of the *rpoB* gene and thus methods that detect point mutations in this region are both sensitive and specific (Telenti *et al.*, 1993; Ramaswamy and Musser, 1998). The most useful methods are direct sequencing from *rpoB* amplicons and commercial kit methods which are available based on a NAA and hybridising stage.

Service Provision

Tuberculosis diagnosis cannot be provided in isolation, and the results obtained should be integrated into the multidisciplinary team discussions required to manage patients with tuberculosis. An overview of the diagnostic process is provided in Figure 10.1.

TREATMENT

The Pre-Antibiotic Era

The association between tuberculosis, poverty and overcrowding led to the theory that the disease could be treated by providing the patient with good food, bed rest and plenty of fresh air. The Sanatorium movement swept Europe and North America in the nineteenth century and provided hope and a cure for many. The isolation imposed on the patients may have helped to reduce the spread of tuberculosis, but there is little evidence that sanatorium treatment significantly contributed to cure (Dormandy, 1999). Before anti-tuberculosis antibiotics became available, surgical treatment proved beneficial with lobectomy and pneumonectomy which removed considerable bacterial load, and in some difficult cases of MDRTB such methods are again used.

The Antibiotic Era

The real revolution in tuberculosis therapy came with the description of streptomycin by Waksman and Schatz (Schatz and Waksman, 1944). They worked on the theory that mycobacteria were soil organisms and that the soil may contain organisms that would produce substances active against them in their competition for the same ecological niche. Attention was focussed on the *Streptomycetes*. The drug was rapidly introduced into clinical practice in the United States, but in the United Kingdom, clinical trials were performed to demonstrate its efficacy.

This was the first example of a placebo-controlled comparative clinical trial, and its unexpected result laid the foundation for the modern drug development paradigm. Although streptomycin monotherapy brought about a rapid clinical improvement, the benefit was transitory and later there was no significant difference in outcome (Crofton and Mitchison, 1948). Moreover, many of the infecting strains had developed resistance. Fortunately, help was at hand: *para*-amino salicylic (PAS) acid was developed by Lehman in Denmark and combination therapy prevented the emergence of resistance (Lehman, 1946). The introduction of isoniazid in 1952 provided another highly bacteridical drug (Middlebrook and Cohn, 1953; Middlebrook and Dressler, 1954). A series of clinical trials performed by the United Kingdom Medical Research Council and the United States Public Health Service concluded that the optimum therapy for pulmonary tuberculosis was with these three drugs for 3 months followed by a further 15 of isoniazid and PAS (Fox and Mitchison, 1975). Many other anti-tuberculosis drugs were introduced in the 1950s including ethambutol and second line agents such as thiacetazone and prothionamide. Pyrazinamide was developed but was soon abandoned due to intolerance to the high doses employed. A crucial trial in Madras demonstrated that the results of domiciliary treatment were as good as chemotherapy given in the context of a sanatorium (Fox, Ellard and Mitchison, 1999). The tuberculosis hospitals began to close, and the focus moved to community-based treatment regimens.

The introduction of rifampicin permitted the duration of therapy to be reduced to 9 months (Furesz and Timball, 1963; Canetti *et al.*, 1968), and the delineation of the role of pyrazinamide in chemotherapy permitted a further reduction to 6 months (Fox and Mitchison, 1975). The base regimen consists of an intensive phase of 2 months of rifampicin, isoniazid and parazinamide followed by a consolidation phase of 4 months of rifampicin and isoniazid. Further attempts to shorten the duration of therapy to 5 or 4 months resulted in unacceptably high relapse rates. Further trials demonstrated that intermittent therapy, given two or three times a week, was as effective as daily treatment. The series of clinical trials leading to the current therapeutic regimen are described in Fox, Ellard and Mitchison (1999).

Directly Observed Therapy Short Course

The recognition of the growing tuberculosis epidemic in the 1990s led to the development of new strategies to deliver anti-tuberculosis chemotherapy more effectively. The directly observed therapy short course (DOTS) strategy was devised. This made use of the known efficacy of chemotherapy in a clinical trial setting where all of the doses of the trial agent are observed. In countries where patients are required to pay for their drugs, the poorest member of the community, most likely to suffer tuberculosis, are likely to stop treatment before bacteriological cure can be achieved. As the patient feels well long before bacteriological cure, this is a major problem in tuberculosis therapy. The homeless and those addicted to drugs often lack social structures or self-discipline to complete a 6-month course of therapy. In DOTS each of the doses of treatment is observed. This may be by a health care worker specifically employed by the tuberculosis-control program, a local pharmacist or a responsible family member. By increasing the number of doses taken, the risk of relapse and resistance is reduced. The DOTS idea depends on the development of effective tuberculosis treatment and control program to deliver directly observed therapy. However, DOTS is not a panacea. It can be difficult to organise and in some countries the results of such programs have been disappointing (Walley *et al.*, 2001).

Antibiotic Resistance

The piecemeal introduction of new chemotherapeutic agents highlighted the importance of preventing the development of resistance (Crofton and Mitchison, 1948; Gillespie, 2002). In any prokaryotic genome, mutations are constantly occurring due to base changes caused by exogenous agents, DNA polymerase errors, deletions, insertions and duplications (David, 1970). For prokaryotes there is a constant rate of spontaneous mutation of 0.0033 mutations/DNA replication that is uniform for a diverse spectrum of organisms (Drake, 1999). The mutation rate for individual genes varies significantly between and within genes, approximately 10–9 for rifampicin and 10–6 for isoniazid (David, 1970; Billington, McHugh and Gillespie, 1999; Gillespie, 2002). The reasons for these variations are uncertain but are thought to be under the influence of local DNA sequence. Thus a patient with tuberculosis with approximately 10^{13} organisms in the body will already have 10 000 rifampicin resistance mutants. To prevent these surviving and coming to dominate the patient's infection the pre-existent mutants are killed by the second or third drug of the combination. It has been assumed that the risk of resistance of an organism developing resistance to two agents is the multiple of the risk of each separately, but the risk of mutants emerging in a patient depends partly on this and the size of bacterial polutations with compartments. Therefore, risk of resistance may be more accurately calculated using the formula $P = 1 - (1 - r)^n$, where P is the probability of drug resistance emerging, r is the probability of drug-resistant mutants and n is the number of bacilli in a lesion usually calculated to be 10^8 per lesion (Shimao, 1987). Where patients have developed extensive cavities, or empyema, the bacterial population at one site may be very much higher, increasing the risk of resistance mutants emerging. Also, poor drug penetration into the fibrous cavities or into empyema can significantly reduce the effective dose and produce the situation where there effectively is monotherapy (Lipsitch and Levin, 1998). It is for these reasons that the emergence of drug resistance is associated with poor adherence to an adequately prescribed regimen, and inadequate regimen, with patients who have multiple cavities or empyema (Iseman and Madsen, 1991; Elliott *et al.*, 1995).

Molecular Mechanism of Resistance

Telenti and colleagues were the first to determine the site of mutation that resulted in rifampicin resistance in *M. tuberculosis*. They used the evidence that *E. coli* became resistant to rifampicin through mutation in the β subunit of the *rpoB* gene and sequenced this gene from a series of epidemiologically unrelated strains. They showed that almost all rifampicin-resistant isolates had mutations in a small region of *rpoB*. Subsequently, further clinical studies indicated that mutations are found in this region in up to 95% of resistant isolates (Telenti *et al.*, 1993). For other antibiotics the resistance mechanisms are more heterogeneous. *Mycobacterium tuberculosis* catalase KatG is a peroxynitratase that generates acyl, acylperoxo and pyridyl radicals of isoniazid (Zhang *et al.*, 1992). These inhibit the β-keto-acyl ACP synthase, KasA which interferes with effective cell-wall synthesis (Sreevatsan *et al.*, 1997a). Resistance emerges by modification of catalase by partial or total gene deletions, point mutations or insertions, which lead to the abolition or diminution of its catalase activity (Zhang *et al.*, 1992). This is the most effective mechanism of resistance, as it is found in more than 80% of isoniazid-resistant strains (Ramaswamy and Musser, 1998). Alternatively, low-level resistance can be caused by point mutations in the regulatory region of *inhA* operon resulting in over expression of *inhA* (Wilson, De Lisle and Collins, 1995). Strains with this mutation have normal mycolic acid synthesis but low-level resistance to isoniazid. Point mutations have also been demonstrated in the regulatory region of *ahpC*, which compensate for the effects of absent or reduced catalase (KatG) function and do not directly result in resistance.

The molecular mechanism of resistance has been described for most of the remaining anti-tuberculosis drugs including pyrazinamide (Scorpio and Zhang, 1996), streptomycin (Honore and Cole, 1994; Cooksey *et al.*, 1996), ethambutol (Belanger *et al.*, 1996) and fluoroquinolones (Cambau *et al.*, 1994). For some of the antibiotics, as many as half of the mechanism of resistance is unexplained.

Drug Toxicity

In most instances, tuberculosis treatment is well tolerated, but the main anti-tuberculosis drugs are associated with significant adverse effects. Rifampicin is associated with acute hepatitis most likely to occur in combination with isoniazid. It induces liver enzymes and can significantly affect the metabolism of other drugs. It interferes with the activity of antiretroviral drugs. At higher doses given in intermittent therapy, patients may suffer a flu-like syndrome (Dickinson *et al.*, 1977). Isoniazid is associated with acute hepatitis caused by the hydrazine radical produced *in vivo*. It is for this reason that hepatitis is more common with isoniazid and rifampicin combination than with either of them alone. Pyrazinamide is associated with hepatic damage and up to 5% of patients may not be able to tolerate the drug. Streptomycin has now largely fallen out of use in part because of the risk of renal and vestibular toxicity. In addition, it must be given parenterally, and this may be a problem in resource-poor countries where the sterility of needles cannot be guaranteed. In the context of drug intolerance or resistance, fluoroquinolones such as ciprofloxacin, ofloxacin or moxifloxacin can be used and are well tolerated and are especially useful in patients with pre-existent hepatic disease (Yew *et al.*, 1992; Kennedy *et al.*, 1993).

Management of Drug-Resistant Tuberculosis

MDRTB is defined as infection with an organism resistant to at least isoniazid and rifampicin. This definition is practical because of the critical role these drugs play in successful treatment. Drug-resistant tuberculosis has a much higher mortality between 30 and 60%. There is an increased need for operative intervention to remove infected cavities or perform pneumonectomy. Thus the management of MDRTB differs significantly from susceptible cases where treatment is protocol driven. In the case of MDRTB each patient must be judged as an individual and the regimen must be tailored to the susceptibilities of the infecting organisms and the presence of other complicating features such as drug toxicity or organ failure. There are some basic rules which govern the construction of a therapeutic regimen to treat unexpected or known resistant isolates; they are:

1. *In vitro* susceptibility should be performed on all isolates.
2. In areas where rates of resistance are greater than 5%, all initial treatment regimens should contain four drugs to reduce the risk that further resistance will emerge.
3. Directly observed therapy should be used to ensure that the regimen is taken correctly.
4. The patient should be prescribed a regimen that contains three drug to which the organism is susceptible.
5. New drugs should only be added to the regimen in pairs and ideally only if the strain is known to be susceptible. Piecemeal addition of single agent is likely to result in additional resistance.
6. Drugs used to treat MDRTB often lack the bactericidal and sterilising efficacy of first-line agents, and thus the duration of treatment will need to be longer for up to 2 years.

Treatment of Smear-Negative Pulmonary Tuberculosis

Attempts have been made to utilise treatment regimens shorter than the standard 6 months. A study performed in Hong Kong of daily treatment with isoniazid, rifampicin, pyrazinamide and streptomycin for 2 or 3 months was associated with a relapse rate of 32 or 13%, respectively, during 5 years of follow-up. A later study of 4-month treatment with an initial 4-drug regimen for 2 months followed by a consolidation phase of 2 months of isoniazid and rifampicin was successful. HIV-infected individuals infected with *M. tuberculosis* are often smear negative. Although shorter regimens can be effective, it is probably advisable to treat all patients with pulmonary tuberculosis with a standard 6-month regimen.

Treatment of Extrapulmonary Tuberculosis

The same basic principles underlie the treatment of extrapulmonary tuberculosis as those discussed above. It should be noted, however, that it may be more difficult for drugs to penetrate into protected sites such as the CSF. There are few controlled clinical trials to inform the duration of treatment, but these suggest that standard therapeutic regimens including the four first-line agents given between 6 and 9 months are adequate. Tuberculous meningitis and tuberculomas provide an exception to this advice, and most authorities suggest treatment durations of between 9 and 12 months. There is considerable trial evidence that corticosteroids have a valuable role to play in the management of tuberculous meningitis and pericarditis. Operative intervention is more often required in extrapulmonary tuberculosis and may be required to drain psoas abscesses, to stabilise the vertebral column in cases of spinal tuberculosis. Emergency surgical intervention may be required to drain tuberculous pus causing pressure on the spinal cord.

TUBERCULOSIS CONTROL

National Infection Control

Early efforts to control tuberculosis were centered on the Sanatorium movement. Segregation of sputum smear-positive patients from the general population may have reduced the risk of transmission. More important was the increasing wealth of the people in industrialised countries providing better nutrition which is though to increase resistance to infection and the emergence of disease. Improved housing and reduced family size also played its part, as there is evidence that the risk of transmission can be related to average room size and the number of individuals sleeping in the same room.

Bovine tuberculosis remained a threat to public health and was tackled by pasteurisation of the milk. In addition, government-sponsored eradication campaigns in most industrialised countries were instituted with compulsory tuberculin testing of cattle and slaughter of those found to be infected. The advent of anti-tuberculosis chemotherapy meant that smear-positive tuberculosis could be identified and rapidly rendered noninfectious through treatment. National programs such as mass – miniaturised radiography assisted case finding and resulted in a decline of 4% per year. It was not just industrialised countries who with effective case finding and treatment could reduce the incidence of disease. Tanzania, one of the poorest countries in the world, developed an efficient tuberculosis control program that brought about a reduction in the number of cases nationally from 25 000 in 1961 to 12 000 in 1985.

Hospital Infection Control

Institutions such as hospitals and prison are high-risk environments for tuberculosis transmission. Individuals are living close proximity and many may be immunocompromised as a result of immunosuppressive therapy or HIV infection. Patients with sputum smear-positive tuberculosis should be managed as an outpatient if this is possible. If hospital admission is required then they should be boarded in a side room until they have received 2 weeks of therapy. Special precautions are required for patients suspected of being infected with a multiple drug-resistant organism. A list of patients who should be considered at high risk of MDRTB is listed in Table 10.1. Specimens for rapid diagnosis should be obtained, including sputum smear, to determine whether the risk of transmission is high, and sputum should be taken for culture. It is in this circumstance that DNA amplification methods

Table 10.1 Initial assessment of patients with suspected tuberculosis

1. Diagnosis – Does the patient have tuberculosis?	
Suspected	Symptoms and signs} persistent cough, night sweats
	Suggestive of tuberculosis} weight loss, fever, CXR changes with or without microbiological/histological support (i.e. smear positive/PPD strongly positive)
Confirmed	Culture positive disease
2. Infectiousness – Is the patient with suspected or confirmed tuberculosis likely to be infectious to others?	
Infectious	(a) Sputum smear positive pulmonary disease
	(b) For suspected pulmonary disease assume infectious until proven otherwise
	(c) Disease of the airway, e.g. laryngeal tuberculosis
	(d) The following nonpulmonary lesions; open abscess or lesions in which the concentration of organisms is high and drainage from the lesion extensive
Potentially infectious	(a) Sputum smear negative pulmonary disease in which one or more cultures positive or culture results not yet known
Non-infectious	(a) Sputum smear negative, culture negative pulmonary disease
	(b) Non-pulmonary disease except those mentioned above
3. Drug resistance or MDRTB – Is the patient with suspected or confirmed tuberculosis likely to have drug-resistant disease?	
Suspected drug resistance	(a) Previous treatment for tuberculosis
	(b) Contact with a person with known drug-resistant disease
	(c) Birth, travel or residence in an area with a high prevalence of MDRTB, e.g. countries in Asia, Africa, Latin America, South & East Europe
	(d) HIV infection
	(e) Failure to respond clinically to a standard treatment regime, e.g. temperature still elevated after 2 weeks treatment
	(f) Poor compliance with therapy
	(g) Prolonged smear or culture positive while on treatment
Confirmed	(a) Resistance to one or more first line anti-tuberculosis drugs confirmed by conventional or molecular techniques

to establish the diagnosis of tuberculosis are of particular importance. Methods that determine the *rpoB* genotype provide a rapid mechanism of determining rifampicin susceptibility with approximately 95% sensitivity (Honore *et al.*, 1993). This can be used as a surrogate marker of multiple drug resistance, and rifampicin monoresistance is rare, being usually accompanied at least by isoniazid resistance. Such patients should be managed in negative-pressure single-side rooms. Health care workers should wear personal protective equipment including impervious gowns and high-efficiency particulate masks. When patients with MDRTB are to undergo procedures likely to generate aerosols, these should be performed in a negative pressure room. A summary of the requirements is found in Table 10.1. Health care workers who are immunosuppressed should not work with smear-positive tuberculosis patients because of their increased susceptibility to infection and disease progression. They should be able to exclude themselves from contact with such patients without prejudice.

Vaccination

The first attempt to make a vaccination against tuberculosis was made by Koch who injected heat-killed bacilli. The intense reaction tuberculin-positive individuals expressed became known as the Koch reaction. This reaction was not accompanied by effective protection. Calmette and Guérin produced a prototypic vaccine by passaging *M. bovis* 231 times through a medium containing glycerol and ox bile. The bacille Calmette–Guérin (BCG) was found to be safe and gained wide acceptance throughout Europe. Many studies have confirmed the efficacy of BCG against tuberculosis especially against miliary forms of the disease such as meningitis (Brewer, 2000). Despite this, the use of BCG remains controvertial. Physicians in the US oppose its use preferring the ability to be able to use the Mantoux reaction to diagnose acute disease. Compilations of the results of clinical trials show a wide variation in the protective efficacy with results ranging from –2 to 70% (Brewer, 2000) with an average effectiveness of approximately 50%. This wide discrepancy may be due to different immune responses to mycobacterial antigens during early life. In a country with high tuberculosis prevalence, exposure to mycobacterial antigens through environmental as well as pathogenic species may mean that BCG cannot provide much additional protection.

The manifest shortcomings of BCG has prompted a widespread search for an alternative vaccine. Research groups are searching the genome for potential new vaccine candidates. Several candidate vaccines are at various stages of clinical development, but all are some years from introduction into routine practice.

REFERENCES

Adjers-Koskela, K. and Katila, M. L. (2003) Susceptibility testing with the manual mycobacteria growth indicator tube (MGIT) and the MGIT 960 system provides rapid and reliable verification of multidrug-resistant tuberculosis. J Clin Microbiol 41, 1235–1239.

Advisory Committee on Dangerous Pathogens. (2001) The management, design and operation of microbiological containment laboratories.

Alland, D., Kalkut, G. E., Moss, A. R. *et al*. (1994) Transmission of tuberculosis in New York City. An analysis by DNA fingerprinting and conventional epidemiologic methods. N Engl J Med 330, 1710–1716.

Anon. (2000) Diagnostic standards and classification of tuberculosis in adults and children. Am J Respir Crit Care Med 161, 1376–1395.

Association of State and Territorial Public Health Directors and Centers for Disease Control. (1995) *Mycobacterium tuberculosis*: assessing your laboratory.

Banner, A. S. (1979) Tuberculosis. Clincal aspects and diagnosis. Arch Intern Med 139, 1387–1390.

Beggs, C. B., Noakes, C. J., Sleigh, P. A. *et al*. (2003) The transmission of tuberculosis in confined spaces: an analytical review of alternative epidemiological models. Int J Tuberc Lung Dis 7, 1015–1026.

Belanger, A. E., Besra, G. S., Ford, M. E. *et al*. (1996) The embAB genes of *Mycobacterium avium* encode an arabinosyl transferase involved in cell wall arabinan biosynthesis that is the target for the antimycobacterial drug ethambutol. Proc Natl Acad Sci U S A 93, 11919–11924.

Bemer, P., Bodmer, T., Munzinger, J. *et al*. (2004) Multicenter evaluation of the MB/BACT system for susceptibility testing of *Mycobacterium tuberculosis*. J Clin Microbiol 42, 1030–1034.

Bhattacharya, M., Dietrich, S., Mosher, L. *et al*. (1998) Cross-contamination of specimens with *Mycobacterium tuberculosis*: clinical significance, causes, and prevention. Am J Clin Pathol 109, 324–330.

Bifani, P. J., Mathema, B., Kurepina, N. E. and Kreiswirth, B. N. (2002) Global dissemination of the *Mycobacterium tuberculosis* W-Beijing family strains. Trends Microbiol 10, 45–52.

Billington, O. J., McHugh, T. D. and Gillespie, S. H. (1999) Physiological cost of rifampin resistance induced *in vitro* in *Mycobacterium tuberculosis*. Antimicrob Agents Chemother 43, 1866–1869.

Blower, S. M., McLean, A. R., Porco, T. C. *et al*. (1995) The intrinsic transmission dynamics of tuberculosis epidemics. Nat Med 1, 815–821.

Braden, C. R. (1995) Infectiousness of a university student with laryngeal and cavitary tuberculosis. Clin Infect Dis 21, 565–570.

Breathnach, A. S., de Ruiter, A., Holdsworth, G. M. *et al*. (1998) An outbreak of multi-drug-resistant tuberculosis in a London teaching hospital. J Hosp Infect 39, 111–117.

Brewer, T. F. (2000) Preventing tuberculosis with bacillus Calmette-Guerin vaccine: a meta-analysis of the literature. Clin Infect Dis 31 (Suppl. 3), S64–S67.

Brosch, R., Gordon, S. V., Marmiesse, M. *et al*. (2002) A new evolutionary scenario for the *Mycobacterium tuberculosis* complex. Proc Natl Acad Sci U S A 99, 3684–3689.

Brosch, R., Pym, A. S., Gordon, S. V. and Cole, S. T. (2001) The evolution of mycobacterial pathogenicity: clues from comparative genomics. Trends Microbiol 9, 452–458.

Cambau, E., Sougakoff, W., Besson, M. *et al*. (1994) Selection of a gyrA mutant of *Mycobacterium tuberculosis* resistant to fluoroquinolones during treatment with ofloxacin. J Infect Dis 170, 479–483.

Canetti, G. (1955) *The Tubercle Bacillus in the Pulmonary Lesion of Man*, Springer, New York, NY.

Canetti, G., Le Lirzin, M., Porven, G. *et al*. (1968) Some comparative aspects of rifampicin and isoniazid. Tubercle 49, 367–376.

Cole, S. T., Brosch, R., Parkhill, J. *et al*. (1998) Deciphering the biology of *Mycobacterium tuberculosis* from the complete genome sequence. Nature 393, 537–544.

Collins, C. H., Grange, J. M. and Yates, M. D. (1997) *Tuberculosis Bacteriology: Organization and Practice*, 2nd edn, Butterworh Heineman, London.

Conaty, S. J., Claxton, A. P., Enoch, D. *et al*. (2005) The clinical interpretation of nucleic acid amplification tests for tuberculosis. Do rapid tests change treatment decisions? J Infect 50, 187–192.

Cooksey, R. C., Morlock, G. P., McQueen, A. *et al*. (1996) Characterization of streptomycin resistance mechanisms among *Mycobacterium tuberculosis* isolates from patients in New York City. Antimicrob Agents Chemother 40, 1186–1188.

Corbett, E. L., Watt, C. J., Walker, N. *et al*. (2003) The growing burden of tuberculosis: global trends and interactions with the HIV epidemic. Arch Intern Med 163, 1009–1021.

Crofton, J. and Mitchison, D. A. (1948) Streptomycin resistance in pulmonary tuberculosis. Br Med J 2, 1009–1015.

Cruickshank, D. B. (1952) Bacteriology, In Modern Practice in Tuberculosis. (ed. T. H. Sellors and J. L. Livingstone) p. 53 Butterworth; London.

Dauendorffer, J. N., Guillemin, I., Aubry, A. *et al*. (2003) Identification of mycobacterial species by PCR sequencing of quinolone resistance-determining regions of DNA gyrase genes. J Clin Microbiol 41, 1311–1315.

David, H. L. (1970) Probability distribution of drug-resistant mutants in unselected populations of *Mycobacterium tuberculosis*. Appl Microbiol 20, 810–814.

Davies, A. P., Newport, L. E., Billington, O. J. and Gillespie, S. H. (1999) Length of time to laboratory diagnosis of *Mycobacterium tuberculosis* infection: comparison of in-house methods with reference laboratory results. J Infect 39, 205–208.

de C. Ramos M., Soini, H., Roscanni, G. C. *et al*. (1999) Extensive cross-contamination of specimens with *Mycobacterium tuberculosis* in a reference laboratory. J Clin Microbiol 37, 916–919.

Dickinson, J. M., Mitchison, D. A., Lee, S. K. *et al*. (1977) Serum rifampicin concentration related to dose size and to the incidence of the 'flu' syndrome during intermittent rifampicin administration. J Antimicrob Chemother 3, 445–452.

Dormandy, T. (1999) *The White Death: A History of Tuberculosis*, The Hambeldon Press, London.

Drake, J. W. (1999) The distribution of rates of spontaneous mutation over viruses, prokaryotes, and eukaryotes. Ann N Y Acad Sci 870, 100–107.

Drobniewski, F. A., Caws, M., Gibson, A. and Young, D. (2003) Modern laboratory diagnosis of tuberculosis. Lancet Infect Dis 3, 141–147.

Dubos, R. and Dubos, J. (1952) *The White Plague: Tuberculosis, Man and Society*, Rutgers University Press, Piscataway, NJ.

Dye, C. (2000) Tuberculosis, 2000–2010: control, but not elimination. Int J Tuberc Lung Dis 4, S146–S152.

Elliott, A. M., Berning, S. E., Iseman, M. D. and Peloquin, C. A. (1995) Failure of drug penetration and acquisition of drug resistance in chronic tuberculous empyema. Tuber Lung Dis 76, 463–467.

Fox, W. and Mitchison, D. A. (1975) Short-course chemotherapy for pulmonary tuberculosis. Am Rev Respir Dis 111, 325–353.

Fox, W., Ellard, G. A. and Mitchison, D. A. (1999) Studies on the treatment of tuberculosis undertaken by the British Medical Research Council tuberculosis units, 1946–1986, with relevant subsequent publications. Int J Tuberc Lung Dis 3, S231–S279.

Furesz, S. and Timball, M. T. (1963) The antibacterial activity of rifamycins. Chemotherapia 7, 200.

Gillespie, S. H. (2002) Evolution of drug resistance in *Mycobacterium tuberculosis*: clinical and molecular perspective. Antimicrob Agents Chemother 46, 267–274.

Glynn, J. R. (1998) Resurgence of tuberculosis and the impact of HIV Infection. Br Med Bull 54, 579–593.

Hobby, G. L., Holman, A. P., Iseman, M. D. and Jones, J. M. (1973) Enumeration of tubercle bacilli in sputum of patients with pulmonary tuberculosis. Antimicrob Agents Chemother 4, 94–104.

Hoge, C. W., Fisher, L., Donnell, H. D. Jr. et al. (1994) Risk factors for transmission of Mycobacterium tuberculosis in a primary school outbreak: lack of racial difference in susceptibility to infection. Am J Epidemiol 139, 520–530.

Holmberg, S. D. (1990) The rise of tuberculosis in America before 1820. Am Rev Respir Dis 142, 1228–1232.

Honore, N. and Cole, S. T. (1994) Streptomycin resistance in mycobacteria. Antimicrob Agents Chemother 38, 238–242.

Honore, N., Perrani, E., Telenti, A. et al. (1993) A simple and rapid technique for the detection of rifampin resistance in Mycobacterium Leprae. Int J Lepr Other Mycobact Dis 61, 600–604.

Indrigo, J., Hunter, R. L. Jr and Actor, J. K. (2003) Cord factor trehalose, 6,6′-dimycolate (TDM) mediates trafficking events during mycobacterial infection of murine macrophages. Microbiology 149, 2049–2059.

Iseman, M. D. and Madsen, L. A. (1991) Chronic tuberculous empyema with bronchopleural fistula resulting in treatment failure and progressive drug resistance. Chest 100, 124–127.

Kaplan, G., Post, F. A., Moreira, A. L. et al. (2003) Mycobacterium tuberculosis growth at the cavity surface: a microenvironment with failed immunity. Infection Immunity 71, 7099–7108.

Kato-Maeda, M., Rhee, J. T., Gingeras, T. R. et al. (2001) Comparing genomes within the species Mycobacterium tuberculosis. Genome Res 11, 547–554.

Kaufmann, S. H. E. (2001) How can immunology contribute to the control of tuberculosis. Nat Immunol 1, 20–30.

Kennedy, N., Fox, R., Uiso, L. et al. (1993) Safety profile of ciprofloxacin during long-term therapy for pulmonary tuberculosis. J Antimicrob Chemother 32, 897–902.

Kim, B. J., Lee, S. H., Lyu, M. A. et al. (1999) Identification of mycobacterial species by comparative sequence analysis of the RNA polymerase gene (rpoB). J Clin Microbiol 37, 1714–1720.

Kuemmerer, J. M. and Comstock, G. W. (1967) Sociologic concomitants of tuberculin sensitivity. Am Rev Respir Dis 96, 885–892.

Lebrun, L., Gonullu, N., Boutros, N. et al. (2003) Use of INNO-LIPA assay for rapid identification of mycobacteria. Diagn Microbiol Infect Dis 46, 151–153.

Lehman, J. (1946) Para-aminosalacylic acid in the treatment of tuberculosis. Lancet i, p. 15.

Lipsitch, M. and Levin, B. R. (1998) Population dynamics of tuberculosis treatment: mathematical models of the roles of non-compliance and bacterial heterogeneity in the evolution of drug resistance. Int J Tuberc Lung Dis 2, 187–199.

Lurie, M. B. (1931) The correlation between the histological changes and fate of living tubercle bacilli in the organs of tuberculous rabbits. J Exp Med 55, 31–56.

Lurie, M. B. (1964) Resistance to tuberculosis: experimental studies in native and acquired defence mechanisms.

Maguire, H., Dale, J. W., McHugh, T. D. et al. (2002) Molecular epidemiology of tuberculosis in London, 1995–7 showing low rate of active transmission. Thorax 57, 617–622.

Mehta, J. B., Dutt, A., Harvill, L. and Mathews, K. M. (1991) Epidemiology of extrapulmonary tuberculosis. A comparative analysis with pre-AIDS era. Chest 99, 1134–1138.

Middlebrook, G. and Cohn, M. L. (1953) Some observations on the pathogenicity of isoniazid resistant tubercle bacilli. Science 118, 297–299.

Middlebrook, G. and Dressler, S. H. (1954) Clinical evaluation of isoniazid. Am Rev Tuberc. 70, 1102–1103.

Moss, A. R., Alland, D., Telzak, E. et al. (1997) A city-wide outbreak of a multiple-drug-resistant strain of Mycobacterium tuberculosis in New York. Int J Tuberc Lung Dis 1, 115–121.

Rad, M. E., Bifani, P., Martin, C. et al. (2003) Mutations in putative mutator genes of Mycobacterium tuberculosis strains of the W-Beijing family. Emerg Infect Dis 9, 838–845.

Ramaswamy, S. and Musser, J. M. (1998) Molecular genetic basis of antimicrobial agent resistance in Mycobacterium tuberculosis: 1998 update. Tuber Lung Dis 79, 3–29.

Rieder, H. L. (1999) Epidemiologic Basis of Tuberculosis Control. International Union Against Tuberculosis and Lung Diseases, Paris.

Riley, R. L., Mills, C. C., O'grady, F. et al. (1962) Infectiousness of air from a tuberculosis ward. Ultraviolet irradiation of infected air: comparative infectiousness of different patients. Am Rev Respir Dis 85, 511–525.

Ruddy, M., McHugh, T. D., Dale, J. W. et al. (2002) Estimation of the rate of unrecognized cross-contamination with Mycobacterium tuberculosis in London microbiology laboratories. J Clin Microbiol 40, 4100–4104.

Russell, D. G. (2001) Mycobacterium tuberculosis: here today, and here tomorrow. Nat Rev Mol Cell Biol 2, 569–577.

Salo, W. L., Aufderheide, A. C., Buikstra, J. and Holcomb, T. A. (1994) Identification of Mycobacterium tuberculosis DNA in a pre-Columbian Peruvian mummy. Proc Natl Acad Sci USA 91, 2091–2094.

Scarparo, C., Piccoli, P., Rigon, A. et al. (2001) Direct identification of mycobacteria from MB/BacT alert 3D bottles: comparative evaluation of two commercial probe assays. J Clin Microbiol 39, 3222–3227.

Scarparo, C., Piccoli, P., Rigon, A. et al. (2002) Evaluation of the BACTEC MGIT, 960 in comparison with BACTEC 460 TB for detection and recovery of mycobacteria from clinical specimens. Diagn Microbiol Infect Dis 44, 157–161.

Schatz, A. and Waksman, S. A. (1944) Effect of streptomycin and other antibiotic substances upon Mycobacterium tuberculosis and related organisms. Proc Soc Exp Biol Med 57, 244–248.

Scorpio, A. and Zhang, Y. (1996) Mutations in pncA, a gene encoding pyrazinamidase/nicotinamidase, cause resistance to the antituberculous drug pyrazinamide in tubercle bacillus. Nat Med 2, 662–667.

Shimao, T. (1987) Drug resistance in tuberculosis control. Tubercle 68, 5–18.

Sreevatsan, S., Pan, X., Stockbauer, K. E. et al. (1997b) Restricted structural gene polymorphism in the Mycobacterium tuberculosis complex indicates evolutionarily recent globaládissemination. Proc Natl Sci USA 94, 9869–9874.

Sreevatsan, S., Pan, X., Zhang, Y. et al. (1997a) Analysis of the oxyR-ahpC region in isoniazid-resistant and -susceptible Mycobacterium tuberculosis complex organisms recovered from diseased humans and animals in diverse localities. Antimicrob Agents Chemother 41, 600–606.

Sultan, L., Nyka, W., Mills, C. et al. (1960) Tuberculosis disseminators. A study of the variability of aerial infectivity of tuberculous patients. Am Rev Respir Dis 82, 358–369.

Supply, P., Mazars, E., Lesjean, S. et al. (2000) Variable human minisatellite-like regions in the Mycobacterium tuberculosis genome. Mol Microbiol 36, 762–771.

Supply, P., Warren, R. M., Banuls, A. L. et al. (2003) Linkage disequilibrium between minisatellite loci supports clonal evolution of Mycobacterium tuberculosis in a high tuberculosis incidence area. Mol Microbiol 47, 529–538.

Telenti, A., Imboden, P., Marchesi, F. et al. (1993) Detection of rifampicin-resistance mutations in Mycobacterium tuberculosis. Lancet 341, 647–650.

Tocque, K., Doherty, M. J., Bellis, M. A. et al. (1998) Tuberculosis notifications in England: the relative effects of deprivation and immigration. Int J Tuberc Lung Dis 2, 213–218.

Tortoli, E. (2003) Impact of genotypic studies on mycobacterial taxonomy: the new mycobacteria of the 1990s. Clin Microbiol Rev 16, 319–354.

van Creval, R., Karyadi, E., Preyers, F. et al. (2001) Increased production of interleukin 4 by CD4+ and CD8+ T cells from patients with tuberculosis is related to the presence of pulmonary cavities. J Infect Dis 181, 1194–1197.

van Crevel, R., Ottenhoff, T. H. and van der Meer, J. W. (2002) Innate immunity to Mycobacterium tuberculosis. Clin Microbiol Rev 15, 294–309.

van Embden, J. D., Cave, M. D., Crawford, J. T. et al. (1993) Strain identification of Mycobacterium tuberculosis by DNA fingerprinting: recommendations for a standardized methodology. J Clin Microbiol 31, 406–409.

Vetter, E., Torgerson, C., Feuker, A. et al. (2001) Comparison of the BACTEC MYCO/F Lytic bottle to the isolator tube, BACTEC Plus Aerobic F/bottle, and BACTEC Anaerobic Lytic/10 bottle and comparison of the BACTEC Plus Aerobic F/bottle to the Isolator tube for recovery of bacteria, mycobacteria, and fungi from blood. J Clin Microbiol 39, 4380–4386.

von Gottberg, A., Sacks, L., Machala, S. and Blumberg, L. (2001) Utility of blood cultures and incidence of mycobacteremia in patients with suspected tuberculosis in a South African infectious disease referral hospital. Int J Tuberc Lung Dis 5, 80–86.

Walley, J. D., Khan, M. A., Newell, J. N. and Khan, M. H. (2001) Effectiveness of the direct observation component of DOTS for tuberculosis: a randomised controlled trial in Pakistan. Lancet 357, 664–669.

Wang, S. X., Sng, L. H. and Tay, L. (2004) Preliminary study on rapid identification of Mycobacterium tuberculosis complex isolates by the BD ProbeTec ET System. J Med Microbiol 53, 57–59.

Wilson, T. M., De Lisle, G. W. and Collins, D. M. (1995) Effect of inhA and katG on isoniazid resistance and virulence of *Mycobacterium bovis*. Mol Microbiol 15, 1009–1015.

Yeager, H. Jr, Lacy, J., Smith, L. R. and LeMaistre, C. A. (1967) Quantitative studies of mycobacterial populations in sputum and saliva. Am Rev Respir Dis 95, 998–1004.

Yew, W. W., Lee, J., Wong, P. C. and Kwan, S. Y. (1992) Tolerance of ofloxacin in the treatment of pulmonary tuberculosis in presence of hepatic dysfunction. Int J Clin Pharmacol Res 12, 173–178.

Zhang, Y., Heym, B., Allen, B. *et al.* (1992) The catalase-peroxidase gene and isoniazid resistance of *Mycobacterium tuberculosis*. Nature 358, 591–593.

11

Non-Tuberculosis Mycobacteria

Stephen H. Gillespie

Centre for Medical Microbiology, Royal Free and University College Medical School, Hampstead Campus, London, UK

INTRODUCTION

Non-tuberculosis mycobacteria are important human pathogens. They were once classified as atypical or anonymous to distinguish them from *Mycobacterium tuberculosis* which is an obligate pathogen (Runyon 1957). They are, for the most part, environmental organisms – something which is characteristic of the genus as a whole. Thus the term is inaccurate as well as unhelpful, and the name non-tuberculosis mycobacteria (NTM) is preferred.

DESCRIPTION OF THE ORGANISM

Taxonomy

The taxonomy of NTM has been transformed by the use of sequence data from the 16S rRNA gene. This gene is approximately 1500 nucleotides containing conserved and hypervariable regions. The sequence of the two hypervariable regions, A and B can be used to speciate organisms. In mycobacteria, region A is the most useful for this purpose, as some distinct species share a sequence in region B. The linear sequence of 16S rRNA folds to give it a number of helices; variation in the sequence in these helices is typical of different groups of species (Tortoli 2003; Tortoli *et al.* 2001). For example, the addition of a cysteine at position 184 in helix 10 is typical of thermotolerant rapid growers. There are more extensive and variable ength insertions at position 455 in helix 18 of between 8 and 14 nucleotides, and these are associated with slow growers. Strains without any insertions are, with a few exceptions, rapid growers (Brown-Elliott and Wallace Jr. 2002; Tortoli 2003; Tortoli *et al.* 2001).

Alternative sequencing targets have been proposed for identification of mycobacteria including heat-shock protein *hsp65* and the gyrase genes (Hafner *et al.* 2004). Mycolic acids are β-hydroxy fatty acids with long side chains which make up a large component of the mycobacterial cell wall. They differ in the number of carbon atoms in the chain and the presence of different functional groups. The mycolic acid pattern of a cell wall is usually typical of a species. This difference can be detected by running the lipids on GLC, TLC or HPLC (Hale, Pfyffer and Salfinger 2001; Tortoli 2003).

One consequence of the interest in mycobacteria as a result of the HIV epidemic coupled with the availability of inexpensive sequencing facilities is the proliferation of new mycobacterial species that have been isolated from human specimens. This chapter will focus on those species which are most commonly isolated in a routine clinical laboratory and will confine a description of the other species to Table 11.1 for slow-growing organisms and Table 11.2 for the rapid growers. Those seeking additional details should refer to (Brown-Elliott and Wallace Jr. 2002; Tortoli *et al.* 2001).

Diagnosis of Non-Tuberculosis Mycobacteria

Isolation of obligate pathogens such as *M. tuberculosis* from a specimen is sufficient for a diagnosis, but for NTM this may not be the case. Many of these organisms are found in the body as commensal organisms or can contaminate laboratory reagents or the specimen from the environment. In many clinical circumstances multiple isolates of NTMs are required before a diagnosis can be confirmed (American Thoracic Society 1997; Hale, Pfyffer and Salfinger 2001; Joint Tuberculosis Committee 2000).

Pulmonary Infection

Pulmonary disease caused by NTM may appear subtly different from that of tuberculosis. Cavities are often thin walled, and there is less surrounding infiltrate. Spread is more contiguous with more marked involvement of the pleura (Kerbiriou *et al.* 2003). Occasionally NTM and notably *M. avium-intracellulare* may cause a single pulmonary nodule (Huang *et al.* 1999; Prince *et al.* 1989) Table 11.3.

Before confirming a diagnosis of NTM infection, it is necessary to exclude alternative diagnoses such as tuberculosis and lung malignancy that is often common in the patients suspected of having these infections. *Mycobacterium kansasii*, *M. xenopi*, *M. malmoense* and the rapid growers are part of the normal flora or can be environmental contaminants; so a single isolate is often insufficient for an unequivocal diagnosis. Thus, to establish a diagnosis of NTM infection, multiple isolates of an NTM are required in specimens from contaminated sites. A single specimen that is smear positive or from bronchoalveolar lavage is more likely to be significant (Joint Tuberculosis Committee 2000). In contrast, a single isolate from a biopsy specimen or from a sterile site can be diagnostic, provided it is supported by compatible histology and laboratory contamination is excluded.

For patients with *M. kansasii*, *M. xenopi* and *M. malmoense* infection sputum is usually an adequate sample. For HIV seronegative individuals infected with *M. avium-intracellulare*, sputum is often negative and brochial lavage samples should be obtained (American Thoracic Society 1997; Hale, Pfyffer and Salfinger 2001; Joint Tuberculosis Committee 2000).

Lymphadenitis

In the industrialised world NTMs are the commonest cause of mycobacterial lymphadenitis (Grange, Yates and Pozniak 1995). It is still important, however, to exclude the possibility of tuberculosis, usually with a DNA amplification technique. The diagnosis of NTM lymphadenitis depends on demonstrating the presence of granulomata histologically in the context of a negative tuberculin test. A single

Principles and Practice of Clinical Bacteriology Second Edition   Editors Stephen H. Gillespie and Peter M. Hawkey

Table 11.1   Summary of characteristic of rarely isolated mycobacteria

Species name	Environmental source	Description	Pathogenic associations
M. bohemicum	Not known	Shares some phenotypic characteristics with MIA complex	Lymphadenitis in immunocompromised patients
M. celatum	Not known	Shares phenotypic characteristics with *M. xenopi*	Disseminated and pulmonary infection in AIDS patients, rarely pulmonary infection or lymphadenitis in immunocompetent patients
M. conspicuum	Not known	Only grows at 37 °C in liquid media	Disseminated infection in severely immunocompromised patients
M. dorcium	Not known	Scotochromogen	Meningitis not confirmed pathogen
M. heckeshornense		Phenotypically and genetically related to *M. xenopi*	Pulmonary cavitation and infiltrates
M. interjectum		Yellow scotochromogen, related to M. *simiae*	Lymphadenitis and chronic lung infection
M. intermedium		A photochromogen related to *M. simiae*	Chronic pulmonary infection in patient with chronic obstructive pulmonary disease
M. kubicae		Strongly acid fast, rod shaped and often bent – related to *M. simiae*	Possible respiratory pathogen
M. lentiflavum		A scotochromogen related to *M. simiae*	Usually non-pathogenic, but lymphadenitis, chronic pulmonary disease, disseminated infection and abscesses described
M. palustre	River water	Yellow scotochromogen related *M. simiae* and *M. kubicae*	Possible cause of lymphadenitis
M. branderi		Non-chromogenic most closely related to *M. celatum*	Ulcerative tenosynovitis and pulmonary infection
M. heidelburgense		Phenotypically similar to *M. malmoense* but genotypically closest to *M. simiae*	Lymphadenitis and pulmonary infection described
M. triplex		Related to *M. simiae*	Disseminated infection in severely immunocompromised patients
M. elephantis		A scotochromogenic rapid grower, unrelated to other rapid growers	Lymphadenitis and possibly pulmonary infection
M. goodie		Rapid grower which shares phenotypic and genotypic similarities with *M. smegmatis*	Traumatic osteomyelitis and chronic pulmonary infection
M. immunogenum		Rapid grower sharing phenotypic and genotypic similarities with *M. abscessus* and *M. chelonei*	Cutaneous infection, keratitis and catheter related infections
M. wolinskyi		Related to *M. smegmatis*	Surgical infections, traumatic cellulitis and osteomyelitis

Table 11.2   Summary of rapidly growing mycobacteria

Group	Species	Disease associations
Mycobacterium fortuitum group	*M. fortuitum*	Localised wound infections, venous catheter infections and surgical infections notably augmentation mammoplasty. Venous catheter infections, a common problem in immunocompromised patients
	M. peregrinum	
	M. mucogenicum	
	M. senegalense	
	M. septicum	
	Third biovariant[a]	
M chelonae-abscessus group	*M. chelonae*	Wound infections, venous catheter infections and post-traumatic or surgical corneal infections
	M. abscessus	Venous catheter infections, chronic lung infection and surgical wound infections
	M. immunogenicum	Venous catheter infection
M. smegmatis group	*M. smegmatis*	Venous catheter infections, wound infections, osteomyelitis and lung infection
	M. goodie	Cellulitis, bursitis, osteomyelitis and lipoid pneumonia
	M. wolinski	Cellulitis and traumatic osteomyelitis

[a] Several species have been described within this taxon: *Mycobacterium boenickei* sp. nov., *M. houstonense* sp. nov., *M. neworleansense* sp. nov. and *M. brisbanense* sp. nov. and *M. porcinum*.

isolate from a biopsy specimen is sufficient to make the diagnosis, although yields may be less than 50% of cases (American Thoracic Society 1997). This may, in part, be due to the methods employed and the presence of fastidious mycobacterial species such as *M. haemophilum* and *M. genavense*.

Cutaneous Infection

Infection of the skin with rapidly growing mycobacteria often follows colonisation and infection of intravascular devices. Thus the significance of rapidly growing mycobacteria in skin samples must be evaluated carefully. Multiple isolates are required in clinical circumstances that

are compatible with the diagnosis for certainty. Diagnostic uncertainty may be reduced if cases may form part of known outbreaks with contaminated injections or prostheses.

For *M. ulcerans* and *M. marinum*, significance is more easy to ascribe, as these species are likely to be isolated from patients with characteristic cutaneous lesions making diagnosis easier.

MYCOBACTERIUM AVIUM-INTRACELLULARE

This mycobacterium, responsible for infection in birds, was first identified in 1890 and given the name *M. avium*. A related organism was identified as the cause of an outbreak of pulmonary infection in

Table 11.3 A summary of the association between non-tuberculosis mycobacteria (NTM) species and clinical infection

Disease association	Species
Pulmonary	*Mycobacterium avium-intracellulare*
	M. kansasii
	M. xenopi
	M. malmoense
	M. szlugai
	M. abscessus
	M. fortuitum
Lymph node	*M. avium-intracellulare*
	M. scrofulaceae
	M. malmoense
Cutaneous	*M. marinum*
	M ulcerans
	M. abscessus
	M. fortuitum
	M. avium-intracellulare
Disseminated	*M. avium-intracellulare*
	M. kansasii
	M. genavense
	M. haemophilum
	M. chelonae
	M. fortuitum
	M. haemophilum
	M. scrofulaceum
	M. celatum
	M. Simiae

a State hospital in Georgia and for some time was known as the Battey bacillus (after the Battey State hospital from which it was isolated). This was given the name *M. intracellulare*.

Epidemiology

Organisms of the *M. avium-intracelluare* complex (MAIC) are widely distributed in nature and have a strong association with acid brown water swamps in the southern USA. *Mycobacterium avium-intracellulare* occurs worldwide where conditions are favorable for its growth: warm temperatures, low pH, low dissolved oxygen, high soluble zinc, high humic acid and high fulvic acid. The availability of these environmental conditions may explain, in part, the geographical variation in the incidence of disease. The organism can be isolated from soil, dust, animals and tap water, and any of these may be the source of infection for immunocompromised patients. The mere isolation of this organism in an environment does not explain the epidemiology on its own, as *M. avium-intracelluare* can be isolated from environmental samples in Congo and Kenya but where MAIC infection is uncommon in AIDS patients.

Before the onset of the HIV, epidemic infections with MAIC were not commonly recognised. Infections, when they were recognised, were mainly those of the respiratory tract and associated with concomitant chronic lung disease or deficient cellular immunity. Its prevalence is increasing (Falkinham III 1996). The predisposing conditions include pneumoconiosis and silicosis due to chronic and long-term exposure to dusts as a result of occupations (e.g. coal mining and farming) (Schaefer *et al.* 1969). For example, in one study, 73% of patients had pre-existing pulmonary disease, 38% smoked and 33% reported alcohol abuse. It has been recognised that MAIC can cause a pulmonary infection in elderly patients with no overt evidence of immunocompromise (Kennedy and Weber 1994; Prince *et al.* 1989).

A chronic lymphadenitis, usually in the cervical and facial region, is the most common presentation in children between 1 and 10 years of age, and MAIC is responsible for the majority of mycobacterial

lymphadenitis in the developed world (Grange, Yates and Pozniak 1995).

Disseminated infection may occur in patients without AIDS, usually associated with malignancy and inherited or iatrogenic immunodeficiency. The HIV initially brought a massive increase in the number of cases of MAIC infection until highly active antiretroviral therapy (HAART) became available. In areas of the developed world where this treatment is available, MAIC infections have become uncommon once again. The main portal of entry is thought to be the respiratory and gastrointestinal tracts. Initially the patient may be colonised, but later disseminated disease with bacteraemia may develop. Pulmonary infection or invasion of other organs, including bone marrow, may take place. Dissemination does not usually take place until the CD4 count has fallen well below 100/mm^3. Colonisation and infection with *M. avium-intracellulare* are associated with a worse prognosis compared with uninfected patients. The outcome can be improved with effective treatment.

In areas where *M. tuberculosis* is endemic, HIV-infected individuals are rarely infected with MAIC. The reasons for this dichotomy of presentation are not well understood.

Clinical Features

Lymphadenitis

The submandibular, pre-auricular, submaxilliary and rarely intraparotid lymph nodes are involved. The infection follows a chronic course with discharge and sinus formation and must be differentiated from other causes of localised lymphadenopathy such as tuberculosis, lymphoma and cat scratch disease.

Respiratory Infection

There are four syndromes recognised: solitary pulmonary nodules, chronic bronchitis or bronchiectasis, a tuberculosis-like syndrome and diffuse pulmonary infiltrates. In patients with bronchitis–bronchiectasis syndrome are usually in the older age group. Diffuse pulmonary infiltrates are usually found in severely immunocompromised individuals in whom a wide range of clinical and radiological features have been described but bacteraemia predominates (Kennedy and Weber 1994; Prince *et al.* 1989).

Disseminated Disease

The usual symptoms are fever and night sweats, diarrhea, abdominal pain and weight loss. On examination, patients are found to be wasted and febrile. Localised disease is less common, but pulmonary nodules, infiltrates or cavitation may be present. Additionally, intra-abdominal abscess, skin infection, osteomyelitis and cervical lymphadenitis may co-exist (Huang *et al.* 1999). At post-mortem many HIV-infected patients have undiagnosed evidence of disseminated MAIC infection.

On clinical examination patients are wasted and have hepatosplenomegaly. They are often anemic with reduced platelets and have elevated alkaline phophatase enzymes. Intra-abdominal lymphadenopathy can be detected by ultrasound or CT scanning.

Management of *M. avium-intracellulare* disease

HIV Seronegative Patients

Pulmonary disease: Older studies of treatment and the natural history of disease show that patients who are symptomatic have progressive disease which is difficult to treat, whereas many of those who were asymptomatic at the time of isolation went on to develop invasive disease (Hunter *et al*. 1981). Isolated pulmonary disease in otherwise healthy women is often difficult to treat, as these patients frequently fail therapy (Prince *et al*. 1989; Huang *et al*. 1999).

Treatment with three drugs including rifampicin, isoniazid with ethambutol were previously thought to give the best results (Hunter *et al*. 1981). Five-year follow-up of patients treated with this regimen showed that 15% of patients failed therapy and 14% relapsed (Anon, 2002). The activity of clarithromycin and quinolones suggest that they may have a role to play, and clinical trials are underway to evaluate this (Research Committee, British Thoracic Society 2001). Open trials suggest that sputum conversion rates of greater than 75% can be achieved with regimens that include a macrolide (Wallace Jr. *et al*. 1994a; Griffith *et al*. 1998; Wallace Jr. *et al*. 2003). Thus, although no comparative clinical trials have yet been reported, macrolides should probably be included in regimens used to treat *M. avium-intracellulare* infections in immunocompetent patients. Treatment can also be given intermittently (Griffith *et al*. 2000). Relapse, when it occurs, may be due not only to treatment failure but to infection with a new and different strains from the environment (Wallace Jr. *et al*. 2003). Some authorities suggest that rifabutin should be the rifamycin of choice for treatment of *M. avium-intracellulare* infection because of its greater *in vitro* activity. However, this drug has a different adverse event profile, and only comparative clinical trials can tell whether the additional activity is gained without increase adverse events (Research Committee, British Thoracic Society 2001).

Lymphadenitis: Surgical excision is essential for diagnosis, as the yield from fine needle aspiration is not complete and there is a considerable risk of sinus formation (American Thoracic Society 1997). Optimal treatment of this condition is surgical excision which has a lower re-operation rate than incision and drainage, curettage or aspiration (Flint *et al*. 2000). Relapse and sinus formation are rare occurring in less than 5% of cases (Rahal *et al*. 2001; Danielides *et al*. 2002). Antimicrobial chemotherapy appears to be unnecessary (American Thoracic Society 1997), although there are reports of successful management with clarithromycin monotherapy (Tunkel and Romaneschi 1995).

HIV seropositive patients: Disseminated *M. avium-intracelluare* infection is a late complication of HIV infection, and since the introduction of HAART, it has become much less common in developed countries, occurring in patients who are untreated or who have been unable to tolerate therapy. The optimal regimen has not yet been established in part, because patients with this infection are at very late stage of their HIV disease where the clinical course is complicated by other opportunistic infections and the complications of HIV itself. In the era of HAART, the management of disseminated *M. avium-intracellulare* infection is underwritten by therapeutic efforts to reduce the HIV viral load, increase the CD4 count and bring about reversal of the immune deficit.

Antibiotics have an important role in reducing bacteremia, and the antibiotics which have been shown to be able to do that include macrolides such as clarithromycin and azithromycin, quinolones such as ciprofloxacin and rifamycins such as rifabutin. The macrolides are highly active and are the cornerstones of all regimens. They are capable of reducing the count of bacteria in the blood when given alone (Hoy *et al*. 1990; Dautzenberg *et al*. 1993). Monotherapy results in the rapid emergence of resistance, and thus combination therapy should be chosen. Clinical trials have supported the superiority of clarithromycin, ethambutol and rifabutin over rifampicin, ethambutol, clofazimine and ciprofloxacin (Shafran *et al*. 1996). A recent comparative trial suggested that lower doses of rifabutin together with ethambutol are more effective than a four-drug regimen of rifampacin, ethambutol, clofazimine and ciprofloxacin while still retaining much of the activity of clarithromycin and rifabutin doses (Shafran *et al*. 1996).

Prophylaxis of *M. avium-intracellulare*: Prophylaxis is necessary to prevent infection in patients with late-stage HIV infection with low CD4 count. Macrolides have been shown to be more effective than

rifabutin which also is an effective agent but is associated with a higher rate of intolerance (Havlir *et al*. 1996; Phillips *et al*. 2002). Ultimately the choice of prophylactic agent will depend on the choice of HAART, as rifabutin interacts with protease inhibitors and patients differ considerably in their ability to tolerate drugs (Cohn *et al*. 2002).

MYCOBACTERIUM KANSASII

Mycobacterium kansasii is a photochromogenic organism that can be isolated from environmental sources. It is one of the commonest NTMs isolated in UK laboratories (Evans *et al*. 1996b).

Epidemiology

Mycobacterium kansasii is one of the most common NTMs isolated making up approximately 5% of significant isolates from surveys (Lillo *et al*. 1990; Evans *et al*. 1996b). Infection is acquired from the environment, and person-to-person spread is thought not to occur. *Mycobacterium kansasii* can be isolated from swimming pools, hot and cold water supplies and storage tanks, and this may be the source of infection in the hospital environment (McSwiggen and Collins). Areas of high and low incidence emphasise the importance of environmental sources in the epidemiology of this infection. Unlike other NTM infections the incidence of *M. kansasii* does not appear to be rising (Breathnach *et al*. 1998; Evans *et al*. 1996b).

Infection is typically found in older patients, and there is a strong male preponderance. A history of cigarette smoking is strongly associated with infection. Chronic lung conditions including pneumoconiosis, treated tuberculosis, chronic obstructive pulmonary disease and bronchiectasis all predispose to *M. kansasii* infection (Corbett *et al*. 1999).

Clinical Feature

The characteristic clinical association is with a chronically progressive pulmonary infection which follows a rather indolent course. The presentation frequently resembles tuberculosis and is often mistaken for it. Cough, sputum production, hemoptysis and the constitutional effects of infection are commonly seen. Pulmonary infection usually arises in patients who have a compromised respiratory tract and who are immunocompetent.

Localised infections are reported but are much less common than in other mycobacteriosis. Lymphadenitis and cutaneous and bone infections have all been reported. Cutaneous infection is usually associated with immunosuppression as is disseminated disease. Some cases of disseminated disease are associated with HIV infection (Breathnach *et al*. 1995). Radiographic abnormalities in patients with *M. kansasii* infection were frequently unilateral. Air-space shadowing involving more than one bronchopulmonary segment and pleural effusions is seen less frequently than in tuberculosis. Cavitation is seen in patients with *M. kansasii* infection as frequently as in those with tuberculosis. Hilar lymphadenopathy and pleural effusions are rare (Evans *et al*. 1996a).

Management of *M. kansasii* Infection

Pulmonary Infections

Mortality rates of *M. kansasii* infection are high, but this is often due to the severe underlying conditions that co-exist in these patients (Jenkins *et al*. 1994). There is widespread agreement that rifampicin is an essential component of successful regimens. Almost all patients become culture negative within 4 months. On the other hand, resistance to this agent or its absence in the regimen is associated with treatment failure (Ahn *et al*. 1983; Banks *et al*. 1984; Pezzia *et al.* 1981). With regimens that contain rifampicin, relapse rates are typically low with figures of between 2.5 and 9% (Ahn *et al.* 1983; Jenkins *et al.* 1994).

The American Thoracic Society recommend a regimen of isoniazid, rifampicin and ethambutol given daily for 18 months with at least 12 months of negative sputum cultures. In patients who are unable to tolerate one of these three drugs, clarithromycin would seem a reasonable alternative, but its effectiveness has not been established by clinical trials (see below). Pyrazinamide has no role to play as in therapy for *M. kansasii* infections, because all isolates are resistant (American Thoracic Society 1997). A prospective clinical trial performed by the British Thoracic Society (BTS) in 173 patients with two sputum cultures positive with NTM showed that *M. kansasii* pulmonary infection responds well to 9 months of treatment with rifampicin and ethambutol, but patients who contract this disease have a high mortality rate from other causes. Isoniazid did not appear to be a necessary part of the regimen (Jenkins *et al.* 1994). Consequently the BTS recommend that 9 months of rifampicin and ethambutol is adequate treatment for most patients, but when there is evidence of compromising conditions treatment can be extended to 15–24 months (Joint Tuberculosis Committee 2000). The use of intermittent drug regimens or short course is not sufficiently studied for advice to be given.

In patients in whom there is an inadequate response, clarithromycin and fluoroquinolones could be used, as these agents are highly active against *M. kansasii* and are likely to be beneficial (Yew *et al.* 1994; Gillespie, Morrissey and Everett 2001), although there is no clinical trial data available. Alternatives include prothionamide and streptomycin could be added (Joint Tuberculosis Committee 2000) but both are associated with frequent adverse events. When rifampicin resistance is present a regimen which includes clarithromycin and ciprofloxacin is associated with a favorable outcome (Wallace Jr. *et al.* 1994b).

Extrapulmonary Infection

The treatment of extrapulmonary disease should probably be similar to the pulmonary regimens. For lymphadenopathy, excision is recommended, as this is the optimal treatment for *M. avium-intracellulare* infection, the commonest cause (see below) (American Thoracic Society 1997; Joint Tuberculosis Committee 2000).

MYCOBACTERIUM MALMOENSE

Mycobacterium malmoense was first isolated from four patients with pulmonary disease present in Malmo, Sweden (Portaels, Larsson and Jenkins 1995). It is closely related to *M. avium-intracellulare* and is a non-chromogenic strain that grows very slowly. The slow growth rate may be the reason that some infections are not diagnosed.

Epidemiology

Like other NTMs *M. malmoense* has been isolated from the environment but only rarely (Portaels, Larsson and Jenkins 1995). It has been isolated from human and animal sources. Most cases have been reported from northern Europe including the UK, but cases have been reported from the Americas. The number of cases is increasing, and this is thought to be due to a real rise in the number of cases not just an improvement in mycobacterial diagnostic methods. *Mycobacterium malmoense* has an affinity for the respiratory tract, and the majority of infections are pulmonary. Patients are predisposed to pulmonary infection by previous pulmonary disease, notably treated tuberculosis. Other associated conditions include lung carcinoma, chronic obstructive airways disease and pneumoconiosis (Jenkins 1981; Falkinham III 1996). Immunosuppression by leukemia or other malignancy can predispose to infection, but *M. malmoense* is rarely associated with HIV. The commonest extrapulmonary presentation is cervical lymphadenitis

which occurs most often in children. Disseminated infection is found most often in severely immunocompromised individuals including those with HIV or leukemia. Asymptomatic colonisation of the respiratory tract is reported and must be distinguished from invasive disease.

Clinical Features

The prodromal period varies from a few weeks to more than a year. Cough, weight loss and hemoptysis are the most prominent symptoms, and the overall picture closely resembles tuberculosis. Radiological examination reveals the presence of cavitation in the majority of patients.

Management of *M. malmoense* Disease

Pulmonary Disease

Rifampicin and ethambutol containing regimens given for 18–24 months are better than those in whom other regimens or shorter durations of treatment were used (Banks *et al.* 1984). The addition of second- or third-line drugs increased the rate at which adverse events were reported without improving the outcome. Surgery has an important role to play in those who are suitable for operation, and chemotherapy should be continued afterwards for at least 18 months. A clinical trial of chemotherapy in *M. malmoense* infection showed that treatment of *M. malmoense* with rifampicin and ethambutol for 2 years is preferable to a regimen that contains isoniazid, although there was a non-significant reduction in the relapse rate for the three drug regimen. However, there is a higher death rate for the three-drug regimen (Joint Tuberculosis Committee 2000). Macrolides and quinolones are active *in vitro* (Yew *et al.* 1994; Gillespie, Morrissey and Everett 2001), and there are some anecdotal reports of treatment response when these agents are used in the management of patients who are highly susceptible to infection (Scmitt *et al.* 1999).

Extrapulmonary Disease

Lymphadenitis is the commonest form of *M. malmoense* extrapulmonary disease, and this syndrome should be treated with excision. Otherwise, extrapulmonary disease should be treated in the same way as pulmonary disease.

MYCOBACTERIUM XENOPI

Mycobacterium xenopi is a slow-growing, non-chromogenic mycobacterium which is frequently isolated from human specimens as a coloniser or as a pathogen. It was first isolated from a skin granuloma of a toad *Xenopus laevus* kept in the laboratory for pregnancy testing. It is regularly isolated from human respiratory specimens, and its significance must be carefully determined.

Epidemiology

In Western Europe it is one of the most common NTMs isolated but much less common in the United States. It has been isolated from hospital hot water supplies, and nosocomial outbreaks have been reported. Person-to-person transmission is not thought to occur but cannot be excluded when there are nosocomial outbreaks. The reservoir of *M. xenopi* in the environment is uncertain, but the organism is isolated in fresh or seawater. Clustering of cases in coastal areas suggests a link with the sea. Birds are highly susceptible to infection, and seabirds have been proposed as a reservoir of infection.

Infection, when it occurs, is most likely to occur in males over middle age, often with a long history of previous lung problems including obstructive airways disease emphysema and healed tuberculosis (Jenkins 1981; Falkinham III 1996). *Mycobacterium xenopi* has also been reported in HIV-infected individuals who are severely immunocompromised (Claydon, Coker and Harris 1991; Kerbiriou *et al*. 2003).

Clinical Features

Isolation of *M. xenopi* is not always associated with disease. Retrospective reviews differ in the proportion of case attributable to the organism, but it has been estimated that half of isolates are not clinically significant. A study of HIV-positive individuals with multiple isolated *M. xenopi* showed that almost all of the patients cleared the organism without specific chemotherapy but with HAART (Kerbiriou *et al*. 2003).

The presentation of pulmonary disease is subacute in most instances. It is almost always associated with cough, but fever is not a prominent sign. Hemoptysis and weight loss may be presenting symptoms. Solitary or multiple nodules or cavitation may be found on radiological examination of the chest (Kerbiriou *et al*. 2003). A syndrome indistinguishable from tuberculosis may also develop. Extrapulmonary disease is rare. Disseminated disease has been reported in HIV-infected patients but is much less common than *M. kansasii* and *M. avium-intracellulare*.

Management of *M. xenopi* Disease

Pulmonary Disease

Myocbacterium xenopi poses many diagnostic and therapeutic problems. In some patients *M. xenopi* may act as a coloniser without causing disease (Smith and Citron 1983; Simor, Salit and Vellend 1984; Jiva *et al*. 1997). Thus it will be present in multiple specimens thus passing the test for significance, although in many such cases it is not responsible for clinical symptoms. In addition, infection with *M. xenopi* is normally indolent with disease developing over a number of years (Banks *et al*. 1984). Thus an isolate in an apparently asymptomatic patient cannot be lightly dismissed, especially in HIV-infected individuals. To overcome the diagnostic difficulty it has been proposed that the criteria for diagnosis of *M. xenopi* infection should be two sputum isolations in the absence of other likely causes of symptoms. (Juffermans *et al*. 1998)

Early studies have suggested that regimens should contain rifampicin and isoniazid together with ethambutol or streptomycin (Costrini *et al*. 1981; Banks *et al*. 1984). A clinical trial suggests that rifampicin and ethambutol is the optimum regimen, although there is a trend to a higher cure rate when isoniazid is added but that the complication rate is increased (Research Committee, British Thoracic Society 2001). In view of the higher complication rate with added isoniazid guidelines suggest that this drug is included only if treatment is failing to render sputum culture negative (Research Committee, British Thoracic Society 2001).

Macrolides and quinolones may have an important role in the treatment of *M. xenopi* infections, as they are active *in vitro* and in animal models (Yew *et al*. 1994; Gillespie, Morrissey and Everett 2001; Lounis *et al*. 2001). There are anecdotal reports of the value of these agents (Scmitt *et al*. 1999). Clinical trial data are not yet available to inform therapeutic choice, but these could rationally be added to treatment in patients who were failing to respond.

The results of medical therapy can be poor, with the 5-year mortality up to 57% (although a minority of these deaths were directly attributed to mycobacterial infection) (Research Committee, British Thoracic Society 2001). Pulmonary resection is often necessary as an adjunct to treatment (Parrot and Grosset 1988). Pulmonary resection may be considered in patients whose disease is localised and who are failing on therapy but otherwise have good pulmonary function.

A study has reported that when these criteria are applied sputum conversion is complete in all but the patients who have incomplete resection (Lang-Lazdunski *et al*. 2001).

MYCOBACTERIUM MARINUM

Epidemiology

Mycobacterium marinum was first recognised as a pathogen of fish. Human infection, when it occurs, usually arises as a result of water contact. Initially most cases were associated with poorly maintained contaminated swimming pools – swimming pool granuloma (Falkinham III 1996). This has now almost completely disappeared due to improved construction and water purification. Most cases seen today arise in association with the maintenance of fish tanks, and fish farming is a risk factor of infection.

Clinical Features

Infection with *M. marinum* causes a chronic granulomatous infiltration of the skin with a similar appearance to sportrichosis (Falkinham III 1996; Casal and Casal 2001). Nodular or pustular lesions are the most common, and erythematous swelling crusting or swelling may occur. The upper limb, especially the hand, is the most common site of infection. The lesions spread locally in the majority of cases after 4–6 weeks.

Mycobacterium marinum infection must be differentiated from other chronic granulomatous skin infections such as tuberculosis, nocardiasis, coccidioidomycosis, histoplasmosis, leishmaniasis, sporotrichosis, leprosy and syphilis.

Management of *M. marinum* Infection

The organisms are often resistant to isoniazid but susceptible to rifampicin, ethambutol and pyrazinamide. Minocycline and other tetracyclines have been used in treatment, but rifampicin and ethambutol appear to be more successful (American Thoracic Society 1997).

MYCOBACTERIUM ULCERANS

Epidemiology

Mycobacterium ulcerans was first isolated from patients in Bairnsdale Australia with necrotising skin ulcers. Buruli ulcer, the lesion caused by this organism, is described in many tropical and subtropical countries in Africa, Central and South America and Southeast Asia (Hayman 1991; Marston *et al*. 1995).

An environmental source for *M. ulcerans* has not been identified, but epidemiological studies indicate that it is to be found near slow-flowing or stagnant water. Infection is most common in individuals under 15 years of age, and males and females are affected equally. The prevalence of infection is highest in villages close to rivers where farming activities occur. Infection appears to occur by direct inoculation into the skin, as wearing protective clothing such as long trousers makes infection less likely (Marston *et al*. 1995). The incidence of infection is rising in some countries of West Africa (van der Werf *et al*. 1999).

Clinical Features

The disease begins as a small subcutaneous swelling increasing in size until the skin is raised. At first it is attached to the skin but not the deep fascia, but as the lesion progresses it extends to involve this layer. The skin over the lesion is at first darker but then loses its pigmentation before becoming necrotic and ulcerating. Some cases have a small central vesicle. Rarely the disease may present in an edematous form mimicking cellulitis. Occasionally the necrosis spreads through the deep fascia involving muscle and bone (Hayman 1993). Ulcers are typically painless, usually found on the lower limbs but may more rarely occur on the face or trunk especially in children. The disease is accompanied by remarkably few systemic symptoms, but occasionally secondary infections resulting in sepsis or tetanus cause severe systemic disease and death. Buruli ulcer should be differentiated from foreign body granulomata, phycomycosis fibroma or fibrosarcoma. Extensive scarring can lead to contractures of the limbs, blindness and other adverse sequelae, which impose a substantial health and economic burden.

Pathology

The ulcer usually has straight or undermined edge with subcutaneous spread, producing nodules and a gelatinous material which is readily removed. An edematous form of the diseases has also been described. Microscopically necrotic tissue lines the ulcer, and multiple acid-fast bacilli can be seen. The bacteria form tangled masses and globular forms where many bacteria are found within macrophages have been described, but this latter term should be avoided as it may give rise to confusion with leprosy (Hayman 1993).

Caseation is not a characteristic of Buruli ulceration. Necrosis is present in the lesion including fat necrosis due to infection of the arterioles serving the fat lobules. Mycobacteria can be found in lipid lacunae. The necrosis is characterised by calcification especially in chronic lesions, although this feature is rarely described in Australian cases. In studies of cases biopsied sequentially, three stages are described: necrosis and tissue degeneration, an organising stage and a healing stage (Hayman 1993).

Management of *M. ulcerans* Infection

Surgical treatment including excision and local debridement is the mainstay of therapy (van der Werf *et al.* 1999). There are few clinical trials of antimicrobial treatment, and these have been disappointing. Patients have been treated with isoniazid and streptomycin or oxytetracycline and dapsone and a combination of rifampicin, minocycline and cotrimoxazole, and these may be beneficial. A significant number of patients are left with residual disability.

Control

Vaccination with BCG does appear to provide some protection against infection. Wearing long trousers may be beneficial by preventing the initial inoculation injury necessary to establish infection.

RAPID GROWING MYCOBACTERIA

Mycobacterium fortuitum and the *M. chelonei* group are the species most frequently encountered in the clinical laboratory. *Mycobacterium fortuitum* was first isolated from a frog and *M. chelonei* from a turtle. The group of rapidly growing organisms is conveniently divided into three taxonomically related groups. *Mycobacterium fortuitum* has three species: *M. fortuitum*, *M. perigrinum* and an unnamed taxon. Additional species are being considered for inclusion. The *M. chelonei* group includes *M. chelonei* sensu stricto, *M. abscessus* and *M. immunogenum*. The smegmatis group includes *M. smegmatis*, *M. goodi* and *M. wolinshyi* (Table 11.2). The classification of this group is evolving rapidly with the increasing use of 16S sequencing for identification (Brown-Elliott and Wallace Jr. 2002; Tortoli 2003).

Epidemiology

The rapidly growing mycobacteria are environmental saprophytes and commensal organisms. They may be found in human specimens without evidence of disease; so care must be taken with ascribing significance to isolates (see above). Infections with rapidly growing mycobacteria are mainly associated with skin, soft tissue and catheters (Brown-Elliott and Wallace Jr. 2002; Wallace Jr. 1989; McWhinney *et al.* 1992). Surgical wounds can become infected, and infection may follow the use of any prosthesis that is infected with these organisms (Ozluer and De'Ambrosis 2001). Outbreaks have been associated with contaminated breast implants, breast augmentation surgery and infected prosthetic devices and intravenous injections and intraperitoneal canulae (Galil *et al.* 1999). Keratitis, ophthalmitis and lymphadenitis have been described (Griffith, Girard and Wallace Jr. 1993; Marin-Casanova *et al.* 2003). Disseminated infection can develop in immunocompromised patients, especially those undergoing remission induction chemotherapy. Pneumonia and disseminated skin infection can also occur in immunocompromised patients (American Thoracic Society 1997).

Pathology

Infection with rapidly growing mycobacteria is characterised histologically by polymorphonuclear leukocytosis in microabscesses. Necrosis is almost invariably present, and granulomatous change with Langerhan's giant cells is found in 80%. Caseation necrosis is rarely reported. Acid-fast bacilli are scanty and may not be seen in as many as two-thirds of biopsies. When organisms are present they are usually found in clumps extracellularly (Sungkanuparph, Sathapatayavongs and Pracharktam 2003a; Sungkanuparph, Sathapatayavongs and Pracharktam 2003b).

Clinical Features

A primary source for infection is not always apparent. Non-surgical skin infection is described in children and young adults. It usually takes the form of cellulitis with abscess formation. The infection follows and indolent path with the patient only seeking medical attention after several weeks. The lesions are red and only mildly tender. The organisms are inoculated into the skin as a result of penetrating trauma, foreign body or pre-existing skin disease. Postoperative infections are often associated with implantation of prostheses including cardiac valves and silicon prostheses used in augmentation mammoplasty. A series of sternotomy wound infections has been reported.

Pulmonary infection is associated with underlying pulmonary disease less often than for other NTMs and may follow a progressive course to death. Cough is the universal presenting symptom, and constitutional symptoms are reported along with progressive disease. Upper lobe infiltrates are most common with most patients developing bilateral disease. Cavitation is present in a minority of patients. Specific underlying diseases are infrequent, but they include previously treated mycobacterial disease, coexistent *M. avium* complex infection, cystic fibrosis and gastroesophageal disorders with chronic vomiting (Griffith, Girard and Wallace Jr. 1993).

Different clinical patterns of disseminated disease have been associated with renal transplantation, renal failure and collagen vascular diseases where skin lesions predominate over organ involvement. In patients with malignancy or defects in cell-mediated immunity, disease is more widespread alongside skin involvement. In this latter situation the mortality is high. Lymphadenitis with rapidly growing mycobacteria has been described in children and has been associated with tooth extraction in some cases.

The outcome of infection with rapidly growing mycobacteria is in large part dependent on the nature of the underlying medical condition. Where there is a severe underlying immune deficit which cannot be reversed, the mortality rate is high.

Management of Infection with Rapid Growers

Pulmonary Disease

More than 80% of cases of pulmonary disease are caused by *M. abscessus* which is the naturally most resistant member of the group of organisms (Griffith, Girard and Wallace Jr. 1993). Treatment of *M. abscessus* infections is often disappointing. Treatment can bring about clinical improvement, but cure is rare. When surgery is technically possible it is recommended (Joint Tuberculosis Committee 2000). Susceptibility testing of rapidly growing mycobacteria is thought to give a good guide to treatment, and regimens should be constructed on the basis of susceptibility test results (Wallace Jr. *et al*. 1985). Regimens should probably include rifampicin, ethambutol and clarithromycin. Fluoroquinolones, sulphonamides, amikacin, cefoxitin and penems may have a role to play in treatment (Wallace Jr. *et al*. 1985; Joint Tuberculosis Committee 2000; Tanaka *et al*. 2002).

Extrapulmonary Disease

Many cases of infection by rapid growers occur in the context of an infected prosthetic device, for example intravenous canulae or other implant. Successful therapy of these catheter-related infections involves removal of the catheter and antimicrobial therapy, usually for 2–4 months (McWhinney *et al*. 1992). Although disease due to *M. fortuitum* may resolve if the catheter is removed, reinsertion of another catheter in a similar location without drug therapy usually results in disease recurrence (as in the above case) (McWhinney *et al*. 1992). Adjunctive therapy should be with ciprofloxacin, amikacin and clarithromycin for up to 4 months. When there is a tunneled line that also has a tissue infection then treatment may need to be extended for 6 months (Brown-Elliott and Wallace Jr. 2002). The oxazolidinone linezolid is active against many mycobacteria including the rapid growers, and there have been several reports of its value in multiple drug regimens, especially when resistant *M. abscessus* is being treated.

Postinjection abscesses should be treated by surgical drainage and clarithromycin for 3–6 months. This advice comes as a result of the experience obtained from a series of outbreaks (Villanueva *et al*. 1997; Galil *et al*. 1999).

Wound infections are one of the most common manifestations of infection with rapidly growing mycobacteria. Infections have often been associated with augmentation mammoplasty and other plastic surgery procedures (Bolan *et al*. 1985; Brown-Elliott and Wallace Jr. 2002; Clegg *et al*. 1983). Therapy depends of the removal of any infected foreign material followed by 6 months of chemotherapy (Morris-Jones *et al*. 2001; Ozluer and De'Ambrosis 2001). Clarithromycin is the main choice with other drugs being added to prevent the emergence of resistance (Wallace Jr. *et al*. 1993).

Disseminated cutaneous infection is mainly with *M. abscessus* usually in patients who are compromised by renal failure or corticosteroid therapy (Wallace Jr. 1989). This is one of the most common presentations of non-pulmonary infection with rapidly growing organisms (Brown-Elliott and Wallace Jr. 2002). Treatment includes drainage of any abscesses that are present coupled with clarithromycin for 6 months together with another agent to which the isolate is susceptible during the first 2 months (Wallace Jr. *et al*. 1993).

MYCOBACTERIUM HAEMOPHILUM

Mycobacterium haemophilum is a fastidious member of the genus which grows more slowly than other species and requires iron-supplemented medium and a lower temperature of incubation (Dawson and Jennis 1980). It was first described in a woman with Hodgkin's disease but has subsequently been described in patients with severe defects in cell-mediated immunity including HIV (Kiehn and White 1994).

Epidemiology

Infection has been described throughout the world. Although almost all of the patients reported had severe immunodeficiency, lymphadenitis has been reported from immunocompetent children. Prior to the HIV pandemic most patients had been treated with immunosuppressive therapy after organ transplantation or were suffering from lymphoma. *Mycobacterium haemophilum* is now a recognised pathogen for HIV-infected individuals. The true extent of infection with *M. haemophilus* is unknown, but the current incidence is probably an underestimate as few laboratories use appropriate media or culture conditions that permit this organism to be isolated. The extent of the underreporting can only be guessed at, but one laboratory found 13 cases in 20 months, equivalent to one third of all cases reported in the world literature up to that point (Straus *et al*. 1994). Little is known of the mode of transmission, but person-to-person transmission is not thought to occur.

Clinical Features

Several syndromes have been associated with *M. haemophilum* including lymphadenitis in immunocompetent children, cutaneous ulceration, bacteremia, infection of bones and joints and infection of lungs in severely immunocompromised individuals.

Cutaneous lesions are the commonest manifestation of *M. haemophilum* infection. The lesions tend to cluster on the extremities often over joints, and this suggests that this distribution is related to the lower temperatures in these areas. Lesions are typically raised violaceous and fluctuant, ranging in size from 0.5 to 2 cm. Later the lesions enlarge and become pustular and may be painful. The appearances must be distinguished from Kaposi's sarcoma which may co-exist in HIV-positive patients. The majority of patients report joint symptoms including tenderness and swelling. Fewer patients have pulmonary infection, but this is associated with a poor outcome even with therapy. Patients also present with septic arthritis or osteomyelitis (Shah *et al*. 2001). Lymphadenitis may arise in the cervical or perihilar region in children, producing a clinical picture similar to infection with *M. avium-intracellulare* (Armstrong *et al*. 1992).

Laboratory Diagnosis

Mycobacterium haemophilum is a fastidious organism that requires an egg-based chocolate agar medium supplemented with ferric ammonium citrate hemoglobin and hemin for optimum growth. The optimum temperature of incubation is 32 °C. As the organism grows slowly, the incubation period should be extended beyond 8 weeks (Dawson and Jennis 1980). There are no standard methods for determining susceptibility, but there are reports that the organism is susceptible to rifabutin, ciprofloxacin, cycloserine and kanamycin. Approximately half are susceptible to rifampicin, but most are resistant to isoniazid and all to pyrazinamide and ethambutol (Saubolle *et al*. 1996).

Management of *M. haemophilum* Infection

There are few clear guidelines for therapy, but the outcome appears to be influenced by the patient's underlying immunosuppression with patient's recover associated with reversal of the immune deficit (Paech *et al*. 2002). The organisms are most susceptible to ciprofloxacin, clarithromycin, rifabutin and rifampin, and regimens that include these drugs have been associated with a successful outcome (Saubolle *et al*. 1996).

OTHER NON-TUBERCULOSIS MYCOBACTERIA

With almost 100 species of mycobacteria described, there is an expanded understanding of the organisms that infect humans. Many of these are isolated in clinical laboratories rarely. Many of these are found in the Tables 11.1 and 11.2. For most there are only anecdotal case reports; so firm recommendations for therapy cannot be given in most instances, but guidance is usually derived by analogy with therapeutic strategies used for the treatment of related species.

Mycobacterium scrofulaceum is closely related to *M. avium* and *M. intraacellulare* and shares many biochemical, phenotypic and environmental characteristics. It has been described in children with lymphadenitis in cutaneous infections and as a cause of cavitary lung disease in patients predisposed by pneumoconiosis (Corbett *et al.* 1999; Primm, Lucero and Falkinham III 2004). Despite its similarity to *M. avium*, it is only rarely isolated in patients with advanced HIV disease.

Mycobacterium simiae is a photochromogenic mycobacterium that was first isolated from feral monkeys imported into the Unites States. *Mycobacterium simiae* shares many similarities with *M. avium* and *M. scrofulaceum* and has been associated with infections in HIV-infected individuals, which resemble *M. avium-intracellulare* disease (Al-Abdely, Revankar and Graybill 2000). Treatment for significant infections should be with a regimen suitable for *M. avium*.

Mycobacterium szulgai is a scotochromogen at 37 °C but a photochromogen at 25 °C. It has been described throughout the world, and the majority of infections are pulmonary or disseminated (Tortoli *et al.* 1998a). In addition to this infections in bursa, tendon sheaths, bones, lymph nodes and skin have been reported (American Thoracic Society 1997). It appears to have an environmental source which has not yet been identified. The main risk factors of infection are cigarette smoking, chronic lung disease and a high alcohol intake. For skin, bursa and bone infection, trauma is the major predisposing condition. Therapy is with isoniazid, rifampicin, streptomycin and ethambutol. Therapy can be amended in light of the results of susceptibility tests. Surgery may be used to complement chemotherapy.

Mycobacterium paratuberculosis is closely related to *M. avium-intracelluare* and mainly causes disease in cattle – Johne's disease. Some authors have linked *M. paratuberculosis* with Crohn's disease, but this link has not been proved (Hermon-Taylor and Bull 2002; Bull *et al.* 2003). The organism is unable to grow in the absence of mycobactin on artificial media.

Mycobacterium genavense is a rarely reported organism found in association with severe immunocompromised including advanced HIV infection. Infections have been reported throughout the world. It grows more slowly than most other mycobacteria and could be discarded if cultures are disposed of in less than 35 days. The source of *M. genavense* infection is unknown (Tortoli *et al.* 1998b).

REFERENCES

Ahn, C. H., Lowell, J. R., Ahn, S. S. *et al.* (1983) Chemotherapy for pulmonary disease due to *Mycobacterium kansasii*: efficacies of some individual drugs. Am Rev Respir Dis 128, 1048–1050.

Al-Abdely, H. M., Revankar, S. G., and Graybill, J. R. (2000) Disseminated *Mycobacterium simiae* infection in patients with. AIDS J Infect 41, 143–147.

American Thoracic Society. (1997) Diagnosis and treatment of disease caused by non-tuberculous mycobacteria. Am J Respir Crit Care Med 156, S1–S25.

Anon. (2002) Pulmonary disease caused by *Mycobacterium avium-intracellulare* in HIV-negative patients: five-year follow-up of patients receiving standardised treatment. Int J Tuberc Lung Dis 6, 628–634.

Armstrong, K. L., James, R. W., Dawson, D. J. *et al.* (1992) *Mycobacterium haemophilum* causing perihilar or cervical lymphadenitis in healthy children. J Pediatr 121, 202–205.

Banks, J., Hunter, A. M., Campbell, I. A. *et al.* (1984) Pulmonary infection with *Mycobacterium kansasii* in Wales, 1970–9: review of treatment and response. Thorax 39, 376–382.

Bolan, G., Reingold, A. L., Carson, L. A. *et al.* (1985) Infections with *Mycobacterium chelonei* in patients receiving dialysis and using processed hemodialyzers. J Infect Dis 152, 1013–1019.

Breathnach, A. S., de Ruiter, A., Holdsworth, G. M. *et al.* (1998) An outbreak of multi-drug-resistant tuberculosis in a London teaching hospital. J Hosp Infect 39, 111–117.

Breathnach, A., Levell, N., Munro, C. *et al.* (1995) Cutaneous *Mycobacterium kansasii* infection: case report and review. Clin Infect Dis 20, 812–817.

Brown-Elliott, B. A. and Wallace, R. J. Jr. (2002) Clinical and taxonomic status of pathogenic nonpigmented or late-pigmenting rapidly growing mycobacteria. Clin Microbiol Rev 15, 716–746.

Bull, T. J., McMinn, E. J., Sidi-Boumedine, K. *et al.* (2003) Detection and verification of *Mycobacterium avium* subsp. *paratuberculosis* in fresh ileo-colonic mucosal biopsy specimens from individuals with and without Crohn's disease. J Clin Microbiol 41, 2915–2923.

Casal, M. and Casal, M. M. (2001) Multicenter study of incidence of *Mycobacterium marinum* in humans in Spain. Int J Tuberc Lung Dis 5, 197–199.

Claydon, E. J., Coker, R. J., and Harris, J. R. (1991) *Mycobacterium malmoense* infection in HIV positive patients. J Infect 23, 191–194.

Clegg, H. W., Foster, M. T., Sanders, W. E. Jr., and Baine, W. B. (1983) Infection due to organisms of the *Mycobacterium fortuitum* complex after augmentation mammaplasty: clinical and epidemiologic features. J Infect Dis 147, 427–433.

Cohn, S. E., Kammann, E., Williams, P. *et al.* (2002) Association of adherence to *Mycobacterium avium* complex prophylaxis and antiretroviral therapy with clinical outcomes in acquired immunodeficiency syndrome. Clin Infect Dis 34, 1129–1136.

Corbett, E. L., Hay, M., Churchyard, G. J. *et al.* (1999) *Mycobacterium kansasii* and *M. scrofulaceum* isolates from HIV-negative South African gold miners: incidence, clinical significance and radiology. Int J Tuberc Lung Dis 3, 501–507.

Costrini, A. M., Mahler, D. A., Gross, W. M. *et al.* (1981) Clinical and roentgenographic features of nosocomial pulmonary disease due to *Mycobacterium xenopi*. Am Rev Respir Dis 123, 104–109.

Danielides, V., Patrikakos, G., Moerman, M. *et al.* (2002) Diagnosis, management and surgical treatment of non-tuberculous mycobacterial head and neck infection in children. ORL J Otorhinolaryngol Relat Spec 64, 284–289.

Dautzenberg, B., Saint Marc, T., Meyohas, M. C. *et al.* (1993) Clarithromycin and other antimicrobial agents in the treatment of disseminated *Mycobacterium avium* infections in patients with acquired immunodeficiency syndrome. Arch Intern Med 153, 368–372.

Dawson, D. J. and Jennis, F. (1980) Mycobacteria with a growth requirement for ferric ammonium citrate, identified as *Mycobacterium haemophilum*. J Clin Microbiol 11, 190–192.

Evans, A. J., Crisp, A. J., Hubbard, R. B. *et al.* (1996a) Pulmonary *Mycobacterium kansasii* infection: comparison of radiological appearances with pulmonary tuberculosis. Thorax 51, 1243–1247.

Evans, S. A., Colville, A., Evans, A. J. *et al.* (1996b) Pulmonary *Mycobacterium kansasii* infection: comparison of the clinical features, treatment and outcome with pulmonary tuberculosis. Thorax 51, 1248–1252.

Falkinham, J. O. III. (1996) Epidemiology of infection by nontuberculous mycobacteria. Clin Microbiol Rev 9, 177–215.

Flint, D., Mahadevan, M., Barber, C. *et al.* (2000) Cervical lymphadenitis due to non-tuberculous mycobacteria: surgical treatment and review. Int J Pediatr Otorhinolaryngol 53, 187–194.

Galil, K., Miller, L. A., Yakrus, M. A. *et al.* (1999) Abscesses due to mycobacterium abscessus linked to injection of unapproved alternative medication. Emerg Infect Dis 5, 681–687.

Gillespie, S. H., Morrissey, I., and Everett, D. (2001) A comparison of the bactericidal activity of quinolone antibiotics in a *Mycobacterium fortuitum* model. J Med Microbiol 50, 565–570.

Grange, J. M., Yates, M. D., and Pozniak, A. (1995) Bacteriologically confirmed non-tuberculous mycobacterial lymphadenitis in south east England: a recent increase in the number of cases. Arch Dis Child 72, 516–517.

Griffith, D. E., Brown, B. A., Cegielski, P. *et al.* (2000) Early results (at 6 months) with intermittent clarithromycin-including regimens for lung disease due to *Mycobacterium avium* complex. Clin Infect Dis 30, 288–292.

Griffith, D. E., Brown, B. A., Murphy, D. T. *et al.* (1998) Initial (6-month) results of three-times-weekly azithromycin in treatment regimens for *Mycobacterium avium* complex lung disease in human immunodeficiency virus-negative patients. J Infect Dis 178, 121–126.

Griffith, D. E., Girard, W. M., and Wallace, R. J. Jr. (1993) Clinical features of pulmonary disease caused by rapidly growing mycobacteria. An analysis of 154 patients. Am Rev Respir Dis 147, 1271–1278.

Hafner, B., Haag, H., Geiss, H. K., and Nolte, O. (2004) Different molecular methods for the identification of rarely isolated non-tuberculous mycobacteria and description of new *hsp65* restriction fragment length polymorphism patterns. Mol Cell Probes 18, 59–65.

Hale, Y. M., Pfyffer, G. E., and Salfinger, M. (2001) Laboratory diagnosis of mycobacterial infections: new tools and lessons learned. Clin Infect Dis 33, 834–846.

Havlir, D. V., Dube, M. P., Sattler, F. R. *et al.* (1996) Prophylaxis against disseminated *Mycobacterium avium* complex with weekly azithromycin, daily rifabutin, or both. California Collaborative Treatment Group. N Engl J Med 335, 392–398.

Hayman, J. (1991) Postulated epidemiology of *Mycobacterium ulcerans* infection. Int J Epidemiol 20, 1093–1098.

Hayman, J. (1993) Out of Africa: observations on the histopathology of *Mycobacterium ulcerans* infection. J Clin Pathol 46, 5–9.

Hermon-Taylor, J. and Bull, T. (2002) Crohn's disease caused by *Mycobacterium avium* subspecies *paratuberculosis*: a public health tragedy whose resolution is long overdue. J Med Microbiol 51, 3–6.

Hoy, J., Mijch, A., Sandland, M. *et al.* (1990) Quadruple-drug therapy for *Mycobacterium avium-intracellulare* bacteremia in AIDS patients. J Infect Dis 161, 801–805.

Huang, J. H., Kao, P. N., Adi, V., and Ruoss, S. J. (1999) *Mycobacterium avium-intracellulare* pulmonary infection in HIV-negative patients without pre-existing lung disease: diagnostic and management limitations. Chest 115, 1033–1040.

Hunter, A. M., Campbell, I. A., Jenkins, P. A., and Smith, P. A. (1981) Treatment of pulmonary infection cuased by mycobacteria of *Mycobacterium avium-intracellulare* complex. Thorax 36, 326–329.

Jenkins, P. A. (1981) The epidemiology of opportunist mycobacterial infections in Wales, 1952–1978. Rev Infect Dis 3, 1021–1023.

Jenkins, P. A., Banks, J., Campbell, I. A., and Smith, A. P. (1994) *Mycobacterium kansasii* pulmonary infection: a prospective study of the results of nine months of treatment with rifampicin and ethambutol. Research Committee, British Thoracic Society. Thorax 49, 442–445.

Jiva, T. M., Jacoby, H. M., Weymouth, L. A. *et al.* (1997) *Mycobacterium xenopi*: innocent bystander or emerging pathogen? Clin Infect Dis 24, 226–232.

Joint Tuberculosis Committee. (2000) Management of opportunist mycobacterial infections: Joint tuberculosis committee guidelines 1997. Thorax 55, 210–218.

Juffermans, N. P., Verbon, A., Danner, S. A. *et al.* (1998) *Mycobacterium xenopi* in HIV-infected patients: an emerging pathogen. AIDS 12, 1661–1666.

Kennedy, T. P. and Weber, D. J. (1994) Nontuberculous mycobacteria. An underappreciated cause of geriatric lung disease. Am J Respir Crit Care Med 149, 1654–1658.

Kerbiriou, L., Ustianowski, A., Johnson, M. A. *et al.* (2003) Human immunodeficiency virus type 1-related pulmonary *Mycobacterium xenopi* infection: a need to treat? Clin Infect Dis 37, 1250–1254.

Kiehn, T. E. and White, M. (1994) *Mycobacterium haemophilum*: an emerging pathogen. Eur J Clin Microbiol Infect Dis 13, 925–931.

Lang-Lazdunski, L., Offredo, C., Pimpec-Barthes, F. *et al.* (2001) Pulmonary resection for *Mycobacterium xenopi* pulmonary infection. Ann Thorac Surg 72, 1877–1882.

Lillo, M., Orengo, S., Cernoch, P., and Harris, R. L. (1990) Pulmonary and disseminated infection due to *Mycobacterium kansasii*: a decade of experience. Rev Infect Dis 12, 760–767.

Lounis, N., Truffot-Pernot, C., Bentoucha, A. *et al.* (2001) Efficacies of clarithromycin regimens against *Mycobacterium xenopi* in mice. Antimicrob Agents Chemother 45, 3229–3230.

Marin-Casanova, P., Calandria Amiguetti, J. L., Garcia-Martos, P. *et al.* (2003) Endophthalmitis caused by *Mycobacterium abscessus*. Eur J Ophthalmol 13, 800–802.

Marston, B. J., Diallo, M. O., Horsburgh, C. R. Jr. *et al.* (1995) Emergence of Buruli ulcer disease in the Daloa region of Cote d'Ivoire. Am J Trop Med Hyg 52, 219–224.

McWhinney, P. H., Yates, M., Prentice, H. G. *et al.* (1992) Infection caused by *Mycobacterium chelonae*: a diagnostic and therapeutic problem in the neutropenic patient. Clin Infect Dis 14, 1208–1212.

Morris-Jones, R., Fletcher, C., Morris-Jones, S. *et al.* (2001) *Mycobacterium abscessus*: a cutaneous infection in a patient on renal replacement therapy. Clin Exp Dermatol 26, 415–418.

Ozluer, S. M. and De'Ambrosis, B. J. (2001) *Mycobacterium abscessus* wound infection. Australas J Dermatol 42, 26–29.

Paech, V., Lorenzen, T., von Krosigk, A. *et al.* (2002) Remission of cutaneous *Mycobacterium haemophilum* infection as a result of antiretroviral therapy in a Human Immunodeficiency Virus-infected patient. Clin Infect Dis 34, 1017–1019.

Parrot, R. G. and Grosset, J. H. (1988) Post-surgical outcome of 57 patients with *Mycobacterium xenopi* pulmonary infection. Tubercle 69, 47–55.

Pezzia, W., Raleigh, J. W., Bailey, M. C. *et al.* (1981) Treatment of pulmonary disease due to *Mycobacterium kansasii*: recent experience with rifampin. Rev Infect Dis 3, 1035–1039.

Phillips, P., Chan, K., Hogg, R. *et al.* (2002) Azithromycin prophylaxis for *Mycobacterium avium* complex during the era of highly active antiretroviral therapy: evaluation of a provincial program. Clin Infect Dis 34, 371–378.

Portaels, F., Larsson, L., and Jenkins, P. A. (1995) Isolation of *Mycobacterium malmoense* from the environment in Zaire. Tuber Lung Dis 76, 160–162.

Primm, T. P., Lucero, C. A., and Falkinham, J. O. III. (2004) Health impacts of environmental mycobacteria. Clin Microbiol Rev 17, 98–106.

Prince, D. S., Peterson, D. D., Steiner, R. M. *et al.* (1989) Infection with *Mycobacterium avium* complex in patients without predisposing conditions. N Engl J Med 321, 863–868.

Rahal, A., Abela, A., Arcand, P. H. *et al.* (2001) Non-tuberculous mycobacterial adenitis of the head and neck in children: experience from a tertiary care pediatric center. Laryngoscope 111, 1791–1796.

Research Committee, British Thoracic Society. (2001) First randomised trial of treatments for pulmonary disease caused by *M avium intracellulare*, *M malmoense*, and *M xenopi* in HIV negative patients: rifampicin, ethambutol and isoniazid versus rifampicin and ethambutol. Thorax 56, 167–172.

Runyon, E. H. (1957) Anonymous mycobacteria in pulmonary disease. Med Clin North America 43, 273–290.

Saubolle, M. A., Kiehn, T. E., White, M. H. *et al.* (1996) *Mycobacterium haemophilum*: microbiology and expanding clinical and geographic spectra of disease in humans. Clin Microbiol Rev 9, 435–447.

Schaefer, W. B., Birn, K. J., Jenkins, P. A., and Marks, J. (1969) Infection with the avian-Battey group of mycobacteria in England and Wales. Br Med J 2, 412–415.

Scmitt, H., Schnitzler, N., Riehl, J. *et al.* (1999) Successful treatment of pulmonary *Mycobacterium xenopi* infection in a natural killer cell-deficient patient with clarithromycin, rifabutin, and sparfloxacin. Clin Infect Dis 29, 120–124.

Shafran, S. D., Singer, J., Zarowny, D. P. *et al.* (1996) A comparison of two regimens for the treatment of *Mycobacterium avium* complex bacteremia in AIDS: rifabutin, ethambutol, and clarithromycin versus rifampin, ethambutol, clofazimine, and ciprofloxacin. Canadian HIV Trials Network Protocol 010 Study Group. N Engl J Med 335, 377–383.

Shah, M. K., Sebti, A., Kiehn, T. E. *et al.* (2001) *Mycobacterium haemophilum* in immunocompromised patients. Clin Infect Dis 33, 330–337.

Simor, A. E. Salit, I. E., and Vellend, H. (1984) The role of *Mycobacterium xenopi* in human disease. Am Rev Respir Dis 129, 435–438.

Smith, M. J. and Citron, K. M. (1983) Clinical review of pulmonary disease caused by *Mycobacterium xenopi*. Thorax 38, 373–377.

Straus, W. L., Ostroff, S. M., Jernigan, D. B. *et al.* (1994) Clinical and epidemiologic characteristics of *Mycobacterium haemophilum*, an emerging pathogen in immunocompromised patients. Ann Intern Med 120, 118–125.

Sungkanuparph, S., Sathapatayavongs, B., and Pracharktam, R. (2003a) Infections with rapidly growing mycobacteria: report 20 cases. Int J Infect Dis 7, 198–205.

Sungkanuparph, S., Sathapatayavongs, B., and Pracharktam, R. (2003b) Rapidly growing mycobacterial infections: spectrum of diseases, antimicrobial susceptibility, pathology and treatment outcomes. J Med Assoc Thai 86, 772–780.

Tanaka, E., Kimoto, T., Tsuyuguchi, K. *et al.* (2002) Successful treatment with faropenem and clarithromycin of pulmonary *Mycobacterium abscessus* infection. J Infect Chemother 8, 252–255.

Tortoli, E. (2003) Impact of genotypic studies on mycobacterial taxonomy: the new mycobacteria of the 1990s. Clin Microbiol Rev 16, 319–354.

Tortoli, E., Bartoloni, A., Bottger, E. C. *et al.* (2001) Burden of unidentifiable mycobacteria in a reference laboratory. J Clin Microbiol 39, 4058–4065.

Tortoli, E., Besozzi, G., Lacchini, C. *et al.* (1998a) Pulmonary infection due to *Mycobacterium szulgai*, case report and review of the literature. Eur Respir J 11, 975–977.

Tortoli, E., Brunello, F., Cagni, A. E. *et al.* (1998b) *Mycobacterium genavense* in AIDS patients, report of 24 cases in Italy and review of the literature. Eur J Epidemiol 14, 219–224.

Tunkel, D. E. and Romaneschi, K. B. (1995) Surgical treatment of cervicofacial non-tuberculous mycobacterial adenitis in children. Laryngoscope 105, 1024–1028.

Villanueva, A., Calderon, R. V., Vargas, B. A. *et al.* (1997) Report on an outbreak of postinjection abscesses due to *Mycobacterium abscessus*, including management with surgery and clarithromycin therapy and comparison of strains by random amplified polymorphic DNA polymerase chain reaction. Clin Infect Dis 24, 1147–1153.

Wallace, R. J. Jr. (1989) The clinical presentation, diagnosis, and therapy of cutaneous and pulmonary infections due to the rapidly growing mycobacteria, *M. fortuitum and M. Chelonae*. Clin Chest Med 10, 419–429.

Wallace, R. J. Jr., Brown, B. A., Griffith, D. E. *et al.* (1994a) Initial clarithromycin monotherapy for *Mycobacterium avium-intracellulare* complex lung disease. Am J Respir Crit Care Med 149, 1335–1341.

Wallace, R. J. Jr., Dunbar, D., Brown, B. A. *et al.* (1994b) Rifampin-resistant *Mycobacterium kansasii*. Clin Infect Dis 18, 736–743.

Wallace, R. J. Jr., Swenson, J. M., Silcox, V. A., and Bullen, M. G. (1985) Treatment of non-pulmonary infections due to *Mycobacterium fortuitum* and *Mycobacterium chelonei* on the basis of *in vitro* susceptibilities. J Infect Dis 152, 200–214.

Wallace, R. J. Jr., Tanner, D., Brennan, P. J., and Brown, B. A. (1993) Clinical trial of clarithromycin for cutaneous (disseminated) infection due to *Mycobacterium chelonae*. Ann Intern Med 119, 482–486.

Wallace, R. J. Jr., Zhang, Y., Brown-Elliott, B. A. *et al.* (2003) Repeat positive cultures in *Mycobacterium intracellulare* lung disease after macrolide therapy represent new infections in patients with nodular bronchiectasis. J Infect Dis 186, 266–273.

van der Werf, T. S., van der Graaf, W. T., Tappero, J. W., and Asiedu, K. (1999) *Mycobacterium ulcerans* infection. Lancet 354, 1013–1018.

Yew, W. W., Piddock, L. J., Li, M. S. *et al.* (1994) *In-vitro* activity of quinolones and macrolides against mycobacteria. J Antimicrob Chemother 34, 343–351.

12

Aerobic Actinomycetes

Stephen H. Gillespie

Centre for Medical Microbiology, Royal Free and University College Medical School, Hampstead Campus, London, UK

INTRODUCTION

The aerobic actinomycetes form a heterogeneous group of organisms only a few of which are human pathogens. They are classified together on morphological criteria: Gram-positive organisms which grow as branching filamentous cells (Plate 7). Although once thought to be fungi imperfecti, they are prokaryotes and there is nothing to suggest that they are higher bacteria. They are closely related to corynebacteria and mycobacteria. Mycobacteria are acid fast by virtue of long chain mycolic acid present in their cell wall. Some strains of corynebacteria contain mycolic acids with a much shorter chain length and may be acid fast when grown under appropriate conditions. Organisms of the genera *Nocardia*, *Rhodococcus*, *Gordona* and *Tsukamurella* possess mycolic acids of intermediate chain length and consequently express a degree of acid fastness.

NOCARDIA SPP.

Introduction

Edmond Nocard, a veterinarian working on the island of Guadeloupe, described a filamentous organism as the cause of bovine farcy. In the following year the organism was characterized and named *Nocardia farcinica*, although later it was recognised that it was a *Mycobacterium* that is the cause of farcy (Nocard 1898). The organism Nocard had isolated was probably *Mycobacterium farcinoges*. The first case of human nocardiasis was reported 1 year later in a patient with pneumonia and a brain abscess: this organism was classified as *Nocardiae asteroides* (Blanchard 1896). *Nocardia transvalensis* was isolated from an African patient with Madura foot (Pijper and Pullinger 1927). More than 25 different species have been described. Nocardiae are regularly but rarely implicated in human infections of immunocompetent patients (Beaman *et al*. 1976). They are increasingly recognised in patients on immunosuppressive therapy and those infected with HIV (Beaman *et al*. 1976; McNeil *et al*. 1990; Miralles 1994; Pintado *et al*. 2003).

DESCRIPTION OF THE ORGANISM

Morphological and Physiological Characteristics

Nocardiae are Gram-positive, aerobic catalase-positive non-motile filamentous bacteria that exhibit branching. The filaments break up into rods, and coccal forms and aerial filaments are always produced. The cell wall contains *meso*-diaminopimelic acid,

arabinose and galactose. It is naturally resistant to lysozyme digestion. The G+C% ranges from 64 to 72 (McNeil *et al*. 1990; Saubolle and Sussland 2003). Like other members of the corynebacteria *Mycobacterium Nocardia* (CMN) group, *Nocardia* spp. contain mycolic acids in the cell wall. The carbon chain length ranges between C_{44} and C_{60} and is responsible for the weak acid fastness these organism exhibit when grown on appropriate lipid containing media (Butler, Kilburn and Kubica 1987; Saubolle and Sussland 2003). Peptidoglycan makes up as little of 25% of the cell-wall mass rising to 45% during the stationary phase of growth (Beaman and Moring 1988). As in mycobacterial cell walls the peptidoglycan is attached to the arabinogalactan polymer by way of a phosphodiester link.

Taxonomy

The nocardiae are phylogenetically related to the *Mycobacterium*, *Corynebacterium*, *Gordona* and *Tsukamurella* as defined by 16S sequencing. There are at least 25 established species described which include (Chun and Goodfellow 1995; Ruimy *et al*. 1996; Roth *et al*. 2003; Wallace *et al*. 1991) *N. asteroides*, *N. carea*, *N. brasiliensis*, *N. pseudobrasiliensis*, *N. veterana*, *N. farcinica*, *N. brevicatena*, *N. otidiscaviarum*, *N. nova*, *N. seriolae*, *N. transvalensis*, *N. N. vacinii* and *N. pseudobrasiliensis*. Some authors suggest that *N. asteroides* should be divided into five species on the basis of 16S sequencing (Roth *et al*. 2003), and new proposed species are added regularly. The species previously named *N. amarae* has been transferred to the genus *Gordona* on the basis of chemical, microbiological and 16S sequences (Goodfellow *et al*. 1994).

Pathogenesis

Experiments with T-cell deficient mice and experience of immuno-compromised patients indicate an important role for T-cell immunity in *Nocardia* infections.

Nocardia Antigens

During the phases of growth the composition of the nocardial cell wall changes significantly and the virulence of *Nocardia* appears to vary with the growth phase of the organism with logarithmically growing filamentous cells being more virulent than stationary phase cells of the same organisms (Beaman and Sugar 1983; Beaman and Moring 1988). The cell wall also contains a number of antigens which have been associated with pathogenicity in mycobacteria: tuberculostearic acid

Principles and Practice of Clinical Bacteriology Second Edition Editors Stephen H. Gillespie and Peter M. Hawkey
© 2006 John Wiley & Sons, Ltd

and trehalose 6–6′ dimycolate (cord factor). The latter substance is thought to play a role in reducing phagosomal–lysosomal fusion (Crowe *et al.* 1994; Spargo *et al.* 1991).

Nocardia possesses a superoxide dismutase which is expressed on the surface of the organism and protects it from the toxic effects of superoxide radicals (Beaman *et al.* 1985). This effect is increased by the action of the organism's catalase. Several studies have reported the presence of toxins including a haemolysin in strains of *N. asteroides*, *N. brasiliensis* and *N. otidiscaviarum*.

Interaction with Phagocytes

Nocardiae are readily phagocytosed by macrophages, and although the majority of organisms are killed, some are able to survive as L-forms where they are able to multiply. This may also explain the reports of relapsing infection after apparently successful therapy. L-forms have also been shown to produce fatal infections in mice.

Virulent *Nocardia asteroides* is able to inhibit phagosomal–lysosomal fusion, and this effect appears to be mediated by cord factor (Spargo *et al.* 1991; Crowe *et al.* 1994). There is also evidence that virulent strains are able to block acidification of the phagosome, whereas avirulent strains are unable to do this. This shows similarity with the behaviour of *Mycobacterium tuberculosis* (see Chapter 10) (Indrigo, Hunter Jr. and Actor 2003). *Nocardia* is also able to use acid phosphatase as a carbon source inhibiting the effect of this toxic macrophage enzyme. *Nocardia* also possess mycolic acids, and these have been implicated in mycobacterial pathogenicity.

EPIDEMIOLOGY

Habitat

The nocardiae are environmental organisms found in soil and vegetation, and they are thought to have a role in the decay of organic plant material. Nocardiae have been isolated from marine environments and freshwater sources including tap water (Beaman and Beaman 1994). *Nocardia asteroides* is more frequently isolated in temperate countries and *N. brasiliensis* in tropical and subtropical regions (Beaman and Beaman 1994). Isolation of *Nocardia* from human specimens may represent colonization, contamination or infection. One study from Australia suggested that as little as 20% of isolates were clinically significant (Georghiou and Blacklock 1992).

Animal Infection

Nocardiae are reported to cause infection in a wide range of animal species including cattle, horses, dogs and pigs. These infections take the form of bovine mastitis which results in reduction in milk production. Equine infections are rarely reported and take the form of bronchopneumonia which may develop into disseminated disease. This syndrome is associated with foals with combined immunodeficiency and adult horses with hyperadrenalism secondary to pituitary tumours. Localized abscesses are also reported rarely.

Human Infection

It is normally assumed that nocardial infections are rare, but a number of studies have suggested that their incidence is underestimated (Beaman *et al.* 1976; Georghiou and Blacklock 1992). With the increasing number of patients who are immunocompromised, the number of cases of *Nocardia* infection is likely to rise. Organisms of the *Nocardia* genus may act as primary pathogens in adults without evidence of immunocompromised or may act as opportunists. Opportunist infections are especially associated with organ transplant-

ation, malignancy, lymphoma sarcoidosis collagen vascular disease or HIV infection (McNeil and Brown 1994).

In immunocompetent patients cutaneous infection is usually caused by *N. asteroides*, *N. brasiliensis* and *N. otidiscavitarum*. There are many cutaneous syndromes described including lymphocutaneous infection, skin abscesses and skin infection associated with dissemination (Georghiou and Blacklock 1992; Sachs 1992; Goodfellow *et al.* 1994). Mycetoma is a chronic granulomatous subcutaneous infection of the foot which occurs in tropical countries and is thought to be associated with walking barefoot, although mycetoma of the hand has also been reported (Beaman and Beaman 1994). The infection can be caused by aerobic actinomycetes including *N. brasiliensis*, *Actinomadura madurae*, *Streptomyces somaliensis* or other nocardiae. Ocular infection is usually secondary to minor corneal trauma or inadequately cleaned contact lenses or ocular surgery. This may lead to keratitis and later to ophthalmitis (Douglas *et al.* 1991). Ocular infection can be seen from a distant site in immunocompromised patients with another infective source.

Colonization of the respiratory tract of immunocompetent patients is not uncommon and may even result in a mild self-limiting infection. *Nocardia asteroides* is implicated in invasive disease which takes the form of a subacute or chronic cavitatory pneumonia (Georghiou and Blacklock 1992). Common predisposing conditions include local compromised of pulmonary defences such as chronic obstructive pulmonary disease, pulmonary fibrosis, healed tuberculosis, bronchiectasis or general immunosuppression such as steroid therapy or HIV infection (Georghiou and Blacklock 1992). Rarely pulmonary nocardiasis can take an acute fulminating course (Neu *et al.* 1967).

Disseminated infection can occur in any patient but more usually presents in immunocompromised patients and is associated with a poor prognosis. It can develop as a late complication of infection with any of the *Nocardia* spp. (Esteban *et al.* 1994; Poonwan *et al.* 1995). Dissemination of the infection may result in secondary lesions in the brain and the eyes. Central nervous system infection takes the form of isolated brain abscess, and nocardial meningitis is very rare (Bross and Gordon 1991).

HIV Infection

Nocardiasis is a relatively rare complication of infection with HIV in comparison with other organisms such as *Pneumocystis jiroveci* and *Toxoplasma gondii* (Pintado *et al.* 2003). This may be because non-T-cell mechanisms are more important for defence against this organism (McNeil and Brown 1994). *Nocardia asteroides* is the species most commonly associated with infection in HIV-infected individuals, but case reports with *N. farcinica* and *N. nova* have been made (Jones *et al.* 2000).

CLINICAL FEATURES

Although species of *Nocardia* may occasionally be found in healthy persons, they are not considered to be part of the normal flora.

Pulmonary Disease

The majority of cases of pulmonary nocardiasis are caused by *N. asteroides*. The first case of pulmonary nocardiasis occurring in a patient in the USA was described by Flexner in 1898. The pathological feature of pulmonary nocardiasis is usually a suppurative lesion, for example, lung abscess, but a granulomatous response or a mixture of these two may occur. The clinical and radiological features are very variable and non-specific, making the diagnosis difficult without the use of invasive techniques. In the normal course of infection, one or more lung abscesses may develop and enlarge to form cavities similar to those seen in pulmonary tuberculosis. Pulmonary nocardiasis may mimic pulmonary malignance, actinomycosis and fungal infection (Menendez *et al.* 1997). *Nocardia brasiliensis* possesses considerably more virulence than *N. asteroides* and may cause primary pulmonary infection in otherwise healthy individuals.

The manifestations within the lungs may vary from a mild, diffuse infiltration to a lobar or multi-lobar consolidation. There may be solitary masses, reticulonodular infiltrates and pleural effusions. In many cases the correct diagnosis is not made until *post mortem* when *Nocardia* can be seen microscopically and grown in culture. From the lung the organisms may spread through the blood stream; there they seem to have a predeliction for the brain and the kidneys but can establish infection anywhere in the body. Unlike actinomycosis or tuberculosis, bone destruction is rare.

Cutaneous Nocardosis

Primary cutaneous nocardosis usually occurs following traumatic introduction of the organisms (usually *N. brasiliensis*) into the skin (Clark *et al*. 1995). The lesions induce pus formation and may develop into cellulitis or pyoderma. Although the infection usually remains local and self-limiting, it may progress and on occasion can spread via the lymphatics to produce a lymphocutaneous lesions (Sachs 1992). This is more likely to occur in *N. brasiliensis* infection (Seddon, Parr and Ellis-Pegler 1995). As local infection can resemble those caused by *S. aureus*, the diagnosis cannot be confirmed without laboratory evidence. Cutaneous infection has most commonly been reported in the USA or Australia and rarely in Europe.

Infections can result in fungating tumour-like masses, termed mycetomas (Figure 12.1). These are chronic subcutaneous infections in which the abscesses extend by destruction of the soft-tissue, sometimes the bone. The infection finally erupts through the skin. In equatorial Africa the mycetomas are caused by *Actinomadua pellitieri* and *Streptomyces somaliensis*, but in Mexico the majority of infections are caused by *N. brasiliensis* (Seddon, Parr and Ellis-Pegler 1995). Mycetomas are usually found in developing countries and rarely in Europe and North America. It is important to differentiate between actinomycotic and eumycotic mycetoma because the latter requires complete excision to effect cure, but the actinomycotic form may respond to aggressive medical treatment and conservative surgical excision.

Madura foot is a chronic granulomatous infection of bones and soft tissue of the result resulting in mycetoma formation and gross deformity. It occurs in the Sudan, North Africa and the west coast of India, principally among those who walk barefoot and are prone to contamination of foot injuries by soil-derived organisms. One of the causative organisms is *N. madurae*, but it is also caused by other nocadiae and fungi.

Systemic Infection

Nocardial lesions in the lungs or elsewhere in the body frequently erode into blood vessels. In systemic nocardiosis the nocardiae behave as pyogenic bacteria, and infection becomes relentlessly progressive. Infection of the central nervous system occurs frequently, but this is often insidious in onset and difficult to diagnose and treat successfully. It often takes the form of a cerebral abscess which may be confused with a pyogenic or fungal abscess or with cerebral toxoplasmosis (McNeil and Brown 1994).

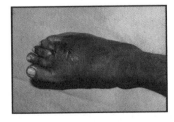

Figure 12.1 Swollen foot of patient with 'Madura foot'.

Clinical Features of Other Nocardiae

Nocardia farcinica is microbiologically related to but distinct from *N. asteroides* and is noted for its propensity to cause serious systemic infection in both normal and immunocompromised hosts and for its natural resistance to multiple antimicrobial agents. *Nocardia farcinica* may cause a variety of clinical presentations including cerebral abscess, pulmonary and cutaneous infection. The similarity between *N. farcinica* and *N. asteroides* means that care must be taken to distinguish these species (Wallace *et al*. 1991; Schiff *et al*. 1993).

Nocardia transvalensis tends to cause pneumonia or disseminated disease in severely immunocompromised patients. It has also been recorded as a cause of mycetoma (McNeil and Brown 1994).

LABORATORY DIAGNOSIS

Nocardiae are difficult to recognize and identify in the routine diagnostic laboratory, and this is made more difficult by the slow growth on primary isolation from clinical samples.

Collection of Specimens

In pulmonary cases specimens of sputum or bronchoalveolar lavage should be collected. Biopsy and autopsy specimens exudate pus, and scrapings from skin lesions should be collected in tightly stoppered bottles and sent to the laboratory without delay. Specimens held in storage or delayed in transit should be kept at 4 °C. Preserving fluids are not necessary, but if the laboratory is at some distance, the specimen should be inoculated onto blood agar locally prior to dispatch.

Direct Examination

A presumptive diagnosis of pulmonary nocardiosis may be made by microscopical examination of sputum or broncho-alveolar lavage. Pus and other exudates should be diluted in sterile water if necessary and examined for the presence of microcolonies. These are best seen in an unstained wet mount between slide and cover slip under greatly reduced light or phase contrast light. In many cases the sputum contains numerous lymphocytes and macrophages, some of which contain pleomorphic Gram-positive and weakly acid-fast bacilli and occasionally extracellular branching filaments. Differential of the microcolonies is difficult and requires specialist attention. Branched hyphal filaments may belong to any number of genera, for example, *Nocardia*, *Actinomadura* and *Streptomyces*.

Nocardiae are Gram positive with a variety of shapes and sizes, but they stain irregularly and their filaments are generally beaded. Acid fastness is variable in *N. asteroides*, *N. brasiliensis* and *N. caviae*, both in clinical specimens and culture. The usual Ziehl–Nielsen procedure may be employed, but the period of decolorization with acid alcohol must not exceed 5–10 s.

Tissue Sections

The microcolonies (granules) show up well in sections stained with haematoxylin and eosin, but this stain often fails to demonstrate other morphological forms of nocardiae. The Grocott-Gomori silver methenamine stains the granules very well, and Gram–Wiegert technique is effective for both granules and filaments.

Culture and Isolation

Nocardiae grow on most standard bacteriological media in 2–30 days. Suitable media include brain heart infusion agar and trypticase soy agar enriched with blood. *Nocardia asteroides* grows well on

Sabouraud's agar incubated at 37 °C in a humid environment, producing the typical wrinkled pigmented forms. Colonies are cream, orange or pink coloured, and their surface may develop dry chalky appearances and the colonies adhere firmly to the medium. Colonies with abundant aerial growth have a cotton wool ball appearance that may resemble many *Streptomyces* spp. The colours are due to various carotenoid-like pigments, and their intensity depends on the specific culture conditions used to grow the organisms.

Recovery of *Nocardia* spp. from mixed cultures is facilitated by the use of selective media. Buffered yeast extract (similar to that used for *Legionella*) containing polymyxin anisomycin and vancomycin and Sabouraud's dextrose agar with chloramphenical has been used for sputum. However, many isolates are susceptible to chloramphenicol. Media using paraffin as the sole carbon source have been shown to be effective as a selective medium for nocardiae. Also as *N. asteroides* grows well at 45 °C, initial incubation at temperatures above 37 °C may help to separate this species from other bacteria. Cultures should be examined at intervals from 2 days to 2 weeks. *Nocardia* spp. grow well on Lowenstein–Jensen medium, produce moist glabrous colonies and must be differentiated from mycobacteria.

Identification

The only constant morphological feature of nocardiae is a tendency for the aerial or vegetative mycelium to fragment. The consistency and composition of the growth medium can affect the growth and stability of both aerial and substrate hyphae. *Nocardia* spp. are Gram positive, weakly acid-fast, non-motile aerobic, catalase positive and oxidase a number of sugars. They are ONPG positive, reduce nitrate and utilize urea. With few exceptions, all *Nocardia* spp. grow in nutrient broth supplemented with lysozyme, whereas most *Rhodococcus*, *Actinomadura* and *Streptomyces* spp. do not. In general, *N. farcinica* is differentiated from *N. asteroides* by the ability of the former to grow and produce acid from rhamnose, to grow on 2,3-butylene glycol and at 45 °C. The Gordon series of tests involving 40 different physiological properties is unable to distinguish between *N. farcinica* and *N. asteroides*. Antibiotic susceptibility tests can assist in speciation as an adjunt to biochemical tests. Differentiation tests are summarized on Table 12.1.

For laboratories that have the facility the most rapid method of species identification is to use 16S rDNA amplification and sequencing (Cloud *et al*. 2004).

Table 12.1 Differentiation of *Nocardia* species

Characteristic	N. asteroides	N. brasiliensis	N. transvalensis	N. farcinica
Decomposition of				
adenine	–	–	–	–
casein	–	+	–	–
hypoxanthine	–	+	+	–
tyrosine	–	+	–	–
xanthine	–	–	+	–
Growth in lysozyme	+	+	+	+
Acetamide utilization	17%			80%
Acid from rhamnose	10%	–	–	80%
Growth at 45 °C for 3 days[a]	43%	–	–	100%
Resistance to:				
cefamandole	5%			93%
tobramycin	17%			100%

% = percentage of 40 strains reacting positive.
[a] See Wallace *et al*. (1991).

Serodiagnostic Tests

There is currently no single serodiagnostic test that is routinely used to identify patients with nocardial infection. Patients infected with nocardiae only develop a minimal antibody response that is non-specific. Infection also occurs in patients who are immunosuppressed, making serology an unattractive method for making a diagnosis.

MANAGEMENT OF INFECTION

There are no specific clinical signs diagnostic for pulmonary nocardosis, and the presentation of disease may run the full spectrum from acute to chronic pulmonary infection. Thus pulmonary infection caused by *Nocardia* spp. is often misdiagnosed as pyogenic infections, tuberculosis, actinomycosis, mycoes of various aetiologies, benign tumours and various forms of neoplasm. Difficulties in diagnosis are compounded when co-existent disease is taken into account, for example, nocarodosis and tuberculosis occur together in 6–30% of cases. In many cases the correct diagnosis is only made when the organisms are visualized on a tissue section isolated in pure culture.

Nocardia spp. show species-specific drug resistance and good identification, and carefully controlled MICs should point the way for directed and prolonged therapy. Sulphonamides or cotrimoxazole are usually the treatment of choice, but many patients, especially those with HIV, are unable to tolerate them and may also be infected with resistant strains (Jones *et al*. 2000). Thus, alternative agents such aminocycline, erythromycin, amikacin or imipenem may have to be used. The prognosis is poor particularly if the organisms have metastisised to other organs in the body. The *in vitro* activities of new quinolones and cephamycins look promising as agents for treatment. Linezolid has been shown to be active *in vitro* and in an experimental model (Gomez-Flores *et al*. 2004). In the absence of consensus on the length of therapy, some authors suggest that a synergistic combination of a β-lactam/β-lactamase inhibitor with ciprofloxacin or amikacin followed by a short course of trimethoprim-sulphamethoxazole may be effective in eradicating nocardial disease and may reduce the need for long-term treatment (Tripodi *et al*. 2001).

PREVENTION AND CONTROL

The widespread distribution of nocardiae in the soil makes control of infection with these organisms impossible. Superficial infections are often the result of local trauma and soil contamination of wounds, and therefore one can suggest that all wounds are adequately cleaned and debrided. The use of prophylactic antibiotics is probably not beneficial, as these organisms are resistant to most of the common oral agents.

Person-to-person transmission is very rare, although there are two reports of nosocomial outbreaks among renal transplant patients.

ACTINOMADURA SPP.

Vincent first isolated an organism which he thought responsible for Madura foot which he named *Streptothrix madurae*. It is now known that many organisms can give rise to this syndrome including other aerobic actinomycetes such as *Nocardia* spp. and fungi. The organism that Vincent described was renamed *Nocardia* and then transferred to *Actinomadura* together with *A. pellettieri* (McNeil and Brown 1994). It has been separated from the other aerobic actinomycetes by 16S rRNA sequencing. The genus is defined on the basis of cell wall chemotype the morphology of its aerial hyphae and the presence of a specific sugar madurose which has been identified as 3-O-methyl-D-galactose. It does not possess mycolic acid in its cell wall and fails to grow in the presence of lysozyme. There are as many as 26 species defined, but only two of these are important in human infections: *A. madurae* and *A. pellettieri*.

Actinomadura is one of the most frequent bacterial causes of Madura foot significantly outnumbering cases caused by *Nocardia*. The peak age of onset is between 16 and 25 years, and it may take up to 10 years for cases to present depending of the species involved (Maiti, Ray and Bandyopadhyay 2002). *Actinomadura* infection has also been associated with infection of long-standing indwelling canulae and wounds, although these are rare. Microscopy of the specimens reveals organisms in the form of branched filaments with short chains of spores. Colonies have a molar tooth appearance after 48 h in culture, and aerial hyphae are sparse and may only be seen after 2 weeks in culture. Speciation is on the basis of biochemical tests and 16S sequencing (McNeil and Brown 1994).

Most isolates of *A. madurae* are susceptible to amikacin and imipenem, and many are resistant to ampicillin. Cephalosporins, trimethoprim suphamethoxazole and penicillins have limited activity (McNeil *et al.* 1990).

DEMATOPHILUS CONGOLENSIS

Dermatophilus congolensis was first recognized as a cattle disease in the Belgian Congo in 1915 by van Saceghem. It is a chronic dermatitis damaging skin wool and more rarely causing foot rot. Humans acquire infection by close contact with infected animals or their products. The clinical spectrum of the disease includes asymptomatic infection, pustular eruption and a pitted keratolysis (Gillum *et al.* 1988; Towersey *et al.* 1993).

The diagnosis of dematophilosis depends on visualizing the organism in wet mounts or specimens stained with methylene blue or Geimsa. Gram stain is not an effective staining method, as some of the detail of the organism is obscured. *Dermatophilus congolensis* may be isolated with difficulty in brain heart infusion agar containing horse blood but may require animal passage. Colonies are small grey white they pit the agar and β-haemolysis may be seen. The organism is motile and catalase positive and hydrolyses casein slowly (McNeil and Brown 1994). The organism is susceptible to penicillin, streptomycin, chloramphenicol, tetracycline, erythromycin, and trimethoprim-sulphamethoxazole.

GORDONA SPP.

Gordona spp. are rarely isolated from human subjects but have been associated with primary cutaneous infection, catheter-related sepsis, pulmonary infection that resembles tuberculosis and rarely brain abscess (Drancourt *et al.* 1994; Pham *et al.* 2003). The genus was defined in 1988 on the basis of 16S rRNA sequence and includes species previously included in the genus *Rhodococcus*. The following species are found in the genus: *G. bronchialis*, *G. rubopertincus*, *G. sputi* and *G. terrae*. *Gordona* spp. have mycolic acid with a shorter chain length than mycobacteria but longer than those of *Rhodococcus*. *Gordona* produces mycobactins under conditions of iron limitation. Wrinkled dry beige colonies grow after 3–7 days of incubation on blood agar. Microscopically, the organisms are small, Gram-positive, weakly acid-fast, beaded bacilli. Species within the genus cannot be fully differentiated on the basis of hydrolysis of amino acids and casein, acid production from sugars or HPLC for mycolic acids (Chun and Goodfellow 1995). Species identification requires the use of molecular methods.

OTHER AEROBIC ACTINOMYCETES

Fewer than 20 cases of serious illness due to *Tsukamurella* have been reported, mostly been ascribed to the species *Tsukamurella paurometabola*. They are frequently misidentified as *Rhodococcus*, or *Corynebacterium* species are most easily identified by 16S rRNA PCR and sequencing. Infection usually occurs in the context of immunosuppression, typically bone marrow transplantation and indwelling venous or CAPD canulae. Treatment is usually successful if associated with removal of intravascular devices and antibiotic therapy (Schwartz *et al.* 2002). *Oerskovia* sp. exist naturally in soil and water, decaying plant material, brewery sewage, aluminum hydroxide gels and grass cuttings, and as they are found in the environment, care must be taken to distinguish colonization and specimen contamination from infection. Reported clinical infection usually occurs in immunocompromised hosts with implantable devices. Most are due to *O. xanthineolytica* and others due to *O. turbata*. A majority of patients required removal of a foreign body as part of their treatment which is usually successful in most cases (Urbina, Gohh and Fischer 2003).

Rothia dentocariosa is a rare cause of endocarditis occuring most frequently in patients with prior heart conditions. Although the clinical course is typically subacute, complications are common, in particular the reported incidence of mycotic aneurysms is as high as 25% (Boudewijns *et al.* 2003). It can be difficult to identify in the laboratory (Von Graevenitz 2004). Penicillin is the treatment of choice, but additional complications may necessitate prompt surgical intervention.

REFERENCES

Beaman, B. L. and Beaman, L. (1994) *Nocardia* species: host-parasite relationships. Clin Microbiol Rev 7,213–264.

Beaman, B. L., Black, C. M., Doughty, F., and Beaman, L. (1985) Role of superoxide dismutase and catalase as determinants of pathogenicity of *Nocardia asteroides*: importance in resistance to microbicidal activities of human polymorphonuclear neutrophils. Infect Immun 47,135–141.

Beaman, B. L., Burnside, J., Edwards, B., and Causey, W. (1976) Nocardial infections in the United States, 1972–1974. J Infect Dis 134,286–289.

Beaman, B. L. and Moring, S. E. (1988) Relationship among cell wall composition, stage of growth, and virulence of *Nocardia asteroids* GUH-2. Infect Immun 56,557–563.

Beaman, B. L. and Sugar, A. M. (1983) *Nocardia* in naturally acquired and experimental infections in animals. J Hyg (Lond) 91,393–419.

Blanchard, R. (1896) Parasites vegetaux á l'exclusion des bacteries, in Traite de pathologie génerale, Vol. 2 (ed. C. Bouchard), Masson et Cie, Paris, France, pp. 811–813.

Boudewijns, M., Magerman, K., Verhaegen, J. *et al.* (2003) *Rothia dentocariosa*, endocarditis and mycotic aneurysms: case report and review of the literature. Clin Microbiol Infect 9,222–229.

Bross, J. E. and Gordon, G. (1991) Nocardial meningitis: case reports and review. Rev Infect Dis 13,160–165.

Butler, W. R., Kilburn, J. O., and Kubica, G. P. (1987) High-performance liquid chromatography analysis of mycolic acids as an aid in laboratory identification of *Rhodococcus* and *Nocardia* species. J Clin Microbiol 25,2126–2131.

Chun, J. and Goodfellow, M. (1995) A phylogenetic analysis of the genus *Nocardia* with 16S rRNA gene sequences. Int J Syst Bacteriol 45,240–245.

Clark, N. M., Braun, D. K., Pasternak, A., and Chenoweth, C. E. (1995) Primary cutaneous *Nocardia otitidiscaviarum* infection: case report and review. Clin Infect Dis 20,1266–1270.

Cloud, J. L., Conville, P. S., Croft, A. *et al.* (2004) Evaluation of partial 16S ribosomal DNA sequencing for identification of *Nocardia* species by using the MicroSeq 500 system with an expanded database. J Clin Microbiol 42,578–584.

Crowe, L. M., Spargo, B. J., Ioneda, T. *et al.* (1994) Interaction of cord factor (alpha, alpha'-trehalose-6,6'-dimycolate) with phospholipids. Biochim Biophys Acta 1194,53–60.

Pham, A. S., De, I., Rolston, K. V. *et al.* (2003) Catheter-related bacteremia caused by the nocardioform actinomycete *Gordonia terrae*. Clin Infect Dis 36,524–527.

Douglas, R. M., Grove, D. I., Elliott, J. *et al.* (1991) Corneal ulceration due to *Nocardia asteroides*. Aust N Z J Ophthalmol 19,317–320.

Drancourt, M., McNeil, M. M., Brown, J. M. *et al.* (1994) Brain abscess due to *Gordona terrae* in an immunocompromised child: case report and review of infections caused by *G. terrae*. Clin Infect Dis 19,258–262.

Esteban, J., Ramos, J. M., Fernandez-Guerrero, M. L., and Soriano, F. (1994) Isolation of *Nocardia* sp. from blood cultures in a teaching hospital. Scand J Infect Dis 26,693–696.

Georghiou, P. R. and Blacklock, Z. M. (1992) Infection with *Nocardia* species in Queensland. A review of 102 clinical isolates. Med J Aust 156,692–697.

Gillum, R. L., Qadri, S. M., Al Ahdal, M. N. *et al.* (1988) Pitted keratolysis: a manifestation of human dermatophilosis. Dermatologica 177,305–308.

Gomez-Flores, A., Welsh, O., Said-Fernandez, S. *et al.* (2004) *In vitro* and *in vivo* activities of antimicrobials against *Nocardia brasiliensis*. Antimicrob Agents Chemother 48,832–837.

Goodfellow, M., Chun, J., Stubbs, S., and Tobili, A. S. (1994) Transfer of *Nocardia amarae* Lechevalier and Lechevalier 1974 to the genus Gordona as Gordona *amarae* comb. nov. Lett Appl Microbiol 19,401–405.

Indrigo, J., Hunter, R. L. Jr., and Actor, J. K. (2003) Cord factor trehalose 6, 6'-dimycolate (TDM) mediates trafficking events during mycobacterial infection of murine macrophages. Microbiology 149,2049–2059.

Jones, N., Khoosal, M., Louw, M., and Karstaedt, A. (2000) Nocardial infection as a complication of HIV in South Africa. J Infect 41,232–239.

Maiti, P. K., Ray, A., and Bandyopadhyay, S. (2002) Epidemiological aspects of mycetoma from a retrospective study of 264 cases in West Bengal. Trop Med Int Health 7,788–792.

McNeil, M. M. and Brown, J. M. (1994) The medically important aerobic actinomycetes: epidemiology and microbiology. Clin Microbiol Rev 7,357–417.

McNeil, M. M., Brown, J. M., Jarvis, W. R., and Ajello, L. (1990) Comparison of species distribution and antimicrobial susceptibility of aerobic actinomycetes from clinical specimens. Rev Infect Dis 12,778–783.

Menendez, R., Cordero, P. J., Santos, M. *et al.* (1997) Pulmonary infection with *Nocardia* species: a report of 10 cases and review. Eur Respir J 10,1542–1546.

Miralles, G. D. (1994) Disseminated *Nocardia farcinica* infection in an AIDS patient. Eur J Clin Microbiol Infect Dis 13,497–500.

Neu, H. C., Silva, M., Hazen, E., and Rosenheim, S. H. (1967) Necrotizing nocardial pneumonitis. Ann Intern Med 66,274–284.

Nocard, M. E. (1898) Note sur la maladie des boeufs de la Guadaloupe sous le nom de farcin. Annales de l'Institut Pasteur 2,293–303.

Pijper, A. and Pullinger, B. D. (1927) South African nocardiasis. J Trop Med Hyg 30,153–156.

Pintado, V., Gomez-Mampaso, E., Cobo, J. *et al.* (2003) Nocardial infection in patients infected with the human immunodeficiency virus. Clin Microbiol Infect 9,716–720.

Poonwan, N., Kusum, M., Mikami, Y. *et al.* (1995) Pathogenic *Nocardia* isolated from clinical specimens including those of AIDS patients in Thailand. Eur J Epidemiol 11,507–512.

Roth, A., Andrees, S., Kroppenstedt, R. M. *et al.* (2003) Phylogeny of the genus Nocardia based on reassessed 16S rRNA gene sequences reveals underspeciation and division of strains classified as *Nocardia asteroides* into three established species and two unnamed taxons. J Clin Microbiol 41,851–856.

Ruimy, R., Riegel, P., Carlotti, A. *et al.* (1996) *Nocardia pseudobrasiliensis* sp. nov., a new species of *Nocardia* which groups bacterial strains previously identified as *Nocardia brasiliensis* and associated with invasive diseases. Int J Syst Bacteriol 46,259–264.

Sachs, M. K. (1992) Lymphocutaneous *Nocardia brasiliensis* infection acquired from a cat scratch: case report and review. Clin Infect Dis 15,710–711.

Saubolle, M. A. and Sussland, D. (2003) Nocardiosis: review of clinical and laboratory experience. J Clin Microbiol 41,4497–4501.

Schiff, T. A., Sanchez, M., Moy, J. *et al.* (1993) Cutaneous nocardiosis caused by Nocardia nova occurring in an HIV-infected individual: a case report and review of the literature. J Acquir Immune Defic Syndr 6,849–851.

Schwartz, M. A., Tabet, S. R., Collier, A. C. *et al.* (2002) Central venous catheter-related bacteremia due to *Tsukamurella* species in the immunocompromised host: a case series and review of the literature. Clin Infect Dis 35, e72–e77.

Seddon, M., Parr, D., and Ellis-Pegler, R. B. (1995) Lymphocutaneous *Nocardia brasiliensis* infection: a case report and review. N Z Med J 108,385–386.

Spargo, B. J., Crowe, L. M., Ioneda, T. *et al.* (1991) Cord factor (alpha, alphatrehalose 6,6'-dimycolate) inhibits fusion between phospholipid vesicles. Proc Natl Acad Sci U S A 88,737–740.

Towersey, L., Martins Ede, C., Londero, A. T. *et al.* (1993) *Dermatophilus congolensis* human infection. J Am Acad Dermatol 29,351–354.

Tripodi, M. F., Adinolfi, L. E., Andreana, A. *et al.* (2001) Treatment of pulmonary nocardiosis in heart-transplant patients: importance of susceptibility studies. Clin Transplant 15,415–420.

Urbina, B. Y., Gohh, R., and Fischer, S. A. (2003) *Oerskovia xanthineolytica* endocarditis in a renal transplant patient: case report and review of the literature. Transpl Infect Dis 5,195–198.

Von Graevenitz, A. (2004) *Rothia dentocariosa*: taxonomy and differential diagnosis. Clin Microbiol Infect 10,399–402.

Wallace, R. J. Jr., Brown, B. A., Tsukamura, M. *et al.* (1991) Clinical and laboratory features of *Nocardia* nova. J Clin Microbiol 29,2407–2411.

Section Three

Gram-Negative Organisms

13

Moraxella catarrhalis and *Kingella kingae*

Alex van Belkum[1] and Cees M. Verduin[2]

[1]*Department of Medical Microbiology & Infectious Diseases, Erasmus University Medical Center Rotterdam EMCR, Rotterdam; and [2]PAMM, Laboratory of Medical Microbiology, Veldhoven, The Netherlands*

INTRODUCTION

Moraxella (Branhamella) catarrhalis, formerly called *Neisseria catarrhalis* or *Micrococcus catarrhalis*, is a Gram-negative, aerobic diplococcus frequently found as a commensal of the upper respiratory tract (Jawetz, Melnich and Adelberg, 1976; Ninane *et al.*, 1977; Johnson, Drew and Roberts, 1981). Over the last 20–30 years, the species has been recognized as a pathogen of the upper respiratory tract in otherwise healthy children and elderly people (Hager *et al.*, 1987; Catlin, 1990; Murphy, 1996; Karalus and Campagnari, 2000). Moreover, *M. catarrhalis* is an important cause of the infective complications of adult chronic obstructive pulmonary disease (COPD) (Hager *et al.*, 1987; Catlin, 1990; Murphy, 1996). In immunocompromised hosts, the bacterium causes pneumonia, endocarditis, septicaemia and meningitis (Doern, 1986). *Moraxella catarrhalis* hospital outbreaks of respiratory disease have been described (Patterson *et al.*, 1988; Richards *et al.*, 1993), establishing the bacterium as a genuine nosocomial pathogen. Because *M. catarrhalis* had long been considered a harmless commensal (Jawetz, Melnich and Adelberg, 1976; Johnson, Drew and Roberts, 1981), knowledge of its antimicrobial resistance profile, pathogenic characteristics and virulence factors has only been acquired recently. Its potential to cause disease, however, is currently undisputed.

Kingella kingae is a member of the HACEK group of microorganisms identified in 1968 as *Moraxella kingii*, named after Elizabeth O. King (*Haemophilus parainfluenzae, Haemophilus aphrophilus, Actinobacillus actinomycetemcomitans, Cardiobacterium hominis, Eikenella corrodens* and *Kingella kingae*). It was renamed in 1974 (see Davies and Maesen, 1986, for a historical review). The bacteria are coccoid, medium-sized (1 μm wide and 2–3 μm in length) Gram-negative rods which are seen as pairs or short chains with rounded or square termini. The organism may be nutritionally fastidious, requiring increased amounts of serum and CO_2 for optimal growth on blood agar medium (Picket *et al.*, 1991). Bactec blood culture bottles are considered an excellent growth medium for this species (Yagupsky *et al.*, 1992). *Kingella kingae* may cause pitting in the agar and generates mild β-haemolysis, and the bacterium sometimes produces a brownish pigment. The bacteria do not grow on MacConkey agar and are usually catalase negative and oxidase positive. They may display twitching motility and are usually highly sensitive to penicillin. Some strains produce β-lactamase. This bacterial species may be mistaken for *Neisseria gonorrhoeae*. There are two additional species within the genus: the non-haemolytic *Kingella oralis* and *Kingella denitrificans*. The organism is a frequent inhabitant of the oropharynx and may be encountered in the urethral tract as well and was until recently considered a rare cause of disease in humans. However, improvements in diagnostic techniques (Lejbkowicz *et al.*, 1999; Yagupsky, 1999b) and an increased awareness amongst clinicians have resulted in increased numbers of reports (Odum and Frederiksen, 1981). This chapter focuses on *M. catarrhalis* in particular, since the body of available literature for this organism is at least twice that of *K. kingae*.

DESCRIPTION OF THE ORGANISM

In the past, *M. catarrhalis* was considered a non-pathogenic member of the resident flora of the nasopharynx. It was one of the species belonging to the so-called non-gonococcal, non-meningococcal neisseriae. The bacterium was first described in 1896 (Frosch and Kolle, 1896) as *M. catarrhalis*. In the early 1960s, it was shown that *M. catarrhalis* in fact comprised two distinct species: *Neisseria cinerea* and *Neisseria catarrhalis* (Berger, 1963). These could be separated on the basis of nitrate and nitrite reduction and tributyrin conversion. Because of the phylogenetic separation between *N. catarrhalis* and the so-called true *Neisseria* species, the bacterium was moved to the genus *Branhamella* in honour of Sara E. Branham (Catlin, 1970). In 1984, *Branhamella catarrhalis* was reassigned to the genus *Moraxella* as *Moraxella (Branhamella) catarrhalis* (Bovre, 1984). Ribosomal DNA sequencing has substantiated the validity of the current taxonomic classification (Enright *et al.*, 1994; Pettersson *et al.*, 1998). Catlin (1991) has proposed the formation of a new family, Branhamaceae, to accommodate the genera *Moraxella* and *Branhamella*. However, comparison of 16S rDNA sequences of *Moraxella* spp. has demonstrated the relatedness of *M. catarrhalis* to *Moraxella lacunata* subsp. *lacunata* and to a 'false' *Neisseria*, *Neisseria ovis* (Enright *et al.*, 1994). Apparently, there is no rationale for a separate *Branhamella* genus. Consequently, at the time of writing, *M. catarrhalis* is the preferred name for this bacterial species.

In 1972, the first paper on the taxon *K. denitrificans* was published (Hollis, Wiggins and Weaver, 1972). The bacterium was called TM-1, because of its recovery on Thayer–Martin medium from throat swabs. Later, the TM-1 group was reported to be similar to saccharolytic *Moraxella* species, now known as *Kingella kingae* (Henriksen and Bøvre, 1976). *Kingella kingae* was previously known as *Moraxella* new species 1, *M. kingii*. In 1976, these species were moved to the new genus *Kingella* (Snell and Lapage, 1976). By the end of the 1980s, the species within the *Kingella* genus were studied for differences in their fatty acid composition. This appeared to be inconclusive for an adequate separation of the species *K. denitrificans* and *K. kingae* (Wallace *et al.*, 1988). Later it was demonstrated that species identification of kingellae could be achieved with biochemical testing. The *Haemophilus–Neisseria* identification panel (American Microscan, Sacramento, CA, USA) was able to distinguish the kingellae from all of the other Gram-negative bacteria tested for (Janda, Bradna and

Ruther, 1989). More definitive identification schemes were those based on DNA–DNA hybridization studies (Tonjum, Bukholm and Bovre, 1989). *Kingella kingae*, *K. denitrificans* and *Kingella indologenes* formed distinct groups, well separated from, for instance, the various *Moraxella* species. However, a little later, *K. indologenes* was moved to the genus *Suttonella* on the basis of its ribosomal RNA sequence (Dewhirst *et al.*, 1990), and the same experimental approach led to the identification of *K. oralis* as a new species in the genus *Kingella* (Dewhirst *et al.*, 1993). The ubiquitous presence of this latter species in the oral cavity was defined shortly thereafter, with 26 of 27 people positive for the organism by culture (Chen, 1996).

In conclusion, both *M. catarrhalis* and *K. kingae* are bacteria belonging to phylum BXII of the proteobacteria (Bergey's Manual Trust website, as per 2000; www.bergeys.org). *Moraxella catarrhalis* is a class III gammaproteobacterium, whereas *K. kingae* belongs to the class II of the betaproteobacteria. *Moraxella catarrhalis* belongs to the Pseudomonales (order VIII), Moraxellaceae (family II) and *Moraxella* (genus I). For *K. kingae*, these are Neisseriales (order IV), Neisseriaceae (family I) and the genus *Kingella* (genus IX).

EPIDEMIOLOGY

Epidemiological studies on *M. catarrhalis* have been difficult to perform, as typing systems have become available only recently and because of the lack of a reliable and simple serological test. Clinical interest in *M. catarrhalis* is only recent, and many laboratories were not used to reporting *M. catarrhalis* as a pathogen. The identification of *M. catarrhalis* from sputa is complicated by the presence of non-pathogenic neisseriae. *Kingella kingae* has been identified as an organism usually present in the oropharyngeal mucosa, and epidemiological studies, defining transmission and assessing the pathogenic potential of these bacteria, are urgently required (Abuamara *et al.*, 2000).

Conventional and Molecular Typing Systems

Several phenotyping strategies have been described for epidemiological typing of *M. catarrhalis*, though none of these has been accepted internationally and no extensive use of this technology has been reported. Serological typing of lipopolysaccharide (LPS) (Vaneechoutte *et al.*, 1990), isoelectric focusing of β-lactamase proteins (Nash *et al.*, 1986) and electrophoretic profiling of outer membrane proteins (OMPs) (Bartos and Murphy, 1988) and biotyping (Denamur *et al.*, 1991; Peiris and Heald, 1992) have also been used. More recently, methods based on nucleic acid polymorphism have become available. Comparison of restriction endonuclease analysis (REA) with phenotyping (Christensen, 1999) has indicated that REA of genomic DNA, including pulsed-field gel electrophoresis (PFGE), may be used successfully for delineating disease outbreaks (Patterson *et al.*, 1988; Kawakami *et al.*, 1994; Vu-Thien *et al.*, 1999; Yano *et al.*, 2000). The use of strain-specific DNA probes has also been documented (Beaulieu *et al.*, 1993; Walker *et al.*, 1998). Studies identified amplified fragment length polymorphism (AFLP) analysis and automated ribotyping as useful typing procedures (Bootsma *et al.*, 2000; Verduin *et al.*, 2000). Moreover, using these two strategies, researchers have found complement-resistant strains of *M. catarrhalis* to form a distinct clonal lineage within the species. These observations are in agreement with those of earlier studies identifying *M. catarrhalis* as a genetically heterogeneous species from which successful clones occasionally proliferate (Enright and McKenzie, 1997). Expansion of such successful types has also been documented during distinct periods of time and in particular geographic regions (Martinez *et al.*, 1999). Frequent horizontal gene transfer seems to occur (Bootsma *et al.*, 1999; Luna *et al.*, 2000), and phenotypic characteristics could be acquired through cross-species gene acquisition.

For *K. kingae*, all of the methods listed could be used, but only immunoblotting, PFGE and ribotyping have been explored in detail (Slonim *et al.*, 1998).

Carriage of *M. catarrhalis* and *K. kingae*

The *M. catarrhalis* carriage rate in children is high (up to 75%) (Van Hare *et al.*, 1987; Faden, Harabuchi and Hong, 1994; Varon *et al.*, 2000). In contrast, the carriage rate of *M. catarrhalis* in healthy adults is only about 1–3% (DiGiovanni *et al.*, 1987; Ejlertsen, 1991). This inverse relationship between age and colonization has been known since 1907 (Arkwright, 1907). The age-dependent development of secretory immunoglobulin A (IgA) could explain the phenomenon. Remarkably, IgG antibody levels do not correlate with the state of colonization or with lower respiratory tract infection (LRTI) with *M. catarrhalis* in children (Ejlertsen *et al.*, 1994). Interestingly, nasopharyngeal carriage rates are significantly higher in winter and autumn than in spring and summer (Van Hare *et al.*, 1987). Monthly or bimonthly sampling of the nasopharynx of children ($n = 120$) by Faden, Harabuchi and Hong (1994) revealed the presence of *M. catarrhalis* in 77.5% of subjects at least once during the first 2 years of life. Furthermore, these authors showed a clear relationship between the frequency of colonization and the development of otitis media. A small Japanese study revealed that colonization in children attending a daycare centre is highly dynamic (Yano *et al.*, 2000). Although clusters of genotypes seemed to persist for periods of 2–6 weeks, frequent changes in the nature of colonizing strains of *M. catarrhalis* were observed. Other authors have described a relationship between the frequency of colonization and the occurrence of upper respiratory infection (Brorson and Malmvall, 1981; Prellner *et al.*, 1984). Klingman *et al.* (1995) investigated the colonization of the respiratory tract of patients with bronchiectasis. Patients were colonized with the same strain for an average of 2.3 months, as determined by restriction fragment length polymorphism (RFLP) patterns, and colonization with a new strain did not correlate with changes in clinical status.

The first large longitudinal carriage study for *K. kingae* was performed by Yagupsky *et al.* (1993, 1995a) in Israel. Children visiting daycare centres had an average carriage rate of 6.1–34.6%, peaking in December and April, nearly 73% having at least one positive culture. These data are similar to those for *Haemophilus influenzae* carriage among children. Frequent transfer of *K. kingae* strains in daycare attendees has been documented using molecular typing. Authors confirmed this hypothesis in a follow-up study (Slonim *et al.*, 1998). Yagupsky, Peled and Katz (2002) highlighted several important features of *K. kingae* carriage. Carriage appeared to be highest in the 0–3-year age group. The carriage rates were equal for males and females. It is striking to note that infection significantly more often occurred in males, for which, at the time of writing, there is no reasonable explanation. The pathogenicity of *K. kingae*, with its seasonal preponderance, could not be explained on the basis of carriage patterns alone.

MORAXELLA CATARRHALIS-MEDIATED DISEASES IN CHILDHOOD

Moraxella catarrhalis is now considered an important pathogen in respiratory tract infections, both in children and in adults with underlying COPD, occasionally causing systemic disease (Doern, Miller and Winn, 1981; Collazos, de Miguel and Ayarza, 1992). Bacteremia caused by *M. catarrhalis* should be considered in febrile children with an immune dysfunction and an upper respiratory tract infection (Abuhammour *et al.*, 1999). In addition, *M. catarrhalis* may be the sole cause of sinusitis, otitis media, tracheitis, bronchitis or pneumonia and, less commonly, ocular infections in children. In children, nasopharyngeal colonization often precedes the development of *M. catarrhalis*-mediated disease (Faden, Harabuchi and Hong, 1994).

Sinusitis

Sinusitis is a very common infection in early childhood, accounting for about 5–10% of upper respiratory tract infections (Wald, 1992a, 1998a, b). It is often underdiagnosed in children because the symptoms are non-specific. In acute sinusitis (where symptoms are present for 10–30 days) and subacute disease (30–120 days), *Streptococcus pneumoniae*, *H. influenzae* and *M. catarrhalis* are the most frequently isolated bacterial pathogens (Bluestone, 1986; Blumer, 1998; Cappelletty, 1998). *Moraxella catarrhalis* accounts for approximately 20% of cases. Interestingly, in children with asthma, the same distribution of bacterial pathogens is found (Wald, 1992b), although Goldenhersh *et al.* (1990) isolated *M. catarrhalis* as the predominant pathogen in chronic sinusitis (symptoms >30 days) in children with respiratory allergy. This would indicate an even more prominent role for *M. catarrhalis* in the aetiology of these widely distributed and highly prevalent infectious diseases.

Otitis Media

Acute otitis media (AOM) is a frequent infection in children: Before the age of 1 year, around 50% of children would have experienced at least one period of AOM. This figure rises to 70% within 3 years (Kabongo, 1989; Klein, 1994). It is the most serious and frequent infection caused by *M. catarrhalis* in children, and therefore, *M. catarrhalis* causes tremendous morbidity, requiring a widespread use of antibiotics, and vaccine development has been suggested (Faden, Harabuchi and Hong, 1994; Berman, 1995; Cohen, 1997; Faden *et al.*, 1997; Froom *et al.*, 1997; Klein, 1999). Since 1980, a marked increase has been reported in the isolation of *M. catarrhalis* from middle-ear exudates (Kovatch, Wald and Michaels, 1983; Shurin *et al.*, 1983; Marchant, 1990; Block, 1997). This increase in *M. catarrhalis* isolation to its present isolation frequency of approximately 15–20% (Patel *et al.*, 1995) has been accompanied by the appearance of β-lactamase-producing strains (90–95%). The exact magnitude of this apparent increase is unclear (Marchant, 1990), since tympanocentesis and culture of middle-ear fluid are not performed routinely. The isolation rates for *M. catarrhalis* might be an underestimation, given the relative anaerobic environment of the middle ear during infection (Anonymous, 1994). In a study using polymerase chain reaction (PCR), *M. catarrhalis* DNA was detected in 46.4% of paediatric chronic middle-ear effusion specimens ($n = 97$), as compared with 54.6% for *H. influenzae* DNA and 29.9% for *S. pneumoniae* DNA (Post *et al.*, 1995). A large percentage (48%) of specimens were PCR positive and culture negative, whereas all culture-positive specimens were also PCR-positive. It is very unlikely that the PCR-positive, yet culture-negative specimens reflect persistence of DNA from old infections (Aul *et al.*, 1998). The severity of symptoms and numbers of bacteria in middle-ear fluid appear to be less for *M. catarrhalis* than for *S. pneumoniae* or *H. influenzae* (Faden *et al.*, 1992).

Lower Respiratory Tract Infections

LRTIs due to *M. catarrhalis* mostly occur in children under the age of 1 year (Boyle *et al.*, 1991). Korppi *et al.* (1992) have investigated the seroconversion to *M. catarrhalis* in patients hospitalized with middle respiratory tract infection (laryngitis, tracheitis and bronchitis) and LRTI. They found seroconversion in only 4 of 76 children (5%) who had *M. catarrhalis*-positive nasopharyngeal aspirate cultures, as compared with 4 of 373 children (1%) who had negative cultures, *M. catarrhalis* being an unlikely pathogen in children. In contrast, several other studies have implicated *M. catarrhalis* as a cause of LRTI in children. *Moraxella catarrhalis* has been isolated in pure culture from secretions obtained by tracheal aspiration in neonates, infants and children with pneumonia (Haddad *et al.*, 1986; Berg and Bartley, 1987; Berner *et al.*, 1996). In a prospective study combining

microbiological and clinical criteria, *M. catarrhalis* was identified as a significant respiratory pathogen in children (Boyle *et al.*, 1991). Both local and systemic antibody responses to *M. catarrhalis* infection have been documented in several studies (Chapman *et al.*, 1985; Black and Wilson, 1988; Goldblatt *et al.*, 1990; Goldblatt, Turner and Levinsky, 1990; Faden, Hong and Murphy, 1992). Pneumonia in children may be complicated by bacteremia with *M. catarrhalis* (Collazos, de Miguel and Ayarza, 1992; Ioannidis *et al.*, 1995; Thorsson, Haraldsdottir and Kristjansson, 1998). *Moraxella catarrhalis* appears to have a role in the causation of LRTI.

Other Infections in Children

Moraxella catarrhalis has been implicated as a cause of bacterial tracheitis in childhood (Wong and Mason, 1987; Brook, 1997; Bernstein, Brilli and Jacobs, 1998), for which preceding viral infection has been considered a significant predisposing factor. Although reports on ocular infections have been rare (Macsai, Hillman and Robin, 1988; Abbott, 1992; Bergren *et al.*, 1993; Weiss, Brinser and Nazar-Stewart, 1993), conjunctivitis and keratitis have also been described.

KINGELLA KINGAE-MEDIATED DISEASES IN CHILDHOOD

Kingella kingae has been identified as one of the emerging causes of childhood infectious disease, with an estimated incidence of more than 25 invasive infections per year per 100 000 children ≤24 months old (Yagupsky *et al.*, 1993; Yagupsky and Dagan, 1997). This bacterial species is capable of causing vaginitis, meningitis (Namnyak, Quinn and Ferguson, 1991; Hay *et al.*, 2002), soft-tissue infection (Rolle *et al.*, 2001), stomatitis (Amir and Yagupsky, 1998), LRTIs (Yagupsky and Dagan, 1997) and endophthalmitis (Carden *et al.*, 1991). However, these infections are rare: most being case reports. In contrast, Centers for Disease Control and Prevention (CDC) evaluations have demonstrated that invasive infection in children caused by *K. kingae* are not so rare. A significant proportion of clinical isolates received during surveillance studies were from invasive disease: 58 of 78 isolates from blood, bone, joints or fluids (Graham *et al.*, 1990).

Bone and Joint Infections

Osteomyelitis due to *K. kingae* has been reported regularly, even in geographically remote regions such as Iceland (Birgisson, Steingrimsson and Gudnason, 1997). As a species, *K. kingae* most frequently causes osteoarticular infections, most commonly involving the bones and joints of the lower extremities (Dodman, Robson and Pincus, 2000). Owing to the fastidious growth requirements of the organism, these infections have been defined as 'hard to detect and treat' (Wildt and Boas, 2001). In contrast with standard radiography, 99mTc bone scans are informative (La Scola, Iorgulescu and Bollini, 1998). The investigators found that all affected children suffered from upper respiratory tract infection or eczema in the month before the development of osteomyelitis. An analysis of children suffering from septic arthritis or (sub)acute osteomyelitis found that *K. kingae* has become one of the most frequently isolated bacteria (Lundy and Kehl, 1998), replacing *H. influenzae*, probably because of successful vaccination. The use of improved automated blood culture has improved the yield of *K. kingae*-positive cultures derived from joint fluid and blood (Yagupsky *et al.*, 1992; Costers *et al.*, 2003).

Endocarditis and Bacteremia

Kingella kingae bacteremia in children is well recognized (Amir, 1992; Yagupsky and Dagan, 1994; Krause and Nimri, 1996;

Moylett *et al.*, 2000). Endocarditis may arise as a primary disease or secondary to hair-cartilage hypoplasia (Ferber *et al.*, 1997) or chickenpox (Waghorn and Cheetham, 1997), neurological symptoms being described well (Wells, Rutter and Donald, 2001). *Kingella kingae* has a particular tropism for the osteoarticular and cardiac sites of infection, being observed at both sites in the same patient (Sarda *et al.*, 1998).

MORAXELLA CATARRHALIS INFECTIONS IN ADULTS

Moraxella catarrhalis has been associated with a variety of clinical syndromes in adults, including nosocomial infections with rare cases of endocarditis (Neumayer *et al.*, 1999; Stefanou *et al.*, 2000).

Laryngitis

Moraxella catarrhalis is the most common bacterial species isolated from adult patients with laryngitis. Schalen *et al.* (1980, 1985) found that of 40 adults suffering from this disease, 22 were infected by *M. catarrhalis* (55%). Among 40 healthy adults, *M. catarrhalis* could not be isolated from the larynx. The exact role of *M. catarrhalis*, either as a commensal or as a pathogen in adult laryngitis, is incompletely understood.

Bronchitis and Pneumonia

Moraxella catarrhalis is a common cause of exacerbations in COPD (McLeod *et al.*, 1986; Davies and Maesen, 1988; Cabello *et al.*, 1997; Eller *et al.*, 1998; Miravitlles *et al.*, 1999). A study has shown *M. catarrhalis* to be the single most frequently isolated bacterium (Sportel *et al.*, 1995). Sarubbi *et al.* (1990) reviewed all respiratory tract cultures (*n* = 16 627) over a period of 42 months and identified *M. catarrhalis* in 2.7% (*n* = 457) of these cultures. *Moraxella catarrhalis* was found to be the second most commonly isolated respiratory tract pathogen after non-typeable *H. influenzae*. These and many other authors (Pollard *et al.*, 1986; Wallace *et al.*, 1989; Dabernat, Avril and Boussougant, 1990; Wood, Johnson and McCormack, 1996) demonstrated striking seasonality, with winter and spring being the periods with the greatest incidence of *M. catarrhalis* isolation. Preceding viral respiratory tract infection could be the cause of *M. catarrhalis* infections, although this hypothesis remains untested (Van Hare and Shurin, 1987; DiGiovanni *et al.*, 1987).

Pneumonia caused by *M. catarrhalis* tends to be a relatively mild disease. It differs from bronchitis by the presence of mostly lower-lobe infiltrates on chest X-rays (Nicotra *et al.*, 1986; Wright and Wallace, 1989). High fever, pleuritic pain and toxic states are uncommon, as are empyema and bacteremia (Pollard *et al.*, 1986). Collazos, de Miguel and Ayarza (1992) reviewed cases of pneumonia due to *M. catarrhalis* which were similar in both characteristics and clinical symptoms to those described for patients with bronchitis or pneumonia without bacteremia, with a mortality rate in bacteremic cases of 13.3%. Ioannidis *et al.* (1995) described the clinical spectrum of *M. catarrhalis* bacteremia in 58 patients. Predisposing factors included neutropenia, malignancy or respiratory impairment, either alone or in combination. Mortality was high (29%) among patients with underlying respiratory disease, and the infection was more severe when the patient was co-infected with other respiratory tract pathogens. *Moraxella catarrhalis* pneumonia often occurs in patients with end-stage pulmonary or malignant disease, with short-term mortality in some patient categories as high as 45% (Wright and Wallace, 1989). Most patients are among the elderly having underlying cardiopulmonary disease or COPD (Barreiro *et al.*, 1992). Many patients appear to be malnourished, smokers or (ex-smokers), and men are at a greater risk than women.

Nosocomial Infections

It has been difficult to confirm the spread of *M. catarrhalis* among hospitalized patients, no reliable typing system and low pathogenicity being the main reasons. Patterson *et al.* (1988) used REA to confirm an outbreak in a hospital. During the investigation of another putative outbreak, immunoblotting with normal human serum was combined with REA to type *M. catarrhalis* strains (Richards *et al.*, 1993). Both methods provided adequate discrimination between strains but were not always in complete agreement (Morgan *et al.*, 1992). Clear vehicles of bacterial dissemination have not yet been identified in the clinical setting. However, Ikram *et al.* (1993) found the nosocomial spread of *M. catarrhalis* to be common, especially in respiratory wards. They also showed considerable contamination of the environment with *M. catarrhalis*, implying possible aerosol-mediated dissemination. Person-to-person transmission (Ikram *et al.*, 1993) or spread from environmental sources (Calder *et al.*, 1986) has been implicated in nosocomial transmission. Nursery schools have been suggested as being the prime sites where frequent exchanges of strains may occur (Yano *et al.*, 2000).

KINGELLA KINGAE INFECTIONS IN ADULTS

Although most infections are diagnosed among the (very) young, rare cases of chorioamnionitis (Maccato *et al.*, 1991), septic arthritis (Esteve *et al.*, 2001) and meningitis (Reekmans *et al.*, 2000) due to *Kingella* spp. have been diagnosed in adults. In AIDS patients, some of the *Kingella* species may cause infection (Minamoto and Sordillo, 1992; Urs *et al.*, 1994). However, most reported *K. kingae* infections in adults involve the circulatory system.

Bacteremia and Endocarditis

Kingella kingae bacteremia may occur in immunocompetent people (Roiz, Peralta and Arjona, 1997). In some cases, underlying diseases have been associated with the development of *K. kingae* vegetations on the mitral valve. The combination of a motile vegetation in a patient with systemic lupus erythematosus (SLE) and anti-phospholipid syndrome is an example of such a complex disease pattern (Wolak *et al.*, 2000). In addition, primary endocarditis with secondary infectious foci in other parts of the body has been described as well (Lewis and Bamford, 2000). Native valve endocarditis may occur (Hassan and Hayek, 1993), but also the colonization of prosthetic valves has been described, with *K. denitrificans* rather than *K. kingae* (Chakraborty, Meigh and Kaye, 1999). This shows that endocarditis due to *Kingella* spp. may be highly diverse in its presentation, which with difficulties in isolation and identification make this a potentially difficult disease to manage.

LABORATORY DIAGNOSIS

Isolation of *M. catarrhalis* from clinical specimens, i.e. sputum, may be complicated by the presence of non-pathogenic *Neisseria*. Although selective agar media are not necessary routinely, they have been used to isolate *M. catarrhalis* with some success. As an example, acetazolamide, which reduces the growth of *Neisseria* species when used under aerobic conditions, and the antimicrobial agents vancomycin, trimethoprim and amphotericin B may be included in an agar medium to inhibit the growth of the normal flora (Vaneechoutte *et al.*, 1988; Doern, 1990). Vancomycin may also be used for selective enrichment (Yagupsky *et al.*, 1995b).

The following criteria unambiguously distinguish *M. catarrhalis* from other bacterial species: Gram stain; colony morphology; lack of pigmentation of the colony on blood agar; oxidase production; DNase production; failure to produce acid from glucose, maltose, sucrose, lactose and fructose; growth at 22 °C on nutrient agar; failure to grow on modified Thayer–Martin (MTM) medium; and, lastly, reduction of nitrate and nitrite (Doern and Morse, 1980; Singh *et al.*, 1997). The Gram stain still plays a crucial role, both in assessing the significance of isolating the bacterium from clinical material (e.g. sputum) and in its subsequent identification. In typical Gram stains, *M. catarrhalis* is a Gram-negative diplococcus, with flattened abutting sides, often de-colouring poorly. Colonies on blood agar are non-haemolytic, round, opaque and convex, with a greyish-white colour. The colony remains intact when pushed across the surface of the agar. The bacteria are oxidase positive and produce DNase. Reduction of nitrate and nitrite and tributyrin hydrolysis are additional differentiating characteristics (Speeleveld *et al.*, 1994). The identity of *M. catarrhalis* is best confirmed by positive reactions in at least three of the criteria.

The kingellae do not produce catalase, and their oxidase reactivity may be feeble, especially when the enzymatic activity is probed with dimethyl-*p*-phenylenediamine instead of the more effective tetramethyl compound (Davies, 1997). When cultured in rich media (e.g. 10–20% ascitic fluid agar), the kingellae all ferment glucose and some may ferment maltose and sucrose. In conventional sugar fermentation tests, however, negative results may be obtained. In comparison with species from the genus *Moraxella*, *Kingella* may be recognized by their ability to reduce nitrite to nitrogen and by their ability to breakdown certain carbohydrates. Both species are able to hydrolyse tributyrin and are able to produce DNA-hydrolysing enzymes. *Kingella kingae* may be misidentified as *Eikenella corrodens*, *Cardiobacterium hominis* and, obviously, *Moraxella* or *Neisseria* spp. Although, in the past, *K. kingae* isolated from eye infections, blood or cerebrospinal fluid (CSF) was often correctly identified (Namnyak, Quinn and Ferguson, 1991; Mollee, Kelly and Tilse, 1992), the correct identification from strains derived from bone and joint infection in both adults and children has been problematic (Chiquita, Elliott and Namyak, 1991; Meis *et al.*, 1992; Yagupsky *et al.*, 1992). Apparently, the choice of culture media for these specimens is critical. When several specialized blood-culture systems were compared, it was shown that all strains were able to grow in BacT/Alert aerobic and BacT/Alert Pedi-BacT bottles produced by Organon Teknika, USA (Host *et al.*, 2000). However, in BacT/Alert FAN anaerobic and Bactec Plus Aerobic media (Becton Dickinson Microbiology Systems, USA), growth was observed for only 63% and 88%, respectively, of the inoculated strains after 12 days.

PCR tests for *M. catarrhalis* have been designed and used clinically, direct detection of *M. catarrhalis* DNA by PCR being concordant with culture and endotoxin detection. However, DNA assays yield significantly more positive results than culture when, for instance, middle-ear effusions are analysed, which suggests superior sensitivity of the DNA amplification assays (Dingman *et al.*, 1998). The clinical relevance of PCR has been validated extensively in the chinchilla model for otitis media. This animal model was instrumental in demonstrating the quick and effective effusion-mediated clearance of DNA and dead *M. catarrhalis* bacteria from the middle-ear cleft, implying that in this case a positive PCR result was indicative of the presence of viable bacteria (Post *et al.*, 1996a). Moreover, PCR has also been reliably used for detecting mixed infections in the same experimental infection model (Bakaletz *et al.*, 1998), thereby validating the applicability of multiplex PCR approaches for detecting mixed bacterial infections (Post *et al.*, 1995; Hendolin *et al.*, 1997). The sensitivity of the PCR tests corresponds to 6 or 7 genome equivalents, making PCR the most sensitive diagnostic assay (Post *et al.*, 1996b). However, the costs and technical demands of PCR make it not appropriate for routine microbiology laboratory testing for this diagnostic application. PCR diagnostic tests have also been developed for *K. kingae* (Stahelin *et al.*, 1998; Yagupsky, 1999a). However, these tests have thus far only been used to corroborate clinical diagnoses for case studies. Extended use, for instance, in epidemiological studies has not been reported yet, and the precise value of PCR tests for *K. kingae* will have to await further evaluation.

ANTIMICROBIAL SUSCEPTIBILITY

Although possessing universal β-lactamase-mediated resistance to penicillins and inherent resistance to trimethoprim, *M. catarrhalis* remains sensitive to most other antibiotics used for treating respiratory infections (Berk and Kalbfleisch, 1996; Hoogkamp-Korstanje *et al.*, 1997; McGregor *et al.*, 1998; Felmingham and Gruneberg, 2000). Strains producing β-lactamase are expected to be resistant to penicillin, ampicillin, amoxicillin and piperacillin (Jorgensen *et al.*, 1990; Fung *et al.*, 1992; Jensen, Schonheyder and Thomsen, 1994; Livermore, 1995; Berk and Kalbfleisch, 1996; Bootsma *et al.*, 1996; Doern *et al.*, 1996).

Kingella kingae still appears to be universally susceptible to penicillin and ampicillin, with minimal inhibitory concentration (MIC) values ranging between 0.5 mg/l and 0.25 mg/l, respectively. Although this was already reported in the early 1990s (Graham *et al.*, 1990), this situation has not changed significantly over the past decade. All isolates are sensitive to gentamicin (MIC ≤1 mg/l) but are inherently resistant to clindamycin. General extrapolation of susceptibility patterns for these organisms is difficult as most *Kingella* papers describe only limited numbers of patient isolates. Meta-analysis has revealed that the kingellae are still universally susceptible to β-lactam antibiotics, aminoglycosides, fluoroquinolones, erythromycin and cotrimoxazole. Significant resistance to the older quinolones, clindamycin, trimethoprim and vancomycin may exist, however (Amir and Shockelford, 1991). Meropenem appears to be more potent than imipenem, although the initial studies performed used suboptimal test systems (Clark and Joyce, 1993). The largest study thus far published on the susceptibility of clinical *K. kingae* isolates revealed that all strains were susceptible to erythromycin, gentamicin, chloramphenicol, tetracycline and ciprofloxacin (Yagupsky, Katz and Peled, 2001). Nearly 40% of all strictly local sets of strains were resistant to clindamycin, and a single isolate was resistant to cotrimoxazole. High levels of resistance to tetracycline have also been observed. These are due to the presence of a large plasmid encoding the TetM protein (Knapp *et al.*, 1988).

β-Lactamase Production in *M. catarrhalis* and *K. kingae*

The first β-lactamase-positive strain was isolated in 1976 (Wallace *et al.*, 1989). By 1980, however, 75% of *M. catarrhalis* isolates from the United States produced β-lactamase (Catlin, 1990; Wallace *et al.*, 1990). Recent studies from Australia, Europe and the United States all noted β-lactamase production in over 90% of isolates (Manninen, Huovinen and Nissinen, 1997; Doern *et al.*, 1999; Fluit *et al.*, 1999; Thornsberry *et al.*, 1999; Walker *et al.*, 2000). Walker *et al.* (2000) investigated trends in antibiotic resistance of *M. catarrhalis* isolates in a single hospital over a 10-year period. During this period, β-lactamase production increased from 30% to 96%. The trend towards resistance to penicillin and ceftriaxone was due to an increased frequency of β-lactamase-positive isolates. In *M. catarrhalis*, two types of β-lactamases are found that are phenotypically identical: the BRO-1 and BRO-2 type. Both are membrane associated, differ by a single amino acid, are chromosomally encoded and are relatively easily transferred from cell to cell. Fortunately, both enzymes are readily inactivated by β-lactamase inhibitors, and all isolates are still susceptible to amoxicillin in combination with clavulanic acid (Hoogkamp-Korstanje *et al.*, 1997; McGregor *et al.*, 1998). BRO-1 is the most common enzyme and confers higher MICs than BRO-2, the difference being attributed to the production of more enzymes as a consequence of higher transcriptional activity. Studies have shown that the β-lactamase of *M. catarrhalis* is lipidated, suggesting a Gram-positive origin (Bootsma *et al.*, 1999). The inferred lack of a genetic barrier between Gram-negative and Gram-positive bacterial species is a reason for clinical concern. β-Lactamase from *M. catarrhalis* is also thought to frustrate penicillin therapy of concomitant infections. This phenomenon is referred to as indirect pathogenicity, and in such circumstances, treatment failures have been reported (Patel et al., 1995). This emphasizes the importance of reporting not only pure but also mixed cultures positive for M. catarrhalis (Wardle, 1986). Strains of K. denitrificans have been demonstrated, on rare occasions, to produce β-lactamases (Minamoto and Sordillo, 1992; Sordillo et al., 1993).

CELL-WALL STRUCTURES OF M. CATARRHALIS

Lipooligosaccharides, Peptidoglycan and Capsule

Moraxella catarrhalis lipooligosaccharide (LOS) is semirough, meaning that it probably contains a single repeating O-antigen (Fomsgaard et al., 1991; Holme et al., 1999). It appears to be antigenically well conserved (Murphy, 1989). This corroborates the observation that it did not serve as a useful basis for a typing system (Doern, 1990). The antigenic specificities of three serotypes that have been identified are caused by differences in terminal sugars of one of the branches. Moreover, a structural overlap was documented with the LPS moieties from species of the Neisseria and Haemophilus genus. The LOS of serotypes B and C contains oligosaccharide chains of variable length. This could be due to phase-variable expression of the biosynthetic genes, as suggested by the presence of tandem repeats (Peak et al., 1996). Another explanation offered is that variations in the activity of enzymes involved in cell-wall assembly result in a different oligosaccharide. LOS is also present in culture supernatants as subcellular elements called blebs, which may facilitate the distribution of LOS in the host. Whether these structures serve some physiological function is, at the time of writing, unknown. LOS (serotype A), once adequately detoxified, can be used as a vaccine when conjugated to a protein carrier (Hu et al., 2000). Increases in anti-LOS IgG are observed upon immunization. The increased levels of antibodies enhanced clearance of bacteria from the lungs of mice after an aerosol-mediated M. catarrhalis infection.

Keller, Gustafson and Keist (1992) suggested that M. catarrhalis has a multilayered peptidoglycan architecture. This peptidoglycan layer was shown to be responsible for the extraordinary capacity of the organism to trigger various functional capacities of macrophages. Secretion of tumour necrosis factor and nitrite metabolism plus the cells' tumoricidal activity were clearly enhanced, which could be a partial explanation for the low virulence of M. catarrhalis. It seems as if peptidoglycan is involved in some sort of suicidal activity, and further studies of the basic mechanisms of this phenomenon are needed. The presence of a polysaccharide capsule has been previously suggested (Ahmed et al., 1991). Capsules are considered to be an important virulence factor in both Gram-positive and Gram-negative bacteria. Unlike the situation in many other bacterial pathogens, the capsule is not detectable when examining colonies of M. catarrhalis on agar plates.

Outer Membrane Proteins

In contrast to other non-enteric Gram-negative bacterial species, different M. catarrhalis strains have OMP profiles that show a high degree of similarity. Murphy and co-workers identified eight major proteins, designated OMPs A through H, ranging from 21 kDa to 98 kDa (Bartos and Murphy, 1988; Murphy and Loeb, 1989; Murphy, 1990). OMPs C and D appeared to be two different but stable forms of the same protein, the CD protein. The CD OMP gene appeared to be strictly conserved among M. catarrhalis strains (Murphy, Kirkham and Lesse, 1993). The CD protein was found to be involved in binding purified human mucin from the nasopharynx and middle ear, but not

to mucin from the saliva and tracheobronchial tract (Bernstein and Reddy, 2000). Antibodies raised in mice enhanced clearance in a pulmonary challenge model (Yang et al., 1997; Murphy et al., 1998). The OMP E antigen appears to be of little immunogenicity, but it does possess universally surface-expressed epitopes when different M. catarrhalis strains are studied (Bhushan et al., 1997). It may have a function in the uptake of nutrients (Murphy et al., 2000).

In addition to OMPs A–H, Klingman and Murphy (1994) described a novel OMP, designated high-molecular-weight OMP (HMW-OMP) or ubiquitous surface protein UspA. UspA has been recently shown to be encoded by two different genes, which share the coding potential for a homologous, internal protein domain at more than 90% amino acid sequence homology (Aebi et al., 1997; Lafontaine et al., 2000). Mutation in the UspA1 gene significantly decreased adherence, and UspA2 is essential for complement resistance (Aebi et al., 1998; Lafontaine et al., 2000). UspA1 protein binds specifically to fibronectin (McMichael et al., 1998), whereas the UspA2 protein preferentially binds to vitronectin. Both proteins are immunogenic in mice, and immunized animals clear bacteria from their lungs rapidly. Genetic studies have shown that intraspecies variability in the genes may be attributed to variation in regions of repetitive DNA (Cope et al., 1999). Moreover, many related genes have been identified in the genomes of a wide variety of bacterial species, suggesting that the proteins serve essential and universally required functions. Although the UspA1/2 antigens, at the time of writing, are the best-studied M. catarrhalis proteins, their vaccine potential still is matter of ongoing investigations (Hays et al., 2003).

Moraxella catarrhalis expresses both transferrin (TbpA and TbpB) and lactoferrin (LbpA and LbpB) receptors on its surface (Bonnah, Yu and Schryvers, 1995). These proteins are partially homologous and provide the cell with the capacity to sequester iron from host carrier proteins (Aebi et al., 1996; Campagnari, Ducey and Rebmann, 1996; Du et al., 1998). The receptors themselves appear to be significant virulence factors, since mutation analysis of the transferrin receptor has demonstrated an impaired growth capacity for the mutated strain. The receptors are also immunogenic and may be interesting vaccine candidates (Chen et al., 1999a; Yu et al., 1999). Molecular knock-out of the tbpB gene revealed that in the presence of a TbpB-specific monoclonal antibody and human complement, the mutant resisted killing as opposed to the wild type, which was rapidly killed (Luke et al., 1999).

Pericellular Structures

The attachment of bacteria to mucosal epithelial cells is often mediated by pili or fimbriae. Some studies have provided evidence for the expression of pili by M. catarrhalis (Marrs and Weir, 1990), whereas others have been unable to demonstrate their presence (Ahmed et al., 1991). Marrs and Weir (1990) found several characteristics that point to the presence of type 4 pili in M. catarrhalis. In addition, electron microscopic data revealed that, besides pili similar to those of type 4, an additional non–type 4 class of pili exists. Pili provide the single pericellular structure for which the kingellae have been well investigated. Initially, type 4 pili were discovered in K. denitrificans. N-Terminal amino acid sequencing of the composing proteins, in conjunction with electron microscopy, DNA transformation patterns and immunoblotting, provided experimental proof of the presence of these elements (Weir and Marrs, 1992). The same authors then also characterized the genes encoding for the building blocks of these pili (Weir, Lee and Marrs, 1996). This included the discovery of DNA-uptake sequences in the vicinity of the genes, similar to those described previously for the gonococcus (Weir, Lee and Marrs, 1997). The presence of pili has not yet been documented for the other Kingella species, but it is considered not unlikely that similar structures will be identified in the near future.

VIRULENCE

In general, the pathogenicity and virulence of a microorganism are defined by five cardinal requirements:

1. binding to and colonization of mucous surfaces;
2. entry into host tissues;
3. multiplication in the *in vivo* environment;
4. interference with host defence mechanisms;
5. damaging the host.

Relatively little is known about the precise virulence traits of *M. catarrhalis*, and for *K. kingae*, essentially nothing is known.

Adherence

It is noteworthy that only a limited number of studies on the precise interaction between *M. catarrhalis* receptors and human antigens have been undertaken. An elegant study was presented by Reddy *et al.* (1997). Using a purified middle-ear mucin glycoprotein, it was shown that only the CD protein of *M. catarrhalis* was capable of establishing a specific interaction with the sialo version of the human protein. A follow-up study from the same laboratory revealed immense heterogeneity in the interaction between upper respiratory tract pathogens and human mucins (Bernstein and Reddy, 2000). The general mechanism of *M. catarrhalis* adherence to host cell surfaces has been studied by Rikitomi, Ahmed and Nagatake (1997). Another study found no differences between source of isolation (blood or lungs) and haemagglutination (Jordan, Berk and Berk, 1990). Furthermore, these investigators showed that attachment was not primarily determined by lectin–carbohydrate interactions. In contrast to the findings of these investigators, an *in vitro* adherence study employing Hep-2 cell cultures demonstrated that strains derived from infections adhere more efficiently than mere colonizers (Fitzgerald *et al.*, 1999b). In addition, experimental periodate treatment suggested that bacterial adherence appears to be mediated by microbial carbohydrate moieties. Adherence of *M. catarrhalis* appeared to be stimulated by neutrophil defensins, peptides with broad-spectrum antimicrobial activity, released from activated neutrophils during inflammation, suggesting that defensin-mediated adherence contributes to the persistence of infection (Gorter *et al.*, 2000). A study by Ahmed *et al.* (2000) investigated the influence of surface charge on adherence. Although bacteria and epithelial cells are both overall negatively charged, interaction between the negatively charged surface of *M. catarrhalis* cells and positively charged subdomains called microplicea on pharyngeal epithelial cells has been documented.

Animal Models

The low virulence of *M. catarrhalis* lack a good animal model. In a murine model designed to study phagocytic responses and clearance mechanisms after endotracheal challenge with *M. catarrhalis*, a high influx of polymorphonuclear neutrophils into the lungs, with clearance in 24–48 h, was noted (Verghese *et al.*, 1990). A similar model uses transoral inoculation of bacteria into the lungs under anaesthesia and operative exposure of the trachea (Unhanand *et al.*, 1992). This model permits an evaluation of the interaction of bacteria with lower respiratory tract epithelium and the precise assessment of pathologic changes in the lungs (Maciver *et al.*, 1993). Kyd *et al.* (1999) used a model for mucosal immunization involving direct inoculation of killed bacteria into the Peyer's patches, followed by an intratracheal booster with dead *M. catarrhalis*. Enhanced clearance of bacteria from the lungs was observed, correlating with higher levels of specific IgA and IgG in serum and bronchoalveolar lavage fluid. Since *M. catarrhalis* inhabits the upper respiratory tract, inhalation models are preferred to intraperitoneal, endotracheal or transoral inoculation models (Hu *et al.*, 1999).

Purulent otitis media could be induced in Sprague–Dawley rats (Westman *et al.*, 1999). This infection progressed in a relatively mild fashion, and upon immunization, a protection rate of 50% or more was induced. Using the same model, a clear increase in the density of goblet cells in the middle ear up to 60 days after inoculation of bacteria was found, suggesting a highly increased mucosal secretory capacity (Caye-Thomasen *et al.*, 2000). In another rat model, inhalation of heat-killed *M. catarrhalis* cells clearly affected the laryngeal mucosa (Jecker *et al.*, 1999), resulting in a clinical syndrome reminiscent of laryngotracheitis in children.

Inoculation of *M. catarrhalis* into the middle ear of chinchillas and gerbils gave rise to effusion, but no live bacteria could be recovered from the middle ear after 24 h (Doyle, 1989). Later studies, however, revealed the feasibility of studying otitis media in the chinchilla model. Chinchillas are not generally available, and PCR would not have reached it current state of clinical applicability without the studies in these animals (Post *et al.*, 1996a; Aul *et al.*, 1998; Bakaletz *et al.*, 1998).

Complement Resistance

In general, rough strains of Gram-negative bacteria, producing LPS devoid of O-specific side chains, are highly susceptible to complement protein C5b-9-mediated killing, whereas smooth strains that synthesize complete LPS are often complement resistant (Taylor, 1995). Since the LPS of *M. catarrhalis* is of the rough type (Fomsgaard *et al.*, 1991), it presumably does not play a major role in complement resistance. However, Zaleski *et al.* (2000) showed that inactivation of a gene involved in the biosynthesis of the galα1–4 galβ1–4 glc LOS epitope results in enhanced susceptibility to serum-mediated killing. Apparently, deviant LOS structures render strains more susceptible to complement attack. Complement resistance may be considered a virulence factor of *M. catarrhalis*: most strains (89%) isolated from LRTIs are resistant to complement-mediated killing, whereas strains from the upper respiratory tract of children are mostly sensitive (58%) (Hol *et al.*, 1993, 1995). Complement-resistant strains inhibit the terminal pathway of complement, i.e. the formation of the membrane attack complex of complement (Verduin *et al.*, 1994a). The binding of human vitronectin, an inhibitor of the terminal pathway of complement, appears to play a crucial role (Verduin *et al.*, 1994b). Another study suggested that OMP *CopB*/ OMP B2 is involved in the resistance of *M. catarrhalis* to killing by normal human serum (Helminen *et al.*, 1993). An isogenic mutant not expressing *CopB* was killed by normal human serum, whereas the wild-type parent strain survived. It has been shown that inactivation of the *CopB* gene inhibits iron acquisition from lactoferrin and transferrin (Aebi *et al.*, 1996). Yet another OMP, OMP E, has also been shown to be involved in complement resistance. A *M. catarrhalis* OMP E knock-out mutant showed a clear increase in serum sensitivity (Murphy *et al.*, 2000). We conclude that complement resistance in *M. catarrhalis* probably is a highly multifactorial process, both from the perspective of both the host and the pathogen.

IMMUNITY

Many aspects of immunity to respiratory tract infections are still unknown, but may include local factors such as mucociliary clearance, aerodynamics, alveolar macrophage activity, complement-mediated killing and surfactant activity. These factors play important roles in host defence against oropharyngeal pathogens (Toews, Hansen and Strieter, 1990). The development of an inflammatory response or the development of specific antibody response may, however, augment these host defence mechanisms. As an example, in COPD patients, local host defence to respiratory pathogens is relatively poor (Ejlertsen, 1991). This points to an important role for local defence mechanisms in non-specific clearance of various bacterial species. Compared to other species, *M. catarrhalis* was cleared relatively slowly from the lungs, and a more pronounced, 400-fold increase in numbers of polymorphonuclear leukocytes in the lungs was observed (Onofrio *et al.*, 1981).

Antibody Responses to Whole *M. catarrhalis* Bacteria

Several authors have investigated antibody responses to *M. catarrhalis* in different patient cohorts. Chapman *et al.* (1985) showed that 90% of adult patients with LRTI had bactericidal antibodies in their convalescent sera; only 37% had bactericidal antibody present in their acute sera. Black and Wilson (1988) obtained similar results. Most children with AOM due to *M. catarrhalis* showed an increase in serum IgG antibody titres (Leinonen *et al.*, 1981; Faden, Hong and Murphy, 1992). In a study in infants with otitis media, a specific IgG response was detected in 10 of 12 children 8 months of age or older, compared with 1 of 6 in younger children. In addition, immunoblotting revealed four immunodominant OMPs: UspA, CopB, TbpB and a protein of 60 kDa, probably OMP CD (Mathers, Leinonen and Goldblatt, 1999).

Antibodies to *M. catarrhalis* are very low or absent in children below the age of 1 year, and the development of an antibody response in children correlates with a decrease in colonization. Antibodies to OMPs of *M. catarrhalis* appear around the age of 4 years (Goldblatt, Turner and Levinsky, 1990; Christensen, 1999). Furthermore, failure to produce significant levels of IgG3 antibodies against *M. catarrhalis* predisposes to infection with the bacterium (Goldblatt *et al.*, 1994).

Moraxella catarrhalis LOS Immunogenicity

It has been reported that antibody responses to these surface structures constitute a major part of the humoral immune response during infection with *M. catarrhalis*. This antibody response is not serotype specific but is directed to common epitopes of the LOS of different *M. catarrhalis* serotypes (Rahman *et al.*, 1995; Oishi *et al.*, 1996). Hence, *M. catarrhalis* LOS may be of interest for evaluation as a possible vaccine candidate (Gu *et al.*, 1998; Hu *et al.*, 2000).

Immunogenicity of Outer Membrane Proteins

OMPs B1, CopB/OMP B2, LbpB, OMP CD, OMP E, OMP G, TbpB and UspA have all been mentioned as potential vaccine candidates. No antibody response to TbpA or LbpA could be detected in convalescent sera of patients with pulmonary infections, limiting their role as vaccine candidates (Yu *et al.*, 1999). Helminen *et al.* (1994) and Sethi, Surface and Murphy (1997) showed that CopB/OMP B2 is a target for antibodies, increasing pulmonary clearance in the mouse. However, certain regions in the protein show interstrain variability; therefore, if this protein is to be developed into a vaccine, only its conserved regions should be targeted. Sethi, Hill and Murphy (1995) found a predominant antibody response to a minor 84-kDa OMP, designated OMP B1. OMP CD does not appear to be an immunodominant antigen, as indicated by an absence of a new antibody response to this protein after exacerbation of *M. catarrhalis* infection in COPD patients. Helminen *et al.* (1994) also presented evidence that UspA may be a target for protective antibodies in humans. Chen *et al.* (1996) immunized mice with purified UspA, and fewer bacteria were isolated from the lungs of immunized mice than from non-immunized controls. The finding that IgG3 is a major contributing factor in the immune response to *M. catarrhalis* was confirmed by a study by Chen *et al.* (1999b) of the immune response of healthy adults and children to UspA1 and UspA2. In a small cohort of children suffering from otitis media, antibodies specific to UspA1/2

proteins were identified (Samukawa *et al.*, 2000a). The amounts of these specific antibodies varied strongly with age (Samukawa *et al.*, 2000b).

Local Antibody Response

Only few investigators have studied the development of antibody responses in middle-ear fluid of children with otitis media (Faden, Hong and Murphy, 1992; Faden, Harabuchi and Hong, 1994; Takada *et al.*, 1998; Karalus and Campagnari, 2000). IgG and IgA appeared to be produced locally in most patients, but also antibodies derived from serum were detected in middle-ear fluids. Faden, Hong and Murphy (1992) showed that middle-ear fluid IgG, IgM and IgA antibody was produced in 100%, 29% and 71% of children, respectively. Of interest, many children with local antibodies in their middle-ear fluid did not develop a systemic antibody response. Local antibodies may play an important role in the recovery and prevention of AOM (Goldblatt, Turner and Levinsky, 1990). In a study focusing on local IgA antibodies to UspA in the nasopharyngeal secretions of children colonized with *M. catarrhalis*, no response was detected (Samukawa *et al.*, 2000b).

Vaccines

The development of vaccines for the prevention of *M. catarrhalis*-mediated disease is, at the time of writing, a hot topic. The most promising vaccine candidates have been recently reviewed by McMichael (2000). The combination of data that are available in today's literature suggests that the development of a *M. catarrhalis* vaccine is well underway: animal models of infection have been developed and described, and several vaccine candidate molecules have been studied for prevalence and genetic conservation among different isolates. Although new candidate molecules are regularly brought forward (Fitzgerald *et al.*, 1999a), relatively little or nothing is known about the optimal routes for vaccine delivery or whether there is a need for adjuvants. A study using a rat model suggests that the mucosal route of delivery for *M. catarrhalis* is more effective than systemic immunization (Kyd and Cripps, 2000).

ACKNOWLEDGEMENT

The authors express their gratitude to all those who directly or indirectly provided inspiration for this chapter. Mrs. Loes van Damme (Erasmus MC, Department of Medical Microbiology) is thanked for providing insight into the classical microbiology of *K. kingae* and its relatives.

REFERENCES

Abbott, M. (1992) Neisseriaceae and *Moraxella* sp.: the role of related microorganisms associated with conjunctivitis in the newborn. Int J STD AIDS 3:212–13.

Abuamara, S., Louis, J.S., Guyard, M.F. *et al.* (2000) *Kingella kingae* osteoarticular infections in children: a report of a series of eight new cases. Arch Pediatr 7:927–32.

Abuhammour, W.M., Abdel-Haq, N.M., Asmar, B.I., and Dajani, A.S. (1999) *Moraxella catarrhalis* bacteremia: a 10-year experience. South Med J 92:1071–4.

Aebi, C., Stone, B., Beucher, M. *et al.* (1996) Expression of the CopB outer membrane protein by *Moraxella catarrhalis* is regulated by iron and affects iron acquisition from transferrin and lactoferrin. Infect Immun 64:2024–30.

Aebi, C., Maciver, I., Latimer, J.L. *et al.* (1997) A protective epitope of *Moraxella catarrhalis* is encoded by two different genes. Infect Immun 65:4367–77.

Aebi, C., Lafontaine, E.R., Cope, L.D. *et al.* (1998) Phenotypic effect of isogenic uspA1 and uspA2 mutations on *Moraxella catarrhalis* 035E. Infect Immun 66:3113–19.

Ahmed, K., Rikitomi, N., Ichinose, A., and Matsumoto, K. (1991) Possible presence of a capsule in *Branhamella catarrhalis*. Microbiol Immunol 35:361–6.

Ahmed, K., Nakagawa, T., Nakano, Y. *et al.* (2000) Attachment of *Moraxella catarrhalis* occurs to the positively charged domains of pharyngeal epithelial cells. Microb Pathog 28:203–9.

Amir, J. (1992) *Kingella kingae* infection in children. Pediatr Infect Dis J 11:339.

Amir, J., and Shockelford, P.G. (1991) *Kingella kingae* intervertebral disc infection. J Clin Microbiol 9:1083–6.

Amir, J., and Yagupsky, P. (1998) Invasive *Kingella kingae* infection associated with stomatitis in children. Pediatr Infect Dis J 17:757–8.

Anonymous (1994) Otitis media bacteriology and immunology. Pediatr Infect Dis J 13:20–2.

Arkwright, J.A. (1907) On the occurrence of the *Micrococcus catarrhalis* in normal and catarrhal noses and its differentiation from the other Gram-negative diplococci. J Hyg (Lond) 7:145–54.

Aul, J.J., Anderson, K.W., Wadowsky, R.M. *et al.* (1998) Comparative evaluation of culture and PCR for the detection and determination of persistence of bacterial strains and DNAs in the *Chinchilla laniger* model of otitis media. Ann Otol Rhinol Laryngol 107:508–13.

Bakaletz, L.O., White, G.J., Post, J.C., and Ehrlich, G.D. (1998) Blinded multiplex PCR analyses of middle ear and nasopharyngeal fluids from chinchilla models of single- and mixed-pathogen-induced otitis media. Clin Diagn Lab Immunol 5:219–24.

Barreiro, B., Esteban, L., Prats, E. *et al.* (1992) *Branhamella catarrhalis* respiratory infections. Eur Respir J 5:675–9.

Bartos, L.C., and Murphy, T.F. (1988) Comparison of the outer membrane proteins of 50 strains of *Branhamella catarrhalis*. J Infect Dis 158:761–5.

Beaulieu, D., Scriver, S., Bergeron, M.G. *et al.* (1993) Epidemiological typing of *Moraxella catarrhalis* by using DNA probes. J Clin Microbiol 31:736–9.

Berg, R.A., and Bartley, D.L. (1987) Pneumonia associated with *Branhamella catarrhalis* in infants. Pediatr Infect Dis J 6:569–73.

Berger, U. (1963) Die anspruchslosen Neisserien. Ergeb Mikrobiol Immunitatsforsch Exp Ther 36:97–167.

Bergren, R.L., Tasman, W.S., Wallace, R.T., and Katz, L.J. (1993) *Branhamella (Moraxella) catarrhalis* endophthalmitis. Arch Ophthalmol 111:1169–70.

Berk, S.L., and Kalbfleisch, J.H. (1996) Antibiotic susceptibility patterns of community-acquired respiratory isolates of *Moraxella catarrhalis* in Western Europe and in the USA. The Alexander Project Collaborative Group. J Antimicrob Chaemother 38:85–96.

Berman, S. (1995) Otitis media in children. N Engl J Med 332:1560–5.

Berner, R., Schumacher, R.F., Brandis, M., and Forster, J. (1996) Colonization and infection with *Moraxella catarrhalis* in childhood. Eur J Clin Microbiol Infect Dis 15:506–9.

Bernstein, J.M., and Reddy, M. (2000) Bacteria–mucin interaction in the upper aerodigestive tract shows striking heterogeneity: implications in otitis media, rhinosinusitis, and pneumonia. Otolaryngol Head Neck Surg 122:514–20.

Bernstein, T., Brilli, R., and Jacobs, B. (1998) Is bacterial tracheitis changing? A 14-month experience in a pediatric intensive care unit. Clin Infect Dis 27:458–62.

Bhushan, R., Kirkham, C., Sethi, S., and Murphy, T.F. (1997) Antigenic characterization and analysis of the human immune response to outer membrane protein E of *Branhamella catarrhalis*. Infect Immun 65:2668–75.

Birgisson, H., Steingrimsson, O., and Gudnason, T. (1997) *Kingella kingae* infections in paediatric patients: 5 cases of septic arthritis, osteomyelitis and bacteremia. Scand J Infect Dis 29:495–8.

Black, A.J., and Wilson, T.S. (1988) Immunoglobulin G (IgG) serological response to *Branhamella catarrhalis* in patients with acute bronchopulmonary infections. J Clin Pathol 41:329–33.

Block, S.L. (1997) Causative pathogens, antibiotic resistance and therapeutic considerations in acute otitis media. Pediatr Infect Dis J 16:449–56.

Bluestone, C.D. (1986) Otitis media and sinusitis in children. Role of *Branhamella catarrhalis*. Drugs 31:132–41.

Blumer, J. (1998) Clinical perspectives on sinusitis and otitis media. Pediatr Infect Dis J 17:S68–72.

Bonnah, R.A., Yu, R., and Schryvers, A.B. (1995) Biochemical analysis of lactoferrin receptors in the Neisseriaceae: identification of a second bacterial lactoferrin receptor protein. Microb Pathog 19:285–97.

Bootsma, H.J., van Dijk, H., Verhoef, J. *et al.* (1996) Molecular characterization of the BRO beta-lactamase of *Moraxella (Branhamella) catarrhalis*. Antimicrob Agents Chaemother 40:966–72.

Bootsma, H.J., Aerts, P.C., Posthuma, G. *et al.* (1999) *Moraxella (Branhamella) catarrhalis* BRO beta-lactamase: a lipoprotein of gram-positive origin? J Bacteriol 181:5090–3.

Bootsma, H.J., van der Heide, H.G., van de Pas, S. *et al.* (2000) Analysis of *Moraxella catarrhalis* by DNA typing: evidence for a distinct subpopulation associated with virulence traits. J Infect Dis 181:1376–87.

Bovre, K. (1984) The genus *Moraxella*. In N.R. Krieg, and J.G. Holt (eds), Bergey's Manual of Systematic Bacteriology, Vol. 1. Williams & Wilkins, Baltimore, MD, pp. 296–303.

Boyle, F.M., Georghiou, P.R., Tilse, M.H., and McCormack, J.G. (1991) *Branhamella (Moraxella) catarrhalis*: pathogenic significance in respiratory infections. Med J Aust 154:592–6.

Brook, I. (1997) Aerobic and anaerobic microbiology of bacterial tracheitis in children. Pediatr Emerg Care 13:16–8.

Brorson, J.E., and Malmvall, B.E. (1981) *Branhamella catarrhalis* and other bacteria in the nasopharynx of children with longstanding cough. Scand J Infect Dis 13:111–3.

Cabello, H., Torres, A., Celis, R. *et al.* (1997) Bacterial colonization of distal airways in healthy subjects and chronic lung disease: a bronchoscopic study. Eur Respir J 10:1137–44.

Calder, M.A., Croughan, M.J., McLeod, D.T., and Ahmad, F. (1986) The incidence and antibiotic susceptibility of *Branhamella catarrhalis* in respiratory infections. Drugs 31:11–6.

Campagnari, A.A., Ducey, T.F., and Rebmann, C.A. (1996) Outer membrane protein B1, an iron-repressible protein conserved in the outer membrane of *Moraxella (Branhamella) catarrhalis*, binds human transferrin. Infect Immun 64:3920–4.

Cappelletty, D. (1998) Microbiology of bacterial respiratory infections. Pediatr Infect Dis J 17:S55–61.

Carden, S.M., Colville, D.J., Gonis, G., and Gilbert, G.L. (1991) *Kingella kingae* endophthalmitis in an infant. Aust N Z J Ophthalmol 19:217–20.

Catlin, B.W. (1970) Transfer of the organism named *Neisseria catarrhalis* to *Branhamella* gen. nov. Int J Syst Bacteriol 20:155–9.

Catlin, B.W. (1990) *Branhamella catarrhalis*: an organism gaining respect as a pathogen. Clin Microbiol Rev 3:293–320.

Catlin, B.W. (1991) Branhamaceae fam. nov., a proposed family to accommodate the genera *Branhamella* and *Moraxella*. Int J Syst Bacteriol 41:320–3.

Caye-Thomasen, P., Hermansson, A., Tos, M., and Prellner, K. (2000) Middle ear secretory capacity after acute otitis media caused by *Streptococcus pneumoniae*, *Moraxella catarrhalis*, non-typeable or type B *Haemophilus influenzae*. A comparative analysis based on goblet cell density. Acta Otolaryngol 543:54–5.

Chakraborty, R.N., Meigh, R.E., and Kaye, G.C. (1999) *Kingella kingae* prosthetic valve endocarditis. Indian Heart J 51:438–9.

Chapman, A.J., Jr., Musher, D.M., Jonsson, S. *et al.* (1985) Development of bactericidal antibody during *Branhamella catarrhalis* infection. J Infect Dis 151:878–82.

Chen, C. (1996) Distribution of a newly described species, *Kingella oralis*, in the human oral cavity. Oral Microbiol Immunol 11:425–7.

Chen, D., McMichael, J.C., VanDerMeid, K.R. *et al.* (1996) Evaluation of purified UspA from *Moraxella catarrhalis* as a vaccine in a murine model after active immunization. Infect Immun 64:1900–5.

Chen, D., McMichael, J.C., VanDerMeid, K.R. *et al.* (1999a) Evaluation of a 74-kDa transferrin-binding protein from *Moraxella (Branhamella) catarrhalis* as a vaccine candidate. Vaccine 18:109–18.

Chen, D., Barniak, V., VanDerMeid, K.R., and McMichael, J.C. (1999b) The levels and bactericidal capacity of antibodies directed against the UspA1 and UspA2 outer membrane proteins of *Moraxella (Branhamella) catarrhalis* in adults and children. Infect Immun 67:1310–6.

Chiquita, P.E., Elliott, J., and Namyak, S.S. (1991) *Kingella kingae* dactylitis in an infant. J Infect 22:102–3.

Christensen, J.J. (1999) *Moraxella (Branhamella) catarrhalis*: clinical, microbiological and immunological features in lower respiratory tract infections. APMIS 88:1–36.

Clark, R.B., and Joyce, S.E. (1993) Activity of meropenem and other antimicrobial agents against uncommon gram-negative organisms. J Antimicrob Chaemother 32:233–7.

Cohen, R. (1997) The antibiotic treatment of acute otitis media and sinusitis in children. Diagn Microbiol Infect Dis 27:35–9.

Collazos, J., de Miguel, J., and Ayarza, R. (1992) *Moraxella catarrhalis* bacteremic pneumonia in adults: two cases and review of the literature. Eur J Clin Microbiol Infect Dis 11:237–40.

Cope, L.D., Lafontaine, E.R., Slaughter, C.A. *et al.* (1999) Characterization of the *Moraxella catarrhalis* uspA1 and uspA2 genes and their encoded products. J Bacteriol 181:4026–34.

Costers, M., Wouters, C., Moens, P., and Verhaegen, J. (2003) Three cases of *Kingella kingae* infection in young children. Eur J Pediatr 162:530–1.

Dabernat, H., Avril, J.L., and Boussougant, Y. (1990) In-vitro activity of cefpodoxime against pathogens responsible for community-acquired respiratory tract infections. J Antimicrob Chaemother 26:1–6.

Davies, B.I. (1997) *Branhamella, Moraxella, Kingella*. In Emmerson, Hawkey, and Gillespie (eds), Principles and Practice of Clinical Bacteriology. Wiley, pp. 251–64.

Davies, B.I., and Maesen, F.P. (1986) Epidemiological and bacteriological findings on *Branhamella catarrhalis* respiratory infections in The Netherlands. Drugs 31:28–33.

Davies, B.I., and Maesen, F.P. (1988) The epidemiology of respiratory tract pathogens in southern Netherlands. Eur Respir J 1:415–20.

Denamur, E., Picard-Pasquier, N., Mura, C. *et al.* (1991) Comparison of molecular epidemiological tools for *Branhamella catarrhalis* typing. Res Microbiol 142:585–9.

Dewhirst, F.E., Paster, B.J., La Fontaine, S., and Rood, J.I. (1990) Transfer of *Kingella indologenes* (Snell and Lapage 1976) to the genus *Suttonella* gen. nov. as *Suttonella indologenes* comb. nov.; transfer of *Bacteroides nodosus* (Beveridge 1941) to the genus *Dichelobacter* gen. nov. as *Dichelobacter nodosus* comb. nov.; and assignment of the genera *Cardiobacterium, Dichelobacter*, and *Suttonella* to Cardiobacteriaceae fam. nov. in the gamma division of the proteobacteria on the basis of 16S rRNA sequence comparisons. Int J Syst Bacteriol 40:426–33.

Dewhirst, F.E., Chen, C.K., Paster, B.J., and Zambon, J.J. (1993) Phylogeny of species in the family Neisseriaceae isolated from human dental plaque and description of *Kingella oralis* sp. nov. Int J Syst Bacteriol 43:490–9.

DiGiovanni, C., Riley, T.V., Hoyne, G.F. *et al.* (1987) Respiratory tract infections due to *Branhamella catarrhalis*: epidemiological data from Western Australia. Epidemiol Infect 99:445–53.

Dingman, J.R., Rayner, M.G., Mishra, S. *et al.* (1998) Correlation between presence of viable bacteria and presence of endotoxin in middle-ear effusions. J Clin Microbiol 36:3417–9.

Dodman, T., Robson, J., and Pincus, D. (2000) *Kingella kingae* infections in children. J Paediatr Child Health 36:87–90.

Doern, G.V. (1986) *Branhamella catarrhalis* – an emerging human pathogen. Diagn Microbiol Infect Dis 4:191–201.

Doern, G.V. (1990) *Branhamella catarrhalis*: phenotypic characteristics. Am J Med 88:33S–5S.

Doern, G.V., and Morse, S.A. (1980) *Branhamella (Neisseria) catarrhalis*: criteria for laboratory identification. J Clin Microbiol 11:193–5.

Doern, G.V., Miller, M.J., and Winn, R.E. (1981) *Branhamella (Neisseria) catarrhalis* systemic disease in humans. Case reports and review of the literature. Arch Intern Med 141:1690–2.

Doern, G.V., Brueggemann, A.B., Pierce, G. *et al.* (1996) Prevalence of antimicrobial resistance among 723 outpatient clinical isolates of *Moraxella catarrhalis* in the United States in 1994 and 1995: results of a 30-center national surveillance study. Antimicrob Agents Chaemother 40:2884–6.

Doern, G.V., Jones, R.N., Pfaller, M.A., and Kugler, K. (1999) *Haemophilus influenzae* and *Moraxella catarrhalis* from patients with community-acquired respiratory tract infections: antimicrobial susceptibility patterns from the SENTRY antimicrobial Surveillance Program (United States and Canada, 1997). Antimicrob Agents Chaemother 43:385–9.

Doyle, W.J. (1989) Animal models of otitis media: other pathogens. Pediatr Infect Dis J 8:S45–7.

Du, R.P., Wang, Q., Yang, Y.P. *et al.* (1998) Cloning and expression of the *Moraxella catarrhalis* lactoferrin receptor genes. Infect Immun 66:3656–65.

Ejlertsen, T. (1991) Pharyngeal carriage of *Moraxella (Branhamella) catarrhalis* in healthy adults. Eur J Clin Microbiol Infect Dis 10:89.

Ejlertsen, T., Thisted, E., Ostergaard, P.A., and Renneberg, J. (1994) Maternal antibodies and acquired serological response to *Moraxella catarrhalis* in children determined by an enzyme-linked immunosorbent assay. Clin Diagn Lab Immunol 1:464–8.

Eller, J., Ede, A., Schaberg, T. *et al.* (1998) Infective exacerbations of chronic bronchitis: relation between bacteriologic etiology and lung function. Chest 113:1542–8.

Enright, M.C., and McKenzie, H. (1997) *Moraxella (Branhamella) catarrhalis* – clinical and molecular aspects of a rediscovered pathogen. J Med Microbiol 46:360–71.

Enright, M.C., Carter, P.E., MacLean, I.A., and McKenzie, H. (1994) Phylogenetic relationships between some members of the genera *Neisseria, Acinetobacter, Moraxella*, and *Kingella* based on partial 16S ribosomal DNA sequence analysis. Int J Syst Bacteriol 44:387–91.

Esteve, V., Porcheret, H., Clerc, D. *et al.* (2001) Septic arthritis due to *Kingella kingae* in an adult. Joint Bone Spine 68:85–6.

Faden, H., Bernstein, J., Brodsky, L. *et al.* (1992) Effect of prior antibiotic treatment on middle ear disease in children. Ann Otol Rhinol Laryngol 101:87–91.

Faden, H., Hong, J., and Murphy, T. (1992) Immune response to outer membrane antigens of *Moraxella catarrhalis* in children with otitis media. Infect Immun 60:3824–9.

Faden, H., Harabuchi, Y., and Hong, J.J. (1994) Epidemiology of *Moraxella catarrhalis* in children during the first 2 years of life: relationship to otitis media. J Infect Dis 169:1312–7.

Faden, H., Duffy, L., Wasielewski, R. *et al.* (1997) Relationship between nasopharyngeal colonization and the development of otitis media in children. Tonawanda/Williamsville Pediatrics. J Infect Dis 175:1440–5.

Felmingham, D., and Gruneberg, R.N. (2000) The Alexander Project 1996–1997: latest susceptibility data from this international study of bacterial pathogens from community-acquired lower respiratory tract infections. J Antimicrob Chaemother 45:191–203.

Ferber, B., Bruckheimer, E., Schlesinger, Y. *et al.* (1997) *Kingella kingae* endocarditis in a child with hair cartilage hypoplasia. Pediatr Cardiol 18:445–6.

Fitzgerald, M., Mulcahy, R., Murphy, S. *et al.* (1999a) Transmission electron microscopy studies of *Moraxella (Branhamella) catarrhalis*. FEMS Immunol Med Microbiol 23:57–66.

Fitzgerald, M., Murphy, S., Mulcahy, R. *et al.* (1999b) Tissue culture adherence and haemagglutination characteristics of *Moraxella (Branhamella) catarrhalis*. FEMS Immunol Med Microbiol 24:105–14.

Fluit, A.C., Schmitz, F.J., Jones, M.E. *et al.* (1999) Antimicrobial resistance among community-acquired pneumonia isolates in Europe: first results from the SENTRY antimicrobial surveillance program 1997. SENTRY Participants Group. Int J Infect Dis 3:153–6.

Fomsgaard, J.S., Fomsgaard, A., Hoiby, N. *et al.* (1991) Comparative immunochemistry of lipopolysaccharides from *Branhamella catarrhalis* strains. Infect Immun 59:3346–9.

Froom, J., Culpepper, L., Jacobs, M. *et al.* (1997) Antimicrobials for acute otitis media? A review from the International Primary Care Network. BMJ 315:98–102.

Frosch, P., and Kolle, W. (1896) Die Mikrokokken. In Flugge (ed.), Die Mikroorganism, 3 ed, vol. 2. Verlag von Vogel, Leipzig, pp. 154–5.

Fung, C.P., Powell, M., Seymour, A. *et al.* (1992) The antimicrobial susceptibility of *Moraxella catarrhalis* isolated in England and Scotland in 1991. J Antimicrob Chaemother 30:47–55.

Goldblatt, D., Seymour, N.D., Levinsky, R.J., and Turner, M.W. (1990) An enzyme-linked immunosorbent assay for the determination of human IgG subclass antibodies directed against *Branhamella catarrhalis*. J Immunol Methods 128:219–25.

Goldblatt, D., Turner, M.W., and Levinsky, R.J. (1990) *Branhamella catarrhalis*: antigenic determinants and the development of the IgG subclass response in childhood. J Infect Dis 162:1128–35.

Goldblatt, D., Scadding, G.K., Lund, V.J. *et al.* (1994) Association of Gm allotypes with the antibody response to the outer membrane proteins of a common upper respiratory tract organism, *Moraxella catarrhalis*. J Immunol 153:5316–20.

Goldenhersh, M.J., Rachelefsky, G.S., Dudley, J. *et al.* (1990) The microbiology of chronic sinus disease in children with respiratory allergy. J Allergy Clin Immunol 85:1030–9.

Gorter, A.D., Hiemstra, P.S., de Bentzmann, S. *et al.* (2000) Stimulation of bacterial adherence by neutrophil defensins varies among bacterial species but not among host cell types. FEMS Immunol Med Microbiol 28:105–11.

Graham, D.R., Brand, J.D., Thornsberry, C. *et al.* (1990) Infections caused by *Moraxella, Moraxella urethralis, Moraxella*-like groups M-5 and M-6, and *Kingella kingae* in the United States, 1953–1980. Rev Infect Dis 12:423–31.

Gu, X.X., Chen, J., Barenkamp, S.J. *et al.* (1998) Synthesis and characterization of lipooligosaccharide-based conjugates as vaccine candidates for *Moraxella (Branhamella) catarrhalis*. Infect Immun 66:1891–7.

Haddad, J., Le Faou, A., Simeoni, U., and Messer, J. (1986) Hospital-acquired bronchopulmonary infection in premature infants due to *Branhamella catarrhalis*. J Hosp Infect 7:301–2.

Hager, H., Verghese, A., Alvarez, S., and Berk, S.L. (1987) *Branhamella catarrhalis* respiratory infections. Rev Infect Dis 9:1140–9.

Hassan, I.J., and Hayek, L. (1993) Endocarditis caused by *Kingella denitrificans*. J Infect 27:291–5.

Hay, F., Chellun, P., Romaru, A. *et al.* (2002) *Kingella kingae*, a rare cause of meningitis. Arch Pediatr 9:37–40.

Hays, J.P., van der Schee, C., Loogman, A. *et al.* (2003) Total genome polymorphism and low frequency of intra genomic variation in the uspA1 and uspA2 genes of *Moraxella catarrhalis* in otitis prone and otitis non-prone children up to two years of age: consequences for vaccine design. Vaccine 21:1118–24.

Helminen, M.E., Maciver, I., Paris, M. *et al.* (1993) A mutation affecting expression of a major outer membrane protein of *Moraxella catarrhalis* alters serum resistance and survival in vivo. J Infect Dis 168:1194–201.

Helminen, M.E., Maciver, I., Latimer, J.L. *et al.* (1994) A large, antigenically conserved protein on the surface of *Moraxella catarrhalis* is a target for protective antibodies. J Infect Dis 170:867–72.

Hendolin, P.H., Markkanen, A., Ylikoski, J., and Wahlfors, J.J. (1997) Use of multiplex PCR for simultaneous detection of four bacterial species in middle ear effusions. J Clin Microbiol 35:2854–8.

Henriksen, S.D., and Bøvre, K. (1976) Transfer of *Kingella kingae* Hendriksen and Bøvre to the genus *Kingella* gen. nov. in the family Neisseriaceae. Int J Syst Bacteriol 26:447–50.

Hol, C., Verduin, C.M., van Dijke, E. *et al.* (1993) Complement resistance in *Branhamella* (*Moraxella*) *catarrhalis*. Lancet 341:1281.

Hol, C., Verduin, C.M., Van Dijke, E.E. *et al.* (1995) Complement resistance is a virulence factor of *Branhamella* (*Moraxella*) *catarrhalis*. FEMS Immunol Med Microbiol 11:207–11.

Hollis, D.G., Wiggins, G.L., and Weaver, R.E. (1972) An unclassified Gram-negative rod isolated from the pharynx on Thayer–Martin medium (selective agar). Appl Microbiol 24:772–7.

Holme, T., Rahman, M., Jansson, P.E., and Widmalm, G. (1999) The lipopolysaccharide of *Moraxella catarrhalis* structural relationships and antigenic properties. Eur J Biochem 265:524–9.

Hoogkamp-Korstanje, J.A., Dirks-Go, S.I., Kabel, P. *et al.* (1997) Multicentre in-vitro evaluation of the susceptibility of *Streptococcus pneumoniae*, *Haemophilus influenzae* and *Moraxella catarrhalis* to ciprofloxacin, clarithromycin, co-amoxiclav and sparfloxacin. J Antimicrob Chaemother 39:411–14.

Host, B., Schumacher, H., Prag, J., and Arpi, M. (2000) Isolation of *Kingella kingae* from synovial fluids using four commercial blood culture bottles. Eur J Clin Microbiol Infect Dis 19:608–11.

Hu, W.G., Chen, J., Collins, F.M., and Gu, X.X. (1999) An aerosol challenge mouse model for *Moraxella catarrhalis*. Vaccine 18:799–804.

Hu, W.G., Chen, J., Battey, J.F., and Gu, X.X. (2000) Enhancement of clearance of bacteria from murine lungs by immunization with detoxified lipooligosaccharide from *Moraxella catarrhalis* conjugated to proteins. Infect Immun 68:4980–5.

Ikram, R.B., Nixon, M., Aitken, J., and Wells, E. (1993) A prospective study of isolation of *Moraxella catarrhalis* in a hospital during the winter months. J Hosp Infect 25:7–14.

Ioannidis, J.P., Worthington, M., Griffiths, J.K., and Snydman, D.R. (1995) Spectrum and significance of bacteremia due to *Moraxella catarrhalis*. Clin Infect Dis 21:390–7.

Janda, W.M., Bradna, J.J., and Ruther, P. (1989) Identification of *Neisseria* spp., *Haemophilus* spp., and other fastidious gram-negative bacteria with the MicroScan *Haemophilus–Neisseria* identification panel. J Clin Microbiol 27:869–73.

Jawetz, E., Melnich, J.L., and Adelberg, E.A. (1976) Review of Medical Microbiology, 12th ed. Lange Medical Publications, Los Altos, CA, p. 183.

Jecker, P., McWilliam, A., Napoli, S. *et al.* (1999) Acute laryngitis in the rat induced by *Moraxella catarrhalis* and *Bordetella pertussis*: number of neutrophils, dendritic cells, and T and B lymphocytes accumulating during infection in the laryngeal mucosa strongly differs in adjacent locations. Pediatr Res 46:760–6.

Jensen, K.T., Schonheyder, H., and Thomsen, V.F. (1994) In vitro activity of beta-lactam and other antimicrobial agents against *Kingella kingae*. J Antimicrob Chaemother 33:635–40.

Johnson, M.A., Drew, W.L., and Roberts, M. (1981) *Branhamella* (*Neisseria*) *catarrhalis* – a lower respiratory tract pathogen? J Clin Microbiol 13:1066–9.

Jordan, K.L., Berk, S.H., and Berk, S.L. (1990) A comparison of serum bactericidal activity and phenotypic characteristics of bacteremic, pneumonia-causing strains, and colonizing strains of *Branhamella catarrhalis*. Am J Med 88:28S–32S.

Jorgensen, J.H., Doern, G.V., Maher, L.A. *et al.* (1990) Antimicrobial resistance among respiratory isolates of *Haemophilus influenzae*, *Moraxella catarrhalis*, and *Streptococcus pneumoniae* in the United States. Antimicrob Agents Chaemother 34:2075–80.

Kabongo, M.L. (1989) *Branhamella catarrhalis* infections. Am Fam Physician 40:34–9.

Karalus, R., and Campagnari, A. (2000) *Moraxella catarrhalis*: a review of an important human mucosal pathogen. Microbes Infect 2:547–59.

Kawakami, Y., Ueno, I., Katsuyama, T. *et al.* (1994) Restriction fragment length polymorphism (RFLP) of genomic DNA of *Moraxella* (*Branhamella*) *catarrhalis* isolates in a hospital. Microbiol Immunol 38:891–5.

Keller, R., Gustafson, J.E., and Keist, R. (1992) The macrophage response to bacteria. Modulation of macrophage functional activity by peptidoglycan from *Moraxella* (*Branhamella*) *catarrhalis*. Clin Exp Immunol 89:384–9.

Klein, J.O. (1994) Lessons from recent studies on the epidemiology of otitis media. Pediatr Infect Dis J 13:1031–4.

Klein, J.O. (1999) Management of acute otitis media in an era of increasing antibiotic resistance. Int J Pediatr Otorhinolaryngol 49:S15–7.

Klingman, K.L., and Murphy, T.F. (1994) Purification and characterization of a high-molecular-weight outer membrane protein of *Moraxella* (*Branhamella*) *catarrhalis*. Infect Immun 62:1150–5.

Klingman, K.L., Pye, A., Murphy, T.F., and Hill, S.L. (1995) Dynamics of respiratory tract colonization by *Branhamella catarrhalis* in bronchiectasis. Am J Respir Crit Care Med 152:1072–8.

Knapp, J.S., Johnson, S.R., Zenilman, J.M. *et al.* (1988) High-level tetracycline resistance resulting from TetM in strains of *Neisseria* spp., *Kingella denitrificans*, and *Eikenella corrodens*. Antimicrob Agents Chaemother 32:765–7.

Korppi, M., Katila, M.L., Jaaskelainen, J., and Leinonen, M. (1992) Role of *Moraxella* (*Branhamella*) *catarrhalis* as a respiratory pathogen in children. Acta Paediatr 81:993–6.

Kovatch, A.L., Wald, E.R., and Michaels, R.H. (1983) Beta-lactamase-producing *Branhamella catarrhalis* causing otitis media in children. J Pediatr 102:261–4.

Krause, I., and Nimri, R. (1996) *Kingella kingae* occult bacteremia in a toddler. Pediatr Infect Dis J 15:557–8.

Kyd, J., and Cripps, A. (2000) Identifying vaccine antigens and assessing delivery systems for the prevention of bacterial infections. J Biotechnol 83:85–90.

Kyd, J., John, A., Cripps, A., and Murphy, T.F. (1999) Investigation of mucosal immunisation in pulmonary clearance of *Moraxella* (*Branhamella*) *catarrhalis*. Vaccine 18:398–406.

La Scola, B., Iorgulescu, I., and Bollini, G. (1998) Five cases of *Kingella kingae* skeletal infection in a French hospital. Eur J Clin Microbiol Infect Dis 17:512–15.

Lafontaine, E.R., Cope, L.D., Aebi, C. *et al.* (2000) The UspA1 protein and a second type of UspA2 protein mediate adherence of *Moraxella catarrhalis* to human epithelial cells in vitro. J Bacteriol 182:1364–73.

Leinonen, M., Luotonen, J., Herva, E. *et al.* (1981) Preliminary serologic evidence for a pathogenic role of *Branhamella catarrhalis*. J Infect Dis 144:570–4.

Lejbkowicz, F., Cohn, L., Hashman, N., and Kassis, I. (1999) Recovery of *Kingella kingae* from blood and synovial fluid of two pediatric patients by using the BacT/Alert system. J Clin Microbiol 37:878.

Lewis, M.B., and Bamford, J.M. (2000) Global aphasia without hemiparesis secondary to *Kingella kingae* endocarditis. Arch Neurol 57:1774–5.

Livermore, D.M. (1995) Beta-lactamases in laboratory and clinical resistance. Clin Microbiol Rev 8:557–84.

Luke, N.R., Russo, T.A., Luther, N., and Campagnari, A.A. (1999) Use of an isogenic mutant constructed in *Moraxella catarrhalis* to identify a protective epitope of outer membrane protein B1 defined by monoclonal antibody 11C6. Infect Immun 67:681–7.

Luna, V.A., Cousin, S., Jr., Whittington, W.L., and Roberts, M.C. (2000) Identification of the conjugative *mef* gene in clinical *Acinetobacter junii* and *Neisseria gonorrhoeae* isolates. Antimicrob Agents Chaemother 44:2503–6.

Lundy, D.W., and Kehl, D.K. (1998) Increasing prevalence of *Kingella kingae* in osteoarticular infections in young children. J Pediatr Orthop 18:262–7.

Maccato, M., McClean, W., Riddle, G., and Faro, S. (1991) Isolation of *Kingella denitrificans* from amniotic fluid in a woman with chorioamnionitis: a case report. J Reprod Med 36:685–7.

Maciver, I., Unhanand, M., McCracken, G.H., and Hansen, E.J. (1993) Effect of immunization of pulmonary clearance of *Moraxella catarrhalis* in an animal model. J Infect Dis 168:469–72.

Macsai, M.S., Hillman, D.S., and Robin, J.B. (1988) *Branhamella* keratitis resistant to penicillin and cephalosporins. Case report. Arch Ophthalmol 106:1506–7.

Manninen, R., Huovinen, P., and Nissinen, A. (1997) Increasing antimicrobial resistance in *Streptococcus pneumoniae*, *Haemophilus influenzae* and *Moraxella catarrhalis* in Finland. J Antimicrob Chaemother 40:387–92.

Marchant, C.D. (1990) Spectrum of disease due to *Branhamella catarrhalis* in children with particular reference to acute otitis media. Am J Med 88:15S–19S.

Marrs, C.F., and Weir, S. (1990) Pili (fimbriae) of *Branhamella* species. Am J Med 88:36S–40S.

Martinez, G., Ahmed, K., Zheng, C.H. *et al.* (1999) DNA restriction patterns produced by pulsed-field gel electrophoresis in *Moraxella catarrhalis* isolated from different geographical areas. Epidemiol Infect 122:417–22.

Mathers, K., Leinonen, M., and Goldblatt, D. (1999) Antibody response to outer membrane proteins of *Moraxella catarrhalis* in children with otitis media. Pediatr Infect Dis J 18:982–8.

McGregor, K., Chang, B.J., Mee, B.J., and Riley, T.V. (1998) *Moraxella catarrhalis*: clinical significance, antimicrobial susceptibility and BRO beta-lactamases. Eur J Clin Microbiol Infect Dis 17:219–34.

McLeod, D.T., Ahmad, F., Capewell, S. *et al.* (1986) Increase in bronchopulmonary infection due to *Branhamella catarrhalis*. BMJ 292:1103–5.

McMichael, J.C. (2000) Progress toward the development of a vaccine to prevent *Moraxella* (*Branhamella*) *catarrhalis* infections. Microbes Infect 2:561–8.

McMichael, J.C., Fiske, M.J., Fredenburg, R.A. *et al.* (1998) Isolation and characterization of two proteins from *Moraxella catarrhalis* that bear a common epitope. Infect Immun 66:4374–81.

Meis, J.F.G.M., Sauerwein, R.W., Gyssens, I.C. *et al.* (1992) *Kingella kingae* intervertebral diskitis in an adult. Clin Infect Dis 15:530–2.

Minamoto, G.Y., and Sordillo, E.M. (1992) *Kingella denitrificans* as a cause of granulomatous disease in a patient with AIDS. Clin Infect Dis 15:1052–3.

Miravitlles, M., Espinosa, C., Fernandez-Laso, E. *et al.* (1999) Relationship between bacterial flora in sputum and functional impairment in patients with acute exacerbations of COPD. Study Group of Bacterial Infection in COPD. Chest 116:40–6.

Mollee, T., Kelly, P., and Tilse, M. (1992) Isolation of *Kingella kingae* from a corneal ulcer. J Clin Microbiol 30:2516–7.

Morgan, M.G., McKenzie, H., Enright, M.C. *et al.* (1992) Use of molecular methods to characterize *Moraxella catarrhalis* strains in a suspected outbreak of nosocomial infection. Eur J Clin Microbiol Infect Dis 11:305–12.

Moylett, E.H., Rossmann, S.N., Epps, H.R., and Demmler, G.J. (2000) Importance of *Kingella kingae* as a pediatric pathogen in the USA. Pediatr Infect Dis J 19:263–5.

Murphy, T.F. (1989) The surface of *Branhamella catarrhalis*: a systematic approach to the surface antigens of an emerging pathogen. Pediatr Infect Dis J 8:S75–7.

Murphy, T.F. (1990) Studies of the outer membrane proteins of *Branhamella catarrhalis*. Am J Med 88:41S–45S.

Murphy, T.F. (1996) *Branhamella catarrhalis*: epidemiology, surface antigenic structure, and immune response. Microbiol Rev 60:267–79.

Murphy, T.F., and Loeb, M.R. (1989) Isolation of the outer membrane of *Branhamella catarrhalis*. Microb Pathog 6:159–74.

Murphy, T.F., Kirkham, C., and Lesse, A.J. (1993) The major heat-modifiable outer membrane protein CD is highly conserved among strains of *Branhamella catarrhalis*. Mol Microbiol 10:87–97.

Murphy, T.F., Kyd, J.M., John, A. *et al.* (1998) Enhancement of pulmonary clearance of *Moraxella* (*Branhamella*) *catarrhalis* following immunization with outer membrane protein CD in a mouse model. J Infect Dis 178:1667–75.

Murphy, T.F., Brauer, A.L., Yuskiw, N., and Hiltke, T.J. (2000) Antigenic structure of outer membrane protein E of *Moraxella catarrhalis* and construction and characterization of mutants. Infect Immun 68:6250–6.

Namnyak, S.S., Quinn, R.J.M., and Ferguson, J.D.M. (1991) *Kingella kingae* meningitis in an infant. J Infect 23:104–6.

Nash, D.R., Wallace, R.J., Steingrube, V.A., and Shurin, P.A. (1986) Isoelectric focusing of beta-lactamases from sputum and middle ear isolates of *Branhamella catarrhalis* recovered in the United States. Drugs 31:48–54.

Neumayer, U., Schmidt, H.K., Mellwig, K.P., and Kleikamp, G. (1999) *Moraxella catarrhalis* endocarditis: report of a case and literature review. J Heart Valve Dis 8:114–7.

Nicotra, B., Rivera, M., Luman, J.I., and Wallace, R.J. (1986) *Branhamella catarrhalis* as a lower respiratory tract pathogen in patients with chronic lung disease. Arch Intern Med 146:890–3.

Ninane, G., Joly, J., Piot, P., and Kraytman, M. (1977) *Branhamella* (*Neisseria*) *catarrhalis* as pathogen. Lancet 2:149.

Odum, L., and Frederiksen, W. (1981) Identification and characterization of *Kingella kingae*. Acta Pathol Microbiol Scand 89:311–5.

Oishi, K., Tanaka, H., Sonoda, F. *et al.* (1996) A monoclonal antibody reactive with a common epitope of *Moraxella* (*Branhamella*) *catarrhalis* lipopolysaccharides. Clin Diagn Lab Immunol 3:351–4.

Onofrio, J.M., Shulkin, A.N., Heidbrink, P.J. *et al.* (1981) Pulmonary clearance and phagocytic cell response to normal pharyngeal flora. Am Rev Respir Dis 123:222–5.

Patel, J.A., Reisner, B., Vizirinia, N. *et al.* (1995) Bacteriologic failure of amoxicillin–clavulanate in treatment of acute otitis media caused by nontypeable *Haemophilus influenzae*. J Pediatr 126:799–806.

Patterson, T.F., Patterson, J.E., Masecar, B.L. *et al.* (1988) A nosocomial outbreak of *Branhamella catarrhalis* confirmed by restriction endonuclease analysis. J Infect Dis 157:996–1001.

Peak, I.R., Jennings, M.P., Hood, D.W. *et al.* (1996) Tetrameric repeat units associated with virulence factor phase variation in *Haemophilus* also occur in *Neisseria* spp. and *Moraxella catarrhalis*. FEMS Microbiol Lett 137:109–14.

Peiris, V., and Heald, J. (1992) Rapid method for differentiating strains of *Branhamella catarrhalis*. J Clin Pathol 45:532–4.

Pettersson, B., Kodjo, A., Ronaghi, M. *et al.* (1998) Phylogeny of the family Moraxellaceae by 16S rDNA sequence analysis, with special emphasis on differentiation of *Moraxella* species. Int J Syst Bacteriol 48:75–89.

Picket, M.J., Hollis, D.G., and Bottone, E.J. (1991) Miscellaneous gram-negative bacteria. In A. Balows (ed.), Manual of Clinical Microbiology, 5th ed. ASM, Washington, DC, pp. 410–28.

Pollard, J.A., Wallace, R.J., Jr., Nash, D.R. *et al.* (1986) Incidence of *Branhamella catarrhalis* in the sputa of patients with chronic lung disease. Drugs 31:103–8.

Post, J.C., Preston, R.A., Aul, J.J. *et al.* (1995) Molecular analysis of bacterial pathogens in otitis media with effusion. JAMA 273:1598–604.

Post, J.C., Aul, J.J., White, G.J. *et al.* (1996a) PCR-based detection of bacterial DNA after antimicrobial treatment is indicative of persistent, viable bacteria in the chinchilla model of otitis media. Am J Otolaryngol 17:106–11.

Post, J.C., White, G.J., Aul, J.J. *et al.* (1996b) Development and validation of a multiplex PCR-based assay for the upper respiratory tract bacterial pathogens *Haemophilus influenzae*, *Streptococcus pneumoniae*, and *Moraxella catarrhalis*. Mol Diagn 1:29–39.

Prellner, K., Christensen, P., Hovelius, B., and Rosen, C. (1984) Nasopharyngeal carriage of bacteria in otitis-prone and non-otitis-prone children in day-care centres. Acta Otolaryngol 98:343–50.

Rahman, M., Holme, T., Jonsson, I., and Krook, A. (1995) Lack of serotype-specific antibody response to lipopolysaccharide antigens of *Moraxella catarrhalis* during lower respiratory tract infection. Eur J Clin Microbiol Infect Dis 14:297–304.

Reddy, M.S., Murphy, T.F., Faden, H.S., and Bernstein, J.M. (1997) Middle ear mucin glycoprotein: purification and interaction with nontypable *Haemophilus influenzae* and *Moraxella catarrhalis*. Otolaryngol Head Neck Surg 116:175–80.

Reekmans, A., Noppen, M., Naessens, A., and Vincken, W. (2000) A rare manifestation of *Kingella kingae* infection. Eur J Intern Med 11:343–4.

Richards, S.J., Greening, A.P., Enright, M.C. *et al.* (1993) Outbreak of *Moraxella catarrhalis* in a respiratory unit. Thorax 48:91–2.

Rikitomi, N., Ahmed, K., and Nagatake, T. (1997) *Moraxella* (*Branhamella*) *catarrhalis* adherence to human bronchial and oropharyngeal cells: the role of adherence in lower respiratory tract infections. Microbiol Immunol 41:487–94.

Roiz, M.P., Peralta, F.G., and Arjona, R. (1997) *Kingella kingae* bacteremia in an immunocompetent adult. J Clin Microbiol 35:1916.

Rolle, U., Schille, R., Hormann, D. *et al.* (2001) Soft tissue infection caused by *Kingella kingae* in a child. J Pediatr Surg 36:946–7.

Samukawa, T., Yamanaka, N., Hollingshead, S. *et al.* (2000a) Immune responses to specific antigens of *Streptococcus pneumoniae* and *Moraxella catarrhalis* in the respiratory tract. Infect Immun 68:1569–73.

Samukawa, T., Yamanaka, N., Hollingshead, S. *et al.* (2000b) Immune response to surface protein A of *Streptococcus pneumoniae* and to high-molecular-weight outer membrane protein A of *Moraxella catarrhalis* in children with acute otitis media. J Infect Dis 181:1842–5.

Sarda, H., Ghazali, D., Thibault, M. *et al.* (1998) Multifocal invasive *Kingella kingae* infection. Arch Pediatr 5:159–62.

Sarubbi, F.A., Myers, J.W., Williams, J.J., and Shell, C.G. (1990) Respiratory infections caused by *Branhamella catarrhalis*. Selected epidemiologic features. Am J Med 88:9S–14S.

Schalen, L., Christensen, P., Kamme, C. *et al.* (1980) High isolation rate of *Branhamella catarrhalis* from the nasopharynx in adults with acute laryngitis. Scand J Infect Dis 12:277–80.

Schalen, L., Christensen, P., Eliasson, I. *et al.* (1985) Inefficacy of penicillin V in acute laryngitis in adults. Evaluation from results of double-blind study. Ann Otol Rhinol Laryngol 94:14–7.

Sethi, S., Hill, S.L., and Murphy, T.F. (1995) Serum antibodies to outer membrane proteins (OMPs) of *Moraxella* (*Branhamella*) *catarrhalis* in patients with bronchiectasis: identification of OMP B1 as an important antigen. Infect Immun 63:1516–20.

Sethi, S., Surface, J.M., and Murphy, T.F. (1997) Antigenic heterogeneity and molecular analysis of CopB of *Moraxella* (*Branhamella*) *catarrhalis*. Infect Immun 65:3666–71.

Shurin, P.A., Marchant, C.D., Kim, C.H. *et al.* (1983) Emergence of beta-lactamase-producing strains of *Branhamella catarrhalis* as important agents of acute otitis media. Pediatr Infect Dis J 2:34–8.

Singh, S., Cisera, K.M., Turnidge, J.D., and Russell, E.G. (1997) Selection of optimum laboratory tests for the identification of *Moraxella catarrhalis*. Pathology 29:206–8.

Slonim, A., Walker, E.S., Mishori, E. *et al.* (1998) Person-to-person transmission of *Kingella kingae* among day care center attendees. J Infect Dis 178:1843–6.

Snell, J.J.S., and Lapage, S.P. (1976) Transfer of some saccharolytic *Moraxella* species to *Kingella* Hendriksen and Bøvre 1976, with descriptions of *Kingella indologenes*, sp. nov. and *Kingella denitrificans* sp. nov. Int J Syst Bacteriol 26:451–8.

Sordillo, E.M., Rendel, M., Sood, R. *et al.* (1993) Septicemia due to beta-lactamase positive *Kingella kingae*. Clin Infect Dis 17:818–9.

Speeleveld, E., Fossepre, J.M., Gordts, B., and Van Landuyt, H.W. (1994) Comparison of three rapid methods, tributyrine, 4-methylumbelliferyl butyrate, and indoxyl acetate, for rapid identification of *Moraxella catarrhalis*. J Clin Microbiol 32:1362–3.

Sportel, J.H., Koeter, G.H., van Altena, R. *et al.* (1995) Relation between beta-lactamase producing bacteria and patient characteristics in chronic obstructive pulmonary disease (COPD). Thorax 50:249–53.

Stahelin, J., Goldenberger, D., Gnehm, H.E., and Altwegg, M. (1998) Polymerase chain reaction diagnosis of *Kingella kingae* arthritis in a young child. Clin Infect Dis 27:1328–9.

Stefanou, J., Agelopoulou, A.V., Sipsas, N.V. *et al.* (2000) *Moraxella catarrhalis* endocarditis: case report and review of the literature. Scand J Infect Dis 32:217–8.

Takada, R., Harabuchi, Y., Himi, T., and Kataura, A. (1998) Antibodies specific to outer membrane antigens of *Moraxella catarrhalis* in sera and middle ear effusions from children with otitis media with effusion. Int J Pediatr Otorhinolaryngol 46:185–95.

Taylor, P.W. (1995) Resistance of bacteria to complement. In J.A. Roth, C.A. Bolin, K.A. Brogden, C. Minion, and M.J. Wannemueller (eds), Virulence of Bacterial Pathogens, 2nd ed. American Society for Microbiology, Washington, DC, pp. 49–64.

Thornsberry, C., Jones, M.E., Hickey, M.L. *et al.* (1999) Resistance surveillance of *Streptococcus pneumoniae*, *Haemophilus influenzae* and *Moraxella catarrhalis* isolated in the United States, 1997–1998. J Antimicrob Chaemother 44:749–59.

Thorsson, B., Haraldsdottir, V., and Kristjansson, M. (1998) *Moraxella catarrhalis* bacteraemia. A report on 3 cases and a review of the literature. Scand J Infect Dis 30:105–9.

Toews, G.B., Hansen, E.J., and Strieter, R.M. (1990) Pulmonary host defenses and oropharyngeal pathogens. Am J Med 88:20S–4S.

Tonjum, T., Bukholm, G., and Bovre, K. (1989) Differentiation of some species of Neisseriaceae and other bacterial groups by DNA–DNA hybridization. APMIS 97:395–405.

Unhanand, M., Maciver, I., Ramilo, O. *et al.* (1992) Pulmonary clearance of *Moraxella catarrhalis* in an animal model. J Infect Dis 165:644–50.

Urs, S., D'Silva, B.S., Jeena, C.P. *et al.* (1994) *Kingella kingae* septicemia in association with HIV disease. Trop Doct 24:127.

Van Hare, G.F., and Shurin, P.A. (1987) The increasing importance of *Branhamella catarrhalis* in respiratory infections. Pediatr Infect Dis J 6:92–4.

Van Hare, G.F., Shurin, P.A., Marchant, C.D. *et al.* (1987) Acute otitis media caused by *Branhamella catarrhalis*: biology and therapy. Rev Infect Dis 9:16–27.

Vaneechoutte, M., Verschraegen, G., Claeys, G., and van den Abeele, A.M. (1988) Selective medium for *Branhamella catarrhalis* with acetazolamide as a specific inhibitor of *Neisseria* spp. J Clin Microbiol 26:2544–8.

Vaneechoutte, M., Verschraegen, G., Claeys, G., and Van Den Abeele, A.M. (1990) Serological typing of *Branhamella catarrhalis* strains on the basis of lipopolysaccharide antigens. J Clin Microbiol 28:182–7.

Varon, E., Levy, C., De La Rocque, F. *et al.* (2000) Impact of antimicrobial therapy on nasopharyngeal carriage of *Streptococcus pneumoniae*, *Haemophilus influenzae*, and *Branhamella catarrhalis* in children with respiratory tract infections. Clin Infect Dis 31:477–81.

Verduin, C.M., Jansze, M., Hol, C. *et al.* (1994a) Differences in complement activation between complement-resistant and complement-sensitive *Moraxella* (*Branhamella*) *catarrhalis* strains occur at the level of membrane attack complex formation. Infect Immun 62:589–95.

Verduin, C.M., Jansze, M., Verhoef, J. *et al.* (1994b) Complement resistance in *Moraxella* (*Branhamella*) *catarrhalis* is mediated by a vitronectin-binding surface protein. Clin Exp Immunol 97:50.

Verduin, C.M., Kools-Sijmons, M., van der Plas, J. *et al.* (2000) Complement-resistant *Moraxella catarrhalis* forms a genetically distinct lineage within the species. FEMS Microbiol Lett 184:1–8.

Verghese, A., Berro, E., Berro, J., and Franzus, B.W. (1990) Pulmonary clearance and phagocytic cell response in a murine model of *Branhamella catarrhalis* infection. J Infect Dis 162:1189–92.

Vu-Thien, H., Dulot, C., Moissenet, D. *et al.* (1999) Comparison of randomly amplified polymorphic DNA analysis and pulsed-field gel electrophoresis for typing of *Moraxella catarrhalis* strains. J Clin Microbiol 37:450–2.

Waghorn, D.J., and Cheetham, C.H. (1997) *Kingella kingae* endocarditis following chickenpox in infancy. Eur J Clin Microbiol Infect Dis 16:944–6.

Wald, E.R. (1992a) Microbiology of acute and chronic sinusitis in children. J Allergy Clin Immunol 90:452–6.

Wald, E.R. (1992b) Sinusitis in children. N Engl J Med 326:319–23.

Wald, E.R. (1998a) Sinusitis. Pediatr Ann 27:811–8.

Wald, E.R. (1998b) Sinusitis overview. Pediatr Ann 27:787–8.

Walker, E.S., Preston, R.A., Post, J.C. *et al.* (1998) Genetic diversity among strains of *Moraxella catarrhalis*: analysis using multiple DNA probes and a single-locus PCR-restriction fragment length polymorphism method. J Clin Microbiol 36:1977–83.

Walker, E.S., Neal, C.L., Laffan, E. *et al.* (2000) Long-term trends in susceptibility of *Moraxella catarrhalis*: a population analysis. J Antimicrob Chaemother 45:175–82.

Wallace, P.L., Hollis, D.G., Weaver, R.E., and Moss, C.W. (1988) Cellular fatty acid composition of *Kingella* species, *Cardiobacterium hominis*, and *Eikenella corrodens*. J Clin Microbiol 26:1592–4.

Wallace, R.J., Steingrube, V.A., Nash, D.R. *et al.* (1989) BRO beta-lactamases of *Branhamella catarrhalis* and *Moraxella* subgenus *Moraxella*, including evidence for chromosomal beta-lactamase transfer by conjugation in *B. catarrhalis*, *M. nonliquefaciens*, and *M. lacunata*. Antimicrob Agents Chaemother 33:1845–54.

Wallace, R.J., Jr., Nash, D.R., and Steingrube, V.A. (1990) Antibiotic susceptibilities and drug resistance in *Moraxella* (*Branhamella*) *catarrhalis*. Am J Med 88:46S–50S.

Wardle, J.K. (1986) *Branhamella catarrhalis* as an indirect pathogen. Drugs 31:93–6.

Weir, S., and Marrs, C.F. (1992) Identification of type 4 pili in *Kingella denitrificans*. Infect Immun 60:3437–41.

Weir, S., Lee, L.W., and Marrs, C.F. (1996) Identification of four complete type 4 pilin genes in a single *Kingella denitrificans* genome. Infect Immun 64:4993–9.

Weir, S., Lee, L.W., and Marrs, C.F. (1997) Type-4 pili of *Kingella denitrificans*. Gene 192:171–6.

Weiss, A., Brinser, J.H., and Nazar-Stewart, V. (1993) Acute conjunctivitis in childhood. J Pediatr 122:10–4.

Wells, L., Rutter, N., and Donald, F. (2001) *Kingella kingae* endocarditis in a sixteen month old child. Pediatr Infect Dis J 20:454–5.

Westman, E., Melhus, A., Hellstrom, S., and Hermansson, A. (1999) *Moraxella catarrhalis*-induced purulent otitis media in the rat middle ear. Structure, protection, and serum antibodies. APMIS 107:737–46.

Wildt, and Boas, M. (2001) *Kingella kingae* osteomyelitis. Ugeskr Laeger 163:6287–8.

Wolak, T., Abu-Shakra, M., Flusser, D. *et al.* (2000) *Kingella* endocarditis and meningitis in a patient with SLE and associated antiphospholipid syndrome. Lupus 9:393–6.

Wong, V.K., and Mason, W.H. (1987) *Branhamella catarrhalis* as a cause of bacterial tracheitis. Pediatr Infect Dis J 6:945–6.

Wood, G.M., Johnson, B.C., and McCormack, J.G. (1996) *Moraxella catarrhalis*: pathogenic significance in respiratory tract infections treated by community practitioners. Clin Infect Dis 22:632–6.

Wright, P.W., and Wallace, R.J. (1989) Pneumonia due to *Moraxella* (*Branhamella*) *catarrhalis*. Semin Respir Infect 4:40–6.

Yagupsky, P. (1999a) Diagnosis of *Kingella kingae* arthritis by polymerase chain reaction analysis. Clin Infect Dis 29:704–5.

Yagupsky, P. (1999b) Use of blood culture systems for isolation of *Kingella kingae* from synovial fluid. J Clin Microbiol 37:3785.

Yagupsky, P., and Dagan, R. (1994) *Kingella kingae* bacteremia in children. Pediatr Infect Dis J 13:1148–9.

Yagupsky, P., and Dagan, R. (1997) *Kingella kingae*: an emerging cause of invasive infections in young children. Clin Infect Dis 24:860–6.

Yagupsky, P., Dagan, R., Howard, C.W. *et al.* (1992) High prevalence of *Kingella kingae* in joint fluid from children with septic arthritis revealed by the BACTEC blood culture system. J Clin Microbiol 30:1278–81.

Yagupsky, P., Dagan, R., Howard, C.B. *et al.* (1993) Clinical features and epidemiology of invasive *Kingella kingae* infections in southern Israel. Pediatrics 92:800–4.

Yagupsky, P., Dagan, R., Prajgrod, F., and Merires, M. (1995a) Respiratory carriage of *Kingella kingae* among healthy children. Pediatr Infect Dis J 14:673–8.

Yagupsky, P., Merires, M., Bahar, J., and Dagan, R. (1995b) Evaluation of novel vancomycin-containing medium for primary isolation of *Kingella kingae* from upper respiratory tract specimens. J Clin Microbiol 33:1426–7.

Yagupsky, P., Katz, O., and Peled, N. (2001) Antibiotic susceptibility of *Kingella kingae* isolates from respiratory carriers and patients with invasive infections. J Antimicrob Chaemother 47:191–3.

Yagupsky, P., Peled, N., and Katz, O. (2002) Epidemiological features of invasive *Kingella kingae* infections and respiratory carriage of the organism. J Clin Microbiol 40:4180–4.

Yang, Y.P., Myers, L.E., McGuinness, U. *et al.* (1997) The major outer membrane protein, CD, extracted from *Moraxella* (*Branhamella*) *catarrhalis* is a potential vaccine antigen that induces bactericidal antibodies. FEMS Immunol Med Microbiol 17:187–99.

Yano, H., Suetake, M., Kuga, A. *et al.* (2000) Pulsed-field gel electrophoresis analysis of nasopharyngeal flora in children attending a day care center. J Clin Microbiol 38:625–9.

Yu, R.H., Bonnah, R.A., Ainsworth, S., and Schryvers, A.B. (1999) Analysis of the immunological responses to transferrin and lactoferrin receptor proteins from *Moraxella catarrhalis*. Infect Immun 67:3793–9.

Zaleski, A., Scheffler, N.K., Densen, P. *et al.* (2000) Lipooligosaccharide P(k) (Galalpha 1–4Galbeta 1–4Glc) epitope of *Moraxella catarrhalis* is a factor in resistance to bactericidal activity mediated by normal human serum. Infect Immun 68:5261–8.

14

Neisseria meningitidis

Dlawer A. A. Ala'Aldeen and David P. J. Turner

Molecular Bacteriology and Immunology Group, Division of Microbiology, University Hospital of Nottingham, Nottingham, UK

INTRODUCTION

Neisseria meningitidis and *N. gonorrhoeae* are the only two recognized obligate human pathogens of the genus *Neisseria*, family Neisseriaceae. *Neisseria* meningitidis (meningococcus) is the most common overall cause of pyogenic meningitis worldwide and is the only bacterium that is capable of generating epidemic outbreaks of meningitis. Vieusseaux was the first to describe an outbreak of an apparently new disease, a cerebrospinal fever, which spread rapidly in and around Geneva in the spring of 1805. Several other outbreaks of similar nature were also recorded over the subsequent decades, but the organism was first discovered in Vienna in 1887 by Anton Weichselbaum. He was able to isolate the causative organism in the meningeal exudate of six cases of cerebrospinal fever and gave it the descriptive name of *Diplococcus intracellular meningitidis*, which was later changed to *Neisseria meningitidis* after the German scientist and clinician, Albert Neisser.

DESCRIPTION OF THE ORGANISM

Cultural Characteristics and Biochemical Reactions

Neisseria meningitidis is a Gram-negative, non-sporing, aflagellate, aerobic diplococcus of approximately 0.8 μm in diameter. On solid media, *N. meningitidis* grows as a transparent, non-haemolytic, non-pigmented (grey) convex colony, approximately 0.5–5 mm in diameter, depending on the length of incubation. The organism is relatively fastidious in its growth requirements, and optimal growth conditions are achieved at 35–37 °C, at pH 7.0–7.4 in a moist environment with 5–10% CO_2. It grows poorly on unenriched media but grows reasonably well on blood, chocolate and Modified New York City agar, and on Mueller Hinton agar without the addition of blood. It can survive and grow slowly at temperatures ranging from 25 to 42 °C. It is best transported on chocolate agar slopes and stored freeze-dried, frozen at −70 °C or in liquid nitrogen.

The organism is oxidase and catalase positive. It is capable of utilizing glucose with no gas formation and, unlike gonococci, it also utilizes maltose, although occasional strains are maltose negative on primary isolation. It does not utilize lactose, sucrose or fructose. Meningococci produce γ-glutamyl aminopeptidase, but not prolyl aminopeptidase or β-galactosidase.

Genome Sequence and Analysis

The genomic sequence of two meningococcal strains (serogroup A and B, respectively) were described in 2000 (Parkhill *et al.*, 2000;

Tettelin *et al.*, 2000). Sequencing of a third (serogroup C) strain has also been completed, but not yet published (http://www.sanger.ac.uk/ Projects/N_meningitidis/seroC.shtml). The genome is approximately 2.2 Mb in length and contains just over 2000 predicted coding regions. The overall G + C content is about 52%, but many coding regions possess a significantly lower G + C content, suggesting that the meningococcus has relatively recently acquired DNA from other organisms (horizontal exchange). Indeed, a number of genes were identified which are homologous to known virulence factors in other organisms. Comparison of the meningococcal and gonococcal genomic sequences shows that the two species are more than 90% homologous at the protein-coding level. One of the most notable and unique features of the genome is the abundance and diversity of repetitive DNA elements. These range from the short (10 bp) neisserial uptake sequence, which is involved in the recognition and uptake of DNA from the environment, to large gene duplications and prophage sequences up to 39 kb in length. Over 250 Correia elements, 156-bp sequences bounded by 26-bp inverted repeats, are present in the meningococcal genome. The function of these sequences, which flank many important virulence-associated genes, is unclear, although a role in genome plasticity has been suggested.

Another key finding derived from the genome sequence data is the presence of more than 60 putative phase-variable (contingency) genes (Snyder, Butcher and Saunders, 2001). The phase-variable nature of these genes is predicted on the basis of the presence of homopolymeric tracts or simple repeat sequences (Henderson, Owen and Nataro, 1999). The length of the homopolymeric tracts or repeat sequences can change frequently as a result of slipped-strand mispairing during replication. This process brings the ATG initiation codon into or out of frame with the remainder of the gene, thus activating or inactivating the gene. Many of the phase-variable genes, for example, the capsule biosynthesis operon, are known to be important virulence determinants. The presence of many phase-variable genes may enhance the ability of the organism to adapt rapidly to new environments within the human host.

Neisseria meningitidis possesses specific DNA sequences that are absent from other *Neisseria* species (Perrin *et al.*, 2002). Regions of DNA that are common to the meningococcus and gonococcus, but not present in the commensal species *N. lactamica*, have also been identified (Perrin, Nassif and Tinsley, 1999). The latter sequences may determine aspects of the life cycle common to the pathogenic *Neisseria* species, such as mucosal colonization and epithelial cell invasion, whereas the former may play a role in the pathogenesis of meningococcal disease – particularly haematogenous dissemination and crossing of the blood-cerebrospinal fluid (B–CSF) barrier. The meningococcal-specific sequences reside in eight chromosomal regions, which range in size from 1.8 to 40 kb (Klee *et al.*, 2000). Five of these regions were shown to be conserved in a representative set of strains and/or carried genes with homologies to previously described virulence factors in other species. For example, one conserved region contained genes homologous to the filamentous haemagglutinin precursor (FhaB) and its accessory protein (FhaC) of *Bordetella pertussis*.

GENERAL EPIDEMIOLOGY

Compulsory notification of meningococcal disease in the UK, imposed in 1912, helped to demonstrate the rise and fall of epidemics. One large peak in reported cases (a few thousand) occurred during the First World War and an even higher one (exceeding 12 000 cases) coincided with the Second World War. Currently, we are experiencing a worldwide epidemic with a clear increase in the number of cases reported in the past two decades. The occurrence of cases is generally unpredictable, but occasional clustering occurs in various geographical areas, particularly during the periods of increased disease activity. The baseline incidence rate between epidemics in the industrialized countries is between one and three cases per 10^5 population compared to 10–25 per 10^5 in underdeveloped countries. Attack rates are highest in those aged under 5 years, and in winter and springtime, possibly due to the prevalence of viral respiratory infections which are thought to act as predisposing factors. In the UK notification of the disease has more than doubled since the early eighties. Until recently, between 2000 and 2500 cases of invasive meningococcal disease were reported annually in England and Wales. Fifty–sixty per cent of cases occur in children less than 5 years of age, of which around 40% are usually in children aged less than 1 year. The highest incidence of cases occurs in the month of January and the lowest in August and September.

In third-world countries more than 310 000 people are thought to suffer from meningococcal diseases annually, of which 35 000 cases would be fatal (Robbins and Freeman, 1988). Over 300 million people live in countries within the meningitis belt of Savannah Africa, which is a vast area extending from Ethiopia in the east to The Gambia in the west and from the Sahara in the north to the tropical rain forest of Central Africa in the south. This belt has expanded since it was first described in 1963 (Lapeyssonie, 1963), and it now includes 15 countries, namely Ethiopia, Sudan, Central African Republic, Chad, Cameroon, Nigeria, Niger, Benin, Togo, Ghana, Burkina Faso, Mali, Guinea, Senegal and The Gambia (Greenwood, 1987). In 1988 and 1989, 80% of meningococcal isolates in the African continent occurred in these 15 countries, and in 1989 more than half the reported cases were from Ethiopia (Riedo, Plikaytis and Broome, 1995). During epidemics, attack rates in some countries of the meningitis belt reach as high as 1000 cases per 10^5 population which contrasts with 5–25 cases per 10^5 in the industrialized countries (Rey, 1991). In many regions of the meningitis belt, epidemics occur in cycles, with intervals between epidemics in the majority of cases less than 12 years (ranging from 2 to 25 years) (Riedo, Plikaytis and Broome, 1995). Epidemics last for approximately 2–4 years (Lapeyssonie, 1963), but in contrast to other areas, cases occur mainly in the hot dry months, possibly due to the influence of dry air on the integrity of the mucosal membranes of the nasopharynx. Major global outbreaks associated with the Hajj pilgrimage have occurred in recent years. Some northern African countries (like Egypt) also experience fluctuating epidemics. Large-scale epidemics have also occurred in many countries of Asia (e.g., Pakistan, India and China) and Central and South America (e.g., Cuba, Chile and Brazil). In Cuba, the attack rate reached levels of more than 50 cases per 10^5 population (Sierra *et al.*, 1991).

Age and Sex

Age-specific rates for meningococcal infection in England and Wales show a sudden rise in the attack rate among infants at around 3 months of age (coinciding with the decline in maternally acquired antibody), which peaks at the age of 6 months (more than 50 cases per 10^5) and remains relatively high for the first 12–24 months of life (Jones,

1995). The attack rate falls dramatically to 10 cases per 10^5 at the age of 2 years and continues to decline until it reaches the overall adult rate of 0.4 cases per 10^5, only to produce a second but smaller peak (less than five cases per 10^5) among the 17–18 year olds. Boys are probably at a slightly higher risk of invasive disease in the first few years of life, whereas among the teenagers this is reversed. It is interesting that during epidemics the average age of patients is older than between epidemics.

Carriage

As an obligate human pathogen, the natural habitat of the meningococcus is the human nasopharynx. Although seasonal variation, temperature and humidity seem to have a clear impact on invasive meningococcal disease, they have very little, if any, effect on the prevalence of carriage. It is interesting that the highest attack rate of invasive meningococcal disease in Europe and United States is in the first year of life, whereas the highest carriage rate is found among teenagers and young adults. While it is difficult to estimate the true carriage rate in different groups in any community at any one time, it is known that several factors may influence carriage in the population. These include factors related to the organism, the host and the environment. Some serogroups of *N. meningitidis*, such as 29E, are regarded of low pathogenicity and are isolated from carriers, whereas serogroups A, B and C are responsible for over 90% of the invasive meningococcal infections worldwide. Carriage is more common in the second and third decades of life, and more common among smokers than non-smokers. The body's immune status, changes in the oropharyngeal flora and concurrent viral upper respiratory tract infections are also believed to influence carriage. Viral infections, particularly influenza A infection, are increasingly blamed for predisposing individuals to meningococcal carriage and invasive disease (Cartwright *et al.*, 1991; Makras *et al.*, 2001). Carriage rates are known to be much higher among family members and close contacts of infected patients, and in closed communities with overcrowded conditions, such as military recruits' training camps, prisons and schools. Baseline carriage rates of 5–15% are considered usual; however, these can fluctuate and reach much higher levels in certain communities.

Carriage rate is shown to be low (<3%) in children under the age of 4 years but can increase up to tenfold among teenagers, young adults and enclosed populations such as military recruits (Cartwright *et al.*, 1987; Blackwell *et al.*, 1990; Caugant *et al.*, 1992; Caugant *et al.*, 1994; Tyski *et al.*, 2001). Carriage among adults above the age of 25 years is usually below 10%.

It is difficult to obtain an estimate of the actual duration of meningococcal carriage, but it has been shown to be as short as several days to as long as several months (Ala'Aldeen *et al.*, 2000; Neal *et al.*, 2000). It is important to emphasize that carriage is a dynamic process that is influenced by numerous factors. In the 1997–1998 academic year, a longitudinal study was carried out on the dynamics of meningococcal carriage and acquisition among first-year students at Nottingham University, UK. Pharyngeal swabs were obtained from 2453 first-year students at the start of the academic year (October), later on during the autumn term and again in March. During freshers week (the first week of term), the carriage rate rose rapidly from 6.9% on day 1 to 11.2% on day 2 and 19% on day 3 to 23.1% on day 4 (Ala'Aldeen *et al.*, 2000; Neal *et al.*, 2000). The average carriage rate during the first week was 12.9%, by November it was 31% and in December it had reached 34.2%. In March the rate was 28%. Of the students who arrived at the University as meningococcal carriers, 44.1% were still positive later on in the autumn term and 57.1% of these had remained persistent carriers at 6 months. These carried the same or different strains at different times. The study revealed a high rate of turnover of meningococcal carriage among students. Of the index carriers who lost carriage during the autumn, 16% were recolonized at 6 months. Of the 344 index non-carriers followed up, 22.1% acquired carriage during the autumn term and another 13.7% by March.

Patients with invasive disease may carry the organisms for days or weeks before becoming ill. Carriage of a particular meningococcal strain does not necessarily protect against colonization or invasion with a homologous or heterologous strain. In the Nottingham study, one student developed invasive meningococcal disease after carrying the same organism for over 7 weeks.

THE CELL ENVELOPE: ANATOMY, HETEROGENEITY AND STRAIN CLASSIFICATION

The Cell Envelope

The meningococcal cell is contained within a permeable polysaccharide capsule which overlays a multilayered envelope (Figure 14.1), very similar to that of *N. gonorrhoeae* and other Gram-negative bacteria. This envelope consists of three layers: an outer membrane layer, which contains proteins, lipooligosaccharides (LOSs) and phospholipids; a thin but rigid peptidoglycan layer; an inner cytoplasmic membrane which contains, among other things, enzymatic components of the electron transport chain and oxidative phosphorylation. Pili, long filamentous protein projections, also extend outwards from the cell envelope. The cell envelope has been extensively studied for pathogenesis, classification, immune response and vaccine development purposes. Many of the prominent components of the cell envelope, particularly the outer membrane components, show wide structural and antigenic diversity which has been exploited for epidemiological purposes in classifying clinical isolates.

Classification Systems of Clinical Isolates

For epidemiological purposes, several classification systems have been developed for *N. meningitidis*. The most widely used and well established is the one that divides strains immunologically into serogroups, serotypes, serosubtypes and immunotypes, based on antigenic differences in their capsules, PorB protein, PorA protein and LOSs, respectively. For example, a serogroup B serotype 4, serosubtype 15 and immunotype 10 is written as B:4:P1.15:L10.

Other classification tools have also been used including multilocus enzyme electrophoresis (MLEE) and multilocus sequence typing (MLST). The MLEE typing (electrotyping, ET) scheme is based on the presence and electrophoretic mobility of iso-forms of cytosolic enzymes whose molecular weights vary between different strains of meningococci. MLEE has enabled subdivision of genetically related strains of serogroup A to subgroups and serogroups B and C to lineages and clonal complexes (Achtman, 1994). Using this method, it was possible to trace the movement of a certain clone (clone III-I) of serogroup A meningococci from China to Nepal in 1983, to India and Pakistan in 1985, to Mecca in 1987 and to the meningitis belt of Africa and other parts of the world by 1989 (Caugant *et al.*, 1987; Moore *et al.*, 1992). Furthermore, using MLEE it was possible to establish that meningococcal populations are genetically highly diverse. However, only a few of the clonal complexes appear to be responsible for most meningococcal disease. For example, strains belonging to ET-5 and ET-37 clonal complexes dominated the disease-causing serogroup B and C strains, respectively, in the UK and the rest of Europe. These dominant complexes are, however,

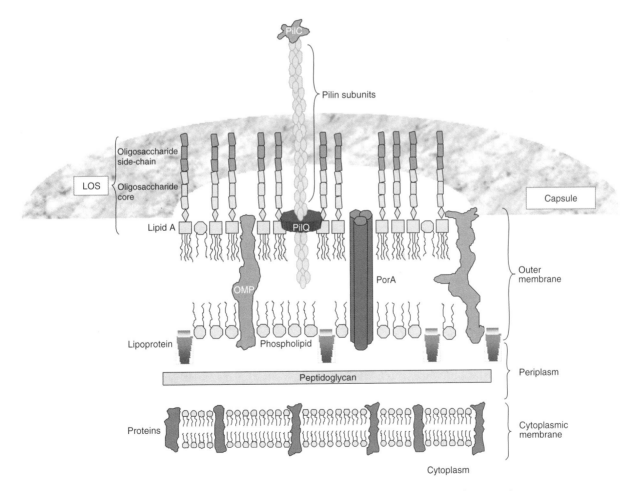

Figure 14.1 Schematic diagram of the meningococcal cell envelope. LOS, lipooligosaccharide; OMP, outer membrane protein.

changing slowly with time, and an increase in the incidence of disease appears to coincide with the introduction of a new clonal complex to a particular community.

With the advent of DNA sequencing it became possible to develop the more precise and less cumbersome MLST scheme (Maiden et al., 1998). In this typing scheme, relatively small DNA fragments from seven meningococcal housekeeping genes are amplified by PCR and sequenced. These genes include abcZ, adk, aroE, fumC, gdh, pdhC and pgm. Sequences of these genes from diverse meningococcal strains representing various clonal complexes are then aligned. Alleles of each gene are assigned numbers, and strains with particular combinations of allelic mixes of all the seven genes are then allocated a sequence type (ST). MLST has a greater discriminatory power than MLEE, otherwise the two methods serve the same purpose.

In addition to typing schemes, there are numerous comparative fingerprinting schemes, some of which can be converted to typing schemes too. Pulse-field gel electrophoresis (PFGE), restriction fragment length polymorphism (RFLP), fluourescent-labelled amplified fragment length polymorphism (fAFLP) (Goulding et al., 2003) and other PCR-based methods have all been used.

Capsular Polysaccharides: the Serogroup Antigens

The meningococcus possesses a phase-variable polysaccharide capsule, as do many non-pathogenic species of Neisseria. The structural and antigenic heterogeneity of the capsular polysaccharide forms the basis of the classification into at least 13 serogroups, namely A, B, C, D, 29E, H, I, J, K, L, W135, X, Y and Z. The serogroup A capsule is composed of mannosamine phosphate, while that of serogroups B, C, Y and W135 contains sialic acid, which is believed to be important in survival and virulence. The capsular polysaccharides of serogroup B and C meningococci are composed of homopolymers of N-acetylneuraminic acid with α-2,8 and α-2,9 linkages, respectively. This minor difference in structure leads to markedly different immunological properties: whereas the latter is immunogenic in humans and generates protective antibodies, the former is poorly immunogenic.

Serogroup A meningococci are now rare in the more developed countries but are the major pathogens in the meningitis belt of Africa and a number of Asian countries, where they are responsible for the epidemics. In the UK, serogroup A was responsible for the epidemics in the first half of the century, whereas in the second half this serogroup became increasingly uncommon and was replaced by serogroup B (up to 70%) and serogroup C (20–30%) with occasional isolates of serogroups A, W135, X, Y and 29E strains. Since the introduction of serogroup C-conjugate vaccine into the childhood immunization programme, this serogroup is now gradually becoming uncommon in the UK. In Europe and the American continent, serogroups B and C are currently responsible for about 85% of the cases, with serogroups Y and W135 accounting for most of the remainder. Serogroups H, I and K have been reported in China. In the United States, until the late 1980s serogroup B meningococci accounted for 50% of the cases and serogroup C for 35%. However, in 1989–1991, the two serogroups were almost equally distributed (46 and 45%, respectively) (Jackson and Wenger, 1993). In Norway in 1975, less than 50% of isolates were serogroup B strains but by 1981 these had reached 89.1%, only to drop again to 71.5% by 1990 (Lystad and Aasen, 1991). Serogroup A strains, however, were responsible for 24% of the cases in Norway in 1975 but by 1990 they had almost disappeared. Serogroup B meningococci also dominate in the current epidemics in Central and South America and in South Australia and New Zealand.

With increased international travel, global dissemination of outbreak-associated strains is common. Typical examples are the international epidemics of serogroup A and W135 meningococcal disease that occurred in association with the Hajj pilgrimage in the late 1980s and 1990s, respectively. The infected pilgrims caused numerous secondary infections in their home countries in different parts of the world, such as

Europe, North America and Asia. Similarly, the intercontinental spread of a particular clone of serogroup B:15:P1.16 (clone ET-5), which was responsible for outbreaks in Norway, Iceland, Denmark, the Netherlands and UK, has also been documented as the cause of outbreaks in South Africa and South America (Caugant et al., 1987).

Until recently, non-groupable strains were considered non-pathogenic. It is now clear that expression of the capsular is phase variable (Snyder, Butcher and Saunders, 2001). This means that all capsulated strains are capable of switching the capsule on and off, unless they loose some of the essential genetic components of the capsule biosynthesis pathways. Temporary loss of capsular expression may allow better affinity between bacterial and host surface molecules and facilitate colonization. Capsule expression appears to be an essential requirement for bacterial survival following invasion of various body compartments, e.g. blood and CSF, and may also be required for detachment and transmission from one host to another.

The PorB and PorA Outer Membrane Proteins: the Serotype and Serosubtype Antigens

The meningococcal outer membrane contains a number of major outer membrane proteins (OMPs). PorB (previously named class 2/3 protein) and PorA (class 1 protein) function as porins, allowing entry of nutrients. Like most surface-exposed meningococcal antigens, they exhibit a large degree of antigenic diversity between different strains, hence their use for the serotype and serosubtype classification of meningococci. Strains of N. meningitidis are subdivided into over 20 different serotypes and subtypes, designated by numbers. However, many isolates cannot be typed or subtyped with the available monoclonal antibodies, and these are designated non-typable and non-subtypable, respectively.

The antigenic structure of serotype and serosubtype antigens changes rapidly in vivo in an individual person and over time in the community. The surface-exposed variable loops of both porins can change antigenicity by insertion or deletion of amino acids or by horizontal transfer of subfragments of their respective genes. These mechanisms will make these genes elusive targets for vaccine development.

Lipooligosaccharide (Endotoxin)

Meningococcal LOS (also called endotoxin) consists of an outer membrane-bound glycolipid which differs from the lipopolysaccharides (LPS) of the Enterobacteriaceae by lacking the repetitive O-side-chains and thus resembles the rough form of LPS (Griffiss et al., 1988). It consists of three main components, namely an oligosaccharide core, a hydrophobic lipid A (the toxic component) and a variable oligosaccharide side-chain (the immunogenic component). Nine glycosyltransferase (lgt) genes, located at three chromosomal loci, are responsible for the biosynthesis of the oligosaccharide side-chains. The structural and immunogenic heterogeneity of the oligosaccharide side-chains, partly due to the phase-variable expression of four of the glycosyltransferases (lgtA, lgtC, lgtD and lgtG), has been used to classify strains into 12 immunotypes using monoclonal antibodies. Immunotypes L1-L8 are predominantly associated with meningococcal serogroups B and C, whereas immunotypes L9-L12 are associated with serogroup A strains. Two of the three meningococcal lgt loci (in contrast to the stable gonococcal lgt loci) are hypervariable genomic regions. Horizontal gene transfer from commensal Neisseria contributes to the genetic diversity of the lgt loci (Zhu et al., 2002).

Meningococcal LOS side-chain variation may be important in meningococcal virulence: commensal nasopharyngeal isolates typically possess a short side-chain, whereas invading isolates from blood and CSF display long side-chains. The longer LOS forms terminate in a galactose residue, which can be modified by sialyltransferase, located in the outer membrane of the cell, using either endogenous or host-derived 5′cytidinemonophospho-N-acetylneuraminc acid (CMP-NANA) as a sialyl donor. LOS sialylation decreases the susceptibility of the organism to antibody- and complement-mediated killing, conferring a survival advantage after invasion of the bloodstream. Conversely, adhesion to nasopharyngeal epithelial cells is suppressed by LOS sialylation, probably as a result of steric hindrance of the interaction of OMPs (such as Opa and Opc) with host-cell ligands. LOS is itself known to be cytotoxic to various host cells and may interact synergistically with pili to cause cell damage. A viable LOS-deficient meningococcal mutant has been generated *in vitro* (Steeghs *et al.*, 1998), which will assist in future studies of LOS–host interactions.

PATHOGENESIS

The pathogenesis of meningococcal disease begins with the transmission of the organism from the nasopharynx of one individual to another, which is believed to occur either by direct physical contact or respiratory droplets. The latter probably only transmit the bacteria to persons in close proximity (about 1 m away), thus explaining the association of meningococcal carriage (and disease) with overcrowding (Nelson, 1996). Close contact with a case of invasive meningococcal disease greatly increases the risk of developing invasive disease (usually within a few days) with the same strain. A small number of pathogenic (sometimes referred to as hypervirulent) meningococcal clones such as ET-5 (ST-32) and ET-37 (ST-11) are responsible for most cases of disease; 90% of strains isolated from carriers are never associated with disease and are considered non-pathogenic (Bevanger *et al.*, 1998). In the overwhelming majority of cases, therefore, the acquisition of a new meningococcal strain results in the asymptomatic colonization of the nasopharyngeal mucosa rather than disease. Even the acquisition of a pathogenic clone causes disease in only 1% of individuals (Caugant *et al.*, 1994). There is thus a strong argument to regard the bacterium as a common respiratory commensal, which rarely causes any deleterious effects to its host. The detection of meningococci in subepithelial layers in the nasopharynx of healthy carriers (Sim *et al.*, 2000) suggests that cellular invasion leading to an intracellular habitat may represent a successful survival mechanism for the organism. Disease may therefore be regarded as an occasional 'accident' resulting from this strategy; the nasopharyngeal transmissibility and colonization/invasion fitness of pathogenic clones having the side-effect of invasive disease (a dead-end event in the biology of the organism) in susceptible persons (Taha *et al.*, 2002). Transient invasion of the bloodstream, which is a hostile environment for bacteria, is probably quite common but goes unrecognized in most cases as the host's immune system clears low-level bacteraemia. Rarely, if the bacteria survive and multiply in the bloodstream, resulting in an unchecked bacteraemia, they may induce septicaemia or proceed to invade meningothelial cells, resulting in meningitis. The predilection of meningococci to invade the meninges is poorly understood but probably involves the same mechanisms utilized to attach to and invade nasopharyngeal and submucosal cells. Metastatic infections at other sites are comparatively rare. The factors that contribute to both fitness and pathogenicity include the polysaccharide capsule, the type IV pilus, OMPs such as Opa and OpcA and secreted proteins such as immunoglobulin A1 (IgA1) protease. Screening of the meningococcal genome sequence has identified several other previously unrecognized proteins, which may also contribute to fitness and pathogenicity.

Polysaccharide Capsule

The ability to express a polysaccharide capsule (analogous to other bacteria that commonly colonize the nasopharynx such as *Haemophilus influenzae* and *Streptococcus pneumoniae*) is believed to be important for the transmission of the meningococcus, protecting the organism from desiccation and enabling aerosol transmission; the absence of a capsule in *N. gonorrhoeae* makes the organism highly susceptible to drying. In the nasopharynx, phase-variants lacking

capsule expression are common, as are strains lacking the capsule biosynthesis operon, indicating that capsule expression is not a prerequisite for colonization (Claus *et al.*, 2002). Indeed, intimate adherence (and subsequent transmigration across the epithelial cell barrier) is enhanced in the absence of the capsule. The transcriptional regulatory protein, CrgA, has been shown to downregulate capsule expression and promote adhesion to epithelial cells (Deghmane *et al.*, 2002). Following invasion, the capsule is an essential virulence factor which protects the organism against phagocytosis and complement-mediated killing (Klein *et al.*, 1996).

Type IV Pili

Analogous to many Gram-negative bacteria, meningococci produce phase- and antigenically variable type IV pili, which act as long-range attachment organelles. Type IV pili are filamentous structures, 6 nm in diameter, which protrude from the cell surface. At least 22 proteins are required for pilus biogenesis and function. The pilus fibre is composed of repeating units of the 18–22-kDa PilE protein, which may be post-translationally glycosylated. A 110-kDa polypeptide, PilC, located at the tip of the pilus, is believed to act as an adhesin. PilC is upregulated by initial interactions with host cells and is essential for obtaining a fully adhesive phenotype. The pilus receptor is thought to be to be CD46 (Kallstrom *et al.*, 1997).

In addition to adhesion, type IV pili are important for other phenotypes including bacterial aggregation, twitching motility and DNA transformation. Pili may also enhance LOS-mediated host-cell cytotoxicity and lead to cortical plaque formation. A role in modulating host-cell signal transduction pathways is postulated, as pili produce transient increases in cytosolic free calcium (Kallstrom *et al.*, 1998).

Opa and OpcA

After the initial pilus-mediated adhesion, further interactions of meningococci with epithelial cells are mediated by opacity OMPs, such as Opa and OpcA. Downregulation of capsule expression and removal of sialic acid from LOS are required for Opa and OpcA to interact with host-cell molecules. Opa are a family of phase-variable basic proteins with a molecular weight of about 28 kDa, each encoded by a distinct *opa* gene. Meningococci usually possess four or five Opa loci. They are surface-exposed and, apart from two hypervariable regions, are fairly constant in sequence. Opa variants bind members of the large CEACAM (formerly CD66) family including CEACAM 1, 3, 5 and 6, which are differentially expressed on host cells. The binding of CEACAM molecules mediates adhesion to and invasion of epithelial cells (Virji *et al.*, 1996). Opa binding of CEACAMs on polymorphonuclear cells leads to phagocytosis.

The OpcA proteins (formerly known as Opc) are another group of OMPs, which, although superficially similar to the Opa proteins in terms of their size and physicochemical properties, share only 22% homology with these proteins (Olyhoek *et al.*, 1991). Transcriptional (phase-variable) regulation leads to zero, intermediate or high-level expression. OpcA promotes adherence of non-piliated meningococci to epithelial cells via cell-surface heparan sulphate proteoglycans (de Vries *et al.*, 1998). OpcA also facilitates adhesion to and invasion of endothelial cells through the binding of vitronectin. The OpcA–vitronectin interaction allows the meningococcus to attach to the integrin $\alpha_v\beta_3$ on the apical surface of the endothelial cells (Virji, Makepeace and Moxon, 1994).

IgA1 Protease and Other Autotransporter Proteins

Meningococci secrete a protease via the autotransporter pathway (type V secretion) that was demonstrated to cleave IgA1 (but not IgA2) at the hinge region, thus generating functionally useless antigen-binding Fab fragments. The functional significance of this cleavage is uncertain, but surface-bound Fab fragments may enhance the resistance of the bacterium to the bactericidal activity of IgG and IgM antibodies. Lamp1, the major integral membrane glycoprotein of lysosomes, is also cleaved by IgA1 protease. Pilus-induced elevation of cytosolic free calcium has been shown to trigger lysosomal exocytosis, which in turn increases the surface levels of Lamp1. Once on the surface, Lamp1 is accessible to cleavage by IgA1 protease secreted by the adherent bacteria (Lin *et al.*, 1997). Reducing the lysosomal levels of Lamp1 enhances the intracellular survival of the organism. Furthermore, enhanced IgA1 protease activity is a feature of pathogenic meningococcal clones (Vitovski, Read and Sayers, 1999).

In addition to IgA1 protease, several other meningococcal proteins are translocated to the outer membrane or secreted via the autotransporter pathway. These include NhhA and App, which show homology to adhesins in *H. influenzae*, and AspA, a subtilisin-like serine protease which shows homology to a virulence determinant in *Bordetella pertussis* (Peak *et al.*, 2000; Abdel-Hadi *et al.*, 2001; Turner, Wooldridge and Ala'Aldeen, 2002). The role of these autotransporter proteins in relation to meningococcal fitness and pathogenesis has yet to be determined.

PATHOPHYSIOLOGY

During meningococcal septicaemia, there are signs and symptoms of circulatory failure, multi-organ dysfunction and coagulopathy. There is increased vascular permeability and vasodilatation that results in capillary leak syndrome with peripheral oedema. Loss of intravascular fluid and plasma proteins results in hypovolaemia and reduced venous return, and hence reduced cardiac output, hypotension and reduced perfusion of vital organs. Systemic hypoxia, acidosis and gross electrolyte and metabolic impairment eventually culminate in multiorgan dysfunction. On the molecular level, the underlying pathophysiology of meningococcal sepsis is complex and involves numerous interactive cascades, including cytokine, chemokine, host-cell receptors and coagulation and complement components. Figure 14.2 summarizes some of these interactive events.

The Role of Endotoxin

The lipid A moiety of the endotoxin is thought to be primarily responsible for septic shock, extensive tissue damage and multi-organ dysfunction through stimulating the release of inflammatory mediators including tumour necrosis factor-α (TNF-α) and a series of interleukins and other cytokines. In addition, meningococcal endotoxin

seems to trigger a number of major intravascular cascade systems, including coagulation, fibrinolysis, complement and kallikrein-kinin (van Deuren, Brandtzaeg and van der Meer, 2000). Although endotoxin is largely membrane-associated, it is shed from live organisms in large quantities via the outer membrane vesicles (blebs), which also contain OMPs. This bleb shedding is known to occur *in vitro* (in broth cultures) and *in vivo*. Increasing levels of circulating endotoxin are associated with increasing seriousness of disease, and very high levels (>700 ng/l) are associated with fulminant septic shock, disseminated intravascular coagulation (DIC) and a high fatality rate (Brandtzaeg, 1995). Endotoxin seems to trigger, directly or indirectly, the release of extremely high levels of monocyte tissue factor activity, as detected in severely septicaemic patients, suggesting that monocytes are responsible for the overwhelming DIC (Osterud and Flaegstad, 1983).

Coagulopathy

Coagulopathy is a common feature of meningococcal septicaemia and is indicated by the presence of: prolonged kaolin partial thromboplastin time, prothrombin time and thrombin time; reduced platelet count, plasma fibrinogen, plasminogen and α-2-antiplasmin; increased fibrin degradation products, fibrinopeptide A and plasminogen activator inhibitor-I; reduced functional levels of coagulation factor V, antithrombin III, protein C, protein S and extrinsic pathway inhibitor (Brandtzaeg, 1995; van Deuren, Brandtzaeg and van der Meer, 2000). Severity of disease is also predicted from markedly increased levels of immunoreactive plasminogen activator inhibitor-I, and figures exceeding 1850 μg/l almost always reflect a fatal outcome (Brandtzaeg, 1995). The fibrinolytic cascade seems to be activated early in the disease, which is then downregulated by increasing levels of plasminogen activator inhibitor-I, resulting in the formation of microthrombi in various body organs, including the skin and the adrenal glands. The uncontrolled and excessive consumption of coagulation factors then results in a haemorrhagic diathesis with disseminated tissue haemorrhages.

Activated Complement Cascades

It has long been established that late complement components are vital for complete protection against invasive meningococcal disease. Both the classical and alternative pathways are important for protection. Those with complement deficiencies, including C2, C3, C4b, C5–9 and properdin, are at high risk of developing meningococcal disease and may suffer more severe disease. Those with late complement component deficiencies are particularly at much greater risk of acquiring meningococcal disease and developing second episodes of disease. It is interesting that these patients suffer from disease at an older age with lower mortality rate, and with serogroup Y being the most frequently isolated serogroup. Whereas men suffering from the X-linked properdin deficiency are at much greater risk of developing severe disseminated meningococcal disease with a very high mortality rate (Densen *et al.*, 1987). This highlights the importance of the early complement components and the initiation of complement activation and, by inference, the role of the alternative pathway in meningococcal lysis.

The complement cascade is triggered during meningococcal infection, probably by the effect of endotoxin alone. In cases of severe septic shock, the alternative pathway seems to dominate over the classical pathway. Although complement-mediated bacteriolysis is an effective protective antibacterial mechanism, it is believed to have some indirect detrimental effect via the release of the C3a, C4a and C5a anaphylatoxins and/or the release of membrane-bound endotoxin which may follow the attack of the membrane-attack complex (Frank, Joiner and Hammer, 1987). Increased levels of C3 activation products and terminal complement complexes are thought to reflect the severity of disease (Brandtzaeg, 1995).

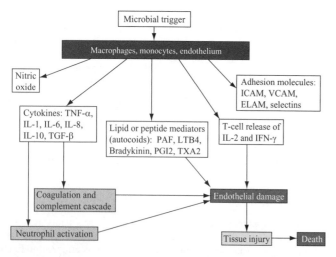

Figure 14.2 Summary of molecular events during meningococcal sepsis.

The Source of Pro-Inflammatory Cytokines

Meningococcal disease is associated with elevated levels of pro-inflammatory cytokines. During septic shock, increased levels of endotoxin, TNF-α, IL-1β, IL-6 and IL-8 are detected in the peripheral blood and reflect disease severity. During meningococcal meningitis without septicaemia, there may be a compartmentalized CSF rise in endotoxin levels associated in about half of the cases with a CSF rise in TNF-α and IL-1β, which all reflect the severity of the meningitis (Brandtzaeg et al., 1992a; Brandtzaeg, Ovsteboo and Kierulf, 1992b). The rise in IL-6 and IL-8 does not seem to be compartmentalized. Epithelial cells of the meninges upregulate the genes for TNF-α, IL-6 and IL-8 in response to exposure to meningococcal cells (Wells et al., 2001). IL-1β, on the other hand, appears to be produced by meningeal macrophages (Garabedian, Lemaigre-Dubreuil and Mariani, 2000).

Host Factors that Affect the Risk and Severity of Meningococcal Disease

In addition to the complement component deficiencies described above, other host factors are known to affect the risk and/or severity of meningococcal disease. Deficiencies in serum mannose-binding lectin, which binds to meningococci and activates complement, predisposes to invasive meningococcal disease, especially in children (Hibberd et al., 1999; Jack et al., 2001). Genetic differences in the ability to express certain cytokines may also impinge on the outcome of infection. For example, persons who are homozygous for an IL-1β gene variant, which results in low-level expression of IL-1β, are prone to more severe disease (Sparling, 2002). Similarly, a polymorphism in the neutrophil receptor for IgG_2 (CD32) has recently been shown to be associated with more severe bacteraemia and sepsis, although it does not predispose to infection per se, nor does it increase the severity of meningitis (Domingo et al., 2002). The coagulation system plays an important role in the outcome of meningococcal disease. Deficiency of either protein C or its cofactor protein S is associated with an increased risk of purpura fulminans (Esmon et al., 1997). Patients with severe meningococcal disease have been shown to have decreased endothelial expression of the two receptors, thrombomodulin and protein C receptor, which are essential for the activation of protein C.

Consequences of the Bacterial Insults

General

The uncontrolled and complicated cascades of organism- or endotoxin-induced events, including inflammatory mediators and coagulopathies, eventually lead to capillary leakage syndrome, circulatory collapse, myocarditis, renal failure, adult respiratory distress syndrome, adrenal haemorrhage, haemorrhagic skin and serosal surface lesions, muscular infarction and other organ dysfunctions. Of these, circulatory failure is the most prominent cause of death in overwhelming meningococcal disease. Myocardial dysfunction is probably due to a combination of many effects, including a direct effect of the bacterium and its endotoxin, inflammatory mediators, impaired coronary artery perfusion, hypoxia, acidosis, hypocalcaemia, hypokalaemia, hypomagnesaemia and hypophosphataemia.

Outcome of disease can be predicted from the initial renal function (e.g. serum creatinine level), and those who develop uraemia or anuria early in the disease will have much poorer prognosis than those who develop only a mild degree of renal impairment. Adrenal haemorrhages, the so-called Waterhouse–Fredrichsen syndrome, which is observed in a large percentage of fatal cases, is not unique to fulminant meningococcal septicaemia, as it occurs in other overwhelming Gram-negative endotoxaemia with DIC. The resulting adrenal insufficiency is an additional factor contributing to the circulatory collapse.

Haemorrhagic skin lesions (petechiae and ecchymoses), which vary in size and severity, reflect the meningococcal predilection for the dermal blood vessels and reflect the severity of DIC and state of septicaemia. There is vascular damage (with or without vasculitis) with endothelial cell injury or death with meningococci identifiable and/or cultivable from lesion biopsies.

The central nervous system

Bacterial endotoxin and peptidoglycan are believed to be mainly responsible, directly or indirectly through inflammatory mediators, for the meningeal inflammation, blood–brain barrier dysfunction and the changes in CSF. These include cellular, biochemical and hydrodynamic changes of the CSF. The combination of increased CSF secretion, impaired CSF absorption and blood–brain barrier dysfunction can lead to the accumulation of CSF, raised intracranial pressure and reduced cerebral blood flow. Increased CSF lactate indicates brain tissue hypoxia. In patients with meningococcal septicaemia and meningitis, the brain tissue hypoxia, which is worsened by the shock-induced underperfusion, vasculitis and DIC, may result in neuronal injury and varying degrees of brain tissue damage.

CLINICAL MANIFESTATIONS

Meningococci are capable of causing a wide range of different clinical syndromes, which can vary in severity from a transient mild sore throat to meningitis or acute meningococcal septicaemia, which can cause death within hours of the appearance of symptoms. Bacteraemia with or without sepsis, meningococcal septicaemia with or without meningitis, meningoencephalitis, chronic meningococcaemia, pneumonia, septic arthritis, pericarditis, myocarditis, endocarditis, conjunctivitis, panophthalmitis, genitourinary tract infection, pelvic infection, peritonitis and proctitis are all among the diseases caused by meningococci. Meningitis and/or septicaemia are by far the most common presentations of disease. It is important to remember that the clinical picture can progress from one end of the spectrum to the other during the course of disease.

Transient Meningococcaemia

Relatively mild culture-positive flu-like illness can present with non-specific symptoms such as raised temperature and joint pain with or without rash and can persist for days or weeks. Such mild bacteraemia may resolve completely without treatment or progress to acute severe disease or persist as chronic meningococcaemia.

Acute Meningococcal Septicaemia With or Without Meningitis

Acute meningococcal disease may present initially with a series of non-specific manifestations which combined may provide indicative diagnostic clues. This may include signs of septicaemia, meningitis or both. Presentation with severe meningococcal septicaemia is less common but deadlier than meningitis.

The majority of cases occur in young children and present with rapid onset of raised temperature with or without vomiting, photophobia, convulsions, skin rash, lethargy, irritability, drowsiness, reluctance to feed, diarrhoea and other more localized signs and symptoms. In older children and adults, headache, restlessness, generalized muscular pain and arthralgia are also noted. Rarely, acute abdominal pain may be the predominant feature. The onset may be preceded by prodromal symptoms mimicking viral upper respiratory tract infections. On examination, the patient is often distressed with temperature ranging from normal to 41 °C.

Meningococcal septicaemia is most often associated with the characteristic meningococcal rash (Plate 8). These lesions, when present, can be macular, maculopapular, petechial, purpuric, ecchymotic or necrotic in nature and can appear anywhere on the body, including the trunk, extremities, pressure sites, palm, sole, face, palate and conjunctivae. Adjacent lesions can coalesce and form confluent purpurae, petechiae, haemorrhagic bullae or even gangrenous necrotic skin and subcutaneous lesions. In severe cases, DIC with intense peripheral vasoconstriction may cause peripheral gangrene, which can involve fingers, toes or even entire limbs.

In severe cases, patients present in septic shock (hypotension which is refractory to fluid resuscitation) with tachycardia, progressively impaired peripheral perfusion, increased respiratory rate, hypoxia, cyanosis, acidosis, electrolyte imbalance, elevated plasma lactate levels, progressive circulatory failure with oliguria or even anuria. Progressive impairment of central perfusion leads to a decline in the level of consciousness and may lead to coma. Poor prognostic signs include extensive and rapidly progressing necrotic skin lesions, shock, multi-organ failure, decreasing level of consciousness and DIC.

Varying degrees of respiratory failure and acute respiratory distress syndrome are found in most patients with meningococcal septicaemic shock. These are caused by various factors, including pulmonary oedema due to capillary leak syndrome, pulmonary vascular occlusion due to DIC and elastase release from neutophils (Donnelly *et al.*, 1995). Reversible oliguria or renal failure occur in severe cases due to hypovolaemia and impaired organ perfusion. In cases of prolonged shock, tubular and cortical necrosis can cause irreversible renal failure. Myoglobinuria, due to muscle necrosis, may further contribute to renal failure (van Deuren *et al.*, 1998).

Meningitis

Signs and symptoms of meningococcal meningitis include general (non-specific) and localized (specific) ones. Patients often remain alert, but signs of cerebral involvement, such as convulsions, declining level of consciousness and coma may be present. Depending on the previous immune status, localized signs may appear early or late in the course. In the young and previously healthy, localized signs of meningeal irritation may occur early in the disease. Unless accompanied by the characteristic rash (present in more than half of the cases), none of these are characteristic to meningococcal meningitis. Most, but not all, patients present with clear signs of meningeal irritation. Of these, neck stiffness is by the far the most important; however, it is present in less than half of the patients (Carpenter and Petersdorf, 1962). In infants and young children, neck stiffness is difficult to detect, but when the child is left to assume the most comfortable position, extension of the neck may be observed. Additional signs include a bulging fontanelle due to increased intracranial pressure, irritability and lying on one side facing away from the source of light (photophobia). A positive Kernig's sign, the inability to extend the knee when the hip is flexed in supine position, is another important and pathognomonic sign of meningeal irritation. An attempt to flex the neck in a sudden movement with the patient in a sitting position with the legs outstretched will induce a reflex flexion of both hips and knees (Brudzinski's sign).

Other Manifestations

Acute or chronic meningococcal septic arthritis is relatively common and may complicate any of the acute meningococcal diseases. It usually affects one or more joints and manifests at any stage of disease (Schaad, 1980). Occasionally, arthritis is the only presenting complaint, which most often involves a single large joint such as the knee or the hip. Meningococci may be seen in Gram stains and isolated from cultures of joint aspirates. An immunoreactive arthritis (or pericarditis) may also occur typically 3–5 days into the illness.

Upper and lower respiratory tract infection, such as otitis media, pharyngitis, bronchitis and pneumonia, can occur as part of disseminated

meningococcal disease or as primary infections. Primary pneumonia is typically caused by serogroup Y meningococci (Koppes, Ellenbogen and Gebhart, 1977), especially in the elderly, and has a good prognosis. Blood or sputum cultures can yield the organism; culture-positive nasopharyngeal swabs can be misleading and are not therefore recommended. Respiratory infections due to meningococci, which respond well to conventional antibiotics, are probably under-diagnosed, as cultures are not routinely obtained in many patients unless empirical treatment fails. Furthermore, meningococci isolated from sputum can easily be dismissed as respiratory *Neisseriae* and not considered as significant.

Similarly, other organs may be attacked by meningococci as part of meningococcal septicaemia or as a primary organ infection. These include pericarditis, myocarditis and endocarditis. Myocarditis is thought to be present in most cases of overwhelming septicaemia. Also, meningococcal urethritis and endocervicitis in sexually active women and proctitis in homosexuals can occur (Faur, Weisburd and Wilson, 1975). Meningococci seem to be responsible for 2% of all bacterial conjunctivitis, and this may proceed to systemic meningococcal disease in nearly one-third of cases (Barquet *et al.*, 1990).

Chronic Meningococcal Disease

Finally, chronic meningococcal disease (chronic meningococcaemia) is a rare but long recognized clinical entity which presents as chronic intermittent high temperature, joint pain and headache with or without skin lesions affecting adults more than children (Benoit, 1963). The condition can last for months and cause a variety of complications. The organism may be isolated from the blood in the first few weeks of the illness.

Morbidity and Mortality

A significant number of the patients who recover end up with permanent neurological sequelae, including mental retardation, cerebellar ataxia, cranial nerve deficits, deafness (due to auditory nerve damage), persistent headache, hydrocephalus, dementia, convulsions, peripheral nerve lesions, subdural empyema and psychoneurological complications.

Mortality rate from meningococcal disease varies between 5 and 70% depending on a number of factors, including the severity of disease, the speed by which it develops, the organs involved, the age and immune status of the patient, the socioeconomic status, the standard of health care and the speed by which the disease is diagnosed and antibiotics administered. An analysis of case fatality rates between 1963 and 1998 in Oxford, UK, showed a rate of 11%, which had not fallen during this period (Goldacre, Roberts and Yeates, 2003). In those patients who present with signs of severe septicaemia, particularly in underdeveloped countries, the mortality rate can reach up to 70% (deMorais *et al.*, 1974). In developed countries, 15–30% of septicaemic patients die, despite treatment in advanced intensive care units. The mortality rate for patients with septicaemic shock remains very high (20–50%), despite state-of-the-art intensive care facilities.

The overall mortality from meningococcal meningitis in Europe and North America is 5–15%, despite the sensitivity of the organism to many antibiotics and the high standard of health care in these two continents (Jones, 1995). Children seem to have a better chance of recovery from meningococcal meningitis than adults. The mortality rate for children under the age of 5 is around 5% compared to 10–15% in adults.

A combination of improvements in initial management of patients, use of a mobile intensive care service and centralization of care in a specialist unit were shown to reduce the overall case-fatality rate (from 23% in 1992/93 to 2% in 1997) in a paediatric population, despite disease severity remaining largely unchanged (Booy *et al.*, 2001). Most deaths occur in the first 2 or 3 days of admission to hospital. Very rapid development of septicaemia and septic shock is a poor prognostic sign; some patients succumb within a few hours of the appearance of the first symptoms.

LABORATORY DIAGNOSIS

Specimens

Cerebrospinal Fluid

Any signs and symptoms suggestive of meningitis should prompt the clinician to consider performing a lumbar puncture without delay. However, this procedure is not an absolute must in all cases of meningococcal disease, and its risks must be weighed against its benefits. In the presence of raised intracranial pressure (ICP), lumbar puncture is associated with fatal cerebral herniation and therefore should not be performed (Mellor, 1992). A mild increase in intracranial pressure is present in virtually all patients with meningococcal meningitis and is not a contraindication for lumbar puncture. However, when clinically evident, it is considered severe enough to prevent lumbar puncture attempts. Even in the presence of markedly increased intracranial pressure, papilloedema may appear late in the process (24–48 h); therefore, more acute signs of raised ICP and incipient cerebral herniation should be looked for: these include severe deterioration in the level of consciousness, bradycardia, abnormal blood pressure, respiratory abnormalities, cranial nerve paralyses (especially 3rd and 6th nerves) and changes in muscle tone or other neurological abnormalities. The role of computerized tomography (CT) scanning to exclude raised ICP prior to lumbar puncture is controversial (Oliver, Shope and Kuhns, 2003). Cerebral herniation is uncommon and may occur without any CT abnormalities; careful clinical examination may be more sensitive in detecting significantly raised ICP. Lumbar puncture should also be avoided in the presence of septicaemic shock, as this may aggravate the circulatory collapse, and it is important to remember that, in cases of meningococcal septicaemia without signs of meningitis, the CSF examination may not reveal any cellular or biochemical abnormality.

In meningitis, depending on white cell counts and bacterial load, the CSF may remain clear or become very cloudy. Two hundred white cells per mm^3 (0.2×10^9 cells/l) can make the CSF look turbid. The leucocyte count (predominantly polymorphonuclear cells) can reach as high as 10 000 per mm^3, as can the bacterial cell count (not measured routinely). Protein content is raised and may even reach levels as high as 10 g/l, and glucose levels fall below 40% of plasma glucose levels or may even be undetectable. Raised lactate dehydrogenase and neuraminidase levels can also be detected in the CSF of patients with meningococcal meningitis.

Direct microscopy of the CSF deposit often provides the first positive diagnostic clue that can guide the choice of antibiotic. The Gram stain may reveal Gram-negative diplococci in more than two-thirds of the cases presenting with meningitis, provided the patient has not received antibiotics prior to lumbar puncture. Diplococci are seen both inside and outside the pus cells. The serogroup of the causative organism, knowledge of which assists in the management of contacts of the index case, can be obtained without delay by probing the CSF deposit with capsular polysaccharide-specific antibodies conjugated to latex or fluorescein isothiocyanate.

Recovery of the organism from the CSF remains the ultimate diagnostic goal. The organism is often isolated in cases with septicaemia without signs of meningitis; CSF parameters may otherwise be normal. Following the recent awareness campaigns in the UK and the increased awareness of General Practitioners (GPs) of meningococcal disease, many patients will have received antibiotics before they arrive in the hospital. This dramatically reduces the chances of recovering *N. meningitidis* from the CSF.

Blood

Blood cultures may yield the organism in up to 60% of cases if taken before antibiotics are administered. This rate is markedly reduced after the first dose of penicillin. *Neisseria meningitidis* may grow in the conventional (simple) broth culture systems without producing visible turbidity; therefore, in all patients with suspected meningococcal disease, blind subculture (usually after 12–18 h) of the broths onto solid media is strongly recommended. If negative, the procedure may be repeated after 48 h of incubation.

Throat and Nasopharynx

Throat and nasopharyngeal culture may also reveal the organism in more than half of the cases, which can support but not necessarily confirm the diagnosis. With the increasing use of antibiotics prior to admission, these specimens may be the only source where isolates can be obtained. Therefore, it is very important that throat swabs are taken by the GP or the hospital doctor as soon as the disease is suspected irrespective of whether the first dose of penicillin is administered. Swabs must be transported to the laboratory as soon as possible in order to maximize the chances of recovery of the organism.

Skin Lesions

Skin lesions may also yield the causative organism. Petechial lesions can be injected with a small volume of sterile saline and aspirated with a hypodermic needle and inoculated into a broth medium. Alternatively, skin biopsies can be Gram stained and processed for culture immediately. In patients with meningococcal septicaemia, the organisms can be recovered from skin biopsies even after the start of antibiotics (van Deuren *et al.*, 1993).

Isolation of *N. meningitidis*

No effort should be spared to isolate the causative organism and, where possible, all the relevant culture specimens should be taken before antibiotics are administered. Recovery and identification of the organism is important for clinical and epidemiological reasons. Isolation of the organisms will confirm the diagnosis, influence the choice of antibiotics, provide useful epidemiological information and boost the confidence of the clinicians in managing meningococcal disease. Meningococci can be isolated from a number of possible sources, including the CSF, blood, throat or nasopharyngeal swabs, skin lesions, joint aspirates, eye swabs, or any other body fluid or tissue specimens.

Identification

Isolates must be identified to species level in the local laboratory and further classified in a reference laboratory. Meningococci must be differentiated from other oxidase-positive, Gram-negative diplococci using biochemical and immunological tests (Barrow and Feltham, 1993). The rapid carbohydrate utilization test, although largely superseded by more convenient tests in many countries, provides fast and reliable means of identification. In this test a heavy bacterial inoculum is suspended in broth containing individual sugars; meningococci will ferment glucose and maltose (but not lactose and sucrose) within an hour or so. γ-Glutamyl aminopeptidase activity can be detected in rapid tests, such as Gonocheck-II (EY Laboratories, Inc., San Mateo, California, USA) where adding meningococcal colonies to the rehydrated substrates will produce a yellow reaction. Isolates from the nasopharynx, throat, conjunctivae, rectum or other sites where gonococci and respiratory *Neisseriae* can also be carried, must be precisely identified. Co-agglutination tests (e.g., Phadebact Monoclonal GC Test, Launch Diagnostics Ltd, Kent, England), using specific anti-gonococcal antibodies conjugated to protein A of non-viable staphylococci, will identify or exclude gonococci but not meningococci.

The serogroup of the organism is identified by detecting the capsular polysaccharide antigen using immunological tests including latex agglutination, coagglutination and counter-immunoelectrophoresis.

It is important to remember that there is always a potential risk of acquiring meningococcal infection through handling live organisms while performing biochemical or antibiotic sensitivity tests. Therefore, any work with live cultures of *N. meningitidis* should be performed in a class 1 safety cabinet.

Detection of Meningococcal Antigens, Antibodies and DNA

In culture-negative CSF, presumptive diagnosis can be obtained by detecting very low levels of meningococcal antigens using various immunological and biochemical techniques. Using specific antibodies, the capsular polysaccharide antigens of serogroups A, C, W135 and Y can be reliably identified in the CSF of more than three quarters of cases. This test is not reliable for serogroup B due to the capsule's poor immunogenicity and its cross-reaction with the capsule of *E. coli* serotype K1, which is another common cause of meningitis in young infants.

High titres of specific anti-meningococcal antibodies in a single specimen of serum, or rising titres in paired sera, can also provide laboratory confirmation of suspected cases when other methodologies prove negative (Clarke *et al.*, 2002). Antibodies are detected by enzyme-linked immunosorbent assay with crude preparations of outer membrane vesicles as antigens (Jones and Kaczmarski, 1995).

Meningococcal DNA can be detected in the CSF and whole blood (collected in bottles with EDTA) during the acute stage of the disease, even after the start of antibiotic therapy. Diagnostic polymerase chain reaction (PCR) tests are now widely used in reference laboratories in most developed countries. In UK, quantitative PCR has demonstrated a direct correlation between bacterial load on admission and disease severity (Hackett *et al.*, 2002a; Hackett *et al.*, 2002b). An increasing proportion of meningococcal cases are now confirmed on the basis of a positive PCR test alone. In the absence of an isolate, PCR can also be used to determine the capsular group, PorB type and PorA subtype of the infecting organism.

Other Investigations

Peripheral white cell count varies from subnormal levels to over 45×10^9 cells/l. In patients presenting with skin rash and septicaemia, clotting parameters must be monitored. In these cases, prothrombin and activated partial thromboplastin time are usually prolonged. Diminished plasma fibrinogen and increased fibrinogen degradation product titre are detected.

Poor prognostic peripheral blood parameters include low white blood cell count ($<10 \times 10^9$ cells/l), thrombocytopenia, grossly abnormal coagulation indices (Algren *et al.*, 1993) and raised serum creatinine.

MANAGEMENT

In the pre-antibiotic era, particularly following the great success of the use of anti-diphtheria antiserum in treating cases of diphtheria, therapeutic administration of immune sera became fashionable for a number of infectious diseases, including meningococcal meningitis. Good results were obtained with anti-meningococcal antisera administered intrathecally, intracisternally or intravenously with clear reduction in mortality rates (Flexner, 1913). The first use of anti-

biotics was in the latter part of 1935, following the discovery of the sulphonamides, which were then replaced by penicillin in the following decade. For the past six decades, penicillin has remained the mainstay of specific anti-meningococcal treatment.

Chemotherapy

The choice of antibiotic treatment varies depending on a number of interrelated factors including those relating to the patient, organism, pharmacokinetics of the antibiotics (CSF penetration), circumstances, epidemiological data, affordability and hospital facilities. The standard approach, particularly in meningitis, is to empirically cover for most of the possible causative organisms using relatively broad-spectrum antibiotics until the diagnosis is confirmed and then switch to penicillin.

The Start of Chemotherapy

Treatment should commence as soon as the disease is first suspected. Parenteral penicillin given by the GP before referral to hospital, or by hospital doctors on arrival, has been shown to reduce the mortality rate (Cartwright *et al.*, 1992; Strang and Pugh, 1992). GPs are recommended to give benzlypenicillin (1.2 g for adult; 600 mg for a child) intramuscularly or intravenously. It is wise, however, to place an intravenous cannula as soon as possible and obtain a throat or nasopharyngeal swab and two sets of blood cultures, one from each arm, before antibiotics are commenced. Initial antibiotic therapy should not be delayed, pending other investigations such as lumbar puncture or CT scanning.

Empirical Therapy

It is widely recommended now that until positive laboratory identification of the organism is made, the patient must be treated empirically with a broad-spectrum bactericidal antibiotic that is effective against most other possible causes of pyogenic meningitis. Recently, third generation cephalosporins, such as cefotaxime (200 mg/kg/day) or ceftriaxone (100 mg/kg/day), have replaced the conventional penicillin and chloramphenicol combination in the more developed countries (Peltola, Anttila and Renkonen, 1989). Although these cephalosporins do not necessarily improve on morbidity or mortality, they offer the advantage of using a single, relatively non-toxic therapeutic agent which penetrates the CSF extremely well. They are effective against *N. meningitidis*, *S. pneumoniae*, *H. influenzae*, *E. coli* and *Streptococcus agalactiae* (serogroup B streptococci), but not against *Listeria monocytogenes*. Furthermore, chloramphenicol-resistant *H. influenzae* are now widespread, and strains of meningococci with reduced sensitivity to penicillin [minimum inhibitory concentration (MIC) ≥ 0.12 mg/l] have emerged in many countries. The mechanism of resistance does not seem to involve β-lactamase production, rather it is believed to be due to the reduced affinity of penicillin for penicillin-binding protein 2, as a result of changes in the nucleotide sequence of its gene, *penA* (Zhang *et al.*, 1990; Saez-Nieto *et al.*, 1992). Until the mid 1980s no such strains were known, and by the end of the decade nearly half of the strains in Spain showed some degree of insensitivity to penicillin, while remaining fully sensitive to the third generation cephalosporins. The clinical significance of meningococci with reduced sensitivity to penicillin is unclear; few treatment failures have been described, and there is no association with a fatal outcome (Trotter *et al.*, 2002). Nevertheless, the increasing frequency of strains less susceptible to penicillin highlights the need for ongoing surveillance.

Cefuroxime, a second generation cephalosporin, and ceftazidime, another third generation cephalosporin, are also effective in treating meningococcal meningitis. However, these two agents are not recommended for empirical treatment, especially knowing that ceftazidime is not very effective against *S. pneumoniae*. The major shortcoming of the cephalosporins is that they are not effective against *L. monocytogenes*, although a relatively uncommon cause of meningitis, and they are comparatively expensive. In many under-developed countries, where cephalosporins are not affordable, the combination of penicillin and chloramphenicol is still used.

Specific Antibiotic Treatment

Penicillin remains the drug of first choice and is highly effective for penicillin-sensitive strains. The drug's penetration into the CSF is poor in the absence of an inflamed blood–CSF barrier; however, when high and frequent (meningitic) doses of penicillin are given, good therapeutic levels are achieved. Premature infants and neonates are given up to 100 mg/kg daily in two divided doses, infants (1–4 weeks) 150 mg/kg daily in three divided doses, children (1 month to 12 years) 180–300 mg/kg daily in 4–6 divided doses and adults 2.4 g every 4 h. The temptation to reduce the dose upon clinical improvement must be avoided because the CSF penetration of penicillin reduces, as the inflamed blood–CSF barrier recovers.

Meningococci are usually highly sensitive to benzylpenicillin (MIC 0.003–0.06 mg/l). Strains with MIC between 0.12 and 1.0 mg/l are considered of reduced sensitivity, although many of these still respond to therapeutic doses of intravenous penicillin. Meningococci are also highly sensitive to many, but not all, other penicillin derivatives such as ampicillin, amoxycillin and antipseudomonal penicillins (MICs <0.06 mg/l). Phenoxymethylpenicillin (Penicillin V) is not very effective (MIC of 0.25 mg/l), and meningococci are highly resistant to the penicillinase-resistant penicillins, such as methicillin (MIC of ≥6.0 mg/l).

All meningococcal strains isolated in 1994 in England and Wales were susceptible to therapeutic doses of penicillin, with 93% of strains showing an MIC of <0.1 mg/l and with the remainder having an MIC of 0.16–0.64 mg/l (Jones and Kaczmarski, 1995). In countries where penicillin-resistant meningococcal strains have not been isolated, the cephalosporins can be replaced with high-dose parenteral benzyl-penicillin as soon as the diagnosis is confirmed. However, irrespective of the country and the prevalence of resistant strains, it is safe practice to test for sensitivity of the organism to penicillin before the antibiotics are switched.

In cases of penicillin allergy or organism's penicillin insensitivity, the cephalosporins should be continued, unless history suggests anaphylaxis. Chloramphenicol, which is bactericidal to meningococci and achieves a high concentration in the CSF, is a good alternative to penicillin in the poorer countries. It is given at doses of 100 mg/kg/day (maximum of 4 g/day) or as single long-acting chloramphenicol injections in oil. The latter is as effective as a 5-day parenteral penicillin course for treating meningococcal meningitis. Single injection of chloramphenicol in oil is not as effective against pneumococcal meningitis and, therefore, should not be used for empirical treatment of pyogenic meningitis.

Antibiotic treatment should be given for a total of 7 days (Radetsky, 1990). Five-day courses have also been successful; however, in the presence of acute complications, treatment is usually extended to 10–14 days. It is interesting that it has been clearly demonstrated that, following antibiotic administration, the increased levels of circulating free endotoxin and endotoxin-activated inflammatory mediator systems decline very rapidly (Brandtzaeg, 1995).

Management of Septic Shock

Management of meningococcal septicaemia is a medical emergency and requires intensive care facilities for the first 48 h until the patient is clinically and physiologically stable. Clinical and laboratory signs of septic shock must be recognized immediately, and those with rapidly progressing disease, rapidly spreading skin rash, increasing tachycardia and tachypnoea, oliguria, low peripheral white blood cell count and signs of DIC require urgent intervention. Hypotension, which may develop late, is not always present on admission, particularly in children, and therefore could be a misleading parameter for impending septic shock (Nadel, Levin and Habibi, 1995). An increasing gap between core and peripheral temperature is a poor prognostic sign suggesting worsening underperfusion.

Hypovolaemia and electrolyte imbalance should be monitored and corrected with fresh frozen plasma or colloid, electrolyte supplements and inotropes as soon as possible. Hypokalaemia, despite acidosis, hypocalcaemia and hypoglycaemia are seen in the early phase of disease and ought to be corrected with care. Dialysis may be necessary, especially if severe oliguria or renal failure is present. It is very important to ensure that a high concentration of oxygen is given in all cases. A child's ventilation can deteriorate very rapidly, and pulmonary oedema and/or respiratory distress syndrome may develop soon after admission. Therefore, all children must be electively ventilated even if no signs of underventilation or respiratory failure are evident (Nadel, Levin and Habibi, 1995). Furthermore, children requiring colloid support of 40–80 ml/kg, or those who are obtunded, should also be electively ventilated. Low doses of heparin are also often given, although this remains controversial. Recent data have supported the use of protein C (or activated protein C) concentrate in patients with septic shock/DIC (Alberio, Lammle and Esmon, 2001). The optimum use and degree of benefit from this therapy remains unclear at present.

Management of Raised Intracranial Pressure

Attempts to correct raised intracranial pressure, if present, are made only after systemic shock is brought under control. In severe meningitis, extensive measures may be needed to reduce intracranial pressure. Electrolyte imbalance, hypovolaemia and cardiac output must be corrected. While colloids are administered to correct the hypovolaemia, subsequent crystalloids are restricted. Other measures such as hyperventilation (maintaining pCO_2 between 3.5 and 4.5 kPa) and careful use of mannitol are also important. Patients must be managed in an intensive care unit and nursed in a quiet room with the head slightly raised above the rest of the body.

The Use of Corticosteroids in Meningitis

Dexamethasone reduces the severity of neurological complications in bacterial meningitis (regardless of the aetiological agent) and is now recommended for both adults and children with suspected meningitis (van de Beek *et al.*, 2003). Treatment should commence with, or a few minutes before, the initiation of antibiotic therapy. Recommended regimens are 10 mg 6-hourly for 4 days (de Gans and van de Beek, 2002) in adults and 0.4 mg/kg 12-hourly for 2 days in children (El Bashir, Laundy and Booy, 2003).

Other Measures

Supportive measures are required to correct any other organ dysfunction and/or other complications. Monitoring cardiac output using echocardiogram is useful. There are a number of potentially therapeutic agents under investigation that may be useful in correcting the pathophysiology of the disease. One of the most promising of these is recombinant amino-terminal fragment of bactericidal/permeability protein (rBPI21). This naturally occurring antimicrobial protein, which is found in neutrophil granules, binds to and neutralizes the effects of endotoxin. In a phase III trial, rBPI21 reduced the incidence of complications from meningococcal disease (Levin *et al.*, 2000).

PREVENTION AND CONTROL

Control of meningococcal outbreaks consists of a number of interrelated measures which include management of individual cases, notification of known and suspected cases, chemoprophylaxis to close contacts, good clinical surveillance, small- or large-scale vaccination and information dissemination. Concerted efforts are required among the hospital physician, the microbiologist, the public health specialist and the GP to prevent secondary cases and control fear and anxiety among case contacts.

In all cases of suspected meningococcal diseases, attempts should be made to obtain positive culture from CSF, blood, throat swab or any other appropriate specimens while the patient is in the hospital, or retrospectively by sending paired sera or other clinical materials for the detection of antigen, antibody or DNA to a reference laboratory. All meningococcal isolates should be sent to such a reference laboratory for a full identification of individual strains. While in the hospital, patients should be isolated for the first 24 h following antibiotic administration. The risk of acquiring disease from patients is minimal in the ward but is greater in the laboratory, particularly for those who handle live cultures (see section on Laboratory Diagnosis).

Notification of all strongly suspected and proven cases of meningococcal disease to the appropriate public health authorities at the earliest opportunity is essential so that close contacts of the index case are identified. Detailed information is needed on full clinical and epidemiological aspects of these cases. In many countries the disease is notifiable by law, and as soon as an outbreak is recognized, an active surveillance must follow. Outbreaks are identified when two or more meningococcal cases are associated in time, place and identity of the meningococcal isolates. Outbreaks can occur within individual families, communities or institutions.

Contacts of the index case are approximately a 1000 times more at risk of developing disease than the rest of the community and are therefore given prophylactic antibiotics to eradicate possible carriage. Close contacts include household and kissing contacts that have been exposed to the index case over the previous 10 days (incubation period is 2–10 days) and also include infants and young children in play groups and nurseries. Nasopharyngeal culture of close contacts is unnecessary, delays antibiotic administration and may have undesirable social implications. It is important to adequately inform close contacts about the disease and the need for compliance with prophylaxis. They must receive clear instructions to seek medical advice immediately, should they develop any signs or symptoms of meningococcal disease. It is interesting that nearly one-third of immediate family contacts of patients carry an organism, which in 90% of cases has the same identity as that of the index strain.

Chemoprophylaxis

Between the 1940s and the 1960s, sulphonamides were the mainstay of treatment and erradication of carriage. However, sulphonamide-resistant strains (MICs ≥10 mg/l) are now prevalent worldwide. Whereas resistance to rifampicin (MIC ≥5 mg/l) is much less common and resistance to ciprofloxacin (MIC <0.02 mg/l) is extremely rare. A 2-day course of 12-hourly oral rifampicin (600 mg in adults, 10 mg/kg in children and 5 mg/kg in infants) is highly effective in up to 95% of cases. However, failure due to lack of compliance, drug resistance or recurrence is known. Rifampicin should not be used during pregnancy and in patients suffering from liver diseases. Also, the drug is expensive, causes discolouration of body fluids and soft contact lenses and interferes with the function of the oral contraceptive pill. Alternatives are single doses of ciprofloxacin (500 mg orally; not in young children) and ceftriaxone (250–500 mg intramuscularly, 125 mg in children under the age of 12 years). The latter, which has been shown to be more effective than rifampicin (Schwartz et al., 1988), offers the additional advantage of being licensed for use in pregnancy and young children. Minocycline (100 mg orally twice daily for 3 days) is also effective; however, it

should not be given to children and pregnant women because of its vestibular toxicity and its ability to discolour the growing teeth.

Prophylactic antibiotics are given simultaneously to all the identified close contacts as soon as the index case is diagnosed. The throat of the index case is also cleared of organisms by giving these antibiotics before discharge from hospital. Random administration of antibiotics to school contacts, which should be coordinated by the department of public health medicine, is only advised in nursery and play-group settings, but not for older children, unless they are close contacts. Except in exceptional circumstances, infants under the age of 3 months are not given prophylaxis. Finally, close contacts of cases caused by serogroup A or C (or rarely W135 or Y) meningococci may be offered the currently available capsular polysaccharide (or serogroup C conjugate) vaccines. A vaccine against serogroup B is not yet available (see section on Vaccines).

Natural Immunity

Protection has been correlated with the presence of bactericidal antibodies (Frasch, 1983), and various lines of evidence highlight the importance of humoral bactericidal activity in host defence against N. meningitidis. For example, the peak incidence of the disease occurs in children under 1 year of age, who, as a group, have few or no bactericidal antibodies. Meningococcal disease is up to 10 000 times more common in terminal complement component-deficient individuals than in normal healthy individuals, with a much higher frequency of second episodes of the illness (Andreoni, Kayhty and Densen, 1993). Passive immunization with antibodies (serum therapy) and active immunization with some capsular polysaccharides (which stimulates antibody response but not T-helper cell response) have been successful in the treatment and prevention of the disease. Further evidence emerged following a study reported by Goldschneider, Gotschlich and Artenstein (1969) where preimmune sera were obtained from new recruits of the American Armed Forces and examined for susceptibility using serum bactericidal activity and immunofluorescence. Unlike the control group, the majority of those who lacked bactericidal activity against prevalent strains acquired meningococci. Thus, bactericidal assays have become established as the best available test to determine the protective ability of specific antisera raised against polysaccharide vaccine candidates. However, it is not certain to what extent the in vitro experimental conditions reflect events occurring in vivo, nor whether the above data apply to infections with all serogroups. The data linking bactericidal antibodies with protection relate primarily to the serogroup A and C polysaccharides but have been extended to include bactericidal antibodies against OMPs.

While the great majority of the studies have focused on the role of serum bactericidal activity in the host defence against meningococcal disease, much less attention has been given to the killing of bacteria by phagocytosis, and therefore little is known about the relative importance of phagocyte-mediated killing of meningococci as compared to serum bactericidal activity. Finally, for most antigens, an efficient humoral immune response resulting in the production of antibodies and the generation of memory response requires help from T lymphocytes. However, T cells respond to peptide antigens associated with molecules of the major histocompatibility complex (HLA in humans) and so will not be stimulated by polysaccharide vaccines. This may explain why the protective efficacy of these vaccines is short lived and ineffective in young children. A way to overcome this problem is to conjugate the polysaccharide containing B-cell epitopes to carrier proteins containing epitopes for recognition by T cells. This may be achieved by using proteins containing T-cell epitopes as carriers in conjugate vaccines.

Ideal Vaccines

An ideal meningococcal vaccine should be safe, offer long-lasting immunity to all age groups, cross-protect against all meningococcal serogroups and serotypes, be given orally or nasally and be easily incorporated into the World Health Organization's Expanded Programme on Immunization. So far no such vaccine has been developed. Nevertheless, there are vaccines available against four out of 13 serogroups of meningococci. These are based on the native capsular polysaccharide of serogroups A, C, Y and W135. However, these unconjugated polysaccharide vaccines offer only a relatively short-lived immunity against their homologous serogroups, and they do not affect carriage rates or interrupt transmission of infection within a community. Furthermore, they offer little protection to children under the age of 2 years, the most vulnerable age group, and hence they are not suitable for including in paediatric immunization programmes. No polysaccharide vaccine yet exists against serogroup B meningococcal disease, and this is seen as a major obstacle to the wider use of meningococcal vaccines in Europe and American continents.

Serogroup A and C Capsular Polysaccharide

The first successful vaccines produced were against serogroup A and C *N. meningitidis* using high-molecular-weight (>100 kDa) capsular polysaccharide from these strains (Gotschlich, Liu and Artenstein, 1969). A series of large-scale field trials were conducted in the 1970s among different age groups in different parts of the world, including Europe, Africa and South America, which showed that the capsular polysaccharide vaccine is effective in controlling epidemics of serogroup A disease in almost all age groups. It soon became clear that antibody responses among infants to the serogroup A and C capsular polysaccharide vaccines depended on a number of factors including the age of the infant, the molecular weight of the antigen, the number of doses of antigen and the prior experience of the infant with naturally occurring antigens cross-reactive with the meningococcal capsular polysaccharide. It became evident that children under the age of 2 years do respond, particularly to serogroup A capsular polysaccharide, with small increases in specific antibodies. The strength of the response and its duration increased with age, and in adults 100% seroconversion was achieved which lasted longer than it would be in children. For example, Reingold *et al.* showed that the serogroup A capsular polysaccharide vaccine efficacy in children vaccinated at less than 4 years of age almost disappeared over the following 3 years, whereas those who were 4 years of age or older when vaccinated showed evidence of vaccine-induced clinical protection for 3 years after vaccination (Reingold *et al.*, 1985).

The serogroup C capsular polysaccharide vaccine, when administered routinely among recruits of the United States army to prevent severe outbreaks, has virtually eradicated the serogroup C meningococcal disease in this population. Subsequent trials and antibody response studies among children confirmed this success among age groups above 2 years but not among children less than 2 years of age. It is interesting from more recent data to show that although anticapsular antibodies and bactericidal activity in adults decline substantially by 2 years following vaccination, both persist at a level significantly above prevaccination levels for up to 10 years (Zangwill *et al.*, 1994).

On the basis of these results, serogroup C capsular polysaccharide vaccines were, until recently, recommended for general use in epidemics except for children under the age of 18 months. In contrast, serogroup A capsular polysaccharide vaccines were recommended to be given during epidemics in all age groups including infants with a booster dose given to those under the age of 18 months. These vaccines were also offered to those who are at increased risk of meningococcal disease, including those with late complement component deficiencies and those with functional or anatomical asplenia. Military recruits in the UK and USA and all would-be Hajj pilgrims were also offered the vaccines. Following recent Hajj-related outbreaks of serogroup W135 meningococci, a certificate of vaccination against serogroups A, C, Y and W135 meningococci is a visa requirement for all pilgrims travelling on the Hajj. It is important that, due to significant shortcomings, these capsular polysaccharide vaccines were not useful for routine immunization of infants.

Glycoconjugate Serogroup A and C Vaccines

The success of linking Hib capsular polysaccharide to a protein carrier which results in a vaccine which induces a thymus-dependent (T-cell dependent) IgG response in young children, and thus immunological memory, encouraged the development of capsular polysaccharide–protein conjugate vaccines for serogroups A and C meningococci. Purified capsular polysaccharides have been conjugated to either tetanus toxoid or to non-toxic cross-reacting mutant of diphtheria toxin (CRM197) and shown to be safe and immunogenic (inducing a memory response) in clinical trials.

The UK was the first country to use meningococcal serogroup C conjugate (MCC) vaccines which were licensed on the basis of immunogenicity and safety data but without a formal efficacy study. It is now increasingly clear that the vaccine is protective in all age groups, including children under the age of 2 years. Meningococcal serogroup C conjugate vaccines were introduced into the UK childhood vaccination programme in late 1999 and scheduled for all under the age of 18 years (Miller, Salisbury and Ramsay, 2001). In 2002, these vaccines were also made available to those aged 20–24 years. Preliminary estimates of efficacy suggest that the vaccine is 88–96% effective against invasive serogroup C meningococcal disease. Occasional, unexplained vaccine failures have been documented in all age groups. More recently, a multicentre collaborative study showed that the MCC vaccine protects against carriage of Group C meningococci (Maiden, Stuart and the UK Meningococcal Carriage Group, 2002).

The success of the MCC vaccines, and the imminent development of similar conjugate vaccines against serogroup A, W135 and Y meningococcal disease, heralds the prospect of effective vaccine prevention of virtually all non-serogroup B invasive meningococcal disease.

Vaccines Against Serogroup B Disease

Serogroup B Capsular Polysaccharide

Serogroup B meningococcal capsular polysaccharide consists of repeated residues of α-(2–8)-linked oligomers of sialic acid which serves as an important virulence factor and protective cell component for the organism. Candidate vaccines based on the native serogroup B polysaccharide induce a transient antibody response of predominantly IgM isotype. This poor immunogenicity of the serogroup B capsular polysaccharide could be due to sensitivity to neuraminidases or immunotolerance of the host due to its similarity to sialic acid moieties in human brain tissues (Finne *et al.*, 1987), which has caused considerable concern regarding the possible induction in humans of adverse autoimmune consequences by administration of serogroup B capsular polysaccharide-based vaccines. Nevertheless, attempts are on going to produce a capsular polysaccharide-based serogroup B vaccine. Attempts have been made to generate anticapsular bactericidal antibodies using mimotopes or using a modified B polysaccharide in which the *N*-acetyl groups at position C-5 of the sialic acid residues are replaced with *N*-propionyl groups. The latter has been conjugated to either tetanus toxoid or to recombinant meningococcal PorB and subjected to extensive preclinical trials (Fusco *et al.*, 1998).

OMPs

In view of the problem associated with the poor immunogenicity of serogroup B capsular polysaccharide, much of the attention has focused on non-capsular antigens, including constitutively expressed and iron-regulated OMPs. It is important to know that recurrence of meningococcal disease is extremely rare in the absence of immunodeficiencies, irrespective of the serogroup of the infecting strain. This indicates that non-capsular antigens can generate long-lasting cross-protective immunity.

Outer Membrane Vesicles

In the past decade, a number of serogroup B meningococcal vaccines based on outer membrane vesicles (some enriched with PorA and PorB) were developed and tested in clinical trials (Bjune *et al.*, 1991; Sierra *et al.*, 1991; Zollinger *et al.*, 1991). Following large-scale placebo-controlled, randomized double-blind trials, only the vaccines produced in Norway and Cuba showed significant protective efficacy. The Norwegian vaccine consisted of OMPs from the Norwegian epidemic strain of *N. meningitidis* (B:15:P1.7.16) and was given to children aged 14–16 years. It produced a point estimate of protective efficacy of 57% after a 30-month follow-up and therefore was considered insufficiently effective for general use (Bjune *et al.*, 1991). The Cuban vaccine consisted of serogroup C capsular polysaccharide mixed with OMPs from a Cuban epidemic strain (B:4:P1.15). When given to children aged 10–16 years, it offered an estimated point efficacy of 83% after 16 months of follow-up (Sierra *et al.*, 1991), and as a result, the vaccine is now incorporated into the routine childhood vaccination programme in Cuba. Although the Cuban trial did not directly address efficacy in children aged less than 10 years, follow-up studies of the mass vaccination have suggested that the overall protective efficacy based on vaccine coverage and incidence of disease in children under 6 years of age is about 93% (Sierra *et al.*, 1991). However, when the Cuban vaccine was tested in a case-control study in Brazil, protective efficacy was reported to vary with age. The vaccine was effective in children aged 4 years and older, but not in younger children (de Moraes *et al.*, 1992). Comparative clinical trials between the Norwegian and Cuban vaccine preparations were later conducted in Iceland and Chile in order to resolve the discrepancies. In these comparative trials the immunogenicity of the two vaccines were thought to be comparable, but the details have not yet been formally published.

It is interesting that PorA and PorB attracted a lot of attention in the 1980s and early 1990s. These proteins show considerable interstrain antigenic variation, hence their use as a basis for the serotyping and subtyping scheme for characterizing strains of *N. meningitidis*. However, they are still considered attractive candidate vaccine antigens because within a particular epidemiological setting the majority of strains causing disease belong to only a limited number of types and subtypes. An obvious drawback of vaccines based entirely on serotype and serosubtype antigens is the fact that the predominant types and subtypes associated with disease in any one area change from time to time. In an attempt to overcome the antigenic variability of PorA, outer membrane vesicle vaccines have been produced which are enriched with a mixture of several (e.g. six; hexavalent) recombinant PorA antigens obtained from prevalent meningococcal serosubtypes. Clinical trials using a hexavalent PorA-based outer membrane vesicle vaccine were largely disappointing. Apart from poor immunogenicity among infants, the vaccine induced strain-specific bactericidal activity and therefore would fail to protect against antigenically diverse meningococci (van der Voort *et al.*, 1997).

Other Approaches

Since the completion of the meningococcal genome sequence project, work on meningococcal vaccines has taken a giant leap forward. Several novel approaches have been pursued, and vaccine development has entered a new energetic race. Pizza *et al.* (2000) screened the entire genome sequence of a serogroup B meningococcal strain *in silico* and identified hundreds of putative genes coding for possible target antigens. They cloned and expressed 350 of these candidate antigens in *E. coli* and examined them for their ability to generate bactericidal antibodies in mice. They then focused on detailed characterization and vaccine assessment of seven of these antigens.

Another approach was to screen expression libraries of meningococcal genomes with human T-helper cells or antibodies and identify T-cell- and B-cell-stimulating antigens. This approach has also identified a number of potently immunogenic vaccine candidates that are important virulence and/or survival factors in the organism. Of these antigens, some

are membrane-bound T-cell immunogens, including TspA (Kizil *et al.*, 1999). Others are surface-exposed or partially secreted proteins, including AutA, App and AspA (Ait-Tahar *et al.*, 2000; Abdel-Hadi *et al.*, 2001; Turner, Wooldridge and Ala'Aldeen, 2002).

Other studies focusing on lipooligosaccharides or their mimotopes are also underway. Given the increased interest in meningococcal vaccines and significant investment by the pharmaceutical industry, there is a great deal of optimism that we may see a series of interesting vaccine trials in the near future, culminating in the development of a protective vaccine against serogroup B meningococcal disease.

REFERENCES

Abdel-Hadi, H., Wooldridge, K.G., Robinson, K. *et al.* (2001) Identification and characterization of App: an immunogenic autotransporter protein of *Neisseria meningitidis. Molecular Microbiology*, **41**, 611–623.

Achtman, M. (1994) Clonal spread of serogroup A meningococci: a paradigm for the analysis of microevolution in bacteria. *Molecular Microbiology*, **11**, 15–22.

Ait-Tahar, K., Wooldridge, K.G., Turner, D.P. *et al.* (2000) Auto-transporter A protein of *Neisseria meningitidis*: a potent CD4+ T-cell and B-cell stimulating antigen detected by expression cloning. *Molecular Microbiology*, **37**, 1094–1105.

Ala'Aldeen, D.A., Neal, K.R., Ait-Tahar, K. *et al.* (2000) Dynamics of meningococcal long-term carriage among University students and their implications for mass vaccination. *Journal of Clinical Microbiology*, **38**, 2311–2316.

Alberio, L., Lammle, B. and Esmon, C.T. (2001) Protein C replacement in severe meningococcemia: rationale and clinical experience. *Clinical Infectious Diseases*, **32**, 1338–1346.

Algren, J.T., Lal, S., Cutliff, S.A. *et al.* (1993) Predictors of outcome in acute meningococcal infection in children. *Critical Care Medicine*, **21**, 447–452.

Andreoni, J., Kayhty, H. and Densen, P. (1993) Vaccination and the role of capsular polysaccharide antibody in prevention of recurrent meningococcal disease in late complement component-deficient individuals. *The Journal of Infectious Diseases*, **168**, 227–231.

Barquet, N., Gasser, I., Domingo, P. *et al.* (1990) Primary meningococcal conjunctivitis: report of 21 patients and review. *Reviews of Infectious Diseases*, **12**, 838–847.

Barrow, G. and Feltham, R.K.A. (1993) *Cowan and Steel's Manual for the Identification of Medical Bacteria*, Cambridge University Press.

Benoit, F.L. (1963) Chronic meningococcaemia: case report and review of the literature. *The American Journal of Medicine*, **35**, 103–112.

Bevanger, L., Bergh, K., Gisnas, G. *et al.* (1998) Identification of nasopharyngeal carriage of an outbreak strain of *Neisseria meningitidis* by pulsed-field gel electrophoresis versus phenotypic methods. *Journal of Medical Microbiology*, **47**, 993–998.

Bjune, G., Hoiby, E.A., Gronnesby, J.K. *et al.* (1991) Effect of outer membrane vesicle vaccine against group B meningococcal disease in Norway. *Lancet*, **338**, 1093–1096.

Blackwell, C.C., Weir, D.M., James, V.S. *et al.* (1990) Secretor status, smoking and carriage of *Neisseria meningitidis. Epidemiological Infection*, **104**, 203–209.

Booy, R., Habibi, P., Nadel, S. *et al.* (2001) Reduction in case fatality rate from meningococcal disease associated with improved healthcare delivery. *Archives of Disease in Childhood*, **85**, 386–390.

Brandtzaeg, P. (1995) Pathogenesis of meningococcal infections, in *Meningococcal Disease* (ed. K. Cartwright), John Wiley and Sons, Ltd, Chichester, pp. 71–114.

Brandtzaeg, P., Halstensen, A., Kierulf, P. *et al.* (1992a) Molecular mechanisms in the compartmentalized inflammatory response presenting as meningococcal meningitis or septic shock. *Microbial Pathogenesis*, **13**, 423–431.

Brandtzaeg, P., Ovsteboo, R. and Kierulf, P. (1992b) Compartmentalization of lipopolysaccharide production correlates with clinical presentation in meningococcal disease. *The Journal of Infectious Diseases*, **166**, 650–652.

Carpenter, R. and Petersdorf, R.G. (1962) The clinical spectrum of bacterial meningitis. *The American Journal of Medicine*, **33**, 262–275.

Cartwright, K., Reilly, S., White, D. *et al.* (1992) Early treatment with parenteral penicillin in meningococcal disease. *BMJ*, **305**, 143–147.

Cartwright, K.A., Jones, D.M., Smith, A.J. *et al.* (1991) Influenza A and meningococcal disease. *Lancet*, **338**, 554–557.

Cartwright, K.A., Stuart, J.M., Jones, D.M. *et al.* (1987) The Stonehouse survey: nasopharyngeal carriage of meningococci and *Neisseria lactamica. Epidemiological Infection*, **99**, 591–601.

Caugant, D.A., Froholm, L.O., Bovre, K. *et al.* (1987) Intercontinental spread of *Neisseria meningitidis* clones of the ET-5 complex. *Antonie Van Leeuwenhoek*, **53**, 389–394.

Caugant, D.A., Hoiby, E.A., Magnus, P. *et al.* (1994) Asymptomatic carriage of *Neisseria meningitidis* in a randomly sampled population. *Journal of Clinical Microbiology*, **32**, 323–330.

Caugant, D.A., Hoiby, E.A., Rosenqvist, E. *et al.* (1992) Transmission of *Neisseria meningitidis* among asymptomatic military recruits and antibody analysis. *Epidemiological Infection*, **109**, 241–253.

Clarke, S.C., Reid, J., Thom, L. *et al.* (2002) Laboratory confirmation of meningococcal disease in Scotland, 1993–9. *Journal of Clinical Pathology*, **55**, 32–36.

Claus, H., Maiden, M.C., Maag, R. *et al.* (2002) Many carried meningococci lack the genes required for capsule synthesis and transport. *Microbiology*, **148**, 1813–1819.

de Gans, J. and van de Beek, D. (2002) Dexamethasone in adults with bacterial meningitis. *The New England Journal of Medicine*, **347**, 1549–1556.

de Moraes, J.C., Perkins, B.A., Camargo, M.C. *et al.* (1992) Protective efficacy of a serogroup B meningococcal vaccine in Sao Paulo, Brazil. *Lancet*, **340**, 1074–1078.

de Vries, F.P., Cole, R., Dankert, J. *et al.* (1998) *Neisseria meningitidis* producing the Opc adhesin binds epithelial cell proteoglycan receptors. *Molecular Microbiology*, **27**, 1203–1212.

Deghmane, A.E., Giorgini, D., Larribe, M. *et al.* (2002) Down-regulation of pili and capsule of *Neisseria meningitidis* upon contact with epithelial cells is mediated by CrgA regulatory protein. *Molecular Microbiology*, **43**, 1555–1564.

deMorais, J.S., Munford, R.S., Risi, J.B. *et al.* (1974) Epidemic disease due to serogroup C *Neisseria meningitidis* in Sao Paulo, Brazil. *The Journal of Infectious Diseases*, **129**, 568–571.

Densen, P., Weiler, J.M., Griffiss, J.M. *et al.* (1987) Familial properdin deficiency and fatal meningococcemia. Correction of the bactericidal defect by vaccination. *The New England Journal of Medicine*, **316**, 922–926.

Domingo, P., Muniz-Diaz, E., Baraldes, M.A. *et al.* (2002) Associations between Fc gamma receptor IIA polymorphisms and the risk and prognosis of meningococcal disease. *The American Journal of Medicine*, **112**, 19–25.

Donnelly, S.C., MacGregor, I., Zamani, A. *et al.* (1995) Plasma elastase levels and the development of the adult respiratory distress syndrome. *American Journal of Respiratory and Critical Care Medicine*, **151**, 1428–1433.

El Bashir, H., Laundy, M. and Booy, R. (2003) Diagnosis and treatment of bacterial meningitis. *Archives of Disease in Childhood*, **88**, 615–620.

Esmon, C.T., Ding, W., Yasuhiro, K. *et al.* (1997) The protein C pathway: new insights. *Thrombosis and Haemostasis*, **78**, 70–74.

Faur, Y.C., Weisburd, M.H. and Wilson, M.E. (1975) Isolation of *Neisseria meningitidis* from the Genito-urinary tract and anal canal. *Journal of Clinical Microbiology*, **2**, 178–182.

Finne, J., Bitter-Suermann, D., Goridis, C. *et al.* (1987) An IgG monoclonal antibody to group B meningococci cross-reacts with developmentally regulated polysialic acid units of glycoproteins in neural and extraneural tissues. *Journal of Immunology*, **138**, 4402–4407.

Flexner, S. (1913) The results of the serum treatment in thirteen hundred cases of epidemic meningitis. *The Journal of Experimental Medicine*, **17**, 553–576.

Frank, M.M., Joiner, K. and Hammer, C. (1987) The function of antibody and complement in the lysis of bacteria. *Reviews of Infectious Diseases*, **9** (Suppl. 5), S537–S545.

Frasch, C.E. (1983) Immunisation against *Neisseria meningitides*, in *Medical Microbiology* (eds C. Easmon and J. Jeljaszewics), Academic Press, London, pp. 115–144.

Fusco, P.C., Michon, F., Laude-Sharp, M. *et al.* (1998) Preclinical studies on a recombinant group B meningococcal porin as a carrier for a novel *Haemophilus influenzae* type b conjugate vaccine. *Vaccine*, **16**, 1842–1849.

Garabedian, B.V., Lemaigre-Dubreuil, Y. and Mariani, J. (2000) Central origin of IL-1beta produced during peripheral inflammation: role of meninges. *Molecular Brain Research*, **75**, 259–263.

Goldacre, M.J., Roberts, S.E. and Yeates, D. (2003) Case fatality rates for meningococcal disease in an English population, 1963–98: database study. *BMJ*, **327**, 596–597.

Goldschneider, I., Gotschlich, E.C. and Artenstein, M.S. (1969) Human immunity to the meningococcus. I. The role of humoral antibodies. *The Journal of Experimental Medicine*, **129**, 1307–1326.

Gotschlich, E.C., Liu, T.Y. and Artenstein, M.S. (1969) Human immunity to the meningococcus. 3. Preparation and immunochemical properties of the group A, group B, and group C meningococcal polysaccharides. *The Journal of Experimental Medicine*, **129**, 1349–1365.

Goulding, J.N., Stanley, J., Olver, W. *et al.* (2003) Independent subsets of amplified fragments from the genome of *Neisseria meningitidis* identify the same invasive clones of ET37 and ET5. *Journal of Medical Microbiology*, **52**, 151–154.

Greenwood, B.M. (1987) The epidemiology of acute bacterial meningitis in tropical Africa, in *Bacterial Meningitis* (eds J.D. Williams and J. Burnie), Academic Press, London, pp. 61–91.

Griffiss, J.M., Schneider, H., Mandrell, R.E. *et al.* (1988) Lipooligosaccharides: the principal glycolipids of the neisserial outer membrane. *Reviews of Infectious Diseases*, **10** (Suppl. 2), S287–S295.

Hackett, S.J., Carrol, E.D., Guiver, M. *et al.* (2002a) Improved case confirmation in meningococcal disease with whole blood Taqman PCR. *Archives of Disease in Childhood*, **86**, 449–452.

Hackett, S.J., Guiver, M., Marsh, J. *et al.* (2002b) Meningococcal bacterial DNA load at presentation correlates with disease severity. *Archives of Disease in Childhood*, **86**, 44–46.

Henderson, I.R., Owen, P. and Nataro, J.P. (1999) Molecular switches – the ON and OFF of bacterial phase variation. *Molecular Microbiology*, **33**, 919–932.

Hibberd, M.L., Sumiya, M., Summerfield, J.A. *et al.* (1999) Association of variants of the gene for mannose-binding lectin with susceptibility to meningococcal disease. Meningococcal Research Group. *Lancet*, **353**, 1049–1053.

Jack, D.L., Jarvis, G.A., Booth, C.L. *et al.* (2001) Mannose-binding lectin accelerates complement activation and increases serum killing of *Neisseria meningitidis* serogroup C. *The Journal of Infectious Diseases*, **184**, 836–845.

Jackson, L.A. and Wenger, J.D. (1993) Laboratory-based surveillance for meningococcal disease in selected areas, United States, 1989–1991. *Morbidity and Mortality Weekly Report*, **42**, 21–30.

Jones, D. (1995) Epidemiology of meningococcal disease in Europe and the USA, in *Meningococcal Disease* (ed. K. Cartwright), Wiley, Chichester, pp. 146–157.

Jones, D.M. and Kaczmarski, E.B. (1995) Meningococcal infections in England and Wales: 1994. *Communicable Disease Report*, **5**, R125–R130.

Kallstrom, H., Islam, M.S., Berggren, P.O. *et al.* (1998) Cell signaling by the type IV pili of pathogenic *Neisseria*. *The Journal of Biological Chemistry*, **273**, 21777–21782.

Kallstrom, H., Liszewski, M.K., Atkinson, J.P. *et al.* (1997) Membrane cofactor protein (MCP or CD46) is a cellular pilus receptor for pathogenic *Neisseria*. *Molecular Microbiology*, **25**, 639–647.

Kizil, G., Todd, I., Atta, M. *et al.* (1999) Identification and characterization of TspA, a major CD4(+) T-cell- and B-cell-stimulating *Neisseria*-specific antigen. *Infection and Immunity*, **67**, 3533–3541.

Klee, S.R., Nassif, X., Kusecek, B. *et al.* (2000) Molecular and biological analysis of eight genetic islands that distinguish *Neisseria meningitidis* from the closely related pathogen *Neisseria gonorrhoeae*. *Infection and Immunity*, **68**, 2082–2095.

Klein, N.J., Ison, C.A., Peakman, M. *et al.* (1996) The influence of capsulation and lipooligosaccharide structure on neutrophil adhesion molecule expression and endothelial injury by *Neisseria meningitidis*. *The Journal of Infectious Diseases*, **173**, 172–179.

Koppes, G.M., Ellenbogen, C. and Gebhart, R.J. (1977) Group Y meningococcal disease in United States Air Force recruits. *The American Journal of Medicine*, **62**, 661–666.

Lapeyssonie, L. (1963) La meningite cerebro-spinale en Arfique. *Bulletin of the World Health Organization*, **28** (Suppl.), 3–114.

Levin, M., Quint, P.A., Goldstein, B. *et al.* (2000) Recombinant bactericidal/permeability-increasing protein (rBPI21) as adjunctive treatment for children with severe meningococcal sepsis: a randomised trial. rBPI21 Meningococcal Sepsis Study Group. *Lancet*, **356**, 961–967.

Lin, L., Ayala, P., Larson, J. *et al.* (1997) The *Neisseria* type 2 IgA1 protease cleaves LAMP1 and promotes survival of bacteria within epithelial cells. *Molecular Microbiology*, **24**, 1083–1094.

Lystad, A. and Aasen, S. (1991) The epidemiology of meningococcal disease in Norway 1975–91. *NIPH Annals*, **14**, 57–65, 65–66 (discussion).

Maiden, M.C., Bygraves, J.A., Feil, E. *et al.* (1998) Multilocus sequence typing: a portable approach to the identification of clones within populations of pathogenic microorganisms. *Proceedings of the National Academy of Sciences of the United States of America*, **95**, 3140–3145.

Maiden, M.C., Stuart, J.M. and the UK Meningococcal Carriage Group. (2002) Carriage of serogroup C meningococci 1 year after meningococcal C conjugate polysaccharide vaccination. *Lancet*, **359**, 1829–1830.

Makras, P., Alexiou-Daniel, S., Antoniadis, A. *et al.* (2001) Outbreak of meningococcal disease after an influenza B epidemic at a Hellenic Air Force recruit training center. *Clin Infect Dis*, **33**, e48–e50.

Mellor, D.H. (1992) The place of computed tomography and lumbar puncture in suspected bacterial meningitis. *Archives of Disease in Childhood*, **67**, 1417–1419.

Miller, E., Salisbury, D. and Ramsay, M. (2001) Planning, registration, and implementation of an immunisation campaign against meningococcal serogroup C disease in the UK: a success story. *Vaccine*, **20** (Suppl. 1), S58–S67.

Moore, P.S., Plikaytis, B.D., Bolan, G.A. *et al.* (1992) Detection of meningitis epidemics in Africa: a population-based analysis. *International Journal of Epidemiology*, **21**, 155–162.

Nadel, S., Levin, M. and Habibi, P. (1995) Treatment of meningococcal disease in childhood, in *Meningococcal Disease* (ed. K. Cartwright), Wiley, Chichester, pp. 207–243.

Neal, K.R., Nguyen-Van-Tam, J.S., Jeffrey, N. *et al.* (2000) Changing carriage rate of *Neisseria meningitidis* among university students during the first week of term: cross sectional study. *BMJ*, **320**, 846–849.

Nelson, J.D. (1996) Jails, microbes, and the three-foot barrier. *The New England Journal of Medicine*, **335**, 885–886.

Oliver, W.J., Shope, T.C. and Kuhns, L.R. (2003) Fatal lumbar puncture: fact versus fiction – an approach to a clinical dilemma. *Pediatrics*, **112**, e174–e176.

Olyhoek, A.J., Sarkari, J., Bopp, M. *et al.* (1991) Cloning and expression in *Escherichia coli* of opc, the gene for an unusual class 5 outer membrane protein from *Neisseria meningitidis* (meningococci/surface antigen). *Microbial Pathogenesis*, **11**, 249–257.

Osterud, B. and Flaegstad, T. (1983) Increased tissue thromboplastin activity in monocytes of patients with meningococcal infection: related to an unfavourable prognosis. *Thrombosis and Haemostasis*, **49**, 5–7.

Parkhill, J., Achtman, M., James, K.D. *et al.* (2000) Complete DNA sequence of a serogroup A strain of *Neisseria meningitidis* Z2491. *Nature*, **404**, 502–506.

Peak, I.R., Srikhanta, Y., Dieckelmann, M. *et al.* (2000) Identification and characterisation of a novel conserved outer membrane protein from *Neisseria meningitidis*. *FEMS Immunology and Medical Microbiology*, **28**, 329–334.

Peltola, H., Anttila, M. and Renkonen, O.V. (1989) Randomised comparison of chloramphenicol, ampicillin, cefotaxime, and ceftriaxone for childhood bacterial meningitis. Finnish Study Group. *Lancet*, **1**, 1281–1287.

Perrin, A., Bonacorsi, S., Carbonnelle, E. *et al.* (2002) Comparative genomics identifies the genetic islands that distinguish *Neisseria meningitidis*, the agent of cerebrospinal meningitis, from other *Neisseria* species. *Infection and Immunity*, **70**, 7063–7072.

Perrin, A., Nassif, X. and Tinsley, C. (1999) Identification of regions of the chromosome of *Neisseria meningitidis* and *Neisseria gonorrhoeae* which are specific to the pathogenic *Neisseria* species. *Infection and Immunity*, **67**, 6119–6129.

Pizza, M., Scarlato, V., Masignani, V. *et al.* (2000) Identification of vaccine candidates against serogroup B meningococcus by whole-genome sequencing. *Science*, **287**, 1816–1820.

Radetsky, M. (1990) Duration of treatment in bacterial meningitis: a historical inquiry. *The Pediatric Infectious Disease Journal*, **9**, 2–9.

Reingold, A.L., Broome, C.V., Hightower, A.W. *et al.* (1985) Age-specific differences in duration of clinical protection after vaccination with meningococcal polysaccharide A vaccine. *Lancet*, **2**, 114–118.

Rey, M. (1991) Improving the early control of outbreaks of meningococcal disease, in *Neisseria 1990* (eds M. Achtman, P. Kohl, C. Marchal *et al.*), Walter de Gruyter, Berlin, pp. 117–122.

Riedo, F.X., Plikaytis, B.D. and Broome, C.V. (1995) Epidemiology and prevention of meningococcal disease. *The Pediatric Infectious Disease Journal*, **14**, 643–657.

Robbins, A. and Freeman, P. (1988) Obstacles to developing vaccines for the Third World. *Scientific American*, **259**, 126–133.

Saez-Nieto, J.A., Lujan, R., Berron, S. *et al.* (1992) Epidemiology and molecular basis of penicillin-resistant *Neisseria meningitidis* in Spain: a 5-year history (1985–1989). *Clinical Infectious Diseases*, **14**, 394–402.

Schaad, U.B. (1980) Arthritis in disease due to *Neisseria meningitidis*. *Reviews of Infectious Diseases*, **2**, 880–888.

Schwartz, B., Al-Tobaiqi, A., Al-Ruwais, A. *et al.* (1988) Comparative efficacy of ceftriaxone and rifampicin in eradicating pharyngeal carriage of group A *Neisseria meningitidis*. *Lancet*, **1**, 1239–1242.

Sierra, G.V., Campa, H.C., Varcacel, N.M. *et al.* (1991) Vaccine against group B *Neisseria meningitidis*: protection trial and mass vaccination results in Cuba. *NIPH Annals*, **14**, 195–207, 208–110 (discussion).

Sim, R.J., Harrison, M.M., Moxon, E.R. *et al.* (2000) Underestimation of meningococci in tonsillar tissue by nasopharyngeal swabbing. *Lancet*, **356**, 1653–1654.

Snyder, L.A., Butcher, S.A. and Saunders, N.J. (2001) Comparative whole-genome analyses reveal over 100 putative phase-variable genes in the pathogenic *Neisseria* spp. *Microbiology*, **147**, 2321–2332.

Sparling, P.F. (2002) A plethora of host factors that determine the outcome of meningococcal infection. *The American Journal of Medicine*, **112**, 72–74.

Steeghs, L., den Hartog, R., den Boer, A. *et al.* (1998) Meningitis bacterium is viable without endotoxin. *Nature*, **392**, 449–450.

Strang, J.R. and Pugh, E.J. (1992) Meningococcal infections: reducing the case fatality rate by giving penicillin before admission to hospital. *BMJ*, **305**, 141–143.

Taha, M.K., Deghmane, A.E., Antignac, A. *et al.* (2002) The duality of virulence and transmissibility in *Neisseria meningitidis*. *Trends in Microbiology*, **10**, 376–382.

Tettelin, H., Saunders, N.J., Heidelberg, J. *et al.* (2000) Complete genome sequence of *Neisseria meningitidis* serogroup B strain MC58. *Science*, **287**, 1809–1815.

Trotter, C.L., Fox, A.J., Ramsay, M.E. *et al.* (2002) Fatal outcome from meningococcal disease – an association with meningococcal phenotype but not with reduced susceptibility to benzylpenicillin. *Journal of Medical Microbiology*, **51**, 855–860.

Turner, D.P., Wooldridge, K.G. and Ala'Aldeen, D.A. (2002) Autotransported serine protease A of *Neisseria meningitidis*: an immunogenic, surface-exposed outer membrane, and secreted protein. *Infection and Immunity*, **70**, 4447–4461.

Tyski, S., Grzybowska, W., Dulny, G. *et al.* (2001) Phenotypical and genotypical characterization of *Neisseria meningitidis* carrier strains isolated from Polish recruits in 1998. *European Journal of Clinical Microbiology and Infectious Diseases*, **20**, 350–353.

van de Beek, D., de Gans, J., McIntyre, P. *et al.* (2003) Corticosteroids in acute bacterial meningitis. *Cochrane Database Syst Rev*, CD004305.

van der Voort, E.R., van Dijken, H., Kuipers, B. *et al.* (1997) Human B- and T-cell responses after immunization with a hexavalent PorA meningococcal outer membrane vesicle vaccine. *Infection and Immunity*, **65**, 5184–5190.

van Deuren, M., Brandtzaeg, P. and van der Meer, J.W. (2000) Update on meningococcal disease with emphasis on pathogenesis and clinical management. *Clinical Microbiology Reviews*, **13**, 144–166.

van Deuren, M., Neeleman, C., Assmann, K.J. *et al.* (1998) Rhabdomyolysis during the subacute stage of meningococcal sepsis. *Clinical Infectious Diseases*, **26**, 214–215.

van Deuren, M., van Dijke, B.J., Koopman, R.J. *et al.* (1993) Rapid diagnosis of acute meningococcal infections by needle aspiration or biopsy of skin lesions. *BMJ*, **306**, 1229–1232.

Virji, M., Makepeace, K., Ferguson, D.J. *et al.* (1996) Carcinoembryonic antigens (CD66) on epithelial cells and neutrophils are receptors for Opa proteins of pathogenic neisseriae. *Molecular Microbiology*, **22**, 941–950.

Virji, M., Makepeace, K. and Moxon, E.R. (1994) Distinct mechanisms of interactions of Opc-expressing meningococci at apical and basolateral surfaces of human endothelial cells; the role of integrins in apical interactions. *Molecular Microbiology*, **14**, 173–184.

Vitovski, S., Read, R.C. and Sayers, J.R. (1999) Invasive isolates of *Neisseria meningitidis* possess enhanced immunoglobulin A1 protease activity compared to colonizing strains. *The FASEB Journal*, **13**, 331–337.

Wells, D.B., Tighe, P.J., Wooldridge, K.G. *et al.* (2001) Differential gene expression during meningeal-meningococcal interaction: evidence for self-defense and early release of cytokines and chemokines. *Infection and Immunity*, **69**, 2718–2722.

Zangwill, K.M., Stout, R.W., Carlone, G.M. *et al.* (1994) Duration of antibody response after meningococcal polysaccharide vaccination in US Air Force personnel. *The Journal of Infectious Diseases*, **169**, 847–852.

Zhang, Q.Y., Jones, D.M., Saez Nieto, J.A. *et al.* (1990) Genetic diversity of penicillin-binding protein 2 genes of penicillin-resistant strains of *Neisseria meningitidis* revealed by fingerprinting of amplified DNA. *Antimicrobial Agents and Chemotherapy*, **34**, 1523–1528.

Zhu, P., Klutch, M.J., Bash, M.C. *et al.* (2002) Genetic diversity of three lgt loci for biosynthesis of lipooligosaccharide (LOS) in *Neisseria* species. *Microbiology*, **148**, 1833–1844.

Zollinger, W.D., Boslego, J., Moran, E. *et al.* (1991) Meningococcal serogroup B vaccine protection trial and follow-up studies in Chile. The Chilean National Committee for Meningococcal Disease. *NIPH Annals*, **14**, 211–212, 213 (discussion).

15

Neisseria gonorrhoeae

Catherine A. Ison

*Sexually Transmitted Bacteria Reference Laboratory, Centre for Infection,
Health Protection Agency, Colindale, London, UK*

DESCRIPTION OF THE ORGANISM

Neisseria gonorrhoeae belongs to the genus *Neisseria* and is an obligate human pathogen with no other natural host. *Neisseria gonorrhoeae* is sexually transmitted, primarily causing infection of the anogenital tract and is always considered a pathogen. This contrasts with *Neisseria meningitidis*, the only other species of the genus considered a primary pathogen, which colonises the upper respiratory tract as a commensal and occasionally invades to cause systemic disease. The other species of *Neisseria*, such as *Neisseria lactamica*, *Neisseria cinerea* and *Moraxella catarrhalis*, are generally considered commensals but have been implicated as causes of infection in immunocompromised patients.

Neisseria spp. are oxidase positive, Gram-negative cocci. The clinically important species *N. gonorrhoeae*, *N. meningitidis*, *N. lactamica*, *N. cinerea* and *M. catarrhalis* are relatively easy to distinguish from the non-pathogenic *Neisseria*. However, differentiation between the non-pathogenic species is more complex, and as a consequence, the taxonomy has been controversial (Knapp, 1988). *Neisseria* has a typical Gram-negative cell envelope, which consists of a cytoplasmic membrane, a thin layer of peptidoglycan and an outer membrane. Many of the major antigens of the cell envelope are shared between *N. gonorrhoeae* and *N. meningitidis* (Table 15.1), with the exception of the capsule which is never expressed by *N. gonorrhoeae* but when expressed by *N. meningitidis* enhances survival in blood.

PATHOGENESIS

Attachment and Invasion

Neisseria gonorrhoeae colonises mucosal surfaces primarily of the lower genital tract and occasionally ascends to the upper genital tract or invades and colonises the blood. For infection to occur, the organism must adhere to the epithelium, enter or invade the host cells, acquire sufficient nutrients from the host to survive and evade the host's immune response. Infection with *N. gonorrhoeae* stimulates an inflammatory response, resulting in a massive infiltration of polymorphs.

If *N. gonorrhoeae* is to attach successfully to the epithelium, it must avoid being swept away by cervical secretions in women and urine in men. In addition, both the bacterial and epithelial cells are negatively charged and, therefore, would naturally repel. Pili, which are hydrophobic surface appendages, are the primary mediators of attachment, firstly, overcoming the electrostatic forces between the bacterial and host cell and then by attachment to the mucosal surface by specific receptors. *Neisseria gonorrhoeae* produces type VI pili that have multiple functions including adhesion and genetic transfer such as transformation (Merz and So, 2000). The host cell receptor for neisserial pili has been identified as CD46, a member of the complement resistance proteins (Kallstrom *et al.*, 1997). Secondary attachment is mediated mainly by the Opa proteins, but lipooligosaccharide (LOS) and the gonococcal porin Por have also been implicated in adhesion to the cell surface.

Opa proteins are a family of phase-variable proteins that confer a tight attachment to the epithelial cell, recognising one of two classes of cellular receptors, the heparan sulphate proteoglycans and distinct carcinoembryonic antigens (CD66), initiating invasion through the epithelial cell (Naumann, Rudel and Meyer, 1999; Hauck and Meyer, 2003). LOS has been described as binding to asialoglycoprotein receptors on the cell surface to aid attachment (Porat *et al.*, 1995) and as having a potential role in invasion (Song *et al.*, 2000). The gonococcal porin, Por, which functions as a channel for essential nutrients, is also associated with increased bacterial invasion (Massari *et al.*, 2003) and transcytosis. Invasion of the epithelial cell follows attachment, which is mediated by Opa proteins and other outer membrane components, through multiple mechanisms, and then the gonococci travel through the epithelial cell by transcytosis to establish colonisation in the subepithelial layer.

Table 15.1 Cell envelope antigens of *Neisseria gonorrhoeae* and their equivalent in *Neisseria meningitidis*

N. gonorrhoeae	N. meningitidis	Description	Function
Por	Class 1 (Por A)	Major (Outer membrane protein) OMP	Porin/Transcytosis
	Class 2/3 (Por B)	Major OMP	Porin/Transcytosis
Rmp	Class 4	Reduction-modified protein	Site for blocking antibody
Opa	Class 5	Family of heat-modifiable proteins	Attachment/Invasion
Pilin	Class 1 pilin	Pilin	Attachment
LOS	LOS	Lipooligosaccharide	Toxic to epithelial cells
Fbp/Frp	Fbp/Frp	Iron binding/ restricted protein	Iron acquisition
Tbp A and B	Tbp A and B	Transferrin-binding protein	Iron acquisition

Principles and Practice of Clinical Bacteriology Second Edition   Editors Stephen H. Gillespie and Peter M. Hawkey

Infection of the epithelial cells induces cytokines, including tumour necrosis factor-α, that initiate an inflammatory response or may lead to the induction of apoptosis and loss of epithelial cells and give the bacterium access to the bloodstream.

There is no animal model for gonorrhoea; therefore, the process of colonisation has been worked out using *ex vivo* models particularly organ culture. These showed that the gonococci adhere to the microvilli of non-ciliated cells and, by producing blebs, cause damage to ciliated cells (McGee *et al.*, 1983; Harvey *et al.*, 1997). A simple model where pili initiate the primary attachment followed by invasion mediated by Opa proteins and transcytosis by Por is probably too simplistic. The interaction of the bacterium and the host cell is more complex, and it is likely that multiple mechanisms play a part in the infection process (Nassif *et al.*, 1999).

Iron Acquisition

Once colonisation has been established, *N. gonorrhoeae* needs to acquire iron to survive. Unlike other organisms that produce siderophores to chelate iron from the environment, *N. gonorrhoeae* has evolved mechanisms to acquire iron directly from human transferrin and lactoferrin (Alcorn and Cohen, 1994). It is postulated that, in the absence of sufficient iron, transcription of transferrin and lactoferrin receptors and transferring-binding proteins is induced in the organism. These receptors interact with human transferrin and lactoferrin, and iron is removed and transported across the bacterial cell membrane into the periplasmic space. During this process, there is a transient association with iron-binding proteins, and then, iron is transported across the cytoplasmic membrane. Other essential nutrients are transported through the porin, which is a polymer of the outer membrane proteins Por and Rmp.

Immune Response

On successful colonisation, the organism elicits an inflammatory and antibody response in the host (Brooks and Lammel, 1989). *Neisseria gonorrhoeae* is known to be ingested by macrophages, and this may be a primary mechanism by which the host eradicates infection. However, evidence that the phagocytosed organisms are killed is inconclusive. Sialylation of LOS (Parsons *et al.*, 1988) and the expression of the Opa proteins have been shown *in vitro* to enhance the ability of the organism to resist phagocytic killing. Antibody to *N. gonorrhoeae*, both immunoglobulin G (IgG) and IgA, is produced in response to infection, both in serum and in local secretions. The function of antibody produced in response to *N. gonorrhoeae* is not known. Systemic antibody may activate the complement pathway and play a role in limiting disseminated infection. However, it is not known whether complement-mediated lysis occurs in mucosal secretions. Antibody in secretions may inhibit attachment or phagocytosis, although there is little evidence that it affords protection against subsequent infection.

Genetic Variation

The ability of *N. gonorrhoeae* to cause repeated infection has been attributed to its ability to exhibit both phase and antigenic variation, by each new infection appearing novel. In phase variation, the control of expression is on or off, whereas in antigenic variation there is a change in the primary sequence, resulting in the expression of a different epitope on the same protein.

Proteins expressed on the cell surface of *N. gonorrhoeae*, such as Pili, Opa proteins and LOS, are genetically variable. In pili, antigenic variation occurs by intragenic recombination between the silent loci (pilS), which are normally not expressed, donating sequences to the pilin expression locus (pilE). The source of the donor sequences may be from within the same bacterial cell, but gonococci may naturally take up species-related DNA released by spontaneous cell lysis, and hence the source may be exogenous. The result of this recombination is that non-functional pili may be produced, resulting in phase variation where expression is switched off. Expression of different antigenic variants of pili is thought to vary the functional activity of the protein, such as adhesion to different host cells (Robertson and Meyer, 1992).

Opa proteins are a family of proteins, encoded by 11 genes, of which zero to multiple forms may be expressed in any clinical isolate. *Opa* genes include 5′ repeats of CTCTT, which are found in each gene and encode the hydrophobic core of the leader peptide. The number of these repeats determines the translational frame of the gene, and if the number of repeats is out of frame, the gene will be switched off and hence controls phase variation (Stern *et al.*, 1986). Antigenic variants may occur through homologous recombination in a similar manner to pili.

The ability of the gonococcus to control and change expression of its major outer membrane antigens enables the organism to appear novel to the host. This plays an important role in pathogenesis. It acts to prolong infections, prevent immunity to reinfection and to change the functional properties of the organism.

EPIDEMIOLOGY

The epidemiology of gonorrhoea has been influenced in the last two decades by two factors: the advent of acquired immunodeficiency syndrome (AIDS) and the development of resistance to first-line therapies such as penicillin and, more recently, ciprofloxacin.

In developing countries, the incidence of gonorrhoea is largely unknown but is considerably greater than that in industrialised countries. The global total of new cases among adults in 1999 was estimated at 62 million by the World Health Organization. There was an estimate of 17 million cases in sub-Saharan Africa and 7.5 million cases in Latin America and The Caribbean compared with 1 million in Western Europe and 1.5 million in North America. The continuing high prevalence results from the numbers of antibiotic-resistant strains circulating and from the lack of resources for effective antibiotics and/or laboratory facilities to aid the diagnosis.

In many industrialised countries, after the advent of AIDS in 1984, there was a decline in the number of cases of gonorrhoea, as well as syphilis. For example, in England and Wales, reported cases of gonorrhoea fell from approximately 50 000 before 1984 to 10 216 infections reported in 1994 (PHLS *et al.*, 2002). The decrease has been attributed to changes in sexual behaviour as a result of education campaigns and fear of a fatal disease and occurred in most groups but was most evident in men who have sex with men (MSM). It is likely that a contribution to this decline was made by good diagnostic tests, effective therapy and good contact tracing. Since 1994, there have been small but sustained increases in gonorrhoea, such that in England and Wales, the number of cases has more than doubled to 24 958 in 2002 (Health Protection Agency *et al.*, 2003). This increase which has been noted for the other bacterial sexually transmitted infections (STIs), syphilis and chlamydial infection, has been explained by a return to unsafe sexual practices, particularly failure to use condoms. This may be true now that AIDS is perceived as a treatable infection, but the ability of the organism to adapt to changes or to interventions with antimicrobial agents may play an important role.

In contrast to chlamydial infection, gonorrhoea is not evenly distributed amongst the population, with the highest rates seen in inner cities and in certain subgroups of the population such as homosexual or bisexual men, young people and ethnic minorities. This may reflect differences in sexual behaviour but may also be due to inequalities in socioeconomic status and access to health care. The greatest burden of infection is usually in young people. In England and Wales, the peak of infection is found among women aged 16–19 years of age and in men of 20–24 years of age. A greater number of men are diagnosed with gonorrhoea than women, although this will be influenced by the high percentage of symptomatic cases in men who will seek care compared to the predominantly asymptomatic and hence undetected infection in women. The ability of the organism to cause infection on a single sexual contact (transmission rate) is unknown but is thought to be greater from infected male to uninfected female (50–60%) than from infected female to uninfected male (35%).

CLINICAL FEATURES

Neisseria gonorrhoeae is primarily an infection of mucosal surfaces (reviewed in Ison and Martin, 2003). Uncomplicated gonococcal infection (UGI) occurs when *N. gonorrhoeae* colonises the columnar epithelium of the cervix, urethra, rectum or pharynx. The cervix is the primary site in women and the urethra in men. Colonisation of the mucosa often stimulates an inflammatory response, resulting in a purulent discharge. In a minority of cases, infection spreads from the lower genital tract to the upper genital tract, which is normally sterile, causing complicated gonococcal infection, which presents as pelvic inflammatory disease (PID) or salpingitis in women and prostatitis or epididymitis in men. Disseminated gonococcal infection (DGI) where *N. gonorrhoeae* invades and colonises the blood after a primary mucosal infection is only rarely encountered. Unlike complicated infection, DGI does not result from untreated or multiple infections but is a distinct entity.

Uncomplicated infection in men usually presents as urethritis, often with dysuria. The discharge is the most common symptom and ranges from scant to copious and purulent. In women, endocervical infection is predominant, resulting in increased secretions, manifesting as a vaginal discharge. Asymptomatic infection does occur in a minority of men (usually <10%) but is more common in women. Over 50% of cases of gonorrhoea in women are asymptomatic and are usually detected by tracing contacts of infected men. Rectal gonorrhoea occurs in homosexual men practising anal intercourse and may result in symptoms that range from mild discomfort to discharge and bleeding. In women, urethral and rectal infection are often the result of contamination with vaginal discharge and only occasionally the result of true colonisation. Pharyngeal infection occurs in patients practising oral intercourse, and there is no evidence that colonisation relates to the presence of a sore throat or tonsillitis.

Complicated gonococcal infection in men is rare because most mucosal infection is symptomatic and usually receives adequate therapy. In women, where infection is often asymptomatic, the organisms ascend from the cervical canal to the fallopian tubes and subsequently to the pelvic cavity, causing pelvic pain and tenderness. These symptoms may also be caused by other STIs, such as chlamydiae, as well as by organisms found as part of the normal vaginal flora. PID caused by *N. gonorrhoeae* is uncommon in industrialised countries but is often found in developing countries where treatment and health care are often inadequate. Infertility may result from PID, and the incidence increases with sequential episodes of infection.

DGI may present as a low-grade fever, followed by a rash and arthritis. DGI is more common in women and in patients with a deficiency in the late components of the complement pathway. It has also been linked to certain nutritional variants of *N. gonorrhoeae* that exhibit hypersusceptibility to penicillin. The interplay between the host and the organism that leads to the dissemination of the infection is not understood. In the 1970s, many cases were reported particularly from Sweden and the west coast of the United States. However, in recent years, it has become increasingly less common (Ison and Martin, 2003).

LABORATORY DIAGNOSIS

The rapid and accurate detection of any STI is crucial to allow suitable therapy to be given and for transmission to be prevented. In an ideal situation, this is achieved on the first occasion the patient attends for care. Historically, microscopy has been used for the presumptive diagnosis, and this is widely used in most parts of the world, even where resources for laboratory diagnosis are poor. In most countries, where laboratory facilities are available, it is normal practice to confirm the diagnosis of gonorrhoea by isolation and identification of *N. gonorrhoeae*. Detection using molecular methods has not been used as widely as for chlamydial infection, largely because existing methods have been considered satisfactory and less expensive. However, the use of nucleic acid amplification tests (NAATs) for chlamydia has highlighted the potential use of a combined approach to both chlamydial and gonococcal infection and the advantage of using non-invasive specimens.

Presumptive Diagnosis

Historically, the presence of intracellular Gram-negative cocci (Figure 15.1) has been used for the presumptive diagnosis of gonorrhoea. In symptomatic men, this has a sensitivity of >95% as compared with culture for *N. gonorrhoeae*. In asymptomatic men and women, the sensitivity is lower (30–50%) Manavi *etal*., 2003. This is probably due to reduced numbers of organisms in asymptomatic infection, which are below the detection limit of microscopy. The sensitivity is also low for rectal smears, but this is due to the large numbers of other Gram-negative organisms colonising the rectum. The sensitivity is highest if the specimen is taken through a proctoscope and if pus is seen. This is a quick and inexpensive test that is easily performed in the clinic and has a high specificity when used by experienced personnel.

Isolation

Specimens for the isolation of *N. gonorrhoeae* may be collected from the urethra, cervix, rectum and oropharynx. In heterosexual men, it is usual to take specimens only from the urethra, whereas for homosexual men, rectal and pharyngeal specimens should be taken, depending on the sexual history. In women, it is essential to sample the cervix, but it may also be helpful to take specimens from the urethra and rectum and the pharynx when appropriate.

Specimens are collected with disposable plastic loops or swabs that are made of material that is non-inhibitory to *N. gonorrhoeae*. Direct inoculation of the specimen onto the isolation medium and incubation at 36 °C until transported to the laboratory is preferred by many workers as giving the highest isolation rates. However, there is little evidence that specimens taken with a swab and placed into transport medium such as Stuart's or Amies', stored at 4 °C and sent to the laboratory as soon as possible give lower rates of isolation. The choice of method is often decided by the proximity of the clinic to the laboratory.

Neisseria gonorrhoeae is a fastidious organism that infects sites that, in some instances, are colonised by many other bacteria as part of the normal flora. An enriched medium is required that supplies nutrients such as essential amino acids, glucose and iron. Chocolatised horse blood agar was used for many years, but it is now more common to use a medium containing GC agar base that is supplemented with lysed horse blood or a growth supplement such as Kellogg's, IsoVitaleX or Vitox. These are modifications of a medium originally described by Thayer and Martin (1966) that contained GC agar base supplemented with haemoglobin. Antibiotics are often added to the primary isolation medium to suppress the growth of other organisms. The cocktail that is most often preferred consists of vancomycin or lincomycin, which inhibit Gram-positive organisms, colistin and trimethoprim, which inhibit other Gram-negative organisms, and nystatin or amphotericin to prevent overgrowth by yeasts (Table 15.2). Antibiotics added to culture media to aid the selection of *N. gonorrhoeae* may also retard the growth of or inhibit certain gonococcal strains. The use of a combination of non-selective and selective media overcomes this problem, but if resources are limited, then a selective medium is preferable.

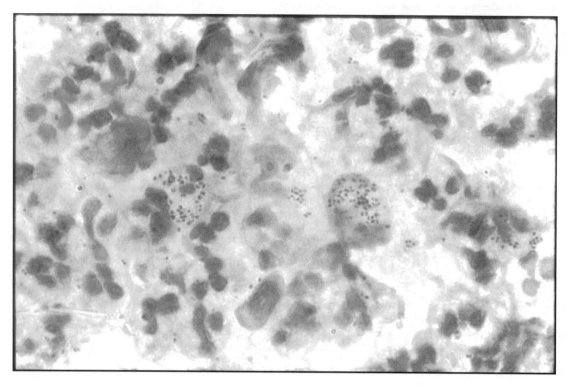

Figure 15.1   Gram-stained urethral smear showing intracellular Gram-negative diplococci.

Table 15.2   Selective agents for the isolation of *Neisseria gonorrhoeae*

Selective agent	Concentration used	Organisms inhibited
Vancomycin	2–4 mg/l	Gram-positive organisms
Lincomycin	1 mg/l	Gram-positive organisms
Colistin	7.5 mg/l	Gram-negative organisms (other than *Neisseria*)
Trimethoprim	5 mg/l	Gram-negative organisms (other than *Neisseria*)
Nystatin	12 500 IU/l	*Candida* spp.
Amphotericin	1–1.5 mg/l	*Candida* spp.

Identification

Identification of *N. gonorrhoeae* may be achieved by using specific reagents that confirm the identity of *N. gonorrhoeae* and eliminate other species of *Neisseria* or by using a battery of tests that will allow speciation into the individual members of the genus. In both instances, the presence of Gram-negative cocci that are oxidase positive (produce cytochrome *c* oxidase) is used as presumptive identification of *Neisseria* spp. Historically, carbohydrate utilisation tests have been used for identifying *N. gonorrhoeae*, which allows both differentiation from other species and speciation of other *Neisseria*. *Neisseria gonorrhoeae* produces acid from glucose alone, whereas *N. meningitidis* also produces acid from maltose and *N. lactamica* from maltose and lactose (Knapp, 1988) (Table 15.3). Conventional carbohydrate utilisation tests use an agar base that supports the growth of *Neisseria* with added carbohydrate. These media require a heavy inoculum of a pure culture and further incubation for 24 h at 36 °C before a result can be obtained. Interpretation of the colour change may be difficult. The prolonged incubation time has been overcome by the use of liquid media or commercially available kits that require between 1 and 4 h of incubation for a result. However, all these tests should be inoculated with a heavy growth of a pure culture, which normally requires at least one subculture from the primary isolation plate. Despite these disadvantages, carbohydrate utilisation has been considered the definitive means of identifying *N. gonorrhoeae*, particularly for medicolegal purposes.

Table 15.3   Identification tests for the pathogenic *Neisseria* spp.

Organism	Glucose	Acid production from			γ-Glutamyl aminopeptidase	Propyl aminopeptidase	Butyrate esterase
		Maltose	Lactose	Sucrose			
N. gonorrhoeae	+	–	–	–	–	+[a]	–
N. meningitidis	+	+	–	–	+	v	–
N. lactamica	+	+	+	–	–	–	–
N. cinerea	–	–	–	–	–	NA	–
M. catarrhalis	–	–	–	–	–	–	+

+, positive; –, negative; v, variable result.
[a] There have been reports of negative strains.

In an attempt to provide a more rapid identification of *N. gonorrhoeae* that is obtained direct from the primary isolation medium, specific antibodies to *N. gonorrhoeae* have been utilised that identify *N. gonorrhoeae* and eliminate other species of *Neisseria* but do not provide a full speciation. Monoclonal antibodies raised to specific epitopes on the two types of the major outer membrane proteins, PIA and PIB (Tam *et al.*, 1982), have been linked either to fluorescein (Microtrak *N. gonorrhoeae* Culture Confirmation Test, Trinity Biotech plc) or to staphylococcal protein A (Phadebact Monoclonal GC Test, Boule). The immunofluorescence reagent has the advantage of requiring little growth but the disadvantage that smears must be examined using a fluorescence microscope. The coagglutination reagent does not need any expensive equipment but requires sufficient growth to provide a cloudy suspension that must be boiled before use. Both the reagents have the advantage that they can be performed on non-viable organisms collected direct from the primary isolation medium. These reagents contain a mixture of specific antibodies to give a reagent that reacts with all strains of *N. gonorrhoeae* but that does not react with other species of *Neisseria*. When mixtures of antibodies are used in this manner, rather than a single antibody to a conserved epitope, there will always be the possibility that some strains will not react. Both of these reagents have been shown to be highly sensitive and specific, and any unexpected negatives can be resolved by using biochemical methods. In laboratories requiring only confirmation of *N. gonorrhoeae*, without identification of the other *Neisseria*, these immunological approaches to identification have largely replaced conventional carbohydrate utilisation tests.

Many laboratories prefer to use commercially available kits which are a combination of both carbohydrate utilisation tests and tests to detect preformed enzymes, particularly the aminopeptidases. These tests should be interpreted with caution because of many reports of gonococci that are unusually prolylaminopeptidase negative, which may lead to the misidentification of *N. gonorrhoeae* as other genera Alexander and Isan, 2005. These are useful for laboratories that isolate *N. gonorrhoeae* occasionally and have little experience of alternative methods. However, these kits are designed to differentiate between the species of *Neisseria* but are most accurate for the clinically important species.

Molecular Detection

Previous attempts at antigen detection for *N. gonorrhoeae* used primarily an immunological approach. The gonococcal antigen test that was evaluated most extensively was an enzyme-linked immunosorbent assay (ELISA) Gonozyme (Abbott Laboratories). This test had a high sensitivity and specificity when used with urethral specimens from symptomatic men, but in women, the sensitivity and specificity were lower, offering little advantage over existing tests. The antibodies used in this test showed some cross-reaction with other *Neisseria* spp., resulting in a lower specificity, which is a particular problem in low-prevalence population where the level of false positives produced may be unacceptably high. Detection of *N. gonorrhoeae* in clinical specimens using DNA-based methods has the same advantage of immunological approaches, in that they are highly sensitive, detect small numbers of organisms and detect non-viable organisms, but molecular detection can also be highly specific (Koumans *et al.*, 1998; Van Doornum *et al.*, 2001; Van Dyck *et al.*, 2001).

Molecular detection for *N. gonorrhoeae* has been achieved using DNA probes and amplification of specific sequences. The use of probes in a hybridisation test (Gene Pace 2) has proved highly sensitive and specific when used with genital specimens (Koumans *et al.*, 1998), but because of the lack of amplification they are likely to be less sensitive than using NAATs. The NAAT that has been most widely evaluated in the ligase chain reaction (LCx, Abbott Laboratory), but this test is no longer available. The two tests currently most often used amplify the DNA by the polymerase chain reaction (PCR) (Cobas Amplicor, Roche) or strand displacement (BD Probtec, Becton Dickensen). All these tests have been shown to be highly sensitive and specific for

use with urethral specimens in men and endocervical specimens in women (Van Dyck *et al.*, 2001). In common with detection of chlamydial infection, NAATs for *N. gonorrhoeae* can be used with urine for men (Palladino *et al.*, 1999; Akduman *et al.*, 2002). However, in women, the sensitivity and specificity is highest using self-taken swabs or tampons and urines are less satisfactory (Van Doornum *et al.*, 2001). The specificity of these tests with all specimens is high, but confirmation of a positive test should be considered when a single test is used to screen low-prevalence populations or if there are medicolegal implications.

The use of molecular testing for gonorrhoea has generally not been as widely used as for the detection of chlamydial infection. The emergence of commercial kits that combine the detection of *N. gonorrhoeae* and chlamydia on the same sample, for relatively little extra cost, may encourage their more extensive use. There is also increasing evidence that NAATs may be more sensitive than culture for detecting *N. gonorrhoeae*, particularly for non-invasive samples. It is likely that the increased sensitivity of NAATs indicates that they are affected less by inadequacies in specimen storage and transport than culture. None of these kits are currently licensed for use with non-genital samples, e.g. from the pharynx or the rectum. The detection of *N. gonorrhoeae* in the presence of commensal *Neisseria* in the pharynx or multiple other bacteria in the rectum requires the use of primers to detect sequences totally specific to *N. gonorrhoeae*. This has proved difficult, and if used on these specimens, it is advisable to use two tests that detect different sequences to reduce the number of false positives. However, culture for *N. gonorrhoeae* in the rectum and the pharynx is less than 100% sensitive, and so NAATs may prove a very useful diagnostic test for these sites in the future.

Typing Methods

The typing of *N. gonorrhoeae* has been used extensively as an epidemiological tool to monitor trends over time, particularly of antibiotic-resistant strains, to study reinfection versus treatment failure and in tracing contacts, such as for medicolegal purposes. Historically, the methods used have been phenotypic in nature and include antimicrobial susceptibility patterns, plasmid profiles, auxotyping and serotyping. Antimicrobial susceptibility patterns and plasmid profiles are limited in their usefulness but do give information on the development and type of resistance (Sarafian and Knapp, 1989).

Auxotyping is differentiation of strains by their nutritional requirements by using a range of chemically defined medium lacking individual amino acids (Catlin, 1973). Many auxotypes have been identified, but four types predominate in most populations – non-requiring (prototrophic or wild type), proline requiring (pro), arginine requiring (arg) and arginine, hypoxanthine and uracil requiring (AHU). AHU-requiring strains have been associated with increased susceptibility to penicillin, DGI and heterosexual gonorrhoea, whereas non-requiring and proline-requiring strains are more resistant to antibiotics. Although auxotyping has revealed some interesting associations, it is technically demanding and lacks sufficient discrimination for most purposes.

Serotyping of *N. gonorrhoeae* is achieved by the pattern of reactivity of a boiled suspension of the gonococcal strain with a panel of monoclonal antibodies linked to staphylococcal protein A, used in a coagglutination technique (Knapp *et al.*, 1984). The panel consists of 12 antibodies directed to specific epitopes on the two types of the major outer membrane proteins, Por, with six antibodies directed at PIA and six at PIB. The pattern of reactivity denotes the serovar. *Neisseria gonorrhoeae* express either PIA or PIB, and it is unusual to encounter hybrids that express both types. Serotyping is simple to perform and inexpensive and is favoured by many diagnostic laboratories. However, the endpoint is the macroscopic examination and grading of the agglutination and is subjective, and interpretation may vary between individuals. The amount and quality of antibody used also influences the result, and hence, the reproducibility of the technique over long time periods is less than 100%. Serotyping has greatly enhanced the ability to discriminate between strains, particularly if used in combination with auxotyping to produce auxotype/serovar (A/S) classes. Problems with the availability of the antibodies in more recent years and the lack of discrimination for some purposes has led to the search for a genotypic method that could offer reproducibility and enhance discrimination. The ideal method would be as simple and easy to perform as serotyping so that it could be used in diagnostic laboratories.

Genotyping methods use the degree of variation in one or more genes to discriminate between strains that are unlinked or to confirm strains from a common source. *Neisseria gonorrhoeae* are known to exhibit a high degree of genetic variation, to be competent for gene exchange at all parts of its life cycle and to be non-clonal. If diversity in variable genes is used, a high degree of discrimination may be achieved, in contrast to a conserved or housekeeping gene(s) where less discrimination would be expected.

A wide range of techniques for genotyping *N. gonorrhoeae* have been described that have examined diversity in the whole genome, number of repetitive sequences or individual genes, resulting in different degrees of discrimination (Ison, 1998). Techniques described that target the whole genome include restriction endonuclease fingerprinting, ribotyping and pulsed-field gel electrophoresis (PFGE) (Unemo *et al.*, 2002a). Amplification by PCR using either random primers or repetitive element sequence-based primers (rep-PCR) and lip typing (Trees *et al.*, 2000) detects variation in the number of repetitive sequences. Techniques that target specific genes include opa typing that examines diversity in the family of 11 *opa* genes (O'Rourke *et al.*, 1995), sequencing or probing of part or the complete *por* gene (Unemo *et al.*, 2002b; McKnew *et al.*, 2003; Martin *et al.*, 2004) and multilocus sequence typing (MLST) (Viscidi and Demma, 2003), which examines variation in seven or more housekeeping genes. All genotypic methods have proved as or more discriminatory than phenotypic methods, although there is variation, in that ribotyping and methods that examine repetitive sequences are less discriminatory and PFGE and opa typing are probably the most discriminatory.

The considerable degree of genetic diversity exhibited by *N. gonorrhoeae* raises questions regarding the validity of monitoring the genotype over time or at different geographical locations because the genotype may vary too quickly to be useful for epidemiological purposes. There are instances, e.g. the emergence of antibiotic-resistant strains, where it would be informative to know whether the same strains are spreading between countries. This may well be best addressed through analysis of the mechanism of resistance, such as comparison of mutations in *gyrA* and *parC* genes in ciprofloxacin-resistant gonococci. Highly discriminatory methods have proved useful for studying sexual networks and outbreaks, which occur over relatively short time periods. In this instance, if diversity in a highly variable gene is used, then it is possible that each gonococcal strain will appear unique, unless linked as part of a short transmission chain. This has been used to detect clusters in populations which were not visible by epidemiological techniques alone (Ward *et al.*, 2000).

Despite considerable progress in genotyping for *N. gonorrhoeae*, a single method that is simple and inexpensive to perform, which could be used in diagnostic laboratories, has not emerged. Many techniques, largely based on the *por* gene either alone or in combination with other genes, are emerging as methods that may be useful for answering specific questions relating to outbreaks. These methods are sequence based and so can be automated and give unambiguous and manageable data. Although the availability of sequencing is increasing in many laboratories and the cost is falling, it is unlikely that it will become used universally. However, routine typing of gonococcal isolates may not give meaningful results, because of the high rate of variation, and it may be better to address specific questions, using the more complex techniques, in a limited number of specialist centres.

MANAGEMENT OF INFECTION

Gonorrhoea is often treated before the susceptibility of the infecting isolate is known, such as after a presumptive diagnosis by microscopy, treatment as a contact of an infected patient or in the absence of a susceptibility test when syndromic treatment is used. It is important to know the patterns of resistance in any geographical location, informed by surveillance programmes, to enable the outcome of therapy to be predicted with reasonable confidence. The accepted dogma is that any first-line therapy should effectively treat >95% of cases.

Penicillin, or ampicillin, was the antibiotic of choice for the treatment of gonorrhoea for many years because it was effective and inexpensive. Resistance emerged slowly and was largely overcome by increasing the dosage. However, isolates of *N. gonorrhoeae* that exhibit plasmid and/or chromosomal resistance to penicillin became disseminated worldwide by the 1980s, presenting a therapeutic problem. National and international guidelines, such as those produced by the World Health Organisation, were changed to recommend the alternatives: ciprofloxacin, 500 mg orally; ceftriaxone, 125 or 250 mg intramuscularly; or the oral cephalosporin, cefixime (400 mg) and spectinomycin 2 g intramuscularly. Ciprofloxacin has been heavily used in many parts of the world, for the treatment of many infectious diseases including gonorrhoea, and resistance emerged in gonococci and has increased to levels above 5%, such that ciprofloxacin is currently being removed as a first-line therapy in some countries.

Ceftriaxone is the preferred alternative but is given by injection, making it a less attractive first-line therapy. Cefixime is currently the only oral third-generation cephalosporin for which there is good efficacy data, but there are concerns regarding its continued availability in some parts of the worlds. Azithromycin is recommended in some guidelines and is often used, but for the treatment of gonorrhoea, the recommended dose is 2 g, which is poorly tolerated, and there is a tendency to use a 1-g dose. However, there is increasing evidence that treatment failure is increasing with both dosages (Young *et al.*, 1997), and azithromycin is not a good alternative to the highly active third-generation cephalosporins.

Antimicrobial Resistance

Neisseria gonorrhoeae is inherently susceptible to most antibiotics, but through continual usage, resistance has emerged (Table 15.4), either by the acquisition of plasmids from other organisms, such as *Haemophilus*, or by the selection of resistant mutants.

Resistance to penicillin may be plasmid or chromosomally mediated. Penicillinase-producing *N. gonorrhoeae* (PPNG) which exhibit high-level plasmid-mediated resistance (minimum inhibitory concentration, MIC, ≥4 mg/l) were first documented in 1976 (Ashford *et al.*, 1976; Percival *et al.*, 1976; Phillips, 1976). Isolates that originated from Africa and Asia were found to carry plasmids of 3.2 MDa and 4.4 MDa, respectively (Roberts *et al.*, 1977), and both encoded for a TEM-1 type β-lactamase, the smaller plasmid having a deletion in a non-functional

Table 15.4 Breakpoints that define categories of antimicrobial resistance in *Neisseria gonorrhoeae*

Strain nomenclature	Strain definition
CMRNG	Penicillin MIC ≥1 mg/l (≥2 mg/l*), tetracycline 2–8 mg/l
PPNG	Penicillinase-producing *N. gonorrhoeae*, tetracycline MIC <16 mg/l
TRNG	Non-penicillinase–producing *N. gonorrhoeae* (non-PPNG), tetracycline ≥16 mg/l
PP/TRNG	PPNG, tetracycline MIC ≥16 mg/l
QRNG	PPNG, non-PPNG, TRNG, PP/TRNG or CMRNG with ciprofloxacin MIC ≥1 mg/l

MIC, minimum inhibitory concentration.
* NCCLS recommended breakpoint.

region of the plasmid (Pagotto *et al.*, 2000). Transfer of penicillinase plasmids may occur between strains of *N. gonorrhoeae* by conjugation, but it requires the presence of a conjugative plasmid of 24.5 MDa to mobilise the transfer (Roberts, 1989). Despite their original geographical associations, PPNG carrying these plasmids have now disseminated worldwide. PPNG carrying plasmids of differing sizes have been reported in more recent years, but these also encode for the TEM-1 type β-lactamase and have not spread so widely.

Chromosomally mediated resistant *N. gonorrhoeae* (CMRNG) exhibit low-level resistance to penicillin and also to tetracycline. Unlike PPNG which produce an enzyme that destroys penicillin, chromosomal resistance is the additive result of mutations at multiple sites which decrease the permeability of the cell envelope and hence forms a barrier against penicillin. Mutations in *penA*, *mtr*, *penB*, *ponA1* and *penC*, acquired in a stepwise manner (Ropp *et al.*, 2002), are necessary to produce a strain that is likely to fail therapy. *penA* encodes for penicillin-binding protein 2 (PBP2), and a resistant isolate exhibits a reduced affinity for penicillin (Brannigan *et al.*, 1990). The *mtr* locus consists of an operon, *mtrRCDE*, which controls susceptibility to a range of hydrophobic antibiotics and detergents, and resistance occurs through overexpression of the *mtrRCDE* efflux pump (Hagman *et al.*, 1995). However, expression of *mtr* requires the presence of the *penB* mutation, which has been correlated to a mutation in the *por* gene (Gill *et al.*, 1998). These three loci confer decreased susceptibility, but mutations in two further loci, *ponA1* (PBP1) (Ropp and Nicholas, 1997) and *penC*, are required for full resistance. The additive effect of these mutations increases the MIC from ≤0.06 mg/l in a susceptible strain to ≥1 mg/l, with an increased chance of therapeutic failure.

Ciprofloxacin, a fluoroquinolone, has been used for over a decade as an alternative to penicillin because it was highly active against *N. gonorrhoeae* and is given orally as a single dose, usually at 500 mg. However, reports of decreased susceptibility appeared soon after its introduction, and therapeutic failure is now common in many parts of the world. The targets for ciprofloxacin are the enzymes DNA gyrase (GryA) and the topoisomerase IV (ParC), which are responsible for supercoiling the bacterial DNA. Resistance occurs through mutations at multiple sites in the quinolone resistance-determining region (QRDR) of these genes (Belland *et al.*, 1994), with the highest level of resistance occurring when mutations are present in both *gyrA* and *parC* (Knapp *et al.*, 1997; Trees *et al.*, 2001). The relationship between the level of resistance and therapeutic failure was initially unknown and has been controversial. There is good agreement that high-level resistance, linked to an increased chance of therapeutic failure, correlates with an MIC ≥1 mg/l. However, decreased susceptibility (MIC 0.12–0.5 mg/l) has also been associated with treatment failure but occurs less often than among isolates exhibiting high-level resistance.

Tetracycline is not commonly used as a first-line treatment for gonorrhoea but is often given concurrently for the treatment of possible chlamydial infection and has been used in parts of the world where more effective therapies are not available or are too expensive. Resistance to tetracycline in *N. gonorrhoeae* may be high level and plasmid mediated (MIC ≥16 mg/l), which is the result of acquisition of the *tetM* determinant into the gonococcal conjugative plasmid (Morse *et al.*, 1986). This plasmid mobilises itself between gonococci and other genera found in the genital tract, and hence, resistance has spread quickly since its first description in 1985. Low-level resistance (MIC ≥2 mg/l) is the result of chromosomal mutations (Johnson and Morse, 1988), and its relationship to therapeutic failure is unclear.

Resistance to spectinomycin, in contrast to other antimicrobial agents, is high level and occurs in a single step, resulting in an increase in MIC from 8 mg/l to 16 mg/l to >256 mg/l, which is invariably resistant to therapy (Easmon *et al.*, 1982; Galimand, Gerbaud and Courvalin, 2000). There have been several documented episodes of spectinomycin-resistant gonorrhoea, but these have not spread widely. Spectinomycin is not commonly used as a first-line therapy and has become difficult to obtain in some countries but may still be particularly useful when resistance to the newer alternative treatments develops.

Therapeutic resistance to ceftriaxone has not been documented, although gonococcal isolates exhibiting reduced susceptibility have been reported (Schwebke *et al.*, 1995), and there are anecdotal reports of resistance. It is likely that there will be increased usage of the third-generation cephalosporins in the coming years as a result of the high levels of resistance to ciprofloxacin in many parts of the world, increasing the selective pressure for resistant mutants to appear. Azithromycin may also be used increasingly for the same reasons, but resistance has been reported (Young, Moyes and McMillan, 1997) and is increasing, and this is unlikely to be a long-term alternative for the treatment of gonorrhoea.

Susceptibility Testing

In vitro susceptibility testing is used to predict the outcome of therapy, detect the emergence of resistance and monitor drifts in susceptibility. This may be achieved by disk diffusion, breakpoint agar dilution technique or determination of the MIC by agar dilution. Susceptibility testing for *N. gonorrhoeae* may be difficult because the organism is fastidious and requires an enriched medium, which may interfere with the action of some of the antibiotics. There are methods recommended by the National Committee for Clinical Laboratory Standards (NCCLS), the World Health Organization (World Health Organization, 1989) and the British Society of Antimicrobial Chemotherapy (BSAC) (King, 2001) among others, which differ largely in medium used and concentration of antibiotic in the disk. Many laboratories use variations of these methodologies, which may be used successfully if an appropriate panel of control strains is used to aid interpretation of the results. The results of *in vitro* susceptibility testing should be correlated with therapeutic outcome wherever possible: susceptible strains having a <5% likelihood of treatment failure, resistant strains having a >15% chance of treatment failure and intermediate resistance or decreased susceptibility having a 5–15% possible treatment failure.

Although many patients with gonorrhoea are treated empirically, it is still common practice to routinely test gonococcal isolates from individual patients to ensure correct patient management. The choice of method is a matter for each laboratory, but disk diffusion is used most often. An alternative method that is appropriate for routine use is the breakpoint agar dilution technique, where a few concentrations of antibiotic are used to categorise strains into susceptible, reduced susceptibility and resistant. This has the advantage of giving a predicted range for the MIC and may be used with a multipoint inoculator, enabling multiple strains to be tested simultaneously. While the breakpoint technique is applicable to laboratories receiving large numbers of gonococcal isolates, for laboratories encountering few isolates or wishing to perform the occasional MIC, the E-test is very useful. The E-test is performed by placing a paper strip containing varying concentrations of antibiotic onto a lawn of bacteria, the resulting zone of inhibition giving an indication of the MIC. This method is commercially available and expensive but is useful and growing in popularity (Biedenbach and Jones, 1996).

Conventional susceptibility testing is reliant on obtaining a viable organism from the patient. However, there is increasing interest in using molecular tests for the diagnosis of gonorrhoea, particularly for screening in primary care settings or where facilities for culture are not available. This will result in fewer organisms being available for testing both for individual patient management and for surveillance programmes; hence, the data that are produced will be less representative. The ideal solution is to develop molecular methods for detecting resistance in *N. gonorrhoeae*, which detect resistance to the main therapeutic agents. The antimicrobial agents currently most commonly used are ciprofloxacin and ceftriaxone or an alternative third-generation cephalosporin. Resistance to ciprofloxacin is known to be the result of mutations in both the *gyrA* and *parC* gene, but the mutations occur at multiple sites on both genes, and it has been difficult to identify mutations found in all strains that would predict therapeutic failure. Levels of ciprofloxacin resistance are increasing, and it is possible that the technical problems of detecting resistance will not be overcome before the drug is withdrawn as a useful therapy. Treatment failure to ceftriaxone is not yet described, although reduced susceptibility is known and has been attributed to the same mutations as chromosomal resistance to penicillin. However, the design of any test to detect therapeutic resistance is difficult until such resistance emerges.

Penicillin, or ampicillin, was used for many years for treating gonorrhoea, but owing to the high prevalence of resistance in many parts of the world it is used infrequently as a first-line therapy unless good surveillance data are available to indicate that the gonococcal population is susceptible. Molecular methods for detecting plasmid-mediated resistance to penicillin by using amplification by PCR (Palmer, Leeming and Turner, 2000) have been described, and the mutations responsible for chromosomal resistance to penicillin are known; therefore, if considered necessary, molecular detection of penicillin resistance is possible. Resistance to other antimicrobial agents may be of interest, particularly for surveillance purposes, such as tetracycline or azithromycin. Amplification methods are in use for plasmid-mediated resistance to tetracycline and to divide the plasmid into two types (Ison, Tekki and Gill, 1993; Xia, Pang and Roberts, 1995), which may be useful for epidemiological studies of antibiotic-resistant gonorrhoea.

It will be a challenge to continue to provide representative surveillance data in the era of NAATs. One approach is to screen for gonorrhoea by using a molecular test and a non-invasive sample and then to recall patients who have a positive test, to enable collection of specimens for the isolation of *N. gonorrhoeae*, which can then be used for susceptibility testing, before the patient is treated.

Surveillance

The efficacy of first-line therapies for gonorrhoea is dependent on good surveillance data giving patterns of susceptibility to inform the choice of an appropriate antimicrobial agent (Ison, Billon and Tapsall, 1998). It is suggested that any therapy used for gonorrhoea should be successful in at least 95% of patients; otherwise, a change in therapy should be considered. There are many established surveillance programmes worldwide, including in the United States (Schwarcz *et al.*, 1990), Canada (Dillon, 1992), Australia (Members of the Australian Gonococcal Surveillance Programme, 1984) and the Netherlands (Van de Laar *et al.*, 1997), that have longitudinal data that are useful to show temporal changes. The World Health Organization initiated a global antimicrobial surveillance programme (GASP) in 1990, but it is currently most active in the Americas and the Caribbean and the Western Pacific Region. More recently, additional programmes have been initiated as resistant gonorrhoea continues to spread. In England and Wales, national surveillance was initiated in 2000 (Paine *et al.*, 2001) and has already provided information regarding the rapid increase in ciprofloxacin resistance above the 5% level (Fenton *et al.*, 2003). This information has been used to inform a change in national guidelines for the first-line treatment of gonorrhoea from ciprofloxacin to ceftriaxone. The WHO WPR (Western Pacific Region) surveillance programme (2001) has demonstrated alarming levels of resistance to penicillin, approaching 100% in some countries and continuing increases in ciprofloxacin resistance.

The challenge for surveillance programmes is to produce comparable data between countries. There are many methodologies used worldwide, but it is unlikely that a common methodology will be agreed upon. The use of a well-defined panel of control strains reflecting different categories of susceptibility and resistance to therapeutic agents, together with regular participation in national and/or international quality assessment programmes, where panels of isolates are circulated for testing, is likely to be a more successful approach.

PREVENTION AND CONTROL

Gonorrhoea is the second most common bacterial STI. Although it is a treatable infection, the number of cases continues to rise. There has been considerable interest in producing a vaccine that may give protection particularly against complicated infection in women, because of the sequelae of PID and infertility, but attempts to induce protection by vaccination have been largely unsuccessful. This is due to the antigenic variability of the major surface antigens such as pili. The first major trial of a gonococcal vaccine in men using purified pili from a single strain was unsuccessful because antibody produced was strain specific (Boslego *et al.*, 1991). Alternative candidates including the major outer membrane protein, Por, and proteins involved in iron acquisition have shown promise but have not been developed for large-scale trials. An additional problem in developing a vaccine has been the lack of knowledge on the host response required to induce protection. Control of gonococcal infection is, therefore, dependent on good access to care, rapid and accurate diagnostic tests and effective therapy. This must be allied to an efficient system to trace and treat sexual contacts. The impact of health education may be seen following the fall in gonorrhoea incidence associated with the publicity surrounding the human immunodeficiency virus (HIV) epidemic.

Gonorrhoea, together with other bacterial STIs, has been linked to the transmission of HIV infection (Fleming and Wasserheit, 1999), and in some studies, aggressive treatment of bacterial STIs has reduced the acquisition of HIV (Grosskurth *et al.*, 2000). It is, therefore, imperative that all efforts be taken to control infections such as gonorrhoea, to aid the control efforts for AIDS. In the absence of a vaccine that provides long-term protection, control is dependent on public health interventions, which in turn require resources. Therefore, in resource-poor settings, control and prevention measures are least effective and the prevalence of gonorrhoea is usually high.

REFERENCES

Akduman, D., Ehret, J.M., Messina, K. *et al.* (2002) Evaluation of a strand displacement amplification (BD Probtec-SDA) for detection of *Neisseria gonorrhoeae* in urine specimens. *Journal of Clinical Microbiology*, **40**, 281–283.

Alcorn, T.M. and Cohen, M.S. (1994) Gonococcal pathogenesis: adaptation and immune evasion in the human host. *Current Opinion on Infectious Diseases*, **7**, 310–316.

Alexander, S.A. and Ison, C.A. (2005) Evaluation of commercial kits for the identification of *Neisseria gonorrhoeae*. *Journal of Medical Microbiology*, **54**, 827–831.

Ashford, W.A., Golash, R.G. and Hemming, V.G. (1976) Penicillinase-producing *Neisseria gonorrhoeae*. *Lancet*, **2**, 657–658.

Belland, R.J., Morrison, S.G., Ison, C. *et al.* (1994) *Neisseria gonorrhoeae* acquires mutations in analogous regions of *gyrA* and *parC* in fluoroquinolone-resistant isolates. *Molecular Microbiology*, **14**, 371–380.

Biedenbach, D.J. and Jones, R.N. (1996) Comparative assessment of Etest for testing susceptibilities of *Neisseria gonorrhoeae* to penicillin, tetracycline, ceftriaxone, cefotaxime, and ciprofloxacin: investigation using 510(k) review criteria, recommended by the Food and Drug Administration. *Journal of Clinical Microbiology*, **34**, 3214–3217.

Boslego, J.W., Tramont, E.C., Chung, R.C. *et al.* (1991) Efficacy trial of a parenteral gonococcal pilus vaccine in men. *Vaccine*, **9**, 154–162.

Brannigan, J.A., Tirodimos, I.A., Zhang, Q.Y. *et al.* (1990) Insertion of an extra amino acid is the main cause of the low affinity of penicillin-binding protein 2 in penicillin resistant strains of *Neisseria gonorrhoeae*. *Molecular Microbiology*, **4**, 913–919.

Brooks, G.F. and Lammel, C.J. (1989) Humoral Immune response to gonococcal infection. *Clinical Microbiology Reviews*, **2**, S5–S10.

Catlin, B.W. (1973) Nutritional profiles of *Neisseria gonorrhoeae*, *Neisseria meningitidis* and *Neisseria lactamica* in chemically defined media and the use of growth requirements for gonococcal typing. *The Journal of Infectious Diseases*, **128**, 178–194.

Dillon, J.R. (1992) National microbiological surveillance of the susceptibility of gonococcal isolates to antimicrobial agents. *Canadian Journal of Infectious Diseases*, **3**, 202–206.

Easmon, C.S.F., Ison, C.A., Bellinger, C.M. *et al.* (1982) Emergence of resistance after spectinomycin treatment for gonorrhoea due to β-lactamase-producing strain of *Neisseria gonorrhoeae*. *British Medical Journal*, **284**, 1604–1605.

Fenton, K.A., Ison, C., Johnson, A.P. *et al.* (2003) Ciprofloxacin resistance in *Neisseria gonorrhoeae* in England and Wales, 2002. *Lancet*, **361**, 1867–1869.

Fleming, D. and Wasserheit, J. (1999) From epidemiological synergy to public health policy and practice: the contribution of other sexually transmitted diseases to sexual transmission of HIV infection. *Sexually Transmitted Infection*, **75**, 3–17.

Galimand, M., Gerbaud, G. and Courvalin, P. (2000) Spectinomycin resistance in *Neisseria* spp. due to mutations in 16S rRNA. *Antimicrobial Agents and Chemotherapy*, **44**, 1365–1366.

Gill, M.J., Simjee, S., Al-Hattawi, K. *et al.* (1998) Gonococcal resistance to b-lactams and tetracycline involves mutation in loop 3 of the porin encoded at the penB locus. *Antimicrobial Agents and Chemotherapy*, **42**, 2799–2803.

Grosskurth, H., Mwijarubi, E., Todd, J. *et al.* (2000) Operational performance of an STD control programme in Mwanza Region, Tanzania. *Sexually Transmitted Infection*, **76**, 426–436.

Hagman, K.E., Pan, W., Spratt, B.G. *et al.* (1995) Resistance of *Neisseria gonorrhoeae* to antimicrobial agents is modulated by the mtrRCDE efflux system. *Microbiology*, **141**, 611–622.

Harvey, H.A., Ketterer, M.R., Preston, A. *et al.* (1997) Ultrastructural analysis of primary human urethral epithelial cell cultures infected with *Neisseria gonorrhoeae*. *Infection and Immunity*, **65**, 2420–2427.

Hauck, C.R. and Meyer, T.F. (2003) 'Small' talk: Opa proteins as mediators of *Neisseria*–host cell communication. *Current Opinion in Microbiology*, **6**, 43–49.

Health Protection Agency, SCIEH, ISD, National Public, Health Service for Wales, CDSC Northern Ireland and the UASSG (2003) *Renewing the Focus. HIV and other sexually transmitted infections in the United Kingdom in 2002.* London: Health Protection Agency. www.hpa.org.uk/infections/topics_az/hiv_and_sti/publications/annual2003/annual2003.pdf (13 July 2005).

Ison, C. and Martin, D. (2003) Gonorrhea. In *Atlas of Sexually Transmitted Diseases and AIDS, 3rd Edition*. Morse, S.A., Ballard, R.C., Holmes, K.K. and Moreland, A.A. (Eds). Mosby (Chapter 6).

Ison, C.A. (1998) Genotyping of *Neisseria gonorrhoeae*. *Current Opinion in Infectious Diseases*, **11**, 43–46.

Ison, C.A., Tekki, N. and Gill, M.J. (1993) Detection of the *tetM* determinant in *Neisseria gonorrhoeae*. *Sexually Transmitted Diseases*, **20**, 329–333.

Ison, C.A., Dillon, J.R. and Tapsall, J. (1998) The epidemiology of global antibiotic resistance among *Neisseria gonorrhoeae* and *Haemophilus ducreyi*. *Lancet*, **351** (Suppl.), 8–11.

Johnson, S.R. and Morse, S.A. (1988) Antibiotic resistance in *Neisseria gonorrhoeae*, genetics and mechanisms of resistance. *Sexually Transmitted Diseases*, **15**, 217–224.

Kallstrom, H., Liszewski, M.K., Atkinson, J.P. and Jonsson, A.B. (1997) Membrane cofactor protein (MCP or CD46) is a cellular pilus receptor for pathogenic *Neisseria*. *Molecular Microbiology*, **25**, 639–647.

King, A. (2001) Recommendations for susceptibility tests on fastidious organisms and those requiring special handling. *The Journal of Antimicrobial Chemotherapy*, **48** (Suppl.), 77–80.

Knapp, J.S. (1988) Historical perspectives and identification of *Neisseria* and related species. *Clinical Microbiology Reviews*, **1**, 415–431.

Knapp, J.S., Tam, M.R., Nowinski, R.C. *et al.* (1984) Serological classification of *Neisseria gonorrhoeae* with use of monoclonal antibodies to gonococcal outer membrane protein I. *The Journal of Infectious Diseases*, **150**, 44–48.

Knapp, J.S., Fox, K.K., Trees, D.L. and Whittington, W.L. (1997) Fluoroquinolone resistance in *Neisseria gonorrhoeae*. *Emerging Infectious Diseases*, **3**, 33–38.

Koumans, E.H., Johnson, R.E., Knapp, J.S. *et al.* (1998) Laboratory testing for *Neisseria gonorrhoeae* by recently introduced nonculture tests: a performance review with clinical and public health considerations. *Clinical Infectious Diseases*, **27**, 1171–1180.

Manavi, K, Young, H. and Clutterbuck, D. (2003) Sensitivity of microscopy for the rapid diagnosis of gonorrhoea in men and women and the role of gonorrhoea serovars. *International Journal of STD & AIDs*, **14**, 390–394.

Martin, I.M.C., Ison, C.A., Aanensen, D.M. *et al.* (2004) Rapid sequence-based identification of gonococcal transmission clusters in a large metropolitan area. *The Journal of Infectious Diseases*, **189**, 1497–1505.

Massari, P., Ram, S., MacLeod, H. and Wetzler, L.M. (2003) The role of porins in neisserial pathogenesis and immunity. *Trends in Microbiology*, **11**, 87–93.

McGee, Z., Stephens, D., Hoffman, L. *et al.* (1983) Mechanisms of mucosal invasion by pathogenic *Neisseria*. *Reviews of Infectious Diseases*, **5**, S708–S714.

McKnew, D.L., Lynn, F., Zenilman, J.M. and Bash, M.C. (2003) Porin variation among clinical isolates of *Neisseria gonorrhoeae* over a 10-year period, as determined by Por variable region typing. *The Journal of Infectious Diseases*, **187**, 1213–1222.

Members of the Australian Gonococcal Surveillance Programme (1984) Penicillin sensitivity of gonococci in Australia: development of Australian gonococcal surveillance programme. *The British Journal of Venereal Diseases*, **60**, 226–230.

Merz, A.J. and So, M. (2000) Interactions of pathogenic neisseriae with epithelial cell membranes. *Annual Review of Cell and Developmental Biology*, **16**, 423–457.

Morse, S.A., Johnson, S.R., Biddle, J.W. *et al.* (1986) High-level tetracycline resistance in *Neisseria gonorrhoeae* is result of acquisition of streptococcal *tetM* determinant. *Antimicrobial Agents and Chemotherapy*, **30**, 664–670.

Nassif, X., Pujol, C., Morand, P. and Eugene, E. (1999) Interactions of pathogenic *Neisseria* with host cells. Is it possible to assemble the puzzle? *Molecular Microbiology*, **32**, 1124–1132.

National Committee for Clinical Laboratory Standards (2003) *Approved Standard: Performance Standards for Antimicrobial Disk Susceptibility Tests, 5th ed.* Document M2-A5. Villanova, PA: National Committee for Clinical Laboratory Standards.

Naumann, M., Rudel, T. and Meyer, T.M. (1999) Host cell interactions and signalling with *Neisseria gonorrhoeae*. *Current Opinion in Microbiology*, **2**, 62–70.

O'Rourke, M., Ison, C.A., Renton, A.R. and Spratt, B.G. (1995) Opa-typing: a high resolution tool for studying the epidemiology of gonorrhea. *Molecular Microbiology*, **17**, 865–875.

Pagotto, F., Aman, A.T., Ng, L.K. *et al.* (2000) Sequence analysis of the family of penicillinase-producing plasmids of *Neisseria gonorrhoeae*. *Plasmid*, **43**, 24–34.

Paine, T.C., Fenton, K.A., Herring, A. *et al.* (2001) GRASP: a new national sentinel surveillance initiative for monitoring gonococcal antimicrobial resistance in England and Wales. *Sexually Transmitted Infections*, **77**, 398–401.

Palladino, S., Pearman, J.W., Kay, I.D. *et al.* (1999) Diagnosis of *Chlamydia trachomatis* and *Neisseria gonorrhoeae*. Genitourinary infections in males by the Amplicor PCR assay of urine. *Diagnostic Microbiology and Infectious Disease*, **33**, 141–146.

Palmer, H.M., Leeming, J.P. and Turner, A. (2000) A multiplex polymerase chain reaction to differentiate beta-lactamase plasmids of *Neisseria gonorrhoeae*. *The Journal of Antimicrobial Chemotherapy*, **45**, 777–782.

Parsons, N.J., Patel, P.V., Tan, E.L. *et al.* (1988) Cytidine 5′-monpohospho-N-acetylneuraminic acid and a low molecular weight factor from human blood cells induce lipopolysaccharide alteration in gonococci when conferring resistance to killing by human serum. *Microbial Pathogenesis*, **5**, 303–309.

Percival, A., Rowlands, J., Corkhill, J.E. *et al.* (1976) Penicillinase-producing gonococci in Liverpool. *Lancet*, **2**, 1379–1382.

Phillips, I. (1976) Beta-lactamase-producing, penicillin-resistant gonococcus. *Lancet*, **2**, 656–657.

PHLS, DHSS&PS, the Scottish ISD(D)5 Collaborative Group (2002) *Sexually transmitted infections in the UK: new episodes seen at Genitourinary Medicine Clinics, 1995–2001.* London: Public Health Laboratory Service. www.hpa.org.uk/infections/topics_az/hiv_and_sti/publications/sti_report2001.pdf (13 July 2005).

Porat, N., Apicella, M.A. and Blake, M.S. (1995) A lipooligosaccharide-binding site on HepG2 cells similar to the gonococcal opacity-associated surface protein Opa. *Infection and Immunity*, **63**, 2164–2172.

Roberts, M., Elwell, L.P. and Falkow, S. (1977) Molecular characterization of two beta-lactamase-specifying plasmids isolated from *Neisseria gonorrhoeae*. *Journal of Bacteriology*, **131**, 557–563.

Roberts, M.C. (1989) Plasmids of *Neisseria gonorrhoeae* and other *Neisseria* species. *Clinical Microbiology Reviews*, **2** (Suppl.), S18–S23.

Robertson, B.D. and Meyer, T.F. (1992) Genetic variation in pathogenic bacteria. *Trends in Genetics*, **8**, 422–427.

Ropp, P.A. and Nicholas, R.A. (1997) Cloning and characterization of the *ponA* gene encoding penicillin-binding protein 1 from *Neisseria gonorrhoeae* and *Neisseria meningitidis*. *Journal of Bacteriology*, **179**, 2783–2787.

Ropp, P.A., Hu, M., Olesky, M. *et al.* (2002) Mutations in ponA, the gene encoding penicillin-binding protein, 1, and a novel locus, penC, are required for high-level chromosomally mediated penicillin resistance in *Neisseria gonorrhoeae*. *Antimicrobial Agents and Chemotherapy*, **46**, 769–777.

Sarafian, S.K. and Knapp, J.S. (1989) Molecular epidemiology. *Clinical Microbiology Reviews*, **2**, S49–S55.

Schwarcz, S.K., Zenilman, J.M., Schnell, D. *et al.* (1990) National surveillance of antimicrobial resistance in *Neisseria gonorrhoeae*. The Gonococcal Isolate Surveillance Project. *Journal of American Medical Association*, **264**, 1413–1417.

Schwebke, J.R., Whittington, W., Rice, R.J. *et al.* (1995) Trends in susceptibility of *Neisseria gonorrhoeae* to ceftriaxone from 1985 through 1991.
Antimicrobial Agents and Chemotherapy, **39**, 917–920.

Song, W., Ma, L., Chen, R. and Stein, D.C. (2000) Role of lipooligosaccharide
in Opa-independent invasion of *Neisseria gonorrhoeae* in human epithelial
cells. *The Journal of Experimental Medicine*, **191**, 949–959.

Stern, A., Brown, M., Nickel, P. and Meyer, T.F. (1986) Opacity genes in
Neisseria gonorrhoeae: control of phase and antigenic variation. *Cell*, **47**,
61–71.

Tam, M.R., Buchanan, T.M., Sandstrom, E.G. *et al.* (1982) Serological classification of *Neisseria gonorrhoeae* with monoclonal antibodies. *Infection
and Immunity*, **36**, 1042–1053.

Thayer, J.D. and Martin, J.E. Jr (1966) Improved medium selective for
cultivation of N. gonorrhoeae and N. meningitidis. *Public Health Report*,
81, 559–562.

Trees, D.L., Schultz, A.J. and Knapp, J.S. (2000) Use of the neisserial lipoprotein
(Lip) for subtyping *Neisseria gonorrhoeae*. *Journal of Clinical Microbiology*,
38, 2914–2916.

Trees, D.L., Sandul, A.L., Neal, S.W. *et al.* (2001) Molecular epidemiology of
Neisseria gonorrhoeae exhibiting decreased susceptibility and resistance to
ciprofloxacin in Hawaii, 1991–1999. *Sexually Transmitted Diseases*, **28**,
309–314.

Unemo, M., Berglund, T., Olcén, P. and Fredlund, H. (2002a) Pulsed-field
gel electrophoresis as an epidemiologic tool for *Neisseria gonorrhoeae*;
identification of clusters within serovars. *Sexually Transmitted Diseases*,
29, 25–31.

Unemo, M., Olcén, P., Berglund, T. *et al.* (2002b) Molecular epidemiology of
Neisseria gonorrhoeae: sequence analysis of the *porB* gene confirms presence
of two circulating strains. *Journal of Clinical Microbiology*, **40**,
3741–3749.

Van de Laar, M.J.W., van Duynhoven, Y.T.P.H., Dessens, M. *et al.* (1997)
Surveillance of antibiotic resistance in *Neisseria gonorrhoeae* in the
Netherlands, 1977–95. *Genitourinary Medicine*, **73**, 510–517.

Van Doornum, G.J., Schouls, L.M., Pijl, A. *et al.* (2001) Comparison between the
LCx Probe system and the COBAS AMPLICOR system for the detection
of *Chlamydia trachomatis* and *Neisseria gonorrhoeae* infections in
patients attending a clinic for sexually transmitted diseases in Amsterdam,
the Netherlands. *Journal of Clinical Microbiology*, **39**, 829–835.

Van Dyck, E., Ieven, M., Pattyn, S. *et al.* (2001) Detection of *Chlamydia
trachomatis* and *Neisseria gonorrhoeae* by enzyme immunoassay, culture
and three nucleic acid amplification tests. *Journal of Clinical Microbiology*,
39, 1751–1756.

Viscidi, R.P. and Demma, J.C. (2003) Genetic diversity of *Neisseria gonorrhoeae* housekeeping genes. *Journal of Clinical Microbiology*, **41**,
197–204.

Ward, H., Ison, C.A., Day, S.E. *et al.* (2000) A prospective social and molecular
investigation of gonococcal transmission. *Lancet*, **356**, 1812–1817.

WHO Western Pacific Region Gonococcal Antimicrobial Surveillance
Programme (2001) Surveillance of antibiotic resistance in *Neisseria
gonorrhoeae* in the WHO Western Pacific Region, 2000. *Communicable
Diseases Intelligence*, **25**, 274–277.

World Health Organization (1989) *Bench-Level Laboratory Manual for
Sexually Transmitted Diseases*. *WHO/VDT/89.443*. Geneva: World Health
Organization.

Xia, M., Pang, Y. and Roberts, M.C. (1995) Detection of two groups of
25.2 MDa TetM plasmids by polymerase chain reaction of the downstream
region. *Molecular and Cellular Probes*, **9**, 327–332.

Young, H., Moyes, A. and McMillan, A. (1997) Azithromycin and erythromycin
resistant *Neisseria gonorrhoeae* following treatment with azithromycin.
International Journal of STD & AIDS, **8**, 299–302.

16

Acinetobacter spp.

Peter Hawkey[1,2] and Eugenie Bergogne-Berezin[3]

[1]*The Medical School, University of Birmingham;* [2]*Health Protection Agency, Birmingham Heartlands Hospital, Birmingham, UK;*
and [3]*Faculty of Medicine Bichat, University Paris 7, Paris, France*

INTRODUCTION

Aerobic non-fermentative Gram-negative bacilli have increasingly become important as human pathogens over the last 20 years. *Pseudomonas aeruginosa* has been well studied, and most other Gram-negative aerobic bacilli such as *Acinetobacter* spp., *Stenotrophomonas* spp., *Flavobacterium* spp. and *Alcaligenes* (*Achromobacter* spp.) have been recognised more recently as pathogens. In a survey of Gram-negative bacilli isolated from 396 intensive care units (ICUs) in the United States, *P. aeruginosa* (19.7% strains) was the second most common isolate after *Escherichia coli*, followed by *Acinetobacter* spp. (5.3%) and *Stenotrophomonas maltophilia* (3.7%) (Quinn, 1998). We have also included *Flavobacterium* and *Achromobacter* as, although these are rare causes of infection, they may cause substantial nosocomial outbreaks. After many years of confusing and erratic taxonomy of these organisms, molecular taxonomic and identification methods have defined new species and genera.

They are saprophytic bacilli, present in the environment (soil, water and phytosphere) and may be recovered from the hospital environment and from colonised or infected patients or from staff, predominantly on the hands. As opportunistic pathogens, these bacilli are responsible for nosocomial infections, mainly severe pneumonia, septicaemia and urinary tract infections (UTIs). Infections caused by them are often difficult to treat because of their frequent multiple resistance to many antibiotics (Gales *et al.*, 2001), which, combined with their persistence in different hospital environments, has resulted in an increased interest in their biology.

ACINETOBACTER

Among Gram-negative bacteria causing nosocomial infections (particularly in ICUs), *Acinetobacter* spp. have attracted increasing attention during the last two decades (Baker and Hawkey, 2004). *Acinetobacter* spp. are implicated in a wide spectrum of infections (e.g. bacteraemia, nosocomial pneumonia, UTIs, secondary meningitis and superinfections in burn patients). One of the most striking features of *Acinetobacter* spp. is their extraordinary ability to express multiple resistance mechanisms against major antibiotic classes: many strains have become resistant to broad-spectrum β-lactams (third-generation cephalosporins, carboxypenicillins and, increasingly, carbapenems). They produce a wide range of aminoglycoside-inactivating enzymes, and many strains are resistant to fluoroquinolones. Among *Acinetobacter* spp., *Acinetobacter baumannii* is the most frequent species involved in nosocomial infections (Seifert *et al.*, 1993) and by far the most resistant to major antibiotic classes. The quest for understanding the epidemiology of *Acinetobacter* spp. has given rise to studies using increasingly sophisticated techniques for typing these organisms isolated from patients and from the hospital environment.

Microbiology

Definition

Acinetobacter spp. are aerobic Gram-negative bacilli commonly present in soil and water as free-living saprophytes. They are also isolated as commensals from skin, throat and various secretions of healthy people, as well as causing human infections (Joly-Guillou and Brunet, 1996; Bergogne-Berezin, 2001). Morphologically, they are aerobic Gram-negative coccobacilli and usually found in diploid formation or chains of variable length. They are not motile, hence their scientific name from the Greek word (ακινετοζ), i.e. 'unable to move', although the cells display a 'twitching motility', presumably because of the presence of polar fimbriae. They are strictly aerobic and grow easily on all common media at temperatures from 20 to 30 °C, with, for most strains, the optimum at 33–35 °C. Growth at 41 and 44 °C occurs for a few species and is a discriminating phenotypic characteristic. The key features for identifying *Acinetobacter* spp. are an oxidase-negative, catalase-positive, indole-negative, nitrate-negative, Gram-negative bacillus. Production, or not, of acid from D-glucose, D-ribose, D-xylose and L-arabinose (utilised oxidatively as carbon sources) had been used in early identification methods for strains formerly designated var. *anitratus* or var. *lwoffii* (Table 16.1). These phenotypic characteristics included in commercial identification systems with 20 biochemical tests (API 20NE, Bio-Merieux) are not considered a sufficiently reliable identification method: more precise and accurate identification requires DNA-based methods, which are increasingly used.

Taxonomy

Acinetobacter spp. have undergone considerable changes in taxonomy over the years: strains were designated *Bacterium anitratum*, *Herellea vaginicola*, *Mima polymorpha*, *Achromobacter*, *Micrococcus calcoaceticus*, *Diplococcus*, B5W and *Cytophaga* (Juni, 1984), until the genus *Acinetobacter* was reclassified in 1986 by Bouvet and Grimont using modern genotypic methods (genetic transformation, DNA–DNA hybridisation and RNA sequence comparison) into genomic species including *A. baumannii* (Bouvet and Grimont, 1986). Since then, a variety of new genotypic methods [ribotyping, tRNA spacer fingerprinting, amplified fragment length polymorphism (AFLP) and amplified ribosomal DNA restriction analysis (ARDRA)]

Table 16.1 Major properties for identifying the most frequently isolated *Acinetobacter* spp.

Biovars	Genus *Acinetobacter*: *Acinetobacter calcoaceticus* (Brisou and Prévot, 1954)[a]										
	A. anitratus (glucose acidified)						*A. lwoffii*[b] (glucose negative)				*A. radioresistens*
Genospecies	*A. calcoaceticus*	*A. baumannii*	Unnamed genospecies	*A. haemolyticus*	*A. junii*	Unnamed genospecies	*A. johnsonii*	*A. lwoffii*	Unnamed genospecies	Unnamed genospecies	
Bouvet and Grimont (1986)	1[d]	2	3	4	5	6	7	8/9	10	11	12[c]
Growth characteristics	No growth at 41 °C	Growth at 44 °C	Growth at 41 °C but not at 44 °C	37 °C	37 °C	37 °C	–	37 °C	37 °C	37 °C	37 °C
Glucose	Positive	Positive	Positive	Positive (52%)	Negative	Positive (66%)	Negative	Negative	Positive	Negative	Negative (acidified)
Haemolysis	–	–	–	+(100%)	–	+(100%)	–	–	–	–	–
Gelatin hydrolysis	–	–	–	+(96%)	–	+(100%)	–	–	–	–	–
Utilisation tests[e]											
β-Alanine	+	+	+	–	–	–	–	–	+	+	–
DL-Lactate assimilation	+[f]	+	–[g]	+	–	+	+	+	+	–	
Citrate	+	+	+	+(91%)	+(82%)	+	+	–	+	+	+
Malonate	+	+	+	–	–	–	–	–	–	–	+

[a] Bergey's Manual (1984).
[b] *A. alcaligenes* is genospecies.
[c] Five other unnamed genospecies have been identified (13–17) (Nowak *et al.*, 1995) and are being studied.
[d] The numbered genospecies are based on DNA–DNA hybridisations.
[e] Eleven other assimilation tests of carbon and energy sources may be used (Bouvet and Grimont, 1986; Gerner-Smidt, 1992).
[f] 100% positive strains.
[g] 0% positive strains.

have identified more than 20 genomic species and new species have been created: *Acinetobacter nannii* (formerly *Acinetobacter calcoaceticus* var. *anitratus* and *Acinetobacter glucidotytica non liquefaciens*), *Acinetobacter haemolyticus*, *Acinetobacter junii*, *Acinetobacter johnsonii* and *Acinetobacter radioresistens*. The definition of *Acinetobacter* is based on phenotypic characteristics (biotyping) and on the identification of genotypic species. The evolution of taxonomy and the main characteristics of 'new' *Acinetobacter* spp. are summarised in Tables 16.1 and 16.2, respectively. Species designation can also include previous designations such as *A. anitratus* for glucose-positive strains, *Acinetobacter lwoffii* for glucose-negative strains. 'Acinetobacter calcoaceticus' was used in relatively recent literature. The ability to produce acid from glucose (oxidation of glucose to produce gluconic acid) is not regarded any more as a characteristic of taxonomic significance.

Genetics

The only complete genome sequence of an *Acinetobacter* that is available are sequence assemblies of *A. calcoaceticus* ADP1 (BD413), from www.genoscope.cns.fr. This strain has been extensively studied as an environmental bacterium capable of degrading a variety of chemicals including aromatic hydrocarbons. This strain is also naturally transformable and has been used to sample the environment for recombinant DNA sequences (Palmen and Hellingwerf, 1997). An *rec*A-independent system for gene amplification has been described in ADP1, which, if present in the clinically significant species, could be important in upregulated chromosomally located antibiotic resistance genes (Reams and Neidle, 2004). Until whole genome sequence information from *A. baumannii* or other clinically significant genomospecies is available, there is only a limited understanding of the genetics of *Acinetobacter*. A survey of predominantly *A. baumannii* from around the world showed that 17 of 25 isolates carried integrons and all but one were class 1, usually carrying aminoglycoside-modifying enzyme genes. The integrons showed similarities to each other, despite being carried by isolates with distinct genotypes (Seward and Towner, 1999).

Pathogenesis of *Acinetobacter* Infection

Virulence of Acinetobacter *spp.*

Although considered originally to be a relatively low-grade pathogen (Juni, 1978), *Acinetobacter* has many putative virulence factors that have been identified (Avril and Mensard, 1991). The presence of a polysaccharide capsule formed of L-rhamnose, D-glucuronic acid and D-mannose renders the cell surface of strains more hydrophilic, hydrophobicity being greater in *A. baumannii* isolated from catheters or tracheal devices (Rosenberg *et al.*, 1983; Kaplan *et al.*, 1985). *Acinetobacter* cells adhere to human epithelial cells probably via the capsule and fimbriae (Rosenberg *et al.*, 1982). The lipopolysaccharide (LPS) of *Acinetobacter* has been shown to be active in *in vitro* assays against eukaryotic cells and to stimulate cytokine release (Garcia *et al.*, 1999). A homologue of the *tra*T gene responsible for serum resistance in *E. coli* has been found in *Acinetobacter* (Montenegro *et al.*, 1985). Lipases (butyrate esterase, caprylate esterase and leucine arylamidase) produced by *Acinetobacter* may damage tissue lipids (Bergogne-Berezin and Towner, 1996). No DNAase, elastase, haemolysin, trypsin or chymotrypsin production has been detected in *Acinetobacter* strains. A novel catechol siderophore has been identified in *A. baumanii* 8399 which came from a patient with nosocomial pneumonia, and the biosynthetic and transport proteins have been characterised (Dorsey *et al.*, 2003). The virulence of *Acinetobacter* has been studied in many animal models of systemic infections (Bergogne-Berezin and Towner, 1996), including an experimental pneumonia model in immunosuppressed mice (Joly-Guillou *et al.*, 1997).

Host Predisposing Factors

Various risk factors predisposing to severe *Acinetobacter* infection have been identified, including the presence of a severe underlying diseases, e.g. malignancy, burns, immunosuppression and major surgery, both elderly and neonatal patients being at increased risk (Bergogne-Berezin and Towner, 1996). The setting for infection is usually the medical/surgical ICU, renal or burns unit, where catheters of all kinds act as portals of entry. The role of antibiotics in predisposing to infection is important as acquisition of multidrug-resistant *Acinetobacter* is significantly associated with prior antibiotic treatment as well as the factors listed above (Smolyakov *et al.*, 2003).

Epidemiology

Strains of *Acinetobacter* are widely distributed in nature and are found in virtually all samples of soil and fresh water when appropriate culture techniques are used (Bergogne-Berezin and Towner, 1996). A range of species have been identified from soil, sewage, plants and food products, although the species associated with human disease are not normally found in such sites.

Human Carriage

Human carriage of *Acinetobacter* has been demonstrated in normal individuals: it forms part of the bacterial flora of the skin and has been found in the axillae, groin and toe webs of normal individuals. *Acinetobacter* colonises the oral cavity, the respiratory tract and the gastrointestinal tract and is found predominantly in moist skin areas (Bergogne-Berezin and Towner, 1996). Studies of the carriage of *Acinetobacter* on human skin and mucous membranes show *A. lwoffii* (47%), *A. johnsonii* (21%) and *A. radioresistens* (12%) and genospecies 3 (11%) to be the most common in both patients and controls. *Acinetobacter baumannii* and genospecies 13, which are the most common nosocomial species, were very rare in controls (6.5% and 1%) (Seifert *et al.*, 1997). A study in Hong Kong found genospecies 3 to be the most common (36%) and *A. lwoffii*/ *A. johnsonii* to be rare (1.6%) in controls from the community, whereas in patients, *A. baumannii* was the most common (35%), emphasising variation in species carried between different geographical locations (Chu *et al.*, 1999). The community is not likely to be a reservoir for *A. baumannii* (identified phenotypically), as suggested by data from a study of hand carriage in New York, where multidrug-resistant strains typical of nosocomial isolates were absent from the community (Zeana *et al.*, 2003).

Sources and Spread in Nosocomial Infections

Sporadic cases of *Acinetobacter* infections are seen in many hospitals and in a variety of patient settings. However, outbreaks of infection caused by endemic strains are being increasingly described, particularly in ICUs. The carriage rate of *Acinetobacter* on the skin of hospitalised patients is significantly higher than the community, and this site has been thought to be an important source (Seifert *et al.*, 1997). This has been postulated to be due to reduced hygiene standards among hospitalised patients and the warm, humid atmosphere of hospital beds, which is supported by the observation that colonisation is more frequent in summer months. However, two recent studies using DNA-based identification techniques have demonstrated that much of this colonisation is due to species of *Acinetobacter* not commonly associated with clinical infection (Seifert *et al.*, 1997; Berlau *et al.*, 1999).

In contrast, very high rates of *A. baumannii* carriage have been described during outbreaks of infection in ICUs. Frequent colonisation of the skin, throat, respiratory and digestive tract has been described in

Table 16.2   Current nomenclature of non-fermentative, Gram-negative bacilli (list limited to those potentially involved in infections) (after Bruckner and Colonna, 1995)

Main groups (genera) as designated in the medical language	Current name	Synonym
Acinetobacter	*Acinetobacter baumannii*	*Acinetobacter anitratus*
	Acinetobacter calcoaceticus	*Acinetobacter anitratus*
		Acinetobacter calcoaceticus spp. *calcoaceticus*
	Acinetobacter haemolyticus	
	Acinetobacter johnsonii	
	Acinetobacter junii	
	Acinetobacter lwoffii	*Acinetobacter anitratus*
		Acinetobacter calcoaceticus spp. *lwoffii*
Alcaligenes	*Agrobacterium tumefaciens*	*Agrobacterium radiobacter* (CDC group Vd-3)
	Alcaligenes faecalis	*Alcaligenes odorans*
		Pseudomonas odorans
	Alcaligenes piechaudii	
	Alcaligenes xylosoxidans subsp. *denitrificans*	*Alcaligenes denitrificans* (CDC group Vc)
	Alcaligenes xylosoxidans subsp. *xylosoxidans*	*Alcaligenes denitrificans* subsp. *xylosoxidans*
		Alcaligenes xylosoxidans (CDC groups IIIa and IIIb)
Burkholderia	*Burkholderia cepacia*	*Pseudomonas cepacia*
		Pseudomonas multivorans
		Pseudomonas kingae (CDC group EO-1)
	Burkholderia gladioli	*Pseudomonas gladioli*
		Pseudomonas marginata
	Burkholderia mallei	*Pseudomonas mallei*
		Actinobacillus mallei
	Burkholderia pickettii	*Pseudomonas pickettii* (CDC groups Va-1 and Va-2)
		Pseudomonas thomasii
	Burkholderia pseudomallei	*Pseudomonas pseudomallei*
Comamonas	*Comamonas acidovorans*	*Pseudomonas acidovorans*
	Comamonas testosteroni	*Pseudomonas testosteroni* (CDC group EF-19)
	Comamonas terrigena	
Chryseobacterium	*Chryseomonas luteola*	*Pseudomonas luteola* (CDC group Ve-1)
	Chryseobacterium gleum	*Flavobacterium gleum* (CDC group IIb)
	Chryseobacterium indologenes	*Flavobacterium indologenes* (CDC group IIb)
	Chryseobacterium meningosepticum	*Flavobacterium meningosepticum* (CDC group IIa)
	Chryseobacterium odoratum (*Myroides odoratus*)	*Flavobacterium odoratum* (CDC group M-4f)
Flavobacterium	*Empedobacter brevis*	*Flavobacterium breve*
	Flavimonas oryzihabitans	*Pseudomonas oryzihabitans* (CDC group Ve-2)
	Flavobacterium sp. group IIe	CDC group IIe
	Flavobacterium sp. group IIh	CDC group IIh
	Flavobacterium sp. group Iii	CDC group Iii
Pseudomonas	*Pseudomonas aeruginosa*	
	Pseudomonas alcaligenes	
	Pseudomonas chlororaphis	*Pseudomonas aureofaciens*
	Pseudomonas delafieldii	
	Pseudomonas fluorescens	
	Pseudomonas mendocina	CDC group Vb-2
	Pseudomonas pertucinogena	*Bordetella pertussis* (rough phase IV)
	Pseudomonas pseudoalcaligenes	*Pseudomonas alcaligenes* biotype B
	Pseudomonas putida	
	Pseudomonas stutzeri	CDC group Vb-1
	Pseudomonas stutzeri-like	CDC group Vb-3
	Pseudomonas sp. group 1	*Pseudomonas denitrificans*
	Pseudomonas-like group 2	CDC group IV-d
Sphingobacterium	*Sphingobacterium mizutaii*	*Flavobacterium mizutaii*
	Sphingobacterium multivorum	*Flavobacterium multivorum* (CDC group IIK-2)
	Sphingobacterium spiritivorum	*Flavobacterium spiritivorum*
		Sphingobacterium versatilis (CDC group IIK-3)
	Sphingobacterium thalpophilum	*Flavobacterium thalpophilum*
	Sphingobacterium yabuuchiae	*Flavobacterium yabuuchiae*
	Sphingomonas paucimobilis	*Pseudomonas paucimobilis* (CDC group IIK 11)
	Weeksella virosa	*Flavobacterium genitale* (CDC group II-f)
	Weeksella zoohelcum	*Flavobacterium*-like group IIj
Stenotrophomonas	*Stenotrophomonas maltophilia*	*Xanthomonas maltophilia*
	Myroides odoratus	*Pseudomonas maltophilia* (CDC group M-4f)

ICU patients. It is likely that respiratory and genitourinary colonisation are the most important sources of nosocomial infection. In one study of adult patients in a French ICU, 33% developed oropharyngeal or rectal carriage of *A. baumannii* and/or *Klebsiella pneumoniae* within a median period of 9 days (Garrouste-Org *et al.*, 1996). Studies of burns patients have demonstrated colonisation rates with *A. baumannii* of over 50%, mainly due to superficial wound colonisation (Fujita, Lilly and Ayliffe, 1982; Sherertz and Sullivan, 1985; Wisplinghoff, Perbix and Seifert, 1999). High rates of colonisation of the respiratory tract have also been described in outbreaks involving mechanically ventilated patients (Allen and Green, 1987; Mah *et al.*, 2001). In one study, 45% of tracheostomies were found to be colonised with *Acinetobacter* spp. (Rosenthal, 1974). Although colonisation of the digestive tract is not generally considered to be as an important source of *Acinetobacter*, several studies have demonstrated significant rectal carriage of *A. baumannii* in ICU patients (Timsit *et al.*, 1993; Corbella *et al.*, 1996).

A wide range of hospital environment sites, both moist and dry, have been shown to be contaminated with *Acinetobacter* spp. (Bergogne-Berezin and Towner, 1996). Materials in contact with colonised patients act as sources after or during admission of the patient, particularly if moist. This has been seen with outbreaks associated with wet contaminated mattresses with breaks in the plastic covers in a burns unit (Sherertz and Sullivan, 1985). *A. baumannii* has been found to be a cause of nosocomial infection amongst cats/dogs and horses in veterinary hospitals, spreading in a very similar fashion to human infections (Boerlin *et al.*, 2001).

Other items of equipments, such as curtains, keyboards, arterial pressure transducers and, particularly, ventilator equipment, may be sources of outbreaks (Cefai *et al.*, 1990; Bergogne-Berezin and Towner, 1996; Baker and Hawkey, 2004). Once the hospital equipment has become contaminated, the ability particularly of the clinically important genospecies (e.g. *A. baumannii*, genospecies 3, 7 and 13) to survive desiccation may be exceptional (Jawad *et al.*, 1996). Genomospecies of *Acinetobacter* associated with nosocomial infection were found to have significantly greater desiccation tolerance, as were freshly isolated strains (Jawad *et al.*, 1996). Bacteria from the equipment may then contaminate the hands of staff and be transmitted to patients. In one outbreak, environmental contamination correlated closely with the number of colonised patients, and improved cleaning was thought to contribute to termination of the outbreak (Denton *et al.*, 2004).

CLINICAL FEATURES OF *ACINETOBACTER* INFECTIONS

Acinetobacter spp. have been implicated in many nosocomial/opportunistic infections. The predominant sites of *Acinetobacter* nosocomial infections have varied with time and local epidemiological factors. In early reports, UTIs predominated in ICUs, but more recently the incidence of UTIs has decreased, possibly in relation to a better care of urinary catheters, whereas the incidence of nosocomial pneumonia has increased significantly (Bergogne-Berezin, 2001). A European survey of causative agents in nosocomial pneumonia carried out in seven countries using the same protocol has established an overall incidence of approximately 10% for *Acinetobacter* (Bergogne-Berezin, 2001). Less common infections such as meningitis, endocarditis, skin and soft-tissue infections, peritonitis or surgical wound infections have been seen as sporadic cases (Bergogne-Berezin and Towner, 1996). *Acinetobacter* species are involved in super-infection in burn patients (Bergogne-Berezin, Joly-Guillou and Vieu, 1987).

Respiratory Tract Infections

Acinetobacter pneumonia does not differ clinically from most severe Gram-negative pneumonias. A combination of clinical signs including fever, neutrophilia, purulent sputum production and appearance of new infiltrates on radiograph or computed tomographic (CT) scan must lead to microbiological investigation. The aetiologic agent is be isolated from bronchial aspirates, bronchial brushings or bronchoalveolar lavage. *Acinetobacter* respiratory tract infections occur predominantly in mechanically ventilated patients (Bergogne-Berezin and Towner, 1996). In the United States, the National Nosocomial Infection Surveillance (NNIS) reports that 4% of all nosocomial pneumonias are caused by *Acinetobacter* (Schaberg, Culver and Gaynes, 1991). In the European Prevalence of Infection in Intensive Care (EPIIC) study, *Acinetobacter* accounted for 10% of all pneumonia. The most severe form is bacteraemic pneumonia, often associated with shock and sepsis occurring in the presence of risk factors (underlying severe pathologies, high APACHE II Acute Physiology And Chronic Health Evaluation score and broad-spectrum antibiotic therapy) (Lortholary *et al.*, 1995; Husni *et al.*, 1999).

Crude mortality rates in *Acinetobacter* pneumonia ranging from 30 to 75% are reported, with the highest rate in ventilated patients (Fagon *et al.*, 1989; Torres *et al.*, 1990). A recent Spanish ICU study demonstrated an attributable mortality of 53% in patients with *A. baumannii* infections (Garcia-Garmendia *et al.*, 1999). When compared with a control group, the excess length of stay was 13 days and the odds ratio for death was **4.0**. Rare cases of community-acquired pneumonia have been reported, being implicated in 10% of patients (particularly alcoholics) in a report from Northern Australia (Anstey, Currie and Withnall, 1992).

Bacteraemia

Risk factors predisposing to bacteraemia and predominant sources of the organism are pneumonias, trauma, surgical procedures, presence of catheters or intravenous lines, dialysis and burns (Chastre *et al.*, 1996; Cisneros *et al.*, 1996; Weinbren *et al.*, 1998; Cisneros and Rodriguez-Bano, 2002). Immunosuppression or respiratory failure at admission increased the risk of bacteraemia by threefold, with increased risk for nosocomial pneumonia. Prolonged hospital stay, previous admission to another hospital, enteral feeding and previous use of third-generation cephalosporins are all identified as risk factors for colonisation/infection with *Acinetobacter* and are therefore indirectly risk factors for *Acinetobacter* bacteraemia (Mulin *et al.*, 1995; Scerpella *et al.*, 1995). Relevant clinical signs to define true bacteraemic episodes were fever, leukocytosis and successive positive blood cultures with the same genotypic isolate of *Acinetobacter* (Darveau *et al.*, 1991; Craig and Ebert, 1992; Decre, Benoit and Joly-Guillou, 1995; Dijkshoorn, 1996; Lai *et al.*, 1999). The incidence of *Acinetobacter* bacteraemia ranks second after pneumonias, and its prognosis is determined by the underlying condition of the patient (Lai *et al.*, 1999). Bacteraemia may be caused by both sporadic strains and epidemic strains that are localised to particular hospitals, as shown by molecular typing in five hospitals studied in a US multicentre study (Wisplinghoff *et al.*, 2000). Outcome depends on the underlying condition of the patient, but septic shock may occur in 20–30% of patients (Seifert, Strate and Pulverer, 1995).

Urinary Tract Infections

Several reviews have attributed to *Acinetobacter* spp. extremely variable rates, ranging from 2 to 61%, of nosocomially acquired UTIs, and recent epidemiological surveys have indicated an overall incidence of up to 30.5% of acquired UTIs (Joly-Guillou *et al.*, 1992; Bergogne-Berezin, 1996). Risk factors for UTIs do not differ from those involved in bacteraemia. The presence of indwelling urinary catheter is an additional factor, and removal of the catheter is recommended as an efficient measure to control *Acinetobacter* bacteraemia.

Skin and Soft-Tissue Infections

Colonisation of burns and traumatic and postoperative wounds with *Acinetobacter* spp. is not uncommon, particularly in ICU settings. Differentiating between wound colonisation and infection may often be difficult. However, cellulitis surrounding intravenous catheter sites caused by *Acinetobacter* has been described (Gervich and Grout, 1985). There have been infrequent reports of extensive soft-tissue necrosis adjacent to colonised wounds (Cisneros *et al.*, 1996) and synergistic necrotising fasciitis in conjunction with *Streptococcus pyogenes* (Allen and Hartman, 2000). *Acinetobacter* colonisation of extremity wounds was frequently observed in casualties of the Vietnam War. In many of these cases, colonisation of wounds was followed 3–5 days later by bacteraemia (Tong, 1972). In several large case series, 4–27% of all *Acinetobacter* bacteraemias occurred as a result of infected surgical or burn wounds (Seifert, Strate and Pulverer, 1995; Cisneros *et al.*, 1996; Wisplinghoff *et al.*, 2000).

Colonisation of burns with *Acinetobacter* spp. is seen in specialist units and in most burns patients when a frequency of isolation similar to *P. aeruginosa* has been reported (Santucci *et al.*, 2003).

Meningitis

Nosocomial meningitis is an infrequent manifestation of *Acinetobacter* infection. Cases of meningitis due to this organism have been reported after neurosurgical procedures, but rare cases of primary meningitis, especially in children, have also occurred. *Acinetobacter* meningitis may result from the introduction of the organism directly into the central nervous system (CNS) following intracranial surgery, myelography or ventriculography as well as after transnasal aspiration of craniopharyngioma or lumbar puncture (Siegman-Igra *et al.*, 1993). The course of the disease has been described as a relatively indolent bacterial meningitis with meningeal signs, high fever, lethargy or headache. However, an acute clinical course may also occur and problems in diagnosis may occur since the organism mimics meningococci or *Haemophilus influenzae*.

Acinetobacter Infections in Children

Recent studies have highlighted the occurrence of paediatric *Acinetobacter* infections, [previously considered rare or misidentified] (McDonald *et al.*, 1998; Mishra *et al.*, 1998; Nagels *et al.*, 1998). Multivariate analysis showed that infants with peripheral intravenous catheters were more likely to develop *Acinetobacter* nosocomial infection, and environmental cultures had revealed that after installation of a new air conditioner, the incidence of cases had increased by airborne dissemination of the bacteria. In a preterm infant ICU, the role of *A. baumannii* in infections was related to mechanical ventilation in 15 preterm infants (McDonald *et al.*, 1998): despite satisfactory conditions of care and hygiene, infants were colonised when ventilation was longer than that in controls, with body temperatures greater than 37 °C for a longer time than controls. However, no fatalities occurred in that particular study.

Surveillance and Control of *Acinetobacter* Nosocomial Infections

With the increasing importance of *Acinetobacter* infection worldwide, the development of local, large national or international surveys has resulted in a better understanding of the epidemiological profile of this organism. The annual incidence of *Acinetobacter* infection increased from 1980 to 1990, reaching up to 560 isolates per year in ICUs only, representing approximately 9–10% of infecting organisms. Since 1991, their number decreased significantly, being as low as 150 isolates per year, representing 0.5% of admissions in ICUs: changes in therapeutic strategies and improved hospital infection control have probably been favourable factors for decreased incidence of *Acinetobacter* infections (Bergogne-Berezin, 2001). An international survey on *Acinetobacter* infection was carried out in 1995–1996 (6 months), collecting data from France, Germany, Spain, Israel, the United Kingdom and South Africa. The results showed that the following:

1. the incidence of *Acinetobacter* infections was higher in large Hospitals (>800 beds) (9–14% of total *Acinetobacter* infections) versus less than 5% in small hospitals (<500 beds);
2. the incidence was higher during prolonged hospital stays (>10 days);
3. infection was rare in paediatric hospitals (unpublished data; presented as a communication at the 4th International Symposium on *Acinetobacter*, Eilat, Israel, 1996).

The application of highly discriminatory molecular typing techniques [AFLP and pulsed-field gel electrophoresis (PFGE)] to examples of three major ribotype groups of isolates from European hospitals confirmed the correlation of ribotypes 1 and 2 with the newly defined clones I and II. A third widely disseminated clone designated III was identified as a new epidemic clone (van Dessel *et al.*, 2004). Analysis of US data collected through the NNIS system between 1987 and 1996 shows a consistent increase in isolates of *Acinetobacter* spp. in the summer months (McDonald, Banerjee and Jarvis, 1999). Interestingly, such a clear seasonal variation was not seen in data from Latin America (Gales *et al.*, 2001).

LABORATORY DIAGNOSIS

Isolation

All the frequently encountered species of *Acinetobacter* grow readily on common laboratory media. In investigating outbreaks, commonly used selective and/or differential media [e.g. MacConkey agar, cystine lactose electrolyte–deficient (CLED) agar] have been made more selective for particular strains with specific antibiotic resistance patterns by the addition of antibiotics (Allen and Green, 1987). Alternatively, a selective medium such as Leeds *Acinetobacter* medium may be used for the selective isolation of most *Acinetobacter* spp. (Jawad *et al.*, 1994). When looking for small numbers in environmental specimens, liquid enrichment in minimal media with vigorous shaking has been useful (Bergogne-Berezin and Towner, 1996). Although *A. baumannii*, genomospecies 3 and 3TU have growth optima of 37 °C, a lower temperature such as 30 °C will ensure that all species are isolated.

Identification

The differentiation of the different genomospecies is not possible using phenotypic characteristics, although some level of discrimination is possible. Some key test findings are summarised in Table 16.1. The great majority of clinical isolates when carefully identified belong to *A. baumannii* (genomospecies 2), with genomospecies 3, *A. johnsonii*, *A. lwoffii* and *A. junii* being found much less frequently (Seifert *et al.*, 1993). Few clinical laboratories have the facilities for the molecular identification of genomospecies and rely on commercial phenotypic systems (e.g. API2ONE [BioMerieux]). Several studies have shown a poor correlation with DNA-based methods (Bernards *et al.*, 1996; Jawad *et al.*, 1998). Ribotyping (Gerner-Smidt, 1992) and RNA sequencer fingerprinting (Janssen and Dijkshoorn, 1996), AFLP (Ehrenstein *et al.*, 1996) and ARDRA should all accurately identify individual genomospecies, particularly differentiating 1, 2, 3 and 3TU.

Strain Typing

Besides sporadic cases of *Acinetobacter* infection, outbreaks of nosocomial infections due to *Acinetobacter* spp. have been widely reported in the medical literature, and they have become an increasing concern in hospitals. To determine the source (environment and/or cross-contamination) and the mode of spread of strains, many typing systems have been developed, from early phenotypic typing (phage typing, serology and bacteriocin typing), to more reliable molecular techniques such as ribotyping (Biendo *et al.*, 1999), analysis of restriction length polymorphisms in chromosomal DNA by PFGE (Gouby *et al.*, 1992; Seifert and Gerner-Smidt, 1995), PCR fingerprinting, ARDRA (Bernards *et al.*, 1997), random amplified polymorphic DNA analysis (RAPD) (Mathai *et al.*, 2001; Levidiotou *et al.*, 2002), AFLP fingerprinting (Janssen and Dijkshoorn, 1996; Koeleman *et al.*, 1997), infrequent restriction site PCR (IRS-PCR) (Wu *et al.*, 2002) and repetitive extragenic palindromic sequence-based PCR (REP-PCR) (Snelling *et al.*, 1996). All of these methods have shown excellent discrimination between clinical isolates, but at present there is no agreed standard method. Moreover, in most clinical laboratories, these techniques are not available. Instead, antibiogram and biotyping are used, although they remain relatively unreliable typing approaches. Improvement in typing methods and their wider adoption is necessary to identify relatedness between strains during outbreaks of infections and in wards with endemic *Acinetobacter* infection, in order to apply suitable measures.

TREATMENT

One problem associated with *Acinetobacter* spp. has been their intrinsic resistance to multiple antibiotics and their particular propensity to rapidly acquire antibiotic resistance. Since 1975, increasing resistance in clinical isolates of *Acinetobacter* spp. has been described. In the most recent surveillance reports, a high incidence of resistance among clinical isolates of *Acinetobacter* spp. has been reported to a range of antibiotics including aminoglycosides, third-generation cephalosporins, fluoroquinolones, extended-spectrum penicillins and monobactams (Hanberger *et al.*, 1999; Gales *et al.*, 2001). In a study conducted in more than 100 ICUs in five European countries, resistance rates to ciprofloxacin, gentamicin, piperacillin and ceftazidime frequently exceeded 50% (Hanberger

et al., 1999). There was a wide variation in resistance patterns detected in different countries, which is likely to reflect both species distribution and differences in the use of antibiotics. A recent study of 595 *Acinetobacter* spp. isolates in the United Kingdom demonstrated that 89% of *A. baumannii* was resistant to ceftazidime and more than 40% showed resistance to ciprofloxacin or gentamicin (Henwood *et al.*, 2002). The carbapenems retain the most activity, but recent reports have demonstrated increasing resistance to both imipenem and meropenem (Afzal-Shah and Livermore, 1998; Da Silva, Leitao and Peixe, 1999). Outbreaks of infection have been described in which sulbactam or colistin are the only antibiotics to which isolates of *A. baumannii* are susceptible (Corbella *et al.*, 2000; Fierobe *et al.*, 2001).

Species of *Acinetobacter* other than *A. baumannii* such as *A. lwoffii*, *A. johnsonii* and *A. junii* are less commonly associated with nosocomial infections and are generally more susceptible to antibiotics (Bergogne-Berezin and Towner, 1996). However, there is some evidence that resistance in these species is increasing.

β-Lactams

β-Lactamase production is the most common resistance mechanism, and *Acinetobacter* spp. produce a wide variety of enzymes (Table 16.3).

Carbapenems

Imipenem remains one of the most efficient drugs in treating *Acinetobacter* infections, and it has been used worldwide. At the time of writing, emergence and spread of imipenem-resistant strains has been observed in outbreaks, particularly when imipenem has been heavily used (Weinbren *et al.*, 1998; Zarrilli *et al.*, 2004). β-Lactams capable of destroying imipenem have been described (Table 16.3). Other mechanisms, particularly loss of outer membrane proteins, have been described as mechanisms for carbapenem resistance (Bou *et al.*, 2000).

Table 16.3 Summary of the properties of some of the β-lactamases identified in *Acinetobacter* spp.

β-Lactamase	Location of gene	Substrate	Properties	References
TEM-1/2	Plasmid	Penicillins	Common in nearly all Gram-negative bacterial. Inhibited by clavulanic acid	Vila *et al.* (1993)
CARB-5	Plasmid	Penicillins	Inhibited by clavulanic acid	Paul *et al.* (1989)
ACE-1 ACE-2 ACE-3 ACE-4	Chromosomal	Cephalosporins	Confers resistance to cephalosporins. Possesses some activity against penicillins. No activity against aztreonam or the broad-spectrum cephalosporins, only ACE-1 hydrolyses cefuroxime	
Amp-C	Chromosomal	Cephalosporins	Inducible β-lactamase. Poorly inhibited by clavulanic acid	Segal, Nelson and Elisha (2004)
PER-1	Plasmid	ESBL[a]	Confers resistance to all β-lactam antibiotics including aztreonam but not cefotoxin. No activity against carbapenems. Inhibited by clavulanic acid	Vahaboglu *et al.* (1997)
VEB-1	Plasmid	ESBL[a]		Poirel *et al.* (2003)
OXA-21 ARI-1 (OXA-23) OXA-24–27	Plasmid	Carbapenems	Class D, non-metalloenzymes with weak carbapenemase activity	Paton *et al.* (1993), Bou, Oliver and Martinez-Beltran (2000), Afzal-Shah, Woodford and Livermore (2001)
IMP-1 IMP-2 IMP-4 IMP-5 VIM-2	Plasmid and integrons	Carbapenems	Metalloenzymes with potent carbapenemase activity. Confers resistance to carbapenems and all β-lactam classes except monobactams	Chu *et al.* (2001), Da Silva *et al.* (2002), Yum *et al.* (2002), Nishio *et al.* (2004)

[a] Extended-spectrum β-lactamase.

Aminoglycosides

Aminoglycosides can be inactivated by a large range of aminoglycoside-modifying enzymes, and none of them (including the most active against *Acinetobacter*, amikacin) is totally immune from enzymatic inactivation (Seward, Lambert and Towner, 1998).

Fluoroquinolones

When first introduced, fluoroquinolones were highly active against *Acinetobacter*, but particularly the acquisition of mutations in the *gryA* and *parC* genes has resulted in a high incidence of fluoroquinolone resistance (Moreau *et al.*, 1996; Brisse *et al.*, 2000). A novel efflux pump has been described in *Acinetobacter* (Magnet, Courvalin and Lambert, 2001).

Other Antibiotics

Tetracycline resistance is mediated by the efflux proteins TetA and B in *A. baumannii* (Guardabassi *et al.*, 2000). Plasmid-mediated resistance to both trimethoprim and chloramphenicol has also been described (Bergogne-Berezin and Towner, 1996).

Therapeutic Options for Treatment of *Acinetobacter* Infection

The therapeutic options for treating *Acinetobacter* nosocomial infection are restricted because of the high levels of antimicrobial resistance. Based on the results of animal models of *Acinetobacter* pneumonia (Joly-Guillou *et al.*, 1997), carbapenems or carboxypenicillins have been used, in most cases in combination with an aminoglycoside, as an empirical treatment for nosocomial pneumonia. Other strategies have used β-lactamase inhibitors, and among them, sulbactam has demonstrated therapeutic efficacy largely because of its intrinsic activity as a single drug against *Acinetobacter* (Rodriguez-Hernandez *et al.*, 2001; Wood *et al.*, 2002). Although sulbactam has been shown to be superior to clavulanic acid and tazobactam, it does not enhance the activity of β-lactams (Higgins *et al.*, 2004).

Single Drug versus Combination Therapy

In vitro studies have shown that bactericidal activities of β-lactams including imipenem and the β-lactamase inhibitor sulbactam are very slow acting, as time-dependent drugs, whereas aminoglycosides, fluoroquinolones and rifampicin are rapidly bactericidal (concentration-dependent drugs) (Xirouchaki and Giamarellou, 1992; Joly-Guillou *et al.*, 1995; Marques *et al.*, 1997). Therefore, combinations of drugs remaining active against *Acinetobacter* spp., such as an aminoglycoside plus a β-lactam (ticarcillin or third-generation cephalosporin or imipenem), could offer synergistic and highly bactericidal effects (Marques *et al.*, 1997). For isolates highly resistant to aminoglycosides and/or fluoroquinolones, non-conventional and/or unusual combinations have been proposed, based on *in vitro* data, and outcomes favourably observed in treating patients (Xirouchaki and Giamarellou, 1992; Joly-Guillou *et al.*, 1995; Bajaksouzian *et al.*, 1997; Marques *et al.*, 1997). Owing to multiple mechanisms of resistance in clinical isolates, each strain must be tested carefully to individual and combined antibiotics, using appropriate *in vitro* techniques, and the choice of a suitable combination should take into account data obtained from animal models (Wolff *et al.*, 1999). Despite increasing development of resistance to ciprofloxacin, options using a new fluoroquinolone, levofloxacin, in combination with imipenem or amikacin in an animal model constitute promising therapeutic strategies. Extrapolation from *in vitro* data is questionable, whereas experimental infection in animal models had a better correlation with outcome.

OTHER AEROBIC GRAM-NEGATIVE BACILLI

INTRODUCTION

There are a large number of aerobic Gram-negative bacilli, other than *Acinetobacter* spp., identified as non-fermenters. They include three genera that are important in medical bacteriology: *Stenotrophomonas*, *Flavobacterium* and *Alcaligenes*. They have undergone many changes in taxonomy in recent years: previously classified on a phenotypic basis (morphology, motility and metabolism), the taxonomy of these bacteria has been revised using genetic information, and the position of the species has been reassigned to different genera on the basis of nucleic acid hybridisation. The current nomenclature of the species belonging to these three genera is summarised in Table 16.2. The other objective of this table is to summarise the previous designations of each species and changes in nomenclature. Table 16.2 includes genera and species of minor clinical importance, but as saprophytic bacteria present in the environment, they are found in water, soil and foodstuffs. In the hospital environment, moist items may be contaminated with such bacteria.

STENOTROPHOMONAS MALTOPHILIA (PREVIOUSLY *XANTHOMONAS*)

Taxonomy

Stenotrophomonas maltophilia (1994) was transferred from the genus *Xanthomonas* identified as *Pseudomonas maltophilia* according to *Bergey's Manual* (Juni, 1984). *Pseudomonas maltophilia* had been previously assigned to the genus *Xanthomonas* on the basis of phenotypic and genotypic characteristics (Swings, De Vos and Van den Mooter, 1983). These bacteria are characterised by the presence of a single or a small number of polar flagella (motile bacteria), frequently pigmented colonies (yellow or yellowish-orange), oxidase-negative reaction and acidified sugars (except for rhamnose or mannitol). They are generally proteolytic. *Stenotrophomonas maltophilia* is the most clinically important species of the genus.

Habitat and Pathogenicity of *S. maltophilia*

Isolated from soil, plants, water and raw milk, this ubiquitous bacterium is frequent in the hospital environment and is an emerging opportunistic pathogen (Denton and Kerr, 1998). It has been isolated from ventilatory equipment (thermal humidifying units), moist respirometers, dialysis fluids and antiseptic solutions. In patients, *S. maltophilia* may be isolated from multiple sites of nosocomial infections: respiratory tract, endocarditis, bacteraemia, meningitis and UTI. *Stenotrophomonas maltophilia* is implicated also in severe cutaneous infections, cellulitis and abscesses. This bacterium produces proteolytic enzymes and other pathogenic extracellular enzymes such as DNAase, RNAase, elastase, lipase, hyaluronidase, mucinase and haemolysin which contribute to the severity of these infections in otherwise immunode-pressed patients in ICUs (Von Gravenitz, 1985; Denton and Kerr, 1998). A high incidence of infection with S. *maltophilia* occurs in patients with cancer, leukaemia or lymphoma, and *S. maltophilia* is increasingly implicated in pulmonary superinfections in cystic fibrosis (CF) patients. A molecular epidemiological study of patients with CF and strains from both the hospital and home environment showed some evidence of cross-infection from the hospital, but most strains carried by patients were unique and probably represented acquisition from water and other home environments (Denton *et al.*, 1998). This hypothesis is supported by the finding that clinical and environmental strains do not separately cluster when examined for intraspecies diversity (Berg, Roskot and Smalla, 1999). *Stenotrophomonas maltophilia* is frequently associated with other bacteria at sites of infection. However, it is increasingly isolated as the sole pathogen because of its high natural resistance to most major antibiotics and to the selective pressure exerted by antibiotic treatments in ICUs (Rolston, Messer and Ho, 1990; Wolff, Brun-Buisson and Lode, 1997; Denton and Kerr, 1998).

Bacteriological Characteristics of *S. maltophilia*

Stenotrophomonas maltophilia does not produce pigments, and colonies are pale yellowish or non-pigmented. Selective media are used for isolating *S. maltophilia* from the environment (Juhnke and des, 1989) or from human specimens with the addition of imipenem (10 mg/l) because of the intrinsic resistance of the organism to carbapenems and/or vancomycin to inhibit Gram-positive bacteria when the specimen is contaminated. The oxidase reaction is negative or weakly and slowly positive. The main biochemical characteristics are summarised in Table 16.4. Four characteristics are important for identifying *S. maltophilia*: it produces lysine decarboxylase, hydrolyses aesculin and gelatin, and gelatin, and requires L-methionine for growth (50 mg/l in synthetic mineral broth) (Swings, De Vos and Van den Mooter, 1983). Other tests – indole, urease and ornithine decarboxylase production – are negative. Identification by more specific tests and/or automated methods has been described (Gilardi, 1985; Kampfer and Dott, 1988; Tenover, Mizuki and Carlson, 1990).

Stenotrophomonas maltophilia Susceptibility/Resistance to Antibiotics

Stenotrophomonas maltophilia is intrinsically resistant to most antibiotics by chromosomally mediated mechanisms. Variation in susceptibility is often seen, even on repeat testing of the same isolate. This is probably due to differential expression of the two chromosomally located β-lactamases L1 and L2 (Avison *et al.*, 2002). L1 is a metalloenzyme responsible for the intrinsic resistance to carbapenems, and L2 is a broad-spectrum cephalosporin (Denton and Kerr, 1998). The availability of Zn^{++} in the media has the effect on the minimum inhibitory concentration (MIC) values for imipenem but not meropenem (Cooke *et al.*, 1996). Production of a metallo-β-lactamase can be detected using a diffusion gradient strip combined with ethylenediaminetetraacetic acid (EDTA) (Etest) (Walsh *et al.*, 2002). Fluoroquinolone resistance is probably due to mutation in *gyrA/parC*, and two efflux pump genes have been described with both fluoroquinolones and β-lactams as substrates (Poole, 2004).

β-Lactams

Stenotrophomonas maltophilia is susceptible only to latamoxef and to combinations of ticarcillin and clavulanic acid or piperacillin plus tazobactam. It is naturally resistant to imipenem and meropenem by production of a carbapenemase. In addition, mutants exhibit high-level expression of constitutive L1 and L2 β-lactamases (Akova, Bonfiglio and Livermore, 1991).

Aminoglycosides

Only a few strains are susceptible to gentamicin, neomycin and kanamycin; although AAC6′ has been described, most resistance is probably due to reduced uptake (Denton and Kerr, 1998).

Fluoroquinolones

Stenotrophomonas maltophilia is generally resistant to quinolones. New agents such as temafloxacin and sparfloxacin seem to be more active than ciprofloxacin. Resistance to quinolones in *S. maltophilia* is generally associated with resistance to chloramphenicol and to doxycycline in more than 50% of cases. This resistance is associated with alteration of outer membrane proteins (Lecso-Bornet *et al.*, 1992).

Other Antibiotics

Clinical strains of *S. maltophilia* are variably susceptible to minocycline, rifampicin and the combination trimethoprim–sulphamethoxazole. Nosocomial infections due to *S. maltophilia* require combination therapy including rifampicin plus fluoroquinolone or ticarcillin plus clavulanic acid, associated with tobramycin or amikacin or even a trimethoprim–sulphamethoxazole combination. The new tetracycline derivative tigecycline shows promising activity (Betriu *et al.*, 2002).

CHRYSEOBACTERIUM (PREVIOUSLY *FLAVOBACTERIUM*)

Definition and Taxonomy

The genus *Chryseobacterium* (*Flavobacterium* spp.) is a large group of non-motile, oxidase-positive (except for *Chryseobacterium odoratum*), Gram-negative, strictly aerobic, non-fermentative bacilli. Species belonging to the genus *Chryseobacterium* are summarised in Table 16.2. The taxonomy of the genus has been evolving (MacMeekin and Shewan, 1978; Bruckner and Colonna, 1995), and two *Flavobacterium*-like species, initially designated groups Iif and Iij, have been transferred recently to the genus *Sphingobacterium* and designated species *Weeksella virosa* and *Weeksella zoohelcum*, respectively, in the new proposed nomenclature (Bruckner and Colonna, 1995). These latter species differ from *Flavobacterium* spp., in that they do not metabolise sugars: they are rarely recovered from human specimens, and unlike *Flavobacterium* spp., they are very susceptible to most antibiotics.

Table 16.4 Main characteristics and tests for identifying aerobic Gram-negative bacilli

Genera	Morphology	Oxidase	G + C%	Characteristics
Acinetobacter	Non-motile, diplobacilli, coccobacilli	−	38–55	No pigment. Glucose acidified by genospecies 1–4 (*A. baumannii, A. calcoaceticus, A. haemolyticus*)
Stenotrophomonas	Motile, polar flagella	− (*S. maltophilia*, negative or weak, slowly positive)	62	Yellowish-orange or lemon (straw) pigment, gelatin hydrolysis
Alcaligenes	Motile, degenerated, peritrichous	+	57–69	No pigment, no gelatine hydrolysis (*A. denitrificans* subsp. *xylosoxidans*: carbohydrate assimilation)
Flavobacterium	Non-motile	+	30–42	Bright yellowish-orange pigment or light yellow, non-diffusible
Burkholderia	Motile, polar lophotrichous, flagella	+ (variable intensity)	67.4–69.5	Diffusible pigment yellow or violet, greenish-yellow or brownish-black (in 10% of strains)
Pseudomonas	Motile, polar monotrichous, flagella	Generally +	57–70	Pigments, bluish-greenish-yellow (*P. aeruginosa*: pyocyanin, pyoverdin; *P. fluorescens*: pyoverdin, greenish; *P. stutzeri*: wrinkled rough colonies, carotenoid pigment)

G + C%, DNA base composition (mol%).

Five other *Flavobacteria* spp. have been renamed as *Sphingobacterium* spp. and *Pseudomonas paucimobilis* is designated *Sphingomonas paucimobilis* (Table 16.2): these 'new species' are of minor clinical significance. Several other taxonomic changes have occurred regarding *Flavobacterium* spp. (Bergogne-Berezin, 1996), and the genus *Flavobacterium* is 'survived' by the transfer to this group of several unnamed species (CDC groups Ve2, Iie, Iih, Iii; *Pseudomonas oryzihabitans* becoming *Flavimonas oryzihabitans*), these species being of low pathogenicity (Esteban *et al.*, 1993) (Table 16.2).

Epidemiology and Pathogenicity

Chryseobacterium spp. are ubiquitous organisms and are found in the hospital environment. Environmental studies have traced their source from contaminated water, ice machines and humidifiers (Pichinoty *et al.*, 1985). Epidemiological markers used for delineating outbreaks of *Flavobacterium meningosepticum* infections were phenotypes of resistance and serology based on the 0 antigenic type. Nine 0 serovars have been identified: A–H and K. The C serovar is isolated in the Far East, whereas serovar G predominates in European countries and has been isolated from clinical samples in ICUs (tracheal aspirations, expectorations and sinuses). New typing procedures, such as ribotyping (Colding *et al.*, 1994), based on molecular biology are being investigated. These bacteria are implicated in various nosocomial infections: *Chryseobacterium meningosepticum* and *Chryseobacterium multivorum* are the predominant species isolated in septicaemia, meningitis and endocarditis (Brunn *et al.*, 1989; Hsueh *et al.*, 1996). Many cases of meningitis due to *F. meningosepticum* have been observed in neonates (Abrahamsen, Finne and Lingaas, 1989) and, infrequently, in immunocompromised patients. In adults, *F. meningosepticum* has been isolated from cases of pneumonia, postoperative bacteraemia and meningitis, usually associated with severe underlying pathologies. Rare cases of community-acquired *C. meningosepticum* meningitis have been cited, and *Chryseobacterium indologenes* has been associated with infections on indwelling devices (Hsueh *et al.*, 1996).

Bacteriology

Chryseobacterium spp. grow between 5 and 30 °C, but the strains isolated from clinical material grow at 35 °C. On nutrient agar, they produce colonies 1–2 mm in diameter, frequently pigmented light yellow or yellowish-orange because of a non-diffusible pigment. The degree of pigmentation may be more pronounced at lower temperatures (15–20 °C). The metabolism is strictly aerobic, and sugars are metabolised by the oxidative pathway [except for *C. odoratum* and *Sphingobacterium multivorum* (Pichinoty *et al.*, 1985) which do not acidify glucose] (Table 16.3). Indole-positive species (*Chryseobacterium breve*, *C. meningosepticum* and *C. indologenes*) are usually strongly proteolytic. Aesculin, citrate and urease tests are variably positive (Bruckner and Colonna, 1995).

Antimicrobial Susceptibility

Chryseobacterium strains isolated from clinical specimens are resistant to many antibiotics: they are generally resistant to all aminoglycosides, third-generation cephalosporins, penicillins (mezlocillin, piperacillin and ticarcillin), aztreonam and imipenem. The most active antibiotics against *C. meningosepticum* are rifampicin, clindamycin (MICs 1–4 mg/l) and ciprofloxacin, which has proven efficacious in treating pneumonia in paediatric patients. Cases of neonatal septicaemia have been treated with clindamycin combined with piperacillin. Other antibiotic combinations have been used in treating *Chryseobacterium* infections (Von Gravenitz, 1985; Raimondi, Moosdeen and Williams, 1986). Susceptibility to β-lactams can be recovered by combining β-lactamase inhibitors with β-lactam antibiotics.

ALCALIGENES

Taxonomy

An early description of *Bacillus alcaligenes faecalis* made by Petrushky in 1896 reported the observation of a Gram-negative motile bacillus, isolated from beer and from human stools. Later, many species have been included in the genus *Alcaligenes* on the basis of DNA–DNA and DNA–rRNA hybridisations. Major taxonomic changes have occurred: *Achromobacter xylosoxidans* has been transferred to the genus *Alcaligenes* as *Alcaligenes xylosoxidans* subsp. *xylosoxidans*. An additional group, *Agrobacterium* spp., has been included in the species *Alcaligenes* and has been isolated in clinical conditions (Popoff *et al.*, 1984) (Table 16.2). All *Alcaligenes* strains are short rods (0.5–2.6 μm), Gram-negative, motile, with one to eight peritrichous (non-polar) flagella, usually described as degenerated. They are oxidase and catalase positive. Other species included in the genus *Alcaligenes* – *Alcaligenes faecalis*, *Alcaligenes piechaudii* and *Alcaligenes xylosoxidans* subsp. *denitrificans* – are not saccharolytic. The only saccharolytic species is *A. xylosoxidans* subsp. *xylosoxidans*. Not all *Alcaligenes* spp. possess specific physiological or biochemical characteristics (Table 16.4), and those most commonly involved in nosocomial infections are *A. faecalis* and *A. xylosoxidans* subsp. *denitrificans*.

Epidemiology and Pathogenicity

Alcaligenes faecalis and *A. denitrificans* are isolated from various environmental sources such as respirators, haemodialysis systems, intravenous solutions and even disinfectants (Duggan *et al.*, 1996; Gomez-Cerezo *et al.*, 2003). They have occasionally been isolated from a variety of human specimens: blood, faeces, sputum, urine, cerebrospinal fluid, wounds, burns and swabs from throat, eyes and ear discharges. *Alcaligenes* strains do not seem to possess any specific virulence determinants, and they are an infrequent cause of hospital-acquired infection in patients with severe underlying disease. Rare cases of peritonitis, pneumonia, bacteraemia or UTI are found in the literature. In many instances, the organism is considered as a coloniser, and its pathogenic role remains somewhat controversial. Nosocomial outbreaks are usually associated with an aqueous source of contamination (Von Gravenitz, 1985; Duggan *et al.*, 1996). Recent findings have underlined that *Alcaligenes* spp. were isolated predominantly from respiratory tract specimens, particularly from sputum of CF patients (Saiman *et al.*, 2001).

Bacteriology

Identification of *Alcaligenes* spp. is based on the main characteristics cited above. In addition, alkalinisation of a series of substrates (malonate, acetamide, allantoin, histidine and itaconate), gelatin hydrolysis (*A. faecalis* and *A. piechaudii*) and several negative characteristics help identify *Alcaligenes* spp., usually by using commercial multitest systems (Kampfer and Dott, 1988; Tenover, Mizuki and Carlson, 1990). Assimilation tests (carbohydrates) and the presence of nitrate and nitrite reductases contribute to the precise identification of species. A combination of substrate assimilation tests and automated ribotyping has been shown to be a rapid and simple approach to the identification of this difficult group of bacteria (Clermont *et al.*, 2001).

Antimicrobial Susceptibility/Resistance

Most strains are resistant to aminoglycosides, chloramphenicol and tetracyclines. They are variably susceptible to trimethoprim–sulphamethoxazole and to newer β-lactams. *Alcaligenes xylosoxidans* subsp. *xylosoxidans* has been shown to be susceptible to ureidopenicillins, latamoxef, imipenem and some fluoroquinolones (ciprofloxacin and ofloxacin) (Mensah, Philippon and Richard, 1989; Philippon *et al.*, 1990; Bizet, Tekaia and Philippon, 1993). There have been several reports of multiple β-lactam resistance in *A. xylosoxidans* to broad-spectrum penicillins, with two resistance phenotypes (Philippon *et al.*, 1990): these phenotypes result from constitutive β-lactamase production. Moreover, three different types of cephalosporinases and the presence of other β-lactamases have been demonstrated. Treatment of infections due to this uncommon opportunistic organism requires combination therapy including extended-spectrum β-lactams (piperacillin and imipenem) and fluoroquinolones (ciprofloxacin and sparfloxacin) or amikacin with fluoroquinolones (levofloxacin) or trimethoprim–sulphamethoxazole.

MISCELLANEOUS AEROBIC GRAM-NEGATIVE BACILLI

Many aerobic, non-fermentative, Gram-negative bacilli which either are poorly characterised as species or have an uncertain positive taxonomy may occasionally be isolated from patients with predisposing medical conditions. Some species have been excluded from the genus *Pseudomonas*, such as *Pseudomonas cepacia* (Goldmann and Klinger, 1986; Rabkin *et al.*, 1989; Werneburg and Monteil, 1989), *Pseudomonas gladioli* and *Pseudomonas pickettii*, and form a new group, designated *Burkholderia* spp. (Yabuuchi *et al.*, 1992), with corresponding species *Burkholderia cepacia*, *Burkholderia pickettii* and *Burkholderia gladioli*, as listed in Table 16.2. These organisms are cited in clinical conditions, with the presence of *B. cepacia* in the hospital environment (Drabick *et al.*, 1996; Pankhurst and Philpott-Howard, 1996) and in patients with CF superinfection (Walters and Smith, 1993; Simpson *et al.*, 1994; Govan and Deretic, 1996; Segonds *et al.*, 1996) or of *B. pickettii* in nosocomial infection (Poty, Denis and Baufine-Ducrocq, 1987). Included recently in the *Burkholderia* genus, *Burkholderia mallei* and *Burkholderia pseudomallei* are environmental organisms predominantly found in Asian countries: they are responsible for infections in animals and may be transmitted in humans, the name of the disease being melioidosis (severe pneumonia with subacute cavitating forms) (Sookpranee *et al.*, 1991). Still classified in the genus *Pseudomonas*, *Pseudomonas stutzeri* is an organism of low pathogenicity and has been isolated in pneumonia and septicaemia (Potvliege *et al.*, 1987). The new group *Comamonas* (Table 16.2) has been formed of *Pseudomonas testosteroni* and *Pseudomonas acidovorans* (Tamaoka, Ha and Komagata, 1978), which are ubiquitous organisms, potentially pathogenic as saprophytic opportunistic species. Listed in the *Alcaligenes* spp., *A. xylosoxidans* (Table 16.2), renamed *Alcaligenes xylosoxidans*, is involved in human infections such as osteomyelitis (Walsh, Klein and Cunha, 1993). Among these multiple Gram-negative bacilli with taxonomic uncertainties, the most frequent and significant is *B. cepacia*, which is increasingly isolated from pulmonary sites of infections, particularly in CF patients. They produce a variety of virulence factors, mainly exoenzymes, and the organism is responsible for rapid pulmonary deterioration in CF patients (Goldmann and Klinger, 1986; Segonds *et al.*, 1996).

REFERENCES

Abrahamsen TG, Finne PH, Lingaas E (1989) *Flavobacterium meningosepticum* infections in a neonatal intensive care unit. Acta Paediatr. Scand., 78: 51–55.

Afzal-Shah M, Livermore DM (1998) Worldwide emergence of carbapenem-resistant *Acinetobacter* spp. J. Antimicrob. Chemother., 41: 576–577.

Afzal-Shah M, Woodford N, Livermore DM (2001) Characterization of OXA-25, OXA-26, and OXA-27, molecular class D beta-lactamases associated with carbapenem resistance in clinical isolates of *Acinetobacter baumannii*. Antimicrob. Agents Chemother., 45: 583–588.

Akova M, Bonfiglio G, Livermore DM (1991) Susceptibility to beta-lactam antibiotics of mutant strains of *Xanthomonas maltophilia* with high- and low-level constitutive expression of L1 and L2 beta-lactamases. J. Med. Microbiol., 35: 208–213.

Allen DM, Hartman BJ (2000) *Acinetobacter* species. In: Mandell, Douglas and Bennett's Principles and Practice of Infectious Diseases. 5th ed. Eds. Mandell GL, Bennett JE, Dolin R. Churchill Livingstone, London; 2339–2344.

Allen KD, Green HT (1987) Hospital outbreak of multi-resistant *Acinetobacter anitratus*: an airborne mode of spread? J. Hosp. Infect., 9: 110–119.

Anstey NM, Currie BJ, Withnall KM (1992) Community-acquired *Acinetobacter pneumonia* in the Northern Territory of Australia. Clin. Infect. Dis., 14: 83–91.

Avison MB, Higgins CS, Ford PJ *et al.* (2002) Differential regulation of L1 and L2 beta-lactamase expression in *Stenotrophomonas maltophilia*. J. Antimicrob. Chemother., 49: 387–389.

Avril J, Mensard LR (1991) Factors influencing the virulence of *Acinetobacter*. In: The Biology of Acinetobacter. Eds. Towner KJ, Bergogne-Berezin E, Fewson CA. Plenum Press, New York; 77–82.

Bajaksouzian S, Visalli MA, Jacobs MR, Appelbaum PC (1997) Activities of levofloxacin, ofloxacin, and ciprofloxacin, alone and in combination with amikacin, against *Acinetobacter*s as determined by checkerboard and time-kill studies. Antimicrob. Agents Chemother., 41: 1073–1076.

Baker N, Hawkey P (2004) The management of resistant *Acinetobacter* infections in the intensive therapy unit. In: Management of Multiple Drug-Resistant Infections. Ed. Gillespie SH. Humana Press, Totowa, NJ; 117–140.

Berg G, Roskot N, Smalla K (1999) Genotypic and phenotypic relationships between clinical and environmental isolates of *Stenotrophomonas maltophilia*. J. Clin. Microbiol., 37: 3594–3600.

Bergogne-Berezin E (1996) Resistance of *Acinetobacter* spp. to antimicrobials. Overview of clinical resistance patterns and therapeutic problems. In: *Acinetobacter* – Microbiology, Epidemiology, Infection, Management. Eds. Bergogne-Berezin E, Joly-Guilou ML, Towner KJ. CRC Press, New York; 133–183.

Bergogne-Berezin E (2001) The increasing role of *Acinetobacter* species as nosocomial pathogens. Curr. Infect. Dis. Rep., 3: 440–444.

Bergogne-Berezin E, Towner KJ (1996) *Acinetobacter* spp. as nosocomial pathogens: microbiological, clinical, and epidemiological features. Clin. Microbiol. Rev., 9: 148–165.

Bergogne-Berezin E, Joly-Guilou ML, Vieu JF (1987) Epidemiology of nosocomial infections due to *Acinetobacter calcoaceticus*. J. Hosp. Infect., 10: 105–113.

Berlau J, Aucken H, Malnick H, Pitt T (1999) Distribution of *Acinetobacter* species on skin of healthy humans. Eur. J. Clin. Microbiol. Infect. Dis., 18: 179–183.

Bernards AT, van der TJ, van Boven CP, Dijkshoorn L (1996) Evaluation of the ability of a commercial system to identify *Acinetobacter* genomic species. Eur. J. Clin. Microbiol. Infect. Dis., 15: 303–308.

Bernards AT, de Beaufort AJ, Dijkshoorn L, van Boven CP (1997) Outbreak of septicaemia in neonates caused by *Acinetobacter junii* investigated by amplified ribosomal DNA restriction analysis (ARDRA) and four typing methods. J. Hosp. Infect., 35: 129–140.

Betriu C, Rodriguez-Avial I, Sanchez BA *et al.* (2002) Comparative in vitro activities of tigecycline (GAR-936) and other antimicrobial agents against Stenotrophomonas maltophilia. J. Antimicrob. Chemother., 50: 758–759.

Biendo M, Laurans G, Lefebvre JF *et al.* (1999) Epidemiological study of an *Acinetobacter baumannii* outbreak by using a combination of antibiotyping and ribotyping. J. Clin. Microbiol., 37: 2170–2175.

Bizet C, Tekaia F, Philippon A (1993) In-vitro susceptibility of *Alcaligenes faecalis* compared with those of other *Alcaligenes* spp. to antimicrobial agents including seven beta-lactams. J. Antimicrob. Chemother., 32: 907–910.

Boerlin P, Eugster S, Gaschen F *et al.* (2001) Transmission of opportunistic pathogens in a veterinary teaching hospital. Vet. Microbiol., 82: 347–359.

Bou G, Cervero G, Dominguez MA *et al.* (2000) Characterization of a nosocomial outbreak caused by a multiresistant *Acinetobacter baumannii* strain with a carbapenem-hydrolyzing enzyme: high-level carbapenem resistance in *A. baumannii* is not due solely to the presence of beta-lactamases. J. Clin. Microbiol., 38: 3299–3305.

Bou G, Oliver A, Martinez-Beltran J (2000) OXA-24, a novel class D beta-lactamase with carbapenemase activity in an *Acinetobacter baumannii* clinical strain. Antimicrob. Agents Chemother., 44: 1556–1561.

Bouvet P, Grimont P (1986) Taxonomy of the genus *Acinetobacter* with the recognition of *Acinetobacter baumannii* sp. nov., *Acinetobacter haemolyticus* sp. nov., *Acinetobacter johnsonii* sp. nov., and *Acinetobacter junii* sp. nov. and emended descriptions of *Acinetobacter calcoaceticus* and *Acinetobacter lwoffii* Int. J. Syst. Bacteriol., 36: 228–240.

Brisse S, Milatovic D, Fluit AC *et al.* (2000) Molecular surveillance of European quinolone-resistant clinical isolates of *Pseudomonas aeruginosa* and *Acinetobacter* spp. using automated ribotyping. J. Clin. Microbiol., 38: 3636–3645.

Bruckner DA, Colonna P (1995) Nomenclature for aerobic and facultative bacteria. Clin. Infect. Dis., 21: 263–272.

Brunn B, Tvenstrup J, Jensen JE *et al.* (1989) *Flavobacterium meningosepticum* infection. Eur. J. Clin. Microbiol. Infect. Dis., 8: 509–514.

Cefai C, Richards J, Gould FK, McPeake P (1990) An outbreak of *Acinetobacter* respiratory tract infection resulting from incomplete disinfection of ventilatory equipment. J. Hosp. Infect., 15: 177–182.

Chastre JJ, Trouillet L, Vaugnet A, Joly-Guillou ML (1996) Nosocomial pneumonia caused by *Acinetobacter* spp. In: *Acinetobacter* – Microbiology, Epidemiology, Infection, Management. Eds. Bergogne-Berezin E, Joly-Guilou ML, Towner KJ. CRC Press, New York; 117–132.

Chu YW, Leung CM, Houang ET *et al.* (1999) Skin carriage of *Acinetobacter*s in Hong Kong. J. Clin. Microbiol., 37: 2962–2967.

Chu YW, Afzal-Shah M, Houang ET *et al.* (2001) IMP-4, a novel metallo-beta-lactamase from nosocomial *Acinetobacter* spp. collected in Hong Kong between 1994 and 1998. Antimicrob. Agents Chemother., 45: 710–714.

Cisneros JM, Rodriguez-Bano J (2002) Nosocomial bacteremia due to *Acinetobacter baumannii*: epidemiology, clinical features and treatment. Clin. Microbiol. Infect., 8: 687–693.

Cisneros JM, Reyes MJ, Pachon J *et al.* (1996) Bacteremia due to *Acinetobacter baumannii*: epidemiology, clinical findings, and prognostic features. Clin. Infect. Dis., 22: 1026–1032.

Clermont D, Harmant C, Bizet C (2001) Identification of strains of *Alcaligenes* and *Agrobacterium* by a polyphasic approach. J. Clin. Microbiol., 39: 3104–3109.

Colding H, Bangsborg J, Fiehn NE *et al.* (1994) Ribotyping for differentiating *Flavobacterium meningosepticum* isolates from clinical and environmental sources. J. Clin. Microbiol., 32: 501–505.

Cooke P, Heritage J, Kerr K *et al.* (1996) Different effects of zinc ions on in vitro susceptibilities of *Stenotrophomonas maltophilia* to imipenem and meropenem. Antimicrob. Agents Chemother., 40: 2909–2910.

Corbella X, Pujol M, Ayats J *et al.* (1996) Relevance of digestive tract colonization in the epidemiology of nosocomial infections due to multiresistant *Acinetobacter baumannii*. Clin. Infect. Dis., 23: 329–334.

Corbella X, Montero A, Pujol M *et al.* (2000) Emergence and rapid spread of carbapenem resistance during a large and sustained hospital outbreak of multiresistant *Acinetobacter baumannii*. J. Clin. Microbiol., 38: 4086–4095.

Craig WA, Ebert SC (1992) Continuous infusion of beta-lactam antibiotics. Antimicrob. Agents Chemother., 36: 2577–2583.

Da Silva GJ, Leitao GJ, Peixe L (1999) Emergence of carbapenem-hydrolyzing enzymes in *Acinetobacter baumannii* clinical isolates. J. Clin. Microbiol., 37: 2109–2110.

Da Silva GJ, Correia M, Vital C *et al.* (2002) Molecular characterization of bla (IMP-5), a new integron-borne metallo-beta-lactamase gene from an *Acinetobacter baumannii* nosocomial isolate in Portugal. FEMS Microbiol. Lett., 215: 33–39.

Darveau RP, Cunningham MD, Seachord CL *et al.* (1991) Beta-lactam antibiotics potentiate magainin 2 antimicrobial activity in vitro and in vivo. Antimicrob. Agents Chemother., 35: 1153–1159.

Decre D, Benoit C, Joly-Guillou ML (1995) In vivo activity and bactericidal activity of levofloxacin (LEVX) alone or in combination with amikacin against *Acinetobacter* spp. in comparison with oxifloxacin (OFL) and ciprofloxacin (CIP). *35th ICAAC*, San Francisco, Abstract E100.

Denton M, Kerr KG (1998) Microbiological and clinical aspects of infection associated with *Stenotrophomonas maltophilia*. Clin. Microbiol. Rev., 11: 57–80.

Denton M, Todd NJ, Kerr KG *et al.* (1998) Molecular epidemiology of *Stenotrophomonas maltophilia* isolated from clinical specimens from patients with cystic fibrosis and associated environmental samples. J. Clin. Microbiol., 36: 1953–1958.

Denton M, Wilcox MH, Parnell P *et al.* (2004) Role of environmental cleaning in controlling an outbreak of *Acinetobacter baumannii* on a neurosurgical intensive care unit. J. Hosp. Infect., 56: 106–110.

Dijkshoorn L (1996) *Acinetobacter* – microbiology. In: *Acinetobacter* – Microbiology, Epidemiology, Infection, Management. Eds. Bergogne-Berezin E, Joly-Guilou ML, Towner KJ. CRC Press, New York; 37–69.

Dorsey CW, Tolmasky ME, Crosa JH, Actis LA (2003) Genetic organization of an *Acinetobacter baumannii* chromosomal region harbouring genes related to siderophore biosynthesis and transport. Microbiology, 149: 1227–1238.

Drabick JA, Gracely EJ, Heidecker GJ, LiPuma JJ (1996) Survival of *Burkholderia cepacia* on environmental surfaces. J. Hosp. Infect., 32: 267–276.

Duggan JM, Goldstein SJ, Chenoweth CE *et al.* (1996) *J. Chemother.* bacteremia: report of four cases and review of the literature. Clin. Infect. Dis., 23: 569–576.

Ehrenstein B, Bernards AT, Dijkshoorn L *et al.* (1996) *Acinetobacter* species identification by using tRNA spacer fingerprinting. J. Clin. Microbiol., 34: 2414–2420.

Esteban J, Valero-Moratalla ML, Alcazar R, Soriano F (1993) Infections due to *Flavimonas oryzihabitans*, case report and literature review. Eur. J. Clin. Microbiol. Infect. Dis., 12: 797–800.

Fagon JY, Chastre J, Domart Y *et al.* (1989) Nosocomial pneumonia in patients receiving continuous mechanical ventilation. Prospective analysis of 52 episodes with use of a protected specimen brush and quantitative culture techniques. Am. Rev. Respir. Dis., 139: 877–884.

Fierobe L, Lucet JC, Decre D *et al.* (2001) An outbreak of imipenem-resistant *Acinetobacter baumannii* in critically ill surgical patients. Infect. Control Hosp. Epidemiol., 22: 35–40.

Fujita K, Lilly HA, Ayliffe GA (1982) Spread of resistant gram-negative bacilli in a burns unit. J. Hosp. Infect., 3: 29–37.

Gales AC, Jones RN, Forward KR *et al.* (2001) Emerging importance of multidrug-resistant *Acinetobacter* species and *Stenotrophomonas maltophilia* as pathogens in seriously ill patients: geographic patterns, epidemiological features, and trends in the SENTRY Antimicrobial Surveillance Program (1997–1999). Clin. Infect. Dis., 32 (Suppl. 2): S104–S113.

Garcia A, Salgado F, Solar H *et al.* (1999) Some immunological properties of lipopolysaccharide from *Acinetobacter baumannii*. J. Med. Microbiol., 48: 479–483.

Garcia-Garmendia JL, Ortiz-Leyba C, Garnacho-Montero J *et al.* (1999) Mortality and the increase in length of stay attributable to the acquisition of *Acinetobacter* in critically ill patients. Crit. Care Med., 27: 1794–1799.

Garrouste-Org O, Marie Rouveau M, Villiers S *et al.* (1996) Secondary carriage with multi-resistant *Acinetobacter baumannii* and *Klebsiella pneumoniae* in an adult ICU population: relationship with nosocomial infections and mortality. J. Hosp. Infect., 34: 279–289.

Gerner-Smidt P (1992) Ribotyping of the *Acinetobacter calcoaceticus–Acinetobacter baumannii* complex. J. Clin. Microbiol., 30: 2680–2685.

Gervich DH, Grout CS (1985) An outbreak of nosocomial *Acinetobacter* infections from humidifiers. Am. J. Infect. Control, 13: 210–215.

Gilardi GL (1985) Cultural and biochemical aspects for identification of glucose non-fermenting Gram-negative rods. In: Non-Fermentative Gram-Negative Rods: Laboratory Identification and Clinical Aspects. Ed. Gilardi GL. Marcel-Dekker, New York; 17–84.

Goldmann DA, Klinger JD (1986) *Pseudomonas cepacia*: biology, mechanisms of virulence, epidemiology. J. Pediatr., 108: 806–812.

Gomez-Cerezo J, Suarez I, Rios JJ *et al.* (2003) *J. Chemother.* bacteremia: a 10-year analysis of 54 cases. Eur. J. Clin. Microbiol. Infect. Dis., 22: 360–363.

Gouby A, Carles-Nurit MJ, Bouziges N *et al.* (1992) Use of pulsed-field gel electrophoresis for investigation of hospital outbreaks of *Acinetobacter baumannii*. J. Clin. Microbiol., 30: 1588–1591.

Govan JR, Deretic V (1996) Microbial pathogenesis in cystic fibrosis: mucoid *Pseudomonas aeruginosa* and *Burkholderia cepacia*. Microbiol. Rev., 60: 539–574.

Guardabassi L, Dijkshoorn L, Collard JM *et al.* (2000) Distribution and in-vitro transfer of tetracycline resistance determinants in clinical and aquatic *Acinetobacter* strains. J. Med. Microbiol., 49: 929–936.

Hanberger H, Garcia-Rodriguez JA, Gobernado M *et al.* (1999) Antibiotic susceptibility among aerobic gram-negative bacilli in intensive care units in 5 European countries. French and Portuguese ICU Study Groups. JAMA, 281: 67–71.

Henwood CJ, Gatward T, Warner M *et al.* (2002) Antibiotic resistance among clinical isolates of *Acinetobacter* in the UK, and in vitro evaluation of tigecycline (GAR-936). J. Antimicrob. Chemother., 49: 479–487.

Higgins PG, Wisplinghoff H, Stefanik D, Seifert H (2004) In vitro activities of the beta-lactamase inhibitors clavulanic acid, sulbactam, and tazobactam alone or in combination with beta-lactams against epidemiologically characterized multidrug-resistant *Acinetobacter baumannii* strains. Antimicrob. Agents Chemother., 48: 1586–1592.

Hsueh PR, Teng LJ, Ho SW *et al.* (1996) Clinical and microbiological characteristics of *Flavobacterium indologenes* infections associated with indwelling devices. J. Clin. Microbiol., 34: 1908–1913.

Husni RN, Goldstein LS, Arroliga AC *et al.* (1999) Risk factors for an outbreak of multi-drug-resistant *Acinetobacter* nosocomial pneumonia among intubated patients. Chest, 115: 1378–1382.

Janssen P, Dijkshoorn L (1996) High resolution DNA fingerprinting of *Acinetobacter* outbreak strains. FEMS Microbiol. Lett., 142: 191–194.

Jawad A, Hawkey PM, Heritage J, Snelling AM (1994) Description of Leeds *Acinetobacter* medium, a new selective and differential medium for isolation of clinically important *Acinetobacter* spp., and comparison with Herellea agar and Holton's agar. J. Clin. Microbiol., 32: 2353–2358.

Jawad A, Heritage J, Snelling AM *et al.* (1996) Influence of relative humidity and suspending menstrua on survival of *Acinetobacter* spp. on dry surfaces. J. Clin. Microbiol., 34: 2881–2887.

Jawad A, Snelling AM, Heritage J, Hawkey PM (1998) Exceptional desiccation tolerance of *Acinetobacter* radioresistens. J. Hosp. Infect., 39: 235–240.

Joly-Guillou ML, Brunet F (1996) Epidemiology of *Acinetobacter* spp. surveillance and management of outbreaks. In: *Acinetobacter* – Microbiology, Epidemiology, Infection, Management. Eds. Bergogne-Berezin E, Joly-Guilou ML, Towner KJ. CRC Press, New York; 71–100.

Joly-Guillou ML, Decre D, Wolfe J, Bergogne-Berezin E (1992) *Acinetobacter* spp. clinical epidemiology in 89 intensive care units. A retrospective study in France during 1991. *2nd International Conference on the Prevention of Infection (CIPI)*, Nice, Abstract Cj1.

Joly-Guillou ML, Decre D, Herrman JL *et al.* (1995) Bactericidal in-vitro activity of beta-lactams and beta-lactamase inhibitors, alone or associated, against clinical strains of *Acinetobacter baumannii*: effect of combination with aminoglycosides. J. Antimicrob. Chemother., 36: 619–629.

Joly-Guillou ML, Wolff M, Pocidalo JJ *et al.* (1997) Use of a new mouse model of *Acinetobacter baumannii* pneumonia to evaluate the postantibiotic effect of imipenem. Antimicrob. Agents Chemother., 41: 345–351.

Juhnke ME, des JE (1989) Selective medium for isolation of *Xanthomonas maltophilia* from soil and rhizosphere environments. Appl. Environ. Microbiol., 55: 747–750.

Juni E (1978) Genetics and physiology of *Acinetobacter*. Annu. Rev. Microbiol., 32: 349–371.

Juni E (1984) Genus III. *Acinetobacter*. Brisou and Prévot 1954. In: Bergey's Manual of Systematic Bacteriology, Vol. 1. Eds. Krieg NR, Hold JG. Williams & Wilkins, Baltimore, MD; 303–307.

Kampfer P, Dott W (1988) Differentiation of some gram-negative glucose nonfermenting bacteria using miniaturized carbon sources assimilation tests. Zentralbl. Bakteriol. Mikrobiol. Hyg.[B], 186: 468–477.

Kaplan N, Rosenberg E, Jann B, Jann K (1985) Structural studies of the capsular polysaccharide of *Acinetobacter calcoaceticus* BD4. Eur. J. Biochem., 152: 453–458.

Koeleman JG, Parlevliet GA, Dijkshoorn L *et al.* (1997) Nosocomial outbreak of multi-resistant *Acinetobacter baumannii* on a surgical ward: epidemiology and risk factors for acquisition. J. Hosp. Infect., 37: 113–123.

Lai SW, Ng KC, Yu WL *et al.* (1999) *Acinetobacter baumannii* bloodstream infection: clinical features and antimicrobial susceptibilities of isolates. Kaohsiung. J. Med. Sci., 15: 406–413.

Lecso-Bornet M, Pierre J, Sarkis-Karam D *et al.* (1992) Susceptibility of *Xanthomonas maltophilia* to six quinolones and study of outer membrane proteins in resistant mutants selected in vitro. Antimicrob. Agents Chemother., 36: 669–671.

Levidiotou S, Galanakis E, Vrioni G *et al.* (2002) A multi-resistant *Acinetobacter baumannii* outbreak in a general intensive care unit. In Vivo, 16: 117–122.

Lortholary O, Fagon JY, Hoi AB *et al.* (1995) Nosocomial acquisition of multiresistant *Acinetobacter baumannii*, risk factors and prognosis. Clin. Infect. Dis., 20: 790–796.

MacMeekin TA, Shewan JM (1978) Taxonomic strategies for *Flavobacterium* and related genera. J. Appl. Bacteriol., 45: 321–322.

Magnet S, Courvalin P, Lambert T (2001) Resistance-nodulation-cell division-type efflux pump involved in aminoglycoside resistance in *Acinetobacter baumannii* strain BM4454. Antimicrob. Agents Chemother., 45: 3375–3380.

Mah MW, Memish ZA, Cunningham G, Bannatyne RM (2001) Outbreak of *Acinetobacter baumannii* in an intensive care unit associated with tracheostomy. Am. J. Infect. Control, 29: 284–288.

Marques MB, Brookings ES, Moser SA *et al.* (1997) Comparative in vitro antimicrobial susceptibilities of nosocomial isolates of *Acinetobacter baumannii* and synergistic activities of nine antimicrobial combinations. Antimicrob. Agents Chemother., 41: 881–885.

Mathai E, Kaufmann ME, Richard VS *et al.* (2001) Typing of *Acinetobacter baumannii* isolated from hospital-acquired respiratory infections in a tertiary care centre in southern India. J. Hosp. Infect., 47: 159–162.

McDonald LC, Walker M, Carson L *et al.* (1998) Outbreak of *Acinetobacter* spp. bloodstream infections in a nursery associated with contaminated aerosols and air conditioners. Pediatr. Infect. Dis. J., 17: 716–722.

McDonald LC, Banerjee SN, Jarvis WR (1999) Seasonal variation of *Acinetobacter* infections: 1987–1996. Nosocomial Infections Surveillance System. Clin. Infect. Dis., 29: 1133–1137.

Mensah K, Philippon A, Richard C (1989) Infections nosocomiales a *Alcaligenes denitrificans* subsp. *xylosoxidans*: sensibilite de 41 souches a 38 antibiotiques. Med. Mal. Inf., 19: 167–172.

Mishra A, Mishra S, Jaganath G *et al.* (1998) *Acinetobacter* sepsis in newborns. Indian Pediatr., 35: 27–32.

Montenegro MA, Bitter-Suermann D, Timmis JK *et al.* (1985) traT gene sequences, serum resistance and pathogenicity-related factors in clinical isolates of *Escherichia coli* and other gram-negative bacteria. J. Gen. Microbiol., 131: 1511–1521.

Moreau NJ, Houot S, Joly-Guillou ML, Bergogne-Berezin E (1996) Characterisation of DNA gyrase and measurement of drug accumulation in clinical isolates of *Acinetobacter baumannii* resistant to fluoroquinolones. J. Antimicrob. Chemother., 38: 1079–1083.

Mulin B, Talon D, Viel JF *et al.* (1995) Risk factors for nosocomial colonization with multiresistant *Acinetobacter baumannii*. Eur. J. Clin. Microbiol. Infect. Dis., 14: 569–576.

Nagels B, Ritter E, Thomas P *et al.* (1998) *Acinetobacter baumannii* colonization in ventilated preterm infants. Eur. J. Clin. Microbiol. Infect. Dis., 17: 37–40.

Nishio H, Komatsu M, Shibata N *et al.* (2004) Metallo-β-Lactamase-producing Gram-negative bacilli: labratory based surveillance in cooperation with 13 clinical laboratories in the Kinki region of Japan. J. Clin. Microbiol., 42: 5256–5263.

Nowak A, Burkiewicz A, Kur J (1995) PCR differentiation of seventeen genospecies of Acinetobacter. FEMS Microbiol. Lett., 15: 181–187.

Palmen R, Hellingwerf KJ (1997) Uptake and processing of DNA by *Acinetobacter calcoaceticus* – a review. Gene, 192: 179–190.

Pankhurst CL, Philpott-Howard J (1996) The environmental risk factors associated with medical and dental equipment in the transmission of *Burkholderia* (*Pseudomonas*) *cepacia* in cystic fibrosis patients. J. Hosp. Infect., 32: 249–255.

Paton R, Miles R, Hood J, Amyes S (1993) ARI-1: beta-lactamase mediated imipenem resistance in *Acinetobacter baumannii*. Int. J. Antimicrob. Agents, 2: 81–88.

Paul G, Joly-Guillou ML, Bergogne-Berezin E *et al.* (1989) Novel carbenicillin-hydrolyzing beta-lactamase (CARB-5) from *Acinetobacter calcoaceticus* var. *anitratus*. FEMS Microbiol. Lett., 50: 45–50.

Philippon A, Mensah K, Fournier G, Freney J (1990) Two resistance phenotypes to beta-lactams of *Alcaligenes denitrificans* subsp. *xylosoxidans* in relation to beta-lactamase types. J. Antimicrob. Chemother., 25: 698–700.

Pichinoty F, Baratti J, Kammoun S *et al.* (1985) [Isolation and description of strains of Flavobacterium multivorum of telluric origin]. Ann. Inst Pasteur Microbiol., 136A: 359–370.

Poirel L, Menuteau O, Agoli N *et al.* (2003) Outbreak of extended-spectrum beta-lactamase VEB-1-producing isolates of *Acinetobacter baumannii* in a French hospital. J. Clin. Microbiol., 41: 3542–3547.

Poole K (2004) Efflux-mediated multiresistance in Gram-negative bacteria. Clin. Microbiol. Infect., 10: 12–26.

Popoff MY, Kersters K, Kiredjian M *et al.* (1984) [Taxonomic position of *Agrobacterium* strains of hospital origin]. Ann. Microbiol. (Paris), 135A: 427–442.

Potvliege C, Jonckheer J, Lenclud C, Hansen W (1987) *Pseudomonas stutzeri* pneumonia and septicemia in a patient with multiple myeloma. J. Clin. Microbiol., 25: 458–459.

Poty F, Denis C, Baufine-Ducrocq H (1987) [Nosocomial *Pseudomonas* pickettii infection. Danger of the use of ion-exchange resins]. Presse Med. 16: 1185–1187.

Quinn JP (1998) Clinical problems posed by multiresistant nonfermenting gram-negative pathogens. Clin. Infect. Dis., 27 (Suppl. 1): S117–S124.

Rabkin CS, Jarvis WR, Anderson RL *et al.* (1989) *Pseudomonas cepacia* typing systems: collaborative study to assess their potential in epidemiologic investigations. Rev. Infect. Dis., 11: 600–607.

Raimondi A, Moosdeen F, Williams JD (1986) Antibiotic resistance pattern of *Flavobacterium meningosepticum*. Eur. J. Clin. Microbiol., 5: 461–463.

Reams AB, Neidle EL (2004) Gene amplification involves site-specific short homology-independent illegitimate recombination in *Acinetobacter* sp. strain ADP1. J. Mol. Biol., 338: 643–656.

Rodriguez-Hernandez MJ, Cuberos L, Pichardo C *et al.* (2001) Sulbactam efficacy in experimental models caused by susceptible and intermediate *Acinetobacter baumannii* strains. J. Antimicrob. Chemother., 47: 479–482.

Rolston KV, Messer M, Ho DH (1990) Comparative in vitro activities of newer quinolones against *Pseudomonas* species and *Xanthomonas maltophilia* isolated from patients with cancer. Antimicrob. Agents Chemother., 34: 1812–1813.

Rosenberg M, Bayer EA, Delarea J, Rosenberg E (1982) Role of thin fimbriae in adherence and growth of *Acinetobacter calcoaceticus* RAG-1 on hexadecane. Appl. Environ. Microbiol., 44: 937.

Rosenberg E, Kaplan N, Pines O *et al.* (1983) Capsular polysaccharides interfere with adherence of *Acinetobacter calcoaceticus* to hydrocarbon. FEMS Microbiol. Lett., 17: 157–160.

Rosenthal SL (1974) Sources of pseudomonas and *Acinetobacter* species found in human culture materials. Am. J. Clin. Pathol., 62: 807–811.

Saiman L, Chen Y, Tabibi S (2001) Identification and antimicrobial susceptability of *Alcaligenes Xylosoxidans* isolated from patients with cystic fibrosis. J. Clin. Microbiol., 39: 3942–3945.

Santucci SG, Gobara S, Santos CR *et al.* (2003) Infections in a burn intensive care unit: experience of seven years. J. Hosp. Infect., 53: 6–13.

Scerpella EG, Wanger AR, Armitige L *et al.* (1995) Nosocomial outbreak caused by a multiresistant clone of *Acinetobacter baumannii*: results of the case–control and molecular epidemiologic investigations. Infect. Control Hosp. Epidemiol., 16: 92–97.

Schaberg DR, Culver DH, Gaynes RP (1991) Major trends in the microbial etiology of nosocomial infection. Am. J. Med., 91: 72S–75S.

Segal H, Nelson EC, Elisha BG (2004) Genetic environment and transcription of ampC in an *Acinetobacter baumannii* clinical isolate. Antimicrob. Agents Chemother., 48: 612–614.

Segonds C, Chabanon G, Couetdic G *et al.* (1996) Epidemiology of pulmonary colonization with *Burkholderia cepacia* in cystic fibrosis patients. The French Observatoire *Burkholderia cepacia* Study Group. Eur. J. Clin. Microbiol. Infect. Dis., 15: 841–842.

Seifert H, Gerner-Smidt P (1995) Comparison of ribotyping and pulsed-field gel electrophoresis for molecular typing of *Acinetobacter* isolates. J. Clin. Microbiol., 33: 1402–1407.

Seifert H, Baginski R, Schulze A, Pulverer G (1993) The distribution of *Acinetobacter* species in clinical culture materials. Zentralbl. Bakteriol., 279: 544–552.

Seifert H, Strate A, Pulverer G (1995) Nosocomial bacteremia due to *Acinetobacter baumannii*. Clinical features, epidemiology, and predictors of mortality. Medicine (Baltimore), 74: 340–349.

Seifert H, Dijkshoorn L, Gerner-Smidt P *et al.* (1997) Distribution of *Acinetobacter* species on human skin: comparison of phenotypic and genotypic identification methods. J. Clin. Microbiol., 35: 2819–2825.

Seward RJ, Towner KJ (1999) Detection of integrons in worldwide nosocomial isolates of *Acinetobacter* spp. Clin. Microbiol. Infect., 5: 308–318.

Seward RJ, Lambert T, Towner KJ (1998) Molecular epidemiology of aminoglycoside resistance in *Acinetobacter* spp. J. Med. Microbiol., 47: 455–462.

Sherertz RJ, Sullivan ML (1985) An outbreak of infections with *Acinetobacter calcoaceticus* in burn patients: contamination of patients' mattresses. J. Infect. Dis., 151: 252–258.

Siegman-Igra Y, Bar-Yosef S, Gorea A, Avram J (1993) Nosocomial *Acinetobacter* meningitis secondary to invasive procedures: report of 25 cases and review. Clin. Infect. Dis., 17: 843–849.

Simpson IN, Finlay J, Winstanley DJ *et al.* (1994) Multi-resistance isolates possessing characteristics of both *Burkholderia (Pseudomonas) cepacia* and *Burkholderia gladioli* from patients with cystic fibrosis. J. Antimicrob. Chemother., 34: 353–361.

Smolyakov R, Borer A, Riesenberg K *et al.* (2003) Nosocomial multi-drug resistant *Acinetobacter baumannii* bloodstream infection: risk factors and outcome with ampicillin-sulbactam treatment. J. Hosp. Infect., 54: 32–38.

Snelling AM, Gerner-Smidt P, Hawkey PM *et al.* (1996) Validation of use of whole-cell repetitive extragenic palindromic sequence-based PCR (REP-PCR) for typing strains belonging to the *Acinetobacter calcoaceticus–Acinetobacter baumannii* complex and application of the method to the investigation of a hospital outbreak. J. Clin. Microbiol., 34: 1193–1202.

Sookpranee T, Sookpranee M, Mellencamp MA, Preheim LC (1991) *Pseudomonas pseudomallei*, a common pathogen in Thailand that is resistant to the bactericidal effects of many antibiotics. Antimicrob. Agents Chemother., 35: 484–489.

Swings J, De Vos P, Van den Mooter M (1983) Transfer of *Pseudomonas maltophilia* Hugh, 1981 to the genus *Xanthomonas* as *Xanthomonas maltophilia* (Hugh 1981) comb. Int. J. Syst. Bacteriol., 33: 409–413.

Tamaoka J, Ha DM, Komagata K (1978) Reclassification of *Pseudomonas acidovorans* Den Dooren De Jong 1926 and *Pseudomonas testosteroni* Marcus and Talalay 1956 as *Comamonas acidovorans* comb. no and *Comamonas testosteroni* comb. no with an emended description of the genus *Comamonas*. Int. J. Syst. Bacteriol., 37: 52–59.

Tenover FC, Mizuki TS, Carlson LG (1990) Evaluation of autoSCAN-W/A automated microbiology system for the identification of non-glucose-fermenting gram-negative bacilli. J. Clin. Microbiol., 28: 1628–1634.

Timsit JF, Garrait V, Misset B *et al.* (1993) The digestive tract is a major site for *Acinetobacter baumannii* colonization in intensive care unit patients. J. Infect. Dis., 168: 1336–1337.

Tong MJ (1972) Septic complications of war wounds. JAMA, 219: 1044–1047.

Torres A, Aznar R, Gatell JM *et al.* (1990) Incidence, risk, and prognosis factors of nosocomial pneumonia in mechanically ventilated patients. Am. Rev. Respir. Dis., 142: 523–528.

Vahaboglu H, Ozturk R, Aygun G *et al.* (1997) Widespread detection of PER-1-type extended-spectrum beta-lactamases among nosocomial *Acinetobacter* and *Pseudomonas aeruginosa* isolates in Turkey: a nationwide multicenter study. Antimicrob. Agents Chemother., 41: 2265–2269.

van Dessel H, Dijkshoorn L, van der RT *et al.* (2004) Identification of a new geographically widespread multiresistant *Acinetobacter baumannii* clone from European hospitals. Res. Microbiol., 155: 105–112.

Vila J, Marcos A, Marco F *et al.* (1993) In vitro antimicrobial production of beta-lactamases, aminoglycoside-modifying enzymes, and chloramphenicol acetyltransferase by and susceptibility of clinical isolates of *Acinetobacter baumannii*. Antimicrob. Agents Chemother., 37: 138–141.

Von Gravenitz A (1985) Ecology, clinical significance and antimicrobial susceptibility of infrequently encountered glucose-nonferminting gram-negative rods. In: Non-Fermentative Gram-Negative Rods: Laboratory Identification and Clinical Aspects. Ed. Gilardi GL. Marcel-Dekker, New York; 181–232.

Walsh RD, Klein NC, Cunha BA (1993) *J. Chemother.* osteomyelitis. Clin. Infect. Dis., 16: 176–178.

Walsh TR, Bolmstrom A, Qwarnstrom A, Gales A (2002) Evaluation of a new Etest for detecting metallo-beta-lactamases in routine clinical testing. J. Clin. Microbiol., 40: 2755–2759.

Walters S, Smith EG (1993) *Pseudomonas cepacia* in cystic fibrosis: transmissibility and its implications. Lancet, 342: 3–4.

Weinbren MJ, Johnson AP, Kaufmann ME, Livermore DM (1998) *Acinetobacter* spp. isolates with reduced susceptibilities to carbapenems in a UK burns unit. J. Antimicrob. Chemother., 41: 574–576.

Werneburg B, Monteil H (1989) New serotypes of *Pseudomonas cepacia*. Res. Microbiol., 140: 17–20.

Wisplinghoff H, Perbix W, Seifert H (1999) Risk factors for nosocomial bloodstream infections due to *Acinetobacter baumannii*: a case-control study of adult burn patients. Clin. Infect. Dis., 28: 59–66.

Wisplinghoff H, Edmond MB, Pfaller MA *et al.* (2000) Nosocomial bloodstream infections caused by *Acinetobacter* species in United States hospitals: clinical features, molecular epidemiology, and antimicrobial susceptibility. Clin. Infect. Dis., 31: 690–697.

Wolff M, Brun-Buisson C, Lode H (1997) The changing epidemiology of severe infections in the ICU. Clin. Microbiol. Infect., 3 (Suppl. 1): S36–S47.

Wolff M, Joly-Guillou ML, Farinotti R, Carbon C (1999) In vivo efficacies of combinations of beta-lactams, beta-lactamase inhibitors, and rifampin against *Acinetobacter baumannii* in a mouse pneumonia model. Antimicrob. Agents Chemother., 43: 1406–1411.

Wood GC, Hanes SD, Croce MA *et al.* (2002) Comparison of ampicillin-sulbactam and imipenem-cilastatin for the treatment of *Acinetobacter* ventilator-associated pneumonia. Clin. Infect. Dis., 34: 1425–1430.

Wu TL, Su LH, Leu HS *et al.* (2002) Molecular epidemiology of nosocomial infection associated with multi-resistant *Acinetobacter baumannii* by infrequent-restriction-site PCR. J. Hosp. Infect., 51: 27–32.

Xirouchaki E, Giamarellou H (1992) In vitro interactions of aminoglycosides with imipenem or ciprofloxacin against aminoglycoside resistant *Acinetobacter baumannii*. J. Chemother., 4: 263–267.

Yabuuchi E, Kosako Y, Oyaizu H *et al.* (1992) Proposal of *Burkholderia* gen. nov. and transfer of seven species of the genus *Pseudomonas* homology group II to the new genus, with the type species *Burkholderia cepacia* (Palleroni and Holmes, 1981) comb. nov. Microbiol. Immunol., 36: 1251–1275.

Yum JH, Yi K, Lee H *et al.* (2002) Molecular characterization of metallo-beta-lactamase-producing *Acinetobacter baumannii* and *Acinetobacter* genomospecies 3 from Korea: identification of two new integrons carrying the bla(VIM-2) gene cassettes. J. Antimicrob. Chemother., 49: 837–840.

Zarrilli R, Crispino M, Bagattini M *et al.* (2004) Molecular epidemiology of sequential outbreaks of *Acinetobacter baumannii* in an intensive care unit shows the emergence of carbapenem resistance. J. Clin. Microbiol., 42: 946–953.

Zeana C, Larson E, Sahni J *et al.* (2003) The epidemiology of multidrug-resistant *Acinetobacter baumannii*: does the community represent a reservoir? Infect. Control Hosp. Epidemiol., 24: 275–279.

17

Haemophilus spp.

Derrick W. Crook[1] and Derek W. Hood[2]

[1]*Department of Clinical Microbiology, John Radcliffe Hospital, Oxford, UK; and* [2]*Molecular Infectious Diseases, Department of Paediatrics, Weatherall Institute of Molecular Medicine, John Radcliffe Hospital, Oxford, UK*

INTRODUCTION

Haemophilus influenzae is the most important cause of human disease among the species belonging to this genus. It is a free-living, human-adapted organism without another reservoir (Kilian, 1976). Its principal habitat is the nasopharynx, but it is found inhabiting other mucosal surfaces including the genital tract and occasionally the intestinal tract. Despite being a fastidious organism which is relatively difficult to grow, it was first isolated and named in 1890 by Pfeiffer who mistakenly thought it was the cause of the contemporaneous influenza pandemic (Turk, 1982). The name *H. influenzae* persisted, even though it was soon recognized that influenza virus caused flu.

Haemophilus influenzae has come to prominence in many unexpected ways. Despite the erroneous suggestion it caused flu, it was soon recognized to be a leading cause of spontaneous childhood meningitis and other life-threatening infections (Pittman, 1930, 1931). It was the first organism where a vaccine made of its polysaccharide capsule conjugated to a carrier protein successfully produced protective immunity in children (Eskola *et al.*, 1990). The impressive success of this vaccine in nearly eliminating childhood *H. influenzae* disease lead the way to the development of further conjugate vaccines to the major childhood pathogens, *Streptococcus pneumonia* and *Neisseria meningitidis*. Restriction enzymes (*Hind*II and *Hind*III), one of the essential tools of recombinant technology, were discovered from studies into the molecular biology of *H. influenzae*, and Werner Arber, Daniel Nathans and Hamilton O. Smith were awarded the Nobel Prize for this discovery. Most recently, *H. influenzae* was the first free-living organism to be whole sequenced using novel methodology developed by Craig Venter and The Institute for Genomic Research which was a crucial building block in making it possible to sequence whole genomes including the human genome (Fleischmann *et al.*, 1995).

DESCRIPTION OF THE ORGANISM

Taxonomy

The full taxonomic root of this bacterium is Phylum *Proteobacteria*, class Gammaproteobacteria, order Pasteurellales, family Pasteurellaceae, genus *Haemophilus* and species *influenzae*. A 16s phylogenetic tree is depicted in Figure 17.1 and shows the relationship of the human members of the genus in relationship to *Escherichia coli*. The closely related, human-adapted *Haemophilus* spp. include *H. parainfluenzae* (the most plentiful colonizing member of this genus), *H. haemolyticus*, *H. aegyptius*, *H. ducreyi*, *H. segnis*, *H. aphrophilus* and *H. paraphrophilus*. Relatively little is known of these other

commensal *Haemophilus* spp., and they will be referred to briefly at the end of the chapter.

Characteristics

Haemophilus influenzae is a small (0.2–0.3 to 0.5–0.8 μm) Gram-negative, nonmotile, coccobacillus (Plate 9). The organism grows well on rich medium and produces 2–3-μm grey translucent colonies after 18–25 h of incubation. It is a fastidious microorganism and has specific growth requirements that are used in routine clinical microbiology for identification. It does not grow on nutrient agar, such as Columbia agar, without the growth supplements X factor (haemin) and V factor (NAD). By contrast *H. parainfluenzae* only requires V factor for growth. Rich medium such as chocolate agar supports the growth of *Haemophilus* spp. and is the preferred solid medium for propagation of the organism. Growth is enhanced in the presence of 5% CO_2. The precise differentiation of human *Haemophilus* spp. as outlined in Table 17.1 is beyond the scope of most diagnostic laboratories; however, satisfactory differentiation can be achieved using X or V dependence and growth characteristics on a blood agar (preferably horse blood) exhibiting presence or absence of haemolysis and satelitism to a staphylococcal streak. Speciation using sequence analysis of ribosomal genes is a better method for speciation of haemophili than biochemical tests; however, this technology is currently available only to some reference laboratories.

Population Genetics

The population of *H. influenzae* has an unusual organization. There is a minority population of capsulate (serotypable) and an abundant population of noncapsulate (nonserotypable) strains consisting of independently evolving lineages. The capsulate strains produce one of six chemically and antigenically distinct polysaccharide capsules, a, b, c, d, e and f. The evolutionary origins of their capsules are unclear; however, the capsulate strains form a relatively nondiverse population of organisms consisting of few lineages compared to the much more diverse population of noncapsulate strains (Plate 10). This suggests that capsule may have been acquired relatively recently in the evolutionary past of *H. influenzae*. The outstanding and dramatic feature arising from this differentiation of *H. influenzae* into capsular types by Margaret Pittman in 1930 was the recognition that childhood invasive disease was caused almost exclusively by type b capsulate strains (Pittman, 1930, 1931). More detailed characterization of the population structure of the species shows that the majority of childhood invasive type b isolates cluster in a relatively small number of related lineages.

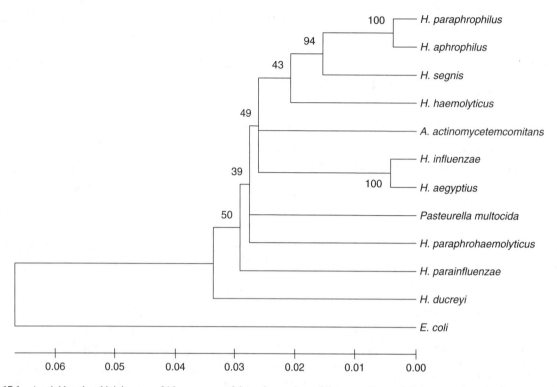

Figure 17.1 A neighbourhood joining tree of 16s sequences of the main members of the genus *Haemophilus* compared to other important related *Pasteurellaceae* and rooted on *Escherichia coli* as an out-group (data were obtained from the RDP website: http://rdp.cme.msu.edu/html/analyses.html). The bootstrap values on the nodes indicate low values (i.e. <70) for many of the tree branches and limited confidence for the topology represented here (tree produced by MEGA version 2.1: http://www.megasoftware.net/). However, this represents an approximate estimate of the relationship between these species.

Table 17.1 Characteristics of *Haemophilus* species

Species	Factor requirement			Fermentation			
	X	V	Haemolysis	Glucose	Sucrose	Lactose	Mannose
H. influenzae	+	+	−	+	−	−	−
H. haemolyticus	+	+	+	+	−	−	−
H. ducreyi	+	−	−	−	−	−	−
H. parainfluenzae	−	+	−	+	+	−	+
H. parahaemolyticus	−	+	+	+	+	−	−
H. segnis	−	+	−	+	+	−	−
H. paraphrophilus	−	+	−	+	+	+	+
H. aphrophilus	−	−	−	+	+	+	+

The appreciation of the population structure of *H. influenzae* described above has been substantially advanced by the development of a sequence-based typing scheme based on multilocus sequence typing (MLST) (Plate 8) (Meats *et al.*, 2003). This is accessible through the Internet (http://haemophilus.mlst.net). MLST and modifications of this scheme for studying the population of *H. influenzae* is giving new insights into the relationships between lineages of this organism. This powerful technique has effectively replaced other typing schemes such as outer membrane protein (OMP) or restriction fragment length polymorphism typing schemes and will facilitate investigation of the changing epidemiology of *H. influenzae*. It is likely that improvements in characterizing *H. influenzae* by MLST will be made by including further and more variable genes in the typing scheme.

PATHOGENICITY

Haemophilus influenzae expresses a number of cell-surface features that are required for establishment of colonization in the nasopharynx

within individual hosts and importantly play a role in the virulence of bacteria in susceptible hosts (virulence factors).

Capsule

Haemophilus influenzae can express one of six antigenically distinct (serotype a–f) polysaccharide capsules that exist loosely associated with the outer surface of the bacterium. Expression of a capsule is considered a prerequisite for *H. influenzae* to be able to cause invasive disease and correspondingly bacteria expressing serotype b capsule account for the majority of invasive disease isolates (Zwahlen *et al.*, 1989). The serotype b capsule consists of a negatively charged phosphodiester-linked linear polymer of disaccharide units of polyribosyl ribitol phosphate (PRP). The genes required for its synthesis are located on a large DNA segment (*cap* locus) that is a compound transposon (Kroll, Loynds and Moxon, 1991). Serum antibody to serotype b capsular polysaccharide is protective, and the Hib vaccine, designed to elicit production of this antibody, is now routinely used in national childhood immunization schemes. The vaccines currently in use contain PRP conjugated covalently to a protein carrier. The proposed roles for capsular polysaccharide in *H. influenzae* are increased resistance to phagocytic host defences in the bloodstream, by interfering with binding of serum components, and increased resistance to desiccation during host-to-host passage.

Lipopolysaccharide

Lipopolysaccharide (LPS) is a complex glycolipid found on the Gram-negative bacterial cell wall that is essential for membrane integrity. In *H. influenzae*, the LPS comprises two distinct, covalently linked regions, the lipid A and the core oligosaccharide (Richards *et al.*, 2001). LPS is an important virulence determinant of *H. influenzae*, playing a key role in the processes of bacterial colonization, persistence and survival in the human host, but also represents an ideal target for bactericidal as well as endotoxin-neutralizing antibodies (Bouchet *et al.*, 2003). The lipid A portion of *H. influenzae* LPS is embedded in the outer membrane and mediates the endotoxic effects of the LPS molecule responsible for some of the pathophysiology associated with severe infection. The host response to *H. influenzae* lipid A is primarily mediated by activation of Toll-like receptor 4 (TLR4), facilitated by binding of LPS-binding protein (LBP) and the host cell receptor, CD14 (Lazou Ahren *et al.*, 2001; Wang *et al.*, 2002). This interaction triggers a cascade of signalling events, resulting in secretion of a variety of potent mediators and cytokines produced primarily by activated macrophages and monocytes. The overproduction of these effector molecules, such as interleukin-1 and tumour necrosis factor-α, contributes to the pathophysiology of endotoxic shock, which can include hypotension, disseminated intravascular coagulopathy and multiorgan failure. The structure of *H. influenzae* oligosaccharide can be characterized in terms of a conserved inner core region and a variable outer core region which can display significantly different patterns of sugars between strains. The oligosaccharide can influence interactions of the bacteria with host cells and components of the host immune system.

Pili and Other Adhesion Proteins

Successful colonization by *H. influenzae* requires that the bacterium binds to host cells, particularly at its normal site of colonization in the nasopharynx. Fundamental to this binding are a number of cell-surface proteins that act as adhesins. Pili (or fimbriae), typically polymeric hair-like structures that extend from the bacterial cell surface, are associated with binding to respiratory epithelial cells. Not all *H. influenzae* express pili, and other nonpilus membrane proteins are involved in the attachment process. At least four other major adhesins have now been identified in Hib and NTHi, including the Hap, HMW1, HMW2 and Hia/Hsf proteins, the distribution of which is highly variable between strains (Kubiet *et al.*, 2000; Hardy, Tudor and St Geme III, 2003).

Phase and Antigenic Variation

Bacterial pathogens must be able to adapt to the changing microenvironments within and between hosts. *Haemophilus influenzae* has evolved a mechanism to alter the expression of cell-surface proteins, or enzymes required for the biosynthesis of the LPS molecule, through the inclusion of microsatellites within the respective genes. These microsatellites, typically composed of tetranucleotide repeats, are unstable and promote random on–off switching of expression of the respective genes within a population (phase variation). This mechanism provides considerable flexibility in the patterns of antigens that can be expressed at any given time within a population of *H. influenzae* (Bayliss, Field and Moxon, 2001).

EPIDEMIOLOGY

Understanding the host–organism relationship is benefited by splitting the consideration of epidemiology into first capsulate *H. influenzae* type b (Hib) and second noncapsulate or nonserotypable *H. influenzae*. An important general characteristic applicable to both categories of the organism is that it is a human-restricted commensal without another reservoir. Also, the emergence of antibiotic resistance is equally applicable to all types of *H. influenzae*.

Haemophilus influenzae type b

Haemophilus influenzae type b (Hib) is a major cause of childhood infectious disease. Understanding the transmission of Hib is important to a better understanding of the distribution and occurrence of disease. Acquisition probably occurs by close bodily contact, and the main source is other children (Barbour *et al.*, 1995). Acquisition is usually followed by carriage where the organism dwells for months harmlessly in the nasopharynx; however, in a few susceptible individuals, acquisition appears to be the point in time when invasive disease commences. Carriage rates vary after birth, and risk factors for carriage are the following: advancing age from birth until 4 years, underdeveloped country, crowding, day-care attendance and presence of siblings. The major preventative factor is Hib immunization which near eliminates carriage and produces a marked herd effect protecting against disease (Barbour *et al.*, 1995).

Disease caused by Hib has been near eliminated by the highly efficacious conjugate Hib vaccine (Peltola, 2000). Consequently, the epidemiology of disease deals with either the prevaccine era in developed countries or those parts of the world where vaccine has not been nationally implemented. It is estimated that as much as 60% of the world's children are unimmunized and therefore are unprotected (Peltola, 2000).

The main disease manifestations of Hib disease are meningitis, primary bacteraemia, pneumonia, epiglottitis and arthritis (Figure 17.2) (Peltola, 2000). The major burden of disease caused by Hib is meningitis. Its relative importance varied or varies between countries or regions of the world. In the USA, before implementation of vaccine, it was the most important cause of childhood meningitis accounting for 80% of cases (Schlech III *et al.*, 1985; Schuchat *et al.*, 1997) (Figure 17.3). In UK, it accounted for approximately 50% of cases in the pre-Hib vaccine era. By contrast, it was a much less prominent cause of meningitis in the meningitis belt of Africa where meningococcal meningitis causes the vast majority of cases.

There are a large number of factors that affect the epidemiology of disease (Frazer, 1982; Broome, 1987). Age is a major factor. Neonates are protected, and disease peaks by 9 months of age and declines to very low levels by 4 years. The age-specific disease incidence is inversely related to serum antibodies to Hib. Male gender is a risk factor of disease. Disease incidence varies by country and ethnic origin. For example, the incidence among Native Americans less than 4 years of age exceeded 150 cases/100 000/year compared to an

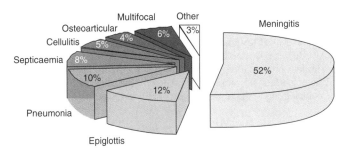

Figure 17.2 A pie chart showing Hib disease manifestations in percentages.

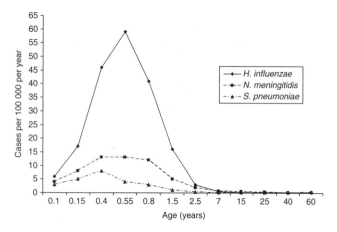

Figure 17.3 Incidence of *H. influenzae*, *N. meningitidis* and *S. pneumoniae* in the USA prior to Hib immunization.

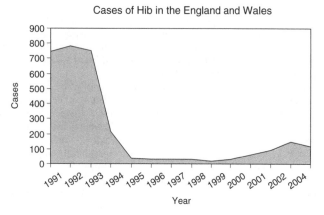

Figure 17.4 Cases of Hib-invasive disease in England and Wales per annum in children aged less than 14 years. The data were obtained from the HPA website http://www.hpa.org.uk

incidence for the whole country of 50 cases/100 000/year (Peltola, 2000). Mortality from Hib meningitis in resource-rich countries was approximately 5%. Long-term morbidity occurred in at least 10% of meningitis cases and consisted of deafness and neurological and learning deficits (Peltola, 2000).

Implementation of Hib-conjugate vaccine has had the largest impact on the epidemiology of disease. It has near disappeared from North America and Europe. There has been a similar dramatic decline in the Gambia, a resource-poor country with a very high incidence before the implementation of a national immunization programme (Peltola, 2000). There has been a striking re-emergence of Hib disease in UK over the past 5 years (Trotter, Ramsay and Slack, 2003) (Figure 17.3). There is compelling data suggesting that Hib disease occurring in Hib-immunized cases was associated with receiving combined Hibacellular pertussis vaccine that is associated with lower Hib antibody levels (McVernon *et al.*, 2003). This formulation was introduced in UK during 2000. As UK disease was increasing before this date and a similar, but smaller, rise has been reported in the Netherlands (Rijkers *et al.*, 2003), it raises the prospect that other factors may also be contributing to the re-emergence of Hib-invasive disease.

Noncapsulate or Nonserotypable *H. influenzae*

Noncapsulate or nonserotypable *H. influenzae* (Nc-hi) is acquired soon after birth, and 20–80% of people of all ages carry the organism.

These organisms rarely cause bacteraemia and meningitis. However, it is prominently associated with otitis media, sinusitis, chronic bronchitis and pneumonia (Murphy and Apicella, 1987; Murphy, 2003). Otitis media is one of the most common childhood diseases, and *H. influenzae* is one of the leading bacterial species cultured from middle ear fluid of cases. In a comprehensive Finnish study where bacteria were grown from middle ear fluid, 23% were *H. influenzae*, 23% were *Moraxella catarrhalis* and 26% were *Streptococcus pneumoniae* (Kilpi *et al.*, 2001). The other diseases are largely conditions of adults. Chronic bronchitis and pneumonia associated with *H. influenzae* occur more frequently with advancing age.

Antibiotic Resistance

Until the early 1970s all strains *of H. influenzae* were susceptible to antibiotics. In 1973 the first isolate resistant to ampicillin and producing β-lactamase was reported. Thereafter, not only did the prevalence of β-lactamase-producing strains rapidly rise in most parts of the world, but strains resistant to tetracycline, chloramphenicol and trimethoprim or multiresistant to the same antibiotics emerged in all parts of the world (Powell, 1988). Prevalence of resistance has remained reasonably constant over the past 10 years and in Europe and ranges from 10 to 20% in most countries (Schito, Debbia and Marchese, 2000). Large integrative and conjugative resistance elements encode most of the ampicillin (TEM β-lactamase producing), tetracycline and chloramphenicol resistance (Dimopoulou *et al.*, 1997; Dimopoulou *et al.*, 2002). Ampicillin resistance also arises from ROB-β-lactamase probably associated with small plasmids. Increasingly, strains are being isolated, which are β-lactamase negative, and the mechanism of resistance in these strains has been shown to be associated with mutations in penicillin-binding protein genes (Ubukata *et al.*, 2001). Trimethoprim resistance arises from mutations in chromosomal dihydrofolate reductase (de Groot *et al.*, 1988).

CLINICAL FEATURES

Hib-Invasive Disease

Clinical features of invasive disease in infants vary from a mild febrile illness, a feature of so-called occult bacteraemia, to fully blown sepsis syndrome with meningitis. Presentation of meningitis in children varies largely depending on age. Infants may present with as little as fever and irritability. Severe cases of all ages present with altered mental status, stiff neck and systemic features of sepsis, which in a few cases may be associated with disseminated intravascular necrosis including a purpuric rash and septicaemic shock. Examination of CSF and blood culture is the main means of identifying the bacterial aetiology. Obtaining CSF from cases must be performed with care, and steps must be taken to ensure that there is no evidence of cerebral oedema, usually by head CT (only in older children).

Epiglottitis is a particularly acute life-threatening manifestation of invasive Hib disease and is a medical emergency. Presentation of this disease manifestation is dramatic and may result in rapid death from sudden respiratory obstruction. Cases usually have general features of sepsis consisting of fever, tachycardia and tachypnoea. Pronounced local signs and symptoms consisting of rapid progression, sore throat, hoarseness, barking cough and stridor are encountered. The epiglottis is inflamed and swollen on visualization, but great care must be taken in examining these cases, as they may suddenly obstruct during examination of the epiglottis.

Pneumonia is a major and relatively unrecognized manifestation of Hib disease. Its recognition came principally from studies in the Gambia (Mulholland *et al.*, 1997). The main features are fever with tachypnoea and other features of respiratory distress such as nasal flaring and intercostal in drawing. Chest examination and X-ray are the key means of diagnosing pneumonia.

Septic arthritis presents with general features of sepsis and local features of arthritis. In a child this is often manifested by unwillingness to use a limb and pain on moving the joint. Examination and culture of joint fluid is the key procedure to make the diagnosis.

A rare manifestation is cellulitis in children. An unusual and specific form of cellulites involving the neck in adults has been described. This presents as a severe septic illness with marked oedema of the skin and an associated violacious rash. The skin is exquisitely tender to palpation.

Many of these disease manifestations have also been associated with cases of invasive disease caused by *H. influenzae* type f. This has become more apparent as the implementation of Hib-conjugate vaccine (Campos *et al.*, 2004).

Noncapsulate *H. influenzae*

The main disease manifestation associated with Nc-hi is otitis media (Murphy, 2003). This is one of the leading childhood illnesses and is associated with irritability and on otoscopy an inflamed tympanic membrane. In a proportion of cases the eardrum will perforate and pus will discharge from the ear. Tympanocentesis is the most reliable means of diagnosing and determining the aetiology of otitis media.

A high proportion of cases of conjunctivitis in children will be caused by Nc-hi. *Haemophilus aegyptius* has a particular proclivity for causing conjunctivitis. Culture of conjunctival swabs is the main test for making a diagnosis. One extreme clinical manifestation of disease caused by *H. aegyptius* is Brazilian purpuric fever, a fulminant septicaemic and often fatal disease.

Sinusitis is another disease manifestation associated with Nc-hi. Sinusitis is principally associated with local features of infection. There is local pain, a sense of pressure in the head and occasionally visible pus draining from the ostea of the sinuses. Occasionally there may be local facial oedema and in very severe cases spread to involve adjacent structures such as the orbit and even the meninges and brain. Diagnosis is based mainly on radiological examination, either skull X-ray with special views or CT scan. There will be opacification of the sinus and often a fluid level. Aspiration, examination and culture of fluid from the affected sinus are the best direct means of determining the aetiology of sinusitis.

Chronic bronchitis and exacerbations of bronchitis are the main disease manifestations associated with Nc-hi in adults. The exact aetiological role of bacteria in chronic bronchitis and exacerbations of bronchitis is not conclusively established. However, *H. influenzae* and *S. pneumonia* are cultured from sputum of up to 50% of cases, and many clinicians believe these bacteria play a role in the disease.

Severe invasive disease may be caused by Nc-hi. Neonatal sepsis has been well described and has features similar to group B streptococcal neonatal sepsis (Wallace Jr. *et al.*, 1983). In adults the main manifestation is pneumonia, particularly among the elderly (Murphy and Apicella, 1987; Murphy, 2003). An unusual manifestation is meningitis, usually in association with previous head trauma, particularly skull fracture (Kaufman *et al.*, 1990).

LABORATORY DIAGNOSIS

Direct examination of CSF and sterile site aspirates using Gram stain is valuable. Pleomorphic Gram-negative coccobacilli are often visible, though it may require careful scrutiny of the sample, as counter staining is poor with neutral red. Some labs use safranin or carbol fuchin to better stain the organism.

Direct culture (whether from a sample or a blood culture bottle) on chocolate agar and incubation in 5–10% CO_2 is the best way of growing *H. influenzae* on solid medium. Blood culture using any of the major commercial systems is effective at growing *H. influenzae*.

Differentiation of *Haemophilus* spp. in the routine laboratory can be satisfactorily achieved by using X, V and X&V requirements. In addition, further confirmation can be achieved by culture on blood agar in the presence of a staphylococcal streak for satelitism and haemolysis.

Antibiotic susceptibility testing using antibiotic discs must be done on media supplemented to support the growth of *Haemophilus* spp. Susceptibility testing by discs is frequently inaccurate, and it is important that disc zone size readings are supplemented by measurement of β-lactamase activity. One simple method is an acidometric paper strip method (Slack, Wheldon and Turk, 1977). Chloramphenicol disc susceptibility is also frequently inaccurate and should be supplemented with an assay for chloramphenicol-acetyl-transferase activity (Azemun *et al.*, 1981).

Direct detection of capsular type b antigen has been used as a means of rapidly diagnosing the presence of Hib in CSF, sterile-site fluid or urine. This is seldom used since the implementation of Hib vaccine. PCR of CSF has been developed for diagnosing *Haemophilus* spp. meningitis (van Ketel, de Wever and van Alphen, 1990; Radstrom *et al.*, 1994; Corless *et al.*, 2001).

Capsular typing of *H. influenzae* can be performed serologically; however, it is subject to error. A PCR-based method has been developed; it overcomes many of the problems of cross-reactions and is used in reference laboratories to characterize strains (Falla *et al.*, 1994).

TREATMENT OF INFECTION

The first-line antibiotic treatment of Hib-invasive disease is a third generation cephalosporin with good CSF penetration. Ceftriaxone and cefotaxime, in meningitis doses and depending on age, are both effective for treating *H. influenzae* meningitis and septicaemia. Cefuroxime must not be used for treating cases of meningitis, as it is less effective than the third generation cephalosporins (Schaad *et al.*, 1990). Adjunctive steroid therapy has been shown to reduce the rate of deafness in children with Hib meningitis (McIntyre *et al.*, 1997). In resource-poor countries, chloramphenicol alone (depending on the prevalence of chloramphenicol resistance) or in combination with ampicillin is effective therapy.

The antibiotic treatment of noninvasive infections such as otitis media, sinusitis and chronic bronchitis is much less clear. There are large meta-analyses that have not demonstrated a clear-cut advantage to administer antibiotics to all cases (Froom *et al.*, 1990; Saint *et al.*, 1995; Del Mar, Glasziou and Hayem, 1997; Glasziou *et al.*, 2004). There is some suggestion that subgroups are benefited by antibiotic treatment. Even with limited data supporting the use of antibiotics, many clinicians administer antibiotics. There are reasonable grounds for giving amoxicillin orally as the first-line. Alternatives include amoxicillin-clavulinate, trimethoprim, tetracycline (for adults only) and quinolones (for adults only).

PREVENTION AND CONTROL

The main preventative measure for controlling Hib disease is PRP-conjugate vaccines. There are a number of highly efficacious conjugate vaccine preparations. They consist of the capsular antigen PRP conjugated to one of the following: tetanus toxoid (PRP-T), outer membrane protein (PRP-OMP) or mutant diphtheria toxoid (PRP-CRM, HbOC). The vaccine is administered as part of childhood immunization in most resource-rich countries, and depending on the country, three doses are given at intervals between 2 and 6 months of age. Many countries give a booster dose at 1–2 years of age.

Antibiotic prophylaxis of close contacts of cases has been used to prevent secondary cases; however, this is only rational where the contacts have not received Hib vaccine. Rifampicin 20 mg/kg orally once a day for 4 days eradicates carriage and is believed to prevent secondary cases.

OTHER *HAEMOPHILUS* SPP.

Haemophilus ducreyi

Haemophilus ducreyi is the aetiological agent of the sexually transmitted disease soft chancre. This is rarely encountered in resource-rich countries. However, it is more prevalent in resource-poor countries and is more particularly associated with prostitution. A major feature of infection is the potentiation of HIV transmission and possible modification of the disease in HIV-infected people (Nasio *et al.*, 1996; Lewis, 2003).

The incubation is approximately 2 weeks. The main disease manifestations are painful genital ulcers from a few mm to 2 cm in size with a defined margin that start as an erythematous papule. The margin may be ragged in some cases and has a grey base which is relatively friable and bleeds. Inguinal lymphadenopathy occurs in most cases and in half they will suppurate and discharge spontaneously.

Diagnosis is essentially clinical. The organism is fastidious and difficult to grow without highly specialized media and prolonged culture. Nucleic acid amplification using PCR offers the best prospect for a reliable laboratory diagnostic test for this infection (Lewis, 2003). Treatment is with antibiotics. The following regimens are recommended by the CDC or WHO: azithromycin 1 gm as a single dose orally, ciprofloxacin 500 mg PO twice a day for 3 days, cephtriaxone 250 mg IM in a single dose, amoxicillin clavulanate (500/125 mg) PO three times a day for 7 days and trimethoprim-sulfamethoxazole (TMP-SMX) two tablets twice a day for 7 days.

Haemophilus parainfluenzae

Haemophilus parainfluenzae is the most abundant colonizing *Haemophilus* spp. It is a well-adapted commensal; all people are colonized soon after birth, and it is rarely associated with disease. Case reports of infective endocarditis, neurosurgical meningitis and prosthetic device infection have been published. It has been isolated from brain and liver abscesses. Its role in lung disease and nephropathies is less clear.

Rare Species

Haemophilus aphrophilus, *H. paraphrophilus* and *H. segnis* are very rare causes of disease. The best recognized disease manifestation is infective endocarditis. However, they account for less than 2% of all the cases of infective endocarditis. Given that these species are slow growing, they may be missed if blood cultures are not incubated for prolonged periods. These species may be rarely found causing brain or lung abscess. They can also be cultured from empyaema fluid.

REFERENCES

Azemun, P., Stull, T., Roberts, M. *et al.* (1981) Rapid detection of chloramphenicol resistance in *Haemophilus influenzae*. *Antimicrob Agents Chemother*, **20**, 168–70.

Barbour, M. L., Mayon-White, R. T., Coles, C. *et al.* (1995) The impact of conjugate vaccine on carriage of *Haemophilus influenzae* type b. *J Infect Dis*, **171**, 93–8.

Bayliss, C. D., Field, D. and Moxon, E. R. (2001) The simple sequence contingency loci of *Haemophilus influenzae* and *Neisseria meningitidis*. *J Clin Invest*, **107**, 657–62.

Bouchet, V., Hood, D. W., Li, J. *et al.* (2003) Host-derived sialic acid is incorporated into *Haemophilus influenzae* lipopolysaccharide and is a major virulence factor in experimental otitis media. *Proc Natl Acad Sci U S A*, **100**, 8898–903.

Broome, C. V. (1987) Epidemiology of *Haemophilus influenzae* type b infections in the United States. *Pediatr Infect Dis J*, **6**, 779–82.

Campos, J., Hernando, M., Roman, F. *et al.* (2004) Analysis of invasive *Haemophilus influenzae* infections after extensive vaccination against *H. influenzae* type b. *J Clin Microbiol*, **42**, 524–9.

Corless, C. E., Guiver, M., Borrow, R. *et al.* (2001) Simultaneous detection of *Neisseria meningitidis*, *Haemophilus influenzae*, and *Streptococcus pneumoniae* in suspected cases of meningitis and septicemia using real-time PCR. *J Clin Microbiol*, **39**, 1553–8.

Del Mar, C., Glasziou, P. and Hayem, M. (1997) Are antibiotics indicated as initial treatment for children with acute otitis media? A meta-analysis. *BMJ*, **314**, 1526–9.

Dimopoulou, I. D., Jordens, J. Z., Legakis, N. J. *et al.* (1997) A molecular analysis of Greek and UK *Haemophilus influenzae* conjugative resistance plasmids. *J Antimicrob Chemother*, **39**, 303–7.

Dimopoulou, I. D., Russell, J. E., Mohd-Zain, Z. *et al.* (2002) Site-specific recombination with the chromosomal tRNA (Leu) gene by the large conjugative Haemophilus resistance plasmid. *Antimicrob Agents Chemother*, **46**, 1602–3.

Eskola, J., Kayhty, H., Takala, A. K. *et al.* (1990) A randomized, prospective field trial of a conjugate vaccine in the protection of infants and young children against invasive *Haemophilus influenzae* type b disease. *N Engl J Med*, **323**, 1381–7.

Falla, T. J., Crook, D. W., Brophy, L. N. *et al.* (1994) PCR for capsular typing of *Haemophilus influenzae*. *J Clin Microbiol*, **32**, 2382–6.

Fleischmann, R. D., Adams, M. D., White, O. *et al.* (1995) Whole-genome random sequencing and assembly of *Haemophilus influenzae* Rd. *Science*, **269**, 496–512.

Frazer, D. W. (1982) In *Haemophilus Influenzae*. (eds S. H. A. W. Snell and F. Peter), Elsevier. Science Publishing Co Inc., New York.

Froom, J., Culpepper, L., Grob, P. *et al.* (1990) Diagnosis and antibiotic treatment of acute otitis media: report from International Primary Care Network. *BMJ*, **300**, 582–6.

Glasziou, P. P., Mar, C. B., Sanders, S. L. *et al.* (2004) Antibiotics for acute otitis media in children. *Cochrane Database Syst Rev*, CD000219.

de Groot, R., Campos, J., Moseley, S. L. *et al.* (1988) Molecular cloning and mechanism of trimethoprim resistance in *Haemophilus influenzae*. *Antimicrob Agents Chemother*, **32**, 477–84.

Hardy, G. G., Tudor, S. M. and St Geme, J. W. III. (2003) The pathogenesis of disease due to nontypeable *Haemophilus influenzae*. *Methods Mol Med*, **71**, 1–28.

Kaufman, B. A., Tunkel, A. R., Pryor, J. C. *et al.* (1990) Meningitis in the neurosurgical patient. *Infect Dis Clin North Am*, **4**, 677–701.

van Ketel, R. J., de Wever, B. and van Alphen, L. (1990) Detection of *Haemophilus influenzae* in cerebrospinal fluids by polymerase chain reaction DNA amplification. *J Med Microbiol*, **33**, 271–6.

Kilian, M. (1976) A taxonomic study of the genus *Haemophilus*, with the proposal of a new species. *J Gen Microbiol*, **93**, 9–62.

Kilpi, T., Herva, E., Kaijalainen, T. *et al.* (2001) Bacteriology of acute otitis media in a cohort of Finnish children followed for the first two years of life. *Pediatr Infect Dis J*, **20**, 654–62.

Kroll, J. S., Loynds, B. M. and Moxon, E. R. (1991) The *Haemophilus influenzae* capsulation gene cluster: a compound transposon. *Mol Microbiol*, **5**, 1549–60.

Kubiet, M., Ramphal, R., Weber, A. *et al.* (2000) Pilus-mediated adherence of *Haemophilus influenzae* to human respiratory mucins. *Infect Immun*, **68**, 3362–7.

Lazou Ahren, I., Bjartell, A., Egesten, A. *et al.* (2001) Lipopolysaccharide-binding protein increases toll-like receptor 4-dependent activation by nontypeable *Haemophilus influenzae*. *J Infect Dis*, **184**, 926–30.

Lewis, D. A. (2003) Chancroid: clinical manifestations, diagnosis, and management. *Sex Transm Infect*, **79**, 68–71.

McIntyre, P. B., Berkey, C. S., King, S. M. *et al.* (1997) Dexamethasone as adjunctive therapy in bacterial meningitis. A meta-analysis of randomized clinical trials since 1988. *JAMA*, **278**, 925–31.

McVernon, J., Andrews, N., Slack, M. P. *et al.* (2003) Risk of vaccine failure after *Haemophilus influenzae* type b (Hib) combination vaccines with acellular pertussis. *Lancet*, **361**, 1521–3.

Meats, E., Feil, E. J., Stringer, S. *et al.* (2003) Characterization of encapsulated and noncapsulated *Haemophilus influenzae* and determination of phylogenetic relationships by multilocus sequence typing. *J Clin Microbiol*, **41**, 1623–36.

Mulholland, K., Hilton, S., Adegbola, R. *et al.* (1997) Randomised trial of *Haemophilus influenzae* type-b tetanus protein conjugate vaccine [corrected] for prevention of pneumonia and meningitis in Gambian infants. *Lancet*, **349**, 1191–7.

Murphy, T. F. (2003) Respiratory infections caused by non-typeable *Haemophilus influenzae*. *Curr Opin Infect Dis*, **16**, 129–34.

Murphy, T. F. and Apicella, M. A. (1987) Nontypable *Haemophilus influenzae*: a review of clinical aspects, surface antigens, and the human immune response to infection. *Rev Infect Dis*, **9**, 1–15.

Nasio, J. M., Nagelkerke, N. J., Mwatha, A. *et al.* (1996) Genital ulcer disease among STD clinic attenders in Nairobi: association with HIV-1 and circumcision status. *Int J STD AIDS*, **7**, 410–4.

Peltola, H. (2000) Worldwide *Haemophilus influenzae* type b disease at the beginning of the 21st century: global analysis of the disease burden 25 years after the use of the polysaccharide vaccine and a decade after the advent of conjugates. *Clin Microbiol Rev*, **13**, 302–17.

Pittman, M. (1930) The "S" and "R" forms of *H. influenzae. Proc Soc Exp Biol Med*, **27**, 299.

Pittman, M. (1931) Variation and type specificity in the bacterial species *H. influenzae. J Exp Med*, **53**, 471–92.

Powell, M. (1988) Antimicrobial resistance in *Haemophilus influenzae. J Med Microbiol*, **27**, 81–7.

Radstrom, P., Backman, A., Qian, N. *et al.* (1994) Detection of bacterial DNA in cerebrospinal fluid by an assay for simultaneous detection of *Neisseria meningitidis, Haemophilus influenzae*, and streptococci using a seminested PCR strategy. *J Clin Microbiol*, **32**, 2738–44.

Richards, J. C., Cox, A. D., Schweda, E. K. *et al.* (2001) Structure and functional genomics of lipopolysaccharide expression in *Haemophilus influenzae. Adv Exp Med Biol*, **491**, 515–24.

Rijkers, G. T., Vermeer-de Bondt, P. E., Spanjaard, L. *et al.* (2003) Return of *Haemophilus influenzae* type b infections. *Lancet*, **361**, 1563–4.

Saint, S., Bent, S., Vittinghoff, E. *et al.* (1995) Antibiotics in chronic obstructive pulmonary disease exacerbations. A meta-analysis. *JAMA*, **273**, 957–60.

Schaad, U. B., Suter, S., Gianella-Borradori, A. *et al.* (1990) A comparison of ceftriaxone and cefuroxime for the treatment of bacterial meningitis in children. *N Engl J Med*, **322**, 141–7.

Schito, G. C., Debbia, E. A. and Marchese, A. (2000) The evolving threat of antibiotic resistance in Europe: new data from the Alexander project. *J Antimicrob Chemother*, **46** (Suppl. T1), 3–9.

Schlech, W. F. III, Ward, J. I., Band, J. D. *et al.* (1985) Bacterial meningitis in the United States, 1978 through 1981. The National Bacterial Meningitis Surveillance Study. *JAMA*, **253**, 1749–54.

Schuchat, A., Robinson, K., Wenger, J. D. *et al.* (1997) Bacterial meningitis in the United States in 1995. Active Surveillance Team. *N Engl J Med*, **337**, 970–6.

Slack, M. P., Wheldon, D. B. and Turk, D. C. (1977) A rapid test for beta-lactamase production by *Haemophilus influenzae. Lancet*, **2**, 906.

Trotter, C. L., Ramsay, M. E. and Slack, M. P. (2003) Rising incidence of *Haemophilus influenzae* type b disease in England and Wales indicates a need for a second catch-up vaccination campaign. *Commun Dis Public Health*, **6**, 55–8.

Turk, D. C. (1982) In *Haemophilus Influenzae* (eds S. H. A. W. Snell and F. Peter), Elsevier Science Publishing Co Inc., New York.

Ubukata, K., Shibasaki, Y., Yamamoto, K. *et al.* (2001) Association of amino acid substitutions in penicillin-binding protein, 3 with beta-lactam resistance in beta-lactamase-negative ampicillin-resistant *Haemophilus influenzae. Antimicrob Agents Chemother*, **45**, 1693–9.

Wallace, R. J. Jr., Baker, C. J., Quinones, F. J. *et al.* (1983) Nontypable *Haemophilus influenzae* (biotype, 4) as a neonatal, maternal, and genital pathogen. *Rev Infect Dis*, **5**, 123–36.

Wang, X., Moser, C., Louboutin, J. P. *et al.* (2002) Toll-like receptor 4 mediates innate immune responses to *Haemophilus influenzae* infection in mouse lung. *J Immunol*, **168**, 810–5.

Zwahlen, A., Kroll, J. S., Rubin, L. G. *et al.* (1989) The molecular basis of pathogenicity in *Haemophilus influenzae*: comparative virulence of genetically-related capsular transformants and correlation with changes at the capsulation locus cap. *Microb Pathog*, **7**, 225–35.

18

Bordetella spp.

Qiushui He[1], Jussi Mertsola[2] and Matti K. Viljanen[3]

[1]*National Public Health Institute, Kiinamyllynkatu;* [2]*Department of Pediatrics; and* [3]*Department of Medical Microbiology, University of Turku, Kiinamyllynkatu, Turku, Finland*

INTRODUCTION

Bacteria of the genus *Bordetella* are small, aerobic and Gram-negative coccobacilli. Currently, the genus consists of seven species: *Bordetella pertussis*, *B. parapertussis*, *B. bronchiseptica*, *B. avium*, *B. hinzii*, *B. holmesii* and *B. trematum*. The species can be distinguished from each other based on their phenotypic characteristics (Table 18.1). All but *B. avium* have been isolated from humans. Of these *Bordetella* species, *B. pertussis*, *B. parapertussis* and *B. bronchiseptica* are the most important human pathogens.

THE GENUS *BORDETELLA*

Bordetella pertussis

Bordetella pertussis was first isolated in 1906 (Bordet and Gengou, 1906) and classified as *Haemophilus pertussis* because of the requirement of blood in laboratory media. However, unlike *Haemophilus*, *B. pertussis* does not need growth factors X (haematin) and V (nicotinamide adenine dinucleotides). From the 1960s on, *Bordetella* was used as a new genus name in honour of Bordet who was the first to report its isolation.

Bordetella pertussis is the most important cause of pertussis (whooping cough) in humans. No animal or environmental reservoir is known for this organism. *Bordetella pertussis* is fastidious and grows slowly (mean generation time 2.3–5 h) (Rowatt, 1957; Hoppe, 1999a). The colonies of *B. pertussis* on culture plates usually become visible after 3–4 days of incubation.

Bordetella parapertussis

Bordetella parapertussis was first described in 1937 (Bradford and Slavin, 1937; Eldering and Kendrick, 1937; Eldering and Kendrick, 1938). *Bordetella parapertussis* usually causes cough with posttussive vomiting that is similar to, but often milder than, that due to *B. pertussis*. Some severe cases of parapertussis have also been reported (Heininger *et al.*, 1994). The incidence of human *B. parapertussis* infections has not been definitely established. Current estimations may be too low (He *et al.*, 1998). *Bordetella pertussis*, *B. parapertussis* and *B. bronchiseptica* share a number of virulence factors and tropism for the ciliated cells of the respiratory tract. Unlike *B. pertussis*, *B. parapertussis* does not secrete pertussis toxin (Table 18.2). *Bordetella parapertussis* is less fastidious than *B. pertussis*. The colonies of *B. parapertussis* on culture plates usually become visible after 2–3 days of incubation.

Table 18.1 Characteristics of *Bordetella* species[a]

Characteristics	*B. pertussis*	*B. parapertussis*	*B. bronchiseptica*	*B. avium*	*B. hinzii*	*B. holmesii*	*B. trematum*
Host	Humans	Humans, sheep	Mammals	Birds, reptiles	Birds, humans	Humans?	Humans?
Disease	Pertussis (whooping cough)	Pertussis (mild)	Various respiratory diseases	Turkey Coryza	Septicaemia in patients with underlying disease	Septicaemia, pertussis-like symptoms	Unknown
Isolation site in humans	Respiratory tract	Respiratory tract	Respiratory tract, blood	–	Respiratory tract, blood	Respiratory tract, blood	Wounds, ear infections
Growth on:							
Blood agar	–	+	+	+	+	+	+
MacConkey agar	–	±	+	+	+	+	+
Oxidase	+	–	+	+	+	–	–
Nitrate reduction	–	–	+	–	–	–	±
Urease production	–	+ (24 h)	+ (4 h)	–	±	–	–
Motility (37 °C) +	–	–	+	+	+		
G + C content (mol%)	67.7	68.1	68.1	61.6	65–67	61.5–62.3	64–65
Genome size (bp)	4 086 186	4 773 551	5 339 179	3732255	ND	ND	ND

+, growth or activity present; –, not present; ±, may or may not be present; ND, not determined.
[a] Adapted from Gerlach *et al.* (2001) and Hoppe (1999a).

Principles and Practice of Clinical Bacteriology Second Edition   Editors Stephen H. Gillespie and Peter M. Hawkey
© 2006 John Wiley & Sons, Ltd

Table 18.2 Virulence factors produced by the three most studied members of the genus *Bordetella*[a]

	B. pertussis	*B. parapertussis*	*B. bronchiseptica*
Pertussis toxin[b]	+	–	–
Filamentous haemagglutinin[b]	+	+	+
Fimbriae[b]	+	+	+
Pertactin[b]	+	+	+
Adenylate cyclase toxin/haemolysin	+	+	+
Dermonecrotic toxin	+	+	+
Tracheal cytotoxin[c]	+	+	+
Lipopolysaccharide	+	+	+
Tracheal colonization factor	+	–	–
Serum resistance locus (BrkA)	+	+	+/–
Flagella	–	–	+

+, produced; –, gene cryptic or absent; +/–, produced by some strains.
[a] Adapted from Smith, Guzmán and Walker (2001).
[b] Included in acellular pertussis vaccines.
[c] Production of tracheal cytotoxin is not regulated by the *bvg* locus.

Bordetella parapertussis has also been found in lambs, in which it causes chronic nonprogressive pneumonia (Hoppe, 1999b). Human and ovine strains have been found to be genetically different (Table 18.3), suggesting that transmission between sheep and humans is unlikely.

Bordetella bronchiseptica

Bordetella bronchiseptica is a common respiratory pathogen in a wide range of mammalian species, including cats, dogs, horses, rabbits and swine (Woolfrey and Moody, 1991). In humans, reports of respiratory disease caused by *B. bronchiseptica* remain rare and usually involve children and immunocompromised patients. As early as in 1926, *B. bronchiseptica* was isolated from a child with signs and symptoms compatible with pertussis (Lautrop, 1960).

Bordetella bronchiseptica shares substantially similar sets of virulence factors with *B. pertussis* and *B. parapertussis*. *Bordetella bronchiseptica* does not produce pertussis toxin (Table 18.2). *Bordetella bronchiseptica* grows faster on culture plates than *B. pertussis* and *B. parapertussis*, and the colonies usually become visible after 24 h of incubation.

Table 18.3 Insertion sequence elements (copy number) in the genomes of *Bordetella* species[a]

Species (hosts)	IS481	IS1001	IS1002
B. pertussis	+ (80–100)	–	+ (4–7)
B. parapertussis (human)	–	+ (20–21)	+ (9)
B. parapertussis (sheep)	–	+ (23–29)	–
B. bronchiseptica (mainly pig)	–	+ (1–7)	–
B. bronchiseptica (guinea pig/horse)	+ (1–2)	–	–
B. bronchiseptica (koala)	–	–	+ (1)
B. holmesii (human)	+ (8–10)	+	–
B. avium	–	–	–
B. hinzii	–	–	–
B. trematum	–	–	–

+, presence of insertion sequence (IS) element; –, absence of IS element.
[a] Adapted from van der Zee *et al.* (1997), Gerlach *et al.* (2001) and Reischl *et al.* (2001).

Other *Bordetella* species

Bordetella avium, first described in 1984 (Kersters *et al.*, 1984), is a respiratory pathogen of birds. *Bordetella avium* infections have not been identified in humans. *Bordetella hinzii* was isolated from the respiratory tracts of birds, where it appears to be nonpathogenic (Vandamme *et al.*, 1995) and has rarely been found in immunocompromised patients. *Bordetella holmesii*, first described in 1995 (Weyant *et al.*, 1995), was originally isolated from blood cultures of young adults. Recently, it has been isolated from the nasopharyngeal samples of schoolchildren with cough (Yih *et al.*, 1999). *Bordetella trematum* has been isolated from wounds and ear infections in humans (Vandamme *et al.*, 1996). *Bordetella petrii* was recently isolated from an anaerobic, dechlorinating bioreactor culture enriched from river sediment. This *Bordetella* species has been proposed as a member of the genus *Bordetella* and is the first member of the genus isolated from the natural environment and capable of anaerobic growth (Von Wintzingerode *et al.*, 2001).

Bordetella genome

The genome sequencing of *B. pertussis*, *B. parapertussis* and *B. bronshiseptica* has been completed (Table 18.1) (http://www.sanger.ac.uk/projects/Microbes/). The sequencing of the *B. avium* genome has been just finished. The representative strains selected for sequencing were Tohama I of *B. pertussis*, strain 12822 of *B. parapertussis* and strain RB50 of *B. bronchiseptica*. Tohama I was isolated in the 1950s in Japan (Kasuga *et al.*, 1954) and was later widely used in laboratory studies and in the production of some acellular pertussis vaccines. *Bordetella parapertussis* strain 12822 was isolated in Germany in 1993 (Heininger *et al.*, 2002). *Bordetella bronchiseptica* strain RB50 was isolated from the naris of a 3-month-old New Zealand White rabbit (Cotter and Miller, 1994). The genome of *B. pertussis* has been found to be significantly smaller than that of *B. bronchiseptica* and *B. parapertussis*. It remains to be shown whether the adaptation of *B. pertussis* to humans as the only host has resulted from the deletion of some genes of the *B. bronchiseptica* strain or whether the adaptation has also required the acquisition of significant amounts of novel genetic material by horizontal transfer (Gerlach *et al.*, 2001). The genome of *B. avium* is also much smaller than that of *B. bronchiseptica*.

Systematics and Evolution

Genetically, *B. pertussis*, *B. parapertussis* and *B. bronchiseptica* are closely related and should be considered subspecies of a single species (*B. bronchiseptica* cluster) with different host adaptation (Musser *et al.*, 1986; Arico *et al.*, 1987; Gerlach *et al.*, 2001). They have probably also diverged from a common ancestor. However, typical differences among the three species have been found. As for virulence factors, pertussis toxin and tracheal colonization factor are only produced by *B. pertussis* (Arico and Rappuoli, 1987; Finn and Stevens, 1995). A large genetic locus required for O antigen biosynthesis has been identified in *B. parapertussis* and *B. bronchiseptica* (Preston *et al.*, 1999). This locus is replaced by an insertion sequence (IS) in *B. pertussis*. Three IS elements have been characterized among the *Bordetella* species (van der Zee *et al.*, 1996b, 1997) (Table 18.3). IS481 and IS1002 have been found in all *B. pertussis* strains, whereas IS1001 has only been found in some *B. bronchiseptica* isolates. Regarding *B. parapertussis*, both IS1002 and IS1001 have been identified in human isolates and only IS1001 in sheep isolates. Human *B. parapertussis* isolates are homogeneous, whereas sheep *B. parapertussis* isolates are heterogeneous (van der Zee *et al.*, 1996b, 1997; Yuk *et al.*, 1998). The IS481 element was identified in *B. holmesii* isolates, whereas none of the three IS elements occur in *B. avium*, *B. hinzii* and *B. trematum* (Gerlach *et al.*, 2001; Reischl *et al.*, 2001).

PATHOGENESIS

The putative virulence factors of *B. pertussis* can be divided into adhesins and toxins. The adhesins facilitate the adherence of the bacterium to host ciliated respiratory epithelial cells and immune effector cells, and the toxins are involved in the clinical disease process by acting locally or systemically. The pathogenesis of infectious diseases caused by *B. pertussis*, *B. parapertussis* and *B. bronchiseptica* has been extensively studied, whereas very little is known about the pathogenesis of infections caused by the other *Bordetella* species. The expression of nearly all adhesins and toxins is positively regulated by the products of the *bvgAS* locus (Matto *et al.*, 2001). BvgA and BvgS comprise a two-component signalling system that mediates transition between three identifiable phases – a virulent (Bvg$^+$) phase, an avirulent (Bvg$^-$) phase and an intermediate (Bvg^i) phase – in response to specific environmental signals (Cotter and Miller, 1997; Matto *et al.*, 2001). The Bvg$^+$ phase has been shown to be necessary and sufficient for respiratory tract colonization. The Bvg^i phase is characterized by the expression of a subset of Bvg-activated factors as well as factors maximally expressed in this phase. The Bvg^i phase is assumed to be involved in aerosol transmission. In the Bvg$^-$ phase, *B. pertussis* and *B. bronchiseptica* express different sets of proteins. The Bvg$^-$ phase of *B. bronchiseptica* has been shown to be necessary and sufficient for survival under nutrient-restricted conditions. The interplay between *B. pertussis* and host cells is a complex process. Using high-density DNA microarrays, a recent *in vitro* study showed that *B. pertussis* induces mucin gene transcription of human bronchial epithelial cells and then counters this defence by using mucin as a binding substrate. The early transcriptional response of these cells to *B. pertussis* is dominated by altered expression of cytokines, DNA-binding proteins and NF-κB-regulated genes (Belcher *et al.*, 2000).

Adhesins

Filamentous Hemagglutinin (FHA)

Filamentous haemagglutinin (FHA) is a 220-kDa surface-associated protein which functions as the major adhesin of *B. pertussis*. The gene for FHA, *fhaB*, has coding potential for a protein of 367 kDa (Delisse-Gathoye *et al.*, 1990; Domenighini *et al.*, 1990). More than one-third of this precursor protein is later cleaved in a complex posttranslational maturation process to unveil the biologically active polypeptide (Locht *et al.*, 1993). FHA contains an Arginine-Glycine-Aspartic acid (Arg-Gly-Asp) peptide sequence, which binds sulfated glycosaminoglycans on cell-surface proteoglycans (Brennan and Shahin, 1996). Deletion of the *FHA* gene in mutant *B. pertussis* strains reduces bacterial adherence to epithelial cells by 70% (Menozzi *et al.*, 1994). A recent *in vitro* study shows that secreted and cell-associated FHA can elicit proinflammatory and proapoptotic responses in human monocytic cells and bronchial epithelial cells, suggesting a previously unrecognized role for FHA in *B. pertussis*–host interaction (Abramson, Kedem and Relman, 2001).

Immunization with FHA protects mice against aerosol challenge with *B. pertussis* (Oda *et al.*, 1984; Sato and Sato, 1984). In humans, addition of FHA to an acellular pertussis vaccine containing only PT increased protection against pertussis in a vaccine efficacy trial (Ad Hoc Group for the Study of Pertussis Vaccines, 1988; Storsaeter and Olin, 1992). FHA is included in all multicomponent acellular pertussis vaccines.

Fimbriae

Fimbriae (FIM), also known as pili and agglutinogens, are long filamentous protrusions that extend from the bacterial cell surface and facilitate binding of the bacterium to host cells. *Bordetella pertussis*

produces two antigenically different fimbriae, named serotype 2 and serotype 3. These are constructed of a major subunit, encoded by *fim2* and *fim3* genes, respectively, and a minor subunit encoded by *fimD* (Willems, van der Heide and Mooi, 1992). Fim2 and Fim3 are proteins of 22 and 22.5 kDa (Steven *et al.*, 1986; Ashworth *et al.*, 1988). Both Fim2 and Fim3 play a role in the attachment of *B. pertussis* to the cells of immune systems, including monocytes, their FimD interacting with the host integrin (Hazenbos *et al.*, 1995; Geuijen *et al.*, 1997).

Purified Fim have been shown to protect mice against respiratory infection with *B. pertussis* (Oda *et al.*, 1985; Ashworth *et al.*, 1988). The protection is serotype specific (Robinson *et al.*, 1989). In humans, the protection provided by whole-cell vaccines is to some extent serotype specific (Preston and Carter, 1992). Acellular vaccines including Fim, in addition to PT, FHA and Prn, are more effective against pertussis than those without Fim (Gustafsson *et al.*, 1996; Olin *et al.*, 1997).

Pertactin

The pertactin (Prn) of *B. pertussis* is a 69-kDa outer membrane protein occurring in all virulent strains. The sizes of Prn molecules produced by *B. parapertussis* and *B. bronchiseptica* are 70 and 68 kDa, respectively (Montaraz, Novotny and Ivanyi, 1985; Li *et al.*, 1991). Prn has been shown to play a role in attachment to and invasion in host cells (Leininger *et al.*, 1991; Brennan and Shahin, 1996). Like FHA, Prn has an Arg-Gly-Asp-binding site interacting with host integrins (Brennan and Shahin, 1996). Prn has two immunodominant regions, 1 and 2, composed of repeating units of five (GGXXP) and three (PQP) amino acids, respectively (Charles *et al.*, 1991). Region 1 is adjacent to the Arg-Gly-Asp motif. The crystal structure of Prn indicates that region 1 forms a loop protruding from a β-sheet (Emsley *et al.*, 1996).

Prn elicits protective immunity in both animals and humans (Brennan *et al.*, 1988; Shahin *et al.*, 1990; Cherry *et al.*, 1998; Storsaeter *et al.*, 1998). In a vaccine efficacy trial, addition of Prn to an acellular vaccine containing PT and FHA provided increased protection against pertussis (Greco *et al.*, 1996; Gustafsson *et al.*, 1996; Olin *et al.*, 1997). Prn is included in most acellular pertussis vaccines.

Recently, divergence was found between the Prn of *B. pertussis* vaccine strains and that of circulating strains (Mooi *et al.*, 1998; Mastrantonio *et al.*, 1999; Mooi *et al.*, 1999; Cassiday *et al.*, 2000; Fry *et al.*, 2001). This polymorphism is essentially limited to region 1. The three most prevalent Prn variants are Prn1–3. Vaccine strains produce Prn1 (Boursaux-Eude *et al.*, 1999; Mooi *et al.*, 2000), whereas Prn2 is the predominant type found in circulating strains. Prn1 and Prn3 are similar, with a difference of only two amino acids, whereas Prn2 has an additional repeat. The biologic importance of this variation remains to be determined. A recent study showed that individuals infected with Prn2 strains have significantly less antibodies to Prn1 than those infected with Prn3 strains or those immunized with a booster dose of acellular vaccines containing Prn1 (He *et al.*, 2003). Moreover, in contrast to vaccine recipients and subjects infected with Prn3 strains, individuals infected with Prn2 strains had hardly any antibodies to the variable region of Prn1.

Polymorphisms have also been reported in the Prn of *B. bronchiseptica*. The variation occurs in both immunodominant regions (Boursaux-Eude and Guiso, 2000; Register, 2001). The Prn of *B. parapertussis* appears to be more conserved than that of *B. pertussis* and *B. bonchiseptica* (Boursaux-Eude and Guiso, 2000).

Toxins

Pertussis Toxin

Pertussis toxin (PT) is a 106-kDa hexameric protein with many biologic activities *in vivo*. It is a member of the A-B bacterial toxin superfamily and consists of five distinct subunits, S1–S5 (Tamura *et al.*, 1982). The molar ratios of the subunits in the native molecule are 1:1:1:2:1. The A portion or S1 subunit has enzymatic activities, catalyzing the transfer of the ADP-ribose moiety of NAD^+ to a family of G proteins which are involved in signal transduction pathways within the host cell. The B oligomer (S2–S5) facilitates attachment of PT to the host cell and delivers the toxic action of the S1 subunit to the cell (Tamura *et al.*, 1982). The genes encoding the PT subunits are clustered in an operon typical of many other bacterial toxins. Genetic analysis has shown that each subunit is translated separately and equipped with an aminoterminal signal sequence that is cleaved during transport to the periplasmic space. The holotoxin is then assembled in the periplasmic space and secreted through the outer membrane into the environment of the bacterium (Locht and Keith, 1986; Nicosia *et al.*, 1986).

Immunization with PT has been found to protect against both intracerebral and aerosol challenges with *B. pertussis*. PT has been included in all acellular pertussis vaccines. Divergence has been found in the PT S1 subunit between *B. pertussis* vaccine strains and circulating strains (Mooi *et al.*, 1998; Mastrantonio *et al.*, 1999; Mooi *et al.*, 1999; Cassiday *et al.*, 2000; Fry *et al.*, 2001). Almost all currently circulating strains in Europe and USA have PTs different from those of vaccine strains. The biologic significance of this antigenic variation is not known.

Of the *Bordetella* species, only *B. pertussis* produces PT. Although *B. parapertussis* and *B. pronchiseptica* contain a PT operon, these organisms cannot produce the toxin because a number of mutations occur in the promoter regions of the genes coding for PT subunits (Arico *et al.*, 1987).

Other Virulence Factors

Adenylate cyclase toxin is a bifunctional protein of the RTX (repeat in toxin) family. The toxin is activated by calmodulin (Ladant *et al.*, 1989) inside the cell, where it catalyses the formation of high levels of cyclic AMP (Greenlee, Andreasen and Storm, 1982). Immunization with adenylate cyclase toxin provides protection against bacterial colonization in animal models (Guiso, Szatanik and Rocancourt, 1991). Serum antibodies against the toxin have been found in humans with pertussis (Arciniega *et al.*, 1991).

BrkA (*Bordetella* resistance to killing) protein is a Bvg-regulated molecule, which mediates the serum resistance and adherence of the bacterium (Fernandez and Weiss, 1994). It shares sequence identity with *B. pertussis* Prn and is a member of the diverse group of proteins found in Gram-negative bacteria secreted via an autotransporter mechanism. Sera taken from acellular vaccine recipients or individuals with a history of a recent infection with *B. pertussis* fail to kill wild-type *B. pertussis* but kill BrkA mutant strains (Oliver and Fernandez, 2001). The BrkA protein prevents killing of *B. pertussis* by the antibody-dependent classical pathway of complement (Barnes and Weiss, 2003).

In Gram-negative bacteria, lipopolysaccharide (LPS) generally comprises lipid A, a core oligosaccharide and a long polysaccharide O antigen. The LPS molecule of *B. pertussis* lacks the O antigen because its genetic locus required for O antigen biosynthesis is replaced by an IS element (Preston *et al.*, 1999). The *B. pertussis* LPS molecule has antigenic and immunomodulating properties. LPS may act in synergy with tracheal cytotoxin or other toxins (Amano, Fukushi and Watanabe, 1990; Watanabe *et al.*, 1990; Schaeffer and Weiss, 2001; Ausiello *et al.*, 2002). In addition, it has endotoxin activities.

Tracheal cytotoxin is a specific fragment of *B. pertussis* cell-wall peptidoglycan that causes ciliostasis and damage to ciliated epithelial cells (Hewlett, 1997). Its toxicity is mediated by intracellular production of interleukin-1 and nitric oxide (Heiss *et al.*, 1994).

Tracheal colonization factor (Tcf) is a virulence factor encoded by the *tcfA* gene (Finn and Stevens, 1995). Tcf is pro-

duced by *B. pertussis* but not by *B. parapertussis* or *B. bronchiseptica*. Its C-terminal half shows more than 50% homology to the C terminus of the precursors of Prns of *Bordetella* species, and its N-terminal half contains three Arg-Gly-Asp motifs. When a strain of *B. pertussis* lacking this protein is used to infect mice with an aerosol challenge, the number of bacteria isolated from the tracheas is only 10% of that obtained with the parental strain (Finn and Stevens, 1995). Polymorphism in *tcfA* gene has been reported (Van Loo *et al.*, 2002).

EPIDEMIOLOGY

Epidemiologic Characteristics of Pertussis

Pertussis is an endemic disease with epidemic peaks occurring every 2–5 years (Cherry, 1984). The seasonal pattern of the disease is not clearly marked (Gordon and Hood, 1951; Cherry, 1984). The changing seasonal profile of pertussis was documented in a recent report of a large outbreak in British Columbia, Canada (Skowronski *et al.*, 2002). During this outbreak in 2000, pertussis reports began to increase in March, peaked in June and markedly dropped in July. However, in the past 10 years, during outbreak and nonoutbreak periods, pertussis reports have begun to increase in July, peaking between late August and November.

Transmission is by aerosol from coughing individuals. Contagiousness is highest during the first 2 weeks of illness. The attack rate among vulnerable individuals is 50–100%, depending on the nature of exposure (Lambert, 1965; Mertsola *et al.*, 1983a; Long, Welkon and Clark, 1990). Pertussis morbidity and mortality is higher in females than in males. Immunized contacts can become transiently colonized and remain asymptomatic or symptomatic (Lambert, 1965; Mertsola *et al.*, 1983a; Long, Welkon and Clark, 1990; He *et al.*, 1994). Long-term carriers of *B. pertussis* have not been identified. Pertussis affects all races equally.

Infections caused by *B. parapertussis* are not rare (Heininger *et al.*, 1994; He, Viljanen *et al.*, 1998). The clinical picture of patients infected with *B. parapertussis* is indistinguishable from that of pertussis. However, the disease tends to be less severe than pertussis and its duration shorter than that for pertussis. Lymphocytosis is not found in parapertussis, as *B. parapertussis* does not produce PT. Dual infections caused by *B. pertussis* and *B. parapertussis* have been reported (Reizenstein *et al.*, 1996; Linnemann and Perry, 1977; He, Viljanen *et al.*, 1998).

Incidence and Mortality

In the prevaccine era, almost all children contracted pertussis. In the United States, pertussis was the leading cause of death from vaccine preventable diseases among children less than 14 years of age (Long, 1997). Since the 1950s, standard pertussis vaccines prepared from killed whole-cell *B. pertussis* have been available and are used in most European countries and the USA. After widespread use of vaccines in these countries, incidence and mortality have dramatically decreased (Trollfors, 1984; Cherry *et al.*, 1988). At present it is estimated that, annually, there are 40 million cases of pertussis and that pertussis causes more than 300 000 deaths (Anonymous, 1996). Most deaths occur in infants too young to receive any or all doses of pertussis vaccine.

Pertussis in Older Children and Adults

Studies from countries with high vaccination coverage have shown that pertussis has shifted from infants to older age groups, indicating that pertussis immunizations and childhood pertussis do not provide lifelong immunity (Nelson, 1978; Mertsola *et al.*, 1983a; Thomas, 1989; He *et al.*, 1994; Grimpel *et al.*, 1996; Skowronski *et al.*, 2002). In recent prospective studies conducted in France, the UK and the USA with high vaccination coverage, the incidence of pertussis among adults ranged from 330 to 508 per 100 000 person-years (Miller et al., 2000; Strebel et al., 2001; Gilberg et al., 2002). However, the reported incidences of pertussis among adults were only 0–4 over the same period. These figures may reflect poor efficacy of pertussis diagnostics in adults. The pool of undiagnosed pertussis cases in older children and adults provides a reservoir for severe infections in unvaccinated infants (Heininger, Stehr and Cherry, 1992; Wirsing von König et al., 1995; Wang et al., 2002). The role of adults as a source of transmission of pertussis in households is stressed by the findings in 16 families of 23 infants dying of pertussis (Wortis et al., 1996). Thirteen (81%) had been exposed by other family members with cough. Adults accounted for at least 46% of those contacts.

Re-Emergence of Pertussis in Immunized Population

In the past 10 years, resurgence of pertussis has been seen in countries with high coverage of vaccinations with whole-cell vaccines (Bass and Wittler, 1994; De Serres et al., 1995; Andrews, Herceg and Roberts, 1997; de Melker et al., 1997; Skowronski et al., 2002). In 1996, a sudden increase in pertussis incidence was reported from the Netherlands (total number of reported cases: 319 in 1995 vs. 2778 in 1996) (de Melker et al., 1997). With epidemic cycles expected every 3–5 years and a recent outbreak in 1994, this increase was unexpected. The reported number of cases has since remained high. In 2000, 4837 cases were reported from the Netherlands (WHO, 2002). Many explanations may be given for the re-emergence of pertussis in countries of good vaccination coverage: better awareness of pertussis by clinicians, improved surveillance systems, use of molecular diagnostic tools, waning vaccine-induced immunity, quality of vaccines used in some countries and adaptation of B. pertussis to vaccine-induced immunity. As stated above, antigenic divergence has been found between B. pertussis vaccine strains and circulating strains in several countries.

CLINICAL MANIFESTATIONS

The incubation period of pertussis ranges from 6 to 20 days. After incubation, the signs and symptoms of the catarrhal stage merge. A typical clinical course of pertussis has three stages: catarrhal, paroxysmal and convalescent. The catarrhal stage usually lasts for less than 2 weeks. The cough of the catarrhal stage becomes increasingly severe and gradually leads to the paroxysmal stage which lasts from 1 to 4 weeks. During the paroxysms, the patient may lean forward with the eyes bulging and tears streaming. Gradually, the face becomes flushed or even cyanotic, and finally, the typical whoop develops after an inspiratory gasp. However, the whoop is not frequently seen in older children or adults. The convalescent stage begins with a decrease in the severity of cough and frequency of paroxysms. The stage can last for several weeks and even months. In an immunized population, mild and atypical clinical cases often occur. Most complications of pertussis, such as pneumonia, pulmonary sequelae, encephalitis and death, occur during the paroxysmal stage. Pertussis is most severe in young infants.

The characteristics of reported cases of pertussis among infants younger than 12 months in the United States were analysed (Tanaka et al., 2003). Of the 18 500 cases reported in 1990s, 91% had paroxysmal cough, 68% had vomiting, 60% had whoop, 56% had apnea, 22% had pneumonia, 1.9% had seizures and 0.3% had encephalopathy. Of these patients, 67% were hospitalized and 0.5% died. Of the 93 fatal cases, 85% had pulmonary infiltrates on admission, 73% had apnea, 41% had pulmonary hypertension, 19% had seizures and 12% had encephalopathy (Vitek et al., 2003).

LABORATORY DIAGNOSIS

The laboratory diagnosis of pertussis is important, particularly in immunized populations, because the signs and symptoms of pertussis

patients are often mild or atypical. The laboratory methods can be generally divided into two main categories: tests detecting the pathogen and its components including culture, polymerase chain reaction (PCR) and the direct fluorescent antibody (DFA) test, and tests assessing the host response to the pathogen, including bacterial agglutination and enzyme immunoassay (EIA). As all methods have limitations in their sensitivity, specificity or practicality, an optimal diagnosis often requires a combination of methods (Onorato and Wassilak, 1987; Müller, Hoppe and Wirsing von König, 1997).

Detection of the Bacterium and Its Components

Bacterial Culture

Culture has remained the gold standard of diagnosis. The specimens should be taken from the posterior nasopharynx, preferably by intranasal aspiration or by swab. Both calcium alginate and Dacron swabs are recommended. Calcium alginate swab is better for culture than Dacron swab (Hoppe and Weiss, 1987). However, Dacron swab on a plastic shaft is preferred for PCR because both the alginate component and the aluminum shaft of calcium alginate swab are found to inhibit a PCR-based assay (Wadowsky et al., 1994). The specimens should be plated immediately onto selective media. Charcoal agar supplemented with 10% horse blood and 40 mg/l of cephalexin is currently the medium of choice (Hoppe, 1999a). Bordetellae are first identified by colony morphology and gram stain. Bordetella parapertussis grows faster than B. pertussis and is oxidase negative, whereas B. pertussis is oxidase positive. In addition, B. parapertussis shows urease activity and pigment formation on tyrosine agar. The two species are also identified by slide agglutination with antisera.

The prevalence of positive cultures is often highest during the first 2 weeks of illness, whereas cultures are seldom positive if cough has lasted for more than 4 weeks. Usually, the sensitivity of culture is less than 50% (higher in unvaccinated infants), and false-negative results are common. Culture plates should be incubated for at least 7 days before being discarded as negative. Extension of incubation of culture plates to 12 days may improve the recovery of Bordetella species (Katzko, Hofmeister and Church, 1996).

Polymerase Chain Reaction

Because of its merits in specificity, sensitivity and rapidity, numerous versions of PCR have been developed for the diagnosis of pertussis. Several genes of B. pertussis have been used as targets of PCR; the two most common of them are the IS481 and the PT promoter region. PCR has proved superior to culture in the analysis of nasopharyngeal specimens. In some countries, the method has been applied to routine diagnosis of pertussis (He et al., 1996).

Recently, LightCycler real-time PCR has been used for diagnosis of pertussis (Kösters et al., 2002; Sloan et al., 2002). The IS481 has been the target gene of these PCRs. The reactions in the LightCycler system occur in closed glass capillaries. The accumulation of the PCR products is followed online, and the products are subsequently identified using fluorescent hybridization probes or dyes binding to double-stranded DNA (SYBR green I). Product identification can be supported by assessing the specific melting characteristics of the PCR product. Therefore, gel electrophoresis or handling of PCR products is not needed, which reduces the risk of contamination frequently associated with post-PCR amplicon manipulation.

Two studies have shown that IS481-based PCR detects B. holmesii in addition to B. pertussis (Loeffelholz et al., 2000; Reischl et al., 2001). Further sequencing and restriction fragment length polymorphism analysis have shown that IS481 is present in B. holmesii. These results and culture of B. holmesii from nasopharyngeal specimens of patients with cough (Yih et al., 1999) suggest that the specificity and predictive value of IS481-based PCR assays for the diagnosis of pertussis may be compromised. The prevalence of *B. holmesii* infections in humans remains to be determined.

PCR has been applied to the diagnostics of parapertussis, and IS1001 has been used as the target gene. PCR has proved specific and superior to culture in the analysis of nasopharyngeal specimens (van der Zee *et al.*, 1996a; He, Viljanen *et al.*, 1998; Kösters *et al.*, 2002).

Direct Fluorescence Antibody Test

DFA has been used since 1960. This test uses polyclonal or monoclonal fluorescein-labelled antibodies to *B. pertussis* for direct detection of the organism in nasopharyngeal specimens (Hoppe, 1999a; McGowan, 2002). It has the advantage of being rapid and independent of viable organisms. Like culture, DFA is likely to be positive in early disease. Although widely used, this test is compromised by inadequate sensitivity and specificity (Friedman, 1988; Halperin, Bortolussi and Wort, 1989; Ewanowich *et al.*, 1993). False-positive results have occurred because of cross-reactions of antibodies used with the organisms of the nasopharyngeal flora as well as errors of the microscopist (Ewanowich *et al.*, 1993). The DFA test should only be used in combination with culture (Müller, Hoppe and Wirsing von König, 1997).

Detection of Serum Pertussis Antibodies

Various methods have been used to assess serum antibody responses to *B. pertussis*. They include functional antibody assays such as agglutination and toxin neutralization as well as EIAs (Goodman, Wort and Jackson, 1981; Granström *et al.*, 1982; Viljanen *et al.*, 1982; Mertsola *et al.*, 1983b; Meade, Mink and Manclark, 1990; Müller, Hoppe and Wirsing von König, 1997). Agglutination assays do not differentiate between antibody isotypes and have been found less sensitive than EIAs (Mertsola *et al.*, 1983b). Toxin neutralization assays are more laborious and less sensitive than EIAs (Granström *et al.*, 1988). In recent years, EIA has thus become the method of choice in the serologic diagnostics of pertussis (Meade, Mink and Manclark, 1990; Müller, Hoppe and Wirsing von König, 1997).

In EIA, purified antigens of *B. pertussis* such as PT, FHA and Prn are preferable because the assays using them as antigens have better specificity than those using crude bacterial extracts (Mertsola *et al.*, 1990; Isacson *et al.*, 1995; Stehr *et al.*, 1998; De Melker *et al.*, 2000). Demonstration of a rise in the concentration of IgG antibodies to *B. pertussis* in paired sera is usually definite proof of the diagnosis of pertussis. Measurement of IgA antibodies may improve diagnostic efficacy even further (Heininger, 2001). Single-specimen analysis is more practical for routine purposes because paired samples are not easy to obtain (Viljanen *et al.*, 1982; De Melker *et al.*, 2000). For this approach, cut-off values based on the analysis of age-matched controls are needed. EIA has been widely used for the study of pertussis in schoolchildren and adults (Miller *et al.*, 2000; Strebel *et al.*, 2001; Gilberg *et al.*, 2002).

In general, the laboratory confirmation of pertussis is difficult in immunized populations. Combinations of culture, PCR and EIA should be used.

Serotyping

Bordetella pertussis strains produce different combinations of the three fimbrial antigens (1, 2 and 3), which are detected by bacterial agglutination with absorbed antisera or with monoclonal antibodies against these antigens (Mooi *et al.*, 2000). The antigen of serotype 1 is part of the lipooligosaccharide. As it does not change, this antigen is not used for serotyping. Strains of *B. pertussis* with three serotypes (types 1, 2; 1, 2, 3; 1, 3) are circulating in human populations and cause

disease. The use of monoclonal antibodies is currently recommended for serotyping (Mooi *et al.*, 2000).

Genotyping

In the past 10 years, several methods have been developed and applied to the genotyping of *B. pertussis*. They include gene typing by sequencing, multilocus sequence typing, restriction fragment length polymorphism (RFLP) and pulsed-field gel electrophoresis (PFGE).

Gene Typing

PCR-based sequencing of the two virulence genes *Prn* and *PT* (S1 subunit) has been successfully used to detect antigenic shifts in *B. pertussis* populations (Mooi *et al.*, 2000). One of the main advantages of gene typing is that sequencing data can be easily compared between laboratories. However, consensus about the genes and the gene segments to be analysed is required for comparisons (Mooi *et al.*, 2000). An alternative to sequencing, based on LightCycler real-time PCR, has been developed for rapid identification of Prn and PT variants (Mäkinen *et al.*, 2001, 2002). The method is especially suitable for large-scale monitoring of gene variation in *B. pertussis* isolates.

Multilocus Sequencing Typing

Multilocus sequencing typing (MLST) is derived from the principle of multilocus enzyme electrophoresis, but it also takes advantage of the speed and simplicity of automated DNA sequencing (Enright and Spratt, 1999). Generally, 450–500-bp internal fragments of seven housekeeping genes are sequenced from each isolate. DNA of this length can be accurately sequenced on both strands using a single pair of primers. In most bacterial pathogens, MLST can show sufficient variation for the identification of many different alleles within the bacterial population (Enright and Spratt, 1999).

Because *B. pertussis* is a very homogeneous species and very little polymorphism has been found in its housekeeping genes, genes encoding for surface proteins have been analysed for typing purposes. A low level of allelic variation was seen when 15 genes coding for surface proteins were studied (Van Loo *et al.*, 2002). In the Dutch isolates from 1949 to 1999, polymorphism is only seen in genes encoding the S1 and S3 subunits of PT, tracheal colonization factor and Prn. Although *Prn* gene is polymorphic, it has not been used for MLST typing, as variation in it is mainly due to insertion or deletion of repeats, a process that is expected to occur relatively frequently. Thus, only genes coding for the PT subunits and tracheal colonization factor are used to define MLST types. When 196 strains isolated from different countries were analysed, nine MLST types were identified (Van Loo *et al.*, 2002).

Restriction Fragment Length Polymorphism

The genome of *B. pertussis* harbours two IS elements, IS481 and IS1002 (Table 18.3). Only four to seven copies of IS1002 occur in the genome, whereas the number of IS481 copies may be 80–100. Because of the high copy number, IS481-based fingerprinting patterns are too complicated even for computer-assisted analysis (van der Zee *et al.*, 1996b). Therefore, IS1002-based RFLP has been used for studying the population structure of *B. pertussis*. In comparison to PFGE, IS1002 fingerprinting is easy to perform and less laborious. However, because of the low copy number of IS1002 in the genome of *B. pertussis*, the discriminatory power of this method does not match that of PFGE (Mooi *et al.*, 2000).

Pulsed-Field Gel Electrophoresis

To construct a restriction map of *B. pertussis* chromosome, Stibitz and Garletts used several rare-cutting restriction enzymes in PFGE (Stibitz and Garletts, 1992). Two enzymes, XbaI and SpeI, cleaved the *B. pertussis* chromosome into 25 and 16 fragments, respectively, suggesting that these two enzymes could be used in the typing of *B. pertussis* isolates by PFGE. Later, several studies have shown that the PFGE using these two enzymes is a very efficient tool in the typing of *B. pertussis* (Khattak, Matthews and Burnie, 1992; Khattak and Matthews, 1993; Nouvellon *et al.*, 1999; Prevost *et al.*, 1999; Brennan *et al.*, 2000; Bisgard *et al.*, 2001; Weber *et al.*, 2001), *B. parapertussis*, *B. bronchiseptica* and *B. holmesii* (Khattak and Matthews, 1993). PFGE has also shown its power in studies assessing the transmission of pertussis in schools and hospitals (Nouvellon *et al.*, 1999; Brennan *et al.*, 2000). An important advantage of PFGE is good comparability of data from different laboratories.

Antimicrobial Sensitivities

The antimicrobial susceptibility tests of *B. pertussis* and *B. parapertussis* have not been standardized (Hoppe and Haug, 1988). Various methods have been used. Testing by broth dilution methods has usually resulted in higher minimal inhibitory concentrations than that by agar dilution methods. Mueller-Hinton broth supplemented with 5% horse blood has been recommended for testing the activity of erythromycin (Hoppe and Tschirner, 1995). Erythromycin-containing plates should be incubated at room temperature for 48 h for *B. parapertussis* and 72 h for *B. pertussis*. The E-test has been evaluated for susceptibility testing of *B. pertussis* and found to be a good alternative to agar dilution testing (Hoppe, 1996).

Erythromycin has been the drug of choice in the treatment and prophylaxis of pertussis, and, until recently, no routine susceptibility testing of *B. pertussis* has been needed, as all strains have been susceptible to this drug. Erythromycin resistance of *B. pertussis* was first recognized in Arizona in 1994. Since then, three additional resistant isolates have been identified in the USA. The rate of erythromycin resistance was found to be <1% when 1030 *B. pertussis* strains were recently isolated and analysed in the USA (Wilson *et al.*, 2002).

The new macrolides show *in vitro* activity against *B. pertussis* similar to that of erythromycin (Hoppe and Eichhorn, 1989). Several quinolones are very active against *B. pertussis* and *B. parapertussis in vitro* (Hoppe and Simon, 1990). Other agents showing good *in vitro* activity against *B. pertussis* are piperacillin and mezlocillin, ceftazidime, cefotaxime and ceftriaxone (Kerr and Preston, 2001).

THERAPY AND CONTROL

Antimicrobial Therapy

Erythromycin is still the standard recommended treatment for pertussis (American Academy of Pediatrics, 2000). Erythromycin estolate is preferred with a dose of 40 mg/kg/day (maximum 2 g/day). It has been shown that erythromycin therapy resulted in elimination of *B. pertussis* from the nasopharynx in 2–7 days (mean 3.6 days), and it is usually considered that 5 days after initiation of the erythromycin treatment the patient is not anymore contagious (Baraff, Wilkins and Wehrle, 1978). The recommended duration of treatment is 14 days. For vaccinated children, 7 days therapy has shown to be effective (Halperin *et al.*, 1997). New macrolides have good *in vitro* activity against *B. pertussis*. Small studies have showed promising results with clarithromycin (10 mg/kg/day, twice a day for 7 days) or azithromycin (10 mg/kg/day, once a day for 5 days) (Aoyama *et al.*, 1996). Cotrimoxazole is considered as an alternative for those who do not tolerate macrolides, and based on *in vitro* results, quinolones could also be an alternative for adults.

PREVENTION AND CONTROL

Chemoprophylaxis

There are several reports showing wide spread of pertussis in families, schools, nursing homes and hospitals. Prompt use of chemoprophylaxis (macrolides with the same dosage as for treatment) in household contacts is effective. The clinical problem usually is that the spread of the infection has already occurred when the index case is diagnosed. The prophylaxis is especially indicated in families with unvaccinated infants.

Use of macrolides in schools or in hospitals is problematic, and some studies indicate that many adults do not complete their prophylaxis. The general value and the practicability of the wide use of prophylaxis in the community are questionable. It is important to increase awareness of pertussis disease in adults and schoolchildren in the population, to start treatment as early as possible in the catarrhal phase of pertussis and to give prophylaxis to those in high risk if infected (infants, other high-risk patients). During outbreak conditions it is important to stress the importance of proper immunizations in children.

Vaccines

Immunization is considered the most effective method of preventing and controlling pertussis in the population. Whole-cell and acellular pertussis vaccines are currently used worldwide.

Whole-Cell Pertussis Vaccines

Shortly after the isolation of *B. pertussis*, attempts were made to develop a vaccine (Madsen, 1933; Lapin, 1943). Since the 1950s, standard pertussis vaccines containing killed whole *B. pertussis* cells have been available in most developed countries (Cherry *et al.*, 1988). The widespread use of these vaccines has been highly successful in decreasing the incidence and mortality of pertussis in these countries (Cherry *et al.*, 1988). However, concern about a possible relationship between immunization and permanent neurologic complications (Kulenkampff, Schwartzman and Wilson, 1974; Miller *et al.*, 1981) resulted in decreased vaccine uptake in several countries, such as the UK, Japan and Sweden. Large pertussis epidemics then emerged in these countries. Although extensive and well-designed studies failed to confirm any evidence of permanent neurologic complications attributable to the whole-cell vaccine (American Academy of Pediatrics Committee on Infectious Diseases, 1991, 1996), extensive efforts have been made to develop less reactogenic acellular vaccines.

Acellular Pertussis Vaccines

Acellular vaccines are based on purified *B. pertussis* antigens. Vaccines containing only PT and FHA were first developed in Japan and introduced into general use as early as 1981 (Sato, Kimura and Fukumi, 1984; Kimura and Kuno-Sakai, 1990). Although extensive use of such vaccines in Japanese children provided evidence of low reactogenicity and satisfactory efficacy (Mortimer *et al.*, 1990), the data showing the safety and efficacy of these vaccines in infants were considered inadequate (Noble *et al.*, 1987). Furthermore, no consensus prevailed on the optimal antigen(s), and their concentrations to be included in acellular vaccines. As a result, manufacturers developed a variety of candidate preparations, 13 of which were evaluated in large phase 1–2 trials in the early 1990s (Decker *et al.*, 1995; Edwards *et al.*, 1995). Acellular vaccines caused much less frequent and less severe adverse reactions than whole-cell vaccines (Decker *et al.*, 1995). Furthermore, most acellular vaccines induced antibody responses to vaccine antigens that equalled or exceeded those induced by whole-cell vaccines (Edwards *et al.*, 1995). In the last decade, nine controlled efficacy trials of acellular vaccines have been carried out in different countries (Hewlett, 1997; Plotkin and Cadoz, 1997). The vaccines tested were either monocomponent ones containing only PT, bicomponent containing PT and FHA, tricomponent containing PT, FHA and Prn or multicomponent containing PT, FHA, Prn and FIM. The vaccines were produced by different manufacturers, and diphtheria–tetanus whole-cell pertussis vaccines or diphtheria–tetanus vaccines served as controls. In most trials, three primary doses were administered to infants less than 6 months of age. The trials showed that acellular vaccines prevent pertussis and cause less adverse reactions than whole-cell vaccines (Cherry, 1997). Because of their low reactogenecity, acellular pertussis vaccines are suitable for booster immunizations of older children and even adults (Tran Minh *et al.*, 1999).

Response to Immunization and Persistence of Immunity

Active and passive immunizations in mice have shown that antibodies to PT, FHA and Prn of *B. pertussis* can confer various degrees of protection against intracerebral or respiratory challenge by the bacterium (Oda *et al.*, 1984; Sato and Sato, 1984; Robinson *et al.*, 1985; Shahin *et al.*, 1990). High levels of antibodies to these components have been found in the convalencent sera of humans with pertussis (Viljanen *et al.*, 1982; Mertsola *et al.*, 1983b; Robertson *et al.*, 1987; He *et al.*, 1993; Nenning *et al.*, 1996; Long, 1997). In the 1940s, high levels of agglutinating antibodies against *B. pertussis* were found to be associated with clinical protection (Miller *et al.*, 1943; Sako, 1947). However, it is now known that the protective effect seen in those studies was partly due to antibodies against FIM and PRN (Mink *et al.*, 1994). In the vaccine efficacy trials, high levels of antibodies against vaccine antigens were reported. Although high levels of antibodies to Prn, FIM or PT in pre-exposure sera are associated with a lower likelihood of pertussis acquisition in households (Cherry *et al.*, 1998; Storsaeter *et al.*, 1998), no single serologic marker or combination of markers has been found to indicate definite protection against pertussis. This suggests that, in addition to antibodies, other arms of the immune system are involved in protection.

Bordetella pertussis was found to be able to survive in mammalian cells, and a role for cell-mediated immunity (CMI) in controlling *B. pertussis* infection was evident from studies of mice. Therefore, CMI induced after immunization and disease was investigated (Zepp *et al.*, 1996; Ausiello *et al.*, 1997; He, Tran Minh *et al.*, 1998; Ryan *et al.*, 1998; Tran Minh *et al.*, 1999). In general, the data obtained so far suggest that immunity conferred by whole-cell vaccines, such as natural immunity, is mediated by T helper 1 cells, whereas both Th1 and Th2 cytokine profiles have been seen in individuals immunized with acellular vaccines.

It is known that neither immunization nor natural infection provides long-term immunity. The protection provided by whole-cell vaccines is short-lived with an estimated duration of 6–12 years. However, the data of protection by acellular vaccines are limited. Preliminary studies show that the efficacy of multicomponent and some monocomponent or bicomponent acellular vaccines may be sustained for several years (Storsaeter and Olin, 1992; Gustafsson *et al.*, 1996; Taranger *et al.*, 1997; Salmaso *et al.*, 1998). The persistence of immunity depends on the vaccine used, the schedule followed, the number of doses used and the epidemiologic background of the study populations.

Immunization Schedules

Pertussis vaccines and immunization programs vary from country to country, as shown in Table 18.4. In most European countries, the recommendation is three diphtheria-tetanus-pertussis (whole-cell or acellular) doses before the age of 7 months and a booster dose at 11–18 months of age. In the United States, the recommendation is

Table 18.4 Current pertussis immunization schedules in European countries[a]

Country	Vaccine used[b]	Months
Austria	DTaP	2, 3, 4, 13
Belgium	DTaP	2, 3, 4, 15
Denmark	DTaP	3, 5, 12
Finland	DTwP	3, 4, 5, 18–24
France	DTaP or DTwP	2, 3, 4, 16
Germany	DTaP	2, 3, 4, 11
Greece	DTwP or DTaP	2, 4, 6, 18–24
Iceland	DTaP	3, 5, 12 or 2, 4, 6
Italy	DTaP	2, 4, 10
Luxembourg	DTaP	2, 3, 4, 11
The Netherlands	DTwP	2, 3, 4, 11
Norway	DTaP	3, 5, 11
Portugal	DTwP	2, 4, 6, 15
Spain	DTwP or DTaP	2, 4, 6, 18–24
Sweden	DTaP	3, 5, 12
Switzerland	DTaP	2, 4, 6, 15
UK	DTwP	2, 3, 4

[a] Adapted from Schmitt *et al.* (2003).
[b] DTaP or DTwP is used in combination with other vaccines.

DTaP for both primary series at 2, 4 and 6 months of age and booster doses at ages 15–18 months and 4–6 years. In Scandinavia, several countries have recently adopted a schedule of 3, 5 and 12 months.

The increased incidence of pertussis in schoolchildren, adolescents and adults stresses the importance of booster immunizations in these populations. In Finland, a second booster dose given at 4–6 years of age has been recently introduced into immunization programs. In France and Germany, a second booster dose given at 11 years of age has already been introduced.

In conclusion, although acellular pertussis vaccines have proved highly effective against pertussis disease, the following important issues remain to be dealt with: 1) the optimal antigens and their quantities in an acellular vaccine – as antigenic divergence with respect to Prn and PT has been found between *B. pertussis* vaccine strains and circulating strains, it remains to be shown whether acellular vaccines provide equal protection against *B. pertussis* representing vaccine and nonvaccine type strains; 2) the rates of waning of pertussis immunity after immunizations with different vaccines; 3) the optimal strategy of boostering (universal – targeted) for the prevention of pertussis in adolescents and adults; 4) the mechanisms of the increase of local reaction often seen after booster immunizations; 5) a better understanding of antigens in terms of interference with reactogenicity and immunogenicity when vaccines are used in combination; 6) provision of effective vaccines for the developing countries at reasonable price.

REFERENCES

Abramson, T., Kedem, H. and Relman, D.A. (2001) Proinflammatory and proapoptotic activities associated with *Bordetella pertussis* filamentous hemagglutinin. *Infection and Immunity*, **69**, 2650–2658.

Ad Hoc Group for the Study of Pertussis Vaccines. (1988) Placebo-controlled trial of two acellular pertussis vaccines in Sweden. *Lancet*, **i**, 955–960.

Amano, K., Fukushi, K. and Watanabe, M. (1990) Biochemical and immunological comparison of lipopolysaccharides from *Bordetella* species. *Journal of General Microbiology*, **136**, 481–487.

American Academy of Pediatrics Committee on Infectious Diseases. (1991) The relationship between pertussis vaccine and brain damage: reassessment. *Pediatrics*, **88**, 397–400.

American Academy of Pediatrics Committee on Infectious Diseases. (1996) The relationship between pertussis vaccine and central nervous system sequelae: continuing assessment. *Pediatrics*, **97**, 279–281.

American Academy of Pediatrics. (2000) Pertussis, in *Red Book: Report on the Committee on Infectious Diseases*, 25th edn (ed. L.K. Pickering), American Academy of Pediatrics, Elk Grove Village, IL, pp. 435–448.

Andrews, R., Herceg, A. and Roberts, C. (1997) Pertussis notifications in Australia, 1991–1997. *Communicable Diseases Intelligence*, **21**, 145–148.

Anonymous. (1996) Pertussis vaccines. *The International Vaccine Institute Newsletter*, **2**, 3–8.

Aoyama, T., Sunakawa, K., Iwata, S. *et al.* (1996) Efficacy of short-term treatment of pertussis with clarithromycin and azithromycin. *The Journal of Pediatrics*, **129**, 761–764.

Arciniega, J.L., Hewlett, E.L., Johnson, F.D. *et al.* (1991) Human serologic response to envelope-associated proteins and adenylate cyclase toxin of *Bordetella pertussis*. *The Journal of Infectious Diseases*, **163**, 135–142.

Arico, B. and Rappuoli, R. (1987) *Bordetella parapertussis* and *Bordetella bronchiseptica* contain transcriptionally silent pertussis toxin genes. *Journal of Bacteriology*, **169**, 2847–2853.

Arico, B., Gross, R., Smida, J. and Rappuoli, R. (1987) Evolutionary relationships in the genus *Bordetella*. *Molecular Microbiology*, **1**, 301–308.

Ashworth, L.A., Robinson, A., Funnell, S. *et al.* (1988) Agglutinogens and fimbriae of *Bordetella pertussis*. *The Tokai Journal of Experimental and Clinical Medicine*, **13**, 203–210.

Ausiello, C.M., Fedele, G., Urbani, F. *et al.* (2002) Native and genetically inactivated pertussis toxins induce human dendritic cell maturation and synergize with lipopolysaccharide in promoting T helper type 1 responses. *The Journal of Infectious Diseases*, **186**, 351–360.

Ausiello, C.M., Urbani, F., la Sala, A. *et al.* (1997) Vaccine- and antigen-dependent type 1 and type 2 cytokine induction after primary vaccination of infants with whole-cell or acellular pertussis vaccines. *Infection and Immunity*, **65**, 2168–2174.

Baraff, L.J., Wilkins, J. and Wehrle, P.F. (1978) The role of antibiotics, immunizations, and adenoviruses in pertussis. *Pediatrics*, **61**, 224–230.

Barnes, M.G. and Weiss, A.A. (2003) Activation of the complement cascade by *Bordetella pertussis*. *FEMS Microbiology Letters*, **220**, 271–275.

Bass, J.W. and Wittler, R.R. (1994) Return of epidemic pertussis in the United States. *Pediatric Infectious Disease Journal*, **13**, 343–345.

Belcher, C.E., Drenkow, J., Kehoe, B. *et al.* (2000) The transcriptional responses of respiratory epithelial cells to *Bordetella pertussis* reveal host defensive and pathogen counter-defensive strategies. *Proceedings of the National Academy of Sciences of the United States of America*, **97**, 13847–13852.

Bisgard, K.M., Christie, C.D., Reising, S.F. *et al.* (2001) Molecular epidemiology of *Bordetella pertussis* by pulsed-field gel electrophoresis profile: Cincinnati, 1989–1996. *The Journal of Infectious Diseases*, **183**, 1360–1367.

Bordet, J. and Gengou, O. (1906) Le microbe de la coqueluche. *Annals de l'Institut Pasteur, Paris*, **20**, 731–741.

Boursaux-Eude, C. and Guiso, N. (2000) Polymorphism of repeated regions of pertactin in *Bordetella pertussis*, *Bordetella parapertussis*, and *Bordetella bronchiseptica*. *Infection and Immunity*, **68**, 4815–4817.

Boursaux-Eude, C., Thiberge, S., Carletti, G. *et al.* (1999) Intranasal murine model of *Bordetella pertussis* infections: II. Sequence variation and protection induced by a tricomponent acellular vaccine. *Vaccine*, **17**, 2651–2660.

Bradford, W.L. and Slavin, B. (1937) An organism resembling *Hemophilus pertussis*. *American Journal of Public Health*, **27**, 1277–1282.

Brennan, M., Strebel, P., George, H. *et al.* (2000) Evidence for transmission of pertussis in schools, Massachusetts, 1996: epidemiologic data supported by pulsed-field gel electrophoresis studies. *The Journal of Infectious Diseases*, **181**, 210–215.

Brennan, M.J. and Shahin, R.D. (1996) Pertussis antigens that abrogate bacterial adherence and elicit immunity. *American Journal of Respiratory and Critical Care Medicine*, **154**, S145–S149.

Brennan, M.J., Li, Z.M., Cowell, J.L. *et al.* (1988) Identification of a, 69-kilodalton nonfimbrial protein as an agglutinogen of *Bordetella pertussis*. *Infection and Immunity*, **56**, 3189–3195.

Cassiday, P., Sanden, G., Heuvelman, K. *et al.* (2000) Polymorphism in *Bordetella pertussis* pertactin and pertussis toxin virulence factors in the United States, 1935–1999. *The Journal of Infectious Diseases*, **182**, 1402–1408.

Charles, I.G., Li, J.L., Roberts, M. *et al.* (1991) Identification and characterization of a protective immunodominant B cell epitopes of pertactin (P.69) from *Bordetella pertussis*. *European Journal of Immunology*, **21**, 1147–1153.

Cherry, J.D. (1984) The epidemiology of pertussis and pertussis immunization in the United Kingdom and the United States: a comparative study. *Current Problems in Pediatrics*, **14**, 1–78.

Cherry, J.D. (1997) Comparative efficacy of acellular pertussis vaccines: an analysis of recent trials. *Pediatric Infectious Disease Journal*, **16**, S90–S96.

Cherry, J.D., Brunell, P.A., Golden, G.S. *et al.* (1988) Report of the task force on pertussis and pertussis immunization-1988. *Pediatrics*, **81** (Suppl.), 939–984.

Cherry, J.D., Gornbein, J., Heininger, U. *et al.* (1998) A search for serologic correlates of immunity to *Bordetella pertussis* cough illness. *Vaccine*, **16**, 1901–1906.

Cotter, P.A. and Miller, J.F. (1994) BvgAS-mediated signal transduction: analysis of phase-locked regulatory mutants of *Bordetella bronchiseptica* in a rabbit model. *Infection and Immunity*, **62**, 3381–3390.

Cotter, P.A. and Miller, J.F. (1997) A mutation in the *Bordetella bronchiseptica bvgS* gene results in reduced virulence and increased resistance to starvation, and identifies a new class of *bvg*-regulated antigens. *Molecular Microbiology*, **24**, 671–685.

de Melker, H.E., Conyn-van Spaendonck, M.A., Rumke, H.C. *et al.* (1997) Pertussis in the Netherlands: an outbreak despite high levels of immunization with whole cell vaccine. *Emerging Infectious Diseases*, **3**, 175–178.

De Melker, H.E., Versteegh, F.G., Conyn-Van Spaendonck, M.A. *et al.* (2000) Specificity and sensitivity of high levels of immunoglobulin G antibodies against pertussis toxin in a single serum sample for diagnosis of infection with *Bordetella pertussis*. *Journal of Clinical Microbiology*, **38**, 800–806.

De Serres, G., Boulianne, N., Douville-Fradet, M. *et al.* (1995) Pertussis in Quebec. Ongoing epidemic since the late 1980s. *Canada Communicable Disease Report*, **21**, 45–48.

Decker, M.D., Edwards, K.M., Steinhoff, M.C. *et al.* (1995) Comparison of 13 acellular pertussis vaccines: adverse reactions. *Pediatrics*, **96**, 557–566.

Delisse-Gathoye, A., Locht, C., Jacob, F. *et al.* (1990) Cloning, partial sequence, expression, and antigenic analysis of the filamentous hemagglutinin gene of *Bordetella pertussis*. *Infection and Immunity*, **58**, 2895–2905.

Domenighini, M., Relman, D., Capiau, C. *et al.* (1990) Genetic characterization of *Bordetella pertussis* filamentous haemagglutinin: a protein processed from an unusually large precursor. *Molecular Microbiology*, **4**, 787–800.

Edwards, K.M., Meade, B.D., Decker, M.D. *et al.* (1995) Comparison of 13 acellular pertussis vaccines: overview and serologic response. *Pediatrics*, **96**, 548–557.

Eldering, G. and Kendrick, P. (1937) A group of cultures resembling both *Bacillus pertussis* and *Bacillus bronchisepticus* but identical with neither [Abstract]. *Journal of Bacteriology*, **33**, 71.

Eldering, G. and Kendrick, P. (1938) *Bacillus para-pertussis*: a species resembling both *Bacillus pertussis* and *Bacillus bronchisepticus* but identical with neither. *Journal of Bacteriology*, **35**, 561–572.

Emsley, P., Charles, I.G., Fairweather, N.F. *et al.* (1996) Structure of *Bordetella pertussis* virulence factors P.69 pertactin. *Nature*, **381**, 90–92.

Enright, M.C. and Spratt, B.G. (1999) Multilocus sequence typing. *Trends in Microbiology*, **7**, 482–487.

Ewanowich, C.A., Chui, L.W., Paranchych, M.G. *et al.* (1993) Major outbreak of pertussis in northern Alberta, Canada: analysis of discrepant direct fluorescent-antibody and culture results by using polymerase chain reaction methodology. *Journal of Clinical Microbiology*, **31**, 1715–1725.

Fernandez, R.C. and Weiss, A.A. (1994) Cloning and sequencing of a *Bordetella pertussis* serum resistance locus. *Infection and Immunity*, **62**, 4727–4738.

Finn, T.M. and Stevens, L.A. (1995) Tracheal colonization factor: a *Bordetella pertussis* secreted virulence determinant. *Molecular Microbiology*, **16**, 625–634.

Friedman, R.L. (1988) Pertussis: the disease and new diagnostic methods. *Clinical Microbiology Reviews*, **1**, 365–376.

Fry, N.K., Neal, S., Harrison, T.G. *et al.* (2001) Genotypic variation in the *Bordetella pertussis* virulence factors pertactin and pertussis toxin in historical and recent clinical isolates in the United Kingdom. *Infection and Immunity*, **69**, 5520–5528.

Gerlach, G., von Wintzingerode, F., Middendorf, B. *et al.* (2001) Evolutionary trends in the genus *Bordetella*. *Microbes and Infection*, **3**, 61–72.

Geuijen, C.A.W., Willems, R.J.L., Bongaerts, M. *et al.* (1997) Role of the *Bordetella pertussis* minor fimbrial subunit FimD in colonization of the mouse respiratory tract. *Infection and Immunity*, **65**, 4222–4228.

Gilberg, S., Njamkepo, E., Parent du Chatelet, I. *et al.* (2002) Evidence of *Bordetella pertussis* infection in adults presenting with persistent cough in a French area with very high whole-cell vaccine coverage. *The Journal of Infectious Diseases*, **186**, 415–418.

Goodman, Y.E., Wort, A.J. and Jackson, F.L. (1981) Enzyme-linked immunosorbent assay for detection of pertussis immunoglobulin A in nasopharyngeal secretions as an indicator of recent infection. *Journal of Clinical Microbiology*, **13**, 286–292.

Gordon, J.E. and Hood, R.I. (1951) Whooping cough and its epidemiological anomalies. *The American Journal of the Medical Sciences*, **222**, 333–361.

Granström, G., Wretlind, B., Salenstedt, C.R. *et al.* (1988) Evaluation of serologic assays for diagnosis of whooping cough. *Journal of Clinical Microbiology*, **26**, 1818–1823.

Granström, M., Granström, G., Lindfors, A. *et al.* (1982) Serologic diagnosis of whooping cough by an enzyme-linked immunosorbent assay using fimbrial hemagglutinin as antigen. *The Journal of Infectious Diseases*, **146**, 741–745.

Greco, D., Salmaso, S., Mastrantonio, P. *et al.* (1996) A controlled trial of two acellular vaccines and one whole-cell vaccine against pertussis. *The New England Journal of Medicine*, **334**, 341–348.

Greenlee, D.V., Andreasen, T.J. and Storm, D.R. (1982) Calcium-independent sitmulation of *Bordetella pertussis* adenylate cyclase toxin by calmodulin. *Biochemistry*, **21**, 2759–2764.

Grimpel, E., Begue, P., Anjak, I. *et al.* (1996) Long-term human serum antibody responses after immunization with whole-cell pertussis vaccine in France. *Clinical and Diagnostic Laboratory Immunology*, **3**, 93–97.

Guiso, N., Szatanik, M. and Rocancourt, M. (1991) Protective activity of *Bordetella* adenylate cyclase-hemolysin against bacterial colonization. *Microbial Pathogenesis*, **11**, 423–431.

Gustafsson, L., Hallander, H.O., Olin, P. *et al.* (1996) A controlled trial of a two-component acellular, a five-component acellular, and a whole-cell pertussis vaccine. *The New England Journal of Medicine*, **334**, 349–355.

Halperin, S.A., Bortolussi, R. and Wort, A.J. (1989) Evaluation of culture, immunofluorescence and serology for the diagnosis of pertussis. *Journal of Clinical Microbiology*, **27**, 752–757.

Halperin, S.A., Bortolussi, R., Langley, J.M. *et al.* (1997) Seven days erythromycin estolate is as effective as fourteen days for the treatment of *Bordetella pertussis* infections. *Pediatrics*, **100**, 65–71.

Hazenbos, W.L., van den Berg, B.M., Geuijen, C.W. *et al.* (1995) Binding of FimD on *Bordetella pertussis* to very late antigen-5 on monocytes activates complement receptor type 3 via protein tyrosine kinases. *Journal of Immunology*, **155**, 3972–3978.

He, Q., Mäkinen, J., Berbers, G. *et al.* (2003) *Bordetella pertussis* protein pertactin induces type-specific antibodies: one possible explanation for the emergence of antigenic variants? *The Journal of Infectious Diseases*, **187**, 1200–1205.

He, Q., Mertsola, J., Himanen, J.P. *et al.* (1993) Evaluation of pooled and individual components of *Bordetella pertussis* as antigens in an enzyme immunoassay for diagnosis of pertussis. *European Journal of Clinical Microbiology & Infectious Diseases*, **12**, 690–695.

He, Q., Schmidt-Schläpfer, G., Just, M. *et al.* (1996) Impact of polymerase chain reaction on clinical pertussis research: Finnish and Swiss experiences. *The Journal of Infectious Diseases*, **174**, 1288–1295.

He, Q., Tran Minh, N.N., Edelman, K. *et al.* (1998) Cytokine mRNA expression and proliferative responses induced by pertussis toxin, filamentous hemagglutinin, and pertactin of *Bordetella pertussis* in the peripheral blood mononuclear cells of infected and immunized schoolchildren and adults. *Infection and Immunity*, **66**, 3796–3801.

He, Q., Viljanen, M.K., Arvilommi, H. *et al.* (1998) Whooping cough caused by *Bordetella pertussis* and *Bordetella parapertussis* in an immunized population. *JAMA*, **280**, 635–637.

He, Q., Viljanen, M.K., Nikkari, S. *et al.* (1994) Outcomes of *Bordetella pertussis* infection in different age groups of an immunized population. *The Journal of Infectious Diseases*, **170**, 873–877.

Heininger, U. (2001) Recent progress in clinical and basic pertussis research. *European Journal of Pediatrics*, **160**, 203–213.

Heininger, U., Cotter, P.A., Fescemyer, H.W. *et al.* (2002) Comparative phenotypic analysis of the *Bordetella parapertussis* isolate chosen for genomic sequencing. *Infection and Immunity*, **70**, 3777–3784.

Heininger, U., Stehr, K. and Cherry, J.D. (1992) Serious pertussis overlooked in infants. *European Journal of Pediatrics*, **151**, 342–343.

Heininger, U., Stehr, K., Schmitt-Grohe, S. *et al.* (1994) Clinical characteristics of illness caused by *Bordetella parapertussis* compared with illness caused by *Bordetella pertussis*. *Pediatric Infectious Disease Journal*, **13**, 306–309.

Heiss, L.N., Lancaster, J.R., Corbett, J.A. *et al.* (1994) Epithelial autotoxity of nitric oxide: role in the respiratory cytopathology of pertussis. *Proceedings of the National Academy of Sciences of the United States of America*, **91**, 267–270.

Hewlett, E.L. (1997) Pertussis: current concepts of pathogenesis and prevention. *Pediatric Infectious Disease Journal*, **16**, S78–S84.

Hoppe, J.E. (1996) Update on epidemiology, diagnosis, and treatment of pertussis. *European Journal of Clinical Microbiology & Infectious Diseases*, **15**, 189–193.

Hoppe, J.E. (1999a) Bordetella, in *Manual of Clinical Microbiology*, 7th edn (eds P.R. Murray, E.J. Baron, M.A. Pfaller, F.C. Tenover and R.H. Yolken), American Society for Microbiology, Washington, USA, pp. 614–624.

Hoppe, J.E. (1999b) Update on respiratory infection caused by *Bordetella parapertussis*. *Pediatric Infectious Disease Journal*, **18**, 375–381.

Hoppe, J.E. and Eichhorn, A. (1989) Activity of new macrolides against *Bordetella pertussis* and *Bordetella parapertussis*. *European Journal of Clinical Microbiology*, **8**, 653–654.

Hoppe, J.E. and Haug, A. (1988) Antimicrobial susceptibility of *Bordetella pertussis* (Part I). *Infection*, **16**, 126–130.

Hoppe, J.E. and Simon, C.G. (1990) *In vitro* susceptibilities of *Bordetella pertussis* and *Bordetella parapertussis* to seven fluoroquinolones. *Antimicrobial Agents and Chemotherapy*, **34**, 2287–2288.

Hoppe, J.E. and Tschirner, T. (1995) Comparison of media for agar dilution susceptibility testing of *Bordetella pertussis* and *Bordetella parapertussis*. *European Journal of Clinical Microbiology & Infectious Diseases*, **14**, 775–779.

Hoppe, J.E. and Weiss, A. (1987) Recovery of *Bordetella pertussis* from four kinds of swabs. *European Journal of Clinical Microbiology*, **6**, 203–205.

Isacson, J., Trollfors, B., Hedvall, G. *et al.* (1995) Response and decline of serum IgG antibodies to pertussis toxin, filamentous hemagglutinin and pertactin in children with pertussis. *Scandinavian Journal of Infectious Diseases*, **27**, 273–277.

Kasuga, T., Nakase, Y., Ukishima, K. *et al.* (1954) Studies on *Haemophilus pertussis*. Part III. Some properties of each phase of *H. pertussis*. *Kitasato Archives of Experimental Medicine*, **27**, 37–48.

Katzko, G., Hofmeister, M. and Church, D. (1996) Extended incubation of culture plates improves recovery of *Bordetella* spp. *Journal of Clinical Microbiology*, **34**, 1563–1564.

Kerr, J.R. and Preston, M.W. (2001) Current pharmacotherapy of pertussis. *Expert Opinion on Pharmacotherapy*, **2**, 1275–1282.

Kersters, K., Hinz, K.H., Hertle, A. *et al.* (1984) *Bordetella avium* sp. Nov, isolated from the respiratory tracts of turkeys and other birds. *International Journal of Systematic Bacteriology*, **34**, 56–70.

Khattak, M.N. and Matthews, R.C. (1993) Genetic relatedness of *Bordetella* species as determined by macrorestriction digests resolved by pulsed-field gel electrophoresis. *International Journal of Systematic Bacteriology*, **43**, 659–664.

Khattak, M.N., Matthews, R.C. and Burnie, J.P. (1992) Is *Bordetella pertussis* clonal? *BMJ*, **304**, 813–815.

Kimura, M. and Kuno-Sakai, H. (1990) Developments in pertussis immunisation in Japan. *Lancet*, **336**, 30–32.

Kösters, K., Reischl, U., Schmetz, J. *et al.* (2002) Real-time LightCycler PCR for detection and discrimination of *Bordetella pertussis* and *Bordetella parapertussis*. *Journal of Clinical Microbiology*, **40**, 1719–1722.

Kulenkampff, M., Schwartzman, J.S. and Wilson, J. (1974) Neurological complications of pertussis inoculation. *Archives of Disease in Childhood*, **49**, 46–49.

Ladant, D., Michelson, S., Sarfati, R. *et al.* (1989) Charaterization of the calmodulin-binding and of the catalytic domains of *Bordetella pertussis* adenylate cyclase. *The Journal of Biological Chemistry*, **264**, 4015–4020.

Lambert, H.J. (1965) Epidemiology of a small pertussis outbreak in Kent County, Michigan. *Public Health Reports*, **80**, 365–369.

Lapin, J.H. (1943) Whooping cough, in (ed. C.C.Thomas), Springfield III.

Lautrop, H. (1960) Laboratory diagnosis of whooping-cough or *Bordetella* infections. *Bulletin of the World Health Organization*, **23**, 15–35.

Leininger, E., Roberts, M., Kenimer, J.G. *et al.* (1991) Pertactin, an Arg-Gly-Asp-containing *Bordetella pertussis* surface protein that promotes adherence of mammalian cells. *Proceedings of the National Academy of Sciences of the United States of America*, **88**, 345–349.

Li, L.J., Dougan, G., Novotny, P. *et al.* (1991) P.70 pertactin, an outer-membrane protein from *Bordetella parapertussis*: cloning, nucleotide sequence and surface expression in *Escherichia coli*. *Molecular Microbiology*, **5**, 409–417.

Linnemann, C.C. and Perry, E.B. (1977) *Bordetella parapertussis*: recent experience and a review of the literature. *American Journal of Diseases of Child*, **131**, 560–563.

Locht, C. and Keith, J.M. (1986) Pertussis toxin gene: nucleotide sequence and genetic organisation. *Science*, **232**, 1258–1263.

Locht, C., Bertin, P., Menozzi, F.D. *et al.* (1993) The filamentous haemagglutinin, a multifaceted adhesion produced by virulent *Bordetella* spp. *Molecular Microbiology*, **9**, 653–660.

Loeffelholz, M.J., Thompson, C.J., Long, K.S. *et al.* (2000) Detection of *Bordetella holmesii* using *Bordetella pertussis* IS481 PCR assay. *Journal of Clinical Microbiology*, **38**, 467.

Long, S.S. (1997) *Bordetella pertussis* (pertussis) and other species, in *Principles and Practice of Pediatric Infectious Diseases* (eds S.S. Long, L.K. Pickering and C.G. Prober), Churchill Livingstone, New York, 976–986.

Long, S.S., Welkon, C.J. and Clark, J.L. (1990) Widespread silent transmission of pertussis in families: antibody correlates of infection and symptomatology. *The Journal of Infectious Diseases*, **161**, 480–486.

Madsen, T. (1933) Vaccination against whooping cough. *JAMA*, **101**, 187–188.

Mäkinen, J., Mertsola, J., Viljanen, M.K. *et al.* (2002) Rapid typing of *Bordetella pertussis* pertussis toxin gene variants by LightCycler real-time PCR and fluorescence resonance energy transfer hybridization probe melting curve analysis. *Journal of Clinical Microbiology*, **40**, 2213–2216.

Mäkinen, J., Viljanen, M.K., Mertsola, J. *et al.* (2001) Rapid identification of *Bordetella pertussis* pertactin gen variants using LightCycler real-time polymerase chain reaction combined with melting curve analysis and gel electrophoresis. *Emerging Infectious Diseases*, **7**, 952–958.

Mastrantonio, P., Spigaglia, P., van Oirschot, H. *et al.* (1999) Antigenic variants in *Bordetella pertussis* strains isolated from vaccinated and unvaccinated children. *Microbiology*, **145**, 2069–2075.

Matto, S., Foreman-Wykert, A.K., Cotter, P.A. *et al.* (2001) Mechanisms of *Bordetella* pathogenesis. *Frontiers in Bioscience*, **6**, e168–e186.

McGowan, K.L. (2002) Diagnostic tests for pertussis: culture vs. DFA vs. PCR. *Clinical Microbiology Newsletter*, **24**, 143–149.

Meade, B.D., Mink, C.M. and Manclark, C.R. (1990) Serodiagnosis of pertussis, in *Proceedings of the Six International Symposium on Pertussis* (ed. C.R. Manclark), DHHS publication no. (FDA) 90–1164, FDA, Bethesda, 322–329.

Menozzi, F.D., Mutombo, R., Renauld, G. *et al.* (1994) Heparin-inhibitable lectin activity of the filamentous hemagglutinin adhesin of *Bordetella pertussis*. *Infection and Immunity*, **62**, 769–778.

Mertsola, J., Ruuskanen, O., Eerola, E. *et al.* (1983a) Intrafamilial spread of pertussis. *The Journal of Pediatrics*, **103**, 359–363.

Mertsola, J., Ruuskanen, O., Kuronen, T. *et al.* (1983b) Serologic diagnosis of pertussis: comparison of enzyme-linked immunosorbent assay and bacterial agglutination. *The Journal of Infectious Diseases*, **147**, 252–257.

Mertsola, J., Ruuskanen, O., Kuronen, T. *et al.* (1990) Serologic diagnosis of pertussis: evaluation of pertussis toxin and other antigens in enzyme-linked immunosorbent assay. *The Journal of Infectious Diseases*, **161**, 996–971.

Miller, D.L., Ross, E.M., Alderslade, R. *et al.* (1981) Pertussis immunisation and serious acute neurological illness in children. *BMJ*, **282**, 1595–1599.

Miller, E., Fleming, D.M., Ashworth, L.A. *et al.* (2000) Serological evidence of pertussis in patients presenting with cough in general practice in Birmingham. *Communicable Disease and Public Health*, **3**, 132–134.

Miller, J.J., Silverberg, R.J., Saito, T.M. *et al.* (1943) An agglutinative reaction for *Hemophilus pertussis*, II: its relation to clinical immunity. *The Journal of Pediatrics*, **22**, 644–651.

Mink, C.M., O'Brien, C.H., Wassilak, S. *et al.* (1994) Isotype and antigen specificity of pertussis agglutinins following whole-cell pertussis vaccination and infection with *Bordetella pertussis*. *Infection and Immunity*, **62**, 1118–1120.

Montaraz, J.A., Novotny, P. and Ivanyi, J. (1985) Identification of a 68-kilodalton protective protein antigen from *Bordetella bronchiseptica*. *Infection and Immunity*, **47**, 744–751.

Mooi, F.R., Hallander, H., Wirsing von König, C.H. *et al.* (2000) Epidemiological typing of *Bordetella pertussis* isolates: recommendations for a standard methodology. *European Journal of Clinical Microbiology & Infectious Diseases*, **19**, 174–181.

Mooi, F.R., He, Q., van Oirschot, H. *et al.* (1999) Variation in the *Bordetella pertussis* virulence factors pertussis toxin and pertactin in vaccine strains and clinical isolates in Finland. *Infection and Immunity*, **67**, 3133–3134.

Mooi, F.R., van Oirschot, H., Heuvelman, K. *et al.* (1998) Polymorphism in the *Bordetella pertussis* virulence factors P.69/pertactin and pertussis toxin in the Netherlands: temporal trends and evidence for vaccine-driven evolution. *Infection and Immunity*, **66**, 670–675.

Mortimer, E.A. Jr., Kimura, M., Cherry, J.D. *et al.* (1990) Protective efficacy of the Takeda acellular pertussis vaccine combined with diphtheria and tetanus toxoids following household exposure of Japanese children. *American Journal of Diseases of Child*, **144**, 899–904.

Müller, F.M., Hoppe, J.E. and Wirsing von König, C.H. (1997) Laboratory diagnosis of pertussis: state of the art in 1997. *Journal of Clinical Microbiology*, **35**, 2435–2443.

Musser, J.M., Hewlett, E.L., Peppler, M.S. *et al.* (1986) Genetic diversity and relationship in populations of *Bordetella* spp. *Journal of Bacteriology*, **166**, 230–237.

Nelson, K.E. (1978) The changing epidemiology of pertussis in young infants. The role of adults as reservoirs of infection. *American Journal of Diseases of Child*, **132**, 371–373.

Nenning, M.E., Shinefield, H.R., Edwards, K.M. *et al.* (1996) Prevalence and incidence of adult pertussis in an urban population. *JAMA*, **275**, 1672–1674.

Nicosia, A., Perugini, M., Franzini, C. *et al.* (1986) Cloning and sequencing of the pertussis toxin genes: operon structure and gene duplication. *Proceedings of the National Academy of Sciences of the United States of America*, **83**, 4631–4635.

Noble, G.R., Bernier, R.H., Esber, E.C. *et al.* (1987) Acellular and whole-cell pertussis vaccines in Japan. Report of a visit by US scientists. *JAMA*, **257**, 1351–1356.

Nouvellon, M., Gehanno, J.F., Pestel-Caron, M. *et al.* (1999) Usefulness of pulsed-field gel electrophoresis in assessing nosocomial transmission of pertussis. *Infection Control and Hospital Epidemiology*, **20**, 758–760.

Oda, M., Cowell, J.L., Burstyn, D.G. *et al.* (1984) Protective activities of the filamentous hemagglutinin and the lymphocytosis-promoting factor of *Bordetella pertussis* in mice. *The Journal of Infectious Diseases*, **150**, 823–833.

Oda, M., Cowell, J.L., Burstyn, D.G. *et al.* (1985) Antibodies to *Bordetella pertussis* in human colostrum and their protective activity against aerosol infection of mice. *Infection and Immunity*, **47**, 441–445.

Olin, P., Rasmussen, F., Gustafsson, L. *et al.* (1997) Randomised controlled trial of two-component, three-component, and five-component acellular pertussis vaccines compared with whole-cell pertussis vaccine. *Lancet*, **350**, 1569–1577.

Oliver, D.C. and Fernandez, R.C. (2001) Antibodies to BrkA augment killing of *Bordetella pertussis*. *Vaccine*, **20**, 235–241.

Onorato, I.M. and Wassilak, S.G. (1987) Laboratory diagnosis of pertussis: the state of the art. *Pediatric Infectious Disease Journal*, **6**, 145–151.

Plotkin, S.A. and Cadoz, M. (1997) The acellular pertussis vaccine trials: an interpretation. *Pediatric Infectious Disease Journal*, **16**, 508–517.

Preston, A., Allen, A.G., Cadisch, J. *et al.* (1999) Genetic basis for lipopolysaccharide O-antigen biosynthesis in *Bordetellae*. *Infection and Immunity*, **67**, 3763–3767.

Preston, N.W. and Carter, E.J. (1992) Serotype specificity of vaccine-induced immunity to pertussis. *Communicable Disease Report. CDR Review*, **2**, R155–R156.

Prevost, G., Freitas, F.L., Stoessel, P. *et al.* (1999) Analysis with a combination of macrorestriction endonucleases reveals a high degree of polymorphism among *Bordetella pertussis* isolates in eastern France. *Journal of Clinical Microbiology*, **37**, 1062–1068.

Register, K.B. (2001) Novel genetic and phenotypic heterogeneity in *Bordetella bronchiseptica* pertactin. *Infection and Immunity*, **69**, 1917–1921.

Reischl, U., Lehn, N., Sanden, G.N. *et al.* (2001) Real-time PCR assay targeting IS481 of *Bordetella pertussis* and molecular basis for detecting *Bordetella holmesii*. *Journal of Clinical Microbiology*, **39**, 1963–1966.

Reizenstein, E., Lindberg, L., Möllby, R. *et al.* (1996) Validation of nested *Bordetella* PCR in a pertussi vaccine trial. *Journal of Clinical Microbiology*, **34**, 810–815.

Robertson, P.W., Goldberg, H., Jarvie, B.H. *et al.* (1987) *Bordetella pertussis* infection: a cause of persistent cough in adults. *The Medical Journal of Australia*, **147**, 522–525.

Robinson, A., Ashworth, L.A., Baskerville, A. *et al.* (1985) Protection against intranasal infection of mice with Bordetella Pertussis. *Developments in Biological Standardization*, **61**, 165–172.

Robinson, A., Gorringe, A.R., Funnell, S.G. *et al.* (1989) Serospecific protection of mice against intranasal infection with *Bordetella pertussis*. *Vaccine*, **7**, 321–324.

Rowatt, E. (1957) The growth of *Bordetella pertussis*: a review. *Journal of General Microbiology*, **17**, 297–326.

Ryan, M., Murphy, G., Ryan, E. *et al.* (1998) Distinct T-cell subtypes induced with whole cell and acellular pertussis vaccines in children. *Immunology*, **93**, 1–10.

Sako, W. (1947) Studies on pertussis immunization. *The Journal of Pediatrics*, **30**, 29–40.

Salmaso, S., Mastrantonio, P., Wassilak, S.G.F. *et al.* (1998) Persistence of protection through 33 months of age provided by immunization in infancy with two three-component acellular pertussis vaccines. *Vaccine*, **13**, 1270–1275.

Sato, H. and Sato, Y. (1984) *Bordetella pertussis* infection in mice: correlation of specific antibodies against two antigens, pertussis toxin and filamentous hemagglutinin with mouse protectivity in an intracerebral or aerosol challenge system. *Infection and Immunity*, **46**, 415–421.

Sato, Y., Kimura, M. and Fukumi, H. (1984) Development of a pertussis component vaccine in Japan. *Lancet*, **1**, 122–126.

Schaeffer, L.M. and Weiss, A.A. (2001) Pertussis toxin and lipopolysaccharide influence phagocytosis of *Bordetella pertussis* by human monocytes. *Infection and Immunity*, **69**, 7635–7641.

Schmitt, H.-J., Booy, R., Weil-Olivier, C. *et al.* (2003) Child vaccination policies in Europe: a report from the summits of independent European vaccination experts. *The Lancet Infectious Diseases*, **3**, 103–108.

Shahin, R.D., Brennan, M.J., Li, Z.M. *et al.* (1990) Characterization of the protective capacity and imunogenicity of the 69-kD outer membrane protein of *Bordetella pertussis*. *The Journal of Experimental Medicine*, **171**, 63–73.

Skowronski, D.M., De Serres, G., MacDonald, D. *et al.* (2002) The changing age and seasonal profile of pertussis in Canada. *The Journal of Infectious Diseases*, **185**, 1448–1453.

Sloan, L.M., Hopkins, M.K., Shawn Mitchell, P. *et al.* (2002) Multiplex Light-Cycler PCR assay for detection and differentiation of *Bordetella pertussis* and *Bordetella parapertussis* in nasopharyngeal specimens. *Journal of Clinical Microbiology*, **40**, 96–100.

Smith, A.M., Guzmán, C.A. and Walker, M.J. (2001) The virulence factors of *Bordetella pertussis*: a matter of control. *FEMS Microbiology Reviews*, **25**, 309–333.

Stehr, K., Cherry, J.D., Heininger, U. *et al.* (1998) A comparative efficacy trial in Germany in infants who received either the Lederle/Takeda acellular pertussis component DTP (DTaP) vaccine, the Lederle whole-cell component DTP vaccine, or DT vaccine. *Pediatrics*, **101**, 1–11.

Steven, A.C., Bisher, M.E., Trus, B.L. *et al.* (1986) Helical structure of *Bordetella pertussis* fimbriae. *Journal of Bacteriology*, **167**, 968–974.

Stibitz, S. and Garletts, T.L. (1992) Derivation of a physical map of the chromosome of *Bordetella pertussis* Tohama I. *Journal of Bacteriology*, **174**, 7770–7777.

Storsaeter, J. and Olin, P. (1992) Relative efficacy of two acellular pertussis vaccines during three years of passive surveillance. *Vaccine*, **10**, 142–144.

Storsaeter, J., Hallander, H.O., Gustafsson, L. *et al.* (1998) Levels of anti-pertussis antibodies related to protection after household exposure to *Bordetella pertussis*. *Vaccine*, **16**, 1907–1916.

Strebel, P., Nordin, J., Edwards, K. *et al.* (2001) Population-based incidence of pertussis among adolescents and adults, Minnesota, 1995–1996. *The Journal of Infectious Diseases*, **183**, 1353–1359.

Tamura, M., Nogimori, K., Murai, S. *et al.* (1982) Subunit structure of islet-activating protein, pertussis toxin, in conformity with the A-B model. *Biochemistry*, **21**, 5516–5522.

Tanaka, M., Vitek, C.R., Brian Pascual, F. *et al.* (2003) Trends in pertussis among infants in the United States, 1980–1999. *JAMA*, **290**, 2968–3975.

Taranger, J., Trollfors, B., Lagergard, T. *et al.* (1997) Unchanged efficacy of a pertussis toxoid vaccine throughout the two years after the third vaccination of infants. *Pediatric Infectious Disease Journal*, **16**, 180–184.

Thomas, M.G. (1989) Epidemiology of pertussis. *Review of Infectious Diseases*, **11**, 255–262.

Tran Minh, N.N., He, Q., Ramalho, A. *et al.* (1999) Acellular vaccines containing reduced quantities of pertussis antigens as a booster in adolescents. *Pediatrics*, **104**, E70.

Trollfors, B. (1984) *Bordetella pertussis* whole cell vaccine-efficacy and toxicity. *Acta Paediatrica Scandinavia*, **73**, 417–425.

van der Zee, A., Agterberg, C., Peeters, M. *et al.* (1996a) A clinical validation of *Bordetella pertussis* and *Bordetella parapertussis* polymerase chain reaction: comparison with culture and serology using samples from patients with suspected whooping cough from a highly immunized population. *The Journal of Infectious Diseases*, **174**, 89–96.

van der Zee, A., Groenendijk, H., Peeters, M. *et al.* (1996b) The differentiation of *Bordetella parapertussis* and *Bordetella bronchiseptica* from humans and animals as determined by DNA polymorphism mediated by two different insertion sequence elements suggests their phylogenetic relationship. *International Journal of Systematic Bacteriology*, **46**, 640–647.

van der Zee, A., Mooi, F., van Embden, J. *et al.* (1997) Molecular evolution and host adaptation of *Bordetella* spp. phylogenetic analysis using multilocus enzyme electrophoresis and typing with three insertion sequences. *Journal of Bacteriology*, **179**, 6609–6617.

Van Loo, I.H.M., Heuvelman, K.J., King, A.J. *et al.* (2002) Multilocus sequence typing of *Bordetella pertussis* based on surface protein genes. *Journal of Clinical Microbiology*, **40**, 1994–2001.

Vandamme, P., Heyndrickx, M., Vancanneyt, M. *et al.* (1996) *Bordetella trematum* sp. nov., isolated from wounds and ear infections in humans, and reassessment of Alcaligenes denitrificans Ruger and Tan 1983. *International Journal of Systematic Bacteriology*, **46**, 849–858.

Vandamme, P., Hommez, J., Vancanneyt, M. *et al.* (1995) *Bordetella hinzii* sp. nov., isolated from poultry and humans. *International Journal of Systematic Bacteriology*, **45**, 37–45.

Viljanen, M.K., Ruuskanen, O., Granberg, C. *et al.* (1982) Serological diagnosis of pertussis: IgM, IgA and IgG antibodies against *Bordetella pertussis* measured by enzyme-linked immunosorbent assay (ELISA). *Scandinavian Journal of Infectious Diseases*, **14**, 117–122.

Vitek, C.R., Brian Pascual, F., Baughman, A.L. and Murphy, T.V. (2003) Increase in deaths from pertussis among young infants in the United States in the 1990s. *Pediatric Infectious Disease Journal*, **22**, 628–634.

Von Wintzingerode, F., Schattke, A., Siddiqui, R.A. *et al.* (2001) *Bordetella petrii* sp. nov., isolated from an anaerobic bioreactor, and emended description of the genus *Bordetella*. *International Journal of Systematic and Evolutionary Microbiology*, **51**, 1257–1265.

Wadowsky, R.M., Laus, S., Libert, T. *et al.* (1994) Inhibition of PCR-based assay for *Bordetella pertussis* by using calcium alginate fiber and aluminum shaft components of a nasopharyngeal swab. *Journal of Clinical Microbiology*, **32**, 1054–1057.

Wang, J., Yang, Y., Li, J. *et al.* (2002) Infantile pertussis rediscovered in China. *Emerging Infectious Diseases*, **8**, 859–861.

Watanabe, M., Takimoto, H., Kumazawa, Y. *et al.* (1990) Biological properties of lipopolysaccharides from *Bordetella* species. *Journal of General Microbiology*, **136**, 489–493.

Weber, C., Boursaux-Eude, C., Coralie, G. *et al.* (2001) Polymorphism of *Bordetella pertussis* isolates circulating for the last 10 years in France, where a single effective whole-cell vaccine has been used for more than 30 years. *Journal of Clinical Microbiology*, **39**, 4396–4403.

Weyant, R.S., Hollis, D.G., Weaver, R.E. *et al.* (1995) *Bordetella holmesii* sp. nov., a new gram-negative species associated with septicemia. *Journal of Clinical Microbiology*, **33**, 1–7.

Willems, R.J., van der Heide, H.G. and Mooi, F.R. (1992) Characterization of a *Bordetella pertussis* fimbrial gene cluster which is located directly downstream of the filamentous haemagglutinin gene. *Molecular Microbiology*, **6**, 2661–2671.

Wilson, K.E., Cassiday, P.K., Popovic, T. *et al.* (2002) *Bordetella pertussis* isolates with a heterogeneous phenotype for erythromycin resistance. *Journal of Clinical Microbiology*, **40**, 2942–2944.

Wirsing von König, C.H., Postels-Multani, S., Bock, H.L. *et al.* (1995) Pertussis in adults: frequency of transmission after household exposure. *Lancet*, **346**, 1326–1329.

Woolfrey, B.F. and Moody, J.A. (1991) Human infections associated with *Bordetella bronchiseptica*. *Clinical Microbiology Reviews*, **4**, 243–255.

World Health Organization. (2002) WHO vaccine-preventable diseases: monitoring system 2002 global summary, WHO/V&B02.20.

Wortis, N., Strebel, P.M., Wharton, M. *et al.* (1996) Pertussis deaths: report of 23 cases in the United States, 1992 and 1993. *Pediatrics*, **97**, 607–612.

Yih, W.K., Silva, E.A., Ida, J. *et al.* (1999) *Bordetella holmesii*-like organisms isolated from Massachusetts patients with pertussis-like symptoms. *Emerging Infectious Diseases*, **5**, 441–443.

Yuk, M.H., Heininger, U., Martinez de Tejada, G. *et al.* (1998) Human but not ovine isolates of *Bordetella parapertussis* are highly clonal as determined by PCR-based RAPD fingerprinting. *Infection*, **26**, 270–273.

Zepp, F., Knuf, M., Habermehl, P. *et al.* (1996) Pertussis-specific cell-mediated immunity in infants after vaccination with a tricomponent acellular pertussis vaccine. *Infection and Immunity*, **64**, 4078–4084.

19

Brucella spp.

Edward J. Young

Department of Internal Medicine, Baylor College of Medicine, Veterans Affairs Medical Center, Houston, TX, USA

INTRODUCTION

Brucellosis is primarily a disease of animals (zoonosis), but it was first recognized for the illness it caused in humans. Jeffery Allen Marston, an assistant surgeon in the Royal Army Medical Corps (RAMC), is credited with the first reliable description of brucellosis based on his experience with the disease. In 1886, David Bruce, another RAMC physician, isolated the organism later called *Brucella melitensis*, the causative agent of Malta fever, as the disease was known (Vassallo, 1992). Between 1904 and 1907, the Mediterranean Fever Commission, established to investigate epidemic brucellosis in Malta, published landmark studies on the nature of the disease (Williams, 1989). In the Third *Report* (1905), Themistocles Zammit, a Maltese physician, reported that native goats were the reservoir of the infection, and in the Seventh *Report* (1907), fresh goat's milk was reported as the vehicle of transmission from animals to humans (Vassallo, 1996).

Meanwhile, in 1895, Bernhard Bang, a Danish physician, isolated *Brucella abortus* from cyetic tissue of cattle suffering from contagious abortions. In 1921, human infection with the agent of Bang's disease was reported in Rhodesia by Bevan and, in 1924, in the United Kingdom by Orpan; however, similarities between the goat and cattle organisms were not immediately apparent. The third major nomen species, *Brucella suis*, was isolated from aborted swine in 1914 by Jacob Traum, a bacteriologist with the US Department of Agriculture (USDA). Keefer reported the first human infection with the swine organism in 1924 in Baltimore, although it was initially believed to be of bovine origin (Spink, 1956). The close taxonomic relationship between the agents of Malta fever and Bang's disease was finally recognized by the work of Alice Evans at the USDA. Evans suggested that the caprine, bovine and swine agents formed a distinct genus, and she proposed the name *Brucella* to honor Bruce. Additional species (*Brucella ovis* and *Brucella neotomae*) were added to the genus, but they do not appear to cause human infection. In 1966, Carmichael and associates isolated *Brucella canis* from kennel-bred dogs suffering from contagious abortion; however, it is a rare cause of human infection.

Since 1994, *Brucella* spp. with phenotypic and phylogenetic characteristics different from those of previously recognized nomen species have been reported from a variety of marine mammals including cetaceans, seals and otters. Initially termed *Brucella maris*, host preferences and DNA polymorphisms indicate the existence of at least two distinct new species (Bricker *et al.*, 2000). The names *Brucella pinnipediae* and *Brucella ceteceae* have been proposed for the seal and cetacean isolates, respectively (González *et al.*, 2002).

Human infection with the marine organism has been reported (Sohn *et al.*, 2003) (Table 19.1).

Table 19.1 Pathogenicity of *Brucella* species and biovars

Nomen species	Biovars	Preferred host	Pathogenicity for humans
B. melitensis	1–3	Goats, sheep	High
B. abortus	1–6, 9	Cattle	Moderate
B. suis	1	Swine	High
	2	Swine	Low
	3	Swine	High
	4	Reindeer/caribou	Moderate
	5	Rodents	High
B. canis	None	Dogs	Low
B. ovis	None	Sheep	None
B. neotomae	None	Desert wood rats	None
B. maris[a]	–	Marine mammals	Low

[a] Pending nomenclature. (Includes *B. pinnipediae* and *B. ceteceae*.)

DESCRIPTION OF THE ORGANISM

Morphology

Brucella are small, Gram-negative coccobacilli that lack capsules, endospores or native plasmids. The cell wall consists of an outer layer of lipopolysaccharide (LPS) protein approximately 9 nm thick. Thin-section electron micrographs reveal an electron-dense layer 3–5 nm thick consisting of a highly cross-linked muramyl–mucopeptide complex associated with lipoproteins. Matrix and porin proteins penetrate the peptidoglycan layer at irregular intervals. A low-density periplasmic space separates the peptidoglycan layer from the cell membrane.

Classification

The genus *Brucella* consists of seven nomen species, some containing several biovars, differentiated on the basis of cultural, metabolic and antigenic characteristics (Tables 19.2 and 19.3). However, taxonomic studies employing DNA and DNA–rRNA hybridization reveal a high degree of homology among nomen species, indicating that the genus is monospecific (Verger *et al.*, 1985; Gándara *et al.*, 2001). Nevertheless, by current convention, the original nomen species classification, based in part on natural host preferences, is retained for epidemiological purposes. In fact, genomic fingerprints and restriction length polymorphisms of conserved loci indicate that the present taxonomy is relevant. For example, the *omp* 2 gene locus is conserved in all *Brucella* sp., and complete sequence analysis of these loci reveals nucleotide differences among nomen species (Ficht *et al.*, 1990). Phylogenetically, *Brucella* sp. appears to have a common origin with

Table 19.2 Characteristics of *Brucella* species and biovars

Nomen species	Biovars	Requirement for CO_2	Production of H_2S	Urease activity	Growth on media containing dyes[a]		Agglutination by monospecific antiserum[b]		
					Thionin	Basic fuchsin	A	M	R
B. melitensis	1	−	−	Variable	+	+		+	−
	2	−	−	Variable	+	+	+	−	−
	3	−	−	Variable	+	+	+	+	−
B. abortus	1	(+)	+	Slow	−	+	+	−	−
	2	(+)	+	Slow	−	−	+	−	−
	3	(+)	+	Slow	+	+	+	−	−
	4	(+)	+	Slow	−	(+)	−	+	−
	5	−	−	Slow	+	+	−	+	−
	6	−	(+)	Slow	−	+	+	−	−
	9	−	−	Slow	+	+	−	+	−
B. suis	1	−	−	Rapid	+	(−)	+	−	−
	2	−	−	Rapid	+	−	+	−	−
	3	−	−	Rapid	+	+	+	−	−
	4	−	−	Rapid	+	(−)	+	+	−
	5	−	−	Rapid	+	−	−	+	−
B. canis		−	−	Rapid	+	−	−	−	+
B. ovis		+	−	–	+	(−)	−	−	+
B. neotomae		−	+	Rapid	−	−	+	−	−
B. maris*		+/−	−	Rapid	+	+	+/−	+/−	−

+, Positive; −, negative; (+), most strains positive; (−), most strains negative.

[a] Dye concentration, 20 μg/ml.

[b] A, monospecific for *B. abortus* A antigen; M, monospecific for *B. melitensis* M antigen; R, monospecific for rough *Brucella* antigen.

free-living, soil-dwelling bacteria. Based on 5S and 16S rRNA sequences, *Brucella* is included in the α-2 subdivision of the Protobacteriaceae, closely related to human pathogens such as *Bartonella* sp. and plant pathogens and symbionts such as *Agrobacterium tumefaciens* and *Sinorhizobium meliloti* (Del Vecchio *et al.*, 2002; Paulsen *et al.*, 2002).

Genetics

Restriction endonuclease techniques have shown that the genome of *B. melitensis* comprises two circular chromosomes of 2.1 Mb and 1.15 Mb (Michaux *et al.*, 1993). The physical maps of the genomes of six nomen species showed only small differences from those of *B. melitensis*, except for a large 640 Kb inversion in the small chromosome of *B. abortus* 544 (Michaux-Charachon *et al.*, 1997). In contrast, strains of the four biovars of *B. suis* varied in chromosome number and size, having either one 3.1 Mb chromosome (biovar 3) or two chromosomes of smaller size (biovars 1, 2 and 4) (Jumas-Bilak *et al.*, 1998). These differences appear to represent rearrangements in three rRNA operons. There is pronounced asymmetry of the genes in different functional categories between the two *Brucella* chromosomes; however, each contains genes that are essential for structural and metabolic integrity, hence qualifying as true chromosomes and not plasmids (Paulsen *et al.*, 2002).

Antigens

Lipopolysaccharide

The major surface component of *Brucella* sp. is LPS consisting of an outer O-polysaccharide chain linked through a core oligosaccharide to lipid A which anchors the complex in the outer cell membrane. Smooth (S) strains contain the complete complex, whereas rough (R) strains lack the O-side chain. The O-antigen contains several epitopes, including the A and M antigens of Wilson and Miles (Wilson and

Miles, 1932). The A epitope predominates in *B. abortus* strains and consists of a linear homopolymer of α-1,2-linked 4-formamido-4, 6-dideoxy-D-mannose (*N*-formyl perosamine) (Perry and Bundle, 1990). In *B. melitensis* strains, the O-chain consists of repeating units of five *N*-formyl perosamine residues, four α-1,2-linked and one α-1,3-linked. A common, or C epitope, is present in all brucellae that accounts for serological cross-reactivity. The presence of perosamine in the *Brucella* LPS also explains the cross-reactivity with other Gram-negative bacteria such as *Escherichia coli* O:157, *Salmonella* O:30, *Vibrio cholerae* and *Yersinia enterocolitica* O:9.

The structure of the core oligosaccharide has not been fully elucidated; however, the presence of mannose and 2-amino-2, 5-dideoxy-D-glucase (quinovosamine) and 2-keto-3-deoxyoctulosonic acid (KDO) is known (Moriyón and López-Goñi, 1998).

The lipid A moiety is a disaccharide of 2,3-diamino-2,3-dideoxy-glucose in α-1,6 linkages, to which are attached a large proportion of long chain fatty acids (>C16), very small amounts of hydroxylated fatty acids, no −OH myristic acid and a proportion of amide- and ester-linked fatty acids (Qureshi *et al.*, 1994). Ethanolamine and arabinosamine, often present in the lipid A of Enterobacteriaceae, have not been found, which may explain differences between *Brucella* and classical endotoxins. For example, *Brucella* endotoxin is only weakly pyrogenic for rabbits, does not elicit the dermal Shwartzman reaction, and is a much less potent inducer of interleukin-1 (IL-1) and TNF-α from human monocytes (Goldstein *et al.*, 1992).

Outer Membrane Proteins

Several outer membrane proteins (OMPs) of brucellae have been identified. They were first classified according to their apparent molecular mass into group 1 (88–94 kDa), group 2 (35–39 kDa) and group 3 (25–31 kDa) proteins. In addition, a lipoprotein similar to the Braun antigen in *E. coli* has been partially characterized. The genes encoding these OMPs have been sequenced, and studies are underway to clarify their precise roles (Moriyón and López-Goñi, 1998; Guzmán-Verri *et al.*, 2002).

Table 19.3 Oxidative metabolism and bacteriophage lysis of *Brucella* species and biovars

Nomen species	L-Alanine	L-Asparagine	L-Glutamic acid	L-Arginine	L-Citrulline	DL-Ornithine	L-Lysine	D-Ribose	D-Xylose	D-Galactose	D-Glucose	i-Erythritol	Lysis at RTD by bacteriophage		
													Tbilisi	Weybridge	Berkeley
B. melitensis 1, 2, 3	+	+	+	–	–	–	–	–	–	–	–	–	NL	NL	L
B. abortus 1–6, 9	+	+	+	–	–	–	–	+	–	+	+	+	L	L	L
B. suis Biovar 1	±	–	–	+	+	+	+	+	+	+	+	+	NL	L	L
Biovar 2	±	±	±	+	+	+	–	+	+	+	+	+	NL	L	L
Biovar 3	±	–	+	+	+	+	+	+	+	–	+	+	NL	L	L
Biovar 4	–	–	+	+	+	+	+	+	+	–	+	+	NL	L	L
Biovar 5	–	+	+	+	+	+	+	+	+	–	+	±	NL	L	L
B. canis	±	–	+	+	+	+	+	+	–	±	+	–	NL	NL	NL
B. ovis	±	+	+	–	–	–	–	±	–	–	–	–	NL	NL	NL
B. neotomae	±	+	+	–	–	–	–	+	–	+	+	+	NL	L	L
B. maris	–	–	+	–	–	–	–	+	+	±	+	+	NL	L	L

+, oxidized by all strains; –, not oxidized by any strain; ±, oxidized by some strains; L, lysis; NL, no lysis at routine test dilution (RTD) by *Brucella* phages.

Periplasmic Proteins

Among the periplasmic proteins that have been identified is a Cu–Zn superoxide dismutase that inactivates reactive oxygen intermediates.

Cytoplasmic Proteins

Several stress proteins such as GroEL, GroES, DnaK and HtrA as well as a bacterioferritin and a lumazine synthetase have been identified. The latter has been proposed as a potential serodiagnostic antigen (Goldbaum *et al.*, 1999); however, it appears to be less sensitive than LPS antigens for this purpose. The L7/L12 ribosomal protein appears to play a role in stimulating T-helper-1 (Th-1)-type cellular immunity.

PATHOGENESIS

Intracellular Survival

Brucella spp. are facultative intracellular pathogens that have evolved mechanisms to evade destruction by phagocytic cells of the host. Although the mechanisms are incompletely understood, several pathogen virulence factors have been proposed. *O*-Polysaccharide is considered important since smooth strains resist phagocytosis and kill better than rough strains. In addition, virulent strains of *B. abortus* contain a Cu–Zn superoxide dismutase enzyme that inhibits reactive oxygen radicals. Two nucleotide-like substances (5′-guanosine monophosphate and adenine) have been recovered from culture filtrates of *B. abortus* that inhibits phagolysosome fusion and suppresses the myeloperoxidase–H_2O_2–halide killing mechanism of neutrophils (Canning, Roth and Deyoe, 1986).

Host Defenses

Intact skin is an effective barrier to the entry of brucellae, but even minute abrasions may permit the ingress of bacteria. The low pH of gastric juice also provides some protection against oral infection, and antacids or histamine blockers appear to increase susceptibility (Steffen, 1977). Normal human serum has moderate anti-brucella activity, and complement opsonizes the organism for phagocytosis. Human neutrophils destroy some brucellae; however, they lack activity against *B. melitensis* (Young, 1985). Organisms that escape killing by neutrophils enter the lymphatics where they localize within organs rich in elements of the reticuloendothelial system (lymph nodes, spleen, liver and bone marrow).

Immune Responses

Innate immunity to *Brucella* infection has been demonstrated in swine and cattle. Macrophages from naturally resistant strains of cattle are better able to control intracellular replication of *B. abortus*, a process that appears to be controlled by the bovine homologue of the *Nramp1* (*natural resistance associated macrophage protein 1*) gene (Feng *et al.*, 1996; Adams and Templeton, 1998).

Acquired immunity involves both humoral and cellular factors, although the latter predominates, as with other intracellular pathogens. The humoral response is characterized by the appearance of immunoglobulin M (IgM) antibodies directed principally against LPS within the first week of infection. This is followed by a switch to IgG synthesis after the second week. Thereafter, both subclasses of Igs increase over time.

Cellular immunity involves T-cell-dependent activation of macrophages, the main cellular reservoir for brucellae. CD4$^+$ T cells appear to play a role in protection against Brucella infection, either by activating CD8$^+$ T cells or by secreting cytokines that mediate macrophage activation (Araya *et al.*, 1989). Among the cytokines involved in activating anti-brucella activity of macrophages, interferon-γ (IFN-γ) is of particular importance (Jiang and Baldwin, 1993). Nevertheless, it is likely that both humoral and cellular immune mechanisms work in concert for recovery from, and resistance to, *Brucella* infection (Casadevall, 2003).

EPIDEMIOLOGY

Brucellosis exists worldwide in domestic and wild animals and is especially prevalent in countries bordering the Mediterranean, throughout the Arabian peninsula, in the Indian subcontinent and in parts of Mexico, Central and South America. In many areas, such as sub-Saharan Africa, brucellosis is known to exist, but the prevalence is unknown, owing to a lack of diagnostic and reporting mechanisms. The importance of enzootic brucellosis in areas adjacent to countries where the disease is almost eliminated is illustrated by the border between Mexico and the United States. In the border states of Texas and California, most cases of human brucellosis originate from unpasteurized dairy products imported from Mexico (Taylor and Perdue, 1989; Chomel *et al.*, 1994). In counties of Texas bordering Mexico, for example, the prevalence of human brucellosis is eight times higher than elsewhere in the United States (Doyle and Bryan, 2000; Fosgate *et al.*, 2002).

Brucella melitensis occurs primarily in goats and sheep, although in some countries, camels, and even cattle, may be important reservoirs. It is the most frequent cause of human brucellosis worldwide. *Brucella abortus* is found primarily in cattle and other bovidae, such as buffalo and yaks. *Brucella suis* biovars 1–3 are found in swine, whereas biovar 4 is limited to reindeer and caribou. Human infection with *B. suis* is a particular risk for abattoir workers and for hunters of feral swine. *Brucella canis* occurs in dogs, especially under conditions of intense breeding. Pathogenicity of *B. canis* for humans is low, and infections generally involve dog breeders or laboratory personnel.

Human-to-human transmission of brucellosis is rare, although cases believed to be sexually transmitted have been reported (Ruben *et al.*, 1991). Few cases of brucellosis have been reported in people with acquired immunodeficiency syndrome (AIDS), which is surprising in view of their susceptibility to other infections (Moreno *et al.*, 1998).

CLINICAL FEATURES

Presentation

The onset of human brucellosis may be acute or insidious, and the clinical manifestations are protean. The disease is generalized and may involve any organ or system of the body. Presenting complaints are numerous and nonspecific, including malaise, anorexia, fatigue, sweats, weight loss, back or joint pains and depression. Objective physical findings are few, notably fever, mild lymphadenopathy and, occasionally, hepatomegaly or splenomegaly. Fever is present in most patients with active infection. When monitored over time without the intervention of antibiotics or antipyretics, the fever pattern is undulating. The importance of a careful history of animal exposure, travel to brucella enzootic countries or the ingestion of 'exotic' foods such as dairy products made from raw milk cannot be overemphasized. People engaged in ranching, animal husbandry, veterinary medicine, abattoir work, hunting game involving feral swine and laboratory medicine should be considered as being at increased risk of brucellosis (Young, 1983; Memish and Mah, 2001).

Nervous System

Depression and mental inattention are common; however, direct invasion of the nervous system occurs in less than 5% of cases. Meningitis or meningoencephalitis is the most common manifestation of neurobrucellosis, and it may be acute or chronic. Other neurological syndromes include peripheral neuropathy, myelitis, radiculitis, brain and epidural abscess and demyelinating and meningovascular syndromes. The presence of brucella antibodies in the cerebrospinal fluid (CSF) is essentially diagnostic. With antimicrobial therapy, the prognosis is usually good; however, severe neurological sequelae have been reported (McLean, Russell and Khan, 1992).

Skeletal System

Osteoarticular involvement is the most common complication of brucellosis, occurring in 20–80% of cases (Colmenero et al., 1996). The axial skeleton is frequently involved, and sacroilitis and spondylitis are common findings (Ariza et al., 1993; Solera et al., 1999). Peripheral joint arthritis primarily involves large, weight-bearing joints such as hips, knees and ankles. Radiographic findings are infrequent, but computed tomography is useful for detecting joint destruction, vertebral osteomyelitis and, rarely, paraspinal abscess. A post-infectious spondyloarthropathy has been described. Treatment is antimicrobial drugs, with surgery reserved for spinal instability, abscess drainage or prosthetic joint infection.

Gastrointestinal System

Brucellosis resembles typhoid fever, in that systemic symptoms predominate over gastrointestinal complaints. Nevertheless, more than two-thirds of patients complain of nausea, anorexia, vomiting, weight loss and abdominal discomfort. Inflammation of Peyer's patches and the ileal and colonic mucosa has been described. The liver is probably always involved; however, liver function tests are usually only mildly elevated. The spectrum of histological lesions of the liver is varied, depending, in part, on the infecting Brucella species. Nonspecific mononuclear cell infiltrates, epithelioid granuloma and abscess have all been reported. Cholecystitis, pancreatitis and spontaneous peritonitis are rare complications.

Cardiovascular System

Endocarditis occurs in less than 2% of cases but accounts for most brucellosis-related deaths. Both native and prosthetic valve endocarditis have been reported. When the diagnosis is delayed, treatment may require a combination of antimicrobial drugs and valve replacement surgery (Jacobs et al., 1990). Myocarditis, pericarditis and mycotic aneurysms have been reported rarely.

Respiratory System

A variety of pulmonary manifestations have been reported, including bronchitis, pneumonia, lung masses, abscess and empyema. Organisms are rarely found by Gram stain of sputum.

Genitourinary System

Brucellae is isolated from urine; however, renal lesions are rare. Interstitial nephritis, glomerulonephritis, pyelonephritis and renal abscess have been described. Orchitis, usually unilateral, is common and must be differentiated from testicular neoplasm (Navarro-Martínez et al., 2001). Abortion is a prominent complication of brucellosis in animals and may occur in humans as well (Khan, Mah and Memish, 2001).

Chronic Infection

Chronic brucellosis occurs when a focus of infection persists in an organ, such as the spleen, kidney or bone marrow, despite antimicrobial therapy. Symptoms may recur over long periods of time and are usually associated with objective signs of infection (e.g. fever), accompanied by persistent elevation of IgG antibodies in the serum. Treatment often requires surgical drainage of the infectious focus. This condition must be differentiated from simple relapse of infection following a course of therapy and usually occurs within a few weeks of discontinuing antibiotics. In addition, few patients experience a delayed recovery from acute brucellosis, in which they have nonspecific symptoms, such as fatigue and malaise, but lack objective signs of disease and do not have elevated IgG antibodies.

Childhood Brucellosis

Once considered rare in children, brucellosis infects people of all ages, especially in areas where B. melitensis is enzootic. The clinical manifestations and complications are similar regardless of the age of the patient.

LABORATORY DIAGNOSIS

Bacteriological Diagnosis

The diagnosis of brucellosis is made with certainty when brucellae are recovered from blood, bone marrow or other tissues. Brucella can be cultured on any high-quality peptone-based media. Growth is enhanced by the addition of blood or serum, but primary isolation may require prolonged (several weeks) incubation. Culture of bone marrow is reported to give a higher yield than the culture of peripheral blood (Gotuzzo et al., 1986). Rapid isolation techniques such as Bactec and BacT/Alert systems have improved recovery rates and decreased the time required for isolation from weeks to days (Yagupsky, 1999). However, findings of rapid bacterial identification systems should be interpreted with caution, since some do not incorporate databases necessary to differentiate Brucella from other bacteria, such as Moraxella phenylpyruvica or Haemophilus sp. Polymerase chain reaction (PCR) and antibody-based antigen detection systems have been devised to demonstrate the presence of organisms in serum or tissues (Morata et al., 2001).

Regardless of the methods used, clinicians should notify the laboratory that brucellosis is suspected, so that cultures are not discarded prematurely, and that appropriate biohazard precautions be taken to protect personnel.

Serological Diagnosis

In the absence of a positive culture, a presumptive diagnosis of brucellosis is made by demonstrating high or rising titers of antibodies to Brucella in the serum. A variety of serological tests have been used to detect antibodies to Brucella; however, the serum agglutination test (SAT) remains the gold standard against which others are compared (Young, 1991). The SAT measures the total quantity of agglutinating antibodies but does not differentiate between Ig isotypes. To determine the titer of IgG antibodies, serum is treated with disulfide-reducing agents (2-mercaptoethanol or dithiothreitol), which destroys the agglutinability of IgM, but does not alter IgG (Buchanan and Faber, 1980). Non-agglutinating antibodies may also be detected by using Coomb's reagent. The SAT using B. abortus S119 antigen can detect antibodies to other smooth species (B. melitensis and B. suis) but not rough organisms such as B. canis. A more sensitive assay that also employs LPS antigen is the enzyme-linked immunoadsorbent assay (ELISA) (Gazapo et al., 1989). Unfortunately, the ELISA antigen is not standardized, making interpretation of results between laboratories difficult. An ELISA employing cytoplasmic antigens has been reported to differentiate between active and inactive disease (Goldbaum, 1999).

Although no single titer is always diagnostic, most patients with active infection have titers ≥1:160, with IgG antibodies predominating. Regardless of the serological test, the results must be analyzed in light of epidemiological, clinical and laboratory information.

MANAGEMENT

Antimicrobial Therapy

Numerous antimicrobial agents have activity *in vitro* against *Brucella*, but a relatively few are effective in treating human brucellosis. The tetracyclines, especially doxycycline, are among the most effective; however, clinical relapse unrelated to antibiotic resistance has led to a preference for combination drug therapy (Young, 2002). Antimicrobial therapy must be continued for a minimum of 6 weeks, and patients should be cautioned to continue therapy even after symptoms have resolved, to assure a cure. Tetracycline (500 mg four times daily by mouth for 6 weeks) plus streptomycin (1 g daily intramuscularly for 2–3 weeks) resulted in the highest cure rate and lowest rate of relapse. Doxycycline (100 mg twice daily by mouth) is now favored over generic tetracycline because of its lower incidence of gastrointestinal upset. Doxycycline in combination with rifampin (600–900 mg daily by mouth), both administered for 45 days, is an acceptable treatment for uncomplicated brucellosis; however, the relapse rate is higher than that with doxycycline plus an aminoglycoside. Gentamicin is as effective as streptomycin; however, no prospective study has compared the two drugs, and the dose and duration of therapy with gentamicin have not been conclusively demonstrated. Most experts recommend gentamicin (5 mg/kg/day) to be administered for 14 days in combination with doxycycline administered for 6 weeks (Solera *et al.*, 1997a). Laboratory studies have shown that the addition of an aminoglycoside to other agents produces a more rapid killing of brucellae *in vitro* (Rubinstein *et al.*, 1991).

The fixed combination of trimethoprim/sulfamethoxazole (co-trimoxazole) has also been used to treat brucellosis, but high rates of relapse have been reported unless the drug is combined with other agents (Ariza *et al.*, 1985; Solera *et al.*, 1997b). Nevertheless, co-trimoxazole has certain advantages, such as in the treatment of pregnant women and children less than 6 years of age, for whom tetracycline compounds are contraindicated because of the risk of staining teeth.

Fluoroquinolone compounds also have activity against *Brucella in vitro*, but treatment is complicated by a high rate of relapse and, possibly, the selection of resistant strains. For this reason, quinolones should generally be used in combination with other drugs such as doxycycline.

Treatment of Complications

The most effective treatments for neurobrucellosis and brucellar endocarditis remain undecided. Many authorities choose doxycycline in combination with two or more other agents, with treatment continued for several months depending on the clinical and laboratory responses. Some third-generation cephalosporins cross the blood–brain barrier well, but they should only be used in combination with other drugs and *in vitro* susceptibility of the infecting strain of brucellae should be assured. Treatment of brucellar endocarditis may require valve replacement, in addition to antimicrobial chemotherapy. Spinal brucellosis generally responds to drug therapy, with surgery being reserved for cases with paraspinal abscess or spinal instability. Prosthesis infection may respond to prolonged drug therapy; however, local relapses are treated with the removal of the prosthetic devise. Treatment of brucellosis in pregnancy remains a problem, but rifampin and co-trimoxazole have been used despite the absence of data assuring their safety in this setting.

Prognosis

Before the advent of effective antimicrobial chemotherapy, brucellosis was a chronic, relapsing, debilitating disease with a mortality of about 2%. Although serious disease can be caused by any *Brucella* species, *B. melitensis* is generally more virulent and accounts for most deaths. With appropriate therapy and follow-up, most patients will recover completely, although a few will experience relapses or prolonged convalescence.

Prevention

The prevention of human brucellosis depends on eliminating the disease in domestic livestock. Vaccines employing live, attenuated strains of *B. abortus* (strains 19 and RB51) or *B. melitensis* (strain Rev 1) have been used successfully to control brucellosis in developed countries; however, they have had little impact in much of the Third World. The threat of *Brucella* as a weapon of biological terrorism could serve as a stimulus to develop vaccines for humans.

REFERENCES

Adams, L.G., Templeton, J.W. (1998) Genetic resistance to bacterial diseases. *Rev Sci Tech Off Intern Epizoot* 17, 200–219.

Araya, L.N., Elzer, P.H., Rowe, G.E. *et al.* (1989) Temporal development of protective cell-mediated and humoral immunity in BALB/C mice infected with *Brucella abortus. J Immunol* 143, 3330–3337.

Ariza, J., Gudiol, F., Pallares, R. *et al.* (1985) Comparative trial of co-trimoxazole versus tetracycline–streptomycin in treating human brucellosis. *J Infect Dis* 152, 1358–1359.

Ariza, J., Pujol, M., Valverde, J. *et al.* (1993) Brucellar sacroiliitis: Findings in 63 episodes and current relevance. *Clin Infect Dis* 16, 761–765.

Bricker, B.J., Ewalt, D.R., MacMillan, A.P. *et al.* (2000) Molecular characterization of *Brucella* strains isolated from marine mammals. *J Clin Microbiol* 38, 1258–1262.

Buchanan, T.M., Faber, L.C. (1980) 2-Mercaptoethanol brucella agglutination test: Usefulness for predicting recovery from brucellosis. *J Clin Microbiol* 11, 691–693.

Canning, P.C., Roth, A., Deyoe, B.L. (1986) Release of 5′-guanosine-monophosphate and adenine by *Brucella abortus* and their role in the intracellular survival of the bacteria. *J Infect Dis* 154, 464–470.

Casadevall, A. (2003) Antibody-mediated immunity against intracellular pathogens: Two-dimensional thinking comes full circle. *Infect Immun* 71, 4225–4228.

Chomel, B.B., DeBess, E.E., Mangiamele, D.M. *et al.* (1994) Changing trends in the epidemiology of human brucellosis in California from 1973 to 1992: A shift toward foodborne transmission. *J Infect Dis* 170, 1216–1223.

Colmenero, J.D., Reguera, J.M., Martos, F. *et al.* (1996) Complications associated with *Brucella melitensis* infection: A study of 530 cases. *Medicine* 75, 195–211.

Del Vecchio, V.G., Kapatral, V., Redkar, R.J. *et al.* (2002) The genome sequence of the facultative intracellular pathogen *Brucella melitensis. Proc Natl Acad Sci U S A* 99, 443–448.

Doyle, T.J., Bryan, R.T. (2000) Infectious disease morbidity in the U.S. region bordering Mexico, 1990–1998. *J Infect Dis* 182, 1503–1510.

Feng, J., Li, Y., Hashad, M. *et al.* (1996) Bovine natural resistance associated macrophage protein 1 (*Nramp 1*) gene. *Genome Res* 6, 956–964.

Ficht, T.A., Bearden, S.W., Sowa, B.A. *et al.* (1990) Genetic variation at the omp 2 porin locus of the brucellae: Species-specific markers. *Mol Microbiol* 4, 1135–1142.

Fosgate, G.T., Carpenter, T.E., Chomel, B.B. *et al.* (2002) Time-space clustering of human brucellosis, California, 1973–1992. *Emerg Infect Dis* 8, 672–678.

Gándara, B., López Merino, A., Rogel, M.A. *et al.* (2001) Limited genetic diversity of *Brucella* spp. *J Clin Microbiol* 39, 235–240.

Gazapo, E., Lahos, J.G., Subiza, J.L. *et al.* (1989) Changes in IgM and IgG antibody concentrations in brucellosis over time: Importance for diagnosis and follow-up. *J Infect Dis* 159, 219–225.

Goldbaum, F.A., Velikovsky, C.A., Baldi, P.C. *et al.* (1999) The 18-kDa cytoplasmic protein of *Brucella* species – an antigen useful for diagnosis – is a lumazine synthase. *J Med Microbiol* 48, 833–839.

Goldstein, J., Hoffman, T., Frasch, C. *et al.* (1992) Lipopolysaccharide (LPS) from *Brucella abortus* is less toxic than that from *Escherichia coli*, suggesting the possible use of *B. abortus* or LPS from *B. abortus* as a carrier in vaccines. *Infect Immun* **60**, 1385–1389.

González, L., Patterson, I.A., Reid, R.J. *et al.* (2002) Chronic meningoencephalitis associated with *Brucella* sp. infection in live-stranded striped dolphins (*Stenella coeruleoalba*). *J Comp Pathol* **126**, 147–152.

Gotuzzo, E., Carrillo, C., Guerra, J. *et al.* (1986) An evaluation of diagnostic methods for brucellosis: The value of bone marrow cultures. *J Infect Dis* **153**, 122–125.

Guzmán-Verri, C., Manterola, L., Sola-Landa, A. *et al.* (2002) The two-component system BvrR/BvrS essential for *Brucella abortus* virulence regulates the expression of outer membrane proteins with counterparts in members of the Rhizobiaceae. *Proc Natl Acad Sci U S A* **99**, 12375–12380.

Jacobs, F., Abramowicz, D., Vereerstraeten, P. *et al.* (1990) Brucellar endocarditis: The role of combined medical and surgical treatment. *Rev Infect Dis* **12**, 740–744.

Jiang, X., Baldwin, C.L. (1993) Effects of cytokines on intracellular growth of *Brucella abortus*. *Infect Immun* **61**, 124–129.

Jumas-Bilak, E., Michaux-Charachon, S., Bourg, G. *et al.* (1998) Differences in chromosome number and genome rearrangements in the genus *Brucella*. *Mol Microbiol* **27**, 99–106.

Khan, M.Y., Mah, M.W., Memish, Z.A. (2001) Brucellosis in pregnant women. *Clin Infect Dis* **32**, 1172–1177.

McLean, D.R., Russell, N., Khan, M.Y. (1992) Neurobrucellosis: Clinical and therapeutic features. *Clin Infect Dis* **15**, 582–590.

Memish, Z.A., Mah, M.W. (2001) Brucellosis in laboratory workers at a Saudi Arabian hospital. *Am J Infect Control* **29**, 48–52.

Michaux, S., Paillisson, J., Carles-Nurit, M.J. *et al.* (1993) Presence of two independent chromosomes in the *Brucella melitensis* 16M genome. *J Bacteriol* **175**, 701–705.

Michaux-Charachon, S., Bourg, G., Jumas-Bilak, E. *et al.* (1997) Genome structure and phylogeny in the genus *Brucella*. *J Bacteriol* **179**, 3244–3249.

Morata, P., Queipo-Ortuño, M.I., Reguera, J.M. *et al.* (2001) Diagnostic yield of a PCR assay in focal complications of brucellosis. *J Clin Microbiol* **39**, 3743–3746.

Moreno, S., Ariza, J., Espinoza, F.J. *et al.* (1998) Brucellosis in patients infected with the human immunodeficiency virus. *Eur J Clin Microbiol Infect Dis* **17**, 319–326.

Moriyón, I., López-Goñi, I. (1998) Structure and properties of the outer membrane of *Brucella abortus* and *Brucella melitensis*. *Int Microbiol* **1**, 19–26.

Navarro-Martínez, A., Solera, J., Corredoira, J. *et al.* (2001) Epididymoorchitis due to *Brucella melitensis*: A retrospective study of 59 patients. *Clin Infect Dis* **33**, 2017–2022.

Paulsen, I.T., Seshadri, R., Nelson, K.E. *et al.* (2002) The *Brucella* genome reveals fundamental similarities between animal and plant pathogens and symbionts. *Proc Natl Acad Sci U S A* **99**, 13148–13153.

Perry, M.B., Bundle, D.R. (1990) Lipopolysaccharide antigens and carbohydrates of *Brucella*. In: Adams, L.G. (ed.), *Advances in Brucellosis Research*. Texas A&M University Press, College Station, TX, pp. 76–88.

Qureshi, N., Takayama, K., Seydel, U. *et al.* (1994) Structural analysis of the lipid A derived from the lipopolysaccharide of *Brucella abortus*. *J Endotoxin Res* **1**, 137–148.

Ruben, B., Band, J.D., Wong, P. *et al.* (1991) Person-to-person transmission of *Brucella melitensis*. *Lancet* **337**, 14–15.

Rubinstein, E., Lang, R., Shasha, B. *et al.* (1991) In vitro susceptibility of *Brucella melitensis* to antibiotics. *Antimicrob Agents Chemother* **35**, 1925–1927.

Sohn, A.H., Probert, W.S., Glaser, C.A. *et al.* (2003) Human neurobrucellosis with intracerebral granuloma caused by a marine mammal *Brucella* spp. *Emerg Infect Dis* **9**, 485–488.

Solera, J., Espinoza, A., Martínez-Alfaro, E. *et al.* (1997a) Treatment of human brucellosis with doxycycline and gentamicin. *Antimicrob Agents Chemother* **41**, 80–84.

Solera, J., Martínez-Alfaro, E., Espinoza, A. (1997b) Recognition and optimum treatment of brucellosis. *Drugs* **53**, 245–256.

Solera, J., Lozano, E., Martínez-Alfaro, E. *et al.* (1999) Brucellar spondylitis: Review of 35 cases and literature survey. *Clin Infect Dis* **29**, 1440–1449.

Spink, W.W. (1956) *The Nature of Brucellosis*. University of Minnesota Press, Minneapolis.

Steffen, R. (1977) Antacids: A risk factor for traveler's brucellosis? *Scand J Infect Dis* **9**, 311–312.

Taylor, J.P., Perdue, J.N. (1989) The changing epidemiology of human brucellosis in Texas, 1977–1986. *Am J Epidemiol* **130**, 160–165.

Vassallo, D.J. (1992) The Corps disease: Brucellosis and its historical association with the Royal Army Medical Corps. *J R Army Med Corps* **138**, 140–150.

Vassallo, D.J. (1996) The saga of brucellosis: Controversy over credit for linking Malta fever with goats' milk. *Lancet* **348**, 804–808.

Verger, J.M., Grimont, F., Grimont, P.A.D. *et al.* (1985) *Brucella* a monospecific genus as shown by desoxyribonucleic acid hybridization. *Int J Syst Bacteriol* **35**, 292–295.

Williams, E. (1989) The Mediterranean Fever Commission: Its origin and achievements. In: Young, E.J., Corbel, M.J. (eds), *Brucellosis: Clinical and Laboratory Aspects*. CRC Press, Boca Raton, FL, pp. 11–23.

Wilson, G.S., Miles, A.A. (1932) The serological differentiation of smooth strains of the *Brucella* group. *Br J Exp Pathol* **13**, 1–13.

Yagupsky, P. (1999) Detection of brucellae in blood culture. *J Clin Microbiol* **37**, 3437–3442.

Young, E.J. (1983) Human brucellosis. *Rev Infect Dis* **5**, 821–842.

Young, E.J. (1991) Serologic diagnosis of human brucellosis: Analysis of 214 cases by agglutination tests and review of the literature. *Rev Infect Dis* **13**, 359–372.

Young, E.J. (2002) *Brucella* species (Brucellosis). In: Yu, V., Weber, R., Raoult, D. (eds), *Antimicrobial Therapy and Vaccines, Vol. 1 Microbes, 2nd ed.* Apple Trees Productions, New York, pp. 121–140.

Young, E.J., Borchert, M., Kretzer, F. *et al.* (1985) Phagocytosis and killing of *Brucella* by human polymorphonuclear leukocytes. *J Infect Dis* **151**, 682–690.

20

Actinobacillus actinomycetemcomitans

Brian Henderson[1] and Derren Ready[2]

[1]*Division of Infection and Immunity, Eastman Dental Institute, University College London,* [2]*Eastman Dental Hospital, University College London Hospitals NHS Trust, London, UK*

INTRODUCTION

'But what if I should tell such people in future that there are more animals living in the scum of the teeth in a man's mouth, than there are men in the whole kingdom'. This is a quote from the 17th century microscopist, Antoni van Leeuwenhoek, the first man to see bacteria. Surprisingly, it is only in very recent years that we have begun to realize how much human morbidity is caused by these animals (oral bacteria) – a fact that is poetically captured in a quote from a recent report of the American Surgeon General about the 'silent epidemic of oral diseases – which disproportionately burdens minorities and the poor' (US Public Health Service, 2000).

The two main oral diseases of humans are caused by bacteria. Caries affects the teeth and the periodontal diseases affect the tissues and supporting structures around the tooth. The periodontal (*perio* – around and *odontos* – tooth) diseases are either acute (gingivitis) or chronic (periodontitis). These diseases cause inflammation of the gums (gingivae) and destruction of the tissues, periodontal ligament and alveolar bone that link the teeth to the jaw bone (Figure 20.1). A number of clinically diagnosable forms of periodontitis exist (Armitage, 1999). The most severe form of periodontitis is localized aggressive periodontitis (LAP), previously called localized juvenile periodontitis,

and the evidence suggests that it is caused by the subject of this chapter, *Actinobacillus actinomycetemcomitans* (Slots and Ting, 1999; Henderson *et al.*, 2003).

DESCRIPTION OF THE ORGANISM

Actinomyces spp. cause a chronic inflammatory lesion, actinomycosis, and a Gram-negative organism isolated from such a lesion by Klinger in 1912 resulted in the name *Bacterium actinomycetum comitans* being coined for this bacterium (Klinger, 1912). Thus begins a history of name changing, starting with *Bacterium comitans* (Lieske, 1921) and *A. actinomycetemcomitans* (Topley and Wilson, 1929). Interest in this organism stemmed from the finding of the bacterium in actinomycotic lesions in which the *Actinomyces* spp. had been eliminated by antibiotics (Holme, 1951). However, subsequent studies revealed that *A. actinomycetemcomitans* is a member of the normal oral microbiota (Heinrich and Pulverer, 1959) and that it is involved in the pathology of periodontitis (Zambon, 1985) and various non-oral infections (Van Winkelhoff and Slots, 1999). The past two decades has seen increasing attention being paid to this organism, and its genome has just been sequenced (Najar, 2002) revealing that this bacterium is closely related to members of the *Pasteurellaceae* such as *Haemophilus influenzae*.

Actinobacillus actinomycetemcomitans is a fastidious, nonmotile, nonsporing, small Gram-negative rod, $0.4–0.5 \times 1.0–1.5\,\mu m$ in size. Microscopically, the cells may appear coccobacillary, particularly if the colonies are 2–3 days old and have been isolated directly from a solid medium. Longer forms may be seen in smears from older cultures or from cells grown in a glucose-containing liquid medium. It is a facultatively anaerobic bacterium that grows best in an aerobic atmosphere supplemented with 5–10% CO_2, although it will grow anaerobically. The colonies of *A. actinomycetemcomitans* on chocolate or blood agar are generally small, with a diameter of 1–2 mm, translucent, often with a rough, irregular morphology (see section on Pathogenicity), dome-shaped and nonhaemolytic and may be adherent to the agar surface (Figure 20.2).

POPULATION STRUCTURE OF *A. ACTINOMYCETEMCOMITANS*

The immunodominant antigen of *A. actinomycetemcomitans* is a high molecular mass *o*-polysaccharide of the lipopolysaccharide (LPS), and six serotypes (a–f) are currently recognized (Kaplan *et al.*, 2001). The serotype a-specific antigen consists solely of 6-deoxyhexose, 6-deoxy-D-talose which is unique among bacteria, and no genes involved in the synthesis of 6-deoxy-D-talan had previously been reported

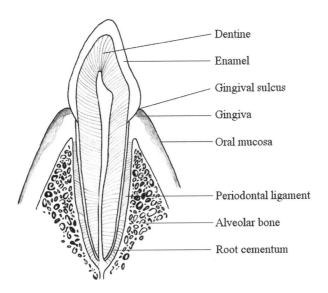

Figure 20.1 A schematic cross-sectional diagram of the human tooth showing the periodontium.

Dentine

Enamel

Gingival sulcus

Gingiva

Oral mucosa

Periodontal ligament

Alveolar bone

Root cementum

Principles and Practice of Clinical Bacteriology Second Edition Editors Stephen H. Gillespie and Peter M. Hawkey

Figure 20.2   Primary isolation of *A. actinomycetemcomitans* on TSBV selective medium showing the characteristic stellate inner structure.

(Shibuya *et al.*, 1991). Investigation of the distribution of *A. actinomycetemcomitans* serotypes and clonotypes in Northern European, Northern African and African-American populations has established that individuals tend to be stably colonized by a single serotype/genotype (Saarela *et al.*, 1999). Serotype b is found at higher frequency both in patients with periodontitis and those with non-oral infections due to this organism than in healthy controls (Lakio *et al.*, 2002). In contrast, serotype c is more often found in periodontally healthy individuals (Lakio *et al.*, 2002). It is proposed that the organism can be grouped into three major phylogenetic lineages consisting of (i) serotype b strains, (ii) serotype c strains and (iii) serotypes a, d, e and f strains (Kaplan *et al.*, 2002a).

Population genetic studies of *A. actinomycetemcomitans* suggest that it is clonal, consisting of genetically distinct subpopulations that correlate with the known serotypes (Poulsen *et al.*, 1994; Haubek *et al.*, 1995). However, it has been proposed that LAP actually represents two different diseases. In Northern European Caucasian populations, LAP is associated with an oligoclonal population of bacteria, none of which had a specific 530-bp deletion from the leukotoxin promoter (see section *A. actinomycetemcomitans toxins*). The simplest interpretation is that LAP in this population is due to *A. actinomycetemcomitans* from the resident microbiota acting as an opportunistic pathogen. More detailed population studies, which included individuals of African origin, revealed the association between LAP in these individuals and a particular serotype b strain isolate with a 530-bp deletion in the leukotoxin promoter. This deletion results in significant enhancement of leukotoxin production (Brogan *et al.*, 1994). DNA fingerprinting and multilocus enzyme electrophoresis suggest that these isolates belong to a single clone, termed JP2 (Haubek *et al.*, 1997), and disease caused by this clone is proposed to be endemic in Morocco (Haubek *et al.*, 2002). However, this interpretation has recently been challenged by Kaplan and coworkers (2002a) and further studies are needed to resolve this issue.

PATHOGENICITY

The disease with which *A. actinomycetemcomitans* is implicated as a pathogen is LAP (Slots and Ting, 1999). This is a curious disease afflicting only the incisors and premolars and causing rapid destruction of the periodontal ligament and alveolar bone which link these teeth to the jawbone (Figure 20.1). As described, this disease is more prevalent in individuals of African origin. Four systems are thought to contribute to the virulence of *A. actinomycetemcomitans*. The first is a novel adhesion system, first discovered in *A. actinomycetemcomitans* and encoded by the tight adherence (tad) locus, more recently coined the 'widespread colonization island' – a modified type IV secretion system now recognized to exist in many bacteria and in *Archaea* (Kachlany *et al.*, 2001; Planet *et al.*, 2003). The second set of virulence determinants are the toxins: leukotoxin and cytolethal distending toxin (CDT). The third system is a collection of exported or cell-surface proteins with immunomodulatory functions and/or ability to stimulate tissue destruction. The final virulence characteristic is the ability to invade host cells by apparently novel mechanisms. It should be noted that in the absence of good animal models, and a paucity of *A. actinomycetemcomitans* isogenic knockouts, these proposed virulence mechanisms are largely speculative.

Adhesion of *A. actinomycetemcomitans*

The rough, adherent colony morphology of *A. actinomycetemcomitans* is associated with a dense collection of surface-attached fibrils (Fine *et al.*, 1999). Using the novel transposon, IS903ϕkan, to generate insertional mutants it was demonstrated that rough colony formation and tight adherence of *A. actinomycetemcomitans* was due to a cluster of novel genes called the *tad* locus (or widespread colonization island) (Planet *et al.*, 2003). The *tad* locus contains 14 genes (Figure 20.3),

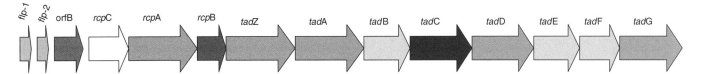

Figure 20.3 The *tad* locus also known as the Widespread Colonization Island of *A. actinomycetemcomitans*. This contains genes encoding for a type IV secretion system that exports the proteins *flp*-1 and 2 onto the cell surface.

12 of which are required for the synthesis, and transport to the cell surface, of the bundled Flp pili which produce the tenacious adhesion of this bacterium. Seven distinct *flp-1* allelic classes have been identified which are highly divergent between classes (Kaplan *et al.*, 2002b). A number of the genes in the *tad* locus appear to be homologues of genes encoding bacterial type II and type IV secretion systems. It is now established that *tad*-related genes are found in both bacterial and Archaeal species including human pathogens such as *Bordetella pertussis*, *H. ducreyi* and *Mycobacterium tuberculosis* (Planet *et al.*, 2003). The *tad* locus explains the tight adherence of *A. actinomycetemcomitans*, and it is postulated that this contributes to virulence. Rough strains show greater colonization of the rat oral cavity than smooth strains (Fine *et al.*, 2001). It has been reported that knockout of the *tadA* gene (an ATPase which energizes secretion of Flp proteins) or the *flp1* and *flp2* genes in *H. ducreyi*, the causative agent of chancroid, has minimal influence on virulence in an animal model (Nika *et al.*, 2002). However, in rats fed *A. actinomycetemcomitans* there was decreased colonization, lowered immune responses and no bone resorption when the bacterium lacked either *tadA* or *flp-1* compared with the response to wild-type organisms (Schreiner *et al.*, 2003). Thus this adhesion system appears to be an important virulence determinant in *A. actinomycetemcomitans*.

Actinobacillus actinomycetemcomitans toxins

Two toxins are produced by this organism. The best studied is an RTX (repeat in toxin) termed leukotoxin (LtxA). The generic RTX toxin operon consists of four genes: *rtx*C, A, B and D in order of transcription. The toxin, RtxA, requires posttranslational modification via lipid acylation, catalysed by RtxC, to be biologically active. Transport proteins, encoded by *rtx*B and *rtx*D, promote movement of the toxin from the cytoplasm to the cell surface (Kachlany *et al.*, 2000; Narayan *et al.*, 2002). Leukotoxin kills lymphoid and myeloid cells from human and some nonhuman primates (Narayan *et al.*, 2002), and the basis of this cell/species selectivity is due to the toxin binding to human cells via the β_2-integrin receptor, LFA-1 (Lally *et al.*, 1997). At low concentrations, LtxA promotes neutrophil degranulation, resulting in the release of the collagenolytic proteinase, matrix metalloproteinase 8 (Claesson *et al.*, 2002), but it also inhibits phagocytosis possibly by increasing intracellular Ca (Taichman *et al.*, 1991). High concentrations of LtxA can cause cell lysis *in vitro* (Karkelian *et al.*, 1998), and it is also a potent inducer of leucocyte apoptosis (Korostoff *et al.*, 1998). There is experimental evidence to support the hypothesis that LtxA directly perturbs mitochondrial function to induce cellular apoptosis (Korostoff *et al.*, 2000).

There are marked strain differences in leukotoxin expression with the highly virulent strain, JP2, producing large amounts of toxin, apparently due to a 528-bp deletion from the promoter (Kolodrubetz *et al.*, 1996).

The *A. actinomycetemcomitans* leukotoxin is unique among the RTX toxins in not being secreted but being associated with the cell membrane (Berthold *et al.*, 1992). It has now been established that adherence to the cell surface is related to possession of a working *tad* locus, as mutations in this locus result in release of the leukotoxin

(Kachlany *et al.*, 2000). Here is an interesting interaction between two of the putative virulence mechanisms of this organism.

A second toxin, CDT, is also produced by *A. actinomycetemcomitans* (Sugai *et al.*, 1998). The toxin is encoded by an operon (*cdt*ABC) which is found in a number of pathogens including *E. coli*, *Shigella* spp., *Campylobacter jejuni* and *H. ducreyi* (Lara-Tejero and Galan, 2002). The toxin inhibits cell-cycle progression in G_2, and intoxicated cells will eventually apoptose. The proposed mechanism is cleavage of chromosomal DNA by CdtB, which has nuclease activity resulting in activation of the DNA damage response and cessation of cell-cycle progression. The entry of CdtB is believed to be aided by the actions of CdtA and CdtC (Lara-Tejero and Galan, 2002). Most *A. actinomycetemcomitans* isolates encode at least one *cdt* gene (Fabris *et al.*, 2002).

The main consequence of the possession of these two toxins is not yet established. The simplest hypothesis is that both toxins are involved in immune evasion (see Henderson and Oyston, 2003 for review of this area). Support for this hypothesis comes from studies showing that CdtB can inhibit human CD4 lymphocyte proliferation and induce apoptosis of this cell population (Shenker *et al.*, 2001).

Exported and Cell-Surface Virulence Factors

There is a growing list of putative virulence factors exported by this bacterium and/or associated with the bacterial surface, and the literature has recently been reviewed (Henderson *et al.*, 2003) and will not be dealt with in detail. Possibly the most interesting of these factors is the molecular chaperone, chaperonin 60. Studies to identify a potent bone-resorbing protein released by cultures of *A. actinomycetemcomitans* concluded that it was chaperonin 60 (Kirby *et al.*, 1995), and immunogold labelling revealed this protein on the bacterial surface (Henderson and Wilson, unpublished). This surface location has been confirmed, and the chaperonin 60 was also reported to be cytotoxic for epithelial cells (Goulhen *et al.*, 1998). Now that the genome of *A. actinomycetemcomitans* has been sequenced, it should be much easier to identify the exported virulence factors of this bacterium using functional proteomics.

Cellular Invasion of *A. actinomycetemcomitans*

A growing number of bacterial pathogens have been found capable of invading nonphagocytic host cells (Wilson, McNab and Henderson, 2002). *Actinobacillus actinomycetemcomitans* – a member of the normal oral microbiota – has been found inside buccal epithelial cells in the majority of subjects surveyed (Rudney, Chen and Sedgewick, 2001), and about a quarter of the clinical isolates of this bacterium examined were invasive (Lepine *et al.*, 1998). Receptors for uptake include the transferrin receptor and the platelet-activating factor receptor (Henderson *et al.*, 2003). There is much that is not understood about cell invasion by *A. actinomycetemcomitans*, and this organism displays several unusual phenomena related to invasion. For example, bacterial replication within host cells is substantially faster than that for the free-living organism (Meyer, Lippmann and Fives-Taylor, 1996). Another unusual aspect of the invasion of epithelial cells by this organism is its rapid exit from the cell after invasion and its ability to move from one cell to another (Meyer, Lippmann and Fives-Taylor, 1996), both of which are dependent on host-cell microtubules (Meyer *et al.*, 1999).

To conclude, *A. actinomycetemcomitans* has an armamentarium of factors and mechanisms which, applying Occam's razor, must contribute to its virulence. Indeed, with so many fascinating evolutionary tricks it is surprising that this organism does not cause more problems. Much work needs to be done to generate panels of isogenic mutants and test them in appropriate animal or human volunteer models to truly identify what behaviour of this organism contributes to virulence and what does not.

EPIDEMIOLOGY

The periodontal diseases are among mankind's most common afflictions. Periodontitis is estimated to have a worldwide prevalence of 10–15%, and it is only because this disease does not kill or maim that it remains so relatively unstudied (Irfan, Dawson and Bissada, 2001). *Actinobacillus actinomycetemcomitans* as well as *Tannerella forsythensis* (previously *Bacteroides forsythus*) and *Porphyromonas gingivalis* was officially designated as aetiological agents of periodontitis in 1996 (American Academy of Periodontology, 1996). *Actinobacillus actinomycetemcomitans* has been shown to be present in up to 95% of periodontal lesions in patients with LAP (Christersson, 1993). However, contrary to these findings other studies have reported a much lower frequency of *A. actinomycetemcomitans* isolation from affected patients (Lopez *et al.*, 1995; Albandar, Brown and Löe, 1997); additionally this organism has been readily isolated from the oral cavity of healthy subjects (Mombelli *et al.*, 1999). These observations suggest that host susceptibility is also an important factor in development of this disease. In a USA study the prevalence of LAP in the adolescent population was shown to be approximately 0.5%, which varied in relation to the subject's racial group, with minority ethnic subjects (African-American and Asian-American) having a prevalence 15 times greater than the Caucasian population (Löe and Brown, 1991). Subjects of African origin have been shown to harbour a particular clone of *A. actinomycetemcomitans* (JP2). The population genetics of this organism has been described in more detail in the section on Population Structure *of A. actinomycetemcomitans*.

An early finding was that LAP is associated with functional abnormalities of neutrophils, such as decreased chemotaxis and phagocytosis, and these abnormalities have been shown to cluster in families (Van Dyke *et al.*, 1985). Mutations in two important neutrophil receptors – the *N*-formylpeptide receptor (Gwinn, Sharma and De Nardin, 1999; Zhang *et al.*, 2003) and the Fc gamma receptor (Fu, Korostoff, Fine and Wilson, 2002) – have recently been suggested to be important in the development of LAP in African-Americans. Familial transfer of *A. actinomycetemcomitans* has been shown to occur between spouses and from parent to child (Greenstein and Lamster, 1997), which may also be an important factor in the high prevalence of disease found in particular families (Stabholz *et al.*, 1998).

CLINICAL FEATURES

Actinobacillus actinomycetemcomitans has been implicated as an important aetiological agent in periodontitis. Periodontitis is an inflammatory disease of the periodontium which occurs in response to bacterial plaque on the adjacent teeth, characterized by gingivitis, destruction of the alveolar bone and periodontal ligament, generally termed loss of attachment. Clinical features of periodontitis may also include oral malodour, gingival bleeding, pain, swelling, gingival recession and tooth mobility and migration. *Actinobacillus*

actinomycetemcomitans is particularly associated with local aggressive periodontitis, in which bone loss is seen around the incisors, molars and the mesial surface of the second molars. The loss of tissues supporting the tooth may be clinically evident due to an apical and lateral migration of the junctional epithelium between the tooth surface and gingival crevice and the development of a periodontal pocket. Progression of the disease can lead to bone destruction along the affected root surface, which presents as either horizontal bone loss with destruction of the entire thickness of the alveolar bone or vertical bone loss where only a portion of the bone is resorbed. If left untreated, periodontitis may lead to acute periodontal abscess and subsequent tooth loss.

Actinobacillus actinomycetemcomitans is also an opportunistic pathogen and has been isolated from patients with actinomycosis, infective endocarditis, osteomyelitis, brain and subcutaneous abscesses and rarely endophthalmitis (Van Winkelhoff and Slots, 1999). Cervicofacial actinomycosis is a chronic infection of the head and neck, most often occurring in males and resulting from infection with *Actinomyces* spp. and often one or more companion organisms including *A. actinomycetemcomitans*, *H. aphrophilus*, *Eikenella corrodens*, *Prevotella* spp. and *Fusobacterium* spp. Early clinical signs include a small hard swelling in the mouth, neck or jaw, followed by the appearance of many small, communicating abscesses, with surrounding granulation and sinuses which discharge sulphur-like granules. Infections may spread to the brain, nearby bones and soft tissues, with occasional pulmonary and abdominal infections reported. Actinomycosis may also affect the tonsils, tongue, cheek or lips after trauma at the affected site.

Actinobacillus actinomycetemcomitans is a rare but important pathogen associated with infective endocarditis and has been reported in patients with abnormal native heart valves and after the implantation of prosthetic heart valves. This bacterium is included in the HACEK (*H. aphrophilus*, *A. actinomycetemcomitans*, *Cardiobacterium hominis*, *E. corrodens* and *Kingella kingae*) group of organisms which are estimated to cause approximately 3% of community-acquired cases of infective endocarditis (Das *et al.*, 1997). As oral bacteria gain entry to the circulatory system by crossing the gingiva, poor oral health is an important risk factor of infective endocarditis. Many dental procedures can also cause oral bacteria to enter the bloodstream, although the incidence of infective endocarditis following dental procedures in at risk patients is extremely low. Clinical features of infective endocarditis may be nonspecific and vary considerably. If not correctly diagnosed, infective endocarditis can be a common cause of pyrexia of unknown origin. General symptoms may include, fever, malaise, weight loss, splinter haemorrhages under the proximal nails, finger clubbing, splenomegaly, anaemia, dyspnoea and cardiac murmur.

LABORATORY DIAGNOSIS

Specimens

Suitable samples for the isolation of *A. actinomycetemcomitans* include dental plaque, tissue biopsy, aspirates and blood samples. Clinical specimens collected from sites within the oral cavity need to be taken carefully to avoid contamination with the normal oral microflora. The use of swabs should therefore be avoided whenever possible, due to the likelihood of contamination with saliva. A needle and syringe can be used for sampling abscesses and lesions. Samples of subgingival plaque collected from patients with periodontitis should be collected by the use of a sterile curette or by paper points after the removal of supragingival plaque. Once the plaque has been sampled it should be immediately placed into a pre-reduced transport medium such as reduced transport fluid (Syed and Loesche, 1972) and taken to the laboratory for processing within 30 min of collection. Infective endocarditis can be diagnosed by the presence of positive blood cultures as routinely used in clinical microbiology. The detection of growth in the blood cultures is generally present within 3–4 days, although blood culture bottles from patients with suspected infective endocarditis should be incubated for a total of 14 days.

Growth Requirement

Clinical samples should be plated onto blood agar or chocolate blood agar supplemented with 5% defibrinated horse blood and should be incubated in an aerobic atmosphere supplemented with 5–10% CO_2. *Actinobacillus actinomycetemcomitans* will also grow anaerobically on fastidious anaerobe agar. Faint colonies may be visible after 24 h of incubation. However, 48–72-h incubation is required for colonies of 1–3 mm to appear. The colonies are generally round, convex and colourless in appearance; they may have central wrinkling and be adherent to the agar surface. Colonies incubated for 4–7 days on a serum-containing medium will develop star-like structures centrally, which can be visualized with a plate microscope. This rough star-like morphology may be lost on subculturing producing smooth isolates which are less adherent to the agar surface. A range of selective media can also be used for the isolation of *A. actinomycetemcomitans* directly from clinical samples. Possibly, the most commonly used is Tryptic Soy-serum-Bacitracin-Vancomycin agar, which contains 10% serum, 75 mg/l of bacitracin and 5 mg/l of vancomycin; the presence of these antibiotics suppresses the growth of Gram-positive bacteria, and the absence of blood eliminates the X and V factor requiring *Haemophilus* species and allows the catalase test to be carried out directly on the colony. Finally the presence of serum allows for the characteristic star-like inner structure colony appearance.

Identification

Table 20.1 summarizes the characteristics of *A. actinomycetemcomitans* used for identification purposes. Adherent colonies with a stellate appearance that are catalase positive and reduce nitrates to nitrite are characteristic of *A. actinomycetemcomitans*. This bacterium ferments glucose, galactose and maltose but does not ferment lactose, sucrose, trehalose, salicin or sorbitol. It produces negative reactions for urease activity, indole production, aesculin hydrolysis and *o*-nitrophenyl-β-D-galactopyranoside. There are several commercial kits available that are able to give a positive identification for *A. actinomycetemcomitans*. Confirmation may be achieved by PCR-based methods using the bacterial small-subunit 16S rRNA gene (16S rDNA). The 16S rDNA contains regions which are highly conserved in bacteria as well as species-specific hypervariable regions. Primers designed within the 16S rDNA have been used to amplify these hypervariable regions found in *A. actinomycetemcomitans* with the production of amplicons of known sizes.

These methods have been shown to be simple to perform, rapid and reliable and suitable for the confirmation of cultured *A. actinomycetemcomitans* or direct identification of this bacterium in clinical samples. More recently a multiplex PCR approach has been used, in which *A. actinomycetemcomitans*-specific primers have been combined with primers specific to other periodontal pathogens for use in a single PCR reaction, allowing the simultaneous detection of a number of periodontal pathogens in a single-patient sample. PCR techniques can also be used for the identification of the JP2 strain by using primers to amplify sequences in the leukotoxin promoter region to identify the deletion found in these JP2 strains.

Antimicrobial Susceptibility

Currently there are no specific recommendations for the susceptibility testing of *A. actinomycetemcomitans* by the British Society for Antimicrobial Chemotherapy. However, several studies have tested this bacterium using modifications of the methods recommended for other fastidious bacteria, using either Muller-Hinton medium, *Haemophilus* test medium or Wilkins-Chalgren medium supplemented with 5% (v/v) defibrinated horse blood. Antibiotic susceptibility has been performed by disk diffusion, E-test and the agar dilution method (in which the antibiotic is incorporated into the medium), with incubation at 37 °C in 5–10% CO_2 for 24–48 h. Isolates of *A. actinomycetemcomitans* are generally susceptible to ciprofloxacin. However, resistance to other antibiotics has been recorded including penicillin (range 10–94%), amoxicillin (0–25%), erythromycin (4–10%) and tetracycline (0–4%) (Slots *et al.*, 1980; Eick, Pfister and Straube, 1999; Madinier *et al.*, 1999; Muller *et al.*, 2002).

TREATMENT AND MANAGEMENT

Periodontal therapy aims to reduce the bacterial load present on the root surface; this may be successfully achieved in the majority of cases by mechanical techniques with or without an antimicrobial adjunct. A standard cycle of periodontal therapy consists of oral hygiene instructions and mechanical debridement of the tooth and root surface. Oral hygiene instruction to control plaque is an essential part of a treatment plan for patients with periodontal diseases. By determining where the majority of the plaque accumulates in a subject's dentition, plaque control can be targeted at problem areas. The removal of supragingival and subgingival dental plaque, calculus from the tooth and root surfaces (scaling) and the removal of surface layers of the root surface (root planning) are normally required to assist healing. In some cases, surgical treatment will also be required. There should be an evaluation of the patient with a review and if necessary reinforcement of oral hygiene. Antimicrobial agents and antibiotics may also be used in treatment. The use of a 0.2% chlorhexidine rinse for

Table 20.1 Identification of *Actinobacillus actinomycetemcomitans*

Colony morphology	Microscopic appearance	Catalase	ONPG	Indole	Urease	Nitrate reduction	Aesculin hydrolysis	Acid from							
								Glucose	Galactose	Maltose	Lactose	Sucrose	Trehalose	Salicin	Sorbitol
Adherent to agar; stellate	Small Gram-negative rods; coccobacilli	+	–	–	–	+	–	+	+	+	–	–	–	–	–

+, positive reaction; –, negative reaction.

2 weeks posttreatment is often recommended to assist with plaque control. Tetracyclines are recommended for refractory periodontal diseases in the UK. However, combination antibiotic therapies are frequently used including metronidazole–amoxycillin and metronidazole–ciprofloxacin.

A proportion of individuals may not respond to the above treatment regimes, and to prevent loss of dentition, a range of therapeutic options are under development. These include the use of bone replacement materials with or without associated bone growth factors. Awaiting on the wings is the use of biologicals such as anti-cytokine antibodies or cytokine-soluble receptors to block the effects of cytokines such as interlukin-1 or tumour necrosis factor. Another treatment modality which is probably to be tested soon is osteoprotegerin or RANK-Fc a natural inhibitor of the cells (osteoclasts) that promote bone resorption (Henderson and Nair, 2003).

REFERENCES

Albandar, J.M., Brown, L.J. and Löe, H. (1997) Putative periodontal pathogens in subgingival plaque of young adults with and without early-onset periodontitis. *Journal of Periodontology* **68**, 973–981.

American Academy of Periodontology. (1996) Consensus report: section on epidemiology. *Annals of Periodontology* **1**, 216–218.

Armitage, G.C. (1999) Development of a classification system for periodontal diseases and conditions. *Annals of Periodontology* **4**, 1–6.

Berthold, P., Forti, D., Kieba, I.R. *et al.* (1992) Electron immunocytochemical localization of *Actinobacillus actinomycetemcomitans* leukotoxin. *Oral Microbiology and Immunology* **7**, 24–27.

Brogan, J.M., Lally, E.T., Poulsen, K. *et al.* (1994) Regulation of *Actinobacillus actinomycetemcomitans* leukotoxin expression: analysis of the promoter regions of leukotoxic and minimally leukotoxic strains. *Infection and Immunity* **62**, 501–508.

Christersson, L.A. (1993) *Actinobacillus actinomycetemcomitans* and localized juvenile periodontitis. Clinical, microbiologic and histologic studies. *Swedish Dental Journal* **90**, 1–46.

Claesson, R., Johansson, A., Belibasakis, G. *et al.* (2002) Release and activation of matrix metalloproteinase 8 from human neutrophils triggered by the leukotoxin of *Actinobacillus actinomycetemcomitans*. *Journal of Periodontology Research* **37**, 353–359.

Das, M., Badley, A.D., Cockerill, F.R. *et al.* (1997) Infective endocarditis caused by HACEK microorganisms. *Annual Review of Medicine* **48**, 25–33.

Eick, S., Pfister, W. and Straube, E. (1999) Antimicrobial susceptibility of anaerobic and capnophilic bacteria isolated from odontogenic abscesses and rapidly progressive periodontitis. *International Journal of Antimicrobial Agents* **12**, 41–46.

Fabris, A.S., DiRienzo, J.M., Wikstrom, M. and Mayer, M.P.A. (2002) Detection of cytolethal distending toxin activity and *cdt* genes in *Actinobacillus actinomycetemcomitans* isolates from geographically diverse populations. *Oral Microbiology and Immunology* **17**, 231–238.

Fine, D.H., Furgang, D., Schreiner, H.C. *et al.* (1999) Phenotypic variation in *Actinobacillus actinomycetemcomitans* during laboratory growth: implications for virulence. *Microbiology* **145**, 1335–1347.

Fine, D.H., Goncharoff, P., Schreiner, H. *et al.* (2001) Colonization and persistence of rough and smooth colony variants of *Actinobacillus actinomycetemcomitans* in the mouths of rats. *Archives of Oral Biology* **46**, 1065–1078.

Fu, Y., Korostoff, J.M., Fine, D.H. and Wilson, M.E. (2002) Fc gamma receptor genes as risk markers for localized aggressive periodontitis in African-Americans. *Journal of Periodontology* **73**, 517–523.

Goulhen, F., Hafezi, A., Uitto, V.-J. *et al.* (1998) Subcellular localization and cytotoxic activity of the GROEL-like protein isolated from *Actinobacillus actinomycetemcomitans*. *Infection and Immunity* **66**, 5307–5313.

Greenstein, G. and Lamster, I. (1997) Bacterial transmission in periodontal diseases: a critical review. *Journal of Periodontology* **68**, 421–431.

Gwinn, M.R., Sharma, A. and De Nardin, E. (1999) Single nucleotide polymorphisms of the N-formyl peptide receptor in localized juvenile periodontitis. *Journal of Periodontology* **70**, 1194–1201.

Haubek, D., Poulsen, K., Asikainen, S. and Kilian, M. (1995) Evidence for absence in Northern Europe of especially virulent clonal types of *Actinobacillus actinomycetemcomitans*. *Journal of Clinical Microbiology* **33**, 395–401.

Haubek, D., DiRienzo, J.M., Tinoco, E.M.B. *et al.* (1997) Racial tropism of a highly toxic clone of *Actinobacillus actinomycetemcomitans* associated with juvenile periodontitis. *Journal of Clinical Microbiology* **35**, 3037–3042.

Haubek, D., Ennibi, O.-K., Abdellaoui, L. *et al.* (2002) Attachment loss in Moroccan early onset periodontitis patients and infection with the JP2-type of *Actinobacillus actinomycetemcomitans*. *Journal of Clinical Periodontology* **29**, 657–660.

Heinrich, S. and Pulverer, G. (1959) Zur atiologie und microbiologie des *Actinobacillus actinomycetemcomitans* unter den begleitbakterien des *Actinomyces israeli*. *Zentralblatt fur Bakteriologie, Parasitenkunde, Infektionskrankheiten und Hygiene* **176**, 91–101.

Henderson, B. and Nair, S.P. (2003) Hard labour: bacterial infection of the skeleton. *Trends in Microbiology* **11**, 570–577.

Henderson, B. and Oyston, P.C. (2003) *Bacterial Evasion of Host Immune Responses*, Cambridge University Press, Cambridge.

Henderson, B., Ward, J.M., Nair, S.P. and Wilson, M. (2003) Molecular pathogenicity of the oral opportunistic pathogen *Actinobacillus actinomycetemcomitans*. *Annual Reviews of Microbiology* **57**, 39–55.

Holme, P. (1951) Studies of the aetiology of human actinomycoses. II. Do the "other microbes" of actinomycosis possess virulence? *Acta Pathologica, Microbiologica et Immunologica Scandinavica* **28**, 391–406.

Irfan, U.M., Dawson, D.V. and Bissada, N.F. (2001) Epidemiology of periodontal disease: a review and clinical perspectives. *Journal of the International Academy of Periodontology* **3**, 14–21.

Kachlany, S.C., Fine, D.H. and Figurski, D.H. (2000b) Secretion of RTX leukotoxin by *Actinobacillus actinomycetemcomitans*. *Infection and Immunity* **68**, 6094–6100.

Kachlany, S.C., Planet, P.J., Bhattacharjee, M.K. *et al.* (2000a) Nonspecific adherence by *Actinobacillus actinomycetemcomitans* requires genes widespread in bacteria and archaea. *Journal of Bacteriology* **182**, 6169–6176.

Kachlany, S.C., Planet, P.J., DeSalle, R. *et al.* (2001) Genes for tight adherence of *Actinobacillus actinomycetemcomitans*: from plaque to plague to pond scum. *Trends in Microbiology* **9**, 429–437.

Kaplan, J.B., Kokeguchi, S., Murayama, Y. and Fine, D.H. (2002b) Sequence diversity in the major fimbrial subunit gene (flp-1) of *Actinobacillus actinomycetemcomitans*. *Oral Microbiology and Immunology* **17**, 354–359.

Kaplan, J.B., Perry, M.B., MacLean, L.L. *et al.* (2001) Structural and genetic analysis of O polysaccharide from *Actinobacillus actinomycetemcomitans* serotype f. *Infection and Immunity* **69**, 5375–5384.

Kaplan, J.B., Schreiner, H.C., Furgang, D. and Fine, D.H. (2002a) Population structure and genetic diversity of *Actinobacillus actinomycetemcomitans* strains isolated from localized juvenile periodontitis patients. *Journal of Clinical Microbiology* **40**, 1181–1187.

Karkelian, D., Lear, J.D., Lally, E.T. and Tanaka, J.C. (1998) Characterization of *Actinobacillus actinomycetemcomitans* leukotoxin pore formation in HL60 cells. *Biochimica et Biophysica Acta* **1406**, 175–187.

Kirby, A.C., Meghji, S., Nair, S.P. *et al.* (1995) The potent bone resorbing mediator of *Actinobacillus actinomycetemcomitans* is homologous to the molecular chaperone GroEL. *The Journal of Clinical Investigation* **96**, 1185–1194.

Klinger, R. (1912) Untersuchungen uber menschliche Actinomykose. *Centralblat fur Bakteriologie* **62**, 191–200.

Kolodrubetz, D., Spitznagel, J., Wang, B. *et al.* (1996) *cis* elements and *trans* factors are both important in strain-specific regulation of the leukotoxin gene in *Actinobacillus actinomycetemcomitans*. *Infection and Immunity* **64**, 3451–3460.

Korostoff, J., Wang, J.-F., Kieba, I. *et al.* (1998) *Actinobacillus actinomycetemcomitans* leukotoxin induces apoptosis in HL-60 cells. *Infection and Immunity* **66**, 4474–4483.

Korostoff, J., Yamaguchi, N., Miller, M. *et al.* (2000) Perturbation of mitochondrial structure and function plays a central role in *Actinobacillus actinomycetemcomitans* leukotoxin-induced apoptosis. *Microbial Pathogenesis* **29**, 267–278.

Lakio, L., Kuula, H., Dogan, B. and Asikainen, S. (2002) *Actinobacillus actinomycetemcomitans* proportion of subgingival bacterial flora in relation to its clonal type. *European Journal of Oral Science* **110**, 212–217.

Lally, E.T., Kieba, I.R., Sato, A. *et al.* (1997) RTX toxins recognize a β2 integrin on the surface of human target cells. *The Journal of Biological Chemistry* **272**, 30463–30469.

Lara-Tejero, M. and Galan, J.E. (2002) Cytolethal distending toxin: limited damage as a strategy to modulate cellular functions. *Trends in Microbiology* **10**, 147–152.

Lepine, G., Caudry, S., DiRienzo, J.M. and Ellen, R.P. (1998) Epithelial cell invasion by *Actinobacillus actinomycetemcomitans* strains from restriction fragment-length polymorphism groups associated with juvenile periodontitis or carrier status. *Oral Microbiology and Immunology* **13**, 341–347.

Lieske, R. (1921) *Morphology und Biologies der Strahlenpilze*, Leipzig, GDR: Borntraeger.

Löe, H. and Brown, L.J. (1991) Early onset periodontitis in the United States of America. *Journal of Periodontology* **62**, 608–616.

Lopez, N.J., Mellado, J.C., Giglio, M.S. and Leighton, G.X. (1995) Occurrence of certain bacterial species and morphotypes in juvenile periodontitis in Chile. *Journal of Periodontology* **66**, 559–567.

Madinier, I.M., Fosse, T.B., Hitzig, C. *et al.* (1999) Resistance profile survey of 50 periodontal strains of *Actinobacillus actinomycetemcomitans*. *Journal of Periodontology* **70**, 888–892.

Meyer, D.H., Lippmann, J.E. and Fives-Taylor, P.M. (1996) Invasion of epithelial cells by *Actinobacillus actinomycetemcomitans*: a dynamic, multistep process. *Infection and Immunity* **64**, 2988–2997.

Meyer, D.H., Rose, J.E., Lippmann, J.E. and Fives-Taylor, P.M. (1999) Microtubules are associated with intracellular movement and spread of the periodontopathogen *Actinobacillus actinomycetemcomitans*. *Infection and Immunity* **67**, 6518–6525.

Mombelli, A., Gmur, R., Lang, N.P. *et al.* (1999) *Actinobacillus actinomycetemcomitans* in Chinese adults. Serotype distribution and analysis of the leukotoxin gene promoter locus. *Journal of Clinical Periodontology* **26**, 505–510.

Muller, H.P., Holderrieth, S., Burkhardt, U. and Hoffler, U. (2002) In vitro antimicrobial susceptibility of oral strains of *Actinobacillus actinomycetemcomitans* to seven antibiotics. *Journal of Clinical Periodontology* **29**, 736–742.

Najar, F.Z. (2002) Sequence and analysis of *Actinobacillus actinomycetemcomitans*, University of Oklahoma, PhD Thesis.

Narayan, S.M., Nagaraja, T.G., Chengappa, M.M. and Stewart, G.C. (2002) Leukotoxins of Gram-negative bacteria. *Veterinary Microbiology* **84**, 337–356.

Nika, J.R., Latimer, J.L., Ward, C.K. *et al.* (2002) *Haemophilus ducreyi* requires the flp gene cluster for microcolony formation in vitro. *Infect Immun* **70**, 2965–2975.

Planet, P.J., Kachlany, S.C., Fine, D.H. *et al.* (2003) The widespread colonisation island of *Actinobacillus actinomycetemcomitans*. *Nature Genetics* **34**, 193–198.

Poulsen, K., Theilade, E., Lally, E.T. *et al.* (1994) Population structure of *Actinobacillus actinomycetemcomitans*: a framework for studies of disease-associated properties. *Microbiology* **140**, 2049–2060.

Rudney, J.D., Chen, R. and Sedgewick, G.J. (2001) Intracellular *Actinobacillus actinomycetemcomitans* and *Porphyromonas gingivalis* in buccal epithelial cells collected from human subjects. *Infection and Immunity* **69**, 700–707.

Saarela, M., Dogan, B., Alaluusua, S. and Asikainen, S. (1999) Persistence of oral colonization by the same *Actinobacillus actinomycetemcomitans* strain(s). *Journal of Periodontology* **70**, 504–509.

Schreiner, H.C., Sinatra, K., Kaplan, J.B. *et al.* (2003) Tight-adherence genes of *Actinobacillus actinomycetemcomitans* are required for virulence in a rat model. *Proceedings of the National Academy of Sciences of the United States of America* **100**, 7295–7300.

Shenker, B.J., Hoffmaster, R.H., Zekavat, A. *et al.* (2001) Induction of apoptosis in human T cells by *Actinobacillus actinomycetemcomitans* cytolethal distending toxin is a consequence of G_2 arrest of the cell cycle. *Journal of Immunology* **167**, 435–441.

Shibuya, N., Amano, K., Azuma, J. *et al.* (1991) 6-Deoxy-D-talan and 6-deoxy-L-talan. Novel serotype-specific polysaccharide antigens from *Actinobacillus actinomycetemcomitans*. *The Journal of Biological Chemistry* **266**, 16318–16323.

Slots, J. and Ting, M. (1999) *Actinobacillus actinomycetemcomitans* and *Porphyromonas gingivalis* in human periodontal disease: occurrence and treatment. *Periodontology 2000* **20**, 82–121.

Slots, J., Evans, R.T., Lobbins, P.M. and Genco, R.J. (1980) In vitro antimicrobial susceptibility of *Actinobacillus actinomycetemcomitans*. *Antimicrobial Agents and Chemotherapy* **18**, 9–12.

Stabholz, A., Mann, J., Agmon, S. and Soskolne, W.A. (1998) The description of a unique population with a very high prevalence of localized juvenile periodontitis. *Journal of Clinical Periodontology* **25**, 872–878.

Sugai, M., Kawamoto, T., Peres, S.Y. *et al.* (1998) The cell cycle-specific growth-inhibitory factor produced by *Actinobacillus actinomycetemcomitans* is a cytolethal distending toxin. *Infection and Immunity* **66**, 5008–5019.

Syed, S.A. and Loesche, W.J. (1972) Survival of human dental plaque flora in various transport media. *Applied Microbiology* **24**, 638–644.

Taichman, N.S., Iwase, M., Lally, E.T. *et al.* (1991) Early changes in cytosolic calcium and membrane potential induced by *Actinobacillus actinomycetemcomitans* leukotoxin in susceptible and resistant cells. *Journal of Immunology* **147**, 3587–3594.

Topley, W.W.C. and Wilson, G.S. (1929) *The Principles of Bacteriology and Immunity*, Edward Arnold, London.

US Public Health Service. (2000) Oral Health in America: A Report of the Surgeon General.

Van Dyke, T.E., Schweinebraten, M., Cianciola, L.J. *et al.* (1985) Neutrophil chemotaxis in families with localized juvenile periodontitis. *Journal of Periodontal Research* **20**, 503–514.

Van Winkelhoff, A.J. and Slots, J. (1999) *Actinobacillus actinomycetemcomitans* and *Porphyromonas gingivalis* in nonoral infections. *Periodontology 2000* **20**, 122–135.

Wilson, M., McNab, R. and Henderson, B. (2002) *Bacterial Disease Mechanisms: An Introduction to Cellular Microbiology*, Cambridge University Press, Cambridge.

Zambon, J.J. (1985) *Actinobacillus actinomycetemcomitans* in human periodontal disease. *Journal of Clinical Periodontology* **12**, 1–20.

Zhang, Y., Syed, R., Uygar, C. *et al.* (2003) Evaluation of human leukocyte N-formylpeptide receptor (FPR1) SNPs in aggressive periodontitis patients. *Genes and Immunity* **4**, 22–29.

21

Francisella tularensis

Petra C. F. Oyston

Defence Science and Technology Laboratory, Salisbury, Wiltshire, UK

INTRODUCTION

Francisella tularensis is the causative agent of tularemia. It is considered a potential bioweapon because of its high infectivity by the aerosol route. There is no vaccine currently available, and antibiotic therapy may need to be protracted. The organism is highly fastidious, requiring rich media for isolation and specialised reagents for positive identification.

DESCRIPTION OF THE ORGANISM

Francisella tularensis was first isolated from ground squirrels dying from a plague-like disease in California in 1911 (McCoy and Chapin, 1912). It is a small (0.2–$0.5\,\mu m \times 0.7$–$1.0\,\mu m$) Gram-negative coccobacillus, non-motile and obligately aerobic. It produces catalase weakly, is oxidase negative and H_2S positive. It produces acid but not gas from glucose, maltose and mannose, although catabolism of carbohydrates is slow (Sjostedt, 2002).

The taxonomy of *F. tularensis* is complex, the species being differentiated into subspecies on the basis of differences in virulence, geographic distribution and natural reservoir (Sjostedt, 2002). The most virulent subspecies, *F. tularensis* subspecies *tularensis* (also referred to as type A), is usually isolated in North America. The less virulent *F. tularensis* subspecies *holarctica* (also called type B) is distributed more widely (Olsufjev and Meshcheryakova, 1983; Eigelsbach and McGann, 1984). Two other subspecies, *novicida* and *mediasiatica*, have also been described but rarely cause disease in humans (Eigelsbach and McGann, 1984; Hollis *et al.*, 1989; Forsman, Sandstrom and Sjostedt, 1994).

Francisella tularensis requires enriched medium for growth such as cysteine glucose blood agar. Centers for Disease Control and Prevention (CDC) guidelines recommend the use of enriched chocolate agar (cysteine heart agar supplemented with 9% heated sheep red blood cells, CHAB) if the pathogen is suspected (Anonymous, 2001). The appearance of individual colonies may require 2–4 days of incubation at $37\,°C$. On CHAB, colonies are 2–4 mm in size, greenish-white, round, smooth and slightly mucoid, whilst on media containing whole blood there is usually a small zone of α-haemolysis surrounding colonies. *Francisella tularensis* grows poorly at $28\,°C$, and this can be exploited to distinguish *F. tularensis* from *Yersinia pestis*, *Francisella philomiragia* and *F. tularensis* subspecies *novicida* – all of which grow well at $28\,°C$. It can be distinguished from other Gram-negative bacteria except *Legionella* by cysteine dependence. Unlike *F. tularensis*, *Legionella* are motile and do not produce H_2O. *Francisella tularensis* does not grow well in liquid media, even when the medium is supplemented with cysteine, and requires a large inoculum to obtain visible growth within 24 h.

PATHOGENICITY

Francisella tularensis is able to infect a wide range of hosts, including humans. *Francisella tularensis* is one of the most highly infectious bacteria known, with an infectious dose in humans of only ten bacteria (McCrumb, 1961). Very few classical virulence factors have been identified for this pathogen. It does not produce secreted exotoxins, and the lipopolysaccharide (LPS) of *F. tularensis* does not exhibit the properties of a classical endotoxin (Sandstrom *et al.*, 1992). The capsule is essential for serum resistance but is not required for survival following phagocytosis (Sandstrom, Lofgren and Tarnvik, 1988).

EPIDEMIOLOGY

The pathogen has only been isolated in the northern hemisphere and most frequently in Scandinavia, northern America, Japan and Russia (Boyce, 1975; Ohara *et al.*, 1991; Berdal *et al.*, 1996; Stewart, 1996; Tarnvik, Sandstrom and Sjostedt, 1996), but has never been isolated in the United Kingdom. It circulates in populations of small rodents-endemic regions, and outbreaks in human populations frequently mirror outbreaks of disease occurring in wild animals (Tarnvik, Sandstrom and Sjostedt, 1996). *Francisella tularensis* is considered to be an intracellular pathogen *in vivo* (Fortier *et al.*, 1994). It is capable of multiplying in a range of cell types, but its primary target appears to be the macrophage. *Francisella tularensis* has also been shown to be able to replicate to high numbers inside amoebae, which may provide an environmental reservoir for the bacterium (Abd *et al.*, 2003).

CLINICAL FEATURES

Tularemia in humans occurs in several forms, depending on the route of infection. Tularemia is a severely debilitating and even fatal disease, especially when caused by *F. tularensis* subspecies *tularensis*, and many cases of disease caused by lower-virulence strains are undiagnosed. The most common form of the disease is ulceroglandular tularemia which usually occurs as a consequence of a bite from an arthropod vector. Cases of ulceroglandular tularemia in hunters and trappers have been known to arise following the handling of infected meat, with infection via cuts or abrasions. After an incubation period of 3–6 days, the patient experiences the sudden onset of flu-like symptoms, especially chills, fever, headache and generalised aches (Evans *et al.*, 1985; Ohara *et al.*, 1991). An ulcer forms at the site of infection, usually on the lower limbs in the cases of tick-borne disease. This ulcer may persist for several months. Disease with

Principles and Practice of Clinical Bacteriology Second Edition   Editors Stephen H. Gillespie and Peter M. Hawkey

similar symptoms but without the appearance of an ulcer is termed glandular tularemia. Bacteria disseminate from the site of infection via the lymphatic system to regional lymph nodes, which then become enlarged to resemble the bubo associated with bubonic plague. Subsequent dissemination spreads the bacteria to other tissues such as the spleen, liver, lungs, kidneys, intestine, central nervous system and skeletal muscles. This bacteremic phase of the infection is transient and occurs relatively early in the infection process. Recovery from the disease can be protracted, taking many months, but the ulcero-glandular form of tularemia is fatal in fewer than 3% of cases. If the initial site of infection is the conjunctiva, oculoglandular tularemia may develop. Ulcers and nodules develop on the conjunctiva, and without treatment, the infection spreads to the local lymph nodes, similar to typical ulceroglandular disease.

Typhoidal tularemia is an acute disease with septicaemia without lymphadenopathy or the appearance of an ulcer. In addition to the generalised symptoms seen in ulceroglandular tularemia, the patient may become delirious and shock may develop. When caused by the highly virulent *F. tularensis* subspecies *tularensis*, typhoidal tularemia is associated with a mortality rate of between 30 and 60% (Evans *et al.*, 1985; Gill and Cunha, 1997).

The ingestion of infected foodstuffs or of bacteria in drinking water results in tularemia infection. Either oropharyngeal or gastrointestinal tularemia develops, depending on the site of colonisation of host tissues. Oropharyngeal tularemia presents as a painful sore throat with enlargement of the tonsils and the formation of a yellow-white pseudomembrane, often accompanied by swollen cervical lymph nodes. Gastrointestinal tularemia produces a range of diseases, from a mild but persistent diarrhea to an acute fatal disease with extensive ulceration of the bowel, depending on the size of the infecting dose.

Pneumonic tularemia is the most acute form of disease. It arises following the inhalation of bacteria, although pneumonic disease occurs as a secondary complication of the other forms of tularemia (Gill and Cunha, 1997). *Francisella tularensis* is highly infectious by the aerosol route, with fewer than ten bacteria being required to cause infection. Following inhalation of the bacteria, there is an incubation period of 3–5 days. Subsequent diagnosis of pneumonic tularemia is difficult because of the symptoms being variable, but patients usually present with an acute flu-like illness with or without clinical pneumonia or pneumonitis. Symptoms include fever, non-productive cough, pharyngitis, pleuritic chest pain and hilar lymphadenopathy. Severe pleuropneumonitis develops with pleural effusions, but chest signs may be absent in some patients. Pneumonic tularemia is rare and usually arises in farmers who have been handling hay which has previously been the site of residence of infected rodents (Stewart, 1996).

LABORATORY DIAGNOSIS

Francisella tularensis can be isolated from ulcer scrapings, lymph node biopsies or sputum. The organism is rarely cultured directly from blood, although this is becoming feasible with the development of sensitive blood culture systems. In inhalational tularemia, the organism can be detected in sputum, pharyngeal washings and pleural fluid.

Because of the difficulty in culturing *F. tularensis*, most cases of tularemia are diagnosed on the basis of clinical picture and/or serology. Serological tests for the diagnosis of *F. tularensis* infection are attractive because diagnostic work involving culture procedures carries a risk of infection for non-vaccinated laboratory staff. The diagnosis of human cases of tularemia is usually confirmed by the demonstration of an antibody response to *F. tularensis*, which occurs about 2 weeks after the onset of the disease (Koskela and Salminen, 1985). The detection of serum antibodies is most frequently achieved by agglutination or using an enzyme-linked immunosorbent assay (ELISA) (Carlsson *et al.*, 1979; Koskela and Salminen, 1985;

Syrjala *et al.*, 1986). A latex agglutination test commercially available from BBL (Becton Dickinson, Franklin Lakes, NJ, USA) has been used to identify individuals with antibodies to *F. tularensis*: reactions at greater than 1:20 dilutions are considered specific and significant. Commercially available antigens are also used with standard tube agglutination tests. A fourfold increase during illness or a single titre of 1:160 or greater is considered diagnostic (Brown *et al.*, 1980). However, while serological tests are frequently used for diagnosis, strains of *F. tularensis* have occasionally been isolated which fail to induce antibodies that agglutinate commercially available *F. tularensis* antigens (Clarridge *et al.*, 1996).

ELISA-based tests can be used to detect the bacteria in clinical samples. For example, a capture ELISA using monoclonal antibodies against *F. tularensis* LPS recognised all strains of *F. tularensis* tested other than those of subspecies *novicida*, with no cross-reactivity with other bacterial species tested (Grunow *et al.*, 2000). An immunochromatographic hand-held assay has also been developed based on a polyclonal and a monoclonal antibody to LPS. This assay is principally designed for field use, being compact, easy to use and gives results in 15 min; however, the relatively low sensitivity means that a negative result does not exclude tularemia (Grunow *et al.*, 2000).

A range of polymerase chain reaction (PCR)-based assays have been reported for the diagnosis of tularemia. These PCR assays use primers directed against genes encoding outer membrane proteins such as *fopA* (Fulop *et al.*, 1996) or the 17 kDa outer membrane lipoprotein (Junhui *et al.*, 1996; Grunow *et al.*, 2000). In a study with lesion swab samples from 40 human cases of ulceroglandular tularemia, a PCR assay was positive for 73% of the samples, whereas bacteria were cultured from only 25% of the samples (Sjostedt *et al.*, 1997). PCR-based tests have the advantage that they are safer than culturing bacteria. Improved methods for isolating and processing tissue or blood samples and for the transport of samples to the laboratory might further improve the sensitivity and utility of the PCR test. Indeed, a specially formulated filter paper designed for sample collection has been proposed for the rapid preparation of template DNA from clinical or from field collected tick vectors. When linked with a TaqMan 5′ nuclease assay or a PCR-enzyme immunoassay, sensitivities of <100 cfu were reported (Higgins *et al.*, 2000).

Molecular methods are also available to discriminate between *Francisella* strains. Long random-sequence oligonucleotide primers and primers specific for repetitive extragenic palindromic (REP) sequences and enterobacterial repetitive intragenic consensus (ERIC) sequences were evaluated. REP-PCR was suitable for identifying strains of *F. tularensis* subspecies *novicida*, but not for separating strains of subspecies *holarctica* and *tularensis*, as the patterns obtained were too similar (Johansson *et al.*, 2000). A one-base difference in the 16S rRNA sequences of subspecies *tularensis* and *holarctica* has been demonstrated, allowing the development of a PCR capable of differentiating the two subspecies (Forsman, Sandstrom and Sjostedt, 1994). PCR amplification of tandem-repeat regions may be highly discriminatory and a useful tool in strain typing (Farlow *et al.*, 2001; Johansson *et al.*, 2001). A specific PCR has been developed which produces amplicons of different lengths (target unknown), which has been used in combination with the 17-kDa lipoprotein PCR and can distinguish *F. tularensis* subspecies *holarctica* from strains of other *F. tularensis* subspecies (Johansson *et al.*, 2000).

MANAGEMENT

The aminoglycosides streptomycin and gentamicin are bactericidal against *F. tularensis* and are currently the drugs of choice for the treatment of tularemia infections. Current PHLS guidelines Lutzomial recommend gentamicin for 10 days. In children, the recommended dose is for

10 days or streptomycin intramuscularly twice daily for 2 weeks. These antimicrobials are not used routinely as monotherapy for the treatment of acute febrile illnesses including pneumonia. Thus, if tularemia is suspected, an aminoglycoside should be added to an appropriate antibiotic regimen and not used as a single agent. Other antibiotics can then be stopped once *F. tularensis* has been confirmed and the antibiotic sensitivity of the organism determined.

The fluoroquinolones have been shown to have good bactericidal activity against *F. tularensis* and are attractive because of their low toxicity and potential for oral therapy. In the UK national guidelines, a regimen of ciprofloxacin intravenously every 12 h, changing to the oral route for 14 days, is suggested for non-pregnant adults. In a recent human epidemic outbreak in Spain, ciprofloxacin was the antibiotic with the lowest rate of therapeutic failure and with the fewest side effects. Ciprofloxacin was also shown to be suitable for treating tularemia in children and in a case of relapse following gentamicin therapy. Tetracycline and chloramphenicol are bacteriostatic against *F. tularensis*, and these antibiotics are associated with high relapse rates, although the chances of relapse are reduced with longer treatment regimens. Owing to the poor penetration of aminoglycosides into CSF, chloramphenicol may be added to streptomycin in cases of meningitis complications. The pathogen is resistant to β-lactam antibiotics, and macrolides are not recommended.

PREVENTION AND CONTROL

Attenuated strains of *F. tularensis* have been used to immunise millions of people against tularemia, especially in the former Soviet Union. The most widely used vaccine strain in recent times is the live vaccine strain (LVS), which was derived from a virulent strain of *F. tularensis* subspecies *holarctica*. In clinical trials in humans, vaccination with LVS by the intradermal route provided either complete protection against a subsequent challenge with fully virulent *F. tularensis* or at least reduced the severity of disease (Saslaw *et al.*, 1961a,b). The LVS vaccine has also been shown to be effective when given by the aerogenic and oral routes of delivery, but the vaccine is usually administered by scarification. However, the precise history of the LVS strain is uncertain and the nature of the attenuation is not currently known. In addition, this strain has been found to be unusual compared with fully virulent *F. tularensis*. For example, the expansion of Vγ9Vδ2 T cells seen in patients with tularemia is not observed following vaccination with LVS (Poquet *et al.*, 1998). Owing to residual virulence in the mouse model and a lack of understanding of the basis of attenuation or the mechanism of induction of protection, this strain is not currently licensed for human use in the United States or Europe.

REFERENCES

Abd, H., Johansson, T., Golovliov, I. *et al.* (2003) Survival and growth of *Francisella tularensis* in *Acanthamoeba castellanii*. *Applied and Environmental Microbiology*, **69**, 600–606.

Anonymous (2001) *Basic Laboratory Protocols for the Presumptive Identification of Francisella tularensis*. Centers for Disease Control and Prevention, Atlanta, GA.

Berdal, B. P., Mehl, R., Meidell, N. K. *et al.* (1996) Field investigations of tularemia in Norway. *FEMS Immunology and Medical Microbiology*, **13**, 191–195.

Boyce, J. M. (1975) Recent trends in the epidemiology of tularemia in the United States. *The Journal of Infectious Diseases*, **131**, 197–199.

Brown, S. L., McKinney, F. T., Klein, G. C. *et al.* (1980) Evaluation of a safranin-O-stained antigen microagglutination test for *Francisella tularensis* antibodies. *Journal of Clinical Microbiology*, **11**, 146–148.

Carlsson, H. E., Lindberg, A. A., Lindberg, G. *et al.* (1979) Enzyme-linked immunosorbents assay for immunological diagnosis of human tularemia. *Journal of Clinical Microbiology*, **10**, 615–621.

Clarridge, J. E., Raich, T. J., Sjostedt, A. *et al.* (1996) Characterization of two unusual clinically significant *Francisella* strains. *Journal of Clinical Microbiology*, **34**, 1995–2000.

Eigelsbach, H. T., McGann, V. G. (1984) Genus *Francisella* Dorofe'ev 1947, in *Bergey's Manual of Systematic Bacteriology, Vol. 1*, Kreig, N. R., Holt, J. G., eds. Williams & Wilkins, Baltimore, MD, pp. 394–399.

Evans, M. E., Gregory, D. W., Schaffner, W. *et al.* (1985) Tularemia: a 30 year experience with 88 cases. *Medicine (Baltimore)*, **64**, 251–269.

Farlow, J., Smith, K. L., Wong, J. *et al.* (2001) *Francisella tularensis* strain typing using multiple-locus, variable-number tandem repeat analysis. *Journal of Clinical Microbiology*, **39**, 3186–3192.

Forsman, M., Sandstrom, G., Sjostedt, A. (1994) Analysis of 16S ribosomal DNA sequences of *Francisella* strains and utilization for determination of the phylogeny of the genus and for identification of strains by PCR. *International Journal of Systematic Bacteriology*, **44**, 38–46.

Fortier, A. H., Green, S. J., Polsinelli, T. *et al.* (1994) Life and death of an intracellular pathogen: *Francisella tularensis* and the macrophage. *Immunology and Serology*, **60**, 349–361.

Fulop, M., Leslie, D. L., Titball, R. (1996) A rapid, highly sensitive method for the detection of *Francisella tularensis* in clinical samples using the polymerase chain reaction. *The American Journal of Tropical Medicine and Hygiene*, **54**, 364–366.

Gill, V., Cunha, B. A. (1997) Tularemia pneumonia. *Seminars in Respiratory Infections*, **12**, 61–67.

Grunow, R., Splettstoesser, W., McDonald, S. *et al.* (2000) Detection of *Francisella tularensis* in biological specimens using a capture enzyme-linked immunosorbent assay, an immunochromatographic handheld assay, and a PCR. *Clinical and Diagnostic Laboratory Immunology*, **7**, 86–90.

Higgins, J. A., Hubalek, Z., Halouzka, J. *et al.* (2000) Detection of *Francisella tularensis* in infected mammals and vectors using a probe-based polymerase chain reaction. *The American Journal of Tropical Medicine*, **62**, 310–318.

Hollis, D. G., Weaver, R. E., Steigerwalt, A. G. *et al.* (1989) *Francisella philomiragia* comb. nov. (formerly *Yersinia philomiragia*) and *Francisella tularensis* biogroup *novicida* (formerly *Francisella novicida*) associated with human disease. *Journal of Clinical Microbiology*, **27**, 1601–1608.

Johansson, A., Ibrahim, A., Goransson, I. *et al.* (2000) Evaluation of PCR-based methods for discrimination of *Francisella* species and subspecies and development of a specific PCR that distinguishes the two major subspecies of *Francisella tularensis*. *Journal of Clinical Microbiology*, **38**, 4180–4185.

Johansson, A., Goransson, I., Larsson, E. *et al.* (2001) Extensive allelic variation among *Francisella tularensis* strains in a short-sequence tandem repeat. *Journal of Clinical Microbiology*, **39**, 3140–3146.

Junhui, Z., Ruifu, Y., Jianchun, L. *et al.* (1996) Detection of *Francisella tularensis* by the polymerase chain reaction. *Journal of Medical Microbiology*, **45**, 477–482.

Koskela, P., Salminen, A. (1985) Humoral immunity against *Francisella tularensis* after natural infection. *Journal of Clinical Microbiology*, **22**, 973–979.

McCoy, G. W., Chapin, C. W. (1912) Further observations on a plague-like disease of rodents with a preliminary note on the causative agent, *Bacterium tularense*. *The Journal of Infectious Diseases*, **10**, 61–72.

McCrumb, F. R. (1961) Aerosol infection of man with *Pasteurella tularensis*. *Bacteriological Reviews*, **25**, 262–267.

Ohara, Y., Sato, T., Fujita, H. *et al.* (1991) Clinical manifestations of tularemia in Japan – analysis of 1355 cases observed between 1924 and 1987. *Infection*, **19**, 14–17.

Olsufjev, N. G., Meshcheryakova, I. S. (1983) Subspecific taxonomy of *Francisella tularensis* McCoy and Chapin 1912. *International Journal of Systematic Bacteriology*, **33**, 872–874.

Poquet, Y., Kroca, M., Halary, F. *et al.* (1998) Expansion of V gamma 9Vdelta2T cells is triggered by *Francisella tularensis*-derived phospho-antigens in tularemia but not after tularemia vaccination. *Infection and Immunity*, **66**, 2107–2114.

Sandstrom, G., Lofgren, S., Tarnvik, A. (1988) A capsule-deficient mutant of *Francisella tularensis* LVS exhibits enhanced sensitivity to killing by serum but diminished sensitivity to killing by polymorphonuclear leukocytes. *Infection and Immunity*, **56**, 1194–1202.

Sandstrom, G., Sjostedt, A., Johansson, T. *et al.* (1992) Immunogenicity and toxicity of lipopolysaccharide from *Francisella tularensis* LVS. *FEMS Microbiology and Immunology*, **5**, 201–210.

Saslaw, S., Eigelsbach, H. T., Prior, J. A. *et al.* (1961a) Tularemia vaccine study. II. Respiratory challenge. *Archives of Internal Medicine*, **107**, 702–714.

Saslaw, S., Eigelsbach, H. T., Wilson, H. E. *et al.* (1961b) Tularemia vaccine study. I. Intracutaneous challenge. *Archives of Internal Medicine*, **107**, 689–701.

Sjostedt, A. (2002) Family XVII. Francisellaceae, genus I. *Francisella*, in *Bergey's Manual of Systematic Bacteriology*, Brenner, D. J., ed. Springer-Verlag, Berlin.

Sjostedt, A., Eriksson, U., Berglund, L. *et al.* (1997) Detection of *Francisella tularensis* in ulcers of patients with tularemia by PCR. *Journal of Clinical Microbiology*, **35**, 1045–1048.

Stewart, S. J. (1996) Tularemia: association with hunting and farming. *FEMS Immunology and Medical Microbiology*, **13**, 197–199.

Syrjala, H., Koskela, P., Ripatti, T. *et al.* (1986) Agglutination and ELISA methods in the diagnosis of tularemia in different severities of the disease. *The Journal of Infectious Diseases*, **153**, 142–145.

Tarnvik, A., Sandstrom, G., Sjostedt, A. (1996) Epidemiological analysis of tularemia in Sweden 1931–1993. *FEMS Immunology and Medical Microbiology*, **13**, 201–204.

22

Rickettsia spp.

James G. Olson[1], Franca R. Jones[2] and Patrick J. Blair[3]

[1]*Virology Department, U.S. Naval Medical Research Center Detachment, Unit 3800, American Embassy, APO AA 34031;*
[2]*National Naval Medical Center, Microbiology Laboratory, Bethesda, MD 20889, USA; and* [3]*US Embassy Jakarta, US NAMRU 2, FPO AP 96520–8132*

INTRODUCTION

Members of the order Rickettsiales are represented by a diverse group of Gram-negative bacteria that are largely intracellular parasites. Although many species are adapted to existence within arthropods, they are frequently capable of infecting vertebrates, including humans. Prior to 1986, eight rickettsioses were recognized. In the last two decades this number has increased to 14 with the advent of improved cell culture and molecular identification techniques (Raoult and Roux, 1997). At least 13 species in five genera are pathogenic for humans. The diseases caused by these microorganisms have had significant roles in the history of civilization. However, other diseases that have had greater epidemic potential and that lack effective interventions have overshadowed rickettsial diseases. The discovery of new groups has sparked a renewed interest of late, and rickettsioses are now seen as emerging infectious diseases that potentially threaten human health in locations throughout the globe.

DESCRIPTION OF THE ORGANISMS

Definition

Rickettsiae, members of the family *Rickettsiacea*, are fastidious Gram-negative, obligate intracellular bacterial organisms that survive within membrane-bound vacuoles (Rikihisa, 1991). Organisms included in the Rickettsiales are distinguished by short, nonflagellated, rod-shaped coccobacilli ranging from 0.8 to 2.0 μm in length and from 0.3 to 0.5 μm in width (Ris and Fox, 1949). All species are characterized by a five-layered outer envelope surrounded by lipopolysaccharide (Schramek, Brezina and Tarasevich, 1976). The genus *Ehrlichia* is made up of three pathogenic species of small obligate intracellular Gram-negative bacteria. Ehrlichiea coccobacilli have been observed to take two forms while inside host cells: a small (0.2–0.4 μm), electron dense form and a large, light form (0.8–1.5 μm) (Popov *et al.*, 1995). Ehrlichiea are surrounded by thin outer and inner membranes, but unlike members of the genus *Rickettsia*, show no thickening of either leaflet or of the outer membrane. Morphologically, members of the genus *Ehrlichia* do not appear to contain significant amounts of peptidoglycan.

Classification

Recent restructuring of the order Rickettsiales, based on genetic differences among rickettsial and ehrlichial species (Weisburg *et al.*,

1989; Brenner *et al.*, 1993), has resulted in the creation of two families within the order: *Rickettsiaceae*, which includes members of the genera *Rickettsia* and *Orientia*, and *Anaplasmataceae*, made up of members of the genera *Ehrlichia*, *Wolbachia*, *Anaplasma* and *Neorickettsia* (Dumler, Barbet and Bekker, 2001). This reclassification has moved former members, most notably *Coxiella*, *Bartonella*, *Grahamella* and *Rochalimae*, from Rickettsiales to different orders. *Coxiella burnetii*, the causative agent of Q fever, is now a member of the family *Coxiellaceae* in the order Legionellales, and *Bartonella*, *Grahamella* and *Rochalimae* have been united in the family *Bartonellaceae*, order Rhizobiales (Comer, Paddock and Childs, 2001). The genus *Rickettsia* is comprised of three genetically and antigenically similar groups, namely spotted fever, typhus and scrub typhus groups.

Molecular Characterization

The previous standards for defining rickettsial species include a variety of direct phenotypic features including morphology and antigenic properties, as well as indirect properties such as geographic distribution, host cell type, animal reservoir and disease characteristics. Classical laboratory methods for identifying different rickettsial family members utilized cross-immunity and protection in guinea pigs in addition to complement fixation (CF) and mouse toxin neutralization tests (Philip *et al.*, 1967). Serologic identification of rickettsial isolates met with some success but was plagued by a high level of cross-reactivity between species. As such, phylogenetic analysis of rickettsial species has been conducted to discern the degree of relatedness among groups and in some cases to define new species following molecular analysis of conserved genes. Initial attempts were conducted to determine cross-hybridization of genomic sequences among different rickettsial species. These techniques were confounded by the necessity to generate large amounts of uncontaminated DNA. The use of smaller quantities of DNA during polymerase chain reaction (PCR) tests has allowed for the identification of species.

PATHOGENESIS

Many rickettsiae infect endothelial cells of the microcirculatory system, especially the capillaries. Damage to the vasculature is the basis for similarities in pathology of rickettsial diseases (Walker and Matern, 1980). The primary cellular lesion results from dilation and destruction of intracellular membranes, particularly the rough endoplasmic reticulum (Silverman, 1984). As endothelial cells die, necrosis of the blood vessels leads to the formation of hyaline thrombi that manifest themselves clinically as petechiae. Gross and microscopic lesions are found in the brain, heart, kidney and lungs.

Unlike the other rickettsiae, infection by the spotted fever group family member (SFG), *Rickettsia rickettsii*, is not confined to the capillary endothelium with the bacterium often found to invade and destroy smooth muscle and vascular endothelium cells of larger vessels. Other SFG rickettsiae may cause interstitial pneumonia. In fatal cases of Rocky Mountain spotted fever (RMSF), the distribution of rickettsiae parallels observed pulmonary vasculitis, suggesting that rickettsial invasion of the microcirculation results in the interstitial pneumonitis. The vasculitis, vascular damage and increased vascular permeability result in alveolar septal congestion and oedema, fibrin formation, macrophage accumulation and haemorrhage and interlobar, septal and pleural effusion (Walker, Crawford and Cain, 1980). Interstitial pneumonitis is more frequent in scrub typhus than in epidemic typhus or spotted fever patients (Allen and Spitz, 1945).

Lesions of the central nervous system (CNS) play an important role in the clinical manifestations of RMSF and typhus. All portions of the brain and spinal cord may be involved, but the midbrain and inferior nucleus are most frequently affected. Focal proliferations of the endothelial and neuroglial cells (once called typhus nodules) are typical. Neurological findings in RMSF are the result of microinfarcts and frequently involve the white matter. Neurological pathology in epidemic and scrub typhus include mononuclear cell meningitis, perivascular cuffing of arteries, focal haemorrhages and degeneration of ganglionic cells (Allen and Spitz, 1945).

Interstitial myocarditis is found in RMSF, epidemic typhus and scrub typhus. RMSF heart lesions are patchy and consist of a mixed mononuclear cell infiltrate. *Rickettsia rickettsii* may be detected in the cardiac vessels by immunofluorescence (Walker, Palletta and Cain, 1980).

The urinary tract is commonly affected in RMSF and typhus. Azotemia is usually prerenal and reflects intravascular volume depletion. Acute renal failure in RMSF is probably due to tubular necrosis that results from systemic hypotension (Allen and Spitz, 1945). Histopathological studies of RMSF, epidemic typhus and scrub typhus have shown that the testes, epididymis, scrotal skin and adrenals may also be involved. Swollen Kupffer cells of the liver are prominent in typhus, and phagocytosis of erythrocytes and inflammatory cells may be noted in the spleen (National Research Council, 1953).

Unlike the *Rickettsiaceae*, members of the family *Anaplasmataceae* primarily infect cells of bone marrow origin. Although ehrlichial agents that infect humans have been classified based on the predominant type of cells they infect, i.e. monocytes [human monocytic ehrlichiosis (HME)] or granulocytes [human granulocytic ehrlichiosis (HGE)], many species have been found in other cell types. Few studies have been performed to investigate the human pathology of ehrlichial disease; what is known is based on sparse biopsy material from individuals who died from ehrlichiosis (Olano and Walker, 2002).

Bone marrow has been the most examined tissue in cases of HME. The most common finding is myeloid hyperplasia, while other findings may include granulomas, diffuse histiocytic aggregates, haemophagocytosis, lymphoid aggregates and plasmocytosis (Olano and Walker, 2002). Cells infected by *E. chaffeensis* include Kupffer cells, splenic macrophages and macrophages present in perivascular lymphohistiocytic infiltrates, which may be present in several organs including the CNS. Interstitial mononuclear cell pneumonia, pulmonary oedema and diffuse alveolar damage may be observed in the lung (Walker and Dumler, 1997). Severe thrombocytopenia can result in haemorrhage in various organs.

In the case of HGE, caused by *Anaplasma phagocytophila*, lymphoid depletion, erythroleucophagogocytosis and apoptosis are observed in the spleen (Olano and Walker, 2002). Lymph nodes may also be affected, showing lymphadenitis early in disease and paracortical hyperplasia later. During the convalescent stage of disease, bone marrow shows normocellularity and paratrabecular and nonparatrabecular lymphoid aggregates. Periportal lymphohistiocytic infiltrates, lobular hepatitis, Kupffer cell hyperplasia and apoptotic hepatocytes may be observed in the liver. In some cases, interstitial pneumonia is observed (Walker and Dumler, 1997).

CLINICAL AND EPIDEMIOLOGICAL FEATURES

The Rickettsiales are often associated with distinct arthropods that in turn serve as vectors for transmission of the bacterium. While a number of rickettsial species exist, distinct groups can be broken out based upon their ability to cause disease in humans (Weiss, 1981). Some of these diseases include RMSF (*R. rickettsii*), epidemic typhus (*R. prowazekii*), murine typhus (*R. typhi*), scrub typhus (*Orientia tsutsugamushi*), boutonneuse fever (*R. conorii*) and rickettsial pox (*R. akari*). Most of the illnesses caused by members of the order Rickettsiales are characterized by acute onset of fever accompanied by nonspecific signs and symptoms. In many diseases a characteristic rash follows the systemic symptoms and may be pathognomonic. Table 22.1 provides a list of agents and a summary of diseases associated with aetiological agents, their geographical distributions and mode of transmission to humans.

Table 22.1 Features of the pathogenic *Rickettsiales* species

Biogroup	Species	Disease in humans	Distribution	Transmission to humans
Spotted fever	*R. rickettsii*	Rocky Mountain spotted fever	Western Hemisphere	Tick bite
	R. conorii	Mediterranean spotted fever; also called Boutonneuse fever	Primarily Mediterranean countries, Africa, India, Southwest Asia	Tick bite
	R. siberica	Siberian tick typhus	Siberia, Mongolia, northern China	Tick bite
	R. australis	Australian tick typhus	Australia	Tick bite
	R. akari	Rickettsial pox	USA, USSR	Mite bite
	R. japonica	Oriental spotted fever	Japan	Presumably tick bite
Typhus	*R. prowazekii*	Louse-borne typhus	Primarily highland areas of South America and Africa	Infected louse faeces
		Recrudescent typhus (Brill-Zinsser disease)	Worldwide: follows distribution of persons with primary infections	Reactivation of latent infection
		Sporadic typhus	USA	Contact with flying squirrels Glaucomys volans
	R. typhi	Flea-borne typhus	Worldwide	Infected flea faeces
Scrub typhus	*O. tsutsugamushi*	Scrub typhus	Asia, northern Australia, Pacific Islands	Chigger bite
Ehrlichiosis	*N. sennetsu*	Sennetsu ehrlichiosis	Japan and Malasya	Trematode in fish
	E. chaffeensis	Human monocytic ehrlichiosis	USA, possibly Africa and Europe	Tick bite
	A. phagocytophila	Human granulocytotropic ehrilichiosis	USA, Europe	Tick bite
	E. ewingii	Human granulocytotropic ehrilichiosis	USA	Tick bite

Rocky Mountain Spotted Fever

RMSF, caused by infection with *Rickettsia rickettsii*, is an acute, seasonal, tick-borne, potentially fatal disease characterized by fever, headache and rash and frequently myalgia and anorexia after an incubation period of between 4 and 14 days. While first reported in the western United States, RMSF has been found in Costa Rica and further down in South America (Fuentes, 1986). Initial symptoms of disease are nonspecific including high fever, headache, muscle pain and occasional nausea, cough and vomiting. Later complications include vascular damage, increased permeability, oedema, haemorrhage, disseminated intravascular coagulation, interstitial pneumonitis, CNS involvement, myocarditis and renal failure (Clements, 1992). The case fatality ratio (CFR) has been reduced to between 3 and 5% since antibiotic therapy became available, but untreated, the CFR ranged between 23 and 70% (Harrell, 1949). Most of the approximately 600 cases reported annually in the United States are located in the southeast and south central states where the principal vector ticks, *Dermacentor variabilis* and *Amblyomma americanum*, are prevalent. In South America *A. cajenense* is a major vector, but other species may also be involved in transmission. The highest incidence occurs in children aged 5–9 years, and males outnumber females by nearly 2–1. Delay in initiating antibiotic therapy and increased age of patients are risk factors that significantly increase complications and death.

Spotted Fevers

A number of other SFG have been described from a wide geographic range. African tick typhus, Japanese tick typhus, Kenya tick typhus, Israel tick typhus, Indian tick typhus, Siberian tick typhus, Australian tick typhus and Oriental tick typhus are caused by additional spotted fever group rickettsiae that are transmitted by ticks. Mediterranean spotted fever (MSF), caused by infection with *R. conorii*, is similar to RMSF but is somewhat milder and is often characterized by a primary necrotic lesion known as an eschar or tache noir, where the site of tick bite has occurred. Among hospitalized patients, the CFR is 2% (Raoult *et al.*, 1986). The tick vector, *Rhipicephalus sanguineus*, is found mostly in Europe. In North America, the best-characterized, tick-borne rickettsial pathogen is *R. rickettsii*. However, a second SFG member with close homology to the Thai tick typhus rickettsia strain TT-118 has been isolated from *A. cajennense* in South Texas (Billings *et al.*, 1998). Evidence of *R. rickettsii* infections has also been documented through South America, primarily due to the tick vector *A. cajenennse* (Ripoll *et al.*, 1999; Martino *et al.*, 2001; Rozental *et al.*, 2002). Interestingly, the degree of virulence and the ability to cause cytopathic injury to human endothelial cells within American strains of *R. rickettsii* have been attributed to polymorphic genetic changes (Eremeeva *et al.*, 2003). Recently a potential new spotted fever group member has been identified in Northeastern Peru (Blair *et al.*, 2004). Demographic and temporal information on vectors and prevalence in humans in the South and Central America should allow for an assessment of risk in these areas.

Louse-Borne Typhus

Louse-borne (epidemic) typhus caused by infection with *R. prowazekii* is an acute febrile disease accompanied by headache, myalgia and rash that is transmitted by the human body louse, *Pediculus humanus*. Complications include interstitial pneumonitis, CNS involvement, myocarditis and acute renal failure. During epidemics, the disease may have CFRs between 10 and 66% (Megaw, 1942; Foster, 1981), depending on the health and nutrition of populations afflicted. Outbreaks of louse-borne typhus occur when pediculosis is wide spread as a result of the disruption of regular hygiene and lack of adequate water supplies for bathing. Typical settings where louse-borne typhus outbreaks occur include refugee camps, prisons and communities

ravaged by war or disaster. An endemic form of the disease occurs in several areas of the world where high altitude and cold weather combine to promote conditions of pediculosis. Of some interest is the evolutionary divergence of *R. prowazekii* from members of the spotted fever group. While both can be traced to a common ancestor some 40–80 million years ago, evidence suggests that *R. prowazekii* has evolved more rapidly, even shedding genetic material while adapting to its specific host (Ogata *et al.*, 2001).

Brill-Zinsser Disease

A recrudescent form of louse-borne typhus (Brill-Zinsser disease) is generally milder than the initial episode but serves as the mechanism for reintroduction of the disease into human populations. Patients who recover from louse-borne typhus may become rickettsemic as a result of diminished immunity due to age, illness or a variety of causes and transmit rickettsiae to their body lice. If infected lice infest susceptible humans they are capable of beginning an epidemic of louse-borne typhus.

Flea-Borne Typhus

Flea-borne (murine) typhus is caused by *R. typhi* infection and is an acute febrile illness typically associated with headache, myalgia, anorexia and rash. While presentation of this disease is mild compared with louse-borne typhus and fatalities are rare (Miller and Beeson, 1946), its importance should not be minimized, as it is endemic across the globe. Indeed, the worldwide distribution of the disease is attributed to the distribution of the rat flea (*Xenopsilla cheopis*). The natural reservoir for typhus are *Rattus rattus* and *R. norvegicus* species and is a major cause of febrile illness in the tropics. Populations that are occupationally exposed to rodents have increased risk of infection. Agricultural workers are often exposed to rodents and their fleas as a result of harvesting crops. A highly sensitive PCR-based assay can detect *R. typhi* in samples of fleas (Webb *et al.*, 1990). Additionally, species-specific monoclonal antibodies have been developed and used successfully to elucidate rickettsial species in flea cryosections (Fang *et al.*, 2003).

Scrub Typhus

Scrub typhus (Tsutsugamushi fever), caused by *O. tsutsugamushi*, is a common febrile disease in South and Southeast Asia. In these areas, more than a billion people are at risk, and roughly one million cases occur each year (Watt *et al.*, 2000). It is characterized by severe headache, myalgia, arthralgia, maculopapular rash and an eschar in about 20% of patients (Berman and Kundin, 1973). Several species of larval mites of the family *Trombiculidae* serve both as vectors and reservoirs of the rickettsiae. The parasitic stage is conducted by a chigger that develops after 7–10 days. The vector is often a member of the order Rodentia, family *Muridae*. Untreated, the CFR is less than 10%, and with antibiotic therapy, fatalities are rare. Populations at risk of acquiring scrub typhus are those who come in contact with mite-infested habitats. Military populations and agricultural workers whose occupations require them to spend time in habitats transitional between forest and field have significant risk of infection. Several strains of *O. tsutsugamushi* cause human illness and infection with one does not confer immunity to subsequent infection with others. *O. tsutsugamushi* has historically been detected by the Weil-Felix serologic test and more recently by molecular and immunological tests.

Rickettsialpox

Rickettsialpox is caused by infection with *R. akari* and is an uncommon febrile illness characterized by sparse, discrete, maculopapular lesions on the face, trunk and extremities that become vesicular. Eschars occur in most (90%) patients and begin as red papules and develop into shallow, punched out ulcers. Transmission of *R. akari* occurs via the mite *Liponyssoides sanguineus* when humans come in contact with premises infested with the house mouse, *Mus musculus*. Most cases in the United States are from urban environs of low socioeconomic status. Cases have been reported from the Ukraine region of the former Soviet Union.

Ehrlichiosis

Human Monocytic Ehrlichiosis

HME in the United States is caused by infection with *E. chaffeensis*. This acute febrile disease is most frequently characterized by headache, myalgia, nausea, arthralgias and malaise (Fishbein, Dawson and Robinson, 1994). Other manifestations of illness may include cough, pharyngitis, lymphadenopathy, diarrhoea, vomiting, abdominal pain and changes in mental status (Fishbein, Dawson and Robinson, 1994; Paddock and Childs, 2003). Laboratory findings include leucopoenia, thrombocytopenia and elevated liver enzymes (Fishbein, Dawson and Robinson, 1994). The disease is potentially life threatening, and complications include disseminated intravascular collapse and death. The CFR is approximately 3%. More than 500 confirmed cases of HME were reported from 1986 to 1997 from all but two states with the majority of cases being reported from the south-central and south-eastern United States (Paddock and Childs, 2003). In 1999 and 2000, 156 and 200 cases of confirmed or probable HME were reported in the United States (CDC, 2002; Paddock and Childs, 2003). More than 70% of cases occur between May and July when vector ticks are actively questing for blood meals (Fishbein, Dawson and Robinson, 1994). Males account for 75% of the HME cases reported. It is not clear whether the difference in incidence between males and females is due to exposure or other factors. Several lines of evidence indicate that *A. americanum* is the vector for *E. chaffeensis* transmission (Anderson *et al.*, 1993; Ewing *et al.*, 1995; Lockhart *et al.*, 1995; Lockhart *et al.*, 1996). The white-tailed deer is the only recognized reservoir for maintaining *E. chaffeensis* in nature, although there are indications that other mammals including goats, dogs, coyotes, red foxes, raccoons and others may serve as hosts (Paddock and Childs, 2003). Risk factors of infection include occupational and recreational exposure to tick-infested areas.

Human Granulocytic Ehrlichiosis

HGE is caused by infection with *A. phagocytophilia* (formerly known as *Ehrlichia phagocytophilia* or *Ehrlichia equi* or the HGE agent). A human granulocytic ehrlichial agent was first identified in 1994 (Chen *et al.*, 1994). Clinical symptoms of HGE are similar to those of HME but to a milder degree (Olano and Walker, 2002). The CFR ranges from 0.5 to 1%. Four hundred and forty-nine cases of HGE were reported as of 1997 (McQuiston *et al.*, 1999). Cases of HGE have been reported in the United States (southern New England, mid-Atlantic, upper Midwest and northern California) and Europe (Blanco and Oteo, 2002; Lotric-Furlan *et al.*, 2003; Ruscio and Cinco, 2003). *Ixodes scapularis* (Bakken and Dumler, 2000), *I. pacificus* (Reubel *et al.*, 1998) and *I. ricinus* (Petrovec *et al.*, 1999) are the vectors responsible for transmission in the eastern and midwestern United States, California and Europe, respectively. The primary reservoir of *A. phagocytophilia* is the white-footed mouse, although other reservoirs may include the whitetail deer, the wood rat, coyotes and cotton mice (Olano and Walker, 2002). Risk factors, seasonality and the predominance of males with infection are similar to those of HME. Another agent that naturally infects dogs, *Ehrlichia ewingii*, has recently been confirmed to cause HGE in four patients who were receiving immunosuppressive therapy, indicating that the infection was likely opportunistic (Buller *et al.*, 1999).

Sennetsu Fever

Neorickettsia sennetsu (formerly known as *Ehrlichia sennetsu*) causes a mononucleosis-like, acute febrile disease characterized by chills, headache, malaise, insomnia, diaphoresis, pharyngitis and anorexia (Misao and Katsuta, 1956). Lymphadenopathy is generalized and characterized by tenderness and is most prominent in the postauricular and posterior cervical areas. Hepatomegaly and splenomegaly occur in about 33% of cases (Tachibana, 1986). The geographic distribution of confirmed cases include Japan and Malaysia. The disease is self-limiting, and no fatalities have been reported. It has been suggested that the primary mode of transmission is from eating fish contaminated with the bacterium (Dumler, Barbet and Bekker, 2001); however, information on potential vectors and reservoirs is lacking. Risk factors of infection are also unknown.

LABORATORY DIAGNOSIS

Laboratory diagnosis of rickettsial diseases is routinely accomplished by serologic assays that detect anti-rickettsial antibodies. Alternative procedures include isolation of rickettsiae from patient tissues and techniques aimed at direct detection of rickettsiae in tissue. However, the later two methods are generally restricted to research laboratories. Although none of the established techniques provides a reliable diagnosis early enough to affect the outcome of the disease, several new rapid diagnostic techniques have shown promise in research laboratories.

Adaptation of some of these new techniques may provide useful tools for early diagnosis of rickettsial diseases in a clinical setting. Specifically, PCR assays for the aetiologic agents of most rickettsial diseases have been developed. These techniques are undoubtedly the most sensitive means by which rickettsiae can be detected directly in clinical specimens. More recently, PCR-based techniques for the genotypic identification of rickettsial agents has been augmented with the addition of RFLP analysis to differentiate species and genotypes and to provide for an estimate of genetic divergences among selected genes (Regnery, Spruill, and Plikaytis, 1991). Additional laboratory techniques that have been used successfully to detect rickettsiae early in the course of infection include detection of antigen in peripheral blood cells and immunoglobulin M (IgM) antibody-capture immunoassays. However, serodiagnosis is still the preferred diagnostic approach, and testing of serum specimens collected during the acute and convalescent phases of illness is recommended. Preferred techniques for diagnosing rickettsial diseases are summarized in Table 22.2.

Serology

Most serologic diagnosis of rickettsial diseases is performed by the indirect immunofluorescence assay (IFA) primarily because of the availability of reagents and the ease and economy with which it can be incorporated into existing antibody screening systems. A variety of other tests have been used for serodiagnosis of rickettsial diseases.

Table 22.2 Preferred laboratory techniques for the diagnosis of rickettsial diseases

Infecting organism	IFA (indirect)	ELISA	Isolation	PCR
R. rickettsii	X	X	X	X
R. conorii	X	X	X	X
R. prowazekii	X	X	X	X
R. typhi	X	X	X	X
O. tsutsugamushi	X		X	
E. chaffeensis	X			X
A. phagocytophila	X			X

Table 22.3 Highlights of various serological techniques for the diagnosis of rickettsial infections

Technique	Minimum positive titre	Time after onset antibody usually detected	Comments
IFA	16–64 depending on investigator and organism	2–3 weeks	Relatively insensitive, requires little antigen; can be used for all rickettsiae and related organisms
CF	8 or 16, depending on investigator	3–4 weeks	Less sensitive than IFA or ELISA but very specific
ELISA	Optical density 0.25>controls	1 week in some instances	IgM capture assay promising for early diagnosis
Latex agglutination	64	1–2 weeks	Lacks sensitivity for late convalescent sera
IHA	40	1–2 weeks	Sensitivity ≥ IFA; more sensitive than CF
Immunoperoxidase	20	7 days	Not evaluated for all rickettsiea; useful in field situations
Microagglutination	≥ to 8	1–2 weeks	Requires considerable antigen; less sensitive than IFA
Weil-Felix	40–320, depending on investigator	2–3 weeks	Lacks sensitivity and specificity

A summary of the more common techniques can be found in Table 22.3, and details can be found in the respective references at the end of this chapter. However, relatively few of these techniques are used regularly by a majority of laboratories, and those that cannot prepare the needed reagents are limited to techniques that utilize commercially available antigens. Some tests compromise sensitivity (latex agglutination and Weil-Felix) for convenience, while other highly sensitive tests [enzyme-linked immunosorbent assay (ELISA)] are cumbersome in clinical settings.

Rickettsioses must be distinguished from several viral and bacterial illnesses, including meningococcaemia, measles, enteroviral exanthems, leptospirosis, typhoid fever, rubella and dengue fever. The interpretation of serologic tests is frequently confounded by the lack of specificity of available rickettsial reagents. Whole rickettsiae are used as antigens for many antibody assays, and as rickettsiae share common antigens (or epitopes) with other bacteria (e.g. *Proteus* species), sera from nonrickettsial patients can react at low titres with rickettsial antigens. This has necessitated the designation of minimum positive titres (Table 22.3) to confirm recent rickettsial infections. As the establishment of minimum positive titres is assigned somewhat arbitrarily, testing of acute- and convalescent-phase serum specimens is recommended.

Recommending a preferred serologic technique is not entirely straightforward. ELISA techniques, particularly IgM-capture assays, are among the most sensitive procedures available for rickettsial diagnosis, but they require large quantities of purified antigens that are unavailable commercially. ELISAs are also quite amenable for the large-scale screening of serum specimens, but this advantage is usually lost for rickettsial diseases, because with the exception of epidemic typhus, rickettsioses usually occur sporadically at a relatively low frequency. A variety of agglutination type assays are in use for serodiagnosis of rickettsial disease. In general these assays require a minimum of specialized equipment but do not compare favourably with the sensitivity of ELISAs or IFA.

Indirect Fluorescent Antibody Assays

The IFA procedure for the rickettsiae is similar to conventional IFA techniques, with inactivated yolk sac or tissue culture suspensions of rickettsiae being used as antigens (Newhouse *et al.*, 1979). The IFA assay has been almost universally successful for diagnosis of disease caused by rickettsiae and rickettsiae-like organisms including those of the typhus group, spotted fever group and scrub typhus. IFA is the gold standard for laboratory diagnosis of ehrlichial diseases (Walker and the Task Force on Consensus Approach for Ehrlichiosis (CAFE), 2000). For all rickettsiae, demonstration of a fourfold or greater increase in antibody titre between acute- and convalescent-phase serum samples is also confirmatory. IFA tests are reasonably sensitive, although the subjectivity of endpoint determinations is an obvious disadvantage. For the majority of patients with ehrlichial disease, titres will be negative in the first week of illness; so obtaining a convalescent sample is generally crucial to confirm diagnosis (Childs *et al.*, 1999). An additional disadvantage is that serologic testing by IFA is generally group reactive, rather than species specific, for the rickettsiae and related organisms.

Complement Fixation

CF tests have been used for many years for the diagnosis of most rickettsial diseases. In general, these tests while specific are less sensitive and more cumbersome than IFA or the agglutination assays and, for these reasons, are no longer used widely. Group-specific soluble antigens from either typhus or spotted fever group rickettsiae are prepared by methods that employ ether extraction. Antibodies reactive with these group-specific soluble antigens generally appear in the patient's serum 10–14 days after onset of disease and may last a year or longer. Species-specific antigens have been utilized to differentiate infections caused by members within the typhus or spotted fever groups. These specific CF assays rely on differential titres obtained with antigens from individual species of *Rickettsia* (Elisberg and Bozeman, 1969). The development of CF assays for diagnosis of ehrlichiosis has not been reported.

Enzyme Immunoassays

A variety of enzyme immunoassays (or ELISAs) have been developed for serologic diagnosis of rickettsial diseases. These tests offer several major advantages over other serologic assays in that they lend themselves well to automation, are not reliant on user interpretation of results and allow large numbers of sera to be run. Disadvantages generally include decreased specificity over some other assays, length of set-up time is not practical for small numbers of sera and the lack of standardization between laboratories may give rise to conflicting results. For these reasons, ELISA is generally limited to being performed in research laboratories.

Generally, the lack of specificity that is associated with rickettsial ELISAs can be avoided by not using whole-cell preparations of rickettsiae as antigens. Enzyme immunoassays that utilize lipopolysaccharide antigens have been successfully developed for spotted fever and typhus group rickettsiae (Jones *et al.*, 1993). The antibody response to this antigen appears to be group specific. Development of ELISAs for diagnosis of ehrlichial disease is still in progress (Paddock and Childs, 2003). Specifically, major antigenic surface proteins of *E. chaffeensis* (Yu *et al.*, 1999; Alleman *et al.*, 2000) and *A. phagocytophilia* (Ijdo *et al.*, 1999; Tajima *et al.*, 2000) are being targeted as antigens for these ELISAs.

Agglutination Assays

The indirect haemagglutination (IHA) test has received only limited evaluations, but its sensitivity is apparently equal to that of IFA. An erythrocyte-sensitizing substance (ESS), which can be obtained from typhus and spotted fever group rickettsiae by alkali extraction, is adsorbed onto sheep or human group erythrocytes, and the coated cells are then used as antigens for simple agglutination tests. The respective ESSs exhibit group-specific antigenic reactivity and appear to be lipopolysaccharide in nature. The convenience of the IHA technique makes it suitable for a clinical setting, although the relatively short shelf life (6 months) of sensitized erythrocytes is a disadvantage (Elisberg and Bozeman, 1969). Nonetheless, it is potentially valuable as a bedside technique, particularly in field situations.

The latex agglutination test has been used with some success in a number of public health laboratories in recent years. Latex spheres are coated with ESS, and the sensitized particles are used as antigens in an agglutination test. The simplicity of the latex agglutination test offers a convenience that is best appreciated at the hospital level, but because it is primarily an IgM assay, it occasionally lacks sensitivity for late-convalescent-phase sera. The convenience of the latex test makes it useful as a bedside diagnostic procedure. It is recommended for that setting over the Weil-Felix test which is still used in similar circumstances.

The Weil-Felix test became popular in the 1920s after it was observed that certain *Proteus* strains would agglutinate early-convalescent-phase sera from patients with suspected rickettsial disease (Weil and Felix, 1916). The test fell into disfavour because it lacks both sensitivity and specificity (Hechemy *et al.*, 1979; Brown *et al.*, 1983), but it has managed to survive because of its convenience in a clinical setting. Despite its simplicity, it does not reliably provide either early or specific diagnosis of rickettsioses.

Diagnostic Criteria

Fourfold rises in titre detected by any technique (Weil-Felix technique excepted) are considered evidence of rickettsial and ehrlichial infections. With single-serum specimens, CF titres of ≥16 in a clinically compatible case are also considered positive. With the IFA test, single titres of 64 or higher are considered of borderline significance for typhus and spotted fever group infections. For *E. chaffeensis*, patients with single titres of 1:64–1:128 are considered probable, while single titres of ≥1:256 are considered confirmed. Single antibody titers of ≥1:80 against *A. phagocytophilia* are considered positive. Although these titres are somewhat higher than those recommended by others (Table 22.2), they usually present no problem in identifying recent infections if appropriately timed sera are available for testing.

The rickettsial species responsible for an infection is difficult to identify by conventional serologic tests. Rickettsiae within either the spotted fever or typhus group are antigenically related, resulting in extensive serologic cross-reactivity. For example, both murine- and louse-borne typhus are endemic in certain areas of the world, and routine IFA tests cannot distinguish between these two infections. Antibody adsorption (Goldwasser and Shepard, 1959) or toxin neutralization (Hamilton, 1945) tests can distinguish patients who have epidemic or murine typhus infections, but only specialty laboratories can perform these tests. In addition, convalescent-phase sera from patients with RMSF or typhus occasionally cross-react with typhus or other spotted fever group rickettsiae, respectively (Ormsbee *et al.*, 1978a). However, such cross-reactions are infrequent, and the heterologous titres are routinely much lower than in homologous reactions. An ELISA utilizing lipopolysaccharide antigens that eliminates cross-reaction between typhus group and spotted fever group infections has been developed (Jones *et al.*, 1993). Cross-reactivity between *E. chaffeensis* and *A. phagocytophilia* occurs in 10–30% of all patients; for this reason it is recommended that patient's serum be tested against antigens from both species (Comer *et al.*, 1999). An IFA assay has been described for HGE using cell-culture-derived antigen that ablates cross-reactivity of serum from patients infected with other human ehrlichial agents (Nicholson *et al.*, 1997).

Determining the specific aetiology of spotted fever group infections can be more or less problematic depending on the country of origin. In the Western Hemisphere, RMSF is by far the most prevalent spotted fever group infection. However, in the United States, RMSF must occasionally be differentiated from rickettsialpox; differences in the respective clinical and epidemiological features are valuable for determining the likely aetiology. Furthermore, in contrast to RMSF, sera from patients with rickettsialpox are uniformly negative in Weil-Felix tests. Recent data from Europe and Asia suggests that the distributions of *R. conorii* and *R. sibirica* may overlap more than previously known and could confound attempts to identify specific aetiologic agents by conventional serologic testing. However, as all rickettsiae are susceptible to the tetracyclines, it is not necessary to identify the specific aetiologic agent to ensure that individual patients receive proper treatment.

The antigenic diversity of the scrub typhus rickettsiae presents a different problem with respect to serologic diagnosis. The most important consideration is an awareness of the antigenic diversity of *O. tsutsugamushi* strains in a given area. Unless an appropriate combination of strains of *O. tsutsugamushi* is included in the battery of test antigens, the titres of some serum specimens could appear falsely low, and some infections could even go undetected (Bourgeois *et al.*, 1977).

Finally, although laboratory testing is central to the diagnosis of rickettsial diseases, clinical findings can contribute to diagnosis, particularly in the face of equivocal serologic results. Although there are no unique pathognomonic features for the rickettsioses, eschars (at the site of arthropod attachment) are useful indicators of MSF and scrub typhus in endemic areas, the vesicular rash of rickettsialpox is unique among the rickettsioses and the triad of fever, headache and rash is a useful indicator of RMSF. Because of the hazards of working with living rickettsiae, isolation attempts are usually limited to situations where the outcome is fatal and postmortem tissues are the only specimens available for testing. Even then, direct fluorescent antibody tests of formalin-fixed, paraffin-embedded tissues are faster and safer than rickettsial isolation for diagnosis.

Culture\Isolation

Primary Isolation

The growth of rickettsial organisms in the laboratory necessitates either living host cells or cell cultures. Isolation and identification of members of the genus *Rickettsia* is not recommended for routine diagnosis of rickettsial diseases. These procedures are only recommended for research laboratories that have biosafety containment level 3 facilities and whose personnel have extensive experience in cultivating rickettsiae. Postmortem tissues are usually contaminated with bacteria; therefore, suspensions of these tissues are inoculated into susceptible animals for attempted isolation of rickettsiae. In most cases, the guinea pig is the animal of choice for rickettsial isolation. They are susceptible to infection by *R. typhi* and *R. rickettsii* (Ormsbee *et al.*, 1978b), which can be monitored by measuring the body temperature. Moreover, male guinea pigs often present with scrotal swelling when infected with *R. typhi* or *R. rickettsii*. Mice are the animal of choice for isolation of *R. akari* and *O. tsutsugamushi*; however, some inbred mice are resistant to infection by scrub typhus rickettsiae (Groves *et al.*, 1980), and outbred mice should be used for isolation of these agents.

Blood specimens are collected from animals that develop overt signs of illness following inoculation; ill animals are then humanely killed, target tissues are harvested aseptically and blood and tissues are repassaged in animals, embryonated eggs or tissue cultures. Gimenez stain (Gimenez, 1964) or direct immunofluorescence may be used to confirm the isolation of rickettsiae. Further passage into tissue culture or embryonated eggs is necessary to increase the yield of organisms. Surviving animals are bled 28 days postinoculation and the sera tested for rickettsial antibodies. Animals that were seronegative before inoculation and are positive after 28 days are considered to have been infected with rickettsiae. Seroconversion in the absence of illness probably indicates the presence of a strain of reduced virulence for the animal of choice.

Successful isolation of both *E. chaffeensis* (Dawson *et al.*, 1991; Dumler *et al.*, 1995; Paddock *et al.*, 1997; Standaert *et al.*, 2000) and *A. phagocytophilia* (Goodman *et al.*, 1996; Klein *et al.*, 1997; Nicholson *et al.*, 1997; Heimer, Tisdale and Dawson, 1998) has been reported, but isolation of *A. phagocytophilia* is more frequent (Olano and Walker, 2002). After inoculation of patient blood onto susceptible cell lines, the resulting isolate must then be identified. The only methods currently available that permit species-level identification of isolates are PCR-based techniques, that amplify the 16S RNA gene, combined with sequencing (Anderson *et al.*, 1991) or hybridization probes (Anderson *et al.*, 1992). Independent confirmation of PCR results by IFA on convalescent serum specimens is desirable.

Propagation of Isolates

Propagation of members of the genus *Rickettsia* is a hazardous procedure that should be attempted only under biosafety level 3 conditions. Cell cultures provide a convenient method for rickettsial propagation. Numerous cell lines have been used successfully for growing rickettsiae; Vero, primary chicken embryo, WI-38, HeLa and virtually any other cell line are suitable for this purpose. Infected animal tissues are homogenized thoroughly in sterile diluent and inoculated onto monolayers of susceptible cells. Cultures are incubated at 35 °C and monitored for up to 14 days. No other special growth conditions are necessary, although it is known that *R. rickettsii*, *R. prowazekii* and *R. typhi* (but not *O. tsutsugamushi*) grow better in an environment with 5% CO_2 than they do when exposed to the 0.2–0.3% CO_2 that is found in atmospheric air (Kopmans-Gargantiel and Wisseman, 1981). Rickettsial growth can be monitored by scrapping monolayers off the flask with an inoculating loop and preparing smears then staining for rickettsiae by the Gimenez technique (Gimenez, 1964) and by the direct fluorescent-antibody procedure, if appropriate. Rickettsiae also grow quite well in the yolk sac of embryonated hen eggs, although this technique is quite labour intensive.

Final confirmation of *Rickettsia* isolation requires that the isolate be morphologically similar to rickettsiae, grow intracellularly, fail to grow on bacteriologic media and react with appropriate immune serum but not with nonimmune sera. Species-level identification of *Rickettsia* can be performed by restriction endonuclease digestion of PCR-amplified products from isolates. Additional details concerning the procedures for rickettsial isolation can be found in the review by Weiss (1981).

Isolation of *E. chaffeensis* involves layering EDTA-anticoagulated blood onto Histopague and removing the leucocyte-rich fraction to flasks containing a semiconfluent or confluent layer of DH82 canine peritoneal macrophages (Paddock *et al.*, 1997; Standaert *et al.*, 2000). Cells are then incubated at 37 °C with 5% CO_2. Detection of morulae is performed by staining using Diff-Quik. *Ehrlichia chaffeensis* has also been culture isolated from cerebrospinal fluid (Standaert *et al.*, 2000). Culture of ehrlichial species homologous to the HGE agent have been performed by inoculating either whole blood (Nicholson *et al.*, 1997) or buffy coat (Lin *et al.*, 2002) onto HL-60 cells (human promyelocytic leukaemia cell line) and incubating the cells at 37 °C with 5% CO_2. Infection is detected by staining of cells.

Direct Detection of Organisms

Immunofluorescence

Rickettsiae have been detected by immunofluorescence in tissue biopsies obtained from the site of tick attachment and from cutaneous lesions (Woodward *et al.*, 1976). Technical improvements in the biopsy assay have resulted in a highly specific assay for confirming MSF. Evaluations of this technique indicate that it will detect *R. rickettsii* or *R. conorii* in about 50% of patients with RMSF or MSF, respectively. All of the factors that contribute to the lack of sensitivity are not known, although most false-negative results were from patients who had received specific antibiotic therapy before the biopsy was performed.

Rickettsiae are also present in circulating endothelial cells. Because of their relatively low number, detection by direct fluorescence microscopy was not feasible until Drancourt and colleagues (Drancourt *et al.*, 1992) utilized monoclonal antibody-coated beads to concentrate circulating endothelial cells from patients with MSF and then detected rickettsiae in the cells by direct immunofluorescence. The procedure had a sensitivity of 66% in initial evaluation. It takes less than 3 h to perform and shows promise as a rapid diagnostic assay for MSF and other rickettsial infections.

Direct fluorescent antibody testing of tissues collected postmortem is a useful approach for the retrospective diagnosis of rickettsial diseases. This technique was first applied to the rickettsiae by Walker and Cain (1978), who successfully detected rickettsiae in kidney tissue of seven of ten patients who had died of suspected RMSF.

Ehrlichial organisms can be seen in intracytoplasmic inclusions, termed morulae, in monocytes and neutrophils, in the cases of HME and HGE, respectively (Olano and Walker, 2002). Use of eosin-azure stains is necessary for visualization (Paddock and Childs, 2003). Although the presence of morulae is diagnostic, it is not sensitive and even when visualized they are detected in less than 5% of circulating leucocytes (Paddock and Childs, 2003).

Polymerase Chain Reaction

Studies by Tzianabos, Anderson and McDade (1989) showed that rickettsial DNA can be detected in blood during rickettsial infections. Using nucleotide primers corresponding to a portion of a defined rickettsial gene, they successfully detected *R. rickettsii* DNA in seven of nine patients with confirmed cases of RMSF by using PCR. Additional testing with clinical specimens indicated that the technique is highly specific but lacks sensitivity; PCR detected rickettsiae best in patients who were seriously ill with RMSF and is of potential importance in the early diagnosis of RMSF and other rickettsial infections. In other work, multiple genes have been utilized to discern subgroups. For instances, the analysis of the 16S rRNA [ribosomal DNA (rRNA)] (Roux and Raoult, 1995) has been used to diverge major groups of rickettsial organisms. Other genes include the common 17-kDa antigen (Tzionabos, Anderson and McDade, 1989), citrate synthase (*gltA*) (Roux *et al.*, 1997), *rOmpA* (Roux, Fournier and Raoult,1996) and *rOmpB* (Roux and Raoult, 2000). These genes contain immunogenic antigens that cross-react well among subgroup members (Raoult and Roux, 1997). More recently, the type 1 signal protein lepB (Rahman *et al.*, 2003) and the 120-kDa antigen (Dash and Jackson, 1998) have been used to characterize SFG members. It seems likely that molecular analysis of rickettsial agents, whether by PCR or restriction fragment length polymorphism (RFLP) analysis, will continue to provide the means by which group members are added to this diverse family. That said, novel new molecular techniques including real-time PCR (Jiang, Temenak and Richards, 2003) and the analysis of intergenic spacer regions in DNA sequences (Vitorino *et al.*, 2003) should also prove useful.

PCR detection of *E. chaffeensis* and *A. phagocytophila* has been performed directly from blood samples using a variety of genes as targets, the most common of which is the gene that encodes 16S rRNA. Recently, Massung and Slater (2003) systematically compared PCR assays for detection of *A. phagocytophilia* using primer sets published in the literature. The highest sensitivity and specificity was obtained using a 16S rRNA-nested assay and a direct assay using primers against the major surface protein of *A. phagocytophila*. Most of the direct assays to amplify 16S rRNA lacked specificity and amplified sequences from *E. chaffeensis* (Massung and Slater, 2003). PCR for the specific detection of *E. chaffeensis* in the blood of patients with ehrlichiosis has been described (Anderson *et al.*, 1992). Blood collected in EDTA is used to prepare DNA template for amplification using *E. chaffeensis*-specific primer pair HE1 and HE3 to the *16S rRNA* gene. The identity of the PCR-amplified products is confirmed with oligonucleotide probe HE4. PCR amplification of *E. chaffeensis* DNA can also be performed using primers against the 120-kDa protein gene and the variable length PCR target (VLPT) (Standaert *et al.*, 2000). The majority of patients with ehrlichiosis do not develop antibody responses within the first week of infection; thus, detection of organisms may be possible by PCR when antibody tests are not yet positive (Childs *et al.*, 1999).

TREATMENT

Antibiotic therapy is, in general, of great benefit to patients infected with rickettsiae. RMSF cases when treated with doxycycline, tetracyclines or chloramphenicol have a CFR much reduced from untreated patients (Harrell, 1949) and show rapid evidence of clinical improvement including defervescence and cessation of constitutional signs and symptoms. Other spotted fevers, most notably MSF, have been successfully treated with erythromycin and ciprofloxacin (Beltran and Herrero, 1992). Tetracyclines, doxycycline and chloramphenicol are the antibiotics of choice for treatment of most rickettsial and ehrlichial illnesses.

PREVENTION AND CONTROL

Prevention of the arthropod-borne diseases caused by rickettsiae-like agents includes a variety of measures that reduce the likelihood of contact between vector and susceptible humans. In general, avoiding exposure to habitats known to be endemic for the diseases and/or infested with vectors of the diseases is prudent advice but not usually practical. Persons who are exposed to vector-infested endemic areas must rely on personal protective measures to minimize contact with potentially infected vectors. Recommendations for prevention of tick-borne rickettsiae and ehrlichiae infections include avoidance of tick-infested habitats and, when that is not feasible, taking personal precautions that prevent tick bite. Tick infestations may be reduced by wearing long trousers that fit tightly around the ankles and shirts that fit tightly around the wrists. Clothing impregnated with permethrin or other repellents may also prevent tick infestation. Early removal of ticks, either before they imbed their mouthparts into the skin or within hours of imbedding, may prevent infection. Similar precautions for preventing louse-borne infections include prevention of louse infestations by regular bathing and washing clothes with soap and hot water. Disinfestation of human populations infested with body lice is an effective measure to contain epidemics. The use of a vaccine containing formalin-killed organisms may also have been a factor in limiting the spread of louse-borne typhus during World War II. Prevention of scrub typhus depends on avoiding the habitat infested by vector chiggers. Elimination of vegetation that supports mites and their small mammal hosts may be an effective means of reducing human infections. Prophylactic administration of antibiotics has been shown to be effective among populations occupationally exposed to risk (Olson *et al.*, 1980). Prevention of flea-borne typhus depends on measures similar to those of plague control and includes the use of insecticides to reduce populations of fleas on rodents, followed by rodent-control programs. The effectiveness of repellent-impregnated clothing, however, remains to be evaluated for the prevention of most diseases.

Immune Mechanisms

Immunity to both spotted fever group and typhus group rickettsial infections (like other intracellular bacteria) involves both the cellular and humoral arms of the immune response. Several vaccines for rickettsial disease have been produced, although none have been completely effective. Vaccines for human ehrlichiosis have not yet been developed.

In the case of spotted fever group rickettsiae, vaccines have been developed but have not been effective in preventing RMSF. Individual antigens have been identified that play a role in eliciting immunity in both the mouse toxin neutralization assay and the guinea-pig model. Two high molecular weight surface proteins, rOmpA (also termed the 190- or 155-kDa antigens based on differing molecular mass estimates) and rOmpB (also termed 135- or 120-kDa antigen) have been shown to contain protective epitopes. Both rOmpA and rOmpB proteins from *R. rickettsii* contain amino acid sequence homology with the species protective antigen of *R. typhi*. These surface proteins show promise as candidates for subunit vaccines (McDonald, Anacker and Garjian, 1987). Attempts to define protection are being augmented by studies to define specific immune responses. Of note has been work to identify CD8 T-cell epitopes against the OmpB gene of *R. conorii* (Li *et al.*, 2003). Protective epitopes for *E. chaffeensis* as yet have not been characterized.

REFERENCES

Alleman, A. R., Barbet, A. F., Bowie, M. V. *et al.* (2000) Expression of a gene encoding the major antigenic protein 2 homolog of *Ehrlichia caffenesis* and potential application for serodiagnosis. *Journal of Clinical Microbiology*, **38**, 3705–3709.

Allen, A. C. and Spitz, S. (1945) A comparative study of the pathology of scrub typhus (tsutsugamushi disease) and other rickettsial diseases. *American Journal of Pathology*, **21**, 603–680.

Anderson, B. E., Dawson, J. E., Jones, D. C. *et al.* (1991) *Ehrlichia chaffeensis*, a new species associated with human ehrlichiosis. *Journal of Clinical Microbiology*, **29**, 2838–2842.

Anderson, B. E., Sims, K. G., Olson, J. G. *et al.* (1993) *Amblyomma americanum*: a potential vector for human ehrlichiosis. *The American Journal of Tropical Medicine and Hygiene*, **49**, 239–244.

Anderson, B. E., Sumner, J. W., Dawson, J. E. *et al.* (1992) Detection of the etiologic agent of human ehrlichiosis by polymerase chain reaction. *Journal of Clinical Microbiology*, **30**, 775–780.

Bakken, J. S. and Dumler, J. S. (2000) Human granulocytic ehrlichiosis. *Clinical Infectious Diseases*, **31**, 554–560.

Beltran, R. R. and Herrero, J. I. H. (1992) Evaluation of ciprofloxacin and doxycycline in the treatment of Mediterranean spotted fever. *European Journal of Clinical Microbiology and Infectious Diseases*, **11**, 427–431.

Berman, S. J. and Kundin, W. D. (1973) Scrub typhus in South Vietnam. *Annals of Internal Medicine*, **79**, 26–30.

Billings, A. N., Yu , X. J., Teel, P. D. and Walker, D. H. (1998) Detection of a spotted fever group rickettsia in *Amblyomma cajennense* (Acari: Ixodidae) in south Texas. *Journal of Medical Entomology*, **35**, 474–478.

Blair, P. J., Moron: C., Schoeler, G. B., *et al.* (2004) Characterization of spatted fever group *Rickettsia* from flea and tick specimens collected in northen Peru. *Journal of Clinical Microbiology,* **42**, 4961–4967.

Blanco, J. R. and Oteo, J. A. (2002) Human granulocytic ehrlichiosis in Europe. *Clinical Microbiology Infections*, **8**, 763–772.

Bourgeois, A. L., Olson, J. G., Fang, R. C. Y. *et al.* (1977) Epidemiological and serological study of scrub typhus among Chinese military in the Pescadores Islands of Taiwan. *Transactions of the Royal Society of Tropical Medicine and Hygiene*, **71**, 338–342.

Brenner, D. J., O'Connor, S. P., Winkler, H. H. and Steigerwalt, A. G. (1993) Proposals to unify the genera *Bartonella* and *Rochalimaea*, with descriptions of *Bartonella quintana* comb. nov., *Bartonella vinsonii* comb. nov., *Bartonella henselae* comb. nov., and *Bartonella elizabethae* comb. nov., and to remove the family *Bartonellaceae* from the order Rickettsiales. *International Journal of Systematic Bacteriology*, **43**, 777–786.

Brown, G. W., Shirai, A., Rogers, C. *et al.* (1983) Diagnostic criteria for scrub typhus: probability values for immunofluorescent antibody and *Proteus* OXK agglutinin titers. *The American Journal of Tropical Medicine and Hygiene*, **32**, 1101–1107.

Buller, R. S., Arens, M., Hmiel, S. P. *et al.* (1999) *Ehrlichia ewingii*, a newly recognized agent of human ehrlichiosis. *The New England Journal of Medicine*, **341**, 148–155.

Centers for Disease Control and Prevention. (2002) Summary of notifiable diseases, United States 2000. *Morbidity and Mortality Weekly Report*, **49**, 1–100.

Chen, S.-M., Dumler, J. S., Bakken, J. S. *et al.* (1994) Identification of a granulocytotropic *Ehrlichia* species as the etiologic agent of human disease. *Journal of Clinical Microbiology*, **32**, 589–595.

Childs, J. E., Sumner, J. W., Nicholson, W. L. *et al.* (1999) Outcome of diagnostic tests using samples from patients with culture-proven human monocytic ehrlichiosis implications for surveillance. *Journal of Clinical Microbiology*, **37**, 2997–3000.

Clements, M. L. (1992) Rocky Mountain spotted fever, in *Infectious Diseases* (eds S. L. Gorbach, J. G. Bartlett and N. R. Blacklow), Saunders, Philadelphia, pp. 1304–1312.

Comer, J. A., Nicholson, W. L., Olson, J. G. *et al.* (1999) Serologic testing for human granulocytic ehrlichiosis at a national referral center. *Journal of Clinical Microbiology*, **37**, 558–564.

Comer, J. A., Paddock, C. D. and Childs, J. E. (2001) Urban zoonoses caused by *Bartonella*, *Coxiella*, *Ehrlichia*, and *Rickettsia* species. *Vector Borne Zoonotic Diseases*, **1**, 91–118.

Dasch, G. A. and Jackson, L. M. (1998) Genetic analysis of isolates of the spotted fever group of rickettsiae belonging to the *R. conorii* complex. *Annals of the New York Academy of Sciences*, **849**, 11–20.

Dawson, J. E., Anderson, B. E., Fishbein, D. B. *et al.* (1991) Isolation and characterization of an *Ehrlichia* sp. from a patient with human ehrlichiosis. *Journal of Clinical Microbiology*, **29**, 2741–2745.

Drancourt, M., George, F., Brouqui, P. *et al.* (1992) Diagnosis of Mediterranean spotted fever by indirect immunofluorescence of *Rickettsia conorii* in circulating endothelial cells isolated with monoclonal antibody-coated immunomagnetic beads. *The Journal of Infectious Diseases*, **166**, 660–663.

Dumler, J. S., Barbet, A. F. and Bekker, C. P. J. (2001) Reorganization of genera in the families *Rickettsiaceae* and *Anaplasmataceae* in the order Rickettsiales: unification of some species of *Ehrlichia* and *Ehrlichia* with *Neorickettsia*, descriptions of six new species combinations and designation of *Ehrlichia equi* and 'HE agent' as subjective synonyms of *Ehrlichia phagocytophila*. *International Journal of Systematic and Evolutionary Microbiology*, **51**, 2145–2165.

Dumler, J. S., Chen, S. M., Asanovich, K. *et al.* (1995) Isolation and characterization of a new strain of *Ehrlichia chaffeensis* from a patient with nearly fatal monocytic ehrlichiosis. *Journal of Clinical Microbiology*, **33**, 1704–1711.

Elisberg, B. L. and Bozeman, F. M. (1969) *Rickettsiae*, in *Diagnostic Procedures for Viral and Rickettsial Infections* (eds E. H. Lennette and N. J. Schmidt), American Public Health Association, New York, pp. 826–868.

Eremeeva, M. E., Klemt, R. M., Santucci-Domotor, L. A. *et al.* (2003) Genetic analysis of isolates of *Rickettsia rickettsii* that differ in virulence. *Annals of the New York Academy of Sciences*, **990**, 717–722.

Ewing, S. A., Dawson, J. E., Kocan, A. A. *et al.* (1995) Experimental transmission of *Ehrlichia chaffeensis* (Rickettsiales: *Ehrlichieae*) among white-tailed deer by *Amblyomma americanum* (Acari: Ixodidae). *Journal of Medical Entomology*, **32**, 368–374.

Fang, R., Fournier, P. E., Houhamdi, L. *et al.* (2003) Detection of *R. felis* and *R. typhi* in fleas using monoclonal antibodies. *Annals of the New York Academy of Sciences*, **990**, 213–220.

Fishbein, D. B., Dawson, J. E. and Robinson, L. E. (1994) Human ehrlichiosis in the United States, 1985–1990. *Annals of International Medicine*, **120**, 736–743.

Foster, G. M. (1981) Typhus disaster in the wake of war: the American-Polish relief expedition, 1919–1920. *Bulletin of Historical Medicine*, **55**, 221–232.

Fuentes, L. (1986) Ecological study of Rocky Mountain spotted fever in Costa Rica. *The American Journal of Tropical Medicine and Hygiene*, **35**, 192–196.

Gimenez, D. F. (1964) Staining rickettsiae in yolk-sac cultures. *Staining Technologies*, **39**, 135–140.

Goldwasser, R. A. and Shepard, C. C. (1959) Fluorescent antibody methods in the differentiation of murine and epidemic typhus fever: specificity changes resulting from previous immunization. *Journal of Immunology*, **82**, 373–380.

Goodman, J. L., Nelson, C., Vitale, B. *et al.* (1996) Direct cultivation of the causative agent of human granulocytic ehrlichiosis. *The New England Journal of Medicine*, **334**, 209–215.

Groves, M. G., Rosenstreich, D. L., Taylor, B. A. *et al.* (1980) Host defenses in experimental scrub typhus: mapping the gene that controls natural resistance in mice. *Journal of Immunology*, **125**, 1395–1399.

Hamilton, H. L. (1945) Specificity of toxic factors associated with epidemic and murine strains of typhus rickettsiae. *The American Journal of Tropical Medicine and Hygiene*, **25**, 391–395.

Harrell, G. T. (1949) Rocky Mountain spotted fever. *Medicine*, **28**, 333–370.

Hechemy, K. E., Stevens, R. W., Sasowski, S. *et al.* (1979) Discrepancies in Weil-Felix and microimmunofluorescence test results for Rocky Mountain spotted fever. *Journal of Clinical Microbiology*, **9**, 292–293.

Heimer, R., Tisdale, D. and Dawson, J. E. (1998) A single tissue culture system for the propagation of the agents of human ehrlichioses. *The American Journal of Tropical Medicine and Hygiene*, **58**, 812–815.

Ijdo, J. W., Wu, C., Magnarelli, L. A. *et al.* (1999) Serodiagnosis of human granulocytic ehrlichiosis by a recombinant HE-44-based enzyme-linked immunosorbent assay. *Journal of Clinical Microbiology*, **37**, 3540–3544.

Jiang, J., Temenak, J. J. and Richards, A. L. (2003) Real-time PCR duplex assay for *Rickettsia prowazekii* and *Borrelia recurrentis*. *Annals of the New York Academy of Sciences*, **990**, 302–310.

Jones, D., Anderson, B., Olson, J. *et al.* (1993) Enzyme-linked immunosorbent assay for detection of human immunoglobulin G to lipopolysaccharide of spotted fever group rickettsiae. *Journal of Clinical Microbiology*, **31**, 138–141.

Klein, M. B., Miller, J. S., Nelson, C. M. *et al.* (1997) Primary bone marrow progenitors of both granulocytic and monocytic lineages are susceptible to infection with the agent of human granulocytic ehrlichiosis. *The Journal of Infectious Diseases*, **176**, 1405–1409.

Kopmans-Gargantiel, A. I. and Wisseman, C. L. Jr. (1981) Differential requirements for enriched atmospheric carbon dioxide content for intracellular growth in cell culture among selected members of the genus *Rickettsia*. *Infection and Immunity*, **31**, 1277–1280.

Li, Z., Diaz-Montero, C. M., Valbuena, G. *et al.* (2003) Identification of CD8 T-lymphocyte epitopes in OmpB of *Rickettsia conorii*. *Infection and Immunity*, **71**, 3920–3926.

Lin, Q., Zhi, N., Ohashi, N. *et al.* (2002) Analysis of sequences and loci of *p44* homologs expressed by *Anaplasma phagocytophila* in acutely infected patients. *Journal of Clinical Microbiology*, **40**, 2981–2988.

Lockhart, J. M., Davidson, W. R., Dawson, J. E. *et al.* (1995) Temporal association of *Amblyomma americanum* with the presence of *Ehrlichia chaffensis* reactive antibodies in white-tailed deer. *Journal of Wildlife Diseases*, **31**, 119–124.

Lockhart, J. M., Davidson, W. R., Stallknecht, D. E. *et al.* (1996) Site-specific geographic association between *Amblyomma americanum* (Acari: Ixodidae) infestations and *Ehrlichia chaffeensis*-reactive (Rickettsiales) antibodies in white-tail deer. *Journal of Medical Entomology*, **33**, 153–158.

Lotric-Furlan, S., Petrovec, M., Avisic-Zupanac, T. *et al.* (2003) Human granulocytic ehrlichiosis in Slovenia. *Annals of the New York Academy of Sciences*, **990**, 279–284.

Martino, O., Orduna, T., Lourtau, L. *et al.* (2001) Spotted fever group rickettsial disease in Argentinean travelers. *Revista da Sociedidc Brasileira de Medicina Tropical*, **34**, 559–562.

Massung, R. F. and Slater, K. G. (2003) Comparison of PCR assays for detection of the agent of human granulocytic ehrlichiosis, *Anaplasma phagocytophilium*. *Journal of Clinical Microbiology*, **41**, 717–722.

McDonald, G. A., Anacker, R. L. and Garjian, K. (1987) Cloned gene of *Rickettsia rickettsii* surface antigen: candidate vaccine for Rocky Mountain spotted fever. *Science*, **235**, 83–85.

McQuiston, J. H., Paddock, C. D., Holman, R. C. *et al.* (1999) The human ehrlichiosis in the United States. *Emerging Infectious Diseases*, **5**, 635–642.

Megaw, J. W. D. (1942) Louse-borne typhus fever. *British Medical Journal*, **ii**, 401–403, 433–435.

Miller, E. S. and Beeson, P. B. (1946) Murine typhus fever. *Medicine*, **25**, 1–15.

Misao, T. and Katsuta, K. (1956) Epidemiology of infectious mononucleosis. *Japan Journal of Clinical and Experimental Medicine*, **33**, 73–82.

National Research Council, Division of Medical Sciences, Committee on Pathology. (1953) Pathology of epidemic typhus: report of fatal cases studied by the United States of America Typhus Commission in Cairo, Egypt during, 1943–1945. *Archives of Pathology*, **56**, 397–435.

Newhouse, V. F., Shepard, C. C., Redus, M. D. *et al.* (1979) A comparison of the complement fixation, indirect fluorescent antibody and microagglutination test for the serological diagnosis of rickettsial diseases. *The American Journal of Tropical Medicine and Hygiene*, **28**, 387–395.

Nicholson, W. L., Comer, J. A., Sumner, J. W. *et al.* (1997) An indirect immunofluorescence assay using a cell culture-derived antigen for detection of antibodies to the agent of human granulocytic ehrlichiosis. *Journal of Clinical Microbiology*, **35**, 1510–1516.

Ogata, H., Audic, S., Renesto-Audiffren, P. *et al.* (2001) Mechanisms of evolution in *Rickettsia conorii* and *R. prowazekii*. *Science*, **293**, 2093–2098.

Olano, J. P. and Walker, D. H. (2002) Human ehrlichioses. *The Medical Clinics of North America*, **86**, 375–392.

Olson, J. G., Bourgeois, A. L., Fang, R. C. Y. *et al.* (1980) Prevention of scrub typhus: Prophylactic administration of doxycycline in a randomized double-blind trial. *The American Journal of Tropical Medicine and Hygiene*, **29**, 989–997.

Ormsbee, R., Peacock, M., Philip, R. *et al.* (1978a) Antigenic relationships between the typhus and spotted fever groups of rickettsiae. *American Journal of Epidemiology*, **108**, 53–59.

Ormsbee, R., Peacock, M., Gerloff, R. *et al.* (1978b) Limits of rickettsial infectivity. *Infection and Immunity*, **19**, 239–245.

Paddock, C. D. and Childs, J. E. (2003) *Ehrlichia chaffeensis*: a prototypical emerging pathogen. *Clinical Microbiology Reviews*, **16**, 37–64.

Paddock, C. D., Sumner, J. W., Shore, G. M. *et al.* (1997) Isolation and characterization of *Ehrlichia chaffeensis* strains from patients with fatal ehrlichiosis. *Journal of Clinical Microbiology*, **35**, 2496–2502.

Petrovec, M., Sumner, J. W., Nicholson, W. L. *et al.* (1999) Identity of ehrlichial DNA sequences derived from *Ioxides ricinus* ticks with those obtained from patients with human granulocytic ehrlichiosis in Slovenia. *Journal of Clinical Microbiology*, **37**, 209–210.

Philip, C. B., Lackman, D. B., Philip, R. N. *et al.* (1967) Serological evidence of rickettsial zoonoses in South American domestic animals. *Acta Medica et Biologica (Niigata)*, **15**, 53–60.

Popov, V. L., Chen, S.-M., Feng, H.-M. *et al.* (1995) Ultrastructural variation of cultured *Ehrlichia chaffeensis*. *Journal of Medical Microbiology*, **43**, 411–421.

Rahman, M. S., Simser, J. A., Macaluso, K. R. and Azad, A. F. (2003) Molecular and functional analysis of the lepB gene, encoding a type I signal peptidase from *Rickettsia rickettsii* and *Rickettsia typhi*. *Journal of Bacteriology*, **185**, 4578–4584.

Raoult, D. and Roux, V. (1997) Rickettsioses as paradigms of new or emerging infectious diseases. *Clinical Microbiology Reviews*, **10**, 694–719.

Raoult, D., Weiller, P. J., Chagnon, A. *et al.* (1986) Mediterranean spotted fever: clinical, laboratory and epidemiological features of 199 cases. *The American Journal of Tropical Medicine and Hygiene*, **35**, 845–850.

Regnery, R. L., Spruill, C. L. and Plikaytis, B. D. (1991) Genotypic identification of rickettsiae and estimation of intraspecies sequence divergence for portions of two rickettsial genes. *Journal of Bacteriology*, **173**, 1576–1589.

Reubel, G. H., Kimsey, R. B., Barlough, J. E. *et al.* (1998) Experimental transmission of *Ehrlichia equi* to horses through naturally infected ticks (*Ixodes pacificus*) from northern California. *Journal of Clinical Microbiology*, **36**, 2131–2134.

Rikihisa, Y. (1991) The tribe Ehrlichiae and ehrlichial diseases. *Clinical Microbiology Reviews*, **4**, 286–308.

Ripoll, C. M., Remondegui, C. E., Ordonez, G. *et al.* (1999) Evidence of rickettsial spotted fever and ehrlichial infections in a subtropical territory of Jujuy, Argentina. *American Journal of Tropical Medical Hygiene*, **61**, 350–354.

Ris, H. and Fox, J. P. (1949) The cytology of rickettsiae. *The Journal of Experimental Medicine*, **89**, 681–686.

Roux, V., Fournier, P. E. and Raoult, D. (1996) Differentiation of spotted fever group rickettsiae by sequencing and analysis of restriction fragment length polymorphism of PCR-amplified DNA of the gene encoding the protein rOmpA. *Journal of Clinical Microbiology*, **34**, 2058–2065.

Roux, V. and Raoult, D. (1995) Phylogenetic analysis of the genus *Rickettsia* by 16S rDNA sequencing. *Research in Microbiology*, **146**, 385–396.

Roux, V. and Raoult, D. (2000) Phylogenetic analysis of members of the genus *Rickettsia* using the gene encoding the outer-membrane protein rOmpB (ompB). *International Journal of Systematic and Evolutionary Microbiology*, **50** Pt 4, 1449–1455.

Roux, V., Rydkina, E., Eremeeva, M. and Raoult, D. (1997) Citrate synthase gene comparison, a new tool for phylogenetic analysis, and its application for the rickettsiae. *International Journal of Systematic Bacteriology*, **47**, 252–261.

Rozental, T., Bustamante, M. C., Amorim, M. *et al.* (2002) Evidence of spotted fever group rickettsiae in state of Rio de Janeiro, Brazil. *Revista do Instituto de Medicina Tropical de São Paulo*, **44**, 155–158.

Ruscio, M. and Cinco, M. (2003) Human granulocytic ehrlichiosis in Italy: first report on two confirmed cases. *Annals of the New York Academy of Sciences*, **990**, 350–352.

Schramek, S. R., Brezina, R. and Tarasevich, I. V. (1976) Isolation of a lipopolysaccharide antigen from *Rickettsia* species. *Acta Virologica*, **20**, 270.

Silverman, D. J. (1984) *Rickettsia rickettsii*-induced cellular injury of human vascular endothelium in vitro. *Infection and Immunity*, **44**, 545–553.

Standaert, S. M., Yu, T., Scott, M. A. *et al.* (2000) Primary isolation of *Ehrlichia chaffeensis* from patients with febrile illnesses: clinical and molecular characteristics. *The Journal of Infectious Diseases*, **181**, 1082–1088.

Tachibana, N. (1986) Sennetsu fever: the disease, diagnosis, and treatment, in *Microbiology* (eds H. Winkler and M. Ristic), American Society for Microbiology, Washington, DC, pp. 205–208.

Tajima, T., Zhi, N., Lin, Q. *et al.* (2000) Comparison of two recombinant major outer membrane proteins of the human granulocytic ehrlichiosis agent for use in an enzyme-linked immunosorbent assay. *Clinical and Diagnostic Laboratory Immunology*, **7**, 652–657.

Tzianabos, T., Anderson, B. E. and McDade, J. E. (1989) Detection of *Rickettsia rickettsii* DNA in clinical specimens by using polymerase chain reaction technology. *Journal of Clinical Microbiology*, **27**, 2866–2868.

Vitorino, L., Ze-Ze, L., Sousa, A. *et al.* (2003) rRNA intergenic spacer regions for phylogenetic analysis of Rickettsia species. *Annals of the New York Academy of Sciences*, **990**, 726–733.

Walker, D. H. and the Task Force on Consensus Approach for Ehrlichiosis (CAFE). (2000) Diagnosing human ehrlichioses: current status and recommendations. *ASM News*, **66**, 287–290.

Walker, D. H. and Cain, B. G. (1978) A method for specific diagnosis of Rocky Mountain spotted fever on fixed, paraffin-embedded tissue by immunofluorescence. *The Journal of Infectious Diseases*, **137**, 206–209.

Walker, D. H., Crawford, G. C. and Cain, B. G. (1980) Rickettsial infection of the pulmonary microcirculation: the basis for interstitial pneumonitis in Rocky Mountain spotted fever. *Human Pathology*, **11**, 263–272.

Walker, D. H. and Dumler, J. S. (1997) Human monocytic and granulocytic ehrlichioses: discovery and diagnosis of emerging tick-borne infections and the critical role of the pathologist. *Archives of Pathology and Laboratory Medicine*, **121**, 785–791.

Walker, D. H. and Matern, W. D. (1980) Rickettsial vasculitis. *American Heart Journal*, **100**, 896–906.

Walker, D. H., Palletta, C. E. and Cain, B. G. (1980) Pathogenesis of myocarditis in Rocky Mountain spotted fever. *Archives of Pathology and Laboratory Medicine*, **104**, 171–174.

Watt, G., Kantipong, P., Jongsakul, K. *et al.* (2000) Doxycycline and rifampicin for mild scrub-typhus infections in northern Thailand: a randomised trial. *Lancet*, 356, 1057–1061.

Webb, L., Carl, M., Malloy, D. C. *et al.* (1990) Detection of murine typhus infection in fleas by using the polymerase chain reaction. *Journal of Clinical Microbiology*, **28**, 530–534.

Weil, E. and Felix, A. (1916) Zur serologischen Diagnosis des Fleckfiebers. *Wiener Klinische Wochenschrift*, **29**, 33–35.

Weisburg, W. G., Dobson, M. E., Samuel, J. E. *et al.* (1989) Phylogenetic diversity of the Rickettsiae. *Journal of Bacteriology*, **171**, 4202–4206.

Weiss, E. (1981) The family *Rickettsiaceae*: human pathogens, in *The Prokaryotes* (ed. M. P. Starr *et al.*), Springer-Verlag, Berlin, pp. 2137–2160.

Woodward, T. E., Pedersen, C. D. Jr., Oster, C. N. *et al.* (1976) Prompt confirmation of Rocky Mountain spotted fever: identification of rickettsiae in skin tissues. *The Journal of Infectious Diseases*, **134**, 297–301.

Yu, X.-J., Corcquet-Valdes, P. A., Cullman, L. C. *et al.* (1999) Comparison of *Ehrlichia chaffeensis* recombinant proteins for serologic diagnosis of human monocytotropic ehrlichiosis. *Journal of Clinical Microbiology*, **37**, 2568–2575.

23

Bartonella spp.

J. M. Rolain and D. Raoult

Unité des Rickettsies, CNRS UMR 6020A, Faculté de Médecine, Université de la Méditerranée, Marseilles, France

INTRODUCTION

Bartonella species are emerging and reemerging bacteria that belong to the α2 subgroup of Proteobacteria (Birtles and Raoult, 1996; Houpikian and Raoult, 2001). The interest in these microorganisms has increased since they are agents of common clinical infectious diseases such as cat-scratch disease (CSD), bacillary angiomatosis or blood culture-negative endocarditis (Maurin and Raoult, 1996; Anderson and Neuman, 1997; Maurin, Birtles and Raoult, 1997; Brouqui and Raoult, 2001). The current knowledge in the field is due essentially to the improvement of new molecular biology methods, especially polymerase chain reaction (PCR) and sequencing (Relman *et al.*, 1990). These bacteria have a unique feature among prokaryotes, the ability to induce angiogenic tumors such as bacillary angiomatosis and verruga peruana (Dehio, 2001). People usually become infected with *Bartonella* species incidentally, with the exception of *Bartonella bacilliformis* and *Bartonella quintana*, as the organisms are normally found in the reservoir hosts, which include animals such as cats and dogs that live in close contact with people (Jacomo, Kelly and Raoult, 2002).

DESCRIPTION OF THE ORGANISM

Historical Background

Until 1990, only two diseases were recognized to be caused by *Bartonella* species: Carrion's disease due to *B. bacilliformis* and trench fever due to *B. quintana* (Karem, Paddock and Regnery, 2000). *Bartonella bacilliformis* was first viewed in 1909 by Alberto Barton in erythrocytes of patients suffering from Oroya fever, the acute form of Carrion's disease in Peru (Ihler, 1996). In 1993, Brenner *et al.* proposed unifying species of the former genus *Rochalimaea* with the genus *Bartonella* based on comparison of 16S rRNA gene sequences. Thus, the species *Rochalimaea quintana* (the agent of trench fever, isolated in 1918) (Strong *et al.*, 1918), *Rochalimaea vinsonii* (isolated in 1946 from small rodents) (Baker, 1946) and *Rochalimaea henselae* (first described in 1992 and the agent of CSD) (Regnery *et al.*, 1992a) have been renamed as *Bartonella quintana*, *Bartonella vinsonii* and *Bartonella henselae*, respectively. Moreover, Birtles *et al.* (1995) proposed unifying the bacteria of the genus *Grahamella* with the genus *Bartonella* based on phylogenetic studies. Numerous other species have been described in the past few years in several animals (Breitschwerdt and Kordick, 2000) (Table 23.1).

Human infections due to *Bartonella* spp. include old and newly characterized diseases. *Bartonella bacilliformis* was reported for the first time in 1913 (Strong *et al.*, 1913), and *B. quintana* was first described during World War I in 1918 (Strong *et al.*, 1918). *Bartonella quintana* is transmitted by the human body louse and was responsible for more than one million deaths during World War I. Sporadic cases have been reported since these periods, especially in homeless patients in United States, France, Russia, Peru and Burundi (Brouqui *et al.*, 1996).

Bartonella henselae was first described in 1992 (Regnery *et al.*, 1992a), and its role in bacillary angiomatosis was demonstrated using molecular biology and PCR sequencing (Relman *et al.*, 1990). In 1983, Stoler *et al.* described the presence of cutaneous angiomatosis lesions in acquired immune deficiency syndrome (AIDS) patients. Small coccobacilli were viewed in histological biopsies by Warthin–Starry staining (Wear *et al.*, 1983). The species *B. quintana* was first identified in cutaneous biopsy samples by using PCR sequencing (Relman *et al.*, 1990). At the same time, a new species was isolated from the blood of a bacteremic patient (Slater *et al.*, 1990) and also in peliosis hepatitis in AIDS patients (Perkocha *et al.*, 1990). Further studies demonstrated that this bacterium was a new species, and this new species was named *B. henselae* (Regnery *et al.*, 1992a). Subsequently, Regnery *et al.*, in 1992, demonstrated that most patients suffering from CSD, disease first described in 1950 by Debré *et al.*, had antibodies directed against *B. henselae* (Regnery *et al.*, 1992a). Subsequent studies have demonstrated the role of *B. henselae* as the etiologic agent of CSD.

In recent years, new *Bartonella* species have been linked to human infections, and thus, the genus *Bartonella* now comprises 21 characterized species, of which 8 are potentially pathogenic in humans: *B. bacilliformis*, *B. quintana*, *B. henselae*, *Bartonella vinsonii* subsp. *berkhoffii* in a patient with endocarditis (Roux *et al.*, 2000), *Bartonella vinsonii* subsp. *arupensis* isolated in a patient with fever and bacteremia (Welch *et al.*, 1999) and in one case of infective endocarditis (unpublished data), *Bartonella grahamii* in a case of retinitis (Kerkhoff *et al.*, 1999), *Bartonella elizabethae* in a case of endocarditis (Daly *et al.*, 1993) and *Bartonella washoensis* in a case of myocarditis (Chang *et al.*, 2001; Kosoy *et al.*, 2003).

Classification – Phylogeny

Phylogenetically, the genus *Bartonella* belongs to the α-subgroup of Proteobacteria, very close to the genus *Brucella*, *Agrobacterium* and *Rhizobium* (Birtles and Raoult, 1996). Bacteria of the genus *Bartonella* are facultative aerobic, intracellular, oxidase and catalase negative, short Gram-negative bacilli. All are nonflagellated organisms, with the exception of flagellated *B. bacilliformis* and *Bartonella clarridgeiae* (Kordick *et al.*, 1997). Members of the genus *Bartonella* possess few phenotypic markers useful for species delineation. Most of the current knowledge about *Bartonella* taxonomy is supported by several genes, namely the 16S rRNA (Birtles *et al.*, 1995), the citrate synthase (*gltA*) (Birtles and Raoult, 1996), the 16S/23S intergenic

Table 23.1   Species of the genus *Bartonella* previously reported: epidemiological and clinical data

Species	Reservoir host	Disease in humans	First description (year)	Vector or detection in arthropods	Detection in erythrocytes
Bartonella bacilliformis	Human	Carrion's disease	1919	Sand fly (*Lutzomia* spp.)	+++
Bartonella talpae	Mole	Unknown		Unknown	
Bartonella peromysci		Unknown	1942	Unknown	
Bartonella quintana	Human	TF, BA, BAC, END	1961	Human body lice, fleas	+
Bartonella vinsonii subsp. *vinsonii*	Rodents	Unknown	1946	Unknown	
Bartonella henselae	Cats	CSD, BA, BAC, END	1990	Fleas (*Ctenocephalides felis*)	++
Bartonella henselae	Rats	END (1 case)	1993	Fleas	
Bartonella grahamii		RET (1 case)	1995	Unknown	
Bartonella taylorii		Unknown	1995	Unknown	
Bartonella doshiae		Unknown	1995	Unknown	
Bartonella clarridgeiae	Cats	Unknown	1995	Fleas (*Ctenocephalides felis*)	++
Bartonella vinsonii subsp. berkhoffii	Dogs	END (1 case)	1995	Fleas and ticks	
Bartonella tribocorum	Rats	Unknown	1998	Unknown	
Bartonella koehlerae	Cats	Unknown	1999	Fleas (supposed vectors)	+
Bartonella alsatica	Rabbit	Unknown	1999	Fleas or ticks	
Bartonella vinsonii subsp. arupensis	Rodents	BAC (1 case)	1999	Unknown	
Bartonella bovis (*weissii*)	Cows	Unknown	2002 (1999)	Unknown	
Bartonella washoensis	Rodents	MYOC (1 case)	2000	Unknown	
Bartonella birtlesii	Rats	Unknown	2000	Unknown	
Bartonella capreoli	Ruminant	Unknown	2002	Unknown	
Bartonella schoenbuchensis	Ruminant	Unknown	2001	Unknown	

BA, bacillary angiomatosis; BAC, bacteremia; CSD, cat-scratch disease; END, endocarditis; MYOC, myocarditis; RET, retinitis; TF, trench fever.

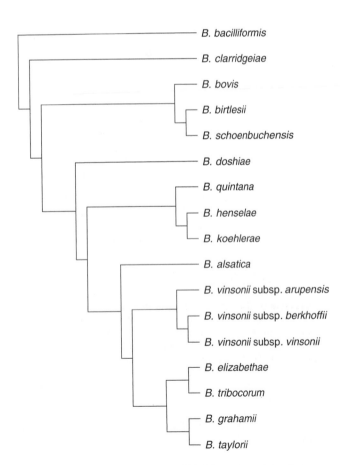

Figure 23.1   Neighbor-joining tree resulting from comparison of sequences of the *groEL* gene of most *Bartonella* spp. identified to date.

spacer region (*ITS*) (Roux and Raoult, 1995), the *rpoB* gene (Renesto *et al.*, 2001) and the 60-kDa heat-shock protein (*groEL*) (Marston, Sumner and Regnery, 1999). A phylogenetic neighbor-joining tree resulting from comparison of sequences of the *groEL* gene of most *Bartonella* spp. identified to date is presented in Figure 23.1. Comparison of phylogenetic organizations obtained from these genes allowed six statistically significant evolutionary clusters to be identified (Houpikian and Raoult, 2001). *Bartonella bacilliformis* and *B. clarridgeiae* appear divergent species. *Bartonella henselae*, *Bartonella koehlerae* and *B. quintana* cluster together, as well as *Bartonella vinsonii* subsp. *vinsonii* and *B. vinsonii* subsp. *berkhoffii*. The fifth group includes bacteria isolated from various rodents, and the sixth group comprises *Bartonella tribocorum*, *B. elizabethae* and *B. grahamii* (Houpikian and Raoult, 2001).

With the exception of 16S rRNA gene, all the sequences studied have good discriminatory powers, but the usefulness of some genes is restricted by high similarity values between some species (La Scola *et al.*, 2003). The only genes with lower interspecies similarity values are *rpoB* and *gltA*. Cutoff sequence similarity values have been proposed for species definition in *Bartonella* genus (La Scola *et al.*, 2003). A new *Bartonella* isolate should be considered as another species if a 327-bp *gltA* fragment shares <96.0% sequence similarity with the validated species and if an 825-bp *rpoB* fragment shares <95.4% sequence similarity with the validated species (La Scola *et al.*, 2003).

PATHOGENESIS

The growth of the bacteria is dependent upon hemin (Myers, Cutler and Wisseman, 1969) and, for some species, CO_2. Isolates can be obtained in rabbit or sheep blood agar plates. Culture on axenic media is difficult since the bacteria are fastidious and culture is usually obtained only after 2–4 weeks of incubation. Subcultures are usually more easy in 3–5 days on agar plates. In primary isolation, colonies are usually rough and become smooth after several passages in agar plates. These bacteria are considered to be facultative intracellular bacteria since target cells in mammals could be either endothelial cells or red blood cells (Dehio, 2001).

In vivo, *B. bacilliformis* has been observed in human erythrocytes in patients suffering from Oroya fever (Pinkerton and Weinman, 1937) as well as *B. quintana* in homeless people (Rolain *et al.*, 2002). Interestingly, *B. quintana* has also been observed in erythroblast cells in bone marrow in bacteremic homeless patients (Rolain *et al.*, 2003b). In cats, *B. henselae* (Kordick and Breitschwerdt, 1995; Rolain *et al.*, 2001) and *B. koehlerae* (Rolain *et al.*, 2003c) have also been observed in erythrocytes. *Bartonella tribocorum* was also observed in erythrocytes of rats (Schulein *et al.*, 2001). It was shown in red blood cells of rats that the bacterial replication of *B. tribocorum* was regulated and stopped at a maximum of 8 bacteria/cell (Schulein *et al.*, 2001). The mechanism responsible for this is unknown, but it prevents hemolysis and allows the persistence of *Bartonella* spp. within erythrocytes (Schulein *et al.*, 2001). Adherence to red blood cells is mediated by the flagella in *B. bacilliformis* (Walker and Winkler, 1981; Scherer, DeBuron-Connors and Minnick, 1993; Rolain *et al.*, 2003g). As the other *Bartonella* do not possess polar flagella, other mechanisms are involved in adhesion to red blood cells, especially the bundle-forming pili of *B. henselae* (Greub and Raoult, 2002). It has been speculated that the binding of *Bartonella* to spectrin may be a first step necessary to alter the erythrocyte membrane for internalizing the bacteria (Buckles and McGinnis, 2000), but a bacterial protein, namely deformin, may also be involved (Mernaugh and Ihler, 1992; Iwaki-Egawa and Ihler, 1997).

Apart from their tropism for red blood cells, a second typical pathogenic feature of *Bartonella* spp. is their ability to trigger angiogenesis. The tropism of these bacteria for endothelial cells seems correlated with the ability to induce angioproliferative lesions: verruga peruana for *B. bacilliformis* and bacillary angiomatosis for *B. henselae* and *B. quintana*. *In vitro* culture of *B. henselae* or *B. quintana* in endothelial cells increases the lifetime of the culture, stimulates cellular multiplication (Palmari *et al.*, 1996) and leads to morphological changes of the cells similar to those observed with angiogenic factors (Koehler *et al.*, 1992; Maeno *et al.*, 1999). This angiogenic effect is mainly due to an increase in the production of vascular endothelial growth factor (VEGF) by the endothelial cells infected with *Bartonella* (Kempf *et al.*, 2001). *In vitro*, *B. bacilliformis*, *B. quintana* and *B. henselae* are cultivated in human endothelial cells (Zbinden, Höchli and Nadal, 1995; Brouqui and Raoult, 1996; Dehio *et al.*, 1997). Isolation of *Bartonella* in these cell-culture systems usually leads to a more rapid isolation in 10–15 days (Koehler *et al.*, 1992). *Bartonella henselae* binds to human endothelial cells by outer membrane proteins (Burgess and Anderson, 1998). The endothelial receptors involved in *Bartonella* adhesion include intercellular adhesion molecule-1 (ICAM-1) (Dehio *et al.*, 1997). Endothelial cells are invaded either by endocytosis or by the way of an invasome (Dehio *et al.*, 1997).

EPIDEMIOLOGY

Bartonella bacilliformis is the agent of Carrion's disease and is transmitted to humans by the sand fly *Lutzomia* in the Peruvian Andes (Peru, Ecuador and Colombia) (Alexander, 1995; Amano *et al.*, 1997). Humans are the only known reservoir of this disease. In these areas, the prevalence of the disease may be high, even though most of the cases are asymptomatic (Herrer, 1990). The disease is also observed in areas where *Lutzomia* is absent, suggesting that other vectors may exist.

Bartonella quintana infections are transmitted by the body louse (Brouqui *et al.*, 1999). Humans are the only known reservoir of *B. quintana* (Maurin and Raoult, 1996), a worldwide species. Transmission of the disease to humans by the body louse (*Pediculus humanis corporis*) has been demonstrated first as trench fever during World War I (Ohl and Spach, 2000). The bacterium has been detected in cat's fleas from France, suggesting that other vectors or animal reservoir may also exist (Rolain *et al.*, 2003d). After World War I, the incidence of trench fever decreased; however, during World War II, the disease reemerged (Foucault *et al.*, 2002). Reports have indicated a reemergence of *B. quintana* infections (bacteremia in homeless people, bacillary angiomatosis and endocarditis) in Europe and the United States (Drancourt *et al.*, 1995; Spach *et al.*, 1995b; Maurin and Raoult, 1996; Foucault *et al.*, 2002). Epidemiological factors significantly correlated with infection by *B. quintana* are poor living conditions (homeless), presence of body lice and alcoholism (Spach *et al.*, 1995b; Brouqui *et al.*, 1996; Koehler *et al.*, 1997). A study in the United States has demonstrated that intravenous drug use is a risk factor for *B. quintana* infections, suggesting that this bacterium could be transmitted intravenously (Comer *et al.*, 1997).

Bartonella henselae, a species first recognized in 1990 (Slater *et al.*, 1990), is the main etiological agent of CSD (Regnery *et al.*, 1992a) and is also responsible for bacillary angiomatosis and peliosis hepatis in immunocompromised (mainly AIDS) patients (Koehler *et al.*, 1997), bacteremia and endocarditis (Anderson and Neuman, 1997). Cats are the main reservoir of *B. henselae*. Bacteremic cats are asymptomatic, and the bacterium is transmitted to the cats via the cat flea *Ctenocephalides felis* (Koehler, Glaser and Tappero, 1994; Chomel *et al.*, 1996). Humans may be contaminated by cat scratches or bites. The role of the cat flea (*C. felis*) as a vector for human transmission has been proposed (Chomel *et al.*, 1996; Anderson and Neuman, 1997). *Bartonella henselae* has also been recently recovered from ticks (Chang *et al.*, 2001; Sanogo *et al.*, 2003), but the role of tick is not known. Contact with cats is the main epidemiological factor associated with CSD (Regnery, Martin and Olson, 1992; Kordick *et al.*, 1995), endocarditis (Raoult *et al.*, 1996, 2003; Fournier *et al.*, 2001) and bacillary angiomatosis (Tappero *et al.*, 1993b; Koehler *et al.*, 1997; Gasquet *et al.*, 1998).

The epidemiology of other pathogenic species of *Bartonella* is not well understood, but the role of several vectors in the transmission of *Bartonella* species is acknowledged since DNA amplification of several species has been reported in ticks (Chang *et al.*, 2001) and fleas (Rolain *et al.*, 2003d).

CLINICAL FEATURES

Carrion's Disease

The disease may present as an acute septicemia with hemolysis, known as Oroya fever (acute form) (Gray *et al.*, 1990; Maguina Vargas, 1998), and a chronic form named as verruga peruana. Asymptomatic acute infections have also been reported (Amano *et al.*, 1997). The acute phase corresponds to a massive invasion of erythrocytes and is responsible for a mortality rate of 40–85% without treatment (Ihler, 1996). In most cases of fatal Oroya fever, death has been a result of secondary infection, primarily with *Salmonella* (Bass, Vincent and Person, 1997b). During this phase, bacteria are easily shown in red blood cells after Giemsa or immunofluorescence staining (Rolain *et al.*, 2003g). Among the endemic population, Carrion's disease is also present as a chronic form called verruga peruana which is characterized by benign cutaneous vascular lesions (Arias-Stella, Lieberman and Erlandson, 1986; Maguina Vargas, 1998). These benign lesions are usually friable and bleed readily (Ihler, 1996).

Trench Fever and Chronic Bacteremia

Trench fever, also known as quintan fever, is characterized by infection of blood by *B. quintana*. The disease is transmitted by the human body louse. The incubation period for the disease is between 15 and 25 days. Clinical manifestations of trench fever range from asymptomatic infection to severe illness. The classical presentation reported among troops during World War I corresponds to a febrile illness of acute onset and periodic nature, often accompanied by severe headache and pain in the long bones of the legs (Maurin and Raoult, 1996).

The term quintan fever refers to the 5-day recurrences most often observed. Although trench fever often results in prolonged disability, no fatalities have been recorded. Few illnesses become chronic, with or without attacks of fever and aching (Byam *et al.*, 1919). Persistent bacteremia has long been a recognized symptom of *B. quintana* infection among troops during the World Wars (Maurin and Raoult, 1996). *Bartonella quintana* has been demonstrated to cause bacteremia in homeless patients (Foucault *et al.*, 2002) and in patients with AIDS (Koehler *et al.*, 2003). In these patients, *B. quintana* may cause acute, severe illnesses but often produces a chronic infection persisting for months with few or no symptoms (Raoult *et al.*, 1994; Drancourt *et al.*, 1996b).

Cat-Scratch Disease

Classical clinical CSD is defined as typical CSD, and other clinical manifestations of infections caused by *B. henselae* have been designed as atypical CSD (Bass, Vincent and Person, 1997a). In typical CSD, nearly all patients give a history of a scratch or bite by and contact with a cat. This is manifested by a regional lymphadenopathy several days after being bitten by a cat. The infection is usually self-limited, but late in the course, in about 10% of patients, the node(s) may suppurate if they are not drained, preferably by needle aspiration. Most patients with typical CSD remain afebrile and are not ill throughout the course of the disease (Carithers, 1985; Margileth, 1993; Bass, Vincent and Person, 1997a). Lymphadenopathy occurs most often in nodes that drain the area where cat scratches occur, most often the axilla, the neck and the groin (Carithers, 1985; Margileth, 1993). Systemic or severe disease complicates CSD in 5–14%, without lymphadenopathy (Carithers, 1985; Margileth, 1993).

Compared with typical CSD patients, atypical CSD patients have prolonged fever (>2 weeks), malaise, fatigue, myalgia and arthralgia, weight loss and splenomegaly. In immunocompromised patients, especially in HIV-infection patients, severe, progressive, disseminated disease may occur (Koehler *et al.*, 2003). Without appropriate therapy, infection spreads systematically to involve virtually all organs, and the outcome is often fatal (Bass, Vincent and Person, 1997a). Visceral manifestation of CSD in children is a possible atypical clinical manifestation of CSD. Other atypical presentation of CSD includes conjunctival granuloma with adjacent preauricular lymphadenitis that evolves and regresses spontaneously over a course of weeks to months (Parinaud's oculoglandular syndrome). This syndrome is the most common form of atypical CSD (about 5%) (Bass, Vincent and Person, 1997a). The disease is transmitted by exposure to infected cats, with inoculation of the organism indirectly into the eye. CSD encephalopathy may occur in 0.17–2% of the cases (Bass, Vincent and Person, 1997a). The onset of neurologic complications varies from a few days to 2 months after the onset of lymphadenopathy. Symptoms include convulsions, persistent headache, malaise and lethargy, lasting for 1 to several weeks. Transient blindness with stellate macular retinitis due to *B. henselae* includes unilateral loss of vision, with central scotoma, optic disc swelling and macular star formation, and there is spontaneous resolution, with complete recovery of vision usually within 1–3 months (Carithers and Margileth, 1991). Both encephalopathy and neuroretinitis are unusual complications of CSD.

Although *B. quintana* is not considered an agent of CSD, the bacterium has been isolated from the blood of patients with lymphadenopathy only (Raoult *et al.*, 1994; Drancourt *et al.*, 1996b). In the cases described by Raoult *et al.* (1994) and Drancourt *et al.* (1996b), the only epidemiologic risk factor identified was the presence of infected cat fleas, and recently, we have found *B. quintana* DNA in cat fleas (Rolain *et al.*, 2003d).

Endocarditis

Bartonella quintana (Spach *et al.*, 1993, 1995a; Drancourt *et al.*, 1995; Raoult *et al.*, 1996, 2003; Fournier *et al.*, 2001), *B. henselae*

(Hadfield *et al.*, 1993; Drancourt *et al.*, 1996a; Raoult *et al.*, 1996, 2003; Fournier *et al.*, 2001), *B. elizabethae* (Daly *et al.*, 1993) and *B. vinsonii* subsp. *berkhoffii* (Roux *et al.*, 2000) have been isolated from patients suffering from bacterial endocarditis. The first case of *Bartonella* endocarditis was reported in an HIV-infected homosexual man in 1993 (Spach *et al.*, 1993). *Bartonella quintana* endocarditis has been reported subsequently in three non-HIV–infected, homeless men in France (Drancourt *et al.*, 1995). All patients required valve replacements because of extensive valvular damages, and pathologic investigation confirmed the diagnosis of endocarditis. Following these reports, the first cases in a series of 22 *Bartonella* endocarditis have been reported, with 5, 4 and 13 cases due to *B. quintana*, *B. henselae* and *Bartonella* spp., respectively (Fournier *et al.*, 2001). *Bartonella quintana* endocarditis can be differentiated from *B. henselae* endocarditis mainly on the basis of associated epidemiologic factors. *Bartonella quintana* endocarditis is most often observed in homeless, chronic alcoholic patients, without previous known valvulopathy, whereas *B. henselae* endocarditis most often occurs in patients with known valvulopathy who had had contact with cats. We have reported a series of 101 cases of *Bartonella* endocarditis, diagnosed by serology, PCR and/or culture (Raoult *et al.*, 2003). In this series, 12 patients died and in 2 patients, the infection relapsed.

Bacillary Angiomatosis and Peliosis Hepatitis

Bacillary angiomatosis is a vascular proliferative disease most often involving the skin but may disseminate to other organs. The disease was first described in HIV-infected patients (Stoler *et al.*, 1983) and organ transplant recipients (Kemper *et al.*, 1990) but may also affect immunocompetent patients (Cockerell *et al.*, 1987). Both *B. quintana* and *B. henselae* are considered etiologic agents of bacillary angiomatosis (Koehler *et al.*, 1992; Spach, 1992). Primary skin lesions correspond to papules, which then increases to form nodules. These lesions may bleed when punctured. The clinical differential diagnosis includes pyogenic granuloma, hemangioma, subcutaneous tumors and Kaposi's sarcoma (Maurin and Raoult, 1996). Cutaneous bacillary angiomatosis lesions may be accompanied by involvement of bone marrow, liver, spleen or lymph nodes (Cockerell *et al.*, 1987; Koehler *et al.*, 1988; Marasco, Lester and Parsonnet, 1989; Milam *et al.*, 1990). Conversely, involvement of the skin does not always occur in bacillary angiomatosis, and patients may only have spleen, liver or lymph node infection (Kemper *et al.*, 1990; Shinella and Alba Greco, 1990).

Peliosis hepatitis is defined as a vascular proliferation of sinusoid hepatic capillaries, resulting in blood-filled spaces in the liver. This disease was first described in patients with tuberculosis and advanced cancers and in association with the use of drugs such as anabolic steroids (Koehler, 1996). *Bartonella henselae* has been recently recognized as an agent of peliosis hepatitis in HIV-infected patients (Perkocha *et al.*, 1990; Maurin, Birtles and Raoult, 1997). Peliosis hepatitis may be considered a visceral manifestation of bacillary angiomatosis, and the term bacillary peliosis has been proposed.

DIAGNOSIS

Because of the nonspecific clinical manifestations, laboratory testing is often necessary to confirm the diagnosis. Diagnosis of infections due to *Bartonella* spp. is often difficult because of the fastidious nature of these bacteria and limitations of serology. Indeed, histological examination of biopsies and PCR, among others, remain essential for the diagnosis of such infections.

Specimen Collection

Several samples are used in establishing the diagnosis of infections due to *Bartonella* spp.: blood, biopsy specimens and arthropods (xenodiagnosis). These specimen collections should be made as early as possible during illness.

Blood can be obtained either on citrate-containing vial for culture or on ethylenediaminetetraacetic acid (EDTA) for PCR diagnosis. EDTA anticoagulant should be avoided for culture because it leads to detachment of the cell monolayer from coverslips. The best anticoagulant for an optimal yield is sodium citrate. Blood should be obtained before antimicrobial therapy. If inoculation of cell culture or PCR process delays for more than 24 h, plasma, buffy coat or whole blood must be frozen at $-70\,°C$ or in liquid nitrogen.

For serologic diagnosis, blood sample should be collected early during the disease and a second sample obtained after 1 or 2 weeks. Serum samples should be stored at $-20\,°C$ or lower for long periods, without degradation of the antibodies. Moreover, blood smears are useful mainly for the diagnosis of Carrion's disease.

Biopsy specimens useful for a diagnosis of CSD are lymph nodes, and those for a diagnosis of endocarditis due to *Bartonella* are cardiac valve and vascular aneurysm or graft.

Bartonellae can be detected by culture, PCR and immunodetection (Rolain *et al.*, 2003d; Sanogo *et al.*, 2003) in various arthropods including ticks, lice and fleas (Rydkina *et al.*, 1999; Chang *et al.*, 2001; Fournier *et al.*, 2002; Rolain *et al.*, 2003d; Sanogo *et al.*, 2003). The arthropod should primarily undergo surface disinfection with iodinated alcohol and then be crushed in medium and inoculated onto a shell vial for culture (Kelly, Raoult and Mason, 1991). Specimen collection of arthropods can be made easily (stored dried in a box) and sent by mail to a reference laboratory for analysis.

Direct Diagnosis

Culture

Bacteria of the genus *Bartonella* may be isolated from blood and biopsies either on axenic medium (Drancourt *et al.*, 1995; Spach *et al.*, 1995b) or in cell-culture systems (Koehler *et al.*, 1992; Zbinden, Höchli and Nadal, 1995). *Bartonella* can be grown on blood agar at $37\,°C$ in a 5% CO_2 atmosphere, except for *B. bacilliformis* which is best grown at $30\,°C$. Primary isolation of the bacteria necessitates 12–15 days of incubation and can be as long as 45 days, and thus, agar plates should be conserved for 2 months (Koehler *et al.*, 1992). Sensitivity can be increased after cell lysis (Koehler *et al.*, 1992; Welch *et al.*, 1992) or after freezing the samples (Brenner *et al.*, 1997). Blood culture systems that are currently used in laboratories support the growth of *Bartonella* (Larson *et al.*, 1994). The two techniques of culture, axenic and with cells, should be used to increase the sensitivity of the isolation (Drancourt *et al.*, 1995). Culture of *Bartonella* spp. may be successful with blood samples from bacteremic patients (La Scola and Raoult, 1999; Foucault *et al.*, 2002) and with cardiac valve specimens of endocarditis patients (Fournier *et al.*, 2001; Raoult *et al.*, 2003) as well as skin, lymph node or other organ biopsy samples from patients with bacillary angiomatosis (La Scola and Raoult, 1999). In contrast, *B. henselae* is rarely reported from patients with CSD (La Scola and Raoult, 1999).

Immunodetection

Immunological detection of *Bartonella* spp. has been achieved in various situations such as bacillary angiomatosis in cutaneous biopsies (Reed *et al.*, 1992), peliosis hepatitis (Reed *et al.*, 1992; Slater and Welch, 1995), CSD in lymph nodes (Slater and Welch, 1995) and in heart valves of patients suffering from *Bartonella* endocarditis (Lepidi, Fournier and Raoult, 2000).

Immunofluorescence detection of *B. henselae* during CSD is performed directly on lymph node smears (Rolain *et al.*, 2003e). Immunofluorescence detection in red blood cells has been reported both in *B. quintana* bacteremic homeless peoples (Rolain *et al.*, 2002) and in patients suffering from Oroya fever (Rolain *et al.*, 2003g).

Molecular Biology

Amplification by PCR of citrate synthase gene (*gltA*) (Drancourt *et al.*, 1995; Birtles and Raoult, 1996), 16S rRNA gene (Regnery *et al.*, 1992a; Raoult *et al.*, 1994), the 16S-23S rRNA *ITS* (Roux and Raoult, 1995), the 60-kDa heat-shock protein gene (*groEL*) (Marston, Sumner and Regnery, 1999) and the *pap31* gene (Zeaiter, Fournier and Raoult, 2002) is currently used for the diagnosis of *Bartonella*-related infections. If the former is amplified, it is further sequenced for identification or diagnosis is confirmed using restriction fragment length polymorphism (RFLP) or hybridization (Norman *et al.*, 1995; Matar *et al.*, 1999; Renesto *et al.*, 2001; Margolis *et al.*, 2003). These techniques are very specific, but their sensitivity may vary according to the type of samples. Thus, the current strategy for the diagnosis of *Bartonella* infections is to use two different target genes (*ITS* gene and *pap31*), and if results of the two PCRs are discordant, use a third gene (*groEL*) (Zeaiter, Fournier and Raoult, 2002). Clinical samples are considered positive only if at least two genes are positive and if sequences obtained give the same identification (Rolain *et al.*, 2003e). Amplification of *Bartonella* spp. DNA in tissue biopsy samples is mainly useful for patients with CSD (lymph node or other organs) (Anderson, Sims and Regnery, 1994; Goral *et al.*, 1994; Zeaiter, Fournier and Raoult, 2002; Rolain *et al.*, 2003a,e), bacillary angiomatosis (angiomatous lesions of skin or other organs) (Relman *et al.*, 1990) or endocarditis (when cardiac valve is removed) (Fournier *et al.*, 2001; Raoult *et al.*, 2003). By using serum samples, we have developed a new one-step nested PCR with the LightCycler (LCN-PCR) for the diagnosis of *Bartonella* endocarditis (Zeaiter *et al.*, 2003). Primers used derived from the riboflavin synthase-encoding gene *ribC*, and specificity of this method was 100% and sensitivity increased to 58.1% (Zeaiter *et al.*, 2003).

Indirect Diagnosis

Serology is the only noninvasive method useful in the diagnosis of *Bartonella* infections, especially for CSD, bacteremia and endocarditis (Maurin, Rolain and Raoult, 2002). Two techniques have been developed for the serology of *Bartonella* infections: immunofluorescence assay (IFA) and enzyme immunoassay (EIA). The major difficulty of these methods is discrepancies in titers obtained among these techniques, depending on the way bacterial antigens are prepared. Indeed, antigens can be prepared either from bacteria cultivated in cells or directly from agar cultures. The IFA was first described for CSD by Regnery *et al.* (1992b) who used *B. henselae* grown in Vero cells as antigens. It remains the most frequently used in-house technique for CSD, with reported sensitivities varying considerably, from nearly 100% to <30% (Regnery *et al.*, 1992b; Sander, Berner and Ruess, 2001; Maurin, Rolain and Raoult, 2002). Commercially prepared antigens are now available for *B. henselae* and *B. quintana* serology (Zbinden *et al.*, 1997; Sander *et al.*, 1998; Harrison and Doshi, 1999). Sensitivity and specificity of one of these tests have been compared with those of our in-house IFA (Maurin, Rolain and Raoult, 2002). The sensitivity of a serological test was greatly influenced by the cutoff values used for each test. At a cutoff titer of 64, the commercial test had higher sensitivity (91.2% versus 52.9%) than our in-house test, in detecting antibodies in CSD patients, but sensitivity was only 87%. Conversely, we have demonstrated that our in-house test was superior to the commercial test for the diagnosis of *Bartonella* endocarditis and bacteremia in homeless patients (Maurin, Rolain and Raoult, 2002). Significant differences in antibody titers were found between the two IFAs, and at a cutoff titer of ≥1024, our in-house test was 100% predictive of *Bartonella* endocarditis (Maurin, Rolain and Raoult, 2002). This was also found to be true in a study showing that IFA had a positive predictive value of 95% for an immunoglobulin G (IgG) titer to *B. henselae* and/or *B. quintana* (≥1:800) in patients with endocarditis (Fournier, Mainardi and Raoult, 2002). False negative is due either to antigenic heterogeneity among *B. henselae* species or to *B. clarridgeiae* (Drancourt *et al.*, 1996a). Conversely, we have found that patients without CSD may have positive serology either with commercial test or with in-house tests, especially in cases of more severe diseases such as mycobacterial infections, lymphoma or Kaposi's sarcoma (Giladi *et al.*, 2001; Ridder *et al.*, 2002; Rolain *et al.*, 2003e). The specificity of serology has been studied, and cross-reactions have been shown either between the different species of *Bartonella* (Dalton *et al.*, 1995; Sander *et al.*, 1998) or between *Bartonella* spp. and *Coxiella burnetii* (La Scola and Raoult, 1996) or *Chlamydia* (Maurin *et al.*, 1997). These cross-reactions should be considered in blood culture-negative endocarditis, where these pathogens are implicated. Moreover, the seroprevalence in a particular geographic area as well as the antigen preparation of the various IFAs used should be considered before interpreting results (Sander, Berner and Ruess, 2001). Even when two different serological tests were coupled, the specificity and the sensitivity were not sufficient for the diagnosis of CSD (Maurin, Rolain and Raoult, 2002). False-negative PCR could be due to a previous infection, with disappearance of the bacteria in the lymph node, whereas antibodies against *B. henselae* are still present, suggesting that the results of PCR strongly depend on the duration of illness (Ridder *et al.*, 2002). Since CSD lymphadenopathy may last for months, it is possible that patients with positive serology and negative PCR had had a previous infection due to *B. henselae* which was not acute at the time of sampling. This may also explain why isolation of the bacteria from lymph nodes is so difficult (La Scola and Raoult, 1999), because bacteria disappear from the node and positive serology reflects a previous infection.

TREATMENT

In vitro Susceptibility to Antibiotics

Antibiotic susceptibility testing of *Bartonella* spp. has been carried out using methods adapted to their fastidious growth and has shown that many antibiotics are bacteriostatic *in vitro* (Maurin *et al.*, 1995; Sobraques *et al.*, 1999; Rolain *et al.*, 2000). *Bartonella* spp. are susceptible to aminoglycosides, chloramphenicol, tetracyclines, macrolides, rifampin, fluoroquinolones and co-trimoxazole (Maurin *et al.*, 1995; Ives *et al.*, 1997). However, minimum inhibitory concentrations (MICs) poorly correlate with the *in vivo* efficacy of antibiotics in patients suffering from *Bartonella*-related infections (Maurin *et al.*, 1995). Failures of treatment using a β-lactam, a macrolide, a tetracycline, rifampin or fluoroquinolone for either disease have been reported (Spach and Koehler, 1998). Relapses are also frequent following withdrawal of treatment. Such discrepancies between *in vitro* and clinical data may be explained by the lack of bactericidal effect of antibiotics against *Bartonella* spp., except for aminoglycosides (Rolain, Maurin and Raoult, 2000). The lack of bactericidal activity of antibiotics was also confirmed in cell-culture models for both *B. henselae* in murine macrophage-like cells (Musso, Drancourt and Raoult, 1995) and *B. quintana* in red blood cells (Rolain *et al.*, 2003f).

Treatment in Humans

Before the antibiotic era, the only available treatment for acute anemia during Oroya fever was blood transfusion, but the effectiveness of this treatment was poor and the mortality rate was high (about 80% of cases) (Schultz, 1968). Now the recommended therapy for Oroya fever is chloramphenicol because this antibiotic is also effective against salmonellosis and other enteropathogenic bacteria that often

complicate the disease (Uertega and Payne, 1955; Cuadra, 1956; Gray *et al.*, 1990; Maguina *et al.*, 2001). Therapeutic failures and persistent bacteremia have been reported when this drug was used, and treatment with chloramphenicol appears not to eliminate the patient's risk of developing the eruptive phase of bartonellosis. Chloramphenicol is effective in most of the patients with few exceptions, and thus, a combination with another antibiotic (especially a β-lactam compound) should be recommended. Cotrimoxazole, macrolides and fluoroquinolones have also been used successfully in some patients and may represent safe alternatives (Maguina Vargas, 1998). Chloramphenicol is not effective for treating verruga peruana (Maguina Vargas, 1998). Since 1975, rifampin has become the drug of choice for treating the eruptive phase of Carrion's disease. However, failures have also been reported with rifampin (Maguina Vargas, 1998). Ciprofloxacin as well as azithromycin has been used successfully in adults with multiple eruptive-phase lesions (Maguina and Gotuzzo, 2000).

CSD typically does not respond to antibiotic therapy (Spaulding and Hennessy, 1960; Margileth, 1992), and patients should be reassured that the adenopathy is benign and that it will subside spontaneously within 2–4 months (Margileth, 2000). Management consists of analgesics for pain, prudent follow-up and drainage when necessary (Margileth, 2000). The only prospective double-blind, placebo-controlled study of treatment for immunocompetent patients with uncomplicated CSD with azithromycin was reported by Bass *et al.* (1998). Although, no significant difference in the rate or degree of resolution between the two groups was found at 30 days, patients treated with azithromycin had a significantly faster reduction of lymph node volume during the first month of azithromycin treatment as compared with that in placebo (Bass *et al.*, 1998). The current recommendation for the moderate-to-severely-ill patient is no treatment. The azithromycin regimen could be an alternative for patients with large, bulky lymphadenopathy (Bass *et al.*, 1998). When CSD lymph nodes suppurate, needle aspiration is probably the best treatment, and the patient usually notes decreased pain within 24–48 h (Margileth, 2000).

Most cases of Trench fever were reported before the antibiotic era. There were no fatal cases of Trench fever during World War, and aspirin was the most effective drug for the pain (Hurst, 1942). However, successful treatment with tetracycline or chloramphenicol has been reported after World War II in some trench fever patients (Bass, Vincent and Person, 1997b). We have demonstrated in a randomized, placebo-controlled clinical trial that patients with chronic bacteremia should be treated with gentamycin for 14 days and with doxycycline for 28 days (Foucault, Raoult and Brouqui, 2003). These results have been confirmed in a randomized, placebo-controlled clinical trial (Foucault, Raoult and Brouqui, 2003). Consequently, patients with chronic *B. quintana* bacteremia should be treated with gentamycin (3 mg/kg intravenously once a day) for 14 days, with doxycycline (200 mg/day per oral once a day) for 28 days (Foucault, Raoult and Brouqui, 2003). Patients with chronic bacteremia should be evaluated for endocarditis. Patients with renal insufficiency, obesity or increased fluid volume should be monitored closely and the dose of gentamicin adjusted with a twice-daily dosing schedule to avoid nephrotoxicity of the drug.

Antibiotic treatment of bacillary angiomatosis and peliosis hepatis has never been studied systematically. The first described patient with bacillary angiomatosis was treated empirically with erythromycin, and the lesion completely resolved (Stoler *et al.*, 1983). Subsequently, erythromycin has become the drug of first choice and has been used successfully to treat many patients with bacillary angiomatosis (Koehler and Tappero, 1993; Tappero *et al.*, 1993a). Doxycycline and fluoroquinolones can be used as alternatives (Slater *et al.*, 1990; Lucey *et al.*, 1992; Koehler and Tappero, 1993). Intravenous administration of erythromycin should be used in patients with severe disease (Koehler and Tappero, 1993). Combination therapy, with the addition of rifampin to either erythromycin or doxycycline, is recommended for immunocompromised patients with acute, life-threatening *Bartonella* infection. Treatment should be given for at least 3 months for bacillary angiomatosis and 4 months for peliosis hepatis to prevent relapses (Krekorian *et al.*, 1990; Szaniawski *et al.*, 1990; Koehler *et al.*, 1992; Cotell and Noskin, 1994; Gazineo *et al.*, 2001; Koehler and Relman, 2002).

Among patients with *Bartonella* endocarditis that we reported, those receiving an aminoglycoside were more likely to fully recover and, if aminoglycoside was prescribed for at least 14 days, more likely to survive, confirming the important role of this antibiotic compound in the treatment of *Bartonella* endocarditis (Raoult *et al.*, 2003). These data strongly suggest the use of aminoglycoside therapy for at least 14 days for cases of *Bartonella* spp. endocarditis, and this antibiotic should be accompanied by doxycycline for 6 weeks (Raoult *et al.*, 2003). Since *B. quintana* bacteremia is optimally treated with doxycycline plus gentamicin (Foucault, Raoult and Brouqui, 2003), in the absence of any prospective study, it is logical that this regimen should be used for endocarditis when *Bartonella* spp. are identified as the causative agent.

PREVENTION

Bartonella quintana infections can be prevented by eradicating body lice (Maurin and Raoult, 1996), and for *B. henselae*, immunocompromised patients should avoid contacts with cats (Koehler, Glaser and Tappero, 1994; Maurin and Raoult, 1996), and also eradicate cat fleas (Koehler, Glaser and Tappero, 1994).

ACKNOWLEDGEMENT

We thank Vijay A.K.B. Gundi for English corrections.

REFERENCES

Alexander B (1995) A review of bartonellosis in Ecuador and Colombia. Am J Trop Med Hyg 52 (4):354–359.

Amano Y, Rumbea J, Knobloch J *et al.* (1997) Bartonellosis in Ecuador: sero-survey and current status of cutaneous verrucous disease. Am J Trop Med Hyg 57 (2):174–179.

Anderson BE, Neuman MA (1997) *Bartonella* spp. as emerging human pathogens. Clin Microbiol Rev 10:203–219.

Anderson B, Sims K, Regnery R (1994) Detection of the *Rochalimaea henselae* DNA in specimens from cat-scratch disease patients by polymerase chain reaction. J Clin Microbiol 32:942–948.

Arias-Stella J, Lieberman PH, Erlandson RA (1986) Histology, immunochemistry, and ultrastructure of the verruga in Carrion's disease. Am J Surg Pathol 10:595–610.

Baker JA (1946) A rickettsial infection in Canadian voles. J Exp Med 84:37–51.

Bass JW, Vincent JF, Person DA (1997a) The expanding spectrum of *Bartonella* infections. II. Cat scratch disease. Pediatr Infect Dis J 16 (2):163–179.

Bass JW, Vincent JM, Person DA (1997b) The expanding spectrum of *Bartonella* infections. I. Bartonellosis and trench fever. Pediatr Infect Dis J 16:2–10.

Bass JW, Freitas BC, Freitas AD *et al.* (1998) Prospective randomized double blind placebo-controlled evaluation of azithromycin for treatment of cat-scratch disease. Pediatr Infect Dis J 17 (6):447–452.

Birtles RJ, Raoult D (1996) Comparison of partial citrate synthase gene (*gltA*) sequences for phylogenetic analysis of *Bartonella* species. Int J Syst Bacteriol 46 (4):891–897.

Birtles RJ, Harrison TG, Saunders NA, Molyneux DH (1995) Proposals to unify the genera *Grahamella* and *Bartonella*, with descriptions of *Bartonella talpae* comb. nov., *Bartonella peromysci* comb. nov., and three new species, *Bartonella grahamii* sp. nov., *Bartonella taylorii* sp. nov., and *Bartonella doshiae* sp. nov. Int J Syst Bacteriol 45 (1):1–8.

Breitschwerdt B, Kordick D (2000) *Bartonella* infection in animals: carriership, reservoir potential, pathogenicity and zoonotic potential for human infection. Clin Microbiol Rev 13:428–438.

Brenner DJ, O'Connor S, Winkler HH, Steigerwalt AG (1993) Proposals to unify the genera *Bartonella* and *Rochalimaea*, with descriptions of *Bartonella quintana* comb. nov., *Bartonella vinsonii* comb. nov., *Bartonella henselae* comb. nov., and *Bartonella elizabethae* comb. nov., and to remove the family Bartonellaceae from the order Rickettsiales. Int J Syst Bacteriol 43:777–786.

Brenner SA, Rooney JA, Manzewitsch P, Regnery RL (1997) Isolation of *Bartonella* (*Rochalimae*) *henselae*: effects of methods of blood collection and handling. J Clin Microbiol 35:544–547.

Brouqui P, Raoult D (1996) *Bartonella quintana* invades and multiplies within endothelial cells *in vitro* and *in vivo* and forms intracellular blebs. Res Microbiol 147:719–731.

Brouqui P, Raoult D (2001) Endocarditis due to rare and fastidious bacteria. Clin Microbiol Rev 14 (1):177–207.

Brouqui P, Houpikian P, Tissot-Dupont H *et al.* (1996) Survey of the seroprevalence of *Bartonella quintana* in homeless people. Clin Infect Dis 23:756–759.

Brouqui P, La Scola B, Roux V, Raoult D (1999) Chronic *Bartonella quintana* bacteremia in homeless patients. N Engl J Med 340 (3):184–189.

Buckles EL, McGinnis HE (2000) Interaction of *Bartonella bacilliformis* with human erythrocyte membrane proteins. Microb Pathog 29 (3):165–174.

Burgess AWO, Anderson BE (1998) Outer membrane proteins of *Bartonella henselae* and their interaction with human endothelial cells. Microb Pathog 25:157–164.

Byam W, Caroll JH, Churchill JH *et al.* (1919) Trench Fever: A Louse-Born Disease. Oxford University Press: London.

Carithers HA (1985) Cat-scratch disease: an overview based on a study of 1200 patients. Am J Dis Child 139:1124–1133.

Carithers HA, Margileth AM (1991) Cat scratch disease. Acute encephalopathy and other neurologic manifestations. Am J Dis Child 145:98–101.

Chang CC, Chomel BB, Kasten RW *et al.* (2001) Molecular evidence of *Bartonella* spp. in questing adult Ixodes pacificus ticks in California. J Clin Microbiol 39 (4):1221–1226.

Chomel BB, Kasten RW, Floyd-Hawkins K *et al.* (1996) Experimental transmission of *Bartonella henselae* by the cat flea. J Clin Microbiol 34 (8):1952–1956.

Cockerell CJ, Whitlow MA, Webster GF, Friedman-Kien AE (1987) Epithelioid angiomatosis: a distinct vascular disorder in patients with the acquired immunodeficiency syndrome or AIDS-related complex. Lancet 2:654–656.

Comer JA, Flynn C, Regnery RL *et al.* (1997) Antibodies to *Bartonella* species in inner-city intravenous drug users in Baltimore, Md. Arch Intern Med 156:2491–2495.

Cotell S, Noskin G (1994) Bacillary angiomatosis. Clinical and histologic features, diagnosis, and treatment. Arch Intern Med 154:524–528.

Cuadra MC (1956) Salmonellosis complications in human bartonellosis. Tex Rep Biol Med 14:97–113.

Dalton MJ, Robinson LE, Cooper J *et al.* (1995) Use of *Bartonella* antigens for serologic diagnosis of cat-scratch disease at a National Referral Center. Arch Intern Med 155:1670–1676.

Daly JS, Worthington MG, Brenner DJ *et al.* (1993) *Rochalimaea elizabethae* sp. nov. isolated from a patient with endocarditis. J Clin Microbiol 31 (4):872–881.

Debré R, Lamy M, Jammet ML *et al.* (1950) La maladie des griffes du chat. Société Médicale des Hôpitaux Paris 66:76–79.

Dehio C (2001) *Bartonella* interactions with endothelial cells and erythrocytes. Trends Microbiol 9:279–285.

Dehio C, Meyer M, Berger J *et al.* (1997) Interaction of *Bartonella henselae* with endothelial cells results in bacterial aggregation on the cell surface and the subsequent engulfment and internalisation of the bacterial aggregate by a unique structure, the invasome. J Cell Sci 110:2141–2154.

Drancourt M, Mainardi JL, Brouqui P *et al.* (1995) *Bartonella* (*Rochalimaea*) *quintana* endocarditis in three homeless men. N Engl J Med 332:419–423.

Drancourt M, Birtles RJ, Chaumentin G *et al.* (1996a) New serotype of *Bartonella henselae* in endocarditis and cat-scratch disease. Lancet 347:441–443.

Drancourt M, Moal V, Brunet P *et al.* (1996b) *Bartonella* (*Rochalimaea*) *quintana* infection in a seronegative hemodialyzed patient. J Clin Microbiol 34:1158–1160.

Foucault C, Barrau K, Brouqui P, Raoult D (2002) *Bartonella quintana* bacteremia among homeless people. Clin Infect Dis 35:684–689.

Foucault C, Raoult D, Brouqui P (2003) Randomized open trial of gentamicin and doxycycline for eradication of *Bartonella quintana* from blood in patients with chronic bacteremia. Antimicrob Agents Chemother 47 (7):2204–2207.

Fournier PE, Lelievre H, Eykyn SJ *et al.* (2001) Epidemiologic and clinical characteristics of *Bartonella quintana* and *Bartonella henselae* endocarditis: a study of 48 patients. Medicine (Baltimore) 80 (4):245–251.

Fournier PE, Mainardi JL, Raoult D (2002) Value of microimmunofluorescence for diagnosis and follow-up of *Bartonella* endocarditis. Clin Diagn Lab Immunol 9 (4):795–801.

Fournier PE, Ndihokubwayo JB, Guidran J *et al.* (2002) Human pathogens in body and head lice. Emerg Infect Dis 8 (12):1515–1518.

Gasquet S, Maurin M, Brouqui P *et al.* (1998) Bacillary angiomatosis in immunocompromised patients: a clinicopathological and microbiological study of seven cases and review of literature. AIDS 12:1793–1803.

Gazineo JL, Trope BM, Maceira JP *et al.* (2001) Bacillary angiomatosis: description of 13 cases reported in five reference centers for AIDS treatment in Rio de Janeiro, Brazil. Rev Inst Rev Inst Med Trop Sao Paulo 43 (1):1–6.

Giladi M, Kletter Y, Avidor B *et al.* (2001) Enzyme immunoassay for the diagnosis of cat-scratch disease defined by polymerase chain reaction. Clin Infect Dis 33 (11):1852–1858.

Goral S, Anderson B, Hager C, Edwards KM (1994) Detection of *Rochalimaea henselae* DNA by polymerase chain reaction from suppurative nodes of children with cat-scratch disease. Pediatr Infect Dis J 13:994–997.

Gray GC, Johnson AA, Thornton SA *et al.* (1990) An epidemic of Oroya fever in the peruvian Andes. Am J Trop Med Hyg 42:215–221.

Greub G, Raoult D (2002) *Bartonella*: new explanations for old diseases. J Med Microbiol 51 (11):915–923.

Hadfield TL, Warren R, Kass M *et al.* (1993) Endocarditis caused by *Rochalimaea henselae*. Hum Pathol 24:1140–1141.

Harrison TG, Doshi N (1999) Serological evidence of *Bartonella* spp. infection in the UK. Epidemiol Infect 123:233–240.

Herrer A (1990) Epidemiologia de la verruga peruana. Gonzales-Magabun: Lima, Peru.

Houpikian P, Raoult D (2001) Molecular phylogeny of the genus *Bartonella*: what is the current knowledge? FEMS Microbiol Lett 200 (1):1–7.

Hurst H (1942) Trench fever. Br Med J 70:318–320.

Ihler GM (1996) *Bartonella bacilliformis*: dangerous pathogen slowly emerging from deep background. FEMS Microbiol Lett 144 (1):1–11.

Ives TJ, Manzewitsch P, Rechery RL *et al.* (1997) *In vitro* susceptibilities of *Bartonella henselae*, *B. quintana*, *B. elizabethae*, *Rickettsia rickettsii*, *R. conorii*, *R. akari*, and *R. prowazekii* to macrolide antibiotics as determined by immunofluorescent–antibody analysis of infected Vero cell monolayers. Antimicrob Agents Chemother 41:578–582.

Iwaki-Egawa S, Ihler GM (1997) Comparison of the abilities of proteins from *Bartonella bacilliformis* and *Bartonella henselae* to deform red cell membranes and to bind to red cell ghost proteins. FEMS Microbiol Lett 157:207–217.

Jacomo V, Kelly PJ, Raoult D (2002) Natural history of *Bartonella* infections (an exception to Koch's postulate). Clin Diagn Lab Immunol 9 (1):8–18.

Karem KL, Paddock CD, Regnery RL (2000) *Bartonella henselae*, *B. quintana*, and *B. bacilliformis*: historical pathogens of emerging significance. Microbes Infect 2 (10):1193–1205.

Kelly PJ, Raoult D, Mason PR (1991) Isolation of spotted fever group rickettsias from triturated ticks using a modification of the centrifugation-shell vial technique. Trans R Soc Trop Med Hyg 85:397–398.

Kemper CA, Lombard CM, Deresinski SC, Tompkins LS (1990) Visceral bacillary epithelioid angiomatosis: possible manifestations of disseminated cat scratch disease in the immunocompromised host: a report of two cases. Am J Med 89:216–222.

Kempf VA, Volkmann B, Schaller M *et al.* (2001) Evidence of a leading role for VEGF in *Bartonella henselae*-induced endothelial cell proliferations. Cell Microbiol 3 (9):623–632.

Kerkhoff FT, Bergmans AMC, van der Zee A, Rothova A (1999) Demonstration of *Bartonella grahamii* DNA in ocular fluids of a patient with neuroretinitis. J Clin Microbiol 37 (12):4034–4038.

Koehler JE (1996) *Bartonella* infections. Adv Pediatr Infect Dis 11:1–27.

Koehler JE, Relman DA (2002) *Bartonella* species. In Yu VL, Weber R, Raoult D (eds). *Antimicrobial Therapy and Vaccines*. Apple Trees Production: New York. 925–931.

Koehler JE, Tappero JW (1993) Bacillary angiomatosis and bacillary peliosis in patients infected with human immunodeficiency virus. Clin Infect Dis 17:612–624.

Koehler JE, Leboit PE, Egbert BM, Berger TG (1988) Cutaneous vascular lesions and disseminated cat-scratch disease in patients with the acquired immunodeficiency syndrome (AIDS) and AIDS-related complex. Ann Intern Med 109:449–455.

Koehler JE, Quinn FD, Berger TG *et al.* (1992) Isolation of *Rochalimaea* species from cutaneous and osseous lesions of bacillary angiomatosis. N Engl J Med 327:1625–1631.

Koehler JE, Glaser CA, Tappero JW (1994) *Rochalimaea henselae* infection: a new zoonosis with the domestic cat as a reservoir. JAMA 271:531–535.

Koehler JE, Sanchez MA, Garrido CS *et al.* (1997) Molecular epidemiology of *Bartonella* infections in patients with bacillary angiomatosis-peliosis. N Engl J Med 337:1876–1883.

Koehler JE, Sanchez MA, Tye S *et al.* (2003) Prevalence of *Bartonella* infection among human immunodeficiency virus-infected patients with fever. Clin Infect Dis 37 (4):559–566.

Kordick DL, Breitschwerdt EB (1995) Intraerythrocytic presence of *Bartonella henselae*. J Clin Microbiol 33:1655–1656.

Kordick DL, Wilson KH, Sexton DJ *et al.* (1995) Prolonged *Bartonella* bacteremia in cats associated with cat-scratch disease patients. J Clin Microbiol 33:3245–3251.

Kordick DL, Hilyard EJ, Hadfield TL *et al.* (1997) *Bartonella clarridgeiae*, a newly recognized zoonotic pathogen causing inoculation papules, fever, and lymphadenopathy (cat scratch disease). J Clin Microbiol 35 (7):1813–1818.

Kosoy M, Murray M, Gilmore RD Jr *et al.* (2003) *Bartonella* strains from ground squirrels are identical to *Bartonella washoensis* isolated from a human patient. J Clin Microbiol 41 (2):645–650.

Krekorian TD, Radner AB, Alcorn JM *et al.* (1990) Biliary obstruction caused by epithelioid angiomatosis in patient with AIDS. Am J Med 89:820–822.

La Scola B, Raoult D (1996) Serological cross-reactions between *Bartonella quintana*, *Bartonella henselae*, and *Coxiella burnetii*. J Clin Microbiol 34:2270–2274.

La Scola B, Raoult D (1999) Culture of *Bartonella quintana* and *Bartonella henselae* from human samples: a 5-year experience (1993 to 1998). J Clin Microbiol 37 (6):1899–1905.

La Scola B, Zeaiter Z, Khamis A, Raoult D (2003) Gene-sequence-based criteria for species definition in bacteriology: the *Bartonella* paradigm. Trends Microbiol 11 (7):318–321.

Larson AM, Dougherty MJ, Nowowiejski DJ *et al.* (1994) Detection of *Bartonella* (*Rochalimaea*) *quintana* by routine acridine orange staining of broth blood cultures. J Clin Microbiol 32 (6):1492–1496.

Lepidi H, Fournier PE, Raoult D (2000) Quantitative analysis of valvular lesions during *Bartonella* endocarditis. Am J Clin Pathol 114 (6):880–889.

Lucey D, Dolan MJ, Moss CW *et al.* (1992) Relapsing illness due to *Rochalimaea henselae* in immunocompetent hosts: implication for therapy and new epidemiological associations. Clin Infect Dis 14:683–688.

Maeno N, Oda H, Yoshiie K *et al.* (1999) Live *Bartonella henselae* enhances endothelial cell proliferation without direct contact. Microb Pathog 27 (6):419–427.

Maguina C, Gotuzzo E (2000) Bartonellosis – new and old. Infect Dis Clin North Am 14 (1):1–22.

Maguina C, Garcia PJ, Gotuzzo E *et al.* (2001) Bartonellosis (Carrion's disease) in the modern era. Clin Infect Dis 33 (6):772–779.

Maguina Vargas C (1998) Bartonellosis O enfermedad de carrion. Nuevos aspectos de una vieja enfermedad. Lima: Peru.

Marasco WA, Lester S, Parsonnet J (1989) Unusual presentation of cat scratch disease in a patient positive for antibody to the human immunodeficiency virus [clinical conference]. Rev Infect Dis 11:793–803.

Margileth AM (1992) Antibiotic therapy for cat scratch disease: clinical study of therapeutic outcome in 268 patients and a review of the literature. Pediatr Infect Dis J 11:474–478.

Margileth AM (1993) Cat scratch disease. Adv Pediatr Infect Dis 8:1–21.

Margileth AM (2000) Recent advances in diagnosis and treatment of cat scratch disease. Curr Infect Dis Rep 2 (2):141–146.

Margolis B, Kuzu I, Herrmann M *et al.* (2003) Rapid polymerase chain reaction-based confirmation of cat scratch disease and *Bartonella henselae* infection. Arch Pathol Lab Med 127 (6):706–710.

Marston EL, Sumner JW, Regnery RL (1999) Evaluation of intraspecies genetic variation within the 60 kDa heat-shock protein gene (*groEL*) of *Bartonella* species. Int J Syst Bacteriol 49:1015–1023.

Matar GM, Koehler JE, Malcolm G *et al.* (1999) Identification of *Bartonella* species directly in clinical specimens by PCR-restriction fragment length polymorphism analysis of a 16S rRNA gene fragment. J Clin Microbiol 37 (12):4045–4047.

Maurin M, Raoult D (1996) *Bartonella* (*Rochalimaea*) *quintana* infections. Clin Microbiol Rev 9:273–292.

Maurin M, Gasquet S, Ducco C, Raoult D (1995) MICs of 28 antibiotic compounds for 14 *Bartonella* (formerly *Rochalimaea*) isolates. Antimicrob Agents Chemother 39:2387–2391.

Maurin M, Birtles RJ, Raoult D (1997) Current knowledge of *Bartonella* species. Eur J Clin Microbiol Infect Dis 16:487–506.

Maurin M, Eb F, Etienne J, Raoult D (1997) Serological cross-reactions between *Bartonella* and *Chlamydia* species: implications for diagnosis. J Clin Microbiol 35 (9):2283–2287.

Maurin M, Rolain JM, Raoult D (2002) Comparison of in-house and commercial slides for detection of immunoglobulins G and M by immunofluorescence against *Bartonella henselae* and *Bartonella quintana*. Clin Diagn Lab Immunol 9 (5):1004–1009.

Mernaugh G, Ihler G (1992) Deformation factor: an extracellular protein synthesized by *Bartonella bacilliformis* that deforms erythrocyte membranes. Infect Immun 60:937–943.

Milam MW, Balerdi MJ, Toney JF *et al.* (1990) Epithelioid angiomatosis secondary to disseminated cat scratch disease involving the bone marrow and skin in a patient with acquired immune deficiency syndrome: a case report. Am J Med 88:180–183.

Musso D, Drancourt M, Raoult D (1995) Lack of bactericidal effect of antibiotics except aminoglycosides on *Bartonella* (*Rochalimaea*) *henselae*. J Antimicrob Chemother 36:101–108.

Myers WF, Cutler LD, Wisseman CL Jr (1969) Role of erythrocytes and serum in the nutrition of *Rickettsia quintana*. J Bacteriol 97:663–666.

Norman AF, Regnery RL, Jameson P *et al.* (1995) Differentiation of *bartonella*-like isolates at the species level by PCR-restriction fragment length polymorphism in the citrate synthase gene. J Clin Microbiol 33 (7):1797–1803.

Ohl ME, Spach DH (2000) *Bartonella quintana* and urban trench fever. J Clin Microbiol 31 (1):131–135.

Palmari J, Teysseire N, Dussert C, Raoult D (1996) Image cytometry and topographical analysis of proliferation of endothelial cells *in vitro* during *Bartonella* (*Rochalimaea*) infection. Anal Cell Pathol 11:13–30.

Perkocha LA, Geaghan SM, Yen JT *et al.* (1990) Clinical and pathological features of bacillary peliosis hepatis in association with human immunodeficiency virus infection. N Engl J Med 323:1581–1586.

Pinkerton H, Weinman D (1937) Carrion's disease. I. Behavior of the etiological agent within cells growing or surviving *in vitro*. Proc Soc Exp Biol Med 37:587–590.

Raoult D, Drancourt M, Carta A, Gastaut JA (1994) *Bartonella* (*Rochalimaea*) *quintana* isolation in patient with chronic adenopathy, lymphopenia, and a cat. Lancet 343:977.

Raoult D, Fournier PE, Drancourt M *et al.* (1996) Diagnosis of 22 new cases of *Bartonella* endocarditis. Ann Intern Med 125:646–652.

Raoult D, Fournier PE, Vandenesch F *et al.* (2003) Outcome and treatment of *bartonella* endocarditis. Arch Intern Med 163 (2):226–230.

Reed JA, Brigati DJ, Flynn SD *et al.* (1992) Immunocytochemical identification of *Rochalimaea henselae* in bacillary (epithelioid) angiomatosis, parenchymal bacillary peliosis, and persistent fever with bacteriema. Am J Surg Pathol 16:650–657.

Regnery R, Martin M, Olson J (1992) Naturally occurring "*Rochalimaea henselae*" infection in domestic cat. Lancet 340:557.

Regnery RL, Anderson BE, Clarridge JE *et al.* (1992a) Characterization of a novel *Rochalimaea* species, *R. henselae* sp. nov., isolated from blood of a febrile, human immunodeficiency virus-positive patient. J Clin Microbiol 30:265–274.

Regnery RL, Olson TG, Perkins BA, Bibb W (1992b) Serological response to *Rochalimaea henselae* antigen in suspected cat-scratch disease. Lancet 339:1443–1445.

Relman DA, Loutit JS, Schmidt TM *et al.* (1990) The agent of bacillary angiomatosis: an approach to the identification of uncultured pathogens. N Engl J Med 323:1573–1580.

Renesto P, Gouvernet J, Drancourt M *et al.* (2001) Use of *rpoB* gene analysis for detection and identification of *Bartonella* species. J Clin Microbiol 39 (2):430–437.

Ridder GJ, Boedeker CC, Technau-Ihling K *et al.* (2002) Role of cat-scratch disease in lymphadenopathy in the head and neck. Clin Infect Dis 35 (6):643–649.

Rolain JM, Maurin M, Bryskier A, Raoult D (2000) *In vitro* activities of telithromycin (HMR 3647) against *Rickettsia rickettsii*, *Rickettsia conorii*, *Rickettsia africae*, *Rickettsia typhi*, *Rickettsia prowazekii*, *Coxiella burnetii*, *Bartonella henselae*, *Bartonella quintana*, *Bartonella bacilliformis*, and *Ehrlichia chaffeensis*. Antimicrob Agents Chemother 44 (5):1391–1393.

Rolain JM, Maurin M, Raoult D (2000) Bactericidal effect of antibiotics on *Bartonella* and *Brucella* spp. clinical implications. J Antimicrob Chemother 46 (5):811–814.

Rolain JM, La Scola B, Liang Z *et al.* (2001) Immunofluorescent detection of intraerythrocytic *Bartonella henselae* in naturally infected cats. J Clin Microbiol 39 (8):2978–2980.

Rolain JM, Foucault C, Guieu R *et al.* (2002) *Bartonella quintana* in human erythrocytes. Lancet 360 (9328):226–228.

Rolain JM, Chanet V, Laurichesse H *et al.* (2003a) Cat scratch disease with vertebral osteomyelitis and spleen abscesses. Ann N Y Acad Sci 990:397–403.

Rolain JM, Foucault C, Brouqui P, Raoult D (2003b) Erythroblast cell as a target for *Bartonella quintana* in bacteremic homeless peoples. Ann N Y Acad Sci 990:485–487.

Rolain JM, Fournier PE, Raoult D, Bonerandi JY (2003c) First isolation and detection by immunofluorescence assay of *Bartonella koehlerae* in erythrocytes from a French cat. J Clin Microbiol 41 (8):4001–4002.

Rolain JM, Franc M, Davoust B, Raoult D (2003d) Molecular detection of *Bartonella quintana*, *B. koehlerae*, *B. henselae*, *B. clarridgeiae*, *Rickettsia felis* and *Wolbachia pipientis* in cat fleas, France. Emerg Infect Dis 9 (3):338–342.

Rolain JM, Gouriet F, Enea M *et al.* (2003e) Detection by immunofluorescence assay of *Bartonella henselae* in lymph nodes from patients with cat scratch disease. Clin Diagn Lab Immunol 10 (4):686–691.

Rolain JM, Maurin M, Mallet MN *et al.* (2003f) Culture and antibiotic susceptibility of *Bartonella quintana* in human erythrocytes. Antimicrob Agents Chemother 47 (2):614–619.

Rolain JM, Novelli S, Ventosilla P *et al.* (2003g) Immunofluorescence detection of *Bartonella bacilliformis* flagella *in vitro* and *in vivo* in human red blood cells as viewed by laser confocal microscopy. Ann N Y Acad Sci 990:581–584.

Roux V, Raoult D (1995) The 16S–23S rRNA intergenic spacer region of *Bartonella* (*Rochalimaea*) species is longer than usually described in other bacteria. Gene 156:107–111.

Roux V, Eykyn SJ, Wyllie S, Raoult D (2000) *Bartonella vinsonii* subsp. *berkhoffii* as an agent of afebrile blood culture-negative endocarditis in a human. J Clin Microbiol 38 (4):1698–1700.

Rydkina EB, Roux V, Gagua EM *et al.* (1999) *Bartonella quintana* in body lice collected from homeless persons in Russia. Emerg Infect Dis 5 (1):176–178.

Sander A, Posselt M, Oberle K, Bredt W (1998) Seroprevalence of antibodies to *Bartonella henselae* in patients with cat scratch disease and in healthy controls: evaluation and comparison of two commercial serological tests. Clin Diagn Lab Immunol 5 (4):486–490.

Sander A, Berner R, Ruess M (2001) Serodiagnosis of cat scratch disease: response to *Bartonella henselae* in children and a review of diagnostic methods. Eur J Clin Microbiol Infect Dis 20 (6):392–401.

Sanogo YO, Zeaiter Z, Caruso G *et al.* (2003) *Bartonella henselae* in *Ixodes ricinus* ticks (Acari: Ixodida) removed from humans, Belluno Province, Italy. Emerg Infect Dis 9 (3):329–332.

Scherer DC, DeBuron-Connors I, Minnick MF (1993) Characterization of *bartonella bacilliformis* flagella and effect of antiflagellin antibodies on invasion of human erythrocytes. Infect Immun 61:4962–4971.

Schulein R, Seubert A, Gille C *et al.* (2001) Invasion and persistent intracellular colonization of erythrocytes. A unique parasitic strategy of the emerging pathogen *Bartonella*. J Exp Med 193 (9):1077–1086.

Schultz MG (1968) A history of bartonellosis (Carrion's disease). Am J Trop Med Hyg 17:503–515.

Shinella RA, Alba Greco M (1990) Bacillary angiomatosis presenting as a soft-tissue tumor without skin involvement. Hum Pathol 21:567–569.

Slater LN, Welch DF (1995) *Rochalimaea* species (recently renamed *Bartonella*). In Mandell GL, Bennett JE, Dolin R (eds). Principles and Practice of Infectious Diseases. Churchill Livingstone. 1741–1747.

Slater LN, Welch DF, Hensel D, Coody DW (1990) A newly recognized fastidious gram-negative pathogen as a cause of fever and bacteremia. N Engl J Med 323:1587–1593.

Sobraques M, Maurin M, Birtles R, Raoult D (1999) *In vitro* susceptibilities of four *Bartonella bacilliformis* strains to 30 antibiotic compounds. Antimicrob Agents Chemother 43 (8):2090–2092.

Spach D (1992) Bacillary angiomatosis. Int J Dermatol 31:19–24.

Spach DH, Koehler JE (1998) *Bartonella*-associated infections. Emerg Infect Dis 12 (1):137–155.

Spach DH, Callis KP, Paauw DS *et al.* (1993) Endocarditis caused by *Rochalimaea quintana* in a patient infected with human immunodeficiency virus. J Clin Microbiol 31:692–694.

Spach DH, Kanter AS, Daniels NA *et al.* (1995a) *Bartonella* (*Rochalimaea*) species as a cause of apparent "culture-negative" endocarditis. Clin Infect Dis 20:1044–1047.

Spach DH, Kanter AS, Dougherty MJ *et al.* (1995b) *Bartonella* (*Rochalimaea*) *quintana* bacteremia in inner-city patients with chronic alcoholism. N Engl J Med 332:424–428.

Spaulding WB, Hennessy JN (1960) Cat scratch disease. A study of eighty-three cases. Am J Med 28:504–509.

Stoler MH, Bonfiglio TA, Steigbigel RT, Pereira M (1983) An atypical subcutaneous infection associated with acquired immune deficiency syndrome. Am J Clin Pathol 80:714–718.

Strong R, Swift H, Opie E *et al.* (1918) Report on progress of trench fever investigations. JAMA 70:1597–1599.

Strong RP, Tyzzer EE, Brues CT *et al.* (1913) Verruga peruana, Oroya fever and Uta. JAMA 61:1713–1716.

Szaniawski WK, Don PC, Bitterman SR, Schachner JR (1990) Epithelioid angiomatosis in patients with AIDS: a report of seven cases and review of the literature. J Am Acad Dermatol 23:41–48.

Tappero JW, Koehler JE, Berger TG *et al.* (1993a) Bacillary angiomatosis and bacillary splenitis in immunocompetent adults. Ann Intern Med 118:363–365.

Tappero JW, Mohle-Boetani JC, Koehler JE *et al.* (1993b) The epidemiology of bacillary angiomatosis and bacillary peliosis. JAMA 269:770–775.

Uertega O, Payne EH (1955) Treatment of the acute febrile phase of Carrion's disease with chloramphenicol. Am J Trop Med Hyg 4:507–511.

Walker TS, Winkler HH (1981) *Bartonella bacilliformis*: colonial types and erythrocyte adherence. Infect Immun 31:480–486.

Wear DJ, Margileth AM, Hadfield TL *et al.* (1983) Cat scratch disease: a bacterial infection. Science 221:1403–1405.

Welch DF, Pickett DA, Slater LN *et al.* (1992) *Rochalimaea henselae* sp. nov., a cause of septicemia, bacillary angiomatosis, and parenchymal bacillary peliosis. J Clin Microbiol 30:275–280.

Welch D, Carrol K, Hofmeister E *et al.* (1999) Isolation of a new subspecies, *Bartonella vinsonii* subsp. *arupensis*, from a cattle rancher: identity with isolates found in conjunction with *Borrelia burgdorferi* and *Babesia microti* among naturally infected mice. J Clin Microbiol 37:2598–2601.

Zbinden R, Höchli M, Nadal D (1995) Intracellular location of *Bartonella henselae* cocultivated with vero cells and used for an indirect fluorescent-antibody test. Clin Diagn Lab Immunol 2:693–695.

Zbinden R, Michael N, Sekulovski M *et al.* (1997) Evaluation of commercial slides for detection of immunoglobulin G against *Bartonella henselae* by indirect immunofluorescence. Eur J Clin Microbiol Infect Dis 16 (9):648–652.

Zeaiter Z, Fournier PE, Raoult D (2002) Genomic variation of *Bartonella henselae* strains detected in lymph nodes of patients with cat scratch disease. J Clin Microbiol 40 (3):1023–1030.

Zeaiter Z, Fournier PE, Greub G, Raoult D (2003) Diagnosis of *Bartonella* endocarditis by a real-time nested PCR assay using serum. J Clin Microbiol 41 (3):919–925.

24

Mycoplasma spp.

Christiane Bébéar, Sabine Pereyre and Cécile M. Bébéar

Laboratoire de Bactériologie, Université Victor Segalen Bordeaux 2, Bordeaux, France

INTRODUCTION

Mycoplasmas are ubiquitous microorganisms which have been known as animal pathogens since the end of the nineteenth century. They are the smallest prokaryotes able to multiply autonomously. Even though the term mycoplasma only applies, taxonomically, to the genus *Mycoplasma*, it is still commonly used to indicate these organisms as a whole. It will be used in this chapter to designate any of the organisms in the class Mollicutes.

The first mycoplasma was isolated in 1898, by Nocard and Roux. It was *Mycoplasma mycoides* spp. *mycoides* the cause of contagious bovine peripneumonia. The first recorded case of a human mycoplasma infection, an abscess of the Bartholin gland, was described in 1937 by Dienes and Edsall. The species isolated was probably *M. hominis*. In 1954 Shepard isolated mycoplasmas from the urogenital tract. These mycoplasmas produced particularly small colonies. They were called T strain, indicating tiny colonies, and are now known as ureaplasmas. It was only in 1962 that Chanock, Hayflick and Barile succeeded in culturing *M. pneumoniae*, the cause of primary atypical pneumonia, in an acellular medium. They then identified the Eaton agent, which had been cultured on chick embryo, as a mycoplasma. Among the potential human pathogens, *M. genitalium* was discovered in 1981 from urethral swabs collected from men with nongonococcal urethritis (NGU) (Taylor-Robinson, 1995). More recently, two species of mycoplasmas were found in HIV-seropositive patients, *M. fermentans*, known elsewhere since 1952, and *M. penetrans* (Lo *et al.*, 1991).

DESCRIPTION OF THE ORGANISMS

Taxonomy and Phylogeny

Mycoplasmas are completely lacking a cell wall. They belong to the class Mollicutes (*mollis cutis*, soft skin). This class consists of four orders, Mycoplasmatales, Entomoplasmatales, Acholeplasmatales and Anaeroplasmatales, which are distinguished on the basis of their natural habitat, their sterol requirements and a certain number of other properties. The class Mollicutes include at least 183 known species.

Taxonomy of Mollicutes has developed a great deal with the introduction of molecular methods and has been recently reviewed (Johansson and Pettersson, 2002). The species isolated from humans belong mainly to the order Mycoplasmatales, family Mycoplasmataceae, that includes two genera, *Mycoplasma* and *Ureaplasma* (Table 24.1). Facultatively anaerobic, they require sterols for growth. Their principal energy sources are sugars and arginine or urea in the case of *Ureaplasma* genus. Thirteen of the 107 known *Mycoplasma* species and one of the seven *Ureaplasma* species, *Ureaplasma urealyticum* have been isolated from humans. *Ureaplasma urealyticum* is an heterogeneous species which has been proposed for division into two separate species, *U. urealyticum* and *U. parvum*, corresponding to the two former biovars, 1 and 2 (Kong *et al.*, 1999). They will be considered together as *Ureaplasma* spp. Other species of *Mycoplasma* and *Ureaplasma* genera are of animal origin. *Acholeplasma laidlawii*, the type species of the genus *Acholeplasma*, and *A. oculi* have been

Table 24.1 Taxonomy and main characteristics of mycoplasmas (Johansson and Pettersson, 2002)

Classification	G + C%	Genome size (kbp)	Sterol requirement	Host
Order I: Mycoplasmatales				
Family I: Mycoplasmataceae				
Genus I: *Mycoplasma* (107 species)	23–40	580–1360	Yes	Humans, animals
Genus II: *Ureaplasma* (7 species)	27–30	750–1170	Yes	Humans, animals
Order II: Entomoplasmatales				
Family I: Entomoplasmataceae				
Genus I: *Entomoplasma*	27–29	790–1140	Yes	Insects, plants
Genus II: *Mesoplasma*	27–30	870–1100	No	Insects, plants
Family II: Spiroplasmataceae				
Genus I: *Spiroplasma*	25–30	780–2240	Yes	Insects, plants
Order III: Acholeplasmatales				
Family I: Acholeplasmataceae				
Genus I: *Acholeplasma*	27–36	1500–1690	No	Animals, some plants and insects
Order IV: Anaeroplasmatales				
Family I: Anaeroplasmataceae				
Genus I: *Anaeroplasma*	29–33	1600	Yes	Bovine and ovine rumen
Genus II: *Asteroleplasma*	40	1600	No	

Principles and Practice of Clinical Bacteriology Second Edition   Editors Stephen H. Gillespie and Peter M. Hawkey
© 2006 John Wiley & Sons, Ltd

described rarely in humans. Phytoplasmas present in insects and plants, closely related to acholeplasmas, cannot be cultivated. The *Entomoplasmatales* are mycoplasmas isolated from insects and plants. The order contains the *Entomoplasma*, *Mesoplasma* and *Spiroplasma* (mycoplasmas with a helical morphology) genera (Johansson and Pettersson, 2002).

Phylogenetically, mycoplasmas are very simple microorganisms. Despite their small size, as demonstrated by 16S rRNA sequence comparison, they are highly evolved bacteria coming from anaerobic clostridia-like Gram-positive ancestors containing a low level of guanine + cytosine. Mycoplasmas probably evolved from these ancestors by successive reduction in the genome and loss of the cell wall. The closest bacterial species to this phylogenetic branch are *Clostridium innocuum* and *C. ramosum* (Woese, 1987). Based on 16S rRNA sequences, noncultivable wall-less parasitic bacteria, which infect animals including primates and are observed in the blood, attaching to erythrocytes, are given the *Candidatus* status in the genus *Mycoplasma*. They were previously classified in the genus *Haemobartonella* and *Eperythrozoon*.

Habitat

Mollicutes are ubiquitous microorganisms. In nature, flowers and plant surfaces constitute an important reservoir. Insects act as vectors and can also suffer from specific mycoplasma-induced diseases. Mycoplasmas are also found in animals, especially animals in intensive production facilities and laboratory animals, and humans.

The 16 species found in humans are listed in Table 24.2. Mycoplasmas can be grouped by the site where they are usually isolated in immunocompetent subjects, the respiratory and genital tracts. Most of the mycoplasmas isolated from the oropharynx are simple commensals (*M. salivarium*, *M. orale*, *M. buccale*, *M. faucium*, *M. lipophilum* and *A. laidlawii*). Only *M. pneumoniae* colonizes the lower respiratory tract and possesses certain pathogenic capacity. Seven species can be considered genital mycoplasmas. Two of these, *Ureaplasma* spp. and *M. hominis* are part of the genital commensal flora of a great number of people (Taylor-Robinson, 1996). Frequency of colonization varies with age, hormonal factors, race, socioeconomic level and sexual activity. It is difficult to evaluate the rate of colonization in the general population, although it is known to be higher in women than in men. It may be as high as 50% vaginally in women for *Ureaplasma* spp.,

while for *M. hominis* it is probably less than 10%. Both species are responsible for human infections (Taylor-Robinson, 1996; Blanchard and Bébéar, 2002). The role of the other species (*M. genitalium*, *M. fermentans* and *M. penetrans*) is much less well known because they are rarely detected by culture. However, the use of polymerase chain reaction (PCR) technology has considerably improved our knowledge of *M. genitalium* occurrence in several diseases. *Mycoplasma spermatophilum* and *M. primatum* have been detected rarely. The natural habitat of *M. pirum* is unknown.

Occasionally, animal mycoplasmas are detected in immunosuppressed hosts. Furthermore, mycoplasmas are frequent contaminants of cell cultures, particularly continuous cell lines. Five species (*M. hyorhinis*, *M. orale*, *M. fermentans*, *M. arginini* and *A. laidlawii*) are responsible for about 95% of contaminations. Their frequency and the possible consequences for the use of these cells necessitate regular surveillance (Drexler and Uphoff, 2000).

PRINCIPAL PROPERTIES

Morphology and Structure

Mycoplasmas are very small organisms (0.2–0.3 μm). When examined by dark field or phase contrast microscopy, they appear to be pleomorphic, coccoid or filamentous, depending on the species and culture conditions. Spiroplasmas are helical in shape. They are not stained by Gram stain. A technique of DNA fluorochrome staining is available for screening cell cultures for mycoplasma contamination but is not specific for mycoplasmas.

Electron microscopy shows that certain species, *M. pneumoniae*, *M. genitalium*, *M. penetrans* and some animal mycoplasmas, have a specialized terminal structure. This slender extremity, the tip, plays an important role in the attachment of mycoplasmas to different substrates, including host cells and in their mobility using a gliding movement.

Mycoplasmas have no cell wall and differ from L-forms by the absence of precursors of peptidoglycan and binding proteins for penicillins. All of the species of the genus *Mycoplasma* and *Ureaplasma* possess a cytoplasmic membrane containing cholesterol.

The genome of mycoplasmas varies in size, from 580 kbp in *M. genitalium*, which was completely sequenced in 1995, to 2200 kbp in *Spiroplasma* spp. The guanine + cytosine content ranges from

Table 24.2 Mycoplasmas isolated in humans (Blanchard and Bébéar, 2002; Waites, Rikihisa and Taylor-Robinson, 2003)

Species	Primary site of colonization		Metabolism of		Pathogenic role[a]
	Oropharynx	Genitourinary tract	Glucose	Arginine	
M. pneumoniae	+	−	+	−	+
M. salivarium	+	−	−	+	−
M. orale	+	−	−	+	−
M. buccale	+	−	−	+	−
M. faucium	+	−	−	+	−
M. lipophilum	+	−	−	+	−
M. hominis	−	+	−	+	+
M. genitalium	−	+	+	−	+
M. fermentans	+	+	+	+	?
M. penetrans	−	+	+	+	?
M. primatum	−	+	−	+	−
M. spermatophilum	−	+	−	+	−
M. pirum	?	?	+	+	−
Ureaplasma spp.[b]	−	+	−	−	+
A. laidlawii	+	−	+	−	−
A. oculi	?	?	+	−	−

[a] In immunocompetent patients: + = proven role; ? = questionable; − = nonpathogenic.

[b] Metabolizes urea, *Ureaplasma* spp. include *U. urealyticum* and *U. parvum*.

23 to 40%. With the exception of acholeplasmas, they use the codon UGA to code for tryptophane. They possess one or two rRNA operons, and their number of tRNAs is reduced. The recent sequencing of the entire genome of eight mycoplasmas, including the human species, *M. genitalium* (Fraser *et al.*, 1995), *M. pneumoniae* (Himmelreich *et al.*, 1996), *M. penetrans* (Sasaki *et al.*, 2002) and *U. parvum* (Glass *et al.*, 2000), has marked a turning point in the molecular genetic analysis of these organisms. As mycoplasmas include the smallest known self-replicating organisms, their molecular study is an important step to define the minimal set of genes essential for life. Furthermore, despite their small size, comparative genome analysis of mycoplasmas shows a considerable variety with many specific adaptations (Dandekar *et al.*, 2002).

Nutritional Requirements

Because of their small number of genes, mycoplasmas have a very limited capacity to synthesize their component parts. This renders them rather fastidious. In order to grow they require nucleic acid precursors (yeast extract) and the cholesterol contained in serum (except for acholeplasmas and other nonhuman species).

The optimum atmospheric conditions are equally variable. *Mycoplasma pneumoniae* and *M. genitalium* are aerobic, but their growth is stimulated in the presence of 5% CO_2. *Mycoplasma hominis* and *Ureaplasma* spp. are indifferent but grow better on agar media in an atmosphere containing 5% CO_2. The optimum temperature for growth is approximately 36–38 °C.

In broth culture, the generation time varies from 1 h for *Ureaplasma* spp. to 6 h for *M. pneumoniae* and more for *M. genitalium*. Broth cultures do not usually become cloudy. On agar media, colonies are slow growing. As they are extremely small (15–300 μm), they must be observed with the help of low-power magnification. The appearance of the colonies is variable but takes on a characteristic fried-egg aspect because the organisms penetrate deeply into the agar in the central region of the colony (Figure 24.1). Colonies of *Ureaplasma* spp. are very small and irregular.

Biochemical Characteristics

Species of mycoplasmas isolated from humans can be grouped as glucose fermenters (including *M. pneumoniae*, *M. genitalium*, *M. fermentans* and *M. penetrans*) and nonfermenters. Nonfermenters such as *M. hominis* draw their energy from the hydrolysis of arginine by the metabolic pathway of arginine dihydrolase. *Mycoplasma*

fermentans and *M. penetrans* use both arginine and glucose. The source of energy of *Ureaplasma* spp. involves the degradation of urea. These characteristics are used to identify mycoplasmas of human origin (Table 24.2).

Mycoplasma pneumoniae is able to reduce 2,3,5-triphenyl-tetrazolium chloride. Hemadsorption and hemagglutination of guinea-pig or chicken erythrocytes and production of peroxides which confer hemolytic abilities are also found in this species.

Antigenic Structure

Membrane antigens probably play an essential role in the host response to infection. *Mycoplasma pneumoniae* possesses a glycolipid which is not completely specific and can be found in diverse tissues, microorganisms, plants and several membrane-bound proteins playing a role in attachment. The P1 protein (170 kDa), found on the narrow extremity, is the major cytadhesin, although other proteins also participate [P30, HMW (high molecular weight) 1–3 and others (Krause, 1998)]. Although variability of cytadhesin P1 has been described, allowing to separate two groups of strains, *M. pneumoniae* is an homogeneous species. *Mycoplasma hominis* is much more heterogeneous. However, no serovar or subspecies could be clearly distinguished among *M. hominis* strains (Ladefoged, 2000). An additional level of antigenic variability, related to variations in the size of antigen proteins, has been described for several species of mycoplasmas, *Ureaplasma* spp. in particular (Zheng *et al.*, 1995). The two different *Ureaplasma* species, include 14 serovars (serovars 2, 4, 5 and 7–13 for *U. urealyticum* and serovars 1, 3, 6 and 14 for *U. parvum*).

PATHOGENESIS

Mycoplasma pneumoniae

The pathogenesis of infection with *M. pneumoniae* has been studied using different models, organ cultures and experimental animal models, hamster inoculated intranasally and chimpanzees. In these animal models, *M. pneumoniae* colonizes the respiratory epithelium diffusely and provokes lesions comparable histopathologically to those observed in humans (perivascular and peribronchiolar infiltration of mononuclear cells). Recently, a murine model was proposed (Hardy *et al.*, 2003).

Two mechanisms contribute to the pathogenesis of *M. pneumoniae* infections, attachment of *M. pneumoniae* to the respiratory epithelium followed by localized cellular lesions and immunopathological disorders which can lead to lesions elsewhere (Razin, Yogev and Naot, 1998). *Mycoplasma pneumoniae* can adhere to a number of substrates, glass, plastic, red blood cells and respiratory epithelial cells. *In vivo*, this attachment permits the mycoplasma to escape the ciliary movement and come into close contact with the cell membrane. This role is confirmed by the loss of pathogenicity for hamster of nonadherent strains. Attachment, which takes place at the narrow extremity of the terminal structure, is mediated by several proteins including P1, the principal cytadhesin involved. The host cells have receptors containing sialic acid. The attachment blocks ciliary action and causes cellular alterations due to the production by *M. pneumoniae* of peroxides and superoxides which also act as a hemolysin.

Immunopathological mechanisms are suspected to be involved in some of the lesions provoked by *M. pneumoniae*. The histological appearance of these lesions, the presence of lesions at distance, from which mycoplasmas are rarely isolated, and the presence of autoantibodies reacting with a variety of host tissues argue in favor of this hypothesis. The formation of autoantibodies reacting with the I antigen of erythrocytes is responsible for production of cold agglutinins.

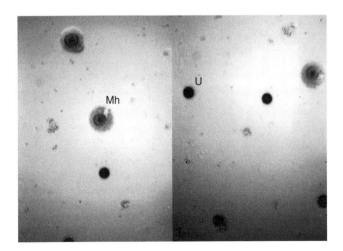

Figure 24.1 Fried-egg colonies of *M. hominis* (Mh) and small granular colonies of *Ureaplasma* spp. (U). Magnification × 170.

Genital Mycoplasmas

The pathogenesis of genital mycoplasma infections is less well known. Various models of genital infections have been developed using different animals. In mice, establishment of *M. hominis* and *Ureaplasma* spp. in the vagina is improved by treating the animals with hormones. Extension of the infection toward the upper genital tract has been observed. In male chimpanzees, intraurethral inoculation of *Ureaplasma* spp. or *M. genitalium* leads to a local leukocyte reaction followed sometimes by dissemination in the blood for *M. genitalium*. Upper genital tract infections (salpingitis and parametritis) have been provoked after inoculation of female monkeys with *M. genitalium* (Taylor-Robinson, 1995, 1996).

The attachment process has been described for the three species, *Ureaplasma* spp., *M. hominis* and *M. genitalium*. This last organism possesses an adhesin, MgPa, which is very similar to P1 of *M. pneumoniae* and facilitates its attachment on epithelial cells. Factors involved in *Ureaplasma* spp. or *M. hominis* attachment are much less known.

Diverse enzymatic activities (urease and IgA1 protease for *Ureaplasma* spp. and phospholipase for *Ureaplasma* spp. and *M. hominis*), the depletion of arginine for *M. hominis*, the production of certain metabolites such as ammonia, explain in part their pathogenicity. It is not known whether one of the two *Ureaplasma* spp. has a particular pathogenic capability.

Although mycoplasmas are generally considered as extracellular organisms, intracellular localization has been reported for several species including *M. genitalium*, *M. hominis*, *M. penetrans* as well as *M. pneumoniae*. This localization may protect the mycoplasma against host defenses and contribute to disease chronicity (Waites, Rikihisa and Taylor-Robinson, 2003).

EPIDEMIOLOGICAL AND CLINICAL CHARACTERISTICS

Respiratory Infections

M. pneumoniae

Mycoplasma pneumoniae causes acute respiratory infections, most frequently in children from 4 years of age to young adults, but it can be observed in younger children and in older persons. The infection is endemic in the population with epidemic peaks every 4–7 years. Surges occurred in Europe in 1987, 1992 and 1999. Although more frequent in cold weather, there is not a clear seasonal trend in frequency. This infection is not very contagious, but spread throughout households is common. The persistence of *M. pneumoniae* in the respiratory tract contributes to the endemic nature of the disease. Other than during epidemics, when it can be found in the oropharynx of healthy subjects, *M. pneumoniae* is not usually found among the commensal flora of the respiratory tract.

Infection often starts as a simple tracheobronchitis. *M. pneumoniae* is the second most common cause of community-acquired pneumonia after *Streptococcus pneumoniae* and is probably responsible for 15–20% of X-ray-proven cases of pneumonia (Foy, 1993). Higher rates have been observed in children (Principi *et al.*, 2001). In its characteristic form, this disease develops as a primary atypical pneumonia with a progressive installation, fever, involvement of the upper respiratory tract, dry cough and marked radiological lesions. It evolves slowly and progressively. The disease is ordinarily mild. Respiratory symptoms alone do not distinguish *M. pneumoniae* from other causes of atypical pneumonia. Association with other symptoms can be a better indication, cutaneous lesions (Stevens Johnson syndrome and multiform erythemia), otitis media, pharyngitis, neurological involvement (meningitis and meningoencephalitis), hemolytic anemia due to the presence of cold hemagglutinins, coagulation problems, arthritis, myocarditis, pericarditis and involvement of the pancreas,

digestive tract and kidneys (Jacobs, 2002). These complications may occur alone and therefore be difficult to associate with their causal agent. Although an autoimmune response is thought to play a role in some extrapulmonary complications, *M. pneumoniae* has been directly isolated or identified by PCR from extrapulmonary sites such as synovial, pericardial and cerebrospinal (CSF) fluids. Its possible role as etiologic or exacerbating factor in bronchial asthma is still debated (Martin *et al.*, 2001).

Other Mycoplasmas

In adults, these are all commensal organisms with the possible exception of *M. fermentans*. A certain number of cases of fulminant infections with a syndrome of respiratory distress with or without systemic involvement have been reported in subjects previously in good health (Lo *et al.*, 1993). In experimental animal models, it has been shown that *M. fermentans* can produce cellular lesions (Stadtländer *et al.*, 1993). *M. fermentans* may be a little known cause of respiratory infections.

Urogenital Tract Infections

In Men

Ureaplasma spp. and *M. genitalium* cause nonchlamydial nongonococcal urethritis (NGU). *Ureaplasma* spp. have long been implicated as a cause of acute NGU on the basis of isolation studies including quantification of the organisms, controlled therapeutic and serological studies and human and animal inoculation studies (Taylor-Robinson, Ainsworth and McCormack, 1999). However, the proportion of cases due to ureaplasmas is unclear and controversial because of the frequent isolation of *Ureaplasma* spp. in the urethra of healthy men, and this ranked them behind *Chlamydia trachomatis* and *M. genitalium* as a cause of acute NGU. Recently, *Ureaplasma* spp. have been implicated in chronic NGU with symptoms or signs re-emerging after treatment (Horner *et al.*, 2001).

Because of the development of PCR, our knowledge on the pathogenic role of *M. genitalium* has significantly increased. *M. genitalium* is strongly associated with acute NGU (Totten *et al.*, 2001), independent of the presence of *C. trachomatis* or ureaplasmas, as reported by a number of studies from different countries. *M. genitalium* might be responsible for 15–25% of acute NGU. It has also been associated with chronic NGU after treatment. *M. genitalium* and *Ureaplasma* spp. have been detected in some cases of sexually acquired reactive arthritis. Evidence that *M. hominis* plays a role in NGU is lacking.

There is still controversy concerning the role of genital mycoplasmas in prostatitis. *M. genitalium* was detected by PCR in 4% of prostatic biopsies from chronic idiopathic prostatitis. There are also a few reports implicating ureaplasmas or *M. hominis* as causes of epididymitis. Nevertheless, the role of mycoplasmas in prostatitis or epididymitis seems to be minimal.

In Women

Mycoplasmas do not cause vaginitis but proliferate in women with bacterial vaginosis (BV), together with other microorganisms such as *Gardnerella vaginalis* and anaerobes. *M. hominis* is strongly associated with BV and is found in the vagina of two-thirds of women with BV in high numbers compared to less than 10% of healthy women. *Ureaplasma* spp. have been associated with BV to a lesser extent and could be responsible for some cases of urethral syndrome in women. *M. genitalium* seems to have no role in BV but has been associated with mucopurulent cervicitis (Manhart *et al.*, 2001).

BV may lead to pelvic inflammatory disease (PID), and *M. hominis* has been isolated from the endometrium and fallopian tubes of 10% of women with salpingitis, accompanied by a significant antibody response. However, PID is a disease of multifactorial etiology, and whether *M. hominis* is a primary agent of PID or acts in association with other bacteria in a context of BV is still controversial. In contrast, there is little evidence that *Ureaplasma* spp. play a similar role in PID. *M. genitalium*, significantly associated with cervicitis and endometritis (Cohen *et al.*, 2002), could be an agent of PID as indicated by serological data and primate inoculation studies.

Furthermore, several studies have reported interdependence between *M. hominis* and *Trichomonas vaginalis* which could be a carrier of the *M. hominis* infection *in vivo* (Taylor-Robinson, 1998).

Reproduction Disorders and Infections in Pregnancy

The role of mycoplasmas in infertility is still unknown. Evidence to incriminate *Ureaplasma* spp. in male and female infertility is sparse. *Ureaplasma* spp. have been reported to decrease sperm motility and alter the spermatozoa morphology. Experimental models showed that ureaplasmas, and recently *M. genitalium*, could adhere to spermatozoa. Whether it is necessary to detect and treat genital mycoplasmas in *in vitro* fertilization is still debatable.

Doubts also remain about the role of genital mycoplasmas in recurrent spontaneous abortion, stillbirth, prematurity or low birth weight (Cassell *et al.*, 1993).

Mycoplasma hominis and *Ureaplasma* spp. were shown to be responsible for some cases of chorioamnionitis and endometritis. They have also been incriminated in post-partum or post-abortum fevers accompanied by isolation of both organisms from the blood (Neman-Simha *et al.*, 1992). In some cases, it could represent the transient invasion of bloodstream following vaginal delivery or it could be a real septicemia leading to a neonatal infection. It should be noticed that *M. hominis* has been implicated more frequently than *Ureaplasma* spp. in post-partum fever.

Disorders of the Urinary Tract

Animal models and human isolation studies have suggested a possible causal association of *Ureaplasma* spp. in the development of infectious stones. The mechanism could be related to the urease activity of ureaplasmas with crystallization of struvite and calcium phosphates in urines (Grenabo, Hedelin and Pettersson, 1988).

Mycoplasma hominis appears to cause a small number of cases of acute pyelonephritis in immunocompetent patients as indicated by the isolation of the microorganisms from the upper urinary tract accompanied by a specific antibody response.

Neonatal Infections

Colonization of neonates by genital mycoplasmas could occur *in utero*, but more frequently at the time of delivery by contact with mycoplasmas from the lower genital tract of the mother. The frequency of colonized newborns could be up to 50% when the mother is colonized. The highest rates concern the low birth weight neonates. Colonization of infants tends to decrease rapidly after 3 months.

Several clinical features such as bacteremia, septicemia, respiratory tract and central nervous system infections have been described for both *Ureaplasma* spp. and *M. hominis* (Waites *et al.*, 1988). Respiratory tract infections are manifested most frequently as pneumonia and in some rare cases as respiratory distress syndromes. Since 1988, several studies have reported the association of colonization of the respiratory tract by *Ureaplasma* spp. with bronchopulmonary dysplasia in low birth weight infants (<1000 g) (Wang *et al.*, 1993). This association has yet to be confirmed, as bronchopulmonary dysplasia is a complex multifactorial pathology. However, if a causality link is confirmed, it could lead to a distinct change in the therapeutic management of this disease.

Both the common genital mycoplasma species could invade the CSF fluid of neonates causing some cases of meningo-encephalitis with neurological damage or death. However, most of the infected newborns had a subclinical meningitis without sequelae (Waites, Rikihisa and Taylor-Robinson, 2003).

Systemic Infections

Mycoplasmas can infect organs other than the respiratory or urogenital tracts. These infections are certainly underestimated, as mycoplasmas are usually considered late in diagnosis when negative results have been observed for other microorganisms or in case of treatment failure.

Extrapulmonary or extragenital infections (mainly septic arthritis) occur frequently in an immunodeficiency state. Mycoplasmas have been identified, by culture or PCR, in about 40% of cases of septic arthritis in patients with hypogammaglobulinemia (Furr, Taylor-Robinson and Webster, 1994). *Ureaplasma* spp. are the most frequently isolated species, followed by *M. hominis*, *M. pneumoniae* and other species in some cases. These arthritis are generally cured successfully by antibiotics, but only with a concomitant treatment of the immuno-suppression.

Other infections by mycoplasmas have often been discovered fortuitously. As *M. hominis* is able to grow on blood agar used in routine culture, it explains why it is much more frequently isolated than the other species (Meyer and Clough, 1993; Blanchard and Bébéar, 2002). *M. hominis* has been demonstrated to be responsible for septicemia, retroperitoneal abscesses and peritonitis, hematoma infections, vascular and catheter-related infections, sternal wound infections associated with mediastinitis after thoracic surgery, prosthetic valve endocarditis, brain abscesses and pneumonia mainly by hematogenous spread. In most of the cases, these infections occurred in immunocompromised or transplanted patients or patients with major disruptions of anatomic barriers or polytraumatisms.

Exceptional cases of human infections due to animal mycoplasmas, *M. arginini* and *M. felis*, have been reported in immunocompromised patients.

The possible role of mycoplasmas in HIV disease has been a matter of debate and was fueled by the ability of these bacteria to promote HIV replication *in vitro* and/or HIV-associated cytopathic effects (Blanchard and Montagnier, 1994). The role of *M. fermentans* and then of *M. penetrans*, a species isolated from the urine of HIV-positive homosexual male, has been suggested in some reports. However, no conclusive evidence has demonstrated none other than an opportunistic role for mycoplasmas in HIV infection. *M. penetrans* has been recently isolated from the blood and the throat of a HIV-negative patient with primary antiphospholipid syndrome.

Recently, another matter of debate concerned the association of *M. fermentans* infection, detected either serologically or by PCR, and chronic fatigue syndrome or Gulf War illness. However, others did not find such an association, and in 2004, no evidence suggest that infection by *M. fermentans* is associated with the development of such diseases (Blanchard and Bébéar, 2002).

DIAGNOSTIC METHODS

Laboratory diagnosis can be performed using direct detection of mycoplasmas, by culture or molecular methods (PCR) or using serology. Both methods are available for *M. pneumoniae*. Direct detection is the only method adapted to genital mycoplasmas (Bébéar *et al.*, 1997; Waites *et al.*, 2001a).

Specimens

Specimens Type

Whatever the sampling method used, it must collect cells to which mycoplasmas have adhered. Most of the specimens suitable for culture can also be used for PCR.

Different types of sampling methods are proposed for the respiratory tract. Sputum specimens are not very useful because they contain too many contaminants. Throat swabs and nasopharyngeal aspirates for young children give good results because of the diffuse nature of the infection. Bronchial brushing, bronchoalveolar lavages, pleural fluid specimens and biopsies can also be taken.

Genital mycoplasmas can be cultured from urethral swabs, first-void urine, semen, prostatic secretions, cervico-vaginal swabs, endometrial biopsies, tubal brushing, pouch of Douglas fluid, amniotic fluid, placenta, endotracheal samples from neonates, etc.

Mycoplasmas can be detected in other samples, CSF, synovial fluid, biopsies, mucocutaneous specimens and blood. However, the classic media used for blood culture contain anticoagulants which act as inhibitors for the growth of mycoplasmas.

Transport and Storage

As soon as specimens are taken, the swabs must be placed in the appropriate transport medium in order to avoid drying. Different media can be used for transport and storage, sucrose phosphate transport medium (2SP) containing 5% fetal bovine serum, without antibiotics and likely to be used for selection of both mycoplasmas and *Chlamydia*, by either culture or PCR, commercial transport media or culture media.

All samples must be transmitted to the laboratory as soon as possible. If they cannot be cultured immediately, they should be stored at 4 °C for a maximum of 48 h and beyond that at −70 °C or in liquid nitrogen.

Culture

Culture of mycoplasmas is relatively simple for certain species such as *M. hominis* and *Ureaplasma* spp., more delicate for *M. pneumoniae*, very fastidious and rarely successful for *M. genitalium* and at a less extent for *M. fermentans* and *M. penetrans*.

Growth Media

The growth media used are complex and rendered selective by the addition of a β-lactam antibiotic (penicillin or ampicillin) and sometimes polymyxin.

For the culture of *M. pneumoniae*, the medium frequently used is a Hayflick modified medium containing heart infusion broth supplemented with 10% (v/v) freshly prepared yeast extract (25% wt/v) and 20% horse serum (Waites *et al.*, 2001a). SP-4 medium is more complex and contains fetal bovine serum. A number of species of mycoplasmas grow better in this medium. Liquid media contain glucose (0.5%) and phenol red (0.002%). The pH ranges from 7.4 to 7.6. *M. hominis* grows on modified Hayflick medium and SP-4 medium, (initial pH around 7.2). The broth form of the two media contains arginine (1%) instead of glucose. *M. hominis* grows also on the medium at pH 6.0 designed for the culture of *Ureaplasma* spp.

Ureaplasma spp. are cultured at pH 6.0, on a medium based on trypticase soy broth, enriched with yeast extract, horse serum, cysteine and urea. The liquid medium contains the pH indicator, phenol red (Waites *et al.*, 2001a).

The medium SP-4 is the best adapted for the growth of fastidious species. It is enriched with glucose for *M. genitalium* and with glucose and sometimes arginine for *M. fermentans*, *M. penetrans* and *M. pirum*.

Whatever the species sought, it is better to use both liquid and agar media (initially or after subculture). Broth cultures should be serially diluted 1:10 from 10^{-1} to 10^{-4}, in order to eliminate any possible inhibitors present in the tissue and for quantitative evaluation. Agar media should be dot inoculated. Liquid specimens (bronchoalveolar lavage and urine) should be centrifuged before plating. Given the fastidious nature of mycoplasmas, quality control of the culture media is essential.

Detection of Growth

In liquid media containing glucose, growth of fermenting species (particularly *M. pneumoniae*) is detected by a change in color of the pH indicator (acidification of the medium). Growth of *Ureaplasma* spp. produces an alcalinization of the liquid medium containing urea as does the growth of *M. hominis* and other such species on broth media containing arginine (Table 24.2). The color change occurs after 18–24 h for *Ureaplasma* spp., 48 h for *M. hominis*, 6–20 days for *M. pneumoniae* and even longer for more fastidious species.

After specimens have been plated and incubated under appropriate conditions for the particular species sought, colonies must be observed with magnification. The appearance varies from species to species. *M. pneumoniae* is often granular, while colonies of *M. hominis* are small (200–300 μm), fried-egg shaped. Colonies of *Ureaplasma* spp. are irregular and very small (15–60 μm) (Figure 24.1). Colonies grown on solid media containing urea and manganous sulfate or calcium choride and are black which makes them easier to see and more difficult to confuse with artifacts such as crystals in the medium. These colonies can be observed directly after culture of the specimen on agar, or as control, after subculture of liquid media which have changed color.

Identification

Mycoplasma pneumoniae is identified on the basis of certain biochemical properties (fermentation of glucose and lack of arginine hydrolysis), hemadsorption or hemagglutination of guinea pig or chicken erythrocytes (absent in commensal respiratory mycoplasmas) and hemolysis. Antigenic identification, the previous reference method, is rarely used in the absence of commercially available immune serum. Furthermore, it is difficult to separate *M. pneumoniae* and *M. genitalium* using this method in that they present closely related antigens and similar biochemical properties. Amplification of genetic material from culture by PCR is an excellent method of identification.

Identification of *Ureaplasma* spp. and *M. hominis* is relatively simple. The change of a color indicator due to metabolic properties and the physical appearance of colonies are the principal methods. It is necessary to control the aspect of colonies because a color change can be caused by the presence of other bacteria or cells. Antigenic identification is very rarely performed, and separation of serovars is not current practice. Identification of the two species, *U. urealyticum* and *U. parvum*, can be performed by PCR (see paragraph 7.3.2).

Different commercial kits have been proposed for the detection of *Ureaplasma* spp. and *M. hominis* from lower genital tract specimens. They give satisfactory results if identification is verified by the physical appearance of colonies on agar.

Interpretation of Culture Results

Isolation of *M. pneumoniae* from a patient with a respiratory infection is an important indicator because this organism is not part of the commensal flora.

Isolation of *Ureaplasma* spp. or *M. hominis* poses a more delicate problem of interpretation. Isolation from normally sterile specimens is significant, but for other specimens where they can be present as commensals, it is useful to make a quantitative evaluation. The criteria proposed as significant for NGU are as follows: 10^4 color-changing unity (CCU)/ml for a urethral swab and 10^3 CCU/ml for first-void urine. The presence of *Ureaplasma* spp. in cervicovaginal swabs is very difficult to interpret because of its natural frequency. *M. hominis* can be found in high numbers ($\geq 10^4$ CCU/ml) during a vaginosis. Its presence can also suggest an infection higher up in the genital tract. In this case, clinical signs and associated bacteriological elements must be taken into account. The presence of mycoplasmas in peripheral neonatal specimens can be due to simple contamination. Isolation from endotracheal specimens in high number ($\geq 10^3$ CCU/ml) is more significant.

Non-Culture Methods

Culture is well adapted to the detection of *M. hominis* and *Ureaplasma* spp. from specimens where they can be found easily, such as urogenital specimens. Nonculture methods are useful for detection of fastidious and slow-growing organisms such as *M. pneumoniae* and *M. genitalium*. They are also adapted to the detection of all mycoplasma species, including *M. hominis* and *Ureaplasma* spp., from some specimens where they are difficult to grow, such as synovial fluids or tissues.

Antigenic detection and DNA probes have been proposed for *M. pneumoniae*. Because of their lack of sensitivity, they have been replaced by PCR. For instance, complete sequences of the 16S rRNA genes are available from almost all the established mycoplasma species and could be selected as target sequences in PCR-based tests. Some examples of target sequences for amplification of several human mycoplasmas will be taken. However, despite many developments in the use of PCR for mycoplasma detection, there is still a major drawback, the lack of standardization of the techniques, related to the lack of evaluated commercially available kit dedicated to mycoplasmas and the use of in-house PCRs.

PCR for M. pneumoniae

PCR detection of *M. pneumoniae* was first proposed 15 years ago (Bernet *et al.*, 1989). Many reports proposing different in-house PCRs in respiratory tract infections have been reviewed recently (Loens *et al.*, 2003). Amplification techniques demonstrated to be highly sensitive and specific and could become the preferred diagnostic procedures for *M. pneumoniae* infection, if available at a reasonnable cost and standardized.

Different extraction methods are proposed. Because of the simple structure of mycoplasmas, sample preparation by freeze boiling could be sufficient (Ieven *et al.*, 1996). Several target regions have been used, 16S rRNA gene, ATPase operon gene, *tuf* gene and P1 adhesin gene. In comparative studies, P1 gene primers were found to give more sensitive results than 16S rRNA gene primers. Nested PCR was also proposed (Abele-Horn *et al.*, 1998; Dorigo-Zetsma *et al.*, 1999). Different techniques can be used for detection of amplification products.

Multiplex PCR tests detecting *M. pneumoniae* and other respiratory pathogens such as *C. pneumoniae* and *Legionella pneumophila* may prove useful, but the major progress will certainly be the development of real-time PCR allowing quantitative detection of *M. pneumoniae* DNA (Ursi *et al.*, 2003). Alternative amplification techniques such as Nucleic Acid Sequence Base Amplification have also been reported (Loens *et al.*, 2003).

PCR can also be used to separate, for epidemiological purpose, the two groups of *M. pneumoniae* strains (Cousin-Allery *et al.*, 2000).

PCR for Other Mycoplasmas

The PCR technique is practically the only method for detection of *M. genitalium* (Jensen *et al.*, 1991; de Barbeyrac *et al.*, 1993). The most frequently used target is *MgPa* adhesin gene. Real-time PCR was recently proposed (Yoshida *et al.*, 2002).

Concerning other mycoplasma species, PCR remains the best method for detection of *M. fermentans*, using amplification of insertion-like elements (Wang *et al.*, 1992) and *M. penetrans*. PCR is potentially interesting for the detection in some specific specimens of *Ureaplasma* spp., using amplification of the urease gene (Blanchard *et al.*, 1993) or other target and allows characterization of the two species *U. urealyticum* and *U. parvum* (Robertson *et al.*, 1993; Kong *et al.*, 2000). PCR is also widely used to detect mycoplasmal contamination of cell cultures (Teyssou *et al.*, 1993; Drexler and Uphoff, 2000).

Serology

M. pneumoniae

Because of their simplicity, serologic tests are very often used to diagnose respiratory infections, mainly in severe illness. Even if PCR or culture are available, *M. pneumoniae* serology gives additional information when showing antibody response in infected patients. Its drawback is that it only allows a retrospective diagnosis. Antibody response to protein and glycolipid antigens of *M. pneumoniae* can be detected after about 1 week of illness, depending on the test, and peeks at 3–6 weeks and then declines gradually (Waites, Rikihisa and Taylor-Robinson, 2003).

Demonstration of seroconversion (fourfold change in titer) between two successive sera collected 2–4 weeks apart, or a high level of antibodies detected by a single or late test, is an important element of diagnosis. The presence of IgM often considered as evidence of acute infection is particularly interesting to detect in children. IgM are less often observed in adults.

Complement fixation (CF) reaction detects antibodies directed against a glycolipid antigen. Thus it was the reference test in the past; it is still valid as long as certain criteria of interpretation are strictly adhered to (seroconversion or a titer of at least 64). Nevertheless, it lacks sensitivity and specificity because of possible cross-reactions with other bacteria, mainly *M. genitalium*, but also with some human tissues or plants. False-positive reactions due to cross-reactive auto-antibodies can be observed in neurological or pancreatic illness, usually at a low level. CF detects IgM and IgG responses but does not differentiate among antibody classes.

Alternative techniques with greater sensitivity are proposed, immunofluorescence assays, particle agglutination assays using latex beads or gelatin and mainly EIAs (Enzyme-linked Immuno Assays) (Waites, Talkington and Bébéar, 2001b). EIAs are the most widely used techniques. Different antigen preparations and a variety of assay conditions (96-well microtiter plate formats, strips of 8-well format and rapid EIAs using qualitative membrane-based test) are available. They can detect IgG and IgM simultaneously or separately.

Cold agglutinins, IgM antibodies to the I antigen of erythrocytes, are found in only 30–50% of *M. pneumoniae* infections. Furthermore, their presence at a significant titer (≥ 64) is not specific of these infections.

Genital Mycoplasmas

Several techniques have been proposed for detection of antibodies during *Ureaplasma* spp. and *M. hominis* infections. The results of these tests were difficult to interpret because of the ubiquity of these species in humans and because of the lack of information concerning the level of immunity found in general population. Efforts to develop serologic tests for *M. genitalium* have been blocked by the problem of antigens in common with *M. pneumoniae*. Serologic tests for genital mycoplasma infections are not commercially available in Europe or US and are not recommended for routine diagnostic.

Antibiotic Susceptibility

Antibiotic Susceptibility Testing

The nutritional requirements and cultural conditions of mycoplasmas are rather distant from the standard conditions recommended for bacteria. Actually no specific guidelines for susceptibility testing of mycoplasmas are available but recommendations have been recently summarized (Waites *et al.*, 2001a).

Standard broth, microbroth and agar dilution methods used for susceptibility testing of bacteria have been adapted for use with mycoplasmas. Agar media are recommended for mycoplasmas which produce colonies easily observed with a binocular microscope (practically all the species of the genus *Mycoplasma*), while broth media are recommended for *Ureaplasma* spp. Initial inoculum should be quantified and contain 10^4–10^5 CCU/ml. After a period of incubation appropriate for the species, the results should be read when growth becomes visible on the control medium containing no antibiotic. The MIC is the lowest concentration of antibiotic inhibiting the growth of colonies or a color change of the pH indicator. These methods, performed with appropriate controls, give useful results for the treatment of mycoplasmal infections.

Susceptibility testing kits, available in Europe, have been adapted only for use with *Ureaplasma* spp. and *M. hominis*. They consist of microwells containing dried antimicrobials, generally in one or two concentrations corresponding to the threshold proposed for conventional bacteria to classify a strain as susceptible, intermediate or resistant. The results obtained with these kits are satisfactory, but the kits should be better used after isolation of the organism than directly on the original specimen.

Generally, agar disk diffusion is not acceptable for mycoplasma susceptibility testing. However, the agar gradient diffusion technique or E-test has been recently adapted to determine fluoroquinolone, tetracycline and macrolide susceptibilities for *M. hominis* and *Ureaplasma* spp. (Waites *et al.*, 2001a).

No breakpoints specific for mycoplasmas are available actually; so it may be preferable to merely report MICs. Some laboratories have adopted the breakpoints used for the interpretation of MICs for other bacteria. MIC values of $\leq 1\,\mu g/ml$ should be considered predictive of effective treatment (Taylor-Robinson and Bébéar, 1997).

Intrinsic Antibiotic Resistance

Two kinds of intrinsic resistance are present in mycoplasmas, one related to the class Mollicutes and one related to the different mycoplasma and ureaplasma species (Bébéar and Bébéar, 2002). Innate antibiotic resistance has been used to isolate mycoplasmas from specimens contaminated with other bacteria and to differentiate mycoplasma species.

The lack of a cell wall makes all the members of the class Mollicutes intrinsically resistant to β-lactams and to all antimicrobials, such as glycopeptides and fosfomycin, which target the cell wall. In addition, mycoplasmas are also resistant to polymyxins, sulfonamides, trimethroprim, nalidixic acid and rifampin. The molecular mechanism of resistance to rifampin investigated in *S. citri* was shown to be a natural mutation of the *rpoB* gene encoding the β subunit of the DNA-dependent RNA polymerase, the target of rifampin, preventing the binding of the antibiotic to its target.

Resistance to the macrolide-lincosamide-streptogramin-ketolide (MLSK) group differs between species.

Mycoplasma pneumoniae and *M. genitalium* are naturally susceptible to all MLSK antibiotics, except to lincomycin which shows modest activity against these two species (Table 24.3). *Ureaplasma* spp. are susceptible to macrolides and ketolides. *Mycoplasma hominis* is resistant to 14-membered ring Macrolides, including erythromycin and its derivatives, and to the 15-membered ring azithromycin but is susceptible to the 16-membered macrolide, josamycin. *Mycoplasma hominis* is susceptible to lincosamides but not to telithromycin. The genetic basis of the intrinsic resistance of *M. hominis* to erythromycin is related to mutations in the peptidyltransferase loop of domain V of 23S rRNA (Furneri *et al.*, 2000; Pereyre *et al.*, 2002).

Activities of Antibiotics Used for Treatment

The number of antibiotics that can be used for treating mycoplasmal infections is limited. The most widely used include tetracyclines, MLSK group and fluoroquinolones. Phenicols and aminoglycosides are kept for the treatment of special cases.

Tetracyclines, MLSK and fluoroquinolones share the advantages of being potentially active against the bacteria which could be associated with mycoplasmas in respiratory and urogenital tract infections. They share also the ability to reach high intracellular concentrations where mycoplasmas could localize. Only fluoroquinolones and ketolides have shown mycoplasmicidal qualities.

A compilation of MIC data obtained with different antimicrobials against human mycoplasmas is shown in Table 24.3. A lot of data are available for *M. pneumoniae*, *M. hominis* and *Ureaplasma* spp. The results for a very small number of strains of *M. genitalium* that are currently available indicate that this species has an antibiotic susceptibility profile similar to that of *M. pneumoniae* (Renaudin, Tully and Bébéar, 1992).

Among the different mechanisms of acquired bacterial resistance, target alterations and efflux mechanisms have been described in mycoplasmas. Both genetic mutations or acquisition of new genes via gene transfer have been documented. Indeed, mycoplasmas are characterized by high mutation frequencies, and transposons carrying antibiotic resistance have been recognized in these bacteria (Roberts, 1992).

Tetracyclines: Acquired resistance to tetracyclines concerns only *Ureaplasma* spp. and *M. hominis*, but not *M. pneumoniae*. High-level resistance to tetracyclines (MICs $\geq 8\,\mu g/ml$) in *M. hominis* and *Ureaplasma* spp. has been associated with the presence of the *tet(M)* determinant (Roberts *et al.*, 1985; Roberts and Kenny, 1986), associated with a conjugative transposon, member of the Tn*1545* family named Tn*916*, widely distributed among urogenital bacteria. The *tet(M)* determinant encodes the TetM protein which protects the ribosome from the action of tetracyclines and confers cross-resistance to all tetracyclines. Glycylcyclines such as tigecycline remain active against tetracycline-resistant strains of *M. hominis* containing *tet(M)* gene but not against *tet(M)*-containing *Ureaplasma* spp. In addition, the *tet(M)* resistance gene is one of the commonly used markers in genetic studies performed in mycoplasmas.

The occurrence of tetracycline resistance in genital mycoplasmas varies geographically and according to prior antimicrobial exposure. It was evaluated to about 10% in patients attending STD clinics in UK and less than 5% in France in 1993 (Bébéar *et al.*, 1993). This resistance rate seems to increase significantly in some regions of France, especially for *M. hominis* (Bébéar *et al.*, personal communication).

MLSK group: Acquired resistance to MLSK appears to remain a rare phenomenon. Some cases of macrolide resistance have been reported during treatment with macrolide of *M. pneumoniae* infection, and resistant strains of *M. pneumoniae* have been obtained *in vitro* by selection in the presence of various macrolides, streptogramins and a ketolide. A variety of mutations could be selected *in vitro*, mutations in the peptidyl transferase loop of domain V in the 23S rRNA, but also point mutations, insertions or deletions in L4 and L22 ribosomal protein genes (Lucier *et al.*, 1995; Pereyre *et al.*, 2004).

Table 24.3 Minimal inhibitory concentration ranges for various antimicrobials against *M. pneumoniae*, *M. genitalium*, *M. hominis* and *Ureaplasma* spp. (adapted from Bébéar *et al.*, 2002; Kenny and Cartwright, 2001)

Antimicrobial	*M. pneumoniae*	*M. genitalium*	*M. hominis*	*Ureaplasma* spp.
Tetracyclines[a], glycylcyclines				
Tetracycline	0.63–0.25	0.06–0.12	0.2–2	0.05–2
Doxycycline	0.02–0.5	≤0.01–0.3	0.03–2	0.02–1
Minocycline	0.06–0.5	≤0.01–0.2	0.03–1	0.06–1
Tigecycline	0.06–0.25	ND	0.125–0.5	1–16
MLSK group				
Erythromycin	≤0.004–0.06	≤0.01	32 to >1000	0.02–4
Roxithromycin	≤0.01–0.03	≤0.01	>16 to >64	0.06–4
Clarithromycin	≤0.004–0.125	≤0.01–0.06	16 to >256	≤0.004–2
Azithromycin	≤0.004–0.01	≤0.01–0.03	4 to >64	0.06–0.5
Josamycin	≤0.01–0.02	0.01–0.02	0.05–2	0.03–4
Spiramycin	≤0.01–0.25	0.12–1	32 to >64	4–32
Midecamycin	≤0.015	ND	0.25	ND
Clindamycin	≤0.008–2	0.2–1	≤0.008–2	0.2–64
Lincomycin	4–8	1–8	0.2–4	8–256
Pristinamycin	0.02–0.05	ND	0.1–0.5	0.1–1
Quinupristin/Dalfopristin	0.008–0.12	0.05	0.03–2	0.03–0.5
Telithromycin	0.0002–0.06	≤0.015	2–32	≤0.015–0.25
Cethromycin	≤0.001–0.016	ND	≤0.008–0.031	≤0.008–0.031
Fluoroquinolones				
Pefloxacin	2	ND	1–4	1–8
Norfloxacin	ND	ND	4–16	4–16
Ciprofloxacin	0.5–2	2	0.5–4	0.1–4
Ofloxacin	0.05–2	1–2	0.5–4	0.2–4
Sparfloxacin	≤0.008–0.5	0.05–0.1	≤0.008–0.1	0.003–1
Levofloxacin	0.5–1	0.5–1	≤0.008–0.5	0.12–2
Trovafloxacin	≤0.008–0.5	0.03–0.06	≤0.008–0.06	≤0.008–0.5
Gatifloxacin	0.06–1	0.12	0.06–0.25	0.25–2
Moxifloxacin	0.06–0.3	0.03	0.06	0.12–0.5
Gemifloxacin	≤0.008–0.12	0.05	0.0025–0.01	≤0.008–0.25
Garenoxacin	0.015–0.12	0.06–0.12	0.008–0.25	0.06–0.25
Chloramphenicol	2–10	0.5–25	4–25	0.4–8
Aminoglycosides				
Gentamicin	4	ND	2–16	0.1–13
New agents				
Linezolid	64–256	ND	2–8	>64
Evernimicin	2–16	ND	1–16	8–16

ND = not determined.

[a] Susceptible strains.

The frequency of acquired macrolide resistance in clinical strains of *M. hominis* and *Ureaplasma* spp. is not known but, like for *M. pneumoniae*, is probably very low. Acquired resistance to 16-membered macrolides and lincosamides has been recently described in a patient repetitively exposed to antibiotics (Pereyre *et al.*, 2002). Neither the methyltransferase enzyme (Erm family) nor the enzyme modifying the antibiotic have been identified so far in mycoplasmas.

Fluoroquinolones: Resistance to fluoroquinolones has been reported only for genital mycoplasmas. High-level resistance was observed in clinical isolates of *M. hominis* (Bébéar *et al.*, 1999) and *Ureaplasma* spp. (Bébéar *et al.*, 2003) obtained from patients treated previously with fluoroquinolones. These strains presented target alterations, located in the quinolone resistance-determining regions of the gyrase *gyrA* gene and of the topoisomerase IV *parC* and *parE* genes. Therapeutic failures of *M. genitalium*-positive NGU have been associated with mutations in the *gyrA* and *parC* genes of *M. genitalium* for patients treated with levofloxacin (Deguchi *et al.*, 2001).

Furthermore, an active efflux system, possibly an ATP-binding cassette-type efflux pump, was reported recently in ethidium bromide (EtBr)-selected strains of *M. hominis* showing a MDR resistance phenotype with increased MICs of ciprofloxacin and EtBr (Raherison *et al.*, 2002).

Other products: High-level resistance to aminoglycosides has been described in strains of *M. fermentans* isolated from cell cultures, but not from human sources. The resistance gene *aacA-aphD* found in the transposon Tn*4001* which confers resistance to gentamicin and the *cat* gene encoding chloramphenicol acetyltransferase are used as markers in genetic studies of mycoplasmas.

TREATMENT, PREVENTION AND CONTROL

M. pneumoniae Infection

The number of evaluations of the *in vivo* activity of antibiotics in infections due to mycoplasmas is low. However, since most of the antimicrobial agents active against mycoplasmas are only bacteriostatic, eradicating mycoplasmas depends on the efficacy of the host immune system.

An experimental model has been used for assessing the efficacy of erythromycin, roxithromycin and pristinamycin in the treatment of respiratory infection caused by intranasal inoculation of *M. pneumoniae* in hamster. Recently, an experimental model of pneumonia was described in mice with *M. pneumoniae* (Hardy *et al.*, 2003). Different clinical trials showed also the efficiency of such treatments for community-acquired pneumonia (Bébéar *et al.*, 1993).

Because of the lack of simple, rapid detection methods, suspected *M. pneumoniae* infections are treated before they can be confirmed.

Macrolides are usually prescribed for patients of all ages. Recently, clinical effectiveness of azithromycin prophylaxis was demonstrated during a *M. pneumoniae* outbreak in a closed setting (Hyde *et al.*, 2001). For adults, fluoroquinolones, which are active *in vitro*, are an interesting alternative, but the number of documented studies is still rather low. The antibiotic treatment can shorten the course of the disease.

Different types of vaccines against *M. pneumoniae* have been tested. None are presently available. The most promising will probably contain the antigenic fraction of purified P1 protein.

Genital Mycoplasma Infections

The treatment of genital mycoplasma infections can occur after demonstration of their role in the infection or in the absence of microbiological proof when they are suspected to be possibly involved. The choice of antibiotic to treat these infections must take into account the specific species isolated or suspected, any other microorganisms isolated or possibly associated and the physical circumstances of the development of the disease.

In adults, the treatment of these infections is the same as for *C. trachomatis* with which they are sometimes associated. Generally, tetracyclines are the antibiotics of first choice. There have been some cases of therapeutic failure related to resistance *in vitro*. The fluoroquinolones are certainly interesting. When these antibiotics are contraindicated (pregnancy and neonatal infections), macrolides are considered first. The activity must be tested *in vitro*. It is sometimes necessary to treat neonates with tetracycline such as when the strain involved is resistant to erythromycin. These cases are very rare. A recent study compared the antibiotic treatment efficacy of tetracyclines and azithromycin in *M. genitalium*-infected adults (Falk, Fredlund and Jensen, 2003). The data suggested that tetracyclines are not sufficient to eradicate *M. genitalium* and randomized controlled treatment trials are needed.

The length and method of administration of the treatment depend on the location of the infection. As most active antibiotics have a bacteriostatic effect on mycoplasmas, the course of treatment must be sufficiently long. Clinical and biological criteria come into play to confirm the efficacy of treatment. Clearing of mycoplasmas is a valid criteria for sites which are normally sterile, but it is no longer reliable for sites where mycoplasmas can be found naturally. It must, nevertheless, be indicated that even though treatment of mycoplasma infections is relatively simple in otherwise healthy subjects, it is much more complicated in immunosuppressed subjects. It is then essential to verify eradication of the organism which can be difficult.

REFERENCES

Abele-Horn M., Busch U., Nitschko H. *et al.* (1998) Molecular approaches to diagnosis of pulmonary diseases due to *Mycoplasma pneumoniae*. *Journal of Clinical Microbiology, 36*, 548–551.

Bébéar C., de Barbeyrac B., Bébéar C.M. *et al.* (1997) New developments in diagnostics and treatment of Mycoplasma infections in humans. *Wiener Klinische Wochenschrift, 109*, 594–599.

Bébéar C., de Barbeyrac B., Dewilde A. *et al.* (1993) Multicenter study of the in vitro sensitivity of genital mycoplasmas to antibiotics. *Pathologie Biologie, 41*, 289–293.

Bébéar C., Dupon M., Renaudin H. and de Barbeyrac B. (1993) Potential improvements in therapeutic options for mycoplasmal respiratory infections. *Clinical Infectious Diseases, 17* (Suppl. 1), S202–S207.

Bébéar C.M. and Bébéar C. (2002) Antimycoplasmal agents, in *Molecular Biology and Pathogenicity of Mycoplasmas* (eds S. Razin and R. Herrmann), Kluwer Academic/Plenum Publishers, New York, pp. 545–566.

Bébéar C.M., Renaudin H., Charron A. *et al.* (2003) DNA gyrase and topoisomerase IV mutations in clinical isolates of *Ureaplasma* spp. and *Mycoplasma hominis* resistant to fluoroquinolones. *Antimicrobial Agents and Chemotherapy, 47*, 3323–3325.

Bébéar C.M., Renaudin J., Charron A. *et al.* (1999) Mutations in the *gyrA*, *parC*, and *parE* genes associated with fluoroquinolone resistance in clinical isolates of *Mycoplasma hominis*. *Antimicrobial Agents and Chemotherapy, 43*, 954–956.

Bernet C., Garret M., de Barbeyrac B. *et al.* (1989) Detection of *Mycoplasma pneumoniae* by using the Polymerase Chain Reaction. *Journal of Clinical Microbiology, 27*, 2492–2496.

Blanchard A. and Bébéar C.M. (2002) Mycoplasma of humans, in *Molecular Biology and Pathogenicity of Mycoplasmas* (eds S. Razin and R. Herrmann), Kluwer Academic/Plenum Publishers, New York, pp. 45–71.

Blanchard A. and Montagnier L. (1994) AIDS – associated mycoplasmas. *Annual Review of Microbiology, 48*, 687–712.

Blanchard A., Hentschel J., Duffy L. *et al.* (1993) Detection of *Ureaplasma urealyticum* by polymerase chain reaction in the urogenital tract of adults, in amniotic fluid, and in the respiratory tract of newborns. *Clinical Infectious Diseases, 17* (Suppl. 1), S148–S153.

Cassell G.H., Waites K.B., Watson H.L. *et al.* (1993) *Ureaplasma urealyticum* intrauterine infection: role in prematurity and disease in newborns. *Clinical Microbiology Review, 6*, 69–87.

Cohen C.R., Manhart L.E., Bukusi E.A. *et al.* (2002) Association between *Mycoplasma genitalium* and acute endometritis. *Lancet, 350*, 765–766.

Cousin-Allery A., Charron A., de Barbeyrac B. *et al.* (2000) Molecular typing of *Mycoplasma pneumoniae* strains by PCR-based methods and pulsed-field gel electrophoresis. Application to french and danish isolates. *Epidemiology and Infection, 124*, 103–111.

Dandekar T., Snel B., Schmidt S. *et al.* (2002) Comparative genome analysis of the Mollicutes, in *Molecular Biology and Pathogenicity of Mycoplasmas* (eds S. Razin and R. Herrmann), Kluwer Academic/Plenum Publishers, New York, pp. 255–278.

de Barbeyrac B., Bernet-Poggi C., Fébrer F. *et al.* (1993) Detection of *Mycoplasma pneumoniae* and *Mycoplasma genitalium* by polymerase chain reaction in clinical samples. *Clinical Infectious Diseases, 17* (Suppl. 1), S83–S89.

Deguchi T., Maeda S., Tamaki M. *et al.* (2001) Analysis of the *gyrA* and *parC* genes of *Mycoplasma genitalium* detected in first-pass urine of men with non-gonococcal urethritis before and after fluoroquinolone treatment. *Journal of Antimicrobial Chemotherapy, 48*, 742–744.

Dorigo-Zetsma J.W., Zaat S.A.J., Vriesema A.J.M. and Dankert J. (1999) Demonstration by a nested PCR for *Mycoplasma pneumoniae* that *M. pneumoniae* load in the throat is higher in patients hospitalized for *M. pneumoniae* infection that in non-hospitalized subjects. *Journal of Medical Microbiology, 48*, 1115–1122.

Drexler H.G. and Uphoff C.C. (2000) Contamination of cell culture, *Mycoplasma*, in *The Leukemia Lymphoma Cell Lines Factsbook* (ed. H.G. Drexler) Academic Press, San Diego, pp. 609–627.

Falk L., Fredlund H. and Jensen J.S. (2003) Tetracycline treatment does not eradicate *Mycoplasma genitalium*. *Sexually Transmitted Diseases, 79*, 318–319.

Foy H.M. (1993) Infections caused by *Mycoplasma pneumoniae* and possible carrier state in different populations of patients. *Clinical Infectious Diseases, 17* (Suppl. 1), S37–S46.

Fraser C.M., Gocayne J.D., White O. *et al.* (1995) The minimal gene complement of *Mycoplasma genitalium*. *Science, 270*, 397–403.

Furneri P.M., Rappazzo G., Musumarra M.P. *et al.* (2000) Genetic basis of natural resistance to erythromycin in *Mycoplasma hominis*. *Journal of Antimicrobial Chemotherapy, 45*, 547–548.

Furr P.M., Taylor-Robinson D. and Webster D.B. (1994) Mycoplasmas and ureaplasmas in patients with hypogammaglobulinaemia and their role in arthritis: microbiological observations over 20 years. *Annals of the Rheumatic Diseases, 53*, 183–187.

Glass J.I., Lafkowitz E.J., Glass J.S. *et al.* (2000) The complete sequence of the mucosal pathogen *Ureaplasma urealyticum*. *Nature, 407*, 757–762.

Grenabo L., Hedelin H. and Pettersson S. (1988) Urinary infection stones caused by *Ureaplasma urealyticum*: a review. *Scandinavian Journal of Infectious Diseases, 53* (Suppl.), 46–49.

Hardy R.D., Rios A.M., Chavez-Bueno S. *et al.* (2003) Antimicrobial and immunologic activities in a murine model of *Mycoplasma pneumoniae*-induced pneumonia. *Antimicrobial Agents and Chemotherapy, 47*, 1614–1620.

Himmelreich R., Hilbert H., Plagens H. *et al.* (1996) Complete sequence analysis of the genome of the bacterium *Mycoplasma pneumoniae*. *Nucleic Acids Research, 24*, 4420–4449.

Horner P., Thomas B., Gilroy C.B. *et al.* (2001) Role of *Mycoplasma genitalium* and *Ureaplasma urealyticum* in acute and chronic nongonococcal urethritis. *Clinical Infectious Diseases, 32*, 995–1003.

Hyde T.B., Gilbert M., Schwartz S.B. *et al.* (2001) Azithromycin prophylaxis during a hospital outbreak of *Mycoplasma pneumoniae* pneumonia. *Journal of Infectious Diseases, 183*, 907–912.

Ieven M., Ursi D., Van Bever H. *et al.* (1996) Detection of *Mycoplasma pneumoniae* by two polymerase chain reactions and role of *M. pneumoniae* in acute respiratory tract infections in pediatric patients. *Journal of Infectious Diseases, 173,* 1445–1452.

Jacobs E. (2002) *Mycoplasma pneumoniae* disease manifestations and epidemiology, in *Molecular Biology and Pathogenicity of Mycoplasmas* (eds S. Razin and R. Herrmann), Kluwer Academic/Plenum Publishers, New York, pp. 519–530.

Jensen J.S., Uldum S.A., Sondegard-Anderson J. *et al.* (1991) Polymerase chain reaction for detection of *Mycoplasma genitalium* in clinical samples. *Journal of Clinical Microbiology, 29,* 46–50.

Johansson K.E. and Pettersson B. (2002) Taxonomy of Mollicutes, in *Molecular Biology and Pathogenicity of Mycoplasmas* (eds S. Razin and R. Herrmann), Kluwer Academic/Plenum Publishers, New York, pp. 1–29.

Kenny G.E. and Cartwright F.D. (2001) Susceptibilities of *Mycoplasma hominis, M. pneumoniae* and *Ureaplasma urealyticum* to GAR-936, dalfopristin, dirithromycin, evernimicin, gatifloxacin, linezolid, moxifloxacin, quinupristin-dalfopristin, and telithromycin compared to their susceptibilities to reference macrolides, tetracyclines and fluoroquinolones. *Antimicrobial Agents and Chemotherapy, 45,* 2604–2608.

Kong F.R., James C., Ma Z.F. *et al.* (1999) Phylogenetic analysis of *Ureaplasma urealyticum* – support for the establishment of a new species, *Ureaplasma parvum. International Journal of Systematic Bacteriology, 49,* 1879–1889.

Kong F.R., Ma Z.F., James C. *et al.* (2000) Species identification and subtyping of *Ureaplasma parvum* and *Ureaplasma urealyticum* using PCR-based assays. *Journal of Clinical Microbiology, 38,* 1175–1179.

Krause D.C. (1998) *Mycoplasma pneumoniae* cythadherence: organization and assembly of the attachment organella. *Trends in Microbiology, 6,* 15–18.

Ladefoged S.A. (2000) Molecular dissection of *Mycoplasma hominis. APMIS: acta pathologica, microbiologica, et immunologica Scandinavica, 108,* 5–45.

Lo S.C., Hayes M.M., Wang R.Y. *et al.* (1991) Newly discovered mycoplasma isolated from patients with HIV. *Lancet, 338,* 1415–1418.

Lo S.C., Wear D.J., Green S.L. *et al.* (1993) Adult respiratory distress syndrome with or without systemic disease associated with infections due to *Mycoplasma fermentans. Clinical Infectious Diseases, 17* (Suppl. 1), S259–S263.

Loens K., Ursi D., Goossens H. and Ieven M. (2003) Molecular diagnosis of *Mycoplasma pneumoniae* respiratory tract infections. *Journal of Clinical Microbiology, 41,* 4915–4923.

Lucier T.S., Heitzman K., Lin S.K. and Hu P.C. (1995) Transition mutations in the 23S rRNA of erythromycin-resistant isolate of *Mycoplasma pneumoniae. Antimicrobial Agents and Chemotherapy, 39,* 2770–2773.

Manhart L.E., Dutro S.M., Holmes K.K. *et al.* (2001) *Mycoplasma genitalium* is associated with mucopurulent cervicitis. *International Journal of STD & AIDS, 12* (Suppl. 2), 69.

Martin R.J., Kraft M., Chu H.W. *et al.* (2001) A link between chronic asthma and chronic infection. *The Journal of Allergy and Clinical Immunology, 107,* 595–601.

Meyer R.D. and Clough W. (1993) Extragenital *Mycoplasma hominis* infections adults: emphasis on immunosuppression. *Clinical Infectious Diseases, 17* (Suppl. 1), S243–S249.

Neman-Simha V., Renaudin H., de Barbeyrac B. *et al.* (1992) Isolation of genital mycoplasmas from blood of febrile obstetrical-gynecologic patients and neonates. *Scandinavian Journal of Infectious Diseases, 24,* 317–321.

Pereyre S., Gonzalez P., de Barbeyrac B. *et al.* (2002) Mutations in 23S rRNA account for intrinsic resistance to macrolides in *Mycoplasma hominis* and *Mycoplasma fermentans* and for acquired resistance to macrolides in *Mycoplasma hominis. Antimicrobial Agents and Chemotherapy, 46,* 3142–3150.

Pereyre S., Guyot C., Renaudin H. *et al.* (2004) *In vitro* selection and characterization of resistance to macrolides and related antibiotics in *Mycoplasma pneumoniae. Antimicrobial Agents and Chemotherapy, 48,* 460–465.

Principi N., Esposito S., Biasi F. *et al.* (2001) Role of *Mycoplasma pneumoniae* and *Chlamydia pneumoniae* in children with community-acquired lower respiratory tract infections. *Clinical Infectious Diseases, 32,* 1281–1289.

Raherison S., Gonzalez P., Renaudin H. *et al.* (2002) Evidence of an active efflux in resistance to ciprofloxacin and to ethidium bromide by *Mycoplasma hominis. Antimicrobial Agents and Chemotherapy, 46,* 672–679.

Razin S., Yogev D. and Naot Y. (1998) Molecular biology and pathogenicity of mycoplasmas. *Microbiology and Molecular Biology Reviews, 62,* 1094–1156.

Renaudin H., Tully J.G. and Bébéar C. (1992) *In vitro* susceptibility of *Mycoplasma genitalium* to antibiotics. *Antimicrobial Agents and Chemotherapy, 36,* 870–872.

Roberts M.C. (1992) Antibiotic resistance, in *Mycoplasmas: Molecular Biology and Pathogenesis* (eds J. Maniloff, R.N. McElhaney, L.R. Finch and J.B. Baseman), American Society for Microbiology, Washington, DC, pp. 513–523.

Roberts M.C. and Kenny G.E. (1986) Dissemination of the *tetM* tetracycline resistant determinant to *Ureaplasma urealyticum. Antimicrobial Agents and Chemotherapy, 29,* 350–352.

Roberts M.C., Koutsky L.A., Holmes K.K. *et al.* (1985) Tetracycline-resistant *Mycoplasma hominis* strains contain streptococcal *tetM* sequences. *Antimicrobial Agents and Chemotherapy, 28,* 141–143.

Robertson J.A., Vekris A., Bébéar C. and Stemke G. (1993) Polymerase chain reaction using 16S rRNA gene sequences distinguishes the two biovars of *Ureaplasma urealyticum. Journal of Clinical Microbiology, 31,* 824–830.

Sasaki Y., Ishikawa J., Yamashita A. *et al.* (2002) The complete genomic sequence of *Mycoplasma penetrans*, an intracellular bacterial pathogen in humans. *Nucleic Acids Research, 30,* 5293–5300.

Stadtländer C.T.K.H., Watson H.L., Simecka J.W. and Cassell G.H. (1993) Cytopathogenicity of *Mycoplasma fermentans* (including strain *incognitus*). *Clinical Infectious Diseases, 17* (Suppl. 1), S289–S301.

Taylor-Robinson D. (1995) The history and role of *Mycoplasma genitalium* in sexually transmitted diseases. *Genitourinary Medicine, 71,* 1–8.

Taylor-Robinson D. (1996) Infections due to species of *Mycoplasma* and *Ureaplasma*: an update. *Clinical Infectious Diseases, 23,* 671–684.

Taylor-Robinson D. (1998) *Mycoplasma hominis* parasitism of *Trichomonas vaginalis. Lancet, 352,* 2022–2023.

Taylor-Robinson D., Ainsworth J.G. and McCormack W.M. (1999) Genital mycoplasmas, in *Sexually Transmitted Diseases*, 3rd edn (eds K.K. Holmes, P.F. Sparling, P.A. Mardh, S.M. Lemon, W.E. Stamm, P. Piot and J.N. Wasserheit), McGraw-Hill, New York, pp. 533–548.

Taylor-Robinson D. and Bébéar C. (1997) Antibiotic susceptibilities of mycoplasmas and treatment of mycoplasmal infections. *The Journal of Antimicrobial Chemotherapy, 40,* 622–630.

Teyssou R., Poutiers F., Saillard C. *et al.* (1993) Detection of mollicute contamination in cell cultures by 16S rDNA amplification. *Molecular and Cellular Probes, 7,* 209–216.

Totten P.A., Schwartz M.A., Sjostrom K.E. *et al.* (2001) Association of *Mycoplasma genitalium* with nongonococcal urethritis in heterosexual men. *Journal of Infectious Diseases, 183,* 269–276.

Ursi D., Dirven K., Loens K. *et al.* (2003) Detection of *Mycoplasma pneumoniae* in respiratory samples by real-time PCR using an inhibition control. *Journal of Microbiological Methods, 55,* 149–153.

Waites K., Rikihisa Y. and Taylor-Robinson D. (2003) *Mycoplasma* and *Ureaplasma*, in *Manual of Clinical Microbiology*, 8th edn (eds P.R. Murray, E.J. Baron, M.A. Pfaller, J.H. Jorgensen and R.M. Yolken), ASM Press, Washington, DC, pp. 972–990.

Waites K.B., Bébéar C.M., Robertson J.A. *et al.* (2001a) Laboratory diagnosis of mycoplasmal infections. Cumitech 34 of American Society for Microbiology (coordinating ed. F.S. Nolte), ASM press, 1–30.

Waites K.B., Rudd P.T., Crouse D.T. *et al.* (1988) Chronic *Ureaplasma urealyticum* and *Mycoplasma hominis* infections of central nervous system in preterm infants. *Lancet, i,* 17–21.

Waites K.B., Talkington D.F. and Bébéar C.M. (2001b) Mycoplasmas, in *Manual of Commercial Methods in Clinical Microbiology* (ed. A.L. Truant), American Society for Microbiology, Washington, DC, pp. 201–224.

Wang E.E., Cassell G.H., Sanchez P.J. *et al.* (1993) *Ureaplasma urealyticum* and chronic lung disease of prematurity: critical appraisal of the literature on causation. *Clinical Infectious Diseases, 17* (Suppl. 1), S112–S116.

Wang R.Y.H., Wu W.S., Dawson M.S. *et al.* (1992) Selective detection of *Mycoplasma fermentans* by polymerase chain reaction and by using a nucleotide sequence within the insertion sequence-like element. *Journal of Clinical Microbiology, 30,* 245–248.

Woese C.R. (1987) Bacterial evolution. *Microbiological Reviews, 51,* 221–271.

Yoshida T., Deguchi T., Ito M. *et al.* (2002) Quantitative detection of *Mycoplasma genitalium* from first-pass urine of men with urethritis and asymptomatic men by real-time PCR. *Journal of Clinical Microbiology, 40,* 1451–1455.

Zheng X., Teng L.J., Watson H.R. *et al.* (1995) Small repeating units within the *Ureaplasma urealyticum* MB antigen gene encode serovar specificity and are associated with antigen size variation. *Infection and Immunity, 63,* 891–898.

25

Chlamydia spp. and Related Organisms

S. J. Furrows[1] and G. L. Ridgway[2]

[1]*Department of Clinical Parasitology, The Hospital of Tropical Diseases; and* [2]*Pathology Department, The London Clinic, London, UK*

INTRODUCTION

Organisms comprising the order Chlamydiales are bacteria adapted to an existence as obligate intracellular parasites of eukaryotic cells. They have an affinity with Gram-negative organisms and infect a wide range of avian, mammalian and invertebrate hosts. The range of clinical syndromes found in human chlamydial infection is wide and includes ocular, respiratory and genital infections.

DESCRIPTION OF THE ORGANISM

Chlamydial Classification

First recognised in 1907, the chlamydiae were initially thought to be viruses and have undergone several reclassifications subsequently.

According to current thinking, the order Chlamydiales contains four defined families: Chlamydiaceae, Parachlamydiaceae, Waddliaceae and Simkaniaceae (Everett, Bush and Andersen, 1999; Bush and Everett, 2001). The important human and animal pathogens belong to the family Chlamydiaceae, which contains two genera: *Chlamydia* spp. and *Chlamydophila* spp. (Figure 25.1).

Chlamydia trachomatis

Chlamydia trachomatis is primarily a human pathogen, causing ocular, urogenital and neonatal infections. Historically, it was subdivided into three strains or biovars: trachoma, lymphogranuloma venereum (LGV) and mouse pneumonitis (Nigg and Eaton, 1944). Serotyping of *C. trachomatis* according to antigenic determinants on the major outer membrane protein (MOMP) subdivides the species

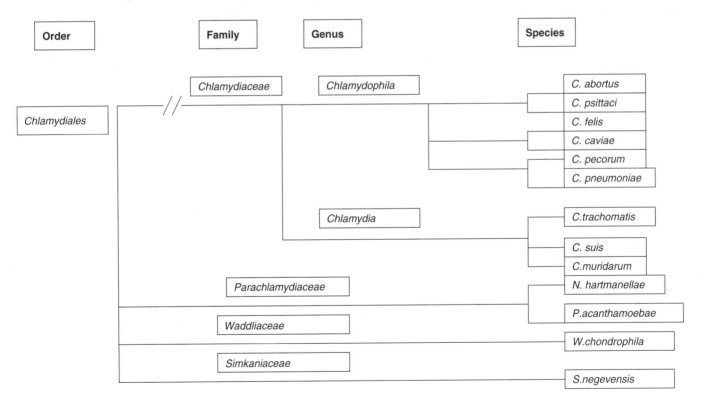

Figure 25.1 From www.chlamydiae.com. Adapted from Bush RM, Everett KDE (2001).

Principles and Practice of Clinical Bacteriology Second Edition   Editors Stephen H. Gillespie and Peter M. Hawkey

into 18 serovars affecting humans (Wang and Grayston, 1991). Serovars A, B, Ba and C are associated primarily with ocular disease; serovars D–K are associated with oculogenital disease, which may be transmitted to neonates; and serovars L1–L3 are associated with LGV.

Chlamydophila pneumoniae

Chlamydophila pneumoniae was originally isolated in 1965 from the conjunctiva of a Taiwanese child participating in a trachoma vaccine trial (Kuo *et al.*, 1986). Subsequent isolation in respiratory illness led to the acronym TWAR (Taiwan acute respiratory) strains. In 1989, TWAR strain was established as a new species of *Chlamydia pneumoniae*, recently reclassified as *C. pneumoniae* on the basis of ribosomal RNA sequence data (Everett, Bush and Andersen, 1999). *Chlamydophila pneumoniae* lacks an animal reservoir and is transmitted from person to person. *Chlamydophila pneumoniae* is a common cause of respiratory infection. More controversially, it has been associated with chronic cardiovascular and neurological illnesses. The exact role, if any, of *C. pneumoniae* in the pathogenesis of these diseases is still unclear.

Chlamydophila psittaci

Chlamydophila psittaci is a diverse group of organisms: at least 11 types are distinguished using microimmunofluorescence, 16S rRNA gene sequencing and polymerase chain reaction (PCR)/sequencing of the MOMP (Andersen, 1997; Takahashi *et al.*, 1997; Bush and Everett, 2001). These strains are thought to differ in virulence, but all are potentially transmissible to humans. *Chlamydophila psittaci* primarily affects birds, and infection has been documented in over 130 avian species (Macfarlane and Macrae, 1983). Infection is spread to humans by the respiratory route, to cause an ornithosis (infection of avian or animal origin) or psittacosis (infection acquired from parrots or related birds). Human-to-human transmission is very rare (Byrom, Walls and Mair, 1979).

Chlamydophila felis

Chlamydophila felis is endemic among housecats worldwide, primarily causing inflammation of feline conjunctiva, rhinitis and respiratory problems. It is recovered from the stomach and reproductive tract. Zoonotic infection of humans with *C. felis* has been reported (Hartley *et al.*, 2001).

Chlamydophila abortus

Chlamydophila abortus strains are endemic among ruminants and colonise the placenta. There are several case reports of miscarriages in women exposed to infected sheep and goats, with the diagnosis confirmed by serology or by PCR (Jorgensen, 1997; Pospischil *et al.*, 2002).

Other chlamydial species include *Chlamydophila caviae*, *Chlamydophila pecorum*, *Chlamydia suis* and *Chlamydia muridarum* (Figure 25.1). These species have not been shown to colonise or infect humans; hence, they are not discussed further in this chapter.

Life Cycle

The life cycle of chlamydial species (including both *Chlamydia* and *Chlamydophila*) is both unique and complex (Figure 25.2). A useful review of the developmental biology is found in Hatch (1999). A metabolically inert extracellular infectious transport particle of 200–300 nm size called the elementary body (EB) alternates with a larger intracellular reproductive particle of 1000 nm size called the

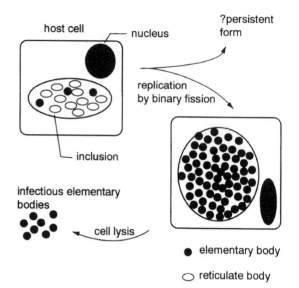

Figure 25.2 Diagram of chlamydial replication. Reproduced by permission of Karin D.E. Everett.

reticulate body (RB). The infective cycle commences when an EB is taken up by potentially phagocytic epithelial cells to form a phagosome. The EB protects itself from destruction within the phagosome by inhibiting fusion of host cell lysosomes (Scidmore, Fischer and Hackstadt, 2003). Its rigid trilaminar cell wall becomes bilaminar as cross-linking sulphide bonds are reduced, allowing compounds required for development and maturity to move in and out of the organism. The dense nucleoid of the EB unfolds and develops into the diffuse DNA characteristic of the RB. This takes place over about 6 h, at which time the transformed RB begins to divide by binary fission. As cell division continues, so the inclusion moves centripetally to enclose partially the host nucleus. At this time, *C. trachomatis*, but not the other species, lays down a glycogen matrix (Moulder, 1991). Division continues over the next 18–20 h, whereupon some of the RBs begin to condense into EBs. Protein synthesis ceases and the dense nucleoid reappears along with the thickened cell wall. Meanwhile, other RBs continue to divide. Inclusions of *C. trachomatis* mature around 48 h. The other species mature somewhat later at around 72 h. The whole process is carefully controlled by the chlamydiae, such that even normal host cell apoptosis is inhibited (Fan *et al.*, 1998; Carratelli *et al.*, 2002). Programmed rupture of the host cell membrane culminates in the release of the EBs into the surrounding environment.

There is evidence that under certain conditions, the cycle of EB/RB/EB development is arrested at the RB stage, leading to persistent (often termed latent) infection (Kutlin, Roblin and Hammerschlag, 1998). This phenomenon is known to be important in non-human and ocular infections and may have importance in chronic genital infection and in the possible role of *C. pneumoniae* in coronary atherosclerosis. The precise mechanism involved is not known, but many factors, such as the action of interferon-γ (IFN-γ), or tryptophan and cysteine deprivation, or antibiotics such as penicillins, may lead to an inhibited developmental cycle.

Genome

The genomes of several of the family Chlamydiaceae have now been sequenced. Published sequences include *C. trachomatis* serotypes B, D and L2 (Stephens, Kalman and Lammel, 1998) and *C. pneumoniae* strains J139 and CUL039 (Kalman *et al.*, 1999). General findings are that the chlamydial genome is smaller than any other prokaryote, with the sole exception of *Mycoplasma* spp. The complete genome of

C. trachomatis serovar D is 1.043 million base pairs, and for *C. pneumoniae* is 1.23 million base pairs. The genome appears relatively stable, with few gene rearrangements. Genes such as *omp1*, which encodes the MOMP, have been identified. Many candidate genes for chlamydial adhesins and for membrane proteins have also been identified. Certain metabolic pathways are missing, including amino acid and purine–pyrimidine biosynthesis, anaerobic fermentation and transformation competence proteins. Their absence may reflect the dependency of chlamydial species on the host cell machinery for the synthesis of these compounds. Interestingly, the necessary genes are present for the synthesis of penicillin-binding proteins, yet these proteins are not expressed. Speculation also exists over the possible presence of genes with eukaryotic homology.

Most strains of *C. trachomatis*, and some strains of *C. psittaci*, contain a plasmid of approximately 7.5 kb (McClenaghan, Honeycombe and Bevan, 1988). The function of the plasmid is unknown, although the extensive conservation between strains implies a critical function. *Chlamydophila pneumoniae* does not contain any extrachromosomal genetic material (Grayston *et al.*, 1989).

Structure and Major Antigens

Lipopolysaccharide

Chlamydiae have a genus-specific lipopolysaccharide (LPS) antigen. This LPS is similar to the rough LPS of certain salmonellae, but it carries a trisaccharide epitope of 3-deoxy-D-manno-octulosonic acid, which is specific to the family Chlamydiaceae. This epitope is immunodominant and is exposed on the surface of RBs and EBs. The LPS is likely to be important in the pathogenesis of chlamydial infection, by induction of tumour necrosis factor-α (TNF-α) and other proinflammatory cytokines, leading to scarring and fibrosis. The potency of chlamydial LPS is considerably less than that of gonococcal or salmonella LPS (Ingalls *et al.*, 1995), which may explain the prevalence of asymptomatic infection.

There are many protein antigens, with genus, species and serotype specificity.

MOMP

The dominant surface antigen is the MOMP, encoded by the *omp1* gene. This 40-kDa protein is the basis of the serological classification of *C. trachomatis* (Wang and Grayston, 1991). The antigenicity of MOMP appears to be less significant in *C. pneumoniae*, and its antigenic variation is much less marked in both *C. pneumoniae* and *C. psittaci* (Carter *et al.*, 1991).

Omc Proteins

Genes termed *omp2* and *omp3* encode two cysteine-rich proteins of the outer membrane complex of approximately 60 kDa and 12–15 kDa, respectively, termed OmcA and OmcB. They are synthesised during the transition of RB to EB, and the S–S-linked complex may provide the cell wall rigidity of the EB (Everett and Hatch, 1995).

Polymorphic Outer Membrane Proteins

Another group of surface-exposed proteins are the polymorphic outer membrane proteins (PMPs). There are at least 9 of these in *C. trachomatis*, and more than 20 in *C. pneumoniae* (Stephens, Kalman and Lammel, 1998; Kalman *et al.*, 1999). Their role is not yet defined.

Type III Secretory System

The chlamydiae possess genes for a type III secretory system, which may contribute to pathogenicity and confer survival advantages (Hueck, 1998; Winstanley and Hart, 2001). The type III secretory system is characteristic of Gram-negative organisms and is surface activated by contact with eukaryotic host cells. The surface of EBs and RBs is characterised by many spike-like, protein-containing projections, which may be associated with this secretory system. The projections on RBs have been shown to penetrate the inclusion membrane. Upon activation of the type III system, chlamydial proteins are injected into the cytoplasm of the host cell. Putative candidate proteins include inclusion membrane (Inc) proteins, which modify the inclusion membrane and subvert host cell signalling, proteins involved in the transport of essential nutrients into the inclusion and proteins involved in the modulation of apoptosis.

Heat Shock Proteins

The 12-, 60- and 75-kDa heat shock proteins are closely related to their counterparts in *Escherichia coli* (GroEL, GroES and DnaK) and also to related mitochondrial proteins in humans (Hsp10, Hsp60 and Hsp70). They are highly conserved chaperonins involved in protein folding. Hsp60 is highly immunogenic and is thought to play a major role in pathogenicity. The mechanism for this relates to an interaction with IFN-γ which, although inhibitory to chlamydiae, may result in the upregulation of Hsp60 at the expense of MOMP. This in turn might lead to a host autoimmune response, leading to the fibrosis and scarring associated with chlamydial infection. Many studies have shown an association between antibodies to chlamydia heat shock

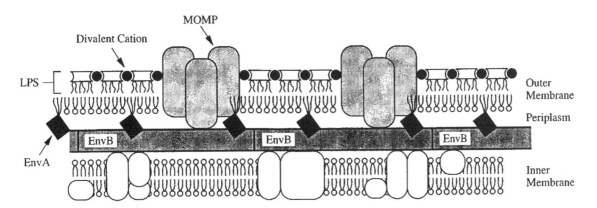

Figure 25.3 Diagram of the cell envelope. Reproduced by permission of Karin D.E. Everett.

proteins and the chronic sequelae of chlamydial infection (Ward, 1999), although this does not prove causation. A study of endemic trachoma in the Gambia found an association between antibodies to chlamydial heat shock proteins and conjunctival scarring (Peeling *et al.*, 1998).

EPIDEMIOLOGY

Chlamydia trachomatis is the commonest cause of bacterial genital infection in both men and women in the developed world and is the major cause of secondary (preventable) infertility. The World Health Organization (WHO) estimates that there are 90 million new cases each year (Gerbase *et al.*, 1998). Prevalence is almost certainly under-reported, not least because many infections are asymptomatic. An apparent greater prevalence in women than men (7:1) probably reflects greater opportunity to screen women than men. Peak prevalence in women is between 16 and 25 years of age. Studies in family practitioner settings in the United Kingdom suggest that the prevalence in this age group is within the range 3–6%. In men, the age of peak prevalence is older, reflecting the higher age of peak sexual activity. Risk factors for infection include multiple sexual partners, recent partner change and low socioeconomic status.

Trachoma, the infection caused by ocular strains of *C. trachomatis* (serotypes A–C), is endemic in North and sub-Saharan Africa, the Middle East, northern Asia, the Far East, Australasia and Latin America. In some villages, there is a greater than 90% attack rate. True figures for prevalence are unknown, with estimates in the range of 600 million people affected, of whom some 6 million will be blinded (Thylefors *et al.*, 1995). Trachoma is spread directly from eye to eye or indirectly by flies that transmit infected secretions from person to person (so called ocular promiscuity).

LGV is confined to the tropics and is sexually transmitted.

Chlamydophila pneumoniae is now recognised as a common respiratory pathogen (Hahn *et al.*, 2002), responsible for between 6 and 20% of community-acquired pneumonia. Humans are the only known reservoir, and transmission is via the respiratory route. Serological studies show that seropositivity rates are low in the under-5s, but rise rapidly during school age, such that by early adulthood, 50% are seropositive. This seropositivity rate increases gradually with age, reaching approximately 75% in the elderly. In addition, studies suggest that *C. pneumoniae* may go on to cause chronic infection of the respiratory tract and, possibly, of other sites such as the vascular system and central nervous system.

Chlamydophila psittaci is carried by animals, classically psittacine birds, and has no human reservoir. Human infection is almost always the result of zoonotic exposure and may occur both sporadically and in outbreaks (Yung and Grayson, 1988). Humans are infected via the respiratory route, particularly via inhaled infected dust in proximity to sick birds. Human-to-human transmission is unusual, except with virulent strains. Psittacosis is now rare, and many earlier references may have been erroneously assigned to *C. psittaci* before the role of *C. pneumoniae* was appreciated: such studies based the diagnosis of psittacosis on positive serology using the genus-specific complement fixation test method, which we now appreciate does not distinguish between *C. psittaci* and *C. pneumoniae*.

PATHOGENICITY

Host immunity is thought to be responsible for the pathogenesis of many of the clinical features of chlamydial infection. The host immune response to chlamydial infection is complex and far from fully understood (Cohen and Brunham, 1999). It is relatively ineffectual against acute infection. Cell-mediated immunity is thought to be important in the hypersensitivity reaction that leads to lasting tissue damage, particularly in recurrent and persistent infections. The host immune response may be considered as follows.

Antibody Response

Antibody to LPS is group specific and forms the basis of the now largely obsolete complement fixation test for chlamydial antibody. Species-specific antibodies of immunoglobulin G (IgG), IgM and IgA classes are produced in serum and in local secretions against MOMP, in response to chlamydial infection. The type and degree of response reflect the site and nature of infection. IgG and IgM levels are generally greater in systemic infections such as LGV or pelvic inflammatory disease (PID). IgM appears early but may persist, particularly in complicated infections. Serotype-specific antibodies are known to be neutralising in cell culture and may also play a part in short-term restriction of chlamydial infection. The role of IgA as a marker of persistent infection, particularly with *C. pneumoniae*, is controversial. In cases of reinfection by *C. pneumoniae*, IgM class antibody may not appear, whilst in recurrent *C. trachomatis* infection early antibody may reflect previously infecting serotypes.

Cellular Response

In acute infection, there is a brisk polymorphonuclear response, and polymorphonuclear leucocytes (PMNLs) are shown *in vitro* to inactivate chlamydiae. However, they have a limited role in the eradication of persistent infection. Resistance to infection and clearance of primary infection are due in large part to T-cell function, with both CD4 and CD8 cells having a protective role. MOMP-specific T-helper 1 (Th1) CD4 cells may have an important role in immunity, while Hsp60-specific Th2 CD4 cells are associated with the pathological sequelae of persistent infection. Much less is known about the role of cytotoxic CD8 cells in chlamydial immunity.

Cytokines

Infection of epithelial mucosal cells with *C. trachomatis* has been shown to generate several cytokines, including interleukin-1α (IL-1α), IL-6, IL-8, GRO-α and granulocyte–macrophage colony-stimulating factor (GM–CSF), which generate and sustain an inflammatory response (Rasmussen *et al.*, 1997). IFNs, in particular IFN-γ, play an important part in the immune response to chlamydial infection by inhibiting intracellular replication at the RB stage through tryptophan depletion. However, this may result in continuing secretions of chlamydial antigens, leading to further sensitisation and also induction of persistent non-replicating infection (Beatty, Morrison and Byrne, 1994). TNF-α is produced and may contribute to the tissue damage resulting from infection.

CLINICAL FEATURES

Chlamydia trachomatis

Chlamydia trachomatis is capable of infecting several different systems, to cause ocular, genital, rheumatological and neonatal infections, in addition to the tropical infection LGV.

Ocular Infections

The most important ocular infection syndrome is trachoma (Mecaskey *et al.*, 2003), caused predominantly by the trachoma serovars A–C. The incubation period is uncertain but is probably within the range 7–14 days. Acute infection presents as a follicular conjunctivitis, with congestion and oedema affecting both the palpebral and bulbar conjunctivae. There is papillary hyperplasia, giving the palpebral conjunctiva a velvety appearance. In hyperendemic areas, infection tends to be more pronounced in the upper lid. Follicles rupture to leave shallow pits termed Herbert's pits. Keratitis develops in the cornea. Recurrent infection leads to limbal scarring and new vessel formation (pannus). Palpebral conjunctival scarring (cicatrisation) leads to in-turning of the eyelashes (entropion), which scrape the bulbar corneal surface (trichiasis). It is the cycle of recurrent infection, with conjunctival scarring and pannus extending over the cornea, which results in impaired vision or blindness.

Clinical diagnosis of trachoma is based on WHO classification. Two or more of the following are required:

1. upper tarsal lymphoid follicles;
2. conjunctival scarring;
3. pannus;
4. limbal follicles or Herbert's pits.

Further staging of disease takes account of the intensity of disease and presence of complications.

Another ocular infection syndrome caused by *C. trachomatis* is adult inclusion conjunctivitis (paratrachoma) (Viswalingam, Wishart and Woodland, 1983). This is a disease of developed countries, characterised by infection with the genital serovars D–K of *C. trachomatis*. The incubation period is 2–3 weeks. Transmission is by contact of eyes with infected genital secretions. An acute follicular conjunctivitis develops, which affects the lower conjunctiva more than the upper. Severe disease presents as diffuse punctuate kerato-conjunctivitis. Pannus and scarring are unusual. The disease, if untreated, runs a fluctuating course before healing without cicatrisation. It is important to enquire about exposure to genital infection, either in the patient or in their partner.

Genital Infections

Genital infection in men presents most commonly as non-gonococcal urethritis (NGU). In most communities, this is far commoner than gonococcal urethritis (GU). *Chlamydia trachomatis* is responsible for 30–50% of cases of NGU: other cases may be caused by *Ureaplasma urealyticum*, *Mycoplasma genitalium*, *Trichomonas vaginalis* or herpes simplex. The incubation period for NGU is between 7 and 14 days (Stamm *et al.*, 1984), compared with 2–5 days for GU. A mucopurulent discharge is followed by dysuria and urethral irritation. Symptoms may be minimal, and some 40% of men with NGU are asymptomatic. NGU and GU may occur simultaneously. Treatment of the GU only, with a single dose of antibiotics ineffective against *C. trachomatis*, will result in post-gonococcal urethritis (PGU). PGU usually appears about 1 week after the treatment of GU, reflecting the difference in incubation period. Symptoms and signs are identical to those of NGU. *Chlamydia trachomatis* is responsible for 80–90% of cases of PGU.

Epididymitis is a recognised complication of chlamydial genital infection in men. Urethritis is usually also present. *Chlamydia trachomatis* is the commonest cause of epididymitis in men aged under 35 years, being responsible for 50% of cases (Berger *et al.*, 1979). Its role in prostatitis remains controversial: Although some investigators have detected *C. trachomatis* in prostatic secretions (Shortliffe, Sellers and Schachter, 1992), the balance of evidence at present does not support a role in either acute or chronic prostatitis. Rectal infection is well documented, particularly in anoreceptive homosexual males, and should be considered in the differential diagnosis of proctitis.

Infection of the female genital tract commences with urethritis and cervicitis and may progress to endometritis, salpingitis and peritoneal involvement (perihepatitis). Approximately 70% of women with chlamydial genital infection are asymptomatic (Cates and Wasserheit, 1991). Many women, if untreated, will go on to develop serious long-term sequelae of infection such as PID, infertility and ectopic pregnancy.

Urethritis alone occurs in some 25% of infected women. Symptoms are absent or mild and include abdominal discomfort and dysuria. The relationship of urethral infection to urethral syndrome (dysuria and frequency with less than 10^5 organisms per millilitre of urine) is controversial. In one study of 59 women with urethral syndrome, 42 also had pyuria and 11 of those 42 were infected with *C. trachomatis* (Stamm *et al.*, 1980).

Cervicitis is the commonest manifestation of chlamydial infection in women. It is frequently asymptomatic, though mucopurulent cervicitis may be apparent on speculum examination. Symptoms, where present, include a mucopurulent vaginal discharge and vaginal bleeding. Chlamydial infection cannot be distinguished on clinical grounds from other causes of vaginal discharge. In prepubertal children, the vaginal epithelium is columnar. With the onset of puberty, the epithelium becomes squamous as the columnar epithelium retreats to the endocervix. In consequence, chlamydial infection of the vaginal epithelium is not seen in adults but occurs in female children.

Endometritis and salpingitis due to chlamydial infection encompass the condition PID. Although PID is the result of a polymicrobial infection, there is no doubt that the most important initiating micro-organism is *C. trachomatis*, followed closely by *Neisseria gonorrhoeae*. The proportion of women with acute PID from whom chlamydia is isolated varies in different populations but is usually between 20 and 40% (Bevan, Ridgway and Rothermel, 2003). Concurrent infection with both organisms is not uncommon. An estimated 8% of women with cervical *C. trachomatis* infection progress to acute salpingitis (Cates and Wasserheit, 1991). Salpingitis is often subclinical but may cause acute illness with fever and pelvic pain. Examination reveals lower abdominal and adnexal tenderness. Tubal abscess may occur. Spread to the peritoneum results in perihepatitis (Fitz–Hugh–Curtis syndrome), an inflammation of the peritoneal covering of the liver, which presents with right upper quadrant pain and may mimic gall bladder disease (Bolton and Darougar, 1983). At laparoscopy, there is reddening of the tubal serosal surface, with intense fibrosis giving rise to friable strands described as violin string adhesions between the liver surface and the abdominal wall (Van Dongen, 1993).

Chlamydial PID is the commonest preventable cause of infertility. Chronic chlamydial infection of the upper genital tract leads to chronic inflammation and scarring, resulting in abdominal pain, secondary infertility and an increased risk of ectopic pregnancy. Some 10% of women are infertile following a single infection, rising to more than 50% after more than two episodes. Although most women with infertility due to tubular disease do not have a history of chlamydial infection or PID, approximately 80% have positive genital chlamydial serology, compared with 20% of women with non-tubal infertility (Osser, Persson and Liedholm, 1989). The risk of ectopic pregnancy is increased by seven to ten times following chlamydial PID.

The effect of chlamydial infection on pregnancy is not well understood. Apart from causing a vaginal discharge consequent on mucopurulent cervicitis, the main concern is with ascending genital tract infection, leading to amnionitis and perinatal endometritis. Acute non-gonococcal salpingitis is unusual in pregnancy. The role of chlamydial infection in premature rupture of membranes, prematurity and neonatal death is not clear. While some reports have reported an association, others have not been able to confirm this. The role of *C. trachomatis* in post-delivery sepsis is even less clear-cut. There is no doubt that chlamydial infection is a risk factor for pelvic sepsis after therapeutic abortion, but less certainty that the organism is an important cause of puerperal sepsis.

Sexually Acquired Reactive Arthritis

This is a rheumatoid seronegative condition which includes non-dysenteric Reiter's syndrome. Symptoms appear 1–4 weeks after lower genital infection and affect the large joints, particularly of the legs, or the sacroiliac joints. Men are more frequently affected than women (10:1). Approximately 1% of men with urethritis develop sexually acquired reactive arthritis (SARA). The classic triad of Reiter's syndrome (urethritis, conjunctivitis and arthritis) occurs in 30–50% (Keat, Thomas and Taylor-Robinson, 1983). In over 90% of patients, more than one joint is involved. Skin involvement may include the penis (circinate balanitis) or keratoderma of the palms and soles.

Two-thirds of patients with SARA carry the HLA-B27 haplotype. The reactive arthritis is thought to be an immune-mediated inflammatory response to an infection at a distant site. *Chlamydia trachomatis* may act as a trigger organism with an enhanced immune response in susceptible individuals but is not responsible for all cases. Resolution usually occurs without specific treatment, but relapse is common.

Neonatal and Childhood Infections

True congenital infection has never been described. Infection is always associated with ruptured membranes. The incidence of transmission from mother to neonate may be higher than previously thought, now that molecular methods of diagnosis are available: one PCR-based study found a vertical transmission rate, from infected mothers to their infants, of 55%, with 45% developing conjunctivitis and 30% pneumonia (Shen, Wu and Liu, 1995). The pharynx, middle ear, rectum and vagina may also be colonised, with a delay of up to 7 months before cultures are positive (Bell *et al.*, 1992). If untreated, the infection may persist for over 2 years. Antenatal screening for *C. trachomatis* has been shown to be effective at reducing neonatal morbidity (Pereira *et al.*, 1990) and is recommended in Centers for Disease Control and Prevention (CDC) guidelines (Centers for Disease Control and Prevention, 2002).

Neonatal inclusion conjunctivitis (ophthalmia neonatorum) due to chlamydial infection is much commoner than gonococcal ophthalmia. The incubation period for chlamydial infection is significantly longer (6–21 days), compared to the 48-h incubation period for gonococcal ophthalmia. The conditions are clinically very similar and may occur together: it is important to ensure that gonococcal ophthalmia does not mask a chlamydial infection, as treatment for the former may not be adequate for the latter. Chlamydial ophthalmia presents with a watery ocular discharge, which may progress to a purulent conjunctivitis with marked periorbital oedema. Because the conjunctiva at birth lacks a lymphoid layer, follicles do not develop initially but may be seen after 3–6 weeks. Unlike trachoma, the lower conjunctival surface is more heavily infected than the upper. Although scarring has been reported, the condition seems to be self-limiting after a relapsing course (Schachter and Dawson, 1978).

Pneumonitis presents within 3 weeks to 3 months of birth. Pneumonitis develops when organisms present in the conjunctiva pass down the nasolacrimal duct into the pharynx. Infection via the eustachian tube may cause otitis media. There is a history of conjunctivitis in approximately half of babies with chlamydial pneumonitis (Tipple, Beem and Saxon, 1979). Most cases of pneumonitis are afebrile (Beem and Saxon, 1977). There may be tachypnoea and a characteristic staccato cough. In severe cases, there may be episodes of apnoea and respiratory failure. On examination, there may be scattered crepitations in the lung fields, but there is good air entry and no wheeze. There are patchy interstitial infiltrates and hyperinflation on chest X-ray. If untreated, infection may persist for weeks or months. There is a peripheral blood eosinophilia, hypergammaglobulinemia and raised anti-chlamydial IgM titres. Pulmonary function tests show an obstructive picture, with significant limitation of expiratory airflow and signs of abnormally elevated volumes of trapped air. There is a possible association between *C. trachomatis* pneumonitis and childhood asthma (Weiss, Newcomb and Beem, 1986). Another study describes a positive association between serological evidence of previous infection with *C. trachomatis* and asthma-related symptoms

(Bjornsson *et al.*, 1996). Chlamydiae may be shed from the rectum in the first year of life, particularly in babies with pneumonia. Such shedding does not occur until 2–3 months after birth, suggesting that it reflects colonisation of the gastrointestinal tract. The long-term sequelae of this colonisation is unknown.

Lymphogranuloma Venereum

LGV is a sexually transmitted disease caused by the L1–L3 serovars of *C. trachomatis*. It is endemic in the tropics, notably Africa, India, Southeast Asia, South America and the Caribbean, and may occur as a sporadic disease elsewhere. As with other chlamydial genital infections, asymptomatic carrier sites, including the female cervix and the rectal mucosa of homosexual men, are important reservoirs.

The primary genital lesion in men is an insignificant and often painless papule, ulcer or vesicle on the penis. Similar lesions may occur on the vulva in women. The lesion appears between 3 and 30 days after infection and heals rapidly without leaving a scar. There is unilateral lymphadenopathy in two-thirds of patients (Schachter and Dawson, 1978). The secondary stage occurs after 1–6 weeks and is characterised by regional lymphadenopathy and systemic symptoms (Schachter and Osaba, 1983). Regional inguinal lymph nodes in the groin become enlarged and tender, with fixation and erythema of the overlying skin. The inflammatory process spreads from the lymph nodes into the surrounding tissue, forming an inflammatory mass. Abscesses within the mass coalesce to form buboes, which may break down to discharge externally as chronic fistulae. Enlargement of inguinal and femoral nodes may lead to a characteristic groove sign. Buboes should not be excised or drained other than by aspiration. They usually subside, but disruption to lymph drainage may occur. In women and anoreceptive men, internal lesions (vaginal, cervical and rectal) may go unnoticed. The lymph node drainage is to the pararectal nodes, and these may discharge via the rectum. Systemic symptoms of this second stage include fever, headache and myalgia. If untreated, the disease may proceed to the third stage, which is more serious in women and homosexual men. In these groups, rectal stricture or rectovaginal and rectal fistulae may occur (Lynch *et al.*, 1999; Papagrigoriadis and Rennie, 1999). Carcinomatous change is not unknown. The vulva, scrotum or penis may become affected by ulceration and oedematous granulomatous hypertrophy ('esthiomene', Greek meaning 'eating away'). Impaired lymphatic drainage may lead to elephantiasis of the vulva or scrotum.

The differential diagnosis for LGV includes other causes of genital ulceration such as syphilis, chancroid and granuloma inguinale. The possibility of dual infection should be borne in mind. Typically, patients with LGV have very high titres of chlamydial antibody, both generic and species specific. The diagnosis of LGV may be confirmed with positive chlamydial serology, isolation of *C. trachomatis* from affected tissue or pus or histopathology. Antigen or nucleic acid amplification test (NAAT) tests may be effective but have not been fully evaluated.

Chlamydophila pneumoniae

The incubation period for infection due to *C. pneumoniae* is about 21 days, which is relatively long for a respiratory pathogen. Most *C. pneumoniae* infections are asymptomatic. Acute infection, when symptomatic, presents most commonly with bronchitis or pneumonia. Typically, there is a biphasic illness characterised by insidious-onset pharyngitis, followed by hoarseness and persistent non-productive cough. Fever, if present, is low grade. Upper respiratory infections, including sinusitis and pharyngitis, may occur either in isolation or in conjunction with a lower respiratory tract infection. Symptoms due to *C. pneumoniae* respiratory infection may persist for weeks or months despite appropriate antibiotic therapy. Radiological changes are variable and indistinguishable from those associated with other aetiological agents. Both lobar and more diffuse changes may be seen. Studies show that up to 56% of chlamydial pneumonias are mixed with other causative agents such as *Streptococcus pneumoniae* and *Mycoplasma pneumoniae* (Kauppinen *et al.*, 1995). It is suggested that the cilostatic effect of *C. pneumoniae* may predispose to invasion by the pneumococcus, leading to a more severe pneumonia.

Chlamydophila pneumoniae is also implicated in chronic respiratory infection. Several studies have highlighted a role in persistent cough syndrome. Chronic *C. pneumoniae* infection is common in patients with chronic bronchitis, as demonstrated by high serum levels of specific IgG and IgA. Studies on chronic obstructive pulmonary disease (COPD) suggest that 4–18% of exacerbations may be associated with *C. pneumoniae*, often as part of a mixed infection. *Chlamydophila pneumoniae* has also been implicated in the development of adult-onset asthma and in exacerbations of both adult and childhood asthma (Hahn, Dodge and Golubjatnikov, 1991). The mainstay of diagnosis in respiratory infections is serology, performed on acute and convalescent serum samples taken at least 3 weeks apart.

Chlamydophila pneumoniae has also been linked with non-respiratory infections, with varying degrees of evidence. A possible association between chronic *C. pneumoniae* infection and chronic coronary heart disease was first described in 1988 (Saikku *et al.*, 1988). Some, but not all, studies have shown persistence of antibodies, demonstrated chlamydia-like particles in atheromatous plaques on electron microscopy and isolated *C. pneumoniae* from atheromatous plaques. These results are not sufficiently conclusive, and animal models for pathogenesis studies are not satisfactory. Antibiotic treatment for atheroma is not currently indicated. Other studies have suggested a causative role for *C. pneumoniae* in neurological diseases including Alzheimer's disease, stroke and multiple sclerosis. Although the organism has occasionally been grown from cerebrospinal fluid (CSF) in some of these disease states, there is no consensus on its significance (Furrows *et al.*, 2004).

Chlamydophila psittaci

Chlamydophila psittaci causes the zoonotic respiratory infections psittacosis or ornithosis (Yung and Grayson, 1988). The incubation period is classically 1–2 weeks but may be up to 4 weeks. Although onset may be insidious, with malaise and limb pains, it is more commonly abrupt, with high fever, rigors and headache. Clinical manifestations are often non-specific. The most common symptom is fever, which occurs in at least 50% of patients. There may be a cough, but sputum is scanty and respiratory distress unusual except in severe disease. The pulse rate is characteristically slow. Physical signs in the chest are usually confined to fine crackles on auscultation, with little or no evidence of consolidation. Extrapulmonary signs and symptoms include epistaxis, pharyngeal oedema, hepatomegaly, splenomegaly, drowsiness, confusion, adenopathy, palatal petechiae, herpes labialis, myalgia and Horder's spots (a pink blanching maculopapular eruption resembling the rose spots of typhoid fever). Complications of infection include hepatitis, meningoencephalitis, pancreatitis, nephritis and endocarditis.

Routine haematological and biochemical investigations are generally unhelpful in diagnosis. The white cell count is normal or only slightly elevated but shows a relative neutropenia in about 25% of cases. Eosinophilia has been seen in convalescence. Liver function tests are mildly abnormal in 50% of cases and may suggest cholestasis. The chest X-ray is abnormal in approximately 75% of cases, with a variety of appearances described such as lobar consolidation, homogenous ground-glass appearance or patchy reticular shadowing. Diagnosis is by acute and convalescent (at least 10 days later) serology. Isolation of the organism from blood or respiratory secretions should not be attempted without proper containment facilities because of the danger to laboratory workers. Much previous epidemiology of psittacosis precedes the recognition of *C. pneumoniae* as a pathogen. This is particularly a problem where diagnosis has been based on the group-reactive complement fixation test, which cannot distinguish between the two diseases.

Other Chlamydial Species

Sporadic zoonotic abortion due to *C. abortus* has been confirmed by genetic analysis of isolates from women who work with sheep. *Chlamydophila felis* is endemic among housecats worldwide, primarily causing inflammation of feline conjunctiva, rhinitis and respiratory problems. Zoonotic infection of humans with *C. felis* involving conjunctivitis, endocarditis and pyelonephritis has been reported (Regan, Dathan and Treharne, 1979; Hartley *et al.*, 2001).

LABORATORY DIAGNOSIS

Several techniques are available for laboratory diagnosis of chlamydial infection. Diagnosis has traditionally been by cell culture, which is expensive and technically demanding. Antigen detection methods and molecular assays are now widely used. Chlamydiae cannot be cultivated in artificial cell-free media.

Specimen Collection

Detection tests are performed on scrapings or swabs from infected sites or on first-catch urine samples. Chlamydial infection is intracellular, so the sample must contain cells that requires firm scraping or swabbing of the site. Samples from the male urethra should be taken by swabbing 2–4 cm within the urethra. For endocervical swabs, the endocervix should first be wiped clear of excess secretions and then swabbed. Conjunctival swabs for trachoma should be taken by everting the affected upper eyelid and swabbing firmly. Specimens from cases of neonatal ophthalmia are taken by firmly swabbing the lower eyelid. Urine samples, used to detect urogenital infection, should be first-catch samples, as these will contain the greatest amount of urethral secretions. It is important to check with the laboratory which test is being used, as the swab material, transport medium and transport conditions vary for each test. In general, cotton-tipped plastic swabs are suitable, but some molecular assays require specially provided swabs. Swabs for culture should be expressed into chlamydial transport medium. The swab should not be left in the medium. Urine samples are not suitable for chlamydial culture. Transport medium should be transported at 4 °C and then stored at −70 °C if not immediately cultured. Swabs for immunofluorescence should be rolled on to a Teflon-coated slide, fixed in methanol and air dried (Table 25.1).

Microscopy

Direct microscopy of a sample will not detect chlamydiae but may provide indirect evidence of infection. In NGU, microscopy of a Gram-stained urethral smear demonstrates greater than 4 polymorphonuclear lymphocytes (PMNL) per high power field (×100 objective) and an absence of intracellular diplococci. A spun first-catch urine deposit will yield greater than 15 PMNL per low power field (×40 objective). Either sample may then be used for antigen detection or NAAT tests for chlamydial infection.

Cell Culture

Cell culture is the traditional method of diagnosis and, although highly specific, is less sensitive than desirable (at best 90% compared with NAATs). It is a time-consuming method, requiring specialised cell-culture equipment, making it impracticable for most diagnostic laboratories.

Table 25.1 Use of routine laboratory tests for diagnosis of chlamydial infections

Clinical presentation	Specimen	Available laboratory test
Cervicitis (*C. trachomatis* D–K)	Endocervical swab	NAAT, EIA, DIF, culture
	First-catch urine	NAAT, EIA, DIF
Pelvic inflammatory disease/Fitz–Hugh–Curtis syndrome (*C. trachomatis* D–K)	Endocervical swab	NAAT, EIA, DIF, culture
	Fallopian/peritoneal swab	NAAT, EIA, DIF, culture
Urethritis (*C. trachomatis* D–K)	Urethral swab	NAAT, EIA, DIF, culture
	First-catch urine	NAAT, EIA, DIF
Lymphogranuloma venereum (*C. trachomatis* L1–L3)	Serum	EIA, MIF, CFT
	Swab of cervix/urethra/rectum/ulcer	Culture, DIF, EIA
	Lymph node aspirate	Culture
Trachoma (*C. trachomatis* A–C)	Conjunctival swab (upper fornix)	NAAT, EIA, DIF, culture
Paratrachoma (*C. trachomatis* D–K)	Conjunctival swab (lower fornix)	NAAT, EIA, DIF, culture
Ophthalmia neonatorum (*C. trachomatis* D–K)	Conjunctival swab (lower fornix)	NAAT, EIA, DIF, culture
Neonatal pneumonia (*C. trachomatis* D–K)	Serum	IgM (EIA, MIF)
	Nasopharyngeal aspirate	NAAT, EIA, DIF, culture
Ornithosis (*C. psittaci*)	Serum	MIF, CFT, WHIF
Community-acquired pneumonia (*C. pneumoniae*)	Serum	MIF
	Respiratory secretions	EIA, DIF

CFT, complement fixation test; DIF, direct immunofluorescence; EIA, enzyme immunoassay; MIF, microimmunofluorescence; NAAT, nucleic acid amplification tests, e.g. polymerase chain reaction, strand displacement assay, transcription-mediated assay, ligase chain reaction (no longer available); WHIF, whole-cell immunofluorescence. Adapted from Blanchard TJ, Mabey DCW (1997), *Principles and Practice of Clinical Bacteriology*, 1st edition.

Many cell lines are suitable for chlamydial culture. McCoy cells (derived from a mouse fibroblast L cell line), HeLa 229, buffalo green monkey and baby hamster kidney (BHK21) are most commonly used to culture *C. trachomatis*. The human fibroblast HL cell line or Hep2 cells are used for the culture of *C. pneumoniae*. *Chlamydophila psittaci* is a category 3 pathogen, and culture should not be attempted in the routine laboratory.

EBs derived from the clinical specimen are added to a monolayer of host cells. The host cells must be in a stationary phase, i.e. not synthesising macromolecules. Classically, host cells were treated with γ-irradiation to achieve stationary phase, but now chemical treatment such as cytochalasin B, idoxyuridine or cycloheximide is used. Cycloheximide has the advantage that host cells do not require pre-treatment. Centrifugal force (300 g) or treatment of the cell sheet with polycations such as diethylaminoethyl (DEAE)–dextran is employed to overcome opposing negative cell-surface charges between the monolayer and the EBs. Culture requires careful choice of water and reagents, including fetal calf serum, balanced salts, glutamine, amino acids and glucose. Cultures are incubated in 10% carbon dioxide for 48–72 h, depending on the chlamydial species (shorter for *C. trachomatis* and longer incubation for *C. pneumoniae* and *C. psittaci*). Cells are then stained with a variety of reagents to demonstrate the presence of inclusions. *Chlamydia trachomatis*, the only species with glycogen-containing inclusions, is stained with glycogen-specific stains such as iodine or periodic acid–Schiff stain. All species stain with Giemsa and inclusions show autofluorescence if viewed by dark ground illumination. The advent of monoclonal antibodies and fluorochromes has resulted in more specific staining of cell sheets. These antibodies are usually raised against either the group-specific LPS or species-specific MOMP (Boman *et al.*, 1997). For optimal isolation, a blind passage is required.

Direct Immunofluorescent Tests

Direct immunofluorescent test (DIF) may be used as a culture confirmation test or for direct detection of EBs in clinical material, particularly from the genital tract and eye (Uyeda *et al.*, 1984). The technology is based upon highly specific monoclonal antibodies, which may be group specific (directed against the LPS and theoretically able to detect all chlamydial species) or species specific (directed against the MOMP of a single species such as *C. trachomatis*). A fluorescence microscope is required. These tests are sensitive but are highly operator dependent: observer error or non-specific fluorescence creates false positives and reduces the specificity of the test. The expense and technical time required for DIF tests make them unsuitable for laboratories with a heavy diagnostic workload.

Enzyme Immunoassays

The development of enzyme immunoassays (EIA) allowed diagnostic facilities to become widely available. Several test formats exist, which use antibodies to detect the presence of an antigen, usually the group LPS, in clinical specimens. EIAs are comparatively cheap and are automated. Sensitivity is estimated at the order of 60–80% compared with that of NAATs (Taylor-Robinson and Tuffrey, 1987). A major problem of EIAs is specificity, and it is mandatory that all positives should be confirmed either with a blocking test (monoclonal anti-LPS used to block a repeated EIA) or with an independent test such as a DIF test. This increases specificity to over 99% (Moncada, Schachter and Bolan, 1990).

Rapid bedside test versions of the EIA have also been developed but have unacceptably low sensitivity when used in low-prevalence populations (Rani *et al.*, 2002).

Nucleic Acid Amplification Technology

Nucleic acid amplification technology (NAAT) has revolutionised the diagnosis of chlamydial infections. High-quality, commercially produced tests are now available for the diagnosis of *C. trachomatis* using PCR, ligase chain reaction (LCR), transcription-mediated amplification (TMA) and strand displacement assay (SDA) (Cherneskey, 2002). These assays amplify either the target nucleic acid or the probe after it has annealed to target nucleic acid. The major targets for NAATs are multiple-copy gene products such as ribosomal RNA or plasmid DNA, as the presence of multiple copies increases the sensitivity of the assay.

The main advantages of NAATs are their high sensitivity and specificity (Van Dyck *et al.*, 2001; Black *et al.*, 2002). They can be automated if required. The exquisite sensitivity of NAATs means that good laboratory practice is paramount if false positives are to be avoided (Furrows and Ridgway, 2001; Verhoeven *et al.*, 2002). NAATs are expensive of reagents, and pooling of samples is recommended by some authorities to reduce unit costs. This approach has some merit in screening scenarios but is likely to be counterproductive in high-prevalence settings.

NAAT assays have demonstrated that the conventional tests, including culture, considerably underestimate the prevalence of *C. trachomatis*. It is now established that cell culture has, at best, a sensitivity of 90% compared with NAATs. The sensitivity of DIF, EIAs and DNA hybridisation tests was always felt to be inferior to that of cell culture in a field setting, giving an overall sensitivity for these tests of only 75%. There is a need to redefine the gold standard for chlamydial diagnosis because of the introduction of NAATs.

Serology

The diagnosis of chlamydial infection by estimating antibody response is far from satisfactory. The traditional complement fixation test, targeting LPS, is group reactive and cannot distinguish between species. Tests based on recombinant LPS in an enzyme-linked immunoadsorbent assay (ELISA) format also tend to be group specific. The technically demanding microimmunofluorescence (MIF) test, targeting the MOMP antigen, is the definitive test. This test is technically very demanding, requiring the use of serovar and species-specific antigens spotted onto slides and incubated with serum in dilutions. Although able to distinguish between species and serovars, the test is not widely available and difficult to standardise.

Chlamydial antibody is widely distributed in the population, much of it related to *C. pneumoniae* respiratory infection. Antibody in acute infection is late in appearing, and hence, paired sera are required at least 3 weeks apart, to demonstrate a significant (greater than fourfold) rise in IgG titre. IgM determination is useful in the diagnosis of neonatal infection and in adult respiratory infection but may not be seen in recurrent respiratory infection in adults. Persistence of serum IgA is regarded by some authorities as evidence of chronic chlamydial infection. Serology (greater than fourfold rise in IgG) may help in the confirmation of chlamydial PID but has no place in the diagnosis of lower chlamydial genital tract infection.

MANAGEMENT

Choice of Antibiotic

The intracellular habitat of chlamydiae restricts the classes of antibiotics that will be effective. In general, chlamydiae are not susceptible to antibiotics that do not penetrate cells efficiently, e.g. the aminoglycosides. The chlamydiae are sensitive to selected groups of antibiotics such as tetracyclines, macrolides and related compounds, rifampicin and more recent fluoroquinolones. The development of newer agents within these groups, such as the long-acting azalide macrolide, azithromycin, has brought new alternatives for therapy compared with traditional antibiotics. Choice of antibiotic, such as for the empirical treatment of genital or respiratory infection, must take into account all the likely causative organisms and their likely susceptibility, site affected, need for topical versus systemic therapy, cost and probable compliance.

β-Lactam antibiotics are not often recommended for routine chlamydial treatment. The chlamydial cell wall does not contain substantial amounts of peptidoglycan, although penicillin-binding proteins are present. Penicillins inhibit the differentiation of RBs to EBs and are reversibly bacteriostatic. Their use *in vivo* may have an inhibitory effect which precludes isolation of chlamydiae from a patient treated with β-lactams. Amoxicillin has a limited role in the treatment of chlamydial genital infection in pregnancy. Patients require careful follow-up to ensure eradication of the organism. Cephalosporins are not active against chlamydiae and should not be used.

The sulphonamides were extensively used to treat *C. trachomatis* before the introduction of tetracyclines. They act by competitively inhibiting the incorporation of *para*-amino benzoic acid in the folic acid pathway. They are now not commonly used, except in some treatment regimens for LGV, because they are poorly bactericidal. *Chlamydophila psittaci* and *C. pneumoniae* are resistant to sulphonamides because they derive their folic acid from the host cell pool. Trimethoprim is not active against chlamydiae.

The tetracyclines are highly active against chlamydiae and are therefore widely used to treat all chlamydial infections. There is considerable experience with their use, but they are not ideal. Gastrointestinal side effects, incompatibility with milk and milk products, multiple dosages and inability to use in pregnancy and for the treatment of paediatric infections are the main limitations. Doxycycline is the favoured preparation, because twice-daily dosage has the advantage over four-times-daily dosage of other forms.

Macrolides are a suitable alternative. There is most experience with erythromycin, but as with the tetracyclines, gastrointestinal side effects may be a problem. Azithromycin, which is an amino-substituted 15-membered azalide macrolide, penetrates extremely well into cells and has a long intracellular half-life. A single 1-g dose orally is effective against uncomplicated *C. trachomatis* genital infections, and a 5-day course of 500 mg daily is effective against chlamydial PID. Other macrolides such as clarithromycin, roxithromycin and josamycin are also effective, but their use has been largely superseded by azithromycin. This antibiotic is the only agent effective against lower genital tract chlamydial infection in a single dose.

Rifampicin is one of the most active antibiotics against chlamydiae *in vitro*. Unfortunately, as with other bacterial species, resistance is easily induced. Rifampicin is occasionally used to treat refractory chlamydial infections such as LGV or psittacosis, usually in combination with erythromycin.

Older fluorinated quinolones, including lomefloxacin, fleroxacin and ciprofloxacin, are moderately active *in vitro* but are not reliable agents *in vivo* and should not be used. Ofloxacin is effective clinically. The more recent quinolones such as gatifloxacin and moxifloxacin are highly active against chlamydiae *in vitro*. The role of fluoroquinolones in the treatment of chlamydial infections is not established owing to the impact of azithromycin on the treatment of acute lower genital tract chlamydial infection. Azithromycin is not as reliable for the treatment of gonorrhoea, and therefore, ofloxacin is favoured for the treatment of PID. Newer fluoroquinolones are being recommended increasingly for the treatment of community-acquired respiratory infections. Their *in vitro* activity against *C. pneumoniae* should ensure that they are clinically effective against respiratory infections with this organism. However, clinical trial data with specific definitive results against *C. pneumoniae* are still required. Resistance to ofloxacin has been demonstrated *in vitro* against *C. trachomatis* but not against *C. pneumoniae* (Morrissey *et al.*, 2002). The clinical significance of this finding is unclear.

Treatment Regimens

Several sets of guidelines on the management of chlamydial infections are available, including US guidelines (Centers for Disease Control and Prevention, 2002), European guidelines (Stary, 2001) and WHO guidelines (2001). WHO guidelines for treatment offer the alternatives of management based on a definite diagnosis or of syndromic management based on the identification of consistent groups of symptoms and easily recognised signs (syndromes). Syndromic management is particularly appropriate for trachoma and for sexually transmitted chlamydial infections in settings where laboratory facilities for diagnosis are limited. Where syndromic management is used, the choice of first-line drugs must cover all common pathogens causing the syndrome. For example, urethritis should be managed with drugs that cover *N. gonorrhoeae* as well as *C. trachomatis*.

Chlamydia trachomatis

Trachoma has been treated traditionally with topical sulphonamides or topical tetracyclines. Topical treatment is of limited use, and systemic oral therapy, usually with oxytetracycline or doxycycline, is preferred. Compliance is a problem, and tetracyclines are contraindicated in young children (Dawson and Schachter, 1985). Populations in hyperendemic areas require retreatment at twice-yearly intervals, in addition to education on hygiene and fly control. Trials of treatment with a single oral dose of azithromycin show 100% compliance and good control of disease for up to 12 months (Mabey *et al.*, 1998).

Genital infections are always treated systemically. For uncomplicated genital tract disease, several options exist: WHO recommend doxycycline for 7 days or a single dose of azithromycin. Alternative regimens include 7-day courses of erythromycin, ofloxacin or tetracycline. Ciprofloxacin is not reliable and should not be used to treat genital chlamydial infections.

Amoxicillin for 7 days is widely used in pregnancy, but careful follow-up is essential to avoid relapse. Erythromycin (if tolerated) is a suitable alternative. Tetracyclines should be avoided during pregnancy and lactation.

Resolution of SARA usually occurs without specific treatment, but relapse is common. A double-blind study comparing lymecycline (a tetracycline) with placebo in patients with reactive arthritis suggested that 3 months of treatment was effective in arthritis associated with chlamydial infection (Lauhio *et al.*, 1991). There is continuing uncertainty about the role, if any, of anti-chlamydial antibiotics in SARA.

Neonatal chlamydial infections are always treated with systemic antibiotics. Topical treatment of chlamydial ophthalmia will not eradicate chlamydiae from other sites such as the respiratory tract. Standard treatment is systemic erythromycin for 2–3 weeks, with local saline bathing if required. Topical silver nitrate (Crédés prophylaxis) is ineffective and may lead to a chemical conjunctivitis.

Early treatment of LGV is essential, before scarring complicates management. Current WHO recommendations are 2 weeks of doxycycline, with erythromycin or tetracycline as alternatives. Rifampicin has been used with good effect but should only be used as part of a multidrug regimen because resistance develops readily. Fluctuant buboes should be aspirated to prevent rupture and sinus tract formation. Scarring does not respond well to therapy.

Chlamydophila pneumoniae

Symptomatic respiratory infections should be treated with a 2–3-week course of a macrolide/azalide or tetracycline. Treatment is usually started empirically before the diagnosis is confirmed. There is no evidence to support longer courses of treatment for chronic or persistent infections, including possible involvement in cardiovascular or neurological disease.

Chlamydophila psittaci

Standard therapy is with oral tetracycline or doxycycline for 3 weeks. An alternative is erythromycin for 3 weeks. Shorter courses may lead to relapse (Schachter and Dawson, 1978). Without treatment, recovery takes 2–3 weeks, and the mortality is up to 20%.

PREVENTION AND CONTROL

Prevention of infection is always preferred to treatment. The vast clinical scope of infections due to chlamydial infection necessitates a different approach for each infection.

Trachoma control is a huge undertaking, involving education on hygiene and fly controls as well as mass treatment programmes, more recently with a single dose of azithromycin. Antibiotic therapy is being combined with other factors, to provide a comprehensive attack on this disease. These treatment modalities are encompassed in the acronym SAFE. This stands for surgery (to eyelids to combat trichiasis), antibiotics (azithromycin), face (encouragement of improved personal hygiene, particularly face washing) and environment (improved provision of latrines and fly control) (Mabey, Solomon and Foster, 2003). Rapid reinfection occurs in endemic areas; therefore, treatment is needed twice a year. The hope is that the reservoir of infection will eventually be eliminated (Mecaskey *et al.*, 2003).

Patients with genital infection should be identified and treated promptly, and contact tracing of partners is an important control measure. There is ongoing debate over the role of screening within groups at risk, particularly in the 16–25-year group. Education about the use of barrier contraceptives and change in sexual behaviour is important.

There are no specific control measures against respiratory infections caused by *C. pneumoniae*. Respiratory infections caused by *C. psittaci* may be avoided by reducing contact with birds, particularly if they appear unwell.

Vaccination

Although there has been over 40 years of active research, an effective vaccine against human chlamydial infection is still not available. However, there is considerable activity in this field. Several approaches are being pursued (Brunham *et al.*, 2000; Stephens, 2000). Firstly, the pulsing of dendritic cells derived from mouse bone marrow cells *ex vivo* with chlamydiae may produce a Th1-type cell-mediated immune response, suggesting the possibility of using chlamydial components to stimulate a response. *In vivo* studies have so far proved disappointing. Vectors such as *Vibrio cholerae* ghosts are showing potential (Eko *et al.*, 1999). Putative DNA vaccines directed against MOMP are showing potential against respiratory but not genital infections in animal experiments. Techniques aimed at stimulating the mucosal response, including gene constructs or locally active gels, to allow slow release of stimulatory antigens and the use of carrier viruses are also promising. Sequencing of the chlamydial genome has identified further potential vaccine candidates. It is likely that, owing to the nature of chlamydial infections, vaccines that stimulate both a Th1 and Th2 response will be required.

A particularly exciting approach has been the development of a DNA vaccine utilising a cytomegalovirus (CMV) promoter virus, coding for MOMP or 60-kDa components, active against *C. pneumoniae* in a murine respiratory model (Penttila *et al.*, 2000).

REFERENCES

Andersen AA (1997) Two new serovars of *Chlamydia psittaci* from North American birds. *J Vet Diagn Invest*; **9**: 159–164.

Beatty WL, Morrison RP, Byrne GI (1994) Persistent chlamydiae: from cell culture to a paradigm for chlamydial pathogenesis. *Microbiol Rev*; **58**: 686–699.

Beem MO, Saxon EM (1977) Respiratory tract colonisation and a distinctive pneumonia syndrome in infants infected with *Chlamydia trachomatis*. *N Engl J Med*; **296**: 306–310.

Bell TA, Stamm WE, Wang SP *et al.* (1992) Chronic *Chlamydia trachomatis* infections in infants. *JAMA*; **267**: 400–402.

Berger RE, Alexander ER, Harnisch JP *et al.* (1979) Etiology, manifestations and therapy of acute epididymitis: prospective study of 50 cases. *J Urol*; **121**: 750–754.

Bevan CD, Ridgway GL, Rothermel CD (2003) Efficacy and safety of azithromycin as monotherapy or combined with metronidazole compared with two standard multidrug regimens for the treatment of acute pelvic inflammatory disease. *J Int Med Res*; **31**: 45–54.

Bjornsson E, Hjelm E, Janson C *et al.* (1996) Serology of chlamydia in relation to asthma and bronchial hyperresponsiveness. *Scand J Infect Dis*; **28**: 63–69.

Black CM, Marrazzo J, Johnson RE *et al.* (2002) Head-to-head multicenter comparison of DNA probe and nucleic acid amplification tests for

Chlamydia trachomatis infection in women performed with an improved reference standard. *J Clin Microbiol*; **40**: 3757–3763.

Bolton JP, Darougar S (1983) Perihepatitis. *Br Med Bull*; **39**: 159–162.

Boman J, Gaydos C, Juto P *et al.* (1997) Failure to detect *Chlamydia trachomatis* in cell culture by using a monoclonal antibody directed against the major outer membrane protein. *J Clin Microbiol*; **35**: 2679–2680.

Brunham RC, Zhang DJ, Yang X, McClarty GM (2000) The potential for vaccine development against chlamydial infection and disease. *J Infect Dis*; **181**(Suppl. 3): S538–S543.

Bush RM, Everett KDE (2001) Molecular evolution of the *Chlamydiaceae*. *Int J Syst Evol Microbiol*; **51**: 203–220.

Byrom NP, Walls J, Mair HJ (1979) Fulminant psittacosis. *Lancet*; **i**: 353–356.

Carratelli CR, Rizzo A, Catania MR *et al.* (2002) *Chlamydia pneumoniae* infections prevent the programmed cell death on THP-1 cell line. *FEMS Microbiol Lett*; **215**: 69–74.

Carter MW, al-Mahdawi SA, Giles IG *et al.* (1991) Nucleotide sequence and taxonomic value of the major outer membrane protein of *Chlamydia pneumoniae* IOL-207. *J Gen Microbiol*; **137**: 465–475.

Cates W Jr, Wasserheit JN (1991) Genital chlamydial infections: epidemiology and reproductive sequelae. *Am J Obstet Gynecol*; **164**: 1771–1781.

Centers for Disease Control and Prevention (2002) Sexually transmitted diseases treatment guidelines—2002. *MMWR Recomm Rep*; **51**(RR-6): 1–80. www.cdc.gov/mmwr/preview/mmwrhtml/rr5106a1.htm.

Cherneskey MA (2002) *Chlamydia trachomatis* diagnostics. *Sex Transm Infect*; **78**: 232–234.

Cohen CR, Brunham RC (1999) Pathogenesis of *Chlamydia* induced pelvic inflammatory disease. *Sex Transm Infect*; **75**: 21–24.

Dawson CR, Schachter J (1985) Strategies for treatment and control of blinding trachoma: cost effectiveness of topical or systemic antibiotics. *Rev Infect Dis*; **7**: 768–773.

Eko FO, Witte A, Huter V *et al.* (1999) New strategies for combination vaccines based on the extended recombinant bacterial ghost system. *Vaccine*; **17**: 1643–1649.

Everett KD, Hatch TP (1995) Architecture of the cell envelope of *Chlamydia psittaci* 6BC. *J Bacteriol*; **177**: 877–882.

Everett KDE, Bush RM, Andersen AA (1999) Amended description of the order Chlamydiales, proposal of Parachlamydiaceae fam. nov. and Simkaniaceae fam. Nov., each containing one monotypic genus, revised taxonomy of the family Chlamydiaceae with description of five new species, and standards for the identification of organisms. *Int J Syst Bacteriol*; **49**: 415–440.

Fan T, Lu H, Hu H *et al.* (1998) Inhibition of apoptosis in chlamydia-infected cells: blockade of mitochondrial cytochrome c release and caspase activation. *J Exp Med*; **187**: 487–496.

Furrows SJ, Ridgway GL (2001) "Good laboratory practice" in diagnostic laboratories using nucleic acid amplification methods. *Clin Microbiol Infect*; **7**: 227–229.

Furrows SJ, Hartley JC, Bell J *et al.* (2004) *Chlamydophila pneumoniae* infection of the central nervous system in patients with multiple sclerosis. *J Neurol Neurosurg Psychiatry*; **75**: 152–154.

Gerbase AC, Rowley JT, Heymann DH *et al.* (1998) Global prevalence and incidence estimates of selected curable STDs. *Sex Transm Infect*; **74**(Suppl. 1): S12–S16.

Grayston JT, Kuo C-C, Campbell LA (1989) *Chlamydia pneumoniae* sp. nov. for *Chlamydia* sp. strain TWAR. *Int J Syst Bacteriol*; **39**: 88–90.

Hahn DL, Dodge RW, Golubjatnikov R (1991) Association of *Chlamydia pneumoniae* (strain TWAR) infection with wheezing, asthmatic bronchitis, and adult-onset asthma. *JAMA*; **266**: 225–230.

Hahn DL, Azenabor AA, Beatty WL, Byrne GI (2002) *Chlamydia pneumoniae* as a respiratory pathogen. *Front Biosci*; **7**: e66–e76.

Hartley JC, Kaye S, Stevenson S *et al.* (2001a) PCR detection and molecular identification of *Chlamydiaceae* species. *J Clin Microbiol*; **39**: 3072–3079.

Hartley JC, Stevenson S, Robinson AJ *et al.* (2001b) Conjunctivitis due to *Chlamydophila felis* (*Chlamydia psittaci* feline pneumonitis agent) acquired from a cat: case report with molecular characterisation of isolates from the patient and cat. *J Infect*; **43**: 7–11.

Hatch TP (1999) Developmental biology. In Stephens RS (Ed.), *Chlamydia: Intracellular Biology, Pathogenesis and Immunity*. ASM Press, Washington, DC. pp. 29–67.

Hueck CJ (1998) Type III protein secretion systems in bacterial pathogens of animals and plants. *Microbiol Mol Bio Rev*; **62**: 379–433.

Ingalls RR, Rice PA, Qureshi N *et al.* (1995) The inflammatory cytokine response to *Chlamydia trachomatis* is endotoxin mediated. *Infect Immun*; **63**: 3125–3130.

Jorgensen DM (1997) Gestational psittacosis in a Montana sheep rancher. *Emerg Infect Dis*; **3**: 191–194.

Kalman S, Mitchell W, Marathe R. *et al.* (1999) Comparative genomes of *Chlamydia pneumoniae* and *C. trachomatis*. *Nat Genet*; **21**: 385–389.

Kauppinen MT, Herva E, Kujala P *et al.* (1995) The etiology of community acquired pneumonia among hospitalised patients during a *Chlamydia pneumoniae* epidemic in Finland. *J Infect Dis*; **172**: 1330–1335.

Keat A, Thomas BJ, Taylor-Robinson D (1983) Chlamydial infection in the aetiology of arthritis. *Br Med Bull*; **39**: 168–174.

Kuo C-C, Chen H-H, Wang SP, Grayston JT (1986) Identification of a new group of *Chlamydia psittaci* strains called TWAR. *J Clin Microbiol*; **24**: 1034–1037.

Kutlin A, Roblin PM, Hammerschlag MR (1998) Continuous *Chlamydia pneumoniae* infection *in vitro*. In Stephens RS *et al.* (Ed.), *Chlamydial Infections. Proceedings of the Ninth International Symposium on Human Chlamydial Infection*. International Chlamydia Symposium, San Francisco, CA.

Lauhio A, Leirisalo-Repo M, Lahdevirta J *et al.* (1991) Double-blind, placebo-controlled study of three-month treatment with lymecycline in reactive arthritis, with special reference to *Chlamydia* arthritis. *Arthritis Rheum*; **34**: 6–14.

Lynch CM, Felder TL, Schwandt RA, Shashy RG (1999) Lymphogranuloma venereum presenting as a rectovaginal fistula. *Infect Dis Obstet Gynecol*; **7**: 199–201.

Mabey D, Bailey R, Faal H *et al.* (1998) Azithromycin in control of trachoma 2. Community based treatment of trachoma with oral azithromycin: a one year follow-up study in the Gambia. In: Stephens RS, Byrne GI, Christiansen G *et al.* (Eds). *Chlamydial Infections. Proceedings of the Ninth International Symposium on Human Chlamydial Infections*. International Chlamydia Symposium, San Francisco, CA. pp. 355–358.

Mabey DC, Solomon AW, Foster H (2003) Trachoma. *Lancet*; **362**: 223–229.

Macfarlane JT, Macrae AD (1983) Psittacosis. *Med Bull*; **39**: 163–167.

McClenaghan M, Honeycombe JR, Bevan BJ (1988) Distribution of plasmid sequences in avian and mammalian strains of *Chlamydia psittaci*. *J Gen Microbiol*; **134**: 559–565.

Mecaskey JW, Knirsch CA, Kumaresan JA, Cook JA (2003) The possibility of eliminating blinding trachoma. *Lancet Infect Dis*; **3**: 728–733.

Moncada J, Schachter J, Bolan G (1990) Confirmatory assay increases specificity of the chlamydiazyne test for *Chlamydia trachomatis* infection of the cervix. *J Clin Microbiol*; **28**: 1770–1773.

Morrissey I, Salman H, Bakker S *et al.* (2002) Serial passage of *Chlamydia* spp. in sub-inhibitory concentrations. *J Antimicrob Chemother*; **49**: 757–761.

Moulder JW (1991) Interaction of chlamydiae and host cells in vitro. *Microbiol Rev*; **55**: 143–190.

Nigg C, Eaton MD (1944) Isolation from normal mice of a pneumotropic virus which forms elementary bodies. *J Exp Med*; **79**: 497.

Osser S, Persson K, Liedholm P (1989) Tubal infertility and silent chlamydial salpingitis. *Human Reprod*; **4**: 280–284.

Papagrigoriadis S, Rennie JA (1999) Lymphogranuloma venereum as a cause of rectal strictures. *Postgrad Med J*; **74**: 168–169.

Peeling RW, Bailey RL, Conway DJ *et al.* (1998) Antibody response to the 60 kDa heat shock protein is associated with scarring trachoma. *J Infect Dis*; **177**: 256–262.

Penttila T, Vuola JM, Puurula V *et al.* (2000) Immunity to *Chlamydia pneumoniae* induced by vaccination with DNA vectors expressing a cytoplasmic protein (Hsp60) or outer membrane proteins (MOMP and Omp2). *Vaccine*; **19**: 1256–1265.

Pereira LH, Embil JA, Haase DA. *et al.* (1990) Cytomegalovirus infection among women attending a sexually transmitted disease clinic: association with clinical symptoms and other sexually transmitted diseases. *Am J Epidemiol*; **131**: 683–692.

Pospischil A, Thoma R, Hilbe M *et al.* (2002) Abortion in humans caused by *Chlamydophila abortus* (*Chlamydia psittaci* serovar 1) [in German]. *Schweiz Arch Tierheilkd*; **144**: 463–466.

Rani R, Corbitt G, Killough R, Curless E (2002) Is there any role for rapid tests for *Chlamydia trachomatis*? *Int J STD AIDS*; **13**: 22–24.

Rasmussen SJ, Eckmann L, Quayle AJ *et al.* (1997) Secretion of proinflammatory cytokines by epithelial cells in response to *Chlamydia* infection suggests a central role for epithelial cells in chlamydial pathogenesis. *J Clin Invest*; **99**: 77–87.

Regan RJ, Dathan JR, Treharne JD (1979) Infective endocarditis with glomerulonephritis associated with cat chlamydia (*C. psittaci*) infection. *Br Heart J*; **42**: 349–352.

Saikku P, Leinonen M, Mattila K *et al*. (1988) Serological evidence of an association of a novel *Chlamydia*, TWAR, with chronic coronary heart disease and acute myocardial infarction. *Lancet*; **2**: 983–986.

Schachter J, Dawson CR (1978) *Human Chlamydial Infections*. PSG Publishing, Littleton, MA. pp. 45–62.

Schachter J, Osaba AO (1983) Lymphogranuloma venereum. *Br Med Bull*; **39**: 151–154.

Schachter J, Grossman M, Sweet RL *et al*. (1986) Prospective study of perinatal transmission of *Chlamydia trachomatis*. *JAMA*; **255**: 3374–3377.

Scidmore MA, Fischer ER, Hackstadt T (2003) Restricted fusion of *Chlamydia trachomatis* vesicles with endocytic compartments during the initial stages of infection. *Infect Immun*; **71**: 973–984.

Shen L, Wu S, Liu G (1995) Study on the perinatal infection caused by *Chlamydia trachomatis* [in Chinese]. *Zhonghua Fu Chan Ke Za Zhi*; **30**: 714–717.

Shortliffe LMD, Sellers RG, Schachter J (1992) The characterisation of the nonbacterial prostatitis: search for an etiology. *J Urol*; **148**: 1461–1466.

Stamm WE, Wagner KF, Amsel R *et al*. (1980) Causes of the acute urethral syndrome in women. *N Engl J Med*; **303**: 409–415.

Stamm WE, Koutsky LA, Bebedetti JK *et al*. (1984) *Chlamydia trachomatis* urethral infections in men. Prevalence, risk factors and clinical manifestations. *Ann Intern Med*; **100**: 47–51.

Stary A (2001) European guideline for the management of chlamydial infection. *Int J STD AIDS*; **12**(S3): 30–33.

Stephens RS (2000) Chlamydial genomics and vaccine antigen discovery. *J Infect Dis*; **181**(Suppl. 3): S521–S523.

Stephens RS, Kalman S, Lammel C (1998) Genome sequence of an obligate intracellular pathogen of humans: *Chlamydia trachomatis*. *Science*; **282**: 754–759.

Takahashi T, Masuda M, Tsuruno T *et al*. (1997) Phylogenetic analyses of *Chlamydia psittaci* strains from birds based on 16S rRNA gene sequence. *J Clin Microbiol*; **35**: 2908–2914.

Taylor-Robinson D, Tuffrey M (1987) Comparison of detection procedures for *Chlamydia trachomatis* including enzyme immunoassays in a mouse model of genital infection. *J Med Microbiol*; **24**: 169–173.

Thylefors B, Negrel AD, Pararajasegaram R, Dadzie KY (1995) Global data on blindness. *Bull World Health Organ*; **73**: 115–121.

Tipple MA, Beem MO, Saxon EM (1979) Clinical characteristics of the afebrile pneumonia associated with *Chlamydia trachomatis* infection in infants less than six months of age. *Pediatrics*; **63**: 192–197.

Uyeda CT, Welborn P, Ellison-Birang N *et al*. (1984) Rapid diagnosis of chlamydial infections with the MicroTrak direct test. *J Clin Microbiol*; **20**: 948–950.

Van Dongen PW (1993) Diagnosis of Fitz–Hugh–Curtis syndrome by ultrasound. *Eur J Obstet Gynecol Reprod Biol*; **50**: 159–162.

Van Dyck E, Ieven M, Pattyn S *et al*. (2001) Detection of *Chlamydia trachomatis* and *Neisseria gonorrhoeae* by enzyme immunoassay, culture, and three nucleic acid amplification tests. *J Clin Microbiol*; **39**: 1751–1756.

Verhoeven V, Ieven M, Meheus A *et al*. (2002) First, do not harm: also an issue in NAA assay diagnostics for chlamydial infection [letter]. *Sex Transm Infect*; 76–77.

Viswalingam ND, Wishart MS, Woodland RM (1983) Adult chlamydial ophthalmia. *Br Med Bull*; **39**: 123–127.

Wang SP, Grayston JT (1991) Three new serovars of *Chlamydia trachomatis*. Da, Ia and L2a. *J Infect Dis*; **163**: 403–405.

Ward ME (1999) Mechanisms of Chlamydia-induced disease. In Stephens, RS (Ed.), *Chlamydia: Intracellular Biology, Pathogenesis and Immunity*. ASM Press, Washington, DC. pp. 171–210.

Weiss SG, Newcomb RW, Beem MO (1986) Pulmonary assessment of children after chlamydial pneumonia of infancy. *J Pediatr*; **108**: 659–664.

Winstanley C, Hart CA (2001) Type III secretion systems and pathogenicity islands. *J Med Microbiol*; **50**: 116–126.

Yung AP, Grayson ML (1988) Psittacosis – a review of 135 cases. *Med J Aust*; **148**: 228–233.

FURTHER READING

Stephens RS (Ed.) (1999) *Chlamydia: Intracellular Biology, Pathogenesis, and Immunity*. ASM Press, Washington, DC.

www.chlamydiae.com is a very useful source of up-to-date information on this group of organisms.

26

Tropheryma whipplei

F. Fenollar and D. Raoult

Unité des Rickettsies, CNRS UMR 6020A, Université de la Méditerranée, Marseilles, France

INTRODUCTION

Tropheryma whipplei is the bacterium responsible for Whipple's disease (Misbah and Mapstone, 2000; Dutly and Altwegg, 2001; Fenollar and Raoult, 2001b; Maiwald and Relman, 2001). An American pathologist George Hoyt Whipple reported the first case in 1907 (Whipple, 1907). This was the case of a fatal disease in an anorexic patient with arthralgias, diarrhea, chronic cough and fever (Whipple, 1907). During the necropsy, abnormalities were principally observed in the intestine, the mesenterium, the heart and the lung (Whipple, 1907). Microorganisms were also detected in tissues by using silver staining. However, because of the observation of fats in the stools, the intestine and the mesenteric glands, a disorder of fat metabolism was suspected (Whipple, 1907) and named intestinal lipodystrophy. Whipple's disease is a rare systemic infection (Misbah and Mapstone, 2000; Dutly and Altwegg, 2001; Fenollar and Raoult, 2001b; Maiwald and Relman, 2001). We have been able to detect *T. whipplei* in biopsies from Dr Whipple's first patient (Dumler, Baisden and Raoult, 2003). The classical symptoms consist of weight loss, abdominal pains, diarrhea, arthralgias and lymphadenopathy (Misbah and Mapstone, 2000; Dutly and Altwegg, 2001; Fenollar and Raoult, 2001b; Maiwald and Relman, 2001). The symptomatology can be very protean and nonspecific. Cardiac, ocular or central nervous involvements can be observed (Misbah and Mapstone, 2000; Dutly and Altwegg, 2001; Fenollar and Raoult, 2001b; Maiwald and Relman, 2001). These manifestations are not always associated with digestive symptoms (Misbah and Mapstone, 2000; Dutly and Altwegg, 2001; Fenollar and Raoult, 2001b; Maiwald and Relman, 2001). For a long time, the diagnosis has been based on a small-bowel biopsy, which allows the observation of positive inclusions by using periodic acid–Schiff (PAS) staining (Misbah and Mapstone, 2000; Dutly and Altwegg, 2001; Fenollar and Raoult, 2001b; Maiwald and Relman, 2001). However, for patients without digestive symptoms, the results of the small-bowel biopsy could be negative, leading to delay in diagnosis. For 10 years, a new tool based on polymerase chain reaction (PCR) targeting the 16S rDNA followed by sequencing has been employed (Wilson *et al.*, 1991; Relman *et al.*, 1992). Since 1999, the culture of the bacterium has allowed new perspectives on the diagnosis (Raoult *et al.*, 2000). The progression of the disease without treatment is always fatal (Misbah and Mapstone, 2000; Dutly and Altwegg, 2001; Fenollar and Raoult, 2001b; Maiwald and Relman, 2001). Even with adequate treatment, relapses are possible (Misbah and Mapstone, 2000; Dutly and Altwegg, 2001; Fenollar and Raoult, 2001b; Maiwald and Relman, 2001). The isolation of *T. whipplei* has opened new perspectives on the characterization of the microorganism (La Scola *et al.*, 2001) and allowed the development of immunohisto-logical tools for the diagnosis (Raoult *et al.*, 2001), the sequencing of the full genome and the introduction of studies on its physiopathology, *in vitro* antibiotic susceptibilities and serology.

DESCRIPTION OF THE ORGANISM

Morphology

When he described the disease for the first time, Whipple had noticed the presence of rod microorganisms $2\,\mu m$ long, mainly in the vacuoles of macrophages of affected tissue (Whipple, 1907). In 1949, Black-Schaffer showed that the macrophages from the intestine and the lymphatic glands of patients with Whipple's disease contained PAS-positive material (Black-Schaffer, 1949). The PAS staining showed purple macrophages, but no bacteria were identified. He concluded that the macrophages must contain mucopolysaccharides. In 1960, a first observation with electron microscopy of structures such as bacteria was performed on an intestinal biopsy (Cohen *et al.*, 1960). This observation was confirmed 1 year later, and a bacterial etiology of Whipple's disease was highly suspected (Chears and Ashworth, 1961; Yardley and Hendrix, 1961). Morphologically, the bacterium develops a characteristically trilamellar ultrastructure including a plasmic membrane surrounded by a thin wall, itself surrounded by a structure similar to a plasmic membrane (Silva, Macedo and Moura Nunes, 1961). This morphology is not typical of Gram-positive or Gram-negative bacteria.

Immunohistochemical Study

Antibodies directed against *Shigella* and *Streptococcus* A, B, C and G, when tested against intestinal biopsies from patients with Whipple's disease, have been shown to present cross-reactivity against the Whipple's disease bacterium (Ectors *et al.*, 1992; Alkan, Beals and Schnitzer, 2001). When using primary antibodies at optimal dilutions, only the cross-reactivity of antibodies directed against *Streptococcus* B persists (Du Boulay, 1982). This staining seems to be due to the presence of rhamnose-containing polysaccharides (Evans and Ali, 1985).

The culture of *T. whipplei* has been an elusive goal for many generations of microbiologists. Several attempts to cultivate *T. whipplei* have failed (Dutly and Altwegg, 2001) but have been reported with bacteria from the genera *Corynebacterium*, *Streptococcus*, *Propionibacterium*, *Bacteroides*, *Brucella*, *Escherichia*, *Clostridium*, *Klebsiella* and *Haemophilus* (Dobbins, 1987b; Dutly and Altwegg, 2001; Fenollar and Raoult, 2001b). In 1997, two strains of *T. whipplei* were reported to have been isolated from cardiac valves of patients with endocarditis (Schoedon *et al.*, 1997). A macrophage system with its microbicidal functions deactivated by a combination of dexamethasone, interleukin-4 (IL-4) and IL-10 was used. Unfortunately, the findings of this work were not confirmed and no isolate could be subcultured (Schoedon *et al.*, 1997). Two years later, a different approach was used for cultivating *T. whipplei*. The bacterium was isolated and established from a strain obtained from a cardiac valve of a patient with Whipple's disease endocarditis. *Tropheryma whipplei* has been propagated in human fibroblasts (HEL) in minimum essential medium (MEM) with 10% fetal serum calf and 2 mM glutamine (Raoult *et al.*, 2000). The use of an appropriate cellular system that can be conserved for several weeks, the high cell-to-bacterium ratio and the step of centrifugation to increase the adhesion of cells to bacteria have probably facilitated the isolation of the bacterium. At first, the bacterium multiplies slowly, and the doubling time has been estimated at 17 days. Currently, this time of multiplication is passed at 36 h (36 h passage). As of yet, culture of the bacterium in axenic medium has not been reported. The multiplication site of the bacterium is controversial. Some authors suggest that the bacterium multiplies in the digestive lumen and that they are phagocytosed and slowly degraded in the macrophages (Dobbins and Ruffin, 1967). *In vitro*, the bacteria apparently multiply in the monocytes with or without IL-4 (Schoedon *et al.*, 1997) (unpublished data). Later, they are released from the infected cells. Inside HeLa cells, *T. whipplei* multiplies actively in an acidic vacuole at pH 5 (Ghigo *et al.*, 2002). This acidic pH may inhibit the antibiotic activity (Street, Donoghue and Neild, 1999) and could then be responsible for the loss of activity of a high number of antibiotic regimens (Dutly and Altwegg, 2001). In a study that combined the use of *in situ* hybridization and confocal microscopy, the ribosomal RNA of the bacterium was localized to the extracellular spaces. This seems to indicate the presence of bacteria with an active metabolism outside the cells (Fredericks and Relman, 2001).

Molecular Characterization

In 1991, Wilson sequenced a fragment of 721 base pairs of the 16S rDNA (Wilson *et al.*, 1991). This sequence contained some ambiguous nucleotides, but it did not appear to be closely related to any known bacteria. One year later, sequencing of 1321 base pairs of the 16S rDNA (representing 90% of the sequence) from the biopsy of a patient and 384 base pairs from four biopsies from four other patients confirmed these data (Relman *et al.*, 1992). The sequence did not correspond to any microorganism characterized before. The bacterium was then named *Tropheryma whippelii*, from the Greek *trophi* (nourish) and *eryma* (barrier), linked to the malabsorption syndrome observed in the disease and in honor of George Hoyt Whipple. Following the culture and the description of the bacterium, and the deposit of a reference isolate, the name was corrected. The bacterium was officially named *Tropheryma whipplei* because the name of Whipple was not correctly Latinized in the original version (La Scola *et al.*, 2001).

Phylogenetic analysis based on the sequence of the 16S rDNA has shown that the bacterium associated with Whipple's disease could be classified among the group of the *Actinomyces* (class Actinobacteria), relatively close to two known species in human pathology, *Actinomyces pyogenes* and *Rothia dentocariosa*, and to bacteria principally found in the environment including the *Cellulomonadaceae* (Relman *et al.*, 1992; Maiwald *et al.*, 1996) (Figure 26.1). The molecular taxonomy classified *T. whipplei* close to bacteria with a high GC% which have in common the soil as a reservoir or a contamination source for humans. An analysis based on the complete sequence of the 16S rDNA (Maiwald *et al.*, 1996) places *T. whipplei* more precisely between the *Cellulomonadaceae* and the *Actinomyces*, comporting group B peptidoglycan. The sequence similarity of the 16S rDNA of *T. whipplei* with those of the two nearest species (*Cellulomonas cellasea* and *Corynebacterium aquaticum*) is only 90%.

The possibility of a second bacterial species being associated with Whipple's disease [Whipple's-disease-associated bacterial organism (WABO)], closer to *Nocardia*, has been suggested following PCR and sequences obtained from an intestinal biopsy from a patient with a clinical and histological Whipple's disease (Harmsen *et al.*, 1994). Significant sequence differences (19 substitutions for 225 amplified and sequenced nucleotides) have been observed by using specific primers (Relman *et al.*, 1992). One case of co-infection with the two variants WABO and *T. whipplei* has been reported from a patient with a cerebral Whipple's disease (Neumann *et al.*, 1997). These data are difficult to interpret because they are based on a very short portion of the 16S rDNA sequence. The possibility of a second bacterial species being responsible for Whipple's disease is not excluded, but more data are necessary. Indeed, a variation of the 16S rDNA sequence of the bacterium responsible for the Whipple's disease cannot be definitely excluded, and this should be investigated in more patients. The variability of the 16S rDNA sequence with a difference of only one position among the 225 nucleotides analyzed has also been observed from two intestinal biopsies from two patients with a clinical and histological Whipple's disease (Harmsen *et al.*, 1994; Maiwald *et al.*, 2000). Phylogenetic analysis of *T. whipplei* based on the 23S and 5S rDNA sequences also places the bacterium in the class of *Actinobacteria*. However, the insufficient number of available sequences does not allow a precise comparison to be made between the analyzed sequences and the different species analyzed using 16S rDNA sequence (Maiwald *et al.*, 2000). *Tropheryma whipplei* shows a degree of heterogeneity that has been proven with the sequencing of the 16S–23S rDNA interregion (Maiwald *et al.*, 1996; Hinrikson, Dutly and Altwegg, 1999) and the 23S rDNA (Hinrikson, Dutly and Altwegg, 2000b). The interregion 16S–23S rDNA is known to be more variable than the structural flanking genes. This zone is considered as a tool for subtyping bacteria in various taxonomic groups (Gürtler and Stanisich, 1996). The sequences of the interregion 16S–23S rDNA and those of 200 nucleotides from the 23S rDNA of *T. whipplei* have been determined (Maiwald *et al.*, 1996). The reported length of the 16S–23S interregion of *T. whipplei* is 294 base pairs, without the coding sequences for the tDNA and the 5S rDNA. These data are comparable to the size and the structure of the 16S–23S interregion of most Gram-positive bacteria with high GC% (Gürtler and Stanisich, 1996). The research of similar sequences of the 16S–23S interregion of *T. whipplei* with those of other *Actinomyces* reveals globally a weak homology (Maiwald *et al.*, 1996). Nevertheless, several small zones with a high homology with other *Actinomyces* are present (Maiwald *et al.*, 1996). The first sequence of the 16S–23S interregion described by Maiwald *et al.* has been confirmed by the detection of homolog sequences in samples of nine Swiss patients with Whipple's disease (Hinrikson, Dutly and Altwegg, 1999). In a later study, the 16S–23S interregion of *T. whipplei* was studied by PCR, followed by sequencing in 38 samples from 28 patients with Whipple's disease (Hinrikson *et al.*, 1999). Sequence analysis revealed the existence of five dimorphic sites constituting three types of spacers (Hinrikson *et al.*, 1999) (Figure 26.2). The more common genotype, i.e. the type 1, corresponds perfectly with the original sequence described of 294 base pairs (Maiwald *et al.*, 1996). By comparison with this reference sequence, the spacer types 2 and 3 are different in only two and five nucleotide positions, respectively. Very recently, three genotypes have been identified among a series of 43 patients (Maiwald *et al.*, 2000). These spacers of types 4, 5 and 6 differ, respectively, in four, three and six nucleotide positions in comparison with type 1. Type 3 was not found in this study. The existence of different types of spacers has been confirmed by sequencing and through analysis with restriction enzymes (Maiwald *et al.*, 2000). The six different types of spacers are not found with the same frequency. Types 1 and 2 are the most common, whereas types 3–6 are only observed sporadically. In addition, the relative frequencies of types 1 and 2 differ between the studies of Hinrikson, Dutly and Altwegg (2000a, b) and Maiwald *et al.* (2000). Currently, differences cannot be attributed to a particular geographic origin because in all the studies most of the patients lived in central Europe. A possible bias could be that the Swiss studies (Hinrikson, Dutly and Altwegg, 2000a)

Figure 26.1 Phylogenetic taxonomy of *Tropheryma whipplei* based on DNA 16S ribosomic sequences.

included not only patients with Whipple's disease but also patients with a positive PCR but with no symptoms of Whipple's disease. When several samples were available for the same patient, the same spacer type was found in each sample from the same patient (Hinrikson *et al.*, 1999; Dutly *et al.*, 2000; Maiwald *et al.*, 2000). Nevertheless, one exception existed – a patient seemed to present a possible double infection (Maiwald *et al.*, 2000). On the basis of these data, each patient seemed then to be infected with only one strain (Gürtler and Stanisich, 1996). The variations found in the 16S–23S interregion raise the question of the possible existence of six different species or subspecies. The 16S rDNA partial sequences determined in

most of the studies have not revealed differences with the reference sequence. However, as for the other bacteria, even a complete identity of 16S rDNA sequences does not prove species identity (Fox, Wisotzkey and Jurtshuk, 1992) but suggests only a very close relation (Stackebrandt and Goebel, 1994). To resolve the problem, DNA–DNA hybridization studies are necessary. Currently, these six types are considered as subtypes of only one species, because of a weak number of variable nucleotides, similar to or less than the variations observed in other species.

For the *Actinobacteria*, one insertion in domain III of the 23S rDNA sequence of approximately 100 nucleotides has been described

Nucleotides of the interregion of the *T. whipplei* 16S–23S DNA. (The base numeration corresponds to those of the reference sequence (X99636). P = position)

```
Type 1:  AAGGAGCTTT TGCGGTTACT TTATAACTTT AAAGTGATAC CGCCATAGTG
Type 2:  ---------- ---------- ---------- ---------- ----------
Type 3:  ---------- ---------- ---------- ---------- ----------
Type 4:  ---------- ---------- ---------- ---------- ----------
Type 5:  ---------- ---------- ---------- ---------- ----------
Type 6:  ---------- ---------- ---------- ---------- ----------

              P56:                                        P86:
Type 1:  CACTGTGATT GTGGTGTTGT TATCACGGTC TGGGTCTGGT CTGTTTTGGG
Type 2:  -----C---- ---------- ---------- ---------- ----------
Type 3:  -----C---- ---------- ---------- ---------- ----------
Type 4:  ---------- ---------- ---------- ---------- ----------
Type 5:  ---------- ---------- ---------- ---------- ----------
Type 6:  -----C---- ---------- ---------- ---------- ----------

         P 94: P98:            P115:      P124:
Type 1:  CACAAGCTCA CGGGAGTGGA ACATTGAATC TGTTCACCGG TTTGTTTACC
Type 2:  ---------- ---------- ---------- ---------- ----------
Type 3:  ---------- ---------- ---------- ----T----- ----------
Type 4:  ----G---T- ---------- -----A---- ---------- ----------
Type 5:  ----G----- ---------- -----A---- ---------- ----------
Type 6:  ---------- ---------- ---------- ----T----- ----------

         P143: P148:      P157:
Type 1:  GGTCTGGGTG CTTATGCACT TGTGTGCACT ATTGGGTTTT GAGAGGCCAG
Type 2:  ---------- ---------- ---------- ---------- ----------
Type 3:  -------C-  -------G-- ---------- ---------- ----------
Type 4:  ---------- ---------- ---------- ---------- ----------
Type 5:  ---------- ---------- ---------- ---------- ----------
Type 6:  ---T----C- -------G-- ---------- ---------- ----------

              P200:
Type 1:  GCTTTACGGG TGGGCAGTGT CCGGCCGTAT TTTGAGAACT GCACAGTGGA
Type 2:  ---------- ---------- ---------- ---------- ----------
Type 3:  ---------- ---------- ---------- ---------- ----------
Type 4:  ---------A ---------- ---------- ---------- ----------
Type 5:  ---------A ---------- ---------- ---------- ----------
Type 6:  ---------- ---------- ---------- ---------- ----------

Type 1:  CGCGAGCATC TTTCTAGACT TAGTCTAGAG TCTGCGCTTA TGCGCAGTAT TTGT
Type 2:  ---------- ---------- ---------- ---------- ---------- ----
Type 3:  ---------- ---------- ---------- ---------- ---------- ----
Type 4:  ---------- ---------- ---------- ---------- ---------- ----
Type 5:  ---------- ---------- ---------- ---------- ---------- ----
Type 6:  ---------- ---------- ---------- ---------- ---------- ----
```

Partial DNA ribosomic 23S (domain III):

```
                        P88:
Type A:  TTTTGCTGTG CTTACGGTAC CCCTTTTACG GTGTGCGGGA TAAGTGTTCA
Type B:  ---------- ---------- ---C------ ---------- ----------
```

Figure 26.2 Schematic representation of the different subspecies of *Tropheryma whipplei*.

and does not seem to be present in any other bacterial group (Roller, Ludwig and Schleifer, 1992). This insertion is more variable between species than the other parts of the 23S rDNA and 16S rDNA sequences. A part of this domain III of *T. whipplei* has been amplified from clinical samples by using primers targeting the flanking region (Hinrikson, Dutly and Altwegg, 2000a). Analysis of the sequence has revealed an insertion of 80 nucleotides (Hinrikson, Dutly and Altwegg, 2000b; Maiwald *et al.*, 2000), confirming the classification of *T. whipplei* in the *Actinobacteria*. Among the 28 studied patients, 27 carried this identical insertion sequence (type A), whereas the other patient had a different insertion sequence (type B), which differed in only one position (Hinrikson, Dutly and Altwegg, 2000a). The similarity of the sequences with those of other *Actinobacteria* was high (greater than 90%) for the region immediately upstream, but only moderate (approximately 70%) for the region downstream of the insertion (Hinrikson, Dutly and Altwegg, 2000b). Compared to other *Actinobacteria*, the insertion sequence itself of *T. whipplei* is smaller (80 nucleotides compared with 86–116 nucleotides), and its similarity

to other sequences is low (Roller, Ludwig and Schleifer, 1992; Hinrikson, Dutly and Altwegg, 2000b; Maiwald *et al.*, 2000). In addition, it does not contain specific elements of the group (Roller, Ludwig and Schleifer, 1992), and more particularly, there is no similarity with the insertion sequences found for the *Cellulomonadaceae* and the *Actinomyces* of group B peptidoglycan, i.e. the groups close to *T. whipplei*, based on 16S rDNA sequence comparisons (Maiwald *et al.*, 1996). Until now, a link between the different genotypes and the symptomatology has not been demonstrated. Some authors suggested that some specific strains may not have a pathogenic role, that others could be responsible for typical Whipple's disease and that others could be responsible for atypical Whipple's disease such as blood culture-negative endocarditis, but this is not yet demonstrated (Maiwald *et al.*, 1998; Hinrikson *et al.*, 1999; Dutly and Altwegg, 2001). Very recently, the genome of *T. whipplei* has been entirely sequenced and deposited in GenBank. *Tropheryma whipplei* has a unique circular chromosome and a small genome size (925 kb), with a GC content of 47%.

EPIDEMIOLOGY OF WHIPPLE'S DISEASE

Whipple's disease is considered rare, and there is no real estimation of the prevalence of the disease. In 1985, Dobbins estimated that 2000 cases have been reported since 1907 (Dobbins, 1987b). The disease seems to be more common in farmers (Dobbins, 1987b). In 1987, an analysis of 696 cases demonstrated that approximately 55% of cases had been reported in Europe and 38% in the north of America (Dobbins, 1987b) and that 97% of the patients were Caucasian (Dobbins, 1987b). The disease has not been frequently reported in Asiatic, black, Indian or Hispanic people. Whipple's disease has been rarely diagnosed in children, but it can be observed in anyone irrespective of the age (Dobbins, 1987b). Few familial cases have been described (brothers, father/daughter), but analysis has not shown particular risk factors (Dobbins, 1987b; Fenollar and Raoult, 2001b). No epidemic or inter-human transmissions have been described (Dutly and Altwegg, 2001). Of note, few grouped cases have been reported (Dobbins, 1987b). Whipple's disease involves principally men with a mean age of 50 years, and there is a man-to-woman sex ratio of 8:1 (Dobbins, 1987b). New studies seem to show a higher prevalence in women than previously supposed (von Herbay et al., 1997b). A prevalence of approximately 26% in HLA-B27-positive patients has been observed in patients with Whipple's disease, i.e. three or four times more frequent than in a normal population (Dobbins, 1987a, b). However, a direct association between an HLA-B27-positive type and a diminution of resistance to infection has never been observed. In addition, these data have not been confirmed in other populations, principally in Italy and in Argentina (Bai et al., 1991; Olivieri et al., 2001). The source of T. whipplei and its transmission are presently unknown. The bacterium seems to be present in the environment: studies based on PCR have shown the presence of T. whipplei DNA in sewage water (Maiwald et al., 1998) and in human stools (Gross, Jung and Zoller, 1999; Dutly and Altwegg, 2001; Maibach, Dutly and Altwegg, 2002). Tropheryma whipplei could therefore be ingested orally (Fenollar and Raoult, 2001b; Maiwald et al., 2001). PCR-based studies have found that the T. whipplei DNA can be detected in saliva, duodenal biopsies, gastric liquids and stools of subjects without Whipple's disease (Ehrbar et al., 1999; Street, Donoghue and Neild, 1999; Dutly et al., 2000). Indeed, a study of 40 healthy subjects has detected the presence of T. whipplei DNA in 35% (Street, Donoghue and Neild, 1999). Another PCR-based study has shown the presence of T. whipplei DNA in 4.8% of duodenal biopsies and 13% of gastric liquids from 105 patients without clinical suspicion of Whipple's disease (Ehrbar et al., 1999). These data suggest that T. whipplei could be a commensal bacterium of the gut. However, these data should be interpreted with caution because they are based on 'home-made' PCR and they may not be confirmed (Dutly and Altwegg, 2001). Indeed, T. whipplei could have a particular geographic distribution. This hypothesis is supported by a study that investigated the presence of T. whipplei DNA in duodenal biopsies and gastric liquid from Swiss and Malaysian patients without clinical suspicion of Whipple's disease (Dutly and Altwegg, 2001). The presence of T. whipplei DNA was detected in 14 of 105 Swiss patients (Ehrbar et al., 1999), whereas none was detected in 108 Malaysian patients (Dutly and Altwegg, 2001). These preliminary data should be substantiated to conclude on the existence of a heterogenicity of geographic distribution for T. whipplei. New PCR-based studies did not find T. whipplei DNA in the duodenal biopsies of control patients. The demonstration of T. whipplei in duodenal biopsies of patients with an extraintestinal Whipple's disease without clinical, endoscopic or histologic evidence of digestive involvement suggests the existence of an intestinal entry for T. whipplei (Müller et al., 1997).

EPIDEMIOLOGY AND PATHOGENESIS

The epidemiology and physiopathology of Whipple's disease are still obscure. The distribution of the disease is not well defined. It is currently difficult to determine any difference between the importance of risk factors (genetic factors and presence of immune defect) and the existence of different strains of T. whipplei in the physiopathology of Whipple's disease.

IMMUNOLOGY AND PATHOLOGY

Different immunological abnormalities have been observed in patients with Whipple's disease, and it was often difficult to make distinctions between the primary and the secondary causes. This is probably due to the small number of patients with Whipple's disease and to the several clinical manifestations. The supposed immunological defect must probably be subtle and specific to T. whipplei. Indeed, patients with Whipple's disease are not usually predisposed to other infections. Only a few cases of co-infections with HIV-positive patients or patients with nocardiosis or giardiasis have been described (Dutly and Altwegg, 2001; Fenollar and Raoult, 2001b). In patients with Whipple's disease, the lamina propria often presents a massive infiltration of macrophages and a relative poverty of infiltration by the plasma cells and the lymphocytes (Maxwell et al., 1968). The number of B lymphocytes is diminished, and the T lymphocytes are lost, suggesting a defect in B cells and/or T cells in response to the causative agent (Dutly and Altwegg, 2001). This could also reflect a secondary loss of the lymphocytes because of lymphangiectasis. A study has found a diminution of immunoglobulin A (IgA) B cells but an augmentation of the number of IgM B cells in the lamina propria (Eck et al., 1997). However, the level of secretory IgA determined from intestinal aspirations is normal (Dobbins, 1981). The humoral response to T. whipplei is not currently measurable in patients. Some studies reported a normal level of IgG and a normal or often diminished level of IgM, whereas the level of IgA appeared slightly increased before treatment and was normal after treatment (Dobbins, 1981, 1987b). For several patients with Whipple's disease, a study has shown a diminution of the level of IgG2 subclass. This subclass of IgG is usually produced in response to infections caused by encapsulated bacteria. Its synthesis is regulated by an immune response based on cellular mediation and interferon-γ (IFN-γ) (Marth et al., 1997). The circulating T-cell population and those of the lamina propria can be characterized by a reduced CD4-to-CD8 ratio of T cells (i.e. augmentation of the number of positive T8 cells), associated with a shift in the T-cell subpopulations to an increased expression of CD45RO, a marker of memory on T cells (Marth et al., 1994). These changes are present in ill patients and also in patients with residual PAS-positive cells but are not observed in patients in complete remission (Marth et al., 1994). These changes are accompanied by a reduced in vitro response of T-cell proliferation to mitogens or other stimuli such as phytohemagglutinin (PHA), concanavalin A (ConA) and anti-CD2 and anti-CD3 antibodies (Dobbins, 1981, 1987b; Marth et al., 1994). These alterations and an impaired delayed-type hypersensitivity are present not only in patients with an acute disease but also in patients with a long period of remission (Dobbins, 1981; Marth et al., 1994). Studies of macrophages in Whipple's disease have shown dysfunctions: they have an incapacity to degrade the bacterial antigens (proteins and DNA derived from E. coli and Streptococcus pyogenes). Even so, phagocytosis and intracellular killing do not seem to be decreased (Bjerknes et al., 1988). In patients with Whipple's disease, this deficiency of macrophages in degrading the intracellular microorganisms is persistent (Bjerknes et al., 1988; Marth et al., 1994). Another study has shown that the intestinal macrophages equally presented reduced phagocytosis (Lukacs, Dobi and Szabo, 1978). A diminution of production of IL-12 and IFN-γ by monocytes has been shown in patients with Whipple's disease, compared to controls, not only during the acute phase but also in the remission phase (Marth et al., 1997). These results, therefore, suggest that the macrophages and not the T cells are directly implicated in the immune defect. IL-12 is a cytokine and a pivotal mediator of cellular immunity.

The granulocytes, the monocytes, the macrophages and the dendritic cells, which are the first cells to encounter a foreign antigen during infection, produce it. IL-12 activates the natural killer (NK) cells plus the T cells and induces the production of IFN-γ by T-helper type 1 cells. The low production of IL-12 lowers IFN-γ production by T cells and consequently controls the microbicidal effect of macrophages. Diminished levels of IL-12p40 have been observed in patients' sera (Marth *et al.*, 2002). These data suggest that Whipple's disease could be associated with a primary defect in the production of IL-12 by monocytes (Ramaiah and Boynton, 1998). In addition, the secretion of this defective cytokine is also associated with low secretions of serum IgG2, a subclass of IgG dependent on IFN-γ. A deficit in IL-12 production is correlated with chronic infections or relapses due to microorganisms of low virulence, such as *Mycobacterium avium* (Levin *et al.*, 1995). An IL-12 defect, but less severe than those observed in patients with Whipple's disease, has been found in two related persons (one sister and one brother of two different patients with Whipple's disease) (Marth *et al.*, 1997). These data suggest the existence of a genetic base to these abnormalities. It has also been demonstrated that patients with Whipple's disease have a reduced expression of CD11b markers, the α-chain of complement receptor 3 (CR3) (Marth *et al.*, 1994). CR3 is a member of the integrin family and facilitates microbial phagocytosis (Marth *et al.*, 1994), playing a role in antigen presentation and mediating the intracellular death induced by the IFN-γ, produced in response to ingested bacteria (Marth *et al.*, 1994). Indeed, a study has reported the case of a patient with Whipple's disease refractory to all antibiotic therapies and with reduced rates of IFN-γ *in vitro*, for whom a treatment with antibiotics and recombinant IFN-γ allowed a cure (Schneider *et al.*, 1998). The recent improvement in the treatment of infections due to intracellular bacteria such as those of the *M. avium* complex has shown that immunotherapy with IFN-γ, which leads to monocyte activation, can have a complementary therapeutic use for patients resistant to antibiotics (Holland *et al.*, 1994).

Two types of granulomatous reactions are observed. In mesenteric adenopathy, granulomas such as those observed in the presence of a lipid-associated foreign body are observed. In the other tissues, epithelioid granulomas with or without giant cells, similar to those observed in sarcoidosis, are observed (Rouillon *et al.*, 1993). The presence of systematic granulomatous inflammation in adenopathy, the liver or spleen in 9% of patients with Whipple's disease leads to a possible confusion with the diagnosis of sarcoidosis (Dobbins, 1987b). Any kidney involvement, such as glomerulonephritis and interstitial or granulomatous nephritis, is not common and usually appears late during the disease (Dobbins, 1987b; Marie, Lecomte and Levesque, 2000). Hypothyroiditis has been rarely described (Dobbins, 1987b). Despite improvements in diagnostic tools, most case reports underline the difficulty of Whipple's disease diagnosis in cases of various and atypical presentations.

CLINICAL FEATURES OF WHIPPLE'S DISEASE

Whipple's disease is often diagnosed late in its evolution. For a long time, Whipple's disease has been considered as a mainly digestive disease. However, it is currently estimated that 15% of diagnosed patients have no digestive symptoms (Misbah *et al.*, 1997). Indeed, the clinical manifestations are several and not specific, depending on the implicated organs (Dutly and Altwegg, 2001; Fenollar and Raoult, 2001b). No major organ is spared by the *T. whipplei* infection. The various clinical manifestations of Whipple's disease are described in Table 26.1. Weight loss, diarrhea, malabsorption and abdominal pains are the most frequent symptoms observed at the time of diagnosis (Marth and Raoult, in press). In very rare cases, constipation is observed instead of diarrhea (Mur Villacampa, 1987). Splenomegaly and hepatomegaly are observed in 15% of patients (Marth, 1999). Very rare cases of hepatitis have also been observed (Schultz *et al.*,

Table 26.1 Clinical manifestations in Whipple's disease

Manifestations
Classic manifestations
Diarrhea
Arthralgias
Weight loss
Abdominal pain
Polyadenopathy
Cutaneous pigmentation
Unknown fever
Sarcoidosis
Central nervous system involvement
Dementia
Abnormal ocular movement
Ophthalmologic manifestations
Uveitis
Cardiovascular manifestations
Blood culture negative endocarditis
Pericarditis
Myocarditis

2002). Ascites has been found in 5% of patients (Ramaiah and Boynton, 1998). Arthralgias are classical symptoms in Whipple's disease. Three-quarters of patients have joint complaints before the diagnosis of the disease (Puéchal, 2001). Arthralgias consist usually in the involvement of peripheral joints and are migratory, non-destroyed and non-deforming (Dobbins, 1987b; Puéchal, 2001). Large joints are usually involved and less frequently the small joints. The diagnosis of rheumatic disease is often considered before the diagnosis of Whipple's disease. Vertebral localization of Whipple's disease is rare, but the diagnosis should be considered in any case of HLA-B27-negative spondylarthropathy (Varvöli *et al.*, 2002) or spondylodiscitis (Altwegg *et al.*, 1996). The involvement of the Whipple's bacterium in a knee prosthesis infection of a patient with systemic Whipple's disease has been confirmed by PCR techniques targeting the 16S rDNA sequence (Frésard, 1996). The neurological involvement of *T. whipplei* infection has been estimated at 6–43% (Maizel, Ruffin and Dobbins, 1970; Dobbins, 1987b; Fleming, Wiesner and Shorter, 1988; Fantry and James, 1995; Vital-Durand *et al.*, 1997; Gerard *et al.*, 2002).

All patients with Whipple's disease have a neurological localization (Dobbins, 1988; von Herbay *et al.*, 1997a; Schnider *et al.*, 2002). This localization may correspond to a sanctuary of the bacterium responsible for relapses of the disease with central nervous system involvement. Currently, clinical neurological symptoms are estimated at more than 15% (Marth and Raoult, in press). These signs may or may not be associated with digestive symptoms (Dobbins, 1987b; Louis *et al.*, 1996; Gerard *et al.*, 2002). The classical signs are facial myoclonus, external ophthalmoplegia and dementia (Dobbins, 1987b; Louis *et al.*, 1996; Gerard *et al.*, 2002). The neurological manifestations are diverse: cephalea, ataxia, hypoacusia, personality changes, seizures, memory loss, somnolence, dysarthria and pyramidal signs have been described (Dobbins, 1987b; Louis *et al.*, 1996; Gerard *et al.*, 2002). Hypothalamic involvements causing insomnia, hypersomnia, polyuria and polydipsia are less common (Dobbins, 1987b; Louis *et al.*, 1996; Gerard *et al.*, 2002). Hypothalamopituitary involvement could be responsible for the deterioration of the secretion of sex hormones and hypogonadism (Di Stefano *et al.*, 1998). More rarely, parkinsonian syndrome, meningitis, myopathies and myoclonus have been observed in Whipple's disease (Louis *et al.*, 1996; Gerard *et al.*, 2002). An abnormal movement called oculo-facio-cervical myorrhythmia, considered by some as pathognomonic of Whipple's disease, has been also described (Louis *et al.*, 1996). Cardiac involvement is observed for at least one-third of patients (Fenollar, Lepidi and Raoult, 2001) and is more and more frequently observed in patients without digestive symptoms (Elkins, Shuman and Pirolo, 1999; Wolfert and Wright, 1999; Fenollar, Lepidi and Raoult, 2001).

The manifestations could be cardiac murmur, pericardiac friction rub, nonspecific electrocardiogram modification, congestive heart failure, sudden death, valvular insufficiency, coronary arteritis, myocardial fibrosis, lymphocystis myocarditis, pericarditis and blood culture-negative endocarditis (McAllister and Fenoglio, 1975; Khairy and Graham, 1996; Fenollar, Lepidi and Raoult, 2001). Hypotension is also frequently observed (McAllister and Fenoglio, 1975; Khairy and Graham, 1996). The ocular involvement is rare, estimated to affect 2–3% of patients with Whipple's disease, and is only rarely observed in patients without digestive symptoms (Dobbins, 1987b). The manifestations usually include blurred vision, uveitis, retinitis, retinal hemorrhages, choroiditis, papillary edema, optical atrophy, retrobulbar neuritis and keratitis (Selsky *et al.*, 1984; Williams *et al.*, 1998; Misbah and Mapstone, 2000; Fenollar and Raoult, 2001b). The neuroocular manifestations include ophthalmoplegia, supranuclear paralysis, nystagmus, myoclonia and ptosis (Adams *et al.*, 1987; Chan, Yannuzzi and Foster, 2001). In 1987, Dobbins estimated that pulmonary involvement could be observed in 30–40% of patients with Whipple's disease (Dobbins, 1987b). Cough is a frequently observed symptom and is found in approximately 50% of patients (Enzinger and Helwig, 1963). Chronic cough, thoracic pain or dyspnea secondary to a pleural or parenchyma involvement or a pulmonary emboli could be observed (Enzinger and Helwig, 1963; Wimberg, Rose and Rappaport, 1978; Dobbins, 1987b). Peripheral lymphadenopathies are observed in half of the patients with Whipple's disease (Alkan, Beals and Schnitzer, 2001). Very rare cases of mediastinal lymphadenopathy have also been described (MacDermott *et al.*, 1997).

Cutaneous involvement is rare and various. Hyperpigmentation is the most common manifestation described in approximately one-third of reported cases (Keren, 1981).

DIAGNOSIS

In the absence of clinical and radiological indicators for establishing a diagnosis of Whipple's disease, laboratory techniques play a crucial role. The different tools are summarized in Figure 26.3 and Table 26.2.

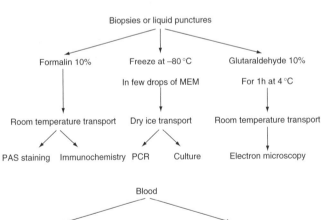

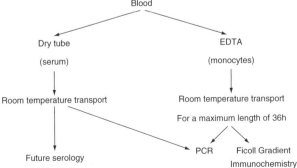

Figure 26.3 Samples, transport means and techniques for the diagnosis of Whipple's disease.

Table 26.2 Diagnostic tools for the diagnosis of Whipple's disease

Technique	Sample	Inconvenience	Advantage
Electron microscopy	Biopsy, aspirate fluid	Necessity to have an apparatus	Unknown sensitivity
Periodic acid–Schiff staining	Biopsy, aspirate fluid	Low specificity	Easy to perform Retrospective
Genomic detection	Biopsy, aspirate fluid Blood	Unknown specificity	Available in most laboratories
Immunodetection	Biopsy, aspirate fluid Blood	Non-commercialized antibodies	Specificity Sensitivity Retrospective
Culture	Biopsy, aspirate fluid, blood	Cell-culture technology Long time for the primary isolation	New strain isolation
Serology	Serum	Low specificity Presently experimental	Noninvasive procedure

Nonspecific Diagnosis

Blood

The blood may show inconstant leucocytosis, hypochromic anemia and a very inconstant thrombocytosis (Dobbins, 1987b).

For unknown reasons, a significant percentage of patients develop hypereosinophilia (Dobbins, 1987b). An inflammatory syndrome can also be observed (Dobbins, 1987b; Dutly and Altwegg, 2001; Fenollar and Raoult, 2001b). Malabsorption signs (hypoalbuminemia, hypocholesterolemia, vitaminic defects and hypocalcemia) could be present (Dobbins, 1987b; Dutly and Altwegg, 2001; Fenollar and Raoult, 2001b). Abnormalities of Ig levels are sometimes present. The fat excretion in stools can be increased (Dobbins, 1987b) and the D-xylose absorption reduced (Dobbins, 1987b).

Histological Diagnosis

Electron Microscopy Diagnosis

Electron microscopy can be performed on various biopsies (Fenollar and Raoult, 2001b). The analysis allows the detection of bacteria in the macrophages (Fenollar and Raoult, 2001b). This technique is not available in most laboratories, and there have been no studies concerning the sensitivity of this technique in comparison with histology or PCR and therefore cannot be considered as an established tool for the diagnosis of the disease.

Specific Staining

Histology was the first tool used for the diagnosis of Whipple's disease. Classically, the diagnosis is performed with a duodenal (or jejunal) biopsy, with the observation, using PAS staining, in the mucosa, of spumous macrophages containing purple material. Owing to a sparse involvement, several duodenal biopsies may be necessary to establish the diagnosis. Since the organisms are localized to the submucosa, the diagnosis could be missed if the biopsies are too superficial (Fenollar and Raoult, 2001b). Depending on the clinical manifestation, other samples could be tested, such as cardiac valves, adenopathies, muscle, lung, liver, spleen, synovial biopsies, synovial liquid, bone marrow or cerebrospinal fluid (CSF) (Fleming, Wiesner and Shorter, 1988; Dutly and Altwegg, 2001; Fenollar and Raoult, 2001b). However, PAS staining, which has been considered for a long time as diagnostic in Whipple's disease, is not completely specific. Patients with infections due to *Actinomyces*, *Mycobacterium avium-intracellulare*, *Rhodococcus equi*, *Bacillus cereus*, *Corynebacterium* sp., *Histoplasma* or *Fungi* could have PAS-positive macrophages in their samples (Dobbins, 1987b; Misbah and Mapstone, 2000; Fenollar and Raoult, 2001b). The diagnosis of Whipple's disease can be excluded when a positive Ziehl–Neelsen staining is obtained. In patients suffering from melanosis coli, histiocytosis, macroglobulinemia and Crohn's disease, the diagnosis can be confused with Whipple's disease (Marth and Raoult, in press; Dobbins, 1987b; Misbah and Mapstone, 2000; Fenollar and Raoult, 2001b). Gastric or rectal biopsies are not adequate for the diagnosis of Whipple's disease. Indeed, weakly PAS-positive macrophages can be observed in the stomach and the rectum of healthy people (Dutly and Altwegg, 2001). Furthermore, mononuclear cells from tissues and synovial liquid may also contain nonspecific PAS-positive cells (O'Duffy *et al.*, 1999). The involvement of lymphatic tissues, gastrointestinal tract and other organs such as the liver or lungs could be associated with the presence of granulomas composed of non-caseous epithelioid cells. Of note, these granulomas associated with Whipple's disease can be PAS negative (Cho *et al.*, 1984). In these latter cases, electron microscopy, immunohistochemistry or PCR is then necessary to confirm the diagnosis of Whipple's disease. These data suggest that the diagnosis of Whipple's disease should not be exclusively based on the presence of PAS-positive macrophages and should not be completely excluded because of the negativity of the PAS staining.

Specific Diagnosis

Immunohistochemical Diagnosis

Immunohistochemistry performed using rabbit polyclonal antibodies directed against *T. whipplei* can be performed on various biopsies (duodenum, cardiac valve, adenopathy and brain), on various liquid puncture (aqueous humor) and blood monocytes (Raoult *et al.*, 2000, 2001; Raoult, Lepidi and Harlé, 2001; Baisden *et al.*, 2002; Lepidi *et al.*, 2002; Dumler, Baisden and Raoult, 2003). The use of polyclonal and monoclonal antibodies (Liang, La Scola and Raoult, 2002) has demonstrated good results. This technique is sensitive and specific for the diagnosis of Whipple's disease. The technique can be performed with previously fixed samples. We stained biopsies from Dr Whipple's patient nearly one century later (Dumler, Baisden and Raoult, 2003). However, it is necessary to possess the specific antibodies, which are not commercially available. This method offers added advantage in terms of specificity over PCR methods, owing to the direct visualization of bacilli and antigens within cells in tissue sections, and offers increased sensitivity and specificity over the traditional PAS method.

Molecular Biology

Because physicians do not often consider Whipple's disease in their differential diagnosis, PCR targeting broad-spectrum 16S rDNA is useful. PCR can be performed on various biopsies (adenopathy, cardiac valve, kidney, brain and synovial) and samples (CSF, articular liquid and aqueous humor blood). PCR on duodenum saliva or stools uses primers targeting different genes of the bacterium (Fenollar and Raoult, 2001a). The hypothesis that the diagnostic PCR of the Whipple's disease could be performed exclusively using DNA extracted from blood has been raised (Lowsky *et al.*, 1994; Marth *et al.*, 1996). However, blood does not appear to be a good sample for this technique because of the likely presence of PCR inhibitors (Lowsky *et al.*, 1994; Marth *et al.*, 1996). For example, Marth *et al.* (1996)

did not succeed in finding some evidence of *T. whipplei* DNA in the blood of four patients with active Whipple's disease (confirmed by a PAS positive on intestinal biopsies). However, this technique has at times contributed to disease diagnosis (Messori *et al.*, 2001; Peters, du Plessi and Humphrey, 2002). It has notably allowed the diagnosis of Whipple's disease in a patient with neurological disease without digestive involvement (Jiménez Carmena *et al.*, 2000). The utility of saliva and stool samples remains to be determined (Gross *et al.*, 1959; Gross, Jung and Zoller, 1999; Dutly *et al.*, 2000; Maibach, Dutly and Altwegg, 2002).

The DNA extraction is a key step, and different protocols have been proposed (Fenollar and Raoult, 2001a). These protocols can be applied using a fresh sample or one fixed in paraffin. The sequence targets presently used for the molecular diagnosis of Whipple's disease are the 16S rDNA, the intergenic region 16S–23S, the 23S rDNA, the *rpoB* gene and the *hsp65* gene (Fenollar and Raoult, 2001a). The availability of the full genome of the bacterium should allow the optimization of the PCR diagnosis. Currently, if an amplification product is obtained by PCR, it should then be confirmed by sequencing. The PCR sensitivity depends on the primers used and on the number of cycles (Fenollar and Raoult, 2001a, b). The amplification of short fragments, compared to long fragments, seems to be more sensitive (Rickman *et al.*, 1995; Lynch *et al.*, 1997; Ramzan *et al.*, 1997). The use of fresh clinical samples obtains better results in comparison with fixed samples comprising partially degraded DNA (Ramzan *et al.*, 1997). Some studies have shown that PCR performed in intestinal samples is more sensitive than the histological studies. A greater sensitivity has also been reported in CSF (von Herbay *et al.*, 1997a). The main problem with PCR is specificity, due notably to the problem of contamination, depending on the technique used and the samples studied. PCR seemed to be a technique full of promise. Positive results have been reported in saliva, gastric liquids and duodenal biopsies in patients without suspicion of Whipple's disease (Ehrbar *et al.*, 1999; Street, Donoghue and Neild, 1999; Dutly *et al.*, 2000), but these results have not been confirmed by others (Maiwald *et al.*, 2001; Fenollar *et al.*, 2002). The utilization of PCR has simplified the laboratory diagnosis, principally for patients with neurological involvement. The PAS staining could necessitate invasive procedures such as a cerebral biopsy to confirm the diagnosis, whereas PCR can be more simply performed on the CSF (Cohen *et al.*, 1996; von Herbay *et al.*, 1997a). The number of patients with atypical Whipple's disease is rising (Rickman *et al.*, 1995; Misbah *et al.*, 1997; Averbuch-Heller *et al.*, 1999; Dutly and Altwegg, 2001; Fenollar and Raoult, 2001b; Akar *et al.*, 2002; Peters, du Plessi and Humphrey, 2002). For a definite molecular diagnosis of Whipple's disease, especially with atypical cases, we suggest that at least two sequenced amplicons from two pairs of primers be obtained from two different genes to avoid false-positive results due to contamination.

Culture

Cell cultures from diverse biopsies, puncture liquids or blood can be performed using human fibroblasts with MEM with 10% fetal calf serum and 2 mM glutamine incubated at 37 °C under a 5% CO_2 atmosphere (La Scola *et al.*, 2001). However, the culture cannot become a tool for the diagnosis unless there is an improvement in culture conditions. One of the major problems is the delay in primary isolation. The first isolate of *T. whipplei* was obtained from a cardiac valve. It demonstrated a cytopathic effect after 2 months. Indeed, the bacterium multiplies slowly in human fibroblasts (MRC5 and HEL cells), HeLa cells and peripheral blood monocytes (Relman *et al.*, 1992). In addition, the culture of the bacterium from contaminated samples such as duodenal biopsies requires the use of a mix of antibiotics (Raoult *et al.*, 2001). Its limitations are the long generation time of these bacteria, the necessity to have qualified personnel and the capacities of cell-culture techniques.

Serology

The recent culture of *T. whipplei* has allowed preliminary testing of serologic techniques, based on indirect immunofluorescence, on a series of patients with Whipple's disease and a control group (Raoult *et al.*, 2000). IgG antibodies greater than or equal to 1/100 were observed so frequently in the control subjects that they were not useful for diagnosis. The presence of IgG in most subjects, independent of the presence or the absence of Whipple's disease, suggests the spread of the microorganism or the frequency of cross-reactivity (Ehrbar *et al.*, 1999; Street, Donoghue and Neild, 1999; Dutly *et al.*, 2000; Hinrikson, Dutly and Altwegg, 2000b). IgM antibodies are more specific. Titers greater than or equal to 1/50 were present in 5 of 7 patients with a classical Whipple's disease, in 2 patients with Whipple's disease endocarditis, in none of 10 patients with an endocarditis other than those of Whipple's disease, in 2 of 9 patients with an autoimmune disease and, lastly, in 1 of 20 blood donors. However, after several subcultures, there was an antigenic modification of the bacterium, leading to a loss of specificity (Marth and Raoult, in press).

For the serological diagnosis of Whipple's disease, another approach has been reported with a recombinant protein *hsp65* by using an enzyme-linked immunoadsorbent assay (ELISA) (Morgenegg *et al.*, 2001). Unfortunately, the results observed for 10 patients with Whipple's disease and 89 control subjects did not show significant differences (Morgenegg *et al.*, 2001). Whipple's disease serology is not presently achievable, but only experimental.

TREATMENT

Whipple's disease was always fatal before the use of antibiotic therapy. However, after the introduction of an antibiotic therapy, a rapid improvement in the patient's clinical state was observed. Diarrhea and fever disappear in less than 1 week, and the arthralgia and the other symptoms improve within several weeks (Marth and Raoult, in press). The neurological disturbances take a longer time to regress. The treatment of Whipple's disease has been established empirically, in the absence of data concerning the *in vitro* antibiotic susceptibilities of *T. whipplei*. The rarity of the disease has notably not allowed randomized trials to determine the best drugs to use and the optimal length of treatment. In 1952, the first antibiotic used, chloramphenicol, was responsible for a regression of the symptoms, with the disappearance of histological abnormalities. Since then, several antibiotic treatments have been prescribed with success, including penicillin, penicillin in combination with streptomycin, ampicillin, chloramphenicol, tetracycline and trimethoprim–sulfamethoxazole (Dobbins, 1987b; Fleming, Wiesner and Shorter, 1988; Vital-Durand *et al.*, 1997). Until 1970, the antibiotic regimen used most frequently was a combination of penicillin and streptomycin using parenteral administration, at the initiation of treatment, followed by per os tetracyclines. However, the evidence of relapses, which could occur late and involve the CNS despite prolonged antibiotic therapy, demonstrated a limitation of tetracycline-based treatment (Dutly and Altwegg, 2001). The treatment has been reoriented toward antibiotics, with good penetration into the brain (Dutly and Altwegg, 2001), and the prescription of trimethoprim–sulfamethoxazole was suggested (Wang and Prober, 1983). The rate of relapse, after antibiotic therapy, is estimated to be from 2 to 33%, depending on the initial treatment (Dutly and Altwegg, 2001). When performing a synthesis of several studies, the rate of relapse can be more precisely evaluated at 2% (1 of 46) for patients treated with trimethoprim–sulfamethoxazole and at 25% (33 of 133) for patients treated with tetracycline (Fleming, Wiesner and Shorter, 1988; Keinath *et al.*, 1985; Feurle and Marth, 1994; Vital-Durand *et al.*, 1997). The rate of clinical relapse seems to be lower after treatment based on trimethoprim–sulfamethoxazole. Currently, the recommended treatment is a per os trimethoprim (160 mg) and sulfamethoxazole (800 mg) two times per day, preceded with an intra-

venous treatment of 14 days based either on streptomycin (1 g/d) and benzylpenicillin (penicillin G: 1.2 million U/day) or on trimethoprim–sulfamethoxazole for a duration of 1–2 years (Keinath *et al.*, 1985; Dobbins, 1987b; Fleming, Wiesner and Shorter, 1988; Feurle and Marth, 1994; Fantry and James, 1995). The optimum length of treatment duration has not been determined. Antibiotic therapies of short duration (few weeks to few months) are thought to be sufficient by some authors (Fleming, Wiesner and Shorter, 1988; Dutly and Altwegg, 2001), and the recommendations for a longer treatment are based on prudence rather than on evidence base. The objective of treatment is to avoid neurological relapses described after short treatment durations (Dutly and Altwegg, 2001). For patients allergic to trimethoprim–sulfamethoxazole or for those with no response to the treatment, chloramphenicol, cefixime or ceftriaxone could be used (Vital-Durand *et al.*, 1997; Dutly and Altwegg, 2001; Fenollar and Raoult, 2001b).

The follow-up strategy for patients with Whipple's disease is not yet established. At the end of the antibiotic therapy, PCR and PAS staining of affected tissues could be useful for detecting a possible relapse (Marth and Strober, 1996; von Herbay *et al.*, 1997a; Dutly and Altwegg, 2001; Fenollar and Raoult, 2001b). If a good clinical response is observed, it is sufficient to follow the patient by performing a duodenal biopsy at 6 months and 1 year after the diagnosis, and the antibiotic treatment can be stopped if no PAS-positive material is observed. In a few cases, where the PAS-positive material persists, treatment must be continued or an alternative therapy could be considered. An increase in PAS-positive material may be an indicator of relapse (Geboes *et al.*, 1992). The results of PCR after the onset of antibiotic treatment can quickly become negative (Müller *et al.*, 1997; Ramzan *et al.*, 1997; Pron *et al.*, 1999). These results are probably linked to the degradation of bacterial DNA before the resolution of the rigid wall of Gram-positive bacteria (von Herbay, Ditton and Maiwald, 1996; von Herbay *et al.*, 1996, 1997a; Petrides *et al.*, 1998). However, the value of the PCR following treatment is controversial (von Herbay, Ditton and Maiwald, 1996; Müller *et al.*, 1997; Ramzan *et al.*, 1997; Brändle *et al.*, 1999). If PCR becomes negative a few weeks after the initiation of the treatment, then it seems to suggest the efficiency of the treatment but not the eradication of the microorganism. The neurological manifestations of Whipple's disease have become the most common symptoms during relapses and have a bad prognosis. The follow-up of patients includes CSF analysis every 6 months.

The recent discovery of a possible defect in cellular immunity in Whipple's disease could lead to a search for new therapeutic approaches. Indeed, it should be possible to combine antibiotic therapy with immunotherapy using IFN-γ. IFN-γ plays an important role in the control of bacterial infection (Gallin *et al.*, 1995). This therapeutic option has yet to be applied. IFN-γ has been used with success in the treatment of a patient with Whipple's disease presenting resistance to presumed efficacious antibiotics (Schneider *et al.*, 1998). The choice of IFN-γ as a therapeutic agent is based on the observation of a deficit in the production of IFN-γ and IL-12 *in vitro* by the monocytes of patients with Whipple's disease (Marth *et al.*, 1997), and it has been employed successfully in the treatment of other infections due to intracellular microorganisms (Holland *et al.*, 1994; Gallin *et al.*, 1995). It is suggested that the utilization of IFN-γ in association with antibiotics could then diminish the number of relapses, but more studies are necessary to prove the real efficacy of the treatment.

REFERENCES

Adams, M., Rhyner, P., Day, J. *et al.* (1987) Whipple's disease confined to the central nervous system. Ann Neurol **21**:104–108.

Akar, Z., Tanriover, N., Tuzgen, S. *et al.* (2002) Intracerebral Whipple disease: unusual location and bone destruction. J Neurosurg **97**:988–991.

Alkan, S., Beals, T. and Schnitzer, B. (2001) Primary diagnosis of Whipple disease manifesting as lymphadenopathy. Use of polymerase chain reaction for detection of *Tropheryma whippeli*. Am J Clin Pathol **116**:898–904.

Altwegg, M., Fleisch-Marx, A., Goldenberger, D. *et al.* (1996) Spondylodiscitis caused by *Tropheryma whippeli*. Schweiz Med Wochenschr **126**:1495–1499.

Averbuch-Heller, L., Paulson, G., Daroff, R. and Leight, J. (1999) Whipple's disease mimicking progressive supranuclear palsy: the diagnostic value of eye movement recording. J Neurol Neurosurg Psychiatry **66**:532–535.

Bai, J., Mota, A., Mauriño, E. *et al.* (1991) Class I and class II HLA antigens in an homogeneous Argentinian population with Whipple's disease: lack of association with HLA-B27. Am J Gastroenterol **86**:992–994.

Baisden, B., Lepidi, H., Raoult, D. *et al.* (2002) Diagnosis of Whipple disease by immunohistochemical analysis. A sensitive and specific method for the detection of *Tropheryma whipplei* (the Whipple Bacillus) in paraffin-embedded tissue. Am J Clin Pathol **118**:742–748.

Bjerknes, R., Odegaard, S., Bjerkvig, R. *et al.* (1988) Whipple's disease—demonstration of a persisting monocyte and macrophage dysfunction. Scand J Gastroenterol **23**:611–619.

Black-Schaffer, B. (1949) Tinctorial demonstration of a glycoprotein in Whipple's disease. Proc Soc Exp Biol Med **72**:225–227.

Brändle, M., Ammann, P., Spinas, G. *et al.* (1999) Relapsing Whipple's disease presenting with hypopituitarism. Clin Endocrinol **50**:339–403.

Chan, R., Yannuzzi, L. and Foster, C. (2001) Ocular Whipple's disease. Earlier definitive diagnosis. Ophthalmology **108**:2225–2231.

Chears, W. and Ashworth, C. (1961) Electron microscopy study of the intestinal mucosa in Whipple's disease—demonstration of encapsulated bacilliform bodies in the lesion. Gastroenterology **41**:129–138.

Cho, C., Linscheer, W., Hirschkorn, M. and Ashutosh, K. (1984) Sarcoidlike granulomas as an early manifestation of Whipple's disease. Gastroenterology **87**:941–947.

Cohen, A., Schimmel, E., Holt, P. and Isselbacher, K. (1960) Ultrastructural abnormalities in Whipple's disease. Proc Soc Exp Biol Med **105**:411–414.

Cohen, L., Berthet, K., Dauga, C. *et al.* (1996) Polymerase chain reaction of cerebrospinal fluid to diagnose Whipple's disease. Lancet **347**:329.

Di Stefano, M., Jorizzo, R., Brusco, G. *et al.* (1998) Bone mass and metabolism in Whipple's disease: the role of hypogonadism. Scand J Gastroenterol **33**:1180–1185.

Dobbins, W. O., III (1981) Is there an immune deficit in Whipple's disease? Dig Dis Sci **26**:247–252.

Dobbins, W. O., III (1987a) HLA antigens in Whipple's disease. Arthritis Rheum **30**:102–105.

Dobbins, W.O., III (1987b) Whipple's Disease. Thomas, Springfield, IL.

Dobbins, W. O., III (1988) Whipple's disease. Mayo Clin Proc **63**:623–628.

Dobbins, W. O., III and Ruffin, J. (1967) A light- and electron-microscopic study of bacterial invasion in Whipple's disease. Am J Pathol **51**:225–242.

Du Boulay, C. (1982) An immunohistochemical study of Whipple's disease using immunoperoxidase technique. Hum Pathol **13**:925–929.

Dumler, J., Baisden, S. and Raoult, D. (2003) Immunodetection of *Tropheryma whipplei* intestinal tissues of Dr Whipple's 1907 patient. N Engl J Med **348**:1411–1412.

Dutly, F. and Altwegg, M. (2001) Whipple's disease and "*Tropheryma whippeli*". Clin Microbiol Rev **14**:561–583.

Dutly, F., Hinrikson, H., Seidel, T. *et al.* (2000) *Tropheryma whippeli* DNA in saliva of patients without Whipple's disease. Infection **28**:219–222.

Eck, M., Kreipe, H., Harmsen, D. and Muller-Hermelink, H. (1997) Invasion and destruction of mucosal plasma cells by *Tropheryma whippeli*. Hum Pathol **28**:1424–1428.

Ectors, N., Geboes, K., Rutgeerts, P. *et al.* (1992) RFD7–RFD9 co-expression by macrophages point to cell–macrophage interaction deficiency in Whipple's disease. Gastroenterology **106**:A676.

Ehrbar, H., Bauerfeind, P., Dutly, F. *et al.* (1999) PCR-positive tests for *Tropheryma whippeli* in patients without Whipple's disease. Lancet **353**:2214.

Elkins, C., Shuman, T. and Pirolo, J. (1999) Cardiac Whipple's disease without digestive symptoms. Ann Thorac Surg **67**:250–251.

Enzinger, F. and Helwig, E. (1963) Whipple's disease: a review of the literature and report fifteen patients. Virchows Arch Pathol Anat **336**:238–269.

Evans, D. and Ali, M. (1985) Immunocytochemistry in the diagnosis of Whipple's disease. J Clin Pathol **38**:372–374.

Fantry, G. and James, S. (1995) Whipple's disease. Dig Dis **13**:108–118.

Fenollar, F. and Raoult, D. (2001a) Molecular techniques in Whipple's disease. Expert Rev Mol Diagn **1**:299–309.

Fenollar, F. and Raoult, D. (2001b) Whipple's disease. Clin Diagn Lab Immunol **8**:1–8.

Fenollar, F., Lepidi, H. and Raoult, D. (2001) Whipple's endocarditis: review of the literature and comparisons with Q fever, *Bartonella* infection, and blood culture-positive endocarditis. Clin Infect Dis **33**:1309–1316.

Fenollar, F., Fournier, P., Gerolami, R. *et al.* (2002) Quantitative detection of *Tropheryma whipplei* DNA by Real-Time PCR. J Clin Microbiol **40**:1119–1120.

Feurle, G. and Marth, T. (1994) An evaluation of antimicrobial treatment for Whipple's disease—tetracycline versus trimethoprim–sulfamethoxazole. Dig Dis Sci **39**:1642–1648.

Fleming, J., Wiesner, R. and Shorter, R. (1988) Whipple's disease: clinical, biochemical, and histopathologic features and assessment of treatment in 19 patients. Mayo Clin Proc **63**:539–551.

Fox, G., Wisotzkey, J. and Jurtshuk, P. (1992) How close is close: 16S rRNA sequence identity may not be sufficient to guarantee species identity. Int J Syst Bacteriol **42**:166–170.

Fredericks, D. and Relman, D. (2001) Localization of *Tropheryma whippeli* rRNA in tissues from patients with Whipple's. J Infect Dis **183**:1229–1237.

Frésard, A. G. C. B. P. (1996) Prosthetic joint infection caused by *Tropheryma whippeli* (Whipple's bacillus). Clin Infect Dis **22**:575–576.

Gallin, J., Farber, J., Holland, S. and Nutman, T. (1995) Interferon-gamma in the management of infectious diseases. Ann Intern Med **123**:216–224.

Geboes, K., Ectors, N., Heidbuchel, H. *et al.* (1992) Whipple's disease: the value of upper gastrointestinal endoscopy for the diagnosis and follow-up. Acta Gastroenterol Belg **55**:209–219.

Gerard, A., Sarrot-Reynaud, F., Liozon, E. *et al.* (2002) Neurologic presentation of Whipple. Report of 12 cases and review of the literature. Medicine **81**:443–457.

Ghigo, E., Capo, C., Aurouze, M. *et al.* (2002) The survival of *Tropheryma whipplei*, the agent of Whipple's disease, requires phagosome acidification. Infect Immun **70**:1501–1506.

Gross, J., Wollaeger, E., Sauer, W. *et al.* (1959) Whipple's disease: report of four cases, including two brothers, with observations on pathologic physiology, diagnosis and treatment. Gastroenterology **36**:65–93.

Gross, M., Jung, C. and Zoller, W. (1999) Detection of *Tropheryma whippeli* (Whipple's disease) in faeces. Ital J Gastroenterol **31**:70–72.

Gürtler, V. and Stanisich, V. (1996) New approaches to typing and identification of bacteria using the 16S–23S rDNA spacer region. Microbiology **142**:3–16.

Harmsen, D., Heesemann, J., Brabletz, T. *et al.* (1994) Heterogeneity among Whipple's-disease-associated bacteria. Lancet **343**:1288.

Hinrikson, H., Dutly, F. and Altwegg, M. (1999) Homogeneity of 16S–23S ribosomal intergenic spacer regions of *Tropheryma whippeli*. Swiss patients with Whipple's disease. J Clin Microbiol **37**:152–156.

Hinrikson, H., Dutly, F., Nair, S. and Altwegg, M. (1999) Detection of three different types of "*Tropheryma whippeli*" directly from clinical specimens by sequencing, single-strand conformation polymorphism (SSCP) analysis and type-specific type PCR of their 16S–23S ribosomal intergenic spacer region. Int J Syst Bacteriol **49**:1701–1706.

Hinrikson, H., Dutly, F. and Altwegg, M. (2000a) Analysis of the actinobacterial insertion in domain III of the 23S rRNA gene of uncultured variants of the bacterium associated with Whipple's disease using broad-range and "*Tropheryma whippeli*"-specific PCR. Int J Syst Evol Microbiol **50**:1007–1011.

Hinrikson, H., Dutly, F. and Altwegg, M. (2000b) Evaluation of a specific nested PCR targeting domain III of the 23S rRNA gene of "*Tropheryma whippeli*" and proposal of a classification system for its molecular variants. J Clin Microbiol **38**:595–599.

Holland, S., Eisenstein, E., Kuhns, D. *et al.* (1994) Treatment of refractory disseminated non-tuberculosis mycobacterial infection with interferon-gamma—a preliminary report. N Engl J Med **330**:1348–1355.

Jiménez Carmena, J., Jiménez, M., Mena, F. *et al.* (2000) Whipple's disease with isolated central nervous system symptomatology diagnosed by molecular identification of *Tropheryma whippeli* in peripheral blood. Neurologia **15**:173–176.

Keinath, R., Merrell, D., Vlietstra, R. and Dobbins, W. I. (1985) Antibiotic treatment and relapse in Whipple's disease. Gastroenterology **88**:1867–1873.

Keren, D. (1981) Whipple's disease: a review emphasizing immunology and microbiology. Crit Rev Clin Lab Sci **14**:75–108.

Khairy, P. and Graham, A. (1996) Whipple's disease and the heart. Can J Cardiol **12**:831–834.

La Scola, B., Fenollar, F., Fournier, P. *et al.* (2001) Description of *Tropheryma whipplei* gen. nov., sp. nov., the Whipple's disease bacillus. Int J Syst Evol Microbiol **51**:1471–1479.

Lepidi, H., Costedoat, N., Piette, J. *et al.* (2002) Immunohistological detection of *Tropheryma whipplei* (Whipple bacillus) in lymph nodes. Am J Med **113**:334–336.

Levin, M., Newport, M., D'Souza, S. *et al.* (1995) Familial disseminated atypical mycobacterial infection in childhood: a human mycobacterial susceptibility gene? Lancet **345**:79.

Liang, Z., La Scola, B. and Raoult, D. (2002) Monoclonal antibodies to immunodominant epitope of *Tropheryma whipplei.* Clin Diagn Lab Immunol **9**:156–159.

Louis, E., Lynch, T., Kaufmann, P. *et al.* (1996) Diagnostic guidelines in central nervous system Whipple's disease. Ann Neurol **40**:561–568.

Lowsky, R., Archer, J., Fyles, G. *et al.* (1994) A diagnosis of Whipple's disease by molecular analysis of peripheral blood. N Engl J Med **331**:1343–1346.

Lukacs, G., Dobi, S. and Szabo, M. (1978) A case of Whipple's disease with repeated operations for ileus and complete cure. Acta Hepatogastroenterol (Stuttg) **25**:238–242.

Lynch, T., Odel, J., Fredericks, D. *et al.* (1997) Polymerase chain reaction-based detection of *Tropheryma whippeli* in central nervous system Whipple's disease. Ann Neurol **42**:120–124.

MacDermott, R., Shepard, J., Graemecook, F. and Bloch, K. (1997) A 59-year-old man with anorexia, weight loss, and a mediastinal mass: Whipple's-disease involving the small intestine and mesenteric and mediastinal lymph nodes. N Engl J Med **337**:1612–1619.

Maibach, R., Dutly, F. and Altwegg, M. (2002) Detection of *Tropheryma whipplei* DNA in feces by PCR using a target capture method. J Clin Microbiol **40**:2466–2471.

Maiwald, M. and Relman, D. (2001) Whipple's disease and *Tropheryma whippeli*: secrets slowly revealed. Clin Infect Dis **32**:457–463.

Maiwald, M., Ditton, H., von Herbay, A. *et al.* (1996) Reassessment of the phylogenetic position of the bacterium associated with Whipple's disease and determination of the 16S–23S ribosomal intergenic spacer sequence. Int J Syst Bacteriol **46**:1078–1082.

Maiwald, M., Schuhmacher, F., Ditton, H. and von Herbay, A. (1998) Environmental occurrence of the Whipple's disease bacterium (*Tropheryma whippeli*). Appl Environ Microbiol **64**:760–762.

Maiwald, M., von Herbay, A., Lepp, P. and Relman, D. (2000) Organization, structure, and variability of the rRNA operon of the Whipple's disease's bacterium (*Tropheryma whippeli*). J Bacteriol **182**:3292–3297.

Maiwald, M., von Herbay, A., Persing, D. *et al.* (2001) *Tropheryma whippeli* DNA is rare in the intestinal mucosa of patients without other evidence of Whipple disease. Ann Intern Med **134**:115–119.

Maizel, H., Ruffin, J. and Dobbins, W. I. (1970) Whipple's disease: a review of 19 patients from one hospital and a review of the literature since 1950. Medicine **72**:343–355.

Marie, I., Lecomte, F. and Levesque, H. (2000) Granulomatous nephritis as the first manifestation of Whipple disease. Ann Intern Med **132**:94–95.

Marth, T. (1999) Whipple's disease. In Mandell, Dolin and Bennett (eds), Principles and Practice of Infectious Disease. Churchill Livingstone, Philadelphia. pp. 1170–1174.

Marth, T. and Raoult, D. (2003) Whipple's disease. Lancet **361**: 239–246.

Marth, T. and Strober, W. (1996) Whipple's disease. Semin Gastrointest Dis **7**:41–48.

Marth, T., Roux, A., von Herbay, A. *et al.* (1994) Persistent reduction of complement receptor 3 a-chain expressing mononuclear blood cells and transient inhibitory serum factors in Whipple's disease. Clin Immunol Immunopathol **72**:217–226.

Marth, T., Fredericks, D., Strober, W. and Relman, D. (1996) Limited role for PCR-based diagnosis of Whipple's disease from peripheral blood mononuclear cells. Lancet **348**:66–67.

Marth, T., Neurath, M., Cuccherini, B. and Strober, W. (1997) Defects of monocyte interleukin-12 production and humoral immunity in Whipple's disease. Gastroenterology **113**:442–448.

Marth, T., Kleen, N., Stallmach, A. *et al.* (2002) Dysregulated peripheral and mucosal Th1/Th2 response in Whipple's disease. Gastroenterology **123**:1468–1477.

Maxwell, J., Ferguson, A., McCay, A. *et al.* (1968) Lymphocytes in Whipple's disease. Lancet **I**:887–889.

McAllister, H. J. and Fenoglio, J. (1975) Cardiac involvement in Whipple's disease. Circulation **52**:152–156.

Messori, A., Di Bella, P., Polonara, G. *et al.* (2001) An unusual spinal presentation of Whipple disease. Am J Neuroradiol **22**:1004–1008.

Misbah, S. and Mapstone, N. (2000) Whipple's disease revisited. J Clin Pathol **53**:750–755.

Misbah, S., Ozols, B., Franks, A. and Mapstone, N. (1997) Whipple's disease without malabsorption: new atypical features. Q J Med **90**:765–772.

Morgenegg, S., Maibach, R., Chaperon, D. *et al.* (2001) Antibodies against recombinant heat shock protein 65 of *Tropheryma whipplei* in patients with and without Whipple's disease. J Microbiol Methods **47**:299–306.

Müller, C., Petermann, D., Stain, C. *et al.* (1997) Whipple's disease: comparison of histology with diagnosis based on polymerase chain reaction in four consecutive cases. Gut **40**:425–427.

Mur Villacampa, M. (1987) Presentacion inusual de la enfermedad de Whipple. Rev Esp Enferm Dig **72**:183.

Neumann, K., Neumann, V., Zierz, S. and Lahl, R. (1997) Coinfection with *Tropheryma whippeli* and a Whipple's disease-associated bacterial organism detected in a patient with central nervous system Whipple's disease. J Clin Microbiol **35**:1645.

O'Duffy, J., Griffing, W., Li, C. *et al.* (1999) Whipple arthritis. Arthritis Rheum **42**:812–817.

Olivieri, I., Brandi, G., Padula, A. *et al.* (2001) Lack of association with spondyloarthritis and HLA-B27 in Italian patients with Whipple's disease. J Rheumatol **28**:1294–1297.

Peters, G., du Plessis, D. and Humphrey, P. (2002) Cerebral Whipple's disease with a stroke-like presentation and cerebrovascular pathology. J Neurol Neurosurg Psychiatry **73**:336–339.

Petrides, P., Müller-Höcker, J., Fredricks, D. and Relman, D. (1998) PCR analysis of *T. whippeli* DNA in a case of Whipple's disease: effect of antibiotics and correlation with histology. Am J Gastroenterol **93**:1579–1582.

Pron, B., Poyart, C., Abachin, T. *et al.* (1999) Diagnosis and follow-up of Whipple's disease by amplification of the 16S rRNA gene of *Tropheryma whippeli.* Eur J Clin Microbiol Infect Dis **18**:62–65.

Puéchal, X. (2001) Whipple's disease and arthritis. Curr Opin Rheumatol **13**:74–79.

Ramaiah, C. and Boynton, R. (1998) Whipple's disease. Gastroenterol Clin North Am **27**:683–695.

Ramzan, N., Loftus, E., Burgart, L. *et al.* (1997) Diagnosis and monitoring of Whipple disease by polymerase chain reaction. Ann Intern Med **126**:520–527.

Raoult, D., Birg, M., La Scola, B. *et al.* (2000) Cultivation of the bacillus of Whipple's disease. N Engl J Med **342**:620–625.

Raoult, D., La Scola, B., Lecocq, P. *et al.* (2001) Culture and immunological detection of *Tropheryma whippeli* from the duodenum of a patient with Whipple disease. JAMA **285**:1039–1043.

Raoult, D., Lepidi, H. and Harlé, J. (2001) *Tropheryma whipplei* circulating in blood monocytes. N Engl J Med **345**:548.

Relman, D., Schmidt, T., MacDermott, R. and Falkow, S. (1992) Identification of the uncultured bacillus of Whipple's disease. N Engl J Med **327**:293–301.

Rickman, L., Freeman, W., Green, W. *et al.* (1995) Uveitis caused by *Tropheryma whippeli* (Whipple's bacillus). Lancet **332**:363–366.

Roller, C., Ludwig, W. and Schleifer, K. (1992) Gram-positive bacteria with a high DNA G + C content are characterized by a common insertion within their 23S rRNA genes. J Gen Microbiol **138**:1167–1175.

Rouillon, A., Menkes, C., Gerster, J. *et al.* (1993) Sarcoid-like forms of Whipple's disease—report of 2 cases. J Rheumatol **20**:1070–1072.

Schneider, T., Stallmach, A., von Herbay, A. *et al.* (1998) Treatment of refractory Whipple disease with interferon-γ. Ann Intern Med **129**:875–877.

Schnider, P., Reisinger, E., Gerschlager, W. *et al.* (2002) Long-term follow-up in cerebral Whipple's disease. Eur J Gastroenterol Hepatol **8**:899–903.

Schoedon, G., Goldenberger, D., Forrer, R. *et al.* (1997) Deactivation of macrophages with interleukin-4 is the key to the isolation of *Tropheryma whippeli.* J Infect Dis **176**:672–677.

Schultz, M., Hartmann, A., Dietmaier, W. *et al.* (2002) Massive steatosis hepatis: an unusual manifestation of Whipple's disease. Am J Gastroenterol **97**:771–772.

Selsky, E., Knox, D., Maumenee, A. and Green, W. (1984) Ocular involvement in Whipple's disease. Retina **4**:103–106.

Silva, M., Macedo, P. and Moura Nunes, J. (1961) Ultrastructure of bacilli and the bacillary origin of the macrophagic inclusions in Whipple's disease. J Gen Microbiol **131**:1001–1013.

Stackebrandt, E. and Goebel, B. (1994) Taxonomic note: a place for DNA–DNA reassociation and 16S rRNA sequence analysis in the present species definition in bacteriology. Int J Syst Bacteriol **44**:846–849.

Street, S., Donoghue, H. and Neild, G. (1999) *Tropheryma whippeli* DNA in saliva of healthy people. Lancet **354**:1178–1179.

Varvöli, C., Buban, T., Szakall, S. *et al.* (2002) Fever of unknown origin with seronegative spondylarthropathy: an atypical manifestation of Whipple's disease. Ann Rheum Dis **61**:377–378.

Vital-Durand, D., Lecomte, C., Cathebras, P. *et al.* (1997) Whipple's disease—clinical review of 52 cases. The SNFMI Research Group On Whipple Disease. Societe Nationale Francaise de Medecine Interne. Medicine **76**:170–184.

von Herbay, A., Ditton, H. and Maiwald, M. (1996) Diagnostic application of a polymerase chain reaction assay for the Whipple's bacterium to intestinal biopsies. Gastroenterology **110**:1735–1743.

von Herbay, A., Maiwald, M., Ditton, H. and Otto, H. (1996) Histology of intestinal Whipple's disease revisited. A study of 48 patients. Virchows Arch **429**:335–343.

von Herbay, A., Ditton, H., Schuhmacher, F. and Maiwald, M. (1997a) Whipple's disease: staging and monitoring by cytology and polymerase chain reaction analysis of cerebral fluid. Gastroenterology **113**:434–441.

von Herbay, A., Otto, H., Stolte, M. *et al.* (1997b) Epidemiology of Whipple's disease in Germany. Analysis of 110 patients diagnosed in 1965–1995. Scand J Gastroenterol **32**:52–57.

Wang, E. and Prober, C. (1983) Ventricular cerebrospinal fluid concentrations of trimethoprim–sulfamethoxazole. J Antimicrob Chemother **11**:385–389.

Whipple, G. (1907) A hitherto undescribed disease characterized anatomically by deposits of fat and fatty acids in the intestinal and mesenteric lymphatic tissues. Bull Johns Hopkins Hosp **18**:382–393.

Williams, J., Edward, D., Tessier, H. et al. (1998) Ocular manifestations of Whipple disease – an atypical presentation. Arch Ophthalmol **116**:1232–1234.

Wilson, K., Blitchington, R., Rothingham, R. F. and Wilson, J. (1991) Phylogeny of the Whipple's-disease associated bacterium. Lancet **338**:474–475.

Wimberg, C., Rose, M. and Rappaport, H. (1978) Whipple's disease of the lung. Am J Med **65**:873–880.

Wolfert, A. and Wright, J. (1999) Whipple's disease presenting as sarcoidosis and valvular heart disease. South Med J **92**:820–825.

Yardley, J. and Hendrix, T. (1961) Combined electron and light microscopy in Whipple's disease-demonstration of 'bacillary bodies' in the intestine. Bull Johns Hopkins Hosp **109**:80–98.

27

Identification of Enterobacteriaceae

Peter M. Hawkey

Division of Immunity and Infection, The Medical School, Edgbaston, Birmingham, UK

The family Enterobacteriacae is the most widely studied family of organisms in the world. They have a worldwide distribution, and while largely found in animals, medical microbiologists are inclined to gain a distorted view of the family as they only encounter species associated with human disease, such as *Escherichia coli*, *Proteus mirabilis*, *Salmonella* spp. and *Klebsiella* spp. The genera *Erwinia* and *Pectobacteria* are major plant pathogens causing blights, wilts and rots in many different crops. *Yersinia ruckeri* is a major pathogen of farmed salmon, and the salmonellae are pathogens of cattle, sheep and poultry. Various species of Enterobacteriaceae have been isolated from the gut contents of animals ranging from fleas to elephants, some being adapted to very specific hosts, e.g. *P. myxofaciens* in the gut of gypsy moth larvae.

The family is numerically important to the medical microbiologist, as they may account for 80% of clinically significant Gram-negative bacilli and about 50% of isolates from cases of septicaemia. The family now has over 20 genera and more than 100 species, of which about 50 are definitely or probably associated with human disease (Farmer *et al.*, 1985). Taxonomy now makes use of formation on the relationship of isolates from examining a wide range of semantides (large information-bearing molecules). These are DNA, both in terms of sequences of genes and DNA/DNA hybridization studies, RNA, particularly 16S rRNA phylogenetic trees, and proteins both as functional enzymes and as electrophoretic cellular protein patterns. The synthesis of this information has led to the development of a clear and scientifically robust taxonomy of the Enterobacteriaceae. These developments are comprehensively and authoritatively reviewed by Farmer and colleagues (1985) in a review which, although now 20 years old, is still the most useful. Changes in taxonomy are often reluctantly accepted by medical microbiologists, but this is an evolving area and much of that which is familiar remains. I urge the reader to embrace the newer names and use them – they will become familiar in time. Although there appears to be a number of unfamiliar genera, such as *Ewingella*, *Obesumbacterium* and *Xenorhabdus*, these organisms are very rarely encountered. In clinical material 99% of isolates are represented by 23 species, whereas the remaining 1% belongs to 74 species (Farmer *et al.*, 1985). The adage from the authors, 'when you hear hoof beats, think horses, not zebras', is indeed true.

There are four reasons why clinical microbiologists may wish to identify microorganisms:

1. to help predict the likely outcome of the infection;
2. to identify potential cross-infection risks and cross-infection, retrospectively;
3. to attempt to predict likely sensitivity to antimicrobials;
4. to obtain research information on new disease associations with microorganisms.

The level to which an identification is made is very much a question of the clinical significance of the cultured bacteria. In the case of a moist, noninflamed surgical wound from which a mixture of faecal bacteria, such as enterococci, *E. coli*, *Klebsiella* spp., *P. mirabilis* and *Pseudomonas aeruginosa* are cultured, identification of the individual isolates of Enterobacteriaceae is inappropriate and a report of mixed faecal flora should be made. Isolates of Enterobacteriaceae from blood cultures, normally sterile sites (e.g., cerebrospinal fluid, peritoneum and deep tissues), infections at different sites requiring linking (e.g., isolates from a blood culture and urine), enteric pathogens (e.g. *Salmonella enteritidis* and *Shigella sonnei*) and if epidemiologically important (e.g., multi-resistant *K. pneumoniae*) should be identified to species level.

With the development of newer technologies the utilization of identification as a predictor of pathogenicity, for example, identification of *Shigella dysenteriae*, can be replaced by identification of the genes responsible for pathogenicity, regardless of the host background. Thus isolates of *E. coli* carrying Shiga toxin genes can be identified in 3 h with a greater sensitivity than biochemical methods (Iijima *et al.*, 2004). Species identification of some Enterobacteriaceae combined with antimicrobial sensitivity data may predict responses to certain agents. This is true of the use of cefotaxime/ceftazidine to treat bacteraemia and pneumonia caused by *Enterobacter* spp. because of the selection of derepressed mutants expressing the chromosomal gene *ampC* encoded by the β-lactamase found in that species (Stearne *et al.*, 2004).

There are other species-specific β-lactamase associations which have been reviewed elsewhere (Livermore, 1995). These predictions are not confined to β-lactams: *Serratia marcescens* was found to be the only species of Enterobacteriaceae that carries the *aac*(6')-Ic that encodes the aminoglycoside-inactivating enzyme AAC(6')-Ic, when 186 strains of 10 species of *Serratia* were probed with a specific DNA probe (Snelling *et al.*, 1993). The enzyme confers broad resistance to aminoglycosides, and aminoglycoside-sensitive strains of *S. marcescens* can express the gene fully after exposure, leading to treatment failure. Some species of Enterobacteriaceae are intrinsically resistant to certain antimicrobials, and this can be used to help in the presumptive identification of isolates in the clinical laboratory. Table 27.1 lists this information.

Suspicion that an isolate cultured from a clinical specimen belongs to the Enterobacteriaceae arises from the following characteristics:

- Gram-negative bacillus $0.5-2 \times 2-4\,\mu m$;
- cytochrome oxidase negative;
- ferments glucose;
- nitrates reduced to nitrites;
- facultative anaerobe.

Some species have characteristic appearances on solid media which may give a clue to identity: swarming – *Proteus* spp.; pronounced mucoid capsule – *Klebsiella/Enterobacter* spp. Definitive identification

Table 27.1   Intrinsic antimicrobial resistance encountered in clinically significant Enterobacteriaceae

Species	Agents to which most strains are resistant	Common chromosomal β-lactamase[a]
Citrobacter diversus	Ampicillin[b], cefuroxime, cephalothin[c], cefoxitin (piperacillin[d], ticarcillin[e])	K1 type[f]
Citrobacter freundii	Ampicillin, cephalothin, cefoxitin (piperacillin, cefotaxime, ceftazidime, aztreonam)	AmpC
Enterobacter spp.	Ampicillin, cephalothin, cefoxitin (piperacillin, cefotaxime, ceftazidime, aztreonam)	AmpC
Klebsiella pneumoniae	Ampicillin, ticarcillin	SHV-1
Klebsiella oxytoca	Ampicillin, ticarcillin (piperacillin, cefuroxime, cefotaxime, aztreonam)[g]	K1
Morganella morganii	Polymyxins, nitrofurantoin, ampicillin, cephalothin, cefuroxime (ticarcillin, piperacillin, cefotaxime, ceftazidime)	AmpC
Proteus mirabilis	Polymyxins, tetracycline, nitrofurantoin	
Proteus vulgaris/penneri	Polymyxins, nitrofurantoin, ampicillin, cefuroxime, cephalothin, cefoxitin (piperacillin, ticarcillin)	K1 type[h]
Providencia stuartii/rettgeri	As *M. morganii*, plus tetracycline except *Prov. stuartii* resistant to gentamicin/tobramycin	AmpC
Serratia marcescens	Polymyxins[i], ampicillin, cephalothin, cefuroxime (ticarcillin, piperacillin, cefotaxime)	AmpC

Resistance to agents shown in parentheses is only seen in hyperproducers of β-lactamase.
[a] Expression of these genes varies; so resistance in individual strains may vary according to induction or mutation to hyperproduction.
[b] Representative of amoxycillin.
[c] Representative of cephalexin, cefazolin and cephradine.
[d] Representative of azlocillin/mezlocillin.
[e] Representative of carbenicillin.
[f] Some strains have AmpC-like enzymes (Jones *et al.*, 1994).
[g] Hyperproduction of K1 only occurs in 10–20% of isolates, rarer than *Enterobacter* spp. AmpC.
[h] Cefuroximases classified by Bush, Jacoby and Medeiros (1995) as 2e.
[i] Frequently forms target zones around disc.

relies on the use of the fermentation of carbohydrate tests and a number of other tests to identify the presence or absence of specific enzymes, e.g. amino acid decarboxylases, urease and phenylalanine deaminase. Microbiologists at the end of the nineteenth century, such as Escherich and Gartner (who discovered *Salmonella enteritidis* in 1888) relied heavily on the use of agglutination tests to differentiate Enterobacteriaceae. It was the work of Theobald Smith and Herbert Durham, at the University of Cambridge, who appreciated the value of carbohydrate fermentation tests in identification (Durham, 1900). Using a range of fermentable vegetable extracts, as well as other tests, Durham recognized ten distinct types which accord broadly with many of the genera now recognized in the Enterobacteriaceae. Initially, laboratories prepared their own identification media, and the choice and quality control were highly individualistic. In the late 1960s, miniaturized, disposable, commercially prepared systems such as the API and Enterotube systems became available. These offered rapid, reliable identification and in the case of the API20E system used freeze-dried reagents that were rehydrated by inoculation with a bacterial suspension. Laboratories relied on either flow charts or tables for translating test results into a species identification (Barrow and Feltham, 1993). The widely used API20E system for the identification of Enterobacteriaceae translates the results from 21 individual tests into an octal number. The tests are arranged in groups of three and the positive and negative results for each are converted into a single octal, representing the results for that group of three tests, as summarized in Table 27.2. A 7 digit number is generated, which can be searched for on a computer database. The database can be derived from a large number of isolates, and a probability of correct identification can be given, taking into account unusual biochemical properties of biotypes of species. The API20E system has been in use for 30 years and, in a recent evaluation, was still found to provide a good level of accuracy of identification (78.7% at 24 h) which although slightly lower than earlier evaluations is acceptable (O'Hara, Rhoden and Miller, 1992).

Because of the special requirements for the identification of enteric pathogens, such as *Salmonella* spp. (Chapter 29) and *Shigella* spp. (Chapter 28), shortened sets of biochemical tests are used to screen isolates. These are described in the relevant chapter and range from single tubes of media, like triple sugar iron agar, to disposable cupule systems to detect preformed enzymes, such as APIZ. The practical details of using these various systems for identification are described in detail elsewhere (Pedler, 2004).

More recently, the principles enshrined in packaged disposable kits have been further developed into automated systems. Generally, these use pre-prepared cartridges of freeze-dried reagents which, after rehydration and inoculation, are incubated and continuously monitored by an automated spectrophotometer. Most systems use conventional biochemical tests with colour changes or release of fluorophores, but growth inhibition by antibiotics and dyes is also included. A number of different approaches have been developed, for example, the Biolog system determines carbon source utilization profiles and the Midi microbial identification system utilizes an automated high-resolution gas chromatography system coupled to a computer (Stager and Davis, 1992). The most widely used systems are VITEK (BioMerieux Marcy I' Etoile, France) and PHOENIX (Becton and Dickinson Franklin Lakes, New Jersey, USA) which combine identification with antimicrobial susceptibility testing. The VITEK2 system incorporates an expert system to match MIC results with identification which had a 95.7% correlation with human review of results, thus greatly reducing laboratory interpretative input (Sanders *et al.*, 2001). The combination of VITEK2 with direct inoculation from an automated blood culture system has the potential to save 1 or 2 days in reporting the antimicrobial sensitivity and identification of Enterobacteriaceae causing septicaemia, although polymicrobial disease will give an erroneous result (Ling, Liu and Cheng, 2003). It is easy for microbiologists to lose sight of the basic biochemical characteristics of the species of the Enterobacteriaceae when using automated systems. Should problems arise, back-up systems should be available and the microbiologist should be aware of the key biochemical properties of the commonly encountered and clinically significant Enterobacteriaceae, which are summarized in Table 27.3.

Table 27.2   Octal conversion of binary code

Binary	Conversion formula	Octal
– – –	0 + 0 + 0 =	0
+ – –	0 + 0 + 1 =	1
– + –	0 + 2 + 1 =	2
+ + –	0 + 2 + 1 =	3
– – +	4 + 0 + 0 =	4
+ – +	4 + 0 + 1 =	5
– + +	4 + 2 + 0 =	6
+ + +	4 + 2 + 1 =	7

Table 27.3 Biochemical reactions of the commonly encountered, clinically significant Enterobacteriaceae (after Farmer *et al.*, 1985)

Species	Indole production	Methyl red	Voges–Proskauer	Citrate (Simmons')	Hydrogen sulphide (TSI)	Urea hydrolysis	Phenylalanine deaminase	Lysine decarboxylase	Arginine dihydrolase	Ornithine decarboxylase	Motility (36 °C)	Gelatine hydrolysis (22 °C)	D-Glucose, gas
Escherichia coli	+	+	–	–	–	–	–	+	V	V	+	–	+
Shigella serogroups A, B and C	V	+	–	–	–	–	–	–	–	–	–	–	–
Shigella sonnei	–	+	–	–	–	–	–	–	–	+	–	–	–
Salmonella, most serotypes	–	+	–	+	+	–	–	+	V	+	+	–	+
Salmonella typhi	–	+	–	–	V	–	–	+	–	–	+	–	–
Salmonella paratyphi A	–	+	–	–	V	–	–	–	V	+	+	–	+
Citrobacter freundii	–	+	–	+	V	V	–	–	V	V	+	–	+
Citrobacter diversus	+	+	–	+	–	V	–	–	V	+	+	–	+
Klebsiella pneumoniae	–	V	+	+	–	+	–	+	–	–	–	–	+
Klebsiella oxytoca	+	V	+	+	–	+	–	+	–	–	–	–	+
Enterobacter aerogenes	–	–	+	+	–	–	–	+	–	+	+	–	+
Enterobacter cloacae	–	–	+	+	–	V	–	–	+	+	+	–	+
Hafnia alvei	–	V	V	V	–	–	–	+	–	+	V	–	+
Serratia liquefaciens	–	–	+	+	–	–	–	+	–	+	+	+	V
Serratia marcescens	–	V	+	+	–	V	–	+	–	+	+	+	V
Proteus mirabilis	–	+	V	V	+	+	+	–	–	+	+	+	+
Proteus vulgaris	+	+	–	V	+	+	+	–	–	–	+	+	V
Providencia rettgeri	+	+	–	+	–	+	+	–	–	–	+	–	V
Providencia stuartii	+	+	–	+	–	V	+	–	–	–	+	–	V
Providencia alcalifaciens	+	+	–	+	–	–	+	–	–	–	+	–	–
Morganella morganii	+	+	–	–	–	+	+	–	–	+	+	–	V
Yersinia enterocolitica	V	+	–	–	–	V	–	–	–	+	+	–	+
Yersinia pestis	–	V	–	–	–	–	–	–	–	–	–	–	–
Yersinia pseudotuberculosis	–	+	–	–	–	+	–	–	–	–	–	–	–

Table 27.3 (Continued)

Species	Lactose fermentation	Sucrose fermentation	D-Mannitol fermentation	Dulcitol fermentation	Adonitol fermentation	D-Sorbitol fermentation	L-Arabinose fermentation	Raffinose fermentation	L-Rhamnose fermentation	D-Xylose fermentation	Melibiose fermentation	DNAse, 25 °C	ONPG
Escherichia coli	+	V	+	V	−	+	+	V	V	+	V	−	+
Shigella serogroups A, B and C	−	−	+	−	−	V	V	V	−	−	V	−	−
Shigella sonnei	−	−	+	−	−	−	+	−	V	−	V	−	+
Salmonella, most serotypes	−	−	+	+	−	+	+	−	+	+	+	−	−
Salmonella typhi	−	−	+	−	−	+	−	−	−	V	−	−	−
Salmonella paratyphi A	−	−	+	+	−	+	+	−	+	+	+	−	−
Citrobacter freundii	V	V	+	V	+	+	+	V	+	+	V	−	+
Citrobacter diversus	V	V	+	V	+	+	+	−	+	+	−	−	+
Klebsiella pneumoniae	+	+	+	V	+	+	+	+	+	+	+	−	+
Klebsiella oxytoca	+	+	+	V	+	+	+	+	+	+	+	−	+
Enterobacter aerogenes	+	+	+	V	+	+	+	+	+	+	+	−	+
Enterobacter cloacae	+	+	+	−	V	+	+	+	+	+	+	−	+
Hafnia alvei	−	V	+	−	−	−	+	−	+	+	−	−	+
Serratia liquefaciens	−	+	+	−	−	+	+	−	−	+	V	V	+
Serratia marcescens	−	+	+	−	V	+	−	−	−	−	−	+	+
Proteus mirabilis	−	V	−	−	−	−	−	−	−	+	−	V	−
Proteus vulgaris	−	+	−	−	−	−	−	−	−	+	−	V	−
Providencia rettgeri	−	V	+	−	+	−	−	−	V	V	−	−	−
Providencia stuartii	−	V	V	−	−	−	−	−	−	−	−	V	−
Providencia alcalifaciens	−	V	−	−	+	−	−	−	−	−	−	−	V
Morganella morganii	−	−	−	−	−	−	−	−	−	−	−	−	−
Yersinia enterocolitica	−	+	+	−	−	+	+	−	−	V	−	−	+
Yersinia pestis	−	−	+	−	−	V	+	−	−	+	V	−	V
Yersinia pseudotuberculosis	−	−	+−	−	−	−	V	V	V	+	V	−	V

ONPG, *o*-nitrophenyl-β-D-glactpyranoside; +, >90% positive; V, 10–90% positive; −, <10% positive, after 48 h incubation at 37 °C.

Bacterial identification by sequencing the 16S rRNA gene is a universal bacterial identification method which has been used in combination with whole-genome DNA/DNA hybridization by bacterial taxonomists. Because the first 500 bp of this highly conserved gene have the greatest heterogeneity, this sequence, which can be generated using PCR and automated DNA sequencing, has been used in the clinical laboratory (particularly for *Mycobacterium* spp.) to identify unknown isolates (Patel, 2001). Consensus primers can be used, and DNA sequenced in-house or commercial systems are available (e.g., MicroSeq – www.appliedbiosystems.com). The technique is extremely powerful as illustrated by the reclassification of the causative agent of granuloma inguinale. 16S rRNA sequencing has reclassified *Calymmatobacterium granulomatis* which cannot be cultured on conventional media, as *Klebsiella granulomatis* (Kharsany *et al.*, 1999). The costs for 16S rRNA identification are relatively high, and a balance must be made between the clinical value of the identification and cost. Individual molecular tests for the rapid identification of some species or Enterobacteriaceae have been developed, for example, *Yersinia pestis* in relation to bioterrorism (McAvin *et al.*, 2003).

REFERENCES

Barrow, G.I. and Feltham, R.K.A. (eds) (1993) *Cowan and Steel's Manual for the Identification of Medical Bacteria*, 3rd edn, Cambridge University Press, Cambridge, UK.

Bush, K., Jacoby, G.A. and Medeiros, A.A. (1995) A functional classification scheme for β-lactamases and its correlation with molecular structures. *Antimicrobial Agents and Chemotherapy*, **39**, 1211–1233.

Durham, H.E. (1900) Some theoretical considerations upon the nature of agglutinins, together with further observations upon *Bacillus typhi abdominalis*, *Bacillus enteritidis*, *Bacillus coli communis*, *Bacillus lactis aeruginosa* and some other bacilli of allied character. *The Journal of Experimental Medicine*, **5**, 354–388.

Farmer, J.J., Davis, B.R., Hickman-Brenner, F.W. *et al.* (1985) Biochemical identification of new species and biogroups of Enterobacteriaceae isolated from clinical specimens. *Journal of Clinical Microbiology*, **21**, 46–76.

Iijima, Y., Asako, N.T., Aihara, M. *et al.* (2004) Improvement in the detection rate of diarrhoeagenic bacteria in human stool specimens by a rapid real-time PCR assay. *Journal of Medical Microbiology*, **53**, 617–622.

Jones, M.E., Avison, M.B., Damidinsuren, E. *et al.* (1994) Heterogeneity at the, β-lactamase structural gene *ampC* amongst *Citrobacter* spp, assessed by polymerase chain reaction analysis: potential for typing at a molecular level. *Journal of Medical Microbiology*, **41**, 209–214.

Kharsany, A.B., Hoosen, A.A., Kiepiela, P. *et al.* (1999) Phylogenetic analysis of *Calymmatobacterium granulomatis* based on 16S rRNA gene sequences. *Journal of Medical Microbiology*, **48**, 841–847.

Ling, T.K., Liu, Z.K. and Cheng, A.F. (2003) Evaluation of the VITEK2 system for rapid direct identification and susceptibility testing of gram-negative bacilli from positive blood cultures. *Journal of Clinical Microbiology*, **41**, 4705–4707.

Livermore, D. (1995) β-Lactamases in laboratory and clinical resistance. *Clinical Microbiology Reviews*, **8**, 557–584.

McAvin, J.C., McConathy, M.A., Rohrer, A.J. *et al.* (2003) A real-time fluorescence polymerase chain reaction assay for the identification of *Yersinia pestis* using a field-deployable thermocycler. *Military Medicine*, **168**, 852–855.

O'Hara, C.M., Rhoden, D.L. and Miller, J.M. (1992) Re-evaluation of the API 20E identification system versus conventional biochemicals for identification of members of the family Enterobacteriaceae: a new look at an old product. *Journal of Clinical Microbiology*, **30**, 123–125.

Patel, J.B. (2001) 16S rRNA gene sequencing for bacterial pathogen identification in the clinical laboratory. *Molecular Diagnosis*, **6**, 313–321.

Pedler, S.J. (2004) Bacteriology of intestinal disease, in *Medical Bacteriology: A Practical Approach*, 2nd edn (eds P.M. Hawkey and D.A. Lewis), Oxford University Press, Oxford, pp. 177–213.

Sanders, C.C., Peyret, M., Moland, E.S. *et al.* (2001) Potential impact of the VITEK 2 system and the Advanced Expert System on the clinical laboratory of a University-based hospital. *Journal of Clinical Microbiology*, **39**, 2379–2385.

Snelling, A.M., Hawkey, P.M., Heritage, J. *et al.* (1993) The use of DNA probe and PCR to examine the distribution of the *aac* (6)-Ic gene in *Serratia marcescens* and other Gram-negative bacteria. *The Journal of Antimicrobial Chemotherapy*, **31**, 841–854.

Stager, C.E. and Davis, J.R. (1992) Automated systems for identification of microorganisms. *Clinical Microbiology Reviews*, **5**, 302–327.

Stearne, L.E., van Boxtel, D., Lemmens, N. *et al.* (2004) Comparative study of the effects of ceftizoxime, piperacillin, and piperacillin-tazobactam concentrations on antibacterial activity and selection of antibiotic-resistant mutants of *Enterobacter cloacae* and *Bacteroides fragilis* in vitro and in vivo in mixed-infection abscesses. *Antimicrobial Agents and Chemotherapy*, **48**, 1688–1698.

28

Escherichia coli and *Shigella* spp.

Christopher L. Baylis[1], Charles W. Penn[2], Nathan M. Thielman[3], Richard L. Guerrant[3], Claire Jenkins[4] and Stephen H. Gillespie[5]

[1]*Campden & Chorleywood Food Research Association (CCFRA), Chipping Campden, Gloucestershire, UK;* [2]*School of Biosciences, University of Birmingham, Edgbaston, Birmingham, UK;* [3]*University of Virginia School of Medicine, Charlottesville, VA, USA;* [4]*Department of Medical Microbiology, Royal Free Hospital NHS Trust, London, UK; and* [5]*Centre for Medical Microbiology, Royal Free and University College Medical School, Hampstead Campus, London, UK*

ESCHERICHIA COLI

INTRODUCTION

Since its first description by Dr Theodore Escherich in 1885, *Escherichia coli* is the most extensively studied bacterial species. Much of our knowledge of bacteria at a molecular and cellular level has been obtained through studies with *E. coli*, particularly using the strain K12, isolated in 1922 from a patient with diphtheria. The complete genome sequence of a strain of *E. coli* K12 has been published, revealing the degree of genome plasticity, as indicated by the presence of phage remnants and insertion elements (Blattner *et al.*, 1997).

DESCRIPTION OF THE ORGANISM

The genus *Escherichia* includes motile and non-motile bacteria which belong to the family Enterobacteriaceae (Edwards and Ewing, 1972). They are Gram-negative, oxidase negative, non-spore-forming, rod-shaped bacteria, facultative anaerobes, which are often motile by peritrichously arranged flagella. They are capable of fermenting a wide variety of carbohydrates with production of both acid and gas, although anaerogenic biotypes exist. Rapid fermentation of lactose is a characteristic feature of many strains, particularly those of *E. coli*, whilst other *Escherichia* species, including strains of enteroinvasive *E. coli* (EIEC), as well as some *E. coli* (metabolically inactive) strains ferment it slowly or fail to utilise this substrate at all. Other genera within the family Enterobacteriaceae (*Klebsiella*, *Enterobacter*, *Serratia* and *Citrobacter*), which share this ability to ferment lactose rapidly (typically within 24 h), are collectively termed coliform bacteria. Besides *E. coli*, other species belonging to the genus *Escherichia* are *Escherichia blattae*, *Escherichia fergusonii*, *Escherichia hermannii* and *Escherichia vulneris*. The species *Escherichia adecarboxylata* has since been assigned to the genus *Lecercia*, and a new species, *Escherichia albertii*, has been described (Abbott *et al.*, 2003). The biochemical characteristics of the genus *Escherichia* are given in Table 28.1.

With the exception of *E. blattae*, which has been isolated from the cockroach intestine, but not from human clinical specimens, the primary habitat of *E. coli* and the other species is the gastrointestinal tract of humans and other warm-blooded animals where they generally exist as harmless commensal organisms. They can also occur in water, food and soil, but this is invariably the result of faecal contamination.

Although most *E. coli* strains are harmless, there are others that cause disease in humans and animals that have evolved to become important pathogens in their own right. Clinically, two distinct types of pathogenic *E. coli* are recognised. One group commonly called extraintestinal pathogenic *E. coli* (ExPEC) includes those *E. coli* associated with newborn meningitis (NBM) or sepsis and urinary tract infections (UTIs). The second group termed intestinal pathogenic *E. coli* (IPEC) includes *E. coli* responsible for a range of distinct classes of diarrhoeal disease.

Escherichia coli is a genetically diverse species that includes many intestinal and extraintestinal pathotypes. Many of these have highly efficient and specialised mechanisms of colonisation and pathogenicity, developed through the acquisition of virulence-associated genes and adaptation to their changing surroundings aided by mutations and natural selection. Many virulence-associated genes are localised on mobile genetic elements such as plasmids, bacteriophages and pathogenicity islands (PAIs). First described in uropathogenic *E. coli*, and now recognised in other pathotypes, notably enteropathogenic *E. coli* (EPEC) and enterohaemorrhagic *E. coli* (EHEC), PAIs are distinct types of genetic element thought to have evolved from mobile genetic elements by horizontal gene transfer. Characterised by their large size (>10 kb), presence of virulence-associated genes and a G+C content that is different from that of the rest of the genome, they are often flanked by repeat structures carrying fragments of other mobile and accessory genetic elements such as plasmids, bacteriophages and insertion sequence (IS). It is this exchange of virulence genes between different bacteria that is largely responsible for the evolution of different bacterial pathotypes, and horizontal transfer plays a major role in the creation of new virulent clones (Johnson, 2002).

Multilocus enzyme electrophoresis (MLEE) and sequencing of the *malate dehydrogenase* gene (*mdh*) indicate that pathogenic strains of *E. coli* have arisen many times and that they do not have a single evolutionary origin within the species (Pupo *et al.*, 1997). Many prophage elements present in the *E. coli* O157 Sakai genome indicate that bacteriophages have played an important role in the emergence of this pathogen and, possibly, other pathotypes of *E. coli* (Ohnishi, Kurokawa and Hayashi, 2001). Although thought to be uncommon, genome rearrangements within species have been reported to be very frequent in human clinical isolates (Hughes, 2000). Comparison of genome size within a species also enables the degree of divergence between strains to be assessed. In strain MG1655 of *E. coli* K12 the chromosome is 4.6 Mb (Blattner *et al.*, 1997), whereas genome sizes of 5.4 Mb and 5.5 Mb have been reported for *E. coli* O157:H7 strain

Table 28.1 Biochemical characteristics associated with different members of the genus *Escherichia*

Test	E. coli	E. coli (inactive)	E. blattae	E. fergusonii	E. hermannii	E. vulneris	E. albertii
ONPG	95	45	0	83	98	100	0
Indole	98	80	0	98	99	0	0
Methyl red	99	95	100	100	100	100	?
Voges-Proskauer	0	0	0	0	0	0	0*
Citrate (Simmons)	1	1	50	17	1	0	0
Lysine decarboxylase	90	40	100	95	6	85	100
Arginine dihydrolase	17	3	0	5	0	30	0
Ornithine decarboxylase	65	20	100	100	100	0	100
Motility	95	5	0	93	99	100	0*
D-Glucose acid	100	100	100	100	100	100	100
D-Glucose gas	95	5	100	95	97	97	100
Yellow pigment	0	0	0	0	98	50	0
Fermentation of							
Lactose	95	25	0	0	45	15	0
Sucrose	50	15	0	0	45	8	0
D-Mannitol	98	93	0	98	100	100	100
Adonitol	5	3	0	98	0	0	0
Cellobiose	2	2	0	96	97	100	0
D-Sorbitol	94	75	0	0	0	1	0
D-Arabitol	5	5	0	0	0	0	0
L-Rhamnose	80	65	100	92	97	93	0

Data from Farmer *et al.* (1985) with the exception of *E. albertii*, which was derived from Abbott *et al.* (2003). Figures represent percentage positive reactions after incubation at 36 °C for 2 days except for those indicated by an asterisk, which were incubated at 35 °C. ONPG, *ortho*-nitrophenyl-β-D-galactopyranoside.

EDL933 (Perna *et al.*, 2001) and the O157:H7 Sakai strain (Hayashi *et al.*, 2001), respectively.

The complete genome sequence of *E. coli* K12 strain MG1655 has provided a benchmark against which other *E. coli* strains have been compared. Comparison of *E. coli* O157:H7 EDL933 against this K12 strain has highlighted that lateral gene transfer in this O157:H7 strain has been more extensive than was initially expected (Perna *et al.*, 2001). DNA sequences found in MG1655 but not EDL933 were designated K islands (KI) and DNA sequences in EDL933 but not MG1655, O islands (OI). There are 177 OI and 234 KI greater than 50 bp in length. Whilst they both share a common 'backbone' sequence of 4.1 Mb (Perna *et al.*, 2001), since the divergence of O157:H7 from K12 about 4.5 million years ago (Reid *et al.*, 2000), this O157:H7 strain has acquired 1387 new genes (Perna *et al.*, 2001). Many of these are putative virulence factors, while others are genes encoding alternative metabolic capacities as well as several prophages (Perna *et al.*, 2002). Whilst the amount of apparent horizontal transfer in MG1655 is not as high as that observed in EDL933, there is evidence that the genome of MG1655 also contains DNA obtained through horizontal transfer (Blattner *et al.*, 1997).

PATHOTYPES OF *E. COLI* AND CLINICAL FEATURES OF INFECTIONS

Extraintestinal Pathogenic *E. coli*

Unlike *E. coli* responsible for diarrhoeal disease and life-threatening infections, ExPEC can either behave as harmless inhabitants of the human intestine or become serious pathogens when they enter the blood, cerebrospinal fluid (CSF) or urinary tract. Certain *E. coli* strains are responsible for classic syndromes such as UTIs, bacteraemia and neonatal meningitis. This has given rise to particular strains, often characterised by specific O:K:H serotypes being classified as uropathogenic *E. coli* (UPEC) or meningitis-associated *E. coli* (MAEC). Some strains differ from commensal *E. coli* by the presence of pathogenicity-associated genes and particular clones of *E. coli* such as strains of O4:H5 which appear to show a greater propensity to cause UTI. The high-affinity iron-uptake system mediated by the hydroxamate siderophore aerobactin, which can be chromosomally or plasmid encoded, is

reported to be particularly prevalent among ExPEC, including those that cause septicemia, pyelonephritis and meningitis (Carbonetti *et al.*, 1986). However, *E. coli* is capable of infecting many anatomical sites and some strains demonstrate pathogenic versatility and the ability to cause infections other than the classic syndromes (Johnson and Russo, 2002).

Urinary Tract Infections

Escherichia coli which generally originate from faeces or the periurethral flora are responsible for most UTIs. After colonising the periurethral area, these organisms may ascend the urinary tract from the urethra meatus or from the insertion of catheters and infect the bladder. In some cases, these organisms continue to progress through the ureters and cause acute pyelonephritis, involving one or both kidneys. Patients with acute pyelonephritis present with a range of symptoms including fever, flank pain and bacteriuria with or without diaphoresis, rigors, groin or abdominal pain and nausea and vomiting. Abdominal tenderness of one or both kidneys may be elicited on examination. In a small proportion of cases, including those with pyelonephritis, the bacteria may spread beyond the urinary tract and enter the blood.

The group of *E. coli*, often called UPEC, which are responsible for acute and chronic UTIs are distinct from the commensal *E. coli* found in the colon of humans and are represented by a few serogroups (O1, O2, O4, O6, O7 and O75). They often possess genes encoding many pathogenicity-associated factors including adhesins, siderophores (i.e. aerobactin), capsule and toxins implicated in UTI pathogenesis. PAIs as a distinct type of genetic element were first described for UPEC strain 536 (O6:K15:H31), which has become a model organism to study ExPEC pathogenesis. The published complete genome sequence of UPEC strain CFT073 reveals the presence of numerous pathogenicity-associated genes in UPEC, especially genes encoding potential fimbrial adhesins, autotransporter iron sequestration systems as well as showing the acquisition of PAIs by horizontal gene transfer (Welch *et al.*, 2002). Unlike *E. coli* O157:H7 and other EHEC strains, the UPEC CFT073 genome contained no genes for type III secretion system, phage or plasmid-encoded virulence genes.

Individual pathogenicity-associated factors found in UPEC strains include adhesins, particularly fimbriae, which facilitate adherence and bacterial colonisation. Many fimbriae, with different host receptor specificity, are expressed by UPEC strains. These include the mannose-resistant P, M, S, F1C and Dr fimbriae, which haemagglutinate erythrocytes in the presence of mannose. The type 1 fimbriae are common among Enterobacteriaceae, although in UPEC, the presence of type 1 fimbriae may increase their virulence for the urinary tract by promoting bacterial persistence and by enhancing the inflammatory response to infection (Wullt, 2003). The earliest described and most commonly associated adhesin in UPEC are P fimbriae, particularly the PapG adhesin. These fimbriae, encoded by the *pap* (pyelonephritis-associated pili) operon, recognise the disaccharide α-D-galactosyl-(1–4)β-D-galactose receptor, which is a very common P blood group antigen. This enables it to bind to red cells, but also to uroepithelial cells in most of the population. There exists a strong relationship between the presence of P fimbriae and severity of infection, especially pyelonephritis. Also in UPEC are P-related sequences (Prs) which are closely related to P fimbriae but possess the F or PrsG adhesin which bind to galactosyl-*N*-acetyl(α1–3)-galactosyl-*N*-acetyl.

Toxins produced by UPEC include haemolysin and cytotoxic necrotizing factor 1 (CNF1) and secreted autotransporter toxin (Sat) which has been shown to have a cytopathic effect on various bladder and kidney cell lines (Bahrani-Mougeot *et al.*, 2002). A high percentage of *E. coli* isolated from patients with pyelonephritis secrete haemo-lysin, which can be plasmid or chromosomally encoded. Additionally, on the chromosome of UPEC strains, the *hly* operon (*hlyCABD*) is often located near the P fimbrial genes on the same PAI. In strain 536, four PAIs have been characterised which carry many pathogenicity-associated genes including two α-haemolysin gene clusters (PAI I$_{536}$ and PAI II$_{536}$), P-related fimbriae (PAI II$_{536}$), S fimbriae (PAI II$_{536}$) and the salmochelin (PAI III$_{536}$) and yersiniabactin (PAI IV$_{536}$) siderophore systems (Dobrindt *et al.*, 2002). The K15 capsule deter-minant of UPEC strain 536 is also found on a PAI (PAI V$_{536}$) (Schneider *et al.*, 2004). Whilst the role of capsular antigens in UTI remains controversial, it is possible that they enable the bacteria to resist phagocytosis and survive in human serum or aid adherence of the bacteria to host cells. Interestingly, PAIs show considerable variability in their composition and structural organisation. Some PAIs show genetic instability, having a tendency to be deleted from the chromo-some, which can be influenced by environmental conditions, whereas others appear to be relatively stable (Middendorf *et al.*, 2004).

Escherichia coli Bacteraemia and Meningitis-Associated *E. coli*

Although *E. coli* are able to colonise and infect the gastrointestinal and urinary tracts of humans, septicaemia remains a relatively rare complication of *E. coli* infections. Nevertheless, *E. coli* is one of the most common Gram-negative bacteria responsible for bacteraemia in humans. Isolates of *E. coli* that infect the bloodstream often possess virulence factors that enable the organisms to circumvent the normal clearance mechanisms and evade the host immune response. These include a range of adhesins (P, S and M), the siderophore aerobactin and haemolysin which are found in other ExPEC. The lipopolysac-charide (LPS) of *E. coli*, as with other Gram-negative bacteria, is an important pathogenicity factor which may cause fatal septic shock and disseminated intravascular coagulation. In response to the presence of *E. coli*, epithelial cells and cells of the host immune system have been shown to secrete many host-cell factors including interleukins (IL), tumour necrosis factor (TNF) and activators of the complement cascade. Epithelial cells are capable of producing a variety of cytokines in response to bacterial stimuli. Epithelial cells originating from the human urinary tract have been shown to produce IL-6, IL-1α and IL-8, whereas human peripheral blood monocytes additionally produce IL-1β and TNF-α in response to the presence of *E. coli* (Agace *et al.*, 1993). Whilst *E. coli* bacteraemia can occur in UTI, especially when the tract is obstructed, studies suggest a relationship between the magnitude of *E. coli* bacteraemia and the development of meningitis.

The most common Gram-negative organism responsible for meningitis during the neonatal period is *E. coli*. Sepsis and NBM are often associated with *E. coli* belonging to a limited number of serotypes, particularly those expressing the K1 capsular antigen (Korhonen *et al.*, 1985), e.g. O83:K1 and O7:K1 and especially O18:K1. Whilst few specific pathogenic determinants have been described for *E. coli* causing NBM, isolates have been shown to possess many factors, some of which are also found in UPEC strains and others that appear to be specific to this group. The polysialic acid homopolymer K1 capsule is believed to increase serum survival by blocking complement activation (Pluschke *et al.*, 1983), and expression of this, together with production of aerobactin, is believed to be important for blood stream dissemination. Adherence is a critical step in the pathogenesis of *E. coli* meningitis. Factors involved in the binding of *E. coli* to brain microvascular endothelial cells (BMECs) include S fimbriae which are also important in UPEC pathogenesis. Subsequent invasion is facilitated by various microbial determinants including invasion of brain endothelium (Ibe) proteins, which may promote the crossing of the blood–brain barrier. A primary determinant of this event is a high density of bacteraemia, but how circulating *E. coli* cross the blood–brain barrier is not fully understood: one mechanism might involve transcytosis through the endothelial cells aided by specific pathogen–host-cell interactions. Another factor which may contribute or enhance invasion of BMEC by *E. coli* is outer membrane protein A (OmpA), which shows structural similarities to *Neisseria* outer membrane protein (Opa) and surface protein PIII (Prasadarao *et al.*, 1996). Another candidate protein necessary for the invasion of BMEC is a novel ibe10 protein found in CSF isolates of *E. coli*, which has been shown to interact with endothelial cells, thus enhancing invasion by *E. coli* cells (Prasadarao *et al.*, 1999). Whilst possession of P fimbriae is important in *E. coli* causing pyelonephritis, it is not thought to be relevant in strains responsible for NBM. Furthermore, whereas α-haemolysin and CNF-1 are less common in NBM strains, *ibe*10 and *sfa* genes are reported to be more commonly associated with meningitis strains compared with blood or commensal *E. coli*.

Intestinal Pathogenic *E. coli*

Six distinct groups have been defined within IPEC commonly associated with intestinal disease: EIEC, enterotoxigenic *E. coli* (ETEC), EPEC, EHEC, enteroaggregative *E. coli* (EAggEC) and the diffusely adherent *E. coli* (DAEC). These diarrhoeagenic *E. coli* are described in more detail, along with aspects of the associated pathogenesis and their epidemiology (Donnenberg, 2002; Kaper, Nataro and Mobley, 2004).

Some of the virulence-associated genes and key features of the different pathotypes are presented in Table 28.2. In the past, patho-genic *E. coli* were defined by their serotype on the basis of somatic O and flagella H antigens and to a lesser extent the K antigens. Although this practice continues, it has become more common to define individual pathotypes based on their pathogenetic characteristics. This is particularly important because of the existence of strains with the same serotype that belong to different pathotypes based on their pathology and complement of specific virulence determinants, which is apparent with some strains of EPEC and EHEC.

Enteroinvasive *E. coli*

Strains belonging to this group are biochemically, genetically and pathogenically closely related to *Shigella* spp. Compared with most *E. coli* strains, EIEC are atypical. They are generally lysine decarbox-ylase negative and non-motile and 70% are unable to ferment lactose (Silva *et al.*, 1980). These characteristics are shared with *Shigella*, although some strains of *Shigella sonnei* can ferment lactose slowly. Genetically, *Shigella* spp. and *E. coli* are very closely related and it has been proposed that *Shigella* spp. have evolved from *E. coli* and

Table 28.2 Summary of pathogenicity-related characteristics of intestinal pathotypes of *E. coli*

Pathotype	Pathogenicity-associated genes or factors	Mechanism	Clinical features
Enteroinvasive *E. coli*	140 MDa plasmid (pINV) chromosomal genes	Bacterial attachment and invasion of colonic enterocytes via endocytosis, multiplication causing host-cell death and inflammation, accompanied by necrosis and ulceration of large bowel	Ulceration of bowel, watery diarrhoea, dysenteric stools (bacillary dysentery)
Enterotoxigenic *E. coli*	Plasmid-encoded CFAs: CFA-I (rigid rod-like fimbriae), CFA-III (bundle-forming group), CFA-II and CFA-IV (flexible fimbriae) and type IV-related longus pili	Colonisation of surface of small-bowel mucosa (CFA I–IV) and production of enterotoxins LTI, LTII, STa, STb	Acute watery diarrhoea, usually without blood mucus or pus
	Labile toxins: LTI, LTII (plasmid encoded)	ADP ribosylation of G proteins → adenylate cylase activation → increased cAMP secretion → reduced Na absorption/Cl secretion → diarrhoea	
	Heat-stable toxins: STa, STb (plasmid and transposon encoded)	Guanylate cyclase (G C-C) activation → increased cGMP secretion → chloride secretion and/or inhibition of NaCl absorption → diarrhoea	
Enteroaggregative *E. coli*	Plasmid (60 MDa) encoded: Aggregative adherence fimbriae (AAF/I and AAF/II), transcriptional regulator (AggR) *E. coli* heat-stable-like toxin-1 (EAST-1) ShET1 is Shigella enterotoxin-1 Plasmid-encoded toxin (Pet)	Adherence and colonisation of intestinal mucosa facilitated by AAF/I and AAF/II Release of toxins and damage to host epithelial cells	Aggregative adherence (AA) phenotype Persistent diarrhoea
Diffusely adherent *E. coli*	Afa/Dr family adhesins (AIDA)	DA phenotype facilitated by surface fimbriae, e.g. F1845 encoded via *daaC* gene, or by other related adhesins which are plasmid or chromosomally encoded	Diffusely adherent (DA) phenotype
	EAST-1 *set* genes (enterotoxins) Possible TTSS with *esc*	Events in pathogenesis remain unclear	Watery diarrhoea, usually without blood
Enteropathogenic *E. coli*	50–70 MDa plasmid (EAF) encodes: bundle-forming pilus (BFP), plasmid-encoded regulator (Per) and LEE-encoded regulator (ler) Chromosomal PAI (LEE)	Localised adherence (LA) via BFP	A/E lesions
	TTSS comprising: intimin (*eaeA*), secreted proteins; Tir, EspA, EspB, EspD, EspF, EspG and mitochondria-associated protein (MAP) EAST-1 Cytolethal distending toxin (CDT)	A/E histopathology; cytoskeletal rearrangement of host epithelial cells involving TTSS Intimate effacing adherence mediated by intimin Destruction of microvilli and interference with host-cell signalling cascades	Acute diarrhoea
Enterohaemorrhagic *E. coli*	Large (60 MDa) plasmid encodes: enterohaemolysin (Ehx), LCT, EspP	A/E histopathology and intimate adherence similar to EPEC. Alternative adherence mechanisms (besides *eae*) known. TTSS aids pathogenesis, toxins (Stx/VT) inhibit protein synthesis of host cells and mediate different pathological effects	Bloody diarrhoea (haemorrhagic colitis)
	Chromosomal PAI (LEE)		Haemolytic uraemic syndrome (HUS) TTP
Verocytotoxin-producing *E. coli*	Chromosomal (prophage) encoded Shiga toxins (Stx)/verocytotoxin (VT) VT1, VT2 and VT2v		

A/E, attaching and effacing; CFA, colonisation factor antigen; EAF, EPEC adherence factor; EAST, enteroaggregative heat-stable toxin; EPEC, enteropathogenic *E. coli*; Esp, *E. coli* secreted protein; LCT, large clostridial toxins; LEE, locus of enterocyte effacement; TTP, thrombotic thrombocytopenic purpura; TTSS, type III protein secretion system.

that they constitute a single species. This has been supported by a recent publication comparing the genome sequence of *Shigella flexneri* 2a with the genomes of *E. coli* K12 and O157 (Jin *et al.*, 2002).

Both *Shigella* and EIEC are responsible for bacillary dysentery, a disease that has had a major impact on society throughout history. Transmission is via the faecal–oral route, with contaminated food and water being the main sources of infection, but person-to-person transmission can also occur. The syndrome produced by EIEC infection can be indistinguishable from *Shigella* infection. In both cases the site

of infection is predominantly the colon. The most common symptom is watery diarrhoea which may precede dysenteric stools containing mucus and blood. In severe cases the bacteria may attack the colonic mucosa, invading epithelial cells, multiplying and causing ulceration of the bowel. In EIEC, genes necessary for invasiveness are carried on a 140-MDa plasmid (pINV). These genes encoding a range of invasion plasmid antigens (IpaA to IpaD) are necessary for strains to be fully pathogenic. Serotypes associated with the EIEC pathotype include O159:H2 and O143:H⁻.

Enterotoxigenic *E. coli*

Strains belonging to the ETEC pathotype are characterised by the production of at least one of two types of enterotoxin: LT (oligomeric heat-labile enterotoxin) and ST (monomeric heat-stable enterotoxin). The LT is classified into type I (LTI) and type II (LTII). LTI is a plasmid-encoded periplasmic protein resembling cholera toxin (CT) expressed by *Vibrio cholerae*, and LTII is chromosomally encoded but is similar to LTI both structurally and in its mode of action (Nair and Takeda, 1997). The pathway used by LT within the host cell is similar to that used by other AB$_5$ toxins including CT and Shiga toxin

(Stx)/verocytotoxin (Sandvig and Van Deurs, 2005; Lencer and Tsai, 2003). These pathways are described in Figure 28.1. In contrast, STs are peptides that stimulate guanylate cyclase in intestinal epithelial cells, resulting in intracellular accumulation of cyclic guanylate monophosphate (cGMP), leading to net fluid secretion and diarrhoea (Figure 28.1). They are classified into two distinct groups (STa and STb) or STI and STII (Nair and Takeda, 1997), and genes for ST are found on plasmids, although some have been found on transposons. Human infections are more commonly associated with STa, which is also found in *Yersinia enterocolitica* and *V. cholerae*, and it shares 50% protein identity with EAST-1 found in EAggEC. Some strains of ETEC

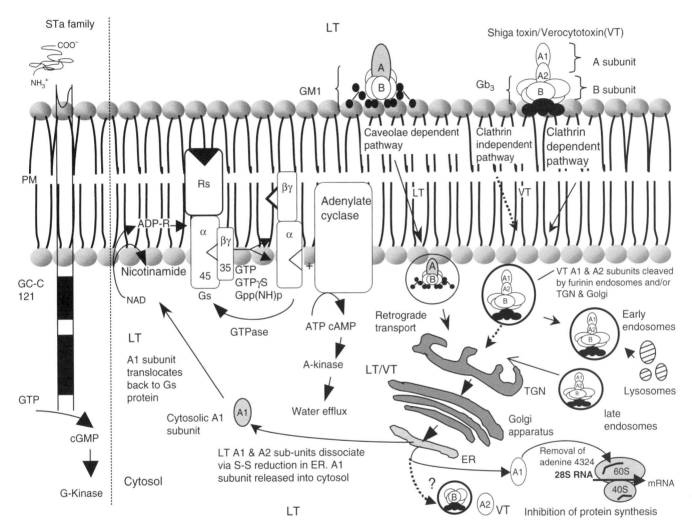

Figure 28.1 Molecular pathogenesis of three *E. coli* toxins. (ST) Binding of the 18–19 amino acid peptide, heat-stable enterotoxin (ST), to an extracellular domain of membrane-spanning enzyme guanylate cyclase (GC-C) increases accumulation of cyclic GMP (cGMP) which activates G kinase. This leads to altered sodium and chloride transport via phosphorylation of membrane proteins, leading to net fluid secretion and diarrhoea. (LT) Analogous to cholera toxin, *E. coli* heat-labile toxin (LT) binds at the cell surface to a monosialganglioside receptor (GM1) via its pentameric B subunit. The single A subunit comprises two functional domains: A1 peptide which exhibits the toxins ADP ribosyltransferase activity and A2 peptide which tethers A and B subunits together and contains the endoplasmic reticulum (ER) targeting motif K(R)DEL. Upon receptor binding the toxin–receptor complex enters the cell via endocytosis. It then carries retrogradely from the plasma membrane (PM) to the trans Golgi network (TGN), Golgi apparatus and the ER where proteolytic cleavage releases the B subunit from the two A subunits, which remain covalently linked via a disulphide bond. This bond is subsequently reduced, and the A1 peptide is released into the cytosol where this enzymatically active peptide translocates back to the PM where it catalyses the dissolution of NAD to nicotinamide and ADP ribose, thus enabling access to its substrate G$_{S\alpha}$ which undergoes toxin-induced ADP ribosylation. This results in the dissociation of the G$_{S\beta\gamma}$ subunit from the G$_{S\alpha}$ subunit, which activates adenylate cyclase which catalyses the formation of cyclic AMP. Raised levels of cAMP elicit a Cl$^-$ secretory response in intestinal crypt epithelial cells and water efflux in the gut. (VT) Most verocytotoxins (VT) bind to a globotriosyl ceramide (Gb$_3$) receptor via the pentameric B subunits of the toxin molecule. The toxin molecule can then be internalised by two distinct mechanisms depending on the cell type. The receptor–toxin complex is endocytosed, via clathrin-coated pits or by another endocytosis mechanism. Cleavage of the A1 and A2 subunits by furin can occur in endosomes, the TGN or in the Golgi. The toxin molecule is transported retrogradely to the Golgi apparatus and ER where ER-resident chaperons and enzymes may facilitate reduction of disulphide bonds between the A1 and A2 subunits. The destination of the B fragment and A2 subunit is not yet fully understood. The A1 subunit is transported into the cytosol where it facilitates the removal of an adenine at position 4324 in the 28S RNA of the 60S ribosomal subunit, resulting in the inhibition of protein synthesis.

have been shown to carry enterotoxin genes formerly characterised in *Shigella* species, for example the ShET1 toxin gene which is described in more detail with other virulence determinants found in EAggEC.

Another characteristic feature of ETEC is their ability to adhere to the intestinal epithelium mediated by adhesive fimbriae, also called *colonisation factors* (CFs), *colonisation factor antigens* (CFAs), *coli surface antigens* (CSAs) or *putative colonisation factors* (PCFs) (Elsinghorst, 2002). Depending on their structure, CFAs can be classified as fimbrial (rigid, filamentous, rod-like structures) or fibrillar (thinner, more flexible) or nonfimbrial if they lack these structures. Like the toxins, the CFAs are also plasmid encoded and associations are reported to exist between the toxin type and CFA carried on a single plasmid. For example, ST has been associated with CFA-I and CFA-IV, and LT and ST together have been associated with CFA-II (Wolf, 1997). The CFs also confer host specificity on the strain because they require the presence of specific host receptors for colonisation to occur. Consequently, ETEC strains are often host specific.

In developing and tropical countries, ETEC are particularly prevalent. They are often associated with travellers' diarrhoea, and in developing countries, ETEC-mediated diarrhoea is primarily a disease of young children, but high prevalence also occurs among infants and older children. The onset is usually abrupt and characterised by watery diarrhoea, usually without blood or mucus, and vomiting is only present in few cases. Whilst ETEC diarrhoea may be mild and self-limiting, in severe cases, dehydration and electrolyte imbalance can become life-threatening.

Infections are acquired through consumption of contaminated food or drink. The organisms colonise the proximal small intestine with the aid of many CFAs. These include those belonging to CFA I group (rigid rod-like fimbriae), CFA-III (bundle-forming group) and CFA-II and CFA-IV (flexible fimbriae). Following colonisation of the intestinal mucosa, localised delivery of enterotoxins occurs, resulting in diarrhoea.

Enteroaggregative *E. coli*

In certain developing countries, EAggEC (or EAEC) are a major cause of chronic infantile diarrhoea and they have also emerged as a cause of diarrhoeal disease in adults and children in developed countries. Currently, EAggEC are defined as *E. coli* strains that do not secrete enterotoxins LT or ST and which adhere to HEp-2 cells in a distinct pattern termed aggregative adherence (AA). This is characterised by the layering of the bacteria in a stacked-brick configuration on the surface of the host cells, which is associated with the carriage of a 60-MDa plasmid (Baudry *et al.*, 1990).

Strains of EAggEC appear to express putative pathogenic determinants, which include aggregative adhesion fimbriae I (AAF/I) and AAF/II that are plasmid encoded, and which have been shown to mediate adhesion to HEp-2 cells by certain strains of EAggEC (Nataro, Steiner and Guerrant, 1998). Expression of *AAF* genes requires the transcriptional regulator AggR. DNA probes for AAF/I and AAF/II have enabled the screening of strains for these surface structures. It has been reported that only a minority of EAggEC strains carry the genes for AAF (Czeczulin *et al.*, 1997), whist genes homologous to AggR are found in most of the strains (Nataro and Steiner, 2002), suggesting that other yet-undiscovered adhesins exist. Some strains of EAggEC are reported to express nonfimbrial surface-exposed proteins which mediate adhesion to HEp-2 cells by charge (Chart, Smith and Rowe, 1995), but again, only a proportion of EAggEC strains adhere by means of these surface proteins.

Toxins that have also been associated with strains of EAggEC include an *E. coli* heat-stable-like enterotoxin termed enteroaggregative heat-stable toxin-1 (EAST-1) and a heat-labile toxin termed plasmid-encoded toxin (Pet). However, in one small study the prevalence of Pet in EAggEC strains was reported to be low (Vila *et al.*, 2000). Furthermore, EAST-1 is not exclusive to EAggEC but is also found in other *E. coli*, notably ETEC, EPEC and EHEC strains, including O157:H7 and other VTEC (Nataro, Steiner and Guerrant,

1998), and has also been reported in DAEC (Vila *et al.*, 2000). However, EAST alone may not be sufficient to cause diarrhoea without the presence of other virulence determinants. One candidate for the regulation of virulence genes in EAggEC is AggR (Sheikh *et al.*, 2002). Many EAggEC strains elaborate the *Shigella* enterotoxin (ShET1), first identified in *Shigella flexneri* 2a strains, and ShET2 found in other *Shigella* strains. In *Shigella flexneri* 2a, ShET1 is encoded by the *set1* chromosomal gene, whereas ShET2 is encoded in the *sen* gene located on the invasion-associated (140 MDa) plasmid (Vargas *et al.*, 1999). The ShET1 toxin gene was reported in ETEC expressing ST and LT, in DAEC and in *E. coli* not belonging to any diarrhoeagenic category, although not all strains carrying the *set* genes expressed the toxin (Vila *et al.*, 2000). ShET1 production was reported to be common in diarrhoea patients, regardless of pathotype, whereas the ShET2 gene (*sen*) was found in EAEC, ETEC-ST and *E. coli*, not belonging to any diarrhoeagenic group.

Pathogenesis is thought to involve adherence to the intestinal mucosa and colonisation, probably facilitated by AAF/I and AAF/II. A functional class of colonisation factor in EAggEC termed dispersin could also aid colonisation by facilitating bacterial dispersal (Sheikh *et al.*, 2002). Colonisation is followed by enhanced mucus production which may promote persistent colonisation, followed by the release of toxins that results in damage to the host's epithelial cells. A flagellar protein in EAEC (FliC–EAEC) has been shown to interact with epithelial cells, leading to IL-8 secretion (Steiner *et al.*, 2000). This protein shows amino acid sequence similarity to a flagellin found in *Shigella dysenteriae*, which, although phenotypically non-motile, have retained some flagellar operons; however, these are commonly non-functional in shigellae (Al Mamun, Tominaga and Enomoto, 1996). Flagella in EAggEC may therefore be responsible for inflammation and could play a role in pathogenesis alone or together with EAEC toxins.

Diffusely Adherent *E. coli*

This relatively recently described pathotype of *E. coli* is characterised by three features: ability to adhere to HEp-2 cells in a diffuse pattern (Nataro and Steiner, 2002), presence of the Afa/Dr family or adhesin involved in diffuse adherence (AIDA) adhesins (Peiffer *et al.*, 2000) and absence of virulence genes typical of other *E. coli* pathotypes. This diffusely adherent (DA) phenotype is believed to be mediated by fimbrial structures. This adhesion pattern is also exhibited by some EPEC strains. One surface fimbria (F1845) encoded via the *daaC* gene has been characterised (Bilge *et al.*, 1989), and this, together with other genetically related adhesins, may provide potential targets for detection and diagnosis of DAEC. Little is known about the clinical characteristics of DAEC, except that they are capable of causing watery diarrhoea without blood and that DAEC strains are able to induce finger-like projections extending from the surface of infected HEp-2 or Caco-2 cells. One recent report has provided evidence of a type III secretion system in DAEC with similar secreted protein (*esc*) genes to those found in EPEC and EHEC (Kyaw *et al.*, 2003). There is evidence that DAECs also produce the EAST toxin which is common in EPEC and EAggEC strains and that they can possess *set* genes which encode enterotoxins found in *Shigella* spp. (Vila *et al.*, 2000). Furthermore, *in vitro* studies show that Afa/Dr DAEC may have a marked proinflammatory effect on colonic epithelial cells, inducing secretion of IL-8 which promotes polymorphonuclear leukocyte (PMNL) migration across the epithelium (Bétis *et al.*, 2003).

Enteropathogenic *E. coli*

The EPECs, together with the closely related EHEC, are perhaps the most extensively studied group of pathogenic *E. coli*. The terminology and the assigned role of specific proteins and putative virulence determinants in EPEC and EHEC pathogenesis have changed. For recent reviews, see Donnenberg (2002) and Kenny (2002).

The first stage in EPEC pathogenesis is localised non-intimate attachment of the organism to the intestinal epithelium via the inducible bundle-forming pilus (BFP) encoded by the *bfp* operon on a 50–70-MDa plasmid, designated the EPEC adherence factor (EAF) plasmid. The BFP, a member of the type IV class of fimbriae produced by other pathogenic bacteria, facilitates interbacterial aggregation, leading to a characteristic local adherence (LA) pattern. EPEC strains not possessing the EAF plasmid are commonly termed atypical EPEC and appear to be closely related to Verocytotoxin-producing *E. coli* (VTEC). Whereas humans appear to be the only reservoir for 'typical EPEC', atypical EPECs have been isolated from both humans and animals.

Adherence of EPECs to the epithelial cells induces a variety of signal transduction pathways in the host cell, mediated by various bacterial proteins. One of the most important characteristics of EPEC is attaching and effacing (A/E) demonstrated histologically, characterised by the local effacement of the microvilli and intimate adherence between the bacterium and the host's epithelial cell membrane. This lesion is often characterised by the polymerisation of actin in the host cell, which results in a pedestal-like structure beneath the adherent bacteria (Knutton *et al.*, 1989). Although the EAF plasmid may be not essential for the formation of A/E lesions, its presence may enhance colonisation (Trabulsi, Keller and Gomes, 2002). A simple diagnostic assay for the A/E lesion and confirmatory test for EPEC is the fluorescence actin staining (FAS) technique (Shariff *et al.*, 1993).

Several genes implicated in A/E lesion formation are located on large (35.6 kb) PAI termed the locus of enterocyte effacement (LEE) (McDaniel *et al.*, 1995). This chromosomal locus encodes four distinct elements involved in EPEC pathogenesis. Firstly, there is the type III protein secretion system (TTSS) which is capable of transporting proteins across the cytoplasmic and outer membrane of the bacterial pathogen and the membrane of the host cell where they interfere with host-cell signalling cascades.

Secondly, there are several secreted proteins, including many *E. coli* secreted proteins (Esp), which serve as substrates for the LEE-encoded TTSS. Thirdly, there is the adhesin intimin, a 94-kDa outer membrane protein encoded by the *E. coli* attaching and effacing (*eaeA*) gene in the LEE which is required for intimate adherence of EPEC to host cells at the site of A/E lesions. This protein shows sequence similarity to invasin found in *Yersinia* (Jerse *et al.*, 1990) and plays an important role in virulence and host-cell invasion, including activation of signal transduction pathways induced by products of the TTSS and subsequent pedestal formation (Phillips *et al.*, 2000). The fourth component, termed the translocated intimin receptor (Tir), is a bacterial protein that acts as the receptor for intimin. This too is translocated into the plasma membrane of the host cell via the bacterial TTSS (Kenny *et al.*, 1997).

Proteins reported to be essential for A/E lesion formation are themselves components of the extracellular translocation apparatus. These include EspA (25 kDa), EspB (38 kDa) and EspD (40 kDa). The EspA protein is a major component of a filamentous sheath-like structure, sometimes called a needle complex (NC) or EspA filament which is a characteristic feature of TTSS (Büttner and Bonas, 2002). This NC connects the bacteria to the host cells to form a pore or translocon through which other LEE-encoded effector proteins are secreted. These translocated effector proteins which include Tir, EspF, EspG and MAP (mitochondria-associated protein) perform specific functions or attack specific targets after introduction into and sub-localisation in the host cell. The translocation apparatus protein EspB (formerly called EaeB) may also serve as an effector protein. Besides these proteins there are also many chaperone proteins encoded by the *ces* genes of the LEE. These interact specifically with certain effector or translocator proteins of the TTSS (Crawford, Blank and Kaper, 2002). Although their exact function is not fully understood, they are believed to play a role in providing stability to the secreted cognate protein or participate in the regulation of the transcription of type III substrates (Feldman and Cornelis, 2003).

Besides the *bfp* operon, the BFP plasmid also encodes other genes involved in the regulation of the LEE. These include the plasmid-encoded regulator (Per) transcriptional activator (Gómez-Duarte and Kaper, 1995) which activates transcription of the chromosomal regulator protein LEE-encoded regulator (ler). This in turn upregulates the transcription of other LEE genes including *EspA*, *EspB* and *EspD* (Mellies *et al.*, 1999) and increases the expression of chromosomal *eae*. ler has been shown to be essential for the formation of A/E lesions in EPEC and EHEC O157:H7, since all the genes known to be important for A/E lesion formation are regulated by *ler*, as well as potentially activating other genes or even the entire LEE (Elliott *et al.*, 2000).

Although there are similarities, it is now known that there are important differences between the regulatory proteins involved in pedestal formation by EPEC and EHEC O157:H7 (Campellone and Leong, 2003). Whereas both EPEC and EHEC Tir bind intimin and focus cytoskeletal rearrangement of the host cell, there are also other regulatory proteins with as-yet-undefined roles in pedestal formation that are present in EPEC but absent from EHEC (Campellone and Leong, 2003). Besides the virulence determinant already described, EPEC strains may also possess genes for the production of EAST-1 and cytolethal distending toxin (CDT), although the role of these toxins in diarrhoeal disease remains unclear and not all EPECs encode these toxins (Blank, Nougayréde and Donnenberg, 2002).

Enterohaemorrhagic and other Verocytotoxin-Producing *E. coli*

EHECs have emerged as one of the most important threats to human health (Kaper, Nataro and Mobley, 2004). Strains belonging to serotype O157:H7 are now regarded as model EHEC based on their phylogenetic profiles and histopathology. Since it was first identified in 1983 (Riley *et al.*, 1983) as the cause of two outbreaks of gastroenteritis in the United States, the previously 'rare serotype' *E. coli* O157:H7 has become responsible for an increasing number of large food- and waterborne outbreaks of haemorrhagic colitis throughout the world including the United States, Canada, United Kingdom and Japan (Kaper and O'Brien, 1998).

The term EHEC was originally used to describe strains causing distinct clinical manifestations such as haemorrhagic colitis (HC) and haemolytic uraemic syndrome (HUS) in humans. EHECs are characterised by the production of a distinct group of prophage-encoded toxins, their ability to cause A/E lesions on epithelial cells using mechanisms similar to those found in EPEC and the possession of a 60-MDa plasmid carrying other virulence determinants that may contribute to pathogenesis. Besides O157:H7 and O157:H⁻, *E. coli* strains traditionally belonging to common EPEC serogroups, e.g. O111 and O26 strains, which have acquired the genes and ability to produce these toxins now also belong to the EHEC group. Genetically, it has been proposed that *E. coli* O157:H7 originally evolved in a step-wise manner from an ancestral strain of EPEC O55:H7 (Feng *et al.*, 1998). The close relationship between O157:H7 and O55:H7 is supported by numerous studies showing genetic similarities between these serotypes (Perna *et al.*, 1998; Shaikh and Tarr, 2003), including evidence that the H7 *fliC* genes of O55:H7 and O157:H7 are almost identical (Reid, Selander and Whittam, 1998). However, *E. coli* O157:H7 and O157:H⁻ strains appear to belong to a unique clonal lineage with various phenotypic characteristics that distinguish these strains from other *E. coli*. These include the lack of β-glucuronidase (GUD) activity and the inability of most strains to ferment sorbitol in 24 h, which have been used as diagnostic features to aid methods for their detection and isolation.

Carriage of the toxin genes is, however, evident in over 200 serotypes of *E. coli*, some of which do not possess the additional virulence genes found in EHEC and have not yet been associated with human infection. Other serotypes have been responsible for sporadic cases of infection and outbreaks (Goldwater and Bettelheim, 1994; Johnson *et al.*, 1996). Therefore, EHECs can be regarded as a subgroup of a larger group of toxin-producing *E. coli* which are collectively termed VTEC or Stx-producing *E. coli* (STEC). These bacteria are present in the intestinal tract of a wide range of domestic and wild animals, but farm animals, particularly cattle and sheep, are major reservoirs of these pathogens (Chapman *et al.*, 1993; Blanco *et al.*, 2001). They have also been isolated from a range of non-ruminant animals, including birds, horses, dogs and cats (Beutin, 1999; Wasteson, 2001). Persistent carriage and host associations between particular serotypes and certain animals exist, but others appear to be carried transiently. VTEC infections have been linked to a diverse range of food types, although many are linked directly or indirectly to animal reservoirs, e.g. undercooked beef, unpasteurised milk and cheese. Infection can also occur via contaminated water, contact with animals and their faeces and secondary person-to-person transmission.

The ability of some *E. coli* strains to produce this range of toxins with potent cytotoxicity against Vero cells was first described in 1977 (Konowalchuk, Spiers and Stavric, 1977). These toxins, termed Verotoxins or Verocytotoxins (VT), were later shown to be structurally and functionally similar to the Stx produced by *S. dysenteriae* 1 (O'Brien and Holmes, 1987). Consequently, they are now commonly called Stx. Two major antigenically distinct forms of toxin, termed VT1 (Stx1) and VT2 (Stx2), are produced by VTEC, and strains may carry genes (*vtx/stx*) for the production of either VT1 or VT2 alone or both together.

Both toxins consist of an enzymatically active A subunit (32 kDa) and a pentameric B subunit (7.7 kDa monomers). The B subunits form a ring which mediates binding to a neutral glycolipid receptor. Most of the VT types recognise and bind to globotriaosylceramide (Gb$_3$) receptors (Lingwood, 1996). These receptors are found in many different eukaryotic cells but are particularly abundant in the cortex of the human kidney. The exception to this is the VT2 variant toxin VT2e that preferentially binds to globotetraosylceramide (Gb$_4$). Structurally, VT1 is identical to Stx from *S. dysenteriae* type 1, differing by only a single amino acid in the A polypeptide (Strockbine *et al.*, 1988).

The VT2 group shows substantial sequence heterogeneity, especially in the B subunit where amino acid exchanges give rise to many VT2 variant forms. These include VT2c and VT2d, which are produced by VTEC from both animals and humans (Piérard *et al.*, 1998; Ramachandran *et al.*, 2001; Eklund, Leino and Siitonen, 2002), VT2e, which is found in VTEC commonly associated with oedema disease in pigs (Wasteson, 2001), and VT2f, which is associated with VTEC isolated from pigeons (Morabito *et al.*, 2001). In contrast, the VT1 family appears homogeneous, although variant forms of the *vtx*1 gene have been recently reported (Zhang *et al.*, 2002; Bürk *et al.*, 2003). Furthermore, expression of *vtx*1 is also increased under iron-restricted conditions, by as much as a fivefold increase in VT1 production (Wagner *et al.*, 2002). Under these conditions, toxin expression but not lysis is stimulated, resulting in the accumulation of intracellular and not extracellular VT1.

Despite detecting antibodies against VT in the serum of patients with HUS, the route taken by VT from the intestine to the kidneys or other target organs remained unknown for some time. Recent studies report that VT is possibly transported in the blood by PMNLs, which have been shown to bind both VT1 and VT2 (te Loo *et al.*, 2001). The pathway used by these toxins, by which they exert their action, is similar to that used by the plant toxin ricin and is shown in Figure 28.1. Although the A subunit appears to be the active component, the B pentamer may also induce apoptosis in certain cell lines independently of A subunit activity (Marcato, Mulvey and Armstrong, 2002).

The effects of VTs on intestinal epithelium cells are not completely understood, but animal models of VTEC infection suggest that, in the gut, VT may participate in both immune activation and induction of apoptosis as well as causing proinflammatory responses by host cells, stimulating the influx of acute inflammatory cells, contributing to intestinal tissue damage (Thorpe *et al.*, 2001). Localised production of the host proinflammatory cytokines, TNF-α and IL-1, may also exacerbate toxin-mediated vascular damage (Tesh, Ramegowda and Samuel, 1994).

Although VT2e is thought to be chromosomally encoded, the *vtx* (*stx*) genes of VTEC are encoded by prophages of the lambda (λ) family. As well as facilitating horizontal transfer of toxin and other virulence-associated genes in *E. coli*, the phage cycle and phage induction play a central role in regulating the expression and production of VT. Toxin production has been linked to prophage induction associated with the activity of RecA protease which is part of the bacterial SOS DNA repair system (Schmidt, 2001). Induction of prophage can occur spontaneously upon exposure to certain agents that damage DNA.

These can include ultraviolet (UV) irradiation, mitomycin C and certain fluoroquinolone antibiotics that can induce both the SOS response and high levels of phage induction with subsequent increased production of VT (Kimmitt, Harwood and Barer, 2000; Teel *et al.*, 2002).

VTECs are responsible for a range of clinical conditions and syndromes in humans. An important feature of VTEC infections, particularly those by *E. coli* O157:H7, is the low infectious dose, which is reported to be as low as less than 100 cells (Willshaw *et al.*, 1994). Besides outbreaks and sporadic infection, asymptomatic carriage is another important feature of these organisms that probably contributes to secondary spread of infection. The incubation period is typically 1–14 days, with diarrhoea being the most common clinical feature. This can be mild, non-bloody or bloody diarrhoea accompanied by severe abdominal cramps. Vomiting may occur in 50% of patients, but fever is uncommon. The presence of visible blood in the stool is one feature strongly associated *E. coli* O157:H7 infections (Slutsker *et al.*, 1997). The quantity of blood in the faeces may be as little as a few streaks, but the condition can progress to HC, which is characterised by bloody diarrhoea in which the stool is almost entirely comprised of blood. This severe condition can be life threatening, especially in patients at the extremes of age.

In its severest form, infection can progress to diseases such as HUS, a cause of acute renal failure in children, characterised by a triad of microangiopathic haemolytic anaemia, thrombocytopenia and acute renal impairment. Approximately 90% of HUS cases are reported to be VTEC associated (Rose and Chant, 1998). Also involving features characteristic of HUS, but with the additional involvement of the central nervous system and neurological problems, is thrombotic thrombocytopenic purpura (TTP) which, although rare, is more common in adults, especially the elderly. Complete HUS/TTP is commonly established 6 days after the onset of the gastrointestinal symptoms. The mortality rate is often higher in infants and the elderly where infection may lead to clinical complications, and mortality rates as high as 35% have been reported (Carter *et al.*, 1987). The number of cases of VTEC infection developing into HUS depends on the age of the patient and is often dependent on the serotype and toxin type expressed by the infecting organism. Significant risk factors for HUS include age less than 15 years or greater than 65 years (Dundas *et al.*, 2001). Severity of infection in humans, particularly the likelihood of HUS and bloody diarrhoea, has been linked to the carriage of *vtx*2 (Eklund, Leino and Siitonen, 2002). Epidemiological evidence from human infections also implicates VT2 as a significant predictive risk factor for HUS. In the longer term, whilst many patients infected with VTEC recover, 15–40% of survivors have renal sequelae (Fitzpatrick, 1999). Production of VT1 is commonly associated with strains found in animals and milder forms of infection.

Besides VT, strains of EHEC and other VTEC carry a varied repertoire of additional putative virulence determinants (Law, 2000) that may or may not play a direct role in pathogenicity. Some genes, e.g. the LEE, have been fully characterised and extensively studied in EPEC as well as VTEC, while the role of others remains unclear. Functional and regulatory differences exist between the LEE of EPEC and EHEC, although the LEE of *E. coli* O157 and other VTEC have the same distinct regions and genes found in EPEC. A strong association between severe human disease, particularly HC and HUS, and the capacity of VTEC to produce intimin-mediated A/E lesions has been reported. However, HUS has occurred in patients infected by *eaeA*-negative VTEC (Paton and Paton, 1998). Other adhesins and attachment mechanisms facilitating cellular attachment besides intimin have been reported (Paton and Paton, 2002; Jenkins *et al.*, 2003), but their role in colonisation and pathogenesis has yet to be fully elucidated.

Most *E. coli* O157:H7 and other EHEC strains possess a large (60 MDa) plasmid (pO157) encoding many putative virulence genes and several IS elements (Burland *et al.*, 1998). Among these is an EHEC-specific plasmid-encoded haemolysin termed enterohaemolysin (Ehx), which is genetically related to α-haemolysin except that it does not show activity against lymphocytes (Bauer and Welch, 1996). Although there is good association (89%) between VT and Ehx production (Beutin *et al.*, 1989), the precise role of Ehx in disease is not fully understood. Several other haemolysins (cytolysins) have been described for *E. coli* from different pathotypes, of which the chromosomally encoded α-haemolysin is probably the best characterised (Beutin, 1991). Also on pO157 are many putative virulence factors, including EHEC catalase peroxidase, sequences with similarity to a family of toxins termed large clostridial toxins (LCT) and an extracellular serine protease (EspP). The latter may contribute to the mucosal haemorrhage in patients with EHEC-associated HC (Karch, Schmidt and Brunder, 1998). DNA sequences homologous to the gene encoding EAST-1 (*astA*) found in EAggEC are also found in the chromosome in many EHEC strains, including O157:H7 as well as other VTEC (Kaper *et al.*, 1998).

LABORATORY METHODS FOR ISOLATION AND DETECTION OF PATHOGENIC *E. COLI*

Whilst most strains of *E. coli* grow well on a range of microbiological culture media, the growth and isolation of some pathogenic strains requires specific methodology. Strains of pathogenic *E. coli* can be phenotypically identical to commensal *E. coli* strains, whereas others may give rise to atypical reactions with particular biochemical tests which aid the identification of *E. coli*. Rapid lactose fermentation remains a key diagnostic feature of media used for the initial isolation or subsequent confirmation of *E. coli*. MacConkey agar and *E. coli* broth are widely used for the initial isolation and confirmation of suspect *E. coli*, respectively. Strains of *E. coli* are commonly distinguished from other faecal coliforms by their ability to grow and produce gas from lactose at 44 °C and indole production from tryptophan. However, these two tests are not always exclusive to *E. coli* as other bacteria, e.g. *Klebsiella*, can give rise to false-positive results. Furthermore, strains of EIEC often ferment lactose slowly or not at all, which together with the absence of indole production and synthesis of lysine decarboxylase, can mean that they are not recognised as *E. coli*.

The presumptive identification of *E. coli* has been improved by the introduction of chromogenic media that provide better diagnostic characteristics mediated by specific enzyme activity which yields colonies of a distinct colour. Most chromogenic substrates used in *E. coli* specific media rely upon the activity of GUD which is prevalent in approximately 95% of *E. coli* strains. A notable exception, however, is *E. coli* O157:H7, which is largely GUD negative. Another common enzyme exploited in chromogenic media is β-D-galactosidase which is responsible for lactose fermentation and common in coliform bacteria, including *E. coli*. Some media contain individual chromogenic substrates to enable specific identification of the target organism, whereas others contain more than one substrate, which enables a differential count or presumptive identification to be made. Fluorogenic substrates that follow the same principle are also available, although they are now less popular because of the requirement to observe plates under long-wave UV light and the problems caused by diffusion of fluorescence through the medium.

Lactose fermentation remains a useful diagnostic feature of media for the isolation of urinary pathogens, including *E. coli*. A good example of such a medium is cysteine lactose electrolyte-deficient (CLED) agar which is used for routine diagnostic urinary bacteriology. This medium is recommended because it is reported that 1.5% of *E. coli* isolated from urine require cysteine and that subsequent broths used for their identification will require cysteine supplementation (McIver and Tapsall, 1990). The non-selective medium CLED supports the growth of a wide range of urinary pathogens whilst preventing swarming of *Proteus* spp. Lactose-fermenting organisms, including *E. coli*, lower the pH of the medium, which turns from green to yellow. The development of chromogenic media combining the basal CLED medium with various chromogenic substrates has the potential to improve presumptive identification of urinary isolates (Fallon *et al.*, 2003).

The correlation between specific O and H antigens with different pathotypes of *E. coli* has led to serotyping being used for identification purposes. Whilst this remains useful for certain pathotypes associated with infection, especially those that comprise limited serotypes

and strains belonging to distinct clonal lineages, this becomes less reliable when the pathogenicity-associated genes are located on mobile genetic elements. Consequently, there is no longer a clear distinction between certain pathotypes of *E. coli* based on serotyping. Detection of specific antigens or toxins associated with a particular *E. coli* pathotype using appropriate immunological methods such as ELISA can improve detection and identification of these strains. This approach has been used to confirm ETEC colonies using a GM1 ganglioside ELISA to detect LT and ST, which compared favourably against a gene probe (Sommerfelt *et al.*, 1988). Numerous commercial assays are available for the detection of the somatic O antigen of *E. coli* O157 and also for the detection of VT from culture supernatants or directly from sample enrichments, thus enabling detection of all VTEC in clinical samples and foods (Bettelheim and Beutin, 2003).

Although phenotypic tests remain useful for the presumptive identification of pathogenic *E. coli*, the introduction of rapid molecular-based technologies has revolutionised clinical diagnosis. Detection and confirmation of specific DNA sequences associated with known pathogenicity-associated genes or conserved regions unique to a particular pathotype can be used to aid confirmation of the presence or identity of these bacteria. DNA probes and techniques such as polymerase chain reaction (PCR) can be applied directly to clinical samples and foods. Alternatively, PCR can be applied directly to suspect colonies to confirm the presence of specific gene sequences. For the detection of ETEC, this has included using non-radioactively labelled oligonucleotide DNA probes and PCR targeted against the genes encoding LT and ST (Yavzori *et al.*, 1998) and EAST-1 (Yamamoto and Echeverria, 1996). Popular targets for probe and PCR-based detection of EPEC strains include the EAF plasmid (Franke *et al.*, 1994) and the gene encoding BFP (*bfpA*) (Gunzburg, Tornieporth and Riley, 1995). Demonstrating AA pattern in the HEp-2 assay is used to confirm the presence of EAggEC, but this has been improved by the development of an EAggEC-specific probe (Baudry *et al.*, 1990). Detection of EAggEC and DAEC strains has been improved by using PCR primers targeted against various plasmid and chromosomally encoded genes associated with adherence and colonisation.

Detection of EHEC strains, particularly *E. coli* O157:H7, has received much attention recently owing to the risks posed by these bacteria and the severity of the infection they cause. Unlike typical *E. coli*, including other VTEC strains, most O157:H7 strains share the inability to ferment sorbitol in 24 h, which is exploited in selective plating media, the most commonly used medium being sorbitol MacConkey agar (SMAC). Since it was described for the differentiation of *E. coli* O157:H7 from other *E. coli* in clinical samples (March and Ratnam, 1986), SMAC and modified versions of this medium have been universally adopted as the medium of choice for the isolation of *E. coli* O157. A limitation of relying solely on the lack of sorbitol fermentation for the isolation of *E. coli* O157 is the existence of sorbitol-fermenting (SF) strains (Fratamico, Buchanan and Cooke, 1993) and O157 strains that display both SF and GUD activity (Gunzer *et al.*, 1992). Alternatives to SMAC for O157 isolation include chromogenic media (Restaino *et al.*, 1999), including some that allow isolation and recognition of other VTEC (Bettelheim, 1998) and blood agar containing washed sheep red cells, which allows recognition of Ehx production (Beutin *et al.*, 1989).

Detection of nucleotide sequences related to the toxin (*stx/vtx*) genes including *vtx2* variants using the PCR technique has become a popular method for detecting VTEC in foods and clinical samples (Lin *et al.*, 1993). DNA probes have also been developed for the confirmation of suspect isolates (Samadpour, Ongerth and Liston, 1994). Alternative targets for DNA probes and PCR include the haemolysin (*hlyA*) gene (Lehmacher *et al.*, 1998), the *E. coli* attaching and effacing (*eae*) gene (Louie *et al.*, 1994), the GUD (*uidA*) gene (Feng, 1993), the 60-MDa plasmid found in O157:H7 and other VTEC (Levine *et al.*, 1987) and the O157 *fliC* (flagellin) gene (Gannon *et al.*, 1997).

Multiplex PCR has been used to simultaneously detect EHEC, EPEC and ETEC in faecal samples from patients with watery diarrhoea, HC and HUS using primers targeted against *eae*, *bfp*, *stx1*, *stx2*, *lt* and *st* (Vidal *et al.*, 2004).

ANTIBIOTIC RESISTANCE

Most *E. coli* are sensitive to antimicrobial agents active against Gram-negative bacteria, although resistance among enteric bacteria, including *E. coli*, has increased markedly over the past 50 years since of the widespread use of antibiotics (Houndt and Ochman, 2000).

Multiple antibiotic resistance can be acquired via plasmids or drug efflux systems. The chromosomal multiple antibiotic resistance locus in *E. coli*, designated *marA*, influences the expression of the *acrAB* efflux pump and other chromosomal genes, resulting in resistance to a range of antibiotics including tetracycline and many unrelated antibiotics including chloramphenicol, β-lactams and nalidixic acid (George and Levy, 1983). Susceptibility of *E. coli* strains to amoxicillin has decreased over recent years owing to the presence of TEM-1 and TEM-2 β-lactamase. The effectiveness of cotrimoxazole and trimethoprim has been reduced by frequent carriage on plasmids and integrons of *dhfr* resistance genes (Yu *et al.*, 2003). Because resistance is high, a fluoroquinolone or nitrofurantoin should be considered for empirical treatment. One study of *E. coli* urinary isolates from females in the United States from 1995 to 2001 revealed resistance rates to be constant for ampicillin (36–37%) and co-trimoxazole (15–17%), with increasing resistance to ciprofloxacin (0.7–2.5%) and low resistance to nitrofurantoin (0.4–0.8%) (Karlowsky *et al.*, 2002), which together with fosfomycin-trometanol remain highly active against urinary Enterobacteriaceae with over 90% of *E. coli* reported to be susceptible (Chomarat, 2000). One prospective study has identified prior UTI is a common risk factor for resistance to different antibiotics use to treat UTI (Sotto *et al.*, 2001).

In a study of the incidence of antibiotic resistance in *E. coli* isolates from blood and CSF between 1991 and 1996 in England and Wales, a significant increase in the incidence of strains resistant to ampicillin and ciprofloxacin was reported (Threlfall *et al.*, 1998). Between 1990 and 1999, ciprofloxacin resistance in *E. coli* from bacteraemias rose from 0.8% to 3.7% in England and Wales (Livermore *et al.*, 2002). Fluoroquinolone resistance is normally acquired by a single mutation in *gyrA*, with high-level resistance probably the result of a second mutation in *gyrA* and *parC*, but mutations in other genes may play a role (Kern *et al.*, 2000). Resistance to β-lactam antibiotics is most frequently due to the production of β-lactamases with some variants able to inactivate the newer cephalosporins. These variants, termed extended-spectrum β-lactamases (ESBL), are predominantly plasmid mediated and arose through single mutations in the progenitor enzymes TEM-1,2 and SHV-1 (Gniadkowski, 2001). Recently, a chromosomal β-lactamase (CTX-M) from *Kluyvera* spp. has become mobilised onto plasmids. Some variants are particularly common in *E. coli* both in the United Kingdom (Woodford *et al.*, 2004) and elsewhere (Munday *et al.*, 2004). Strains of EHEC O157 have developed resistance to antibiotics (Meng *et al.*, 1998), but their use in the treatment of EHEC infections remains controversial. The spread of ESBLs, together with the emergence of broad-spectrum resistance among *E. coli* populations, will pose a significant clinical problem in the future.

MANAGEMENT OF *E. COLI* INFECTION

Most infections caused by *E. coli* respond to treatment with appropriate antibiotics, although higher mortality rates are associated with septicaemia, especially in immunocompromised and elderly patients. In cases of UTI, antibiotic treatment must often be started before antimicrobial susceptibilities are known, and if CTX-M ESBL-producing strains are common, a carbapenem is the preferred choice. Antimicrobial susceptibility testing is important both to confirm whether the correct empirical choice has been made and to detect resistance. As *E. coli* is generally spread via the faecal–oral route, IPEC infections can be prevented by good hygiene and the provision of clean water and good food handling practices, especially avoiding cross-contamination from raw to cooked foods and ensuring adequate cooking and proper storage of foods. This also applies to hospitals where handwashing by nursing staff and environmental cleaning are important in controlling outbreaks.

In cases of VTEC infection, early recognition of the disease and appropriate clinical management of the patient are critical. Patients with acute gastroenteritis should receive supportive measures that ensure maintenance of fluid and electrolyte balance. Antimotility and antidiarrhoeal agents may be inappropriate as these may lead to the progression of the disease to HUS/TTP along with the use of certain antibiotics (Todd, Duncan and Coia, 2001). Previous reports have warned that antibiotic treatment of children with *E. coli* O157:H7 infection increases the risk of HUS (Wong *et al.*, 2000). This observation conflicts with the findings of a retrospective study of the 1996 Sakai City outbreak which reported that early use of fosfomycin reduced the risk of HUS in children (Takeda *et al.*, 1998). However, *in vitro* evidence suggests that treatment with four quinolones increases the risk of toxin release (Kimmitt, Harwood and Barer, 1999) and that certain fluoroquinolone antibiotics such as ciprofloxacin induce VT prophages and enhance toxin production by VTEC (Zhang *et al.*, 2000). Patients who develop HUS are commonly treated by dialysis, and in extreme cases renal transplantation may be necessary. Current clinical management relies on maintenance of the patient and supportive care. Various novel approaches, including vaccines and other experimental therapies, have been developed and are undergoing evaluation in clinical trials (Kaper and O'Brien, 1998), although currently no alternative approaches have been routinely adopted for the treatment of VTEC infections.

SHIGELLA

INTRODUCTION

The disease caused by *Shigella* spp., dysentery, was recognised by Hippocrates as a condition characterised by the frequent passage of stools containing blood and mucus. He also recognised some aspects of the epidemiology, noting that after a dry winter and a wet spring the number of cases became more frequent in the summer. It was not until 1875 when the cause of amoebic dysentery was discovered and later when Shiga isolated the organism that was later known as *Shigella dysenteriae* that the two dysenteric illnesses could be clearly differentiated.

Dysentery has always played an important part in human history, influencing the course of military campaigns. Although the mortality rate among soldiers with bacillary dysentery was approximately 2.5%, the high attack rate meant that as many men died of this cause as the effects of battle. At crucial times, many soldiers were infected and incapable of performing military duties. When people are herded together in gaols or on ships, epidemic dysentery may sweep through the population with devastating effect. Dysenteric disease remains an ever-present danger in the refugee camps of modern times.

The first of the genus to be identified was *S. dysenteriae*, which for many years was known as Shiga's bacillus (Shiga, 1898). *Shigella flexneri* was originally described by Flexner in 1900 (Flexner, 1990). *Shigella sonnei* was first isolated in 1904, but it was not until 1915 that its pathogenic potential was recognised by Sonne (Sonne, 1915).

DESCRIPTION OF THE ORGANISM

Shigellae are small Gram-negative rods, non- or late lactose fermenters, fermenting sugars without gas production. Although non-motile using conventional tests, recent studies have shown the presence of flagellar genes and expression of a motile phenotype under certain conditions (Giron, 1995). For details of the biochemical features of *Shigella* spp., see Table 28.3.

Shigellae do not survive as well as salmonellas in clinical specimens and so should be plated onto isolation medium as quickly as possible. Some selective media, deoxycholate citrate agar (DCA) or Wilson and Blair designed for the isolation of Salmonella, can be too inhibitory for Shigella; thus, it is useful to use a non-inhibitory medium such as MacConkey, Salmonella–Shigella (SS) or xylose–lysine–deoxycholate (XLD) in parallel (see below).

CLASSIFICATION

The genus *Shigella* is part of the Enterobacteriaceae, which have been classified together on the basis of the disease they cause. The shigellae have been shown to be closely related to *E. coli* by DNA hybridisation, isoenzyme analysis and the presence of Shiga-like toxin (Brenner *et al.*, 1969; Crosa *et al.*, 1973; Goullet, 1980). Recently, the full genome of *Shigella flexneri* 2a was sequenced and comparisons with *E. coli* K12 showed that *Shigella* were phylogenetically indistinguishable from *E. coli* (Jin *et al.*, 2002; Wei *et al.*, 2003). It has been suggested that *Shigella* spp. should be reclassified and new nomenclature introduced to clarify the relationship between these two species (Lan and Reeves, 2002). They are most closely related to the EIEC, which also are capable of producing a dysenteric illness.

The genus *Shigella* contains four species: *S. dysenteriae*, *S. flexneri*, *Shigella boydii* and *S. sonnei*. The species are differentiated on the basis of simple biochemical tests and serology of their LPSs: *S. dysenteriae* are non-mannitol fermenters and their O-polysaccharide LPS is unrelated antigenically to the other shigellas. *Shigella flexneri* and *S. boydii* ferment mannitol, but the latter are antigenically distinct from *S. flexneri*. All *S. flexneri* O-specific polysaccharides of the LPS antigens have a common rhamnose-containing tetrasaccharide (Robbins, Chu and Schneerson, 1992). Included among the *S. boydii* are variants, some of which are capable of producing gas from glucose. *Shigella sonnei* differs from the other members of the genus in that it is a late lactose fermenter, which can be detected with ONPG, unlike other members of the genus which are non-lactase fermenting (Tabe 28.3). The O-side chain is composed of a repeating disaccharide containing an unusual monosaccharide altruronic acid (Robbins, Chu and Schneerson, 1992). This antigen is identical to that of *Plesiomonas shigelloides* 017 which make up most human isolates (see Chapter 16) and is the source of the cross-reactions between these two species.

The O-polysaccharide of the LPS is used to divide the species into serotypes as follows: *S. dysenteriae* into 12 serotypes, *S. flexneri* into 13, *S. boydii* into 18 and a single serotype of *S. sonnei*.

PATHOGENESIS

Shigella spp. are largely limited to mucosal infection of the distal ileum and colon; intestinal perforation, although rare, has been reported (Azad, Islam and Butler, 1986); toxic megacolon may occur

in up to 3% of patients in developing countries (Bennish, 1991). Bacteraemia is the most important lethal complication (Struelens *et al.*, 1990); meningitis and pneumonia due to general spread are rare (Bennish, 1991). Seizures are common especially with *S. dysenteriae* infection, and these resemble febrile seizures, although they occur in children over 5 years old (Rawashdeh, Ababneh and Shurman, 1994; Thapa *et al.*, 1995).

To establish an infection, *Shigella* must invade the enteric epithelium via the basolateral epithelial surface. The organisms gain access to the basolateral side via M-cells, the specialised antigen-presenting cells of the lymphoid follicles (Wassef, Keren and Mailloux, 1989; Perdomo *et al.*, 1994). Subsequent killing of macrophages in the submucosa leads to the release of cytokines (Zychlinsky, Prevost and Sansonetti, 1992; Zychlinsky *et al.*, 1994), which attracts many PMNLs. A path for *Shigella* invasion is opened via the tight junctions which are broken down by the PMNLs (Perdomo, Gounon and Sansonetti, 1994) (Figure 12b.1). Recent research has shown that *S. flexneri* can also invade the colonic epithelial layer by manipulation of the epithelial cell tight junction proteins in an M-cell/PMN-independent manner (Sakaguchi *et al.*, 2002).

Once inside the submucosa, *S. flexneri* makes contact with the basolateral membrane of the epithelial cells (Tran Van Nhieu *et al.*, 1999). Shigellae have a virulence plasmid that encodes two loci required for invasion: the *ipa* locus and the *mxi-spa* locus (Jennison and Verma, 2004). The *ipa* operon encodes the invasion plasmid antigens IpA, IpaB, IpaC and IpaD, whereas the *mxi-spa* operon encodes the components of a type III secretion system that delivers the Ipa proteins from the bacterial cytoplasm to the cytoplasmic membrane or cytosol of the host cell (Menard, Sansonetti and Parsot, 1993; Menard *et al.*, 1994). Bacterial proteins are secreted into the epithelial cell's cytoplasm through a pore formed by IpaB and IpaC (Blocker *et al.*, 1999; Harrington *et al.*, 2003). IpaC and IpgD polymerise actin within the host cell and create cell-surface extensions which form around the bacterium, inducing the epithelial cell to take up *S. flexneri* into a vacuole (Bourdet-Sicard *et al.*, 1999; Tran Van Nhieu *et al.*, 1999; Niebuhr *et al.*, 2002). IpaB, and possibly IpaC, lyses the vacuole, releasing the bacteria into the epithelial cytoplasm, where they replicate (High *et al.*, 1992; De Geyter *et al.*, 1997). The epithelial cells lyse, probably because of the host inflammatory response to the replicating bacteria (Jennison and Verma, 2004).

Shigella exploits the host-cell actin via the outer membrane protein, IcsA, to move through the host-cell cytoplasm and into adjacent epithelial cells (Bernardini *et al.*, 1989; Purdy, Hong and Payne, 2002). It does this by creating a polymerisation actin tail behind the bacterium, by expressing IcsA on one pole of the bacterium, and propelling *S. flexneri* through the cytoplasm until it contacts the cytoplasmic membrane. The force of the contact creates a protrusion into the neighbouring epithelial cell. Both membranes are lysed, thus releasing *S. flexneri* into the neighbouring epithelial cell (Jennison and Verma, 2004).

Shigella dysenteriae type 1 produces a potent exotoxin (Stx) which enhances local vascular damage. It is structurally closely related to the Shiga-like toxins of *E. coli*. It produces fluid accumulation in the rabbit ileal loop model but does not appear to be necessary for intracellular killing of mucosal epithelial cells. Systemic distribution of the toxin results in microangiopathic renal damage with subsequent development of the HUS (Obrig *et al.*, 1988). The Stx is a bipartite molecule with an enzymatic A subunit activated by proteolytic cleavage and a multimeric receptor-binding B subunit (O'Brien *et al.*, 1992). The binding subunits, like CT, bind to a host-cell-surface glycolipid, in this case Gb_3, which possesses a Gala-1,4-Gal as the carbohydrate moiety. Stx exerts its lethal effect by inhibiting the host-cell 60S ribosomal subunit by cleaving a specific adenosine residue of the 28S rRNA. Recent studies have shown that Stx can cause apoptosis in certain cell types (Cherla, Lee and Tesh, 2003).

Table 28.3 Typical biochemical reactions of *Shigella* spp. and *E. coli*

	Indole	LDC	ODC	Motility	Glucose with gas	ONPG	D-Mannitol
S. sonnei	−	−	+	−	−	+	+
S. dysenteriae	−[a]	−	−	−	−	−	−
S. boydii	−	−	−	−	−	−	+
S. flexneri	−	−	−	−	−	−	+
E. coli	+	+	+	+	+	+	+

LDC, lysine decarboxylase; ODC, ornithine decarboxylase; ONPG, *ortho*-nitrophenyl-β-D-galactopyranoside.
[a] *Shigella dysenteriae* type 2 is positive.

EPIDEMIOLOGY

It is estimated that there are 164.7 million cases of bacillary dysentery annually, of which 163.2 million are in developed countries and 1.5 million in industrialised countries (Kotloff et al., 1999). Approximately 1.1 million people die from shigellosis each year – 61% of these are children under 5 years old (Kotloff et al., 1999). Infection is spread by the faecal–oral route with great ease, as the infective dose is low (10^2–10^3 cfu). Shigella spp. are obligate human pathogens and do not infect other hosts, although experimental infection can be achieved in primates. Most cases of bacillary dysentery spread from person to person, and this may occur rapidly, especially in closed communities and when individuals are brought together in large numbers and sanitary arrangements are inadequate. Dysentery is a disease of poverty, and the incidence of infection can be correlated with poor housing and sanitary facilities and when large populations are suddenly displaced through war or natural disaster. Epidemics of disease may be both water and food borne and can be associated with faecally contaminated wells.

Shigellosis is a disease of children under 5 years old in developing countries, reaching a peak at between 18 and 23 months (Henry, 1991). Disease is more severe in children who are malnourished (Bennish, 1991). In one study, all of the fatal cases were severely malnourished (Thapa et al., 1995). Patients suffered convulsions, bacteraemia, renal failure, intestinal perforation and toxic megacolon (Rawashdeh, Ababneh and Shurman, 1994; Thapa et al., 1995). Shigella dysentery has a significant effect on childhood mortality and on growth in excess of dysenteric illnesses of other aetiologies (Bennish, 1991; Henry, 1991).

In developing countries, S. flexneri and S. dysenteriae are most frequently isolated, whereas S. sonnei and S. flexneri predominate in developed countries (Kotloff et al., 1999). The change in pattern of infection occurred in industrialised countries more than 50 years ago and resulted in a fall in mortality from dysentery as the more virulent forms became less common (Figure 12b.2). Haemolytic uraemic syndrome is associated with infection with S. dysenteriae.

Between 2000 and 2004, the annual number of cases per 100 000 population per year in the United Kingdom varied between 1.6 and 1.1 (data from the Health Protection Agency, Communicable Disease Surveillance Centre, London). A study of infectious intestinal disease among cases in the community in England, carried out between August 1993 and January 1996, showed that Shigella spp. was isolated in 0.8% of faecal samples from cases and none of the controls (Tompkins et al., 1999). However, in many industrialised countries the scale of bacillary dysentery is not clearly established as many cases do not come to medical attention or are neither investigated bacteriologically nor notified to public health authorities. Infection is more common in children than in adults. Shigella is an important cause of travellers' diarrhoea, especially in individuals returned from developing countries (Vila et al., 1994). Data from the Health Protection Agency on Shigella isolates received between April and June 2004 showed that 24% of all reports of Shigella infection recorded recent travel abroad, 63% of which specified travel to the Indian subcontinent, 19 of those were due to S. sonnei (Health Protection Agency, 2004).

CLINICAL FEATURES

Dysentery is an infection, not a toxaemia, and symptoms result from changes in the bowel wall, caused by multiplication of the organisms after ingestion. The incubation period appears to be 2–3 days on average.

Clinical features vary markedly among the different types of shigellae, sonne dysentery being the mildest. Diarrhoea is frequently the only symptom in sonne dysentery and consists of several loose stools in the first 24 h. This acute stage passes off rapidly, and in most cases, by the second day, the condition largely subsides. From then on, the frequent bowel movement is more of a social inconvenience. In some, the infection can be a mild and trivial affair; for others the

symptoms are more severe and vomiting can lead to dehydration, especially among the young. Fever is usually absent, but sometimes, at the onset, there is a sharp rise in temperature and general signs and symptoms may suggest meningeal involvement. Isolation of Shigella from blood and CSF specimens is rare (Struelens et al., 1990; Langman, 1996). Shigella septicaemia is most common in infants and in people with malnutrition or in the immunocompromised (Huebner et al., 1993; Huskins et al., 1994). Data published by Gupta et al. (2004) identified infants and elderly patients (>80 years) as having the highest rates of Shigella bacteraemia. Shigella dysenteriae, S. flexneri and S. boydii were isolated from blood cultures more frequently than S. sonnei, raising the possibility that these serotypes may be more invasive (Gupta et al., 2004).

Abdominal pain is not usually a prominent symptom, although the illness may mimic appendicitis or even intussusception in babies. In some severe cases, the onset may be sudden, with vomiting, headache, rigors, severe colic and exhausting diarrhoea. This may lead to dehydration, tetany and meningeal signs.

While the illness caused by S. flexneri and S. dysenteriae may be no worse than sonne dysentery, in most cases the patient is more acutely ill with constitutional upset. Diarrhoea is severe and persistent. Faecal material soon gives way to mucus and blood, and it is some days before faecal material returns. Abdominal pain and tenderness are frequent features, and the patient is toxic and febrile. The pulse is rapid and weak, and the patient becomes feeble, the skin punched, the tongue coated and urine scanty. The patient suffers from great thirst and cramps in the limbs; confusion or delirium may ensue. Generally, symptoms gradually subside over a period of 10–14 days, but relapses do occur, with a flare-up of the dysenteric diarrhoea and rapid death. Fulminant choleraic or gangrenous forms are usually due to S. dysenteriae. Bacteraemia, pneumonia, meningitis, seizures and HUS are recognised complications.

During the acute stage of the disease, Shigella organisms are excreted in large numbers in the faeces, but during recovery the numbers fall, although the organism may remain in the faeces for several weeks after the symptoms have subsided.

LABORATORY DIAGNOSIS

The organisms are usually present in large numbers in the intestinal mucus or the faeces in the early stages. Freshly passed stools should be examined, although rectal swabs showing marked faecal staining may be used. If the specimen includes blood and mucus, these should be plated directly. When faeces are kept alkaline, shigellae may survive for days, but in acid stools they die in a few hours. If there is likely to be much of a delay, it may be useful to collect faeces into buffered 30% glycerol saline solution. Nothing is to be gained by culturing urine or blood, since these are invariably negative. Numerous waterborne epidemics are on record, and in several instances, shigellas have been isolated from water itself.

Microscopic examination of the mucus in the early stages of acute bacillary dysentery shows a marked predominance of polymorphonuclear cells and red cells.

Faecal material is plated out directly onto deoxycholate citrate agar (DCA), Salmonella–Shigella (SS) agar or xylose–lysine–deoxycholate (XLD) medium. Shigellae appear as small colourless or slightly pink colonies on DCA and as a pink or red colonies with, in some cases, a pink or yellow periphery on XLD. A few strains grow poorly on inhibitory media, and it is advisable to use MacConkey agar and to examine any non-lactose-fermenting colonies after overnight incubation. A preliminary identification can often be made by slide agglutination with specific antisera from the growth on the primary plate. In any event, full biochemical and serological tests must be performed on subcultures that have been checked for purity.

The biochemical identification of Shigella is complicated by the similarity of some strains of other genera, in particular strains of Hafnia,

Providencia, Aeromonas and atypical *E. coli*; non-lactose-fermenting or anaerogenic strains of *E. coli* are a common problem.

The PCR can be used to identify *Shigellae* either from a cultured isolate or directly from faeces. The *ipa*H gene is a good target for PCR as it is a multiple-copy element found on the large invasion plasmid (pINV) and the chromosome (Islam *et al.*, 1998; Dutta *et al.*, 2001). This PCR test also detects EIEC, as this group of diarrhoeagenic *E. coli* also harbours pINV. Molecular typing techniques, such as pulsed-field gel electrophoresis (PFGE), can be used to investigate outbreaks of *Shigella* infection and to trace the source of infection (Chen *et al.*, 2003; Surdeanu *et al.*, 2003).

Safety

Most potential isolates belong to Advisory Committee on Dangerous Pathogens (ACDP) Group 2, and good laboratory practice is all that is required. Should *S. dysenteriae* be suspected, then all manipulations should be carried out in a class I microbiological safety cabinet in the containment level 3 laboratory. Since the infectious dose is between 10 and 100 cfu, great care must be taken to avoid environmental contamination. Gloves should be worn and the bench surfaces disinfected (e.g. with clear phenol) before removing the used gloves, followed by careful handwashing. Aerosolisation is also a problem – avoid splashes.

Antimicrobial Sensitivities

The earliest reports described *Shigella* spp. as susceptible to sulphonamides, but within a few years resistant strains began to appear. Between 1947 and 1950, antibiotic-resistant strains became predominant in many countries. In 1992, it was recommended that should antibiotic therapy be needed for the treatment of shigellosis in developed countries, children should be treated with ampicillin or trimethoprim plus sulphamethoxazole and adults should be treated with one of the fluoroquinolone antimicrobials (Bennish and Salam, 1992). However, 46% of isolates of *S. dysenteriae*, *S. flexneri* and *S. boydii* received by the reference laboratory at the Health Protection Agency in London between 1995 and 1996 were resistant to ampicillin and trimethoprim (Cheasty *et al.*, 1998). Fifteen per cent of *S. sonnei* isolates were resistant to both these antimicrobials and 50% were resistant to one or the other (Cheasty *et al.*, 1998). Resistance to fluoroquinolones has been rarely reported, and nearly all *Shigella* isolates are susceptible to these agents (Ashkenazi *et al.*, 2003). Therefore, if treatment must be commenced before laboratory-based tests are available, it is still recommended that fluoroquinolone antibiotics, such as ciprofloxacin, should be used to treat adults but that nalidixic acid should be used to treat children (Cheasty *et al.*, 1998; Ashkenazi *et al.*, 2003; Phavichitr and Catto-Smith, 2003). Fluoroquinolones are not currently approved for children.

Most cases of shigellosis, especially those caused by *S. sonnei*, are mild and do not require antibiotic therapy. Symptomatic treatment with the maintenance of hydration is all that is required. However, treatment with a suitable antibiotic is necessary in the very young, the aged or the debilitated and in severe infections. Because of the high incidence of the antibiotic resistance among shigellae, it is necessary to determine the resistance pattern of the strains before starting treatment where possible.

MANAGEMENT OF INFECTION

Oral Rehydration Therapy

Oral rehydration therapy (ORT) is the cornerstone of worldwide efforts to reduce mortality from acute diarrhoea, and while toxaemia plays a major role in the severity of illness in shigellosis, dehydration has to be corrected. The World Health Organization (WHO)/United Nations Children's Fund (UNICEF) oral rehydration formula (ORS) contains glucose and sodium in a molar ratio of 1.2:1. Potassium chloride is added to replace potassium lost in the stool. Trisodium citrate dihydrate (or sodium bicarbonate) corrects metabolic acidosis caused by faecal loss of bicarbonate. ORT has proved to be a simple, safe and cost-effective means of preventing and managing dehydration in the community (Bhattacharya and Sur, 2003).

When patients, especially babies and elderly persons, are acutely ill, there is much loss of fluids and salts in vomit and in stools, and replacement of fluids may need to be controlled by electrolyte estimations. Intravenous administration seldom needs to be continued for more than 24 h, but children suffering from HUS may require dialysis. Patients with tetany may respond to intravenous normal saline alone or with the addition of 20% calcium gluconate.

On clinical grounds, antibiotics are not required for mild or moderate cases; this applies to virtually every case of dysentery seen in Britain, where 94% of cases are due to *S. sonnei* (Crowley, Ryan and Wall, 1997). In severe form of dysentery due to *S. dysenteriae* and *S. flexneri* where shigellae have penetrated intestinal epithelial cells in large numbers, the use of antibiotics such as ciprofloxacin, pivmecillinam or ceftriaxone (Bennish and Salam, 1992) to kill these intracellular organisms may prevent severe destruction of the intestinal wall. Persistent hiccough or the passage of sloughs in the stools are ominous signs. The choleraic and gangrenous types of bacillary dysentery are almost invariably fatal.

Biotherapy is the use of live organisms or their metabolic products for treatment. The most commonly available sources are fermented milk products such as yoghurt, kefir and buttermilk. Unless pasteurised, these products contain live bacteria, and they have been used for the prevention and treatment of gastrointestinal complaints. Kefir is traditionally made from goats' milk by the addition of kefir granules consisting of *Lactobacillus* spp. and a yeast (*Torula kefir*). *Saccharomyces boulardii* is a non-pathogenic yeast which is sometimes used to treat diarrhoea. It does not multiply in the gut, and there is no permanent colonisation of the bowel (Roffe, 1996).

Antimicrobial Therapy

During the last decade, quinolones such as norfloxacin, ciprofloxacin, ofloxacin and fleroxacin have emerged as drugs of choice for the treatment of various bacterial enteric infections, including shigellosis (Bhattacharya and Sur, 2003). Controlled trials have shown that quinolones in varying regimens, from a single dose (Salam *et al.*, 1994) to 5 days of treatment, significantly reduce the intensity and severity of traveller's diarrhoea as well as shigellosis (Wistrom and Norrby, 1995). Quinolone resistance is presently uncommon among shigellae, but it is inevitable that resistance will develop from increased usage of these agents.

Antimotility drugs such as diphenoxylate (Lomotil) are not recommended in the treatment of diarrhoea, although loperamide, a synthetic antidiarrhoeal agent, has been shown to decrease the number of unformed stools and shortens the duration of diarrhoea caused by *Shigella* in adults treated with ciprofloxacin (Murphy *et al.*, 1993; Bhattacharya and Sur, 2003).

INFECTION CONTROL

The mild and often fleeting nature of the clinical illness associated with sonne dysentery means that patients, often children, remain ambulant and are free to continue their daily activities and act as dispersers of the organism. When the arrangements for the disposal of human faeces are crude or non-existent, heavy contamination of the environment is inevitable, and even with modern sanitation the disease can spread easily and rapidly if there is a break in hygiene standards. The dose of *Shigella* needed to cause infection is very small (less than 200 organisms), and given moist, alkaline, cool and shady conditions, large numbers will survive for many days. Even on dried linen, organisms will survive in the dark for long periods. Environmental contamination, particularly in and around toilet facilities, remains an important focus for continuing outbreaks of dysentery.

Shigellae are sensitive to heat and are killed in 1 hour at 55 °C. At laboratory temperatures below this level, they survive for varying periods of time, and unless strict care is exercised, a laboratory bench will remain a rich source of contamination. Shigellae are readily killed by disinfectants: in 6 h by 0.5% clear phenolics or in 15–30 min by 1% phenol. A handwash of 1% benzalkonium chloride will kill *S. sonnei* in 1 min in a dilution of 1:64. A solution of 3% (1:30) is a useful hand decontamination fluid, which is safe to use on children's hands and has been used to control outbreaks of sonne dysentery in preschool institutions.

Drinking water should be protected and kept well away from toilet facilities as shigella can survive in tap or sterilised water for as long as 4–6 weeks. Waterborne outbreaks tend to be explosive, especially in developing countries. Handwashing, although very important, cannot be guaranteed to remove all dysentery bacilli from hands, which readily become recolonised from inanimate objects.

Given the low infectious dose, cases with diarrhoea should stay at home. Infection is rarely spread in the absence of diarrhoea, and cases with no diarrhoea need not be excluded from work or school. Exceptions to this include those working in the food industry, health-care or nursery staff and children at nursery, who should be excluded or isolated until free from diarrhoea and three consecutive faeces specimens collected at intervals of at least 24 h are negative (PHLS Salmonella Sub-committee, 1990).

Spread by Personal Contact

Sonne dysentery spreads by direct contact and is largely a disease of young children. This reflects the intimate nature of contact between young children and decreases with age as the standard of hygiene increases. Symptomless excreters play a much less important role in the spread of disease than in the acute case. Although splashing occurs with the contamination of the toilet area, door-handles and so on, it is likely that hand-to-hand contact facilitates spread among young children. Likewise, staff-to-patient and patient-to-staff transmission can occur.

Adequate toilet facilities in schools and nurseries, supervised handwashing for young children and separate nappy changing areas in nurseries are simple preventative measures to minimise transmission (Crowley, Ryan and Wall, 1997). Children without symptoms can be allowed to stay at school, provided their hygiene is supervised. The use of antibiotics prophylactically in an attempt to limit the spread should be avoided.

Vaccines

Serum antibodies do not seem to be protective against intestinal shigella infection. Killed shigellae induced humoral immunity only, and research and clinical trials are underway with live attenuated strains. The growing understanding of the mechanisms of pathogenicity involved in *Shigellosis* has contributed to improvements in vaccine development. Mutations in the *S. flexneri* chromosome and the virulence plasmid (pINV) have been used to generate non-invasive live vaccine strains (Jennison and Verma, 2004). Invasive vaccine strains, attenuated by mutations in either virulence or metabolic genes, have also been produced and have been shown to induce a strong immune response (Kotloff *et al.*, 1996). *Shigella* LPS can be complexed to prote-osomes and delivered intranasally, producing a safe subunit vaccine that can also induce a strong serum response (Fries *et al.*, 2001).

Immunity to *S. flexneri* is serotype specific, and vaccination against one serotype will only provide protection to infection by the homologous serotype. Studies have shown that mixing many different serotypes into a vaccine cocktail can result in a cross-reactive immune response (Noriga *et al.*, 1999). Novel approaches include the development of a vaccine strain able to express the O-antigen of more than one serotype (Guan and Verma, 1998).

REFERENCES

Abbott, S. L., O'Connor, J., Robin, T. *et al.* (2003) "Biochemical properties of a newly described *Escherichia* species, *Escherichia albertii*." *Journal of Clinical Microbiology* **41**: 4852–4854.

Agace, W., Hedges, S., Andersson, U. *et al.* (1993) "Selective cytokine production by epithelial cells following exposure to *Escherichia coli*." *Infection and Immunity* **61**: 602–609.

Al Mamun, A. A. M., Tominaga, A. and Enomoto, M. (1996) "Detection and characterization of the flagellar master operon in the four *Shigella* subgroups." *Journal of Bacteriology* **178**: 3722–3726.

Ashkenazi, S., Levy, I., Kazaronovski, V. and Samara, Z. (2003) "Growing antimicrobial resistance of *Shigella* isolates." *The Journal of Antimicrobial Chemotherapy* **51**: 427–429.

Azad, M. A., Islam, M. and Butler, T. (1986) "Colonic perforation in *Shigella dysenteriae* 1 infection." *The Pediatric Infectious Disease Journal* **1**: 103–104.

Bahrani-Mougeot, F., Gunther, N. W. IV, Donnenberg, M. S. and Mobley, H. L. T. (2002) "Uropathogenic *Escherichia coli*." In *Escherichia coli: Virulence Mechanisms of a Versatile Pathogen*. M. S. Donnenberg (ed.). London: Academic Press, 239–268.

Baudry, B., Savarino, S. J., Vial, P. *et al.* (1990) "A sensitive and specific DNA probe to identify enteroaggregative *Escherichia coli*, a recently discovered diarrheal pathogen." *The Journal of Infectious Diseases* **161**: 1249–1251.

Bauer, M. E. and Welch, R. A. (1996) "Characterization of an RTX toxin from enterohemorrhagic *Escherichia coli* O157:H7." *Infection and Immunity* **64**: 167–175.

Bennish, M. L. (1991) "Mortality due to shigellosis: community and hospital data." *Reviews of Infectious Diseases* **13** (Suppl. 4): S219–S224.

Bennish, M. L. and Salam, M. A. (1992) "Rethinking options for the treatment of shigellosis." *The Journal of Antimicrobial Chemotherapy* **30**: 243–247.

Bernardini, M. L., Mounier, J. d'Hauteville, H. *et al.* (1989) "Identification of icsA, a plasmid locus of *Shigella flexneri* that governs bacterial intra- and intercellular spread through interaction with F-actin." *Proceedings of the National Academy of Sciences of the United States of America* **86**: 3867–3871.

Bétis, F., Brest, P., Hofman, V. *et al.* (2003) "The Afa/Dr adhesins of diffusely adhering *Escherichia coli* stimulate interleukin-8 secretion, activate mitogen-activated protein kinases, and promote polymorphonuclear transepithelial migration in T84 polarized epithelial cells." *Infection and Immunity* **71**: 1068–1074.

Bettelheim, K. A. (1998) "Reliability of CHROMagar O157 for the detection of enterohaemorrhagic *Escherichia coli* (EHEC) O157 but not EHEC belonging to other serogroups." *Journal of Applied Microbiology* **85**: 425–428.

Bettelheim, K. A. and Beutin, L. (2003) "Rapid laboratory identification and characterisation of verocytotoxigenic (shiga toxin producing) *Escherichia coli* (VTEC/STEC)." *Journal of Applied Microbiology* **95**: 205–217.

Beutin, L. (1991) "The different hemolysins of *Escherichia coli*." *Medical Microbiology and Immunology* **180**: 167–182.

Beutin, L. (1999) "*Escherichia coli* as a pathogen in dogs and cats." *Veterinary Research* **30**: 285–298.

Beutin, L., Montenegro, M. A., Orskov, I. *et al.* (1989) "Close association of verotoxin (Shiga-like toxin) production with enterohemolysin production in strains of *Escherichia coli*." *Journal of Clinical Microbiology* **27**: 2559–2564.

Bhattacharya, S. K. and Sur, D. (2003) "An evaluation of current shigellosis treatment." *Expert Opinion on Pharmacotherapy* **4**: 1315–1320.

Bilge, S. S., Clausen, C. R., Lau, W. and Moseley, S. L. (1989) "Molecular characterization of a fimbrial adhesin, F1845, mediating diffuse adherence of diarrhea-associated *Escherichia coli* to HEp-2 cells." *Journal of Bacteriology* **171**: 4281–4289.

Blanco, J., Blanco, M., Blanco, J. E. *et al.* (2001) "Epidemiology of verocyto-toxigenic *Escherichia coli* (VTEC) in ruminants." In *Verocytotoxigenic E. coli.* G. Duffy, P. Garvey and D. A. McDowell (eds). Trumbull, CT: Food & Nutrition Press, 113–148.

Blank, E. T., Nougayréde, J. P. and Donnenberg, M. S. (2002) "*Enteropathogenic Escherichia coli.*" In *Escherichia coli: Virulence Mechanisms of a Versatile Pathogen.* M. S. Donnenberg (ed.). London: Academic Press, 81–118.

Blattner, F. R., Plunkett, G., Bloch, C. A. *et al.* (1997) "The complete genome sequence of *Escherichia coli* K12." *Science* **277**: 1453–1474.

Blocker, A., Gounon, P., Larquet, E. *et al.* (1999) "The tripartite type III secreton of *Shigella flexneri* inserts IpaB and IpaC into host membranes." *The Journal of Cell Biology* **147**: 683–693.

Bourdet-Sicard, R., Rudiger, M., Jockusch, B. M. *et al.* (1999) "Binding of the *Shigella* protein IpaA to vinculin induces F-actin depolymerization." *The EMBO Journal* **18**: 5853–5862.

Brenner, D. J., Fanning, G. R., Johnson, K. E. *et al.* (1969) "Polynucleotide sequence relationships among members of Enterobacteriaceae." *Journal of Bacteriology* **98**: 637–650.

Bürk, C., Dietrich, R., Acar, G. *et al.* (2003) "Identification and characterization of a new variant of Shiga toxin 1 in *Escherichia coli* ONT:H19 of bovine origin." *Journal of Clinical Microbiology* **41**: 2106–2112.

Burland, V., Shao, Y., Perna, N. T. *et al.* (1998) "The complete DNA sequence and analysis of the large virulence plasmid of *Escherichia coli* O157:H7." *Nucleic Acids Research* **26**: 4196–4204.

Büttner, D. and Bonas, U. (2002) "Port of entry—the type III secretion translocon." *Trends in Microbiology* **10**: 186–192.

Campellone, K. G. and Leong, J. M. (2003) "Tails of two tirs: actin pedestal formation by enteropathogenic *E. coli* and enterohemorrhagic *E. coli* O157:H7." *Current Opinion in Microbiology* **6**: 82–90.

Carbonetti, N. H., Boonchai, S., Parry, S. H. *et al.* (1986) "Aerobactin-mediated iron uptake by *Escherichia coli* isolates from human extraintestinal infections." *Infection and Immunity* **51**: 966–968.

Carter, A. O., Borczyk, A. A., Carlson, J. A. *et al.* (1987) "A severe outbreak of *Escherichia coli* O157:H7 associated hemorrhagic colitis in a nursing home." *The New England Journal of Medicine* **317**: 1496–1500.

Chapman, P. A., Siddons, C. A., Wright, D. J. *et al.* (1993) "Cattle as a possible source of verocytotoxin-producing *Escherichia coli* O157 infections in man." *Epidemiology and Infection* **111**: 439–447.

Chart, H., Smith, H. R. and Rowe, B. (1995) "Enteroaggregative strains of *Escherichia coli* belonging to serotypes O126:H27 and O44:H18 express antigenically similar 18 kDa outer membrane-associated proteins." *FEMS Microbiology Letters* **132**: 17–22.

Cheasty, T., Skinner, J. A., Rowe, B. and Threlfall, E. J. (1998) "Increasing incidence of antibiotic resistance in Shigellas from humans in England and Wales: recommendations for therapy." *Microbial Drug Resistance* **4**: 57–60.

Chen, J. H., Chiou, C. S., Chen, P. C. *et al.* (2003) "Molecular epidemiology of *Shigella* in a Taiwan township during 1996–2000." *Journal of Clinical Microbiology* **41**: 3078–3088.

Cherla, R. P., Lee, S. and Tesh, V. L. (2003) "Shiga toxins and apoptosis." *FEMS Microbiology Letters* **228**: 159–166.

Chomarat, M. (2000) "Resistance of bacteria in urinary tract infections." *International Journal of Antimicrobial Agents* **16**: 483–487.

Crawford, J. A., Blank, E. T. and Kaper, J. B. (2002) "The LEE-encoded type III secretion system in EPEC and EHEC: assembly, function and regulation." In *Escherichia coli: Virulence Mechanisms of a Versatile Pathogen.* M. S. Donnenberg (ed.). London: Academic Press, 337–359.

Crosa, J. H., Brenner, D. J., Ewing, W. H. and Falkow, S. (1973) "Molecular relationships among the Salmonelleae." *Journal of Bacteriology* **115**: 307–315.

Crowley, D. S., Ryan, M. J. and Wall, P. G. (1997) "Gastroenteritis in children under 5 years of age in England and Wales." *Communicable Disease Report* **7**: R82–R86.

Czeczulin, J. R., Balepur, S., Hicks, S. *et al.* (1997) "Aggregative adherence fimbria II, a second fimbrial antigen mediating aggregative adherence in enteroaggregative *Escherichia coli.*" *Infection and Immunity* **65**: 4135–4145.

De Geyter, C., Vogt, B., Benjelloun-Touimi, Z. *et al.* (1997) "Purification of IpaC, a protein involved in entry of *Shigella flexneri* into epithelial cells and characterization of its interaction with lipid membranes." *FEBS Letters* **400**: 149–154.

Dobrindt, U., Blum-Oehler, G., Nagy, G. *et al.* (2002) "Genetic structure and distribution of four pathogenicity islands (PAI I_{536} to PAI IV_{536}) of uropathogenic *Escherichia coli* strain 536." *Infection and Immunity* **70**: 6365–6372.

Donnenberg, M. S. (ed.) (2002) *Escherichia coli: Virulence Mechanisms of a Versatile Pathogen.* London: Academic Press.

Dundas, S., Todd, W. T., Stewart, A. I. *et al.* (2001) "The central Scotland *Escherichia coli* O157:H7 outbreak: risk factors for the hemolytic uremic syndrome and death among hospitalized patients." *Clinical Infectious Diseases* **33**: 923–931.

Dutta, S., Chatterjee, A., Dutta, P. *et al.* (2001) "Sensitivity and performance characteristics of a direct PCR with stool samples in comparison to conventional techniques for diagnosis of *Shigella* and enteroinvasive *Escherichia coli* infection in children with acute diarrhoea in Calcutta, India." *Journal of Medical Microbiology* **50**: 667–674.

Edwards, P. R. and Ewing, W. H. (1972) *Identification of Enterobacteriaceae.* Minneapolis, MN: Burgess Publishing.

Eklund, M., Leino, K. and Siitonen, A. (2002) "Clinical *Escherichia coli* strains carrying *stx* genes: *stx* variants and *stx*-positive virulence profiles." *Journal of Clinical Microbiology* **40**: 4585–4593.

Elliott, S. J., Sperandio, V., Giron, J. A. *et al.* (2000). "The locus of entero-cyte effacement (LEE)-encoded regulator controls expression of both LEE- and non-LEE-encoded virulence factors in enteropathogenic and enterohemorrhagic *Escherichia coli.*" *Infection and Immunity* **68**: 6115–6126.

Elsinghorst, E. A. (2002) "Enterotoxigenic *Escherichia coli.*" In *Escherichia coli: Virulence Mechanisms of a Versatile Pathogen.* M. S. Donnenberg (ed.). London: Academic Press, 155–187.

Fallon, D., Ackland, G., Andrews, N. *et al.* (2003) "A comparison of the performance of commercially available chromogenic agars for the isolation and presumptive identification of organisms from urine." *Journal of Clinical Pathology* **56**: 608–612.

Farmer, J. J., III, Davis, B. R., Hickman-Brenner, F. W. *et al.* (1985) "Biochemical identification of new species and biogroups of Enterobacteriaceae isolated from clinical specimens." *Journal of Clinical Microbiology* **21**: 46–76.

Feldman, M. F. and Cornelis, G. R. (2003) "The multitalented type III chaperones: all you can do with 15 kDa." *FEMS Microbiology Letters* **219**: 151–158.

Feng, P. (1993) "Identification of *Escherichia coli* serotype O157:H7 by DNA probe specific for an allele of *uid* A gene." *Molecular and Cellular Probes* **7**: 151–154.

Feng, P., Lampel, K. A., Karch, H. and Whittam, T. S. (1998) "Genotypic and phenotypic changes in the emergence of *Escherichia coli* O157:H7." *The Journal of Infectious Diseases* **177**: 1750–1753.

Fitzpatrick, M. (1999) "Haemolytic uraemic syndrome and *E coli* O157." *British Medical Journal* **318**: 684–685.

Flexner, S. (1990) "On the etiology of tropical dysentery." *Bulletin of the Johns Hopkins Hospital* **11**: 231–4252.

Franke, J., Franke, S., Schmidt, H. *et al.* (1994) "Nucleotide sequence analysis of enteropathogenic *Escherichia coli* (EPEC) adherence factor probe and development of PCR for rapid detection of EPEC harboring virulence plasmids." *Journal of Clinical Microbiology* **32**: 2460–2463.

Fratamico, P. M., Buchanan, R. L. and Cooke, P. H. (1993) "Virulence of an *Escherichia coli* O157:H7 sorbitol-positive mutant." *Applied and Environmental Microbiology* **59**: 4245–4252.

Fries, L. F., Montemarano, A. D., Mallet, C. P. *et al.* (2001) "Safety and immunogenicity of a proteosome-*Shigella flexneri* 2a lipopolysaccharide vaccine administered intranasally to healthy adults." *Infection and Immunity* **69**: 4545–4553.

Gannon, V. P., D'Souza, S., Graham, T. *et al.* (1997) "Use of the flagellar H7 gene as a target in multiplex PCR assays and improved specificity in identification of enterohemorrhagic *Escherichia coli* strains." *Journal of Clinical Microbiology* **35**: 656–662.

George, A. M. and Levy, S. B. (1983) "Amplifiable resistance to tetracycline, chloramphenicol, and other antibiotics in *Escherichia coli*: involvement of a non-plasmid-determined efflux of tetracycline." *Journal of Bacteriology* **155**: 531–540.

Giron, J. A. (1995) "Expression of flagella and motility by Shigella." *Molecular Microbiology* **18**: 63–75.

Gniadkowski, M. (2001) "Evolution and epidemiology of extended-spectrum beta-lactamases (ESBLs) and ESBL-producing microorganisms." *Clinical Microbiology and Infection* **7**: 597–608.

Goldwater, P. N. and Bettelheim, K. A. (1994) "The role of enterohaemor-rhagic *E. coli* serotypes other than O157:H7 as causes of disease." In *Recent Advances in Verocytotoxin-Producing Escherichia coli Infections.* M. A. Karmali and A. G. Goglio (eds). Amsterdam: Elsevier Science, 57–60.

Gómez-Duarte, O. G. and Kaper, J. B. (1995) "A plasmid-encoded regulatory region activates chromosomal *eaeA* expression in enteropathogenic *Escherichia coli*." *Infection and Immunity* **63**: 1767–1776.

Goullet, P. (1980) "Esterase electrophoretic pattern relatedness between *Shigella* species and *Escherichia coli*." *Journal of General Microbiology* **117**: 493–500.

Guan, S. and Verma, N. K. (1998) "Serotype conversion of a *Shigella flexneri* candidate vaccine strain via a novel site specific chromosome-integration system." *FEMS Microbiology Letters* **166**: 79–87.

Gunzburg, S. T., Tornieporth, N. G. and Riley, L. W. (1995) "Identification of enteropathogenic *Escherichia coli* by PCR-based detection of the bundle-forming pilus gene." *Journal of Clinical Microbiology* **33**: 1375–1377.

Gunzer, F., Böhm, H., Russmann, H. *et al.* (1992) "Molecular detection of sorbitol-fermenting *Escherichia coli* O157 in patients with hemolytic–uremic syndrome." *Journal of Clinical Microbiology* **30**: 1807–1810.

Gupta, A., Polyak, C. S., Bishop, R. D. *et al.* (2004) "Laboratory-confirmed shigellosis in the United States, 1989–2002: epidemiologic trends and patterns." *Clinical Infectious Diseases* **38**: 1372–1377.

Harrington, A. T., Hearn, P. D., Picking, W. L. *et al.* (2003) "Structural characterization of the N terminus of IpaC from *Shigella flexneri*." *Infection and Immunity* **71**: 1255–1264.

Hayashi, T., Makino, K., Ohnishi, M. *et al.* (2001) "Complete genome sequence of enterohemorrhagic *Escherichia coli* O157:H7 and genomic comparison with a laboratory strain K-12." *DNA Research* **8**: 11–22.

Health Protection Agency (2004) "Travel Health. Imported Infections, England and Wales: April to June 2004." *CDR Weekly* **14**(38). www.hpa.org.uk/cdr/archives/2004/cdr3204.pdf (21 July 2005).

Henry, F. J. (1991) "The epidemiologic importance of dysentery in communities." *Reviews of Infectious Diseases* **13** (Suppl. 4): S238–S244.

High, N., Mounier, J., Prevost, M. C. and Sansonetti, P. J. (1992) "IpaB of *Shigella flexneri* causes entry into epithelial cells and escape from the phagocytic vacuole." *The EMBO Journal* **11**: 1991–1999.

Houndt, T. and Ochman, H. (2000) "Long-term shifts in patterns of antibiotic resistance in enteric bacteria." *Applied and Environmental Microbiology* **66**: 5406–5409.

Huebner, J., Czerwenka, W., Gruner, E. and von Graevenitz, A. (1993) "Shigellemia in AIDS patients: case report and review of the literature." *Infection* **21**: 122–124.

Hughes, D. (2000) "Evaluating genome dynamics: the constraints of rearrangements within bacterial genomics." http://genomebiology.com/2000/1/6/reviews/0006 (8 December 2000).

Huskins, W. C., Griffiths, J. K., Faruque, A. S. and Bennish, M. L. (1994) "Shigellosis in neonates and young infants." *The Journal of Pediatrics* **125**: 14–22.

Islam, M. S., Hossain, M. S., Hasan, M. K. *et al.* (1998) "Detection of Shigellae from stools of dysentery patients by culture and polymerase chain reaction techniques." *Journal of Diarrhoeal Diseases Research* **16**: 248–251.

Jenkins, C., Perry, N. T., Cheasty, T. *et al.* (2003) "Distribution of the *saa* gene in strains of Shiga toxin-producing *Escherichia coli* of human and bovine origins." *Journal of Clinical Microbiology* **41**: 1775–1778.

Jennison, A. V. and Verma, N. K. (2004) "*Shigella flexneri* infection: pathogenesis and vaccine development." *FEMS Microbiology Reviews* **28**: 43–58.

Jerse, A. E., Yu, J., Tall, B. D. and Kaper, J. B. (1990) "A genetic locus of enteropathogenic *Escherichia coli* necessary for the production of attaching and effacing lesions on tissue culture cells." *Proceedings of the National Academy of Sciences of the United States of America* **87**: 7839–7843.

Jin, Q., Yuan, Z., Xu, J. *et al.* (2002) "Genome sequence of *Shigella flexneri* 2a: insights into pathogenicity through comparison with genomes of *Escherichia coli* K12 and O157." *Nucleic Acids Research* **30**: 4432–4441.

Johnson, J. R. (2002) "Evolution of pathogenic *Escherichia coli*." In *Escherichia coli: Virulence Mechanisms of a Versatile Pathogen*. M. S. Donnenberg (ed.). London: Academic Press, 55–77.

Johnson, J. R. and Russo, T. A. (2002) "Uropathogenic *Escherichia coli* as agents of diverse non-urinary tract extraintestinal infections." *The Journal of Infectious Diseases* **186**: 859–864.

Johnson, R. P., Clarke, R. C., Wilson, J. B. *et al.* (1996) "Growing concerns and recent outbreaks involving non-O157:H7 serotypes of verocytotoxigenic *Escherichia coli*." *Journal of Food Protection* **59**: 1112–1122.

Kaper, J. B. and O'Brien, A. D. (1998) *Escherichia coli O157:H7 and Other Shiga Toxin-Producing E. coli Strains*. Washington, DC: ASM Press.

Kaper, J. B., Elliott, S., Sperandio, V. *et al.* (1998) "Attaching-and-effacing intestinal histopathology and the locus of enterocyte effacement." In *Escherichia coli O157:H7 and Other Shiga Toxin-Producing E. coli Strains*. J. B. Kaper and A. D. O'Brien (eds). Washington, DC: ASM Press, 163–182.

Kaper, J. B., Nataro, J. P. and Mobley, H. L. (2004) "Pathogenic *Escherichia coli*." *Nature Reviews. Microbiology* **2**: 123–140.

Karch, H., Schmidt, H. and Brunder, W. (1998) "Plasmid-encoded determinants of *Escherichia coli* O157:H7." In *Escherichia coli O157:H7 and Other Shiga Toxin-Producing E. coli Strains*. J. B. Kaper and A. D. O'Brien (eds). Washington, DC: ASM Press, 183–194.

Karlowsky, J. A., Kelly, L. J., Thornsberry, C. *et al.* (2002) "Trends in antimicrobial resistance among urinary tract infection isolates of *Escherichia coli* from female outpatients in the United States." *Antimicrobial Agents and Chemotherapy* **46**: 2540–2545.

Kenny, B. (2002) "Enteropathogenic *Escherichia coli* (EPEC)—a crafty subversive little bug." *Microbiology* **148**: 1967–1978.

Kenny, B., DeVinney, R., Stein, M. *et al.* (1997) "Enteropathogenic *E. coli* (EPEC) transfers its receptor for intimate adherence into mammalian cells." *Cell* **91**: 511–520.

Kern, W. V., Oethinger, M., Jellin-Ritter, A. S. and Levy, S. B. (2000) "Non-target gene mutations in the development of fluoroquinolone resistance in *Escherichia coli*." *Antimicrobial Agents and Chemotherapy* **44**: 814–820.

Kimmitt, P. T., Harwood, C. R. and Barer, M. R. (1999) "Induction of type 2 Shiga toxin synthesis in *Escherichia coli* O157 by 4-quinolones." *Lancet* **353**: 1588–1589.

Kimmitt, P. T., Harwood, C. R. and Barer, M. R. (2000) "Toxin gene expression by shiga toxin-producing *Escherichia coli*: the role of antibiotics and the bacterial SOS response." *Emerging Infectious Diseases* **6**: 458–465.

Knutton, S., Baldwin, T., Williams, P. H. and McNeish, A. S. (1989) "Actin accumulation at sites of bacterial adhesion to tissue culture cells: basis of a new diagnostic test for enteropathogenic and enterohemorrhagic *Escherichia coli*." *Infection and Immunity* **57**: 1290–1298.

Konowalchuk, J., Spiers, J. I. and Stavric, S. (1977) "Vero response to a cytotoxin of *Escherichia coli*." *Infection and Immunity* **18**: 775–779.

Korhonen, T. K., Valtonen, M. V., Parkkinen, J. *et al.* (1985) "Serotypes, hemolysin production, and receptor recognition of *Escherichia coli* strains associated with neonatal sepsis and meningitis." *Infection and Immunity* **48**: 486–491.

Kotloff, K. L., Noriega, F., Losonsky, G. A. *et al.* (1996) "Safety, immunogenicity, and transmissibility in humans of CVD 1203, a live oral *Shigella flexneri* 2a vaccine candidate attenuated by deletions in aroA and virG." *Infection and Immunity* **64**: 4542–4548.

Kotloff, K. L., Winickoff, J. P., Ivanoff, B. *et al.* (1999) "Global burden of *Shigella* infections: implications for vaccine development and implementation of control strategies." *Bulletin of the World Health Organization* **77**: 651–665.

Kyaw, C. M., De Araujo, C. R., Lima, M. R. *et al.* (2003) "Evidence for the presence of a type III secretion system in diffusely adhering *Escherichia coli* (DAEC)." *Infection, Genetics and Evolution* **3**: 111–117.

Lan, R. and Reeves, P. R. (2002) "*Escherichia coli* in disguise: molecular origins of Shigella." *Microbes and Infection* **4**: 1125–1132.

Langman, G. (1996) "*Shigella sonnei* meningitis." *South African Medical Journal* **86**: 91–92.

Law, D. (2000) "Virulence factors of *Escherichia coli* O157 and other Shiga toxin-producing *E. coli*." *Journal of Applied Microbiology* **88**: 729–745.

Lehmacher, A., Meier, H., Aleksic, S. and Bockemuhl, J. (1998) "Detection of hemolysin variants of Shiga toxin-producing *Escherichia coli* by PCR and culture on vancomycin-cefixime-cefsulodin blood agar." *Applied and Environmental Microbiology* **64**: 2449–2453.

Lencer, W. I. and Tsai, B. (2003) "The intracellular voyage of cholera toxin: going retro." *Trends in Biochemical Sciences* **28**: 639–645.

Levine, M. M., Xu, J. G., Kaper, J. B. *et al.* (1987) "A DNA probe to identify enterohemorrhagic *Escherichia coli* of O157:H7 and other serotypes that cause hemorrhagic colitis and hemolytic uremic syndrome." *The Journal of Infectious Diseases* **156**: 175–182.

Lin, Z., Kurazono, H., Yamasaki, S. and Takeda, Y. (1993) "Detection of various variant verotoxin genes in *Escherichia coli* by polymerase chain reaction." *Microbiology and Immunology* **37**: 543–548.

Lingwood, C. A. (1996) "Role of verotoxin receptors in pathogenesis." *Trends in Microbiology* **4**: 147–153.

Livermore, D. M., James, D., Reacher, M. *et al.* (2002) "Trends in fluoroquinolone (ciprofloxacin) resistance in Enterobacteriaceae from bacteremias, England and Wales, 1990–1999." *Emerging Infectious Diseases* **8**: 473–478.

Louie, M., de Azavedo, J., Clarke, R. *et al.* (1994) "Sequence heterogeneity of the *eae* gene and detection of verotoxin-producing *Escherichia coli* using serotype-specific primers." *Epidemiology and Infection* **112**: 449–461.

Marcato, P., Mulvey, G. and Armstrong, G. D. (2002) "Cloned shiga toxin 2 B subunit induces apoptosis in Ramos Burkitt's lymphoma B cells." *Infection and Immunity* **70**: 1279–1286.

March, S. B. and Ratnam, S. (1986) "Sorbitol–MacConkey medium for detection of *Escherichia coli* O157:H7 associated with hemorrhagic colitis." *Journal of Clinical Microbiology* **23**: 869–872.

McDaniel, T. K., Jarvis, K. G., Donnenberg, M. S. and Kaper, J. B. (1995) "A genetic locus of enterocyte effacement conserved among diverse enterobacterial pathogens." *Proceedings of the National Academy of Sciences of the United States of America* **92**: 1664–1668.

McIver, C. J. and Tapsall, J. W. (1990) "Assessment of conventional and commercial methods for identification of clinical isolates of cysteine-requiring strains of *Escherichia coli* and *Klebsiella* species." *Journal of Clinical Microbiology* **28**: 1947–1951.

Mellies, J. L., Elliott, S. J., Sperandio, V. *et al.* (1999) "The Per regulon of enteropathogenic *Escherichia coli*: identification of a regulatory cascade and a novel transcriptional activator, the locus of enterocyte effacement (LEE)-encoded regulator (Ler)." *Molecular Microbiology* **33**: 296–306.

Menard, R., Sansonetti, P. J. and Parsot, C. (1993) "Nonpolar mutagenesis of the *ipa* genes defines IpaB, IpaC, and IpaD as effectors of *Shigella flexneri* entry into epithelial cells." *Journal of Bacteriology* **175**: 5899–5906.

Menard, R., Sansonetti, P., Parsot, C. and Vasselon, T. (1994) "Extracellular association and cytoplasmic partitioning of the IpaB and IpaC invasins of *S. flexneri*." *Cell* **79**: 515–525.

Meng, J., Zhao, S., Doyle, M. P. and Joseph, S. W. (1998) "Antibiotic resistance of *Escherichia coli* O157:H7 and O157:NM isolated from animals, food, and humans." *Journal of Food Protection* **61**: 1511–1514.

Middendorf, B., Hochhut, B., Leipold, K. *et al.* (2004) "Instability of pathogenicity islands in uropathogenic *Escherichia coli* 536." *Journal of Bacteriology* **186**: 3086–3096.

Morabito, S., Dell'Omo, G., Agrimi, U. *et al.* (2001) "Detection and characterization of shiga toxin-producing *Escherichia coli* in feral pigeons." *Veterinary Microbiology* **82**: 275–283.

Munday, C. J., Xiong, J., Li, C. *et al.* (2004) "Dissemination of CTX-M type beta-lactamases in Enterobacteriaceae isolates in the People's Republic of China." *International Journal of Antimicrobial Agents* **23**: 175–180.

Murphy, G. S., Bodhidatta, L., Echeverria, P. *et al.* (1993) "Ciprofloxacin and loperamide in the treatment of bacillary dysentery." *Annals of Internal Medicine* **118**: 582–586.

Nair, G. B. and Takeda, Y. (1997) "The heat-labile and heat-stable enterotoxins of *Escherichia coli*." In *Escherichia coli: Mechanisms of Virulence*. M. Sussman (ed.). Cambridge: Cambridge University Press, 237–256.

Nataro, J. P. and Steiner, T. (2002) "Enteroaggregative and diffusely adherent *Escherichia coli*." In *Escherichia coli: Virulence Mechanisms of a Versatile Pathogen*. M. S. Donnenberg (ed.). London: Academic Press, 189–207.

Nataro, J. P., Steiner, T. and Guerrant, R. L. (1998) "Enteroaggregative *Escherichia coli*." *Emerging Infectious Diseases* **4**: 251–261.

Newman, C. P. S. (1993) "Surveillance and control of *Shigella sonnei* infection." *Communicable Disease Report* **3**: R63–R70.

Niebuhr, K., Giuriato, S., Pedron, T. *et al.* (2002) "Conversion of PtdIns(4,5)P(2) into PtdIns(5)P by the *S. flexneri* effector IpgD reorganizes host cell morphology." *The EMBO Journal* **21**: 5069–5078.

Noriga, F. R., Liao, F. M., Maneval, D. R. *et al.* (1999) "Strategy for cross-protection among *Shigella flexneri* serotypes." *Infection and Immunity* **67**: 782–788.

O'Brien, A. D. and Holmes, R. K. (1987) "Shiga and shiga-like toxins." *Microbiological Reviews* **51**: 206–220.

O'Brien, A. D., Tesh, V. L., Donohue-Rolfe, A. *et al.* (1992) "Shiga toxin: biochemistry, genetics, mode of action, and role in pathogenesis." *Current Topics in Microbiology and Immunology* **180**: 65–94.

Obrig, T. G., Del Vecchio, P. J., Brown, J. E. *et al.* (1988) "Direct cytotoxic action of Shiga toxin on human vascular endothelial cells." *Infection and Immunity* **56**: 2373–2378.

Ohnishi, M., Kurokawa, K. and Hayashi, T. (2001) "Diversification of *Escherichia coli* genomes: are bacteriophages the major contributors?" *Trends in Microbiology* **9**: 481–485.

Paton, A. W. and Paton, J. C. (2002) "Direct detection and characterisation of shiga toxigenic *Escherichia coli* by multiplex PCR for *stx*$_1$, *stx*$_2$, *eae*, *ehxA* and *saa*." *Journal of Clinical Microbiology* **40**: 271–274.

Paton, J. C. and Paton, A. W. (1998) "Pathogenesis and diagnosis of shiga toxin-producing *Escherichia coli* infections." *Clinical Microbiology Reviews* **11**: 450–479.

Peiffer, I., Blanc-Potard, A. B., Bernet-Camard, M. F. *et al.* (2000) "Afa/Dr diffusely adhering *Escherichia coli* C1845 infection promotes selective injuries in the junctional domain of polarized human intestinal Caco-2/TC7 cells." *Infection and Immunity* **68**: 3431–3442.

Perdomo, J. J., Gounon, P. and Sansonetti, P. J. (1994) "Polymorphonuclear leukocyte transmigration promotes invasion of colonic epithelial monolayer by *Shigella flexneri*." *The Journal of Clinical Investigation* **93**: 633–643.

Perdomo, O. J., Cavaillon, J. M., Huerre, M. *et al.* (1994) "Acute inflammation causes epithelial invasion and mucosal destruction in experimental shigellosis." *The Journal of Experimental Medicine* **180**: 1307–1319.

Perna, N. T., Mayhew, G. F., Posfai, G. *et al.* (1998) "Molecular evolution of a pathogenicity island from enterohemorrhagic *Escherichia coli* O157:H7." *Infection and Immunity* **66**: 3810–3817.

Perna, N. T., Plunkett, G., Burland, V. *et al.* (2001) "Genome sequence of enterohaemorrhagic *Escherichia coli* O157:H7." *Nature* **409**: 529–533.

Perna, N. T., Glasner, J. D., Burland, V. and Plunckett, G. III (2002) "The genomes of *Escherichia coli* K12 and pathogenic *E. coli*." In *Escherichia coli: Virulence Mechanisms of a Versatile Pathogen*. M. S. Donnenberg (ed.). London: Academic Press, 3–53.

Phavichitr, N. and Catto-Smith, A. (2003) "Acute gastroenteritis in children: what role for antibacterials?" *Paediatric Drugs* **5**: 279–290.

Phillips, A. D., Girón, J., Dougan, G. and Frankel, G. (2000) "Intimin from enteropathogenic *Escherichia coli* mediates remodelling of the eukaryotic cell surface." *Microbiology* **146**: 1333–1344.

PHLS Salmonella Sub-committee (1990) "Notes on the control of human sources of gastrointestinal infections and bacterial intoxications in the United Kingdom." *Communicable Disease Report* Suppl. 1.

Piérard, D., Muyldermans, G., Moriau, L. *et al.* (1998) "Identification of new verocytotoxin type 2 variant B-subunit genes in human and animal *Escherichia coli* isolates." *Journal of Clinical Microbiology* **36**: 3317–3322.

Pluschke, G., Mayden, J., Achtman, M. and Levine, R. P. (1983) "Role of the capsule and the O antigen in resistance of O18:K1 *Escherichia coli* to complement-mediated killing." *Infection and Immunity* **42**: 907–913.

Prasadarao, N. V., Wass, C. A., Huang, S.-H. and Kim, K. S. (1999) "Identification and characterization of a novel Ibe10 binding protein that contributes to *Escherichia coli* invasion of brain microvascular endothelial cells." *Infection and Immunity* **67**: 1131–1138.

Prasadarao, N. V., Wass, C. A., Weiser, J. N. *et al.* (1996) "Outer membrane protein A of *Escherichia coli* contributes to invasion of brain microvascular endothelial cells." *Infection and Immunity* **64**: 146–153.

Pupo, G. M., Karaolis, D. K., Lan, R. and Reeves, P. R. (1997) "Evolutionary relationships among pathogenic and nonpathogenic *Escherichia coli* strains inferred from multilocus enzyme electrophoresis and *mdh* sequence studies." *Infection and Immunity* **65**: 2685–2692.

Purdy, G. E., Hong, M. and Payne, S. M. (2002) "*Shigella flexneri* DegP facilitates IcsA surface expression and is required for efficient intercellular spread." *Infection and Immunity* **70**: 6355–6364.

Ramachandran, V., Hornitzky, M. A., Bettelheim, K. A. *et al.* (2001) "The common ovine shiga toxin 2-containing *Escherichia coli* serotypes and human isolates of the same serotype possess a Stx2d toxin type." *Journal of Clinical Microbiology* **39**: 1932–1937.

Rawashdeh, M. O., Ababneh, A. M. and Shurman, A. A. (1994) "Shigellosis in Jordanian children: a clinico-epidemiologic prospective study and susceptibility to antibiotics." *Journal of Tropical Pediatrics* **40**: 355–359.

Reid, S. D., Herbelin, C. J., Bumbaugh, A. C. *et al.* (2000) "Parallel evolution of virulence in pathogenic *Escherichia coli*." *International Journal of Food Microbiology* **58**: 73–82.

Reid, S. D., Selander, R. K. and Whittam, T. S. (1998) "Sequence diversity of flagellin (*fliC*) alleles in pathogenic *Escherichia coli*." *Journal of Bacteriology* **181**: 153–160.

Restaino, L., Frampton, E. W., Turner, K. M. and Allison, D. R. (1999) "A chromogenic plating medium for isolating *Escherichia coli* O157:H7 from beef." *Letters in Applied Microbiology* **29**: 26–30.

Riley, L. W., Remis, R. S., Helgerson, S. D. *et al.* (1983) "Hemorrhagic colitis associated with a rare *Escherichia coli* serotype." *The New England Journal of Medicine* **308**: 681–685.

Robbins, J. B., Chu, C. and Schneerson, R. (1992) "Hypothesis for vaccine development: protective immunity to enteric diseases caused by nontyphoidal salmonellae and shigellae may be conferred by serum IgG antibodies to the O-specific polysaccharide of their lipopolysaccharides." *Clinical Infectious Diseases* **15**: 346–361.

Roffe, C. (1996) "Biotherapy for antibiotic-associated and other diarrhoeas." *The Journal of Infection* **32**: 1–10.

Rose, P. and Chant, I. (1998) "Hematology of hemolytic–uremic syndrome." In *Escherichia coli O157:H7 and Other Shiga Toxin-Producing E. coli Strains*. J. B. Kaper and A. D. O'Brien (eds). Washington, DC: ASM Press, 293–301.

Sakaguchi, T., Kohler, H., Gu, X. *et al.* (2002) "*Shigella flexneri* regulates tight junction-associated proteins in human intestinal epithelial cells." *Cell Microbiology* **4**: 367–381.

Salam, I., Katelaris, P., Leigh-Smith, S. and Farthing, M. J. (1994) "Randomised trial of single-dose ciprofloxacin for travellers' diarrhoea." *Lancet* **344**: 1537–1539.

Samadpour, M., Ongerth, J. E. and Liston, J. (1994) "Development and evaluation of oligonucleotide DNA probes for detection and genogrouping of shiga-like toxin producing *Escherichia coli*." *Journal of Food Protection* **57**: 399–402.

Sandvig, K. and Van Deurs, B. (2002) "Transport of Protein toxins into cells: pathways used by ricin, cholea toxin and shiga toxin." FEBS *Letters* **529**: 49–53.

Schmidt, H. (2001) "Shiga-toxin-converting bacteriophages." *Research in Microbiology* **152**: 687–695.

Schneider, G., Dobrindt, U., Bruggemann, H. *et al.* (2004) "The pathogenicity island-associated K15 capsule determinant exhibits a novel genetic structure and correlates with virulence in uropathogenic *Escherichia coli* strain 536." *Infection and Immunity* **72**: 5993–6001.

Shaikh, N. and Tarr, P. I. (2003) "*Escherichia coli* O157:H7 Shiga toxin-encoding bacteriophages: integrations, excisions, truncations, and evolutionary implications." *Journal of Bacteriology* **185**: 3596–3605.

Shariff, M., Bhan, M., Knutton, S. *et al.* (1993) "Evaluation of the fluorescence actin staining test for detection of enteropathogenic *Escherichia coli*." *Journal of Clinical Microbiology* **31**: 386–389.

Sheikh, J., Czeczulin, J. R., Harrington, S. *et al.* (2002) "A novel dispersin protein in enteroaggregative *Escherichia coli*." *The Journal of Clinical Investigation* **110**: 1329–1337.

Shiga, K. (1898) "Ueber den dysenteriebacillus (*Bacillus dysenteriae*)." *Zentralblatt der Bakteriologie* **24**: 817–828.

Silva, R. M., Regina, M., Toledo, F. and Trabulsi, L. R. (1980) "Biochemical and cultural characteristics of invasive *Escherichia coli*." *Journal of Clinical Microbiology* **11**: 441–444.

Slutsker, L., Ries, A. A., Greene, K. D. *et al.* (1997) "*Escherichia coli* O157:H7 diarrhea in the United States: clinical and epidemiologic features." *Annals of Internal Medicine* **126**: 505–513.

Sommerfelt, H., Svennerholm, A. M., Kalland, K. H. *et al.* (1988) "Comparative study of colony hybridization with synthetic oligonucleotide probes and enzyme-linked immunosorbent assay for identification of enterotoxigenic *Escherichia coli*." *Journal of Clinical Microbiology* **26**: 530–534.

Sonne, C. (1915) "Ueber die Bacteriologie der giftarmen Dysenteribacillen (Paradysenteribacille)." *Zentralblatt Fur Bakteriologie I, Originale Abteilung* **75**: 408–456.

Sotto, A., De Boever, C. M., Fabbro-Peray, P. *et al.* (2001) "Risk factors for antibiotic-resistant *Escherichia coli* isolated from hospitalized patients with urinary tract infections: a prospective study." *Journal of Clinical Microbiology* **39**: 438–444.

Steiner, T. S., Nataro, J. P., Poteet-Smith, C. E. *et al.* (2000) "Enteroaggregative *Escherichia coli* expresses a novel flagellin that causes IL-8 release from intestinal epithelial cells." *The Journal of Clinical Investigation* **105**: 1769–1777.

Strockbine, N. A., Jackson, M. P., Sung, L. M. *et al.* (1988) "Cloning and sequencing of the genes for shiga toxin from *Shigella dysenteriae* type 1." *Journal of Bacteriology* **170**: 1116–1122.

Struelens, M. J., Mondal, G., Roberts, M. and Williams, P. H. (1990) "Role of bacterial and host factors in the pathogenesis of *Shigella* septicemia." *European Journal of Clinical Microbiology & Infectious Diseases* **9**: 337–344.

Surdeanu, M., Ciudin, L., Pencu, E. and Straut, M. (2003) "Comparative study of three different DNA fingerprint techniques for molecular typing of *Shigella flexneri* strains isolated in Romania." *European Journal of Epidemiology* **18**: 703–710.

Takeda, T., Yoshino, K., Uchida, H. *et al.* (1998) "Early use of fosfomycin for shigella toxin-producing *Escherichia coli* O157 infection reduces the risk of hemolytic–uremic syndrome." In *Escherichia coli O157:H7 and Other Shiga Toxin-Producing E. coli Strains*. J. B. Kaper and A. D. O'Brien (eds). Washington, DC: ASM Press, 385–387.

te Loo, D. M., van Hinsbergh, V. W. M., van den Heuvel, L. P. W. J. and Monnens, L. A. H. (2001) "Detection of verocytotoxin bound to circulating polymorphonuclear leukocytes of patients with hemolytic uremic syndrome." *Journal of the American Society of Nephrology* **12**: 800–806.

Teel, L. D., Melton-Celsa, A. R., Schmitt, C. K. and O'Brien, A. D. (2002) "One of two copies of the gene for the activatable shiga toxin type 2d in *Escherichia coli* O91:H21 strain B2F1 is associated with an inducible bacteriophage." *Infection and Immunity* **70**: 4282–4291.

Tesh, V. L., Ramegowda, B. and Samuel, J. E. (1994) "Purified Shiga-like toxins induce expression of proinflammatory cytokines from murine peritoneal macrophages." *Infection and Immunity* **62**: 5085–5094.

Thapa, B. R., Ventkateswarlu, K., Malik, A. K. and Panigrahi, D. (1995) "Shigellosis in children from north India: a clinicopathological study." *Journal of Tropical Pediatrics* **41**: 303–307.

Thorpe, C. M., Smith, W. E., Hurley, B. P. and Acheson, D. W. K. (2001) "Shiga toxins induce, superinduce, and stabilize a variety of C-X-C chemokine mRNAs in intestinal epithelial cells, resulting in increased chemokine expression." *Infection and Immunity* **69**: 6140–6147.

Threlfall, E. J., Cheasty, T., Graham, A. and Rowe, B. (1998) "Antibiotic resistance in *Escherichia coli* isolated from blood and cerebrospinal fluid: a 6-year study of isolates from patients in England and Wales." *International Journal of Antimicrobial Agents* **9**: 201–205.

Todd, W. T. A., Duncan, S. and Coia, J. (2001) "Clinical management of *E. coli* O157 infection." In *Verocytotoxigenic E. coli*. G. Duffy, P. Garvey and D. A. McDowell (eds). Trumbull, CT: Food & Nutrition Press, 393–420.

Tompkins, D. S., Hudson, M. J., Smith, H. R. *et al.* (1999) "A study of infectious intestinal disease in England: microbiological findings in cases and controls." *Communicable Disease and Public Health* **2**: 108–113.

Trabulsi, L. R., Keller, R. and Gomes, T. A. (2002) "Typical and atypical enteropathogenic *Escherichia coli*." *Emerging Infectious Diseases* **8**: 508–513.

Tran Van Nhieu, G., Caron, E., Hall, A. and Sansonetti, P. J. (1999) "IpaC induces actin polymerization and filopodia formation during *Shigella* entry into epithelial cells." *The EMBO Journal* **18**: 3249–3262.

Tran Van Nhieu, G., Bourdet-Sicard, R., Dumenil, G. *et al.* (2000) "Bacterial signals and cell responses during *Shigella* entry into epithelial cells." *Cell Microbiology* **2**: 187–193.

Vargas, M., Gascon, J., De Anta, M. T. J. and Vila, J. (1999) "Prevalence of *Shigella* enterotoxins 1 and 2 among Shigella strains isolated from patients with traveler's diarrhea." *Journal of Clinical Microbiology* **37**: 3608–3611.

Vidal, R., Vidal, M., Lagos, R. *et al.* (2004) "Multiplex PCR for diagnosis of enteric infections associated with diarrheagenic *Escherichia coli*." *Journal of Clinical Microbiology* **42**: 1787–1789.

Vila, J., Gascon, J., Abdalla, S. *et al.* (1994) "Antimicrobial resistance of Shigella isolates causing traveler's diarrhea." *Antimicrobial Agents and Chemotherapy* **38**: 2668–2670.

Vila, J., Vargas, M., Henderson, I. R. *et al.* (2000) "Enteroaggregative *Escherichia coli* virulence factors in traveler's diarrhea strains." *The Journal of Infectious Diseases* **182**: 1780–1783.

Wagner, P. L., Livny, J., Neely, M. N. *et al.* (2002) "Bacteriophage control of shiga toxin 1 production and release by *Escherichia coli*." *Molecular Microbiology* **44**: 957–970.

Wassef, J. S., Keren, D. F. and Mailloux, J. L. (1989) "Role of M cells in initial antigen uptake and in ulcer formation in the rabbit intestinal loop model of shigellosis." *Infection and Immunity* **57**: 858–863.

Wasteson, Y. (2001) "Epidemiology of VTEC in non-ruminant animals." In *Verocytotoxigenic E. coli*. G. Duffy, P. Garvey and D. A. McDowell (eds). Trumbull, CT: Food & Nutrition Press, 149–160.

Wei, J., Goldberg, M. B., Burland, V. *et al.* (2003) "Complete genomic sequence and comparative genomics of *Shigella flexneri* serotype 2a strain 2457T." *Infection and Immunity* **71**: 2775–2786.

Welch, R. A., Burland, V., Plunkett, G. 3rd, *et al.* (2002) "Extensive mosaic structure revealed by the complete genome sequence of uropathogenic *Escherichia coli*." *Proceedings of the National Academy of Sciences of the United States of America* **99**: 17020–17024.

Willshaw, G. A., Thirlwell, J., Jones, A. P. *et al.* (1994) "Vero cytotoxin-producing *Escherichia coli* O157 in beefburgers linked to an outbreak of diarrhoea, haemorrhagic colitis and haemolytic uraemic syndrome in Britain." *Letters in Applied Microbiology* **19**: 304–307.

Wistrom, J. and Norrby, S. R. (1995) "Fluoroquinolones and bacterial enteritis, when and for whom?" *The Journal of Antimicrobial Chemotherapy* **36**: 23–39.

Wolf, M. K. (1997) "Occurrence, distribution and associations of O and H serogroups, colonization factor antigens and toxins of enterotoxigenic *Escherichia coli*." *Clinical Microbiology Reviews* **10**: 569–584.

Wong, C. S., Jelacic, S., Habeeb, R. L. *et al.* (2000) "The risk of the hemolytic-uremic syndrome after antibiotic treatment of *Escherichia coli* O157:H7 infections [see comments]." *The New England Journal of Medicine* **342**: 1930–1936.

Woodford, N., Ward, M. E., Kaufmann, M. E. *et al.* (2004) "Community and hospital spread of *Escherichia coli* producing CTX-M extended-spectrum beta-lactamases in the UK." *The Journal of Antimicrobial Chemotherapy* **54**: 735–743.

Wullt, B. (2003) "The role of P fimbriae for *Escherichia coli* establishment and mucosal inflammation in the human urinary tract." *International Journal of Antimicrobial Agents* **21**: 605–621.

Yamamoto, T. and Echeverria, P. (1996) "Detection of the enteroaggregative *Escherichia coli* heat-stable enterotoxin 1 gene sequences in enterotoxigenic *E. coli* strains pathogenic for humans." *Infection and Immunity* **64**: 1441–1445.

Yavzori, M., Porath, N., Ochana, O. *et al.* (1998) "Detection of enterotoxigenic *Escherichia coli* in stool specimens by polymerase chain reaction." *Diagnostic Microbiology and Infectious Disease* **31**: 503–509.

Yu, H. S., Lee, J. C., Kang, M. Y. *et al.* (2003) "Changes in gene cassettes of class1 integrons among *Eischerichia Coli* isolates from urine specimens collected in Korea during the last two decades." *Journal of Clinical Microbiology* **41**: 5429–5433.

Zhang, W., Bielaszewska, M., Kuczius, T. and Karch, H. (2002) "Identification, characterization, and distribution of a Shiga toxin 1 gene variant (stx_{1c}) in *Escherichia coli* strains isolated from humans." *Journal of Clinical Microbiology* **40**: 1441–1446.

Zhang, X., McDaniel, A. D., Wolf, L. E. *et al.* (2000) "Quinolone antibiotics induce Shiga toxin-encoding bacteriophages, toxin production, and death in mice." *The Journal of Infectious Diseases* **181**: 664–670.

Zychlinsky, A., Prevost, M. C. and Sansonetti, P. J. (1992) "*Shigella flexneri* induces apoptosis in infected macrophages." *Nature* **358**: 167–169.

Zychlinsky, A., Fitting, C., Cavaillon, J. M. and Sansonetti, P. J. (1994) "Interleukin 1 is released by murine macrophages during apoptosis induced by *Shigella flexneri*." *The Journal of Clinical Investigation* **94**: 1328–1332.

29

Salmonella spp.

Claire Jenkins[1] and Stephen H. Gillespie[2]

[1]*Department of Medical Microbiology, Royal Free Hospital NHS Trust, London, UK; and,* [2]*Centre for Medical Microbiology, Royal Free and University College Medical School, Hampstead Campus, London, UK*

INTRODUCTION

Infection with organisms of the *Salmonella* spp. is an important public health problem throughout the world. Non-typhoidal Salmonellosis is a common cause of food-borne infection, and typhoid continues to exact a considerable death toll in developing countries.

DESCRIPTION OF THE ORGANISM

Classification

Salmonella are fermentative facultatively anaerobic, oxidase-negative Gram-negative rods. They are generally motile, aergenic, nonlactose fermenting, urease negative, citrate utilizing and acetyl methyl carbinol negative.

Taxonomists, it appears, cut their teeth on the Enterobacteriaceae. They were simple to cultivate and submitted themselves readily to biochemical testing (Barrow and Feltham, 1992). The consequence of this early enthusiasm is a proliferation of genera and species which in the modern age, dominated as it is by the results of molecular studies, would never be dignified with genus or species designation. Comparisons of the *Salmonella* genome with other enteric bacteria, such as *Escherichia coli*, highlight large regions of DNA showing a high degree of conservation between species (70–80% of the chromosome) (McCelland *et al.*, 2001) – sufficiently similar to cause them to be classified as the same species. Although this has been proposed, it is unlikely to happen as the genus *Salmonella* has an important clinical identity giving the name utility in day-to-day microbiological practice.

Historically, the classification of *Salmonella* was based on host specificity, serotyping the lipopolysaccharide (LPS) and flagella antigens and phage typing, resulting in more than 2500 different, often imaginatively named, serovars (Kauffmann, 1957). However, molecular methods, such as DNA hybridization, protein isozyme analysis, and DNA sequence similarity, show that *Salmonella* serovars are sufficiently similar at the genomic level to be regarded as the same species. The genus *Salmonella* consists of two species, *Salmonella enterica* and *S. bongori*. Within the species *S. enterica* there are six subspecies differentiated by biochemical variations, namely *enterica*, *salamae*, *arizonae*, *diarizonae*, *houtenae* and *indica* (Barrow and Feltham, 1992, Old and Threlfall, 1998). Using this system the designation for *S. enteritidis* phage type (PT4) is *S. enterica* subsp. *enterica* serotype Enteritidis PT4, although this can be abbreviated to *S. enteritidis* (Old and Threlfall, 1998). Serovars within the subspecies

other than subsp. *enterica* are designated by their antigenic formulae, for example, *S. enterica* subsp. *arizonae* 61:k:1,5,7.

Population Structure of *Salmonella* spp.

The population structure of *Salmonella* spp. has been studied by a number of techniques including multilocus sequence typing (Kidgell *et al.*, 2002), genome sequencing (McCelland *et al.*, 2001, Wain *et al.*, 2002) and DNA microarrays. Comparison of the housekeeping genes, involved in DNA replication and central metabolism, indicate that the *Salmonella* serovars are very similar (97.6–99.5%). Despite their overall similarity, comparative genomics show that each serovar has many insertions and deletions relative to other serovars (Edwards, Olsen and Maloy, 2002). These unique regions represent 10–12% of their 5-Mb genome and probably encode gene products responsible for the ability of the serovars to infect different hosts (for example, *S. enterica* subsp. *arizonae* 61:k:1,5,7 infect sheep) and cause a range of diseases (for example, *S. typhi*).

PATHOGENICITY

Salmonella cause localized infection of the gastrointestinal tract but can also multiply in the reticuloendothelial system resulting in systemic infection and death. *Salmonella* invade host cells by subverting host-cell signal transduction pathways and promoting cytoskeletal rearrangements. This results in the uptake of the microorganism via large vesicles or macropinosomes. To avoid the intracellular environment of the host, *Salmonella* invade macrophages, where they proliferate within a membrane-bound vacuole.

Non-*Salmonella typhi* serotypes

Invasion

Salmonella cross the small intestine epithelium by translocating through the M cells of the Peyer's patches, resulting in the destruction of the M cells and the adjacent epithelium. The *Salmonella* genome contains five large insertions, or pathogenicity islands, absent from *E. coli*. Most of the genes controlling entry into the M cells are located on *Salmonella* pathogenicity island-1 (SPI-1) and encode a type III secretion system responsible for exporting proteins from the bacterium to the host-cell cytosol. Four proteins encoded by genes on SPI-1, SopE, SipA, SipC and StpP are injected into the host cell in this way and, once inside, initiate cytoskeletal rearrangements, membrane ruffling and, ultimately, bacterial internalization.

Intracellular Survival

Once inside the macrophage, *Salmonella* remains inside a phagosome or *Salmonella*-containing vacuole (SCV). Proteins encoded for by genes on SPI-2, such as SpiC, inhibit fusion of the SCV with the microbiocidal contents of the host-cell lysosomes and endosomes. Other SPI-2 proteins may interfere with the intracellular trafficking of the host cell to increase the chances of survival of *Salmonella* within the phagosome. For example, SPI-2 proteins may reduce the amounts of NADPH oxidase in the macrophage which is essential for the production of microbiocidal compounds, such as reactive oxygen and nitrogen intermediates.

Virulence Plasmids

Jones *et al.* (1982) originally identified a cryptic plasmid capable of conferring a virulent phenotype on *S. typhimurium*. It is now recognized that non-typhoid salmonellae associated with intestinal invasion – *S. dublin*, *S. cholerasuis*, *S. gallinarum-pullorum* and in particular *S. virchow* – possess plasmids of differing sizes which contain genes important for invasive infection. All these plasmids possess *spv* (Salmonella plasmid virulence) genes contained on an 8-kb regulon. The regulon consists of the positive regulator *spv*R and four structural genes: *spv*A, *spv*B, *spv*C and *spv*D. Efficient transcription of SpvR depends on an alternative σ factor RpoS which is encoded on the chromosome. The activity of the RpoS increases, as the organisms enter the post-expontential phase of growth (Abe *et al.*, 1994). The expression of the *spv* genes is regulated by the growth phase of the bacteria: during early exponential growth in rich media these genes are not expressed, but the synthesis of SpvA begins in the transition period between late logarithmic and early post-exponential phase growth. The other proteins are expressed later. It appears that nutrient starvation, iron limitation and pH rather than cell density, is the stimulus for spv expression.

The *spv* genes enhance the growth of *Salmonella* strains within the macrophages and non-professional phagocytic cells (Heffernan *et al.*, 1987, Eriksson *et al.*, 2003). The nature of the initial signal for it's induction is unknown.

Salmonella typhi

Natural History of Infection

Although *S. typhimurium* can cause systemic disease in the human host, infection is usually localized in the intestinal epithelium (Ohl and Miller, 2001). *Salmonella typhi* is not typically associated with acute diarrhoea, suggesting that the initial interaction between this serovar and the human gut is less inflammatory. After ingestion, *S. typhi* must pass through the stomach, which provides a considerable barrier to infection. Organisms cannot be cultured from the stomach 30 min after ingestion, and antacids and H_2 antagonists reduce the number of organisms required to initiate infection. Bacteria will multiply in the intestine, but stools are only intermittently positive to culture during the first week, and even after experimental infection with 10^5–10^7 cfu, the presence of organisms in the stools during the first 5 days did not necessarily indicate that illness would develop (Hornick *et al.*, 1970). They are transported from the small intestine to the mesenteric lymph nodes via the M cells of the Peyer's patch and then disseminated widely via the lymphatics and bloodstream to grow intracellularly within the cells of the reticuloendothelial system. This time corresponds roughly to the incubation period. After approximately 1 week there is secondary bacteraemia, and typhoid bacilli reinvade the gut via the liver and gall bladder, settling in Peyer's patches and causing inflammation, ulceration and necrosis. This may lead to the complications of perforation or haemorrhage.

The Virulence of S. typhi

The complete genome sequence of a multiple-drug resistant *S. typhi* CT18 was published in 2001 (Parkhill *et al.*, 2001). Scattered along the backbone of the conserved core DNA are regions unique to *S. typhi*, for example, SPI-7, a large pathogenicity island containing genes that encode for Vi polysaccharide production (Wain *et al.*, 2002). Almost all virulent *S. typhi* possess a Vi antigen – a polysaccharide composed of *N*-actylglucosamine uronic acid. Organisms lacking the antigen require a much higher dose to infect; antibodies to Vi are detected in patients recovering from typhoid, and when Vi antigen is used as a vaccine, protective immunity is conferred.

Also present on the *S. typhi* genome are more than 200 pseudogenes, stretches of DNA that encode gene-like sequence but have been inactivated, usually by single-base mutations. One hundred and forty-five *S. typhi* pseudogenes are present as active genes in *S. typhimurium*, a Salmonella serovar that has a wider host range and causes different symptoms in humans. The genome of *S. typhi* may have undergone degeneration to facilitate a specialized association with humans and that by reducing the options for routes of invasion into human tissues this pathogen favours a route that promotes the likelihood of systemic spread (Wain *et al.*, 2002).

The macrophage has an important role to play in the pathogenesis of typhoid fever. This was recognized as long ago as 1891 by Mallory, who stated, 'the typhoid bacillus produces a mild diffusible *toxine*, partly within the intestinal tract, and partly within the blood and organs of the body. The *toxine* produced proliferation of the endothelial cella which acquire for a time malignant properties. The new formed cells are epithelioid in character, have irregular, lightly stained, eccentrically situated nuclei, abundant sharply defined acidophilic protoplasm and are characterized by marked phagocytic properties' (Mallory, 1898). In volunteer experiments, patients were made tolerant of endotoxin by repeated injection of small doses to the extent that they could tolerate 2.5 μg with no ill effect. In spite of this the volunteers became ill when challenged with *S. typhi*, exhibiting the characteristic clinical features of the disease (Greisman, Hornick and Woodward, 1964). Like other members of the Enterobacteriaceae, *S. typhi* and other salmonellas possess a LPS with a lipid A core. Stimulation of cells of the monocyte lineage results in the release of tumour necrosis factor-α and other cytokines. This gives rise to the fever, chills and rigors common in the clinical presentation and explains the phenomena so carefully recorded by previous workers.

LPSs also assist salmonellae by defending the organism against the effects of the complement pathway. The long 'O' side chain on the molecule means that activated complement (membrane attack complex) is unable to damage the outer membrane. The *rck* gene, encoded on the *S. typhimurium* virulence plasmid, codes for a protein, which acts to prevent the formation and insertion of fully polymerized tubular C9 complexes into the outer membrane.

EPIDEMIOLOGY

Non-Typhoid Salmonellosis

Examples of the way in which changes in animal husbandry practice and food preparation habits affect the incidence are illustrated by the trends in laboratory reporting of some of the major gastrointestinal pathogens in England and Wales in Figure 29.1. The incidence of salmonellosis reached its highest level in 1997, when over 32 000 cases were reported. The 12.5% increase in laboratory-confirmed cases between 1996 and 1997 represented a sharp up-turn after a period of relative stability since the early 1990s. Much of the increase in 1997 was accounted by a resurgence of cases of *S. enteritidis* PT4 which is closely associated with eggs and poultry (PHLS, 1999) changes in farming practices, and a move to more industrial-scale farm size and

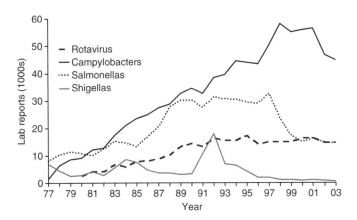

Figure 29.1 Trends in laboratory reporting of some of the major gastrointestinal pathogens in England and Wales (data from the Health Protection Agency).

feeding make infection more likely to occur and difficult to eradicate when it does.

Since 1997 the incidence of this phage type has continued to decline and, in 2001, *S. enteritidis* PT4 accounted for less than 50% of all cases for the first time since 1983 (Wall and Ward, 1999). The reason for this dramatic fall in the incidence of salmonellosis is mainly attributed to vaccination of poultry flocks (Ward *et al.*, 2000). The vaccine is based on *S. enteritidis* PT4, and very high proportion of flocks has been vaccinated. Improvements in the microbiological quality of food at all stages in the food chain from the point of production to the point of consumption, the so-called farm-to-fork approach, and the adoption of hazard analysis critical control point (HACCP) systems (O'Brien *et al.*, 1998) have also contributed to the decline in salmonellosis.

Although the incidence of *S. enteritidis* PT4 has declined, since autumn 2002, there has been an increase in infections caused by other *S. enteritidis* phage types associated with raw shell eggs imported from outside the UK. *Salmonella* spp. were not recovered from UK eggs bearing a quality mark (Lion mark) (PHLS, 2002a).

Infective Dose

The stomach acts as a barrier to *Salmonella* infection; consequently patients with achlorhydria, or who are taking antacids or H_2 antagonists, the young and the very old, are more susceptible to infection. Similarly ingestion of organisms in food, especially fatty food, is more likely to result in infection, as the organisms are protected from deleterious effects of gastric acid. This can be seen in outbreaks of *Salmonella* infection associated with chocolate (Linnane, Roberts and Mannion, 2002, PHLS, 2002b). Endoscopy with a contaminated instrument can also circumvent this natural defence (Spach, Silverstein and Stamm, 1993). A large number of experimental studies have been performed to determine the infective dose, and the lowest dose causing infection varied from 10^5 to 10^{10} (Blaser and Newman, 1982). An exception to these relatively high doses was a single study where patients were infected with *S. sofia* and *S. bovis-morbificans* by the nasal route, and the minimum infective dose was only 25 organisms (Mackenzie and Livingstone, 1968).

Patients of all ages are affected by salmonellosis, but the main burden of disease falls on the elderly and those who are immunocompromised.

Outbreaks in Hospitals and Institutions

Outbreaks of salmonellosis may occur in many situations, but individuals in institutional care, for example, hospitals and old people's homes, are particularly vulnerable and the mortality risks are significantly higher. Outbreaks characteristically take two forms, food borne or person-to-person spread. In a study carried out in England and Wales, between 1992 and 2000, food-borne transmission was reported in just 25 (1.8%) of 1396 hospital outbreaks investigated compared to 1212 (86%) that were caused by person-to-person spread (Meakins *et al.*, 2003). If a case of salmonellosis is admitted to the ward, *Salmonella* spreads rapidly if good control measures are not in place, and cases continue to rise sporadically over a number of days. Nursing and medical staff play a role in perpetuating such an outbreak through inadequate infection control practices or because they become infected and become asymptomatic carriers. Meakins *et al.* (2003) reported that *Salmonella* was the cause of 3.7% of hospital outbreaks.

Although person-to-person spread is the most common mode of transmission associated with hospital outbreaks, point-source outbreaks caused by the ingestion of a contaminated food item by a large number of patients can be a significant problem. In 1988, the Chief Medical Officer advised that raw shell eggs should be replaced with pasteurized eggs in recipes in institutions with high-risk groups. Despite this advice, a large nosocomial outbreak of *S. enteritidis* PT 6a occurred at a London hospital in 2002. The source was thought to be raw shell eggs from the hospital kitchen (PHLS, 2002c).

Typhoid

The incidence of typhoid is falling worldwide due to improvements in public health, such as provision of clean water and good sewage systems, but it still remains a major threat to human health. The World Health Organization estimates there are 17 million typhoid cases annually and that these infections are associated with about 600 000 deaths (Pang *et al.*, 1998). Typhoid is predominantly a disease of the developing world: the incidence in the Far East is approximately 1000/100 000 (Sinha *et al.*, 1999, Lin *et al.*, 2000). In some developing countries the majority of cases are reported to be in the 5–14 year age group, whereas other reports show a more even distribution but confirm that typhoid is an uncommon but serious infection in patients over 30 (Butler *et al.*, 1991, Lin *et al.*, 2000). The annual numbers of cases of typhoid in England and Wales between 1980 and 2001 are shown in Figure 29.2 (PHLS, 2002d). In industrialized countries the majority of *S. typhi* infections are acquired abroad (Figure 29.2). Infection is acquired by ingestion of contaminated food and water or contact with a patient or carrier of the disease. It is restricted in host range to human beings, and there is no known

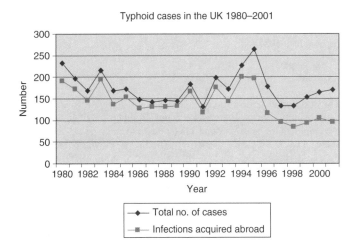

Figure 29.2 Cases of typhoid in England and Wales between 1980 and 2001, including those infections acquired abroad (data from the Health Protection Agency).

animal reservoir. In addition, *S. typhi* have caused laboratory-acquired infections.

CLINICAL FEATURES

Infection with organisms of the genus *Salmonella* results in three main syndromes: infection localized to the intestine, often known as salmonellosis, invasive or bacteraemic disease and enteric fever.

Salmonellosis

Symptoms of salmonellosis can develop in as little as 6 h and usually appear within 24 h after ingestion of the organisms. Patients complain of nausea, vomiting and diarrhoea. Fever and abdominal pain are also common, but the degree of each of these symptoms varies considerably between patients. In salmonellosis, patients complain of symptoms related to enteritidis and dysentery; blood in the stool is not uncommon. Abdominal pain is usually mild or moderate but can be sufficiently severe to mimic an acute abdomen in some cases. After symptoms subside patients continue to excrete organisms in their stools for up to 3 months. Approximately 1–3% may excrete organisms for more than a year.

In community outbreaks the majority of patients will suffer no serious effects but a small proportion, usually the elderly, will suffer invasive disease with a significant mortality. This is more likely to occur with Blegdam, Bredeney, Cholerasuis, Dublin, Enteritidis, Panama and Virchow serotypes (Old and Threlfall, 1998). Bacteraemia may result in the seeding of the viscera, meninges, bones, joints and serous cavities (Hohmann, 2001). Infection of the bones can result in an acute osteomyelitis. This is common in patients with sickle cell disease and may be very difficult to treat (Workman *et al.*, 1996).

Certain Salmonella serotypes, particularly Enteritidis, Typhimurium and Dublin (Levine *et al.*, 1991), have become an important problem among those infected with HIV, who are estimated to be between 20 and 100 times more at risk of infection than the general population (Gruenewald, Blum and Chan, 1994, Angulo and Swerdlow, 1995). Salmonella septicaemia usually occurs late in the course of HIV disease in industrialized countries. In Africa, Salmonella septicaemia is one of the leading causes of treatable and preventable death in HIV-infected individuals (Gilks *et al.*, 1990). Bacteraemia in HIV-positive patients in Malawi was associated with an inpatient mortality of 47% and a 1-year mortality of 77% (Gordon *et al.*, 2002). Recurrent *Salmonella* septicaemia is rare in immunocompetent patients but is a common problem in HIV-infected individuals, especially in those who do not complete an adequate course of antimicrobials (Hart *et al.*, 2000, Gordon *et al.*, 2002). It has been considered an AIDS-defining illness since 1987.

Typhoid

The incubation period for typhoid is longer than that for salmonellosis, usually 1 week. Historical accounts of typhoid written before the development of antimicrobial chemotherapy emphasized four main stages, lasting roughly 1 week each. The first week was characterized by rising fever, the second by rose spots, abdominal pain and splenomegaly and the third by the abdominal complications of haemorrhage or perforation followed by recovery in the fourth week (Stuart and Pullen, 1946).

The disease classically begins with an intermittent fever pattern which becomes more sustained over the first few days of infection. Most patients report fever or rigors (Parry *et al.*, 2002) and headache; a smaller proportion report nausea, vomiting, abdominal cramps and cough.

Patients can present paradoxically with either diarrhoea [more common in young children and HIV patients (Parry *et al.*, 2002)]

or constipation. Given the diversity of these clinical symptoms and the lack of a strong clinical pattern, patients should be suspected of having typhoid with an appropriate history or ingestion of suspected food.

On examination a fever is usually present, and careful recording may demonstrate the stepwise progression characteristics of this syndrome. The abdomen may be tender, and approximately half of patients will have a palpable liver and/or spleen. There is relative brachycardia. Careful inspection of the skin may reveal the presence of rose spots which characteristically begin around the umbilicus, but this sign is difficult to elucidate in patients with dark skin. Signs in the chest or meningismus may suggest an alternative diagnosis, emphasizing the fact that clinical diagnosis of typhoid may be difficult. Paratyphoid fever shares many of these clinical features.

In countries where schistosomiasis is endemic, retained eggs may be a source for recurrent typhoid infection as it is thought that *S. typhi* binds to the egg surface and later multiplies to cause a relapse.

Many infections may mimic typhoid, and it is important to exclude malaria in those with an appropriate travel history. A detailed travel history should be taken to generate a full differential diagnosis which might include the conditions listed in Table 29.1.

Complications

Complications occur in 10–15% of patients and are particularly probable in patients who have been ill for more than 2 weeks (Parry *et al.*, 2002). The case fatality rate with antimicrobial chemotherapy is less than 5% but is higher in children under the age of 1 year and older adults (Butler *et al.*, 1991). In young children, seizures may occur as a result of fever, hypoglycaemia or electrolyte imbalance. The most important complications are intestinal perforation or haemorrhage. The pathophysiological background to these complications has already been described. Intestinal perforation results in release of colonic bacteria into the peritoneum with consequent peritonitis. Haemorrhage results from severe ulceration of Peyer's patches. In the period before antimicrobial therapy, the mortality of intestinal perforation was approximately 90%. The appearance of chloramphenicol reduced this to around 60%. Delay in administering antimicrobial treatment is the most important factor associated with high mortality. Currently, the average case fatality rate is less than 1% but varies from less than 2% in Vietnam to 30–50% in Papua New Guinea (Parry *et al.*, 2002). Relapse occurs in 5–10% of patients, and reinfection can also occur. The latter can be distinguished from relapse by molecular typing (Wain *et al.*, 1999). Asymptomatic carriage of *S. typhi* occurs in 10% of convalescing patients, and 1–4% become long-term carriers (Parry *et al.*, 2002). Typhoid fever in pregnancy may lead to miscarriage, although antimicrobial therapy has made this outcome less common (Seoud *et al.*, 1988).

Typhoid may be complicated by acute pneumonia causing diagnostic confusion. Meningitis, cholecystitis and hepatitis are rare

Table 29.1 Differential diagnosis of typhoid

Malaria
Yersinia enterocolitica, Y. pseudotuberculosis infection
Campylobater infections
Amoebiasis
Brucellosis
Tuberculosis
Plague
intestinal anthrax
Melioidosis
Oroya fever
Rat-bite fever
Leptospirosis
Relapsing fever
Rickettsial infections

complications but should be considered when the appropriate clinical features are present. Like other salmonella infections, osteomyelitis or other localized infection may develop.

LABORATORY DIAGNOSIS

Specimens

A multiplicity of specimens is submitted for the isolation of *Salmonella*, including specimens of food and sewage, in addition to specimens from patients. Salmonellosis is most conveniently diagnosed by culture of diarrhoeal stool. Three stools should be submitted for culture to maximize the clinical yield. When patients are febrile or enteric fever is suspected, blood culture should be performed.

In enteric fever, culture of the stool may be negative in up to half of patients during the first week of infection but is more likely to be positive from the second week of illness. A positive isolate of *S. typhi* from stool alone is not diagnostic in the absence of characteristic clinical features, as up to 10% of patients become chronic carriers after acute infection. A definitive diagnosis is made by culture of enteric fever from a sterile site. Blood culture is positive in up to 75% of patients, but a higher yield is obtained by culturing bone marrow, with the added advantage that a positive culture can usually be obtained even when the patient has commenced antimicrobial therapy. *Salmonella typhi* can be isolated from urine, especially after the second week of illness. Chronic excretion of this organism in the urine can cause diagnostic confusion in a patient being investigated for urinary tract infection. The organism can also be isolated from jejunal juice obtained from a duodenal string test (Hoffman *et al.*, 1984). Positive cultures can be obtained from skin snips of rose spots, although in practice it is rarely necessary to perform this technique. Clot culture has a venerable place in the mythology of microbiology, enabling a bacterial culture and serological test to be performed on a single sample. It should now not be performed, as it has clearly been shown that routine clot culture is not cost effective if a sensitive blood culture method is used (Duthie and French, 1990).

The diagnosis of *Salmonella* spp. can be conveniently divided into four phases: initial isolation, screening identification, definitive identification and typing.

Initial Isolation

The problem with isolating *Salmonella* spp. from stools is in ensuring adequate sensitivity and selection (Anonymous, 1998). Salmonella will grow on strongly inhibitory media but *Shigella* spp., which are often sought at the same time, are more readily inhibited. Thus a compromise must be made. In order to maximize diagnostic yield a fluid enrichment medium, such as selenite F (Leifson, 1936), is usually inoculated in parallel with the plates, allowing these to be subcultured if the plates show no suspicious colonies. Selenite F is inhibitory to enterococci and *E. coli* due to the presence of sodium selenite. Salmonellae are able to grow continuously in this medium, and although the other species will slowly begin to multiply, useful selective enrichment does occur. Care must be taken with the quality control of this medium, as the degree of inhibition can vary widely and may reduce the diagnostic yield. The inhibition is most pronounced at low oxygen tension; so the medium should be poured into universal bottles to a depth greater than 5 cm (Gillespie, 1994). Rappaport, Konforti and Navon (1956) developed a magnesium chloride-malachite green enrichment broth, which they showed was more effective than Selenite F for the isolation of a wide range of *Salmonella* spp. This medium is not suitable for the isolation of *S. typhi* (Anderson and Kennedy, 1965) and is not often used in the clinical diagnostic laboratory, although it is frequently used, with modifications, for the isolation of *Salmonella* spp. from food (Fricker, 1987).

Deoxycholate citrate agar (DCA) is highly selective but may inhibit some shigellae. The growth of coliforms and Gram-positive organisms is strongly inhibited due to sodium citrate and sodium deoxycholate in the medium. Lactose and an indicator system are included, and H_2S-producing organisms will reduce the ferric ammonium citrate in the medium to produce a black centre to the colonies. Typically, Salmonellae produce pale colonies with a black centre, but this appearance can be mimicked by *Proteus* spp. and some *Citrobacter* organisms. Xylose lysine deoxycholate agar is a popular medium, and as it is less inhibitory than DCA, it is often used in parallel to optimize the yield of shigellae. The medium contains three sugars – xylose, lactose and sucrose – together with lysine. The colour changes that occur in the colonies depend on the biochemical reactions of the organisms. Organisms that do not ferment the sugars or decarboxylate the lysine will appear as colourless colonies, and those that ferment two or more sugars will produce bright yellow colonies. Organisms, including *Salmonella*, which ferment xylose only and decarboxylate lysine (with the exception of *S. paratyphi* A) and produce H_2S will appear as red colonies with a black centre. *Salmonella* spp. producing little or no H_2S, e.g. *S. typhi* and *S. pullorum*, will appear as red colonies with or without black centres. Red colonies can also be produced by *Proteus* spp. and *Pseudomonas* spp. A number of chromogenic agars are available, designed for the improved identification of salmonellae from faecal samples (Perez *et al.*, 2003), but the cost usually prohibits their use routinely.

Wilson and Blair medium is a more reliable modification of the original bismuth sulphite agar. It is highly inhibitory to other species and is particularly useful for the isolation of *S. typhi* from heavily contaminated specimens. Selection depends on the presence of brilliant green, sodium sulphite and bismuth ammonium citrate. *Salmonella typhi* has a characteristic silvery sheen with an adjacent brown-black zone in the agar.

Screening Identification

A large number of suspect colonies must be examined to maximize the diagnostic yield. To limit the number of full biochemical tests that must be performed, a screening test is performed to exclude organisms such as *Citrobacter* and *Proteus*. Several techniques are used, including short sugar series, the Kligler iron medium and Kohn's tubes. Commercial screening tests such as the APIZ are also available. Kligler's is the most popular of the composite media; it contains sucrose and an excess of lactose, ferric ammonium citrate and a suitable indicator. Organisms that ferment glucose but not lactose or sucrose, such as *Salmonella* spp., are inoculated; there is an initial acid production that turns the medium yellow, but under the aerobic conditions of the slope, this reverts to alkaline. When lactose is also fermented there is sufficient acid so that the medium does not revert. The medium is blackened by H_2S-producing organisms, and gas produced by sugar fermentation disrupts the medium. This can be used along with indole and urease tests (*Salmonella* spp. do not produce indole or hydrolyse urea), and together the results are used to select those organisms which should be definitely identified. The growth on the slope may be used for serological testing. Alternatively, commercial tests use the substrate for several biochemical reactions which when inoculated with a heavy suspension of organisms change colour rapidly, due to the action of preformed enzymes.

Definitive Identification

The definitive identification of *Salmonella* spp. depends on obtaining biochemical and serological confirmation of the diagnosis. The techniques employed for biochemical identifications are similar to those already described (see Hawkey, 1997). Serological identification depends on determining which flagella or H antigens (from the German word *Hauch*, meaning mist) and LPS or somatic O antigens (from the

German *Ohne hauch*, meaning without mist) are present. These are usually identified using a bank of polyclonal and specific rabbit antisera raised against boiled organisms by simple bacterial agglutination. Further identification is then made by reference to the Kauffman and White scheme (Old and Threlfall, 1998). Negative O agglutination may be found, and this may be due to the presence of an unusual or new serotype. Negative agglutination may also occur due to the presence of a capsular antigen (e.g. Vi) and can be overcome by heating the culture for 1 h at 100 °C. The flagellar antigens are diphasic in many serotypes, e.g. *S. typhimurium* produces antigenic specificity 'i' in phase 1 and '1,2' in phase 2. Phase 1 antigens are specific, whereas phase 2 antigens may share antigens with other serotypes. When salmonellae are found to have organisms in the nonspecific phase, variation can be induced by cultivation in semisolid agar containing antiserum against phase 2, which selects for phase 1 (Gillespie, 1994). Phage typing systems, based on host specificity of bacteriophages, have been developed for organisms most commonly implicated in food-borne outbreaks. Phage types are not indicative of clonality because phage-type conversions may result from the acquisition of both plasmids and bacteriophages. However, phage typing proved invaluable during the pandemic of *S. enteritidis* in the late 1980s and early 1990s (Old and Threlfall, 1998).

Molecular Methods for Detecting and Typing Salmonellae

Molecular techniques for the detection of *Salmonella* spp., such as PCR, have considerable advantages in terms of specificity, speed and standardization over the conventional methodologies described above. However, it is difficult to perform PCR directly on faecal samples due to the presence of inhibitory substances and large quantities of bacterial DNA other than the target DNA (Wilson, 1997). DNA extraction from faeces can be improved by pretreating the sample with polyvinyl pyrrolidone (McLauchlin *et al.*, 2000). Commercial kits for the detection of *Salmonella* spp. by real-time PCR, such as the RealArt™ *Salmonella* TM PCR Kit (Artus GmbH, Hamburg, Germany), are available but are not used routinely. Amar *et al.* (2004) found that culture and PCR methods used for the detection of *Salmonella* from clinical faecal samples were of similar sensitivity. However, using culture, results are available within 2–3 days, whereas those obtained by real-time PCR assays can be available within 3 h, which can be advantageous for rapid intervention and appropriate treatment.

For many diagnostic purposes and control of infection, serotyping and phage typing are sufficient. For adequate surveillance and determining the source of food-borne infections and outbreaks in hospitals and institutions, more sophisticated typing methods must be employed. Plasmid typing is based on the numbers and molecular weight of plasmids found in many *Salmonella* spp. and can be used for strain differentiation within serotypes during outbreak investigations (Threlfall *et al.*, 1994). For salmonella serotypes that do not carry plasmids, chromosomally based restriction fragment length polymorphism methods, such as ribotyping and insertion sequence (IS200) typing, can be used. IS200 typing has been shown to be more useful for certain serotypes (e.g. *S. infantis* and *S. heidelberg*) than other (e.g. *S. enteritidis*) (Stanley, Jones and Threlfall, 1991). Pulsed field gel electrophoresis and amplified fragment length polymorphism each have the advantage of providing fingerprints of the whole genome and, at present, are the most commonly used typing methods in the investigation of outbreaks (Ward *et al.*, 2002, Lan *et al.*, 2003).

Laboratory Diagnosis of Typhoid

Widal Test

The Widal test uses O and H antigens from the *S. typhi* and *S. paratyphi* in a simple bacterial agglutination test to aid the diagnosis of enteric fever. Patients infected with *S. typhi* produce serum antibodies to the O and H antigens of this pathogen, and the detection of these specific antibodies forms the basis of this test, the standardized protocol for which was established in 1950 (Felix, 1950). Although simple to perform, the Widal test is difficult to interpret, requiring a detailed knowledge of the patient's medical, travel and vaccination history. Acute and convalescent serum should be obtained and a positive diagnosis made on the basis of a fourfold rise or fall in *S. typhi* O and H antibodies. Certain patients may have high H antibody titres due to previous infection or vaccination. The O antibody concentrations fall about 6 months after previous exposure to typhoid. Cross-reacting antibodies produced by exposure to other serotypes may be a problem, but this is usually identified by including a nonspecific salmonella antigen preparation in the test battery. In addition, patients with typhoid may mount no detectable antibody response or demonstrable rise in antibody titre (Parry *et al.*, 2002). If paired sera are not available a single serum may be valuable if it yields an antibody concentration significantly in excess of the community norm for the test. Thus a patient from an industrialized country with a titre of 180 was likely to have acute infection, whereas this figure would be equivocal for a patient likely to have previous exposure.

ELISA tests and immunoblotting with LPS prepared from strains of *S. typhi* have been evaluated for the detection of anti-*S. typhi* LPS IgM serum antibodies in infected patients and have compared favourably with the Widal test (Chart, 1995). However, the determination of a cut-off value that clearly delineates antibody-positive sera from antibody-negative sera remains a problem. PCR tests targeting the flagellin genes to detect *S. typhi* directly in the blood have been developed but are not widely used (Song *et al.*, 1993). Other molecular markers of *S. typhi*, including a distinctive IS200 insertion element present in multiple copies in the *S. typhi* genome, are under investigation as potential diagnostic targets (Calva *et al.*, 1997).

Antigen Detection

The O, H and Vi antigens may be detected in the blood and urine of patients with typhoid. Co-agglutination, counter-immunoelectro-immunophoresis and ELISA techniques for their detection have been described (Barrett *et al.*, 1982, Shetty, Srinivasa and Bhat, 1985). Sensitivity of over 90% has been reported but, although valuable, these techniques cannot replace culture with the increasingly unpredictable susceptibility pattern of enteric fever organisms.

TREATMENT

Salmonella spp. are susceptible to a range of antimicrobial agents, including chloramphenicol, 4-fluoroquinolonoes, ampicillin, trimethoprim, aminoglycosides and third generation cephalosporins. Salmonellosis rarely requires systemic antimicrobials unless infection is complicated by bacteraemia or localized infection, such as oesteomylitis. The importance of resistance and the serotypes in which it occurs differs from country to country. In the UK, from 1975 onwards there has been an increase in multiresistant (resistance to four or more antibiotics) *S. typhimurium*, including reduced susceptibility to ciprofloxacin (MIC 0.250 mg/l), which is the drug of choice for the treatment of invasive salmonellosis. In developing countries, multiresistance is more common (Old and Threlfall, 1998).

Traditionally choramphenicol is the first-line agent for the treatment of typhoid and remains an appropriate treatment for typhoid in areas where the bacterium is still susceptible and resources for drug expenditure are few. The 4-fluroquinolones, including ciprofloxacin and ofloxacin, are now the first choice agent. Antibiotic trials have shown that these drugs are safe in all age groups and rapidly effective (3–7 days) (Gotuzzo and Carrillo, 1994, White and Parry, 1996). The cost of fluroquinolones and the potential emergence of resistance are two problems associated with this treatment. Full fluroquinolone resistance is rare, although quinolone-resistant strains are often resistant to other antibiotics making the choice of drugs limited. In patients with quinolone-resistant *S. typhi* infections, treatment with 20 mg of ofloxacin per kg of body weight for 10–14 days has been successful in 90–95% of patients (Wain *et al.*, 1997). Cefotaxime, ceftriaxone and cefoperazone have excellent *in vitro* activity against *S. typhi* and other salmonellae and have acceptable efficacy in the treatment of typhoid fever. Only intravenous formulations are available. Cefotaxime is given 1 g three times daily (in children: 200 mg/kg daily in divided doses) for 14 days. Ceftriaxone has an advantage of only requiring a single dose daily. The cephalosporins are not active against many MDR strains, and this limits its use in empirical treatment when resistant typhoid is likely.

Azithromycin

Animal models have shown that azithromycin is highly effective against *S. typhi* and non-typhoid *Salmonella*. It has now been shown to be effective in a series of open and randomized control trials. Oral administration is an aid to administration, and the results of clinical studies demonstrate that it is as effective as chloramphenicol, cefriaxone and ciprofloxacin. It was also effective in cases of MDR typhoid. This provides a useful alternative for the management of children with uncomplicated typhoid in developing countries (Butler *et al.*, 1999, Girgis *et al.*, 1999, Frenck *et al.*, 2000).

Multiple drug-resistant *S. typhi* was first described in Mexico in the 1960s, probably arising from uncontrolled antibiotic use (Anderson and Smith, 1972, Olarte and Galindo, 1973). It is now commonplace in *S. typhi* and non-typhoid *Salmonella* in the Indian subcontinent (Rowe, Ward and Threlfall, 1997), Latin America, Africa and countries in the Far East (Bhutta *et al.*, 1991, Coovadia *et al.*, 1992, Rasaily *et al.*, 1994). Resistance to chloramphenicol, ampicillin and trimethoprim is generally plasmid encoded, although additional chromosomal resistance to ciprofloxacin has been reported (Rowe, Ward and Threlfall, 1995). Because of the efficacy and low relapse and carrier rates associated with their use, the 4-quinolone drugs are now the drugs of choice in the treatment of adult typhoid, certainly in areas where multiresistant typhoid fever has been reported (Ferreccio *et al.*, 1988, DuPont, 1993). However, because of its cheapness, chloramphenicol will continue to be used in other areas where the local strains are sensitive, although azithromycin may in the future be a useful alternative especially in children.

In children with possible multiresistant typhoid, a third generation cephalosporin, e.g. cefotaxime will be the preferred drug if 4-quinolone drugs are to be avoided. However, their cost and the need for intravenous administration are significant disadvantages, particularly in the developing countries, and ciprofloxacin is being used increasingly in children with typhoid. Azithromycin is another potential alternative that combines oral availability with intracellular penetration and activity against the pathogen.

Treatment of Chronic Carriers

Chronic carriage is common in patients recovering from typhoid who have not been treated with antibiotics. Chloramphenicol therapy reduces chronic carriage rates to less than 10%, and chronic carriage rates after 4-fluroquinolone therapy are thought to be less than 2% (Parry *et al.*, 2002). In some circumstances, eradication of chronic carriage might be necessary. Before embarking on this course the real benefits to the patient in the context of the risk posed to public health should be carefully considered. Thus treatment of a single asymptomatic person living alone might not be necessary, but a food handler with no other means of employment would require it. Chronic carriage may be eradicated by further courses of specific antimicrobial therapy, and clinicians vary widely in the regimes they recommend.

PREVENTION AND CONTROL

The provision of safe drinking water, effective sewage disposal and hygienic food preparation have contributed to the control of typhoid in developing countries. Mass immunization has been used successfully in some areas (Bodhidatta *et al.*, 1987). In developed countries, most cases are the result of travel to an endemic area, and travellers should take suitable precautions, such as boiling drinking water and cooking food thoroughly (Parry *et al.*, 2002).

Vaccination

In 1896, Almoth Wright described the first use of a typhoid vaccine given to two officers in the Indian Medical Service. Initial attempts at large-scale vaccination of soldiers in India and in the Boer War led to controversy about its efficacy, which continued over the next 50 years. It was not until the 1960s that controlled field trials were performed (Woodward, 1980, Engels *et al.*, 1998). These studies were with heat- or phenol-killed organisms and the most successful demonstrated protection of 65–88% for up to 7 years.

The first effective oral typhoid vaccine was made by attenuating a virulent strain of *S. typhi* Ty2, yielding the strain Ty2A after radiation (Germanier and Furer, 1975). It has been shown to be effective in a number of field trials and produces approximately 70% protection, lasting for up to 7 years (Levine *et al.*, 1987). The genome has multiple mutations, and the organism was originally thought to be attenuated because of specific mutations in the galE enzyme essential for production of the O polysaccharide. Volunteer studies with another *S. typhi* strain with a single galE mutation was not attenuated and caused disease, suggesting that the mechanism of attenuation is elsewhere. The vaccine is not very immunogenic and is given in four doses – one capsule on alternate days – and requires a booster every 5 years. It is a live, attenuated vaccine and therefore should not be administered to immunocompromised patients or patients taking antibiotics (Parry *et al.*, 2002).

An alternative vaccine approach is to use purified Vi antigen, which is both a virulence factor and a protective antigen. This idea had been proposed many years ago, but unpromising results from early field trials discouraged its use (Woodward, 1980). The Vi antigen is thought to prevent antibodies binding to the O antigen, enabling *S. typhi* to survive in the blood, and is also associated with the inhibition of complement-mediated lysis and phagocytosis. Immunization with the Vi antigen stimulates protective antibody responses against typhoid fever (Garmory, Brown and Titball, 2002). The vaccine can be given to adults and children over the age of 2 as a single dose of 25 µg administered intramuscularly. Two to three year boosters are required (Parry *et al.*, 2002). Like Ty2A approximately 70% protection is afforded for up to 7 years (Acharya *et al.*, 1987). It has the advantage that it is inexpensive to produce by a semisynthetic procedure using pectin from citrus fruit. A new Vi vaccine conjugated to a nontoxic recombinant *Pseudomonas aeruginosa* exotoxin A (rEPA) with the potential to be immunogenic in infants under the age of 2 was evaluated recently in Vietnam. The protective efficacy was 91.5% (Lin *et al.*, 2001).

Another approach is to use strains of *Salmonella* with defined mutations in metabolic pathways. These strains arose out of a need to produce attenuated organisms for teaching purposes. It was only later that their potential as vaccine candidates was recognized (Hoiseth and Stocker, 1981). Work has continued in developing auxotrophic mutants, harbouring defined mutations in genes coding for enzymes in the prechorismate biosynthetic pathway (*aro* mutants). These strains are unable to scavenge essential aromatic metabolites such as *para*-aminobenzoic acid. As these substances are not available in mammalian tissues the organism is attenuated. As attenuation comes through metabolic handicap, these mutants should also be attenuated in immunocompromised populations. The most characterized strains are

CVD906 and CVD908 which are mutants of Ty2 and ISP 1820, respectively, in which *aro*C and *aro*D mutations have been induced. Of these, CVD908 is the most promising vaccine candidate. In phase II clinical trials with CVD908, LPS-specific IgA-secreting cells were detected in 92 and 100% low-dose and high-dose vaccines, and serum LPS-specific IgG was detected in 46 and 49%, respectively. There were few side-effects associated with the vaccine (Tacket *et al.*, 2000). Further studies are also being carried out on CVD 909, another potential vaccine candidate, which consists of CVD908 constitutively expressing the Vi antigen (Wang *et al.*, 2000).

Many other attenuation strategies are under investigation. Promising results have been shown in trials involving strain χ4073. This strains has mutations in adenylate cyclase *cya* and camp receptor *crp* that affect the expression of genes involved in carbohydrate and amino acid metabolism. Following inoculation with a single dose of up to 5×10^8 cfu of χ4073, most healthy volunteers developed a specific serum antibody response and specific antibody-secreting cells to *S. typhi* and the vaccine was well tolerated (Tacket *et al.*, 1997). In addition, a *phoP/phoQ* mutant constructed from strains Ty800, a derivative of *S. typhi* Ty2, has been shown to be safe and immunogenic (Hohmann *et al.*, 1996). Further studies are also being conducted on CVD915, a *gua*BA deletion mutant of *S. typhi* Ty2 (Wang *et al.*, 2001).

Attenuated *S. enterica* strains have also been manipulated to carry foreign genes, such as the tetanus toxin fragment (TetC), opening the prospect for oral vaccination against a wide range of diseases (Garmory, Brown and Titball, 2002).

Prevention of Salmonellosis in the Food Industry

Central to the prevention of transmission of salmonellosis via food is the cooperation of farmers, veterinarians, abattoir workers and agriculture ministries. Several codes of practice for the control of salmonellas in chickens have been in operation in the UK since 1993, and there have been many improvements in the poultry industry in infection control and hygiene at breeding sites. In 1994, vaccination against *S. enteritidis* started in breeder flocks and in 1998 in layer flocks (Ward *et al.*, 2000). A European Commission Green Paper, published in 1997, stressed that the responsibility for food safety rests with the food industry and proposed that HACCP systems be adopted by the food industry (European Commission Green Paper, 1997). HACCP is an analytical tool for the systematic assessment of the many steps on the production, processing, packaging and kitchen preparation of food and the identification of steps critical to the safety of the product, also referred to as the stable-to-table or farm-to-fork strategy (WHO, 1997, O'Brien *et al.*, 1998). Adequate education and training should be provided for workers at all levels in the food industry. Similarly good practice in food preparation facilities is necessary to minimize the risks of transmission. The Food Standards Agency recommends that food businesses should use pasteurized egg rather than raw egg, in products that will not be cooked or only lightly cooked before eating (Food Standards Agency, 2002). Coupled with these measures, all outbreaks should be investigated using epidemiological techniques both as a local and national surveillance programme. The final defence against salmonellosis must be the consumer both in preparation of food at home and in the exercise of choice in purchase of commercially prepared food. In kitchens and food preparation areas where ordinary eggs are being used, good food hygiene practices are important to avoid the risk of cross-contamination (Food Standards Agency, 2002).

Prevention of Salmonellosis in the Hospital Environment

Salmonellosis can be a particular problem in hospitals and similar institutions because of the presence of elderly or immunocompromised patients. Common-source outbreaks require urgent investigation of the patients to determine the presence of a pathogen and to provide specimens for typing and also the food which should be investigated to indicate the source and breakdown in the hygienic arrangements. During a large *S. enteritidis* outbreak at a London hospital, control measures included removal of a range of high-risk foods from the patients' menus, including all raw shell egg-based products, meals prepared from raw chicken and sandwiches prepared on site (PHLS, 2002c). For a 2-week period, all food production was halted on site and cold food was bought in from external suppliers as individual pre-packed items. Good food and personal hygiene practices among all staff was reinforced, training was increased and monitoring at kitchen, ward and departmental level was carried out. A trust-wide hand washing campaign was also introduced (PHLS, 2002c).

As salmonellosis is transmitted by the faecal–oral route, it can be readily contained by isolation of the source in a single room with enteric precautions.

REFERENCES

Abe, A., Matsui, H., Danbara, H. *et al.* (1994) Regulation of spvR gene expression of *Salmonella* virulence plasmid pKDSC50 in *Salmonella choleraesuis* serovar Choleraesuis. *Molecular Microbiology*, **12**, 779–787.

Acharya, I.L., Lowe, C.U., Thapa, R. *et al.* (1987) Prevention of typhoid fever in Nepal with the Vi capsular polysaccharide of *Salmonella typhi*. A preliminary report. *The New England Journal of Medicine*, **317**, 1101–1014.

Amar, C.F., East, C., Maclure, E. *et al.* (2004) Blinded application of microscopy, bacteriological culture, immunoassays and PCR to detect gastrointestinal pathogens from faecal samples of patients with community-acquired diarrhoea. *European Journal of Clinical Microbiology Infectious Diseases*, **23**, 529–534.

Anderson, E.S. and Smith, H.R. (1972) Chloramphenicol resistance in the typhoid bacillus. *British Medical Journal*, **3**, 329–331.

Anderson, K. and Kennedy, H. (1965) Comparison of selective media for the isolation of salmonellae. *Journal of Clinical Pathology*, **18**, 747–749.

Angulo, F.J. and Swerdlow, D.L. (1995) Bacterial enteric infections in persons infected with human immunodeficiency virus. *Clinical Infectious Diseases*, **21** (Suppl. 1), S84–S93.

Anonymous. (1998) PHLS Standard operating procedure for the investigation of faeces specimens for bacterial pathogens, in *PHLS Standard Operating Procedures*, Technical services, PHLS Head Quarters, London.

Anserson, E.S. and Smith, H.R. (1972) Chloramphenicol resistance in the typhoid bacillus. *British Medical Journal*, **3**, 329–331.

Barrett, T.J., Snyder, J.D., Blake, P.A. and Feeley, J.C. (1982) Enzyme-linked immunosorbent assay for detection of *Salmonella typhi* Vi antigen in urine from typhoid patients. *Journal of Clinical Microbiology*, **15**, 235–237.

Barrow, G.I. and Feltham, R.K.A. (1992) *Cowan and Steel's Manual for the Identification of Medical Bacteria*, 3rd edn, Cambridge University Press, Cambridge, UK.

Bhutta, Z.A., Naqvi, S.H., Razzaq, R.A. and Farooqui, B.J. (1991) Multidrug-resistant typhoid in children: presentation and clinical features. *Reviews of Infectious Diseases*, **13**, 832–836.

Blaser, M.J. and Newman, L.S. (1982) A review of human salmonellosis: I. Infective dose. *Reviews of Infectious Diseases*, **4**, 1096–1106.

Bodhidatta, L., Taylor, D.N., Thisyakorn, U. and Echeverria, P. (1987) Control of typhoid fever in Bangkok, Thailand, by annual immunization of school children with parenteral typhoid vaccine. *Reviews of Infectious Disease*, **9**, 841–845.

Butler, T., Islam, A., Kabir, I. and Jones, P.K. (1991) Patterns of morbidity and mortality in typhoid fever dependent on age and gender: review of 552 hospitalized patients with diarrhea. *Reviews of Infectious Diseases*, **13**, 85–90.

Butler, T., Sridhar, C.B., Daga, M.K. *et al.* (1999) Treatment of typhoid fever with azithromycin versus chloramphenicol in a randomized multicentre trial in India. *The Journal of Antibicrobial Chemotherapy*, **44** (2), 243–250.

Calva, E., Ordonez, L.G., Fernandez-Mora, M. *et al.* (1997) Distinctive IS200 insertion between gyrA and rcsC genes in *Salmonella typhi*. *Journal of Clinical Microbiology*, **35**, 3048–3053.

Chart, H. (1995) Detection of antibody responses to infection with *Salmonella*. *Serodiagnosis, Immunotherapy and Infectious Disease*, **7**, 34–39.

Coovadia, Y.M., Gathiram, V., Bhamjee, A. *et al.* (1992) An outbreak of multiresistant *Salmonella typhi* in South Africa. *The Quarterly Journal of Medicine*, **82**, 91–100.

DuPont, H.L. (1993) Quinolones in *Salmonella typhi* infection. *Drugs*, **45** (Suppl. 3), 119–124.

Duthie, R. and French, G.L. (1990) Comparison of methods for the diagnosis of typhoid fever. *Journal of Clinical Pathology*, **43**, 863–865.

Edwards, R.A., Olsen, G.J. and Maloy, S.R. (2002) Comparative genomics of closely related salmonellae. *Trends in Microbiology*, **10**, 94–99.

Engels, E.A., Falagas, M.E., Lau, J. and Bennish, M.L. (1998) Typhoid fever vaccines: a metaanalysis of studies of efficacy and toxicity. *BMJ*, **316**, 110–116.

Eriksson, S., Lucchini, S., Thompson, A. *et al.* (2003) Unravelling the biology of macrophage infection by gene expression profiling of intracellular *Slamonella enterica. Molecular Microbiology*, **47**, 103–118.

European Commission Green Paper. (1997) The general principles of food law in the European Union. Brussels: EC (Com(97)176).

Felix, A. (1950) Standardization of diagnostic agglutination tests. *Bulletin of the World Health Organization*, **2**, 643–649.

Ferreccio, C., Morris, J.G. Jr., Valdivieso, C. *et al.* (1988) Efficacy of ciprofloxacin in the treatment of chronic typhoid carriers. *The Journal of Infectious Diseases*, **157** (6), 1235–1239.

Food Standards Agency. (2002) *Salmonella* outbreaks prompt agency to issue hygiene alert (Press Release). London: FSA, 15 October 2002. www.food.gov.uk/news/newsarchive/salmonellaoutbreaknews.

Frenck, R.W. Jr., Nakhla, I., Sultan, Y. *et al.* (2000) Azithromycin versus ceftriaxone for the treatment of uncomplicated typhoid fever in children. *Clinical Infectious Diseases*, **31** (5), 1134–1138.

Fricker, C.R. (1987) The isolation of salmonellas and campylobacters. *The Journal of Applied Bacteriology*, **63**, 99–116.

Garmory, H.S., Brown, K.A. and Titball, R.W. (2002) *Salmonella* vaccines for use in humans: present and future perspectives. *FEMS Microbiology Reviews*, **26**, 339–353.

Germanier, R. and Furer, E. (1975) Isolation and characterization of Gal E mutant Ty21a of *Salmonella typhi*: a candidate strains for a live, oral typhoid vaccine. *The Journal of Infectious Diseases*, **131**, 553–558.

Gilks, C.F., Brindle, R.J., Otieno, L.S. *et al.* (1990) Life-threatening bacteraemia in HIV-1 seropositive adults admitted to hospital in Nairobi, Kenya. *Lancet*, **36**, 545–549.

Gillespie, S.H. (1994) Examination of faeces for bacterial pathogens, in *Medical Microbiology Illustrated* (ed. S.H. Gillespie), Butterworth-Heinemann, Oxford, pp. 192–210.

Girgis, N.I., Butler, T., Frenck, R.W. *et al.* (1999) Azithromycin versus ciprofloxacin for treatment of uncomplicated typhoid fever in a randomized trial in Egypt that included patients with multidrug resistance. *Antimicrobial Agents and Chemotherapy*, **43** (6), 1441–1444.

Gordon, M.A., Banda, H.T., Gondwe, M. *et al.* (2002) Non-typhoidal salmonella bacteraemia among HIV-infected Malawian adults: high mortality and frequent recrudescence. *AIDS*, **16**, 1633–1641.

Gotuzzo, E. and Carrillo, C. (1994) Quinolones in typhoid fever. *Infectious Diseases in Clinical Practice*, **3**, 345–351.

Greisman, S.E., Hornick, R.B. and Woodward, T.E. (1964) The role of endotoxin during typhoid fever and tularemia in man. 3. Hyperreactivity to endotoxin during infection. *The Journal of Clinical Investigation*, **43**, 1747–1757.

Gruenewald, R., Blum, S. and Chan, J. (1994) Relationship between human immunodeficiency virus infection and salmonellosis in, 20- to 59-year-old residents of New York City. *Clinical Infectious Diseases*, **18**, 358–363.

Hart, C.A., Beeching, N.J., Duerden, B.I. *et al.* (2000) Infections in AIDS. *Journal of Medical Microbiology*, **49**, 947–967.

Hawkey, P.M. (1997) Identification of Enterobacteriaceae, in *Principles and Practice of Clinical Bacteriology*, 1st edn (eds A.M. Emmerson, P.M. Hawkey and S.H. Gillespie), John Wiley & Sons Ltd, Chichester, p. 367.

Heffernan, E.J., Fierer, J., Chikami, G. and Guiney, D. (1987) Natural history of oral Salmonella dublin infection in BALB/c mice: effect of an 80-kilobase-pair plasmid on virulence. *The Journal of Infectious Diseases*, **155**, 1254–1259.

Hoffman, S.L., Punjabi, N.H., Rockhill, R.C. *et al.* (1984) Duodenal string-capsule culture compared with bone-marrow, blood, and rectal-swab cultures for diagnosing typhoid and paratyphoid fever. *The Journal of Infectious Diseases*, **149**, 157–161.

Hohmann, E. (2001) Non-typhoidal salmonellosis. *Clinical Infectious Diseases*, **32**, 263–269.

Hohmann, E.L., Oletta, C.A., Killeen, K.P. and Miller, S.I. (1996) *phoP/phoQ*-deleted *Salmonella typhi* (Ty 800) is a safe and immunogenic single-dose typhoid fever vaccine in volunteers. *The Journal of Infectious Diseases*, **173**, 1408–1414.

Hoiseth, S.K. and Stocker, B.A. (1981) Aromatic-dependent *Salmonella typhimurium* are non-virulent and effective as live vaccines. *Nature*, **291**, 238–239.

Hornick, R.B., Greisman, S.E., Woodward, T.E. *et al.* (1970) Typhoid fever: pathogenesis and immunologic control. *The New England Journal of Medicine*, **283**, 739–746.

Jones, G.W., Rabert, D.K., Svinarich, D.M. and Whitfield, H.J. (1982) Association of adhesive, invasive, and virulent phenotypes of *Salmonella typhimurium* with autonomous 60-megadalton plasmids. *Infection and Immunity*, **38**, 476–486.

Kauffmann, F. (1957) Das Kauffmann-White Schema, in *Ergebnisse der Mikrobiologie, Immunitatsforschung und Experimentellen Therapie*, Vol 30 (ed. F. Kauffmann) Springer Verlag, Berlin-Gottngen-heidelberg, p. 160.

Kidgell, C., Reichard, U., Wain, J. *et al.* (2002) *Salmonella typhi*, the causative agent of typhoid fever is approximately 50,000 years old. *Infection, Genetics and Evolution*, **2**, 39–45.

Lan, R., Davison, A.M., Reeves, P.R. and Ward, L.R. (2003) AFLP analysis of Salmonella enterica serovar Typhimurium isolates of phage types DT 9 and DT 135: diversity within phage types and its epidemiological significance. *Microbes and Infection*, **5**, 841–850.

Leifson, E. (1936) New selenite enrichment media for the isolation of typhoid and partyphoid (*Salmonella*) bacilli. *American Journal of Hygiene*, **24**, 423–432.

Levine, M.M., Ferreccio, C., Black, R.E. and Germanier, R. (1987) Large-scale field trial of Ty21a live oral typhoid vaccine in enteric-coated capsule formulation. *Lancet*, **1**, 1049–1052.

Levine, W.C., Buehler, J.W., Bean, N.H. and Tauxe, R.V. (1991) Epidemiology of nontyphoidal *Salmonella* bacteremia during the human immunodeficiency virus epidemic. *The Journal of Infectious Diseases*, **164**, 81–87.

Lin, F.Y., Ho, V.A., Khiem, H.B. *et al.* (2001) The efficacy of a Salmonella typhi Vi conjugate vaccine in two-to-five-year-old children. *The New England Journal of Medicine*, **344**, 1263–1269.

Lin, F.Y., Vo, A.H., Phan, V.B. *et al.* (2000) The epidemiology of typhoid fever in the Dong Thap Province, Mekong Delta region of Vietnam. *The American Journal of Tropical Medicine and Hygiene*, **62**, 644–648.

Linnane, E., Roberts, R.J. and Mannion, P.T. (2002) An outbreak of Salmonella enteritidis phage type 34a infection in primary school children: the use of visual aids and food preferences to overcome recall bias in a case control study. *Epidemiology and Infection*, **129**, 35–39.

Mackenzie, C.R. and Livingstone, D.J. (1968) Salmonellae in fish and food. *South African Medical Journal*, **42**, 999–1003.

Mallory, F.B. (1898) A histological study of typhoid fever. *The Journal of Experimental Medicine*, **3**, 611–638.

McCelland, M., Sanderson, K.E., Spieth, J. *et al.* (2001) Complete genome sequence of *Salmonella enterica* serovar Typhimurium LT2. *Nature*, **413**, 848–852.

McLauchlin, J., Amar, C., Pedraza-Diaz, S. and Nichols, G.L. (2000) Molecular epidemiological analysis of Cryptosporidium spp. in the United Kingdom: results of genotyping Cryptosporidium spp. in 1,705 fecal samples from humans and 105 fecal samples from livestock animals. *Journal of Clinical Microbiology*, **38**, 3984–3990.

Meakins, S.M., Adak, G.K., Lopman, B.A. and O'Brien, S.J. (2003) General outbreaks of infectious intestinal disease (IID) in hospitals, England and Wales, 1992–2000. *The Journal of Hospital Infection*, **53**, 1–5.

O'Brien, S., Rooney, R., Stanwell-Smith, R. and Handysides, S. (1998) Taking control of infectious intestinal disease. *Communicable Disease and Public Health*, **1**, 144–145.

Ohl, M.E. and Miller, S.I. (2001) *Salmonella*: a model for bacterial pathogenesis. *Annual Reviews in Medicine*, **52**, 259–274.

Olarte, J. and Galindo, E. (1973) *Salmonella typhi* resistant to chloroamphenicol, ampicillin and other antimicrobial agents: strains isolated during an extensive typhoid fever epidemic in Mexico. *Antimicrobial Agents and Chemotherapy*, **4**, 597–601.

Old, D. and Threlfall, E.J. (1998) Salmonella, in *Topley and Wilson's Microbiology and Microbial Infections*, 9th edn (eds L. Collier, A. Balows and M. Sussman), Oxford University Press, New York, Vol. 2, pp. 969–998.

Pang, T., Levine, M.M., Ivanoff, B. *et al.* (1998) Typhoid fever – important issues still remain. *Trends in Microbiology*, **6**, 131–133.

Parkhill, J., Dougan, G., James, K.D. *et al.* (2001) Complete genome sequence of *a multiple drug resistant* Salmonella enterica serovar Typhi CT18. *Nature*, **413**, 848–852.

Parry, C.M., Hein, T.T., Dougan, G. *et al.* (2002) Typhoid fever. *The New England Journal of Medicine*, **347**, 1770–1781.

Perez, J.M., Cavalli, P., Roure, C. *et al.* (2003) Comparison of four chromogenic media and Hektoen agar for detection and presumptive identification of *Salmonella* strains in human stools. *Journal of Clinical Microbiology*, **41**, 1130–1134.

PHLS. (1999) The rise and fall of *salmonella*? *Communicable Disease Report. CDR Weekly*, **9** (14).

PHLS. (2002a) Public health investigation of *Salmonella* Enteriditis in raw shell eggs. *Communicable Disease Report. CDR Weekly*, **12** (50).

PHLS. (2002b) An international outbreak of *Salmonella* Oranienburg infection. *Communicable Disease Report. CDR Weekly*, **12** (2).

PHLS. (2002c) Salmonella Enteritidis outbreak in a London hospital – update. *Communicable Disease Report. CDR Weekly*, **12** (48).

PHLS. (2002d) Laboratory reports of typhoid and paratyphoid, England and Wales: 1980 to 2001. *Communicable Disease Report. CDR Weekly*, **12** (7).

Rappaport, F., Konforti, N. and Navon, B. (1956) A new enrichment medium for certain salmonellae. *Journal of Clinical Microbiology*, **9**, 261–266.

Rasaily, R., Dutta, P., Saha, M.R. *et al.* (1994) Multi-drug resistant typhoid fever in hospitalised children. Clinical, bacteriological and epidemiological profiles. *European Journal of Epidemiology*, **10**, 41–46.

Rowe, B., Ward, L.R. and Threlfall, E.J. (1995) Ciprofloxacin-resistant *Salmonella typhi* in the UK. *Lancet*, **346**, 1302.

Rowe, B., Ward, L.R. and Threlfall, E.J. (1997) Multiresistant *Salmonella typhi* – a world-wide epidemic. *Clinical Infectious Diseases*, **24** (Suppl. 1), S106–S109.

Seoud, M., Saade, G., Uwaydah, M. and Azoury, R. (1988) Typhoid fever in pregnancy. *Obstetrics and Gynaecology*, **71**, 711–714.

Shetty, N.P., Srinivasa, H. and Bhat, P. (1985) Coagglutination and counter immunoelectrophoresis in the rapid diagnosis of typhoid fever. *American Journal of Clinical Pathology*, **84**, 80–84.

Sinha, A., Sazawal, S., Kumar, R. *et al.* (1999) Typhoid fever in children aged less than 5 years. *Lancet*, **354**, 734–737.

Song, J.H., Cho, H., Park, M.Y. *et al.* (1993) Detection of *Salmonella typhi* in the blood of patients with typhoid fever by polymerase chain reaction. *Journal of Clinical Microbiology*, **31**, 1439–1443.

Spach, D.H., Silverstein, F.E. and Stamm, W.E. (1993) Transmission of infection by gastrointestinal endoscopy and bronchoscopy. *Annals of International Medicine*, **118**, 117–128.

Stanley, J., Jones, C.S. and Threlfall, E.J. (1991) Evolutionary lines among *Salmonella enteritidis* phage types are identified by insertion sequence IS200 distribution. *FEMS Microbiology Letters*, **82**, 83–90.

Stuart, B.M. and Pullen, R.L. (1946) Typhoid: clinical analysis of three hundred and sixty cases. *Archives of Internal Medicine*, **78**, 629–661.

Tacket, C.O., Sztein, M.B., Losonsky, G.A. *et al.* (1997) Safety of live oral *Salmonella typhi* vaccine strains with deletions in htrA and aroC aroD and immune response in humans. *Infection and Immunity*, **65**, 452–456.

Tacket, C.O., Sztein, M.B., Wasserman, S.S. *et al.* (2000) Phase 2 clinical trial of attenuated *Salmonella enterica* serovar typhi oral live vector vaccine CVD 908-htrA in U.S. volunteers. *Infection and Immunity*, **68**, 1196–1201.

Threlfall, E.J., Hampton, M.D., Chart, H. and Rowe, B. (1994) Use of plasmid profile typing for surveillance of *Salmonella enteritidis* phage type 4 from humans, poultry and eggs. *Epidemiology*, **112**, 25–31.

Wain, J., Hien, T.T., Connerton, P. *et al.* (1999) Molecular typing of multiple-antibiotic-resistant *Salmonella enterica* serovar Typhi from Vietnam: application to acute and relapse cases of typhoid fever. *Journal of Clinical Microbiology*, **37**, 2466–2472.

Wain, J., Hoa, N.T., Chinh, N.T. *et al.* (1997) Quinolone-resistant *Salmonella typhi* in Vietnam: molecular basis of resistance and clinical response to treatment. *Clinical Infectious Diseases*, **25**, 1404–1410.

Wain, J., House, D., Parkhill, J. *et al.* (2002) Unlocking the genome of the human typhoid bacillus. *The Lancet Infectious Diseases*, **2**, 163–170.

Wall, P.G. and Ward, L.R. (1999) Epidemiology of *Salmonella enterica* serovar Enteritidis phage type 4 in England and Wales, in *Salmonella Enterica Serovar Enteritidis in Humans and Animals: Epidemiology, Pathogenesis and Control* (ed. A.M. Saeed), Iowa State University Press, Iowa, pp. 19–15.

Wang, J.Y., Noriega, F.R., Galen, J.E. *et al.* (2000) Constitutive expression of the Vi polysaccharide capsular antigen in attenuated *Salmonella enterica* serovar typhi oral vaccine strain CVD 909. *Infection and Immunity*, **68**, 4647–4652.

Wang, J.Y., Pasetti, M.F., Noriega, F.R. *et al.* (2001) Construction, genotypic and phenotypic characterization, and immunogenicity of attenuated δ-guaBA *Salmonella enterica* serovar Typhi strain CVD 915. *Infection and Immunity*, **69**, 4734–4741.

Ward, L.R., Maguire, C., Hampton, M.D. *et al.* (2002) Collaborative investigation of an outbreak of *Salmonella enterica* serotype Newport in England and Wales in 2001 associated with ready-to-eat salad vegetables. *Communicable Disease and Public Health*, **5**, 301–304.

Ward, L.R., Threlfall, J., Smith, H.R. and O'Brien, S. (2000) *Salmonella enteritidis* Epidemic. *Science*, **287**, 1753–1754.

White, N.J. and Parry, C.M. (1996) The treatment of typhoid fever. *Current Opinion in Infectious Diseases*, **9**, 298–302.

WHO. (1997) Introducing the Hazard Analysis Critical Control Point System. Geneva: (DOC WHO/FSF/FOS/97.2).

Wilson, I.G. (1997) Inhibition and facilitation of nucleic acid amplification. *Applied and Environmental Microbiology*, **63**, 3741–3751.

Woodward, W.E. (1980) Volunteer studies of typhoid fever and vaccines. *Transactions of the Royal Society of Tropical Medicine and Hygiene*, **74**, 553–556.

Workman, M.R., Philpott-Howard, J., Bragman, S. *et al.* (1996) Emergence of ciprofloxacin resistance during treatment of Salmonella osteomyelitis in three patients with sickle cell disease. *The Journal of Infection*, **32**, 27–32.

30

Klebsiella, Citrobacter, Enterobacter and *Serratia* spp.

C. Anthony Hart

Medical Microbiology Department, Royal Liverpool Hospital, Liverpool, UK

INTRODUCTION

The four genera *Klebsiella, Citrobacter, Enterobacter* and *Serratia* are all members of the family Enterobacteriaceae but exhibit different degrees of relatedness to each other and to other members of the family. For example, *Serratia* spp. are 25% related to the other three genera and *Enterobacter* spp. (except *E. aerogenes*) are 45% related to *Klebsiella*. Interestingly *Citrobacter freundii* and *C. diversus* are only 50% related to each other (Brenner, 1984). The phenotypic characteristics of the genera are summarized in Table 30.1. They are all oxidase-negative rods that are facultative anaerobes.

All the four genera are associated with nosocomial infections and infection in the immunocompromised host. However, *Klebsiella* spp. are much more frequently associated with infection, being responsible for 3% of community-acquired and 9% of hospital-acquired cases of septicaemia in one hospital (Eykyn, Gransden and Phillips, 1990). They are estimated to be responsible for 16.7 infectious episodes per 10 000 hospital discharges (Jarvis *et al.*, 1985). The other three genera were together responsible for 0.9% of community-acquired and 4% of nosocomial episodes of septicaemia.

KLEBSIELLA SPECIES

The genus *Klebsiella* is named after the German bacteriologist Edwin Klebs (1834–1913). In the UK literature several other species have been differentiated on biochemical properties (Table 30.2), including *K. pneumoniae* (Friedlander's bacillus), *K. aerogenes, K. ozaenae* and *K. rhinoscleromatis* (Orskov, 1984). On the basis of DNA homology studies, these should all be included in the species *K. pneumoniae*. However, to maintain continuity with previous literature, the old names are used to designate subspecies. For example, *K. aerogenes* is *K. pneumoniae* subsp. *aerogenes* and Friedlander's bacillus is *K. pneumoniae* subsp. *pneumoniae*. Other species in the genus include *K. granulomatis* (see Chapter 00) and *K. oxytoca* (Jain, Radsak and Mannheim, 1974). *Klebsiella ornitholytica, K. planticola* and *K. terrigena* have been reclassified as species in the new genus *Raoultella* (Drancourt *et al.*, 2001).

DESCRIPTION OF THE ORGANISM

Klebsiella spp. are straight bacilli, $0.3–1\,\mu m \times 0.6–6\,\mu m$, not motile and most isolates possess a thick polysaccharide capsule and fimbriae. Their DNA mol. % G+C is 53–58.

They are facultative anaerobes that have no particular growth factor requirement. They hydrolyse sugars fermentatively, mostly with release of carbon dioxide rather than hydrogen. *Klebsiella* spp. usually hydrolyse urea and most can utilize citrate and glucose as sole carbon sources. *Klebsiella oxytoca* differs from *K. pneumoniae* in being indole positive and able to liquefy gelatin, degrade pectate and utilize gentisate and *m*-hydroxybenzoate. All are lactose fermenters and those that are the most efficient carry *lac* genes on both chromosome and plasmid.

Table 30.1 Biochemical characteristics of the genera of *Klebsiella, Citrobacter, Enterobacter* and *Serratia*

	Klebsiella spp.	*Citrobacter* spp.	*Enterobacter* spp.	*Serratia* spp.
Motile	−	+	+	+
Indole	V (*K. oxytoca*)	+	−	−
Methyl red	V	+	−	V
Voges-Proskauer	V	−	+	+
H₂S production	−	+	−	−
Arginine decarboxylase	−	+	V	−
Lysine decarboxylase	+	−	V	+
Ornithine decarboxylase	−	V	+	V
Gelatinase	−	−	+	+
Urease	+	V	V	+
Acid from				
Adonitol	+	V	V	V
Inositol	+	−	V	+

+, ≥90% strains positive; V, 10–90% strains positive; −, ≤ 10% strains positive.

Table 30.2   Biochemical characteristics of *Klebsiella* spp

| | *Klebsiella pneumoniae* | | | |
	subsp. *pneumoniae*	subsp. *aerogenes*	subsp. *rhinoscleromatis*	*K. oxytoca*
Indole production	−	−	−	+
Pectate degradation	−	−	−	+
β-Galactosidase	+	+	−	+
Growth at 10 °C	−	−	−	+
Gentisate utilization	−	−	−	+
Methyl red	+	−	+	−
Voges-Proskauer	+	−	−	+
Urease	+	+	−	+
Gas from glucose	+	+	−	+
Growth in KCN	−	+	+	+
Growth on Simmon's citrate	+	+	−	+
Acid from dulcitol	+	V	−	V
Acid from lactose	+	+	−	+
Lysine decarboxylase	+	+	−	+
Histamine assimilation	−	−	−	−
Melezitose assimilation	−	−	−	+ (75%)

+, ≥90% strains positive; V, 10–50% strains positive; −, ≤10% strains positive.

PATHOGENICITY

Apart from rhinoscleroma, ozaena and granuloma inguinale, infections with *K. pneumoniae* are nosocomial and opportunistic. Rhinoscleroma is a chronic granulomatous infection of the upper airways, although contiguous skin and the trachea may also be affected. Lesions are characterized by a submucosal infiltrate of plasma cells and histiocytes called Mickulicz cells. These are foamy macrophages with vacuoles containing *K. pneumoniae* subsp. *rhinoscleromatis*. The infective dose, portal of entry and incubation period remain unknown. It is not clear how a bacterium with a hydrophilic capsule resistant to phagocytosis enters macrophages and survives (Hart and Rao, 2000).

The pathogenesis and pathogenic determinants of nosocomial and opportunistic *K. pneumoniae* infections are shown in Figure 30.2. Most of our information on transmission of *K. pneumoniae* comes from studies of outbreaks of nosocomial infection (Podschun and Ullmann, 1998). Reservoirs of *K. pneumoniae* tend to be the intestinal tract of other hospitalized patients (Le Frock, Ellis and Weinstein, 1979; Hart and Gibson, 1982; Kerver *et al.*, 1987), although other sites such as the oropharynx, skin (the nearer the perineum the heavier the colonization) and vagina may be carriage sites (Hart and Gibson, 1982). Although *K. pneumoniae* can be transmitted via surgical instruments, urinary catheters, food (Donowitz *et al.*, 1981) and even expressed breast milk (Cooke *et al.*, 1980), hands are by far the most important route (Knittle, Eitzman and Baer, 1975; Casewell and Phillips, 1977). *Klebsiella* spp. appear to be able to survive on skin better than, for example, *Escherichia coli* or *Citrobacter* spp. (Hart, Gibson and Buckles, 1981; Casewell and Desai, 1983), perhaps because of the hydrophilic capsule. This has been disputed by others (Fryklund, Tullus and Burman, 1995).

In most cases infection is preceded by intestinal colonization which, it is presumed, is mediated by fimbriae. Although there is considerable heterogeneity among the mannose-inhibitable type I fimbriae, the receptor-binding moiety (29 kDa subunit) is highly conserved (Abraham *et al.*, 1988). These fimbriae appear to be important in attachment to respiratory and urinary tract mucosal cells (Williams and Tomas, 1990). Some *K. pneumoniae* possess plasmid-encoded proteins that mediate autoagglutination and attachment to plastic surfaces (Denoya, Trevisan and Zorzopoulos, 1986).

Klebsiella pneumoniae becomes established in other mucosal sites by migration and can enter the blood by translocation. Once in the bloodstream or in tissues it must evade the host's non-specific immune system. There is no doubt that the hydrophilic capsule inhibits phagocytic uptake of bacteria (Denoya, Trevisan and Zorzopoulos, 1986). Heavily encapsulated *K. pneumoniae* are more virulent than less encapsulated strains in experimentally infected mice (Williams *et al.*, 1983). Specific antibody can opsonize the bacteria and facilitate phagocytosis, but the ability of virulent *K. pneumoniae* to shed capsular material may render this process less efficient. The capsule may also render bacteria resistant to opsonization and killing by the alternative complement cascade (Domenico, Johanson and Straus, 1982). The presence of long O-polysaccharide chains on lipopolysaccharides (LPSs) contributes greatly to resistance to serum killing (Merino *et al.*, 1992), perhaps by sterically hindering the binding of the complement membrane attack complex to the outer membrane. However in murine models it is clear that the capsular polysaccharides are more important than LPS O-side chains in modulating complement deposition (Cortes *et al.*, 2002a).

Once in tissues the iron scavengers, in particular enterochelin, allow the bacteria to grow even when iron is scarce (Tarkkanen *et al.*, 1992). The presence of antibiotic resistance plasmids is also of assistance in allowing *Klebsiella* to grow in the tissues of patients receiving antibiotics. The tissue damage by *Klebsiella* is probably because of both release of toxic molecules by the bacteria and the host's response, in particular frustrated activated phagocytes releasing toxic oxygen radicals. The major bacterial determinant is endotoxin released either as LPSs alone or in complexes with capsular polysaccharide (Straus, 1987; Williams and Tomas, 1990). Global analysis of *K. pneumoniae* virulence-associated genes has been undertaken using transposon muta-genesis assays (Cortes *et al.*, 2002b; Struve, Forestier and Krogfelt, 2003). Mutations in genes involved in core LPS biosynthesis (*waaL*, *waaE* and *wbbO*), synthesis of cell membrane and surface structures (*plsX*, *surA* and *ompA*) and in protein and histone biosynthesis (*tufA* and *hupA*) all inhibited colonization of intestine and urinary tract, whereas mutations in genes encoding GDP-fucose synthesis (involved in capsule synthesis) and fimbria synthesis inhibited urinary tract colonization alone (Struve, Forestier and Krogfelt, 2003). Disruption of *htrA* (a heat-shock–induced serine protease) rendered *K. pneumoniae* much more sensitive to temperature, oxidative stress and complement-mediated killing (Cortes *et al.*, 2002b).

Capsule

A key characteristic of *Klebsiella* spp. is the production of mucoid colonies when grown on solid media. This is enhanced when the media contain an excess of carbohydrate. The mucoid character is due to the thick polysaccharide capsule (Figure 30.1) which absorbs a large amount of water. The capsular material may also diffuse freely into the surrounding medium as extracapsular polysaccharide. There are at least 82 different capsular (K) serotypes and subtypes with similar antigenicity but different polysaccharide backbones have been described (Allen *et al.*, 1987b). Despite this antigenic diversity, the range of monosaccharide units is limited to L-fucose, L-rhamnose, D-mannose, D-glucose, D-galactose, D-glucuronic acid or D-galacturonic acid, some with *O*-acetyl and pyruvate ketal groups (Sutherland, 1985). *Klebsiella pneumoniae* subsp. *pneumoniae* is predominantly serotype 3, as is *K. pneumoniae* subsp. *rhinoscleromatis*. *Klebsiella pneumoniae* subsp. *ozaenae* encompasses serogroups 3, 4, 5 and 6, but each of the serogroups has been associated with *K. pneumoniae* subsp. *aerogenes*.

The genetic control of *Klebsiella* capsular synthesis is complex. A 29-kb chromosomal fragment from *K. pneumoniae* K2 Chedid was able to induce K2 capsular biosynthesis in capsuleless mutants of *K. pneumoniae*, but an extra plasmid-encoded *rmp*A gene was required for its expression in *E. coli* (Arakawa *et al.*, 1991). *rmp*A2 that is encoded on a 200-kb plasmid in bacteraemic isolates of *K. pneumoniae* is a trans-activator of capsular polysaccharide biosynthesis (Lai, Peng and Chang, 2003). Two genes *rcs*A (regulation of capsule synthesis) and *rcs*B have been cloned from the chromosome of *K. pneumoniae* subsp. *aerogenes* K21 that cause expression of a mucoid character in *E. coli*. This, however, is due to activation of synthesis of colanic acid which is antigenically similar to the K21 capsular polysaccharide (Allen, Hart and Saunders, 1987a). Capsular polysaccharide biosynthesis is greater under nitrogen than carbon limitation and increased at lower (approximately 30 °C) temperatures (Mengistu, Edwards and Saunders, 1994). The initial biosynthesis takes place in the cytoplasm and utilizes UTP and ATP. The assembled polysaccharide is transported across the inner membrane using an undecaprenyl phosphate (P-C55) carrier (Troy, 1979). How it traverses the periplasmic space

and outer membrane and forms the capsule is unclear. Group II capsules of *E. coli* are homopolymers of a single monosaccharide (Jann and Jann, 1990). These are transported across the periplasmic space utilizing the so-called traffic ATPases or ABC (ATP-binding cassette) transporters (Ames and Joshi, 1990; Pavelka, Wright and Silver, 1991). A further set of polypeptides are required for translocation of these group II *E. coli* capsular polysaccharides across the outer membrane and to tether them in place (Boulnois and Roberts, 1990). Unfortunately the capsules of *Klebsiella* spp. are more closely related to those of group I *E. coli*; for example, the *Klebsiellar* K5 and K54 antigens are structurally identical to *E. coli* K55 and K28 respectively (Altman and Dutton, 1985). The capsular polysaccharides appear to form fibres protruding radially from the bacterial surface (Figure 30.1). The presence of the capsule imparts a net negative charge to the bacterium and render it highly hydrophilic.

Adhesins

Klebsiella spp. possess both fimbrial and non-fimbrial adhesins (Pruzzo *et al.*, 1989; Tarkkanen *et al.*, 1992). The fimbriae can be of type 1 or type 3 (mannose-resistant haemagglutination of tanned erythrocytes only), and occasionally P fimbriae.

A mannose-inhibitable non-fimbrial adhesin has been found on unencapsulated *Klebsiella* spp. (Pruzzo *et al.*, 1989), as has a plasmid-encoded non-fimbrial adhesin (29 kDa) which facilitates adhesion of *Klebsiella* to epithelial cells (Darfeuille-Michaud *et al.*, 1992). The situation is reversed for certain *K. pneumoniae* capsular serotypes that have mannose disaccharide units in their polysaccharide (Athamna *et al.*, 1991). Here the disaccharide acts as the receptor (rather than adhesin) for a mannose/*N*-acetylglucosamine-specific lectin on the surface of macrophages.

Lipopolysaccharide

The repertoire of O-antigens on *K. pneumoniae* LPS is limited to eight (O types 1, 3–5, 7–9 and 12), with O:1 being the most common.

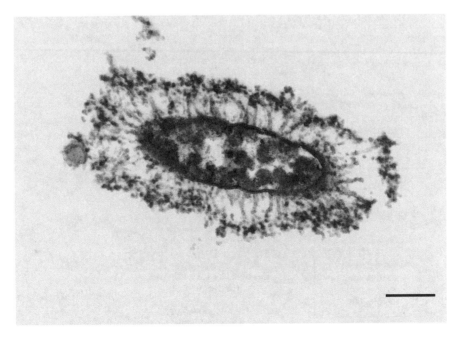

Figure 30.1 Thin-section electron micrograph of *Klebsiella pneumoniae* stained with ruthenium red to demonstrate the capsule. Bar = 1 μm.

Other Toxins

Klebsiella pneumoniae strains carrying genes for, or expressing heat-labile and heat-stable enterotoxins have been described (Betley, Miller and Mekalanos, 1986). Their clinical importance is unclear (Figure 30.2). A channel-forming bacteriocin (microcin E492) from *K. pneumoniae* is able to induce apoptosis in human cells (Hetz *et al.*, 2002).

Iron Scavenging

Klebsiella pneumoniae is able to scavenge iron from its surrounding medium using either enterochelin (phenolte siderophore), which is detected more frequently in pathogenic isolates, or aerobactin (hydroxamate siderophore).

NATURAL HABITAT

Although *Klebsiella* spp. are listed as part of the enteric flora in healthy individuals; if they are present it is only in small numbers ($<10^2$ cfu/g faeces). Premature babies and especially those in neonatal intensive care units (Tullus *et al.*, 1988; Hart, 1993), hospitalized patients even when moderately (Le Frock, Ellis and Weinstein, 1979) or severely ill (Kerver *et al.*, 1987), those with chronic disease and elderly individuals are relatively easily colonized by *Klebsiella* spp. This colonization may also extend to the oropharynx (Mackowiak, Martin and Smith, 1979). Administration of antimicrobials that affect the colonization resistance (Van der Waaij, 1982) of the intestinal tract also predisposes to colonization by *Klebsiella* and other Enterobacteriaceae. In one study it was found that the half-life for carriage of aminoglycoside-resistant *K. pneumoniae*, but not of *K. oxytoca* in elderly patients, was 100 days and some subjects excreted bacteria for over 200 days (Hart and Gibson, 1982). *Klebsiella* spp. can be part of

the normal flora of a variety of other animals and are widely distributed in the inanimate environment. A recent study found *Klebsiella* spp. in 53% of 208 samples of surface water (streams, lakes and the Baltic Sea) and *K. pneumoniae* was the species most frequently isolated (Podschun *et al.*, 2001).

EPIDEMIOLOGY

Klebsiella spp. are important nosocomial pathogens worldwide. In a UK prevalence survey, *Klebsiella* spp. were responsible for 8.3% of urinary tract infection, 4% of wound infections and 3.5% of pneumonias (Meers *et al.*, 1980). In the USA the importance of *K. pneumoniae* as a cause of nosocomial bacteraemia has declined from 9.1% of cases in 1975 to 4.5% in 1985–1989 (Pittet, 1993). In neonatal intensive care units *Klebsiella* spp. appear in the top three or four pathogens both in developed (Gladstone *et al.*, 1990) and developing countries (Rajab and De Louvois, 1990). Explosive life-threatening outbreaks of septicaemia due to multidrug-resistant *Klebsiella* spp. occur with monotonous regularity (Donowitz *et al.*, 1981; Morgan, Hart and Cooke, 1984; Fryklund, Tullus and Burman, 1995). In this setting septicaemia usually follows intestinal colonization with *Klebsiella* spp. which, it is suggested, can result from use of ampicillin rather than cephalosporins in the unit (Tullus and Burman, 1989). Outbreaks of infection are a feature in compromised patients, often with multidrug-resistant organisms (Casewell *et al.*, 1977; Curie *et al.*, 1978). *Klebsiella* spp. have been reported to be responsible for 14% of cases of bacteraemia associated with intravascular devices (Eykyn, Gransden and Phillips, 1990).

Rhinoscleroma is found in Eastern Europe, and occasionally in Italy, Switzerland, Spain and southern France. It is also endemic in Africa, Latin America and parts of the Middle and Far East. It is commoner in women than men and usually presents in early adulthood. It is associated with poor and crowded living conditions with poor sanitation.

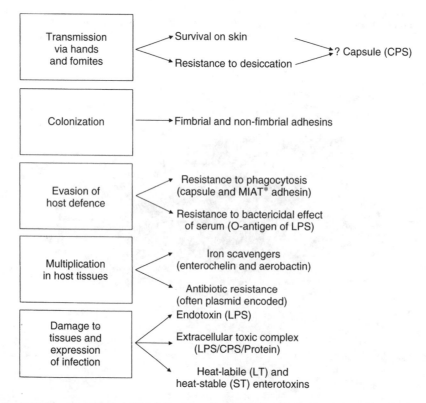

Figure 30.2 Pathogenesis of nosocomial and opportunistic *Klebsiella pneumoniae* infection. MIAT*, mannose-inhibitable adhesin/phage T7 receptor.

Pneumonia due to *K. pneumoniae* subsp. *pneumoniae* is an uncommon occurrence and generally occurs as a community-acquired infection in elderly debilitated males.

CLINICAL FEATURES

Nosocomial *Klebsiella* infections can occur at almost any site, and the clinical features do not differ from those caused by any of the Enterobacteriaceae. Although *K. pneumoniae* subsp. *pneumoniae* is historically associated with pneumonia, a worldwide study of bacteraemic pneumonia found Friedlander's bacillus is less than 1% of cases (Ko *et al.*, 2002). In addition a distinctive syndrome of *K. pneumoniae* bacteraemia in association with community-acquired liver abscesses meningitis or endophthalmitis was found, the first being found almost exclusively in Taiwan (Ko *et al.*, 2002). Risk factors for community-acquired *K. pneumoniae* bacteraemia include alcoholic liver disease and diabetes mellitus (Tsay *et al.*, 2002).

Rhinoscleroma begins as a painless chronic inflammatory swelling which causes nasal or respiratory tract obstruction. The nasal lesions enlarge locally to produce the so-called Hebra nose, which is grossly distorted and splayed. Local spread and local metastatic foci, often with lymph node involvement, are frequently described. Ozaena is seldom seen nowadays and there are doubts over its status as a disease entity.

DIAGNOSIS

The nosocomial *Klebsiella* spp. are simple to detect by standard cultural techniques, and their isolation from normally sterile sites or in significant numbers from sites with a normal flora provides evidence of infection.

The diagnosis of rhinoscleroma depends upon the typical histopathological appearance (Mickulicz cells) which can be made more specific by using anti-capsular antibody (Meyer *et al.*, 1983). The differential diagnosis of rhinoscleroma includes espundia, rhinosporidiosis, leprosy, yaws and malignancy.

Antimicrobial Susceptibility

Klebsiella spp. seem to have a particular propensity to acquire, build up and maintain antibiotic resistance plasmids (Hart, Gibson and Buckles, 1981), especially if nosocomially acquired (Tsay *et al.*, 2002). Most *Klebsiella* are intrinsically resistant to ampicillin, but acquisition of plasmid-encoded resistance has resulted in outbreaks of nosocomial infection. In the 1970s and 1980s outbreaks of infection due to gentamicin- and tobramycin (but not amikacin)-resistant *K. pneumoniae* were reported in increasing numbers (Casewell *et al.*, 1977; Curie *et al.*, 1978; Morgan, Hart and Cooke, 1984). With the introduction of second- and third-generation cephalosporins, isolates of *K. pneumoniae* resistant to cefuroxime, ceftazidime and cefotaxime soon emerged. Initially this was chromosomally encoded (Hart and Percival, 1982), but subsequently plasmid-encoded extended-spectrum β-lactamases caused large outbreaks of infection especially in adult intensive care units (Jacoby and Medeiros, 1991; Hibbert-Rogers *et al.*, 1994). In general, these extended-spectrum β-spectrum lactamases have arisen by evolution of *TEM* and *SHV* genes (Hibbert-Rogers *et al.*, 1994). Resistance to fluoroquinolones can be high; for example, in one French hospital 97% of β-lactamase-resistant *K. pneumoniae* were resistant to ciprofloxacin (Bergogne-Bérézin, Decré and Joly-Guillon, 1993). It appears that *Klebsiella* spp. do not remain predictably susceptible to a new antibiotic beyond a relatively short 'honeymoon period'.

Epidemiological Typing Methods

The major typing method for *Klebsiella* spp. is capsular serotyping. Although this was originally done by the Quellung reaction, other techniques such as immunofluorescence (Murcia and Rubin, 1979) or counter-current immunoelectrophoresis (Palfreyman, 1978) are more readily applicable. An analysis of *K. pneumoniae* isolated from blood showed that K2 (8.9% of isolates), K21 (7.8%) and K55 (4.8%) were the most common serotypes and that 25 serotypes accounted for 70% of the isolates (Cryz *et al.*, 1986). In a UK survey, K21 (42% of isolates) and K2 (16.5%) were the most common isolates from nosocomial infections (Casewell and Talsania, 1979). Other methods available for typing *Klebsiella* spp. are biotyping (Simoons-Smit *et al.*, 1987), antimicrobial susceptibility patterns, Klebecin typing (Edmondson and Cooke, 1979) and a variety of molecular methods including pulsed-field gel electrophoresis (PFGE) of digested chromosomal DNA (Gouby *et al.*, 1994). A recent comparison of PFGE versus O:K serotyping has concluded that both were highly discriminative but PFGE was more effective (Hansen *et al.*, 2002).

TREATMENT AND PREVENTION

Because *K. pneumoniae* can be resistant to many different antimicrobial agents, empirical therapy must be based on knowledge of local susceptibility patterns. Prevention of nosocomial infection requires adherence to infection-control policies, and handwashing is an important part of this. Passive immunization with anticapsular polysaccharides has had some effect in treating and preventing *Klebsiella* infection in burns patients (see references in Cryz *et al.*, 1986).

Klebsiella pneumoniae subsp. *rhinoscleromatis* is sensitive *in vitro* to many antibiotics, including streptomycin, cotrimoxazole, gentamicin and tobramycin. However, ciprofloxacin (Borgstein, Sada and Cortes, 1993) or ofloxacin appear to be the drugs of choice but they must be given for long periods (months). Surgical removal of lesions will also be necessary.

CITROBACTER SPECIES

The genus *Citrobacter* was first named in 1932, but until recently the name had not gained full acceptance (Sakazaki, 1984). Species in the present genus have been previously named *Levinia amalonatica*, *E. freundii*, *Salmonella hormaechii*, *S. ballerup* and the Ballerup/Bethesda group. Currently 11 species are recognized: *Citrobacter koseri* (formerly *C. diversus*), *C. freundii*, *C. amalonaticus*, *C. farmeri*, *C. youngae*, *C. braakii*, *C. gillenii*, *C. muraliniae*, *C. werkmanii*, *C. sedlakii* and *C. rodentium* (O'Hara, Roman and Miller, 1995).

DESCRIPTION OF THE ORGANISM

Citrobacter spp. are rods ($1 \mu m \times 2$–$6 \mu m$), usually non-encapsulated, but some *C. freundii* strains produce Vi capsular antigen (Houng *et al.*, 1992). They are motile by means of peritrichous flagella and some possess class 1 fimbriae (Abraham *et al.*, 1988). All are lactose fermenters but some isolates of *C. freundii* are late lactose fermenters. This together with their ability to produce H_2S in the butt of triple sugar or Kligler's iron-agar has led to their being confused with *Salmonella* spp. but *C. freundii* is indole positive. At least 42 O-antigens, one of which cross-reacts with *E. coli* 0157 (Bettelheim *et al.*, 1993), and 90 H-antigens are expressed by the genus. The mol. % G+C of their DNA is 51–52.

PATHOGENESIS

The virulence determinants of *Citrobacter* spp. are little understood. It is presumed that they colonize the intestine and other sites using fimbriae. Production of capsule (Vi) and *E. coli* 0157 lipopolysaccharide is more related to confusing microbiologists than causing disease. *Citrobacter rodentium* and some strains of *C. freundii* carry the *eae* (enteropathogenic *E. coli* attaching and effacing) gene and are associated with murine colonic hyperplasia (Frankel et al., 1994). Recently an outbreak of gastroenteritis and haemolytic uraemic syndrome was associated with C. freundii producing verotoxin 2 (Tschäpe et al., 1995). Expression of a 32-kDa outer membrane protein was found among isolates of C. koseri causing neonatal meningitis (Li et al., 1990). Citrobacter koseri is easily engulfed by human macrophages, but survives phagolysome fusion and replicates within them (Townsend et al., 2003).

EPIDEMIOLOGY AND INFECTIONS

Normal Habitat

Citrobacter spp. are found in the intestinal tract of humans and other animals and are widely distributed in the environment. Citrobacter koseri appears to become established with some ease in the neonatal intestine and even in the mother's or nurse's intestinal flora (Williams et al., 1984; Goering et al., 1992). In contrast, spontaneous loss of intestinal carriage of other Citrobacter spp. that were multidrug resistant occurred more readily than that of Klebsiella spp. or E. coli (Hart and Gibson, 1982).

Citrobacter spp. are an infrequent cause of bacteraemia; in a UK survey 0.7% of episodes were caused by C. freundii and C. koseri equally (Eykyn, Gransden and Phillips, 1990), and they do not appear in the 'top 10' bacteraemic isolates from 1975 to 1989 in the USA (Pittet, 1993). Citrobacter koseri does, however, appear to be an important neonatal pathogen (Williams et al., 1984; Goering et al., 1992). In common with other Enterobacteriaceae, Citrobacter spp. can cause meningitis, bacteraemia pneumonia, urinary tract infection and wound infections. Although C. koseri is predominantly a neonatal pathogen, occasional cases of meningitis are reported in adults. Following C. koseri neonatal meningitis, abscess formation is common and the long-term outlook is poor with seizures, paralysis and mental retardation in 50% of cases (Straussberg, Harel and Amir, 2001).

Most infections are sporadic and the reservoir for Citrobacter spp. is the intestinal tract (Chen et al., 2002). Spread is most likely to be by hands, although Citrobacter spp. survive less well on skin than Klebsiella or Enterobacter spp. (Hart, Gibson and Buckles, 1981). For epidemiological purposes Citrobacter spp. have been differentiated by biotype, O-serotyping, plasmid profiles (Williams et al., 1984; Goering et al., 1992), outer membrane protein profiles (Tschäpe et al., 1995) and analysis of chromosomal DNA (Morris et al., 1986).

DIAGNOSIS

Citrobacter spp. can be easily grown from most infected sites. One of the major pitfalls in the laboratory is distinguishing them from Salmonella spp. and E. coli.

Antimicrobial Susceptibility

Antibiotic susceptibilities are not entirely predictable, but in one survey all isolates of C. freundii and C. koseri were sensitive to ciprofloxacin and imipenem, and over 50% were sensitive to gentamicin and trimethoprim (Reeves, Bywater and Holt, 1993). A particular feature of Citrobacter spp. is the carriage of inducible cephalosporinases (AmpC β-lactamases) and the frequency of the emergence of the de-repressed mutants overproducing such enzymes (Stapleton, Shannon and Phillips, 1995). These enzymes are chromosomally encoded, not inhibited by clavulanate, able to hydrolyse third-generation cephalosporins and are induced by their substrates. They have been detected in strains isolated well before the 'antibiotic era' (Barlow and Hall, 2002). Approximately 10% of hospitalized children were found to be excreting Citrobacter spp. expressing these enzymes (Berkowitz and Metchock, 1995).

TREATMENT AND PREVENTION

Antimicrobial chemotherapy will depend upon the local susceptibility pattern. However, treatment with fluorinated quinolones, imipenem or amikacin is likely to be successful depending on the site of infection. Prevention is by appropriate infection-control measures (Williams et al., 1984).

ENTEROBACTER SPECIES

Enterobacter cloacae was first described as Bacterium cloacae and as Cloaca cloacae but renamed E. cloacae when the genus was established (Richard, 1984).

Enterobacter spp. are biochemically active and E. aerogenes is similar to K. pneumoniae. Enterobacter spp. differ from Klebsiella spp. in being motile and from Serratia spp. by being negative for lipase, Tween 80 esterase and DNAase. Currently there are 15 species in the genus: E. cloacae, E. aerogenes (syn K. mobilis), E. agglomerans (now Pantoea agglomerans), E. gergoviae, E. sakazakii, E. cowanii, E. hormaechei, E. taylorae, E. asburiae, E. intermedius, E. amnigenus, E. dissolvens, E. kobei, E. pyrinus and E. nimipressuralis. The role of the last seven species in human infection is unclear. Enterobacter amnigenus and E. intermediums are environmental bacteria that do not grow at 41 °C.

DESCRIPTION OF THE ORGANISM

Enterobacter spp. are 0.6–1.0 μm × 1.2–3.0 μm long, motile by means of peritrichous flagella and possess class 1 fimbriae (Abraham et al., 1988). They ferment glucose with production of acid and gas and are methyl red negative and Voges-Proskauer positive. Although the optimal growth temperature is 30 °C, most clinical isolates grow well at 37 °C. Enterobacter cloacae has 53 O-antigens and 56 H-antigens. Approximately 80% of E. aerogenes isolates are encapsulated. The capsule is usually thinner than that of Klebsiella but shares some antigens (e.g. K68, K26 and K59). Enterobacter (Pantoea) agglomerans and E. sakazakii produce a yellow diffusible pigment at 20 °C on agar, the mol. % DNA + C content is 52–60. Analysis of a collection of 206 strains of E. cloacae from a variety of sources indicated that although phenotypically similar they were genetically quite diverse (Hoffmann and Roggenkamp, 2003).

PATHOGENICITY

Enterobacter spp. are important nosocomial and opportunistic pathogens. There is little information on their virulence determinants. It is presumed that they adhere to mucosal surfaces using fimbriae.

EPIDEMIOLOGY

Normal Habitat

Enterobacter spp. are found in soil and water but E. cloacae and E. aerogenes can be a minority component of the intestinal flora of human and other animals and in sewage (up to 10^7 cfu/g).

Enterobacter spp. and in particular E. cloacae (the most common) and E. aerogenes are associated with sporadic and occasionally epidemic nosocomial infection (Gaston, 1988). Enterobacter sakazakii and E. hormaechei have been particularly associated with neonatal sepsis (Arseni et al., 1987; da Silva et al., 2002), and E. taylorae is a rare cause of nosocomial infection. Enterobacter spp. are responsible for 1.5% (Eykyn, Gransden and Phillips, 1990) to 6% (Pittet, 1993) of bacteraemic episodes in adults. Common source outbreaks have been linked to enteral feeds and dextrose infusions (Maki et al., 1976), probably due to an ability to grow in concentrated glucose solutions. Hospital cross-infection can produce outbreaks, for example, by contamination of urinals (Mummery, Rowe and Gross, 1974) or by transfer on hands in neonatal units (Haertl and Bandlow, 1993; Ahmet, Houang and Hurley, 1995). *Enterobacter* spp. as well as *Klebsiella* spp. survive on hands (Hart, Gibson and Buckles, 1981).

For epidemiological purposes, *Enterobacter* spp. have been subdivided by biotype, andibiogram, cloacin typing, phage typing (Gaston, 1988), plasmid profile, pyrolysis mass spectrometry (Ahmet, Houang and Hurley, 1995) and a variety of molecular biological techniques including PFGE (Haertl and Bandlow, 1993) and repetitive element polymerase chain reaction (Georghiou *et al.*, 1995).

ANTIMICROBIAL SUSCEPTIBILITY

All isolates of *E. cloacae* and *E. aerogenes* are resistant to cephradine, cefuroxime and amoxycillin, but most UK isolates remain sensitive to cefotaxime, ceftazidime, cefpirome, imipenem, gentamicin and ciprofloxacin (Reeves, Bywater and Holt, 1993). The presence of plasmid-encoded resistance is not uncommon. Like *Citrobacter* spp., *Enterobacter* spp. have the capacity to produce both inducible and de-repressed chromosomally encoded cephalosporinases. Recently strains of *E. aerogenes* resistant to imipenem have emerged (de Champs *et al.*, 1993). These were found to hyperexpress a chromosomally encoded cephalosporinase, a variety of plasmid-encoded β-lactamases (TEM types) and to have lost an outer membrane (porin) protein. Other species such as *E. amnigenus*, *E. gergorviae* and *E. sakazakii* appear less resistant (Stock and Wiedemann, 2002).

TREATMENT AND PREVENTION

Treatment depends upon the local antimicrobial susceptibility patterns and prevention is by adherence to appropriate control of infection procedures. In a carefully controlled study bacteraemia due to *Enterobacter* spp. did not adversely affect the outcome in ICU patients provided there was appropriate antimicrobial chemotherapy (Blot, Vandewoude and Colardyn, 2003).

SERRATIA SPECIES

The type species, *Serratia marcescens*, was first described by Bizio in 1823 as a cause of red discoloration of cornmeal porridge: 'bleeding polenta'. The red pigment produced by *S. marcescens* is water insoluble, not light fast and is called prodigiosin (Yu, 1979). It has been responsible for the appearance of 'blood' on foodstuffs throughout history. Its presence in sputum has also led to a misdiagnosis of bronchiectasis or even bronchial carcinoma. *Serratia marcescens* has also been used extensively as a biological tracer. The earliest experiment was in the House of Commons in 1906 and involved gargling with a solution of *S. marcescens* and then declaiming to an audience of agar plates (doubtless more attentive and receptive than the usual audience!).

Serratia spp. are indole negative, produce lecithinase and lipase and are more likely to be gelatinase and DNAase positive than the other Enterobacteriaceae (Grimont and Grimont, 1984). The genus contains 10 species, namely *S. marcescens*, *S. liquefaciens*, *S. rubidaea* (previously *S. marinorubra*), *S. ficaria*, *S. fonticola*, *S. odorifera*, *S. entomophilia*, *S. plymuthica*, *S. grimesii* and *S. proteomaculans* subsp. *quinovora*. Of these, *S. marcescens*, *S. liquefaciens* and *S. rubidaea* are most often associated with human infection. *Serratia fonticola* is found in water. *Serratia ficaria* shuttles between fig tree and fig wasp but can cause human infection, and *S. plymuthica* and *S. proteomaculans* subsp. *quinovora* are rare causes of human disease.

DESCRIPTION OF THE ORGANISM

Serratia spp. are straight rods 0.5–0.8 µm×0.9–2.0 µm. Most are motile by means of peritrichous flagella. In a survey of respiratory tract isolates, over 17% of *Serratia* spp. possessed type 3 fimbriae (Hornick *et al.*, 1991). Mucoid colonies of *S. plymuthica* occur regularly but most other *Serratia* spp. are non-capsulate. *Serratia odorifera* does possess a microcapsule that cross-reacts with *Klebsiella* K4 or K68 antisera. There are 21 somatic antigens (01–021), 25 flagellar (H1–H25) antigens and 16–50% of strains have the serotype O14:H12. Prodigiosin is produced by two biogroups of *S. marcescens* and most *S. rubidaea* and *S. plymuthica* between 20 and 35 °C. Colonies of these bacteria are completely red or have a red centre or margin. Most species produce a fishy-urinary odour. *Serratia odorifera* and *S. ficaria* smell musty. All *Serratia* spp. are Voges-Proskauer positive (except for 40% of *S. plymuthica*) and hydrolyse tributyrin; mol. % DNA G + C content is 52–60. Recently a spore-forming strain of *S. marcescens* subsp. *sakuensis* has been isolated from domestic wastewater (Ajithkumar *et al.*, 2003).

PATHOGENICITY

Other than class 3 fimbriae no other virulence determinants have been ascribed to *Serratia* spp. However, their ability to survive in disinfectant solution, to grow at relatively low temperatures and to adhere to plastics can provide a reservoir for infection. They are not part of the adult normal flora but can colonize premature neonates (Christensen *et al.*, 1982; Newport *et al.*, 1985).

Prodigiosin production seems unrelated to pathogenicity since up to 80% of clinical isolates are non-producers, despite the fact that it is being developed as an anti-cancer agent (Manderville, 2001). *Serratia marcescens* produces a pore-forming haemolysin, which allows the bacterium to acquire haemoglobin and haem using an extracellular scavenger protein (Letoffe, Ghigo and Wandersman, 1994). In addition to the haemolysis and iron scavenger protein, transposon mutagenesis studies identified LPS biosynthetic enzymes as important virulence determinants (Kurz *et al.*, 2003). *In vitro*, a metalloprotease secreted by *S. marcescens* has been shown to contribute to the bacterium's cytotoxicity (Marty *et al.*, 2002).

EPIDEMIOLOGY AND INFECTIONS

Normal Habitat

Serratia spp. are widely distributed in the environment (water, soil and plants) and can be found in the rodent gut (Grimont and Grimont, 1984) and occasionally in the human intestinal tract. *Serratia* spp. appear to colonize the premature neonate's intestinal tract when in intensive care but are rarely found in staff or mothers (Christensen *et al.*, 1982; Newport *et al.*, 1985).

Serratia spp. are nosocomial and opportunist pathogens, *S. marcescens* being the most important. The inanimate environment, especially when moist, is the major reservoir of *S. marcescens* as is the premature neonate's intestinal tract (Christensen *et al.*, 1982; Newport *et al.*, 1985). Transmission has been by fomites such as disinfectant and cleansing solutions, contact lenses (Ahanotu and Ahearn, 2002), hand lotions, urinary catheters, catheter hubs, parenteral fluids, mechanical respirators and fibre-optic bronchoscopes (Yu, 1979). Spread appears to be via hands, although *Serratia* spp. tend to be infrequently detected on hands (Christensen *et al.*, 1982; Newport *et al.*, 1985). Community-acquired *S. plymuthica*, *S. ficaria* and *S. proteamaculans* subsp. *quinovora* infections have resulted by contamination from soil and plant material.

Infections can be sporadic or epidemic. *Serratia* spp. is a relatively rare cause of bacteraemia, causing 0.4% of cases in one large UK hospital (Eykyn, Gransden and Phillips, 1990). In the USA in 1975, *Serratia* spp. were responsible for 3.8% of cases of nosocomial bacteraemia but since then have not appeared in the top 10 isolates (Pittet, 1993). Bacteraemia in premature neonates can be particularly devastating (Berger *et al.*, 2002).

Epidemiological typing methods for *Serratia* spp. include serotyping (Newport *et al.*, 1985), bacteriocin susceptibility, phage typing and biotyping (Grimont and Grimont, 1984), antibiograms (Christensen *et al.*, 1982) and a variety of molecular techniques including polymerase chain reaction of repetitive intergenic consensus sequences (Lin *et al.*, 1994).

CLINICAL FEATURES

Serratia spp. can cause urinary tract infection, pneumonia, meningitis (rarely), endophthalmitis, bacteraemia and wound infections. One particularly striking 'non-illness' is the red diaper syndrome due to excretion of *S. marcescens* in the infant stool (Yu, 1979).

DIAGNOSIS

Serratia spp. are easily cultured on most media from infected sites. Some strains of *S. plymuthica* do not grow well at 37 °C.

Antimicrobial Susceptibility

As for *Enterobacter* spp., *Serratia* spp. are generally resistant to cefuroxime, cephradine and amoxycillin but sensitive to ceftazidime, imipenem, cefpirome and ciprofloxacin (Reeves, Bywater and Holt, 1993). Susceptibility to cefotaxime and gentamicin is less predictable. *Serratia* spp. are also relatively resistant to disinfectants, *S. marcescens* being the most resistant and *S. plymuthica* the least. *Serratia* spp. can also express chromosomally encoded cephalosporinases (*AmpC*). The rarer *Serratia* spp. are naturally resistant to cefazolin, cefuroxime and rifampicin but usually sensitive to cotrimoxazole, piperacillin, carbapenems and fluoroquinolones (Stock *et al.*, 2003).

Serratia spp. are able to acquire and maintain resistance plasmids; incompatibility groups B, C, F1, F_{11}, H_2, M, N, P and W have been detected (Grimont and Grimont, 1984). Plasmids appear to be able to move from genus to genus, and one American hospital outbreak of infection due to gentamicin-resistant *S. marcescens* was followed by another due to similarly resistant *K. pneumoniae* carrying the same plasmid (Thomas *et al.*, 1977). Recently plasmid-encoded resistance to imipenem has been described in clinical isolates of *S. marcescens* (Ito *et al.*, 1995).

TREATMENT AND PREVENTION

Specific treatment depends upon the susceptibility of the isolate but most are still susceptible to ciprofloxacin, imipenem and ceftazidime. Prevention is by appropriate infection-control measures. Especially care must be taken with storage of disinfectants, plastic blood containers and contact lenses.

REFERENCES

Abraham, S.N., Sun, D., Dale, J.B. and Beachey, E.H. (1988) Conservation of the D-mannose-adhesion protein among type 1 fimbriated members of the family Enterobacteriaceae. *Nature*, **336**, 682–684.

Ahanotu, E.N. and Ahearn, D.G. (2002) Association of *Pseumonas aeruginosa* and *Serratia marsescens* with extended-wear soft contact lenses in asymptomatic patients. *Contact Lens Association of Ophthalmologists*, **28**, 157–159.

Ahmet, Z., Houang, E. and Hurley, R. (1995) Pyrolysis mass spectrometry of cephalosporin-resistant *Enterobacter cloacae*. *Journal of Hospital Infection*, **31**, 99–104.

Ajithkumar, B., Ajithkumar, V.P., Iriye, R. *et al.* (2003) Spore-forming *Serratia marcescens* subsp. *Sakuensis* subsp. *nov.*, isolated from a domestic wastewater treatment tank. *International Journal of Systematic and Evolutionary Microbiology*, **53**, 253–258.

Allen, P., Hart, C.A. and Saunders, J.R. (1987a) Isolation of *Klebsiella* and characterization of two *rcs* genes that activate colanic acid capsular biosynthesis in *Escherichia coli*. *Journal of General Microbiology*, **133**, 331–340.

Allen, P.M., Williams, J.M., Hart, C.A. and Saunders, J.R. (1987b) Identification of two chemical types of K21 capsular polysaccharide from *Klebsiellae*. *Journal of General Microbiology*, **133**, 1365–1370.

Altman, E. and Dutton, G.G.S. (1985) Chemical and structural analysis of the capsular polysaccharide from *Escherichia coli* O9:K28(A):H-(K28 Antigen). *Carbohydrate Research*, **178**, 293–303.

Ames, G.F.L. and Joshi, A.K. (1990) Energy coupling in bacteria and periplasmic permeases. *Journal of Bacteriology*, **171**, 4133–4137.

Arakawa, Y., Ohta, M., Wacharotayankun, R. *et al.* (1991) Biosynthesis of *Klebsiella* K2 capsular polysaccharide in *Escherichia coli* HB101 requires the functions of *rmp*A and chromosomal *cps* gene cluster of the virulent strain *Klebsiella pneumoniae* Chedid (O1:K2). *Infection and Immunity*, **59**, 2043–2050.

Arseni, A., Malamon-Ladas, E., Koutsia, C. *et al.* (1987) Outbreak of colonization of neonates with *Enterobacter sakazakii*. *Journal of Hospital Infection*, **9**, 143–150.

Athamna, A., Ofek, I., Keisari, Y. *et al.* (1991) Lectinophagocytosis and encapsulated *Klebsiella pneumoniae* mediated by surface lectins of guinea-pig alveolar macrophages and human monocyte derived macrophages. *Infection and Immunity*, **59**, 1673–1682.

Barlow, M. and Hall, B.G. (2002) Origin and evolution of the AmpC β-lactamases of *Citrobacter freundii*. *Antimicrobial Agents and Chemotherapy*, **46**, 1190–1198.

Berger, A., Rohrmeister, K., Haiden, N. *et al.* (2002) *Serratia marcescens* in the neonatal intensive care unit: re-emphasis of the potentially devastating sequelae. *Wiener Kliniche Wochenschrift*, **114**, 1017–1022.

Bergogne-Bérézin, E., Decré, D. and Joly-Guillon, M.L. (1993) Opportunistic nosocomial multiply resistant bacterial infections: their treatment and prevention. *Journal of Antimicrobial Chemotherapy*, **32** (Suppl. A), 39–48.

Berkowitz, F.E. and Metchock, B. (1995) Third generation cephalosporin-resistant gram negative bacilli in the feces of hospitalized children. *Pediatric Infectious Disease Journal*, **14**, 97–100.

Betley, M.J., Miller, V.L. and Mekalanos, J.J. (1986) Genetics of bacterial enterotoxins. *Annual Review of Microbiology*, **40**, 577–605.

Bettelheim, K.A., Evangelidis, H., Pearce, J.C. *et al.* (1993) Isolation of a *Citrobacter freundii* strain which carries the *Escherichia coli* 0157 antigen. *Journal of Clinical Microbiology*, **31**, 760–761.

Blot, S.I., Vandewoude, K.H. and Colardyn, F.A. (2003) Evaluation of outcome in critically ill patients with nosocomial enterobacter bacteremia: results of a matched cohort study. *Chest*, **123**, 1208–1213.

Borgstein, J., Sada, E. and Cortes, R. (1993) Ciprofloxacin for rhinoscleroma and ozaena. *Lancet*, **342**, 122.

Boulnois, G.J. and Roberts, I.S. (1990) Genetics of capsular polysaccharide production in bacteria. *Current Topics in Microbiology*, **150**, 1–18.

Brenner, D.J. (1984) Family 1: Enterobacteriaceae. In *Bergey's Manual of Systematic Bacteriology*, Vol. 1 (eds N.R. Krieg and J.G. Holt), pp. 408–420. Williams & Wilkins, Baltimore.

Casewell, M.W. and Desai, N. (1983) Survival of multi-resistant and other Gram negative bacilli on finger-tips. *Journal of Hospital Infection*, **4**, 350–360.

Casewell, M.W. and Phillips, I. (1977) Hands as a route of transmission for *Klebsiella* species. *British Medical Journal*, **ii**, 1315–1317.

Casewell, M. and Talsania, H.G. (1979) Predominance of certain *Klebsiella* capsular types in hospitals in the United Kingdom. *Journal of Infection*, **1**, 77–79.

Casewell, M.W., Dalton, M.T., Webster, M. and Phillips, I. (1977) Gentamicin-resistant *Klebsiella aerogenes* in a urological ward. *Lancet*, **ii**, 444–446.

de Champs, C., Henquell, C., Guelon, D. *et al.* (1993) Clinical and bacteriological study of nosocomial infections due to *Enterobacter aerogenes* resistant to imipenem. *Journal of Clinical Microbiology*, **31**, 123–127.

Chen, Y.S., Wong, W.V. and Fung, C.P. *et al.* (2002) Clinical features and antimicrobial susceptibility trends in *Citrobacter freundii* bacteremia. *Journal of Microbiology, Immunology and Infection*, **35**, 109–114.

Christensen, G.D., Korones, S.B., Reed, L. *et al.* (1982) Epidemic *Serratia marcescens* in a neonatal intensive care unit: importance of the gastrointestinal tract as reservoir. *Infection Contrology*, **3**, 127–133.

Cooke, E.M., Sazegar, T., Edmondson, A.S. *et al.* (1980) *Klebsiella* species in hospital food and kitchens: a source of organisms in the bowel of patients. *Journal of Hygiene*, **84**, 97–101.

Cortes, G., de Astorza, B., Benedi, V.J. and Alberti, S. (2002a) Role of the *htrA* gene in *Klebsiella pneumoniae* virulence. *Infection and Immunity*, **70**, 4772–4776.

Cortes, G., Borrell, N., de Astorza, B. *et al.* (2002b) Molecular analysis of the contribution of the capsular polysaccharide and the lipopolysaccharide O side chain to the virulence of *Klebsiella pneumoniae* in a murine model of pneumonia. *Infection and Immunity*, **70**, 2583–2590.

Cryz, S.J., Mortimer, P.M., Mansfield, V. and Germainer, R. (1986) Sero-epidemiology of *Klebsiella* bacteremic isolates and implications for vaccine development. *Journal of Clinical Microbiology*, **23**, 687–690.

Curie, K., Speller, D.C.E., Simpson, R.A. *et al.* (1978) A hospital epidemic caused by a gentamicin-resistant *Klebsiella aerogenes. Journal of Hygiene*, **80**, 115–123.

Darfeuille-Michaud, A., Jallat, C., Aubel, D. *et al.* (1992) R-plasmid encoded adhesive factor in *Klebsiella pneumoniae* strains responsible for human nosocomial infections. *Infection and Immunity*, **60**, 44–55.

Denoya, C.D., Trevisan, A.R. and Zorzopoulos, J. (1986) Adherence of multi-resistant strains of *Klebsiella pneumoniae* to cerebrospinal fluid shunts: correlation with plasmid content. *Journal of Medical Microbiology*, **21**, 225–231.

Domenico, P., Johanson, W.G. and Straus, D.C. (1982) Lobar pneumonia in rats produced by *Klebsiella pneumoniae. Infection and Immunity*, **37**, 327–355.

Donowitz, L.G., Marsik, F.J., Fisher, K.A. and Wenzel, R.P. (1981) Contaminated breast milk: a source of *Klebsiella* bacteremia in a newborn intensive care unit. *Journal of Infectious Diseases*, **3**, 716–720.

Drancourt, M., Bollet, C., Carta, C. and Rousselier, P. (2001) Phylogenetic analyses of *Klebsiella* species delineate *Klebsiella* and *Raoultella* gen. nov., with description of *Raoultella ornitholytica* comb.nov., *Raoultella terrigena* comb.nov. and *Raoultella planticola* comb.nov. *International Journal of Systematic and Evolutionary Microbiology*, **51**, 925–932.

Edmondson, A.S. and Cooke, E.M. (1979) The development and assessment of a bacteriocin typing method for *Klebsiella. Journal of Hygiene*, **82**, 207–223.

Eykyn, S.J., Gransden, W.R. and Phillips, I. (1990) The causative organisms of septicaemia and their epidemiology. *Journal of Antimicrobial Chemotherapy*, **25** (Suppl. c), 41–58.

Frankel, G., Candy, D., Everest, P. and Dougan, G. (1994) Characterization of the C-terminal domains of intimin-like proteins of enteropathogenic and enterohemorrhagic *Escherichia coli*, *Citrobacter freundii* and *Hafnia alvei. Infection and Immunity*, **62**, 1835–1842.

Fryklund, B., Tullus, K. and Burman, L.G. (1995) Survival on skin and surfaces of epidemic and non-epidemic strains of enterobacteria from neonatal special care units. *Journal of Hospital Infection*, **29**, 201–208.

Gaston, M.A. (1988) *Enterobacter*: an emerging nosocomial pathogen. *Journal of Hospital Infection*, **11**, 197–208.

Georghiou, P.R., Hanill, R.J., Wright, C.E. *et al.* (1995) Molecular epidemiology of infections due to *Enterobacter aerogenes*: identification of hospital outbreak-associated strains by molecular techniques. *Clinical Infectious Diseases*, **20**, 84–94.

Gladstone, I.M., Ehrenkranz, A.R.A., Edberg, S.C. and Baltimore, R.S. (1990) A ten year review of neonatal sepsis and comparisons with previous, 50-year experience. *Pediatric Infectious Disease Journal*, **9**, 819–825.

Goering, R.V., Ehrenkranz, N.J., Sanders, C.C. and Sander, W.E. (1992) Long term epidemiological analysis of *Citrobacter diversus* in a neonatal intensive care unit. *Pediatric Infectious Disease Journal*, **11**, 99–104.

Gouby, A., Neuwirth, C., Bourg, G. *et al.* (1994) Epidemiological study by pulsed-field gel electrophoresis of a outbreak of extended-spectrum β-lactamase producing *Klebsiella pneumoniae* in a geriatric hospital. *Journal of Clinical Microbiology*, **32**, 301–305.

Grimont, P.A.D. and Grimont, F. (1984) Genus VIII: *Serratia*. In *Bergey's Manual of Systematic Bacteriology*, Vol. 1 (eds N.R. Krieg and J.G. Holt), pp. 477–494. Williams & Wilkins, Baltimore.

Haertl, R. and Bandlow, G. (1993) Epidemiological fingerprinting of *Enterobacter cloacae* by small-fragment restriction endonuclease analysis and pulsed-field gel electrophoresis of genomic restriction fragments. *Journal of Clinical Microbiology*, **31**, 128–133.

Hansen, D.S., Skov, R., Benedi, J.V. *et al.* (2002) Klebsiella typing: pulsed-field gel electrophoresis (PFGE) in comparison with O:K – serotyping. *Clinical Microbiology and Infection*, **8**, 397–404.

Hart, C.A. (1993) *Klebsiellae* and neonates. *Journal of Hospital Infection*, **23**, 83–86.

Hart, C.A. and Gibson, M.F. (1982) Comparative epidemiology of gentamicin-resistant enterobacteria: persistence of carriage and infection. *Journal of Clinical Pathology*, **35**, 452–457.

Hart, C.A. and Percival, A. (1982) Resistance to cephalosporins among gentamicin-resistant *Klebsiellae. Journal of Antimicrobial Chemotherapy*, **9**, 275–286.

Hart, C.A. and Rao, S.K. (1999) Donovanosis. *Journal of Medical Microbiology*, **48**, 707–709.

Hart, C.A., Gibson, M.R. and Buckles, A. (1981) Variation in skin and environmental survival of hospital gentamicin-resistant enterobacteria. *Journal of Hygiene*, **87**, 277–285.

Hetz, C., Bono, M.R., Barros, L.F. and Lagos, R. (2002) Microcin E492, a channel-forming bacteriocin from *Klebsiella pneumoniae*, induces apoptosis in some human cell lines. *Proceedings of the National Academy of Sciences USA*, **99**, 2696–2701.

Hibbert-Rogers, L.C.F., Heritage, J., Todd, N. and Hawkey, P. (1994) Convergent evolution of TEM-26: a β-lactamase with extended spectrum activity. *Journal of Antimicrobial Chemotherapy*, **33**, 707–720.

Hoffmann, H. and Roggenkamp, A. (2003) Population genetics of the nomenspecies *Enterobacter cloacae. Applied and Environmental Microbiology*, **69**, 5306–5318.

Hornick, D.B., Allen, B.C., Horn, M.A. and Clegg, S. (1991) Fimbrial types among respiratory isolates belonging to the family Enterobacteriaceae. *Journal of Clinical Microbiology*, **21**, 1795–1800.

Houng, H.S., Noon, K.F., Ou, J.T. and Barron, L.S. (1992) Expression of Vi antigen in *Escherichia coli* K12: characterization of ViaB from *Citrobacter freundii* and identity of ViaA with KesB. *Journal of Bacteriology*, **174**, 5910–5915.

Ito, H., Arakawa, Y., Ohsuka, S. *et al.* (1995) Plasmid mediated dissemination of the metallo-β-lactamase gene *bla*IMP among clinically isolated strains of *Serratia marcescens. Antimicrobial Agents and Chemotherapy*, **39**, 824–829.

Jacoby, G.A. and Medeiros, A.A. (1991) More extended spectrum β-lactamases. *Antimicrobial Agents and Chemotherapy*, **35**, 1697–1704.

Jain, K., Radsak, K. and Mannheim, W. (1974) Differentiation of the *Oxytocum* group from *Klebsiella* by deoxyribonucleic acid hybridisation. *International Journal of Systematic Bacteriology*, **24**, 402–407.

Jann, B. and Jann, K. (1990) Structure and biosynthesis of capsular antigen of *Escherichia coli. Current Topics in Microbiology and Immunology*, **150**, 19–42.

Jarvis, W.R., Munn, V.P., Highsmith, A.K. *et al.* (1985) The epidemiology of nosocomial infection caused by *Klebsiella pneumoniae. Infection Contrology*, **6**, 68–74.

Kerver, A.J., Rommes, J.H., Mevissen-Verhage, E.A.E. *et al.* (1987) Colonization and infection in surgical intensive care patients: a prospective study. *Intensive Care Medicine*, **13**, 347–351.

Knittle, M.A., Eitzman, D.V. and Baer, H. (1975) Role of hand contamination in the epidemiology of gram-negative nosocomial infections. *Journal of Pediatrics*, **86**, 433–437.

Ko, W.C., Paterson, D.L., Sagnimeni, A.J. *et al.* (2002) Community-acquired *Klebsiella pneumoniae* bacteremia: global differences in clinical patterns. *Emerging Infectious Diseases*, **8**, 160–166.

Kurz, C.L., Chauvet, S., Andros, E. *et al.* (2003) Virulence factors of the human opportunistic pathogen *Serratia marcescens* identified by *in vivo* screening. *The EMBO Journal*, **22**, 1451–1460.

Lai, Y.-C., Peng, H.-L. and Chang, H.-Y. (2003) RmpA2, an activator of capsule biosynthesis in *Klebsiella pneumoniae* CG-43, regulates K2 *cps* gene expression at the transcriptional level. *Journal of Bacteriology*, **185**, 788–800.

Le Frock, J.L., Ellis, C.A. and Weinstein, L. (1979) Impact of hospitalisation on the aerobic microflora. *American Journal of Medical Science*, **277**, 269–274.

Letoffe, S., Ghigo, J.M. and Wandersman, C. (1994) Iron acquisition from heme and hemoglobin by *Serratia marcescens* extracellular protein. *Proceedings of the National Academy of Sciences USA*, **91**, 9876–9880.

Li, J., Musser, J.M., Beltran, P. *et al.* (1990) Genotypic heterogeneity of strains of *Citrobacter diversus* expressing a 32-kilodalton outer membrane protein associated with neonatal meningitis. *Journal of Clinical Microbiology*, **28**, 1760–1765.

Lin, P.Y., Lau, Y.J., Hu, B.S. *et al.* (1994) Use of PCR to study the epidemiology of *Serratia marcescens* isolates in nosocomial infection. *Journal of Clinical Microbiology*, **32**, 1935–1938.

Mackowiak, P.A., Martin, R.M. and Smith, J.W. (1979) The role of bacterial interference in the increased prevalence of Gram-negative bacilli among alcoholics and diabetics. *American Review of Respiratory Disease*, **120**, 589–593.

Maki, D.G., Rhame, F.S., Mackel, D.C. and Bennet, J.V. (1976) Nationwide epidemic of septicaemia caused by contaminated intravenous products. 1. Epidemiological and clinical features. *American Journal of Medicine*, **60**, 471–485.

Manderville, R.A. (2001) Synthesis, proton-affinity and anti-cancer properties of the prodigiosin-group natural products. *Current Medicinal Chemistry and Anti-Cancer Agents*, **1**, 195–218.

Marty, K.B., Williams, C.L., Guynn, L.J. *et al.* (2002) Characterization of a cytotoxic factor in culture filtrates of *Serratia macescens*. *Infection and Immunity*, **70**, 1121–1128.

Meers, P.D., Ayliffe, G.A.J., Emmerson, A.M. *et al.* (1980) Report on the National Survey of Infection in Hospitals, 1980. *Journal of Hospital Infection* **2** (Suppl.), 1–48.

Mengistu, Y., Edwards, C. and Saunders, J.R. (1994) Continuous culture studies on the synthesis of capsular polysaccharide by *Klebsiella pneumoniae* K1. *Journal of Applied Bacteriology*, **76**, 424–430.

Merino, S., Camprubi, S., Alberts, S. *et al.* (1992) Mechanism of *Klebsiella pneumoniae* resistance to complement-mediated killing. *Infection and Immunity*, **60**, 2529–2535.

Meyer, P.R., Shum, T.K., Becker, T.S. and Taylor, C.R. (1983) Scleroma (rhinoscleroma). A histologic immunohistochemical study with bacteriologic correlates. *Archives of Pathology and Laboratory Medicine*, **107**, 377–383.

Morgan, M.E.I., Hart, C.A. and Cooke, R.W.I. (1984) *Klebsiella* infection in a neonatal intensive care unit: role of bacteriological surveillance. *Journal of Hospital Infection*, **5**, 377–385.

Morris, J.G., Lin, F.Y.C., Morrison, C.B. *et al.* (1986) Molecular epidemiology of neonatal meningitis due to *Citrobacter diversus*: a study of isolates from hospitals in Maryland. *Journal of Infectious Diseases*, **154**, 409–414.

Mummery, R.V., Rowe, B. and Gross, R.J. (1974) Urinary tract infections due to atypical *Enterobacter cloacae*. *Lancet*, **ii**, 1333.

Murcia, A. and Rubin, S.J. (1979) Reproducibility of an indirect immunofluorescent-antibody technique for capsular serotyping of *Klebsiella pneumoniae*. *Journal of Clinical Microbiology*, **9**, 208–213.

Newport, M.T., John, J.F., Michel, Y.M. and Levkoff, A.H. (1985) Endemic *Serratia marcescens* infection in a neonatal intensive care nursery associated with gastrointestinal colonization. *Pediatric Infectious Diseases*, **4**, 160–167.

O'Hara, C.M., Roman, S.B. and Miller, J.M. (1995) Ability of commercial identification systems to identify newly recognized species of *Citrobacter*. *Journal of Clinical Microbiology*, **33**, 242–245.

Orskov, I. (1984) Genus V: *Klebsiella*. In *Bergey's Manual of Systematic Bacteriology*, Vol. 1 (eds N.R. Krieg and J.G. Holt), pp. 461–465. Williams & Wilkins, Baltimore.

Palfreyman, J.M. (1978) *Klebsiella* serotyping by counter-current immunoelectrophoresis. *Journal of Hygiene*, **81**, 219–225.

Pavelka, M.S., Wright, L.F. and Silver, R.P. (1991) Identification of two genes *kps*M and *kps*T in region 3 of the polysialic acid gene cluster of *Escherichia coli* K1. *Journal of Bacteriology*, **173**, 4603–4610.

Pittet, D. (1993) Nosocomial bloodstream infections. In *Prevention and Control of Nosocomial Infections*, 2nd edn (ed. R.P. Wenzel), pp. 512–555. Williams & Wilkins, Baltimore.

Podschun, R. and Ullmann, U. (1998) *Klebsiella* spp. as nosocomial pathogens: epidemiology, taxonomy, typing methods and pathogenicity factors. *Clinical Microbiology Reviews*, **11**, 589–603.

Podschun, R., Pietsch, S., Heller, C. and Ullmann, U. (2001) Incidence of *Klebsiella* species in surface waters and their expression of virulence factors. *Applied and Environmental Microbiology*, **67**, 3325–3327.

Pruzzo, C., Guzman, C.A., Caegari, L. and Satta, G. (1989) Impairment of phagocytosis by the *Klebsiella pneumoniae* mannose-inhibitable adhesin-T7 receptor. *Infection and Immunity*, **57**, 975–982.

Rajab, A. and De Louvois, J. (1990) Survey of infection in babies at Khoula hospital Oman. *Annals of Tropical Paediatrics*, **10**, 39–43.

Reeves, D.S., Bywater, M.J. and Holt, H.A. (1993) The activity of cefpirome and ten other antimicrobial agents against 2858 clinical isolates collected from 20 centres. *Journal of Antimicrobial Chemotherapy*, **31**, 345–362.

Richard, C. (1984) Genus VI: *Enterobacter*. In *Bergey's Manual of Systematic Bacteriology*, Vol. 1 (eds N.R. Krieg and J.G. Holt), pp. 465–469. Williams & Watkins, Baltimore.

Sahly, H., Kekow, J., Podshun, R. *et al.* (1994) Comparison of the antibody responses to the 77 *Klebsiellar* types in analysing spondylitis and various rheumatic diseases. *Infection and Immunity*, **62**, 4838–4843.

Sakazaki, R. (1984) Genus IV: *Citrobacter*. In *Bergey's Manual of Systemic Bacteriology*, Vol. 1 (eds N.R. Krieg and N.G. Holt), pp. 458–461. Williams & Wilkins, Baltimore.

da Silva, C.L., Miranda, L.E., Moreira, B.M. *et al.* (2002) *Enterobacter hormaechei* bloodstream infection at three neonatal intensive care units in Brazil. *Pediatric Infectious Disease Journal*, **21**, 175–177.

Simoons-Smit, A.M., Verweij-van Vught, A.M.J.J., De Vries, P.M.J.M. and Maclaren, D.M. (1987) Comparison of biochemical and serological typing results and antimicrobial susceptibility patterns in the epidemiological investigation of *Klebsiella* spp. *Epidemiology and Infection*, **99**, 625–634.

Stapleton, P., Shannon, K. and Phillips, I. (1995) The ability of β-lactam antibiotics to select mutants with depressed β-lactamase synthesis from *Citrobacter freundii*. *Journal of Antimicrobial Chemotherapy*, **36**, 483–496.

Stock, I. and Wiedemann, B. (2002) Natural antibiotic susceptibility of *Enterobacter amnigenus*, *Enterobacter cancerogenus*, *Enterobacter gergoviae* and *Enterobacter sakazakii* strains. *Clinical Microbiology and Infection*, **8**, 564–578.

Stock, I., Burak, S., Sherwood, K.J. *et al.* (2003) Natural antimicrobial susceptibilities of 'unusual' *Serratia* species: *S. ficaria*, *S. fonticola*, *S. odorifera*, *S. plymuthica* and *S. rubidaea*. *Journal of Antimicrobial Chemotherapy*, **51**, 865–885.

Straus, D.C. (1987) Production of an extracellular toxic complex by various strains of *Klebsiella pneumoniae*. *Infection and Immunity*, **55**, 44–48.

Straussberg, R., Harel, L. and Amir, J. (2001) Long-term outcome of neonatal *Citrobacter koseri* (*diversus*) meningitis treated with imipenem/meropenem and surgical drainage. *Infection*, **29**, 280–282.

Struve, C., Forestier, C. and Krogfelt, K.A. (2003) Application of a novel multi-screening signature-tagged mutagenesis assay for identification of *Klebsiella pneumoniae* genes essential in colonization and infection. *Microbiology*, **149**, 167–176.

Sutherland, I.W. (1985) Biosynthesis and composition of gram-negative bacterial extracellular and wall polysaccharides. *Annual Review of Microbiology*, **39**, 243–270.

Tarkkanen, A.M., Allen, B.C., Williams, P.H. *et al.* (1992) Fimbriation, capsulation and iron scavenging systems of *Klebsiella* strains associated with human urinary tract infection. *Infection and Immunity*, **60**, 1187–1192.

Thomas, F.E., Jackson, R.T., Melly, A. and Alford, R.H. (1977) Sequential hospital-wide outbreaks of resistant serratia and *Klebsiella* species. *Archives of International Medicine*, **137**, 581–584.

Townsend, S.M., Pollack, H.A., Gonzalez-Gomez, I. *et al.* (2003) *Citrobacter koseri* brain abscess in the neonatal rat: survival and replication within human and rat macrophages. *Infection and Immunity*, **71**, 5871–5880.

Troy, F.A. (1979) The chemistry and biosynthesis of selected bacterial capsular polymers. *Annual Review of Microbiology*, **33**, 519–560.

Tsay, R.-W., Siu, L.K., Fung, C.-P. and Chang, F.-Y. (2002) Characteristics of bacteremia between community-acquired and nosocomial *Klebsiella pneumoniae* infection. *Archives of International Medicine*, **162**, 1021–1027.

Tschäpe, H., Prager, R., Streckel, W. *et al.* (1995) Verotoxinogenic *Citrobacter freundii* associated with severe gastroenteritis and cases of haemolytic uraemic syndrome in a nursery school: green butter as the infection source. *Epidemiology and Infection*, **114**, 441–450.

Tullus, K. and Burman, L.G. (1989) Ecological effect of ampicillin and cefuroxime in neonatal units. *Lancet*, **i**, 1405–1407.

Tullus, K., Berglund, B., Fryklund, B. *et al.* (1988) Epidemiology of faecal strains of the family Enterobacteriaceae in 22 neonatal units and influence of antibiotic policy. *Journal of Clinical Microbiology*, **26**, 1166–1170.

Van der Waaij, D. (1982) Colonization resistance of the digestive tract: clinical consequences and implications. *Journal of Antimicrobial Chemotherapy*, **10**, 263–270.

Williams, P. and Tomas, J.M. (1990) The pathogenicity of *Klebsiella pneumoniae*. *Reviews of Infectious Diseases*, **1**, 196–204.

Williams, P., Lambert, P.A., Brown, M.R.W. and Jones, R.J. (1983) The role of the O and K antigens in determining the resistance of *Klebsiella aerogenes* to serum killing and phagocytosis. *Journal of General Microbiology*, **129**, 2181–2191.

Williams, W.W., Mariano, J., Spurrier, M. *et al.* (1984) Nosocomial meningitis due to *Citrobacter diversus* in neonates: new aspects of the epidemiology. *Journal of Infectious Diseases*, **150**, 229–236.

Yu, V.L. (1979) *Serratia marcescens*: historical perspective and clinical review. *New England Journal of Medicine*, **300**, 887–893.

31

Donovanosis and *Klebsiella* spp.

John Richens

Centre for Sexual Health and HIV Research, Royal Free and University College Medical School, London, UK

INTRODUCTION

Donovanosis is the name given to an ulcerative tropical sexually transmitted infection that was first described in India by McLeod (1881). Charles Donovan was the first to describe the presence of intracellular inclusions (now know as Donovan bodies) as the characteristic pathological feature (Donovan, 1905). The disease has been known by a confusing variety of different names. Among experts, donovanosis is the preferred term, whilst Index Medicus retains granuloma inguinale as its major medical subject heading. Donovan bodies have to be distinguished from the protozoal Leishman-Donovan bodies that characterize leishmaniasis which were also described by Charles Donovan at much the same time. Donovanosis is one of the least common sexually transmitted infections, and it has declined markedly in many formerly endemic areas (Gupta and Kumar, 2002). It has been targeted for eradication in Australia (Bowden and Savage, 1998a), and arguments have been put forward for global eradication (O'Farrell, 1995).

DESCRIPTION OF THE ORGANISM

The Gram-negative intracellular coccobacilli associated with donovanosis were first cultured successfully in chick embryo yolk sac by Anderson in the United States (Anderson, 1943). Subsequently, very few successful isolations were reported, but a resurgence in interest in the organism developed when successful culture efforts were reported by one group in Durban, employing human peripheral blood mononuclear cells (Kharsany *et al.*, 1997), and a second group employing HEp 2 cells in Darwin (Carter *et al.*, 1997). Both groups have subsequently scrutinized the genetic material of their cultured organisms, confirming the long-held hypotheses that the organism is closely related to *Klebsiella* species. The Darwin group have put forward a case for renaming the organism *Klebsiella granulomatis* (Carter *et al.*, 1999), citing 98.8% homology of the rRNA gene to *Klebsiella pneumoniae* and 99.8% homology to *Klebsiella rhinoscleromatis* (the latter organism causes a histologically analogous intracellular infection, rhinoscleroma, in the upper respiratory tract). They also reported 99.7% and 99.8% homology between the *PhoE* genes in *Calymmatobacterium* and *K. pneumoniae* and *K. rhinoscleromatis*, respectively. The Durban group have argued that the differences from *Klebsiella* are great enough to justify retaining the separate species name, *Calymmatobacterium* (Kharsany *et al.*, 1999). The ultrastructural features of the organisms in tissue and in culture have been recently described by Kharsany *et al.* (1998) and are typical of Gram-negative bacteria.

PATHOGENESIS

The histopathological features of established infection comprise the presence of intracellular bacteria within histiocytes, surrounded by a dense plasma cell infiltrate, ulceration of skin or mucosa and local epithelial hyperplasia (Sehgal, Shyamprasad and Beohar, 1984). Infection spreads readily to local lymph nodes where a notable characteristic is the breakdown of overlying skin to form further ulcers. Haematogenous dissemination is a rare event, most likely to occur in pregnant women with untreated lesions of the cervix (Paterson, 1998). Involvement of bone and viscera may occur in these cases. The intracellular location of bacteria may be linked to the lack of the sucrose transporter gene *ScrA* found in extracellular *Klebsiella* species (Hart and Rao, 1999). An intriguing detailed histopathological and ultrastructural study of the behaviour of donovanosis within the epidermis has been recently undertaken by Ramdial *et al.* (2000).

EPIDEMIOLOGY

The disease is most frequently encountered in India, Brazil, Papua New Guinea and parts of South Africa. It is on the verge of eradication from the Australian aboriginal population and has already disappeared from the southern United States where it used to be common. The association of the disease with risky sexual behaviour and the anogenital location of lesions point strongly to sexual transmission as the dominant mode of spread, although unconvincing efforts have been made to argue the importance of non-sexual transmission (Goldberg, 1964). Mother-to-child transmission may occur during delivery (Bowden *et al.*, 2000).

CLINICAL FEATURES

The classic presentation is a painless beefy-red ulcer in the anogenital region, developing 3–40 days after exposure (Clarke, 1947). Hypertrophic lesions standing out from the level of surrounding skin are often seen. As lesions advance, a pungent odour often develops. Single or multiple lesions may occur. When the infection reaches local lymph nodes, the picture ranges from painless adenopathy, to subcutaneous abscess (often referred to as a pseudobubo), to ulcerative lesions arising in the overlying skin. Primary lesions of the mouth with extension to cervical nodes may occur (Veeranna and Raghu, 2002). In females, primary lesions may occur on the cervix (Arnell and Potekin, 1940) and infection may extend upwards into the uterus and tubes. Complications include the development of lymphoedema,

Principles and Practice of Clinical Bacteriology Second Edition   Editors Stephen H. Gillespie and Peter M. Hawkey

scarring and the late development of squamous carcinoma. Destructive lesions result in genital deformity including penile autoamputation. Rare cases of perinatal transmission have been described, often involving the ears when infectious material is forced into the external auditory meatus from the cervix during delivery (Bowden *et al.*, 2000). The extraordinary diversity of possible clinical presentations is well illustrated in a recent letter from Australia describing donovanosis presenting as psoas abscess, perinephric abscess, intraoral lesions, cervical adenopathy, infertility and spinal cord compression (Mein, Anstey and Bowden, 1999). Haematogenous spread of infection to bone, joints, liver and spleen has been observed in several patients, especially following delivery in women with untreated cervical lesions of donovanosis (Cherny *et al.*, 1957; Paterson, 1998). In these circumstances, the infection may cause death (Pund and McInnes, 1944). In two studies of donovanosis in HIV-coinfected patients, one from South Africa showed no influence of HIV on clinical outcomes (Hoosen *et al.*, 1996), while the other, from India, showed delayed response to treatment with erythromycin in HIV-infected patients (Jamkhedar *et al.*, 1998).

DIAGNOSIS

The differential diagnosis of donovanosis ulceration includes syphilis, chancroid and genital amoebiasis. Inguinal lesions have to be distinguished from lymphogranuloma venereum, tuberculosis, lymphoma and filariasis. Lesions of the cervix and glans penis are not infrequently mistaken for carcinoma, and it is not unknown for Donovan bodies to be missed on haematoxylin and eosin-stained biopsies, only to be recognized in resection specimens after radical surgery (Rajam and Rangiah, 1953; Hoosen *et al.*, 1990).

The mainstay of diagnosis is the identification of Donovan bodies in material collected from lesions (Richens, 1991). Such materials are collected by gently abrading the lesion to make a smear or by taking punch or snip biopsies. Donovan bodies show up well with Giemsa, Wright's or Leishman stains. They are not clearly seen with haematoxylin and eosin in biopsy material which is better processed with Giemsa, silver stains or a slow overnight Giemsa (Sehgal and Jain, 1987). Donovan bodies appear as pleomorphic coccobacilli $1-2\,\mu\mathrm{m} \times 0.5-0.7\,\mu\mathrm{m}$, clustered in groups within thin-walled vacuoles of macrophages. They may need to be distinguished carefully from the types of inclusions found in conditions such as leishmaniasis, histoplasmosis, rhinoscleroma and malakoplakia. Culture is not established for routine diagnosis but could be considered by laboratories able to maintain HEp 2 cells (Carter *et al.*, 1997). Careful decontamination of biopsy material with amikacin, vancomycin and metronidazole is recommended (Kharsany *et al.*, 1997). Polymerase chain reaction (PCR) tools for diagnosis have been developed but are not commercially available (Carter and Kemp, 2000). A serological test for donovanosis with 100% sensitivity and 98% specificity has been described, but it is not currently in use (Freinkel *et al.*, 1992). All patients with donovanosis should be offered screening for other sexually transmitted infections and for HIV.

TREATMENT

Recent experience in Australia with azithromycin suggests that this antibiotic is superior to alternatives, although direct comparisons with alternatives have not been conducted (Bowden *et al.*, 1996; Mein *et al.*, 1996; Bowden and Savage, 1998b). Azithromycin may be administered on a daily or weekly basis with equivalent good results (Bowden *et al.*, 1996). Published guidelines (Richens, 1999; Sexually Transmitted Diseases Treatment Guidelines, 2002) for the management of donovanosis also recommend doxycycline, fluorinated quinolones, ceftriaxone and erythromycin as alternatives. Antibiotic treatment should be continued until lesions have fully healed. A change

of antibiotic therapy or addition of a second agent should be considered in patients not responding. Operative procedures may be required for patients with deforming genital lesions (Bozbora *et al.*, 1998). Metronidazole is useful as adjunctive therapy for patients with especially malodorous lesions.

CONTROL

Partner management should be undertaken whenever donovanosis is diagnosed. The recent Australian campaign to eradicate donovanosis provides a useful model for other endemic areas. Key features of this campaign were setting up a co-ordinating structure, dialogue with aboriginal community leaders, education for the community and for health workers, provision of drugs and diagnostic support, improved surveillance, operational research and liaison with neighbouring Papua New Guinea which contains an important reservoir of infection.

REFERENCES

Anderson, K. (1943) The cultivation from granuloma inguinale of a microorganism having the characteristics of Donovan bodies in the yolk sac of chick embryos. *Science*, 9, 560–1.

Arnell, R.E. and Potekin, J.S. (1940) Granuloma inguinale (granuloma venereum) of the cervix. An analysis of thirty-eight cases. *Am J Obstet Gynecol*, 39, 626–35.

Bowden, F.J. and Savage, J. (1998a) Is the eradication of donovanosis possible in Australia? [editorial]. *Aust N Z J Public Health*, 22, 7–9.

Bowden, F.J. and Savage, J. (1998b) Donovanosis treatment with azithromycin [letter, comment]. *Int J STD AIDS*, 9, 61–2.

Bowden, F.J., Mein, J., Plunkett, C. and Bastian, I. (1996) Pilot study of azithromycin in the treatment of genital donovanosis. *Genitourin Med*, 72, 17–9.

Bowden, F.J., Bright, A., Rode, J.W. and Brewster, D. (2000) Donovanosis causing cervical lymphadenopathy in a five-month-old boy. *Pediatr Infect Dis J*, 19, 167–9.

Bozbora, A., Erbil, Y., Berber, E., Ozarmagan, S. and Ozarmagan, G. (1998) Surgical treatment of granuloma inguinale. *Br J Dermatol*, 138, 1079–81.

Carter, J.S. and Kemp, D.J. (2000) A colorimetric detection system for *Calymmatobacterium granulomatis*. *Sex Transm Infect*, 76, 134–6.

Carter, J., Hutton, S., Sriprakash, K.S. *et al.* (1997) Culture of the causative organism of donovanosis (*Calymmatobacterium granulomatis*) in HEp-2 cells. *J Clin Microbiol*, 35, 2915–7.

Carter, J.S. *et al.* (1999) Phylogenetic evidence for reclassification of *Calymmatobacterium granulomatis* as *Klebsiella granulomatis* comb. Nov. *Int J Syst Bacteriol*, 49, 1695–700.

Cherny, W.B.V., Jones, C.P., Peete, C.H. and Durham, N.C. (1957) Disseminated granuloma inguinale and its relationship to granuloma of the cervix and pregnancy. *Am J Obstet Gynecol*, 74, 597–605.

Clarke, C.W. (1947) Notes on the epidemiology of granuloma inguinale. *J Vener Dis Inf*, 28, 189.

Donovan, C. (1905) Ulcerating granuloma of the pudenda. *Ind Med Gaz*, 40, 414.

Freinkel, A.L., Dangor, Y., Koomhof, H.J. and Ballard, R.C. (1992) A serological test for granuloma inguinale. *Genitourin Med*, 68(4), 269–72.

Goldberg, J. (1964) Studies on granuloma inguinale: VII. Some epidemiological considerations of the disease. *Br J Vener Dis*, 40, 140–5.

Gupta, S. and Kumar, B. (2002) Donovanosis in India: declining fast? *Int J STD AIDS*, 13(4), 277.

Hart, C.A. and Rao, S.K. (1999) Donovanosis. *J Med Microbiol*, 48, 707–9.

Hoosen, A.A., Draper, G., Moodley, J. and Cooper, K. (1990) Granuloma inguinale of the cervix: a carcinoma look-alike. *Genitourin Med*, 66(5), 380–2.

Hoosen, A.A., Mphatsoe, M., Kharsany, A.B. *et al.* (1996) Granuloma inguinale in association with pregnancy and HIV infection. *Int J Gynaecol Obstet*, 53(2), 133–8.

Jamkhedar, P.P., Hira, S.K., Shroff, H.J. and Lanjewar, D.N. (1998) Clinico-epidemiologic features of granuloma inguinale in the era of acquired immune deficiency syndrome. *Sex Transm Dis*, 25(4), 196–200.

Kharsany, A.B., Hoosen, A.A., Kiepiela, P. *et al.* (1997) Growth and cultural characteristics of *Calymmatobacterium granulomatis* – the aetiological agent of granuloma inguinale (Donovanosis). *J Med Microbiol*, 46, 579–85.

Kharsany, A.B., Hoosen, A.A., Naicker, T. *et al.* (1998) Ultrastructure of *Calymmatobacterium granulomatis*: comparison of culture with tissue biopsy specimens. *J Med Microbiol*, 47(12), 1069–73.

Kharsany, A.B., Hoosen, A.A., Kiepala, P. *et al.* (1999) Phylogenetic analysis of *Calymmatobacterium granulomatis* based on 16S sequences. *J Med Microbiol*, 48, 841–7.

McLeod, K. (1881) Précis of operations performed in the wards of the first surgeon, Medical College Hospital, during the year 1881. *Ind Med Gaz*, 17, 113–23.

Mein, J.K., Anstey, N.M. and Bowden, F.J. (1999) Missing the diagnosis of donovanosis in northern Australia [letter]. *Med J Aust*, 170, 48.

Mein, J., Bastian, I., Guthridge, S. *et al.* (1996) Donovanosis: sequelae of severe disease and successful azithromycin treatment. *Int J STD AIDS*, 7, 448–51.

O'Farrell, N. (1995) Global eradication of donovanosis: an opportunity for limiting the spread of HIV-1 infection. *Genitourin Med*, 71, 27–31.

Paterson, D.L. (1998) Disseminated donovanosis (granuloma inguinale) causing spinal cord compression: case report and review of donovanosis involving bone. *Clin Infect Dis*, 26, 379–83.

Pund, E.R. and McInnes, G.F. (1944) Granuloma venereum: a cause of death. Report of six fatal cases. *Clinics*, 3, 221–34.

Rajam, R.V. and Rangiah, P.N. (1953) Granuloma venereum and its relationship to epidermoid carcinoma. *Ind J Vener Dis Dermatol*, 19, 1–19.

Ramdial, P.K., Kharsany, A.B., Reddy, R. and Chetty, R. (2000) Transepithelial elimination of cutaneous vulval granuloma inguinale. *J Cutan Pathol*, 27(10), 493–9.

Richens, J. (1991) The diagnosis and treatment of donovanosis (granuloma inguinale). *Genitourin Med*, 67, 441–52.

Richens, J. (1999) National guideline for the management of donovanosis (granuloma inguinale). Clinical Effectiveness Group (Association of Genitourinary Medicine and the Medical Society for the Study of Venereal Diseases). *Sex Transm Infect*, 75(Suppl 1), S38–9.

Sehgal, V.N. and Jain, M.K. (1987) Tissue section Donovan bodies – identification through slow Giemsa (overnight) technique. *Dermatologica*, 174, 228–31.

Sehgal, V.N., Shyamprasad, A.L. and Beohar, P.C. (1984) The histopathological diagnosis of donovanosis. *Br J Vener Dis*, 60, 45–7.

Sexually Transmitted Diseases Treatment Guidelines (2002) Centers for Disease Control and Prevention. *MMWR Recomm Rep, 51(RR-6)*, 1–78.

Veeranna, S. and Raghu, T.Y. (2002) Oral donovanosis. *Int J STD AIDS*, 13, 855–6.

32

Proteus, Providencia and *Morganella* spp.

Division of Immunity and Infection, The Medical School, Edgbaston, Birmingham B15 2TT, UK

TAXONOMY AND GENERAL DESCRIPTION OF THE GENERA

These genera have been grouped as the tribe Proteeae as they can oxidatively deaminate phenylalanine and tryptophan. Other characteristics which suggest an oxidase-negative non-lactose–fermenting bacterium may belong to this tribe are swarming (Proteus), motility, methyl red (MR)$^+$/Voges–Proskauer$^-$, H$_2$S positive, urease production (except *Providencia*), protease/lipase production (*Proteus*) and resistance to polymxyins.

DNA homology studies have shown that *Proteus morganii*, as it was previously known, has but 20% homology with the other members of the Proteeae (Brenner *et al.*, 1978), and in consequence the separate genus *Morganella* was created and existing and new species redistributed amongst *Proteus* and *Providencia*. Because of the similarities in pathogenicity and natural habitat, clinical microbiologists do still tend to group these bacteria together. *Proteus mirabilis* and *Proteus vulgaris* were originally described in 1885, and although the taxonomic status of *P. mirabilis* has been stable, *P. vulgaris* has undergone some changes. DNA hybridization studies by Hickman *et al.* (1982) suggested that *P. vulgaris* can be divided into three groups: (i) biogroup (BG) 1 strains being named as *Proteus penneri*, which is also defined as genomospecies 1; (ii) BG 2 strains are genetically distinct (genomospecies 2) and the name *P. vulgaris* has been reassigned to this species; (iii) BG 3 strains contain the genomospecies 3, 4, 5 and 6, of which genomospecies 3 has been named *P. hauseri*, the rest being unnamed (O'Hara *et al.*, 2000b). *Proteus myxofaciens* is not a human pathogen and is only reported to have been isolated from gypsy moth larvae (Farmer *et al.*, 1985). The bacterium, previously known as *Proteus morganii*, described by Morgan in 1906 from the faeces of children with diarrhoea, was reassigned to its own genus *Morganella* (Brenner *et al.*, 1978). Subsequently two subspecies, *morganii* and *sibonii*, have been characterized (O'Hara, Brenner and Miller, 2000a; O'Hara *et al.*, 2000b).

There are currently five recognized species in the genus *Providencia*, which was named after Providence, Rhode Island, where C.A. Stuart worked on these bacteria. The new species *Providencia rustigianii* and *Providencia heimbachae* were originally grouped within *Providencia alcalifaciens*, but neither are found in human clinical specimens (Muller *et al.*, 1986). The remaining species are easily distinguished (Table 32.1). The occurrence of plasmid-encoded urease activity in some strains of *Providencia stuartii*, which were previously confused with *Proteus* (now *Providencia*) *rettgeri* should be noted (Farmer *et al.*, 1985). Full descriptions of the species have recently been reviewed (O'Hara, Brenner and Miller, 2000a).

The biochemical characteristics useful in differentiating the genera of the tribe Proteeae are summarized in Table 32.1 (in bold type).

A rapid identification scheme for the common species, using six media, has also been described (Senior, 1997). One of the most striking features of the genus *Proteus* in culture is their ability to swarm on solid media. Short motile cells at the edge of the growing colony differentiate into very elongated hyperflagellated swarm cells. These align along their axes and by coordinated flagella action swarm as a two-dimensional raft within a film of hydrated polysaccharide. Migration ceases periodically with increased cell septation and reduced flagellation, giving rise to the bulls-eye appearance (Figure 32.1). Transposon mutagenesis has shown that many differentiation and flagellar genes are involved in swarming. A sugar transferase gene (cmfA) and a membrane-spanning protein influencing cell shape (ccmA) have also been shown to influence swarming (Hay *et al.*, 1999). Putrescine has recently been shown to act as an extracellular signal for effective swarming, putrescine-repressing activity of the SpeB enzyme which is involved in the synthesis of putrescine (Sturgill and Rather, 2004).

All members of the genus possess flagellae (usually peritrichous) and are motile with the exception of some strains of *M. morganii*. The flagellae of *P. mirabilis* are variable in morphology.

The major antigenic determinates of the Proteeae are somatic (O) and flagellar (H) antigens. Typing schemes for detecting O antigens have been described in detail for most of the species (Penner, 1981).

PATHOGENICITY OF PROTEEAE

A detailed review of potential virulence factors in *Proteus* ssp. has been published (Rozalski, Sidorczyk and Kotelko, 1997). It has long been recognized that the possession of urease (a nickel metallo-enzyme) by a number of the members of the Proteeae represents a likely pathogenicity factor (Mobley and Hausinger, 1989). A wide range of biochemically and genetically distinct ureases are found amongst all of the urease-positive members of the Proteeae, many of which have now been cloned and sequenced. The gene in *P. mirabilis* is chromosomal and consists of the structural genes ureABC; ureD is upstream and ureEFG downstream, ureDEFG encode proteins that are involved in nickel ion insertion and enzyme assembly. The ureR gene lies upstream of ureD and is a positive transcriptional regulator required for both basal and induced urease activity, which is an essential virulence factor in urinary tract infections (Dattelbaum *et al.*, 2003). This gene cluster interestingly shows considerable homology and organizational similarities to urease gene clusters on plasmids in *Escherichia coli*, *Prov. stuartii* and *Salmonella cubana*, suggesting common routes of evolution (D'Orazio and Collins, 1993). The formation of urinary stones results from the pH rise that occurs in urine due to the release of ammonium ions from urea as a result of activity of bacterial ureases. The stones are a mixture of struvite

Principles and Practice of Clinical Bacteriology Second Edition Editors Stephen H. Gillespie and Peter M. Hawkey
© 2006 John Wiley & Sons, Ltd

Table 32.1 Key biochemical characteristics of the species belonging to the genera *Proteus*, *Providencia* and *Morganella*

Test or property[a]	*P. mirabilis*	*P. myxofaciens*[b]	*P. penneri*	*P. vulgaris*[c] BG 2	*P. vulgaris*[c] BG 3	*M. morganii*[d]	*Prov. alcalifaciens*	*Prov. rustiganii*	*Prov. heimbachae*	*Prov. stuarti*	*Prov. rettgeri*
Indole	−[e]	−	−	+	+	+	+	+	−	+	+
Citrate	V[e]	V	−	−	−	−	+	V	V	+	+
H$_2$S	+[e]	−	V	+	+	−	−	−	−	−	−
Urease	+	+	+	+	+	+	−	−	−	V[f]	+
Ornithine decarboxylase	+	−	−	−	−	+	−	−	−	−	−
Gelatinase	+	+	V	+	+	−	−	+	−	−	−
Lipase	+	+	V	V	V	−	−	−	−	−	−
Swarming[g]	+	+	+	+	+	−	−	−	−	−	−
Fermentation of											
Mannose	−	−	−	−	−	+	+	+	+	+	+
Maltose	−	+	+	+	+	−	−	−	−	−	−
Xylose	+	−	+	+	+	−	−	−	−	−	−
Salicin	−	−	−	+	−	−	−	−	−	−	V
Inositol	−	−	−	−	−	−	−	−	V	+[h]	+
Adonitol	−	−	−	−	−	−	+	−	−	−[h]	+
Arabitol	−	−	−	−	−	−	−	−	+	−	+
Trehalose	+	+	V	−	V	V[i]	−	−	−	+	−
Galactose							−	+	+	+	+
Rhamnose	−	−	−	−	−	−	−	−	+	−	V
Aesculin hydrolysis	−	−	−	+	−	−	−	−	−	−	V

[a] Tests in bold type are useful in differentiating the three genera.
[b] Only occurs as pathogen of gypsy moth larvae.
[c] BG, biogroup.
[d] Biogroup 1 strains are lysine decarboxylase positive (plasmid mediated) non-motile and glycerol positive.
[e] +, 90–100% positive; V, 10–90% positive; −, 0–10% positive.
[f] Plasmid-encoded property.
[g] 37 °C on nutrient or blood agar.
[h] BG 4 insitol negative and BF6 adonitol positive, both rare.
[i] Subsp. *morganii* is negative, subsp. *sibonii* is positive.

(MgNH$_4$ PO$_4$·6H$_2$O) and carbonate apatite (Ca$_{10}$(PO4)$_6$ CO$_3$) which precipitate at pH < 6.5 (Mobley and Hausinger, 1989). It has been observed that *P. mirabilis* is associated with blocked urinary catheters rather than other urease-producing bacteria (Mobley and Warren, 1987). The capsular polysaccharide from *P. mirabilis* enhances struvite formation, possibly by weakly concentrating Mg^{2+} because of its structure and anionic nature (Dumanski *et al.*, 1994). The most frequently encountered member of the Proteeae in infected colonized urinary catheters is *Prov. stuartii* (Mobley *et al.*, 1988). It appears this is not due to increased entry of *Prov. stuartii* into the catheter system but due to the ability of *Prov. stuartii* to adhere to catheter material via mannose-resistant *Klebsiella*-like haemagglutinins (MR/K), whereas strains only found in patients for a short duration express mannose-sensitive haemagglutinins (Mobley *et al.*, 1988). The formation of crystalline encrustations in uretheral catheters does not only require that the bacteria produce urease as do *P. mirabilis*, *P. vulgaris* and *Prov. rettgeni* but that the urine pH rises above 8.0, these species producing encrustations in a model whereas urease-positive *M. morganii*, *K. pneumoniae* and *P. aeruginosa* did not (Stickler *et al.*, 1998). *Proteus mirabilis* also expresses MR/K-type fimbriae which probably contribute to catheter adherence, but also produce *P. mirabilis* fimbriae (PMF), uroepithelial cell adhesin (UCA), ambient temperature fimbriae and mannose-resistant/*Proteus*-like fimbriae (MR/P) (Rozalski, Sidorczyk and Kotelko, 1997). The association of MR/P fimbriae and the development of pyelonephritis was made by Silverblatt and Ofek in 1978 (cited in Rozalski, Sidorczyk and Kotelko, 1997), and the relevant gene has been cloned. The use of mrpA$^-$ mutants in a mouse model of ascending urinary tract infection confirms the important role of MR/P fimbriae in the pathogenesis of acute pyelonephritis (Rozalski, Sidorczyk and Kotelko, 1997). PMF have been shown to play a role in the development of both bladder colonization and kidney colonization (Zunino *et al.*, 2003). Recently the gene encoding UCA has been cloned, and amino acid sequence homology was found between a range of different pilins in other bacteria, such as *Haemophilus*, *E. coli* pilins F17, G and IC (Cook *et al.*, 1995). UCA seems to be distinct from MR/P and may be responsible for uroepithelial cell adhesion. The use of a Tn5-based

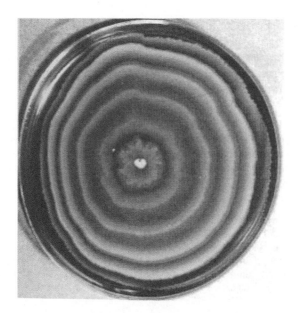

Figure 32.1 Swarming culture of *Proteus mirabilis* (blood agar, 22 °C incubated for 40 h) demonstrating the 'bulls-eye' appearance.

transposon mutagenesis system in *P. mirabilis* to produce swarm cell-negative mutants in the mouse model has shown the importance of hyperflagellated swarm cells in establishing ascending infections of the urinary tract (Allison *et al.*, 1994). During swarm cell differentiation a number of virulence factors are produced, including an extracellular metalloprotease with activity against IgA (Senior, Loomes and Kerr, 1991). Molecular analysis of the *zapA* gene, which encodes a metalloprotease (serralysin type) is expressed only in swarmer cells; *zapA⁻* mutants failed to degrade IgA and had decreased survival in the mouse UTI model (Walker *et al.*, 1999).

Siderophores are essential for bacteria to acquire iron, particularly when multiplying in an animal host. The Proteeae are unique amongst Enterobacteriaceae in possessing amino acid deaminases which result in the production of α-keto and α-hydroxycarboxylic acids, such as phenylpyruvic and indolylpyruvic acids; these in turn, bind actively to iron and have been demonstrated to have siderophore activity (Drechsel *et al.*, 1993).

A study of traveller's diarrhoea in UK patients demonstrated a strong association with the isolation of *Prov. alcalifaciens* from faeces (Haynes and Hawkey, 1989). A case-control study in Bangladesh also demonstrated a highly significant association with diarrhoea in children (Albert, Faruque and Mahalanabis, 1998). There has also been a report of *Prov. alcalifaciens* with evidence for its aetiological role in causing a large outbreak of food poisoning in Japan (Murata *et al.*, 2001). It has further been shown that strains of *Prov. alcalifaciens* isolated from the faeces of patient suffering from diarrhoea in Bangladesh are capable of invading Hep-2 cell monolayers, with actin condensation (Albert *et al.*, 1995). Subsequently some strains have been shown to be resistant to complement and translocate from the gut (Vieira, Koh and Guth, 2003).

EPIDEMIOLOGY

Habitat

Most of the Proteeae are widely distributed in nature, being frequently found in the faeces of animals and humans as well as in associated materials such as decomposing meat and sewage. All of the clinically significant species were found in a study of faecally contaminated calf bedding, bovine meat being suggested as a possible route of colonization of the human gut (Hawkey, McCormick and Simpson, 1986). Faeces have been reported as a source for nosocomial infections caused by *P. mirabilis* (Burke *et al.*, 1971), *M. morganii* (Williams *et al.*, 1983) and *Prov. stuartii* (Hawkey *et al.*, 1982). Maternal vaginal carriage of *P. mirabilis* (probably secondary to faecal carriage) has been described as a significant source of neonatal infections (Bingen *et al.*, 1993). The carriage of *P. mirabilis* in the prepuce of 22% of uncircumcised and 1.7% of circumcised male infants is thought to explain the occurrence of *P. mirabilis* infections in male babies (Glennon *et al.*, 1988).

Proteus mirabilis and less frequently *P. vulgaris*, possibly due to its low carriage rate in faeces (Senior and Leslie, 1986), are associated with urinary tract infections and the infected urine, particularly when patients have an indewelling catheter, can be a source for cross-infection; infected urine can therefore be regarded as a habitat for Proteeae (Hickman and Farmer, 1976; Warren, 1986). Similarly *Prov. stuartii* (Hawkey *et al.*, 1982) and *M. morganii* (McDermott and Mylotte, 1984) nosocomial infection can be acquired from infected urine, although in the case of the last organism infected wounds may be a significant source (Williams *et al.*, 1983).

Providencia alcalifaciens can be isolated from both human and animal faeces and has been assumed in the past not to be a human pathogen. However, early workers such as C.A. Stuart and Patricia Carpenter (Central Public Health Laboratory, London) thought it could cause diarrhoeal illness (Penner, 1981). Recently good evidence for an enteropathogenic role has emerged (see above).

HUMAN INFECTIONS

The most common infection caused by Proteeae is infection of the urinary tract. *Proteus mirabilis* is by far the most common species of Proteeae isolated from the urinary tract most frequently being nosocomial (de Champs *et al.*, 2000) and is the third most frequently isolated species of Enterobacteriaceae in the clinical laboratory (Hickman and Farmer, 1976). Whilst the association of Proteeae with urinary tract infection, particularly the anatomically abnormal or catheterized urinary tract (*P. mirabilis* and *Prov. stuartii* in particular), is very well documented (Warren *et al.*, 1982; Hawkey, 1984; Ehrenkranz *et al.*, 1989), detailed prospective epidemiological studies using strain typing are rare. It seems likely that faecal carriage can be a source for subsequent infection of the abnormal tract (Hawkey *et al.*, 1982). The colonized groin area of geriatric patients has also been proposed as a source for a Proteeae (Ehrenkranz *et al.*, 1989), although this needs to be proven by typing studies. Cross-infection of both multiple and an endemic strain amongst catheterized patients in the hospital setting via hands and fomites (e.g. jugs and specific gravity cylinders) has been documented particularly for *Prov. stuartii* (Hawkey *et al.*, 1982; Rahav *et al.*, 1994). In neonatal units, the vaginal carriage of *P. mirabilis* (Bingen *et al.*, 1993) and rectal/vaginal carnage in a nurse (Burke *et al.*, 1971) and probable subsequent spread via hands have led to serious outbreaks of septicaemia, meningitis and omphalitis. Cross-infection by *M. morganii* is rare, but episodes in which infected wounds co-infected with *P. mirabilis* have acted as a source with subsequent transfer to other patients have been described and documented with strain typing (Williams *et al.*, 1983). In a review of 19 cases of *M. morganii* bacteraemia the most common source was found to be an infected wound, and extensive use of cephalosporins (to which *M. morganii* is often resistant) was thought to be an important factor leading to the development of infection (McDermott and Mylotte, 1984). The source and route of transmission of *Prov. alcalifaciens* in human diarrhoeal illness have not been investigated, but are likely to be via food (Murata *et al.*, 2001) or asymptomatic faecal carriage leading to transient hand carriage.

EPIDEMIOLOGICAL TYPING METHODS

A wide range of epidemiological typing methods have been applied to Proteeae, the oldest being O serotyping (Penner, 1981) which has been replaced by molecular methods. In the case of the swarming *Proteus* spp. use has been made for many years of the Dienes test. Originally described by Dienes in 1946, a line of demarcation appears between two swarming strains if they are different, whereas identical strains swarm into one another (Figure 32.2). The phenomenon appears to be largely due to both sensitivity to and production of bacteriocins (Senior, 1977). It is an easily applied method and has been shown to have a similar index of discrimination to ribotyping and PFGE (Pfaller *et al.*, 2000).

Bacteriocin typing schemes have been described for Proteeae (Senior, 1977), as has Bacteriophage typing (Hickman and Farmer, 1976). In a study of all species of Proteeae a wide range of ribotype patterns were seen, suggesting that this method would be a sensitive typing method (Pignato *et al.*, 1999). A recent comparison of ribotyping and arbitrarily primed polymerase chain reaction typing (AP-PCR) in investigating *P. mirabilis* cross-infection in a maternity hospital concluded that AP-PCR was equivalent to ribotyping and easier to perform (Bingen *et al.*, 1993). Ribotyping has been successfully applied to an outbreak of *Prov. stuartii* catheter infection, suggesting multiple sources of infecting bacteria (Rahav *et al.*, 1994). Tandem tetramer-based microsatellite typing has been suggested to have a high discrimination index when applied to *P. mirabilis* (Cieslikowski *et al.*, 2003). PFGE has been applied to *P. mirabilis* typing (Sabbuba, Mahenthiralingam and Stickler, 2003).

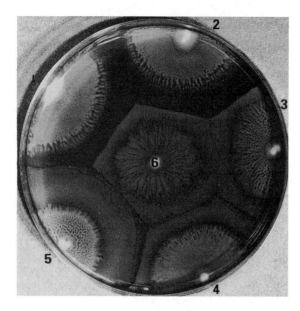

Figure 32.2 Diene's test, lines of compatibility are seen between isolates 1 and 2; 3 and 6 indicating that they are indistinguishable but different from each other. Isolates 4 and 5 are incompatible and different from the other two groups of isolates (1 and 2; 3 and 6). Blood agar, incubated for 48 h at 22 °C.

CLINICAL MANIFESTATIONS

The urinary tract is the most commonly infected site, although infection of the anatomically normal and uncatheterized tract is rare. Pyelonephritis and urinary stone formation are common features of infection with *P. mirabilis* for the reasons described above. *Providencia stuartii*, *Proteus mirabilis* and *M. morganii* are the most frequently found species in the catheterized urinary tract; often more than one species of Proteeae is present and usually mixed with other bacterial species. *Providencia stuartii* is the species which persists best; in a prospective study of bacteriuria in 20 catheterized patients, the mean duration of new episodes was 10.4 weeks for *Prov. stuartii*, 5.5 weeks for *P. mirabilis* and 2.9 weeks for enterococci (Warren *et al.*, 1982).

The frequent presence of *Prov. stuartii* in catheter bags has been shown to explain the condition in which the bag turns deep purple in colour, termed the purple urine bag syndrome. Some strains of *Prov. stuartii* possess an indoxyl suphatase which degrades high levels of urinary indoxyl sulphate (usually in patients with bacterial overgrowth of the small bowel) to indoxyl, which condenses to form the insoluble dyes indirubin (red) and indigo (blue). These dyes are soluble in the plasticizers of the bag wall and stain it purple (Dealler Hawkey and Millar, 1988).

Proteus mirabilis has been reported as causing a number of other infections, such as bacteraemia arising from the infected/colonized urinary tract (Muder *et al.*, 1992; de Champs *et al.*, 2000), nosocomial wound infections (Chow *et al.*, 1979; Williams *et al.*, 1983; de Champs *et al.*, 2000) and meningitis in neonates arising from infected umbilical stumps (Burke *et al.*, 1971). It is also found with other bacteria in brain abscess derived from middle ear infections. Chronic colonization of the ear may cause local infections and local spread results in its isolation from brain abscesses. Other members of the Proteeae, notably *M. morganii*, are associated with soft-tissue infections, such as wound infections (Tucci and Isenberg, 1981; Williams *et al.*, 1983; McDermott and Mylotte, 1984). In studies of bacteriuria *P. mirabilis* is the most frequently reported of the Proteeae

(McDermott and Mylotte, 1984). *Proteus vulgaris* and *P. penneri* cause a spectrum of infections similar to *P. mirabilis* but are somewhat rarer. In long-term care and geriatric facilities, where patients with urinary catheters are common, *Prov. stuartii* is the most frequent cause of bacteremia (Hawkey, 1984; Muder *et al.*, 1992). *Providencia stuartii* has been noted to cause nosocomial pneumonia and wound and burn infections, but these are rare (Hawkey, 1984). The role of Proteeae in diarrhoeal disease has long been debated (Penner, 1981). Definitive evidence for the enteropathogenicity of *Prov. alcalifaciens* has now emerged (see above).

LABORATORY DIAGNOSIS

Isolation

All members of the Proteeae likely to be encountered in clinical material are capable of growth on commonly used laboratory media (nutrient, blood and MacConkey's agar) incubated at 37 °C in room air. A number of selective media have been developed for the isolation of Proteeae (Penner, 1981). PIM agar, which has been used in a number of studies, utilizes clindamycin and colistin as the selective agents and tryptophan/tyrosine as differential agents (Hawkey, McCormick and Simpson, 1986).

Identification of Proteeae is generally reliably achieved using one of the recognized commercial identification systems, either automated or manual and chromogenic agars are useful for urinary isolates (Filius *et al.*, 2003). It is worth noting that *Prov. stuartii* may pose problems and be misidentified by most systems if supplementary tests are not included (Cornaglia *et al.*, 1988). A modified MacConkey's agar which detects phosphatase activity has been shown to alleviate this problem (Thaller *et al.*, 1992). The biochemical characteristics of the Proteeae species are summarized in Table 32.1. In the case of blood culture isolates, the pathogenic status of those isolates is clear; however, from colonized wounds and catheter urines the clinical condition of the patient is paramount in deciding the level of identification and reporting. One study has shown that the Proteeae may be selectively under-reported by some microbiology laboratories from catheter urine specimens, with the potential for missed treatment of bacteraemic patients (Damron *et al.*, 1986).

RESISTANCE TO ANTIMICROBIALS AND CHEMOTHERAPY OF PROTEEAE

β-Lactam antibiotics are widely used to treat infections caused by bacteria belonging to the family Enterobacteriacae. In the case of the Proteeae, *P. mirabilis*, unlike the other species [except *Prov. alcalifaciens*, which is generally very susceptible to antibiotics (Penner, 1981)], does not appear to produce a chromosomal cephalosporinase or AmpC type penicillinase (Livermore, 1995). In consequence, many cases of infection caused by *P. mirabilis* can often be treated with ampicillin or older cephalosporins such as cephradine or cephalexin. *Providencia stuartii* carries a number of intrinsic plasmid-mediated antibiotic resistance determinants and may be difficult to treat (Hawkey, 1984; Warren, 1986). The extended-spectrum cephalosporins are suitable for treating infections caused by the *Providencia* spp., *M. morganii* and *P. vulgaris*, although plasmid-mediated extended-spectrum β-lactamases are being increasingly reported from the Proteeae particularly *P. stuartii* (Tumbarello *et al.*, 2004) and *P. miratilis* (de Champs *et al.*, 2000). The TEM family of β-lactamases is found in most of the species of Proteeae at varying frequencies, often dictated by local usage of ampicillins (Hawkey, 1984). Imipenem resistance is very rare in the Proteeae with the exception of occasional isolates of *P. mirabilis*, when altered penicillin-binding proteins appear to be responsible (Neuwirth *et al.*, 1995), although the carbapenemase IMP-1 has been reported from Japan in a *P. vulgaris* isolate (Arakawa *et al.*, 2000).

Most isolates of Proteeae are susceptible to aminoglycosides, provided none of the plasmid-mediated aminoglycoside-inactivating enzymes are present [e.g. APH(3′)]. *Providencia stuartii* and probably some strains of *Prov. rettgeri* carry a chromosomally encoded gentamicin 2′-*N*-acetyl-transferase (AAC2′) which is not always expressed. The enzyme probably plays a role in the maintenance of peptidoglycan structure (Franklin and Clarke, 2001). Commonly used antibiotics for the treatment of urinary tract infections caused by Proteeae are trimethoprim and fluoroquinolones, such as ciprofloxacin, which are effectively subject to sensitivity testing of individual isolates as resistance to these agents is increasing. Caution should be observed in using chloramphenicol to treat brain abscesses caused by *P. mirabilis* as most strains carry an inducible chloramphenicol acetylase gene (Charles *et al.*, 1985) which will result in treatment failure. *Providencia stuartii* has been regarded by some workers to be resistant to the disinfectant chlorhexidine. It has been suggested that chlorhexidine inhibits membrane-bound ATPase, and this may provide an understanding for resistance, although a study of *Prov. stuartii* mutants resistant to high levels of chlorhexidine failed to confirm this (Chopra, Johnson and Bennett, 1987). Proteeae are usually intrinsically resistant to the polymyxins and nitrofurantoins, possibly due to particular properties of their bacterial membranes.

An interesting alternative to antibiotic treatment for UTIs caused by *P. mirabilis* has been suggested following the successful protection of mice from infection by vaccination with purified mannose – resistant, *Proteus*-like fimbriae (Li *et al.*, 2004).

REFERENCES

Albert, M. J., Ansaruzzaman, M., Bhuiyan, N. A. *et al.* (1995) Characteristics of invasion of HEp-2 cells by *Providencia alcalifaciens*. *Journal of Medical Microbiology*, **42**, 186–190.

Albert, M. J., Faruque, A. S., and Mahalanabis, D. (1998) Association of *Providencia alcalifaciens* with diarrhea in children. *Journal of Clinical Microbiology*, **36**, 1433–1435.

Allison, C., Emody, L., Coleman, N. *et al.* (1994) The role of swarm cell differentiation and multicellular migration in the uropathogenicity of *Proteus mirabilis*. *Journal of Infectious Diseases*, **169**, 1155–1158.

Arakawa, Y., Shibata, N., Shibayama, K. *et al.* (2000) Convenient test for screening metallo-beta-lactamase-producing gram-negative bacteria by using thiol compounds. *Journal of Clinical Microbiology*, **38**, 40–43.

Bingen, E., Boissinot, C., Desjardins, P. *et al.* (1993) Arbitrarily primed polymerase chain reaction provides rapid differentiation of Proteus mirabilis isolates from a pediatric hospital. *Journal of Clinical Microbiology*, **31**, 1055–1059.

Brenner, D. J., Farmer, J. J. III, and Fanning, G. R. (1978) Deoxyribonucleic acid relatedness of *Proteus* and *Providencia* species. *International Journal of Systematic Bacteriology*, **28**, 269–282.

Burke, J. P., Ingall, D., Klein, J. O. *et al.* (1971) *Proteus mirabilis* infections in a hospital nursery traced to a human carrier. *New England Journal of Medicine*, **284**, 115–121.

Charles, I. G., Harford, S., Brookfield, J. F. *et al.* (1985) Resistance to chloramphenicol in *Proteus mirabilis* by expression of a chromosomal gene for chloramphenicol acetyltransferase. *Journal of Bacteriology*, **164**, 114–122.

Chopra, I., Johnson, S. C., and Bennett, P. M. (1987) Inhibition of *Providencia stuartii* cell envelope enzymes by chlorhexidine. *Journal of Antimicrobial Chemotherapy*, **19**, 743–751.

Chow, A. W., Taylor, P. R., Yoshikawa, T. T. *et al.* (1979) A nosocomial outbreak of infections due to multiply resistant *Proteus mirabilis*: role of intestinal colonization as a major reservoir. *Journal of Infectious Diseases*, **139**, 621–627.

Cieslikowski, T., Gradecka, D., Mielczarek, M. *et al.* (2003) Tandem tetramer-based microsatellite fingerprinting for typing of Proteus mirabilis strains. *Journal of Clinical Microbiology*, **41**, 1673–1680.

Cook, S. W., Mody, N., Valle, J. *et al.* (1995) Molecular cloning of *Proteus mirabilis* uroepithelial cell adherence (uca) genes. *Infection and Immunity*, **63**, 2082–2086.

Cornaglia, G., Dainelli, B., Berlutti, F. *et al.* (1988) Commercial identification systems often fail to identify *Providencia stuartii*. *Journal of Clinical Microbiology*, **26**, 323–327.

de Champs, C., Bonnet, R., Sirot, D. *et al.* (2000) Clinical relevance of *Proteus mirabilis* in hospital patients: a two year survey. *Journal of Antimicrobial Chemotherapy*, **45**, 537–539.

D'Orazio, S. E. and Collins, C. M. (1993) Characterization of a plasmid-encoded urease gene cluster found in members of the family Enterobacteriaceae. *Journal of Bacteriology*, **175**, 1860–1864.

Damron, D. J., Warren, J. W., Chippendale, G. R. *et al.* (1986) Do clinical microbiology laboratories report complete bacteriology in urine from patients with long-term urinary catheters? *Journal of Clinical Microbiology*, **24**, 400–404.

Dattelbaum, J. D., Lockatell, C. V., Johnson, D. E. *et al.* (2003) UreR, the transcriptional activator of the *Proteus mirabilis* urease gene cluster, is required for urease activity and virulence in experimental urinary tract infections. *Infection and Immunity*, **71**, 1026–1030.

Dealler, S. F., Hawkey, P. M., and Millar, M. R. (1988) Enzymatic degradation of urinary indoxyl sulfate by *Providencia stuartii* and *Klebsiella pneumoniae* causes the purple urine bag syndrome. *Journal of Clinical Microbiology*, **26**, 2152–2156.

Drechsel, H., Thieken, A., Reissbrodt, R. *et al.* (1993) Alpha-keto acids are novel siderophores in the genera *Proteus*, *Providencia*, and *Morganella* and are produced by amino acid deaminases. *Journal of Bacteriology*, **175**, 2727–2733.

Dumanski, A. J., Hedelin, H., Edin-Liljegren, A. *et al.* (1994) Unique ability of the *Proteus mirabilis* capsule to enhance mineral growth in infectious urinary calculi. *Infection and Immunity*, **62**, 2998–3003.

Ehrenkranz, N. J., Alfonso, B. C., Eckert, D. G. *et al.* (1989) Proteeae species bacteriuria accompanying Proteeae species groin skin carriage in geriatric outpatients. *Journal of Clinical Microbiology*, **27**, 1988–1991.

Farmer, J. J. III, Davis, B. R., Hickman-Brenner, F. W. *et al.* (1985) Biochemical identification of new species and biogroups of Enterobacteriaceae isolated from clinical specimens. *Journal of Clinical Microbiology*, **21**, 46–76.

Filius, P. M., van Netten, D., Roovers, P. J. *et al.* (2003) Comparative evaluation of three chromogenic agars for detection and rapid identification of aerobic Gram-negative bacteria in the normal intestinal microflora. *Clinical Microbiology and Infection*, **9**, 912–918.

Franklin, K. and Clarke, A. J. (2001) Overexpression and characterization of the chromosomal aminoglycoside, 2′-*N*-acetyltransferase of *Providencia stuartii*. *Antimicrobial Agents and Chemotherapy*, **45**, 2238–2244.

Glennon, J., Ryan, P. J., Keane, C. T. *et al.* (1988) Circumcision and periurethral carriage of *Proteus mirabilis* in boys. *Archives of Disease in Childhood*, **63**, 556–557.

Hawkey, P. M. (1984) *Providencia stuartii*: a review of a multiply antibiotic-resistant bacterium. *Journal of Antimicrobial Chemotherapy*, **13**, 209–226.

Hawkey, P. M., McCormick, A., and Simpson, R. A. (1986) Selective and differential medium for the primary isolation of members of the Proteeae. *Journal of Clinical Microbiology*, **23**, 600–603.

Hawkey, P. M., Penner, J. L., Potten, M. R. *et al.* (1982) Prospective survey of fecal, urinary tract, and environmental colonization by *Providencia stuartii* in two geriatric wards. *Journal of Clinical Microbiology*, **16**, 422–426.

Hay, N. A., Tipper, D. J., Gygi, D. *et al.* (1999) A novel membrane protein influencing cell shape and multicellular swarming of *Proteus mirabilis*. *Journal of Bacteriology*, **181**, 2008–2016.

Haynes, J. and Hawkey, P. M. (1989) *Providencia alcalifaciens* and travellers' diarrhoea. *British Medical Journal*, **299**, 94–95.

Hickman, F. W. and Farmer, J. J. III (1976) Differentiation of *Proteus mirabilis* by bacteriophage typing and the Dienes reaction. *Journal of Clinical Microbiology*, **3**, 350–358.

Hickman, F. W., Steigerwalt, A. G., Farmer, J. J. III *et al.* (1982) Identification of *Proteus penneri* sp. nov., formerly known as *Proteus vulgaris* indole negative or as *Proteus vulgaris* biogroup, 1. *Journal of Clinical Microbiology*, **15**, 1097–1102.

Li, X., Lockatell, C. V., Johnson, D. E. *et al.* (2004) Development of an intranasal vaccine to prevent urinary tract infection by *Proteus mirabilis*. *Infection and Immunity*, **72**, 66–75.

Livermore, D. M. (1995) beta-Lactamases in laboratory and clinical resistance. *Clinical Microbiology Reviews*, **8**, 557–584.

McDermott, C. and Mylotte, J. M. (1984) *Morganella morganii*: epidemiology of bacteremic disease. *Infection Control*, **5**, 131–137.

Mobley, H. L. and Hausinger, R. P. (1989) Microbial ureases: significance, regulation, and molecular characterization. *Microbiological Reviews*, **53**, 85–108.

Mobley, H. L. and Warren, J. W. (1987) Urease-positive bacteriuria and obstruction of long-term urinary catheters. *Journal of Clinical Microbiology*, **25**, 2216–2217.

Mobley, H. L., Chippendale, G. R., Tenney, J. H. *et al.* (1988) MR/K hemagglutination of *Providencia stuartii* correlates with adherence to catheters and with persistence in catheter-associated bacteriuria. *Journal of Infectious Diseases*, **157**, 264–271.

Muder, R. R., Brennen, C., Wagener, M. M. *et al.* (1992) Bacteremia in a long-term-care facility: a five-year prospective study of, 163 consecutive episodes. *Clinical Infectious Diseases*, **14**, 647–654.

Muller, H. E., O'Hara, C. M., Fanning, G. R. *et al.* (1986) *Providencia heimbache*, a new species of Enterobacteriaceae isolated from animals. *International Journal of Systematic Bacteriology*, **36**, 252–256.

Murata, T., Iida, T., Shiomi, Y. *et al.* (2001) A large outbreak of foodborne infection attributed to *Providencia alcalifaciens. Journal of Infectious Diseases*, **184**, 1050–1055.

Neuwirth, C., Siebor, E., Duez, J. M. *et al.* (1995) Imipenem resistance in clinical isolates of *Proteus mirabilis* associated with alterations in penicillin-binding proteins. *Journal of Antimicrobial Chemotherapy*, **36**, 335–342.

O'Hara, C. M., Brenner, F. W., and Miller, J. M. (2000a) Classification, identification, and clinical significance of *Proteus, Providencia*, and *Morganella. Clinical Microbiology Reviews*, **13**, 534–546.

O'Hara, C. M., Brenner, F. W., Steigerwalt, A. G. *et al.* (2000b) Classification of *Proteus vulgaris* biogroup, 3 with recognition of *Proteus hauseri* sp. nov., nom. rev. and unnamed Proteus genomospecies 4, 5 and 6. *International Journal of Systematic Bacteriology*, **50**, 1869–1875.

Penner, J. L. (1981) The tribe Proteeae. In *The Pokaryotes: A Handbook on Habitats, Isolation and Identification of Bacteria* (eds M. P. Starr, H. Stolp, H. G. Truper, A. Balows and H. G. Schlegel). Springer-Verlag, Heidelberg, 1204–1224.

Pfaller, M. A., Mujeeb, I., Hollis, R. J. *et al.* (2000) Evaluation of the discriminatory powers of the Dienes test and ribotyping as typing methods for *Proteus mirabilis. Journal of Clinical Microbiology*, **38**, 1077–1080.

Pignato, S., Giammanco, G. M., Grimont, F. *et al.* (1999) Molecular characterization of the genera *Proteus, Morganella*, and *Providencia* by ribotyping. *Journal of Clinical Microbiology*, **37**, 2840–2847.

Rahav, G., Pinco, E., Silbaq, F. *et al.* (1994) Molecular epidemiology of catheter-associated bacteriuria in nursing home patients. *Journal of Clinical Microbiology*, **32**, 1031–1034.

Rozalski, A., Sidorczyk, Z., and Kotelko, K. (1997) Potential virulence factors of Proteus bacilli. *Microbiological and Molecular Biological Reviews*, **61**, 65–89.

Sabbuba, N. A., Mahenthiralingam, E., and Stickler, D. J. (2003) Molecular epidemiology of *Proteus mirabilis* infections of the catheterized urinary tract. *Journal of Clinical Microbiology*, **41**, 4961–4965.

Senior, B. W. (1977) The Dienes phenomenon: identification of the determinants of compatibility. *Journal of General Microbiology*, **102**, 235–244.

Senior, B. W. (1997) Media and tests to simplify the recognition and identification of members of the Proteeae. *Journal of Medical Microbiology*, **46**, 39–44.

Senior, B. W. and Leslie, D. L. (1986) Rare occurrence of *Proteus vulgaris* in faeces: a reason for its rare association with urinary tract infections. *Journal of Medical Microbiology*, **21**, 139–144.

Senior, B. W., Loomes, L. M., and Kerr, M. A. (1991) The production and activity in vivo of *Proteus mirabilis* IgA protease in infections of the urinary tract. *Journal of Medical Microbiology*, **35**, 203–207.

Stickler, D., Morris, N., Moreno, M. C. *et al.* (1998) Studies on the formation of crystalline bacterial biofilms on urethral catheters. *European Journal of Clinical Microbiology and Infectious Disease*, **17**, 649–652.

Sturgill, G. and Rather, P. N. (2004) Evidence that putrescine acts as an extracellular signal required for swarming in *Proteus mirabilis. Molecular Microbiology*, **51**, 437–446.

Thaller, M. C., Berlutti, F., Pantanella, F. *et al.* (1992) Modified MacConkey medium which allows simple and reliable identification of *Providencia stuartii. Journal of Clinical Microbiology*, **30**, 2054–2057.

Tucci, V. and Isenberg, H. D. (1981) Hospital cluster epidemic with *Morganella morganii. Journal of Clinical Microbiology*, **14**, 563–566.

Tumbarello, M., Citton, R., Spanu, T. *et al.* (2004) ESBL-producing multidrug-resistant *Providencia stuartii* infections in a university hospital. *Journal of Antimicrobial Chemotherapy*, **53**, 277–282.

Vieira, A. B., Koh, I. H., and Guth, B. E. (2003) *Providencia alcalifaciens* strains translocate from the gastrointestinal tract and are resistant to lytic activity of serum complement. *Journal of Medical Microbiology*, **52**, 633–636.

Walker, K. E., Moghaddame-Jafari, S., Lockatell, C. V. *et al.* (1999) ZapA, the IgA-degrading metalloprotease of *Proteus mirabilis*, is a virulence factor expressed specifically in swarmer cells. *Molecular Microbiology*, **32**, 825–836.

Warren, J. W. (1986) *Providencia stuartii*: a common cause of antibiotic-resistant bacteriuria in patients with long-term indwelling catheters. *Reviews of Infectious Diseases*, **8**, 61–67.

Warren, J. W., Tenney, J. H., Hoopes, J. M. *et al.* (1982) A prospective microbiologic study of bacteriuria in patients with chronic indwelling urethral catheters. *Journal of Infectious Diseases*, **146**, 719–723.

Williams, E. W., Hawkey, P. M., Penner, J. L. *et al.* (1983) Serious nosocomial infection caused by *Morganella morganii* and *Proteus mirabilis* in a cardiac surgery unit. *Journal of Clinical Microbiology*, **18**, 5–9.

Zunino, P., Sosa, V., Allen, A. G. *et al.* (2003) *Proteus mirabilis* fimbriae (PMF) are important for both bladder and kidney colonization in mice. *Microbiology*, **149**, 3231–3237.

33

Yersinia spp.

M. B. Prentice

Department of Microbiology, University College Cork, Cork, Ireland

INTRODUCTION

The genus *Yersinia* contains three species that are well-established human pathogens. *Yersinia pestis* is responsible for bubonic plague, an often-fulminant systemic zoonosis. *Yersinia pseudotuberculosis* and *Y. enterocolitica* cause yersiniosis, a self-limiting gastrointestinal illness that may have serious complications in special circumstances. As for many other bacterial pathogens, genome-sequencing projects are now revealing the genetic basis for these phenotypic differences.

DESCRIPTION OF THE ORGANISMS

Taxonomy

Yersinia pestis was isolated for the first time by Alexandre Yersin in 1894 in Hong Kong (Yersin, 1894), following the spread of infection from mainland China. *Yersinia pseudotuberculosis* was first reported in 1883 from caseating lesions in rodents and was for many years classified with *Yersinia pestis* in the genus *Pasteurella* (Bercovier and Mollaret, 1984). *Yersinia enterocolitica* was first reported as a *Y. pseudotuberculosis*-like organism isolated from enteric specimens and skin lesions of patients in the United States in 1939 (Schleifstein and Coleman, 1939) and later named *Bacterium enterocoliticum*. In the 1960s European microbiologists identified another heterogeneous group of bacteria resembling but distinct from *Yersinia* (then *Pasteurella*) *pseudotuberculosis* and termed them *Pasteurella X*. In 1964 Frederiksen pointed out similarities between *Pasteurella X* and *Bacterium enterocoliticum* and proposed the new species name *Yersinia enterocolitica* to cover this biochemically disparate group (Frederiksen, 1964). *Yersinia intermedia*, *Y. frederiksenii*, *Y. kristensenii*, *Y. aldovae* *Y. mollaretti* and *Y. bercovieri* have subsequently been split off from *Y. enterocolitica sensu stricto* within the group of *Y. enterocolitica*-like organisms on biochemical and genetic grounds. They are of doubtful pathogenicity and little clinical significance.

Apart from *Y. ruckeri*, a fish pathogen, DNA relatedness among *Yersinia* spp. is at least 40%. The molar G + C ratio of *Yersinia* is 46–50%, consistent with that for Enterobacteriaceae (Bercovier and Mollaret, 1984). Multi-locus sequence typing (MLST) of housekeeping genes suggests that *Y. pestis* is actually a clone derived from *Y. pseudotuberculosis* (Achtman *et al.*, 1999). At least two strains of *Y. pestis* have now been sequenced (Parkhill *et al.*, 2001; Deng *et al.*, 2002), allowing *Y. pestis* genomic array construction. Microarray-based gene complement comparison between *Y. pestis* and *Y. pseudotuberculosis* strains confirms the very close relationship shown on MLST typing (Hinchliffe *et al.*, 2003). Current *Y. pestis* strains form a very homogeneous group that is estimated to have emerged 1500–20 000 years ago (Achtman *et al.*, 1999).

In this short period of evolutionary time, *Y. pestis* has evolved the ability to colonise an insect vector (the flea) and establish a transmission cycle between mammalian hosts by novel subcutaneous and pneumonic routes of infection.

Organism Morphology and Growth

Gram-negative coccobacilli are short rods with rounded ends 1–3 μm long and 0.5–0.8 μm in diameter (Bercovier and Mollaret, 1984: 67). They are non-sporing, and only *Y. pestis* strains (grown *in vivo* or at 37 °C) form a capsule (Brubaker, 1991; Perry and Fetherston, 1997). *Yersinia pestis* in particular shows bipolar staining and pleomorphism. *Yersinia pestis* is non-motile but all other species are motile at 25 °C and non-motile at 37 °C. All species are facultative anaerobes and produce catalase but not oxidase.

All yersiniae grow on nutrient agar forming smaller colonies than other Enterobacteriaceae. All have a growth optimum of less than 37 °C (25–29 °C) and grow over a range of 4–42 °C. *Yersinia pestis* requires 48-h incubation to produce 1 mm colonies at 25–37 °C, whereas the enteropathogenic yersiniae produce visible colonies after 24 h. *Yersinia pestis* colonies are opaque, smooth and round with irregular edges, whereas other *Yersinia* spp. rapidly produce smooth-edged colonies with an elevated centre. Yersiniae are non-haemolytic on blood agar. Although *Y. enterocolitica* grows well on MacConkey agar and other enteric media, *Y. pestis* and *Y. pseudotuberculosis* grow poorly on MacConkey agar on primary isolation (however, laboratory-adapted *Y. pseudotuberculosis* strains may flourish) (Mair and Fox, 1986). Iron is required for the growth of all yersiniae and is important in pathogenicity.

Normal Habitat

Yersinia pestis is primarily a rodent pathogen usually transmitted by the bite of an infected flea and therefore has two main habitats: (i) in the stomach or proventriculus of various flea species at ambient temperature or (ii) in the blood stream or tissues of a rodent host at body temperature (Perry and Fetherston, 1997). There may be an inanimate reservoir in the soil of rodent burrows for some months after their inhabitants have died (Mollaret, 1963), but *Y. pestis* has exacting nutritional requirements (Brubaker, 1991) that make it less well equipped to survive in the environment than the other *Yersinia* spp. *Yersinia pseudotuberculosis* and *Y. enterocolitica* are spread by the enteric route, and their habitat is either in the host's gut or its associated lymphatic tissue, or in the food and water by which the host is

Principles and Practice of Clinical Bacteriology Second Edition   Editors Stephen H. Gillespie and Peter M. Hawkey

infected. *Yersinia enterocolitica* strains lacking virulence determinants and other avirulent *Yersinia* spp. are frequently found in the environment (including human foods) and can be isolated in equal frequency from the stools of asymptomatic and symptomatic humans (Van Noyen *et al.*, 1981). Outbreaks of disease related to specific food or water supplies contaminated with fully virulent *Y. enterocolitica* are relatively rare but well recognised (Cover and Aber, 1989; Lee *et al.*, 1990; Jones, 2003).

PATHOGENICITY

Yersinia pestis must cause disseminated infection with bacteraemia of at least 10^8 bacteria per ml in the mammalian host for transmission by a flea vector (Brubaker, 2003) or the respiratory route. In contrast, *Y. pseudotuberculosis* and *Y. enterocolitica* are more likely to be transmitted if the host remains alive to spread the infection by chronic faecal excretion. The main virulence factors are listed in Table 33.1. All the pathogenic yersiniae share a tropism for lymphoid tissue and an ability to avoid the non-specific immune response of the host (Brubaker, 1991, 2003). Although *Y. pestis* and *Y. pseudotuberculosis* grow well in macrophage cell lines (Pujol and Bliska, 2003), in animal models of disease they are predominantly found in extracellular sites in the later stages of systemic infection.

Factors Unique to *Y. pestis*

Yersinia pestis possesses two plasmids not found in other *Yersinia* spp. that produce virulence factors and other factors required for flea colonisation: the 100-kb pFra plasmid (sometimes called pMT1) (Prentice *et al.*, 2001) and the 9.5-kb pPst (or pPCP1) plasmid (Table 33.1).

pFra encodes a phospholipase D enzyme Ymt (originally called *Yersinia* murine toxin because of its action on some strains of mice), essential for survival of *Y. pestis* in the flea midgut (Hinnebusch

et al., 2002). It also encodes a polypeptide [Fraction 1 (F1-antigen)] which is exported on to the cell surface, forming a fibrillar capsule (Zavialov *et al.*, 2003) that protects against phagocytosis by macrophages (Du, Rosqvist and Forsberg, 2002).

The small pPst plasmid carries the *pla* gene, encoding the Pla surface protease which activates mammalian plasminogen and degrades complement (Sodeinde *et al.*, 1992), as well as adhering to the extracellular matrix component laminin (Lahteenmaki *et al.*, 1998). The localised, uncontrolled proteolysis resulting from these activities promotes disseminated *Y. pestis* infection from the subcutaneous site of the fleabite. Disseminated infection and bacteraemia with a Gram-negative organism carries the risk of causing sudden host death by endotoxic shock, mainly due to cell wall lipid A. Shortening host life is unhelpful for maintaining the plague transmission cycle (Keeling and Gilligan, 2000) and *Y. pestis* remodels its lipid A to be less acylated and less toxigenic at 37 °C (Kawahara *et al.*, 2002), potentially prolonging host survival with a high-grade bacteraemia.

Virulent *Y. pestis* strains possess the property of pigmentation: colonies acquire brown colouration on haemin-containing media by storing haemin in the outer membrane. White non-pigmented *Y. pestis* mutants arise spontaneously on plating a fully virulent strain on haemin media and are of greatly reduced virulence in mice, but virulence is restored in mice that are overloaded with iron or haemin (Jackson and Burrows, 1956). In *Y. pestis*, the pigmented phenotype is genetically linked with other iron metabolism-related virulence factors (Perry and Fetherston, 1997; Buchrieser *et al.*, 1999). The haemin storage locus essential for the pigmentation phenotype is required for the blocking of the flea gut (Hinnebusch, Perry and Schwan, 1996; Darby *et al.*, 2002; Joshua *et al.*, 2003), apparently by the production of a biofilm (Darby *et al.*, 2002; Joshua *et al.*, 2003).

Factors Shared by *Y. pestis* and *Y. pseudotuberculosis*

The pH 6 antigen is a fibrillar protein expressed on the surface of *Y. pestis* grown at pH 6.7 or less at above 37 °C. Its expression is

Table 33.1 Pathogenicity factors of *Yersinia*

Yersinia spp.	Host/vector virulence factor	Gene	Gene location	Action
Y. pestis only	Fraction 1 antigen (Du, Rosqvist and Forsberg, 2002)	*caf1*	100 kb pFra	Inhibits macrophage phagocytosis
	Phospholipase D/murine toxin (Hinnebusch *et al.*, 2002)	*ymt*	100 kb pFra	Promotes *Y. pestis* survival in flea midgut
	Plasminogen activator (Sodeinde *et al.*, 1992)	*pla*	9.5 kb pPst plasmid	Dissemination from subcutaneous injection
Y. pestis and *Y. pseudotuberculosis*	pH 6 antigen (Lindler and Tall, 1993)	*psaA*	Chromosomal	Fibrillar surface protein, binds host lipoproteins
All pathogenic yersiniae	Low calcium response (Cornelis *et al.*, 1998)	*yadA, yopH, lcrV* etc	70 kb Lcr/pYV plasmid	Type III secretion system and adhesin
	Lipopolysaccharide (Darwin and Miller, 1999; Skurnik, Peippo and Ervela, 2000; Karlyshev *et al.*, 2001)	*rfb*	Chromosomal	Endotoxin activity
				Serum resistance (enteropathogenic *Yersinia* only: *Y. pestis* is rough)
	Pigmentation (Hinnebusch, Perry and Schwan, 1996; Darby *et al.*, 2002; Joshua *et al.*, 2003)	*hms* locus	Chromosomal	Biofilm formation, essential for *Y. pestis* flea blocking, present in enteropathogens
Mouse-lethal strains of all three species	Yersiniabactin synthesis (Carniel, Guilvout and Prentice, 1996)	*irp2*	Chromosomal	Siderophore present in all strains of *Yersinia* spp. that are highly virulent for mice (numerous other siderophores in all *Yersinia*)
Enteropathogenic yersiniae only	Inv protein	*inv*	Chromosomal	Invasion of mammalian epithelial cells
	Ail protein (Miller *et al.*, 1989; Pierson and Falkow, 1990) (*inv* inactivated in *Y. pestis*)	*ail*		
Y. enterocolitica only	Myf antigen (Iriarte *et al.*, 1993)	*myfA*	Chromosomal	Homologous to pH 6 antigen i.e. fimbrial structure
	ST toxin (Ramamurthy *et al.*, 1997)	*yst*	Chromosomal	Resembles *E. coli* ST toxin

associated with a more rapidly progressive illness after *Y. pestis* infection in mice (Lindler and Tall, 1993). It may act as a 'stealth cloak', coating the organism with host antigens by binding lipoprotein (Makoveichuk *et al.*, 2003).

Factors Common to all Pathogenic Yersiniae

The pYV (plasmid associated with *Yersinia* virulence) or Lcr (low calcium response) plasmid is a 70-kb plasmid essential for virulence in all the three pathogenic species. It was originally recognised to regulate a complex reaction in which *in vitro* vegetative growth is restricted when yersiniae are grown at 37 °C in low calcium conditions (<2.5 mM Ca^{2+}). At the same time a series of proteins encoded by the virulence plasmid are produced and, in some cases, secreted in quantity. The growth restriction is reversed either by reduction of temperature to 26 °C or by the addition of Ca^{2+}. The proteins produced include Yops (originally an acronym for *Yersinia* outer membrane proteins: they are now known not to be all membrane anchored) and the adhesin YadA.

The plasmid-encoded Yop virulon was one of the first recognised bacterial type III secretion systems, injecting paralysing bacterial proteins into host immune response cells, thus allowing pathogenic *Yersinia* spp. to avoid phagocytosis and remain outside host cells. This phenotype is in contrast to that produced by chromosomally encoded type III secretion systems in *Salmonella*, where the SPI-1 (*Salmonella* pathogenicity island I)-encoded system facilitates bacterial invasion of epithelial cells and the SPI-2-encoded system facilitates the replication of intracellular bacteria within membrane-bound *Salmonella*-containing vacuoles (Waterman and Holden, 2003). There are now known to be chromosomal type III secretion systems in *Yersinia*, but their function is not yet defined (Haller *et al.*, 2000; Parkhill *et al.*, 2001). The virulon consists of the Yop proteins and a type III secretion apparatus, called Ysc.

The Ysc apparatus is composed of 25 proteins (Figures 33.1 and 33.2). There are two groups of Yops (Figure 33.2): intracellular effectors (YopE, YopH, YpkA/YopO, YopP/YopJ, YopM and YopT) and those (YopB, YopD and LcrV) forming the molecular syringe or translocation apparatus to deliver the effectors into the eukaryotic cells, penetrating the plasma membrane (Cornelis *et al.*, 1998) (Figure 33.1). YopE, YopT, YopO and YopH disrupt the host actin cytoskeleton. YopH

was the first bacterial protein recognised to be a tyrosine phosphatase (Guan and Dixon, 1990) and has multiple disruptive effects on macrophage, neutrophil and lymphocyte function (Cornelis *et al.*, 1998; Hamid *et al.*, 1999; Yao *et al.*, 1999). YopP (called YopJ in *Y. pseudotuberculosis* and *Y. pestis*) causes macrophage-specific apoptosis (Monack *et al.*, 1997).

In *Y. pestis*, unlike the enteropathogenic *Yersinia* spp., all freely excreted Yops apart from Lcr V (the cytoplasmic V antigen) are hydrolysed after excretion by the outer membrane plasminogen activator Pla encoded by the Pst plasmid. Also in contrast to the findings in animal models of enteropathogenic *Yersinia* infection, large amounts of LcrV accumulate in animals infected with *Y. pestis* (Brubaker, 2003). LcrV has long been recognised as a protective antigen when included in experimental plague vaccines, and it has recently been found to act as an immunosuppressant, decreasing the production of the proinflammatory cytokines interferon-γ and tumour necrosis factor-α by upregulating host interleukin-10 (Brubaker, 2003). Effector Yops are still delivered via the *Y. pestis* injectisome to paralyse and lyse host phagocytic cells, even with active Pla hydrolysing free Yops, and the overall effect of this combined assault on the non-specific immune system is to allow uncontrolled multiplication of *Y. pestis* in extravascular buboes (Brubaker, 2003).

In vitro regulation by Ca^{2+} was originally thought to mean that Yops would be induced by the calcium-restricted intracellular environment of the host after invasion had occurred, but it is now known that host cell contact in the presence of Ca^{2+} can trigger Yop expression (Pettersson *et al.*, 1996).

YadA forms a fibrillar matrix on the surface of the enteropathogenic yersiniae, making them adherent to a variety of molecules on eukaryotic cells and promoting intestinal colonisation in *Y. enterocolitica* (but not apparently in *Y. pseudotuberculosis*) (Cornelis *et al.*, 1998). It is produced at 37 °C irrespective of the Ca^{2+} concentration. It is the prototype of a family of adhesins seen in a variety of Enterobacteriaceae (Hoiczyk *et al.*, 2000). In *Y. pestis*, the *yadA* gene is inactivated by a point mutation but several other homologues of *yadA* are present on the chromosome (Parkhill *et al.*, 2001).

Yersinia pestis lipopolysaccharide (LPS) is rough (lacks extended O-group chains) (Brubaker, 1991; Skurnik, Peippo and Ervela, 2000). The enteropathogenic yersiniae produce longer O-group side chains corresponding to the different serotypes.

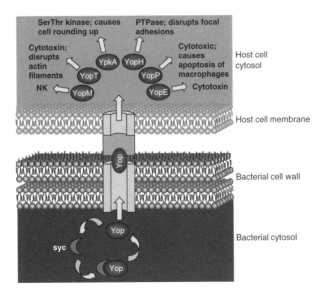

Figure 33.1 Components of the *Yersinia* plasmid-specified type III secretion system. Cartoon copyright Professor R.W. Titball; reproduced with permission.

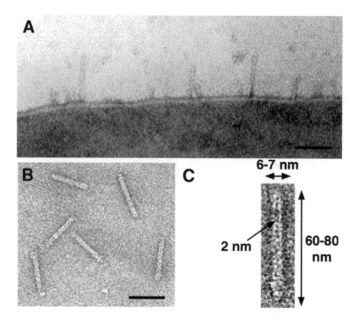

Figure 33.2 The needle-like injectisome of the *Y. enterocolitica* type III secretion system Ysc. Electron microscopy shows needles on the *Y. enterocolitica* cell surface (A) and after cell-free isolation (B). All scale bars are 50 nm. At higher magnification (C), the 2-nm wide hollow centre of the needles is visible. Modified from Hoiczyk E. and Blobel G. (2001) Proc Natl Acad Sci USA **98**(8):4669–74; reproduced with permission. Copyright (2001) National Academy of Sciences, USA.

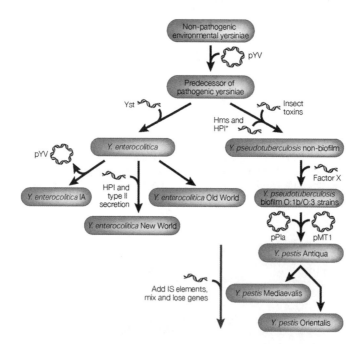

Figure 33.3 Simplified model of *Yersinia* spp. evolution based on the present knowledge of genome data. The non-pathogenic yersiniae gain the virulence plasmid pYV to form the predecessor of pathogenic yersiniae. *Yersinia enterocolitica* diverges from *Y. pseudotuberculosis* and forms three lineages: 1A, Old World and New World. *Yersinia pseudotuberculosis* gains the ability to parasitise insects and form biofilms in hosts before evolving into *Y. pestis* through the acquisition of the plasmids pPla (pPst) and pMT1 (pFra), genome mixing and decay. Hms, haemin storage; HPI and HPI*, high-pathogenicity islands; IS, insertion sequence. Reproduced with permission from Nat Rev Microbiol [Wren B. W. (2003). Nat Rev Microbiol **1**(1):55–64]. Copyright (2003) Macmillan Magazines Ltd.

Factors in Mouse-Lethal Yersiniae Belonging to all Three Species

The siderophore Yersiniabactin is synthesised by genes located on a chromosomal high-pathogenicity island present in *Yersinia* and apparently transmissible between different Enterobacteriaceae (Carniel, Guilvout and Prentice, 1996; Schubert *et al.*, 1998). The term high pathogenicity refers to the fact that island was shown to be present in all strains of *Yersinia* spp. that are highly virulent for mice or humans but absent from strains with lower virulence (de Almeida *et al.*, 1993). Besides Yersiniabactin, there are multiple iron and iron-containing compound uptake systems in *Y. pestis* and other *Yersinia* spp. (Perry and Fetherston, 1997). Enteropathogenic low-virulence *Y. enterocolitica* strains unable to produce yersiniabactin can utilise siderophores produced by other bacteria (e.g. desferrioxamine B) and these substances potentiate their virulence (Perry, 1993).

Factors in Enteropathogenic Yersiniae

The chromosomal genes *inv* (invasion: encodes *Yersinia* invasin) and *ail* (attachment invasion locus) when introduced into non-invasive *Escherichia coli* confer the ability of *Y. enterocolitica* strains to penetrate cultured human epithelial cells. Only pathogenic serotypes of *Y. enterocolitica* produce these two proteins (Miller *et al.*, 1989; Pierson and Falkow, 1990). Invasin specifically binds host cell β_1-chain integrin receptors on the surface of intestinal M cells (Marra and Isberg, 1997). Synthesis of Inv is maximal at 28 °C and that of Ail at 37 °C, suggesting they act at different phases of the infective process. *Yersinia pseudotuberculosis* also produces Inv but Ail is not apparently important for adhesion to mammalian cells in *Y. pseudotuberculosis* (Yang *et al.*, 1996). The *inv* gene is disrupted in *Y. pestis*, but the *ail* gene is apparently intact (Parkhill *et al.*, 2001).

Although, as stated above, *Y. pestis* produced rough LPS, lacking an O-antigen, both *Y. enterocolitica* and *Y. pseudotuberculosis* require O-antigens for pathogenicity (Darwin and Miller, 1999; Karlyshev *et al.*, 2001).

Pathogenic *Y. enterocolitica* strains produce a family of heat-stable enterotoxins which have the same method of action as the heat-stable toxin STa of *E. coli* (Ramamurthy *et al.*, 1997) and a fibrillar adhesin Myf (mucoid *Yersinia* factor) (Iriarte *et al.*, 1993), which resembles ph6 antigen seen in *Y. pseudotuberculosis* and *Y. pestis*. The contribution of these gene products to pathogenicity is controversial.

A possible schema for *Yersinia* evolution, linking these pathogenicity factors with overall genome-sequence data is shown in Figure 33.3.

EPIDEMIOLOGY

Y. pestis

From historical descriptions of disease and current areas of endemicity, *Y. pestis* has been linked with three pandemics (Plate 11). The first began in the reign of the Roman Emperor Justinian in AD 541 (Perry and Fetherston, 1997). The second pandemic corresponds to the Black Death, which killed up to one-third of the European population between 1347 and 1352 (Ziegler, 1982). The third (current) *Y. pestis* pandemic followed the spread of infection from mainland China to Hong Kong in 1894 (Yersin, 1894). Most mortality in this pandemic was seen in India (White, 1918) and China (Teh, 1922) in the late nineteenth and early twentieth century, when millions died. It has been suggested that each pandemic is associated with a different biovar of *Y. pestis*, distinguished by a different *in vitro* phenotype of glycerol fermentation and nitrate reduction (Devignat, 1951).

The aetiology of the Black Death is still controversial, with some authors disputing that *Y. pestis* could have caused this syndrome (Scott and Duncan, 2001; Cohn, 2002). Recently, it has been claimed that ancient DNA evidence from Marseilles, France, proves the association of *Y. pestis* with the Black Death beyond any doubt (Drancourt *et al.*, 1998; Raoult *et al.*, 2000). This evidence has not been reproducible in ancient DNA studies of human remains from other well-documented Black Death burial sites (Gilbert *et al.*, 2004).

Yersinia pestis has been isolated from all continents and plague is enzootic in Africa, North and South America and Asia (Middle East, Far East and countries of the ex-USSR), following the third pandemic (Butler, 1983; Christie and Corbel, 1990; Perry and Fetherston, 1997). Many different combinations of rodent and flea vector species have been described. Classically, efficient flea vectors feeding on an infected host permit the establishment of *Y. pestis* in the proventriculus. Subsequent 'blocking' of the gastrointestinal tract with an overgrowth of *Y. pestis* and regurgitation of infected blood introduces the pathogen into the bloodstream of a new host. Sylvatic plague with a cycle of infection in rodents outside urban areas is the rule. Between 1987 and 2001, 36 876 cases of plague with 2847 deaths were reported to the World Health Organization. Most of the cases were from Africa and

Asia. The United States, which contains a very large enzootic focus, contributed only 125 human cases (12 fatalities) over this period, probably because the areas involved are rural and largely uninhabited. Urban plague requires contact with infected fleas from urban rodents. Worldwide the urban black rat *Rattus rattus* is the most important reservoir and *Xenopsylla cheopis* (the oriental rat flea) the most efficient vector for human disease (Butler, 1983; Christie and Corbel, 1990). Transmission can also rarely occur from human to human by the human flea *Pulex irritans* or by the respiratory route from human to human or domestic animal to human (Doll *et al.*, 1994; World Health Organization, 1999).

Yersinia pseudotuberculosis also causes a zoonosis of wild and domestic animals with human as an incidental host (Ljungberg *et al.*, 1995; Sanford, 1995). Disease incidence is highest in the winter months and results from contact with sick or asymptomatic animals or ingestion of food or water contaminated with their excreta. *Yersinia enterocolitica* exists worldwide in the environment as commensal biotype 1A strains, but the disease-causing serotypes are more frequently identified with a specific host. [There is some evidence that a subset of biotype IA strains are host associated and have some pathogenicity factors that are active *in vitro* (Tennant, Grant and Robins-Browne, 2003).] Serotype O:9 and O:3 strains are the predominant types associated with diarrhoea in Northern Europe, and the disease is more common in autumn/winter (Tauxe *et al.*, 1987; Cover and Aber, 1989; Prentice, Cope and Swann, 1991). Human disease is linked to asymptomatic infection in pigs and consumption of pork products (Tauxe *et al.*, 1987). In Belgium (Verhaegen *et al.*, 1998) and Scandinavia (Cover and Aber, 1989) and New Zealand (Fenwick and McCarthy, 1995) *Y. enterocolitica* is a common cause of diarrhoeal illness. In Germany, 40% of blood donors have antibodies to *Yersinia* Yop proteins (similar in all pathogenic *Yersinia* spp.), despite relatively infrequent isolation of *Y. enterocolitica* and almost no *Y. pseudotuberculosis* cases (Neubauer and Sprague, 2003), suggesting a high level of asymptomatic infection with *Y. enterocolitica*. In the United Kingdom, serological and culture evidence of infection is less frequently seen than in Germany (Prentice, Cope and Swann, 1991). In the United States, sporadic infection is uncommonly reported but outbreaks are well recognised, some also associated with pork products (Lee *et al.*, 1990; Jones, 2003) and O:3 strains are now more commonly isolated than the highly pathogenic O:8 biotype 1B strains which used to predominate (Lee *et al.*, 1990).

CLINICAL FEATURES

Plague

Lymphadenitis in the regional lymph nodes serving the area of the fleabite produces a tense tender swelling or bubo after 2–6 days incubation in cases of bubonic plague. Fever and prostration is usual and progression to septicaemic plague with meningitis or secondary pneumonia can occur. Pneumonic plague may result in transmission of infection by aerosol. Primary septicaemia can occur in the absence of a bubo. Untreated bubonic plague has a mortality of 50–90% and untreated meningitis or septicaemia is usually fatal (Tieh *et al.*, 1948; Butler, 1983).

Yersiniosis

Yersinia pseudotuberculosis causes mesenteric adenitis and chronic diarrhoea in animals, which resolve spontaneously or progress to a fatal septicaemia with widespread deposition of caseating lesions in the lymphatic tissue (Fukushima *et al.*, 1994; Martins, Bauab and Falcao, 1998). Humans develop a mesenteric adenitis that may simulate appendicitis (pseudoappendicitis) (Attwood *et al.*, 1987). With improvements in water supply hygiene, human *Y. pseudotuberculosis* infection has become rarer (Fukushima *et al.*, 1994). Far-eastern scarlatiniform fever is a childhood form of *Y. pseudotuberculosis* infection reported from far-eastern Russia, characterised by a desquamating rash, arthralgia and polyarthritis (Kuznetsov, 1974). A similar childhood syndrome has also been associated with *Y. pseudotuberculosis* in Japan, but not Western countries, possibly reflecting the restriction of superantigen-producing strains of *Y. pseudotuberculosis* to the Far East (Fukushima *et al.*, 2001).

Most *Y. enterocolitica* infections are in children of under 5 years of age (Prentice, Cope and Swann, 1991; Abdel-Haq *et al.*, 2000). Following an incubation period of 1–11 days (Cover and Aber, 1989), *Y. enterocolitica* causes an enteritis [containing frank blood in 25–50% of cases (Marks *et al.*, 1980; Abdel-Haq *et al.*, 2000) and occasionally pseudoappendicitis (as a result of terminal ileitis rather than mesenteric adenitis)]. Infection can occasionally be chronic and relapsing (Hoogkamp-Korstanje, de Koning and Heeseman, 1988). Pseudoappendicitis is more common in young adults. Spontaneous resolution is the norm, but certain groups such as iron-overloaded patients (e.g. thalassaemia, haemochromatosis and iron-supplement overdose), those with liver disease or diabetes and the elderly are at increased risk of a septicaemic illness with a fatal outcome (Cover and Aber, 1989; Prentice, Cope and Swann, 1991). *Yersinia enterocolitica* may be a more common cause of mycotic aneurysm than *Salmonella* spp. in some populations (Prentice *et al.*, 1993). Medical interventions associated with systemic disease include the use of desferrioxamine (iron-chelation therapy in thalassaemia) and blood transfusion. The latter occurs when a blood donor has asymptomatic bacteraemia at the time of donation and depends on the ability of *Y. enterocolitica* to grow at 4 °C in stored blood (Prentice, 1992; Hogman and Engstrand, 1996). Immunological complications are important especially in Northern Europe where HLA-B27 is frequent (Hermann *et al.*, 1993). Reactive arthritis follows several weeks after diarrhoea with other complications such as erythema nodosum and glomerulonephritis (van der Heijden *et al.*, 1997). *Yersinia enterocolitica* antibodies are found in autoimmune thyroid disease patients in these countries but it is not thought that *Y. enterocolitica* is a major factor in triggering thyroid disease (Toivanen and Toivanen, 1994).

LABORATORY DIAGNOSIS

Staining, Culture and Antigen Detection

Plague

In known endemic areas the clinical features described above are well recognised, and a presumptive diagnosis may be made on the appearance of bipolar staining Gram-negative rods from a bubo aspirate, CSF or sputum. Bipolar staining is more prominent with Wright–Giemsa staining than Gram staining. Rapid diagnosis is provided by detection of *Y. pestis* F1-antigen by immunofluorescence in clinical material. A promising rapid antigen-detection test in the form of an F1-antibody–containing dipstick has been shown to be a sensitive and specific assay in field conditions on a variety of clinical specimens (sputum, bubo aspirate and csf) in a developing country (Chanteau *et al.*, 2003). A polymerase chain reaction (PCR) assay based on the *Y. pestis*-specific *pla* gene (Hinnebusch and Schwan, 1993) located on a high copy number plasmid is probably the most successful of several molecular targets tested and has been adapted to a real-time 5′-nuclease assay (Higgins *et al.*, 1998).

Culture of non-contaminated samples (blood and lymph nodes) with routine biochemical identification is required for confirmation of rapid identification methods and sensitivity testing. It should be remembered that *Y. pestis* is potentially dangerous to laboratory workers (a Hazard Group 3 or P3 pathogen), and all samples sent from suspected cases should be labelled as high risk. Regrettably, we must now consider that identification of a patient with *Y. pestis* infection, particularly pulmonary disease, in a non-endemic country with no history of travel to a plague endemic zone suggests deliberate release of the organism (Inglesby *et al.*, 2000). In the UK, the reference laboratory for all *Yersinia* spp. including *Y. pestis* is HPA-colindale Laboratory of Enteric Pathogens (61 Colindale Avenue, London NW9 5HT. Tel.: (+ 44) 020 8200 4400x3173; Fax: (+ 44) 020 8905 9929).

In the United Kingdom, the response to any covert or overt release of *Y. pestis*, as for other bioterrorism pathogens, would be coordinated by the Health Protection Agency and up-to-date information is available from their Web site. A comprehensive Plague Manual (World Health Organization, 1999) can be downloaded from the World Health Organization's Web site. Extensive information on plague diagnosis, treatment and control is also available from the Centers for Disease Control. Strict federal anti-bioterrorism legislation now applies to all laboratories and individuals culturing or holding *Y. pestis* strains in any circumstances in the United States (American Society for Microbiology).

Yersiniosis

Isolation from clinical material free from contamination (blood culture and lymph nodes) is by culture on conventional media (blood agar and nutrient agar) incubated at 25 and 37 °C. Growth occurs in conventional and proprietary blood culture broths, but prolonged incubation may be required for primary isolation. For isolation from faeces, routine enteric media such as MacConkey, desoxycholate or SS agar incubated at 25–30 °C for 48 h can be used. A selective medium specifically for yersiniae, cefsulodin-irgasan-novobiocin (CIN) agar, incubated at 30 °C, is commercially available (Oxoid, Basingstoke, UK). Cold enrichment by incubation in phosphate-buffered saline at 4 °C for 3 weeks is extremely effective at encouraging the overgrowth of yersiniae in mixed culture with other Enterobacteriaceae. This is impractical in a routine diagnostic service and may promote the isolation of commensal yersiniae (Van Noyen *et al.*, 1981). *Yersinia pseudotuberculosis* is rarely isolated from faeces culture even in proven cases of human infection, unless an outbreak has been recognised, and diagnosis is usually serological (Mair and Fox, 1986). Twenty-five different published PCR methods for detection of plasmid or chromosomal *Y. enterocolitica* target genes in clinical specimens or foodstuffs are listed in a recent review (Fredriksson-Ahomaa and Korkeala, 2003). In all comparisons, PCR was more sensitive than culture (Fredriksson-Ahomaa and Korkeala, 2003), and widespread use of PCR may increase the apparent incidence of *Y. enterocolitica*.

Serology

A passive haemagglutination test using *Y. pestis* F1-antigen (unique to *Y. pestis*) is available in countries where plague is endemic (World Health Organization, 1999). Whole-cell agglutination antibody is diagnostic for infection with serogroups I, III and V of *Y. pseudotuberculosis*, but groups II and IV share O-antigen sugars with group B and D salmonellae (Mair and Fox, 1986). High titres may be obtained soon after the onset of symptoms. *Yersinia enterocolitica*-agglutinating antibodies are more likely to be found in convalescence (Mair and Fox, 1986). *Yersinia enterocolitica* O:9 strains share polysaccharide antigens with *Brucella abortus* (Weynants *et al.*, 1996), and cross-reactions between O:5, 27 strains and *E. coli* also occur (Chart, Okubadejo and Rowe, 1992).

Identification

Table 33.2 gives the main reactions distinguishing the species. Other genus-specific characteristics have been mentioned under the general description above. The API-20E kit (bioMérieux, Marcy I' Étiole, France) incubated at 28 °C reliably identifies *Y. enterocolitica* and *Y. pseudotuberculosis* (Sharma *et al.*, 1990; Neubauer *et al.*, 1998). Further biochemical characterisation of *Y. enterocolitica*-like strains into five biovars and different species is possible (Wauters, Kandolo

Table 33.2 Characteristics of *Yersinia* spp.

Test	Y. pestis	Y. pseudotuberculosis	Y. enterocolitica	Y. intermedia	Y. frederiksenii	Y. kristensenii
Motility (37 °C)	–	–	–	–	–	–
Motility (25 °C)	–	+	+	+	+	+
Ornithine decarboxylase	–	–	+	+	+	+
Lysine decarboxylase	–	–	–	–	–	–
Urease	–	+	+	+	+	+
Gelatinase	–	–	–	–	–	–
Voges–Proskauer (25 °C)	–	–	+	+	+	–
Indole	–	–	d	+	+	d
Simmon's citrate	–	–	–	+	d	–
Acid production						
Glucose	+	+	+	+	+	+
Rhamnose	–	+	–	+	+	–
Sucrose	–	–	+	+	+	–
Melibiose	d	+	–	+	–	–
Sorbitol	–	–	+	+	+	+

Biochemical tests incubated at 28 °C for 3 days unless indicated. All catalase positive, oxidase negative. None forms H_2S in TSI medium. +, 90% or more of strains positive; –, 90% or more of strains negative; d, 11–89% strains positive [adapted from Bercovier and Mollaret (1984)].

and Janssens, 1987). Biotyping and serotyping of *Yersinia* spp. is a reference laboratory service, offered in the United Kingdom by the laboratory mentioned above for *Y. pestis* identification. In the clinical laboratory, dealing with fresh isolates, the possession of the pYV-related property of pigmentation and small colony growth on Congo Red-magnesium oxalate agar (CR-MOX) at 37 °C gives useful information about clinical significance (Farmer *et al.*, 1992). As a plasmid-related property, this may be lost on repeated subculture, but simultaneous testing for non-plasmid–derived factors associated with virulent serotypes (such as the absence of pyrazinamidase and the inability to ferment salicin or aesculin) overcomes this (Farmer *et al.*, 1992). The main serotypes associated with disease in Europe are O:9, O:3 or O:5, 27. These are of low virulence with little capacity to cause systemic disease outside the risk groups outlined above. Biotype 1B and serotype O:8 strains mainly found in North America have the capacity to cause severe disease in non-iron–overloaded hosts.

ANTIMICROBIAL SENSITIVITIES

Yersinia pestis clinical isolates are generally sensitive *in vitro* to most agents active against Gram-negative bacteria. Antibiotics active *in vitro* include streptomycin, ciprofloxacin, chloramphenicol, tetracycline, ampicillin, cotrimoxazole and cephalothin. A strain resistant to multiple antimicrobials was first reported from Madagascar in 1997 (Galimand *et al.*, 1997), and subsequently a different strain resistant to the first-line antibiotic streptomycin was also identified (Guiyoule *et al.*, 2001). Worryingly, both plasmids responsible for these resistance patterns were self-transferrable to other bacteria.

A similar generally sensitive spectrum is seen for *Y. pseudotuberculosis* (Soriano and Vega, 1982), but *Y. enterocolitica* strains express one or two chromosomal β-lactamases: a non-inducible broad-spectrum β-lactamase (enzyme A) and an inducible cephalosporinase (known as enzyme B, strictly speaking a class C enzyme) (Pham *et al.*, 2000). Despite the presence of an inducible cephalosporinase in *Y. enterocolitica*, the newer third-generation cephalosporins, as well as quinolone and penam antimicrobials, are active *in vitro* against all *Yersinia* species (Soriano and Vega, 1982; Pham *et al.*, 2000) including *Y. pestis* (Smith *et al.*, 1995; Frean *et al.*, 1996).

MANAGEMENT OF INFECTION

Streptomycin is traditionally regarded as the most effective treatment for plague at a dosage of 1 g twice daily (30 mg/kg/per day) for 10 days (Butler, 1983). Retrospective case review suggests the more readily available aminoglycoside gentamicin is as effective as streptomycin in the treatment of human plague, when given at standard doses for severe sepsis (Boulanger *et al.*, 2004). Oral chloramphenicol is recommended for plague meningitis. In a mouse septicaemia model, third-generation cephalosporins and quinolones were as effective as streptomycin and tetracycline (Bonacorsi *et al.*, 1994). However in a mouse model of pneumonic plague, β-lactam antibiotics were less effective than aminoglycosides and quinolones (Byrne *et al.*, 1998). A consensus view of treatment for pneumonic plague resulting from biological weapon attack suggests streptomycin, gentamicin, tetracycline or fluoroquinolones may be effective (Inglesby *et al.*, 2000).

Yersiniosis is usually a self-limiting condition but for chronic or relapsing illness with *Y. enterocolitica*, focal disease outside the gastrointestinal tract or septicaemia, cotrimoxazole or ciprofloxacin has been recommended (Hoogkamp-Korstanje, 1987; Gayraud *et al.*, 1993; Crowe, Ashford and Ispahani, 1996). First-generation cephalosporins and amoxicillin/clavulanate are not effective as sole therapy in septicaemia (Gayraud *et al.*, 1993) and fluoroquinolones are probably the treatment of choice. *Yersinia pseudotuberculosis* is treated with similar agents to those effective against *Y. enterocolitica*, and as β-lactamases are absent, ampicillin can also be used (Van Noyen *et al.*, 1995; Martins, Bauab and Falcao, 1998).

PREVENTION AND CONTROL

The control of plague is based on reducing the likelihood of people being bitten by infected fleas or exposed to aerosols from humans or animals with plague pneumonia. A detailed summary of the international and patient-centred measures in force to prevent the spread of plague from and within enzootic areas is given in the WHO Plague Manual mentioned above. Bubonic plague is an internationally quarantinable disease. Current vaccines are restricted in application by local and systemic reactions and short duration of protective immunity. They are indicated only for those working with live *Y. pestis* in laboratories, or likely to be in contact with rats and fleas in endemic areas (Butler, 1983; Titball and Williamson, 2001). A new subunit vaccine is in clinical trials (Titball and Williamson, 2001). Person to person spread of yersiniosis has been described but is rarely demonstrated. Most documented outbreaks have been food or water borne (Tauxe *et al.*, 1987; Cover and Aber, 1989; Prentice, Cope and Swann, 1991). Control is based on the avoidance of consumption of contaminated food and water. The food industry has to be particularly alert to avoid contamination of products whose safety is assured by refrigeration with psychrotrophic yersiniae able to grow at 4 °C.

REFERENCES

Abdel-Haq, N. M., Asmar, B. I., Abuhammour, W. M. and Brown, W. J. (2000) *Yersinia enterocolitica* infection in children. Pediatr Infect Dis J **19**:954–8.

Achtman, M., Zurth, K., Morelli, G. *et al.* (1999) *Yersinia pestis*, the cause of plague, is a recently emerged clone of *Yersinia pseudotuberculosis*. Proc Natl Acad Sci USA **96**:14043–8.

de Almeida, A. M. P., Guiyoule, A., Guilvout, I. *et al.* (1993) Chromosomal *irp2* gene in *Yersinia*: distribution, expression, deletion and impact on virulence. Microb Pathog **14**:9–21.

American Society for Microbiology. Biological Weapons Resources Center. www.asm.org/Policy/index.asp?bid = 520 (9 April 2004).

Attwood, S. E., Mealy, K., Cafferkey, M. T. *et al.* (1987) *Yersinia* infection and acute abdominal pain. Lancet **1**:529–33.

Bercovier, H. and Mollaret, H. H. (1984) Genus XIV. Yersinia, p. 498–506. *In* N. R. Krieg (ed.), Bergey's Manual of Systematic Bacteriology, Vol. 1. Williams & Wilkins, Baltimore.

Bonacorsi, S. P., Scavizzi, M. R., Guiyoule, A. *et al.* (1994) Assessment of a fluoroquinolone, 3 beta-Lactams, 2 aminoglycosides, and a cycline in treatment of murine *Yersinia pestis* infection. Antimicrobial Agents Chemother **38**:481–6.

Boulanger, L. L., Ettestad, P., Fogarty, J. D. *et al.* (2004) Gentamicin and tetracyclines for the treatment of human plague: review of 75 cases in new Mexico, 1985–99. Clin Infect Dis **38**:663–9.

Brubaker, R. R. (1991) Factors promoting acute and chronic diseases caused by yersiniae. Clin Microbiol Rev **4**:309–24.

Brubaker, R. R. (2003) Interleukin-10 and inhibition of innate immunity to Yersiniae: roles of Yops and LcrV (V antigen). Infect Immun **71**:3673–81.

Buchrieser, C., Rusniok, C., Frangeul, L. *et al.* (1999) The 102-kilobase pgm locus of *Yersinia pestis*: sequence analysis and comparison of selected regions among different *Yersinia pestis* and *Yersinia pseudotuberculosis* strains. Infect Immun **67**:4851–61.

Butler, T. (1983) Plague and Other *Yersinia* infections. Plenum Press, New York.

Byrne, W. R., Welkos, S. L., Pitt, M. L., Davis, K. J. *et al.* (1998) Antibiotic treatment of experimental pneumonic plague in mice. Antimicrobial Agents Chemother **42**:675–81.

Carniel, E., Guilvout, I. and Prentice, M. (1996) Characterization of a large chromosomal 'high-pathogenicity island' in biotype 1B *Yersinia enterocolitica*. J Bacteriol **178**:6743–51.

Centers for Disease Control. Emergency preparedness & response. www.bt.cdc.gov/agent/plague/index.asp (9 April 2004).

Chanteau, S., Rahalison, L., Ralafiarisoa, L. *et al.* (2003) Development and testing of a rapid diagnostic test for bubonic and pneumonic plague. Lancet **361**:211–6.

Chart, H., Okubadejo, O. A. and Rowe, B. (1992) The serological relationship between *Escherichia coli* O157 and *Yersinia enterocolitica* O9 using sera from patients with brucellosis. Epidemiol Infect **108**:77–85.

Christie, A. B. and Corbel, M. J. (1990) Plague and other yersinial diseases, p. 399–421. *In* G. R. Smith and C. S. F. Easmon (eds), Bacterial Diseases, 8th edn, Vol. 3. Edward Arnold, London.

Cohn, S. (2002) The Black Death Transformed: Disease and Culture in Early Renaissance Europe. Arnold, London.

Cornelis, G. R., Boland, A., Boyd, A. P. *et al.* (1998) The virulence plasmid of Yersinia, an antihost genome. Microbiol Mol Biol Rev **62**:1315–52.

Cover, T. L. and Aber, R. C. (1989) *Yersinia enterocolitica*. N Engl J Med **321**:16–24.

Crowe, M., Ashford, K. and Ispahani, P. (1996) Clinical features and antibiotic treatment of septic arthritis and osteomyelitis due to *Yersinia enterocolitica*. J Med Microbiol **45**:302–9.

Darby, C., Hsu, J. W., Ghori, N. and Falkow, S. (2002) *Caenorhabditis elegans*: plague bacteria biofilm blocks food intake. Nature **417**:243–4.

Darwin, A. J. and Miller, V. L. (1999) Identification of *Yersinia enterocolitica* genes affecting survival in an animal host using signature-tagged transposon mutagenesis. Mol Microbiol **32**:51–62.

Deng, W., Burland, V., Plunkett, G. III, *et al.* (2002) Genome sequence of *Yersinia pestis* KIM. J Bacteriol **184**:4601–11.

Devignat, R. (1951) Variétés de l'espèce *Pasteurella pestis*. Nouvelle hypothèse. Bull OMS **4**:247–63.

Doll, J. M., Zeitz, P. S., Ettestad, P. *et al.* (1994) Cat-transmitted fatal pneumonic plague in a person who traveled from Colorado to Arizona. Am J Trop Med Hyg **51**:109–14.

Drancourt, M., Aboudharam, G., Signoli, M. *et al.* (1998) Detection of 400-year-old *Yersinia pestis* DNA in human dental pulp: an approach to the diagnosis of ancient septicemia. Proc Natl Acad Sci USA **95**:12637–40.

Du, Y., Rosqvist, R. and Forsberg, A. (2002) Role of fraction 1 antigen of *Yersinia pestis* in inhibition of phagocytosis. Infect Immun **70**:1453–60.

Farmer, J. J., Carter, G. P., Miller, V. L. *et al.* (1992) Pyrazinamidase, CR-MOX agar, salicin fermentation-esculin hydrolysis, and d-xylose fermentation for identifying pathogenic serotypes of *Yersinia enterocolitica*. J Clin Microbiol **30**:2589–94.

Fenwick, S. G. and McCarthy, M. D. (1995) *Yersinia enterocolitica* is a common cause of gastroenteritis in Auckland. NZ Med J **108**:269–71.

Frean, J. A., Arntzen, L., Capper, T. *et al.* (1996) In vitro activities of 14 antibiotics against 100 human isolates of *Yersinia pestis* from a southern African plague focus. Antimicrobial Agents Chemother **40**:2646–7.

Frederiksen, W. (1964) Presented at the Proceedings of the 14th Scandinavian Congress of Pathology and Microbiology, Oslo.

Fredriksson-Ahomaa, M. and Korkeala, H. (2003) Low occurrence of pathogenic *Yersinia enterocolitica* in clinical, food, and environmental samples: a methodological problem. Clin Microbiol Rev **16**:220–9.

Fukushima, H., Gomyoda, M., Kaneko, S. *et al.* (1994) Restriction endonuclease analysis of virulence plasmids for molecular epidemiology of *Yersinia pseudotuberculosis* infections. J Clin Microbiol **32**:1410–3.

Fukushima, H., Matsuda, Y., Seki, R. *et al.* (2001) Geographical heterogeneity between Far Eastern and Western countries in prevalence of the virulence plasmid, the superantigen *Yersinia pseudotuberculosis*-derived mitogen, and the high-pathogenicity island among *Yersinia pseudotuberculosis* strains. J Clin Microbiol **39**:3541–7.

Galimand, M., Guiyoule, A., Gerbaud, G. *et al.* (1997) Multidrug resistance in *Yersinia pestis* mediated by a transferable plasmid. N Engl J Med **337**:677–80.

Gayraud, M., Scavizzi, M. R., Mollaret, H. H. *et al.* (1993) Antibiotic treatment of *Yersinia enterocolitica* septicemia: a retrospective review of 43 cases. Clin Infect Dis **17**:405–10.

Gilbert, M. T., Cuccui, J., White, W. *et al.* (2004) Absence of *Yersinia pestis*-specific DNA in human teeth from five European excavations of putative plague victims. Microbiology **150**:341–54.

Guan, K. L. and Dixon, J. E. (1990) Protein tyrosine phosphatase activity of an essential virulence determinant in *Yersinia*. Science **249**:553–6.

Guiyoule, A., Gerbaud, G., Buchrieser, C. *et al.* (2001) Transferable plasmid-mediated resistance to streptomycin in a clinical isolate of *Yersinia pestis*. Emerg Infect Dis **7**:43–8.

Haller, J. C., Carlson, S., Pederson, K. J. and Pierson, D. E. (2000) A chromosomally encoded type III secretion pathway in *Yersinia enterocolitica* is important in virulence. Mol Microbiol **36**:1436–46.

Hamid, N., Gustavsson, A., Andersson, K. *et al.* (1999) YopH dephosphorylates Cas and Fyn-binding protein in macrophages. Microb Pathog **27**:231–42.

Health Protection Agency, UK. Deliberate release homepage. www.hpa.org.uk/infections/topics_az/deliberate_release/menu.htm (9 April 2004).

Hermann, E., Yu, D. T., Meyer zum Buschenfelde, K. H. and Fleischer, B. (1993) HLA-B27-restricted CD8 T cells derived from synovial fluids of patients with reactive arthritis and ankylosing spondylitis. Lancet **342**:646–50.

Higgins, J. A., Ezzell, J., Hinnebusch, B. J. *et al.* (1998) 5′ nuclease PCR assay to detect *Yersinia pestis*. J Clin Microbiol **36**:2284–8.

Hinchliffe, S. J., Isherwood, K. E., Stabler, R. A. *et al.* (2003) Application of DNA microarrays to study the evolutionary genomics of *Yersinia pestis* and *Yersinia pseudotuberculosis*. Genome Res **13**:2018–29.

Hinnebusch, J. and Schwan, T. G. (1993) New method for plague surveillance using polymerase chain reaction to detect *Yersinia pestis* in fleas. J Clin Microbiol **31**:1511–4.

Hinnebusch, B., Perry, R. and Schwan, T. (1996) Role of the *Yersinia pestis* hemin storage (*hms*) locus in the transmission of plague by fleas. Science **273**:367–70.

Hinnebusch, B. J., Rudolph, A. E., Cherepanov, P., Dixon, J. E., Schwan, T. G. and Forsberg, A. (2002) Role of *Yersinia* murine toxin in survival of *Yersinia pestis* in the midgut of the flea vector. Science **296**:733–5.

Hogman, C. F. and Engstrand, L. (1996) Factors affecting growth of *Yersinia enterocolitica* in cellular blood products. Transfus Med Rev **10**:259–75.

Hoiczyk, E., Roggenkamp, A., Reichenbecher, M. *et al.* (2000) Structure and sequence analysis of *Yersinia* YadA and *Moraxella* UspAs reveal a novel class of adhesins. EMBO J **19**:5989–99.

Hoogkamp-Korstanje, J. A. A. (1987) Antibiotics in *Yersinia enterocolitica* infections. J Antimicrob Chemother **20**:123–31.

Hoogkamp-Korstanje, J. A. A., de Koning, J. and Heeseman, J. (1988) Persistence of *Yersinia enterocolitica* in man. Infection **16**:81–5.

Inglesby, T. V., Dennis, D. T., Henderson, D. A. *et al.* (2000) Plague as a biological weapon: medical and public health management. Working Group on Civilian Biodefense. JAMA **283**:2281–90.

Iriarte, M., Vanooteghem, J. C., Delor, I. *et al.* (1993) The Myf fibrillae of *Yersinia enterocolitica*. Mol Microbiol **9**:507–20.

Jackson, S. and Burrows, T. (1956) The virulence enhancing effect of iron on non-pigmented mutants of virulent strains of *Pasteurella pestis*. Br J Exp Pathol **37**:577–83.

Jones, T. F. (2003) From pig to pacifier: chitterling-associated yersiniosis outbreak among black infants. Emerg Infect Dis **9**:1007–9.

Joshua, G. W., Karlyshev, A. V., Smith, M. P. *et al.* (2003) A *Caenorhabditis elegans* model of *Yersinia* infection: biofilm formation on a biotic surface. Microbiology **149**:3221–9.

Karlyshev, A. V., Oyston, P. C., Williams, K. *et al.* (2001) Application of high-density array-based signature-tagged mutagenesis to discover novel *Yersinia* virulence-associated genes. Infect Immun **69**:7810–9.

Kawahara, K., Tsukano, H., Watanabe, H. *et al.* (2002) Modification of the structure and activity of lipid A in *Yersinia pestis* lipopolysaccharide by growth temperature. Infect Immun 70:4092–8.

Keeling, M. J. and Gilligan, C. A. (2000) Metapopulation dynamics of bubonic plague. Nature **407**:903–6.

Kuznetsov, V. G. (1974) (Clinical laboratory and geographical epidemiological parallels in Far East scarlatiniform fever and Izumi fever) Russian language text. Voen Med Zh **11**:45–9.

Lahteenmaki, K., Virkola, R., Saren, A. *et al.* (1998) Expression of plasminogen activator pla of *Yersinia pestis* enhances bacterial attachment to the mammalian extracellular matrix. Infect Immun **66**:5755–62.

Lee, L. A., Gerber, A. R., Lonsway, D. R. *et al.* (1990) *Yersinia enterocolitica* O:3 infections in infants and children, associated with the household preparation of chitterlings. N Engl J Med **322**:984–7.

Lindler, L. E. and Tall, B. E. (1993) *Yersinia pestis* pH 6 antigen forms fimbriae and is induced by intracellular association with macrophages. Mol Microbiol **8**:311–24.

Ljungberg, P., Valtonen, M., Harjola, V. P. *et al.* (1995) Report of four cases of *Yersinia pseudotuberculosis* septicemia and a literature review. Eur J Clin Microbiol Infect Dis **14**:804–10.

Mair, N. and Fox, E. (1986) *Yersiniosis*. Public Health Laboratory Service, London.

Makoveichuk, E., Cherepanov, P., Lundberg, S. *et al.* (2003) pH6 antigen of *Yersinia pestis* interacts with plasma lipoproteins and cell membranes. J Lipid Res **44**:320–30.

Marks, M. I., Pai, C. H., Lafleur, L. *et al.* (1980) *Yersinia enterocolitica* gastroenteritis: a prospective study of clinical, bacteriologic, and epidemiologic features. J Pediatr **96**:26–31.

Marra, A. and Isberg, R. R. (1997) Invasin-dependent and invasin-independent pathways for translocation of *Yersinia pseudotuberculosis* across the Peyer's patch intestinal epithelium. Infect Immun **65**:3412–21.

Martins, C. H., Bauab, T. M. and Falcao, D. P. (1998) Characteristics of *Yersinia pseudotuberculosis* isolated from animals in Brazil. J Appl Microbiol **85**:703–7.

Miller, V. L., Farmer, J. J. III, Hill, W. E. and Falkow, S. (1989) The ail locus is found uniquely in *Yersinia enterocolitica* serotypes commonly associated with disease. Infect Immun **57**:121–31.

Mollaret, H. H. (1963) [Experimental preservation of plague in soil]. Bull Soc Pathol Exot Filiales **56**:1168–82.

Monack, D. M., Mecsas, J., Ghori, N. and Falkow, S. (1997) *Yersinia* signals macrophages to undergo apoptosis and YopJ is necessary for this cell death. Proc Natl Acad Sci USA **94**:10385–90.

Neubauer, H. K. and Sprague, L. D. (2003) Epidemiology and diagnostics of Yersinia-infections. Adv Exp Med Biol **529**:431–8.

Neubauer, H., Sauer, T., Becker, H. *et al.* (1998) Comparison of systems for identification and differentiation of species within the genus *Yersinia*. J Clin Microbiol **36**:3366–8.

Parkhill, J., Wren, B. W., Thomson, N. R. *et al.* (2001) Genome sequence of *Yersinia pestis*, the causative agent of plague. Nature **413**:523–7.

Perry, R. D. (1993) Acquisition and storage of inorganic iron and hemin by the yersiniae. Trends Microbiol **1**:142–7.

Perry, R. D. and Fetherston, J. D. (1997) *Yersinia pestis* – etiologic agent of plague. Clin Microbiol Rev **10**:35–66.

Pettersson, J., Nordfelth, R., Bergman, T. *et al.* (1996) Modulation of virulence factor expression by pathogen target cell contact. Science **273**:1231–3.

Pham, J. N., Bell, S. M., Martin, L. and Carniel, E. (2000) The beta-lactamases and beta-lactam antibiotic susceptibility of *Yersinia enterocolitica*. J Antimicrob Chemother **46**:951–7.

Pierson, D. E. and Falkow, S. (1990) Nonpathogenic isolates of *Yersinia enterocolitica* do not contain functional inv-homologous sequences. Infect Immun **58**:1059–64.

Prentice, M. B. (1992) *Yersinia enterocolitica* and blood transfusion. Br Med J **305**:663–4.

Prentice, M. B., Cope, D. and Swann, R. A. (1991) The epidemiology of *Yersinia enterocolitica* infection in the British Isles 1983–1988. Contrib Microbiol Immunol **12**:17–25.

Prentice, M. B., Fortineau, N., Lambert, T. *et al.* (1993) *Yersinia enterocolitica* and mycotic aneurysm. Lancet **341**:1535–6.

Prentice, M. B., James, K. D., Parkhill, J. *et al.* (2001) *Yersinia pestis* pFra shows biovar-specific differences and recent common ancestry with a *Salmonella enterica* serovar typhi plasmid. J Bacteriol **183**:2586–94.

Pujol, C. and Bliska, J. B. (2003) The ability to replicate in macrophages is conserved between *Yersinia pestis* and *Yersinia pseudotuberculosis*. Infect Immun **71**:5892–9.

Ramamurthy, T., Yoshino, K., Huang, X. *et al.* (1997) The novel heat-stable enterotoxin subtype gene (*ystB*) of *Yersinia enterocolitica*: nucleotide sequence and distribution of the *yst* genes. Microb Pathog **23**:189–200.

Raoult, D., Aboudharam, G., Crubezy, E. *et al.* (2000) Molecular identification by 'suicide PCR' of *Yersinia pestis* as the agent of medieval black death. Proc Natl Acad Sci USA **97**:12800–3.

Sanford, S. E. (1995) Outbreaks of yersiniosis caused by *Yersinia pseudotuberculosis* in farmed cervids. J Vet Diagn Invest **7**:78–81.

Schleifstein, J. I. and Coleman, M. B. (1939) An unidentified microorganism resembling *B. lignieri* and *Past. pseudotuberculosis*, and pathogenic for man. N Y State J Med **39**:1749–53.

Schubert, S., Rakin, A., Karch, H. *et al.* (1998) Prevalence of the 'high-pathogenicity island' of Yersinia species among *Escherichia coli* strains that are pathogenic to humans. Infect Immun **66**:480–5.

Scott, S. and Duncan, C. (2001) Biology of Plagues: Evidence from Historical Populations. Cambridge University Press, Cambridge.

Sharma, N. K., Doyle, P. W., Gerbasi, S. A. and Jessop, J. H. (1990) Identification of *Yersinia* species by the API 20E. J Clin Microbiol **28**:1443–4.

Skurnik, M., Peippo, A. and Ervela, E. (2000) Characterization of the O-antigen gene clusters of *Yersinia pseudotuberculosis* and the cryptic O-antigen gene cluster of *Yersinia pestis* shows that the plague bacillus is most closely related to and has evolved from *Y. pseudotuberculosis* serotype O:1b. Mol Microbiol **37**:316–30.

Smith, M. D., Vinh, D. X., Nguyen, T. T. *et al.* (1995) In vitro antimicrobial susceptibilities of strains of *Yersinia pestis*. Antimicrobial Agents Chemother **39**:2153–4.

Sodeinde, O. A., Subrahmanyam, Y. V., Stark, K. *et al.* (1992) A surface protease and the invasive character of plague. Science **258**:1004–7.

Soriano, F. and Vega, J. (1982) The susceptibility of *Yersinia* to eleven antimicrobials. J Antimicrobial Chemother **10**:543–7.

Tauxe, R. V., Vandepitte, J., Wauters, G. *et al.* (1987) *Yersinia enterocolitica* infections and pork: the missing link. Lancet **1**:1129–32.

Teh, W. L. (1992) plague in the orient with special reference to the Manchurian outbreaks. J Hyg (Cambridge) **21**:62–76.

Tennant, S. M., Grant, T. H. and Robins-Browne, R. M. (2003) Pathogenicity of *Yersinia enterocolitica* biotype 1A. FEMS Immunol Med Microbiol **38**:127–37.

Tieh, T., Landauer, E., Miyagawa, F. *et al.* (1948) Primary pneumonic plague in Mukden, 1946, and report of 39 cases with three recoveries. J Infect Dis **82**:52–8.

Titball, R. W. and Williamson, E. D. (2001) Vaccination against bubonic and pneumonic plague. Vaccine **19**:4175–84.

Toivanen, P. and Toivanen, A. (1994) Does *Yersinia* induce autoimmunity? Int Arch Allergy Immunol **104**:107–11.

van der Heijden, I. M., Res, P. C., Wilbrink, B., Leow, A., Breedveld, F. C., Heesemann, J. and Tak, P. P. (1997) *Yersinia enterocolitica*: a cause of chronic polyarthritis. Clin Infect Dis **25**:831–7.

Van Noyen, R., Selderslaghs, R., Bogaerts, A. *et al.* (1995) *Yersinia pseudotuberculosis* in stool from patients in a regional Belgian hospital. Contrib Microbiol Immunol **13**:19–24.

Van Noyen, R., Vandepitte, J., Wauters, G. and Selderslaghs, R. (1981) *Yersinia enterocolitica*: its isolation by cold enrichment from patients and healthy subjects. J Clin Pathol **34**:1052–6.

Verhaegen, J., Charlier, J., Lemmens, P. *et al.* (1998) Surveillance of human *Yersinia enterocolitica* infections in Belgium: 1967–1996. Clin Infect Dis **27**:59–64.

Waterman, S. R. and Holden, D. W. (2003) Functions and effectors of the *Salmonella* pathogenicity island 2 type III secretion system. Cell Microbiol **5**:501–11.

Wauters, G., Kandolo, K. and Janssens, K. (1987) Revised biogrouping scheme of *Yersinia enterocolitica*. Contrib Microbiol Immunol **9**:14–21.

Weynants, V., Tibor, A., Denoel, P. A. *et al.* (1996) Infection of cattle with *Yersinia enterocolitica* O:9 a cause of the false-positive serological reactions in bovine brucellosis diagnostic tests. Vet Microbiol **48**:101–12.

White, F. (1918) Twenty years of plague in India, with special reference to the outbreak of 1917–18. Indian J Med Res 6:190.

World Health Organization (1999) Plague manual: epidemiology, distribution, surveillance and control. www.who.int/csr/resources/publications/plague/ WHO_CDS_CSR_EDC_99_2_EN/en/ (9 April 2004).

World Health Organization (2003) Human plague in 2000 and 2001. Wkly Epidemiol Rec **78**:130–5.

Wu, L.-T. (1922) Plague in the orient with special reference to the Manchurian outbreaks. J Hyg (Cambridge) **26**:62–76.

Yang, Y. X., Merriam, J. J., Mueller, J. P. and Isberg, R. R. (1996) The psa locus is responsible for thermoinducible binding of *Yersinia pseudotuberculosis* to cultured cells. Infect Immun **64**:2483–9.

Yao, T., Mecsas, J., Healy, J. I., Falkow, S. and Chien, Y. (1999) Suppression of T and B lymphocyte activation by a *Yersinia pseudotuberculosis* virulence factor, *yopH*. J Exp Med **190**:1343–50.

Yersin, A. (1894) La peste bubonique à Hong Kong. Ann Inst Pasteur **8**:662–7.

Zavialov, A. V., Berglund, J., Pudney, A. F. *et al.* (2003) Structure and biogenesis of the capsular F1 antigen from *Yersinia pestis*: preserved folding energy drives fiber formation. Cell **113**:587–96.

Ziegler, P. (1982) The Black Death. Penguin, London.

34

Vibrio spp.

Tom Cheasty

Laboratory of Enteric Pathogens, Centre for Infection, Colindale Avenue, London, UK

INTRODUCTION

A number of major texts have been written about cholera and cholera vibrios, and this chapter provides an outline for the clinical microbiologist (Pollitzer, 1959; Barua and Burrows, 1974; Colwell, 1984; Barua and Greenough, 1992; Wachsmuth, Blake and Olsvik, 1994). The genus *Vibrio* (Pacini, 1854) is among the oldest still recognized by bacteriology. Taxonomy was reviewed by Baumann, Furniss and Lee (1984) who listed 20 species, of which 12 had been isolated from clinical material. More recently, Farmer and Hickman-Brenner (1992) listed 34 species. However, of these only 12 had been isolated from clinical material. These studies illustrate the way in which our knowledge of the genus *Vibrio* has advanced. Until recently, *Vibrio* consisted of a heterogeneous and poorly described group of organisms. However, a stricter definition of the genus, coupled with the transfer of some members to other well-understood groups, has allowed the taxonomy to be clarified. Some, such as the organisms now known as *Campylobacter*, have been transferred to their own genera.

Vibrios are fermentative, Gram-negative bacteria that are oxidase positive. They are part of a widespread group of organisms, many of which are common in the environment. In clinical bacteriology we focus our attention mainly upon *Vibrio cholerae*, the cause of cholera, and *V. parahaemolyticus*, a major cause of food poisoning. However, it is important to remember that they are not the only vibrios that can cause human disease and indeed there are very many other vibrios that are not associated with human disease.

The genus as currently understood is comparatively well characterized and consists of oxidase positive fermentative organisms. All members are facultative anaerobes and grow well on artificial culture media. It is usual to speak of these organisms as belonging to two groups: those isolated from clinical specimens and the marine vibrios. However, this distinction is by no means clear.

All vibrios grow better with the addition of sodium chloride to culture media. An absolute requirement for sodium is not restricted to vibrios of marine origin. However, many of those isolated from clinical material are thought to originate from marine sources. Of the 12 vibrios isolated from clinical specimens, most have also been associated with marine environments. They have been isolated from food poisoning-associated consumption of seafoods and from wound infections resulting from contamination with a variety of marine materials (Table 34.1).

DESCRIPTION OF THE ORGANISM

The Family Vibrionaceae

The genus *Vibrio* is part of the family Vibrionaceae. The family also includes the genera *Photobacterium*, *Aeromonas* and *Plesiomonas*.

Table 34.1 Examples of disease syndromes associated with vibrios

Species	Enteritis	Wound infection	Ear infection	Septicaemia
V. cholerae O1, O139	+			
V. cholerae non-O1	+	+	+	+
V. mimicus	+		+	
V. fluvialis	+			
V. parahaemolyticus	+			+
V. vulnificus		+		+
V. damsela		+		+
V. metschnikovii		+		
V. alginolyticus		+	+	+

Members of the family are usually oxidase positive and have polar flagella for motility. Peritrichous flagella may occasionally be produced.

The organisms are rigid Gram-negative rods, motile by polar flagella in liquid media, but some species produce peritrichous flagella on solid media. They are chemoorganotrophs; metabolism is both oxidative and fermentative. They utilize glucose fermentatively and are catalase positive; most species are oxidase positive and reduce nitrate to nitrite. They are facultative anaerobes but do not have exacting nutritional requirements.

It is not certain that this family is a natural grouping in an evolutionary sense. It arose out of the wish to separate these organisms from the Enterobacteriaceae. DNA homology studies suggest *Vibrio* and *Photobacterium* may be closely related. However, *Aeromonas* is more distantly related and may in fact be closer to *Pasteurella multocida* and the Enterobacteriaceae than to vibrios, and Colwell, Macdonell and De Ley (1986) proposed the removal of *Aeromonas* from the family Vibrionaceae to a new family Aeromonadaceae. It is only in recent years that the definition of the genus *Vibrio* has become more delineated.

The Genus *Vibrio*

Historically, taxonomically and medically the genus *Vibrio* focuses on the organisms that cause cholera. These are largely members of the species *V. cholerae*; other species may occasionally cause clinically similar illness. Advances in molecular biology, immunology and bacterial taxonomy mean that *V. cholerae* is now among the best understood of bacterial species. The taxonomy of the organisms, the virulence mechanisms by means of which it causes disease and the means of treating apparent cholera have all been thoroughly investigated (see pathogenesis of cholera).

Principles and Practice of Clinical Bacteriology Second Edition   Editors Stephen H. Gillespie and Peter M. Hawkey

Genus Definition

The organisms are short asporogenous rods, with axis curved or straight, $0.5 \times 3\,\mu m$. They are single or occasionally united in S shapes or spirals. They are motile by a single polar flagellum which is sheathed in certain if not all species. Some species produce lateral (peritrichous) flagella on solid media. Carbohydrates are fermented with the production of acid and no gas. A wide range of extracellular enzymes is produced, including amylase, chitinase, DNAase, gelatinase, lecithinase and lipase. They grow on simple mineral media. NaCl stimulates growth with an optimum of 1–3%. Some strains require NaCl for growth. Temperature optima range from 18 to 37 °C and pH range 6.0–9.0. They are sensitive to O/129 (2,4-diamino-6, 7-diisopropylpterdine). G + C content of DNA is 40–50 mol%.

Non-Cholera Vibrios

The non-cholera Vibrios are found in aquatic environments worldwide and tend to be found more frequently in association with the warmer coastal waters and water temperatures of 18 °C or more (Kaspar and Tamplin, 1993). They are readily found in and on a wide variety of seafood including shellfish, particularly oysters. Only a small number of the many species apart from *V. cholerae* are associated with human disease (Figure 34.1), and of these the most significant are *V. parahaemolyticus*, *V. mimicus* and *V. vulnificus*. Because these bacteria are natural inhabitants of the aquatic environment, they create a continuous threat to the health of humans and the safety of food.

Vibrio parahaemolyticus

Vibrio parahaemolyticus is a well-known cause of diarrhoeal disease, symptoms include diarrhoea (which can be bloody), abdominal cramps, nausea and fever. Hughes *et al.* (1978) reported an outbreak of 'dysentery-like' disease following a fish meal in Bangladesh. Also in susceptible hosts *V. parahaemolyticus* can also cause severe

wound, soft-tissue and septicaemic infections, and Daniels *et al.* (2000) reported a fatality rate of 29% in septicaemia cases in the USA.

Vibrio vulnificus

Vibrio vulnificus comprises three biotypes, with biotype 1 (indole and ornithine decarboxylase positive) the most common found in clinical and environmental isolates and biotype 2 (indole and ornithine decarboxylase negative) a pathogen only for eels. *Vibrio vulnificus* is predominantly associated with causing severe wound infections, septicaemia and occasionally diarrhoea. Most infections are associated with the consumption of raw shellfish and in particular oysters (Strom and Paranjpye, 2000). In the USA *V. vulnificus* is considered to be the most common cause of serious vibrio infection with a reported fatality rate of more than 50% in cases of primary septicaemia (Klontz *et al.*, 1998).

Antibiotics used for the treatment of extraintestinal infections with both *V. vulnificus* and *V. parahaemolyticus* include tetracyclines and the newer fluoroquinolones. Rehydration is the key to diarrhoeal episodes caused by these vibrios.

For the laboratory isolation of these non-cholera vibrios, from cases of extraintestinal infection blood agar or other non-selective media are adequate whereas for the isolation from faeces, TCBS is the agar of choice.

Other Non-cholera Vibrios

The remaining non-cholera vibrios including *V. alginolyticus*, *V. damsela*, *V. fluvialis* and *V. metschnikovii* are mainly considered to be opportunistic pathogens. Infections caused by these species although small in number are commonly associated with contamination of wounds, incidents of trauma and ear infections following exposure to the aquatic environment (West, 1989; Hornstrup and Gahrn-Hansen, 1993). *Vibrio fluvialis* has also been reported as an occasional cause of gastroenteritis.

PATHOGENESIS OF CHOLERA

Explanations of the pathogenesis of infection usually focus on interaction of the organism with its host. In this context, cholera is a comparatively simple disease. *Vibrio cholerae* enters the intestine and attaches to the mucosal surface. When attached, it produces cholera toxin (CT), which attaches to the intestinal mucosal cells. An important function of mucosal cells is the control of ion transport; under normal conditions the net flow of ions is from the lumen to the tissue, which results in a net uptake of water. CT disrupts ion fluxes without causing apparent damage to the mucosa (Gangarosa *et al.*, 1960). CT decreases the net flow of sodium into the tissue. This produces a net flow of chloride and water out of the tissue into the lumen, causing the massive diarrhoea and electrolyte imbalance that is associated with this infection.

What a bacterium can do to its host is only part of its life as a pathogen. An organism must be transmitted between hosts and establish itself in new hosts. The ability of an organism to survive in the external environment and to adapt to the internal environment provided by the host is an important aspect of its pathogenic armamentarium. When considered in this way, the pathogenesis of cholera and the ecology of *V. cholerae* are much more complex.

Transmission

Until the 1970s it was believed that the ecology of *V. cholerae* was restricted to its role as a human pathogen. That is to say, if infection of the human host could be eliminated the major ecological niche of *V. cholerae* would have been eliminated and the organism would no longer be able to survive. It is now clear that the human host represents only a small compartment of the ecology of the vibrio. Thus, even if

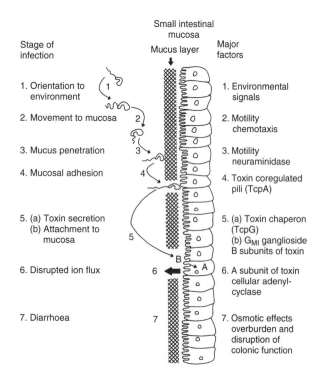

Figure 34.1 The pathogenesis of cholera

the human infection were eliminated, this would not materially influence the survival of the species. It is now clear that *V. cholerae*, including strains belonging to serogroups O1 and O139, is a member of the autochthonous flora of brackish waters. It is able to persist and grow in a variety of aquatic environments in association with a variety of aquatic plants, zooplankton and blue-green algae independently of infected humans. In some circumstances, it may be a member of the biofilm flora found on submerged surfaces. This ability to colonize many different bodies of water and to associate with different members of microbiota undoubtedly contributes to the patterns of endemic cholera. This environmental survival was originally investigated in the search for an explanation for seasonal patterns of cholera in endemic areas. However, it has much wider implications than this (Colwell, 1984; West, 1989; Drasar, 1992; Pascual *et al.*, 2000). In addition to water, contaminated food also plays an important part in the transmission of *V. cholerae*.

Infective Dose

When the interaction of a pathogen with a host is considered, this is often described as though it were a unitary process, i.e. that a bacterium and a person come into contact and this results in disease. The situation is much more complicated than this. For *V. cholerae* we are fortunate in having the results of volunteer studies in which the numbers of bacteria required to initiate infection in healthy adults were examined. The numbers of bacteria involved in the initiation of infection under these conditions were much larger than might have been expected for a major epidemic disease (Table 34.2) (Cash *et al.*, 1974; Levine *et al.*, 1982, 1984; Suntharasarnai *et al.*, 1992). These results can be interpreted in several ways. It may be that they indicate the scale of the importance of host factors in initiating infection. Thus, for example, neutralization of gastric acid by bicarbonate decreases the numbers of bacteria required to initiate infection from 10^8 to 10^5. Bacterial factors may also be important. The numbers of organisms required may reflect the size of bacterial population needed to ensure that the cells responsive to environmental signals are available. Thus, it may be that only a small fraction of laboratory-grown cells are immediately responsive to the environmental signals transmitted in the intestinal environment.

This would mean that not all cells were equally able to initiate the transcriptional regulation of those genes required for expression of virulence. It is also important to realize that these volunteer studies do not reflect the situation during an epidemic. Studies of epidemics caused by Classical and El Tor biotypes of *V. cholerae* suggest that in an epidemic most people are exposed to the organism. However, although many become temporary carriers, few develop clinical disease. The case–carrier ratio is 1 : 50 for the Classical biotype and 1:90 for the El Tor biotype. It is not certain that these differences are significant – they may simply represent the different methodologies used in the studies.

Table 34.2 Human challenge studies with virulent *V. cholerae* inoculum 10^5 bacteria Stomach acid neutralized

	Classical biotype		El Tor biotype	
	Inaba 569B	Ogawa 395	Inaba N16961	Ogawa E7946
No response	4	3	0	1
Positive stool culture	6	0	3	2
Positive stool culture diarrhoea	28	12	25	6
Positive stool culture severe diarrhoea (cholera)	14	10	10	1
Total number of subjects	52	25	38	10

Virulence Factors

Since the description of CT and the demonstration that the small intestinal mucosa remains intact in cholera, there has been a major resurgence of interest in the virulence factors of cholera vibrios. Prior to these discoveries, it was believed that the destruction of mucus by mucinase was the main virulence factor and that this caused destruction of the intestinal mucosa. A number of virulence factors have been described. Not all are firmly established as factors important in the disease. Some have been demonstrated in experimental systems, others have been invoked to explain the side effects caused by genetically manipulated candidate cholera vaccine strains in some early studies. A general schema for the initiation of clinical infection is set out in Figure 34.1. All strains with the ability to cause cholera possess a set of virulence genes required for pathogenesis in humans. These genes are found in clusters and include genes for CT, the toxin-coregulated pilus (TCP), a colonization factor, and ToxR, a regulatory protein that coregulates the expression of both CT and TCP. These virulence genes or their homologues have also been found in environmental strains comprising a variety of serogroups and may thus provide an important environmental virulence gene pool (Chakraborty *et al.*, 2000).

Cholera Toxin

CT is the most important of the virulence factors produced by *V. cholerae*. CT is encoded by a lysogenic, filamentous bacteriophage CTXΦ which is found as a prophage in the bacterial chromosome. Typically the CTXΦ genome has two functionally distinct regions (i) the core which is 4.6-kb and encodes CT as well as other virion structural gene-encoded proteins; and (ii) the 2.5-kb RS2 region that encodes the regulation, replication and integration functions of the CTXΦ genome. Strains of *V. cholerae* that do not produce CT do not produce disease in human volunteers (Levine *et al.*, 1982). Such strains may cause a milder form of diarrhoea because of the presence of other toxins (Cash *et al.*, 1974; Levine *et al.*, 1988). This has proved a particular problem with candidate genetically engineered vaccine strains.

CT consists of a number of subunits: one A or enzymatic subunit and five identical B or binding subunits. The molecular weight of the A subunit is 27 kDa and that of each of the B subunits is 11.7 kDa. The genes that code for the toxin subunits are part of the same operon. These genes are part of a coordinated regulated system that also controls the expression of TCP and other virulence factors (DiRita *et al.*, 1991). The expression of these virulence factors is modulated by environmental influences. Environmental regulation of this system seems to be mediated by the toxR protein which spans the cytoplasmic membrane (DiRita, 1992). The system is further complicated by the fact that toxR is itself environmentally regulated and its expression at 37 °C may be downregulated in favour of the heat-shock response (Figure 34.2) (Parsot and Mekalanos, 1990).

Although the details of the regulatory system are not fully understood, it is certain that when *V. cholerae* colonize the intestine, the toxin is produced. It seems that the A and B subunits are produced separately and then assembled in the bacterial cytoplasm. TcpG is a chaperon, which is involved in the assembly of toxin-coregulated pili and may be involved in this process. Following assembly, the CT is transported to the periplasmic space and excreted by the bacterium. This is in contrast to the situation with heat-labile toxin of *Escherichia coli*, cells of which seem to have to lyse to release the toxin. When released into the intestine, CT binds to mucosal cells via the host Gm, ganglioside. The B subunits bind to the ganglioside and form a pseudoporin in the cell membrane, thus enabling the A subunit to enter the cell. Prior to insertion into the cell, the A subunit is nicked probably in the small intestinal environment after release by the bacterium. This nicking converts the A subunit into two smaller proteins, A1 and A2. It is the A1 subunit that enters the cells via the pore formed by the B subunits. The A subunit ADP ribosolates a membrane protein called G.

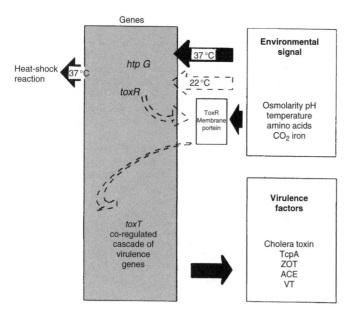

Figure 34.2 Environmental control of virulence in *V. cholerae*.

This G protein regulates the activity of the host cell adenyl cyclase and thus determines the level of cyclic AMP in the host cells. This brings about an uncontrollable rise in cyclic AMP levels. This has a variety of effects, but in this context the most important is to alter the activities of sodium and chloride membrane transport systems (DiRita, 1992). These changes result in a disruption of the biodirectional ion fluxes in the mucosal membrane. The overall effect of this is to increase the relative concentrations of sodium and chloride ions in the lumen. Diarrhoea then results as a direct effect of the osmotic imbalance. It is the osmotic imbalance which prevents, or at least much reduces, the active absorption of fluid that causes the diarrhoea.

CT may not be produced in large enough amounts to produce the diarrhoea in all individuals. The efflux of fluids must be sufficient to overcome the absorptive capacities of the colon. It is important to remember that the result of CT is a physiological lesion. Cells remain normal and are capable of carrying out normal functions, including absorption of salts and water. It is merely that the membrane fluxes from the cell to lumen are much exaggerated. It is for this reason that rehydration together with a source of energy can correct the physiological disturbance by amplifying and reinforcing the ability of the membrane to absorb and reabsorb fluid.

Toxin-Coregulated Pili

Attachment or adhesion to the intestinal mucosa is an essential part of the virulence mechanism of intestinal pathogens. *Vibrio cholerae* produces a variety of haemagglutinins and pili which had been postulated to be important in virulence (Freter and Jones, 1976; Jones and Freter, 1976). But these do not occur uniformly in all virulent strains. Toxin-coregulated pili were not demonstrated in spite of many investigations, because of the conditions under which test organisms were grown. Only when strains grown on solid media were investigated, were pili obvious. Since their discovery, it has been demonstrated that their production is regulated by the same cascade mechanism as is important for the production of toxin (Kaufman *et al.*, 1993; Pearson *et al.*, 1993). Thus, virulent *V. cholerae* produce TCP, a type IV pilus, which is considered to be the most important colonization factor because it has been demonstrated to be necessary for intestinal colonization in both animal models and human challenge studies, and a

toxin that is important for the regulation of ion fluxes at the same time, when the correct environmental conditions trigger the transcription of the virulence genes (Parsot and Mekalanos, 1990; DiRita, 1992). The genes for TCP are a component of the vibrio pathogenicity island, which is distinct from the CTXΦ genome. Biogenesis of TCP requires the products of a number of genes (the *tcp* gene cluster) located adjacent to the *tcpA* gene on the chromosome. The *tcpA* gene encodes for the major subunit of TCP (Ogierman *et al.*, 1993; Kovach, Shaffer and Peterson, 1996). How toxin-coregulated pili bring about attachment of *V. cholerae* to the mucosal cells is still not fully understood. Kirn *et al.* (2000) have proposed a multistep model in which TCP-mediated bacterial interactions promote microcolony formation, resulting in the local accumulation of bacteria on the epithelial surface. Most research has concentrated on the study of the regulation of the genes responsible for their production. Nevertheless, their importance in virulence has been established in volunteer studies. The strains that do not produce toxin-coregulated pili are avirulent in human volunteers (Herrington *et al.*, 1988).

Other Virulence Factors

Vibrio cholerae produces a number of other toxins in addition to CT. The existence of these toxins was first recognized when human volunteer studies were undertaken using mutagenized and biochemically engineered strains of *V. cholerae* that did not produce CT. Initial studies with live or vaccine strains showed that these strains that did not produce entire CT still produced symptoms. A number of toxins have been described, including the zona occludens toxin (ZOT) (Fasano *et al.*, 1991; Baudry *et al.*, 1992). This toxin disrupts the tight junctions between mucosal cells. In effect, this increases the surface area of the mucosa available for ion and water loss. Such losses normally occur through the brush border, but the action of this toxin exposes the sides of the cells and this may increase the efficiency of the virulence. Another enterotoxin, or accessory cholera enterotoxin (ACE), has also been described. Both ZOT and Ace proteins have been shown to have other cellular functions; Ace is associated with phage packaging and secretion whilst ZOT is needed for CTXΦ assembly (Waldor *et al.*, 1996).

Vibrio cholerae has also been reported to produce verocell toxin, which is analogous to the toxin produced by *Shigella dysenteriae*1. *Vibrio cholerae* also produces haemolysin. The role of these additional toxins in the pathogenesis of cholera is not well understood. The variety of toxins produced by *V. cholerae* probably reflects its undoubted ability to produce a number of extracellular enzymes. These may well be important for the bacteria in non-intestinal environments such as aquatic systems, as part of the biofilm communities and in association with aquatic microflora and fauna. The production of extracellular enzymes and toxins may help the organism to digest a variety of substances and to increase the range of nutrients available to it.

In 2000 Heidelberg *et al.* published the entire genome sequence of *V. cholerae* O1 biotype El Tor (N16961). The genome was shown to consist of two circular chromosomes, one large (2 961 146 bp) and one small (1 072 314 bp). Most of the genes for growth and pathogenicity are located on the large chromosome, whilst on the small chromosome are located components of essential regulatory and metabolic pathways. The role(s) of these two chromosomes remains an enigma. It is possible that large chromosome genes (chrI) adapts the organism for growth in the human intestine, whereas the small chromosome (chrII) genes are essential for survival in environmental niches (Schoolnik and Yildiz, 2000).

Vibrio cholerae Non-O1, Non-O139 (Non-cholera Vibrios, Non-O1 *V. cholerae*)

Some of these strains also appear to produce CT together with ZOT (Karasawa *et al.*, 1993). Many others appear to produce cytolysins that are not cholera like (McCardell, Madden and Shah, 1985). A few produce a toxin similar to the heat-stable toxin of *E. coli* and some produce a shigella-like toxin (O'Brien *et al.*, 1984). Others produce no detectable enterotoxin at all and are probably non-pathogenic. Much research is still needed.

V. mimicus

These organisms were originally considered to be sucrose-negative *V. cholerae*. Some strains produce CT. Others produce an enterotoxin similar to the heat-stable toxin produced by non-O1 *V. cholerae* and haemolysins similar to that produced by *V. parahaemolyticus* (Ramamurthy *et al.*, 1994).

V. parahaemolyticus

The factors necessary for pathogenesis have not yet been established. However, shigella-like toxin is produced (O'Brien *et al.*, 1984). An enterotoxin has been purified, but it is not certain that this is responsible for the symptoms of *V. parahaemolyticus* food poisoning. The haemolysin is also thought to act as a virulence factor. Studies of this have revealed a regulatory system analogous to the toxR system in *V. cholerae* (Lin *et al.*, 1993).

EPIDEMIOLOGY OF CHOLERA

Cholera is a pandemic disease par excellence. Although it has undoubtedly been common in India since records began, and has spread to neighbouring countries on frequent occasions, the great days of cholera came after the resumption of world trade following the Napoleonic wars. Whether widespread cholera had occurred previously is not known. Writings of Europeans in India had from the earliest days of this intercourse made reference to this disease. Epidemics undoubtedly occurred among Europeans especially European troops stationed in India (Orton, 1831; MacPherson, 1872; MacNamara, 1876). Until 1817, spread was probably limited but in that year the first pandemic started and reached Europe. This was rapidly followed by the second pandemic, which reached England through the port of Sunderland in 1831. It is notable that Snow, who later became famous for the elucidation of the water-borne transmission of cholera (Snow, 1849, 1855), was working as a doctor's apprentice in Newcastle at about this time. Early nineteenth-century Europe was ravaged by a series of epidemics of cholera. We do not know for certain what caused these epidemics, although it is assumed to be *V. cholerae*. It was not until the fifth pandemic that Koch (1884) succeeded in isolating and identifying the organism. Even when the organism had been isolated, there were many who did not believe that it was the cause of the disease, though few went so far as Pettenkofer von (1892) in their wish to demonstrate that *V. cholerae* was non-pathogenic. He drank a culture and did not get cholera. As we now know, this probably reflects the relative resistance of some people to cholera vibrios. Comparable studies in healthy volunteers have not always been successful in inducing disease. Pandemics of cholera occur worldwide; we are now in the seventh pandemic, and with the emergence of *V. cholerae* O139 have been waiting to see if we are about to see the beginning of the eighth pandemic (Table 34.3).

However, for most practical purposes it is the local epidemics in endemic areas that are of the greatest practical importance. Historically it is likely that endemic cholera has been restricted to the Ganges delta. However, in the recent pandemic – the seventh pandemic – many other endemic areas have been detected. How an area becomes an endemic focus is unclear; the answers probably relate to aspects of the ecology of *V. cholerae* that we do not fully understand. In endemic areas successive, usually seasonally related, epidemics of cholera occur.

Historically, cholera has been considered to be a water-borne disease (Feachem, 1981, 1982; Feachem, Miller and Drasar, 1981).

Table 34.3 The chronology of cholera

At home	To 1817	Endemic on the Indian Subcontinent	
The historic pandemics	1817–1823	First pandemic	John Snow and the Broad Street Pump (1854)
	1829–1850	Second pandemic	
	1852–1860	Third pandemic	
	1863–1879	Fourth pandemic	
The classical pandemics	1881–1896	Fifth pandemic	Robert Koch and the comma bacillus (1884)
	1899–1923	Sixth pandemic	
The El Tor pandemic	1961 to date	Seventh pandemic	De and Chatterje cholera toxin (1953)
O139 Bengal	1992 to date		

There is considerable evidence that water-borne transmission of cholera is very important. However, in an epidemic situation, food-borne transmission is also very significant. *Vibrio cholerae* has often been isolated from seafood and shellfish, and the vibrios are able to grow well in cooked rice and on other foods. However, external contamination of foods may be equally important. In the recent outbreak in Chile, cholera was transmitted by vegetables irrigated with untreated sewage. Similar transmission related to melons was seen in Peru (Crowcroft, 1994). Thus, both food and water can act as vehicles for transmission (Glass *et al.*, 1991; Risks of Transmission of Cholera by Food, 1991). Poor sanitation is undoubtedly important for maintaining transmission during epidemics. This probably explains why secondary transmission is rare when cases are imported into countries with good sanitation. Investigations of other outbreaks in an epidemic situation may demonstrate what vehicle is important.

Major questions remain about the epidemiology of cholera. Two questions are of particular importance:

1. How does cholera survive in the inter-epidemic periods in endemic areas? That is, why are there cholera seasons?

There is linked to this a subsidiary question:

1a. How do new endemic areas develop?
2. How do pandemics arise?

Recent studies on the survival of *V. cholerae* have provided some clues to both of these questions. However, neither can be regarded as settled. *Vibrio cholerae* is now accepted as being a normal member of the autochthonous flora of natural aqueous habitats. A variety of members of the aquatic flora and fauna have been suggested as likely to produce the right habitat. These include the plants, crustaceans, amoebae and various forms of phytoplankton (Thorn, Warhurst and Drasar, 1992; Islam, Drasar and Sack, 1993, 1994). These studies assume that the vibrios persist as active members of a biofilm macrobiota. It is likely that the occurrence of starved and viable, but not culturable, forms (VBNC) is also important. These studies are particularly pertinent when considering the maintenance of endemic foci and the development of new endemic foci. If we postulate that in endemic areas vibrios persist as part of the normal flora of a number of bodies of water it can be readily seen how local epidemics will arise. Seasonal variation can be explained in terms of variations in the aquatic microbiota, e.g. algal blooms linked to an increase in the numbers of *V. cholerae* in water. Further, colonization of lakes and watercourses in non-endemic areas may result in the development of new endemic areas and the range of areas in which cholera is endemic. Endemic foci associated with colonized water sources probably occur in Bangladesh (Tamplin *et al.*, 1990), Louisiana (Blake *et al.*, 1980) in parts of Africa (Glass *et al.*, 1991) and perhaps in Australia (Desmarchelier and Reichelt, 1981). It is possible that during the current South American outbreak, new endemic foci have been established.

It has long been assumed that pandemics result from the spread of strains of *V. cholerae* from a single source. The Ganges delta has been considered to be the ancestral home of vibrios, and it has long been believed that pandemics spread from this area. Under this theory, a new, highly virulent strain of *V. cholerae* would develop in an endemic focus and then spread widely. However, application of ribotyping systems to cholera vibrios isolated during the seventh pandemic casts doubts on this dogma.

A variety of ribotypes have been demonstrated, each in part restricted to a particular area (Koblavi, Grimont and Grimont, 1990). Thus, different types of *V. cholerae* El Tor occur on the Indian subcontinent, Africa and in South America. It has been suggested that this diversity of types represents the acquisition of virulence factors by *V. cholerae* normally resident in aquatic environments in the different areas. Studies on isolates from the epidemic caused by *V. cholerae* O139 suggest a single type is responsible for all the cases so far studied.

The reasons for the mechanisms of pandemic spread remain obscure. In the most recent pandemics rapid travel of persons by air and the movement of ballast water by ships can be seen as a means of moving local microbiotas around the world and possibly enhancing the range for colonization of *V. cholerae*. However, in past pandemics this did not occur. Carriers in cholera, in the long-term sense that they occur in typhoid, are very rare and, thus, spread by humans, until the advent of rapid air transport does not seem to be an adequate explanation. Although the epidemiology of cholera has been studied since before the advent of microbiology and the pioneering studies of John Snow established many of the principles of epidemiology, our increased knowledge of the survival of the vibrio, its pathogenicity and the advent of molecular typing methods have not made our understanding that much clearer. For example, cholera had not been seen in South America for nearly 100 years, and it was assumed for some reason it would not occur there and perhaps the population was not susceptible. The recent extension of the seventh pandemic to South America has made us realize that there is no area of the world in which cholera cannot become established. The advent of the new strain of *V. cholerae* O139 has added a further question: how is and by what mechanism, do new pandemic occur during the history of cholera? Of the pandemics for which a cause is known, three were caused by different organisms.

CLINICAL FEATURES OF CHOLERA

The presentation of cholera is diverse, with many asymptomatic cases and others with severe symptoms. The incubation period varies from >1 day to 5 days, symptoms are frequently sudden and the infection can be rapidly progressive. Patients may lose as much as 100% of their body weight in diarrhoea over 3 to 4 days, bringing about death in a few hours if not adequately treated, but a less fulminant course is usual. All of the symptoms arise as a result of the dehydration and electrolyte disturbance consequent on the loss of water and salts into the gut lumen. Diarrhoea is severe, but abdominal pain is usually absent. Vomiting is often present in the early stages of cholera. Rarely ileus may occur on presentation and is associated with shock, and this may mimic acute intestinal obstruction. The state of consciousness is altered but the patient can be roused to lucidity. There is generalized weakness and cardiac arrhythmias may arise.

Other species are associated with enteritis and wound or ear infection after contamination arising in marine environments. In some cases septicaemia may result. Examples of species associated with these syndromes are given in Table 34.1.

LABORATORY DIAGNOSIS

Collection and Transport of Specimens

Many vibrios have been isolated from samples routinely submitted to clinical laboratories. When a suspected outbreak of cholera is being investigated, such samples may not be sufficient. The drying and acidification of specimens during their transport to the laboratory may prove to be a problem. Wherever possible samples should be dealt with a minimum delay. Alkaline peptone water used for the selective enrichment of vibrios can also be used for transport. This can act as an enrichment transport system. An alternative enrichment transport system using Tellurite taurocholate peptone broth has been used successfully in Bangladesh. Of the routinely available transport media, Cary and Blair is the most satisfactory and may maintain the viability of vibrios for up to 4 weeks. In the absence of a suitable transport system blotting paper can be impregnated with liquid stool and packed in plastic bags. If these do not dry out, the vibrios will remain viable for about 4 weeks.

A rapid presumptive diagnosis can be achieved by mixing a drop of liquid faeces with a drop of specific antiserum. In patients with cholera the characteristic darting movement of the vibrios is inhibited.

Isolation

Vibrios can be isolated on normal laboratory media. Specialist media have been devised for the isolation of *V. cholerae*, and these may be applicable to the isolation of other organisms. There are doubts about the cost-effectiveness of employing specialized media for the isolation of vibrios, unless cholera is suspected. When a specialized medium, such as thiosulphate citrate bile salts sucrose agar (TCBS), has been employed in parallel with other media in routine clinical practice vibrios have been isolated, when present, on a variety of media. TCBS is extremely useful for isolating *V. cholerae*. But it should be remembered that some vibrio species do not grow well on TCBS. These include *V. metschnikovii*, *V. hollisae* and *V. cincinnatiensis*. Growth of *V. damsela* may also be reduced. The results obtained with TCBS produced by different manufacturers may be different, and each batch should be quality controlled to ensure the isolation characteristics are understood.

Isolation Methods

A simple schema for the isolation of *V. cholerae* is set out in Table 34.4. Although a number of vibrios have been isolated from clinical material, not all are of equal importance. In the routine laboratory, it is important that the system in place will ensure that *V. cholerae* is recognized if it occurs.

Different isolation techniques are necessary for halophilic and for non-halophilic vibrios. Enrichment methods designed principally for the isolation of *V. parahaemolyticus*, for example, are not ideal for the non-halophilic vibrios. If cholera vibrios are being looked for, the alkaline peptone water should not contain salt. This reduces overgrowth by other halophilic vibrios. It is also important to keep the pH alkaline: at a lower pH *Aeromonas* may so overgrow the vibrios that they cannot be isolated. This is particularly so with freshwater samples. Only one species of *Vibrio* is likely to be significant in any one clinical specimen. A method that permits isolation of both non-halophilic (*V. cholerae*) and halophilic (*V. parahaemolyticus*) vibrios is thus needed.

Table 34.4 Detection of *V. cholerae* in the laboratory

Day 0	Streak inoculate selective (TCBS) and non-selective media
	Inoculate alkaline peptone water
Incubate	
Day 1	Suspicious colonies – > agglutination with *V. cholerae* O1 antiserum
	Subculture for purity
	Alkaline peptone water
	Streak inoculate to selective and non-selective media incubate
Day 2	Confirm agglutination results with pure culture; set up identification tests
	Test suspicious colonies isolated from alkaline peptone water

If the presence of a vibrio is suspected in patients with loose motions, about 5–10 ml of faecal material should be collected in a disposable container. The specimen should reach the laboratory as soon as possible. In the laboratory, inoculate about 2 g of faeces into 20 ml alkaline peptone water (peptone 10 g, NaCl 10 g, distilled water 1 l, pH 8.6) and incubate for 5–8 h at 37 °C. This will enrich for halophilic vibrios. For *V. cholerae* alkaline peptone water without NaCl is used. At the same time, TCBS is streak inoculated with a heavy load of faecal material.

After 5–8 h incubation from the first alkaline peptone water inoculate, a new alkaline peptone water preparation with 1 ml of fluid from the top of the alkaline peptone water and a second TCBS plate with a loopful from the top and incubate overnight. Look for typical colonies on both TCBS plates (yellow or green colonies about 2 mm in diameter). Inoculate a third TCBS plate from the alkaline peptone water. If no colonies occur on this plate after overnight incubation the sample is probably negative. This schema can be adapted for other specimens.

Identification

Suspicious colonies are streak inoculated onto nutrient agar as a purity check. Vibrios may be identified by the tests summarized in Table 34.5. A small number of basic tests are needed to identify at the genus level.

Oxidase

Colonies grown on nutrient agar should be tested using filter paper impregnated with 1% tetramethyl *p*-phenylene-diamine dihydrochloride. The oxidase test on growth from a plate containing a carbohydrate that is fermented by the organism can give a false negative.

0/129 Sensitivity

This is available as the phosphate (2,4-diamino-6,7-diisopropylpteridine phosphate) from BDH, product number 44169. Disks are prepared containing 10 and 150 Mg and the test performed as for a disk diffusion sensitivity method on nutrient agar – not specialized sensitivity testing agars.

CLED

Cystine-lactose-electrolyte-deficient medium was designed to prevent swarming of proteus. Being electrolyte deficient it is useful for distinguishing the non-halophilic vibrios (e.g. *V. cholerae*), which can grow on it, from the halophilic ones (e.g. *V. parahaemolyticus*), which cannot.

O/F Oxidation/Fermentation (tests)

Media for these tests should have the sodium chloride concentration raised to 1%, which will support good growth of all the pathogenic *Vibrio* spp.

Vibrios can be identified using commercial identification systems such as API 20E or standard laboratory tests when standard peptone water sugars are used. Results are more reliable if these are supplemented by 1% NaCl. Indeed, 1% NaCl is essential as a growth factor for most of the clinically important vibrios. Exceptions to this are *V. cholerae* and *V. mimicus*. O'Hara *et al.* (2003) evaluated six commercially available systems for the identification of the family Vibrionaceae and reported that none of the six systems tested were able to identify isolates to the species level with an accuracy of more than 81%, indicating that care must be taken when using these six commercial kits for the identification of *Vibrio* species.

V. cholerae O1, O139, Non-O1, Non-O139

The definition of the species *V. cholerae* and the relationship of the species to the cholera vibrios that cause cholera has long been a source of confusion. Rather like *E. coli*, *V. cholerae* is a large and diverse species, not all members of which are pathogens. In the past it was customary to speak of cholera vibrios, non-cholera vibrios (NCVs) and non-agglutinable vibrios (NAGs). It is now clear that these organisms are all members of the same species. Taxonomic studies including DNA–DNA homology have shown that *V. cholerae* is a single, relatively homogeneous, closely related species (Citarella and Colwell, 1970). Like other Gram-negative bacteria, *V. cholerae* has a complex cell wall which includes lipopolysaccharides, the carbohydrate components of which are designated as O antigens; there are now 206 somatic antigens in the *V. cholerae* serotyping scheme established by Shimada (Shimada *et al.*, 1994; Faruque and Nair, 2002).

Table 34.5 The 12 *Vibrio* species that are found in human clinical specimens

Test	Percentage positive* for											
	V. cholerae	*V. mimicus*	*V. metschnikovii*	*V. cincinnatiensis*	*V. hollisae*	*V. danisela*	*V. fluvialis*	*V. furnissii*	*V. alginolyticus*	*V. parahaemolyticus*	*V. vulnificus*	*V. carchariae*
D-Glucose, acid production	100	100	100	100	100	100	100	100	100	100	100	50
D-Glucose, gas production	0	0	0	0	0	10	0	100	0	0	0	0
Arginine, Moeller (1% NaCl)	0	0	60	0	0	95	95	100	0	0	0	0
Lysine, Moeller (1% NaCl)	99	100	35	60	0	50	0	0	99	100	99	100
Ornithine, Moeller (1% NaCl)	99	99	0	0	0	0	0	0	50	95	55	0
myo-nositol	0	0	40	100	0	0	0	0	0	0	0	0
Sucrose	100	0	100	100	0	5	100	100	99	1	15	50
Nitrate → nitrate	99	100	0	100	100	100	100	100	100	100	100	100
Oxidase	100	100	0	100	100	95	100	100	100	100	100	100
Growth in nutrient broth with												
0% NaCl	100	100	0	0	0	0	0	0	0	0	0	0
1% NaCl	100	100	100	100	99	100	99	99	99	100	99	100
Isolated from												
1 Faeces	+++	++	−	−	+	−	+	+	±	++	±	−
2 Other samples	±	+	+	+	−	+	−	−	+	±	+	+

* After 48 h incubation at 35 °C.
Most of the positive reactions occur during the first 24 h. NaCl 1% is added to the standard medium to enhance growth. Percentages approximate for guidance only. Boxes indicate useful tests.

Most of the strains of *V. cholerae* isolated from cases of cholera belong to the O1 serogroup. Within this serogroup are further subdivisions, three serogroups Inaba, Ogawa and Hikojima. There are also two biotypes, Classical and El Tor. Each biotype may include members of all serotypes. It should be remembered that not all O1 strains cause cholera. Members of serogroups O2 to O206 may cause diarrhoea but are not thought to cause epidemic or endemic cholera. Although serogroups O2 to O206 (excluding O139) are generally recognized to be less pathogenic than the Classical or El Tor biotypes, they can cause symptoms very similar or identical to those of cholera and may cause outbreaks of enteritis. Indeed, they are more commonly isolated in Britain from cases of enteritis than the El Tor or Classical biotypes. Only *V. cholerae* O1 is reportable to the World Health Organisation as cholera. Isolation of the Classical biotype is much less common and the current, seventh, cholera pandemic is caused by the El Tor biotype. Members of the recently isolated O139 serogroup cause a disease clinically indistinguishable from cholera and have been responsible for epidemics (Albert *et al.*, 1993; Bhattacharya *et al.*, 1993; Ramamurthy *et al.*, 1993; Cheasty and Rowe, 1994).

Strains of *V. cholerae* O1 can be divided into the two biotypes on the basis of the tests set out in Table 34.6.

Serotyping

All strains of *V. cholerae* share a common H antigen. There are at least 206 O groups in the extended serotyping scheme of Shimada. Apart from O1 and O139 there is no link between the O group and pathogenicity. The O1 antiserum is available commercially as *V. cholerae* 'polyvalent' antiserum. Subtyping of O1 strains may be done by testing for agglutination with absorbed Inaba and Ogawa antisera (Table 34.7).

Rapid Identification

Rapid methods for the detection of both *V. cholerae* O1 and O139 are available using microscopy for immobilization studies (Beneson *et al.*, 1964) and immunoassays (Hasan *et al.*, 1994; Qadri *et al.*, 1995). Nato *et al.* (2003) described a one-step immunochromatographic dipstick tests for the detection of *V. cholerae* O1 and O139 in stool samples. The dipsticks represent the first rapid test which has been successfully used to diagnose cholera from rectal swabs, and this would immensely improve surveillance for cholera, especially in remote settings (Bhuiyan *et al.*, 2003). Molecular methods such as PCR and DNA probes for the identification of *V. cholerae* directly from both clinical and environmental samples are also available; however their use in areas of epidemic cholera may not be practicable.

Typing

Further subtyping/molecular analysis of strains has also been reported using such techniques as pulsed-field gel electrophoresis (PFGE), multilocus enzyme elecrophoresis (MLEE) and multilocus sequence typing (MLST). Studies using MLEE have grouped the toxigenic O1 El Tor biotype strains into four major electrophoretic types (ET), the Australian type, the Gulf coast type, the seventh pandemic type and the Latin American type, designated ET1–ET4 respectively (Sack *et al.*, 2004). Kotetishvili, Stine and Chen (2003) reported that when investigating strains of *V. cholerae* O1 and O139, MLST using three housekeeping genes *gyrB*, *pgm* and *recA* and three virulence-associated genes *tcpA*, *ctxA* and *ctxB* was more discriminatory than PFGE and additionally provided a measure of phylogenetic relatedness.

Vibrio mimicus

Previously *V. mimicus* had been identified as *V. cholerae*. Organisms that are unable to ferment sucrose produce green colonies on TCBS but which are otherwise phenotypically similar to *V. cholerae* have been isolated from shellfish, brackish water and stools of human beings with diarrhoea in many parts of the world. Studies on more than 50 such organisms showed that they have DNA relatedness to each other at species level but are related only distantly to *V. cholerae* (DNA–DNA homology 20–50%). Thus these bacteria would appear to be distinct species (Davis *et al.*, 1981). *Vibrio mimicus* shares O group serotypes with *V. cholerae*. Only some strains produce enterotoxin.

Vibrio parahaemolyticus

These vibrios are halophilic, i.e. they have an obligate requirement for sodium chloride. *Vibrio parahaemolyticus* is widely distributed in warm coastal and estuarine waters, requires sodium for growth and has been found in gastroenteritis outbreaks associated with the eating of seafoods throughout the world. This vibrio was first identified in Japan as a cause of food poisoning associated with shrimps. Further outbreaks have been associated with the eating of crabs, prawns and other seafoods. In the current *V. parahaemolyticus* pandemic involving eight countries, there is an association with three predominant serotypes O1:K?, O3:K6 and O4:K68 (Bhuiyan *et al.*, 2002).

TREATMENT

The nature of the life-threatening symptoms of cholera results from massive loss of water and salts and without effective treatment the case-fatality rate in cases of severe cholera can be as high as 50%. All other effects are probably consequent on the dehydration. Effective therapy concentrates on the replacement of fluids and prevention of further loss. Both oral and intravenous rehydration have their place in the management of cholera. Many cases of cholera, especially if diagnosed early, can be managed by oral rehydration. However, the volumes of fluid lost in severe cases may be very large. Especially if diagnosis has been delayed, replacement of adequate volumes of fluid by mouth may prove difficult. Fluid losses of up to 20 l per day have been reported. For these reasons, the use of intravenous rehydration methods may prove very valuable and should not be excluded.

Rehydration therapy alone will lead to recovery from cholera (Guidelines for Cholera Control, 1986). However, it may be necessary

Table 34.6 The biotypes of cholera vibrios (*V. cholerae* O1)

	Biovar	
	Classical	El Tor
O1 antiserum	+	+
Voges-Proskauer reaction	−	+
Haemolysis of sheep erythrocytes	−	+
Chick red cell agglutination	−	+
Polymyxin 50 i.u.	S	R
Classical phage IV	S	R
El Tor phage 5	R	S

Table 34.7 Serotyping of *V. cholerae*

	O antigen	Type antigen	
V. cholerae	O:1	Inaba	Ogawa
Subtype			
Inaba	+	+	−
Ogawa	+	−	+
Hikojima	+	+	+
V. cholerae non-O1 (non-cholera vibrios (NCVs), non-agglutinable vibrios (NAGs), non-O1 *V. cholerae*)	O2–O206		
O139: Bengal cholera	O:139		

to rehydrate for quite a long time. It must be remembered that CT becomes permanently bound to mucosal cells, and thus ion fluxes are probably unbalanced throughout the life of that cell. For this reason, antibiotics are usually added to the treatment regimen. The empirical evidence is that they reduce the time during which rehydration therapy has to be continued. Antibiotics also reduce the load of *V. cholerae* in the intestine. Elimination of the organism by antibiotics prevents the release and binding of more CT, and thus only the limited amount of CT already released before antibiotic therapy is commenced continues to act on the intestine.

Though there is a role for antibiotics in the treatment of cholera, there is no role for antimotility agents. Such agents increase the retention of both vibrios and their attendant CT in the gut, and probably exacerbate the physiological effects. When using antibiotics it is important to remember that *V. cholerae*, like other bacteria, is able to develop antibiotic resistance. *Vibrio cholerae* is also able to acquire antibiotic resistance plasmids and, although they are not as stable hosts of plasmids as are some other organisms, this can be of considerable practical importance (Glass *et al.*, 1983). A number of antimicrobial agents have been used in endemic situations. These include tetracycline, doxycycline, furazolidone, trimethoprim-sulphamethoxazole and norfloxacin. Indeed, extensive use of tetracycline has resulted in many of the endemic strains becoming tetracycline resistant, and since the late 1970s there has appeared in Africa, the Indian subcontinent and South America, strains of multidrug resistant *V. cholerae* (Mhalu, Mmari and Ijumba, 1979; Threlfall, Rowe and Huq, 1980; Threlfall, Said and Rowe, 1993). Thus, although empirical therapy may be necessary, the performance of antibiotic resistance tests should also be a priority. *Vibrio cholerae* strains isolated from individuals undergoing antibiotic therapy should be maintained on media containing the antibiotics. It may be helpful to use a medium containing antibiotics in general use in isolation procedures. This would be in addition to standard media. Such procedures may be helpful in obtaining a more accurate picture of the effective antibiotic resistances in the population of *V. cholerae* in that many plasmids are unstable in *V. cholerae* in the absence of the selective pressure from antibiotics for their maintenance.

PREVENTION AND CONTROL

Cholera Vaccines

The first candidate cholera vaccine was produced within a year of the first isolation of *V. cholerae* by Koch. Since then, a variety of vaccines have been developed and used. The discovery of CT (De, 1959) provided a further stimulus to vaccine development. All types of vaccine have been tried. These include live parenteral vaccines, killed parenteral vaccines, oral killed vaccines, oral subunit vaccines and live oral vaccines. Oral vaccines include those developed by chemical mutagenesis such as Texas Star; those produced by using the techniques of molecular biology to disarm fully virulent strains; and those that use the ability of other organisms such as *Salmonella typhi* type 21A, which has an expression vector for genes encoding relevant *V. cholerae* antigens.

Conventional parenteral vaccines have been of little use in cholera. The level of protection afforded is low, perhaps 40–80% protection during the first 2 to 3 months after administration but no significant protection 6 months after. There is some evidence suggesting that the vaccine may be more effective when administered to the inhabitants of highly endemic areas. However, the vaccine is not at all satisfactory, and the World Health Organisation has recommended that its use be discontinued. The advances in our understanding of the pathogenesis of cholera reawaken interest in oral vaccines. The isolation of the infection from systemic immune systems, and the lack of success from parenteral vaccines, made cholera a prime target for the development of oral vaccines. Studies have focused on the development of a

live oral vaccine, although killed vaccines have also been investigated. Considerable advances have been made and several oral candidate vaccines have been developed and shown to protect in volunteer studies.

The first new vaccine to reach the field test stage was a killed oral vaccine (Clemens *et al.*, 1987). Killed vaccines were tested in an extensive field trial in Bangladesh (Clemens *et al.*, 1986). The basic vaccine consisted of a mixture of heat-killed and formalin-killed cells of *V. cholerae* O1 of both the Ogawa and Inaba subtypes. In this trial, the basic vaccine was administered alone or in combination with the B subunit of CT, and *E. coli* K12 preparation was administered as a control. These vaccines provided good, although short-term, protection against *V. cholerae*, both in the field trial and in adult volunteers. These vaccines needed to be administered and boosted on two occasions by further vaccine. This would have disadvantages for large-scale use. However, their success demonstrated the possibility of a successful vaccination campaign using an oral vaccine.

The use of a live oral vaccine should overcome the problem of the need for multiple vaccine administration. Several such candidate vaccine strains have been developed and tested. The prototype strain of this approach was the vaccine strain Texas Star SR (Levine *et al.*, 1984). This strain, isolated after nitrosoguanidine mutagenesis, produces B subunits only. Texas Star may be considered as establishing the principles by which the development and testing of candidate vaccines were governed. More recent strains have shown greater promise. A number of genetically engineered strains have been developed and tested (Table 34.8) (Levine *et al.*, 1988). Those engineered with attenuated *V. cholerae* strain CVD103 HgR have shown the greatest promise. This strain is a derivative of CVD103 which was itself derived from wildtype O1 classical Inaba strain 56913. Strain 569B is pathogenic for volunteers. A gene coding for mercury resistance was introduced into CVD103 to produce CVD103 HgR. This marker enables us to differentiate vaccine strain from wildtype *V. cholerae* O1. In volunteer studies the efficacy was about 60%. This is better than currently available vaccines. In some groups, effective efficacy was much higher than this: up to 100% in some instances. The vaccine has undergone volunteer studies or clinical trials in the USA (Kotloff *et al.*, 1992), Switzerland, Thailand, Indonesia (Suharyono *et al.*, 1992), Peru (Guttuzzo *et al.*, 1993), Chile and Costa Rica. Apart from initial volunteer studies, most of these have been randomized placebo-controlled double-blind trials. As mentioned above, the vaccine shows good protective efficacy but, perhaps equally important in terms of its large-scale application, there are no side effects. Occasional subjects reported diarrhoea, but this was of no greater incidence than in the control groups. Studies have been undertaken in both adults and children. We now have a vaccine

Table 34.8 Candidate vaccine strains of *V. cholerae*

Strain number	Toxigenic status	Number	Parent strain	
			Biotype	Serotype
Texas Star SR	A-13′	3083	EI Tor	Ogawa
J13K 70	A-B⁻	N 16961	EI Tor	Inaba
CVD) 1O1	A-B~	395	Classical	Ogawa
CVD 1O2[a]	A-B⁺	CVD 1O1	Classical	Ogawa
CVD 103	A-B′	569 B	Classical	Inaba
CVD 103 Hg[b]	A-B~	CVD 103	Classical	Inaba
CM 104	A-B⁻[d]	JBK 70	EI Tor	Inaba
CVD) 105	A-B⁺[d]	CVD 1O1	Classical	Ogawa
CVD 109	A-B⁻[e]	E7946	EI Tor	Ogawa
CVD 110⁻[c]	A-B′	CVD 109	EI Tor	Ogawa

[a] Thymine-dependent mutant.
[b] Hg resistance inserted into hylA locus.
[c] Hg resistance and CtxB inserted into parent HigA locus.
[d] Genes encoding EI Tor haemolysin deleted.
[e] ACE ZOT and Ctx genes deleted.

produced by the methods of molecular biology that is a significant improvement on vaccines produced by other methods and has considerable potential for use in control of cholera. It is likely that future generations of engineered vaccines will be even more useful but, at the present, the claims that molecular biology are able to produce a single-dose, live oral vaccine providing significant immunity after oral administration seem to have been amply fulfilled (Levine and Kaper, 1993).

It is worthwhile noting briefly a few considerations about immunity to cholera. It has long been assumed that the production of local intestinal immunity is what protects against infection. However, the best laboratory correlate with immunity is circulating vibriocidal antibody. The reasons for this are not clear. Nonetheless, field studies have shown that vibriocidal antibody in populations is the best indicator of the susceptibility of individuals to cholera, and further, such vibriocidal antibody is produced as a result of stimulation by effective vaccines (Wasserman *et al.*, 1994). Antibody production is not the only important factor in determining susceptibility to cholera. There is undoubtedly a genetic element. Studies in Bangladesh have indicated that among those developing clinical cholera, there is an excess of persons with blood group O (Levine *et al.*, 1979). The reason for this has not been fully clarified, although it has been suggested that H substance might act as a receptor for the fucose-resistant adhesin of *V. cholerae*.

Two oral vaccines (Dukoral and Orochol) are currently commercially available, Dukoral consisted of heat- and formalin-inactivated *V. cholerae* O1 Inaba and Classical biotype strains in association with recombinant CT B subunit (rCTB). This vaccine claims protection against cholera caused by *V. cholerae* O1 and diarrhoea caused by enterotoxigenic *E. coli*. The second oral vaccine Orochol contains an avirulent mutant *V. cholerae* strain (CVD103 HgR). Other oral vaccines currently under development may well bring further benefits.

REFERENCES

Albert MJ, Siddique AK, Islam MS *et al.* (1993) Large outbreak of clinical cholera due to *Vibrio cholerae* non-O1 in Bangladesh (Letter). *Lancet*, **341**, 704.

Barua D and Burrows W (eds) (1974) *Cholera*. Saunders, Philadelphia.

Barua D and Greenough WB (eds) (1992) *Cholera* (Current Topics in Infectious Diseases). Plenum, New York.

Baudry B, Fasano A, Ketley LM *et al.* (1992) Cloning of a gene (ZOT) encoding a new toxin produced by *Vibrio cholerae*. *Infection and Immunity*, **60**, 428–434.

Baumann P, Furniss AL and Lee JV (1984) Section 5. Facultatively anaerobic Gram-negative rods. In *Bergey's Manual of Systematic Bacteriology*, Vol. 1 Holt JG and Krieg NR (eds), Williams & Wilkins, Baltimore, pp. 518–538.

Benenson AS, Islam MR, Greenough WB III (1964) Rapid identification of *Vibrio cholerae* by dark-field microscopy. *Bulletin World Health Organization*, **30**, 827–831.

Bhattacharya SK, Bhattacharya MF, Balakrish Nair G *et al.* (1993) Clinical profile of acute diarrhoea cases infected with the new epidemic strain of *Vibrio cholerae* O139: designation of the disease as cholera. *Journal of Infection*, **27**, 11–15.

Bhuiyan NA, Ansaruzzaman M, Kamruzzaman M *et al.* (2002) Prevalence of the pandemic genotype of *Vibrio parahaemolyticus* in Dhaka, Bangladesh, and significance of its distribution across different serotypes. *Journal of Clinical Microbiology*, **40**, 284–286.

Bhuiyan NA, Qadri F, Faruque AS *et al.* (2003) Use of dipsticks for rapid diagnosis of cholera caused by *Vibrio cholerae* O1 and O139 from rectal swabs. *Journal of Clinical Microbiology*, **41**, 3939–3941.

Blake PA, Allegra DT, Snyder JD *et al.* (1980) Cholera: a possible endemic focus in the United States. *New England Journal of Medicine*, **302**, 305–309.

Cash RA, Music SI, Libonati P *et al.* (1974) Response of man to infection with *Vibrio cholerae* 1. Clinical, serologic, and bacteriologic responses to a known inoculum. *Journal of Infectious Diseases*, **129**, 45–52.

Chakraborty S, Mukhopadhyay AK, Bhadra RK *et al.* (2000) Virulence genes in environmental strains of *Vibrio cholerae*. *Applied Environmental Microbiology*, **66**, 4022–4028.

Cheasty T and Rowe B (1994) New cholera strains. *PHLS Microbiology Digest*, **11**, 73–76.

Citarella RV and Colwell RR (1970) Polyphasic taxonomy of the genus *Vibrio*. *Journal of Bacteriology*, **104**, 434–442.

Clemens JD, Harris JR, Khan MR *et al.* (1986) Field trial of oral cholera vaccines in Bangladesh. *Lancet*, **ii**, 124–127.

Clemens JD, Stanton BF, Chakraborty J *et al.* (1987) B subunit-whole cell and whole cell-only oral vaccines against cholera: studies on reactogenicity and immunogenicity. *Journal of Infectious Diseases*, **155**, 79–85.

Colwell RR (ed.) (1984) *Vibrios in the Environment*. Wiley, New York.

Colwell RR, Macdonell MT and De Ley J (1986) Proposal to recognize the family *Aeromonadaceae* fam. nov. *International Journal of Systematic Bacteriology*, **36**, 473–477.

Crowcroft NS (1994) Cholera: current epidemiology. In *Communicable Disease Report Review*, **4**, Review Number 13:R157–164. Public Health Laboratory Service, London.

Daniels NA, MacKinnon L, Bishop R *et al.* (2000) *Vibrio parahaemolyticus* infections in the United States, 1973–1998. *Journal of Infectious Diseases*, **181**, 1661–1666.

Davis BR, Fanney GR, Maddon JM *et al.* (1981) Characterisation of biochemically atypical *Vibrio cholerae* strain and designation of a new pathogenic species *Vibrio mimicus*. *Journal of Clinical Microbiology*, **14**, 631–639.

De SN (1959) Enterotoxicity of bacteria-free culture filtrates of *Vibrio cholerae*. *Nature*, **183**, 1533.

Desmarchelier PM and Reichelt JL (1981) Phenotypic characterization of clinical and environmental isolates of *Vibrio cholerae* from Australia. *Current Microbiology*, **5**, 123–127.

DiRita VJ (1992) Co-ordinate expression of virulence genes by ToxR in *Vibrio cholerae*. *Molecular Microbiology*, **6**, 451–458.

DiRita VJ, Parsot C, Jander G *et al.* (1991) Regulatory cascade controls virulence in *Vibrio cholerae*. *Proceedings of the National Academy of Sciences of the USA*, **88**, 5403–5407.

Drasar BS (1992) Pathogenesis and ecology: the case of cholera. *Journal of Tropical Medicine and Hygiene*, **95**, 365–372.

Farmer JJ and Hickman-Brenner FW (1992) The Genera *Vibrio* and *Photobacterium*. In *The Prokaryotes*, 2nd edn, Vol. 3 Balows A, Tresper HG and Dworkin M (eds) Springer-Verlag, New York, pp. 2952–3011.

Faruque SM and Nair GB (2002) Molecular ecology of toxigenic *Vibrio cholerae*. *Microbiology and Immunology*, **46**, 59–66.

Fasano A, Baudry B, Pumplin DW *et al.* (1991) *Vibrio cholerae* produces a second enterotoxin which affects intestinal tight junctions. *Proceedings of the National Academy of Science of the USA*, **88**, 5242–5246.

Feachem RG (1981) Environmental aspects of cholera epidemiology I. A review of selected reports of endemic and epidemic situations during 1961–1980. *Tropical Diseases Bulletin of the Bureau of Hygiene and Tropical Diseases*, **78**, 675–698.

Feachem RG (1982) Environmental aspects of cholera epidemiology III. Transmission and control. *Tropical Diseases Bulletin of the Bureau of Hygiene and Tropical Diseases*, **79**, 1–47.

Feachem R, Miller C and Drasar B (1981) Environmental aspects of cholera epidemiology II. Occurrence and survival of *Vibrio cholerae* in the environment. *Tropical Diseases Bulletin of the Bureau of Hygiene and Tropical Diseases*, **78**, 865–880.

Freter R and Jones GW (1976) Adhesive properties of *Vibrio cholerae*: nature of the interaction with intact mucosal surfaces. *Infection and Immunity*, **14**, 246–256.

Gangarosa EJ, Beisel WR, Benyajati C *et al.* (1960) The nature of the gastrointestinal lesion in Asiatic cholera and its relation to pathogenesis: a biopsy study. *American Journal of Tropical Medicine and Hygiene*, **9**, 125–135.

Glass RI, Claeson M, Blake PA *et al.* (1991) Cholera in Africa: lessons on transmission and control for Latin America. *Lancet*, **338**, 791–795.

Glass RI, Huq MI, Lee JV *et al.* (1983) Plasmid-borne multiple drug resistance in *Vibrio cholerae* serogroup O1, biotype El Tor: evidence for a point-source outbreak in Bangladesh. *Journal of Infectious Diseases*, **147**, 204–209.

Guttuzzo E, Butron B, Seas C *et al.* (1993) Safety, immunogenicity, and excretion pattern of single dose live oral cholera vaccine CVD 103-HgR in Peruvian adults of high and low socio-economic levels. *Infection and Immunity*, **61**, 3994–3997.

Guidelines for Cholera Control (1986) World Health Organization: Programme for Control of Diarrhoeal Diseases, 80.4 Rev 1.

Hasan JA, Huq A, Tamplin ML *et al.* (1994) A novel kit for rapid detection of *Vibrio cholerae* O1. *Journal of Clinical Microbiology*, **32**, 249–252.

Heidelberg JF, Eisen JA, Nelson WC *et al.* (2000) DNA sequence of both chromosomes of the cholera pathogen *Vibrio cholerae*. *Nature*, **406**, 477–483.

Herrington DA, Hall RH, Losonsky G *et al.* (1988) Toxin, toxin-coregulated pili, and the *toxR* regulon are essential for *Vibrio cholerae* pathogenesis in humans. *Journal of Experimental Medicine*, **168**, 1487–1492.

Hornstrup MK and Gahrn-Hansen B (1993) Extraintestinal infections caused by *Vibrio parahaemolyticus* and *Vibrio alginolyticus* in a Danish county, 1987–1992. *Scandinavian Journal of Infectious Diseases*, **25**, 735–740.

Hughes JM, Boyce JM, Aleem AR *et al.* (1978) *Vibrio parahaemolyticus* enterocolitis in Bangladesh: report of an outbreak. *American Journal of Tropical Medicine and Hygiene*, **1**, 106–112.

Islam MS, Drasar BS and Sack RB (1993) The aquatic environment as a reservoir of *Vibrio cholerae*: a review. *Journal of Diarrhoeal Diseases Research*, **11**, 197–206.

Islam MS, Drasar BS and Sack RB (1994) The aquatic flora and fauna as reservoirs of *Vibrio cholerae*: a review. *Journal of Diarrhoeal Diseases Research*, **12**, 87–96.

Jones GW and Freter R (1976) Adhesive properties of *Vibrio cholerae*: nature of the interaction with isolated rabbit brush border membranes and human erythrocytes. *Infection and Immunity*, **14**, 240–245.

Karasawa T, Mihara T, Kurazono H *et al.* (1993) Distribution of the *zot* (zonula occludens toxin) gene among strains of *Vibrio cholerae* O1 an α non- O1. *FEMS Microbiology Letters*, **106**, 143–145.

Kaspar CW and Tamplin ML (1993) Effects of temperature and salinity on the survival of *Vibrio vulnificus* in seawater and shellfish. *Applied Environmental Microbiology*, **59**, 2425–2429.

Kaufman MR, Shaw CE, Jones ID *et al.* (1993) Biogenesis and regulation of the *Vibrio cholerae* toxin-coregulated pilus: analogies to other virulence factor secretory systems. *Gene*, **126**, 43–49.

Kirn TJ, Lafferty MJ, Sandoe CM and Taylor RK (2000) Delineation of pilin domains required for bacterial association into microcolonies and intestinal colonization by *Vibrio cholerae*. *Molecular Microbiology*, **35**, 896–910.

Klontz KC, Lieb S, Schreiber M *et al.* (1998) Syndromes of *Vibrio vulnificus* infections. Clinical and epidemiologic features in Florida cases, 1981–1987. *Annals of Internal Medicine*, **15**, 318–323.

Koblavi S, Grimont F and Grimont PA (1990) Clonal diversity of *Vibrio cholerae* O1 evidenced by rRNA gene restriction patterns. *Research in Microbiology*, **141**, 645–657.

Koch R (1884) Ueber die Cholerabacterian. *Deutsche Medizinische Wochenschrift*, **10**, 725–728.

Kotetishvili M, Stine OC and Chen Y (2003) Multilocus sequence typing has better discriminatory ability for typing *Vibrio cholerae* than does pulsedfield gel electrophoresis and provides a measure of phylogenetic relatedness. *Journal of Clinical Microbiology*, **41**, 2191–2196.

Kotloff KL, Wasserman SS, O'Donnell S *et al.* (1992) Safety and immunogenicity in North Americans of a single dose of live oral cholera vaccine CVD 103-HgR: results of a randomized, placebo-controlled, double-blind crossover trial. *Infection and Immunity*, **60**, 4430–4432.

Kovach ME, Shaffer MD and Peterson KM (1996) A putative integrase gene defines the distal end of a large cluster of ToxR-regulated colonization genes in *Vibrio cholerae*. *Microbiology*, **142**, 2165–2174.

Levine MM and Kaper JB (1993) Live oral vaccines against cholera: an update. *Vaccine*, **11**, 207–212.

Levine MM, Black RE, Clements ML *et al.* (1982) The pathogenicity of nonenterotoxigenic *Vibrio cholerae* serogroup O1 biotype El Tor isolated from sewage water in Brazil. *Journal of Infectious Disease*, **145**, 296–299.

Levine MM, Black RE, Clements ML *et al.* (1984) Evaluation in humans of attenuated *Vibrio cholerae* El Tor Ogawa strain Texas Star-SR as a live oral vaccine. *Infection and Immunity*, **43**, 515–522.

Levine MM, Kaper JB, Herrington D *et al.* (1988) Volunteer studies of deletion mutants of *Vibrio cholerae* O1 prepared by recombinant techniques. *Infection and Immunity*, **56**, 161–167.

Levine MM, Nalin DR, Rennels MB *et al.* (1979) Genetic susceptibility to cholera. *Annals of Human Biology*, **6**, 369–374.

Lin Z, Kumagai K, Baba K *et al.* (1993) *Vibrio parahaemolyticus* has a homolog of the *Vibrio cholerae* toxRS operon that mediates environmentally induced regulation of the thermostable direct hemolysin gene. *Journal of Bacteriology*, **175**, 3844–3855.

MacNamara C (1876) *A History of Asiatic Cholera*. Macmillan, London.

MacPherson J (1872) *Annals of Cholera from the Earliest Periods to the Year 1817*. Ranken, London.

McCardell BA, Madden JM and Shah DB (1985) Isolation and characterization of a cytolysin produced by *Vibrio cholerae* serogroup non-O1. *Canadian Journal of Microbiology*, **31**, 711–720.

Mhalu FS, Mmari PW and Ijumba J (1979) Rapid emergence of El Tor *Vibrio cholerae* resistant to antimicrobial agents during first six months of fourth cholera epidemic in Tanzania. *Lancet*, **17**, 345–347.

Nato F, Boutonnier A, Rajerison M *et al.* (2003) One-step immunochromatographic dipstick tests for rapid detection of *Vibrio cholerae* O1 and O139 in stool samples. *Clinical Diagnostic Laboratory Immunology*, **10**, 476–478.

O'Brien AD, Chen ME, Holmes RX *et al.* (1984) Environmental and human isolates of *Vibrio cholerae* and *Vibrio parahaemolyticus* produce a *Shigella dysenteriae* 1 (Shiga)-like cytotoxin. *Lancet*, **i**, 77–78.

O'Hara CM, Sowers EG, Bopp CA *et al.* (2003) Accuracy of six commercially available systems for identification of members of the family *Vibrionaceae*. *Journal of Clinical Microbiology*, **41**, 5654–5659.

Ogierman MA, Zabihi S, Mourtzios L and Manning PA (1993) Genetic organization and sequence of the promoter-distal region of the tcp gene cluster of *Vibrio cholerae*. *Gene*, **126**, 51–60.

Orton R (1831) *The Epidemic Cholera of India: Madras*. Burgess & Hill, London.

Pacini F (1854) Osservazione microscopiche e duduzioni patalogiche sul cholera asiatico. *Gazzette Medicale de Italiana Toscana Firenze*, **6**, 405–412.

Parsot C and Mekalanos JJ (1990) Expression of ToxR, the transcriptional activator of the virulence factors in *Vibrio cholerae*, is modulated by the heat shock response. *Proceedings of the National Academy of Sciences of the USA*, **87**, 9898–9902.

Pascual M, Rodo X, Ellner SP *et al.* (2000) Cholera dynamics and El Nino-Southern Oscillation. *Science*, **289**, 1766–1769.

Pearson GDN, Woods A, Chiang SL *et al.* (1993) CTX genetic element encodes a site-specific recombination system and an intestinal colonization factor. *Proceedings of the National Academy of Sciences of the USA*, **90**, 3750–3754.

Pettenkofer von M (1892) Ueber Cholera unt Berficksichtigung der jiingsten Cholera-Epidemic in Hamburg. *Miinchener Medizische Wochenschrift*, **46**, 345–348.

Pollitzer R (1959) *Cholera*. WHO, Geneva.

Qadri F, Hasan JA, Hossain *et al.* (1995) Evaluation of the monoclonal antibody-based kit Bengal SMART for rapid detection of *Vibrio cholerae* O139 synonym Bengal in stool samples. *Journal of Clinical Microbiology*, **33**, 732–734.

Ramamurthy T, Albert ML, Huq A *et al.* (1994) *Vibrio mimicus* with multiple toxin types isolated from human and environmental sources. *Journal of Medical Microbiology*, **40**, 194–196.

Ramamurthy T, Garg S, Sharma R *et al.* (1993) Emergence of novel strain of *Vibrio cholerae* with epidemic potential in southern and eastern India (Letter). *Lancet*, **341**, 703–704.

Risks of Transmission of Cholera by Food (1991) Health Programs Development, Veterinary Public Health Program. World Health Organization, Washington, DC.

Sack DA, Sack RB, Nair GB and Siddique AK (2004) Cholera. *Lancet*, **17**, 223–233.

Schoolnik GK and Yildiz FH (2000) The complete genome sequence of *Vibrio cholerae*: a tale of two chromosomes and of two lifestyles. *Genome Biology*, **1**, 1016.1–1016.3.

Shimada T, Arakawa E, Itoh K *et al.* (1994) Extended serotyping scheme for *V. cholerae*. *Current Microbiology*, **28**, 175–178.

Snow J (1849) *On Mode of Communication of Cholera*. John Churchill, London.

Snow J (1855) *On the Mode of Communication of Cholera*, 2nd edn. John Churchill, London.

Strom MS and Paranjpye RN (2000) Epidemiology and pathogenesis of *Vibrio vulnificus*. *Microbes and Infection*, **2**, 177–188.

Suharyono, Simanjuntak C, Witham N *et al.* (1992) Safety and immunogenicity of single-dose live oral cholera vaccine CVD 103-HgR in 5–9-year-old Indonesian children. *Lancet*, **340**, 689–694.

Suntharasarnai P, Migasena S, Vongsthongsri U *et al.* (1992) Clinical and bacteriological studies of El Tor cholera after ingestion of known inocula in Thai volunteers. *Vaccine*, **10**, 502–505.

Tamplin ML, Gauzens AL, Huq A *et al.* (1990) Attachment of *Vibrio cholerae* serogroup O1 to zooplankton and phytoplankton of Bangladesh waters. *Applied and Environmental Microbiology*, **56**, 1977–1980.

Thorn S, Warhurst D and Drasar BS (1992) Association of *Vibrio cholerae* with fresh water amoebae. *Journal of Medical Microbiology*, **36**, 303–306.

Threlfall EJ, Rowe B and Huq I (1980) Plasmid-encoded multiple antibiotic resistance in *Vibrio cholerae* El Tor from Bangladesh. *Lancet*, **7**, 1247–1248.

Threlfall EJ, Said B and Rowe B (1993) Emergence of multiple drug resistance in *Vibrio cholerae* O1 El Tor from Ecuador. *Lancet*, **6**, 1173.

Wachsmuth K, Blake PA and Olsvik O (eds) (1994) *Vibrio cholerae and Cholera: Molecular to Global Perspectives*. American Society for Microbiology, Washington, DC.

Waldor MK and Mekalanos JJ (1996) Lysogenic conversion by a filamentous phage encoding cholera toxin. *Science*, **272**, 1910–1914.

Wasserman SS, Losonsky GA, Noriega F *et al.* (1994) Kinetics of the vibriocidal antibody response to live oral cholera vaccines. *Vaccine*, **12**, 1000–1003.

West PA (1989) The human pathogenic vibrios: a public health update with environmental perspectives. *Epidemiology and Infection*, **103**, 1–34.

35

Aeromonas and *Plesiomonas* spp.

Alpana Bose

Centre for Medical Microbiology, Royal Free and University College Medical School, London, UK

INTRODUCTION

The aeromonads and plesiomonads were originally placed in the family Vibrionaceae, which also includes *Vibrio*, *Photobacterium* and *Enhydrobacter*. However phylogenetic studies have indicated that *Aeromonas* is not closely related to vibrios and thus has been transferred to a separate family, Aeromonadaceae (Colwell, MacDonell and De Ley, 1986). It has also been proposed that *Plesiomonas* being closer to the genus *Proteus* should belong to the family Enterobacteriaceae (Ruimy *et al.*, 1994; Janda, 1998).

DESCRIPTION OF ORGANISM

Aeromonas

Members of the genus *Aeromonas* are Gram-negative, non-spore-forming straight rods, which occur singly, in pairs or short chains. They are facultative anaerobes, being both catalase and oxidase positive. The aeromonads break down carbohydrates with the production of acid or acid and gas. Most of the mesophilic species within this genus are motile and have a single polar flagellum. Swarming motility with the production of lateral flagella has also recently been described (Kirov *et al.*, 2002). Although strains of *A. salmonicida* are capable of producing lateral flagella they are non-motile. This is thought to be as a result of inactivation of the *lafA* (flagellin gene), by transposase 8 (IS3 family) (Merino *et al.*, 2003).

The aeromonads can grow at range of temperatures from 5 to 44 °C. The optimum temperature for growth is 22–28 °C. Most isolates of clinical significance will grow readily at 37 °C. The pH range for growth is 5.5–9.0. Growth is inhibited in 6% salt broth. The G + C content of DNA is 57–63%.

Taxonomy

The genus *Aeromonas* was proposed by Kluvyer and van Neil in 1936 (Popoff, 1984). Since then it has undergone a number of taxonomic and nomenclature changes.

Phenotypic Classification: A phenotypic classification system was put forward on the basis of a numerical study by Popoff and Vernon (1976) who divided the aeromonads into two groups. The generally motile mesophiles which grow at 35–37 °C were known as *A. hydrophilia*. The subspecies within this group were known as *hydrophilia*, *anaerogenes* and *sobria*. These are now known as *A. hydrophilia*, *A. caviae* and *A. sobria* respectively and are responsible for the majority of human infections. The non-motile psychrophiles, primarily fish pathogens, and growing at 22–28 °C, were known as *A. salmonicida*. Further phenotypic distinctions could be made on the basis of motility, production of melanin-like pigment on tyrosine agar and ability to produce indole. Just over a decade later Hickman-Brenner described two further mesophilic phenospecies, *A. veronii* (Hickman-Brenner *et al.*, 1987) and *A. shubertii* (Hickman-Brenner *et al.*, 1988). Since then many other species have been described such as *A. trota*, *A. jandaei*, *A. allosaccharophilia*, *A. bestiarum*, *A. enchelaiae* and *A. popoffi* based on biochemical and genetic analysis (Janda, 1998).

Genotypic Classification: DNA hybridisation studies initially established 12 genomic species or hybridisation groups (HG). Each of the phenotypes was found to contain several genomic species which could be differentiated in some cases by biochemical tests. This has caused considerable confusion, for example, HG3 has been named *A. salmonicida* (Janda, 1991). This genomic species contains the psychrophilic, non-motile strains of phenospecies *A. salmonicida* and the mesophilic strains of phenospecies *A. hydrophilia*. Therefore it is important to specify the system used to classify the species. The discrepancy between the different modes of classification can be further exemplified by considering HG8 and HG10. Although described independently with different phenotypes they are genetically identical. The rules of taxonomic nomenclature mean that the genomic species is known as *A. veronii*. However the phenotypic characteristic is used to describe the biotype: thus HG8 is known as *A. veronii biotype sobria* and HG10 is referred to as *A. veronii biotype veronii* (Hickman-Brenner *et al.*, 1987; Janda, 1991).

Abbott *et al.* (1992) proposed a new classification scheme that aimed to correlate the genomic species and phenospecies (Table 35.1). This scheme includes 13 genospecies ordered in four complexes or phenospecies.

Table 35.1 Hybridisation groups, mesophilic genomic and phenotypic species of *Aeromonas* spp.

Hybridisation group	Genomic species	Phenospecies or complex
1	*hydrophilia*	*hydrophilia*
2	Not classified	
3	*salmonicida*	
4	*caviae*	*caviae*
5A, 5B	Media	
6	*eucreonophilia*	
7	*sobria*	*sobria*
8	*veronii biotype sobria*	
9	*jandaei*	
10	*veronii biotype veronii*	
11	Not classified	
12	*shubertii*	*shubertii-trota*
13	*trota*	

Four new species were subsequently described: *A. allosaccharophila* (Martinez-Murcia *et al.*, 1992), *A. encheleia* (Esteve, Gutiérrez and Ventosa, 1995), *A. bestiarum* (Ali *et al.*, 1996) and *A. popoffii* (Huys *et al.*, 1997). Five new species were subsequently described: *A. allosaccharophila* (Martinez-Murcia *et al.*, 1992), *A. encheleia* (Esteve, Gutiérrez and Ventosa, 1995), *A. bestiarum* (Ali *et al.*, 1996), *A. popoffii* (Huys *et al.*, 1997) and *A. culicicola* (Pidiyar *et al.*, 2002). Two additional species, *ichthiosmia* and *enteropelogenes*, appear to be subjective synonyms of previously published species (Janda and Abbott, 1998). Recently a new species, *A. molluscorum* sp. nov., isolated from the bivalve mollusc has been proposed. 16S rRNA gene sequence analysis showed that the strains were similar to *A. encheleia*. Biochemical tests that distinguish this strain from other *Aeromonas* spp. include their negative reactions in tests for indole production, lysine decarboxylase, gas from glucose and starch hydrolysis (Minana-Galbis *et al.*, 2004).

Thus, at least 14 species are considered in the most recent classification (Janda and Abbott, 1998); among them, six are considered pathogen for human, while the rest are non-pathogenic or 'environmental' on the basis of the frequency of isolation from clinical or environmental sources.

Plesiomonas

Like aeromonads the plesiomonads are Gram-negative, non-spore-forming straight rods. However, the plesiomonads are more pleomorphic occurring singly, in pairs, short chains or as long filamentous forms. They are also facultative anaerobes, being catalase and oxidase positive. Carbohydrate is broken down with the production of acid only. The plesiomonads are motile and have several polar flagella.

The optimum temperature for growth is 30–37 °C, although plesiomonads will grow at a range of temperatures between 8 and 44 °C. The pH range for growth is 5–7.7 and growth is inhibited by 6% salt broth. The G + C content of DNA is 51%.

Taxonomy: These organisms were initially described by Ferguson and Henderson in 1947 and were known as C27. The C27 organisms were initially placed in the genus *Pseudomonas* as '*Pseudomonas shigelloides*'. They were subsequently transferred to the genus *Aeromonas* as '*A. shigelloides*'. Habs and Shubert proposed the name *Plesiomonas* in 1962 and the organisms were given their own genus. The name for this genus was chosen from the Greek word for 'neighbour' as it was thought that the organism was closely related to *Aeromonas*. However, *Plesiomonas* spp. have been found to be more closely related to the genus *Proteus* in the family Enterobacteriaceae. *Plesiomonas shigelloides* is the only species in the genus. This species name was chosen, as a minority of strains share a common O-antigen with *Shigella sonnei*. Despite this, the genus *Plesiomonas* still resides in the family Vibrionaceae.

The major characteristics of the genus *Aeromonas* and *Plesiomonas* are summarised in Table 35.2.

PATHOGENESIS

Aeromonas

The pathogenicity of *Aeromonas* spp. is mainly related to tissue adherence and toxin production.

S-Layer

Tissue adherence is mediated by the S-layer composed of single-surface array protein of about 50 kDa. The S-layer can be found in the fish pathogen *A. salmonicida*. Human pathogens *A. veronii biotype sobria* and *A. hydrophilia* have also been found to have an S-layer; however, the molecular weight of the protein is 52–58 kDa and sequence analysis shows no homology at the amino terminus

Table 35.2 A summary of the major characteristics of the genera *Aeromonas* and *Plesiomonas*

Characteristics	Aeromonas	Plesiomonas
Gram stain	Gram-negative straight rod	Gram-negative straight rod
Oxygen requirement	Facultative anaerobe	Facultative anaerobe
Biochemical reactions[a]	Oxidase positive	Oxidase positive
	Catalase positive	Catalase positive
	Break down carbohydrates producing acid or acid and gas	Break down carbohydrates producing acid only
Flagella	Single flagellum	Several flagella
Temperature range	5–44 °C	8–44 °C
Optimum temperature	22–28 °C	30–37 °C
pH range	5.5–9.0	5–7.7
Growth in 6% salt broth	Inhibited	Inhibited

[a] See Table 35.4 for further biochemical reactions.

(Janda *et al.*, 1987; Dooley and Trust, 1988). The S-layer is thought to promote colonisation of the organisms to gut mucosa. In addition, the S-layer confers surface hydrophobicity, resistance to complement-mediated lysis and enhances interaction with macrophages. The S-layer appears to be anchored by lipopolysaccharide (LPS) in *A. hydrophilia* and *A. veronii biotype sobria*.

Fimbriae and Adhesins

Mesophilic strains possess fimbriae (filamentous) and outer membrane, S-layer synonymous adhesins (non-filamentous) with haemagglutinating activity which promote tissue adherence. The production of adhesins is increased in liquid culture at low temperatures (5 °C) (Ho, Sohel and Schoolnik, 1992; Gosling, 1996).

Capsular Layer

During growth in glucose-rich media strains of *A. hydrophilia* and *A. veronii* have been shown to produce a capsular polysaccharide which may contribute to increased virulence of these strains (Martinez *et al.*, 1995).

Extracellular Proteins

The toxins produced by *Aeromonas* spp. include cytotoxins including aerolysin, phospholipases, haemolysins and enterotoxins some of which have haemolytic activity. The mechanism of action of the enterotoxin is similar to that of *Vibrio cholerae* (Merino, Camprubi and Tomàs, 1992; Merino *et al.*, 1995; Tanoue *et al.*, 2005). *Aeromonas* spp. can also produce other substances such as amylase, chitinase, lipase and nuclease that can be released by the organisms and act as extracellular virulence factors (Janda, 1991). In addition, exoenzymes such as proteases have been implicated as virulence factors, as species deficient in proteases show reduced virulence. It has been shown that proteases are also required for the activation of other virulence factors such as aerolysin which form pores in the cell membrane leading to cell lysis (Howard and Buckley, 1985; Abrami *et al.*, 1998).

Endotoxin

Endotoxin or LPS of *Aeromonas* spp. has been implicated as an important virulence factor. It has been shown that strains that possess

O-antigen LPS secrete more exotoxin than strains that do not have O-antigen LPS. The possession of O-antigen LPS has been reported to be temperature dependent with some strains that have LPS becoming more virulent when grown at low temperatures (Merino, Camprubi and Tomàs, 1992). The O-antigen LPS is also used to serotype *Aeromonas* strains (Sakazaki and Shimada, 1984).

Siderophores

Almost all strains of *Aeromonas* spp. produce siderophores. In other Gram-negative organisms siderophores have been associated with establishment of infection. *Aeromonas hydrophilia*, *A. caviae* and some strains of *A. sobria* have been shown to produce a new siderophore known as amonabactin (Barghouthi *et al.*, 1989). This is thought to enhance growth in iron-deficient media.

Plesiomonas

Research into elucidating virulence factors of *P. shigelloides* has focused mainly on investigating enteropathogenic mechanisms. The production of enterotoxin in the rabbit ileal loop model has been demonstrated. Both heat-labile and heat-stable enterotoxins have been characterised. The heat-stable enterotoxin is genetically different to that of *Escherichia coli* and *V. cholerae* (Matthews, Douglas and Guiney, 1988). Cytolysins that may cause tissue destruction or inhibition of the normal bacterial gut flora and thus have a role in enhancing colonisation or invasion of epithelial cells have also been identified (Abbott, Kokka and Janda, 1991). Other potential virulence factors such as haemolysin, elastin and plasmids have been detected but their pathogenic significance is uncertain.

EPIDEMIOLOGY

Aeromonas

Bacteria of the genera *Aeromonas* are found in abundance in freshwater, saltwater and chlorinated drinking water. Counts tend to rise with increasing water temperatures, and in the deeper more anoxic layers there may be more than 10^5 per gram mud and greater than 10^5 per litre of surface water. In summer they form majority of the aerobic flora. The organisms can also be isolated from fish, birds and reptiles, raw foods such as fresh vegetables, meat and ready-to-eat products (Hanninen and Siitonen, 1995; Blair, McMahon and McDowell, 1999). More recently a rare species, *Aeromonas culicicola*, has been isolated from the midgut of female *Culex quinquefasciatus* and *Aedes aegyptii* mosquitoes (Pidiyar *et al.*, 2002) and drinking water supply in Spain (Figueras *et al.*, 2005).

Members of the Aeromonadacaeae have been implicated in cases of human infection. Aeromonads are also important in veterinary medicine causing an acute or chronic haemorrhagic septicaemia, furunculosis, in fish. They have also been implicated as food-spoilage organisms.

Plesiomonas

The only member of the genus *Plesiomonas*, *P. shigelloides*, is found in aquatic environments primarily freshwater and estuaries within tropical and temperate climates but can also be isolated from seawater in summer months. The organism forms part of the normal bacterial flora of fish, reptiles and amphibians. It has been isolated sporadically from mammals such as wolves, cheetah and Black lemur (Jagger, 2000). *Plesiomonas shigellodes* has been implicated as an uncommon cause of diarrhoea in cats. In humans, extragastrointestinal infection occurs mainly in neonates or in patients with underlying chronic diseases or who are immunocompromised. Diarrhoeal illness may also occur in immunocompetent individuals.

CLINICAL FEATURES

Aeromonas

Despite the explosion in the number of new species only a few have been implicated as human pathogens. Clinically significant strains belong to *A. hydrophilia* (HG1), *A. cavaie* (HG4) and *A. veronii biotype sobria* (HG8). These account for over 85% of all clinical isolates and are the causative agents of a wide range of extragastrointestinal and systemic infections. In addition, *A. veronii biotype veronii* (HG10), *A. jandaie* (HG9) and *A. shubertii* (HG12) have also been isolated albeit less frequently from clinical samples (Janda, 1991). The remainder of the *Aeromonas* species have been mainly isolated from environmental sources including water, fish, birds, other animals and industrial sources. The pathogenic potential of newer *Aeromonas* species is yet to be determined. Almost all of the recently described mesophilic species have been isolated from faeces and may reflect transient carriage as a consequence of their proliferation in water supplies in the summer months (Janda, 1991). *Aeromonas* species are recovered most commonly from faecal samples followed by samples from wound infections.

Gastroenteritis

There has been considerable debate regarding the enteropathogenic potential of *Aeromonas* species. The clinical presentation is said to vary tremendously, but most cases have been described as watery diarrhoea, associated with vomiting and low-grade pyrexia. The duration of symptoms is less than a week, although they can persist for longer in a minority of cases. In addition, there have also been a few cases of bloody diarrhoea (Gracey, Burke and Robinson, 1982). Evidence in favour of *Aeromonas* spp. as an aetiological agent of gastroenteritis comes from numerous reports linking infection with *Aeromonas* spp. and gastroenteritis. The discovery of enterotoxin produced by many *Aeromonas* strains provided a mechanism by which *Aeromonas* spp. could act as an enteropathogen.

In a Finnish study it was found that *Aeromonas* spp. was the third most common enteropathogen isolated, after *Campylobacter* and *Salmonella*. Furthermore 96% of patients with faecal isolates of *Aeromonas* spp. had gastroenteritis. *Aeromonas caviae* (41%), *A. hydrophila* (27%) and *A. veronii biotype sobria* (22%) were the most frequent isolates (Rautelin *et al.*, 1995). There have also been a number of case reports describing diarrhoeal illness associated with the isolation of *Aeromonas* spp. as well as serological evidence of infection. Champsaur *et al.* (1982) described a case of a patient with 'rice-water' diarrhoea. The stool was culture negative for *V. cholerae* and enterotoxigenic *E. coli* but was positive for *A. sobria*. This strain produced enterotoxin, proteolysin, cytolysin, hemolysin and a cell-rounding factor. Acute and convalescent phase sera showed an increase in neutralising antibodies to enterotoxin, cytolysin and hemolysin. Further evidence supporting a causative role of *Aeromonas* spp. in diarrhoeal disease comes from case reports where there has been improvement in diarrhoea with antibiotics active against *Aeromonas* spp. and resolution of abnormal bowel pathology with the disappearance of the organism from the stool (Roberts, Parenti and Albert, 1987).

Evidence against the view that *Aeromonas* spp. is a human enteropathogen comes from case–control studies that have shown that three main species *A. hydrophilia*, *A. veronii* and *A. caviae* could be isolated from symptomatic individuals and healthy controls. Most isolates of *A. hydrophilia* and *A. veronii* have been shown *in vitro* to express a cytolytic enterotoxin, β-haemolysin. In addition, human volunteer studies have yielded inconclusive results. In a study by Morgan *et al.* (1985), of fifty-seven healthy adults who were given doses of up to 10^{10} cfu of strains of *A. hydrophilia* which produced both cytotoxin and enterotoxin, only two developed mild diarrhoea. It was suggested that the lack of infectivity was secondary to immunity in the adult population as a result of exposure to strains in the local water supply. The observation that episodes of travellers diarrhoea in Peace Corp volunteers in Thailand were associated with *Aeromonas* spp. in faecal samples whereas the indigenous population from whom *Aeromonas* spp. was also isolated remained asymptomatic seemed to corroborate this view (Pitarangsi *et al.*, 1982). However, there have been reports that suggest that over 40% of individuals with diarrhoea and *Aeromonas* spp. isolated from the stool may have been infected with another recognised pathogen or have a non-infective disease such as bowel malignancy or inflammatory bowel disease (Millership *et al.*, 1986). This together with the lack of an animal model that could be used to fulfil Koch's postulates has failed to convince sceptics.

However, it is difficult to ignore the mounting evidence outlined above, suggesting that *Aeromonas* spp. do have an enteropathogenic role. Several outbreaks have also been described. These have been mainly in long-term care facilities (Bloom and Bottone, 1990) and child daycare centres (de la Morena *et al.*, 1993). Therefore, an alternative view is emerging that certain strains can sometimes produce gastrointestinal disease in predisposed individuals provided that there is a high enough infective dose.

Musculoskeletal and Wound Infections

Aeromonas spp. have been isolated either in pure culture or more often as a mixed culture together with enteric organisms such as *Bacteroides* spp., *Enterococci* and *Clostridia* from clinical samples taken from sites of musculoskeletal and wound infections. Although most clinically significant *Aeromonas* spp. have been recovered from such sites, strains of *A. hydrophilia* account for the majority of clinical isolates. There is a wide spectrum of clinical presentation ranging from mild superficial skin infection to more severe deep infections such as septic arthritis and osteomyelitis (Janda and Abbott, 1998).

Musculoskeletal and wound infections occur most commonly in healthy individuals who participate in recreational or occupational aquatic activities. *Aeromonas* infection is in many cases secondary to penetrating injuries to the lower limbs in aquatic environments. Patients often give a history of striking a submerged object barefoot whilst walking along the banks of a river or lake (Voss, Rhodes and Johnson, 1992). These infections arise almost exclusively as a result of exposure to freshwater as opposed to seawater. This observation is difficult to explain as the bacterial densities in both uncontaminated freshwater and seawater are similar (approximately 10^2 cfu/ml) and *Aeromonas* spp. have been isolated from marine crustaceans (Holmes, Niccolls and Sartory, 1996). Recently an outbreak of *A. hydrophilia* wound infection in people playing 'mud-football' has been reported in Western Australia. The field had been prepared using water from a near-by river (Vally *et al.*, 2004).

The other important environmental source of infection with *Aeromonas* spp. is soil. In a review of 32 infections of the foot, at least 33% were related to trauma secondary to inanimate objects such as broken glass and nails contaminated with soil (Wakabongo, 1995). *Aeromonas* infections as a consequence of severe crush injuries causing compound fractures or extensive burns have also been associated with soil-borne pathogens. In some cases osteomyelitis can occur several months after the event (Voss, Rhodes and Johnson, 1992).

In a minority of patients no history of trauma can be elicited. However these patients have underlying disorders such as chronic ulcers, chronic vascular insufficiency, malignancy and diabetes mellitus (Voss, Rhodes and Johnson, 1992).

Medicinal leech therapy has also been implicated as a significant albeit rare source of infection. Leeches are used by many Plastic Surgeons and other surgical specialists to relieve venous congestion and reduce oedema over skin flaps. *Aeromonas veronii biotype sobria* forms part of the normal flora of the leech and is thought to break down ingested erythrocytes and thus aid digestion (Dabb, Malone and

Leverett, 1992). Prophylactic antibiotics are recommended for treatment with medicinal leeches as up to one-fifth of such treatment regimens are associated with Gram-negative infections many of which are secondary to *Aeromonas* spp. (Mercer *et al.*, 1987). The clinical presentation following infection is varied and can range from mild wound infection to myonecrosis and bacteraemia. Recently a case of *Aeromonas veronii biotype sobria* meningitis associated with the use of leech therapy has been reported. The leeches were used to salvage a skin flap after surgical excision of a glomus tumour of the central nervous system. The authors advised caution in the use of leech therapy in close proximity to the central nervous system (Ouderkirk *et al.*, 2004).

Hospital-acquired *Aeromonas*-associated wound infections have also been described, but these are more likely to arise as a result of translocation of endogenous gut flora secondary to intervention. There have been rare cases of indwelling catheter-related *Aeromonas* bacteraemia and wound infections post laparotomy (Janda and Abbott, 1998). *Aeromonas* has also been recovered from hospital water supplies. There have been observations of a seasonal variation in nosocomial *A. hydrophilia* infection and in the number of *Aeromonas* spp. in hospital water supplies. A high prevalence of *A. hydrophilia* infection in the summer months correlated with maximal proliferation rates in hospital storage tanks (Picard and Goullet, 1986). Other studies have failed to demonstrate an epidemiological link. Millership, Stephenson and Tabaqchali (1988) investigated the occurrence of *Aeromonas* spp. in hospital water supplies. The isolation rate among hospital patients was 6% compared to 3.6% in individuals from the community. When molecular typing of the isolates was performed, no relationship could be established between water and clinical samples.

Bacteraemia

Aeromonas species that have been isolated from blood cultures include *A. hydrophilia*, *A. veronii* (both biotypes), *A. jandaie* and *A. cavaie* (Janda *et al.*, 1994). There have also been less-frequent case reports of *A. shubertii* bacteraemia (Hickman-Brenner *et al.*, 1988; Abbott *et al.*, 1998). The most common groups of patients who are susceptible to *Aeromonas*-associated bacteraemia are immunocompromised adults and infants under the age of 2 with underlying medical conditions. In these groups of patients *Aeromonas* bacteraemia can occur as a result of haematogenous translocation of the organisms from their gastrointestinal tracts. There are a variety of premorbid conditions that are associated with an increased risk of *Aeromonas* bacteraemia. These include pancreatitis, trauma, anaemia, gastrointestinal disorders, respiratory and cardiac abnormalities. However, in most cases *Aeromonas* bacteraemia is associated with underlying malignancy, hepatobiliary disease and diabetes mellitus. Mortality rates can be as high as 50% in these groups (Janda and Abbott, 1996). Clinical presentations and prognostic factors of 104 cases of monomicrobial *Aeromonas* bacteraemia were analysed by Ko *et al.* (2000) in Taiwan. The majority of infections occurred in patients with hepatic cirrhosis and malignancy. The crude fatality rate at 2 weeks was 32% and the outcome was worse in patients who had secondary bacteraemia where a source of infection could be identified and those with more severe illness at presentation such as septic shock.

The fatality rate is greater than 90% in the subgroup of patients who develop bacteraemia as a result of severe wound infection and myonecrosis secondary to trauma. Infection occurs as a result of exogenous sources such as exposure to freshwater (Janda and Abbott, 1996). In patients who develop bacteraemia secondary to burns the mortality rate approaches 67% (Ko and Chuang, 1995). There is also a smaller subgroup of adults in whom no underlying premorbid condition can be found. These patients do not have a history of exposure to freshwater and the portal of entry cannot be determined (Janda and Abbott, 1996).

Polymicrobial bacteraemias with members of the Enterobacteriaceae family, *Pseudomonas* spp., *Streptococci* spp. and *Enterococci* spp., have also been reported. Some surveys have found that certain *Aeromonas* species are more likely to participate in polymicrobial bacteraemia in patients with certain underlying medical conditions, for example, in patients with underlying malignancy. In these patients the frequency of polymicrobial bacteraemia involving *A. caviae* is much greater than the other species (Janda *et al.*, 1994).

Respiratory Tract Infections

Although rare, *Aeromonas* spp. have been associated with infections of the upper and lower respiratory tract such as epiglotitis, pharyngitis, pneumonia, empyema and pulmonary abscess formation. Immuno-competent individuals can acquire infection from aspirating freshwater associated with swimming, near drowning or other accidents. Patients with underlying medical problems or who are immunocompromised develop pulmonary infection as a result of haematogenous dissemination from their gastrointestinal tract. Respiratory infections can be associated with concomitant *Aeromonas* bacteraemia (Janda and Abbott, 1998).

The mortality rate is approximately 50% and is not related to the presence of underlying disease. Factors indicative of poor prognosis include pneumonia, haemoptysis, rapidly progressive infection and bacteraemia (Goncalves *et al.*, 1992).

Peritonitis

Peritonitis is an uncommon but serious consequence of *Aeromonas* spp. infection or colonisation with a mortality rate of up to 60%. The majority of cases are associated with bacteraemia. *Aeromonas hydrophilia* is predominantly isolated although the other clinically significant *Aeromonas* spp. have also been implicated. In a review of 34 cases, *Aeromonas*-associated peritonitis occurred primarily in patients with underlying chronic hepatic impairment; this was followed by patients with chronic renal failure undergoing chronic ambulatory peritoneal dialysis and those with intestinal perforation (Munoz *et al.*, 1994).

Haemolytic Uraemic Syndrome

Haemolytic uraemic syndrome (HUS) associated with *A. hydrophila* enterocolitis has been reported. This extremely rare condition occurred in a 23-month-old infant who developed HUS 6 days after an episode of abdominal pain and bloody diarrhoea. *Aeromonas hydrophila* was isolated from the patient's stool and produced cytotoxin against Vero cells. A rising titre of cytotoxin-neutralising antibody was demonstrated in the serum (Bogdanovic *et al.*, 1991). In a subsequent study 82 cases of HUS were investigated. There were two cases in which a link to recent *A. hydrophilia* infection could be established. It was felt that *Aeromonas* spp. should be recognised as a sporadic cause of HUS secondary to a diarrhoeal illness (Robson, Leung and Trevenen, 1992). The cytotoxin produced by *A. hydrophilia* has been shown to be genetically and antigenically unrelated to the shiga toxin produced by *E. coli* 0157:H7 (Janda, 1991).

Other Infections

In addition there have been documented cases of meningitis (Ouderkirk *et al.*, 2004) and ocular infections including corneal ulceration (Puri *et al.*, 2003), endophthalmitis (Lee, O'Hogan and Dal Pra, 1997) and keratitis associated with contact lens wear (Pinna *et al.*, 2004).

Plesiomonas

Plesiomonas shigelloides has been implicated as an uncommon cause of diarrhoea in humans. Extragastrointestinal manifestations are rare.

Gastroenteritis

Plesiomonas shigelloides has been isolated from stools of patients with diarrhoea. The incidence is higher in tropical and subtropical regions such as the Indian subcontinent and Southeast Asia. In Europe, America and Canada cases of travellers diarrhoea have also been reported. Patients with *P. shigelloides*-associated gastroenteritis present with a variety of symptoms associated with diarrhoea including abdominal pain, tenesmus, nausea, vomiting, lethargy, rigors and headache. The incubation period can range from 1 to 9 days. There have been reports of large-volume secretory diarrhoea as well as of bloody mucoid diarrhoea as a result of invasive disease. Infection can occur in immunocompetent individuals, and the duration of illness in untreated cases is approximately 11 days. In some cases, the bacterium can be isolated from stool samples more than 2 months following the initial diarrhoeal episode suggesting chronic infection. However, there have also been reports of asymptomatic carriage of *P. shigelloides*, and volunteer studies have failed to establish an aetiological link between *P. shigelloides* and gastroenteritis. Evidence to support the view that *P. shigelloides* can act as an enteropathogen comes predominantly from a number of published case reports and case–control studies (Jagger, 2000).

Infection with *P. shigelloides* is acquired mainly through aquatic sources and is often associated with contamination of water sources with sewage. The incidence is higher in the summer months reflecting the increased proliferation of the organisms in warmer temperatures. Infection can also arise as a result of inadequately treated drinking water. This was thought to be the source of an outbreak of *P. shigelloides*-associated gastroenteritis in Japan affecting 978 people (Tsukamoto *et al.*, 1978). In addition there have been reports of *P. shigelloides*-associated diarrhoea in individuals who have been swimming or participating in other recreational activities in fresh- and seawater (Soweid and Clarkston, 1995).

Infection with *P. shigelloides* can also occur as a consequence of eating fish, shellfish and crustaceans. It is thought that inadequate refrigeration or cooking contributes to food-borne infection.

Bacteraemia and Other Infections

Bacteraemia is a rare consequence of infection with *P. shigelloides*. In many cases the source is unknown. The majority of affected individuals are either neonates or immunocompromised adults. The mortality rate can be as high as 62% (Lee *et al.*, 1996).

Other extragastrointestinal infections that have been described include meningitis, endophthalmitis, cellulitis, wound infections and cholecystitis in immunocompromised individuals. Cases of osteomy-elitis and septicaemia have also been described in immunocompetent patients (Ingram, Morrison and Levitz, 1987). Recently two cases of peritonitis in patients undergoing chronic ambulatory renal dialysis have been described (Woo *et al.*, 2004).

LABORATORY DIAGNOSIS

Aeromonas spp. and *Plesiomonas* spp. grow well on 5% sheep blood, chocolate and MacConkey blood agars. They also grow well in the broth of blood culture systems and in thioglycollate or brain–heart infusion broths. *Plesiomonas* spp. does not grow so well on thiosulphate citrate bile sucrose media.

On 5% sheep blood agar *Aeromonas* spp. colonies are large, round, raised and opaque. Most colonies are β-haemolytic except *A. caviae*, which is non-haemolytic. In comparison, colonies of *P. shigelloides* appear smooth, shiny, opaque and are generally non-haemolytic.

On MacConkey agar *Aeromonas* spp. and *Plesiomonas* spp. can be lactose fermenters or non-lactose fermenters, although *Plesiomonas* spp. does not ferment lactose on most enteric media.

Selective techniques are often needed to isolate *Aeromonas* spp. and *Plesiomonas* spp. from mixed cultures. Enteric media may inhibit some members of *Aeromonas* spp., and thus isolation of the bacterium from stool cultures can sometimes be difficult. Either cefsulodin-irgasan-novobiocin agar or blood agar that contains ampicillin (10 or 30 µg/ml) can be used as a selective medium (Janda, Abbott and Carnahan, 1995). The plates can be examined for growth after 24 h. Selective media for the isolation of *Plesiomonas* spp. can be used and includes bile peptone broth and trypticase soy broth with ampicillin (Rahim and Kay, 1988). *Plesiomonas* spp. grow well at 35 °C producing colonies within 24 h.

The oxidase test distinguishes *Aeromonas* spp. and *Plesiomonas* spp. (positive) from other Enterobacteriaceae (negative).

The string test can be used to distinguish *Vibrios* spp. from *Aeromonas* spp. and *Plesiomonas* spp. When the organisms are emulsified in 0.5% sodium deoxycholate, the cells of *Vibrios* spp. are lysed and this can be detected by the subsequent release of DNA which can be pulled into a string using an inoculating loop. The cells of *Aeromonas* spp. and *Plesiomonas* spp. are unaffected.

The vibriostatic test using 0/129 (2,4-diamino-6,7-diisopropylpteridine) impregnated discs has also been used to separate the vibrios which are susceptible from other oxidase-positive glucose fermenters (resistant). However, recently strains of *V. cholerae* 01 have been found to be resistant to 0/129 and hence the dependability of this test is becoming questionable.

Some commercial kits are unable to distinguish between *Aeromonas* spp. and *Vibrios* spp. Hence, the identification of *Aeromonas* spp. and *P. shigelloides* should be confirmed using conventional biochemical or serotyping methods. The biochemical and biophysical features that may be used to differentiate members of the family Vibrionaceae are summarised in Table 35.3. Differentiation between *Aeromonas* spp. and *Plesiomonas* spp. requires further biochemical analysis (Table 35.4).

Most commercial identification systems fail to accurately identify aeromonads to a species level. Thus many systems proceed no further than identification to genus level reporting an *Aeromonas* isolate as 'Aeromonas species' or 'Aeromonas hydrophilia complex'. To solve this problem various methods have been used to identify mesophilic *Aeromonas* spp. to a species level (Table 35.4). For example, *A. hydrophilia* is motile, catalase positive, urease negative and converts nitrate to nitrite.

Abbott *et al.* have recently developed phenotypic identification schemes to species level for *Aeromonas* spp. based on initial Moeller decarboxylase and dihydrolase reactions. Several new tests have also been described to help separate members of the *A. cavaiae* complex (*A. caviae*, *A. media* and *A. eucreonophilia*) which included utilisation of citrate, DL-lactate, urocanic acid and fermentation of glucose 1-phosphate, glucose 6-phosphate, lactulose and D-mannose (Abbott, Cheung, Janda, 2003).

PCR-based 16S and 23S rRNA gene analysis have also been used to identify *P. shigelloides* and *Aeromonas* spp. to genomic species level (Ruimy *et al.*, 1994; Gonzalez-Rey *et al.*, 2000; Laganowska and Kaznowski, 2004). In addition the rpoD sequence that encodes sigma factor and the gyrB sequence, a gene that encodes the B-subunit of DNA gyrase, have been utilised to establish phylogenetic relationships between *Aeromonas* species. The results corroborate current taxanomic divisions (Yanez *et al.*, 2003; Soler *et al.*, 2004). Furthermore, gyrB sequence analysis was superior to 16S rRNA at identifying different *Aeromonas* species and has therefore been proposed as a useful marker for identification of *Aeromonas* strains (Yanez *et al.*, 2003).

MANAGEMENT

The management of extragastrointestinal infections secondary to *Aeromonas* spp. or *Plesiomonas* spp. relies on prompt diagnosis and treatment with antibiotics to which the organisms are susceptible. Most gastrointestinal infections are short lived and therefore the majority of patients do not require antibiotics.

Aeromonas

A recent study investigated the antibiotic susceptibility pattern of *A. veronii biotype sobria*, *A. caviae*, *A. jandaei* and *A. hydrophila* isolated from faecal specimens from patients with gastroenteritis. All strains were resistant to ampicillin but susceptible to cefotaxime, ciprofloxacin and nalidixic acid. The susceptibility to chloramphenicol, tetracycline and trimethoprim-sulfamethoxazole varied (Vila *et al.*, 2003). *Aeromonas* spp. can possess a conjugative plasmid that confers multiple antibiotic resistance (Chang and Bolton, 1987). *Aeromonas* spp.

Table 35.3 Differentiation of the genera within the family Vibrionaceae (adapted from Blair, McMahon and McDowell, 1999)

Characteristics	Aeromonas	Plesiomonas	Vibrio	Photobacterium	Enhydrobacter
Motility	+[b]	+	+	+	−
Na+ stimulates/required for growth	−	−	+	+	−
Sensitivity to 0/129	−	+[b]	+[b]	+	−
Lipase	+[b]	−	+[b]	Variable	+
D-Mannitol utilisation	+[b]	−	+[b]	−	−

[b] Most species/strains positive.

Table 35.4 Differentiation between *Aeromonas* spp. and *Plesiomonas* spp.

Species	Oxidase	Fermentation of				Lysine decarboxylase	Arginine dihydrolase	Ornithine decarboxylase
		Glucose (gas)	Lactose	Sucrose	Myo-inositol			
A. hydrophilia	+	+	−	+	−	+	+	−
A. caviae	+	−	+	+	−	−	+	−
A. veronii biotype veronii	+	+	−	+	−	+	−	+
A. jandaei	+	+	−	−	−	+	+	−
A. shubertii	+	−	−	−	−	+	+	−
P. shigelloides	+	−	−	−	+	+	+	+

+, >90% of strains positive; −, >90% of strains negative.

can also produce chromosomal β-lactamases such as Bush group 2d penicillinase, group 1 cephalosporinase and a metallocarbapenemase (Walsh *et al.*, 1995). Unlike other known carbapenamases the metallocarbapenemse found in *Aeromonas* spp. have narrow substrate profile and only hydrolyse carbapenams (Rossolini, Walsh and Amicosante, 1996). Despite this, many isolates of the clinically significant *Aeromonas* spp., with the exception of *A. jandaei* and *A. veronii biotype veronii*, remain susceptible to imipenem, with a minimal inhibitory concentrations observed (Overman and Janda, 1999). In addition to strains in which β-lactamase production may be induced, derepressed muntants for β-lactamase production have been isolated from clinical and environmental specimens (Walsh *et al.*, 1997).

Plesiomonas

P. shigelloides strains are usually susceptible to second- and third-generation cephalosporins, nalidixic acid, quinolones, co-trimoxazole, chloramphenicol and nitrofurantoin. Resistance to aminoglycosides is variable. Most *P. shigelloides* strains produce β-lactamase and are therefore resistant to all penicillins; however these isolates are susceptible to penicillins in combination with β-lactamase inhibitors.

Stock and Wiedemann investigated the natural susceptibility of 74 *P. shigelloides* strains isolated from humans, water and animals to a range of antibiotics. *Plesiomonas* strains were sensitive or displayed intermediate sensitivity to tetracyclines, several aminoglycosides, aminopenicillins in combination with β-lactamase inhibitors, all cephalosporins (except cefoperazone, ceftazidime and cefepime), carbapenems, aztreonam, quinolones, trimethoprim, sulfamethoxazole, azithromycin, chloramphenicol, nitrofurantoin and fosfomycin. Uniform resistance was found to all penicillins, roxithromycin, clarithromycin, lincosamides, streptogramins, glycopeptides and fusidic acid. Variable resistance to streptomycin, erythromycin and rifampicin was detected (Stock and Wiedemann, 2001).

Treatment of *P. shigelloides*-associated infection should be guided by sensitivity testing of the isolate. Quinolones may be used to treat *P. shigelloides*-associated gastroenteritis and in some cases antibiotic treatment has been shown to decrease the duration of diarrhoeal illness.

PREVENTION AND CONTROL

There have been no reports of person-to-person spread of infection; therefore special infection-control procedures are unnecessary. Proper cooking of food will prevent most food-borne infections as the vegetative cells and many of the exoenzymes and exotoxins are destroyed by heat challenge. However both *Aeromonas* spp. and *Plesiomonas* spp. have been shown to produce heat-stable exotoxins.

REFERENCES

Abbott, S. L., Cheung, W. K. W., Janda, J. M. (2003). The genus *Aeromonas*: biochemical characteristics, atypical reactions, and phenotypic identification schemes. J Clin Microbiol, 41, 2348–2357.

Abbott, S. L., Cheung, W. K. W., Kroske-Bystrom, S. *et al.* (1992) Identification of *Aeromonas* strains to the genospecies level in the clinical laboratory. J Clin Microbiol, 30, 1262–1266.

Abbott, S. L., Kokka, R. P., Janda, J. M. (1991) Laboratory investigations on the low pathogenic potential of *Plesiomonas shigelloides*. J Clin Microbiol, 29, 148–153.

Abbott, S. L., Seli, L. S., Catino, M. *et al.* (1998) Misidentification of unusual species as a member of the genus *Vibrio*: a continuing problem. J Clin Microbiol, 36, 1103–1104.

Abrami, L., Fivaz, M., Decroly, E. *et al.* (1998) The pore-forming toxin proaerolysin is activated by furin. J Biol Chem, 273, 32656–32661.

Ali, A., Carnahan, A. M., Altwegg, M. *et al.* (1996) *Aeromonas bestiarum* sp. nov. (formerly genomospecies DNA group 2 *A. hydrophila*), a new species isolated from non-human sources. Med Microbiol Lett, 5, 156.

Barghouthi, S., Young, R., Olsen, M. O. J. *et al.* (1989) Amonabactin, a novel tryphtophan- or phenylalanine-containing phenolate siderophore in *Aeromonas hydrophilia*. J Bacteriol, 171, 1811–1816.

Blair, I. S., McMahon, M. A. S., McDowell D. A. (1999) Aeromonas. An Introduction, p. 25–30. In: Richard, C. A. B. Robinson, K., Pradip Patel, eds. Encyclopaedia of food microbiology,1st ed., Vol.1, Academic Press.

Bloom, H. G., Bottone, E. J. (1990) *Aeromonas hydrophilia* diarrhoea in long term care setting. J Am Geriatr Soc, 38, 804–806.

Bogdanovic, R., Cobeljic, M., Markovic, V. *et al.* (1991) Haemolytic uraemic syndrome associated with *Aeromonas hydrophilia* enterocolitis. Pediatr Nephrol, 5, 293–295.

Champsaur, H., Andremont, A., Mathieu, D. *et al.* (1982) Cholera like illness due to *Aeromonas sobria*. J Infect Dis, 145, 248–254.

Chang, B. J., Bolton, S. M. (1987) Plasmids and resistance to antimicrobial agents in *Aeromonas sobria* and *Aeromonas hydrophilia* clinical isolates. Antimicrobial Agents Chemother, 31, 1281–1282.

Colwell, R. R., MacDonell, M. T., De Ley, J. (1986) Proposal to recognise the family *Aeromonadaceae* fam. Nov. Int J Syst Bacteriol, 36, 473–477.

Dabb, R. W., Malone, J. M., Leverett, L. C. (1992) The use of medicinal leeches in the salvage of skin flaps with venous congestion. Ann Plast Surg, 29, 250–256.

de la Morena, M. L., Van, R., Singh, K. *et al.* (1993) Diarrhoea associated with *Aeromonas* spp. in children in day care centres. J Infect Dis, 168, 215–218.

Dooley, J. S. G., Trust, T. J. (1988) Surface protein composition of *Aeromonas hydrophila* strains virulent for fish: identification of a surface array protein. J Bacteriol, 170, 499–506.

Esteve, C., Gutiérrez, M. C., Ventosa, A. (1995) *Aeromonas encheleia* sp. nov., isolated from European eels. Int J Syst Bacteriol, 45, 462–466.

Figueras, M. J., Suarez-Franquet, A., Chacon, M. R. *et al.* (2005) First record of the rare species *Aeromonas culicicola* from a drinking water supply. Appl Environ Microbiol, 71, 538–541.

Goncalves, J. R., Brum, G., Fernandes, A. *et al.* (1992) *Aeromonas hydrophilia* fulminant pneumonia in a fit young man. Thorax, 47, 482–483.

Gonzalez-Rey, C., Svenson, S. B., Bravo, L. *et al.* (2000) Specific detection of *Plesiomonas shigelloides* isolated from aquatic environments, animals and human diarrhoeal cases by 23S rRNA gene. FEMS Immunol Med Microbiol, 29, 107–113.

Gosling, P. J. (1996) Pathogenic mechanisms. In: Austin, B. *et al.*, eds. The genus *Aeromonas*. London: Wiley, 245–265.

Gracey, M., Burke, V., Robinson, J. (1982) *Aeromonas* associated gastroenteritis. Lancet, ii, 1304–1306.

Hanninen, M. L., Siitonen, A. (1995) Distribution of *Aeromonas* phenospecies and genospecies among strains isolated from water, foods or from human clinical samples. Epidemiol Infect, 115, 39–50.

Hickman-Brenner, F. W., MacDonald, K. L., Steiger-Walt, A. G. *et al.* (1987) *Aeromonas veronii*, a new ornithine decarboxylase-positive species that may cause diarrhoea. J Clin Microbiol, 25, 900–906.

Hickman-Brenner, F. W., Fanning, G. R., Arduino, M. J. *et al.* (1988) *Aeromonas shubertii*, a new mannitol-negative species found in human clinical specimens. J Clin Microbiol, 26, 1561–1564.

Ho, A. S. Y., Sohel, I., Schoolnik, G. K. (1992) Cloning and characterization of fxp, the flexible pilin gene of *Aeromonas hydrophila*. Mol Microbiol, 6, 2725–2732.

Holmes, P., Niccolls, L. M., Sartory, D. P. (1996) The ecology of mesophilic *Aeromonas* in the aquatic environment. In: Austin, B. Altwegg, M. Gosling, P. J. Joseph, S., eds. The genus *Aeromonas*. Chichester, England: John Wiley & Sons, 127–150.

Howard, S. P., Buckley, J. T. (1985) Activation of the hole-forming toxin aerolysin by extracellular processing. J Bacteriol, 163, 336–340.

Huys, G., Kämpfer, P., Altwegg, M. *et al.* (1997) *Aeromonas popoffii* sp. nov., a mesophilic bacterium isolated from drinking water production plants and reservoirs. Int J Syst Bacteriol, 47, 1165–1171.

Ingram, C. W., Morrison, A. J., Levitz, R. E. (1987) Gastroenteritis, sepsis, osteomyelitis caused by *Plesiomonas shigelloides* in an immunocompetent host: case report and review of the literature. J Clin Microbiol, 25, 1791–1793.

Jagger, T. D. (2000) *Pleisomonas shigelloides* – a veterinary perspective. Rev Infect Dis, 2, 199–210.

Janda, J. M. (1991) Recent advances in the study of the taxonomy, pathogenicity, and infectious syndromes associated with the genus *Aeromonas*. Clin Microbiol Rev, 4, 397–410.

Janda, J. M. (1998) *Vibrio, Aeromonas* and *Plesiomonas*. In: Collier, L., Balowsa, A., Sussman, M., eds. Topley and Wilson's microbiology and microbial infections. Systemic bacteriology, Vol. 2, Arnold, 1065–1089.

Janda, J. M., Abbott, S. L. (1996) Human pathogens. In: Austin, B. Altwegg, M. Gosling, P. J. Joseph, S., eds. The genus *Aeromonas*. Chichester, England: John Wiley & Sons, 151–173.

Janda, J. M., Abbott, S. L. (1998) Evolving concepts regarding the genus *Aeromonas*: an expanding panorama of species, disease presentations, and unanswered questions. Clin Infect Dis, 27, 332–344.

Janda, J. M., Abbott, S. L., Carnahan, A. M. (1995) *Aeromonas* and *Plesiomonas*. In: Murray, P. R. Baron, E. J. Pfaller, M. A. *et al.*, eds. Manual of clinical microbiology. Washington DC: American Society for Microbiology, 477–482.

Janda, J. M., Gutherz, L. S., Kokka, R. P. *et al.* (1994) *Aeromonas* species in septicaemia: laboratory characteristics and clinical observations. Clin Infect Dis, 19, 77–83.

Janda, J. M., Oshiro, L. S., Abbott, S. L. *et al.* (1987) Virulence markers of mesophilic *Aeromonas*: association of autoagglutination phenomenon with mouse pathogenicity and the presence of a cell-associated layer. Infect Immun, 55, 3070–3077.

Kirov, S. M., Tassell, B. C., Semmler, A. B. T. *et al.* (2002) Lateral flagella and swarming motility in *Aeromonas* species. J Bacteriol, 184, 547–555.

Ko, W. C., Chuang, Y. C. (1995) *Aeromonas* bacteraemia: review of 59 episodes. Clin Infect Dis, 20, 1298–1304.

Ko, W. C., Lee, H. C., Chuang, Y. C. *et al.* (2000) Clinical features and therapeutic implications of 104 episodes of monobacterial *Aeromonas* bacteraemia. J Infect Dis, 40, 267–273.

Laganowska, M., Kaznowski, A. (2004) Restriction fragment length polymorphism of 16S–23S rDNA intergenic spacer of *Aeromonas* spp. Syst Appl Microbiol, 27, 549–557.

Lee, L. R., O'Hogan, S., Dal Pra, M. (1997) *Aeromonas sobria* endophthalmitis. Aust N Z J Ophthalmol, 25, 299–300.

Lee, A. C. W., Yeun, K. Y., Ha, S., Y. *et al.* (1996) *Plesiomonas shigelloides* septicaemia: case report and literature review. Pediatr Hematol Oncol, 13, 265–269.

Martinez, M. J., Simon-Pujol, D., Congregado, F. *et al.* (1995) The presence of capsular polysaccharide in mesophilic *Aeromonas hydrophila* serotypes O:11 and O:34. FEMS Microbiol Lett, 128, 69–73.

Martinez-Murcia, A. J., Esteve, C., Garay, E. *et al.* (1992) *Aeromonas allosaccharophila* sp. nov., a new mesophilic member of the genus *Aeromonas*. FEMS Microbiol Lett, 70, 199–205.

Matthews, B. G., Douglas, H., Guiney, D. G. (1988) Production of a heat stable enterotoxin by *Plesiomonas shigelloides*. Microb Pathog, 5, 207–213.

Mercer, N. S., Beere, D. M., Bornemisza, A. J. *et al.* (1987) Medicinal leeches as sources of wound infection. BMJ, 294, 937–938.

Merino, S., Camprubi, S., Tomàs, J. M. (1992) Effect of growth temperature on outer membrane components and virulence of *Aeromonas hydrophila* strains of serotype O:34. Infect Immun, 60, 4343–4349.

Merino, S., Gavín, R., Vilches, S. *et al.* (2003) A colonization factor (production of lateral flagella) of mesophilic *Aeromonas* spp. is inactive in *Aeromonas salmonicida* strains. Appl Environ Microbiol, 69, 663–667.

Merino, S., Rubires, X., Knochel, S. *et al.* (1995) Emerging pathogens: *Aeromonas* spp. Int J Food Microbiol, 28, 157–168.

Millership, S. E., Barer, M. R., Tabaqchali, S. (1986) Toxin production by *Aeromonas* spp. from different sources. J Med Microbiol, 22, 311–314.

Millership, S. E., Stephenson, J. R., Tabaqchali, S. (1988) Epidemiology of *Aeromonas* species in a hospital. J Hosp Infect, 11, 169–175.

Minana-Galbis, D., Farfan, M., Fuste, M. C. *et al.* (2004) *Aeromonas molluscorum* sp. nov., isolated from bivalve molluscs. Int J Syst Evol Microbiol, 54, 2073–2078.

Morgan, D. R., Johnson, P. C., DuPont, H. L. *et al.* (1985) Lack of correlation between known virulence properties of *Aeromonas hydrophilia* and enteropathogenicity for humans. Infect Immun, 50, 62–65.

Munoz, P., Fernandez-Baca, V., Pelaez, T. *et al.* (1994) *Aeromonas peritonitis*. Clin Infect Dis, 18, 32–37.

Ouderkirk, J. P., Bekhor, D., Turett, G. S. *et al.* (2004) *Aeromonas meningitis* complicating medicinal leech therapy. Clin Infect Dis, 38, 36–37.

Overman, T. L., Janda, J. M. (1999) Antimicrobial susceptibility patterns of *Aeromonas jandaei*, *A. schubertii*, *A. trota*, *A. veronii biotype veronii*. J Clin Microbiol, 37, 706–708.

Picard, B., Goullet, Ph (1986) Seasonal prevalence of nosocomial *Aeromonas hydrophilia* infection related to *Aeromonas* in hospital water. J Hosp Infect, 10, 152–155.

Pidiyar, V., Kaznowski, A., Badri, N. N. *et al.* (2002) *Aeromonas culicicola* sp. nov., from the midgut of *Culex quinquefasciatus*. Int J Syst Evol Microbiol, 52, 1723–1728.

Pinna, A., Sechi, L. A., Zanetti, S. *et al.* (2004) *Aeromonas cavaie* keratitis associated with contact lens wear. Ophthalmology, 111, 348–351.

Pitarangsi, C., Echeverria, P., Whitmire, R. *et al.* (1982) Enteropathogenicity of *Aeromonas hydrophilia* and *Plesiomonas shigelloides*: prevalence among individuals with and without diarrhoea in Thailand. Infect Immun, 35, 666–673.

Popoff, M. (1984) *Aeromonas*. In: Holt, J. G., ed. Bergey's manual of determinative bacteriology, Vol. 1, pp. 545–548.

Popoff, M., Vernon, M. (1976) A taxonomic study of *Aeromonas hydrophilia–Aeromonas punctata* group. J Gen Microbiol, 94, 11–22.

Puri, P., Bansal, V., Dinakaran, S. *et al.* (2003) *Aeromonas sobria* corneal ulcer. Eye, 17, 104–105.

Rahim, Z., Kay, B. A. (1988) Enrichment for *Plesiomonas shigelloides* from stool. J Clin Microbiol, 26, 789–790.

Rautelin, H., Sivonen, A., Kuikka, A. *et al.* (1995) The role of *Aeromonas* isolated from faeces of Finnish patients. Scand J Infect Dis, 27, 207–210.

Roberts, I. M., Parenti, D. M., Albert, M. B. (1987) *Aeromonas hydrophilia*-associated colitis in a male homosexual. Arch Intern Med, 147, 1502–1503.

Robson, W. L. M., Leung, A. K. C., Trevenen, C. L. (1992) Haemolytic uraemic syndrome associated with *Aeromonas hydrophilia* enterocolitis. Pediatr Nephrol, 6, 221–222.

Rossolini, G. M., Walsh, T., Amicosante, G. (1996) The *Aeromonas* metallo-beta-lactamases: genetics, enzymology and contribution to drug resistance. Microb Drug Resist, 2, 245–252.

Ruimy, R., Breittmayar, V., Elbaze, P. *et al.* (1994) Phylogenetic analysis and assessment of the genera *Vibrio*, *Photobacterium*, *Aeromonas* and *Plesiomonas* from small subunit rRNA sequences. Int J Syst Bacteriol, 44, 416–426.

Sakazaki, R., Shimada, T., (1984) O-serogrouping scheme for mesophilic *Aeromonas* strains. Jpn J Med Sci Biol, 37, 247–255.

Soler, L., Yanez, M. A., Chacon, M. R. *et al.* (2004) Phylogenetic analysis of the genus *Aeromonas* based on two housekeeping genes. Int J Syst Evol Microbiol, 54, 1511–1519.

Soweid, A. M., Clarkston, W. K. (1995) *Plesiomonas shigelloides*: an unusual cause of diarrhoea in cats. Am J Gastroenterol, 90, 2235–2236.

Stock, I., Wiedemann, B. (2001) Natural antimicrobial susceptibilities of *Plesiomonas shigelloides* strains. J Antimicrob Chemother, 48, 803–811.

Tanoue, N., Takahashi, A., Okamoto, K. *et al.* (2005) A pore-forming toxin produced by *Aeromonas sobria* activates cAMP-dependent Cl(−) secretory pathways to cause diarrhoea. FEMS Microbiol Lett, 242, 195–201.

Tsukamoto, T., Kinoshita, Y., Shimada, *et al.* (1978) Two epidemics of diarrhoeal disease possibly caused by *Plesiomonas shigelloides*. J Hyg Camb, 80, 275–280.

Vally, H., Whittle, A., Cameron, S. *et al.* (2004) Outbreak of *Aeromonas hydrophila* wound infections associated with mud football. Clin Infect Dis, 38, 1084–1089.

Vila, J., Ruiz, J., Gallardo, F. *et al.* (2003) *Aeromonas* spp. and travellers diarrhoea: clinical features and antimicrobial resistance. Emerg Infect Dis, 9, 552–555.

Voss, L. M., Rhodes, K. H., Johnson, K. A. (1992) Musculoskeletal and soft tissue *Aeromonas* infection: an environmental disease. Mayo Clin Proc, 67, 422–427.

Wakabongo, M. (1995) *Aeromonas* as agents of infection of the foot. J Am Podiatr Med Assoc, 85, 505–508.

Walsh, T. R., Payne, D. J., MacGowan, A. P. *et al.* (1995) A clinical isolate of *Aeromonas sobria* with three chromosomally mediated inducible beta-lactamases: a cephalosporinase, a penicillinase and a third enzyme displaying carbapenamase activity. J Antimicrob Chemother, 35, 271–279.

Walsh, T. R., Stunt, R. A., Nabi, J. A. *et al.* (1997) Distribution and expression of beta-lactamase genes among *Aeromonas* spp. J Antimicrob Chemother, 40, 171–178.

Woo, P. C., Lau, S. K., Wong, S. S. *et al.* (2004) Two cases of continuous ambulatory peritoneal dialysis associated peritonitis due to *Plesiomonas shigelloides*. J Clin Microbiol, 42, 933–935.

Yanez, M. A., Catalan, V., Apraiz, D. *et al.* (2003) Phylogenetic analysis of members of the genus *Aeromonas* based on gyrB gene sequences. Int J Syst Evol Microbiol, 53, 875–883.

36

Pseudomonas and *Burkholderia* spp.

Tyrone L. Pitt[1] and Andrew J. H. Simpson[2]

[1]*Centre for Infection, Health Protection Agency, London; and* [2]*Biological Sciences, Defence Science and Technology Laboratory, Wiltshire, UK*

INTRODUCTION

Early Descriptions

In 1850, Sédillot, a French military surgeon, observed the formation of blue pus in the wound dressings of injured soldiers. Fordos also documented this, in 1860, but it was not until 1882 that Gessard described the organism responsible for the pigmentation, which he named *Bacillus pyocyaneus*. In 1900, Migula adopted the generic name *Pseudomonas* (Greek: 'pseudes', false; 'monas', unit) and called the species *Pseudomonas pyocyanea*. The epithet *aeruginosa* (Latin: aeruginosus, full of copper rust; i.e. green) became widely used and is now the approved species name. (For early references, see Forkner, 1960.)

Taxonomy

Up to 1984, over 100 species were included in the genus *Pseudomonas*. Many of these are plant pathogens, and their names reflect their primary hosts. The genus was subdivided into five groups based on rRNA homology (Palleroni, 1984), and today, only rRNA group I remains in the genus *Pseudomonas* (Table 36.1). It comprises 57 species defined by 16S rRNA phylogenetic analysis and includes the fluorescent species *Pseudomonas aeruginosa*, *Pseudomonas putida* and *Pseudomonas fluorescens* as well as the non-fluorescent *Pseudomonas stutzeri* (Anzai *et al.*, 2000). Organisms that were previously categorized as rRNA group II, 'the pseudomallei group', were transferred to the genus *Burkholderia* by Yabuuchi *et al.* (1992) and include the *Burkholderia cepacia* complex, *Burkholderia mallei*, *Burkholderia pseudomallei* and many plant pathogens. rRNA group III was formerly *Comamonas acidovorans* but is now incorporated into the genus *Delftia* (Wen *et al.*, 1999). Members of rRNA groups IV and V have been reclassified, respectively, as *Brevundimonas* and *Stenotrophomonas*.

DESCRIPTION OF ORGANISMS

The diversity of the pseudomonads makes it difficult to consider their properties as a single group. Furthermore, the bulk of the literature pertaining to *Pseudomonas* concerns specifically *P. aeruginosa*, and less so *B. cepacia* and *B. pseudomallei*, and the literature is relatively sparse for other species. As a result, this chapter deals with each species independently to avoid generalizations that might be inaccurate.

Table 36.1 Key tests for medically important pseudomonads

rRNA homology group		Species	Distinguishing features
I *Pseudomonas*	Fluorescent on King's B agar	*Pseudomonas aeruginosa* (oxidase+)	Arginine+, grows at 42 °C not 5 °C
		Pseudomonas fluorescens (oxidase+)	Arginine+, grows at 5 °C not 42 °C
		Pseudomonas putida (oxidase+)	Arginine+, grows at 5 °C
	Non-fluorescent	*Pseudomonas alcaligenes* (oxidase+)	Glucose−, motile+
		Pseudomonas pseudoalcaligenes (oxidase+)	Glucose−, fructose+
		Pseudomonas stutzeri (oxidase+)	NO_3+, arginine−, maltose+
II *Burkholderia*		*Burkholderia cepacia* (oxidase+)	Arginine−, lysine+, R-colistin, gentamicin
		Burkholderia pseudomallei (oxidase+)	Arginine+, lysine−, grows at 42 °C, R-colistin, gentamicin
		Burkholderia mallei (oxidase±)	Arginine+, motile−
		Ralstonia pickettii (oxidase+)	NO_3+, arginine−
		Burkholderia gladioli (oxidase−)	Lactose−, lysine−
III *Delftia*		*Delftia acidovorans* (oxidase+)	Asaccharolytic
IV *Brevundimonas*		*Brevundimonas diminuta* (oxidase+)	NO_3−, arginine−
		Brevundimonas vesicularis (oxidase+)	Aesculin hydrolysis
V *Stenotrophomonas*		*Stenotrophomonas maltophilia* (oxidase−)	Maltose+, DNAase+, R-imipenem, gentamicin

Principles and Practice of Clinical Bacteriology Second Edition   Editors Stephen H. Gillespie and Peter M. Hawkey

Cultural Characteristics of *P. aeruginosa*

Pigments

Pseudomonas aeruginosa grows well on simple bacteriological media, and most strains elaborate the blue phenazine pigment pyocyanin and fluorescein (yellow), which together impart the characteristic blue-green coloration to agar cultures. The production of pyocyanin is unique to the species, and this is enhanced by culture on King's A medium (King *et al.*, 1954), which contains potassium and magnesium salts in sufficient concentration to suppress fluorescein production. The latter, also called pyoverdine, is optimally produced on King's B medium, which contains less of these salts, and is visible under ultraviolet illumination. *Pseudomonas chlororaphis* may be confused with *P. aeruginosa* as it can grow as green colonies on King's A agar. Two other uncommon pigments are pyorubrin (port wine) and the brownish-black pigment pyomelanin, which is best demonstrated on tyrosine agar. This pigment should not be confused with the brown discoloration of blood agar by some weakly haemolytic strains.

Colonial Forms and Biochemistry

On agar culture at 37 °C, most isolates form flat diffuse colonies with an irregular edge (type 1), and some are coliform in appearance (type 2). Other colony types are generally rare. They include the dry, pepper-corn (type 3), mucoid (type 4), rugose (type 5) and dwarf colonies characteristic of type 6. Many cultures produce an ester-like odour owing to the formation of trimethylamine.

Pseudomonas aeruginosa is a Gram-negative rod of variable length, motile by means of a single polar flagellum, molecular oxygen being necessary for motility. Anaerobic growth is possible only in the presence of an alternative terminal electron acceptor such as nitrate or arginine. Motility tests in conventional Craigie agar tubes are therefore inappropriate. It grows over a wide temperature range (10–44 °C), but growth is optimal at around 35 °C. It will not grow at 4 °C, which distinguishes it from the psychrophilic species *P. putida* and *P. fluorescens*, but isolates will grow over three successive subcultures at 42 °C, while the latter do not.

In keeping with other pseudomonads, most strains utilize a range of single organic compounds as energy sources. The species metabolizes glucose and other sugars by an oxidative pathway. Acid is formed in peptone–water sugars by young cultures, but this is neutralized by alkali released by the breakdown of peptone. Sugar utilization is therefore best demonstrated in an ammonium salt-based medium. All strains produce a cytochrome oxidase enzyme that is detected in the oxidase test with Kovac's reagent. Other common features of *P. aeruginosa* are reduction of nitrate to nitrogen gas and the formation of ammonia from arginine.

PATHOGENICITY

Virulence-Associated Factors

Pseudomonas aeruginosa produces a wide array of cell-associated and secreted factors, some of which have been closely linked with disease-producing potential. These include exotoxin A (ETA), phospholipase, proteases, pyocyanin, pili, flagella and lipopolysaccharide (LPS). Nevertheless, in the absence of impairment of host defences, the species has low intrinsic virulence for humans and, to cause disease, must be introduced into the tissues, or systemically, in sufficient numbers to overwhelm the host defences. This feature probably explains some of the conflicting experimental evidence concerning the role of different virulence factors. Several animal models such as the burned mouse, neutropenic mouse (induced with cyclophosphamide or anti-neutrophil monoclonal antibodies) and the murine corneal scratch model have been used by different investigators. Certain isolates infect a broad range of host organisms from plants (*Arabidopsis*), nematode worms (*Caenorhabditis elegans*) to humans, and a universal genetic mechanism underlying the pathogenic process has been proposed (He *et al.*, 2004).

Like many Gram-negative bacteria, *P. aeruginosa* utilizes a type III secretion system that enables the organism to secrete and inject virulence factors into the cytosol of the host eukaryotic cells. The type III secretion system of *P. aeruginosa* transports four effector proteins (ExoS, ExoT, ExoU and ExoY), and the genes encoding these are present in nearly all clinical and environmental isolates, although individual isolates and populations from distinct disease sites differ in their effector gene complement (Feltman *et al.*, 2001). An additional protein, PcrV, which is a structural analogue of the *Yersinia* V antigen LcrV, is involved in the translocational process of the type III secretion system. Production of PcrV has been associated with strains from acute infections, and the risk of mortality was reported to be sixfold higher in patients infected with strains producing ExoS, ExoT, ExoU or PcrV, suggesting that these strains have increased virulence and that their presence is a predictor of poor clinical outcome (Roy-Burman *et al.*, 2001).

Adherence

Injury to the epithelial mucosa by viral infection or previous colonization by other bacteria predisposes tissue to the attachment of *P. aeruginosa*. The organism grows naturally as a biofilm either on a surface or attached to large glycoproteins such as those found in respiratory mucins. Adherence properties vary from strain to strain and the attachment matrix used, but four main groups of adhesins have been identified which mediate the attachment of *P. aeruginosa* to host tissues and mucins. They are pili, mucin-binding outer membrane (OM) protein F, surface lectins and mucoid exopolysaccharide (alginate).

The contribution of each class of adhesin to the attachment of strains of *P. aeruginosa* to different tissues is difficult to quantify. However, nearly all strains produce pili under favourable conditions. The main pilus adhesin is the type IV pilus, and they are responsible for motility of the bacteria across a solid surface (twitching motility) and for the binding and entry of some bacteriophages. They are the protein polymers (pilin) of a single gene *pilA*, but their assembly and function involves a large number of genes. Many studies have shown that pili are necessary for strain virulence at several sites of infection, and they may also be involved in cytotoxicity. *Pseudomonas aeruginosa* synthesizes two surface lectins termed PA-IL, binding to galactose, and PA-IIL (binds to fucose), which appear to function as adhesins similar to the fimbrial adhesins of uropathogenic *Escherichia coli*. They interact with the ABO(H) and P blood group glycosphingolipid antigens and, through this, contribute to tissue infectivity and pathogenesis of the species (Gilboa-Garber *et al.*, 1994).

Transcription of alginate biosynthetic genes is induced upon attachment of the pseudomonas cell to a surface, and increased alginate production ensues. The consortium of alginate and bacterial cells (microcolonies) is the basis of the biofilm mode of growth and protects the bacteria from the external environment, complement and antibiotics. Flagella and type IV pili do not appear to be necessary for biofilm formation, but they have a role in biofilm development (Klausen *et al.*, 2003). Exotoxin S also functions as an adhesin and is found on the cell surface, where it interacts with glycosphingolipid receptors.

Flagella

The loss of flagella by strains is strongly correlated with a decrease in virulence in the experimental animal. Flagella are therefore considered to be a significant virulence factor, in that they promote chemotaxis and motility during invasion of the tissues. The flagellin protein binds to respiratory mucins, and this facilitates the clearance of the organism from the airways via the normal mucociliary escalator. They can also function as ligands for macrophages and other phagocytic cells. Thus, although they are important in the establishment of respiratory tract infection and may tether the cell to epithelial membranes, flagella may also contribute to the host clearance mechanisms and the elimination of the organism (Feldman *et al.*, 1998). Most isolates of *P. aeruginosa* from patients with cystic fibrosis (CF) lack flagella and are generally resistant to macrophage phagocytosis, and so they may promote the establishment of chronic infection in these patients.

Lipopolysaccharide

The cell envelope of *P. aeruginosa*, as with other Gram-negative bacteria, is composed of an external unit membrane (OM), a layer of peptidoglycan and an inner cytoplasmic membrane. The latter is mainly composed of phospholipids with randomly intercalated molecules of proteins. *Pseudomonas aeruginosa* produces two chemically distinct types of LPS, a serotype specific (B-band) and a common antigen (A-band). The B-band LPS consists of three basic units: lipid A, core polysaccharide and O-specific side chain which determines the O-serotype of the organism. A-band LPS is antigenically uniform (Lam *et al.*, 1989). The relative expression of A-band and B-band LPS in *P. aeruginosa* may enable alteration of surface characteristics to survive under extreme conditions or to evade the host immune response. The levels of the two LPS species are influenced by the dissolved oxygen tensions, with the B-band variety being mostly absent from cells grown in reduced oxygen concentrations, which may explain the loss of O-serotype of isolates recovered from the near-anaerobic conditions found in biofilms of *P. aeruginosa* growing in the CF patient's lungs (Worlitzsch *et al.*, 2002).

The LPS of *P. aeruginosa* has a lower intrinsic toxicity for mammalian cells compared with the LPS of the Enterobacteriaceae. It plays a major role in protecting the cell from the complement-mediated bactericidal action of normal human serum. Most isolates of *P. aeruginosa* from clinical samples are resistant to serum, and the highest frequency of resistance is found in isolates from blood. However, isolates from chronic infection, in particular CF and bronchiectasis, and, less frequently, persistent urinary tract infection in paraplegics, are often sensitive to serum. LPS-deficient mutants are invariably serum sensitive and avirulent in the experimental animal. Furthermore, antibody to LPS is protective not only in animal models but also in humans. Evidence suggests that antibodies specific to O-antigens of *P. aeruginosa*, particularly the high-molecular-weight O-polysaccharide, can protect against mucosal surface colonization by the organism, and this is achieved through circulating antibody alone rather than by induction of local immunoglobulin A (IgA) antibodies (Pier, Meluleni and Goldberg, 1995).

Extracellular Polysaccharides

All strains of *P. aeruginosa* elaborate an extracellular slime that aggregates loosely around the cell. More than 50% of the dry weight of slime is composed of polysaccharide, chiefly of the sugars glucose, rhamnose and mannose, with galactosamine, glucuronic acid, nucleic acids (20%) and proteins as a minor fraction (Brown, Foster and Clamp, 1969). The alginate polysaccharide is hyper-produced by mucoid strains of *P. aeruginosa* and is composed of β-1,4-linked D-mannuronic acid and L-guluronic acid, the ratio of which confers the degree of viscosity on the polymer and hence the physical appearance of the mucoid colony. Mucoid strains are almost exclusively isolated from the sputum of patients with CF or those with bronchiectasis, but few strains (<2%) from other sources may grow mucoid on first isolation. The frequency of these forms may be higher for isolates from the urine of spinal injury patients who have had repeated urinary tract infections. Isolates elaborate mucoid exopolysaccharide best on glycerol-containing agar but often throw off non-mucoid colonies on subculture. Increased oxygenation of the medium favours the mucoid phenotype. Conversion of a non-mucoid strain to mucoid can be engineered by exposure of the organism to hydrogen peroxide and oxygen by-products of polymorphonuclear leucocytes (Mathee *et al.*, 1999). Alginate production by other species is rare, but *Azotobacter vinelandii* and some pseudomonads (*P. fluorescens*, *P. putida* and *Pseudomonas mendocina*) can manufacture the polymer.

Exoenzymes

The lethal toxin of *P. aeruginosa*, ETA, is produced by approximately 90% of the species and is closely related to diphtheria toxin. Both toxins catalyse the transfer of the ADP-ribosyl moiety of oxidized NAD (NAD^+) to elongation factor 2 (EF2). This reaction inactivates EF2, terminates peptide chain elongation and leads to the inhibition of protein synthesis and cell death. The toxin is composed of two fragments. Fragment A is of molecular weight 21 000 Da and has enzymic activity. Fragment B is 37 000 Da and acts as the binding moiety. ETA is optimally produced in iron-limited conditions and is encoded by a single copy of a structural gene, *toxA*, and regulated by two genes *regA* and *regB* on an operon (Wick *et al.*, 1990). It is highly toxic for mammalian cells (LD_{50} for mice <0.01 mg/kg) and enters the cell via a receptor-mediated endocytic pathway. ETA also has activity as a T-cell mitogen and interleukin-1 (IL-1) inducer (Nicas and Iglewski, 1986).

Pseudomonas aeruginosa produces another ADP-ribosyl transferase, exoenzyme S. It catalyses the transfer of ADP-ribose from NAD^+ to other eukaryotic proteins, and it is more resistant to heat and is not neutralized by antibodies to ETA. Exoenzyme S activity is also partially destroyed by the reducing agents urea and dithiothreitol, which potentiate ETA activity. Strains lacking only the *exoS* gene are less virulent in animal models, but 94% and 80% of CF and non-CF isolates, respectively, are reported to carry the gene (Lanotte *et al.*, 2004).

Two distinct haemolysins have been identified in *P. aeruginosa*: a heat-labile phospholipase C (PLC) and a heat-stable rhamnolipid. PLC is believed to be exclusive to the species and is found in almost all strains. It degrades phospholipids containing quaternary ammonium groups such as phosphatidylcholine, the major phospholipid component of lung surfactant. PLC is therefore a potential virulence factor particularly for strains colonizing the human lung. Rhamnolipid also degrades phospholipids because of its detergent action and, in high concentrations, attracts and lyses leucocytes and has ciliostatic action (Woods and Vasil, 1994).

Most strains of *P. aeruginosa* produce a range of proteolytic enzymes that are active against a variety of substrates such as gelatin, casein, elastin, collagen and fibrin. Three classes of protease have been identified: a general protease, elastase and alkaline protease. They are distinguishable by their optimum pH of activity, substrate specificity and physical properties (Woods and Vasil, 1994). The general protease is lysine specific and has a limited range of activity. Alkaline protease has not been widely studied but has activity against a wider range of substrates and has been implicated in corneal damage in *P. aeruginosa* eye infections. Elastase degrades many proteins such as human lung elastin, Igs, complement factors, basement membrane constituents and mucins. Elastolytic activity is the result of at least two proteins, LasA and LasB, under the control of the gene *lasR*. Elastase production by strains is controlled, at the translational level, by the levels of zinc and iron in the environment. Exoenzyme production has been associated with a pathogenic role in *P. aeruginosa*. A recent study of a large number of CF isolates found that isolates from patients with good clinical status often had higher elastase and neuraminidase activities than isolates from patients with worse clinical scores. In contrast, PLC activity was higher in isolates from patients with poor clinical status (Lanotte *et al.*, 2003).

Most *P. aeruginosa* organisms produce a cytotoxin that is localized to the periplasmic space as an inactive form, but is activated by proteases, including an endogenous elastase. Cytotoxic activity may be linked to the observed lack of leucocytes during sepsis due to *P. aeruginosa* and the fall in the granulocyte count in some patients. Experimental studies of the action of the cytotoxin in a leukopenic mouse model showed that challenge with a high cytotoxin-producing strain caused earlier and a significantly greater mortality than that observed for a cytotoxin-deficient strain. Moreover, pretreatment of mice with a cytotoxin-specific antibody decreased the mortality following challenge with the cytotoxic strain (Baltch *et al.*, 1994).

Siderophores

To obtain iron, an essential requirement for the establishment and maintenance of bacterial infection, *P. aeruginosa* excretes iron chelators such as pyochelin and the fluorescent yellow-green pigment pyoverdine (Poole and McKay, 2003). It is also able to use many heterologous siderophores of fungal and bacterial origin, and together, they play a key role in the growth and virulence of the species in iron-limited conditions. Pyochelin is poorly water soluble and of low molecular weight (325 Da), whereas pyoverdine (approximately 1500 Da) is highly soluble in water and more active. Three structurally different pyoverdine compounds have been identified in *P. aeruginosa* by isoelectric focusing (De Vos *et al.*, 2001).

Pyocyanin

The phenazine pigment pyocyanin and its colourless precursor 1-hydroxyphenazine are highly toxic for human respiratory epithelial cilia. They induce, at low concentrations, ciliary dyskinesis and later disrupt the integrity of the epithelial surface. Pyocyanin causes a fall in intracellular cyclic adenosine mono phosphate (amp), and ciliary beat can be restored by agents such as the β_2-adrenoceptor agonists salmeterol and isoprenaline. Pyocyanin is bactericidal for many species including *E. coli*, *Staphylococcus aureus* and *Mycobacterium smegmatis*. Other activities of pyocyanin include the production of reactive nitrogen intermediates and promotion of elastase–anti-elastase imbalance by increasing the release of neutrophil elastase and enhancing the oxidative inactivation of α_1-protease inhibitor (Ran *et al.*, 2003).

Quorum Sensing

Pseudomonas aeruginosa produces small diffusible signal molecules, acyl homoserine lactones (AHLs), during growth in an environment, and these molecules can be detected by surrounding organisms. When the concentration of AHLs accumulates in the environment because of increasing numbers of bacteria, intracellular levels of AHLs are then sufficiently high to induce the activation of transcriptional regulators, thus enabling the bacteria to act as a community in the coordinated regulation of gene expression. This process is known as quorum sensing, and two different systems have been described: the *las* system and the *rhl* system (Smith and Iglewski, 2003). These systems are the target for therapeutic compounds in the hope that the latter will attenuate the virulence of the organism and potentially assist the host in clearing the infection. Cell-to-cell signalling is considered an essential requirement for the establishment of biofilms, and AHLs have been detected in the lungs of CF patients infected with *P. aeruginosa* (Favre-Bonté *et al.*, 2002).

EPIDEMIOLOGY

Pseudomonas aeruginosa is widespread in the natural environment and can adapt to habitats ranging from surface waters to disinfectants. It can multiply in distilled water, presumably by the utilization of gaseous dissolved nutrients, but is rarely isolated from sea water (except from sewage outfalls and polluted river estuaries). It is not associated with disease in fish. The species is present in soil and the rhizosphere and, as a result, may be frequently recovered from fresh vegetables and plants (Rhame, 1980). In the hospital environment, sinks, taps and drains are invariably colonized by *P. aeruginosa* and other pseudomonads. In contrast, the domestic household environment is rarely contaminated, with the exception of washing machines, where the organism may reside in water traps.

Natural carriage by humans is infrequent. In healthy subjects, faecal carriage rates vary between 2% and 10% (possibly higher in vegetarians). Faecal colonization appears to be transient in healthy people, and there is a rapid turnover of strain types. *Pseudomonas aeruginosa* dies rapidly on dry healthy skin, but in conditions of superhydration, such as in divers undergoing long-term saturation dives, the frequency of skin colonization is increased and accompanied by infections, particularly otitis externa (Alcock, 1977).

Clinical and environmental strains of *P. aeruginosa* form a coherent taxonomic group with few distinguishing characteristics. Environmental isolates are able to utilize gasoline supplied in the gas phase as the sole carbon source, but clinical strains do not share this property (Foght *et al.*, 1996).

Despite the apparent ubiquity of *P. aeruginosa* in the natural environment and the vast array of potential virulence factors, the incidence of community-acquired infections in healthy subjects is relatively low. However, in the hospital environment, particularly in immunosuppressed, debilitated and burns patients, the incidence of *P. aeruginosa* infection is high. The frequency of faecal carriage is sharply increased in hospitalized patients (Stoodley and Thom, 1970), possibly because of ingestion of contaminated food and the widespread use of antibiotics. The salivary carriage rate is similar in hospital patients and normal controls (approximately 5%), but skin colonization in burns patients may reach 80% by the ninth day following the burn (Holder, 1977).

Almost any type of hospital equipment or utensil has been implicated as a reservoir for *P. aeruginosa*, including disinfectants, antiseptics, intravenous fluids and eyewash solutions. These sources may serve as foci for the dissemination of the organism in common-source outbreaks, and this usually is the result of poor sterilization. Widespread contamination of the inanimate ward environment such as sink traps and taps has been cited by many workers as a potential infection hazard particularly for the contamination of the hands of personnel, but not all studies have supported this conclusion (Beck-Sagué, Baneijee and Jands,1994). Many surveys have concluded that *P. aeruginosa* accounts for approximately 10% of all hospital-acquired infections, but its prevalence in various body sites and hospital units varies from study to study. It is among the most frequent causes of hospital-acquired pneumonia, and prevalence rates of 15–25% have been reported from some centres. The species was by far the most common, and dangerous, microorganism recovered from burn wounds in the 1960s and 1970s. Today, the number of serious infections in burns patients caused by *P. aeruginosa* has declined considerably (in the developed world), owing to improvements in burn management and treatment.

INFECTIONS DUE TO *P. AERUGINOSA*

Skin and Eye

Pseudomonas aeruginosa rarely causes disease in the healthy individual unless introduced into the tissues in relatively large numbers. Prolonged exposure to contaminated water [$>10^6$ colony-forming units (CFU)/mL] may lead to skin infections such as folliculitis, soggy dermatitis of the interdigital spaces or otitis externa. Folliculitis due to *P. aeruginosa* is characterized by a diffuse maculopapular or vesiculopustular rash and usually follows immersion in contaminated swimming pools, spas and hot tubs (Schlech *et al.*, 1986). The condition is usually self-limiting, but topical treatment or, rarely, systemic antibiotics may be required. *Pseudomonas aeruginosa* is a common cause of superficial otitis externa, but it may also cause an invasive form, sometimes in diabetics, which fails to respond to topical antibiotic therapy. This malignant infection can lead to erosion through the external auditory canal and involvement of cranial nerves, with a mortality of 20%.

Pseudomonas aeruginosa is probably the most devastating of bacterial pathogens for the human eye, in particular the cornea, where it produces ulceration or keratitis. Other infections include conjunctivitis, endophthalmitis and orbital cellulitis. The most common sources of the organism are contact lens fluids, ocular medications and mascara which have been become contaminated with *P. aeruginosa* (Zloty and Belin, 1994). Septicaemic spread of the organism from an orbital focus accompanied by pneumonia and ecthyma gangrenosum has been described in an immunocompromised individual (Maccheron *et al.*, 2004).

Burns

Pseudomonas aeruginosa is a common cause of infections in burns, established through colonization of the burn wound by the patient's own flora or from the environment. Most fatalities (usually arising from septicaemia) are associated with full-thickness burns, and there is a strong correlation with the percentage area of the burn (Smith, 1994). The most commonly used anti-pseudomonal topical agent is silver sulphadiazine, which is most effective when used prophylactically. The addition of cerium increases its antibacterial properties as well as its persistence in the eschar. Topical silver nitrate also reduces mortality in patients with extensive burns, and methylated sulphonamide (mafenide) cream has good burn eschar penetration. Patients with burns infected with *P. aeruginosa* have an increased mortality rate and longer hospital stays compared to non-infected patients. They also have an increased number of surgical procedures and higher associated antibiotic costs (Tredget *et al.*, 2004).

Wounds

Pseudomonas aeruginosa is a common isolate from surgical wounds, and its frequency is related to the site and extent of the surgery and the underlying clinical state of the patient. Data from the National Nosocomial Infection Surveillance (NNIS) survey in the United States showed an incidence in surgical wounds of about 7% in ward areas compared with 10% in intensive care patients (Beck-Sagué , Baneijee and Jands, 1994).

Bones and Joints

Bones and joints can be infected either by direct injection in puncture wounds (rare) or by haematogenous spread in intravenous drug abusers and diabetic subjects, resulting in chronic osteomyelitis. Removal of the focus as well as the surrounding soft tissue is necessary if antibiotic therapy is to be successful. Septic arthritis is unusual, and many patients have a history of drug abuse or a predisposing condition (Mader *et al.*, 1994). Infection of the temporal bone by *P. aeruginosa* (mastoiditis) is a common complication of otitis media, particularly in children.

Blood and Cerebrospinal Fluid

Reports of *P. aeruginosa* bacteraemia are relatively rare in the early literature, but owing to the increased susceptibility of the hospitalized population, frequencies of *P. aeruginosa* bacteraemia today vary from 5% to 20% in many studies and septicaemia is most common in the immunocompromised, in particular granulocytopenic patients and the elderly. The mortality rates associated with *P. aeruginosa* bacteraemia range from 17% to 78%, and attributable mortality has been estimated at 34–48%, although it is often difficult to distinguish mortality due directly to bacteraemia from mortality due to underlying disease. A study by Blot *et al.* (2003) of matched cohorts in 53 intensive care unit (ICU) patients with *P. aeruginosa* bacteraemia found that patients with the organism had a higher incidence of acute respiratory failure, haemodynamic instability and longer ICU stay and ventilator dependence with a 62% mortality rate compared with 47% for controls. Attributable mortality was therefore 15%. Multivariate survival analysis found the APACHE II score to be the only variable independently associated with mortality. A minority of patients exhibit ecthyma gangrenosum, and if untreated, necrosis of deep tissue can occur. The primary sources of bacteraemia are the gastrointestinal tract, the respiratory tract and skin. Early treatment is critical, particularly in the immunocompromised, and combination therapy with an aminoglycoside and a β-lactam is usual.

Infection of the central nervous system (CNS) with *P. aeruginosa* is uncommon. Nevertheless, the risk is increased for neonates, for those undergoing neurosurgery or transplantation or for those with a chronic underlying condition. *Pseudomonas aeruginosa* may be isolated from a minority of intracranial abscesses, but the incidence rises in patients with chronic otitis media and mastoiditis. Nosocomial transmission can occur in neurosurgical ICUs, particularly in patients with percutaneous external ventricular catheters to monitor cerebrospinal fluid pressure. The application of a sterile occlusive dressing may decrease the risk of pseudomonas ventriculitis in these patients (Trick *et al.*, 2000).

Urinary Tract

Primary urinary tract infection acquired in the community because of *P. aeruginosa* is rare except in those patients with anatomical abnormalities and spinal cord injuries. The great majority of urinary tract infections are of nosocomial origin and are invariably the consequence of long-term catheterization. Rates of approximately 5% have been reported. Haematogenous spread from a primary focus is uncommon. Paraplegic patients in institutional care are especially at risk, and reservoirs of contaminated urine such as drainage bottles and bed pans are probably the major sources for the organism. The infection is usually resolved by the removal of the catheter or other predisposing factor, but prostatic infections in the presence of calculi are exceedingly difficult to treat (Kunin, 1994).

Respiratory Tract (Non-Cystic Fibrosis)

Community-acquired pneumonia due to *P. aeruginosa* is rare and usually presents in patients with an underlying disease such as malignancy or chronic pulmonary disease or with a history of intravenous drug abuse. In intubated hospitalized patients, the organism frequently colonizes the lower respiratory tract, particularly those with a tracheostomy who have been exposed to contaminated inhalation equipment. Few of these patients progress to pneumonia. A large recent study reported that the incidence of *P. aeruginosa* was 7%, with a mortality of 28%, which is higher than that reported by other surveys (Arancibia *et al.*, 2002). This study also showed that fluoroquinolone therapy was associated with the emergence of multiresistant strains, and the presence of an underlying disease affecting immune function was linked to the presence of such strains. In most patients without anatomical abnormalities or prolonged immunosuppression, infection is relatively transient and specific treatment is not indicated. A diagnosis of ventilator-associated pneumonia is problematic, with variable sensitivity and specificity for the range of methods employed. The reader is referred to the comprehensive review of Chastre and Fagon (2002) for further detail.

Elevation of the pro-inflammatory cytokine IL-1 is believed to be associated with the outcome of *P. aeruginosa* pneumonia, and an absence of or reduction in IL-1 improves the host defence against the organism. In contrast, IL-18 (gamma interferon) appears to impair the host response to *P. aeruginosa* (Schultz *et al.*, 2003).

Pseudomonas aeruginosa is commonly recovered from the lower airways of patients with non-CF bronchiectasis, and in a minority, a chronic infected state is established, which is seldom relieved by therapy. Isolates from these patients may exhibit some of the features normally associated with CF, such as mucoid alginate production and colonial variation. Diffuse panbronchiolitis is an idiopathic inflammatory condition largely restricted to Japan that is characterized by suppurative and obstructive airway disease. Many of these patients are co-infected by *P. aeruginosa* and *Haemophilus influenzae*, and mortality is high if left untreated. Erythromycin is the drug of choice, although its effectiveness appears to be due to mechanisms other than antibacterial activity (Koyama and Geddes, 1997).

There have been many reports of *P. aeruginosa* respiratory infections in acquired immune deficiency syndrome (AIDS) patients. Most infections appear to be community acquired, and the organism is often present in pure culture. Morbidity and mortality are difficult to define accurately as many patients have advanced human immunodeficiency virus (HIV) disease, but mortality may reach 40%. Risk factors identified include advanced HIV immunosuppression and use of systemic *Pneumocystis* prophylaxis and/or broad-spectrum antibiotics.

Cystic Fibrosis

Until the 1970s, most CF patients who succumbed to bacterial infection were infected with *S. aureus*, and *P. aeruginosa* infection was relatively rare. Since that time, there has been an inexorable rise in the number of patients who harbour *P. aeruginosa*. Today, in some centres, as many as 30% of children aged 2–5 years and 80% of adult patients are colonized by or infected with this organism. Its presence and persistence in the lungs correlates with clinical decline, and until recently, it was seldom eradicated once colonization was established. The gut does not appear to be the principal source of the organism, as respiratory tract colonization often precedes the recovery of the organism from the faeces of CF patients. Oropharyngeal swabs have relatively poor sensitivity (46%) for *P. aeruginosa* in the lower airway but are quite specific for the organism. Thus, a negative oropharyngeal culture indicates that isolation of the organism from the lower airway is unlikely, but a positive culture does not predict lower airway infection. Similar findings were shown for bronchoscopic sampling of the lower airways to detect early infection in neonates (Saiman, 2004).

As *P. aeruginosa* is ubiquitous, CF patients may acquire it from a variety of sources in the environment. This is corroborated by the fact that patients are, with the exception of siblings, usually colonized by genetically different strains. Most individuals are colonized by a single strain and often retain this strain for their lifetime, but a minority may harbour two, three or more different strain genotypes. There is a reported risk of cross-infection with *P. aeruginosa* among CF patients, and this is highest when non-colonized patients come into contact with chronically infected patients in hospitals, CF centres or summer holiday camps (Govan and Nelson, 1992). A recent survey indicated that cross-infection with highly transmissible strains has occurred both within and across different CF centres in England and Wales. The study found that about 20% of all strains examined from many centres fell into two prevalent genotypes (Scott and Pitt, 2004).

An explanation for the apparent predilection of *P. aeruginosa* for the CF respiratory tract was proposed by Pier (2002). In Pier's (2002) scenario, *P. aeruginosa* may be killed early in the lung by antimicrobial factors such as host cationic peptides (defensins) and nitric oxide or bound by the CF transmembrane conductance regulator (CFTR). This serves as an epithelial cell receptor for *P. aeruginosa*, and binding occurs through the conserved core oligosaccharides of the bacterial OM. Binding of CFTR to *P. aeruginosa* results in the internalization of the bacterium by the epithelial cells, and these cells are shed into the airway surface liquid and removed by the mucociliary escalator. The airway surface fluid of CF patients is dehydrated because of CFTR dysfunction, which combined with hypersecretion of abnormal mucus and a slow ciliary escalator provides the conditions where persistent colonization and subsequent infection can occur. Increased numbers of asialo GM_1 molecules, a receptor for various bacteria, have been reported on CF epithelia, and *P. aeruginosa* apparently binds to this compound via surface structures such as pili and flagella; however, the specificity of this mechanism has been questioned by Schroeder *et al.* (2001). It has been suggested that *P. aeruginosa* grows in a biofilm in the viscous mucus layer in near-anaerobic conditions, and this triggers the emergence of mucoid cell types. However, growth of *P. aeruginosa* biofilms in the presence of reactive oxygen species such as hydrogen peroxide can also trigger the emergence of mucoid cells (Ratjen and Doring, 2003). The selection of *P. aeruginosa* in the CF lung is therefore probably multifactorial, but it is evident that a cycle of inflammation and tissue damage is established and accompanied by an extended and exaggerated response to bacterial and viral invaders. This heightened inflammatory state over time results in an environment where high bacterial loads (often in excess of 10^8 CFU/g of sputum) prove refractory to lung clearance mechanisms and antimicrobial therapy.

There is pronounced antibody production in response to colonization by *P. aeruginosa* in CF patients. This humoral response is associated with a poor prognosis. IgA and IgG antibodies are significantly elevated in CF compared with normal subjects, while IgM is not. Precipitating antibodies against *P. aeruginosa* are significantly elevated in at least 30% of patients, and these antibodies are always detectable when the mucoid phenotype of *P. aeruginosa* is present in the sputum. Indeed, the rise in precipitins is much more evident with chronically colonized patients as opposed to those with intermittent colonization.

Pier *et al.* (1987) suggested that the survival of CF patients to adulthood without being colonized by mucoid *P. aeruginosa* was related to the presence of specific serum antibodies against the mucoid exopolysaccharide. These antibodies were termed opsonic killing and were significantly more common in older CF patients not colonized by *P. aeruginosa*. Chronically colonized patients also had high titres of these antibodies, but they were not specific for the exopolysaccharide. This suggested that the opsonic-killing antibodies against alginate protected patients against chronic colonization with mucoid *P. aeruginosa*. An enzyme-linked immunoadsorbent assay (ELISA) which measures antibody to the lipid A, core oligosaccharide and O-polysaccharide of *P. aeruginosa* LPS revealed that the systemic antibody response is increased to all parts of the molecule (Kronborg *et al.*, 1992). Furthermore, antibodies in sputum are mainly anti-lipid A and anti-O-polysaccharide of the IgG and IgA isotypes. The determination of serum antibody by ELISA is of value, particularly in paediatric CF patients, where it may help detect *P. aeruginosa* infection at an early stage and aid the differentiation between early infection and colonization as well as monitor disease progression. In general, positive serum IgG antibody titres do not predate isolation of *P. aeruginosa* from sputum in CF patients but in some are present soon after acquisition of the organism. A positive titre indicates significant exposure to *P. aeruginosa* and may be useful to detect infection in sputum-negative patients and possibly indicate the effect of early treatment. It is paradoxical, however, that CF patients with high titres of antibodies, which supposedly protect against infection, are often those who are chronically colonized by *P. aeruginosa*. It is clear, therefore, that elevated antibodies against *P. aeruginosa*, which may protect against septicaemia, are insufficient to eliminate the bacterium from the lungs of CF patients. West *et al.* (2002) claimed that a positive serum antibody response to *P. aeruginosa* antigens (ETA, a cell lysate and elastase) preceded culture positivity by 6–12 months and argued that such diagnostic markers provided an opportunity for transforming the treatment strategy of neonates with CF. This work has yet to be independently confirmed by others.

Many isolates of *P. aeruginosa* from CF have an altered LPS structure in their cell envelope. This involves loss of all or part of the polysaccharide side chain (O-antigen) and results in the absence of a specific O-serotype reaction for most strains (Hancock *et al.*, 1983). By electron microscopy, LPS-deficient strains exhibit outer layers characterized by atypical extrusions (lipophilic blebs) from the cell wall. These strains are also invariably serum sensitive, which requires the action of both the classical and alternative complement pathways. Both LPS-deficient and LPS-complete strains can activate complement, but only in the deficient form is the terminal complement complex able to anchor stably in the OM. Serum-sensitive strains should not survive in the bloodstream, and this is consistent with the extremely rare finding of *P. aeruginosa* septicaemia in CF patients.

Many isolates of *P. aeruginosa* from CF patients are auxotrophic (i.e. require specific growth factors). The most common requirement is for methionine, but other amino acids such as leucine, arginine or ornithine are less commonly required. Auxotrophic and prototrophic (wild-type) *P. aeruginosa* isolates colonizing the same CF patient invariably constitute an isogenic group, and it is most likely that auxotrophs are selected from the prototrophic population during the course of pulmonary infection (Barth and Pitt, 1995).

LABORATORY DIAGNOSIS AND STRAIN TYPING

Media

Pseudomonas aeruginosa grows readily on simple media. It is relatively resistant to quaternary ammonium compounds, in particular cetyltrimethyl ammonium chloride (cetrimide) and benzalkonium chloride. Cetrimide resistance has been exploited as a selective feature for the isolation and presumptive identification of the species from clinical specimens, and cetrimide-containing media combined with nalidixic acid are commercially available. Irgasan is also selective for *P. aeruginosa*. However, a minority of isolates from CF patients may exhibit hypersensitivity to both compounds and thus fail to grow on these media. Enrichment in acetamide broth has been recommended for the isolation of *P. aeruginosa* from faeces.

Typing Methods

That *P. aeruginosa* is one of the most diverse of bacterial pathogens is not surprising, given its metabolic versatility and its ability to adapt to a wide variety of habitats. The species is characterized by a superficially clonal population structure, with frequent genetic recombination leading to the emergence of highly successful epidemic clones (Pirnay *et al.*, 2002). No single method is sufficient for the unequivocal identification of strain types of *P. aeruginosa* in epidemiological studies, and combinations of two or more methods are recommended to ensure efficient discrimination between strains. Many of the early methods used for typing such as pyocin typing, phage typing and serotyping are seldom performed today and have been largely replaced by molecular techniques to define genotypes. For details of the early methods, see the first edition of this book.

Pseudomonas aeruginosa is serologically heterogeneous, and a scheme of 17 O-antigen groups is recognized internationally (Liu *et al.*, 1983). The scheme was subsequently extended to 20 serotypes. Isolates are typed by slide agglutination with live cultures from agar plates or in microtitre trays with boiled suspensions. The percentage typeability of O-serotyping for most isolates of *P. aeruginosa* (excluding CF isolates) is approximately 90%, and the reproducibility of the technique is high. Two serotypes, O6 and O11, predominate in clinical material (20% and 15%, respectively), with O11 being particularly frequent in hospital outbreaks. The loss of serotype specificity in the LPS of CF isolates drastically reduces their typeability with O-antisera, and usually less than 20% of these isolates are able to be serotyped. Many CF isolates exhibit polyagglutination with antisera to unrelated serotypes. At the time of writing, O-typing antisera can be obtained commercially from the Institut Pasteur, Paris, France and Denka-Seiken, Tokyo, Japan.

Rare cutting restriction endonucleases, in particular *Spe*I and *Xba*I, have been widely utilized for DNA macrorestriction analysis and for generating fragments in the size range of 50–2000 kb, which are separated by pulsed-field gel electrophoresis (PFGE). Isolates with approximately 80% genomic similarity by the Dice coefficient (differ by three or fewer bands) are considered to be of the same clone, while four-to-seven-band difference implies probable relatedness. PFGE has excellent discriminatory power for *P. aeruginosa* and is the gold standard technique against which other molecular techniques are compared (Grundmann *et al.*, 1995). Amplified fragment length polymorphism (AFLP) is a PCR-based technique that randomly samples a small fraction of the entire genome and has been successfully applied to the typing of *P. aeruginosa* and gives a similar level of discrimination as PFGE (Speijer *et al.*, 1999). Other approaches to genetic typing of strains include ribotyping with *Pvu*II restriction, specific gene probes (Grundmann *et al.*, 1995), various DNA targets by PCR (repetitive elements, consensus sequences and random priming) and DNA sequencing of specific genes encoding OM proteins (Pirnay *et al.*, 2002).

ANTIBIOTIC RESISTANCE AND THERAPY

Compared with the Enterobacteriaceae, *P. aeruginosa* is relatively resistant to many antibiotics. However, there are many compounds with good-to-excellent activity against the species. They include the semisynthetic penicillins (e.g. ticarcillin), ureidopenicillins (piperacillin), carboxypenicillins (carbenicillin), third-generation cephalosporins (ceftazidime), carbapenems (imipenem and meropenem), monobactams (aztreonam), aminoglycosides (gentamicin and amikacin), fluoroquinolones (ciprofloxacin) and polymyxins (polymyxin B and colistin). To improve outcomes, an aminoglycoside has often been combined previously in treatment regimens with a β-lactam, third-generation cephalosporin, monobactam or carbapenem. This requirement for combination therapy has been challenged in several studies (Paul *et al.*, 2004). Monotherapy with an appropriate antibiotic may be as effective if given early and in suitable doses.

A variety of antibiotic resistance mechanisms exist in *P. aeruginosa*. These include the following:

1. low cell-wall permeability conferring intrinsic resistance;
2. the production of extracellular chromosomal and plasmid-mediated β-lactamases, aminoglycosidases and cephalosporinases;
3. an alteration in antibiotic-binding protein sites;
4. an active efflux mechanism which pumps out antibiotic from the cell.

The development of resistance to β-lactams *in vivo* during treatment is often the result of derepression of chromosomal β-lactamase expression. Some antibiotics, in particular cefoxitin, are potent inducers of these enzymes and may select out subpopulations resistant to other cephalosporins such as ceftazidime. Resistance emerging from therapy with β-lactam antibiotics may also be due to the modification of the penicillin-binding proteins of *P. aeruginosa*. Most impermeability mediated resistance is due to an efflux pump mechanism, MexAB–OprM, which removes a very wide range of antimicrobials including β-lactams, chloramphenicol, fluoroquinolones, macrolides and various dyes and detergents. Indeed, a combination of upregulated efflux, loss of the major porin OprD and impermeability to aminoglycosides confers resistance to every known anti-pseudomonal drug class except the polymyxins (Livermore, 2002).

Pseudomonas aeruginosa from the natural environment and from non-CF patients are generally susceptible to the above-mentioned agents. Resistance rates have changed little over the last two decades and today remain around 12% for these agents (Henwood *et al.*, 2001). Similar studies from Europe have also shown little variation in resistance rates; however, in some parts of southern Europe, Southeast Asia and South America, resistance rates may be as high as 50% for many anti-pseudomonal agents. Nevertheless, a survey of isolates from British CF patients attending regional treatment centres revealed disturbingly high levels of resistance. It showed that only 38% of isolates were susceptible to all of the six agents tested but almost 50% were resistant to gentamicin, 39% to ceftazidime, 32% to piperacillin, 30% to ciprofloxacin, 10% to tobramycin and 3% to colistin. Approximately 40% were resistant to two or more compounds (combinations of ceftazidime, gentamicin, piperacillin or ciprofloxacin). These resistance rates were generally similar to those reported for CF patients in the United States and, less so, in Germany (Pitt *et al.*, 2003). Despite this background of resistance, these and other antibiotics continue to be used with clinical success in CF patients. Part of the explanation for this is that the minimum inhibitory concentration (MIC) of some antibiotics may vary by 80-fold within the same specimen and the result obtained is greatly influenced by the colony chosen for test. To overcome this, mixed morphotype testing has been advocated, and studies have shown good agreement between multiple colonies tested singly and mixed morphotypes. Morlin *et al.* (1994) examined the correlation of the MIC of an agent for mixed colonies with the MIC for the most resistant morphotypes in sputum samples and found that the former correctly predicted the highest MICs for isolated colonies. On 90% of occasions, the MIC for mixed morphotypes predicted susceptibility, but for resistance it was only just over half of the cases.

Many antimicrobial compounds, some of which have negligible anti-pseudomonal activity, can, at concentrations as low as 1/20th and 1/50th of their MIC for a particular organism, suppress the synthesis of pseudomonas aggressins, such as elastase, phospholipase and exotoxins, and in this way reduce the inflammatory state in the lung and improve lung function. The MIC of macrolide antibiotics for *P. aeruginosa* exceeds their achievable peak serum concentrations, but clinical experience supports their anti-inflammatory action in the CF lung (Saiman *et al.*, 2003).

In the course of infection in the CF lung, *P. aeruginosa* accumulates many mutations, and these are expressed in the variety of phenotypes of the organisms recovered from sputum. Recently, hypermutable strains of *P. aeruginosa*, which have mutations in genes controlling their DNA repair systems, were reported to be exclusive to CF, and approximately one-third of CF patients in a survey in Madrid were shown to harbour these strains (Oliver *et al.*, 2000). Mutation rates of 100- to 1000-fold higher than the normal mutation rate (approximately $1 \times 10^{-8}/10^{-9}$) were found, and antibiotic resistance was most associated with hypermutating strains. The clinical significance of hypermutators over non-mutator strains is unknown, but there is speculation over whether antibiotic therapy drives the process of the selection of these strains. However, it is noteworthy that while organisms with more resistance are most likely to be hypermutators, some hypermutators may exhibit little or no resistance to anti-pseudomonal agents.

A recent study that analysed data from a placebo group in an inhaled tobramycin trial found that the *in vitro* susceptibility tests for the most resistant strains did not correlate with clinical improvement and questioned the validity of standard susceptibility methods to guide therapy in CF patients (Smith *et al.*, 2003). This was addressed by Moskowitz *et al.* (2004) who speculated that the response to antibiotic exposure of stationary-phase, microaerophilic *P. aeruginosa* growing as biofilms in the clinical laboratory may better predict the response of CF patients to antimicrobial therapy. They devised an assay that gave reproducible results and found that biofilm inhibitory concentrations (BICs) were much higher than the corresponding conventionally determined MIC for the β-lactam antibiotics but were similar for meropenem, ciprofloxacin and the aminoglycosides. Azithromycin had good anti-biofilm activity, but all isolates were uniformly resistant in the standard method. Clinical trials are clearly needed to compare the efficacy of therapy based on biofilm susceptibility testing with standard testing.

CF patients may require long-term maintenance antibiotics and additional treatment of acute exacerbations of infection. The primary aim for long-term maintenance antibiotic therapy against *P. aeruginosa* in CF is to reduce the severity and frequency of acute exacerbations. Eradication of the organism is difficult, but there are occasional reports of success with different regimens. Early aggressive antibiotic treatment is widely practised using a combination of inhaled colistin with oral ciprofloxacin or inhaled colistin or inhaled tobramycin alone. Long-term follow-up of patients treated with inhaled tobramycin (80 mg twice a day for 1 year after initial colonization) showed that it not only postponed chronic infection but also led to eradication of the bacteria from the sputum in 14 of 15 treated patients (Ratjen, Doring and Nikolaizik, 2001). Aminoglycosides are suitable agents for nebulization, and there is increasing evidence of their efficacy in CF, although not in other groups of patients.

PREVENTION AND CONTROL

Vaccines

Pseudomonas aeruginosa expresses many antigenic surface structures and extracellular products, many of which have been investigated for their potential to induce protective antibodies. Native LPS is poorly immunogenic, and it is best bound to a carrier protein such as ETA or tetanus toxoid, to stimulate an antibody response in the subject. An alternative approach is the use of high-molecular-weight polysaccharide polymers from culture supernatants that are highly antigenic and free of the toxicity of native LPS.

Many clinical trials with *O*-polysaccharide-based vaccines have been reported, and the results vary with the type and quality of the preparations (Cryz, 1994). The first whole-cell vaccines reduced mortality rates in burns patients when compared with controls, but it was not until the development of heptavalent LPS vaccine, Pseudogen, that large-scale studies in different patient groups were undertaken. The vaccine gave little or no protection to cancer patients against *P. aeruginosa* and was also ineffective in intensive care patients and in CF. Another LPS-based vaccine, PEV-01, entered clinical trials, and initial results in burns patients were promising, although this has been questioned on the grounds of poor study design (Cryz, 1994). An octavalent O-polysaccharide–toxin A conjugate vaccine was shown to be safe in volunteers and patient groups and elicited functional anti-LPS and antitoxin A antibodies following immunization. Lang *et al.* (2004) reported a 10-year retrospective analysis of clinical outcomes with this vaccine in a cohort of 30 CF children with mean age 7 years, 26 of whom were followed up. The patients were given yearly vaccine boosters, and results were compared with a closely matched control group who did not receive the vaccine. The time to infection with *P. aeruginosa* was longer in the vaccinated group compared with controls, and fewer vaccinated patients than controls became chronically infected (32% versus 72%; $P < 0.001$). This was associated with better preservation of lung function. Vaccinated patients gained more weight during the study period, a possible indication of an improved overall health status.

Pseudomonas alginate is poorly immunogenic, but both the high-molecular-weight polysaccharide moiety and alginate combined with toxin A elicit opsonizing antibodies in humans. It has been shown that the larger the polymer size, the more immunogenic is the alginate vaccine. At a dose of 100 pg of alginate, long-lasting opsonic antibodies are formed in humans which enhance the deposition of C3 onto mucoid cells and mediate the killing of the challenge strain (Garner, Desjardins and Pier, 1990).

Recent research in immunotherapy has focused on the use of DNA vaccines in which animals are immunized with plasmid DNA that expresses bacterial proteins in eukaryotic cells. DNA immunization with flagellin B, OM protein F and ETA has been shown to be effective at protecting animals challenged with virulent strains of

P. aeruginosa. Variation in effectiveness is probably a consequence of the choice of bacterial protein and the route of administration. The type III secretion system proteins have also been investigated for their potential as vaccine targets. Active and passive immunization with a purified type III translocation protein, PcrV, enhanced survival of mice challenged with a lethal dose of *P. aeruginosa* (see review by Holder, 2004).

Normal human serum contains *P. aeruginosa*-specific antibodies, and in the past, pools of normal IgG have proved successful for the reduction of septicaemia and mortality of burns patients infected with *P. aeruginosa*. There were therefore great expectations for passive immunotherapy with hyperimmune plasma, but it proved disappointing in clinical use. Many aspects of passive immunotherapy such as their prophylactic value, the type of patient most amenable to treatment, the duration of treatment and the most reliable clinical indicators of therapy remain contentious. Both murine and human monoclonal antibodies have been prepared against specific LPS, and other antigens of *P. aeruginosa*, including toxin A, have performed with varying success in animal challenge models (Holder, 2004).

OTHER *PSEUDOMONAS* SPP.

Pseudomonas fluorescens and *P. putida* are both fluorescent pseudomonads. They grow poorly at 37 °C, and most strains will grow at 4 °C. This property contributes to the fact that these species, in particular *P. fluorescens*, are able to multiply in stored refrigerated blood and blood products and may give rise to fatal reactions when injected intravenously owing to the release of endotoxin. They are both of low pathogenicity in humans but have been isolated from urine, faeces, sputum and occasionally the blood of immunosuppressed patients. Hospital outbreaks of infection are rare, but *P. fluorescens* is often recovered from hospital sites such as floors and sink traps, in environmental screens.

Pseudomonas stutzeri has a characteristic appearance on agar and colonies may grow 'rough', 'smooth' or intermediate even in pure cultures. The rough colonies may be confused with *B. pseudomallei*, and older cultures may appear light brown in colour. The species is actively denitrifying, and large volumes of gas are produced in nitrate broth. Most strains will grow at 41 °C. *Pseudomonas stutzeri* has been widely reported from hospital-acquired infections, but these are usually self-limiting. Isolates are sensitive to β-lactams, colistin and gentamicin.

BURKHOLDERIA SPP.

BURKHOLDERIA CEPACIA COMPLEX

Taxonomy

This species was first described by Burkholder in 1950 as the agent of slippery skin in onions, and the genus *Burkholderia* is named after him. Subsequent to this first description, it became known as '*Pseudomonas multivorans*' and '*Pseudomonas kingii*', but later the epithet *Pseudomonas cepacia* was adopted. However, Yabuuchi *et al.* (1992) proposed the transfer of seven species of rRNA homology group II, including *P. cepacia*, to the new genus *Burkholderia*. *Burkholderia cepacia* is not a homogeneous species but forms a complex of phenotypically similar, genetically distinct, genomovars within the genus *Burkholderia*. It was the pioneering work of Vandamme *et al.* (1997) using polyphasic taxonomic methods that revealed a complex of five genomovars among strains isolated from various clinical and environmental sources. Today, the *B. cepacia* complex comprises nine genomovars: *B. cepacia* (I), *Burkholderia multivorans* (II), *Burkholderia cenocepacia* (III), *Burkholderia stabilis* (IV), *Burkholderia vietnamiensis* (V), *Burkholderia dolosa* (VI), *Burkholderia ambifaria* (VII), *Burkholderia anthina* (VIII) and *Burkholderia pyrrocinia* (IX).

The individual genomovars cannot be identified by their phenotypic properties or reactions in biochemical tests. Classification of *B. cepacia* complex clinical isolates to genomovar level is best achieved by the amplification of the *recA* gene to assign the isolate to the complex. Restriction fragment length polymorphism (RFLP) analysis with *Hae*III restriction enzyme is then used to generate genomovar-specific patterns (Mahenthiralingam *et al.*, 2000), and these are then confirmed with specific PCR tests.

PATHOGENICITY

Strains of *B. cepacia* from animal or plant sources are relatively avirulent (LD_{50} 1×10^8 CFU) for normal mice, but they persist in high numbers in a burned mouse model. The species produces a variety of extracellular putative virulence factors that may contribute to its pathogenicity for plant and animal hosts. These include lipases, haemolysins and proteases. Invasion of host cells and intracellular survival are fundamental to its establishment of disease. *Burkholderia cepacia* invades cultured respiratory epithelial cells and progresses into the cytoplasm where they are enclosed by membrane-bound vacuoles. It is also able to penetrate, survive and grow in *Acanthamoeba*, which may suggest a symbiotic relationship between the bacteria and free-living amoebae, with the latter perhaps acting as a reservoir for the organism and even possibly serving as a vehicle for its transmission to CF patients.

Using immunohistological techniques, Sajjan *et al.* (2001) were able to localize *B. cepacia* in the lungs of CF patients with cepacia syndrome in the exudates within bronchial lumens and between airway epithelial cells. The organisms were diffusely distributed in patients with chronic infections, but in acute infections the distribution was more focal, with the bacteria located on injured airway surfaces and in abscesses. In the progression from the chronic state to invasive infection, *B. cepacia* migrates across the epithelial barrier to invade the lung parenchyma and capillaries, thereby initiating septicaemia.

EPIDEMIOLOGY

Nosocomial Infections

A common feature of infections in humans with *B. cepacia* is the association of infusion of contaminated fluids. Numerous nosocomial outbreaks have been documented involving contamination of antiseptics, infusion fluids, tubing for irrigation and pressure-monitoring devices, but in general, outbreaks are rare and can usually be curtailed by the recognition and removal of the source. Infections also tend not to be severe, and systemic sepsis is uncommon. Patient-to-patient spread is unusual, and prolonged carriage by individuals (with the exception of CF) has not been documented. An association between *B. cepacia* and chronic granulomatous disease has been reported (O'Neil *et al.*, 1986). Phagocytic cells from these patients do not produce peroxides necessary for the intracellular killing of bacteria.

Cystic Fibrosis

Burkholderia cepacia infection was first recognized in CF patients in the mid-1970s, but it was 10 years later when its prevalence rose to nearly 40% of patients in a major centre in Toronto, Canada (Isles *et al.*, 1984), and elsewhere in the United States. The clinical outcome of *B. cepacia* infection in CF patients may vary from asymptomatic carriage through a gradual but accelerated decline to a rapidly fatal fulminant septicaemia in a minority of cases, the latter condition being termed cepacia syndrome. The contribution of *B. cepacia* to clinical deterioration and lung damage in most patients is unclear. Some early studies reported no significant difference in inflammatory markers and lung function between *B. cepacia* colonized and non-colonized patients, but more recently, a clear excess of mortality has been demonstrated in patients colonized with an epidemic strain of *B. cepacia* (Ledson *et al.*, 2002).

The frequency of *B. cepacia* complex organisms among all CF patients is generally low (approximately 5% of patients in UK centres), but this may increase dramatically if transmissible strains are implicated. Infections are more common in adult (>15 years) than paediatric patients. Segregation of *B. cepacia*-positive CF patients from others who do not harbour the organism is now widely practised, and this, combined with efficient infection control measures, reduces acquisition of the organism by new patients. However, a 6-year review of *B. cepacia* genotypes from patients referred to a North American transplantation centre showed that patient-to-patient spread was rare, and no evidence was found that they represented a major source of *B. cepacia* strains found in the resident CF clinic population (Heath *et al.*, 2002). This may be due to the fact that almost all the strains examined lacked the *cblA* gene which is almost unique to strains of the ET12 clonal lineage. This highly transmissible lineage probably originated in Ontario and spread through northeastern centres in the United States to the United Kingdom, apparently carried by CF patients returning from summer camp holidays, and at its peak was found in almost one-third of United Kingdom and Ireland patients colonized with *B. cepacia* (Pitt *et al.*, 1996).

Burkholderia cepacia can be transmitted between CF patients both in and out of hospitals (Govan *et al.*, 1993) and between non-CF and CF patients. Guidelines for the management and control of *B. cepacia* infections in patients and their carers have been published by CF associations in Europe and North America and have been reviewed recently (Saiman and Siegel, 2004). Apart from practical infection control advice, they also highlight the importance of regular sensitive and specific bacteriological screening to establish the colonization status of patients.

Most *B. cepacia* complex isolates from CF patients are of genomovar III (60–70%). *Burkholderia multivorans* (genomovar II) is the second most common, with reported frequencies of approximately 5–37%; the other genomovars are uncommon (Mahenthiralingam, Baldwin and Vandamme, 2002). Accordingly, most outbreak isolates are genomovar III. Although genomovar II strains may occasionally be transmitted between patients, most strains appear to be genotypically unique. Some genomovar III strains may superinfect patients already colonized with other genomovars. This is clinically important, as the former are universally associated with clinical decline and a higher mortality. Patients infected with *B. cepacia* complex organisms present an additional problem in their selection for heart–lung transplantation as long-term survival is influenced by genomovar status. Outcomes are variable, with some 1-year mortality rates of between 50% and 100%, but survival rates greater than 75% have also been reported. However, the highest mortality is associated with genomovar III-colonized patients (De Soyza *et al.*, 2001).

Three transmissibility genetic factors have so far been described in *B. cepacia*: *cbl*A which encodes for thick bundles of peritrichous (cable) fimbriae, an insertion sequence hybrid IS1356/IS402 and the *B. cepacia* epidemic strain marker (BCESM), a 1.4-kb sequence containing an open-reading frame with homology to transcriptional regulatory genes. These markers have strong association with the prevalent epidemic strain (ET 12) in the United Kingdom, and some authors have suggested that strict segregation of patients might only be appropriate for those harbouring this genotype (Clode *et al.*, 2000).

Burkholderia cepacia can be detected by air sampling after physiotherapy of CF patients has taken place and may persist for short periods after the room is vacated. Some strains of *B. cepacia* are able to survive in respiratory droplets on environmental surfaces, and it is also a common contaminant of dental water systems. It is relatively common in the domestic environment but is infrequent in salads and fresh foods. The risk posed by the natural environment as a reservoir for infection of CF patients is therefore probably low, but the patho-genic potential of environmental strains is unclear. A comparison of the traits related to virulence and transmissibility in large numbers of isolates from Italian CF patients with isolates from the maize rhizosphere (Bevivino *et al.*, 2002) found that genomovar III isolates comprised most of the strains in both groups. The clinical strains were more resistant to antibiotics, but most strains were equally as able as environmental strains to rot onion tissue. There is a potential danger to CF patients from the widespread application of strains for biodegradation and biocontrol purposes to the agricultural environment. Most rhizosphere isolates have closest affinity with *Burkholderia* spp. with known bioremediation and biocontrol capabilities and are unrelated to plant or human pathogenic strains. Nevertheless, the finding of a genomovar III strain from agricultural soil in the United States by LiPuma *et al.* (2002) suggests that human pathogenic strains are not necessarily distinct from environmental strains and that patients may acquire them on occasion direct from the environment.

LABORATORY DIAGNOSIS

Cultural Characteristics

Burkholderia cepacia complex organisms are slender motile rods. They grow well aerobically on nutrient agar but often prefer temperatures of 25–35 °C for 48 h for optimal growth. Most strains grow at 41 °C but not at 42 °C, and none grows at 4 °C. Stored cultures survive poorly on refrigerated agar slopes, and viability is better maintained as suspensions in sterile tap water and in frozen suspensions in glycerol on beads. Colonies on nutrient agar are opaque and vary in appearance from greyish-white, through yellow to a reddish-purple or deep brown colour. Members of the complex do not form ammonia from arginine and are lysine and ornithine decarboxylase positive. The oxidase reaction is slow and nitrate is not reduced; some strains produce a melanin-like pigment on tyrosine agar. Automated biochemical identification assays, such as Vitek and Microscan, give a reasonable level of identification, but the API 20NE system is reliable for identifying members of the complex, except *B. stabilis*.

Selective media are necessary to optimize isolation of the species from clinical and environmental specimens. Most selective media exploit the biochemical diversity and intrinsic resistance of the complex to antibacterial agents. They include an oxidation–fermentation agar base supplemented with lactose, polymyxin B and bacitracin; a commercially available agar containing crystal violet, bile salts, ticarcillin and polymyxin B; and a medium containing the selective agents 9-chloro-9-(4-diethylaminophenyl)-10-phenylacridan and polymyxin B. Henry *et al.* (1999) reported that a medium containing lactose and sucrose in an enriched casein and yeast extract with polymyxin, gentamicin and vancomycin was superior to oxidation–fermentation agar and the commercially available selective medium for the recovery of *B. cepacia* from CF sputum. For the optimal isolation of *B. cepacia*, sputum should be liquefied and diluted in saline or water and selective agar plates incubated at 37 °C for 48 h and then at room temperature for up to 5 days. Many pseudomonads and other polymyxin-resistant Gram-negative species are able to grow on *B. cepacia* selective media, and the identity of isolates should be confirmed by additional tests. A liquid enrichment medium (Malka broth) supplemented with polymyxin B is recommended for isolating the organism from environmental swabs and soils (Butler *et al.*, 1995).

Typing Methods

Many typing methods have been described for *B. cepacia*. They include biotyping, serotyping of O and H antigens, bacteriocin production/sensitivity, plasmid profiling, chromosomal DNA analysis and multilocus enzyme electrophoresis (Wilkinson and Pitt, 1995). The methods vary in their discriminatory power, reproducibility and typeability, and therefore, a combination of them is recommended. Ribotyping using the restriction enzyme *Eco*RI has been widely applied to epidemiological studies of *B. cepacia*, and at least 50 different patterns were identified among strains from British CF patients (Pitt *et al.*, 1996). Various PCR methods have been used to detect polymorphisms in the intergenic spacer regions of rRNA, random DNA targets, flagellin genes and insertion sequences as a means of strain identification. PFGE of *Xba*I chromosomal DNA digests offers the highest level of discrimination. Multilocus restriction typing (MLRT) which indexes variation at five chromosomal loci by restriction analysis of PCR-amplified genes was described by Coeyne and LiPuma (2002). Most clusters delineated by MLRT corresponded closely to those defined by PFGE.

Antibiotic Susceptibility and Treatment

Burkholderia cepacia is intrinsically very resistant to many antimicrobial agents, and resistance is even more common in clinical isolates from CF patients who have previously received antibiotics. *Burkholderia cepacia* is generally resistant to the antibiotics active against *P. aeruginosa*. Nearly all strains are resistant to the aminoglycosides, polymyxin, ticarcillin and azlocillin, while variable sensitivity is shown to aztreonam, ciprofloxacin, tetracyclines and carbapenems. A proportion of strains remain relatively susceptible to trimethoprim–sulphamethoxazole and chloramphenicol. Pitt *et al.* (1996) found that about three-quarters of CF isolates were susceptible to ceftazidime, piperacillin (+tazobactam) and meropenem but usually at only high breakpoint levels. However, the MICs for individual antimicrobials vary widely within genomovars (Nzula, Vandamme and Govan, 2002). Combinations of two, three or even four agents may show *in vitro* synergy against *B. cepacia*, but Manno *et al.* (2003) found that less than half of *B. cepacia* complex isolates tested were susceptible to combinations of ciprofloxacin–piperacillin and rifampicin–ceftazidime. They also showed that rates of antagonism for cotrimoxazole and chloramphenicol in combination with β-lactams were higher than those observed for ciprofloxacin plus β-lactams. Interpretation of these synergy tests is difficult; data are lacking on the relationship with clinical response. As a result, particularly where patients have been treated previously, combination antibiotic regimens are often tailored to individual patients (Jones, Dodd and Webb, 2001; Conway *et al.*, 2003).

Resistance to β-lactams is a consequence of a combination of low permeability of the OM and inducible β-lactamases. β-Lactam resistance may also be mediated by a class A β-lactamase very similar to the chromosomal β-lactamase of *Klebsiella oxytoca* (Trepanier, Prince and Huletsky, 1997). Multidrug resistance efflux pumps have been described and may contribute to high-level resistance to tetracycline and nalidixic acid. Plasmids can be detected in a minority of *B. cepacia* strains, but their contribution to overall resistance is unknown. Conjugative transfer of plasmid DNA from *P. aeruginosa* to *B. cepacia* and from *B. cepacia* transconjugants to other strains of the same species has been demonstrated.

BURKHOLDERIA PSEUDOMALLEI

INTRODUCTION

Burkholderia pseudomallei is the causative agent of melioidosis, which is a glanders-like disease of humans and animals. The species has been known by many names in the past and was for a long time classified among the genus *Pseudomonas* but was transferred to *Burkholderia* by Yabuuchi *et al.* (1992). *Burkholderia pseudomallei* and *B. mallei* formed a distinct group within *Pseudomonas* RNA group II because of their phenotypic and genetic similarities; they share 92–94% DNA–DNA homology and have identical 16S rRNA sequences.

EPIDEMIOLOGY

Burkholderia pseudomallei is a free-living saprophyte in soil and water in tropical areas where melioidosis is endemic, mainly Southeast Asia and northern Australia. Sporadic cases have also been reported from other parts of the world including the Indian subcontinent, the Pacific Islands, Central Africa, Caribbean islands and Central and South America (Dance, 1991). The organism persists in soil during the dry season and spreads through the surface with the rains; the number of bacteria in the soil probably affects the incidence of melioidosis in these areas. It has been recovered from a wide variety of sources within endemic areas, but the relationship between environmental contamination and the incidence of melioidosis is unclear. Furthermore, the species is able to survive well in conditions of nutrient depletion, and this may be relevant to its persistence in the environment.

Infection is acquired through wounds and skin abrasions and by inhalation and rarely through ingestion. The organism may also enter the body through mucosal membranes of the eye and nose. Human-to-human transmission is considered to be extremely rare. Venereal spread of melioidosis has been described in humans, and intrauterine and mammary infection in goats. Early studies postulated that an insect vector was involved in the transmission of the organism, but this is now refuted. Although *B. pseudomallei* may infect a wide range of animals and birds, transmission of infection from animal to humans is rare, and cross-infection between people is very uncommon. There have been occasional reports of infection through contaminated injections as well as accidental exposure in the laboratory.

Regular contact with soil and water appears to be the primary risk factor for acquiring melioidosis, hence the frequency of the disease among rice farmers in Southeast Asia and Aboriginals in Australia. Several thousand cases are thought to occur each year in Thailand, and melioidosis accounts for approximately 18% of community-acquired septicaemia in endemic areas of the country (Suputtamongkol *et al.*, 1994a). In Thailand, the peak incidence of infection is in the 40-to-60-year age group, but all ages are affected with a male:female ratio of 3:2. Most (75%) cases present during the rainy season, but incubation periods varying from 2 days to 26 years have been documented. Particular risk factors for developing disease are the presence of other underlying diseases such as diabetes mellitus, renal failure and alcoholism (White, 2003).

PATHOGENESIS AND INFECTION

Early work suggested that *B. pseudomallei* produces two heat-labile toxic constituents, both of which are lethal for mice when injected intraperitoneally but only one of which is dermonecrotic. Proteins with cytotoxic activity of approximately 31 kDa and 3 kDa have been purified from culture filtrates of isolates of *B. pseudomallei* from different sources, but it is unclear whether these toxins represent those described originally. An acidic rhamnolipid of 762 Da was also identified in heated culture filtrates of *B. pseudomallei* with toxicity for various cell lines, and it also lysed mouse, sheep and human erythrocytes (Häußler *et al.*, 2003).

Most strains of *B. pseudomallei* produce protease, lipase, PLC and haemolysin but not elastase. A type II secretion gene cluster controls the secretion of protease, lipase and PLC, but there is no direct correlation between the level of protease activity and the virulence of *B. pseudomallei* for mice. Another cluster of genes has been identified in the species which is homologous to, and in the same order as, a group of genes in *Ralstonia solanacearum* which in the latter are located in a type III secretion system-associated pathogenicity island. These genes were present only in the ara⁻ phenotype and absent from ara⁺ strains (see *Burkholderia thailandensis*), suggesting that they have a role in virulence (Winstanley and Hart, 2000). The species obtains iron from serum and cells through a hydroxamate siderophore, malleobactin, which is optimally produced under iron-deficient conditions.

As with *B. cepacia*, *B. pseudomallei* adheres to, penetrates and survives in a variety of mammalian cells. Structurally intact bacteria can be demonstrated within macrophages in experimental melioidosis that, despite an augmented cellular immune response, results in intracellular survival and growth. Phagocytes containing *B. pseudomallei* remain viable and retain their capacity to produce an oxidative burst for the first hour of incubation. The occurrence of latency and the tendency for melioidosis patients to relapse are also consistent with the organism surviving intracellularly. Harley *et al.* (1998) investigated intracellular survival of the species by studying the uptake of a test strain by fresh mouse peritoneal macrophages and cultured cells by electron microscopy. They found that *B. pseudomallei* appeared initially in phagosomes with intact membranes, but fusion with lysosomes was not observed. Loss of the membrane occurred rapidly, and the organisms escaped into the cytoplasm. However, fusion of phagocytic and non-phagocytic cells occurs when infected with *B. pseudomallei*, leading to the formation of multinucleated giant cells. The latter have been reported in the tissues of patients with melioidosis (Wong, Puthucheary and Vadivelu, 1995). *Burkholderia pseudomallei* has a specific high-affinity binding site for human insulin, and this feature has been postulated to be related to the predisposing influence of diabetes mellitus for melioidosis.

Presentation and Diagnosis of Melioidosis

Melioidosis varies greatly in clinical presentation (White, 2003). Patients may be asymptomatic but seropositive, present with mild localized infections, or may have a fulminating rapidly fatal septicaemia. The disease is characterized by pneumonia (either primary or secondary) and the widespread development of abscesses, particularly in the skin, liver, spleen, lung and prostate gland. Large joints are often involved. Among hospitalized adult cases, approximately 60% will have positive blood cultures. Children often present with localized, unilateral parotid gland swelling. Chronic suppurative infections also occur, and the disease may be easily mistaken for tuberculosis. The overall in-hospital mortality remains high, at approximately 40%.

Latency and relapse are also prominent features of this disease (Chaowagul *et al.*, 1993). A chronic state is established in many patients after acute infection, and relapse may occur after the apparent cure or remission of primary infection. The definitive diagnosis of melioidosis is made by the isolation of *B. pseudomallei* from tissues or body fluids, such as blood, urine and sputum. The organism is relatively easy to isolate on conventional media, and selective broth-enrichment media increase diagnostic sensitivity. Every attempt to aspirate samples of pus should be made. In Gram-stained smears, the organism may be present in only small numbers and bipolar staining may not be observed. Culture of throat swabs has a high predictive value (Wuthiekanun *et al.*, 2001). An immunofluorescence antibody test, ELISA latex agglutination test and PCR have all been used for rapid diagnosis. Serodiagnosis of melioidosis is based on the demonstration of a rising antibody titre to reference strains of *B. pseudomallei*. The indirect haemagglutination assay (IHA) is probably the most widely used, but high background titres in non-infected patients from endemic areas reduce its effectiveness. IHA titres of 40–80 in non-infected adults in endemic areas are not unusual, but titres of 640 or above are considered strongly indicative or diagnostic of the disease. ELISA tests for detecting both specific IgG and IgM antibody to *B. pseudomallei* have been proposed, with claims of high specificity and sensitivity. The ELISA is comparable in performance with an IgG immunofluorescence assay and is more sensitive than the IHA test.

LABORATORY DIAGNOSIS

Both *B. pseudomallei* and *B. mallei* are classified, by the UK Advisory Committee on Dangerous Pathogens (ACDP), as Hazard Group 3 organisms, and therefore, their culture and manipulation presents a risk to laboratory personnel. All procedures involving live cultures must be performed within a Laboratory Containment Level 3 facility. Both organisms are considered to be potential biological warfare agents.

Burkholderia pseudomallei grows well on nutrient agar at an optimum temperature of 37 °C (range 15–43 °C). Colonial morphology is best observed after 48 h at 37 °C, although growth is visible after 24 h. Colonies may be smooth and opaque or wrinkled; the latter is enhanced by glycerol. The growth has a characteristic musty odour (**sniffing of cultures is discouraged**), and on blood agar there may be weak haemolysis. Variation between rough and smooth colonial forms is frequent, and purified colonies may not breed true on subculture. Cells exhibiting bipolar staining are apparently associated with the rough colonial form. Mucoid isolates are rare. No extracellular pigments are formed. Colonial morphology is best observed on Ashdown's medium, which is a simple agar containing crystal violet, glycerol and gentamicin (Ashdown, 1979). Colonies concentrate the dye from this medium and grow rugose or smooth.

Burkholderia pseudomallei oxidizes glucose in Hugh and Leifson's medium, produces ammonia from arginine and degrades poly-β-hydroxybutyrate, starch, gelatin and Tween 80. Dance *et al.* (1989b) recommend that an oxidase-positive Gram-negative rod showing bipolar or irregular staining should be tested for resistance to colistin 10 μg and gentamicin 10 μg discs and subcultured on Ashdown's medium and incubated for 48–72 h. Presumptive isolates from this medium (resistant to both antibiotics) should be screened by API 20NE, which is highly efficient for confirmation of the identity of *B. pseudomallei*, although false identification of some other pseudomonads may occur. *Burkholderia pseudomallei* may be misidentified as *B. mallei*, *B. cepacia*, *P. stutzeri* or *Flavobacterium* spp.; *B. cepacia* is arginine negative and lysine decarboxylase positive, and *B. mallei* is not motile and arginine negative. A latex agglutination test can be used to confirm the presence of the LPS antigen of *B. pseudomallei*, which is conserved throughout the species.

Various PCR assays targeting 16S or 23S rDNA and other DNA sequences have been described for the identification of *B. pseudomallei* either in pure culture or in clinical specimens. They vary considerably in their performance characteristics but generally are disappointing in their specificity, although reported sensitivities range from 40% to 100% with spiked samples. The protocol of Hagen *et al.* (2002) is probably the most efficient for the rapid and specific detection of *B. pseudomallei*. They used a seminested PCR assay targeting a genus-specific sequence of the ribosomal protein subunit 21 (rpsU) and a nested PCR to amplify the filament-forming flagellin gene (*fliC*). The combination of PCR and sequencing of the amplicons resulted in high sensitivity and specificity in fixed tissue sections and avoided the need for culture of the organism.

Typing Methods

Until the advent of molecular typing techniques, relatively few methods were applied to the epidemiological type identification of *B. pseudomallei*. There is little, if any, serological heterogeneity to be exploited for antigenic typing schemes. The immunodominant epitope of the OM, LPS, is structurally and antigenically conserved, and there is minimal variation in OM and cellular proteins. The flagella are antigenic, but no H-typing scheme has been described. A capsule-like exopolysaccharide has been identified, but this is also antigenically uniform throughout the species.

Ribotyping using *Bam*HI digests of chromosomal DNA was the first molecular typing method used for studies of the epidemiology of *B. pseudomallei*. A survey of 350 isolates of *B. pseudomallei* recovered over 71 years from 23 countries revealed 44 ribotypes. Two ribotypes together accounted for almost half of the collection, and one of these was statistically associated with an Asian origin (Pitt, Trakulsomboon and Dance, 2000). PFGE of *Xba*I DNA macrorestriction digests gives optimal discrimination between strains and has been utilized for many studies.

Other techniques applied to strain typing include random amplified polymorphic DNA (RAPD) analysis, multilocus enzyme electrophoresis and multilocus sequence typing (MLST). The latter technique has similar discriminatory power to PFGE (Godoy et al., 2003).

ANTIBIOTIC SUSCEPTIBILITY AND TREATMENT

In general, B. pseudomallei is susceptible in vitro to carbapenems (imipenem and meropenem), anti-pseudomonal penicillins (such as piperacillin and azlocillin), tetracyclines such as doxycycline, amoxicillin plus clavulanic acid (co-amoxiclav), ticarcillin plus clavulanic acid, ceftazidime, ceftriaxone, cefotaxime, aztreonam, cefoperazone plus sulbactam, chloramphenicol and trimethoprim–sulphamethoxazole. Quinolones such as ofloxacin and ciprofloxacin have intermediate activity. Burkholderia pseudomallei is resistant to penicillins, aminopenicillins and many early cephalosporins, including cefuroxime. It is also resistant to macrolides, clindamycin and rifampicin. Aminoglycoside resistance is uniform, although 60% of strains may be sensitive to kanamycin. Sulphamethoxazole and trimethoprim sensitivity tests may be unreliable because of poorly defined endpoints (Dance et al., 1989a). Resistance to ampicillin is mediated by a clavulanate-susceptible β-lactamase, three different mechanisms of β-lactam resistance being identified in B. pseudomallei (Godfrey et al., 1991).

High-dose ceftazidime, with or without additional co-trimoxazole, has been shown to halve the mortality of severe melioidosis compared with the previously used conventional regimen (see below) and is the drug of choice for severe infections. Co-amoxiclav (Augmentin) is a good alternative (Suputtamongkol et al., 1994b), particularly for empirical treatment of severe sepsis in endemic areas, because of its broad spectrum of activity. Imipenem plus cilastatin was shown to be equivalent to ceftazidime alone in a single large trial (Simpson et al., 1999), and meropenem might be expected to have equivalent efficacy. Cefotaxime and ceftriaxone are poorly active in vivo despite good in vitro activity (Chaowagul et al., 1999). Before 1987, the conventional treatment for melioidosis was chloramphenicol, doxycycline and trimethoprim–sulphamethoxazole, in combination. This regimen, given intravenously, was associated with poor outcomes in acute disease. However, it may be the only option in severe β-lactam allergy.

Prolonged oral eradication therapy, following successful treatment of acute disease, is required to reliably limit disease relapse. The four-drug conventional regimen has been shown to be superior to co-amoxiclav for this purpose and remains the eradication regimen of choice. Doxycycline, as a single agent, and fluoroquinolones (ciprofloxacin and ofloxacin) have each proved to be unsatisfactory (White, 2003). The need for chloramphenicol in the oral eradication regimen has been questioned (Chetchotisakd et al., 2001) and is the subject of further clinical trials. Trimethoprim–sulphamethoxazole alone has been used successfully in Australia (Currie et al., 2000), but as yet there are no clinical trial data to support its use. For children and pregnant women, prolonged courses of co-amoxiclav are recommended for eradication.

BURKHOLDERIA THAILANDENSIS

In recent years, it has become apparent that some environmental isolates of B. pseudomallei are of a different phenotype to those from disease. This phenotype was first characterized by their ability to assimilate arabinose; in contrast, all clinical isolates were unable to assimilate this sugar. These newly defined strains, termed ara$^+$, were non-virulent in mice compared with the highly virulent ara$^-$ clinical strains. They were also distinguishable by ribotype and had sufficient differences in 16S rRNA encoding genes to allow Brett et al. (1998) to propose the new species name of Burkholderia thailandensis. Isolates from the environment should therefore be screened for arabinose assimilation for the presumptive identification of this species. At the time of writing, there

has been only one unconfirmed report of human infection because of this organism, and its clinical relevance remains in doubt. Antibiotic susceptibility patterns are similar to those of B. pseudomallei.

BURKHOLDERIA MALLEI

Burkholderia mallei is an obligate animal parasite (particularly of equine species) that was at one time widespread in the world. It is exceedingly rare today and restricted to a few foci in Asia, Africa and the Middle East. Most interest in this organism now arises from its theoretical potential for use as a biological warfare agent.

Cells are straight or slightly curved rods arranged singly, in pairs end to end, in parallel bundles or in Chinese-letter form; they may stain irregularly owing to granular inclusions. It does not produce a capsule and flagella are absent. Burkholderia mallei is aerobic and on nutrient agar forms smooth, grey translucent colonies 0.5–1 mm in diameter in 18 h at 37 °C. It is non-motile, does not grow on MacConkey or cetrimide agar, does not form pigments and is catalase positive, and few are oxidase positive. All strains reduce nitrate but not nitrite. Gelatin hydrolysis, urease activity and production of lecithinase is variable; almost all hydrolyse tyrosine, but without the formation of pigment. All strains break down arginine but not lysine or ornithine.

Data on the antibiotic susceptibilities of B. mallei are scanty. The few strains reported appear to be sensitive to sulphonamides and usually also to streptomycin, tetracycline and novobiocin; some are sensitive to chloramphenicol. Kenny et al. (1999) examined 17 strains from reference culture collections and reported the following MICs in broth dilution tests for 90% of the strains tested: >64 μg/mL for ampicillin, 16 μg/mL for piperacillin, 8 μg/mL for amoxicillin–clavulanate, >64 μg/mL for cefuroxime, 8 μg/mL for ceftazidime, 0.25 μg/mL for imipenem, >64 μg/mL for chloramphenicol, 2 μg/mL for doxycycline, 8 μg/mL for ofloxacin, 8 μg/mL for ciprofloxacin, 0.5 μg/mL for gentamicin, 4 μg/mL for azithromycin, 16 μg/mL for rifampin, >64 μg/mL for sulphamethoxazole, 32 μg/mL for trimethoprim and >64 μg/mL for co-trimoxazole. They concluded that the superior tissue penetration of azithromycin over gentamicin made this a possible candidate antibiotic for treating glanders, but there is virtually no clinical experience with this agent.

There are no contemporary accounts in the literature of glanders in humans with the exception of the report of a laboratory-acquired case in a military researcher working with B. mallei (Srinivasan et al., 2001). The individual was diabetic and had not worn gloves when handling the live organism. His illness closely resembled melioidosis, with initial axillary lymphadenopathy followed by a systemic illness with liver and splenic abscesses. He was effectively cured by treatment with imipenem plus doxycycline followed by azithromycin plus doxycycline for a total of 6 months. In view of the lack of experience of treating glanders in recent times, treatment of cases should follow regimens that are suitable for melioidosis.

BURKHOLDERIA GLADIOLI

Burkholderia gladioli was previously known as Pseudomonas marginata but is closely related to B. cepacia. Strains grow readily on polymyxin-containing selective media, and they can be differentiated from B. cepacia by their negative reactions for oxidase enzyme and utilization of maltose and lactose. The species may also be confused with Oligella ureolytica because of its variable urea-hydrolysing properties. It is most often isolated as an opportunist in the CF lung where it may cause a primary invasive infection, but it has also been recovered from patients with chronic granulomatous disease. Multiresistant isolates with shared phenotypic properties of B. cepacia and B. gladioli have been reported from CF patients by Baxter et al. (1997) who found the 'Edinburgh epidemic strain' of B. cepacia to give equivocal reactions in biochemical tests and fatty acid analysis. A PCR assay for the species based on 23S rDNA sequences was shown by Whitby *et al.* (2000) to have a sensitivity of 96% and to be 100% specific. Human clinical cases have been reviewed previously (Graves *et al.*, 1997).

OTHER PSEUDOMONADS

STENOTROPHOMONAS MALTOPHILIA

The work of Swings *et al.* (1983) placed strains of the species *Pseudomonas maltophilia* into the genus *Xanthomonas*. However, many strains of the former did not form the bright-yellow colonies of many *Xanthomonas* strains, and hence a new genus *Stenotrophomonas* (a unit feeding on few substrates) was proposed by Palleroni and Bradbury (1993) to accommodate the single species *maltophilia*. *Stenotrophomonas maltophilia* forms opaque grey-green yellowish colonies on nutrient agar at 37 °C and often has a lavender hue on blood agar. It has several polar flagella and gives a variable weak oxidase reaction. Most strains require methionine for growth. Maltose, glucose and xylose are utilized in ammonium salt-based media and nitrate is reduced.

Stenotrophomonas maltophilia is widely distributed in the natural world and has been isolated from various environments (soil, water and milk). After *P. aeruginosa*, it is probably the most frequently isolated pseudomonad in the clinical laboratory. It is an occasional cause of bacteraemia, endocarditis and pneumonia and has been recovered from many other disease states. Nevertheless, the clinical significance of *S. maltophilia* is unclear. Although it can cause severe and fatal infections, in many patients, particularly those with wound infections, infection is often trivial and self-limiting. However, mechanically ventilated patients receiving antimicrobials in the Intensive Therapy Unit (ITU) and neutropenic patients are at increased risk of *S. maltophilia* infection or colonization. Establishing the importance of isolates from ITU patients can be very difficult, particularly if the organism is a common isolate from patients in that unit. Contamination of monitoring equipment possibly contributes to its spread, although carriage on the hands of hospital workers in outbreaks is uncommon. The hospital sources from which the organism has been isolated include sinks, nebulizers, transducers, disinfectants, defrost water baths and ice-making machines. *Stenotrophomonas maltophilia* has been reported as the most common contaminant identified from embryos and semen samples stored in liquid nitrogen, and the organism was further shown to suppress significantly fertilization and embryonic development *in vitro* (Bielanski *et al.*, 2003). Human carriage is variable; faecal carriage rates have been reported to be as high as 33% in patients with haematological malignancy but only 3% in a control patient group (Denton and Kerr, 1998).

Stenotrophomonas maltophilia has been increasingly isolated from the sputum of CF patients. Talmaciu *et al.* (2000) reported that since 1993 the incidence of *S. maltophilia*-positive patients increased from 2.8% to 6.2% in 1997. Their data suggest that colonized patients had worse growth parameters, clinical scores and spirometric values than *S. maltophilia*-negative patients, and there appeared to be a higher incidence of colonization with *S. aureus* among those patients who grew *S. maltophilia*. Considerable variation has been reported in the prevalence of the species in CF centres worldwide, 19% in a British centre and over 30% in a Spanish centre (Denton and Kerr, 1998), and this may be attributable to, among other factors, the wider use of carbapenems and longer duration of hospitalization of patients. Cross-infection between patients does not appear to be significant. *Stenotrophomonas maltophilia* will often grow reasonably well on *B. cepacia* selective media containing polymyxin B.

Many typing methods have been applied to epidemiological studies of *S. maltophilia*. O-Serotyping is poorly discriminatory owing to the disproportionate frequencies of three of the 31 types described, and the technique has not been widely used. Other methods proposed

for typing include protein profiles, bacteriocins and pyrolysis mass spectrometry. Ribotyping is highly discriminatory for the species, and PFGE of *Xba*I chromosomal DNA digests is able to resolve genetically distinct strains. Arbitrarily primed PCR is slightly less discriminatory than PFGE but offers the advantage of speed and less labour. Typing of isolates from apparent outbreaks usually reveals the presence of multiple strains, with small clusters of patients sharing the same strain. Indeed, the diversity of genotypes identified in *S. maltophilia* incidents argues against clonal spread of resistant hospital strains, and the marked genetic distance between strains may lead to further examination of the homogeneity of the species.

Stenotrophomonas maltophilia is intrinsically resistant to imipenem owing to the production of a zinc-dependent carbapenemase, which may explain in part its prevalence in intensive care and neutropenic patients. Most strains are sensitive to co-trimoxazole, doxycycline, minocycline, ticarcillin/clavulanic acid, cefotaxime and ceftazidime. Co-trimoxazole is the antibiotic of choice for therapy. Synergy, *in vitro*, has been observed for combinations of co-trimoxazole with many agents, including carbenicillin, rifampicin and gentamicin. There may be significant differences in sensitivity test results of strains after incubation at different temperatures. Strains are more resistant to aminoglycosides at 30 °C and more sensitive to colistin at 37 °C. There is also considerable discrepancy between disc diffusion susceptibility values for strains and broth dilution MICs (Denton and Kerr, 1998). This makes the interpretation and reporting of susceptibility tests results problematic.

RALSTONIA SPP.

Ralstonia pickettii is a non-pigmented pseudomonad, formerly a member of the genus *Burkholderia*, which grows well at 37 °C and less so at 41 °C. The species is synonymous with the organism earlier described as *Pseudomonas thomasii* and resembles *B. cepacia* biochemically, as it does not attack arginine. Most strains are resistant to aminoglycosides and colistin but are generally susceptible to chloramphenicol, co-trimoxazole and cephalosporins. *Ralstonia pickettii* is an occasional cause of nosocomial infections in hospital patients and has been recovered from the ward environment as well as contaminated antiseptic and disinfectant solutions. It has also been associated with pseudobacteraemia. Three other species of *Ralstonia*, *Ralstonia gilardii*, *Ralstonia paucula* and *Ralstonia mannitolilytica*, have rarely been associated with human infections (Coeyne, Vandamme and Lipuma, 2002).

BREVUNDIMONAS SPP.

Brevundimonas diminuta and *Brevundimonas vesicularis* are closely related species that were previously classified as rRNA homology group IV within the genus *Pseudomonas*. They grow rather slowly on ordinary nutrient media and require pantothenate, biotin and cyanocobalamin for optimal growth; *B. diminuta* also requires cysteine or methionine. *B. vesicularis* gives only a weak oxidase reaction, forms yellow or orange colonies and acidifies glucose and maltose in ammonium salt media. Both species are rare in pathological specimens and of doubtful clinical significance, although occasional serious infections may occur.

DELFTIA ACIDOVORANS

Delftia acidovorans was previously classified as a member of the genus *Comamonas* (Wen *et al.*, 1999). It has been isolated on many occasions from hospital patients and their environment and is a common contaminant of dental unit water lines and has been occasionally reported as a cause of eye infections. The organism is able to degrade polyurethane, a widely used industrial polymer, and utilize certain herbicides. Some isolates are able to grow on polymyxin-containing agar and may be misidentified as *B. cepacia* in sputum specimens from CF patients. It acidifies mannitol but not ethanol, glucose or maltose in ammonium salt media and is generally asaccharolytic. Isolates are usually resistant to gentamicin and susceptible to the ureidopenicillins, tetracyclines, quinolones and co-trimoxazole.

PANDORAEA SPP.

These organisms were originally classified as CDC weak oxidizer group 2 but were recently assigned to the genus *Pandoraea*, which is a close relative of *Burkholderia*. They are nonfermentative Gram-negative rods isolated predominantly from soil and the sputum of CF patients. Four species have been named, *Pandoraea apista*, *Pandoraea pulmonicola*, *Pandoraea pnomenusa* and *Pandoraea sputorum*; *Burkholderia norimbergensis* has also been reclassified as a new *Pandoraea* species. Members may be differentiated from each other by specific PCR assays (Coeyne *et al.*, 2001). The few isolates tested were resistant to ampicillin, extended-spectrum cephalosporins and aminoglycosides but varied in their susceptibility to ciprofloxacin.

REFERENCES

Alcock SR (1977) Acute otitis externa in divers working in the North Sea: a microbiological survey of seven saturation dives. *The Journal of Hygiene*, **78**, 395–409.

Anzai Y, Kim H, Park J-Y *et al.* (2000) Phylogenetic affiliation of the pseudomonads based on 16S rRNA sequence. *International Journal of Systematic and Evolutionary Microbiology*, **50**, 1563–1589.

Arancibia F, Bauer TT, Ewig S *et al.* (2002) Community-acquired pneumonia due to Gram-negative bacteria and *Pseudomonas aeruginosa*. *Archives of Internal Medicine*, **162**, 1849–1858.

Ashdown LR (1979) An improved screening technique for isolation of *Pseudomonas pseudomallei* from clinical specimens. *Pathology*, **11**, 293–297.

Baltch AL, Smith RP, Franke M *et al.* (1994) *Pseudomonas aeruginosa* cytotoxin as a pathogenicity factor in a systemic infection of leukopenic mice. *Toxicon*, **32**, 27–34.

Barth AL and Pitt TL (1995) Auxotrophic variants of *Pseudomonas aeruginosa* are selected from prototrophic wild-type strains in respiratory infections in cystic fibrosis patients. *Journal of Clinical Microbiology*, **33**, 37–40.

Baxter IA, Lambert PA and Simpson IN (1997) Isolation from clinical sources of *Burkholderia cepacia* possessing characteristics of *Burkholderia gladioli*. *The Journal of Antimicrobial Chemotherapy*, **39**, 169–175.

Beck-Sagué CM, Banerjee SN and Jands WR (1994) Epidemiology and control of *Pseudomonas aeruginosa* in U.S. hospitals. In *Pseudomonas aeruginosa Infections and Treatment*, Baltch AL and Smith RP (eds). Marcel Dekker, New York, pp. 51–71.

Bevivino A, Dalmastri C, Tabacchioni S *et al.* (2002) *Burkholderia cepacia* complex bacteria from clinical and environmental sources in Italy: genomovar status and distribution of traits related to virulence and transmissibility. *Journal of Clinical Microbiology*, **40**, 846–851.

Bielanski A, Bergeron H, Lau PC and Devenish J (2003) Microbial contamination of embryos and semen during long term banking in liquid nitrogen. *Cryobiology*, **46**, 146–152.

Blot S, Vandewoude K, Hoste E and Colardyn F (2003) Reappraisal of attributable mortality in critically ill patients with nosocomial bacteraemia involving *Pseudomonas aeruginosa*. *The Journal of Hospital Infection*, **53**, 18–24.

Brett PJ, DeShazer D and Woods DE (1998) *Burkholderia thailandensis* sp. nov., a *Burkholderia pseudomallei*-like species. *International Journal of Systematic Bacteriology*, **48**, 317–320.

Brown MRW, Foster JHS and Clamp JR (1969) Composition of *Pseudomonas aeruginosa* slime. *The Biochemical Journal*, **112**, 521–525.

Butler SL, Doherty CJ, Hughes JE *et al.* (1995) *Burkholderia cepacia* and cystic fibrosis: do natural environments present a potential hazard? *Journal of Clinical Microbiology*, **33**, 1001–1004.

Chaowagul W, White NJ, Dance DAB *et al.* (1989) Melioidosis: a major cause of community-acquired septicemia in Northeastern Thailand. *The Journal of Infectious Diseases*, **159**, 890–899.

Chaowagul W, Suputtamongkol Y, Dance DAB *et al.* (1993) Relapse in melioidosis: incidence and risk factors. *The Journal of Infectious Diseases*, **168**, 1181–1185.

Chaowagul W, Simpson AJH, Suputtamongkol Y and White NJ (1999) Empirical cephalosporin treatment of melioidosis. *Clinical Infectious Diseases*, **28**, 1328.

Chastre J and Fagon J-Y (2002) Ventilator-associated pneumonia. *American Journal of Respiratory and Critical Care Medicine*, **165**, 867–903.

Chetchotisakd P, Chaowagul W, Mootsikapun P *et al.* (2001) Maintenance therapy of melioidosis with ciprofloxacin plus azithromycin compared with cotrimoxazole plus doxycycline. *The American Journal of Tropical Medicine and Hygiene*, **64**, 24–27.

Clode FE, Kaufmann ME, Malnick H and Pitt TL (2000) Distribution of genes encoding putative transmissibility factors among epidemic and non-epidemic strains of *Burkholderia cepacia* from cystic fibrosis patients in the United Kingdom. *Journal of Clinical Microbiology*, **38**, 1763–1766.

Coeyne T and LiPuma JJ (2002) Multilocus restriction typing: a novel tool for studying global epidemiology of *Burkholderia cepacia* complex infection in cystic fibrosis. *The Journal of Infectious Diseases*, **185**, 1454–1462.

Coeyne T, Liu L, Vandamme P and LiPuma JJ (2001) Identification of *Pandoraea* species by 16S ribosomal DNA-based PCR assays. *Journal of Clinical Microbiology*, **39**, 4452–4455.

Coeyne T, Vandamme P and LiPuma JJ (2002) Infection by *Ralstonia* species in cystic fibrosis patients: identification of *R. pickettii* and *R. mannitolilytica* by polymerase chain reaction. *Emerging Infectious Diseases*, **8**, 692–696.

Conway SP, Brownlee KG, Denton M and Peckham DG (2003) Antibiotic treatment of multidrug-resistant organisms in cystic fibrosis. *American Journal of Respiratory Medicine*, **2**, 321–332.

Cryz SJ (1994) Vaccines, immunoglobulins, and monoclonal antibodies for the prevention and treatment of *Pseudomonas aeruginosa* infections. In *Pseudomonas aeruginosa Infections and Treatment*. Baltch AL and Smith RP (eds). Marcel Dekker, New York, pp. 519–545.

Currie BJ, Fisher DA, Howard DM *et al.* (2000) Endemic melioidosis in tropical northern Australia: a 10-year prospective study and review of the literature. *Clinical Infectious Diseases*, **31**, 981–986.

Dance DAB (1991) Melioidosis: the tip of the iceberg? *Clinical Microbiology Reviews*, **4**, 52–60.

Dance DAB, Wuthiekanun V, Chaowagul W and White NJ (1989a) The antimicrobial susceptibility of *Pseudomonas pseudomallei*: emergence of resistance in vitro and during treatment. *The Journal of Antimicrobial Chemotherapy*, **24**, 295–309.

Dance DAB, Wuthiekanun V, Naigowit P and White NJ (1989b) Identification of *Pseudomonas pseudomallei* in clinical practice: use of simple screening tests and API 2ONE. *Journal of Clinical Pathology*, **42**, 645–648.

De Soyza A, McDowell A, Archer L *et al.* (2001) *Burkholderia cepacia* complex genomovars and pulmonary transplantation outcomes in patients with cystic fibrosis. *Lancet*, **358**, 1780–1781.

De Vos D, De Chial M, Cochez C *et al.* (2001) Study of pyoverdine type and production by *Pseudomonas aeruginosa* isolated from cystic fibrosis patients. *Archives of Microbiology*, **175**, 384–388.

Denton M and Kerr KG (1998) Microbiological and clinical aspects of infection associated with *Stenotrophomonas maltophilia*. *Clinical Microbiology Reviews*, **11**, 57–80.

Favre-Bonté S, Pache JC, Robert J *et al.* (2002) Detection of *Pseudomonas aeruginosa* cell-to-cell signals in lung tissue of cystic fibrosis patients. *Microbial Pathogenesis*, **32**, 143–147.

Feldman M, Bryan R, Rajan S *et al.* (1998) Role of flagella in pathogenesis of *Pseudomonas aeruginosa* pulmonary infection. *Infection and Immunity*, **66**, 43–51.

Feltman H, Schulert G, Khan S *et al.* (2001) Prevalence of type III secretion genes in clinical and environmental isolates of *Pseudomonas aeruginosa*. *Microbiology*, **147**, 2659–2669.

Foght JM, Westlake DWS, Johnson WM and Ridgway HF (1996) Environmental gasoline-utilizing isolates and clinical isolates of *Pseudomonas aeruginosa* are taxonomically indistinguishable by chemotaxonomic and molecular techniques. *Microbiology*, **142**, 2333–2340.

Forkner CE (1960) *Pseudomonas aeruginosa* infections. In *Modern Medical Monographs*. Grune & Stratton, New York, No. 22, pp. 1–5.

Garner CV, Desjardins D and Pier GB (1990) Immunogenic properties of *Pseudomonas aeruginosa* mucoid exopolysaccharide. *Infection and Immunity*, **58**, 1835–1842.

Gilboa-Garber N, Sudakevitz D, Sheffi M *et al.* (1994) PA-I and PA-II lectin interactions with the ABO(H) and P blood group glycosphingolipid antigens may contribute to the broad spectrum adherence of *Pseudomonas aeruginosa* to human tissues in secondary infection. *Glycoconjugate Journal*, **11**, 414–417.

Godfrey AJ, Wong S, Dance DAB *et al.* (1991) *Pseudomonas pseudomallei* resistance to β-lactams due to alterations in the chromosomally encoded beta-lactamases. *Antimicrobial Agents and Chemotherapy*, **35**, 1635–1640.

Godoy D, Randle G, Simpson AJ *et al.* (2003) Multilocus sequence typing and evolutionary relationships among the causative agents of melioidosis and glanders, *Burkholderia pseudomallei* and *Burkholderia mallei*. *Journal of Clinical Microbiology*, **41**, 2068–2079.

Govan JRW and Nelson JW (1992) Microbiology of lung infection in cystic fibrosis. *British Medical Bulletin*, **48**, 912–930.

Govan JRW, Brown PH, Maddison J *et al.* (1993) Evidence for transmission of *Pseudomonas cepacia* by social contact in cystic fibrosis. *Lancet*, **342**, 15–19.

Graves M, Robin T, Chipman AM *et al.* (1997) Four additional cases of *Burkholderia gladioli* infection with microbiological correlates and review. *Clinical Infectious Diseases*, **25**, 838–842.

Grundmann H, Schneider C, Hartung D *et al.* (1995) Discriminatory power of three DNA-based typing techniques for *Pseudomonas aeruginosa*. *Journal of Clinical Microbiology*, **33**, 528–534.

Hagen RM, Gauthier YP, Sprague LD *et al.* (2002) Strategies for PCR based detection of *Burkholderia pseudomallei* DNA in paraffin wax embedded tissues. *Molecular Pathology*, **55**, 398–400.

Hancock REW, Mutharia LM, Chan L *et al.* (1983) *Pseudomonas aeruginosa* isolates from cystic fibrosis: a class of serum sensitive, nontypable strains deficient in lipopolysaccharide O side chains. *Infection and Immunity*, **42**, 170–177.

Harley VS, Dance DAB, Drasar BS and Tovey G (1998) An ultrastructural study of the phagocytosis of *Burkholderia pseudomallei*. *Microbios*, **94**, 35–45.

Häußler S, Rohde M, von Neuhoff N *et al.* (2003) Structural and functional cellular changes induced by *Burkholderia pseudomallei* rhamnolipid. *Infection and Immunity*, **71**, 2970–2975.

He J, Baldini RL, Deziel E *et al.* (2004) The broad host range pathogen *Pseudomonas aeruginosa* strain PA 14 carries two pathogenicity islands harbouring plant and animal virulence genes. *Proceedings of the National Academy of Sciences of the United States of America*, **101**, 2530–2535.

Heath DG, Hohneker K, Carriker C *et al.* (2002) Six-year molecular analysis of *Burkholderia cepacia* complex isolates among cystic fibrosis patients at a referral center for lung transplantation. *Journal of Clinical Microbiology*, **40**, 1188–1193.

Henry DA, Campbell M, McGimpsey C *et al.* (1999) Comparison of isolation media for recovery of *Burkholderia cepacia* complex from respiratory secretions of patients with cystic fibrosis. *Journal of Clinical Microbiology*, **37**, 1004–1007.

Henwood CJ, Livermore DM, James D *et al.* (2001) Antimicrobial susceptibility of *Pseudomonas aeruginosa*: results of a UK survey and evaluation of the British Society for Antimicrobial Chemotherapy disc susceptibility test. *The Journal of Antimicrobial Chemotherapy*, **47**, 789–799.

Holder IA (1977) Epidemiology of *Pseudomonas aeruginosa* in a burn hospital. In *Pseudomonas aeruginosa: Ecological Aspects and Patient Colonization*. Young VM (ed.). Raven Press, New York, pp. 77–95.

Holder IA (2004) Pseudomonas immunotherapy: a historical overview. *Vaccine*, **22**, 831–839.

Isles A, Maclusky I, Corey M *et al.* (1984) *Pseudomonas cepacia* infection in cystic fibrosis: an emerging problem. *The Journal of Pediatrics*, **104**, 206–210.

Jones AM, Dodd ME and Webb AK (2001) *Burkholderia cepacia*: current clinical issues, environmental controversies and ethical dilemmas. *The European Respiratory Journal*, **17**, 295–301.

Kenny DJ, Russell P, Rogers D *et al.* (1999) In vitro susceptibilities of *Burkholderia mallei* in comparison to those of other pathogenic *Burkholderia* spp. *Antimicrobial Agents and Chemotherapy*, **43**, 2773–2775.

King EO, Ward MK and Raney DA (1954) Two simple media for the demonstration of pyocyanin and fluorescein. *The Journal of Laboratory and Clinical Medicine*, **44**, 301–307.

Klausen M, Heydorn A, Ragas P *et al.* (2003) Biofilm formation by *Pseudomonas aeruginosa* wild type, flagella and type IV pili mutants. *Molecular Microbiology*, **48**, 1511–1524.

Koyama H and Geddes DM (1997) Erythromycin and diffuse panbronchiolitis. *Thorax*, **52**, 915–918.

Kronborg G, Fomsgaard A, Galanos G *et al.* (1992) Antibody response to lipid A, core, and O-sugars of the *Pseudomonas aeruginosa* lipopolysaccharide in chronically infected cystic fibrosis patients. *Journal of Clinical Microbiology*, **30**, 1848–1855.

Kunin CM (1994) Infections of the urinary tract due to *Pseudomonas aeruginosa*. In *Pseudomonas aeruginosa Infections and Treatment*. Baltch AL and Smith RP (eds). Marcel Dekker, New York, pp. 237–256.

Lam MYC, McGroarty EJ, Kropinski AM *et al.* (1989) Occurrence of a common lipopolysaccharide antigen in standard and clinical strains of *Pseudomonas aeruginosa*. *Journal of Clinical Microbiology*, **27**, 962–967.

Lang AB, Rudeberg A, Schoni MH *et al.* (2004) Vaccination of cystic fibrosis patients against *Pseudomonas aeruginosa* reduces the proportion of patients infected and delays time to infection. *The Pediatric Infectious Disease Journal*, **23**, 504–510.

Lanotte P, Mereghetti L, Lejeune B *et al.* (2003) *Pseudomonas aeruginosa* and cystic fibrosis: correlation between exoenzyme production and patient's clinical state. *Pediatric Pulmonology*, **36**, 405–412.

Lanotte P, Watt S, Mereghetti L *et al.* (2004) Genetic features of *Pseudomonas aeruginosa* isolates from cystic fibrosis patients compared with those of isolates from other origins. *Journal of Medical Microbiology*, **53**, 73–81.

Ledson MJ, Gallagher MJ, Jackson M *et al.* (2002) Outcome of *Burkholderia cepacia* colonisation in an adult cystic fibrosis centre. *Thorax*, **57**, 142–145.

LiPuma JJ, Spilker T, Coeyne T and Gonzalez CF (2002) An epidemic *Burkholderia cepacia* complex strain identified in soil. *Lancet*, **359**, 2002–2003.

Liu PV, Matsumoto H, Kusama H and Bergan T (1983) Survey of heat-stable, major somatic antigens of *Pseudomonas aeruginosa*. *International Journal of Systematic Bacteriology*, **33**, 256–264.

Livermore DM (2002) Multiple mechanisms of antimicrobial resistance in *Pseudomonas aeruginosa*: our worst nightmare? *Clinical Infectious Diseases*, **34**, 634–640.

Maccheron LJ, Groeneveld ER, Ohlrich SJ *et al.* (2004) Orbital cellulitis, panophthalmitis, and ecthyma gangrenosum in an immunocompromised host with pseudomonas septicaemia. *American Journal of Ophthalmology*, **137**, 176–178.

Mader JT, Vibhagool A, Mader J and Calhoun JH (1994) *Pseudomonas aeruginosa* bone and joint infections. In *Pseudomonas aeruginosa Infections and Treatment*. Baltch AL and Smith RP (eds). Marcel Dekker, New York, pp. 293–326.

Mahenthiralingam E, Bischof J, Byrne SK *et al.* (2000) DNA-based diagnostic approaches for the identification of *Burkholderia cepacia* complex, *Burkholderia vietnamiensis*, *Burkholderia multivorans*, *Burkholderia stabilis*, and *Burkholderia cepacia* genomovars I and III. *Journal of Clinical Microbiology*, **38**, 3165–3173.

Mahenthiralingam E, Baldwin A and Vandamme P (2002) *Burkholderia cepacia* complex infection in patients with cystic fibrosis. *Journal of Medical Microbiology*, **51**, 533–538.

Manno G, Ugolotti E, Belli ML *et al.* (2003) Use of the E test to assess synergy of antibiotic combinations against isolates of *Burkholderia cepacia-*complex from patients with cystic fibrosis. *European Journal of Clinical Microbiology & Infectious Diseases*, **22**, 28–34.

Mathee K, Ciofu O, Sternberg C *et al.* (1999) Mucoid conversion of *Pseudomonas aeruginosa* by hydrogen peroxide: a mechanism for virulence activation in the cystic fibrosis lung. *Microbiology*, **145**, 1349–1357.

Morlin GL, Hedges DL, Smith AL and Burns JL (1994) Accuracy and cost of antibiotic susceptibility testing of mixed morphotypes of *Pseudomonas aeruginosa*. *Journal of Clinical Microbiology*, **32**, 1027–1030.

Moskowitz SM, Foster JM, Emerson J and Burns JL (2004) Clinically feasible biofilm susceptibility assay for isolates of *Pseudomonas aeruginosa* from patients with cystic fibrosis. *Journal of Clinical Microbiology*, **42**, 1915–1922.

Nicas TI and Iglewski BH (1986) Toxins and virulence factors of *Pseudomonas aeruginosa*. In *The Bacteria, Vol. X*. Nicas TI and Iglewski BH (eds). Academic Press, New York, pp. 195–213.

Nzula S, Vandamme P and Govan JRW (2002) Influence of taxonomic status on the in vitro antimicrobial susceptibility of the *Burkholderia cepacia* complex. *The Journal of Antimicrobial Chemotherapy*, **50**, 261–269.

O'Neil K, Herman JH, Modlin JF *et al.* (1986) *Pseudomonas cepacia*: an emerging pathogen in chronic granulomatous disease. *The Journal of Pediatrics*, **108**, 940–942.

Oliver A, Canton R, Campo P *et al.* (2000) High frequency of hypermutable *Pseudomonas aeruginosa* in cystic fibrosis lung infection. *Science*, **288**, 1251–1254.

Palleroni NJ (1984) Genus I. *Pseudomonas* Migula 1894. In *Bergey's Manual of Systematic Bacteriology*. Krieg NR and Holt JG (eds). Williams & Wilkins, Baltimore, pp. 141–199.

Palleroni NJ, Bradbury JF (1993) *Stenotrophomonas*, a new bacterial genus for *Xanthomonas maltophilia* (Hugh 1980) Swings *et al.* 1983. *International Journal of Systematic Bacteriology*, **43**, 606–609.

Paul M, Benuri-Silbiger I, Soares-Weiser K and Leibovici L (2004) Beta lactam monotherapy versus beta lactam-aminoglycoside combination therapy for sepsis in immunocompetent patients: systematic review and meta-analysis of randomised trials. *British Medical Journal*, **328**, 668.

Pier GB (2002) *CFTR* mutations and host susceptibility to *Pseudomonas aeruginosa* lung infection. *Current Opinion in Microbiology*, **5**, 81–86.

Pier GB, Saunders JM, Ames P *et al.* (1987) Opsonophagocytic killing antibody to *Pseudomonas aeruginosa* mucoid exopolysaccharide in older non-colonized patients with cystic fibrosis. *The New England Journal of Medicine*, **317**, 793–798.

Pier GB, Meluleni L and Goldberg JB (1995) Clearance of *Pseudomonas aeruginosa* from the murine gastrointestinal tract is effectively mediated by O-antigen specific circulating antibodies. *Infection and Immunity*, **63**, 2818–2825.

Pirnay J-P, De Vos D, Cochez C *et al.* (2002) *Pseudomonas aeruginosa* displays an epidemic population structure. *Environmental Microbiology*, **4**, 898–911.

Pitt TL, Kaufmann ME, Patel PS *et al.* (1996) Type characterisation and antibiotic susceptibility of *Burkholderia (Pseudomonas) cepacia* isolates from patients with cystic fibrosis in the United Kingdom and the Republic of Ireland. *Journal of Medical Microbiology*, **44**, 203–210.

Pitt TL, Trakulsomboon S and Dance DAB (2000) Molecular phylogeny of *Burkholderia pseudomallei. Acta Tropica*, **74**, 181–185.

Pitt TL, Sparrow M, Warner M and Stefanidou M (2003) Survey of resistance of *Pseudomonas aeruginosa* from UK patients with cystic fibrosis to six commonly prescribed antimicrobial agents. *Thorax*, **58**, 794–796.

Poole K and McKay GA (2003) Iron acquisition and its control in *Pseudomonas aeruginosa*: many roads lead to Rome. *Frontiers in Bioscience*, **8**, 661–686.

Ran H, Hassett DJ and Lau GW (2003) Human targets of *Pseudomonas aeruginosa* pyocyanin. *Proceedings of the National Academy of Sciences of the United States of America*, **100**, 14315–14320.

Ratjen F and Doring G (2003) Cystic fibrosis. *Lancet*, **361**, 681–689.

Ratjen F, Doring G and Nikolaizik W (2001) Eradication of *Pseudomonas aeruginosa* with inhaled tobramycin in patients with cystic fibrosis. *Lancet*, **358**, 983–984.

Rhame FS (1980) The ecology and epidemiology of *Pseudomonas aeruginosa*. In *Pseudomonas aeruginosa: the Organism, Diseases It Causes, and Their Treatment*. Sabath LD (ed.). Hans Huber, Bern, pp. 31–51.

Roy-Burman A, Savel RH, Racine S *et al.* (2001) Type III protein secretion is associated with death in lower respiratory and systemic *Pseudomonas aeruginosa* infections. *The Journal of Infectious Diseases*, **183**, 1767–1774.

Saiman L (2004) Microbiology of early CF lung disease. *Paediatric Respiratory Reviews*, **5** (Suppl. A), S367–S369.

Saiman L and Siegel J (2004) Infection control in cystic fibrosis. *Clinical Microbiology Reviews*, **17**, 57–71.

Saiman L, Marshall BC, Mayer-Hamblett N *et al.* (2003) Azithromycin in patients with cystic fibrosis chronically infected with *Pseudomonas aeruginosa*: a randomized trial. *The Journal of the American Medical Association*, **290**, 1749–1756.

Sajjan U, Corey M, Humar A *et al.* (2001) Immunolocalisation of *Burkholderia cepacia* in the lungs of cystic fibrosis patients. *Journal of Medical Microbiology*, **50**, 535–546.

Schlech WF, Simonsen N, Sumarah R and Martin RS (1986) Nosocomial outbreak of *Pseudomonas aeruginosa* folliculitis associated with a physiotherapy pool. *Canadian Medical Association Journal*, **134**, 909–913.

Schroeder TH, Zaidi TS and Pier GB (2001) Lack of adherence of clinical isolates of *Pseudomonas aeruginosa* to asialo GM$_1$ on epithelial cells. *Infection and Immunity*, **69**, 719–729.

Schultz MJ, Knapp S, Florquin S *et al.* (2003) Interleukin-18 impairs the pulmonary host response to *Pseudomonas aeruginosa*. *Infection and Immunity*, **71**, 1630–1634.

Scott FW and Pitt TL (2004) Identification and characterization of transmissible *Pseudomonas aeruginosa* strains in cystic fibrosis in England and Wales. *Journal of Medical Microbiology*, **53**, 609–615.

Simpson AJH, Suputtamongkol Y, Smith MD *et al.* (1999) Comparison of imipenem and ceftazidime as therapy for severe melioidosis. *Clinical Infectious Diseases*, **29**, 381–387.

Smith AL, Fiel SB, Mayer-Hamblett N *et al.* (2003) Lack of association between the in vitro antibiotic susceptibility testing of *Pseudomonas aeruginosa* isolates and clinical response to parenteral antibiotic administration in cystic fibrosis. *Chest*, **123**, 1495–1502.

Smith RP (1994) Skin and soft tissue infections due to *Pseudomonas aeruginosa*. In *Pseudomonas aeruginosa Infections and Treatment*. Baltch AL and Smith RP (eds). Marcel Dekker, New York, pp. 327–369.

Smith RS and Iglewski BH (2003) *Pseudomonas aeruginosa* quorum sensing as a potential antimicrobial target. *The Journal of Clinical Investigation*, **112**, 1460–1465.

Speijer H, Savelkoul PH, Bonten MJ *et al.* (1999) Application of different genotyping methods for *Pseudomonas aeruginosa* in a setting of endemicity in an intensive care unit. *Journal of Clinical Microbiology*, **37**, 3654–3661.

Srinivasan A, Kraus CN, DeShazer D *et al.* (2001) Glanders in a military research microbiologist. *The New England Journal of Medicine*, **345**, 256–258.

Stoodley BJ and Thom BT (1970) Observations on the intestinal carriage of *Pseudomonas aeruginosa. Journal of Medical Microbiology*, **3**, 367–375.

Suputtamongkol Y, Hall AJ, Dance DAB *et al.* (1994a) The epidemiology of melioidosis in Ubon Ratchatani, northeast Thailand. *International Journal of Epidemiology*, **23**, 1082–1090.

Suputtamongkol Y, Rajchanuwong A, Chaowagul W *et al.* (1994b) Ceftazidime vs. amoxicillin/clavulanate in the treatment of severe melioidosis. *Clinical Infectious Diseases*, **19**, 846–853.

Swings J, De Vos P, Van den Mooter M and De Ley J (1983) Transfer of *Pseudomonas maltophilia* Hugh 1981 to the genus *Xanthomonas* as *Xanthomonas maltophilia* (Hugh 1981) comb. nov. *International Journal of Systematic Bacteriology*, **33**, 409–413.

Talmaciu I, Varlotta L, Mortensen J and Schidlow DV (2000) Risk factors for emergence of *Stenotrophomonas maltophilia* in cystic fibrosis. *Pediatric Pulmonology*, **30**, 10–15.

Tredget EE, Shankowsky HA, Rennie R *et al.* (2004) *Pseudomonas* infections in the thermally injured patient. *Burns*, **30**, 3–26.

Trepanier S, Prince A and Huletsky A (1997) Characterization of the *penA* and *penR* genes of *Burkholderia cepacia* 249 which encode the chromosomal class A penicillinase and its LysR-type transcriptional regulator. *Antimicrobial Agents and Chemotherapy*, **41**, 2399–2405.

Trick WE, Kioski CM, Howard KM *et al.* (2000) Outbreak of *Pseudomonas aeruginosa* ventriculitis among patients in a neurosurgical intensive care unit. *Infection Control and Hospital Epidemiology*, **21**, 204–208.

Vandamme P, Holmes B, Vancanneyt M *et al.* (1997) Occurrence of multiple genomovars of *Burkholderia cepacia* in cystic fibrosis patients and proposal of *Burkholderia multivorans* sp. nov. *International Journal of Systematic Bacteriology*, **47**, 1188–1200.

Wen A, Fegan M, Hayward C *et al.* (1999) Phylogenetic relationships among members of the Comamonadaceae, and description of *Delftia acidovorans* (den Dooren de Jong 1926 and Tamaoka *et al.* 1987) gen. nov., comb. nov. *International Journal of Systematic Bacteriology*, **49**, 567–576.

West SHE, Zeng L, Lee BL *et al.* (2002) Respiratory infections with *Pseudomonas aeruginosa* in children with cystic fibrosis. *The Journal of the American Medical Association*, **287**, 2958–2967.

Whitby PW, Pope LC, Carter KB *et al.* (2000) Species-specific PCR as a tool for the identification of *Burkholderia gladioli. Journal of Clinical Microbiology*, **38**, 282–285.

White NJ (2003) Melioidosis. *Lancet*, **361**, 1715–1722.

Wick MJ, Frank DW, Storey DJ and Iglewski BH (1990) Identification of *regB*, a gene required for optimal exotoxin A yields in *Pseudomonas aeruginosa*. *Molecular Microbiology*, **4**, 489–497.

Wilkinson SG and Pitt TL (1995) *Burkholderia (Pseudomonas) cepacia*: surface chemistry and typing methods: pathogenicity and resistance. *Reviews in Medical Microbiology*, **6**, 1–17.

Winstanley C and Hart CA (2000) Presence of type III secretion genes in *Burkholderia pseudomallei* correlates with ara$^-$ phenotypes. *Journal of Clinical Microbiology*, **38**, 883–885.

Wong KT, Puthucheary SD and Vadivelu J (1995) The histopathology of human melioidosis. *Histopathology*, **26**, 51–55.

Woods DE and Vasil ML (1994) Pathogenesis of *Pseudomonas aeruginosa* infections. In *Pseudomonas aeruginosa Infections and Treatment*. Baltch AL and Smith RP (eds). Marcel Dekker, New York, pp. 21–50.

Worlitzsch D, Tarran R, Ulrich M *et al.* (2002) Effects of reduced mucus oxygen concentration in airway *Pseudomonas* infections of cystic fibrosis patients. *The Journal of Clinical Investigation*, **109**, 317–325.

Wuthiekanun V, Suputtamongkoi Y, Simpson AJH *et al.* (2001) Value of throat swab in diagnosis of melioidosis. *Journal of Clinical Microbiology*, **39**, 3801–3802.

Yabuuchi E, Kosako Y, Oyaizu H *et al.* (1992) Proposal of *Burkholderia* gen. nov. and transfer of seven species of the genus *Pseudomonas* homology group II to the new genus, with the type species *Burkholderia cepacia* (Palleroni and Holmes 1981) comb. nov. *Microbiology and Immunology*, **36**, 1251–1275.

Zloty P and Belin MW (1994) Ocular infections caused by *Pseudomonas aeruginosa*. In *Pseudomonas aeruginosa Infections and Treatment*. Baltch AL and Smith RP (eds). Marcel Dekker, New York, pp. 371–399.

37

Legionella spp.

T. G. Harrison

Health Protection Agency, Systemic Infection Laboratory, Centre for Infection, London, UK

INTRODUCTION

In July 1976 a dramatic outbreak of an acute, febrile respiratory illness occurred in the city of Philadelphia among approximately 4400 Legionnaires (veterans of the US army) who had gathered to attend the 58th convention of the Pennsylvania Branch of the American Legion. There were a total of 182 convention-associated cases of whom 29 died and, as most of the victims were military veterans, the outbreak aroused considerable media interest. The newspapers used colourful names to describe the illness: 'Philly Killer', 'Legion Fever', 'Legion Malady' and 'Legionnaires' disease (LD)'. Surprisingly the latter of these was adopted by the scientific community, and this name has, in part, helped to perpetuate media interest in this uncommon form of pneumonia ever since.

The search for the cause of the Philadelphia outbreak was exhaustive, and only after more than 6 months of investigations was the isolation of the causative organism reported (McDade *et al.*, 1977). In order to isolate the organism, these investigators had inoculated guinea pigs with autopsy material from an LD patient. Tissues from the guinea pigs that developed fever were used to inoculate embryonated hens' eggs. Using these rickettsial isolation techniques McDade and colleagues (1977) were able to demonstrate the presence of Gram-negative bacilli in preparations of the infected eggs. Convalescent sera from patients (91%) identified in the outbreak reacted with antigens prepared from these bacteria in an indirect immunofluorescent antibody test. These results were consistent with the view that the bacterium was the aetiological agent of LD.

Investigators at the Centers for Disease Control, Atlanta recognised epidemiological similarities between the Philadelphia outbreak, where an air-conditioning system was implicated as the source of infection, and several earlier outbreaks of illness including one in Washington DC during 1965, one in Pontiac, Michigan during 1968 and one in the same Philadelphia hotel in 1974 which occurred amongst delegates to a convention of the Independent Order of Odd Fellows. Retrospective examination of stored sera collected from patients during these outbreaks revealed that these outbreaks had also been caused by this organism (McDade *et al.*, 1977; Terranova, Cohen and Fraser, 1978). Thus although newly recognised, LD was clearly not a new illness. Outbreaks of LD have now been recognised to have occurred since the 1940s.

The realisation that a hitherto unrecognised bacterium was responsible for an illness with significant morbidity and mortality led to period of intense interest in all aspects of the organism. It was quickly appreciated that in addition to air-conditioning systems, water-distribution systems were also important sources of infection. It was also established that legionellae occur widely in both natural and artificial aquatic environments. As knowledge of the habitat expanded so did the number of legionellae-like organisms identified.

DESCRIPTION OF THE ORGANISM

Taxonomy

The causative agent of LD was initially designated the Legionnaires' Disease Bacterium (McDade *et al.*, 1977). DNA/DNA homology studies on the few clinical and environmental isolates then available revealed that they were members of a distinct species, and the family Legionellaceae and the genus *Legionella* were subsequently defined in 1979 for a single species *Legionella pneumophila* (Brenner, Steigerwalt and McDade, 1979).

The decision, taken at that time, to create a new family was based on phenotypic evidence: *L. pneumophila* was distinct from other Gram-negative bacteria in nutritional requirements and cell-wall composition. Since this time 48 other species of *Legionella* have been identified (Park *et al.*, 2003) (Table 37.1), and the validity of the family Legionellaceae has been substantiated both phenotypically and phylogenetically. However classification at the genus level has been controversial. The generally held view is that all the species are phenotypically very similar and hence should all be placed into the genus *Legionella*. On the basis of limited phenotypic evidence (e.g. colony autofluorescence) and a different interpretation of DNA homology data, some workers maintain that the family should be divided into three genera: *Legionella*, *Fluoribacter* and *Tatlockia*. Alternative names have been formally proposed and consequently *L. bozemanii*, *L. dumoffi*, *L. gormanii*, *L. micdadei* and *L. maceachernii* may be found cited in the literature as *Fluoribacter bozemanae*, *F. dumoffi*, *F. gormanii*, *Tatlockia micdadei* and *T. maceachernii* respectively (Garrity, Brown and Vickers, 1980; Fox and Brown, 1993).

More recent phylogenetic studies (Harrison and Saunders, 1994; Ratcliff *et al.*, 1997) have again demonstrated that the species of the family Legionellaceae are monophyletic, showing no significant discrete divisions. Furthermore if the family were to be divided on the basis suggested by Fox and Brown (1993) many genera with only one or two species would be created. The phenotypic data do not support this view, as although identification of members of the genus is relatively simple, distinguishing between many of the species is very difficult and often requires the application of molecular techniques (Table 37.1).

Definition

Legionellae are nutritionally fastidious, pleomorphic Gram-negative, non-spore–forming organisms. *In vivo* they are short rods or coccobacilli measuring 0.3–0.9 μm in width and 2.0–6.0 μm in length; on artificial media they may be seen as rods or occasionally as filamentous 6.0–20.0 μm or more in length. They are motile with a single monopolar

Table 37.1 Members of the family Legionellaceae

Legionella species	Multiple serogroups	Causes human illness[a]
L. adelaidensis		−
L. anisa	+	Sporadic and an outbreak of Pontiac Fever
L. beliardensis		−
L. birminghamensis		−
L. bozemanii	2	+	Sporadic and nosocomial cases
L. brunensis		−
L. busanensis		−
L. cherrii		−
L. cincinnatiensis		+
L. drozanskii[1]		−
L. dumoffii		+	Sporadic cases and a single documented outbreak
L. erythra	2[b]	−
L. fairfieldensis		−
L. fallonii[c]		−
L. feeleii	2	+
L. geestiana		−
L. gormanii		+
L. gratiana		−
L. gresilensis		−
L. hackeliae	2	+
L. israelensis		−
L. jamestowniensis		−
L. jordanis		+
L. lansingensis		+
L. londiniensis		−
L. longbeachae	2	+
L. lytica[c]		+
L. maceachernii		+
L. micdadei		+
L. moravica		−
L. nautarum		−
L. oakridgensis		+
L. parisiensis		−
L. pneumophila[d]	15	++
L. quateirensis		−
L. quinlivanii	2	−
L. rowbothamii[c]		−
L. rubrilucens		−
L. sainthelensi	2[e]	+
L. santicrusis		−
L. shakespearei		−
L. spiritensis	2	−
L. steigerwaltii		−
L. taurinensis		−
L. tucsonensis		+
L. wadsworthii		−
L. waltersii		−
L. worsleiensis		−
L. genomospecies 1		−

[a] Evidenced by isolation of the organism, unless otherwise indicated.

[b] L. erythra serogroup 2 is serologically indistinguishable from L. rubrilucens.

[c] This species was originally designated as LLAP (see Growth Requirements for details).

[d] L. pneumophila comprises three subspecies: subsp. pneumophila; subsp. fraseri and subsp. pascullei.

[e] L. sainthelensi Sgp 2 is serologically indistinguishable from L. santicrucis.

flagellum although non-motile strains do occur. Branched-chain fatty acids predominate in the cell wall which also contains major amounts of ubiquinones with more than ten isoprene units in the side chain. L-Cysteine and iron salts are required for growth in vitro, carbohydrates are not fermented or oxidised and nitrate is not reduced to nitrite. Legionella are urease negative, catalase positive and give variable results in the oxidase test. The legionellae have a genome size of approximately 2.5×10^9 Da (3.9 Mb) and a guanine-plus-cytosine content of 38–52 mol%, the exception being L. gresilensis, which has a reported G + C content of 33 mol%.

PATHOGENICITY

As discussed below legionellae are primary pathogens of protozoans, which can from time to time give rise to infection in humans and possibly other animals. It appears that the pathogenic mechanisms that have evolved to enable legionellae to survive in their protozoal host also enable them to multiple in mammalian alveolar macrophages.

Infection in humans follows inhalation of fine aerosol (<5 μm particle size) containing organisms that are both viable and virulent. Once within the alveoli the legionellae are taken up by alveolar macrophages where they multiply. Two modes of entry into macrophages have been observed. The first termed 'coiling phagocytosis', in which long phagocytic pseudopods coil round the organism, was extensively studied by Horwitz and colleagues (Horwitz and Silverstein, 1980; Horwitz, 1983; Horwitz and Maxfield, 1984; Shuman and Horwitz, 1996). Although initially thought to be highly significant, workers have since reported that other strains of L. pneumophila are taken up by conventional phagocytosis (Rechnitzer and Blom, 1989). Phagocytosis is mediated by a three-component, receptor ligand-acceptor molecule, system comprising complement receptors on phagocytes (CR1 and CR3), fragments of the complement component C3 fixed on the bacterial surface as ligand, and the major outer membrane protein (MOMP), which acts as the C3-acceptor molecule, in the bacterial outer membrane (Payne and Horwitz, 1987). During phagocytosis the plasma membrane is remodelled with components of the phagocyte being sorted into, or out of, the nascent phagosome: the CR3 receptors are concentrated within the phagosome, while class I and class II MHC molecules are excluded from it. This mode of entry appears to limit the oxidative burst of the phagocyte and hence enhances intracellular survival of the legionellae. A less well-characterised opsonin-independent mode of entry has also been reported (Stone and Abu Kwaik, 1998). Once within the macrophage legionellae are able to reprogram the phagocyte endocytic pathway to escaping killing by inhibiting acidification of the nascent phagosome and preventing phagosome/lysosome fusion. Subsequently the phagosome recruits smooth vesicles, mitochondria and rough endoplasmic reticulum, and within 4–8 h a ribosome-lined replicative vacuole is formed. Multiplication of the legionellae proceeds by binary fission with a doubling time of approximately 2 h. During the late replicative phase the legionella-containing phagosome merges with lysosomes, but replication continues until the host cell is packed with organisms and finally disrupts releasing bacteria to infect further host cells. The actual killing and lysis of the phagocytes appears to occur in a biphasic manner. In the early stages of infection legionellae induce apoptosis, a process that is mediated by the activation of caspase-3 (Gao and Abu Kwaik, 1999), but once bacterial replication ceases, the legionellae cause necrosis of their host cell probably by inducing a pore-forming toxin (Alli et al., 2000).

The host limits infection primarily by cell-mediated immune mechanisms. Once activated by cytokines macrophages are poor hosts for legionellae; phagocytic activity is reduced by about 50%, possibly through decreased expression of CR1 and CR3, and intracellular iron is decreased by downregulation of the transferrin receptor expression (Bellinger-Kawahara and Horwitz, 1990). Studies, using polymorphonuclear leucocyte (PMN)-depleted guinea pigs, suggest that PMNs also play an important role in host defence. Although this observation is consistent with the large numbers of PMNs seen in histology specimens from LD patients, in vitro studies have not been able to elucidate the mechanism by which this occurs. Although a strong humoral response accompanies infection, this appears to provide little protection and possibly may be detrimental, promoting the uptake of L. pneumophila by macrophages (Horwitz and Silverstein, 1981).

Virulence Factors

In recent years a number of genetic loci have been identified in *L. pneumophila* that are required for its intracellular survival. Many, but not all, of these loci are necessary for the infection process in both mammalian and protozoan cells. The first virulence-associated gene of *L. pneumophila* to be identified was the macrophage infectivity potentiator (*mip*) gene that encodes a 24-kD protein surface protein (Mip). Mip is a prokaryotic homologue of the FK506-binding proteins and exhibits peptidyl-prolyl *cis/trans* isomerase (PPIase) activity. *mip*-like genes have been detected in all other species of legionella and some other bacteria. Cianciotto *et al.* (1990), using a pair of isogenic strains (one of which was deficient in Mip), demonstrated that Mip was required for the full expression of virulence by *L. pneumophila*. It appears that Mip is required either for optimal internalisation of the legionellae by macrophages or for resistance to the bactericidal mechanisms active shortly after phagocytosis. It is, however, noteworthy that although much less efficiently, the Mip-deficient mutant did multiple in macrophages and did produce illness in guinea pigs. Clearly Mip can only contribute in part to the virulence of a strain.

The Dot/Icm system of *L. pneumophila* is encoded by genes located in two 20-kb regions of the chromosome which may, at least in some strains, constitute a single pathogenicity island (Vogel and Isberg, 1999; Brassinga *et al.*, 2003). The *dot* (defective in organelle trafficking) and *icm* (intracellular multiplication) loci were independently identified by two different laboratories so the majority of the 24 genes so far identified have two designations. The Dot/Icm system shares homology with type IV transporters of other pathogenic bacteria such as *Bordetella pertussis*, *Agrobacterium tumefaciens*, *Helicobacter pylori* and *Brucella* species. Type IV systems, which are thought to have evolved from systems involved in conjugative transfer of DNA, are critical for the pathogenic process as they export important virulence factors across bacterial membranes. Hilbi *et al.* (2001) showed that the Dot/Icm system is required for stimulation of uptake of the bacteria upon contact with the host cell. Two proteins, DotA and RalF, which are secreted by Dot/Icm have been identified. DotA previously thought to be an inner membrane protein may actually be a pore-forming protein that is inserted into the host membrane (Nagai and Roy, 2001). RalF is required for the localisation of the ADP-ribosylation factor on phagosomes containing *L. pneumophila* (Nagai *et al.*, 2002).

In addition to those mentioned above, a number of other genes have been investigated. *Legionella pneumophila* contains a type II secretion system that is required for the expression of type IV pili (Liles, Viswanathan and Cianciotto, 1998). The system is known to transport several molecules including the 38- to 42-kD Zn-metalloprotease. This enzyme, variously called tissue-destructive protease, zinc metalloprotease and major secretory protein, has been studied in detail and has been shown to be cytotoxic and haemolytic. Experimentally it causes pulmonary lesions which closely resemble those seen in infection (Conlan, Williams and Ashworth, 1988; Szeto and Shuman, 1990; Williams *et al.*, 1993). The type II secretion system is dependent on the *pilBCD* genes that are involved in the biogenesis of temperature-regulated type IV pili (Stone and Abu Kwaik, 1998). Two genes have been studied in detail: *pil*E (pilin protein) and *pil*D (prepilin peptidase). The *pil*E gene/pilin is not required for intracellular growth but may be involved in attachment to the host cells. However the prepilin peptidase is essential for pilus production and type II secretion of proteins. Other genes that promote various stages of intracellular infection, but which are less well studied included the *pmi* (protozoan macrophage infectivity) gene and *mil* (macrophage-specific infectivity) gene.

Cellular Constituents

The cell wall of legionellae is typical of a Gram-negative organism being trilaminar with an outer membrane, peptidoglycan layer and cytoplasmic membrane; however the composition of the cell wall is unusual.

The subunits of peptidoglycan are extensively cross-linked, producing a highly stable layer which confers some resistance to degradation by lysozyme (Amano and Williams, 1983). The lipopolysaccharide (LPS) composition of legionella has also been shown to be unusual, the monosaccharide and the fatty-acid composition being particularly complex in structure (Sonesson *et al.*, 1989). The LPS comprises a Lipid A part with 16–24 amide-linked 3-hydroxy fatty acids, 5–11 ester-linked non-hydroxy fatty acids and long-chain ester-linked (ω-1)-oxo fatty acids and (ω-1)-hydroxy fatty acids. These features appear to be unique and characteristic of *Legionella* species (Sonesson *et al.*, 1994a,b). It is thought that these unusually long–branched-chain fatty acids may inhibit binding of the CD14 LPS receptor on macrophages, thus reducing the endotoxic activity of *Legionella* LPS by approximately 1000-fold compared to that of salmonella LPS (Wong *et al.*, 1979; Neumeister *et al.*, 1998). The LPS core oligosaccharide lacks the heptose of classic LPS but contains abundant 6-deoxy sugars and is highly o-acetylated. The O-chain polysaccharide of the *L. pneumophila* serogroup 1 LPS is an α(2–4)-linked homopolymer of the 5-*N*-acetimidoyl-7-*N*-acetyl derivative of 5,7-diamino-3,5,7, 9-tetradeoxynon-2-ulosonic acid, termed legionaminic acid.

Legionella contain large amounts (>80%) of branched-chain fatty acids and only small amounts of hydroxy acids. Although qualitatively similar, the fatty-acid profiles of some *Legionella* species show marked and consistent quantitative differences. In addition all species contain unusual members of the ubiquinone group of respiratory isoprenoid quinones which contain nine to 15 isoprenyl unit side chains. The combination of fatty-acid profiling and ubiquinone analysis has proved a valuable, although not routine, tool for the identification of legionellae (Wait, 1988; Diogo *et al.*, 1999).

As noted above the *L. pneumophila* LPS is unusual in composition and has little endotoxic activity in comparison with typical LPS. However it has been suggested that differences in LPS structure between strains might account for differences in virulence in humans. Monoclonal antibodies (mAbs) have been used extensively to differentiate between strains of *L. pneumophila* serogroup 1. Dournon *et al.* (1988) were the first to notice that strains (designated MAB2[+]) which reacted with one such mAb, directed against an LPS epitope (Petitjean *et al.*, 1990), were more likely to be isolated from patients than were strains which did not react (MAB2[-] strains). This has led to the widely held view that MAB2[+] strains are more virulent than MAB2[-] strains. There is some evidence that MAB2[-] strains infect human monocytes less well than do MAB2[+] strains; however some recent studies have not confirmed this (Edelstein and Edelstein, 1993). An alternative suggestion is that the LPS of mAb2-positive strains is more hydrophobic than mAb2-negative LPS, resulting in better survival of mAb2-positive strains in aerosols which, in turn, makes it more likely that they will be inhaled in a viable state by a susceptible individual than will be mAb2-negative strains. Data from two studies support this hypothesis. First, Dennis and Lee(1988) showed that mAb2-positive strains survived significantly better than did mAb2-negative strains in a laboratory-generated aerosol held at a suboptimal relative humidity of 60%. Second, Luck *et al.* (2001) isolated an mAb2-negative mutant from an mAb2-positive wildtype parent strain. Analysis showed that mAb2-negative strain had a point mutation in the active site of O-acetyltransferase and consequently its LPS contained legionaminic acid that lacked 8-O-acetyl groups: earlier work by the same group had shown that increased 8-O-acetylation of legionella LPS resulted in increased hydrophobicity of the organism (Zahringer *et al.*, 1995).

EPIDEMIOLOGY

Normal Habitat

Extensive environmental studies indicate that legionellae are ubiquitous organisms occurring over a wide geographical area in river mud, streams, lakes, rivers, warm spa water, water in hydrothermal areas, numerous other natural bodies of water and, most recently, subterrestrial groundwater sediments (Fliermans *et al.*, 1981; Verissimo *et al.*, 1991; Fliermans and Tyndall, 1993). Legionellae have also been isolated from diverse artificial habitats including air-conditioning systems, potable water supplies, ornamental fountains and plumbing fixtures and fittings in hospital, shops and homes (Tobin, Swann and Bartlett, 1981; Colbourne *et al.*, 1988; Redd *et al.*, 1990; Hlady *et al.*, 1993).

Early studies indicated that legionellae could be detected by immunofluorescence (IF) in most natural bodies of water examined but could only be isolated and cultured from a few of these (Fliermans *et al.*, 1981). Initially, because of the questionable specificity of the reagents used in the IF, only the isolation rate was considered to be an accurate indication of the likely prevalence of legionellae. The isolation rate is, in part, dependent on the temperature of the body of water examined: legionellae can be readily isolated and cultured from water between about 20 and 40 °C but are rarely isolated from colder sources. It is now known that at low temperatures, in common with many environmental organisms, legionellae enter a 'dormant state' where they remain viable but non-culturable (VBNC) (Colbourne *et al.*, 1988; Steinert *et al.*, 1997; Ohno *et al.*, 2003). Hence isolation rates probably give a poor estimate of the true prevalence of legionellae which are very clearly widely distributed, being found in almost all natural bodies of fresh water although they may be present at only low numbers and possibly in a VBNC state.

The ubiquitous nature of legionellae, even in low-nutrient environments, appears to contrast with exacting requirements for culture and isolation of the organism in the laboratory. The explanation seems to be that naturally legionellae grow in association with other microorganisms. Tison *et al.* (1980) first demonstrated that *L. pneumophila* could grow in mineral salts medium at 45 °C in association with cyanobacteria (*Fischerella* sp.). In the same year Rowbotham (1980) observed that when legionellas were phagocytosed by *Acanthamoeba* trophozoites *in vitro*, they could survive and multiple within the amoebal vacuole. It has subsequently been shown that other protozoa such as species of *Hartmanella*, *Naegleria* and *Valkampfia*, *Tetrahymena* and *Cyclidium* can also be parasitised (Breiman *et al.*, 1990), and it is now widely acknowledged that these host–parasite relationships are central to the propagation and distribution of legionellae within the environment. Indeed it has been suggested that legionellae may not be free-living bacteria *per se* (Fields, 1996). This close association has several potential benefits for legionellae such as the provision of adequate concentrations of nutrients in otherwise nutrient poor conditions, the protection from adverse external conditions within, for example, amoebal cysts (Kilvington and Price, 1990) and possibly a mode of distribution with the amoebal cyst acting as a carrier.

In common with most other aquatic bacteria legionellae colonise surfaces at the aqueous/solid interface. Here together with their protozoal hosts, other bacteria and algae, they form a complex consortium of organisms loosely termed 'biofilm'. It is believed that biofilm formation provides nutritional advantages and protection from adverse environmental influences to its constituent organisms compared to their planktonic counterparts.

In humans legionella infection can vary in severity from a mild flu-like illness to acute life-threatening pneumonia. There is no evidence of person-to-person spread, infection being acquired primarily by inhalation of aerosol containing virulent organisms. The pneumonic form of the disease, LD, is well characterised, but the mild non-pneumonic form is ill defined and encompasses any condition that is not LD but where there is some evidence of legionella infection. The term 'Pontiac Fever' is sometimes used interchangeably with non-pneumonic legionellosis; however it is more appropriately applied specifically for those cases characterised by a self-limiting, flu-like illness with a high attack rate and short incubation period. The data presented below relate only to LD; that is where there is clear evidence of pneumonia and the criteria for laboratory diagnosis have been met (Anonymous, 1990).

Incidence and Risk Factors

Cases of LD have been reported from countries throughout the world. However reported incidence varies considerably ranging, for example, from 0.0 to 34.1 cases per million population in 2002 for 31 European countries (Joseph, 2004). This variation may in part be a reflection of genuine differences in incidence but is probably due to considerable under-ascertainment in many countries.

In England and Wales surveillance is organised through a voluntary reporting scheme where about 300 laboratories report cases to the HPA Communicable Disease Surveillance Centre (CDSC). In the 23 years from 1980 to 2002 a total of 4391 cases of LD were reported to CDSC, an average of 191 cases per year (range 112–367) giving an incidence of approximately 3.6 cases per million per year. However prospective studies in the UK and elsewhere indicate that about 2.0–5.0% of cases of all pneumonia cases requiring admission to hospital are LD (Woodhead *et al.*, 1986; Marston *et al.*, 1997; Lim *et al.*, 2001). This equates to between 2000 and 5000 cases per annum in England and Wales, suggesting that 10% or less of LD cases are recognised and reported each year. Estimates from the United States suggest that <5% of LD cases are reported to their national surveillance scheme (Benin, Benson and Besser, 2002).

LD is seen in patients of all ages (1–92 years) but is most frequent in adults over 40 years old and very rare in children. There are many more cases among males than among females (ratio for England and Wales of 2.8:1). Significant risk factors include a compromised immune system, recent surgery, congestive heart failure, chronic bronchitis, liver cirrhosis, renal insufficiency and heavy smoking. Not surprisingly, therefore, in certain patient groups, such as organ transplant patients, the incidence is much higher than 5%, and LD may even be the main cause of lung infections. The overall mortality is 13% for England and Wales, but this can be much higher in immunocompromised patients. Recent data suggest that overall the rate is falling in some countries. In the United States the average mortality was 20% between 1980 and 1998 falling from a high of 34% in 1985 to 11.5% in 1998 (Benin, Benson and Besser, 2002). Similarly in Europe mortality fell from 13% in 1998 to 6% in 2002. While there is some evidence that the decline is due to improvements in speed of diagnosis, and hence the initiation of appropriate therapy (Heath, Grove and Looke, 1996; Formica *et al.*, 2000), improved ascertainment of less-severe cases probably accounts for most of the fall (Benin, Benson and Besser, 2002). In support of this view mortality rates appear to be low in outbreaks of LD that are investigated using urinary antigen assays compared with those seen in outbreaks where this sensitive and rapid technology was not used. For example in the Central London outbreak of 1988 the mortality was 3/79 (3.8%) (Westminster Action Committee, 1988) and in the Murcia outbreak of 2001, the largest to date, the mortality was only 5/449 (1.1%) (Garcia-Fulgueiras *et al.*, 2003). In contrast in the Stafford Hospital outbreak, which occurred in 1985 before urine assays were available, 22/68 (32%) patients died (O'Mahony *et al.*, 1990) and in the 1999 'Flower show' outbreak in the Netherlands 21/188 (11%) patients died. In this latter case although urinary antigen testing was available it was rarely used in the Netherlands prior to the outbreak (Den Boer *et al.*, 2002).

A seasonal variation in the incidence of LD has been recognised in many parts of the world, with a peak incidence in late summer and early autumn in both the Northern and Southern hemispheres (Joseph *et al.*, 1999; Benin, Benson and Besser, 2002). However there have been several major outbreaks during the winter and early spring following periods of unseasonably warm weather (Westminster Action Committee, 1988; Watson *et al.*, 1994). Although highly publicised, outbreaks of LD are uncommon and only account for 25% of cases. There is, however, some evidence to suggest that many of the remaining 75% of cases are associated with a common source of infection although this is not recognised at the time (Bhopal *et al.*, 1991). Cooling towers have been identified as the source in most of the major outbreaks of LD, but overall water-distribution systems are the most frequently implicated sources.

Travel-Associated LD

Travel-associated LD is now recognised as a significant problem worldwide. In the UK almost half the annual cases of LD, both sporadic and epidemic, are travel associated, the majority being linked to Mediterranean holiday resorts. Where investigations have been undertaken, hotel water systems have been most frequently identified as the likely source. The incidence of cases appears to be highest in those countries and areas where the tourist industry is still developing or where preventative action is poorly enforced. Although countries such as Spain, with long-established tourist industries, account for the highest numbers of cases, the actual rate of infection has declined from about six per million visitors in 1989–1991 to about three per million visitors in 2002. This contrasts with countries such as Turkey where the rate of infection has remained high at >10 per million visitors over the same period.

Nosocomially Acquired LD

Hospitals often have large and complex water-distribution systems and a population of susceptible patients. It is not therefore surprising that nosocomially acquired LD is a significant problem, accounting for 888 (9%) of cases in Europe in 2000–2002 (Joseph, 2004). Although overall this represents only a small number of cases annually the mortality in this group of patients is disproportionately high at approximately 30% (Benin, Benson and Besser, 2002). A second feature of nosocomial LD is that very much larger proportion of infections is caused by non-*L. pneumophila* serogroup 1 strains than is reported overall. Both these observations probably reflect the immuno-compromised status of most patients in the nosocomial group.

Legionella spp. Able to Cause Human Infection

To date 21 species have been reported as pathogenic for humans (Table 37.1). However infections caused by species other than *L. pneumophila* are rare and almost always seen in patients who are immunocompromised (Dournon, 1988; Benin, Benson and Besser, 2002; Muder and Yu, 2002). One exception to this is *L. longbeachae* serogroup 1 which is now recognised to be a major cause of community-acquired pneumonia in Australia. Between 1996 and 2000 42% of notified cases of LD in Australia were caused by *L. longbeachae* (Li, O'Brien and Guest, 2002). The reservoir for this organism is thought to be potting mixes prepared from composted indigenous wood where it is found growing in association with amoebae (Hughes and Steele, 1994). Cases of *L. longbeachae* infection associated with potting composts have recently been reported outside Australia (Anonymous, 2000).

The majority of sporadic, and almost all outbreak-associated, cases of LD are caused by a subset of so-called 'virulent' strains of *L. pneumophila* serogroup 1. These strains are characterised by their reaction with a particular mAb and are referred to as MAB2[+] (Dournon *et al.*, 1988) or MAb3/1[+] (Helbig *et al.*, 2002), depending on the origin of the antibodies being used. These antibodies recognise a virulence-associated epitope on the LPS of *L. pneumophila* serogroup 1 strains. Strains of *L. pneumophila* serogroup 1 that are MAB2[−] or MAB3/1[−], together with strains of other *L. pneumophila* serogroups, account for most remaining cases (Joseph, 2004). With a few exceptions *Legionella* species other than *L. pneumophila* are only seen as the cause of rare sporadic cases of LD or persistent endemic problems in large hospitals (Muder and Yu, 2002). It does, however, seem probable that under the appropriate circumstances any *Legionella* species could cause infection as no consistent differences have been observed

between the virulence traits of the various species examined to date (O'Connell, Dhand and Cianciotto, 1996; Alli *et al.*, 2003).

CLINICAL FEATURES

Legionnaires' Disease

Although overall an uncommon form of pneumonia, in the context of severe community-acquired pneumonias, LD is much more significant, probably being the second most common form accounting for 14–37% of cases (Hubbard, Mathur and Macfarlane, 1993). LD has no special features that clearly distinguish it from other pneumonias, but the clinical picture is usually suggestive of the diagnosis (Mayaud and Dournon, 1988).

The onset of LD is generally more insidious than in typical pneumococcal pneumonia, with fever, headaches and malaise being the first manifestations. In about half of the cases gastrointestinal abnormalities are present. Respiratory signs often only appear later in the course of the disease. A cough is rare and if present it is generally non-productive early in the illness although later small amounts of purulent, and sometimes bloody sputum, may be expectorated. Lung examination usually reveals abnormalities at this stage of the disease, with at least rales and very often lung consolidation being evident. A small pleural effusion, sometimes suspected from the presence of thoracic pains, is present in about one-third of patients but is difficult to recognise on clinical examination. Progression of the lung infection may be rapid and be responsible for increasing dyspnoea and acute respiratory failure (Mayaud and Dournon, 1988).

A frequently reported but non-specific feature of LD is the presence in many cases of gastrointestinal symptoms or neurological signs. In the series studied by Mayaud and colleagues (1984), diarrhoea was present in about 50% cases and appeared early in the course of the disease. Nausea, vomiting or right lower abdominal pain with tenderness were each seen in 20–30% patients. Nervous system manifestations, which were dominated by alterations of the mental status such as disorientation and confusion, were found in about 25% cases.

Routine laboratory findings are not particularly helpful. Liver abnormalities are frequently found with elevated serum levels of transaminases and/or alkaline phosphatase or, in some cases, of bilirubin. Proteinuria, haematuria and renal insufficiency are infrequent in patients receiving early treatment. Elevation of creatinine phosphokinase and aldolase blood levels, suggestive of muscle damage, is seen in some patients.

The prognosis for patients treated with appropriate antibiotics within 7 days of onset is good and they usually recover even if they are immunocompromised (Mayaud *et al.*, 1984). In contrast the prognosis is poor in patients not treated before acute respiratory failure and shock develop. This is usually late in the course of the illness. In such cases rhabdomyolysis, acute renal failure, pancytopenia, disseminated intravascular coagulation, icterus or coma are not infrequent (Mayaud *et al.*, 1984). However, in most instances patients die from the respiratory distress syndrome and not from the extrapulmonary manifestations of the disease.

Pontiac Fever

Pontiac fever is an acute, non-pneumonic, flu-like illness which is usually self-limiting (Glick *et al.*, 1978). The attack rate is usually very high (>90%) and the disease affects previously healthy, and often young, individuals. The incubation period is short, varying from a few hours to 2–3 days, and the illness resolves spontaneously usually within 2–6 days. Outbreaks of Pontiac fever have been reported to be caused by *L. pneumophila*, *L. feeleii*, *L. micdadei* and *L. anisa* (Glick *et al.*, 1978; Goldberg *et al.*, 1989). Although the subject of considerable speculation, the explanation for the differences between LD and Pontiac fever remains unclear.

DIAGNOSTIC METHODS

Although in most instances the clinical picture of LD allows sufficient suspicion of the diagnosis for appropriate treatment to be initiated evidence from laboratory tests are essential for a definitive diagnosis. There are three approaches to the laboratory diagnosis of legionella infections: (i) culture of the causative organism; (ii) demonstration of the organism, its components or products in clinical specimens; and (iii) demonstration of significant levels of antibody directed against the organism in patient sera.

Specimens for Culture

Culture of legionellae from clinical specimens is relatively easy and, in situations where clinical awareness is high, has a sensitivity of 50–80% (Dournon, 1988; Winn, 1993). Furthermore diagnosis by culture and isolation has several advantages over any other approach. First, isolation of a *Legionella* sp. provides definitive proof of the diagnosis as colonisation without infection has not been demonstrated (Bridge and Edelstein, 1983). Second, isolation may be the only way to establish a diagnosis, for example if the patient does not produce antibodies or if there are no reagents available to detect the infecting species/serogroup. Third, isolation of the infecting strain allows subtyping to be undertaken, providing valuable epidemiological data for the control and prevention of further cases of infection.

Given these clear advantages it is perhaps surprising that overall only 12% of cases reported to CDSC between 1980 and 2002 were confirmed by isolation of the organism. The explanation for this disappointing observation is not entirely clear but may in part be due to the fact that the diagnosis of legionellosis is often only considered after both the initiation of antibiotic therapy and the failure to identify more common pathogens.

Specimens for culture should be taken as early as possible and ideally before antibiotic treatment is initiated. Lower respiratory tract specimens are most likely to yield positive results, but sputa are quite satisfactory provided the specimens are pre-treated (Dournon, 1988). The pre-treatments used to reduce contaminants exploit the property that legionellae are generally more tolerant to heating and resistant to low pH exposure than are other organisms found in the respiratory tract. Isolation is usually attempted using buffered charcoal yeast extract (BCYE) agar with and without selective antibiotics; dilution of the specimen may also help. Many combinations of antibiotics have been tried but for the culture of strains from clinical specimens a mixture of polymixin, vancomycin, cefamandole and anisomycin has proved very effective.

Microscopic and Colonial Appearance

A bacterium which is Gram-negative, catalase positive and grows on complete BCYE agar but not on the same medium lacking supplemental L-cysteine can be presumptively identified as a *Legionella* sp. Colonies usually first appear after 3–6 days of incubation but may be visible in about 36 h from specimens where large numbers of legionellae are present with few contaminating organisms. Growth tends to be delayed by about 12–24 h where selective agars are used.

Colonies of legionella are convex, have an entire edge, glisten and have a characteristic granular or 'ground glass' appearance which is most pronounced in young colonies. Colony colour varies considerably and in large part depends upon the thickness and formulation of growth medium used. In general however coloration varies from blue/green when the colonies are first visible to pink/purple as they grow larger. As the colonies age they become less characteristic, being white/grey and smoother; however the pink/purple coloration can still be seen at their edges. As the name suggests the colonies of *L. erythra* have a slight red coloration.

Almost all species of *Legionella* are weakly motile having one polar flagellum. Flagella can be easily demonstrated by one of the many described 'silver-plating stains', based on Fontana's stain for spirochaetes. Furthermore the flagella from all species share common antigens and can be visualised using a suitable anti-flagellar serum. Expression of flagellae appears to depend on the growth phase (most abundant in late log-phase), the temperature of incubation and the 'passage' history of the isolate.

Growth Requirements

Legionellae will grow in the temperature range of 29–40 °C with an optimum of 35 °C. However they will withstand considerably higher temperatures, of 50 °C and above, for a considerable time (>30 min), a characteristic which is exploited in their culture and isolation from clinical and environmental specimens (Dennis, 1988). They grow well aerobically on BCYE agar, particularly when supplemented with alpha-ketoglutarate. The function of the charcoal in the medium seems to prevent superoxide formation. Amino acids such as arginine, threonine, methionine, serine, isoleucine, valine and cystine form the major sources of energy for the organism. Legionellae show an absolute requirement for L-cysteine, although *L. oakridgensis* appears not to require L-cysteine as a culture medium supplement after primary isolation. Trace metals such as calcium, cobalt, copper, magnesium, manganese, nickel, vanadium and zinc enhance growth. Although legionellae do not appear to require iron in greater amounts than do other organisms, the presence of soluble iron stimulates growth.

Studies by Rowbotham (1993) have shown that some legionellae originally described as *Legionella*-like amoebal pathogens (LLAPs) can be obtained from environmental and clinical samples by co-cultivation with amoebae but cannot be isolated using standard legionella culture media such as BCYE. Subsequent work has shown that these LLAPs are *Legionella* spp. and most can be grown, albeit poorly, on BCYE provided a reduced incubation temperature of 30 °C is used (Adeleke *et al.*, 2001).

Differentiation of Species

There are a number of simply demonstrated phenotypic characteristics that allow the Legionellaceae to be subdivided into groups of species, or in some cases into individual species (Table 37.2). As in most cases these characteristics have been determined using small numbers of strains of each species, or even a single isolate, the stability of these phenotypic markers is uncertain. Identification of an isolate should, therefore, always be confirmed by an alternative method such as chemical analysis of cellular components, serological identification or molecular analysis (Table 37.2).

Antigenic Properties

Legionellae can be subdivided into serogroups (Sgp) by their reaction with hyperimmune antisera-containing antibodies directed against the somatic LPS or O-antigens. To date 70 serogroups have been recognised among the 49 species, although in the cases of *L. erythra* Sgp 2 and *L. sainthelensi* Sgp 2 these are indistinguishable from *L. rubrilucens* and *L. santicrucis* respectively (Table 37.1). There is considerable antigenic overlap between some of the serogroups, particularly within the blue–white autofluorescent species complex, and therefore hyperimmune antisera require extensive absorption to render them serogroup specific. There is also significant antigenic heterogeneity within many serogroups. Thus strains of one serogroup may react to different degrees with a serogroup-specific antiserum. Thomason and Bibb (1984) examined this phenomenon in detail using five cross-absorbed polyclonal antisera and 176 *L. pneumophila* Sgp 1 strains. They identified 17 subsets that fell broadly into three subgroups. Serogroup heterogeneity is most clearly revealed using mAbs, which identify type-specific epitopes. Panels of such mAbs

Table 37.2 Summary of phenotypic characteristics and biochemical reactions helpful in differentiating between *Legionella* species (modified from Harrison and Saunders, 1994)

Test	Result	Relevant species	Comment
Subculture onto BCYE Subculture onto BCYE-Cys	Supports growth Does not support growth	All species[a]	Blood agar is not a suitable substitute for BCYE-Cys, particularly for environmental specimens, as non-legionellae may grow on BCYE but not on blood agar
Gram's stain	Gram-negative	All species	Legionellae will readily counterstain with 0.1% basic fuchsin
Catalase test	Positive	All species (*L. pneumophila* only weakly)	Some species posses a peroxidase rather than a catalase but they still give a positive result in the test if 3% H_2O_2 is used
Oxidase test	Negative	All species	Positive results are sometimes recorded but this is probably due to contamination with BCYE medium
Hippurate hydrolysis	Positive Negative	*L. pneumophila* positive All other species negative	Some strains, particularly of subsp. fraseri, are said to be negative in this test
Colony autofluorescence under long-wavelength UV (approximately 365 nm)	Red fluorescence	*L. erythra*, *L. taurinensis* and *L. rubrilucens*	Red fluorescence may only be seen after prolonged or reduced temperature (approximately 30 °C) incubation
	Blue–white fluorescence	*L. anisa*, *L. bozemanii*, *L. cherrii*, *L. dumoffii*, *L. gormanii*, *L. gratiana*, *L. parisiensis*, *L. steigerwaltii*, *L. tucsonensis*	Many of these species are phenotypically and antigenically very similar to others in the group
	No fluorescence	All other species	
Bromocresol-purple spot test	Positive Negative	*L. micdadei*, *L. maceachernii* positive. All other species negative	These species are serologically distinct

[a] Except some of the so-called LLAPs which only grow very poorly on BCYE (see *Growth Requirements* for details).

have been used extensively to subgroup strains of *L. pneumophila* Sgp 1 for epidemiologic purposes (Helbig *et al.*, 2002).

In addition to O-antigens legionella have flagellar antigens. As noted above flagella from all species so far investigated share common antigens, a property that can be useful in the identification of new or unknown strains of legionella. It has also been shown that legionella flagella can be subtyped using panels of mAbs (Saunders and Harrison, 1988).

Alternatives to Culture

Despite the many advantages of culture the time taken to obtain results by this method is measured in days. In contrast the direct demonstration of legionella antigen or nucleic acids in clinical specimens can be achieved within a few minutes of specimen collection. Furthermore a diagnosis may be established by visualisation or detection of the organism in tissue even when they are no longer viable, after antibiotic therapy or retrospectively in fixed tissues.

Microscopy

Bacteriological stains such as Gimenez and modified Warthin–Starry have been used successfully to detect legionellae in clinical material. However such stains may reveal any bacterial species and hence a specific diagnosis cannot be established. Immunofluorescence using rabbit hyperimmune antisera has been used to diagnose LD since legionellae were first recognised, but initially the poor sensitivity and, particularly, the poor specificity of this method severely limited its use. Subsequently the problem of specificity has been overcome by using reagents derived from mAbs, the most widely used of these being directed against the *L. pneumophila* MOMP (Gosting *et al.*, 1984). This reagent reacts with all the serogroups of *L. pneumophila* and hence obviates the need for repeated testing of a specimen with several different

antisera. The clinical utility of this reagent has been evaluated (Edelstein *et al.*, 1985), and the specificity has been found to be excellent allowing a diagnosis to be confidently established very rapidly. The sensitivity is, however, still considerably lower than that of culture.

Antigen Detection

It is now well established that the use of enzyme immunoassays (EIAs) for the detection of *L. pneumophila* antigen in urine allows diagnoses of LD to be established early in the course of infection (Birtles *et al.*, 1990; Hackman *et al.*, 1996; Harrison *et al.*, 1998). Until the early 1990s the use of such assays was restricted to those few reference laboratories able to develop and maintain in-house EIAs (Sathapatayavongs *et al.*, 1982; Birtles *et al.*, 1990), but commercially produced kits are now widely available. Currently available kits have been shown to be highly specific and reliable in routine use (Dominguez *et al.*, 1999; Harrison and Doshi, 2001), and legionella urinary antigen detection has become the single most widely used diagnostic methodology. The limitation to this method is the poor sensitivity that the kits have for LD caused by any legionellae other than the 'virulent' (mAb2-positive) *L. pneumophila* serogroup 1 strains. Practically this means that while the sensitivity is approximately 90% for the diagnosis of community-acquired LD, it is <50% for nosocomially acquired LD (Helbig *et al.*, 2003).

Attempts to overcome this limitation led to the development of a 'broad-spectrum' EIA that uses hyperimmune serum raised against a pool of *Legionella* species and serogroups to capture the soluble antigen (Tang and Toma, 1986). Unfortunately this method of producing the capture antiserum is not amenable to standardisation and this methodology has not been translated into a commercially assay. An alternative approach is to use reagents prepared against defined *L. pneumophila*, or preferably *Legionella* genus, common antigens. While the successful detection of common protein antigens such as momp or mip in urine has not been reported, preliminary studies suggest that a *Legionella* genus common 19-kDa peptidoglycan-associated lipoprotein target is detectable in the urine of experimentally infected guinea pigs (Kim *et al.*, 2003). Whether or not this is also true for humans with a natural infection remains to be seen.

Molecular Methods

Methods for the detection of legionella nucleic acid have been available since the early 1980s when Kohne and colleagues (1984) developed a radiolabelled DNA probe complementary to regions of the ribosomal RNAs of all 20 *Legionella* species known at the time. A commercial nucleic acid hybridisation test, based on this work, and intended for the detection of legionelae in clinical samples, was introduced soon after. This kit was shown to be reasonably sensitive performing similarly to IF, but was less sensitive than culture (Pasculle *et al.*, 1989; Finkelstein *et al.*, 1993). Furthermore questions concerning the kit's specificity (Laussucq *et al.*, 1988) and its use of radioisotopes limited the kit's utility and it was subsequently withdrawn from the market.

The theoretical sensitivity of PCR (i.e. detection of one copy of target sequence) quickly led to widespread use as an alternative to direct probing, which typically requires large number of organisms. Furthermore, as alternatives to radioisotope labelling are now available which have at least equivalent sensitivity and none of the inherent disadvantages, PCR is now seen as the molecular method of choice, promising advantages of both specificity and sensitivity. Development of PCR-based assays for legionellae has focused on three main targets. Assays designed to detect *L. pneumophila* typically use primers directed against the *mip* gene, while those targeting *Legionella* species use primers for either the 5S rRNA or 16S rRNA genes (Starnbach, Falkow and Tompkins, 1989; Mahbubani *et al.*, 1990; Lisby and Dessau, 1994; Jonas *et al.*, 1995; Fry and Harrison, 1998; Ballard *et al.*, 2000). Where lower respiratory tract samples are used, PCR appears to be at least as sensitive as culture (Jaulhac *et al.*, 1992; Jonas *et al.*, 1995; Cloud *et al.*, 2000). However this does not seem to be the case for detection of legionellae in environmental samples, where the sensitivity is consistently lower than culture (Ballard *et al.*, 2000; Riffard *et al.*, 2003). The difference is probably due to the concentration steps, typically a 100-fold, which are integral to the processing of environmental samples. *Legionella* DNA has been detected in non-respiratory samples such as serum and urine and leucocyte samples, but with a lower sensitivity than culture (Lindsay, Abraham and Fallon, 1994; Murdoch *et al.*, 1996; Helbig *et al.*, 1999). At present the status of PCR for the diagnosis of LD is much the same as applied to urinary antigen detection in 1980s, which is confined to a few specialist and reference laboratories able to use it in conjunction with the full range of other diagnostic approaches. Until validated commercial kits are available, the real impact of PCR will not be known.

Diagnosis by Estimation of Antibody Levels

For more than 20 years the primary method for the diagnosis of LD was by serology and although now displaced by urinary antigen testing, estimation of antibody levels against killed preparations of legionellae is still a very commonly employed method for the diagnosis of LD. It is an important tool for case finding during outbreak investigations and for late or retrospective diagnosis. However this approach does not often allow a diagnosis to be established during acute illness, and hence has little value in patient management or in the early detection of outbreaks.

Historically the indirect immunofluorescent antibody test (IFAT) was the method used in 1977 to establish that *L. pneumophila* was the causative agent of LD (McDade *et al.*, 1977). Since this time several

variants of this test have been comprehensively evaluated for the diagnosis of infection caused by *L. pneumophila* serogroup 1, and the IFAT became the reference test for LD diagnosis (Wilkinson, Cruce and Broome, 1981; Harrison and Taylor, 1988). The test used in the UK and parts of Europe until the late 1990s employed a formolised yolk-sac antigen and FITC-labelled anti-human conjugate capable of detecting IgG, IgM and IgA. This test was both sensitive (approximately 80%) and was with a high specificity (>99%) (Bornstein *et al.*, 1987; Harrison and Taylor, 1988), the latter being particularly important in a disease of low prevalence such as LD, if a positive result is to have any real significance. Reagents for this IFAT are no longer available, but various commercially produced IFAT reagents, typically using agar-grown antigens, can be obtained. Reagents for IFATs to detect infection caused by species and serogroups other than *L. pneumophila* serogroup 1 are also commercially available; however it should be emphasised that the sensitivity and specificity of IFAT using these reagents are largely unknown.

In addition to the IFAT several other methods for antibody detection such as microagglutination (Farshy, Klein and Feeley, 1978; Harrison and Taylor, 1982) and indirect haemagglutination (Edson *et al.*, 1979; Lennette *et al.*, 1979) have been devised. As EIAs are widely used in microbiology several have been developed for *L. pneumophila* (Wreghitt, Nagington and Gray, 1982; Herbrink *et al.*, 1983; Barka, Tomasi and Stadtsbaeder, 1986). Reagent manufacturer's market diagnostic kits using this methodology, but these typically incorporate antigens prepared from pools of *L. pneumophila* serogroups 1–6. None of these assays use purified antigens and hence the EIA plates are coated with a complex mixture of bacterial proteins and LPS. The use of such crude pooled antigens may in part be responsible for the poor performance of these assays (Elder *et al.*, 1983; Harrison and Taylor, 1988; Malan *et al.*, 2003).

MANAGEMENT

Legionellae are susceptible to a wide range of antibiotics *in vitro*, but these findings are not necessarily predictive of their *in vivo* efficacy. As legionella are facultative intracellular organisms, some antibiotics that are active *in vitro* are not active *in vivo* because their intracellular penetration is poor. Furthermore the media used to culture legionellae inhibit the activity of some antibiotics, making a valid determination of MICs and MBCs difficult (Edelstein and Meyer, 1980; Dowling, McDevitt and Pasculle, 1984).

The antimicrobial agents that show the best activity against legionellae are those that are concentrated within the phagocytic cells such as macrolides and quinolones, and these antibiotics form the basis of recent treatment guidelines (Bartlett *et al.*, 2000; British Thoracic Society, 2001). Although erythromycin is the traditional drug of choice, studies reveal that it is only inhibitory, and reversibly so, in its activity against intracellular legionellae (Horwitz and Silverstein, 1983). In addition it is often poorly tolerated by many patients and treatment failures are not infrequent (Edelstein, 1998). In contrast azithromycin is bactericidal even at quite low concentrations (Edelstein and Edelstein, 1991), is highly effective in preventing disease in an aerosol-infected guinea pig model (Fitzgeorge, Lever and Baskerville, 1993) and is well tolerated by patients. Azithromycin is only licensed for oral use in some countries and is not suitable for treatment of severe legionellosis in this form. Although clarithromycin is significantly less active than azithromycin, it is more active than erythromycin and hence is recommended in some guidelines (British Thoracic Society, 2001). Most fluoroquinolones are more active than erythromycin *in vitro*, and some have been shown to be highly effective in both animal and cell-culture models of infection (Dournon *et al.*, 1986; Rajagopalan-Levasseur *et al.*, 1990; Edelstein *et al.*, 1996).

PREVENTION AND CONTROL OF INFECTION

Five essential elements must be in place if legionellae are to give rise to infection: (i) A virulent strain must be present in an environmental reservoir; (ii) conditions must be such that it can multiple to significant numbers; (iii) a mechanism must exist for creating an aerosol containing the organisms from the reservoir; (iv) the aerosol must be disseminated and finally a susceptible human must inhale the aerosol. Prevention of infection depends upon breaking this 'chain of causation'.

As discussed above legionellae can be found in low numbers in almost all aquatic environments. It is probably not possible to eradicate legionellae from such a system because of regrowth from residual biofilm or reseeding from other environmental sources. However it is feasible to ensure that the numbers of organisms remain at a low level, and it is the aim of most preventative procedures to break this link of the chain. Legionellae grow best in water at temperatures between 20 and 40 °C, particularly where there is an accumulation of organic and inorganic material. Thus the strategy is to keep the systems clean by regular maintenance, to keep cold water at below 20 °C and hot water above 50 °C, to avoid the use of certain materials in the construction of water systems and, where appropriate, to use a programme of treatment with biocides and other chemicals.

Aerosols can be generated from many sources such as cooling towers, showers, taps, nebulisers and other domestic and industrial equipment. The use of effective drift eliminators in cooling towers can considerably reduce the release of aerosol. However in some instances, for example showers, there is little that can be done but to avoid their use. Where the population at risk is particularly susceptible (e.g. transplant patients), this approach is sometimes taken. Approved codes of practice and guidelines detailing all aspects of design, maintenance and running of water systems and cooling towers have been provided from a number of governmental and professional bodies and much excellent advice is now available (Health and Safety Commission, 2000).

REFERENCES

Adeleke, A. A., Fields, B. S., Benson, R. F., Daneshvar, M. I., Pruckler, J. M., Ratcliff, R. M., Harrison, T. G., Weyant, R. S., Birtles, R. J., Raoult, D., & Halablab, M. A. (2001). "*Legionella drozanskii* sp. nov., *Legionella rowbothamii* sp. nov. and *Legionella fallonii* sp. nov.: three unusual new *Legionella* species", *International Journal of Systematic & Evolutionary Microbiology*, vol. 51, Pt 3, pp. 1151–1160.

Alli, O. A., Gao, L. Y., Pedersen, L. L., Zink, S., Radulic, M., Doric, M., & Abu, K. Y. (2000). "Temporal pore formation-mediated egress from macrophages and alveolar epithelial cells by *Legionella pneumophila*", *Infection & Immunity*, vol. 68, no. 11, pp. 6431–6440.

Alli, O. A., Zink, S., von Lackum, N. K., & Abu-Kwaik, Y. (2003). "Comparative assessment of virulence traits in *Legionella* spp", *Microbiology*, vol. 149, Pt 3, pp. 631–641.

Amano, K., & Williams, J. C. (1983). "Partial characterization of peptidoglycan-associated proteins of *Legionella pneumophila*", *Journal of Biochemistry*, vol. 94, no. 2, pp. 601–606.

Anonymous (1990). "Epidemiology, prevention and control of legionellosis: memorandum from a WHO meeting [Review] [55 refs]", *Bulletin of the World Health Organization*, vol. 68, no. 2, pp. 155–164.

Anonymous (2000). "Legionnaires' disease associated with potting soil – California, Oregon, and Washington, May–June 2000", *Morbidity and Mortality Weekly Report*, vol. 49, no. 34, pp. 777–778.

Ballard, A. L., Fry, N. K., Chan, L., Surman, S. B., Lee, J. V., Harrison, T. G., & Towner, K. J. (2000). "Detection of *Legionella pneumophila* using a real-time PCR hybridization assay", *Journal of Clinical Microbiology*, vol. 38, no. 11, pp. 4215–4288.

Barka, N., Tomasi, J. P., & Stadtsbaeder, S. (1986). "ELISA using whole *Legionella pneumophila* cell as antigen. Comparison between monovalent and polyvalent antigens for the serodiagnosis of human legionellosis", *Journal of Immunological Methods*, vol. 93, no. 1, pp. 77–81.

Bartlett, J. G., Dowell, S. F., Mandell, L. A., File, J. T., Musher, D. M., & Fine, M. J. (2000). "Practice guidelines for the management of community-acquired pneumonia in adults. Infectious Diseases Society of America", *Clinical Infectious Diseases*, vol. 31, no. 2, pp. 347–382.

Bellinger-Kawahara, C. & Horwitz, M. A. (1990). "Complement component C3 fixes selectively to the major outer membrane protein (MOMP) of *Legionella pneumophila* and mediates phagocytosis of liposome-MOMP complexes by human monocytes", *Journal of Experimental Medicine*, vol. 172, no. 4, pp. 1201–1210.

Benin, A. L., Benson, R. F., & Besser, R. E. (2002). "Trends in legionnaires disease, 1980–1998: declining mortality and new patterns of diagnosis", *Clinical Infectious Diseases*, vol. 35, no. 9, pp. 1039–1046.

Bhopal, R. S., Fallon, R. J., Buist, E. C., Black, R. J., & Urquhart, J. D. (1991). "Proximity of the home to a cooling tower and risk of non-outbreak Legionnaires' disease", *BMJ*, vol. 302, no. 6773, pp. 378–383.

Birtles, R. J., Harrison, T. G., Samuel, D., & Taylor, A. G. (1990). "Evaluation of urinary antigen ELISA for diagnosing *Legionella pneumophila* serogroup 1 infection", *Journal of Clinical Pathology*, vol. 43, no. 8, pp. 685–690.

Bornstein, N., Fleurette, J., Bebear, C., & Chabanon, G. (1987). "Bacteriological and serological diagnosis of community-acquired acute pneumonia, specially Legionnaire's disease. Multicentric prospective study of 274 hospitalized patients", *Zentralblatt Fur Bakteriologie, Mikrobiologie, Und Hygiene – Series A, Medical Microbiology, Infectious Diseases, Virology, Parasitology*, vol. 264, no. 1–2, pp. 93–101.

Brassinga, A. K., Hiltz, M. F., Sisson, G. R., Morash, M. G., Hill, N., Garduno, E., Edelstein, P. H., Garduno, R. A., & Hoffman, P. S. (2003). "A 65-kilobase pathogenicity island is unique to Philadelphia-1 strains of *Legionella pneumophila*", *Journal of Bacteriology*, vol. 185, no. 15, pp. 4630–4637.

Breiman, R. F., Fields, B. S., Sanden, G. N., Volmer, L., Meier, A., & Spika, J. S. (1990). "Association of shower use with Legionnaires' disease. Possible role of amoebae", *JAMA*, vol. 263, no. 21, pp. 2924–2926.

Brenner, D. J., Steigerwalt, A. G., & McDade, J. E. (1979). "Classification of the Legionnaires' disease bacterium: *Legionella pneumophila*, genus novum, species nova, of the family Legionellaceae, familia nova", *Annals of Internal Medicine*, vol. 90, no. 4, pp. 656–658.

Bridge, J. A. & Edelstein, P. H. (1983). "Oropharyngeal colonization with *Legionella pneumophila*", *Journal of Clinical Microbiology*, vol. 18, no. 5, pp. 1108–1112.

British Thoracic Society (2001). "BTS guidelines for the management of community acquired pneumonia in adults", *Thorax*, vol. 56, Suppl. 4, pp. IV1–64.

Cianciotto, N. P., Eisenstein, B. I., Mody, C. H., & Engleberg, N. C. (1990). "A mutation in the *mip* gene results in an attenuation of *Legionella pneumophila* virulence", *Journal of Infectious Diseases*, vol. 162, no. 1, pp. 121–126.

Cloud, J. L., Carroll, K. C., Pixton, P., Erali, M., & Hillyard, D. R. (2000). "Detection of *Legionella* species in respiratory specimens using PCR with sequencing confirmation", *Journal of Clinical Microbiology*, vol. 38, no. 5, pp. 1709–1712.

Colbourne, J. S., Dennis, P. J., Trew, R. M., Berry, C., & Vesey, G. (1988). "*Legionella* and public water supplies", *Water Science Technology*, vol. 20, pp. 5–10.

Conlan, J. W., Williams, A., & Ashworth, L. A. (1988). "In vivo production of a tissue-destructive protease by *Legionella pneumophila* in the lungs of experimentally infected guinea-pigs", *Journal of General Microbiology*, vol. 134, Pt 1, pp. 143–149.

Den Boer, J. W., Yzerman, E. P., Schellekens, J., Lettinga, K. D., Boshuizen, H. C., Van Steenbergen, J. E., Bosman, A., Van den Hof, S., Van Vliet, H. A., Peeters, M. F., van Ketel, R. J., Speelman, P., Kool, J. L., & Conyn-Van Spaendonck, M. A. (2002). "A large outbreak of Legionnaires' disease at a flower show, the Netherlands, 1999", *Emerging Infectious Diseases*, vol. 8, no. 1, pp. 37–43.

Dennis, P. J. (1988). "Isolation of legionellae from environmental specimens" in *A laboratory manual for legionella*, T. G. Harrison & A. G. Taylor, eds, Chichester: John Wiley and Sons, pp. 31–44.

Dennis, P. J. & Lee, J. V. (1988). "Differences in aerosol survival between pathogenic and non-pathogenic strains of *Legionella pneumophila* serogroup 1", *The Journal of Applied Bacteriology*, vol. 65, pp. 135–141.

Diogo, A., Verissimo, A., Nobre, M. F., & da Costa, M. S. (1999). "Usefulness of fatty acid composition for differentiation of *Legionella* species", *Journal of Clinical Microbiology*, vol. 37, no. 7, pp. 2248–2544.

Dominguez, J., Gali, N., Matas, L., Pedroso, P., Hernandez, A., Padilla, E., & Ausina, V. (1999). "Evaluation of a rapid immunochromatographic assay for the detection of legionella antigen in urine samples", *European Journal of Clinical Microbiology & Infectious Diseases*, vol. 18, no. 12, pp. 896–898.

Dournon, E. (1988). "Isolation of legionellae from clinical specimens" in *A laboratory manual for legionella*, T. G. Harrison & A. G. Taylor, eds, Chichester: John Wiley and Sons, pp. 13–30.

Dournon, E., Bibb, W. F., Rajagopalan, P., Desplaces, N., & McKinney, R. M. (1988). "Monoclonal antibody reactivity as a virulence marker for *Legionella pneumophila* serogroup 1 strains", *Journal of Infectious Diseases*, vol. 157, no. 3, pp. 496–501.

Dournon, E., Rajagopalan, P., Vilde, J. L., & Pocidalo, J. J. (1986). "Efficacy of pefloxacin in comparison with erythromycin in the treatment of experimental guinea pig legionellosis", *Journal of Antimicrobial Chemotherapy*, vol. 17, Suppl. B, pp. 41–48.

Dowling, J. N., McDevitt, D. A., & Pasculle, A. W. (1984). "Disk diffusion antimicrobial susceptibility testing of members of the family Legionellaceae including erythromycin-resistant variants of *Legionella micdadei*", *Journal of Clinical Microbiology*, vol. 19, no. 6, pp. 723–729.

Edelstein, P. H. (1998). "Antimicrobial chemotherapy for Legionnaires disease: time for a change", *Annals of Internal Medicine*, vol. 129, no. 4, pp. 328–330.

Edelstein, P. H. & Edelstein, M. A. (1991). "In vitro activity of azithromycin against clinical isolates of *Legionella* species", *Antimicrobial Agents & Chemotherapy*, vol. 35, no. 1, pp. 180–181.

Edelstein, P. H. & Edelstein, M. A. (1993). "Intracellular growth of *Legionella pneumophila* serogroup 1 monoclonal antibody type 2 positive and negative bacteria", *Epidemiology and Infection*, vol. 111, no. 3, pp. 499–502.

Edelstein, P. H., & Meyer, R. D. (1980). "Susceptibility of *Legionella pneumophila* to twenty antimicrobial agents", *Antimicrobial Agents & Chemotherapy*, vol. 18, no. 3, pp. 403–408.

Edelstein, P. H., Beer, K. B., Sturge, J. C., Watson, A. J., & Goldstein, L. C. (1985). "Clinical utility of a monoclonal direct fluorescent reagent specific for *Legionella pneumophila*: comparative study with other reagents", *Journal of Clinical Microbiology*, vol. 22, no. 3, pp. 419–421.

Edelstein, P. H., Edelstein, M. A., Ren, J., Polzer, R., & Gladue, R. P. (1996). "Activity of trovafloxacin (CP-99,219) against *Legionella* isolates: in vitro activity, intracellular accumulation and killing in macrophages, and pharmacokinetics and treatment of guinea pigs with *L. pneumophila* pneumonia", *Antimicrobial Agents & Chemotherapy*, vol. 40, no. 2, pp. 314–319.

Edson, D. C., Stiefel, H. E., Wentworth, B. B., & Wilson, D. L. (1979). "Prevalence of antibodies to Legionnaires' disease. A seroepidemiologic survey of Michigan residents using the hemagglutination test", *Annals of Internal Medicine*, vol. 90, no. 4, pp. 691–693.

Elder, E. M., Brown, A., Remington, J. S., Shonnard, J., & Naot, Y. (1983). "Microenzyme-linked immunosorbent assay for detection of immunoglobulin G and immunoglobulin M antibodies to *Legionella pneumophila*", *Journal of Clinical Microbiology*, vol. 17, no. 1, pp. 112–121.

Farshy, C. E., Klein, G. C., & Feeley, J. C. (1978). "Detection of antibodies to legionnaires disease organism by microagglutination and micro-enzyme-linked immunosorbent assay tests", *Journal of Clinical Microbiology*, vol. 7, no. 4, pp. 327–331.

Fields, B. S. (1996). "The molecular ecology of legionellae", *Trends in Microbiology*, vol. 4, no. 7, pp. 286–290.

Finkelstein, R., Brown, P., Palutke, W. A., Wentworth, B. B., Geiger, J. G., Bostic, G. D., & Sobel, J. D. (1993). "Diagnostic efficacy of a DNA probe in pneumonia caused by *Legionella* species", *Journal of Medical Microbiology*, vol. 38, no. 3, pp. 183–186.

Fitzgeorge, R. B., Lever, S., & Baskerville, A. (1993). "A comparison of the efficacy of azithromycin and clarithromycin in oral therapy of experimental airborne Legionnaires' disease", *Journal of Antimicrobial Chemotherapy*, vol. 31, Suppl. E, pp. 171–176.

Fliermans, C. B. & Tyndall, R. L. (1993). "Association of *Legionella pneumophila* with natural ecosystem", in *Legionella: Current status and emerging perspectives*, J. M. Barbaree, R. Breiman, & A. P. Dufour, eds, Washington, DC: American Society for Microbiology, pp. 284–285.

Fliermans, C. B., Cherry, W. B., Orrison, L. H., Smith, S. J., Tison, D. L., & Pope, D. H. (1981). "Ecological distribution of *Legionella pneumophila*", *Applied & Environmental Microbiology*, vol. 41, no. 1, pp. 9–16.

Formica, N., Tallis, G., Zwolak, B., Camie, J., Beers, M., Hogg, G., Ryan, N., & Yates, M. (2000). "Legionnaires' disease outbreak: Victoria's largest identified outbreak", *Communicable Diseases Intelligence*, vol. 24, no. 7, pp. 199–2022.

Fox, K. F. & Brown, A. (1993). "Properties of the genus *Tatlockia*. Differentiation of *Tatlockia* (*Legionella*) *maceachernii* and *micdadei* from each other and from other legionellae", *Canadian Journal of Microbiology*, vol. 39, no. 5, pp. 486–491.

Fry, N. K. & Harrison, T. G. (1998). "Diagnosis and epidemiology of infections caused by *Legionella* spp", in *Molecular bacteriology: protocols and clinical applications*, N. Woodford & A. P. Johnson, eds, Totowa: Humana Press, pp. 213–241.

Gao, L. Y. & Abu Kwaik, Y. (1999). "Activation of caspase 3 during *Legionella pneumophila*-induced apoptosis", *Infection and Immunity*, vol. 67, no. 9, pp. 4886–4944.

Garcia-Fulgueiras, A., Navarro, C., Fenoll, D., Garcia, J., Gonzales-Diego, P., Jimenez-Bunuales, T., Rodrigues, M., & Lopez, R. (2003). "Legionniares' disease outbreak in Murcia, Spain", *Emerging Infectious Diseases*, vol. 9, no. 8, p. 915–921.

Garrity, G. M., Brown, A., & Vickers, R. M. (1980). "*Tatlockia* and *Fluoribacter*. Two new genera of organisms resembling *Legionella pneumophila*", *International Journal of Systematic Bacteriology*, vol. 30, pp. 609–614.

Glick, T. H., Gregg, M. B., Berman, B., Mallison, G., Rhodes, W. W. Jr., & Kassanoff, I. (1978). "Pontiac fever. An epidemic of unknown etiology in a health department: I. Clinical and epidemiologic aspects", *American Journal of Epidemiology*, vol. 107, no. 2, pp. 149–160.

Goldberg, D. J., Wrench, J. G., Collier, P. W., Emslie, J. A., Fallon, R. J., Forbes, G. I., McKay, T. M., Macpherson, A. C., Markwick, T. A., & Reid, D. (1989). "Lochgoilhead fever: outbreak of non-pneumonic legionellosis due to *Legionella micdadei*", *Lancet*, vol. 1, no. 8633, pp. 316–318.

Gosting, L. H., Cabrian, K., Sturge, J. C., & Goldstein, L. C. (1984). "Identification of a species-specific antigen in *Legionella pneumophila* by a monoclonal antibody", *Journal of Clinical Microbiology*, vol. 20, no. 6, pp. 1031–1035.

Hackman, B. A., Plouffe, J. F., Benson, R. F., Fields, B. S., & Breiman, R. F. (1996). "Comparison of Binax Legionella Urinary Antigen EIA kit with Binax RIA Urinary Antigen kit for detection of *Legionella pneumophila* serogroup 1 antigen", *Journal of Clinical Microbiology*, vol. 34, no. 6, pp. 1579–1580.

Harrison, T. G. & Doshi, N. (2001). "Evaluation of the Bartels Legionella Urinary Antigen enzyme immunoassay", *European Journal of Clinical Microbiology & Infectious Diseases*, vol. 20, no. 10, pp. 738–740.

Harrison, T. G. & Saunders, N. A. (1994). "Taxonomy and typing of legionellae", *Reviews in Medical Microbiology*, vol. 5, no. 2, pp. 79–90.

Harrison, T. G. & Taylor, A. G. (1982). "A rapid microagglutination test for the diagnosis of *Legionella pneumophila* (serogroup 1) infection", *Journal of Clinical Pathology*, vol. 35, no. 9, pp. 1028–1031.

Harrison, T. G. & Taylor, A. G. (1988). "The diagnosis of Legionnaires' disease by estimation of antibody levels, in *A laboratory manual for legionella*, T. G. Harrison & A. G. Taylor, eds, Chichester: John Wiley and Sons, pp. 113–135.

Harrison, T., Uldum, S., Alexiou-Daniel, S., Bangsborg, J., Bernander, S., Drasar, V., Etienne, J., Helbig, J., Lindsay, D., Lochman, I., Marques, T., de Ory, F., Tartakovskii, I., Wewalka, G., & Fehrenbach, F. (1998). "A multicenter evaluation of the Biotest legionella urinary antigen EIA", *Clinical Microbiology and Infection*, vol. 4, no. 7, pp. 359–365.

Health and Safety Commission (2000). *Legionnaires' disease. The control of legionella bacteria in water systems: Approved Code of Practice and Guidance*. London: HMSO.

Heath, C. H., Grove, D. I., & Looke, D. F. (1996). "Delay in appropriate therapy of *Legionella pneumonia* associated with increased mortality", *European Journal of Clinical Microbiology & Infectious Diseases*, vol. 15, no. 4, pp. 286–290.

Helbig, J. H., Bernander, S., Castellani, P. M., Etienne, J., Gaia, V., Lauwers, S., Lindsay, D., Luck, P. C., Marques, T., Mentula, S., Peeters, M. F., Pelaz, C., Struelens, M., Uldum, S. A., Wewalka, G., & Harrison, T. G. (2002). "Pan-European study on culture-proven Legionnaires' disease: distribution of *Legionella pneumophila* serogroups and monoclonal subgroups", *European Journal of Clinical Microbiology & Infectious Diseases*, vol. 21, no. 10, pp. 710–716.

Helbig, J. H., Engelstadter, T., Maiwald, M., Uldum, S. A., Witzleb, W., & Luck, P. C. (1999). "Diagnostic relevance of the detection of Legionella DNA in urine samples by the polymerase chain reaction", *European Journal of Clinical Microbiology & Infectious Diseases*, vol. 18, no. 10, pp. 716–722.

Helbig, J. H., Uldum, S. A., Bernander, S., Luck, P. C., Wewalka, G., Abraham, B., Gaia, V., & Harrison, T. G. (2003). "Clinical utility of urinary antigen detection for diagnosis of community-acquired, travel-associated, and nosocomial legionnaires' disease", *Journal of Clinical Microbiology*, vol. 41, no. 2, pp. 838–840.

Herbrink, P., Meenhorst, P. L., Groothuis, D. G., Munckhof, H. V., Bax, R., Meijer, C. J., & Lindeman, J. (1983). "Detection of antibodies against *Legionella pneumophila* serogroups 1–6 and the Leiden-1 strain by micro ELISA and immunofluorescence assay", *Journal of Clinical Pathology*, vol. 36, no. 11, pp. 1246–1252.

Hilbi, H., Segal, G., & Shuman, H. A. (2001). "Icm/dot-dependent upregulation of phagocytosis by *Legionella pneumophila*", *Molecular Microbiology*, vol. 42, no. 3, pp. 603–617.

Hlady, W. G., Mullen, R. C., Mintz, C. S., Shelton, B. G., Hopkins, R. S., & Daikos, G. L. (1993). "Outbreak of Legionnaire's disease linked to a decorative fountain by molecular epidemiology", *American Journal of Epidemiology*, vol. 138, no. 8, pp. 555–562.

Horwitz, M. A. (1983). "Formation of a novel phagosome by the Legionnaires' disease bacterium (*Legionella pneumophila*) in human monocytes", *Journal of Experimental Medicine*, vol. 158, no. 4, pp. 1319–1331.

Horwitz, M. A. & Maxfield, F. R. (1984). "*Legionella pneumophila* inhibits acidification of its phagosome in human monocytes", *Journal of Cell Biology*, vol. 99, no. 6, pp. 1936–1943.

Horwitz, M. A. & Silverstein, S. C. (1980). "Legionnaires' disease bacterium (*Legionella pneumophila*) multiples intracellularly in human monocytes", *Journal of Clinical Investigation*, vol. 66, no. 3, pp. 441–450.

Horwitz, M. A. & Silverstein, S. C. (1981). "Activated human monocytes inhibit the intracellular multiplication of Legionnaires' disease bacteria", *Journal of Experimental Medicine*, vol. 154, no. 5, pp. 1618–1635.

Horwitz, M. A. & Silverstein, S. C. (1983). "Intracellular multiplication of Legionnaires' disease bacteria (*Legionella pneumophila*) in human monocytes is reversibly inhibited by erythromycin and rifampin", *Journal of Clinical Investigation*, vol. 71, no. 1, pp. 15–26.

Hubbard, R. B., Mathur, R. M., & Macfarlane, J. T. (1993). "Severe community-acquired legionella pneumonia: treatment, complications and outcome", *Quarterly Journal of Medicine*, vol. 86, no. 5, pp. 327–332.

Hughes, M. S. & Steele, T. W. (1994). "Occurrence and distribution of *Legionella* species in composted plant materials", *Applied & Environmental Microbiology*, vol. 60, no. 6, pp. 2003–2005.

Jaulhac, B., Nowicki, M., Bornstein, N., Meunier, O., Prevost, G., Piemont, Y., Fleurette, J., & Monteil, H. (1992). "Detection of *Legionella* spp. in bronchoalveolar lavage fluids by DNA amplification", *Journal of Clinical Microbiology*, vol. 30, no. 4, pp. 920–924.

Jonas, D., Rosenbaum, A., Weyrich, S., & Bhakdi, S. (1995). "Enzyme-linked immunoassay for detection of PCR-amplified DNA of legionellae in bronchoalveolar fluid", *Journal of Clinical Microbiology*, vol. 33, no. 5, pp. 1247–1252.

Joseph, C. (on behalf of the European Working Group for Legionella Infections) (2004). "Legionnaires' disease in Europe 2000–2002", *Epidemiology and Infection*, vol. 132, pp. 417–424.

Joseph, C. A., Harrison, T. G., Ilijic-Car, D., & Bartlett, C. L. (1999). "Legionnaires' disease in residents of England and Wales: 1998", *Communicable Disease & Public Health*, vol. 2, no. 4, pp. 280–284.

Kilvington, S. & Price, J. (1990). "Survival of *Legionella pneumophila* within cysts of *Acanthamoeba polyphaga* following chlorine exposure", *Journal of Applied Bacteriology*, vol. 68, no. 5, pp. 519–525.

Kim, M. J., Sohn, J. W., Park, D. W., Park, S. C., & Chun, B. C. (2003). "Characterization of a lipoprotein common to *Legionella* species as a urinary broad-spectrum antigen for diagnosis of Legionnaires' disease", *Journal of Clinical Microbiology*, vol. 41, no. 7, pp. 2974–2979.

Kohne, D. E., Stein, P. E., Brenner, D. J. (1984). "Nucleic acid probe specific for members of the genus Legionella", in Legionella: *Proceedings of the 2nd International Symposium*, C. Thornsberry *et al.*, eds, Washington: ASM, pp. 107–108.

Laussucq, S., Schuster, D., Alexander, W. J., Thacker, W. L., Wilkinson, H. W., & Spika, J. S. (1988). "False-positive DNA probe test for *Legionella* species associated with a cluster of respiratory illnesses", *Journal of Clinical Microbiology*, vol. 26, no. 8, pp. 1442–1444.

Lennette, D. A., Lennette, E. T., Wentworth, B. B., French, M. L., & Lattimer, G. L. (1979). "Serology of Legionnaires disease: comparison of indirect fluorescent antibody, immune adherence hemagglutination, and indirect hemagglutination tests", *Journal of Clinical Microbiology*, vol. 10, no. 6, pp. 876–879.

Li, J. S., O'Brien, E. D., & Guest, C. (2002). "A review of national legionellosis surveillance in Australia, 1991–2000", *Communicable Diseases Intelligence*, vol. 26, no. 3, pp. 461–468.

Liles, M. R., Viswanathan, V. K., & Cianciotto, N. P. (1998). "Identification and temperature regulation of *Legionella pneumophila* genes involved in type IV pilus biogenesis and type II protein secretion", *Infection & Immunity*, vol. 66, no. 4, pp. 1776–1782.

Lim, W. S., Macfarlane, J. T., Boswell, T. C., Harrison, T. G., Rose, D., Leinonen, M., & Saikku, P. (2001). "Study of community acquired pneumonia aetiology (SCAPA) in adults admitted to hospital: implications for management guidelines", *Thorax*, vol. 56, no. 4, pp. 296–301.

Lindsay, D. S., Abraham, W. H., & Fallon, R. J. (1994). "Detection of *mip* gene by PCR for diagnosis of Legionnaires' disease", *Journal of Clinical Microbiology*, vol. 32, no. 12, pp. 3068–3069.

Lisby, G. & Dessau, R. (1994). "Construction of a DNA amplification assay for detection of *Legionella* species in clinical samples", *European Journal of Clinical Microbiology & Infectious Diseases*, vol. 13, no. 3, pp. 225–231.

Luck, P. C., Freier, T., Steudel, C., Knirel, Y. A., Luneberg, E., Zahringer, U., Helbig, J. H. (2001). "A point mutation in the active site of *Legionella pneumophila* O-acetyltransferase results in modified lipopolysaccharide but does not influence virulence", *Internal Journal of Medical Microbiology*, vol. 291, no. 4, pp. 345–352.

Mahbubani, M. H., Bej, A. K., Miller, R., Haff, L., DiCesare, J., & Atlas, R. M. (1990). "Detection of *Legionella* with polymerase chain reaction and gene probe methods", *Molecular & Cellular Probes*, vol. 4, no. 3, pp. 175–187.

Malan, A. K., Martins, T. B., Jaskowski, T. D., Hill, H. R., & Litwin, C. M. (2003). "Comparison of two commercial enzyme-linked immunosorbent assays with an immunofluorescence assay for detection of *Legionella pneumophila* types 1–6", *Journal of Clinical Microbiology*, vol. 41, no. 7, pp. 3060–3063.

Marston, B. J., Plouffe, J. F., File, T. M. J., Hackman, B. A., Salstrom, S. J., Lipman, H. B., Kolczak, M. S., & Breiman, R. F. (1997). "Incidence of community-acquired pneumonia requiring hospitalization. Results of a population-based active surveillance study in Ohio. The Community-Based Pneumonia Incidence Study Group", *Archives of International Medicine*, vol. 157, no. 15, pp. 1709–1718.

Mayaud, C. & Dournon, E. (1988). "Clinical features of Legionnaires' disease", in *A laboratory manual for legionella*, T. G. Harrison & A. G. Taylor, eds, Chichester: John Wiley and Sons pp. 5–11.

Mayaud, C., Carette, M. F., Dournon, E., Bure, A., Francois, T., Akoun, G. (1984). "Clinical features and prognosis of severe pneumonia caused by *Legionella pneumophila*", *Proceedings of the 2nd International Symposium*, C. Thornsberry *et al.*, eds, Washington: ASM, pp. 11–12.

McDade, J. E., Shepard, C. C., Fraser, D. W., Tsai, T. R., Redus, M. A., & Dowdle, W. R. (1977). "Legionnaires' disease: isolation of a bacterium and demonstration of its role in other respiratory disease", *New England Journal of Medicine*, vol. 297, no. 22, pp. 1197–1203.

Muder, R. R. & Yu, V. L. (2002). "Infection due to *Legionella* species other than *L. pneumophila*", *Clinical Infectious Diseases*, vol. 35, no. 8, pp. 990–998.

Murdoch, D. R., Walford, E. J., Jennings, L. C., Light, G. J., Schousboe, M. I., Chereshsky, A. Y., Chambers, S. T., & Town, G. I. (1996). "Use of the polymerase chain reaction to detect *Legionella* DNA in urine and serum samples from patients with pneumonia", *Clinical Infectious Diseases*, vol. 23, no. 3, pp. 475–480.

Nagai, H. & Roy, C. R. (2001). "The DotA protein from *Legionella pneumophila* is secreted by a novel process that requires the Dot/Icm transporter", *EMBO Journal*, vol. 20, no. 21, pp. 5962–5970.

Nagai, H., Kagan, J. C., Zhu, X., Kahn, R. A., & Roy, C. R. (2002). "A bacterial guanine nucleotide exchange factor activates ARF on *Legionella* phagosomes", *Science*, vol. 295, no. 5555, pp. 679–682.

Neumeister, B., Faigle, M., Sommer, M., Zahringer, U., Stelter, F., Menzel, R., Schutt, C., & Northoff, H. (1998). "Low endotoxic potential of *Legionella pneumophila* lipopolysaccharide due to failure of interaction with the monocyte lipopolysaccharide receptor CD14", *Infection & Immunity*, vol. 66, no. 9, pp. 4151–4177.

O'Connell, W. A., Dhand, L., & Cianciotto, N. P. (1996). "Infection of macrophage-like cells by *Legionella* species that have not been associated with disease", *Infection & Immunity*, vol. 64, no. 10, pp. 4381–4384.

O'Mahony, M. C., Stanwell-Smith, R. E., Tillett, H. E., Harper, D., Hutchison, J. G., Farrell, I. D., Hutchinson, D. N., Lee, J. V., Dennis, P. J., Duggal, H. V. *et al.* (1990). "The Stafford outbreak of Legionnaires' disease", *Epidemiology & Infection*, vol. 104, no. 3, pp. 361–380.

Ohno, A., Kato, N., Yamada, K., & Yamaguchi, K. (2003). "Factors influencing survival of *Legionella pneumophila* serotype 1 in hot spring water and tap water", *Applied & Environmental Microbiology*, vol. 69, no. 5, pp. 2540–2547.

Park, M. Y., Ko, K. S., Lee, H. K., Park, M. S., & Kook, Y. H. (2003). "*Legionella busanensis* sp. nov., isolated from cooling tower water in Korea", *International Journal of Systematic & Evolutionary Microbiology*, vol. 53, Pt 1, pp. 77–80.

Pasculle, A. W., Veto, G. E., Krystofiak, S., McKelvey, K., & Vrsalovic, K. (1989). "Laboratory and clinical evaluation of a commercial DNA probe for the detection of *Legionella* spp", *Journal of Clinical Microbiology*, vol. 27, no. 10, pp. 2350–2358.

Payne, N. R. & Horwitz, M. A. (1987). "Phagocytosis of *Legionella pneumophila* is mediated by human monocyte complement receptors", *Journal of Experimental Medicine*, vol. 166, no. 5, pp. 1377–1389.

Petitjean, F., Dournon, E., Strosberg, A. D., & Hoebeke, J. (1990). "Isolation, purification and partial analysis of the lipopolysaccharide antigenic determinant recognized by a monoclonal antibody to *Legionella pneumophila* serogroup 1", *Research in Microbiology*, vol. 141, no. 9, pp. 1077–1094.

Rajagopalan-Levasseur, P., Dournon, E., Dameron, G., Vilde, J. L., & Pocidalo, J. J. (1990). "Comparative postantibacterial activities of pefloxacin, ciprofloxacin, and ofloxacin against intracellular multiplication of *Legionella pneumophila* serogroup 1", *Antimicrobial Agents & Chemotherapy*, vol. 34, no. 9, pp. 1733–1738.

Ratcliff, R. M., Donnellan, S. C., Lanser, J. A., Manning, P. A., & Heuzenroeder, M. W. (1997). "Interspecies sequence differences in the Mip protein from the genus *Legionella*: implications for function and evolutionary relatedness", *Molecular Microbiology*, vol. 25, no. 6, pp. 1149–1158.

Rechnitzer, C. & Blom, J. (1989). "Engulfment of the Philadelphia strain of *Legionella pneumophila* within pseudopod coils in human phagocytes. Comparison with other *Legionella* strains and species", *APMIS*, vol. 97, no. 2, pp. 105–114.

Redd, S. C., Lin, F. Y., Fields, B. S., Biscoe, J., Plikaytis, B. B., Powers, P., Patel, J., Lim, B. P., Joseph, J. M., Devadason, C. *et al*. (1990). "A rural outbreak of Legionnaires' disease linked to visiting a retail store", *American Journal of Public Health*, vol. 80, no. 4, pp. 431–434.

Riffard, S., Douglass, S., Brooks, T., Springthorpe, S., Filion, L. G., & Sattar, S. A. (2003). "Occurrence of *Legionella* in groundwater: an ecological study", *Water Science & Technology*, vol. 43, no. 12, pp. 99–102.

Rowbotham, T. J. (1980). "Preliminary report on the pathogenicity of *Legionella pneumophila* for freshwater and soil amoebae", *Journal of Clinical Pathology*, vol. 33, no. 12, pp. 1179–1183.

Rowbotham, T. J. (1993). "Legionella-like amoebal pathogens", in *Legionella: current status and emerging perspectives*, J. M. Barbaree, R. F. Breiman, & A. P. Dufour, eds, Washington: ASM, pp. 137–140.

Sathapatayavongs, B., Kohler, R. B., Wheat, L. J., White, A., Winn, W. C. Jr, Girod, J. C., & Edelstein, P. H. (1982). "Rapid diagnosis of Legionnaires' disease by urinary antigen detection. Comparison of ELISA and radioimmunoassay", *American Journal of Medicine*, vol. 72, no. 4, pp. 576–582.

Saunders, N. A. & Harrison, T. G. (1988). "The application of nucleic acid probes and monoclonal antibodies to the investigation of legionella infections", in *A laboratory manual for legionella*, T. G. Harrison & A. G. Taylor, eds, Chichester: John Wiley and Sons, pp. 137–153.

Shuman, H. A. & Horwitz, M. A. (1996). "*Legionella pneumophila* invasion of mononuclear phagocytes", *Current Topics in Microbiology & Immunology*, vol. 209, pp. 99–112.

Sonesson, A., Jantzen, E., Bryn, K., Larsson, L., & Eng, J. (1989). "Chemical composition of a lipopolysaccharide from *Legionella pneumophila*", *Archives of Microbiology*, vol. 153, no. 1, pp. 72–78.

Sonesson, A., Jantzen, E., Tangen, T., & Zahringer, U. (1994a) "Chemical composition of lipopolysaccharides from *Legionella bozemanii* and *Legionella longbeachae*", *Archives of Microbiology*, vol. 162, no. 4, pp. 215–221.

Sonesson, A., Jantzen, E., Tangen, T., & Zahringer, U. (1994b) "Lipopolysaccharides of *Legionella erythra* and *Legionella oakridgensis*", *Canadian Journal of Microbiology*, vol. 40, no. 8, pp. 666–671.

Starnbach, M. N., Falkow, S., & Tompkins, L. S. (1989). "Species-specific detection of *Legionella pneumophila* in water by DNA amplification and hybridization", *Journal of Clinical Microbiology*, vol. 27, no. 6, pp. 1257–1261.

Steinert, M., Emody, L., Amann, R., & Hacker, J. (1997). "Resuscitation of viable but nonculturable *Legionella pneumophila* Philadelphia JR32 by *Acanthamoeba castellanii*", *Applied & Environmental Microbiology*, vol. 63, no. 5, pp. 2047–2053.

Stone, B. J. & Abu Kwaik, Y. (1998). "Expression of multiple pili by *Legionella pneumophila*: identification and characterization of a type IV pilin gene and its role in adherence to mammalian and protozoan cells", *Infection & Immunity*, vol. 66, no. 4, pp. 1768–1775.

Szeto, L. & Shuman, H. A. (1990). "The *Legionella pneumophila* major secretory protein, a protease, is not required for intracellular growth or cell killing", *Infection & Immunity*, vol. 58, no. 8, pp. 2585–2592.

Tang, P. W. & Toma, S. (1986). "Broad-spectrum enzyme-linked immuno-sorbent assay for detection of *Legionella* soluble antigens", *Journal of Clinical Microbiology*, vol. 24, no. 4, pp. 556–558.

Terranova, W., Cohen, M. L., & Fraser, D. W. (1978). "1974 outbreak of Legionnaires' disease diagnosed in 1977. Clinical and epidemiological features", *Lancet*, vol. 2, no. 8081, pp. 122–124.

Thomason, B. M. & Bibb, W. F. (1984). "Use of absorbed antisera for demonstration of antigenic variation among strains of *Legionella pneumophila* serogroup 1", *Journal of Clinical Microbiology*, vol. 19, no. 6, pp. 794–797.

Tison, D. L., Pope, D. H., Cherry, W. B., & Fliermans, C. B. (1980). "Growth of *Legionella pneumophila* in association with blue-green algae (cyanobacteria)", *Applied & Environmental Microbiology*, vol. 39, no. 2, pp. 456–459.

Tobin, J. O., Swann, R. A., & Bartlett, C. L. (1981). "Isolation of *Legionella pneumophila* from water systems: methods and preliminary results", *British Medical Journal Clinical Research Ed*, vol. 282, no. 6263, pp. 515–517.

Verissimo, A., Marrao, G., da Silva, F. G., & da Costa, M. S. (1991). "Distribution of *Legionella* spp. in hydrothermal areas in continental Portugal and the island of Sao Miguel, Azores", *Applied & Environmental Microbiology*, vol. 57, no. 10, pp. 2921–2927.

Vogel, J. P. & Isberg, R. R. (1999). "Cell biology of *Legionella pneumophila*", *Current Opinion in Microbiology*, vol. 1, pp. 30–44.

Wait, R. (1988). "Confirmation of the identity of legionellae by whole cell fatty-acid and isoprenoid quinone profiles", in *A laboratory manual for legionella*, T. G. Harrison & A. G. Taylor, eds, Chichester: John Wiley and Sons, pp. 69–101.

Watson, J. M., Mitchell, E., Gabbay, J., Maguire, H., Boyle, M., Bruce, J., Tomlinson, M., Lee, J., Harrison, T. G., Uttley, A. *et al*. (1994). "Piccadilly Circus legionnaires' disease outbreak", *Journal of Public Health Medicine*, vol. 16, no. 3, pp. 341–347.

Westminster Action Committee (1988). *Broadcasting House Legionnaires' Disease*.

Wilkinson, H. W., Cruce, D. D., & Broome, C. V. (1981). "Validation of *Legionella pneumophila* indirect immunofluorescence assay with epidemic sera", *Journal of Clinical Microbiology*, vol. 13, no. 1, pp. 139–146.

Williams, A., Rechnitzer, C., Lever, M. S., & Fitzgeorge, R. B. (1993). "Intracellular production of *Legionella pneumophila* tissue-destructive protease in alveolar macrophages", in *Legionella: current status and emerging perspectives*, J. M. Barbaree, R. F. Breiman, & A. P. Dufour, eds, Washington: ASM, pp. 88–90.

Winn, W. C. J. (1993). "*Legionella* and the clinical microbiologist", *Infectious Disease Clinics of North America*, vol. 2, pp. 377–922.

Wong, K. H., Moss, C. W., Hochstein, D. H., Arko, R. J., & Schalla, W. O. (1979). " 'Endotoxicity' of the Legionnaires' disease bacterium", *Annals of Internal Medicine*, vol. 90, no. 4, pp. 624–627.

Woodhead, M. A., Macfarlane, J. T., Macrae, A. D., & Pugh, S. F. (1986). "The rise and fall of Legionnaires' disease in Nottingham", *Journal of Infection*, vol. 13, no. 3, pp. 293–296.

Wreghitt, T. G., Nagington, J., & Gray, J. (1982). "An ELISA test for the detection of antibodies to *Legionella pneumophila*", *Journal of Clinical Pathology*, vol. 35, no. 6, pp. 657–660.

Zahringer, U., Knirel, Y. A., Lindner, B., Helbig, J. H., Sonesson, A., Marre, R., & Rietschel, E. T. (1995). "The lipopolysaccharide of *Legionella pneumophila* serogroup 1 (strain Philadelphia 1): chemical structure and biological significance", *Progress in Clinical Biology Research*, vol. 392, pp. 113–139.

38

Coxiella burnetii

James G. Olson[1], Franca R. Jones[2] and Patrick J. Blair[3]

[1]*Virology Department, U.S. Naval Medical Research Center Detachment, Unit 3800, American Embassy, APO AA 34031;*
[2]*National Naval Medical Center, Microbiology Laboratory, Bethesda, MD 20889, USA; and* [3]*US Embassy Jakarta, US NAMRU 2,*
FPO AP 96520-8132

INTRODUCTION

Coxiella burnetii is the causative agent of Q fever. Although traditionally classified as a rickettsial disease, *Coxiella* is a member of the family *Coxiellaceae* in the order Legionellales (http://www.cme.msu.edu/bergeys/april2001-genus.pdf). Although not often considered a significant public health threat, *C. burnetii* has received renewed attention as a potential bioterrorism agent due to its remarkable stability outside the host, its low infectious dose, and – perhaps most pertinent – its capability to be transmitted by the aerosol route.

DESCRIPTION OF THE ORGANISM

Definition

Coxiella burnetii is an obligate, intracellular Gram-negative bacterium ranging from 0.4 to 1 μm in length and 0.2 to 0.4 μm in width (Maurin and Raoult, 1999). Entry of *C. burnetii* into host cells is a passive process that results in the fusion of membrane-bound vesicles containing *C. burnetii* organisms with lysosomes (Baca and Paretsky, 1983; Baca, Klassen and Aragon, 1993; Heinzen *et al.*, 1996). *Coxiella burnetii* thrives within the acidic environment of phagolysosomes (Hackstadt and Williams, 1981, 1983; Zuerner and Thompson, 1983; Chen *et al.*, 1990). Like *Chlamydia*, *Coxiella* have a unique intracellular cycle consisting of two distinct morphological forms of the bacteria. First, small cell variants (SCVs) are 204–450 nm rods and have dense material in the periplasm. These forms are not metabolic but are highly infectious and resistant to many environmental conditions. The acidic environment of the phagolysosome appears to trigger the conversion of SCVs to LCVs (large cell variants; 2 μm in length), which are metabolically active and appear to form spores that develop to SCVs that are then released from the cell to restart the infectious cycle (McCaul, 1991). *Coxiella* are not flagellated.

Molecular Characterization

Although *C. burnetii* has been known historically as a member of the order Rickettsiales, partly based on its inability to grow in axenic medium, sequencing of the 16S rRNA gene (Stein *et al.*, 1993) has allowed for the reclassification of this organism into the gamma subdivision of Proteobacteria order Legionellales, family *Coxiellaceae*. On the basis of a combination of different molecular techniques, including DNA–DNA hybridization (Vodkin, Williams and Stephenson, 1986), restriction fragment length polymorphism (Hendrix, Samuel and Mallavia, 1991), and pulsed-field gel electrophoresis (Heinzen *et al.*,

1990), it has been demonstrated that considerable variation exists between the genomes of *C. burnetii* isolates. It does appear, however, that isolates tend to cluster depending on geographic location (Maurin and Raoult, 1999).

CLINICAL FEATURES, PATHOGENESIS, AND EPIDEMIOLOGY

Acute Q fever is characterized by two clinical presentations, atypical pneumonia and hepatitis. The primary mode of transmission is from inhalation of contaminated aerosols (Marrie, 1990). It is estimated that only between 1 and 10 bacteria are necessary to cause infection (Norlander, 2000). The specific route of transmission appears to correlate with the subsequent clinical presentation (Marrie *et al.*, 1996). For example, in Nova Scotia, Canada, Switzerland, and northern Spain, the primary manifestation of Q fever is pneumonia, presumably caused by inhalation of aerosols (Marrie *et al.*, 1988; Norlander, 2000). In contrast, in France and southern Spain, acute Q fever manifests as granulomatous hepatitis, which is assumed to result from the drinking of contaminated milk (Fishbein and Raoult, 1992). Acute Q fever is an acute febrile disease commonly associated with sudden onset of headache, chills, myalgia, arthralgia, photophobia, lymphadenopathy, conjunctivitis, nausea or vomiting, diarrhea and pharyngitis. Rash is rare and abnormal X-ray findings are present in 50% of cases. Although Q fever is rarely fatal (CFR approximately 1%), it may become a chronic illness that lasts longer than 6 months and has a higher rate of mortality (Sawyer, Fishbein and McDade, 1987). While chronic disease can afflict multiple organ systems, Q fever typically manifests itself as endocarditis and is often diagnosed among patients who have preexisting heart valve disease (Turck *et al.*, 1976; Ellis, Smith and Moffat, 1982).

Microscopically, the pathology of Q fever is similar to bacterial pneumonia and can be characterized by severe interalveolar and a patchy, focally necrotizing haemorrhagic pneumonia that involves the alveolar-lining cells (Perin, 1949). Necrotizing bronchitis and bronchiolitis may be present. *Coxiella burnetii* is found in histiocytes of the alveolar exudate. Histiocytic hyperplasia is found in mediastinal lymph nodes, spleen, and adrenals. Hepatocellular damage in acute Q fever consists of granulomatous changes in the lobules and occasional involvement of the portal areas (Srigley *et al.*, 1985). Granulomas consist of non-distinctive focal histiocytic and mixed inflammatory cell infiltrates with multinucleated giant cells.

Q fever is distributed worldwide and is often transmitted from infectious aerosols in animal tissues or products and occasionally from unpasteurized milk. The disease is only rarely reported in the United States. New Zealand is the only country where Q fever has not

Principles and Practice of Clinical Bacteriology Second Edition   Editors Stephen H. Gillespie and Peter M. Hawkey
© 2006 John Wiley & Sons, Ltd

been reported (Hilbink *et al.*, 1993). Naturally infected ticks may play a role in maintenance and transmission of *C. burnetii* among animal hosts, but probably play little or no role in transmission to humans (Eklund, Parker and Lackman, 1947). Occupational exposure to infected livestock, primarily cattle, sheep, and goats is a major risk factor. Abattoir workers, sheep shearers, and wool gatherers have the highest rates of infection and disease (Bernard *et al.*, 1982). Persons who have contact with sheep, particularly fetuses and birth products, have the greatest risk. There have been several reports of transmission of Q fever from pets to humans, likely due to human handling of litter, birth products, cats (Kosatsky, 1984), or dogs (Laughlin *et al.*, 1991; Buhariwalla, Cann and Marrie, 1996).

LABORATORY DIAGNOSIS

Q fever must be distinguished from several viral and bacterial illnesses, including meningococcemia, measles, enteroviral exanthems, leptospirosis, typhoid fever, rubella, dengue fever, and Legionnaires' disease. Laboratory diagnosis of Q fever is routinely accomplished by serologic assays that detect anti-*C. burnetii* antibodies. Alternative procedures include isolation of *C. burnetii* from patient blood samples or tissue biopsies and techniques aimed at direct detection of organisms in these specimens. However, serologic methods are more widely used by clinical laboratories. Because antibody levels to *C. burnetii* are not detectable until 2–3 weeks following presentation of symptoms, serology cannot typically provide a diagnosis early enough to affect the outcome of disease. More recently, nested polymerase chain reactions (PCRs) have successfully identified *C. burnetii* in clinical and ectoparasite samples and PCR is now a readily utilized technique for rapid identification (Berri, Arricau-Bouvery and Rodolakis, 2003; Zhang *et al.*, 1998).

Serology

Indirect Fluorescent Antibody (IFA) Assays

IFA remains the gold standard for diagnosis of Q fever (Peacock *et al.*, 1983; Tissot-Dupont, Thirion and Raoult, 1994). IgG, IgM, and IgA antibodies can be measured against both *C. burnetii* phase I (obtained from spleens of infected mice) and phase II (obtained from *C. burnetii* grown in cell culture) antigens (Tissot-Dupont, Thirion and Raoult, 1994). Although seroconversion can be detected between 7 and 15 days after symptoms appear, 90% of patients will seroconvert by the third week. Diagnosis generally depends upon the demonstration of fourfold or greater increase in antibody titer between the acute- and convalescent-phase serum samples. Single serum specimens, however, can be used to make a diagnosis in many cases (Table 38.1). Although cross-reactivity has been described to occur between *C. burnetii* and *Legionella* or *Bartonella*, differential diagnosis can be made when antibody titers against both phase I and II antigens are obtained (Maurin and Raoult, 1999). IFA tests are reasonably sensitive; however, the subjectivity of endpoint determinations is an obvious disadvantage.

Table 38.1 Microimmunofluorescence cut-off values for Q fever diagnosis using a single serum specimen

Phase II antibody titer		Phase I antibody titer (IgG)	Interpretation
IgG	IgM		
≤100			Active Q fever improbable
≥200	≥50		Acute Q fever (100% predictive)
		≥1:800	Chronic Q fever (98% predictive)
		≥1:1600	Chronic Q fever (100% predictive)

Adapted from Maurin and Raoult (1999).

Complement Fixation

The CF assay for diagnosis of Q fever utilizes an extract of either phase I or phase II antigens of *C. burnetii* (Field, Hunt and Murphy, 1983; Herr *et al.*, 1985). By the fourth week after the onset of symptoms from Q fever, more than 90% of patients have antibodies that fix complement in the presence of phase II antigen (Elisberg and Bozeman, 1969). Antigenic phase variation must be considered for the interpretation of serologic results for Q fever. In acute, self-limited Q fever infections, antibodies to the phase II antigen appear first and dominate the humoral immune response. A CF titer ≥40 indicates acute Q fever (Guigno *et al.*, 1992). With chronic Q fever infections, however, phase I titers eventually equal or exceed phase II titers; thus a CF titer ≥200 indicates chronic Q fever (Peter *et al.*, 1992). The detection of antibodies that fix complement in the presence of phase I CF antigen has been useful in recognition of subacute Q fever endocarditis. The CF assay has the disadvantages that it is more time consuming than IFA, and seroconversion is detected almost 1 week later than IFA.

Other Serologic Techniques

Several other serologic methods have been described for diagnosis of Q fever including enzyme-linked immunosorbent assay (ELISA), microagglutination, dot blotting, western blotting, and radioimmunoassay (Maurin and Raoult, 1999). ELISA tests are purported to be more sensitive than IFA and CF (Peter *et al.*, 1988; Cowley *et al.*, 1992; Uhaa *et al.*, 1994; Field *et al.*, 2000, 2002), but they are not widely used in the clinical setting due to decreased specificity over some other assays, and because the length of setup time is not practical for small numbers of sera. Furthermore, the lack of standardization of homemade ELISA between laboratories may give rise to conflicting results. A new ELISA, available commercially to measure IgG and IgM, can be standardized between laboratories, and is more sensitive than IFA or CF (Field *et al.*, 2000, 2002). Samples positive by ELISA, however, should be confirmed using IFA for enhanced specificity. ELISA may be useful for large-scale serosurveys. Although microagglutination may be sensitive enough to detect antibodies soon after symptoms appear, a large amount of antigen is required for its use (Kazar *et al.*, 1981).

Bacterial Isolation and Culture

Many laboratory-acquired cases of Q fever have been reported; thus attempted isolation and culture of *C. burnetii* are only recommended for research laboratories that have personnel with experience in cultivating *Coxiellae* in addition to biosafety containment level-3 facilities. The guinea pig remains the animal of choice for *C. burnetii* isolation, although mice and embryonated eggs are also used (Perin and Bengston, 1942; Ormsbee, 1952; Williams, Thomas and Peacock, 1986). Clinical specimens from infected humans are injected intraperitoneally into guinea pigs (Maurin and Raoult, 1999). Infection is monitored by measuring body temperature, animals are killed 5–8 days later, then spleens are extracted for isolation of *C. burnetii*. Spleen extracts are inoculated into embryonated eggs to propagate the bacteria.

Coxiella burnetii can infect monocytes, macrophages, and a wide variety of cell lines (Baca and Paretsky, 1983; Norlander, 2000). Several different human specimens, including blood, cerebrospinal fluid, cardiac valve, liver biopsy, and birth products can be inoculated onto cell cultures, typically grown in shell vials, which are then centrifuged to increase bacterial adherence, and incubated at 37 °C with 5% CO_2 and for 5–7 days (Maurin and Raoult, 1999). Although *Coxiellae* have typical Gram-negative cell walls, they do not stain with the Gram procedure. Growth can be observed with staining using the Gimenez technique (Gimenez, 1964) or by indirect immunofluorescence using monoclonal or polyclonal anti-*C. burnetii* antibodies.

More recently, green fluorescent protein has served as a marker for detecting transformed pathogens inside of host eukaryotic cells (Lukacova *et al.*, 1999).

Direct Detection of Organisms

Immunofluorescence

Coxiella burnetii have been detected by immunofluorescence in tissue biopsies obtained from patients with Q fever endocarditis. Biopsy specimens can be analyzed fresh or after fixation in formalin and paraffin embedding. Detection of organisms can be carried out using immunohistochemistry (Brouqui, Dumler and Raoult, 1994), capture ELISA/ELIFA (Thiele, Karo and Krauss, 1992), or immunofluorescence/immunoelectron microscopy with monoclonal or polyclonal antibodies (McCaul and Williams, 1990; Thiele, Karo and Krauss, 1992).

Polymerase Chain Reaction

PCR detection of *C. burnetii* DNA has been reported from clinical specimens (Stein and Raoult, 1992). Current PCR primers are designed to amplify the genes encoding 16S rRNA, superoxide dismutase, and glutathione. The addition of nested primers have allowed for the successful amplification of these genes from clinical samples (Berri, Arricau-Bouvery and Rodolakis, 2003; Zhang *et al.*, 1998).

TREATMENT

Antibiotic therapy is, in general, of great benefit to patients infected with *C. burnetii* especially when initiated within the first three days of illness. Acute Q fever is typically treated successfully with doxycycline, or other quinolones (Hoover, Vodkin and Williams, 1992; Maurin and Raoult, 1999). Other antimicrobials, including erythromycin or chloramphenicol, may be useful (Clark and Lennette, 1952; D'Angelo and Hetherington, 1979). Chronic Q fever, on the other hand, requires combination therapy of both doxycycline and ciprofloxacin for long periods up to years to prevent episodic illness (Maurin and Raoult, 1999; Raoult, 1993).

PREVENTION

Q fever can be prevented by avoidance of contact with potentially infectious animal tissues and products. Occupationally exposed persons may reduce their risk of infection with *C. burnetii* by wearing respirators that prevent aerosol infections. Q fever vaccines for human use are killed, purified whole-cell preparations of phase I *C. burnetii* that contain lipopolysaccharide–protein complex antigens (Kazar and Rehacek, 1987; Williams *et al.*, 1993). A vaccine used in Australia is highly effective in preventing illness among abattoir workers (Marmion *et al.*, 1984). Individuals at high risk of acquiring Q fever should consider vaccination, however, emergence of Q fever has been noted in populations not typically considered at risk, notably individuals in urban areas exposed to pets during parturition.

REFERENCES

Babudieri, B. (1959) Q fever: a zoonosis. *Adv. Vet. Sci.* 5:81.

Baca, O.G., and Paretsky, D. (1983) Q fever and *Coxiella burnetii*: a model for host-parasite interactions. *Microbiol. Rev.* 46:127–149.

Baca, O.G., Klassen, D.A., and Aragon, A.S. (1993) Entry of *Coxiella burnetii* into host cells. *Acta Virol.* 37:143–155.

Bernard, K.W., Parham, G.L., Winkler, W.G., *et al.* (1982) Q fever control measures: recommendations for research facilities using sheep. *Infect. Control.* 3:461–465.

Berri, M., Arricau-Bouvery, N., and Rodolakis, A. (2003) PCR-based detection of *Coxiella burnetii* from clinical samples. *Methods Mol. Biol.* 216:153–161.

Brouqui, P., Dumler, J.S., and Raoult, D. (1994) Immunohistologic demonstration of *Coxiella burnetii* in the valves of patients with Q fever endocarditis. *Am. J. Med.* 97:451–458.

Buhariwalla, F., Cann, B., and Marrie, T.J. (1996) A dog related outbreak of Q fever. *Clin. Infect. Dis.* 23:753–755.

Chen, S-Y., Vodkin, M., Thompson, H.A., and Williams, J.C. (1990) Isolated *Coxiella burnetii* synthesizes DNA during acid activation in the absence of host cells. *J. Gen. Microbiol.* 136:89–96.

Clark, W.H., and Lennette, E.H. (1952) Treatment of Q fever with antibiotics. *Ann. N Y Acad. Sci.* 55:1004–1006.

Cowley, R., Fernandez, F., Freemantle, W., and Rutter, D. (1992) Enzyme immunoassay for Q fever: comparison with complement fixation and immunofluorescence tests and dot immunoblotting. *J. Clin. Microbiol.* 30:2451–2455.

D'Angelo, L.J., and Hetherington, R. (1979) Q fever treated with erythromycin. *Br. Med. J.* 4: 305–306.

Eklund, C.M., Parker, R.R., and Lackman, D.B. (1947) A case of Q fever probably contracted by exposure to ticks in nature. *Public Health Rep.* 62:1413–1416.

Elisberg, B.L., and Bozeman, F.M. (1969) Rickettsiae. p. 826–868. In Diagnostic procedures for viral and Rickettsial Infections. Lennette E.H., and Schmidt, N.J. (eds). American Public Health Association. NY.

Ellis, M.E., Smith, C.C., and Moffat, M.A.J. (1982) Chronic or fatal Q fever infection: a review of 16 patients seen in north-east Scotland (1967–80). *Q. J. Med.* (new series) 205:54–66.

Field, P.R., Hunt, J.G., and Murphy, A.M. (1983) Detection and persistence of specific IgM antibody to *Coxiella burnetii* by enzyme-linked immunosorbent assay: a comparison with immunofluorescence and complement fixation tests. *J. Infect. Dis.* 148:477–487.

Field, P.R., Mitchell, J.L., Santiago, A., *et al.* (2000) Comparison of a commercial enzyme-linked immunosorbent assay with immunofluorescence and complement fixation tests for detection of *Coxiella burnetii* (Q fever) immunoglobulin M. *J. Clin. Microbiol.* 38:1645–1647.

Field, P.R., Santiago, A., Chan, S-W., *et al.* (2002) Evaluation of a novel commercial enzyme-linked immunosorbent assay for detecting *Coxiella burnetii*-specific immunoglobulin G for Q fever prevaccination screening and diagnosis. *J. Clin. Microbiol.* 40:3526–3529.

Fishbein, D.B., and Raoult, D. (1992) A cluster of *Coxiella burnetii* infections associated with exposure to vaccinated goats and their unpasteurized dairy products. *Am. J. Trop. Med. Hyg.* 47:35–40.

Gimenez, D.F. (1964) Staining Rickettsiae in yolk-sac cultures. *Stain Technol.* 39:135–140.

Guigno, D., Compland, B., Spith, E.G., *et al.* (1992) Primary humoral antibody response to *Coxiella burnetii*, the causative agent of Q fever. *J. Clin. Microbiol.* 30:1958–1967.

Hackstadt, T., and Williams, J.C. (1981) Biochemical stratagem for obligate parasitism of eukaryotic cells by *Coxiella burnetii. Proc. Natl. Acad. Sci. U S A* 78:3240–3244.

Hackstadt, T., and Williams, J.C. (1983) pH dependence of the *Coxiella burnetii* glutamate transport system. *J. Bacteriol.* 154:598–603.

Heinzen, R.A., Scidmore, M.A., Rockey, D.D., and Hackstadt, T. (1996) Differential interaction with endocytic and exocytic pathways, distinguish parasitophorous vacuoles of *Coxiella burnetii* and *Chlamydia trachomatis*. *Infect. Immun.* 64:796–809.

Heinzen, R.A., Stiegler, G.L., Whiting, L.L., *et al.* (1990) Use of pulsed field electrophoresis to differentiate *Coxiella burnetii* strains. *Ann. N Y Acad. Sci.* 590:504–513.

Hendrix, L.R., Samuel, J.E., and Mallavia, L.P. (1991) Differentiation of *Coxiella burnetii* isolates by analysis of restriction endonuclease-digested DNA separated by SDS-PAGE. *J. Gen. Microbiol.* 13:269–276.

Herr, S., Huchzermeyer, H.F., Te Brugge, L.A., *et al.* (1985) The use of single complement fixation test technique in bovine brucellosis, Johne disease, dourine, equine piroplasmosis and Q fever serology. *Onderstepoort J. Vet. Res.* 52:279–282.

Hilbink, F., Penrose, M., Kovácová, E., and Kazár, J. (1993) Q fever is absent from New Zealand. *Int. J. Epidemiol.* 22:945–949.

Hoover, T.A., Vodkin, M.H., and Williams, J. (1992) A *Coxiella burnetii* repeated DNA element resembling a bacterial insertion sequence. *J. Bacteriol.* 174:5540–5548.

Kazar, J., and Rehacek, J. (1987) Q fever vaccines: present status and application in man. *Zentbl. Bakteriol. Mikrobiol. Hyg.* 267:74–78.

Kazar, J., Brezina, R., Schramek, S., *et al.* (1981) Suitability of the microagglutination test for detection of post-infection and post-vaccination Q fever antibodies in human sera. *Acta Virol.* 25:235–240.

Kosatsky, T. (1984) Household outbreak of Q-fever pneumonia related to a parturient cat. *Lancet* ii:1447–1449.

Laughlin, T., Wang, D., Williams, J., and Marrie, T.J. (1991) Q fever: from deer to dog to man. *Lancet* 337:676–677.

Lukacova, M., Valkova, D., Quevedo, D.M., *et al.* (1999) Green fluorescent protein as a detection marker for *Coxiella burnetii* transformation. *FEMS Microbiol. Lett.* 175:255–260.

Marmion, B.P., Ormsbee, R.A., Kyrkou, M., *et al.* (1984) Vaccine prophylaxis of abattoir-associated Q fever. *Lancet* ii:1411–1414.

Marrie, T.J. (1990) Epidemiology of Q fever. p. 49–70. In T.J. Marrie (ed.), *Q fever*, Vol. 1. *The disease*. CRC Press, Inc., Boca Raton, FL.

Marrie, T.J., Durant, H., Williams, J.C., *et al.* (1988) Exposure to parturient cats is a risk factor for acquisition of Q fever in Maritime Canada. *J. Infect. Dis.* 158:101–108.

Marrie, T.J., Stein, A., Janigan, D., and Raoult, D. (1996) Route of infection determines the clinical manifestations of acute Q fever. *J. Infect. Dis.* 173:484–487.

Maurin, M., and Raoult, D. (1999) Q fever. *Clin. Microbiol. Rev.* 12:518–553.

McCaul, T.F. (1991) The development cycle of *Coxiella burnetii*. p. 223–258. In J.C. Williams, and H.A. Thompson (eds), *Q Fever: the biology of Coxiella Burnetii*. CRC Press, Inc., Boca Raton, FL.

McCaul, T.F., and Williams, J.C. (1990) Localization of DNA in *Coxiella burnetii* by post-embedding immunoelectron microscopy. *Ann. N Y Acad. Sci.* 590:136–147.

Norlander, L. (2000) Q fever epidemiology and pathogenesis. *Microbes Infect.* 2(4):417–424.

Ormsbee, R.A. (1952) The growth of *Coxiella burnetii* in embryonated eggs. *J. Bacteriol.* 63:73.

Peacock, M.G., Philip, R.N., Williams, J.C., and Faulkner, R.S. (1983) Serological evaluation of Q fever in humans: enhanced phase I titers of immunoglobulins G and A are diagnostic for Q fever endocarditis. *Infect. Immun.* 41:1089–1098.

Perin, T.L. (1949) Histopathologic observations in a fatal case of Q fever. *Arch. Pathol.* 47:361–365.

Perin, T.K., and Bengston, I.A. (1942) The histopathology of experimental Q fever in mice. *Public Health Rep.* 57:790–794.

Peter, O., Dupuis, G., Bee, D., *et al.* (1988) Enzyme-linked immunoabsorbent assay for diagnosis of chronic Q fever. *J. Clin. Microbiol.* 26:1978–1982.

Peter, O., Flepp, M., Bestetti, G., *et al.* (1992) Q fever endocarditis: diagnostic approaches and monitoring of therapeutic effects. *Clin. Investig.* 70:932–937.

Raoult, D. (1993) Treatment of Q fever. *Antimicrobial Agents Chemother.* 37:1733–1736.

Sawyer, L.A., Fishbein, D.B., and McDade, J.E. (1987) Q fever: current concepts. *Rev. Infect. Dis.* 9:935–946.

Srigley, J.R., Vellend, H., Palmer, N., *et al.* (1985) Q fever: The liver and bone marrow pathology. *Am. J. Surg. Pathol.* 9:752–758.

Stein, A., and Raoult, D. (1992) Detection of *Coxiella burnetii* by DNA amplification using polymerase chain reaction. *J. Clin. Microbiol.* 30:2462–2466.

Stein, A., Saunders, N.A., Taylor, A.G., and Raoult, D. (1993) Phylogenic homogenicity of *Coxiella* strains as determined by 16S ribosomal RNA sequencing. *FEMS Microbiol. Lett.* 113:244–339.

Thiele, D., Karo, M., and Krauss, H. (1992) Monoclonal antibody based capture ELISA/ELIFA for detection of *Coxiella burnetii* in clinical specimens. *Eur. J. Epidemiol.* 8:568–574.

Tissot-Dupont, H., Thirion, X., and Raoult, D. (1994) Q fever serology: cutoff determination for microimmunofluorescence. *Clin. Diagn. Lab. Immunol.* I:189–196.

Turck, W.P.G., Howitt, G., Turnberg, L.A., *et al.* (1976) Chronic Q fever. *Q. J. Med.* (new series) 45:193–217.

Uhaa, I.J., Fishbein, D.B., Olson, J.G., *et al.* (1994) Evaluation of specificity of indirect enzyme-linked immunosorbent assay for diagnosis of human Q fever. *J. Clin. Microbiol.* 32:1560–1565.

Vodkin, M.H., Williams, J.C., and Stephenson, E.H. (1986) Genetic heterogeneity among isolates of *Coxiella burnetii*. *J. Gen. Microbiol.* 132:455.

Williams, J.C., Peacock, M.G., Waag, D.M., *et al.* (1993) Vaccines against coxiellosis and Q fever. Development of a chloroform–methanol residue subunit of phase I *Coxiella burnetii* for the immunization of animals. *Ann. N Y Acad. Sci.* 653:88–111.

Williams, J.C., Thomas, L.A., and Peacock, M.G. (1986) Humoral immune response to Q fever: enzyme-linked immunosorbent assay antibody response to *Coxiella burnetii* in experimentally infected guinea pigs. *J. Clin. Microbiol.* 24:935–939.

Zhang, G.Q., Hotta, A., Mizutani, M., *et al.* (1998) Direct identification of *Coxiella burnetii* plasmids in human sera by nested PCR. *J. Clin. Microbiol.* 36:2210–2213.

Zuerner, R.I., and Thompson, A. (1983) Protein synthesis by intact *Coxiella burnetii* cells. *J. Bacteriol.* 156:186–191.

Section Four

Spiral Bacteria

39

Leptospira spp.

P. N. Levett

Saskatchewan Health, Provincial Laboratory, Regina, Saskatchewan, Canada

INTRODUCTION

Leptospirosis is a zoonosis caused by infection with pathogenic strains of the genus *Leptospira*. The most severe manifestation of leptospirosis is the syndrome of multiorgan infections known as Weil's disease, first described by Adolf Weil in Heidelberg in 1886. Weil described an infectious disease with jaundice and nephritis. Earlier descriptions of diseases that were probably leptospirosis were reviewed recently (Faine *et al.*, 1999). The aetiology of leptospirosis was demonstrated independently in 1915 by Inada and Ido, who detected in the blood of Japanese miners with infectious jaundice both spirochaetes and specific antibodies, and by Uhlenhuth and Fromme, who detected spirochaetes in the blood of German soldiers afflicted by 'French disease' while in the trenches (Everard, 1996). The importance of occupation as a risk factor was recognized early. The role of the rat as a source of human infection was discovered in 1917, while the potential for leptospiral disease in dogs and in livestock was not recognized until the 1930s and 1940s.

CLASSIFICATION

Definition

Leptospires are tightly coiled spirochaetes, usually $0.1\,\mu m \times 6–20\,\mu m$, but occasional cultures may contain much longer cells. The cells have pointed ends, either or both of which are usually bent into a distinctive hook (Figure 39.1). Morphologically all leptospires are indistinguishable. They have Gram-negative cell walls and may be stained using carbol fuchsin counterstain (Faine *et al.*, 1999).

Leptospires are obligate aerobes with an optimum growth temperature of 28–30 °C. They grow in simple media enriched with vitamins (vitamins B_2 and B_{12} are growth factors), long-chain fatty acids and ammonium salts (Johnson and Faine, 1984). Growth of leptospires is often slow on first isolation and cultures are retained for up to 13 weeks before being discarded, but pure subcultures in liquid media usually grow within 10–14 days.

The complete genome sequences of several strains of *Leptospira* have been determined and the data are leading to important discoveries regarding gene transfer between *Leptospira* serovars (Haake *et al.*, 2004; Nascimento *et al.*, 2004).

Species

Traditionally the genus was divided into two species, *Leptospira interrogans*, comprising all pathogenic strains, and *L. biflexa*, containing the saprophytic strains isolated from the environment (Johnson and Faine, 1984). The phenotypic classification of leptospires has been replaced by a genotypic one, in which a total of 16 species of *Leptospira* are defined by DNA hybridization studies (Yasuda *et al.*, 1987; Ramadass *et al.*, 1992; Pérolat *et al.*, 1998; Brenner *et al.*, 1999). Pathogenic and non-pathogenic serovars occur within the same species (Table 39.1); in addition there are several examples of serovars which exhibit sufficient genetic diversity to be classified into different species. There is evidence of significant horizontal transfer between species (Haake *et al.*, 2004). The phenotypic characteristics previously used to differentiate *L. interrogans* sensu lato and *L. biflexa* sensu lato are no longer useful. At present neither serogroup nor serovar reliably predicts the species of *Leptospira*. DNA hybridization is available in relatively few research laboratories. Leptospiral species can be identified by 16S rRNA gene sequencing, but the serological classification of pathogenic leptospires remains the principal method of identifying isolates as pathogens.

Serovars

Leptospires are divided into numerous serovars by agglutination and cross-agglutinin adsorption (Johnson and Faine, 1984; Kmety and Dikken, 1993). Antigenically related serovars are grouped into serogroups. More than 60 serovars of *L. biflexa* sensu lato are recognized (Johnson and Faine, 1984), and within *L. interrogans* sensu lato over 230 serovars are organized into 23 serogroups (Kmety and

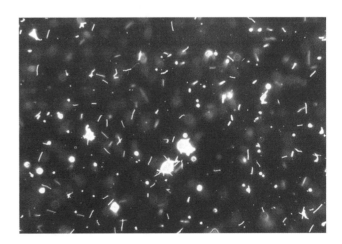

Figure 39.1 Leptospires viewed by darkfield microscopy (Public Health Image Library).

Table 39.1   Named species of *Leptospira* and distribution of serogroups

Species	Includes some serovars of serogroup
L. alexanderi	Manhao Hebdomadis[a] Javanica[a] Mini[a]
L. biflexa[b]	Semaranga[a, c]
L. borgpetersenii	Javanica Ballum Hebdomadis[a] Sejroe[a] Tarassovi[a] Mini
L. fainei	Hurstbridge
L. inadai[b]	Lyme
L. interrogans	Icterohaemorrahgiae[a] Canicola Pomona[a] Australis[a] Autumnalis[a] Pyrogenes[a] Grippotyphosa[a] Djasiman Hebdomadis[a] Sejroe[a] Bataviae
L. kirschneri	Grippotyphosa[a] Autumnalis[a] Cynopteri Hebdomadis[a] Australis[a] Pomona[a]
L. meyeri[b]	Ranarum Semaranga[a, c]
L. noguchii	Panama Autumnalis[a] Pyrogenes[a] Louisiana
L. santarosai	Shermanii Hebdomadis[a] Tarassovi[a]
L. weilii	Celledoni Icterohaemorrhagiae[a] Sarmin
L. wolbachii[b]	Codice[c]

[a] Serovars of these serogroups are found within two or more genospecies.
[b] *L. biflexa*, *L. inadai* and *L. wolbachii* are species which consist currently of non-pathogenic strains only. *L. meyeri* consists of both pathogenic strains and non-pathogenic strains.
[c] These serogroups comprise non-pathogenic leptospires.

Dikken, 1993). Further serovars have been isolated but have yet to be validly published.

EPIDEMIOLOGY

Geographical Distribution

Leptospirosis is considered to be one of the most widespread zoonoses (World Health Organization, 1999). Human infection is acquired by direct or indirect contact with infected animal tissues or urine. The incidence of leptospirosis is much higher in warm-climate countries than in temperate regions. This is due largely to longer survival of leptospires in damp soil with high humidity at ambient temperatures of 25–30 °C, and at a slightly alkaline pH. However, in most warm-climate countries opportunities for exposure to infected animals, whether domesticated or wild, are greater and other avocational activities further magnify the risk of exposure. In tropical regions there are more potential reservoir animals and almost invariably there are more serovars isolated from animals and humans. Conversely, in temperate regions in developed countries there are relatively few reservoir animals and thus less diversity among serovars.

Routes of Transmission

Leptospires usually enter the body through the mucous membranes of the upper respiratory tract or the conjunctivae, or through abraded skin, following exposure to contaminated water, infected urine or animal tissues. Water-borne transmission has been documented; numerous point-source outbreaks have been reported (Levett, 2001). Ingestion of water or immersion in water is commonly identified as a risk factor, suggesting that the oral mucosae and the conjunctivae are readily penetrated by leptospires.

Life Cycle

Animals, including man, can be divided into maintenance hosts or accidental hosts. A maintenance host is a species in which infection is endemic, usually transferred from animal to animal by direct contact. Often such infections occur while young animals are still in the nest. Other animals, such as humans, may become infected by indirect contact with the maintenance host. The most important maintenance hosts are rodents and other small mammals, which serve as reservoirs of infection for domestic animals, dogs and humans. The extent to which infection is transmitted depends upon many factors, which include climate, population densities and the degree of contact between maintenance and accidental hosts. Some serovars are host adapted; for example, the brown rat (*Rattus norvegicus*) is the ubiquitous carrier of serovar Icterohaemorrhagiae. Domestic animals may also be maintenance hosts; dairy cattle may harbour serovar Hardjo, while pigs may carry serovars Pomona, Tarassovi or Bratislava. Distinct variations in maintenance hosts and their associated serovars occur throughout the world and it is necessary to understand the local epidemiology in order to control the disease in humans.

Occupational and Recreational Exposure

Occupation remains a significant risk factor for humans, although protective measures have reduced the incidence of occupational disease in many countries. Direct contact with infected animals accounts for most infections in farmers, veterinarians, abattoir workers and meat inspectors, while indirect contact is important for sewer workers, miners, soldiers, septic tank cleaners, fish farmers, rice-field workers and sugar-cane cutters. In developed countries there has been a change in epidemiology of leptospirosis in recent years, with a shift away from occupational exposures and towards recreational exposures. Increasingly the disease occurs in tourists who have taken part in adventure tourism activities in tropical regions, which have often involved exposure to fresh water (Haake *et al.*, 2002).

CLINICAL FEATURES

The spectrum of symptoms caused by leptospiral infection is extremely broad; the classical syndrome of Weil's disease represents only the most severe presentation. Formerly it was considered that distinct clinical syndromes were associated with specific serovars. However, this association has been disproven. In humans, severe leptospirosis is frequently, but not invariably, caused by serovars of the Icterohaemorrhagiae serogroup. The specific serovars involved depend largely on the geographical location and the ecology of local maintenance hosts.

The clinical presentation of leptospirosis is biphasic (Figure 39.2), the acute, or septicaemic, phase lasting about a week, and being followed by the immune phase, characterized by antibody production and excretion of leptospires in the urine (Turner, 1967). Most of the complications of leptospirosis are associated with localization of leptospires within the tissues during the immune phase, and thus occur during the second week of the illness. In severe disease the two phases may not be apparent.

The great majority of infections are either subclinical or of very mild severity and will probably not be brought to medical attention. A smaller proportion of infections, but the overwhelming majority of the recognized cases, will present with a febrile illness of sudden onset, the symptoms of which include chills, headache, myalgia, abdominal pain, conjunctival suffusion and less often a skin rash (Table 39.2). This anicteric syndrome usually lasts for about a week, and its resolution coincides with the

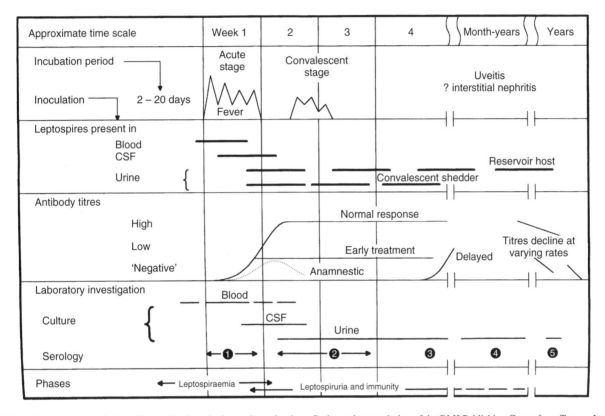

Figure 39.2 Biphasic nature of clinical leptospirosis and relevant investigations. Redrawn by permission of the BMJ Publishing Group from Turner, L. H., *BMJ* 1969; i: 231–235.

Table 39.2 Symptoms on admission in 88 patients with severe leptospirosis (Edwards *et al.*, 1990) and 150 anicteric patients (Berman *et al.*, 1973)

Symptom	Prevalence in leptospirosis (%)	
	Severe	Anicteric
Jaundice	95	1.5
Fever	85	97
Anorexia	85	—
Headache	76	98
Conjunctival suffusion	54	42
Vomiting	50	33
Myalgia	49	79
Abdominal pain	43	28
Nausea	37	41
Dehydration	37	—
Cough	32	20
Hepatomegaly	27	15
Lymphadenopathy	21	21
Diarrhoea	14	29
Rash	2	7

appearance of antibodies. The fever may be biphasic and may recur after a remission of 3 to 4 days. The headache is often intense, with retroorbital pain and photophobia. Aseptic meningitis may be found in ≤25% of all leptospirosis cases and may account for a significant minority of all causes of aseptic meningitis. Mortality is almost nil in anicteric leptospirosis.

The differential diagnosis includes common viral infections, such as influenza, and in the tropics, dengue, in addition to the bacterial causes of PUO, such as typhoid and brucellosis. A comprehensive list of other conditions that may be mimicked by leptospirosis would include malaria, viral hepatitis, rickettsiosis, mononucleosis and HIV seroconversion (Turner, 1967; Levett, 2001).

Icteric leptospirosis is a much more severe disease, in which the clinical course is often very rapidly progressive. Severe cases often present late in the course of the disease, and this contributes to the high mortality rate, which ranges between 5 and 15%. Between 5 and 10% of all patients with leptospirosis have the icteric form of the disease (Heath, Alexander and Galton, 1965). The jaundice occurring in leptospirosis is not associated with hepatocellular necrosis and liver function returns to normal after recovery. Serum bilirubin levels may be high and may take several weeks to fall to within normal limits. There are moderate rises in transaminase levels and minor elevation of the alkaline phosphatase level usually occurs.

The complications of severe leptospirosis emphasize the multi-systemic nature of the disease. Leptospirosis is a common cause of acute renal failure (ARF), which occurs in 16–40% of cases (Ramachandran *et al.*, 1976; Winearls *et al.*, 1984; Edwards *et al.*, 1990). Thrombocytopenia (platelet count $<100 \times 10^9$/l) occurs in ≥50% of cases and is a significant predictor for the development of ARF (Edwards, Nicholson and Everard, 1982). However, thrombocytopenia in leptospirosis is transient and does not result from disseminated intravascular coagulation (Edwards *et al.*, 1986). Adult respiratory distress syndrome (Ramachandran and Perera, 1977) is another common complication. Pulmonary haemorrhage has become more commonly recognized in recent years (Im *et al.*, 1986; Zaki, Shieh, and the Epidemic Working Group, 1996; Yersin *et al.*, 2000; Carvalho and Bethlem, 2002). Serum amylase levels are often raised significantly, but pancreatitis is not a common finding (Edwards and Everard, 1991).

Fatality in severe leptospirosis often follows cardiac arrest, resulting from the development of interstitial myocarditis, pericarditis and cardiac arrhythmias (Ramachandran, 1975; Lee *et al.*, 1986). These complications occur in ≤50% of patients (Areán, 1957; de Brito *et al.*, 1987) and are more common in older patients.

Anterior uveitis, usually bilateral, occurs after recovery from the acute illness in a minority of cases (Rathinam *et al.*, 1997). Uveitis may present weeks, months or occasionally years after the acute stage. The incidence of ocular complications is variable, but this probably reflects the long time-scale over which they may occur (Mancel *et al.*, 1999). In the United States the incidence was estimated at 3% (Heath, Alexander and Galton, 1965). Rare complications include cerebrovascular accidents, rhabdomyolysis and Guillain–Barré syndrome (Levett, 2001). Leptospirosis during pregnancy has been shown to cause fetal infection and death, but only a handful of cases have been reported (Levett, 2001).

LABORATORY DIAGNOSIS

Leptospiral diagnosis depends heavily upon serological detection of specific antibodies, and effective serological assays have been developed in recent years. A definitive identification of the infecting serovar is possible only if an isolate is obtained from blood, urine, CSF or other specimen during the acute illness. The selection of appropriate specimens is dependent on the stage of the disease, and therefore an adequate history is essential, with particular attention to the duration of symptoms.

Microscopy

Blood, urine, CSF and other specimens may be examined directly by darkground microscopy. However, both the sensitivity and specificity of microscopy may be unacceptably low (Vijayachari *et al.*, 2001), since many artefacts may be mistaken for leptospires.

Histopathology

Spirochaetes can be demonstrated by silver staining, using stains such as the modified Wartin–Starry. This approach is not specific for leptospires. A more recent approach is the use of immunohistochemical staining using pooled antisera against a combination of serovars (Zaki, Shieh, and the Epidemic Working Group, 1996). Using this approach the morphology of the spirochaetes in tissue sections can be seen readily (Plate 12).

Culture

Leptospires grow in simple media, to which either serum or albumin and Tween have been added (Faine *et al.*, 1999). Numerous media containing serum have been described, including those of Fletcher, Korthof, Stuart and Ellinghausen (Turner, 1970). Pooled rabbit serum, rich in vitamin B_{12}, is usually employed. The most widely used medium for isolation of human leptospires is EMJH (Ellinghausen and McCullough, 1965; Johnson and Harris, 1967). This is a serum-free medium, containing Tween 80 and bovine serum albumin fraction V. Growth of contaminants can be inhibited by the addition of 5-fluorouracil (Johnson and Rogers, 1964).

During the first (leptospiraemic) phase of the disease, cultures of blood in EMJH are made at the patient's bedside, by injecting one drop of blood into the bottle. Urine is an appropriate specimen for culture from the start of the second week of illness (Figure 39.2). Survival of leptospires in human urine is limited, and hence urine samples should be processed as soon as possible by neutralization and centrifugation at 1500 *g* for 30 min, after which the pellet is resuspended in phosphate-buffered saline and inoculated into semisolid medium containing 5-fluorouracil. Specimens that may be contaminated, such as urine, or contaminated cultures, may be filtered using a 0.22 μm filter before inoculation. Cultures are incubated at 28–30 °C. Low-power (×100–200) darkground microscopy is used to examine cultures weekly for up to 13 weeks before cultures are discarded. After obtaining pure subcultures, isolates are identified using a combination of molecular and serologic methods. Since some serovars are distributed within several species, identification of the species by 16S rRNA gene sequencing is valuable. Serovar identification can be performed by cross-agglutinin absorption testing, available in only a few reference laboratories (Dikken and Kmety, 1978), by agglutination with a panel of monoclonal antibodies (Terpstra, 1992), or by molecular typing. Although identification of isolates is not helpful for patient management, it is essential for understanding of the epidemiology in a region and the design of appropriate public health interventions.

Molecular Typing

Pulsed-field gel electrophoresis (PFGE) has become the standard molecular method for identification of serovars (Herrmann *et al.*, 1992). Earlier approaches using restriction endonuclease digestion of genomic DNA led to extremely complex patterns, which were difficult to interpret. Numerous PCR-based methods have been applied to epidemiological characterization of isolates from different sources. Of particular significance is the potential for some methods (such as PCR-REA and LSSP-PCR) which retain the use of specific primers, to be applied directly to DNA derived from clinical material, without the necessity of first isolating leptospires in culture.

Serology

Serology provides the means of diagnosis in most cases of leptospirosis. However, antibodies do not appear in the blood until late in the first week of the disease (Figure 39.2) and the diagnosis is thus often retrospective.

Microscopic Agglutination Test

The reference method for serological diagnosis of leptospirosis is the microscopic agglutination test (MAT), performed on paired sera, usually taken a minimum of 3 days apart. Patients' sera are mixed with live antigen suspensions of leptospiral serovars. The range of antigens used should include all serogroups and all locally common serovars. After incubation the serum/antigen mixtures are examined by darkfield microscopy for agglutination and the titres are determined.

Interpretation of the MAT is complicated by the high degree of cross-reaction that occurs between different serogroups, especially in acute-phase samples. This is to some extent predictable, thus patients often have similar titres to all serovars of an individual serogroup, but 'paradoxical' reactions, in which the highest titres are detected to a serogroup unrelated to the infecting one, are also common (Figure 39.3).

Acute infection is suggested by a single titre of ≥800 and confirmed by a fourfold or greater rise in titre between paired sera. Titres following acute infection may be extremely high (≥25 600) and may take months, or even years, to fall to low levels (Cumberland *et al.*, 2001). Often, it becomes possible to distinguish a predominant serogroup only months after infection, as cross-reacting titres decline at different rates. It is important, therefore, to examine several sera taken at monthly intervals after the acute disease, in order to determine the presumptive infecting serogroup (Levett, 2003). Some patients are found to have serological evidence of previous infection with a different leptospiral serovar. In these cases, serological diagnosis is complicated further by the 'anamnestic response', in which the first rise in antibody titre is usually directed against the infecting serovar from the previous exposure.

The MAT is a complex test to control, perform and interpret. An international proficiency testing scheme has been developed and has produced a positive effect on MAT testing in many countries (Chappel *et al.*, 2004).

Other Serological Tests

Because of the complexity of the MAT, rapid screening tests for leptospiral antibodies have been developed as an aid to presumptive diagnosis. Such tests are genus specific and the methods used have included complement fixation, slide agglutination, counter immunoelectrophoresis, indirect haemagglutination, microcapsule agglutination, latex agglutination, lateral flow and ELISA in a variety of formats (Levett, 2001). Many methods have been described but relatively few have found widespread use in diagnostic laboratories.

Several methods for the use of ELISA have been described (Adler *et al.*, 1980; Terpstra, Ligthart and Schoone, 1985), using different serovars or combinations of serovars as antigens. The IgM antibodies produced in leptospirosis are broadly cross-reactive, but titres vary slightly depending upon the antigen used in the assay and the infecting serovar. In acute infection, IgM antibodies reach very high titres (≥1280 is quite common). Lower titres may be non-specific and a second specimen is required to detect a rise in titre. A high IgM titre also indicates the necessity for a convalescent specimen to be taken 10–14 days after the acute sample. In either case a fourfold or greater rise in titre of IgM is considered diagnostic. Several IgM-ELISA assays are available commercially. IgM-dipstick assays and a latex agglutination assay have been evaluated (Smits *et al.*, 1999, 2001; Levett *et al.*, 2001) and have the advantage of requiring only very basic laboratory equipment.

Molecular Methods of Diagnosis

Several primer sets have been described for use in PCR, but few have been evaluated using either human (Merien, Baranton and Pérolat, 1995; Brown *et al.*, 1995) or veterinary (Wagenaar *et al.*, 2000; Harkin, Roshto and Sullivan, 2003) clinical samples. PCR has been found useful in confirming a diagnosis of leptospirosis, particularly in severe disease, when patients may die before seroconversion occurs (Brown *et al.*, 2003). Blood, urine, CSF, aqueous humor, dialysate fluid and postmortem tissues have all been used as sources of leptospiral DNA (Merien *et al.*, 1993; Brown *et al.*, 1995). Real-time quantitative assays (Smythe *et al.*, 2002; Levett *et al.*, 2005; Merien *et al.*, 2005) appear to offer an increase in sensitivity over earlier methods.

MANAGEMENT

Treatment of leptospirosis is dependent upon the severity and duration of symptoms at the time of presentation. Patients with mild, flu-like symptoms require only symptomatic treatment, but should be cautioned to seek further medical help if they become jaundiced. Patients who present with more severe anicteric leptospirosis will require hospital admission and close observation. If the headache is particularly severe a lumbar puncture usually produces a dramatic improvement.

The management of icteric leptospirosis requires admission of the patient to the ICU initially. Patients with pre-renal azotaemia can be rehydrated for the first 2–3 days while their renal function is observed, but patients in ARF require dialysis as a matter of urgency (Nicholson *et al.*, 1989). This may be accomplished by peritoneal dialysis or by haemodialysis. Cardiac monitoring is also necessary during the first few days after admission.

The role of antibiotics in the management of leptospirosis is still contentious, despite numerous reports of their use. A major difficulty in assessing the efficacy of antibiotic treatment results from the late presentation of many patients with severe disease, after the leptospires have localized in the tissues. Few well-designed and controlled studies have been reported (Guidugli, Castro and Atallah, 2003).

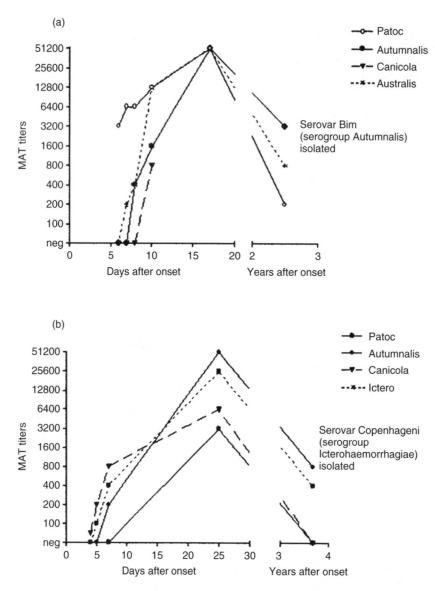

Figure 39.3 Paradoxical immune response to acute infection with serovar Bim, in which the presumptive serogroup (Autumnalis) was identified during follow-up (a) and with serovar Copenhageni, in which serogroup Icterohaemorrhagiae was never identified as the predominant serogroup (b). Reproduced by permission of ASM Press, from Levett, P. N., Clinical Microbiology Reviews, 2001; 14: 296–326.

Doxycycline (100 mg bds for 7 days) was shown to reduce the duration and severity of illness in anicteric leptospirosis by an average of 2 days (McClain *et al.*, 1984). Patients with severe disease were excluded from this study.

Three randomized trials of penicillin produced conflicting results. One study included 42 patients with severe leptospirosis, of whom 19 were jaundiced (Watt *et al.*, 1988). The mean duration of illness before admission was 9 days. No patients required dialysis and there were no deaths. IV penicillin was given at a dose of 6 MU/day for 7 days and was found to halve the duration of fever. A second study included 79 patients with icteric leptospirosis, of whom four died (Edwards *et al.*, 1988). Patients in the treatment group received IV penicillin at a dose of 8 MU/day for 5 days. The mean duration of illness prior to admission was 6 days. No difference was observed between treatment and control groups in outcome or duration of the illness. In the largest study, of 253 patients, the mean duration of

illness prior to admission was 6 days and 94% of the subjects were icteric (Costa *et al.*, 2003). The dose of penicillin was 6 MU/day for 7 days and there was no significant difference in mortality between treated and untreated groups in this study.

Recently, prospective open-label trials comparing penicillin with ceftriaxone (Panaphut *et al.*, 2003) or penicillin with cefotaxime or doxycycline (Suputtamongkol *et al.*, 2004) have shown these four antibiotics to be equally effective. Less than half of the subjects in these studies were jaundiced, and most patients had received some antibiotics before hospital admission.

A common finding in some of these studies has been the prevention of leptospiruria or a significant reduction in its duration (McClain *et al.*, 1984; Edwards *et al.*, 1988; Watt *et al.*, 1988). This finding alone is sufficient justification for antibiotic use, but any antibiotic treatment should be started as early as possible and should be regarded as an adjunct to other therapeutic approaches.

CONTROL

Control and prevention of leptospirosis in humans is inextricably linked with the control of animal reservoirs. The most efficient approaches to control in a given geographical location depend upon the local epidemiology. The epidemiology of leptospirosis may be altered markedly by changing agricultural and cultural practices. As an example, recent large outbreaks of leptospirosis have occurred in recreational athletes from several countries after competitive events in the USA and in Borneo (Morgan *et al.*, 2002; Sejvar *et al.*, 2003), associated with prolonged exposure to fresh water. The necessity for continued surveillance of human and veterinary leptospirosis should be emphasized.

The use of protective clothing has markedly reduced the incidence of leptospirosis associated with many occupations. Chemoprophylaxis with doxycycline (200 mg once weekly) is effective if the risk exposure is predicted beforehand (Takafuji *et al.*, 1984).

Immunization of animals may be practised in regions where livestock animals are important reservoirs, both for economic reasons and to prevent infection of humans. Current vaccines are killed suspensions of whole cells grown in protein-free medium. Usually two doses are given, followed by annual booster injections. Cattle are often immunized against serovars Hardjo and Pomona, while pigs may be immunized against Pomona, Tarassovi, Bratislava, Grippotyphosa and Icterohaemorrhagiae, depending upon the geographical region. Canine vaccines usually include Icterohaemorrhagiae and Canicola, but in North America serovars Grippotyphosa and Pomona are included. Vaccines are used in human populations at high risk of acquiring leptospirosis. A trivalent vaccine is used in Cuba, containing serovars Icterohaemorrhagiae, Pomona and Canicola (Martínez Sánchez *et al.*, 1998). Widespread use of vaccines has been restricted to China, Japan and some other SE Asian countries and has been effective in controlling epidemics, particularly in rice-field workers.

Immunity to leptospirosis was long thought to result solely from the humoral response to lipopolysaccharide (Adler and Faine, 1977). Vaccination with killed bacterins induces low titres of antibody and requires the administration of booster doses at frequent intervals. Immunity generated by such vaccines is relatively serovar specific. In animals, immunization protects against disease but rarely prevents infection and renal colonization and thus does not effectively prevent immunized animals from becoming carriers. Immunized dogs have served as reservoirs for human infection (Feigin *et al.*, 1973; Everard *et al.*, 1987).

More recent evidence suggests that cell-mediated immunity is also important, both in humans and animals (Naiman *et al.*, 2001; Klimpel, Matthias and Vinetz, 2003). A monovalent bovine Hardjo vaccine induced a T-helper 1 immune response (Naiman *et al.*, 2002) and protected against both renal colonization and urinary shedding (Bolin and Alt, 2001).

REFERENCES

Adler, B. and Faine, S. (1977) Host immunological mechanisms in the resistance of mice to leptospiral infections. *Infection and Immunity* **17**: 67–72.

Adler, B., Murphy, A. M., Locarnini, S. A. and Faine, S. (1980) Detection of specific anti-leptospiral immunoglobulins M and G in human serum by solid-phase enzyme-linked immunosorbent assay. *Journal of Clinical Microbiology* **11**: 452–457.

Areán, V. M. (1957) Leptospiral myocarditis. *Laboratory Investigation* **6**: 462–471.

Berman, S. J., Tsai, C. C., Holmes, K. K., *et al.* (1973) Sporadic anicteric leptospirosis in South Vietnam. A study in 150 patients. *Annals of Internal Medicine* **79**: 167–173.

Bolin, C. A. and Alt, D. P. (2001) Use of a monovalent leptospiral vaccine to prevent renal colonization and urinary shedding in cattle exposed to Leptospira borgpetersenii serovar hardjo. *American Journal of Veterinary Research* **62**: 995–1000.

Brenner, D. J., Kaufmann, A. F., Sulzer, K. R. *et al.* (1999) Further determination of DNA relatedness between serogroups and serovars in the family Leptospiraceae with a proposal for *Leptospira alexanderi* sp. nov. and four new *Leptospira* genomospecies. *International Journal of Systematic Bacteriology* **49**: 839–858.

Brown, P. D., Carrington, D. G., Gravekamp, C., *et al.* (2003) Direct detection of leptospiral material in human postmortem samples. *Research in Microbiology* **154**: 581–586.

Brown, P. D., Gravekamp, C., Carrington, D. G. *et al.* (1995) Evaluation of the polymerase chain reaction for early diagnosis of leptospirosis. *Journal of Medical Microbiology* **43**: 110–114.

Carvalho, C. R. and Bethlem, E. P. (2002) Pulmonary complications of leptospirosis. *Clinics in Chest Medicine* **23**: 469–478.

Chappel, R. J., Goris, M. G., Palmer, M. F. and Hartskeerl, R. A. (2004) Impact of proficiency testing on results of the microscopic agglutination test for diagnosis of leptospirosis. *Journal of Clinical Microbiology* **42**: 5484–5488.

Costa, E., Lopes, A. A., Sacramento, E. *et al.* (2003) Penicillin at the late stage of leptospirosis: a randomized controlled trial. *Revista do Instituto de Medicina Tropical de São Paulo* **45**: 141–145.

Cumberland, P. C., Everard, C. O. R., Wheeler, J. G. and Levett, P. N. (2001) Persistence of anti-leptospiral IgM, IgG and agglutinating antibodies in patients presenting with acute febrile illness in Barbados 1979–1989. *European Journal of Epidemiology* **17**: 601–608.

de Brito, T., Morais, C. F., Yasuda, P. H. *et al.* (1987) Cardiovascular involvement in human and experimental leptospirosis: pathologic findings and immunohistochemical detection of leptospiral antigen. *Annals of Tropical Medicine and Parasitology* **81**: 207–214.

Dikken, H. and Kmety, E. (1978) Serological typing methods of leptospires. In *Methods in Microbiology*. T. Bergan and J. R. Norris. Academic Press, London. **11**: 259–307.

Edwards, C. N. and Everard, C. O. R. (1991) Hyperamylasemia and pancreatitis in leptospirosis. *The American Journal of Gastroenterology* **86**: 1665–1668.

Edwards, C. N., Nicholson, G. D. and Everard, C. O. R. (1982) Thrombocytopenia in leptospirosis. *American Journal of Tropical Medicine and Hygiene* **31**: 827–829.

Edwards, C. N., Nicholson, G. D., Hassell, T. A. *et al.* (1986) Thrombocytopenia in leptospirosis: the absence of evidence for disseminated intravascular coagulation. *American Journal of Tropical Medicine and Hygiene* **35**: 352–354.

Edwards, C. N., Nicholson, G. D., Hassell, T. A. *et al.* (1988) Penicillin therapy in icteric leptospirosis. *American Journal of Tropical Medicine and Hygiene* **39**: 388–390.

Edwards, C. N., Nicholson, G. D., Hassell, T. A. *et al.* (1990) Leptospirosis in Barbados: a clinical study. *The West Indian Medical Journal* **39**: 27–34.

Ellinghausen, H. C. and McCullough, W. G. (1965) Nutrition of *Leptospira pomona* and growth of 13 other serotypes: fractionation of oleic albumin complex and a medium of bovine albumin and polysorbate 80. *American Journal of Veterinary Research* **26**: 45–51.

Everard, J. D. (1996) Leptospirosis. In *The Wellcome Trust Illustrated History of Tropical Diseases*. F. E. G. Cox. The Wellcome Trust, London: 111–119, 416–418.

Everard, C. O. R., Jones, C. J., Inniss, V. A., *et al.* (1987) Leptospirosis in dogs on Barbados. *Israel Journal of Veterinary Medicine* **43**: 288–295.

Faine, S., Adler, B., Bolin, C. and Perolat, P. (1999) *Leptospira and leptospirosis*. Melbourne, MedSci.

Feigin, R. D., Lobes, L. A., Anderson, D. and Pickering, L. (1973) Human leptospirosis from immunized dogs. *Annals of Internal Medicine* **79**: 777–785.

Guidugli, F., Castro, A. A. and Atallah, A. N. (2003) Antibiotics for treating leptospirosis (Cochrane Review). In *Cochrane Library, Issue 3*. Update Software, Oxford.

Haake, D. A., Dundoo, M., Cader, R. *et al.* (2002) Leptospirosis, water sports, and chemoprophylaxis. *Clinical Infectious Diseases* **34**: e40–43.

Haake, D. A., Suchard, M. A., Kelley, M. M. *et al.* (2004) Molecular evolution and mosaicism of leptospiral outer membrane proteins involves horizontal DNA transfer. *Journal of Bacteriology* **186**: 2818–2828.

Harkin, K. R., Roshto, Y. M. and Sullivan, J. T. (2003) Clinical application of a polymerase chain reaction assay for diagnosis of leptospirosis in dogs. *Journal of the American Veterinary Medical Association* **222**: 1224–1229.

Heath, C. W., Alexander, A. D. and Galton, M. M. (1965) Leptospirosis in the United States: 1949–1961. *New England Journal of Medicine* 273: 857–864, 915–922.

Herrmann, J. L., Bellenger, E., Perolat, P. *et al.* (1992) Pulsed-field gel electrophoresis of *Not*I digests of leptospiral DNA: a new rapid method of serovar identification. *Journal of Clinical Microbiology* **30**: 1696–1702.

Im, J.-G., Yeon, K. M., Han, M. C. *et al.* (1986) Pulmonary manifestations of leptospirosis. *Journal of Korean Radiological Society* **22**: 49–56.

Johnson, R. C. and Faine, S. (1984) *Leptospira*. In *Bergey's Manual of Systematic Bacteriology*. N. R. Krieg and J. G. Holt. Williams & Wilkins, Baltimore. 1: 62–67.

Johnson, R. C. and Harris, V. G. (1967) Differentiation of pathogenic and saprophytic leptospires. 1. Growth at low temperatures. *Journal of Bacteriology* **94**: 27–31.

Johnson, R. C. and Rogers, P. (1964) 5-Fluorouracil as a selective agent for growth of leptospirae. *Journal of Bacteriology* **87**: 422–426.

Klimpel, G. R., Matthias, M. A. and Vinetz, J. M. (2003) *Leptospira interrogans* activation of human peripheral blood mononuclear cells: preferential expansion of TCR gammadelta(+) T cells vs TCR alphabeta(+) T cells. *Journal of Immunology* **171**: 1447–1455.

Kmety, E. and Dikken, H. (1993) *Classification of the Species Leptospira Interrogans and History of its Serovars*. Groningen, University Press Groningen.

Lee, M. G., Char, G., Dianzumba, S. and Prussia, P. (1986) Cardiac involvement in severe leptospirosis. *West Indian Medical Journal* **35**: 295–300.

Levett, P. N. (2001) Leptospirosis. *Clinical Microbiology Reviews* **14**: 296–326.

Levett, P. N. (2003) Usefulness of serologic analysis as a predictor of the infecting serovar in patients with severe leptospirosis. *Clinical Infectious Diseases* **36**: 447–452.

Levett, P. N., Branch, S. L., Whittington, C. U. *et al.* (2001) Two methods for rapid serological diagnosis of acute leptospirosis. *Clinical and Diagnostic Laboratory Immunology* **8**: 349–351.

Levett, P. N., Morey, R. E., Galloway, R. L. *et al.* (2005) Detection of pathogenic leptospires by real-time quantitative PCR. *Journal of Medical Microbiology* **54**: 45–49.

Mancel, E., Merien, F., Pesenti, L. *et al.* (1999) Clinical aspects of ocular leptospirosis in New Caledonia (South Pacific). *Australian and New Zealand Journal of Ophthalmology* **27**: 380–386.

Martínez Sánchez, R., Obregon Fuentes, A. M., Perez Sierra, A. *et al.* (1998) The reactogenicity and immunogenicity of the first Cuban vaccine against human leptospirosis. *Revista Cubana de Medicina Tropical* **50**: 159–166.

McClain, J. B. L., Ballou, W. R., Harrison, S. M. and Steinweg, D. L. (1984) Doxycycline therapy for leptospirosis. *Annals of Internal Medicine* **100**: 696–698.

Merien, F., Baranton, G. and Pérolat, P. (1995) Comparison of polymerase chain reaction with microagglutination test and culture for diagnosis of leptospirosis. *Journal of Infectious Diseases* **172**: 281–285.

Merien, F., Pérolat, P., Mancel, E. *et al.* (1993) Detection of leptospiral DNA by polymerase chain reaction in aqueous humor of a patient with unilateral uveitis. *Journal of Infectious Diseases* **168**: 1335–1336.

Merien, F., Portoni, D., Bourch, P. (2005) A rapid and quantitative method for the detection of Leptospira species in human leptospirosis. *FEMS Microbiology Letters* **249**: 139–147.

Morgan, J., Bornstein, S. L., Karpati, A. M. *et al.* (2002) Outbreak of Leptospirosis among Triathlon participants and community residents in Springfield, Illinois, 1998. *Clinical Infectious Diseases* **34**: 1593–1599.

Naiman, B. M., Alt, D., Bolin, C. A. *et al.* (2001) Protective killed *Leptospira borgpetersenii* vaccine induces potent Th1 immunity comprising responses by CD4 and gammadelta T lymphocytes. *Infection and Immunity* **69**: 7550–7558.

Naiman, B. M., Blumerman, S., Alt, D. *et al.* (2002) Evaluation of type 1 immune response in naive and vaccinated animals following challenge with *Leptospira borgpetersenii* serovar *hardjo*: involvement of WC1(+) gammadelta and CD4 T cells. *Infection and Immunity* **70**: 6147–6157.

Nascimento, A. L., Ko, A. I., Martins, E. A. *et al.* (2004) Comparative genomics of two *Leptospira interrogans* serovars reveals novel insights into physiology and pathogenesis. *Journal of Bacteriology* **186**: 2164–2172.

Nicholson, G. D., Edwards, C. N., Hassell, T. A. *et al.* (1989) Urinary diagnostic indices in the management of leptospirosis. *West Indian Medical Journal* **38**: 33–38.

Panaphut, T., Domrongkitchaiporn, S., Vibhagool, A. *et al.* (2003) Ceftriaxone compared with sodium penicillin G for treatment of severe leptospirosis. *Clinical Infectious Diseases* **36**: 1507–1513.

Pérolat, P., Chappel, R. J., Adler, B. *et al.* (1998) *Leptospira fainei* sp. nov., isolated from pigs in Australia. *International Journal of Systematic Bacteriology* **48**: 851–858.

Ramachandran, S. (1975) Electrocardiographic abnormalities in leptospirosis. *Journal of Tropical Medicine and Hygiene* **78**: 210–213.

Ramachandran, S. and Perera, M. V. F. (1977) Cardiac and pulmonary involvement in leptospirosis. *Transactions of the Royal Society of Tropical Medicine and Hygiene* **71**: 56–59.

Ramachandran, S., Rajapakse, C. N. A., Perera, M. V. F. and Yoganathan, M. (1976) Patterns of acute renal failure in leptospirosis. *Journal of Tropical Medicine and Hygiene* **79**: 158–160.

Ramadass, P., Jarvis, B. D. W., Corner, R. J. *et al.* (1992) Genetic characterization of pathogenic *Leptospira* species by DNA hybridization. *International Journal of Systematic Bacteriology* **42**: 215–219.

Rathinam, S. R., Rathnam, S., Selvaraj, S. *et al.* (1997) Uveitis associated with an epidemic outbreak of leptospirosis. *American Journal of Ophthalmology* **124**: 71–79.

Sejvar, J., Bancroft, E., Winthrop, K. *et al.* (2003) Leptospirosis in 'Eco-Challenge' athletes, Malaysian Borneo, 2000. *Emerging Infectious Diseases* **9**: 702–707.

Smits, H. L., Ananyina, Y. V., Chereshsky, A. *et al.* (1999) International multicenter evaluation of the clinical utility of a dipstick assay for detection of *Leptospira*-specific immunoglobulin M antibodies in human serum specimens. *Journal of Clinical Microbiology* **37**: 2904–2909.

Smits, H. L., Chee, H. D., Eapen, C. K. *et al.* (2001) Latex based, rapid and easy assay for human leptospirosis in a single test format. *Tropical Medicine and International Health* **6**: 114–118.

Smythe, L. D., Smith, I. L., Smith, G. A. *et al.* (2002) A quantitative PCR (TaqMan) assay for pathogenic *Leptospira* spp. *BMC Infectious Diseases* **2**: 13.

Suputtamongkol, Y., Niwattayakul, K., Suttinont, C. *et al.* (2004) An open, randomized, controlled trial of penicillin, doxycycline, and cefotaxime for patients with severe leptospirosis. *Clinical Infectious Diseases* **39**: 1417–1424.

Takafuji, E. T., Kirkpatrick, J. W., Miller, R. N. *et al.* (1984) An efficacy trial of doxycycline chemoprophylaxis against leptospirosis. *The New England Journal of Medicine* **310**: 497–500.

Terpstra, W. J. (1992) Typing leptospira from the perspective of a reference laboratory. *Acta Leidensia* **60**: 79–87.

Terpstra, W. J., Ligthart, G. S. and Schoone, G. J. (1985) ELISA for the detection of specific IgM and IgG in human leptospirosis. *Journal of General Microbiology* **131**: 377–385.

Turner, L. H. (1967) Leptospirosis I. *Transactions of the Royal Society of Tropical Medicine and Hygiene* **61**: 842–855.

Turner, L. H. (1970) Leptospirosis III. Maintenance, isolation and demonstration of leptospires. *Transactions of the Royal Society of Tropical Medicine and Hygiene* **64**: 623–646.

Vijayachari, P., Sugunan, A. P., Umapathi, T. and Sehgal, S. C. (2001) Evaluation of darkground microscopy as a rapid diagnostic procedure in leptospirosis. *Indian Journal of Medical Research* **114**: 54–58.

Wagenaar, J., Zuerner, R., Alt, D. and Bolin, C. A. (2000) Comparison of polymerase chain reaction assays with bacteriologic culture, immunofluorescence, and nucleic acid hybridization for detection of *Leptospira borgpetersenii* serovar *hardjo* in urine of cattle. *American Journal of Veterinary Research* **61**: 316–320.

Watt, G., Padre, L. P., Tuazon, M. L. *et al.* (1988) Placebo-controlled trial of intravenous penicillin for severe and late leptospirosis. *Lancet* **i**: 433–435.

Winearls, C. G., Chan, L., Coghlan, J. D. *et al.* (1984) Acute renal failure due to leptospirosis: clinical features and outcome in six cases. *Quarterly Journal of Medicine* **53**: 487–495.

World Health Organization (1999) Leptospirosis worldwide, 1999. *Weekly Epidemiological Record* **74**: 237–242.

Yasuda, P. H., Steigerwalt, A. G., Sulzer, K. R. *et al.* (1987) Deoxyribonucleic acid relatedness between serogroups and serovars in the family *Leptospiraceae* with proposals for seven new *Leptospira* species. *International Journal of Systematic Bacteriology* **37**: 407–415.

Yersin, C., Bovet, P., Mérien, F. *et al.* (2000) Pulmonary haemorrhage as a predominant cause of death in leptospirosis in Seychelles. *Transactions of the Royal Society of Tropical Medicine and Hygiene* **94**: 71–76.

Zaki, S. R., Shieh, W.-J. and the Epidemic Working Group (1996) Leptospirosis associated with outbreak of acute febrile illness and pulmonary haemorrhage, Nicaragua, 1995. *Lancet* **347**: 535.

40

Helicobacter spp. and Related Organisms

Peter J. Jenks

Department of Microbiology, Plymouth Hospitals, Derriford Hospital, Plymouth, UK

INTRODUCTION

Helicobacter pylori is a Gram-negative, microaerobic, spiral bacterium that colonizes the stomachs of approximately half the world's population (Dunn, Cohen and Blaser, 1997). The bacterium is uniquely adapted for survival within the gastric mucosa, and colonization usually persists for years or even decades. Infection with *H. pylori* is associated with chronic gastritis and peptic ulceration, and the bacterium is also considered a risk factor for developing gastric adenocarcinoma and mucosa-associated lymphoid tissue (MALT) lymphoma (Blaser, 1990; Parsonnet *et al.*, 1991; Parsonnet *et al.*, 1994). The recognition of an etiologic link between *H. pylori* and severe gastroduodenal disease has revolutionized the management of these conditions and provided a potential strategy for the prevention of gastric cancer. Evidence is beginning to emerge of a link between other *Helicobacter* species and human disease conditions.

THE GENUS *HELICOBACTER*

The existence of gastric spiral bacteria was first noted over 100 years ago, well before the recognition of any association with gastric disease, and there are early reports of such organisms in the stomachs of dogs, cats and humans (Salomon, 1896; Lim, 1920). The simultaneous occurrence of bacteria and peptic ulceration was also described well before the isolation of *H. pylori* (Rosenow and Sandford, 1915), and in 1975 a relationship was reported between the migration of polymorphonuclear leukocytes through the gastric mucosa and the presence of spiral bacteria in the gastric epithelium (Steer, 1975). In 1983, Robin Warren and Barry Marshall reported that curved bacteria were often present in gastric biopsy specimens submitted for histological examination (Marshall, 1983; Warren, 1983), and this group went on to isolate and fulfil Koch's postulates for this '*pyloric Campylobacter*' (Marshall *et al.*, 1985). Further analysis, particularly of cellular fatty acids, menaquinones, morphology and biochemical capabilities, suggested that this organism was not a *Campylobacter*, and in 1989 it was placed in a phylogenetically distinct genus, *Helicobacter*, with *H. pylori* as the type species (Goodwin *et al.*, 1989). Analysis of the 16S rRNA sequence relationships of members of this genus reveals two large groups of *Helicobacter* species (Table 40.1). The first group contains *H. pylori*, *Helicobacter heilmannii* and other *Helicobacter* species that colonize gastric mucosa, and the second more diverse group contains the enteric, biliary and blood-associated *Helicobacter* species (reviewed in Solnick and Schauer, 2001).

Table 40.1 *Helicobacter* species and their hosts

Helicobacter species	Natural host(s)
Gastric	
'*Candidatus* Helicobacter bovis'	Cattle
'*Candidatus Helicobacter suis*'	Pig
Helicobacter acinonychis	Cheetah
Helicobacter bizzozeronii	Dog
Helicobacter cetorum	Dolphin, whale
Helicobacter felis	Cat, dog
Helicobacter heilmannii	Human, nonhuman primate
Helicobacter mustelae	Ferret
Helicobacter nemestrinae	Pigtailed macaque
Helicobacter pylori	Human, rhesus macaque
Helicobacter salomonis	Dog
'*Helicobacter suncus*'	House musk shrew
Enterohepatic	
Helicobacter aurati	Hamster
Helicobacter bilis	Mouse, dog, human
'*Helicobacter canadensis*'	Human
Helicobacter canis	Dog, human
Helicobacter cholecystus	Hamster
Helicobacter cinaedi	Human, hamster
Helicobacter sp. cotton-top	Cotton-top tamarin
Helicobacter fennelliae	Human
'*Helicobacter ganmani*'	Mouse
Helicobacter hepaticus	Mouse
'*Helicobacter mainz*'	Human
Helicobacter marmotae	Woodchuck, cat
'*Helicobacter mesocricetorum*'	Hamster
Helicobacter muridarum	Mouse, rat
Helicobacter pametensis	Tern, pig
Helicobacter pullorum	Chicken, human
'*Helicobacter rappini*'	Sheep, human
Helicobacter sp. flexispira	Pig, sheep, dog, human
Helicobacter rodentium	Mouse
Helicobacter trogontum	Rat
'*Helicobacter typhlonius*'	Mouse
Helicobacter westmeadii	Human
Helicobacter winghamensis	Human

EPIDEMIOLOGY

Helicobacter pylori is the most common persistent bacterial infection of humans, infecting between 70 and 90% of the population of developing countries and 25–50% of the population of developed countries

(Dunn, Cohen and Blaser, 1997). Most infections are acquired before the age of 10 years, with males and females infected at approximately the same rates, and there is an inverse relationship between the incidence of infection and socioeconomic status. The only natural reservoir of *H. pylori* is the human stomach, and although it is possible to experimentally infect other animal species, there is no evidence that these act as important sources of transmission of the bacterium. The fastidious nutritional requirements of *H. pylori* mean that it is unable to grow or survive for prolonged periods outside the human host. Definite proof of the exact mode of transmission of *H. pylori* is lacking, but person-to-person transmission via gastric–oral and/or faecal–oral route is considered the most likely means by which the organism is spread. Intrafamilial transmission seems particularly important, with the mother playing a key role in transmitting the agent to the child. The organism can be cultivated from vomitus and, occasionally, from saliva and faeces, indicating that the organism is potentially transmissible during episodes of gastrointestinal illness, particularly with vomiting (Parsonnet, Shmuely and Haggerty, 1999). It is possible that transmission occurs more frequently in the acute phase of the infection when the organism may be excreted in higher numbers. Detection of *H. pylori* in dental plaque is most likely to reflect regurgitation of the bacterium into the mouth, and there is little evidence for long-term survival of the bacterium within this ecological niche. Iatrogenic transmission of *H. pylori* via contaminated endoscopes has been largely eliminated by the rigorous implementation of disinfection procedures.

HELICOBACTER SPECIES AND HUMAN DISEASE

All patients infected with *H. pylori* develop chronic superficial gastritis, a condition that is asymptomatic in most cases (Blaser, 1990). In a proportion of patients infection is complicated by severe gastroduodenal disease, and *H. pylori* is now recognized to cause approximately 90% of duodenal and 70% of gastric ulcers (Figure 40.1). Epidemiological evidence of a link with adenocarcinoma (both intestinal and diffuse histological types) of the distal stomach and gastric MALT lymphoma has resulted in the classification of the organism as a group I carcinogen (IARC Working Group on the Evaluation of Carcinogenic Risks to Humans, 1994). In the developing world infection is also linked with chronic diarrhoea, malnutrition and poor growth, as well as predisposing to enteric infections, including typhoid fever and cholera (Frenck and Clemens, 2003). Numerous studies have attempted to identify an association between infection with *H. pylori* and extradigestive diseases, particularly atherosclerosis and autoimmune disorders. Interpretation of such studies is often difficult owing to the presence of multiple confounding factors, and it remains unclear whether such associations are causal or occasional. Enterohepatic *Helicobacter* species of clinical interest include *Helicobacter cinaedi*, *Helicobacter fennelliae*, '*Helicobacter mainz*' and '*Helicobacter westmeadii*', which are particularly associated with gastroenteritis and bacteraemia in human immunodeficiency virus (HIV)-positive homosexual males. The recent isolation of various *Helicobacter* species from bile and gallbladder tissue of patients with conditions ranging from chronic cholecystitis and primary sclerosing cholangitis to gallbladder and primary hepatic cancer suggests a possible role of these micro-organisms in certain hepatobiliary disorders (reviewed in Leong and Sung, 2002). Again, many of these studies have been small and lacked well-matched controls, but it is plausible that some *Helicobacter* species are able to survive in this niche and cause human disease.

MICROBIOLOGY OF *H. PYLORI*

Helicobacter pylori is a spiral, Gram-negative bacterium that is motile by means of four to eight sheathed flagella (Figure 40.2). The bacterium is an obligate microaerobe, requiring 5–7% O_2 and 5–10% CO_2, and grows optimally in a humid atmosphere at 37 °C on a variety of basal media supplemented with 5–10% horse or sheep blood. *Helicobacter pylori* will also grow in liquid media such as brain–heart infusion (BHI), brucella or Isosensitest broths supplemented with 5–10% fetal calf serum or 1–2% cyclodextrin. A selective supplement containing vancomycin, trimethoprim, cefsulodin (or polymyxin B) and amphotericin B is frequently added to media used for the isolation of *H. pylori* from clinical specimens and is useful for reducing contamination by other organisms. Sterile saline is a suitable medium for the transport of gastric biopsy specimens. More complex media, such as Stuart's medium or supplemented BHI, may improve recovery rates where prolonged transportation times are unavoidable. Although growth is frequently visible after 3–5 days, primary isolation may require incubation for up to 10 days.

After prolonged culture on solid or liquid medium, the normal helical bacillary morphology of the bacterium changes to predominantly coccoid forms, and it has been postulated that these represent a dormant stage important for survival in a hostile environment. However, numerous studies have failed to convincingly demonstrate that coccoid forms can either regenerate into spiral cells or are

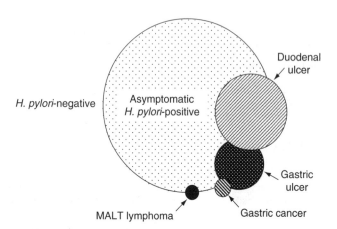

Figure 40.1 Representation of gastroduodenal diseases associated with *H. pylori* infection. MALT, mucosa-associated lymphoid tissue.

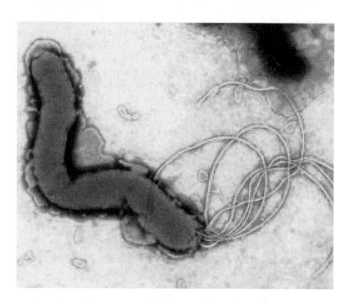

Figure 40.2 Electron micrograph of *H. pylori* showing spiral morphology and sheathed unipolar flagella.

themselves infectious agents. Furthermore, most evidence suggests that the conversion of *H. pylori* from bacillary to the coccoid form is a passive process and that the coccoid form is the morphological manifestation of cell death (Kusters *et al.*, 1997).

Helicobacter pylori possesses urease, oxidase and catalase enzyme activities but is unable to hydrolyse hippurate or reduce nitrate (Goodwin and Armstrong, 1990). The organism is resistant to nalidixic acid (30 µg) and sensitive to cephalothin (30 µg). *Helicobacter pylori* is able to metabolize a variety of substrates via a specialized series of pathways that are different from those found in other bacteria, and detailed description of the biochemical and metabolic pathways of *H. pylori* can be found in excellent reviews of these subjects (Kelly, 1998; Doig *et al.*, 1999; Marais *et al.*, 1999). The presence of incomplete citric acid and other biosynthetic pathways is likely to explain the complex nutritional requirements of *H. pylori*, and its microaerophilic nature is probably due to the involvement of oxygen-sensitive enzymes in central metabolic pathways. The ability of *H. pylori* to use molecular hydrogen, present as a by-product of fermentation by normal colonic flora, as an energy-yielding substrate has been recently shown to facilitate survival within the stomach (Olson and Maier, 2002).

In response to the sequestration and storage of available iron by the host, *H. pylori* has evolved many mechanisms for scavenging iron molecules from host-borne sources including three heme-binding outer membrane proteins (Worst, Otto and Graaf, 1995) and a 70-kDa lactoferrin-binding protein (Dhaenens, Szczebara and Husson, 1997). Gene homologues of uptake systems for ferrous iron (*feo*), ferric iron (*ceu*) and iron–citrate complexes (*fec*) have also been identified in the complete genome sequences of *H. pylori* strain 26695 (Tomb *et al.*, 1997; Alm *et al.*, 1999). Once iron enters the bacterium, the ferritin Pfr binds non-heme iron and forms iron-containing paracrystalline inclusion bodies that allow storage and protects against iron toxicity (Frazier *et al.*, 1993; Bereswill *et al.*, 1998).

GENOME, GENOMICS AND GENETIC DIVERSITY

The size of the circular *H. pylori* chromosome varies from 1.6 to 1.73 Mb, with a G + C composition of between 34.1 and 37.5 mol%. The genomes of two strains of *H. pylori*, 26695 and J99, have been sequenced and have provided new insights into the unique lifestyle and pathogenesis of this bacterium (Tomb *et al.*, 1997; Alm *et al.*, 1999). Like other bacteria that have evolved to colonize a relatively restricted environmental niche, *H. pylori* has a small genome containing a minimal set of metabolic genes. Features that may be particularly relevant to understanding the biology of *H. pylori* include the absence of several key regulatory proteins and biosynthetic enzymes and the presence of a unique combination of virulence factors and multiple DNA-restriction and -modification enzymes. Comparison of the two completed genomes has revealed that around 7% of genes are specific to each strain, with almost half of these genes being clustered in a single hypervariable region, known as the plasticity zone. While the function of 58% of gene products is either known or may be predicted by sequence homology, 18.5% gene products have orthologues of unknown function in other species and 23.5% of proteins are *H. pylori* specific (Alm *et al.*, 1999). Although a recent comparative analysis of the *H. pylori* genomes has reduced the number of hypothetical proteins further (Boneca *et al.*, 2003), one of the major challenges facing researchers is to assign function to and characterize these currently unclassified genes, many of which are likely to be novel determinants important in survival and colonization of the stomach. One approach to this problem has been to use the complete genome sequence to perform *in silico* structural and functional predictions using methods such as fold assignments for genome proteins (Pawlowski *et al.*, 1999) and protein–protein interaction screens (Rain *et al.*, 2001). Signature-tagged mutagenesis, which utilizes negative selection to identify genes that may be important for

pathogenesis, has also provided information on previously unknown colonization factors of *H. pylori*, including a collagenase (Kavermann *et al.*, 2003). Global analysis of gene expression using whole-genome microarray-based technology (reviewed in Thompson and de Reuse, 2002) and comparative proteome analysis (Jungblut *et al.*, 2000) are also providing important clues to the function of genes with unknown function. The goal of characterizing these genes would be the selection of candidates suitable for the rational development of new therapeutic agents that may provide alternatives to antibiotic-based therapies in the future.

Because *H. pylori* is naturally competent for transformation, it is able to acquire new genetic material by horizontal transfer (Nedenskov-Sorensen, Bukholm and Bovre, 1990). *Helicobacter pylori* exhibits a panmictic (free recombinational) population structure that results from frequent genetic recombination during mixed colonization by unrelated strains (Suerbaum *et al.*, 1998). Gene transfer between *H. pylori* strains is common and can generate novel subtypes that exhibit changes in important virulence markers, such as the *cag* pathogenicity island (Kersulyte, Chalkauskas and Berg, 1999). The persistent accumulation of *de novo* mutations also contributes to *H. pylori* genetic variation and evolution (Wang, Humayun and Taylor, 1999), and this process appears to be facilitated by the absence of certain DNA repair systems, such as components of the mismatch repair (MMR) system (Labigne and Jenks, 2001). The ability to generate such an extraordinary degree of genetic diversity has been shown to facilitate adaptation of *H. pylori* to environmental challenges within the stomach and contribute to the longevity of infection (Loughlin *et al.*, 2003). Approximately 40% of *H. pylori* strains contain plasmids varying in size from 1.5 to 23.3 kb. Although virulence factors have not been identified on plasmids, these mobile genetic elements appear to participate in recombinational exchange with the chromosome, so contributing to the extensive genetic heterogeneity characteristic of *H. pylori* (De Ungria *et al.*, 1999).

GENE REGULATION

Despite advances in our knowledge of the factors that are essential for colonization and the induction of mucosal damage, little is known of the molecular mechanisms that regulate the coordinated expression of genes responsible for these functions. Sequence analysis of the genome sequences has revealed a low abundance of regulatory networks when compared with other Gram-negative bacteria (Tomb *et al.*, 1997; Alm *et al.*, 1999). *Helicobacter pylori* possesses only three sigma factors RpoD (σ^{80}), RpoN (σ^{54}) and FliA (σ^{28}) and lacks homologues of the stress-response sigma factors, RpoS (σ^{38}) and RpoH (σ^{32}). There are relatively few two-component regulatory systems, with only four histidine kinases with their cognate response regulators as well as two orphan response regulators (Beier and Frank, 2000). Of the handful of other transcriptional regulators, HspR represses the expression of the major chaperones of *H. pylori* (Spohn and Scarlato, 1999), the ferric uptake regulator (Fur) functions in the regulation of iron homeostasis (Delany *et al.*, 2001) as well as acid resistance and the regulation of the amidase and formamidase enzymes, and NikR largely mediates the nickel-responsive induction of urease (van Vliet *et al.*, 2002).

The paucity of regulatory systems may reflect the fact that *H. pylori* survives in the relatively restricted niche of the human gastric mucosa, with an apparent lack of competition from other micro-organisms. However, the human stomach is far from a stable habitat, experiencing considerable fluctuations in nutrient availability, reactive oxygen species, pH and temperature, and a certain degree of adaptability would be expected to allow successful and persistent colonization. Furthermore, it is clear that *H. pylori* is able to sense and respond to diverse environmental stimuli despite the absence of classical global regulators.

PATHOGENESIS

Colonization of the Gastric Mucosa

Motility

Motility in *H. pylori* is conferred by four to eight unipolar flagella which allow it to remain highly motile in the viscous environment of the mucus layer overlying the gastric mucosa. The bacterium requires flagellum-dependent motility to survive and multiply in this environmental niche as well as to penetrate the mucus layer to reach the surface of the epithelial cells. The flagellar filaments are composed of polymers of a major (FlaA) and a minor flagellin (FlaB), and the expression of both subunits is necessary for full motility and colonization (Eaton *et al.*, 1996). Each flagellar filament is connected to the basal body via a flagellar hook, and the flagella are enclosed in a membranous sheath which is thought to shield the inner filament and protect it from acid-induced damage (Geis *et al.*, 1993). Export and assembly of flagellar proteins in *H. pylori* is mediated by a highly conserved family of proteins which are implicated in the export of flagellar proteins and other virulence factors in many bacterial species. In contrast, the regulation of flagellar gene expression in *H. pylori* differs from that in model organisms such as *Escherichia coli* and can be ordered into three hierarchical classes that respond to many stimuli including growth phase and changes in environmental conditions (Colland *et al.*, 2001; Josenhans *et al.*, 2002; Thompson *et al.*, 2003).

Mechanisms of Acid Tolerance

Acid resistance is central to the pathogenesis of *H. pylori*, and a series of complex physiological adaptations accompany changes in environmental pH (McGowan *et al.*, 2003; Merrell *et al.*, 2003). The urease enzyme, which catalyses the hydrolysis of urea to carbon dioxide and ammonium, provides *H. pylori* with its most important defence against gastric acidity. *Helicobacter pylori* urease accounts for up to 6% of total cellular proteins of the organism, and the generation of ammonia by ureolysis creates a neutral microenvironment around the bacterium that protects it from the acid pH of the stomach (Goodwin, Armstrong and Marshall, 1986). The relatively high affinity constant (K_m) of *H. pylori* for urea, coupled with its high rate of activity, means that the enzyme remains catalytically efficient despite the low intragastric concentration of urea (Dunn *et al.*, 1990). The urease apoenzyme of *H. pylori* is composed of a hexapolymeric structure consisting of the two structural subunits, UreA (26 kDa) and UreB (61 kDa) in a 1:1 molar ratio, to which are bound two nickel ions per monomer (Dunn *et al.*, 1990). As well as encoding the two structural subunits, the urease gene complex also encodes an acid-gated urea channel (UreI) and the accessory assembly proteins (UreE–H). The latter interact with the apoenzyme and incorporate nickel ions into the urease complex, ensuring that the enzyme remains catalytically active under conditions of low pH (Labigne, Cussac and Courcoux, 1991). Because nickel is an essential cofactor of urease activity, *H. pylori* has evolved many systems for the incorporation and storage of nickel ions within the bacterial cell, and the expression and activity of *H. pylori* urease are upregulated by increased availability of this cofactor through the regulatory protein NikR (van Vliet *et al.*, 2002). Urease is necessary for colonization of the gastric mucosa, and its essential role appears to extend beyond that of simple acid protection (Eaton and Krakowka, 1994). Other physiologically important functions of urease include the provision of a nitrogen source via production of ammonia, chemotactic motility and the promotion of nutrient release from the gastric mucosa through the induction of epithelial cell inflammation and apoptosis. Other systems also contribute to ammonia production in *H. pylori*, including an aliphatic amidase (Skouloubris, Labigne and DeReuse, 1997) and an arginase, which synthesizes urea from L-arginine,

providing an intracellular pool of substrate for the urease enzyme (McGee *et al.*, 1999).

The product of another member of the urease gene cluster, UreI, is an integral cytoplasmic membrane protein that is essential for survival of the *H. pylori* in an acid environment (Skouloubris *et al.*, 1998). UreI functions as a proton-gated transmembrane channel that regulates the passage of urea across the inner membrane into the cytoplasm (Weeks *et al.*, 2000). The UreI protein forms a membrane complex with the urease enzyme catalytic subunits (UreA/B), which facilitates rapid delivery and hydrolysis of urea to ammonia (Voland *et al.*, 2003). In acid conditions, the urea channel opens and, because of increased substrate delivery, cytoplasmic urease activity increases to produce more ammonia. Delivery of ammonia to the periplasmic space neutralizes the immediate environment of the bacterium and maintains the proton motive force across the inner cytoplasmic membrane due to the formation of ammonium ions. As the pH rises, UreI-mediated urea transport is switched off, resulting in the lowering of cytoplasmic urease activity, thereby preventing lethal alkalinization of the cytoplasm owing to excess production of ammonia.

Adhesion

The ability to bind to epithelial surfaces is important for many bacterial pathogens that colonize mucosal surfaces. Once they escape the gastric fluid and penetrate the gastric mucus, most *H. pylori* remain within the mucus layer, but a certain proportion (~2%) are found in intimate contact with the surface of the underlying epithelial cells (Hessey *et al.*, 1990). Interaction of *H. pylori* with mucosal cells can be associated with the formation of adherence pedestals reminiscent of enteropathogenic *E. coli* (Dytoc *et al.*, 1993), and *H. pylori* adherence to gastric epithelial cells may result in several cellular events, including modulation of cell signalling cascades, rearrangement of the host cytoskeleton and cytokine release (Segal *et al.*, 1999).

Many specific adhesins and cognate host-cell receptors have been implicated in the association of *H. pylori* with the surface of gastric epithelial cells. The best-characterized adhesins of *H. pylori* are members of a superfamily of about 30 genes encoding putative *Helicobacter* outer membrane proteins (Hops). The blood group antigen-binding adhesin (BabA2) belongs to this superfamily of outer membrane proteins and binds to the fucosylated Lewis[b] blood group antigen (Ilver *et al.*, 1998). Two other Hops, encoded by the *alpAB* (adherence-associated lipoprotein) gene locus, also mediate specific adherence of *H. pylori* to human gastric tissue, probably via a different receptor (Odenbreit *et al.*, 1999). High levels of sialylated glycoconjugates are associated with *H. pylori* infection, and the formation of sialyl-Lewis x glycosphingolipids in the gastric epithelium is induced by *H. pylori* infection (Mahdavi *et al.*, 2002). By mediating binding to sialyl-Lexis x antigens via the sialic acid-binding adhesin (SabA), *H. pylori* is able to exploit inflammation-activated domains of sialylated epithelium and complement baseline adhesion to other host receptors (Mahdavi *et al.*, 2002). This spectrum of adhesin–receptor interactions contributes to the chronicity of *H. pylori* infection and represents an adaptation to human genetic diversity and the host defences (Mahdavi *et al.*, 2002).

Resistance to Oxidative Stress

Helicobacter pylori has adapted to survive in an environment which is bathed in oxygen-derived free radicals generated from the bacterium's own metabolism and the inflammatory defences of the host. The ability of *H. pylori* to neutralize reactive oxygen species confers resistance against phagocytic killing (Ramarao, Gary-Owen and

Meyer, 2000) and is essential for long-term survival in the murine gastric mucosa (Harris *et al.*, 2003). *Helicobacter pylori* produces many detoxifying enzymes that protect against the effects of reactive oxygen species, including catalase, superoxide dismutase and alkyl hydroperoxide reductase. These antioxidant enzyme-encoding genes are upregulated in response to oxidative stress conditions, despite the fact that *H. pylori* lacks both the RpoS sigma factor and the oxidatively activated transcriptional regulators OxyR and SoxRS (Jenks, unpublished data). The *H. pylori* neutrophil-activating protein (NapA) was originally described as a promoter of neutrophil adhesion to endothelial cells (Yoshida *et al.*, 1993) and has significant homology to bacteriferritins and the *E. coli* DNA-binding protein from starved cells Dps (Evans *et al.*, 1995). The latter is a non-specific DNA-binding protein that is induced by stress conditions and protects DNA from oxidative damage by sequestering iron that may otherwise generate free radicals through the Fenton reaction. Evidence that NapA has a similar DNA-protective role in *H. pylori* is provided by the observation that *napA* mutants are more susceptible to oxidative stress and display increased oxidant-induced genomic DNA damage (Jenks, unpublished data). The fact that NapA is also able to induce inflammatory reactions and tissue damage suggests that it has evolved additional functions that contribute to the pathological process.

Induction of Pathological Changes

Despite being non-invasive, *H. pylori* is capable of inducing severe pathological changes in the gastric mucosa of the host. The generation of mucosal lesions by *H. pylori* is a result of a complex series of interactions between bacterial factors that cause direct or indirect tissue damage and the host response that is determined by genetic and environmental factors. Strains expressing high levels of vacuolating cytotoxin (VacA), the *cag* pathogenicity island and a functional BabA2 correlate with more severe disease (Covacci *et al.*, 1999). While the presence of these and other markers signals an increased predisposition to the development of severe pathology, many factors, particularly geographic variation in the genotype of dominant circulating strains, limit their value as indicators of pathogenic strains. Caution is therefore required in applying these disease associations in different human populations.

Vacuolating Cytotoxin

Helicobacter pylori secretes a VacA that induces the formation of vacuoles in the cytoplasm of eukaryotic cells (Figure 40.3) (Leunk *et al.*, 1988). Although the *vacA* gene is present in almost all strains of *H. pylori*, isolates differ in vacuolating activity because of variability in the coding region of the signal sequence and the middle region of the *vacA* gene. Certain alleles of the signal sequence correlate with higher expression of active toxin, are associated with the greater gastric inflammation and are more commonly isolated from individuals with peptic ulcer disease (Atherton *et al.*, 1995). Mosaicism of the middle region determines targeting and internalization of the toxin but does not affect activity once the toxin enters the host-cell cytoplasm (Pagliaccia *et al.*, 1998). The precursor form of VacA has a molecular weight of 140 kDa and undergoes amino- and carboxy-terminal processing to yield a mature secreted toxin of approximately 87 kDa. Mature VacA contains amino- and carboxyl-terminal domains (p37 and p58, respectively), both of which are essential for toxin activity (Ye, Willhite and Blanke, 1999). Release of VacA from *H. pylori* occurs both by a specific secretion pathway and by budding of outer membrane vesicles, with the latter mechanism representing a potential alternative pathway for the delivery of other bacterial components to the gastric mucosa (Fiocca *et al.*, 1999). Once secreted, VacA monomers oligomerize into inactive rosettes comprising of 12 or 14 subunits (Cover *et al.*, 1994). Acidification enhances cytotoxic activity and results in the disruption of VacA oligomers into monomeric components that insert into lipid membranes to form hexameric, anion-conductive channels. These channels increase paracellular epithelial permeability to low-molecular-weight molecules, and this selective permeabilization may increase the supply of essential nutrients, including urea, therefore favouring growth *in vivo* (Papini *et al.*, 1998; Tombola *et al.*, 2001). VacA channels are actively endocytosed and result in the formation of vacuoles which are a hybrid between the lysosomal and late endosomal compartments of eukaryotic cells (Papini *et al.*, 1994). In a separate pathway distinct from vacuole formation, VacA is also capable of binding to receptor-type protein tyrosine phosphatase (PtprZ) which mediates multiple intracellular signalling pathways that regulate membrane trafficking, cytoskeletal rearrangements and cellular adhesion (Yahiro *et al.*, 1999). Alterations in the activity of PtprZ, and hence the rate of dephosphorylation of host-cell target proteins, may result in epithelial cell detachment from the underlying basement membrane, exposing the lamina propria to the damaging effects of gastric acid and leading to ulcer formation (Fujikawa *et al.*, 2003; Peek, 2003). Interaction of VacA with PtprZ may also interfere with signal transduction pathways known to be essential for cell proliferation and ulcer healing (Peek, 2003).

VacA has many other biological effects including alteration of tight-junction permeability and induction of gastric epithelial cell apoptosis. The toxin inhibits antigen presentation and processing within the gastric mucosa (Molinari *et al.*, 1998) and may also prevent phagosome maturation (Zheng and Jones, 2003), effects that would contribute to the evasion of the host immune response. Oral administration of mice with VacA leads to mucosal degeneration and epithelial cell injury (Telford *et al.*, 1994), and co-culture experiments have indicated that the toxin is important during the early stages of infection of the host (Salama *et al.*, 2001).

Cytotoxin-Associated Gene A and the cag Pathogenicity Island

Cytotoxin-associated gene A (CagA) was first described as an immunodominant antigen with a molecular mass of 120 kDa (Crabtree *et al.*, 1991) and was subsequently found to be expressed by most VacA-producing *H. pylori* strains (Covacci *et al.*, 1993). More recently, the multigenic locus upstream of *cagA* was found to have the typical features of a pathogenicity island (PAI) that has been acquired by *H. pylori* relatively recently in its evolution (Censini *et al.*, 1996). The *cag* PAI encodes a type IV secretion system, a syringe-like structure that serves to inject effector molecules into the host cell, allowing the bacterium to modulate certain aspects of the host cell's metabolism, including cytoskeletal rearrangements, host-cell morphological

(A) 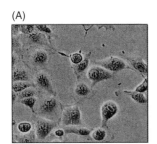   (B)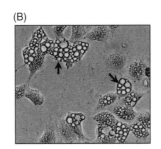

Figure 40.3 Vacuolating cytotoxin-induced vacuolation of rabbit kidney epithelial (RK13) cells observed at ×50 magnification. (A) Epithelial cells after the addition of liquid broth alone (negative control). (B) Epithelial cells after the addition of liquid broth culture supernatants of a VacA-producing strain of *H. pylori*, with examples of vacuolation indicated by arrows. These images were kindly provided by Darren Letley.

changes and expression of proto-oncogenes. Microscopic studies have recently confirmed the presence of sheathed, central hollow needle-shaped organelle at the surface of *H. pylori* (Rohde *et al.*, 2003) (Figure 40.4). A functional type IV secretion system translocates CagA directly into gastric epithelial cells (Segal *et al.*, 1999; Asahi *et al.*, 2000; Odenbreit *et al.*, 2000; Stein, Rappuoli and Covacci, 2000) where it localizes on the inner surface of the plasma membrane and is tyrosine phosphorylated by the Src family of protein tyrosine kinases (Selbach *et al.*, 2002; Stein *et al.*, 2002). CagA tyrosine phosphorylation sites (EPIYA motifs) are located in the C-terminal half of CagA, and the number of motifs determines the phosphorylation level of individual CagA molecules. Phosphorylated CagA then binds the Src homology 2-containing tyrosine phosphatase, SHP-2, and stimulates its phosphatase activity, resulting in the dephosphorylation of various host-cell proteins (Puls, Fischer and Haas, 2002). SHP-2 phosphatase plays an important role in the regulation of intracellular signal transduction, stimulating mitogenic signalling and mediating the adhesion, migration and division of cells through actin reorganization (reviewed in Hatakeyama, 2003). By stimulating SHP-2, CagA may deregulate cell growth and induce proliferation of gastric epithelial cells, ultimately leading to malignant transformation (Hatakeyama, 2003). Polymorphism in the number of CagA tyrosine phosphorylation sites may determine the biological activity of CagA and the associated risk of gastric carcinoma, since proteins with more EPIYA motifs undergo greater tyrosine phosphorylation, exhibit increased SHP-2 binding and induce greater phenotypic changes on host cells (Higashi *et al.*, 2002).

A further important function of the *cag* PAI is the induction of proinflammatory chemokines such as interleukin-8 (IL-8), Rantes and epithelial neutrophil-activating peptide 78 ENA-78, from gastric epithelial cells via the stress kinase pathway (Naumann *et al.*, 1999). Chemokine induction occurs independently of CagA translocation, and distinct genes of the *cag* PAI are essential for these two processes (Fischer *et al.*, 2001). The host cellular responses that lead to the activation of immunomodulatory chemokines may result from direct activation of the stress kinase pathway by the type IV system itself (Foryst-Ludwig and Naumann, 2000) or by type IV apparatus-mediated translocation of other bacterial effector molecules, such as peptidoglycan, into the host cell (Viala *et al.*, 2003). The significance of the potential for interplay between *cag* PAI-mediated activation of gastric epithelial cell kinases and blockade of related cell signalling pathways by VacA is currently unknown. However, it appears that different bacterial factors modify the intensity of the host inflammatory immune response, therefore contributing to the persistence of *H. pylori* and the degree of induced mucosal damage.

Lipopolysaccharide

Helicobacter pylori lipopolysaccharide (LPS) contains various modifications that allow the bacterium to moderate the degree of inflammation induced and which may contribute to its persistence in the gastric mucosa and disease causation. The LPS of *H. pylori* has lower biological activity and immunogenicity than the LPS of Enterobacteriaceae (Muotiala *et al.*, 1992). The glycoprotein moiety of *H. pylori* LPS O-side chains contains antigenic domains identical to the mono- and di-fucosylated Lewis[x] and Lewis[y] and also H-type 1 blood group antigens that are also expressed at the surface of gastric parietal cells, in mucin and by the gastric H^+K^+-ATPase. Most strains of *H. pylori* produce either Lewis[x] or Lewis[y], but certain strains express both these carbohydrate antigens (Monteiro *et al.*, 1998). Antigenic mimicry between *H. pylori* LPS and human gastric epithelial cell surface glycoforms may result in immune tolerance against *H. pylori* antigens via downregulation of the T helper (Th) type 1 response. In chronic infection, immune cross-reactivity may ultimately result in the induction of autoantibodies that recognize mucosal epitopes and contribute to the development of chronic active gastritis (Negrini *et al.*, 1991; Appelmelk *et al.*, 1996). Lewis antigen expression is not a stable trait, and LPS displays a high frequency of phase variation which is mediated by length alteration of homopolymeric tracts by slipped-strand mispairing (Appelmelk *et al.*, 1998). This increased heterogeneity of *H. pylori* is another factor in allowing adaptation to microenvironmental changes and evasion of the immune response.

Other biological properties attributed to *H. pylori* LPS include the inhibition of glycosylation and sulphation of gastric mucus, which may impede its protective function and increase the vulnerability of the epithelial surface to gastric acidity (Moran, 1996). The LPS of certain *H. pylori* strains has also been shown to induce secretion of pepsinogen, and the increased proteolytic activity of pepsin may be implicated in *H. pylori*-induced mucosal damage (Young *et al.*, 1992).

DIAGNOSIS OF *H. PYLORI*

Diagnosis of *H. pylori* infection may be established by many invasive and non-invasive techniques. Culture of the organism from multiple endoscopic gastric mucosal biopsies taken from the antrum and corpus is the most specific method and offers the distinct advantage of allowing further investigations, such as susceptibility testing and typing, to be performed. Definitive identification of the organism relies on the typical 'seagull' morphology of the Gram film and positive oxidase, catalase and rapid urease tests. Histological examination of at least two biopsies is as sensitive as culture for the detection of *H. pylori* and is also relatively specific. Stains for the recognition of the organism include Warthin–Starry, Giemsa and Creosyl violet, and sensitivity can be improved by the use of immunostaining with an anti-*H. pylori* antibody. The biopsy urease test detects the presence of urease activity in gastric biopsies by using a broth that contains urea and a pH indicator. This test is frequently performed in the endoscopy suite and is rapid, sensitive and cheap. The polymerase chain reaction (PCR) has been used for the detection of *H. pylori* in various samples but has not been widely adopted for routine diagnostic use.

Non-invasive methods include the urea breath test, which is semi-quantitative and is frequently used to check eradication after treatment. The urea breath test is based on the release of carbon dioxide from urea by the highly active urease enzyme. Urea labelled with either radioactive ^{14}C or non-radioactive ^{13}C is given as a test meal, and the labelled carbon dioxide is exhaled and detected by scintillation counting or mass spectroscopy. Laboratory testing for serum IgG by enzyme-linked immunoadsorbent assay is frequently used for screening for *H. pylori* infection. Other non-invasive tests include the detection of antibody in saliva, which is less sensitive then serum testing, and *H. pylori* faecal antigen detection by enzyme immunoassay

(A)

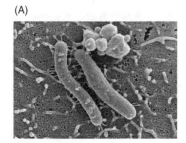

(B)

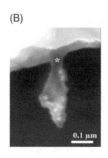

0.1 μm

Figure 40.4 Structure of the *H. pylori cag* type IV secretion system. (A) Surface filamentous structures of *H. pylori* are induced after contact with gastric epithelial (AGS) cells. (B) Immunogold labelling of surface-exposed needle-shaped organelle at the surface of the bacterium. The asterisk indicates the central, hollow structure. These images were kindly provided by Manfred Rohde and Rainer Haas.

(EIA), which is easy to perform and may be particularly useful in the screening of children.

Susceptibility Testing

Until recently, methods of susceptibility testing of *H. pylori* suffered from a lack of consistency and conflicting results were frequently found when different techniques were compared. This combined with cost implications and the convenience of non-culture-based diagnostic tests means that laboratory susceptibility testing is rarely performed before empirical therapy is commenced. The recent trend towards standardization of testing is designed to improve agreement in reporting and should mean that culture and susceptibility testing of *H. pylori* can be performed in most hospital laboratories. Agar dilution is generally regarded as the gold standard method of susceptibility testing for *H. pylori* and is highly reproducible, but laborious and time-consuming (National Committee for Clinical Laboratory Standards, 1999). The epsilometer (E-test) is frequently used since it is easier to perform and has good reproducibility (Cederbrant, Kahlmeter and Ljungh, 1993). Disk diffusion has been shown to be reliable in *H. pylori* when test methods are well controlled and, as with the E-test, allows visualization of resistant subpopulations within the zone of inhibition. When performed appropriately, there is a good correlation with all methods for the testing of clarithromycin, tetracycline and amoxicillin, and disk testing would appear an adequate choice for these agents. Testing for metronidazole susceptibility is more difficult since the distribution of MICs is continuous, resistant subpopulations are frequently present and results are more easily affected by test conditions. Although the E-test is usually recommended, even this method may overestimate resistance (Mégraud *et al.*, 1999), and more than one testing method may be appropriate, particularly for strains close to the breakpoint. Continued refinement of protocols and participation in quality control schemes is important for improving the reproducibility of testing and will allow the implementation of community surveillance, both to monitor the prevalence of resistant strains and to guide empirical treatment based on local resistance patterns. Basing individual therapy on pretreatment laboratory susceptibility data significantly improves the eradication rate (Toracchio *et al.*, 2000). As the proportion of patients colonized with resistant strains continues to rise, routine susceptibility testing would allow a more rational approach to antibiotic selection, extending the useful life of these agents for the treatment of not only *H. pylori* but other infections as well.

A rapid and useful alternative to conventional susceptibility testing is to use molecular methods to detect resistance markers, permitting resistance testing without the need for culture. Because resistance to metronidazole is associated with multiple changes within *rdxA* and possibly other reductase-encoding genes, it has not been possible to develop simple molecular assays capable of detecting nitroimidazole resistance in *H. pylori*. Although many genotype-based methods, usually based on PCR, have been developed to identify the limited number of point mutations that cause macrolide resistance, these are yet to be widely adopted by routine diagnostic microbiology laboratories.

TREATMENT OF *H. PYLORI*

Eradication of *H. pylori* from the gastric and duodenal mucosa of infected patients is the most important goal in the management of peptic ulcer disease and other *H. pylori*-associated conditions (Jenks, 2002). Although the bacterium appears susceptible to many antibiotics *in vitro*, relatively few are effective in clinical practice. This is partly because certain antibiotics are unable to achieve therapeutic concentration in the gastric mucosa, and the slow rate of growth and metabolism of the organism also contribute to this relative

in vivo resistance. *Helicobacter pylori* is therefore difficult to eradicate, and effective treatment requires multidrug regimens, consisting of two antibiotics (usually selected from clarithromycin, metronidazole, amoxicillin and tetracycline), combined with acid suppressants and/or bismuth compounds (Malfertheiner *et al.*, 2002). Although several controlled clinical trials have shown that current first-line regimens are effective in most patients, a significant proportion fail therapy in clinical trials and success rates are frequently as low as 70% in everyday clinical practice. Several factors adversely affect treatment outcome, including advanced age, smoking, high pretreatment intragastric bacterial load, bacterial genotype, poor patient compliance due to the high incidence of side effects associated with multidrug regimens and host genetic polymorphisms of the cytochrome P450 isoenzymes that are specifically involved in the metabolism of proton-pump inhibitors. However, as happens for many infectious diseases, resistance to the antibiotic component of the regimen is the most important cause of treatment failure, and increased use of antibiotics has resulted in a significant rise in both the prevalence of resistance and multiresistant strains. Although resistance to amoxicillin and tetracycline remains relatively rare, 10–50% of clinical strains isolated in western Europe are resistant to metronidazole and up to 20% are resistant to clarithromycin (Mégraud, 1998). Resistance to either the macrolide or amoxicillin component of a regimen correlates closely with reduced therapeutic efficacy. Although there have been conflicting reports concerning the clinical impact of imidazole resistance in *H. pylori*, recent meta-analyses have demonstrated that resistance to this class of antibiotics is also an important predictor of treatment failure (Houben *et al.*, 1999; Van der Wouden *et al.*, 1999). How to respond to the worldwide increase in the prevalence of antibiotic resistance is the most important challenge facing the clinical management of patients with *H. pylori* infection. The solution will involve safeguarding the already limited number of treatment options as well as identifying novel molecular targets with potential for therapeutic intervention.

Resistance to Antimicrobial Agents

The antimicrobial toxicity of metronidazole is dependent on reduction of its nitro moiety to the nitro anion radical and other DNA-damaging compounds, including nitroso and hydroxylamine derivatives. In the vast majority of *H. pylori* strains, resistance to metronidazole is associated with mutational inactivation of the *rdxA* gene, which encodes an oxygen-insensitive NADPH nitroreductase involved in the reductive activation of metronidazole (Goodwin *et al.*, 1998; Jenks, Ferrero and Labigne, 1999). Other reductase-encoding genes, including *frxA* (which encodes NADPH flavin oxidoreductase) and *fdxB* (which encodes ferredoxin-like protein), are occasionally associated with resistance to metronidazole (Jeong *et al.*, 2000; Kwon *et al.*, 2000). Although resistance has been reported to be associated with inactivation of *frxA* alone, the most important role of these genes is to mediate the transition to high-level resistance once inactivation of the *rdxA* gene has occurred (Jeong *et al.*, 2000).

Most clarithromycin-resistant *H. pylori* strains emerge after spontaneous mutation of the 23S rRNA gene and selection of mutants after exposure to the drug (Versalovic *et al.*, 1996). Stable amoxicillin resistance in *H. pylori* is not mediated through β-lactamase production but results from point mutations within the *penicillin-binding protein (PBP) 1A* gene (Gerrits *et al.*, 2002). Nonstable amoxicillin resistance, which is also associated with a marked reduction in treatment efficacy, is probably due to a decreased expression of the penicillin-binding protein PBP 4 (Dore, Graham and Sepulveda, 1999). The small number of tetracycline-resistant *H. pylori* strains that have been studied contain a mutation within the 16S rRNA gene that alters the affinity of the ribosome for the antibiotic (Trieber and Taylor, 2002). Although an initial analysis of membrane-associated systems suggested that active efflux does not play a role in the intrinsic resistance of *H. pylori* (Bina *et al.*, 2000), recent studies have suggested that altered membrane permeability contributes to resistance to both β-lactams and tetracycline (Dailidiene *et al.*, 2002; Kwon *et al.*, 2003).

PREVENTION AND ERADICATION OF *H. PYLORI* BY VACCINES

Currently recommended *H. pylori* eradication therapies are relatively expensive, and the impact of antibiotic resistance is likely to further reduce the efficacy of currently available regimens. Although novel antimicrobial agents may facilitate treatment in the future, only a vaccine will have a global impact on the prevalence of *H. pylori* infection and, hence, on *H. pylori*-associated morbidity and mortality (Suerbaum and Josenhans, 1999). Evaluation of the cost-effectiveness of vaccination against *H. pylori* in developed countries found that it had the potential to deliver the highest health benefit of all vaccines tested and could significantly reduce health care expenditure (Rupnow *et al.*, 1999). Although the implementation of public health measures might prove more effective in developing countries (Rupnow *et al.*, 1999), the introduction of a vaccine in those regions with the highest disease burden would represent a major advance in the management of *H. pylori* infection.

Helicobacter pylori has developed a spectrum of strategies to resist host defence mechanisms and establish persistent gastric infection. By modulating cytokine production, *H. pylori* is able to bias Th cell differentiation, resulting in an ineffective immune response that contributes to both persistence and gastric disease processes (reviewed in Zevering, Jacob and Meyer, 1999). Although the mechanisms of immune subversion are unclear, it is known that protection can occur in a strictly Th1-dependent fashion and that the immunosuppressive cytokine IL-10 downregulates protective host responses and facilitates persistent colonization (Figure 40.5). It is attractive to envisage an immune-based strategy that would reprogramme the natural tendency of *H. pylori* to subvert the host response and establish a persistent infection, to one that elicits a protective response resulting in eradication.

Experiments using animal models of infection and a variety of *H. pylori* antigens have demonstrated the feasibility of developing both prophylactic and therapeutic vaccines against *H. pylori* infection. These have included the urease holoenzyme and its subunits, the HspB (GroEL) and HspA (GroES) heat shock proteins, purified VacA cytotoxin, CagA, NapA and catalase (reviewed in Suerbaum and Josenhans, 1999). The observation that these antigens may serve as effective vaccines against *H. pylori*, despite being located within the cytoplasm, appears to be explained by their adsorption onto the cell surface during bacterial lysis (Phadnis *et al.*, 1996). Unfortunately, in

human clinical trials, these antigens either have not been successful in preventing infection or have failed to induce sterilizing immunity (reviewed in Prinz, Hafsi and Voland, 2003). Homology searches of the complete genome sequence of *H. pylori* and the application of techniques that allow genome-wide screening for immunogenic proteins may facilitate the identification of novel antigens for inclusion in prophylactic and/or therapeutic subunit vaccines against *H. pylori*. Ultimately, improved results are dependent on an improved understanding of the host immune response to infection as well as the use of multiple protective antigens, modification of the route and mode of administration and development of less toxic and more potent adjuvants.

REFERENCES

Alm RA, Ling LS, Moir DT *et al.* (1999) Genomic-sequence comparison of two unrelated isolates of the human gastric pathogen *Helicobacter pylori*. *Nature*, **397**, 176–180.

Appelmelk BJ, Simoons-Smit I, Negrini R *et al.* (1996) Potential role of molecular mimicry between *Helicobacter pylori* lipopolysaccharide and host Lewis blood group antigens in autoimmunity. *Infection and Immunity*, **64**, 2031–2040.

Appelmelk BJ, Shiberu B, Trinks C *et al.* (1998) Phase variation in *Helicobacter pylori* lipopolysaccharide. *Infection and Immunity*, **66**, 70–76.

Asahi M, Azuma T, Ito S *et al.* (2000) *Helicobacter pylori* CagA can be tyrosine phosphorylated in gastric epithelial cells. *The Journal of Experimental Medicine*, **191**, 593–602.

Atherton JC, Cao P, Peek RM *et al.* (1995) Mosaicism in vacuolating cytotoxin alleles of *Helicobacter pylori*: association of specific vacA types with cytotoxin production and peptic ulceration. *The Journal of Biological Chemistry*, **270**, 1771–1777.

Barnard FM, Loughlin MF, Fainberg HP *et al.* (2004) Global regulation of virulence and the stress response by CsrA in the highly adapted human gastric pathogen *Helicobacter pylori*. *Molecular Microbiology*, **51**, 15–32.

Beier D and Frank R (2000) Molecular characterization of two-component systems of *Helicobacter pylori*. *Journal of Bacteriology*, **182**, 2068–2076.

Bereswill S, Waidner U, Odenbreit S *et al.* (1998) Structural, functional and mutational analysis of the *pfr* gene encoding a ferritin from *Helicobacter pylori*. *Microbiology*, **144**, 2505–2516.

Bina JE, Alm RA, Uria-Nickelsen M *et al.* (2000) *Helicobacter pylori* uptake and efflux: basis for intrinsic susceptibility to antibiotics in vitro. *Antimicrobial Agents and Chemotherapy*, **44**, 248–254.

Blaser MJ (1990) *Helicobacter pylori* and the pathogenesis of gastroduodenal inflammation. *The Journal of Infectious Diseases*, **161**, 626–633.

Boneca IG, de Reuse H, Epinat JC *et al.* (2003) A revised annotation and comparative analysis of *Helicobacter pylori* genomes. *Nucleic Acids Research*, **31**, 1704–1714.

Cederbrant G, Kahlmeter G and Ljungh A (1993) The E test for antimicrobial susceptibility testing of *Helicobacter pylori*. *The Journal of Antimicrobial Chemotherapy*, **31**, 65–71.

Censini S, Lange C, Xiang Z *et al.* (1996) *cag*, a pathogenicity island of *Helicobacter pylori*, encodes type I-specific and disease-associated virulence factors. *Proceedings of the National Academy of Sciences of the United States of America*, **93**, 14648–14653.

Colland F, Rain JC, Gounon P *et al.* (2001) Identification of the *Helicobacter pylori* anti-sigma28 factor. *Molecular Microbiology*, **41**, 477–487.

Covacci A, Censini S, Bugnoli M *et al.* (1993) Molecular characterization of the 128-kDa immunodominant antigen of *Helicobacter pylori* associated with cytotoxicity and duodenal ulcer. *Proceedings of the National Academy of Sciences of the United States of America*, **90**, 5791–5795.

Covacci A, Telford JL, Del Giudice J *et al.* (1999) *Helicobacter pylori* virulence and genetic geography. *Science*, **284**, 1328–1333.

Cover TL, Tummuru MK, Cao P *et al.* (1994) Divergence of genetic sequences for the vacuolating cytotoxin among *Helicobacter pylori* strains. *The Journal of Biological Chemistry*, **269**, 10566–10573.

Crabtree JE, Taylor JD, Wyatt JI *et al.* (1991) Mucosal IgA recognition of *Helicobacter pylori* 120 kDa protein, peptic ulceration, and gastric pathology. *Lancet*, **338**, 332–335.

Dailidiene D, Bertoli MT, Miciuleviciene J *et al.* (2002) Emergence of tetracycline resistance in *Helicobacter pylori*: multiple mutational changes in 16S ribosomal DNA and other genetic loci. *Antimicrobial Agents and Chemotherapy*, **46**, 3940–3946.

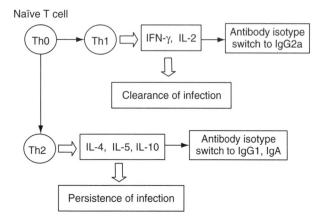

Figure 40.5 Proposed immune correlates of persistence and protection in *H. pylori* infection. IFN-γ, interferon-γ; Ig, immunoglobulin; IL, interleukin; Th, T helper.

De Ungria MC, Kolesnikow T, Cox PT and Lee A (1999) Molecular characterization and interstrain variability of pHPS1, a plasmid isolated from the Sydney Strain (SS1) of *Helicobacter pylori*. *Plasmid*, **41**, 97–109.

Delany I, Spohn G, Rappuoli R and Scarlato V (2001) The Fur repressor controls transcription of iron-activated and -repressed genes in *Helicobacter pylori*. *Molecular Microbiology*, **42**, 1297–1309.

Dhaenens L, Szczebara F and Husson MO (1997) Identification, characterization and immunogenicity of the lactoferrin-binding protein from *Helicobacter pylori*. *Infection and Immunity*, **65**, 514–518.

Doig P, de Jonge BL, Alm RA *et al.* (1999) *Helicobacter pylori* physiology predicted from genomic comparison of two strains. *Microbiology and Molecular Biology Reviews*, **63**, 675–707.

Dore MP, Graham DY and Sepulveda AR (1999) Different penicillin-binding profiles in amoxicillin-resistant *Helicobacter pylori*. *Helicobacter*, **4**, 154–161.

Dunn BE, Campbell GP, Pérez-Pérez GI and Blaser MJ (1990) Purification and characterization of urease from *Helicobacter pylori*. *The Journal of Biological Chemistry*, **265**, 3464–3469.

Dunn BE, Cohen H and Blaser MJ (1997) *Helicobacter pylori*. *Clinical Microbiology Reviews*, **10**, 720–741.

Dytoc M, Gold B, Louie M *et al.* (1993) Comparison of *Helicobacter pylori* and attaching-effacing *Escherichia coli* adhesion to eukaryotic cells. *Infection and Immunity*, **61**, 448–456.

Eaton KA and Krakowka S (1994) Effect of gastric pH on urease-dependent colonization of gnotobiotic piglets by *Helicobacter pylori*. *Infection and Immunity*, **62**, 3604–3607.

Eaton KA, Suerbaum S, Josenhans C and Krakowka S (1996) Colonization of gnotobiotic piglets by *Helicobacter pylori* deficient in two flagellin genes. *Infection and Immunity*, **64**, 2445–2448.

Evans DJ, Evans DG, Takemura T *et al.* (1995) Characterization of a *Helicobacter pylori* neutrophil-activating protein. *Infection and Immunity*, **63**, 2213–2220.

Fiocca R, Necchi V, Sommi P *et al.* (1999) Release of *Helicobacter pylori* vacuolating cytotoxin by both a specific secretion pathway and budding of outer membrane vesicles. Uptake of released toxin and vesicles by gastric epithelium. *The Journal of Pathology*, **188**, 220–226.

Fischer W, Püls J, Buhrdorf R *et al.* (2001) Systematic mutagenesis of the *Helicobacter pylori cag* pathogenicity island: essential genes for CagA translocation in host cells and induction of interleukin-8. *Molecular Microbiology*, **42**, 1337–1348.

Foryst-Ludwig A and Naumann M (2000) PAK1 activates the NIK-IKK NF-kappaB pathway and proinflammatory cytokines in *H. pylori*-infection. *The Journal of Biological Chemistry*, **275**, 39779–39785.

Frazier BA, Pfeifer JD, Russell DG *et al.* (1993) Paracrystalline inclusions of a novel ferritin containing nonheme iron, produced by the human gastric pathogen *Helicobacter pylori*: evidence for a third class of ferritins. *Journal of Bacteriology*, **175**, 966–972.

Frenck RW and Clemens J (2003) Helicobacter in the developing world. *Microbes and Infection*, **5**, 705–713.

Fujikawa A, Shirasaka D, Yamamoto S *et al.* (2003) Mice deficient in protein tyrosine phosphatase receptor type Z are resistant to gastric ulcer induction by VacA of *Helicobacter pylori*. *Nature Genetics*, **33**, 375–381.

Geis G, Suerbaum S, Forstoff B *et al.* (1993) Ultrastructure and biochemical studies of the flagellar sheath of *Helicobacter pylori*. *Journal of Medical Microbiology*, **38**, 371–377.

Gerrits MM, Schuijffel D, vanZwet AA *et al.* JG (2002) Alterations in penicillin-binding protein 1A confer resistance to beta-lactam antibiotics in *Helicobacter pylori*. *Antimicrobial Agents and Chemotherapy*, **46**, 2229–2233.

Goodwin A, Kersulyte D, Sisson G *et al.* (1998) Metronidazole resistance in *Helicobacter pylori* is due to null mutations in a gene (*rdxA*) that encodes an oxygen-insensitive NADPH nitroreductase. *Molecular Microbiology*, **28**, 383–393.

Goodwin CS and Armstrong JA (1990) Microbiological aspects of *Helicobacter pylori* (*Campylobacter pylori*). *European Journal of Clinical Microbiology*, **9**, 1–13.

Goodwin CS, Armstrong JA and Marshall BJ (1986) *Campylobacter pyloridis*, gastritis and peptic ulceration. *Journal of Clinical Pathology*, **39**, 353–365.

Goodwin CS, Armstrong JA, Chilvers T *et al.* (1989) Transfer of *Campylobacter pylori* and *Campylobacter mustelae* to *Helicobacter* gen. nov. as *Helicobacter pylori* comb. nov. and *Helicobacter mustelae* comb. nov. respectively. *International Journal of Systematic Bacteriology*, **39**, 397–405.

Harris AG, Wilson JE, Danon SJ *et al.* (2003) Catalase (KatA) and KatA-associated protein (KapA) are essential to persistent colonization in the *Helicobacter pylori* SS1 mouse model. *Microbiology*, **149**, 665–672.

Hatakeyama M (2003) *Helicobacter pylori* CagA—a potential bacterial oncoprotein that functionally mimics the mammalian Gab family of adaptor proteins. *Microbes and Infection*, **5**, 143–150.

Hessey SJ, Spencer J, Wyatt JI *et al.* (1990) Bacterial adhesion and disease activity in *Helicobacter*-associated chronic gastritis. *Gut*, **31**, 134–138.

Higashi H, Tsutsumi R, Fujiyata A *et al.* (2002) Biological activity of the *Helicobacter pylori* virulence factor CagA is determined by variation in the tyrosine phosphorylation sites. *Proceedings of the National Academy of Sciences of the United States of America*, **99**, 14428–14433.

Houben MH, Van De Beek D, Hensen EF *et al.* (1999) A systematic review of *Helicobacter pylori* eradication therapy – the impact of antimicrobial resistance on eradication rates. *Alimentary Pharmacology & Therapeutics*, **13**, 1047–1055.

IARC Working Group on the Evaluation of Carcinogenic Risks to Humans (1994) *Helicobacter pylori*. In *Schistosomes, Liver Flukes, and Helicobacter pylori: Views and Expert Opinions of an IARC Working Group on the Evaluation of Carcinogenic Risks to Humans*. International Agency for Research on Cancer, Lyon, pp. 177–240.

Ilver D, Arnqvist A, Ögren J *et al.* (1998) *Helicobacter pylori* adhesin binding fucosylated histo-blood group antigens revealed by retagging. *Science*, **279**, 373–377.

Jenks PJ (2002) Causes of failure of eradication of *Helicobacter pylori*. *British Medical Journal*, **325**, 3–4.

Jenks PJ, Ferrero RL and Labigne A (1999) The role of the *rdxA* gene in the evolution of metronidazole resistance in *Helicobacter pylori*. *The Journal of Antimicrobial Chemotherapy*, **43**, 753–758.

Jeong JY, Mukhopadhyay AK, Dailidiene D *et al.* (2000) Sequential inactivation of *rdxA* (HP0954) and *frxA* (HP0642) nitroreductase genes causes moderate and high-level metronidazole resistance in *Helicobacter pylori*. *Journal of Bacteriology*, **182**, 5082–5090.

Josenhans C, Niehus E, Amersbach S *et al.* (2002) Functional characterization of the antagonistic flagellar late regulators FliA and FlgM of *Helicobacter pylori* and their effects on the *H. pylori* transcriptome. *Molecular Microbiology*, **43**, 307–322.

Jungblut PR, Bumann D, Haas G *et al.* (2000) Comparative proteome analysis of *Helicobacter pylori*. *Molecular Microbiology*, **36**, 710–725.

Kavermann H, Burns BP, Angermüller K *et al.* (2003) Identification and characterization of *Helicobacter pylori* genes essential for gastric colonization. *The Journal of Experimental Medicine*, **197**, 813–822.

Kelly DJ (1998) The physiology and metabolism of the human gastric pathogen *Helicobacter pylori*. *Advances in Microbial Physiology*, **40**, 137–189.

Kersulyte D, Chalkauskas H and Berg DE (1999) Emergence of recombinant strains of *Helicobacter pylori* during human infection. *Molecular Microbiology*, **31**, 31–43.

Kusters JG, Gerrits MM, Strijp JA and Vandenbrouke-Grauls CM (1997) Coccoid forms of *Helicobacter pylori* are the morphological manifestation of cell death. *Infection and Immunity*, **65**, 3672–3679.

Kwon DH, Kato M, El-Zaatari FA *et al.* (2000) Frame-shift mutations in NAD(P)H flavin oxidoreductase encoding gene (*frxA*) from metronidazole resistant *Helicobacter pylori* ATCC43504 and its involvement in metronidazole resistance. *FEMS Microbiology Letters*, **188**, 197–202.

Kwon DH, Dore MP, Kim JJ *et al.* (2003) High-level beta-lactam resistance associated with multidrug resistance in *Helicobacter pylori*. *Antimicrobial Agents and Chemotherapy*, **47**, 2169–2178.

Labigne A and Jenks PJ (2001) Mutagenesis. In *Helicobacter pylori: Physiology and Genetics*. Mobley HL, Mendz GL and Hazell SL (eds). American Society for Microbiology, Washington, DC, pp. 335–344.

Labigne A, Cussac V and Courcoux P (1991) Shuttle cloning and nucleotide sequence of *Helicobacter pylori* genes responsible for urease activity. *Journal of Bacteriology*, **173**, 1920–1931.

Leong RW and Sung JJ (2002) *Helicobacter* species and hepatobiliary disease. *Alimentary Pharmacology & Therapeutics*, **16**, 1037–1045.

Leunk RD, Johnson PT, David BC *et al.* (1988) Cytotoxin activity in broth-culture filtrates of *Campylobacter pylori*. *Journal of Medical Microbiology*, **26**, 93–99.

Lim RK (1920) A parasitic spiral organism in the stomach of the cat. *Parasitology*, **12**, 108–112.

Loughlin MF, Barnard FM, Jenkins D *et al.* (2003) *Helicobacter pylori* mutants defective in RuvC Holliday junction resolvase display reduced macrophage survival and spontaneous clearance from the murine gastric mucosa. *Infection and Immunity*, **71**, 2022–2031.

Mahdavi J, Sondén B, Hurtig M *et al.* (2002) *Helicobacter pylori* SabA adhesin in persistent infection and chronic inflammation. *Science*, **297**, 573–578.

Malfertheiner P, Mégraud F, O'Morain C et al. (2002) Current concepts in the management of Helicobacter pylori infection—the Maastricht 2–2000 Consensus Report. Alimentary Pharmacology & Therapeutics, **16**, 167–180.

Marais A, Mendz GL, Hazell SL and Mégraud F (1999) Metabolism and genetics of Helicobacter pylori: the genomic era. Microbiology and Molecular Biology Reviews, **63**, 642–674.

Marshall BJ (1983) Unidentified curved bacilli on gastric epithelium in chronic active gastritis. Lancet, **i**, 1273–1275.

Marshall BJ, Armstrong JA, McGechie DB and Glancy RJ (1985) Attempt to fulfill Koch's postulates for pyloric Campylobacter. The Medical Journal of Australia, **142**, 436–439.

McGee DJ, Radcliff FJ, Mendz GL et al. (1999) Helicobacter pylori rocF is required for arginase activity and acid protection in vitro but is not essential for colonization of mice or for urease activity. Journal of Bacteriology, **181**, 7314–7322.

McGowan CC, Necheva AS, Forsyth MH et al. (2003) Promoter analysis of Helicobacter pylori genes with enhanced expression at low pH. Molecular Microbiology, **48**, 1225–1239.

Mégraud F (1998) Epidemiology and mechanism of antibiotic resistance in Helicobacter pylori. Gastroenterology, **115**, 1278–1282.

Mégraud F, Lehn N, Lind T et al. (1999) Antimicrobial susceptibility testing of Helicobacter pylori in a large multicenter trial: the MACH2 study. Antimicrobial Agents and Chemotherapy, **43**, 2747–2752.

Merrell DS, Goodrich ML, Otto G et al. (2003) pH-regulated gene expression of the gastric pathogen Helicobacter pylori. Infection and Immunity, **71**, 3529–3539.

Molinari M, Salio M, Galli C et al. (1998) Selective inhibition of li-dependent antigen processing by Helicobacter pylori toxin VacA. The Journal of Experimental Medicine, **187**, 135–140.

Monteiro MA, Chan KH, Rasko DA et al. (1998) Simultaneous expression of type 1 and type 2 Lewis blood group antigens by Helicobacter pylori lipopolysaccharides. The Journal of Biological Chemistry, **273**, 11533–11543.

Moran AP (1996) The role of lipopolysaccharide in Helicobacter pylori pathogenesis. Alimentary Pharmacology & Therapeutics, **10** (Suppl. 1), 39–50.

Muotiala A, Helander IM, Pyhälä L et al. (1992) Low biological activity of Helicobacter pylori lipopolysaccharide. Infection and Immunity, **60**, 1714–1716.

National Committee for Clinical Laboratory Standards (1999) Performance Standards for Antimicrobial Susceptibility Testing—Ninth Informational Supplement: Document M100–S9. National Committee for Clinical Laboratory Standards, Villanova, PA.

Naumann M, Wessler S, Bartsch C et al. (1999) Activation of activator protein 1 and stress response kinases in epithelial cells colonized by Helicobacter pylori encoding the cag pathogenicity island. The Journal of Biological Chemistry, **274**, 31655–31662.

Nedenskov-Sorensen P, Bukholm G and Bovre K (1990) Natural competence for genetic transformation in Campylobacter pylori. The Journal of Infectious Diseases, **161**, 365–366.

Negrini R, Lisato L, Zanella I et al. (1991) Helicobacter pylori infection induces antibodies cross-reacting with human gastric mucosa. Gastroenterology, **101**, 437–445.

Odenbreit S, Till M, Hofreuter D et al. (1999) Genetic and functional characterization of the alpAB gene locus essential for the adhesion of Helicobacter pylori to human gastric tissue. Molecular Microbiology, **31**, 1537–1548.

Odenbreit S, Püls J, Sedlmaier B et al. (2000) Translocation of Helicobacter pylori CagA into gastric epithelial cells by type IV secretion. Science, **287**, 1497–1500.

Olson JW and Maier RJ (2002) Molecular hydrogen as an energy source for Helicobacter pylori. Science, **298**, 1788–1790.

Pagliaccia C, de Bernard M, Lupetti P et al. (1998) The m2 form of the Helicobacter pylori cytotoxin has cell type-specific vacuolating activity. Proceedings of the National Academy of Sciences of the United States of America, **95**, 10212–10217.

Papini E, de Bernard M, Milia E et al. (1994) Cellular vacuoles induced by Helicobacter pylori originate from late endosomal compartments. Proceedings of the National Academy of Sciences of the United States of America, **91**, 9720–9724.

Papini E, Satin B, Norais N et al. (1998) Selective increase of the permeability of polarized epithelial cell monolayers by Helicobacter pylori vacuolating cytotoxin. The Journal of Clinical Investigation, **102**, 813–820.

Parsonnet J, Friedman GD, Vandersteed DP et al. (1991) Helicobacter pylori infection and the risk of gastric cancer. The New England Journal of Medicine, **325**, 1127–1129.

Parsonnet J, Hansen S, Rodriguez L et al. (1994) Helicobacter pylori infection and gastric lymphoma. The New England Journal of Medicine, **330**, 1267–1271.

Parsonnet J, Shmuely H and Haggerty T (1999) Fecal and oral shedding of Helicobacter pylori from healthy infected adults. Journal of the American Medical Association, **282**, 2240–2245.

Pawlowski K, Zhang B, Rychlewski L and Godzik A (1999) The Helicobacter pylori genome: from sequence analysis to structural and functional predictions. Proteins, **36**, 20–30.

Peek RM (2003) Intoxicated cells and stomach ulcers. Nature Genetics, **33**, 328–330.

Phadnis SH, Parlow MH, Levy M et al. (1996) Surface localization of Helicobacter pylori urease and a heat shock protein homolog requires bacterial autolysis. Infection and Immunity, **64**, 905–912.

Prinz C, Hafsi N and Voland P (2003) Helicobacter pylori virulence factors and the host immune response: implications for therapeutic vaccination. Trends in Microbiology, **11**, 134–138.

Puls J, Fischer W and Haas R (2002) Activation of Helicobacter pylori CagA by tyrosine phosphorylation is essential for dephosphorylation of host cell proteins in gastric epithelial cells. Molecular Microbiology, **43**, 961–969.

Rain JC, Selig L, De Reuse H et al. (2001) The protein–protein interaction map of Helicobacter pylori. Nature, **409**, 211–215.

Ramarao N, Gary-Owen SD and Meyer TF (2000) Helicobacter pylori induces but survives the extracellular release of oxygen radicals from professional phagocytes using its catalase activity. Molecular Microbiology, **38**, 103–113.

Rohde M, Püls J, Buhrdorf R et al. (2003) A novel sheathed surface organelle of the Helicobacter pylori cag type IV secretion system. Molecular Microbiology, **49**, 219–234.

Rosenow EC and Sandford AH (1915) The bacteriology of ulcer of the stomach and duodenum in man. The Journal of Infectious Diseases, **17**, 210–226.

Rupnow MF, Owens DK, Shachter R and Parsonett J (1999) Helicobacter pylori vaccine development and use: a cost-effectiveness analysis using the Institute of Medicine Methodology. Helicobacter, **4**, 272–280.

Salama NR, Otto G, Tompkins L and Falkow S (2001) Vacuolating cytotoxin of Helicobacter pylori plays a role during colonization in a mouse model of infection. Infection and Immunity, **69**, 730–736.

Salomon H (1896) Ueber das spirillum des saugetiermagens und sein verhalten zu den gelegzellen. Zentralblatt fur Bakteriologie, **19**, 433–441.

Segal ED, Cha J, Lo J et al. (1999) Altered states: involvement of phosphorylated CagA in the induction of host cellular growth changes by Helicobacter pylori. Proceedings of the National Academy of Sciences of the United States of America, **96**, 14559–14564.

Selbach M, Moese S, Hauch CR et al. (2002) Src is the kinase of the Helicobacter pylori CagA protein in vitro and in vivo. The Journal of Biological Chemistry, **277**, 6775–6778.

Skouloubris S, Labigne A and DeReuse H (1997) Identification and characterization of an aliphatic amidase of Helicobacter pylori. Molecular Microbiology, **25**, 989–998.

Skouloubris S, Thiberge JM, Labigne A and Reuse HD (1998) The Helicobacter pylori UreI protein is not involved in urease activity but is essential for bacterial survival in vivo. Infection and Immunity, **66**, 4517–4521.

Solnick JV and Schauer DB (2001) Emergence of diverse Helicobacter species in the pathogenesis of gastric and enterohepatic diseases. Clinical Microbiology Reviews, **14**, 59–97.

Spohn G and Scarlato V (1999) The autoregulatory HspR repressor protein governs chaperone gene transcription in Helicobacter pylori. Molecular Microbiology, **34**, 663–674.

Steer HW (1975) Ultrastructure of cell migration through the gastric epithelium and its relationship to bacteria. Journal of Clinical Pathology, **28**, 639–646.

Stein M, Rappuoli R and Covacci A (2000) Tyrosine phosphorylation of the Helicobacter pylori CagA antigen after cag-driven host cell translocation. Proceedings of the National Academy of Sciences of the United States of America, **97**, 1263–1268.

Stein M, Bagnoli F, Halenbach R et al. (2002) c-Src/Lyn kinases activate Helicobacter pylori CagA through tyrosine phosphorylation of the EPIYA motifs. Molecular Microbiology, **43**, 971–980.

Suerbaum S and Josenhans C (1999) Virulence factors of H. pylori: implications for vaccine development. Molecular Medicine Today, **5**, 32–39.

Suerbaum S, Smith JM, Bapumia K *et al.* (1998) Free recombination within *Helicobacter pylori. Proceedings of the National Academy of Sciences of the United States of America*, **95**, 12619–12624.

Telford JL, Ghiara P, Dell'Orco M *et al.* (1994) Gene structure of the *Helicobacter pylori* cytotoxin and evidence of its key role in gastric disease. *The Journal of Experimental Medicine*, **179**, 1653–1658.

Thompson LJ and de Reuse H (2002) Genomics of *Helicobacter pylori. Helicobacter*, **7** (Suppl. 1), 1–7.

Thompson LJ, Merrell DS, Neilan BA *et al.* (2003) Gene expression profiling of *Helicobacter pylori* reveals a growth-phase-dependent switch in virulence gene expression. *Infection and Immunity*, **71**, 2643–2655.

Tomb JF, White O, Kervalage AR *et al.* (1997) The complete genome sequence of the gastric pathogen *Helicobacter pylori. Nature*, **388**, 539–547.

Tombola F, Morbiato L, Del Giudice G *et al.* (2001) The *Helicobacter pylori* VacA toxin is a urea permease that promotes urea diffusion across epithelia. *The Journal of Clinical Investigation*, **108**, 803–804.

Toracchio S, Cellini L, Di Campli E *et al.* (2000) Role of antimicrobial susceptibility testing on efficacy of triple therapy in *Helicobacter pylori* eradication. *Alimentary Pharmacology & Therapeutics*, **14**, 1639–1643.

Trieber CA and Taylor DE (2002) Mutations in the 16S rRNA genes of *Helicobacter pylori* mediate resistance to tetracycline. *Journal of Bacteriology*, **184**, 2131–2140.

Van der Wouden EJ, Thijs JC, van Zwet AA *et al.* (1999) The influence of in vitro nitroimidazole resistance on the efficacy of nitroimidazole containing anti-*Helicobacter pylori* regimens: a meta-analysis. *The American Journal of Gastroenterology*, **94**, 1751–1759.

van Vliet AH, Poppelaars SW, Davies BJ *et al.* (2002) NikR mediates nickel-responsiveness transcriptional induction of urease expression in *Helicobacter pylori. Infection and Immunity*, **70**, 2846–2852.

Versalovic J, Shortridge D, Kibler K *et al.* (1996) Mutations in 23S rRNA are associated with clarithromycin resistance in *Helicobacter pylori. Antimicrobial Agents and Chemotherapy*, **40**, 477–480.

Viala J, Chaput C, Boneca IG *et al.* (2003) Nod1-dependent proinflammatory responses to *Helicobacter pylori* infection in gastric epithelial cells. *Gastroenterology*, **124** (Suppl. 1), **A-43**.

Voland P, Weeks DL, Marcus EA *et al.* (2003) Interactions among the seven *Helicobacter pylori* proteins encoded by the urease gene cluster. *American Journal of Physiology. Gastrointestinal and Liver Physiology*, **284**, G96–G106.

Wang G, Humayun MZ and Taylor DE (1999) Mutation as an origin of genetic variability in *Helicobacter pylori. Trends in Microbiology*, **7**, 488–493.

Warren JR (1983) Unidentified curved bacilli on gastric epithelium in chronic active gastritis. *Lancet*, **i**, 1273–1275.

Weeks DL, Eskandari S, Scott DR and Sachs G (2000) A H +-gated urea channel: the link between *Helicobacter pylori* urease and gastric colonization. *Science*, **287**, 482–485.

Worst DJ, Otto BR and Graaf J (1995) Iron-repressible outer membrane proteins of *Helicobacter pylori* involved in heme uptake. *Infection and Immunity*, **63**, 4161–4165.

Yahiro K, Niidome T, Kimura M *et al.* (1999) Activation of *Helicobacter pylori* VacA toxin by alkaline or acid conditions increases its binding to a 250-kDa receptor protein-tyrosine phosphatase beta. *The Journal of Biological Chemistry*, **274**, 36693–36699.

Ye D, Willhite DC and Blanke SR (1999) Identification of the minimal intracellular vacuolating domain of the *Helicobacter pylori* vacuolating cytotoxin. *The Journal of Biological Chemistry*, **274**, 9277–9282.

Yoshida N, Granger DN, Evans DJ *et al.* (1993) Mechanisms involved in *Helicobacter pylori*-induced inflammation. *Gastroenterology*, **105**, 1431–1440.

Young GO, Stemmet N, Lastovica A *et al.* (1992) *Helicobacter pylori* lipopolysaccharide stimulates gastric mucosal pepsinogen secretion. *Alimentary Pharmacology & Therapeutics*, **6**, 169–177.

Zevering Y, Jacob L and Meyer TF (1999) Naturally acquired human immune responses against *Helicobacter pylori* and implications for vaccine development. *Gut*, **45**, 465–474.

Zheng PW and Jones NL (2003) *Helicobacter pylori* strains expressing the vacuolating cytotoxin interrupt phagosome maturation in macrophages by recruiting and retaining TACO (coronin 1) protein. *Cellular Microbiology*, **5**, 25–40.

41

Campylobacter and *Arcobacter* spp.

Diane E. Taylor[1] and Monika Keelan[2]

[1]*Department of Medical Microbiology and Immunology; and* [2]*Department of Laboratory Medicine & Pathology, Division of Medical Laboratory Science, University of Alberta, Edmonton, Alberta, Canada*

INTRODUCTION

The epsilon subdivision of Proteobacteria includes the family Campylobacteraceae, which is comprised of genera *Campylobacter*, *Arcobacter* and, more recently, genus *Sulfurospirillum*, as well as the generically misclassified *Bacteroides ureolyticus* (Vandamme, 2000). The complex taxonomy of Gram-negative spiral-shaped bacteria is continuously evolving with new information acquired from phylogenetic studies (Dekeyser *et al.*, 1972; Butzler *et al.*, 1973; Skirrow, 1977; Vandamme and De Ley, 1991; Trust *et al.*, 1994; Ursing, Lior and Owen, 1994; On, 1996; Vandamme *et al.*, 1997). Genus *Suflurospirillum* are free-living *Campylobacter*-like organisms that reduce sulfur and are isolated from mud, and *Bacteroides ureolyticus* are non-motile rods that resemble *Campylobacter* that have been isolated from soft-tissue ulcers, periodontal disease and urethritis (Vandamme, 2000). These newer additions to the Campylobacteracea family are not discussed in detail in this chapter.

Taxonomic issues resolved through studies that exploit DNA–rRNA hybridization, DNA–DNA hybridization and 16S rRNA sequencing concluded that the genus *Helicobacter* is too unrelated to *Campylobacter* and *Arcobacter* to group them together (Romaniuk *et al.*, 1987; Vandamme *et al.*, 1991). Identification methods for Campylobacteria have been extensively reviewed (On, 1996). Quantitative DNA–DNA hybridization demonstrated that the proposed species *C. hyoilei* actually represents additional strains of *C. coli*; therefore its usage is to be discouraged in favour of *C. coli* (Vandamme *et al.*, 1997; Vandamme and On, 2001). Phenotypic characteristics also provide support for differentiation of bacterial species (Thompson *et al.*, 1988; Vandamme *et al.*, 1991, 1992b), although identification of Campylobacteraceae is problematic due to the relative biochemical inactivity and fastidious growth requirements (On, 1996, 2001). *Campylobacter sputorum* is emended to include three biovars, *sputorum*, *fecalis* and *paraureolyticus*, with previously defined biovar *bubulus* now considered to be biovar *sputorum* (On, Atabay and Corry, 1999; Vandamme and On, 2001). Extensive 16S rRNA gene sequence diversity in *C. hyointestinalis* confirms the taxonomic position of the two different subspecies previously described – *C. hyointestinalis* subsp. *hyointestinalis* and *C. hyointestinalis* subsp. *lawsonii* (On and Holmes, 1995; Harrington and On, 1999). Two new *Campylobacter* spp. isolated from humans have also been characterized: *C. lanienae* sp. nov. and *C. hominus* sp. nov. (Logan *et al.*, 2000; Lawson *et al.*, 2001).

Values of less than 97.0% 16S rRNA similarity between two organisms suggest they represent distinct species (Stackebrandt and Goebel, 1994). As an alternative to 16S rRNA gene sequencing, the *recA* gene and restriction fragment length polymorphism (RFLP) analysis of the 23S rRNA gene may possibly resolve conflicts in highly divergent 16S rRNA genes of certain species (On, 2000).

DNA–DNA hybridization is generally accepted to be superior to 16S rRNA phylogenetic analysis for establishing relationships between closely related taxa (Stackebrandt and Goebel, 1994). A bacterial species is defined as a group of strains (including the type strain) sharing 70% or greater DNA–DNA relatedness with 5C or less ΔT_m (Wayne *et al.*, 1987). Amplified fragment length polymorphisms (AFLP) analysis is a high-resolution, whole-genome fingerprinting method that is able to concurrently identify taxonomic and epidemiological relationships among *Campylobacter* species (On and Harrington, 2000). A polyphasic taxonomic approach that examines the overall phenotypic, ecologic and genotypic consistencies is considered to be essential for the accurate classification and identification of bacteria within *Proteobacteria* (Stackebrandt and Goebel, 1994; Vandamme *et al.*, 1996; Harrington and On, 1999).

The species of genera *Campylobacter* and *Arcobacter* are remarkably diverse in their natural habitats and ability to produce disease. Habitats include the human oral cavity, human and/or animal intestinal tracts, the genital tracts of cattle and the sediments in salt marshes. Diseases include diarrhoea, bacteremia, septic abortion, infertility and possibly diseases of the oral cavity. The procedures for isolation and culture vary considerably with the species with respect to temperature, atmospheric conditions and media composition. Unfortunately, the phenotypic characteristics of Campylobacteria do not reflect their diversity, making it difficult to readily differentiate the genera and species. These identification and classification problems remain a challenge for the clinical laboratory.

PROPERTIES OF THE FAMILY CAMPYLOBACTERACEAE

Detection and identification of *Campylobacter* and *Arcobacter* species is a challenging task due to cross-species similarities in phenotype and general biochemical unreactivity. Most species are curved, S-shaped or spiral Gram-negative rods. Four species (*C. showae*, *C. rectus*, *C. gracilis* and *C. hominus* sp. nov.) produce only straight rods. Campylobacteria may stain poorly with Gram counterstains other than carbol or basic fuchsin. The cells vary in length from 0.5 to 5 μm and in width from 0.2 to 0.8 μm. With the exception of *C. gracilis* and *C. hominus* sp. nov. which are non-motile, most of the species have a darting type of motility produced by the action of unsheathed flagella. There may be a single polar flagellum or bipolar flagella in all species except *C. showae*, which has a cluster of two to five flagella at one end. Spores are not produced but spherical coccal forms have been observed in some species in older cultures or under unfavourable growth conditions. A characteristic common to all Campylobacteraceae is their ability to grow under microaerobic conditions, although some also grow under aerobic or anaerobic conditions. Growth is optimal between 30 and 37 °C. None of the species ferment or oxidize carbohydrates and all species utilize organic amino acids or tricarboxylic acid intermediates as sources of carbon. The DNA base ratio ranges between 27 and 47%. All species of the family except *A. nitrofigilis* are either commensal or pathogens of humans or animals. *Arcobacter nitrofigilis* is unique in the family in its ability to fix nitrogen and in its association with the roots of halophytic plants. A list of the species is provided in Table 41.1.

The Genus *Campylobacter*

Although the bacteria of the genus *Campylobacter* are the most common bacterial cause of enteritis worldwide, they are largely unfamiliar to the general public although their importance to human health is as great, or greater, than the well-known *Salmonella*, *Shigella* and verotoxin-producing *Escherichia coli*. Campylobacteria first identified with human illness were referred to 'related vibrios' to distinguish them from species of *Vibrio*, e.g. *Vibrio cholerae* (King, 1957, 1962). A convincing association of these organisms with human diarrhoea was demonstrated a decade later (Dekeyser *et al.*, 1972; Butzler *et al.*, 1973). Their significance to human health, however, was not established until a medium was designed for routine isolation of the organisms (Skirrow, 1977).

Table 41.1 Classification of Campylobacteraceae

Genus *Campylobacter*

Group 1. Thermophilic enteropathogenic species
    C. jejuni subsp. *jejuni*
    C. jejuni subsp. *doylei*
    C. coli[a]
    C. lari
    C. upsaliensis
    C. helveticus
Group 2. Animal pathogens and commensals that may infect humans
    C. fetus subsp. *fetus*
    C. fetus subsp. *venerealis*
    C. hyointestinalis subsp. *hyointestinalis*
    C. hyointestinalis subsp. *lawsonii*
    C. sputorum biovar *sputorum*[b]
    C. sputorum biovar *faecalis*
    C. sputorum biovar *paraureolyticus*
    C. mucosalis
    C. lanienae sp. nov.
Group 3. Species associated with periodontal disease
    C. concisus
    C. curvus
    C. rectus
    C. showae
    C. gracilis
    C. hominus sp. nov.
Genus *Arcobacter*
    A. nitrofigilis
    A. cryaerophilus
    A. butzleri
    A. skirrowii
Genus *Sulfurospirillum*
    S. arcachonense
    S. arsenophilum
    S. barnesii
    S. deleyianum
    [Bacteroides] ureolyticus

The type species of genus Campylobacter and genus *Arcobacter* are *C. fetus* subsp. *fetus* and *A. nitrofigilis*.
[a] *C. hyoilei* is synonymous with *C. coli*, but its name usage is discouraged.
[b] *C. sputorum* biovar *sputorum* incorporates biovar *bubulus*.

Campylobacteraceae can be conveniently separated into three groups on the basis of phenotypic characteristics (Table 41.1). The first and most important group to human health consists of five thermophilic enteropathogenic species. In this group are *C. jejuni* with two subspecies, *C. coli*, *C. lari*, *C. upsaliensis* and *C. helveticus*. A second group includes four species well known to veterinary microbiologists because of their frequent occurrence as pathogens or commensals in farm livestock, although some species have been shown to be pathogenic for humans. Included are *C. fetus* with two subspecies, *C. hyointestinalis* with two subspecies, *C. sputorum* with three biovars and *C. mucosalis*. A third group, found in human periodontal disease, consists of *C. concisus*, *C. curvas*, *C. rectus*, *C. showae* and *C. gracilis*. Two new species, *C. lanienae* sp. nov. and *C. hominus* sp. nov., have been grouped according to their similarity with other *Campylobacter* species.

Genetics and Pathogenicity

Several molecular methods have been specifically designed for use with *Campylobacter*. DNA has been successfully introduced into *C. jejuni* subsp. *jejuni* by natural transformation, electrotransformation or conjugation (Taylor, 1992a). However, not all strains appeared to be competent and plasmid DNA transformed with low frequency regardless of the technique, although conjugation is fairly efficient. Many chromosomal genes have been stably expressed in *E. coli*, including amino acid biosynthesis genes, ribosomal RNA genes and transfer RNA genes. One of the genes (*flbA*) involved in flagellar biogenesis in *C. jejuni* subsp. *jejuni* was cloned, although expression in *E. coli* was not detected (Miller, Pesci and Pickett, 1993). All of these genes are general housekeeping genes and appear to be highly conserved across species boundaries. Several shuttle vector series are available for *Campylobacter* genes and may be used for cloning, expression systems, sequence analysis and for site-specific mutagenesis (Taylor, 1992a; Yao *et al.*, 1993). These vectors were not functional for genes from *C. hyointestinalis* and hence a separate vector has been developed (Waterman, Hackett and Manning, 1993).

Since *C. jejuni* subsp. *jejuni* genome has been sequenced (Parkhill *et al.*, 2000), it has become possible to decipher some of the functions of the bacteria by examining the content and expression of the genome (Dorrell *et al.*, 2001). At least 21% of genes in the NCTC11168 appear to be dispensable as they are absent or highly divergent. A total of 1300 *C. jejuni* core genes present are associated with metabolic, biosynthetic, cellular and regulatory processes, as well as many virulence determinants. All of these genes are general housekeeping genes and appear to be highly conserved across species boundaries. Most recently, comparative genome analysis of 18 *C. jejuni* from different sources using whole-genome DNA microarrays has revealed that 16.3% of the genes present in the sequenced strain, NCTC11168, were either absent or highly variable in sequence (Pearson *et al.*, 2003). Seven clusters of variable sequences comprising 50% of the variable gene pool were found, suggesting these clusters were lost or acquired as a group during evolution. These clusters differ from variable gene clusters in islands and pathogenicity islands because their G + C content is not different from the genome and not associated with mobile elements for DNA transfer. PR1 contains genes utilized in alternative electron transfer, which may play a role in oxygen-restricted environments. Three of the gene clusters (PR4, 5 and 6) are involved with the production of surface structures (lipo-oligosaccharides, flagella and capsule) and may be important in the avoidance of innate and adaptive immune responses in the host. The remaining clusters contain proteins of the outer and periplasmic membranes which may play a role in *Campylobacter* adaptation to different ecological niches.

The factors determining the clinical outcome of infection with *C. jejuni* are likely dependent upon the infecting strain as well as on the susceptibility of the host. Colonization of human intestine with *C. jejuni* results in the expression of several putative virulence factors, yet the mechanisms of pathogenesis are unclear (Van Vliet and Ketley, 2001). Colonization may be influenced by the expression of lipopolysaccharides (LPS) on the bacterial cell surface. A mutation in the *galE* gene involved in the synthesis of LPS reduced the ability of *C. jejuni* to adhere and invade INT 407 cells, suggesting that LPS is a virulence factor for *C. jejuni* infection, although the *C. jejuni* was still able to colonize chickens (Fry *et al.*, 2000). Flagella are an important determinant in virulence since colonization of intestinal tracts does not occur with non-motile mutants of *C. jejuni* (Wassenaar, Fry and van der Zeijst, 1995). Internalization of *C. jejuni* into mammalian cells requires the secretion of invasion proteins (Konkel *et al.*, 1999). *Campylobacter jejuni* also produces cytolethal distending toxin which interferes with the cell cycle (Whitehouse *et al.*, 1998). Porins may affect the ability of *Campylobacter* to survive transmission from animals to humans (Bacon, Johnson and Rodgers, 1999; Dedieu, Pages and Bolla, 2002). The presence of plasmids in *C. jejuni* may play an important role in the pathogenesis associated with infection. A transmissible plasmid encodes resistance to tetracycline via the Tet(O) determinant in *C. jejuni* (Taylor *et al.*, 1981; Connell *et al.*, 2003). Another plasmid, pVir, isolated from *C. jejuni* 81–176 (Bacon *et al.*, 2000, 2002), encodes proteins with homology to components of a type IV secretion system (including *virB11*) associated with virulence in other bacterial pathogens (Christie, 2001). Two recent reports also cite the presence of *virB11* in human and animal isolates of *C. jejuni* (Bang *et al.*, 2003; Datta, Niwa and Itoh, 2003). The pVir plasmid enhances the ability of *C. jejuni* 81–176 to adhere and invade intestinal epithelial cells in culture (Bacon *et al.*, 2002). Continued study on the prevalence of putative virulence genes will provide insight into the adaptability of *Campylobacter* species and their association with the development of disease (Bang *et al.*, 2003; Datta, Niwa and Itoh, 2003).

Flagellar Genes and Phase Variation

Several attempts have been made to clone toxin genes as well as membrane proteins from *C. jejuni* subsp. *jejuni*, but none of these efforts has resulted in stable, entire clones (Calva *et al.*, 1989; Taylor, 1992a). The flagella biosynthesize genes that have been cloned from *C. jejuni* subsp. *jejuni*; however, no expression of gene products in *E. coli* has ever been detected. Both *C. jejuni* subsp. *jejuni* and *C. coli* produce two flagellin genes, *flaA* and *flaB*, which have strong sequence homology but differ in their regulatory mechanisms (Guerry *et al.*, 1992; Taylor, 1992a). Although both protein products are present in the flagella, mutations in the *flaA* gene induce the synthesis of truncated flagella, which results in only partial motility, whereas *flaB* mutations do not alter the length of the flagella or the motility of the cell. Recombination of the flagellin gene locus and horizontal gene transfer have been demonstrated and may provide a mechanism by which the bacteria evades the immune response or adapts to its environment (Wassenaar, Fry and van der Zeijst, 1995). Phase variation in *Campylobacter* spp. occurs relatively frequently *in vitro* and the transition from aflagellate to flagellate has also been observed *in vivo* by passage through the rabbit intestine (Caldwell *et al.*, 1985). The genetic mechanism for this is unknown but may involve turning off or on of the expression of the *flaA* gene. *Campylobacter* spp. are also able to reversibly express flagella of different antigenic specificities and, at least in some cases, this corresponds to the production of multiple flagellins and to a DNA rearrangement (Guerry *et al.*, 1992). A recombination event within the hypervariable sequences in the internal region of the *fla* gene may account for this rearrangement, but the exact mechanism is unknown.

Surface Proteins

Virulence-related surface layer proteins (S proteins) from *C. fetus* have been cloned (*sapA* genes) and will express in *E. coli* (Blaser and Gotschlich, 1990). It was shown that wildtype and S mutants possessed several *sapA* homologues but that only the mutants lacked expression sequences (Tummuru and Blaser, 1993). Comparison of *sapA* sequences from a pair of variant strains suggested that antigenic variation had occurred as a result of site-specific recombination of unique coding regions. The occurrence of Guillain–Barré syndrome following infection with *C. jejuni* (in particular serotype 0:19) has provoked much interest. It has been found that lipopolysaccharides extracted from *C. jejuni* have core oligosaccharides with terminal structures resembling human gangliosides (Aspinall *et al.*, 1994). Genes involved in LPS biosynthesis have been cloned (Klena, Gray and Konkel, 1998; Wood *et al.*, 1999). Both O- and N-glycosylation pathways are present in *Campylobacter*, which will facilitate the study of glycoprotein synthesis and the role of these molecules in pathogenesis (Szymanski *et al.*, 2003). A functional N-glycosylation pathway has recently been transferred from *C. jejuni* to *E. coli* (Wacker *et al.*, 2003). Alterations in protein glycosylation in *C. jejuni* may affect the immunogenicity of the proteins (Young *et al.*, 2002). *Campylobacter hyolei* (*coli*) LOS have been expressed in *E. coli* (Korolik *et al.*, 1997). Phase variation of lipooligosaccharide may also affect invasiveness of *C. jejuni* through alterations in the biosynthesis of the ganglioside GM1-like LOS structure (Linton *et al.*, 2000; Gilbert *et al.*, 2002; Guerry *et al.*, 2002).

Difficulties that have been experienced during the cloning of *Campylobacter* genes are numerous and have included both the failure to express the gene in *E. coli* and the instability of the genes in this host. The possible reasons for this may include differences in methylation between species, differences in promoter sequences used, the high A–T content of *Campylobacter* DNA and/or the lack of accessory proteins that might enable expression. As well, *C. hyointestinalis* was shown to have a restriction barrier against DNA from other *Campylobacter* spp. (Waterman, Hackett and Manning, 1993).

Campylobacter jejuni is a highly plastic organism as demonstrated by the number of hypervariable regions in the genome (Pearson *et al.*, 2003). These regions may be involved in the colonization of *C. jejuni* in animals and in lateral gene transfer among strains, which may have implications regarding the epidemiology of *C. jejuni* human infections. Increased understanding of the host–pathogen interaction is mediated by the identification of a unique protein glycosylation system and plasmid-encoded components of a type IV secretion system (Linton *et al.*, 2000).

Thermophilic Enteropathogenic Campylobacteria: Clinical Significance, Occurrence and Epidemiology

Thermophilic enteropathogenic species, with few exceptions, are isolated from the intestinal tracts of humans and/or animals at 42–43 °C. *Campylobacter jejuni* subsp. *jejuni* is recognized as the most important cause of *Campylobacter* enteritis (campylobacteriosis) in humans occurring in 89–93% of the cases. *Campylobacter coli* is most closely related to *C. jejuni* and is the cause of 7–10% of the human cases, while *C. lari* is the cause of only 0.1–0.2% (Healing, Greenwood and Pearson, 1992). The frequency of infections in the human population due to these three species is difficult to estimate precisely, but incidences ranging from 6 to 7 per 100 000 to 87 per 100 000 have been reported in the USA and the UK (Skirrow and Blaser, 1992). Most infections are due to sporadic cases: exposure to raw meat, a pet with diarrhoea and drinking untreated water, family outbreaks and larger community outbreaks arising from contaminated food, water or unpasteurized milk (Adak *et al.*, 1995; Sopwith *et al.*, 2003). It has also been implicated in travellers' diarrhoea. In developing countries it is the paediatric population in which the largest number of cases occur. Incidences as high as 40 000 per 100 000 have been reported and most of these occur in children 2 years old or younger. The adult population in the underdeveloped world is much less affected, presumably owing to the development of immunity during childhood exposure. Isolates of these three thermophilic species are rarely recovered from healthy humans in developed countries, but they can be obtained almost as frequently from healthy children as from children with diarrhoea in underdeveloped countries (Taylor, 1992b).

The campylobacteriosis produced in humans by these three thermophilic species varies from a mild short-term diarrhoea of 2 days to a severe form that may last a week or longer. The diarrhoea is watery and may contain blood. Vomiting only occurs in a small percentage of cases. The diarrhoea may be accompanied by severe pain, leading to suspicions of acute appendicitis, cholecystitis or peritonitis. Complications are infrequent but include septicaemia, reactive arthritis, Guillain–Barré syndrome, erythema nodosum and meningitis.

C. jejuni subsp. jejuni

In addition to its role as a major cause of human enteritis, C. jejuni subsp. jejuni also causes diarrhoea in cattle and has been isolated from dogs, cats and non-human primates with diarrhoea. This species has also been implicated as the cause of mastitis in cattle and abortion in sheep. In chickens, turkeys, pigeons, crows and gulls and other wild birds, C. jejuni subsp. jejuni constitutes part of normal intestinal flora. Other sources of isolates include hamsters, flies and mushrooms. Since there are such a large number of potential contributors to the contamination of natural waters, it is not surprising that large outbreaks have been traced to unchlorinated water supplies. Unpasteurized milk has also been implicated in a number of milk-borne outbreaks. However, the majority of the human cases can be traced to poultry, particularly to the consumption of chickens that have been incompletely cooked or to cookware, utensils and cutting boards that have been contaminated during the preparation of the chicken. These cases are, for the most part, sporadic but they constitute the majority of the human cases. Transmission of C. jejuni subsp. jejuni from cats and dogs has been reported to be the cause of sporadic cases, usually limited to a small number of associated individuals often in the same family (Tauxe, 1992). Occasionally complications of C. jejuni infection result in the development of more severe neurological and autoimmune illnesses such as Guillain–Barré, Miller–Fisher and Reiter's syndromes (Nachamkin, Allos and Ho, 1998).

Differentiation of C. jejuni subsp. jejuni from C. coli is commonly achieved with the hippurate hydrolysis test, in which C. coli is negative. Difficulty arises when some C. jejuni strains are also unable to hydrolyse hippurate (Totten et al., 1987). Additional tests such as growth on a minimal media are useful (Table 41.2).

Epidemiological investigations have confirmed that campylobacteriosis is a zoonotic disease. The use of both phenotyping and genotyping methods is required to achieve greater discrimination among subtype isolates of C. jejuni from humans and animals (Wassenaar and Newell, 2000). Tracing infections to animal sources has been accomplished mostly through the use of serotyping systems used mainly by reference laboratories (Penner, 1988). One system distinguishes strains on the basis of differences in specificities in thermostable antigens that are known to be lipopolysaccharide (Penner system) and the other employs differences in thermolabile antigens to distinguish between strains (Lior system). Serotyping of heat-stable antigens is widely accepted as a basis for reference typing of strains (Frost et al., 1998). Although serotyping is a widely used method that may be useful in certain epidemiological studies, up to 20% of human and veterinary strains may be non-typable (Neilsen, Engberg and Mogens, 1997). Serotyping has been reported to be useful when used in combination with genotypic methods such as pulsed-field gel electrophoresis (PFGE) or flagellar typing by PCR-RFLP-fla (Birkenhead et al., 1993). Genotyping methods have the advantage to be able to type strains that cannot be typed using phenotypic methods because genotyping methods measure stable chromosomal differences. In contrast phenotyping methods measure characteristics (i.e. antigens) that may not be stably expressed. PCR-RFLP-fla typing is a simple, inexpensive method that has been successfully applied to animals and humans (Newell et al., 2000). In the past, PFGE was reported to have superior discriminatory potential, compared to other molecular typing methods for C. jejuni, although some strains may not be typable due to the inability of commonly used enzymes to digest the DNA (Newell et al., 2000). Random amplification of polymorphic DNA (RAPD) typing is less sensitive to genetic instability than fla typing, but is not very reproducible with complex banding patterns (Madden, Moran and Scates, 1996). Flagellar gene sequencing may also be used to

Table 41.2 Characteristics for differentiating thermophilic enteropathogenic species of Campylobacter

Phenotypic characteristics	C. jejuni		C. coli	C. lari	C. upsaliensis	C. helveticus
	subsp. jejuni	subsp. doylei				
Catalase activity	+	v	+	+	−	−
Nitrate reduction	+	−	+	+	+	+
Selenite reduction					+	−
Hydrolysis of						
Hippurate	+	+	−	−	−	−
Indoxyl acetate	+	+	+	−	+	+
Growth						
at 42 °C	+	−	+	+	v[a]	+
at 25 °C	−	−	−	−	−	−
on minimal media	−	−	+	−	−	−
on potato starch					+	−
Anaerobic growth with trimethylamine N-oxide	−	−	−	+	−	−
Susceptibility[b] to						
Nalidixic acid	S	S	S	v	S	S
Cephalothin	R	S	R	R	v	S

Test reactions: +, present in >90% of strains examined; −, present in <11% of strains examined; v, variable strain-dependent reactions. Sources of data include Goossens and Butzler (1992), Stanley et al. (1992), Steele and Owen (1988), Tenover and Fennell (1992), On and Holmes (1995), On (1996) and Vandamme (2000).

[a] At least 80% of strains examined have this characteristic.

[b] Susceptibility to 30 μg disk of antibiotic: S, susceptible; R, resistant; v, variable strain dependent.

discriminate isolates of *C. jejuni* (Meinersmann *et al.*, 1997; Petersen and On, 2000). RAPD with PCR-RFLP-*iam* (invasion-associated marker) typing is a recent method by which invasive strains may be differentiated from non-invasive strains (Carvalho *et al.*, 2001). RFLP and PFGE have been successfully used to identify nosocomial transmission of *C. jejuni* (Llovo *et al.*, 2003). Other techniques to distinguish strains include plasmid profile determinations, phage typing, analysis of proteins from outer membranes, multilocus enzyme electrophoresis and ribotyping. Multilocus sequence typing (MLST) has recently been proposed to be a useful strategy for diverse bacterial populations that have a weakly clonal population structure such as *Campylobacter* (Dingle *et al.*, 2001, 2002). MLST is based on the sequence typing of seven housekeeping genes, which evolve slowly due to the stabilizing selection for the conservation of metabolic function. A set of reference isolates for examining clonal complexes of *C. jejuni* have been established (Wareing *et al.*, 2003). Whole-genome typing performed on digests of the *Campylobacter* genome using microarrays (Kim *et al.*, 2002) examines the entire genome, not just the housekeeping genes, and therefore is likely to be more discriminatory and more relevant than other typing strategies. Whole-genome typing by amplified fragment length polymorphism (AFLP) fingerprinting is reported to provide a similar level of discrimination as PFGE, although more recent reports suggest that AFLP is a superior method that facilitates concurrent species identification and relationships among *Campylobacter* strains (On and Harrington, 2000). Future studies must directly compare genomotyping, PCR-RFLP and AFLP with standard methods of serotyping and PFGE currently used in reference public health laboratories.

The molecular techniques are generally too complex for routine use and future research needs to be directed towards simplification.

C. jejuni subsp. *doylei*

The subspecies *C. jejuni* subsp. *doylei*, has been proposed for groups of bacteria isolated from faecal samples of paediatric patients with diarrhoea in Australia and South Africa and from gastric biopsies of adults in West Germany and the UK (Steele and Owen, 1988). These bacteria, previously referred to as atypical *C. jejuni* or as GCLO-2, grow optimally at 35–37 °C and poorly or not at all at 42 °C. They do not reduce nitrate to nitrite and approximately 20% do not produce catalase. The isolates are susceptible to cephalothin and generally also to nalidixic acid. Because of their preference for a growth temperature of 35–37 °C and their susceptibility to cephalothin, they are not likely to be isolated with procedures used routinely for *C. jejuni* subsp. *jejuni* and other thermophiles. The bacteria produce non-haemolytic, pinpoint colonies after incubation for 5 to 6 days at 37 °C. In this respect and in their inability to reduce nitrate or grow at 42 °C, they resemble *Helicobacter fennelliae* (formerly *C. fennelliae*), with which they may be confused. Unlike *C. jejuni* subsp. *doylei*, however, the colonies of *H. fennelliae* have a distinct bleach-like odour and a tendency to swarm (Tenover and Fennell, 1992). The pathologenesis mechanism of *C. jejuni* subsp. *doylei* isolates has not yet been demonstrated.

C. coli

Campylobacter coli is a significant cause of human enteritis but its incidence is much lower than that of *C. jejuni* subsp. *jejuni*. This may be due to under-reporting arising from the inhibition of some *C. coli* strains by the antibiotics in the isolation media (Tenover and Fennell, 1992). Incidence varies from one area to the next as shown by the higher proportion of human enteritis cases in developing countries caused by *C. coli* (Skirrow and Blaser, 1992). *Campylobacter coli* is well known as the species that is most frequently found in healthy pigs, but it also occurs in the intestines of cattle, sheep and chickens. Many strains of *C. coli* can be serotyped on the basis of

their thermostable and thermolabile antigens. Although most strains of *C. coli* are susceptible to nalidixic acid, recent reports have drawn attention to the development of resistance to this antibiotic. In such cases they may be misclassified as *C. lari*, which are characteristically resistant to nalidixic acid. Molecular methods are often required to differentiate *C. coli* from *C. jejuni*, since some *C. jejuni*, like *C. coli*, are hippurate negative (Totten *et al.*, 1987).

C. lari

Organisms of this species were first isolated from seagulls and referred to as nalidixic acid-resistant thermophilic *Campylobacter* (NARTC). Susceptibility to nalidixic acid is dependent upon the strain. They have also been obtained from other species of birds, and from dogs, monkeys and sheep (Tenover and Fennell, 1992). *Campylobacter lari* is also a cause of human enteritis but such cases are rare relative to *C. jejuni* subsp. *jejuni*. Some strains of *C. lari* are capable of producing urease, a characteristic not shared by any other species of the genus. The ability to grow anaerobically in 0.1% trimethylamine-*N*-oxide hydrochloride is a key distinguishing feature, as well as an inability to hydrolyse indole acetate permits its differentiation from other species of the thermophilic enteropathogenic group of the genus.

C. upsaliensis

In studies to investigate transmission of Campylobacteria from dogs to humans, a number of strains were isolated from dogs that differed from the typical thermophilic enteropathogenic species by testing negative or weakly positive for catalase production. They were known as the catalase-negative or weak (CNW) group of strains before it was recognized that they were representatives of a new species. This species, named *C. upsaliensis*, consists of strains that are generally more susceptible to antibiotics and have a wider diversity in their susceptibilities than do the other members of the thermophilic group. For effective isolation, the filtration technique and media without antibiotics are recommended. Adoption of these measures has permitted the isolation of *C. upsaliensis* from patients with gastroenteritis and bacteremia, providing a convincing line of evidence that they are causes of diarrhoea in humans (Goossens and Butzler, 1992). *Campylobacter upsaliensis* has been isolated from a patient with diarrhoea (Taylor, Hiratsuka and Mueller, 1989), and also from the faeces of asymptomatic and diarrhoeic cats and dogs (Sandstedt and Ursing, 1991).

Isolates of *C. upsaliensis* are readily differentiated from other thermophilic species by a negative or weak reaction for catalase, and by their inability to hydrolyse hippurate or grow anaerobically in 0.1% trimethylamine *N*-oxide hydrochloride. *Campylobacter upsaliensis* isolates are usually susceptible to both cephalothin and nalidixic acid. They most closely resemble *C. helveticus* strains, from which they can be differentiated by their positive reactions for reduction of selenite and growth on potato starch.

C. helveticus

The most recently defined species of the thermophilic enteropathogenic group is *C. helveticus*. It was discovered during studies on the detection of *C. upsaliensis* in cats and dogs with gastroenteritis (Stanley *et al.*, 1992). The species reflects the phenotypic characteristics of *C. upsaliensis* but has a homology level of only 35% in DNA–DNA hybridizations and less homology with other species of the thermophilic group. Its differentiation from *C. upsaliensis* is based on its inability to reduce selenite or grow on potato starch. It is differentiated from the other thermophilic species by its lack of catalase production. On blood agar, *C. helveticus* produces smooth, flat colonies with a characteristic blue-green hue and a watery spreading appearance.

Approximately half of the Campylobacters isolated from cats belong to *C. helveticus* and the species is present in both healthy cats and in cats with gastroenteritis, but only 2% of the Campylobacters isolated from dogs belong to this species.

Laboratory Diagnosis

Procedures for Isolating Campylobacteria

Bacteria of the genus *Campylobacter* have a characteristic morphology and a darting type of motility that permits their identification by direct examination of broth suspensions of faeces. However, *C. jejuni* subsp. *jejuni* cannot be distinguished from *C. coli* by this procedure and the test is considerably less sensitive than isolation by culture.

A method used in early studies utilized the supernatant from a saline suspension of faeces which is drawn up in a syringe and then forced through a 0.65 µm filter. The filtered fluid, which contains the filter-passing Campylobacteria, is inoculated on a solid isolation medium (Dekeyser *et al.*, 1972). It has been replaced by direct plating on a selective isolation medium, the first of which was described in 1977 (Skirrow, 1977). The Skirrow medium consists of a blood agar base in which is incorporated lysed horse blood, vancomycin, trimethoprim and polymyxin B. Blood contains catalase, peroxidase and superoxide dismutase which are believed to remove toxic oxygen derivatives that inhibit Campylobacteria, while the three antibiotics inhibit the growth of other enteric bacteria. The Skirrow medium is immensely successful in isolating *C. jejuni* subsp. *jejuni* but some faecal flora are not inhibited and some species of *Campylobacter* are inhibited. It is best to avoid cephalothin in media as it may inhibit some Campylobacters. Semisolid motility agar is blood free (Goossens and Butzler, 1991) and, like Skirrow media, is effective at isolating Campylobacteria at 42 °C, but neither is effective at 37 °C (Endtz *et al.*, 1991). This has prompted investigations of other media formulations to overcome these problems. Different cephalosporins have been examined for their inhibition of faecal bacteria, and amphotericin B and cycloheximide have been used to inhibit yeasts. Activated charcoal has been used as a replacement for blood in blood-free media (charcoal-selective media) (Karmali *et al.*, 1986). Such media are generally less expensive and have an advantage in areas of the world where regular supplies of sterile horse or sheep blood are not readily available. One blood-free selective medium developed to replace blood with charcoal is referred to as modified charcoal cefoperazone deoxycholate agar (CCDA medium) (Hutchinson and Bolton, 1984). It also contains ferrous sulfate and sodium pyruvate to enhance aerotolerance and growth of the bacteria.

A simpler method of filtering faecal suspensions requires six to eight drops of faecal suspension to be applied to a membrane filter (0.4 or 0.65 µm pore size) placed on a solid base medium with blood. The filter is removed and discarded 30 min after the application of the suspension. Only about 10% of the Campylobacteria in the sample pass through the filter to produce colonies, thereby limiting detectability of these organisms in faeces to those samples with more than 10^5 cfu per gram (Goossens and Butzler, 1992). Some authors advocate that both inoculation on a selective medium and filtration on a non-selective medium be performed to enhance recovery of the thermophilic species. A charcoal-based media with cefoperazone, amphotericin and teichoplainin has been used for the primary isolation of *C. upsaliensis* (Aspinall *et al.*, 1996). Considering that some Campylobacters are susceptible to the antibiotics in media, they should be avoided to prevent favouring fast-growing isolates, or altering the true distribution of different Campylobacters present in the specimen (Silley, 2003).

To isolate *Campylobacter* spp. from water, milk or food, pre-enrichment steps are recommended. Enrichment is usually unnecessary for fresh stool samples but should be considered for samples that have been stored or maintained in transport media. A number of enrichment media have been described and tested, but agreement is lacking on the choice of an enrichment medium. Those that have been examined include Campy-thio broth, alkaline peptone water, Doyle and Roman enrichment broth, Lovett's enrichment broth and the Preston semi-solid enrichment medium. It should be noted that these media have been tested primarily for improving isolation of *C. jejuni* (Goossens and Butzler, 1992; Tenover and Fennell, 1992), but may in fact be artificially be selecting for this species (Silley, 2003).

Since Campylobacteria are microaerophilic it is essential that they be cultured in an atmosphere with reduced oxygen concentration. The recommended gaseous mixture is 5% oxygen, 10% carbon dioxide and 85% nitrogen for most *Campylobacter* species. The most convenient method of achieving such an atmosphere is with commercially available gas-generating envelopes that are activated in the anaerobic jars. Hydrogen has been shown to enhance growth of some species and to be essential for microaerobic growth of *C. mucosalis*, *C. hyointestinalis* subsp. *lawsonii*, *C. concisus*, *C. curvus*, *C. rectus*, *C. showae*, *C. gracilis* and *C. hominus* sp. nov. (Goossens and Butzler, 1992; Vandamme, 2000). These hydrogen-requiring species need 6% oxygen, 6% carbon dioxide, 3% hydrogen and 85% nitrogen. Gas-generating envelopes that produce an atmosphere of approximately 10% hydrogen, 10% carbon dioxide and 80% nitrogen are available but it should be noted that the hydrogen may be hazardous (Tenover and Fennell, 1992). A commercially available supplement consisting of equal concentrations of ferrous sulfate, sodium metabisulfite and sodium pyruvate may be incorporated in the isolation medium to enhance aerotolerance and growth. The use of this supplement is recommended by many workers and is particularly important in laboratories that use the burning candle to reduce the concentration of oxygen.

If more than a 2-h delay will elapse prior to culturing the specimens, they are best placed in Cary–Blair transport media (Wang *et al.*, 1983). Specimens should be inoculated to two selective plates, one with blood and one without, and incubated at 42 °C under microaerobic conditions since most Campylobacteria will grow, apart from *C. fetus*. However, *Campylobacter* and *Arcobacter* species grow well at 37 °C on selective media, such as CCDA. Culture should be made at both 37 and 42 °C or repeated at 37 °C, if negative at 42 °C (Nachamkin, 1995). Since blood culture isolation is difficult, if curved Gram-negative rods are present, it is best to subculture to non-selective media at 37 °C under microaerobic conditions. Plates should be incubated for 72 h before they may be considered negative.

Identification

Accurate identification of *Campylobacter* species is necessary to collect useful surveillance data for subsequent risk assessment studies and development of effective interventions to prevent the spread of Campylobacteria. Unfortunately, the biochemical inertness of Campylobacteria prevents effective discrimination of the species by phenotypic methods, such as tolerance to different temperatures and to agents that inhibit growth. However, tests for oxidase, catalase, hippuricase, urease, indoxyl acetate hydrolysis, nitrate and nitrite reductases, hydrogen sulfide production, susceptibility to nalidixic acid and cephalothin and growth at 15, 25 and 42 °C are the most reliable for species differentiation. Tests for tolerance to 1% glycine, 3.5% sodium chloride and 0.1% trimethylamine *N*-oxide are less reproducible.

The experienced worker may be aided by slight differences in colony morphology on isolation media. Colonies produced by *C. jejuni* subsp. *jejuni* are grey, moist, flat and spreading after 42 h of incubation at 42 °C. They are often found to spread along the lines of inoculation. *Campylobacter coli* colonies tend to be creamy-grey, moist and more discrete than those of *C. jejuni* subsp. *jejuni*. Colonies of *C. lari* are generally grey and discrete but more variable, resembling either *C. jejuni* subsp. *jejuni* or *C. coli* colonies.

Any colonies isolated from selective media under microaerobic conditions that are oxidase positive, Gram-negative curved, S-shaped or spiral rods 0.2–0.9 mm wide and 0.5–5 mm long may be reported as *Campylobacter* species (Nachamkin, Engberg and Aarestrup, 2000). The ability of many *C. jejuni* to hydrolyse hippurate allows its differentiation from *C. coli*, which lacks this ability. Speciation becomes more difficult for *C. jejuni* strains that are unable to hydrolyse hippurate (Totten *et al.*, 1987). A list of reactions by which the thermophilic species may be differentiated are summarized in Table 41.2.

Computerized schemes utilizing probability matrices for phenotypic characteristics (On and Holmes, 1995; On, 1996) in combination with other methods, including molecular methods, have been developed to improve the identification of species within the Campylobacteraceae family. These methods encompass a variety of techniques including enzymology (Elharrif and Megraud, 1986), serology (Hebert *et al.*, 1983), cellular fatty-acid profiles (Goodwin *et al.*, 1989), electrophoretic patterns (Costas *et al.*, 1987), ribotyping (Fitzgerald, Owen and Stanely, 1996; Smith *et al.*, 1998), DNA–DNA hybridization (Romaniuk and Trust, 1989; Stucki *et al.*, 1995), PCR-DNA fingerprinting (Giesendorf *et al.*, 1994; van Doorn *et al.*, 1997; Vandamme *et al.*, 1997), PCR-RFLP analysis (Cardarelli-Leite *et al.*, 1996; Marshall *et al.*, 1999) PCR-ELISA (Sails *et al.*, 2001; Kulkarni *et al.*, 2002), PFGE (Yan, Chang and Taylor, 1991; Ribot *et al.*, 2001; Wu *et al.*, 2002; Llovo *et al.*, 2003) and AFLP (On and Harrington, 2000; Duim *et al.*, 2001; Moreno *et al.*, 2002; Schouls *et al.*, 2003).

No single PCR method is yet able to identify all the species of *Campylobacter*, so it will likely be more effective to use a polyphasic approach (On, 1996; On and Jordan, 2003). PCR protocols are based on a variety of genes such as 16S rRNA, 23S rRNA, *ceuE*, *flaA*, *mapA* and *glyA* (Oyofo *et al.*, 1992; Eyers *et al.*, 1993; Gonzalez *et al.*, 1997; Al Rashid *et al.*, 2000), or some utilize random primers (On and Jordan, 2003). While several oligo probes and PCR methods are able to identify several *Campylobacter* species, only the combined PCR-oligo hybridization strategy using species-specific oligo probes based on the highly conserved *glyA* gene demonstrated sufficient sensitivity and specificity to distinguish between *C. jejuni*, *C. coli*, *C. lari*, *C. upsaliensis* and *A. butzleri* (Al Rashid *et al.*, 2000). A real-time PCR-biprobe assay is reported to accurately identify *C. jejuni*/*C. coli*, *C. lari*, *C. upsaliensis*, *C. hyointestinalis*, *C. fetus* as well as *C. helveticus* and *C. lanienae* from isolates and faecal DNA (Logan *et al.*, 2001). A standardized PCR-based detection assay, using a single select primer pair, has been developed and may be useful in large-scale screening programs for the most common food-borne *Campylobacter* species: *C. jejuni*, *C. coli* and *C. lari* (Lübeck *et al.* 2003a,b). Oligonucleotide microarrays have recently been developed that can differentiate *C. jejuni*, *C. coli*, *C. lari* and *C. upsaliensis* based on PCR amplification of specific regions in five target genes followed by microarray analysis of amplified DNA (*fur*, *glyA*, *cdtABC*, *ceuB-C* and *fliY*), with confirmation by *hipO* or *ask* PCR (Volokhov *et al.*, 2003).

Antibiotic Susceptibility

The US Food and Drug Administration analysed data from an international multicenter study of seven laboratory sites and recommended to the National Committee for Clinical Laboratory Standards (NCCLS) subcommittees on antimicrobial susceptibility testing (AST) and veterinary antimicrobial susceptibility testing that the AST method for *Campylobacter* be agar dilution (McDermott and Walker, 2003). NCCLS document M31-A2 containing the testing method and the tentative quality-control ranges and document M7-A6 with M100-S13 supplemental tables were made available in January 2003. The study was validated using isolates of *C. jejuni* subsp. *jejuni*,

C. jejuni subsp. *doylei*, *C. coli*, *C. fetus* and *C. lari*. The quality-control organism is *C. jejuni* isolate ATCC 33560. Tentative quality-control ranges have been proposed for five antibiotics: ciprofloxacin, doxycycline, gentamicin, meropenem and tetracycline. Minimal inhibitory concentrations (MICs) obtained by traditional agar dilution susceptibility testing on Mueller–Hinton media supplemented with 5% horse or sheep blood incubated at 35 °C for 24–48 h under microaerobic conditions have been reported to generally correlate well with the E-test method (Engberg *et al.*, 2001), although MICs at low or high concentrations do not correlate that well. The relationship between MICs and clinical outcome for *Campylobacter* has not yet been established.

Originally, Campylobacters were universally resistant to penicillins, cephalosporins (except broad spectrum), trimethoprim, sulfamethoxazole, rifampicin and vancomycin and susceptible to erythromycin, fluoroquinolones, tetracyclines, aminoglycosides and clindamycin (Nachamkin, Engberg and Aarestrup, 2000). Since 1989, antimicrobial resistance to the latter group of drugs has been rising at different rates throughout the world (Nachamkin, Engberg and Aarestrup, 2000).

The mechanisms of antibiotic resistance in *Campylobacter* have been reviewed recently (Trieber and Taylor, 2000). In a Canadian study, tetracycline resistance more than doubled over a 10-year period to a level of 55% (Gaudreau and Gilbert, 1998). Tetracycline resistance in *Campylobacter* is mediated by Tet(O), a ribosomal protection protein that inhibits tetracycline interaction with the ribosome (Connell *et al.*, 2003). The tetracycline-resistance gene (*tetO*) appears to be closely related to the *tetM* gene found in *Streptococcus* sp. (Taylor, 1984, 1992a), and both protein products function as GTPase proteins (Grewel, Manarathu and Taylor, 1993). In the absence of Tet(O) tetracycline interrupts protein synthesis by binding the ribosome and inhibiting the ribosomal elongation rate, producing short chains (Trieber and Taylor, 2000). Mutations in the *miaA* gene product and mutations in the ribosomal protein S12 (encoded by *rpsL*) alter translation accuracy which in turn affects Tet(O)-mediated tetracycline resistance (Taylor *et al.*, 1998). Antibiotic resistance genes, usually carried on plasmids, have been cloned and will express in *E. coli*. High levels of tetracycline resistance are usually associated with the Tet(O) determinant on a self-transmissible plasmid, but it may also be found on the chromosome, although it is unknown whether this is mediated by a transposon or through host-cell recombination (Taylor *et al.*, 1981; Taylor and Courvalin, 1988; Trieber and Taylor, 2000).

Chloramphenicol resistance in *Campylobacter* spp. occurs rarely and is mediated either through mutation of the 23S rRNA or the chloramphenicol-resistance determinant (*cat*) that is closely related to genes specifying a chloramphenicol transferase in *Clostridium* spp. (Taylor, 1984, 1992a; Spahn and Prescott, 1996).

The level of resistance to β-lactam antibiotics varies with antibiotic. Resistance to penicillin and cephalosporins is common, less common with ampicillin, rarely seen with amoxacillin and not at all with imipenem (Trieber and Taylor, 2000). For this reason, imipenem is used only to treat the most severe systemic infections and the other β-lactam antibiotics avoided. Resistance to β-lactams occurs by modification of a porin to reduce antibiotic uptake, alterations of penicillin-binding protein or production of β-lactamase. The addition of clavulanate, a β-lactamase inhibitor, to β-lactam antibiotics results in total susceptibility *in vitro* (Tajada *et al.*, 1996).

Trimethoprim resistance is intrinsic to Campylobacters due to the presence of either *dfr1* and *dfr9*, genes encoding dihydrofolate reductase (Taylor, 1992b). As a result, trimethoprim is often used in selective media for Campylobacters.

Aminoglycoside resistance occurs by modification of the antibiotic by one of three enzymes: aminoglycoside phosphotranserases (APH), aminoglycoside adenyltransferases (AAD) and acetyltransferases. Kanamycin resistance is due to the production of APH, which is encoded by three different genes: *aphA-1*, which is chromosomally mediated and found among the Enterobacteriaceae; *aphA-3*, which is found in Gram-positive cocci; and *aphA-7*, which is thought to be indigenous to *Campylobacter* (Taylor, 1992a). Streptomycin and spectinomycin resistance occurs with AAD, encoded by *aadA* and *aadE* genes (Björk, 1995). Streptothricin resistance in *C. coli* arises from antibiotic modification by acetyltransferase encoded by *sat4* (Jacob *et al.*, 1994). Aminoglycoside resistance may also arise from chromosomal mutations of ribosomal proteins and rRNA. Although spontaneous chromosomal streptomycin resistance in *C. coli* could be transformed to another *C. coli* by natural transformation, the nature of the mutation has not yet been characterized (Wang and Taylor, 1990; Trieber and Taylor, 2000).

Macrolide resistance in *Campylobacter* is very low and thought to be chromosomally mediated (Taylor, 1992b). Macrolides (erythromycin, azithromycin and clarithromycin) are protein synthesis inhibitors that reversibly bind the ribosome to a site that may include 23S rRNA and proteins (Spahn and Prescott, 1996). Erythromycin-resistance determinants have been identified in *C. rectus* as rRNA methylases (ErmB, GrmC, GrmQ and GrmFS) which methylate an adenine in the 23S rRNA and prevent binding of erythromycin (Roe, Weinberg and Roberts, 1995). In contrast, in *C. jejuni* and *C. coli*, the mechanism of erythromycin resistance is likely due to mutations at the 23S rRNA genes which result in decreased ability of erythromycin to bind the ribosomes (Trieber and Taylor, 1999).

Campylobacter spp. are generally highly susceptible to fluoroquinolones (ciprofloxacin, ofloxacin and norfloxacin), antibiotics derived from nalidixic acid and target type II topoisomerases, DNA gyrase and topoisomerase IV (Drlica and Zhao, 1997). Fluoroquinolones form a complex with the gyrase and the DNA to inhibit DNA replication and transcription and do not actually bind directly to the gyrase or the DNA (Willmott and Maxwell, 1993). Resistance to fluoroquinolones in *C. jejuni* arises from a chromosomal mutation genes encoding either gyrase, *gyrA* (Thre-86 to Ile, Asp-90 to Ala and Ala-70 to Thr) and/or topoisomerase IV, *parC* (Arg-139). Both human and veterinary use of fluoroquinolones may result in resistance in *Campylobacter* (Wang, Huang, Taylor, 1993; Piddock, 1995). The incidence of resistance to fluoroquinolones in *C. jejuni* and *C. coli* is on the rise, reaching levels as high as 88% in Spain (Trieber and Taylor, 2000). Resistance to nalidixic acid in *C. jejuni* and *C. coli* usually results in cross-resistance to fluoroquinolones but is not true for *C. fetus* which is intrinsically resistant to nalidixic acid (Taylor, Ng and Lior, 1985). A multidrug-resistance efflux pump has been characterized in *C. jejuni* and is a three-gene operon encoded on the chromosome as *CmeA*, *CmeB* and *CmeC* (Lin, Michel and Zhang, 2002; Pumbwe and Piddock, 2002).

Treatment

Most commonly, *Campylobacter* infections manifest as an acute diarrhoeal disease. General treatment for *Campylobacter* infections is supportive and symptomatic, with antibiotics being reserved for those with prolonged serious symptoms (abdominal pain, diarrhoea and blood in the stool), such as observed in immunocompromised persons, and in pregnancy in which the fetus may be affected (Allos and Blaser, 1995). Erythromycin is the drug of choice to treat *C. jejuni* infections in humans. Up to 68.4% of *C. coli* isolates may be resistant to erythromycin (Nachamkin, Engberg and Aarestrup, 2000). Alternatively, ciprofloxacin, gentamicin and tetracycline may be used to treat *Campylobacter* enteritis (Lucey *et al.*, 2000; Wilson *et al.*, 2000). Antimicrobial agents used for the treatment of infection in chickens and cattle are not currently monitored in the industry, making it difficult to assess levels of antimicrobial use. Increasing resistance to quinolones has been reported and may be associated with multidrug resistance on farms producing poultry and beef (Engberg *et al.*, 2001). Recent studies of human and animal *C. jejuni* isolates reported increasing resistance to ciprofloxacin, tetracycline and erythromycin in isolates collected in 2001 as compared with several

years previously (Gaudreau and Gilbert, 2003; Luber *et al.*, 2003). Multidrug resistance is associated with an efflux pump (Lin *et al.*, 2003). Ciprofloxacin is used to treat human infections, while enrofloxacin is used for veterinary use (Luo *et al.*, 2003). Ciprofloxacin-resistant mutants can be isolated from chickens after only 1 day of enrofloxacin treatment, and these isolates are also cross-resistant to enrofloxacin and nalidixic acid. Resistance was shown to be due to a mutation in *gyrA* and suggests that *Campylobacter* is hypermutable under the selective pressure of fluoroquinolones (Luo *et al.*, 2003).

Prevention

As infection with *Campylobacter* spp. is a zoonosis, control is largely directed against removing the source of infection. The consumption of unchlorinated water is a well-recognized source. Contamination of drinking water for poultry may be a significant source of intestinal colonization in poultry flocks, and control measures can lead to reduced carriage (Healing, Greenwood and Pearson, 1992). Strict hygiene and decontamination technologies are needed to assure the safety of meat (Moorhead and Dykes, 2002). Adequate cooking of poultry is probably the single most important control measure. Pasteurization of milk is also important, as is the protection of doorstep bottles of milk from pecking by birds (Healing, Greenwood and Pearson, 1992). Improved implementation of hazard analysis of critical control points and consumer food-safety education efforts are necessary to prevent transmission of food-borne pathogens such as *Campylobacter* (Zhao *et al.*, 2001).

ANIMAL PATHOGENS AND COMMENSAL CAMPYLOBACTERIA THAT MAY CAUSE OPPORTUNISTIC INFECTIONS IN HUMANS

The four species within this group (*C. fetus*, *C. hyointestinalis*, *C. sputorum* and *C. mucosalis*) are, with the exception of *C. sputorum* biovar *sputorum*, either pathogens or commensals of farm animals. Human infections are uncommon and those that have been reported are usually opportunistic in predisposed patients. There is considerable diversity among the species in their habitats, pathogenicity, atmospheric conditions for growth and biochemical reactions (Table 41.3), although genetic analysis has shown them to be a closely related group (Thompson *et al.*, 1988; Vandamme *et al.*, 1991).

Although *C. hyointestinalis*, *C. coli* and *C. mucosalis* are readily isolated from porcine intestines, they have not been shown to have a direct effect in the development of intestinal disease. Some reports suggest that porcine diseases may be caused by intracellular *Campylobacter*-like organisms that cannot be cultured (McOrist, Boict and Lawson, 1989; Gebhart *et al.*, 1991). However, it is interesting to note that *C. hyointestinalis* and *C. mucosalis* occur far more frequently in piglets with intestinal lesions than in healthy animals, but the basis for this observation is not clear.

The filtration technique optimizes recovery of these Campylobacteria from stools and helps to ascertain their incidence in human infection along with antibiotic-free media, when incubated under microaerobic conditions at 35–37 °C. Campylobacteria, which are smaller than most enteric bacteria, pass through the filters and produce colonies, while the larger enteric bacteria are excluded.

C. fetus

Bacteria now classified as *C. fetus* have been known since 1913, when they were believed to belong to the genus *Vibrio* (Tenover and Fennell, 1992). Currently, the species consists of two groups classified as *C. fetus* subsp. *fetus* and *C. fetus* subsp. *venerealis*. *Sma*I-based macrorestriction digests, or PCR, allow for differentiation of the two subspecies (Hum *et al.*, 1997). The assignment of the two groups to

Table 41.3 Differentiation of *C. fetus*, *C. hyointestinalis*, *C. sputorum*, *C. mucosalis* and *C. lanienae* sp. nov.

Phenotypic characteristics	C. fetus		C. hyointestinalis			C. sputorum		C. mucosalis	C. lanienae sp. nov.
	subsp. *fetus*	subsp. *venerealis*	subsp. *hyointestinalis*	subsp. *lawsonii*	biovar *sputorum*	biovar *fecalis*	biovar *paraureolyticus*		
Catalase activity	+	v[a]	+	+	−	+	+	−	+
Urease activity	−	−	+	+	−	−	+	−	−
Nitrate reduction	+	+	+	+	+	+	+	−	+
Nitrite reduction	−	−	−	−	+	+	+	+	+
H_2S production	−	−	+	+	+	+	+	+	−
Growth									
at 42 °C	v	−	+	+	+	+	+	+	+
at 25 °C	+	+	v	−	−	−	−	−	−
Growth in the presence of									
4% NaCl		−	−	−	−	v	−	−	−
1% glycine	+	−	+	v	+	+	+	+	−
H_2 requirement	−	−	−	+	−	−	−	+	
Susceptibility[b] to									
Nalidixic acid	R	v	R	R	S	R	R	v[a]	R
Cephalothin	S	S	v	S	S	S	S	S	S

Test reactions: +, present in >90% of strains examined; −, present in <11% of strains examined; v, variable strain-dependent reactions. Sources of data include Goossens and Butzler (1992), Gebhart *et al.* (1985), Roop *et al.* (1985), Tenover and Fennell (1992), On and Holmes (1995), On (1996) and Vandamme (2000).
[a] At least 80% of strains examined have this characteristic.
[b] Susceptibility to 30 µg disk of antibiotic: S, susceptible; R, resistant; v, variable strain dependent.

the taxonomic status of subspecies is based more on convenience than on differences that can be demonstrated by either DNA hybridization or phenotypic characteristics that are used to identify and classify Campylobacteria. The subspecies are, however, quite distinct in their habitats and in the sites of the host where they produce infections. The major habitat of *C. fetus* subsp. *fetus* is the intestinal tract of cattle and sheep, and isolations of the organisms from healthy animals are not uncommon and may cause spontaneous abortion (Skirrow, 1994). The bacteria may be acquired from food or water that has been contaminated with bacteria from faeces, aborted fetuses or from vaginal discharges of infected animals. After colonization of the intestine, the organisms may spread via the bloodstream to the placenta, for which they have a high affinity, and multiply in the developing fetus, which is then generally aborted.

In contrast, the normal habitat of *C. fetus* subsp. *venerealis* is the genital tract of cattle and sheep.

Human infections caused by *C. fetus* subsp. *fetus* are rare and usually occur in immunocompromised or debilitated patients. Gastro-enteritis, endocarditis and septicaemia have been reported with complications arising from septicaemia that include meningitis, endocarditis, infection of the fetus and abortion, pleuropericardial effusion, cellulitis as well as abscesses from most parts of the body (On, 1996; Ganeshram *et al.*, 2000; Peetermans *et al.*, 2000; Briedis *et al.*, 2002; Tremblay, Gaudreau and Lorange, 2003). Some of the intestinal isolates have been recovered at 42 °C, but most strains of the species cannot tolerate this temperature, preferring 35–37 °C, and are also susceptible to cephalothin, the most common selective agent for culture. Since many *C. fetus* subsp. *fetus* strains would not be isolated, accurate determination of the incidence of this subspecies in human diarrhoea is not available.

The involvement of *C. fetus* subsp. *venerealis* in human disease is even much less frequent than is *C. fetus* subsp. *fetus* (Garcia, Eaglesome and Rigby, 1983).

C. hyointestinalis

A group of bacteria isolated from pigs with proliferative ileitis (Gebhart *et al.*, 1983) was confirmed to be a separate species and was named *C. hyointestinalis* (Gebhart *et al.*, 1985). This bacteria is comprised of two subspecies, *C. hyointestinalis* subsp. *hyointestinalis* and *C. hyointestinalis* subsp. *lawsonii*, isolated from primarily enteric sources and pig stomachs, respectively (On and Homes 1995). *Campylobacter hyointestinalis* has also been isolated from stomachs or intestines of hamsters, cattle and deer, as well as from human faeces (Gebhart *et al.*, 1985; Hill, Thomas and Mackenzie, 1987; Harrington and On, 1999). Isolates obtained from a homosexual male (Fennell *et al.*, 1986) first implicated the species as another aetiological agent of human diarrhoea. In subsequent studies, isolates from both male and female patients with diarrhoea have been reported (Edmonds *et al.*, 1987), but isolates have also been obtained from asymptomatic individuals (Salama, Garcia and Taylor, 1992; On and Harrington, 2001) and the suggestion has been raised that *C. hyointestinalis* may be an opportunistic enteropathogen in humans (Minet, Grosbois and Megraud, 1988).

Campylobacter hyointestinalis is most closely related to *C. fetus* subsp. *fetus* (Vandamme *et al.*, 1991) and possesses a similar colony morphology and other phenotypic characteristics (Table 41.2). *Campylobacter hyointestinalis* grows optimally at 35–37 °C, less well at 42 °C and not at all at 25 °C. All strains grow microaerobically in the presence of hydrogen, and only a few do not require hydrogen. The colonies are yellow and do not swarm on moist media or show haemolysis (Gebhart *et al.*, 1985). The species is differentiated from *C. fetus* by its production of hydrogen sulfide in triple sugar iron medium and by its growth in trimethylamine *N*-oxide hydrochloride under anaerobic conditions. The number of reports of isolations of *C. hyointestinalis* from humans has not increased substantially since the first report of a human isolate (Fennell *et al.*, 1986), and thus its frequency as a pathogen or commensal in the human population remains uncertain. A species-specific PCR test for *C. hyointestinalis* has been described for the detection, but not differentiation of both subspecies (Linton, Owen and Stanley, 1996). The existence of polymorphic sites may, however, compromise the specificity of the PCR test. Subspecies of *C. hyointestinalis* may be differentiated by *Sma*I-derived macrorestriction profiles (On and Vandamme, 1997). Recovery of *C. hyointestinalis* from clinical specimens is often unsuccessful because the bacteria do not grow optimally at 42–43 °C and they may be susceptible to selective antibiotics.

C. sputorum

The species consists of three biovars (biotypes): *C. sputorum* biovar *sputorum*, *C. sputorum* biovar *faecalis* and *C. sputorum* biovar *paraureolyticus* differentiated on the basis of their habitats and simple biochemical tests (catalase positive and urease negative) (catalase negative and urease negative) and (catalase negative and urease positive), respectively (On *et al.*, 1998; Vandamme and On, 2001). Although there are reports of an isolate of *C. sputorum* biovar *sputorum* from a human leg abscess and another isolate from the faeces of an infant with diarrhoea the organisms of this biovar are believed to be commensals because they can be isolated from the oral cavity, faeces of healthy humans and reproductive tract of cattle (Karmali and Skirrow, 1984). *Campylobacter sputorum* biovar *fecalis* was originally believed to represent a separate species and was first classified as *Vibrio fecalis*, but it was shown to constitute a separate biogroup within the species *C. sputorum* in DNA-relatedness studies (Roop *et al.*, 1985). Organisms of this biovar have been isolated from sheep faeces, bovine semen and the bovine vagina, but evidence for a pathogenic role has not been forthcoming. *Campylobacter sputorum* biovar *paraureolyticus* has been isolated from the faeces of cattle and also from humans with diarrhoea, yet no pathological role is understood.

The morphology of the bacterial cells of the three biovars is not as tightly coiled as are those of *C. jejuni* subsp. *jejuni* (Tenover and Fennell, 1992). Isolates of each of the biovars grow microerophically at 42 °C but prefer 37 °C, and do not grow at 25 °C. The colonies are small (1–2.5 mm in diameter), grey in colour and do not swarm on moist agar. Some show low levels of alpha-haemolysis. The isolates reduce nitrate and nitrite, produce hydrogen sulfide in triple sugar iron medium and are susceptible to cephalothin.

C. mucosalis

Campylobacter mucosalis (formerly classified as *C. sputorum* subsp. *mucosalis*) is of interest primarily to veterinarians (no human infections reported) because of its occurrence in the intestinal mucosa of pigs with intestinal adenomatosis, necrotic enteritis, regional ileitis and proliferative haemorrhagic enteropathy (Roop *et al.*, 1985).

Organisms of *C. mucosalis* are noted for their yellowish colour and ability to grow at temperatures from 25 to 42 °C. Hydrogen is necessary for microaerophilic growth, a requirement that permits differentiation from *C. fetus*, *C. hyointestinalis*, *C. sputorum* and species of the thermophilic enteropathogenic group. The four species associated with periodontal diseases in humans also require hydrogen for microaerophilic growth, but they can be differentiated from *C. mucosalis* on the basis of their inability to grow at 25 °C.

Campylobacter Species Associated with Periodontal Disease

Five species are included in this group: *C. concisus*, *C. curvus*, *C. rectus*, *C. showae* and *C. gracilis*. Two species, *C. concisus* and *C. rectus*, were described as a result of a taxonomic study on bacteria associated with periodontal disease (Tanner *et al.*, 1981). *Campylobacter curvus* was first assigned to the genus *Wolinella* with the specific name 'curva' to reflect the shape of the cells. *Campylobacter rectus* was previously designated *Wolinella recta*, with its cells the shape of plump, straight rods. The surface of *C. rectus* is covered with an array of hexagonal, packed macromolecular subunits (Tanner *et al.*, 1981). By DNA–rRNA hybridization and immunotyping, both strains more closely relate to the genus *Campylobacter* than to the type species of the genus *Wolinella* (Vandamme *et al.*, 1991). With corrections in the spelling of the epithets, the two species were therefore transferred from *Wolinella* to the genus *Campylobacter* (Vandamme *et al.*, 1991). *Campylobacter gracilis* was originally classified as *Bacteroides gracilis*, due to its description as a non-motile, non-spore–forming anaerobic

Gram-negative rod with a formate- and fumarate-requiring metabolism (Tanner *et al.*, 1981). Later analysis and comparison of the cellular fatty acids, respiratory quinones and proteins of *B. gracilis* with campylobacters resulted in the reclassification and transfer of this species to genus *Campylobacter* (Vandamme *et al.*, 1995). *Campylobacter gracilis* is the only oxidase-negative *Campylobacter* species.

Both *C. gracilis* and *C. rectus* have been detected in endodontic infections (Gomes, Lilley and Drucker, 1996; Le Goff *et al.*, 1997; Sundqvist *et al.*, 1998; Siqueira *et al.*, 2000b, 2001b). The prevalence of campylobacters associated with primary endodontic infections has likely been underestimated due to the difficulty in traditional phenotype-based species identification. Molecular methods, such as nested PCR technology, have increased the sensitivity and specificity for species-specific detection of *C. gracilis* and *C. rectus* (Jordan *et al.*, 2001), at rates of 21.1 and 23.3% respectively, in endodontal infections (Siqueira and Rôças, 2003). *Campylobacter rectus* has also been associated with appendicitis, while *C. gracilis* is associated with serious deep-tissue infections.

Campylobacter concisus and *C. curvus* have been isolated from human gingival crevices and associated with gingivitis, periodontitis and periodontosis (Tanner *et al.*, 1981; Tanner, Listgarten and Ebersole, 1984) and also from blood, oesophagus, stomach and duodenum (Goossens and Butzler, 1992). Both species have been recovered from stool and blood samples of healthy children and adults and those with diarrhoea (Vandamme *et al.*, 1989; Lindblom *et al.*, 1995; Van Etterijck *et al.*, 1996).

DNA–DNA hybridization tests, 16S rRNA sequence data and biochemical reactions typical of Campylobacteria supported the proposal for creating a new species, *C. showae*, for another group of strains isolated from infected root canals, dental plaque and human gingival crevices of apparently healthy adults (Etoh *et al.*, 1993). The species differs from others of the genus in having rod-shaped bacteria with two to five flagella at one end of the cell. It is differentiated from *C. concisus*, *C. rectus* and *C. curvus* in its ability to produce catalase. While *C. concisus* and *C. curvus* are true microaerophils, *C. rectus*, *C. gracilis* and *C. showae* grow optimally under anaerobic conditions (On, 1996). Additional characteristics for differentiating these species are summarized in Table 41.4.

Newly Characterized Campylobacter Species

Campylobacter lanienae sp. nov. was recently identified in the stools of healthy abattoir workers and pigs and found to be most closely related to *C. hyointestinalis* (Logan *et al.*, 2000; Sasaki *et al.*, 2003). Cells are Gram-negative slender, slightly spiral rods with curved ends and single bipolar unsheathed flagella. *Campylobacter lanienae* is microaerophilic and grows weakly under anaerobic conditions. Other characteristics are summarized in Table 41.3.

Campylobacter hominus sp. nov. was recently identified in faecal samples from humans with diarrhoea and also those without clinical illness using 16S rRNA sequences (Lawson *et al.*, 2001). Phylogenetically, *C. hominus* sp. nov. is related to *C. gracilis* and *C. sputorum*. Anaerobic conditions, hydrogen and FAA medium (containing vitamin K, haemin, cysteine hydrochloride and L-arginine) and 37 °C are required for culture of *C. hominus* sp. nov. Cells are Gram-negative, straight, non-motile, non-spore–forming rods. Flagella are absent, but irregular fimbriae-like structures are present (Table 41.4).

THE GENUS ARCOBACTER

The genus *Arcobacter* is classified with rRNA superfamily VI of the Proteobacteria and includes four species: *A. nitrofigilis* and *A. cryaerophilus* (formerly in the genus *Campylobacter*), *A. butzleri* and *A. skirrowii* (Vandamme *et al.*, 1991, 1992b). Although *Arcobacter* spp. resemble Campylobacteria, differences in a number of phenotypic traits warrant their placement in a separate genus. The name

Table 41.4 Differentiation of *C. concisus*, *C. rectus*, *C. curvus*, *C. showae*, *C. gracilis* and *C. hominus* sp. nov.

Characteristics	C. concisus	C. rectus	C. curvus	C. showae	C. gracilis	C. hominus sp. nov.
Catalase activity	−	v	−	+	v	−
Urease activity	−	−	−	−	−	−
Indoxyl acetate hydrolysis	−	+	v	v	v	−
Nitrite reduction	+	+	+	+	+	
Nitrate reduction	v	+	+	+	v	v[a]
H₂S production		+	v	+	−	−
Growth						
at 42 °C	v	v	v	v	v	−
at 37 °C	+	+	+	+	+	+
at 25 °C	−	−	−	−	−	−
Growth in the presence of						
4% NaCl		−	−	−		−
1% glycine	v	+	+	v	+	+
0.05% NaF	+	−	−	+		+
Hydrogen requirement	+	+	+	+	+	+
Susceptibility[b] to						
Nalidixic acid	v	v[c]	R	S	v	v[a]
Cephalothin	R	S	S	S	S	S
Cell morphology						
Helical curved	+	−	+	−	−	−
Rod shaped	−	+	−	+	+	+
Number of flagella	1	1	1	2–5	0	0

Test reactions: +, present in >90% of strains examined; −, present in <11% of strains examined; v, variable strain-dependent reactions.
Sources of data include Etoh *et al*. (1993), Goossens and Butzler (1992), Tanner *et al*. (1981), Tanner, Listgarten and Ebersole (1984), On and Holmes (1995), On (1996) and Vandamme (2000).

[a] 38–66% of strains examined have this characteristic.

[b] Susceptibility to 30 μg disk of antibiotic: S, susceptible; R, resistant; v, variable strain dependent.

[c] At least 80% of strains examined have this characteristic.

Arcobacter was selected to suggest a bow shape, but, like Campylobacteria, they are usually curved, S-shaped or helical cells. Arcobacteria are differentiated from Campylobacteria by two main phenotypic characteristics: the ability to grow at 25 °C and aerotolerance. Although aerotolerance is not usually observed on initial isolation, it does emerge upon subculture.

The four *Arcobacter* spp. grow over a diverse range of temperatures and atmospheric conditions, but all species are able to grow aerobically at 30 °C, anaerobically at 35–37 °C and microaerophilically, with or without hydrogen, at 25–30 °C. Optimal growth occurs under microaerobic conditions with oxygen concentrations of 3–10%. *Arcobacter nitrofigilis* and *A. craerophilus* are capable of growth at 15 °C or at lower temperatures, but only the occasional strain of *A. cryaerophilus* grows at 42 °C, in contrast to 67% of *A. butzleri* and 33% of *A. skirrowii* strains that grow at this temperature.

Tolerance tests have been used to characterize *Arcobacter* species, but these tests are generally unsatisfactory for species identification and classification because of the diversity of the reactions among the four species and among the strains of each species. Some of the tolerance tests advocated include tests for growth in sodium chloride (in concentrations ranging from 1.0 to 3.5%), growth in bile (1.0%), in triphenyltetrazolium chloride (0.04–0.1%), sodium selenite (0.1%), cadmium chloride (2.5 and 2.0 μm disks), glucose (8%), brilliant green (0.004–0.003%) and in glycine (1.0%). MacConkey agar is the only selective media that has value as a differentiating test because it separates *A. butzleri* strains, which grow on the medium, from *A. skirrowii* strains, which do not. Virtually all *Arcobacter* strains (92–100%) are susceptible to nalidixic acid, but there is much diversity among the species in susceptibility to cephalothin.

Useful phenotypic characteristics for the identification of *Arcobacter* spp. are summarized in Table 41.5. All *Arcobacter* strains produce oxidase, hydrolyse indoxyl acetate and catalase, although isolates of *A. butzleri* characteristically exhibit a weak catalase reaction. Virtually all strains of *A. nitrofigilis*, *A. butzleri* and *A. skirrowii* and 70% of the *A. cryaerophilus* strains have DNase activity. All strains are resistant to the pteridine vibriostatic agent 0/129. *Arcobacter*

nitrofigilis strains are the least tolerant of the Arcobacteria to glycine, but are the most tolerant to NaCl, with some strains able to grow at a concentration of NaCl as high as 7.0% (McClung, Patriquin and Davis, 1983). The production of brown pigment from tryptophan is also unique to *A. nitrofigilis*, and the production of urease by strains belonging to other species of *Arcobacter* has not been reported (McClung, Patriquin and Davis, 1983). *Arcobacter nitrofigilis* is capable of nitrogen fixation and a test for nitrogenase is critical for a complete identification, but the test is not routinely provided in the clinical laboratory (Hardy *et al*., 1968). Another criterion for differentiating the species of *Arcobacter* is colony appearance (Vandamme *et al*., 1992b). On charcoal medium white colonies are produced by *A. nitrofigilis*, beige-to-yellow colonies by *A. cryaerophilus*, white-to-beige by *A. butzleri* and greyish colonies by *A. skirrowii*.

Correct identification of *Arcobacter* spp. is difficult with the use of conventional phenotypic characterization due to the limited number of species-specific characteristics and discrepancies between different workers (Table 41.5). More sophisticated techniques such as gas–liquid chromatography and molecular DNA methods are able to absolutely identify species with genus *Arcobacter*. Cellular fatty-acid profiles obtained by gas–liquid chromatography are unique to each *Arcobacter* spp. and are consistent with groupings produced from the results of DNA–rRNA and DNA–DNA hybridizations (Vandamme *et al*., 1992b). The increased availability of DNA-based methods offer a more rapid and specific approach to detection and identification of *Arcobacter* spp. than conventional phenotypic-based methods. PCR-based methods using DNA probes may be used for routine identification of genus *Arcobacter* and specifically *A. butzleri* (Wesley *et al*., 1995; Harmon and Wesley, 1996, 1997; Hurtado and Owen, 1997; Gonzalez *et al*., 2000). *Arcobacter* species have been differentiated and identified using a PCR-based method with two distinct DNA amplifications at different annealing temperatures (Bastijns *et al*., 1995). More recently, a multiplex PCR assay using five primers targeting the 16S and 23S rRNA genes has been developed for the rapid detection and identification of *A. butzler*, *A. cryaerophilus* and *A. skirrowii* after 24 h

Table 41.5   Characteristics for differentiating species of genus *Arcobacter*

Characteristics	*A. nitrofigilis*	*A. cryaerophilus*[a]	*A. butzleri*	*A. skirrowii*
Catalase activity	+	+	v	+
Nitrite reduction	+	v	+	+
Nitrate reduction	+	+	+	+
DNase activity	+	v	v	+
Nitrogenase activity	+	−	−	−
Urease activity	+	−	−	−
Growth				
at 42 °C	−	−	v	v
at 37 °C	−	+	+	+
at 25 °C	+	+	+	+
on MacConkey agar	−	v	v	−
Growth in the presence of				
4% NaCl	+	−	−	+
1% glycine	−	−	−	−
1% bile	−	v	v	−
8% glucose	+	+	+	v
Production of brown pigment from tryptophan	+	−	−	−
Alpha hemolysis	−	−	−	+

Test reactions: +, present in >90% of strains examined; −, present in <11% of strains examined; v, variable strain-dependent reactions. Data compiled from Kiehlbauch *et al.* (1991), McClung, Patriquin and Davis (1983), Neill *et al.* (1985), Vandamme *et al.* (1991, 1992b), On and Holmes (1995), On (1996) and Vandamme (2000).
[a] Including subgroup 1 and subgroup 2.

cultivation in *Arcobacter* broth supplemented with cefoperazone, amphotericin and teicoplanin (Houf *et al.*, 2000).

A high prevalence of Arcobacters has been reported in Danish pig abortions, with more than one species present including 40.7% *A. cryaerophilus*, 37% *A. skirrowii* and 22.2% unclassified (On *et al.*, 2002).

SPECIES OF THE GENUS *ARCOBACTER*

A. nitrofigilis

Arcobacter nitrofigilis is the type species of the genus (Vandamme *et al.*, 1991), but it comprises the most unique group in the genus. The bacteria are microaerophilic, nitrogen fixing and associated with the roots of a halophytic plant, *Spartina alterniflora*, which has as its habitat the sediments of salt marshes where the oxygen content is at reduced levels (McClung and Patriquin, 1980). It is not surprising, therefore, that in the laboratory it is capable of growth at low temperatures and in media with high concentrations of sodium chloride. The observation that cells suspended in distilled water undergo rapid lysis suggests that viability is dependent on the presence of salt. Bacteria of this species are thus unlikely to be isolated in clinical laboratories that set the temperature for isolation at 37 or 42 °C. There is no known significance with respect to disease production in humans or animals.

A. cryaerophilus

During investigations of spiral-shaped organisms from aborted bovine fetuses, an unusual group of strains was isolated on *Leptospira* medium under microaerobic conditions at 30 °C. On subculture, the bacteria were culturable under aerobic, anaerobic and microaerobic conditions (Ellis *et al.*, 1977). The requirement for the two-step isolation procedure and the ability to grow in air at low temperatures are unique attributes of the strains. They were first assigned to a newly created species in the genus *Campylobacter* with the epithet 'cryaerophila' to reflect both their characteristic growth at low temperatures and their aerotolerance, although it was recognized that the group displayed considerable heterogeneity and that, as a group,

they were more distantly related to *Campylobacter* spp. than were the *Campylobacter* spp. to each other (Neill *et al.*, 1985). DNA–rRNA hybridization experiments showed that they clustered along with *C. nitrofigilis* in a separate group that did not include any of the other *Campylobacter* spp. (Vandamme *et al.*, 1991). The relationships were confirmed by DNA–DNA hybridization and were in agreement with the 16S rRNA homology groups determined by others (Thompson *et al.*, 1988). In accordance with these results, the species (accompanied by a correction of the spelling of the epithet) was transferred along with *C. nitrofigilis* to the newly described genus *Arcobacter* (Vandamme *et al.*, 1991).

Early collection of *A. cryaerophilus* isolates originated mostly from bovine, ovine and porcine fetal tissue and is associated with spontaneous abortions (On *et al.*, 2002). In a collection of *A. cryaerophilus* strains made at the Centers for Disease Control (CDC) in Atlanta Georgia, two isolates were from human blood cultures and one from the stool of a patient with severe gastroenteritis (Kiehlbauch *et al.*, 1991). Although no medical significance has been linked to these human isolates, it does suggest that with adoption by clinical laboratories of appropriate isolation techniques, the occurrence of *A. cryaerophilus* in human specimens could be shown to be much higher than is now suspected.

A. butzleri

Through a study of aerotolerant spiral-shaped bacteria at the CDC, a group of strains was discovered on the basis of DNA–DNA hybridizations to be distinct from bacteria now classified as *A. cryaerophilus*. A new species was proposed with the epithet '*butzleri*' and assigned to the genus *Campylobacter*, although the authors recognized that the new species would be placed in genus *Arcobacter* when the proposal for the new genus was accepted (Kiehlbauch *et al.*, 1991).

Arcobacter butzleri is the species most commonly associated with human infections (Pugina *et al.*, 1991; Taylor *et al.*, 1991; Vandamme *et al.*, 1992a; Lerner, Brumberger and Preac-Mursic, 1994). The first indication that *A. butzleri* strains were associated with human enteritis was through the report of a faecal isolate from a patient who experienced diarrhoea and abdominal pain. The isolate was identified as *C. cryaerophitus* (Tee *et al.*, 1988), but in subsequent investigations at the CDC it was confirmed that it belonged to *A. butzleri* (Kiehlbauch *et al.*, 1991). Of the 63 other *A. butzleri* strains under investigation at the CDC, approximately 75% were also from human cases of diarrhoea (Kiehlbauch *et al.*, 1991). Evidence for a pathogenic role for *A. butzleri* mounted with the report of an outbreak of abdominal cramps without diarrhoea in a nursery and primary school in Italy (Vandamme *et al.*, 1992a). Fourteen isolates of *A. butzleri* from this outbreak were demonstrated to be epidemiologically linked through the application of PCR-mediated DNA fingerprinting (Vandamme *et al.*, 1993). *Arcobacter butzleri* is found not only in humans but also in a variety of animals. Among the CDC isolates, 10 were from non-human primates, three from porcine tissue samples, one from a bovine stomach and one from an ostrich yolk sac (Kiehlbauch *et al.*, 1991; Anderson *et al.*, 1993). Seven *A. butzleri* strains have also been recovered from infant monkeys (Russell *et al.*, 1992). *Arcobacter butzleri* has been isolated from food, such as poultry (Festy *et al.*, 1993) and ground pork (Collins, Wesley and Murano, 1994), as well as a variety of water sources, including ground water (Rice *et al.*, 1999), river water (Musmanno *et al.*, 1997), canal water (Dhamabutra, Kamol-Rathanakul and Pienthaweechai, 1992) and a drinking water reservoir (Jacob, Lior and Feuerpfeil, 1993).

Authors with experience in investigating Arcobacteria, particularly *A. butzleri*, stress the importance of improving the isolation procedure. *Arcobacter butzleri* is not often isolated with media and cultural conditions appropriate for *C. jejuni* subsp. *jejuni*.

A. skirrowii

This species is the most recent addition to the genus *Arcobacter*. It was described for a group of aerotolerant strains that were initially believed to belong to *A. cryaerophilus* and have been isolated from animals (Vandamme *et al.*, 1992b, 1996).

Future research should address the seasonal nature to *Campylobacter* infections and the variety of environmental sources that may harbour this bacteria. It is not known why *Campylobacter* infections peak in late May–June each year, or why *Campylobacters* cannot multiply without a warm host, since other food-poisoning related bacteria do not need a host (Jones, 2001).

REFERENCES

Adak, G.K., Cowden, J.M., Nicholas, S., Evan, U.S. (1995) The Public Health Laboratory Service national case control study of primary indigenous sporadic cases of *Campylobacter* infection. *Epidemiology and Infection*, 115, 1522.

Al Rashid, S.T., Dakuna, I., Louie, H. *et al.* (2000) Identification of *Campylobacter jejuni*, *C. coli*, *C. lari*, *C. upsaliensis*, *Arcobacter butzleri*, and *A. butzleri*-like species based on the *glyA* gene. *Journal of Clinical Microbiology*, 38, 1488–1494.

Allos, B.M., Blaser, M.J. (1995) *Campylobocter jejuni* and the expanding spectrum of related infections. *Clinical Infectious Diseases*, 20, 1092–1101.

Anderson, K.F., Kiehlbauch, J.A., Anderson, D.C. *et al.* (1993) *Arcobacter (Campylobacter) butzleri*-associated diarrheal illness in a nonhuman primate population. *Infection and Immunity*, 61, 2220–2223.

Aspinall, G.O., Fugimoto, S., McDonald, A.E. *et al.* (1994) Lipopolysaccharides from *Campylobacter jejuni* associated with Guillain–Barré syndrome patients mimic human gangliosides in structure. *Infection and Immunity*, 62, 2122–2125.

Aspinall, G.O., Wareing, D.R.A., Hayward, P.G., Hutchinson, D.N. (1996) A comparison of a new campylobacter selective medium (CAT) with membrane filtration for the isolation of thermophilic campylobacters including *Campylobacter upsaliensis*. *Journal of Applied Bacteriology*, 80, 645–650.

Bacon, D.J., Alm, R.A., Burr, D.H. *et al.* (2000) Involvement of a plasmid in virulence of *Campylobacter jejuni* 81–178. *Infection and Immunity*, 68, 4384–4390.

Bacon, D.J., Alm, R.A., Hu, L. *et al.* (2002) DNA sequence and mutational analyses of the pVir plasmid of *Campylobacter jejuni* 81–176. *Infection and Immunity*, 70, 6242–6250.

Bacon, D.J., Johnson, W.M., Rodgers, F.G. (1999) Identification and characterisation of a cytotoxic porin-lipopolysaccharide complex from *Campylobacter jejuni*. *Journal of Medical Microbiology*, 48, 139–148.

Bang, D.D., Nielsen, E.M., Scheutz, F. *et al.* (2003) PCR detection of seven virulence and toxin genes of *Campylobacter jejuni* and *Campylobacter coli* isolates from Danish pigs and cattle and cytolethal distending toxin production of the isolates. *Journal of Applied Microbiology*, 94, 1003–1014.

Bastijns, K., Cartuyvels, D., Chapelle, S. *et al.* (1995) A variable 23S rDNA region is a useful discriminating target for genus-specific and species-specific PCR amplification in *Arcobacter* species. *Systematic Applied Microbiology*, 18, 353–356.

Birkenhead, D., Hawkey, P.M., Heritage, J. *et al.* (1993) PCR for the detection and typing of campylobacters. *Letters in Applied Microbiology*, 17, 235–237.

Björk, G.R. (1995) Genetic dissection of synthesis and function of modified nucleosides in bacterial transfer RNA. *Progress in Nucleic Acid Research*, 50, 263–338.

Blaser, M.J., Gotschlich, E.G. (1990) Surface array protein of *Campylobacter fetus*: cloning and gene structure. *Journal of Biological Chemistry*, 265, 14529–14535.

Briedis, D.J., Khamessan, A., McLaughlin, R.W. *et al.* (2002) Isolation of *Campylobacter fetus* subsp. *fetus* from a patient with cellulitis. *Journal of Clinical Microbiology*, 40, 4792–4796.

Butzler, J.-P., Dekeyser, P., Detrain, M., DeHaen, F. (1973) Related vibrios in stools. *Journal of Pediatrics*, 82, 493–495.

Caldwell, M.B., Guerry, P., Lee, E.C. *et al.* (1985) Reversible expression of flagella in *Campylobacter jejuni*. *Infection and Immunity*, 50, 941–943.

Calva, E., Torres, J., Vazquez, M. *et al.* (1989) *Campylobacter jejuni* chromosomal sequences that hybridize to *Vibrio cholerae* and *Escherichia coli* LT enterotoxin genes. *Gene*, 75, 243–251.

Cardarelli-Leite, P., Blom, K., Patton, C.M. *et al.* (1996) Rapid identification of *Campylobacter* species by restriction fragment length polymorphism analysis of a PCR-amplified fragment of the gene coding for 16S rRNA. *Journal of Clinical Microbiology*, 34, 62–67.

Carvalho, A.C.T., Ruiz-Palacios, G.M., Ramos-Cervantes, P. *et al.* (2001) Molecular characterization of invasive and noninvasive *Campylobacter jejuni* and *Campylobacter coli* isolates. *Journal of Clinical Microbiology*, 39, 1353–1359.

Christie, P.J. (2001) Type IV secretion: intercellular transfer of macromolecules by systems ancestrally related to conjugations machines. *Molecular Microbiology*, 40, 294–305.

Collins, C.I., Wesley, I.V., Murano, E.A. (1994) Incidence of *Arcobacter* spp. in ground pork. *Diary, Food Environment Sanitation*, 14, 603.

Connell, S.R., Tracz, D.M., Nierhaus, K.H. *et al.* (2003) Ribosomal protection proteins and their mechanism of tetracycline resistance. *Antimicrobial Agents and Chemotherapy*, 47, 3675–3681.

Costas, M., Owen, R.J., Jackman, P.J.M. (1987) Classification of *Campylobacter sputorum* and allied campylobacters based on numerical analysis of electrophoretic protein patterns. *Systematic Applied Microbiology*, 9, 125–131.

Datta, S., Niwa, H., Itoh, K. (2003) Prevalence of 11 pathogenic genes of *Campylobacter jejuni* by PCR in strains isolated from humans, poultry meat and broiler and bovine feces. *Journal of Medical Microbiology*, 52, 345–348.

Dedieu, L., Pages, J.M., Bolla, J.M. (2002) Environmental regulation of *Campylobacter jejuni* major outer membrane protein porin expression in *Escherichia coli* monitored by using green fluorescent protein. *Applied Environmental Microbiology*, 68, 4209–4215.

Dekeyser, P., Goussuins-Deirain, M., Butzler, J.-P., Stemon, J. (1972) Acute enteritis due to related *Vibrio*: first positive stool cultures. *Journal of Infectious Diseases*, 125, 390–393.

Dhamabutra, N., Kamol-Rathanakul, P., Pienthaweechai, K. (1992) Isolation of campylobacters from the canals of Bangkok metropolitan area. Journal of the Medical Association of Thailand, 75 (6), 350–364.

Dingle, K.E., Colles, F.M., Ure, R. *et al.* (2002) Molecular characterization of *Campylobacter jejuni* clones: a basis for epidemiologic investigation. *Emerging Infectious Diseases*, 8, 949–955.

Dingle, K.E., Colles, F.M., Wareing, D.R.A. *et al.* (2001) Multilocus sequence typing system for *Campylobacter jejuni*. *Journal of Clinical Microbiology*, 39, 14–23.

Dorrell, N., Mangan, J.A., Laing, K.G. *et al.* (2001) Whole genome comparison of *Campylobacter jejuni* human isolates using a low-cost microarray reveals extensive genetic diversity. *Genome Research*, 11, 1706–1715.

Drlica, K., Zhao, X. (1997) DNA gyrase, topoisomerase IV and the 4-quinolones. *Microbiology and Molecular Biology Reviews*, 61, 377–392.

Duim, B., Vandamme, P.A., Rigter, A. *et al.* (2001) Differentiation of *Campylobacter* species by AFLP fingerprinting. *Microbiology*, 147, 2729–2937.

Edmonds, P., Patton, C.M., Griffin, P.M. *et al.* (1987) *Campylobacter hyointestinalis* associated with human gastrointestinal disease in the United States. *Journal of Clinical Microbiology*, 25 (4), 685–691.

Elharrif, A., Megraud, F. (1986) Characterization of thermophilic *Campylobacter*. II. Enzymatic profiles. *Current Microbiology*, 13, 317–322.

Ellis, W.A., Neill, S.I., O'Brien, J.J. *et al.* (1977) Isolation of Spirillum/vibrio-like organisms from bovine fetuses. *Veterinary Record*, 100, 451–452.

Endtz, H.P., Ruijs, G.J., van Klingeren, B. *et al.* (1991) Comparison of six media, including a semisolid agar, for the isolation of various *Campylobacter* species from stool specimens. *Journal of Clinical Microbiology*, 29, 1007–1010.

Engberg, J., Aarestrup, F.M., Taylor, D.E. *et al.* (2001) Quinolone and macrolide resistance in *Campylobacter jejuni* and *C. coli*: resistance mechanisms and trends in human isolates. *Emerging Infectious Disease*, 7, 24–34.

Etoh, Y., Dewhirst, F.E., Paster, B.J. *et al.* (1993) *Campylobacter showae* sp. nov. isolated from the human oral cavity. *International Journal of Systematic Bacteriology*, 43, 631–639.

Eyers, M., Chapelle, S., Van Camp, G. *et al.* (1993) Discrimination among thermophilic *Campylobacter* species by polymerase chain reaction and amplification of 23S rRNA gene fragments. *Journal of Clinical Microbiology*, 31, 3340–3343.

Fennell, C.L., Rompalo, A.M., Totten, P.M. *et al.* (1986) Isolation of *Campylobacter hyointestinalis* from a human. *Journal of Clinical Microbiology*, 24, 146–148.

Festy, B., Squinazi, F., Marin, M. *et al.* (1993) Poultry meat and waters as the possible sources of *Arcobacter butzleri* associated human disease in Paris, France. *Acta Gastro-Enterologica Belgica*, 56, 35.

Fitzgerald, C., Owen, R.J., Stanely, J. (1996) Comprehensive ribotyping scheme for heat-stable serotypes of *Campylobacter jejuni*. *Journal of Clinical Microbiology*, 34, 265–269.

Frost, J.A., Oza, A.N., Thwaites, R.T., Rowe, B. (1998) Serotyping scheme for *Campylobacter jejuni* and *Campylobacter coli* based on direct agglutination of heat stable antigens. *Journal of Clinical Microbiology*, 36, 335–339.

Fry, B.N., Feng, S., Chen, Y.Y. *et al.* (2000) The galE gene of *Campylobacter jejuni* is involved in lipopolysaccharide synthesis and virulence. *Infection and Immunity*, 68, 2594–2601.

Ganeshram, K.N., Ross, A., Cowell, R.P.W. *et al.* (2000) Recurring febrile illness in a slaughterhouse worker. *Postgraduate Medical Journal*, 76, 790–794.

Garcia, M.M., Eaglesome, M.D., Rigby, C. (1983) Campylobacters important in veterinary medicine. *Veterinary Bulletin*, 53, 793–818.

Gaudreau, C., Gilbert, H. (1998) Antimicrobial resistance of clinical strains of *Campylobacter jejuni* subsp. *jejuni* isolated from 1985 to 1997 in Quebec, Canada. *Antimicrobial Agents and Chemotherapy*, 42, 2106–2108.

Gaudreau, C., Gilbert, H. (2003) Antimicrobial resistance of *Campylobacter jejuni* subsp. *jejuni* strains isolated from humans in 1998–2001 in Montréal, Canada. *Antimicrobial Agents and Chemotherapy*, 47, 2027–2029.

Gebhart, C.J., Edmonds, P., Ward, E.G. *et al.* (1985) *Campylobacter hyointestinalis* sp. nov.: a new species of *Campylobacter* found in the intestines of pigs and other animals. *Journal of Clinical Microbiology*, 1, 715–720.

Gebhart, C.J., Lin, G.F., McOrist, S.M. *et al.* (1991) Cloned DNA probes specific for the intracellular *Campylobacter*-like organisms of porcine proliferative enteritis. *Journal of Clinical Microbiology*, 29, 1011–1015.

Gebhart, C.J., Ward, G.H., Chang, K., Kurtz, H.J. (1983) '*Campylobacter hyointestinalis*' (new species) isolated from swine with lesions of proliferative ileitis. *American Journal of Veterinary Research*, 44, 361–367.

Giesendorf, B.A.J., Goossens, H., Niesters, H.G.M. *et al.* (1994) PCR-mediated DNA fingerprinting for epidemiological studies on *Campylobacter* spp. *Journal of Medical Microbiology*, 40, 141–147.

Gilbert, M., Karwaski, M.F., Bernatchez, S. *et al.* (2002) The genetic bases for the variation in the lipo-oligosaccharide of the mucosal pathogen, *Campylobacter jejuni*. Biosynthesis of sialylated ganglioside mimics in the core oligosaccharide. *Biological Chemistry*, 27, 327–337.

Gomes, B.P.F.A., Lilley, J.D., Drucker, D.B. (1996) Clinical significance of dental root canal microflora. *Journal of Dentistry*, 24, 47–55.

Gonzalez, I., Garcia, T., Antolin, A. *et al.* (2000) Development of a combined PCR-culture technique for the rapid detection of *Arcobacter* spp. in chicken meat. *Letters in Applied Microbiology*, 30, 207–212.

Gonzalez, I., Grant, K.A., Richardson, P.T. *et al.* (1997) Specific identification of the enteropathogens *Campylobacter jejuni* and *Campylobacter coli* by using a PCR test based on the *ceuE* gene encoding a putative virulence determinant. *Journal of Clinical Microbiology*, 35, 759–763.

Goodwin, C.S., McConnell, W., McCulloch, R.K. *et al.* (1989) Cellular fatty acid composition of *Campylobacter pylori* from primates and ferrets compared with those of other campylobacters. *Journal of Clinical Microbiology*, 27, 938–943.

Goossens, H., Butzler, J.P. (1991) Isolation of *Campylobacter* spp. from stool specimens with a semisolid medium. *Journal of Clinical Microbiology*, 29, 2681–2682.

Goossens, H., Butzler, J.-P. (1992) Isolation and identification of *Campylobacter* spp. In: *Campylobacter jejuni Current Status and Future Trends* (eds I. Nachamikin, M. J. Blaser, L. S. Tompkins). Ch II. American Society for Microbiology, Washington, DC.

Grewel, J., Manarathu, E.K., Taylor, D.E. (1993) Effect of mutational alteration of asn-128 in the putative GTP-binding domain of tetracycline resistance determinant *tet(0)* from *Campylobacter jejuni*. *Antimicrobial Agents and Chemotherapy*, 37, 2645–2649.

Guerry, P., Aim, R.A., Power, M.E., Trust, T.J. (1992) Molecular and structural analysis of *Campylobacter* flagellin. In: *Campylobacter jejuni: Current Status and Future Trends* (eds I. Nachamkin, M. J. Blaser, L. Tompkins), pp. 267–281. *Proceedings of the NIH Symposium on Campylobacter*. ASM Publications, Washington, DC.

Guerry, P.P., Szymanski, C.M., Prendergast, M.M. *et al.* (2002) Phase variation of *Campylobacter jejuni* 81–176 lipooligosaccharide affects ganglioside mimicry and invasiveness in vitro. *Infection and Immunity*, 770, 787–793.

Hardy, R.W.F., Holsten, R.D., Jackson, E.K., Burns, R.C. (1968) The acetylene-ethylene assay for nitrogen fixation: laboratory and field evaluation. *Plant Physiology*, 43, 1185–1207.

Harmon, K., Wesley, I. (1996) Identification of *Arcobacter* isolates by PCR. *Letters in Applied Microbiology*, 23, 241–244.

Harmon, K., Wesley, I. (1997) Multiplex PCR for the identification of Arcobacter and differentiation of *Arcobacter butzleri* from other *Arcobacters*. *Veterinary Microbiology*, 58, 215–227.

Harrington, C.S., On, S.L.W. (1999) Extensive 16S rRNA gene sequence diversity in *Campylobacter hyointestinalis* strains: taxonomic and applied implications. *International Journal of Systematic Bacteriology*, 49, 1171–1175.

Healing, T.D., Greenwood, M.H., Pearson, A.D. (1992) Campylobacters and enteritis. *Reviews in Medical Microbiology*, 3, 159–167.

Hebert, G.A., Hollis, D.G., Weaver, R.E. *et al.* (1983) Serogroups of *Campylobacter jejuni*, *Campylobacter coli*, and *Campylobacter fetus* defined by direct immunofluorescence. *Journal of Clinical Microbiology*, 17, 529–538.

Hill, B.D., Thomas, R.J., Mackenzie, A.R. (1987) *Campylobacter hyointestinalis*-associated enteritis in Moluccan rusa deer (*Cervus timorensis* subsp. *Moluccensis*). *Journal of Comparative Pathology*, 97 (6), 687–694.

Houf, K., Tutenel, A., De Zutter, L. *et al.* (2000) Development of a multiplex PCR assay for the simultaneous detection and identification of *Arcobacter butzleri*, *Arcobacter cryaerophilus* and *Arcobacter skirrowii*. *FEMS Microbiology Letters*, 193, 89–94.

Hum, S., Quinn, K., Brunner, J., On, S.L.W. (1997) Evaluation of a PCR assay for identification and differentiation of *Campylobacter* subspecies. *Australian Veterinary Journal*, 75, 827–831.

Hurtado, A., Owen, R. (1997) A molecular scheme based on 23S rRNA gene polymorphisms for rapid identification of *Campylobacter* and *Arcobacter* species. *Journal of Clinical Microbiology*, 35, 2401–2404.

Hutchinson, D.N., Bolton, F.J. (1984) Improved blood free selective medium for the isolation of *Campylobacter jejuni* from faecal specimens. *Journal of Clinical Pathology*, 37, 956–957.

Jacob, J., Evers, S., Bischoff, K. *et al.* (1994) Characterization of the *sat4* gene encoding a streptothricin acetyltransferase in *Campylobacter coli* BE/G4. *FEMS Microbiology Letters*, 120, 13–18.

Jacob, J., Lior, H., Feuerpfeil, I. (1993) Isolation of *Arcobacter butzleri* from a drinking water reservoir in Eastern Germany. *Zentralbl Hyg Umweltmed*, 193 (6), 557–562.

Jones, K. (2001) The *Campylobacter* conundrum. *Trends in Microbiology*, 9, 365–366.

Jordan, R.C.K., Daniels, T.E., Greenspan, J.S., Regezi, J.A. (2001) Advanced diagnostic methods in oral and maxillofacial pathology. Part 1. Molecular methods. *Oral Surgery, Oral Medicine, Oral Pathology, Oral Radiology, and Endodontics*, 92, 650–669.

Karmali, M.A., Skirrow, M.B. (1984) Taxonomy of the genus *Campylobacter*. In: *Campylobacter Infection in Man and Animals* (ed. J.-P. Butzler), pp. 1–20. CRC Publications, Boca Raton, Florida.

Karmali, M.A., Simor, A.E., Roscoe, M. *et al.* (1986) Evaluation of a blood-free, charcoal-based, selective medium for the isolation of *Campylobacter* organisms from feces. *Journal of Clinical Microbiology*, 23, 456–459.

Kiehlbauch, J.A., Brenner, D.J., Nicholson, M.A. *et al.* (1991) '*Campylobacter butzleri*' sp. nov. isolated from humans and animals with diarrhoeal illness. *Journal of Clinical Microbiology*, 29, 376–385.

Kim, C.C., Joyce, E.A., Chan, K., Falkow, S. (2002) Improved analytical methods for microarray-based genome-composition analysis. *Genome Biology*. 3, RESEARCH0065.

King, E.O. (1957) Human infections with *Vibrio fetus* and a closely related vibrio. *Journal of Infectious Diseases*, 101, 119–128.

King, E.O. (1962) The laboratory recognition of *Vibrio fetus* and a closely related *Vibrio* isolated from cases of human vibriosis. *Annals of the New York Academy of Sciences*, 98, 700–711.

Klena, J.D., Gray, S.A., Konkel, M.E. (1998) Cloning, sequencing, and characterization of the lipopolysaccharide biosynthetic enzyme heptosyltransferase I gene (*waaC*) from *Campylobacter jejuni* and *Campylobacter coli*. *Gene*, 222, 177–185.

Konkel, M.E., Kim, B.J., Rivera-Amill, V., Garvis, S.G. (1999) Bacterial secreted proteins are required for the internalization of *Campylobacter jejuni* into cultured mammalian cells. *Molecular Microbiology*, 32, 691–701.

Korolik, V., Fry, B.N., Alderton, M.R. *et al.* (1997) Expression of *Campylobacter hyoilei* lipo-oligosaccharide (LOS) antigens in *Escherichia coli*. *Microbiology*, 143, 3481–3489.

Kulkarni, S.P., Lever, S., Logan, J.M. *et al.* (2002) Detection of *Campylobacter* species: a comparison of culture and polymerase chain reaction based methods. *Journal of Clinical Pathology*, 55, 749–753.

Lawson, A.J., On, S.L.W., Logan, J.M.J., Stanley, J. (2001) Isolation and characterization of a new species, *Campylobacter hominis* sp. nov. from the human gastrointestinal tract. *International Journal of Systematic Evolutionary Biology*, 51, 651–660.

Le Goff, A., Bunetel, L., Mouton, C., Bonnaure-Mallet, M. (1997) Evaluation of root canal bacteria and their antimicrobial susceptibility in teeth with necrotic pulp. *Oral Microbiology and Immunology*, 12, 318–322.

Lerner, J., Brumberger, V., Preac-Mursic, V. (1994) Severe diarrhea associated with *Arcobacter butzleri*. *European Journal of Clinical Microbiology and Infectious Disease*, 13, 660–662.

Lin, J., Michel, L.O., Zhang, Q. (2002) CmeABC functions as a multidrug efflux system in *Campylobacter jejuni*. *Antimicrobial Agents and Chemotherapy*, 46, 2124–2131.

Lin, J., Sahin, O., Michel, L.O., Zhang, Q. (2003) Critical role of multidrug efflux pump CmeABC in bile resistance and in vivo colonization of *Campylobacter jejuni*. *Infection and Immunity*, 71 (8), 4250–4259.

Lindblom, G.B., Sjogren, E., Hansson-Westerberg, J., Kaijser, B. (1995) *Campylobacter upsaliensis*, *C. sputorum* and *C. concisus* as common causes of diarrhoea in Swedish children. *Scandinavian Journal of Infectious Diseases*, 27 (2), 187–188.

Linton, D., Gilbert, M., Hitchen, P.G. *et al.* (2000) Phase variation of a beta-1,3 galactosyltransferase involved in generation of the ganglioside GM1-like lipo-oligosaccharide of *Campylobacter jejuni*. *Molecular Microbiology*, 37, 501–514.

Linton, D., Owen, R.J., Stanley, J. (1996) Rapid identification by PCR of the genus *Campylobacter* and of five *Campylobacter* species enteropathogenic for man and animals. *Research in Microbiology*, 147 (9), 707–718.

Llovo, J., Mateo, E., Munoz, A. *et al.* (2003) Molecular typing of *Campylobacter jejuni* isolates involved in a neonatal outbreak indicates nosocomial transmission. *Journal of Clinical Microbiology*, 41, 3926–3928.

Logan, J.M., Burnens, A., Linton, D. *et al.* (2000) *Campylobacter lanienae* sp. nov., a new species isolated from workers in an abattoir. *International Journal of Systematic Evolutionary Biology*, 50, 865–872.

Logan, J.M.J., Edwards, K.J., Saunders, N.A., Stanley, J. (2001) Rapid identification of *Campylobacter* spp. By melting peak analysis of biprobes in real-time PCR. *Journal of Clinical Microbiology*, 39, 2227–2232.

Lübeck, P.S., Cook, N., Wagner, M. *et al.* (2003b) Toward an international standard for PCR-based detection of food-borne thermotolerant Campylobacters: validation in a multicenter collaborative trial. *Applied Environmental Microbiology*, 69, 5670–5672.

Lübeck, P.S., Wolffs, P., On, S.L. *et al.* (2003a) Toward an international standard for PCR-based detection of food-borne thermotolerant Campylobacters: assay development and analytical validation. *Applied and Environmental Microbiology*, 69, 5664–5669.

Luber, P., Wagner, J., Hahn, H., Bartelt, E. (2003) Antimicrobial resistance in *Campylobacter jejuni* and *Campylobacter coli* strains isolated in 1991 and 2001–2002 from poultry and humans in Berlin, Germany. *Antimicrobial Agents and Chemotherapy*, 47 (12), 3825–3830.

Lucey, B., Feurer, C., Greer, P. *et al.* (2000) Antimicrobial resistance profiling and DNA Amplification Fingerprinting (DAF) of thermophilic *Campylobacter* spp. in human, poultry and porcine samples from the Cork region of Ireland. *Journal of Applied Microbiology*, 89 (5), 727–734.

Luo, N., Sahin, O., Lin, J. *et al.* (2003) In vivo selection of Campylobacter isolates with high levels of fluoroquinolone resistance associated with gyrA mutations and the function of the CmeABC efflux pump. *Antimicrobial Agents and Chemotherapy*, 47, 390–394.

Madden, R.H., Moran, L., Scates, P. (1996) Sub-typing of animal and human *Campylobacter* spp. using RAPD. *Letters in Applied Microbiology*, 23, 167–170.

Marshall, S.M., Melito, P.L., Woodward, D.L. *et al.* (1999) Rapid identification of *Campylobacter*, *Arcobacter*, and *Helicobacter* isolates by PCR-restriction fragment length polymorphism analysis of the 16S rRNA gene. *Journal of Clinical Microbiology*, 37, 4158–4160.

McClung, C.R., Patriquin, D.G. (1980) Isolation of nitrogen-fixing *Campylobacter* species from the roots of *Sparlina alterniflora* Loisel. *Canadian Journal of Microbiology*, 26, 881–886.

McClung, C.R., Patriquin, D.G., Davis, R.E. (1983) *Campylobacter nitrofigilis* sp. nov., a nitrogen-fixing *alterniflora*. *International Journal of Systematic Bacteriology*, 33, 605–612.

McDermott, P.F., Walker, R.D. (2003) Standardizing antimicrobial susceptibility testing of *Campylobacter* species. *Journal of Clinical Microbiology*, 41, 1810.

McOrist, S., Boict, R., Lawson, G.H.K. (1989) Antigenic analysis of *Campylobacter* species and an intracellular *Campylobacter*-like organism associated with porcine proliferative enteropathies. *Infection and Immunity*, 57, 957–962.

Meinersmann, R.J., Helsel, L.O., Fields, P.I., Hiett, K.L. (1997) Discrimination of *Campylobacter jejuni* by *fla* gene sequencing. *Journal of Clinical Microbiology*, 35, 2810–2814.

Miller, S., Pesci, E.G., Pickett, C.L. (1993) A *Campylobacter jejuni* homolog of the LcrD/FlbE family of proteins is necessary for flagellar biogenesis. *Infection and Immunity*, 61, 2930–2936.

Minet, J., Grosbois, B., Megraud, F. (1988) *Campylobacter hyointestinalis*: an opportunistic enteropathogen. *Journal of Clinical Microbiology*, 26, 2659–2660.

Moorhead, S.M., Dykes, G.A. (2002) Survival of *Campylobacter jejuni* on beef trimmings during freezing and frozen storage. *Letters in Applied Microbiology*, 34 (1), 72–76.

Moreno, Y., Ferrus, M.A., Vanoostende, A. *et al.* (2002) Comparison of 23S polymerase chain reaction-restriction fragment length polymorphism and amplified fragment length polymorphism techniques as typing systems for thermophilic campylobacters. *FEMS Microbiology Letters*, 211, 97–103.

Musmanno, R.A., Russi, M., Lior, H., Figura, N. (1997) In vitro virulence factors of *Arcobacter butzleri* strains isolated from superficial water samples. *New Microbiology*, 20 (1), 63–68.

Nachamkin, I. (1995) *Campylobacter* and *Arcobacter*. In: *Manual of Clinical Microbiology*, 6th edn (ed. P. R. Murray), Ch. 37. Washington, DC.

Nachamkin, I., Allos, B.M., Ho, T. (1998) *Campylobacter* species and Guillain–Barré syndrome. *Clinical Microbiology Reviews*, 11, 555–567.

Nachamkin, I., Engberg, J., Aarestrup, F.M. (2000) Diagnosis and antimicrobial susceptibility of *Campylobacter* species. In: *Campylobacter*, 2nd edn (eds I. Nachamkin, M.J. Blaser), pp. 45–66. ASM Press, Washington, DC.

Neill, S.D., Campbell, J.N., O'Brien, J.J., Weatherup, S.T.C., Ellis, W.A. (1985) Taxonomic position of *Campylobacter cryaerophila* sp. nov. *International Journal of Systematic Bacteriology*, 35, 342–356.

Neilsen, E.M., Engberg, J., Mogens, M. (1997) Distribution of serotypes of *Campylobacter jejuni* and *C. coli* from Danish patients, poultry, cattle and swine. *FEMS Immunology and Medical Microbiology*, 19, 47–56.

Newell, D.G., Frost, J.A., Duim, B. *et al.* (2000) New developments in the subtyping of *Campylobacter* species. In: *Campylobacter*, 2nd edn (eds I. Nachamkin, M.J. Blaser), pp. 27–44. ASM Press, Washington, DC.

On, S.L.W. (1996) Identification methods for campylobacters, helicobacters, and related organisms. *Clinical Microbiology Reviews*, 9, 405–422.

On, S.L.W. (2000) International Committee on Systematic Bacteriology Subcommittee on the taxonomy of *Campylobacter* and related bacteria. *International Journal of Systematic and Evolutionary Microbiology*, 50, 1401–1403.

On, S.L.W. (2001) Taxonomy of *Campylobacter, Arcobacter, Helicobacter* and related bacteria: current status, future prospects and immediate concerns. *Symposium Series (Society of Applied Microbiology)*, 90, 405–422.

On, S.L.W., Harrington, C.S. (2000) Identification of taxonomic and epidemiological relationships among *Campylobacter* species by numerical analysis of AFLP profiles. *FEMS Microbiology Letters*, 193, 161–169.

On, S.L.W., Harrington, C.S. (2001) Evaluation of numerical analysis of PFGE-DNA profiles for differentiating *Campylobacter fetus* subspecies by comparison with phenotypic, PCR, and 16S rDNA sequencing methods. *Journal of Applied Microbiology*, 90, 285–293.

On, S.L.W., Holmes, B. (1995) Classification and identification of campylobacters, helicobacters and allied taxa by numerical analysis of phenotypic tests. *Systematic and Applied Microbiology*, 18, 374–390.

On, S.L.W., Jordan, P.J. (2003) Evaluation of 11 PCR assays for species-level identification of *Campylobacter jejuni* and *Campylobacter coli*. *Journal of Clinical Microbiology*, 41, 330–336.

On, S.L.W., Vandamme, P. (1997) Identification and epidemiological typing of *Campylobacter hyointestinalis* subspecies by phenotypic and genotypic methods and description of novel subgroups. *Systematic and Applied Microbiology*, 20, 238–247.

On, S.L., Atabay, H.I., Corry, J.E. (1999) Clonality of *Campylobacter sputorum* bv. *paraureolyticus* determined by macrorestriction profiling and biotyping, and evidence for long-term persistent infection in cattle. *Epidemiology and Infection*, 122 (1), 175–182.

On, S.L.W., Atabay, H.I., Corry, J.E.L. *et al.* (1998) Emended description of *Campylobacter sputorum* and revision of its infrasubspecific (biovar) divisions, including *C. sputorum* biovar *paraureolyticus*, a urease-producing variant from cattle and humans. *International Journal of Systematic Bacteriology*, 48, 195–206.

On, S.L.W., Jensen, T.K., Bille-Hansen, V. *et al.* (2002) Prevalence and diversity of *Arcobacter* spp. isolated from the internal organs of spontaneous porcine abortions in Denmark. *Veterinary Microbiology*, 85, 159–167.

Oyofo, B.A., Thornton, S.A., Burr, D.H. *et al.* (1992) Specific detection of *Campylobacter jejuni* and *Campylobacter coli* using polymerase chain reaction. *Journal of Clinical Microbiology*, 30, 2613–2619.

Parkhill, J., Wren, B.W., Mungall, K. *et al.* (2000) The genome sequence of the food-borne pathogen *Campylobacter jejuni* reveals hypervariable sequences. *Nature*, 403, 665–668.

Pearson, B.M., Pin, C., Wright, J. *et al.* (2003) Comparative genome analysis of *Campylobacter jejuni* using whole genome DNA microarrays. *FEBS Letters*, 554, 224–230.

Peetermans, W.E., De Man, F., Moerman, P., Van de Werf, F. (2000) Fatal prosthetic valve endocarditis due to Campylobacter fetus. The British Infection Society (Online May 22).

Penner, J.L. (1988) The genus *Campylobacter*: a decade of progress. *Clinical Microbiology Reviews*, 1, 157–172.

Petersen, L., On, S.L. (2000) Efficacy of flagellin gene typing for epidemiological studies of *Campylobacter jejuni* in poultry estimated by comparison with macrorestriction profiling. *Letters in Applied Microbiology*, 31, 14–19.

Piddock, L.J.V. (1995) Quinolone resistance and *Campylobacter* spp. *Journal of Antimicrobial Chemotherapy*, 36, 891–898.

Pugina, P., Benzi, G., Lauwers, S. *et al.* (1991) An outbreak of *Arcobacter (Campylobacter) butzleri* in Italy. *Microbiol Ecology in Health Disease Supplement*, 4, S94.

Pumbwe, L., Piddock, L.J.V. (2002) Identification and molecular characterisation of CmeB, a *Campylobacter jejuni* multidrug efflux pump. *FEMS Microbiology Letters*, 206, 185–189.

Ribot, E.M., Fitzgerald, C., Kubota, K. *et al.* (2001) Rapid pulsed-field gel electrophoresis protocol for subtyping of *Campylobacter jejuni*. *Journal of Clinical Microbiology*, 39, 1889–1894.

Rice, E.W., Rodgers, M.R., Wesley, I.V. *et al.* (1999) Isolation of *Arcobacter butzleri* from ground water. Letters in Applied Microbiology, 28 (1), 31–35.

Roe, D.E., Weinberg, A., Roberts, M.C. (1995) Characterization of erythromycin resistance in *Campylobacter (Wolinella) rectus*. *Clinical Infectious Diseases*, 20 (Suppl. 2), S370–S371.

Romaniuk, P.J., Trust, T.J. (1989) Rapid identification of *Campylobacter* species using oligonucleotide probes to 16S ribosomal RNA. *Molecular Cellular Probes*, 3, 133–142.

Romaniuk, P.J., Zoltowska, B., Trust, T.J. *et al.* (1987) *Campylobacter pylori*, the spiral bacterium associated with human gastritis, is not a true *Campylobacter* sp. *Bacteriology*, 169, 2137–2141.

Roop, R.M., Smibert, P.M., Johnson, J.L., Krieg, N.R. (1985) *Campylobacter mucosalis* (Lawson, Leaver, Pettigrew, and Rowland, 1981) comb. nov. amended description. *International Journal of Systematic bacteriology*, 35, 189–192.

Russell, R.G., Kiehibauch, J.A., Gebban, C.J., DeTolla, L.J. (1992) Uncommon *Campylobacter* housed in a nursery. *Journal of Clinical Microbiology*, 30, 3024–3027.

Sails, A.D., Fox. A.J., Bolton, F.J. *et al.* (2001) Development of a PCR ELISA assay for the identification of *Campylobacter jejuni* and *Campylobacter coli. Molecular and Cellular Probes*, 15, 291–300.

Salama, S.M., Garcia, M.M., Taylor, D.E. (1992) Differentiation of the subspecies of *Campylobacter fetus* by genomic sizing. *International Journal of Systematic Bacteriology*, 42, 446–450.

Sandstedt, K., Ursing, J. (1991) Description of *Campylobacter upsaliensis* sp. nov. previously known as the CNW group system. *Applied Microbiology*, 14, 39–45.

Sasaki, Y., Fujisawa, T., Ogikubo, K. *et al.* (2003) Characterization of *Campylobacter lanienae* from pig feces. *Journal of Veterinary Medical Sciences*, 65, 129–131.

Schouls, L.M., Reulen, S., Duim, B. *et al.* (2003) Comparative genotyping of *Campylobacter jejuni* by amplified fragment length polymorphism, multilocus sequence typing, and short repeat sequencing: strain diversity, host range, and recombination. *Journal of Clinical Microbiology*, 41, 15–26.

Silley, P. (2003) *Campylobacter* and fluoroquinolones: a bias data set? *Environmental Microbiology*, 5, 219–230.

Siqueira, F.R. Jr, Rôças, I.N. (2003) *Campylobacter gracilis* and *Campylobacter rectus* in primary endodontic infections. *International Endodontic Journal*, 36, 174–180.

Siqueira, F.R. Jr, Rôças, I.N., Souto, R. *et al.* (2000b) Checkerboard DNA-DNA hybridization analysis of endodontic infections. *Oral Surgery, Oral Medicine, Oral Pathology, Oral Radiology, and Endodontics*, 89, 744–748.

Siqueira, F.R. Jr, Rôças, I.N., Souto, R. *et al.* (2001b) Microbiological evaluation of acute periradicular abscesses by DNA-DNA hybridization. *Oral Surgery, Oral Medicine, Oral Pathology, Oral Radiology, and Endodontics*, 92, 451–457.

Skirrow, M.B. (1977) *Campylobacter* enteritis: a 'new' disease. *British Medical Journal*, 11, 9–11.

Skirrow, M.B. (1994) Diseases due to *Campylobacter, Helicobacter* and related bacteria. *Journal of Comparative Pathology*, 111 (2), 113–149.

Skirrow, M.B., Blaser, M.J. (1992) Clinical and epidemiologic considerations. In: *Campylobacter jejuni Current Status and Future Trends* (eds I. Nachamkin, M.J. Blaser, L.S. Tompkins), pp. 3–8. American Society for Microbiology, Washington, DC.

Smith, S.I., Olukoya, D.K., Fox, A.J., Coker, A.O. (1998) Ribosomal RNA gene restriction fragment diversity amongst Penner serotypes of *Campylobacter jejuni* and *C. coli. Zeitschrift Naturforsch*, 544, 65–68.

Sopwith, W., Ashton, M., Frost, J.A. *et al.* (2003) Enhanced surveillance of *Campylobacter* infection in the north west of England 1997-1999. *Journal of Infection*, 46, 35–45.

Spahn, C.M.T., Prescott, C.D. (1996) Throwing a spanner in the works: antibiotic and the translation apparatus. *Journal of Molecular Medicine*, 74, 423–439.

Stackebrandt, E., Goebel, B.M. (1994) Cultural and phylogenetic analysis of mixed microbial populations found in natural and commercial bioleaching environments. *Applied and Environmental Microbiology*, 60 (5): 1614–1621.

Stanley, J., Bumens, A.P., Linton, D. *et al.* (1992) *Campylobacter helveticus* sp. nov., a new thermophilic species from domestic animals: characterization and cloning of a species-specific DNA probe. *Journal of General Microbiology*, 138, 2293–2303.

Steele, T.W., Owen, R.J. (1988) *Campylobacter jejuni* subsp. *doylei* subsp. nov., a subspecies of nitrate-negative campylobacters isolated from human clinical specimens. *International Journal of Systematic Bacteriology*, 38, 316–318.

Stucki, U., Frey, J., Jicolet, J., Burnens, A.P. (1995) Identification of *Campylobacter jejuni* on the basis of a species-specific gene that encodes a membrane protein. *Journal of Clinical Microbiology*, 33, 855–859.

Sundqvist, G., Figdor, D., Persson, S., Sjögren, U. (1998) Microbiologic analysis of teeth with failed endodontic treatment and the outcome of conservative re-treatment. *Oral Surgery, Oral Medicine, Oral Pathology, Oral Radiology, and Endodontics*, 85, 68–93.

Szymanski, C.M., Logan, S.M., Linton, D., Wren, B.W. (2003) Campylobacter – a tale of two protein glycosylation systems. *Trends in Microbiology*, 11, 233–238.

Tajada, P., Gomez-Garces, J.-I., Alos, J.-I. *et al.* (1996) Antimicrobial susceptibilities of *Campylobacter jejuni* and *Campylobacter coli* to 12 β-lactams and combination with β-lactamase inhibitors. *Antimicrobial Agents and Chemotherapies*, 40, 1924–1925.

Tanner, A.C.R., Badger, S., Lai, C.-H. *et al.* (1981) *Wolinella* General nov., *Wolinella succinogenes* (*Vibrio succinogenes* Wolin *et al.*) comb. nov. and description of *Bacteroides gracilis* sp. nov., *Wolinella recta* sp. nov., *Campylobacter concisus* sp. nov. and *Eikenella corrodens* from humans with periodontal disease. *International Journal of Systematic Bacteriology*, 31, 432–445.

Tanner, A.C.R., Listgarten, M.A., Ebersole, J.L. (1984) *Wolinella curva* sp. nov., 'Vibrio succinogenes' of human origin. *International Journal of Systematic Bacteriology*, 34, 275–282.

Tauxe, R.V. (1992) Epidemiology of *Campylobacter jejuni* infections in the United States and other industrialized nations. In: *Campylobacter jejuni Current Status and Future Trends* (eds I. Nachamkin, M.J. Blaser, L.S. Tompkins), Ch. 2. American Society for Microbiology, Washington, DC.

Taylor, D.E. (1984) Plasmids from *Campylobacter*. In: *Campylobacter Infection in Man and Animals* (ed. J.P. Butzler), pp. 87–96. CRC Publications, Florida.

Taylor, D.E. (1992a) Genetics of *Campylobacter* and *Helicobacter*. *Annual Reviews in Microbiology*, 46, 35–64.

Taylor, D.E. (1992b) Antimicrobial resistance of *Campylobacter jejuni* and *Campylobacter coli* to tetracycline, chloramphenicol and erythromycin. In: *Campylobacter jejuni: Current Status and Future Trends* (eds I. Nachamkin, M.J. Blaser, L.S. Tompkins), pp. 74–86. American Society for Microbiology, Washington, DC.

Taylor, D.E., Courvalin, P. (1988) Mechanism of antibiotic resistance in *Campylobacter* species. *Antimicrobial Agents and Chemotherapy*, 32, 1107–1112.

Taylor, D.E., De Grandis, S.A., Karmali, M.A., Fleming, P.C. (1981) Transmissible plasmids from *Campylobacter jejuni*. *Antimicrobial Agents and Chemotherapy*, 19, 831–835.

Taylor, D.E., Hiratsuka, K., Mueller, L. (1989) Isolation and characterization of catalase-negative and catalase-weak strains of *Campylobacter* species, including 'Campylobacter upasaliensis' from humans with gastroenteritis. *Journal of Clinical Microbiology*, 72, 2042–2045.

Taylor, D.N., Kiehlbauch, J.A., Tee, W. *et al.* (1991) Isolation of group 2 aerotolerant *Campylobacter* species from Thai children with diarrhea. *Journal of Infectious Diseases*, 163, 1062–1067.

Taylor, D.E., Ng, L.-K., Lior, H. (1985) Susceptibility of *Campylobacter* species to nalidixic acid, enoxacin, and other DNA gyrase inhibitors. *Antimicrobial Agents and Chemotherapy*, 28, 708–710.

Taylor, D.E., Trieber, C.A., Trescher, G., Bekkering, M. (1998) Host mutations (*miaA* and *rpsL*) reduce tetracycline resistance mediated by Tet(O) and Tet(M). *Antimicrobial Agents and Chemotherapy*, 42, 59–64.

Tee, W., Baird, K., Dyall-Smith, M., Dwyer, B. (1988) *Campylobacter cryaerophila* isolated from a human. *Journal of Clinical Microbiology*, 26, 2469–2473.

Tenover, F.C., Fennell, C.L. (1992) The genera *Campylobacter* and *Helicobacter*. In: *A Handbook on the Biology of Bacteria: Ecophysiology, Isolation, Identification, Application* (eds A. Balows, H.G. Truper, M. Dworkin *et al.*), pp. 3488–3511. Springer-Verlag, New York.

Thompson, L.M. III, Smibert, R.M., Johnson, J.L., Krieg, N.R. (1988) Phylogenetic study of the genus *Campylobacter*. *International Journal of Systematic Bacteriology*, 38, 190–200.

Totten, P.A., Patton, C.M., Tenover, F.C. *et al.* (1987) Prevalence and characterization of hippurate-negative *Campylobacter jejuni* in King County, Washington. *Journal of Clinical Microbiology*, 25, 1747–1752.

Tremblay, C., Gaudreau, C., Lorange, M. (2003) Epidemiology and antimicrobial susceptibilities of 111 *Campylobacter fetus* subsp. *fetus* strains isolated in Quebec, Canada, from 1983 to 2000. *Journal of Clinical Microbiology*, 41, 463–466.

Trieber, C.A., Taylor, D.E. (1999) Erythromycin resistance in *Campylobacter*, p. 3. In: *Proceedings of the 10th International Workshop on Campylobacter, Helicobacter and Related Organisms, Baltimore, MD*.

Trieber, C.A., Taylor, D.E. (2000) Mechanisms of antibiotic resistance in *Campylobacter*. In: *Campylobacter*, 2nd edn (eds I. Nachamkin, M.J. Blaser), pp. 45–66. ASM Press, Washington, DC.

Trust, T.J., Logan, S.M., Gustafson, C.E. *et al.* (1994) Phylogenetic and molecular characterization of a 23S rRNA gene positions the genus *Campylobacter* in the epsilon subdivision of the *Proteobacteria* and shows that the presence of transcribed spacers is common in *Campylobacter* spp. *Journal of Bacteriology*, 176, 4597–4609.

Tummuru, M.K.R., Blaser, M.J. (1993) Rearrangement of *sap A*. homologs with conserved and variable regions in *Campylobacter fetus*. *Proceedings of the National Academy of Sciences USA*, 90, 7265–7269.

Ursing, J.B., Lior, H., Owen, R.J. (1994) Proposal of minimal standards for describing new species of the family *Campylobacteracae*. *International Journal of Systemic Bacteriology*, 44, 842–845.

van Doorn, L.-J., Giesendorf, B.A.J., Bax, R. *et al.* (1997) Molecular discrimination between *Campylobacter jejuni*, *Campylobacter coli*, *Campylobacter lari*, and *Campylobacter upsaliensis* by polymerase chain reaction based on a novel putative GTPase gene. *Molecular Cellular Probes*, 11, 177–185.

Van Etterijck, R., Breynaert, J., Revets, H. *et al.* (1996) Isolation of *Campylobacter concisus* from feces of children with and without diarrhea. *Journal of Clinical Microbiology*, 34 (9), 2304–2306.

Van Vliet, A.H.M., Ketley, J.M. (2001) Pathogenesis of enteric *Campylobacter* infection. *Journal of Applied Microbiology*, 90, 45S–56S.

Vandamme, P. (2000) Taxonomy of the family Campylobacteraceae. In: *Campylobacter*, 2nd edn (eds I. Nachamkin, M.J. Blaser), pp. 3–26. ASM Press, Washington, D.C.

Vandamme, P., De Ley, J. (1991) Proposal for a new family, *Campylobacteraceae*. *International Journal of Systematic Bacteriology*, 41, 451–455.

Vandamme, P., On. S. (2001) Recommendations of the subcommittee on the taxonomy of *Campylobacter* and related bacteria. *International Journal of Systematic Evolutionary Microbiology*, 51, 719–721.

Vandamme, P., Daneshvar, M.I., Dewhirst, F.E. *et al.* (1995) Chemotaxonomic analyses of *Bacteroides gracilis* and *Bacteroides ureolyticus* and reclassification of *B. gracilis* as *Campylobacter gracilis* comb. nov. *International Journal of Systematic Bacteriology*, 45, 145–152.

Vandamme, P., Falsen, E., Pot, B. *et al.* (1989) Identification of EF group 22 campylobacters from gastroenteritis cases as *Campylobacter concisus*. *Journal of Clinical Microbiology*, 27 (8), 1775–1781.

Vandamme, P., Falsen, E., Rossau, R. *et al.* (1991) Revision of *Campylobacter*, *Helicobacter* and *Wolinella* taxonomy: emendation of generic descriptions and proposal of *Arcobacter* General nov. *International Journal of Systematic Bacteriology*, 41, 88–103.

Vandamme, P., Giesendorf, B.A.J., van Belkum, A. *et al.* (1993) Discrimination of epidemic and sporadic isolates of *Arcobacter butzleri* by polymerase chain reaction-mediated DNA finger printing. *Journal of Clinical Microbiology*, 31, 3317–3319.

Vandamme, P., Pot, B., Gillis, M. *et al.* (1996) Polyphasic taxonomy, a consensus approach to bacterial classification. *Microbiology Reviews*, 60, 407–438.

Vandamme, P., Pugina, P., Benzi, G. *et al.* (1992a) Outbreak of recurrent abdominal cramps associated with *Arcobacter butzleri* in an Italian School. *Journal of Clinical Microbiology*, 30, 2335–2337.

Vandamme, P., van Doorn, L.-J., Al Rashid, S.T. *et al.* (1997) *Campylobacter hyoilei* Alderton *et al.*, 1995 and *Campylobacter coli* Veron and Chatelaine 1973 are subjective synonyms. *International Journal of Systematic Bacteriology*, 47, 1055–1060.

Vandamme, P., Vancanneyt, M., Pot, B. *et al.* (1992b) Polyphasic taxonomic study of the emended genus *Arcobacter* with *Arcobacter butzleri* comb. nov. and *Arcobacter skirrowii* sp. nov., an aerotolerant bacterium isolated from veterinary specimens. *International Journal of Systematic Bacteriology*, 42, 344–356.

Volokhov, D., Chizhikov, V., Chumakov, K., Rasooly, A. (2003) Microarray-based identification of thermophilic *Campylobacter jejuni*, *C. coli*, *C. lari*, and *C. upsaliensis*. *Journal of Clinical Microbiology*, 41, 4071–4080.

Wacker, M., Linton, D., Hitchen, P.G. *et al.* (2003) N-linked glycosylation in *Campylobacter jejuni* and its functional transfer into *E. coli*. *Science*, 298, 1790–1793.

Wang, Y., Taylor, D.E. (1990) Natural transformation in *Campylobacter* species. Journal of Bacteriology, 172 (2), 949–955.

Wang, Y., Huang, W.M., Taylor, D.E. (1993) Cloning and nucleotide sequence of the *Campylobacter jejuni gyrA* gene and characterization of quinolones resistance mutations. *Antimicrobial Agents and Chemotherapy*, 37, 457–463.

Wang, W.-L., Reller, L.B., Smallwood, B. *et al.* (1983) Evaluation of transport media for *Campylobacter jejuni* in human fecal specimens. *Journal of Clinical Microbiology*, 18, 803–807.

Wareing, D.R.A., Ure, R., Colles, F.M. *et al.* (2003) Reference isolates for the clonal complexes of *Campylobacter jejuni*. *Letters in Applied Microbiology*, 36, 106–110.

Wassenaar, T.M., Fry, B.N., van der Zeijst, B.A. (1995) Variation of the flagellin gene locus of *Campylobacter jejuni* by recombination and horizontal gene transfer. *Microbiology*, 141, 95–101.

Wassenaar, T.M., Newell, D.G. (2000) Genotyping of *Campylobacter* spp. *Applied Environmental Microbiology*, 66, 1–9.

Waterman, S.R., Hackett, J., Manning, P.A. (1993) Isolation of a restriction-less mutant and development of a shuttle vector for the genetic analysis of *Campylobacter hyointestinalis*. *Gene*, 125, 19–24.

Wayne, L.G., Brenner, D.J., Colwell, R.R. *et al.* (1987) Report of the ad hoc committee on reconciliation of approaches to bacterial systematics. *International Journal of Systematic Bacteriology*, 37, 463–464.

Wesley, I.V., Schroeder-Tucker, L., Baetz, A.L. *et al.* (1995) *Arcobacter*-specific and *Arcobacter butzleri*-specific 16S rRNA-based DNA probes (1995). *Journal of Clinical Microbiology*, 33, 1691–1698.

Whitehouse, C.A., Balbo, P.B., Pesci, E.C. *et al.* (1998) *Campylobacter jejuni* cytolethal distending toxin causes a G2-phase cell cycle block. *Infection and Immunity*, 66, 1934–1940.

Willmott, C.J.R., Maxwell, A. (1993) A single point mutation in the DNA gyrase A protein greatly reduces binding of fluoroquinolones to the gyrase–DNA complex. *Antimicrobial Agents and Chemotherapy*, 37, 126–127.

Wilson, D.L., Abner, S.R., Newman, T.C. *et al.* (2000) Identification of ciprofloxacin-resistant *Campylobacter jejuni* by use of a fluorogenic PCR assay. *Journal of Clinical Microbiology*, 38 (11), 3971–3978.

Wood, A.C., Oldfield, N.J., O'Dwyer, C.A., Ketley, J.M. (1999) Cloning, mutation and distribution of a putative lipopolysaccharide biosynthesis locus in *Campylobacter jejuni*. *Microbiology*, 145, 379–388.

Wu, T.L., Su, L.H., Chia, J.H. *et al.* (2002) Molecular epidemiology of nalidixic acid-resistant *Campylobacter* isolates from humans and poultry by pulsed-field gel electrophoresis and flagellin gene analysis. *Epidemiology Infection*, 129, 227–231.

Yan, W., Chang, N., Taylor, D.E. (1991) Pulsed-field gel electrophoresis of *Campylobacter jejuni* and *Campylobacter coli* genomic DNA and its epidemiologic application. *Journal of Infectious Diseases*, 163, 1068–1072.

Yao, R., Aim, R., Trust, T.J., Guerry, P. (1993) Construction of new *Campylobacter* cloning vectors and a new mutational *cat* cassette. *Gene*, 130, 127–130.

Young, N.M., Brisson, J.-R., Kelly, J. *et al.* (2002) Structure of the *N*-linked glycan present on multiple glycoproteins in the gram-negative bacterium, *Campylobacter jejuni*. *Journal of Biology Chemistry*, 277, 42530–42539.

Zhao, C., Ge, B., De Villena, J. *et al.* (2001) Prevalence of *Campylobacter* spp., *Escherichia coli*, and *Salmonella* serovars in retail chicken, turkey, pork, and beef from the Greater Washington, D.C., area. *Applied and Environmental Microbiology*, 67 (12), 5431–5436.

42

Treponemes

Andrew J. L. Turner

Manchester Medical Microbiology Partnership, Department of Clinical Virology, Manchester Royal Infirmary, Oxford Road, Manchester, UK

INTRODUCTION

Treponemes are spirochaetes, spiral bacteria, which cause four human diseases: syphilis and the endemic treponematoses. The treponematoses are distinguished on the basis of epidemiological characteristics and clinical manifestations. *Treponema pallidum* subspecies *pallidum* causes syphilis; *Treponema pallidum* subspecies *endemicum* causes endemic syphilis or bejel; *Treponema pallidum* subspecies *pertenue* causes yaws and *Treponema carateum* causes pinta (Table 42.1). Syphilis is transmitted sexually and can affect the foetus and central nervous system, whereas central nervous system involvement and congenital transmission do not occur in endemic treponemal infection.

Despite the continued efficacy of penicillin therapy and concerted public health programmes, the treponematoses remain important. WHO has estimated that 2.5 million people are affected by the endemic treponematoses (WHO, 1998) and that there were approximately 12 million new cases of infectious syphilis in adults worldwide in 1999 (WHO, 2001). Syphilis is endemic in many parts of Africa and Southeast Asia, and there have been major epidemics in the former Soviet Union and Eastern Europe recently. In contrast, rates of infectious syphilis in the United States in 2000 were the lowest since reporting began; this decline in syphilis rates following an epidemic in the late 1980s and the concentration of the syphilis in a relatively small number of areas were among the reasons that prompted the development of a national plan to eliminate syphilis from the United States (St Louis and Wasserheit, 1998). Nevertheless, there have been numerous outbreaks of syphilis since the late 1990s in the USA and in other countries. These outbreaks have often occurred among men who have sex with men and have been characterised by high rates of co-infection with HIV, highlighting concerns about interactions between syphilis and HIV infection.

Syphilis also remains topical because of the recent publication of the genome sequence of *T. pallidum* (Fraser *et al.*, 1998) and the impact of the application of genomics and proteomics on the understanding of this hitherto enigmatic organism (Norris and Weinstock, 2000).

DESCRIPTION OF THE ORGANISM

The following sections concentrate on *T. pallidum*; the other treponemes are not discussed in detail.

Morphology

Treponema pallidum is spiral shaped, varying from 0.16 to 0.20 µm in diameter and 5–15 µm in length. Each organism typically has between 6 and 14 spiral turns and pointed ends, lacking the hook shape of some commensal human spirochaetes. As a result of its small size, it is difficult to see by conventional light microscopy; unstained organisms can be visualised by darkground microscopy and this can be used for diagnosis. Ultrastructurally, *T. pallidum* resembles Gram-negative bacteria, with a concentric arrangement of outer membrane, periplasmic space, peptidoglycan layer, inner (cytoplasmic) membrane and protoplasmic cylinder (Figure 42.1). However, the *T. pallidum* outer membrane differs from that of Gram-negative bacteria in that it lacks lipopolysaccharide (LPS); this has been confirmed by the lack of genes encoding LPS biosynthesis in the genome (Fraser *et al.*, 1998). It also has few transmembrane proteins; freeze-fracture studies have shown that the outer membrane of *T. pallidum* contains approximately 100-fold fewer transmembrane proteins than other spirochaetes and Gram-negative bacteria (Radolf, Norgard and Schulz, 1989). The cytoplasmic membrane appears to be more like that of Gram-negative bacteria and contains high concentrations of intramembranous particles which appear to include many of the abundant and highly immunogenic lipoproteins of *T. pallidum*, including Tpp 47, Tpp17, Tpp15 and possibly glycerophosphodiester phosphodiesterase,

Table 42.1 Characteristics of pathogenic Treponemes[a]

Organism	Disease	Distribution	Pathogenicity in humans	Pathogenicity in animals
Treponema pallidum subsp. *pallidum*	Syphilis	Worldwide	Highly invasive; CNS and congenital infection	Rabbit, primate, guinea pig and hamster
Treponema pallidum subsp. *endemicum*	Bejel	North Africa, Near East and Southern Europe	Moderately invasive	Rabbit and hamster
Treponema pallidum subsp. *pertenue*	Yaws	Central Africa, South America and Indonesia	Moderately invasive	Rabbit and hamster
Treponema carateum	Pinta	Central and South America	Not invasive	Primate only

[a]Compiled from Norris *et al.* (2001) and Antal *et al.* (2002).

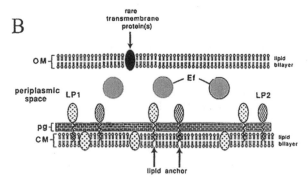

Figure 42.1 *T. pallidum* ultrastructure. (A) Membrane architecture of *T. pallidum* subspecies *pallidum* as demonstrated by freeze-fracture electron microscopy. The figure shows the paucity of OM transmembrane proteins. (B) Proposed molecular architecture of *T. pallidum*. The OM is depicted as having rare transmembrane proteins, while major immunogens are lipoproteins (LP1 and LP2) anchored to the periplasmic leaflet of the OM. The bacterium is also shown with periplasmic endoflagellum (Ef) and peptidoglycan (pg)-cytoplasmic membrane (CM) complex. Adapted with permission from Cox DL *et al.* (1992) *Infection and Immunity* **60** (3), 1076–1083, as modified from Microbes and Infection Volume 4, Salazar JC *et al.*, page 1134, copyright (2002) with permission from Elsevier.

although this is controversial. It has been shown that Tpp47, the major 47 kDa lipoprotein of *T. pallidum*, is a penicillin-binding protein with carboxypeptidase activity (Weigel, Radolf and Norgard, 1994) and this suggests that the other lipoproteins associated with the cytoplasmic membrane may also be involved in peptidoglycan biosynthesis. A further difference from Gram-negative bacteria is that the endoflagella lie entirely within the periplasmic space. There are usually three endoflagella originating from each end of the organism; they run the length of the organism and overlap in the centre with those from the opposite end. They are responsible for motility, which is important in pathogenesis because of its role in tissue invasion and in dissemination; they are also important because they are highly antigenic, stimulating a strong early antibody response that persists throughout infection and is thus a potential target for serological tests.

Physiology and Cultivation

Although limited multiplication can be obtained by co-cultivation with rabbit epithelial cells (Fieldsteel, Cox and Moeckli, 1981), *T. pallidum* has not been continuously cultured *in vitro* and this has hindered understanding of the organism. Until recently, it was necessary to propagate the organism by intratesticular inoculation of rabbits for research. Most research has been done on the Nichols strain; modern isolates of *T. pallidum* appear to be closely related to this strain even though it has been maintained in rabbits since it was originally isolated from the cerebrospinal fluid (CSF) of a patient with neurosyphilis in 1912. *Treponema pallidum* is an obligate human parasite; analysis of the genome sequence has confirmed earlier data that indicated *T. pallidum* has minimal metabolic activity and biosynthetic ability. It does have an extensive set of transport proteins though, which would allow it to scavenge essential nutrients from its

human host (Norris and Weinstock, 2000). *Treponema pallidum* is microaerophilic; its sensitivity to oxygen appears to be due to its lack of superoxide dismutase, catalase and other oxygen radical scavengers. Again, this is confirmed by the genome sequence, which suggests no obvious mechanism for protecting against oxygen toxicity, other than NADH oxidase. It appears similar to lactic acid bacteria in its relationship with oxygen, which may provide further insights into *T. pallidum* (Norris, Cox and Weinstock, 2001). It has a relatively long generation time of 30–33 h, which may be partly responsible for the difficulty of *in vitro* culture and is also of therapeutic significance.

Phylogeny

The treponemes are classified in the genus *Treponema* within the Spirochaetaceae family of the order Spirochaetales; the order also includes the genera *Borrelia* and *Leptospira*. Spirochaetes are characterised by spiral morphology, motility and the periplasmic location of their flagella. The pathogenic treponemes are virtually indistinguishable in terms of DNA homology, morphology, protein content and physiology and until recently the only reliable criterion for differentiating them was their pattern of pathogenesis in humans and in experimentally infected animals (Table 42.1). On the basis of DNA homology and clinical observations, three of the human treponemal pathogens are now classified as subspecies of *Treponema pallidum*: *Treponema pallidum* subspecies *pallidum*, *Treponema pallidum* subspecies *endemicum* and *Treponema pallidum* subspecies *pertenue*; *Treponema carateum* is classified as a separate species. However, it was recently reported that it was possible to differentiate *Treponema pallidum* subspecies *pallidum* from subspecies *endemicum* and *pertenue* on the basis of differences in the flanking regions of the 15-kDa lipoprotein gene *tpp 15* (Tp0171) (Centurion-Lara *et al.*, 1998). Other differences have also been described, most notably in the *T. pallidum* repeat (*tpr*) genes.

Molecular Biology and Antigenicity

Developments in analytical techniques have enabled considerable advances in the understanding of *T. pallidum*. The application of methods including protein gel electrophoresis, monoclonal antibody technology and gene cloning enabled characterisation of the major antigens (Norris and the Treponema Pallidum Polypeptide Research Group, 1993). The recent sequencing of the *T. pallidum* genome (Fraser *et al.*, 1998) has allowed prediction of gene sequences and functions. This analysis can include assessment of missing functions, which is particularly relevant in *T. pallidum* given its limited metabolic capabilities.

The genome consists of a single circular chromosome of approximately 1000 kb pairs and no extrachromosomal elements; it is thus one of the smallest prokaryotic genomes containing 1041 open-reading frames (ORFs). Biological functions have been predicted for more than 50% of these ORFs; a further 17% match hypothetical proteins from other species; the remaining 28% did not have significant similarity to other known sequences, which is typical of most bacterial genomes that have been sequenced. The *T. pallidum* genome appears to lack transposable elements; furthermore there are few sequence differences between *T. pallidum* subspecies and strains, suggesting either that the variants evolved recently or that they are vary stable in terms of DNA content (Norris and Weinstock, 2000). The fact that *T. pallidum* has so far not developed or acquired resistance to penicillin may be related to this apparent genetic stability.

The *T. pallidum* repeat (*tpr*) genes, which encode 12 *T. pallidum* repeat proteins, or Tpr proteins (TprA to TprL), do exhibit interstrain and intrastrain differences though. The *tpr* genes appear to be treponeme specific. The genome sequence suggests similarities to the *T. denticola* major sheath protein (Msp). The *T. denticola* Msp is thought to be a surface-exposed protein which has porin activity and is involved in cell attachment, although this is controversial (Caimano *et al.*, 1999). Nevertheless, Centurion-Lara *et al.* (1999) reported that rabbits immunised with a recombinant fragment of one of the Tpr proteins, TprK, developed opsonic antibody and exhibited reduced lesion severity when challenged with *T. pallidum*, consistent with a surface localisation for TprK. These findings led them to suggest that variation in the expression of Tpr antigens may be important in the evasion of the immune response. Although this was challenged by Hazlett *et al.* (2001), the evidence of a protective effect of immunisation with TprK proteins has been confirmed (Morgan, Lukehart and Van Voorhis, 2002) and extended by demonstrating that protection correlates with the specificity of antibodies to the variable region of the TprK protein used as an immunogen (Morgan, Lukehart and Van Voorhis, 2003). Furthermore, the same group has reported increased diversity among TprK sequences upon passage, suggesting a possible mechanism for evasion of the host immune response, which could contribute to persistence and re-infection (LaFond *et al.*, 2003).

Recently, to circumvent the difficulties of obtaining *T. pallidum* DNA, a large insert bacterial artificial chromosome (BAC) library was constructed in *Escherichia coli* using the entire *T. pallidum* subspecies *pallidum* genome (Smajs *et al.*, 2002). The BAC library covered 97% of the total predicted *T. pallidum* ORFs. Twelve protein products were identified, seven of which corresponded to previously identified *T. pallidum* polypeptides, indicating that at least some *T. pallidum* genes are expressed in *E. coli* and that this approach can be used for functional studies. This BAC library was subsequently used to produce PCR products, encoding 96% of the total predicted *T. pallidum* ORFs, which were cloned into a variety of plasmids, permitting their rapid conversion into a series of other functional vectors (McKevitt *et al.*, 2003). In total, 85 of 248 predicted proteins were expressed in *E. coli*, 12 of which were recognised by rabbit anti-sera; seven of the 12 had previously been identified as *T. pallidum* antigens, the remaining five represent novel antigens. These two studies clearly illustrate the potential of functional genomics and proteomics for utilising the *T. pallidum* genome sequence data to identify antigenic proteins of potential significance in the development of diagnostic tests and, possibly, vaccines.

PATHOGENICITY

The pathogenesis of treponemal infection is poorly understood. In syphilis, infection is believed to occur following inoculation of *T. pallidum* through breaks in the skin or mucous membranes. An early inoculation study in human volunteers showed that as few as 10 organisms may lead to infection (Magnuson *et al.*, 1956). Local multiplication of the organism leads to the formation of a lesion, the chancre, at the site of inoculation, within days or weeks of exposure. Dissemination occurs during the primary stage, resulting in systemic infection: early animal studies demonstrated that the organisms appear in lymph nodes within minutes and disseminate widely within hours. *Treponema pallidum* is extremely invasive, rapidly attaching to mammalian cell surfaces and penetrating endothelial junctions and tissue layers (Thomas *et al.*, 1988; Riviere, Thomas and Cobb, 1989). Most of the pathology is apparently due to the host immune response (Sell and Norris, 1983). Invasiveness may also be important in establishing latent infection; *T. pallidum* may be able to penetrate to areas that are inaccessible to the immune response and hence avoid elimination. The immune response to *T. pallidum* has recently been reviewed in detail (Salazar, Hazlett and Radolf, 2002).

Treponema pallidum has limited toxigenic capabilities. The outer membrane lacks LPS and contains very few transmembrane proteins. This is thought to explain the poor antigenicity of the organism (Cox *et al.*, 1992) and has led to the hypothesis that the organism acts as a 'stealth' organism by minimising the number of surface-localised targets that can be recognised by the host immune system.

An analysis of the *T. pallidum* genome sequence for potential virulence factors by comparison of predicted coding sequences with sequences in databases (Weinstock *et al.*, 1998) identified 67 genes of potential pathogenic significance, including the *tpr* genes and genes encoding putative haemolysins, regulators, polysaccharide biosynthesis proteins and possible surface proteins; less than one-third of these genes had been identified previously, further emphasising the value of genome sequencing in identifying areas for future study. A comparative analysis of the predicted gene products of *T. pallidum* and *Borrelia burgdorferi* (Subrahmanian, Koonin and Aravind, 2000) identified proteins that had previously been undetected in these organisms, such as von Willibrand A factor (vWA) domain-containing proteins, which may be involved in pathogenesis; spirochaete vWA proteins are likely to be involved in adhesion and could be therapeutic targets.

EPIDEMIOLOGY

Syphilis is usually transmitted by the sexual route; vertical transmission also occurs, generally *in utero*. Other routes of transmission, such as blood transfusion and sharp injury, have been described but are uncommon now. The risk of sexual transmission has been estimated to be about 60% per partnership (Garnett *et al.*, 1997). The endemic treponematoses are transmitted by direct person to person contact and possibly, for bejel, via drinking vessels (Antal, Lukehart and Meheus, 2002).

The incidence of syphilis in most industrialised countries fell during the 1980s, following the advent of HIV/AIDS. However, there was an epidemic in the United States between the late 1980s and early 1990s. It mainly affected heterosexual black Americans and was associated with a dramatic increase in the use of crack cocaine, the exchange of sex for money and drugs, decreased effectiveness of STD prevention and control programmes, increasing poverty and urban decay (Nakashima *et al.*, 1996). The adult epidemic was paralleled by an increase in congenital syphilis, with more than 3000 cases annually around the time of the peak of the adult epidemic in 1990. With renewed public health efforts, the incidence of early syphilis declined and in 2000 it was the lowest since reporting began in 1941. However, it increased again in 2001 and 2002 but only in men; increases in rates of syphilis have been reported among men who have sex with men in several US cities and rates of other STDs have also increased in this group, reflecting a general increase in unsafe sexual behaviour concomitant with the availability of highly active anti-retroviral therapy (Katz *et al.*, 2002).

Similar increases have been reported elsewhere. In the United Kingdom, the incidence of infectious syphilis in men and women more than doubled between 1998 and 2000. This was partly due to several localised outbreaks, notably in Bristol, Manchester and London, involving homosexual men and heterosexual men and women; most infections were acquired in the United Kingdom. There were several common features to these outbreaks including high rates of partner change and of anonymous contacts; unprotected oral sex reported as the sole high-risk sexual practice; illicit drug use and HIV co-infection (Doherty *et al.*, 2002). The number of new diagnoses in the UK in 2002 was the highest since 1984 with a high proportion of cases among men who have sex with men and high rate of co-infection with HIV.

There were also dramatic increases in the incidence of syphilis in the former Soviet Union in the 1990s; in the Russian Federation, the notification rate increased 62-fold between 1988 and 1996, probably for a combination of reasons including poverty, unemployment, a decline in government-funded public health services and increased population movement with the opening of national borders (Tichonova *et al.*, 1997). Even though syphilis rates declined in the late 1990s, this was at least partly due to reduced case-finding and a shift from public sector provision to private healthcare services, which are less likely to report cases (Reidner, Dehne and Gromyko, 2000). As in the earlier epidemic in the USA, the adult epidemic in the Russian Federation was accompanied by epidemic of congenital syphilis, which increased 26-fold between 1991 and 1999. Late presentation for antenatal care, or a complete lack of it, resulting in no treatment or inadequate treatment, contributed to the fact that more than 60% of infected mothers delivered an infant with congenital syphilis and to high rates of foetal or infant death (Tikhonova *et al.*, 2003).

Nevertheless, the treponematoses principally affect the developing world: WHO estimated that 4 million of a total of 12 million new cases of syphilis in adults in 1999 occurred in South and Southeast Asia, with a further 4 million estimated to have occurred in sub-Saharan Africa and 3 million in Latin America and the Caribbean (WHO Department of HIV/AIDS and STDs, 2001). Likewise, the endemic treponematoses predominantly affect children under 15 years of age in the remote rural communities in the tropics, with bejel extending to the adjacent arid areas and some temperate zones. The burden of disease that they cause was substantially reduced by the global endemic treponematoses control programme conducted by WHO in the 1950s and 1960s, but the subsequent integration of the programme with local health services was ineffective, leading to a resurgence of endemic treponematoses in many areas. As a result, endemic foci remain in parts of Africa, Southeast Asia, the Pacific Islands and South America, with about 460 000 of the estimated 2 million current worldwide cases thought to be infectious (Antal, Lukehart and Meheus, 2002).

CLINICAL FEATURES

Syphilis

Syphilis is classified as acquired or congenital, with both being divided into early and late stages. Early acquired syphilis includes primary, secondary and early latent syphilis and late acquired syphilis includes late latent and tertiary syphilis.

Primary syphilis is characterised by an ulcer, or chancre, at the site of inoculation and regional lymphadenopathy. The incubation period is typically about 3 weeks but varies from 10 to 90 days. The chancre is classically in the anogenital region, single, painless and indurated, with a clean base that discharges serum. However, chancres may be atypical: they may be multiple, painful, purulent, destructive and extragenital.

Secondary syphilis usually develops 4–8 weeks after the primary lesion has healed, although this may vary from 2 weeks to 6 months. Up to one-third of patients still have a chancre. The lesions of secondary syphilis are generalised, affecting skin and mucous membranes. There is usually a rash, often affecting the palms and soles, condyloma lata (papular lesions which have coalesced in warm moist areas), mucocutaneaous lesions, fever and generalised lymphadenopathy. Less common manifestations include anterior uveitis, meningitis, cranial nerve palsies, hepatitis, splenomegaly, periostitis and glomerulonephritis.

Primary and secondary syphilis resolve spontaneously and the infection then enters the latent stage, which is characterised by asymptomatic infection and positive serology. It is classified into early and late stages on the basis of the risk of recurrence; untreated, about 25% patients relapse and develop recurrent mucocutaneous lesions, which are infectious. About 90% recurrences occur during the first year, 94% occur within 2 years and the rest occur within 4 years of infection (Clark and Danbolt, 1955). Early latent syphilis is defined as the first 2 years following infection by WHO, as it is in the UK, whereas in the USA it is defined as the first year following infection.

After a variable period, usually many years, about one-third of untreated patients will develop tertiary syphilis, with about 15% developing gummatous (benign tertiary) syphilis, about 10% developing cardiovascular syphilis and the remaining 10% developing neurosyphilis. The latter two stages are sometimes classified as quaternary syphilis. Gummas are granulomatous lesions. They can occur in any organ but most commonly affect bone and skin; they can be locally destructive, so the term benign can be misleading. Cardiovascular syphilis usually affects large vessels, classically the ascending aorta when it may lead to aortic regurgitation, angina and aortic aneurysm. Neurosyphilis may either be asymptomatic, which is characterised by abnormal CSF findings but no associated neurological signs or symptoms, or symptomatic. In fact, central nervous system involvement can occur at any time following primary infection and is not confined to tertiary syphilis. It can present in many ways, which may overlap; in the pre-penicillin era, the most common manifestations were meningovascular and parenchymatous neurosyphilis. Parenchymatous neurosyphilis has become uncommon since the advent of antimicrobial therapy; it may present as tabes dorsalis, a syndrome due to dorsal column loss, or general paresis of the insane, a progressive dementia.

Syphilis and HIV Infection

It has been suggested that HIV alters the natural history and clinical presentation of syphilis. A greater risk of persistent chancres, more rapid progression to secondary syphilis, with ulcerative skin lesions as well as an increased risk of neurosyphilis have been described (Hutchinson *et al.*, 1994), but most patients with coincident HIV present with typical manifestations of syphilis.

Congenital Syphilis

Unlike the endemic treponematoses, syphilis can cause vertical infection; most infections are thought to occur *in utero*, although they may occasionally occur perinatally. The consequences include stillbirth, neonatal death or congenital infection, though some neonates escape infection. The outcome depends on the stage of maternal infection and the stage of pregnancy at which transmission occurs; it is worse if the mother has early syphilis but, in contrast to sexual transmission, vertical transmission can occur even with late latent infection. Transmission can largely be prevented by treatment before the 16th week of pregnancy (Fiumara, 1975). The clinical manifestations are divided into early congenital syphilis, occurring within the first 2 years of life, and late congenital syphilis which includes all aspects of the disease occurring after the second year of life. Early congenital syphilis may be symptomatic at birth, which is associated with a worse prognosis, or it may present later, usually within the first 5 weeks of life. Late congenital syphilis is often diagnosed incidentally, on the basis of positive serology, although it can lead to significant morbidity. The paediatric aspects of treponemal infection have been comprehensively reviewed recently (Parish, 2000).

The Endemic Treponematoses

The clinical manifestations of the endemic treponematoses are also divided into early and late stages; the early stage is considered to be infectious, or potentially infectious, and lasts up to 5 years from infection. The endemic treponematoses have a relapsing clinical course and prominent cutaneous manifestations. In pinta, this is the only clinical manifestation; in yaws and endemic syphilis, the mucous membranes and bones can also be affected. In contrast to syphilis, congenital disease and CNS involvement do not occur and late sequelae are usually limited to skin, bone cartilage and mucosal surfaces. Further details can be found in a recent review (Antal, Lukehart and Meheus, 2002).

Oral Treponemes

The oral treponemes, particularly *T. denticola*, are associated with periodontal disease. Difficulty in culturing oral treponemes and the complex nature of the oral microflora have hindered their study. However, the *T. denticola* genome has been sequenced recently (Seshadri *et al.*, 2004). It is considerably larger than that of *T. pallidum*; in part, this appears to be because of expansion of the genome and this may have pathogenic significance, although some of the additional genes are of unknown function. As with *T. pallidum*, this new information, together with the application of current molecular methods, for example quantitative PCR (Yoshida *et al.*, 2004), should lead to greater understanding of the pathogenic role of the oral treponemes.

LABORATORY DIAGNOSIS

The diagnosis of treponemal infection is based on a combination of clinical evaluation, direct detection of organisms and serological testing. Direct detection of treponemes, either by microscopy or PCR, is particularly important in early primary infection before antibody is detectable, but at all other stages of infection serology is still the mainstay of routine laboratory diagnosis. Detailed descriptions of the individual tests and their performance characteristics at different stages of infection can be found in several comprehensive reviews (Young, 1992; Larsen, Steiner and Rudolph, 1995; Wicher, Horowitz and Wicher, 1999).

Direct Detection of *T. pallidum*

Microscopy

Darkground microscopy is the traditional method for direct detection of *T. pallidum* in lesion exudates; it can provide rapid results, with identification of the organism by its characteristic morphology and motility; yet there are many limitations to the technique and it requires technical expertise which is no longer widespread. Immunofluorescence using fixed smears of lesion material, or tissue specimens, has several advantages and is of comparable sensitivity; neither technique differentiates between the pathogenic treponemes. Polyclonal antisera were originally used for immunofluorescence but, despite absorption with non-pathogenic treponemes, the results were non-specific. More recently described monoclonal antibodies are more specific but the reagents are not widely available.

PCR

PCR offers potential for direct detection of treponemes in lesion material, as well as other specimens which can present diagnostic difficulties, such as CSF. Several different assays have been described, using a variety of targets and test formats. The most commonly described target is the 47-kDa protein gene (Tpp 47); other approaches include reverse transcriptase PCR for 16S rRNA and multiplexing to allow detection of other causes of genital ulcers: *Haemophilus ducrei* and herpes simplex virus. Early reports suggested a lack of sensitivity; concerns have also been expressed about specificity, as the function of many of the target genes is unknown. This has stimulated efforts to develop novel *T. pallidum*-specific targets (Liu *et al.*, 2001). Although it has been a useful research technique, PCR has yet to find a role in clinical laboratories. However, a recent comparison of PCR with conventional diagnostic methods suggests that it may have a useful role (Palmer *et al.*, 2003); additional developments, such as real-time PCR, should enhance this potential.

Rabbit Infectivity Testing

The rabbit infectivity test (RIT) is still used as the 'gold standard' for evaluating methods such as PCR for detecting organisms in clinical specimens. The sensitivity of RIT approaches 100%, but it is too impractical a technique for routine use.

Serology

Treponemal antibody tests are classified into two groups: non-treponemal tests, which detect the so-called non-specific antibody, and treponemal tests which detect specific treponemal antibody. Non-treponemal tests include the Venereal Diseases Research Laboratory (VDRL) and rapid plasma reagin (RPR) tests. Treponemal tests include the *T. pallidum* haemagglutination test (TPHA), the *T. pallidum* particle agglutination test (TPPA), the fluorescent treponemal antibody-absorbed test (FTA-abs) and most treponemal antibody enzyme immunoassays (EIAs). The first detectable antibody response is treponemal IgM antibody, towards the end of the second week of infection; IgG antibody is detectable at about 4 weeks. Serology cannot distinguish between syphilis and the endemic treponematoses.

Serology has several applications: it is used for screening low-risk populations, for example routine screening in pregnancy or screening blood and organ/tissue donors; for screening high-risk populations, such as genitourinary medicine clinic attenders; for diagnostic testing of patients whose history of clinical features are consistent with syphilis and to assess the stage of infection and monitor the response to therapy. The approach varies: in the USA and some European countries non-treponemal tests are used for screening; some European countries screen with a TPHA alone whereas in the UK, a combination of a non-treponemal test and a treponemal test, typically either a VDRL or an RPR test in parallel with a TPHA test, has been used. Recently, however, treponemal antigen-based EIAs have become the most widely used screening test (Young, 2000). Guidance for UK laboratories on the use of serological tests (Egglestone and Turner, 2000) highlights the importance of confirming a reactive screening test result by additional testing, ideally using a methodologically independent test; for example, a TPHA would be suitable to confirm a reactive EIA screening test. The guidance also refers to the importance of testing a second specimen to confirm the results and the correct identification of the first specimen.

Non-Treponemal Tests

These are flocculation tests based on the use of cardiolipin, a phospholipid antigen. The VDRL is read microscopically; the RPR test is a modification of the VDRL test that can be read macroscopically and hence is more suitable to larger workloads. Both tests are cheap, simple and quick to perform. They detect IgG and IgM antibodies; antibody becomes detectable about 3–5 weeks after the acquisition of infection. Sensitivity is reported to be about 90% in primary syphilis and up to 100% in the secondary stage. However, false-negative reactions occur in the presence of high titres of antibody (the prozone phenomenon) and false-positive reactions are seen in various acute and chronic conditions in the absence of syphilis (biological false-positive reactions). If performed quantitatively, the non-treponemal tests can be used to assess the stage of infection and the response to treatment.

Treponemal Tests

The FTA-abs is an indirect immunofluorescence test. It becomes reactive around the third week of infection and in primary syphilis has a sensitivity of about 85–100%. Although it has been regarded as the 'gold-standard' treponemal test, it is subjective and poorly standardised and is becoming less widely used, at least in the UK, as its limitations are increasingly recognised (Young, 2000). The TPHA is based on the agglutination of red cells coated with *T. pallidum* antigen. It is less sensitive than the FTA in early primary syphilis but it is more specific. The TPPA, which is based on the use of gelatin particles rather than red cells, appears to be more sensitive and more specific than the TPHA (Pope *et al.*, 2000) and it is replacing it as a confirmatory test.

There are now numerous commercially available treponemal EIAs, based on different formats and antigens and capable of detecting treponemal IgG, or IgG and IgM. Many are now based on recombinant antigens, such as one or more of the lipoprotein antigens Tpp15, Tpp17 and Tpp47. IgM-specific assays are also commercially available. There are few published comparative evaluations, but in one study the sensitivity of EIA in early primary syphilis was reported as up to 77% for combined IgG and IgM assays, compared to 87% for an IgM-specific EIA (Schmidt, Edjlalipour and Luger, 2000). An earlier study had reported a sensitivity of 93% in primary syphilis for an IgM-specific EIA (Lefevre, Bertrand and Bauriad, 1990); the assay is an antibody capture assay and more specific than the FTA-abs IgM test. It is recommended that an IgM test should be used in all cases of suspected primary infection, regardless of the screening test results (Young, 2000). IgM testing is also useful in congenital infection: serial testing is recommended because of the possibility that specific IgM may not be detectable at birth even in infected infants.

Developments in serodiagnosis are likely to follow the sequencing of the *T. pallidum* genome. For example, a putative outer membrane protein (Tp0453) appears to be highly sensitive for early syphilis (Van Voorhis *et al.*, 2003); surface-exposed antigens have theoretical advantages for the diagnosis of all stages of infection, but further work will be required to confirm this potential.

Immunoblots

Western blots and a line immunoassay for treponemal antibody are now commercially available; the line assay uses recombinant and synthetic polypeptide antigens and has been proposed as a confirmatory test (Hagedorn *et al.*, 2002), although this requires further evaluation.

Rapid Tests

Several simple, rapid tests for treponemal antibody are now commercially available; most are in the format of an immunochromatographic strip. An evaluation of these assays was recently published by the WHO Sexually Transmitted Diseases Diagnostics Initiative (WHO, 2003). Sensitivity, compared to TPHA or TPPA testing, varied from 85 to 98%. The rapid assays are potentially suitable for non-laboratory use in the developing world, and they may also have a role as point of care tests elsewhere.

Testing CSF

Laboratory confirmation of neurosyphilis remains problematic. CDC guidelines highlight the importance of considering serological tests, abnormalities of CSF cell count or protein or a reactive VDRL test on CSF. A reactive VDRL test on CSF, in the absence of blood contamination of the CSF, is considered diagnostic of syphilis (CDC, 2002), but false-negative VDRL results can occur. In contrast, a negative treponemal test result on CSF is generally accepted as excluding neurosyphilis. Various indices have been suggested for detecting intrathecal synthesis of treponemal antibody in order to confirm a diagnosis of neurosyphilis (for further details, see Young, 1992); however, routine experience of them is limited and none has been widely adopted.

Response to Treatment

The non-treponemal tests tend to become negative after treatment, particularly in early syphilis and, contrary to some teaching, this also occurs with the treponemal tests (Romanowski *et al.*, 1991). The rate of decline in non-treponemal test titres depends on the stage of infection, the initial titre and the history of previous syphilis. Titres decline approximately fourfold within 6 months in patients with primary or secondary syphilis; in early latent syphilis, it may take 12 months for a fourfold decline in titre to occur (Romanowski *et al.*, 1991). In late syphilis, the non-treponemal tests may remain positive at low titres for many years, the so-called 'serofast syphilis'. IgM testing is less useful than non-treponemal testing because IgM reactivity may persist for many months; quantitative non-treponemal tests are still preferred (Egglestone and Turner, 2000).

Molecular Typing

A PCR-based typing system for *T. pallidum* subspecies *pallidum* has been described (Pillay *et al.*, 1998), on the basis of a combination of size heterogeneity within the acidic repeat protein (*arp*) gene and RFLP patterns of three members of the *tpr* gene family (*tpr* E, G and J). This technique has been applied to genital specimens and blood samples from the USA, Madagascar and South Africa; in South Africa 35 subtypes were identified, with 82 of 161 (51%) specimens belonging to four subtypes (Pillay *et al.*, 2002). This pattern, with a limited number of predominant subtypes together with some strain diversity, suggests that this typing system may be useful in epidemiological studies.

TREATMENT

Penicillin remains the treatment of choice for syphilis and the endemic treponematoses. The efficacy of penicillin was established before randomised clinical trails became routine practice, so recommended regimes are based on a combination of data derived from rabbit models, extensive clinical experience, case series and limited clinical trials. The preparation used (benzathine, aqueous procaine or aqueous crystalline), the dose and duration of treatment depend on the stage and clinical manifestations of disease. The aim is to achieve treponemicidal levels of antimicrobial, which for penicillin is greater than 0.018 mg/l, for at least 7 days to cover a number of division times (30–33 h) of treponemes, with a subtreponemicidal interval of not more than 24–30 h, in early syphilis. A longer duration of treatment is used in late syphilis when the organisms are thought to divide more slowly. Parenteral rather than oral therapy has been the standard approach because of better bioavailability and because treatment is supervised. Recommendations on the preparation used vary; US guidelines are largely based on the use of benzathine penicillin (CDC, 2002); in the UK, although benzathine penicillin is also recommended, procaine penicillin is the treatment of choice (Goh and French, 1999); the WHO recommends benzathine penicillin as first-line therapy with procaine penicillin as an alternative (WHO Department of HIV/AIDS and STDs, 2001). These respective guidelines should be referred to for further details of management, including alternatives in penicillin allergy, treatment in pregnancy, treatment of HIV co-infected patients and of neurosyphilis, follow-up and partner notification.

PREVENTION AND CONTROL

The aims of control programmes are similar for syphilis and the endemic treponematoses: to interrupt transmission, to reduce the duration of infection and to prevent the development of complications of infection. For syphilis, this requires a range of well-established measures, including health education; promotion of safer sex;

provision of accessible, effective and acceptable care and early detection and treatment through case finding, screening and partner notification. Syphilis among women in the USA has largely been controlled by conventional methods and rates declined for the twelfth consecutive year in 2002. In contrast, the current outbreaks of syphilis among men who have sex with men are proving difficult to control and are probably the main obstacles to the current national plan to eliminate syphilis from the United States. One reason for this difficulty may be the high proportion of anonymous contacts, which makes conventional partner notification ineffective; however, evidence that the Internet is used increasingly as the initial means of meeting sexual partners has prompted public health authorities also to use the Internet for partner notification, with some success (CDC, 2004b). The need for additional novel strategies for outbreak control is recognised but they need to be developed and evaluated. Screening in venues where transmission is likely to occur is advocated in the UK (Doherty *et al.*, 2002) but is of unproven benefit. Targeted mass treatment with azithromycin failed to control an outbreak amongst heterosexuals. In Vancouver, Canada (Rekart *et al.*, 2003), a recent report of azithromycin treatment failures has cast doubt on its potentially valuable role (MMWR, 2004). The importance of antenatal screening is well recognised though, including additional testing in the third trimester and at delivery in areas with a high incidence of syphilis (CDC, 2002).

Clearly, there are even greater challenges in the developing world, both for syphilis and the endemic treponematoses. The control measures required for the endemic treponematoses include community screening programmes combined with mass treatment; adequate diagnosis of symptomatic patients with contact tracing and treatment; health education and improved hygiene. These activities need to be underpinned by effective surveillance and logistic support. Efforts to control the endemic treponematoses stalled in the late 1970s, leading to resurgent disease in many areas. Like syphilis, the endemic treponematoses have many of the characteristics that make them susceptible to elimination and the failure to eliminate them is more because of political and management issues than technical difficulties (Antal, Lukehart and Meheus, 2002).

There have been major advances in the understanding of the treponemes since Schaudinn and Hoffman first identified *Spirochaeta pallida* (*T. pallidum*) in syphilitic lesions in 1905, particularly in recent years, and there are realistic prospects for further advances, building on the information provided by the sequencing of the *T. pallidum* genome.

REFERENCES

Antal GM, Lukehart SA, and Meheus AZ (2002) The endemic treponematoses. *Microbes and Infection*, **4**, 83–94.

Caimano MJ, Bourell KW, Bannister TD *et al.* (1999) The *Treponema denticola* major sheath protein is predominantly periplasmic and had only limited surface exposure. *Infection and Immunity*, **67**, 4072–4083.

CDC (2002) Sexually transmitted diseases treatment guidelines. *MMWR*, **51** (No. RR-6): 19–30.

CDC (2004a) Brief report: azithromycin treatment failures in syphilis infections – San Francisco, California, 2002–2003. *MMWR*, **53**, 197–198.

CDC (2004b) Using the Internet for partner notification of sexually transmitted diseases – Los Angeles County, California, 2003. *MMWR*, **53**, 129–131.

Centurion-Lara A, Castro C, Barrett L *et al.* (1999) *Treponema pallidum* major sheath protein homologue Tpr K is a target of opsonic antibody and the protective immune response. *Journal of Experimental Medicine*, **189**, 647–656.

Centurion-Lara A, Castro C, Castillo R *et al.* (1998) The flanking region sequences of the 15-kDa lipoprotein gene differentiate pathogenic treponemes. *The Journal of Infectious Diseases*, **177**, 1036–1040.

Clark EG, and Danbolt N (1955) The Oslo study of the natural history of untreated syphilis: an epidemiologic investigation based on a re-study of the Boeck-Brusgard material. A review and appraisal. *Journal of Chronic Diseases*, **2**, 311–344.

Cox DL, Chang PO, McDowall AW, and Radolf JD (1992) The outer membrane, not a coat of host proteins, limits antigenicity of virulent *Treponema pallidum*. *Infection and Immunity*, **60**, 1076–1083.

Doherty L, Fenton KA, Jones J *et al.* (2002) Syphilis: old problem, new strategy. *British Medical Journal*, **325**, 153–156.

Egglestone SI, and Turner AJL for the PHLS Syphilis Serology Working Group (2000) Serological diagnosis of syphilis. *Communicable Disease and Public Health*, **3**, 158–162.

Fieldsteel AH, Cox DL, and Moeckli RA (1981) Cultivation of virulent *Treponema pallidum* in tissue culture. *Infection and Immunity*, **32**, 908–915.

Fiumara NJ (1975) Syphilis in newborn children. *Clinical Obstetrics and Gynaecology*, **18**, 183–189.

Fraser CM, Norris SJ, Weinstock GM *et al.* (1998) Complete genome sequence of *Treponema pallidum*, the syphilis spirochaete. *Science*, **281**, 375–388.

Garnett GP, Aral SO, Hoyle DV *et al.* (1997) The natural history of syphilis. Implications for the transmission dynamics and control of infection. *Sexually Transmitted Diseases*, **24**, 185–200.

Goh B, and French P (1999) National guidelines for the management of early and late syphilis. In: Radcliffe K, Ahmed-Jushef I, and Cowan F. UK national guidelines on sexually transmitted infections and closely related conditions. *Sexually Transmitted Infection*, **75**, Suppl. 1: S29–S37.

Hagedorn H-J, Kraminer-Hagedorn A, De Bosschere K *et al.* (2002) Evaluation of INNO-LIA syphilis assay as a confirmatory test for syphilis. *Journal of Clinical Microbiology*, **40**, 973–978.

Hazlett KRO, Selatti TT, Nguyen DL *et al.* (2001) The TprK protein of *Treponema pallidum* is periplasmic and is not a target of opsonic antibody or protective immunity. *Journal of Experimental Medicine*, **193**, 1015–1026.

Hutchinson CM, Hook EW III, Shepherd M *et al.* (1994) Altered clinical presentation of early syphilis in patients with human immunodeficiency virus infection. *Annals of International Medicine*, **121**, 94–100.

Katz MH, Schwarcz SK, Kellogg TA *et al.* (2002) Impact of highly active antiretroviral treatment on HIV sero-incidence among men who have sex with men: San Francisco. *American Journal of Public Health*, **92**, 388–394.

LaFond RE, Centurion-Lara A, Godornes C *et al.* (2003) Sequence diversity of *Treponema pallidum* subsp. *pallidum* tprK in human syphilis lesions and rabbit-propagated isolates. *Journal of Bacteriology*, **185**, 6262–6268.

Larsen SA, Steiner BM, and Rudolph AH (1995) Laboratory diagnosis and interpretation of tests for syphilis. *Clinical Microbiology Reviews*, **8**, 1–21.

Lefevre J-C, Bertrand M-A, and Bauriad R (1990) Evaluation of the Captia enzyme immunoassays for detection of immunoglobulins G and M to *Treponema pallidum* in syphilis. *Journal of Clinical Microbiology*, **28**, 1704–1707.

Liu H, Rodes B, Chen C-Y, and Steiner B (2001) New tests for syphilis: rational design of a PCR method for detection of *Treponema pallidum* in clinical specimens using unique regions of the DNA polymerase I gene. *Journal of Clinical Microbiology*, **39**, 1941–1946.

Magnuson HJ, Thomas EW, Olansky S *et al.* (1956) Inoculation syphilis in human volunteers. *Medicine*, **35**, 33–82.

McKevitt M, Patel K, Smajs D *et al.* (2003) Systematic cloning of *Treponema pallidum* open reading frames for protein expression and antigen discovery. *Genome Research*, **13**, 1665–1674.

Morgan CA, Lukehart SA, and Van Voorhis WC (2002) Immunization with the N-terminal portion of *Treponema pallidum* repeat protein K attenuates syphilitic lesion development in the rabbit model. *Infection and Immunity*, **70**, 6811–6816.

Morgan CA, Lukehart SA, and Van Voorhis WC (2003) Protection against syphilis correlates with specificity of antibodies to the variable regions of *Treponema pallidum* repeat protein K. *Infection and Immunity*, **71**, 5605–5612.

Nakashima AK, Rolfs R, Flock M *et al.* (1996) Epidemiology of syphilis in the United States, 1941–1993. *Sexually Transmitted Diseases*, **23**, 16–23.

Norris SJ and the Treponema Pallidum Polypeptide Research Group (1993) Polypeptides of *Treponema pallidum*: progress towards understanding of their structural, functional and immunological roles. *Microbiological Reviews*, **57**, 750–779.

Norris SJ, Cox DL, and Weinstock GM (2001) Biology of *Treponema pallidum*: correlation of functional activities with genome sequence data. *Journal of Molecular Microbiology and Biotechnology*, **3**, 37–62.

Norris SJ, and Weinstock GM (2000) The genome sequence of *Treponema pallidum*, the syphilis spirochaete: will clinicians benefit? *Current Opinion in Infectious Diseases*, **13**, 29–36.

Palmer HM, Higgins SP, Herring AJ, and Kingston MA (2003) *Sexually Transmitted Infections*, **79**, 479–483.

Parish JL (2000) Treponemal infections in the paediatric population. *Clinics in Dermatology*, **18**, 687–700.

Pillay A, Liu H, Chen CY *et al.* (1998) Molecular subtyping of *Treponema pallidum* subspecies *pallidum*. *Sexually Transmitted Diseases*, **25**, 408–414.

Pillay A, Liu H, Ebrahim S *et al.* (2002) Molecular typing of *Treponema pallidum* in South Africa: cross-sectional studies. *Journal of Clinical Microbiology*, **40**, 256–258.

Pope V, Fears M, Morrill WE *et al.* (2000) Comparison of the serodia *Treponema pallidum* particle agglutination, Captia Syphilis-G and SpiroTek Reagin II tests with standard test techniques for diagnosis of syphilis. *Journal of Clinical Microbiology*, **38**, 2543–2545.

Radolf JD, Norgard MV, and Schulz WW (1989) Outer membrane ultrastructure explains the limited antigenicity of virulent *Treponema pallidum*. *Proceedings of the National Academy of Science*, **86**, 2051–2055.

Reidner G, Dehne KL, and Gromyko A (2000) Recent declines in reported syphilis rates in eastern Europe and central Asia: are the epidemics over? *Sexually Transmitted Infections*, **76**, 363–365.

Rekart ML, Patrick DM, Chakraborty B *et al.* (2003) Targeted mass treatment for syphilis with oral azithromycin. *Lancet*, **361**, 313–314.

Riviere GR, Thomas DD, and Cobb CM (1989) An in vitro model of *Treponema pallidum* invasiveness. *Infection and Immunity*, **57**, 2267–2271.

Romanowski B, Sutherland R, Fick GH *et al.* (1991) Serologic response to treatment of infectious syphilis. *Annals of International Medicine*, **114**, 1005–1009.

Salazar JC, Hazlett KRO, and Radolf JD (2002) The immune response to infection with *Treponema pallidum*, the stealth pathogen. *Microbes and Infection*, **4**, 1133–1140.

Schmidt BL, Edjlalipour M, and Luger A (2000) Comparative evaluation of nine different enzyme-linked immunosorbent assays for determination of antibodies against *Treponema pallidum* in patients with primary syphilis. *Journal of Clinical Microbiology*, **38**, 1279–1282.

Sell S, and Norris SJ (1983) The biology, pathology and immunology of syphilis. *International Reviews of Experimental Pathology*, **24**, 204–276.

Seshadri R, Myers GS, Tettelin H *et al.* (2004) Comparison of the genome of the oral pathogen *Treponema denticola* with other spirochaete genomes. *Proceedings of the National Academy of Sciences*, **101**, 5646–5651.

Smajs D, McKevitt M, Wang L *et al.* (2002) BAC library of *T. pallidum* DNA in *E. coli*. *Genome Research*, **12**, 515–522.

St Louis ME, and Wasserheit JN (1998) Elimination of syphilis in the United States. *Science*, **281**, 353–354.

Subrahmanian G, Koonin EV, and Aravind L (2000) Comparative genome analysis of the pathogenic spirochaetes *Treponema pallidum* and *Borrelia burgdorferi*. *Infection and Immunity*, **68**, 1633–1648.

Thomas DD, Navab M, Haake DA *et al.* (1988) *Treponema pallidum* invades intercellular junctions of endothelial cell monolayers. *Proceedings of the National Academy of Science*, **85**, 3608–3612.

Tichonova L, Borisenko K, Ward D *et al.* (1997) Epidemics of syphilis in the Russian Federation: trends, and origins and priorities for control. *Lancet*, **350**, 210–213.

Tikhonova L, Salakhov E, Southwick K *et al.* (2003) Congenital syphilis in the Russian Federation: magnitude, determinants and consequences. *Sexually Transmitted Infections*, **79**, 106–110.

Van Voorhis WC, Barrett LK, Lukehart SA *et al.* (2003) Serodiagnosis of syphilis: antibodies to recombinant Tp0453, Tp92 and Gpd proteins are sensitive and specific indicators of infection by *Treponema pallidum*. *Journal of Clinical Microbiology*, **41**, 3668–3674.

Weigel LM, Radolf JD, and Norgard MV (1994) The 47 kDa major lipoprotein immunogen of *Treponema pallidum* is a penicillin-binding protein with carboxy peptidase activity. *Proceedings of the National Academy of Science*, **91**, 11611–11615.

Weinstock GM, Hardham JM, McLeod MP *et al.* (1998) The genome of *Treponema pallidum*: new light on the agent of syphilis. *FEMS Microbiology Reviews*, **22**, 323–332.

WHO (1998) The World Health Report 1998 – Life in the 21st Century; A vision for all. World Health Organization, Geneva, Switzerland.

WHO (2001) Sexually transmitted infections management guidelines 2001. http://www.who.int/docstore/hiv/STIManagementguidelines/who_hiv_aids_2001.01

WHO (2003) Laboratory-based evaluation of rapid syphilis diagnostics. World Health Organisation, Geneva, Switzerland.

WHO Department of HIV/AIDS and STDs (2001) Global prevalence and incidence of selected curable sexually transmitted infections. Syphilis estimates, 1999. World Health Organisation, Geneva, Switzerland.

Wicher K, Horowitz HW, and Wicher V (1999) Laboratory methods of diagnosis of syphilis for the beginning of the third millennium. *Microbes and Infection*, **1**, 1035–1049.

Yoshida A, Kawada M, Suzuki N *et al.* (2004) TaqMan real-time polymerase chain reaction assay for correlation of *Treponema denticola* numbers with the severity of periodontal disease. *Oral Microbiology and Immunology*, **19**, 196–200.

Young H (1992) Syphilis: new diagnostic directions. *International Journal of STD and AIDS*, **3**, 391–413.

Young H (2000) Guidelines for serological testing for syphilis. *Sexually Transmitted Infections*, **76**, 403–405.

43

Borrelia spp.

Sudha Pabbatireddy and Benjamin J. Luft

Department of Medicine, Division of Infectious Diseases, Stony Brook, NY, USA

HISTORY

Lyme disease is a multisystem, vector-borne, zoonotic infectious disease caused by *Borrelia burgdorferi*: a thin, spiral, motile, extracellular bacterium belonging to the family Spirochaetaceae. The hallmark of infection, erythema migrans (EM), an annular erythematous skin lesion that usually appears at the site of the tick bite was first described by Afzelius in 1909. In the 1930s, Hellerstrom linked EM with the development of meningitis and in the early 1940s Bannwarth associated EM with a syndrome consisting of painful radiculopathy, limb pain, and cerebrospinal fluid pleocytosis. The French physicians Garin and Bujadoux reported the first case of a neurologic manifestation, meningoradiculoneuritis (Kristoferitsch, 1989). The first suggestion that this was a spirochetal infection came in 1948 when Lenhoff, a Swedish pathologist, observed spirochete-like structures in skin biopsies from EM lesions. In 1976 Allen Steere and colleagues noted an association between EM and a cluster of patients with knee arthritis in Old Lyme, Connecticut (Steere *et al.*, 1977b) and described a syndrome of Lyme arthritis. In the subsequent decade, Steere and colleagues defined the multisystem nature of the disease and modified the name to Lyme disease.

The first isolate of this disease-causing spirochete was only obtained in 1981 when Burgdorfer and colleagues demonstrated a new spirochete in *Ixodes* ticks collected on Shelter Island, NY (Burgdorfer *et al.*, 1982). In 1982, spirochetes were identified in the midgut of the adult deer tick, *Ixodes dammini*, and given the name *Borrelia burgdorferi*. Finally, conclusive evidence that *B. burgdorferi* caused Lyme disease came in 1984 when spirochetes were cultured from the blood of patients with EM, from the rash lesion itself, and from the cerebrospinal fluid of a patient with meningoencephalitis and history of prior EM.

DESCRIPTION OF THE ORGANISM

Borrelia are members of the Spirochaetaceae that also include the treponemes. They are helical organisms between 8 and 30 μm long and 0.2–0.5 μm wide. They have up to 10 loose spirals and are actively motile via flagella that wind around the cells. The organisms are readily killed by ultraviolet light and desiccation.

The *Borrelia* have a complex relationship with their vector hose. For *Borrelia burgdorferi* and related species *Borrelia garinii* and *Borrelia afzelii* the vector is the Ixotid tick. For the 15 species responsible for tick-borne relapsing fever the vectors are soft ticks of the *Ornithodorus* genus and for *B. recurrentis* it is the human body louse *Pediculus humanus*.

PATHOGENESIS

Antigenic Structure

Borrelia are flexible helical cells comprised of a protoplasmic cylinder surrounded by a cell membrane, seven to 11 periplasmic flagella and an outer membrane that is loosely associated with the underlying structures. *Borrelia burgdorferi* B31, a tick (*Ixodes scapularis*) isolate from Shelter Island, NY (Burgdorfer *et al.*, 1982), is the type strain of the species *B. burgdorferi* sensu stricto. The DNA sequence of *B. burgdorferi* type strain B31 was published in 1997 and contains 950-kb linear chromosome, nine linear plasmids and 12 circular plasmids (Barbour and Hayes, 1986; Fraser *et al.*, 1997). The remarkable aspect of the *B. burgdorferi* genome is its large number of sequences for predicted or known lipoproteins, including the plasmid-encoded outer-surface proteins (Osp) A through F. These and other differentially expressed outer-surface proteins presumably help the spirochete adapt to and survive in markedly different arthropod and mammalian environments (de Silva and Fikrig, 1997). In addition, during early, disseminated infection, a surface-exposed lipoprotein, called VlsE, has been reported to undergo extensive antigenic variation (Zhang and Norris, 1998). The organism has few proteins with biosynthetic activity and apparently depends on the host for most of its nutritional requirement. Furthermore, the high concentration of outer-surface lipoproteins encoded by the plasmids (Casjens *et al.*, 2000) in conjunction with the variability of plasmid sequences supports the view that these sequences are directly involved in parasite–host interactions and therefore are more than likely to be key virulence factors.

Two major outer-surface proteins, OspA and OspB, are encoded by the same 50-kb linear plasmid and share 56% sequence homology (Dworkin *et al.*, 1998). Along with lipoprotein 6.6, these proteins are expressed primarily in the arthropod phase of the enzootic cycle (Le, 1980; Rawlings, 1995). OspA has an antiparallel beta strand topology with a middle nonglobular region connecting globular N- and C-terminal domains (Sanford, 1976). During tick feeding, as the spirochete migrates from the midgut to the salivary gland, OspC, which is encoded by a 26-kb circular plasmid, is upregulated (Horton and Blaser, 1985). OspE and OspF are immunogenic in some patients (Galloway *et al.*, 1977). A number of proteins have been identified that are differentially expressed by the spirochete during early infection, including decorin-binding proteins A and B (DbpA and DbpB) (Beutler and Munford, 1996) and a 47-kD fibronectin-binding protein (Probert and Johnson, 1998). Additional spirochetal proteins include a 39-kD *Borrelia* membrane protein A (BmpA) (Simpson, Schrumpf and Schwan, 1990), the 41-kD flagellar antigen (Coleman and

Benach, 1989), 60- and 73-kD heat–shock proteins (Hansen *et al.*, 1988; Anzola, Luft and Gorgane, 1992), a 66-kD integral outer-membrane protein that may function as a porin (Skare *et al.*, 1997) and a 93-kD antigen that is a part of the protoplasmic cylinder (Luft *et al.*, 1992).

Genetic Variation

Globally, about 10 genomic species of the *B. burgdorferi* sensu lato species complex have been recognized (Balmelli and Piffaretti, 1996). Most *Borrelia* genomic species have restricted geographical distribution, a consequence of geographic isolation (Price, 1980), and possible adaptation to the local enzootic transmission cycle of ticks and vertebrate hosts (Baranton, Marti Ras and Postic, 1998; Baranton *et al.*, 2001). In the Northeast, where over 90% of Lyme disease cases in the US occur annually *B. burgdorferi* sensu stricto is the predominant genospecies. A survey of *Borrelia* carried by *I. scapularis* ticks along the East coast concluded that *Borrelia* and its tick vector share a common evolutionary history of co-expansion within this geographic region (Qiu *et al.*, 2002).

Borrelia strains isolated from local tick samples in the Northeast are highly diverse. The *Borrelia* local diversity is reflected by high sequence variation of ospC alleles, infection of single ticks by multiple ospA and ospC strains as well as by a nearly even allelic frequency distribution of these alleles at each locality (Guttman *et al.*, 1996; Wang *et al.*, 1999; Qiu *et al.*, 2002). Previous studies have shown considerable genetic variation in *B. burgdoferi* (Wilske *et al.*, 1988; Boerlin *et al.*, 1992; Wallich *et al.*, 1992). These and other studies have subdivided this species into a species complex, which includes *B. burgdorferi* sensu stricto, *B. afzellii*, *B. japonica* (Baranton *et al.*, 1992; Kawabata, Masuzawa and Yanagihara, 1993) and several other groups (Valsangiacomo *et al.*, 1993; Postic *et al.*, 1994; Balmelli and Piffaretti, 1995). The first three species, *B. burgdorferi*, *B. afzelli* and *B. garinii*, are the only ones that have been isolated from humans, and each has been shown to be associated with a set of symptoms in chronic disease (van Dam *et al.*, 1993; Balmelli and Piffaretti, 1995). In the United States, *B. burgdorferi* and three new genospecies have been isolated (*B. andesonii*, *B. bisettii* and *B. lonestari*), while in Europe, all three pathogenic genospecies and two of the others are known to occur. One genospecies (*B. turdae*) has been found only in Japan, while another (*B. sinica*) appears to be restricted to China. So far, *B. burgdorferi* is the only genospecies that has been found in both North America and Europe, which is consistent with the recent migration and geographic range expansion of this species.

The evolutionary cause of high local diversity of *Borrelia* is most likely the host-driven, immune-mediated selection for antigenic variation (Qiu *et al.*, 1997; Wang *et al.*, 2002), surface lipoproteins like OspC, which has been shown to be a key component in parasite–host interaction in animal models (Schwan, 1996), are the most likely targets of such selection.

The B31 genome encodes 136 putative lipoproteins; 38 of these putative lipoproteins are chromosomally encoded while 98 are plasmid encoded. It is, however, unknown how many of these putative lipoproteins are actually expressed or how they differ in their degree of immunoreactivity. Even though the chromosomal genes and ospA are quite homogeneous within *B. burgdorferi*, ospC is highly polymorphic within and between the various genospecies (Dykhuizen *et al.*, 1993; Thiesen *et al.*, 1995). In nature humans are a dead-end hosts for *B. burgdorferi*, i.e., humans do not play a prominent role in the life cycle of this organism. Therefore, the spirochete is not adapted to humans and, as a consequence, many genospecies seem not to be human pathogens. The extensive genetic and antigenic diversity of OspC in all three pathogenic genospecies of *B. burgdorferi* sensu lato has led us to lead to chronic infection and, therefore, be used as a marker to discover new protective antigens (Thiesen *et al.*, 1993). We have shown that alleles of OspC could be clustered into 19 major groups or types designated A–S based on DNA sequence homology (Wang *et al.*, 1999). Sequence variation within a major group is <1% but 15% across all the groups. We have also shown certain OspC alleles are tightly linked to both infectivity and invasiveness. Of the 19 major groups, only four, designated A, B, I and K, contain invasive clones and cause infections of skin and extracutaneous sites, while the others are non-pathogenic for humans or infect only the skin (Seinost *et al.*, 1999). The analysis of ospC gene variability has recently been extended to the other pathogenic *Borrelia* species, *B. garinii* and *B. afzelii*. In that study, only two groups in *B. afzelii* and four groups in *B. garinii* caused invasive disease (Baranton *et al.*, 2001). Thus the ospC gene can be used to predict infectivity and pathogenesis in humans. Wormser and colleagues assessed the pathogenicity of *B. burgdorferi* clinical isolates to determine any potential association between severity of disease and genotype of the infecting *B. burgdorferi*. Isolates were characterized by RFLP analysis and grouped into three rDNA spacer + RFLP genotypes. A highly significant association was found between specific *B. burgdorferi* genotype and hematogenous dissemination in patients with early Lyme disease and disease severity in a murine model of Lyme borreliosis (Wormser *et al.*, 1999).

Despite availability of the B31 genome sequence, the pathogenesis of Lyme disease is not well elucidated; however, considerable anecdotal evidence indicates that *Borrelia* plasmids are important in pathogenesis. *In vitro* passage of *B. burgdorferi* is associated with loss of plasmids (Barbour, 1988). Plasmid loss after 10–17 passages is also coupled with decreased infectivity in mice and changes in spirochetal protein expression (Johnson, Marek and Kodner, 1984; Moody, Barthold and Tergwilliger, 1990). High- and low-infectivity phenotypes of *B. burgdorferi* coexist in an uncloned population, but the proportion of high-infectivity phenotypes decreases with serial *in vitro* passage (Norris *et al.*, 1995). Plasmid profiles vary among certain *B. burgdorferi* sensu lato strains (including *B. burgdorferi* sensu stricto, *B. garinii* and *B. afzelii*) as well as within high- and low-passaged cultures; this variance is associated with infectivity of *B. burgdorferi* in mice. Two recent studies have suggested that plasmids Lp25 and Lp28-1 encode virulence factors involved in *B. burgdorferi* pathogenicity (Purser and Norris, 2000; Labandeira-Rey and Skare, 2001). However, the association between the loss of specific plasmid and loss of infectivity has not been universally observed (McDowell *et al.*, 2001). Clearly, more complex mechanisms independent of plasmid loss appear to be at play. These may include loss of specific genes, small-scale mutations or rearrangements in essential genes and/or differences in gene-expression patterns among clonal populations.

EPIDEMIOLOGY

In the United States, surveillance for Lyme disease was begun by the Centers for Disease Control (CDC) in 1982. Since that time the number of reported cases has increased dramatically. The Council of State and Territorial Epidemiologists designated Lyme disease as a nationally notifiable disease in January 1991. In 2002, 23 763 cases were reported to the CDC, making Lyme disease the most common vector-borne disease in the United States (CDC, 2004). As many as 90% of Lyme disease cases may go unreported. The infection is reported in 49 of the 50 US states, but most cases occur in the Northeastern Midwest and North Central regions of the United States. Nine states account for more than 90% of the reported cases, with Connecticut leading the group. The other states are Rhode Island, New York, Pennsylvania, Delaware, New Jersey, Maryland, Massachusetts and Wisconsin (Melski, 2000; CDC, 2001). The rising incidence of Lyme disease in recent years in the United States may be explained by multiple factors, including an increase in the numbers of Ixodid ticks, the outward migration of residential areas into previously rural woodlands (habitats favored by Ixodid ticks and their hosts), rising deer population and increased recognition by patients and clinicians.

The infecting organism, *B. burgdorferi*, is maintained in and transmitted by ticks of the *Ixodes ricinus* complex, including *I. scapularis* in northeast and northcentral USA, *I. pacificus* on the west coast of the USA, *Ixodes ricinus* in Europe and *I. persulcatus* in Asia (Nadelman and Wormser, 1998). In Europe, three genospecies of the *B. burgdorferi* sensu lato complex are pathogenic, including *B. burgdorferi* sensu stricto, *B. garinii*, and *B. afzelii*. *Borrelia burgdorferi* is the only pathogenic species in North America. The varying relative distribution of these genospecies from region to region throughout Europe and Asia may account for the relative variability of disease syndromes associated with Lyme disease. In the United States, the majority of affected individuals have had symptoms of the illness (Steere *et al.*, 1986; Lastavica *et al.*, 1989), whereas in Europe, the majority have been asymptomatic. Of 346 people who were studied in the highly endemic area of Liso, Sweden, 41 (12%) had symptoms of the illness and 89 (26%) had evidence of subclinical infection (Fahrer *et al.*, 1991). In a serosurvey of 950 Swiss orienteers, 26% had detectable IgG antibodies to *B. burgdorferi*, but only 2–3% had a past history of definite or probable Lyme borreliosis (Gustafson *et al.*, 1992). The ages of patients range from 2 to 88 years (median, 28 years), and the sex ratio is nearly 1:1 (Steere *et al.*, 1983a). Age-adjusted attack rates show a bimodal distribution, with the greatest risk of acquiring the illness in children and middle-aged adults. Additional borrelial species that have been identified are *B. valasiana* and *B. lusitaniae*; although the role of the former in promoting human disease is not entirely clear, it has been detected by polymerase chain reaction (PCR) in skin biopsies of a few patients with symptomatic acute and chronic skin disease. However, the organism has not been cultures from these lesions (van Dam *et al.*, 1993; O'Connell *et al.*, 1998; Weber, 2001).

In field studies in Connecticut and New York, *B. burgdorferi* has been found in 10–50% of nymphal and adult *I. scapularis* (Magnarelli *et al.*, 1976; Bosler *et al.*, 1983). Although *B. burgdorferi* has been demonstrated in mosquitoes and deer flies (Magnarelli and Anderson, 1988), however only ticks of the *I. ricinus* complex seem to be important in the transmission of the spirochete to humans. The seasonal variation of onset in temperate climatic zones is explained by the ecology of the predominant tick vectors. Among the Ixodid tick vectors, the life cycle and feeding habits of *I. scapularis* are best understood. This tick has a three-stage life cycle (larva, nymph and adult) that spans 2 years. Larvae hatch from fertilized eggs in late spring and feed once for 2 or more days in midsummer. Preferred hosts include a broad range of small mammals. The next spring they molt into nymphs, and feed again from 3 or 4 days, with the same host range. After this second blood meal, the nymphs molt into adults. Adult *I. scapularis* has a narrower host range, with a preference for deer. Mating occurs on deer, and the female deposits her eggs and the cycle begins anew (Spielman *et al.*, 1985). During their 2-year life cycle, ticks typically feed once during each of the three stages, usually the late summer for larval ticks, the following spring for nymphs, and autumn for the adults. *Ixodes scapularis* nymphs appear to be the most important vector for transmission of *B. burgdorferi*. According to laboratory studies, a minimum of 36–48 h of attachment of the tick is required for transmission. In the United States, most cases involving *B. burgdorferi* occur between May and August, which corresponds with increased outdoor human and nymphal tick activity.

Lyme disease affects all age groups and both sexes, and transplacental transmission of *B. burgdorferi* has been reported, but seems to be very infrequent (Schlesinger *et al.*, 1985; Weber *et al.*, 1988).

Markowitz *et al.* (1986) retrospectively reviewed 19 cases of Lyme disease during pregnancy and noted adverse fetal outcomes in five cases.

The risk of infection in a given area depends largely on the density of these ticks as well as on their feeding habits and animal hosts, which have evolved differently in different locations. In the northeastern and north central United States, *I. scapularis* ticks are abundant, and a highly efficient cycle of *B. burgdorferi* transmission

occurs between immature larval and nymphal *I. scapularis* ticks and white-footed mice (Spielman, 1994), resulting in high rates of infection in nymphal ticks and a high a frequency of Lyme disease in humans during the late spring and summer months (Orloski *et al.*, 2000). The proliferation of deer, which are the preferred host of the adult tick, was a major factor in the emergence of epidemic Lyme disease in the northeastern United States during the late twentieth century.

Ixodes scapularis is a vector for the agents of human granulocytic ehrlichiosis and *Babesia microtti*. *Ixodes scapularis* and other ticks in the *I. ricinus* complex may transmit multiple pathogens. The proportion of *I. scapularis* or *I. ricinus* ticks coinfected with both *B. burgdorferi* sensu lato and *Anaplasma phagocytophila* is generally low, ranging from <1 to 6% in six geographic areas. A higher prevalence of tick coinfection (26%) has been reported in Westchester county, NY. The proportion of *Ixodes* ticks coinfected with *B. burgdorferi* and *Babesia microti* has ranged from 2% in NJ to 19% on Nantucket Island, Massachusetts. Among patients with a confirmed tick-borne infection, coinfection rates as high as 39% have been reported. The most commonly recognized coinfection in most of the eastern United States is Lyme borreliosis and babesiosis, accounting for approximately 80% of coinfections (Krause *et al.*, 1996).

An enzootic cycle of infection is maintained through passage of *B. burgdorferi* back and forth between ticks and their hosts. Infected nymphal ticks transmit *B. burgdorferi* to mice, which serve as a reservoir from which uninfected larvae may acquire infecting organisms. In this manner, a high rate of infection can be maintained in the tick population when the organism, ticks, mice, and deer are all present in the environment.

The establishment and maintenance of an enzootic cycle requires a competent reservoir host in addition to a tick vector. Variation in vector–host relationships provide the primary explanation for wide regional variation in the rate of infection in the tick population in California (Brown and Lane, 1992), southeast (Oliver *et al.*, 1993) and northeast United States.

On the west coast of the United States the nymphal stage of *I. pacificus* prefers to feed on lizards rather than rodents, and lizards are not susceptible to infection with *B. burgdorferi* (Manweiler, Lane and Tempelis, 1992). There, the spirochete is maintained in nature in a horizontal cycle between the dusky-footed woodrat *Neotoma fiscipes* and *I. neotomae*, a tick that does not feed on humans (Brown and Lane, 1992). The infection range of *I. pacificus* ticks is only 1–3%, and the number of cases on the West Coast is much less than that on the East Coast.

In Europe, Lyme borreliosis is widely established in forested areas. The highest reported frequencies of the disease are in middle Europe and Scandinavia, particularly in Germany, Austria, Slovenia, and Sweden (Stanek *et al.*, 1993). The infection is also found in Russia, China, Japan, and the UK.

CLINICAL MANIFESTATIONS

Lyme disease is a progressive infectious disease with a wide array of clinical manifestations. Infection begins locally in the skin after *B. burgdorferi* is inoculated by a feeding tick. In the majority of individuals, the initial sign of infection is the development of EM (Luft and Dattwyler, 1989). Even at this early phase of infection the clinical expression of the disease is highly variable. Some individuals are relatively asymptomatic, while others develop fever, arthralgias, myalgias, conjunctivitis, meningismus, or multifocal EM lesions, and still others develop more dramatic signs of infection, including acute meningitis, myocarditis with or without conduction block abnormalities, hepatitis, myositis, or frank arthritis. Up to 50% of infected individuals will go on to develop manifestations of late disease if not treated in the acute phase of the infection. In chronic phase of the illness, localized inflammatory processes may occur in one or more organ systems, particularly the nervous and the musculoskeletal systems.

Furthermore, it may be important to know the chronicity of the infection since the pathogenesis of the disease and the response to therapy may be different in patients with acute versus chronic disease. In general, three stages of the illness – early localized disease, early disseminated disease, and persisting late disease can be distinguished.

Coincident Infections

It is significant to note that other zoonotic diseases transmitted by *Ixodes* ticks may occur concomitantly with *B. burgdorferi*. Babesiosis and HGE can occur independently or together with Lyme disease. The first coinfection with *B. burgdorferi* and *B. microti* was reported in 1983 (Grunwaldt, Barbour and Benach, 1983) and the first coinfection of *B. burgdorferi* and *A. phagocytophila* in 1997 (Nadelman *et al.*, 1997). Coinfection may be common because 2.2–26% *I. scapularis* ticks may be coinfected with *B. burgdorferi* and *A. phagocytophila* (Belongia *et al.*, 1999). A higher prevalence of tick coinfection (26%) has been reported in Westchester County, NY. The proportion of *Ixodes* ticks coinfected with *B. burgdorferi* and *Babesia microti* ranged from 2% in New Jersey to 19% on Nantucket Island, Massachusetts. In humans, cross-sectional seroprevalence studies have found markers of dual infection in 9–26% of patients with a tick-borne infection, but such studies often fail to distinguish simultaneous coinfection from sequential infections. In a report of a fatal case of babesiosis, the autopsy revealed significant histopathological evidence of a fulminant myocarditis due to *B. burgdorferi* (Marcus, Steere and Duray, 1985). Among patients with a confirmed tick-borne infection, coinfection rates as high as 39% have been reported. The most commonly recognized coinfection in most of the eastern United States is Lyme borreliosis and babesiosis, accounting for approximately 80% of coinfections. Coinfections can modify the immune response and alter the severity of arthritis in animal models. DNA extracts from 701 ticks collected in 15 localities were examined by PCR for the simultaneous detection of these three pathogens. Overall, 14% were infected with *A. phagocytophilum* followed by 12.4% with *B. burgdorferi* and 2.3% with *B. microti*. In total, the percentage of infected females (32.9%) was 2.4 times higher than in males (13.7%) and 3.2 times higher than in nymphs (10.3%). Among adult ticks (n = 303), 8.3% were dually infected with *A. phagocytophilum* and *B. burgdorferi*, 2.0% with the agent of human anaplasmosis and *B. microti*, and 0.3% with borreliae and *B. microti* (Stanczak *et al.*, 2004). A recent population retrospective survey in Nantucket Island, Massachusetts, found that serological evidence of exposure to babesiosis was not associated with increased severity of acute Lyme disease (Wang *et al.*, 2000).

Russian investigators, Putkonen *et al.* in 1962, have reported a relationship between tick-borne encephalitis due to flaviviruses and EM. Since Lyme borreliosis can be associated with a meningitis or meningoencephalitis, it is not clear whether these particular cases of tick-borne encephalitis associated with EM are due to a dual infection of *B. burgdorferi* and flavivirus or solely of infection with *B. burgdorferi* (Kristoferitsch, 1989). In Europe, species of *Babesia* and *Ehrlichia* are present in Irizinus, which is also a vector for flavivirus causing tick-borne encephalitis (Telford *et al.*, 1997). Furthermore, Lyme borreliosis should be considered in all patients suspected of arboviral encephalitis (Edlinger, Rodhain and Perez, 1985; Rodhain and Edlinger, 1987). Recently, Bartonella henselae has been described as a novel coinfecting agent in patients whose CNS is infected with *B. burgdorferi*.

Skin

Erythema Migrans

The most common manifestation of early localized Lyme disease is EM occurring in up to 85% of patients and is frequently located around the knees, axilla, and in the groins (Barbour and Hayes, 1986;

de Silva and Fikrig, 1997; Fraser *et al.*, 1997). It generally appears at the site of a tick bite three to 30 days (typically within seven to 10 days) after the bite. EM usually begins as a red macule or papule, which expands over the course of days to weeks, presumably as the spirochetes spread centrifugally through the skin. Secondary cutaneous lesions without tick bites may develop after hematogenous spread of spirochetes. Local symptoms include pruritus, tenderness, or paresthesias but are generally rare or absent in secondary lesions. The EM may also appear as target lesion with variable degrees of central clearing and may occasionally have vesicular or necrotic areas in the center. In one US study, spirochetes were cultures from plasma samples in 50% of patients with EM (Wormser *et al.*, 2000a). Spreading through skin and other tissue matrixes may be facilitated by the binding of human plasminogen and its activators to the surface of the spirochete (Coleman *et al.*, 1997). During the dissemination and homing of specific sites, the organism attaches to certain host integrins (Coburn, Leong and Erban, 1993), matrix glycosaminoglycans (Guo *et al.*, 1998), and extracellular-matrix proteins (Probert and Johnson, 1998). Spirochetal lipoproteins, which bind the CD14 molecule and Toll-like receptors on macrophages, are potent activators of the innate immune response, leading to the production of macrophage-derived inflammatory cytokines (Hirschfeld *et al.*, 1999).

In humans, infiltrates of macrophages and T cells in EM lesions express messenger RNA for both inflammatory and anti-inflammatory cytokines (Muellegger *et al.*, 2000). Particularly in disseminated infection, adaptive T-cell and B-cell responses in lymph nodes lead to the production of antibody against many components of the organism (Krause *et al.*, 1991; Akin *et al.*, 1999). It has been shown that the epidermal Langerhans' cells are invaded by *B. burgdorferi* in early Lyme borreliosis. In addition, reduced numbers of CD1a-positive Langerhans' cells were found in EM lesions. Systemic complaints in patients with EM are more common in USA than in Europe (McAlister *et al.*, 1989) secondary to differences in genospecies or concurrent infection with babesiosis or ehrlichiosis (Nadelman *et al.*, 1997). The most frequent symptoms of culture-confirmed cases include fatigue (54%), myalgia (44%), arthralgia (44%), headache (42%), fever and/or chills (39%), and stiffness of the neck (35%) (Nadelman *et al.*, 1996). In an observational cohort study in 10 endemic states, 118 patients with microbiologically confirmed EM presented a median of 3 days after symptom onset. Early EM commonly had homogeneous or central redness rather than a peripheral erythema with partial central clearing. The most common associated symptoms were low-grade fever, headache, neck stiffness, arthralgia, myalgia, or fatigue. By convalescence, 65% of patients had positive IgM or IgG antibody responses to *B. burgdorferi*. Most patients responded promptly to antibiotic treatment (Hassler and Maiwald, 1994). Subsequent episodes of EM have been reported in patients who received appropriate antimicrobial therapy for an initial episode, while primary failure of antibiotic treatment is rare, about 0.14% according to Maraspin, Lotric-Furlan and Strle (2002).

Lymphocytoma

Borrelial lymphocytoma (BL), which is characterized by a dense lymphocytic infiltrate in the dermis or subcutaneous tissue, has been associated with borrelial infection, is the least common manifestation of Lyme disease (5%), and occurs more often in children than in adults (Hovmark, Asbrink and Olsson, 1986). It is usually caused by *B. garinii* and *B. afzelii* and is seen more frequently in Europe than in the USA. It is a tumor-like nodule that typically appears on the ear lobe, nipple, or scrotum (van Dam *et al.*, 1991; Picken *et al.*, 1997b). The lymphocytoma may occur with other manifestations of infection such as meningitis, choroiditis, arthritis, or acrodermatitis atrophicans (Asbrink and Hovmark, 1988). Histopathologically, it may be difficult to differentiate the lymphocytoma from lymphoma. Although some patients with BL may be seronegative, *B. burgdorferi* (Bb) IgG and/or

IgM antibodies are found in the serum of 80% of all BL patients. Direct detection of Bb or Bb-specific DNA in lesional skin by culture or PCR is helpful addition to the diagnosis. Thirty-six cases of borrelial lymphocytoma were detected during the period 1986–1990 in Slovenia led to complete recovery within an average of 3 weeks in all cases with antibiotic therapy (Strle *et al.*, 1992).

Acrodermatitis Chronicum Atrophicans (ACA)

In European patients, especially elderly women with *B. afzelii* infection, a chronic, slowly progressive skin condition called acrodermatitis chronica atrophicans may develop on sun-exposed acral surfaces. The organism has been cultured from such lesions as long as 10 years after the onset of the disease (Asbrink and Hovmark, 1985). These lesions may be preceded by EM and may represent a late or chronic stage of infection. Early lesions have erythematous nodules or plaques with central clearing and involve the extensor areas of the extremities or joints. Later lesions become atrophic and poikilodermatous resembling scleroderma or lichen sclerosus et atrophicus.

In one study, the infiltrates of T cells and macrophages in these lesions had a restricted cytokine profile, with little or no production of interferon-γ (Aberer, Klade and Hobisch, 1991), which may explain in part why the immune response is ineffective in eradicating the spirochete. The CD1a-positive Langerhans' cells were normal or slightly raised in these lesions. Major histocompatibility complex class II molecules on Langerhans' cells were downregulated (Aberer, Koszik and Silberer, 1997). While this downregulation may protect from presentation of autoantigens, it may also cause the impaired capacities of Langerhans' cells to eliminate *B. burgdorferi* and, thus, permit the chronicity of the infection. Alternatively, downregulation of several *B. burgdorferi* genes and insufficient killing of *Borrelia* by complement and antibody has been reported (Liang, Nelson and Fikrig, 2002).

Superficial and deep perivascular and interstitial infiltrate of plasma cells, lymphocytes, and histiocytes. Later lesions show sclerosis with loss of adnexa and elastic fibers. Concomitant with the ACA, neurological (Hopf, 1975; Kristoferitsch *et al.*, 1988) and rheumatological (Herzer *et al.*, 1986) evidence of *B. burgdorferi* infection may be present. In approximately one-third of patients, there is evidence of a polyneuropathy. Patients complain of paraesthesias, hyperaesthesias, weakness, and muscle cramps. There have been reports of profound fatigue and weight loss associated with ACA. They may also have severe rheumatological involvement, with subluxation of the small joints of the hands and feet, arthritis of the large joints, and periosteal thickening of the bones. Bursitis, epicondylitis, and tendonitis have also been reported to occur with ACA. At times it may be difficult to differentiate between the pain of arthritis and the significant neuritis associated with this disease process.

Carditis

Within several weeks after the onset of disease, about 4–10% of untreated patients have acute cardiac involvement – most commonly fluctuating degrees of atrioventricular block, occasionally acute myopericarditis or mild left ventricular dysfunction, and rarely cardiomegaly or fatal pancarditis (Steere, 1989). Diffuse T-wave changes, ST-segment depression, and arrhythmias also are frequently observed. Steere *et al.* (1980) first described the cardiovascular manifestations of Lyme disease in 1980 with a report of 20 North American cases. Patients may complain of dizziness, palpitations, dyspnea, chest pain, or syncope (van der Linde, 1991; van der Linde and Ballmer, 1993). In a mouse model, cardiac infiltrates of macrophages and T cells secrete inflammatory cytokines (Kelleher Doyle *et al.*, 1998). In these mice, the killing of spirochetes through cellular immune mechanisms seems to be the dominant factor in the resolution of the cardiac lesion. In mice with severe combined immunodeficiency, immune serum resolves arthritis but not carditis (Barthold *et al.*, 1997).

In Europe, *B. burgdorferi* has been isolated from endomyocardial biopsy samples from several patients with chronic dilated cardiomyopathy (Stanek *et al.*, 1990; Lardieri *et al.*, 1993). However, this complication has not been observed in the United States (Sonnesyn *et al.*, 1995). A Dutch group noted improvement in left ventricular ejection fraction in eight of nine *B. burgdorferi*-seropositive patients with idiopathic dilated cardiomyopathy who were treated with antibiotics (Gasser *et al.*, 1992). Further studies are warranted to clarify the role that *B. burgdorferi* in chronic and acute congestive heart failure. Gallium and anti-myosin Indium scans have been abnormal in individual cases with myocarditis. Although temporary pacing is frequently required for complete heart block, permanent pacing is rarely needed (McAlister *et al.*, 1989). The prognosis of the acute cardiac manifestations of Lyme disease is good.

Joint Involvement

In the USA, arthritis is the predominant manifestation of disseminated *B. burgdorferi* infection, with about 60% of untreated patients developing joint manifestations usually weeks to years after the initial infection. Popliteal (Baker's) cysts are common, and – if present – dactylitis and Achilles tendon involvement can be helpful as diagnostic signs if reactive arthritis, psoriatic arthritis, and spondyloarthropathies have been ruled out. Months after the onset of illness, patients begin to have intermittent episodes of joint swelling and pain, primarily in large joints, occasionally the temporomandibular joint (Steere, 1989). Musculoskeletal features include arthralgia; intermittent episodes of migratory arthritis, usually monoarthritis or asymmetric oligoarthritis; and chronic arthritis, usually of the knees (Steere, Schoen and Taylor, 1987). In an observational cohort study of 15 patients with Lyme arthritis or fibromyalgia or both, symptoms of Lyme arthritis resolved with antibiotic therapy, whereas symptoms of fibromyalgia persisted in 14 of 15 patients treated with antibiotic therapy (Dinerman and Steere, 1992). The development of articular manifestations was initially described in a cohort of 55 patients evaluated for EM before the role of antibiotic therapy in Lyme disease was identified, 80% experienced some sort of articular manifestation over the course of the mean 6-year follow-up (Steere, Schoen and Taylor, 1987). Approximately 20% of patients had brief episodes of arthralgias which involved multiple joints in a migratory fashion and 50% of patients had EM with or followed by intermittent arthritis. Ten percent of patients experienced chronic Lyme arthritis affecting one to three joints. A number of these patients share a B-cell alloantigen, HLA-DW2, but did not have the immunologic features of rheumatoid arthritis (Steere *et al.*, 1979).

In Wurzburg, Germany, an epidemiologic study suggested a yearly incidence rate of one case per 100 total population of chronic manifestations of Lyme disease with a higher prevalence of Lyme arthritis than of neuroborreliosis. Because of the increased frequency of certain HLA-DR4 and DRB1 alleles in these patients, an autoimmune mechanism has been proposed (Steere *et al.*, 2001). Synovia from affected patients shows synovial hypertrophy, vascular proliferation, and a marked infiltration of mononuclear cells, sometimes with pseudolymphoid follicles that are reminiscent of peripheral lymph nodes (Steere, Schoen and Taylor, 1987; Steere, Duray and Butcher, 1988). Synovial fluid analyses typically show a slightly increased protein and white cell count of a median of 24 000/μl with a polymorphonuclear predominance and a normal glucose level (Steere *et al.*, 1977a). During attacks of arthritis, innate immune responses to *B. burgdorferi* lipoproteins are found, and γ/δ T cells in joint fluid may aid in this response (Vincent *et al.*, 1998). Furthermore, patients with Lyme arthritis usually have higher *Borrelia*-specific antibody titers than patients with any other manifestations of the illness, including late neuroborreliosis (Akin *et al.*, 1999).

While most patients respond favorably to antibiotic therapy, about 10% of adults and fewer than 5% of children with Lyme arthritis develop persisting inflammatory joint disease lasting for more than 1 year, which may eventually lead to joint destruction. In about 10% of patients, particularly those with HLA-DRB1*0401 or related alleles, the arthritis persists in the knees for months or even years (Akin *et al.*, 1999). This postinfectious syndrome occurs more frequently in patients whose symptoms are suggestive of early dissemination of the spirochete to the nervous system, particularly if treatment is delayed (Shadick *et al.*, 1999; Kalish *et al.*, 2001). Autoimmunity may develop within the inflammatory milieu of affected joints in these patients because of molecular mimicry between an immunodominant T-cell epitope of OspA of *B. burgdorferi* and human lymphocyte function-associated antigen-1 (LFA-1$\alpha_{332-342}$), an adhesion molecule that is highly expressed on T cells in synovium (Gross *et al.*, 1998; Steere *et al.*, 2001). T cells that react to OspA$_{(65-173)}$ are concentrated in the joints of these patients (Meyer *et al.*, 2000). When LFA-1$\alpha_{332-342}$ is processed and presented by HLA-DRB1*0401 molecules, this self-peptide may behave as partial agonist. Treatment-resistant and treatment-responsive arthritis differ primarily in the cellular and humoral immune responses to OspA (Kalish, Leong and Steere, 1993; Chen *et al.*, 1999). In a retrospective study of 88 patients with Lyme arthritis, Nocton *et al.* (1994) showed the presence of *B. burgdorferi* DNA in 85% of patients with active Lyme arthritis and in none of the control patients. Because culture of synovial fluid is not a practical method by which the diagnosis of Lyme arthritis can be made, serologic tests and PCR are useful methods to determine whether live spirochetes are present within joint fluid. Enzyme-linked immunosorbent assay (ELISA) and Western blot testing for anti-*B. burgdorferi* antibodies within the synovial fluid is an accurate and reliable method of proving that an arthritis is really caused by Lyme. Severe acute arthritis develops in C3H/HeJ mice that are infected with *B. burgdorferi* (Barthold *et al.*, 1990), antibody against a 37-kD arthritis-resolving protein of the spirochete is critical in the resolution of the arthritis (Feng, Hodzic and Barthold, 2000). In contrast arthritis does not develop in *B. burgdorferi*-infected C57BL/6 mice (Barthold *et al.*, 1990).

Nervous System

All the neurological syndromes associated with *B. burgdorferi* infection can occur without previous EM (Reik, Burgdorfer and Donaldson, 1986). Lyme borreliosis is associated with both acute and chronic neurological abnormalities, affecting both the central and peripheral nervous systems. Clinical data support the hypothesis that, in most individuals, *B. burgdorferi* invades the nervous system early in the course of the infection. The most compelling evidence for this is the frequency of complaints referable to the nervous system in patients with EM. In one series of 314 patients with EM, 80% of the patients had malaise, fatigue, and lethargy; 64% had headache; and 48% complained of a stiff neck (Steere *et al.*, 1983b). Additional evidence of the early invasion of the nervous system is shown by the finding that approximately 12–15% of patients develop one or more of these three acute disorders – meningitis, cranial neuritis, or painful radiculitis – within the first 3 months after infection (Pachner and Steere, 1985). In Europe, the frequency of neuroborreliosis seems higher, potentially due to the greater neurotropism of *B. garinii*, which has not been isolated in North America. In Europe, *B. garinii* may also occasionally cause chronic encephalomyelitis, characterized by spastic paraparesis, cranial neuropathy, or cognitive impairment with marked intrathecal production of antibodies against the spirochete (Oschmann *et al.*, 1998). Cranial and spinal cord magnetic resonance imaging usually show diffuse white matter lesions.

The distinguishing feature for patients who develop meningitis is the presence of CSF abnormalities, a mild pleocytosis largely consisting of polymorphonuclear leukocytes or mononuclear cells,

a modest elevation of CSF protein, and a normal CSF glucose. Papilloedema and increased CSF pressure can occur, resulting in a syndrome indistinguishable from pseudotumor cerebri (Raucher *et al.*, 1985). Local antibody production against *B. burgdorferi* and MRI abnormalities have been observed in the more severely affected patients. In the absence of local antibody production, the diagnosis of *B. burgdorferi* CNS infection is questionable although has been demonstrated to occur.

Localized CNS processes have been associated with *B. burgdorferi* infection (Broderick, Sandok and Mertz, 1987). These may present as acute myelitis (Rousseau *et al.*, 1986), localized encephalitis (Feder, Zalneraitis and Reik, 1988), or cerebellar ataxia. The Headache of disseminated Lyme disease tends to be persistent and may worsen with time. Mild-to-moderate neck stiffness and photophobia may occur and continue for several weeks. Meningoencephalitis may be a prominent feature, manifesting as difficulty with memory and concentration and emotional lability (Oschmann *et al.*, 1998; Coyle *et al.*, 1999). Lyme meningitis typically occurs in the fall, at the same time as the peak incidence of enteroviral meningitis. Clues to Lyme meningitis, which otherwise mimics aseptic viral meningitis, are an associated EM, facial nerve palsy occurring during the early dissemination stage most often involves the facial nerve. One unusual case of *B. burgdorferi*-induced meningoencephalitis has been reported, where a T-cell clone recovered from the cerebrospinal fluid responded to spirochetal epitopes and to autoantigens (Hemmer *et al.*, 1999).

Cranial neuropathies may occur with or without evidence of meningitis. Any cranial nerve may be affected by Lyme disease, but seventh nerve, both unilateral and bilateral (Clark *et al.*, 1985; Oschmann *et al.*, 1998), involvement is by far the most frequent, occurring in up to 10% of patients with this infection. One recent review of 43 cases of bilateral seventh nerve palsy in a nonendemic area found no patients with Lyme disease (Keane, 1994). By contrast, studies in highly endemic areas indicate that Lyme disease may account for up to 25% of cases of facial nerve palsy. Halperin and Golightly (1992) concluded that Lyme disease may be responsible for approximately 25% of new onset Bell's palsy in an endemic area, with the palsy sometimes developing before positive serologic testing. No antibody was found in the CSF of those patients tested, which supported this being a peripheral neuropathy.

Lyme disease can cause a painful radiculitis manifested by neuropathic symptoms such as numbness, tingling, and burning. This radiculoneuropathy may affect the limbs or trunk. Fifty percent of the patients have associated cranial nerve palsies. The peripheral nerve damage in Lyme disease is usually an axonopathy, rather than a demyelinating syndrome (Hansen and Lebech, 1992). Acute or subacute myelitis can occur, associated with spastic paraparesis and CSF pleocytosis (Logigian, Kaplan and Steere, 1990). Mononeuritis and a Guillain–Barre-like syndrome have rarely been reported. Acute painful radiculoneuritis is the most striking early dissemination neurologic syndrome. The most common late stage neurologic syndrome in the United States is a subtle Lyme encephalopathy, which seems to represent CNS infection. Less-common neurologic manifestations include sudden sensorineural hearing loss, cerebellitis, intracranial aneurysm, and myelitis (Finizia, Jonsson and Hanner, 2001).

Meningopolyneuritis (Garin–Bujadoux, Bannwarth's syndrome) is the most dramatic of the peripheral nervous system abnormalities (Henriksson *et al.*, 1986). Concomitant encephalitis and myelitis occur in over 20% of patients. In these patients, long-term sequelae such as spastic paraparesis and neurogenic bladder persist after appropriate therapy. Sensitive neurophysiological techniques reveal that more than 40% of patients with late disease have demonstrated abnormalities (Halperin *et al.*, 1987).

Some patients report persistent symptoms after the recommended antibiotic treatment (Bujak, Weinstein and Dornbush, 1996), which has been termed post-Lyme disease syndrome, post-treatment chronic Lyme disease (PTCLD), or chronic Lyme disease. Nonspecific symptoms, such as fatigue, cognitive impairment, headache, arthralgias, and myalgias (Asch *et al.*, 1994; Shadick *et al.*, 1994; Seltzer *et al.*, 2000) have been reported. Patients with PTCLD report more symptoms of depression than healthy control subjects (Fallon *et al.*, 1992; Kaplan *et al.*, 1999) and patients who have recovered from Lyme disease (Bujak, Weinstein and Dornbush, 1996). The severe fatigue may simply be an indication of chronic CNS infection and, in many cases, the most recognizable symptom of a mild chronic encephalopathy of months to years duration (Bensch, Olsen and Hagberg, 1987). In one study, patients with Lyme disease were evaluated 10–20 years after their original infection (Kalish *et al.*, 2001). Among the facial nerve palsy patients, those who had not received antibiotics were more likely to have arthralgias and sleep difficulties, mild residual deficits, and physical limitations compared with patients who had received treatment. In up to 10% of untreated patients, *B. burgdorferi* may cause chronic neuroborreliosis sometimes after long periods of latent infection (stage 3), thus making correct diagnosis very difficult (Logigian, Kaplan and Steere, 1990). Neither single PET of the brain nor neuropsychological testes of memory have sufficient specificity to be helpful in the diagnosis.

Long-term follow-up of patients with culture-confirmed Lyme disease showed excellent outcomes for the 96 cases who were followed for a mean of 4.9 ± 2.9 years. Other studies have reported frequent somatic complaints in controls (Kaplan *et al.*, 1992; Gaudino, Coyle and Krupp, 1997; Barr *et al.*, 1999) such as fatigue (15 to 43%), joint pain (20 to 27%), muscle aches (19 to 57%), paresthesias (13 to 27%), difficulty concentrating (3 to 13%), and headaches (16 to 20%). As with many chronic illnesses, patients with Lyme disease with persistent physical symptoms, such as pain, are also often experiencing greater emotional distress, including depression and anxiety, than otherwise healthy people, although they do not always meet the clinical criteria for psychopathology. Furthermore, in a double-blind, placebo-controlled study to improve the outcomes of patients with chronic Lyme disease (Klempner *et al.*, 2001) who had persistent subjective symptoms did not improve outcomes with both parenteral (1 month) and oral antibiotics (2 months) directed at *B. burgdorferi* infection. In an observational study of 129 patients indicate that for patients who have been previously treated with recommended antibiotics for Lyme disease and with self-reported neurocognitive symptoms who do not show clear evidence of persisting *Borrelia* infection or evidence of cognitive impairment on objective testing, there is no efficacy to additional antibiotic therapy (Halperin *et al.*, 1988; Logigian, Kaplan and Steere, 1999).

Other Organ Systems

Borrelia burgdorferi will localize to other organ systems. Lyme disease may be characterized by myositis, liver, and spleen involvement. Myositis is characteristically localized near an involved joint or localized neuropathy. Histopathologic examination in 1993 study by Reimers, de Koning and Neubert (1993), revealed focal nodular myositis or interstitial myositis, interstitial lymphohistiocytic infiltrates with plasma cells often localized around small endomysial vessels. Nuclear imaging with gallium-67 may be useful for detection. Patients with EM have been observed to have liver function test abnormalities in the absence of symptomatic hepatitis (Kazakoff, Sinusas and Macchia, 1993). Lymphoplasmocellular infiltrates have been seen in liver and spleen (Duray, 1989). Splenomegaly has been noted infrequently (Nelson and Nemcek, 1992). Necrotizing eosinophilic lymphadenitis was described in a 2-year-old girl with chronic arthritis and positive Lyme serology by ELISA. Although the spirochete is known to induce glomerulonephritis in animals, a recent case of membranoproliferative glomerulonephritis has been reported in which the diagnosis was confirmed serologically, and the patient though was dialysis dependent for a protracted period of 5 months, regained normal urine volume with treatment (Kirmizis *et al.*, 2004). Cases of myositis, panniculitis, osteomyelitis, uveitis and pneumonitis due to

B. burgdorferi have been reported (Steere *et al.*, 1985a; Kramer *et al.*, 1986; Atlas *et al.*, 1988; Kirsch *et al.*, 1988). A nonspecific follicular conjunctivitis may occur in as many as 10% of patients during the flu-like illness of Lyme disease (Mombaerts, Maudgal and Knockaert, 1991). Periorbital edema, episcleritis, photophobia and subconjunctival hemorrhages have been noted (Mombaerts, Maudgal and Knockaert, 1991). Interstitial and ulcerative keratitis have also been observed (deLuise and O'Leary, 1991). It has also been reported to cause vitritis, anterior uveitis (Schubert, 1994). Optic neuritis and perioptic neuritis associated with Lyme disease have been described in both Europe and the United States (Farris and Webb, 1988). Seventh nerve paresis can lead to neurotrophic keratitis. In 38 cases reported by Pachner and Steere (1985), one-half of the patients had facial palsy, a third of these with bilateral involvement. Optic disc edema can occur as a result of optic neuritis or increased intracranial pressure. Pupillary changes can occur, including the development of a Robertson Pupil (Schechter, 1986).

DIAGNOSIS

Until the discovery of the causative agent in 1982, the clinical diagnosis of Lyme borreliosis was based solely on the recognition of EM, either historically or by direct observation. In the absence of EM, the demonstration of an immune response to *B. burgdorferi* in an appropriate clinical setting forms the basis on which most diagnoses are made. In patients in the United States, the diagnosis is usually based on the recognition of the characteristic clinical findings, a history of exposure in an area where the disease is endemic, and except in patients with EM, an antibody response to *B. burgdorferi* by ELISA and Western blotting, interpreted according to the criteria of the Centers for Disease Control and Prevention and Association of State and Territorial Public Health Laboratory Directors (CDC, 1990, 1995).

Microscopy

Borrelia burgdorferi, a loosely coiled spirocheate which is approximately 200 µm wide and 10–30 µm long, is difficult to visualize under bright-field conditions, but is readily visible in phase contrast or dark field. The organism is Gram negative and can also be stained with acridine orange, Giemsa, or by fluorescent antibody technique. Silver stains, either Warthin–Starry or a modified Dieterle stain, have proved to be successful in identifying spirochaetes in fixed, paraffin-embedded tissue. Apart from skin biopsy, spirochaetes have been observed in other tissues, such as the myocardium, synovium, and the nervous system, but the yields from them have been very poor because of low numbers of spirochaetes.

Cultural Methods

In patients presenting with EM, culture is virtually 100% specific and seems to be more sensitive (57–86%) (Lebech, 2002) than serology (about 50% in the USA and less than 50% in Europe) (Asbrink, Hovmark and Olsson, 1986). The culture of *B. burgdorferi* from specimens in Barbour–Stonner–Kelly medium permits a definitive diagnosis. Positive cultures have been obtained only early in the illness, primarily from biopsy samples (Berger *et al.*, 1992) of EM lesions, less often from plasma samples (Wormser *et al.*, 2000a), and only occasionally from cerebrospinal fluid samples in patients with meningitis (Coyle *et al.*, 1999).

Later in the infection, PCR testing is greatly superior to culture in the detection of *B. burgdorferi* in joint fluid (Nocton *et al.*, 1994). *Borrelia burgdorferi* has not been isolated from the cerebrospinal fluid of patients with chronic neuroborreliosis, and *B. burgdorferi* DNA has been detected in CSF samples in only a small number of such patients (Nocton *et al.*, 1996). The Lyme urine antigen test, which has given unreliable results (Klempner *et al.*, 2001), should not be used to support the diagnosis of Lyme disease.

Immunological Tests

Immune Response

In order to understand the difficulties involved in the interpretation of the serological tests for Lyme disease, it is important to understand the kinetics of the immune response to this organism. Infected individuals develop an early and vigorous T-cell response to *B. burgdorferi* (Dattwyler *et al.*, 1988b); the B cell response evolves more slowly, frequently over the course of months. IgM responses against individual antigens appear first, followed by IgG response. Within 3 or 4 weeks after the onset of infection a rise in IgM against one or more spiro-chaetal antigens can be detected in most individuals. The IgM response usually peaks after 6 to 8 weeks and then gradually declines. Early in the course of infection, the IgM response is mainly directed against a 41-kD flagella-associated antigen (Luft, Bosler and Dattwyler, 1992) and outer-surface protein C (OspC). However, epitopes on this 41-kD antigen are cross-reactive with epitopes found on flagella of other borrelias and treponemes. The 41-kD antigen of *B. burgdorferi* has been shown to have a high degree of homology to the 33-kD protein of *Treponema pallidum* (Luft and Dattwyler, 1989). Antibodies against this antigen and 60-kD antigen also does not seem to be specific. Humoral responses to other antigens gradually develop as the disease progresses. Although there are antigenic differences between North America and Europe, antibodies against one or more of the major protein antigens (OspC, 31, 34, 60, or 66-kD) develop as the infection continues. Craft *et al.* (1986) found that, in untreated patients, specific IgM antibodies to the 34-kD OspB antigen developed later in the course of the infection. This could explain also why IgM titers may remain elevated late in the course of the infection. Specific IgG and IgA responses gradually increase during the second and third months of infection and, once established, may remain detectable for years.

Cross Reactions

Some epitopes of antigenic components of *B. burgdorferi*, such as proteins with molecular masses of 41 and 60 kD, are common to *T. pallidum*, oral treponemes, and even *Escherichia coli* (Russell *et al.*, 1984; Magnarelli *et al.*, 1990). False-positive reactions have been reported when sera from patients who were diagnosed with juvenile rheumatoid arthritis, rheumatoid arthritis, systemic lupus erythematous, infectious mononucleosis, and subacute bacterial endocarditis were analyzed for antibodies to *B. burgdorferi* (Zemel, 1992; Kaell *et al.*, 1993).

Currently Used Assays

Serologic assays for Lyme disease were first used in 1983. Sero-diagnostic tests are insensitive during the first several weeks of infection. In the United States, approximately 20–30% of patients have positive responses, usually of the IgM isotype, during this period (Dressler *et al.*, 1993; Engstrom, Shoop and Johnson, 1995), but by convalescence 2 to 4 weeks later, about 70–80% have seroreactivity, even after antibiotic treatment. After 1 month, the majority of patients with active infection have IgG antibody responses (Dressler *et al.*, 1993). In persons who have been ill for longer than 1 month, a positive IgM test alone is likely to represent a false-positive result; therefore, such a response should not be used to support the diagnosis after the first month of infection. In patients with acute neuroborreliosis, especially those with meningitis, the intrathecal production of IgM, IgG, or IgA

antibody against *B. burgdorferi* may often be demonstrated by antibody-capture enzyme immunoassay (Steere *et al.*, 1990), but this test is less often positive in those with chronic neuroborreliosis.

Immunofluorescence and ELISA are the two most commonly used methods for the detection of antibodies to *B. burgdorferi*. ELISAs are most sensitive than IFAs and offer the advantage of having the capability of screening large numbers of patient samples more easily. Although potentially useful and constantly improving, *B. burgdorferi* antibody tests have limited sensitivity (primarily in early disease) and specificity and have not been standardized to date (Steere *et al.*, 1993; Tugwell *et al.*, 1997; Jespersen *et al.*, 2002). These limitations have led to erroneous diagnoses and may have contributed to fundamental misunderstandings of Lyme borreliosis. Therefore, a straightforward two-step serological approach has been proposed (Stanek *et al.*, 1996; CDC, 1997) by CDC in 1994. A positive or equivocal first test, usually an ELISA or indirect immunofluorescence assay (IFA), is followed by an immunoblot test on the same serum sample (Norman *et al.*, 1996), which can detect IgM and IgG antibodies to individual *B. burgdorferi* antigens.

Criteria for Immunoblot Interpretation: Positive IgM blot: presence of two of three significant bands (23-kD(OspC), 39, and 41 kD). Positive IgG blot: presence of five of 10 significant bands: 18 kD, OspC, 28, 30, 39, 41, 45, 58, 66, and 93 kD. If the Western blot is negative, the reactive ELISA or IFA very probably was a false-positive result. Neither ELISA nor immunoblot permits detection of fourfold rises in antibody titer (seroconversion). However, antibodies to *B. burgdorferi* flagella antigens decrease markedly after successful treatment compared with 271 ± 115 SD days after unsuccessful therapy (Panelius *et al.*, 1999). Some individuals lack diagnostic levels of specific antibody in their serum, yet have neurological involvement and diagnostic levels of antibody in their CSF (Dattwyler *et al.*, 1988b; Stiernstedt *et al.*, 1988) as *B. burgdorferi* reaching this immunologically privileged site remain viable and induce a local immune response. The demonstration of intrathecal production of anti-*B. burgdorferi* antibody may be demonstrated by using the formula:

$$\frac{\text{Anti-}B.\ burgdorferi\ \text{titer CSF} \times \text{Serum IgG concentration}}{\text{Anti-}B.\ burgdorferi\ \text{titer serum} \times \text{CSF IgG concentration}}$$

If the ratio is greater than 1, localized production of anti-*Borrelia* antibodies has occurred.

More recently, sensitive ELISAs using recombinant chimeric *Borrelia* proteins of different types, as well as those devoid of OsPA, have been introduced for serological detection of *B. burgdorferi* antibodies (Gomes-Solecki *et al.*, 2000; Gomes-Solecki *et al.*, 2002) which helps distinct between individuals that have been vaccinated with OspA. Furthermore, epitope mapping of the OspA has been introduced by using a panel of monoclonal antibodies (Kharitonenkov *et al.*, 2002). After antibiotic treatment, antibody titers fall slowly, but IgG and even IgM responses may persist for many years after treatment (Kalish *et al.*, 2001). Thus, even an IgM response cannot be interpreted as a demonstration of recent infection or reinfection unless the appropriate clinical characteristics are present.

Other Serological Tests

An ELISA assay has been developed based on a conserved immuno-dominant portion of the variable surface antigen (VlsE) which can be used in vaccinated patients. The *B. burgdorferi* C6 peptide antibody test (Lyme C6 EIA) identifies antibodies to a newly discovered conserved peptide called C6, which is a component of the VlsE of *B. burgdorferi*. The Lyme C6 EIA is important because it detects both IgM and IgG antibodies in patients with Lyme disease but not in vaccines. The Lyme borreliacidal antibody test (BAT) has been proposed as a sensitive and highly specific diagnostic test (Callister *et al.*, 1993). Agger and Case (1997) compared the BAT, indirect immunofluorescence assay (IFA), and ELISA for detection of antibodies to *B. burgdorferi* in 307 patients who had had previous serological tests for Lyme disease. The BAT was less sensitive in active Lyme disease (11%) and did not correlate with the conventional antibody assays.

Schutzer and colleagues have utilized *Borrelia*-specific immunoassays to detect antibody and antigen in patients with active Lyme disease. Brunner and Sigal (2000) have recently confirmed these results and reported that the test can demonstrate *B. burgdorferi* antigen in otherwise seronegative patients. Of 168 patients who fulfilled the CDC criteria for Lyme disease, Lyme immune complexes were found in 25 of 26 patients with early disease, 13 of 13 patients with neuroborreliosis and in 105 of 107 patients who were seropositive and had clinically defined EM. In only two of 147 control patients were Lyme immune complexes, and IgM in immune complexes may be a marker of active infection (Brunner and Sigal, 2000).

MOLECULAR DIAGNOSIS

Specific PCR methods of diagnosis have been described for Lyme disease in skin biopsy. Although possessing the potential to provide a more rapid diagnosis some studies suggest that it is not more sensitive than culture *Borrelia* because the organisms appear to be unevenly distributed throughout biopsy (Picken *et al.*, 1997a). However, others suggest that the sensitivity is superior (Lebech, 2002). The most frequent and serious manifestation of disseminated Lyme borreliosis is neuroborreliosis. In patients with very early neuroborreliosis (<2 weeks), still being negative for specific intrathecal antibody synthesis, a positive PCR was more frequent than in patients with longer disease duration. PCR can be used as a diagnostic aid in these patients. However, in general the measurement of specific intrathecal antibody production in patients with neuroborreliosis was superior to PCR (Lebech, 2002). New oligonucleotide primers based on *B. burgdorferi* sensu lato 16S rRNA gene sequences have been designed for the detection of the bacteria in blood, cerebrospinal, and synovial fluid specimens with PCR. The procedure has been shown to be a rapid and sensitive method for specimens derived from patients with the clinical symptoms of Lyme borreliosis. It can be utilized for both basic research and routine laboratory diagnosis (Chmielewski *et al.*, 2003).

TREATMENT

The primary goals of therapy for Lyme disease are the control of inflammation and the eradication of the infection. Lyme disease is most responsive to antibiotics early in the course of the disease. In one report, the leading reason for failure to respond to antibiotic therapy for Lyme disease was incorrect diagnosis (Sigal, 1990). About 10% of patients with early Lyme disease experience a Jarisch–Herxheimer-like reaction (higher fever, redder rash, or greater pain) during the first 24 h of antibiotic therapy (Steere, 1989). If antibiotic therapy is initiated early in the course of Lyme disease, EM typically resolves promptly and later stage disease is prevented (Steere *et al.*, 1983c; Dattwyler *et al.*, 1990). Early localized infection, limited to a single skin lesion, with mild or no systemic symptoms, is uniformly responsive to short-course oral antibiotic therapy with a number of agents. Of the antibiotics studied to date, amoxicillin 500 mg daily for 8 h, doxycycline 100 mg daily for 12 h, or cefuroxsime axetil 500 mg daily for 12 h have been the most effective for this stage of disease (Massarotti *et al.*, 1992; Nadelman *et al.*, 1992). In a prospective trial, 72 adults with EM were randomized to treatment with amoxicillin plus probenecid or doxycycline (Luft *et al.*, 1996). Both drugs were equally effective; mild fatigue and arthralgia were the only post-treatment complaints, and they resolved within 6 months. No patient required further antimicrobial therapy for Lyme disease. An advantage of doxycycline is its efficacy against the agent of human

granulocytic ehrlichiosis, a possible coinfecting agent. Amoxicillin should be used in children and pregnant women. The optimal duration of therapy is addressed in a randomized, double-blind trial in 180 patients with EM compared 10 days of doxycycline to one dose of intravenous ceftriaxone followed by 10 days of doxycycline. There was no difference among the groups in outcome at the follow-up at 3, 12, and 30 months (Nadelman *et al.*, 1992). Another study suggested that 14 days of antibiotic therapy is adequate for the treatment of EM (Nadelman *et al.*, 1997). In multicenter studies of patients with EM, similar results were obtained with doxycycline, amoxicillin, and cefuroxime axetil, and more than 90% of patients had satisfactory outcomes (Dattwyler *et al.*, 1990; Nadelman *et al.*, 1992). Intravenous ceftriaxone, although effective, was not superior to oral agents as long as the patient did not have objective evidence of neurologic involvement in early disseminated disease with multiple EM lesions and arthralgias (Dattwyler *et al.*, 1997). The first-generation cephalosporins were ineffective (Nadelman *et al.*, 1992).

Although carditis resolves spontaneously, in patients with atrioventricular nodal block with a PR interval greater than 0.3 s, therapy with one of the intravenous regimens for at least part of the course and cardiac monitoring are recommended, but the insertion of a permanent pacemaker is not necessary (Lopez *et al.*, 1991). Facial palsy itself resolves completely or nearly completely in nearly all patients (121 of 122 patients in one series) (Clark *et al.*, 1985). Patients who have facial palsy should undergo a careful neurologic evaluation, including a CSF examination. If facial palsy is the only clinical abnormality and CSF is normal, current practice is to administer oral antibiotics for 21–30 days, a practice that has resulted in favorable outcomes. In one small series, intravenous ceftriaxone for 14 days was superior to a 10-day course of penicillin (Dattwyler *et al.*, 1988a). Most experts prefer a 30-day course of treatment, however, because of the occasional occurrence of late neurologic relapses after shorter courses of therapy. It is not necessary to document clearing of all CSF abnormalities before discontinuation of therapy because clearing of inflammation may lag behind bacteriologic cure. Experience with doxycycline is limited (Dotevall *et al.*, 1988), one patient who had severe neurologic involvement, unresponsive to penicillin responded favorably to chloramphenicol (Diringer, Halperin and Dattwyler, 1987) Arthritis may respond to either oral (Steere *et al.*, 1994) or parenteral (Steere *et al.*, 1985; Roberts *et al.*, 1995) antibiotic therapy, but antibiotic failures occur with either approach. Despite treatment with either oral or intravenous antibiotic therapy, about 10% of patients in the United States have persistent joint inflammation for months or even several years after 2 months or more of oral or 1 month or more of intravenous antibiotic therapy (Steere *et al.*, 1994). If patients have persistent arthritis despite this treatment and if the results of PCR testing of joint fluid are negative, they may be treated with anti-inflammatory agents or arthroscopic synovectomy. In one report, only one-half of patients who had late neurologic symptoms showed either resolution or sustained improvement after 6 months of follow-up after a 2-week course of ceftriaxone (Logigian, Kaplan and Steere, 1990). Those who did not respond, however, did not show progressive worsening. Long-term follow-up of this patient group is essential. In a European trial ceftriaxone and penicillin for 2 weeks were both effective for later neurologic involvement (Hassler *et al.*, 1990). *Borrelia burgdorferi* has not been linked statistically to congenital anomalies, and no increased risk of an adverse outcome of pregnancy has been associated with asymptomatic seropositivity (Williams *et al.*, 1988) or history of previous Lyme disease. It is appropriate to maintain a lower threshold for institution of aggressive antibiotic therapy for suspected Lyme disease during pregnancy, but women should be reassured that no cases of fetal Lyme disease have occurred with currently recommended antibiotic regimen.

After appropriately treated Lyme disease, a small percentage of patients continue to have subjective symptoms – primarily musculoskeletal pain, neurocognitive difficulties, or fatigue – that may last for years. This disabling syndrome, which is sometimes called 'chronic Lyme disease' is similar to chronic fatigue syndrome or fibromyalgia (Dinerman and Steere, 1992). However, in a large study, the frequency of symptoms of pain and fatigue was no greater in patients who had had Lyme disease than in age-matched subjects who had not had this infection (Seltzer *et al.*, 2000). In a study of patients with post-Lyme disease syndrome who received either intravenous ceftriaxone for 30 days followed by oral doxycycline for 60 days or intravenous and oral placebo preparations for the same duration, there were no significant differences between the groups in the percentage of patients who said that their symptoms had improved, gotten worse, or stayed the same (Klempner *et al.*, 2001). Treatment with the standard antibiotics is generally successful for each of the stages of Lyme disease. Some cases of reinfection with a different strain of *B. burgdorferi* have been documented (Asch *et al.*, 1994; Shadick *et al.*, 1999). In one retrospective cohort study form Nantucket, patients with a history of treated Lyme disease were more likely than controls to report subjective complaints without objective evidence (Krupp *et al.*, 2003). There is no evidence that treatment of asymptomatic seropositive individuals is beneficial.

PREVENTION AND CONTROL

Primary prevention strategies will help reduce Lyme disease cases, and some strategies may also prevent other tick-borne illnesses, including babesiosis and granulocytic ehrlichiosis in the United States and tick-borne encephalitis in Europe. The first line of defense is avoidance of tick-infested habitats, use of personal protective measures (e.g. repellents and protective clothing), and checking for and removing attached ticks, and modifications of landscapes in or near residential areas (CDC, 1999). Surveys in such areas have indicated that less than 54% of people tuck their pants into their socks to prevent Lyme disease, 38–79% use repellents, but 79–93% check themselves for ticks (Herrington *et al.*, 1997). Tick control (burning or removing vegetation, acaricide use, and deer elimination) reduces *I. scapularis* populations by up to 94%, and acaricide application to wildlife decreases nymphal *I. scapularis* populations by up to 83%. The effect of these strategies on incidence of Lyme disease in humans is unknown. Studies show that only 40–50% of adults take precautions against tick bites even when they are aware of Lyme disease (Herrington *et al.*, 1997).

In addition, a vaccine for Lyme disease (LYMErix), consisting of recombinant outer-surface protein A (rOspA) for persons aged 15–70 years in the United States, was commercially available in 1998. Results of a large-scale, randomized, controlled (Phase III) trial of safety and efficacy of LYMErix in disease-endemic areas of the northeastern and northcentral United States indicate that the vaccine is safe with an efficacy rate of 76% when administered on a three-dose schedule of 0, 1, and 12 months. The efficacy of the vaccine in preventing asymptomatic infection was 83% in the first year and 100% in the second year (Steere *et al.*, 1998). But Shadick *et al.* (2001) demonstrated that the vaccination against Lyme disease appears only to be economically attractive for individuals who have a seasonal probability of *B. burgdorferi* infection of greater than 1%. Researchers have proposed that an autoimmune reaction might develop within the joints of some Lyme arthritis patients as a result of molecular mimicry between the dominant T-cell epitope of OspA and human leukocyte function-associated antigen 1. Vaccine against Lyme disease was withdrawn from the market in February 2002 because of low sales. The other controversial issue regarding prevention is the prophylactic treatment of Ixodid tick bite with antibiotics.

In 1992, Shapiro *et al.* (1992) demonstrated that even in an area in which Lyme disease is endemic, the risk of infection with *B. burgdorferi* after a recognized deer-tick bit is so low that prophylactic antimicrobial treatment is not routinely indicated. According to Magid *et al.* (1992) empirical treatment of patients with tick bites is indicated when the probability of *B. burgdorferi* infection after a bite is 0.036 or

higher, and this treatment may be preferred when the infection ranges from 0.01 to 0.035. When the probability of infection after a tick bite is less than 0.01, empirical therapy is not warranted. A recent randomized clinical trial showed that a single 200-mg dose of doxycycline administered within 72 h after a recognized *I. scapularis* bite had an efficacy of 87% (95% confidence interval, 25–98%) in preventing EM (Nadelman *et al.*, 2001). None of the people treated with antibiotics in this or any of the previous clinical trials had asymptomatic *B. burgdorferi* infection or any late manifestations of Lyme disease during follow-up periods that ranged from 6 weeks to 3 years.

Many clinicians argue that this practice is unwarranted any may promote the development of antibiotic resistance (Shapiro, 2001). A meta-analysis of several prospective, randomized trials did not indicate that antimicrobial prophylaxis is effective (Wormser *et al.*, 2000b). The cooperative engagement of public health officials, clinicians, and residents of the affected communities will make it possible to reduce the incidence of Lyme disease.

PATHOGENESIS OF RELAPSING FEVER

Borrelia infect the blood stream during infection when large numbers of organisms can be present up to 100000/ml. As a result of the immune pressure against the surface protein variable major protein the organism undergoes antigenic variation mediated by intermolecular and intramolecular recombination. This allows further rounds of bacterial multiplication to occur and clinical relapse. Up to 30 different antigenic variants can be produced but as time progresses the organism reverts to earlier types and this may allow eradication of disease.

CLINICAL MANIFESTATIONS OF RELAPSING FEVER

The clinical features of relapsing fever are similar but the case fatality rate is very different being 4–40% for louse borne disease and less that 5% for tick-borne infection. The initial febrile stage lasts for just under a week stopping abruptly. Subsequent febrile relapses last for 2 or 3 days with an interval between relapses of just over a week. Most patients suffer only a few relapses but more than 12 may occur. Louse-borne relapsing fever is usually associated with a single relapse, whereas multiple relapses are the rule with tick-borne disease. Clinically, the patient may have a cough and examination is likely to reveal hepatosplenomegaly. Hemorrhage is common but is rarely severe. Myocarditis with its associated arrhythmias is the most common cause of death followed by cerebral hemorrhage and hepatic failure.

DIAGNOSIS

The diagnosis of relapsing infection is usually established on epidemiological grounds during an epidemic. In sporadic cases the differential diagnosis includes malaria, typhoid, leptospirosis, rat bite fever, and dengue. The definitive diagnosis is established by detecting *Borrelia* in the peripheral blood of febrile patients. This is achieved by examination of wet blood films or darkground microscopy. Organisms are rarely seen in patient specimens during non-febrile patients. Serodiagnosis detecting the present of specific antibodies to a glycerophosphodiester phosphodiesterase (Porcella *et al.*, 2000).

REFERENCES

Aberer E, Klade H, and Hobisch G (1991) A clinical, histological, and immunohistochemical comparison of acrodermatitis chronica atrophicans and morphea. Am J Dermatopathol 13:34–41.
Aberer E, Koszik F, and Silberer M (1997) Why is chronic Lyme borreliosis chronic? Clin Infect Dis 25:S64–70.

Agger WA and Case KL (1997) Clinical comparison of borreliacidal antibody test with indirect immunofluorescence and enzyme-linked immunosorbent assays for diagnosis of Lyme disease. Mayo Clin Proc 72:510–4.

Akin E, McHugh GL, Flavell RA et al. (1999b) The immunoglobulin (IgG) antibody response to OspA and OspB correlates with severe and prolonged Lyme arthritis and the IgG response to P35 correlates with mild and brief arthritis. Infect Immun 67:173–81.

Anzola J, Luft BJ, Gorgone G et al. (1992) Borrelia burgdorferi HSP70 homolog: Characterization of an immunoreactive stress protein. Infect Immun 60:3704.

Asbrink E and Hovmark A (1985) Successful cultivation of spirochetes from skin lesions of patients with erythema chronicum migrans Afzelius and acrodermatitis chronica atrophicans. Acta Pathol Microbiol Immunol Scand [B] 93:161–3.

Asbrink E and Hovmark A (1988) Early and late cutaneous manifestations in Ixodes-borne borreliosis. Ann N Y Acad Sci 539:4–15.

Asbrink E, Hovmark A, and Olsson I (1986) Clinical manifestations of acrodermatitis chronica atrophicans in 50 Swedish patients. Zentralbl Bakteriol Mikrobiol Hyg [A] 263:253–61.

Asch ES, Bujak DI, Weiss M et al. (1994) Lyme disease: An infectious and postinfectious syndrome. J Rheumatol 21:454–61.

Atlas E, Novak S, Duray PH et al. (1988) Lyme myositis: muscle invasion by Borrelia burgdorferi. Ann Intern Med 109:245–6.

Balmelli T and Piffaretti JC (1995) Association between different clinical manifestations of Lyme disease and different species of Borrelia burgdorferi sensu lato. Res Microbiol 146:329–40.

Balmelli T and Piffaretti JC (1996) Analysis of the genetic polymorphism of Borrelia burgdorferi sensu lato by multilocus enzyme electrophoresis. Int J Syst Bacteriol 46:167–72.

Baranton G, Marti Ras N, and Postic D (1998) Molecular epidemiology of the etiological agents of Lyme borreliosis. Wien Klin Wochenschr 110:850–5.

Baranton G, Postic D, Saint Girons I et al. (1992) Delineation of Borrelia burgdorferi sensu stricto, Borrelia garinii sp. nov., and group VS 461 associated with Lyme borreliosis. Int J Syst Bacteriol 42:378–83.

Baranton G, Seionost G, Theodore D et al. (2001) Distinct levels of genetic diversity of Borrelia burgdorferi are associated with different aspects of pathogenicity. Res Microbiol 154:149–56.

Barbour AG (1988) Plasmad análisis of Borrelia burgdorferi, the Lyme disease agent. J Clin Microbiol 26:475–8.

Barbour AG and Hayes SF (1986) Biology of Borrelia species. Microbiol Rev 50:381–400.

Barr WB, Rastogi R, Ravdin L, and Hilton E (1999) Relations among indexes of memory disturbance and depression in patients with Lyme borreliosis. Appl Neuropsychol 6:12–8.

Barthold SW, Beck DS, Hansen GM et al. (1990) Lyme borreliosis in selected strains and ages of laboratory mice. J Infect Dis 162:133–8.

Barthold SW, Feng S, Bockenstedt LK et al. (1997) Protective and arthritis-resolving activity in sera of mice infected with Borrelia burgdorferi. Clin Infect Dis 25(Suppl. 1):S9–17.

Belongia EA, Reed KD, Mitchell PD et al. (1999) Clinical and epidemiological features of early Lyme disease and human granulocytic ehrlichiosis in Wisconsin. Clin Infect Dis 29:1472–7.

Bensch J, Olsen P, and Hagberg L (1987) Destructive chronic borrelia meningoencephalitis in a child untreated for 15 years. Scand J Infect Dis 19:667–700.

Berger BW, Johnson RC, Kodner C, and Coleman L (1992) Cultivation of Borrelia burgdorferi from erythema migrans lesions and perilesional skin. J Clin Microbiol 30:359–61.

Beutler B and Munford RS (1996) Tumor necrosis factor and the Jarisch–Herxheimer reaction. N Engl J Med 335:347.

Boerlin P, Peter O, Bretz AG et al. (1992) Population genetic analysis of Borrelia burgdorferi isolates by multilocus enzyme electrophoresis. Infect Immun 60:1677–83.

Bosler EM, Coleman JL, Benach JL et al. (1983) Natural distribution of the Ixodes dammini spirochetes. Science 220:321.

Broderick JP, Sandok B, and Mertz LE (1987) Focal encephalitis in a young woman 6 years after the onset of Lyme disease: Tertiary Lyme disease? Mayo Clin Proc 62:313–6.

Brown RN and Lane RS (1992) Lyme disease in California: A novel enzootic transmission cycle of Borrelia burgdorferi. Science 256:1439–42.

Brunner M and Sigal LH (2000) Immune complexes from serum of patients with Lyme disease contain Borrelia burgdorferi antigen and antigen-specific antibodies. Potential use for improved testing. J Infect Dis 182:534–9.

Bujak DI, Weinstein A, and Dornbush RL (1996) Clinical and neurocognitive features of the post-Lyme syndrome. J Rheumatol 23:1393–397.

Burgdorfer W, Barbour AG, Hayes S et al. (1982) Lyme disease: a tick-borne spirochetosis? Science 216:1317–9.

CDC (1990) Case definitions for public health surveillance. MMWR Morb Mortal Wkly Rep 39:1–43.

CDC (1995) Recommendations for test performance and interpretation from the Second National Conference on Serologic Diagnosis of Lyme Disease. MMWR Morb Mortal Wkly Rep 44:590–1.

CDC (1997) Case definitions for infectious conditions under public health surveillance: Lyme disease. MMWR Morb Mortal Wkly Rep 46:20–1.

CDC (1999) Recommendations for the use of Lyme disease vaccine. (1999) Recommendations of the Advisory Committee on Immunization Practices (ACIP). MMWR Morb Mortal Wkly Rep 48 (1–17):21. [Erratum, MMWR Morb Mortal Wkly Rep 1999; 48: 833.]

CDC (2001) Lyme Disease – United States, 1999. Morbidity Mortality Weekly Report 50:181–4.

CDC (2004) Lyme disease – United States, 2001–2002. Morbidity and Mortality Weekly Report. (2004) 53(17):365–9.

Callister SM, Schell RF, Case KL et al. (1993) Characterization of the borreliacidal antibody response to Borrelia burgdorferi in humans: a serodiagnostic test. J Infect Dis 167:158–64.

Casjens S, Palmer N, van Vugt R et al. (2000) A bacterial genome in flux: the twelve linear and nine circular extrachromosomal DNAs in an infectious isolate of the Lyme disease spirochete Borrelia burgdorferi. Mol Microbiol 35:490–516.

Chen J, Field JA, Glickstein L et al. (1999) Association of antibiotic treatment-resistant Lyme arthritis with T cell responses to dominant epitopes of outer surface protein A of Borrelia burgdorferi. Arthritis Rheum 42:1813–22.

Chmielewski T, Fiett J, Gniadkowski M, and Tylewska-Wierzbanowska S (2003) Improvement in the laboratory recognition of Lyme borreliosis with the combination of culture and PCR methods. Mol Diagn 7:155–62.

Clark JR, Carlson R, Sasaki CT et al. (1985) Facial paralysis in Lyme disease. Laryngoscope 95:1341–5.

Coburn J, Leong JM, and Erban JK (1993) Integrin $\alpha_{IIb}\beta_3$ mediates binding of the Lyme disease agent Borrelia burgdorferi to human platelets. Proc Natl Acad Sci USA 90:7059–63.

Coleman JL and Benach JL (1989) Identification and characterization of an endoflagellar antigen of Borrelia burgdorferi. J Clin Invest 84:322.

Coleman JL, Gebbia JA, Piesman J et al. (1997) Plasminogen is required for efficient dissemination of B. burgdorferi in ticks and for enhancement of spirochetemia in mice. Cell 89:1111–9.

Coyle PK, Goodman JL, Krupp LB et al. (1999) Lyme Disease: Continuum: Lifelong Learning in Neurology, Vol. 5. No. 4. Part A. Philadelphia, Lippincott: Williams & Wilkins.

Craft JE, Fisher DK, Shimamoto GT et al. (1986) Antigens of Borrelia burgdorferi recognized during Lyme disease: appearance of an immunoglobulin in response and expansion of immunoglobulin G response late in the illness. J Clin Invest 78:934.

Dattwyler RJ, Luft BJ, Kunkel MJ et al. (1997) Ceftriaxone compared with doxycycline for the treatment of acute disseminated Lyme disease. N Engl J Med 337:289–94.

Dattwyler RJ, Volkman DJ, Connaty SM et al. (1990) Amoxicillin plus probenecid versus doxycycline for treatment of erythema migrans borreliosis. Lancet 336:1404–6.

Dattwyler RJ, Volkman DJ, Halperin JJ et al. (1988b) Specific immune responses in Lyme borreliosis: characterization of T cell and B cell responses to Borrelia burgdorferi. Ann N Y Acad Sci 539:93–102.

Dattwyler RJ, Volkman DJ, Halperin JJ, and Luft BJ (1988a) Treatment of late Lyme borreliosis – randomized comparison of ceftriaxone and penicillin. Lancet 331:1191–4.

deLuise VP and O'Leary MJ (1991) Peripheral ulcerative keratitis related to Lyme disease. (Letter) Am J Ophthalmol 111:244–5.

de Silva AM and Fikrig E (1997) Arthropod- and host-specific gene expression by Borrelia burgdorferi. J Clin Invest 99:377–9.

Dinerman H and Steere AC (1992) Lyme disease associated with fibromyalgia. Ann Intern Med 117:281–5.

Diringer MN, Halperin JJ, and Dattwyler RJ (1987) Lyme meningoencephalitis: report of a severe, penicillin-resistant case. Arthritis Rheum 30:705–8.

Dotevall L, Alestig K, Hanner P et al. (1988) The use of doxycycline in nervous system Borrelia burgdorferi infection. Scand J Infect Dis 53:74–9.

Dressler F, Whalen JA, Reinhardt BN, and Steere AC (1993) Western blotting in the serodiagnosis of Lyme disease. J Infect Dis 167:392–400.

Duray PH (1989) Histopathology of clinical phases of human Lyme disease. Rheum Dis Clin North Am 15:691–710.

Dworkin MS, Anderson DE Jr, Schwan TG et al. (1998) Tick-borne relapsing fever in the northwestern United States and southwestern Canada. Clin Infect Dis 26:122.

Dykhuizen DE, Polin DS, Dunn JJ et al. (1993) Borrelia burgdorferi is clonal: Implications for taxonomy and vaccine development. Proc Natl Acad Sci USA 90:10163–7.

Edlinger E, Rodhain F, and Perez C (1985) Lyme disease in patients previously suspected of arbovirus infection. Lancet ii:93.

Engstrom SM, Shoop E, and Johnson RC (1995) Immunoblot interpretation criteria for serodiagnosis of early Lyme disease. J Clin Microbiol 33:419–27.

Fahrer H, van der Linden S, Sauvain MJ et al. (1991) The prevalence and incidence of clinical and asymptomatic Lyme borreliosis in a population at risk. J Infect Dis 163:305.

Fallon BA, Nields JA, Burrascano JJ et al. (1992) The neuropsychiatric manifestations of Lyme borreliosis. Psychiatr Q 63:95–117.

Farris BK and Webb RM (1988) Lyme disease and optic neuritis. J Clin Neuroophthalmol 8:73–8.

Feder HM Jr, Zalneraitis E, and Reik L Jr (1988) Lyme disease: acute focal meningoencephalitis in a child. Pediatrics 82:931–4.

Feng S, Hodzic E, and Barthold SW (2000) Lyme arthritis resolution with antiserum to a 37-kilodalton Borrelia burgdorferi protein. Infect Immun 68:4169–73.

Finizia C, Jonsson R, and Hanner P (2001) Serum and cerebrospinal fluid pathology in patients with sudden sensorineural hearing loss. Acta Otolaryngol 121:823–30.

Fraser CM, Casjens S, Huang WM et al. (1997) Genomic sequence of a Lyme disease spirochete, Borrelia burgdorferi. Nature 390:580–6.

Galloway RE, Levin J, Butler T et al. (1977) Activation of protein mediators of inflammation and evidence for endotoxemia in Borrelia recurrentis infection. Am J Med 63:933–8.

Gasser R, Dusleag J, Reisinger E et al. (1992) Reversal by ceftriaxone of dilated cardiomyopathy Borrelia burgdorferi infection. Lancet 339:1174–5.

Gaudino EA, Coyle PK, and Krupp LB (1997) Post-Lyme syndrome and chronic fatigue syndrome. Arch Neurol 54:1372–6.

Gomes-Solecki MJ, Dunn JJ, Luft BJ et al. (2000) Recombinant chimeric Borrelia proteins for diagnosis of Lyme disease. J Clin Microbiol 38:2530–5.

Gomes-Solecki MJ, Wormser GP, Schriefer M et al. (2002) Recombinant assay for serodiagnosis of Lyme disease regardless of OspA vaccination status. J Clin Microbiol 40:193–7.

Gross DM, Forsthuber T, Tary-Lehmann M et al. (1998) Identification of LFA-1 as a candidate autoantigen in treatment-resistant Lyme arthritis. Science 281:703–6.

Grunwaldt E, Barbour AG, and Benach JL (1983) Simultaneous occurrence of babesiosis and Lyme disease. N Engl J Med 308:1166.

Guo BP, Brown EL, Dorward DW et al. (1998) Decorin-binding adhesins from Borrelia burgdorferi. Mol Microbiol 30:711–23.

Gustafson R, Svenungsson B, Forsgren M et al. (1992) Two-year survey of the incidence of Lyme borreliosis and tick-borne encephalitis in a high-risk population in Sweden. Eur J Clin Microbiol Infect Dis 11:894.

Guttman DS, Wang PW, Wang I-N et al. (1996) Multiple infections of Ixodes scapularis ticks by Borrelia burgdoferi as revealed by single-strand conformation polymorphism analysis. J Clin Microbiol 34:652–6.

Halperin JJ and Golightly M (1992) Lyme borreliosis in Bell's palsy. Long Island Neuroborreliosis Collaborative Study Group Neurology 42(7):1268–70.

Halperin JJ, Little BW, and Coyle PK et al. (1987) Lyme disease: Cause of a treatable peripheral neuropathy. Neurology 37:1700–6.

Halperin JJ, Pass HL, Anand AK et al. (1988) Nervous system abnormalities in Lyme disease. Ann N Y Acad Sci 539:24–34.

Hansen K and Lebech A-M (1992) The clinical and epidemiologic profile of Lyme neuroborreliosis in Denmark 1985–1990. Brain 115:399.

Hansen K, Bangsborg JM, and Fjordvang H et al. (1988) Immunochemical characterization of and isolation of the gene for a Borrelia burgdorferi immunodominant 60-kilodalton antigen common to a wide range of bacteria. Infect Immun 56:2047.

Hassler D and Maiwald M (1994) Reinfection with Borrelia burgdorferi in an immunocompetent patient. Dtsch Med Wochenschr 119:338–42.

Hassler D, Zoller L, Haude M et al. (1990) Cefotaxime versus penicillin in the late stage of Lyme disease-prospective, randomized therapeutic study. Infection 18:16–20.

Hemmer B, Gran B, Zhao Y et al. (1999) Identification of candidate T-cell epitopes and molecular mimics in chronic Lyme disease. Nat Med 5:1375–82.

Henriksson A, Link H, Cruz M et al. (1986) Immunoglobulin abnormalities in CSF and blood over the course of lymphocytic meningoradiculitis (Bannwarth's syndrome). Ann Neurol 20:337–45.

Herrington JE Jr, Campbell GL, Bailey RE et al. (1997) Predisposing factors for individuals' Lyme disease prevention practices: Connecticut, Maine, Montana. Am J Public Health 87:2035–8.

Herzer P, Wilske B, Preac-Mursic V et al. (1986) Lyme arthritis: Clinical features, serological, and radiographic findings of cases in Germany. Klin Wochenschr 64:206–15.

Hirschfeld M, Kirschning CJ, Schwander R et al. (1999) Inflammatory signaling by Borrelia burgdorferi lipoproteins is mediated by toll-like receptor 2. J Immunol 163:2382–6.

Hopf HC (1975) Peripheral neuropathy in acrodermatitis chronica atrophicans (Herxheimer). J Neurol Neurosurg Psychiatry 38:452–8.

Horton JM and Blaser MJ (1985) The spectrum of relapsing fever in the Rocky Mountains. Arch Intern Med 145:871–5.

Hovmark A, Asbrink E, and Olsson I (1986) The spirochetal etiology of lymphadenosis benigna cutis solitaria. Acta Derm Venereol 66:479–84.

Jespersen DJ, Smith TF, Rosenblatt JE, and Cockerill FR III (2002) Comparison of the Borrelia DotBlot G, MarDx, and VIDAS enzyme immunoassays for detecting immunoglobulin G antibodies to Borrelia burgdorferi in human serum. J Clin Microbiol 40:4782–4.

Johnson RC, Marek N, and Kodner C (1984) Infection of Syrian hamsters with Lyme disease spirochetes. J Clin Microbiol 20:1099–101.

Kaell AT, Redecha PR, and Elkon KB et al. (1993) Occurrence of antibodies to Borrelia burgdorferi in patients with nonspirochetal subacute bacterial endocarditis. Ann Intern Med 119:1079–83.

Kalish RA, Kaplan RF, Taylor E et al. (2001) Evaluation of study patients with Lyme disease, 10–20-year follow-up. J Infect Dis 183:453–60.

Kalish RA, Leong JM, and Steere AC (1993) Association of treatment-resistant chronic Lyme arthritis with HLA-DR4 and antibody reactivity to OspA and OspB of Borrelia burgdorferi. Infect Immun 61:2774–9.

Kaplan RF, Jones-Woodward L, Workman K et al. (1999) Neuropsychological deficits in Lyme disease patients with and without other evidence of central nervous system pathology. Appl Neuropsychol 6:3–11.

Kaplan RF, Meadows ME, Vincent LC et al. (1992) Memory impairment and depression in patients with Lyme encephalopathy: comparison with fibromyalgia and nonpsychotically depressed patients. Neurology 42:1263–7.

Kawabata H, Masuzawa T, and Yanagihara Y (1993) Genomic analysis of Borrelia japonica sp. nov isolated from Ixodes ovatus in Japan. Microbiol Immunol 37:843–8.

Kazakoff MA, Sinusas K, and Macchia C (1993) Liver function test abnormalities in early Lyme disease. Arch Fam Med 2:409–13.

Keane JR (1994) Bilateral seventh nerve palsy: analysis of 43 cases and review of the literature. Neurology 44(7):1198–202.

Kelleher Doyle M, Telford SR III, Criscione L et al. (1998) Cytokines in murine Lyme carditis: Th1 cytokine expression follows expression of proinflammatory cytokines in a susceptible mouse strain. J Infect Dis 177:242–6.

Kharitonenkov IG, Pomelova VG, Bucher DJ et al. (2002) Epitope mapping of the outer surface protein A (OspA) of the spirochete Borrelia burgdorferi using a panel of monoclonal antibodies and lanthanide competition fluoroimmunoassay. Biochemistry (Mosc) 67:640–50.

Kirmizis D, Efstratiadis G, Economidiour D III et al. (2004) MPGN secondary to Lyme disease. Am J Kidney Dis 43:544–51.

Kirsch M, Ruben FL, Steere AC et al. (1988) Fatal adult respiratory distress syndrome in a patient with Lyme disease. JAMA 259:2737–9.

Klempner MS, Hu LT, Evans J et al. (2001) Two controlled trials of antibiotic treatment in patients with persistent symptoms and a history of Lyme disease. N Engl J Med 345:85–92.

Klempner MS, Schmid C, Hu L et al. (2001) Intralaboratory reliability of serologic and urine testing for Lyme disease. Am J Med 110:217–9.

Kramer N, Rickert RR, Brodkin RH et al. (1986) Septal panniculitis as a manifestation of Lyme disease. Am J Med 81:149–52.

Krause A, Brade V, Schoerner C et al. (1991) T cell proliferation induced by Borrelia burgdorferi in patients with Lyme borreliosis: autologous serum required for optimum stimulation. Arthritis Rheum 34:393–402.

Krause PJ, Telford SR III, Spielman A et al. (1996) Concurrent Lyme disease and babesiosis. Evidence for increased severity and duration of illness. JAMA 275:1657.

Kristoferitsch W (1989) Lyme borreliosis in Europe. Rheum Dis Clin North Am 15:767–74.

Kristoferitsch W, Sluga E, Graf M *et al.* (1988) Neuropathy associated with acrodermatitis crónica atrophicans: Clinical and morphological features. Ann N Y Acad Sci 539:35–45.

Krupp LB, Hyman LG, Grimson R *et al.* (2003) Study and treatment of post-Lyme disease (STOP-LD): a randomized double masked clinical trial. Neurology 60:1923.

Labandeira-Rey M and Skare JT (2001) Decreased infectivity in *Borrelia burgdorferi* strain B31 is associated with loss of either linear plasmid 25 or 28–1. Infect Immun 69:446–55.

Lardieri G, Salvi A, Camerini F *et al.* (1993) Isolation of *Borrelia burgdorferi* from myocardium. Lancet 342:490.

Lastavica CC, Wilson ML, Berardi VP *et al.* (1989) Rapid emergence of a focal epidemic of Lyme disease in coastal Massachusetts. N Engl J Med 320:133.

Le CT (1980) Tickborne relapsing fever in children. Pediatrics 66:963–6.

Lebech AM (2002) Polymerase chain reaction in diagnosis of *Borrelia burgdorferi* infections and studies on taxonomic classification. APMIS (Suppl. 105):1–40.

Liang FT, Nelson FK, and Fikrig E (2002) Molecular adaptation of *Borrelia burgdorferi* in the murine host. J Exp Med 196:275–80.

Logigian EL, Kaplan RF, and Steere AC (1990) Chronic neurologic manifestations of Lyme disease. N Engl J Med 323:1438.

Logigian EL, Kaplan RF, and Steere AC (1999) Successful treatment of Lyme encephalopathy with intravenous ceftriaxone. J Infect Dis 180:377–83.

Lopez AJ, O'Keefe P, Morrissey M, and Pickleman J (1991) Ceftriaxone-induced cholelithiasis. Ann Intern Med 115:712.

Luft BJ and Dattwyler RJ (1989) Lyme borreliosis problems in diagnosis and treatment. Curr Clin Top Infect Dis 11:56–81.

Luft BJ, Bosler EM, and Dattwyler RJ (1992) Diagnosis of Lyme Borreliosis. In: Molecular and Immunologic Approaches. Cold Spring Harbor: Cold Spring Harbor Press, 317–24.

Luft BJ, Dattwyler RJ, Johnson RC *et al.* (1996) Azithromycin compared with amoxicillin in the treatment of erythema migrans: a double-blind, randomized, controlled trial. Ann Intern Med 124:785.

Luft BJ, Mudri S, Jiang W *et al.* (1992) The 93-kilodalton protein of *Borrelia burgdorferi*: An immunodominant protoplasmic cylinder antigen. Infect Immun 60:4309.

Magid D, Schwartz B, Craft J, and Schwartz JS (1992) Prevention of Lyme disease after tick bites. A cost-effectiveness analysis. N Engl J Med 327:534–41.

Magnarelli LA and Anderson JF (1988) Ticks and biting insects infected with the etiologic agent of Lyme disease, *Borrelia burgdorferi*. J Clin Microbiol 26:1482.

Magnarelli LA, Anderson JF, Apperson CS *et al.* (1976) Spirochetes in ticks and antibodies to *Borrelia burgdorferi* in white-tailed deer from Connecticut, New York State, and North Carolina. J Wildl Dis 22:178.

Magnarelli LA, Miller JN, Anderson JF, and Riviere GR (1990) Cross-reactivity of non-specific treponemal antibody in serologic tests for Lyme disease. J Clin Microbiol 28:1276–9.

Manweiler SA, Lane RS, and Tempelis CH (1992) The western fence lizard *Sceloporus occidentalis*: Evidence of field exposure to *Borrelia burgdorferi* in relation to infestation by *Ixodes pacificus* (Acari: Ixodidae). Am J Trop Med Hyg 47:328.

Maraspin V, Lotric-Furlan S, and Strle F (2002) Development of erythema migrans in spite of treatment with antibiotics after a tick bite. Wien Klin Wochenschr 114:616–9.

Marcus LC, Steere AC, and Duray PH (1985) Fatal pancarditis in a patient with coexistent Lyme disease and Babesiosis. Ann Intern Med 103:374–6.

Markowitz LE, Steere AC, Benach JL *et al.* (1986) Lyme disease during pregnancy. JAMA 225:3394–6.

Massarotti EM, Luger SW, Rahn DW *et al.* (1992) Treatment of early Lyme disease. Am J Med 92:396–403.

McAlister HF, Klementowicz PT, Andrews C *et al.* (1989) Lyme carditis: an important cause of reversible heart block. Ann Intern Med 110:339–45.

McDowell JV, Sung SY, Labandeira-Rey M *et al.* (2001) Analysis of mechanisms associated with loss of infectivity of clonal populations of *Borrelia burgdorferi* B31M1. Infect Immun 69:3670–7.

Melski JW (2000) Lyme borreliosis. Semin Cutan Med Surg 19:10–8.

Meyer AL, Trollmo C, Crawford F *et al.* (2000) Direct enumeration of *Borrelia*-reactive CD4 T cells ex vivo by using MHC class II tetramers. Proc Natl Acad Sci USA 97:11433–8.

Mombaerts IM, Maudgal PC, and Knockaert D (1991) Bilateral follicular conjunctivitis as a manifestation of Lyme disease. Am J Ophthalmol 112:96–7.

Moody KD, Barthold S, and Tergwilliger GA (1990) Lyme borreliosis in laboratory animals: effect of host species and in vitro passage of *Borrelia burgdorferi*. Am. J Trop Med Hyg 43:87–92.

Muellegger RR, McHugh G, Ruthazer R *et al.* (2000) Differential expression of cytokine mRNA in skin specimens from patients with erythema migrans or acrodermatitis chronica atrophicans. J Invest Dermatol 115:1115–23.

Nadelman RB and Wormser GP (1998) Lyme borreliosis. Lancet 352:557–65.

Nadelman RB, Horowitz HW, Hsieh TC *et al.* (1997) Simultaneous human granulocytic ehrlichiosis and Lyme borreliosis. N Engl J Med 337:27–30.

Nadelman RB, Luger SW, Frank E *et al.* (1992) Comparison of cefuroxime axetil and doxycycline in the treatment of early Lyme disease. Ann Intern Med 117:273.

Nadelman RB, Nowakowski J, Forseter G *et al.* (1996) The clinical spectrum of early Lyme borreliosis in patients with culture-confirmed erythema migrans. Am J Med 100:502–8.

Nadelman RB, Nowakowski J, Fish D *et al.* (2001) Prophylaxis with single-dose doxycycline for the prevention of Lyme disease after an *Ixodes scapularis* tick bite. N Engl J Med 345:79–84.

Nelson JA and Nemcek AA (1992) Vesicular rash, radicular pain, and splenomegaly in a patient with Lyme borreliosis. Clin Infect Dis 15:180–1.

Nichol G, Dennis DT, Steere AC *et al.* (1998) Test-treatment strategies for patients suspected of having Lyme disease: a cost-effectiveness analysis. Ann Intern Med 128:37–48 (Abstract).

Nocton JJ, Bloom BJ, Rutledge BJ *et al.* (1996) Detection of *Borrelia burgdorferi* DNA by polymerase chain reaction in cerebrospinal fluid in patients with Lyme neuroborreliosis. J Infect Dis 174:623–7.

Nocton JJ, Dressler F, Rutledge BJ *et al.* (1994) Detection of *Borrelia burgdorferi* DNA by polymerase chain reaction in synovial fluid from patients with Lyme arthritis. N Engl J Med 330:229–34.

Norman GL, Antig JM, Bigaignon G, and Hogrefe WR (1996) Serodiagnosis of Lyme borreliosis by *Borrelia burgdorferi* sensu stricto, *B. garinii*, and *B. afzelii* western blots (immunoblots). J Clin Microbiol 34:1732–8.

Norris SJ, Howell JK, Garza SA *et al.* (1995) High- and low-infectivity phenotypes of clonal populations of in vitro- cultured *Borrelia burgdorferi*. Infect Immun 63:2206–12.

O'Connell S, Granstrom M, Gray JS, and Stanek G (1998) Epidemiology of European Lyme borreliosis. Zentralbl Bakteriol 287:229–40.

Oliver JH, Chandler FW, Luttrell MP *et al.* (1993) Isolation and transmission of the Lyme disease spirochete from the southeastern United States. Proc Natl Acad Sci USA 90:7371–5.

Orloski KA, Hayes EB, Campbell GL, and Dennis DT (2000) Surveillance for Lyme disease – United States, 1992–98. MMWR CDC Surveill Summ 49(SS–3):1–11.

Oschmann P, Dorndorf W, Hornig C *et al.* (1998) Stages and syndromes of neuroborreliosis. J Neurol 245:262–72.

Pachner AR and Steere A (1985) The triad of neurologic manifestations of Lyme disease: meningitis, cranial neuritis, and radiculoneuritis. Neurology 35:47–53.

Panelius J, Seppala I, Granlund H *et al.* (1999) Evaluation of treatment responses in late Lyme borreliosis on the basis of antibody decrease during the follow-up period. Eur J Clin Microbiol Infect Dis 18:621–9.

Picken MM, Picken RN, Han D *et al.* (1997a) A two year prospective study to compare culture and polymerase chain reaction amplification for the detection and diagnosis of Lyme borreliosis. Mol Pathol 50(4):186–93.

Picken RN, Strle F, Ruzic-Sabljic E *et al.* (1997b) Molecular subtyping of *Borrelia burgdorferi* sensu lato isolates from five patients with solitary lymphocytoma. J Invest Dermatol 108:92–7.

Porcella SF, Raffel SJ, Schrumpf ME *et al.* (2000) Serodiagnosis of Louse-Borne relapsing fever with glycerophosphodiester phosphodiesterase (GlpQ) from *Borrelia recurrentis*. J Clin Microbiol 38(10):3561–71.

Postic D, Assous MV, Grimont PAD, and Baranton G (1994) Diversity of *Borrelia burgdorferi* sensu lato evidenced by restriction fragment length polymorphism of rrf(5S) ffl(23S) intergenic spacer amplicons. Int J Syst Bacteriol 44:743–52.

Price PW (1980) Evolutionary biology of parasites. Monogr Popul Biol 15:1–237.

Probert WS and Johnson BJB (1998) Identification of a 47 kDa fibronectin-binding protein expressed by *Borrelia burgdorferi* isolate B31. Mol Microbiol 30:1003–15.

Purser JE and Norris SJ (2000) Correlation between plasmid content and infectivity in *Borrelia burgdorferi*. Proc Natl Acad Sci USA 97:13865–70.

Qiu WG, Bosler EM, Campbell JR *et al.* (1997) A population genetic study of *Borrelia burgdorferi* sensu stricto from Eastern Long Island, New York, suggested frequency-dependent selection, gene flow and host adaptation. Hereditas 127:203–16.

Qiu WG, Dykhuizen DE, Acosta MS, and Luft BJ (2002) Geographic uniformity of the Lyme disease spirochete (*Borrelia burgdorferi*) and its shared history with tick vector (*Ixodes scapularis*) in the Northeastern United States. Genetics 160:833–49.

Raucher HS, Kaufman DM, Goldfarb J *et al.* (1985) Pseudotumor cerebri and Lyme disease: a new association. J Pediatr 107:931–3.

Rawlings JA (1995) An overview of tick-borne relapsing fever with emphasis on outbreaks in Texas. Tex Med 91:56.

Reik L, Burgdorfer W, and Donaldson JO (1986) Neurologic abnormalities in Lyme disease without erythema chronicum migrans. Am J Med 12:475–6.

Reimers CD, de Koning J, and Neubert U (1993) *Borrelia burgdorferi* byositis. Report of eight patients. J Neurol 240:278–83.

Roberts ED, Bohm RP, Cogswell FB *et al.* (1995) Chronic Lyme disease in the Rhesus monkey. Lab Invest 72:146–60.

Rodhain F and Edlinger E (1987) Serodiagnostic of erythema chronicum migrans (Lyme disease) in cases initially suspected as caused by arboviruses. Zentralbl Bakteriol Mikrobiol Hyg [A] 263:425–6.

Rousseau JJ, Lust C, Zangerle PF *et al.* (1986) Acute transverse myelitis as presenting neurologic feature of Lyme disease (Letter). Lancet 221:1222–3.

Russell H, Sampson JS, Schmid GP *et al.* (1984) Enzyme linked immunosorbent assay and indirect immunofluorescence assay for Lyme disease. J Infect Dis 149:465–70.

Sanford JP (1976) Relapsing fever-treatment and control. In: Johnson, RC ed. The Biology of Parasitic Spirochetes. New York: Academic Press, 389–94.

Schechter SL (1986) Lyme disease associated with optic neuropathy. Am J Med 81:143–5.

Schlesinger PA, Duray PH, Burke BA *et al.* (1985) Maternal-fetal transmission of the Lyme disease spirochete, *Borrelia burgdorferi*. Ann Intern Med 103:67–9.

Schubert HD (1994) Cytologically proven seronegative Lyme choroiditis and vitritis. Retina 14:39–42.

Schwan TG (1996) Ticks and borrelia: Model systems for investigating pathogen arthropod interactions. Infect Agents Dis 5:167–81.

Seinost G, Dykhuizen DE, Dattywyler RJ *et al.* (1999) Four clones of *Borrelia burgdorferi* sensu stricto cause invasive infection in humans. Infect Immun 67:3518–24.

Seltzer EG, Gerber MA, Cartter ML *et al.* (2000) Long-term outcomes of persons with Lyme disease. JAMA 283:609–11.

Shadick NA, Liang MH, Phillips CB *et al.* (2001) The cost-effectiveness of vaccination against Lyme disease. Arch Intern Med 161:554–61.

Shadick NA, Phillips CB, Logigian EL *et al.* (1994) The long-term clinical outcomes of Lyme disease: a population-based retrospective cohort study. Ann Intern Med 121:560–7.

Shadick NA, Phillips CB, Sangha O *et al.* (1999) Musculoskeletal and neurologic outcomes in patients with previously treated Lyme disease. Ann Intern Med 131:919–26.

Shapiro ED (2001) Doxycycline for tick bites: not for everyone. N Engl J Med 345:133–4.

Shapiro ED, Gerber MA, Holabird NB *et al.* (1992) A controlled trial of antimicrobial prophylaxis for Lyme disease after deer-tick bites. N Engl J Med 327(25):1769–73.

Sigal LH (1990) Summary of the first 100 patients seen at a Lyme disease referral center. Am J Med 88:577–81.

Simpson WJ, Schrumpf ME, and Schwan TG (1990) Reactivity of human Lyme borreliosis sera with a 39-kilodalton antigen specific to *Borrelia burgdorferi*. J Clin Microbiol 28:1329.

Skare JT, Mirzabekov TA, Shang ES *et al.* (1997) The Om66 (p66) protein is a *Borrelia burgdorferi* porin. Infect Immun 65:3654.

Sonnesyn SW, Diehl SC, Johnson RC *et al.* (1995) A prospective study of the seroprevalence of *Borrelia burgdorferi* infection in patients with severe heart failure. Am J Cardiol 76:97–100.

Spielman A (1994) The emergence of Lyme disease and human babesiosis in a changing environment. Ann N Y Acad Sci 740:146–56.

Spielman A, Wilson ML, Levine JF, and Piesman J (1985) Ecology of *Ixodes dammini*-borne human babesiosis and Lyme disease. Annu Rev Entomol 30:439–60.

Stanczak J, Gabre RM, Kruminis-Lozowska W *et al.* (2004) *Ixodes ricinus* as a vector of *Borrelia burgdorferi* sensu lato, *Anaplasma phagocytophilum* and *Babesia microti* in urban and suburban forests. Ann Agric Environ Med 11(1):109–14.

Stanek G, Klein J, Bittner R, and Glogar D (1990) Isolation of *Borrelia burgdorferi* from the myocardium of a patient with longstanding cardiomyopathy. N Engl J Med 322:249–52.

Stanek G, O'Connell S, Cimmino M *et al.* (1996) European Union concerted action on risk assessment in Lyme borreliosis: clinical case definitions for Lyme borreliosis. Wien Klin Wochenschr 108:741–7.

Stanek G, Satz N, Strle F, and Wilske B (1993) Epidemiology of Lyme borreliosis. In: Weber K, Burgdorfer W, eds. Aspects of Lyme Borreliosis. Berlin, Germany: Springer-Verlag, 358–70.

Steere AC (1989) Lyme disease. N Engl J Med 321:586–96.

Steere AC, Bartenhagen NH, Craft JE *et al.* (1983a) The early clinical manifestations of Lyme disease. Ann Intern Med 99:76.

Steere AC, Batsford WP, Weinberg M *et al.* (1980) Lyme carditis: Cardiac abnormalities of Lyme disease. Ann Intern Med 93:8–16.

Steere AC, Berardi VP, Weeks KE *et al.* (1990) Evaluation of the intrathecal antibody response to *Borrelia burgdorferi* as a diagnostic test for Lyme neuroborreliosis. J Infect Dis 161:1203–9.

Steere AC, Duray PH, and Butcher EC (1988) Spirochetal antigens and lymphoid cell surface markers in Lyme synovitis. Comparison with rheumatoid synovium and tonsillar lymphoid tissue. Arthritis Rheum 31:487–95.

Steere AC, Duray PH, Kauffmann DJH *et al.* (1985a) Unilateral blindness caused by infection with the Lyme disease spirochete, *Borrelia burgdorferi*. Ann Intern Med 103:382–4.

Steere AC, Gibofsky A, Pattarroyo ME *et al.* (1979) Chronic Lyme arthritis: clinical and immunogenetic differentiation from rheumatoid arthritis. Ann Intern Med 90:286–91.

Steere AC, Green J, Schoen RT *et al.* (1985b) Successful parenteral antibiotic therapy of established Lyme arthritis. N Engl J Med 312:869–74.

Steere AC, Gross D, Meyer AL, and Huber BT (2001) Autoimmune mechanisms in antibiotic treatment-resistant Lyme arthritis. J Autoimmun 16:263–8.

Steere AC, Hutchinson GJ, Craft J *et al.* (1983b) The early clinical manifestations of Lyme disease. Ann Intern Med 99:76–82.

Steere AC, Hutchinson GJ, Rahn DW *et al.* (1983c) Treatment of the early manifestations of Lyme disease. Ann Intern Med 99:22–6.

Steere AC, Levin RE, and Molloy PJ *et al.* (1994) Treatment of Lyme arthritis. Arthritis Rheum 37:878–88.

Steere AC, Malawista SE, Hardin JA *et al.* (1977a) Erythema chronicum migrans and Lyme arthritis. The enlarging clinical spectrum. Ann Intern Med 86:685–98.

Steere AC, Malawista SE, Snydman DR *et al.* (1977b) Lyme arthritis: an epidemic of oligoarticular arthritis in children and adults in three Connecticut communities. Arthritis Rheum 20:7–17.

Steere AC, Schoen RT, and Taylor E (1987) The clinical evolution of Lyme arthritis. Ann Intern Med 107:725–31.

Steere AC, Sikand VK, Meurice F, and Parenti DL (1998) Vaccination against Lyme disease with recombinant *Borrelia burgdorferi* outer-surface lipoprotein A with adjuvant. N Engl J Med 339:209–15.

Steere AC, Taylor E, McHugh GL, and Logigian EL (1993) The overdiagnosis of Lyme disease. JAMA 269:1812–6.

Steere AC, Taylor E, Wilson ML *et al.* (1986) Longitudinal assessment of the clinical and epidemiologic features of Lyme disease in a defined population. J Infect Dis 154:295.

Stiernstedt G, Gustafsson R, Karlsson M *et al.* (1988) Clinical manifestations and diagnosis of neuroborreliosis. Ann N Y Acad Sci 539:46–55.

Strle F, Pleterski-Rigler D, Stanek G *et al.* (1992) Solitary borrelial lymphocytoma: report of 36 cases. Department of Infectious Diseases, University Medical Centre Ljubljana, Slovenia. Infection 20(4): 201–6.

Telford SR III, Armstrong PM, Katavolos P *et al.* (1997) A new tick-borne encephalitis-like virus infecting New England deer ticks, *Ixodes dammini*. Emerg Infec Dis 3:165–70.

Thiesen M, Borre M, Mathiesen MJ *et al.* (1995) Evolution of the *Borrelia burgdorferi* outer surface protein OspC. J Bacteriol 177:3036–44.

Thiesen M, Frederiksen B, Lebech AM *et al.* (1993) Polymorphism in the OspC gene of *Borrelia burgdorferi* and immunoreactivity of OspC protein: implication for taxonomy and for use of OspC protein as a diagnostic agent. J Clin Microbiol 31:2570–6.

Tugwell P, Dennis DT, Weinstein A *et al.* (1997) Laboratory evaluation in the diagnosis of Lyme disease. Ann Intern Med 127:1109–23.

Valsangiacomo C, Balmelli T, Piffaretti JC *et al.* (1993) Different genospecies of *Borrelia burgdorferi* are associated with distinct clinical manifestations of Lyme borreliosis. Clin Infect Dis 17:708–17.

van Dam AP, Kuiper H, Vos K *et al.* (1991) Differential genospecies of *Borrelia burgdorferi* are associated with distinct clinical manifestations of Lyme diseases. CID 123:603–6.

van Dam AP, Kuiper H, Vos K *et al.* (1993) Different genospecies of *Borrelia burgdorferi* are associated with distinct clinical manifestations of Lyme borreliosis. Clin Infect Dis 17:708–17.

van der Linde MR (1991) Lyme carditis: clinical characteristics of 105 cases. Scand J Infect Dis 77:81–4.

van der Linde MR and Ballmer PE (1993) Lyme carditis. In: Weber K, Burgdorfer W, Schier ZG, eds. Aspects of Lyme Borreliosis. Berlin: Springer-Verlag, 131–45.

Vincent MS, Roessner K, Sellati T *et al.* (1998) Lyme arthritis synovial γδ T cells respond to *Borrelia burgdorferi* lipoproteins and lipidated hexapeptides. J Immunol 161:5762–71.

Wallich R, Helmes C, Schaible UE *et al.* (1992) Evaluation of genetic divergence among *Borrelia burgdorferi* isolates by use of OspA fla, HSP60, and HSP70 gene probes. Infect Immun 60 (4856–66):.

Wang I-N, Dykhuizen DE, Qui W *et al.* (1999) Genetic diversity of OspC in a local population of *Borrelia burgdorferi* sensu stricto. Genetics 151:15–30.

Wang TJ, Liang MH, Sangha O *et al.* (2000) Coexposure to *Borrelia burgdorferi* and *Babesia microti* does not worsen the long-term outcome of Lyme disease. Clin Infect Dis 31:1149–54.

Wang G, Ojaimi C, Wu H *et al.* (2002) Disease severity in a murine model of Lyme borreliosis associated with the genotype of the infecting *Borrelia burgdorferi* sensu stricto strain. J Infect Dis 186:782–91.

Weber K (2001) Aspects of Lyme borreliosis in Europe. Eur J Clin Microbiol Infect Dis 20:6–13.

Weber K, Bratzke HJ, Neubert U *et al.* (1988) *Borrelia burgdorferi* in a new born despite oral penicillin for Lyme borreliosis during pregnancy. Pediatr Infect Dis J 7:286–9.

Williams CL, Benach JL, Curran AS *et al.* (1988) Lyme disease during pregnancy: a cord blood serosurvey. Ann N Y Acad Sci 539:504–6.

Wilske B, Preac-Mursic V, Schier ZG *et al.* (1988) Antigenic variability of *Borrelia burgdorferi*. Ann N Y Acad Sci 539:126–43.

Wormser GP, Bittker S, Cooper D, Nowakowski J, Nadelman RB, and Pavia C (2000a) Comparison of the yields of blood cultures using serum or plasma from patients with early Lyme disease. J Clin Microbiol 38:1648–50.

Wormser GP, Liveris D, Nowakowski J *et al.* (1999) Association of specific subtypes of *Borrelia burgdorferi* with hematogenous dissemination in early Lyme disease. J Infect Dis 180:720–5.

Wormser GP, Nadelman RB, Dattwyler RJ *et al.* (2000b) Practice guidelines for the treatment of Lyme disease. Clin Infect Dis 31(Suppl. 1):1–14.

Zemel LS (1992) Lyme disease. A paediatric perspective. J Rheumatol 19(Suppl. 34):1–13.

Zhang J-R and Norris SJ (1998) Genetic variation of the *Borrelia burgdorferi* gene vlsE involves cassette-specific, segmental gene conversion. Infect Immun 66:3698–704.

Section Five

Obligate Anaerobic Bacteria

44

Anaerobic Cocci

D. A. Murdoch

Department of Medical Microbiology, Royal Free Hospital NHS Trust, London, UK

INTRODUCTION

Watt and Jack (1977) published a useful definition of anaerobic cocci as 'cocci that grow well under satisfactory conditions of anaerobiosis and do not grow on suitable solid media in 10% CO_2 in air even after incubation for 7 days at 37 °C'. Anaerobic cocci are a major component of the normal flora and are commonly isolated from human infections (Finegold and George, 1989; Murdoch, 1998), but remarkably little is known about their role in pathogenesis. One reason is that they are usually isolated from polymicrobial infections; until recently, it has been very difficult to identify the more important pathogens. In addition, classification has been unsatisfactory and there have been few identification methods available for routine laboratories. In the last 15 years, there have been several major developments; the classification is finally on a more secure, genomically based footing; with the introduction of preformed enzyme kits, identification is much easier; and molecular techniques can be applied to examine the complex flora of the gingival crevice, genito-urinary and gastro-intestinal tracts. There has also been considerable interest in the pathogenicity of two of the most prominent members of the group, *Finegoldia magna* and *Micromonas micros*.

GRAM-POSITIVE ANAEROBIC COCCI (GPAC)

Classification

'Anaerobic streptococci' were frequently described from human infections in the early part of the 20th century, but the classification was confused by a profusion of poorly defined species (Hare, 1967). The term 'anaerobic streptococcus' is now meaningless and redundant, yet regrettably it is still often used. Rogosa (1971) brought some order to a chaotic situation with the description of the family Peptococcaceae, based on the obligately anaerobic nature of the organisms and their Gram-stained morphology; this led to the removal of microaerophilic species such as *Peptostreptococcus intermedius* to the genus *Streptococcus*. When the Approved List of Bacterial Names was published (Skerman, MacGowan and Sneath, 1980), the Peptococcaceae comprised five genera, *Peptococcus*, *Peptostreptococcus*, *Ruminococcus*, *Coprococcus* and *Sarcina*. With the introduction of taxonomic methods based on nucleic acid techniques, first using DNA-DNA hybridisation and later 16S rRNA sequencing, nine new species were described (Ezaki *et al.*, 1983; Li *et al.*, 1992; Murdoch *et al.*, 1997). They were all assigned to the genus *Peptostreptococcus*, which became a placement of convenience.

Recently, there have been major changes (Table 44.1). As a result of 16S rRNA sequence data, there has been a radical revision of the classification and a recognition that the microscopic morphology does not always reflect an organism's genetic affinities.

The long-standing division of clinically significant strains into the genera *Peptococcus* and *Peptostreptococcus* is now obsolete. The genus *Peptococcus*, with a G+C ratio of 50–51 mol% (Ezaki, Oyaizu and Yabuuchi, 1992), is only very distantly related to other Gram-positive anaerobic cocci (GPAC – a useful term in the diagnostic laboratory), most of which have G+C ratios of 28–37 mol%. The only species now in the genus *Peptococcus* is *Pc. niger*, which is very rarely isolated from human clinical specimens. The type species of the genus *Peptostreptococcus* is *Peptostreptococcus anaerobius*, which has been shown by 16S rRNA sequence analysis to be located in Clostridial cluster XI in the scheme of Collins *et al.* (1994). Collins *et al.* assigned most species in the genus *Peptostreptococcus* to Cluster XIII, phylogenetically distant from *P. anaerobius*. With the removal of these species from the genus *Peptostreptococcus* (Murdoch *et al.*, 2000), six new genera have been described. Murdoch and Shah (1999) constructed the genus *Finegoldia* for *Peptostreptococcus magnus* and the genus *Micromonas* for *Peptostreptococcus micros*. Species of *Finegoldia* form acetic acid as their terminal volatile fatty acid (VFA), are strongly proteolytic but only weakly saccharolytic; *F. magna*, the species most commonly isolated from human clinical material, is the only species at present in this genus. Species of *Micromonas* form acetic acid as their terminal VFA, are very strongly proteolytic and asaccharolytic; *M. micros*, another species of clinical importance, is the only species at present in this genus. The recent construction of a genetic map for *Finegoldia*, which disclosed a chromosome of size 1.9 Mb and a megaplasmid of 200 kb (Todo *et al.*, 2002), supported its reclassification.

The classification of the butyrate-producing GPAC has long been confused. Ezaki *et al.* (2001) recently separated the group into two new genera, *Anaerococcus* and *Peptoniphilus*. The saccharolytic genus *Anaerococcus* includes six species, some frequently isolated from human clinical material, and the asaccharolytic genus *Peptoniphilus* five, including *P. asaccharolyticus*, one of the most important species. However, the taxonomy particularly of *Anaerococcus* is incomplete, with several species awaiting description, and the identification criteria for both genera are still unsatisfactory. Ezaki *et al.* also created the genus *Gallicola* for *Peptostreptococcus barnesae*, a non-fermenting species apparently isolated only from chicken faeces.

Further rationalisations have included the reclassification of *Peptostreptococcus heliotrinreducens* in a new genus, *Slackia*, with an organism previously classified as *Eubacterium exiguum*; these species are now assigned to the family Coriobacteriaceae in the high G+C Gram-positive phylum (Wade *et al.*, 1999). They are closely related to *Atopobium parvulum*, an obligately anaerobic coccus originally

Principles and Practice of Clinical Bacteriology Second Edition Editors Stephen H. Gillespie and Peter M. Hawkey

Table 44.1 Changes to genus *Peptococcus* and genus *Peptostreptococcus*, 1974–2003

Rogosa (1974)	Holdeman Moore, Johnson and Moore (1986)	Ezaki, Oyaizu and Yabuuchi (1992)	2003
Pc. niger (T)	*Pc. niger*	*Pc. niger*	*Pc. niger*
Pc. activus			
Pc. aerogenes			
Pc. anaerobius			
Pc. asaccharolyticus			
Pc. constellatus			
P. anaerobius (T)	*P. anaerobius*	*P. anaerobius*	*P. anaerobius*
	P. asaccharolyticus	*P. asaccharolyticus*	*Pp. asaccharolyticus* (T)
		P. barnesae	*Gallicola barnesae* (T)
			Pp. harei
	P. heliotrinreducens	*P. heliotrinreducens*	*Slackia heliotrinireducens*
		P. hydrogenalis	*Anaerococcus hydrogenalis*
	P. indolicus	*P. indolicus*	*Pp. indolicus*
			Pp. ivorii
			Pp. lacrimalis
			Anaerococcus lactolyticus
P. lanceolatus			
	P. magnus	*P. magnus*	*Finegoldia magnus* (T)
P. micros	*P. micros*	*P. micros*	*Micromonas micros* (T)
			Anaerococcus octavius
P. parvulus	–		*Atopobium parvulum*
	P. prevotii	*P. prevotii*	*Anaerococcus prevotii* (T)
P. productus	*P. productus*	*P. productus*	*Ruminococcus productus*
	P. tetradius	*P. tetradius*	*Anaerococcus tetradius*
			Anaerococcus vaginalis

P., *Peptostreptococcus*; *Pc.*, *Peptococcus*; *Pp.*, *Peptoniphilus*; (T) denotes type species of genus.

described as *Peptostreptococcus parvulus* but placed until recently in the genus *Streptococcus*.

Other anaerobic species of Gram-positive cocci are rarely isolated from human clinical material and have been very little studied. The genus *Sarcina*, which forms spores but is morphologically a coccus, is located in Cluster I and is closely related to *Clostridium perfringens*. The genus *Ruminococcus* is known to be heterogeneous; the type species, *R. flavefaciens*, is placed in Cluster IV and is only distantly related to other ruminococci. A group of other ruminococci are assigned to Cluster XIVa, including *R. productus*, a strongly saccharolytic species placed until recently in the genus *Peptostreptococcus*. The genus *Coprococcus* also falls into Cluster XIVa.

It is likely that future reviews of GPAC will not consider *P. anaerobius* with former members of the genus *Peptostreptococcus*. As these radical taxonomic revisions are very recent, it is much simpler at this stage to consider these organisms together – keeping in mind that they are genetically and therefore phenotypically very diverse. The following abbreviations will be used: *P.* for *Peptostreptococcus*, *Pc.* for *Peptococcus* and *Pp.* for *Peptoniphilus*.

Aerotolerance, Specimen Transport, Laboratory Isolation and Maintenance

Anaerobes involved in human infections are usually relatively (sometimes remarkably) tolerant of oxygen; they require an anaerobic atmosphere in which to multiply, but many can survive prolonged periods in an aerobic atmosphere. Tally *et al.* (1975) reported that 14 clinical strains (not speciated) all survived exposure to atmospheric oxygen for 8 h; nine (63%) survived more than 72 h. When four fresh clinical isolates of *F. magna* and *M. micros* were exposed to air (Murdoch, 1998), 1% of cells were still viable after 48 h. Whilst many GPAC are moderately aerotolerant, it is likely that intestinal organisms such as *Coprococcus* and *Ruminococcus* are much more sensitive to oxygen; unfortunately, many of these organisms have been little studied.

The best specimens for culturing obligate anaerobes are aspirates or tissue specimens. The best means of maximising recovery rates are anaerobic transport systems, which are essential if the specimens sent are swabs. It is important that specimens reach the laboratory as quickly as possible and are not allowed to dry out. Specimens should be processed as soon as possible and must be incubated in an anaerobic atmosphere that includes 10% CO_2 and a palladium catalyst to remove traces of oxygen.

Very little is known about the nutritional requirements of GPAC. As they are phylogenetically diverse, generalisations are hazardous. (Heginbottom, Fitzgerald and Wade, 1990) compared several commercial non-selective solid media and reported that Fastidious Anaerobe Agar (Lab M, Bury, UK) consistently gave the best growth. Supplements such as vitamin K, haemin (Holdeman Moore, Johnson and Moore, 1986) and sodium oleate (Tween 80; final concentration 0.02%) may enhance the growth of some species (Hare, 1967), but none of them appear to be essential. A blood-free medium, Gifu anaerobic medium (GAM; Nissui, Tokyo, Japan), can support the growth well (Ezaki, Oyaizu and Yabuuchi, 1992). For liquid media, there is a great variation in the quality of chopped meat broths, with some commercial products appearing not to support growth at all; other products have maintained fastidious GPAC for several years at room temperature on the open bench, providing a very convenient method of storage (Murdoch, 1998). The best methods of long-term storage are either in liquid nitrogen or lyophilisation in the early stationary phase of growth in a medium containing less than 0.2% fermentable carbohydrate (Holdeman Moore, Johnson and Moore, 1986).

Most strains of *Ruminococcus*, *Coprococcus* and *Sarcina* are considerably less tolerant of oxygen and are unlikely to grow in the conditions encountered in routine laboratories (Ezaki, Oyaizu and Yabuuchi, 1992). Carbohydrates are essential for the growth of ruminococci and sarcinae and stimulate the growth of coprococci (Ezaki, Oyaizu and Yabuuchi, 1992). Excellent guidelines for their isolation and culture are available (Bryant, 1986; Ezaki, Oyaizu and Yabuuchi, 1992).

There are no selective media for GPAC as a group. As GPAC are heterogeneous, a single medium is unlikely to support the growth of all representatives yet be reasonably selective. Most antibiotics, including neomycin and polymyxin, have some activity against some strains of GPAC (Murdoch, 1998). Nalidixic acid–Tween blood agar (nalidixic acid 10 mg/l and 0.1% Tween 80) has been reported to give better isolation rates than neomycin blood agar (neomycin 75 mg/l) (Wren, 1980). Turng *et al.* (1996) have described *P. micros* medium, a selective and differential medium for *P. micros*, which contains colistin–nalidixic acid agar (CNA; Difco) supplemented with glutathione and lead acetate. Strains of *P. micros* use the reduced form of glutathione to form hydrogen sulphide, which reacts with lead acetate to form a black precipitate under the colony. The development of selective media for genetically homogeneous groups, e.g. *F. magna* or *Anaerococcus*, would be realistic and valuable. As most clinical isolates grow relatively slowly on standard media and are usually present in mixed culture, they are frequently overgrown by more robust organisms, and many clinical studies therefore give a falsely low estimate of their frequency.

Laboratory Identification

The lack of an adequate identification scheme undoubtedly held back studies of GPAC for many years (Watt and Jack, 1977). Older schemes (Holdeman Moore, Johnson and Moore, 1986) required equipment such as gas-liquid chromatography (GLC), which was rarely available in routine diagnostic laboratories. However, now that identification is based on preformed enzyme profile (PEP) kits (Murdoch, 1998), which can be used in most diagnostic laboratories, well over 90% of clinical strains can be placed in a phylogenetically valid genus and probably 80% can be identified to the species level. As the relative clinical importance of different species of GPAC is little known, all microbiologists are urged to identify the strains they isolate as fully as they can.

The first stage of identification is to decide to which general group an apparently anaerobic isolate belongs; this can be surprisingly difficult. Some Gram-positive anaerobes, such as *P. asaccharolyticus*, retain Gram's stain poorly, and colonies can appear Gram-negative after 48 h incubation. A simple trick is to test susceptibility to vancomycin, using a 5 µg disc, to which strains of GPAC are sensitive and Gram negatives are usually resistant. A rapid method using a 3% solution of potassium hydroxide (KOH) can also be used; the solution usually becomes viscous and mucoid in the presence of Gram-negative bacteria (Halebian *et al.*, 1981). GPAC must be distinguished from micro-aerophilic organisms, such as many strains of streptococci, *Staphylococcus saccharolyticus* and *Gemella morbillorum*. Streptococci and gemellae are strongly saccharolytic and produce large quantities of lactic acid, which can be detected by GLC. It is effective and simpler to test susceptibility to metronidazole (Watt and Jack, 1977); a 5 µg disc is applied to the edge of an inoculum (not to the centre; some strains of GPAC are extremely sensitive and may not grow at all) and susceptibility read within 48 h. Metronidazole-resistant strains of GPAC appear to be very rare. The decision whether an organism is a rod or a coccus can be very subjective; the cell morphology of *P. anaerobius*, in particular, is very variable. Similarly, strains of *Slackia exigua*, previously identified as *P. heliotrinreducens* (Murdoch and Mitchelmore, 1989), can be coccobacillary; they are easily confused with strains of *M. micros*, and they can be isolated from similar (and sometimes highly significant) clinical specimens (Murdoch, 1998). *Slackia* and *Micromonas* can be distinguished by their PEP (Wade *et al.*, 1999).

Second-stage identification of GPAC relied for many years on the standard biochemical tests used for anaerobes, notably carbohydrate fermentation reactions (Table 44.2). These tests are useful for clostridia and *Bacteroides* but are not appropriate for GPAC, most of which appear not to use carbohydrates, much less ferment them.

Nevertheless, demonstration that an organism is strongly saccharolytic may be a useful way of identifying organisms with which GPAC may be confused, e.g. *Staphylococcus saccharolyticus* and streptococci. Members of the genus *Anaerococcus* are able to ferment carbohydrates, though sometimes weakly (Ezaki *et al.*, 2001); the classification and characterisation of this group are still unclear, and it would be worth evaluating their carbohydrate fermentation reactions using commercial kits as a potential method of identification.

GLC was a mainstay of anaerobic identification for many years, but few laboratories still maintain the equipment. It is of limited use for GPAC (Table 44.3). It divides them into an 'acetic group' (the clinically relevant groups being *Finegoldia*, *Micromonas* and *Slackia*); a large butyrate group, which is heterogeneous and still inadequately characterised; and a few species that produce longer-chain fatty acids, notably isocaproate (*P. anaerobius*), isovalerate (*Pp. ivorii*) and *n*-caproate (*A. octavius*). Production of non-volatile fatty acids by GPAC has been little studied; streptococci form large quantities of lactic acid.

Most GPAC use the products of protein decomposition as a major energy source. Ezaki and Yabuuchi (1985) demonstrated that most strains of clinical importance had consistent amidase and oligopeptidase activities; individual species had consistent PEPs that could be used to distinguish between them. Several of these tests have been incorporated into commercially available PEP kits. The use of PEPs has provided a simple, powerful and widely available means of identification; for instance, *F. magna* and *M. micros*, which had previously only been separable by the size of the cell, are reliably differentiated by several tests available in commercial kits. PEPs have been discriminatory enough to distinguish undescribed groups of GPAC from recognised taxa, such as *Pp. ivorii* and *A. octavius*. Wilson *et al.* (2000) reported that PEPs were valuable for characterising strains in the newly described genus *Anaerococcus*. However, it has become increasingly clear that they will rarely discriminate reliably within the new genus *Peptoniphilus*. Further, it must be emphasised that a heavy inoculum of the test organism (MacFarland standard 4 or more) is required; using a lighter inoculum will lead to misidentification. It is often necessary to inoculate several plates to harvest a sufficient quantity of the test organism. Finally, though they are marketed as 'rapid' kits, the manufacturers neglect to mention the 48 h period required to grow sufficient of the organism for inoculating the kit. The commercially available kits are designed to identify as wide a range of anaerobes as possible; they include many tests of little relevance to GPAC. Development of kits specifically for GPAC on an in-house basis would be valuable.

Few other techniques have been shown to be of use in the routine laboratory. There is one important exception; strains of *P. anaerobius* are sensitive to sodium polyanethol sulphonate (SPS) or liquoid, whereas almost all other strains of GPAC are resistant (Graves, Morello and Kocka, 1974). SPS discs are available, making this a simple and reliable test.

Nucleic acid-based techniques have been adapted to the identification of GPAC with considerable success, although they are at present too expensive for routine use. Dymock *et al.* (1996) used conserved random amplification of 16S rRNA genes by means of eubacterial primers to detect a non-culturable organism related to *M. micros* from a dento-alveolar abscess. Wang *et al.* (1997) described a polymerase chain reaction (PCR)-based technique for the detection of ruminococci in faecal specimens. PCR methods for the identification of *M. micros* and *P. anaerobius* have also been published; they were valuable for their detection in polymicrobial samples from dental specimens (Riggio, Lennon and Smith, 2001; Riggio and Lennon, 2002; Siqueira *et al.*, 2003). Hill *et al.* (2002) recently used amplification of the 16S–23S intergenic spacer region for differentiation of a group of reference and clinical strains. They reported considerable heterogeneity within described butyrate-forming species. A set of multiplex PCR probes have been validated against the known sequences of recognised species of GPAC and found to be superior to identification

Table 44.2 Differential characteristics of GPAC

| Species (strains examined) | Terminal VFA | Production of | | | | Carbohydrate fermentation reactions | | | | | Saccharolytic and proteolytic enzymes | | | | | | | | | | |
		Indole	Urease	ALP	ADH	Glucose	Lactose	Raffinose	Ribose	Mannose	α-GAL	β-GAL	α-GLU	β-GUR	ArgA	ProA	PheA	LeuA	PyrA	TyrA	HisA
F. magnus	A	−	−	d	d	−/w	−	−	−	−	−	−	−	−	+	−	−	+	+	−/w	−/w
M. micros	A	−	−	+	−	−	−	−	−	−	−	−	−	−	+	+	+	+	+	+	+
*S. heliotrinreducens**	A	−	−	−	+	+	+	+	d	d	+	−	+	−	d	+	+	+	−	w	w
*P. productus**	A	w	−	−	−	+	+	+	d	d	+	−	+	−	d	+	+	+	−	−	−
*G. barnesae**	A(b)	d	−	+	−	−	−	−	−	d	−	−	−	−	−	−	−	−	−	d	w
P. asaccharolyticus	B	+	−	−	−	−	−	−	−	−	−	−	−	−	+	−	+	d	−	w	+
*P. indolicus**	B	−	−	+	−	−	−	−	−	−	−	−	−	−	+	+	+	+	−	w	−
*P. ivorii**	IV	d	−	−	−	−	−	−	−	−	−	−	−	−	+	+	−	−	−	−	−
P. harei	B	−	+	−	−	−	−	−	−	−	−	−	−	−	+	−	+	−/w	−	w	+
*P. lacrimalis**	B	−	−	−	−	−	−	−	−	−	−	−	−	+	+	−	+	−	+	d	+
A. prevotii (type strain)	B	−	+	−	−	−	−	+	+	+	+	−	+	+	+	−	−	−	−	w	+
A. tetradius (type strain)	B	−	+	−	−	+	−	−	−	+	−	−	+	+	+	−	w	+	w	w	w
prevotii/tetradius group	B	−	?d	d	−	d	−	d	d	d	d	d	+	−	+	−	d	d	d	d	d
A. hydrogenalis	B	+	d	−/w	−	+	+	+	−	+	−	−	d	−	−	−	d	−	w	−	−
*A. lactolyticus**	B	−	+	−	−	+	+	−	+	+	−	+	−	−	+	−	−	−	−	−	−
*A. octavius**	C	?d	−	−	+	+	−	−	+	d	−	−	−	−	−	+	−	−	w	−	−
A. vaginalis	B	−	−	−/w	d	+	+	d	+	+	−	w/+	−/w	−	+	−	+	+	w	−	+
'*β-GAL*' group	B	+	−	w	d	+	+	−/w	+	+	−	d	−/w	−	+	−	−	+	+	−	−
'*trisimilis*' group*	B	−	−	d	−	+	w	−/w	−/w	+	−	−	+	−	−	+	−	−	+	−	−
P. anaerobius	IC	−	−	−	−	+	−	−	+	w	−	−	−	−	+	+	+	−	+	−	−
*Pc. niger**	C	−	−	−	−	−	−	−	−	w	−	−	−	−	−	−	−	−	−	−	−

A, acetate; ADH, arginine dihydrolase; ALP, alkaline phosphatase; ArgA, arginine arylamidase (AMD); B, butyrate; C, *n*-caproate; HisA, histidine AMD; IC, isocaproate; IV, isovalerate; LeuA, leucine AMD; PheA, phenylalanine AMD; ProA, proline AMD; PyrA, pyroglutamyl AMD; TyrA, tyrosine AMD; VFA, volatile fatty acid; α-GAL, α-galactosidase; β-GAL, β-galactosidase; α-GLU, α-glucosidase; β-GUR, β-glucuronidase.
A., *Anaerococcus*; F., *Finegoldia*; G., *Gallicola*; M., *Micromonas*; P., *Peptostreptococcus*; Pc., *Peptococcus*; Pp., *Peptoniphilus*; S., *Slackia*; −, >90% negative; w, weakly positive; +, >90% positive; d, different reactions. It must be stressed that the reactions of many groups (asterisked) have been inferred from a very limited number of isolates (less than ten) and therefore must be considered provisional.
Data on carbohydrate fermentation reactions from Holdeman Moore, Johnson and Moore (1986) and Murdoch and Mitchelmore (1991).
Data on production of VFAs, indole, urease, ALP, ADH and saccharolytic and proteolytic enzymes mainly from Murdoch and Mitchelmore (1991) and Murdoch (1998), using ATB 32A system (API-bioMérieux).

Table 44.3 Present classification of GPAC based on 16S rRNA sequence analysis

Genus	Species			Major energy source	
	Name	Terminal VFA	G + C content (mol%)	Carbohydrates	Proteins
Cluster I					
Clostridium	*ventriculi*	A	28–31	+ +	–
(at present *Sarcina*)	*maxima*	AB	28–30	+ +	–
Cluster IV					
Ruminococcus	*flavefaciens* (T)	A	39–44	+ +	–
Cluster XI					
Peptostreptococcus	*anaerobius* (T)	IC (IV)	33–34	w	w
Cluster XIII					
Finegoldia	*magnus* (T)	A	32–34	– /w	+ +
Gallicola	*barnesae* (T)	A(b)	34–35	–	w
Micromonas	*micros* (T)	A	28–29	–	++
Anaerococcus	*prevotii* (T)	B	29–33	+	+
	tetradius	B	30–32	+ +	+
	hydrogenalis	B	30–31	+ +	w
	lactolyticus	B	34	+ +	+
	vaginalis	B	28–30	+	+
	octavius	C	26–31	+	w
Peptoniphilus	*asaccharolyticus* (T)	B	30–34	–	+
	harei	B	25	–	+
	indolicus	B	32–34	–	+
	ivorii	IV	28	–	w
	lacrimalis	B	30–31	–	+ +
Cluster XIVa					
(at present *Ruminococcus*)	*productus*	A	44–45	+ +	–
	torques	A	40–42	+ +	–
Coprococcus	*eutactus* (T)	B	39–42	+ +	–
Not in clostridial subphylum					
Peptococcus	*niger* (T)	C	50–51	–	–
Atopobium	*parvulum*	A	46	+ +	w
Slackia	*exigua* (T)	A	60–64	–	+ +
	heliotrinireducens	A	61	–	+ +

–, negative; +, positive; + +, strongly positive; A, acetate; B, butyrate; C, *n*-caproate; IC, isocaproate; IV, isovalerate; NK, not known; *R.*, *Ruminococcus*; *S.*, *Sarcina*; (T), type species of genus; VFA, volatile fatty acid; w, weakly positive.
Data on G + C contents from Ezaki *et al.* (1983), Li *et al.* (1992), Murdoch *et al.* (1997) and Wade *et al.* (1999).
Data on production of VFAs and major energy sources from Holdeman Moore, Johnson and Moore (1986) and Murdoch and Mitchelmore (1991).

by PEP (Song *et al.*, 2003). A range of probes for the detection and enumeration of anaerobes in human faeces have recently been extended to include *Ruminococcus* and related species (Harmsen *et al.*, 2002).

Suggested Laboratory Identification Scheme for GPAC

Using the following scheme, experienced laboratory workers should be able to place at least 80% of clinical strains in the correct genus 48 h after the isolation of individual colonies.

Primary Isolation (First Stage)

Specimens should be processed and placed in an anaerobic environment as rapidly as possible. Enriched blood agar should be used for culture, preferably Fastidious Anaerobe Agar (Lab M). Test for susceptibility to metronidazole by placing a 5 µg disc on the primary inoculum, on the edge of the plate and well away from secondary streaking; this positioning is necessary to prevent metronidazole suppressing growth of very sensitive organisms.

Examine the primary plate after incubation for 48 h and 5 days; plates should not be removed from their anaerobic environment at 24 h unless essential (e.g. possible clostridial infections). It is recommended that plates should not be discarded before 5 days, as many anaerobes grow very slowly – particularly if the plates are old.

When the primary plate is examined, experienced workers can use the Gram stain, colonial morphology and smell to make a presumptive identification of most strains (25–50% isolates) of *P. anaerobius*, *M. micros* and *Pp. asaccharolyticus*. Unfortunately, characteristic features for *F. magna*, the commonest species in human specimens, have not yet been described.

Subculture (Second Stage)

Subculture onto enriched blood agar, which should be as fresh as possible and compatible with PEP kits. Several plates must be used for slowly growing GPAC (colony diameter <1 m after incubation for 48 h), otherwise it will be impossible to harvest enough organisms to inoculate the PEP kit. One plate must be streaked so that single colonies are present. Also inoculate a plate to detect capnophilic organisms, incubating it in an aerobic atmosphere containing 6% CO_2. When GLC is available, a broth culture may be inoculated.

After incubation for 48 h, a PEP kit can be inoculated and read (Table 44.2). It is essential to use an adequate inoculum for the kit (MacFarland standard 4 as a minimum). Strains of *P. anaerobius*, *F. magna* and *M. micros* should be identified without difficulty. The profiles of *A. hydrogenalis* and *A. octavius* should be distinctive enough to allow species level identification, but most strains of *Anaerococcus* can only be identified to the genus level. Likewise, *Pp. ivorii* and *Pp. lacrimalis* can be fully identified using their PEP, but cell and colonial morphology are at present important for speciation of the *asaccharolyticus/hareii/indolicus* group.

Inspect cell and colonial morphology again, checking that they are consistent with PEP identification. If the PEP does not give a clear identification and GLC is available, examine for VFA production; GLC of unidentified strains will usually place them in the butyrate group, which contains a number of undescribed species. If the strain ferments any sugars in the PEP kit and forms butyrate, it can probably be assigned to the genus *Anaerococcus*.

After 5 days, it is good practice to check both aerobic and anaerobic plates again (recording all observations), particularly if the PEP does not give a clear identification (often because of mixed culture). Colonial morphology and smell are most distinctive after 5 days and help distinguish *Pp. harei* from *Pp. asaccharolyticus*. However, by 5 days, many strains (of species such as *F. magna* and *A. tetradius*) will show marked heterogeneity in their colonial morphology, making assessment more difficult.

It is regrettable that a significant proportion of clinical isolates, perhaps 10–20%, cannot yet be identified to phylogenetically valid species. If they are cultured from important specimens, they should be sent to a reference laboratory for further investigation.

Modifications

1. When presumptive strains of *P. anaerobius* are subcultured, an SPS disc should be placed on the primary streaking; the plate can be inspected after 18 h. Organisms susceptible to SPS with appropriate cell and colonial morphology do not need further characterisation. Some strains of *Pp. ivorii* are sensitive to SPS, but their cell morphology is coccoid and regular rather than coccobacillary and pleomorphic, and they do not form α-glucosidase or the VFA isocaproic acid.
2. When the cell morphology from the primary plate suggests a very decolourised strain of GPAC, for example *Pp. asaccharolyticus* or *Pp. indolicus*, place a vancomycin 5 μg disc on the primary inoculum to check the Gram reaction; GPAC are sensitive whereas most Gram-negative anaerobes are resistant.

Normal Flora

Although little studied, GPAC are found on most or all body surfaces (Finegold and George, 1989; Murdoch, 1998). They are major components of the normal flora of the mouth, upper respiratory and gastro-intestinal tracts and female genito-urinary system and are also present (though less commonly) on the skin, certain species having a predilection for specific anatomical sites.

GPAC are a major constituent of the oral flora and are found in plaque and the gingival sulcus; the dominant species is *M. micros*. Many species of GPAC are found in the gut, but they are often poorly characterised; they include *Coprococcus*, *Ruminococcus* and *Sarcina* (particularly in vegetarians; Crowther, 1971). Using optimal culture techniques, Finegold, Attebery and Sutter (1974) showed that '*Peptostreptococcus*' *productus* (now placed in the genus *Ruminococcus*) is one of the commonest organisms in the gastro-intestinal flora, which has been confirmed using 16S rRNA probes (Harmsen *et al.*, 2002). The significance of GPAC in the vaginal ecology is unclear, though large numbers of many species are found in the female genito-urinary tract. The GPAC present on the skin appear to have been little studied; *F. magna*, *P. asaccharolyticus* and *A. vaginalis* are likely to be present, as they have been isolated from superficial wound infections.

Clinical Importance

Comprehensive surveys of anaerobic pathogens have reported that GPAC are second only to *Bacteroides* and account for 25–30% of all isolates from human clinical specimens. Their clinical significance has been underestimated, because most infections involving GPAC are polymicrobial, and the technical problems of classification and identification. Most reports of isolation in pure culture refer to *F. magna* (Murdoch, 1998), but there are also reports of *P. anaerobius*, *Pp. asaccharolyticus*, *Pp. indolicus*, *Pp. harei*, *M. micros* and *A. vaginalis*.

Oral and Respiratory Tract Infections

GPAC have been isolated from a wide variety of infections involving the upper respiratory tract, but usually in mixed growth (Murdoch, 1998). Most dental infections are polymicrobial but are dominated by anaerobic organisms. *Micromonas micros*, a strictly anaerobic, strongly proteolytic organism, being well suited to the subgingival ecosystem (Shah and Gharbia, 1995); it has been implicated in periodontitis and various more serious intra-oral infections, but causation is very difficult to prove in a microbial flora as complex as that of the subgingival plaque (Socransky *et al.*, 1991). Species clusters of potential pathogens, including *M. micros*, *Prevotella melaninogenica* and *P. gingivalis*, have been associated with dental disease (Socransky *et al.*, 1988). *Micromonas micros* has been isolated from endodontic and peritonsillar abscesses (Mitchelmore *et al.*, 1995) and complications such as retropharyngeal and brain abscesses (Murdoch, Mitchelmore and Tabaqchali, 1988b). Chronic sinusitis yields a mainly anaerobic flora, of which *F. magna* and *M. micros* are frequent constituents (Brook, Yocum and Frazier, 1996). Anaerobic infections of the lower respiratory tract are probably much commoner than generally realised (Bartlett, 1993); in the days when transtracheal aspiration was an acceptable technique, GPAC were recognised as one of the major pathogenic groups. They were isolated mainly from cases of aspiration pneumonia, lung abscess and empyema, in association with *Prevotella* spp., *Fusobacterium nucleatum* and microaerophilic streptococci (Bartlett, Gorbach and Finegold, 1974).

Intra-Abdominal Infections

GPAC are usually said to be less common in serious abdominal infections than *Bacteroides* spp., but they predominated in a recent series of retroperitoneal abscesses (Brook and Frazier, 1998a), as well as to identify intra-abdominal isolates to the species level. GPAC have been the commonest anaerobes in two series of liver and spleen abscess (Sabbaj, Sutter and Finegold, 1972; Brook and Frazier, 1998b).

Infections of the Genito-Urinary Tract

GPAC are commonly implicated in gynaecological and obstetric sepsis, but their precise role is still unclear largely because of problems with identification and a wide range of species being present (Murdoch, 1998). Schottmüller in 1910 was the first to associate puerperal sepsis (now rare in developed countries) with 'anaerobic streptococci'; several later papers established their importance in this condition. Complications included abscess formation and infertility, but septicaemia and metastatic abscess formation were rare, in contrast to *Streptococcus pyogenes* (Murdoch, 1998).

Bacterial vaginosis (BV) is now recognised to predispose to various gynaecological and obstetric problems. The place of GPAC in complications associated with BV has been less studied than for anaerobic Gram-negative rods such as *Prevotella*. Hillier *et al.* (1993) compared the vaginal flora of groups of pregnant women with normal flora and BV; they reported that women with BV were more likely to have counts of GPAC $>10^5$/ml, with significantly increased counts of '*P. tetradius*' but not other species of GPAC. High carriage rates of '*P. asaccharolyticus*', '*P. prevotii*' and '*P. tetradius*' have been reported from the vaginal flora of non-pregnant women with BV. There has recently been great interest in the role of BV in various obstetric complications; GPAC have been reported from studies of chorioamni-onitis, post-Caesarean endometritis and preterm labour and delivery.

Laparoscopic studies of the upper genital tract flora of women hospitalised for pelvic inflammatory disease have revealed that GPAC are often present (Soper *et al.*, 1994). However, their precise pathogenic role is unclear although they are one of the commonest groups of organisms isolated from tubo-ovarian abscesses, along with *Escherichia coli* and *Bacteroides fragilis* (Landers and Sweet, 1983).

It is clear that the female genito-urinary tract (in health and disease) hosts a complex flora and that many of the GPAC present cannot yet be adequately characterised (in either phylogenetic or phenotypic terms). When a polymicrobial flora is being analysed, traditional identification methods that use biochemical identification methods such as PEPs are very laborious; this is a field where analysis by molecular probes will be very effective. But, as many of the organisms present appear to belong to the poorly studied genera *Anaerococcus* and *Peptoniphilus*, further taxonomic studies of these groups will be needed before their role in the female genito-urinary tract is properly understood.

Superficial and Soft-Tissue Infections

GPAC are frequently isolated from superficial and soft-tissue infections and post-operative surgical wounds (Brook, 1988; Murdoch, Mitchelmore and Tabaqchali, 1994). The classic study by Bourgault, Rosenblatt and Fitzgerald (1980) established the pre-eminence of *F. magna*.

Brook and Frazier (1990) emphasised the importance of the location of the infection in determining the organisms involved; they reported that superficial infections of the trunk most often yielded *F. magna*, but *Pp. asaccharolyticus* and *P. anaerobius* were commonest in perirectal abscesses, infections of the leg and external genitalia. They reported that GPAC were the commonest anaerobic group in cases of cellulitis, necrotising fasciitis, mediastinitis, pyomyositis and other soft-tissue infections. There appears to be a strong association between GPAC and infections of the feet, particularly in diabetics (Wheat *et al.*, 1986), and non-puerperal breast abscesses (Edmiston *et al.*, 1990), when *F. magna* is again the commonest species isolated.

Musculoskeletal Infections

Anaerobic septic arthritis, formerly a very rare condition, has now become much more frequent, being most often seen as a complication of hip and knee joint replacements. GPAC are the commonest pathogens; they cause a low-grade infection that is frequently missed or diagnosed late, so a high index of suspicion by both clinician and laboratory is necessary. Bourgault, Rosenblatt and Fitzgerald (1980) isolated *F. magna* from 32 bone and joint infections (14%) from 222 positive specimens; they recovered *F. magna* in pure culture from 18 patients (15 with foreign bodies). GPAC have also been associated with various orthopaedic infections including post-traumatic joint infections, traumatic fractures of long bones and osteomyelitis associated with sinus, ear or dental infections. Diabetic foot infections linked to peripheral vascular disease are often complicated by osteomyelitis; *F. magna* is the commonest anaerobe (Wheat *et al.*, 1986).

Septicaemia and CNS Infections

GPAC are only occasionally isolated from blood cultures (Brook, 1989); cases of bacteraemia are probably missed as cultures of some species, including *F. magna* and *P. anaerobius*, may remain signal negative in commercial systems (van der Vorm *et al.*, 2000). Obstetric patients, particularly after septic abortion, appear to be at highest risk (Topiel and Simon, 1986). Analyses of anaerobic septicaemia have not reported deaths linked to GPAC, in contrast to cases yielding clostridia or *Bacteroides* group organisms.

GPAC very rarely cause meningitis, polymicrobial abscesses being the commonest condition (Finegold and George, 1989). Direct spread can occur from infections of the middle ear, sinuses and mouth (usually teeth), and haematogenous seeding can result from pleuro-pulmonary infections or, occasionally, from cardiovascular abnormalities. Murdoch, Mitchelmore and Tabaqchali (1988b) described three brain abscesses that yielded *M. micros*, fusobacteria, '*Bacteroides*' spp. (*sensu lato*) and micro-aerophilic streptococci.

Pathogenesis and Virulence Factors

This area is little studied although there has been some work to identify virulence factors, particularly for *F. magna* and *M. micros*, and several studies have attempted to define the relative importance of GPAC in polymicrobial infections.

Most species of GPAC are isolated from mixed infections with facultative organisms and/or other anaerobes; a synergistic interaction appears likely but it is technically difficult to evaluate the relative importance of the different components. In an elegant series of experiments from the pre-antibiotic era, Smith (1930) demonstrated the principle of microbial synergy in the development of anaerobic abscesses. He used as his model the development of anaerobic lung infections in mice and rabbits. He injected pus into the trachea and cultured the resulting lung abscesses, isolating 17 different organisms. He then injected pure cultures of each organism back into fresh animals, without producing lung abscesses, but then tested combinations of organisms. He found that a combination of an 'anaerobic streptococcus', an anaerobic spirochaete (probably *Fusobacterium*) and an anaerobic Gram-negative rod was necessary to reproduce the disease. Unfortunately, we do not know which organisms he was studying, but they bear a remarkable resemblance to those from a series of polymicrobial deep organ abscesses described by Murdoch, Mitchelmore and Tabaqchali (1988b) that yielded *M. micros*. Brook and co-workers (Brook and Walker, 1984) developed a subcutaneous abscess model in the mouse. They examined a range of bacteria in pure and mixed growth, testing their ability to induce abscesses, enhance the growth of bacterial components and increase mortality. They showed that GPAC could interact with facultative and anaerobic bacteria, including *Staphylococcus aureus*, *Pseudomonas aeruginosa* and several species of streptococci and *Enterobacteriaceae*, in causing abscess formation, but they did not report clear differences between the species of GPAC studied. Possible mechanisms for microbial synergy have included mutual protection from phagocytosis (Ingham *et al.*, 1977) and lowering of oxidation/reduction potentials in host tissue.

Many anaerobes form potential virulence factors, but this characteristic appears less marked for GPAC than, for instance, clostridia. Brook (1987) showed that capsule formation was an important virulence mechanism, capsulate organisms being able to seed more successfully to distant organs. Freshly isolated strains have been shown to produce gelatinase, collagenase and hyaluronidase (Steffen and Hentges, 1981) and DNAase, RNAase, coagulase and haemolysins (Marshall and Kaufman, 1981), but these studies did not find a clear correlation between the species and the potential virulence factor.

More recently, there has been interest in a range of virulence factors formed by *F. magna* which is the most studied GPAC. Krepel *et al.* (1992) reported that clinical strains of *F. magna* from different sites varied in their ability to form a range of potentially significant enzymes; they found that strains of *F. magna* from non-puerperal breast abscesses and diabetic foot infections were more likely to form collagenase and gelatinase than strains from intra-abdominal sepsis. A specific protein receptor on the cell wall described by Myhre (1984) can bind significant amounts of human serum albumin; this receptor was not found in a range of other anaerobic pathogens such as *B. fragilis*. Some strains of *F. magna* form Protein L, a cell surface protein that binds to the κ light-chain variable domain of human and many mammalian immunoglobulin molecules (Nilson *et al.*, 1992). Protein

L can trigger the release of histamine and leukotriene C4 from mast cells and basophilic granules (Patella *et al.*, 1990) and can act as an immunoglobulin superantigen (Genovese *et al.*, 2000). The isolation from the vaginal flora of strains of *F. magna* expressing Protein L has been associated with the presence of BV (Kastern *et al.*, 1990). de Château and Björck (1994) described a mosaic albumin-binding protein, Protein PAB, which shows substantial relatedness to Protein L and a virulence factor elaborated by some β-haemolytic streptococci, Protein G.

Micromonas micros is of interest to oral microbiologists, as strains are frequently isolated from intra-oral infections and their complications. Murdoch, Mitchelmore and Tabaqchali (1988a) noted the intrinsically strong proteolytic activity of *M. micros* and suggested that this could be important in abscess formation. Carlsson, Larsen and Edlund (1993) examined the ability of oral bacteria to form the cytotoxin hydrogen sulphide from glutathione, a tripeptide involved in the intracellular defences against reactive oxygen metabolites; *M. micros* was the most active organism of 37 species studied. van Dalen *et al.* (1993) described a rough morphotype of *M. micros* present in plaque samples that differs from the 'normal' smooth morphotype in its haemolytic ability and hydrophilicity; it possesses long fibrillar structures that appeared to support cellular aggregation. Rough and smooth morphotypes cluster separately when examined by genetic or electrophoretic techniques (Kremer *et al.*, 1997). Both types show synergism with oral strains of *Prevotella* in their ability to form abscesses in a mouse model (van Dalen *et al.*, 1998).

Antibiotic Susceptibility

Unfortunately, most reports on anaerobic susceptibility to antibiotics have presented data for GPAC as a group rather than for individual species; considering their genetic diversity, this information must now be considered of little value. When GPAC have been identified to the species level, differences in susceptibility patterns have been reported, but these are rarely consistent between reports; it is not clear whether these differences are due to true geographical variations, methodological differences or, indeed, misidentification.

Most workers have reported that almost all GPAC are susceptible to metronidazole (Finegold and George, 1989); resistance has frequently been reported (Summanen *et al.*, 1993), but it is possible that some resistant strains were micro-aerophilic streptococci, as their identity has often not been reported. Some strains in the *prevotii/tetradius* group are partially or even completely resistant by disc testing (Murdoch, 1998). Strains of *Pp. asaccharolyticus*, *F. magna* and *M. micros* have been reported to be exquisitely susceptible to metronidazole, but two of the nine strains of *P. anaerobius* tested were of intermediate susceptibility (MIC 2–4 mg/l) (Bowker *et al.*, 1996).

Peptoniphilus asaccharolyticus, *F. magna* and *M. micros* are almost always susceptible to penicillins, but some strains of *P. anaerobius* are much more resistant (Reig and Baquero, 1994; Bowker *et al.*, 1996). Cefoxitin predictably is the most active cephalosporin, other extended spectrum cephalosporins having higher MICs (Wexler and Finegold, 1988). GPAC appear to be very sensitive to carbapenems. Occasional strains of *P. anaerobius* are resistant to penicillin G, cefoxitin and cefotaxime, although β-lactamase production by GPAC has never been demonstrated (Garcia-Rodriguez, Garcia-Sanchez and Munoz-Bellido, 1995).

The macrolides erythromycin, clarithromycin and azithromycin are not active enough to be recommended as first-line therapy. Clindamycin is widely used as anti-anaerobic treatment, but high rates of resistance have often been reported; this may be a result of local use, as there appears to be considerable geographical variation within individual countries (e.g. Spain and the United States). Reig, Moreno and Baquero (1992) and Sanchez, Jones and Croco (1992) have described inducible macrolide-lincosamide resistance in several species of GPAC; they recommended that erythromycin susceptibility be assessed

before individual isolates were reported as sensitive to clindamycin. There are few data on glycopeptides, but resistance has not been reported. Resistance to chloramphenicol, mediated via acetyl transferase activity, has occasionally been reported. The older quinolones such as ciprofloxacin are moderately effective, but the newer agents such as moxifloxacin are extremely active with an MIC_{90} for *Peptostreptococcus* spp. of 1 mg/l (Fung-Tomc *et al.*, 2000).

Treatment with a penicillin or metronidazole (preferred because of better penetration) will usually be effective first-line therapy. A β-lactamase inhibitor will add little to penicillin therapy for GPAC, though in mixed infections it may be essential for associated organisms. Susceptibility to clindamycin varies widely; erythromycin susceptibility should simultaneously be tested. Cephalosporins are usually and carbapenems always effective.

Future studies must present data by species, in particular *P. anaerobius* is genetically so distantly related to other species of GPAC that its data should at least be presented separately.

Characteristics of the More Significant Species of GPAC

Collins Cluster XI

Peptostreptococcus anaerobius: This is now the only species in the genus *Peptostreptococcus* (Murdoch *et al.*, 2000); it was assigned by Collins *et al.* (1994) to *Clostridium* cluster XI with ten species at present in the genus *Clostridium*. It is phenotypically very different from the species with which it was formerly classified (Murdoch, 1998): the cell morphology is highly pleomorphic but usually coccobacillary; it forms distinctive colonies with a sickly sweet smell; proteolytic activity is feeble; several VFAs are formed, notably isocaproic acid; and strains are sensitive to SPS. A molecular identification method based on PCR has been reported (Riggio and Lennon, 2002). *Peptostreptococcus anaerobius* is part of the gut flora and is often grown from clinical specimens, but rarely in pure culture. It is regularly isolated from gastro-intestinal and high vaginal infections, usually abscesses, with other components of the faecal flora such as coliforms and other anaerobes. Its role in causing infections is unclear. It appears not to be involved in periodontitis or associated diseases (Riggio and Lennon, 2002), though the data are contradictory (Murdoch, 1998). Some strains are resistant to a range of β-lactams, probably mediated via modified penicillin-binding proteins.

Collins Cluster XIII

Finegoldia magna: Clinically the most important species in the group, *F. magna* can undoubtedly act as a primary pathogen, causing a range of infections from infected sebaceous cysts to life-threatening infections (Murdoch, 1998). 16S rRNA data place *F. magna* in *Clostridium* cluster XIII (Collins *et al.*, 1994) but in a genus of its own (Murdoch and Shah, 1999). Both cell and colonial morphology are variable (Murdoch and Mitchelmore, 1991), but the Gram-stained appearance always reveals clumps of cocci. Saccharolytic activity is very weak but proteolytic activity is strong (Murdoch and Mitchelmore, 1991); the only VFA formed is acetate. Strains are easily distinguished from those of *M. micros* by their PEP. There are contradictory reports about its role in the normal flora; it is probably present on the skin and in the urogenital tract and possibly in the faecal flora and mouth. It is one of the most common anaerobic pathogens and is frequently recovered in pure culture from a wide range of serious infections, some of which have been fatal. Bourgault, Rosenblatt and Fitzgerald (1980) associated *F. magna* with soft-tissue and bone/joint infections; prosthetic joints are particularly at risk. It can be isolated from abscesses from many sites and appears to have a predilection for diabetic foot infections (Wheat *et al.*, 1986). Unlike other GPAC, it is usually co-cultured with facultative organisms rather than other obligate anaerobes. The few reports on antibiotic susceptibility are not always consistent, but they indicate that most or all strains are sensitive to metronidazole and β-lactams and some strains are resistant to clindamycin (Sanchez, Jones and Croco, 1992).

Micromonas micros: *Micromonas micros* is the only species at present in its genus. Cells are smaller than those of *F. magna* and are usually arranged in chains; the colonial morphology of the smooth morphotype is usually distinctive, with domed, glistening white colonies surrounded by a yellow-brown halo of discoloured agar (Murdoch and Mitchelmore, 1991). A rough morphotype described from gingival plaque forms dry, white, haemolytic colonies with crinkled edges (van Dalen *et al.*, 1993). Strains of *M. micros* are asaccharolytic but very strongly proteolytic; the only VFA formed is acetic acid. The PEP, supported by cell size, should distinguish isolates of *F. magna*; strains of *Slackia exigua* can be more difficult to separate. A selective medium (Turng *et al.*, 1996) and PCR-based methods for detection in polymicrobial dental specimens (Riggio, Lennon and Smith, 2001; Siqueira *et al.*, 2003) have been reported.

Micromonas micros is found in the gingival crevice and the gastro-intestinal tract; it may be part of the normal vaginal flora (Murdoch, 1998). It has been linked with other oral pathogens in the causation of periodontitis and peritonsillar abscesses (Mitchelmore *et al.*, 1995). It has been isolated from a wide range of extra-oral sites, usually in association with other anaerobes, in a mixed anaerobic abscess flora, typically consisting of micro-aerophilic streptococci ('*S. milleri*'), *Prevotella*, *M. micros* and fusobacteria (Murdoch, Mitchelmore and Tabaqchali, 1988b). Reports on antibiotic susceptibility are conflicting but most strains appear highly sensitive, and resistance to metronidazole and cefoxitin has been reported in strains from periodontitis (Rams *et al.*, 1991).

Peptoniphilus asaccharolyticus: *Peptoniphilus asaccharolyticus* is one of the commonest species of GPAC in clinical specimens (Murdoch, Mitchelmore and Tabaqchali, 1994), but with the radical changes in classification and identification; many older reports referring to *Pp. asaccharolyticus* undoubtedly included strains now assigned to other species, notably *A. hydrogenalis*. An asaccharolytic and proteolytic species that forms butyric acid as its terminal VFA, it was made the type species of the new genus *Peptoniphilus* by Ezaki *et al.* (2001); however, laboratory methods for its separation from two other species in the genus, *Pp. indolicus* and *Pp. hareii*, are at present inadequate. As the commercial PEP kits in use at present do not separate these species reliably, undue weight must be placed on the subjective criteria of cell and colonial morphology. *Peptoniphilus asaccharolyticus* has been reported as part of the normal flora of the gastro-intestinal and genito-urinary tracts, causing infections at the latter site (Brook, 1988). There are few data on pathogenicity or susceptibility to antibiotics; Brook (1987) reported that capsule formation was an important virulence determinant.

Anaerococcus vaginalis: This saccharolytic, butyrate-producing species was described as *Peptostreptococcus vaginalis* by Li *et al.* (1992) and re-assigned by Ezaki *et al.* (2001) to the genus *Anaerococcus*. It is easily identified by its distinctive PEP (Murdoch, 1998; Wilson *et al.*, 2000). It appears to be part of the normal vaginal flora and causes gynaecological infections. It accounted for 6% of all strains of GPAC in one survey (Murdoch, Mitchelmore and Tabaqchali, 1994). Murdoch and Kelly (1997) reported its isolation in pure culture and suggested that a large proportion of the organisms identified in previous surveys as *P. prevotii* actually refer to this species. It has been isolated from superficial sites (from leg ulcers with *S. aureus*), and recently from cerebral abscesses (Wilson *et al.*, 2000). There appear so far to be no studies of its pathogenicity, but the range and severity of the infections from which it has been isolated indicate that it is one of the more significant species of GPAC.

Clostridium Cluster I

Sarcina: This genus of strict anaerobes is highly unusual in that the cell morphology which is Gram-positive cocci arranged in 'packets' of eight that can form spores. 16S rRNA data have shown that the two species, *S. maxima* and *S. ventriculi*, fall into *Clostridium* cluster I of Collins *et al.* and are closely related to *C. perfringens*; they are therefore phylogenetically distant from 'true' GPAC. The G + C ratio is 28–31 mol%. *Sarcinae* require carbohydrates to ferment and are common soil organisms, being able to thrive in acid environments as low as pH 1 (Canale-Parola, 1986; Ezaki, Oyaizu and Yabuuchi, 1992). They are part of the faecal flora, particularly of vegetarians (Crowther, 1971; Finegold, Attebery and Sutter, 1974) but are rarely isolated from human pathological specimens.

Clostridium Clusters IV and XIVa

Ruminococcus: This genus is a genetically heterogeneous collection of strictly anaerobic, strongly fermentative organisms. The type species, *R. flavefaciens*, is a member of Cluster IV but many other species are assigned to Cluster XIVa, including *R. productus*, previously classified in the genus *Peptostreptococcus*. Cell morphology is variable, depending on the species. Most species do not use peptides and require carbohydrates for their growth; some can digest cellulose. The major VFA formed is acetic acid. The atmospheric requirements for their growth are stringent (Bryant, 1986), which may explain why they are very rarely isolated from human clinical material. Nevertheless, they are major constituents of the gastro-intestinal flora (Finegold, Attebery and Sutter, 1974; Harmsen *et al.*, 2002).

Coprococcus: This little studied genus (similar to ruminococci) was proposed by Holdeman and Moore (1974) for organisms isolated from the human faecal flora.

Coriobacteriaceae (High G + C Gram-Positive Phylum)

Slackia exigua: Nucleic acid sequence data have shown that *Slackia* is not placed in the clostridial radiation and is closely related to *Atopobium* in the high G + C Gram-positive phylum (Wade *et al.*, 1999). This species fulfils the definition of a GPAC, as the Gram-stained appearance reveals tiny cocci or coccobacilli, and strains are strictly anaerobic (probably why they are rarely isolated from clinical material), forming very slowly growing colonies. *Slackia exigua* is strongly proteolytic and forms no VFAs other than acetate; strains can most easily be distinguished from *M. micros* and *A. parvulum* by their PEP. It is frequently isolated from periodontitis and periapical infections. Strains initially identified as *Peptostreptococcus heliotrinreducens* have been isolated from polymicrobial deep organ abscesses where they were very similar to *M. micros* (Murdoch and Mitchelmore, 1989). Antibiotic susceptibility data are limited, but metronidazole and β-lactams were effective against the extra-oral strains reported.

Atopobium: This genus of obligately anaerobic Gram-positive cocci includes *A. parvulum*, which was initially placed in the genus *Peptostreptococcus*; it was then reclassified in *Streptococcus* because its G + C ratio is 46 mol% and it produces large quantities of lactic acid. *Atopobium parvulum* is strongly saccharolytic but weakly proteolytic. It is probably part of the normal flora of the gastro-intestinal tract and has been isolated from abdominal abscesses and wounds (Hardie, 1986). It is sensitive to the same range of antibiotics as other GPAC, including metronidazole.

GRAM-NEGATIVE ANAEROBIC COCCI

These organisms are rarely cultured from human infections and have been much less studied; however, they are part of the normal human flora, for which reason alone they deserve further investigation.

Veillonella

The genus was described by Prevot in 1933; subsequently using DNA–DNA hybridisation techniques, Mays *et al.* (1982) recognised seven species, but the authors were unable to differentiate them on phenotypic characteristics. The G + C content is 40–44 mol%. The cell morphology consists of small cocci arranged in pairs or short chains. Colonies grow slowly on normal laboratory media, their growth being encouraged by factors such as lactate, pyruvate and nitrate. All species form the VFAs acetic and proprionic acid and reduce nitrate to nitrite, some form catalase. Otherwise they are rarely reactive in carbohydrate fermentation reactions or PEP kits, making differentiation within the genus difficult. A method of identification using RFLP analysis of PCR-amplified 16S rRNA has been described (Sato *et al.*, 1997). A red fluorescence under long-wave UV light formerly recommended for picking out colonies in mixed culture depends on the medium used and is unreliable (Brazier and Riley, 1988). Little is known about their potential pathogenicity; they contain a lipopolysaccharide that is a potential virulence factor. *Veillonellae* are part of the normal flora of the mouth and can reach high counts in the saliva. A recent study using 16S rRNA probes reported that they were minor but regular constituents of human faeces (Harmsen *et al.*, 2002). They have been isolated from oral infections but are rarely found in infections outside the mouth. Brook (1996) cultured them from 4% of paediatric specimens received, the most frequent sites being abscesses, aspiration pneumonias, burns, bites and sinuses (Bhatti and Frank, 2000). There are few data on their susceptibility to antibiotics; most strains tested have been sensitive to metronidazole, β-lactams and clindamycin, but resistant to vancomycin and aminoglycosides.

Acidaminococcus

The sole species at present in the genus, *A. fermentans*, does not ferment sugars but uses amino acids as its energy source. The G + C content is 56 mol%. The VFAs formed are acetic and butyric acids. It has been isolated from the human gastro-intestinal flora and, on very rare occasions, from human pathological specimens.

Megasphaera

The only species at present in the genus, *M. elsdenii*, forms very large cocci (2.5 μm diameter) arranged in pairs or chains. It ferments carbohydrates, lactate and pyruvate and forms several VFAs, with *n*-caproic acid as the major metabolite. It is found in the stomachs of ruminants and occasionally of man. Human infections appear to be very rare.

REFERENCES

Bartlett JG (1993) Anaerobic bacterial infections of the lung and pleural space. *Clinical Infectious Diseases*, **16**, S248–S255.
Bartlett JG, Gorbach SL and Finegold SM (1974) The bacteriology of aspiration pneumonia. *The American Journal of Medicine*, **56**, 202–207.
Bhatti MA and Frank MO (2000) *Veillonella parvula* meningitis: case report and review of *Veillonella* infections. *Clinical Infectious Diseases*, **31**, 839–840.
Bourgault A-M, Rosenblatt JE and Fitzgerald RH (1980) *Peptococcus magnus*: a significant human pathogen. *Annals of Internal Medicine*, **93**, 244–248.
Bowker KE, Wootton M, Holt HA *et al.* (1996) The in-vitro activity of trovafloxacin and nine other antimicrobials against, 413 anaerobic bacteria. *The Journal of Antimicrobial Chemotherapy*, **38**, 271–281.
Brazier JS and Riley TV (1988) UV red fluorescence of *Veillonella* spp. *Journal of Clinical Microbiology*, **26**, 383–384.
Brook I (1987) Bacteraemia and seeding of capsulate *Bacteroides* spp. and anaerobic cocci. *Journal of Medical Microbiology*, **23**, 61–67.
Brook I (1988) Recovery of anaerobic bacteria from clinical specimens in 12 years at two military hospitals. *Journal of Clinical Microbiology*, **26**, 1181–1188.
Brook I (1989) Anaerobic bacterial bacteremia: 12-year experience in two military hospitals. *The Journal of Infectious Diseases*, **160**, 1071–1075.
Brook I (1996) *Veillonella* infections in children. *Journal of Clinical Microbiology*, **34**, 1283–1285.
Brook I and Frazier EH (1998a) Aerobic and anaerobic microbiology of retroperitoneal abscesses. *Clinical Infectious Diseases*, **26**, 938–941.
Brook I and Frazier EH (1998b) Microbiology of liver and spleen abscesses. *Journal of Medical Microbiology*, **47**, 1075–1080.
Brook I and Frazier EH (1990) Aerobic and anerobic bacteriology of wounds and cutaneous abscesses. *Archives of Surgery*, **125**, 1445–1451.
Brook I and Frazier EH (1993) Anaerobic osteomyelitis and arthritis in a military hospital. *The American Journal of Medicine*, **94**, 221–228.
Brook I and Walker RI (1984) Pathogenicity of anaerobic gram-positive cocci. *Infection and Immunity*, **45**, 320–324.
Brook I and Walker RI (1985) The role of encapsulation in the pathogenesis of anaerobic gram-positive cocci. *Canadian Journal of Microbiology*, **31**, 176–180.
Brook I, Yocum P and Frazier EH (1996) Bacteriology and β-lactamase activity in acute and chronic maxillary sinusitis. *Archives of Otolaryngology, Head and Neck Surgery*, **122**, 418–422.
Bryant MP (1986) Genus *Ruminococcus*. In *Bergey's manual of systematic bacteriology*, vol. 2, Mair NS, Sharpe ME and Holt JG (eds). Williams & Wilkins, Baltimore, pp. 1093–1097.
Canale-Parola E (1970) Biology of the sugar-fermenting sarcinae. *Bacteriology Review*, **34**, 82–97.
Canale-Parola E (1986) Genus *Sarcina*. In *Bergey's manual of systematic bacteriology*, vol. 2, Mair NS, Sharpe ME and Holt JG (eds). Williams & Wilkins, Baltimore, pp. 1100–1103.
Carlsson J, Larsen JT and Edlund M-B (1993) *Peptostreptococcus micros* has a uniquely high capacity to form hydrogen sulfide from glutathione. *Oral Microbiology and Immunology*, **8**, 42–45.
de Château M and Björck L (1994) Protein PAB, a mosaic albumin-binding bacterial protein representing the first contemporary example of module shuffling. *The Journal of Biological Chemistry*, **269**, 12147–12151.
Collins MD, Lawson PA, Willems A *et al.* (1994) The phylogeny of the genus *Clostridium*: proposal of five new genera and eleven new species combinations. *International Journal of Systematic Bacteriology*, **44**, 812–826.
Crowther JS (1971) *Sarcina ventriculi* in human faeces. *Journal of Medical Microbiology*, **4**, 343–349.
van Dalen PJ, van Steenbergen TJM, Cowan MM *et al.* (1993) Description of two morphotypes of *Peptostreptococcus micros*. *International Journal of Systematic Bacteriology*, **43**, 787–793.
van Dalen PJ, van Deutekom-Mulder EC, de Graaff J and van Steenbergen JM (1998) Pathogenicity of *Peptostreptococcus micros* morphotypes and *Prevotella* species in pure and mixed culture. *Journal of Medical Microbiology*, **47**, 135–140.
Dymock D, Weightman AJ, Scully C and Wade WG (1996) Molecular analysis of microflora associated with dentoalveolar abscesses. *Journal of Clinical Microbiology*, **34**, 537–542.
Edmiston CE, Walker AP, Krepel CJ and Gohr C (1990) The non-puerperal breast infection: aerobic and anaerobic microbial recovery from acute and chronic disease. *The Journal of Infectious Diseases*, **162**, 695–699.
Ezaki T and Yabuuchi E (1985) Oligopeptidase activity of gram-positive anaerobic cocci used for rapid identification. *Journal of General Applied Microbiology*, **31**, 255–266.
Ezaki T, Yamamoto N, Ninomiya *et al.* (1983) Transfer of *Peptococcus indolicus*, *Peptococcus asaccharolyticus*, *Peptococcus prevotii* and *Peptococcus magnus* to the genus *Peptostreptococcus* and proposal of *Peptostreptococcus tetradius* sp. nov. *International Journal of Systematic Bacteriology*, **33**, 683–698.
Ezaki T, Oyaizu H and Yabuuchi E (1992) The anaerobic gram-positive cocci. In *The prokaryotes*, Balows A, Truper HG, Dworkin M *et al.* (eds). Springer-Verlag KG, Berlin, pp. 1879–1892.
Ezaki T, Kawamura Y, Li N *et al.* (2001) Proposal of the genera *Anaerococcus* gen. nov., *Peptoniphilus* gen. nov. and *Gallicola* gen. nov. for members of the genus *Peptostreptococcus*. *International Journal of Systematic and Evolutionary Microbiology*, **51**, 1521–1528.
Finegold SM and George WL (1989) *Anaerobic infections in humans*. Academic Press, New York.
Finegold SM, Attebery HR and Sutter VL (1974) Effect of diet on human fecal flora: comparison of Japanese and American diets. *The American Journal of Clinical Nutrition*, **27**, 1456–1469.
Fung-Tomc JC, Minassian B, Kolek B *et al.* (2000) Antibacterial spectrum of a novel des-fluoro(6)quinolone, BMS-284756. *Antimicrobial Agents and Chemotherapy*, **44**, 3351–3356.
Garcia-Rodriguez JA, Garcia-Sanchez JE and Munoz-Bellido JL (1995) Antimicrobial resistance in anaerobic bacteria: current situation. *Anaerobe*, **1**, 69–80.

Genovese A, Bouvet J-P, Florio G *et al.* (2000) Bacterial immunoglobulin superantigen proteins A and L activate human heart mast cells by interacting and immunoglobulin E. *Infection and Immunity*, **68**, 5517–5524.

Graves MH, Morello JA and Kocka FE (1974) Sodium polyanethol sulfonate sensitivity of anaerobic cocci. *Applied Microbiology*, **27**, 1131–1133.

Halebian S, Harris B, Finegold SM and Rolfe RD (1981) Rapid method that aids in distinguishing gram-positive from gram-negative anaerobic bacteria. *Journal of Clinical Microbiology*, **13**, 444–448.

Hardie JM (1986) Anaerobic streptococci. In *Bergey's manual of systematic bacteriology*, vol. 2, Sneath PHA, Mair NS, Sharpe ME and Holt JG (eds). Williams & Wilkins, Baltimore, pp. 1066–1068.

Hare R (1967) The anaerobic cocci. In *Recent Advances in Medical Microbiology*, Waterson, AP (ed.). Churchill, London, pp. 285–317.

Harmsen HJM, Raangs GC, He T *et al.* (2002) Extensive set of 16S rRNA-based probes for detection of bacteria in human feces. *Applied and Environmental Microbiology*, **68**, 2982–2990.

Heginbottom M, Fitzgerald TC and Wade WG (1990) Comparison of solid media for cultivation of anaerobes. *Journal of Clinical Pathology*, **43**, 253–256.

Hill KE, Davies CE, Wilson MJ *et al.* (2002) Heterogeneity within the gram-positive anaerobic cocci demonstrated by analysis of 16S–23S intergenic ribosomal RNA polymorphisms. *Journal of Medical Microbiology*, **51**, 949–957.

Hillier SL, Krohn MA, Rabe LK *et al.* (1993) The normal vaginal flora, H_2O_2-producing lactobacilli, and bacterial vaginosis in pregnant women. *Clinical Infectious Diseases*, **16**, S273–S281.

Holdeman LV and Moore WEC (1974) New genus, *Coprococcus*, twelve new species, and emended descriptions of four previously described species of bacteria from human feces. *International Journal of Systematic Bacteriology*, **24**, 260–277.

Holdeman Moore LV, Johnson JL and Moore WEC (1986) Genus *Peptostreptococcus*. In *Bergey's manual of systematic bacteriology*, vol. 2, Sneath PHA, Mair NS, Sharpe ME and Holt JG (eds). Williams & Wilkins, Baltimore, pp. 1083–1092.

Ingham HR, Sisson PR, Tharagonnet D *et al.* (1977) Inhibition of phagocytosis in vitro by obligate anaerobes. *Lancet*, **ii**, 1252–1254.

Kremer BHA, Magee JT, van Dalen PJ and van Steenbergen TJM (1997) Characterization of smooth and rough morphotypes of *Peptostreptococcus micros*. *International Journal of Systematic Bacteriology*, **47**, 363–368.

Krepel CJ, Gohr CM, Walker AP *et al.* (1992) Enzymatically active *Peptostreptococcus magnus*: association with site of infection. *Journal of Clinical Microbiology*, **30**, 2330–2334.

Landers DV and Sweet RL (1983) Tubo-ovarian abscess: contemporary approach to management. *Reviews in Infectious Diseases*, **5**, 876–884.

Li N, Hashimoto Y, Adnan S *et al.* (1992) Three new species of the genus *Peptostreptococcus* isolated from humans: *Peptostreptococcus vaginalis* sp. nov., *Peptostreptococcus lacrimalis* sp. nov. and *Peptostreptococcus lactolyticus* sp. nov. *International Journal of Systematic Bacteriology*, **42**, 602–605.

Marshall R and Kaufman AK (1981) Production of deoxyribonuclease, ribonuclease, coagulase and hemolysins by anaerobic gram positive cocci. *Journal of Clinical Microbiology*, **13**, 787–788.

Mays TD, Holdeman LV, Moore WEC *et al.* (1982) Taxonomy of the genus *Veillonella* Prevot. *International Journal of Systematic Bacteriology*, **32**, 28–36.

Mitchelmore IJ, Prior AJ, Montgomery PQ and Tabaqchali S (1995) Microbiological features and pathogenesis of peritonsillar abscesses. *European Journal of Clinical Microbiology and Infectious Diseases*, **14**, 870–877.

Murdoch DA (1998) Gram-positive anaerobic cocci. *Clinical Microbiology Reviews*, **11**, 3–43.

Murdoch DA and Kelly M (1997) *Peptostreptococcus prevotii* or *Peptostreptococcus vaginalis*? Identification and clinical importance of a new species of *Peptostreptococcus*. *Anaerobe*, **3**, 23–26.

Murdoch DA and Mitchelmore IJ (1989) Isolation of *Peptostreptococcus heliotrinreducens* from human polymicrobial abscesses. *Letters in Applied Microbiology*, **9**, 223–225.

Murdoch DA and Mitchelmore IJ (1991) The laboratory identification of gram-positive anaerobic cocci. *Journal of Medical Microbiology*, **34**, 295–308.

Murdoch DA and Shah HN (1999) Reclassification of *Peptostreptococcus magnus* (Prevot, 1993) Holdeman and Moore 1972 as *Finegoldia magna* comb. nov. and *Peptostreptococcus micros* (Prevot 1933) Smith 1957 as *Micromonas micros* comb. nov. *Anaerobe*, **5**, 555–559.

Murdoch DA, Mitchelmore IJ and Tabaqchali S (1988a) Identification of gram-positive anaerobic cocci by use of systems for detecting preformed enzymes. *Journal of Medical Microbiology*, **25**, 289–293.

Murdoch DA, Mitchelmore IJ and Tabaqchali S (1988b) *Peptostreptococcus micros* in polymicrobial abscesses. *Lancet*, **i**, 594.

Murdoch DA, Mitchelmore IJ and Tabaqchali S (1994) The clinical importance of gram-positive anaearobic cocci isolated at St. Bartholomew's Hospital, London in 1987. *Journal of Medical Microbiology*, **41**, 36–44.

Murdoch DA, Collins MD, Willems A *et al.* (1997) Description of three new species of the genus *Peptostreptococcus* from human clinical specimens: *Peptostreptococcus harei* sp. nov., *Peptostreptococcus ivorii* sp. nov. and *Peptostreptococcus octavius* sp. nov. *International Journal of Systematic Bacteriology*, **47**, 781–787.

Murdoch DA, Shah HN, Gharbia SE and Rajendram D (2000) Proposal to restrict the genus *Peptostreptococcus* (Kluyver & van Niel, 1936) to *Peptostreptococcus anaerobius*. *Anaerobe*, **6**, 257–260.

Myhre EB (1984) Surface receptors for human serum albumin in *Peptococcus magnus* strains. *Journal of Medical Microbiology*, **18**, 189–195.

Nilson BH, Solomon A, Björck L and Akerstrom B (1992) Protein L from *Peptostreptococcus magnus* binds to the κ light chain variable domain. *Journal of Biological Chemotherapy*, **267**, 2234–2239.

Patella V, Casolaro V, Björck L and Marone G (1990) Protein L: a bacterial immunoglobulin-binding protein that activates human basophils and mast cells. *Journal of Immunology*, **145**, 3054–3061.

Reig M and Baquero F (1994) Antibacterial activity of clavulanate and tazobactam on *Peptostreptococcus* species. *The Journal of Antimicrobial Chemotherapy*, **33**, 358–359.

Reig M, Moreno A and Baquero F (1992) Resistance of *Peptostreptococcus* spp. to macrolides and lincosamides: inductible and constitutive phenotypes. *Antimicrobial Agents and Chemotherapy*, **36**, 662–664.

Riggio MP and Lennon A (2002) Development of a PCR assay specific for *Peptostreptococcus anaerobius*. *Journal of Medical Microbiology*, **51**, 1097–1101.

Riggio MP, Lennon A and Smith (2001) Detection of *Peptostreptococcus micros* DNA in clinical samples by PCR. *Journal of Medical Microbiology*, **50**, 249–254.

Rogosa M (1971) Peptococcaceae, a new family to include the gram-positive, anaerobic cocci of the genera *Peptococcus*, *Peptostreptococcus*, and *Ruminococcus*. *International Journal of Systematic Bacteriology*, **21**, 234–237.

Rogosa M (1974) Family III *Peptococcaceae*. In *Bergey's manual of determinative bacteriology*, 8th edn, Buchanan RE and Gibbons NE (eds). Williams & Wilkins, Baltimore, pp. 517–527.

Sabbaj J, Sutter VL and Finegold SM (1972) Anaerobic pyogenic liver abscess. *Annals of Internal Medicine*, **77**, 629–638.

Sanchez ML, Jones RN and Croco JL (1992) Use of the E-test to assess macrolide-lincosamide resistance patterns among *Peptostreptococcus* species. *Antimicrobial Newsletter*, **8**, 45–49.

Sato T, Matsuyama J, Sato M and Hoshino E (1997) Differentiation of *Veillonella atypica*, *Veillonella dispar* and *Veillonella parvula* using restriction fragment-length polymorphism analysis of, 16S rDNA amplified by polymerase chain reaction. *Oral Microbology and Immunology*, **12**, 350–353.

Shah HN and Gharbia SE (1995) The biochemical milieu of the host in the selection of anaerobic species in the oral cavity. *Clinical Infectious Diseases*, **20**, S291–S300.

Siqueira JF, Rocas IN, Andrade AFB and de Uzeda M (2003) *Peptostreptococcus micros* in primary endodontic infections as detected by 16S rDNA-based polymerase chain reaction. *Journal of Endodontics*, **29**, 111–113.

Skerman VBD, MacGowan V and Sneath PHA (1980) Approved lists of bacterial names. *International Journal of Systematic Bacteriology*, **30**, 225–240.

Smith DT (1930) Fusospirochetal disease of the lungs produced with cultures from Vincent's angina. *The Journal of Infectious Diseases*, **46**, 303–310.

Socransky SS, Haffajee AD, Dzink JL and Hillman JD (1988) Associations between microbial species in subgingival plaque samples. *Oral Microbiology and Immunology*, **3**, 1–7.

Socransky SS, Haffajee AD, Smith C and Dibart S (1991) Relation of counts of microbial species to clinical status at the sampled site. *Journal of Clinical Periodontology*, **18**, 766–775.

Song Y, Liu C, McTeague M *et al.* (2003) Rapid identification of gram-positive anaerobic coccal (GPAC) species originally in the genus *Peptostreptococcus* by multiplex PCR assays using genus- and species-specific primers. *Microbiology*, **149**, 1419–1427.

Soper DE, Brockwell NJ, Dalton HP and Johnson D (1994) Observations concerning the microbial etiology of acute salpingitis. *American Journal of Obstetrics and Gynecology*, **170**, 1008–1017.

Steffen EK and Hentges DJ (1981) Hydrolytic enzymes of anaerobic bacteria isolated from human infections. *Journal of Clinical Microbiology*, **14**, 153–156.

Summanen P, Baron EJ, Citron DM *et al.* (1993) *Wadsworth anaerobic bacteriology manual*, 5th edn. Star Publishing, Belmont, California.

Tally FP, Stewart PR, Sutter VL and Rosenblatt JE (1975) Oxygen tolerance of fresh clinical anaerobic bacteria. *Journal of Clinical Microbiology*, **1**, 161–164.

Todo K, Goto T, Miyamoto K and Akimoto S (2002) Physical and genetic map of the *Finegoldia magna* (formerly *Peptostreptococcus magnus*) ATTC 29328 genome. *FEMS Microbiology Letters*, **210**, 33–37.

Topiel SM and Simon GL (1986) Peptococcaceae bacteremia. *Diagnostic Microbiology and Infectious Diseases*, **4**, 109–117.

Turng BF, Guthmiller JM, Minah GE and Falkler WA (1996) Development and evaluation of a selective and differential medium for the primary isolation of *Peptostreptococcus micros*. *Oral Microbiology and Immunology*, **11**, 356–361.

van der Vorm ER, Dondorp AM, van Ketel RJ and Dankert J (2000) Apparent culture-negative prosthetic valve endocarditis caused by *Peptostreptococcus magnus*. *Journal of Clinical Microbiology*, **38**, 4640–4642.

Wade WG, Downes J, Dymock D *et al.* (1999) The family *Coriobacteriaceae*: reclassification of *Eubacterium exiguum* (Poco *et al.*, 1996) and *Peptostreptococcus heliotrinreducens* (Lanigan 1976) as *Slackia exigua* gen. nov., comb. nov. and *Slackia heliotrinireducens* gen. nov., comb. nov., and *Eubacterium lentum* (Prevot 1938) as *Eggerthella lenta* gen. nov., comb. nov. *International Journal of Systematic Bacteriology*, **49**, 595–600.

Watt B and Jack EP (1977) What are anaerobic cocci? *Journal of Medical Microbiology*, **10**, 461–468.

Wheat LJ, Allen SD, Henry M *et al.* (1986) Diabetic foot infections. *Archives of Internal Medicine*, **146**, 1935–1940.

Wilson MJ, Hall V, Brazier J and Lewis MAO (2000) Evaluation of a phenotypic scheme for the identification of 'butyrate-producing' *Peptostreptococcus* species. *Journal of Medical Microbiology*, **49**, 747–751.

Wren MWD (1980) Multiple selective media for the isolation of anaerobic bacteria from clinical specimens. *Journal of Clinical Pathology*, **33**, 61–65.

45

Non-Sporing Gram-Negative Anaerobes

Sheila Patrick[1] and Brian I. Duerden[2]

[1]*Department of Microbiology and Immunobiology, School of Medicine, Queen's University of Belfast, Belfast; and* [2]*Anaerobe Reference Laboratory, Department of Medical Microbiology, University of Wales College of Medicine, and National Public Health Service for Wales, Heath Park, Cardiff, UK*

INTRODUCTION

The major Gram-negative non-sporing anaerobic bacteria of medical interest are classified within the families Bacteroidaceae and Fusobacteriaceae. They are a major component of the normal microbiota of mucosal surfaces of man and the whole range of animals from invertebrates such as termites to primates and are also important pathogens, usually in infections related to those colonised mucosal sites. Their contribution to human infection was first recognised in a report of pulmonary gangrene and appendicitis by Veillon and Zuber from the Institute Pasteur in 1898 although organisms now classified as *Fusobacterium necrophorum* had already been shown to cause calf 'diphtheria' (by Loeffler in 1884), liver abscesses in cattle and facial necrosis in rabbits. In 1861, Pasteur himself had introduced the term anaerobe for microbes that would grow only in the absence of free oxygen. Before 1900, the role of non-sporing anaerobes in suppurative diseases of the middle ear, mastoid and sinuses, pulmonary gangrene and genital tract infection was established by the group of researchers at the Faculté de Medicin de Paris. However, with a few notable exceptions, non-sporing anaerobes were largely ignored in mainstream medical microbiology for most of the 20th century, until the 'anaerobic renaissance' of the 1970s when improved laboratory methods for reliable routine anaerobic microbiology became more widely available. The findings of the early workers were then rediscovered, and the widespread role of these organisms in infections of the abdomen, perineum and genitalia, mouth and respiratory tract, and compromised soft tissues at other sites became more generally recognised (Finegold, 1994).

DESCRIPTION OF THE ORGANISMS

Development of Taxonomy

Veillon and Zuber named their isolates *Bacillus fragilis, Bacillus fusiformis* and others, and the current classification dates from 1919 when Castellani and Chalmers proposed the name *Bacteroides* for all obligately anaerobic, non-sporing bacilli; this definition was modified to exclude Gram-positive organisms by Weiss and Rettger in 1937. Knorr introduced the name *Fusobacterium* in 1922 for the spindle-shaped Gram-negative anaerobes. Despite various subsequent proposals for new genus names, all of which were subsequently deemed to be invalid, these two genera were the basis of the classification presented in the eighth (1974) edition of Bergey's Manual of Determinative Bacteriology, with the addition of the single-species genus *Leptotrichia*

(*L. buccalis*) for fusiform organisms that produce lactic acid rather than butyric acid as their major metabolic product (Holdeman and Moore, 1974). The genus *Bacteroides* contained five groups of species: 1. *B. fragilis*, divided into five sub-species; 2. bile-sensitive non-pigmented species such as *B. oralis*; 3. *B. melaninogenicus* including all strains that produced black- or brown-pigmented colonies on lysed blood agar and divided into three sub-species; 4. non-saccharolytic non-pigmented species; and 5. other, unrelated saccharolytic species. This was the starting point for the rapid growth of clinical and taxonomic studies with Gram-negative anaerobes (Duerden, 1983).

It should be remembered that although these bacteria are Gram negative, they are not closely related taxonomically to facultative and aerobic Gram-negative bacteria generally encountered in infections such as the Enterobacteriaceae and the pseudomonads. *Bacteroides* and the related *Porphyromonas* and *Prevotella* spp. belong to the Bacteroidetes/Chlorobi superphylum. This group diverged early in evolutionary terms from other eubacteria, and this is thought to have occurred well before the divergence of the Gram-positive bacteria from the phylum that contains Gram-negative species such as *Escherichia coli*, the Proteobacteria. Similarly, the fusobacteria are within their own separate phylum and have similarities at the genome sequence level with low mol% G + C Gram-positive bacteria.

The early taxonomic studies on these bacteria were based upon conventional cultural and biochemical tests, which were difficult to perform reliably and to interpret in anaerobic growth conditions. The use of gas–liquid chromatography (GLC), introduced into this area of work in the late 1960s, was a significant advance in showing the possession of distinct metabolic pathways giving reproducible patterns of fatty acid end products. A recent description of GLC methodology for anaerobes is detailed in the Wadswoth-KTL anaerobic bacteriology manual (Jousimies-Somer *et al.*, 2002). However, the significant advances since 1980 have been based upon a combination of chemotaxonomy and genetic analysis in particular, for example, phylogenetic 16S rRNA gene sequencing. rRNA gene sequencing is now a standard method for the validation of culture identity, and suitable methodology is detailed in, for example, Stubbs *et al.* (2000). A comprehensive taxonomic listing of old and new nomenclature can be found in Jousimies-Somer and Summanen (2002). The reader is directed to the following publications for detailed information on taxonomic history and recent updates (Shah and Garbia, 1991; Jousimies-Somer, 1997; Shah, Gharbia and Duerden, 1998; Jousimies-Somer and Summanen, 2002). The chemotaxonomic characters of the main genera are summarized in Table 45.1.

Principles and Practice of Clinical Bacteriology Second Edition Editors Stephen H. Gillespie and Peter M. Hawkey

Table 45.1   Chemotaxonomic characters of the main genera of Gram-negative anaerobes of clinical importance

Character	Bacteroides	Bilophila	Prevotella	Porphyromonas	Fusobacterium	B. ureolyticus
Major products	Acetate, succinate	Acetate succinate	Acetate succinate	Butyrate	Butyrate	Acetate, propionate
Dehydrogenases	G-6-PDH, 6-PDGH, MDH, GDH	...	MDH, GDH	MDH, GDH	GDA	...
DNA G + C (mol%)	40–48	39–40	40–50	48–54	27–32	28
Peptidoglycan diamino acid	meso-DAP	...	DAP	DAP, lysine	DAP, lanthionine	...
Fatty acids	Straight saturated anteiso- and iso-methyl branched		Straight saturated anteiso- and iso-methyl branched	iso-methyl branched	Straight chain, monounsaturated	...
Quinones	MK-10, MK-11		MK-10, MK-11, MK-13	MK-9	–	...
Sphingolipids	+	...	–	–	–	–
Urease	–	+	–	–	–	+

6-PGDH, 6-phosphogluconate dehydrogenase; DAP, diaminopimelic acid; G-6-PDH, glucose-6-phosphate dehydrogenase; GDH, glutamate dehydrogenase; MDH, malate dehydrogenase.

Bacteroides

The sub-species of *B. fragilis* were given species status, and several new species were added to this fragilis group of intestinal *Bacteroides*. As other species have been removed to new genera, the genus *Bacteroides sensu stricto* should now encompass only those species of the fragilis group. It was also recognised that a single species *B. melaninogenicus*, which had been described by Oliver and Wherry in 1921, could not include organisms as diverse as the saccharolytic subsp. *intermedius* and subsp. *melaninogenicus* and the non-saccharolytic but highly proteolytic subsp. *asaccharolyticus*. *Bacteroides asaccharolyticus* was first separated as a distinct species and then assigned to a new genus, *Porphyromonas*, together with related isolates from diseases of the mouth, *P. gingivalis* and *P. endodontalis*; further new species of human and animal origin were added subsequently. The saccharolytic pigmented species were shown to share many properties with the group 2 non-pigmented organisms that were mostly of oral or genital tract origin. These became the 'melaninogenicus-oralis group' and were then assigned to a new genus as *Prevotella* spp. The other '*Bacteroides*' species that do not belong to these three main genera of human isolates have mostly been assigned to various new genera although a few remain with the inappropriate name '*Bacteroides*' while awaiting reclassification. With the exception of '*B.*' *ureolyticus* and *Bilophila wadsworthia*, these are of little concern in medical microbiology.

Fusobacterium

The genus *Fusobacterium* is now defined in biochemical and genetic terms as encompassing Gram-negative non-sporing anaerobes with DNA G + C content of 27–33 mol% that produce butyric and propionic acids as major metabolic products. Many that conform with this description do not exhibit the classical fusiform appearance. The more significant members of the genus in human disease are *F. necrophorum* and *F. nucleatum*.

General Properties

All members of the families Bacteroidaceae and Fusobacteriaceae are obligately anaerobic, Gram-negative bacilli, although many are highly pleomorphic and microscopic appearances range from cocco-bacillary to long filamentous forms. Their susceptibility to oxygen varies; some, such as *B. fragilis*, can remain viable (sub-cultivable) after exposure to oxygen for many hours after growth on agar media, whereas some others cannot withstand exposure for more than a few minutes. Most of those of concern in medical microbiology are at the more tolerant end of this spectrum and can be manipulated on the open bench, although the speed of regrowth and percentage viability

are generally better if bacteria are handled in an anaerobic cabinet without exposure to air. Anaerobic conditions for growth are provided by incubation in anaerobic jars, or cabinets or other small-scale culture systems that exclude air and have a chemical system for removing traces of oxygen. The most commonly used gas mixture is H_2 10%, CO_2 10%, N_2 80% for cabinets, or jars that incorporate a palladium catalyst to ensure the removal of remaining oxygen by reaction with the hydrogen. The species assigned to the main genera of are listed in Table 45.2. The genera *Bacteroides*, *Prevotella* and *Porphyromonas* all fall taxonomically within the class Bacteroides, whereas the genera *Fusobacterium* and *Leptotrichia* are within the class Fusobacteria. Details of the taxonomic status and classification of bacteria, including anaerobes, can be found at the NCBI taxonomy web site http://www.ncbi.nlm.nih.gov/Taxonomy/taxonomyhome.html/.

Bacteroides

Bacteroides fragilis and other members of the genus *Bacteroides sensu stricto* are indistinguishable by microscopy or by colonial appearance. They are small, moderately pleomorphic, non-motile Gram-negative bacilli; cocco-bacilli are common but long filaments

Table 45.2   Gram-negative anaerobic bacteria of clinical significance

Bacteroides	Prevotella	Porphyromonas	Fusobacterium
B. fragilis	Pr. melaninogenica*	P. asaccharolytica	F. necrophorum
B. distasonis	Pr. denticola*	P. gingivalis	F. nucleatum
B. ovatus	Pr. loescheii*	P. endodontalis	F. varium
B. thetaiotaomicron	Pr. intermedia*	P. canontiae**	F. pseudonecrophorum
B. vulgatus	Pr. nigrescens*		F. naviforme
B. uniformis	Pr. corporis*	P. macacae	F. mortiferum
B. eggerthii	Pr. tannerae*	P. levii	F. russii
B. caccae	Pr. pallens*	P. gulae	F. ulcerans
B. merdae			
B. stercoris	Pr. buccae		
B. capillosus	Pr. oris		
B. coagulans	Pr. buccalis		
B. putredinis	Pr. oralis		
	Pr. dentalis		
	Pr. enoeca		
('B.' splanchnicus)	Pr. veroralis		
	Pr. heparinolytica		
...	Pr. oulora		
	Pr. zoogleoformans		
Bilophila wadsworthia	Pr. bivia		
	Pr. disiens		
...			
'B.' ureolyticus			

*Black- or brown-pigmented *Prevotella* spp.
**Non-pigmented *Porphyromonas* spp.

are rare. They grow well on conventional laboratory media containing blood to form circular, low convex, smooth, shiny, translucent or semi-opaque, grey colonies, 1–3 mm in diameter, that are often moist or frankly mucoid after incubation for 24–28 h. Most strains are non-haemolytic, but a few are slightly haemolytic and a small proportion are distinctly β-haemolytic. They are strongly saccharolytic organisms, producing acid from a range of carbohydrates, and moderately proteolytic. The main end products of metabolism detected using GLC are acetic and succinic acids, with smaller amounts of various other acids but not n-butyric or lactic acids. Growth is generally stimulated by whole bile but inhibited by sodium deoxycholate and isolates are tolerant of the bile salt sodium taurocholate. *Bacteroides splanchnicus* is an intestinal *Bacteroides* that differs from the other true *Bacteroides* spp. in producing propionic and n-butyric acids and in genetic and chemotaxonomic characters; it probably represents a distinct genus.

Prevotella

The various species of *Prevotella* cannot be distinguished by cell morphology. Most are short pleomorphic Gram-negative bacilli, often with many cocco-bacillary forms. They are divided into pigmented and non-pigmented groups by the ability of some to assimilate haemoglobin from blood in the medium, converting it to protohaemin which accumulates in the cells, making the colonies dark brown or black after incubation for several days. Most pigmented strains are haemolytic, and pigmentation depends upon lysis of the red cells in the medium and develops most rapidly on lysed blood agar. Pigmentation can vary from almost black to pale brown. Colonies of the non-pigmented species are indistinguishable from each other and very similar to those of the *Bacteroides* spp. – 1–2 mm in diameter, circular, shiny, convex and light buff or grey coloured; most are non-haemolytic. *Prevotella* spp. are moderately to strongly saccharolytic. All produce acid from glucose and various other carbohydrates and, like *Bacteroides* spp., their major metabolic products are acetic and succinic acids; n-butyric acid is not produced. Most species are moderately to strongly proteolytic. They are inhibited by bile and bile salts.

Porphyromonas

Formerly known as the asaccharolytic *B. melaninogenicus* strains, *Porphyromonas* spp. are similar to pigmented *Prevotella* spp. in microscopic and colony morphology. They are small, mainly cocco-bacillary organisms with few long filaments. Growth is slower than with *Bacteroides* spp. and some *Prevotella* spp.; small colonies may be visible after incubation for 24 h but many do not appear until 48 h when they are <1 mm in diameter, smooth, shiny and grey. Dark brown or black pigmentation develops after 3–7 days on blood agar, more rapidly on lysed blood agar. After 4–5 days, they are 1–2 mm in diameter and haemolytic – lysis of the cells being essential for pigmentation. Most strains have a characteristically strong, putrid smell, even more noticeable than those of other non-sporing anaerobes. They do not produce acid from glucose or other carbohydrates but their metabolism, as with all anaerobes, is fermentative. The major metabolic end product is n-butyric acid. Like *Prevotella* spp., *Porphyromonas* spp. are inhibited by bile or bile salts. They are vigorously proteolytic, and this is considered to be a factor in their virulence.

Fusobacterium

Fusobacteria are anaerobic Gram-negative bacilli of varied size and morphology. Classical fusobacteria, such as *F. nucleatum* are fairly long (4–6 μm), regular bacilli with tapered or pointed ends. Spheroplasts and spindle forms are common. Others (e.g. some strains of *F. necrophorum*) are pleomorphic with a mixture of long, filamentous organisms with rounded ends and many short or even cocco-bacillary forms and numerous spheroplasts and L-forms. Yet others are mostly cocco-bacillary. Colonial appearance is also varied, but most fusobacteria produce moderate to large colonies, 1–3 mm in diameter, generally with an irregular or dentate edge. They vary from translucent to granular and opaque. Their metabolism varies; carbohydrates are fermented only feebly or not at all but most species are proteolytic and the more virulent species such as *F. necrophorum* produce a range of enzymes that can break down tissue components (see p. 536 Tissue damage line 50–62). The major metabolic products of fusobacteria, which help define the genus, are n-butyric and propionic acids, the latter produced by deamination of threonine. The G + C content of fusobacteria is low at 27–33 mol%, and these and *Leptotrichia* are phylogenetically distinct from the Bacteroidaceae.

Leptotrichia

The species *L. buccalis* may be synonymous with the original fusiform organism described by Vincent. It has typical, long fusobacterial cells, 5–15 μm long, many with pointed ends. Its colonies are lobate or convoluted, opaque and grey/yellow in colour. Although it is within the Fusobacteriaceae, it differs from *Fusobacterium* spp. in that it produces acid from several sugars and the only major metabolic product is lactic acid, with a small amount of acetic acid; n-butyric acid is not produced. Complete 16S rDNA sequencing and DNA-DNA hybridisation of strains initially assigned to *L. buccalis* reveals sufficient genetic diversity to warrant the designation of five different species (including *L. buccalis*) within this group (Eribe *et al.*, 2004).

Bilophila

The single species *Bil. wadsworthia* is a slow-growing, small Gram-negative bacillus that is stimulated by bile, requires pyruvate for growth and is asaccharolytic and urease positive (Baron *et al.*, 1992). It has no genetic homology with *Bacteroides* spp. and is classified within the Desulfovibrionaceae in the phylum Proteobacteria (Laue, Denger and Cook, 1997).

'Bacteroides' ureolyticus

This slender Gram-negative anaerobe typically forms pitting colonies on agar media (hence its former name *B. corrodens*). It is asaccharolytic, strongly proteolytic and urease positive. Its G + C content of 28 mol% clearly sets it apart from the other genera of Bacteroidaceae. Genotypically it is related to the Campylobacteraceae, in the phylum Proteobacteria (Jousimies-Somer, 1997). It is likely that it will be renamed once further similar bacteria have been isolated and studied.

Growth Conditions

As well as requiring anaerobic conditions, most Gram-negative anaerobic bacilli require enriched media for growth; none grows readily on minimal media or unsupplemented nutrient agar. Most require a source of haemin for synthesis of the menaquinones that are essential components of their anaerobic electron transport systems. Some *Prevotella* and *Porphyromonas* spp. also require menadione (vitamin K). Media for optimal growth comprise a rich nutrient base, e.g. Fastidious Anaerobe Agar (Lab M), Columbia agar base (Oxoid), Brucella agar or Wilkins Chalgren agar, with 5–10% blood plus haemin and menadione. *Bacteroides* spp. and some of the more vigorous *Prevotella* spp. produce reasonable growth after 24 h, but other species grow more slowly and require undisturbed anaerobic incubation for 48–72 h, or even longer, for recognisable growth. Details of the growth conditions for anaerobes of clinical importance can be found in the following references (Moore and Moore, 1977; Summanen *et al.*, 1993; Jousimies-Somer *et al.*, 2002).

NORMAL HUMAN MICROBIOTA

All the Bacteroides and Fusobacteria associated with man are part of the normal resident microbiota of the gastrointestinal, oral or genitourinary mucosae where they form a major part of the complex and interdependent ecosystems. However, the species found at the three sites are quite distinct and each species appears to have a distinct ecological niche (Drasar and Duerden, 1991).

Gastrointestinal Tract

This is the normal habitat of all the species of *Bacteroides sensu stricto* (*B. fragilis* group). They are one of the predominant groups of organisms in the faecal and colonic microbiota, having populations of approximately 10^{12} cfu/g wet weight of faeces and outnumbering the facultative enterobacteria by at least 1000:1. Not all species are equally common. The main components of the normal faecal microbiota are *B. vulgatus*, *B. distasonis*, *B. thetaiotaomicron* and *B. uniformis*. Within the gut, *Bacteroides* spp. contribute to the degradation of ingested material, in particular heterologous polysaccharides in plant material, that the mammalian system is incapable of degrading. This process results in the production of short-chain fatty acids, which once absorbed from the gut are potential substrates for energy metabolism in the intestinal mucosal epithelium. The potential for intimate interaction between the normal microbiota and the human host is illustrated by *B. thetaiotaomicron*. Within the environment of the small intestine, *B. thetaiotaomicron* appears to induce villus cells to fucosylate their cell surface glycoconjugates. *Bacteroides thetaiotaomicron* also secretes an α-fucosidase which cleaves the fucose from the villus cell glycoconjugates, thus releasing the fucose, which then becomes a potential carbon and energy source for the bacterium (Hooper *et al.*, 1999). Studies of the adherent mucosal microbiota indicate considerable variation amongst individuals. Mucosal biopsies of the proximal colon and rectum of 12 individuals showed that in different people one or two species of *Bacteroides* predominated by culture (Poxton *et al.*, 1997). For example, *B. fragilis* was the major isolate in one individual, the second most common isolate in three individuals and not detected in six, whereas *B. vulgatus* was the most common isolate in four individuals. Although *B. fragilis* accounts for fewer than 10% of faecal isolates and was not detected in some of the biopsy patients, it is the principal pathogen in gut-associated sepsis (approximately 75% of isolates from abdominal infections). This indicates that *B. fragilis* has a greater pathogenic potential than the other *Bacteroides* spp. *Bil. wadsworthia* is also a common but not dominant member of the faecal microbiota. The gastrointestinal tract is probably also the normal habitat of *P. asaccharolytica* and '*B.*' *ureolyticus*, but these are present only in small numbers.

Genitourinary Tract

Gram-negative anaerobes are part of the normal microbiota of the vagina although present in relatively small numbers (approximately 10^6 cfu/g of secretions) and outnumbered by Gram-positive species such as lactobacilli. Most isolates belong to the genus *Prevotella*, principally *Pr. bivia* and *Pr. disiens* which are uncommon elsewhere in the body; the pigmented species *Pr. melaninogenica* is also present, but in smaller numbers and less frequently (Drasar and Duerden, 1991).

Mouth

The oral microbiota have long been recognised as a source of a wide range of Gram-negative anaerobic species. The main habitat for these organisms in the mouth is the gingival crevice where both *Prevotella* and *Fusobacterium* spp. make a major contribution to the complex microbiota. The most commonly reported isolates are *Pr. melaninogenica*,

Pr. oralis and other non-pigmented *Prevotella* spp. and fusobacteria, mainly *F. nucleatum*. *Prevotella intermedia* is also isolated from the normal gingival microbiota but, like *B. fragilis* in the colon, represents a much smaller proportion of the total count of *Prevotella* spp. than do other species. *Porphyromonas gingivalis* is rarely isolated from healthy gingivae (Drasar and Duerden, 1991).

PATHOGENICITY

A wide range of Bacteroidaceae and Fusobacteriaceae have been isolated at one time or another from infections of man but a high proportion of these infections are caused by only a few, more virulent species. Three main features are common to most of these infections: 1. the source of infection is the endogenous microbiota of the patient's own gastrointestinal, oropharyngeal or genitourinary mucosa; 2. alterations of the host tissue, e.g. trauma and/or hypoxia, provide suitable conditions for the development of secondary opportunist anaerobic infections; and 3. the infections are generally polymicrobial, often involving mixtures of several anaerobic and facultative species acting synergically to cause damage. The initiation of infection generally depends on host factors, but even in such opportunist situations, some species show particular pathogenic potential not evident for the majority of related species from the same normal habitat. Thus, most species have some pathogenic capability but the majority of serious anaerobic infections are caused by the small number of more virulent species (Duerden, 1994).

The types of infection generally associated with Gram-negative anaerobes are listed in Table 45.3. The species most commonly isolated and considered to be significant pathogens in these situations are: *B. fragilis* in abdominal infections; *Pr. intermedia*, *P. gingivalis*, and *F. nucleatum* in periodontal disease and other infections related to the mouth; *P. asaccharolytica* and '*B.*' *ureolyticus* in superficial necrotising infections; and *F. necrophorum* in the invasive disease necrobacillosis (Lemmiere's disease). *Fusobacterium necrophorum* (including subsp. *necrophorum*, formerly Biovar A, and subsp. *funduliforme*, formerly Biovar B) has more of the characteristics of a virulent primary pathogen than other anaerobic species; typically, it causes severe purulent tonsillitis with pseudomembrane formation and lymph node involvement, septicaemia and metastatic abscess formation. It may be fatal even for a previously healthy person.

To establish an infection, bacteria must attach to target cells (generally mucosal or epithelial cells), invade the tissues, establish themselves by multiplying at the site of infection and avoiding elimination by the

Table 45.3 Gram-negative anaerobic bacteria isolated from the normal microbiota and from infections

Normal microbiota	Infections
Faeces	Abdominal
B. vulgatus	*B. fragilis*
B. distasonis	*Bil. wadsworthia*
B. thetaiotaomicron	*B. thetaiotaomicron*
B. uniformis	
Vagina	Genito-urinary
Pr. disiens	*B. fragilis*
Pr. bivia	*Prevotella* spp.
Pr. melaninogenica	*P. asaccharolytica*
	'*B.*' *ureolyticus*
Mouth	Head and neck
Pr. melaninogenica	*P. gingivalis*
Pr. intermedia	*Pr. intermedia*
Pr. ovalis	*F. nucleatum*
F. nucleatum	*F. necrophorum*
	Superficial
	P. asaccharolytica
	'*B.*' *ureolyticus*
	B. fragilis
	F. ulcerans

host's defence mechanisms and cause damage both to local tissues and, in systemic infections, to the whole patient. A factor that may impact on each of these events is within-strain surface variation in the form of phase and antigenic variation. This type of variation is mediated by genomic rearrangements and is reversible within a single strain (Patrick and Larkin, 1995). Multiple variants may arise during the progression of infection of a single individual or indeed during one subculture in the laboratory. The more virulent Gram-negative anaerobes exhibit virulence factors that help elicit the different stages of infection. A review of the major potential virulence factors of *Bacteroides fragilis* can be found in Patrick (2002). Whole-genome sequencing programmes have added considerably to our knowledge and understanding of pathogenic bacteria and will continue to do so in years to come. Gram-negative non-sporing anaerobe whole-genome sequencing programmes are listed in Table 45.4.

Surface Variation

A single strain of *B. fragilis* may have the capability to reversibly produce three different encapsulating surface structures, namely a large capsule or small capsule, both visible by light microscopy, or an electron-dense layer (EDL) outwith the outer membrane, visible by electron microscopy. In addition, polysaccharides associated with the EDL and large capsulate bacteria (but which do not appear to form the bulk of the large capsule structure) exhibit within-strain antigenic variation (Lutton *et al.*, 1991). This is mediated, at least in part, by invertible promoter regions upstream of the polysaccharide biosynthesis operons. The invertible regions are bound by inverted repeats of 30 or 32 bp in length with striking similarity to the *Salmonella typhimurium* **H** flagellar antigen **i**nversion **cross**-over (*hix*) recombination sites of the invertible hin region (Patrick *et al.*, 2003). In *B. fragilis* NCTC 9343, there are ten different polysaccharide biosynthesis operons. The genes within these operons indicate that some may be similar to enterobacterial 'O-antigen' rather than classical capsules. Analysis of the entire genome has revealed a further 16 regions with similar sequence characteristics, suggesting that these may also be active invertible promoters. Most of these are upstream of conserved putative secreted and outer membrane proteins of unknown function. Also there is evidence of large scale (5–120 kb) DNA inversion events within the genome (Cerdeno-Tarraga *et al.*, 2005).

Whether this phenomenal variation relates to the increased frequency of association of *B. fragilis* with infection when compared with other gastrointestinal tract *Bacteroides* or survival in the gastrointestinal tract remains to be determined.

Adhesion

Bacteroides fragilis can attach to both host cells and components of the extracellular matrix; however, studies of adhesion have rarely taken into account the considerable within-strain surface variation

p. 535 Surface Variation line 19–41. Haemagglutination and attachment to host cells in *B. fragilis* is demonstrable in EDL-enriched populations (Patrick *et al.*, 1988). Sodium periodate treatment of the bacteria abolishes haemagglutination, which suggests that saccharides are involved. Populations of strain NCTC 9343 enriched for either a large or small capsule do not haemagglutinate (Patrick *et al.*, 1996). As recent clinical isolates vary considerably in the proportion of bacteria that express the different capsules, this probably accounts for the variation in adhesion observed with different strains. Furthermore, the EDL/non-capsulate bacteria release extracellular vesicles which by themselves will cause haemagglutination. Despite reports of fimbrial expression by *B. fragilis*, fimbrial genes have not been reliably identified, although a gene with similarity to a putative non-fimbrial adhesin described in *Pr. intermedia* is present in *B. fragilis* NCTC 9343 (Cerdeno-Tarraga *et al.*, 2005). Studies of the attachment to components of the extracellular matrix have generally not taken into account the within-strain variability and the potential attachment of bacterial extracellular components; however, there is clear evidence that *Bacteroides* spp. can attach to some host components. Examples include fibronectin, collagen type 1 and vitronectin.

Periodontal pathogens also exhibit adhesive properties. *Fusobacterium nucleatum*, *P. gingivalis* and *Pr. intermedia* adhere to crevicular epithelium and cause haemagglutination. It seems likely that, as with *B. fragilis*, both proteinaceous haemagglutinins and surface saccharides are involved. More than ten genes encoding haemagglutinin-related proteins have been identified in the *P. gingivalis* genome sequence (Nelson *et al.*, 2003). Another aspect of adhesion that may be important in polymicrobial infections is the ability of bacterial species to coaggregate. These coaggregates may be important in creating appropriate conditions for the symbiotic metabolism and pathogenic interaction characteristic of anaerobic infections. This feature is manifest by oral *Prevotella* and *Porphyromonas* spp. together with streptococci, actinomyces and other oral species.

Fusobacterium necrophorum haemagglutinates erythrocytes from a range of species, including humans, and some will also aggregate human platelets. It is suggested that this causes intravascular coagulation which contributes to the establishment of infection (Hagelskjaer Kristensen and Prag, 2000).

Invasion

Most anaerobic pathogens are not primarily invasive. Initiation of infection depends on initial damage due to trauma, hypoxia, neoplasia or some other alteration to provide the route of entry for the anaerobes. *Fusobacterium necrophorum* is an exception to this rule, as although it may be present as part of the normal microbiota, in some patients it penetrates the mucosa. The reasons for this are unknown but may relate to underlying bacterial or viral pharyngitis leading to a reduction in the host defence at the mucosal surface (Hagelskjaer Kristensen and Prag, 2000).

Table 45.4 Complete genome sequencing projects of Gram-negative anaerobes of clinical importance

Species	Strain	Genome size (Mb)	Reference
Bacteroides fragilis	NCTC 9343	5.21	Cerdeno-Tarraga *et al.* (2005)
	638R	5.44	In progress, The Wellcome Trust Sanger Institute *http://www.sanger.ac.uk*_NCBI Accession No. NC_003228
	YCH 46	5.27	Kuwahara *et al.* (2004)
Bacteroides thetaiotaomicron	VPI-5482	6.2	Xu *et al.* (2003)
Fusobacterium nucleatum subsp. *vincentii*	ATCC 49256	2.12	Kapatral *et al.* (2003)
Fusobacterium nucleatum subsp. *nucleatum*	ATCC 25586	2.17	Kapatral *et al.* (2002)
Porphyromonas gingivalis	W83	2.34	Nelson *et al.* (2003)
Prevotella intermedia	17	3.33	In progress, The Institute for Genomic Research *http://www.tigr.org*_NCBI Accession No. NC_003441
Tannerella forsythensis	ATCC 43037	Unfinished	In progress, The Institute for Genomic Research *http://www.tigr.org*_NCBI Accession No. NC_003915

Establishing Infection

Once the initial damage has allowed the anaerobes to penetrate the tissues, they must establish a focus of infection by multiplying and avoiding elimination by the host's defence mechanisms. Most anaerobic infections are polymicrobial, and the metabolic interdependency of the bacterial mixtures involved (both anaerobic and facultative species) is important to their establishment in the tissues, the satisfaction of their nutritional requirements and the expression of their synergic pathogenicity (e.g. Rotstein, Kao and Houston, 1989); thus, the virulence of anaerobic species is a reflection of their ability to exploit a compromised host environment. Tissue damage and necrosis, a reduction in blood supply leading to hypoxia and the presence of a blood clot or foreign body or substance (especially $CaCl_2$) create conditions appropriate for anaerobic growth. The capacity of facultative species such as *E. coli* to consume oxygen may help create reduced conditions favourable to the growth of anaerobes. *Bacteroides fragilis* has a cluster of genes, the aerotolerance operon, whose products aid its survival on exposure to oxygen. A similar cluster had been identified in the *P. gingivalis* genome. In addition, *B. fragilis* can grow in the presence of nanomolar concentrations of oxygen and this is linked to the presence of a cytochrome *bd* oxidase (Baughn and Malamy, 2004).

Gram-negative anaerobes require several growth factors and nutrients produced by damaged host tissues or by other bacteria acting in synergy. For example, *B. fragilis* uses haemoglobin and haemaglobin–haptoglobin complexes as sources of iron and the porphyrin component. It has an iron-repressible haem-binding outer membrane protein involved in the uptake of this essential growth factor (Otto *et al.*, 1996). *Porphyromonas gingivalis* demetalates protohaem to protophorphyrin, providing for its iron needs. Novel putative haemolysin genes with sequence similarity have been identified in *B. fragilis*, *Pr. melaninogenica*, *Pr. intermedia* and *P. gingivalis*. Hydrolytic enzymes and proteases produced by *P. gingivalis* (e.g. gingipains), *P. asaccharolytica*, *Pr. intermedia* and *Pr. denticola* release nutrients from growth factors for these species themselves and for other members of the polymicrobial ecosystems. Some of the proteases of *P. gingivalis* are subject to the post-translational addition of oligosaccharides within the catalytic domain (Gallagher *et al.*, 2003). *Bacteroides fragilis* produces a serine-thiol-like protease that hydrolyses the α-chain of fibrinogen (Chen *et al.*, 1995). As soluble fibrinogen is converted to insoluble fibrin as part of the normal blood-clot formation and wound healing, this enzyme has the potential to slow clot formation. It also produces heparinase and condroitin sulphatase, which hydrolyse heparin and condroitin sulphate, allowing their use as nutrients. The activities of these enzymes are likely to encompass tissue destruction and, where cells and molecules of the immune system are attacked, evasion of the immune response (see p. 536 Evasion of the host defence line 16–34 and Tissue damage line 38–61).

Evasion of the Host Defence

The host response to anaerobic infection includes phagocytosis and opsonisation and killing by serum immunoglobulin and complement. The virulent species of Gram-negative anaerobe have various means of avoiding and resisting these defence mechanisms. *Bacteroides fragilis* and some black-pigmented, Gram-negative anaerobes produce various extracellular polysaccharides that may form encapsulating structures, not all which are visible by light microscopy (see p. 535 Surface Variation line 21–24). In *B. fragilis* some of these various components have been implicated in adhesion (Patrick *et al.*, 1988), protection against phagocytosis, resistance to killing mediated by the alternative complement pathway (Reid and Patrick, 1984) and the induction of abscess formation (Tzianabos *et al.*, 1993).

Bacteroides fragilis also inhibits macrophage migration and impairs the phagocytosis of other species involved in the polymicrobial infections. Succinic and other short-chain fatty acid metabolic products have also been shown to inhibit chemotaxis, phagocytosis and intracellular killing by phagocytic cells (Rotstein *et al.*, 1989). *Porphyromonas gingivalis* (Laine and von Winkelhoff, 1998) also produces a capsule that protects against phagocytosis and intracellular killing and generates metabolic products that compete with chemotactic peptides, heat-labile opsonins and complement components to block chemotactic receptors on polymorphs. Abscess-derived neutrophils are reported to harbour viable bacteria and be less efficient at killing *B. fragilis* than neutrophils derived from either peritoneal aspirates or peripheral blood (Finlay-Jones *et al.*, 1991). The potential for excreted extracellular polysaccharide and extracellular outer membrane vesicles to mop up opsonin, activate complement and interact with phagocytes, thus diverting the host defence from the bacterial cell, should not be overlooked. Proteases that degrade cytokines, produced by for example *P. gingivalis*, may also interfere with the immune response including phagocytic function.

Fusobacterium necrophorum isolated from animal infection secrete a well-characterised leukotoxin with specific toxicity for cattle and sheep neutrophils (Narayanan *et al.*, 2002); however, it is not clear whether isolates from human infections secrete similar molecules with specificity for human neutrophils.

Several virulent anaerobic species generate products that inhibit or destroy the humoral components of the host's defences. The black-pigmented species produce proteolytic enzymes active against immunoglobulins and complement and most anaerobic bacteria produce soluble metabolites that are leukotoxic, inhibit chemotaxis and damage mucosal cells.

Neuraminidases, which cleave sialic acid from oligosaccharides on host cell glycoproteins and glycolipids, produced by *B. fragilis* could play a key role in virulence (Godoy *et al.*, 1993). There are more than 20 known naturally occurring sialic acids formed by various substitutions and additions to neuraminic acid (Schauer, 1985). There is growing evidence that these sugar residues are involved in the biological activities of the host cells and molecules, in particular in relation to immune system function.

Tissue Damage

Once infection is established, several virulence factors appear to act in combination to produce damage that is manifest as tissue necrosis and abscess formation. Some products of anaerobic metabolism are toxic to mammalian cells: volatile fatty acids and sulphur compounds (e.g. H_2S and amines). Most pathogenic anaerobes also produce extracellular enzymes that hydrolyse tissue components and are thought to play a significant role in pathogenesis (e.g. Rudek and Haque, 1976). *Bacteroides fragilis*, *P. gingivalis*, *P. asaccharolytica*, *Pr. intermedia*, *Pr. melaninogenica* and *Pr. denticola* produce hyaluronidase, condroitin sulphatase, heparinase and a range of enzymes that hydrolyse carbohydrates. All of these enzymes play dual roles, causing tissue damage and providing nutrients for the infecting microbes. *Fusobacterium necrophorum* produces membrane-damaging lipases that may be related to haemolytic activity (Hagelskjaer Kristensen and Prag, 2000).

Proteolytic activity is associated with the strongly proteolytic *Porphyromonas* spp. and some *Prevotella* spp. Proteases, such as the trypsin-like cysteine proteases (gingipains) of *P. gingivalis* (Curtis *et al.* 2001), are thought to be important in the destruction of gingival tissue and of the collagen bridges in the gingival crevice in periodontal disease. Proteases may also be important in the contribution of *P. asaccharolytica* and '*B.*' *ureolyticus* to tissue damage in ulcerative and gangrenous lesions such as genital and perineal ulcers, decubitus and varicose lesions and diabetic gangrene.

Systemic Inflammatory Response Syndrome

Systemic inflammatory response syndrome (SIRS) or 'sepsis' is the most common cause of death in patients who are critically ill in non-coronary intensive care units. Endotoxin, containing lipopolysac-charide (LPS), is one bacterial component that is known to play a central role. In patients where there is no obvious bacterial septi-caemia, it is thought that the source of the endotoxin is the normal intestinal microbiota. *Bacteroides fragilis* LPS is 5000-fold less toxic in a mouse lethality model than enterobacterial LPS, but given the high numbers of *Bacteroides* spp. in the intestinal tract, where entero-bacteria may be outnumbered by up to 1000–1, even allowing for the lower toxicity, it is likely that the numerical dominance of *Bacter-oides* LPS means that it plays some role in endotoxic shock and SIRS (Delahooke, Barclay and Poxton, 1995a, b). LPS triggers inflamma-tory events via the Toll-like receptors (TLRs). LPS from *E. coli* and *Salmonella* spp. interacts with TLR4, whereas LPS from *P. gingivalis* interacts with TLR2, resulting in a less efficient production of proinflammatory cytokines (Erridge, Bennett-Guerrero and Poxton, 2002; Netea *et al.*, 2002). It is now known that the mutation in C3H/HeJ mice that are hyporesponsive to enteric LPS but react with *Bacter-oides* spp. LPS (Delahooke, Barclay and Poxton, 1995b) is in the TLR4 gene. The differences in the reactivity of the different LPS molecules appear to relate to their chemical composition, the number, length and composition of the fatty acid chains (or acyl groups) attached to the diglucosamine backbone of the LPSs being different. The lipid A of *P. gingivalis* is less conical in shape than that of *E. coli*. In addition, enterobacterial lipid A diglucosamine is bisphosphor-ylated whereas in *B. fragilis* it is monophosphorylated; the distal glucosamine residue of the lipid A molecule is lacking in a phosphate group. Also, the core polysaccharide sugar keto-deoxyoctonate (KDO) is phosphorylated which renders it undetectable in the standard KDO thiobarbituric acid assay (Beckmann *et al.*, 1989; Fujiwara *et al.*, 1990).

Whether or not the LPS of *Bacteroides* spp. possesses or is lacking in an O-antigen similar to that found in the enterobacteria is subject to controversy. In the literature, there are reports that extended repeating O-antigen is absent in *Bacteroides* (Lindberg *et al.*, 1990; Comstock *et al.*, 1999). The suggestion is that LPS is more like the lipo-oligosac-charides of, for example, *Nesseria meningitidis* (Jennings *et al.*, 1999) or 'rough' type mutants of the enterobacteria. Other publications, however, quite clearly illustrate silver-stained LPS polyacrylamide gel electrophoresis (PAGE) profiles with ladder patterns characteristic of smooth LPS in *B. fragilis* (Poxton and Brown, 1986) and *B. vulg-atus* (Delahooke, Barclay and Poxton, 1995a). The controversy over the presence or absence of an 'O-antigen' component in *Bacteroides* spp. can be explained by the use of different chemical extraction procedures by different laboratories and the variation in the expres-sion of the polysaccharide biosynthesis operons. The *in silico* identity of the genes within the polysaccharide biosynthesis operons, which are controlled by invertible promoters (see p. 535 Surface Variation line 26-28), suggests that some of these are related to the O-antigens of other bacteria (Cerdeno-Tarraga *et al.*, 2005). The LPS of *P. gingivalis* appears to play a significant role in the pathogenesis of periodontal disease, in which (like the LPS of *B. fragilis*) it reduces the opsonic activity of serum, stimulates gingival inflammation, increases the secretion of collagenase from host cells and reduces collagen formation, and induces localised bone resorption around the tooth root. These effects are attributable to the release of biologically active agents, including the cytokines interleukin-1 and tumour necrosis factor from host cells.

Fusobacterium necrophorum is an exception, as its LPS contains classical KDO and displays endotoxic activity similar to that of enterobacterial LPS, particularly during the septicaemic phase of human necrobacillosis.

Enterotoxin

Enterotoxigenic *B. fragilis* were first described as a cause of acute watery diarrhoeal disease in animals and subsequently humans, in particular children (Myers *et al.*, 1987). The enterotoxin is a zinc-dependent metalloprotease of approximately 20 kDa. The possible relationship of the enterotoxin to the virulence of *B. fragilis* in other types of infection remains to be proven as only between 9 and 26% of isolates harbour the gene encoding the toxin (Patrick, 2002).

CLINICAL FEATURES OF ANAEROBIC INFECTIONS

Infections with Gram-negative anaerobes are principally endogenous in source and related to the body sites where these microbes are part of the normal microbiota. Several features are typical of most anaerobic infections: they are necrotising or gangrenous conditions in tissues rendered susceptible by trauma, reduced blood supply and poor oxygenation, often in the presence of a foreign body or blood clot, and producing copious amounts of foul-smelling pus; thrombophlebitis of surrounding blood vessels is common, enhancing the anaerobic condi-tions. The characteristic smell of anaerobic infections is caused by the volatile end products of anaerobic metabolism. Infections are rarely 'pure', i.e. monobacterial. Anaerobic species are usually present in association with other anaerobic or facultative species in synergic mixtures (see p. 534 pathogenicity line 17–19), but the anaerobes appear to be the principal causes of tissue damage and abscess formation in these mixed infections. From a clinical perspective, infections may be divided into five broad groups: 1. those derived from the gastrointestinal tract, 2. genito-urinary infections in men and women, 3. infections of the head and neck related to the oral microbiota, 4. infections of other soft tissues, and 5. bacteraemia (Finegold and George, 1989; Duerden, 1990).

Gut-Associated Infections

Bacteroides spp. represent a large proportion of the normal microbiota of the lower intestinal tract and are the main cause of serious sepsis associated with surgery, injury, perforation or other underlying abnor-mality of the large intestine. *Bacteroides* spp. are the main compon-ents of the microbiota of post-operative abdominal wound infections, peritonitis and intra-abdominal abscesses (appendix, diverticular or paracolic, pelvic, sub-phrenic etc.). *Bacteroides fragilis* is the most common species isolated from these infections, representing about 75% of the *Bacteroides* isolates in contrast with between 4 and 13% of *Bacteroides* isolates from the faecal microbiota. Thus, *B. fragilis* clearly appears to have particular pathogenic potential and a much greater capacity to cause infection than the other *Bacteroides* spp., as described in the pathogenicity. Amongst the remaining 25% of isolates, *B. thetaiotaomicron* is the most common, with smaller numbers of *B. distasonis* and *B. vulgatus* (Patrick, 2002). The species other than *B. fragilis* are usually found in mixtures of multiple anaerobic species, often representing gross faecal soiling of the tissues. *Bilophila wadsworthia* is associated with gangrenous appendicitis and other intra-abdominal sepsis.

Gram-negative anaerobes are predominant components of the mixed microbiota of peri-anal, pilonidal and perineal abscesses (Duerden, 1991). *Bacteroides fragilis* is again the most common species isolated, but *P. asaccharolytica* and '*B.*' *ureolyticus* appear more frequently in these infections than in abdominal infections, as they do in other superficial soft-tissue infections such as sebaceous abscesses and decubitus or diabetic ulcers (Duerden, Bennett and Faulkner, 1982) (see p. 539 Superficial soft-tissue infections).

Liver abscesses form a specific clinical group of intra-abdominal abscesses. They are generally polymicrobial and anaerobes, particularly *Bacteroides* spp. and anaerobic cocci often predominate. The route of infection is usually via the portal venous system, and the common underlying causes are colorectal malignancy, ulcerative colitis, Crohn's disease and other intra-peritoneal abscesses. The most common species isolated is, again, *B. fragilis*, and *Bacteroides* bacter-aemia is a well-recognised complication. Anaerobes are responsible for only a minority of biliary tract infections, being restricted mainly to obstructive empyema of the gall bladder.

Commonly associated facultative bacteria include *E. coli* and *Streptococcus* spp., in particular *Strep. milleri*, which forms pinpoint colonies and with a characteristic smell.

Genito-Urinary Infections

Gram-negative anaerobes are significant pathogens in a wide range of infections of the genito-urinary tract in both men and women (Duerden, 1991). In men, the infections are principally ulcers and abscesses of the external genitalia and perineum, but in women there are also infections of the vagina and of the uterus and deep pelvic tissues.

In local abscesses associated with the secretory glands in women, e.g. Bartholin's and Skein's abscesses, obstruction of the ducts leads to a build up of secretions and infection with a mixed bacterial microbiota, predominantly anaerobic, that is similar to the microbiota of sebaceous cysts and pilonidal abscesses; *P. asaccharolytica* is a common and important isolate. Anaerobes are also frequent isolates from genital ulcers, a common problem in both men and women. They are probably not primary causes (genital herpes or trauma are common primary diagnoses) but whatever the cause of the initial damage, anaerobes form a major part of the microbiota of established ulcers and contribute to the progressive tissue damage. The term genital ulceration covers a range of superficial necrotising conditions, from erosive balanitis/balanoposthitis in men and superficial labial ulceration in women, through deep, spreading ulcers, typically with undermined edges, to the more severe forms of synergic gangrene. In all groups, regardless of initiating factors, anaerobic bacteria are the predominant cultivable bacteria once the superficial debris is removed. Gram-negative anaerobes are not normally present on the external genitalia but are the most common isolates from infected ulcers. Although *Bacteroides* spp. may be present in some cases, the most common apparently significant isolates are the asaccharolytic and strongly proteolytic *P. asaccharolytica* and '*B.*' *ureolyticus* which have a particular association with superficial necrotising and ulcerative lesions at various body sites. *Prevotella* spp., principally *Pr. intermedia* (which appears to be the most virulent of the *Prevotella* spp.), are also common isolates – being less common than *P. asaccharolytica* in men but about equally common in women, in whom *Prevotella* spp. are part of the normal vaginal microbiota (Masfari, Kinghorn and Duerden, 1983; Masfari, Duerden and Kinghorn, 1986).

Vaginal discharge is a common complaint of women attending Genito-urinary Medicine clinics and anaerobes are involved in many cases (Duerden, 1991). The most common condition is bacterial (anaerobic) vaginosis in which a disturbance of the normal vaginal microbiota with the loss of predominant lactobacilli and their replacement by increased numbers of Gram-negative anaerobes results in a discharge with an offensive, fishy smell. With the proliferation of *Prevotella* spp., *Gardnerella vaginalis* and *Mobiluncus* spp., the vaginal pH rises, the lactate concentration falls and the amounts of succinate, acetate, propionate and butyrate increase, together with volatile amines that cause the smell. These same metabolites may induce the excessive secretion from the vaginal mucosa, but the factors that initiate the condition, other than its clear association with sexual activity and a relationship with the presence of seminal fluid in the vagina, are still not clear. The other main causes of vaginal discharge – gonorrhoea, chlamydia and trichomonas infection – also cause a major disturbance of the normal microbiota with significant increases in the numbers of Gram-negative anaerobes (Masfari, Duerden and Kinghorn, 1986).

Ascending infections of the uterus and pelvis with vaginal anaerobes (mainly *Prevotella* spp.) or with *B. fragilis* or *P. asaccharolytica* are the cause of serious gynaecological sepsis. Some of these infections are the classical infective complications of pregnancy, parturition or abortion such as post-partum or post-abortal uterine infections in which retained products of conception provide ideal conditions for anaerobes to proliferate. With the control of streptococcal puerperal infection, anaerobes have become the most common cause of post-partum sepsis. The same microbes also contribute to deep pelvic infections

unrelated to pregnancy – endometritis, parametritis, tubo-ovarian and pelvic abscesses, generally grouped together as pelvic inflammatory disease – and wound infections and abscesses complicating gynaecological surgery. Laboratory confirmation of the cause of pelvic sepsis is difficult without surgical exploration; laparoscopy does not provide adequate samples for bacteriological culture but evidence from cases that have required open surgery supports the role of anaerobes in these infections (Kinghorn, Duerden and Hafiz, 1986). In gynaecological surgery, the underlying pathology has often resulted in colonisation of the usually sterile deep sites and the incidence of wound infection, before antibiotic prophylaxis was routine, was about 20% and most were anaerobic, with both the vaginal *Prevotella* spp. and *B. fragilis* being common isolates.

Fusobacterium nucleatum, along with *Mobiluncus curtisii*, has been recovered from the amniotic fluid of women with preterm delivery (Citron, 2002).

Infections with Oral Anaerobes

The Gram-negative anaerobes that are part of the normal microbiota of the gingival crevice cause suppurative infections of the gingiva and immediate surrounding tissues related to dental problems but can also cause various abscesses and soft-tissue infections throughout the head, neck and chest (Hardie, 1991).

Gingivitis and periodontal disease are probably the most common anaerobic infections, if not the most common of all infections worldwide, and are the most important cause of tooth loss. Acute ulcerative gingivitis (also known as Vincent's angina, Plaut-Vincent's infection or trench mouth) was one of the first anaerobic infections of man to be recognised. It is associated with poor oral hygiene, malnutrition and general debility and is now seen as a complication of AIDS. Clinically, there is pain, haemorrhage, inflammation and destruction of gum tissue and a foul odour. Spirochaetes and fusiform organisms are seen readily in stained films of the exudate and they may play a role in the disease, but the oral Gram-negative anaerobes, principally *Prevotella* and *Porphyromonas* spp., are also present in large numbers and may be more important in the pathogenesis of the condition. In other forms of periodontal disease, *P. gingivalis* and *Pr. intermedia* have been implicated as significant pathogens in rapidly progressive disease. *Fusobacterium nucleatum* and non-pigmented *Prevotella* spp. are also present in large numbers in periodontal disease and may have a pathogenic role (Tanner and Stillman, 1993).

The oral anaerobes are also important in dental abscesses, root canal infections (a particular association of *P. endodontalis*) and soft-tissue abscesses, e.g. buccal and pharyngeal abscesses. In the normal healthy state, these anaerobes do not colonise other sites in the head and neck and do not cause acute primary infections of the throat, middle ear, mastoid or sinuses. However, when pre-existing damage or prolonged infection leads to the development of chronic otitis media, mastoiditis or sinusitis, these compromised air passages become infected with a mixture of oral streptococci and anaerobes such as *Prevotella* spp. These anaerobes are also the most common cause of brain abscess, especially those originating from chronic otitis media, mastoiditis or sinusitis; cholesteatoma is an important predisposing factor. Infection spreads by direct extension, often with localised osteomyelitis and thrombosis of the lateral sinus. Dental infection may also give rise to brain abscess either by direct extension or by haematogenous spread (Ingham and Sisson, 1991). Similarly, the oral anaerobes can cause lung abscesses as a result of aspiration of organisms from the mouth when there is already some abnormality such as obstruction due to a malignant tumour or an inhaled foreign body (Civen *et al.*, 1995).

Fusobacterium necrophorum subsp. *necrophorum* (previously *F. necrophorum* Biovar A) differs from almost all other Gram-negative anaerobes in being a primary pathogen capable of causing disease in previously healthy people, although there may be synergic involvement of other viruses or bacteria that cause pharyngitis in the initial invasion of the mucosa (Hagelskjaer Kristensen and Prag, 2000).

Epstein–Barr virus infection (infectious mononucleosis) may be a predisposing factor.

Classically, the infection (necrobacillosis or Lemmiere's syndrome) begins as a suppurative tonsillitis with pus formation and a pseudomembrane, which spreads to involve the local lymph nodes in the neck and from there to invade the bloodstream causing septicaemia and disseminating the organisms widely in the tissues to form multiple soft-tissue abscesses at many sites in the body (liver, lung, kidney etc.). It may also occur in children with otitis media, adults with tooth infections and spread from other foci such as sinusitis or mastoiditis in all ages. *Fusobacterium necrophorum* may also cause necrobacillosis arising from sites other than the head and neck in elderly patients with other predisposing disease. In patients with anaerobic sepsis arising from the oropharynx, *F. necrophorum* should immediately be considered as a potential cause, as it is the most frequently isolated species. The classical symptoms of Lemmiere's syndrome are oropharyngeal pain, neck swelling, pulmonary symptoms, arthralgia and fever. It should be considered in patients with severe sepsis and pulmonary symptoms after acute pharyngotonsillar infection. Unilateral suppurative thrombophlebitis of the internal jugular vein is characteristic and can be diagnosed by ultrasonography, axial CT scan or magnetic resonance angiography of the neck. Multiple bilateral necrotic infiltration of the lungs with pleural effusion, empyema and/or pulmonary abscesses is reported in up to 85% of cases (Hagelskjaer Kristensen and Prag, 2000).

Lemmiere's syndrome was first described in the pre-antibiotic era. It became very uncommon during the second half of the 20th century, possibly due to widespread and indiscriminate use of antibiotics for upper respiratory tract infections, but increasing numbers of cases have been reported in the United Kingdom during the past decade (Brazier et al., 2002) and also recently in the United States (Ramirez et al., 2003).

Bone and Joint Infection

Anaerobic bacteria, often along with facultative species, may be associated with both osteomyelitis and septic arthritis. *Bacteroides fragilis* was the predominant anaerobic isolate from osteomyelitis in a 10-year study. In addition, *Prevotella*, *Porphyromonas*, *Fusobacterium* and other *Bacteroides* spp. were isolated. *Bacteroides fragilis* was associated most frequently with infections of the hands and feet where there was underlying vascular disease or neuropathy, whereas *Prevotella*, *Porphyromonas* and *Fusobacterium* were more frequently recovered from skull and bite infections (Brook and Frazier, 1993). Gram-negative non-sporing anaerobes are rarely isolated from prosthetic joint-associated infection, where *Propionibacterium acnes* is the most frequently isolated anaerobe (Tunney et al., 1998). *Bacteroides fragilis* and *Fusobacterium* spp., however, may cause septic arthritis. In approximately 50% of instances this arises from haematogenous spread from a distant focus of infection.

Superficial Soft-Tissue Infections

As well as infections related to the normal carriage sites of Gram-negative anaerobes, another group of soft-tissue infections occurs at sites not immediately adjacent to the mucosae but in tissue damaged by trauma or hypoxia, or where secretory glands become blocked. These infections include sebaceous and pilonidal abscesses, breast abscesses in non-lactating women and infection secondary to inadequate perfusion and oxygenation of the skin and subcutaneous tissues – decubitus and varicose ulcers, diabetic gangrene etc. (Adriaans, Drasar and Duerden, 1991). Although a wide range of anaerobes and facultative species can be isolated from these lesions, there is a particular association between progressively destructive infections and the presence of *P. asaccharolytica* and/or '*B.*' *ureolyticus* (Duerden, Bennett and Faulkner, 1982). There is a specific association of *F. ulcerans* with tropical ulcers – deep eroding ulcers, usually of the lower limb, occurring across many tropical regions (Adriaans Hay and Drasar, 1987; Adriaans et al., 1987). Since the initial description

of this bacterium, further isolates have not been reported. This may have arisen from a phenotypic similarity to *F. varium* (a gastrointestinal species). Many commercial identification kits lack a test for nitrate reduction, a distinguishing feature of *F. ulcerans*. Therefore, an incorrect identification of *F. varium* is generated by the kit database (Citron, 2002).

Bacteraemia

Primary bacteraemia with Gram-negative anaerobes is extremely rare. It is an integral part of necrobacillosis in which *F. necrophorum* invades the bloodstream as part of the spread of infection from the primary site in the throat (see p. 538 Infections with oral anaerobes line 3–5). In most other cases, the bacteraemia is secondary to localised infections at some site in the body, e.g. intra-abdominal abscess and pelvic sepsis, may be the first indication of some serious underlying condition such as a brain abscess (Goldstein, 1996). Bacteraemia with a *Bacteroides* spp. may be the first indication of a malignant tumour of the colon, rectum or cervix. The most common anaerobic species isolated from blood cultures is *B. fragilis*. Mortality and morbidity of *Bacteroides* bacteraemia depends generally upon the underlying cause. A matched-pair study in the United States reported a potential mortality of up to 19% for *B. fragilis* group bacteraemia (Redondo et al., 1995).

LABORATORY DIAGNOSES

The site and nature of the infection (e.g. an abdominal abscess or a necrotising or gangrenous ulcer) may give a clear indication that anaerobes are likely to be involved, and the presence of a foul-smelling exudate or pus is the strongest clinical indication of an anaerobic infection. The diagnosis is confirmed by careful anaerobic culture methods and the identification of the isolates obtained. Because these infections are often associated with sites that have a complex normal bacterial microbiota (often including the organisms that may be the putative pathogens), and because anaerobes are, by their nature, more or less susceptible to exposure to oxygen, care is needed in specimen collection and transport to try to ensure that the organisms sought do not die in transit to the laboratory and that any cultures obtained represent the microbiota at the infected site and not contamination from the normal mucosal microbiota (Drasar and Duerden, 1991; Summanen et al., 1993). Information on all aspects of laboratory diagnosis of anaerobic bacteria and detailed identification protocols can be found in the Wadsworth-KTL Anaerobic Bacteriology Manual (Jousimies-Somer et al., 2002).

Specimen

The most reliable specimens are pus or exudates from the depths of an open lesion or a closed lesion. Specimens such as sputum collected from sites with a normal mucosal microbiota are not suitable for anaerobic culture. Swabs used traditionally to sample wounds and exudates are not ideal for anaerobes that suffer from desiccation and exposure to oxygen, as well as being entrapped in the interstices of the swab. Whenever possible, pus should be aspirated into a sterile container and delivered promptly to the laboratory (Finegold, 1995).

Transport

Direct plating of a sample at the patient's bedside and immediate anaerobic incubation are likely to give the best results, but this is an unattainable ideal in most clinical situations. If pus, or a piece of necrotic tissue obtained by biopsy, is the specimen, it should be transported to the laboratory without delay in a sterile container. Sealed containers with an oxygen-free atmosphere, or from which oxygen can be removed by a simple chemical reaction, are available for optimal transport of such specimens. Pus is generally regarded as its own best transport medium. If there is no alternative to the use of swabs, these should be placed in semi-solid anaerobic transport medium to prevent both desiccation and the toxic effects of oxygen.

Laboratory Examination

Direct microscopy of pus samples may be helpful in some cases; most of the Gram-negative anaerobes are small, pleomorphic organisms that are difficult to see in direct smears. The most useful immediate examination is direct GLC to detect the presence of the volatile fatty acid products of anaerobic metabolism (Jousimies-Somer *et al.*, 2002). This will not, generally, give any indication of the specific identity of the anaerobes present, but the demonstration of a mixture of short-chain fatty acids confirms the presence of anaerobic bacteria.

Isolation of Gram-negative anaerobes is by anaerobic culture on selective and non-selective media. Anaerobic culture conditions can be provided in sealed anaerobic chambers equipped with air locks and functioning as anaerobic incubators as well as workstations, or in anaerobic jars. The anaerobic atmosphere generally used in cabinets is N_2 80%, H_2 10% and CO_2 10%, as the growth of many anaerobes is stimulated by CO_2. The same gas mixture can be used in jars operated by an evacuation and refill system and equipped with room-temperature-active catalysts (e.g. palladium) to remove any remaining oxygen. Alternatively, jars can be used with sachets that generate H_2 and CO_2 when water is added and depend upon the H_2 to remove all the oxygen from the jar with the help of the catalyst. There are many variations on the basic anaerobic jar methodology but all must be carefully controlled to ensure that as much as possible of the O_2 is removed and that air cannot leak back into the system.

Many anaerobes grow more slowly than most common aerobic pathogens, and anaerobic cultures must be incubated for at least 48 h without exposure to oxygen and for an overall minimum of 72–96 h; however, some of the more virulent anaerobes (e.g. *Clostridium perfringens* and *B. fragilis*) produce acceptable growth in 24 h, and most laboratories would not wish to delay the diagnosis by 24 h. If the use of duplicate sets of plates where anaerobic jars are in use (one for examination after 24 h and the other to remain undisturbed for 72–96 h) is considered too extravagant of resources, a useful compromise is for the one set of plates to be examined quickly after 24 h and then re-incubated after as little exposure to air as possible for another undisturbed 48–72 h. This difficulty is overcome by the use of anaerobic cabinets in which the initial examination can be done without removing the plates from the anaerobic atmosphere.

Various media have been recommended for the optimal growth of Gram-negative anaerobes. All comprise a rich nutrient base with added blood. *Bacteroides fragilis* grows well on most formulations of blood agar, but better growth of a wider range of more fastidious species needs more specially enriched media. Most species grow more quickly and produce larger colonies on media such as Fastidious

Anaerobe Agar, Brucella agar or BM medium with ingredients such as proteose peptone, trypticase and yeast extract. Menadione and haemin enhance the growth of some species, especially of *Prevotella*; other demanding strains such as '*B.*' *ureolyticus* require formate and fumarate for good growth and *Bil. wadsworthia* has an absolute requirement for pyruvate. The use of lysed blood assists in the early recognition of pigment production.

As anaerobic bacteria are generally present in mixed infections, their isolation is aided by the use of selective media containing antibiotics inactive against anaerobes. Aminoglycosides have been widely used for many years. The neomycin blood agar developed for the selective isolation of clostridia is too inhibitory for many Gram-negative anaerobes, and kanamycin (75 mg/l) gives better results. An alternative that may be preferred is the use of nalidixic acid (10 mg/l). When it is desirable to eliminate Gram-positive bacteria, the addition of vancomycin (2.5 mg/l) provides a medium highly selective for *Bacteroides* spp. Selective media should never be used without parallel non-selective cultures and growth on the selective media generally requires extended incubation.

Identification of Gram-Negative Anaerobes

Detailed identification to species level of all isolates of Gram-negative anaerobes is beyond the scope of most diagnostic laboratories and would require excessive commitment of resources. They are significant pathogens, however, and certain species either show greater virulence than most of their group (e.g. *B. fragilis*) or are associated with particular types of infections (e.g. *F. necrophorum*, *P. gingivalis* and *P. asaccharolytica*). It is important that primary diagnostic laboratories can at least determine the presence of the main pathogens and identify other significant isolates at least to genus level, for more detailed study in a reference or research laboratory when appropriate.

Preliminary allocation of isolates to the main genera can be made by a relatively simple set of tests based upon colonial and microscopic morphology, growth in the presence of bile (20% in broth or agar medium) or sodium taurocholate (disc tolerance test), disc resistance tests with antibiotics (neomycin 1 mg, kanamycin 1 mg, penicillin 1 or 2 units, rifampicin 15 µg, colistin 10 µg and vancomycin 5 µg) catalase and urease production (Table 45.4), supported by GLC if available.

More detailed identification depends upon sets of biochemical tests with carbohydrate and other substrates and detection of particular enzyme activities. Some of these can be performed with commercial kits (especially the enzymes tests), and these can be useful for the identification of the more common species isolated from clinical specimens. However, most of these systems suffer from the attempt by most manufacturers to produce a single kit for the whole range of anaerobes – Gram-positive and Gram-negative, cocci and bacilli. Tables 45.5–45.9 summarize the sets of tests that enable the identification of most clinically significant Gram-negative anaerobes. The

Table 45.5 Preliminary identification of Gram-negative anaerobic bacteria of clinical importance

	Bacteroides	*Bilophila*	*Prevotella*	*Porphyromonas*	*Fusobacterium*	*Leptotrichia*	'*B.*' *ureolyticus*
Growth in bile (20%)	+	+	–	–	+/–	+/–	–
Tolerance of taurocholate	+	+	–	–	+/–	+/–	–
Antibiotic disk resistance tests							
Neomycin (1 mg)	R	S	S	S	S	S	S
Kanamycin (1 mg)	R	S	R	R	S	S	S
Penicillin (1–2 units)	R	R	S/R	S/R	S	S	S
Rifampicin (15 µg)	S	S	S	S	R/S	S	R
Colistin (10 µg)	R	S	S/R	R	S	R	S
Vancomycin (5 µg)	R	R	R	S	R	S	R
Urease	–	+	–	–	–	–	+
GLC	Ac, Su	Ac, Su	Ac, Su	*n*-Bu	*n*-Bu	Pro, Lac	Ac, Pro

Ac, acetic; Lac, lactic; *n*-Bu, *n*-butyric; Pro, propionic; Su, succinic.

Table 45.6 Identification of *Bacteroides* spp.

	B. fragilis	B. vulgatus	B. distasonis	B. merdae	B. caccae	B. ovatus	B. thetaiotaomicron	B. eggerthii	B. uniformis	B. stercoris	B. splanchnicus
Indole	–	–	–	–	–	+	+	+	+	+	+
Aesculin hydrolysis	+	–/+	+	+	+	+	+	+	+	+	+
Fermentation of											
glucose	+	+	+	+	+	+	+	+	+	+	+
lactose	+	+	+	+	+	+	+	+	+	+	–
sucrose	+	+	+/–	+	+	+	+	–	+	+	–
rhamnose	–	+	+/–	+	+	+	+	+	–	+	–
trehalose	–	–	+	–	–	+	–	–	–	–	–
mannitol	–	–	–	–	–	+	–	–	–	–	–
arabinose	–	+	+	+	+	+	+	+	+	–	+
salicin	–	–	+	–	–	+	–	+	+	–	+
xylan	–	–	–	–	–	+	–	+	+/–	+/–	–
α-fucosidase	+	+	–	–	+	+	+	–	+	+/–	+

Table 45.7 Identification of *Prevotella* spp.

	Pr. buccae	Pr. oris	Pr. zoogleoformans	Pr. heparinolytica	Pr. oralis	Pr. veroralis	Pr. buccalis	Pr. oulora	Pr. bivia	Pr. disiens	Pr. melaninogenica	Pr. denticola	Pr. loescheii	Pr. intermedia*	Pr. nigrescens*	Pr. corporis
Pigment	–	–	–	–	–	–	–	–	–	–	+	+	+	+	+	+
Indole	–	–	–	+	–	–	–	–	–	–	–	–	+/–	+	+	–
Aesculin hydrolysis	+	+	+	+	+	+	+	+/–	–	–	–	+	+/–	–	–	–
Hippurate hydrolysis	–	–	–	–	–	–	–	–	+	+	...	...	...	...	...	...
α-Fucosidase	–	+	+	+	+	+	+	–	+	–	+	+	+	+	+	+
Fermentation of																
glucose	+	+	+	+	+	+	+	+	+	+	+	+	+	+	+	+
sucrose	+	+	+	+	+	+	+	+	–	–	+	+	+	+	+	–
lactose	+	+	+	+	+	+	–	+	+	–	+	+	+	–	–	–
xylose	+	+	+	+	–	–	–	–	–	–	–	–	–	–	–	–
arabinose	+	+	+	+	–	+	+	–	–	–	–	–	–	–	–	–
cellobiose	+	+	+	+	+	–	+	–	–	–	–	–	+	–	–	–
salicin	+	+	+	+	+	–	–	–	–	–	–	–	–	–	–	–
xylan	+	–	–	+/–	–	–	–	–	–	–	...	...	...	...	...	...
inulin	+	+	+	+/–	–	+	+/–	–	–	–	...	...	...	...	...	...

The only phenotypic distinguishing test is multi-focus enzyme electrophoresis for glutamate and malate dehydrogenases; *Pr. nigrescens* has slower mobilities with both enzymes.

Table 45.8 Identification of pigmented human strains of *P.* spp.

	P. asaccharolytica	*P. gingivalis*	*P. endodontalis*
Pigment	+	+	+
Indole	+	+	+
Trypsin-like activity	–	+	–
Phenylacetic acid production	–	+	–
α-Fucosidase	+	–	–

media and methods for performing these tests are varied, and details may be found in manuals or monographs devoted specifically to anaerobic microbiology – the Wadsworth Anaerobic Bacteriology Manual sixth edition (Jouseimies-Somer *et al.*, 2002), the Virginia Polytechnic Institute Anaerobe Laboratory Manual (Moore and Moore, 1977) or Anaerobes in Human Disease (Wren, 1991). Baron and Citron (1997) provide an outline of a cost-effective protocol.

Non-Culture Detection Methods

The ability to identify reliably obligately anaerobic bacteria directly in clinical samples without the need for culture is particularly attractive. Despite the best efforts of clinical bacteriologists to ensure the handling and transit of samples under anaerobic conditions, a high proportion of infections may be missed as a result of loss of viability (Patrick *et al.*, 1995).

Techniques have been developed in some research laboratories for the direct detection of specific pathogenic anaerobes (*B. fragilis*, *P. asaccharolytica*, *P. gingivalis*, *Pr. intermedia* and *Bil. wadsworthia*) by DNA probes and by polymerase chain reaction (PCR) technology, but as yet, there are no non-culture detection methods for anaerobes in general use in clinical diagnostic laboratories. PCR finger printing identification (e.g. amplified ribosomal DNA restriction analysis) of pure culture isolates has resulted in good identification and differentiation of *Bacteroides*, *Prevotella* and *Porphyromonas* spp. of the *fragilis* group, as has a multiplex PCR assay based on 16S rRNA, the 16S–23S rRNA intergenic spacer region and part of the 23S rRNA gene (Liu *et al.*, 2003). PCR-based assays for the detection of enterotoxigenic *B. fragilis* have also proved reliable. Microarrays enabling the simultaneous detection of multiple genes and therefore multiple bacterial species have been applied to studies of the faecal microbiota. Fluorescence *in situ* hybridisation of 16S rRNA is possible with pure cultures of *B. fragilis*; however, such techniques must be approached with caution if they are to be applied directly to clinical samples, as the penetration of the probe into the bacterial cell is dependent on the nature of encapsulating surface structures.

It is likely, however, that accurate and specific multiplex and quantitative (real-time) PCR diagnostic tests or microarray-based technologies will ultimately become part of routine diagnosis once inherent problems of sensitivity in clinical specimens and potential for contamination from exogenous DNA are resolved.

Antibiotic Susceptibility of Gram-Negative Anaerobes

Approaches to susceptibility testing of Gram-negative anaerobes have been debated amongst microbiologists and infectious disease physicians

for many years. A major consideration is whether susceptibility testing should be, as for most aerobes, concentrated in the primary laboratories doing individual susceptibility tests on clinical isolates from their patients or whether strains should be collected for batch testing in recognised specialist centres while individual clinical treatment is based upon accumulated data generated by that batch testing. For aerobes, laboratories test their significant isolates and clinicians expect prompt susceptibility data on their own patients' organisms. This has not been the case with anaerobes because of three perceptions: 1. anaerobic investigations are slower – initial culture may take several days and susceptibility tests at least 2 days more, so that individual results are irrelevant to patient management; 2. methods for susceptibility testing of anaerobes are unreliable, especially when performed on an individual basis, and batch testing in specialist centre gives more reliable results; and 3. the susceptibility of anaerobes is predictable so that clinical treatment can be based upon the batch data with confidence.

These premises were the basis for susceptibility testing of anaerobes having been neglected in primary laboratories and concentrated in specialist centres that receive clinical isolates from a wide range of sources. However, there is an increasing view from clinicians and microbiologists that the same standard of laboratory data is required in anaerobic infections as in aerobic infections. Testing of individual isolates is supported by four changed perceptions: 1. antibiotic susceptibility in anaerobes is variable and patterns differ in different places and at different times; 2. modern laboratory methodology should enable the prompt isolation of the more common anaerobic pathogens; 3. methods for susceptibility testing of anaerobes are available that can be used in primary laboratories; and 4. the most important reason, patients managed with access to accurate and specific laboratory data on their own anaerobic isolates recover more rapidly with fewer complications.

Both approaches are necessary. Primary laboratories need to develop adequate susceptibility tests on anaerobes for first-choice agents while reference laboratories should enhance their activities in collecting strains from a wide range of sources and batch testing to provide the best general advice for clinicians in choosing empirical therapy and devising antibiotic policies.

Methods

The selection of methods for susceptibility testing of anaerobes has been a major difficulty. The two basic methodologies for rapidly growing, relatively non-fastidious aerobes – disc susceptibility tests on solid media and breakpoint determination in liquid or solid media – are not readily adaptable to anaerobic work, are difficult to standardise and do not give reliably reproducible results with anaerobes. Anaerobes tend to grow more slowly than many common aerobes, and the balance between antimicrobial effect, bacterial growth and antibiotic degradation varies much more, affecting disc susceptibility methods in particular. Many anaerobes need more enriched media than are used for susceptibility testing of aerobes; they are more difficult to standardise, and there are interactions between media components and the antibiotics. The presence of fermentable nutrients often results in a very significant drop in pH in the test medium which can have a major impact of the results obtained with pH-sensitive antibiotics. Similarly, many anaerobes require CO_2 for growth and all

Table 45.9 Identification of *Fusobacterium* spp.

	F. nucleatum	*F. necrophorum*	*F. pseudonecrophorum*	*F. russi*	*F. naviforme*	*F. mortiferum*	*F. varium*	*F. gonidiaformans*	*F. ulcerans*
Indole	+	+	+	–	+	–	+/–	+	–
Lipase	–	+	–	–	–	–	–/+	–	–
Aesculin hydrolysis	–	–	–	–	–	+	–	–	–
Propionate from									
lactate	–	+	+	–	–	–	–	–	–
threonine	+	+	+	–	–	+	+	+	+

standard anaerobic gas mixtures contain CO_2 which again lowers the pH (Watt and Brown, 1985). Many anaerobes do not grow well from small inocula, and semi-confluent growth is difficult to achieve. A further practical difficulty arises because the MICs of several agents for anaerobes cluster near the recommended breakpoints for the agents, so that minor variations in method and inevitable margins of error in any technique can give variable results in terms of clinical interpretation and recommendations. Therefore, disc susceptibility tests for anaerobes have been regarded as unreliable for other than the fast-growing anaerobes. Recommended methods in the past have been based upon agar dilution with multipoint inoculation or broth micro-dilution or macro-dilution methods. Some of these have not been entirely satisfactory and are not appropriate for testing individual clinical isolates, hence the methodology has exerted pressure in favour of batch testing in reference laboratories. The E test (AB Biodisc, Denmark) provides reliable susceptibility testing with MIC determination for individual isolates. A plastic strip coated with an antibiotic gradient on one side and with an MIC interpretation scale on the other is placed on a seeded plate as in disc susceptibility testing. After incubation, the edge of the zone of inhibition reaches the scale at a point equivalent to the MIC. This shares some of the disadvantages of disc methods, but experience has shown it to be generally reliable for a wide range of anaerobes (Wexler, 1993; Duerden, 1995; Jouseimies-Somer et al., 2002).

It is likely that molecular detection methods for antibiotic resistance genes will be incorporated into molecular identification protocols. A PCR-restriction fragment length polymorphism analysis identification method for Bacteroides spp. that incorporates nim gene detection has proved to be rapid and accurate (Stubbs et al., 2000).

Antibiotic Susceptibility

The susceptibility of Gram-negative anaerobes varies considerably both between and within the major genera. Metronidazole has been the mainstay of therapy and prophylaxis for 20 years, and it is re-assuring that resistance amongst Gram-negative anaerobes has remained relatively low (Dubreuil, 1996; Snydman et al., 1996), so that it remains the drug of choice for many of these infections. Its use has, however, been extended to the treatment of peptic ulcer disease with the recognition of the involvement of Helicobacter pylori (Freeman et al., 1997). Whether this wider use of metronidazole is leading to the development of increased resistance in Bacteroides of the normal microbiota is not proven, but metronidazole resistance has become more widely recognised in Bacteroides spp. in the last 5 years. Of the eight metronidazole resistance genes described (nimA-G), all but one (nimB) can be found on potentially mobile genetic elements in Bacteroides spp. Within the United Kingdom, referral of isolates to the Anaerobe Reference Laboratory has shown the incidence of metronidazole resistance in Bacteroides spp. increasing to around 7% since 1995 (Brazier, Stubbs and Duerden, 1999).

A number of different types of mobile genetic elements, including self-transmissible large conjugative transposons and plasmids as well as mobilisable plasmids and transposons, have been identified in the Gram-negative anaerobes (Sebald, 1994; Smith, Tribble and Bayley, 1998). There is evidence of horizontal transfer of antibiotic resistance genes via these mobile elements and the potential transfer of these to facultative species (Vedantam and Hecht, 2003). In the case of the large conjugative transposons that carry a tetracycline resistance gene, exposure to tetracycline, even at sub-inhibitory concentrations, increases the frequency of conjugation (Patrick, 2002). The normal microbiota may therefore represent a potential reservoir of transmissible antibiotic resistance genes.

Bacteroides spp. isolates of B. fragilis and other Bacteroides spp. are resistant to benzylpenicillin and ampicillin/amoxicillin due to the production of an amp-C type β-lactamase (Wang, Fast and Benkovic, 1999); in almost all cases this resistance is overcome by the use of a combination of ampicillin/amoxicillin with the β-lactamase inhibitors clavulanate or tazobactam. Bacteroides spp. are also resistant to many other β-lactam agents. Cefoxitin has been widely used in the United States for anaerobic infections because it has greater activity against Bacteroides spp. than most β-lactam agents, but results of susceptibility tests are variable with up to 40% of isolates giving MIC values around or above the recommended breakpoint of 32 mg/l. Most isolates in the United Kingdom remain susceptible to imipenem, but resistance due to the production of a metallo-β-lactamase, not affected by clavulanate or tazobactam, and encoded by the cfiA (ccrA) gene does occur and may become a more significant problem in the future. With the inhibitors of protein synthesis, many strains are resistant to erythromycin and tetracycline but most are susceptible to clindamycin. Like the Bacteroides spp., Bil. wadsworthia produces a β-lactamase and is resistant to most β-lactam agents.

Porphyromonas and Prevotella

Although the genera are generally more susceptible to antimicrobial agents than the Bacteroides spp., results are still unpredictable. In the past, they were considered to be generally penicillin sensitive, but one-third or more of clinical isolates, especially of Prevotella spp., are resistant due to β-lactamase production. Thus, they are susceptible to amoxicillin/ampicillin and clavulanate combinations. Most are susceptible to cefoxitin and highly susceptible to imipenem. They are generally susceptible to erythromycin and clindamycin, but up to half of clinical isolates are resistant to tetracycline and up to 15% are resistant to ciprofloxacin and clindamycin.

Fusobacterium

Most Fusobacterium spp. isolates are highly susceptible to a wide range of agents including penicillin and all the β-lactam agents, but resistance mediated by β-lactamase production is increasing. Ceftizoxime, ciprofloxacin and erythromycin are poorly active and up to 30% of strains may be resistant to clindamycin and tetracycline (Finegold and Jousimies-Somer, 1997).

MANAGEMENT OF ANAEROBIC INFECTIONS

Because of the multi-factorial nature of infection with Gram-negative anaerobes (p. 537–539 Clinical features of anaerobic infections) – mixed infections, tissue necrosis, inadequate blood supply/oxygenation, patient already compromised by underlying disease – the management of these infections requires a combination of approaches. Antibiotic treatment is important but is rarely effective if used alone. The choice is generally between metronidazole, a β-lactam agent such as amoxicillin-clavulanate or imipenem, and clindamycin. The recognition of the presence of aerobic or facultative organisms may also need to be addressed either by adding an agent effective against, e.g. E. coli to the metronidazole, or by choosing an agent effective against both (e.g. imipenem).

It is axiomatic in medical microbiology that antibiotics alone are ineffective, as treatment for abscesses and many of the anaerobic infections have a component of tissue necrosis and abscess formation. Debridement of necrotic tissue and drainage of pus are essential to the effective management of these infections. An interesting exception is disseminated F. necrophorum infection with metastatic abscess formation; this responds well to treatment with penicillin or other antibiotics alone.

Many of the patients who develop anaerobic infections are seriously ill, e.g. peritonitis following major abdominal surgery, and may be developing sepsis syndrome due to the anaerobes or other components of the mixed infection. These patients clearly need intensive supportive therapy for respiratory and circulatory functions.

PREVENTION AND CONTROL

As most infections with Gram-negative anaerobes are endogenous, cross infection is not a significant problem with these patients, although isolation may be appropriate because of the severity of the infection and, on some occasions, because of the intense malodour created by the anaerobes which can be embarrassing and distressing to the patient and unpleasant and disturbing for others.

Prevention of these infections depends upon prophylactic measures that can be taken to avoid creating the conditions that allow anaerobes to become established. Most of these measures relate to surgical practice in abdominal, gynaecological or oral-facial surgery – i.e. in areas where anaerobes predominate in the normal microbiota. Prevention of post-operative anaerobic infection in these situations depends upon good surgical practice, reduction in the bacterial challenge and the use of appropriate prophylactic antibiotics. Surgical skill goes a long way towards preventing anaerobic infections by not leaving non-viable tissue or foreign bodies at the operation site, ensuring adequate perfusion and oxygenation of the tissues that are left, preventing leakage from the viscus or the accumulation of fluid and closing off any potential spaces where infection can develop. Thus, the 'soil' is not conducive to anaerobic infection. The bacterial challenge can be reduced in part by physical removal of the normal microbiota. This is particularly important in the use of laxatives, enemas or even colonic irrigation to reduce the volume of faeces in the colon before abdominal surgery. A further important prophylactic measure is the use of prophylactic antibiotics per-operatively. These have made a major impact in reducing the incidence of post-operative infection during the last three decades. Since the advent of specific prophylaxis encompassing anaerobic organisms, the incidence of post-operative infection in abdominal and gynaecological surgery has fallen dramatically, e.g. from 35 to 5–10% for major colonic surgery, 20 to <5% for appendicectomy, 25 to 5% for hysterectomy (Willis and Fiddian, 1983). The aim of prophylaxis is to have high levels of appropriate antibiotics in the tissues at the time of the operation to prevent any implanted organisms becoming established. Prophylaxis in this way is not designed to destroy the bacteria in their normal habitat, e.g. in the gastrointestinal lumen. The choice of agent must include effective anti-anaerobe activity – usually by including metronidazole or amoxicillin/clavulanate – plus activity against other organisms likely to be present. Thus, in abdominal surgery, an aminoglycoside (e.g. gentamicin) or a cephalosporin (e.g. cefuroxime or cefotaxime) may be added to the metronidazole. In some circumstances, amoxicillin/clavulanate may fulfil both functions. For effective prophylaxis and to avoid disturbing the normal microbiota or selecting resistant strains before surgery, the antimicrobial agents should be given by intravenous injection at the induction of anaesthesia. For most operations, a single dose at this stage is sufficient to give good protection. In some regimens, a second dose after 4 h or at the completion of the operation has been recommended in more complex and longer operations. In any case, the whole period of prophylaxis should not exceed 24 h. If infection is already present at operation and more prolonged use of antibiotics is clinically necessary, this becomes a therapeutic course for established infection and not a prophylactic measure (Willis, 1991; Keighley, 1992).

Mobiluncus

The genus *Mobiluncus* represents a group of slender, curved, rapidly motile, Gram-variable anaerobic rods found in the vagina and associated with *Gardnerella vaginalis* and *Prevotella* spp. in bacterial (anaerobic) vaginosis (Duerden, 1991). It is a member of the Actinomycetaceae and the phylum of high mol% G+C Gram-positive bacteria, the Actinobacteria, and has a G+C content of 52–56 mol%.

Two morphologically distinct types of *Mobiluncus* are recognised and represent two separate species (Tiveljung, Forsum and Monstein, 1996). *Mobiluncus curtisii* strains have short Gram-positive or Gram-variable cells and are much more active in biochemical tests – they hydrolyse arginine and hippurate, reduce nitrate and produce β-galactosidase. *Mobiluncus mulieris* gives negative reactions in these tests and has much longer cells that appear Gram negative. However, both have a cell wall that is essentially of Gram-positive structure, but the peptidoglycan layer is thin, probably resulting in the variable staining pattern. Both species grow well anaerobically but will grow in 5% O_2 and growth is enhanced by CO_2. Although they may be present in large numbers in vaginosis and were described in vaginal discharge by Curtis in 1915, their role in disease is not clear. PCR detection indicates that, whereas *M. curtisii* is frequently associated with bacterial vaginosis, *M. mulieris* may also be detected in apparently healthy women (Schwebke and Lawing, 2001). *Mobiluncus* spp. are always present along with other potential pathogens such as *G. vaginalis* and *B. fragilis* in vaginosis and may have a synergic role (Spiegel, 1987). Treatment with metronidazole removes the anaerobes from the synergic mixture, and vaginosis usually responds well to short-course treatment with agents active against anaerobes, such as metronidazole to which *Mobiluncus* spp. are often resistant (Jones *et al.*, 1985). In addition, as there is a relationship with the deposition of semen in the vagina, the use of a condom will prevent the development of vaginosis (Duerden, 1991).

REFERENCES

Adriaans B, Hay R and Drasar BS (1987) The infectious aetiology of tropical ulcer – a study of the role of anaerobic bacteria. *The British Journal of Dermatology* **116**, 31–37.

Adriaans B, Hay R, Lucas S and Robinson DC (1987) Light and electron microscopic features of tropical ulcer. *Journal of Clinical Pathology* **40**, 1231–1234.

Adriaans B, Drasar BS and Duerden BI (1991) Superficial ulceration and necrosis. In *Anaerobes in Human Disease*, Duerden BI and Drasar BS (eds), Edward Arnold, London, pp. 287–298.

Baron EJ and Citron D (1997) Anaerobe identification flowchart using minimal laboratory resources. *Clinical Infectious Diseases* **25 Suppl. 2**, S143–S146.

Baron EJ, Curren M, Henderson G *et al.* (1992) *Bilophila wadsworthia* isolates from clinical specimens. *Journal of Clinical Microbiology* **30**, 1882–1884.

Baughn AB and Malamy MH (2004) The strict anaerobe *Bacteroides fragilis* grows in and benefits from nanomolar concentrations of oxygen. *Nature* **427**, 441–444.

Beckmann I, van Eijk HG, Meisel-Mikolajczyk F and Wallenburg HC (1989) Detection of 2-keto-3-deoxyoctonate in endotoxins isolated from six reference strains of the *Bacteroides fragilis* group. *The International Journal of Biochemistry* **21**, 661–666.

Brazier JS, Stubbs SL and Duerden BI (1999) Metronidazole resistance among clinical isolates belonging to the *Bacteroides fragilis* group: time to be concerned? *The Journal of Antimicrobial Chemotherapy* **44**, 580–581.

Brazier JS, Hall V, Yusuf E and Duerden BI (2002) *Fusobacterium necrophorum* infections in England and Wales, 1990–2000. *Journal of Medical Microbiology* **51**, 69–72.

Brook I and Frazier EH (1993) Anaerobic osteomyelitis and arthritis in a military hospital: a 10-year experience. *The American Journal of Medicine* **94**, 21–28.

Cerdeno-Tarraga AM, Patrick S, Crossman LC *et al.* (2005) Extensive DNA inversions on the *Bacteroides fragilies* genome control variable gene expression. *Science* **307**, 1463–1465.

Chen Y, Kinouchi T, Kataoka K *et al.* (1995) Purification and characterizaton of a fibrinogen-degrading protease in *Bacteroides fragilis* strain YCH46. *Microbiology and Immunology* **39**, 967–977.

Citron DM (2002) Update on the taxonomy and clinical aspects of the genus *Fusobacterium*. *Clincal Infectious Diseases* **35S**, S22–S27.

Comstock LE, Coyne MJ, Tzianabos AO *et al.* (1999) Analysis of a capsular polysaccharide biosynthesis locus of *Bacteroides fragilis*. *Infection and Immunity* **67**, 3525–3532.

Curtis MA, Aduse-Opoku J, Rangarajan M (1995) Cysteine proteases of *Porphyromonas gingivalis*. *Critical Reviews in Oral Biology and Medicine* **12**, 192–216.

Civen R, Jousimies-Somer H, Marina M, *et al*. (1995) A retrospective review of cases of anaerobic empyema and update of bacteriology. *Clinical Infectious Diseases* **20**, S224–S229.

Delahooke DM, Barclay GR and Poxton IR (1995a) A re-appraisal of the biological activity of bacteroides LPS. *Journal of Medical Microbiology* **42**, 102–112.

Delahooke DM, Barclay GR and Poxton IR (1995b) Tumor necrosis factor induction by an aqueous phenol-extracted lipopolysaccharide complex from *Bacteroides* species. *Infection and Immunity* **63**, 840–846.

Drasar BS and Duerden BI (1991) Anaerobes in the normal flora of man. In *Anaerobes in Human Disease*, Duerden BI and Drasar BS (eds), Edward Arnold, London, pp. 162–179.

Dubreuil L (1996) Antibiotic susceptibility patterns of the *Bacteroides fragilis* group. *Medecine et Maladies Infectieuses* **26**, 196–207.

Duerden BI (1983) The Bacteroidaceae: *Bacteroides*, *Fusobacterium*, and *Leptotrichia*. In *Topley and Wilson's Principles of Bacteriology, Virology and Immunity*, 7th edn, vol. 2, Wilson GS Miles AA and Parker MT (eds), Edward Arnold, London, pp. 114–136.

Duerden BI (1990) Infections due to gram-negative anaerobic bacilli. In *Topley and Wilson's Principles of Bacteriology, Virology and Immunity*, 8th edn, vol. 3, Parker MT and Collier LH (eds), Edward Arnold, London, pp. 287–305.

Duerden BI (1991) Anaerobes in genitourinary infections. In *Anaerobes in Human Disease*, Duerden BI and Drasar BS (eds), Edward Arnold, London, pp. 224–244.

Duerden BI (1994) Virulence factors in anaerobes. *Clinical Infectious Disease* **18 Suppl. 4**, S253–S259.

Duerden BI (1995) The role of the Reference Laboratory in susceptibility testing of anaerobes and a survey of isolates referred from laboratories in England and Wales in, 1993–94. *Clinical Infectious Diseases* **20**, S180–S186.

Duerden BI, Bennett KW and Faulkner J (1982) Isolation of *Bacteroides ureolyticus (B. corrodens)* from clinical infections. *Journal of Clinical Pathology* **35**, 309–312.

Eribe ER, Paster BJ, Caugant DA *et al*. (2004) Genetic diversity of *Leptotrichia* and description of *Leptotrichia goodfellowii* sp. nov., *Leptotrichia hofstadii* sp. nov., *Leptotrichia shahii* sp. nov. and Leptotrichia wadei sp. nov. *International Journal of Systematic and Evolutionary Microbiology* **54**, 583–592.

Erridge C, Bennett-Guerrero E and Poxton IR (2002) Structure and function of lipopolysaccharides. *Microbes and Infection* **4**, 837–851.

Finegold SM (1994) Review of early research on anaerobes. *Clinical Infectious Diseases* **18 Suppl. 4**, S249–S249.

Finegold SM (1995) Anaerobic bacteria: general concepts. In *Mandell, Douglas and Bennett's Principles and Practice of Infectious Diseases*, 4th edn., Mandell GL Bennett JE and Dolin R (eds), Churchill Livingstone, New York, pp. 2156–2173.

Finegold SM and George LW (1989) *Anaerobic Infections in Humans*. Academic Press, San Diego.

Finegold SM and Jousimies-Somer H (1997) Recently described clinically important anaerobic bacteria: medical aspects. *Clinical Infectious Diseases* **25 Suppl. 2**, S88–S93.

Finlay-Jones JJ, Hart PH, Spencer LK *et al*. (1991) Bacterial killing in vitro by abscess-derived neutrophils. *Journal of Medical Microbiology* **34**, 73–81.

Freeman CD, Klutman NE, Lamp KC (1997) Metronidazole. A Therapeutic review and update. *Drugs* **54**, 679–708.

Fujiwara T, Ogawa T, Sobue S and Hamada S (1990) Chemical, immunobiological and antigenic characterizations of lipopolysaccharides from *Bacteroides gingivalis* strains. *Journal of General Microbiology* **136**, 319–326.

Gallagher A, Aduse-Opoku J, Rangarajan M *et al*. (2003) Glycosylation of the Arg-gingipains of *Porphyromonas gingivalis* and comparison with glycoconjugate structure and synthesis in other bacteria. *Current Protein and Peptide Science* **4**, 427–441.

Godoy VG, Dallas MM, Russo TA and Malamy MH (1993) A role for *Bacteroides fragilis* neuraminidase in bacterial growth in two model systems. *Infection and Immunity* **61**, 4415–4426.

Goldstein EJ (1996) Anaerobic bacteremia. *Clinical Infectious Diseases* **23** Suppl. I, S97–S101.

Hagelskjaer Kristensen L and Prag J (2000) Human necrobacillosis, with emphasis on Lemierre's syndrome. *Clinical Infectious Diseases* **31**, 524–532.

Hardie JM (1991) Dental and oral infection. In *Anaerobes in Human Disease*, Duerden BI and Drasar BS (eds), Edward Arnold, London, pp. 245–267.

Holdeman LV and Moore WEC (1974) Gram-negative anaerobic bacteria. In *Bergey's Manual of Determinative Bacteriology*, 8th edn, Buchanan RE and Gibbons NE (eds), Williams & Wilkins, Baltimore, p. 231.

Hooper LV, Xu J, Falk PG *et al*. (1999) A molecular sensor that allows a gut commensal to control its nutrient foundation in a competitive ecosystem. *Proceedings of the National Academy of Science United States of America* **96**, 9833–9838.

Ingham HR and Sisson PR (1991) Respiratory, ENT and CNS infections. In *Anaerobes in Human Disease*, Duerden BI and Drasar BS (eds), Edward Arnold, London, pp. 268–286.

Jennings MP, Srikhanta YN, Moxon ER *et al*. (1999) The genetic basis of the phase variation repertoire of lipopolysaccharide immunotypes in *Neisseria meningitidis*. *Microbiology* **145**, 3013–3021.

Jones BM, Geary I, Alawattagama AA *et al*. (1985) *In-vitro* and *in-vivo* activity of metronidazole against *Gardnerella vaginalis*, *Bacteroides* spp. and *Mobiluncus* spp. in bacterial vaginosis. *The Journal of Antimicrobial Chemotherapy* **16**, 189–197.

Jousimies-Somer H (1997) Recently described clinically important anaerobic bacteria: taxonomic aspects and update. *Clinical Infectious Diseases* **25 Suppl. 2**, S78–S87.

Jousimies-Somer H and Summanen P (2002) Recent taxonomic changes and terminology update of clinically significant anaerobic gram-negative bacteria (excluding spirochetes). *Clinical Infectious Diseases* **35 Suppl. 1**, S17–S21.

Jousimies-Somer H, Summanen P, Citron D *et al*. (2002) *Wadsworth-KTL Anaerobic Bacteriology Manual*, 6th edn. Star Publishing, Belmont.

Kapatral V, Anderson I, Ivanova N *et al*. (2002) Genome sequence and analysis of the oral bacterium *Fusobacterium nucleatum* strain ATCC 25586. *Journal of Bacteriology* **184**, 2005–2018.

Kapatral V, Ivanova N, Anderson I *et al*. (2003) Genome analysis of *Fusobacterium nucleatum* sub spp. *vincentii* and its comparison with the genome of *F. nucleatum* ATCC, 25586. *Genome Research* **13**, 1180–1189.

Keighley MRB (1992) Anaerobes in abdominal surgery. In *Medical and Environmental Aspects of Anaerobes*, Duerden BI, Brazier JS, Seddon SV and Wade WG (eds), Wrighton Biomedical Publishing Ltd, Petersfield, pp. 24–30.

Kinghorn GR, Duerden BI and Hafiz S (1986) Clinical and microbiological investigation of women with acute salpingitis and their consorts. *British Journal of Obstetrics and Gynaecology* **93**, 869–880.

Kuwahara T, Yamashita A, Hirakawa H *et al*. (2004) Genomic analysis of *Bacteroides fragilis* reveals extensive DNA inversions regulating cell surface adaptation. *Proceedings of the National Academy of Sciences of the USA* **101(41)**, 14919–14924.

Laine ML and van Winkelhoff AJ (1998) Virulence of six capsular serotypes of *Porphyromonas gingivalis* in a mouse model. *Oral Microbiology and Immunology* **13**, 322–325.

Laue H, Denger K and Cook AM (1997) Taurine reduction in anaerobic respiration of *Bilophila wadsworthia* RZATAU. *Applied and Environmental Microbiology* **63**, 2016–2021.

Lindberg AA, Weintraub A, Zahringer U and Rietschel ET (1990) Structure-activity relationships in lipopolysaccharides of *Bacteroides fragilis*. *Reviews of Infectious Diseases* **12 Suppl. 2**, S133–S141.

Liu C, Song Y, McTeague M *et al*. (2003) Rapid identification of the species of the *Bacteroides fragilis* group by multiplex PCR assays using group- and species-specific primers. *FEMS Microbiology Letters* **222**, 9–16.

Lutton DA, Patrick S, Crockard AD *et al*. (1991) Flow cytometric analysis of within-strain variation in polysaccharide expression by *Bacteroides fragilis* by use of murine monoclonal antibodies. *Journal of Medical Microbiology* **35**, 229–237.

Masfari AN, Kinghorn GR and Duerden BI (1983) Anaerobic genitourinary infections in men. *British Journal of Venereal Diseases* **59**, 255–259.

Masfari AN, Duerden BI and Kinghorn GR (1986) Quantitative studies of vaginal bacteria. *Genitourinary Medicine* **62**, 265–273.

Moore LVH and Moore WEC (1977) *VPI Anaerobe Laboratory Manual*, 4th edn. Virginia Polytechnic Institute and State University, Blacksburg, Updates 1987, 1991.

Myers LL, Shoop DS, Stackhouse LL *et al*. (1987) Isolation of enterotoxigenic *Bacteroides fragilis* from humans with diarrhea. *Journal of Clinical Microbiology* **25**, 2330–2333.

Narayanan SK, Nagaraja TG, Chengappa MM and Stewart GC (2002) Leukotoxins of gram-negative bacteria. *Veterinary Microbiology* **84**, 337–356.

Nelson KE, Fleischmann RD, DeBoy RT *et al*. (2003) Complete genome sequence of the oral pathogenic bacterium *Porphyromonas gingivalis* strain W83. *Journal of Bacteriology* **185**, 5591–5601.

Netea MG, van Deuren M, Kullberg BJ *et al.* (2002) Does the shape of lipid A determine the interaction of LPS with Toll-like receptors? *Trends in Immunology* **23**, 135–139.

Otto BR, Kusters JG, Luirink J *et al.* (1996) Molecular characterization of a heme-binding protein of *Bacteroides fragilis* BE1. *Infection and Immunity* **64**, 4345–4350.

Patrick S (2002) *Bacteroides.* In *Molecular Medical Microbiology*, Sussman M (ed.), Academic Press, London, pp. 1921–1948.

Patrick S and Larkin MJ (1995) Genetic variation and regulation of virulence determinants. In *Immunological and Molecular Aspects of Bacterial Virulence.* J Wiley, London, pp. 192–249.

Patrick S, Coffey A, Emmerson AM and Larkin MJ (1988) The relationship between cell surface structure expression and haemagglutination in *Bacteroides fragilis. FEMS Microbiology Letters* **50**, 67–71.

Patrick S, Stewart LD, Damani N *et al.* (1995) Immunological detection of *Bacteroides fragilis* in clinical samples. *Journal of Medical Microbiology* **43**, 99–109.

Patrick S, McKenna JP, O'Hagan S and Dermott E (1996) A comparison of the haemagglutinating and enzymic activities of *Bacteroides fragilis* whole cells and outer membrane vesicles. *Microbial Pathogenesis* **20**, 191–202.

Patrick S, Parkhill J, McCoy LJ *et al.* (2003) Multiple inverted DNA repeats of *Bacteroides fragilis* that control polysaccharide antigenic variation are similar to the hin region inverted repeats of *Salmonella typhimurium. Microbiology* **149**, 915–924.

Poxton IR and Brown R (1986) Immunochemistry of the surface carbohydrate antigens of *Bacteroides fragilis* and definition of a common antigen. *Journal of General Microbiology* **132**, 2475–2481.

Poxton IR, Brown R, Sawyerr A and Ferguson A (1997) Mucosa-associated bacterial flora of the human colon. *Journal of Medical Microbiology* **46**, 85–91.

Ramirez S, Hild TG, Rudolph CN *et al.* (2003) Increased diagnosis of Lemierre syndrome and other *Fusobacterium necrophorum* infections at a children's hospital. *Pediatrics* **112**, 380–385.

Redondo MC, Arbo MD, Grindlinger J and Snydman DR (1995) Attributable mortality of bacteremia associated with the *Bacteroides fragilis* group. *Clinical Infectious Diseases* **20**, 1492–1496.

Reid JH and Patrick S (1984) Phagocytic and serum killing of capsulate and non-capsulate *Bacteroides fragilis. Journal of Medical Microbiology* **17**, 247–257.

Rotstein OD, Kao J and Houston K (1989) Reciprocal synergy between *Escherichia coli* and *Bacteroides fragilis* in an intraabdominal infection model. *Journal of Medical Microbiology* 29, 269–276.

Rotstein OD, Vittorini T, Kao J *et al.* (1989) A soluble *Bacteroides* by-product impairs phagocytic killing of *Escherichia coli* by neutrophils. *Infection and Immunity* **57**, 745–753.

Rudek W and Haque R (1976) Extracellular enzymes of the genus *Bacteroides. Journal of Clinical Microbiology* **4**, 458–460.

Schauer R (1985) Sialic acids and their role as biological masks. *Trends in Biochemical Sciences* **10**, 357–360.

Schwebke JR and Lawing LF (2001) Prevalence of *Mobiluncus* spp. among women with and without bacterial vaginosis as detected by polymerase chain reaction. *Sexually Transmitted Diseases* **28**, 195–199.

Sebald M (1994) Genetic basis for antibiotic resistance in anaerobes. *Clinical Infectious Diseases* **18 Suppl. 4**, S297–S304.

Shah HN and Garbia SE (1991) *Bacteroides* and *Fusobacterium*: classification and relationships to other bacteria. In *Anaerobes in Human Disease*, Duerden BI and Drasar BS (eds), Edward Arnold, London, pp. 62–84.

Shah HN, Gharbia SE and Duerden BI (1998) *Bacteroides, Prevotella, Porphyoromonas.* In *Topley and Wilson's Microbiology and Microbial Infections*, 9th edn, vol. 2, Collier L, Balows A and Sussman M (eds), Edward Arnold, London, pp. 1305–1330.

Smith CJ, Tribble GD and Bayley DP (1998) Genetic elements of *Bacteroides* species: a moving story. *Plasmid* **40**, 19–29.

Snydman DR, McDermott L, Cuchural GJ *et al.* (1996) Analysis of trends in antimicrobial resistance patterns among clinical isolates of *Bacteroides fragilis* group species from 1990 to 1994. *Clinical Infectious Diseases* **23 Suppl. 1**, S54–S65.

Spiegel CA (1987) New developments in the etiology and pathogenesis of bacterial vaginosis. *Advances in Experimental Biology* **224**, 127–134.

Stubbs SLJ, Brazier JS, Talbot PR and Duerden BI (2000) PCR-restriction fragment length polymorphism analysis for identification of *Bacteroides* spp. and characterisation of nitroimidazole resistance genes. *Journal of Clinical Microbiology* **38**, 3209–3213.

Summanen P, Baron EJ, Citron DM *et al.* (1993) *Wadsworth Anaerobic Bacteriology Manual*, 5th edn. Star Publishing Co., Belmont.

Tanner A and Stillman N (1993) Oral and dental infections with anaerobic bacteria: clinical features, predominant pathogens and treatment. *Clinical Infectious Diseases* **16**, S304–S309.

Tiveljung U, Forsum and Monstein HJ (1996) Classification of the genus Mobiluncus based on comparative partial, 16S rRNA gene analysis. *International Journal of Systematic Bacteriology* **46**, 332–336.

Tunney MM, Patrick S, Gorman SP *et al.* (1998) Improved detection of infection in hip replacements: a currently underestimated problem. *The Journal of Bone and Joint Surgery [Br]* **80-B**, 568–572.

Tzianabos AO, Onderdonk AB, Rosner B *et al.* (1993) Structural features of polysaccharides that induce intra-abdominal abscesses. *Science* **262**, 416–419.

Vedantam G and Hecht DW (2003) Antibiotics and anaerobes of gut origin. *Current Opinion in Microbiology* **6**, 457–461.

Wang Z, Fast W and Benkovic SJ (1999) On the mechanism of the metallo-beta-lactamase from *Bacteroides fragilis. Biochemistry* **38**, 10013–10023.

Watt B and Brown FV (1985) Effect of the growth of anaerobic bacteria on the surface pH of solid media. *Journal of Clinical Pathology* **38**, 565–569.

Wexler HM (1993) Susceptibility testing of anaerobic bacteria – the state of the art. *Clinical Infectious Diseases* **16 Suppl. 4**, S328–S333.

Willis AT (1991) Abdominal sepsis. In *Anaerobes in Human Disease*, Duerden BI and Drasar BS (eds), Edward Arnold, London, pp. 197–223.

Willis AT and Fiddian RV (1983) Metronidazole in the prevention of anaerobic infection. *Surgery* **93**, 174–179.

Wren MWD (1991) Laboratory diagnosis of anaerobic infections. In *Anaerobes in Human Disease*, Duerden BI and Drasar BS (eds), Edward Arnold, London, pp. 180–196.

Xu J, Bjursell MK, Himrod J *et al.* (2003) A genomic view of the human-*Bacteroides thetaiotaomicron* symbiosis. *Science* **299**, 2074–2076.

46

Clostridium difficile

Mark H. Wilcox

Department of Microbiology, Leeds General Infirmary & University of Leeds, Old Medical School, Leeds, UK

INTRODUCTION

Antibiotic overprescription is associated with the development of resistance. During the last two decades, another major drawback of antibiotic administration, namely *Clostridium difficile* infection, has emerged. Following *C. difficile* acquisition, the bacterium may simply be excreted by the host or asymptomatic colonisation of gut, diarrhoea, colitis, pseudomembranous colitis (PMC) and/or death may occur. These conditions may blur, and it is often unclear which is the exact clinical diagnosis in a symptomatic patient. PMC is a specific term that should be reserved for the severe end of the spectrum of *C. difficile* infection. Sigmoidoscopy with biopsy for histology is required for an accurate diagnosis.

HISTORICAL ASPECTS

Hall and O'Toole (1935) first described a Gram-positive, sporing, obligate anaerobe isolated from the faeces of neonates and named it *Bacillus difficile*, considering the problems they endured with the organism. Snyder (1937) noted that the lethal effects of broth culture filtrates of the organism could be prevented by an antiserum to *B. difficile*. It was postulated that the toxins of the organism might be involved in life-threatening infections in the newborn. The link between antibiotics and PMC has been apparent since the 1950s, but *Staphylococcus aureus* was presumed to be the infective agent. However, increasing numbers of reports implicating clindamycin and PMC began to appear in the 1970s (Tedesco, Barton and Alpers, 1974), and an alternative pathogen was implied, given the excellent activity of clindamycin against *S. aureus*. PMC was shown to have an infective aetiology by experiments cross-infecting hamsters with caecal contents. A toxin produced by a clostridial-like organism was shown to induce PMC, and disease was prevented by *Clostridium sordellii* antitoxin (Rifkin, Fekety and Silva, 1977; Bartlett *et al.*, 1978). However, it was *C. difficile* as opposed to *C. sordellii* that was recovered from the caeca of diseased animals, and Larson *et al.* (1978) provided evidence linking *C. difficile* with histologically proven PMC in both humans and hamsters.

DESCRIPTION OF THE ORGANISM

Clostridium difficile is a Gram-positive, sporing, obligate anaerobe. Colonial morphology can be variable, depending on media and length of incubation and, hence to the inexperienced, is not a good marker. Distinctive colonies of *C. difficile* growing on the selective medium cycloserine–cefoxitin fructose agar (CCFA) are approximately 4 mm in diameter, yellow and ground glass-like and have a slightly filamentous edge. However, the odour associated with colonies is very distinctive and is typically referred to as like elephant or horse manure. *Clostridium difficile* colonies fluoresce yellow-green under long-wave (365 nm) ultraviolet illumination (when grown on blood-based selective media), but this property is medium dependent and is not unique to *C. difficile*. As older colony cultures sporulate, they tend to become whiter and the fluorescence fades. Also, colonial fluorescence cannot be screened for on original formulations of CCFA as the neutral red pH indicator autofluoresces. Agar formulations using an egg yolk agar base have a theoretical advantage over blood agar-based media, as *C. difficile* produces neither lecithinase nor lipase, whereas other clostridia commonly present in the gut such as *Clostridium perfringens* and *Clostridium bifermentans/C. sordellii* (and which occasionally grow on the selective medium) produce lecithinase and thus can easily be recognised.

Gas–liquid chromatographic methods to detect volatile metabolites indicative of *C. difficile* such as isovaleric acid, isocaproic acid or *p*-cresol in the faeces have been investigated, but the complexity of this approach means that it is unsuitable for use in routine laboratories. Clostridial spores are relatively resistant to heat, solvents and alcohols, enzymes, detergents, disinfectants and ultraviolet and ionising radiation. Control of sporulation is poorly understood in *C. difficile* compared with that seen in *Bacillus subtilis*, although sporulation genes are relatively conserved in *Clostridium* spp. and *Bacillus* spp. *sigK* genes encode an RNA polymerase sigma factor that is essential for sporulation. Notably, *C. difficile* does not readily sporulate on agars containing selective agents, and therefore, colonies will become non-viable if plates are left in air for prolonged periods. In contrast, colonies usually sporulate heavily after incubation for 72 h on non-selective agars and hence will survive prolonged exposure to air. The spore-bearing properties of *C. difficile* can be used to select for the organism before culture, e.g. by alcohol shock (Bartley and Dowell, 1991). Also, addition of bile salts, such as sodium cholate or pure grade taurocholate (Buggy, Wilson and Fekety, 1983), or lysozyme (Wilcox, Fawley and Parnell, 2000) to media enhances the recovery of spores by inducing spore germination from environmental samples or from faeces, after alcohol shock.

Diarrhoea-associated isolates produce usually both toxin A (conventionally called an enterotoxin because of its activity in a rabbit ileal loop assay) and toxin B (classically known as a cytotoxin). Some isolates (not associated with diarrhoeal disease) are non-toxigenic, and a small but ill-defined proportion produce toxin B but not toxin A (A–B+). A recent report described a *C. difficile* strain that produces toxin A and an aberrant toxin B (Mehlig *et al.*, 2001).

PATHOGENICITY

Aetiology of Antibiotic-Associated Infective Diarrhoea

For *C. difficile* to colonise the gut of a normal individual, the resident flora, which are usually inhibitory to its establishment, must be altered qualitatively and quantitatively (Borriello and Barclay, 1986; Borriello, 1998). Anaerobic gut bacteria are believed to be crucial to colonisation resistance, but the precise components involved have not been defined. There is *in vitro* evidence that the faecal flora of elderly patients are less inhibitory to the growth of *C. difficile* (Borriello and Barclay, 1986). Most commonly, gut flora are disrupted following exposure to antibiotics. Occasionally, colonisation resistance may also be impaired by antineoplastic chemotherapy, which may have additional deleterious effects on the gut mucosa.

Clostridium difficile toxin can be found in faecal samples from almost all patients with PMC. In contrast, toxin is present in 30% of patients with antibiotic-associated colitis and usually less than 5% of patients with antibiotic-associated diarrhoea (AAD). Corresponding figures for *C. difficile* culture-positive faeces are 95%, 60% and 20–40% (Wilcox and Spencer, 1992). Hence, a *C. difficile* culture-positive result is a less specific finding than the presence of toxin. Also, it is very likely that *C. difficile* is not the only pathogen that causes AAD (see *Epidemiology*).

Clostridium difficile Toxins

Clostridium difficile usually produces two toxins, in 1:1 molar ratios, classically called toxin A and toxin B. These are often called an enterotoxin and cytotoxin, respectively, although both are detected in the standard diagnostic cytotoxin assay. Opinions differ on the contribution of each of these toxins to the disease state. It is likely that both have a role in the gut mucosal inflammation that typifies *C. difficile* infection (Borriello, 1998). It is currently unclear how toxin A–B+ isolates cause disease if no functional toxin A is produced. It is possible that the toxin B produced by such strains has a broader specificity and is more active than that present in fully functional stains (Wilkins and Lyerly, 2003).

Clostridia produce various toxins that cause alterations in the actin cytoskeleton. The toxins have been classified into at least three groups. Two of them, the C3-like toxins and the binary toxins, are ADP ribosyltransferases. The large clostridial toxins (toxins A and B) are glucosyltransferases and are the best-studied *C. difficile* virulence determinants, causing cell death by disrupting the actin cytoskeleton, inducing the production of inflammatory mediators and disrupting epithelial cell tight-junction proteins (Nusrat *et al.*, 2001). The genes coding for *C. difficile* toxins A and B and minor toxins C–E are sited in a chromosomal pathogenicity locus. This locus was highly stable in 50 toxigenic *C. difficile* strains, whereas non-toxigenic isolates lacked the unit (Cohen, Tang and Silva, 2000). Also, isolates with a defective pathogenicity locus were still able to cause clinical disease.

Clostridium difficile strains that produce toxin B but not toxin A (toxin A–B+) have been reported from many countries. In the United Kingdom, approximately 3% of isolates are *C. difficile* toxin A–B+ (Brazier, Stubbs and Duerden, 1999). There is now good evidence that *C. difficile* toxin A–B+ strains are virulent. Several reports have shown that *C. difficile* toxin A–B+ strains can cause symptomatic disease and even endoscopically confirmed severe PMC (Limaye *et al.*, 2000). Furthermore, in a Canadian hospital, 31% of 91 cases of *C. difficile* infection occurring in a 12-month period (including two deaths) were caused by a toxin A–B+ strain (Alfa *et al.*, 2000). In view of these observations, toxin A detection alone is insufficient for the laboratory diagnosis of *C. difficile* infection. Toxin A–B+ isolates have a truncated *toxA* gene that contains a stop codon, which means that functional toxin A is not produced. The truncated gene also does not code for the immunodominant repeating units (i.e. the binding portion) recognised by antibodies used in toxin A-specific enzyme immunoassays (EIAs) (Kato *et al.*, 1999; Moncrief *et al.*, 2000).

A binary toxin (an actin-specific ADP ribosyltransferase) similar to *C. perfringens* ĩ-toxin (which is an enterotoxin) has also been described in *C. difficile*. The virulence potential of *C. difficile* binary toxin is unknown, and to date, it appears to be limited to strains that have altered toxin A and B genes. Approximately 6% of toxigenic isolates of *C. difficile* referred to the Anaerobe Reference Unit from UK hospitals have binary toxin genes (*cdtA* and *cdtB*) genes (Stubbs *et al.*, 2000). However, although binary toxin-positive isolates are widespread, the common UK hospital-associated *C. difficile* ribotypes do not possess binary toxin, implying that it does not play a major role in the aetiology of *C. difficile* diarrhoea. Other potential virulence determinants in *C. difficile* are reviewed elsewhere (Borriello, 1998).

Epidemic C. difficile Strains

Some *C. difficile* strains are clearly more virulent than others, and there is emerging evidence for the presence of several epidemic *C. difficile* strains. Brazier and colleagues originally identified 116 *C. difficile* polymerase chain reaction (PCR) ribotypes but noted that the prevalence of ribotype I in hospitalised patients was much higher than from individuals in the community (Stubbs *et al.*, 1999). PCR ribotype I is known to be widely distributed and accounts for 58% of strains submitted to the Anaerobe Reference Unit from UK hospitals (Brazier, 2001). It is now apparent that there are subtypes of PCR ribotype I, as determined by random amplified polymorphic DNA (RAPD) fingerprinting (Fawley, Freeman and Wilcox, 2003). There is no good evidence that differences in virulence between strains correlate with the amount of toxin(s) produced. However, the persistence of one or a few strain types within many institutions implies either that some strains survive (sporulate) and spread more effectively between patients or that other virulence characteristics make them more successful at colonisation and/or causing infection. Notably, PCR ribotype I has been shown to sporulate more than other *C. difficile* strains (Wilcox and Fawley, 2000).

Other virulence determinants in *C. difficile* remain poorly characterised. It is unclear whether *C. difficile* flagellae or other putative adhesins mediate bacterial adherence to eukaryotic cells. The significance of an outer cell coat (S-layer) is similarly uncertain.

Immune Response to C. difficile

Following acquisition of *C. difficile* in the hospital setting, only about half of patients develop diarrhoea (Kyne *et al.*, 2000). Furthermore, a proportion of these will experience multiple symptomatic episodes. It has proven difficult to delineate the influence of host factors in *C. difficile* infection, given the frailty and medical complexity of typically affected patients. It is now clear, however, that host immune response to *C. difficile* plays a critical role in determining disease occurrence and recurrence. *Clostridium difficile* toxins A and B share 63% amino acid sequence homology but very little antigenic similarity. Population prevalence studies have detected antibodies against toxins A and B in the serum of approximately 70% of individuals (Johnson, 1997), as might be expected from an organism that is widely distributed in the environment. Early studies found no correlation between antibody titre and toxin-neutralising ability. There is some evidence that the capacity of sera to neutralise toxin decreases with age, despite higher levels of antibody (Viscidi *et al.*, 1983; Warny *et al.*, 1994).

Two key recent studies have significantly advanced knowledge about host humoral response in *C. difficile* infection. In the first of these, serial antibody levels were measured in hospitalised patients who were receiving antibiotics (Kyne *et al.*, 2000). Fourteen per cent of the patients were colonised with *C. difficile* on admission, and a further 17% had nosocomial acquisition. Notably, almost half (44%) of all the patients colonised with *C. difficile* remained asymptomatic carriers, and yet antibody levels at baseline were similar in those who did and did not later became colonised. The explanation for this observation was that after colonisation those patients who became asymptomatic carriers had significantly greater increases in serum immunoglobulin G (IgG) anti-toxin A than did those who had C. difficile diarrhoea. This also explains why patients who are already colonised when admitted to hospital are significantly less likely to develop symptomatic infection (Shim et al., 1998). The second study on humoral immunity to C. difficile diarrhoea found that patients who had only one as opposed to multiple episodes of C. difficile diarrhoea had significantly higher levels of both serum IgM and IgG anti-toxin A (Kyne et al., 2001). Hence, patients who can mount a serum IgG response to toxin A have relative protection both against a first episode of C. difficile disease and recurrent symptoms. It is unknown why individuals can or cannot mount a protective humoral response against C. difficile. Furthermore, the relative importance of initiating antibiotic, gut flora disturbance, strain type and host antibody response in determining the risk of C. difficile infection has not been elucidated.

EPIDEMIOLOGY

Clostridium difficile is found as part of normal large-bowel flora in approximately 2–4% of normal young adults. This percentage rises with age, and the elderly have colonisation rates of 10–20%, depending on recent antibiotic exposure and the time spent in an institution, i.e. the likelihood of exposure to C. difficile (Viscidi, Willey and Bartlett, 1981). It is still not known whether the carriage rate of C. difficile, particularly in the asymptomatic elderly, is increasing. Vegetative forms of C. difficile die rapidly on exposure to air, but spores are produced when the bacterium encounters unfavourable conditions. Clostridial spores can survive for many months and possibly years and are the most likely infective mode by which C. difficile is spread. Interestingly, rates of colonisation in neonates, who have a less complex gut flora, may be very high (up to 70%, depending on the degree of spread of the bacterium within individual units), although babies are very seldom symptomatic. Carriage rates in children approximate to those of adults by the age of 2–3 years. Clostridium difficile has also been isolated from farmyard and domestic animals such as horses, cows, pigs, dogs and cats, but there is no evidence that C. difficile infection is a zoonosis.

The incidence of C. difficile diarrhoea has increased since the 1980s, and now there are approximately 44 000 laboratory reports of C. difficile per year in England and Wales. This figure compares with approximately 13 000 and 42 000 reports of salmonella and campylobacter in 2004. Recent reports suggest the emergence of a new epidemic virulent C. difficile strain (type 027) in Canada, USA and, most recently, the UK. A recent US study found that between 1989 and 2000, the incidence of C. difficile infection increased from 0.68% to 1.2% of hospitalised patients, with a corresponding doubling increase (1.6–3.2%) in the subset who developed life-threatening symptoms (Dallal et al., 2002). It is unclear what is the true distribution of clinical disease represented by laboratory reports of C. difficile detection, and some undoubtedly represent asymptomatically colonised individuals. Conversely, given the variability of testing protocols, underdiagnosis also appears to be commonplace. Over 80% of C. difficile infection cases occur in hospitalised patients aged over 65 years, and the frail elderly are more prone to severe C. difficile infection. Although C. difficile diarrhoea is more common in older patients, increasing age itself appears not to be a risk factor for the severity of C. difficile infection (Kyne et al., 1999). Significant risk factors for severe C. difficile diarrhoea included functional disability, cognitive impairment, recent endoscopy and, possibly, enteral tube feeding. Clostridium difficile infection appears to be a marker, but not a cause per se, of increased risk of dying (Wilcox et al., 1996; Kyne et al., 2002).

Clostridium difficile infection is generally underdiagnosed in the community setting. A Swedish study found that 42% of cases of C. difficile infection presented in the community, and half of these did not have a history of hospitalisation within the previous month (Karlstrom et al., 1998). The Intestinal Infectious Disease survey in England identified C. difficile as the third most common cause of intestinal infectious disease (IID) in patients aged greater than 75 years seen by general practitioners (Tompkins et al., 1999). Clearly, the distinction between nosocomial and community-acquired C. difficile diarrhoea is increasingly blurred by the overlap between primary and secondary healthcare provision.

It has been recently estimated that C. difficile infection accounts for more than $1.1 billion in healthcare costs in the United States per annum (Kyne et al., 2002). A UK case–control study found that C. difficile diarrhoea occurred approximately 14 days before routine discharge and was associated with a high (37%) clinical failure rate of treatment. Although many remain non-quantifiable, estimated costs (in 1996) attributable to C. difficile infection were approximately £4000 per case, 94% of which were due to an increased duration of hospital stay by an average of 3 weeks per patient (Wilcox et al., 1996).

Clostridium difficile in the Hospital Environment

Cross-infection with C. difficile is believed to be common and is compounded by widespread environmental contamination (20–70% of sampled sites), which is especially pronounced if a patient has explosive diarrhoea. An infected patient excretes more than 10^5 C. difficile per gram of faeces, and spores may survive on contaminated surfaces for months (Mulligan et al., 1979; Fekety et al., 1981). In one study, one-third of environmental samples taken from hospital rooms occupied by patients with PMC were culture positive for C. difficile, compared with 1.3% of samples from control rooms. Environmental contamination with C. difficile spores has been demonstrated in 34–58% of sites in hospital wards (Samore et al., 1996; Fawley and Wilcox, 2001), including after detergent-based cleaning (Verity et al., 2001). Commodes, bed frames, sluice rooms and toilet floors are the most frequently contaminated sites. Other potential sources of C. difficile contamination include nurses' uniforms, blood pressure cuffs and thermometers. Spores are easily carried on the hands of healthcare workers, facilitating spread, and are relatively resistant to many disinfectants, including alcohol-based products. As the prevalence of C. difficile in the environment increases, so does the contamination rate on the hands of staff working there (Samore et al., 1996; Fawley and Wilcox, 2001). There may also be an association between the level of environmental contamination and the number of cases of C. difficile infection, but it remains unknown whether this is a cause or consequence of diarrhoea. DNA fingerprinting studies identified an endemic strain that caused a cluster of six cases of diarrhoea, but this was only isolated in the environment after the sixth symptomatic patient (Fawley and Wilcox, 2001). This implies that patient-to-patient or staff-to-patient spread predominates.

Other Causes of AAD

Clostridium difficile is the most commonly identified pathogen causing hospital-acquired infective AAD. Most AAD cases are, however, C. difficile toxin/culture negative, and the cause(s) of these remains unclear. Enterotoxigenic C. perfringens may be pathogenic in some cases of AAD (Borriello et al., 1985; Modi and Wilcox, 2001). In hospitalised patients with suspected AAD, 8% of faecal samples were positive for C. perfringens enterotoxin, 16% positive for C. difficile cytotoxin and 2% positive for both C. perfringens and C. difficile toxins (Asha and Wilcox, 2002). Culture and PCR results confirmed most enzyme-linked immunoadsorbent assay (ELISA) results. Approximately 5% of C. perfringens type A isolates carry the cpe gene that encodes the C. perfringens enterotoxin. It is intriguing that C. perfringens isolates from cases of AAD differ from food-poisoning isolates in the location of the enterotoxin gene (cpe). cpe is found in extrachromosomal DNA in isolates from AAD cases but is chromosomal in those causing food poisoning (Collie and McClane,

1998). This difference may be because AAD isolates acquire *cpe* from resident *C. perfringens*.

Staphylococcus aureus was considered as a prime cause of post-antibiotic diarrhoea, until the link was described between *C. difficile* and AAD in the late 1970s. However, a 2-year prospective study reexamined the role of *S. aureus* as a possible cause of AAD (Gravet *et al.*, 1999). One case of *S. aureus* AAD was found for every five cases of *C. difficile* AAD. Ninety-two per cent of implicated *S. aureus* isolates were methicillin resistant, and of the 47 cases described, 9 developed bacteraemia. Blood culture and stool isolates were indistinguishable by DNA fingerprinting. Significantly more cases than controls received a fluoroquinolone, and significantly fewer received amoxicillin. Also, significantly more AAD *S. aureus* isolates produced specific leucotoxins and staphylococcal enterotoxin A than did random *S. aureus* strains. This relatively high prevalence of *S. aureus* AAD could not be confirmed in a recent UK study of over 4000 faecal samples from patients with nosocomial diarrhoea (unpublished data).

CLINICAL FEATURES

Clinical Presentation

Clostridium difficile infection is typified by diarrhoea starting usually within a few days of starting antibiotics, but antimicrobial therapy taken 1–2 months ago can still predispose to infection. Very occasional cases occur where no recent antibiotic consumption can be identified, and it is possible that antimicrobial substances in food are responsible. Diarrhoeal faeces usually have a foul odour and, in the absence of PMC, are rarely bloody. Depending on the extent of colitis, abdominal pain with or without a pyrexia may occur. It is not uncommon for diarrhoea, particularly in the elderly, to cause electrolyte disturbances, hypoalbuminaemia and/or paralytic ileus. If pancolitis develops, this may lead to toxic megacolon, perforation and secondary endotoxic shock. Patients with *C. difficile* infection often have raised peripheral white blood cell (WBC) counts. Wanahita, Goldsmith and Musher (2002) found that of 400 inpatients with WBC counts $\geq 15 \times 10^9$/L, infection was documented in 207 (53%) and 16% of these had confirmed *C. difficile* infection. When patients with haematological malignancy were excluded, a quarter of patients with WBC counts $\geq 30 \times 10^9$/L had *C. difficile* infection. *Clostridium difficile* has been reported to cause very occasional extraintestinal infections typically in patients with bowel disease.

Multiple recurrences of diarrhoea following treatment of *C. difficile* infection are common, and subsequent treatment poses a therapeutic dilemma. Recurrence is a clinical definition (associated with toxin-positive faeces), i.e. a symptom-free period followed by diarrhoea with or without other symptoms. It should be noted that recurrence is sometimes overdiagnosed, notably in studies that included patients who were symptom free for only 1 day.

Patients with PMC tend to have more marked systemic symptoms. If pancolitis develops, this may lead to toxic megacolon, perforation and secondary endotoxic shock. There is a high mortality associated with these complications. In some cases, e.g. toxic megacolon, reduced diarrhoea may occur. Clinicians must be alert to this possibility and to the potential need for urgent surgical intervention in a patient with worsening abdominal and/or systemic signs. Patients with *C. difficile* infection often have raised peripheral WBC counts.

Antimicrobials Predisposing to *C. difficile* Infection

Simplistically, broad-spectrum, particularly anti-anaerobic, agents should be those most associated with *C. difficile* infection, according to colonisation resistance theory. However, *in vivo* antimicrobial activity, and therefore probably the risk of *C. difficile* infection, can be markedly influenced by factors such as drug penetration into the large gut lumen, specific and non-specific antibiotic binding and gut pH and redox potential. It is extremely difficult to determine accurately the risk of developing *C. difficile* infection associated with specific antibiotics. Most studies in this area to date are flawed because of the use of incorrect control groups, bias and inadequate control of confounding and/or small sample sizes (Wilcox, 2001; Thomas, Stevenson and Riley, 2003). Many factors may affect colonisation resistance and therefore potentially distort the risk associated with particular antibiotics. Also, affected patients are frequently exposed to multiple antibiotics and indeed often receive combinations of antimicrobial agents. Importantly, an antibiotic may be assumed to have a high *C. difficile* risk because it was used during an outbreak of *C. difficile* infection.

Accepting the inadequacies of many studies, some antibiotics and antibiotic classes have been consistently associated with increased risks of *C. difficile* infection, whereas others are conspicuous by the absence of reports linking them to *C. difficile* diarrhoea (Table 46.1). The main predisposing factor for *C. difficile* diarrhoea is antibiotic therapy, notably with cephalosporins (in particular third-generation cephalosporins such as cefotaxime and ceftriaxone), clindamycin and broad-spectrum penicillins. Golledge, McKenzie and Riley (1989) found that over a 1-year period, 5% of patients treated with extended-spectrum cephalosporins developed *C. difficile* diarrhoea. A similar study by de Lalla *et al.* (1989) showed that although all cephalosporins were statistically more likely to induce *C. difficile* infection, there was a significantly higher risk of diarrhoea with third-generation as opposed to first-generation cephalosporins. Aronsson, Möllkby and Nord (1985) found that cephalosporins were implicated in *C. difficile* infection 40 times more often than narrow-spectrum penicillins. Duration of antimicrobial therapy is an important factor in assessing the risk of *C. difficile* infection. A large prospective study of patients in five Swedish hospitals demonstrated that patients treated for fewer than 3 days had a significantly lower incidence of *C. difficile* infection than those receiving longer courses of antibiotic therapy (Wistrom *et al.*, 2001). Macrolides, especially clarithromycin, have been cited as a risk factor for *C. difficile* infection, although confounding factors are difficult to exclude, particularly as hospitalised patients often these agents with other antibiotics.

Anti-pseudomonal penicillins such as piperacillin and ticarcillin (with or without β-lactamase inhibitors) appear to have a low propensity to induce *C. difficile* infection, which is interesting given their deleterious effects on endogenous aerobic and anaerobic gut flora. Anand *et al.* (1994) found that 51 cases of *C. difficile* infection were associated with 40 000 doses of third-generation cephalosporins, and yet none were seen following 62 000 doses of ticarcillin–clavulanate ($P = 0.0001$). In a prospective, geriatric ward crossover study, there was a highly significant difference in the rates of colonisation and

Table 46.1 Antibiotic-related risk of *C. difficile* infection

High risk
Cephalosporins
Clindamycin
Medium risk
Ampicillin/amoxicillin
Co-trimoxazole
Macrolides
Tetracyclines
Low risk
Aminoglycosides
Metronidazole
Anti-pseudomonal penicillins ± β-lactamase inhibitor
Fluoroquinolones
Rifampicin
Vancomycin

These subdivisions may be biased because of the relative frequency of use of agents and the complexity of interaction between antibiotic exposure, *C. difficile* and other factors.

C. difficile diarrhoea in cefotaxime as opposed to piperacillin–tazobactam-treated patients (relative risk of diarrhoea 7.4, 95% CI = 1.7–33) (Settle *et al.*, 1998). The reasons why anti-pseudomonal penicillins rarely appear to induce *C. difficile* infection compared with cephalosporins are uncertain but possibly relate to lower-than-expected faecal concentrations of antibiotic and lack of induction of toxin production.

Aminoglycosides have reduced propensity to induce *C. difficile* infection, probably due to their lack of effect on the endogenous anaerobic gut bacteria. So far, there has been little evidence linking fluoroquinolones with *C. difficile* infection. Older fluoroquinolones (including ciprofloxacin) have poor anti-anaerobic activity, whereas new agents such as moxifloxacin and gatifloxacin have significantly increased activity. In a retrospective case–control study of 27 patients, prior exposure to cephalosporins and use of ciprofloxacin were found to be significant risk factors for *C. difficile* infection in a multivariate analysis (Yip *et al.*, 2001). However, a much larger prospective study showed that the incidence of *C. difficile* diarrhoea was significantly lower in patients treated with levofloxacin compared with those treated with cefuroxime (2.2% vs. 9.2%, P < 0.0001) and was similar to those receiving amoxicillin (1.8%, P = 0.7) (Gopal Rao, Mahankali Rao and Starke, 2003). Other antibiotics considered to have a low risk of inducing *C. difficile* infection include penicillin, trimethoprim, rifampicin, fusidic acid and nitrofurantoin. There are also very few published reports associating co-amoxiclav with *C. difficile* infection, despite widespread usage of this combination antimicrobial agent.

Although *C. difficile* infection is most commonly associated with antibiotic treatment, studies have shown that short-term perioperative prophylaxis with antimicrobial agents can also predispose to *C. difficile* diarrhoea. In a retrospective case–control study, significantly fewer patients who developed *C. difficile* diarrhoea than controls received antibiotic prophylaxis for an appropriate duration (i.e. ≤24 h). Length of hospital stay was also significantly longer in the toxin-positive group (16.5 days vs. 10.2 days, P < 0.05), suggesting that overlong antibiotic prophylaxis was detrimental in terms of both diarrhoea and patient stay (Kreisel *et al.*, 1995).

LABORATORY DIAGNOSIS

Many diseases and drugs are associated with the symptom of diarrhoea, and indeed, antibiotics can cause diarrhoea by means that do not involve *C. difficile*, e.g. erythromycin-induced stimulation of gut peristalsis. It is therefore necessary to use the microbiology laboratory to confirm a suspected diagnosis of *C. difficile* infection. There is a great variability both in the criteria used by microbiology laboratories to decide which faecal specimens to test and in the methods employed to detect *C. difficile*. At one extreme, some only test for *C. difficile* if it is specifically requested. At the other extreme, others look for the bacterium/toxins on all specimens received. Both approaches have their merits, but unless some degree of selectivity is operated, there is a real risk of overdiagnosing the infection, given the existence of asymptomatic carriers. In the hospital setting, some laboratories operate a 3-day rule whereby all faecal samples from patients admitted at least 3 days previously are examined for *C. difficile* but not for other enteropathogens such as salmonellae and shigellae (Wood, 2001).

Clostridium difficile Culture

There are two main alternatives for the laboratory diagnosis of *C. difficile* infection. Firstly, detection of the bacterium by culture, which is relatively slow (taking up to 5 days) and is less specific than toxin testing, particularly given the existence of asymptomatic carriage, especially in the elderly, and occasional non-toxigenic (and therefore non-pathogenic) strains. Most laboratories now detect *C. difficile* by toxin testing, and the proportion of centres using culture continues to decline. For those laboratories performing both approaches, there is no consensus on how to interpret the occasional specimen that yields a

toxin-negative result and yet is culture positive. Culture is, however, required if typing is to be performed to establish the epidemiology of strains present in a ward or hospital.

Agar formulations using an egg yolk agar base (CCEY) have a theoretical advantage over blood agar-based media, as *C. difficile* produces neither lecithinase nor lipase, while other clostridia commonly present in the gut such as *C. perfringens* and *C. bifermentans/C. sordellii* (and which occasionally grow on the selective medium) produce lecithinase and thus can easily be recognised. Commercially available formulations of CCEY are available, and these are optimised for *C. difficile* isolation and recognition. Putative *C. difficile* colonies can be identified by the characteristics noted above and by using antigen (e.g. Microgen, UK) or enzyme detection kits.

Clostridium difficile Toxin Detection

Detection of one or both of the toxins of *C. difficile* remains the gold standard test with which other diagnostic tests are compared in evaluations of sensitivity and specificity. The presence of cytotoxin usually causes a rounding-up cytopathic effect (CPE), depending on the cell line used. Both toxins A and B can cause a CPE, although the latter is approximately 1000-fold more potent in this assay. Assay for *C. difficile* cytotoxins can be performed using many different commonly used cell lines, including Vero, HEp2, monkey kidney, HeLa, fibroblast and Chinese hamster ovary cell lines. It is disputed which cell line(s) is the most sensitive, but Vero cell lines are probably most commonly used in Europe. A centrifuged, filtered faecal filtrate is inoculated onto the cell monolayer, which is examined after overnight incubation and again after 48 h. Non-specific CPEs are detected by using a control cell monolayer containing either *C. difficile* or *C. sordellii* antitoxin that neutralises *C. difficile* cytotoxic activity. Disadvantages of the cytotoxin technique include a lack of method standardisation, a need to maintain a cell-culture line and the relative slowness of the test (24–48 h) compared with EIA (1–2 h). However, this delay may be beneficial in avoiding treatment of cases of *C. difficile* diarrhoea that resolve spontaneously. It is also possible to detect 75% of cytotoxin-positive results after incubation for only 6 h (Settle and Wilcox, 1996).

Many kit-based methods (primarily EIAs) to detect toxins A ± B are available. These are quicker and simpler to perform than the cytotoxin technique but tend to suffer from reduced sensitivity. Numerous studies have compared the performance of different kits with each other and with other methods, but no meta-analysis has been performed to determine with confidence the superiority of any particular test (Brazier, 1998). A recent study comparing six different rapid methods with the cytotoxin technique on over 1000 faecal samples (Turgeon *et al.*, 2003) confirmed earlier findings that such kits are about 80–90% sensitive and 90–99% specific. Assays that target both toxins tend to perform better and have the advantage of being able to detect the toxin A–B+ strains. Triage® is a membrane immunoassay for the simultaneous but distinct detection of toxin A and glutamate dehydrogenase. Triage® has been shown to have a high negative predictive value, suggesting that it might be a useful screening test (Barbut *et al.*, 2000; Turgeon *et al.*, 2003). In general, nucleic acid amplification techniques have not offered any significant benefit to date compared with the above approaches for the diagnosis of *C. difficile* diarrhoea, primarily because of the high clinical accuracy associated with direct toxin detection methods.

Histopathology

The histological appearances of large-bowel tissue obtained via sigmoidoscopy remain the optimum way of distinguishing between the different pathological conditions associated with *C. difficile*, although in about 10% of patients evidence of colitis is found only in the caecum or transverse colon (Fekety, 1994). In approximately 25% of individuals, stools may continue to be culture and toxin positive for some time after symptoms have resolved (Bartlett, 1984; Fekety *et al.*, 1984), and follow-up stools from asymptomatic patients should not be sent for toxin assay.

Clostridium difficile Typing

Numerous typing schemes have been proposed for *C. difficile* (Brazier, 2001; Cohen, Tang and Silva, 2001). These differ markedly in ease of use and interpretation and, importantly, in their discriminatory power (Table 46.2). The three most extensively studied are serogrouping (used predominantly in France and Belgium; Delmée, Homel and Wauters, 1985), restriction endonuclease analysis (REA) (used in United States; Kuijper *et al.*, 1987) and ribotyping (used in United Kingdom; Stubbs *et al.*, 1999). Serogrouping is simple to perform and has moderate discriminatory power but is limited by availability of sera. REA has good discriminatory power but produces highly complicated DNA banding patterns that can be difficult to interpret. Ribotyping requires molecular biology expertise but is highly discriminatory and produces simple banding patterns. RAPD and pulsed-field gel electrophoresis (PFGE) have shown potentially very high discriminatory power but have problems associated with reproducibility and demanding methodology, respectively (Fawley, Freeman and Wilcox, 2003). Some strains, such as *C. difficile* serogroup G isolates (PCR ribotype I), have proved to be difficult to type by PFGE, presumably due to high endonuclease activity, although a modified method appears to have overcome the DNA degradation (Fawley and Wilcox, 2002). In general, standardisation and simplification of *C. difficile* typing nomenclature schemes is needed.

MANAGEMENT OF INFECTION

Treatment

There are few proven therapeutic regimens for *C. difficile* infection. First-line treatment should, where possible, involve discontinuation of the precipitating antibiotic(s), although in many patients this is not possible. Specific treatment is indicated when the patient has a systemic illness with evidence of colonic inflammation, PMC or persistent symptoms, despite stopping the precipitating antibiotic. Spontaneous symptomatic resolution may occur in approximately a quarter of cases of *C. difficile* infection. In reality, most patients are commenced on either metronidazole or vancomycin when the infection is diagnosed and before precipitating antibiotic(s) is ceased. Precipitating antibiotic(s) can be substituted with lower-risk agents, but this approach is of unproven benefit. *Clostridium difficile*

infection is treated preferably with oral metronidazole (400 mg or 500 mg three times daily) for 7–10 days. Oral vancomycin (125 mg four times daily) is a more expensive alterative and risks selecting for glycopeptide-resistant enterococci. There is no statistically proven difference between metronidazole and vancomycin when overall response or recurrence rates are examined (Teasley *et al.*, 1983; Wilcox and Spencer, 1992). The mean duration of symptoms is marginally shorter (mean 1.6 days) following vancomycin compared with metronidazole administration (Wilcox and Howe, 1995). Either antibiotic can be administered via the nasogastric route if they cannot be tolerated orally. For intravenous treatment, both antibiotics should be administered together because of the unpredictable colonic concentrations of each agent. Reports of reduced susceptibility of *C. difficile* to metronidazole and vancomycin are very uncommon, but both intermediate (8–16 mg/L) and high-level resistance (>64 mg/L) to metronidazole have been described (Pelaez *et al.*, 2002). Disturbingly high rates of intermediate resistance to metronidazole (6.3%) and vancomycin (3.1%) were reported in Spain in 2002 (Pelaez *et al.*, 2002), but only one metronidazole-resistant isolate has been reported in the United Kingdom (Brazier *et al.*, 2001).

Recurrence

Symptomatic recurrences following treatment of *C. difficile* infection are common (up to 37% of cases), and subsequent treatment poses a therapeutic dilemma (Wilcox *et al.*, 1996). DNA fingerprinting studies have shown that relapses are often re-infections with different strains (O'Neill, Beaman and Riley, 1991; Wilcox *et al.*, 1998). Hence, the common practice of switching metronidazole to vancomycin and vice versa in patients with symptomatic recurrences is illogical, and a first recurrence should be re-treated with oral metronidazole. Courses of 4–6 weeks with tapering and pulsed doses of vancomycin have been used in theory to first kill the vegetative bacteria, to allow spores to germinate and later to kill them (Tedesco, Gordon and Fortson, 1985). Controlled studies of this approach have not been performed. Success of this approach may be due to the prolonged antibiotic course, preventing reacquisition of *C. difficile*, while colonisation resistance by bowel flora remains ineffectual against opportunistic bacterial colonisation. Use of anion exchange resins to bind *C. difficile* toxins is not a practical option because it is cumbersome to use and vancomycin can bind to these agents *in vitro* (Taylor and Bartlett, 1980).

Biotherapy

There is much interest in the biotherapeutic and immunological approaches for the management of *C. difficile* infection. *Saccharomyces*

Table 46.2 Comparison of techniques used to type *C. difficile*

Typing method	Typeability	Reproducibility	Discriminatory power	Interpretation	Performance (number of types/groups)
Antimicrobial susceptibility	All	Fair	Poor	Easy	Easy
Serotyping	Most	Good	Fair (19)	Easy	Easy
Phage/bacteriocin	Few	Fair	Poor	Easy	Easy
Sodium dodecyl sulphate polyacrylamide gel electrophoresis	All	Good	Good (17)	Easy	Moderate
Immunoblot	All	Good	Good	Easy	Moderate
Plasmid profile	Few	Fair	Poor	Moderate	Moderate
Toxinotype	All	Good	Poor (11)	Easy	Moderate
Pyrolysis mass spectrometry	All	Fair	Excellent	Easy	Difficult
Restriction endonuclease analysis	All	Good	Good (55)	Difficult	Moderate
Restriction fragment length polymorphism	All	Good	Poor	Moderate	Moderate
AP-PCR	All	Good	Excellent	Easy	Easy
Random amplified polymorphic DNA	All	Fair	Excellent	Easy	Easy
PCR ribotyping	All	Excellent	Excellent (116)	Easy	Easy
Pulsed-field gel electrophoresis	Most if not all	Excellent	Excellent	Moderate	Difficult

Modified from Cohen, Tang and Silva (2001).

boulardii in particular, which prevents binding of toxin A in a rat ileal model (Pothoulakis *et al.*, 1993), has been extensively examined and is commercially available as a freeze-dried preparation. There have been several placebo-controlled, double-blind trials examining the role of *S. boulardii* in preventing or treating AAD. Early prophylaxis studies investigated the effect of *S. boulardii* given to patients receiving antibiotics. There was a reduced incidence of diarrhoea within 2 weeks of cessation of antibiotic therapy, but diarrhoea was not significantly reduced in the group of patients with *C. difficile*, and neither acquisition of *C. difficile* nor production of toxin was reduced (Adam, Barret and Barrett-Bellet, 1977; Surawicz, Elmer and Speelman, 1989). In a large, double-blind, placebo-controlled study of *S. boulardii* for the treatment of *C. difficile* infection, *S. boulardii* did not reduce subsequent recurrences in patients with a first episode of *C. difficile* diarrhoea. In patients experiencing recurrent diarrhoea, there was a significant decrease in recurrences in those receiving *S. boulardii* compared with placebo (McFarland *et al.*, 1994). The reason for such a discrepancy in response is unclear.

An area of potential concern with biotherapy is the risks associated with the administration of live micro-organisms, particularly to frail elderly patients with inflamed gut mucosae. Cases of fungaemia have been reported in both immunocompromised and immunocompetent patients following administration of *S. boulardii*, highlighting the potential virulence of this yeast in humans (Fredenucci *et al.*, 1998; Niault *et al.*, 1999). Furthermore, *S. boulardii* isolates obtained from varied commercial sources were shown to differ in their virulence in murine models of systemic infection, suggesting a lack of uniformity of yeast strains in such preparations (McCullough *et al.*, 1998). Also, commercially available *S. boulardii* strains had moderate virulence compared with *Saccharomyces cerevisiae* (baker's yeast) in these models.

Biotherapy studies in AAD have been usually open and contained few subjects, and thus, the true efficacy of most, if not all of the, regimens described to date is uncertain. Small series of patients have been treated for, or received prophylaxis against, AAD with biotherapy, including *Lactobacillus acidophilus*, *Lactobacillus GG*, *Enterococcus faecium* SF 68, non-toxigenic *C. difficile*, yoghurt or Brewer's yeast. Trials evaluating the role of *Lactobacillus GG* in acute diarrhoea have reported conflicting results (Gorbach, Chang and Goldin, 1987; Siitonen *et al.*, 1990; Thomas *et al.*, 2001). Notably, a recent large, randomised, double-blind, placebo-controlled trial found neither a decrease in the rate of diarrhoea nor a decrease in the rate of *C. difficile* infection in patients receiving *Lactobacillus GG* (Thomas *et al.*, 2001). Rectal biotherapy and rectal infusion of faeces have been used occasionally in *C. difficile* infection to circumvent gastric acid-mediated degradation of potential probiotics. However, this route of administration is often impractical and unpleasant to administer.

Toxin-Binding Polymer

A high-molecular-weight toxin-binding polymer, GT160–246, is undergoing investigation for treatment of *C. difficile* diarrhoea. It has no direct antimicrobial activity but binds toxins A and B. It has been shown to protect *C. difficile*-infected hamsters from mortality, although not from initial development of colitis (Kurtz *et al.*, 2001). The treatment has been shown to be well tolerated in phase 1 trials and may offer a promising alternative to the current limited choice of proven therapies.

PREVENTION AND CONTROL

Antibacterial Prophylaxis Against *C. difficile* Infection

Antibacterial prophylaxis against *C. difficile* infection is not a practical option as oral antibiotics can adversely affect the normal gut flora and increase the potential period of susceptibility to colonisation by *C. difficile*. A study of prophylaxis with oral metronidazole or vancomycin against *C. difficile* infection concluded that neither was effective at preventing asymptomatic faecal excretion of *C. difficile* (Johnson *et al.*, 1992). Two studies have claimed evidence that metronidazole, given as surgical prophylaxis to surgical patients, may reduce the risk of postoperative *C. difficile* infection (Gerding *et al.*, 1990; Cleary *et al.*, 1998). However, in one of these, the comparator group with more cases of *C. difficile* infection received clindamycin (a high-risk antibiotic) instead of metronidazole as anti-anaerobic cover (Gerding *et al.*, 1990). In the other study, a non-significant difference was seen in the incidence of *C. difficile* colonisation in patients receiving a non-standard prophylaxis regimen (Cleary *et al.*, 1998). Thus, there is no sound evidence of a benefit of using metronidazole to prevent *C. difficile* infection.

Immunotherapy

In the mid 1990s, *C. difficile* toxin-neutralising antibodies were found to be present in nine batches from three commercial (United States) Ig suppliers (Salcedo *et al.*, 1997). There have been several case reports of patients with *C. difficile* infection unresponsive to standard antimicrobial therapy who were successfully treated with intravenous Ig, but no comparative studies of Ig have been performed. Diarrhoea was prevented in clindamycin-treated, *C. difficile*-exposed hamsters when given prophylaxis with bovine colostral Ig concentrate (Lyerly *et al.*, 1991). Bovine Ig concentrate has been reported to inhibit the cytotoxicity and enterotoxicity of *C. difficile* toxin (Kelly *et al.*, 1996), and human faeces were shown to contain neutralising antitoxin activity after oral administration of this preparation (Kelly *et al.*, 1997). However, current concerns about bovine spongiform encephalopathy are likely to undermine further progress with this preparation in humans.

Vaccination

Vaccines preventing *C. difficile* infection in the elderly could potentially revolutionise current therapeutic options. A vaccine containing both *C. difficile* A and B toxoids has been shown to be safe and immunogenic in healthy volunteers, producing increased serum IgG and faecal IgA anti-toxin levels in excess of those associated with protection in clinical studies (Kotloff *et al.*, 2001; Aboudola *et al.*, 2003). It remains unclear, however, whether this approach will be effective in the main target population, particularly since infection in the elderly may be attributable to a compromised immune response.

Hospital Infection Control

The fundamental principle in the control of *C. difficile* infection is the control of antimicrobial prescribing, and there are numerous examples of restrictive antibiotic policies associated with reduction in rates of *C. difficile*-associated diarrhoea (CDAD) (Pear *et al.*, 1994; Ludlam *et al.*, 1999). Despite the recognition of *C. difficile* infection and its well-known link with antibiotic use, incidence of infection may still increase, and constant feedback to prescribers may be required (Stone *et al.*, 2000).

Nosocomial transmission of *C. difficile* is believed to occur primarily via the contaminated hands of healthcare workers or via environmental (including healthcare equipment) contamination and, less commonly, by direct patient-to-patient spread. Symptomatic patients should be isolated or cohort nursed, and hand hygiene is a key intervention to prevent nosocomial spread of *C. difficile* infection. Much of the evidence, however, is observational, although some studies have demonstrated the effectiveness of gloves (Johnson *et al.*, 1990), antiseptic soaps and alcohol-based hand-rub solutions (Bettin *et al.*, 1994;

Lucet *et al.*, 2002). Contamination of healthcare workers' hands can lead to and result from contamination of the environment. Thus, the prevalence of healthcare worker hand contamination with *C. difficile* correlates with the level of environmental contamination (Samore *et al.*, 1996; Wilcox *et al.*, 2003). However, proof that reducing environmental *C. difficile* can decrease the incidence of infection is lacking.

The optimum method for decontaminating hospital environments contaminated with *C. difficile* remains controversial. Hypochlorite was found to be more effective than a quaternary ammonium solution in a bone marrow transplant (BMT) unit, but not in an ITU or a general medical ward (Mayfield *et al.*, 2000). The same workers reported a reduction in *C. difficile* diarrhoea rates in the BMT unit after use of the hypochlorite solution, but no reduction on the other wards studied. When the use of ammonium solution was restarted, rates of *C. difficile* diarrhoea rose, suggesting that hypochlorite solution is effective at reducing the risk of infection in high-risk clinical areas. However, environmental *C. difficile* prevalence was not measured and antibiotic use was altered during the study period. An outbreak of *C. difficile* infection ended following introduction of disinfection with unbuffered hypochlorite (500 ppm available chlorine), and surface contamination with *C. difficile* decreased to 21% of initial levels (Kaatz *et al.*, 1988). Phosphate-buffered hypochlorite (1600 ppm available chlorine, pH 7.6) was found to be more effective at reducing environmental *C. difficile* levels (98% reduction in surface contamination). Wilcox and Fawley (2000) found that some non-chlorine–based hospital cleaning agents, but not chlorine-based cleaning products, caused increased sporulation of *C. difficile in vitro*. Hence, the choice of environmental cleaning agents may in fact increase the persistence of organisms and lead to increased risk of infection. In a recent crossover study on elderly medicine wards, hypochlorite-based cleaning was associated with a reduction in both *C. difficile* environmental prevalence and *C. difficile* infection incidence in one of the two study wards (Wilcox *et al.*, 2003).

REFERENCES

Aboudola S, Kotloff KL, Kyne L *et al.* (2003) *Clostridium difficile* vaccine and serum immunoglobulin G antibody response to toxin A. *Infect Immun* 71, 1608–10.

Adam J, Barret A, Barrett-Bellet C (1977) Essais cliniques controles en double insu de l'ultra-levure lyophilisee. Etude multicentrique par 25 medecins de 388 cas. *Gaz Med Fr* 84, 2072–8.

Alfa MJ, Kabani A, Lyerly D *et al.* (2000) Characterization of a toxin A-negative, toxin B-positive strain of *Clostridium difficile* responsible for a nosocomial outbreak of *Clostridium difficile*-associated diarrhea. *J Clin Microbiol* 38, 2706–14.

Anand A, Bashey B, Mir T, Glatt AE (1994) Epidemiology, clinical manifestations and outcome of *Clostridium difficile*-associated diarrhoea. *Am J Gastroenterol* 89, 519–23.

Aronsson B, Möllby R, Nord CE (1985) Antimicrobial agents and *Clostridium difficile* in acute enteric disease: epidemiological data from Sweden. 1980–1982. *J Infect Dis* 151, 476–81.

Asha N, Wilcox MH (2002) Laboratory diagnosis of *Clostridium perfringens* antibiotic associated diarrhoea. *J Med Microbiol* 51, 891–4.

Barbut F, Lalande V, Daprey G *et al.* (2000) Usefulness of simultaneous toxin A and glutamate dehydrogenase for the diagnosis of *Clostridium difficile*-associated disease. *Eur J Clin Microbiol Infect Dis* 19, 481–4.

Bartlett JG (1984) Treatment of antibiotic-associated pseudomembranous colitis. *Rev Infect Dis* 6 (Suppl 1), S235–41.

Bartlett JG, Chang TW, Gurwith M *et al.* (1978) Antibiotic-associated pseudomembranous colitis due to toxin-producing clostridia. *N Engl J Med* 298, 531–4.

Bartley SL, Dowell VR Jr (1991) Comparison of media for the isolation of *Clostridium difficile* from faecal specimens. *Lab Med* 22, 335–8.

Bettin K, Clabots C, Mathie P *et al.* (1994) Effectiveness of liquid soap vs. chlorhexidine gluconate for the removal of *Clostridium difficile* from bare hands and gloved hands. *Infect Control Hosp Epidemiol* 15, 697–702.

Borriello SP (1998) Pathogenesis of *Clostridium difficile* infection. *J Antimicrob Chemother* 41 (Suppl C), 13–9.

Borriello SP, Barclay FE (1986) An *in-vitro* model of colonisation resistance to *Clostridium difficile* infection. *J Med Microbiol* 21, 299–309.

Borriello SP, Barclay FE, Welch AR *et al.* (1985) Epidemiology of diarrhoea caused by enterotoxigenic *Clostridium perfringens*. *J Med Microbiol* 20, 363–72.

Brazier J (2001) Typing of *Clostridium difficile*. *Clin Microbiol Infect* 7, 428–31.

Brazier JS (1998) The diagnosis of *Clostridium difficile*-associated disease. *J Antimicrob Chemother* 41 (Suppl C), 29–40.

Brazier JS, Stubbs SL, Duerden BI (1999) Prevalence of toxin A negative/B positive *Clostridium difficile* strains. *J Hosp Infect* 42, 248–9.

Brazier JS, Fawley WN, Freeman J, Wilcox MH (2001) Reduced susceptibility of *Clostridium difficile* to metronidazole. *J Antimicrob Chemother* 48, 741–2.

Buggy BP, Wilson KH, Fekety R (1983) Comparison of methods for recovery of *Clostridium difficile* from an environmental source. *J Clin Microbiol* 18, 348–52.

Cleary RK, Grossman R, Fernandez FB *et al.* (1998) Metronidazole may inhibit intestinal colonisation with *Clostridium difficile*. *Dis Colon Rectum* 41, 464–7.

Cohen SH, Tang YJ, Silva J Jr (2000). Analysis of the pathogenicity locus in *Clostridium difficile* strains. *J Infect Dis* 181, 659–63.

Cohen SH, Tang YJ, Silva J Jr (2001) Molecular typing methods for the epidemiological identification of *Clostridium difficile* strains. *Expert Rev Mol Diagn* 1, 61–70.

Collie RE, McClane BA (1998) Evidence that the enterotoxin gene can be episomal in *Clostridium perfringens* isolates associated with non-food-borne human gastrointestinal diseases. *J Clin Microbiol* 36, 30–6.

Dallal RM, Harbrecht BG, Boujoukas AJ *et al.* (2002) Fulminant *Clostridium difficile*: an underappreciated and increasing cause of death and complications. *Ann Surg* 235, 363–72.

de Lalla F, Privitera G, Ortisi G *et al.* (1989) Third generation cephalosporins as a risk factor for *Clostridium difficile*-associated disease: a four-year survey in a general hospital. *J Antimicrob Chemother* 23, 623–31.

Delmée M, Homel M, Wauters G (1985) Serogrouping of *Clostridium difficile* strains by slide agglutination. *J Clin Microbiol* 21, 323–7.

Fawley WN, Wilcox MH (2001) Molecular typing of endemic *Clostridium difficile* infection. *Epidemiol Infect* 126, 343–50.

Fawley WN, Wilcox MH (2002) Pulsed-field gel electrophoresis can yield DNA fingerprints of degradation-susceptible *Clostridium difficile* strains. *J Clin Microbiol* 40, 3546–7.

Fawley WN, Freeman J, Wilcox MH (2003) Evidence to support the existence of subgroups within the UK epidemic *Clostridium difficile* strain (PCR ribotype 1). *J Hosp Infect* 54, 74–7.

Fekety R (1994) Antibiotic-associated colitis. In: Mandell GL, Douglas RG, Bennett JE (eds), *Principles and Practice of Infectious Diseases (4th edition)*. Churchill Livingstone, 978–87.

Fekety R, Kim KH, Batts DH *et al.* (1981) Epidemiology of antibiotic-associated colitis; isolation of *Clostridium difficile* from the hospital environment. *Am J Med* 70, 906–8.

Fekety R, Silva J, Buggy B, Deery HG (1984) Treatment of antibiotic associated colitis with vancomycin. *J Antimicrob Chemother* 14 (Suppl D), 97–102.

Fredenucci I, Chomarat M, Boucaud C, Flandrois JP (1998) *Saccharomyces boulardii* fungemia in a patient receiving Ultra-levure therapy. *Clin Infect Dis* 27, 222–3.

Gerding DN, Olson MM, Johnson S *et al.* (1990) *Clostridium difficile* diarrhea and colonisation after treatment with abdominal infection regimens containing clindamycin or metronidazole. *Am J Surg* 159, 212–7.

Golledge CL, McKenzie T, Riley TV (1989) Extended spectrum cephalosporins and *Clostridium difficile*. *J Antimicrob Chemother* 23, 929–31.

Gopal Rao G, Mahankali Rao CS, Starke I (2003) *Clostridium difficile*-associated diarrhoea in patients with community-acquired lower respiratory infection being treated with levofloxacin compared with beta-lactam-based therapy. *J Antimicrob Chemother* 51, 697–701.

Gorbach SI, Chang TW, Goldin B (1987) Successful treatment of relapsing *Clostridium difficile* colitis with *Lactobacillus GG*. *Lancet* ii, 1519.

Gravet A, Rondeau M, Harf-Monteil C *et al.* (1999) Predominant *Staphylococcus aureus* isolated from antibiotic-associated diarrhea is clinically relevant and produces enterotoxin A and the bicomponent toxin LukE-LukD. *J Clin Microbiol* 37, 4012–9.

Hall IC, O'Toole E (1935) Intestinal flora in newborn infants with a description of a new pathogenic anaerobe, *Bacillus difficilis*. *Am J Dis Child* 49, 390–402.

Johnson S (1997) Antibody responses to clostridial infection in humans. *Clin Infect Dis* 25 (Suppl 2), S173–7.

Johnson S, Gerding DN, Olson MM *et al.* (1990) Prospective, controlled study of vinyl glove use to interrupt *Clostridium difficile* nosocomial transmission. *Am J Med* 88, 137–40.

Johnson S, Homann SR, Bettin KM *et al.* (1992) Treatment of asymptomatic *Clostridium difficile* carriers (fecal excretors) with vancomycin or metronidazole. A randomized, placebo-controlled trial. *Ann Intern Med* 117, 297–302.

Kaatz GW, Gitlin SD, Schaberg DR *et al.* (1988) Acquisition of *Clostridium difficile* from the hospital environment. *Am J Epidemiol* 127, 1289–93.

Karlstrom O, Fryklund B, Tullus K, Burman LG (1998) A prospective nationwide study of *Clostridium difficile*-associated diarrhea in Sweden. The Swedish C difficile Study Group. *Clin Infect Dis* 26, 141–5.

Kato H, Kato N, Katow S *et al.* (1999) Deletions in the repeating sequences of the toxin A gene of toxin A-negative, toxin B-positive *Clostridium difficile* strains. *FEMS Microbiol Lett* 175, 197–203.

Kelly CP, Pothoulakis C, Vavva F *et al.* (1996) Anti-*C. difficile* bovine immunoglobulin concentrate inhibits cytotoxicity and enterotoxicity of *C. difficile* toxins. *Antimicrob Agents Chemother* 40, 373–9.

Kelly CP, Chetham S, Keates S *et al.* (1997) Survival of anti-*C. difficile* bovine immunoglobulin concentrate in the human gastrointestinal tract. *Antimicrob Agents Chemother* 41, 236–41.

Kotloff KL, Wasserman SS, Losonsky GA *et al.* (2001) Safety and immunogenicity of increasing doses of a *Clostridium difficile* toxoid vaccine administered to healthy adults. *Infect Immun* 69, 988–95.

Kreisel D, Savel TC, Silver AL, Cunningham JD (1995) Surgical antibiotic prophylaxis and *Clostridium difficile* toxin positivity. *Arch Surg* 130, 989–93.

Kuijper EJ, Oudbier JH, Stuifbergen WN *et al.* (1987) Application of whole-cell DNA restriction endonuclease profiles to the epidemiology of *Clostridium difficile*-induced diarrhea. *J Clin Microbiol* 25, 751–3.

Kurtz CB, Cannon EP, Brezzani A *et al.* (2001) GT160–246, a toxin binding polymer for treatment of *Clostridium difficile* colitis. *Antimicrob Agents Chemother* 45, 2340.

Kyne L, Merry C, O'Connell B *et al.* (1999) Factors associated with prolonged symptoms and severe disease due to *Clostridium difficile*. *Age Ageing* 28, 107–13.

Kyne L, Warny M, Qamar A, Kelly CP (2000) Asymptomatic carriage of *Clostridium difficile* and serum levels of IgG antibody against toxin A. *N Engl J Med* 342, 390–7.

Kyne L, Warny M, Qamar A, Kelly CP (2001) Association between antibody response to toxin A and protection against recurrent *Clostridium difficile* infection. *Lancet* 357, 189–93.

Kyne L, Hamel MB, Polavaram R, Kelly CP (2002) Health care costs and mortality associated with nosocomial diarrhoea due to *Clostridium difficile*. *Clin Infect Dis* 34, 346–53.

Larson HE, Price AB, Honour P, Borriello SP (1978) *Clostridium difficile* and the aetiology of pseudomembranous colitis. *Lancet* i, 1063–6.

Limaye AP, Turgeon DK, Cookson BT, Fritsche TR (2000) Pseudomembranous colitis caused by a toxin A(–) B(+) strain of *Clostridium difficile*. *J Clin Microbiol* 38, 1696–7.

Lucet JC, Rigaud MP, Mentre F *et al.* (2002) Hand contamination before and after different hand hygiene techniques: a randomised clinical trial. *J Hosp Infect* 50, 276–80.

Ludlam H, Brown N, Sule O *et al.* (1999) An antibiotic policy associated with reduced risk of *Clostridium difficile*-associated diarrhoea. *Age Ageing* 28, 578–80.

Lyerly DM, Bostwick EF, Binion SB, Wilkins TD (1991) Passive immunisation of hamsters against disease caused by *Clostridium difficile* by use of a bovine immunoglobulin G concentrate. *Infect Immun* 59, 2215–8.

Mayfield JL, Leet T, Miller J, Mundy LM (2000) Environmental control to reduce transmission of *Clostridium difficile*. *Clin Infect Dis* 31, 995–1000.

McCullough MJ, Clemons KV, McCluster JH, Stevens DA (1998) Species identification and virulence attributes of *Saccharomyces boulardii* (nom. inval.). *J Clin Microbiol* 36, 2613–7.

McFarland LV, Surawicz CM, Greenberg RN *et al.* (1994) A randomized placebo-controlled trial of *Saccharomyces boulardii* in combination with standard antibiotics for *Clostridium difficile* disease. *JAMA* 271, 1913–8.

Mehlig M, Moos M, Braun V *et al.* (2001) Variant toxin B and a functional toxin A produced by *Clostridium difficile* C34. *FEMS Microbiol Lett* 198, 171–6.

Modi N, Wilcox MH (2001) Evidence for antibiotic induced *Clostridium perfringens* diarrhoea. *J Clin Pathol* 54, 748–51.

Moncrief JS, Zheng L, Neville LM, Lyerly DM (2000) Genetic characterization of toxin A-negative, toxin B-positive *Clostridium difficile* isolates by PCR. *J Clin Microbiol* 38, 3072–5.

Mulligan ME, Rolfe RD, Finegold SM, George WL (1979) Contamination of a hospital environment by *Clostridium difficile*. *Curr Microbiol* 3, 173–5.

Niault M, Thomas F, Prost J *et al.* (1999) Fungemia due to *Saccharomyces* species in a patient treated with enteral *Saccharomyces boulardii*. *Clin Infect Dis* 28, 930.

Nusrat A, von Eichel-Streiber C, Turner JR *et al.* (2001) *Clostridium difficile* toxins disrupt epithelial barrier function by altering membrane microdomain localization of tight junction proteins. *Infect Immun* 69, 1329–36.

O'Neill GL, Beaman MH, Riley TV (1991) Relapse versus reinfection with *Clostridium difficile*. *Epidemiol Infect* 107, 627–35.

Pear SM, Williamson TH, Bettin KM *et al.* (1994) Decrease in nosocomial *Clostridium difficile*-associated diarrhoea by restricting clindamycin use. *Ann Intern Med* 120, 272–7.

Pelaez T, Alcala L, Alonso R *et al.* (2002) Reassessment of *Clostridium difficile* susceptibility to metronidazole and vancomycin. *Antimicrob Agents Chemother* 46, 1647–50.

Pothoulakis C, Kelly CP, Joshi MA *et al.* (1993) *Saccharomyces boulardii* inhibits *Clostridium difficile* toxin A binding and enterotoxicity in rat ileum. *Gastroenterology* 104, 1108–15.

Rifkin GD, Fekety FR, Silva J Jr (1977) Antibiotic-induced colitis implication of a toxin neutralised by *Clostridium sordellii* antitoxin. *Lancet* ii, 1103–6.

Salcedo J, Keates S, Pothoulakis C *et al.* (1997) Intravenous immunoglobulin therapy for severe *Clostridium difficile* colitis. *GUT* 41, 366–70.

Samore MH, Venkataraman L, DeGirolami PC *et al.* (1996) Clinical and molecular epidemiology of sporadic and clustered cases of nosocomial *Clostridium difficile* diarrhoea. *Am J Med* 100, 32–40.

Settle CD, Wilcox MH (1996) Rapid detection of *Clostridium difficile* toxin by gold standard versus kit methods. In: *Federation of Infection Societies Third Conference*, Manchester. Abstract P7.16.

Settle CD, Wilcox MH, Fawley WN *et al.* (1998) Prospective study of the risk of *Clostridium difficile* diarrhoea in elderly patients following treatment with cefotaxime or piperacillin–tazobactam. *Aliment Pharmacol Ther* 12, 1217–23.

Shim JK, Johnson S, Samore MH *et al.* (1998) Primary symptomless colonisation by *Clostridium difficile* and decreased risk of subsequent diarrhoea. *Lancet* 351, 633–6.

Siitonen S, Vapaatalo H, Salminen S *et al.* (1990) Effect of *Lactobacillus GG* yoghurt in prevention of antibiotic associated diarrhea. *Ann Med* 22, 57–9.

Snyder ML (1937) Further studies on *Bacillus difficilis* (Hall & O'Toole). *J Infect Dis* 60, 223–31.

Stone S, Kibbler C, How A, Balstrini A (2000) Feedback is necessary in strategies to reduce hospital acquired infection. *BMJ* 321, 302–3.

Stubbs SL, Brazier JS, O'Neill GL, Duerden BI (1999) PCR targeted to the 16S–23S rRNA gene intergenic spacer region of *Clostridium difficile* and construction of a library consisting of 116 different PCR ribotypes. *J Clin Microbiol* 37, 461–3.

Stubbs S, Rupnik M, Gibert M *et al.* (2000) Production of actin-specific ADP–ribosyltransferase (binary toxin) by strains of *Clostridium difficile*. *FEMS Microbiol Lett* 186, 307–12.

Surawicz CM, Elmer GM, Speelman P (1989) Prevention of antibiotic association diarrhea by *Saccharomyces boulardii*: a prospective study. *Gastroenterology* 96, 981–8.

Taylor NS, Bartlett JG (1980) Binding of *Clostridium difficile* cytotoxin and vancomycin by anion exchange resins. *J Infect Dis* 141, 92–7.

Teasley DG, Gerding DN, Olson MM *et al.* (1983) Prospective randomized trial of metronidazole versus vancomycin for *Clostridium difficile* associated diarrhoea and colitis. *Lancet* ii, 1043–6.

Tedesco FJ, Barton RW, Alpers DH (1974) Clindamycin-associated colitis. A prospective study. *Ann Intern Med* 81, 429–33.

Tedesco FJ, Gordon D, Fortson WC (1985) Approach to patients with multiple relapses of antibiotic associated pseudomembranous colitis. *Am J Gastroenterol* 80, 867–8.

Thomas MR, Litin SC, Osmon DR *et al.* (2001) Lack of effect of *Lactobacillus GG* on antibiotic-associated diarrhea: a randomized, placebo-controlled trial. *Mayo Clin Proc* 76, 883–9.

Thomas C, Stevenson M, Riley TV (2003) Antibiotics and hospital-acquired *Clostridium difficile*-associated diarrhoea: a systematic review. *J Antimicrob Chemother* 51, 1339–50.

Tompkins DS, Hudson MJ, Smith HR *et al.* (1999) A study of infectious intestinal disease in England: microbiological findings in cases and controls. *Commun Dis Public Health* 2, 108–13.

Turgeon DK, Novicki TJ, Quick J *et al.* (2003) Six rapid tests for direct detection of *Clostridium difficile* and its toxins in fecal samples compared with the fibroblast cytotoxicity assay. *J Clin Microbiol* 41, 667–70.

Verity P, Wilcox MH, Fawley W, Parnell P (2001) Prospective evaluation of environmental contamination by *Clostridium difficile* in isolation side rooms. *J Hosp Infect* 49, 204–9.

Viscidi R, Willey S, Bartlett JG (1981) Isolation rates and toxigenic potential of *Clostridium difficile* isolates from various patient populations. *Gastroenterology* 81, 5–9.

Viscidi R, Laughon BE, Yolken R *et al.* (1983) Serum antibody response to toxins A and B of *Clostridium difficile*. *J Infect Dis* 148, 93–100.

Wanahita A, Goldsmith EA, Musher DM (2002) Conditions associated with leukocytosis in a tertiary care hospital, with particular attention to the role of infection caused by *Clostridium difficile*. *Clin Infect Dis* 34, 1585–92.

Warny M, Vaerman JP, Avesani V, Delmee M (1994) Human antibody response to *Clostridium difficile* toxin A in relation to clinical course of infection. *Infect Immun* 62, 384–9.

Wilcox MH (2001) Clarithromycin and risk of *Clostridium difficile*-associated diarrhoea. *J Antimicrob Chemother* 47, 358–9.

Wilcox MH, Fawley WN (2000) Hospital disinfectants and spore formation by *Clostridium difficile*. *Lancet* 356, 1324.

Wilcox MH, Howe R (1995) Diarrhoea caused by *Clostridium difficile*: response time for treatment with metronidazole and vancomycin. *J Antimicrob Chemother* 36, 673–9.

Wilcox MH, Spencer RC (1992) *Clostridium difficile* infection: responses, relapses and re-infections. *J Hosp Infect* 22, 85–92.

Wilcox MH, Cunnliffe JG, Trundle C, Redpath C (1996) Financial burden of hospital-acquired *Clostridium difficile* infection. *J Hosp Infect* 34, 23–30.

Wilcox MH, Fawley WN, Settle CD, Davidson A (1998) Recurrence of symptoms in *Clostridium difficile* infection – relapse or reinfection? *J Hosp Infect* 38, 93–100.

Wilcox MH, Fawley WN, Parnell P (2000) Value of lysozyme agar incorporation and alkaline thioglycollate exposure for the environmental recovery of *Clostridium difficile*. *J Hosp Infect* 44, 65–9.

Wilcox MH, Fawley WN, Wigglesworth N *et al.* (2003) Comparison of effect of detergent versus hypochlorite cleaning on environmental contamination and incidence of *Clostridium difficile* infection. *J Hosp Infect* 54, 109–14.

Wilkins TD, Lyerly DM (2003) *Clostridium difficile* testing: after 20 years, still challenging. *J Clin Microbiol* 41, 531–4.

Wistrom J, Norrby SR, Myhre EB *et al.* (2001) Frequency of antibiotic-associated diarrhoea in 2462 antibiotic-treated hospitalized patients: a prospective study. *J Antimicrob Chemother* 47, 43–50.

Wong SS, Woo PC, Luk WK, Yuen KY (1999) Susceptibility testing of *Clostridium difficile* against metronidazole and vancomycin by disk diffusion and E-test. *Diagn Microbiol Infect Dis* 34, 1–6.

Wood M (2001) When stool cultures from adult inpatients are appropriate. *Lancet* 357, 901.

Yip C, Loeb M, Salama S *et al.* (2001) Quinolone use as a risk factor for nosocomial *Clostridium difficile*-associated diarrhea. *Infect Control Hosp Epidemiol* 22, 572–5.

47

Other *Clostridium* spp.

Ian R. Poxton

Medical Microbiology, University of Edinburgh Medical School, Edinburgh, UK

INTRODUCTION

The Gram-positive, anaerobic, spore-forming rod-shaped bacteria – the clostridia – are ubiquitous. They are found throughout nature – in soils, fresh water and marine sediments, decaying vegetation and animal matter, sewage and the gastrointestinal (GI) tract of vertebrates and invertebrates. Saprophytic forms have a major role in recycling organic matter, while other forms inhabit the GI tract of animals in commensal associations. A few infamous species are well recognised as significant opportunist pathogens causing severe wound infections and GI disease.

DESCRIPTION OF THE GENUS *CLOSTRIDIUM*

Almost 100 different clostridial species are currently recognised. Table 47.1 summarises the phenotypic characteristics of the more commonly encountered species. All form oval or spherical spores that often distend the cell, but spores in some species –, e.g. *C. perfringens* – are difficult to demonstrate in the laboratory. Cell shape is straight or sometimes slightly curved rods of variable length and width. They all have typical Gram-positive cell envelope morphology and most stain Gram positive. However, some only show this characteristic in young cultures, with a few species tending always to appear Gram negative. Most species are obligate anaerobes, but there is a degree of oxygen tolerance, with some species, most notably *C. tertium*, being able to grow in air. Those that can grow in air do not sporulate under aerobic conditions, and their growth is not as profuse as in oxygen-free conditions. There are non-motile and motile species, the latter usually with peritrichous flagella. The %G+C of the DNA of clostridia varies considerably, ranging from 22 to 55%. However, there are two main clusters with %G+C of 22–34 and 40–55, and it was once proposed to separate these into two different genera (Sneath, 1986). A recent discussion paper by Finegold, Song and Liu (2002) highlights many of the current problems and anomalies in the classification of clostridia. As in most, if not all, other groups of bacteria, the naming and classification of the clostridia must be a compromise between that based on the most accurate and modern polyphasic taxonomy and the need to communicate names that can be remembered and mean something to clinicians.

The optimum growth conditions of clostridia reflect their diverse habitats, with optimum temperatures between 15 and 69 °C. Energy production is usually by fermentation of carbohydrates or peptides with end products of metabolism being a mixture of alcohols and volatile and non-volatile fatty acids. These are often used in identification by gas chromatography and typically are responsible for the noxious smells produced by cultures (or wounds) in which clostridia are growing. Those species pathogenic for human and other mammals typically produce one or more potent exotoxins.

Table 47.1 *Clostridium* spp. that may be commonly encountered in the diagnostic laboratory – a brief guide to their phenotype

Species	Growth in air	Position/shape of spore	Pathogen/normal flora/saprophyte	Haemolysis	Lecithinase production	Lipase production
C. bifermentans	N	C/ST O	NF/S	y	Y	N
C. butyricum	N	C/ST O	S/NF	N	N	N
C. botulinum						
Types A, B and F	N	ST O b	P/S	Y	N	Y
Types C and D	N	C/ST O	P/S	Y	y	Y
C. chauvoei	N	ST/C O B	P (animal)	Y	N	N
C. difficile	N	ST/T O	P/NF	N	N	N
C. novyi	N	ST/C O	P	Yd	Y	Y
C. perfringens	N	ST/C O	P/NF	YD	Y	N
C. septicum	N	ST/C O B	P/NF	Y	N	N
C. sordellii	N	C O	NF	y	Y	N
C. sporogenes	N	ST O b	NF	y	y	Y
C. tertium	G	T O	S/NF	y	N	N
C. tetani	N	T/ST S	P/NF/S	Y	N	N

Notes: N, no growth; G, growth; T, terminal; ST, subterminal; C, central; O, oval; S, spherical; B, bulging; P, pathogen; NF, normal flora; S, saprophyte; Y, yes; D, double zone; N, no; Y, yes; N, no; Y, yes; N, no. Upper case: always/nearly always, lower case sometimes.

Principles and Practice of Clinical Bacteriology Second Edition   Editors Stephen H. Gillespie and Peter M. Hawkey

Overview of the Structure of Clostridia and their Virulence Factors

Clostridia are typically Gram positive in structure, with a cell envelope consisting of a cytoplasmic membrane and a peptidoglycan-containing cell wall with *meso*-diaminopimelic acid as the diamino acid of the peptidoglycan side chain. Many species are capsulate, but this character is not of major importance either as a virulence factor or in classification, although capsular antigens were once used to serotype strains of *C. perfringens*. The resistant exospores that members of the genus invariably produce are crucial for survival in hostile environments. All soils and dusts throughout the world are likely to contain spores of clostridia. The various exotoxins and extracellular enzymes produced by clostridia are the main virulence factors of clostridia. Traditionally the toxins are referred to by Greek letters such as alpha and beta. However, it should be stressed that the alpha-toxin of one species may be very different from the alpha-toxin of another species.

The Pathogenic Clostridia

Of the 100 or so species only a handful are commonly found causing infections in human and domestic animal, with a further 10 or so found rarely. Many of the infections arise from contamination of wounds with soil or faeces, and it is sometimes difficult to know if an isolate from a polymicrobial infection is playing a pathogenic role or if it is simply a contaminant.

PATHOGENICITY

The pathogenic mechanisms of clostridia are invariably associated with the exotoxins that they produce. In the past the first test to see whether a clostridial isolate was pathogenic was to inoculate some culture supernatant into a mouse to see if it was lethal. Clostridial exotoxins can be categorised according to several different criteria, such as the effects they have on the whole animal, the cell or at the biochemical level. Many have enzymatic effects, while others have immunomodulating effects, some may be superantigens, while others have an unknown action. Table 47.2 summarises the main groups of

toxins and their effects. Details of pathogenesis will be described more fully for each species in later sections. Toxins often work in concert, and the tissue destruction characteristic of so many clostridial wound infections is a result of one or more exotoxins and perhaps some extracellular enzymes produced by the bacteria. It is often difficult to differentiate between what are exotoxins and extracellular enzymes. A possible way of defining them is that a toxin if injected in pure form is capable of producing pathology, while the extracellular enzymes cannot, and are dependent on true toxins to initiate the pathology. In practice of course this does not matter as it is the result that is important.

Table 47.3 lists the common pathogenic species and the exotoxins that they produce and summarises their role in infection. With the exception of *C. difficile*, which is covered in detail in Chapter 46, the major clostridial pathogens will be described in some detail later.

Table 47.2   Some of the better characterised clostridial exotoxins

Exotoxin type	Effect	Examples
Phospholipase C	Breakdown of cell membranes and tissue destruction	*C. perfringens* alpha-toxin
Neurotoxins	Cleavage of peptides involved in neurotransmission	Botulinum and tetanus toxins
Enterotoxin	Diarrhoea. Apoptosis or necrosis of villus cells and induction of proinflammatory cytokines	*C. perfringens* enterotoxin
Perfringolysin	Binds to cholesterol in membranes and causes cell lysis	*C. perfringens* theta-toxin (perfringolysin O)
Affecting cytoskeleton: (i) Glycosylation of proteins from the Rho family that regulate actin polymerisation (ii) ADP-ribosylation of actin monomers	Dissociation of the actin filaments in the cytoskeleton inducing cell rounding, cell dysfunctions, changes in cell permeability and disruption of intercellular junction	(i) The large clostridial toxins from *C. difficile* (toxins A and B), *C. sordellii* and *C. novyi* (ii) The clostridial binary toxins: *C. perfringens* iota, *C. botulinum* C2 and *C. difficile* binary

Table 47.3   The pathogenic clostridia, their clinical spectrum and major toxins

Species	Major toxins	Infections they cause
C. perfringens	Alpha (phospholipase-C), beta, epsilon iota. Enterotoxin	Gas gangrene, food poisoning and other GI infections: pigbel, enteritis necroticans, antibiotic associated diarrhoea
C. tetani	Tetanospasmin (TeNT)	Tetanus
C. botulinum	Botulinum neurotoxins (BoNT) A, B, C1, D, E and F and C2 exotoxin	Botulism (classical food-borne) Infant botulism Wound botulism Equine dysautonomia
C. novyi	Type A: alpha (lethal, necrotising), gamma (phospholipase C) and epsilon (lipase). Type B: alpha-toxin, beta toxin (lecithinase)	Gas gangrene – in mixed infections with *C. perfringens* and *C. septicum.* Other oedematous wound infections – previously in war wounds, but more recently a cause of death in injecting drug users
C. difficile	Toxin A, toxin B, ADP-ribosylating toxin	Pseudomembranous (antibiotic associated) colitis and antibiotic associated diarrhoea
C. septicum	Lethal, haemolytic and necrotising alpha-toxin	Gas gangrene – in mixed infections with *C. perfringens* and *C. novyi*. A lethal disseminating disease: 'malignant oedema' in cattle and 'braxy' in sheep. Similar diseases in other domestic and wild animals
C. sordellii	Lethal toxin – immunologically cross-reactive with *C. difficile* toxin B	Various soft-tissue infections in human and other animals
C. chauvoei	Beta toxin (deoxyribonuclease) Gamma toxin (hyaluronidase)	Blackleg in cattle and sheep not a pathogen of human
C. barati *C. carnis* *C. fallax* *C. haemolyticum* *C. histolyticum* *C. limosum* *C. ramosum* *C. sporogenes* *C. tertium*	Poorly characterised or unknown	Isolated infrequently from soft-tissue infections with other bacteria – primary role in pathogenesis or contaminant uncertain in most cases

GENERAL METHODS FOR THE ISOLATION OF *CLOSTRIDIUM* SPECIES

Most isolation procedures make use of the fact that clostridia produce heat- and alcohol-resistant spores. Treatment of 70–80 °C for 10 min or addition of an equal volume of 95% or absolute ethanol to a specimen for a few minutes prior to culturing is often used. Many different enrichment and selective media exist for clostridia. Reinforced clostridial agar is one of the most commonly used general plate media. Many clostridia, including *C. perfringens, C. novyi, C. sordellii, C. bifermentans* and some *C. botulinum* are identified by their production of lipases and/or lecithinases. Egg yolk, a rich source of lecithin, is often incorporated into plate media to permit easy identification. A zone of opalescence extending beyond the colony is indicative of lecithinases production, while restricted opalescence under a colony with an associated 'pearly layer' over the colony indicates lipase activity. Proteolytic clostridia produce clear zones in egg-yolk-containing agar. The media of Lowbury and Lilly and Willis and Hobbs are examples of egg-yolk agars. Full details of recipes, including instruction for making egg-yolk suspensions (which is also available commercially), are in Collee, Brown and Poxton (1996).

Spore formation, a diagnostic feature of clostridia, is a very variable characteristic in laboratory cultures. Many species sporulate readily in broth or on plates, and spore numbers usually increase with the age of the culture. However, one species in particular, *C. perfringens*, is notoriously difficult to make sporulate in the laboratory. Several media, including those of Ellner, Duncan and Strong, Phillip, and an alkaline egg-yolk agar (all detailed in Collee, Brown and Poxton 1996), have been developed to encourage sporulation in *C. perfringens*, but none is absolutely effective. In nature, *C. perfringens* readily sporulates as it passes into the small intestine – an event associated with the production of the enterotoxin, the crucial stage in the pathogenesis of perfringens food poisoning (see p. 560).

CLOSTRIDIUM PERFRINGENS

Clostridium perfringens is a relatively large Gram-positive, rod-shaped bacterium (4–$6 \times 1\,\mu m$). In usual laboratory culture it rarely forms spores (see above for special media). There are five types of *C. perfringens* recognised and are classified as types A–E based on the different combinations of lethal toxins they produce (Table 47.4). Food poisoning strains typically produce heat-resistant spores and an enterotoxin that is produced on sporulation. All types of *C. perfringens* produce alpha-toxin, a phospholipase C (lecithinase) that cleaves phosphatidyl choline into phosphoryl choline and an insoluble diglyceride. This results in opalescence in agars containing egg yolk and is the basis for the Nagler reaction where the reaction can be inhibited by

specific anti-alpha-toxin (Collee, Brown and Poxton, 1996). Many strains produce a haemolysin that, together with the alpha-toxin, gives rise to a typical double zone of haemolysis: a clear inner zone and an outer zone of incomplete lysis, on blood agar. Colonies are usually large and smooth and cells are often capsulate.

Clostridium perfringens is an anaerobe, but may grow microaerobically. In optimum growth conditions at a temperature of between 37 and 45 °C, it can grow at a high growth rate with a generation times of 15 min or less.

Clostridium perfringens is ubiquitous – found in the soil, dust and the GI tract of human and other animals. In the GI tract it is usually considered a minor component of the normal flora, or in transient passage. It usually numbers about 10^{3-4} organisms per gram of faeces in humans.

As a pathogen, *C. perfringens* is found in two major conditions: in anaerobic wounds it is the key agent causing gas gangrene (often associated with other clostridia or other bacteria), and much more commonly it is responsible for a type of food poisoning.

Clinical Features

Gas Gangrene

Classically this was a consequence of war wounds associated with contamination of complex deep wounds with soil and faeces. It came to real prominence in the trenches of the First World War in France and Belgium – where the organism was encountered commonly in the soil. However, in other theatres of war, such as the desert and Gallipoli, where the organism is uncommon in soil, gangrene was much less common (Willis and Smith, 1990). More typically these days gangrene is an uncommon complication of diabetic ulcers and other wounds associated with poor tissue perfusion.

The crucial events in initiating pathogenesis are tissue destruction such that the environment of the wound is anoxic and contamination with the organism – most likely in the form of spores. Other clostridial species as well as some facultative organisms may also contribute to the pathology. The incubation period ranges from a few hours to a week, depending on injury and soiling. Pain is an early event, with progressive oedema and swelling of the wound with mild-to-moderate fever. A profuse serous exudate often with blood is produced, and this is followed by bubbles of gas appearing in the discharge. The skin is marbled and the tissues become crepitant. Without surgical intervention the patient becomes extremely toxaemic and shocked, and sudden death is usually due to circulatory failure. Once symptoms have commenced the only successful treatment is by radical and extensive surgical excision of all affected tissue. The use of systemic antitoxin is probably not useful, almost certainly because the antibodies are unable to reach the affected tissue. Hyperbaric oxygen treatment remains controversial and the literature abounds with doubts about its use (Wang *et al.*, 2003).

Prevention is through thorough cleansing and debridement of any likely wound with antibiotic treatment – drugs should include anti-anaerobe antibiotics such as metronidazole, an aminoglycoside and a beta-lactam – to cover against the non-clostridial components of what is often a polymicrobial infection.

The alpha-toxin (phospholipase C) and to a lesser extent the theta toxin are primarily responsible for most of the observed pathology (Awad *et al.*, 1995; Rood, 1998). Moreover, the proteases produced by the bacteria are likely to contribute to the spreading pathology.

Diagnosis of gangrene is primarily a clinical one, as clostridia often contaminate wounds. However, isolation and identification of *C. perfringens* and other clostridia, including *C. novyi, C. septicum* and *C. histolyticum*, from suspected cases of gangrene is helpful.

Table 47.4 Classification of types of *C. perfringens* – by patterns of major toxins produced

Type	Where found	Major lethal toxins			
		α	β	ε	ι
A	Gas gangrene Food poisoning	+	–	–	–
B	Lamb dysentery	+	+	+	–
C	Enteritis necroticans in human. Enteritis in other animals	+	+	–	–
D	Enterotoxaemia and pulpy kidney disease in sheep	+	–	+	–
E	Possible pathogen of sheep and cattle	+	–	–	+

Clostridium perfringens Food Poisoning and Other Forms of GI Disease

The symptoms of acute diarrhoea with severe abdominal pain 8–22 h after ingestion of meat stews are likely to indicate food poisoning caused by *C. perfringens*. Type A strains of *C. perfringens*, typically those producing heat-resistant spores, produce an enterotoxin that is responsible for the symptoms. These strains are often poorly haemolytic compared to the strains found in gangrene and are sometimes referred to as type A2. However, other strains are sometimes involved.

A review of over 1500 cases of *C. perfringens* food poisoning in the UK between 1970 and 1996 (Brett and Gilbert, 1997) showed that meat, poultry or their products were implicated in 97% of outbreaks. 'Preparation of food too far in advance; inadequate cooling; storage at ambient temperature and inadequate reheating' were the major contributory factors in outbreaks. Outbreaks are typically large with more than 1000 individuals involved being not uncommon. In the Brett and Gilbert (1997) studies there were 28 associated deaths from more than 36 000 cases. The usual sequence of events begins with warm, bulk-cooked meat stews or casseroles. These may already contain heat-resistant spores that have survived the cooking, or it has become contaminated with spores from dust. The cooked meat (cf. cooked-meat broth – a common medium for culturing anaerobes) provides the ideal environment for the germination of the spores and subsequent growth of *C. perfringens*. Cell division can occur every 10 min. The resultant vegetative bacterial cell culture is then ingested. On entering the small intestine these vegetative cells sporulate and coincidently release an enterotoxin. This enterotoxin, which appears to be carried on a transposable element (Brynestad, Synstad and Granum, 1997), acts on the villus cells and, depending on the dose, brings about their apoptosis or oncosis (necrosis) (Chakrabarti, Zhou and McClane, 2003). It has been proposed that the enterotoxin has some superantigen properties as it stimulates the production of some proinflammatory cytokines. However, this does not seem to be the case (Wallace *et al.*, 1999). The combined result of cell death and induction of mediators is reduced absorption with some fluid secretion and a painful watery diarrhoea. This form of food poisoning is relatively common and as it is often associated with large outbreaks – due to the bulk cooking of food – it is one of the most commonly reported types, despite its short duration (1–2 days) and relatively trivial symptoms. Treatment is not usually necessary except in the young and very old where dehydration might require correction.

Detection of excessive numbers of *C. perfringens* in faeces can be used in diagnosis: median counts of 8.5×10^6/g in cases of diarrhoea compared to 1.5×10^4/g in healthy people (Poxton *et al.*, 1996). Although typing is possible, utilising methods such as pulse-field gel electrophoresis (Maslanka *et al.*, 1999), it is not commonly done. It should also be recognised that more than one type of *C. perfringens* can be in the same sample of faeces or food, and several colonies must be selected from each specimen. Commercial enzyme immunoassays (EIA) are available for the detection of *C. perfringens* enterotoxin, e.g. that produced by Techlab Inc., Blacksburg, Virginia. It appears not as sensitive as the in-house EIA used by the UK Health Protection Agency Food Microbiology Laboratory, but it is specific and recommended for investigation of outbreaks (Forward, Tompkins and Brett, 2003).

Enterotoxin-producing *C. perfringens* can also be responsible for outbreaks of antibiotic associated diarrhoea that appears very similar to that caused by *C. difficile* (Borriello *et al.*, 1985). The enterotoxin EIA kit mentioned above is probably the best method for diagnosing this condition and allowing differentiation from *C. difficile* diarrhoea (Asha and Wilcox, 2002).

Various forms of necrotic enteritis or enteritis necroticans have been reported in human and other animals – including chickens – and is thought to be often caused by type C strains of *C. perfringens*. Necrotising jejunitis or 'Pigbel' was once common in the natives of Papua New Guinea. This invariably followed feasting on pork and sweet potatoes (which contain protease inhibitors). The pathogenesis appears to be due to the production of high amounts of beta toxin by the *C. perfringens* in the presence of protein-rich pork and protease inhibitors. Normally the toxin would be expected to be destroyed by trypsin, but during the feasting with excessive protein and the protease inhibitor, the trypsin activity is insufficient and pathology occurs. In severely protein-malnourished children in the Developing World a similar condition exists.

CLOSTRIDIUM NOVYI

Cells of *C. novyi* are typically large Gram-positive rods ($5–8 \times 1 \mu m$) with oval central or subterminal spores. It is a rare member of faecal flora of human, but may be more common in other animals.

This species is closely related to type *C. botulinum* and phenotypically (biochemically) is indistinguishable – except that it does not produce the C1 botulinum neurotoxin. Instead it produces one of four toxins or 'soluble antigens' that define the types A–D. The most common type A strains produce an alpha-toxin that is encoded by a lysogenic bacteriophage similar to the toxin-encoding phages of *C. botulinum* types C and D. The three toxin types can be interconverted by curing of one bacteriophage and reinfecting with a different one (Eklund and Poysky, 1974; Eklund *et al.*, 1974).

Clostridium novyi strains are extremely strict anaerobes, increasing in strictness from A to D. They are therefore difficult to isolate except in laboratories using exacting anaerobiosis. However, types B, C and D are not usually associated with disease in human.

Clinical Features

Clostridium novyi type A strains are associated with wound and soft-tissue infections. Its former name *C. oedematiens* was descriptive of the oedematous lesions it produced. It was typically found in soiled wounds, classically war wounds, and sometimes in farm animals. In the past 50 years it has been extremely uncommon, rarely being encountered in anaerobic/gangrenous wounds. However, *C. novyi* infections have recently re-emerged from an unexpected source. In Scotland between April and August 2000, 60 injecting drug users (IDUs) presented with severe skin and muscle damage and 23 died of systemic shock and *C. novyi* type A was isolated from several cases. IDUs, especially those with no veins left in which to inject, have been injecting heroin directly into muscle or skin (muscle- or skin-popping). An epidemiological study which explored the outbreak revealed temporally related cases in England and the Republic of Ireland, and the source of the infection appeared to be a batch of purer than normal heroin contaminated with spores of *C. novyi* type A. The higher concentration of heroin necessitated more citric acid to be added to it to help it dissolve. This appears to have caused more tissue destruction than normal allowing a better environment for the germination of spores and growth of the clostridia. A series of papers have been published describing the outbreak in the same issue of the *Journal of Medical Microbiology* (Brazier *et al.*, 2002; Jones *et al.*, 2002; McGuigan *et al.*, 2002; McLauchlin *et al.*, 2002a,b). Other clostridia, including *C. botulinum* and *C. tetani*, have also been associated with injection of heroin (see pp. 562–563).

Clostridium novyi strains produce several toxins, but it is likely that the alpha-toxin is the key virulence determinant. It is one of a family of large clostridial cytotoxins (Von Eichel-Streiber *et al.*, 1996), with a mass of 250 kDa. *In vivo* it is lethal and causes oedema. *In vitro* it is a potent cytotoxin, acting on cells by modifying small GTP-binding proteins. It shares a great deal of sequence homology with toxins A and B of *C. difficile* and the lethal toxin from *C. sordellii*. However, unlike these toxins it uses UDP-*N*-acetylglucosamine as the source of the sugar moiety it transfers to the their protein substrates (Busch *et al.*, 2000).

Laboratory Diagnosis of *C. novyi* Infection, Patient Management and Prevention

Laboratory diagnosis is by observation of typical lipase- and lecithinase-positive colonies on egg-yolk-containing agar – due to the gamma and epsilon toxins, respectively. On horse blood agar, discrete colonies show a double zone of haemolysis – a narrow inner zone of complete haemolysis and an outer zone of partial haemolysis. Older, spreading colonies do not tend to show this effect. As mentioned earlier, *C. novyi* is one of the strictest anaerobes, and good anaerobic technique is an absolute necessity.

Patient management is similar to that for gas gangrene. Radical surgical intervention is always crucial, with appropriate antibiotic therapy. Prevention is through thorough cleansing of contaminated wounds. Vaccines have been developed for some *C. novyi* infections of animals (Willis and Smith, 1990).

NEUROTOXIC CLOSTRIDIA

Clostridium botulinum

Clostridium botulinum is not a single species in the accepted sense – it is more a collection of different phenotypes of anaerobic, Gram-positive, spore-forming bacteria only related to each other by the powerful neurotoxins with similar pharmacological actions which all produce. They are denoted as types A, B, C, D, E and F, with a type G that was renamed *C. argentinense*, and perhaps will be named *C. botulinum* type G again (Finegold, Song and Liu, 2002). Types F and G are rarely encountered, and for most practical purposes types A–E are the only types to be concerned about. The neurotoxins are not antigenically related in that antiserum to one type does not neutralise another type, but they do share some common active sites.

Type A and proteolytic type B strains are lipase positive and lecithinase negative, and closely resemble *C. sporogenes*, both biochemically and antigenically. Non-proteolytic B and E strains share antigens with *C. perfringens* and some other clostridia including *C. chauvoei*, *C. septicum* and *C. tertium*. Types C and D are similar to each other, and almost identical to *C. novyi*: they are lecithinase and lipase positive, and as mentioned above are interconvertible by infection with the appropriate toxin-encoding bacteriophage (Poxton, 1984; Poxton and Byrne, 1984; Hunter and Poxton, 2002). Carrying of genes encoding virulence factors by bacteriophages is not uncommon in bacterial pathogens (e.g. cholera toxin, the shiga toxin of enterohaemorrhagic *E. coli* and diphtheria toxin in *Corynebacterium diphtheriae* are examples and are discussed elsewhere in this book). However, the interconversion of these type III clostridia by curing and reinfection may be unique. Whether this happens in nature is not known.

Types of *C. botulinum* are widely distributed in nature – soils, lake sediments, vegetables, GI tract of mammals, birds and fish, but they are best known for their role in the uncommon, but extremely dangerous form of food poisoning/intoxication botulism. Botulinum toxins are also considered as prime candidates for use as biological weapons. Toxicoinfections with *C. botulinum* cause infant botulism and wound botulism in human and shaker–foal syndrome and equine grass sickness in horses.

The Action of Botulinum Toxins

The mechanism of action of all botulinum toxins is similar pharmacologically, despite them being antigenically different. Their prime action is to block neurotransmitter release at cholinergic nerve terminals. They are bipartite toxins with heavy (H) and light (L) chains. The H chain is the binding part, while the L is the active component and is a zinc-dependent endopeptidase, targeting SNARE proteins, which are involved in the exocytosis of neurotransmitters at the synapse. Each different toxin acts by proteolysis of specific proteins forming synaptic vesicle docking and fusion complex: e.g. cleavage of synaptobrevin by B, D and F toxins, cleavage of SNAP-25 by C1 and syntaxin by A and E.

Toxins A, B and E are chromosomally encoded, while C1 and D are carried on temperate bacteriophages. This association between toxin-encoding bacteriophage and bacterium in *C. botulinum* types C and D, which permits interconversion as discussed earlier, is sometimes referred to as pseudolysogeny. It appears the association is very unstable and the phage is easily lost. From our recent experience it appears that sporulation results in phage loss. This means that isolation of pure toxin-producing cultures by conventional streak plating from a sporulating culture is difficult.

It may be that in nature there is a 'parental' non-toxin-producing bacterium which can be readily infected by a toxin-encoding bacteriophage that confers it with an advantage in its natural environment. However, until we understand the reasons why the toxins are required by these bacteria in their normal habitats we will not be able to understand the pressures selecting them.

Classical Food-Borne Botulism

Classical food-borne botulism results from ingestion of preformed toxin in food, and symptoms are flaccid paralysis. The disease is often fatal unless modern supportive intensive therapy is available. It is not restricted to human, but can be found in domestic animals (where it is known as forage poisoning). Botulinum toxin is the most powerful natural toxin, where less than 1 µg is the lethal dose for a human. Human disease is usually associated with types A, B and E, while the avian form is usually type C or D, the toxins of the latter types being poorly absorbed through the human gut mucosa. Forage poisoning in domestic animals is usually from A, B and sometimes C or D.

Food-borne botulism is generally not common, but in some parts of the world it is much more than others. Outbreaks are usually associated with home-canned or bottled vegetables or meat, with the very rare occasions where commercially canned food is implicated – usually due to a flaw in the can. In Northern and Northwestern Europe it is extremely uncommon with outbreaks occurring only once in a decade or so. In Southern and Eastern Europe it is more common: for example between 1965 and 1990 in Poitier, France there were 108 cases investigated (Roblot *et al.* 1994). In North America cases occur with some regularity and a recent outbreak in Texas is a typical example (Kalluri *et al.*, 2003). This outbreak affected 15 people out of 38 attending a church supper. Patients presented over a 4-day period with cranial neuropathy and progressive peripheral muscle weakness. Botulinum toxin type A was detected in stool specimens from nine patients. The intoxication was shown to be associated with a chilli dish prepared from frozen chillis that had been stored inadequately eaten 4 days before the first patients presented. Of the 10 patients who received medical care, six had severe neurological impairment and required mechanical ventilation. There were no fatalities.

Toxicoinfectious Botulism

Toxicoinfectious botulism describes the condition where viable bacteria enter the body, either through a wound or into the gut, and if conditions are appropriate, they grow and produce toxin locally. This toxin is absorbed and may produce systemic effects similar to food-borne botulism. Wound botulism, an extremely rare condition, was associated with war wounds. However, similar to the *C. novyi* infections in IDUs, wound botulism has recently been seen to increase in this unexpected patient group.

Prior to 1999 no reports of wound botulism had been notified to the Communicable Diseases Surveillance Centre of the Public Health Laboratory Service of England and Wales (PHLS; now the Health Protection Agency: HPA). Then in 2000 six cases of wound botulism were reported, followed by four in 2001 and 13 in 2002. All were from IDUs employing the practice of skin or muscle popping.

Infant botulism results in a spectrum of disease from 'floppy baby syndrome' to sudden infant death (cot death). Spores of *C. botulinum* get into the immature infant gut where they germinate and the vegetative cells multiply producing toxin which is absorbed through the gut wall. The symptoms are the result of systemic spread of the toxin to cholinergic nerve terminals. Infant botulism has been reported from many countries of the world; however it is extremely rare in many of them (Arnon, 1986). However, in some locations, where the custom is to supplement milk feeds with food in which botulinum spores may be present, it is more common. This is true in the southwest of the USA where honey has been fed to young babies.

Recently we have identified another form of toxicoinfectious botulism. This is equine dysautonomia or equine grass sickness. Here we hypothesise that toxin is produced by *C. botulinum* type C in the gut of the horse and the enteric neurones are destroyed – possibly by the combined action of BoNT/C (C1 neurotoxin), and C2 cytotoxin (Hunter, Miller and Poxton, 1999) causing the characteristic stasis of the GI tract. Dysautonomias with clinical signs and pathological findings similar to equine grass sickness have been recognised in rabbits and hares, and recently we have hypothesised that feline dysautonomia may have *C. botulinum* as the aetiological agent (Nunn *et al.*, 2004).

Laboratory Diagnosis, Patient Management and Prevention

Traditional laboratory diagnosis is by demonstration of specific toxin by toxin-neutralisation tests in mice. Samples can include faeces, food, vomit, gastric fluid, tissue, serum or an environmental sample. Very few laboratories are able, or have ethical permission, to perform this test. Detailed instructions for the test are in Poxton *et al.* (1996). Some recent developments for detection of the toxin include ELISA, which is of much lower sensitivity, and detection of the specific enzymatic activity of the specific toxin. This latter approach is still experimental, but may eventually prove extremely useful. Methods for detecting the organism range from selective culture (Dezfulian *et al.*, 1981) which only seems to work for types A, B and E, to detection of toxin genes by PCR (Lindstrom *et al.*, 2001). These methods only prove the presence of toxigenic bacteria and do not prove that toxin has been produced. If, based on clinical findings, there is suspicion that botulism is a possibility then detection of genes by PCR must be considered as confirmation, especially if the specimen was food. As previously noted several clostridia are phenotypically very similar to *C. botulinum* strains, and proof of identification can only be based on detection of biologically active toxin or the genes encoding it.

Generally, the types of botulism encountered reflect the predominant types of *C. botulinum* present in the local environment. Type E is often associated with marine environments.

Prevention is largely through adherence to good hygiene during food preparation and storage. A pentavalent toxoid vaccine is available from CDC for at-risk groups such as laboratory workers and the military. Its efficacy is however debated, although several botulinum toxoid vaccines of proven use are available for animals.

Treatment of food-borne botulism involves emergency medical intensive care support. The relatively slow incubation period for the disease (due to time taken for toxin to locate to nerve terminals), and the possibility of outbreaks from a common food source, means that antitoxins are often of potential use. Possibly exposed individuals may be offered antitoxin for prophylaxis. Once the toxin has bound, antitoxin is probably of little use for reversing the systems, but systemic antitoxins do prevent further binding of any remaining free toxin and may shorten the course of the disease. It is only for the most common types of botulism – that caused by A and B – that antitoxins are available, and most are of equine origin. However, infant botulism

is treated in the USA under the recommendations of the California Department of Health Services Infant Botulism Treatment and Prevention Program (IBTPP). Botulism Immune Globulin (BIG) – prepared from vaccinated individuals – was officially licensed by the US Food and Drug Administration on 23 October 2003 for the treatment of infant botulism types A and B under the proprietary name of BabyBIG™. Infant botulism is not usually fatal, but prompt treatment with BabyBIG™ may reduce the length of time for recovery and babies may not need mechanical ventilation. Full details for its use are given on the web site of the American Society of Health-System Pharmacists at: (http://www.ashp.org/news/ShowArticle.cfm?id=3637).

CLOSTRIDIUM TETANI

Clostridium tetani has several characteristics in common with *C. botulinum*, but with none of the heterogeneity of the latter species. It is a strictly anaerobic, rod-shaped bacterium ($0.5-1 \times 2-5\,\mu m$), growing singly or in pairs at temperatures between 14 and 43 °C. Cells can sometimes be long and filamentous, and become Gram-negative after about 24 h of incubation. Most strains are motile – by means of peritrichous flagella – and spores typically are round and terminal and their diameter can be up to twice that of the vegetative cell giving sporulating cells a 'drumstick' appearance. *Clostridium tetani* grows well on blood agar, and colonies show a fine branching appearance, which after 48–72 h of incubation can extend over the whole surface of the plate, reminiscent of shigellae, and may not be apparent when given a cursory examination.

Clostridium tetani is not routinely typed, but there are recognised differences in flagellar types. It produces two toxins, the tetanus neurotoxin (TeNT, tetanospasmin) and the haemolysin tetanolysin.

Tetanus and its Pathogenesis

Classical tetanus results from *C. tetani* spores entering a wound in which the redox potential is low enough for their germination and subsequent growth – typically in a deep puncture wound contaminated with soil or animal faeces. The organism is found in the GI tracts of various animals and is ubiquitous in soils throughout the world where it may have a role as a saprophyte. The TeNT is produced within the wound and from there it enters the systemic circulation and travels to the synapses causing spastic paralysis of the voluntary muscles (see last paragraph of this section). The lethal dose for human is approximately 1 ng/kg, making it second only to botulinum toxin as the most powerful toxin known. Unless treated by modern intensive care it is invariably fatal.

Clinical symptoms usually appear 7 and 10 days after the wound is infected, but can be as short as 3 days and as long as 30 days (Collee, Brown and Poxton, 1996). The first symptoms are often trismus – where the muscles of the jaw are affected – giving rise to its common name 'Lockjaw'. Other early symptoms include pain and stiffness, which develops into muscle rigidity and later becomes generalised. External stimuli such as a sudden noise can trigger spasms. Spasms of the larynx may occur and be life threatening. Autonomic problems occur in more severe cases, and these begin a day or two after the muscle spasms. Clinical signs include pyrexia, sweating, hypertension and cyanosis of the digits. Tetanus might be mistaken for strychnine poisoning, convulsions, brain infections or haemorrhage, psychiatric disorders and even local oral disease. However, clinical diagnosis is relatively straightforward, but there are very few clinicians with experience of tetanus. This is because it is extremely uncommon to encounter this disease in the Developed World as it is so well controlled by active vaccination programmes (see Laboratory Diagnosis, Patient Management and Prevention of Tetanus). However, there are an estimated 400 000 cases throughout the world each year, mainly in neonates – following infection of the umbilicus.

As has already been described for *C. novyi* and *C. botulinum*, *C. tetani* is also causing recent problems in IDUs. Between July and November 2003, seven clinically diagnosed cases of tetanus were reported in IDUs in England and Wales (CDR, 2003). Sources of *C. tetani* could be contamination of the drugs themselves, or the equipment used, or conceivably the skin.

Despite the symptoms being very different, the pathogenesis of tetanus is very similar to that of botulism and TeNT is very similar to the BoNTs (see p. 561). Both are zinc-dependent endopeptidases that target synaptic members of the SNARE proteins, which are involved in the exocytosis of neurotransmitters at the synapse. Although they have identical intracellular mechanism of action, TeNT is transported retrogradely through motor neurons and inhibits exocytosis in inhibitory interneurons, compared to the BoNTs which block neurotransmitter release at the neuromuscular junction. The resultant primary symptom of tetanus is the characteristic spastic paralysis of the voluntary muscles, compared to the flaccid paralysis of botulism (Verastegui *et al.*, 2002).

Laboratory Diagnosis, Patient Management and Prevention of Tetanus

Laboratory diagnosis is similar to that for other clostridial infections. In addition a Gram-stained smear of wound exudates may show the presence of the typical rods with bulging spores; however, this is not diagnostic. Wound samples are cultured anaerobically on blood agar and in cooked-meat broth. The wound culture usually produces a mixture of anaerobes, but the *C. tetani* can be picked pure by examining the plate through a hand lens and sampling from the spreading edge of growth – utilising the growth characteristics described earlier. Heat selection of spores should be attempted, but it should be noted that some strains produce heat-sensitive spores. Full details of laboratory isolation and identification are given in Collee, Brown and Poxton (1996).

Tetanus is preventable with toxoid vaccine. It is one of the safest and most effective vaccines currently in use. It is usually administered along with diphtheria, with or without pertussis, in babies. Later in life it is usually boosted jointly with diphtheria toxoid, although it can be administered alone. Boosters are recommended every 10 years. If an individual is thought to be at risk of developing tetanus following a soiled wound and because they have not having been previously vaccinated, or have not completed the three-dose primary course, or they are of unknown vaccine status, then postexposure treatment can be given in the form of an immune globulin preparation. These days this is usually of human origin (prepared from vaccinees), but in the past it was from hyperimmunised horses. In a manner similar to botulism, the rationale for passive therapy after symptoms begin is to prevent further binding of toxin. However, in the case of tetanus, antibiotic therapy is also recommended if there is still evidence of an infected wound (metronidazole or benzyl penicillin), along with thorough wound cleansing.

OTHER CLOSTRIDIA

The following clostridia are uncommon causes of disease in human and are only covered briefly.

Clostridium septicum and *C. chauvoei*

These are two closely related species, both of which cause disease in farm animal. It is only *C. septicum* that is considered a pathogen of humans. It is found in polymicrobial infections in gangrene and in systemic infections associated with malignancies or immunosuppression where it has a variable clinical presentation and is associated with a high mortality. Patients with colorectal cancer had septicaemia and 'vague abdominal symptoms' (Chew and Lubowski, 2001). There have also been reports of secondary *C. septicum* infections following *Escherichia coli* O157 haemolytic uraemic syndrome – perhaps as a zoonosis (Barnham and Weightman, 1998).

Clostridium sordellii

This is another clostridial species infrequently associated with wound infections and gangrene. *Clostridium sordellii* can also be responsible for a rare and rapidly fatal postpartum endometritis in previously healthy young women (Rorbye, Petersen and Nilas, 2000). Recently there have been reports of *C. sordellii* infection associated with allograft musculoskeletal tissue transplantation using tissues removed from cadavers (Malinin *et al.*, 2003).

GENOMICS OF CLOSTRIDIA

The genomes of several species of *Clostridium* are being, or have been sequenced, some because they are important pathogens and others because they may have important uses in biotechnology. *Clostridium tetani* is an example of the former. The tetanus toxin and a collegenase have been shown to be encoded on a 74 kbp plasmid, while other virulence factors such as surface-layer adhesion proteins have been identified, some of which are unique to *C. tetani* (Bruggemann *et al.*, 2003). The genome of *C. perfringens* (Shimizu *et al.*, 2002) has allowed identification of enzymes for anaerobic fermentation leading to gas production, but none for the tricarboxylic acid cycle or respiratory chain. More than 20 newly identified genes have been proposed as coding for virulence factors as well as providing a method for detecting the two-component VirR/VirS regulon that co-ordinately regulates the pathogenicity of *C. perfringens* (Banu *et al.*, 2000). Genomics is perhaps most useful when the pathogenesis of an organism still has many unknowns. Certainly our understanding of *C. perfringens* has been advanced. However, it is probably the recently sequenced genome of *C. difficile* that will further our knowledge of a *Clostridium* most – but that is covered in its own chapter.

SUMMARY

Clostridia are typically thought of as dangerous, exotoxin-producing pathogens which are rarely encountered in modern medicine. Formerly they were found in dirty war and postoperative wounds in the preantibiotic era where they caused gangrene and tetanus. These conditions have largely disappeared from the Developed World. However, some species such as *C. perfringens* continue to cause food poisoning, while others, despite being rare, do cause life-threatening infections. Since the 1970s there has been increasing awareness of *C. difficile* and its real increasing incidence (covered in Chapter 46). It is probably true to say that *C. difficile* is the only species of *Clostridium* (or even anaerobe) that is a real concern to Public Health in the Developed World. However, over the past decade several clostridial infections have appeared when not expected: the occurrence of botulism, tetanus and *C. novyi* infections was unexpected in IDUs. These together with the possible association of *C. septicum* with malignancies and the potential for botulinum toxins as biological warfare agents indicates that we must keep closely aware of these important pathogens.

REFERENCES

Arnon SS (1986) Infant botulism – anticipating the 2nd decade. *Journal of Infectious Diseases* 154: 201–206.

Asha NJ, Wilcox MH (2002) Laboratory diagnosis of *Clostridium perfringens* antibiotic-associated diarrhoea. *Journal of Medical Microbiology* 51: 891–894.

Awad MM, Bryant AE, Stevens DL, Rood JI (1995) Virulence studies on chromosomal alpha-toxin and theta-toxin mutants constructed by allelic exchange provide genetic-evidence for the essential role of alpha-toxin in *Clostridium perfringens*-mediated gas-gangrene. *Molecular Microbiology* 15: 191–202.

Banu S, Ohtani K, Yaguchi H *et al.* (2000) Identification of novel VirR/VirS-regulated genes in *Clostridium perfringens*. *Molecular Microbiology* 35: 854–864.

Barnham M, Weightman N (1998) *Clostridium septicum* infection and hemolytic uremic syndrome. *Emerging Infectious Diseases* 4: 321–324.

Borriello SP, Barclay FE, Welch AR *et al.* (1985) Epidemiology of diarrhoea caused by entero-toxigenic *Clostridium perfringens*. *Journal of Medical Microbiology* 20: 363–372.

Brazier JS, Duerden BI, Hall V *et al.* (2002) Isolation and identification of *Clostridium* spp. from infections associated with the injection of drugs: experiences of a microbiological investigation team. *Journal of Medical Microbiology* 51: 985–989.

Brett MM, Gilbert RJ (1997) 1525 outbreaks of *Clostridium perfringens* food poisoning, 1970–1996. *Reviews in Medical Microbiology* 8 (Suppl. 1): S64–S65.

Bruggemann H, Baumer S, Fricke WF *et al.* (2003) The genome sequence of *Clostridium tetani*, the causative agent of tetanus disease. *Proceedings of the National Academy of Sciences of the United States of America* 100: 1316–1321.

Brynestad S, Synstad B, Granum PE (1997) The *Clostridium perfringens* enterotoxin gene is on a transposable element in type A human food poisoning strains. *Microbiology* 143: 2109–2115.

Busch C, Schömig K, Hofmann F, Aktories K (2000) Characterization of the catalytic domain of *Clostridium novyi* alpha-toxin infection and immunity 68: 6378–6383.

CDR (2003) Cluster of cases of tetanus in injecting drug users in England: update. *CDR Wkly (Online)* http://www.hpa.org.uk/cdr/PDFfiles/2003/cdr4803.pdf (28 November 2003 13 (48): news).

Chakrabarti G, Zhou X, McClane BA (2003) Death pathways activated in CaCo-2 cells by *Clostridium perfringens* enterotoxin. *Infection and Immunity* 71: 4260–4270.

Chew SSB, Lubowski DZ (2001) *Clostridium septicum* and malignancy. *ANZ Journal of Surgery* 71: 647–649.

Collee JG, Brown R, Poxton IR (1996) Chapter 31 in Mackie & McCartney's Practical Medical Microbiology Collee JG, Fraser AG, Marmion BP, Simmons A (eds). Churchill Livingstone: Edinburgh pp. 521–535.

Dezfulian M, McCroskey LM, Hatheway CL, Dowell VR (1981) Selective medium for the isolation of *Clostridium botulinum* from human feces. *Journal of Clinical Microbiology* 13: 526–531.

Eklund MW, Poysky FT (1974) Interconversion of type C and D strains of *Clostridium botulinum* by specific bacteriophages. *Applied Microbiology* 27: 251–258.

Eklund MW, Poysky FT, Meyers JA, Pelroy GA (1974) Interspecies conversion of *Clostridium botulinum* type C to *Clostridium novyi* type A by bacteriophage. *Science* 186: 456–458.

Finegold SM, Song Y, Liu C (2002) Taxonomy – general comments and update on taxonomy of clostridia and anaerobic cocci. *Anaerobe* 8: 283–285.

Forward LJ, Tompkins DS, Brett MM (2003) Detection of *Clostridium difficile* cytotoxin and *Clostridium perfringens* enterotoxin in cases of diarrhoea in the community. *Journal of Medical Microbiology* 52: 753–757.

Hunter LC, Poxton IR (2002) *Clostridium botulinum* types C and D and the closely related *Clostridium novyi*. *Reviews in Medical Microbiology* 13: 75–90.

Hunter LC, Miller JK, Poxton IR (1999) The association of *Clostridium botulinum* type C with equine grass sickness: a toxicoinfection? *Equine Veterinary Journal* 31: 492–499.

Jones JA, Salmon JE, Djuretic T *et al.* (2002) An outbreak of serious illness and death among injecting drug users in England during 2000. *Journal of Medical Microbiology* 51: 978–984.

Kalluri P, Crowe C, Reller M *et al.* (2003) An outbreak of foodborne botulism associated with food sold at a salvage store in Texas. *Clinical Infectious Diseases* 37: 1490–1495.

Lindstrom M, Keto R, Markkula A *et al.* (2001) Multiplex PCR assay for detection and identification of *Clostridium botulinum* types A, B, E, and F

in food and fecal material. *Applied and Environmental Microbiology* 67: 5694–5699.

Malinin TI, Buck BE, Temple HT, Martinez OV, Fox WP (2003) Incidence of clostridial contamination in donors' musculoskeletal tissue. *Journal of Bone and Joint Surgery. British Volume* 85: 1051–1054.

Maslanka SE, Kerr JG, Williams G *et al.* (1999) Molecular subtyping of *Clostridium perfringens* by pulsed-field gel electrophoresis to facilitate food-borne-disease outbreak investigations. *Journal of Clinical Microbiology* 37: 2209–2214.

McGuigan CG, Penrice GM, Gruer L *et al.* (2002) Lethal outbreak of infection with *Clostridium novyi* type A and other spore-forming organisms in Scottish injecting drug users. *Journal of Medical Microbiology* 51: 971–977.

McLauchlin J, Mithani V, Bolton FJ *et al.* (2002b) An investigation into the microflora of heroin. *Journal of Medical Microbiology* 51: 1001–1008.

McLauchlin J, Salmon JE, Ahmed S *et al.* (2002a) Amplified fragment length polymorphism (AFLP) analysis of *Clostridium novyi*, *C. perfringens* and *Bacillus cereus* isolated from injecting drug users during 2000. *Journal of Medical Microbiology* 51: 990–1000.

Nunn F, Cave TA, Knottenbelt C, Poxton IR (2004) Evidence that Key–Gaskell Syndrome is a toxico-infection caused by *Clostridium botulinum* type C/D. *Veterinary Record* 115: 111–115.

Poxton IR (1984) Demonstration of the common antigens of *Clostridium botulinum*, *C. sporogenes* and *C. novyi* by an enzyme-linked immunosorbent assay and electroblot transfer. *Journal of General Microbiology* 130: 975–981.

Poxton IR, Byrne MD (1984) The demonstration of shared antigens in the genus *Clostridium* by an enzyme-linked immunosorbent assay. *Journal of Medical Microbiology* 17: 171–176.

Poxton IR, Brown R, Fraser AG, Collee JG (1996) Chapter 32 in Mackie & McCartney's Practical Medical Microbiology Collee JG, Fraser AG, Marmion BP, Simmons A (eds). Churchill Livingstone: Edinburgh pp. 536–547.

Roblot P, Roblot F, Fauchere JL *et al.* (1994) Retrospective study of 108 cases of botulism in Poitiers, France. *Journal of Medical Microbiology* 40: 379–384.

Rood JI (1998) Virulence genes of *Clostridium perfringens*. *Annual Review of Microbiology* 52: 333–360.

Rorbye C, Petersen IS, Nilas L (2000) Postpartum *Clostridium sordellii* infection associated with fatal toxic shock syndrome. *Acta Obstetrica Gynaecologica Scandinavica* 79: 1134–1135.

Shimizu T, Ohtani K, Hirakawa H *et al.* (2002) Complete genome sequence of *Clostridium perfringens*, an anaerobic flesh-eater. *Proceedings of the National Academy of Sciences of the United States of America* 99: 996–1001.

Sneath PHA (1986) Endospore-forming Gram-positive rods and cocci. Section 13 in Vol. 2 of *Bergey's Manual of Systematic Bacteriology* Sneath, PHA, Mair, NS, Sharpe, ME, Holt, JG (eds). Baltimore: Williams & Wilkins.

Verastegui C, Lalli G, Bohnert S, Meunier FA, Schiavo G (2002) Clostridial neurotoxins. *Journal of Toxicology – Toxin Reviews* 21 (3): 203–227.

Von Eichel-Streiber C, Boquet P, Sauerborn M, Thelestam M (1996) Large clostridial cytotoxins: a family of glycosyltransferases modifying small GTP-binding proteins. *Trends in Microbiology* 4: 375–382.

Wallace FM, Mach AS, Keller AM, Lindsay JA (1999) Evidence for *Clostridium perfringens* enterotoxin (CPE) inducing a mitogenic and cytokine response in vitro and a cytokine response in vivo. *Current Microbiology* 38: 96–100.

Wang C, Schwaitzberg S, Berliner E, Zarin DA, Lau J (2003) Hyperbaric oxygen for treating wounds: a systematic review of the literature. *Archives of Surgery* 138: 272–279; discussion 280.

Willis AT, Smith GR (1990) Gas gangrene and other clostridial infections of man and animals in *Topley and Wilson's Principles of Bacteriology, Virology and Immunity*, 8th edn Vol. 3, *Bacterial Diseases* Smith, GR, Easmon, CSF (eds) pp. 307–329.

48

Anaerobic Actinomycetes and Related Organisms

Val Hall

Anaerobe Reference Laboratory, National Public Health Service for Wales, Microbiology Cardiff, University Hospital of Wales, Heath Park, Cardiff, UK

INTRODUCTION

Actinomyces species are facultatively anaerobic Gram-positive bacilli, endogenous to mucous membranes of humans and animals. Some members of the genus cause actinomycosis, a chronic, subacute suppurating granulomatous infection. This disease occurs most commonly in the cervicofacial area but may affect organs of the thorax, abdomen or pelvis and, rarely, the central nervous system or limbs. Advanced cases of actinomycosis are uncommon in the developed world but may be severely debilitating, disfiguring and potentially fatal. Aggressive therapy is required. The clinical diagnosis of actinomycosis and laboratory isolation and identification of the causative agents are notoriously problematic.

Actinomyces species also occur in polymicrobial superficial soft-tissue abscesses and may play an important role in the development of dental plaque, periodontal diseases and caries. The clinical significance of their association with intrauterine contraceptive devices has been much debated. Recently, a plethora of novel *Actinomyces* species has been isolated from clinical and veterinary sources. The natural habitats and pathogenic potential of these species have yet to be established.

DESCRIPTION OF THE ORGANISMS

General Description

Actinomyces are facultatively anaerobic, Gram-positive, irregularly staining bacilli that are non-acid-fast, non-sporing and non-motile. Growth is enhanced by the presence of blood or serum and CO_2. Anaerobic conditions are favoured but some species can be cultured aerobically or in air plus 5% CO_2. Members of the genus are fermentative and produce acetic, formic, lactic and succinic acids but not propionic acid as endproducts of glucose metabolism. They are indole negative, catalase negative or positive and rarely (weakly) proteolytic. Cell-wall peptidoglycan contains alanine, glutamic acid and lysine with either aspartic acid or ornithine. Diaminopimelic acid and mycolic acids are absent. Cell-wall carbohydrates include glucose, galactose and rhamnose, but not arabinose. Major cellular fatty acids are myristic (tetradecanoic, 14:0), palmitic (hexadecanoic, 16:0), stearic (octadecanoic, 18:0) and oleic acids (octadecenoic, 18:1*cis*-9). The G + C ratio of DNA is 55–71 mol%. Optimal growth temperature is 35–37 °C.

Classification

Current taxonomy is based upon phylogenetic trees constructed from 16S rRNA sequence comparisons and supported by data derived from biochemical characteristics, whole-cell protein profiles and other diverse methods. The genus *Actinomyces* has been shown to comprise at least three subgroups that may represent separate genera (Schaal *et al.*, 1999). Closely related genera are *Arcanobacterium*, *Actinobaculum*, *Mobiluncus* and the recently described *Varibaculum*. Genetic, serological and phenotypic diversity has been demonstrated within some species. These characteristics are particularly variable within the *Actinomyces naeslundii/A. viscosus* complex, wherein clear species distinctions are difficult to define. Johnson *et al.* (1990) restricted *A. viscosus* to strains of animal origin, formerly designated *A. viscosus* serotype I, and designated two 'Genospecies' of *A. naeslundii*. Genospecies 1 comprises strains of *A. naeslundii* serotype I, while *A. naeslundii* Genospecies 2 includes *A. naeslundii* serotypes II and III plus *A. viscosus* serotype II (human strains). Other strains, closely related to these may form additional Genospecies. The term Genospecies was used because, whilst isolates formed genetically distinct taxa, they could not be reliably differentiated in phenotypic tests.

As of early 2004, the genus *Actinomyces* comprises 31 species, 13 of which have been reported only from animal sources (Tables 48.1 and 48.2). Advances in isolation techniques and molecular detection methods are likely to lead to the valid publication of additional species. The 'classically described' species, associated with actinomycosis, and their principal host-species are listed in Table 48.1. *Propionibacterium propionicum* is included in this table and throughout the chapter as this species also causes actinomycosis and resembles *Actinomyces israelii* both morphologically and phenotypically. Table 48.2 lists species described more

Table 48.1 The 'classically described' *Actinomyces* species and *P. propionicum*

Species	Former name(s)	Principal host species
A. bovis		Cattle
A. israelii	*A. israelii* serotype I	Humans and animals
A. gerencseriae	*A. israelii* serotype II	Humans and animals
A. naeslundii Genospecies 1	*A. naeslundii* serotype I	Humans
A. naeslundii	*A. naeslundii* serotypes II and III	Humans
Genospecies 2	*A. viscosus* serotype II	
A. viscosus	*Odontomyces viscosus* *A. viscosus* serotype I	Animals
A. odontolyticus	*Actinobacterium meyeri*	Humans
A. meyeri	*Actinomyces propionicus*	Humans
P. propionicum	*Arachnia propionica*	Humans

Table 48.2 Recently described *Actinomyces* species

Actinomyces species	Principal host species	Reference(s)
A. denticolens	Cattle, ?humans	Dent and Williams (1984a)
A. howellii	Cattle	Dent and Williams (1984b)
A. hordeovulneris	Dogs	Buchanan *et al.* (1984)
A. slackii	Cattle	Dent and Williams (1986)
A. georgiae	Humans	Johnson *et al.* (1990)
A. hyovaginalis	Pigs	Collins *et al.* (1993)
A. neuii subspp. *neuii* and *anitratus*	Humans	Funke *et al.* (1994)
A. radingae	Humans	Wust *et al.* (1995), emended Vandamme *et al.* (1998)
A. turicensis	Humans	Wust *et al.* (1995), emended Vandamme *et al.* (1998)
A. europaeus	Humans	Funke *et al.* (1997)
A. graevenitzii	Humans	Pascual Ramos *et al.* (1997)
A. bowdenii	Dogs, cats	Pascual *et al.* (1999)
A. canis	Dogs	Hoyles *et al.* (2000)
A. radicidentis	Humans	Collins *et al.* (2000)
A. urogenitalis	Humans	Nikolaitchouk *et al.* (2000)
A. suimastitidis	Pigs	Hoyles *et al.* (2001a)
A. catuli	Dogs	Hoyles *et al.* (2001b)
A. marimammalium	Marine mammals	Hoyles *et al.* (2001c)
A. funkei	Humans	Lawson *et al.* (2001)
A. coleocanis	Dogs	Hoyles *et al.* (2002)
A. cardiffensis	Humans	Hall *et al.* (2002)
A. vaccimaxillae	Cattle	Hall *et al.* (2003a)
A. oricola	Humans	Hall *et al.* (2003b)
A. nasicola	Humans	Hall *et al.* (2003c)

recently and of less certain pathogenicity. Early reports of *Actinomyces bovis* from human infections are unverified and probably represent misidentifications. The organisms now classified as *A. viscosus* (formerly *A. viscosus* serotype I) appear to be specific to animals; reports of *A. viscosus* from humans probably represent *A. naeslundii* Genospecies 2. A recent study wherein *Actinomyces* species were identified by a genotypic method (ARDRA) detected five strains that resembled *A. denticolens* in 475 strains isolated from human clinical sources, but failed to detect any of the other animal-associated species (Hall *et al.*, 2001b). This suggests that these species occur rarely or not at all in clinical material and, consequently, they will not be considered further in this chapter. They are *A. howellii*, *A. hordeovulneris*, *A. slackii*, *A. hyovaginalis*, *A. bowdenii*, *A. canis*, *A. marimammalium*, *A. catuli*, *A. suimastitidis*, *A. coleocanis*, *A. vaccimaxillae* and the aforementioned *A. bovis* and *A. viscosus*.

Morphology

Members of the genus demonstrate considerable variation in both cell and colony morphologies. This may lead to difficulties in recognition of putative *Actinomyces* on primary isolation media. However, to the experienced eye, cell and colony appearance can be a useful aid to species-level identifications, increasing confidence in results obtained by biochemical or genotypic tests. Morphologies of *Actinomyces* spp. that occur in human clinical specimens and *P. propionicum* are listed in Table 48.3.

In clinical material, the fine, filamentous, beaded cells of *A. israelii*, *A. gerencseriae*, *A. oricola* and *P. propionicum* may be mistaken for chains of streptococci. On primary isolation, colonies of these species

Table 48.3 Cell and colony morphology of *Actinomyces* spp. from human sources and *P. propionicum*

Species	Cells[a]	Colonies[b]	Comments
A. israelii	Fine, filamentous, beaded, branching	White-to-cream, breadcrumb or molar tooth, gritty, pitting	Slow growing Old colonies may become pink
A. gerensceriae	Fine, filamentous, beaded, branching	Bright white, breadcrumb or molar tooth, pitting and softer than *A. israelii*	Slow growing
A. naeslundii	Medium rods, may have clubbed ends, some branching	White, cream or pinkish, smooth, convex, entire edged	Occasional rough forms occur
A. odontolyticus	Diphtheroid like, some clubbed ends	Cream-to-red, smooth, convex, entire edged	Old colonies may be dark brown
A. meyeri	Short, fine rods	Small, white, smooth, convex, entire edged	Slow growing
A. denticolens like	Diphtheroid like	White-to-pink, heaped or molar tooth, pitting	
A. georgiae	Diphtheroid like	White or cream, smooth, convex, entire edged	
A. neuii subsp. *neuii* and *anitratus*	Diphtheroid like	White or cream, smooth, convex, entire edged	
A. radingae	Cocco-bacilli	Grey-to-white, semitranslucent, smooth, low convex, entire edged	
A. turicensis	Cocco-bacilli	Grey, semitranslucent, smooth, low convex, entire edged	
A. europaeus	Cocco-bacilli	Whitish, semitranslucent, smooth, low convex, entire edged	
A. graevenitzii	Medium length, thick, granular, branched	White, pronounced molar tooth or smooth, convex	Red fluorescence. Rough and smooth forms occur together Old colonies may be dark brown
A. radicidentis	Coccoid	Cream-to-pink, smooth, convex, entire edged	Old colonies may become red
A. urogenitalis	Cocco-bacilli	Cream-to-pink, with darker rings, smooth, convex, entire edged	Old colonies may become red
A. funkei	Fine, filamentous, beaded, branching	Grey, semitranslucent, opaque centre (fried-egg), low convex, entire edged	
A. cardiffensis	Fine, filamentous, beaded, branching, ends may be clubbed	Cream-to-pink, smooth, convex, entire edged	
A. nasicola	Diphtheroid like or cocco-bacilli may be branched	White or grey, smooth, convex, entire edged	
A. oricola	Diphtheroid to filamentous, some branching	White, breadcrumb, pitting	
P. propionicum	Fine, filamentous, beaded, branching or coccoid	Off-white to buff, breadcrumb, gritty, pitting, or smooth, convex, entire edged	Red fluorescence. Rough and smooth forms occur together

[a] Cells of all species are Gram positive or Gram variable.
[b] All colonies ≤1 mm diameter.

develop very slowly and are obligately anaerobic but, on subculture, organisms usually grow more rapidly and become more aerotolerant. Colonies, particularly those of *A. israelii*, may be very hard and grit like, and it may be necessary to crush them between two microscope slides in order to prepare a film suitable for Gram's staining. This characteristic may cause difficulties in obtaining smooth suspensions for susceptibility testing and for inoculation of identification test substrates. Other organisms that sometimes produce breadcrumb-like colonies and, therefore, may be mistaken for *Actinomyces* spp. include *Bifidobacterium* spp., *Streptococcus mutans* and microaerophilic streptococci.

The red pigmentation of mature *Actinomyces odontolyticus* colonies is a well-known distinguishing characteristic but it must be noted that the recently described species *A. radicidentis*, *A. urogenitalis* and *A. graevenitzii* may also produce pink-to-red colonies, particularly in old cultures. Young cultures of the above-named species, and *A. naeslundii*, *A. georgiae*, *A. neuii*, *A. cardiffensis* and *A. nasicola* form smooth, white or cream colonies, may grow more rapidly than *A. israelii* and may grow well in air or in air plus CO_2. Consequently, it is important to differentiate them from *Corynebacterium* spp. The failure of most *Actinomyces* spp. to produce catalase aids identification but *A. neuii*, *A. radicidentis* and some strains of *A. naeslundii* are catalase positive. *Actinomyces radicidentis* is similar to staphylococci in its cell and colony morphology and catalase production, but can be differentiated in biochemical tests and by endproducts of glucose metabolism.

The grey, semitranslucent colony morphology, growth within 5 days and relative aerotolerance of *A. turicensis* are useful aids to distinguish this species from the phenotypically similar *A. meyeri*, which grows slowly and only anaerobically, producing tiny, opaque, white colonies. *Actinomyces turicensis* and other rapidly growing, aerotolerant species with similar morphology (*A. radingae*, *A. europaeus* and *A. funkei*) may be mistaken for non-haemolytic streptococci, lactobacilli or other commensal organisms.

PATHOGENESIS

Actinomyces spp. appear to be unable to invade intact mucous membranes. Trauma such as dental diseases, infection with other organisms, surgery or penetration of mucosa by a foreign body commonly precede infection, but in some cases the portal of entry is not apparent.

The species that cause actinomycosis characteristically form *in vivo* dense microcolonies of cells with protruding clubbed ends comprising both bacterial and host material and surrounded by neutrophils and foamy macrophages. Clearly, the organisms within these clusters are protected from both host defences and antimicrobial agents. As the organisms slowly multiply, the host mounts a 'foreign-body' response and the characteristic chronic abscesses and sinus tracts of advanced disease develop. Doubtless, the other bacterial species invariably present in such lesions influence pathogenicity and may produce specific toxins or agents active against host cells or against antibiotics. *Actinomyces* spp. are not known to produce specific toxins. Humoral antibody responses to infection are not reliably produced; therefore serological tests for actinomycosis have not been developed.

In the development of dental plaque and intraoral diseases, the abilities of *Actinomyces* spp. to co-aggregate with other bacteria and to adhere to mammalian cells are important factors. Their fermentative properties may also provide essential nutrients for other pathogenic species, e.g. succinate for *Porphyromonas* spp., and may regulate pH.

Epidemiology

The 'classically-described' *Actinomyces* spp. and *P. propionicum* are known to be early colonisers of the mucous membranes, especially specific niches within the oral cavity. They may occur also in the gastrointestinal and genitourinary tracts in smaller numbers.

Little is known of the natural habitats of many recently described species. However, *A. graevenitzii* has been found to be an early coloniser of the mouth and *A. urogenitalis* appears to be aptly named (Sarkonen *et al.*, 2000; Hall *et al.*, 2001b). The many reported isolates of *A. turicensis* associate this species with the genitourinary and gastrointestinal tracts and lipid-rich areas of the skin (Sabbe *et al.*, 1999; Hall *et al.*, 2001b). *Actinomyces radingae*, *A. funkei* and *A. europaeus* appear to occupy similar habitats, and *A. neuii* is probably a commensal of the skin. *Actinomyces cardiffensis* has been isolated from sites similar to those of *A. israelii* but has not been proven to cause actinomycosis (Hall *et al.*, 2002).

Association with Intrauterine Contraceptive Devices (IUCDs)

In 1976, *Actinomyces*-like organisms (ALO) were first reported in Papanicolaou-stained cervical smears of IUCD users and were confirmed as *A. israelii* by direct immunofluorescence (Gupta, Hollander and Frost, 1976). Numerous subsequent studies employing combinations of cytological examination, cultivation and immunofluorescence, have endeavoured to elucidate the prevalence and clinical significance of ALO. All studies confirmed a strong association of ALO with IUCD usage. However, other findings have been highly variable, due in part to differing methodologies, variations in study populations and criteria, types of IUCD and other confounding factors. Accuracy of detection by Papanicolaou smears has been questioned. Confirmation by immunofluorescence has been recommended but specific antisera are not commercially available. Cultivation from urogenital sites has a poor record of recovery, largely due to the heavy concomitant flora.

Reported detection rates for ALO or confirmed *Actinomyces* spp. in IUCD users are 1.6–50%, averaging approximately 20%. Several authors reported increasing prevalence with duration of IUCD use, becoming significant after 2–8 years in different studies. Some noted significantly higher incidence in users of plastic devices over those with copper-containing IUCDs, possibly due to the known bactericidal effect of copper. However, both findings have been disputed and may be biased by the recommendation that copper IUCDs are replaced every 2 years, whilst plastic IUCDs have a longer life span. A study of the new generation of hormone-releasing IUCDs found ALO in 2.9% of users of a levonorgestrel-releasing IUCD and in 20% of users of a copper-containing IUCD (Merki-Feld *et al.*, 2000).

Clinical significance of ALO or confirmed *Actinomyces* spp. in the genital tract, IUCD or endometrium has proved difficult to assess. Indisputably, worldwide, overt pelvic actinomycosis remains extremely rare, whilst vast numbers of IUCDs are in use and perhaps 20% are colonised by ALO. Hence, actinomyces appear to occur as commensals and rare opportunist pathogens. However, IUCD usage is associated with increased incidence of pelvic inflammatory disease and other less severe pelvic symptoms, and the role played by actinomyces in these entities has not been established.

The potential, though remote, risk of serious infection (and perhaps, fears of litigation) led many authors to recommend removal of any ALO-associated IUCD, with or without antimicrobial cover, and replacement of the device after 1–2 months, when follow-up smears demonstrated absence of ALO. Other authors recommended removal only in symptomatic patients. The Clinical and Scientific Committee of the British Faculty of Family Planning and Reproductive Healthcare has produced concise recommendations (Cayley *et al.*, 1998). In summary these are: in symptomatic patients with ALO, remove the IUCD, offer antimicrobial therapy and review symptoms; in asymptomatic women with ALO, counsel regarding risks and symptoms, advise consultation if symptoms occur, arrange 6-monthly follow-up, and either leave the device in place or remove or replace it.

ALO have been associated with other foreign bodies of the genital tract or uterus, including vaginal pessaries and forgotten tampons (Bhagavan and Gupta, 1978). Species isolated from IUCDs are *A. israelii*, *A. gerencseriae*, *A. naeslundii*, *A. odontolyticus*, *A. meyeri*,

A. cardiffensis, A. europaeus, A. funkei, A. neuii, A. radingae, A. turicensis, A. urogenitalis, P. propionicum and *A. denticolens*-like strains (various studies and PHLS Anaerobe Reference Unit records).

CLINICAL FEATURES

Actinomycosis

Actinomycosis is a subacute to chronic disease, characterised by abscesses with hard fibrous walls enclosing collections of white or yellow pus that may contain 'sulphur granules' or 'drusen' (microcolonies of the *Actinomyces* spp.). In advanced cases, sinuses develop and discharge at the skin surface or to internal sites. Central suppurative foci regress and cicatrise, producing very disfiguring lesions. Infection is usually endogenous, and concomitant bacteria, frequently other members of the oral microflora, are almost invariably isolated. Acute lesions are usually painful but chronic forms are often painless. The disease progresses over months or years, invading surrounding tissues, regardless of fascia and organ boundaries, and forming multiple indurated abscesses, cavities and fistulae. Osteitis and osteomyelitis occur rarely in human infections, though bone involvement is common in bovine cases. Spontaneous remissions and relapses are characteristic, and the disease may become recalcitrant or eventually fatal if ineffectively treated. Indeed, in the pre-antibiotic era, mortality was high, and 902 deaths were documented in UK for the period 1916–1935 (Cope, 1938). An average of 25 deaths per annum were reported in the USA for the years 1949–1969 (Slack and Gerencser, 1975).

Actinomycosis can affect any part of the body but occurs most often in the cervicofacial region (60% cases). Thoracic and abdominal/pelvic infections account for approximately 15 and 20% of cases, respectively. Rare, exogenous sources of infection are fistfight injuries to the hand (punch actinomycosis), human bites and, possibly, animal bites. Predisposing factors include diabetes, dental diseases and chronic lung diseases. Persons of any age may be affected, and various studies have reported a male to female preponderance of between 2:1 and 4:1. Slack and Gerencser (1975) reviewed the incidence of infection and concluded that 'Actinomycosis occurs throughout the world and that it is neither a rare nor a common disease'. The principal agents of human actinomycosis are *A. israelii* and *A. gerencseriae*. Less frequently, *A. naeslundii*, *A. odontolyticus*, *A. meyeri* or *P. propionicum* are isolated.

Cervicofacial Actinomycosis

Given the oral habitat of the causative organisms, it is not surprising that the majority of actinomycotic lesions occur in the face and neck regions. Infection may be preceded by poor oral hygiene, dental and periodontal diseases, tooth extraction and other dental procedures, jaw fracture, tonsillitis or damage to mucosa by foreign bodies such as fish bones. However, predisposing or causative factors are not always determinable. The principal site of infection is tissue adjacent to the mandible. Other sites include the cheek, chin, maxillary area, neck, mastoid, sinuses, parotid and thyroid glands, tongue, lips, ears and nasal septum. Lesions near the mandibular joint may result in trismus. The primary lesion is often a hard, red and painful 'boil-like' swelling. Untreated, the disease progresses slowly as described above, resulting in characteristic multiple indurated abscesses at various stages of development, discharging sinuses and scarring. Secondary spread to the brain, thorax, other organs or bloodstream may be life threatening.

Thoracic, Abdominal and Pelvic Actinomycosis

The usual source of thoracic actinomycosis is aspirated oral microflora. Consequently, alcoholics are at increased risk of disease. Other predisposing conditions include dentogingival disease, emphysema and bronchiectasis. Less commonly, the disease may extend directly from cervicofacial or abdominal lesions. Abdominal and some pelvic infections arise from damage to the gastrointestinal mucosa, such as may result from surgery, ruptured appendix or diverticula, neoplasia, ulcerative diseases or accidental trauma. Primary hepatic abscesses may be seeded with gastrointestinal flora via the portal vein. As discussed above, colonisation of IUCDs and other foreign bodies in the uterus and cervix is another source of pelvic infection. Causative organisms are as listed for cervicofacial actinomycosis. However, *A. meyeri* is particularly associated with thoracic infections and has a tendency to disseminate to cerebral, muscular and cutaneous sites.

Thoracic infection commonly develops within the lungs but may spread to the pleura, ribs, intercostal tissue and chest wall. Involvement of the latter results in chronic draining sinuses and stigmata characteristic of actinomycosis. Other thoracic presentations include pericarditis, tracheal, paravertebral and psoas abscesses.

Abdominal/pelvic infection may involve the gut wall, bladder, kidney, liver, spleen, fallopian tubes, ovaries, uterus and anorectal abscesses. In chronic disease, fistulous abdominal wall abscesses may develop. In all the abovementioned sites, early symptoms are frequently vague and non-specific. Differential diagnosis commonly includes carcinoma and tuberculosis, and less commonly, Crohn's disease or other ulcerative conditions of the gastrointestinal tract. Correct diagnosis is often made only at a late stage when sinuses are apparent, obstruction of major organs occurs or surgery for suspected neoplasm is performed. By this stage, there may be extensive tissue damage, the disease is difficult to treat, and may become life threatening.

The published literature abounds with cases of thoracic, abdominal and pelvic actinomycosis that mimicked neoplasia, correct diagnosis being made only peri- or postoperatively, or at postmortem examination. Weese and Smith (1975) reported on 57 cases of which only four were diagnosed as actinomycosis upon admission. Misdiagnoses included carcinoma and tuberculosis. In 181 cases of actinomycosis in military personnel studied by Brown (1973), only nineteen were diagnosed correctly by the attending physician, and five of eight cases with pulmonary involvement were thought to have tuberculosis. Other diagnoses included histoplasmosis, nocardiosis, maduramycosis, sporotrichosis and carcinoma. Shannon, Wightman and Carey (1995) described pulmonary actinomycosis as 'a master of disguise' and reported three cases with initial diagnoses of malignant neoplasia. In each case correct diagnoses were made histologically, subsequent to lobectomy or pneumonectomy. Similarly, Williams, Lamb and Lewis-Jones (1990) described actinomycosis of the abdomen and pelvis as 'the great imitator' and warned 'beware the IUCD'. They reported two cases of pelvic actinomycosis, in patients with IUCDs. Both had experienced low-grade symptoms followed by acute exacerbations. One was diagnosed as pelvic malignancy, the other as spastic colon. Both required colostomies, and one suffered relapse after several months. In this patient, actinomycosis was diagnosed only when *A. israelii* was isolated from a metastatic groin abscess. The authors stressed the need to consider actinomycosis when pelvic neoplasia or inflammatory disease is suspected in patients with IUCDs. They concluded that radiology, particularly computed tomography (CT), may be important both in diagnosis and monitoring of treatment efficacy (Williams, Lamb and Lewis-Jones, 1990). CT-guided core needle biopsy may be a valuable aid to diagnosis and may obviate surgical intervention in some cases (Lee *et al.*, 2000). Fine-needle aspiration cytology has been used to confirm diagnosis early in cervicofacial disease, and to collect material for microbiological examination (Nespolon and Moore, 1994).

Infections of the CNS, Bones and Soft tissues

Actinomycotic infections of the central nervous system (CNS), bones and skin are rare. CNS infections arise from direct extension of infection of the head, neck, thorax or abdomen, or by haematogenous spread.

The most common form of CNS involvement is cerebral abscess (75%), predominantly in the temporal or frontal lobes. Infections may be polymicrobial, commonly involving members of the oral microflora. Chronic low-grade meningitis is another presentation and may be misdiagnosed as tuberculous meningitis. Abscesses of the spinal cord, vertebrae or paravertebral regions may follow surgery. *Actinomyces meyeri* appears to be associated most often with such cases.

Osseous involvement results in periostitis, osteomyelitis and bone destruction. Infection usually arises by direct extension from soft tissues; hence the mandible, ribs and spine are the most frequent sites involved. However, infection of the long bones occurs occasionally. Two case reports indicate the potential severity of this manifestation and difficulties in treatment (Macfarlane, Tucker and Kemp, 1993; Vandevelde, Jenkins and Hardy, 1995). In both patients (female, 19 and male, 22 years), infection originated from penetrating injuries to the foot, sustained on farms. Chronic infection of bone, joint and soft tissues occurred over 41 and 11 years, respectively, and after numerous courses of prolonged therapy and minor surgical interventions, both underwent below-knee amputations. In the latter case, where the causative organism was *A. naeslundii*, this apparently resulted in a cure. However, in the former case, disease progressed to eventually involve the femur, hip joint, spine, psoas muscle, perirenal space, pelvis and abdominal wall. At the time of reporting, significant improvement in symptoms had been achieved by prolonged ciprofloxacin therapy (Macfarlane, Tucker and Kemp, 1993). In this patient, the causative organism was not identified to species level. Notably, in both these severe cases, the source of the infecting organisms may have been exogenous, and from a farm environment. Wust *et al.* (2000) reported infection of a hip prosthesis with *A. naeslundii*. This patient responded well to appropriate antimicrobial therapy.

Cutaneous actinomycosis presents as subcutaneous or muscular abscesses and usually occurs by dissemination from 'classic' sites of infection such as the lung. However, cases with no apparent primary focus have been recorded. Cultures are often unsuccessful, but *A. meyeri* has a strong association with disseminated disease.

Lacrimal Canaliculitis

In lacrimal canaliculitis, the tear duct becomes obstructed by concretions of *Actinomyces* spp. or *P. propionicum* and concomitant oral or nasal microflora. Purulent exudate from the puncta and chronic conjunctivitis may occur. Infection may be undiagnosed for months or years and may respond poorly to topical antimicrobial therapy alone. Surgical evacuation of concretions, or manipulation of these through the puncta is usually necessary and these are valuable specimens for laboratory confirmation of causative organisms. The disease occurs mainly in females over 40 years of age, and *P. propionicum* is strongly associated (Brazier and Hall, 1993).

Bacteraemia

Bacteraemia and subacute endocarditis are rarely caused by actinomyces. Cases due to *A. naeslundii*, *A. viscosus* (presumably *A. naeslundii* Genospecies 2), *A. meyeri*, *A. odontolyticus*, *A. israelii*, *A. turicensis* and *A. funkei* have been reported. Some of these have occurred in injecting drug users and possible sources of infection include skin or oral flora. Transient *Actinomyces* bacteraemia following toothbrushing and endodontic therapy has been demonstrated (Silver, Martin and McBride, 1979; Debelian, Olsen and Tronstad, 1995).

Superficial Soft-Tissue Infections

Actinomyces spp. have been isolated from a variety of superficial soft-tissue infections, usually in conjunction with other bacteria. Their clinical significance in these polymicrobial infections is often difficult to establish. Several of the recently described *Actinomyces* spp. have been isolated from such sources, but have not been implicated in 'classic' actinomycosis. Some species appear to be associated with specific sites or disease entities. Notably, *A. turicensis* has been isolated frequently from pilonidal sinuses, perianal and decubital abscesses, balanitis, omphalitis and ear, nose and throat infections (Wust *et al.*, 1995; Vandamme *et al.*, 1998; Sabbe *et al.*, 1999; Hall *et al.*, 2001b). *Actinomyces radingae* and *A. europaeus* have been isolated from various soft-tissue infections, though numbers of isolates studied are small. Likewise, *A. neuii* is frequently isolated from abscesses, including breast abscesses, in association with mixed anaerobic flora. A catalase-negative isolate of *A. neuii* subsp. *neuii* has been reported in an infected mammary prosthesis (Brunner *et al.*, 2000).

Dental Caries and Periodontal Diseases

As the development and progression of dental caries and periodontal diseases involves complex interactions of numerous member species of the oral microflora, the precise roles played by individual species are difficult to ascertain. *Actinomyces* spp. are numerically significant components of plaque and were formerly thought to be associated with dental caries, but later research indicated that mutans streptococci and lactobacilli were more significant (Marsh and Martin, 1992). Demineralisation of tooth enamel by acids produced from carbohydrate metabolism may initiate caries. Whether genospecies of *A. naeslundii* have a disease-causing or protective role continues to be debated (Bowden *et al.*, 1999; Morou-Bermudez and Burne, 2000), and the presence of *A. odontolyticus* may affect the progression of lesions. Periapical actinomycosis is considered to be extremely rare, but a review of the literature suggests this occurs more frequently than previously thought (Sakellariou, 1996). Diagnosis was aided by histological examination of periapical lesions. The recently described species *A. radicidentis* has been isolated in pure culture and, to date exclusively, from infected root canals (Collins *et al.*, 2000).

The organisms associated with periodontal diseases are principally Gram-negative obligate anaerobes, including *Porphyromonas gingivalis*, *Prevotella intermedia* and *Actinobacillus actinomycetemcomitans*. The latter is named for its frequent association with actinomycetes. Human strains of *A. naeslundii* have been shown to cause destructive periodontal disease in animal models. Conversely, later research indicated that *Actinomyces* spp. were more frequently isolated from healthy or non-progressive sites than from sites with active periodontal disease (Moore *et al.*, 1987; Dzink, Socransky and Haffajee, 1988), and a beneficial role was suggested. However, more recent studies have shown associations of *A. naeslundii* with active periodontal sites (Tanner *et al.*, 1998) and *A. meyeri* with destructive periodontitis (Moore and Moore, 1994). Thus, evidence of pathogenicity remains contradictory, but *Actinomyces* spp. appear to play an important ancillary role by maintaining a favourable environment for causative agents of caries and periodontal diseases. Furthermore, the adhesive and co-aggregative properties of *Actinomyces* spp. are important in the development of dental plaque, where filamentous organisms form the basis of 'corn-cob' structures of bacteria. Adherence to buccal epithelia is facilitated by binding to host galactosyl ligands, and fimbriae of *A. naeslundii* mediate binding of cells to host surfaces and bacteria via acidic proline-rich proteins and statherin (Marsh and Martin, 1992; Hallberg *et al.*, 1998). *Actinomyces odontolyticus* has tongue-specific binding properties (Hallberg *et al.*, 1998).

LABORATORY DIAGNOSIS

Appropriate Specimens

Clinical awareness, appropriate specimens and laboratory expertise are important in confirming a diagnosis of actinomycosis. Where neoplasia is suspected, entire perioperative specimens are often placed in formalin and are, therefore, unsuitable for microbiological investigations. Whilst histological examination is valuable, it is non-specific and cannot predict antimicrobial susceptibilities; specific confirmation rests currently upon microbiological evidence. However, in cases where symptoms are non-specific, or other infectious processes are suspected, the patient may have received multiple courses of antibiotics prior to surgery, biopsy or aspiration of clinical material, and culture-failure may result. Clearly, late or misdiagnosis of actinomycosis can have catastrophic consequences for the patient, whereas early detection and appropriate antimicrobial therapy may effect a complete cure and may prevent much suffering, disfigurement, drastic surgery, infertility, recrudescence or death.

For confirmation of clinical actinomycosis, optimal specimens are pus or sinus exudate. These may be examined for sulphur granules, and by microbiological staining and culture. During specimen collection, care should be taken to minimise contamination of material with skin or oral flora. Other appropriate specimens include biopsy material, bronchial secretions obtained by transtracheal aspiration and concretions from lacrimal canaliculi. The latter may be placed directly onto suitable agar media and microscope slides at the bedside, but these must be transferred to the laboratory immediately. For microbiological investigations, biopsy material should be placed in sterile saline.

Swabs are less successful for isolation of *Actinomyces* spp. but may be the only option in cases where frank pus cannot be collected, and from superficial soft-tissue infections. Dry swabs are unsuitable, particularly for maintenance of fastidious concomitant organisms. Rather, swabs in transport media, based on formulations of Stewart or Amies, should be used.

IUCDs and vaginal pessaries may be cultured, but isolation of *Actinomyces* spp. from these devices and associated material is not necessarily indicative of clinical infection. Furthermore, removal of the device via the genital tract may introduce organisms from the local flora. Similarly, data obtained from intraoral sites must be interpreted with caution as isolates may represent endogenous flora. Also for this reason, expectorated sputum is unsuitable for diagnosis of pulmonary infection.

For investigation of endocarditis, several sets of anaerobic blood cultures should be collected and processed in line with local laboratory policies for such cases, including prolonged incubation and terminal subculture.

Ideally, specimens should be collected prior to commencement of antimicrobial therapy. All specimens should be transported to the laboratory without delay, and be examined promptly. Accompanying documentation should request specific examination for actinomyces as extended incubation on specifically selective media is necessary for isolation of the causative organisms.

Detection of Sulphur Granules

Sulphur granules or 'drusen' are yellowish, irregular, gritty particles that are sometimes present in pus or sinus exudate from human and bovine cases of actinomycosis. The granules may be very sparse, or so numerous as to give exudate the appearance of semolina. They are composed of central masses of bacterial filaments, radiating peripherally, and with hyaline, clubbed ends. Clubs are composed of a polysaccharide and protein complex, presumed to be derived from the bacteria, and a variety of salts and polypeptides, thought to result from interaction with host cells (Slack and Gerencser, 1975). There is no evidence that they contain sulphur.

Sulphur granules are valuable aids to diagnosis of actinomycosis. However, they must be differentiated from similar particles that may result from aggregations of other bacteria or host material. The latter are usually soft and easily emulsified on a microscope slide. Conversely, actinomycotic granules are usually hard and impossible to emulsify with a bacteriological loop. They may be crushed between two slides and stained by Gram's and modified Ziehl–Neelsen methods. If available, specific fluorescent antibody techniques may be

applied. Granules may be seen in fixed tissue sections and may be similarly stained.

Sulphur granules are virtually pure concentrations of *Actinomyces* spp. and organisms within may be protected from effects of antimicrobial therapy and host defences. Hence, culture of washed granules greatly enhances isolation of the causative organism from polymicrobial infections. A thin layer of pus or exudate should be examined in a sterile Petri dish, and any granules teased out and washed free of cellular debris in sterile saline. The granules should sink and remain intact. These may be aspirated with a sterile Pasteur pipette for culture on appropriate media. Appropriate safety precautions should be taken for examination of potentially hazardous specimens, such as those where differential diagnosis includes tuberculosis. Pseudosulphur granules, composed of synthetic material may be found in association with IUCDs.

Direct Microscopy

Gram's staining of sulphur granules reveals characteristic masses of Gram-positive beaded, branched filaments. Unlike aggregates of *Nocardia* spp., *Actinomyces* spp. are non-acid-fast when stained by modified Ziehl–Neelsen (Kinyoun) method.

In fixed tissue sections, haematoxylin–eosin staining may reveal granules composed of eosinophilic clubs surrounding basophilic filaments, and the whole usually surrounded by proliferations of polymorphonuclear leucocytes and foamy macrophages. Confirmation of aetiology is obtained by Grocott's methenamine silver stain and Gram's stain (Brown–Brenn modification).

In Gram-stained smears prepared from pus, mucosal secretions, blood cultures or swabs, *Actinomyces* spp. may appear as individual Gram-positive beaded and branched filaments, loose aggregations or spider-like microcolonies, or as short diphtheroidal rods. Other bacteria are often present, and the actinomycetes may be very sparse, even in overt actinomycosis. Therefore, the finding of characteristic filaments in a clinical specimen should arouse suspicion of actinomycosis, but is not pathognomic, but their absence does not rule out infection.

Culture Media

Actinomyces spp. may be grown on various nutrient-rich solid or fluid media. Addition of serum or whole blood (rabbit, sheep or horse blood 5%) enhances growth. Columbia blood agar (Oxoid, UK) can be used, but fastidious anaerobe agar (BioConnections, UK) or anaerobe basal agar (Oxoid) are superior for the cultivation of fastidious anaerobes and *Actinomyces* spp.

Frequently, in clinical material, large numbers of concomitant bacteria are present with *Actinomyces* spp. Broth cultures are rarely advantageous, as less-demanding organisms usually proliferate rapidly, to the detriment of actinomycetes. A fluid medium sufficiently selective to overcome this problem has yet to be discovered. However, pure isolates may be cultivated in thioglycollate broth or fastidious anaerobe broth (BioConnections), with or without cooked meat granules. Light inoculation of broth media with organisms that form gritty colonies (e.g. *A. israelii*) may result in growth of discrete 'cloud-like' colonies, whilst the surrounding broth remains clear and apparently sterile. This phenomenon has technical implications for subsequent manipulations such as subculture and susceptibility testing and may be a reason for culture-failure of enrichment broths of clinical material.

Isolation of slow-growing actinomycetes from polymicrobial infections demands selective media. However, *Actinomyces* spp. are susceptible to a wide range of antimicrobial agents and may fail to grow on media containing neomycin, kanamycin or gentamicin, commonly used for isolation of obligate anaerobes. Several media incorporating cadmium sulphate (15–20 mg/l) have been developed specifically for isolation of *A. naeslundii* from dental plaque.

However, other species of *Actinomyces*, including *A. israelii*, may be inhibited by these media and they have not been evaluated for the isolation of recently described species.

For isolation of *Actinomyces* spp. from clinical specimens, inherent resistance to metronidazole has been exploited. Traynor *et al.* (1981) described a medium incorporating metronidazole (2.5 mg/l) in Columbia base with horse blood (10%). Lewis *et al.* (1995) reduced the horse blood to 5% in their MMBA medium and improved the selectivity of Traynor's formula by addition of mupiricin (128 mg/l). Brazier and Hall (1997) recommended metronidazole (10 mg/l), nalidixic acid (30 mg/l) and horse blood (7%) in fastidious anaerobe agar. Subsequently, however, this concentration of nalidixic acid has been found to inhibit some strains of *Actinomyces* and 15 mg/l nalidixic acid is now preferred. These media are semiselective, inhibiting obligate anaerobes and most Gram-negative bacteria but unable to suppress growth of other Gram-positive non-sporing bacilli and, except MMBA, staphylococci, streptococci and enterococci. However, reduction of the overall bacterial load, combined with good spreading technique, improves the chances of successful isolation of *Actinomyces*.

Application of commercial antimicrobial discs to agar plates may enhance selectivity when special media are not readily available, and may be useful in obtaining pure cultures. This procedure is cheap, simple and flexible, as inhibitory agents such as colistin sulphate, nalidixic acid and/or metronidazole can be chosen to best suppress the flora of the individual specimen. Ciprofloxacin (1 mg/l) discs may be used to inhibit staphylococci (personal observation), but higher concentrations of this agent may inhibit *Actinomyces* spp. Efficacy of selective media for isolation of recently described *Actinomyces* spp. has not been established.

Isolation Techniques

Clinical material should be cultured on an actinomyces selective agar and on a range of media suitable for isolation of other expected pathogens, choice depending upon source of material. *Actinomyces* spp. vary considerably in oxygen tolerance, but growth of most isolates is enhanced by CO_2 and all grow under anaerobic conditions. *Actinomyces naeslundii*, *A. viscosus* and *A. odontolyticus* may grow rapidly and equally well under aerobic or anaerobic conditions. Other species, particularly *A. israelii*, *A. gerencseriae*, *A. meyeri*, *A. cardiffensis* and *A. oricola* grow only anaerobically and very slowly on primary isolation. Aerotolerance and speed of growth may both increase upon repeated subculture. However, primary plates should be incubated anaerobically for 10–14 days, under conditions that avoid desiccation of media and examined at 2–3, 7 and 10–14 days. Anaerobic chambers or jars or the pyrogallol–sodium carbonate method may be used. A major advantage of using anaerobic chambers is that interim examination of cultures can be performed without exposure to air. Subcultures and other manipulations of colonies may be performed on the open bench provided that cultures are returned to anaerobic conditions promptly.

Gram's stain should be performed on colonies consistent with *Actinomyces* spp. or *P. propionicum* (Table 48.3). A hand-lens or stereo microscope are valuable aids for examination and for selection of single colonies from mixed cultures. Blood-containing plates may be examined under long wave (365 nm) ultraviolet light for the bright red fluorescence of *P. propionicum*, *A. graevenitzii* and some *Bifidobacterium* spp. Colonies of other *Actinomyces* spp. occasionally fluoresce weakly (personal observations). This simple test is a useful adjunct but is not specific as many other anaerobic bacteria demonstrate red (or green) fluorescence.

Suspicious colonies should be subcultured to three non-selective plates; one incubated anaerobically to obtain a pure culture; and the others in air and air plus CO_2 5%, to assess atmospheric requirements. A pure culture is essential for confirmation of identity and for susceptibility testing.

Direct Detection by Specific Probes

For pathogens that are slow growing or difficult to isolate, direct detection in clinical material is clearly advantageous. For the 'classic' *Actinomyces* spp. and *P. propionicum*, direct or indirect immunofluorescence and, more recently, DNA probes have been applied for this purpose.

Fluorescent antibody techniques, originally developed for serological grouping of *Actinomyces* spp., have been applied successfully to fresh clinical material, formalin-fixed tissue, histological sections, primary microbiological cultures and pure isolates (Slack and Gerencser, 1975). These authors used serotype-specific antibodies, raised in rabbits, conjugated with fluorescein isothiocyanate and, where necessary, absorbed with cross-reacting antigens. They claimed this procedure to be the most rapid and reliable means of identifying *Actinomyces* and *P. propionicum*, but recommended that, where possible, conventional isolation and identification are also performed.

Similar methods, using species-specific antibodies, became popular in the 1970s to early 1990s, particularly for confirmation of identity of ALO seen in cervical smears and associated with IUCDs (Gupta, Erozan and Frost, 1978; Leslie and Garland, 1991). However, several problems are associated with the technique and it remains a research tool. Antisera have not been produced commercially and antibody production is non-standardised, laborious, complex and involves laboratory animals. The process is further complicated by the well-documented occurrence of cross-reactions, commonly between serotypes and between species, and occasionally between genera (Slack and Gerencser, 1975). Furthermore, in the light of current knowledge regarding antigenic heterogeneity within species, particularly in *A. naeslundii* and *A. viscosus*, it is possible that antisera raised from single or few strains may demonstrate strain or subspecies specificity, rather than species specificity.

Given the complexity of current taxonomy of the genus, it is clear that large numbers of antisera would be necessary to detect all species of *Actinomyces* now known to occur in clinical material, and the method would be impractical for routine use. Efficacy for detection of recently described species has not been reported.

DNA probes species specific for the 'classic' *Actinomyces* spp. have been applied to supra- and subgingival plaque samples (Ximénez-Fyvie, Haffajee and Socransky, 2000) and to infected root canals (Tang *et al.*, 2003). DNA cloning and sequencing, combined with checkerboard DNA hybridisation for some species has been applied to various intraoral niches to elucidate bacterial species associated with caries and in health (Paster *et al.*, 2001; Kazor *et al.*, 2003).

Multiplex fluorescent *in situ* hybridisation (FISH) successfully detected *A. naeslundii* and other five bacteria in a simulated oral biofilm (Thurnheer, Gmür and Guggenheim, 2004). This technique appears promising for the direct detection of *Actinomyces* spp. from clinical material. However, many species-specific probes would be required and, given the intraspecies genetic diversity within this genus, specificity and sensitivity may be problematic. The complexities of these techniques render them suitable only for research purposes at present. However, rapid advances in new technologies, including DNA microarrays, may bring such techniques into routine clinical use in the future.

Identification Procedures

Gas–Liquid Chromatography (GLC) for Products of Glucose Metabolism

The volatile and non-volatile fatty acids formed as endproducts of glucose metabolism are a stable characteristic of obligate and facultative anaerobes. The presence and approximate amounts (major or minor) of specific acids differentiate many anaerobic non-sporing Gram-positive bacilli to genus level. *Actinomyces* produce minor-to-moderate amounts of acetic acid and major amounts of lactic and succinic acids, but no propionic acid. Members of *Actinobaculum, Arcanobacterium* and *Corynebacterium* produce similar endproducts, with or without lactic acid. *Propionibacterium* spp. produce major amounts of acetic and propionic acids and may produce isovaleric, lactic and succinic acids. Bifidobacteria produce acetic and lactic acids as major products, but no succinic acid. Lactobacilli produce only lactic acid (major). *Eubacterium* spp., the related *Eggerthella* and *Slackia*, produce various volatile acids, or none at all, and may produce lactic acid but not succinic. Susceptibility of these genera and clostridia to metronidazole (5 μg disc) differentiates them from other non-sporing Gram-positive bacilli.

Volatile and non-volatile fatty acids may be detected in broth cultures by the GLC methods described in the Anaerobe Laboratory Manual (Holdeman, Cato and Moore, 1977), or a modification thereof. Background levels of acids in uninoculated broth should be determined for comparison.

Conventional Biochemical Tests

Identification of *Actinomyces* to species level is traditionally based on differentiation in carbohydrate fermentation tests, hydrolysis of aesculin, urea and starch, and supplementary reactions including production of indole, catalase and nitrate reductase. Tests have been performed in various formats including broths, agar slopes and stabs, and in various basal media. Prolonged anaerobic incubation may be necessary and, in order to obtain adequate growth, serum is commonly added to broth or sloped media.

The agar plate method of Phillips (1976) was developed specifically for testing of fastidious anaerobes. A series of blood agar plates, each containing a single carbohydrate source, is inoculated and incubated anaerobically until good growth is visible. Acids formed by fermentation are demonstrated by addition of a pH indicator to an agar plug of culture. This method, though time-consuming, is sensitive, well controlled, and practical for testing small batches

of isolates and is, therefore, recommended for testing of *Actinomyces* spp.

Rapid results may be obtained with commercially available tablets (Rosco Diagnostic tablets, BioConnections, UK) that detect carbohydrate fermentation or preformed proteolytic enzymes by chromogenic changes after 4 h of aerobic incubation in suspensions of test organisms. This method is cheap, simple and flexible as the range of tests performed can be tailored to individual needs. However, a heavy inoculum of a fresh (<48 h) culture is essential. A scheme for presumptive identification of *Actinomyces* spp. and related genera using such tests has been developed (Sarkonen *et al.*, 2001).

Despite the widespread use of conventional biochemical tests, identification of *Actinomyces* spp. by these methods is notoriously problematic. Variation in results may occur due to varying methodologies, in particular the choice of basal medium and, possibly, atmospheric conditions. Technical problems include slow growth, obtaining smooth suspensions of colonies and differing results upon repeat testing of individual strains. Interpretation of results is also problematic. For some species, conventional tests differentiate poorly. This may be due, in part, to the heterogeneous nature of some species as previously described, and some anomalies may be clarified by recent or future taxonomic changes. An example is the differentiation of *A. israelii* and *A. gerencseriae* by failure of the latter to ferment arabinose; obsolete identification schemes would list fermentation of arabinose by *A. israelii* (including *A. gerencseriae*) as variable. Differentiating reactions listed in schemes by various authors are not always in agreement, possibly due to differing methodologies used, or the different strains from which data were obtained. Furthermore, most published schemes are now obsolete in respect of current taxonomy. Table 48.4 lists biochemical reactions that may aid the identification to species level of *Actinomyces* spp. from human sources. However, the abovementioned limitations must be heeded, and interpretation should be based on a consensus of reactions together with consistent cell and colony morphology. Recent descriptions of species should be consulted (see references listed in Table 48.2).

Table 48.4 Biochemical reactions that may aid identification of *Actinomyces* spp. from human sources

	car[a]	den	eur	fun	geo	ger	gra	isr	mey	nae	nas	neu	odo	ori	rac	rag	tur	uro
Growth in air	+/–[b]	+/–	+/–	+	+	+/–	+	+/–	0	+	+/–	+	+/–	0	+	+	+	+
Growth air and CO$_2$	+/–	+	+	+	+	+/–	+	+/–	+/–	+	+/–	+	+	+/–	+	+	+	+
Pigment	–/pink	–/pink	0	0	0	0	0	0	0	–/pink	0	0	red/–	0	–/red	0	0	–/red
Nitrate reduction	+	+	+/–	+/–	+/–	+/–	0	+/–	0	+	0	V	+	+	+/–	0	0	+
Urease	0	0	0	0	0	0	0	0	+/–	+/–	0	0	0	0	+/–	0	0	0
Catalase	0	0	0	0	0	0	0	0	0	+/–	0	+	0	0	+	0	0	0
Aesculin hydrolysis	0	+	+/–	0	+/–	+	0	+	0	+	0	0	+/–	+/–	+	+	+/–	+
Starch hydrolysis	0	0	0	nd	+/–	+/–	nd	0	0	+/–	0	nd	+/–	0	nd	nd	nd	nd
Amygdalin	0	+/–	0	nd	+/–	nd	0	+	+/–	+/–	0	nd	0	+/–	nd	0	0	nd
Arabinose	0	0	0	0	+/–	0	0	+	+/–	0	0	0	+/–	0	0	+/–	+/–	+/–
Cellobiose	0	+/–	0	nd	+	0	0	+	0	+/–	+/–	+/–	0	+	nd	+/–	+/–	nd
Glucose	+	+	+	+	+	+	+	+	+	+	+/–	+	+	+	+	+	+	+
Lactose	0	+	0	+/–	+	+	+	+	+	+	0	+	+	0	+	+	0	+
Mannitol	0	+	0	0	+/–	+	0	+/–	0	0	0	+	0	+/–	+	0	0	+/–
Raffinose	0	+	0	0	0	+	nd	+	0	+	0	+	0	+/–	+	+/–	+/–	+
Ribose	+	+	+/–	+/–	+	+	+	+	+	+/–	0	+	+/–	+/–	+	+	+	+/–
Salicin	0	+	0	nd	+/–	+	0	+	0	+/–	0	+/–	+/–	+	nd	+	0	nd
Sucrose	+	+	+/–	+	+	+	+	+	+	+	0	+	+	+	+	+	+	+
Trehalose	0	+	+/–	0	+	+	0	+	0	+	0	+	0	+	+	+/–	+	+
Xylose	0	0	0	+	+	+	0	+	+	0	0	+	+/–	+/–	nd	+	+	+
Pyrazinamidase	0	+	0	+/–	nd	nd	nd	nd	nd	nd	+	+	nd	+/–	+	+/–	0	0
β-Galactosidase	0	+	+	+/–	nd	nd	+	+	0	+/–	+	+	0	+/–	+	+	0	+
α-Glucosidase	+	+	+	+	nd	nd	0	+	+	+/–	+	+	0	+	+	+	+	+
β-NAG[c]	0	0	0	+/–	nd	nd	+	0	0	0	+	+/–	0	0	0	+	0	+

[a] Abbreviations for *Actinomyces* spp.: car, *A. cardiffensis*; den, *A. denticolens* like; eur, *A. europaeus*; fun, *A. funkei*; geo, *A. georgiae*; ger, *A. gerencseriae*; gra, *A. graevenitzii*; isr, *A. israelii*; mey, *A. meyeri*; nae, *A. naeslundii*; nas, *A. nasicola*; neu, *A. neuii*; odo, *A. odontolyticus*; ori, *A. oricola*; rac, *A. radicidentis*; rag, *A. radingae*; tur, *A. turicensis*; uro, *A. urogenitalis*.
[b] Reactions: +, positive; +/–, weak or variable; 0, negative; nd, no data; V, *A. neuii* subspp. *neuii* positive, subspp. *anitratus* negative.
[c] β-NAG, *N*-acetyl-β-glucosaminidase.

Commercial Identification Kits

Commercial kit-form systems (e.g. API rapid ID32A, bioMérieux, France; RapID ANA II, Innovative Diagnostic systems, USA) for identification of anaerobes are used widely in UK clinical laboratories. In these systems, a suspension of fresh culture is added to a panel of freeze-dried substrates and, after aerobic incubation for 4 h, preformed enzymes involved in carbohydrate fermentation or proteolysis are detected by chromogenic reactions. Tests are scored numerically and the resulting code is compared with those in the manufacturer's database. Advantages of kits are ease of use, standardisation of substrates, long shelf-life and rapid results. Disadvantages, for anaerobes in general, include difficulties in interpretation of weak reactions, omission from databases of some clinically relevant species and variable results dependent on culture medium from which the isolate is taken. Furthermore, for aerobic organisms, various kits are available, with tests appropriate to a specific genus, or related genera, e.g. *Streptococcus*, Enterobacteriaceae. However, single kits are marketed for identification of all anaerobes and consequently, for some genera, only a limited number of tests in the panel is relevant. Problems specific to *Actinomyces* spp. include obtaining a smooth suspension for inoculation, lack of specificity of results and inadequate databases. Four commercial systems were recently evaluated specifically for identification of *Actinomyces* spp. and closely related species, and performance was found to be universally poor, particularly for recently described species (Santala *et al.*, 2004).

Despite the limitations listed above, combinations of API-ZYM, API rapid ID32Strep and API CORYNE kits (bioMérieux, France) have proved useful in classification and identification of recently described *Actinomyces* spp. Results may be interpreted with reference to the most recent species descriptions. Sabbe *et al.* (1999) recommend the API CORYNE kit for clear discrimination of *A. turicensis* from other *Actinomyces* spp.

Specific Antibody Techniques

Fluorescent antibody techniques, including immunodiffusion using antisera raised against whole cells or soluble antigens and cell-wall agglutination tests have been applied to mixed cultures and pure isolates of *Actinomyces* spp. In each of these techniques, unavailability of sera, lack of standardisation, cross-reactions and non-specificity have been problematic. Therefore, these methods have not been widely used for identification of *Actinomyces* spp.

Chemotaxonomic Approaches

Combinations of chemotaxonomic and molecular methods are valuable tools in classification of *Actinomyces* spp. However, the specialist expertise and equipment required for performance of tests and analysis of data have ensured that these methods have not gained favour for routine identification. Methods commonly employed for classification are whole-cell protein profiling by sodium dodecyl sulphate–polyacrylamide gel electrophoresis (SDS-PAGE) and analysis of cell-wall constituents such as amino acids, sugars, fatty acids, phospholipids and menaquinones. In each of these methods, findings may vary with culture conditions, particularly media composition. Therefore, results for individual isolates can only be compared with those for strains grown under similar conditions, and standardisation of procedures is important.

Molecular Methods

In recent years, identification methods based on detection of variations at the genetic level have become increasingly practical and, for genera such as *Actinomyces* that are difficult to speciate in conventional tests, have greatly improved the accuracy of identification. In general, the 16S rRNA gene is a popular target for genetic probes because it is mostly highly conserved but contains several, often species-specific, variable regions and occurs in high-copy numbers in each cell. Also, as nucleotide sequence variations within this gene form the mainstay of modern taxonomy, identification methods based on detection of such variations are directly linked to current classification.

Prior to the widespread use of PCR, Barsotti *et al.* (1994) digested total genomic DNA with the endonuclease *Bst*EII to detect restriction fragment length polymorphisms (RFLP) and were able to differentiate *A. bovis*, *A. gerencseriae*, *A. israelii*, *A. meyeri*, *A. odontolyticus* and *Arcanobacterium pyogenes* to species. However, *A. naeslundii* and *A. viscosus* were not clearly distinguishable.

Methodology has been refined and simplified by PCR amplification of specific DNA prior to enzymatic digestion. The technique, known as PCR-RFLP or amplified ribosomal DNA restriction analysis (ARDRA), has been adapted to differentiate species of many genera. In brief, DNA extracted from a pure cultured isolate is used as the template for PCR amplification with specific primers (commonly targeting the almost complete 16S rRNA gene). The amplicon is digested by one or more endonucleases, selected for their ability to cleave the DNA at species-specific variable regions. Resulting DNA fragments are separated by gel electrophoresis, alongside molecular size markers, and are visualised by ethidium bromide staining. Banding patterns are compared with those of reference strains, either visually or by computer software.

For human oral *Actinomyces* spp., two small studies found PCR-RFLP with the endonuclease *Mnl*I to be discriminatory (Sato *et al.*, 1998; Ruby *et al.*, 2002). However, in studies of over 600 human clinical and veterinary strains, banding patterns produced from separate digestions with endonucleases *Hae*III and *Hpa*II were analysed. This method discriminates to species or subspecies level, all currently recognised species of *Actinomyces*, *Arcanobacterium*, *Actinobaculum*, *Varibaculum*, *Mobiluncus*, *Gardnerella*, *Propionibacterium* and many species of *Bifidobacterium*, *Lactobacillus* and other non-sporing Gram-positive bacilli (Hall, Lewis-Evans and Duerden, 2001a; Hall *et al.*, 2001b). The method is practical and cost-effective and is now in routine use for the identification of these organisms at the PHLS Anaerobe Reference Unit.

Antimicrobial Susceptibility Testing

Disc Diffusion Tests

In the United Kingdom, the antimicrobial susceptibilities of clinically significant bacteria are commonly determined by disc diffusion tests, but this method was developed for the testing of rapid-growing aerobic organisms. Where organisms have a longer than average lag phase of growth, the antibiotic diffuses further into the agar before active growth begins and, hence, produces exceptionally large zones of inhibition. Furthermore, there are few recommended anaerobic control organisms, and these do not adequately represent all clinically significant genera. Consequently, disc diffusion methods are not well suited to testing of slow-growing and fastidious organisms such as *Actinomyces* spp. but nevertheless, for reasons of economy and convenience, are commonly used methods. The special methodological problems pertaining to the testing of Actinomycetes, including difficulties in obtaining a uniform inoculum, have been discussed by Schaal and Pape (1980). It is recommended that, if disc diffusion methods are used for fastidious organisms, a maximum of four discs are placed on a standard Petri dish, and any unusual resistance is confirmed by other methods.

Determination of MICs

Determination of minimum inhibitory concentrations (MICs) of antimicrobial agents by agar dilution is the 'gold standard' method for fastidious anaerobes. This method is suitable for the testing of large numbers of isolates, but the labour-intensity of agar preparation renders it impractical for testing of occasional isolates. In the USA, the National Committee for Clinical Laboratory Standards guidelines recommend the use of Brucella agar supplemented with vitamin K and haemin for testing of anaerobes. It should be noted that Wilkins–Chalgren agar does not support good growth of some anaerobes and interpretation of endpoints can be problematic, particularly where hazy trailing endpoints occur.

Broth dilution methods for determination of MICs are not recommended for fastidious organisms, and the granular or 'microcolony' growth of some *Actinomyces* spp. in broth cultures is a further contraindication for use of this method with these organisms.

Conversely, the Etest method (AB-Biodisk, Sweden) is easy to use, accurate and practical for susceptibility testing of individual clinical isolates, including fastidious and slow-growing organisms, though relatively expensive in comparison with disc diffusion testing. The formation of an immediate and stable antimicrobial gradient overcomes problems encountered with slow-growing organisms in the disc diffusion method, and the Etest appears to be less sensitive to variations in inoculum concentration. However, results are still subject to variations in media constituents, quality of growth and incubation time, and to excess moisture on agar surfaces. Clear endpoints are usually readable after incubation for 20–48 h. In an evaluation of 50 strains of *Actinomyces* and closely related genera, results obtained by Etest correlated well with those obtained in the agar dilution method (Hall and Talbot, 1997).

Surveillance-Based Therapy

Alternatively, treatment regimens may be based on surveillance of representative clinical isolates or on reported clinical responses rather than the testing of individual strains. This approach is attractive and is commonly used in cases of actinomycosis, where diagnosis is often clinical or histological, the causative agent is frequently not isolated, and antimicrobial susceptibility patterns are stable and highly predictable. However, for data to retain relevance, it is important to monitor adequate numbers of fresh representative isolates and/or clinical outcomes periodically, and to establish efficacy of new antimicrobial agents and combination therapies.

Monitoring of MICs by Etest at the PHLS Anaerobe Reference Unit confirms that the long-established *Actinomyces* spp. and *P. propionicum* remain exquisitely sensitive to commonly used antimicrobial agents. Typical ranges of MICs were: benzyl penicillin <0.004–0.094 mg/l, amoxycillin <0.016–0.064 mg/l, tetracycline 0.125–2.0 mg/l, erythromycin <0.016–0.25 mg/l, cefuroxime <0.016–0.5 mg/l, clindamycin <0.016–0.5 mg/l and imipenem <0.002–0.064 mg/l. However, for *A. odontolyticus* MICs were generally higher though still within the therapeutic range, e.g. penicillin 0.032–0.5 mg/l, amoxycillin 0.064–1.0 mg/l, imipenem 0.064–0.5 mg/l, and one strain, isolated from a pleural fluid, was resistant to erythromycin with an MIC of >256 mg/l.

For recently described species, few data have been reported to date. Monitoring at the PHLS Anaerobe Reference Unit indicates that most isolates are susceptible to penicillin, erythromycin and tetracycline, but some exceptions occur and empirical data cannot be relied upon. Notably, 13 of 42 *A. turicensis* strains and one of 10 *A. funkei* strains were resistant to tetracycline, some strains having MICs of >256 mg/l. Five strains of *A. turicensis* showed reduced susceptibility or resistance to erythromycin. Also, high-level resistance to erythromycin (MIC > 256 mg/l) occurred in three of seven strains of *A. europaeus* and two of five strains of *A. radingae* examined. The eight strains of *A. urogenitalis* tested were all clinically susceptible to penicillin, but their MICs of 0.5–1.0 mg/l were considerably higher than those of 'classic' *Actinomyces* spp. In many cases, the pathogenicity and, therefore, clinical relevance of antimicrobial resistance in these species is questionable. However, one isolate of *A. turicensis* with MICs of 0.19 mg/l for penicillin, 3.0 mg/l for erythromycin and 48 mg/l for tetracycline was isolated from a blood culture of a 31-year-old intravenous drug abuser with an injection-site abscess and pneumonia.

TREATMENT

Aggressive antimicrobial therapy is essential for treatment of actinomycosis. Additionally, depending upon the site and severity of disease, surgical intervention, ranging from drainage of pus to excision of organs may be required. Historically, prognosis was poor, even when radical surgery was performed. However, treatment options greatly improved with the availability of penicillin and, to a lesser extent, sulphonamides. As reported above, *Actinomyces* spp. and *P. propionicum* are usually susceptible to a wide range of antimicrobial agents, including penicillins, cephalosporins, chloramphenicol, tetracyclines, erythromycin, clindamycin and imipenem. Conversely, fluoroquinolones, aztreonam, aminoglycosides and metronidazole have poor activity.

Although *Actinomyces* are exquisitely sensitive *in vitro* to many agents, bacteria within microcolonies (sulphur granules) and indurated abscesses are protected from their action. Therefore, high-dosage and long-term therapy is often necessary, and the pharmacological properties of penicillin make it the agent of choice in most cases. Alternatives for penicillin-allergic patients include tetracycline, erythromycin, clindamycin or cephalosporins. Therapy aimed at concomitant bacteria may be necessary, particularly where *Actinobacillus actinomycetemcomitans* or β-lactamase producers are present. Schaal (1998) recommends amoxycillin/clavulanic acid for cervicofacial and thoracic cases, and amoxycillin/clavulanic acid with metronidazole and an aminoglycoside or imipenem for abdominal cases. However, Smego and Foglia (1998) claim, controversially, that regimens targeting only *Actinomyces* sp. are usually curative, and cover against secondary organisms is unnecessary.

Typical regimens for straightforward cervicofacial infection are: oral penicillin V or doxycycline, 100 mg twice daily for 2 months, without surgical intervention (Smego and Foglia, 1998); or amoxycillin (2 g)/clavulanic acid (0.2 g) three times daily for 1 week, followed by 1 week at half this dose (Schaal, 1998). For more complicated infections, Martin (1984) recommends parenteral penicillin G, 10–20 million units daily for 4–6 weeks, followed by oral penicillin V 2–4 g daily for 6–12 months. For pelvic actinomycosis, without perforation, Hamid *et al.* (2000) suggest high-dose intravenous penicillin and metronidazole until biological parameters are within normal limits, then oral therapy for 1 month prior to surgery to excise residual abscess and relieve any urinary or bowel compression. Surgical procedures would usually entail bilateral salpingectomy or bilateral salpingoophorectomy and total hysterectomy.

Clinical response to prolonged ciprofloxacin therapy (750 mg twice daily for 6 months, 375 mg for 4 years) in a severe and very long-standing case of abdominal/pelvic infection has been reported (Macfarlane, Tucker and Kemp, 1993). However, *in vitro* activity of this agent is poor.

In all presentations of actinomycosis, early diagnosis greatly improves prognosis and, despite the armamentarium of modern medicine, inadequate treatment or late diagnosis may lead to years of suffering or even death from the disease or associated complications. Recurrence or secondary foci of disease are not uncommon; therefore all patients should be followed for long periods. IUCD users should be assessed regularly and managed as recommended by the British Faculty of Family Planning and Reproductive Healthcare (Cayley *et al.*, 1998).

After dental surgery, prophylaxis is advised for those at risk of cardiac infection, and early treatment of postsurgical inflammation may prevent development of cervicofacial actinomycosis.

REFERENCES

Barsotti O, Decoret D, Benay G *et al.* (1994) rRNA gene restriction patterns as possible taxonomic tools for the genus *Actinomyces*. *Zentalblatt Für Bakteriologie*, **281**, 433–441.

Bhagavan BS and Gupta PK (1978) Genital actinomycosis and intrauterine contraceptive devices. *Human Pathology*, **9**, 567–578.

Bowden GHW, Nolette N, Ryding H and Cleghorn BM (1999) The diversity and distribution of the predominant ribotypes of *Actinomyces naeslundii* genospecies 1 and 2 in samples from enamel and from healthy and carious root surfaces of teeth. *Journal of Dental Research*, **78**, 1800–1809.

Brazier JS and Hall V (1993) *Propionibacterium propionicum* and infections of the lacrimal apparatus. *Clinical Infectious Diseases*, **17**, 892–893.

Brazier JS and Hall V (1997) *Actinomyces*. In *Principles and Practice of Clinical Bacteriology*, Emmerson AM, Hawkey PM and Gillespie SH (eds), Wiley, UK, pp. 625–639.

Brown JR (1973) Human actinomycosis. *Human Pathology*, **4**, 319–330.

Brunner S, Graf S, Riegel P and Altwegg M (2000) Catalase-negative *Actinomyces neuii* subsp. *neuii* isolated from an infected mammary prosthesis. *International Journal of Medical Microbiology*, **290**, 285–287.

Buchanan AM, Scott JL, Gerencser MA *et al.* (1984) *Actinomyces hordeovulneris* sp. nov., an agent of canine actinomycosis. *International Journal of Systematic Bacteriology*, **34**, 439–443.

Cayley J, Fotherby K, Guillebaud J *et al.* (1998) Recommendations for clinical practice: *Actinomyces* like organisms and intrauterine contraceptives. *British Journal of Family Planning*, **23**, 137–138.

Collins MD, Hoyles L, Kalfas S *et al.* (2000) Characterization of *Actinomyces* isolates from infected root canals of teeth: description of *Actinomyces radicidentis* sp. nov. *Journal of Clinical Microbiology*, **38**, 3399–3403.

Collins MD, Stubbs S, Hommez J and Devriese LA (1993) Molecular taxonomic studies of *Actinomyces*-like bacteria isolated from purulent lesions in pigs and description of *Actinomyces hyovaginalis* sp. nov. *International Journal of Systematic Bacteriology*, **43**, 471–473.

Cope VZ (1938) *Actinomycosis*, Oxford University Press, London.

Debelian GJ, Olsen I and Tronstad L (1995) Bacteremia in conjunction with endodontic therapy. *Endodontics and Dental Traumatology*, **11**, 142–149.

Dent VE and Williams RAD (1984a) *Actinomyces denticolens* Dent and Williams sp. nov.: a new species from the dental plaque of cattle. *Journal of Applied Bacteriology*, **56**, 183–192.

Dent VE and Williams RAD (1984b) *Actinomyces howellii*, a new species from the dental plaque of dairy cattle. *International Journal of Systematic Bacteriology*, **34**, 316–320.

Dent VE and Williams RAD (1986) *Actinomyces slackii* sp. nov. from dental plaque of dairy cattle. *International Journal of Systematic Bacteriology*, **36**, 392–395.

Dzink JL, Socransky SS and Haffajee AD (1988) The predominant cultivable microbiota of active and inactive lesions of destructive periodontal diseases. *Journal of Clinical Periodontology*, **15**, 316–323.

Funke G, Alvarez N, Pascual C *et al.* (1997) *Actinomyces europaeus* sp. nov., isolated from human clinical specimens. *International Journal of Systematic Bacteriology*, **47**, 687–692.

Funke G, Stubbs S, von Graevenitz A and Collins MD (1994) Assignment of human-derived CDC group, 1 coryneform bacteria and CDC 1-like coryneform bacteria to the genus *Actinomyces* as *Actinomyces neuii* subp. *neuii* sp. nov., subsp. nov. and *Actinomyces neuii* subsp. *anitratus* subsp. nov. *International Journal of Systematic Bacteriology* **44**, 167–171.

Gupta PK, Erozan YS and Frost JK (1978) Actinomycetes and the IUCD: an update. *Acta Cytologica*, **22**, 281–282.

Gupta PK, Hollander DH and Frost JK (1976) Actinomycetes in cervicovaginal smears: an association with IUD usage. *Acta Cytologica*, **20**, 295–297.

Hall V and Talbot P (1997) Evaluation of the E-test for susceptibility testing of *Actinomyces* species. *Reviews in Medical Microbiology*, **8**(Suppl. 1), S83.

Hall V, Collins MD, Hutson R, Falsen E and Duerden BI (2002) *Actinomyces cardiffensis* sp. nov., from human clinical sources. *Journal of Clinical Microbiology*, **40**, 3427–3431.

Hall V, Collins MD, Hutson R *et al.* (2003a) *Actinomyces vaccimaxillae* sp. nov., from the jaw of a cow. *International Journal of Systematic and Evolutionary Microbiology*, **53**, 603–606.

Hall V, Collins MD, Hutson R *et al.* (2003b) *Actinomyces oricola* sp. nov., from a human dental abscess. *International Journal of Systematic and Evolutionary Microbiology*, **53**, 1515–1518.

Hall V, Collins MD, Lawson PA. BI (2003c) *Actinomyces nasicola* sp. nov., isolated from a human nose. *International Journal of Systematic and Evolutionary Microbiology*, **53**, 1445–1448.

Hall V, Lewis-Evans T and Duerden BI (2001a) Identification of actinomyces, propionibacteria, lactobacilli and bifidobacteria by amplified 16S rDNA restriction analysis. *Anaerobe*, **7**, 55–57.

Hall V, Talbot PR, Stubbs SL and Duerden BI (2001b) Identification of clinical isolates of *Actinomyces* species by amplified, 16S ribosomal DNA restriction analysis. *Journal of Clinical Microbiology*, **39**, 3555–3562.

Hallberg K, Hammarström K-J, Falsen E *et al.* (1998) *Actinomyces naeslundii* genospecies 1 and 2 express different binding specificities to *N*-acetyl-β-D-galactosamine, whereas *Actinomyces odontolyticus* expresses a different binding specificity in colonizing the human mouth. *Oral Microbiology and Immunology*, **13**, 327–336.

Hamid D, Baldauf JJ, Cuenin C and Ritter J (2000) Treatment strategy for pelvic actinomycosis: case report and review of the literature. *European Journal of Obstetrics, Gynecology and Reproductive Biology*, **89**, 197–200.

Holdeman LV, Cato EP and Moore WEC (eds) (1977) *Anaerobe Laboratory Manual*, 4th edition, Virginia Polytechnic Institute and State University, Blacksburg, USA.

Hoyles L, Falsen E, Foster G *et al.* (2000) *Actinomyces canis* sp. nov., isolated from dogs. *International Journal of Systematic and Evolutionary Microbiology*, **50**, 1547–1551.

Hoyles L, Falsen E, Foster G and Collins MD (2002) *Actinomyces coleocanis* sp. nov., from the vagina of a dog. *International Journal of Systematic and Evolutionary Microbiology*, **52**, 1201–1203.

Hoyles L, Falsen E, Holmstrom G *et al.* (2001a) *Actinomyces suimastitidis* sp. nov., isolated from pig mastitis. *International Journal of Systematic and Evolutionary Microbiology*, **51**, 1323–1326.

Hoyles L, Falsen E, Pascual C *et al.* (2001b) *Actinomyces catuli* sp. nov., from dogs. *International Journal of Systematic and Evolutionary Microbiology*, **51**, 679–682.

Hoyles L, Pascual C, Falsen E *et al.* (2001c) *Actinomyces marimammalium* sp. nov., from marine animals. *International Journal of Systematic and Evolutionary Microbiology*, **51**, 151–156.

Johnson JL, Moore LVH, Kaneko B and Moore WEC (1990) *Actinomyces georgiae* sp. nov., *Actinomyces gerencseriae* sp. nov., designation of two genospecies of *Actinomyces naeslundii*, and inclusion of A. *naeslundii* serotypes II and III and *Actinomyces viscosus* serotype II in A. *naeslundii* genospecies 2. *International Journal of Systematic Bacteriology*, **40**, 273–286.

Kazor CE, Mitchell PM, Lee AM *et al.* (2003) Diversity of bacterial populations on the tongue dorsa of patients with halitosis and healthy patients. *Journal of Clinical Microbiology*, **41**, 558–563.

Lawson PA, Nikolaitchouk N, Falsen E, Westling K and Collins MD (2001) *Actinomyces funkei* sp. nov., isolated from human clinical specimens. *International Journal of Systematic and Evolutionary Microbiology*, **51**, 853–855.

Lee YC, Min D, Holcomb K *et al.* (2000) Computed tomography guided core needle biopsy diagnosis of pelvic actinomycosis. *Gynecological Oncology*, **79**, 318–323.

Leslie DE and Garland SM (1991) Comparison of immunofluorescence and culture for the detection of *Actinomyces israelii* in wearers of intrauterine contraceptive devices. *Journal of Medical Microbiology*, **35**, 224–228.

Lewis R, McKenzie D, Bagg J and Dickie A (1995) Experience with a novel selective medium for isolation of *Actinomyces* spp. from medical and dental specimens. *Journal of Clinical Microbiology*, **33**, 1613–1616.

Macfarlane DJ, Tucker LG and Kemp RJ (1993) Treatment of recalcitrant actinomycosis with ciprofloxacin. *Journal of Infection*, **27**, 177–180.

Marsh P and Martin M (1992) *Oral Microbiology*, Chapman and Hall, London.

Martin M (1984) The use of oral amoxicillin for the treatment of actinomycosis. *British Dental Journal*, **156**, 252–254.

Merki-Feld GS, Lebeda E, Hogg B and Keller PJ (2000) The incidence of actinomyces-like organisms in Papanicolaou-stained smears of copper- and levonorgestrel-releasing intrauterine devices. *Contraception*, **61**, 365–368.

Moore WEC and Moore LVH (1994) The bacteria of periodontal diseases. *Periodontology 2000*, **5**, 66–77.

Moore LV, Moore WEC, Cato EP *et al.* (1987) Bacteriology of human gingivitis. *Journal of Dental Research*, **66**, 989–995.

Morou-Bermudez E and Burne RA (2000) Analysis of urease expression in *Actinomyces naeslundii* WVU45. *Infection and Immunity*, **68**, 6670–6676.

Nespolon W and Moore C (1994) Diagnosis of cervicofacial actinomycosis by fine needle aspiration cytology. *Australian Journal of Medical Science*, **15**, 21–22.

Nikolaitchouk N, Hoyles L, Falsen E *et al.* (2000) Characterization of *Actinomyces* isolates from samples from the human urogenital tract: description of *Actinomyces urogenitalis* sp. nov. *International Journal of Systematic and Evolutionary Microbiology*, **50**, 1649–1654.

Pascual C, Foster G, Falsen E *et al.* (1999) *Actinomyces bowdenii* sp. nov., isolated from canine and feline clinical specimens. *International Journal of Systematic Bacteriology*, **49**, 1873–1877.

Pascual Ramos C, Falsen E, Alvarez N *et al.* (1997) *Actinomyces graevenitzii* sp. nov. isolated from human clinical specimens. *International Journal of Systematic Bacteriology*, **47**, 885–888.

Paster BJ, Boches SK, Galvin JL *et al.* (2001) Bacterial diversity in human subgingival plaque. *Journal of Bacteriology*, **183**, 3770–3783.

Phillips KD (1976) A simple and sensitive technique for determining the fermentation reactions of non-sporing anaerobes. *Journal of Applied Bacteriology*, **41**, 325–328.

Ruby JD, Li Y, Luo Y and Caulfield PW (2002) Genetic characterization of the oral *Actinomyces*. *Archives of Oral Biology*, **47**, 457–463.

Sabbe LJM, Van de Merwe D, Schouls L *et al.* (1999) Clinical spectrum of infections due to the newly described *Actinomyces* species *A. turicensis*, *A. radingae*, and *A. europaeus*. *Journal of Clinical Microbiology*, **37**, 8–13.

Sakellariou PL (1996) Periapical actinomycosis: report of a case and review of the literature. *Endodontics and Dental Traumatology*, **12**, 151–154.

Santala AM, Sarkonen N, Hall V *et al.* (2004) Evaluation of four commercial test systems for identification of *Actinomyces* and some closely related species. *Journal of Clinical Microbiology*, **42**, 418–420.

Sarkonen N, Könönen E, Summanen P *et al.* (2000) Oral colonization with *Actinomyces* species in infants by two years of age. *Journal of Dental Research*, **79**, 864–867.

Sarkonen N, Könönen E, Summanen P *et al.* (2001) Phenotypic identification of *Actinomyces* and related species isolated from human sources. *Journal of Clinical Microbiology*, **39**, 3955–3961.

Sato T, Matsuyama J, Takahashi N *et al.* (1998) Differentiation of oral *Actinomyces* species by 16S ribosomal DNA polymerase chain reaction-restriction fragment length polymorphism. *Archives of Oral Biology*, **43**, 247–252.

Schaal KP (1998) Actinomycoses, Actinobacillosis and related diseases. In *Topley and Wilson's Microbiology and Microbial Infections*, 9th edition, Hausler WJ and Sussman M (eds), Arnold, London, Vol. 3, pp. 777–798.

Schaal KP and Pape W (1980) Special methodological problems in antibiotic susceptibility testing of fermentative actinomycetes. *Infection*, **8**, S176–S182.

Schaal KP, Crecelius A, Schumacher G and Yassin AA (1999) Towards a new taxonomic structure of the genus *Actinomyces* and related bacteria. *Nova Acta Leopoldina NF 80*, **312**, 83–91.

Shannon HM, Wightman AJA and Carey FA (1995) Pulmonary actinomycosis – a master of disguise. *Journal of Infection*, **31**, 165–169.

Silver JG, Martin AW and McBride BC (1979) Experimental transient bacteraemias in human subjects with clinically healthy gingivae. *Journal of Clinical Periodontology*, **6**, 33–36.

Slack JM and Gerencser MA (1975) *Actinomyces, Filamentous Bacteria. Biology and Pathogenicity*, Burgess, Minneapolis.

Smego RA and Foglia G (1998) Actinomycosis. *Clinical Infectious Diseases*, **26**, 1255–1261.

Tang G, Samaranayake LP, Yip HK *et al.* (2003) Direct detection of *Actinomyces* spp. from infected root canals in a Chinese population: a study using PCR-based, oligonucleotide–DNA hybridization technique. *Journal of Dentistry*, **31**, 559–568.

Tanner A, Maiden MF, Macuch PJ *et al.* (1998) Microbiota of health, gingivitis and initial periodontitis. *Journal of Clinical Periodontology*, **25**, 85–88.

Thurnheer T, Gmür R and Guggenheim B (2004) Multiplex FISH analysis of a six-species bacterial biofilm. *Journal of Microbiological Methods*, **56**, 37–47.

Traynor RM, Parratt D, Duguid HLD and Duncan ID (1981) Isolation of actinomycetes from cervical specimens. *Journal of Clinical Pathology*, **34**, 914–916.

Vandamme P, Falsen E, Vancanneyt M *et al.* (1998) Characterization of *Actinomyces turicensis* and *Actinomyces radingae* strains from human clinical samples. *International Journal of Systematic Bacteriology*, **48**, 503–510.

Vandevelde AG, Jenkins SG and Hardy PR (1995) Sclerosing osteomyelitis and *Actinomyces naeslundii* infection of surrounding tissues. *Clinical Infectious Diseases*, **20**, 1037–1039.

Weese WC and Smith IM (1975) A study of fifty seven cases of actinomycosis over a thirty six-year period. *Archives of International Medicine*, **135**, 1562–1568.

Williams CE, Lamb GHR and Lewis-Jones HG (1990) Pelvic actinomycosis: beware the intrauterine contraceptive device. *British Journal of Radiology*, **63**, 134–137.

Wust J, Steiger U, Vuong H and Zbinden R (2000) Infection of a hip prosthesis by *Actinomyces naeslundii*. *Journal of Clinical Microbiology*, **38**, 929–930.

Wust J, Stubbs S, Weiss N *et al.* (1995) Assignment of *Actinomyces pyogenes*-like (CDC coryneform group E) bacteria to the genus *Actinomyces* as *Actinomyces radingae* sp. nov. and *Actinomyces turicensis* sp. nov. *Letters in Applied Microbiology*, **20**, 76–81.

Ximénez-Fyvie LA, Haffajee AD and Socransky SS (2000) Microbial composition of supra- and subgingival plaque in subjects with adult periodontitis. *Journal of Clinical Periodontology*, **27**, 722–732.

Index